$368.95 per copy (in United States).
Price subject to change without prior notice.
0209

RSMeans

Facilities Construction Cost Data

24th Annual Edition

NEW FOR 2009!

Line items found in this cost data that have the icon shown above are green.

The identified items fall into a broad definition of what is considered green.

Please see page ix for more information.

2009

Senior Editor
Melville J. Mossman, PE
Stephen C. Plotner

Contributing Editors
Christopher Babbitt
Ted Baker
Barbara Balboni
Robert A. Bastoni
John H. Chiang, PE
Gary W. Christensen
David G. Drain, PE
Cheryl Elsmore
Robert J. Kuchta
Robert C. McNichols
Jeannene D. Murphy
Eugene R. Spencer
Marshall J. Stetson
Phillip R. Waier, PE

Senior Vice President & General Manager
John Ware

Vice President of Operations
Dave Walsh

Vice President of Sales & Marketing
Sev Ritchie

Marketing Director
John M. Shea

Director of Product Development
Thomas J. Dion

Engineering Manager
Bob Mewis, CCC

Production Manager
Michael Kokernak

Production Coordinator
Jill Goodman

Technical Support
Jonathan Forgit
Mary Lou Geary
Roger Hancock
Gary L. Hoitt
Genevieve Medeiros
Debbie Panarelli
Paula Reale-Camelio
Kathryn S. Rodriguez
Sheryl A. Rose

Book & Cover Design
Norman R. Forgit

R.S. Means Company, Inc.
Construction Publishers & Consultants
63 Smiths Lane
Kingston, MA 02364-3008
(781) 422-5000

Printed in the United States of America.
ISSN 1075-0789
ISBN 978-0-87629-152-8

PRINTED WITH SOY INK

This book is printed on recycled paper (10% PCW cover, 20% PCW text), using soy-based printing ink. This book is recyclable.

Foreword

Our Mission

Since 1942, RSMeans has been actively engaged in construction cost publishing and consulting throughout North America.

Today, over 60 years after RSMeans began, our primary objective remains the same: to provide you, the construction and facilities professional, with the most current and comprehensive construction cost data possible.

Whether you are a contractor, an owner, an architect, an engineer, a facilities manager, or anyone else who needs a reliable construction cost estimate, you'll find this publication to be a highly useful and necessary tool.

With the constant flow of new construction methods and materials today, it's difficult to find the time to look at and evaluate all the different construction cost possibilities. In addition, because labor and material costs keep changing, last year's cost information is not a reliable basis for today's estimate or budget.

That's why so many construction professionals turn to RSMeans. We keep track of the costs for you, along with a wide range of other key information, from city cost indexes . . . to productivity rates . . . to crew composition . . . to contractor's overhead and profit rates.

RSMeans performs these functions by collecting data from all facets of the industry and organizing it in a format that is instantly accessible to you. From the preliminary budget to the detailed unit price estimate, you'll find the data in this book useful for all phases of construction cost determination.

The Staff, the Organization, and Our Services

When you purchase one of RSMeans' publications, you are, in effect, hiring the services of a full-time staff of construction and engineering professionals.

Our thoroughly experienced and highly qualified staff works daily at collecting, analyzing, and disseminating comprehensive cost information for your needs. These staff members have years of practical construction experience and engineering training prior to joining the firm. As a result, you can count on them not only for the cost figures, but also for additional background reference information that will help you create a realistic estimate.

The RSMeans organization is always prepared to help you solve construction problems through its four major divisions: Construction and Cost Data Publishing, Electronic Products and Services, Consulting and Business Solutions, and Professional Development Services.

Besides a full array of construction cost estimating books, RSMeans also publishes a number of other reference works for the construction industry. Subjects include construction estimating and project and business management; special topics such as HVAC, roofing, plumbing, and hazardous waste remediation; and a library of facility management references.

In addition, you can access all of our construction cost data electronically using *Means CostWorks®* CD or on the Web using Means CostWorks.com.

What's more, you can increase your knowledge and improve your construction estimating and management performance with an RSMeans Construction Seminar or In-House Training Program. These two-day seminar programs offer unparalleled opportunities for everyone in your organization to get updated on a wide variety of construction-related issues.

RSMeans is also a worldwide provider of construction cost management and analysis services for commercial and government owners.

In short, RSMeans can provide you with the tools and expertise for constructing accurate and dependable construction estimates and budgets in a variety of ways.

Robert Snow Means Established a Tradition of Quality That Continues Today

Robert Snow Means spent years building RSMeans, making certain he always delivered a quality product.

Today, at RSMeans, we do more than talk about the quality of our data and the usefulness of our books. We stand behind all of our data, from historical cost indexes to construction materials and techniques to current costs.

If you have any questions about our products or services, please call us toll-free at 1-800-334-3509. Our customer service representatives will be happy to assist you. You can also visit our Web site at www.rsmeans.com.

Table of Contents

Related RSMeans Products and Services

The engineers at RSMeans, suggest the following products and services as companion information resources to *Facilities Construction Cost Data:*

Construction Cost Data Books

Interior Cost Data 2009
Building Construction Cost Data 2009
Facilities Maintenance & Repair Cost Data 2009

Reference Books

Facilities Planning and Relocation
Life Cycle Costing for Facilities
Cost Planning & Estimating for Facilities Maintenance
Job Order Contracting
Estimating Building Costs
Estimating Handbook, 2nd Edition
ADA Compliance Pricing Guide, 2nd Edition
Building Security: Strategies & Costs

Seminars and In-House Training

Facilities Maintenance & Repair Estimating
Repair & Remodeling Estimating
Unit Price Estimating
Means CostWorks™ Training
Means Data for Job Order Contracting (JOC)

RSMeans on the Internet

Visit RSMeans at **www.rsmeans.com.** The site contains useful interactive cost and reference material. Request or download **FREE** estimating software demos. Visit our bookstore for convenient ordering and to learn more about new publications and companion products.

RSMeans Electronic Data

Get the information found in RSMeans cost books electronically on *Means CostWorks®* CD or on the Web at MeansCostworks.com

RSMeans Business Solutions

Engineers and analysts offer research studies, benchmark analysis, predictive cost modeling, analytics, job order contracting, and real property management consultation, as well as custom-designed, Web-based dashboards and calculators that apply RCD/RSMeans extensive databases. Clients include federal government agencies, architects, construction management firms, and institutional organizations, such as school systems, health care facilities, associations, and corporations.

RSMeans for Job Order Contracting (JOC)

Best practice JOC tools for cost estimating and project management to help streamline delivery processes for renovation projects. Renovation is a $147 billion market in the U.S., and includes projects in school districts, municipalities, health care facilities, colleges and universities, and corporations.

- RSMeans Engineers consult in contracting methods and conduct JOC Facility Audits
- JOCWorks™ Software (Basic, Advanced, PRO levels)
- RSMeans Job Order Contracting Cost Data for the entire U.S.

Construction Costs for Software Applications

Over 25 unit price and assemblies cost databases are available through a number of leading estimating and facilities management software providers (listed below). For more information see the "Other RSMeans Products" pages at the back of this publication.

MeansData™ is also available to federal, state, and local government agencies as multi-year, multi-seat licenses.

- 4Clicks-Solutions, LLC
- ArenaSoft Estimating
- Beck Technology
- BSD – Building Systems Design, Inc.
- CMS – Construction Management Software
- Corecon Technologies, Inc.
- CorVet Systems
- Estimating Systems, Inc.
- Maximus Asset Solutions
- Sage Timberline Office
- US Cost, Inc.
- VFA – Vanderweil Facility Advisers
- WinEstimator, Inc.

How the Book Is Built: An Overview

The Construction Specifications Institute (CSI) and Construction Specifications Canada (CSC) have produced the 2004 edition of MasterFormat, a system of titles and numbers used extensively to organize construction information.

All unit price data in the RSMeans cost data books is now arranged in the 50-division MasterFormat 2004 system.

A Powerful Construction Tool

You have in your hands one of the most powerful construction tools available today. A successful project is built on the foundation of an accurate and dependable estimate. This book will enable you to construct just such an estimate.

For the casual user the book is designed to be:

- quickly and easily understood so you can get right to your estimate.
- filled with valuable information so you can understand the necessary factors that go into the cost estimate.

For the regular user, the book is designed to be:

- a handy desk reference that can be quickly referred to for key costs.
- a comprehensive, fully reliable source of current construction costs and productivity rates so you'll be prepared to estimate any project.
- a source book for preliminary project cost, product selections, and alternate materials and methods.

To meet all of these requirements we have organized the book into the following clearly defined sections.

New: Quick Start

See our new "Quick Start" instructions on the following page.

Estimating with RSMeans Unit Price Cost Data

Please see these steps to complete an estimate using RSMeans unit price cost data.

How To Use the Book: The Details

This section contains an in-depth explanation of how the book is arranged . . . and how you can use it to determine a reliable construction cost estimate. It includes information about how we develop our cost figures and how to completely prepare your estimate.

Unit Price Section

All unit price cost data has been divided into the 50 divisions according to the MasterFormat system of classification and numbering. For a listing of these divisions and an outline of their subdivisions, see the Unit Price Section Table of Contents.

Estimating tips are included at the beginning of each division.

Assemblies Section

The cost data in this section has been organized in an "Assemblies" format. These assemblies are the functional elements of a building and are arranged according to the 7 divisions of the UNIFORMAT II classification system. For a complete explanation of a typical "Assemblies" page, see "How To Use the Assemblies Cost Tables."

Reference Section

This section includes information on Equipment Rental Costs, Crew Listings, Historical Cost Indexes, City Cost Indexes, Location Factors, Reference Tables, Change Orders, Square Foot Costs, and a listing of Abbreviations.

Equipment Rental Costs: This section contains the average costs to rent and operate hundreds of pieces of construction equipment.

Crew Listings: This section lists all the crews referenced in the book. For the purposes of this book, a crew is composed of more than one trade classification and/or the addition of power equipment to any trade classification. Power equipment is included in the cost of the crew. Costs are shown both with the bare labor rates and with the installing contractor's overhead and profit added. For each, the total crew cost per eight-hour day and the composite cost per labor-hour are listed.

Historical Cost Indexes: These indexes provide you with data to adjust construction costs over time.

City Cost Indexes: All costs in this book are U.S. national averages. Costs vary because of the regional economy. You can adjust costs by CSI Division to over 700 locations throughout the U.S. and Canada by using the data in this section.

Location Factors: You can adjust total project costs to over 900 locations throughout the U.S. and Canada by using the data in this section.

Reference Tables: At the beginning of selected major classifications in the Unit Price and Assemblies sections are "reference numbers" shown in a shaded box. These numbers refer you to related information in the Reference Section. In this section, you'll find reference tables, explanations, estimating information that supports how we develop the unit price data, technical data, and estimating procedures.

Change Orders: This section includes information on the factors that influence the pricing of change orders.

Square Foot Costs: This section contains costs for 59 different building types that allow you to make a rough estimate for the overall cost of a project or its major components.

Abbreviations: A listing of abbreviations used throughout this book, along with the terms they represent, is included in this section.

Index

A comprehensive listing of all terms and subjects in this book will help you quickly find what you need when you are not sure where it falls in MasterFormat.

The Scope of This Book

This book is designed to be as comprehensive and as easy to use as possible. To that end we have made certain assumptions and limited its scope in two key ways:

1. We have established material prices based on a national average.
2. We have computed labor costs based on a 30-city national average of union wage rates.

For a more detailed explanation of how the cost data is developed, see "How To Use the Book: The Details."

Project Size/Type

The material prices in Means data cost books are "contractor's prices." They are the prices that contractors can expect to pay at the lumberyards, suppliers/distributers warehouses, etc. Small orders of speciality items would be higher than the costs shown, while very large orders, such as truckload lots, would be less. The variation would depend on the size, timing, and negotiating power of the contractor. The labor costs are primarily for new construction or major renovations rather than repairs or minor alterations. **With reasonable exercise of judgment, the figures can be used for any building work.**

Absolute Essentials for a Quick Start

If you feel you are ready to use this book and don't think you will need the detailed instructions that begin on the following page, this Absolute Essentials for a Quick Start page is for you. These steps will allow you to get started estimating in a matter of minutes.

1 Scope

Think through the project that you will be estimating and identify the many individual work tasks that will need to be covered in your estimate.

2 Quantify

Determine the number of units that will be required for each work task that you identified.

3 Pricing

Locate individual Unit Price line items that match the work tasks you identified. The Unit Price Section Table of Contents that begins on page 1 and the Index in the back of the book will help you find these line items.

4 Multiply

Multiply the Total Incl O&P cost for a Unit Price line item in the book by your quantity for that item. The price you calculate will be an estimate for a completed item of work performed by a subcontractor. Keep adding line items in this manner to build your estimate.

5 Project Overhead

Include project overhead items in your estimate. These items are needed to make the job run and are typically, but not always, provided by the General Contractor. They can be found in Division 1. An alternate method of estimating project overhead costs is to apply a percentage of the total project cost.

Include rented tools not included in crews, waste, rubbish handling and cleanup.

6 Estimate Summary

Include General Contractor's markup on subcontractors, General Contractor's office overhead and profit, and sales tax on materials and equipment.

Adjust your estimate to the project's location by using the City Cost Indexes or Location Factors found in the Reference Section.

Editors' Note: We urge you to spend time reading and understanding the supporting material in the front of this book. An accurate estimate requires experience, knowledge and careful calculation. The more you know about how we at RSMeans developed the data, the more accurate your estimate will be. In addition, it is important to take into consideration the reference material in the back of the book such as Equipment Listings, Crew Listings, City Cost Indexes, Location Factors and Reference Numbers.

Estimating with RSMeans Unit Price Cost Data

Following these steps will allow you to complete an accurate estimate using RSMeans Unit Price cost data.

1 Scope out the project

- Identify the individual work tasks that will need to be covered in your estimate.
- The Unit Price data inside this book has been divided into 44 Divisions according to the CSI MasterFormat2004 – their titles are listed on the back cover of your book.
- Think through the project that you will be estimating and identify those CSI Divisions that you will need to use in your estimate.
- The Unit Price Section Table of Contents that begins on page 1 may also be helpful when scoping out your project.
- Experienced estimators find it helpful to begin an estimate with Division 2, estimating Division 1 after the full scope of the project is known.

2 Quantify

- Determine the number of units that will be required for each work task that you previously identified.
- Experienced estimators will include an allowance for waste in their quantities (waste is not included in RSMeans Unit Price line items unless so stated).

3 Price the quantities

- Use the Unit Price Table of Contents, and the Index, to locate the first individual Unit Price line item for your estimate.
- Reference Numbers indicated within a Unit Price section refer to additional information that you may find useful.
- The crew will tell you who is performing the work for that task. Crew codes are expanded in the Crew Listings in the Reference Section to include all trades and equipment that comprise the crew.
- The Daily Output is the amount of work the crew is expected to do in one day.
- The Labor-Hours value is the amount of time it will take for the average crew member to install one unit of measure.
- The abbreviated Unit designation indicates the unit of measure upon which the crew, productivity and prices are based.
- Bare Costs are shown for materials, labor, and equipment needed to complete the Unit Price line item. Bare costs do not include waste, project overhead, payroll insurance, payroll taxes, main office overhead, or profit.
- The Total Incl O&P cost is the billing rate or invoice amount of the installing contractor or subcontractor who performs the work for the Unit Price line item.

4 Multiply

- Multiply the total number of units needed for your project by the Total Incl O&P cost for the Unit Price line item.
- Be careful that the final unit of measure for your quantity of units matches the unit of measure in the Unit column in the book.
- The price you calculate will be an estimate for a completed item of work.
- Keep scoping individual tasks, determining the number of units required for those tasks, matching them up with individual Unit Price line items in the book, and multiplying quantities by Total Incl O&P costs. In this manner keep building your estimate.
- An estimate completed to this point in this manner will be priced as if a subcontractor, or set of subcontractors, will perform the work. The estimate does not yet include Project Overhead or Estimate Summary components such as General Contractor markups on subcontracted and self-performed work, General Contractor office overhead and profit, contingency, and location factor.

5 Project Overhead

- Include project overhead items from Division 1 – General Requirements.
- These are items that will be needed to make the job run. These are typically, but not always, provided by the General Contractor. They include, but are not limited to, such items as field personnel, insurance, performance bond, permits, testing, temporary utilities, field office and storage facilities, temporary scaffolding and platforms, equipment mobilization and demobilization, temporary roads and sidewalks, winter protection, temporary barricades and fencing, temporary security, temporary signs, field engineering and layout, final cleaning and commissioning.
- These items should be scoped, quantified, matched to individual Unit Price line items in Division 1, priced and added to your estimate.
- An alternate method of estimating project overhead costs is to apply a percentage of the total project cost, usually 5% to 15% with an average of 10% (see General Conditions, page ix).
- Include other project related expenses in your estimate such as:
    - Rented equipment not itemized in the Crew Listings.
    - Rubbish handling throughout the project (see 02 41 19.23).

6 Estimate Summary

- Includes sales tax on materials and equipment.
  Note: Sales tax must be added for materials in subcontracted work.
- Include the General Contractor's markup on self-performed work, usually 5% to 15% with an average of 10%.
- Include the General Contractor's markup on subcontracted work, usually 5% to 15% with an average of 10%.
- Include General Contractor's main office overhead and profit.
- RSMeans gives general guidelines on the General Contractor's main office overhead (see section 01 31 13.50 and Reference Number R013113-50).
- RSMeans gives no guidance on the General Contractor's profit.
- Markups will depend on the size of the General Contractor's operations, his projected annual revenue, the level of risk he is taking on, and on the level of competitiveness in the local area and for this project in particular.
- Include a contingency, usually 3% to 5%.
- Adjust your estimate to the project's location by using the City Cost Indexes or the Location Factors in the Reference Section.
- Look at the rules on the pages for How to Use the City Cost Indexes to see how to apply the Indexes for your location.
- When the proper Index or Factor has been identified for the project's location, convert it to a multiplier by dividing it by 100, then multiply that multiplier by your estimate total cost. Your original estimate total cost will now be adjusted up or down from the national average to a total that is appropriate for your location.

Editors' Notes:

1) *We urge you to spend time reading and understanding the supporting material in the front of this book. An accurate estimate requires experience, knowledge, and careful calculation. The more you know about how we at RSMeans developed the data, the more accurate your estimate will be. In addition, it is important to take into consideration the reference material in the back of the book such as Equipment Listings, Crew Listings, City Cost Indexes, Location Factors, and Reference Numbers.*
2) *Contractors who are bidding or are involved in JOC, DOC, SABER, or IDIQ type contracts are cautioned that workers' compensation Insurance, federal and state payroll taxes, waste, project supervision, project overhead, main office overhead, and profit are not included in bare costs. Your coefficient or multiplier must cover these costs.*

How to Use the Book: The Details

What's Behind the Numbers? The Development of Cost Data

The staff at RSMeans continuously monitors developments in the construction industry in order to ensure reliable, thorough, and up-to-date cost information.

While *overall* construction costs may vary relative to general economic conditions, price fluctuations within the industry are dependent upon many factors. Individual price variations may, in fact, be opposite to overall economic trends. Therefore, costs are continually monitored and complete updates are published yearly. Also, new items are frequently added in response to changes in materials and methods.

Costs—$ (U.S.)

All costs represent U.S. national averages and are given in U.S. dollars. The RSMeans City Cost Indexes can be used to adjust costs to a particular location. The City Cost Indexes for Canada can be used to adjust U.S. national averages to local costs in Canadian dollars. No exchange rate conversion is necessary.

G The processes or products identified by the green symbol in this publication have been determined to be environmentally responsible and/or resource-efficient solely by the RSMeans engineering staff. The inclusion of the green symbol does not represent compliance with any specific industry association or standard.

Material Costs

The RSMeans staff contacts manufacturers, dealers, distributors, and contractors all across the U.S. and Canada to determine national average material costs. If you have access to current material costs for your specific location, you may wish to make adjustments to reflect differences from the national average. Included within material costs are fasteners for a normal installation. RSMeans engineers use manufacturers' recommendations, written specifications, and/or standard construction practice for size and spacing of fasteners. Adjustments to material costs may be required for your specific application or location. Material costs do not include sales tax.

Labor Costs

Labor costs are based on the average of wage rates from 30 major U.S. cities. Rates are determined from labor union agreements or prevailing wages for construction trades for the current year. Rates, along with overhead and profit markups, are listed on the inside back cover of this book.

- If wage rates in your area vary from those used in this book, or if rate increases are expected within a given year, labor costs should be adjusted accordingly.

Labor costs reflect productivity based on actual working conditions. In addition to actual installation, these figures include time spent during a normal weekday on tasks such as, material receiving and handling, mobilization at site, site movement, breaks, and cleanup.

Productivity data is developed over an extended period so as not to be influenced by abnormal variations and reflects a typical average.

Equipment Costs

Equipment costs include not only rental, but also operating costs for equipment under normal use. The operating costs include parts and labor for routine servicing such as repair and replacement of pumps, filters, and worn lines. Normal operating expendables, such as fuel, lubricants, tires, and electricity (where applicable), are also included. Extraordinary operating expendables with highly variable wear patterns, such as diamond bits and blades, are excluded. These costs are included under materials. Equipment rental rates are obtained from industry sources throughout North America—contractors, suppliers, dealers, manufacturers, and distributors.

Equipment costs do not include operators' wages; nor do they include the cost to move equipment to a job site (mobilization) or from a job site (demobilization).

Equipment Cost/Day—The cost of power equipment required for each crew is included in the Crew Listings in the Reference Section (small tools that are considered as essential everyday tools are not listed out separately). The Crew Listings itemize specialized tools and heavy equipment along with labor trades. The daily cost of itemized equipment included in a crew is based on dividing the weekly bare rental rate by 5 (number of working days per week) and then adding the hourly operating cost times 8 (the number of hours per day). This Equipment Cost/Day is shown in the last column of the Equipment Rental Cost pages in the Reference Section.

Mobilization/Demobilization—The cost to move construction equipment from an equipment yard or rental company to the job site and back again is not included in equipment costs. Mobilization (to the site) and demobilization (from the site) costs can be found in the Unit Price Section. If a piece of equipment is already at the job site, it is not appropriate to utilize mob./demob. costs again in an estimate.

General Conditions

Cost data in this book is presented in two ways: Bare Costs and Total Cost including O&P (Overhead and Profit). General Conditions, when applicable, should also be added to the Total Cost including O&P. The costs for General Conditions are listed in Division 1 of the Unit Price Section and the Reference Section of this book. General Conditions for the *Installing Contractor* may range from 0% to 10% of the Total Cost including O&P. For the *General* or *Prime Contractor*, costs for General Conditions may range from 5% to 15% of the Total Cost including O&P, with a figure of 10% as the most typical allowance.

Overhead and Profit

Total Cost including O&P for the *Installing Contractor* is shown in the last column on both the Unit Price and the Assemblies pages of this book. This figure is the sum of the bare material cost plus 10% for profit, the bare labor cost plus total overhead and profit, and the bare equipment cost plus 10% for profit. Details for the calculation of Overhead and Profit on labor are shown on the inside back cover and in the Reference Section of this book. (See the "How to Use the Unit Price Pages" for an example of this calculation.)

Factors Affecting Costs

Costs can vary depending upon a number of variables. Here's how we have handled the main factors affecting costs.

Quality—The prices for materials and the workmanship upon which productivity is based represent sound construction work. They are also in line with U.S. government specifications.

Overtime—We have made no allowance for overtime. If you anticipate premium time or work beyond normal working hours, be sure to make an appropriate adjustment to your labor costs.

Productivity—The productivity, daily output, and labor-hour figures for each line item are based on working an eight-hour day in daylight hours in moderate temperatures. For work that extends beyond normal work hours or is performed under adverse conditions, productivity may decrease. (See the section in "How To Use the Unit Price Pages" for more on productivity.)

Size of Project—The size, scope of work, and type of construction project will have a significant impact on cost. Economies of scale can reduce costs for large projects. Unit costs can often run higher for small projects. Costs in this book are intended for the size and type of project as previously described in "How the Book Is Built: An Overview." Costs for projects of a significantly different size or type should be adjusted accordingly.

Location—Material prices in this book are for metropolitan areas. However, in dense urban areas, traffic and site storage limitations may increase costs. Beyond a 20-mile radius of large cities, extra trucking or transportation charges may also increase the material costs slightly. On the other hand, lower wage rates may be in effect. Be sure to consider both of these factors when preparing an estimate, particularly if the job site is located in a central city or remote rural location.

In addition, highly specialized subcontract items may require travel and per-diem expenses for mechanics.

Other Factors—

- season of year
- contractor management
- weather conditions
- local union restrictions
- building code requirements
- availability of:
  - adequate energy
  - skilled labor
  - building materials
- owner's special requirements/restrictions
- safety requirements
- environmental considerations

Unpredictable Factors—General business conditions influence "in-place" costs of all items. Substitute materials and construction methods may have to be employed. These may affect the installed cost and/or life cycle costs. Such factors may be difficult to evaluate and cannot necessarily be predicted on the basis of the job's location in a particular section of the country. Thus, where these factors apply, you may find significant but unavoidable cost variations for which you will have to apply a measure of judgment to your estimate.

Rounding of Costs

In general, all unit prices in excess of $5.00 have been rounded to make them easier to use and still maintain adequate precision of the results. The rounding rules we have chosen are in the following table.

Prices from ...	Rounded to the nearest ...
$.01 to $5.00	$.01
$5.01 to $20.00	$.05
$20.01 to $100.00	$.50
$100.01 to $300.00	$1.00
$300.01 to $1,000.00	$5.00
$1,000.01 to $10,000.00	$25.00
$10,000.01 to $50,000.00	$100.00
$50,000.01 and above	$500.00

Important Estimating Considerations

One reason for listing a job size "minimum" is to ensure that the construction craftsmen are productive for 8 hours a day on a continuous basis. Otherwise, tasks that require only 6 to 7 hours could be billed as an 8-hour day, thus providing an erroneous productivity estimate. The "productivity," or daily output, of each craftsman includes mobilization and cleanup time, break time and plan layout time, as well as an allowance to carry stock from the storage trailer or location on the job site up to 200′ into the building and to the first or second floor. If material has to be transported over greater distances or to higher floors, an additional allowance should be considered by the estimator. An allowance has also been included in the piping and fittings installation time for a leak check and minor tightening.

Equipment installation time includes the following applicable items: positioning, leveling and securing the unit in place, connecting all associated piping, ducts, vents, etc., which shall have been estimated separately, connecting to an adjacent power source, filling/bleeding, startup, adjusting the controls up and down to ensure proper response, setting the integral controls/valves/regulators/thermostats for proper operation (does not include external building type control systems, DDC systems, etc.), explaining/training owner's operator, and warranty. A reasonable breakdown of the labor costs is as follows:

1.	Movement into building, installation/ setting of equipment	35%
2.	Connection to piping/duct/power, etc.	25%
3.	Filling/flushing/cleaning/touchup, etc.	15%
4.	Startup/running adjustments	5%
5.	Training owner's rep.	5%
6.	Warranty/call back/service	15%

Note that cranes or other lifting equipment are not included on any lines in the Mechanical divisions. For example, if a crane is required to lift a heavy piece of pipe into place high above a gym floor, or to put a rooftop unit on the roof of a four-story building, etc., it must be added. Due to the potential for extreme variation—from nothing additional required to a major crane or helicopter—we feel that including a nominal amount for "lifting contingency" would be useless and detract from the accuracy of the estimate. When using equipment rental from RSMeans, remember to include the cost of the operator(s).

Final Checklist

Estimating can be a straightforward process provided you remember the basics. Here's a checklist of some of the steps you should remember to complete before finalizing your estimate.

Did you remember to . . .

- factor in the City Cost Index for your locale?
- take into consideration which items have been marked up and by how much?
- mark up the entire estimate sufficiently for your purposes?
- read the background information on techniques and technical matters that could impact your project time span and cost?
- include all components of your project in the final estimate?
- make use of Minimum Labor/Equipment Charges for Small Quantities (see the following page for more details)?
- double check your figures for accuracy?
- call RSMeans if you have any questions about your estimate or the data you've found in our publications?

Remember, RSMeans stands behind its publications. If you have any questions about your estimate . . . about the costs you've used from our books . . . or even about the technical aspects of the job that may affect your estimate, feel free to call the RSMeans editors at 1-800-334-3509.

Using Minimum Labor/Equipment Charges for Small Quantities

Estimating small construction or repair tasks often creates situations in which the quantity of work to be performed is very small. When this occurs, the labor and/or equipment costs to perform the work may be too low to allow for the crew to get to the job, receive instructions, find materials, get set up, perform the work, clean up, and get to the next job. In these situations, the estimator should compare the developed labor and/or equipment costs for performing the work (e.g., quantity x labor and/or equipment costs) with the *"minimum labor/equipment charge"* within that Unit Price section of the book.(These "minimum labor/equipment charge" line items appear only in the *Facilities Construction Cost Data and the Repair & Remodeling Cost Data* titles).

If the labor and/or equipment costs developed by the estimator are LOWER THAN the *"minimum labor/equipment charge"* listed at the bottom of specific sections of Unit Price costs, the estimator should adjust the developed costs upward to the *"minimum labor/equipment charge."* The proper use of a *"minimum labor/equipment charge"* results in having enough money in the estimate to cover the contractor's higher cost of performing a very small amount of work during a partial workday.

A *"minimum labor/equipment charge"* should be used only when the task being estimated is the only task the crew will perform at the job site that day. If, however, the crew will be able to perform other tasks at the job site that day, the use of a *"minimum labor/equipment charge"* is not appropriate.

08 52 Wood Windows

08 52 10 – Plain Wood Windows

08 52 10.40 Casement Window			Crew	Daily Output	Labor-Hours	Unit	2009 Bare Costs Material	Labor	Equipment	Total	Total Incl O&P
0010	**CASEMENT WINDOW,** including frame, screen and grilles										
0100	Avg. quality, bldrs. model, 2'-0" x 3'-0" H, dbl. insulated glass	G	1 Carp	10	.800	Ea.	276	32		308	360
0150	Low E glass	G		10	.800		225	32		257	300
0200	2'-0" x 4'-6" high, double insulated glass	G		9	.889		345	35.50		380.50	440
0250	Low E glass	G		9	.889		263	35.50		298.50	350
0300	2'-4" x 6'-0" high, double insulated glass	G		8	1		380	40		420	485
0350	Low E glass	G		8	1		465	40		505	580
0522	Vinyl clad, premium, double insulated glass, 2'-0" x 3'-0"	G		10	.800		266	32		298	345
0524	2'-0" x 4'-0"	G		9	.889		310	35.50		345.50	400
0525	2'-0" x 5'-0"	G		8	1		355	40		395	455
0528	2'-0" x 6'-0"	G		8	1		355	40		395	455
8100	Metal clad, deluxe, dbl. insul. glass, 2'-0" x 3'-0" high	G		10	.800		226	32		258	300
9000	Minimum labor/equipment charge		1 Carp	3	2.667	Job		107		107	176

Example:

Establish the bid price to install two casement windows. Assume installation of 2' x 4' metal clad windows with insulating glass [Unit Price line number 08 52 10.40 0200], and that this is the only task this crew will perform at the job site that day.

Solution:

Step One —Develop the Bare Labor Cost for this task:

Bare Labor Cost = 2 windows @ $35.50/each = $71.00

Step Two —Evaluate the *"minimum labor/equipment charge"* for this Unit Price section against the developed Bare Labor Cost for this task:

"Minimum labor/equipment charge" = $107.00 (compare with $71.00)

Step Three —Choose to adjust the developed labor cost upward to the *"minimum labor/equipment charge."*

Step Four —Develop the bid price for this task (including O&P):

Add together the marked-up Bare Material Cost for this task and the marked-up *"minimum labor/equipment charge"* for this Unit Price section.

2 x ($345.00 + 10%) + ($107.00 + 65.1%)

= 2 x ($345.00 + $34.50) + ($107.00 + $69.66)

= 2 x ($379.50) + $176.66

= $759.00 + $176.66

= $935.66

ANSWER: $935.66 is the correct bid price to use. This sum takes into consideration the Material Cost (with 10% for profit) for these two windows, plus the *"minimum labor/equipment charge"* (with O&P included) for this section of the Unit Price book.

Unit Price Section

Table of Contents

Table of Contents (cont.)

Table of Contents (cont.)

How to Use the Unit Price Pages

The following is a detailed explanation of a sample entry in the Unit Price Section. Next to each bold number below is the described item with the appropriate component of the sample entry following in parentheses. Some prices are listed as bare costs; others as costs that include overhead and profit of the installing contractor. In most cases, if the work is to be subcontracted, the general contractor will need to add an additional markup (RSMeans suggests using 10%) to the figures in the column "Total Incl. O&P."

1 *Division Number/Title (03 30/Cast-In-Place Concrete)*

Use the Unit Price Section Table of Contents to locate specific items. The sections are classified according to the CSI MasterFormat 2004 system.

2 *Line Numbers (03 30 53.40 3920)*

Each unit price line item has been assigned a unique 12-digit code based on the CSI MasterFormat classification.

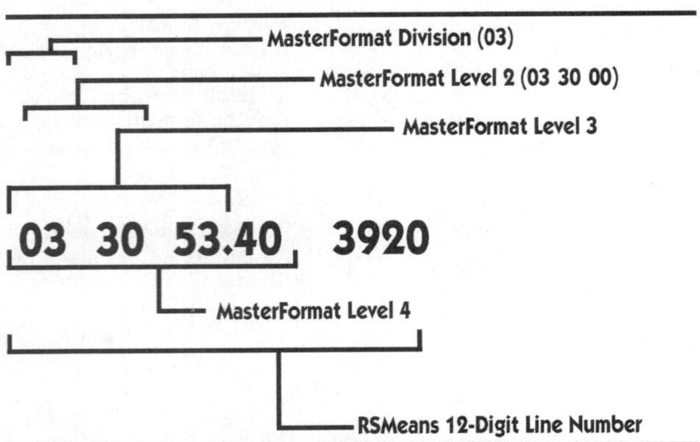

3 *Description (Concrete In Place, etc.)*

Each line item is described in detail. Sub-items and additional sizes are indented beneath the appropriate line items. The first line or two after the main item (in boldface) may contain descriptive information that pertains to all line items beneath this boldface listing.

4 *Reference Number Information*

R033053 -50 You'll see reference numbers shown in shaded boxes at the beginning of some sections. These refer to related items in the Reference Section, visually identified by a vertical gray bar on the page edges.

The relation may be an estimating procedure that should be read before estimating, or technical information.

The "R" designates the Reference Section. The numbers refer to the MasterFormat 2004 classification system.

It is strongly recommended that you review all reference numbers that appear within the section in which you are working.

Note: Not all reference numbers appear in all RSMeans publications.

03 30 Cast-In-Place Concrete

03 30 53 – Miscellaneous Cast-In-Place Concrete

03 30 53.40 Concrete In Place		R033053-50	Crew	Daily Output	Labor-Hours	Unit	2009 Bare Costs Material	Labor	Equipment	Total	Total Incl O&P
3570	6' x 6' x 8" thick			14	3.429		188	135	1.85	324.85	430
3580	8' x 8' x 10" thick			8	6		395	236	3.24	634.24	830
3590	10' x 10' x 12" thick			5	9.600		655	380	5.20	1,040.20	1,350
3800	Footings, spread under 1 C.Y.		C-14C	28	4	C.Y.	209	153	.92	362.92	485
3825	1 C.Y to 5 C.Y.			43	2.605		221	99.50	.60	321.10	410
3850	Over 5 C.Y.			75	1.493		198	57	.34	255.34	310
3900	Footings, strip, 18" x 9", unreinforced		C-14L	40	2.400		120	89.50	.65	210.15	280
3920	18" x 9", reinforced		C-14C	35	3.200		157	122	.74	279.74	375
3925	20" x 10", unreinforced		C-14L	45	[illegible]		117	79.50	.58	197.08	260
3930	20" x 10", reinforced		C-14C	[illegible]	[illegible]		147	107	.64	254.64	[illegible]
3935	24" x 12", unreinforced		C-14L	55	1.745		116	65	.47	181.47	[illegible]
3940	24" x 12", reinforced		C-14C	48	2.333		147	89.50	.54	237.04	310
3945	36" x 12", unreinforced		C-14L	70	1.371		113	51	.37	164.37	208
3950	36" x 12", reinforced		C-14C	60	1.867		141	71.50	.43	212.93	273

5 *Crew (C-14C)*

The "Crew" column designates the typical trade or crew used to install the item. If an installation can be accomplished by one trade and requires no power equipment, that trade and the number of workers are listed (for example, "2 Carpenters"). If an installation requires a composite crew, a crew code designation is listed (for example, "C-14C"). You'll find full details on all composite crews in the Crew Listings.

- For a complete list of all trades utilized in this book and their abbreviations, see the inside back cover.

Crews

Crew No.	Bare Costs		Incl. Subs O & P		Cost Per Labor-Hour	
Crew C-14C	Hr.	Daily	Hr.	Daily	Bare Costs	Incl. O&P
1 Carpenter Foreman (out)	$41.95	$335.60	$69.25	$554.00	$38.25	$63.24
6 Carpenters	39.95	1917.60	65.95	3165.60		
2 Rodmen (reinf.)	44.55	712.80	75.95	1215.20		
4 Laborers	31.60	1011.20	52.15	1668.80		
1 Cement Finisher	38.30	306.40	59.90	479.20		
1 Gas Engine Vibrator		26.00		28.60	.23	.26
112 L.H., Daily Totals		$4309.60		$7111.40	$38.48	$63.49

6 *Productivity: Daily Output (35.0)/ Labor-Hours (3.20)*

The "Daily Output" represents the typical number of units the designated crew will install in a normal 8-hour day. To find out the number of days the given crew would require to complete the installation, divide your quantity by the daily output. For example:

Quantity	÷	Daily Output	=	Duration
100 C.Y.	÷	35.0/ Crew Day	=	2.86 Crew Days

The "Labor-Hours" figure represents the number of labor-hours required to install one unit of work. To find out the number of labor-hours required for your particular task, multiply the quantity of the item times the number of labor-hours shown. For example:

Quantity	x	Productivity Rate	=	Duration
100 C.Y.	x	3.20 Labor-Hours/ C.Y.	=	320 Labor-Hours

7 *Unit (C.Y.)*

The abbreviated designation indicates the unit of measure upon which the price, production, and crew are based (C.Y. = Cubic Yard). For a complete listing of abbreviations, refer to the Abbreviations Listing in the Reference Section of this book.

8 *Bare Costs:*

Mat. (Bare Material Cost) (157)

The unit material cost is the "bare" material cost with no overhead or profit included. Costs shown reflect national average material prices for January of the current year and include delivery to the job site. No sales taxes are included.

Labor (122)

The unit labor cost is derived by multiplying bare labor-hour costs for Crew C-14C by labor-hour units. The bare labor-hour cost is found in the Crew Section under C-14C. (If a trade is listed, the hourly labor cost—the wage rate—is found on the inside back cover.)

Labor-Hour Cost Crew C-14C	x	Labor-Hour Units	=	Labor
$38.25	x	3.20	=	$122.00

Equip. (Equipment) (.74)

Costs for equipment included in each crew are listed in the Crew Listings in the Reference Section. Costs for small tools that are considered as essential everyday tools are covered in the "Overhead" column on the "Installing Contractor's Overhead and Profit" table that lists labor trades, base rates and markups. The unit equipment cost displayed here is derived by multiplying the bare equipment hourly cost in the Crew Listing by the unit Labor-Hours value.

Equipment Cost Crew C-14C	x	Labor-Hour Units	=	Equip.
$.23	x	3.20	=	$.74

Total (279.74)

The total of the bare costs is the arithmetic total of the three previous columns: mat., labor, and equip.

Material	+	Labor	+	Equip.	=	Total
$157	+	$122	+	$.74	=	$279.74

9 *Total Costs Including O&P*

This figure is the sum of the bare material cost plus 10% for profit; the bare labor cost plus total overhead and profit (per the inside back cover or, if a crew is listed, from the crew listings); and the bare equipment cost plus 10% for profit.

Material is Bare Material Cost + 10% = 157 + 15.70	=	$172.70
Labor for Crew C-14C = Labor-Hour Cost (63.24) x Labor-Hour Units (3.20)	=	$202.37
Equip. is Bare Equip. Cost + 10% = .74 + .07	=	$ .81
Total (Rounded)	=	$ 375

Estimating Tips

01 20 00 Price and Payment Procedures

- Allowances that should be added to estimates to cover contingencies and job conditions that are not included in the national average material and labor costs are shown in section 01 21.
- When estimating historic preservation projects (depending on the condition of the existing structure and the owner's requirements), a 15%-20% contingency or allowance is recommended, regardless of the stage of the drawings.

01 30 00 Administrative Requirements

- Before determining a final cost estimate, it is a good practice to review all the items listed in Subdivisions 01 31 and 01 32 to make final adjustments for items that may need customizing to specific job conditions.
- Requirements for initial and periodic submittals can represent a significant cost to the General Requirements of a job. Thoroughly check the submittal specifications when estimating a project to determine any costs that should be included.

01 40 00 Quality Requirements

- All projects will require some degree of Quality Control. This cost is not included in the unit cost of construction listed in each division. Depending upon the terms of the contract, the various costs of inspection and testing can be the responsibility of either the owner or the contractor. Be sure to include the required costs in your estimate.

01 50 00 Temporary Facilities and Controls

- Barricades, access roads, safety nets, scaffolding, security, and many more requirements for the execution of a safe project are elements of direct cost. These costs can easily be overlooked when preparing an estimate. When looking through the major classifications of this subdivision, determine which items apply to each division in your estimate.
- Construction Equipment Rental Costs can be found in the Reference Section in section 01 54 33. Operators' wages are not included in equipment rental costs.
- Equipment mobilization and demobilization costs are not included in equipment rental costs and must be considered separately in section 01 54 36.50.
- The cost of small tools provided by the installing contractor for his workers is covered in the "Overhead" column on the "Installing Contractor's Overhead and Profit" table that lists labor trades, base rates and markups and, therefore, is included in the "Total Incl. O&P" cost of any Unit Price line item. For those users who are constrained by contract terms to use only bare costs, there are two line items in section 01 54 39.70 for small tools as a percentage of bare labor cost. If some of those users are further constrained by contract terms to refrain from using line items from Division 1, you are advised to cover the cost of small tools within your coefficient or multiplier.

01 70 00 Execution and Closeout Requirements

- When preparing an estimate, thoroughly read the specifications to determine the requirements for Contract Closeout. Final cleaning, record documentation, operation and maintenance data, warranties and bonds, and spare parts and maintenance materials can all be elements of cost for the completion of a contract. Do not overlook these in your estimate.

Reference Numbers

Reference numbers are shown in shaded boxes at the beginning of some major classifications. These numbers refer to related items in the Reference Section. The reference information may be an estimating procedure, an alternate pricing method, or technical information.

Note: Not all subdivisions listed here necessarily appear in this publication.

01 11 Summary of Work

01 11 31 – Professional Consultants

01 11 31.10 Architectural Fees

	01 11 31.10 Architectural Fees	Crew	Daily Output	Labor-Hours	Unit	Material	Labor	Equipment	Total	Total Incl O&P
						2009 Bare Costs				
0010	ARCHITECTURAL FEES R011110-10									
0020	For new construction									
0060	Minimum				Project					4.90%
0090	Maximum									16%
0100	For alteration work, to $500,000, add to new construction fee									50%
0150	Over $500,000, add to new construction fee									25%
2000	For "Greening" of building G									3%

01 11 31.20 Construction Management Fees

	01 11 31.20 Construction Management Fees	Crew	Daily Output	Labor-Hours	Unit	Material	Labor	Equipment	Total	Total Incl O&P
0010	CONSTRUCTION MANAGEMENT FEES									
0060	For work to $100,000				Project					10%
0070	To $250,000									9%
0090	To $1,000,000									6%
0100	To $5,000,000									5%
0110	To $10,000,000									4%
0300	$50,000,000 job, minimum									2.50%
0350	Maximum									4%

01 11 31.30 Engineering Fees

	01 11 31.30 Engineering Fees	Crew	Daily Output	Labor-Hours	Unit	Material	Labor	Equipment	Total	Total Incl O&P
0010	ENGINEERING FEES R011110-30									
0020	Educational planning consultant, minimum				Project					.50%
0100	Maximum				"					2.50%
0200	Electrical, minimum				Contrct					4.10%
0300	Maximum									10.10%
0400	Elevator & conveying systems, minimum									2.50%
0500	Maximum									5%
0600	Food service & kitchen equipment, minimum									8%
0700	Maximum									12%
0800	Landscaping & site development, minimum									2.50%
0900	Maximum									6%
1000	Mechanical (plumbing & HVAC), minimum									4.10%
1100	Maximum									10.10%
1200	Structural, minimum				Project					1%
1300	Maximum				"					2.50%
4000	Consultant, using DOE software energy analysis, small bldg, min				SF Flr.					.25
4010	Maximum									.45
4020	Medium building, minimum									.15
4030	Maximum									.35
4040	Large building, minimum									.05
4050	Maximum									.25

01 11 31.50 Models

	01 11 31.50 Models	Crew	Daily Output	Labor-Hours	Unit	Material	Labor	Equipment	Total	Total Incl O&P
0010	MODELS									
0020	Cardboard & paper, 1 building, minimum				Ea.	700			700	770
0050	Maximum					1,600			1,600	1,750
0100	2 buildings, minimum					935			935	1,025
0150	Maximum					2,125			2,125	2,325
0200	Plexiglass and metal, basic layout				SF Flr.	.06			.06	.07
0210	Including equipment and personnel				"	.31			.31	.34
0300	Site plan layout, minimum				Ea.	1,350			1,350	1,475
0350	Maximum				"	2,250			2,250	2,475

01 11 31.75 Renderings

	01 11 31.75 Renderings	Crew	Daily Output	Labor-Hours	Unit	Material	Labor	Equipment	Total	Total Incl O&P
0010	**RENDERINGS** Color, matted, 20" x 30", eye level,									
0020	1 building, minimum				Ea.	1,950			1,950	2,150
0050	Average					2,775			2,775	3,075

01 11 Summary of Work

01 11 31 – Professional Consultants

01 11 31.75 Renderings		Crew	Daily Output	Labor-Hours	Unit	2009 Bare Costs Material	Labor	Equipment	Total	Total Incl O&P
0100	Maximum				Ea.	4,450			4,450	4,900
1000	5 buildings, minimum					3,900			3,900	4,300
1100	Maximum					7,800			7,800	8,575
2000	Aerial perspective, color, 1 building, minimum					2,775			2,775	3,075
2100	Maximum					7,800			7,800	8,575
3000	5 buildings, minimum					5,600			5,600	6,175
3100	Maximum				↓	11,100			11,100	12,200

01 21 Allowances

01 21 16 – Contingency Allowances

01 21 16.50 Contingencies

		Crew	Daily Output	Labor-Hours	Unit	Material	Labor	Equipment	Total	Total Incl O&P
0010	**CONTINGENCIES**, Add to estimate									
0020	Conceptual stage				Project					25%
0050	Schematic stage									20%
0100	Preliminary working drawing stage (Design Dev.)									15%
0150	Final working drawing stage				↓					8%

01 21 53 – Factors Allowance

01 21 53.50 Factors

		Crew	Daily Output	Labor-Hours	Unit	Material	Labor	Equipment	Total	Total Incl O&P
0010	**FACTORS** Cost adjustments R012153-10									
0100	Add to construction costs for particular job requirements									
0500	Cut & patch to match existing construction, add, minimum				Costs	2%	3%			
0550	Maximum					5%	9%			
0800	Dust protection, add, minimum					1%	2%			
0850	Maximum					4%	11%			
1100	Equipment usage curtailment, add, minimum					1%	1%			
1150	Maximum					3%	10%			
1400	Material handling & storage limitation, add, minimum					1%	1%			
1450	Maximum					6%	7%			
1700	Protection of existing work, add, minimum					2%	2%			
1750	Maximum					5%	7%			
2000	Shift work requirements, add, minimum						5%			
2050	Maximum						30%			
2300	Temporary shoring and bracing, add, minimum					2%	5%			
2350	Maximum					5%	12%			
2400	Work inside prisons and high security areas, add, minimum						30%			
2450	Maximum				↓		50%			

01 21 55 – Job Conditions Allowance

01 21 55.50 Job Conditions

		Crew	Daily Output	Labor-Hours	Unit	Material	Labor	Equipment	Total	Total Incl O&P
0010	**JOB CONDITIONS** Modifications to applicable									
0020	cost summaries									
0100	Economic conditions, favorable, deduct				Project				2%	2%
0200	Unfavorable, add								5%	5%
0300	Hoisting conditions, favorable, deduct								2%	2%
0400	Unfavorable, add								5%	5%
0500	General Contractor management, experienced, deduct								2%	2%
0600	Inexperienced, add								10%	10%
0700	Labor availability, surplus, deduct								1%	1%
0800	Shortage, add								10%	10%
0900	Material storage area, available, deduct								1%	1%
1000	Not available, add				↓				2%	2%

01 21 Allowances

01 21 55 – Job Conditions Allowance

01 21 55.50 Job Conditions

	01 21 55.50 Job Conditions	Crew	Daily Output	Labor-Hours	Unit	Material	Labor (2009 Bare Costs)	Equipment	Total	Total Incl O&P
1100	Subcontractor availability, surplus, deduct				Project				5%	5%
1200	Shortage, add								12%	12%
1300	Work space, available, deduct								2%	2%
1400	Not available, add								5%	5%

01 21 57 – Overtime Allowance

01 21 57.50 Overtime

	01 21 57.50 Overtime	Crew	Daily Output	Labor-Hours	Unit	Material	Labor	Equipment	Total	Total Incl O&P
0010	**OVERTIME** for early completion of projects or where R012909-90									
0020	labor shortages exist, add to usual labor, up to				Costs		100%			

01 21 61 – Cost Indexes

01 21 61.20 Historical Cost Indexes

	01 21 61.20 Historical Cost Indexes	Crew	Daily Output	Labor-Hours	Unit	Material	Labor	Equipment	Total	Total Incl O&P
0010	**HISTORICAL COST INDEXES** (See Reference Section)									

01 21 63 – Taxes

01 21 63.10 Taxes

	01 21 63.10 Taxes	Crew	Daily Output	Labor-Hours	Unit	Material	Labor	Equipment	Total	Total Incl O&P
0010	**TAXES** R012909-80									
0020	Sales tax, State, average				%	4.91%				
0050	Maximum R012909-85					7.25%				
0200	Social Security, on first $102,000 of wages						7.65%			
0300	Unemployment, combined Federal and State, minimum						.80%			
0350	Average						6.20%			
0400	Maximum						11.76%			

01 31 Project Management and Coordination

01 31 13 – Project Coordination

01 31 13.20 Field Personnel

	01 31 13.20 Field Personnel	Crew	Daily Output	Labor-Hours	Unit	Material	Labor	Equipment	Total	Total Incl O&P
0010	**FIELD PERSONNEL**									
0020	Clerk, average				Week		380		380	590
0100	Field engineer, minimum						895		895	1,375
0120	Average						1,165		1,165	1,800
0140	Maximum						1,350		1,350	2,100
0160	General purpose laborer, average						1,250		1,250	1,925
0180	Project manager, minimum						1,650		1,650	2,550
0200	Average						1,925		1,925	2,975
0220	Maximum						2,175		2,175	3,375
0240	Superintendent, minimum						1,600		1,600	2,475
0260	Average						1,775		1,775	2,750
0280	Maximum						2,025		2,025	3,125
0290	Timekeeper, average						1,040		1,040	1,600

01 31 13.30 Insurance

	01 31 13.30 Insurance	Crew	Daily Output	Labor-Hours	Unit	Material	Labor	Equipment	Total	Total Incl O&P
0010	**INSURANCE** R013113-40									
0020	Builders risk, standard, minimum				Job					.24%
0050	Maximum R013113-60									.64%
0200	All-risk type, minimum									.25%
0250	Maximum									.62%
0400	Contractor's equipment floater, minimum				Value					.50%
0450	Maximum				"					1.50%
0800	Workers' compensation & employer's liability, average									
0850	by trade, carpentry, general				Payroll		17.80%			
0900	Clerical						.58%			
0950	Concrete						14.58%			

01 31 Project Management and Coordination

01 31 13 – Project Coordination

01 31 13.30 Insurance		Crew	Daily Output	Labor-Hours	Unit	Material	2009 Bare Costs Labor	Equipment	Total	Total Incl O&P
1000	Electrical				Payroll		6.46%			
1050	Excavation						10.01%			
1100	Glazing						13.89%			
1150	Insulation						14.44%			
1200	Lathing						10.63%			
1250	Masonry						14.37%			
1300	Painting & decorating						12.49%			
1350	Pile driving						20.09%			
1400	Plastering						13.60%			
1450	Plumbing						7.84%			
1500	Roofing						31.18%			
1550	Sheet metal work (HVAC)						11.57%			
1600	Steel erection, structural						37.94%			
1650	Tile work, interior ceramic						9.13%			
1700	Waterproofing, brush or hand caulking						6.69%			
1800	Wrecking						34.13%			
2000	Range of 35 trades in 50 states, excl. wrecking, min.						2.09%			
2100	Average						15.50%			
2200	Maximum						168.17%			

01 31 13.40 Main Office Expense

		Crew	Daily Output	Labor-Hours	Unit	Material	Labor	Equipment	Total	Total Incl O&P
0010	**MAIN OFFICE EXPENSE** Average for General Contractors R013113-50									
0020	As a percentage of their annual volume									
0030	Annual volume to $300,000, minimum				% Vol.				20%	
0040	Maximum								30%	
0060	To $500,000, minimum								17%	
0070	Maximum								22%	
0080	To $1,000,000, minimum								16%	
0090	Maximum								19%	
0110	To $3,000,000, minimum								14%	
0120	Maximum								16%	
0130	To $5,000,000, minimum								8%	
0140	Maximum								10%	

01 31 13.50 General Contractor's Mark-Up

		Crew	Daily Output	Labor-Hours	Unit	Material	Labor	Equipment	Total	Total Incl O&P
0010	**GENERAL CONTRACTOR'S MARK-UP** on Change Orders									
0200	Extra work, by subcontractors, add				%					10%
0250	By General Contractor, add									15%
0400	Omitted work, by subcontractors, deduct all but									5%
0450	By General Contractor, deduct all but									7.50%
0600	Overtime work, by subcontractors, add									15%
0650	By General Contractor, add									10%

01 31 13.70 Overhead

		Crew	Daily Output	Labor-Hours	Unit	Material	Labor	Equipment	Total	Total Incl O&P
0010	**OVERHEAD** As a percent of installing contractors direct costs R013113-50									
0040	Includes an allowance for home office expenses, FICA,									
0060	Risk & public liability insur. and unemploy.; minimum				%				5%	
0080	Average								15%	
0100	Maximum								30%	
0120	With profit allowance, by size of project; under $50,000								40%	
0140	$50,000 to $100,000								35%	
0160	$100,000 to $500,000								25%	
0180	$500,000 to $1,000,000								20%	
0200	Mark-up on general contractors total incl. O&P for									
0220	Handling subcontract; minimum				%					5%

01 31 Project Management and Coordination

01 31 13 – Project Coordination

01 31 13.70 Overhead		Crew	Daily Output	Labor-Hours	Unit	Material	2009 Bare Costs Labor	Equipment	Total	Total Incl O&P
0240	Average				%					10%
0260	Maximum				↓					15%

01 31 13.90 Performance Bond

		Crew	Daily Output	Labor-Hours	Unit	Material	Labor	Equipment	Total	Total Incl O&P
0010	PERFORMANCE BOND R013113-80									
0020	For buildings, minimum				Job					.60%
0100	Maximum				"					2.50%

01 32 Construction Progress Documentation

01 32 13 – Scheduling of work

01 32 13.50 Scheduling

		Crew	Daily Output	Labor-Hours	Unit	Material	Labor	Equipment	Total	Total Incl O&P
0010	SCHEDULING									
0020	Critical path, as % of architectural fee, minimum				%					.50%
0100	Maximum				"					1%
0300	Computer-update, micro, no plots, minimum				Ea.				455	500
0400	Including plots, maximum				"				1,455	1,600
0600	Rule of thumb, CPM scheduling, small job ($10 Million)				Job					.05%
0650	Large job ($50 Million +)									.03%
0700	Including cost control, small job									.08%
0750	Large job				↓					.04%

01 32 33 – Photographic Documentation

01 32 33.50 Photographs

		Crew	Daily Output	Labor-Hours	Unit	Material	Labor	Equipment	Total	Total Incl O&P
0010	PHOTOGRAPHS									
0020	8" x 10", 4 shots, 2 prints ea., std. mounting				Set	475			475	520
0100	Hinged linen mounts					530			530	580
0200	8" x 10", 4 shots, 2 prints each, in color					415			415	460
0300	For I.D. slugs, add to all above					5.30			5.30	5.85
0500	Aerial photos, initial fly-over, 6 shots, 1 print ea., 8" x 10"					825			825	905
0550	11" x 14" prints					1,000			1,000	1,100
0600	16" x 20" prints					1,175			1,175	1,300
0700	For full color prints, add					40%				40%
0750	Add for traffic control area				↓	294			294	325
0900	For over 30 miles from airport, add per				Mile	5.30			5.30	5.85
1000	Vertical photography, 4 to 6 shots with									
1010	different scales, 1 print each				Set	1,100			1,100	1,200
1500	Time lapse equipment, camera and projector, buy					3,775			3,775	4,175
1550	Rent per month				↓	565			565	620
1700	Cameraman and film, including processing, B.&W.				Day	1,375			1,375	1,525
1720	Color				"	1,375			1,375	1,525

01 41 Regulatory Requirements

01 41 26 – Permits

01 41 26.50 Permits		Crew	Daily Output	Labor-Hours	Unit	2009 Bare Costs Material	Labor	Equipment	Total	Total Incl O&P
0010	**PERMITS**									
0020	Rule of thumb, most cities, minimum				Job					.50%
0100	Maximum				"					2%

01 45 Quality Control

01 45 23 – Testing and Inspecting Services

01 45 23.50 Testing

		Crew	Daily Output	Labor-Hours	Unit	Material	Labor	Equipment	Total	Total Incl O&P
0010	**TESTING** and Inspecting Services									
0015	For concrete building costing $1,000,000, minimum				Project				4,725	5,200
0020	Maximum								38,000	41,800
0050	Steel building, minimum								4,727	5,200
0070	Maximum								14,818	16,300
0100	For building costing, $10,000,000, minimum								30,091	33,100
0150	Maximum								48,182	53,000
0200	Asphalt testing, compressive strength Marshall stability, set of 3				Ea.				145	165
0220	Density, set of 3								86	95
0250	Extraction, individual tests on sample								136	150
0300	Penetration								41	45
0350	Mix design, 5 specimens								182	200
0360	Additional specimen								36	40
0400	Specific gravity								41	45
0420	Swell test								64	70
0450	Water effect and cohesion, set of 6								182	200
0470	Water effect and plastic flow								64	70
0600	Concrete testing, aggregates, abrasion, ASTM C 131								136	150
0650	Absorption, ASTM C 127								42	46
0800	Petrographic analysis, ASTM C 295								773	850
0900	Specific gravity, ASTM C 127								50	55
1000	Sieve analysis, washed, ASTM C 136								59	65
1050	Unwashed								59	65
1200	Sulfate soundness								114	125
1300	Weight per cubic foot								36	40
1500	Cement, physical tests, ASTM C 150								318	350
1600	Chemical tests, ASTM C 150								245	270
1800	Compressive test, cylinder, delivered to lab, ASTM C 39								12	13
1900	Picked up by lab, minimum								14	15
1950	Average								18	20
2000	Maximum								27	30
2200	Compressive strength, cores (not incl. drilling), ASTM C 42								36	40
2250	Core drilling, 4" diameter (plus technician)				Inch				23	25
2260	Technician for core drilling				Hr.				45	50
2300	Patching core holes				Ea.				22	24
2400	Drying shrinkage at 28 days								236	260
2500	Flexural test beams, ASTM C 78								59	65
2600	Mix design, one batch mix								259	285
2650	Added trial batches								120	132
2800	Modulus of elasticity, ASTM C 469								164	180
2900	Tensile test, cylinders, ASTM C 496								45	50
3000	Water-Cement ratio curve, 3 batches								141	155
3100	4 batches								186	205
3300	Masonry testing, absorption, per 5 brick, ASTM C 67								45	50

01 45 Quality Control

01 45 23 – Testing and Inspecting Services

01 45 23.50 Testing		Crew	Daily Output	Labor-Hours	Unit	2009 Bare Costs Material	Labor	Equipment	Total	Total Incl O&P
3350	Chemical resistance, per 2 brick				Ea.				50	55
3400	Compressive strength, per 5 brick, ASTM C 67								68	75
3420	Efflorescence, per 5 brick, ASTM C 67								68	75
3440	Imperviousness, per 5 brick								87	96
3470	Modulus of rupture, per 5 brick								86	95
3500	Moisture, block only								32	35
3550	Mortar, compressive strength, set of 3								23	25
4100	Reinforcing steel, bend test								55	61
4200	Tensile test, up to #8 bar								36	40
4220	#9 to #11 bar								41	45
4240	#14 bar and larger								64	70
4400	Soil testing, Atterberg limits, liquid and plastic limits								59	65
4510	Hydrometer analysis								109	120
4530	Specific gravity, ASTM D 354								44	48
4600	Sieve analysis, washed, ASTM D 422								55	60
4700	Unwashed, ASTM D 422								59	65
4710	Consolidation test (ASTM D2435), minimum								250	275
4715	Maximum								432	475
4720	Density and classification of undisturbed sample								73	80
4735	Soil density, nuclear method, ASTM D2922								35	38.67
4740	Sand cone method ASTM D1556								27	30.17
4750	Moisture content, ASTM D 2216								9	10
4780	Permeability test, double ring infiltrometer								500	550
4800	Permeability, var. or constant head, undist., ASTM D 2434								227	250
4850	Recompacted								250	275
4900	Proctor compaction, 4" standard mold, ASTM D 698								123	135
4950	6" modified mold								68	75
5100	Shear tests, triaxial, minimum								409	450
5150	Maximum								545	600
5300	Direct shear, minimum, ASTM D 3080								318	350
5350	Maximum								409	450
5550	Technician for inspection, per day, earthwork								210	231
5570	Concrete								245	270
5650	Bolting								268	295
5750	Roofing								244	256
5790	Welding								257	283
5820	Non-destructive metal testing, dye penetrant				Day				310	341
5840	Magnetic particle								310	341
5860	Radiography								450	495
5880	Ultrasonic								309	340
6000	Welding certification, minimum				Ea.				91	100
6100	Maximum				"				250	275
7000	Underground storage tank									
7500	Volumetric tightness test ,<=12,000 gal				Ea.				435	478
7510	<=30,000 gal				"				613	675
7600	Vadose zone (soil gas) sampling, 10-40 samples, min.				Day				1,364	1,500
7610	Maximum				"				2,273	2,500
7700	Ground water monitoring incl. drilling 3 wells, min.				Total				4,545	5,000
7710	Maximum				"				6,364	7,000
8000	X-ray concrete slabs				Ea.				182	200
9000	Thermographic testing, for bldg envelope heat loss, average 2,000 S.F. G				"					500

01 51 Temporary Utilities

01 51 13 – Temporary Electricity

01 51 13.80 Temporary Utilities		Crew	Daily Output	Labor-Hours	Unit	Material	2009 Bare Costs Labor	Equipment	Total	Total Incl O&P
0010	**TEMPORARY UTILITIES**									
0100	Heat, incl. fuel and operation, per week, 12 hrs. per day	1 Skwk	100	.080	CSF Flr	27	3.27		30.27	35
0200	24 hrs. per day	"	60	.133		52	5.45		57.45	66.50
0350	Lighting, incl. service lamps, wiring & outlets, minimum	1 Elec	34	.235		2.63	11.05		13.68	19.90
0360	Maximum	"	17	.471		5.70	22		27.70	40.50
0400	Power for temp lighting only, per month, min/month 6.6 KWH								.75	.83
0450	Maximum/month 23.6 KWH								2.85	3.14
0600	Power for job duration incl. elevator, etc., minimum								47	51.70
0650	Maximum				↓				110	121
0700	Temporary construction water bill per mo. average				Month	62			62	68
1000	Toilet, portable, see Equip. Rental 01 54 33 in Reference Section									

01 52 Construction Facilities

01 52 13 – Field Offices and Sheds

01 52 13.20 Office and Storage Space

		Crew	Daily Output	Labor-Hours	Unit	Material	Labor	Equipment	Total	Total Incl O&P
0010	**OFFICE AND STORAGE SPACE**									
0020	Trailer, furnished, no hookups, 20′ x 8′, buy	2 Skwk	1	16	Ea.	8,200	655		8,855	10,100
0250	Rent per month					163			163	179
0300	32′ x 8′, buy	2 Skwk	.70	22.857		12,200	935		13,135	15,000
0350	Rent per month					200			200	220
0400	50′ x 10′, buy	2 Skwk	.60	26.667		23,200	1,100		24,300	27,400
0450	Rent per month					281			281	310
0500	50′ x 12′, buy	2 Skwk	.50	32		27,900	1,300		29,200	32,800
0550	Rent per month					375			375	410
0700	For air conditioning, rent per month, add				↓	41			41	45
0800	For delivery, add per mile				Mile	4.50			4.50	4.95
1000	Portable buildings, prefab, on skids, economy, 8′ x 8′	2 Carp	265	.060	S.F.	85	2.41		87.41	97.50
1100	Deluxe, 8′ x 12′	"	150	.107	"	95	4.26		99.26	112
1200	Storage boxes, 20′ x 8′, buy	2 Skwk	1.80	8.889	Ea.	4,675	365		5,040	5,750
1250	Rent per month					72			72	79
1300	40′ x 8′, buy	2 Skwk	1.40	11.429		6,400	465		6,865	7,800
1350	Rent per month				↓	99			99	109
5000	Air supported structures, see Div. 13 31 13.13									

01 52 13.40 Field Office Expense

		Crew	Daily Output	Labor-Hours	Unit	Material	Labor	Equipment	Total	Total Incl O&P
0010	**FIELD OFFICE EXPENSE**									
0100	Office equipment rental average				Month	155			155	171
0120	Office supplies, average				"	85			85	93.50
0125	Office trailer rental, see Div. 01 52 13.20									
0140	Telephone bill; avg. bill/month incl. long dist.				Month	80			80	88
0160	Lights & HVAC				"	150			150	165

01 54 Construction Aids

01 54 09 – Protection Equipment

01 54 09.50 Personnel Protective Equipment		Crew	Daily Output	Labor-Hours	Unit	2009 Bare Costs Material	Labor	Equipment	Total	Total Incl O&P
0010	**PERSONNEL PROTECTIVE EQUIPMENT**									
0015	Hazardous waste protection									
0020	Respirator mask only, full face, silicone				Ea.	223			223	245
0030	Half face, silicone					33			33	36.50
0040	Respirator cartridges, 2 req'd/mask, dust or asbestos					5.30			5.30	5.85
0050	Chemical vapor					4.69			4.69	5.15
0060	Combination vapor and dust					9.70			9.70	10.65
0100	Emergency escape breathing apparatus, 5 min					465			465	510
0110	10 min					500			500	550
0150	Self contained breathing apparatus with full face piece, 30 min					1,750			1,750	1,925
0160	60 min					2,925			2,925	3,225
0200	Encapsulating suits, limited use, level A					905			905	995
0210	Level B					270			270	297
0300	Over boots, latex				Pr.	6.35			6.35	7
0310	PVC					21.50			21.50	24
0320	Neoprene					41.50			41.50	46
0400	Gloves, nitrile/PVC					21			21	23.50
0410	Neoprene coated					24			24	26.50

01 54 09.60 Safety Nets

		Crew	Daily Output	Labor-Hours	Unit	Material	Labor	Equipment	Total	Total Incl O&P
0010	**SAFETY NETS**									
0020	No supports, stock sizes, nylon, 4" mesh				S.F.	1.10			1.10	1.21
0100	Polypropylene, 6" mesh					1.59			1.59	1.75
0200	Small mesh debris nets, 1/4" & 3/4" mesh, stock sizes					.74			.74	.81
0220	Combined 4" mesh and 1/4" mesh, stock sizes					2.05			2.05	2.26
0300	Monthly rental, 4" mesh, stock sizes, 1st month					.50			.50	.55
0320	2nd month rental					.25			.25	.28
0340	Maximum rental/year					1.15			1.15	1.27

01 54 16 – Temporary Hoists

01 54 16.50 Weekly Forklift Crew

		Crew	Daily Output	Labor-Hours	Unit	Material	Labor	Equipment	Total	Total Incl O&P
0010	**WEEKLY FORKLIFT CREW**									
0100	All-terrain forklift, 45' lift, 35' reach, 9000 lb. capacity	A-3P	.20	40	Month		1,550	2,225	3,775	4,900

01 54 19 – Temporary Cranes

01 54 19.50 Daily Crane Crews

		Crew	Daily Output	Labor-Hours	Unit	Material	Labor	Equipment	Total	Total Incl O&P
0010	**DAILY CRANE CREWS** for small jobs, portal to portal									
0100	12-ton truck-mounted hydraulic crane	A-3H	1	8	Day		340	1,025	1,365	1,650
0200	25-ton	A-3I	1	8			340	1,050	1,390	1,675
0300	40-ton	A-3J	1	8			340	1,050	1,390	1,700
0400	55-ton	A-3K	1	16			635	1,625	2,260	2,800
0500	80-ton	A-3L	1	16			635	1,850	2,485	3,025
0600	100-ton	A-3M	1	16			635	2,525	3,160	3,800

01 54 19.60 Monthly Tower Crane Crew

		Crew	Daily Output	Labor-Hours	Unit	Material	Labor	Equipment	Total	Total Incl O&P
0010	**MONTHLY TOWER CRANE CREW**, excludes concrete footing									
0100	Static tower crane, 130' high, 106' jib, 6200 lb. capaccity	A-3N	.05	176	Month		7,500	22,500	30,000	36,500

01 54 23 – Temporary Scaffolding and Platforms

01 54 23.60 Pump Staging

		Crew	Daily Output	Labor-Hours	Unit	Material	Labor	Equipment	Total	Total Incl O&P
0010	**PUMP STAGING**, Aluminum R015423-20									
0200	24' long pole section, buy				Ea.	415			415	460
0300	18' long pole section, buy					325			325	355
0400	12' long pole section, buy					218			218	240
0500	6' long pole section, buy					115			115	126

01 54 23.60 Pump Staging		Crew	Daily Output	Labor-Hours	Unit	2009 Bare Costs Material	Labor	Equipment	Total	Total Incl O&P
0600	6′ long splice joint section, buy				Ea.	85.50			85.50	94
0700	Pump jack, buy					139			139	153
0900	Foldable brace, buy					56.50			56.50	62.50
1000	Workbench/back safety rail support, buy					73.50			73.50	81
1100	Scaffolding planks/workbench, 14″ wide x 24′ long, buy					680			680	750
1200	Plank end safety rail, buy					370			370	405
1250	Safety net, 22′ long, buy					335			335	365
1300	System in place, 50′ working height, per use based on 50 uses	2 Carp	84.80	.189	C.S.F.	6.40	7.55		13.95	19.50
1400	100 uses		84.80	.189		3.21	7.55		10.76	16
1500	150 uses		84.80	.189		2.15	7.55		9.70	14.80

01 54 23.70 Scaffolding		Crew	Daily Output	Labor-Hours	Unit	Material	Labor	Equipment	Total	Total Incl O&P
0010	**SCAFFOLDING** R015423-10									
0015	Steel tube, regular, no plank, labor only to erect & dismantle									
0090	Building exterior, wall face, 1 to 5 stories, 6′-4″ x 5′ frames	3 Carp	8	3	C.S.F.		120		120	198
0200	6 to 12 stories	4 Carp	8	4			160		160	264
0301	13 to 20 stories	5 Clab	8	5			158		158	261
0460	Building interior, wall face area, up to 16′ high	3 Carp	12	2			80		80	132
0560	16′ to 40′ high		10	2.400			96		96	158
0800	Building interior floor area, up to 30′ high		150	.160	C.C.F.		6.40		6.40	10.55
0900	Over 30′ high	4 Carp	160	.200	″		8		8	13.20
0906	Complete system for face of walls, no plank, material only rent/mo				C.S.F.	35.50			35.50	39
0908	Interior spaces, no plank, material only rent/mo				C.C.F.	3.40			3.40	3.74
0910	Steel tubular, heavy duty shoring, buy									
0920	Frames 5′ high 2′ wide				Ea.	93			93	102
0925	5′ high 4′ wide					105			105	116
0930	6′ high 2′ wide					106			106	117
0935	6′ high 4′ wide					124			124	136
0940	Accessories									
0945	Cross braces				Ea.	18			18	19.80
0950	U-head, 8″ x 8″					21.50			21.50	24
0955	J-head, 4″ x 8″					15.80			15.80	17.40
0960	Base plate, 8″ x 8″					17.60			17.60	19.35
0965	Leveling jack					38			38	41.50
1000	Steel tubular, regular, buy									
1100	Frames 3′ high 5′ wide				Ea.	71			71	78
1150	5′ high 5′ wide					83			83	91.50
1200	6′-4″ high 5′ wide					104			104	114
1350	7′-6″ high 6′ wide					179			179	197
1500	Accessories cross braces					20.50			20.50	22.50
1550	Guardrail post					18.60			18.60	20.50
1600	Guardrail 7′ section					9.90			9.90	10.90
1650	Screw jacks & plates					27			27	29.50
1700	Sidearm brackets					43			43	47.50
1750	8″ casters					35.50			35.50	39
1800	Plank 2″ x 10″ x 16′-0″					51			51	56
1900	Stairway section					325			325	360
1910	Stairway starter bar					36.50			36.50	40
1920	Stairway inside handrail					66			66	72.50
1930	Stairway outside handrail					98			98	108
1940	Walk-thru frame guardrail					47			47	51.50
2000	Steel tubular, regular, rent/mo.									
2100	Frames 3′ high 5′ wide				Ea.	5			5	5.50

01 54 23 – Temporary Scaffolding and Platforms

01 54 23.70 Scaffolding		Crew	Daily Output	Labor-Hours	Unit	2009 Bare Costs Material	Labor	Equipment	Total	Total Incl O&P
2150	5′ high 5′ wide				Ea.	5			5	5.50
2200	6′-4″ high 5′ wide					5.05			5.05	5.55
2250	7′-6″ high 6′ wide					7			7	7.70
2500	Accessories, cross braces					1			1	1.10
2550	Guardrail post					1			1	1.10
2600	Guardrail 7′ section					1			1	1.10
2650	Screw jacks & plates					2			2	2.20
2700	Sidearm brackets					2			2	2.20
2750	8″ casters					8			8	8.80
2800	Outrigger for rolling tower					3			3	3.30
2850	Plank 2″ x 10″ x 16′-0″					6			6	6.60
2900	Stairway section					40			40	44
2940	Walk-thru frame guardrail				↓	2.50			2.50	2.75
3000	Steel tubular, heavy duty shoring, rent/mo.									
3250	5′ high 2′ & 4′ wide				Ea.	5			5	5.50
3300	6′ high 2′ & 4′ wide					5			5	5.50
3500	Accessories, cross braces					1			1	1.10
3600	U - head, 8″ x 8″					1			1	1.10
3650	J - head, 4″ x 8″					1			1	1.10
3700	Base plate, 8″ x 8″					1			1	1.10
3750	Leveling jack				↓	2			2	2.20
4000	Scaffolding, stl tubular, reg, no plank, labor only to erect & dismantle									
4100	Building exterior 2 stories	3 Carp	8	3	C.S.F.		120		120	198
4150	4 stories	″	8	3			120		120	198
4200	6 stories	4 Carp	8	4			160		160	264
4250	8 stories		8	4			160		160	264
4300	10 stories		8	4			160		160	264
4350	12 stories	↓	8	4	↓		160		160	264
5700	Planks, 2x10x16′-0″, labor only to erect & remove to 50′ H	3 Carp	72	.333	Ea.		13.30		13.30	22
5800	Over 50′ high	4 Carp	80	.400	″		16		16	26.50
6000	Heavy duty shoring for elevated slab forms to 8′-2″ high, floor area									
6100	Labor only to erect & dismantle	4 Carp	16	2	C.S.F.		80		80	132
6110	Materials only, rent.mo				″	29.50			29.50	32.50
6500	To 14′-8″ high									
6600	Labor only to erect & dismantle	4 Carp	10	3.200	C.S.F.		128		128	211
6610	Materials only, rent/mo				″	43			43	47.50

01 54 23.75 Scaffolding Specialties		Crew	Daily Output	Labor-Hours	Unit	Material	Labor	Equipment	Total	Total Incl O&P
0010	**SCAFFOLDING SPECIALTIES**									
1200	Sidewalk bridge, heavy duty steel posts & beams, including									
1210	parapet protection & waterproofing									
1220	8′ to 10′ wide, 2 posts	3 Carp	15	1.600	L.F.	35	64		99	145
1230	3 posts	″	10	2.400	″	54	96		150	218
1500	Sidewalk bridge using tubular steel									
1510	scaffold frames, including planking	3 Carp	45	.533	L.F.	5.60	21.50		27.10	41
1600	For 2 uses per month, deduct from all above					50%				
1700	For 1 use every 2 months, add to all above					100%				
1900	Catwalks, 20″ wide, no guardrails, 7′ span, buy				Ea.	145			145	160
2000	10′ span, buy					203			203	223
3720	Putlog, standard, 8′ span, with hangers, buy					75			75	82.50
3730	Rent per month					10			10	11
3750	12′ span, buy					113			113	124
3755	Rent per month				↓	15			15	16.50

01 54 Construction Aids

01 54 23 – Temporary Scaffolding and Platforms

	01 54 23.75 Scaffolding Specialties	Crew	Daily Output	Labor-Hours	Unit	Material	2009 Bare Costs Labor	Equipment	Total	Total Incl O&P
3760	Trussed type, 16′ span, buy				Ea.	260			260	286
3770	Rent per month					20			20	22
3790	22′ span, buy					310			310	345
3795	Rent per month					30			30	33
3800	Rolling ladders with handrails, 30″ wide, buy, 2 step					214			214	235
4000	7 step					655			655	720
4050	10 step					910			910	1,000
4100	Rolling towers, buy, 5′ wide, 7′ long, 10′ high					1,350			1,350	1,500
4200	For 5′ high added sections, to buy, add					207			207	228
4300	Complete incl. wheels, railings, outriggers,									
4350	21′ high, to buy				Ea.	2,250			2,250	2,475
4400	Rent/month = 5% of purchase cost				″	7.50			7.50	8.25

	01 54 23.80 Staging Aids	Crew	Daily Output	Labor-Hours	Unit	Material	Labor	Equipment	Total	Total Incl O&P
0010	**STAGING AIDS** and fall protection equipment									
0100	Sidewall staging bracket, tubular, buy				Ea.	43			43	47.50
0110	Cost each per day, based on 250 days use				Day	.17			.17	.19
0200	Guard post, buy				Ea.	22.50			22.50	25
0210	Cost each per day, based on 250 days use				Day	.09			.09	.10
0300	End guard chains, buy per pair				Pair	31			31	34
0310	Cost per set per day, based on 250 days use				Day	.17			.17	.19
1000	Roof shingling bracket, steel, buy				Ea.	9.35			9.35	10.25
1010	Cost each per day, based on 250 days use				Day	.04			.04	.04
1100	Wood bracket, buy				Ea.	16.60			16.60	18.25
1110	Cost each per day, based on 250 days use				Day	.07			.07	.07
2000	Ladder jack, aluminum, buy per pair				Pair	108			108	119
2010	Cost per pair per day, based on 250 days use				Day	.43			.43	.48
2100	Steel siderail jack, buy per pair				Pair	122			122	134
2110	Cost per pair per day, based on 250 days use				Day	.49			.49	.54
3000	Laminated wood plank, 2″x10″x16′, buy				Ea.	51			51	56
3010	Cost each per day, based on 250 days use				Day	.20			.20	.22
3100	Aluminum scaffolding plank, 20″ wide x 24′ long, buy				Ea.	890			890	975
3110	Cost each per day, based on 250 days use				Day	3.55			3.55	3.91
4000	Nylon full body harness, lanyard and rope grab				Ea.	215			215	236
4010	Cost each per day, based on 250 days use				Day	.86			.86	.95
4100	Rope for safety line, 5/8″ x 100′ nylon, buy				Ea.	70			70	77
4110	Cost each per day, based on 250 days use				Day	.28			.28	.31
4200	Permanent U-Bolt roof anchor, buy				Ea.	35			35	38.50
4300	Temporary (one use) roof ridge anchor, buy				″	28.50			28.50	31
5000	Installation (setup and removal) of staging aids									
5010	Sidewall staging bracket	2 Carp	64	.250	Ea.		10		10	16.50
5020	Guard post with 2 wood rails	″	64	.250			10		10	16.50
5030	End guard chains, set	1 Carp	64	.125			4.99		4.99	8.25
5100	Roof shingling bracket		96	.083			3.33		3.33	5.50
5200	Ladder jack		64	.125			4.99		4.99	8.25
5300	Wood plank, 2x10x16′	2 Carp	80	.200			8		8	13.20
5310	Aluminum scaffold plank, 20″ x 24′	″	40	.400			16		16	26.50
5410	Safety rope	1 Carp	40	.200			8		8	13.20
5420	Permanent U-Bolt roof anchor (install only)	2 Carp	40	.400			16		16	26.50
5430	Temporary roof ridge anchor (install only)	1 Carp	64	.125			4.99		4.99	8.25

01 54 26 – Temporary Swing Staging

	01 54 26.50 Swing Staging	Crew	Daily Output	Labor-Hours	Unit	Material	Labor	Equipment	Total	Total Incl O&P
0010	**SWING STAGING**, 500 lb cap., 2′ wide to 24′ long, hand operated									

01 54 Construction Aids

01 54 26 – Temporary Swing Staging

01 54 26.50 Swing Staging		Crew	Daily Output	Labor-Hours	Unit	Material	2009 Bare Costs Labor	Equipment	Total	Total Incl O&P
0020	steel cable type, with 60′ cables, buy				Ea.	4,050			4,050	4,450
0030	Rent per month				"	405			405	445
0600	Lightweight (not for masons) 24′ long for 150′ height,									
0610	manual type, buy				Ea.	4,150			4,150	4,575
0620	Rent per month					415			415	455
0700	Powered, electric or air, to 150′ high, buy					13,300			13,300	14,700
0710	Rent per month					935			935	1,025
0780	To 300′ high, buy					13,500			13,500	14,800
0800	Rent per month					945			945	1,025
1000	Bosun's chair or work basket 3′ x 3.5′, to 300′ high, electric, buy					8,625			8,625	9,500
1010	Rent per month				↓	605			605	665
2200	Move swing staging (setup and remove)	E-4	2	16	Move		725	67	792	1,425

01 54 36 – Equipment Mobilization

01 54 36.50 Mobilization or Demob.		Crew	Daily Output	Labor-Hours	Unit	Material	Labor	Equipment	Total	Total Incl O&P
0010	**MOBILIZATION OR DEMOB.** (One or the other, unless noted) R015433-10									
0015	Up to 25 mi haul dist (50 mi RT for mob/demob crew)									
0020	Dozer, loader, backhoe, excav., grader, paver, roller, 70 to 150 H.P.	B-34N	4	2	Ea.		64	117	181	233
0100	Above 150 HP	B-34K	3	2.667			85	261	346	430
0300	Scraper, towed type (incl. tractor), 6 C.Y. capacity		3	2.667			85	261	346	430
0400	10 C.Y.		2.50	3.200			102	315	417	515
0600	Self-propelled scraper, 15 C.Y.		2.50	3.200			102	315	417	515
0700	24 C.Y.		2	4			128	390	518	640
0900	Shovel or dragline, 3/4 C.Y.		3.60	2.222			71	218	289	355
1000	1-1/2 C.Y.	↓	3	2.667			85	261	346	430
1100	Small equipment, placed in rear of, or towed by pickup truck	A-3A	8	1			31	16.05	47.05	68.50
1150	Equip up to 70 HP, on flatbed trailer behind pickup truck	A-3D	4	2			62	57.50	119.50	166
2000	Mob & demob truck-mounted crane up to 75 ton, driver only	1 Eqhv	3.60	2.222			94.50		94.50	149
2100	Crane, truck-mounted, over 75 ton	A-3E	2.50	6.400			238	51.50	289.50	435
2200	Crawler-mounted, up to 75 ton	A-3F	2	8			298	405	703	920
2300	Over 75 ton	A-3G	1.50	10.667	↓		395	610	1,005	1,300
2500	For each additional 5 miles haul distance, add						10%	10%		
3000	For large pieces of equipment, allow for assembly/knockdown									
3001	For mob/demob of vibrofloatation equip, see Div. 31 45 13.10									
3100	For mob/demob of micro-tunneling equip, see Div. 33 05 23.19									
3200	For mob/demob of pile driving equip, see Div. 31 62 19.10									
3300	For mob/demob of caisson drilling equip, see Div. 31 63 26.13									

01 54 39 – Construction Equipment

01 54 39.70 Small Tools		Crew	Daily Output	Labor-Hours	Unit	Material	Labor	Equipment	Total	Total Incl O&P
0010	**SMALL TOOLS** R013113-50									
0020	As % of contractor's bare labor cost for project, minimum				Total		.50%			
0100	Maximum				"		2%			

01 55 Vehicular Access and Parking

01 55 23 – Temporary Roads

01 55 23.50 Roads and Sidewalks		Crew	Daily Output	Labor-Hours	Unit	2009 Bare Costs Material	Labor	Equipment	Total	Total Incl O&P
0010	**ROADS AND SIDEWALKS** Temporary									
0050	Roads, gravel fill, no surfacing, 4" gravel depth	B-14	715	.067	S.Y.	4.97	2.23	.41	7.61	9.55
0100	8" gravel depth	"	615	.073	"	9.95	2.59	.48	13.02	15.70
1000	Ramp, 3/4" plywood on 2" x 6" joists, 16" O.C.	2 Carp	300	.053	S.F.	1.35	2.13		3.48	5
1100	On 2" x 10" joists, 16" O.C.	"	275	.053	"	1.82	2.32		4.14	5.85

01 56 Temporary Barriers and Enclosures

01 56 13 – Temporary Air Barriers

01 56 13.60 Tarpaulins

		Crew	Daily Output	Labor-Hours	Unit	Material	Labor	Equipment	Total	Total Incl O&P
0010	**TARPAULINS**									
0020	Cotton duck, 10 oz. to 13.13 oz. per S.Y., minimum				S.F.	.54			.54	.59
0050	Maximum					.53			.53	.58
0100	Polyvinyl coated nylon, 14 oz. to 18 oz., minimum					.48			.48	.53
0150	Maximum					.68			.68	.75
0200	Reinforced polyethylene 3 mils thick, white					.15			.15	.17
0300	4 mils thick, white, clear or black					.20			.20	.22
0400	5.5 mils thick, clear					.21			.21	.23
0500	White, fire retardant					.35			.35	.39
0600	7.5 mils, oil resistant, fire retardant					.40			.40	.44
0700	8.5 mils, black					.53			.53	.58
0710	Woven polyethylene, 6 mils thick					.35			.35	.39
0720	Steel reinforced polyethylene, 4 mils thick					.53			.53	.58
0730	Polyester reinforced w/ integral fastening system 11 mils thick					1.07			1.07	1.18
0740	Mylar polyester, non-reinforced, 7 mils thick				↓	1.17			1.17	1.29

01 56 13.90 Winter Protection

		Crew	Daily Output	Labor-Hours	Unit	Material	Labor	Equipment	Total	Total Incl O&P
0010	**WINTER PROTECTION**									
0100	Framing to close openings	2 Clab	750	.021	S.F.	.39	.67		1.06	1.54
0200	Tarpaulins hung over scaffolding, 8 uses, not incl. scaffolding		1500	.011		.25	.34		.59	.84
0250	Tarpaulin polyester reinf. w/ integral fastening system 11 mils thick		1600	.010		.80	.32		1.12	1.40
0300	Prefab fiberglass panels, steel frame, 8 uses	↓	1200	.013	↓	.85	.42		1.27	1.64

01 56 23 – Temporary Barricades

01 56 23.10 Barricades

		Crew	Daily Output	Labor-Hours	Unit	Material	Labor	Equipment	Total	Total Incl O&P
0010	**BARRICADES**									
0020	5' high, 3 rail @ 2" x 8", fixed	2 Carp	20	.800	L.F.	4.99	32		36.99	58.50
0150	Movable	"	30	.533	"	4.12	21.50		25.62	39.50
0300	Stock units, 6' high, 8' wide, plain, buy				Ea.	435			435	480
0350	With reflective tape, buy				"	525			525	580
0400	Break-a-way 3" PVC pipe barricade									
0410	with 3 ea. 1' x 4' reflectorized panels, buy				Ea.	305			305	335
0500	Plywood with steel legs, 32" wide					72			72	79
0600	Telescoping Christmas tree, 9' high, 5 flags, buy					122			122	134
0800	Traffic cones, PVC, 18" high					8.40			8.40	9.25
0850	28" high				↓	14.20			14.20	15.65
1000	Guardrail, wooden, 3' high, 1" x 6", on 2" x 4" posts	2 Carp	200	.080	L.F.	1.02	3.20		4.22	6.40
1100	2" x 6", on 4" x 4" posts	"	165	.097		2.16	3.87		6.03	8.75
1200	Portable metal with base pads, buy					19.75			19.75	21.50
1250	Typical installation, assume 10 reuses	2 Carp	600	.027	↓	2.50	1.07		3.57	4.51
1300	Barricade tape, polyethylene, 7 mil, 3" wide x 500' long roll				Ea.	25			25	27.50
5000	Barricades, see Div. 01 54 33.40									

01 56 Temporary Barriers and Enclosures

01 56 26 – Temporary Fencing

01 56 26.50 Temporary Fencing		Crew	Daily Output	Labor-Hours	Unit	2009 Bare Costs Material	Labor	Equipment	Total	Total Incl O&P
0010	TEMPORARY FENCING									
0020	Chain link, 11 ga, 5′ high	2 Clab	400	.040	L.F.	7.25	1.26		8.51	10.10
0100	6′ high		300	.053		7.75	1.69		9.44	11.35
0200	Rented chain link, 6′ high, to 1000′ (up to 12 mo.)		400	.040		2.69	1.26		3.95	5.05
0250	Over 1000′ (up to 12 mo.)		300	.053		2.59	1.69		4.28	5.65
0350	Plywood, painted, 2″ x 4″ frame, 4′ high	A-4	135	.178		5.05	6.80		11.85	16.70
0400	4″ x 4″ frame, 8′ high	″	110	.218		9.85	8.35		18.20	24.50
0500	Wire mesh on 4″ x 4″ posts, 4′ high	2 Carp	100	.160		9.85	6.40		16.25	21.50
0550	8′ high	″	80	.200		15.05	8		23.05	30

01 56 29 – Temporary Protective Walkways

01 56 29.50 Protection

		Crew	Daily Output	Labor-Hours	Unit	Material	Labor	Equipment	Total	Total Incl O&P
0010	PROTECTION									
0020	Stair tread, 2″ x 12″ planks, 1 use	1 Carp	75	.107	Tread	4.16	4.26		8.42	11.60
0100	Exterior plywood, 1/2″ thick, 1 use		65	.123		1.39	4.92		6.31	9.60
0200	3/4″ thick, 1 use		60	.133		2.49	5.35		7.84	11.55
2200	Sidewalks, 2″ x 12″ planks, 2 uses		350	.023	S.F.	.69	.91		1.60	2.27
2300	Exterior plywood, 2 uses, 1/2″ thick		750	.011		.23	.43		.66	.95
2500	3/4″ thick		600	.013		.42	.53		.95	1.34

01 56 32 – Temporary Security

01 56 32.50 Watchman

		Crew	Daily Output	Labor-Hours	Unit	Material	Labor	Equipment	Total	Total Incl O&P
0010	WATCHMAN									
0020	Service, monthly basis, uniformed person, minimum				Hr.				25	27.50
0100	Maximum								45.45	50
0200	Person and command dog, minimum								31	34
0300	Maximum								54.55	60
0500	Sentry dog, leased, with job patrol (yard dog), 1 dog				Week				290	319
0600	2 dogs				″				390	429
0800	Purchase, trained sentry dog, minimum				Ea.				1,364	1,500
0900	Maximum				″				2,727	3,000

01 58 Project Identification

01 58 13 – Temporary Project Signage

01 58 13.50 Signs

		Crew	Daily Output	Labor-Hours	Unit	Material	Labor	Equipment	Total	Total Incl O&P
0010	SIGNS									
0020	High intensity reflectorized, no posts, buy				S.F.	21			21	23

01 71 Examination and Preparation

01 71 23 – Field Engineering

01 71 23.13 Construction Layout

		Crew	Daily Output	Labor-Hours	Unit	Material	Labor	Equipment	Total	Total Incl O&P
0010	CONSTRUCTION LAYOUT									
1100	Crew for layout of building, trenching or pipe laying, 2 person crew	A-6	1	16	Day		635	70	705	1,100
1200	3 person crew	A-7	1	24			1,025	70	1,095	1,750
1400	Crew for roadway layout, 4 person crew	A-8	1	32			1,350	70	1,420	2,250

01 71 23.19 Surveyor Stakes

		Crew	Daily Output	Labor-Hours	Unit	Material	Labor	Equipment	Total	Total Incl O&P
0010	SURVEYOR STAKES									
0020	Hardwood, 1″ x 1″ x 48″ long				C	58			58	64
0100	2″ x 2″ x 18″ long					66			66	72.50

01 71 Examination and Preparation

01 71 23 – Field Engineering

01 71 23.19 Surveyor Stakes		Crew	Daily Output	Labor-Hours	Unit	2009 Bare Costs Material	Labor	Equipment	Total	Total Incl O&P
0150	2″ x 2″ x 24″ long				C	110			110	121
0200	2″ x 2″ x 30″ long				↓	74			74	81.50

01 74 Cleaning and Waste Management

01 74 13 – Progress Cleaning

01 74 13.20 Cleaning Up

		Crew	Daily Output	Labor-Hours	Unit	Material	Labor	Equipment	Total	Total Incl O&P
0010	**CLEANING UP**									
0020	After job completion, allow, minimum				Job					.30%
0040	Maximum				″					1%
0042	Rubbish removal, see Div. 02 41 19.23									
0052	Cleanup of floor area, continuous, per day, during const.	A-5	16	1.125	M.S.F.	1.70	35.50	3.04	40.24	63.50
0100	Final by GC at end of job	″	11.50	1.565	″	2.71	49.50	4.23	56.44	89

01 91 Commissioning

01 91 13 – General Commissioning Requirements

01 91 13.50 Commissioning

		Crew	Daily Output	Labor-Hours	Unit	Material	Labor	Equipment	Total	Total Incl O&P
0010	**COMMISSIONING** Including documentation of design intent									
0100	performance verification, O&M, training, minimum				Project					.50%
0150	Maximum				″					.75%

01 93 Facility Maintenance

01 93 08 – Facility Maintenance Equipment

01 93 08.50 Equipment

		Crew	Daily Output	Labor-Hours	Unit	Material	Labor	Equipment	Total	Total Incl O&P
0010	**EQUIPMENT**, Purchase									
0090	Carpet care equipment									
0110	Dual motor vac, 1 H.P., 16″ brush				Ea.	825			825	910
0120	Upright vacuum, 12″ brush					310			310	345
0130	14″ brush					375			375	415
0140	Soil extractor, hot water 5′ wand, 12″ head					1,975			1,975	2,175
0150	Dry foam, 13″ brush					1,475			1,475	1,625
0160	24″ brush				↓	2,075			2,075	2,300
0240	Floor care equipment									
0260	Polishing, buffing, waxing machine, 175 RPM									
0270	.33 H.P., 11″ dia. brush				Ea.	565			565	620
0280	1 H.P., 16″ dia. brush					970			970	1,075
0290	18″ dia. brush					1,025			1,025	1,125
0300	1.5 H.P., 20″ dia. brush					1,275			1,275	1,400
0310	24″ dia. brush				↓	1,125			1,125	1,250
0330	Scrubber, automatic, 2 stage 1 H.P. vacuum motor									
0340	20″ dia. brush				Ea.	6,225			6,225	6,850
0350	28″ dia. brush				″	9,425			9,425	10,400
1200	Plumbing maintenance equipment									
1220	Kinetic water ram				Ea.	203			203	223
1240	Cable pipe snake, 104 ft., self feed, electric, .5 H.P.				″	2,050			2,050	2,275
1500	Specialty equipment									
1550	Litterbuggy, 8 H.P. gasoline				Ea.	2,500			2,500	2,750
1560	Propane				↓	2,675			2,675	2,925

01 93 Facility Maintenance

01 93 08 – Facility Maintenance Equipment

01 93 08.50 Equipment		Crew	Daily Output	Labor-Hours	Unit	2009 Bare Costs Material	Labor	Equipment	Total	Total Incl O&P
1570	Pressure cleaner, hot water, minimum				Ea.	2,950			2,950	3,250
1580	Maximum					5,350			5,350	5,875
1590	Vacuum cleaners, steel canister									
1600	Dry only, two stage, 1 H.P.				Ea.	305			305	335
1610	Wet/dry, two stage, 2 H.P.					400			400	440
1620	4 H.P.					690			690	755
1630	Wet/dry, three stage, 1.5 H.P.					340			340	375
1670	Squeegee wet only, 2 stage, 1 H.P.					765			765	840
1680	2 H.P.					1,425			1,425	1,550
1690	Asbestos and hazardous waste dry vacs, 1 H.P.					1,800			1,800	1,975
1700	2 H.P.					1,850			1,850	2,025
1710	Three stage, 1 H.P.					2,175			2,175	2,400
1800	Upholstery soil extractor, dry-foam					2,350			2,350	2,575
1900	Snow removal equipment									
1920	Thrower, 3 H.P., 20", single stage, gas				Ea.	575			575	635
1930	Electric start					680			680	750
1940	5 H.P., 24", 2 stage, gas					1,000			1,000	1,100
1950	11 H.P., 36"					2,225			2,225	2,450
1970	Trash receptacles and closures; see Div. 12 93 23.20									
2000	Tube cleaning equip., for heat exchangers, condensers, evap.									
2010	Tubes/pipes 1/4"-1", 115v, .5 H.P. drive				Ea.	1,525			1,525	1,675
2020	Air powered					1,600			1,600	1,750
2040	1" and up, 115 volt, 1 H.P. drive					1,700			1,700	1,875

01 93 09 – Facility Equipment

01 93 09.50 Moving Equipment		Crew	Daily Output	Labor-Hours	Unit	Material	Labor	Equipment	Total	Total Incl O&P
0010	**MOVING EQUIPMENT**, Remove and reset, 100' distance,									
0020	No obstructions, no assembly or leveling unless noted									
0100	Annealing furnace, 24' overall	B-67	4	4	Ea.		161	70.50	231.50	330
0200	Annealing oven, small		14	1.143			46	20	66	94.50
0240	Very large		1	16			645	281	926	1,325
0400	Band saw, small		12	1.333			54	23.50	77.50	111
0440	Large		8	2			80.50	35	115.50	166
0500	Blue print copy machine		7	2.286			92	40	132	189
0600	Bonding mill, 6"		7	2.286			92	40	132	189
0620	12"		6	2.667			108	47	155	221
0640	18"		4	4			161	70.50	231.50	330
0660	24"		2	8			325	141	466	660
0700	Boring machine (jig)	B-68	7	3.429			140	40	180	263
0800	Bridgeport mill, standard	B-67	14	1.143			46	20	66	94.50
1000	Calibrator, 6 unit	"	14	1.143			46	20	66	94.50
1100	Comparitor, bench top	2 Clab	14	1.143			36		36	59.50
1140	Floor mounted	B-67	7	2.286			92	40	132	189
1200	Computer, desk top	2 Clab	25	.640			20		20	33.50
1300	Copy machine	"	25	.640			20		20	33.50
1500	Deflasher	B-67	14	1.143			46	20	66	94.50
1600	Degreaser, small		14	1.143			46	20	66	94.50
1640	Large 24' overall		1	16			645	281	926	1,325
1700	Desk with chair	2 Clab	25	.640			20		20	33.50
1800	Dial press	B-67	7	2.286			92	40	132	189
1900	Drafting table	2 Clab	14	1.143			36		36	59.50
2000	Drill press, bench top	"	14	1.143			36		36	59.50
2040	Floor mounted	B-67	14	1.143			46	20	66	94.50

01 93 09.50 Moving Equipment		Crew	Daily Output	Labor-Hours	Unit	Material	2009 Bare Costs Labor	Equipment	Total	Total Incl O&P
2080	Industrial radial	B-67	7	2.286	Ea.		92	40	132	189
2100	Dust collector, portable	2 Clab	25	.640			20		20	33.50
2140	Stationary, small	B-67	7	2.286			92	40	132	189
2180	Stationary, large	"	2	8			325	141	466	660
2300	Electric discharge machine	B-68	7	3.429	↓		140	40	180	263
2400	Environmental chamber walls, including assembly	4 Clab	18	1.778	L.F.		56		56	92.50
2600	File cabinet	2 Clab	25	.640	Ea.		20		20	33.50
2800	Grinder/sander, pedestal mount	B-67	14	1.143			46	20	66	94.50
3000	Hack saw, power	2 Clab	24	.667			21		21	35
3100	Hydraulic press	B-67	14	1.143			46	20	66	94.50
3500	Laminar flow tables	"	14	1.143			46	20	66	94.50
3600	Lathe, bench	2 Clab	14	1.143			36		36	59.50
3640	6"	B-67	14	1.143			46	20	66	94.50
3680	10"		13	1.231			49.50	21.50	71	102
3720	12"		12	1.333			54	23.50	77.50	111
4000	Milling machine		8	2			80.50	35	115.50	166
4100	Molding press, 25 ton		5	3.200			129	56.50	185.50	265
4140	60 ton		4	4			161	70.50	231.50	330
4180	100 ton		2	8			325	141	466	660
4220	150 ton		1.50	10.667			430	188	618	880
4260	200 ton		1	16			645	281	926	1,325
4300	300 ton		.75	21.333			860	375	1,235	1,750
4700	Oil pot stand		14	1.143			46	20	66	94.50
5000	Press, 10 ton		14	1.143			46	20	66	94.50
5040	15 ton		12	1.333			54	23.50	77.50	111
5080	20 ton		10	1.600			64.50	28	92.50	132
5120	30 ton		8	2			80.50	35	115.50	166
5160	45 ton		6	2.667			108	47	155	221
5200	60 ton		4	4			161	70.50	231.50	330
5240	75 ton		2.50	6.400			258	113	371	530
5280	100 ton	↓	2	8	↓		325	141	466	660
5500	Raised floor, including assembly	2 Carp	250	.064	S.F.		2.56		2.56	4.22
5600	Rolling mill, 6"	B-67	7	2.286	Ea.		92	40	132	189
5640	9"		6	2.667			108	47	155	221
5680	12"		4	4			161	70.50	231.50	330
5720	13"		3.50	4.571			184	80.50	264.50	380
5760	18"		2	8			325	141	466	660
5800	25"		1	16			645	281	926	1,325
6000	Sander, floor stand		14	1.143			46	20	66	94.50
6100	Screw machine		7	2.286			92	40	132	189
6200	Shaper, 16"		14	1.143			46	20	66	94.50
6300	Shear, power assist		4	4			161	70.50	231.50	330
6400	Slitter, 6"		14	1.143			46	20	66	94.50
6440	8"		13	1.231			49.50	21.50	71	102
6480	10"		12	1.333			54	23.50	77.50	111
6520	12"		11	1.455			58.50	25.50	84	120
6560	16"		10	1.600			64.50	28	92.50	132
6600	20"		8	2			80.50	35	115.50	166
6640	24"		6	2.667			108	47	155	221
6800	Snag and tap machine		7	2.286			92	40	132	189
6900	Solder machine (auto)	↓	7	2.286			92	40	132	189
7000	Storage cabinet metal, small	2 Clab	36	.444			14.05		14.05	23
7040	Large	↓	25	.640	↓		20		20	33.50

01 93 09 – Facility Equipment

01 93 09.50 Moving Equipment		Crew	Daily Output	Labor-Hours	Unit	Material	2009 Bare Costs Labor	Equipment	Total	Total Incl O&P
7100	Storage rack open, small	2 Clab	14	1.143	Ea.		36		36	59.50
7140	Large		7	2.286			72		72	119
7200	Surface bench, small	B-67	14	1.143			46	20	66	94.50
7240	Large	"	5	3.200			129	56.50	185.50	265
7300	Surface grinder, large wet	B-68	5	4.800			196	56.50	252.50	365
7500	Time check machine	2 Clab	14	1.143			36		36	59.50
8000	Welder, 30 KVA (bench)		14	1.143			36		36	59.50
8100	Work bench with chair		25	.640			20		20	33.50

01 93 13 – Facility Maintenance Procedures

01 93 13.03 Concrete Facilities Maintenance

		Crew	Daily Output	Labor-Hours	Unit	Material	Labor	Equipment	Total	Total Incl O&P
0010	**CONCRETE FACILITIES MAINTENANCE**									
0200	Epoxy floor - resurface concrete floor with aggregated epoxy									
0210	Clean, patch, prime, sand/agg. mix, power trowel, sealer									
0220	Minimum cost				S.F.					4.50
0230	Maximum cost				"					7.50
1000	Patching concrete									
1010	Floors, 1/4" thick, small areas, regular grout	1 Cefi	170	.047	S.F.	1.05	1.80		2.85	3.98
1015	Epoxy grout	"	100	.080	"	8.40	3.06		11.46	14.05
1020	Walls, including chipping, cleaning and epoxy grout									
1025	Minimum	1 Cefi	65	.123	S.F.	3.50	4.71		8.21	11.20
1030	Average		50	.160		8.15	6.15		14.30	18.55
1035	Maximum		40	.200		11.40	7.65		19.05	24.50
1037	Stair cleaning, sweep	1 Clab	50	.160	Flight		5.05		5.05	8.35
1039	Mop or scrub	"	5	1.600	"	.10	50.50		50.60	83.50
1040	Sealing concrete floor, 2 coats, oil or urethane base, min. cost				S.F.				.50	.60
1050	Maximum cost				"				.75	.90
1080	Sweep and damp mop concrete floor	1 Clab	20	.400	M.S.F.	2.29	12.65		14.94	23.50

01 93 13.04 Masonry Facilities Maintenance

		Crew	Daily Output	Labor-Hours	Unit	Material	Labor	Equipment	Total	Total Incl O&P
0010	**MASONRY FACILITIES MAINTENANCE**									
0040	Caulking masonry, no staging included									
0050	Re-caulk only, oil base, 1/2" x 1/2"	1 Bric	225	.036	L.F.	.53	1.44		1.97	2.91
0060	Butyl		205	.039		.34	1.58		1.92	2.93
0070	Poly sulfide		200	.040		.70	1.62		2.32	3.39
0080	Silicone		195	.041		.53	1.66		2.19	3.28
0100	Cut out and re-caulk, oil base		145	.055		.53	2.23		2.76	4.19
0110	Butyl		130	.062		.34	2.49		2.83	4.40
0120	Polysulfide		125	.064		.35	2.59		2.94	4.58
0130	Silicone		125	.064		.53	2.59		3.12	4.78
0800	Cleaning masonry, no staging included									
0850	Chemical cleaning, brush and wash, minimum cost	D-1	800	.020	S.F.	.03	.73		.76	1.21
0860	Average cost		400	.040		.04	1.45		1.49	2.39
0870	Maximum cost		330	.048		.07	1.76		1.83	2.93
0880	High pressure water only, minimum cost	B-9	2000	.020			.64	.10	.74	1.17
0900	Average cost		1500	.027			.85	.13	.98	1.55
0910	Maximum cost		1000	.040			1.28	.19	1.47	2.32
0920	High pressure water and chemical, minimum cost		1800	.022		.03	.71	.11	.85	1.32
0930	Average cost		1200	.033		.04	1.07	.16	1.27	1.98
0940	Maximum cost		800	.050		.07	1.60	.24	1.91	2.98
0950	Sandblast, wet system, minimum cost		1700	.024		.22	.75	.11	1.08	1.60
0960	Average cost		1100	.036		.27	1.16	.17	1.60	2.41
0970	Maximum cost		700	.057		.33	1.83	.27	2.43	3.68
0980	Dry system, minimum cost		3000	.013		.22	.43	.06	.71	1.01

01 93 Facility Maintenance

01 93 13 – Facility Maintenance Procedures

01 93 13.04 Masonry Facilities Maintenance		Crew	Daily Output	Labor-Hours	Unit	Material	Labor	Equipment	Total	Total Incl O&P
						2009 Bare Costs				
0990	Average cost	B-9	1750	.023	S.F.	.27	.73	.11	1.11	1.63
1000	Maximum cost		1500	.027		.33	.85	.13	1.31	1.91
1010	For walnut shells, add					.37			.37	.40
1020	For corn chips, add					.27			.27	.30
1030	Steam cleaning, minimum cost	B-9	3000	.013			.43	.06	.49	.77
1040	Average cost		2500	.016			.51	.08	.59	.93
1050	Maximum cost		1500	.027			.85	.13	.98	1.55
1060	Add for masking doors & windows									.80
1070	Add for pedestrian protection				Job					10%
1080	Add for wire cut face brick				S.F.					.32
3000	Pointing masonry									
3030	Cut and repoint brick, hard mortar, running bond	1 Bric	80	.100	S.F.	.23	4.05		4.28	6.80
3040	Common bond		77	.104		.23	4.21		4.44	7.05
3050	Flemish bond		70	.114		.24	4.63		4.87	7.75
3060	English bond		65	.123		.24	4.98		5.22	8.30
3070	Soft old mortar, running bond		100	.080		.25	3.24		3.49	5.55
3080	Common bond		96	.083		.25	3.37		3.62	5.75
3090	Flemish bond		90	.089		.26	3.60		3.86	6.10
3100	English bond		82	.098		.26	3.95		4.21	6.70
3110	Stonework, hard mortar		140	.057	L.F.	.35	2.31		2.66	4.13
3120	Soft old mortar		160	.050	"	.39	2.03		2.42	3.71
3130	Repoint, mask and grout method, running bond		95	.084	S.F.	.35	3.41		3.76	5.90
3140	Common bond		90	.089		.35	3.60		3.95	6.20
3150	Flemish bond		86	.093		.35	3.77		4.12	6.50
3160	English bond		77	.104		.35	4.21		4.56	7.20
3170	Scrub coat, sand grout on walls, minimum		120	.067		.35	2.70		3.05	4.76
3180	Maximum		98	.082		.38	3.31		3.69	5.75

01 93 13.05 Metals Facilities Maintenance

0010	**METALS FACILITIES MAINTENANCE**									
2000	Steel surface treat., sand blast & paint, see Div. 05 01 10.51									
2010	Steel siding painting, see Div. 09 91 13									

01 93 13.07 Moisture-Thermal Control Facilities Maintenance

0010	**MOISTURE-THERMAL CONTROL FACILITIES MAINTENANCE**									
0100	Caulking and sealants see Div. 07 92 10									
0110	Caulking around exterior doors and windows, silicone	1 Carp	400	.020	L.F.	.13	.80		.93	1.47
2000	Roofing & siding demolition see Div. 07 05 05.10									
2500	Single-ply or built-up roofing maintenance									
2510	Walk roof, clean drain trap, pick-up trash									
2520	Allow per year, minimum				S.F.					.02
2530	Maximum				"					.06
3000	Roof non-destructive testing, infrared analysis system									
3040	Minimum cost				S.F.					.04
3050	Maximum cost				"					.06

01 93 13.08 Door and Window Facilities Maintenance

0010	**DOOR & WINDOW FACILITIES MAINTENANCE**									
0200	Caulking & sealants see Div. 07 92 10									
0250	Cylinder dead bolt lock set, repair	1 Skwk	11	.727	Ea.		29.50		29.50	48.50
0300	Demolition, windows see Div. 08 05 05.20									
0500	Door closer, overhaul	1 Skwk	5	1.600	Ea.		65.50		65.50	106
0510	Repair concealed door check or closer		4	2			31.50		81.50	133
0520	Adjust		26	.303			12.55		12.55	20.50
0530	Replace, see Div. 08 71 20.31									

01 93 13 – Facility Maintenance Procedures

01 93 13.08 Door and Window Facilities Maintenance		Crew	Daily Output	Labor-Hours	Unit	Material	2009 Bare Costs Labor	Equipment	Total	Total Incl O&P
1500	Remove and reset window, minimum cost (labor)	1 Carp	4	2	Ea.		80		80	132
1510	Average cost		2	4			160		160	264
1520	Maximum cost		1.33	6.015			240		240	395
1900	Washing door, 3′ x 7′, damp wipe	1 Clab	145	.055		.18	1.74		1.92	3.08
2000	Washing windows, at ground level w/ladder, sponge & squeegee									
2010	Both sides, 4′ x 7′ 2 pane	1 Clab	60	.133	Ea.	.21	4.21		4.42	7.20
2020	4′ x 6′, 4 pane		55	.145		.18	4.60		4.78	7.80
2030	3.5′ x 5.5′, 8 pane		45	.178		.15	5.60		5.75	9.40
2040	2.5′ x 5.7′, 12 pane		40	.200		.11	6.30		6.41	10.55
2050	4′ x 6′, 16 pane		35	.229		.18	7.20		7.38	12.10
2060	4′ x 7′, industrial, 20 pane		30	.267		.21	8.45		8.66	14.15
2070	6′ x 7′, austral casement, 6 pane		35	.229		.32	7.20		7.52	12.25
2080	Clear glass partition, 8 sq. ft. per unit		3	2.667	M.S.F.	3.79	84.50		88.29	143
2090	Opaque, 20 sq. ft. per unit		3	2.667	"	3.79	84.50		88.29	143
2100	Alternate pricing method by window area to be washed									
2110	Minimum productivity	1 Clab	1	8	M.S.F.	3.79	253		256.79	420
2120	Average productivity		2.50	3.200		3.79	101		104.79	171
2130	Maximum productivity		4	2		3.79	63		66.79	108
2200	Weatherstripping see Div. 08 75 30.10									

01 93 13.09 Finishes, Facilities Maintenance

		Crew	Daily Output	Labor-Hours	Unit	Material	Labor	Equipment	Total	Total Incl O&P
0010	**FINISHES, FACILITIES MAINTENANCE** R019313-10									
0100	Ceiling maintenance									
0120	Acoustical tile cleaning, chemical R019313-20				S.F.					.25
0130	Clean and apply acoustical coating				"					.60
0400	Floor maintenance R019313-30									
0500	Carpet fiber sealant, anti-soilant, commercial				S.F.					.28
0510	Residential				"					.37
0530	Carpet cleaning, vacuum, dry pick-up									
0540	Unobstructed	1 Clab	30	.267	M.S.F.		8.45		8.45	13.90
0550	Obstructed		22	.364			11.50		11.50	18.95
0560	Wet pick-up, unobstructed		21	.381		8.55	12.05		20.60	29.50
0570	Obstructed		14	.571		8.55	18.05		26.60	39.50
0580	Steam clean, minimum		9760	.001	S.F.		.03		.03	.04
0590	Maximum cost		3250	.002	"		.08		.08	.13
0700	Composition, resilient or wood flooring									
0720	Dust mop, unobstructed	1 Clab	60	.133	M.S.F.		4.21		4.21	6.95
0730	Obstructed		35	.229			7.20		7.20	11.90
0740	Damp mop, unobstructed		26	.308		3.13	9.70		12.83	19.50
0750	Obstructed		16	.500		3.13	15.80		18.93	29.50
0780	Hand scrub, unobstructed		1.80	4.444		3.13	140		143.13	235
0790	Obstructed		1.30	6.154		3.13	194		197.13	325
0800	Machine scrub, unobstructed		17	.471		3.13	14.85		17.98	28
0810	Obstructed		12	.667		3.13	21		24.13	38.50
0820	Machine polish, unobstructed		28	.286			9.05		9.05	14.90
0830	Obstructed		14	.571			18.05		18.05	30
0850	Refinish old wood floors, minimum	1 Carp	400	.020	S.F.	.02	.80		.82	1.34
0860	Maximum cost	"	130	.062	"	.03	2.46		2.49	4.09
0870	Strip & rewax/polish, unobstructed	1 Clab	4	2	M.S.F.	20	63		83	126
0880	Obstructed		3	2.667		20	84.50		104.50	161
0890	Sweeping, unobstructed		47	.170			5.40		5.40	8.90
0900	Obstructed		35	.229			7.20		7.20	11.90
0950	Painting concrete or wood floors, see Div. 09 91 23.52									

01 93 13.09 Finishes, Facilities Maintenance

	01 93 13.09 Finishes, Facilities Maintenance	Crew	Daily Output	Labor-Hours	Unit	2009 Bare Costs Material	Labor	Equipment	Total	Total Incl O&P
1200	Paint removal, see Div. 02 83 19.26									
1300	Scrape after fire damage, see Div. 09 91 03.41									
2000	Wall maintenance									
2400	Washing enamel finish walls with mild cleanser	1 Clab	16	.500	M.S.F.	4.25	15.80		20.05	30.50

01 93 13.10 Specialties, Facilities Maintenance

	01 93 13.10 Specialties, Facilities Maintenance	Crew	Daily Output	Labor-Hours	Unit	Material	Labor	Equipment	Total	Total Incl O&P
0010	**SPECIALTIES, FACILITIES MAINTENANCE**									
0100	Bathroom accessories									
0120	Accessory, replacement see Div. 10 28 13.13									
0130	All purpose cleaner, concentrate				Gal.	17			17	18.70
0150	Clean mirror, 36" x 24"	1 Clab	840	.010	Ea.	.03	.30		.33	.53
0160	48" x 24"		630	.013		.03	.40		.43	.70
0170	72" x 24"		420	.019		.05	.60		.65	1.05
0200	General cleaning of fixtures (basins, water closets, urinals)									
0210	Including shelves, partitions and dispenser servicing	1 Clab	80	.100	Fixture	.03	3.16		3.19	5.25
1500	Partition replacement, hospital, office, toilet see Div. 10 21 13.13									

01 93 13.11 Architectural Equipment, Facilities Maintenance

	01 93 13.11 Architectural Equipment, Facilities Maintenance	Crew	Daily Output	Labor-Hours	Unit	Material	Labor	Equipment	Total	Total Incl O&P
0010	**ARCHITECTURAL EQUIPMENT, FACILITIES MAINTENANCE**									
0100	Bird control needle strips									
0110	Anchor clip mounted	2 Clab	500	.032	L.F.	5.45	1.01		6.46	7.65
0120	Adhesive mounted	"	550	.029	"	5.45	.92		6.37	7.50
0500	Laundry equipment repair and maintenance									
0520	Dryer, domestic, remove or reinstall basket	1 Skwk	6	1.333	Ea.		54.50		54.50	88.50
0530	Commercial, disassemble, clean & reassemble		2	4			163		163	266
0550	Extractor, replace belts (2)		6	1.333			54.50		54.50	88.50
0560	Replace pressure pad		2	4			163		163	266
0570	Flatwork ironer, replace feed ribbons		2	4			163		163	266
0580	Press, overhaul air system		4	2			81.50		81.50	133
0590	Tumbler, extractor, replace bearings and seal		2	4			163		163	266
0600	Washer, domestic, replace bearings and seal		3	2.667			109		109	177
0610	Remove or reinstall tub		4	2			81.50		81.50	133
0700	Medical and ward equipment repair and maintenance									
0710	Aspirator-overhaul	1 Skwk	6	1.333	Ea.		54.50		54.50	88.50
0720	Autoclave-overhaul		4	2	"		81.50		81.50	133
0730	Bed rails, assemble or repair		10	.800	Pr.		32.50		32.50	53
0770	Cart, medication, ice, etc. - replace wheels		6	1.333	Carton		54.50		54.50	88.50
0780	Replace rubber bumper		6	1.333	"		54.50		54.50	88.50
0800	Centrifuge, repair or replace timer or switch		11	.727	Ea.		29.50		29.50	48.50
0850	Litter, attach safety belt		16	.500			20.50		20.50	33.50
0880	Operating table, overhaul and lubricate		6	1.333			54.50		54.50	88.50
0900	Oxygen tent apparatus, overhaul		2	4			163		163	266
0950	Still, clean		3	2.667			109		109	177
0970	Striker saw, overhaul		5	1.600			65.50		65.50	106
1100	Wheelchair, complete overhaul		3	2.667			109		109	177
1110	Replace wheel		11	.727			29.50		29.50	48.50
2000	Other equipment repair and maintenance									
2040	Cart or dolly, assemble or disassemble axles and wheels	1 Skwk	6	1.333	Carton		54.50		54.50	88.50
2090	Forge and harden tool		13	.615	Ea.		25		25	41
3100	Office furniture, repair		7	1.143			46.50		46.50	76
3120	Portable power tool, assemble or disassemble		6	1.333			54.50		54.50	88.50
3130	Repair or replace part		6	1.333			54.50		54.50	88.50
3180	Replace tool handle		16	.500			20.50		20.50	33.50
3200	Sharpen hand tool		16	.500			20.50		20.50	33.50

01 93 13 – Facility Maintenance Procedures

01 93 13.12 Furnishings Facilities Maintenance

	01 93 13.12 Furnishings Facilities Maintenance	Crew	Daily Output	Labor-Hours	Unit	2009 Bare Costs Material	Labor	Equipment	Total	Total Incl O&P
0010	**FURNISHINGS FACILITIES MAINTENANCE**									
1000	Reupholster, side chair	1 Skwk	2.50	3.200	Ea.		131		131	213
1040	Tubular frame chair seat and back		3	2.667			109		109	177
1080	Sofa		2	4			163		163	266
1500	Upholstery fabric protection/cleaner, minimum cost				S.Y.					10
1510	Maximum cost				"					26
1540	Vacuum, large divan	1 Clab	80	.100	Ea.		3.16		3.16	5.20

01 93 13.14 Conveying Systems Facilities Maintenance

	01 93 13.14 Conveying Systems Facilities Maintenance	Crew	Daily Output	Labor-Hours	Unit	Material	Labor	Equipment	Total	Total Incl O&P
0010	**CONVEYING SYSTEMS FACILITIES MAINTENANCE**									
0900	Elevator, general cleaning, vacuum, dust, mop, polish	1 Clab	16	.500	Ea.	.77	15.80		16.57	27
0920	Motor generator set, service	1 Elec	8	1			47		47	72.50
0930	Repair, minor		2	4			188		188	289
0940	Limit switch, repair		11	.727			34		34	52.50
0950	Troubleshoot, main control panel	2 Elec	2.50	6.400			300		300	465
1000	Escalator cleaning, dusting, polishing	1 Clab	20	.400	Flight	.85	12.65		13.50	22
1500	Vehicle maintenance, grease, light vehicle	1 Skwk	20	.400	Ea.		16.35		16.35	26.50
1510	Heavy vehicle		7	1.143			46.50		46.50	76
1520	Wash, light vehicle		8	1		.58	41		41.58	67
1530	Heavy vehicle		6	1.333		1.04	54.50		55.54	89.50
1540	Change oil and filter, light vehicle		10	.800		18	32.50		50.50	73
1570	Rotate tires, light vehicle		18	.444			18.15		18.15	29.50

01 93 13.15 Mechanical Facilities Maintenance

	01 93 13.15 Mechanical Facilities Maintenance	Crew	Daily Output	Labor-Hours	Unit	Material	Labor	Equipment	Total	Total Incl O&P
0010	**MECHANICAL FACILITIES MAINTENANCE**									
0100	Air conditioning system maintenance									
0130	Belt, replace	1 Stpi	15	.533	Ea.		26.50		26.50	41
0170	Fan, clean		16	.500			24.50		24.50	38.50
0180	Filter, remove, clean, replace		12	.667			33		33	51
0190	Flexible coupling alignment, inspect		40	.200			9.85		9.85	15.30
0200	Gas leak locate and repair		4	2			98.50		98.50	153
0250	Pump packing gland, remove and replace		11	.727			36		36	55.50
0270	Tighten		32	.250			12.35		12.35	19.15
0290	Pump disassemble and assemble		4	2			98.50		98.50	153
0300	Air pressure regulator, disassemble, clean, assemble	1 Skwk	4	2			81.50		81.50	133
0310	Repair or replace part	"	6	1.333			54.50		54.50	88.50
0320	Purging system	1 Stpi	16	.500			24.50		24.50	38.50
0400	Compressor, air, remove or install fan wheel	1 Skwk	20	.400			16.35		16.35	26.50
0410	Disassemble or assemble 2 cylinder, 2 stage		4	2			81.50		81.50	133
0420	4 cylinder, 4 stage		1	8			325		325	530
0430	Repair or replace part		2	4			163		163	266
0700	Demolition, for mech. demolition see Div. 23 05 05.10 or 22 05 05.10									
0800	Ductwork, clean									
0810	Rectangular									
0820	6" G	1 Shee	187.50	.043	L.F.		2.01		2.01	3.20
0830	8" G		140.63	.057			2.69		2.69	4.27
0840	10" G		112.50	.071			3.36		3.36	5.35
0850	12" G		93.75	.085			4.03		4.03	6.40
0860	14" G		80.36	.100			4.70		4.70	7.45
0870	16" G		70.31	.114			5.35		5.35	8.55
0900	Round									
0910	4" G	1 Shee	358.10	.022	L.F.		1.05		1.05	1.68
0920	6" G		238.73	.034			1.58		1.58	2.51
0930	8" G		179.05	.045			2.11		2.11	3.35

01 93 Facility Maintenance

01 93 13 – Facility Maintenance Procedures

01 93 13.15 Mechanical Facilities Maintenance		Crew	Daily Output	Labor-Hours	Unit	Material	2009 Bare Costs Labor	Equipment	Total	Total Incl O&P
0940	10″ G	1 Shee	143.24	.056	L.F.		2.64		2.64	4.19
0950	12″ G		119.37	.067			3.16		3.16	5.05
0960	16″ G		89.52	.089			4.22		4.22	6.70
1200	Fire protection equipment									
1220	Fire hydrant, replace	Q-1	3	5.333	Ea.		234		234	365
1230	Service, lubricate, inspect, flush, clean	1 Plum	7	1.143			55.50		55.50	86.50
1240	Test	"	11	.727			35.50		35.50	55
1310	Inspect valves, pressure, nozzle,	1 Spri	4	2	System		96.50		96.50	149
1500	Instrumentation preventive maintenance, rule of thumb									
1510	Annual cost per instrument for calibration & trouble shooting									
1520	Includes flow meters, analytical transmitters, flow control									
1530	chemical feed systems, recorders and distrib. control sys.				Ea.					325
1800	Plumbing fixtures, for installation see Div. 22 41 00									
1840	Open drain with snake	1 Plum	13	.615	Ea.		30		30	46.50
1850	Open drain with toilet auger		16	.500			24.50		24.50	38
1860	Grease trap, clean		2	4			195		195	300
1870	Plaster trap, clean		6	1.333			65		65	101
1900	Relief valve, test and adjust	1 Stpi	20	.400			19.75		19.75	30.50
2000	Repair or replace-steam trap		8	1			49.50		49.50	76.50
2020	Y-type or bell strainer		6	1.333			66		66	102
2040	Water trap or vacuum breaker, screwed joints		13	.615			30.50		30.50	47
2100	Steam specialties, clean									
2120	Air separator with automatic trap, 1″ fittings	1 Stpi	12	.667	Ea.		33		33	51
2130	Bucket trap, 2″ pipe		7	1.143			56.50		56.50	87.50
2150	Drip leg, 2″ fitting		45	.178			8.75		8.75	13.60
2200	Thermodynamic trap, 1″ fittings		50	.160			7.90		7.90	12.25
2210	Thermostatic		65	.123			6.05		6.05	9.40
2240	Screen and seat in y-type strainer, plug type.		25	.320			15.80		15.80	24.50
2242	Screen and seat in y-type strainer, flange type.		12	.667			33		33	51
2500	Valve, replace broken handwheel		24	.333			16.45		16.45	25.50
3000	Valve, overhaul, regulator, relief, flushometer, mixing									
3040	Cold water, gas	1 Stpi	5	1.600	Ea.		79		79	122
3050	Hot water, steam		3	2.667			132		132	204
3080	Globe, gate, check up to 4″ cold water, gas		10	.800			39.50		39.50	61
3090	Hot water, steam		5	1.600			79		79	122
3100	Over 4″ ID hot or cold line		1.40	5.714			282		282	435
3120	Remove and replace, gate, globe or check up to 4″	Q-5	6	2.667			118		118	184
3130	Over 4″	"	2	8			355		355	550
3150	Repack up to 4″	1 Stpi	13	.615			30.50		30.50	47
3160	Over 4″	"	4	2			93.50		98.50	153

01 93 13.16 Electrical Facilities Maintenance		Crew	Daily Output	Labor-Hours	Unit	Material	Labor	Equipment	Total	Total Incl O&P
0010	**ELECTRICAL FACILITIES MAINTENANCE**									
0700	Cathodic protection systems									
0720	Check and adjust reading on rectifier	1 Elec	20	.400	Ea.		13.80		18.80	29
0730	Check pipe to soil potential		20	.400			13.80		18.80	29
0740	Replace lead connection		4	2			94		94	145
0800	Control device, install		5.70	1.404			66		66	101
0810	Disassemble, clean and reinstall		7	1.143			53.50		53.50	82.50
0820	Replace		10.70	.748			35		35	54
0830	Trouble shoot		10	.800			37.50		37.50	58
0900	Demolition, for electrical demolition see Div. 26 05 05.10									
1000	Distribution systems and equipment install or repair a breaker									

01 93 Facility Maintenance

01 93 13 – Facility Maintenance Procedures

01 93 13.16 Electrical Facilities Maintenance		Crew	Daily Output	Labor-Hours	Unit	Material	2009 Bare Costs Labor	Equipment	Total	Total Incl O&P
1010	In power panels up to 200 amps	1 Elec	7	1.143	Ea.		53.50		53.50	82.50
1020	Over 200 amps		2	4			188		188	289
1030	Reset breaker or replace fuse		20	.400			18.80		18.80	29
1100	Megger test MCC (each stack)		4	2			94		94	145
1110	MCC vacuum and clean (each stack)		5.30	1.509			71		71	109
2500	Remove and replace or maint, road fixture & lamp		3	2.667		845	125		970	1,125
2510	Fluorescent fixture		7	1.143		72	53.50		125.50	162
2515	Relamp (fluor.) facility area each tube		60	.133		6.15	6.25		12.40	16.40
2518	Fluorescent fixture, clean (area)		44	.182		.10	8.55		8.65	13.25
2520	Incandescent fixture		11	.727		59	34		93	117
2530	Lamp (incandescent or fluorescent)		60	.133		3.47	6.25		9.72	13.45
2535	Replace cord in socket lamp		13	.615		2.20	29		31.20	47
2540	Ballast electronic type for two tubes		8	1		35.50	47		82.50	112
2541	Starter		30	.267		1.70	12.55		14.25	21
2545	Replace other lighting parts		11	.727		16.60	34		50.60	71
2550	Switch		11	.727		7.40	34		41.40	60.50
2555	Receptacle		11	.727		9.95	34		43.95	63.50
2560	Floodlight		4	2		281	94		375	455
2570	Christmas lighting, indoor, per string		16	.500			23.50		23.50	36
2580	Outdoor		13	.615			29		29	44.50
2590	Test battery operated emergency lights		40	.200			9.40		9.40	14.45
2600	Repair and replace component in communication system		6	1.333		53.50	62.50		116	155
2700	Repair misc. appliances (incl. clocks, vent fan, blower, etc.)		6	1.333			62.50		62.50	96.50
2710	Reset clocks & timers		50	.160			7.50		7.50	11.55
2720	Adjust time delay relays		16	.500			23.50		23.50	36
2730	Test specific gravity of lead-acid batteries	↓	80	.100	↓		4.70		4.70	7.25
3000	Motors and generators									
3020	Disassemble, clean and reinstall motor, up to 1/4 HP	1 Elec	4	2	Ea.		94		94	145
3030	Up to 3/4 HP		3	2.667			125		125	193
3040	Up to 10 HP		2	4			188		188	289
3050	Replace part, up to 1/4 HP		6	1.333			62.50		62.50	96.50
3060	Up to 3/4 HP		4	2			94		94	145
3070	Up to 10 HP		3	2.667			125		125	193
3080	Megger test motor windings		5.33	1.501			70.50		70.50	109
3082	Motor vibration check		16	.500			23.50		23.50	36
3084	Oil motor bearings		25	.320			15.05		15.05	23
3086	Run test emergency generator for 30 minutes		11	.727			34		34	52.50
3090	Rewind motor, up to 1/4 HP		3	2.667			125		125	193
3100	Up to 3/4 HP		2	4			188		188	289
3110	Up to 10 HP		1.50	5.333			251		251	385
3150	Generator, repair or replace part		4	2			94		94	145
3160	Repair DC generator		2	4			188		188	289
4000	Stub pole, install or remove		3	2.667			125		125	193
4500	Transformer maintenance up to 15 kVA	↓	2.70	2.963	↓		139		139	214

Estimating Tips

02 30 00 Subsurface Investigation

In preparing estimates on structures involving earthwork or foundations, all information concerning soil characteristics should be obtained. Look particularly for hazardous waste, evidence of prior dumping of debris, and previous stream beds.

02 41 00 Demolition and Structure Moving

The costs shown for selective demolition do not include rubbish handling or disposal. These items should be estimated separately using RSMeans data or other sources.

- Historic preservation often requires that the contractor remove materials from the existing structure, rehab them, and replace them. The estimator must be aware of any related measures and precautions that must be taken when doing selective demolition and cutting and patching. Requirements may include special handling and storage, as well as security.
- In addition to Subdivision 02 41 00, you can find selective demolition items in each division. Example: Roofing demolition is in Division 7.

02 80 00 Hazardous Material Disposal/ Remediation

This subdivision includes information on hazardous waste handling, asbestos remediation, lead remediation, and mold remediation. See reference R028213-20 and R028319-60 for further guidance in using these unit price lines.

Reference Numbers

Reference numbers are shown in shaded boxes at the beginning of some major classifications. These numbers refer to related items in the Reference Section. The reference information may be an estimating procedure, an alternate pricing method, or technical information.

Note: Not all subdivisions listed here necessarily appear in this publication.

02 21 Surveys

02 21 13 – Site Surveys

02 21 13.09 Topographical Surveys		Crew	Daily Output	Labor-Hours	Unit	Material	Labor	Equipment	Total	Total Incl O&P
						2009 Bare Costs				
0010	TOPOGRAPHICAL SURVEYS									
0020	Topographical surveying, conventional, minimum	A-7	3.30	7.273	Acre	18	315	21	354	550
0100	Maximum	A-8	.60	53.333	"	55	2,250	117	2,422	3,800

02 21 13.13 Boundary and Survey Markers

		Crew	Daily Output	Labor-Hours	Unit	Material	Labor	Equipment	Total	Total Incl O&P
0010	BOUNDARY AND SURVEY MARKERS									
0300	Lot location and lines, large quantities, minimum	A-7	2	12	Acre	32	520	35	587	910
0320	Average	"	1.25	19.200		51	830	56	937	1,450
0400	Small quantities, maximum	A-8	1	32		68	1,350	70	1,488	2,325
0600	Monuments, 3′ long	A-7	10	2.400	Ea.	34	104	7	145	212
0800	Property lines, perimeter, cleared land	"	1000	.024	L.F.	.03	1.04	.07	1.14	1.78

02 21 13.16 Aerial Surveys

		Crew	Daily Output	Labor-Hours	Unit	Material	Labor	Equipment	Total	Total Incl O&P
0010	AERIAL SURVEYS									
1500	Aerial surveying, including ground control, minimum fee, 10 acres				Total					6,270
1510	100 acres									10,500
1550	From existing photography, deduct									1,510
1600	2′ contours, 10 acres				Acre					510
1650	20 acres									350
1800	50 acres									105
1850	100 acres									95
2150	For 1′ contours and									
2160	dense urban areas, add to above				Acre					40%
3000	Inertial guidance system for									
3010	locating coordinates, rent per day				Ea.					4,400

02 32 Geotechnical Investigations

02 32 13 – Subsurface Drilling and Sampling

02 32 13.10 Boring and Exploratory Drilling

		Crew	Daily Output	Labor-Hours	Unit	Material	Labor	Equipment	Total	Total Incl O&P
0010	BORING AND EXPLORATORY DRILLING									
0020	Borings, initial field stake out & determination of elevations	A-6	1	16	Day		635	70	705	1,100
0100	Drawings showing boring details				Total		255		255	372
0200	Report and recommendations from P.E.						580		580	847
0300	Mobilization and demobilization, minimum	B-55	4	6			188	237	425	570
0350	For over 100 miles, per added mile		450	.053	Mile		1.67	2.11	3.78	5.10
0600	Auger holes in earth, no samples, 2-1/2″ diameter		78.60	.305	L.F.		9.60	12.10	21.70	29
0650	4″ diameter		67.50	.356			11.15	14.05	25.20	34
0800	Cased borings in earth, with samples, 2-1/2″ diameter		55.50	.432		20	13.55	17.10	50.65	63.50
0850	4″ diameter		32.60	.736		31.50	23	29	83.50	105
1000	Drilling in rock, "BX" core, no sampling	B-56	34.90	.458			16.20	37.50	53.70	67
1050	With casing & sampling		31.70	.505		20	17.85	41	78.85	95.50
1200	"NX" core, no sampling		25.92	.617			22	50	72	90
1250	With casing and sampling		25	.640		24.50	22.50	52	99	121
1400	Borings, earth, drill rig and crew with truck mounted auger	B-55	1	24	Day		755	950	1,705	2,300
1450	Rock using crawler type drill	B-56	1	16	"		565	1,300	1,865	2,325
1500	For inner city borings add, minimum									10%
1510	Maximum									20%

02 32 19 – Exploratory Excavations

02 32 19.10 Test Pits

		Crew	Daily Output	Labor-Hours	Unit	Material	Labor	Equipment	Total	Total Incl O&P
0010	TEST PITS									
0020	Hand digging, light soil	1 Clab	4.50	1.778	C.Y.		56		56	92.50
0100	Heavy soil	"	2.50	3.200			101		101	167

02 32 Geotechnical Investigations

02 32 19 – Exploratory Excavations

02 32 19.10 Test Pits		Crew	Daily Output	Labor-Hours	Unit	Material	2009 Bare Costs Labor	Equipment	Total	Total Incl O&P
0120	Loader-backhoe, light soil	B-11M	28	.571	C.Y.		21	12.30	33.30	47
0130	Heavy soil	"	20	.800			29	17.25	46.25	66
1000	Subsurface exploration, mobilization				Mile				8.25	9.50
1010	Difficult access for rig, add				Hr.				165	190
1020	Auger borings, drill rig, incl. samples				L.F.				19	21.85
1030	Hand auger								28	32.20
1050	Drill and sample every 5′, split spoon								24.75	28.45
1060	Extra samples				Ea.				33.50	38.55

02 41 Demolition

02 41 13 – Selective Site Demolition

02 41 13.15 Hydrodemolition

		Crew	Daily Output	Labor-Hours	Unit	Material	Labor	Equipment	Total	Total Incl O&P
0010	**HYDRODEMOLITION** R024119-10									
0015	Hydrodemolition, concrete pavement, 4000 PSI, 2″ depth	B-5	500	.112	S.F.		3.88	2.36	6.24	8.90
0120	4″ depth		450	.124			4.31	2.63	6.94	9.90
0130	6″ depth		400	.140			4.85	2.95	7.80	11.15
0410	6000 PSI, 2″ depth		410	.137			4.74	2.88	7.62	10.85
0420	4″ depth		350	.160			5.55	3.38	8.93	12.70
0430	6″ depth		300	.187			6.45	3.94	10.39	14.85
0510	8000 PSI, 2″ depth		330	.170			5.90	3.58	9.48	13.50
0520	4″ depth		280	.200			6.95	4.22	11.17	15.90
0530	6″ depth		240	.233			8.10	4.92	13.02	18.55

02 41 13.17 Demolish, Remove Pavement and Curb

		Crew	Daily Output	Labor-Hours	Unit	Material	Labor	Equipment	Total	Total Incl O&P
0010	**DEMOLISH, REMOVE PAVEMENT AND CURB** R024119-10									
5010	Pavement removal, bituminous roads, 3″ thick	B-38	690	.058	S.Y.		2.05	1.51	3.56	4.99
5050	4″ to 6″ thick		420	.095			3.38	2.49	5.87	8.20
5100	Bituminous driveways		640	.063			2.22	1.63	3.85	5.40
5200	Concrete to 6″ thick, hydraulic hammer, mesh reinforced		255	.157			5.55	4.10	9.65	13.50
5300	Rod reinforced		200	.200			7.10	5.20	12.30	17.20
5400	Concrete, 7″ to 24″ thick, plain		33	1.212	C.Y.		43	31.50	74.50	105
5500	Reinforced		24	1.667	"		59	43.50	102.50	144
5590	Minimum labor/equipment charge		6	6.667	Job		236	174	410	570
5600	With hand held air equipment, bituminous, to 6″ thick	B-39	1900	.025	S.F.		.84	.10	.94	1.48
5700	Concrete to 6″ thick, no reinforcing		1600	.030			1	.12	1.12	1.76
5800	Mesh reinforced		1400	.034			1.14	.14	1.28	2.01
5900	Rod reinforced		765	.063			2.08	.25	2.33	3.68
5990	Minimum labor/equipment charge	B-38	6	6.667	Job		236	174	410	570
6000	Curbs, concrete, plain	B-6	360	.067	L.F.		2.27	.82	3.09	4.58
6100	Reinforced		275	.087			2.97	1.07	4.04	6
6200	Granite		360	.067			2.27	.82	3.09	4.58
6300	Bituminous		528	.045			1.55	.56	2.11	3.12
6390	Minimum labor/equipment charge		6	4	Job		136	49	185	275
6500	Site demo, berms under 4″ in height, bituminous		528	.045	L.F.		1.55	.56	2.11	3.12
6600	4″ or over in height		300	.080	"		2.73	.98	3.71	5.50

02 41 13.20 Selective Demo, Highway Guard Rails & Barriers

		Crew	Daily Output	Labor-Hours	Unit	Material	Labor	Equipment	Total	Total Incl O&P
0010	**SELECTIVE DEMOLITION, HIGHWAY GUARD RAILS & BARRIERS**									
0100	Guard rail, corrugated steel	B-6	600	.040	L.F.		1.36	.49	1.85	2.75
0200	End sections		40	.600	Ea.		20.50	7.35	27.85	41
0300	Wrap around		40	.600	"		20.50	7.35	27.85	41
0400	Timber 4″ x 8″		600	.040	L.F.		1.36	.49	1.85	2.75

02 41 Demolition

02 41 13 – Selective Site Demolition

02 41 13.20 Selective Demo, Highway Guard Rails & Barriers		Crew	Daily Output	Labor-Hours	Unit	Material	2009 Bare Costs Labor	Equipment	Total	Total Incl O&P
0500	Three 3/4″ cables	B-6	600	.040	L.F.		1.36	.49	1.85	2.75
0600	Wood posts		240	.100	Ea.		3.41	1.22	4.63	6.90
0700	Guide rail, 6″ x 6″ box beam	B-80B	120	.267	L.F.		8.90	1.87	10.77	16.60
0800	Median barrier, 6″ x 8″ box beam		240	.133			4.46	.94	5.40	8.30
0850	Precast concrete 3′-6″ high x 2′ wide		300	.107			3.57	.75	4.32	6.60
0900	Impact barrier, UTMCD, barrel type	B-16	60	.533	Ea.		17.15	8.90	26.05	38.50
1000	Resilient guide fence and light shield 6′ high	″	120	.267	L.F.		8.60	4.44	13.04	19.05
1100	Concrete posts, 6′-5″ triangular	B-6	200	.120	Ea.		4.09	1.47	5.56	8.25
1200	Speed bumps 10 1/2 x 2 1/4 x 48 ″		300	.080	L.F.		2.73	.98	3.71	5.50
1300	Pavement marking channelizing		200	.120	Ea.		4.09	1.47	5.56	8.25
1400	Barrier and curb delineators		300	.080			2.73	.98	3.71	5.50
1500	Rumble strips 24″ x 3 1/2″ x1/2″		150	.160			5.45	1.96	7.41	11

02 41 13.33 Minor Site Demolition		Crew	Daily Output	Labor-Hours	Unit	Material	Labor	Equipment	Total	Total Incl O&P
0010	**MINOR SITE DEMOLITION** R024119-10									
0015	No hauling, abandon catch basin or manhole	B-6	7	3.429	Ea.		117	42	159	235
0020	Remove existing catch basin or manhole, masonry		4	6			204	73.50	277.50	410
0030	Catch basin or manhole frames and covers, stored		13	1.846			63	22.50	85.50	127
0040	Remove and reset		7	3.429			117	42	159	235
0100	Roadside delineators, remove only	B-80	175	.183			6.20	3.74	9.94	14.15
0110	Remove and reset	″	100	.320			10.80	6.55	17.35	25
0400	Minimum labor/equipment charge	B-6	4	6	Job		204	73.50	277.50	410
0800	Guiderail, corrugated steel, remove only	B-80A	100	.240	L.F.		7.60	2.43	10.03	15.15
0850	Remove and reset	″	40	.600	″		18.95	6.05	25	38
0860	Guide posts, remove only	B-80B	120	.267	Ea.		8.90	1.87	10.77	16.60
0870	Remove and reset	B-55	50	.480	″		15.05	19	34.05	46
0890	Minimum labor/equipment charge	2 Clab	4	4	Job		126		126	209
0900	Hydrants, fire, remove only	B-21A	5	8	Ea.		315	120	435	625
0950	Remove and reset	″	2	20	″		780	300	1,080	1,575
0990	Minimum labor/equipment charge	2 Plum	2	8	Job		390		390	605
1000	Masonry walls, block, solid	B-5	1800	.031	C.F.		1.08	.66	1.74	2.47
1200	Brick, solid		900	.062			2.16	1.31	3.47	4.94
1400	Stone, with mortar		900	.062			2.16	1.31	3.47	4.94
1500	Dry set		1500	.037			1.29	.79	2.08	2.97
1600	Median barrier, precast concrete, remove and store	B-3	430	.112	L.F.		3.76	4.78	8.54	11.40
1610	Remove and reset	″	390	.123	″		4.15	5.25	9.40	12.55
1650	Minimum labor/equipment charge	A-1	4	2	Job		63	15.80	78.80	121
2900	Pipe removal, sewer/water, no excavation, 12″ diameter	B-6	175	.137	L.F.		4.67	1.68	6.35	9.45
2930	15″-18″ diameter		150	.160			5.45	1.96	7.41	11
2960	21″-24″ diameter		120	.200			6.80	2.45	9.25	13.75
3000	27″-36″ diameter		90	.267			9.10	3.26	12.36	18.35
3200	Steel, welded connections, 4″ diameter		160	.150			5.10	1.84	6.94	10.30
3300	10″ diameter		80	.300			10.20	3.67	13.87	20.50
3390	Minimum labor/equipment charge		3	8	Job		273	98	371	550
3500	Railroad track removal, ties and track	B-13	330	.170	L.F.		5.80	2.39	8.19	12.10
3600	Ballast	B-14	500	.096	C.Y.		3.19	.59	3.78	5.85
3700	Remove and re-install, ties & track using new bolts & spikes		50	.960	L.F.		32	5.90	37.90	58.50
3800	Turnouts using new bolts and spikes		1	48	Ea.		1,600	294	1,894	2,925
3890	Minimum labor/equipment charge		5	9.600	Job		320	59	379	585
4000	Sidewalk removal, bituminous, 2-1/2″ thick	B-6	325	.074	S.Y.		2.52	.90	3.42	5.05
4050	Brick, set in mortar		185	.130			4.42	1.59	6.01	8.90
4100	Concrete, plain, 4″		160	.150			5.10	1.84	6.94	10.30
4200	Mesh reinforced		150	.160			5.45	1.96	7.41	11

02 41 13 – Selective Site Demolition

02 41 13.33 Minor Site Demolition		Crew	Daily Output	Labor-Hours	Unit	Material	2009 Bare Costs Labor	Equipment	Total	Total Incl O&P
4290	Minimum labor/equipment charge	B-39	12	4	Job		133	15.95	148.95	235
6850	Runways, remove rubber skid marks, 4-6 passes	B-59A	35	.686	M.S.F.	33	22	12.10	67.10	85.50
6860	6-10 passes	"	35	.686	"	49.50	22	12.10	83.60	103

02 41 13.34 Selective Demolition, Utility Materials

		Crew	Daily Output	Labor-Hours	Unit	Material	Labor	Equipment	Total	Total Incl O&P
0010	SELECTIVE DEMOLITION, UTILITY MATERIALS R024119-10									
0015	Excludes excavation									
0020	See other utility items in Div. 02 41 13.33									
0100	Fire Hydrant extensions	B-20	14	1.714	Ea.		60.50		60.50	99.50
0200	Precast Utility boxes up to 8′ x 14′ x 7′	B-13	2	28			955	395	1,350	1,975
0300	Handholes and meter pits	B-6	2	12			410	147	557	825
0400	Utility valves 4″-12″	B-20	4	6			212		212	350
0500	14″-24″	B-21	2	14			510	64.50	574.50	900

02 41 13.36 Selective Demolition, Utility Valves and Accessories

		Crew	Daily Output	Labor-Hours	Unit	Material	Labor	Equipment	Total	Total Incl O&P
0010	SELECTIVE DEMOLITION, UTILITY VALVES & ACCESSORIES									
0015	Excludes excavation									
0100	Utility valves 4″-12″ diam.	B-20	4	6	Ea.		212		212	350
0200	14″-24″ diam.	B-21	2	14			510	64.50	574.50	900
0300	Crosses 4″-12″	B-20	8	3			106		106	174
0400	14″-24″	B-21	4	7			255	32.50	287.50	450
0500	Utility cut-in valves 4″-12″ diam.	B-20	20	1.200			42.50		42.50	69.50
0600	Curb boxes	"	20	1.200			42.50		42.50	69.50

02 41 13.38 Selective Demo., Water & Sewer Piping & Fittings

		Crew	Daily Output	Labor-Hours	Unit	Material	Labor	Equipment	Total	Total Incl O&P
0010	SELECTIVE DEMOLITION, WATER & SEWER PIPING AND FITTINGS									
0015	Excludes excavation									
0090	Concrete pipe 4″-10″ dia	B-6	250	.096	L.F.		3.27	1.18	4.45	6.60
0100	42″-48″ diameter	B-13B	96	.583			19.95	11.70	31.65	45.50
0200	60″-84″ diameter	"	80	.700			24	14.05	38.05	54.50
0300	96″ diameter	B-13C	80	.700			24	20	44	61
0400	108″-144″ diameter	"	64	.875			30	25	55	76
0450	Concrete fittings 12″ diameter	B-6	24	1	Ea.		34	12.25	46.25	69
0480	Concrete end pieces 12″ diameter		200	.120	L.F.		4.09	1.47	5.56	8.25
0485	15″ diameter		150	.160			5.45	1.96	7.41	11
0490	18″ diameter		150	.160			5.45	1.96	7.41	11
0500	24″ -36″ diameter		100	.240			8.20	2.94	11.14	16.50
0600	Concrete fittings 24″-36″ diameter		12	2	Ea.		68	24.50	92.50	138
0700	48″-84″ diameter	B-13B	12	4.667			160	94	254	360
0800	96″ diameter	"	8	7			239	141	380	545
0900	108″-144″ diameter	B-13C	4	14			480	405	885	1,225
1000	Ductile iron pipe 4″ diameter	B-21B	200	.200	L.F.		6.85	3.84	10.69	15.40
1100	6″-12″ diameter		175	.229			7.80	4.39	12.19	17.60
1200	14″-24″ diameter		120	.333			11.40	6.40	17.80	25.50
1300	Ductile iron fittings 4″-12″ diameter		24	1.667	Ea.		57	32	89	128
1400	14″-16″ diameter		18	2.222			76	42.50	118.50	171
1500	18″-24″ diameter		12	3.333			114	64	178	257
1600	Plastic pipe 3/4″-4″ diameter	B-20	700	.034	L.F.		1.21		1.21	1.99
1700	6″-8″ diameter		500	.048			1.70		1.70	2.79
1800	10″-18″ diameter		300	.080			2.83		2.83	4.64
1900	20″-36″ diameter		200	.120			4.24		4.24	6.95
1910	42″-48″ diameter		180	.133			4.71		4.71	7.75
1920	54″-60″ diameter		160	.150			5.30		5.30	8.70
2000	Plastic fittings 4″- 8″ diameter	B-6	75	.320	Ea.		10.90	3.92	14.82	22
2100	10″-14″ diameter		50	.480			16.35	5.90	22.25	33

02 41 Demolition

02 41 13 – Selective Site Demolition

02 41 13.38 Selective Demo., Water & Sewer Piping & Fittings		Crew	Daily Output	Labor-Hours	Unit	Material	2009 Bare Costs Labor	Equipment	Total	Total Incl O&P
2200	16"-24" diameter	B-6	20	1.200	Ea.		41	14.70	55.70	82.50
2210	30"-36" diameter		15	1.600			54.50	19.60	74.10	110
2220	42 "- 48" diameter		12	2			68	24.50	92.50	138
2300	Copper pipe 3/4"-2" diameter	Q-1	500	.032	L.F.		1.40		1.40	2.18
2400	2 1/2" - 3" diameter		300	.053			2.34		2.34	3.63
2500	4"- 6" diameter		200	.080			3.51		3.51	5.45
2600	Copper fittings 3/4"- 2" diameter		15	1.067	Ea.		47		47	72.50
2700	Cast iron pipe 4" diameter		200	.080	L.F.		3.51		3.51	5.45
2800	5"- 6" diameter	Q-2	200	.120			5.45		5.45	8.45
2900	8"- 12" diameter	Q-3	200	.160			7.45		7.45	11.50
3000	Cast iron fittings 4" diameter	Q-1	30	.533	Ea.		23.50		23.50	36.50
3100	5"- 6" diameter	Q-2	30	.800			36.50		36.50	56.50
3200	8"- 15" diameter	Q-3	30	1.067			49.50		49.50	77
3300	Vent cast iron pipe 4"- 8" diameter	Q-1	200	.080	L.F.		3.51		3.51	5.45
3400	10"- 15" diameter	Q-3	200	.160	"		7.45		7.45	11.50
3500	Vent cast iron fittings 4"- 8" diameter	Q-2	30	.800	Ea.		36.50		36.50	56.50
3600	10"- 15" diameter	"	20	1.200	"		54.50		54.50	84.50

02 41 13.40 Selective Demolition, Metal Drainage Piping

		Crew	Daily Output	Labor-Hours	Unit	Material	Labor	Equipment	Total	Total Incl O&P
0010	**SELECTIVE DEMOLITION, METAL DRAINAGE PIPING**									
0015	Excludes excavation									
0100	CMP pipe, aluminum, 6"-10" dia	B-21	800	.035	L.F.		1.27	.16	1.43	2.26
0110	12" dia		600	.047			1.70	.22	1.92	3.01
0120	18" dia		600	.047			1.70	.22	1.92	3.01
0140	Steel, 6"-10" dia		800	.035			1.27	.16	1.43	2.26
0150	12" dia		600	.047			1.70	.22	1.92	3.01
0160	18" dia		400	.070			2.55	.32	2.87	4.50
0170	24" dia	B-13	300	.187			6.40	2.63	9.03	13.25
0180	30" - 36" dia		250	.224			7.65	3.16	10.81	15.90
0190	48" - 60" dia		200	.280			9.55	3.95	13.50	19.90
0200	72" dia	B-13B	100	.560			19.15	11.25	30.40	43.50
0210	CMP end sections, steel, 10"-18" dia	B-21	40	.700	Ea.		25.50	3.23	28.73	45
0220	24"-36" dia	B-13	30	1.867			64	26.50	90.50	133
0230	48" dia		20	2.800			95.50	39.50	135	200
0240	60" dia		10	5.600			191	79	270	395
0250	72" dia	B-13B	10	5.600			191	113	304	435
0260	CMP fittings, 8"-12" dia	B-21	60	.467			17	2.15	19.15	30
0270	18" dia	"	40	.700			25.50	3.23	28.73	45
0280	24"-48" dia	B-13	30	1.867			64	26.50	90.50	133
0290	60" dia	"	20	2.800			95.50	39.50	135	200
0300	72" dia	B-13B	10	5.600			191	113	304	435
0310	Oval arch 17" x 13",21" x 15", 15-18" equivalent	B-21	400	.070	L.F.		2.55	.32	2.87	4.50
0320	28" x 20", 24" equivalent	B-13	300	.187			6.40	2.63	9.03	13.25
0330	35" x 24",42" x 29", 30-36" equivalent		250	.224			7.65	3.16	10.81	15.90
0340	49" x 33",57" x 38", 42-48" equivalent		200	.280			9.55	3.95	13.50	19.90
0350	Oval arch 17" x 13" end piece, 15" equivalent	B-21	40	.700	Ea.		25.50	3.23	28.73	45
0360	42" x 29" end piece, 36" equivalent	B-13	30	1.867	"		64	26.50	90.50	133

02 41 13.42 Selective Demolition, Manholes and Catch Basins

		Crew	Daily Output	Labor-Hours	Unit	Material	Labor	Equipment	Total	Total Incl O&P
0010	**SELECTIVE DEMOLITION, MANHOLES & CATCH BASINS**									
0015	Excludes excavation									
0100	Manholes, precast or brick over 8' deep	B-6	8	3	V.L.F.		102	36.50	138.50	207
0200	Cast in place 4'-8' deep	B-9	127	.315	SF Face		10.10	1.51	11.61	18.30
0300	Over 8' deep	"	100	.400	"		12.80	1.92	14.72	23

02 41 Demolition

02 41 13 – Selective Site Demolition

	02 41 13.42 Selective Demolition, Manholes and Catch Basins	Crew	Daily Output	Labor-Hours	Unit	Material	2009 Bare Costs Labor	Equipment	Total	Total Incl O&P
0400	Top, precast, 8" thick, 4'-6' dia	B-6	8	3	Ea.		102	36.50	138.50	207
0500	Steps	1 Clab	60	.133	"		4.21		4.21	6.95

	02 41 13.43 Selective Demolition, Box Culvert	Crew	Daily Output	Labor-Hours	Unit	Material	Labor	Equipment	Total	Total Incl O&P
0010	SELECTIVE DEMOLITION, BOX CULVERT									
0015	Excludes excavation									
0100	Box culvert 8' x 6' x3' to 8' x 8' x 8'	B-69	300	.160	L.F.		5.55	4.32	9.87	13.75
0200	8' x 10' x 3' to 8' x 12' x 8'	"	200	.240	"		8.30	6.50	14.80	20.50

	02 41 13.44 Selective Demolition, Septic Tanks and Related Components	Crew	Daily Output	Labor-Hours	Unit	Material	Labor	Equipment	Total	Total Incl O&P
0010	SELECTIVE DEMOLITION, SEPTIC TANKS & RELATED COMPONENTS									
0020	Excludes excavation									
0100	Septic tanks, precast, 1000-1250 gal	B-21	8	3.500	Ea.		127	16.15	143.15	226
0200	1500 gal		7	4			146	18.45	164.45	258
0300	2000-2500 gal		5	5.600			204	26	230	360
0400	4000 gal	↓	4	7			255	32.50	287.50	450
0500	Precast, 5000 gal, 4 piece	B-13	3	18.667			640	263	903	1,325
0600	15,000 gal	B-13B	1.70	32.941			1,125	660	1,785	2,550
0700	25,000 gal		1.10	50.909			1,750	1,025	2,775	3,950
0800	40,000 gal	↓	.80	70			2,400	1,400	3,800	5,450
0900	Precast, 50,000 gal, 5 piece	B-13C	.60	93.333			3,200	2,675	5,875	8,125
1000	Cast-in-place, 75,000 gal	B-9	.06	666			21,300	3,200	24,500	38,700
1100	100,000 gal	"	.05	800			25,600	3,825	29,425	46,400
1200	HDPE, 1000 gal	B-21	9	3.111			113	14.35	127.35	201
1300	1500 gal		8	3.500			127	16.15	143.15	226
1400	Galley, 4' x 4' x 4'	↓	16	1.750			63.50	8.05	71.55	113
1500	Distribution boxes, concrete, 7 outlets	2 Clab	16	1			31.50		31.50	52
1600	9 outlets	"	8	2			63		63	104
1700	Leaching chambers 13' x 3'-7" x 1'-4", standard	B-13	16	3.500			120	49.50	169.50	249
1800	8' x 4' x 1'-6", heavy duty		14	4			137	56.50	193.50	284
1900	13' x 3'-9" x 1'-6"		12	4.667			160	66	226	330
2100	20' x 4' x 1'-6"	↓	5	11.200			385	158	543	795
2200	Leaching pit 6'-6" x 6' deep	B-21	5	5.600			204	26	230	360
2300	6'-6" x 8' deep		4	7			255	32.50	287.50	450
2400	8' x 6' deep H20		4	7			255	32.50	287.50	450
2500	8' x 8' deep H20		3	9.333			340	43	383	605
2600	Velocity reducing pit, precast 6' x 3' deep	↓	4.70	5.957	↓		217	27.50	244.50	385

	02 41 13.46 Selective Demolition, Steel Pipe With Insulation	Crew	Daily Output	Labor-Hours	Unit	Material	Labor	Equipment	Total	Total Incl O&P
0010	SELECTIVE DEMOLITION, STEEL PIPE WITH INSULATION									
0020	Excludes excavation									
0100	Steel pipe, with insulation, 3/4"-4"	B-1A	400	.060	L.F.		1.94	.46	2.40	3.71
0200	5"-10"	B-1B	360	.089			3.10	2.65	5.75	7.95
0300	12"-16"		240	.133			4.65	3.97	8.62	11.90
0400	18"-24"		160	.200			6.95	5.95	12.90	17.90
0450	26"-36"	↓	100	.320	↓		11.15	9.55	20.70	28.50
0500	Steel gland seal, with insulation, 3/4"-4"	B-1A	100	.240	Ea.		7.75	1.85	9.60	14.85
0600	5"-10"	B-1B	75	.427			14.85	12.70	27.55	38
0700	12"-16"		60	.533			18.60	15.90	34.50	47.50
0800	18"-24"		50	.640			22.50	19.10	41.60	57.50
0850	26"-36"	↓	40	.800			28	24	52	71.50
0900	Demo steel fittings with insulation 3/4"-4"	B-1A	60	.400			12.90	3.09	15.99	25
1000	5"-10"	B-1B	40	.800			28	24	52	71.50
1100	12"-16"		30	1.067			37	32	69	95.50
1200	18"-24"	↓	20	1.600	↓		55.50	47.50	103	143

02 41 Demolition

02 41 13 – Selective Site Demolition

02 41 13.46 Selective Demolition, Steel Pipe With Insulation		Crew	Daily Output	Labor-Hours	Unit	Material	2009 Bare Costs Labor	Equipment	Total	Total Incl O&P
1300	26"-36"	B-1B	15	2.133	Ea.		74.50	63.50	138	191
1400	Steel pipe anchors, 5"-10"		40	.800			28	24	52	71.50
1500	12"-16"		30	1.067			37	32	69	95.50
1600	18"-24"		20	1.600			55.50	47.50	103	143
1700	26"-36"		15	2.133			74.50	63.50	138	191

02 41 13.48 Selective Demolition, Gasoline Containment Piping

		Crew	Daily Output	Labor-Hours	Unit	Material	Labor	Equipment	Total	Total Incl O&P
0010	**SELECTIVE DEMOLITION, GASOLINE CONTAINMENT PIPING**									
0020	Excludes excavation									
0030	Excludes environmental site remediation									
0100	Gasoline plastic primary containment piping 2" to 4"	Q-6	800	.030	L.F.		1.38		1.38	2.14
0200	Fittings 2" to 4"		40	.600	Ea.		27.50		27.50	43
0300	Gasoline plastic secondary containment piping 3" to 6"		800	.030	L.F.		1.38		1.38	2.14
0400	Fittings 3" to 6"		40	.600	Ea.		27.50		27.50	43

02 41 13.50 Selective Demolition, Natural Gas, PE Pipe

		Crew	Daily Output	Labor-Hours	Unit	Material	Labor	Equipment	Total	Total Incl O&P
0010	**SELECTIVE DEMOLITION, NATURAL GAS, PE PIPE**									
0020	Excludes excavation									
0100	Natural gas coils, PE, 1 1/4" to 3"	Q-6	800	.030	L.F.		1.38		1.38	2.14
0200	Joints, 40', PE, 3 – 4"		800	.030			1.38		1.38	2.14
0300	6 – 8"		600	.040			1.84		1.84	2.86

02 41 13.51 Selective Demolition, Natural Gas, Steel Pipe

		Crew	Daily Output	Labor-Hours	Unit	Material	Labor	Equipment	Total	Total Incl O&P
0010	**SELECTIVE DEMOLITION, NATURAL GAS, STEEL PIPE**									
0020	Excludes excavation									
0100	Natural gas steel pipe 1- 4"	B-1A	800	.030	L.F.		.97	.23	1.20	1.85
0200	5 – 10"	B-1B	360	.089			3.10	2.65	5.75	7.95
0300	12 – 16"		240	.133			4.65	3.97	8.62	11.90
0400	18 – 24"		160	.200			6.95	5.95	12.90	17.90
0500	Natural gas steel fittings 1 – 4"	B-1A	160	.150	Ea.		4.84	1.16	6	9.25
0600	5 – 10"	B-1B	160	.200			6.95	5.95	12.90	17.90
0700	12 – 16"		108	.296			10.30	8.85	19.15	26.50
0800	18 – 24"		70	.457			15.95	13.65	29.60	41

02 41 13.52 Selective Demo, Natural Gas, Valves, Fittings, Regulators

		Crew	Daily Output	Labor-Hours	Unit	Material	Labor	Equipment	Total	Total Incl O&P
0010	**SELECTIVE DEMO, NATURAL GAS, VALVES, FITTINGS, REGULATORS**									
0100	Gas stops 1 1/4"-2"	1 Plum	22	.364	Ea.		17.75		17.75	27.50
0200	Gas regulator 1 1/2"-2"	"	22	.364			17.75		17.75	27.50
0300	3 – 4"	Q-1	22	.727			32		32	49.50
0400	Gas plug valve, 3/4 – 2"	1 Plum	22	.364			17.75		17.75	27.50
0500	2 1/2 – 3"	Q-1	10	1.600			70		70	109

02 41 13.54 Selective Demolition, Electric Ducts and Fittings

		Crew	Daily Output	Labor-Hours	Unit	Material	Labor	Equipment	Total	Total Incl O&P
0010	**SELECTIVE DEMOLITION, ELECTRIC DUCTS & FITTINGS**									
0020	Excludes excavation									
0100	Plastic conduit, 1/2" - 2"	1 Elec	600	.013	L.F.		.63		.63	.96
0200	3" - 6"	2 Elec	400	.040	"		1.88		1.88	2.89
0300	Fittings, 1/2" - 2"	1 Elec	50	.160	Ea.		7.50		7.50	11.55
0400	3" - 6"	"	40	.200	"		9.40		9.40	14.45

02 41 13.56 Selective Demolition, Electric Duct Banks

		Crew	Daily Output	Labor-Hours	Unit	Material	Labor	Equipment	Total	Total Incl O&P
0010	**SELECTIVE DEMOLITION, ELECTRIC DUCT BANKS**									
0020	Excludes excavation									
0100	Hand holes sized to 4' x 4' x 4'	R-3	7	2.857	Ea.		132	18.45	150.45	225
0200	Manholes sized to 6' x 10' x 7'	B-13	6	9.333	"		320	132	452	665
0300	Conduit 1 @ 2" diameter, EB plastic, no concrete	2 Elec	1000	.016	L.F.		.75		.75	1.16
0400	2 @ 2" diameter		500	.032			1.50		1.50	2.31

02 41 13.56 Selective Demolition, Electric Duct Banks		Crew	Daily Output	Labor-Hours	Unit	Material	2009 Bare Costs Labor	Equipment	Total	Total Incl O&P
0500	4 @ 2" diameter	2 Elec	250	.064	L.F.		3.01		3.01	4.63
0600	1 @ 3" diameter		800	.020			.94		.94	1.45
0700	2 @ 3" diameter		400	.040			1.88		1.88	2.89
0800	4 @ 3" diameter		200	.080			3.76		3.76	5.80
0900	1 @ 4" diameter		800	.020			.94		.94	1.45
1000	2 @ 4" diameter		400	.040			1.88		1.88	2.89
1100	4 @ 4" diameter		200	.080			3.76		3.76	5.80
1200	6 @ 4" diameter		100	.160			7.50		7.50	11.55
1300	1 @ 5" diameter		500	.032			1.50		1.50	2.31
1400	2 @ 5" diameter		250	.064			3.01		3.01	4.63
1500	4 @ 5" diameter		160	.100			4.70		4.70	7.25
1600	6 @ 5" diameter		100	.160			7.50		7.50	11.55
1700	1 @ 6" diameter		500	.032			1.50		1.50	2.31
1800	2 @ 6" diameter		250	.064			3.01		3.01	4.63
1900	4 @ 6" diameter		160	.100			4.70		4.70	7.25
2000	6 @ 6" diameter		100	.160			7.50		7.50	11.55
2100	Conduit 1 EB plastic, with concrete, 0.92 CF/LF	B-9	150	.267			8.55	1.28	9.83	15.50
2200	2 EB plastic 1.52 CF/LF		100	.400			12.80	1.92	14.72	23
2300	2 x 2 EB plastic 2.51 CF/LF		80	.500			16	2.40	18.40	29
2400	2 x 3 EB plastic 3.56 CF/LF		60	.667			21.50	3.19	24.69	38.50
2500	Conduit 2 @ 2" diameter, steel, no concrete	2 Elec	400	.040			1.88		1.88	2.89
2600	4 @ 2" diameter		200	.080			3.76		3.76	5.80
2700	2 @ 3" diameter		200	.080			3.76		3.76	5.80
2800	4 @ 3" diameter		100	.160			7.50		7.50	11.55
2900	2 @ 4" diameter		200	.080			3.76		3.76	5.80
3000	4 @ 4" diameter		100	.160			7.50		7.50	11.55
3100	6 @ 4" diameter		50	.320			15.05		15.05	23
3200	2 @ 5" diameter		120	.133			6.25		6.25	9.65
3300	4 @ 5" diameter		60	.267			12.55		12.55	19.30
3400	6 @ 5" diameter		40	.400			18.80		18.80	29
3500	2 @ 6" diameter		120	.133			6.25		6.25	9.65
3600	4 @ 6" diameter		60	.267			12.55		12.55	19.30
3700	6 @ 6" diameter		40	.400			18.80		18.80	29
3800	Conduit 2 steel, with concrete, 1.52 CF/LF	B-9	80	.500			16	2.40	18.40	29
3900	2 x 2 EB steel 2.51 CF/LF		60	.667			21.50	3.19	24.69	38.50
4000	2 x 3 steel 3.56 CF/LF		50	.800			25.50	3.83	29.33	46.50
4100	Conduit fittings, PVC type EB, 2" - 3"	1 Elec	30	.267	Ea.		12.55		12.55	19.30
4200	4" - 6"	"	20	.400	"		18.80		18.80	29

02 41 13.60 Selective Demolition Fencing		Crew	Daily Output	Labor-Hours	Unit	Material	Labor	Equipment	Total	Total Incl O&P
0010	**SELECTIVE DEMOLITION FENCING** R024119-10									
0700	Snow fence, 4' high	B-6	1000	.024	L.F.		.82	.29	1.11	1.65
1600	Fencing, barbed wire, 3 strand	2 Clab	430	.037			1.18		1.18	1.94
1650	5 strand	"	280	.057			1.81		1.81	2.98
1700	Chain link, posts & fabric, 8' to 10' high, remove only	B-6	445	.054			1.84	.66	2.50	3.71
1750	Remove and reset	"	70	.343			11.70	4.20	15.90	23.50

02 41 13.62 Selective Demo., Chain Link Fences & Gates		Crew	Daily Output	Labor-Hours	Unit	Material	Labor	Equipment	Total	Total Incl O&P
0010	**SELECTIVE DEMOLITION, CHAIN LINK FENCES & GATES**									
0100	Chain link, gates, 3' - 4' width	B-6	30	.800	Ea.		27.50	9.80	37.30	55
0200	10-12' width		16	1.500			51	18.35	69.35	103
0300	14' width		15	1.600			54.50	19.60	74.10	110
0400	20' width		10	2.400			82	29.50	111.50	166
0500	18' width with overhead & cantilever		80	.300	L.F.		10.20	3.67	13.87	20.50

02 41 13.62 Selective Demo., Chain Link Fences & Gates

		Crew	Daily Output	Labor-Hours	Unit	Material	2009 Bare Costs Labor	Equipment	Total	Total Incl O&P
0510	Sliding	B-6	80	.300	L.F.		10.20	3.67	13.87	20.50
0520	Cantilever to 40′ wide	↓	80	.300	↓		10.20	3.67	13.87	20.50
0530	Motor operators	2 Skwk	1	16	Ea.		655		655	1,075
0540	Transmitter systems	"	15	1.067	"		43.50		43.50	71
0600	Chain link, fence, 5′ high	B-6	890	.027	L.F.		.92	.33	1.25	1.85
0650	3-4′ high		1000	.024			.82	.29	1.11	1.65
0675	12′ high	↓	400	.060	↓		2.04	.73	2.77	4.13
0800	Chain link, fence, braces	B-1	2000	.012	Ea.		.39		.39	.64
0900	Privacy slats	"	2000	.012			.39		.39	.64
1000	Fence posts, steel, in concrete	B-6	80	.300	↓		10.20	3.67	13.87	20.50
1100	Fence fabric & accessories, fabric to 8′ high		800	.030	L.F.		1.02	.37	1.39	2.06
1200	Barbed wire		5000	.005	"		.16	.06	.22	.33
1300	Extension arms & eye tops		300	.080	Ea.		2.73	.98	3.71	5.50
1400	Fence rails		2000	.012	L.F.		.41	.15	.56	.82
1500	Reinforcing wire	↓	5000	.005	"		.16	.06	.22	.33

02 41 13.64 Selective Demolition, Vinyl Fences and Gates

		Crew	Daily Output	Labor-Hours	Unit	Material	Labor	Equipment	Total	Total Incl O&P
0010	**SELECTIVE DEMOLITION, VINYL FENCES & GATES**									
0100	Vinyl fence up to 6 ′ high	B-6	1000	.024	L.F.		.82	.29	1.11	1.65
0200	Gates, up to 6′ high	"	40	.600	Ea.		20.50	7.35	27.85	41

02 41 13.66 Selective Demolition, Misc Metal Fences and Gates

		Crew	Daily Output	Labor-Hours	Unit	Material	Labor	Equipment	Total	Total Incl O&P
0010	**SELECTIVE DEMOLITION, MISC METAL FENCES & GATES**									
0100	Misc steel mesh fences, 4′ - 6′ high	B-6	600	.040	L.F.		1.36	.49	1.85	2.75
0200	Kennels, 6′ - 12′ long	2 Clab	8	2	Ea.		63		63	104
0300	Tops, 6′ - 12′ long	"	30	.533	"		16.85		16.85	28
0400	Security fences, 12′ - 16′ high	B-6	100	.240	L.F.		8.20	2.94	11.14	16.50
0500	Metal tubular picket fences 4′ - 6′ high		500	.048	"		1.64	.59	2.23	3.30
0600	Gates 3′ - 4′ wide	↓	20	1.200	Ea.		41	14.70	55.70	82.50

02 41 13.68 Selective Demolition, Wood Fences and Gates

		Crew	Daily Output	Labor-Hours	Unit	Material	Labor	Equipment	Total	Total Incl O&P
0010	**SELECTIVE DEMOLITION, WOOD FENCES & GATES**									
0100	Wood fence gates 3′ - 4′ wide	2 Clab	20	.800	Ea.		25.50		25.50	41.50
0200	Wood fence, open rail, to 4′ high		2560	.006	L.F.		.20		.20	.33
0300	to 8′ high		368	.043	"		1.37		1.37	2.27
0400	Post, in concrete	↓	50	.320	Ea.		10.10		10.10	16.70

02 41 13.70 Selective Demolition, Rip-Rap and Rock Lining

		Crew	Daily Output	Labor-Hours	Unit	Material	Labor	Equipment	Total	Total Incl O&P
0010	**SELECTIVE DEMOLITION, RIP-RAP & ROCK LINING** R024119-10									
0100	Slope protection, broken stone	B-13	62	.903	C.Y.		31	12.75	43.75	64
0200	3/8 to 1/4 CY pieces		60	.933	S.Y.		32	13.15	45.15	66.50
0300	18 inch depth		60	.933	"		32	13.15	45.15	66.50
0400	Dumped stone		93	.602	Ton		20.50	8.50	29	43
0500	Gabions, 6 – 12 inches deep		60	.933	S.Y.		32	13.15	45.15	66.50
0600	18 – 36 inches deep	↓	30	1.867	"		64	26.50	90.50	133

02 41 13.72 Selective Demo., Shore Protect/Mooring Struct.

		Crew	Daily Output	Labor-Hours	Unit	Material	Labor	Equipment	Total	Total Incl O&P
0010	**SELECTIVE DEMOLITION, SHORE PROTECT/MOORING STRUCTURES**									
0100	Breakwaters, bulkheads, concrete, maximum	B-9	12	3.333	L.F.		107	15.95	122.95	194
0200	Breakwaters, bulkheads, concrete, 12′, minimum		10	4			128	19.15	147.15	232
0300	Maximum	↓	9	4.444			142	21.50	163.50	259
0400	Steel, from shore	B-40B	54	.889			31	24.50	55.50	77
0500	from barge	B-76A	30	2.133	↓		72.50	62	134.50	186
0600	Jetties, docks, floating	B-21B	600	.067	S.F.		2.28	1.28	3.56	5.15
0700	Pier supported, 3″ - 4″ decking		300	.133			4.56	2.56	7.12	10.25
0800	Floating, prefab, small boat, minimum	↓	600	.067	↓		2.28	1.28	3.56	5.15

02 41 Demolition

02 41 13 – Selective Site Demolition

02 41 13.72 Selective Demo., Shore Protect/Mooring Struct.		Crew	Daily Output	Labor-Hours	Unit	Material	2009 Bare Costs Labor	Equipment	Total	Total Incl O&P
0900	Maximum	B-21B	300	.133	S.F.		4.56	2.56	7.12	10.25
1000	Floating, prefab, per slip, minimum		3	13.333	Ea.		455	256	711	1,025

02 41 13.74 Selective Demolition, Piles

		Crew	Daily Output	Labor-Hours	Unit	Material	Labor	Equipment	Total	Total Incl O&P
0010	**SELECTIVE DEMOLITION, PILES**									
0100	Cast in place piles, corrugated, 8" - 10"	B-19	600	.107	V.L.F.		4.22	2.94	7.16	10.15
0200	12"-14"		500	.128			5.05	3.53	8.58	12.20
0300	16"		400	.160			6.35	4.41	10.76	15.20
0400	fluted, 12"		600	.107			4.22	2.94	7.16	10.15
0500	14"-18"		500	.128			5.05	3.53	8.58	12.20
0600	end bearing, 12 "		600	.107			4.22	2.94	7.16	10.15
0700	14"-18"		500	.128			5.05	3.53	8.58	12.20
0800	Precast prestressed piles, 12"-14" dia		700	.091			3.62	2.52	6.14	8.65
0900	16"-24" dia		600	.107			4.22	2.94	7.16	10.15
1000	36"-66" dia		300	.213			8.45	5.90	14.35	20.50
1100	10"-14" thick		600	.107			4.22	2.94	7.16	10.15
1200	16"-24" thick		500	.128			5.05	3.53	8.58	12.20
1300	Pressure grouted pile, 5"		150	.427			16.85	11.75	28.60	40.50
1400	Steel piles, 8"-12" tip		600	.107			4.22	2.94	7.16	10.15
1500	H sections HP8 to HP12		600	.107			4.22	2.94	7.16	10.15
1600	HP14	B-19A	600	.107			4.22	3.82	8.04	11.10
1700	Steel pipe piles, 8"-12"	B-19	600	.107			4.22	2.94	7.16	10.15
1800	14"-18" plain		500	.128			5.05	3.53	8.58	12.20
1900	concrete filled, 14"-18"		400	.160			6.35	4.41	10.76	15.20
2000	Timber piles to 14" dia		600	.107			4.22	2.94	7.16	10.15

02 41 13.76 Selective Demolition, Water Wells

		Crew	Daily Output	Labor-Hours	Unit	Material	Labor	Equipment	Total	Total Incl O&P
0010	**SELECTIVE DEMOLITION, WATER WELLS**									
0100	Well, 40' deep with casing & gravel pack, 24"-36" dia	B-23	.25	160	Ea.		5,125	11,100	16,225	20,700
0200	Riser pipe, 1 1/4 ", for observation well	"	300	.133	L.F.		4.27	9.25	13.52	17.25
0300	Pump, 1/2 to 5 HP up to 100' depth	Q-1	3	5.333	Ea.		234		234	365
0400	Up to 150' well 25 HP pump	Q-22	1.50	10.667			470	515	985	1,300
0500	Up to 500' well 30 HP pump	"	1	16			700	770	1,470	1,950
0600	Well screen 2" to 8"	B-23	300	.133	V.L.F.		4.27	9.25	13.52	17.25
0700	10" to 16"		200	.200			6.40	13.90	20.30	26
0800	18" to 26"		150	.267			8.55	18.50	27.05	34.50
0900	Slotted PVC for wells 1 1/4 "-8"		600	.067			2.13	4.63	6.76	8.60
1000	Well screen and casing 6" to 16"		300	.133			4.27	9.25	13.52	17.25
1100	18" to 26"		150	.267			8.55	18.50	27.05	34.50
1200	30" to 36"		100	.400			12.80	28	40.80	51.50

02 41 13.78 Selective Demolition, Radio Towers

		Crew	Daily Output	Labor-Hours	Unit	Material	Labor	Equipment	Total	Total Incl O&P
0010	**SELECTIVE DEMOLITION, RADIO TOWERS**									
0100	Radio tower,guyed, 50'	2 Skwk	16	1	Ea.		41		41	66.50
0200	190', 40 Lb section	K-2	.70	34.286			1,400	345	1,745	2,900
0300	200', 70 Lb section		.70	34.286			1,400	345	1,745	2,900
0400	300', 70 Lb section		.40	60			2,450	605	3,055	5,075
0500	270', 90 Lb section		.40	60			2,450	605	3,055	5,075
0600	400'		.30	80			3,250	810	4,060	6,775
0700	Self supported, 60'		.90	26.667			1,075	270	1,345	2,250
0800	120'		.80	30			1,225	305	1,530	2,525
0900	190'		.40	60			2,450	605	3,055	5,075

02 41 13.80 Selective Demo., Utility Poles and Cross Arms

		Crew	Daily Output	Labor-Hours	Unit	Material	Labor	Equipment	Total	Total Incl O&P
0010	**SELECTIVE DEMOLITION, UTILITY POLES & CROSS ARMS**									
0100	Utility poles, wood, 20' - 30' high	R-3	6	3.333	Ea.		154	21.50	175.50	262

02 41 Demolition

02 41 13 – Selective Site Demolition

02 41 13.80 Selective Demo., Utility Poles and Cross Arms		Crew	Daily Output	Labor-Hours	Unit	Material	2009 Bare Costs Labor	Equipment	Total	Total Incl O&P
0200	35′ - 45′ high	R-3	5	4	Ea.		185	26	211	315
0300	Cross arms, wood,4′ - 6′ long	1 Elec	5	1.600			75		75	116

02 41 13.82 Selective Removal, Pavement Lines and Markings		Crew	Daily Output	Labor-Hours	Unit	Material	Labor	Equipment	Total	Total Incl O&P
0010	**SELECTIVE REMOVAL, PAVEMENT LINES & MARKINGS**									
0015	Does not include traffic control costs									
0100	Remove permanent painted traffic lines and markings	B-78A	500	.016	C.L.F.		.62	1.66	2.28	2.80
0200	Temporary traffic line tape	2 Clab	1500	.011	L.F.		.34		.34	.56
0300	Thermoplastic traffic lines and markings	B-79A	500	.024	C.L.F.		.94	2.54	3.48	4.26
0400	Painted pavement markings	B-78B	500	.036	S.F.		1.17	.64	1.81	2.62

02 41 13.84 Selective Demolition, Walks, Steps and Pavers		Crew	Daily Output	Labor-Hours	Unit	Material	Labor	Equipment	Total	Total Incl O&P
0010	**SELECTIVE DEMOLITION, WALKS, STEPS AND PAVERS**									
0100	Splash blocks	1 Clab	300	.027	S.F.		.84		.84	1.39
0200	Tree grates	"	50	.160	Ea.		5.05		5.05	8.35
0300	Walks, limestone pavers	2 Clab	150	.107	S.F.		3.37		3.37	5.55
0400	Redwood sections		600	.027			.84		.84	1.39
0500	Redwood planks		480	.033			1.05		1.05	1.74
0600	Shale paver		300	.053			1.69		1.69	2.78
0700	Tile thinset paver		675	.024			.75		.75	1.24
0800	Wood round	B-1	350	.069	Ea.		2.21		2.21	3.65
0900	Asphalt block	2 Clab	450	.036	S.F.		1.12		1.12	1.85
1000	Bluestone		450	.036			1.12		1.12	1.85
1100	Slate, 1″ or thinner		675	.024			.75		.75	1.24
1200	Granite blocks		300	.053			1.69		1.69	2.78
1300	Precast patio blocks		450	.036			1.12		1.12	1.85
1400	Planter blocks		600	.027			.84		.84	1.39
1500	Brick paving, dry set		300	.053			1.69		1.69	2.78
1600	Mortar set		180	.089			2.81		2.81	4.64
1700	Dry set on edge		240	.067			2.11		2.11	3.48
1800	Steps, brick		200	.080	L.F.		2.53		2.53	4.17
1900	Railroad tie		150	.107			3.37		3.37	5.55
2000	Bluestone		180	.089			2.81		2.81	4.64
2100	Wood/steel edging for steps		1000	.016			.51		.51	.83
2200	Timber or railroad tie edging for steps		400	.040			1.26		1.26	2.09

02 41 13.86 Selective Demolition, Athletic Surfaces		Crew	Daily Output	Labor-Hours	Unit	Material	Labor	Equipment	Total	Total Incl O&P
0010	**SELECTIVE DEMOLITION, ATHLETIC SURFACES**									
0100	Synthetic grass	2 Clab	2000	.008	S.F.		.25		.25	.42
0200	Surface coat latex rubber	"	2000	.008	"		.25		.25	.42
0300	Tennis court posts	B-11C	16	1	Ea.		36.50	18.35	54.85	78.50

02 41 13.88 Selective Demolition, Lawn Sprinkler Systems		Crew	Daily Output	Labor-Hours	Unit	Material	Labor	Equipment	Total	Total Incl O&P
0010	**SELECTIVE DEMOLITION, LAWN SPRINKLER SYSTEMS**									
0100	Golf course sprinkler system, 9 hole	4 Skwk	.10	320	Ea.		13,100		13,100	21,300
0200	Sprinkler system, 24′ diam. @ 15′ O.C., per head	B-20	110	.218	Head		7.70		7.70	12.65
0300	60′ diam. @ 24′ O.C., per head	"	52	.462	"		16.30		16.30	27
0400	Sprinkler heads, plastic	2 Skwk	150	.107	Ea.		4.36		4.36	7.10
0500	Impact circle pattern, 28-76′ diam.		75	.213			8.70		8.70	14.20
0600	Pop-up, 42′-76′ diam.		50	.320			13.05		13.05	21.50
0700	39′ - 99′ diameter		50	.320			13.05		13.05	21.50
0800	Sprinkler valves		40	.400			16.35		16.35	26.50
0900	Valve boxes		40	.400			16.35		16.35	26.50
1000	Controls		2	8			325		325	530
1100	Backflow preventer		4	4			163		163	266
1200	Vacuum breaker		4	4			163		163	266

02 41 Demolition

02 41 13 – Selective Site Demolition

02 41 13.90 Selective Demolition, Retaining Walls

	02 41 13.90 Selective Demolition, Retaining Walls	Crew	Daily Output	Labor-Hours	Unit	Material	2009 Bare Costs Labor	Equipment	Total	Total Incl O&P
0010	**SELECTIVE DEMOLITION, RETAINING WALLS**									
0100	Concrete retaining wall, 6′ high, no reinforcing	B-9	12.70	3.150	L.F.		101	15.10	116.10	183
0200	8′ high		10	4			128	19.15	147.15	232
0300	10′ high		7.80	5.128			164	24.50	188.50	298
0400	With reinforcing, 6′ high		11.50	3.478			111	16.65	127.65	202
0500	8′ high		9	4.444			142	21.50	163.50	259
0600	10′ high		7	5.714			183	27.50	210.50	330
0700	20′ high		4	10			320	48	368	585
0800	Concrete cribbing, 12′ high, open/closed face		126	.317	S.F.		10.15	1.52	11.67	18.40
0900	Interlocking segmental retaining wall	B-62	800	.030			1.02	.19	1.21	1.87
1000	Wall caps	"	600	.040			1.36	.25	1.61	2.48
1100	Metal bin retaining wall, 10′ wide, 4-12′ high	B-13	1200	.047			1.60	.66	2.26	3.31
1200	10′ wide, 16-28′ high		1000	.056			1.91	.79	2.70	3.98
1300	Stone filled gabions, 6′ x 3′ x 1′		170	.329	Ea.		11.25	4.64	15.89	23.50
1400	6′ x 3′ x 1′-6″		75	.747			25.50	10.55	36.05	53
1500	6′ x 3′ x 3′		25	2.240			76.50	31.50	108	159
1600	9′ x 3′ x 1′		75	.747			25.50	10.55	36.05	53
1700	9′ x 3′ x 1′-6″		33	1.697			58	24	82	121
1800	9′ x 3′ x 3′		12	4.667			160	66	226	330
1900	12′ x 3′ x 1′		42	1.333			45.50	18.80	64.30	94.50
2000	12′ x 3′ x 1′-6″		20	2.800			95.50	39.50	135	200
2100	12′ x 3′ x 3′		6	9.333			320	132	452	665

02 41 13.92 Selective Demolition, Parking Appurtenances

	02 41 13.92 Selective Demolition, Parking Appurtenances	Crew	Daily Output	Labor-Hours	Unit	Material	2009 Bare Costs Labor	Equipment	Total	Total Incl O&P
0010	**SELECTIVE DEMOLITION, PARKING APPURTENANCES**									
0100	Bumper rails, garage, 6″ wide	B-6	300	.080	L.F.		2.73	.98	3.71	5.50
0200	12″ channel rail		300	.080			2.73	.98	3.71	5.50
0300	Parking bumper, timber		1000	.024			.82	.29	1.11	1.65
0400	Folding, with locks	B-1	100	.240	Ea.		7.75		7.75	12.80
0500	Flexible fixed garage stanchion	B-6	150	.160			5.45	1.96	7.41	11
0600	Wheel stops, precast concrete		120	.200			6.80	2.45	9.25	13.75
0700	Thermoplastic		120	.200			6.80	2.45	9.25	13.75
0800	Pipe bollards, 6″ - 12″ dia		80	.300			10.20	3.67	13.87	20.50

02 41 13.93 Selective Demolition, Site Furnishings

	02 41 13.93 Selective Demolition, Site Furnishings	Crew	Daily Output	Labor-Hours	Unit	Material	2009 Bare Costs Labor	Equipment	Total	Total Incl O&P
0010	**SELECTIVE DEMOLITION, SITE FURNISHINGS**									
0100	Benches, all types	2 Clab	10	1.600	Ea.		50.50		50.50	83.50
0200	Trash receptacles, all types		80	.200			6.30		6.30	10.45
0300	Trash enclosures, steel or wood		10	1.600			50.50		50.50	83.50

02 41 13.94 Selective Demolition, Athletic Screening

	02 41 13.94 Selective Demolition, Athletic Screening	Crew	Daily Output	Labor-Hours	Unit	Material	2009 Bare Costs Labor	Equipment	Total	Total Incl O&P
0010	**SELECTIVE DEMOLITION, ATHLETIC SCREENING**									
0020	See other items in Div. 02 41 13.60 & 02 41 13.62									
0100	Baseball backstops	B-6	2	12	Ea.		410	147	557	825
0200	Basketball goal, single		6	4			136	49	185	275
0300	Double		4	6			204	73.50	277.50	410
0400	Tennis wire mesh, pair ends		5	4.800	Set		164	59	223	330
0500	Enclosed court		3	8	Ea.		273	98	371	550
0600	Wood or masonry handball court		1	24	"		820	294	1,114	1,650

02 41 13.95 Selective Demo., Athletic/Playground Equip.

	02 41 13.95 Selective Demo., Athletic/Playground Equip.	Crew	Daily Output	Labor-Hours	Unit	Material	2009 Bare Costs Labor	Equipment	Total	Total Incl O&P
0010	**SELECTIVE DEMO., ATHLETIC/PLAYGROUND EQUIPMENT**									
0020	See other items in Div. 02 41 13.96									
0100	Bike rack	B-6	32	.750	Ea.		25.50	9.20	34.70	51.50
0200	Climber arch		16	1.500			51	18.35	69.35	103

02 41 13.95 Selective Demo., Athletic/Playground Equip.

02 41 13.95 Selective Demo., Athletic/Playground Equip.		Crew	Daily Output	Labor-Hours	Unit	Material	2009 Bare Costs Labor	Equipment	Total	Total Incl O&P
0300	Fitness trail, 9 to 10 stations wood or metal	B-6	.50	48	Ea.		1,625	590	2,215	3,300
0400	16 to 20 stations wood or metal		.25	96			3,275	1,175	4,450	6,600
0500	Monkey bars 14′ long		8	3			102	36.50	138.50	207
0600	Parallel bars 10′ long		8	3			102	36.50	138.50	207
0700	Post tether ball		32	.750			25.50	9.20	34.70	51.50
0800	Poles 10′-6″ long		32	.750			25.50	9.20	34.70	51.50
0900	Pole ground sockets		80	.300			10.20	3.67	13.87	20.50
1000	See-saw, 2 units		12	2			68	24.50	92.50	138
1100	4 units		10	2.400			82	29.50	111.50	166
1200	6 units		8	3			102	36.50	138.50	207
1300	Shelter, fiberglass, 3 person		10	2.400			82	29.50	111.50	166
1400	Slides, 12′ long		8	3			102	36.50	138.50	207
1500	20′ long		6	4			136	49	185	275
1600	Swings, 4 seats		6	4			136	49	185	275
1700	8 seats		4	6			204	73.50	277.50	410
1800	Whirlers, 8 – 10′ diameter		6	4			136	49	185	275
1900	Football or football/soccer goal posts		3	8	Pair		273	98	371	550
2000	Soccer goal posts		4	6	″		204	73.50	277.50	410
2100	Playground surfacing, 4 ″ depth	2 Clab	900	.018	S.F.		.56		.56	.93
2200	2″ topping	″	1800	.009	″		.28		.28	.46
2300	Platform/paddle tennis court, alum. deck & frame	B-1	.20	120	Court		3,875		3,875	6,400
2400	Aluminum deck and wood frame		.25	96	″		3,100		3,100	5,100
2500	Heater		2	12	Ea.		385		385	640
2600	Wood deck & frame		.25	96	Court		3,100		3,100	5,100
2700	Wood deck & steel frame		.25	96			3,100		3,100	5,100
2800	Steel deck & wood frame		.25	96			3,100		3,100	5,100

02 41 13.96 Selective Demo., Modular Playground Equipment

		Crew	Daily Output	Labor-Hours	Unit	Material	Labor	Equipment	Total	Total Incl O&P
0010	SELECTIVE DEMOLITION, MODULAR PLAYGROUND EQUIPMENT									
0100	Modular playground, 48″ deck, square or triangle shape	B-1	2	12	Ea.		385		385	640
0200	Various posts		18	1.333			43		43	71
0300	Roof 54″ x 54″		18	1.333			43		43	71
0400	Wheel chair transfer module		6	4			129		129	213
0500	Guardrail pipe		90	.267	L.F.		8.60		8.60	14.20
0600	Steps, 3 risers		16	1.500	Ea.		48.50		48.50	80
0700	Activity panel crawl through		8	3			97		97	160
0800	Alphabet/spelling panel		8	3			97		97	160
0900	Crawl tunnel, each 4′ section		8	3			97		97	160
1000	Slide tunnel		16	1.500			48.50		48.50	80
1100	Spiral slide tunnel		12	2			64.50		64.50	107
1200	Ladder, horizontal or vertical		10	2.400			77.50		77.50	128
1300	Corkscrew climber		6	4			129		129	213
1400	Fire pole		12	2			64.50		64.50	107
1500	Bridge, ring chamber		8	3			97		97	160
1600	Bridge, suspension		8	3			97		97	160

02 41 13.97 Selective Demolition, Traffic Light Systems

		Crew	Daily Output	Labor-Hours	Unit	Material	Labor	Equipment	Total	Total Incl O&P
0010	SELECTIVE DEMOLITION, TRAFFIC LIGHT SYSTEMS									
0100	Traffic signal, pedestrian cross walk	R-2	.50	112	Total		4,675	515	5,190	7,875
0200	8 signal intersection, 2 each direction	″	.25	224			9,325	1,025	10,350	15,700
0300	Each traffic phase controller	L-9	3	12			435		435	735
0400	Each semi-actuated detector		1.50	24			870		870	1,475
0500	Each fully-actuated detector		1.50	24			870		870	1,475
0600	Pedestrian push button system		2	18			655		655	1,100

02 41 Demolition

02 41 13 – Selective Site Demolition

02 41 13.97 Selective Demolition, Traffic Light Systems		Crew	Daily Output	Labor-Hours	Unit	Material	2009 Bare Costs Labor	Equipment	Total	Total Incl O&P
0650	Each optically programmed head	L-9	3	12	Total		435		435	735
0700	School signal flashing system	↓	1	36	Signal		1,300		1,300	2,200
0800	Complete traffic light intersection system, without lane control	R-2	.25	224	Ea.		9,325	1,025	10,350	15,700
0900	With lane control		.20	280			11,700	1,300	13,000	19,700
1000	Traffic light, each left turn protection system	↓	.66	84.848	↓		3,525	390	3,915	5,950

02 41 13.98 Select. Demo., Sod, Edging, Planters & Tree Guying

		Crew	Daily Output	Labor-Hours	Unit	Material	Labor	Equipment	Total	Total Incl O&P
0010	**SELECTIVE DEMOLITION, SOD, EDGING, PLANTERS & TREE GUYING**									
0100	Remove sod 1" deep, 3 M.S.F. or greater, level ground	B-63	40	1	M.S.F.		33	3.73	36.73	58
0200	Less than 3 M.S.F., level ground		27	1.431			49	5.55	54.55	86
0300	3 M.S.F. or greater, sloped ground		12	3.333			110	12.45	122.45	194
0400	Less than 3 M.S.F., sloped ground	↓	8	5	↓		165	18.65	183.65	291
0500	Edging, aluminum	B-1	800	.030	L.F.		.97		.97	1.60
0600	Brick horizontal		400	.060			1.94		1.94	3.20
0700	Brick vertical		200	.120			3.87		3.87	6.40
0800	Corrugated aluminum		1200	.020			.65		.65	1.07
0900	Granite 5" x 16"		600	.040			1.29		1.29	2.13
1000	Polyethylene grass barrier		1200	.020			.65		.65	1.07
1100	Precast 2" x 8" x 16"		800	.030			.97		.97	1.60
1200	Railroad ties		200	.120			3.87		3.87	6.40
1300	Wood		400	.060			1.94		1.94	3.20
1400	Steel strips including stakes	↓	600	.040	↓		1.29		1.29	2.13
1500	Planters, concrete, 48" dia	2 Clab	30	.533	Ea.		16.85		16.85	28
1600	Concrete 7' dia		20	.800			25.50		25.50	41.50
1700	Fiberglass 36" dia		30	.533			16.85		16.85	28
1800	60" dia		20	.800			25.50		25.50	41.50
1900	24" square		30	.533			16.85		16.85	28
2000	48" square		30	.533			16.85		16.85	28
2100	Wood 48" square		30	.533			16.85		16.85	28
2200	48" dia		20	.800			25.50		25.50	41.50
2300	72" dia		20	.800			25.50		25.50	41.50
2400	Planter/bench, fiberglass 72" - 96" square		10	1.600			50.50		50.50	83.50
2500	Wood, 72" square		10	1.600			50.50		50.50	83.50
2600	Tree guying, stakes, 3" - 4" caliper		40	.400			12.65		12.65	21
2700	anchors, 3" caliper		40	.400			12.65		12.65	21
2800	3" - 6" caliper		30	.533			16.85		16.85	28
2900	6" caliper		20	.800			25.50		25.50	41.50
3000	8" caliper	↓	15	1.067	↓		33.50		33.50	55.50

02 41 16 – Structure Demolition

02 41 16.13 Building Demolition

		Crew	Daily Output	Labor-Hours	Unit	Material	Labor	Equipment	Total	Total Incl O&P
0010	**BUILDING DEMOLITION** Large urban projects, incl. 20 mi. haul R024119-10									
0011	No foundation or dump fees, C.F. is vol. of building standing									
0020	Steel	B-8	21500	.003	C.F.		.10	.13	.23	.32
0050	Concrete		15300	.004			.15	.19	.34	.44
0080	Masonry		20100	.003			.11	.14	.25	.34
0100	Mixture of types, average	↓	20100	.003			.11	.14	.25	.34
0500	Small bldgs, or single bldgs, no salvage included, steel	B-3	14800	.003			.11	.14	.25	.33
0600	Concrete		11300	.004			.14	.18	.32	.43
0650	Masonry		14800	.003			.11	.14	.25	.33
0700	Wood	↓	14800	.003	↓		.11	.14	.25	.33
1000	Single family, one story house, wood, minimum				Ea.				3,870	4,260
1020	Maximum								6,650	7,315
1200	Two family, two story house, wood, minimum				↓				5,080	5,590

02 41 16 – Structure Demolition

02 41 16.13 Building Demolition		Crew	Daily Output	Labor-Hours	Unit	Material	2009 Bare Costs Labor	Equipment	Total	Total Incl O&P
1220	Maximum				Ea.				9,680	10,650
1300	Three family, three story house, wood, minimum								6,660	7,325
1320	Maximum								11,735	12,910
5000	For buildings with no interior walls, deduct								50%	

02 41 16.17 Building Demolition Footings and Foundations

		Crew	Daily Output	Labor-Hours	Unit	Material	Labor	Equipment	Total	Total Incl O&P
0010	**BUILDING DEMOLITION FOOTINGS AND FOUNDATIONS** R024119-10									
0200	Floors, concrete slab on grade,									
0240	4″ thick, plain concrete	B-9	500	.080	S.F.		2.56	.38	2.94	4.64
0280	Reinforced, wire mesh		470	.085			2.72	.41	3.13	4.94
0300	Rods		400	.100			3.20	.48	3.68	5.85
0400	6″ thick, plain concrete		375	.107			3.41	.51	3.92	6.20
0420	Reinforced, wire mesh		340	.118			3.76	.56	4.32	6.80
0440	Rods		300	.133			4.27	.64	4.91	7.75
1000	Footings, concrete, 1′ thick, 2′ wide	B-5	300	.187	L.F.		6.45	3.94	10.39	14.85
1080	1′-6″ thick, 2′ wide		250	.224			7.75	4.73	12.48	17.80
1120	3′ wide		200	.280			9.70	5.90	15.60	22.50
1140	2′ thick, 3′ wide		175	.320			11.10	6.75	17.85	25.50
1200	Average reinforcing, add								10%	10%
1220	Heavy reinforcing, add								20%	20%
2000	Walls, block, 4″ thick	1 Clab	180	.044	S.F.		1.40		1.40	2.32
2040	6″ thick		170	.047			1.49		1.49	2.45
2080	8″ thick		150	.053			1.69		1.69	2.78
2100	12″ thick		150	.053			1.69		1.69	2.78
2200	For horizontal reinforcing, add								10%	10%
2220	For vertical reinforcing, add								20%	20%
2400	Concrete, plain concrete, 6″ thick	B-9	160	.250			8	1.20	9.20	14.50
2420	8″ thick		140	.286			9.15	1.37	10.52	16.60
2440	10″ thick		120	.333			10.65	1.60	12.25	19.35
2500	12″ thick		100	.400			12.80	1.92	14.72	23
2600	For average reinforcing, add								10%	10%
2620	For heavy reinforcing, add								20%	20%
9000	Minimum labor/equipment charge	A-1	2	4	Job		126	31.50	157.50	244

02 41 16.33 Bridge Demolition

		Crew	Daily Output	Labor-Hours	Unit	Material	Labor	Equipment	Total	Total Incl O&P
0010	**BRIDGE DEMOLITION**									
0100	Bridges, pedestrian, precast, 60′ to 150′ long	B-21C	250	.224	S.F.		7.65	7.70	15.35	21
0200	Steel, 50′ to 160′ long, 8′ to 10′ wide	″	500	.112			3.83	3.85	7.68	10.45
0300	Laminated wood,80′ m to 130′ long	C-12	300	.160			6.30	2.56	8.86	13.10

02 41 19 – Selective Structure Demolition

02 41 19.13 Selective Building Demolition

		Crew	Daily Output	Labor-Hours	Unit	Material	Labor	Equipment	Total	Total Incl O&P
0010	**SELECTIVE BUILDING DEMOLITION**									
0020	Costs related to selective demolition of specific building components									
0025	are included under Common Work Results (XX 05 00)									
0030	in the component's appropriate division.									

02 41 19.16 Selective Demolition, Cutout

		Crew	Daily Output	Labor-Hours	Unit	Material	Labor	Equipment	Total	Total Incl O&P
0010	**SELECTIVE DEMOLITION, CUTOUT** R024119-10									
0020	Concrete, elev. slab, light reinforcement, under 6 C.F.	B-9	65	.615	C.F.		19.70	2.95	22.65	35.50
0050	Light reinforcing, over 6 C.F.		75	.533	″		17.05	2.55	19.60	31
0200	Slab on grade to 6″ thick, not reinforced, under 8 S.F.		85	.471	S.F.		15.05	2.25	17.30	27.50
0250	8 – 16 S.F.		175	.229	″		7.30	1.09	8.39	13.25
0255	For over 16 S.F. see Div. 02 41 16.17 0400									
0600	Walls, not reinforced, under 6 C.F.	B-9	60	.667	C.F.		21.50	3.19	24.69	38.50

02 41 Demolition

02 41 19 – Selective Structure Demolition

02 41 19.16 Selective Demolition, Cutout		Crew	Daily Output	Labor-Hours	Unit	Material	2009 Bare Costs Labor	Equipment	Total	Total Incl O&P
0650	6 – 12 C.F.	B-9	80	.500	C.F.		16	2.40	18.40	29
0655	For over 12 C.F. see Div. 02 41 16.17 2500									
1000	Concrete, elevated slab, bar reinforced, under 6 C.F.	B-9	45	.889	C.F.		28.50	4.26	32.76	51.50
1050	Bar reinforced, over 6 C.F.		50	.800	″		25.50	3.83	29.33	46.50
1200	Slab on grade to 6″ thick, bar reinforced, under 8 S.F.		75	.533	S.F.		17.05	2.55	19.60	31
1250	8 – 16 S.F.		150	.267	″		8.55	1.28	9.83	15.50
1255	For over 16 S.F. see Div. 02 41 16.17 0440									
1400	Walls, bar reinforced, under 6 C.F.	B-9	50	.800	C.F.		25.50	3.83	29.33	46.50
1450	6 – 12 C.F.	″	70	.571	″		18.30	2.74	21.04	33
1455	For over 12 C.F. see Div. 02 41 16.17 2500 and 2600									
2000	Brick, to 4 S.F. opening, not including toothing									
2040	4″ thick	B-9	30	1.333	Ea.		42.50	6.40	48.90	77.50
2060	8″ thick		18	2.222			71	10.65	81.65	129
2080	12″ thick		10	4			128	19.15	147.15	232
2400	Concrete block, to 4 S.F. opening, 2″ thick		35	1.143			36.50	5.45	41.95	66.50
2420	4″ thick		30	1.333			42.50	6.40	48.90	77.50
2440	8″ thick		27	1.481			47.50	7.10	54.60	86
2460	12″ thick		24	1.667			53.50	8	61.50	97
2600	Gypsum block, to 4 S.F. opening, 2″ thick		80	.500			16	2.40	18.40	29
2620	4″ thick		70	.571			18.30	2.74	21.04	33
2640	8″ thick		55	.727			23.50	3.48	26.98	42.50
2800	Terra cotta, to 4 S.F. opening, 4″ thick		70	.571			18.30	2.74	21.04	33
2840	8″ thick		65	.615			19.70	2.95	22.65	35.50
2880	12″ thick		50	.800			25.50	3.83	29.33	46.50
4000	For toothing masonry, see Div. 04 01 20.50									
6000	Walls, interior, not including re-framing,									
6010	openings to 5 S.F.									
6100	Drywall to 5/8″ thick	1 Clab	24	.333	Ea.		10.55		10.55	17.40
6200	Paneling to 3/4″ thick		20	.400			12.65		12.65	21
6300	Plaster, on gypsum lath		20	.400			12.65		12.65	21
6340	On wire lath		14	.571			18.05		18.05	30
7000	Wood frame, not including re-framing, openings to 5 S.F.									
7200	Floors, sheathing and flooring to 2″ thick	1 Clab	5	1.600	Ea.		50.50		50.50	83.50
7310	Roofs, sheathing to 1″ thick, not including roofing		6	1.333			42		42	69.50
7410	Walls, sheathing to 1″ thick, not including siding		7	1.143			36		36	59.50
8500	Minimum labor/equipment charge		4	2	Job		63		63	104

02 41 19.18 Selective Demolition, Disposal Only		Crew	Daily Output	Labor-Hours	Unit	Material	Labor	Equipment	Total	Total Incl O&P
0010	**SELECTIVE DEMOLITION, DISPOSAL ONLY** R024119-10									
0015	Urban bldg w/salvage value allowed									
0020	Including loading and 5 mile haul to dump									
0200	Steel frame	B-3	430	.112	C.Y.		3.76	4.78	8.54	11.40
0300	Concrete frame		365	.132			4.43	5.65	10.08	13.40
0400	Masonry construction		445	.108			3.63	4.62	8.25	11
0500	Wood frame		247	.194			6.55	8.30	14.85	19.80

02 41 19.19 Selective Demolition, Dump Charges		Crew	Daily Output	Labor-Hours	Unit	Material	Labor	Equipment	Total	Total Incl O&P
0010	**SELECTIVE DEMOLITION, DUMP CHARGES** R024119-10									
0020	Dump charges, typical urban city, tipping fees only									
0100	Building construction materials				Ton	100			100	110
0200	Trees, brush, lumber					75			75	82.50
0300	Rubbish only					90			90	99
0500	Reclamation station, usual charge					100			100	110

02 41 Demolition

02 41 19 – Selective Structure Demolition

02 41 19.21 Selective Demolition, Gutting

Line	Description	Crew	Daily Output	Labor-Hours	Unit	2009 Bare Costs Material	Labor	Equipment	Total	Total Incl O&P
0010	SELECTIVE DEMOLITION, GUTTING R024119-10									
0020	Building interior, including disposal, dumpster fees not included									
0500	Residential building									
0560	Minimum	B-16	400	.080	SF Flr.		2.58	1.33	3.91	5.70
0580	Maximum	"	360	.089	"		2.86	1.48	4.34	6.35
0900	Commercial building									
1000	Minimum	B-16	350	.091	SF Flr.		2.94	1.52	4.46	6.55
1020	Maximum		250	.128	"		4.12	2.13	6.25	9.15
3000	Minimum labor/equipment charge		4	8	Job		258	133	391	570

02 41 19.23 Selective Demolition, Rubbish Handling

Line	Description	Crew	Daily Output	Labor-Hours	Unit	Material	Labor	Equipment	Total	Total Incl O&P
0010	SELECTIVE DEMOLITION, RUBBISH HANDLING R024119-10									
0020	The following are to be added to the demolition prices									
0400	Chute, circular, prefabricated steel, 18" diameter	B-1	40	.600	L.F.	47	19.35		66.35	84
0440	30" diameter	"	30	.800	"	48.50	26		74.50	96
0725	Dumpster, weekly rental, 1 dump/week, 20 C.Y. capacity (8 Tons) R024119-20				Week	775			775	852.50
0800	30 C.Y. capacity (10 Tons)					1,000			1,000	1,100
0840	40 C.Y. capacity (13 Tons)					1,300			1,300	1,430
0900	Alternate pricing for dumpsters									
0910	Delivery, average for all sizes				Ea.	75			75	82.50
0920	Haul, average for all sizes					220			220	242
0930	Rent per day, average for all sizes					10			10	11
0940	Rent per month, average for all sizes					75			75	82.50
0950	Disposal fee per ton, average for all sizes				Ton	75			75	82.50
1000	Dust partition, 6 mil polyethylene, 1" x 3" frame	2 Carp	2000	.008	S.F.	.44	.32		.76	1.02
1080	2" x 4" frame	"	2000	.008	"	.27	.32		.59	.83
2000	Load, haul, and dump, 50' haul	2 Clab	24	.667	C.Y.		21		21	35
2040	100' haul		16.50	.970			30.50		30.50	50.50
2080	Over 100' haul, add per 100 L.F.		35.50	.451			14.25		14.25	23.50
2120	In elevators, per 10 floors, add		140	.114			3.61		3.61	5.95
3000	Loading & trucking, including 2 mile haul, chute loaded	B-16	45	.711			23	11.85	34.85	50.50
3040	Hand loading truck, 50' haul	"	48	.667			21.50	11.10	32.60	47.50
3080	Machine loading truck	B-17	120	.267			8.95	5.15	14.10	20.50
5000	Haul, per mile, up to 8 C.Y. truck	B-34B	1165	.007			.22	.46	.68	.86
5100	Over 8 C.Y. truck	"	1550	.005			.16	.34	.50	.65

02 41 19.25 Selective Demolition, Saw Cutting

Line	Description	Crew	Daily Output	Labor-Hours	Unit	Material	Labor	Equipment	Total	Total Incl O&P
0010	SELECTIVE DEMOLITION, SAW CUTTING R024119-10									
0015	Asphalt, up to 3" deep	B-89	1050	.015	L.F.	.40	.53	.38	1.31	1.72
0020	Each additional inch of depth	"	1800	.009		.08	.31	.22	.61	.84
1200	Masonry walls, hydraulic saw, brick, per inch of depth	B-89B	300	.053		.42	1.87	2.28	4.57	5.95
1220	Block walls, solid, per inch of depth	"	250	.064		.42	2.24	2.74	5.40	7.05
2000	Brick or masonry w/hand held saw, per inch of depth	A-1	125	.064		.34	2.02	.51	2.87	4.27
5000	Wood sheathing to 1" thick, on walls	1 Carp	200	.040			1.60		1.60	2.64
5020	On roof	"	250	.032			1.28		1.28	2.11
9000	Minimum labor/equipment charge	A-1	2	4	Job		126	31.50	157.50	244

02 41 19.27 Selective Demolition, Torch Cutting

Line	Description	Crew	Daily Output	Labor-Hours	Unit	Material	Labor	Equipment	Total	Total Incl O&P
0010	SELECTIVE DEMOLITION, TORCH CUTTING R024119-10									
0020	Steel, 1" thick plate	1 Clab	360	.022	L.F.	.22	.70		.92	1.40
0040	1" diameter bar	"	210	.038	Ea.		1.20		1.20	1.99
1000	Oxygen lance cutting, reinforced concrete walls									
1040	12" to 16" thick walls	1 Clab	10	.800	L.F.		25.50		25.50	41.50
1080	24" thick walls	"	6	1.333	"		42		42	69.50

02 41 Demolition

02 41 19 – Selective Structure Demolition

02 41 19.27 Selective Demolition, Torch Cutting			Crew	Daily Output	Labor-Hours	Unit	Material	2009 Bare Costs Labor	Equipment	Total	Total Incl O&P
1090	Minimum labor/equipment charge		E-25	2	4	Job		187	46.50	233.50	395
1100	See Div. 05 05 21.10										

02 43 Structure Moving

02 43 13 – Structure Relocation

02 43 13.13 Building Relocation

			Crew	Daily Output	Labor-Hours	Unit	Material	Labor	Equipment	Total	Total Incl O&P
0010	**BUILDING RELOCATION**										
0011	One day move, up to 24′ wide										
0020	Reset on new foundation, patch & hook-up, average move					Total					11,765
0040	Wood or steel frame bldg., based on ground floor area	G	B-4	185	.259	S.F.		8.30	2.31	10.61	16.25
0060	Masonry bldg., based on ground floor area	G	"	137	.350			11.20	3.13	14.33	22
0200	For 24′ to 42′ wide, add					↓					15%
0220	For each additional day on road, add	G	B-4	1	48	Day		1,525	430	1,955	3,000
0240	Construct new basement, move building, 1 day										
0300	move, patch & hook-up, based on ground floor area	G	B-3	155	.310	S.F.	9	10.45	13.25	32.70	41.50

02 58 Snow Control

02 58 13 – Snow Fencing

02 58 13.10 Snow Fencing System

		Crew	Daily Output	Labor-Hours	Unit	Material	Labor	Equipment	Total	Total Incl O&P
0010	**SNOW FENCING SYSTEM**									
7001	Snow fence on steel posts 10′ O.C., 4′ high	B-1	500	.048	L.F.	2.70	1.55		4.25	5.55

02 65 Underground Storage Tank Removal

02 65 10 – Underground Tank and Contaminated Soil Removal

02 65 10.30 Removal of Underground Storage Tanks

			Crew	Daily Output	Labor-Hours	Unit	Material	Labor	Equipment	Total	Total Incl O&P
0010	**REMOVAL OF UNDERGROUND STORAGE TANKS**	R026510-20									
0011	Petroleum storage tanks, non-leaking										
0100	Excavate & load onto trailer										
0110	3000 gal. to 5000 gal. tank	G	B-14	4	12	Ea.		400	73.50	473.50	730
0120	6000 gal to 8000 gal tank	G	B-3A	3	13.333			445	289	734	1,050
0130	9000 gal to 12000 gal tank	G	"	2	20	↓		670	435	1,105	1,575
0190	Known leaking tank, add					%				100%	100%
0200	Remove sludge, water and remaining product from tank bottom										
0201	of tank with vacuum truck										
0300	3000 gal to 5000 gal tank	G	A-13	5	1.600	Ea.		62.50	152	214.50	266
0310	6000 gal to 8000 gal tank	G		4	2			78	190	268	330
0320	9000 gal to 12000 gal tank	G	↓	3	2.667	↓		104	253	357	440
0390	Dispose of sludge off-site, average					Gal.				6.60	7.25
0400	Insert inert solid CO_2 "dry ice" into tank										
0401	For cleaning/transporting tanks (1.5 lbs./100 gal. cap)	G	1 Clab	500	.016	Lb.	1.69	.51		2.20	2.69
0403	Insert solid carbon dioxide, 1.5 lbs./100 gal.	G	"	400	.020	"	1.69	.63		2.32	2.90
0503	Disconnect and remove piping	G	1 Plum	160	.050	L.F.		2.44		2.44	3.78
0603	Transfer liquids, 10% of volume	G	"	1600	.005	Gal.		.24		.24	.38
0703	Cut accessway into underground storage tank	G	1 Clab	5.33	1.501	Ea.		47.50		47.50	78.50
0813	Remove sludge,wash and wipe tank, 500 gal	G	1 Plum	8	1			49		49	75.50
0823	3,000 gal	G		6.67	1.199			58.50		58.50	90.50
0833	5,000 gal	G	↓	6.15	1.301	↓		63.50		63.50	98.50

02 65 Underground Storage Tank Removal

02 65 10 – Underground Tank and Contaminated Soil Removal

02 65 10.30 Removal of Underground Storage Tanks			Crew	Daily Output	Labor-Hours	Unit	Material	2009 Bare Costs Labor	Equipment	Total	Total Incl O&P
0843	8,000 gal	G	1 Plum	5.33	1.501	Ea.		73		73	113
0853	10,000 gal	G		4.57	1.751			85.50		85.50	132
0863	12,000 gal	G		4.21	1.900			92.50		92.50	144
1020	Haul tank to certified salvage dump, 100 miles round trip										
1023	3000 gal. to 5000 gal. tank					Ea.				725	798
1026	6000 gal. to 8000 gal. tank									860	946
1029	9,000 gal. to 12,000 gal. tank									1,160	1,276
1100	Disposal of contaminated soil to landfill										
1110	Minimum					C.Y.				130.70	144
1111	Maximum					"				392	431
1120	Disposal of contaminated soil to										
1121	bituminous concrete batch plant										
1130	Minimum					C.Y.				65.30	72
1131	Maximum					"				130.70	144
1203	Excavate, pull, & load tank, backfill hole, 8,000 gal +	G	B-12C	.50	32	Ea.		1,175	2,375	3,550	4,500
1213	Haul tank to certified dump, 100 miles rt, 8,000 gal +	G	B-34K	1	8			256	785	1,041	1,275
1223	Excavate, pull, & load tank, backfill hole, 500 gal	G	B-11C	1	16			585	294	879	1,275
1233	Excavate, pull, & load tank, backfill hole, 3-5,000 gal	G	B-11M	.50	32			1,175	690	1,865	2,625
1243	Haul tank to certified dump, 100 miles rt, 500 gal	G	B-34L	1	8			310	194	504	705
2010	Decontamination of soil on site incl poly tarp on top/bottom										
2011	Soil containment berm, and chemical treatment										
2020	Minimum	G	B-11C	100	.160	C.Y.	6.90	5.85	2.94	15.69	20
2021	Maximum	G	"	100	.160		9	5.85	2.94	17.79	22.50
2050	Disposal of decontaminated soil, minimum									78.20	86
2055	Maximum									160.40	176.50

02 82 Asbestos Remediation

02 82 13 – Asbestos Abatement

02 82 13.41 Asbestos Abatement Equip.

			Crew	Daily Output	Labor-Hours	Unit	Material	Labor	Equipment	Total	Total Incl O&P
0010	**ASBESTOS ABATEMENT EQUIP.**	R028213-20									
0011	Equipment and supplies, buy										
0200	Air filtration device, 2000 C.F.M.					Ea.	795			795	875
0250	Large volume air sampling pump, minimum						305			305	335
0260	Maximum						455			455	500
0300	Airless sprayer unit, 2 gun						3,125			3,125	3,425
0350	Light stand, 500 watt						72.50			72.50	80
0400	Personal respirators										
0410	Negative pressure, 1/2 face, dual operation, min.					Ea.	25.50			25.50	28
0420	Maximum						31			31	34.50
0450	P.A.P.R., full face, minimum						124			124	137
0460	Maximum						150			150	165
0470	Supplied air, full face, incl. air line, minimum						157			157	172
0480	Maximum						239			239	263
0500	Personnel sampling pump, minimum						201			201	221
1500	Power panel, 20 unit, incl. G.F.I.						1,100			1,100	1,200
1600	Shower unit, including pump and filters						1,075			1,075	1,200
1700	Supplied air system (type C)						9,800			9,800	10,800
1750	Vacuum cleaner, HEPA, 16 gal., stainless steel, wet/dry						675			675	745
1760	55 gallon						1,700			1,700	1,875
1800	Vacuum loader, 9-18 ton/hr						90,000			90,000	99,000
1900	Water atomizer unit, including 55 gal. drum						230			230	253

02 82 Asbestos Remediation

02 82 13 – Asbestos Abatement

02 82 13.41 Asbestos Abatement Equip.		Crew	Daily Output	Labor-Hours	Unit	2009 Bare Costs Material	Labor	Equipment	Total	Total Incl O&P
2000	Worker protection, whole body, foot, head cover & gloves, plastic				Ea.	6.80			6.80	7.45
2500	Respirator, single use					10.90			10.90	12
2550	Cartridge for respirator					12.30			12.30	13.55
2570	Glove bag, 7 mil, 50″ x 64″					9.20			9.20	10.10
2580	10 mil, 44″ x 60″					74.50			74.50	82
3000	HEPA vacuum for work area, minimum					1,325			1,325	1,475
3050	Maximum					3,100			3,100	3,425
6000	Disposable polyethylene bags, 6 mil, 3 C.F.					1.24			1.24	1.36
6300	Disposable fiber drums, 3 C.F.					12			12	13.20
6400	Pressure sensitive caution labels, 3″ x 5″					1.19			1.19	1.31
6450	11″ x 17″					7.05			7.05	7.75
6500	Negative air machine, 1800 C.F.M.				↓	770			770	845

02 82 13.42 Preparation of Asbestos Containment Area		Crew	Daily Output	Labor-Hours	Unit	Material	Labor	Equipment	Total	Total Incl O&P
0010	**PREPARATION OF ASBESTOS CONTAINMENT AREA**									
0100	Pre-cleaning, HEPA vacuum and wet wipe, flat surfaces	A-9	12000	.005	S.F.	.02	.24		.26	.40
0200	Protect carpeted area, 2 layers 6 mil poly on 3/4″ plywood	″	1000	.064		1.50	2.83		4.33	6.20
0300	Separation barrier, 2″ x 4″ @ 16″, 1/2″ plywood ea. side, 8′ high	2 Carp	400	.040		1.25	1.60		2.85	4.02
0310	12′ high		320	.050		1.40	2		3.40	4.84
0320	16′ high		200	.080		1.50	3.20		4.70	6.95
0400	Personnel decontam. chamber, 2″ x 4″ @ 16″, 3/4″ ply ea. side		280	.057		2.50	2.28		4.78	6.50
0450	Waste decontam. chamber, 2″ x 4″ studs @ 16″, 3/4″ ply ea. side	↓	360	.044	↓	3	1.78		4.78	6.25
0500	Cover surfaces with polyethylene sheeting									
0501	Including glue and tape									
0550	Floors, each layer, 6 mil	A-9	8000	.008	S.F.	.06	.35		.41	.63
0551	4 mil		9000	.007		.04	.31		.35	.55
0560	Walls, each layer, 6 mil		6000	.011		.06	.47		.53	.82
0561	4 mil	↓	7000	.009	↓	.04	.40		.44	.69
0570	For heights above 12′, add						20%			
0575	For heights above 20′, add						30%			
0580	For fire retardant poly, add					100%				
0590	For large open areas, deduct					10%	20%			
0600	Seal floor penetrations with foam firestop to 36 Sq. In.	2 Carp	200	.080	Ea.	1,275	3.20		1,278.20	1,400
0610	36 Sq. In. to 72 Sq. In.		125	.128		2,550	5.10		2,555.10	2,825
0615	72 Sq. In. to 144 Sq. In.		80	.200		5,125	8		5,133	5,650
0620	Wall penetrations, to 36 square inches		180	.089		1,275	3.55		1,278.55	1,400
0630	36 Sq. In. to 72 Sq. In.		100	.160		2,550	6.40		2,556.40	2,825
0640	72 Sq. In. to 144 Sq. In.	↓	60	.267	↓	5,125	10.65		5,135.65	5,650
0800	Caulk seams with latex	1 Carp	230	.035	L.F.	.15	1.39		1.54	2.46
0900	Set up neg. air machine, 1-2k C.F.M. /25 M.C.F. volume	1 Asbe	4.30	1.860	Ea.		82		82	133

02 82 13.43 Bulk Asbestos Removal		Crew	Daily Output	Labor-Hours	Unit	Material	Labor	Equipment	Total	Total Incl O&P
0010	**BULK ASBESTOS REMOVAL**									
0020	Includes disposable tools and 2 suits and 1respirator filter/day/worker									
0100	Beams, W 10 x 19	A-9	235	.272	L.F.	.88	12.05		12.93	20.50
0110	W 12 x 22		210	.305		.99	13.45		14.44	23
0120	W 14 x 26		180	.356		1.15	15.70		16.85	27
0130	W 16 x 31		160	.400		1.29	17.65		18.94	30
0140	W 18 x 40		140	.457		1.48	20		21.48	34
0150	W 24 x 55		110	.582		1.88	25.50		27.38	43.50
0160	W 30 x 108		85	.753		2.43	33.50		35.93	56.50
0170	W 36 x 150		72	.889	↓	2.87	39.50		42.37	66.50
0200	Boiler insulation	↓	480	.133	S.F.	.50	5.90		6.40	10.05
0210	With metal lath, add				%				50%	

02 82 Asbestos Remediation

02 82 13 – Asbestos Abatement

02 82 13.43 Bulk Asbestos Removal		Crew	Daily Output	Labor-Hours	Unit	2009 Bare Costs Material	Labor	Equipment	Total	Total Incl O&P
0300	Boiler breeching or flue insulation	A-9	520	.123	S.F.	.40	5.45		5.85	9.25
0310	For active boiler, add				%				100%	
0400	Duct or AHU insulation	A-10B	440	.073	S.F.	.24	3.22		3.46	5.45
0500	Duct vibration isolation joints, up to 24 Sq. In. duct	A-9	56	1.143	Ea.	3.69	50.50		54.19	85.50
0520	25 Sq. In. to 48 Sq. In. duct		48	1.333		4.31	59		63.31	99.50
0530	49 Sq. In. to 76 Sq. In. duct		40	1.600		5.15	70.50		75.65	120
0600	Pipe insulation, air cell type, up to 4" diameter pipe		900	.071	L.F.	.23	3.14		3.37	5.35
0610	4" to 8" diameter pipe		800	.080		.26	3.53		3.79	6
0620	10" to 12" diameter pipe		700	.091		.30	4.04		4.34	6.90
0630	14" to 16" diameter pipe		550	.116		.38	5.15		5.53	8.70
0650	Over 16" diameter pipe		650	.098	S.F.	.32	4.35		4.67	7.40
0700	With glove bag up to 3" diameter pipe		200	.320	L.F.	2.15	14.15		16.30	25.50
1000	Pipe fitting insulation up to 4" diameter pipe		320	.200	Ea.	.65	8.85		9.50	15
1100	6" to 8" diameter pipe		304	.211		.68	9.30		9.98	15.80
1110	10" to 12" diameter pipe		192	.333		1.08	14.70		15.78	25
1120	14" to 16" diameter pipe		128	.500		1.62	22		23.62	37.50
1130	Over 16" diameter pipe		176	.364	S.F.	1.18	16.05		17.23	27.50
1200	With glove bag, up to 8" diameter pipe		75	.853	L.F.	6.55	37.50		44.05	68
2000	Scrape foam fireproofing from flat surface		2400	.027	S.F.	.09	1.18		1.27	1.99
2100	Irregular surfaces		1200	.053		.17	2.36		2.53	4
3000	Remove cementitious material from flat surface		1800	.036		.11	1.57		1.68	2.67
3100	Irregular surface		1000	.064		.15	2.83		2.98	4.73
4000	Scrape acoustical coating/fireproofing, from ceiling		3200	.020		.06	.88		.94	1.50
5000	Remove VAT from floor by hand		2400	.027		.09	1.18		1.27	1.99
5100	By machine	A-11	4800	.013		.04	.59	.01	.64	1.01
5150	For 2 layers, add				%				50%	
6000	Remove contaminated soil from crawl space by hand	A-9	400	.160	C.F.	.52	7.05		7.57	11.95
6100	With large production vacuum loader	A-12	700	.091	"	.30	4.04	1.08	5.42	8.05
7000	Radiator backing, not including radiator removal	A-9	1200	.053	S.F.	.17	2.36		2.53	4
8000	Cement-asbestos transite board	2 Asbe	1000	.016		.14	.71		.85	1.30
8100	Transite shingle siding	A-10B	750	.043		.22	1.89		2.11	3.29
8200	Shingle roofing	"	2000	.016		.08	.71		.79	1.23
8250	Built-up, no gravel, non-friable	B-2	1400	.029		.08	.91		.99	1.60
8300	Asbestos millboard	2 Asbe	1000	.016		.08	.71		.79	1.23
9000	For type C (supplied air) respirator equipment, add				%					10%

02 82 13.44 Demolition In Asbestos Contaminated Area		Crew	Daily Output	Labor-Hours	Unit	Material	Labor	Equipment	Total	Total Incl O&P
0010	**DEMOLITION IN ASBESTOS CONTAMINATED AREA**									
0200	Ceiling, including suspension system, plaster and lath	A-9	2100	.030	S.F.	.10	1.35		1.45	2.29
0210	Finished plaster, leaving wire lath		585	.109		.35	4.83		5.18	8.20
0220	Suspended acoustical tile		3500	.018		.06	.81		.87	1.38
0230	Concealed tile grid system		3000	.021		.07	.94		1.01	1.60
0240	Metal pan grid system		1500	.043		.14	1.88		2.02	3.20
0250	Gypsum board		2500	.026		.08	1.13		1.21	1.92
0260	Lighting fixtures up to 2' x 4'		72	.889	Ea.	2.87	39.50		42.37	66.50
0400	Partitions, non load bearing									
0410	Plaster, lath, and studs	A-9	690	.093	S.F.	.87	4.10		4.97	7.55
0450	Gypsum board and studs	"	1390	.046	"	.15	2.03		2.18	3.45
9000	For type C (supplied air) respirator equipment, add				%					10%

02 82 13.45 OSHA Testing		Crew	Daily Output	Labor-Hours	Unit	Material	Labor	Equipment	Total	Total Incl O&P
0010	**OSHA TESTING**									
0100	Certified technician, minimum				Day				240	264
0110	Maximum				"				320	352

02 82 Asbestos Remediation

02 82 13 – Asbestos Abatement

02 82 13.45 OSHA Testing

	02 82 13.45 OSHA Testing	Crew	Daily Output	Labor-Hours	Unit	2009 Bare Costs Material	Labor	Equipment	Total	Total Incl O&P
0200	Personal sampling, PCM analysis, NIOSH 7400, minimum	1 Asbe	8	1	Ea.	2.75	44		46.75	74.50
0210	Maximum	"	4	2	"	3	88		91	146
0300	Industrial hygienist, minimum				Day				320	352
0310	Maximum				"				480	528
1000	Cleaned area samples	1 Asbe	8	1	Ea.	2.63	44		46.63	74.50
1100	PCM air sample analysis, NIOSH 7400, minimum		8	1		30.50	44		74.50	106
1110	Maximum		4	2		3.20	88		91.20	147
1200	TEM air sample analysis, NIOSH 7402, minimum								100	125
1210	Maximum								400	500

02 82 13.46 Decontamination of Asbestos Containment Area

	02 82 13.46 Decontamination of Asbestos Containment Area	Crew	Daily Output	Labor-Hours	Unit	Material	Labor	Equipment	Total	Total Incl O&P
0010	DECONTAMINATION OF ASBESTOS CONTAINMENT AREA									
0100	Spray exposed substrate with surfactant (bridging)									
0200	Flat surfaces	A-9	6000	.011	S.F.	.36	.47		.83	1.16
0250	Irregular surfaces		4000	.016	"	.31	.71		1.02	1.48
0300	Pipes, beams, and columns		2000	.032	L.F.	.56	1.41		1.97	2.90
1000	Spray encapsulate polyethylene sheeting		8000	.008	S.F.	.31	.35		.66	.91
1100	Roll down polyethylene sheeting		8000	.008	"		.35		.35	.57
1500	Bag polyethylene sheeting		400	.160	Ea.	.77	7.05		7.82	12.25
2000	Fine clean exposed substrate, with nylon brush		2400	.027	S.F.		1.18		1.18	1.90
2500	Wet wipe substrate		4800	.013			.59		.59	.95
2600	Vacuum surfaces, fine brush		6400	.010			.44		.44	.71
3000	Structural demolition									
3100	Wood stud walls	A-9	2800	.023	S.F.		1.01		1.01	1.63
3500	Window manifolds, not incl. window replacement		4200	.015			.67		.67	1.09
3600	Plywood carpet protection		2000	.032			1.41		1.41	2.28
4000	Remove custom decontamination facility	A-10A	8	3	Ea.	15	133		148	232
4100	Remove portable decontamination facility	3 Asbe	12	2	"	12.75	88		100.75	157
5000	HEPA vacuum, shampoo carpeting	A-9	4800	.013	S.F.	.05	.59		.64	1.01
9000	Final cleaning of protected surfaces	A-10A	8000	.003	"		.13		.13	.21

02 82 13.47 Asbestos Waste Pkg., Handling, and Disp.

	02 82 13.47 Asbestos Waste Pkg., Handling, and Disp.	Crew	Daily Output	Labor-Hours	Unit	Material	Labor	Equipment	Total	Total Incl O&P
0010	ASBESTOS WASTE PACKAGING, HANDLING, AND DISPOSAL									
0100	Collect and bag bulk material, 3 C.F. bags, by hand	A-9	400	.160	Ea.	1.24	7.05		8.29	12.75
0200	Large production vacuum loader	A-12	880	.073		.80	3.21	.86	4.87	7.05
1000	Double bag and decontaminate	A-9	960	.067		1.24	2.94		4.18	6.10
2000	Containerize bagged material in drums, per 3 C.F. drum	"	800	.080		12	3.53		15.53	18.90
3000	Cart bags 50′ to dumpster	2 Asbe	400	.040			1.76		1.76	2.85
5000	Disposal charges, not including haul, minimum				C.Y.				50	55
5020	Maximum				"				170	187
5100	Remove refrigerant from system	1 Plum	40	.200	Lb.		9.75		9.75	15.10
9000	For type C (supplied air) respirator equipment, add				%					10%

02 82 13.48 Asbestos Encapsulation With Sealants

	02 82 13.48 Asbestos Encapsulation With Sealants	Crew	Daily Output	Labor-Hours	Unit	Material	Labor	Equipment	Total	Total Incl O&P
0010	ASBESTOS ENCAPSULATION WITH SEALANTS									
0100	Ceilings and walls, minimum	A-9	21000	.003	S.F.	.27	.13		.40	.52
0110	Maximum		10600	.006		.42	.27		.69	.89
0200	Columns and beams, minimum		13300	.005		.27	.21		.48	.64
0210	Maximum		5325	.012		.47	.53		1	1.38
0300	Pipes to 12″ diameter including minor repairs, minimum		800	.030	L.F.	.37	3.53		3.90	6.10
0310	Maximum		400	.160	"	1.04	7.05		8.09	12.55

02 83 Lead Remediation

02 83 19 – Lead-Based Paint Remediation

02 83 19.23 Encapsulation of Lead-Based Paint

		Crew	Daily Output	Labor-Hours	Unit	Material	2009 Bare Costs Labor	Equipment	Total	Total Incl O&P
0010	**ENCAPSULATION OF LEAD-BASED PAINT**									
0020	Interior, brushwork, trim, under 6"	1 Pord	240	.033	L.F.	2.46	1.17		3.63	4.58
0030	6" to 12" wide		180	.044		3.28	1.56		4.84	6.10
0040	Balustrades		300	.027		1.98	.94		2.92	3.68
0050	Pipe to 4" diameter		500	.016		1.19	.56		1.75	2.21
0060	To 8" diameter		375	.021		1.57	.75		2.32	2.93
0070	To 12" diameter		250	.032		2.36	1.13		3.49	4.40
0080	To 16" diameter		170	.047		3.47	1.66		5.13	6.45
0090	Cabinets, ornate design		200	.040	S.F.	2.97	1.41		4.38	5.50
0100	Simple design		250	.032	"	2.36	1.13		3.49	4.40
0110	Doors, 3' x 7', both sides, incl. frame & trim									
0120	Flush	1 Pord	6	1.333	Ea.	30.50	47		77.50	109
0130	French, 10-15 lite		3	2.667		6.05	94		100.05	157
0140	Panel		4	2		36.50	70.50		107	153
0150	Louvered		2.75	2.909		33.50	102		135.50	201
0160	Windows, per interior side, per 15 S.F.									
0170	1 to 6 lite	1 Pord	14	.571	Ea.	21	20		41	55
0180	7 to 10 lite		7.50	1.067		23	37.50		60.50	85.50
0190	12 lite		5.75	1.391		31	49		80	113
0200	Radiators		8	1		73.50	35		108.50	138
0210	Grilles, vents		275	.029	S.F.	2.15	1.02		3.17	4.01
0220	Walls, roller, drywall or plaster		1000	.008		.59	.28		.87	1.10
0230	With spunbonded reinforcing fabric		720	.011		.68	.39		1.07	1.37
0240	Wood		800	.010		.74	.35		1.09	1.37
0250	Ceilings, roller, drywall or plaster		900	.009		.68	.31		.99	1.25
0260	Wood		700	.011		.84	.40		1.24	1.56
0270	Exterior, brushwork, gutters and downspouts		300	.027	L.F.	1.98	.94		2.92	3.68
0280	Columns		400	.020	S.F.	1.47	.70		2.17	2.75
0290	Spray, siding		600	.013	"	.99	.47		1.46	1.84
0300	Miscellaneous									
0310	Electrical conduit, brushwork, to 2" diameter	1 Pord	500	.016	L.F.	1.19	.56		1.75	2.21
0320	Brick, block or concrete, spray		500	.016	S.F.	1.19	.56		1.75	2.21
0330	Steel, flat surfaces and tanks to 12"		500	.016		1.19	.56		1.75	2.21
0340	Beams, brushwork		400	.020		1.47	.70		2.17	2.75
0350	Trusses		400	.020		1.47	.70		2.17	2.75

02 83 19.26 Removal of Lead-Based Paint

		Crew	Daily Output	Labor-Hours	Unit	Material	2009 Bare Costs Labor	Equipment	Total	Total Incl O&P
0010	**REMOVAL OF LEAD-BASED PAINT** R028319-60									
0011	By chemicals, per application									
0050	Baseboard, to 6" wide	1 Pord	64	.125	L.F.	.66	4.40		5.06	7.80
0070	To 12" wide		32	.250	"	1.32	8.80		10.12	15.50
0200	Balustrades, one side		28	.286	S.F.	1.32	10.05		11.37	17.50
1400	Cabinets, simple design		32	.250		1.32	8.80		10.12	15.50
1420	Ornate design		25	.320		1.32	11.25		12.57	19.45
1600	Cornice, simple design		60	.133		1.32	4.69		6.01	8.95
1620	Ornate design		20	.400		5.05	14.10		19.15	28
2800	Doors, one side, flush		84	.095		1	3.35		4.35	6.45
2820	Two panel		80	.100		1.32	3.52		4.84	7.10
2840	Four panel		45	.178		1.32	6.25		7.57	11.45
2880	For trim, one side, add		64	.125	L.F.	.66	4.40		5.06	7.80
3000	Fence, picket, one side		30	.267	S.F.	1.32	9.40		10.72	16.45
3200	Grilles, one side, simple design		30	.267		1.32	9.40		10.72	16.45
3220	Ornate design		25	.320		1.32	11.25		12.57	19.45

02 83 Lead Remediation

02 83 19 – Lead-Based Paint Remediation

02 83 19.26 Removal of Lead-Based Paint		Crew	Daily Output	Labor-Hours	Unit	Material	Labor	Equipment	Total	Total Incl O&P
						2009 Bare Costs				
4400	Pipes, to 4" diameter	1 Pord	90	.039	L.F.	1	3.13		4.13	6.10
4420	To 8" diameter		50	.160		2	5.65		7.65	11.20
4440	To 12" diameter		36	.222		3	7.80		10.80	15.80
4460	To 16" diameter		20	.400		4	14.10		18.10	27
4500	For hangers, add		40	.200	Ea.	2.54	7.05		9.59	14.05
4800	Siding		90	.039	S.F.	1.17	3.13		4.30	6.30
5000	Trusses, open		55	.145	SF Face	1.85	5.10		6.95	10.25
6200	Windows, one side only, double hung, 1/1 light, 24" x 48" high		4	2	Ea.	25.50	70.50		96	141
6220	30" x 60" high		3	2.667		34	94		128	188
6240	36" x 72" high		2.50	3.200		41	113		154	225
6280	40" x 80" high		2	4		51	141		192	281
6400	Colonial window, 6/6 light, 24" x 48" high		2	4		51	141		192	282
6420	30" x 60" high		1.50	5.333		68	188		256	375
6440	36" x 72" high		1	8		102	282		384	560
6480	40" x 80" high		1	8		102	282		384	560
6600	8/8 light, 24" x 48" high		2	4		51	141		192	282
6620	40" x 80" high		1	8		102	282		384	560
6800	12/12 light, 24" x 48" high		1	8		102	282		384	560
6820	40" x 80" high		.75	10.667		136	375		511	750
6840	Window frame & trim items, included in pricing above									
9000	Minimum labor/equipment charge	1 Pord	3	2.667	Job		94		94	150

02 85 Mold Remediation

02 85 16 – Mold Remediation Preparation and Containment

02 85 16.40 Mold Remediation Plans and Methods

		Crew	Daily Output	Labor-Hours	Unit	Material	Labor	Equipment	Total	Total Incl O&P
0010	**MOLD REMEDIATION PLANS AND METHODS**									
0020	Initial inspection, average 3 bedroom home				Total				250	275
0030	Average 5 bedroom home								340	374
0040	Testing, air sample each								250	275
0050	Swab sample								150	165
0060	Tape sample								150	165
0070	Post remediation air test								250	275
0080	Mold abatement plan, average 3 bedroom home								1,300	1,430
0090	Average 5 bedroom home								1,580	1,740
0100	Packup & removal of contents, average 3 bedroom home, excl storage								10,500	11,550
0110	Average 5 bedroom home, excl storage								26,400	29,000
0120	For demolition in mold contaminated areas, see Div. 02 85 33.50									
0130	For personal protection equipment, see Div. 02 82 13.41									

02 85 16.50 Preparation of Mold Containment Area

		Crew	Daily Output	Labor-Hours	Unit	Material	Labor	Equipment	Total	Total Incl O&P
0010	**PREPARATION OF MOLD CONTAINMENT AREA**									
0100	Pre-cleaning, HEPA vacuum and wet wipe, flat surfaces	A-9	12000	.005	S.F.	.02	.24		.26	.40
0300	Separation barrier, 2" x 4" @ 16", 1/2" plywood ea. side, 8' high	2 Carp	400	.040		1.25	1.60		2.85	4.02
0310	12' high		320	.050		1.40	2		3.40	4.84
0320	16' high		200	.080		1.50	3.20		4.70	6.95
0400	Personnel decontam. chamber, 2" x 4" @ 16", 3/4" ply ea. side		280	.057		2.50	2.28		4.78	6.50
0450	Waste decontam. chamber, 2" x 4" studs @ 16", 3/4" ply each side		360	.044		3	1.78		4.78	6.25
0500	Cover surfaces with polyethylene sheeting									
0501	Including glue and tape									
0550	Floors, each layer, 6 mil	A-9	8000	.008	S.F.	.09	.35		.44	.67
0551	4 mil		9000	.007		.05	.31		.36	.57
0560	Walls, each layer, 6 mil		6000	.011		.09	.47		.56	.86

02 85 Mold Remediation

02 85 16 – Mold Remediation Preparation and Containment

02 85 16.50 Preparation of Mold Containment Area		Crew	Daily Output	Labor-Hours	Unit	Material	2009 Bare Costs Labor	Equipment	Total	Total Incl O&P
0561	4 mil	A-9	7000	.009	S.F.	.07	.40		.47	.73
0570	For heights above 12′, add						20%			
0575	For heights above 20′, add						30%			
0580	For fire retardant poly, add					100%				
0590	For large open areas, deduct					10%	20%			
0600	Seal floor penetrations with foam firestop to 36 sq in	2 Carp	200	.080	Ea.	1,275	3.20		1,278.20	1,400
0610	36 sq in to 72 sq in		125	.128		2,550	5.10		2,555.10	2,825
0615	72 sq in to 144 sq in		80	.200		5,125	8		5,133	5,650
0620	Wall penetrations, to 36 square inches		180	.089		1,275	3.55		1,278.55	1,400
0630	36 sq. in. to 72 sq. in.		100	.160		2,550	6.40		2,556.40	2,825
0640	72 Sq. in. to 144 sq. in.	↓	60	.267	↓	5,125	10.65		5,135.65	5,650
0800	Caulk seams with latex caulk	1 Carp	230	.035	L.F.	.15	1.39		1.54	2.46
0900	Set up neg. air machine, 1-2k C.F.M. /25 M.C.F. volume	1 Asbe	4.30	1.860	Ea.		82		82	133

02 85 33 – Removal and Disposal of Materials with Mold

02 85 33.50 Demolition in Mold Contaminated Area		Crew	Daily Output	Labor-Hours	Unit	Material	Labor	Equipment	Total	Total Incl O&P
0010	**DEMOLITION IN MOLD CONTAMINATED AREA**									
0200	Ceiling, including suspension system, plaster and lath	A-9	2100	.030	S.F.	.10	1.35		1.45	2.29
0210	Finished plaster, leaving wire lath		585	.109		.35	4.83		5.18	8.20
0220	Suspended acoustical tile		3500	.018		.06	.81		.87	1.38
0230	Concealed tile grid system		3000	.021		.07	.94		1.01	1.60
0240	Metal pan grid system		1500	.043		.14	1.88		2.02	3.20
0250	Gypsum board		2500	.026		.08	1.13		1.21	1.92
0255	Plywood		2500	.026	↓	.08	1.13		1.21	1.92
0260	Lighting fixtures up to 2′ x 4′	↓	72	.889	Ea.	2.87	39.50		42.37	66.50
0400	Partitions, non load bearing									
0410	Plaster, lath, and studs	A-9	690	.093	S.F.	.87	4.10		4.97	7.55
0450	Gypsum board and studs		1390	.046		.15	2.03		2.18	3.45
0465	Carpet & pad		1390	.046	↓	.15	2.03		2.18	3.45
0600	Pipe insulation, air cell type, up to 4″ diameter pipe		900	.071	L.F.	.23	3.14		3.37	5.35
0610	4″ to 8″ diameter pipe		800	.080		.26	3.53		3.79	6
0620	10″ to 12″ diameter pipe		700	.091		.30	4.04		4.34	6.90
0630	14″ to 16″ diameter pipe		550	.116	↓	.38	5.15		5.53	8.70
0650	Over 16″ diameter pipe	↓	650	.098	S.F.	.32	4.35		4.67	7.40
9000	For type C (supplied air) respirator equipment, add				%					10%

Estimating Tips

General

- Carefully check all the plans and specifications. Concrete often appears on drawings other than structural drawings, including mechanical and electrical drawings for equipment pads. The cost of cutting and patching is often difficult to estimate. See Subdivision 03 81 for Concrete Cutting, Subdivision 02 41 19.16 for Cutout Demolition, Subdivision 03 05 05.10 for Concrete Demolition, and Subdivision 02 41 19.23 for Rubbish Handling (handling, loading and hauling of debris).
- Always obtain concrete prices from suppliers near the job site. A volume discount can often be negotiated, depending upon competition in the area. Remember to add for waste, particularly for slabs and footings on grade.

03 10 00 Concrete Forming and Accessories

- A primary cost for concrete construction is forming. Most jobs today are constructed with prefabricated forms. The selection of the forms best suited for the job and the total square feet of forms required for efficient concrete forming and placing are key elements in estimating concrete construction. Enough forms must be available for erection to make efficient use of the concrete placing equipment and crew.
- Concrete accessories for forming and placing depend upon the systems used. Study the plans and specifications to ensure that all special accessory requirements have been included in the cost estimate, such as anchor bolts, inserts, and hangers.
- Included within costs for forms-in-place are all necessary bracing and shoring.

03 20 00 Concrete Reinforcing

- Ascertain that the reinforcing steel supplier has included all accessories, cutting, bending, and an allowance for lapping, splicing, and waste. A good rule of thumb is 10% for lapping, splicing, and waste. Also, 10% waste should be allowed for welded wire fabric.
- The unit price items in the subdivision for Reinforcing In Place include the labor to install accessories such as beam and slab bolsters, high chairs, and bar ties and tie wire. The material cost for these accessories is not included; they may be obtained from the Accessories Division.

03 30 00 Cast-In-Place Concrete

- When estimating structural concrete, pay particular attention to requirements for concrete additives, curing methods, and surface treatments. Special consideration for climate, hot or cold, must be included in your estimate. Be sure to include requirements for concrete placing equipment, and concrete finishing.
- For accurate concrete estimating, the estimator must consider each of the following major components individually: forms, reinforcing steel, ready-mix concrete, placement of the concrete, and finishing of the top surface. For faster estimating, Subdivision 03 30 53.40 for Concrete-In-Place can be used; here, various items of concrete work are presented that include the costs of all five major components (unless specifically stated otherwise).

03 40 00 Precast Concrete

03 50 00 Cast Decks and Underlayment

- The cost of hauling precast concrete structural members is often an important factor. For this reason, it is important to get a quote from the nearest supplier. It may become economically feasible to set up precasting beds on the site if the hauling costs are prohibitive.

Reference Numbers

Reference numbers are shown in shaded boxes at the beginning of some major classifications. These numbers refer to related items in the Reference Section. The reference information may be an estimating procedure, an alternate pricing method, or technical information.

Note: Not all subdivisions listed here necessarily appear in this publication.

03 05 Common Work Results for Concrete

03 05 05 – Selective Concrete Demolition

03 05 05.10 Selective Demolition, Concrete		Crew	Daily Output	Labor-Hours	Unit	Material	2009 Bare Costs Labor	Equipment	Total	Total Incl O&P
0010	**SELECTIVE DEMOLITION, CONCRETE** R024119-10									
0012	Excludes saw cutting, torch cutting, loading or hauling									
0050	Break up into small pieces, minimum reinforcing	B-9	24	1.667	C.Y.		53.50	8	61.50	97
0060	Average reinforcing		16	2.500			80	12	92	145
0070	Maximum reinforcing		8	5			160	24	184	291
0150	Remove whole pieces, up to 2 tons per piece	E-18	36	1.111	Ea.		49.50	30	79.50	122
0160	2 – 5 tons per piece		30	1.333			59	36	95	147
0170	5 – 10 tons per piece		24	1.667			74	45	119	183
0180	10 – 15 tons per piece		18	2.222			98.50	60	158.50	244
0250	Precast unit embedded in masonry, up to 1 CF	D-1	16	1			36.50		36.50	59
0260	1 – 2 CF		12	1.333			48.50		48.50	78.50
0270	2 – 5 CF		10	1.600			58		58	94
0280	5 – 10 CF		8	2			72.50		72.50	118
1910	Minimum labor/equipment charge	B-9	2	20	Job		640	96	736	1,150

03 05 13 – Basic Concrete Materials

03 05 13.20 Concrete Admixtures and Surface Treatments

		Crew	Daily Output	Labor-Hours	Unit	Material	Labor	Equipment	Total	Total Incl O&P
0010	**CONCRETE ADMIXTURES AND SURFACE TREATMENTS**									
0040	Abrasives, aluminum oxide, over 20 tons				Lb.	1.63			1.63	1.79
0050	1 to 20 tons					1.68			1.68	1.85
0070	Under 1 ton					1.75			1.75	1.93
0100	Silicon carbide, black, over 20 tons					2.23			2.23	2.45
0110	1 to 20 tons					2.34			2.34	2.58
0120	Under 1 ton					2.44			2.44	2.68
0200	Air entraining agent, .7 to 1.5 oz. per bag, 55 gallon drum				Gal.	9.95			9.95	10.95
0220	5 gallon pail					12.50			12.50	13.75
0300	Bonding agent, acrylic latex, 250 S.F. per gallon, 5 gallon pail					18.50			18.50	20.50
0320	Epoxy resin, 80 S.F. per gallon, 3.5 gallon unit					45			45	49.50
0400	Calcium chloride, 50 lb. bags, TL lots				Ton	525			525	580
0420	Less than truckload lots				Bag	17.75			17.75	19.55
0500	Carbon black, liquid, 2 to 8 lbs. per bag of cement				Lb.	5.90			5.90	6.45
0600	Colors, integral, 2 to 10 lb. per bag of cement, minimum					2.41			2.41	2.65
0610	Average					3.40			3.40	3.74
0620	Maximum					4.80			4.80	5.30
0920	Dustproofing compound, 250 SF/gal, 5 gallon pail				Gal.	6.80			6.80	7.45
1010	Epoxy based, 125 SF/Gal, 5 gallon pail				"	45.50			45.50	50
1100	Hardeners, metallic, 55 lb. bags, natural (grey)				Lb.	.93			.93	1.02
1200	Colors					1.32			1.32	1.45
1300	Non-metallic, 55 lb. bags, natural grey					.37			.37	.41
1320	Colors					.66			.66	.72
1550	Release agent, for tilt slabs, 5 gallon pail				Gal.	15.10			15.10	16.60
1570	For forms, 5 gallon pail					9.35			9.35	10.30
1590	Concrete release agent for forms, 100% biodegradeable, zero VOC, 5 gal pail G					13.20			13.20	14.50
1595	55 gallon drum G					10.25			10.25	11.30
1600	Sealer, hardener and dustproofer, epoxy-based, 125 SF/gal, 5 gallon, min					45.50			45.50	50
1620	5 gallon pail, maximum					50			50	55
1630	Sealer, solvent-based, 250 SF/gal, 55 gallon drum					15.25			15.25	16.75
1640	5 gallon pail					17.40			17.40	19.15
1650	Sealer, water based, 350 S.F., 55 gallon drum					18.30			18.30	20
1660	5 gallon pail					21			21	23
1900	Set retarder, 100 SF/gallon, 5 gallon pail					18.35			18.35	20
2000	Waterproofing, integral 1 lb. per bag of cement				Lb.	1.38			1.38	1.52
2100	Powdered metallic, 40 lbs. per 100 S.F., minimum					1.63			1.63	1.79

03 05 13 – Basic Concrete Materials

03 05 13.20 Concrete Admixtures and Surface Treatments		Crew	Daily Output	Labor-Hours	Unit	2009 Bare Costs Material	Labor	Equipment	Total	Total Incl O&P
2120	Maximum				Lb.	2.28			2.28	2.51
3000	For integral colors, 2500 psi (5 bag mix)									
3100	Red, yellow or brown, 1.8 lb. per bag, add				C.Y.	21.50			21.50	24
3200	9.4 lb. per bag, add					113			113	125
3400	Black, 1.8 lb. per bag, add					30.50			30.50	33.50
3500	7.5 lb. per bag, add					128			128	140
3700	Green, 1.8 lb. per bag, add					43			43	47.50
3800	7.5 lb. per bag, add					180			180	198
6000	Concrete ready-mix additives, recycled coal fly ash, mixed at plant G				Ton	52			52	57
6010	Recycled blast furnace slag, mixed at plant G				"	90			90	99

03 05 13.25 Aggregate		Crew	Daily Output	Labor-Hours	Unit	Material	Labor	Equipment	Total	Total Incl O&P
0010	**AGGREGATE**									
0020	Expanded shale, C.L. lots, 52 lb. per C.F., minimum G				Ton	39.50			39.50	43.50
0050	Maximum G				"	53			53	58
0100	Lightweight vermiculite or perlite, 4 C.F. bag, C.L. lots G				Bag	7.25			7.25	7.95
0150	L.C.L. lots G				"	8.05			8.05	8.85
0250	Sand & stone, loaded at pit, crushed bank gravel				Ton	19.55			19.55	21.50
0350	Sand, washed, for concrete					17.80			17.80	19.55
0400	For plaster or brick					17.80			17.80	19.55
0450	Stone, 3/4" to 1-1/2"					14.10			14.10	15.50
0470	Round, river stone					32.50			32.50	35.50
0500	3/8" roofing stone & 1/2" pea stone					21.50			21.50	24
0550	For trucking 10 miles (5 mile R/T), add to the above	B-34B	117	.063			2.18	4.56	6.74	8.60
0600	30 miles (15 mi R/T), add to the above	"	72	.111			3.55	7.40	10.95	13.95
0850	Sand & stone, loaded at pit, crushed bank gravel				C.Y.	27.50			27.50	30
0950	Sand, washed, for concrete					24.50			24.50	27
1000	For plaster or brick					24.50			24.50	27
1050	Stone, 3/4" to 1-1/2"					27			27	29.50
1055	Round, river stone					30.50			30.50	33.50
1100	3/8" roofing stone & 1/2" pea stone					39			39	43
1150	For trucking 10 miles (5 mi R/T), add to the above	B-34B	78	.103			3.28	6.85	10.13	12.85
1200	30 miles (15 mi R/T), add to the above	"	48	.167			5.35	11.10	16.45	21
1310	Quartz chips, 50 lb. bags				Cwt.	18.70			18.70	20.50
1330	Silica chips, 50 lb. bags					10.25			10.25	11.30
1410	White marble, 3/8" to 1/2", 50 lb. bags					15.45			15.45	17
1430	3/4"				Ton	80			80	88

03 05 13.30 Cement		Crew	Daily Output	Labor-Hours	Unit	Material	Labor	Equipment	Total	Total Incl O&P
0010	**CEMENT**									
0240	Portland, Type I/II, TL lots, 94 lb bags				Bag	10.65			10.65	11.70
0250	LTL/LCL lots				"	11.65			11.65	12.85
0300	Trucked in bulk, per cwt				Cwt.	6.95			6.95	7.60
0400	Type III, high early strength, TL lots, 94 lb bags				Bag	12.30			12.30	13.55
0420	L.T.L. or L.C.L. lots					12.40			12.40	13.65
0500	White, type III, high early strength, T.L. or C.L. lots, bags					24.50			24.50	27
0520	L.T.L. or L.C.L. lots					25			25	27.50
0600	White, type I, T.L. or C.L. lots, bags					24			24	26.50
0620	L.T.L. or L.C.L. lots					25.50			25.50	28

03 05 13.80 Waterproofing and Dampproofing		Crew	Daily Output	Labor-Hours	Unit	Material	Labor	Equipment	Total	Total Incl O&P
0010	**WATERPROOFING AND DAMPPROOFING**									
0050	Integral waterproofing, add to cost of regular concrete				C.Y.	8.30			8.30	9.10

03 05 13.85 Winter Protection		Crew	Daily Output	Labor-Hours	Unit	Material	Labor	Equipment	Total	Total Incl O&P
0010	**WINTER PROTECTION**									

03 05 Common Work Results for Concrete

03 05 13 – Basic Concrete Materials

03 05 13.85 Winter Protection		Crew	Daily Output	Labor-Hours	Unit	Material	Labor	Equipment	Total	Total Incl O&P
						2009 Bare Costs				
0012	For heated ready mix, add, minimum				C.Y.	5.25			5.25	5.80
0050	Maximum				″	6.60			6.60	7.25
0100	Temporary heat to protect concrete, 24 hours, minimum	2 Clab	50	.320	M.S.F.	520	10.10		530.10	590
0150	Maximum	″	25	.640	″	680	20		700	780
0200	Temporary shelter for slab on grade, wood frame/polyethylene sheeting									
0201	Build or remove, minimum	2 Carp	10	1.600	M.S.F.	245	64		309	375
0210	Maximum	″	3	5.333	″	293	213		506	670
0500	Electrically heated pads, 110 volts, 15 watts per S.F., buy				S.F.	6.10			6.10	6.70
0600	20 watts per S.F., buy					8.10			8.10	8.90
0710	Electrically, heated pads, 15 watts/S.F., 20 uses, minimum					.22			.22	.24
0800	Maximum					.36			.36	.40

03 11 Concrete Forming

03 11 13 – Structural Cast-In-Place Concrete Forming

03 11 13.13 Concrete Slip Forming

		Crew	Daily Output	Labor-Hours	Unit	Material	Labor	Equipment	Total	Total Incl O&P
0010	**CONCRETE SLIP FORMING**									
0020	Silos, minimum	C-17E	3885	.021	SFCA	3.75	.85	.02	4.62	5.55
0050	Maximum		1095	.073		5.50	3.01	.08	8.59	11.05
1000	Buildings, minimum		3660	.022		2.45	.90	.02	3.37	4.20
1050	Maximum		875	.091		4.50	3.77	.10	8.37	11.20

03 11 13.20 Forms In Place, Beams and Girders

		Crew	Daily Output	Labor-Hours	Unit	Material	Labor	Equipment	Total	Total Incl O&P
0010	**FORMS IN PLACE, BEAMS AND GIRDERS**									
0500	Exterior spandrel, job-built plywood, 12″ wide, 1 use R031113-60	C-2	225	.213	SFCA	3.71	8.30		12.01	17.80
0550	2 use		275	.175		1.98	6.80		8.78	13.35
0650	4 use		310	.155		1.20	6		7.20	11.30
1000	18″ wide, 1 use		250	.192		3.01	7.45		10.46	15.65
1050	2 use		275	.175		1.66	6.80		8.46	13
1100	3 use		305	.157		1.21	6.10		7.31	11.45
1150	4 use		315	.152		.99	5.95		6.94	10.90
1500	24″ wide, 1 use		265	.181		2.75	7.05		9.80	14.70
1550	2 use		290	.166		1.54	6.45		7.99	12.35
1600	3 use		315	.152		1.10	5.95		7.05	11
1650	4 use		325	.148		.89	5.75		6.64	10.50
2000	Interior beam, job-built plywood, 12″ wide, 1 use		300	.160		4.30	6.20		10.50	15
2050	2 use		340	.141		2.14	5.50		7.64	11.40
2100	3 use		364	.132		1.72	5.15		6.87	10.35
2150	4 use		377	.127		1.40	4.95		6.35	9.70
2500	24″ wide, 1 use		320	.150		2.80	5.85		8.65	12.75
2550	2 use		365	.132		1.57	5.10		6.67	10.20
2600	3 use		385	.125		1.11	4.85		5.96	9.20
2650	4 use		395	.122		.90	4.73		5.63	8.80
3000	Encasing steel beam, hung, job-built plywood, 1 use		325	.148		3.71	5.75		9.46	13.60
3050	2 use		390	.123		2.04	4.79		6.83	10.15
3100	3 use		415	.116		1.48	4.50		5.98	9.10
3150	4 use		430	.112		1.21	4.34		5.55	8.50
3500	Bottoms only, to 30″ wide, job-built plywood, 1 use		230	.209		4.12	8.10		12.22	17.95
3550	2 use		265	.181		2.31	7.05		9.36	14.20
3600	3 use		280	.171		1.65	6.65		8.30	12.80
3650	4 use		290	.166		1.34	6.45		7.79	12.10
4000	Sides only, vertical, 36″ high, job-built plywood, 1 use		335	.143		9.20	5.55		14.75	19.30
4050	2 use		405	.119		5.05	4.61		9.66	13.15

03 11 13.20 Forms In Place, Beams and Girders		Crew	Daily Output	Labor-Hours	Unit	2009 Bare Costs Material	2009 Bare Costs Labor	2009 Bare Costs Equipment	2009 Bare Costs Total	Total Incl O&P
4100	3 use	C-2	430	.112	SFCA	3.68	4.34		8.02	11.20
4150	4 use		445	.108		2.99	4.20		7.19	10.25
4500	Sloped sides, 36" high, 1 use		305	.157		8.25	6.10		14.35	19.20
4550	2 use		370	.130		4.57	5.05		9.62	13.40
4600	3 use		405	.119		3.30	4.61		7.91	11.25
4650	4 use		425	.113		2.68	4.39		7.07	10.20
5000	Upstanding beams, 36" high, 1 use		225	.213		12.55	8.30		20.85	27.50
5050	2 use		255	.188		6.90	7.30		14.20	19.70
5100	3 use		275	.175		5	6.80		11.80	16.75
5150	4 use		280	.171		4.08	6.65		10.73	15.50
9000	Minimum labor/equipment charge	2 Carp	2	8	Job		320		320	530

03 11 13.25 Forms In Place, Columns		Crew	Daily Output	Labor-Hours	Unit	Material	Labor	Equipment	Total	Total Incl O&P
0010	**FORMS IN PLACE, COLUMNS**									
0500	Round fiberglass, 4 use per mo., rent, 12" diameter	C-1	160	.200	L.F.	7.10	7.55		14.65	20.50
0550	16" diameter R031113-60		150	.213		8.40	8.10		16.50	22.50
0600	18" diameter		140	.229		9.45	8.65		18.10	24.50
0650	24" diameter		135	.237		11.75	8.95		20.70	28
0700	28" diameter		130	.246		13.10	9.30		22.40	30
0800	30" diameter		125	.256		13.70	9.70		23.40	31
0850	36" diameter		120	.267		18.25	10.10		28.35	36.50
1500	Round fiber tube, recycled paper, 1 use, 8" diameter G		155	.206		1.39	7.80		9.19	14.45
1550	10" diameter G		155	.206		1.98	7.80		9.78	15.10
1600	12" diameter G		150	.213		2.33	8.10		10.43	15.90
1650	14" diameter G		145	.221		3.21	8.35		11.56	17.35
1700	16" diameter G		140	.229		4	8.65		12.65	18.70
1720	18" diameter G		140	.229		4.68	8.65		13.33	19.45
1750	20" diameter G		135	.237		6.05	8.95		15	21.50
1800	24" diameter G		130	.246		7.95	9.30		17.25	24
1850	30" diameter G		125	.256		11.65	9.70		21.35	29
1900	36" diameter G		115	.278		14.90	10.55		25.45	34
1950	42" diameter G		100	.320		35	12.10		47.10	58.50
2000	48" diameter G		85	.376		41.50	14.25		55.75	69.50
2200	For seamless type, add					15%				
3000	Round, steel, 4 use per mo., rent, regular duty, 14" diameter G	C-1	145	.221	L.F.	15.15	8.35		23.50	30.50
3050	16" diameter G		125	.256		15.50	9.70		25.20	33
3100	Heavy duty, 20" diameter G		105	.305		17	11.55		28.55	38
3150	24" diameter G		85	.376		18.65	14.25		32.90	44
3200	30" diameter G		70	.457		21.50	17.30		38.80	52
3250	36" diameter G		60	.533		23	20		43	59
3300	48" diameter G		50	.640		68	24		92	115
3350	60" diameter G		45	.711		42	27		69	90.50
4500	For second and succeeding months, deduct					50%				
5000	Job-built plywood, 8" x 8" columns, 1 use	C-1	165	.194	SFCA	2.41	7.35		9.76	14.75
5050	2 use		195	.164		1.37	6.20		7.57	11.75
5100	3 use		210	.152		.96	5.75		6.71	10.55
5150	4 use		215	.149		.79	5.65		6.44	10.15
5500	12" x 12" columns, 1 use		180	.178		2.27	6.75		9.02	13.60
5550	2 use		210	.152		1.25	5.75		7	10.85
5600	3 use		220	.145		.91	5.50		6.41	10.10
5650	4 use		225	.142		.74	5.40		6.14	9.70
6000	16" x 16" columns, 1 use		185	.173		2.24	6.55		8.79	13.25
6050	2 use		215	.149		1.19	5.65		6.84	10.60

03 11 13.25 Forms In Place, Columns

03 11 13.25 Forms In Place, Columns		Crew	Daily Output	Labor-Hours	Unit	2009 Bare Costs Material	2009 Bare Costs Labor	2009 Bare Costs Equipment	2009 Bare Costs Total	Total Incl O&P
6100	3 use	C-1	230	.139	SFCA	.90	5.25		6.15	9.70
6150	4 use		235	.136		.73	5.15		5.88	9.30
6500	24" x 24" columns, 1 use		190	.168		2.49	6.40		8.89	13.30
6550	2 use		216	.148		1.37	5.60		6.97	10.75
6600	3 use		230	.139		.99	5.25		6.24	9.80
6650	4 use		238	.134		.81	5.10		5.91	9.30
7000	36" x 36" columns, 1 use		200	.160		1.81	6.05		7.86	12
7050	2 use		230	.139		1.03	5.25		6.28	9.85
7100	3 use		245	.131		.73	4.94		5.67	8.95
7150	4 use		250	.128		.59	4.85		5.44	8.65
7400	Steel framed plywood, based on 50 uses of purchased									
7420	forms, and 4 uses of bracing lumber									
7500	8" x 8" column	C-1	340	.094	SFCA	4.53	3.56		8.09	10.90
7550	10" x 10"		350	.091		3.99	3.46		7.45	10.10
7600	12" x 12"		370	.086		3.38	3.27		6.65	9.10
7650	16" x 16"		400	.080		2.63	3.03		5.66	7.90
7700	20" x 20"		420	.076		2.35	2.88		5.23	7.35
7750	24" x 24"		440	.073		1.67	2.75		4.42	6.40
7755	30" x 30"		440	.073		2.14	2.75		4.89	6.90
7760	36" x 36"		460	.070		1.90	2.63		4.53	6.45
9000	Minimum labor/equipment charge	2 Carp	2	8	Job		320		320	530

03 11 13.30 Forms In Place, Culvert

03 11 13.30 Forms In Place, Culvert		Crew	Daily Output	Labor-Hours	Unit	Material	Labor	Equipment	Total	Total Incl O&P
0010	**FORMS IN PLACE, CULVERT**									
0015	5' to 8' square or rectangular, 1 use	C-1	170	.188	SFCA	6.10	7.15		13.25	18.45
0050	2 use R031113-60		180	.178		3.90	6.75		10.65	15.40
0100	3 use		190	.168		3.16	6.40		9.56	14.05
0150	4 use		200	.160		2.79	6.05		8.84	13.05

03 11 13.35 Forms In Place, Elevated Slabs

03 11 13.35 Forms In Place, Elevated Slabs		Crew	Daily Output	Labor-Hours	Unit	Material	Labor	Equipment	Total	Total Incl O&P
0010	**FORMS IN PLACE, ELEVATED SLABS**									
1000	Flat plate, job-built plywood, to 15' high, 1 use	C-2	470	.102	S.F.	4.53	3.97		8.50	11.55
1050	2 use R031113-60		520	.092		2.49	3.59		6.08	8.70
1100	3 use		545	.088		1.81	3.43		5.24	7.65
1150	4 use		560	.086		1.47	3.33		4.80	7.10
1500	15' to 20' high ceilings, 4 use		495	.097		1.67	3.77		5.44	8.10
1600	21' to 35' high ceilings, 4 use		450	.107		2.55	4.15		6.70	9.65
2000	Flat slab, drop panels, job-built plywood, to 15' high, 1 use		449	.107		4.77	4.16		8.93	12.10
2050	2 use		509	.094		2.62	3.67		6.29	8.95
2100	3 use		532	.090		1.91	3.51		5.42	7.90
2150	4 use		544	.088		1.55	3.43		4.98	7.35
2250	15' to 20' high ceilings, 4 use		480	.100		5.10	3.89		8.99	12
2350	20' to 35' high ceilings, 4 use		435	.110		5.95	4.29		10.24	13.65
3000	Floor slab hung from steel beams, 1 use		485	.099		3.58	3.85		7.43	10.30
3050	2 use		535	.090		3.09	3.49		6.58	9.15
3100	3 use		550	.087		2.92	3.39		6.31	8.80
3150	4 use		565	.085		2.84	3.30		6.14	8.55
3500	Floor slab, with 1-way joist pans, 1 use		415	.116		6.50	4.50		11	14.60
3550	2 use		445	.108		4.33	4.20		8.53	11.70
3600	3 use		475	.101		3.61	3.93		7.54	10.50
3650	4 use		500	.096		3.26	3.73		6.99	9.75
4500	With 2-way joist domes, 1 use		405	.119		6.25	4.61		10.86	14.45
4520	2 use		450	.107		4.09	4.15		8.24	11.35
4530	3 use		460	.104		3.38	4.06		7.44	10.40

03 11 Concrete Forming

03 11 13 – Structural Cast-In-Place Concrete Forming

03 11 13.35 Forms In Place, Elevated Slabs		Crew	Daily Output	Labor-Hours	Unit	Material	2009 Bare Costs Labor	Equipment	Total	Total Incl O&P
4550	4 use	C-2	470	.102	S.F.	3.02	3.97		6.99	9.85
5000	Box out for slab openings, over 16" deep, 1 use		190	.253	SFCA	5	9.80		14.80	21.50
5050	2 use		240	.200	"	2.75	7.80		10.55	15.90
5500	Shallow slab box outs, to 10 S.F.		42	1.143	Ea.	9.50	44.50		54	84
5550	Over 10 S.F. (use perimeter)		600	.080	L.F.	1.27	3.11		4.38	6.55
6000	Bulkhead forms for slab, with keyway, 1 use, 2 piece		500	.096		2.30	3.73		6.03	8.70
6100	3 piece (see also edge forms)		460	.104		2.36	4.06		6.42	9.30
6200	Slab bulkhead form, 4-1/2" high, exp metal, w/ keyway & stakes G	C-1	1200	.027		2.78	1.01		3.79	4.73
6210	5-1/2" high G		1100	.029		3.23	1.10		4.33	5.35
6215	7-1/2" high G		960	.033		4.26	1.26		5.52	6.75
6220	9-1/2" high G		840	.038		4.84	1.44		6.28	7.70
6500	Curb forms, wood, 6" to 12" high, on elevated slabs, 1 use		180	.178	SFCA	1.27	6.75		8.02	12.50
6550	2 use		205	.156		.70	5.90		6.60	10.50
6600	3 use		220	.145		.51	5.50		6.01	9.65
6650	4 use		225	.142		.41	5.40		5.81	9.35
7000	Edge forms to 6" high, on elevated slab, 4 use		500	.064	L.F.	.17	2.42		2.59	4.19
7070	7" to 12" high, 1 use		162	.198	SFCA	.98	7.50		8.48	13.45
7080	2 use		198	.162		.54	6.10		6.64	10.70
7090	3 use		222	.144		.39	5.45		5.84	9.45
7101	4 use		350	.091		.17	3.46		3.63	5.90
7500	Depressed area forms to 12" high, 4 use		300	.107	L.F.	1.12	4.04		5.16	7.90
7550	12" to 24" high, 4 use		175	.183		1.52	6.90		8.42	13.10
8000	Perimeter deck and rail for elevated slabs, straight		90	.356		13.70	13.45		27.15	37
8050	Curved		65	.492		18.75	18.65		37.40	51.50
8500	Void forms, round fiber, 3" diameter G		450	.071		.97	2.69		3.66	5.50
8550	4" diameter G		425	.075		1.44	2.85		4.29	6.30
8600	6" diameter G		400	.080		2.25	3.03		5.28	7.50
8650	8" diameter G		375	.085		3.85	3.23		7.08	9.60
8700	10" diameter G		350	.091		2.40	3.46		5.86	8.35
8750	12" diameter G		300	.107		3.05	4.04		7.09	10
9000	Minimum labor/equipment charge	2 Carp	2	8	Job		320		320	530

03 11 13.40 Forms In Place, Equipment Foundations		Crew	Daily Output	Labor-Hours	Unit	Material	Labor	Equipment	Total	Total Incl O&P
0010	**FORMS IN PLACE, EQUIPMENT FOUNDATIONS**									
0020	1 use	C-2	160	.300	SFCA	3.60	11.65		15.25	23
0050	2 use R031113-60		190	.253		1.98	9.80		11.78	18.40
0100	3 use		200	.240		1.44	9.35		10.79	17
0150	4 use		205	.234		1.17	9.10		10.27	16.35
9000	Minimum labor/equipment charge	1 Carp	3	2.667	Job		107		107	176

03 11 13.45 Forms In Place, Footings		Crew	Daily Output	Labor-Hours	Unit	Material	Labor	Equipment	Total	Total Incl O&P
0010	**FORMS IN PLACE, FOOTINGS**									
0020	Continuous wall, plywood, 1 use	C-1	375	.085	SFCA	7.45	3.23		10.68	13.55
0050	2 use R031113-60		440	.073		4.10	2.75		6.85	9.05
0100	3 use		470	.068		2.98	2.58		5.56	7.55
0150	4 use		485	.066		2.42	2.50		4.92	6.80
0500	Dowel supports for footings or beams, 1 use		500	.064	L.F.	.65	2.42		3.07	4.72
1000	Integral starter wall, to 4" high, 1 use		400	.080		.78	3.03		3.81	5.85
1500	Keyway, 4 use, tapered wood, 2" x 4"	1 Carp	530	.015		.16	.60		.76	1.17
1550	2" x 6"		500	.016		.26	.64		.90	1.35
2000	Tapered plastic		530	.015		.53	.60		1.13	1.58
2250	For keyway hung from supports, add		150	.053		.65	2.13		2.78	4.24
3000	Pile cap, square or rectangular, job-built plywood, 1 use	C-1	290	.110	SFCA	2.46	4.18		6.64	9.60
3050	2 use		346	.092		1.35	3.50		4.85	7.30

03 11 Concrete Forming

03 11 13 – Structural Cast-In-Place Concrete Forming

03 11 13.45 Forms In Place, Footings

	03 11 13.45 Forms In Place, Footings	Crew	Daily Output	Labor-Hours	Unit	2009 Bare Costs Material	Labor	Equipment	Total	Total Incl O&P
3100	3 use	C-1	371	.086	SFCA	.99	3.27		4.26	6.50
3150	4 use		383	.084		.80	3.16		3.96	6.10
4000	Triangular or hexagonal, 1 use		225	.142		2.86	5.40		8.26	12.05
4050	2 use		280	.114		1.58	4.33		5.91	8.90
4100	3 use		305	.105		1.15	3.97		5.12	7.80
4150	4 use		315	.102		.93	3.85		4.78	7.35
5000	Spread footings, job-built lumber, 1 use		305	.105		2.16	3.97		6.13	8.95
5050	2 use		371	.086		1.20	3.27		4.47	6.70
5100	3 use		401	.080		.86	3.02		3.88	5.95
5150	4 use		414	.077		.70	2.93		3.63	5.60
6000	Supports for dowels, plinths or templates, 2′ x 2′ footing		25	1.280	Ea.	5.40	48.50		53.90	86
6050	4′ x 4′ footing		22	1.455		10.80	55		65.80	103
6100	8′ x 8′ footing		20	1.600		21.50	60.50		82	124
6150	12′ x 12′ footing		17	1.882		26	71.50		97.50	147
7000	Plinths, job-built plywood, 1 use		250	.128	SFCA	2.73	4.85		7.58	11
7100	4 use		270	.119	"	.90	4.49		5.39	8.40
9000	Minimum labor/equipment charge	1 Carp	3	2.667	Job		107		107	176

03 11 13.47 Forms In Place, Gas Station Forms

		Crew	Daily Output	Labor-Hours	Unit	Material	Labor	Equipment	Total	Total Incl O&P
0010	**FORMS IN PLACE, GAS STATION FORMS**									
0050	Curb fascia, with template, 12 ga. steel, left in place, 9″ high G	1 Carp	50	.160	L.F.	15.70	6.40		22.10	28
1000	Sign or light bases, 18″ diameter, 9″ high G		9	.889	Ea.	99	35.50		134.50	168
1050	30″ diameter, 13″ high G		8	1	"	157	40		197	239
1990	Minimum labor/equipment charge		2	4	Job		160		160	264
2000	Island forms, 10′ long, 9″ high, 3′- 6″ wide G	C-1	10	3.200	Ea.	440	121		561	685
2050	4′ wide G		9	3.556		455	135		590	720
2500	20′ long, 9″ high, 4′ wide G		6	5.333		730	202		932	1,125
2550	5′ wide G		5	6.400		760	242		1,002	1,225
9000	Minimum labor/equipment charge	1 Carp	3	2.667	Job		107		107	176

03 11 13.50 Forms In Place, Grade Beam

		Crew	Daily Output	Labor-Hours	Unit	Material	Labor	Equipment	Total	Total Incl O&P
0010	**FORMS IN PLACE, GRADE BEAM**									
0020	Job-built plywood, 1 use	C-2	530	.091	SFCA	3.25	3.52		6.77	9.40
0050	2 use R031113-60		580	.083		1.79	3.22		5.01	7.25
0100	3 use		600	.080		1.30	3.11		4.41	6.60
0150	4 use		605	.079		1.06	3.09		4.15	6.25
9000	Minimum labor/equipment charge	2 Carp	2	8	Job		320		320	530

03 11 13.55 Forms In Place, Mat Foundation

		Crew	Daily Output	Labor-Hours	Unit	Material	Labor	Equipment	Total	Total Incl O&P
0010	**FORMS IN PLACE, MAT FOUNDATION**									
0020	Job-built plywood, 1 use	C-2	290	.166	SFCA	2.94	6.45		9.39	13.90
0050	2 use R031113-60		310	.155		1.19	6		7.19	11.25
0100	3 use		330	.145		.75	5.65		6.40	10.20
0120	4 use		350	.137		.70	5.35		6.05	9.55

03 11 13.65 Forms In Place, Slab On Grade

		Crew	Daily Output	Labor-Hours	Unit	Material	Labor	Equipment	Total	Total Incl O&P
0010	**FORMS IN PLACE, SLAB ON GRADE**									
1000	Bulkhead forms w/keyway, wood, 6″ high, 1 use	C-1	510	.063	L.F.	.84	2.38		3.22	4.84
1050	2 uses R031113-60		400	.080		.46	3.03		3.49	5.50
1100	4 uses		350	.091		.27	3.46		3.73	6
1400	Bulkhead form for slab, 4-1/2″ high, exp metal, incl keyway & stakes G		1200	.027		2.78	1.01		3.79	4.73
1410	5-1/2″ high G		1100	.029		3.23	1.10		4.33	5.35
1420	7-1/2″ high G		960	.033		4.26	1.26		5.52	6.75
1430	9-1/2″ high G		840	.038		4.84	1.44		6.28	7.70
2000	Curb forms, wood, 6″ to 12″ high, on grade, 1 use		215	.149	SFCA	2.83	5.65		8.48	12.40
2050	2 use		250	.128		1.57	4.85		6.42	9.75

03 11 13 – Structural Cast-In-Place Concrete Forming

03 11 13.65 Forms In Place, Slab On Grade		Crew	Daily Output	Labor-Hours	Unit	Material	2009 Bare Costs Labor	Equipment	Total	Total Incl O&P
2100	3 use	C-1	265	.121	SFCA	1.13	4.57		5.70	8.80
2150	4 use		275	.116		.92	4.41		5.33	8.25
3000	Edge forms, wood, 4 use, on grade, to 6" high		600	.053	L.F.	.38	2.02		2.40	3.75
3050	7" to 12" high		435	.074	SFCA	.74	2.79		3.53	5.40
3500	For depressed slabs, 4 use, to 12" high		300	.107	L.F.	.56	4.04		4.60	7.25
3550	To 24" high		175	.183		.76	6.90		7.66	12.30
4000	For slab blockouts, to 12" high, 1 use		200	.160		.63	6.05		6.68	10.70
4050	To 24" high, 1 use		120	.267		.80	10.10		10.90	17.55
4100	Plastic (extruded), to 6" high, multiple use, on grade		800	.040		5.50	1.51		7.01	8.55
5000	Screed, 24 ga. metal key joint, see Div. 03 15 05.25									
5020	Wood, incl. wood stakes, 1" x 3"	C-1	900	.036	L.F.	.68	1.35		2.03	2.97
5050	2" x 4"		900	.036	"	.63	1.35		1.98	2.91
6000	Trench forms in floor, wood, 1 use		160	.200	SFCA	1.48	7.55		9.03	14.10
6050	2 use		175	.183		.81	6.90		7.71	12.35
6100	3 use		180	.178		.59	6.75		7.34	11.75
6150	4 use		185	.173		.48	6.55		7.03	11.35
8760	Void form, corrugated fiberboard, 6" x 12", 10' long G		240	.133	S.F.	.84	5.05		5.89	9.25
9000	Minimum labor/equipment charge	1 Carp	2	4	Job		160		160	264

03 11 13.85 Forms In Place, Walls		Crew	Daily Output	Labor-Hours	Unit	Material	Labor	Equipment	Total	Total Incl O&P
0010	**FORMS IN PLACE, WALLS**									
0100	Box out for wall openings, to 16" thick, to 10 S.F.	C-2	24	2	Ea.	21.50	78		99.50	152
0150	Over 10 S.F. (use perimeter)	"	280	.171	L.F.	1.78	6.65		8.43	12.95
0250	Brick shelf, 4" w, add to wall forms, use wall area abv shelf									
0260	1 use R031113-60	C-2	240	.200	SFCA	1.87	7.80		9.67	14.90
0300	2 use		275	.175		1.03	6.80		7.83	12.35
0350	4 use		300	.160		.75	6.20		6.95	11.05
0500	Bulkhead, wood with keyway, 1 use, 2 piece		265	.181	L.F.	1.68	7.05		8.73	13.50
0600	Bulkhead forms with keyway, 1 piece expanded metal, 8" wall G	C-1	1000	.032		4.26	1.21		5.47	6.70
0610	10" wall G		800	.040		4.84	1.51		6.35	7.80
0620	12" wall G		525	.061		5.80	2.31		8.11	10.20
0700	Buttress, to 8' high, 1 use	C-2	350	.137	SFCA	7.05	5.35		12.40	16.55
0750	2 use		430	.112		3.89	4.34		8.23	11.45
0800	3 use		460	.104		2.83	4.06		6.89	9.80
0850	4 use		480	.100		2.33	3.89		6.22	8.95
1000	Corbel or haunch, to 12" wide, add to wall forms, 1 use		150	.320	L.F.	1.95	12.45		14.40	22.50
1050	2 use		170	.282		1.07	11		12.07	19.35
1100	3 use		175	.274		.78	10.65		11.43	18.45
1150	4 use		180	.267		.63	10.35		10.98	17.80
2000	Wall, job-built plywood, to 8' high, 1 use		370	.130	SFCA	2.62	5.05		7.67	11.25
2050	2 use		435	.110		1.61	4.29		5.90	8.85
2100	3 use		495	.097		1.17	3.77		4.94	7.55
2150	4 use		505	.095		.95	3.70		4.65	7.15
2400	Over 8' to 16' high, 1 use		280	.171		9.40	6.65		16.05	21.50
2450	2 use		345	.139		1.34	5.40		6.74	10.40
2500	3 use		375	.128		.96	4.98		5.94	9.25
2550	4 use		395	.122		.78	4.73		5.51	8.65
2700	Over 16' high, 1 use		235	.204		2.71	7.95		10.66	16.10
2750	2 use		290	.166		1.49	6.45		7.94	12.30
2800	3 use		315	.152		1.08	5.95		7.03	11
2850	4 use		330	.145		.88	5.65		6.53	10.30
3000	For architectural finish, add		1820	.026		2.72	1.03		3.75	4.68
4000	Radial, smooth curved, job-built plywood, 1 use		245	.196		2.57	7.60		10.17	15.45

03 11 13.85 Forms In Place, Walls		Crew	Daily Output	Labor-Hours	Unit	Material	2009 Bare Costs Labor	Equipment	Total	Total Incl O&P
4050	2 use	C-2	300	.160	SFCA	1.41	6.20		7.61	11.80
4100	3 use		325	.148		1.03	5.75		6.78	10.65
4150	4 use		335	.143		.85	5.55		6.40	10.15
4200	Below grade, job-built plywood, 1 use		225	.213		4.52	8.30		12.82	18.70
4210	2 use		225	.213		2.52	8.30		10.82	16.45
4220	3 use		225	.213		1.94	8.30		10.24	15.85
4230	4 use		225	.213		1.46	8.30		9.76	15.30
4300	Curved, 2′ chords, job-built plywood, 1 use		290	.166		2.09	6.45		8.54	12.95
4350	2 use		355	.135		1.15	5.25		6.40	9.95
4400	3 use		385	.125		.84	4.85		5.69	8.90
4450	4 use		400	.120		.68	4.67		5.35	8.45
4500	Over 8′ high, 1 use		290	.166		.83	6.45		7.28	11.55
4525	2 use		355	.135		.46	5.25		5.71	9.20
4550	3 use		385	.125		.33	4.85		5.18	8.35
4575	4 use		400	.120		.27	4.67		4.94	8
4600	Retaining wall, battered, job-built plyw'd, to 8′ high, 1 use		300	.160		2	6.20		8.20	12.45
4650	2 use		355	.135		1.10	5.25		6.35	9.90
4700	3 use		375	.128		.80	4.98		5.78	9.10
4750	4 use		390	.123		.50	4.79		5.29	8.45
4900	Over 8′ to 16′ high, 1 use		240	.200		2.14	7.80		9.94	15.20
4950	2 use		295	.163		1.17	6.35		7.52	11.75
5000	3 use		305	.157		.85	6.10		6.95	11.05
5050	4 use		320	.150		.69	5.85		6.54	10.40
5500	For gang wall forming, 192 S.F. sections, deduct					10%	10%			
5550	384 S.F. sections, deduct					20%	20%			
5750	Liners for forms (add to wall forms), A.B.S. plastic									
5800	Aged wood, 4″ wide, 1 use	1 Carp	256	.031	SFCA	3.24	1.25		4.49	5.60
5820	2 use		256	.031		1.78	1.25		3.03	4.02
5830	3 use		256	.031		1.30	1.25		2.55	3.49
5840	4 use		256	.031		1.05	1.25		2.30	3.22
5900	Fractured rope rib, 1 use		192	.042		4.71	1.66		6.37	7.95
5925	2 use		192	.042		2.59	1.66		4.25	5.60
5950	3 use		192	.042		1.88	1.66		3.54	4.82
6000	4 use		192	.042		1.53	1.66		3.19	4.43
6100	Ribbed, 3/4″ deep x 1-1/2″ O.C., 1 use		224	.036		2.72	1.43		4.15	5.35
6125	2 use		224	.036		1.50	1.43		2.93	4.01
6150	3 use		224	.036		1.09	1.43		2.52	3.56
6200	4 use		224	.036		.88	1.43		2.31	3.33
6300	Rustic brick pattern, 1 use		224	.036		3.24	1.43		4.67	5.90
6325	2 use		224	.036		1.78	1.43		3.21	4.32
6350	3 use		224	.036		1.30	1.43		2.73	3.79
6400	4 use		224	.036		1.05	1.43		2.48	3.52
6500	3/8″ striated, random, 1 use		224	.036		2.12	1.43		3.55	4.69
6525	2 use		224	.036		1.17	1.43		2.60	3.64
6550	3 use		224	.036		.85	1.43		2.28	3.29
6600	4 use		224	.036		.69	1.43		2.12	3.12
6850	Random vertical rustication, 1 use		384	.021		6.15	.83		6.98	8.10
6900	2 use		384	.021		3.38	.83		4.21	5.10
6925	3 use		384	.021		2.46	.83		3.29	4.07
6950	4 use		384	.021		2	.83		2.83	3.57
7050	Wood, beveled edge, 3/4″ deep, 1 use		384	.021	L.F.	.21	.83		1.04	1.60
7100	1″ deep, 1 use		384	.021	″	.34	.83		1.17	1.74
7200	4″ wide aged cedar, 1 use		256	.031	SFCA	3.24	1.25		4.49	5.60

03 11 Concrete Forming

03 11 13 – Structural Cast-In-Place Concrete Forming

03 11 13.85 Forms In Place, Walls		Crew	Daily Output	Labor-Hours	Unit	Material	2009 Bare Costs Labor	Equipment	Total	Total Incl O&P
7300	4" variable depth rough cedar	1 Carp	224	.036	SFCA	4.71	1.43		6.14	7.55
7500	Lintel or sill forms, 1 use		30	.267		2.66	10.65		13.31	20.50
7520	2 use		34	.235		1.46	9.40		10.86	17.10
7540	3 use		36	.222		1.06	8.90		9.96	15.80
7560	4 use		37	.216		.86	8.65		9.51	15.20
7800	Modular prefabricated plywood, based on 20 uses of purchased									
7820	forms, and 4 uses of bracing lumber									
7860	To 8' high	C-2	800	.060	SFCA	1.47	2.33		3.80	5.45
8060	Over 8' to 16' high		600	.080		1.51	3.11		4.62	6.80
8600	Pilasters, 1 use		270	.178		3.07	6.90		9.97	14.80
8620	2 use		330	.145		1.69	5.65		7.34	11.20
8640	3 use		370	.130		1.23	5.05		6.28	9.70
8660	4 use		385	.125		1	4.85		5.85	9.10
9010	Steel framed plywood, based on 50 uses of purchased									
9020	forms, and 4 uses of bracing lumber									
9060	To 8' high	C-2	600	.080	SFCA	.83	3.11		3.94	6.05
9260	Over 8' to 16' high		450	.107		.83	4.15		4.98	7.75
9460	Over 16' to 20' high		400	.120		.83	4.67		5.50	8.60
9475	For elevated walls, add						10%			
9900	Minimum labor/equipment charge	2 Carp	2	8	Job		320		320	530

03 11 19 – Insulating Concrete Forming

03 11 19.10 Insulating Forms, Left In Place		Crew	Daily Output	Labor-Hours	Unit	Material	Labor	Equipment	Total	Total Incl O&P
0010	**INSULATING FORMS, LEFT IN PLACE**									
0020	S.F. is for exterior face, but includes forms for both faces									
2000	4" wall, straight block, 16" x 48" (5.33 SF) G	2 Carp	90	.178	Ea.	15.75	7.10		22.85	29
2010	90 corner block, exterior 16" x 38" x 22" (6.67 SF) G		75	.213		18.45	8.50		26.95	34.50
2020	45 corner block, exterior 16" x 34" x 18" (5.78 SF) G		75	.213		17.05	8.50		25.55	33
2100	6" wall, straight block, 16" x 48" (5.33 SF) G		90	.178		16.05	7.10		23.15	29.50
2110	90 corner block, exterior 16" x 32" x 24" (6.22 SF) G		75	.213		18.45	8.50		26.95	34.50
2120	45 corner block, exterior 16" x 26" x 18" (4.89 SF) G		75	.213		16.85	8.50		25.35	32.50
2130	Brick ledge block, 16" x 48" (5.33 SF) G		80	.200		19.65	8		27.65	34.50
2140	Taper top block, 16" x 48" (5.33 SF) G		80	.200		17.55	8		25.55	32.50
2200	8" wall, straight block, 16" x 48" (5.33 SF) G		90	.178		17.05	7.10		24.15	30.50
2210	90 corner block, exterior 16" x 34" x 26" (6.67 SF) G		75	.213		19.60	8.50		28.10	35.50
2220	45 corner block, exterior 16" x 28" x 20" (5.33 SF) G		75	.213		17.75	8.50		26.25	33.50
2230	Brick ledge block, 16" x 48" (5.33 SF) G		80	.200		20.50	8		28.50	35.50
2240	Taper top block, 16" x 48" (5.33 SF) G		80	.200		18.55	8		26.55	33.50
03 11 19.60 Roof Deck Form Boards										
0010	**ROOF DECK FORM BOARDS** R051223-50									
0050	Non-asbestos fiber cement, 1/8" thick	C-13	2950	.008	S.F.	2.67	.35	.05	3.07	3.62
0070	1/4" thick		2950	.008		3.03	.35	.05	3.43	4.01
0100	Fiberglass, 1" thick, economy		2700	.009		3.25	.38	.05	3.68	4.32
0500	Wood fiber, 1" thick G		2700	.009		4.62	.38	.05	5.05	5.85

03 11 23 – Permanent Stair Forming

03 11 23.75 Forms In Place, Stairs		Crew	Daily Output	Labor-Hours	Unit	Material	Labor	Equipment	Total	Total Incl O&P
0010	**FORMS IN PLACE, STAIRS**									
0015	(Slant length x width), 1 use	C-2	165	.291	S.F.	5.10	11.30		16.40	24.50
0050	2 use R031113-60		170	.282		3.10	11		14.10	21.50
0100	3 use		180	.267		2.44	10.35		12.79	19.80
0150	4 use		190	.253		2.11	9.80		11.91	18.50
1000	Alternate pricing method (1.0 L.F./S.F.), 1 use		100	.480	LF Rsr	5.10	18.65		23.75	36.50

03 11 Concrete Forming

03 11 23 – Permanent Stair Forming

03 11 23.75 Forms In Place, Stairs		Crew	Daily Output	Labor-Hours	Unit	Material	Labor	Equipment	Total	Total Incl O&P
1050	2 use	C-2	105	.457	LF Rsr	3.10	17.80		20.90	33
1100	3 use		110	.436		2.44	16.95		19.39	30.50
1150	4 use		115	.417	↓	2.11	16.25		18.36	29.50
2000	Stairs, cast on sloping ground (length x width), 1 use		220	.218	S.F.	1.75	8.50		10.25	15.90
2025	2 use		232	.207		.96	8.05		9.01	14.35
2050	3 use		244	.197		.70	7.65		8.35	13.40
2100	4 use	↓	256	.188	↓	.57	7.30		7.87	12.65

(Material, Labor, Equipment and Total columns: 2009 Bare Costs)

03 15 Concrete Accessories

03 15 05 – Concrete Forming Accessories

03 15 05.12 Chamfer Strips		Crew	Daily Output	Labor-Hours	Unit	Material	Labor	Equipment	Total	Total Incl O&P
0010	**CHAMFER STRIPS**									
2000	Polyvinyl chloride, 1/2" wide with leg	1 Carp	535	.015	L.F.	.48	.60		1.08	1.52
2200	3/4" wide with leg		525	.015		.58	.61		1.19	1.65
2400	1" radius with leg		515	.016		.67	.62		1.29	1.76
2800	2" radius with leg		500	.016		3.17	.64		3.81	4.55
5000	Wood, 1/2" wide		535	.015		.22	.60		.82	1.23
5200	3/4" wide		525	.015		.21	.61		.82	1.24
5400	1" wide	↓	515	.016	↓	.34	.62		.96	1.39

03 15 05.25 Expansion Joints		Crew	Daily Output	Labor-Hours	Unit	Material	Labor	Equipment	Total	Total Incl O&P
0010	**EXPANSION JOINTS**									
0020	Keyed, cold, 24 ga., incl. stakes, 3-1/2" high G	1 Carp	200	.040	L.F.	2.44	1.60		4.04	5.30
0050	4-1/2" high G		200	.040		2.78	1.60		4.38	5.70
0100	5-1/2" high G		195	.041		3.23	1.64		4.87	6.25
0150	7-1/2" high G		190	.042		4.26	1.68		5.94	7.45
0160	9-1/2" high G	↓	185	.043		4.84	1.73		6.57	8.15
0300	Poured asphalt, plain, 1/2" x 1"	1 Clab	450	.018		.51	.56		1.07	1.49
0350	1" x 2"		400	.020		2.05	.63		2.68	3.30
0500	Neoprene, liquid, cold applied, 1/2" x 1"		450	.018		2.21	.56		2.77	3.36
0550	1" x 2"		400	.020		8.85	.63		9.48	10.75
0700	Polyurethane, poured, 2 part, 1/2" x 1"		400	.020		1.40	.63		2.03	2.58
0750	1" x 2"		350	.023		5.60	.72		6.32	7.35
0900	Rubberized asphalt, hot or cold applied, 1/2" x 1"		450	.018		.76	.56		1.32	1.77
0950	1" x 2"		400	.020		3.05	.63		3.68	4.40
1100	Hot applied, fuel resistant, 1/2" x 1"		450	.018		1.14	.56		1.70	2.18
1150	1" x 2"	↓	400	.020		4.58	.63		5.21	6.10
2000	Premolded, bituminous fiber, 1/2" x 6"	1 Carp	375	.021		.40	.85		1.25	1.85
2050	1" x 12"		300	.027		1.91	1.07		2.98	3.86
2140	Concrete expansion joint, recycled paper and fiber, 1/2" x 6" G		390	.021		.35	.82		1.17	1.73
2150	1/2" x 12" G		360	.022		.69	.89		1.58	2.23
2250	Cork with resin binder, 1/2" x 6"		375	.021		1.81	.85		2.66	3.40
2300	1" x 12"		300	.027		7.55	1.07		8.62	10.05
2500	Neoprene sponge, closed cell, 1/2" x 6"		375	.021		1.88	.85		2.73	3.48
2550	1" x 12"		300	.027		7.05	1.07		8.12	9.50
2750	Polyethylene foam, 1/2" x 6"		375	.021		.59	.85		1.44	2.06
2800	1" x 12"		300	.027		1.92	1.07		2.99	3.87
3000	Polyethylene backer rod, 3/8" diameter		460	.017		.06	.69		.75	1.21
3050	3/4" diameter		460	.017		.09	.69		.78	1.25
3100	1" diameter		460	.017		.16	.69		.85	1.33
3500	Polyurethane foam, with polybutylene, 1/2" x 1/2"		475	.017		.95	.67		1.62	2.16
3550	1" x 1"	↓	450	.018	↓	2.40	.71		3.11	3.81

03 15 05.25 Expansion Joints

03 15 05.25 Expansion Joints			Crew	Daily Output	Labor-Hours	Unit	Material	2009 Bare Costs Labor	Equipment	Total	Total Incl O&P
3750	Polyurethane foam, regular, closed cell, 1/2" x 6"		1 Carp	375	.021	L.F.	.70	.85		1.55	2.18
3800	1" x 12"			300	.027		2.50	1.07		3.57	4.51
4000	Polyvinyl chloride foam, closed cell, 1/2" x 6"			375	.021		1.90	.85		2.75	3.50
4050	1" x 12"			300	.027		6.55	1.07		7.62	8.95
4250	Rubber, gray sponge, 1/2" x 6"			375	.021		1.60	.85		2.45	3.17
4300	1" x 12"		↓	300	.027	↓	5.75	1.07		6.82	8.10
4500	Lead wool for joints					Lb.	8.80			8.80	9.70
5000	For installation in walls, add							75%			
5250	For installation in boxouts, add							25%			

03 15 05.70 Shores

03 15 05.70 Shores			Crew	Daily Output	Labor-Hours	Unit	Material	Labor	Equipment	Total	Total Incl O&P
0010	**SHORES**										
0020	Erect and strip, by hand, horizontal members										
0500	Aluminum joists and stringers	G	2 Carp	60	.267	Ea.		10.65		10.65	17.60
0600	Steel, adjustable beams	G		45	.356			14.20		14.20	23.50
0700	Wood joists			50	.320			12.80		12.80	21
0800	Wood stringers			30	.533			21.50		21.50	35
1000	Vertical members to 10' high	G		55	.291			11.60		11.60	19.20
1050	To 13' high	G		50	.320			12.80		12.80	21
1100	To 16' high	G		45	.356	↓		14.20		14.20	23.50
1500	Reshoring	G	↓	1400	.011	S.F.	.50	.46		.96	1.30
1600	Flying truss system	G	C-17D	9600	.009	SFCA		.36	.07	.43	.66
1760	Horizontal, aluminum joists, 6-1/4" high x 5' to 21' span, buy	G				L.F.	36			36	39.50
1770	Beams, 7-1/4" high x 4' to 30' span	G				"	47.50			47.50	52.50
1810	Horizontal, steel beam, adjustable, 4' to 7' span, buy	G				Ea.	330			330	360
1830	6' to 10' span	G					330			330	360
1920	9' to 15' span	G					545			545	600
1940	12' to 20' span	G				↓	650			650	715
1970	Steel stringer, W8x10, 4' to 16' span, buy	G				L.F.	25.50			25.50	28
3000	Rent for job duration, aluminum joist @ 2' O.C., per mo	G				SF Flr.	.90			.90	.99
3050	Steel W8x10	G					.64			.64	.70
3060	Steel adjustable	G				↓	.82			.82	.90
3500	#1 post shore, steel, 5'-7" to 9'-6" high, 10000# cap., buy	G				Ea.	410			410	450
3550	#2 post shore, 7'-3" to 12'-10" high, 7800# capacity	G					465			465	515
3600	#3 post shore, 8'-10" to 16'-1" high, 3800# capacity	G				↓	540			540	595
5010	Frame shoring systems, steel, 12000#/leg, buy										
5040	Frame, 2' wide x 6' high	G				Ea.	106			106	117
5250	X-brace	G					18			18	19.80
5550	Base plate	G					17.60			17.60	19.35
5600	Screw jack	G					38			38	41.50
5650	U-head, 8" x 8"	G				↓	21.50			21.50	24

03 15 05.85 Stair Tread Inserts

03 15 05.85 Stair Tread Inserts			Crew	Daily Output	Labor-Hours	Unit	Material	Labor	Equipment	Total	Total Incl O&P
0010	**STAIR TREAD INSERTS**										
0015	Cast iron, abrasive, 3" wide	G	1 Carp	90	.089	L.F.	10.65	3.55		14.20	17.55
0020	4" wide	G		80	.100		14.15	4		18.15	22
0040	6" wide	G		75	.107		21.50	4.26		25.76	30.50
0050	9" wide	G		70	.114		32	4.57		36.57	42.50
0100	12" wide	G	↓	65	.123		42.50	4.92		47.42	55
0300	Cast aluminum, compared to cast iron, deduct						10%				
0500	Extruded aluminum safety tread, 3" wide	G	1 Carp	75	.107		12.10	4.26		16.36	20.50
0550	4" wide	G		75	.107		16.15	4.26		20.41	25
0600	6" wide	G		75	.107		24	4.26		28.26	33.50
0650	9" wide to resurface stairs	G	↓	70	.114	↓	36.50	4.57		41.07	47.50

03 15 Concrete Accessories

03 15 05 – Concrete Forming Accessories

03 15 05.85 Stair Tread Inserts		Crew	Daily Output	Labor-Hours	Unit	Material	2009 Bare Costs Labor	Equipment	Total	Total Incl O&P
1700	Cement fill for pan-type metal treads, plain	1 Cefi	115	.070	S.F.	1.87	2.66		4.53	6.25
1750	Non-slip	"	100	.080	"	2.06	3.06		5.12	7.05

03 15 13 – Waterstops

03 15 13.50 Waterstops

		Crew	Daily Output	Labor-Hours	Unit	Material	Labor	Equipment	Total	Total Incl O&P
0010	**WATERSTOPS**, PVC and Rubber									
0020	PVC, ribbed 3/16" thick, 4" wide	1 Carp	155	.052	L.F.	1.08	2.06		3.14	4.59
0050	6" wide		145	.055		1.80	2.20		4	5.60
0500	With center bulb, 6" wide, 3/16" thick		135	.059		1.52	2.37		3.89	5.60
0550	3/8" thick		130	.062		3.20	2.46		5.66	7.60
0600	9" wide x 3/8" thick		125	.064		4.50	2.56		7.06	9.15
0800	Dumbbell type, 6" wide, 3/16" thick		150	.053		1.48	2.13		3.61	5.15
0850	3/8" thick		145	.055		3.16	2.20		5.36	7.10
1000	9" wide, 3/8" thick, plain		130	.062		4.46	2.46		6.92	8.95
1050	Center bulb		130	.062		6.80	2.46		9.26	11.50
1250	Ribbed type, split, 3/16" thick, 6" wide		145	.055		1.83	2.20		4.03	5.65
1300	3/8" thick		130	.062		3.93	2.46		6.39	8.40
2000	Rubber, flat dumbbell, 3/8" thick, 6" wide		145	.055		8.65	2.20		10.85	13.15
2050	9" wide		135	.059		12.65	2.37		15.02	17.85
2500	Flat dumbbell split, 3/8" thick, 6" wide		145	.055		1.83	2.20		4.03	5.65
2550	9" wide		135	.059		3.93	2.37		6.30	8.25
3000	Center bulb, 1/4" thick, 6" wide		145	.055		7.15	2.20		9.35	11.55
3050	9" wide		135	.059		14.10	2.37		16.47	19.40
3500	Center bulb split, 3/8" thick, 6" wide		145	.055		9.30	2.20		11.50	13.90
3550	9" wide		135	.059		15.55	2.37		17.92	21
5000	Waterstop fittings, rubber, flat									
5010	Dumbbell or center bulb, 3/8" thick,									
5200	Field union, 6" wide	1 Carp	50	.160	Ea.	28	6.40		34.40	41
5250	9" wide		50	.160		33	6.40		39.40	47
5500	Flat cross, 6" wide		30	.267		58	10.65		68.65	81
5550	9" wide		30	.267		83	10.65		93.65	109
6000	Flat tee, 6" wide		30	.267		53.50	10.65		64.15	76.50
6050	9" wide		30	.267		70.50	10.65		81.15	95.50
6500	Flat ell, 6" wide		40	.200		48.50	8		56.50	66.50
6550	9" wide		40	.200		65.50	8		73.50	85
7000	Vertical tee, 6" wide		25	.320		47	12.80		59.80	72.50
7050	9" wide		25	.320		60	12.80		72.80	87
7500	Vertical ell, 6" wide		35	.229		41.50	9.15		50.65	61
7550	9" wide		35	.229		57.50	9.15		66.65	78

03 21 Reinforcing Steel

03 21 05 – Reinforcing Steel Accessories

03 21 05.10 Rebar Accessories

		Crew	Daily Output	Labor-Hours	Unit	Material	Labor	Equipment	Total	Total Incl O&P
0010	**REBAR ACCESSORIES**									
0030	Steel & plastic made from recycled materials									
0100	Beam bolsters (BB), lower, 1-1/2" high, plain steel G				C.L.F.	100			100	110
0102	Galvanized G					120			120	132
0104	Stainless tipped legs G					515			515	565
0106	Plastic tipped legs G					172			172	189
0108	Epoxy dipped G					233			233	256
0110	2" high, plain G					127			127	140

03 21 Reinforcing Steel

03 21 05 – Reinforcing Steel Accessories

03 21 05.10 Rebar Accessories			Crew	Daily Output	Labor-Hours	Unit	2009 Bare Costs Material	Labor	Equipment	Total	Total Incl O&P
0120	Galvanized	G				C.L.F.	152			152	167
0140	Stainless tipped legs	G					545			545	595
0160	Plastic tipped legs	G					176			176	193
0162	Epoxy dipped	G					248			248	272
0200	Upper (BBU), 1-1/2" high, plain steel	G					281			281	310
0210	3" high	G					505			505	555
0300	Beam bolster with plate (BBP), 1-1/2" high, plain steel	G					330			330	360
0310	3" high	G					520			520	575
0500	Slab bolsters, continuous (SB), 1" high, plain steel	G					86.50			86.50	95
0502	Galvanized	G					95.50			95.50	105
0504	Stainless tipped legs	G					500			500	555
0506	Plastic tipped legs	G					118			118	129
0510	2" high, plain steel	G					100			100	110
0515	Galvanized	G					118			118	129
0520	Stainless tipped legs	G					515			515	565
0525	Plastic tipped legs	G					130			130	143
0530	For bolsters with wire runners (SBR), add	G					145			145	159
0540	For bolsters with plates (SBP), add	G					277			277	305
0700	Bag ties, 16 ga., plain, 4" long	G				C	1.35			1.35	1.49
0710	5" long	G					1.46			1.46	1.61
0720	6" long	G					1.46			1.46	1.61
0730	7" long	G					1.56			1.56	1.72
1200	High chairs, individual (HC), 3" high, plain steel	G					121			121	133
1202	Galvanized	G					170			170	186
1204	Stainless tipped legs	G					535			535	590
1206	Plastic tipped legs	G					160			160	176
1210	5" high, plain	G					185			185	204
1212	Galvanized	G					231			231	254
1214	Stainless tipped legs	G					600			600	660
1216	Plastic tipped legs	G					231			231	254
1220	8" high, plain	G					400			400	440
1222	Galvanized	G					510			510	560
1224	Stainless tipped legs	G					815			815	900
1226	Plastic tipped legs	G					505			505	555
1230	12" high, plain	G					835			835	915
1232	Galvanized	G					1,000			1,000	1,100
1234	Stainless tipped legs	G					1,250			1,250	1,375
1236	Plastic tipped legs	G					940			940	1,025
1400	Individual high chairs, with plate (HCP), 5" high	G					480			480	530
1410	8" high	G					730			730	800
1500	Bar chair (BC), 1-1/2" high, plain steel	G					61.50			61.50	67.50
1520	Galvanized	G					72			72	79
1530	Stainless tipped legs	G					475			475	525
1540	Plastic tipped legs	G					126			126	138
1550	Joist chair (JC), 3/4" x 5", plain steel	G					83			83	91.50
1580	Galvanized	G					100			100	110
1600	Stainless tipped legs	G					500			500	550
1620	Plastic tipped legs	G					144			144	158
1630	Epoxy dipped	G					325			325	355
1700	Continuous high chairs (CHC), legs 8" O.C., 4" high, plain steel	G				C.L.F.	160			160	176
1705	Galvanized	G					187			187	206
1710	Stainless tipped legs	G					575			575	635
1715	Plastic tipped legs	G					199			199	219

03 21 Reinforcing Steel

03 21 05 – Reinforcing Steel Accessories

03 21 05.10 Rebar Accessories			Crew	Daily Output	Labor-Hours	Unit	2009 Bare Costs Material	Labor	Equipment	Total	Total Incl O&P
1718	Epoxy dipped	G				C.L.F.	283			283	310
1720	6″ high, plain	G					231			231	254
1725	Galvanized	G					315			315	350
1730	Stainless tipped legs	G					645			645	710
1735	Plastic tipped legs	G					315			315	350
1738	Epoxy dipped	G					385			385	420
1740	8″ high, plain	G					335			335	370
1745	Galvanized	G					405			405	445
1750	Stainless tipped legs	G					750			750	825
1755	Plastic tipped legs	G					440			440	485
1758	Epoxy dipped	G					585			585	640
1900	For continuous bottom wire runners, add	G					172			172	189
1940	For continuous bottom plate, add	G					335			335	370
2100	Paper tubing, 4′ lengths, for #4 bar	G					87			87	96
2120	For #6 bar	G					103			103	114
2200	Screed chair base, 1/2″ coil thread diam., 2-1/2″ high, plain steel	G				C	395			395	435
2210	Galvanized	G					420			420	460
2220	5-1/2″ high, plain	G					480			480	530
2250	Galvanized	G					505			505	555
2300	3/4″ coil thread diam., 2-1/2″ high, plain steel	G					505			505	555
2310	Galvanized	G					675			675	740
2320	5-1/2″ high, plain steel	G					615			615	680
2350	Galvanized	G					865			865	950
2400	Screed holder, 1/2″ coil thread diam. for pipe screed, plain steel, 6″ long	G					420			420	460
2420	12″ long	G					695			695	760
2500	3/4″ coil thread diam. for pipe screed, plain steel, 6″ long	G					755			755	830
2520	12″ long	G					1,125			1,125	1,225
2700	Screw anchor for bolts, plain steel, 3/4″ diameter x 4″ long	G					475			475	525
2720	1″ diameter x 6″ long	G					815			815	895
2740	1-1/2″ diameter x 8″ long	G					1,375			1,375	1,500
2800	Screw anchor eye bolts, 3/4″ x 7″ long	G					6,125			6,125	6,725
2820	1″ x 9″ long	G					12,100			12,100	13,300
2840	1-1/2″ x 14″ long	G					30,600			30,600	33,600
2900	Screw anchor bolts, 3/4″ x 9″ long	G					2,125			2,125	2,350
2920	1″ x 12″ long	G					4,125			4,125	4,525
3001	Slab lifting inserts, single pickup, galv, 3/4″ diam., 5″ high	G					2,000			2,000	2,200
3010	6″ high	G					2,175			2,175	2,375
3030	7″ high	G					2,300			2,300	2,525
3100	1″ diameter, 5-1/2″ high	G					2,325			2,325	2,550
3120	7″ high	G					2,450			2,450	2,700
3200	Double pickup lifting inserts, 1″ diameter, 5-1/2″ high	G					4,725			4,725	5,200
3220	7″ high	G					5,175			5,175	5,700
3330	1-1/2″ diameter, 5-1/2″ high	G					6,200			6,200	6,825
3800	Subgrade chairs, #4 bar head, 3-1/2″ high	G					78			78	86
3850	12″ high	G					136			136	150
3900	#6 bar head, 3-1/2″ high	G					78			78	86
3950	12″ high	G					136			136	150
4200	Subgrade stakes, no nail holes, 3/4″ diameter, 12″ long	G					365			365	405
4250	24″ long	G					570			570	625
4300	7/8″ diameter, 12″ long	G					530			530	580
4350	24″ long	G					745			745	820
4500	Tie wire, 16 ga. annealed steel	G				Cwt.	174			174	191
4550	Tie wire holder, plastic case	G				Ea.	49			49	54

03 21 Reinforcing Steel

03 21 05 – Reinforcing Steel Accessories

03 21 05.10 Rebar Accessories			Crew	Daily Output	Labor-Hours	Unit	Material	2009 Bare Costs Labor	Equipment	Total	Total Incl O&P
4600	Aluminum case	G				Ea.	76			76	83.50

03 21 05.75 Splicing Reinforcing Bars

			Crew	Daily Output	Labor-Hours	Unit	Material	2009 Bare Costs Labor	Equipment	Total	Total Incl O&P
0010	**SPLICING REINFORCING BARS**										
0020	Including holding bars in place while splicing										
0100	Standard, self-aligning type, taper threaded, #4 bars	G	C-25	190	.168	Ea.	4.55	5.90		10.45	15.20
0105	#5 bars	G		170	.188		5.55	6.60		12.15	17.50
0110	#6 bars	G		150	.213		6.40	7.45		13.85	20
0120	#7 bars	G		130	.246		7.45	8.60		16.05	23
0300	#8 bars	G		115	.278		12.90	9.75		22.65	31
0305	#9 bars	G	C-5	105	.533		14.10	23	7.50	44.60	62.50
0310	#10 bars	G		95	.589		15.70	25.50	8.30	49.50	69.50
0320	#11 bars	G		85	.659		16.70	28.50	9.30	54.50	76.50
0330	#14 bars	G		65	.862		25	37.50	12.15	74.65	103
0340	#18 bars	G		45	1.244		38	54	17.55	109.55	152
0500	Transition self-aligning, taper threaded, #18-14	G		45	1.244		40.50	54	17.55	112.05	154
0510	#18-11	G		45	1.244		41.50	54	17.55	113.05	156
0520	#14-11	G		65	.862		29.50	37.50	12.15	79.15	108
0540	#11-10	G		85	.659		23.50	28.50	9.30	61.30	84
0550	#10-9	G		95	.589		22	25.50	8.30	55.80	76
0560	#9-8	G	C-25	105	.305		21	10.65		31.65	41.50
0580	#8-7	G		115	.278		19.55	9.75		29.30	38.50
0590	#7-6	G		130	.246		19.05	8.60		27.65	36
0600	Position coupler for curved bars, taper threaded, #4 bars	G		160	.200		22	7		29	36
0610	#5 bars	G		145	.221		23	7.70		30.70	38.50
0620	#6 bars	G		130	.246		26	8.60		34.60	44
0630	#7 bars	G		110	.291		38.50	10.15		48.65	60
0640	#8 bars	G		100	.320		40	11.20		51.20	63.50
0650	#9 bars	G	C-5	90	.622		43.50	27	8.75	79.25	103
0660	#10 bars	G		80	.700		47	30.50	9.85	87.35	113
0670	#11 bars	G		70	.800		49	35	11.30	95.30	124
0680	#14 bars	G		55	1.018		61	44	14.35	119.35	157
0690	#18 bars	G		40	1.400		88	61	19.75	168.75	220
0700	Transition position coupler for curved bars, taper threaded, #18-14	G		40	1.400		111	61	19.75	191.75	246
0710	#18-11	G		40	1.400		112	61	19.75	192.75	247
0720	#14-11	G		55	1.018		95	44	14.35	153.35	194
0730	#11-10	G		70	.800		54.50	35	11.30	100.80	130
0740	#10-9	G		80	.700		52	30.50	9.85	92.35	119
0750	#9-8	G	C-25	90	.356		48	12.45		60.45	74.50
0760	#8-7	G		100	.320		44	11.20		55.20	68
0770	#7-6	G		110	.291		42.50	10.15		52.65	64.50
0800	Sleeve type w/ grout filler, for precast concrete, #6 bars	G		72	.444		23.50	15.55		39.05	53
0802	#7 bars	G		64	.500		25.50	17.50		43	58.50
0805	#8 bars	G		56	.571		29	19.95		48.95	66.50
0807	#9 bars	G		48	.667		31.50	23.50		55	75.50
0810	#10 bars	G	C-5	40	1.400		41	61	19.75	121.75	169
0900	#11 bars	G		32	1.750		49	76	24.50	149.50	208
0920	#14 bars	G		24	2.333		79.50	101	33	213.50	293
1000	Sleeve type w/ ferrous filler, for critical structures, #6 bars	G	C-25	72	.444		41	15.55		56.55	72.50
1210	#7 bars	G		64	.500		42	17.50		59.50	76.50
1220	#8 bars	G		56	.571		44	19.95		63.95	83
1230	#9 bars	G	C-5	48	1.167		45	50.50	16.45	111.95	153
1240	#10 bars	G		40	1.400		48	61	19.75	128.75	177

03 21 Reinforcing Steel

03 21 05 – Reinforcing Steel Accessories

03 21 05.75 Splicing Reinforcing Bars			Crew	Daily Output	Labor-Hours	Unit	Material	Labor (2009 Bare Costs)	Equipment	Total	Total Incl O&P
1250	#11 bars	G	C-5	32	1.750	Ea.	58	76	24.50	158.50	218
1260	#14 bars	G		24	2.333		73	101	33	207	285
1270	#18 bars	G	↓	16	3.500		106	152	49.50	307.50	425
2000	Weldable half coupler, taper threaded, #4 bars	G	E-16	120	.133		7.10	6.10	1.12	14.32	20.50
2100	#5 bars	G		112	.143		8.40	6.55	1.20	16.15	22.50
2200	#6 bars	G		104	.154		13.35	7.05	1.29	21.69	29
2300	#7 bars	G		96	.167		15.50	7.60	1.40	24.50	32.50
2400	#8 bars	G		88	.182		16.15	8.30	1.53	25.98	35
2500	#9 bars	G		80	.200		17.80	9.15	1.68	28.63	38.50
2600	#10 bars	G		72	.222		18.20	10.15	1.86	30.21	41
2700	#11 bars	G		64	.250		19.45	11.45	2.10	33	45
2800	#14 bars	G		56	.286		22.50	13.05	2.40	37.95	51
2900	#18 bars	G	↓	48	.333	↓	36.50	15.25	2.80	54.55	71

03 21 10 – Uncoated Reinforcing Steel

03 21 10.60 Reinforcing In Place

			Crew	Daily Output	Labor-Hours	Unit	Material	Labor	Equipment	Total	Total Incl O&P
0015	**REINFORCING IN PLACE** A615 Grade 60, incl. access. labor R032110-10										
0030	Made from recycled materials										
0102	Beams & Girders, #3 to #7	G	4 Rodm	3200	.010	Lb.	.81	.45		1.26	1.65
0152	#8 to #18	G		5400	.006		.81	.26		1.07	1.34
0202	Columns, #3 to #7	G		3000	.011		.81	.48		1.29	1.70
0252	#8 to #18	G		4600	.007	↓	.81	.31		1.12	1.42
0300	Spirals, hot rolled, 8″ to 15″ diameter	G		2.20	14.545	Ton	2,000	650		2,650	3,300
0320	15″ to 24″ diameter	G		2.20	14.545		1,925	650		2,575	3,200
0330	24″ to 36″ diameter	G		2.30	13.913		1,825	620		2,445	3,050
0340	36″ to 48″ diameter	G		2.40	13.333		1,725	595		2,320	2,925
0360	48″ to 64″ diameter	G		2.50	12.800		1,925	570		2,495	3,075
0380	64″ to 84″ diameter	G		2.60	12.308		2,000	550		2,550	3,125
0390	84″ to 96″ diameter	G		2.70	11.852	↓	2,100	530		2,630	3,200
0402	Elevated slabs, #4 to #7 R032110-80	G		5800	.006	Lb.	.84	.25		1.09	1.35
0502	Footings, #4 to #7	G		4200	.008		.81	.34		1.15	1.47
0552	#8 to #18	G		7200	.004		.81	.20		1.01	1.23
0602	Slab on grade, #3 to #7	G		4200	.008		.78	.34		1.12	1.43
0702	Walls, #3 to #7	G		6000	.005		.81	.24		1.05	1.29
0752	#8 to #18	G	↓	8000	.004	↓	.81	.18		.99	1.19
0900	Use the following for a rough estimate guide										
1000	Typical in place, average, under 10 ton job, #3 to #7	G	4 Rodm	1.80	17.778	Ton	1,600	790		2,390	3,125
1010	#8 to #18	G		2.70	11.852		1,625	530		2,155	2,700
1050	10 – 50 ton job, #3 to #7	G		2.10	15.238		1,575	680		2,255	2,875
1060	#8 to #18	G		3	10.667		1,600	475		2,075	2,575
1100	50 – 100 ton job, #3 - #7	G		2.20	14.545		1,525	650		2,175	2,800
1110	#8 to #18	G		3.10	10.323		1,575	460		2,035	2,500
1150	Over 100 ton job, #3 - #7	G		2.30	13.913		1,525	620		2,145	2,725
1160	#8 - #18	G	↓	3.20	10		1,550	445		1,995	2,450
1200	High strength steel, Grade 75, #14 bars only, add	G					153			153	168
2000	Unloading & sorting, add to above		C-5	100	.560			24.50	7.90	32.40	49
2200	Crane cost for handling, add to above, minimum			135	.415			18	5.85	23.85	36.50
2210	Average			92	.609			26.50	8.60	35.10	53.50
2220	Maximum		↓	35	1.600	↓		69.50	22.50	92	141
2350	For stainless steel rebar, add						300%				
2400	Dowels, 2 feet long, deformed, #3	G	2 Rodm	520	.031	Ea.	.64	1.37		2.01	3.05
2410	#4	G		480	.033		1.14	1.48		2.62	3.78
2420	#5	G	↓	435	.037	↓	1.78	1.64		3.42	4.75

03 21 Reinforcing Steel

03 21 10 – Uncoated Reinforcing Steel

03 21 10.60 Reinforcing In Place		Crew	Daily Output	Labor-Hours	Unit	Material	2009 Bare Costs Labor	Equipment	Total	Total Incl O&P
2430	#6 G	2 Rodm	360	.044	Ea.	2.56	1.98		4.54	6.20
2450	Longer and heavier dowels, add G		725	.022	Lb.	.85	.98		1.83	2.62
2500	Smooth dowels, 12″ long, 1/4″ or 3/8″ diameter G		140	.114	Ea.	1.21	5.10		6.31	10.05
2520	5/8″ diameter G		125	.128		2.11	5.70		7.81	12
2530	3/4″ diameter G		110	.145		2.62	6.50		9.12	13.95
2600	Dowel sleeves for CIP concrete, 2-part system									
2610	Sleeve base, plastic, for 5/8″ smooth dowel sleeve, fasten to edge form	1 Rodm	200	.040	Ea.	.52	1.78		2.30	3.61
2615	Sleeve, plastic, 12″ long, for 5/8″ smooth dowel, snap onto base		400	.020		1.03	.89		1.92	2.65
2620	Sleeve base, for 3/4″ smooth dowel sleeve		175	.046		.49	2.04		2.53	4.01
2625	Sleeve, 12″ long, for 3/4″ smooth dowel		350	.023		1.08	1.02		2.10	2.93
2630	Sleeve base, for 1″ smooth dowel sleeve		150	.053		.64	2.38		3.02	4.75
2635	Sleeve, 12″ long, for 1″ smooth dowel		300	.027		1.84	1.19		3.03	4.05
2700	Dowel caps, visual warning only, plastic, #3 to #8	2 Rodm	800	.020		.51	.89		1.40	2.08
2720	#8 to #18		750	.021		1.08	.95		2.03	2.81
2750	Impalement protective, plastic, #4 to #9		800	.020		1.95	.89		2.84	3.67
9000	Minimum labor/equipment charge	1 Rodm	4	2	Job		89		89	152

03 21 10.70 Glass Fiber Reinforced Polymer Bars

		Crew	Daily Output	Labor-Hours	Unit	Material	Labor	Equipment	Total	Total Incl O&P
0010	**GLASS FIBER REINFORCED POLYMER BARS**									
0050	#2 bar, .043 lbs/ ft.				L.F.	.35			.35	.39
0100	#3 bar, .092 lbs/ ft.					.48			.48	.53
0150	#4 bar, .160 lbs/ ft.					.70			.70	.77
0200	#5 bar, .258 lbs/ ft.					.98			.98	1.08
0250	#6 bar, .372 lbs/ ft.					1.35			1.35	1.49
0300	#7 bar, .497 lbs/ ft.					1.70			1.70	1.87
0350	#8 bar, .620 lbs/ ft.					2.25			2.25	2.48
0400	#9 bar, .800 lbs/ ft.					2.90			2.90	3.19
0450	#10 bar, 1.08 lbs/ ft.					3.45			3.45	3.80
0500	For Bends, add per bend				Ea.	1			1	1.10

03 21 13 – Galvanized Reinforcing Steel

03 21 13.10 Galvanized Reinforcing

		Crew	Daily Output	Labor-Hours	Unit	Material	Labor	Equipment	Total	Total Incl O&P
0010	**GALVANIZED REINFORCING**									
0150	Add to uncoated reinforcing price for galvanizing				Ton	1,150			1,150	1,250

03 21 16 – Epoxy-Coated Reinforcing Steel

03 21 16.10 Epoxy-Coated Reinforcing

		Crew	Daily Output	Labor-Hours	Unit	Material	Labor	Equipment	Total	Total Incl O&P
0010	**EPOXY-COATED REINFORCING**									
0100	Add to uncoated reinforcing price for coating with epoxy				Ton	865			865	950

03 22 Welded Wire Fabric Reinforcing

03 22 05 – Uncoated Welded Wire Fabric

03 22 05.50 Welded Wire Fabric

		Crew	Daily Output	Labor-Hours	Unit	Material	Labor	Equipment	Total	Total Incl O&P
0010	**WELDED WIRE FABRIC** ASTM A185									
0030	Made from recycled materials									
0050	Sheets									
0100	6 x 6 - W1.4 x W1.4 (10 x 10) 21 lb. per C.S.F. G	2 Rodm	35	.457	C.S.F.	18.05	20.50		38.55	54.50
0200	6 x 6 - W2.1 x W2.1 (8 x 8) 30 lb. per C.S.F. G		31	.516		26.50	23		49.50	68
0300	6 x 6 - W2.9 x W2.9 (6 x 6) 42 lb. per C.S.F. G		29	.552		32.50	24.50		57	78
0400	6 x 6 - W4 x W4 (4 x 4) 58 lb. per C.S.F. G		27	.593		46.50	26.50		73	96
0500	4 x 4 - W1.4 x W1.4 (10 x 10) 31 lb. per C.S.F. G		31	.516		26.50	23		49.50	68.50
0600	4 x 4 - W2.1 x W2.1 (8 x 8) 44 lb. per C.S.F. G		29	.552		38	24.50		62.50	83.50
0650	4 x 4 - W2.9 x W2.9 (6 x 6) 61 lb. per C.S.F. G		27	.593		50.50	26.50		77	101

03 22 Welded Wire Fabric Reinforcing

03 22 05 – Uncoated Welded Wire Fabric

03 22 05.50 Welded Wire Fabric			Crew	Daily Output	Labor-Hours	Unit	Material (2009 Bare Costs)	Labor (2009 Bare Costs)	Equipment (2009 Bare Costs)	Total (2009 Bare Costs)	Total Incl O&P
0700	4 x 4 - W4 x W4 (4 x 4) 85 lb. per C.S.F.	G	2 Rodm	25	.640	C.S.F.	65.50	28.50		94	121
0750	Rolls										
0800	2 x 2 - #14 galv., 21 lb/C.S.F., beam & column wrap	G	2 Rodm	6.50	2.462	C.S.F.	66	110		176	260
0900	2 x 2 - #12 galv. for gunite reinforcing	G	"	6.50	2.462	"	65.50	110		175.50	260
9000	Minimum labor/equipment charge		1 Rodm	4	2	Job		89		89	152

03 23 Stressing Tendons

03 23 05 – Prestressing Tendons

03 23 05.50 Prestressing Steel

			Crew	Daily Output	Labor-Hours	Unit	Material	Labor	Equipment	Total	Total Incl O&P
0010	PRESTRESSING STEEL										
0100	Grouted strand, post-tensioned in field, 50' span, 100 kip	G	C-3	1200	.053	Lb.	2.33	2.18	.09	4.60	6.30
0150	300 kip	G		2700	.024		1.05	.97	.04	2.06	2.83
0300	100' span, 100 kip	G		1700	.038		2.33	1.54	.06	3.93	5.20
0350	300 kip	G		3200	.020		2.02	.82	.03	2.87	3.63
0500	200' span, 100 kip	G		2700	.024		2.32	.97	.04	3.33	4.23
0550	300 kip	G		3500	.018		2.01	.75	.03	2.79	3.49
0800	Grouted bars, 50' span, 42 kip	G		2600	.025		1.25	1.01	.04	2.30	3.11
0850	143 kip	G		3200	.020		1.21	.82	.03	2.06	2.74
1000	75' span, 42 kip	G		3200	.020		1.26	.82	.03	2.11	2.80
1050	143 kip	G	↓	4200	.015		1.12	.62	.02	1.76	2.31
1200	Ungrouted strand, 50' span, 100 kip	G	C-4	1275	.025		.62	1.13	.02	1.77	2.63
1250	300 kip	G		1475	.022		.62	.98	.02	1.62	2.37
1400	100' span, 100 kip	G		1500	.021		.62	.96	.02	1.60	2.34
1450	300 kip	G		1650	.019		.62	.87	.02	1.51	2.19
1600	200' span, 100 kip	G		1500	.021		.62	.96	.02	1.60	2.34
1650	300 kip	G		1700	.019		.62	.85	.02	1.49	2.15
1800	Ungrouted bars, 50' span, 42 kip	G		1400	.023		.78	1.03	.02	1.83	2.63
1850	143 kip	G		1700	.019		.78	.85	.02	1.65	2.32
2000	75' span, 42 kip	G		1800	.018		.78	.80	.02	1.60	2.24
2050	143 kip	G		2200	.015		.78	.66	.01	1.45	1.98
2220	Ungrouted single strand, 100' slab, 25 kip	G		1200	.027		.62	1.20	.02	1.84	2.76
2250	35 kip	G	↓	1475	.022	↓	.62	.98	.02	1.62	2.37

03 24 Fibrous Reinforcing

03 24 05 – Reinforcing Fibers

03 24 05.30 Synthetic Fibers

			Crew	Daily Output	Labor-Hours	Unit	Material	Labor	Equipment	Total	Total Incl O&P
0010	SYNTHETIC FIBERS										
0100	Synthetic fibers, add to concrete					Lb.	4.43			4.43	4.87
0110	1-1/2 lb. per C.Y.					C.Y.	6.85			6.85	7.55

03 24 05.70 Steel Fibers

			Crew	Daily Output	Labor-Hours	Unit	Material	Labor	Equipment	Total	Total Incl O&P
0010	STEEL FIBERS										
0150	Steel fibers, add to concrete	G				Lb.	.70			.70	.77
0155	25 lb. per C.Y.	G				C.Y.	17.50			17.50	19.25
0160	50 lb. per C.Y.	G					35			35	38.50
0170	75 lb. per C.Y.	G					54			54	59.50
0180	100 lb. per C.Y.	G				↓	70			70	77

03 30 53 – Miscellaneous Cast-In-Place Concrete

03 30 53.40 Concrete In Place			Crew	Daily Output	Labor-Hours	Unit	2009 Bare Costs Material	Labor	Equipment	Total	Total Incl O&P
0010	**CONCRETE IN PLACE**										
0020	Including forms (4 uses), reinforcing steel, concrete, placement,										
0050	and finishing unless otherwise indicated	R033053-50									
0300	Beams, 5 kip per L.F., 10′ span		C-14A	15.62	12.804	C.Y.	400	515	49	964	1,350
0350	25′ span		"	18.55	10.782		430	430	41	901	1,225
0500	Chimney foundations, industrial, minimum		C-14C	32.22	3.476		166	133	.80	299.80	405
0510	Maximum	R033105-80	"	23.71	4.724		203	181	1.09	385.09	525
0700	Columns, square, 12″ x 12″, minimum reinforcing		C-14A	11.96	16.722		400	670	64	1,134	1,600
0720	Average reinforcing			10.13	19.743		720	790	75.50	1,585.50	2,175
0740	Maximum reinforcing			9.03	22.143		1,150	890	84.50	2,124.50	2,850
0800	16″ x 16″, minimum reinforcing			16.22	12.330		315	495	47	857	1,225
0820	Average reinforcing			12.57	15.911		625	640	61	1,326	1,800
0840	Maximum reinforcing			10.25	19.512		1,025	780	74.50	1,879.50	2,525
0900	24″ x 24″, minimum reinforcing			23.66	8.453		275	340	32.50	647.50	895
0920	Average reinforcing			17.71	11.293		570	455	43	1,068	1,425
0940	Maximum reinforcing			14.15	14.134		965	565	54	1,584	2,050
1000	36″ x 36″, minimum reinforcing			33.69	5.936		249	233	22.50	509.50	695
1020	Average reinforcing			23.32	8.576		505	345	33	883	1,150
1040	Maximum reinforcing			17.82	11.223		910	450	43	1,403	1,800
1200	16″ diameter, minimum reinforcing			31.49	6.351		315	255	24.50	594.50	790
1220	Average reinforcing			19.12	10.460		635	420	40	1,095	1,450
1240	Maximum reinforcing			13.77	14.524		1,025	580	55.50	1,660.50	2,150
1300	20″ diameter, minimum reinforcing			41.04	4.873		315	195	18.60	528.60	690
1320	Average reinforcing			24.05	8.316		610	335	32	977	1,250
1340	Maximum reinforcing			17.01	11.758		1,025	470	45	1,540	1,950
1400	24″ diameter, minimum reinforcing			51.85	3.857		294	155	14.75	463.75	595
1420	Average reinforcing			27.06	7.391		620	296	28	944	1,200
1440	Maximum reinforcing			18.29	10.935		1,000	440	42	1,482	1,875
1500	36″ diameter, minimum reinforcing			75.04	2.665		310	107	10.20	427.20	530
1520	Average reinforcing			37.49	5.335		595	214	20.50	829.50	1,025
1540	Maximum reinforcing		↓	22.84	8.757		990	350	33.50	1,373.50	1,725
1900	Elevated slabs, flat slab with drops, 125 psf Sup. Load, 20′ span		C-14B	38.45	5.410		299	216	19.85	534.85	705
1950	30′ span			50.99	4.079		340	163	14.95	517.95	660
2100	Flat plate, 125 psf Sup. Load, 15′ span			30.24	6.878		278	275	25	578	790
2150	25′ span			49.60	4.194		300	168	15.40	483.40	625
2300	Waffle const., 30″ domes, 125 psf Sup. Load, 20′ span			37.07	5.611		267	225	20.50	512.50	685
2350	30′ span			44.07	4.720		255	189	17.30	461.30	610
2500	One way joists, 30″ pans, 125 psf Sup. Load, 15′ span			27.38	7.597		297	305	28	630	855
2550	25′ span			31.15	6.677		294	267	24.50	585.50	790
2700	One way beam & slab, 125 psf Sup. Load, 15′ span			20.59	10.102		305	405	37	747	1,050
2750	25′ span			28.36	7.334		297	294	27	618	840
2900	Two way beam & slab, 125 psf Sup. Load, 15′ span			24.04	8.652		296	345	32	673	930
2950	25′ span		↓	35.87	5.799	↓	256	232	21.50	509.50	690
3100	Elevated slabs including finish, not										
3110	including forms or reinforcing										
3150	Regular concrete, 4″ slab		C-8	2613	.021	S.F.	1.36	.75	.28	2.39	3.01
3200	6″ slab			2585	.022		2.01	.76	.29	3.06	3.75
3250	2-1/2″ thick floor fill			2685	.021		.87	.73	.27	1.87	2.44
3300	Lightweight, 110# per C.F., 2-1/2″ thick floor fill			2585	.022		1.17	.76	.29	2.22	2.83
3400	Cellular concrete, 1-5/8″ fill, under 5000 S.F.			2000	.028		.78	.99	.37	2.14	2.86
3450	Over 10,000 S.F.			2200	.025		.75	.90	.34	1.99	2.63
3500	Add per floor for 3 to 6 stories high			31800	.002			.06	.02	.08	.13
3520	For 7 to 20 stories high		↓	21200	.003	↓		.09	.03	.12	.19

03 30 53 – Miscellaneous Cast-In-Place Concrete

03 30 53.40 Concrete In Place		Crew	Daily Output	Labor-Hours	Unit	2009 Bare Costs Material	Labor	Equipment	Total	Total Incl O&P
3540	Equipment pad, 3′ x 3′ x 6″ thick	C-14H	45	1.067	Ea.	58	42	.58	100.58	134
3550	4′ x 4′ x 6″ thick		30	1.600		84	63	.86	147.86	197
3560	5′ x 5′ x 8″ thick		18	2.667		140	105	1.44	246.44	330
3570	6′ x 6′ x 8″ thick		14	3.429		188	135	1.85	324.85	430
3580	8′ x 8′ x 10″ thick		8	6		395	236	3.24	634.24	830
3590	10′ x 10′ x 12″ thick		5	9.600		655	380	5.20	1,040.20	1,350
3800	Footings, spread under 1 C.Y.	C-14C	28	4	C.Y.	209	153	.92	362.92	485
3825	1 C.Y to 5 C.Y.		43	2.605		221	99.50	.60	321.10	410
3850	Over 5 C.Y.		75	1.493		198	57	.34	255.34	310
3900	Footings, strip, 18″ x 9″, unreinforced	C-14L	40	2.400		120	89.50	.65	210.15	280
3920	18″ x 9″, reinforced	C-14C	35	3.200		157	122	.74	279.74	375
3925	20″ x 10″, unreinforced	C-14L	45	2.133		117	79.50	.58	197.08	260
3930	20″ x 10″, reinforced	C-14C	40	2.800		147	107	.64	254.64	340
3935	24″ x 12″, unreinforced	C-14L	55	1.745		116	65	.47	181.47	236
3940	24″ x 12″, reinforced	C-14C	48	2.333		147	89.50	.54	237.04	310
3945	36″ x 12″, unreinforced	C-14L	70	1.371		113	51	.37	164.37	208
3950	36″ x 12″, reinforced	C-14C	60	1.867		141	71.50	.43	212.93	273
4000	Foundation mat, under 10 C.Y.		38.67	2.896		255	111	.67	366.67	465
4050	Over 20 C.Y.		56.40	1.986		218	76	.46	294.46	365
4200	Wall, free-standing, 8″ thick, 8′ high	C-14D	45.83	4.364		195	173	16.65	384.65	520
4250	14′ high		27.26	7.337		251	291	28	570	785
4260	12″ thick, 8′ high		64.32	3.109		175	124	11.90	310.90	410
4270	14′ high		40.01	4.999		191	199	19.10	409.10	555
4300	15″ thick, 8′ high		80.02	2.499		166	99.50	9.55	275.05	355
4350	12′ high		51.26	3.902		166	155	14.90	335.90	455
4500	18′ high		48.85	4.094		191	163	15.65	369.65	495
4520	Handicap access ramp, railing both sides, 3′ wide	C-14H	14.58	3.292	L.F.	300	130	1.78	431.78	545
4525	5′ wide		12.22	3.928		310	155	2.12	467.12	595
4530	With 6″ curb and rails both sides, 3′ wide		8.55	5.614		310	221	3.03	534.03	710
4535	5′ wide		7.31	6.566		315	259	3.55	577.55	775
4650	Slab on grade, not including finish, 4″ thick	C-14E	60.75	1.449	C.Y.	127	57	.43	184.43	235
4700	6″ thick	″	92	.957	″	121	37.50	.29	158.79	196
4751	Slab on grade, incl. troweled finish, not incl. forms									
4760	or reinforcing, over 10,000 S.F., 4″ thick	C-14F	3425	.021	S.F.	1.35	.76	.01	2.12	2.70
4820	6″ thick		3350	.021		1.97	.78	.01	2.76	3.42
4840	8″ thick		3184	.023		2.70	.82	.01	3.53	4.27
4900	12″ thick		2734	.026		4.04	.96	.01	5.01	6
4950	15″ thick		2505	.029		5.10	1.04	.01	6.15	7.25
5000	Slab on grade, incl. textured finish, not incl. forms									
5001	or reinforcing, 4″ thick	C-14G	2873	.019	S.F.	1.31	.70	.01	2.02	2.56
5010	6″ thick		2590	.022		2.05	.77	.01	2.83	3.50
5020	8″ thick		2320	.024		2.68	.86	.01	3.55	4.33
5200	Lift slab in place above the foundation, incl. forms,									
5210	reinforcing, concrete and columns, minimum	C-14B	2113	.098	S.F.	7.45	3.94	.36	11.75	15.10
5250	Average		1650	.126		8.15	5.05	.46	13.66	17.80
5300	Maximum		1500	.139		8.85	5.55	.51	14.91	19.45
5500	Lightweight, ready mix, including screed finish only,									
5510	not including forms or reinforcing									
5550	1:4 for structural roof decks	C-14B	260	.800	C.Y.	138	32	2.94	172.94	208
5600	1:6 for ground slab with radiant heat	C-14F	92	.783		143	28.50	.28	171.78	202
5650	1:3:2 with sand aggregate, roof deck	C-14B	260	.800		132	32	2.94	166.94	201
5700	Ground slab	C-14F	107	.673		132	24.50	.24	156.74	184
5900	Pile caps, incl. forms and reinf., sq. or rect., under 5 C.Y.	C-14C	54.14	2.069		172	79	.48	251.48	320

03 30 Cast-In-Place Concrete

03 30 53 – Miscellaneous Cast-In-Place Concrete

03 30 53.40 Concrete In Place		Crew	Daily Output	Labor-Hours	Unit	2009 Bare Costs Material	Labor	Equipment	Total	Total Incl O&P
5950	Over 10 C.Y.	C-14C	75	1.493	C.Y.	167	57	.34	224.34	279
6000	Triangular or hexagonal, under 5 C.Y.		53	2.113		127	81	.49	208.49	275
6050	Over 10 C.Y.		85	1.318		152	50.50	.30	202.80	251
6200	Retaining walls, gravity, 4′ high see Div. 32 32	C-14D	66.20	3.021		153	120	11.55	284.55	380
6250	10′ high		125	1.600		146	63.50	6.10	215.60	273
6300	Cantilever, level backfill loading, 8′ high		70	2.857		175	113	10.90	298.90	390
6350	16′ high		91	2.198		166	87.50	8.40	261.90	335
6800	Stairs, not including safety treads, free standing, 3′-6″ wide	C-14H	83	.578	LF Nose	6.20	23	.31	29.51	44.50
6850	Cast on ground		125	.384	″	5	15.10	.21	20.31	30.50
7000	Stair landings, free standing		200	.240	S.F.	5.25	9.45	.13	14.83	21.50
7050	Cast on ground		475	.101	″	4	3.98	.05	8.03	11
9000	Minimum labor/equipment charge	2 Carp	1	16	Job		640		640	1,050

03 31 Structural Concrete

03 31 05 – Normal Weight Structural Concrete

03 31 05.30 Concrete, Field Mix		Crew	Daily Output	Labor-Hours	Unit	Material	Labor	Equipment	Total	Total Incl O&P
0010	**CONCRETE, FIELD MIX**									
0015	FOB forms 2250 psi				C.Y.	90			90	99
0020	3000 psi				″	94			94	103

03 31 05.35 Normal Weight Concrete, Ready Mix		Crew	Daily Output	Labor-Hours	Unit	Material	Labor	Equipment	Total	Total Incl O&P
0010	**NORMAL WEIGHT CONCRETE, READY MIX**, delivered									
0012	Includes local aggregate, sand, Portland cement, and water									
0015	Excludes all additives and treatments									
0020	2000 psi				C.Y.	97			97	107
0100	2500 psi					98.50			98.50	108
0150	3000 psi					101			101	111
0200	3500 psi					104			104	114
0300	4000 psi					106			106	116
0350	4500 psi					109			109	120
0400	5000 psi					111			111	122
0411	6000 psi					127			127	139
0412	8000 psi					206			206	227
0413	10,000 psi					293			293	320
0414	12,000 psi					355			355	390
1000	For high early strength cement, add					10%				
1010	For structural lightweight with regular sand, add					25%				
1300	For winter concrete (hot water), add					5.25			5.25	5.80
1400	For hot weather concrete (ice), add					7.20			7.20	7.90
1410	For mid-range water reducer, add					3.95			3.95	4.35
1420	For high-range water reducer/superplasticizer, add					5.85			5.85	6.40
1430	For retarder, add					2			2	2.20
1440	For non-Chloride accelerator, add					4.75			4.75	5.25
1450	For Chloride accelerator, per 1%, add					2.75			2.75	3.03
1460	For fiber reinforcing, synthetic (1 Lb./C.Y.), add					6			6	6.60
1500	For Saturday delivery, add					5.50			5.50	6.05
1510	For truck holding/waiting time past 1st hour per load, add				Hr.	92			92	101
1520	For short load (less than 4 C.Y.), add per load				Ea.	100			100	110
2000	For all lightweight aggregate, add				C.Y.	45%				
4000	Flowable fill: ash, cement, aggregate, water									
4100	40 – 80 psi				C.Y.	87			87	95.50
4150	Structural: ash, cement, aggregate, water & sand									

03 31 Structural Concrete

03 31 05 – Normal Weight Structural Concrete

03 31 05.35 Normal Weight Concrete, Ready Mix		Crew	Daily Output	Labor-Hours	Unit	Material	2009 Bare Costs Labor	Equipment	Total	Total Incl O&P
4200	50 psi				C.Y.	87			87	95.50
4250	140 psi					90.50			90.50	99.50
4300	500 psi					94			94	103
4350	1000 psi					99.50			99.50	109

03 31 05.70 Placing Concrete

		Crew	Daily Output	Labor-Hours	Unit	Material	Labor	Equipment	Total	Total Incl O&P
0010	**PLACING CONCRETE** R033105-70									
0020	Includes labor and equipment to place and vibrate									
0050	Beams, elevated, small beams, pumped	C-20	60	1.067	C.Y.		36	13.15	49.15	73.50
0100	With crane and bucket	C-7	45	1.600			55	26.50	81.50	118
0200	Large beams, pumped	C-20	90	.711			24	8.80	32.80	48.50
0250	With crane and bucket	C-7	65	1.108			38	18.40	56.40	82
0400	Columns, square or round, 12" thick, pumped	C-20	60	1.067			36	13.15	49.15	73.50
0450	With crane and bucket	C-7	40	1.800			61.50	30	91.50	133
0600	18" thick, pumped	C-20	90	.711			24	8.80	32.80	48.50
0650	With crane and bucket	C-7	55	1.309			45	22	67	96.50
0800	24" thick, pumped	C-20	92	.696			23.50	8.60	32.10	48
0850	With crane and bucket	C-7	70	1.029			35	17.10	52.10	76
1000	36" thick, pumped	C-20	140	.457			15.50	5.65	21.15	31
1050	With crane and bucket	C-7	100	.720			24.50	11.95	36.45	53
1400	Elevated slabs, less than 6" thick, pumped	C-20	140	.457			15.50	5.65	21.15	31
1450	With crane and bucket	C-7	95	.758			26	12.60	38.60	56
1500	6" to 10" thick, pumped	C-20	160	.400			13.55	4.94	18.49	27.50
1550	With crane and bucket	C-7	110	.655			22.50	10.90	33.40	48.50
1600	Slabs over 10" thick, pumped	C-20	180	.356			12.05	4.39	16.44	24.50
1650	With crane and bucket	C-7	130	.554			18.95	9.20	28.15	40.50
1900	Footings, continuous, shallow, direct chute	C-6	120	.400			13.20	.43	13.63	22
1950	Pumped	C-20	150	.427			14.45	5.25	19.70	29.50
2000	With crane and bucket	C-7	90	.800			27.50	13.30	40.80	59
2100	Footings, continuous, deep, direct chute	C-6	140	.343			11.35	.37	11.72	18.90
2150	Pumped	C-20	160	.400			13.55	4.94	18.49	27.50
2200	With crane and bucket	C-7	110	.655			22.50	10.90	33.40	48.50
2400	Footings, spread, under 1 C.Y., direct chute	C-6	55	.873			29	.94	29.94	48
2450	Pumped	C-20	65	.985			33.50	12.15	45.65	68
2500	With crane and bucket	C-7	45	1.600			55	26.50	81.50	118
2600	Over 5 C.Y., direct chute	C-6	120	.400			13.20	.43	13.63	22
2650	Pumped	C-20	150	.427			14.45	5.25	19.70	29.50
2700	With crane and bucket	C-7	100	.720			24.50	11.95	36.45	53
2900	Foundation mats, over 20 C.Y., direct chute	C-6	350	.137			4.53	.15	4.68	7.55
2950	Pumped	C-20	400	.160			5.45	1.97	7.42	10.95
3000	With crane and bucket	C-7	300	.240			8.20	3.99	12.19	17.70
3200	Grade beams, direct chute	C-6	150	.320			10.60	.35	10.95	17.70
3250	Pumped	C-20	180	.356			12.05	4.39	16.44	24.50
3300	With crane and bucket	C-7	120	.600			20.50	10	30.50	44.50
3500	High rise, for more than 5 stories, pumped, add per story	C-20	2100	.030			1.03	.38	1.41	2.09
3510	With crane and bucket, add per story	C-7	2100	.034			1.17	.57	1.74	2.53
3700	Pile caps, under 5 C.Y., direct chute	C-6	90	.533			17.65	.58	18.23	29.50
3750	Pumped	C-20	110	.582			19.75	7.20	26.95	40
3800	With crane and bucket	C-7	80	.900			31	14.95	45.95	66.50
3850	Pile cap, 5 C.Y. to 10 C.Y., direct chute	C-6	175	.274			9.05	.30	9.35	15.15
3900	Pumped	C-20	200	.320			10.85	3.95	14.80	22
3950	With crane and bucket	C-7	150	.480			16.45	8	24.45	35.50
4000	Over 10 C.Y., direct chute	C-6	215	.223			7.40	.24	7.64	12.30

03 31 05 – Normal Weight Structural Concrete

03 31 05.70 Placing Concrete		Crew	Daily Output	Labor-Hours	Unit	Material	2009 Bare Costs Labor	Equipment	Total	Total Incl O&P
4050	Pumped	C-20	240	.267	C.Y.		9.05	3.29	12.34	18.30
4100	With crane and bucket	C-7	185	.389			13.30	6.45	19.75	28.50
4300	Slab on grade, up to 6" thick, direct chute	C-6	110	.436			14.40	.47	14.87	24
4350	Pumped	C-20	130	.492			16.70	6.10	22.80	33.50
4400	With crane and bucket	C-7	110	.655			22.50	10.90	33.40	48.50
4600	Over 6" thick, direct chute	C-6	165	.291			9.60	.31	9.91	16.05
4650	Pumped	C-20	185	.346			11.75	4.27	16.02	24
4700	With crane and bucket	C-7	145	.497			17	8.25	25.25	36.50
4900	Walls, 8" thick, direct chute	C-6	90	.533			17.65	.58	18.23	29.50
4950	Pumped	C-20	100	.640			21.50	7.90	29.40	44
5000	With crane and bucket	C-7	80	.900			31	14.95	45.95	66.50
5050	12" thick, direct chute	C-6	100	.480			15.85	.52	16.37	26.50
5100	Pumped	C-20	110	.582			19.75	7.20	26.95	40
5200	With crane and bucket	C-7	90	.800			27.50	13.30	40.80	59
5300	15" thick, direct chute	C-6	105	.457			15.10	.49	15.59	25
5350	Pumped	C-20	120	.533			18.10	6.60	24.70	37
5400	With crane and bucket	C-7	95	.758			26	12.60	38.60	56
5600	Wheeled concrete dumping, add to placing costs above									
5610	Walking cart, 50' haul, add	C-18	32	.281	C.Y.		8.95	1.72	10.67	16.65
5620	150' haul, add		24	.375			11.95	2.30	14.25	22
5700	250' haul, add		18	.500			15.90	3.07	18.97	30
5800	Riding cart, 50' haul, add	C-19	80	.113			3.58	1.08	4.66	7.10
5810	150' haul, add		60	.150			4.77	1.44	6.21	9.50
5900	250' haul, add		45	.200			6.35	1.92	8.27	12.60
9000	Minimum labor/equipment charge	C-6	2	24	Job		795	26	821	1,325

03 35 Concrete Finishing

03 35 29 – Tooled Concrete Finishing

03 35 29.30 Finishing Floors

		Crew	Daily Output	Labor-Hours	Unit	Material	Labor	Equipment	Total	Total Incl O&P
0010	**FINISHING FLOORS**									
0020	Manual screed finish	C-10	4800	.005	S.F.		.18		.18	.29
0100	Manual screed and bull float		4000	.006			.22		.22	.34
0125	Manual screed, bull float, manual float		2000	.012			.43		.43	.69
0150	Manual screed, bull float, manual float & broom finish		1850	.013			.47		.47	.74
0200	Manual screed, bull float, manual float, manual steel trowel		1265	.019			.68		.68	1.09
0250	Manual screed, bull float, machine float & trowel (walk-behind)	C-10C	1715	.014			.50	.02	.52	.83
0300	Power screed, bull float, machine float & trowel (walk-behind)	C-10D	2400	.010			.36	.05	.41	.62
0350	Power screed, bull float, machine float & trowel (ride-on)	C-10E	4000	.006			.22	.06	.28	.41
0400	Integral topping and finish, using 1:1:2 mix, 3/16" thick	C-10B	1000	.040		.10	1.37	.24	1.71	2.58
0450	1/2" thick		950	.042		.26	1.44	.25	1.95	2.89
0500	3/4" thick		850	.047		.39	1.61	.28	2.28	3.34
0600	1" thick		750	.053		.52	1.83	.32	2.67	3.87
0800	Granolithic topping, laid after, 1:1:1-1/2 mix, 1/2" thick		590	.068		.29	2.32	.41	3.02	4.52
0820	3/4" thick		580	.069		.43	2.36	.41	3.20	4.74
0850	1" thick		575	.070		.57	2.38	.42	3.37	4.93
0950	2" thick		500	.080		1.15	2.74	.48	4.37	6.20
1200	Heavy duty, 1:1:2, 3/4" thick, preshrunk, gray, 20 MSF		320	.125		.73	4.29	.75	5.77	8.55
1300	100 MSF		380	.105		.39	3.61	.63	4.63	6.90
1600	Exposed local aggregate finish, minimum	1 Cefi	625	.013		.22	.49		.71	1.02
1650	Maximum		465	.017		.66	.66		1.32	1.76
1800	Floor abrasives, .25 psf, aluminum oxide		850	.009		.44	.36		.80	1.04

03 35 Concrete Finishing

03 35 29 – Tooled Concrete Finishing

03 35 29.30 Finishing Floors

	03 35 29.30 Finishing Floors	Crew	Daily Output	Labor-Hours	Unit	2009 Bare Costs Material	Labor	Equipment	Total	Total Incl O&P
1850	Silicon carbide	1 Cefi	850	.009	S.F.	.61	.36		.97	1.23
2000	Floor hardeners, metallic, light service, .50 psf, add		850	.009		.51	.36		.87	1.12
2050	Medium service, .75 psf		750	.011		.76	.41		1.17	1.47
2100	Heavy service, 1.0 psf		650	.012		1.01	.47		1.48	1.85
2150	Extra heavy, 1.5 psf		575	.014		1.52	.53		2.05	2.50
2300	Non-metallic, light service, .50 psf		850	.009		.23	.36		.59	.81
2350	Medium service, .75 psf		750	.011		.34	.41		.75	1.02
2400	Heavy service, 1.00 psf		650	.012		.45	.47		.92	1.24
2450	Extra heavy, 1.50 psf		575	.014		.68	.53		1.21	1.58
2800	Trap rock wearing surface for monolithic floors									
2810	2.0 psf	C-10B	1250	.032	S.F.	.03	1.10	.19	1.32	2.01
3000	Floor coloring, dusted on, minimum (0.6 psf), add to above	1 Cefi	1300	.006		.43	.24		.67	.84
3050	Maximum (1.0 psf), add to above	"	625	.013		.71	.49		1.20	1.55
3100	Colored powder only				Lb.	.71			.71	.78
3600	1/2" topping using 0.6 psf powdered color	C-10B	590	.068	S.F.	4.86	2.32	.41	7.59	9.55
3650	1/2" topping using 1.0 psf powdered color	"	590	.068		5.15	2.32	.41	7.88	9.85
3800	Dustproofing, solvent-based, 1 coat	1 Cefi	1900	.004		.17	.16		.33	.44
3850	2 coats		1300	.006		.61	.24		.85	1.05
4000	Epoxy-based, 1 coat		1500	.005		.15	.20		.35	.48
4050	2 coats		1500	.005		.29	.20		.49	.64
4400	Stair finish, float		275	.029			1.11		1.11	1.74
4500	Steel trowel finish		200	.040			1.53		1.53	2.40
4600	Silicon carbide finish, .25 psf		150	.053		.44	2.04		2.48	3.67
9100	Minimum labor/equipment charge	C-10	2	12	Job		435		435	690

03 35 29.35 Control Joints, Saw Cut

	03 35 29.35 Control Joints, Saw Cut	Crew	Daily Output	Labor-Hours	Unit	Material	Labor	Equipment	Total	Total Incl O&P
0010	**CONTROL JOINTS, SAW CUT**									
0100	Sawcut in green concrete									
0120	1" depth	C-27	2000	.008	L.F.	.07	.31	.07	.45	.63
0140	1-1/2" depth		1800	.009		.10	.34	.08	.52	.73
0160	2" depth		1600	.010		.13	.38	.09	.60	.85
0200	Clean out control joint of debris	C-28	6000	.001			.05		.05	.08
0300	Joint sealant									
0320	Backer rod, polyethylene, 1/4" diameter	1 Cefi	460	.017	L.F.	.04	.67		.71	1.09
0340	Sealant, polyurethane									
0360	1/4" x 1/4" (308 LF/Gal)	1 Cefi	270	.030	L.F.	.18	1.13		1.31	1.96
0380	1/4" x 1/2" (154 LF/Gal)	"	255	.031	"	.35	1.20		1.55	2.27

03 35 29.60 Finishing Walls

	03 35 29.60 Finishing Walls	Crew	Daily Output	Labor-Hours	Unit	Material	Labor	Equipment	Total	Total Incl O&P
0010	**FINISHING WALLS**									
0020	Break ties and patch voids	1 Cefi	540	.015	S.F.	.03	.57		.60	.92
0050	Burlap rub with grout		450	.018		.03	.68		.71	1.10
0100	Carborundum rub, dry		270	.030			1.13		1.13	1.77
0150	Wet rub		175	.046			1.75		1.75	2.74
0300	Bush hammer, green concrete	B-39	1000	.048			1.59	.19	1.78	2.81
0350	Cured concrete	"	650	.074			2.45	.29	2.74	4.33
0600	Float finish, 1/16" thick	1 Cefi	300	.027		.27	1.02		1.29	1.89
0700	Sandblast, light penetration	E-11	1100	.029		.70	1.04	.18	1.92	2.80
0750	Heavy penetration	"	375	.085		1.41	3.04	.52	4.97	7.45
0850	Grind form fins flush	1 Clab	700	.011	L.F.		.36		.36	.60
9000	Minimum labor/equipment charge	C-10	2	12	Job		435		435	690

03 35 Concrete Finishing

03 35 33 – Stamped Concrete Finishing

03 35 33.50 Slab Texture Stamping		Crew	Daily Output	Labor-Hours	Unit	Material	2009 Bare Costs Labor	Equipment	Total	Total Incl O&P
0010	**SLAB TEXTURE STAMPING**									
0020	Buy re-usable stamp, small (2 S.F.)				Ea.	82.50			82.50	91
0030	Medium (4 S.F.)					165			165	182
0120	Large (8 S.F.)					330			330	365
0200	Commonly used chemicals for texture systems									
0210	Hardener, colored powder				S.F.	.53			.53	.59
0220	Release agent, colored powder					.09			.09	.10
0230	Curing & sealing compound, solvent based					.06			.06	.06
0300	Broadcasting hardener & release agent, stamping, tooling	3 Cefi	1000	.024			.92		.92	1.44

03 37 Specialty Placed Concrete

03 37 13 – Shotcrete

03 37 13.30 Gunite (Dry-Mix)

Line	Description	Crew	Daily Output	Labor-Hours	Unit	Material	Labor	Equipment	Total	Total Incl O&P
0010	**GUNITE (DRY-MIX)**									
0020	Applied in 1" layers, no mesh included	C-8	2000	.028	S.F.	.32	.99	.37	1.68	2.35
0100	Mesh for gunite 2 x 2, #12, to 3" thick	2 Rodm	800	.020		.66	.89		1.55	2.24
0150	Over 3" thick	"	500	.032		.66	1.43		2.09	3.15
0300	Typical in place, including mesh, 2" thick, minimum	C-16	1000	.072		1.30	2.68	.74	4.72	6.65
0350	Maximum		500	.144		1.30	5.35	1.48	8.13	11.85
0500	4" thick, minimum		750	.096		1.94	3.58	.98	6.50	9.05
0550	Maximum		350	.206		1.94	7.65	2.11	11.70	17
0900	Prepare old walls, no scaffolding, minimum	C-10	1000	.024			.87		.87	1.38
0950	Maximum	"	275	.087			3.15		3.15	5
1100	For high finish requirement or close tolerance, add, minimum						50%			
1150	Maximum						110%			
9000	Minimum labor/equipment charge	C-10	1	24	Job		865		865	1,375

03 39 Concrete Curing

03 39 13 – Water Concrete Curing

03 39 13.50 Water Curing

Line	Description	Crew	Daily Output	Labor-Hours	Unit	Material	Labor	Equipment	Total	Total Incl O&P
0010	**WATER CURING**									
0015	With burlap, 4 uses assumed, 7.5 oz.	2 Clab	55	.291	C.S.F.	8.65	9.20		17.85	24.50
0100	10 oz.	"	55	.291	"	15.55	9.20		24.75	32.50
0400	Curing blankets, 1" to 2" thick, buy, minimum				S.F.	.32			.32	.35
0450	Maximum				"	.48			.48	.53
9000	Minimum labor/equipment charge	1 Clab	5	1.600	Job		50.50		50.50	83.50

03 39 23 – Membrane Concrete Curing

03 39 23.13 Chemical Compound Membrane Concrete Curing

Line	Description	Crew	Daily Output	Labor-Hours	Unit	Material	Labor	Equipment	Total	Total Incl O&P
0010	**CHEMICAL COMPOUND MEMBRANE CONCRETE CURING**									
0300	Sprayed membrane curing compound	2 Clab	95	.168	C.S.F.	5.25	5.30		10.55	14.55
0700	Curing compound, solvent based, 400 S.F./gal, 55 gallon lots				Gal.	15.25			15.25	16.75
0720	5 gallon lots					17.40			17.40	19.15
0800	Curing compound, water based, 250 S.F./gal, 55 gallon lots					18.30			18.30	20
0820	5 gallon lots					21			21	23

03 39 23.23 Sheet Membrane Concrete Curing

Line	Description	Crew	Daily Output	Labor-Hours	Unit	Material	Labor	Equipment	Total	Total Incl O&P
0010	**SHEET MEMBRANE CONCRETE CURING**									
0200	Curing blanket, burlap/poly, 2-ply	2 Clab	70	.229	C.S.F.	15.95	7.20		23.15	29.50

03 41 05 – Precast Concrete Members

03 41 05.10 Precast Beams		Crew	Daily Output	Labor-Hours	Unit	Material	Labor	Equipment	Total	Total Incl O&P
						2009 Bare Costs				
0010	PRECAST BEAMS									
0011	L-shaped, 20′ span, 12″ x 20″	C-11	32	2.250	Ea.	1,750	98.50	59	1,907.50	2,175
1000	Inverted tee beams, add to above, small beams					15%				
1050	Large beams					20%				
1200	Rectangular, 20′ span, 12″ x 20″	C-11	32	2.250		1,300	98.50	59	1,457.50	1,675
1250	18″ x 36″		24	3		1,475	131	79	1,685	1,950
1300	24″ x 44″		22	3.273		1,850	143	86	2,079	2,375
1400	30′ span, 12″ x 36″		24	3		2,000	131	79	2,210	2,525
1450	18″ x 44″		20	3.600		2,500	158	94.50	2,752.50	3,125
1500	24″ x 52″		16	4.500		3,225	197	118	3,540	4,025
1600	40′ span, 12″ x 52″		20	3.600		3,700	158	94.50	3,952.50	4,425
1650	18″ x 52″		16	4.500		4,400	197	118	4,715	5,325
1700	24″ x 52″		12	6		4,850	263	158	5,271	5,975
2000	"T" shaped, 20′ span, 12″ x 20″		32	2.250		2,200	98.50	59	2,357.50	2,675
2050	18″ x 36″		24	3		2,500	131	79	2,710	3,100
2100	24″ x 44″		22	3.273		3,150	143	86	3,379	3,800
2200	30′ span, 12″ x 36″		24	3		3,400	131	79	3,610	4,050
2250	18″ x 44″		20	3.600		4,275	158	94.50	4,527.50	5,075
2300	24″ x 52″		16	4.500		5,475	197	118	5,790	6,500
2500	40′ span, 12″ x 52″		20	3.600		6,275	158	94.50	6,527.50	7,275
2550	18″ x 52″		16	4.500		7,475	197	118	7,790	8,700
2600	24″ x 52″		12	6		8,225	263	158	8,646	9,700

03 41 05.15 Precast Columns

		Crew	Daily Output	Labor-Hours	Unit	Material	Labor	Equipment	Total	Total Incl O&P
0010	PRECAST COLUMNS									
0020	Rectangular to 12′ high, small columns	C-11	120	.600	L.F.	106	26.50	15.75	148.25	180
0050	Large columns		96	.750		184	33	19.70	236.70	283
0300	24′ high, small columns		192	.375		106	16.45	9.85	132.30	156
0350	Large columns		144	.500		184	22	13.15	219.15	256
0700	24′ high, 1 haunch, 12″ x 12″		32	2.250	Ea.	2,550	98.50	59	2,707.50	3,075
0800	20″ x 20″		28	2.571	"	4,450	113	67.50	4,630.50	5,175

03 41 05.25 Precast Joists

		Crew	Daily Output	Labor-Hours	Unit	Material	Labor	Equipment	Total	Total Incl O&P
0010	PRECAST JOISTS									
0015	40 psf L.L., 6″ deep for 12′ spans	C-12	600	.080	L.F.	17.65	3.15	1.28	22.08	26
0050	8″ deep for 16′ spans		575	.083		29.50	3.28	1.34	34.12	39.50
0100	10″ deep for 20′ spans		550	.087		51.50	3.43	1.40	56.33	63.50
0150	12″ deep for 24′ spans		525	.091		70.50	3.60	1.46	75.56	85

03 41 13 – Precast Concrete Hollow Core Planks

03 41 13.50 Precast Slab Planks

		Crew	Daily Output	Labor-Hours	Unit	Material	Labor	Equipment	Total	Total Incl O&P
0010	PRECAST SLAB PLANKS									
0020	Prestressed roof/floor members, grouted, solid, 4″ thick	C-11	2400	.030	S.F.	5.70	1.31	.79	7.80	9.50
0050	6″ thick		2800	.026		6.65	1.13	.67	8.45	10.10
0100	Hollow, 8″ thick		3200	.023		7.35	.99	.59	8.93	10.45
0150	10″ thick		3600	.020		7.75	.88	.53	9.16	10.65
0200	12″ thick		4000	.018		8.35	.79	.47	9.61	11.10

03 41 16 – Precast Concrete Slabs

03 41 16.20 Precast Concrete Channel Slabs

		Crew	Daily Output	Labor-Hours	Unit	Material	Labor	Equipment	Total	Total Incl O&P
0010	PRECAST CONCRETE CHANNEL SLABS									
0335	Lightweight concrete channel slab, long runs, 2-3/4″ thick	C-12	1575	.030	S.F.	6.75	1.20	.49	8.44	9.95
0375	3-3/4″ thick		1550	.031		7	1.22	.50	8.72	10.25
0475	4-3/4″ thick		1525	.031		7.65	1.24	.50	9.39	11.05
1275	Short pieces, 2-3/4″ thick		785	.061		10.15	2.41	.98	13.54	16.15

03 41 Precast Structural Concrete

03 41 16 – Precast Concrete Slabs

	03 41 16.20 Precast Concrete Channel Slabs	Crew	Daily Output	Labor-Hours	Unit	Material	2009 Bare Costs Labor	Equipment	Total	Total Incl O&P
1375	3-3/4" thick	C-12	770	.062	S.F.	10.50	2.45	1	13.95	16.65
1475	4-3/4" thick	↓	762	.063	↓	11.50	2.48	1.01	14.99	17.80
	03 41 16.50 Precast Lightweight Concrete Plank									
0010	**PRECAST LIGHTWEIGHT CONCRETE PLANK**									
0015	Lightweight plank, nailable, T&G, 2" thick	C-12	1800	.027	S.F.	6.15	1.05	.43	7.63	8.95
0150	For premium ceiling finish, add				"	10%				
0200	For sloping roofs, slope over 4 in 12, add						25%			
0250	Slope over 6 in 12, add						150%			

03 41 23 – Precast Concrete Stairs

	03 41 23.50 Precast Stairs	Crew	Daily Output	Labor-Hours	Unit	Material	Labor	Equipment	Total	Total Incl O&P
0010	**PRECAST STAIRS**									
0020	Precast concrete treads on steel stringers, 3' wide	C-12	75	.640	Riser	137	25	10.25	172.25	203
0300	Front entrance, 5' wide with 48" platform, 2 risers		16	3	Flight	450	118	48	616	740
0350	5 risers		12	4		705	157	64	926	1,100
0500	6' wide, 2 risers		15	3.200		505	126	51.50	682.50	820
0550	5 risers		11	4.364		785	172	70	1,027	1,225
0700	7' wide, 2 risers		14	3.429		610	135	55	800	950
0750	5 risers	↓	10	4.800		1,000	189	77	1,266	1,500
1200	Basement entrance stairs, steel bulkhead doors, minimum	B-51	22	2.182		1,350	69.50	8.85	1,428.35	1,600
1250	Maximum	"	11	4.364	↓	2,250	139	17.65	2,406.65	2,725

03 41 33 – Precast Structural Pretensioned Concrete

	03 41 33.60 Tees	Crew	Daily Output	Labor-Hours	Unit	Material	Labor	Equipment	Total	Total Incl O&P
0010	**TEES**									
0020	Quad tee, short spans, roof	C-11	7200	.010	S.F.	6.90	.44	.26	7.60	8.70
0050	Floor		7200	.010		6.90	.44	.26	7.60	8.70
0200	Double tee, floor members, 60' span		8400	.009		8.60	.38	.23	9.21	10.35
0250	80' span		8000	.009		10.10	.39	.24	10.73	12.05
0300	Roof members, 30' span		4800	.015		7.30	.66	.39	8.35	9.60
0350	50' span		6400	.011		8.35	.49	.30	9.14	10.35
0400	Wall members, up to 55' high		3600	.020		11.65	.88	.53	13.06	14.95
0500	Single tee roof members, 40' span		3200	.023		9.20	.99	.59	10.78	12.55
0550	80' span		5120	.014		11.30	.62	.37	12.29	13.90
0600	100' span		6000	.012		17	.53	.32	17.85	20
0650	120' span	↓	6000	.012	↓	18.50	.53	.32	19.35	22
1000	Double tees, floor members									
1100	Lightweight, 20" x 8' wide, 45' span	C-11	20	3.600	Ea.	2,925	158	94.50	3,177.50	3,600
1150	24" x 8' wide, 50' span		18	4		3,200	175	105	3,480	3,925
1200	32" x 10' wide, 60' span		16	4.500		5,400	197	118	5,715	6,400
1250	Standard weight, 12" x 8' wide, 20' span		22	3.273		1,100	143	86	1,329	1,550
1300	16" x 8' wide, 25' span		20	3.600		1,475	158	94.50	1,727.50	1,975
1350	18" x 8' wide, 30' span		20	3.600		1,850	158	94.50	2,102.50	2,425
1400	20" x 8' wide, 45' span		18	4		2,100	175	105	2,380	2,725
1450	24" x 8' wide, 50' span		16	4.500		2,700	197	118	3,015	3,450
1500	32" x 10' wide, 60' span	↓	14	5.143	↓	4,950	225	135	5,310	6,000
2000	Roof members									
2050	Lightweight, 20" x 8' wide, 40' span	C-11	20	3.600	Ea.	2,475	158	94.50	2,727.50	3,100
2100	24" x 8' wide, 50' span		18	4		3,300	175	105	3,580	4,075
2150	32" x 10' wide, 60' span		16	4.500		5,450	197	118	5,765	6,475
2200	Standard weight, 12" x 8' wide, 30' span		22	3.273		1,650	143	86	1,879	2,175
2250	16" x 8' wide, 30' span		20	3.600		1,725	158	94.50	1,977.50	2,275
2300	18" x 8' wide, 30' span	↓	20	3.600	↓	1,925	158	94.50	2,177.50	2,500

03 41 Precast Structural Concrete

03 41 33 – Precast Structural Pretensioned Concrete

03 41 33.60 Tees		Crew	Daily Output	Labor-Hours	Unit	2009 Bare Costs Material	Labor	Equipment	Total	Total Incl O&P
2350	20" x 8' wide, 40' span	C-11	18	4	Ea.	1,975	175	105	2,255	2,600
2400	24" x 8' wide, 50' span		16	4.500		2,650	197	118	2,965	3,375
2450	32" x 10' wide, 60' span		14	5.143		4,625	225	135	4,985	5,650

03 41 36 – Precast Structural Post-Tensioned Concrete

03 41 36.50 Post-Tensioned Jobs

		Crew	Daily Output	Labor-Hours	Unit	Material	Labor	Equipment	Total	Total Incl O&P
0010	**POST-TENSIONED JOBS**									
0100	Post-tensioned in place, small job	C-17B	8.50	9.647	C.Y.	1,175	400	39.50	1,614.50	2,000
0200	Large job	"	10	8.200	"	885	340	33.50	1,258.50	1,550

03 45 Precast Architectural Concrete

03 45 13 – Faced Architectural Precast Concrete

03 45 13.50 Precast Wall Panels

		Crew	Daily Output	Labor-Hours	Unit	Material	Labor	Equipment	Total	Total Incl O&P
0010	**PRECAST WALL PANELS**									
0050	Uninsulated 4" thick, smooth gray									
0150	Low rise, 4' x 8'x 4" thick	C-11	320	.225	S.F.	15.10	9.85	5.90	30.85	41
0210	8'x 8', 4" thick		576	.125		14.95	5.50	3.28	23.73	30
0250	8'x 16'x 4" thick		1024	.070		14.80	3.08	1.85	19.73	24
0600	High rise, 4' x 8' x 4" thick		288	.250		15.10	10.95	6.55	32.60	43.50
0650	8' x 8'x 4" thick		512	.141		14.95	6.15	3.69	24.79	31.50
0700	8' x 16' x 4" thick		768	.094		14.80	4.11	2.46	21.37	26.50
0750	20' x 10', 6" thick, smooth gray		1400	.051		25.50	2.25	1.35	29.10	33.50
0800	Insulated panel, 2" polystyrene, add					1.04			1.04	1.14
0850	2" urethane, add					.81			.81	.89
1200	Finishes, white, add					2.83			2.83	3.11
1250	Exposed aggregate, add					1.69			1.69	1.86
1300	Granite faced, domestic, add					28			28	30.50
2200	Fiberglass reinforced cement with urethane core									
2210	R20, 8' x 8', minimum	E-2	750	.075	S.F.	21.50	3.25	2.32	27.07	32.50
2220	Maximum	"	600	.093	"	24	4.06	2.90	30.96	37

03 47 Site-Cast Concrete

03 47 13 – Tilt-Up Concrete

03 47 13.50 Tilt-Up Wall Panels

		Crew	Daily Output	Labor-Hours	Unit	Material	Labor	Equipment	Total	Total Incl O&P
0010	**TILT-UP WALL PANELS**									
0015	Wall panel construction, walls only, 5-1/2" thick	C-14	1600	.090	S.F.	9.40	3.51	.81	13.72	17.05
0100	7-1/2" thick		1550	.093		11	3.62	.84	15.46	18.95
0500	Walls and columns, 5-1/2" thick walls, 12" x 12" columns		1565	.092		14.55	3.59	.83	18.97	23
0550	7-1/2" thick wall, 12" x 12" columns		1370	.105		16.90	4.10	.95	21.95	26.50
0800	Columns only, site precast, minimum		200	.720	L.F.	24	28	6.50	58.50	80
0850	Maximum		105	1.371	"	28.50	53.50	12.35	94.35	133

03 48 Precast Concrete Specialties

03 48 43 – Precast Concrete Trim

03 48 43.40 Precast Lintels		Crew	Daily Output	Labor-Hours	Unit	Material	2009 Bare Costs Labor	Equipment	Total	Total Incl O&P
0010	**PRECAST LINTELS**									
0800	Precast concrete, 4" wide, 8" high, to 5' long	D-10	28	1.143	Ea.	25.50	45	21.50	92	124
0850	5'-12' long		24	1.333		82.50	52.50	25	160	203
1000	6" wide, 8" high, to 5' long		26	1.231		39.50	48.50	23	111	147
1050	5'-12' long		22	1.455		99.50	57.50	27.50	184.50	232
1200	8" wide, 8" high, to 5' long		24	1.333		47.50	52.50	25	125	165
1250	5'-12' long		20	1.600		135	63	30	228	283
1400	10" wide, 8" high U-Shape, to 14' long		18	1.773		209	70	33.50	312.50	380
1450	12" wide, 8" high U-Shape, to 19' long		16	2		275	79	37.50	391.50	475

03 48 43.90 Precast Window Sills		Crew	Daily Output	Labor-Hours	Unit	Material	Labor	Equipment	Total	Total Incl O&P
0010	**PRECAST WINDOW SILLS**									
0600	Precast concrete, 4" tapers to 3", 9" wide	D-1	70	.229	L.F.	11.75	8.30		20.05	26.50
0650	11" wide		60	.267		15.60	9.70		25.30	33
0700	13" wide, 3 1/2" tapers to 2 1/2", 12" wall		50	.320		15	11.65		26.65	35.50

03 51 Cast Roof Decks

03 51 13 – Cementitious Wood Fiber Decks

03 51 13.50 Cementitious/Wood Fiber Planks		Crew	Daily Output	Labor-Hours	Unit	Material	Labor	Equipment	Total	Total Incl O&P
0010	**CEMENTITIOUS/WOOD FIBER PLANKS** R051223-50									
0050	Plank, beveled edge, 1" thick	2 Carp	1000	.016	S.F.	2.10	.64		2.74	3.37
0100	1-1/2" thick		975	.016		2.41	.66		3.07	3.73
0150	T & G, 2" thick		950	.017		2.69	.67		3.36	4.07
0200	2-1/2" thick		925	.017		3.02	.69		3.71	4.46
0250	3" thick		900	.018		3.49	.71		4.20	5
0300	3-1/2" thick		875	.018		4.94	.73		5.67	6.65
0350	4" thick		850	.019		5.50	.75		6.25	7.30
1000	Bulb tee, sub-purlin and grout, 6' span, add	E-1	5000	.005		2.18	.21	.03	2.42	2.80
1100	8' span	"	4200	.006		2.47	.25	.03	2.75	3.20

03 52 Lightweight Concrete Roof Insulation

03 52 16 – Lightweight Insulating Concrete

03 52 16.13 Lightweight Cellular Insulating Concrete		Crew	Daily Output	Labor-Hours	Unit	Material	Labor	Equipment	Total	Total Incl O&P
0010	**LIGHTWEIGHT CELLULAR INSULATING CONCRETE**									
0020	Portland cement and foaming agent G	C-8	50	1.120	C.Y.	118	39.50	14.75	172.25	210

03 52 16.16 Lightweight Aggregate Insulating Concrete		Crew	Daily Output	Labor-Hours	Unit	Material	Labor	Equipment	Total	Total Incl O&P
0010	**LIGHTWEIGHT AGGREGATE INSULATING CONCRETE**									
0100	Poured vermiculite or perlite, field mix,									
0110	1:6 field mix G	C-8	50	1.120	C.Y.	119	39.50	14.75	173.25	210
0200	Ready mix, 1:6 mix, roof fill, 2" thick G		10000	.006	S.F.	.66	.20	.07	.93	1.12
0250	3" thick G		7700	.007		.99	.26	.10	1.35	1.61
0400	Expanded volcanic glass rock, with binder, minimum G	2 Carp	1500	.011		.40	.43		.83	1.14
0450	Maximum G	"	1200	.013		1.20	.53		1.73	2.20

03 54 Cast Underlayment

03 54 13 – Gypsum Cement Underlayment

03 54 13.50 Gypsum Underlayment		Crew	Daily Output	Labor-Hours	Unit	2009 Bare Costs Material	Labor	Equipment	Total	Total Incl O&P
0010	**GYPSUM UNDERLAYMENT**									
1000	Poured gypsum, 2" thick	C-8	6000	.009	S.F.	1.75	.33	.12	2.20	2.60
1100	3" thick	"	4800	.012	"	2.63	.41	.15	3.19	3.72

03 54 16 – Hydraulic Cement Underlayment

03 54 16.50 Cement Underlayment

		Crew	Daily Output	Labor-Hours	Unit	Material	Labor	Equipment	Total	Total Incl O&P
0010	**CEMENT UNDERLAYMENT**									
2510	Underlayment, P.C based self-leveling, 4100 psi, pumped, 1/4"	C-8	20000	.003	S.F.	1.44	.10	.04	1.58	1.78
2520	1/2"		19000	.003		2.87	.10	.04	3.01	3.37
2530	3/4"		18000	.003		4.31	.11	.04	4.46	4.97
2540	1"		17000	.003		5.75	.12	.04	5.91	6.55
2550	1-1/2"		15000	.004		8.60	.13	.05	8.78	9.75
2560	Hand mix, 1/2"	C-18	4000	.002		2.87	.07	.01	2.95	3.30
2610	Topping, P.C. based self-level/dry 6100 psi, pumped, 1/4"	C-8	20000	.003		2.21	.10	.04	2.35	2.64
2620	1/2"		19000	.003		4.43	.10	.04	4.57	5.10
2630	3/4"		18000	.003		6.65	.11	.04	6.80	7.55
2660	1"		17000	.003		8.85	.12	.04	9.01	10
2670	1-1/2"		15000	.004		13.30	.13	.05	13.48	14.85
2680	Hand mix, 1/2"	C-18	4000	.002		4.43	.07	.01	4.51	5

03 62 Non-Shrink Grouting

03 62 13 – Non-Metallic Non-Shrink Grouting

03 62 13.50 Grout, Non-Metallic Non-shrink

		Crew	Daily Output	Labor-Hours	Unit	Material	Labor	Equipment	Total	Total Incl O&P
0010	**GROUT, NON-METALLIC NON-SHRINK**									
0300	Non-shrink, non-metallic, 1" deep	1 Cefi	35	.229	S.F.	7.15	8.75		15.90	21.50
0350	2" deep	"	25	.320	"	14.30	12.25		26.55	35

03 62 16 – Metallic Non-Shrink Grouting

03 62 16.50 Grout, Metallic Non-Shrink

		Crew	Daily Output	Labor-Hours	Unit	Material	Labor	Equipment	Total	Total Incl O&P
0010	**GROUT, METALLIC NON-SHRINK**									
0020	Column & machine bases, non-shrink, metallic, 1" deep	1 Cefi	35	.229	S.F.	8.90	8.75		17.65	23.50
0050	2" deep	"	25	.320	"	17.85	12.25		30.10	39

03 63 Epoxy Grouting

03 63 05 – Grouting of Dowels and Fasteners

03 63 05.10 Epoxy Only

		Crew	Daily Output	Labor-Hours	Unit	Material	Labor	Equipment	Total	Total Incl O&P
0010	**EPOXY ONLY**									
1500	Chemical anchoring, epoxy cartridge, excludes layout, drilling, fastener									
1530	For fastener 3/4" diam. x 6" embedment	2 Skwk	72	.222	Ea.	3.54	9.10		12.64	18.70
1535	1" diam. x 8" embedment		66	.242		5.30	9.90		15.20	22
1540	1-1/4" diam. x 10" embedment		60	.267		10.60	10.90		21.50	29.50
1545	1-3/4" diam. x 12" embedment		54	.296		17.70	12.10		29.80	39
1550	14" embedment		48	.333		21	13.60		34.60	45.50
1555	2" diam. x 12" embedment		42	.381		28.50	15.55		44.05	56.50
1560	18" embedment		32	.500		35.50	20.50		56	72.50

03 64 Injection Grouting

03 64 23 – Epoxy Injection Grouting

03 64 23.10 Crack Repair		Crew	Daily Output	Labor-Hours	Unit	Material	2009 Bare Costs Labor	Equipment	Total	Total Incl O&P
0010	CRACK REPAIR									
0100	Epoxy injection, 1/8" wide, 12" deep	B-9	80	.500	L.F.	6.50	16	2.40	24.90	36.50
0110	1/4" wide, 12" deep		60	.667		13	21.50	3.19	37.69	53
0200	Latex injection, 1/8" wide, 12" deep		100	.400		2.28	12.80	1.92	17	25.50
0210	1/4" wide, 12" deep		75	.533		4.57	17.05	2.55	24.17	36

03 81 Concrete Cutting

03 81 13 – Flat Concrete Sawing

03 81 13.50 Concrete Floor/Slab Cutting

		Crew	Daily Output	Labor-Hours	Unit	Material	Labor	Equipment	Total	Total Incl O&P
0010	CONCRETE FLOOR/SLAB CUTTING									
0400	Concrete slabs, mesh reinforcing, up to 3" deep	B-89	980	.016	L.F.	.49	.57	.41	1.47	1.91
0420	Each additional inch of depth	"	1600	.010	"	.16	.35	.25	.76	1.02

03 81 16 – Track Mounted Concrete Wall Sawing

03 81 16.50 Concrete Wall Cutting

		Crew	Daily Output	Labor-Hours	Unit	Material	Labor	Equipment	Total	Total Incl O&P
0010	CONCRETE WALL CUTTING									
0800	Concrete walls, hydraulic saw, plain, per inch of depth	B-89B	250	.064	L.F.	.44	2.24	2.74	5.42	7.10
0820	Rod reinforcing, per inch of depth	"	150	.107	"	.62	3.73	4.56	8.91	11.70

03 82 Concrete Boring

03 82 13 – Concrete Core Drilling

03 82 13.10 Core Drilling

		Crew	Daily Output	Labor-Hours	Unit	Material	Labor	Equipment	Total	Total Incl O&P
0010	CORE DRILLING									
0020	Reinf. conc slab, up to 6" thick, incl. bit, layout & set up									
0100	1" diameter core	B-89A	17	.941	Ea.	3.45	34	6.50	43.95	67
0150	Each added inch thick in same hole, add		1440	.011		.63	.40	.08	1.11	1.43
0300	3" diameter core		16	1		7.40	36	6.90	50.30	75.50
0350	Each added inch thick in same hole, add		720	.022		1.38	.81	.15	2.34	3.01
0500	4" diameter core		15	1.067		7.40	38.50	7.35	53.25	80
0550	Each added inch thick in same hole, add		480	.033		1.81	1.21	.23	3.25	4.22
0700	6" diameter core		14	1.143		12.20	41.50	7.85	61.55	90
0750	Each added inch thick in same hole, add		360	.044		2.17	1.61	.31	4.09	5.35
0900	8" diameter core		13	1.231		16.70	44.50	8.50	69.70	101
0950	Each added inch thick in same hole, add		288	.056		2.88	2.01	.38	5.27	6.90
1100	10" diameter core		12	1.333		22	48.50	9.20	79.70	113
1150	Each added inch thick in same hole, add		240	.067		3.62	2.42	.46	6.50	8.45
1300	12" diameter core		11	1.455		27	52.50	10	89.50	128
1350	Each added inch thick in same hole, add		206	.078		4.35	2.81	.54	7.70	10
1500	14" diameter core		10	1.600		33	58	11	102	143
1550	Each added inch thick in same hole, add		180	.089		5.85	3.22	.61	9.68	12.35
1700	18" diameter core		9	1.778		42.50	64.50	12.25	119.25	165
1750	Each added inch thick in same hole, add		144	.111		7.60	4.03	.77	12.40	15.80
1760	For horizontal holes, add to above						20%	20%		
1770	Prestressed hollow core plank, 8" thick									
1780	1" diameter core	B-89A	17.50	.914	Ea.	2.30	33	6.30	41.60	63.50
1790	Each added inch thick in same hole, add		3840	.004		.46	.15	.03	.64	.79
1800	3" diameter core		17	.941		4.81	34	6.50	45.31	68.50
1810	Each added inch thick in same hole, add		1920	.008		.83	.30	.06	1.19	1.46
1820	4" diameter core		16.50	.970		6.60	35	6.70	48.30	72
1830	Each added inch thick in same hole, add		1280	.013		1.21	.45	.09	1.75	2.16

03 82 Concrete Boring

03 82 13 – Concrete Core Drilling

03 82 13.10 Core Drilling		Crew	Daily Output	Labor-Hours	Unit	2009 Bare Costs Material	Labor	Equipment	Total	Total Incl O&P
1840	6″ diameter core	B-89A	15.50	1.032	Ea.	8.10	37.50	7.10	52.70	77.50
1850	Each added inch thick in same hole, add		960	.017		1.38	.60	.11	2.09	2.64
1860	8″ diameter core		15	1.067		11.50	38.50	7.35	57.35	84.50
1870	Each added inch thick in same hole, add		768	.021		1.93	.75	.14	2.82	3.52
1880	10″ diameter core		14	1.143		15.25	41.50	7.85	64.60	93.50
1890	Each added inch thick in same hole, add		640	.025		2.01	.91	.17	3.09	3.88
1900	12″ diameter core		13.50	1.185		17.85	43	8.15	69	99
1910	Each added inch thick in same hole, add		548	.029		3.22	1.06	.20	4.48	5.50
1999	Drilling, core, minimum labor/equipment charge		4	4	Job		145	27.50	172.50	268
3010	Bits for core drill, diamond, premium, 1″ diameter				Ea.	117			117	129
3020	3″ diameter					288			288	315
3040	4″ diameter					320			320	350
3050	6″ diameter					510			510	565
3080	8″ diameter					705			705	775
3120	12″ diameter					1,050			1,050	1,150
3180	18″ diameter					1,900			1,900	2,075
3240	24″ diameter					2,600			2,600	2,875

03 82 16 – Concrete Drilling

03 82 16.10 Concrete Impact Drilling		Crew	Daily Output	Labor-Hours	Unit	Material	Labor	Equipment	Total	Total Incl O&P
0010	**CONCRETE IMPACT DRILLING**									
0050	Up to 4″ deep in conc/brick floor/wall, incl. bit & layout, no anchor									
0100	Holes, 1/4″ diameter	1 Carp	75	.107	Ea.	.06	4.26		4.32	7.10
0150	For each additional inch of depth, add		430	.019		.01	.74		.75	1.25
0200	3/8″ diameter		63	.127		.05	5.05		5.10	8.40
0250	For each additional inch of depth, add		340	.024		.01	.94		.95	1.56
0300	1/2″ diameter		50	.160		.06	6.40		6.46	10.60
0350	For each additional inch of depth, add		250	.032		.01	1.28		1.29	2.13
0400	5/8″ diameter		48	.167		.08	6.65		6.73	11.10
0450	For each additional inch of depth, add		240	.033		.02	1.33		1.35	2.22
0500	3/4″ diameter		45	.178		.10	7.10		7.20	11.80
0550	For each additional inch of depth, add		220	.036		.03	1.45		1.48	2.43
0600	7/8″ diameter		43	.186		.12	7.45		7.57	12.40
0650	For each additional inch of depth, add		210	.038		.03	1.52		1.55	2.54
0700	1″ diameter		40	.200		.14	8		8.14	13.35
0750	For each additional inch of depth, add		190	.042		.04	1.68		1.72	2.82
0800	1-1/4″ diameter		38	.211		.20	8.40		8.60	14.10
0850	For each additional inch of depth, add		180	.044		.05	1.78		1.83	2.99
0900	1-1/2″ diameter		35	.229		.29	9.15		9.44	15.35
0950	For each additional inch of depth, add		165	.048		.07	1.94		2.01	3.28
1000	For ceiling installations, add						40%			

Estimating Tips

04 05 00 Common Work Results for Masonry

- The terms *mortar* and *grout* are often used interchangeably, and incorrectly. Mortar is used to bed masonry units, seal the entry of air and moisture, provide architectural appearance, and allow for size variations in the units. Grout is used primarily in reinforced masonry construction and is used to bond the masonry to the reinforcing steel. Common mortar types are M(2500 psi), S(1800 psi), N(750 psi), and O(350 psi), and conform to ASTM C270. Grout is either fine or coarse and conforms to ASTM C476, and in-place strengths generally exceed 2500 psi. Mortar and grout are different components of masonry construction and are placed by entirely different methods. An estimator should be aware of their unique uses and costs.
- Waste, specifically the loss/droppings of mortar and the breakage of brick and block, is included in all masonry assemblies in this division. A factor of 25% is added for mortar and 3% for brick and concrete masonry units.
- Scaffolding or staging is not included in any of the Division 4 costs. Refer to Subdivision 01 54 23 for scaffolding and staging costs.

04 20 00 Unit Masonry

- The most common types of unit masonry are brick and concrete masonry. The major classifications of brick are building brick (ASTM C62), facing brick (ASTM C216), glazed brick, fire brick, and pavers. Many varieties of texture and appearance can exist within these classifications, and the estimator would be wise to check local custom and availability within the project area. For repair and remodeling jobs, matching the existing brick may be the most important criteria.
- Brick and concrete block are priced by the piece and then converted into a price per square foot of wall. Openings less than two square feet are generally ignored by the estimator because any savings in units used is offset by the cutting and trimming required.
- It is often difficult and expensive to find and purchase small lots of historic brick. Costs can vary widely. Many design issues affect costs, selection of mortar mix, and repairs or replacement of masonry materials. Cleaning techniques must be reflected in the estimate.
- All masonry walls, whether interior or exterior, require bracing. The cost of bracing walls during construction should be included by the estimator, and this bracing must remain in place until permanent bracing is complete. Permanent bracing of masonry walls is accomplished by masonry itself, in the form of pilasters or abutting wall corners, or by anchoring the walls to the structural frame. Accessories in the form of anchors, anchor slots, and ties are used, but their supply and installation can be by different trades. For instance, anchor slots on spandrel beams and columns are supplied and welded in place by the steel fabricator, but the ties from the slots into the masonry are installed by the bricklayer. Regardless of the installation method, the estimator must be certain that these accessories are accounted for in pricing.

Reference Numbers

Reference numbers are shown in shaded boxes at the beginning of some major classifications. These numbers refer to related items in the Reference Section. The reference information may be an estimating procedure, an alternate pricing method, or technical information.

Note: Not all subdivisions listed here necessarily appear in this publication.

04 01 Maintenance of Masonry

04 01 20 – Maintenance of Unit Masonry

04 01 20.20 Pointing Masonry		Crew	Daily Output	Labor-Hours	Unit	Material	2009 Bare Costs Labor	Equipment	Total	Total Incl O&P
0010	**POINTING MASONRY**									
0300	Cut and repoint brick, hard mortar, running bond	1 Bric	80	.100	S.F.	.53	4.05		4.58	7.15
0320	Common bond		77	.104		.53	4.21		4.74	7.40
0360	Flemish bond		70	.114		.56	4.63		5.19	8.10
0400	English bond		65	.123		.56	4.98		5.54	8.65
0600	Soft old mortar, running bond		100	.080		.53	3.24		3.77	5.85
0620	Common bond		96	.083		.53	3.37		3.90	6.05
0640	Flemish bond		90	.089		.56	3.60		4.16	6.40
0680	English bond		82	.098		.56	3.95		4.51	7
0700	Stonework, hard mortar		140	.057	L.F.	.70	2.31		3.01	4.51
0720	Soft old mortar		160	.050	"	.70	2.03		2.73	4.05
1000	Repoint, mask and grout method, running bond		95	.084	S.F.	.70	3.41		4.11	6.25
1020	Common bond		90	.089		.70	3.60		4.30	6.55
1040	Flemish bond		86	.093		.74	3.77		4.51	6.90
1060	English bond		77	.104		.74	4.21		4.95	7.60
2000	Scrub coat, sand grout on walls, minimum		120	.067		2.84	2.70		5.54	7.50
2020	Maximum		98	.082		3.94	3.31		7.25	9.70
9000	Minimum labor/equipment charge		3	2.667	Job		108		108	175

04 01 20.40 Sawing Masonry		Crew	Daily Output	Labor-Hours	Unit	Material	Labor	Equipment	Total	Total Incl O&P
0010	**SAWING MASONRY**									
0050	Brick or block by hand, per inch depth	A-1	125	.064	L.F.		2.02	.51	2.53	3.90

04 01 20.50 Toothing Masonry		Crew	Daily Output	Labor-Hours	Unit	Material	Labor	Equipment	Total	Total Incl O&P
0010	**TOOTHING MASONRY**									
0500	Brickwork, soft old mortar	1 Clab	40	.200	V.L.F.		6.30		6.30	10.45
0520	Hard mortar		30	.267			8.45		8.45	13.90
0700	Blockwork, soft old mortar		70	.114			3.61		3.61	5.95
0720	Hard mortar		50	.160			5.05		5.05	8.35
9000	Minimum labor/equipment charge		4	2	Job		63		63	104

04 01 30 – Unit Masonry Cleaning

04 01 30.20 Cleaning Masonry		Crew	Daily Output	Labor-Hours	Unit	Material	Labor	Equipment	Total	Total Incl O&P
0010	**CLEANING MASONRY**									
0200	Chemical cleaning, new construction, brush and wash, minimum	D-1	1000	.016	S.F.	.05	.58		.63	.99
0220	Average		800	.020		.07	.73		.80	1.26
0240	Maximum		600	.027		.09	.97		1.06	1.67
0260	Light restoration, minimum		800	.020		.07	.73		.80	1.25
0270	Average		400	.040		.10	1.45		1.55	2.46
0280	Maximum		330	.048		.13	1.76		1.89	3
0300	Heavy restoration, minimum		600	.027		.09	.97		1.06	1.67
0310	Average		400	.040		.14	1.45		1.59	2.51
0320	Maximum		250	.064		.19	2.33		2.52	3.97
0400	High pressure water only, minimum	B-9	2000	.020			.64	.10	.74	1.17
0420	Average		1500	.027			.85	.13	.98	1.55
0440	Maximum		1000	.040			1.28	.19	1.47	2.32
0800	High pressure water and chemical, minimum		1800	.022		.08	.71	.11	.90	1.38
0820	Average		1200	.033		.12	1.07	.16	1.35	2.07
0840	Maximum		800	.050		.16	1.60	.24	2	3.08
1200	Sandblast, wet system, minimum		1750	.023		.47	.73	.11	1.31	1.85
1220	Average		1100	.036		.70	1.16	.17	2.03	2.88
1240	Maximum		700	.057		.94	1.83	.27	3.04	4.35
1400	Dry system, minimum		2500	.016		.47	.51	.08	1.06	1.45
1420	Average		1750	.023		.70	.73	.11	1.54	2.10
1440	Maximum		1000	.040		.94	1.28	.19	2.41	3.35

04 01 Maintenance of Masonry

04 01 30 – Unit Masonry Cleaning

04 01 30.20 Cleaning Masonry		Crew	Daily Output	Labor-Hours	Unit	2009 Bare Costs Material	Labor	Equipment	Total	Total Incl O&P
1800	For walnut shells, add				S.F.	.50			.50	.55
1820	For corn chips, add					.50			.50	.55
2000	Steam cleaning, minimum	B-9	3000	.013			.43	.06	.49	.77
2020	Average		2500	.016			.51	.08	.59	.93
2040	Maximum		1500	.027			.85	.13	.98	1.55
4000	Add for masking doors and windows	1 Clab	800	.010		.06	.32		.38	.59
4200	Add for pedestrian protection				Job					10%
9000	Minimum labor/equipment charge	D-4	2	16	"		575	63.50	638.50	995

04 01 30.60 Brick Washing		Crew	Daily Output	Labor-Hours	Unit	Material	Labor	Equipment	Total	Total Incl O&P
0010	**BRICK WASHING**									
0012	Acid cleanser, smooth brick surface	1 Bric	560	.014	S.F.	.05	.58		.63	1
0050	Rough brick		400	.020		.07	.81		.88	1.39
0060	Stone, acid wash		600	.013		.08	.54		.62	.96
1000	Muriatic acid, price per gallon in 5 gallon lots				Gal.	10.50			10.50	11.55

04 05 Common Work Results for Masonry

04 05 05 – Selective Masonry Demolition

04 05 05.10 Selective Demolition		Crew	Daily Output	Labor-Hours	Unit	Material	Labor	Equipment	Total	Total Incl O&P
0010	**SELECTIVE DEMOLITION** R024119-10									
0200	Bond beams, 8" block with #4 bar	2 Clab	32	.500	L.F.		15.80		15.80	26
0300	Concrete block walls, unreinforced, 2" thick		1200	.013	S.F.		.42		.42	.70
0310	4" thick		1150	.014			.44		.44	.73
0320	6" thick		1100	.015			.46		.46	.76
0330	8" thick		1050	.015			.48		.48	.79
0340	10" thick		1000	.016			.51		.51	.83
0360	12" thick		950	.017			.53		.53	.88
0380	Reinforced alternate courses, 2" thick		1130	.014			.45		.45	.74
0390	4" thick		1080	.015			.47		.47	.77
0400	6" thick		1035	.015			.49		.49	.81
0410	8" thick		990	.016			.51		.51	.84
0420	10" thick		940	.017			.54		.54	.89
0430	12" thick		890	.018			.57		.57	.94
0440	Reinforced alternate courses & vertically 48" OC, 4" thick		900	.018			.56		.56	.93
0450	6" thick		850	.019			.59		.59	.98
0460	8" thick		800	.020			.63		.63	1.04
0480	10" thick		750	.021			.67		.67	1.11
0490	12" thick		700	.023			.72		.72	1.19
1000	Chimney, 16" x 16", soft old mortar	1 Clab	55	.145	C.F.		4.60		4.60	7.60
1020	Hard mortar		40	.200			6.30		6.30	10.45
1030	16" x 20", soft old mortar		55	.145			4.60		4.60	7.60
1040	Hard mortar		40	.200			6.30		6.30	10.45
1050	16" x 24", soft old mortar		55	.145			4.60		4.60	7.60
1060	Hard mortar		40	.200			6.30		6.30	10.45
1080	20" x 20", soft old mortar		55	.145			4.60		4.60	7.60
1100	Hard mortar		40	.200			6.30		6.30	10.45
1110	20" x 24", soft old mortar		55	.145			4.60		4.60	7.60
1120	Hard mortar		40	.200			6.30		6.30	10.45
1140	20" x 32", soft old mortar		55	.145			4.60		4.60	7.60
1160	Hard mortar		40	.200			6.30		6.30	10.45
1200	48" x 48", soft old mortar		55	.145			4.60		4.60	7.60
1220	Hard mortar		40	.200			6.30		6.30	10.45

04 05 Common Work Results for Masonry

04 05 05 – Selective Masonry Demolition

04 05 05.10 Selective Demolition		Crew	Daily Output	Labor-Hours	Unit	Material	2009 Bare Costs Labor	Equipment	Total	Total Incl O&P
1250	Metal, high temp steel jacket, 24" diameter	E-2	130	.431	V.L.F.		18.75	13.40	32.15	48.50
1260	60" diameter	"	60	.933			40.50	29	69.50	105
1280	Flue lining, up to 12" x 12"	1 Clab	200	.040			1.26		1.26	2.09
1282	Up to 24" x 24"		150	.053			1.69		1.69	2.78
2000	Columns, 8" x 8", soft old mortar		48	.167			5.25		5.25	8.70
2020	Hard mortar		40	.200			6.30		6.30	10.45
2060	16" x 16", soft old mortar		16	.500			15.80		15.80	26
2100	Hard mortar		14	.571			18.05		18.05	30
2140	24" x 24", soft old mortar		8	1			31.50		31.50	52
2160	Hard mortar		6	1.333			42		42	69.50
2200	36" x 36", soft old mortar		4	2			63		63	104
2220	Hard mortar		3	2.667			84.50		84.50	139
2230	Alternate pricing method, soft old mortar		30	.267	C.F.		8.45		8.45	13.90
2240	Hard mortar		23	.348	"		11		11	18.15
3000	Copings, precast or masonry, to 8" wide									
3020	Soft old mortar	1 Clab	180	.044	L.F.		1.40		1.40	2.32
3040	Hard mortar	"	160	.050	"		1.58		1.58	2.61
3100	To 12" wide									
3120	Soft old mortar	1 Clab	160	.050	L.F.		1.58		1.58	2.61
3140	Hard mortar	"	140	.057	"		1.81		1.81	2.98
4000	Fireplace, brick, 30" x 24" opening									
4020	Soft old mortar	1 Clab	2	4	Ea.		126		126	209
4040	Hard mortar		1.25	6.400			202		202	335
4100	Stone, soft old mortar		1.50	5.333			169		169	278
4120	Hard mortar		1	8			253		253	415
5000	Veneers, brick, soft old mortar		140	.057	S.F.		1.81		1.81	2.98
5020	Hard mortar		125	.064			2.02		2.02	3.34
5100	Granite and marble, 2" thick		180	.044			1.40		1.40	2.32
5120	4" thick		170	.047			1.49		1.49	2.45
5140	Stone, 4" thick		180	.044			1.40		1.40	2.32
5160	8" thick		175	.046			1.44		1.44	2.38
5400	Alternate pricing method, stone, 4" thick		60	.133	C.F.		4.21		4.21	6.95
5420	8" thick		85	.094	"		2.97		2.97	4.91
9000	Minimum labor/equipment charge		2	4	Job		126		126	209

04 05 13 – Masonry Mortaring

04 05 13.10 Cement

		Crew	Daily Output	Labor-Hours	Unit	Material	Labor	Equipment	Total	Total Incl O&P
0010	**CEMENT**									
0100	Masonry, 70 lb. bag, T.L. lots				Bag	9.05			9.05	9.95
0150	L.T.L. lots					9.60			9.60	10.55
0200	White, 70 lb. bag, T.L. lots					15.05			15.05	16.55
0250	L.T.L. lots					16.05			16.05	17.65

04 05 13.20 Lime

		Crew	Daily Output	Labor-Hours	Unit	Material	Labor	Equipment	Total	Total Incl O&P
0010	**LIME**									
0020	Masons, hydrated, 50 lb. bag, T.L. lots				Bag	8.95			8.95	9.85
0050	L.T.L. lots					9.30			9.30	10.25
0200	Finish, double hydrated, 50 lb. bag, T.L. lots					9.90			9.90	10.90
0250	L.T.L. lots					11			11	12.10

04 05 13.23 Surface Bonding Masonry Mortaring

		Crew	Daily Output	Labor-Hours	Unit	Material	Labor	Equipment	Total	Total Incl O&P
0010	**SURFACE BONDING MASONRY MORTARING**									
0020	Gray or white colors, not incl. block work	1 Bric	540	.015	S.F.	.10	.60		.70	1.08

04 05 13 – Masonry Mortaring

04 05 13.30 Mortar

04 05 13.30 Mortar		Crew	Daily Output	Labor-Hours	Unit	Material	2009 Bare Costs Labor	Equipment	Total	Total Incl O&P
0010	**MORTAR** R042110-50									
0100	Type M, 1:1:6 mix	1 Brhe	143	.056	C.F.	4.75	1.80		6.55	8.10
0200	Type N, 1:3 mix		143	.056		4.29	1.80		6.09	7.65
0300	Type O, 1:3 mix		143	.056		6.45	1.80		8.25	10
0400	Type PM, 1:1:6 mix, 2500 psi		143	.056		6.85	1.80		8.65	10.45
0500	Type S, 1/2:1:4 mix	↓	143	.056	↓	7.25	1.80		9.05	10.90
2000	With portland cement and lime									
2100	Type M, 1:1/4:3 mix	1 Brhe	143	.056	C.F.	8.45	1.80		10.25	12.20
2200	Type N, 1:1:6 mix, 750 psi		143	.056		6.80	1.80		8.60	10.35
2300	Type O, 1:2:9 mix (Pointing Mortar)		143	.056		7.50	1.80		9.30	11.15
2400	Type PL, 1:1/2:4 mix, 2500 psi		143	.056		7.45	1.80		9.25	11.10
2500	Type K, 1:3:12 mix, 75 psi		143	.056		6.45	1.80		8.25	10
2600	Type S, 1:1/2:4 mix, 1800 psi	↓	143	.056		9.15	1.80		10.95	12.95
2650	Pre-mixed, type S or N					4.94			4.94	5.45
2700	Mortar for glass block	1 Brhe	143	.056	↓	10.20	1.80		12	14.10
2900	Mortar for Fire Brick, 80 lb. bag, T.L. Lots				Bag	25.50			25.50	28

04 05 13.91 Masonry Restoration Mortaring

0010	**MASONRY RESTORATION MORTARING**									
0020	Masonry restoration mix				Lb.	1.97			1.97	2.17
0050	White				"	2.27			2.27	2.50

04 05 13.93 Mortar Pigments

0010	**MORTAR PIGMENTS**, 50 lb. bags (2 bags per M bricks)									
0020	Color admixture, range 2 to 10 lb. per bag of cement, minimum				Lb.	4.50			4.50	4.95
0050	Average					5.40			5.40	5.95
0100	Maximum				↓	10.45			10.45	11.50

04 05 13.95 Sand

0010	**SAND**, screened and washed at pit									
0020	For mortar, per ton				Ton	17.80			17.80	19.55
0050	With 10 mile haul					26			26	28.50
0100	With 30 mile haul				↓	36			36	40
0200	Screened and washed, at the pit				C.Y.	24.50			24.50	27
0250	With 10 mile haul					36			36	40
0300	With 30 mile haul				↓	50			50	55.50

04 05 13.98 Mortar Admixtures

0010	**MORTAR ADMIXTURES**									
0020	Waterproofing admixture, per quart (1 qt. to 2 bags of masonry cement)				Qt.	9.50			9.50	10.45

04 05 16 – Masonry Grouting

04 05 16.30 Grouting

0010	**GROUTING**									
0011	Bond beams & lintels, 8" deep, 6" thick, 0.15 C.F. per L.F.	D-4	1480	.022	L.F.	.64	.78	.09	1.51	2.04
0020	8" thick, 0.2 C.F. per L.F.		1400	.023		.98	.82	.09	1.89	2.50
0050	10" thick, 0.25 C.F. per L.F.		1200	.027		1.06	.96	.11	2.13	2.83
0060	12" thick, 0.3 C.F. per L.F.	↓	1040	.031	↓	1.28	1.11	.12	2.51	3.31
0200	Concrete block cores, solid, 4" thk., by hand, 0.067 C.F./S.F. of wall	D-8	1100	.036	S.F.	.28	1.35		1.63	2.50
0210	6" thick, pumped, 0.175 C.F. per S.F.	D-4	720	.044		.74	1.60	.18	2.52	3.58
0250	8" thick, pumped, 0.258 C.F. per S.F.		680	.047		1.10	1.69	.19	2.98	4.14
0300	10" thick, pumped, 0.340 C.F. per S.F.		660	.048		1.45	1.74	.19	3.38	4.60
0350	12" thick, pumped, 0.422 C.F. per S.F.		640	.050		1.79	1.80	.20	3.79	5.10
0500	Cavity walls, 2" space, pumped, 0.167 C.F./S.F. of wall		1700	.019		.71	.68	.07	1.46	1.95
0550	3" space, 0.250 C.F./S.F.	↓	1200	.027	↓	1.06	.96	.11	2.13	2.83

04 05 Common Work Results for Masonry

04 05 16 – Masonry Grouting

04 05 16.30 Grouting		Crew	Daily Output	Labor-Hours	Unit	2009 Bare Costs Material	Labor	Equipment	Total	Total Incl O&P
0600	4" space, 0.333 C.F. per S.F.	D-4	1150	.028	S.F.	1.42	1	.11	2.53	3.29
0700	6" space, 0.500 C.F. per S.F.		800	.040	↓	2.13	1.44	.16	3.73	4.82
0800	Door frames, 3' x 7' opening, 2.5 C.F. per opening		60	.533	Opng.	10.65	19.20	2.11	31.96	45
0850	6' x 7' opening, 3.5 C.F. per opening		45	.711	"	14.90	25.50	2.82	43.22	60.50
2000	Grout, C476, for bond beams, lintels and CMU cores	↓	350	.091	C.F.	4.25	3.29	.36	7.90	10.40
9000	Minimum labor/equipment charge	1 Bric	2	4	Job		162		162	262

04 05 19 – Masonry Anchorage and Reinforcing

04 05 19.05 Anchor Bolts		Crew	Daily Output	Labor-Hours	Unit	Material	Labor	Equipment	Total	Total Incl O&P
0010	**ANCHOR BOLTS**									
0020	Hooked, with nut and washer, 1/2" diam., 8" long	1 Bric	200	.040	Ea.	.83	1.62		2.45	3.53
0030	12" long		190	.042		1.26	1.71		2.97	4.15
0040	5/8" diameter, 8" long		180	.044		1.13	1.80		2.93	4.15
0050	12" long		170	.047		1.24	1.91		3.15	4.45
0060	3/4" diameter, 8" long		160	.050		1.86	2.03		3.89	5.35
0070	12" long	↓	150	.053	↓	2.33	2.16		4.49	6.05

04 05 19.16 Masonry Anchors		Crew	Daily Output	Labor-Hours	Unit	Material	Labor	Equipment	Total	Total Incl O&P
0010	**MASONRY ANCHORS**									
0020	For brick veneer, galv., corrugated, 7/8" x 7", 22 Ga.	1 Bric	10.50	.762	C	9.80	31		40.80	61
0100	24 Ga.		10.50	.762		9.15	31		40.15	60
0150	16 Ga.		10.50	.762		24	31		55	76.50
0200	Buck anchors, galv., corrugated, 16 gauge, 2" bend, 8" x 2"		10.50	.762		134	31		165	197
0250	8" x 3"	↓	10.50	.762	↓	138	31		169	202
0300	Adjustable, rectangular, 4-1/8" wide									
0350	Anchor and tie, 3/16" wire, mill galv.									
0400	2-3/4" eye, 3-1/4" tie	1 Bric	1.05	7.619	M	415	310		725	960
0500	4-3/4" tie		1.05	7.619		465	310		775	1,000
0520	5-1/2" tie		1.05	7.619		490	310		800	1,050
0550	4-3/4" eye, 3-1/4" tie		1.05	7.619		460	310		770	1,000
0570	4-3/4" tie		1.05	7.619		525	310		835	1,075
0580	5-1/2" tie		1.05	7.619	↓	550	310		860	1,100
0680	1/4" diameter		10.50	.762	C	39.50	31		70.50	93.50
0850	8" long, 3/16" diameter		10.50	.762		24	31		55	76
0855	1/4" diameter		10.50	.762		46.50	31		77.50	101
1000	Rectangular type, galvanized, 1/4" diameter, 2" x 6"		10.50	.762		67	31		98	124
1050	4" x 6"		10.50	.762		83.50	31		114.50	142
1100	3/16" diameter, 2" x 6"		10.50	.762		37.50	31		68.50	91.50
1150	4" x 6"	↓	10.50	.762	↓	39	31		70	93
1200	Mesh wall tie, 1/2" mesh, hot dip galvanized									
1400	16 gauge, 12" long, 3" wide	1 Bric	9	.889	C	87.50	36		123.50	155
1420	6" wide		9	.889		130	36		166	201
1440	12" wide		8.50	.941		208	38		246	291
1500	Rigid partition anchors, plain, 8" long, 1" x 1/8"		10.50	.762		103	31		134	163
1550	1" x 1/4"		10.50	.762		170	31		201	237
1580	1-1/2" x 1/8"		10.50	.762		138	31		169	201
1600	1-1/2" x 1/4"		10.50	.762		280	31		311	360
1650	2" x 1/8"		10.50	.762		178	31		209	246
1700	2" x 1/4"	↓	10.50	.762	↓	325	31		356	410
2000	Column flange ties, wire, galvanized									
2300	3/16" diameter, up to 3" wide	1 Bric	10.50	.762	C	81.50	31		112.50	140
2350	To 5" wide		10.50	.762		89	31		120	148
2400	To 7" wide		10.50	.762		95	31		126	155
2600	To 9" wide	↓	10.50	.762	↓	103	31		134	163

04 05 19 – Masonry Anchorage and Reinforcing

04 05 19.16 Masonry Anchors		Crew	Daily Output	Labor-Hours	Unit	2009 Bare Costs Material	Labor	Equipment	Total	Total Incl O&P
2650	1/4" diameter, up to 3" wide	1 Bric	10.50	.762	C	111	31		142	172
2700	To 5" wide		10.50	.762		122	31		153	184
2800	To 7" wide		10.50	.762		132	31		163	195
2850	To 9" wide		10.50	.762		144	31		175	208
2900	For hot dip galvanized, add					35%				
4000	Channel slots, 1-3/8" x 1/2" x 8"									
4100	12 gauge, plain	1 Bric	10.50	.762	C	238	31		269	310
4150	16 gauge, galvanized	"	10.50	.762	"	136	31		167	199
4200	Channel slot anchors									
4300	16 gauge, galvanized, 1-1/4" x 3-1/2"				C	52			52	57.50
4350	1-1/4" x 5-1/2"					65			65	72
4400	1-1/4" x 7-1/2"					74.50			74.50	82
4500	1/8" plain, 1-1/4" x 3-1/2"					100			100	110
4550	1-1/4" x 5-1/2"					107			107	117
4600	1-1/4" x 7-1/2"					117			117	128
4700	For corrugation, add					54			54	59
4750	For hot dip galvanized, add					35%				
5000	Dowels									
5100	Plain, 1/4" diameter, 3" long				C	38.50			38.50	42.50
5150	4" long					43			43	47
5200	6" long					52.50			52.50	57.50
5300	3/8" diameter, 3" long					51			51	56
5350	4" long					63.50			63.50	70
5400	6" long					73			73	80.50
5500	1/2" diameter, 3" long					74.50			74.50	82
5550	4" long					88.50			88.50	97
5600	6" long					117			117	129
5700	5/8" diameter, 3" long					102			102	112
5750	4" long					126			126	138
5800	6" long					173			173	190
6000	3/4" diameter, 3" long					127			127	140
6100	4" long					163			163	179
6150	6" long					230			230	254
6300	For hot dip galvanized, add					35%				

04 05 19.26 Masonry Reinforcing Bars		Crew	Daily Output	Labor-Hours	Unit	Material	Labor	Equipment	Total	Total Incl O&P
0010	**MASONRY REINFORCING BARS** R040519-50									
0015	Steel bars A615, placed horiz., #3 & #4 bars	1 Bric	450	.018	Lb.	.78	.72		1.50	2.01
0020	#5 & #6 bars		800	.010		.78	.41		1.19	1.51
0050	Placed vertical, #3 & #4 bars		350	.023		.78	.93		1.71	2.35
0060	#5 & #6 bars		650	.012		.78	.50		1.28	1.66
0500	Joint reinforcing, ladder type, mill std galvanized									
0600	9 ga. sides, 9 ga. ties, 4" wall	1 Bric	30	.267	C.L.F.	11.20	10.80		22	30
0650	6" wall		30	.267		12.75	10.80		23.55	31.50
0700	8" wall		25	.320		13.10	12.95		26.05	35.50
0750	10" wall		20	.400		13.95	16.20		30.15	41.50
0800	12" wall		20	.400		14.75	16.20		30.95	42
1000	Truss type									
1100	9 ga. sides, 9 ga. ties, 4" wall	1 Bric	30	.267	C.L.F.	16.35	10.80		27.15	35.50
1150	6" wall		30	.267		17.25	10.80		28.05	36.50
1200	8" wall		25	.320		17.95	12.95		30.90	41
1250	10" wall		20	.400		18.95	16.20		35.15	47
1300	12" wall		20	.400		20	16.20		36.20	48

04 05 Common Work Results for Masonry

04 05 19 – Masonry Anchorage and Reinforcing

04 05 19.26 Masonry Reinforcing Bars

	04 05 19.26 Masonry Reinforcing Bars	Crew	Daily Output	Labor-Hours	Unit	2009 Bare Costs Material	Labor	Equipment	Total	Total Incl O&P
1500	3/16" sides, 9 ga. ties, 4" wall	1 Bric	30	.267	C.L.F.	21	10.80		31.80	41
1550	6" wall		30	.267		21	10.80		31.80	40.50
1600	8" wall		25	.320		21.50	12.95		34.45	44.50
1650	10" wall		20	.400		22.50	16.20		38.70	50.50
1700	12" wall		20	.400		26	16.20		42.20	54.50
2000	3/16" sides, 3/16" ties, 4" wall		30	.267		24	10.80		34.80	44
2050	6" wall		30	.267		24.50	10.80		35.30	44
2100	8" wall		25	.320		25	12.95		37.95	48.50
2150	10" wall		20	.400		26.50	16.20		42.70	55
2200	12" wall		20	.400		31	16.20		47.20	60
2500	Cavity truss type, galvanized									
2600	9 ga. sides, 9 ga. ties, 4" wall	1 Bric	25	.320	C.L.F.	29.50	12.95		42.45	53.50
2650	6" wall		25	.320		30	12.95		42.95	53.50
2700	8" wall		20	.400		30.50	16.20		46.70	59.50
2750	10" wall		15	.533		32	21.50		53.50	70
2800	12" wall		15	.533		33	21.50		54.50	71.50
3000	3/16" sides, 9 ga. ties, 4" wall		25	.320		38	12.95		50.95	63
3050	6" wall		25	.320		38.50	12.95		51.45	63.50
3100	8" wall		20	.400		39.50	16.20		55.70	69.50
3150	10" wall		15	.533		40	21.50		61.50	79
3200	12" wall		15	.533		42	21.50		63.50	81
3500	For hot dip galvanizing, add					80%				

04 05 23 – Masonry Accessories

04 05 23.13 Control Joint

	04 05 23.13 Control Joint	Crew	Daily Output	Labor-Hours	Unit	Material	Labor	Equipment	Total	Total Incl O&P
0010	**CONTROL JOINT**									
0020	Rubber, for double wythe 8" minimum wall (Brick/CMU)	1 Bric	400	.020	L.F.	1.90	.81		2.71	3.40
0025	"T" shaped		320	.025		1.46	1.01		2.47	3.25
0030	Cross-shaped for CMU units		280	.029		2	1.16		3.16	4.07
0050	PVC, for double wythe 8" minimum wall (Brick/CMU)		400	.020		1.18	.81		1.99	2.61
0120	"T" shaped		320	.025		1.07	1.01		2.08	2.82
0160	Cross-shaped for CMU units		280	.029		1.23	1.16		2.39	3.22

04 05 23.19 Vent Box

	04 05 23.19 Vent Box	Crew	Daily Output	Labor-Hours	Unit	Material	Labor	Equipment	Total	Total Incl O&P
0010	**VENT BOX**									
0020	Extruded aluminum, 4" deep, 2-3/8" x 8-1/8"	1 Bric	30	.267	Ea.	30.50	10.80		41.30	51
0050	5" x 8-1/8"		25	.320		40	12.95		52.95	65
0100	2-1/4" x 25"		25	.320		69.50	12.95		82.45	97.50
0150	5" x 16-1/2"		22	.364		58.50	14.75		73.25	88
0200	6" x 16-1/2"		22	.364		81	14.75		95.75	113
0250	7-3/4" x 16-1/2"		20	.400		69	16.20		85.20	102
0400	For baked enamel finish, add					35%				
0500	For cast aluminum, painted, add					60%				
1000	Stainless steel ventilators, 6" x 6"	1 Bric	25	.320		126	12.95		138.95	159
1050	8" x 8"		24	.333		132	13.50		145.50	168
1100	12" x 12"		23	.348		153	14.10		167.10	191
1150	12" x 6"		24	.333		134	13.50		147.50	169
1200	Foundation block vent, galv., 1-1/4" thk, 8" high, 16" long, no damper		30	.267		24.50	10.80		35.30	44.50
1250	For damper, add					8.20			8.20	9

04 05 23.95 Wall Plugs

	04 05 23.95 Wall Plugs	Crew	Daily Output	Labor-Hours	Unit	Material	Labor	Equipment	Total	Total Incl O&P
0010	**WALL PLUGS** (for nailing to brickwork)									
0020	26 ga., galvanized, plain	1 Bric	10.50	.762	C	31	31		62	84
0050	Wood filled	"	10.50	.762	"	104	31		135	164

04 21 13.13 Brick Veneer Masonry

04 21 13.13 Brick Veneer Masonry		Crew	Daily Output	Labor-Hours	Unit	Material	Labor	Equipment	Total	Total Incl O&P
						2009 Bare Costs				
0010	**BRICK VENEER MASONRY**, T.L. lots, excl. scaff., grout & reinforcing									
0015	Material costs incl. 3% brick and 25% mortar waste									
0020	Standard, select common, 4" x 2-2/3" x 8" (6.75/S.F.)	D-8	1.50	26.667	M	535	990		1,525	2,175
0050	Red, 4" x 2-2/3" x 8", running bond	D-8	1.50	26.667	M	840	990		1,830	2,525
0100	Full header every 6th course (7.88/S.F.)	D-8	1.45	27.586	M	840	1,025		1,865	2,575
0150	English, full header every 2nd course (10.13/S.F.)	D-8	1.40	28.571	M	835	1,050		1,885	2,650
0200	Flemish, alternate header every course (9.00/S.F.)	D-8	1.40	28.571	M	840	1,050		1,890	2,650
0250	Flemish, alt. header every 6th course (7.13/S.F.)	D-8	1.45	27.586	M	840	1,025		1,865	2,575
0300	Full headers throughout (13.50/S.F.)	D-8	1.40	28.571	M	835	1,050		1,885	2,650
0350	Rowlock course (13.50/S.F.)	D-8	1.35	29.630	M	835	1,100		1,935	2,700
0400	Rowlock stretcher (4.50/S.F.)	D-8	1.40	28.571	M	845	1,050		1,895	2,650
0450	Soldier course (6.75/S.F.)	D-8	1.40	28.571	M	840	1,050		1,890	2,650
0500	Sailor course (4.50/S.F.)	D-8	1.30	30.769	M	845	1,150		1,995	2,775
0601	Buff or gray face, running bond, (6.75/S.F.)	D-8	1.50	26.667	M	840	990		1,830	2,525
0700	Glazed face, 4" x 2-2/3" x 8", running bond	D-8	1.40	28.571	M	1,725	1,050		2,775	3,625
0750	Full header every 6th course (7.88/S.F.)	D-8	1.35	29.630	M	1,650	1,100		2,750	3,600
1000	Jumbo, 6" x 4" x 12",(3.00/S.F.)	D-8	1.30	30.769	M	1,775	1,150		2,925	3,800
1051	Norman, 4" x 2-2/3" x 12" (4.50/S.F.)	D-8	1.45	27.586	M	1,100	1,025		2,125	2,875
1100	Norwegian, 4" x 3-1/5" x 12" (3.75/S.F.)	D-8	1.40	28.571	M	1,150	1,050		2,200	2,975
1150	Economy, 4" x 4" x 8" (4.50 per S.F.)	D-8	1.40	28.571	M	940	1,050		1,990	2,750
1201	Engineer, 4" x 3-1/5" x 8", (5.63/S.F.)	D-8	1.45	27.586	M	615	1,025		1,640	2,325
1251	Roman, 4" x 2" x 12", (6.00/S.F.)	D-8	1.50	26.667	M	980	990		1,970	2,675
1300	S.C.R. 6" x 2-2/3" x 12" (4.50/S.F.)	D-8	1.40	28.571	M	1,175	1,050		2,225	3,025
1350	Utility, 4" x 4" x 12" (3.00/S.F.)	D-8	1.35	29.630	M	1,500	1,100		2,600	3,425
1360	For less than truck load lots, add				M	15			15	16.50
1400	For battered walls, add						30%			
1450	For corbels, add						75%			
1500	For curved walls, add						30%			
1550	For pits and trenches, deduct						20%			
1999	Alternate method of figuring by square foot									
2000	Standard, sel. common, 4" x 2-2/3" x 8", (6.75/S.F.)	D-8	230	.174	S.F.	3.60	6.45		10.05	14.40
2020	Standard, red, 4" x 2-2/3" x 8", running bond (6.75/SF)	D-8	220	.182	S.F.	5.65	6.75		12.40	17.20
2050	Full header every 6th course (7.88/S.F.)	D-8	185	.216	S.F.	6.60	8.05		14.65	20.50
2100	English, full header every 2nd course (10.13/S.F.)	D-8	140	.286	S.F.	8.50	10.60		19.10	26.50
2150	Flemish, alternate header every course (9.00/S.F.)	D-8	150	.267	S.F.	7.55	9.90		17.45	24.50
2200	Flemish, alt. header every 6th course (7.13/S.F.)	D-8	205	.195	S.F.	6	7.25		13.25	18.35
2250	Full headers throughout (13.50/S.F.)	D-8	105	.381	S.F.	11.30	14.15		25.45	35.50
2300	Rowlock course (13.50/S.F.)	D-8	100	.400	S.F.	11.30	14.85		26.15	36.50
2350	Rowlock stretcher (4.50/S.F.)	D-8	310	.129	S.F.	3.80	4.79		8.59	11.95
2400	Soldier course (6.75/S.F.)	D-8	200	.200	S.F.	5.65	7.45		13.10	18.25
2450	Sailor course (4.50/S.F.)	D-8	290	.138	S.F.	3.80	5.15		8.95	12.50
2600	Buff or gray face, running bond, (6.75/S.F.)	D-8	220	.182	S.F.	6	6.75		12.75	17.55
2700	Glazed face brick, running bond	D-8	210	.190	S.F.	11.15	7.10		18.25	23.50
2750	Full header every 6th course (7.88/S.F.)	D-8	170	.235	S.F.	13	8.75		21.75	28.50
3000	Jumbo, 6" x 4" x 12" running bond (3.00/S.F.)	D-8	435	.092	S.F.	4.86	3.42		8.28	10.90
3050	Norman, 4" x 2-2/3" x 12" running bond, (4.5/S.F.)	D-8	320	.125	S.F.	5.65	4.65		10.30	13.75
3100	Norwegian, 4" x 3-1/5" x 12" (3.75/S.F.)	D-8	375	.107	S.F.	4.14	3.96		8.10	10.95
3150	Economy, 4" x 4" x 8" (4.50/S.F.)	D-8	310	.129	S.F.	4.18	4.79		8.97	12.35
3200	Engineer, 4" x 3-1/5" x 8" (5.63/S.F.)	D-8	260	.154	S.F.	3.46	5.70		9.16	13.05
3250	Roman, 4" x 2" x 12" (6.00/S.F.)	D-8	250	.160	S.F.	5.75	5.95		11.70	15.90
3300	SCR, 6" x 2-2/3" x 12" (4.50/S.F.)	D-8	310	.129	S.F.	5.25	4.79		10.04	13.50
3350	Utility, 4" x 4" x 12" (3.00/S.F.)	D-8	450	.089	S.F.	4.40	3.30		7.70	10.20
3400	For cavity wall construction, add						15%			

04 21 Clay Unit Masonry

04 21 13 – Brick Masonry

04 21 13.13 Brick Veneer Masonry		Crew	Daily Output	Labor-Hours	Unit	2009 Bare Costs Material	Labor	Equipment	Total	Total Incl O&P
3450	For stacked bond, add						10%			
3500	For interior veneer construction, add						15%			
3550	For curved walls, add						30%			
9000	Minimum labor/equipment charge	D-1	2	8	Job		291		291	470

04 21 13.15 Chimney

		Crew	Daily Output	Labor-Hours	Unit	Material	Labor	Equipment	Total	Total Incl O&P
0010	**CHIMNEY**, excludes foundation, scaffolding, grout and reinforcing									
0100	Brick, 16" x 16", 8" flue	D-1	18.20	.879	V.L.F.	20.50	32		52.50	74.50
0150	16" x 20" with one 8" x 12" flue		16	1		32.50	36.50		69	95
0200	16" x 24" with two 8" x 8" flues		14	1.143		47.50	41.50		89	120
0250	20" x 20" with one 12" x 12" flue		13.70	1.168		38.50	42.50		81	111
0300	20" x 24" with two 8" x 12" flues		12	1.333		54	48.50		102.50	138
0350	20" x 32" with two 12" x 12" flues		10	1.600		67.50	58		125.50	169
1800	Metal, high temp. steel jacket, factory lining, 24" diam.	E-2	65	.862		210	37.50	27	274.50	330
1900	60" diameter	"	30	1.867		765	81.50	58	904.50	1,050
2100	Poured concrete, brick lining, 200' high x 10' diam.					7,275			7,275	8,000
2800	500' x 20' diameter					12,700			12,700	14,000

04 21 13.18 Columns

		Crew	Daily Output	Labor-Hours	Unit	Material	Labor	Equipment	Total	Total Incl O&P
0010	**COLUMNS**, solid, excludes scaffolding, grout and reinforcing									
0050	Brick, 8" x 8", 9 brick per VLF	D-1	56	.286	V.L.F.	7.35	10.40		17.75	25
0100	12" x 8", 13.5 brick per VLF		37	.432		11.05	15.70		26.75	37.50
0200	12" x 12", 20 brick per VLF		25	.640		16.35	23.50		39.85	55.50
0300	16" x 12", 27 brick per VLF		19	.842		22	30.50		52.50	74
0400	16" x 16", 36 brick per VLF		14	1.143		29.50	41.50		71	99.50
0500	20" x 16", 45 brick per VLF		11	1.455		37	53		90	126
0600	20" x 20", 56 brick per VLF		9	1.778		46	64.50		110.50	155
0700	24" x 20", 68 brick per VLF		7	2.286		55.50	83		138.50	195
0800	24" x 24", 81 brick per VLF		6	2.667		66	97		163	230
1000	36" x 36", 182 brick per VLF		3	5.333		149	194		343	480
9000	Minimum labor/equipment charge		2	8	Job		291		291	470

04 21 13.30 Oversized Brick

		Crew	Daily Output	Labor-Hours	Unit	Material	Labor	Equipment	Total	Total Incl O&P
0010	**OVERSIZED BRICK**, excludes scaffolding, grout and reinforcing									
0100	Veneer, 4" x 2.25" x 16"	D-8	387	.103	S.F.	5.05	3.84		8.89	11.75
0105	4" x 2.75" x 16"		412	.097		4.55	3.61		8.16	10.85
0110	4" x 4" x 16"		460	.087		4.48	3.23		7.71	10.20
0120	4" x 8" x 16"		533	.075		5.50	2.79		8.29	10.55
0125	Loadbearing, 6" x 4" x 16", grouted and reinforced		387	.103		8.10	3.84		11.94	15.10
0130	8" x 4" x 16", grouted and reinforced		327	.122		8.25	4.55		12.80	16.40
0135	6" x 8" x 16", grouted and reinforced		440	.091		7.45	3.38		10.83	13.65
0140	8" x 8" x 16", grouted and reinforced		400	.100		8.35	3.72		12.07	15.20
0145	Curtainwall / reinforced veneer, 6" x 4" x 16"		387	.103		12.05	3.84		15.89	19.45
0150	8" x 4" x 16"		327	.122		14.55	4.55		19.10	23.50
0155	6" x 8" x 16"		440	.091		12.15	3.38		15.53	18.80
0160	8" x 8" x 16"		400	.100		14.65	3.72		18.37	22
0200	For 1 to 3 slots in face, add					25%				
0210	For 4 to 7 slots in face, add					15%				
0220	For bond beams, add					20%				
0230	For bullnose shapes, add					20%				
0240	For open end knockout, add					10%				
0250	For white or gray color group, add					10%				
0260	For 135 degree corner, add					250%				

04 21 13.35 Common Building Brick

		Crew	Daily Output	Labor-Hours	Unit	Material	Labor	Equipment	Total	Total Incl O&P
0010	**COMMON BUILDING BRICK**, C62, TL lots, material only R042110-20									

04 21 Clay Unit Masonry

04 21 13 – Brick Masonry

04 21 13.35 Common Building Brick		Crew	Daily Output	Labor-Hours	Unit	2009 Bare Costs Material	Labor	Equipment	Total	Total Incl O&P
0020	Standard, minimum				M	340			340	375
0050	Average (select)				"	430			430	475

04 21 13.40 Structural Brick		Crew	Daily Output	Labor-Hours	Unit	Material	Labor	Equipment	Total	Total Incl O&P
0010	**STRUCTURAL BRICK** C652, Grade SW, incl. mortar, scaffolding not incl.									
0100	Standard unit, 4-5/8" x 2-3/4" x 9-5/8"	D-8	245	.163	S.F.	5.15	6.05		11.20	15.50
0120	Bond beam		225	.178		9.65	6.60		16.25	21.50
0140	V cut bond beam		225	.178		10.45	6.60		17.05	22
0160	Stretcher quoin, 5-5/8" x 2-3/4" x 9-5/8"		245	.163		10.10	6.05		16.15	21
0180	Corner quoin		245	.163		11.30	6.05		17.35	22.50
0200	Corner, 45 deg, 4-5/8" x 2-3/4" x 10-7/16"		235	.170		11.70	6.35		18.05	23

04 21 13.45 Face Brick		Crew	Daily Output	Labor-Hours	Unit	Material	Labor	Equipment	Total	Total Incl O&P
0010	**FACE BRICK** Material Only, C216, TL lots R042110-20									
0300	Standard modular, 4" x 2-2/3" x 8", minimum				M	730			730	805
0350	Maximum R042110-50					835			835	920
0450	Economy, 4" x 4" x 8", minimum					815			815	895
0500	Maximum					1,050			1,050	1,150
0510	Economy, 4" x 4" x 12", minimum					1,200			1,200	1,325
0520	Maximum					1,550			1,550	1,725
0550	Jumbo, 6" x 4" x 12", minimum					1,475			1,475	1,625
0600	Maximum					1,875			1,875	2,075
0610	Jumbo, 8" x 4" x 12", minimum					1,475			1,475	1,625
0620	Maximum					1,875			1,875	2,075
0650	Norwegian, 4" x 3-1/5" x 12", minimum					980			980	1,075
0700	Maximum					1,275			1,275	1,400
0710	Norwegian, 6" x 3-1/5" x 12", minimum					1,425			1,425	1,575
0720	Maximum					1,875			1,875	2,050
0850	Standard glazed, plain colors, 4" x 2-2/3" x 8", minimum					1,525			1,525	1,675
0900	Maximum					1,975			1,975	2,175
1000	Deep trim shades, 4" x 2-2/3" x 8", minimum					1,600			1,600	1,750
1050	Maximum					1,900			1,900	2,100
1080	Jumbo utility, 4" x 4" x 12"					1,325			1,325	1,450
1120	4" x 8" x 8"					1,550			1,550	1,700
1140	4" x 8" x 16"					3,100			3,100	3,400
1260	Engineer, 4" x 3-1/5" x 8", minimum					510			510	560
1270	Maximum					625			625	690
1350	King, 4" x 2-3/4" x 10", minimum					525			525	575
1360	Maximum					560			560	620
1400	Norman, 4" x 2-3/4" x 12"					525			525	575
1450	Roman, 4" x 2" x 12"					525			525	575
1500	SCR, 6" x 2-2/3" x 12"					525			525	575
1550	Double, 4" x 5-1/3" x 8"					525			525	575
1600	Triple, 4" x 5-1/3" x 12"					525			525	575
1770	Standard modular, double glazed, 4" x 2-2/3" x 8"					1,775			1,775	1,950
1850	Jumbo, colored glazed ceramic, 6" x 4" x 12"					2,400			2,400	2,650
2050	Jumbo utility, glazed, 4" x 4" x 12"					4,050			4,050	4,450
2100	4" x 8" x 8"					4,725			4,725	5,200
2150	4" x 16" x 8"					6,200			6,200	6,825
2170	For less than truck load lots, add					15			15	16.50
2180	For buff or gray brick, add					16			16	17.60
3050	Used brick, minimum					395			395	430
3100	Maximum					375			375	415
3150	Add for brick to match existing work, minimum					5%				

04 21 Clay Unit Masonry

04 21 13 – Brick Masonry

04 21 13.45 Face Brick		Crew	Daily Output	Labor-Hours	Unit	2009 Bare Costs Material	Labor	Equipment	Total	Total Incl O&P
3200	Maximum				M	50%				

04 21 26 – Glazed Structural Clay Tile Masonry

04 21 26.10 Structural Facing Tile

		Crew	Daily Output	Labor-Hours	Unit	Material	Labor	Equipment	Total	Total Incl O&P
0010	**STRUCTURAL FACING TILE**, std. colors, excl. scaffolding, grout, reinforcing									
0020	6T series, 5-1/3" x 12", 2.3 pieces per S.F., glazed 1 side, 2" thick	D-8	225	.178	S.F.	8.45	6.60		15.05	20
0100	4" thick		220	.182		11.05	6.75		17.80	23
0150	Glazed 2 sides		195	.205		14.15	7.60		21.75	28
0250	6" thick		210	.190		16.60	7.10		23.70	29.50
0300	Glazed 2 sides		185	.216		19.45	8.05		27.50	34.50
0400	8" thick		180	.222		22	8.25		30.25	37.50
0500	Special shapes, group 1		400	.100	Ea.	6.85	3.72		10.57	13.55
0550	Group 2		375	.107		10.95	3.96		14.91	18.45
0600	Group 3		350	.114		14.45	4.25		18.70	23
0650	Group 4		325	.123		30	4.57		34.57	40.50
0700	Group 5		300	.133		35.50	4.95		40.45	47.50
0750	Group 6		275	.145		49	5.40		54.40	63
1000	Fire rated, 4" thick, 1 hr. rating		210	.190	S.F.	15.95	7.10		23.05	29
1300	Acoustic, 4" thick		210	.190	"	34	7.10		41.10	49
1400	For designer colors, add					25%				
2000	8W series, 8" x 16", 1.125 pieces per S.F.									
2050	2" thick, glazed 1 side	D-8	360	.111	S.F.	9.35	4.13		13.48	16.95
2100	4" thick, glazed 1 side		345	.116		12	4.31		16.31	20
2150	Glazed 2 sides		325	.123		13.45	4.57		18.02	22
2200	6" thick, glazed 1 side		330	.121		21	4.50		25.50	31
2250	8" thick, glazed 1 side		310	.129		21.50	4.79		26.29	31.50
2500	Special shapes, group 1		300	.133	Ea.	12.95	4.95		17.90	22.50
2550	Group 2		280	.143		17.40	5.30		22.70	27.50
2600	Group 3		260	.154		18.55	5.70		24.25	30
2650	Group 4		250	.160		37	5.95		42.95	50
2700	Group 5		240	.167		36	6.20		42.20	49.50
2750	Group 6		230	.174		77.50	6.45		83.95	96
3000	4" thick, glazed 1 side		345	.116	S.F.	11.10	4.31		15.41	19.15
3100	Acoustic, 4" thick		345	.116	"	13.95	4.31		18.26	22.50
3120	4W series, 8" x 8", 2.25 pieces per S.F.									
3125	2" thick, glazed 1 side	D-8	360	.111	S.F.	9.75	4.13		13.88	17.40
3130	4" thick, glazed 1 side		345	.116		10.60	4.31		14.91	18.60
3135	Glazed 2 sides		325	.123		14.45	4.57		19.02	23.50
3140	6" thick, glazed 1 side		330	.121		14.90	4.50		19.40	23.50
3150	8" thick, glazed 1 side		310	.129		21.50	4.79		26.29	32
3155	Special shapes, group I		300	.133	Ea.	6.95	4.95		11.90	15.65
3160	Group II		280	.143	"	7.90	5.30		13.20	17.30
3200	For designer colors, add					25%				
3300	For epoxy mortar joints, add				S.F.	1.58			1.58	1.74
9000	Minimum labor/equipment charge	D-1	2	8	Job		291		291	470

04 21 29 – Terra Cotta Masonry

04 21 29.10 Terra Cotta Masonry Components

		Crew	Daily Output	Labor-Hours	Unit	Material	Labor	Equipment	Total	Total Incl O&P
0010	**TERRA COTTA MASONRY COMPONENTS**									
0020	Coping, split type, not glazed, 9" wide	D-1	90	.178	L.F.	10.40	6.45		16.85	22
0100	13" wide		80	.200		15.75	7.25		23	29
0200	Coping, split type, glazed, 9" wide		90	.178		17.40	6.45		23.85	29.50
0250	13" wide		80	.200		22.50	7.25		29.75	37
0500	Partition or back-up blocks, scored, in C.L. lots									

04 21 Clay Unit Masonry

04 21 29 – Terra Cotta Masonry

04 21 29.10 Terra Cotta Masonry Components		Crew	Daily Output	Labor-Hours	Unit	2009 Bare Costs Material	Labor	Equipment	Total	Total Incl O&P
0700	Non-load bearing 12" x 12", 3" thick, special order	D-8	550	.073	S.F.	19.30	2.70		22	25.50
0750	4" thick, standard		500	.080		7.10	2.97		10.07	12.60
0800	6" thick		450	.089		8.70	3.30		12	14.90
0850	8" thick		400	.100		11.05	3.72		14.77	18.20
1000	Load bearing, 12" x 12", 4" thick, in walls		500	.080		6.95	2.97		9.92	12.45
1050	In floors		750	.053		6.95	1.98		8.93	10.85
1200	6" thick, in walls		450	.089		8.70	3.30		12	14.90
1250	In floors		675	.059		8.70	2.20		10.90	13.10
1400	8" thick, in walls		400	.100		11.05	3.72		14.77	18.20
1450	In floors		575	.070		11.05	2.59		13.64	16.40
1600	10" thick, in walls, special order		350	.114		14.50	4.25		18.75	23
1650	In floors, special order		500	.080		14.50	2.97		17.47	21
1800	12" thick, in walls, special order		300	.133		28	4.95		32.95	38.50
1850	In floors, special order		450	.089		28	3.30		31.30	36
2000	For reinforcing with steel rods, add to above					15%	5%			
2100	For smooth tile instead of scored, add					3.74			3.74	4.11
2200	For L.C.L. quantities, add					10%	10%			
9000	Minimum labor/equipment charge	D-1	2	8	Job		291		291	470

04 21 29.20 Terra Cotta Tile		Crew	Daily Output	Labor-Hours	Unit	Material	Labor	Equipment	Total	Total Incl O&P
0010	**TERRA COTTA TILE**, on walls, dry set, 1/2" thick									
0100	Square, hexagonal or lattice shapes, unglazed	1 Tilf	135	.059	S.F.	6.35	2.26		8.61	10.50
0300	Glazed, plain colors		130	.062		8.75	2.34		11.09	13.25
0400	Intense colors		125	.064		10.20	2.44		12.64	15

04 22 Concrete Unit Masonry

04 22 10 – Concrete Masonry Units

04 22 10.10 Concrete Block		Crew	Daily Output	Labor-Hours	Unit	Material	Labor	Equipment	Total	Total Incl O&P
0010	**CONCRETE BLOCK** Material Only R042210-20									
0020	2" x 8" x 16" solid, normal-weight, 2,000 psi				Ea.	.93			.93	1.02
0050	3,500 psi R042110-50					.88			.88	.97
0100	5,000 psi					1.09			1.09	1.20
0150	Lightweight, std.					1.06			1.06	1.17
0300	3" x 8" x 16" solid, normal-weight, 2000 psi					1.12			1.12	1.23
0350	3,500 psi					1.55			1.55	1.71
0400	5,000 psi					1.65			1.65	1.82
0450	Lightweight, std.					.94			.94	1.03
0600	4" x 8" x 16" hollow, normal-weight, 2000 psi					.97			.97	1.07
0650	3,500 psi					1.04			1.04	1.14
0700	5000 psi					1.24			1.24	1.36
0750	Lightweight, std.					1.22			1.22	1.34
1300	Solid, normal-weight, 2,000 psi					1.29			1.29	1.42
1350	3,500 psi					1.57			1.57	1.73
1400	5,000 psi					1.74			1.74	1.91
1450	Lightweight, std.					1.74			1.74	1.91
1600	6" x 8" x 16" hollow, normal-weight, 2,000 psi					1.39			1.39	1.53
1650	3,500 psi					1.48			1.48	1.63
1700	5,000 psi					1.60			1.60	1.76
1750	Lightweight, std.					1.65			1.65	1.82
2300	Solid, normal-weight, 2,000 psi					1.38			1.38	1.52
2350	3,500 psi					1.71			1.71	1.88
2400	5,000 psi					2.15			2.15	2.37

04 22 10.10 Concrete Block

	04 22 10.10 Concrete Block	Crew	Daily Output	Labor-Hours	Unit	2009 Bare Costs Material	Labor	Equipment	Total	Total Incl O&P
2450	Lightweight, std.				Ea.	1.94			1.94	2.13
2600	8" x 8" x 16" hollow, normal-weight, 2000 psi					1.37			1.37	1.51
2650	3500 psi					1.64			1.64	1.80
2700	5,000 psi					1.95			1.95	2.15
2750	Lightweight, std.					2			2	2.20
3200	Solid, normal-weight, 2,000 psi					2.30			2.30	2.53
3250	3,500 psi					2.62			2.62	2.88
3300	5,000 psi					2.97			2.97	3.27
3350	Lightweight, std.					2.63			2.63	2.89
3400	10" x 8" x 16" hollow, normal-weight, 2000 psi					1.37			1.37	1.51
3410	3500 psi					1.64			1.64	1.80
3420	5,000 psi					1.95			1.95	2.15
3430	Lightweight, std.					2			2	2.20
3480	Solid, normal-weight, 2,000 psi					2.30			2.30	2.53
3490	3,500 psi					2.62			2.62	2.88
3500	5,000 psi					2.97			2.97	3.27
3510	Lightweight, std.					2.63			2.63	2.89
3600	12" x 8" x 16" hollow, normal-weight, 2000 psi					2.22			2.22	2.44
3650	3,500 psi					2.23			2.23	2.45
3700	5,000 psi					2.62			2.62	2.88
3750	Lightweight, std.					3.02			3.02	3.32
4300	Solid, normal-weight, 2,000 psi					3.55			3.55	3.91
4350	3,500 psi					3.15			3.15	3.47
4400	5,000 psi					4.47			4.47	4.92
4500	Lightweight, std.				↓	4.75			4.75	5.25

04 22 10.11 Autoclave Aerated Concrete Block

	04 22 10.11 Autoclave Aerated Concrete Block	Crew	Daily Output	Labor-Hours	Unit	Material	Labor	Equipment	Total	Total Incl O&P
0010	**AUTOCLAVE AERATED CONCRETE BLOCK**, excl. scaffolding, grout & reinforcing									
0050	Solid, 4" x 12" x 24", incl mortar G	D-8	600	.067	S.F.	1.60	2.48		4.08	5.75
0060	6" x 12" x 24" R042110-50 G		600	.067		2.17	2.48		4.65	6.40
0070	8" x 8" x 24" G		575	.070		2.80	2.59		5.39	7.25
0080	10" x 12" x 24" G		575	.070		3.59	2.59		6.18	8.15
0090	12" x 12" x 24" G	↓	550	.073	↓	4.23	2.70		6.93	9

04 22 10.12 Chimney Block

	04 22 10.12 Chimney Block	Crew	Daily Output	Labor-Hours	Unit	Material	Labor	Equipment	Total	Total Incl O&P
0010	**CHIMNEY BLOCK**, excludes scaffolding, grout and reinforcing									
0220	1 piece, with 8" x 8" flue, 16" x 16"	D-1	28	.571	V.L.F.	16.95	21		37.95	52
0230	2 piece, 16" x 16"		26	.615		14.45	22.50		36.95	52
0240	2 piece, with 8" x 12" flue, 16" x 20"	↓	24	.667	↓	28.50	24		52.50	70.50

04 22 10.14 Concrete Block, Back-Up

	04 22 10.14 Concrete Block, Back-Up	Crew	Daily Output	Labor-Hours	Unit	Material	Labor	Equipment	Total	Total Incl O&P
0010	**CONCRETE BLOCK, BACK-UP**, C90, 2000 psi R042210-20									
0020	Normal weight, 8" x 16" units, tooled joint 1 side									
0050	Not-reinforced, 2000 psi, 2" thick R042110-50	D-8	475	.084	S.F.	1.36	3.13		4.49	6.55
0200	4" thick		460	.087		1.52	3.23		4.75	6.90
0300	6" thick		440	.091		2.12	3.38		5.50	7.80
0350	8" thick		400	.100		2.23	3.72		5.95	8.45
0400	10" thick	↓	330	.121		3.03	4.50		7.53	10.65
0450	12" thick	D-9	310	.155		3.44	5.65		9.09	12.90
1000	Reinforced, alternate courses, 4" thick	D-8	450	.089		1.64	3.30		4.94	7.15
1100	6" thick		430	.093		2.25	3.46		5.71	8.10
1150	8" thick		395	.101		2.37	3.76		6.13	8.70
1200	10" thick	↓	320	.125		3.18	4.65		7.83	11
1250	12" thick	D-9	300	.160		3.59	5.80		9.39	13.35
2000	Lightweight, not reinforced, 4" thick	D-8	460	.087	↓	1.80	3.23		5.03	7.25

04 22 Concrete Unit Masonry

04 22 10 – Concrete Masonry Units

04 22 10.14 Concrete Block, Back-Up		Crew	Daily Output	Labor-Hours	Unit	Material	Labor	Equipment	Total	Total Incl O&P
						2009 Bare Costs				
2100	6" thick	D-8	445	.090	S.F.	2.42	3.34		5.76	8.05
2150	8" thick		435	.092		2.94	3.42		6.36	8.80
2200	10" thick		410	.098		3.81	3.63		7.44	10.05
2250	12" thick	D-9	390	.123		4.35	4.47		8.82	12.05
3000	Reinforced, alternate courses, 4" thick	D-8	450	.089		1.93	3.30		5.23	7.45
3100	6" thick		430	.093		2.54	3.46		6	8.40
3150	8" thick		420	.095		3.08	3.54		6.62	9.10
3200	10" thick		400	.100		3.96	3.72		7.68	10.35
3250	12" thick	D-9	380	.126		4.50	4.59		9.09	12.35
9000	Minimum labor/equipment charge	D-1	2	8	Job		291		291	470

04 22 10.16 Concrete Block, Bond Beam		Crew	Daily Output	Labor-Hours	Unit	Material	Labor	Equipment	Total	Total Incl O&P
0010	**CONCRETE BLOCK, BOND BEAM**, C90, 2000 psi									
0020	Not including grout or reinforcing									
0125	Regular block, 6" thick	D-8	584	.068	L.F.	2.21	2.55		4.76	6.55
0130	8" high, 8" thick	"	565	.071		2.51	2.63		5.14	7
0150	12" thick	D-9	510	.094		3.43	3.42		6.85	9.35
0525	Lightweight, 6" thick	D-8	592	.068		2.53	2.51		5.04	6.85
0530	8" high, 8" thick	"	575	.070		3.06	2.59		5.65	7.55
0550	12" thick	D-9	520	.092		4.12	3.35		7.47	9.95
2000	Including grout and 2 #5 bars									
2100	Regular block, 8" high, 8" thick	D-8	300	.133	L.F.	5.15	4.95		10.10	13.65
2150	12" thick	D-9	250	.192		6.60	7		13.60	18.60
2500	Lightweight, 8" high, 8" thick	D-8	305	.131		5.70	4.87		10.57	14.15
2550	12" thick	D-9	255	.188		7.30	6.85		14.15	19.10
9000	Minimum labor/equipment charge	D-1	2	8	Job		291		291	470

04 22 10.18 Concrete Block, Column		Crew	Daily Output	Labor-Hours	Unit	Material	Labor	Equipment	Total	Total Incl O&P
0010	**CONCRETE BLOCK, COLUMN** or pilaster									
0050	Including vertical reinforcing (4-#4 bars) and grout									
0160	1 piece unit, 16" x 16"	D-1	26	.615	V.L.F.	18.65	22.50		41.15	56.50
0170	2 piece units, 16" x 20"		24	.667		24.50	24		48.50	66
0180	20" x 20"		22	.727		35.50	26.50		62	81.50
0190	22" x 24"		18	.889		49.50	32.50		82	107
0200	20" x 32"		14	1.143		53.50	41.50		95	126

04 22 10.19 Concrete Block, Insulation Inserts		Crew	Daily Output	Labor-Hours	Unit	Material	Labor	Equipment	Total	Total Incl O&P
0010	**CONCRETE BLOCK, INSULATION INSERTS**									
0100	Styrofoam, plant installed, add to block prices									
0200	8" x 16" units, 6" thick				S.F.	1.81			1.81	1.99
0250	8" thick					1.81			1.81	1.99
0300	10" thick					2.13			2.13	2.34
0350	12" thick					2.24			2.24	2.46
0500	8" x 8" units, 8" thick					1.49			1.49	1.64
0550	12" thick					1.81			1.81	1.99

04 22 10.23 Concrete Block, Decorative		Crew	Daily Output	Labor-Hours	Unit	Material	Labor	Equipment	Total	Total Incl O&P
0010	**CONCRETE BLOCK, DECORATIVE**, C90, 2000 psi									
0020	Embossed, simulated brick face									
0100	8" x 16" units, 4" thick	D-8	400	.100	S.F.	3.51	3.72		7.23	9.85
0200	8" thick		340	.118		4.83	4.37		9.20	12.35
0250	12" thick		300	.133		6.35	4.95		11.30	15
0400	Embossed both sides									
0500	8" thick	D-8	300	.133	S.F.	5.40	4.95		10.35	13.95
0550	12" thick	"	275	.145	"	6.85	5.40		12.25	16.25
1000	Fluted high strength									

04 22 10 – Concrete Masonry Units

04 22 10.23 Concrete Block, Decorative		Crew	Daily Output	Labor-Hours	Unit	2009 Bare Costs Material	Labor	Equipment	Total	Total Incl O&P
1100	8" x 16" x 4" thick, flutes 1 side,	D-8	345	.116	S.F.	4.16	4.31		8.47	11.55
1150	Flutes 2 sides		335	.119		5.05	4.44		9.49	12.75
1200	8" thick	↓	300	.133		6.55	4.95		11.50	15.20
1250	For special colors, add				↓	.41			.41	.45
1400	Deep grooved, smooth face									
1450	8" x 16" x 4" thick	D-8	345	.116	S.F.	2.71	4.31		7.02	9.95
1500	8" thick	"	300	.133	"	4.67	4.95		9.62	13.15
2000	Formblock, incl. inserts & reinforcing									
2100	8" x 16" x 8" thick	D-8	345	.116	S.F.	4.75	4.31		9.06	12.15
2150	12" thick	"	310	.129	"	5.90	4.79		10.69	14.25
2500	Ground face									
2600	8" x 16" x 4" thick	D-8	345	.116	S.F.	4.17	4.31		8.48	11.55
2650	6" thick		325	.123		5.20	4.57		9.77	13.10
2700	8" thick	↓	300	.133		6.75	4.95		11.70	15.40
2750	12" thick	D-9	265	.181	↓	7.20	6.60		13.80	18.55
2900	For special colors, add, minimum					15%				
2950	For special colors, add, maximum					45%				
4000	Slump block									
4100	4" face height x 16" x 4" thick	D-1	165	.097	S.F.	3.95	3.52		7.47	10.05
4150	6" thick		160	.100		5.40	3.63		9.03	11.85
4200	8" thick		155	.103		6.05	3.75		9.80	12.70
4250	10" thick		140	.114		10.90	4.15		15.05	18.70
4300	12" thick		130	.123		11.40	4.47		15.87	19.80
4400	6" face height x 16" x 6" thick		155	.103		4.74	3.75		8.49	11.25
4450	8" thick		150	.107		6.50	3.88		10.38	13.40
4500	10" thick		130	.123		9.60	4.47		14.07	17.85
4550	12" thick	↓	120	.133	↓	10.35	4.84		15.19	19.25
5000	Split rib profile units, 1" deep ribs, 8 ribs									
5100	8" x 16" x 4" thick	D-8	345	.116	S.F.	3.15	4.31		7.46	10.40
5150	6" thick		325	.123		3.62	4.57		8.19	11.40
5200	8" thick	↓	300	.133		4.13	4.95		9.08	12.55
5250	12" thick	D-9	275	.175		4.91	6.35		11.26	15.65
5400	For special deeper colors, 4" thick, add					1.12			1.12	1.23
5450	12" thick, add					.96			.96	1.06
5600	For white, 4" thick, add					1.12			1.12	1.23
5650	6" thick, add					1.12			1.12	1.23
5700	8" thick, add					1.04			1.04	1.14
5750	12" thick, add				↓	.96			.96	1.06
6000	Split face									
6100	8" x 16" x 4" thick	D-8	350	.114	S.F.	2.82	4.25		7.07	9.95
6150	6" thick		325	.123		3.27	4.57		7.84	11
6200	8" thick	↓	300	.133		3.70	4.95		8.65	12.05
6250	12" thick	D-9	270	.178		4.56	6.45		11.01	15.45
6300	For scored, add					.32			.32	.35
6400	For special deeper colors, 4" thick, add					.64			.64	.71
6450	6" thick, add					.64			.64	.71
6500	8" thick, add					.58			.58	.63
6550	12" thick, add					.49			.49	.53
6650	For white, 4" thick, add					1.12			1.12	1.23
6700	6" thick, add					1.12			1.12	1.23
6750	8" thick, add					1.04			1.04	1.14
6800	12" thick, add				↓	.96			.96	1.06
7000	Scored ground face, 2 to 5 scores									

04 22 Concrete Unit Masonry

04 22 10 – Concrete Masonry Units

04 22 10.23 Concrete Block, Decorative		Crew	Daily Output	Labor-Hours	Unit	2009 Bare Costs Material	Labor	Equipment	Total	Total Incl O&P
7100	8" x 16" x 4" thick	D-8	340	.118	S.F.	4.90	4.37		9.27	12.45
7150	6" thick		310	.129		5.60	4.79		10.39	13.90
7200	8" thick	↓	290	.138		6.60	5.15		11.75	15.55
7250	12" thick	D-9	265	.181	↓	8.20	6.60		14.80	19.65
8000	Hexagonal face profile units, 8" x 16" units									
8100	4" thick, hollow	D-8	340	.118	S.F.	3.12	4.37		7.49	10.50
8200	Solid		340	.118		4.03	4.37		8.40	11.50
8300	6" thick, hollow		310	.129		3.49	4.79		8.28	11.60
8350	8" thick, hollow	↓	290	.138	↓	4.73	5.15		9.88	13.50
8500	For stacked bond, add						26%			
8550	For high rise construction, add per story	D-8	67.80	.590	M.S.F.		22		22	35.50
8600	For scored block, add					10%				
8650	For honed or ground face, per face, add				Ea.	.38			.38	.42
8700	For honed or ground end, per end, add				"	2.98			2.98	3.28
8750	For bullnose block, add					10%				
8800	For special color, add					13%				
9000	Minimum labor/equipment charge	D-1	2	8	Job		291		291	470

04 22 10.24 Concrete Block, Exterior		Crew	Daily Output	Labor-Hours	Unit	Material	Labor	Equipment	Total	Total Incl O&P
0010	**CONCRETE BLOCK, EXTERIOR**, C90, 2000 psi									
0020	Reinforced alt courses, tooled joints 2 sides									
0100	Normal weight, 8" x 16" x 6" thick	D-8	395	.101	S.F.	2.49	3.76		6.25	8.85
0200	8" thick		360	.111		3.65	4.13		7.78	10.70
0250	10" thick	↓	290	.138		4.27	5.15		9.42	13
0300	12" thick	D-9	250	.192		4.52	7		11.52	16.25
0500	Lightweight, 8" x 16" x 6" thick	D-8	450	.089		2.75	3.30		6.05	8.40
0600	8" thick		430	.093		3.67	3.46		7.13	9.65
0650	10" thick	↓	395	.101		4.31	3.76		8.07	10.85
0700	12" thick	D-9	350	.137	↓	6.30	4.98		11.28	15
9000	Minimum labor/equipment charge	D-1	2	8	Job		291		291	470

04 22 10.26 Concrete Block Foundation Wall		Crew	Daily Output	Labor-Hours	Unit	Material	Labor	Equipment	Total	Total Incl O&P
0010	**CONCRETE BLOCK FOUNDATION WALL**, C90 / C145									
0050	Normal-weight, cut joints, horiz joint reinf, no vert reinf									
0200	Hollow, 8" x 16" x 6" thick	D-8	455	.088	S.F.	2.56	3.27		5.83	8.10
0250	8" thick		425	.094		2.68	3.50		6.18	8.60
0300	10" thick	↓	350	.114		3.49	4.25		7.74	10.70
0350	12" thick	D-9	300	.160		3.91	5.80		9.71	13.70
0500	Solid, 8" x 16" block, 6" thick	D-8	440	.091		2.55	3.38		5.93	8.25
0550	8" thick	"	415	.096		3.73	3.58		7.31	9.90
0600	12" thick	D-9	350	.137	↓	5.40	4.98		10.38	14
1000	Reinforced, #4 vert @ 48"									
1100	Hollow, 8" x 16" block, 4" thick	D-8	455	.088	S.F.	2.53	3.27		5.80	8.10
1125	6" thick		445	.090		3.46	3.34		6.80	9.20
1150	8" thick		415	.096		3.93	3.58		7.51	10.10
1200	10" thick	↓	340	.118		5.10	4.37		9.47	12.65
1250	12" thick	D-9	290	.166		5.85	6		11.85	16.15
1500	Solid, 8" x 16" block, 6" thick	D-8	430	.093		2.55	3.46		6.01	8.40
1600	8" thick	"	405	.099		3.73	3.67		7.40	10.05
1650	12" thick	D-9	340	.141	↓	5.40	5.15		10.55	14.25
9000	Minimum labor/equipment charge	D-1	2	8	Job		291		291	470

04 22 10.28 Concrete Block, High Strength		Crew	Daily Output	Labor-Hours	Unit	Material	Labor	Equipment	Total	Total Incl O&P
0010	**CONCRETE BLOCK, HIGH STRENGTH** R042210-20									
0050	Hollow, reinforced alternate courses, 8" x 16" units									

04 22 Concrete Unit Masonry

04 22 10 – Concrete Masonry Units

04 22 10.28 Concrete Block, High Strength		Crew	Daily Output	Labor-Hours	Unit	Material (2009 Bare Costs)	Labor (2009 Bare Costs)	Equipment (2009 Bare Costs)	Total (2009 Bare Costs)	Total Incl O&P
0200	3500 psi, 4" thick	D-8	440	.091	S.F.	2.15	3.38		5.53	7.80
0250	6" thick		395	.101		2.41	3.76		6.17	8.75
0300	8" thick		360	.111		3.57	4.13		7.70	10.65
0350	12" thick	D-9	250	.192		4.40	7		11.40	16.15
0500	5000 psi, 4" thick	D-8	440	.091		2.39	3.38		5.77	8.10
0550	6" thick		395	.101		2.99	3.76		6.75	9.40
0600	8" thick		360	.111		4.05	4.13		8.18	11.15
0650	12" thick	D-9	300	.160		6	5.80		11.80	16
1000	For 75% solid block, add					30%				
1050	For 100% solid block, add					50%				

04 22 10.30 Concrete Block, Interlocking

Code	Description	Crew	Daily Output	Labor-Hours	Unit	Material	Labor	Equipment	Total	Total Incl O&P
0010	**CONCRETE BLOCK, INTERLOCKING**									
0100	Not including grout or reinforcing									
0200	8" x 16" units, 2,000 psi, 8" thick	D-1	245	.065	S.F.	2.69	2.37		5.06	6.80
0300	12" thick		220	.073		3.98	2.64		6.62	8.65
0350	16" thick		185	.086		6	3.14		9.14	11.70
0400	Including grout & reinforcing, 8" thick	D-4	245	.131		7.60	4.70	.52	12.82	16.45
0450	12" thick		220	.145		9.20	5.25	.58	15.03	19.15
0500	16" thick		185	.173		11.55	6.20	.69	18.44	23.50
9000	Minimum labor/equipment charge	D-1	2	8	Job		291		291	470

04 22 10.32 Concrete Block, Lintels

Code	Description	Crew	Daily Output	Labor-Hours	Unit	Material	Labor	Equipment	Total	Total Incl O&P
0010	**CONCRETE BLOCK, LINTELS**, C90, normal weight									
0100	Including grout and horizontal reinforcing									
0200	8" x 8" x 8", 1 #4 bar	D-4	300	.107	L.F.	4.86	3.84	.42	9.12	11.95
0250	2 #4 bars		295	.108		5.20	3.90	.43	9.53	12.45
0400	8" x 16" x 8", 1 #4 bar		275	.116		3.36	4.18	.46	8	10.90
0450	2 #4 bars		270	.119		3.71	4.26	.47	8.44	11.45
1000	12" x 8" x 8", 1 #4 bar		275	.116		6.75	4.18	.46	11.39	14.65
1100	2 #4 bars		270	.119		7.10	4.26	.47	11.83	15.20
1150	2 #5 bars		270	.119		7.50	4.26	.47	12.23	15.60
1200	2 #6 bars		265	.121		7.95	4.34	.48	12.77	16.25
1500	12" x 16" x 8", 1 #4 bar		250	.128		10.65	4.60	.51	15.76	19.65
1600	2 #3 bars		245	.131		10.70	4.70	.52	15.92	19.85
1650	2 #4 bars		245	.131		11	4.70	.52	16.22	20
1700	2 #5 bars		240	.133		11.40	4.79	.53	16.72	21
9000	Minimum labor/equipment charge	D-1	2	8	Job		291		291	470

04 22 10.34 Concrete Block, Partitions

Code	Description	Crew	Daily Output	Labor-Hours	Unit	Material	Labor	Equipment	Total	Total Incl O&P
0010	**CONCRETE BLOCK, PARTITIONS**, excludes scaffolding R042210-20									
0100	Acoustical slotted block									
0200	4" thick, type A-1	D-8	315	.127	S.F.	3.73	4.72		8.45	11.75
0210	8" thick		275	.145		4.71	5.40		10.11	13.95
0250	8" thick, type Q		275	.145		8.30	5.40		13.70	17.90
0260	4" thick, type RSC		315	.127		5.80	4.72		10.52	14.05
0270	6" thick		295	.136		5.80	5.05		10.85	14.55
0280	8" thick		275	.145		5.80	5.40		11.20	15.15
0290	12" thick		250	.160		5.80	5.95		11.75	16
0300	8" thick, type RSR		275	.145		5.80	5.40		11.20	15.15
0400	8" thick, type RSC/RF		275	.145		6.50	5.40		11.90	15.90
0410	10" thick		260	.154		7	5.70		12.70	16.95
0500	NRC .60 type R, 8" thick		265	.151		7.95	5.60		13.55	17.80
0600	NRC .65 type RR, 8" thick		265	.151		7.30	5.60		12.90	17.10
0700	NRC .65 type 4R-RF, 8" thick		265	.151		8.20	5.60		13.80	18.10

04 22 10.34 Concrete Block, Partitions		Crew	Daily Output	Labor-Hours	Unit	Material	2009 Bare Costs Labor	Equipment	Total	Total Incl O&P
0710	NRC .70 type R, 12" thick	D-8	245	.163	S.F.	8.60	6.05		14.65	19.30
1000	Lightweight block, tooled joints, 2 sides, hollow									
1100	Not reinforced, 8" x 16" x 4" thick	D-8	440	.091	S.F.	1.72	3.38		5.10	7.35
1150	6" thick		410	.098		2.33	3.63		5.96	8.40
1200	8" thick		385	.104		2.86	3.86		6.72	9.40
1250	10" thick		370	.108		3.73	4.02		7.75	10.60
1300	12" thick	D-9	350	.137		4.26	4.98		9.24	12.75
1500	Reinforced alternate courses, 4" thick	D-8	435	.092		1.80	3.42		5.22	7.55
1600	6" thick		405	.099		2.42	3.67		6.09	8.60
1650	8" thick		380	.105		2.95	3.91		6.86	9.60
1700	10" thick		365	.110		3.83	4.07		7.90	10.80
1750	12" thick	D-9	345	.139		4.37	5.05		9.42	12.95
2000	Not reinforced, 8" x 24" x 4" thick, hollow		460	.104		1.25	3.79		5.04	7.55
2100	6" thick		440	.109		1.68	3.96		5.64	8.25
2150	8" thick		415	.116		2.10	4.20		6.30	9.10
2200	10" thick		385	.125		2.72	4.53		7.25	10.35
2250	12" thick		365	.132		3.11	4.78		7.89	11.15
2400	Reinforced alternate courses, 4" thick		455	.105		1.34	3.83		5.17	7.65
2500	6" thick		435	.110		1.80	4.01		5.81	8.50
2550	8" thick		410	.117		2.19	4.25		6.44	9.30
2600	10" thick		380	.126		2.82	4.59		7.41	10.50
2650	12" thick		360	.133		3.22	4.84		8.06	11.40
2800	Solid, not reinforced, 8" x 16" x 2" thick	D-8	440	.091		1.42	3.38		4.80	7
2900	4" thick		420	.095		2.30	3.54		5.84	8.25
2950	6" thick		390	.103		2.66	3.81		6.47	9.05
3000	8" thick		365	.110		3.57	4.07		7.64	10.50
3050	10" thick		350	.114		4.20	4.25		8.45	11.45
3100	12" thick	D-9	330	.145		6.20	5.30		11.50	15.40
3300	Solid, reinforced alternate courses, 4" thick	D-8	415	.096		2.39	3.58		5.97	8.45
3400	6" thick		385	.104		2.88	3.86		6.74	9.40
3450	8" thick		360	.111		3.67	4.13		7.80	10.75
3500	10" thick		345	.116		4.31	4.31		8.62	11.70
3550	12" thick	D-9	325	.148		6.30	5.35		11.65	15.65
4000	Regular block, tooled joints, 2 sides, hollow									
4100	Not reinforced, 8" x 16" x 4" thick	D-8	430	.093	S.F.	1.43	3.46		4.89	7.20
4150	6" thick		400	.100		2.03	3.72		5.75	8.25
4200	8" thick		375	.107		2.14	3.96		6.10	8.75
4250	10" thick		360	.111		2.95	4.13		7.08	9.95
4300	12" thick	D-9	340	.141		3.35	5.15		8.50	12
4500	Reinforced alternate courses, 8" x 16" x 4" thick	D-8	425	.094		1.56	3.50		5.06	7.35
4550	6" thick		395	.101		2.16	3.76		5.92	8.50
4600	8" thick		370	.108		2.28	4.02		6.30	9
4650	10" thick		355	.113		3.37	4.19		7.56	10.45
4700	12" thick	D-9	335	.143		3.50	5.20		8.70	12.25
4900	Solid, not reinforced, 2" thick	D-8	435	.092		1.31	3.42		4.73	7
5000	3" thick		430	.093		1.56	3.46		5.02	7.30
5050	4" thick		415	.096		1.80	3.58		5.38	7.80
5100	6" thick		385	.104		2.02	3.86		5.88	8.45
5150	8" thick		360	.111		3.19	4.13		7.32	10.20
5200	12" thick	D-9	325	.148		4.86	5.35		10.21	14.05
5500	Solid, reinforced alternate courses, 4" thick	D-8	420	.095		1.88	3.54		5.42	7.75
5550	6" thick		380	.105		2.12	3.91		6.03	8.70
5600	8" thick		355	.113		3.29	4.19		7.48	10.35

04 22 Concrete Unit Masonry

04 22 10 – Concrete Masonry Units

04 22 10.34 Concrete Block, Partitions		Crew	Daily Output	Labor-Hours	Unit	2009 Bare Costs Material	Labor	Equipment	Total	Total Incl O&P
5650	12″ thick	D-9	320	.150	S.F.	4.40	5.45		9.85	13.65
9000	Minimum labor/equipment charge	D-1	2	8	Job		291		291	470

04 22 10.38 Concrete Brick

		Crew	Daily Output	Labor-Hours	Unit	Material	Labor	Equipment	Total	Total Incl O&P
0010	**CONCRETE BRICK**, C55, grade N, type 1									
0100	Regular, 4 x 2-1/4 x 8	D-8	660	.061	Ea.	.42	2.25		2.67	4.10
0125	Rusticated, 4 x 2-1/4 x 8		660	.061		.47	2.25		2.72	4.16
0150	Frog, 4 x 2-1/4 x 8		660	.061		.45	2.25		2.70	4.14
0200	Double, 4 x 4-7/8 x 8		535	.075		.74	2.78		3.52	5.30

04 22 10.42 Concrete Block, Screen Block

		Crew	Daily Output	Labor-Hours	Unit	Material	Labor	Equipment	Total	Total Incl O&P
0010	**CONCRETE BLOCK, SCREEN BLOCK**									
0200	8″ x 16″, 4″ thick	D-8	330	.121	S.F.	1.96	4.50		6.46	9.45
0300	8″ thick		270	.148		2.98	5.50		8.48	12.20
0350	12″ x 12″, 4″ thick		290	.138		2.21	5.15		7.36	10.75
0500	8″ thick		250	.160		3.04	5.95		8.99	12.95

04 22 10.44 Glazed Concrete Block

		Crew	Daily Output	Labor-Hours	Unit	Material	Labor	Equipment	Total	Total Incl O&P
0010	**GLAZED CONCRETE BLOCK** C744									
0100	Single face, 8″ x 16″ units, 2″ thick	D-8	360	.111	S.F.	8.25	4.13		12.38	15.80
0200	4″ thick		345	.116		8.50	4.31		12.81	16.30
0250	6″ thick		330	.121		9.15	4.50		13.65	17.35
0300	8″ thick		310	.129		9.20	4.79		13.99	17.90
0350	10″ thick		295	.136		11.10	5.05		16.15	20.50
0400	12″ thick	D-9	280	.171		11.75	6.25		18	23
0700	Double face, 8″ x 16″ units, 4″ thick	D-8	340	.118		12.20	4.37		16.57	20.50
0750	6″ thick		320	.125		14.50	4.65		19.15	23.50
0800	8″ thick		300	.133		15.15	4.95		20.10	24.50
1000	Jambs, bullnose or square, single face, 8″ x 16″, 2″ thick		315	.127	Ea.	17.50	4.72		22.22	27
1050	4″ thick		285	.140	″	20.50	5.20		25.70	31
1200	Caps, bullnose or square, 8″ x 16″, 2″ thick		420	.095	L.F.	14.15	3.54		17.69	21.50
1250	4″ thick		380	.105		15.95	3.91		19.86	24
1500	Cove base, 8″ x 16″, 2″ thick		315	.127		11	4.72		15.72	19.75
1550	4″ thick		285	.140		7.15	5.20		12.35	16.30
1600	6″ thick		265	.151		7.70	5.60		13.30	17.50
1650	8″ thick		245	.163		8.10	6.05		14.15	18.70
9000	Minimum labor/equipment charge	D-1	2	8	Job		291		291	470

04 23 Glass Unit Masonry

04 23 13 – Vertical Glass Unit Masonry

04 23 13.10 Glass Block

		Crew	Daily Output	Labor-Hours	Unit	Material	Labor	Equipment	Total	Total Incl O&P
0010	**GLASS BLOCK**									
0100	Plain, 4″ thick, under 1,000 S.F., 6″ x 6″	D-8	115	.348	S.F.	20.50	12.95		33.45	43.50
0150	8″ x 8″		160	.250		14.05	9.30		23.35	30.50
0160	end block		160	.250		39.50	9.30		48.80	58
0170	90 deg corner		160	.250		39	9.30		48.30	58
0180	45 deg corner		160	.250		19.15	9.30		28.45	36
0210	4″ x 8″		160	.250		10.25	9.30		19.55	26.50
0220	6″ x 8″		160	.250		11	9.30		20.30	27
0300	1,000 to 5,000 S.F., 6″ x 6″		135	.296		20	11		31	40
0350	8″ x 8″		190	.211		13.75	7.80		21.55	28
0400	12″ x 12″		215	.186		15.70	6.90		22.60	28.50
0410	4″ x 8″		215	.186		4.43	6.90		11.33	16.05

04 23 Glass Unit Masonry

04 23 13 – Vertical Glass Unit Masonry

04 23 13.10 Glass Block		Crew	Daily Output	Labor-Hours	Unit	Material	2009 Bare Costs Labor	Equipment	Total	Total Incl O&P
0420	6" x 8"	D-8	215	.185	S.F.	4.79	6.90		11.69	16.45
0500	Over 5,000 S.F., 6" x 6"		145	.276		19.50	10.25		29.75	38
0550	8" x 8"		215	.185		13.35	6.90		20.25	26
0600	12" x 12"		240	.167		15.20	6.20		21.40	27
0610	4" x 8"		240	.167		4.33	6.20		10.53	14.75
0620	6" x 8"		240	.167		4.74	6.20		10.94	15.20
0700	For solar reflective blocks, add					100%				
1000	Thinline, plain, 3-1/8" thick, under 1,000 S.F., 6" x 6"	D-8	115	.348	S.F.	15.05	12.95		28	37.50
1050	8" x 8"		160	.250		9.35	9.30		18.65	25.50
1200	Over 5,000 S.F., 6" x 6"		145	.276		16.30	10.25		26.55	34.50
4000	Accessories									
4100	Anchors, 20 ga. galv., 1-3/4" wide x 24" long				Ea.	2.95			2.95	3.25
4200	Emulsion asphalt				Gal.	8.10			8.10	8.90
4300	Expansion joint, fiberglass				L.F.	.67			.67	.74
4400	Steel mesh, double galvanized				"	.94			.94	1.03
9000	Minimum labor/equipment charge	D-1	2	8	Job		291		291	470

04 24 Adobe Unit Masonry

04 24 16 – Manufactured Adobe Unit Masonry

04 24 16.06 Adobe Brick		Crew	Daily Output	Labor-Hours	Unit	Material	Labor	Equipment	Total	Total Incl O&P
0010	**ADOBE BRICK**, Semi-stabilized, with cement mortar									
0060	Brick, 10" x 4" x 14", 2.6/S.F. G	D-8	560	.071	S.F.	3.54	2.65		6.19	8.20
0080	12" x 4" x 16", 2.3/S.F. G		580	.069		4.50	2.56		7.06	9.10
0100	10" x 4" x 16", 2.3/S.F. G		590	.068		4.27	2.52		6.79	8.75
0120	8" x 4" x 16", 2.3/S.F. G		560	.071		3.34	2.65		5.99	7.95
0140	4" x 4" x 16", 2.3/S.F. G		540	.074		2.55	2.75		5.30	7.25
0160	6" x 4" x 16", 2.3/S.F. G		540	.074		2.58	2.75		5.33	7.30
0180	4" x 4" x 12", 3.0/S.F. G		520	.077		2.29	2.86		5.15	7.15
0200	8" x 4" x 12", 3.0/S.F. G		520	.077		2.52	2.86		5.38	7.40

04 25 Unit Masonry Panels

04 25 20 – Pre-Fabricated Masonry Panels

04 25 20.10 Brick and Epoxy Mortar Panels		Crew	Daily Output	Labor-Hours	Unit	Material	Labor	Equipment	Total	Total Incl O&P
0010	**BRICK AND EPOXY MORTAR PANELS**									
0020	Prefabricated brick & epoxy mortar, 4" thick, minimum	C-11	775	.093	S.F.	9.25	4.07	2.44	15.76	20
0100	Maximum	"	500	.144		10.75	6.30	3.78	20.83	27.50
0200	For 2" concrete back-up, add					50%				
0300	For 1" urethane & 3" concrete back-up, add					70%				

04 27 Multiple-Wythe Unit Masonry

04 27 10 – Multiple-Wythe Masonry

04 27 10.10 Cornices		Crew	Daily Output	Labor-Hours	Unit	2009 Bare Costs Material	Labor	Equipment	Total	Total Incl O&P
0010	**CORNICES**									
0110	Face bricks, 12 brick/S.F., minimum	D-1	30	.533	SF Face	5.95	19.40		25.35	38
0150	15 brick/S.F., maximum		23	.696	"	7.15	25.50		32.65	49
9000	Minimum labor/equipment charge		1.50	10.667	Job		390		390	625

04 27 10.30 Brick Walls		Crew	Daily Output	Labor-Hours	Unit	Material	Labor	Equipment	Total	Total Incl O&P
0010	**BRICK WALLS**, including mortar, excludes scaffolding R042110-20									
0020	Estimating by number of brick									
0140	Face brick, 4" thick wall, 6.75 brick/S.F.	D-8	1.45	27.586	M	830	1,025		1,855	2,575
0150	Common brick, 4" thick wall, 6.75 brick/S.F. R042110-50		1.60	25		430	930		1,360	1,975
0204	8" thick, 13.50 bricks per S.F.		1.80	22.222		445	825		1,270	1,825
0250	12" thick, 20.25 bricks per S.F.		1.90	21.053		450	780		1,230	1,775
0304	16" thick, 27.00 bricks per S.F.		2	20		455	745		1,200	1,700
0500	Reinforced, face brick, 4" thick wall, 6.75 brick/S.F.		1.40	28.571		870	1,050		1,920	2,675
0520	Common brick, 4" thick wall, 6.75 brick/S.F.		1.55	25.806		470	960		1,430	2,075
0550	8" thick, 13.50 bricks per S.F.		1.75	22.857		485	850		1,335	1,900
0600	12" thick, 20.25 bricks per S.F.		1.85	21.622		490	805		1,295	1,850
0650	16" thick, 27.00 bricks per S.F.		1.95	20.513		495	760		1,255	1,775
0790	Alternate method of figuring by square foot									
0800	Face brick, 4" thick wall, 6.75 brick/S.F.	D-8	215	.186	S.F.	5.60	6.90		12.50	17.35
0850	Common brick, 4" thick wall, 6.75 brick/S.F.		240	.167		2.90	6.20		9.10	13.20
0900	8" thick, 13.50 bricks per S.F.		135	.296		6.05	11		17.05	24.50
1000	12" thick, 20.25 bricks per S.F.		95	.421		9.10	15.65		24.75	35.50
1050	16" thick, 27.00 bricks per S.F.		75	.533		12.35	19.80		32.15	45.50
1200	Reinforced, face brick, 4" thick wall, 6.75 brick/S.F.		210	.190		5.85	7.10		12.95	17.90
1220	Common brick, 4" thick wall, 6.75 brick/S.F.		235	.170		3.16	6.35		9.51	13.75
1250	8" thick, 13.50 bricks per S.F.		130	.308		6.55	11.45		18	25.50
1300	12" thick, 20.25 bricks per S.F.		90	.444		9.90	16.50		26.40	37.50
1350	16" thick, 27.00 bricks per S.F.		70	.571		13.40	21		34.40	49
9000	Minimum labor/equipment charge	D-1	2	8	Job		291		291	470

04 27 10.40 Steps		Crew	Daily Output	Labor-Hours	Unit	Material	Labor	Equipment	Total	Total Incl O&P
0010	**STEPS**									
0012	Entry steps, select common brick	D-1	.30	53.333	M	430	1,950		2,380	3,600

04 41 Dry-Placed Stone

04 41 10 – Dry Placed Stone

04 41 10.10 Rough Stone Wall		Crew	Daily Output	Labor-Hours	Unit	Material	Labor	Equipment	Total	Total Incl O&P
0011	**ROUGH STONE WALL**, Dry									
0100	Random fieldstone, under 18" thick G	D-12	60	.533	C.F.	11.15	19.65		30.80	44.50
0150	Over 18" thick G	"	63	.508	"	13.40	18.70		32.10	45.50
0500	Field stone veneer G	D-8	120	.333	S.F.	9.60	12.40		22	30.50
0510	Valley stone veneer G		120	.333		9.60	12.40		22	30.50
0520	River stone veneer G		120	.333		9.60	12.40		22	30.50
0600	Rubble stone walls, in mortar bed, up to 18" thick G	D-11	75	.320	C.F.	13.50	12.30		25.80	34.50
9000	Minimum labor/equipment charge	D-1	2	8	Job		291		291	470

04 43 Stone Masonry

04 43 10 – Masonry with Natural and Processed Stone

04 43 10.05 Ashlar Veneer

	04 43 10.05 Ashlar Veneer	Crew	Daily Output	Labor-Hours	Unit	Material	2009 Bare Costs Labor	Equipment	Total	Total Incl O&P
0011	**ASHLAR VENEER** 4" + or - thk, random or random rectangular									
0150	Sawn face, split joints, low priced stone	D-8	140	.285	S.F.	9.40	10.60		20	27.50
0200	Medium priced stone		130	.303		11.25	11.45		22.70	31
0300	High priced stone		120	.333		15.05	12.40		27.45	36.50
0600	Seam face, split joints, medium price stone		125	.320		7.55	11.90		19.45	27.50
0700	High price stone		120	.333		15	12.40		27.40	36.50
1000	Split or rock face, split joints, medium price stone		125	.320		10.80	11.90		22.70	31
1100	High price stone	↓	120	.333	↓	15	12.40		27.40	36.50

04 43 10.10 Bluestone

	04 43 10.10 Bluestone	Crew	Daily Output	Labor-Hours	Unit	Material	Labor	Equipment	Total	Total Incl O&P
0010	**BLUESTONE**, cut to size									
0100	Paving, natural cleft, to 4', 1" thick	D-8	150	.267	S.F.	9.85	9.90		19.75	27
0150	1-1/2" thick		145	.276		9.85	10.25		20.10	27.50
0200	Smooth finish, 1" thick		150	.267		9.85	9.90		19.75	27
0250	1-1/2" thick		145	.276		9.85	10.25		20.10	27.50
0300	Thermal finish, 1" thick		150	.267		9.85	9.90		19.75	27
0350	1-1/2" thick	↓	145	.276	↓	9.85	10.25		20.10	27.50
0500	Sills, natural cleft, 10" wide to 6' long, 1-1/2" thick	D-11	70	.343	L.F.	13.35	13.15		26.50	36
0550	2" thick		63	.381		14.60	14.60		29.20	39.50
0600	Smooth finish, 1-1/2" thick		70	.343		25	13.15		38.15	49
0650	2" thick		63	.381		30	14.60		44.60	56.50
0800	Thermal finish, 1-1/2" thick		70	.343		30	13.15		43.15	54.50
0850	2" thick	↓	63	.381		35	14.60		49.60	62
1000	Stair treads, natural cleft, 12" wide, 6' long, 1-1/2" thick	D-10	115	.278		30	10.95	5.25	46.20	56.50
1050	2" thick		105	.305		30	12	5.75	47.75	58.50
1100	Smooth finish, 1-1/2" thick		115	.278		15	10.95	5.25	31.20	40
1150	2" thick		105	.305		15	12	5.75	32.75	42
1300	Thermal finish		115	.278		15	10.95	5.25	31.20	40
1350	2" thick	↓	105	.305	↓	15	12	5.75	32.75	42
2000	Coping, finished top & 2 sides, 12" to 6'									
2100	Natural cleft, 1-1/2" thick	D-10	115	.278	L.F.	40	10.95	5.25	56.20	67.50
2150	2" thick		105	.305		40	12	5.75	57.75	69.50
2200	Smooth finish, 1-1/2" thick		115	.278		40	10.95	5.25	56.20	67.50
2250	2" thick		105	.305		40	12	5.75	57.75	69.50
2300	Thermal finish, 1-1/2" thick		115	.278		45	10.95	5.25	61.20	73
2350	2" thick	↓	105	.305	↓	45	12	5.75	62.75	75
9000	Minimum labor/equipment charge	D-1	2.50	6.400	Job		233		233	375

04 43 10.45 Granite

	04 43 10.45 Granite	Crew	Daily Output	Labor-Hours	Unit	Material	Labor	Equipment	Total	Total Incl O&P
0010	**GRANITE**, cut to size									
0050	Veneer, polished face, 3/4" to 1-1/2" thick									
0150	Low price, gray, light gray, etc.	D-10	130	.246	S.F.	25.50	9.70	4.63	39.83	48.50
0180	Medium price, pink, brown, etc.		130	.246		28	9.70	4.63	42.33	51
0220	High price, red, black, etc.	↓	130	.246	↓	40	9.70	4.63	54.33	64.50
0300	1-1/2" to 2-1/2" thick, veneer									
0350	Low price, gray, light gray, etc.	D-10	130	.246	S.F.	27.50	9.70	4.63	41.83	50.50
0500	Medium price, pink, brown, etc.		130	.246		32	9.70	4.63	46.33	55.50
0550	High price, red, black, etc.	↓	130	.246	↓	49.50	9.70	4.63	63.83	75
0700	2-1/2" to 4" thick, veneer									
0750	Low price, gray, light gray, etc.	D-10	110	.291	S.F.	36.50	11.45	5.45	53.40	64.50
0850	Medium price, pink, brown, etc.		110	.291		42	11.45	5.45	58.90	70.50
0950	High price, red, black, etc.	↓	110	.291		60	11.45	5.45	76.90	90.50
1000	For bush hammered finish, deduct					5%				
1050	Coarse rubbed finish, deduct				↓	10%				

04 43 Stone Masonry

04 43 10 – Masonry with Natural and Processed Stone

04 43 10.45 Granite

	04 43 10.45 Granite	Crew	Daily Output	Labor-Hours	Unit	Material (2009 Bare Costs)	Labor (2009 Bare Costs)	Equipment (2009 Bare Costs)	Total (2009 Bare Costs)	Total Incl O&P
1100	Honed finish, deduct				S.F.	5%				
1150	Thermal finish, deduct					18%				
1800	Carving or bas-relief, from templates or plaster molds									
1850	Minimum	D-10	80	.400	C.F.	168	15.75	7.50	191.25	219
1900	Maximum	"	80	.400	"	490	15.75	7.50	513.25	575
2000	Intricate or hand finished pieces									
2010	Mouldings, radius cuts, bullnose edges, etc.									
2050	Add, minimum					30%				
2100	Add, maximum					300%				
2500	Steps, copings, etc., finished on more than one surface									
2550	Minimum	D-10	50	.640	C.F.	91	25	12.05	128.05	154
2600	Maximum	"	50	.640	"	146	25	12.05	183.05	214
2700	Pavers, Belgian block, 8"-13" long, 4"-6" wide, 4"-6" deep	D-11	120	.200	S.F.	25.50	7.70		33.20	40.50
2800	Pavers, 4" x 4" x 4" blocks, split face and joints									
2850	Minimum	D-11	80	.300	S.F.	12.35	11.50		23.85	32
2900	Maximum	"	80	.300	"	27.50	11.50		39	49
3000	Pavers, 4" x 4" x 4", thermal face, sawn joints									
3050	Minimum	D-11	65	.369	S.F.	24.50	14.15		38.65	49.50
3100	Maximum	"	65	.369		32	14.15		46.15	58.50
4000	Soffits, 2" thick, minimum	D-13	35	1.371		37.50	52.50	17.20	107.20	145
4100	Maximum		35	1.371		86.50	52.50	17.20	156.20	199
4200	4" thick, minimum		35	1.371		59.50	52.50	17.20	129.20	169
4300	Maximum		35	1.371		113	52.50	17.20	182.70	228
9000	Minimum labor/equipment charge	D-1	2	8	Job		291		291	470

04 43 10.50 Lightweight Natural Stone

	04 43 10.50 Lightweight Natural Stone	Crew	Daily Output	Labor-Hours	Unit	Material	Labor	Equipment	Total	Total Incl O&P
0011	**LIGHTWEIGHT NATURAL STONE** Lava type									
0100	Veneer, rubble face, sawed back, irregular shapes G	D-10	130	.246	S.F.	6.30	9.70	4.63	20.63	27.50
0200	Sawed face and back, irregular shapes G	"	130	.246	"	6.30	9.70	4.63	20.63	27.50

04 43 10.55 Limestone

	04 43 10.55 Limestone	Crew	Daily Output	Labor-Hours	Unit	Material	Labor	Equipment	Total	Total Incl O&P
0010	**LIMESTONE**, cut to size									
0020	Veneer facing panels									
0750	5" thick, 5' x 14' panels	D-10	275	.116	S.F.	52.50	4.59	2.19	59.28	67.50
1000	Sugarcube finish, 2" Thick, 3' x 5' panels		275	.116		35	4.59	2.19	41.78	48.50
1050	3" Thick, 4' x 9' panels		275	.116		38	4.59	2.19	44.78	52
1200	4" Thick, 5' x 11' panels		275	.116		41	4.59	2.19	47.78	55
1400	Sugarcube, textured finish, 4-1/2" thick, 5' x 12'		275	.116		45	4.59	2.19	51.78	59.50
1450	5" thick, 5' x 14' panels		275	.116		53.50	4.59	2.19	60.28	68.50
2000	Coping, sugarcube finish, top & 2 sides		30	1.067	C.F.	58.50	42	20	120.50	154
2100	Sills, lintels, jambs, trim, stops, sugarcube finish, average		20	1.600		58.50	63	30	151.50	199
2150	Detailed		20	1.600		58.50	63	30	151.50	199
2300	Steps, extra hard, 14" wide, 6" rise		50	.640	L.F.	35	25	12.05	72.05	92.50
3000	Quoins, plain finish, 6"x12"x12"	D-12	25	1.280	Ea.	58.50	47		105.50	141
3050	6"x16"x24"	"	25	1.280	"	78	47		125	162
9000	Minimum labor/equipment charge	D-1	2	8	Job		291		291	470

04 43 10.60 Marble

	04 43 10.60 Marble	Crew	Daily Output	Labor-Hours	Unit	Material	Labor	Equipment	Total	Total Incl O&P
0011	**MARBLE**, ashlar, split face, 4" + or - thick, random									
0040	Lengths 1' to 4' & heights 2" to 7-1/2", average	D-8	175	.229	S.F.	16.90	8.50		25.40	32.50
0100	Base, polished, 3/4" or 7/8" thick, polished, 6" high	D-10	65	.492	L.F.	15.35	19.40	9.25	44	58
0300	Carvings or bas relief, from templates, average		80	.400	S.F.	137	15.75	7.50	160.25	185
0350	Maximum		80	.400	"	320	15.75	7.50	343.25	385
0600	Columns, cornices, mouldings, etc.									
0650	Hand or special machine cut, average	D-10	35	.914	C.F.	137	36	17.20	190.20	228

04 43 Stone Masonry

04 43 10 – Masonry with Natural and Processed Stone

04 43 10.60 Marble		Crew	Daily Output	Labor-Hours	Unit	Material	2009 Bare Costs Labor	Equipment	Total	Total Incl O&P
0700	Maximum	D-10	35	.914	C.F.	294	36	17.20	347.20	400
1000	Facing, polished finish, cut to size, 3/4" to 7/8" thick									
1050	Average	D-10	130	.246	S.F.	22.50	9.70	4.63	36.83	45.50
1100	Maximum		130	.246		52.50	9.70	4.63	66.83	78
1300	1-1/4" thick, average		125	.256		32.50	10.10	4.82	47.42	57.50
1350	Maximum		125	.256		65.50	10.10	4.82	80.42	93.50
1500	2" thick, average		120	.267		38	10.50	5	53.50	64
1550	Maximum		120	.267		65.50	10.50	5	81	94.50
1700	Rubbed finish, cut to size, 4" thick									
1740	Average	D-10	100	.320	S.F.	38	12.60	6	56.60	68.50
1780	Maximum	"	100	.320	"	73	12.60	6	91.60	108
2500	Flooring, polished tiles, 12" x 12" x 3/8" thick									
2510	Thin set, average	D-11	90	.267	S.F.	11.35	10.25		21.60	29
2600	Maximum		90	.267		174	10.25		184.25	209
2700	Mortar bed, average		65	.369		10.15	14.15		24.30	34
2740	Maximum		65	.369		174	14.15		188.15	215
2780	Travertine, 3/8" thick, average	D-10	130	.246		14.40	9.70	4.63	28.73	36.50
2790	Maximum	"	130	.246		35.50	9.70	4.63	49.83	59.50
2800	Patio tile, non-slip, 1/2" thick, flame finish	D-11	75	.320		14.70	12.30		27	36
2900	Shower or toilet partitions, 7/8" thick partitions									
3050	3/4" or 1-1/4" thick stiles, polished 2 sides, average	D-11	75	.320	S.F.	44	12.30		56.30	68
3201	Soffits, add to above prices				"	20%	100%			
3210	Stairs, risers, 7/8" thick x 6" high	D-10	115	.278	L.F.	14.80	10.95	5.25	31	39.50
3360	Treads, 12" wide x 1-1/4" thick	"	115	.278	"	21.50	10.95	5.25	37.70	47
3500	Thresholds, 3' long, 7/8" thick, 4" to 5" wide, plain	D-12	24	1.333	Ea.	16.30	49		65.30	97.50
3550	Beveled		24	1.333	"	18.95	49		67.95	101
3700	Window stools, polished, 7/8" thick, 5" wide		85	.376	L.F.	15.20	13.85		29.05	39.50
9000	Minimum labor/equipment charge	D-1	2	8	Job		291		291	470

04 43 10.75 Sandstone or Brownstone		Crew	Daily Output	Labor-Hours	Unit	Material	Labor	Equipment	Total	Total Incl O&P
0011	**SANDSTONE OR BROWNSTONE**									
0100	Sawed face veneer, 2-1/2" thick, to 2' x 4' panels	D-10	130	.246	S.F.	17.25	9.70	4.63	31.58	39.50
0150	4' thick, to 3'-6" x 8' panels		100	.320		17.25	12.60	6	35.85	46
0300	Split face, random sizes		100	.320		12.40	12.60	6	31	41
0350	Cut stone trim (limestone)									
0360	Ribbon stone, 4" thick, 5' pieces	D-8	120	.333	Ea.	152	12.40		164.40	187
0370	Cove stone, 4" thick, 5' pieces		105	.381		153	14.15		167.15	191
0380	Cornice stone, 10" to 12" wide		90	.444		189	16.50		205.50	234
0390	Band stone, 4" thick, 5' pieces		145	.276		97.50	10.25		107.75	124
0410	Window and door trim, 3" to 4" wide		160	.250		82.50	9.30		91.80	106
0420	Key stone, 18" long		60	.667		87	25		112	136
9000	Minimum labor/equipment charge	D-1	2.50	6.400	Job		233		233	375

04 43 10.80 Slate		Crew	Daily Output	Labor-Hours	Unit	Material	Labor	Equipment	Total	Total Incl O&P
0010	**SLATE**									
0040	Pennsylvania - blue gray to black									
0050	Vermont - unfading green, mottled green & purple, gray & purple									
0100	Virginia, blue black									
0200	Exterior paving, natural cleft, 1" thick									
0250	6" x 6" Pennsylvania	D-12	100	.320	S.F.	6.75	11.80		18.55	26.50
0300	Vermont		100	.320		12.50	11.80		24.30	33
0350	Virginia		100	.320		14.10	11.80		25.90	34.50
0500	24" x 24", Pennsylvania		120	.267		13	9.80		22.80	30
0550	Vermont		120	.267		31	9.80		40.80	50

04 43 10 – Masonry with Natural and Processed Stone

04 43 10.80 Slate		Crew	Daily Output	Labor-Hours	Unit	2009 Bare Costs Material	Labor	Equipment	Total	Total Incl O&P
0600	Virginia	D-12	120	.267	S.F.	20.50	9.80		30.30	38.50
0700	18" x 30" Pennsylvania		120	.267		14.75	9.80		24.55	32
0750	Vermont		120	.267		31	9.80		40.80	50
0800	Virginia		120	.267		18.20	9.80		28	36
1000	Interior flooring, natural cleft, 1/2" thick									
1100	6" x 6" Pennsylvania	D-12	100	.320	S.F.	4	11.80		15.80	23.50
1150	Vermont		100	.320		11	11.80		22.80	31
1200	Virginia		100	.320		11.15	11.80		22.95	31.50
1300	24" x 24" Pennsylvania		120	.267		7.75	9.80		17.55	24.50
1350	Vermont		120	.267		25	9.80		34.80	43.50
1400	Virginia		120	.267		14.75	9.80		24.55	32
1500	18" x 24" Pennsylvania		120	.267		7.75	9.80		17.55	24.50
1550	Vermont		120	.267		16	9.80		25.80	33.50
1600	Virginia		120	.267		14.95	9.80		24.75	32.50
2000	Facing panels, 1-1/4" thick, to 4' x 4' panels									
2100	Natural cleft finish, Pennsylvania	D-10	180	.178	S.F.	34	7	3.34	44.34	52
2110	Vermont		180	.178		27	7	3.34	37.34	45
2120	Virginia		180	.178		33	7	3.34	43.34	51.50
2150	Sand rubbed finish, surface, add					10.15			10.15	11.20
2200	Honed finish, add					7.35			7.35	8.10
2500	Ribbon, natural cleft finish, 1" thick, to 9 S.F.	D-10	80	.400		13.10	15.75	7.50	36.35	48
2550	Sand rubbed finish		80	.400		17.75	15.75	7.50	41	53.50
2600	Honed finish		80	.400		16.50	15.75	7.50	39.75	52
2700	1-1/2" thick		78	.410		17.05	16.20	7.70	40.95	53.50
2750	Sand rubbed finish		78	.410		22.50	16.20	7.70	46.40	59.50
2800	Honed finish		78	.410		21.50	16.20	7.70	45.40	58
2850	2" thick		76	.421		20.50	16.60	7.90	45	57.50
2900	Sand rubbed finish		76	.421		28.50	16.60	7.90	53	66
2950	Honed finish		76	.421		26	16.60	7.90	50.50	63.50
3500	Stair treads, sand finish, 1" thick x 12" wide									
3550	Under 3 L.F.	D-10	85	.376	L.F.	22	14.85	7.10	43.95	56
3600	3 L.F. to 6 L.F.	"	120	.267	"	24	10.50	5	39.50	48.50
3700	Ribbon, sand finish, 1" thick x 12" wide									
3750	To 6 L.F.	D-10	120	.267	L.F.	20	10.50	5	35.50	44.50
4000	Stools or sills, sand finish, 1" thick, 6" wide	D-12	160	.200		11.50	7.35		18.85	24.50
4100	Honed finish		160	.200		11	7.35		18.35	24
4200	10" wide		90	.356		17.75	13.10		30.85	40.50
4250	Honed finish		90	.356		16.50	13.10		29.60	39
4400	2" thick, 6" wide		140	.229		18.50	8.40		26.90	34
4450	Honed finish		140	.229		18.50	8.40		26.90	34
4600	10" wide		90	.356		29	13.10		42.10	53
4650	Honed finish		90	.356		27.50	13.10		40.60	51.50
4800	For lengths over 3', add					25%				
9000	Minimum labor/equipment charge	D-1	2.50	6.400	Job		233		233	375

04 43 10.85 Window Sill		Crew	Daily Output	Labor-Hours	Unit	Material	Labor	Equipment	Total	Total Incl O&P
0010	**WINDOW SILL**									
0020	Bluestone, thermal top, 10" wide, 1-1/2" thick	D-1	85	.188	S.F.	16.05	6.85		22.90	28.50
0050	2" thick		75	.213	"	18.75	7.75		26.50	33
0100	Cut stone, 5" x 8" plain		48	.333	L.F.	11.90	12.10		24	32.50
0200	Face brick on edge, brick, 8" wide		80	.200		2.53	7.25		9.78	14.55
0400	Marble, 9" wide, 1" thick		85	.188		8.50	6.85		15.35	20.50
0900	Slate, colored, unfading, honed, 12" wide, 1" thick		85	.188		17.10	6.85		23.95	30

04 43 Stone Masonry

04 43 10 – Masonry with Natural and Processed Stone

04 43 10.85 Window Sill		Crew	Daily Output	Labor-Hours	Unit	2009 Bare Costs Material	Labor	Equipment	Total	Total Incl O&P
0950	2" thick	D-1	70	.229	L.F.	24	8.30		32.30	39.50
9000	Minimum labor/equipment charge	1 Bric	2	4	Job		162		162	262

04 51 Flue Liner Masonry

04 51 10 – Clay Flue Lining

04 51 10.10 Flue Lining

		Crew	Daily Output	Labor-Hours	Unit	Material	Labor	Equipment	Total	Total Incl O&P
0010	**FLUE LINING**, including mortar									
0020	8" x 8"	D-1	125	.128	V.L.F.	4.43	4.65		9.08	12.35
0100	8" x 12"		103	.155		6.90	5.65		12.55	16.75
0200	12" x 12"		93	.172		8.85	6.25		15.10	19.80
0300	12" x 18"		84	.190		17.25	6.90		24.15	30
0400	18" x 18"		75	.213		22	7.75		29.75	36.50
0500	20" x 20"		66	.242		34	8.80		42.80	52
0600	24" x 24"		56	.286		43.50	10.40		53.90	65
1000	Round, 18" diameter		66	.242		30	8.80		38.80	47.50
1100	24" diameter		47	.340		62.50	12.35		74.85	89

04 54 Refractory Brick Masonry

04 54 10 – Refractory Brick Work

04 54 10.10 Fire Brick

		Crew	Daily Output	Labor-Hours	Unit	Material	Labor	Equipment	Total	Total Incl O&P
0010	**FIRE BRICK**									
0012	Low duty, 2000° F, 9" x 2-1/2" x 4-1/2"	D-1	.60	26.667	M	1,375	970		2,345	3,100
0050	High duty, 3000° F	"	.60	26.667	"	2,100	970		3,070	3,900

04 54 10.20 Fire Clay

		Crew	Daily Output	Labor-Hours	Unit	Material	Labor	Equipment	Total	Total Incl O&P
0010	**FIRE CLAY**									
0020	Gray, high duty, 100 lb. bag				Bag	59			59	65
0050	100 lb. drum, premixed (400 brick per drum)				Drum	73			73	80.50

04 57 Masonry Fireplaces

04 57 10 – Brick or Stone Fireplaces

04 57 10.10 Fireplace

		Crew	Daily Output	Labor-Hours	Unit	Material	Labor	Equipment	Total	Total Incl O&P
0010	**FIREPLACE**									
0100	Brick fireplace, not incl. foundations or chimneys									
0110	30" x 29" opening, incl. chamber, plain brickwork	D-1	.40	40	Ea.	535	1,450		1,985	2,950
0200	Fireplace box only (110 brick)	"	2	8	"	176	291		467	665
0300	For elaborate brickwork and details, add					35%	35%			
0400	For hearth, brick & stone, add	D-1	2	8	Ea.	197	291		488	685
0410	For steel angle, damper, cleanouts, add		4	4		138	145		283	385
0600	Plain brickwork, incl. metal circulator		.50	32		1,025	1,175		2,200	3,000
0800	Face brick only, standard size, 8" x 2-2/3" x 4"		.30	53.333	M	535	1,950		2,485	3,725
0900	Stone fireplace, fieldstone, add				SF Face	14.05			14.05	15.50
1000	Cut stone, add				"	15.50			15.50	17.05
9000	Minimum labor/equipment charge	D-1	2	8	Job		291		291	470

04 71 Manufactured Brick Masonry

04 71 10 – Simulated or Manufactured Brick

04 71 10.10 Simulated Brick		Crew	Daily Output	Labor-Hours	Unit	2009 Bare Costs Material	Labor	Equipment	Total	Total Incl O&P
0010	**SIMULATED BRICK**									
0020	Aluminum, baked on colors	1 Carp	200	.040	S.F.	3.53	1.60		5.13	6.50
0050	Fiberglass panels		200	.040		3.87	1.60		5.47	6.90
0100	Urethane pieces cemented in mastic		150	.053		6.70	2.13		8.83	10.85
0150	Vinyl siding panels		200	.040		2.67	1.60		4.27	5.60
0160	Cement base, brick, incl. mastic	D-1	100	.160		4.22	5.80		10.02	14.05
0170	Corner		50	.320	V.L.F.	10.55	11.65		22.20	30.50
0180	Stone face, incl. mastic		100	.160	S.F.	10.20	5.80		16	20.50
0190	Corner		50	.320	V.L.F.	11.25	11.65		22.90	31

04 72 Cast Stone Masonry

04 72 10 – Cast Stone Masonry Features

04 72 10.10 Coping		Crew	Daily Output	Labor-Hours	Unit	Material	Labor	Equipment	Total	Total Incl O&P
0010	**COPING**, stock units									
0050	Precast concrete, 10″ wide, 4″ tapers to 3-1/2″, 8″ wall	D-1	75	.213	L.F.	17.40	7.75		25.15	31.50
0100	12″ wide, 3-1/2″ tapers to 3″, 10″ wall		70	.229		17.40	8.30		25.70	32.50
0110	14″ wide, 4″ tapers to 3-1/2″, 12″ wall		65	.246		20.50	8.95		29.45	37
0150	16″ wide, 4″ tapers to 3-1/2″, 14″ wall		60	.267		17.10	9.70		26.80	34.50
0250	Precast concrete corners		40	.400	Ea.	28.50	14.55		43.05	54.50
0300	Limestone for 12″ wall, 4″ thick		90	.178	L.F.	15.35	6.45		21.80	27.50
0350	6″ thick		80	.200		23	7.25		30.25	37.50
0500	Marble, to 4″ thick, no wash, 9″ wide		90	.178		23	6.45		29.45	36
0550	12″ wide		80	.200		35	7.25		42.25	50.50
0700	Terra cotta, 9″ wide		90	.178		5.60	6.45		12.05	16.60
0750	12″ wide		80	.200		9.20	7.25		16.45	22
0800	Aluminum, for 12″ wall		80	.200		13.30	7.25		20.55	26.50
9000	Minimum labor/equipment charge		2	8	Job		291		291	470

04 72 20 – Cultured Stone Veneer

04 72 20.10 Cultured Stone Veneer Components		Crew	Daily Output	Labor-Hours	Unit	Material	Labor	Equipment	Total	Total Incl O&P
0010	**CULTURED STONE VENEER COMPONENTS**									
0110	On wood frame and sheathing substrate, random sized cobbles, corner stones	D-8	70	.571	V.L.F.	12	21		33	47.50
0120	Field stones		140	.286	S.F.	8.70	10.60		19.30	26.50
0130	Random sized flats, corner stones		70	.571	V.L.F.	11.80	21		32.80	47.50
0140	Field stones		140	.286	S.F.	9.90	10.60		20.50	28
0150	Horizontal lined ledgestones, corner stones		75	.533	V.L.F.	12	19.80		31.80	45
0160	Field stones		150	.267	S.F.	8.70	9.90		18.60	25.50
0170	Random shaped flats, corner stones		65	.615	V.L.F.	12	23		35	50
0180	Field stones		150	.267	S.F.	8.70	9.90		18.60	25.50
0190	Random shaped / textured face, corner stones		65	.615	V.L.F.	12	23		35	50
0200	Field stones		130	.308	S.F.	8.70	11.45		20.15	28
0210	Random shaped river rock, corner stones		65	.615	V.L.F.	12	23		35	50
0220	Field stones		130	.308	S.F.	8.70	11.45		20.15	28
0240	On concrete or CMU substrate, random sized cobbles, corner stones		70	.571	V.L.F.	11.25	21		32.25	47
0250	Field stones		140	.286	S.F.	8.30	10.60		18.90	26.50
0260	Random sized flats, corner stones		70	.571	V.L.F.	11	21		32	46.50
0270	Field stones		140	.286	S.F.	9.50	10.60		20.10	27.50
0280	Horizontal lined ledgestones, corner stones		75	.533	V.L.F.	11.25	19.80		31.05	44.50
0290	Field stones		150	.267	S.F.	8.30	9.90		18.20	25
0300	Random shaped flats, corner stones		70	.571	V.L.F.	11.25	21		32.25	47
0310	Field stones		140	.286	S.F.	8.30	10.60		18.90	26.50

04 72 Cast Stone Masonry

04 72 20 – Cultured Stone Veneer

04 72 20.10 Cultured Stone Veneer Components		Crew	Daily Output	Labor-Hours	Unit	2009 Bare Costs Material	Labor	Equipment	Total	Total Incl O&P
0320	Random shaped / textured face, corner stones	D-8	65	.615	V.L.F.	11.25	23		34.25	49.50
0330	Field stones		130	.308	S.F.	8.30	11.45		19.75	27.50
0340	Random shaped river rock, corner stones		65	.615	V.L.F.	11.25	23		34.25	49.50
0350	Field stones	↓	130	.308	S.F.	8.30	11.45		19.75	27.50
0360	Cultured stone veneer, #15 felt weather resistant barrier	1 Clab	3700	.002	Sq.	4.77	.07		4.84	5.35
0370	Expanded metal lath, diamond, 2.5 lb/sy, galvanized	1 Lath	85	.094	S.Y.	3.47	3.35		6.82	9.10
0390	Water table or window sill, 18" long	1 Bric	80	.100	Ea.	47.50	4.05		51.55	58.50

04 73 Manufactured Stone Masonry

04 73 20 – Simulated or Manufactured Stone

04 73 20.10 Simulated Stone		Crew	Daily Output	Labor-Hours	Unit	Material	Labor	Equipment	Total	Total Incl O&P
0010	**SIMULATED STONE**									
0100	Insulated fiberglass panels, 5/8" ply backer	L-4	200	.120	S.F.	11.40	4.48		15.88	19.80

Division Notes

	CREW	DAILY OUTPUT	LABOR-HOURS	UNIT	2009 BARE COSTS MAT.	LABOR	EQUIP.	TOTAL	TOTAL INCL O&P

Estimating Tips

05 05 00 Common Work Results for Metals

- Nuts, bolts, washers, connection angles, and plates can add a significant amount to both the tonnage of a structural steel job and the estimated cost. As a rule of thumb, add 10% to the total weight to account for these accessories.
- Type 2 steel construction, commonly referred to as "simple construction," consists generally of field-bolted connections with lateral bracing supplied by other elements of the building, such as masonry walls or x-bracing. The estimator should be aware, however, that shop connections may be accomplished by welding or bolting. The method may be particular to the fabrication shop and may have an impact on the estimated cost.

05 12 23 Structural Steel

- Steel items can be obtained from two sources: a fabrication shop or a metals service center. Fabrication shops can fabricate items under more controlled conditions than can crews in the field. They are also more efficient and can produce items more economically. Metal service centers serve as a source of long mill shapes to both fabrication shops and contractors.
- Most line items in this structural steel subdivision, and most items in 05 50 00 Metal Fabrications, are indicated as being shop fabricated. The bare material cost for these shop fabricated items is the "Invoice Cost" from the shop and includes the mill base price of steel plus mill extras, transportation to the shop, shop drawings and detailing where warranted, shop fabrication and handling, sandblasting and a shop coat of primer paint, all necessary structural bolts, and delivery to the job site. The bare labor cost and bare equipment cost for these shop fabricated items is for field installation or erection.
- Line items in Subdivision 05 12 23.40 Lightweight Framing, and other items scattered in Division 5, are indicated as being field fabricated. The bare material cost for these field fabricated items is the "Invoice Cost" from the metals service center and includes the mill base price of steel plus mill extras, transportation to the metals service center, material handling, and delivery of long lengths of mill shapes to the job site. Material costs for structural bolts and welding rods should be added to the estimate. The bare labor cost and bare equipment cost for these items is for both field fabrication and field installation or erection, and include time for cutting, welding and drilling in the fabricated metal items. Drilling into concrete and fasteners to fasten field fabricated items to other work are not included and should be added to the estimate.

05 20 00 Steel Joist Framing

- In any given project the total weight of open web steel joists is determined by the loads to be supported and the design. However, economies can be realized in minimizing the amount of labor used to place the joists. This is done by maximizing the joist spacing, and therefore minimizing the number of joists required to be installed on the job. Certain spacings and locations may be required by the design, but in other cases maximizing the spacing and keeping it as uniform as possible will keep the costs down.

05 30 00 Steel Decking

- The takeoff and estimating of metal deck involves more than simply the area of the floor or roof and the type of deck specified or shown on the drawings. Many different sizes and types of openings may exist. Small openings for individual pipes or conduits may be drilled after the floor/roof is installed, but larger openings may require special deck lengths as well as reinforcing or structural support. The estimator should determine who will be supplying this reinforcing. Additionally, some deck terminations are part of the deck package, such as screed angles and pour stops, and others will be part of the steel contract, such as angles attached to structural members and cast-in-place angles and plates. The estimator must ensure that all pieces are accounted for in the complete estimate.

05 50 00 Metal Fabrications

- The most economical steel stairs are those that use common materials, standard details, and most importantly, a uniform and relatively simple method of field assembly. Commonly available A36 channels and plates are very good choices for the main stringers of the stairs, as are angles and tees for the carrier members. Risers and treads are usually made by specialty shops, and it is most economical to use a typical detail in as many places as possible. The stairs should be pre-assembled and shipped directly to the site. The field connections should be simple and straightforward to be accomplished efficiently, and with minimum equipment and labor.

Reference Numbers

Reference numbers are shown in shaded boxes at the beginning of some major classifications. These numbers refer to related items in the Reference Section. The reference information may be an estimating procedure, an alternate pricing method, or technical information. *Note:* Not all subdivisions listed here necessarily appear in this publication.

05 01 Maintenance of Metals

05 01 10 – Maintenance of Structural Metal Framing

05 01 10.51 Cleaning of Structural Metal Framing		Crew	Daily Output	Labor-Hours	Unit	2009 Bare Costs Material	Labor	Equipment	Total	Total Incl O&P
0010	**CLEANING OF STRUCTURAL METAL FRAMING**									
6125	Steel surface treatments, PDCA guidelines									
6170	Wire brush, hand (SSPC-SP2)	1 Psst	400	.020	S.F.	.02	.72		.74	1.40
6180	Power tool (SSPC-SP3)	″	700	.011		.06	.41		.47	.85
6215	Pressure washing, 2800-6000 S.F./day	1 Pord	10000	.001			.03		.03	.05
6220	Steam cleaning, 2800-4000 S.F./day		2000	.004			.14		.14	.23
6225	Water blasting		2500	.003			.11		.11	.18
6230	Brush-off blast (SSPC-SP7)	E-11	1750	.018		.23	.65	.11	.99	1.53
6235	Com'l blast (SSPC-SP6), loose scale, fine pwder rust, 2.0 #/S.F. sand		1200	.027		.47	.95	.16	1.58	2.38
6240	Tight mill scale, little/no rust, 3.0 #/S.F. sand		1000	.032		.70	1.14	.20	2.04	3
6245	Exist coat blistered/pitted, 4.0 #/S.F. sand		875	.037		.94	1.30	.22	2.46	3.58
6250	Exist coat badly pitted/nodules, 6.7 #/S.F. sand		825	.039		1.57	1.38	.24	3.19	4.43
6255	Near white blast (SSPC-SP10), loose scale, fine rust, 5.6 #/S.F. sand		450	.071		1.31	2.54	.44	4.29	6.40
6260	Tight mill scale, little/no rust, 6.9 #/S.F. sand		325	.098		1.62	3.51	.60	5.73	8.65
6265	Exist coat blistered/pitted, 9.0 #/S.F. sand		225	.142		2.11	5.05	.87	8.03	12.25
6270	Exist coat badly pitted/nodules, 11.3 #/S.F. sand		150	.213		2.65	7.60	1.31	11.56	17.75

05 05 Common Work Results for Metals

05 05 05 – Selective Metals Demolition

05 05 05.10 Selective Demolition, Metals

		Crew	Daily Output	Labor-Hours	Unit	Material	Labor	Equipment	Total	Total Incl O&P
0010	**SELECTIVE DEMOLITION, METALS** R024119-10									
0015	Excludes shores, bracing, cutting, loading, hauling, dumping									
0020	Remove nuts only up to 3/4″ diameter	1 Sswk	480	.017	Ea.		.75		.75	1.38
0030	7/8″ to 1-1/4″ diameter		240	.033			1.49		1.49	2.76
0040	1-3/8″ to 2″ diameter		160	.050			2.24		2.24	4.14
0060	Unbolt and remove structural bolts up to 3/4″ diameter		240	.033			1.49		1.49	2.76
0070	7/8″ to 2″ diameter		160	.050			2.24		2.24	4.14
0140	Light weight framing members, remove whole or cut up, up to 20 lb		240	.033			1.49		1.49	2.76
0150	21 – 40 lb	2 Sswk	210	.076			3.41		3.41	6.30
0160	41 – 80 lb	3 Sswk	180	.133			5.95		5.95	11.05
0170	81 – 120 lb	4 Sswk	150	.213			9.55		9.55	17.65
0230	Structural members, remove whole or cut up, up to 500 lb	E-19	48	.500			21.50	22.50	44	63.50
0240	1/4 – 2 tons	E-18	36	1.111			49.50	30	79.50	122
0250	2 – 5 tons	E-24	30	1.067			47	26.50	73.50	113
0260	5 – 10 tons	E-20	24	2.667			117	55.50	172.50	269
0270	10 – 15 tons	E-2	18	3.111			135	96.50	231.50	345
0340	Fabricated item, remove whole or cut up, up to 20 lb	1 Sswk	96	.083			3.72		3.72	6.90
0350	21 – 40 lb	2 Sswk	84	.190			8.50		8.50	15.75
0360	41 – 80 lb	3 Sswk	72	.333			14.90		14.90	27.50
0370	81 – 120 lb	4 Sswk	60	.533			24		24	44
0380	121 – 500 lb	E-19	48	.500			21.50	22.50	44	63.50
0390	501 – 1000 lb	″	36	.667			29	30	59	84.50
2950	Minimum labor/equipment charge	B-13	2	28	Job		955	395	1,350	1,975

05 05 13 – Shop-Applied Coatings for Metal

05 05 13.50 Paints and Protective Coatings

		Crew	Daily Output	Labor-Hours	Unit	Material	Labor	Equipment	Total	Total Incl O&P
0010	**PAINTS AND PROTECTIVE COATINGS**									
5900	Galvanizing structural steel in shop, under 1 ton				Ton	905			905	995
5950	1 ton to 20 tons					795			795	875
6000	Over 20 tons					735			735	810

05 05 21 – Fastening Methods for Metal

05 05 21.10 Cutting Steel

Line	05 05 21.10 Cutting Steel	Crew	Daily Output	Labor-Hours	Unit	Material	2009 Bare Costs Labor	Equipment	Total	Total Incl O&P
0010	**CUTTING STEEL**									
0020	Hand burning, incl. preparation, torch cutting & grinding, no staging									
0100	Steel to 1/2″ thick	E-25	320	.025	L.F.		1.17	.29	1.46	2.48
0150	3/4″ thick		260	.031			1.44	.36	1.80	3.05
0200	1″ thick		200	.040			1.87	.46	2.33	3.97
9000	Minimum labor/equipment charge		2	4	Job		187	46.50	233.50	395

05 05 21.15 Drilling Steel

Line	05 05 21.15 Drilling Steel	Crew	Daily Output	Labor-Hours	Unit	Material	Labor	Equipment	Total	Total Incl O&P
0010	**DRILLING STEEL**									
1910	Drilling & layout for steel, up to 1/4″ deep, no anchor									
1920	Holes, 1/4″ diameter	1 Sswk	112	.071	Ea.	.08	3.19		3.27	6
1925	For each additional 1/4″ depth, add		336	.024		.08	1.06		1.14	2.06
1930	3/8″ diameter		104	.077		.09	3.44		3.53	6.45
1935	For each additional 1/4″ depth, add		312	.026		.09	1.15		1.24	2.22
1940	1/2″ diameter		96	.083		.10	3.72		3.82	7
1945	For each additional 1/4″ depth, add		288	.028		.10	1.24		1.34	2.41
1950	5/8″ diameter		88	.091		.13	4.06		4.19	7.70
1955	For each additional 1/4″ depth, add		264	.030		.13	1.35		1.48	2.66
1960	3/4″ diameter		80	.100		.15	4.47		4.62	8.45
1965	For each additional 1/4″ depth, add		240	.033		.15	1.49		1.64	2.93
1970	7/8″ diameter		72	.111		.18	4.97		5.15	9.40
1975	For each additional 1/4″ depth, add		216	.037		.18	1.66		1.84	3.27
1980	1″ diameter		64	.125		.20	5.60		5.80	10.55
1985	For each additional 1/4″ depth, add		192	.042		.20	1.86		2.06	3.67
1990	For drilling up, add						40%			
2000	Minimum labor/equipment charge	1 Carp	4	2	Job		80		80	132

05 05 21.90 Welding Steel

Line	05 05 21.90 Welding Steel	Crew	Daily Output	Labor-Hours	Unit	Material	Labor	Equipment	Total	Total Incl O&P
0010	**WELDING STEEL**, Structural R050521-20									
0020	Field welding, 1/8″ E6011, cost per welder, no oper. engr	E-14	8	1	Hr.	4.73	46.50	16.80	68.03	110
0200	With 1/2 operating engineer	E-13	8	1.500		4.73	66	16.75	87.48	141
0300	With 1 operating engineer	E-12	8	2		4.73	86	16.80	107.53	172
0500	With no operating engineer, 2# weld rod per ton	E-14	8	1	Ton	4.73	46.50	16.80	68.03	110
0600	8# E6011 per ton	″	2	4		18.90	187	67	272.90	440
0800	With one operating engineer per welder, 2# E6011 per ton	E-12	8	2		4.73	86	16.80	107.53	172
0900	8# E6011 per ton	″	2	8		18.90	345	67	430.90	685
1200	Continuous fillet, stick welding, incl. equipment									
1300	Single pass, 1/8″ thick, 0.1#/L.F.	E-14	150	.053	L.F.	.24	2.49	.89	3.62	5.85
1400	3/16″ thick, 0.2#/L.F.		75	.107		.47	4.98	1.79	7.24	11.75
1500	1/4″ thick, 0.3#/L.F.		50	.160		.71	7.45	2.68	10.84	17.60
1610	5/16″ thick, 0.4#/L.F.		38	.211		.95	9.85	3.53	14.33	23
1800	3 passes, 3/8″ thick, 0.5#/L.F.		30	.267		1.18	12.45	4.47	18.10	29
2010	4 passes, 1/2″ thick, 0.7#/L.F.		22	.364		1.66	17	6.10	24.76	40
2200	5 to 6 passes, 3/4″ thick, 1.3#/L.F.		12	.667		3.07	31	11.20	45.27	73
2400	8 to 11 passes, 1″ thick, 2.4#/L.F.		6	1.333		5.65	62.50	22.50	90.65	146
2600	For all position welding, add, minimum						20%			
2700	Maximum						300%			
2900	For semi-automatic welding, deduct, minimum						5%			
3000	Maximum						15%			
4000	Cleaning and welding plates, bars, or rods									
4010	to existing beams, columns, or trusses	E-14	12	.667	L.F.	1.18	31	11.20	43.38	71
9000	Minimum labor/equipment charge	″	4	2	Job		93.50	33.50	127	210

05 05 23.05 Anchor Bolts			Crew	Daily Output	Labor-Hours	Unit	2009 Bare Costs Material	Labor	Equipment	Total	Total Incl O&P
0010	**ANCHOR BOLTS**										
0015	Made from recycled materials										
0100	J-type, incl. hex nut & washer, 1/2" diameter x 6" long	G	2 Carp	70	.229	Ea.	1.01	9.15		10.16	16.15
0110	12" long	G		65	.246		1.26	9.85		11.11	17.65
0120	18" long	G		60	.267		1.64	10.65		12.29	19.40
0130	3/4" diameter x 8" long	G		50	.320		1.86	12.80		14.66	23
0140	12" long	G		45	.356		2.33	14.20		16.53	26
0150	18" long	G		40	.400		3.03	16		19.03	30
0160	1" diameter x 12" long	G		35	.457		6.90	18.25		25.15	37.50
0170	18" long	G		30	.533		8.30	21.50		29.80	44
0180	24" long	G		25	.640		10.15	25.50		35.65	53
0190	36" long	G		20	.800		13.90	32		45.90	68.50
0200	1-1/2" diameter x 18" long	G		22	.727		15.80	29		44.80	65.50
0210	24" long	G		16	1		18.80	40		58.80	86.50
0300	L-type, incl. hex nut & washer, 3/4" diameter x 12" long	G		45	.356		2.08	14.20		16.28	26
0310	18" long	G		40	.400		2.67	16		18.67	29.50
0320	24" long	G		35	.457		3.25	18.25		21.50	33.50
0330	30" long	G		30	.533		4.13	21.50		25.63	39.50
0340	36" long	G		25	.640		4.71	25.50		30.21	47
0350	1" diameter x 12" long	G		35	.457		4.23	18.25		22.48	34.50
0360	18" long	G		30	.533		5.25	21.50		26.75	41
0370	24" long	G		25	.640		6.50	25.50		32	49
0380	30" long	G		23	.696		7.65	28		35.65	54.50
0390	36" long	G		20	.800		8.80	32		40.80	62.50
0400	42" long	G		18	.889		10.65	35.50		46.15	70.50
0410	48" long	G		15	1.067		11.95	42.50		54.45	83.50
0420	1-1/4" diameter x 18" long	G		25	.640		6.80	25.50		32.30	49.50
0430	24" long	G		22	.727		8.10	29		37.10	57
0440	30" long	G		20	.800		9.35	32		41.35	63.50
0450	36" long	G		18	.889		10.65	35.50		46.15	70
0460	42" long	G		16	1		12.05	40		52.05	79.50
0470	48" long	G		14	1.143		13.75	45.50		59.25	90.50
0480	54" long	G		12	1.333		16.20	53.50		69.70	106
0490	60" long	G		10	1.600		17.80	64		81.80	126
0500	1-1/2" diameter x 18" long	G		22	.727		10.50	29		39.50	59.50
0510	24" long	G		19	.842		12.25	33.50		45.75	69
0520	30" long	G		17	.941		13.80	37.50		51.30	77
0530	36" long	G		16	1		15.90	40		55.90	83.50
0540	42" long	G		15	1.067		18.10	42.50		60.60	90.50
0550	48" long	G		13	1.231		20.50	49		69.50	104
0560	54" long	G		11	1.455		25	58		83	124
0570	60" long	G		9	1.778		27	71		98	147
0580	1-3/4" diameter x 18" long	G		20	.800		15.80	32		47.80	70.50
0590	24" long	G		18	.889		18.55	35.50		54.05	79
0600	30" long	G		17	.941		21.50	37.50		59	86
0610	36" long	G		16	1		24.50	40		64.50	93
0620	42" long	G		14	1.143		27.50	45.50		73	106
0630	48" long	G		12	1.333		30.50	53.50		84	122
0640	54" long	G		10	1.600		37.50	64		101.50	148
0650	60" long	G		8	2		41	80		121	177
0660	2" diameter x 24" long	G		17	.941		23.50	37.50		61	88
0670	30" long	G		15	1.067		26.50	42.50		69	100

05 05 23 – Metal Fastenings

05 05 23.05 Anchor Bolts			Crew	Daily Output	Labor-Hours	Unit	Material	Labor	Equipment	Total	Total Incl O&P
0680	36" long	G	2 Carp	13	1.231	Ea.	29.50	49		78.50	113
0690	42" long	G		11	1.455		32.50	58		90.50	132
0700	48" long	G		10	1.600		37.50	64		101.50	148
0710	54" long	G		9	1.778		44.50	71		115.50	166
0720	60" long	G		8	2		48	80		128	185
0730	66" long	G		7	2.286		51.50	91.50		143	208
0740	72" long	G		6	2.667		56.50	107		163.50	238
0990	For galvanized, add						75%				

(2009 Bare Costs: Material, Labor, Equipment, Total)

05 05 23.10 Bolts and Hex Nuts			Crew	Daily Output	Labor-Hours	Unit	Material	Labor	Equipment	Total	Total Incl O&P
0010	**BOLTS & HEX NUTS**, Steel, A307										
0100	1/4" diameter, 1/2" long	G	1 Sswk	140	.057	Ea.	.10	2.55		2.65	4.84
0200	1" long	G		140	.057		.12	2.55		2.67	4.86
0300	2" long	G		130	.062		.15	2.75		2.90	5.25
0400	3" long	G		130	.062		.22	2.75		2.97	5.35
0500	4" long	G		120	.067		.24	2.98		3.22	5.75
0600	3/8" diameter, 1" long	G		130	.062		.19	2.75		2.94	5.30
0700	2" long	G		130	.062		.24	2.75		2.99	5.35
0800	3" long	G		120	.067		.31	2.98		3.29	5.85
0900	4" long	G		120	.067		.39	2.98		3.37	5.95
1000	5" long	G		115	.070		.48	3.11		3.59	6.25
1100	1/2" diameter, 1-1/2" long	G		120	.067		.47	2.98		3.45	6
1200	2" long	G		120	.067		.52	2.98		3.50	6.10
1300	4" long	G		115	.070		.79	3.11		3.90	6.60
1400	6" long	G		110	.073		1.07	3.25		4.32	7.20
1500	8" long	G		105	.076		1.38	3.41		4.79	7.80
1600	5/8" diameter, 1-1/2" long	G		120	.067		1.09	2.98		4.07	6.70
1700	2" long	G		120	.067		1.18	2.98		4.16	6.80
1800	4" long	G		115	.070		1.64	3.11		4.75	7.55
1900	6" long	G		110	.073		2.07	3.25		5.32	8.30
2000	8" long	G		105	.076		3	3.41		6.41	9.60
2100	10" long	G		100	.080		3.74	3.58		7.32	10.70
2200	3/4" diameter, 2" long	G		120	.067		1.54	2.98		4.52	7.20
2300	4" long	G		110	.073		2.17	3.25		5.42	8.40
2400	6" long	G		105	.076		2.76	3.41		6.17	9.35
2500	8" long	G		95	.084		4.12	3.76		7.88	11.50
2600	10" long	G		85	.094		5.40	4.21		9.61	13.70
2700	12" long	G		80	.100		6.30	4.47		10.77	15.20
2800	1" diameter, 3" long	G		105	.076		4.49	3.41		7.90	11.25
2900	6" long	G		90	.089		7	3.97		10.97	15.05
3000	12" long	G		75	.107		13.30	4.77		18.07	23.50
3100	For galvanized, add						75%				
3200	For stainless, add						350%				

05 05 23.15 Chemical Anchors		Crew	Daily Output	Labor-Hours	Unit	Material	Labor	Equipment	Total	Total Incl O&P
0010	**CHEMICAL ANCHORS**									
0020	Includes layout & drilling									
1430	Chemical anchor, w/rod & epoxy cartridge, 3/4" diam. x 9-1/2" long	B-89A	27	.593	Ea.	7.55	21.50	4.08	33.13	48
1435	1" diameter x 11-3/4" long		24	.667		12.80	24	4.59	41.39	58.50
1440	1-1/4" diameter x 14" long		21	.762		23.50	27.50	5.25	56.25	77
1445	1-3/4" diameter x 15" long		20	.800		49.50	29	5.50	84	108
1450	18" long		17	.941		59	34	6.50	99.50	128
1455	2" diameter x 18" long		16	1		78.50	36	6.90	121.40	154
1460	24" long		15	1.067		102	38.50	7.35	147.85	184

05 05 23 – Metal Fastenings

05 05 23.20 Expansion Anchors			Crew	Daily Output	Labor-Hours	Unit	Material	2009 Bare Costs Labor	Equipment	Total	Total Incl O&P
0010	**EXPANSION ANCHORS**										
0100	Anchors for concrete, brick or stone, no layout and drilling										
0200	Expansion shields, zinc, 1/4" diameter, 1-5/16" long, single	G	1 Carp	90	.089	Ea.	.28	3.55		3.83	6.15
0300	1-3/8" long, double	G		85	.094		.33	3.76		4.09	6.55
0400	3/8" diameter, 1-1/2" long, single	G		85	.094		.49	3.76		4.25	6.75
0500	2" long, double	G		80	.100		1.10	4		5.10	7.80
0600	1/2" diameter, 2-1/16" long, single	G		80	.100		1.11	4		5.11	7.80
0700	2-1/2" long, double	G		75	.107		1.40	4.26		5.66	8.60
0800	5/8" diameter, 2-5/8" long, single	G		75	.107		1.52	4.26		5.78	8.70
0900	2-3/4" long, double	G		70	.114		2.14	4.57		6.71	9.90
1000	3/4" diameter, 2-3/4" long, single	G		70	.114		1.99	4.57		6.56	9.75
1100	3-15/16" long, double	G		65	.123		3.36	4.92		8.28	11.80
1600	Self-drilling anchor, snap-off, for 3/8" diameter bolt	G		23	.348		1.46	13.90		15.36	24.50
1900	3/4" diameter bolt	G		16	.500		6.30	20		26.30	40
2100	Hollow wall anchors for gypsum wall board, plaster or tile										
2300	1/8" diameter, short	G	1 Carp	160	.050	Ea.	.18	2		2.18	3.50
2400	Long	G		150	.053		.21	2.13		2.34	3.75
2500	3/16" diameter, short	G		150	.053		.36	2.13		2.49	3.92
2600	Long	G		140	.057		.47	2.28		2.75	4.29
2700	1/4" diameter, short	G		140	.057		.45	2.28		2.73	4.27
2800	Long	G		130	.062		.49	2.46		2.95	4.60
3000	Toggle bolts, bright steel, 1/8" diameter, 2" long	G		85	.094		.17	3.76		3.93	6.40
3100	4" long	G		80	.100		.20	4		4.20	6.80
3200	3/16" diameter, 3" long	G		80	.100		.22	4		4.22	6.85
3300	6" long	G		75	.107		.31	4.26		4.57	7.40
3400	1/4" diameter, 3" long	G		75	.107		.30	4.26		4.56	7.40
3500	6" long	G		70	.114		.42	4.57		4.99	8
3600	3/8" diameter, 3" long	G		70	.114		.68	4.57		5.25	8.30
3700	6" long	G		60	.133		1.13	5.35		6.48	10.05
3800	1/2" diameter, 4" long	G		60	.133		1.58	5.35		6.93	10.55
3900	6" long	G		50	.160		1.86	6.40		8.26	12.60
4000	Nailing anchors										
4100	Nylon nailing anchor, 1/4" diameter, 1" long		1 Carp	3.20	2.500	C	9.75	100		109.75	176
4200	1-1/2" long			2.80	2.857		11.20	114		125.20	200
4300	2" long			2.40	3.333		14.25	133		147.25	236
4400	Metal nailing anchor, 1/4" diameter, 1" long	G		3.20	2.500		14.75	100		114.75	181
4500	1-1/2" long	G		2.80	2.857		19.55	114		133.55	210
4600	2" long	G		2.40	3.333		24.50	133		157.50	247
5000	Screw anchors for concrete, masonry,										
5100	stone & tile, no layout or drilling included										
5700	Lag screw shields, 1/4" diameter, short	G	1 Carp	90	.089	Ea.	.33	3.55		3.88	6.20
5800	Long	G		85	.094		.33	3.76		4.09	6.55
5900	3/8" diameter, short	G		85	.094		.61	3.76		4.37	6.85
6000	Long	G		80	.100		.64	4		4.64	7.30
6100	1/2" diameter, short	G		80	.100		.89	4		4.89	7.60
6200	Long	G		75	.107		1.06	4.26		5.32	8.20
6300	5/8" diameter, short	G		70	.114		1.47	4.57		6.04	9.15
6400	Long	G		65	.123		1.79	4.92		6.71	10.05
6600	Lead, #6 & #8, 3/4" long	G		260	.031		.10	1.23		1.33	2.14
6700	#10 - #14, 1-1/2" long	G		200	.040		.22	1.60		1.82	2.88
6800	#16 & #18, 1-1/2" long	G		160	.050		.28	2		2.28	3.61
6900	Plastic, #6 & #8, 3/4" long			260	.031		.03	1.23		1.26	2.06

05 05 23 – Metal Fastenings

05 05 23.20 Expansion Anchors		Crew	Daily Output	Labor-Hours	Unit	Material	Labor	Equipment	Total	Total Incl O&P
7000	#8 & #10, 7/8" long	1 Carp	240	.033	Ea.	.03	1.33		1.36	2.23
7100	#10 & #12, 1" long		220	.036		.04	1.45		1.49	2.44
7200	#14 & #16, 1-1/2" long		160	.050		.06	2		2.06	3.37
8000	Wedge anchors, not including layout or drilling									
8050	Carbon steel, 1/4" diameter, 1-3/4" long G	1 Carp	150	.053	Ea.	.25	2.13		2.38	3.79
8100	3-1/4" long G		140	.057		.32	2.28		2.60	4.13
8150	3/8" diameter, 2-1/4" long G		145	.055		.32	2.20		2.52	4
8200	5" long G		140	.057		.57	2.28		2.85	4.40
8250	1/2" diameter, 2-3/4" long G		140	.057		.97	2.28		3.25	4.84
8300	7" long G		125	.064		1.66	2.56		4.22	6.05
8350	5/8" diameter, 3-1/2" long G		130	.062		1	2.46		3.46	5.15
8400	8-1/2" long G		115	.070		2.12	2.78		4.90	6.90
8450	3/4" diameter, 4-1/4" long G		115	.070		1.66	2.78		4.44	6.40
8500	10" long G		95	.084		3.76	3.36		7.12	9.70
8550	1" diameter, 6" long G		100	.080		6	3.20		9.20	11.90
8575	9" long G		85	.094		7.80	3.76		11.56	14.80
8600	12" long G		75	.107		8.45	4.26		12.71	16.35
8650	1-1/4" diameter, 9" long G		70	.114		19.45	4.57		24.02	29
8700	12" long G		60	.133		25	5.35		30.35	36.50
8750	For type 303 stainless steel, add					350%				
8800	For type 316 stainless steel, add					450%				
8950	Self-drilling concrete screw, hex washer head, 3/16" diam. x 1-3/4" long G	1 Carp	300	.027	Ea.	.16	1.07		1.23	1.94
8960	2-1/4" long G		250	.032		.18	1.28		1.46	2.31
8970	Phillips flat head, 3/16" diam. x 1-3/4" long G		300	.027		.15	1.07		1.22	1.93
8980	2-1/4" long G		250	.032		.17	1.28		1.45	2.30
9000	Minimum labor/equipment charge		4	2	Job		80		80	132

05 05 23.25 High Strength Bolts		Crew	Daily Output	Labor-Hours	Unit	Material	Labor	Equipment	Total	Total Incl O&P
0010	**HIGH STRENGTH BOLTS** R050523-10									
0020	A325 Type 1, structural steel, bolt-nut-washer set									
0100	1/2" diameter x 1-1/2" long G	1 Sswk	130	.062	Ea.	.78	2.75		3.53	5.95
0120	2" long G		125	.064		.83	2.86		3.69	6.20
0150	3" long G		120	.067		1.07	2.98		4.05	6.70
0170	5/8" diameter x 1-1/2" long G		125	.064		1.22	2.86		4.08	6.65
0180	2" long G		120	.067		1.29	2.98		4.27	6.90
0190	3" long G		115	.070		1.51	3.11		4.62	7.40
0200	3/4" diameter x 2" long G		120	.067		1.79	2.98		4.77	7.45
0220	3" long G		115	.070		2.06	3.11		5.17	8
0250	4" long G		110	.073		2.42	3.25		5.67	8.65
0300	6" long G		105	.076		3	3.41		6.41	9.60
0350	8" long G		95	.084		5.50	3.76		9.26	13
0360	7/8" diameter x 2" long G		115	.070		2.68	3.11		5.79	8.70
0365	3" long G		110	.073		3.06	3.25		6.31	9.35
0370	4" long G		105	.076		3.56	3.41		6.97	10.20
0380	6" long G		100	.080		4.36	3.58		7.94	11.40
0390	8" long G		90	.089		6.55	3.97		10.52	14.60
0400	1" diameter x 2" long G		105	.076		3.70	3.41		7.11	10.35
0420	3" long G		100	.080		4.05	3.58		7.63	11.05
0450	4" long G		95	.084		4.47	3.76		8.23	11.85
0500	6" long G		90	.089		5.65	3.97		9.62	13.55
0550	8" long G		85	.094		9.05	4.21		13.26	17.80
0600	1-1/4" diameter x 3" long G		85	.094		8.40	4.21		12.61	17.05
0650	4" long G		80	.100		3.81	4.47		8.28	12.50

05 05 Common Work Results for Metals

05 05 23 – Metal Fastenings

05 05 23.25 High Strength Bolts

			Crew	Daily Output	Labor-Hours	Unit	2009 Bare Costs Material	Labor	Equipment	Total	Total Incl O&P
0700	6" long	G	1 Sswk	75	.107	Ea.	10.95	4.77		15.72	21
0750	8" long	G	↓	70	.114	↓	13.30	5.10		18.40	24
1020	A490, bolt-nut-washer set										
1170	5/8" diameter x 1-1/2" long	G	1 Sswk	125	.064	Ea.	1.75	2.86		4.61	7.25
1180	2" long	G		120	.067		1.99	2.98		4.97	7.70
1190	3" long	G		115	.070		2.33	3.11		5.44	8.30
1200	3/4" diameter x 2" long	G		120	.067		2.31	2.98		5.29	8.05
1220	3" long	G		115	.070		2.63	3.11		5.74	8.65
1250	4" long	G		110	.073		2.97	3.25		6.22	9.25
1300	6" long	G		105	.076		4.10	3.41		7.51	10.80
1350	8" long	G		95	.084		6.55	3.76		10.31	14.15
1360	7/8" diameter x 2" long	G		115	.070		3.15	3.11		6.26	9.20
1365	3" long	G		110	.073		3.56	3.25		6.81	9.90
1370	4" long	G		105	.076		4.22	3.41		7.63	10.95
1380	6" long	G		100	.080		5.60	3.58		9.18	12.75
1390	8" long	G		90	.089		7.80	3.97		11.77	15.90
1400	1" diameter x 2" long	G		105	.076		4.26	3.41		7.67	11
1420	3" long	G		100	.080		4.89	3.58		8.47	11.95
1450	4" long	G		95	.084		5.45	3.76		9.21	12.90
1500	6" long	G		90	.089		6.85	3.97		10.82	14.85
1550	8" long	G		85	.094		10.05	4.21		14.26	18.85
1600	1-1/4" diameter x 3" long	G		85	.094		9.80	4.21		14.01	18.60
1650	4" long	G		80	.100		10.85	4.47		15.32	20.50
1700	6" long	G		75	.107		13.95	4.77		18.72	24
1750	8" long	G	↓	70	.114	↓	17.35	5.10		22.45	28.50

05 05 23.30 Lag Screws

			Crew	Daily Output	Labor-Hours	Unit	Material	Labor	Equipment	Total	Total Incl O&P
0010	**LAG SCREWS**										
0020	Steel, 1/4" diameter, 2" long	G	1 Carp	200	.040	Ea.	.08	1.60		1.68	2.73
0100	3/8" diameter, 3" long	G		150	.053		.26	2.13		2.39	3.81
0200	1/2" diameter, 3" long	G		130	.062		.45	2.46		2.91	4.56
0300	5/8" diameter, 3" long	G	↓	120	.067	↓	1.12	2.66		3.78	5.65

05 05 23.35 Machine Screws

			Crew	Daily Output	Labor-Hours	Unit	Material	Labor	Equipment	Total	Total Incl O&P
0010	**MACHINE SCREWS**										
0020	Steel, round head, #8 x 1" long	G	1 Carp	4.80	1.667	C	2.91	66.50		69.41	113
0110	#8 x 2" long	G		2.40	3.333		4.46	133		137.46	225
0200	#10 x 1" long	G		4	2		3.44	80		83.44	136
0300	#10 x 2" long	G	↓	2	4	↓	5.85	160		165.85	270

05 05 23.40 Machinery Anchors

			Crew	Daily Output	Labor-Hours	Unit	Material	Labor	Equipment	Total	Total Incl O&P
0010	**MACHINERY ANCHORS**, heavy duty, incl. sleeve, floating base nut,										
0020	lower stud & coupling nut, fiber plug, connecting stud, washer & nut.										
0030	For flush mounted embedment in poured concrete heavy equip. pads.										
0200	Stud & bolt, 1/2" diameter	G	E-16	40	.400	Ea.	67	18.30	3.36	88.66	112
0300	5/8" diameter	G		35	.457		74.50	21	3.84	99.34	125
0500	3/4" diameter	G		30	.533		86	24.50	4.47	114.97	144
0600	7/8" diameter	G		25	.640		93.50	29.50	5.35	128.35	163
0800	1" diameter	G		20	.800		98.50	36.50	6.70	141.70	183
0900	1-1/4" diameter	G	↓	15	1.067	↓	131	49	8.95	188.95	244

05 05 23.50 Powder Actuated Tools and Fasteners

			Crew	Daily Output	Labor-Hours	Unit	Material	Labor	Equipment	Total	Total Incl O&P
0010	**POWDER ACTUATED TOOLS & FASTENERS**										
0020	Stud driver, .22 caliber, buy, minimum					Ea.	224			224	246
0100	Maximum					"	360			360	395
0300	Powder charges for above, low velocity					C	8.65			8.65	9.50

05 05 23 – Metal Fastenings

05 05 23.50 Powder Actuated Tools and Fasteners			Crew	Daily Output	Labor-Hours	Unit	Material	Labor	Equipment	Total	Total Incl O&P
							2009 Bare Costs				
0400	Standard velocity					C	12.30			12.30	13.55
0600	Drive pins & studs, 1/4″ & 3/8″ diam., to 3″ long, minimum	G	1 Carp	4.80	1.667		2.88	66.50		69.38	113
0700	Maximum	G	″	4	2		11.25	80		91.25	144
0800	Pneumatic stud driver for 1/8″ diameter studs					Ea.	1,550			1,550	1,725
0900	Drive pins for above, 1/2″ to 3/4″ long	G	1 Carp	1	8	M	355	320		675	925

05 05 23.55 Rivets

			Crew	Daily Output	Labor-Hours	Unit	Material	Labor	Equipment	Total	Total Incl O&P
0010	**RIVETS**										
0100	Aluminum rivet & mandrel, 1/2″ grip length x 1/8″ diameter	G	1 Carp	4.80	1.667	C	6.15	66.50		72.65	117
0200	3/16″ diameter	G		4	2		9.45	80		89.45	142
0300	Aluminum rivet, steel mandrel, 1/8″ diameter	G		4.80	1.667		9.40	66.50		75.90	120
0400	3/16″ diameter	G		4	2		8.60	80		88.60	141
0500	Copper rivet, steel mandrel, 1/8″ diameter	G		4.80	1.667		19.90	66.50		86.40	132
0600	Monel rivet, steel mandrel, 1/8″ diameter	G		4.80	1.667		57.50	66.50		124	173
0700	3/16″ diameter	G		4	2		165	80		245	315
0800	Stainless rivet & mandrel, 1/8″ diameter	G		4.80	1.667		30	66.50		96.50	143
0900	3/16″ diameter	G		4	2		56.50	80		136.50	194
1000	Stainless rivet, steel mandrel, 1/8″ diameter	G		4.80	1.667		23.50	66.50		90	136
1100	3/16″ diameter	G		4	2		42.50	80		122.50	179
1200	Steel rivet and mandrel, 1/8″ diameter	G		4.80	1.667		10.45	66.50		76.95	121
1300	3/16″ diameter	G		4	2		15.70	80		95.70	149
1400	Hand riveting tool, minimum					Ea.	143			143	157
1500	Maximum						273			273	300
1600	Power riveting tool, minimum						1,025			1,025	1,125
1700	Maximum						2,600			2,600	2,850

05 05 23.80 Vibration and Bearing Pads

		Crew	Daily Output	Labor-Hours	Unit	Material	Labor	Equipment	Total	Total Incl O&P
0010	**VIBRATION & BEARING PADS**									
0300	Laminated synthetic rubber impregnated cotton duck, 1/2″ thick	2 Sswk	24	.667	S.F.	75	30		105	138
0400	1″ thick		20	.800		151	36		187	232
0600	Neoprene bearing pads, 1/2″ thick		24	.667		28.50	30		58.50	86
0700	1″ thick		20	.800		58	36		94	130
0900	Fabric reinforced neoprene, 5000 psi, 1/2″ thick		24	.667		12.70	30		42.70	69
1000	1″ thick		20	.800		25.50	36		61.50	94
1200	Felt surfaced vinyl pads, cork and sisal, 5/8″ thick		24	.667		32.50	30		62.50	91
1300	1″ thick		20	.800		59.50	36		95.50	132
1500	Teflon bonded to 10 ga. carbon steel, 1/32″ layer		24	.667		56.50	30		86.50	117
1600	3/32″ layer		24	.667		84.50	30		114.50	148
1800	Bonded to 10 ga. stainless steel, 1/32″ layer		24	.667		100	30		130	165
1900	3/32″ layer		24	.667		130	30		160	198
2100	Circular machine leveling pad & stud				Kip	8.20			8.20	9

05 05 23.85 Weld Shear Connectors

			Crew	Daily Output	Labor-Hours	Unit	Material	Labor	Equipment	Total	Total Incl O&P
0010	**WELD SHEAR CONNECTORS**										
0020	3/4″ diameter, 3-3/16″ long	G	E-10	960	.017	Ea.	.50	.76	.39	1.65	2.39
0030	3-3/8″ long	G		950	.017		.52	.77	.40	1.69	2.45
0200	3-7/8″ long	G		945	.017		.56	.77	.40	1.73	2.49
0300	4-3/16″ long	G		935	.017		.59	.78	.40	1.77	2.54
0500	4-7/8″ long	G		930	.017		.66	.79	.41	1.86	2.64
0600	5-3/16″ long	G		920	.017		.69	.79	.41	1.89	2.68
0800	5-3/8″ long	G		910	.018		.69	.80	.41	1.90	2.71
0900	6-3/16″ long	G		905	.018		.76	.81	.42	1.99	2.80
1000	7-3/16″ long	G		895	.018		.95	.82	.42	2.19	3.01
1100	8-3/16″ long	G		890	.018		1.03	.82	.42	2.27	3.13
1500	7/8″ diameter, 3-11/16″ long	G		920	.017		.81	.79	.41	2.01	2.81

05 05 23 – Metal Fastenings

05 05 23.85 Weld Shear Connectors			Crew	Daily Output	Labor-Hours	Unit	2009 Bare Costs Material	Labor	Equipment	Total	Total Incl O&P
1600	4-3/16" long	G	E-10	910	.018	Ea.	.86	.80	.41	2.07	2.90
1700	5-3/16" long	G		905	.018		.98	.81	.42	2.21	3.04
1800	6-3/16" long	G		895	.018		1.10	.82	.42	2.34	3.18
1900	7-3/16" long	G		890	.018		1.22	.82	.42	2.46	3.33
2000	8-3/16" long	G	↓	880	.018	↓	1.33	.83	.43	2.59	3.47
9000	Minimum labor/equipment charge		1 Sswk	2	4	Job		179		179	330

05 05 23.87 Weld Studs			Crew	Daily Output	Labor-Hours	Unit	Material	Labor	Equipment	Total	Total Incl O&P
0010	**WELD STUDS**										
0020	1/4" diameter, 2-11/16" long	G	E-10	1120	.014	Ea.	.32	.65	.34	1.31	1.93
0100	4-1/8" long	G		1080	.015		.30	.68	.35	1.33	1.96
0200	3/8" diameter, 4-1/8" long	G		1080	.015		.35	.68	.35	1.38	2.02
0300	6-1/8" long	G		1040	.015		.45	.70	.36	1.51	2.20
0400	1/2" diameter, 2-1/8" long	G		1040	.015		.33	.70	.36	1.39	2.06
0500	3-1/8" long	G		1025	.016		.40	.71	.37	1.48	2.16
0600	4-1/8" long	G		1010	.016		.47	.72	.37	1.56	2.27
0700	5-5/16" long	G		990	.016		.58	.74	.38	1.70	2.43
0800	6-1/8" long	G		975	.016		.63	.75	.39	1.77	2.51
0900	8-1/8" long	G		960	.017		.89	.76	.39	2.04	2.81
1000	5/8" diameter, 2-11/16" long	G		1000	.016		.57	.73	.38	1.68	2.39
1010	4-3/16" long	G		990	.016		.71	.74	.38	1.83	2.57
1100	6-9/16" long	G		975	.016		.92	.75	.39	2.06	2.83
1200	8-3/16" long	G	↓	960	.017	↓	1.23	.76	.39	2.38	3.20
9000	Minimum labor/equipment charge		1 Sswk	2	4	Job		179		179	330

05 05 23.90 Welding Rod		Crew	Daily Output	Labor-Hours	Unit	Material	Labor	Equipment	Total	Total Incl O&P
0010	**WELDING ROD**									
0020	Steel, type 6011, 1/8" diam., less than 500#				Lb.	2.36			2.36	2.60
0100	500# to 2,000#					2.13			2.13	2.34
0200	2,000# to 5,000#					2			2	2.20
0300	5/32" diameter, less than 500#					2.28			2.28	2.50
0310	500# to 2,000#					2.05			2.05	2.26
0320	2,000# to 5,000#					1.93			1.93	2.12
0400	3/16" diam., less than 500#					2.26			2.26	2.49
0500	500# to 2,000#					2.04			2.04	2.24
0600	2,000# to 5,000#					1.92			1.92	2.11
0620	Steel, type 6010, 1/8" diam., less than 500#					2.05			2.05	2.26
0630	500# to 2,000#					1.85			1.85	2.04
0640	2,000# to 5,000#					1.74			1.74	1.91
0650	Steel, type 7018 Low Hydrogen, 1/8" diam., less than 500#					2.39			2.39	2.63
0660	500# to 2,000#					2.15			2.15	2.37
0670	2,000# to 5,000#					2.02			2.02	2.22
0700	Steel, type 7024 Jet Weld, 1/8" diam., less than 500#					2.26			2.26	2.49
0710	500# to 2,000#					2.04			2.04	2.24
0720	2,000# to 5,000#					1.92			1.92	2.11
1550	Aluminum, type 4043 TIG, 1/8" diam., less than 10#					4.15			4.15	4.57
1560	10# to 60#					3.74			3.74	4.11
1570	Over 60#					3.52			3.52	3.87
1600	Aluminum, type 5356 TIG, 1/8" diam., less than 10#					4.33			4.33	4.76
1610	10# to 60#					3.90			3.90	4.29
1620	Over 60#					3.67			3.67	4.03
1900	Cast iron, type 8 Nickel, 1/8" diam., less than 500#					29.50			29.50	32.50
1910	500# to 1,000#					26.50			26.50	29
1920	Over 1,000#				↓	25			25	27.50

05 05 Common Work Results for Metals

05 05 23 – Metal Fastenings

05 05 23.90 Welding Rod		Crew	Daily Output	Labor-Hours	Unit	Material	2009 Bare Costs Labor	Equipment	Total	Total Incl O&P
2000	Stainless steel, type 316/316L, 1/8″ diam., less than 500#				Lb.	7.85			7.85	8.65
2100	500# to 1000#					7.10			7.10	7.80
2220	Over 1000#					6.65			6.65	7.35

05 12 Structural Steel Framing

05 12 23 – Structural Steel for Buildings

05 12 23.05 Canopy Framing

		Crew	Daily Output	Labor-Hours	Unit	Material	Labor	Equipment	Total	Total Incl O&P
0010	CANOPY FRAMING									
0020	6″ and 8″ members, shop fabricated G	E-4	3000	.011	Lb.	1.80	.48	.04	2.32	2.92
9000	Minimum labor/equipment charge	1 Sswk	1	8	Job		360		360	660

05 12 23.10 Ceiling Supports

		Crew	Daily Output	Labor-Hours	Unit	Material	Labor	Equipment	Total	Total Incl O&P
0010	CEILING SUPPORTS									
1000	Entrance door/folding partition supports, shop fabricated G	E-4	60	.533	L.F.	30	24	2.23	56.23	80
1100	Linear accelerator door supports G		14	2.286		137	103	9.60	249.60	350
1200	Lintels or shelf angles, hung, exterior hot dipped galv. G		267	.120		20.50	5.40	.50	26.40	33
1250	Two coats primer paint instead of galv. G		267	.120		17.75	5.40	.50	23.65	30
1400	Monitor support, ceiling hung, expansion bolted G		4	8	Ea.	475	360	33.50	868.50	1,225
1450	Hung from pre-set inserts G		6	5.333		510	241	22.50	773.50	1,025
1600	Motor supports for overhead doors G		4	8		242	360	33.50	635.50	975
1700	Partition support for heavy folding partitions, without pocket G		24	1.333	L.F.	68.50	60.50	5.60	134.60	193
1750	Supports at pocket only G		12	2.667		137	121	11.15	269.15	385
2000	Rolling grilles & fire door supports G		34	.941		58.50	42.50	3.94	104.94	148
2100	Spider-leg light supports, expansion bolted to ceiling slab G		8	4	Ea.	195	181	16.75	392.75	570
2150	Hung from pre-set inserts G		12	2.667	″	210	121	11.15	342.15	465
2400	Toilet partition support G		36	.889	L.F.	68.50	40	3.72	112.22	154
2500	X-ray travel gantry support G		12	2.667	″	234	121	11.15	366.15	490

05 12 23.15 Columns, Lightweight

		Crew	Daily Output	Labor-Hours	Unit	Material	Labor	Equipment	Total	Total Incl O&P
0010	COLUMNS, LIGHTWEIGHT									
1000	Lightweight units (lally), 3-1/2″ diameter	E-2	780	.072	L.F.	6.10	3.13	2.23	11.46	14.70
1050	4″ diameter	″	900	.062	″	8.95	2.71	1.93	13.59	16.75
5800	Adjustable jack post, 8′ maximum height, 2-3/4″ diameter G				Ea.	33			33	36.50
5850	4″ diameter G				″	53			53	58

05 12 23.17 Columns, Structural

		Crew	Daily Output	Labor-Hours	Unit	Material	Labor	Equipment	Total	Total Incl O&P
0010	COLUMNS, STRUCTURAL R051223-20									
0015	Made from recycled materials									
0020	Shop fab'd for 100-ton, 1-2 story project, bolted connections									
0800	Steel, concrete filled, extra strong pipe, 3-1/2″ diameter	E-2	660	.085	L.F.	49.50	3.70	2.64	55.84	64
0830	4″ diameter		780	.072		55	3.13	2.23	60.36	68.50
0890	5″ diameter		1020	.055		65.50	2.39	1.71	69.60	78.50
0930	6″ diameter		1200	.047		87	2.03	1.45	90.48	101
0940	8″ diameter		1100	.051		87	2.22	1.58	90.80	102
1100	For galvanizing, add				Lb.	.40			.40	.44
1300	For web ties, angles, etc., add per added lb.	1 Sswk	945	.008		1.50	.38		1.88	2.35
1500	Steel pipe, extra strong, no concrete, 3″ to 5″ diameter G	E-2	16000	.004		1.50	.15	.11	1.76	2.04
1600	6″ to 12″ diameter G		14000	.004		1.50	.17	.12	1.79	2.10
1700	Steel pipe, extra strong, no concrete, 3″ diameter x 12′-0″ G		60	.933	Ea.	185	40.50	29	254.50	310
1750	4″ diameter x 12′-0″ G		58	.966		270	42	30	342	405
1800	6″ diameter x 12′-0″ G		54	1.037		515	45	32	592	680
1850	8″ diameter x 14′-0″ G		50	1.120		910	49	35	994	1,125
1900	10″ diameter x 16′-0″ G		48	1.167		1,325	51	36.50	1,412.50	1,575

05 12 23.17 Columns, Structural

	05 12 23.17 Columns, Structural		Crew	Daily Output	Labor-Hours	Unit	Material	2009 Bare Costs Labor	Equipment	Total	Total Incl O&P
1950	12" diameter x 18'-0"	G	E-2	45	1.244	Ea.	1,775	54	38.50	1,867.50	2,100
3300	Structural tubing, square, A500GrB, 4" to 6" square, light section	G		11270	.005	Lb.	1.50	.22	.15	1.87	2.21
3600	Heavy section	G		32000	.002	"	1.50	.08	.05	1.63	1.85
4000	Concrete filled, add					L.F.	4.04			4.04	4.45
4500	Structural tubing, sq, 4" x 4" x 1/4" x 12'-0"	G	E-2	58	.966	Ea.	248	42	30	320	380
4550	6" x 6" x 1/4" x 12'-0"	G		54	1.037		405	45	32	482	560
4600	8" x 8" x 3/8" x 14'-0"	G		50	1.120		880	49	35	964	1,100
4650	10" x 10" x 1/2" x 16'-0"	G		48	1.167		1,625	51	36.50	1,712.50	1,925
5100	Structural tubing, rect, 5" to 6" wide, light section	G		8000	.007	Lb.	1.50	.30	.22	2.02	2.43
5200	Heavy section	G		12000	.005		1.50	.20	.15	1.85	2.17
5300	7" to 10" wide, light section	G		15000	.004		1.50	.16	.12	1.78	2.07
5400	Heavy section	G		18000	.003		1.50	.14	.10	1.74	2
5500	Structural tubing, rect, 5" x 3" x 1/4" x 12'-0"	G		58	.966	Ea.	240	42	30	312	370
5550	6" x 4" x 5/16" x 12'-0"	G		54	1.037		375	45	32	452	530
5600	8" x 4" x 3/8" x 12'-0"	G		54	1.037		550	45	32	627	715
5650	10" x 6" x 3/8" x 14'-0"	G		50	1.120		880	49	35	964	1,100
5700	12" x 8" x 1/2" x 16'-0"	G		48	1.167		1,625	51	36.50	1,712.50	1,900
6800	W Shape, A992 steel, 2 tier, W8 x 24	G		1080	.052	L.F.	39.50	2.26	1.61	43.37	49.50
6850	W8 x 31	G		1080	.052		51	2.26	1.61	54.87	62.50
6900	W8 x 48	G		1032	.054		79	2.36	1.69	83.05	93
6950	W8 x 67	G		984	.057		111	2.48	1.77	115.25	128
7000	W10 x 45	G		1032	.054		74.50	2.36	1.69	78.55	87.50
7050	W10 x 68	G		984	.057		112	2.48	1.77	116.25	129
7100	W10 x 112	G		960	.058		185	2.54	1.81	189.35	210
7150	W12 x 50	G		1032	.054		82.50	2.36	1.69	86.55	97
7200	W12 x 87	G		984	.057		144	2.48	1.77	148.25	164
7250	W12 x 120	G		960	.058		198	2.54	1.81	202.35	225
7300	W12 x 190	G		912	.061		315	2.67	1.91	319.58	350
7350	W14 x 74	G		984	.057		122	2.48	1.77	126.25	140
7400	W14 x 120	G		960	.058		198	2.54	1.81	202.35	225
7450	W14 x 176	G		912	.061		290	2.67	1.91	294.58	325
8090	For projects 75 to 99 tons, add					All	10%				
8092	50 to 74 tons, add						20%				
8094	25 to 49 tons, add						30%	10%			
8096	10 to 24 tons, add						50%	25%			
8098	2 to 9 tons, add						75%	50%			
8099	Less than 2 tons, add						100%	100%			
9000	Minimum labor/equipment charge		1 Sswk	1	8	Job		360		360	660

05 12 23.20 Curb Edging

	05 12 23.20 Curb Edging		Crew	Daily Output	Labor-Hours	Unit	Material	Labor	Equipment	Total	Total Incl O&P
0010	**CURB EDGING**										
0020	Steel angle w/anchors, shop fabricated, on forms, 1" x 1", 0.8#/L.F.	G	E-4	350	.091	L.F.	1.76	4.13	.38	6.27	10
0100	2" x 2" angles, 3.92#/L.F.	G		330	.097		7.40	4.38	.41	12.19	16.65
0200	3" x 3" angles, 6.1#/L.F.	G		300	.107		11.70	4.82	.45	16.97	22.50
0300	4" x 4" angles, 8.2#/L.F.	G		275	.116		15.50	5.25	.49	21.24	27.50
1000	6" x 4" angles, 12.3#/L.F.	G		250	.128		23	5.80	.54	29.34	36.50
1050	Steel channels with anchors, on forms, 3" channel, 5#/L.F.	G		290	.110		9.30	4.99	.46	14.75	20
1100	4" channel, 5.4#/L.F.	G		270	.119		10.05	5.35	.50	15.90	21.50
1200	6" channel, 8.2#/L.F.	G		255	.125		15.50	5.65	.53	21.68	28
1300	8" channel, 11.5#/L.F.	G		225	.142		21.50	6.45	.60	28.55	36
1400	10" channel, 15.3#/L.F.	G		180	.178		28.50	8.05	.74	37.29	46.50
1500	12" channel, 20.7#/L.F.	G		140	.229		38	10.35	.96	49.31	62
2000	For curved edging, add						35%	10%			

05 12 Structural Steel Framing

05 12 23 – Structural Steel for Buildings

05 12 23.20 Curb Edging			Crew	Daily Output	Labor-Hours	Unit	Material	2009 Bare Costs Labor	Equipment	Total	Total Incl O&P
9000	Minimum labor/equipment charge		E-4	4	8	Job		360	33.50	393.50	705

05 12 23.40 Lightweight Framing			Crew	Daily Output	Labor-Hours	Unit	Material	Labor	Equipment	Total	Total Incl O&P
0010	**LIGHTWEIGHT FRAMING** R051223-35										
0015	Made from recycled materials										
0200	For load-bearing steel studs see Div. 05 41 13.30										
0400	Angle framing, field fabricated, 4" and larger R051223-45	G	E-3	440	.055	Lb.	.87	2.47	.30	3.64	5.90
0450	Less than 4" angles	G		265	.091	"	.90	4.11	.51	5.52	9.15
0460	1/2" x 1/2" x 1/8"	G		200	.120	L.F.	.18	5.45	.67	6.30	11.05
0462	3/4" x 3/4" x 1/8"	G		160	.150		.50	6.80	.84	8.14	14.05
0464	1" x 1" x 1/8"	G		135	.178		.72	8.05	.99	9.76	16.85
0466	1-1/4" x 1-1/4" x 3/16"	G		115	.209		1.33	9.45	1.17	11.95	20.50
0468	1-1/2" x 1-1/2" x 3/16"	G		100	.240		1.62	10.90	1.34	13.86	23.50
0470	2" x 2" x 1/4"	G		90	.267		2.87	12.10	1.49	16.46	27.50
0472	2-1/2" x 2-1/2" x 1/4"	G		72	.333		3.69	15.10	1.86	20.65	34
0474	3" x 2" x 3/8"	G		65	.369		5.30	16.75	2.06	24.11	39
0476	3" x 3" x 3/8"	G		57	.421		6.50	19.10	2.35	27.95	45
0600	Channel framing, field fabricated, 8" and larger	G		500	.048	Lb.	.90	2.18	.27	3.35	5.30
0650	Less than 8" channels	G		335	.072	"	.90	3.25	.40	4.55	7.45
0660	C2 x 1.78	G		115	.209	L.F.	1.60	9.45	1.17	12.22	20.50
0662	C3 x 4.1	G		80	.300		3.69	13.60	1.68	18.97	31
0664	C4 x 5.4	G		66	.364		4.86	16.50	2.03	23.39	38
0666	C5 x 6.7	G		57	.421		6.05	19.10	2.35	27.50	44.50
0668	C6 x 8.2	G		55	.436		7.15	19.80	2.44	29.39	47
0670	C7 x 9.8	G		40	.600		8.80	27	3.35	39.15	64
0672	C8 x 11.5	G		36	.667		10.35	30.50	3.73	44.58	71.50
0710	Structural bar tee, field fabricated, 3/4" x 3/4" x 1/8"	G		160	.150		.50	6.80	.84	8.14	14.05
0712	1" x 1" x 1/8"	G		135	.178		.72	8.05	.99	9.76	16.85
0714	1-1/2" x 1-1/2" x 1/4"	G		114	.211		2.11	9.55	1.18	12.84	21.50
0716	2" x 2" x 1/4"	G		89	.270		2.87	12.25	1.51	16.63	27.50
0718	2-1/2" x 2-1/2" x 3/8"	G		72	.333		5.30	15.10	1.86	22.26	36
0720	3" x 3" x 3/8"	G		57	.421		6.50	19.10	2.35	27.95	45
0730	Structural zee, field fabricated, 1-1/4" x 1-3/4" x 1-3/4"	G		114	.211		.68	9.55	1.18	11.41	19.75
0732	2-11/16" x 3" x 2-11/16"	G		114	.211		1.60	9.55	1.18	12.33	21
0734	3-1/16" x 4" x 3-1/16"	G		133	.180		2.42	8.20	1.01	11.63	18.90
0736	3-1/4" x 5" x 3-1/4"	G		133	.180		3.30	8.20	1.01	12.51	19.90
0738	3-1/2" x 6" x 3-1/2"	G		160	.150		4.98	6.80	.84	12.62	18.95
0740	Junior beam, field fabricated, 3"	G		80	.300		5.15	13.60	1.68	20.43	32.50
0742	4"	G		72	.333		6.95	15.10	1.86	23.91	37.50
0744	5"	G		67	.358		9	16.25	2	27.25	42
0746	6"	G		62	.387		11.25	17.55	2.16	30.96	47.50
0748	7"	G		57	.421		13.75	19.10	2.35	35.20	53
0750	8"	G		53	.453		16.55	20.50	2.53	39.58	59
1000	Continuous slotted channel framing system, shop fab, minimum	G	2 Sswk	2400	.007	Lb.	4.65	.30		4.95	5.65
1200	Maximum	G	"	1600	.010		5.25	.45		5.70	6.65
1250	Plate & bar stock for reinforcing beams and trusses	G					1.65			1.65	1.82
1300	Cross bracing, rods, shop fabricated, 3/4" diameter	G	E-3	700	.034		1.80	1.56	.19	3.55	5.05
1310	7/8" diameter	G		850	.028		1.80	1.28	.16	3.24	4.52
1320	1" diameter	G		1000	.024		1.80	1.09	.13	3.02	4.15
1330	Angle, 5" x 5" x 3/8"	G		2800	.009		1.80	.39	.05	2.24	2.75
1350	Hanging lintels, shop fabricated, average	G		850	.028		1.80	1.28	.16	3.24	4.52
1380	Roof frames, shop fabricated, 3'-0" square, 5' span	G	E-2	4200	.013		1.80	.58	.41	2.79	3.47
1400	Tie rod, not upset, 1-1/2" to 4" diameter, with turnbuckle	G	2 Sswk	800	.020		1.95	.89		2.84	3.81

05 12 Structural Steel Framing

05 12 23 – Structural Steel for Buildings

05 12 23.40 Lightweight Framing

Line	Description		Crew	Daily Output	Labor-Hours	Unit	2009 Bare Costs Material	Labor	Equipment	Total	Total Incl O&P
1420	No turnbuckle	G	2 Sswk	700	.023	Lb.	1.88	1.02		2.90	3.95
1500	Upset, 1-3/4" to 4" diameter, with turnbuckle	G		800	.020		1.95	.89		2.84	3.81
1520	No turnbuckle	G		700	.023		1.88	1.02		2.90	3.95
9000	Minimum labor/equipment charge			2	8	Job		360		360	660

05 12 23.45 Lintels

Line	Description		Crew	Daily Output	Labor-Hours	Unit	Material	Labor	Equipment	Total	Total Incl O&P
0010	**LINTELS**										
0015	Made from recycled materials										
0020	Plain steel angles, shop fabricated, under 500 lb.	G	1 Bric	550	.015	Lb.	1.16	.59		1.75	2.22
0100	500 to 1000 lb.	G		640	.013		1.13	.51		1.64	2.06
0200	1,000 to 2,000 lb.	G		640	.013		1.10	.51		1.61	2.02
0300	2,000 to 4,000 lb.	G		640	.013		1.07	.51		1.58	1.99
0500	For built-up angles and plates, add to above	G					.38			.38	.41
0700	For engineering, add to above						.15			.15	.17
0900	For galvanizing, add to above, under 500 lb.						.50			.50	.55
0950	500 to 2,000 lb.						.45			.45	.50
1000	Over 2,000 lb.						.40			.40	.44
2000	Steel angles, 3-1/2" x 3", 1/4" thick, 2'-6" long	G	1 Bric	47	.170	Ea.	16.20	6.90		23.10	29
2100	4'-6" long	G		26	.308		29	12.45		41.45	52
2500	3-1/2" x 3-1/2" x 5/16", 5'-0" long	G		18	.444		43	18		61	76.50
2600	4" x 3-1/2", 1/4" thick, 5'-0" long	G		21	.381		37	15.45		52.45	66
2700	9'-0" long	G		12	.667		67	27		94	117
2800	4" x 3-1/2" x 5/16", 7'-0" long	G		12	.667		64.50	27		91.50	115
2900	5" x 3-1/2" x 5/16", 10'-0" long	G		8	1		104	40.50		144.50	181
9000	Minimum labor/equipment charge			4	2	Job		81		81	131

05 12 23.60 Pipe Support Framing

Line	Description		Crew	Daily Output	Labor-Hours	Unit	Material	Labor	Equipment	Total	Total Incl O&P
0010	**PIPE SUPPORT FRAMING**										
0020	Under 10#/L.F., shop fabricated	G	E-4	3900	.008	Lb.	2.01	.37	.03	2.41	2.94
0200	10.1 to 15#/L.F.	G		4300	.007		1.98	.34	.03	2.35	2.83
0400	15.1 to 20#/L.F.	G		4800	.007		1.95	.30	.03	2.28	2.74
0600	Over 20#/L.F.	G		5400	.006		1.92	.27	.02	2.21	2.64

05 12 23.65 Plates

Line	Description		Crew	Daily Output	Labor-Hours	Unit	Material	Labor	Equipment	Total	Total Incl O&P
0010	**PLATES**	R051223-80									
0015	Made from recycled materials										
0020	For connections & stiffener plates, shop fabricated										
0050	1/8" thick (5.1 Lb./S.F.)	G				S.F.	7.65			7.65	8.40
0100	1/4" thick (10.2 Lb./S.F.)	G					15.30			15.30	16.85
0300	3/8" thick (15.3 Lb./S.F.)	G					23			23	25.50
0400	1/2" thick (20.4 Lb./S.F.)	G					30.50			30.50	33.50
0450	3/4" thick (30.6 Lb./S.F.)	G					46			46	50.50
0500	1" thick (40.8 Lb.S.F.)	G					61			61	67.50
2000	Steel plate, warehouse prices, no shop fabrication										
2100	1/4" thick (10.2 Lb./S.F.)	G				S.F.	11.20			11.20	12.35

05 12 23.70 Stressed Skin Steel Roof and Ceiling System

Line	Description		Crew	Daily Output	Labor-Hours	Unit	Material	Labor	Equipment	Total	Total Incl O&P
0010	**STRESSED SKIN STEEL ROOF & CEILING SYSTEM**										
0020	Double panel flat roof, spans to 100'	G	E-2	1150	.049	S.F.	12	2.12	1.51	15.63	18.65
0100	Double panel convex roof, spans to 200'	G		960	.058		19.50	2.54	1.81	23.85	28
0200	Double panel arched roof, spans to 300'	G		760	.074		30	3.21	2.29	35.50	41

05 12 23.75 Structural Steel Members

Line	Description		Crew	Daily Output	Labor-Hours	Unit	Material	Labor	Equipment	Total	Total Incl O&P
0010	**STRUCTURAL STEEL MEMBERS**	R051223-10									
0015	Made from recycled materials										
0020	Shop fab'd for 100-ton, 1-2 story project, bolted connections										

05 12 Structural Steel Framing

05 12 23 – Structural Steel for Buildings

05 12 23.75 Structural Steel Members			Crew	Daily Output	Labor-Hours	Unit	2009 Bare Costs Material	Labor	Equipment	Total	Total Incl O&P
0102	W 6 x 9 R051223-15	G	E-2	600	.093	L.F.	14.85	4.06	2.90	21.81	27
0302	W 8 x 10	G		600	.093		16.50	4.06	2.90	23.46	28.50
0502	x 31	G		550	.102		51	4.43	3.17	58.60	68
0702	W 10 x 22	G		600	.093		36.50	4.06	2.90	43.46	50.50
0902	x 49	G		550	.102		81	4.43	3.17	88.60	100
1102	W 12 x 16	G		880	.064		26.50	2.77	1.98	31.25	36
1302	x 22	G		880	.064		36.50	2.77	1.98	41.25	47
1502	x 26	G		880	.064		43	2.77	1.98	47.75	54
1702	x 72	G		640	.088		119	3.81	2.72	125.53	141
1902	W 14 x 26	G		990	.057		43	2.46	1.76	47.22	53.50
2102	x 30	G		900	.062		49.50	2.71	1.93	54.14	61.50
2302	x 34	G		810	.069		56	3.01	2.15	61.16	69
2502	x 120	G		720	.078		198	3.39	2.42	203.81	227
2702	W 16 x 26	G		1000	.056		43	2.44	1.74	47.18	53.50
2902	x 31	G		900	.062		51	2.71	1.93	55.64	63.50
3102	x 40	G		800	.070		66	3.05	2.18	71.23	80.50
3302	W 18 x 35	G	E-5	960	.083		58	3.67	1.95	63.62	72.50
3502	x 40	G		960	.083		66	3.67	1.95	71.62	81.50
3702	x 50	G		912	.088		82.50	3.87	2.06	88.43	100
3902	x 55	G		912	.088		91	3.87	2.06	96.93	109
4102	W 21 x 44	G		1064	.075		72.50	3.32	1.76	77.58	88
4302	x 50	G		1064	.075		82.50	3.32	1.76	87.58	99
4502	x 62	G		1036	.077		102	3.41	1.81	107.22	121
4702	x 68	G		1036	.077		112	3.41	1.81	117.22	131
4902	W 24 x 55	G		1110	.072		91	3.18	1.69	95.87	108
5102	x 62	G		1110	.072		102	3.18	1.69	106.87	121
5302	x 68	G		1110	.072		112	3.18	1.69	116.87	131
5502	x 76	G		1110	.072		125	3.18	1.69	129.87	146
5702	x 84	G		1080	.074		139	3.27	1.74	144.01	160
5902	W 27 x 94	G		1190	.067		155	2.96	1.58	159.54	178
6102	W 30 x 99	G		1200	.067		163	2.94	1.56	167.50	187
6302	x 108	G		1200	.067		178	2.94	1.56	182.50	203
6502	x 116	G		1160	.069		191	3.04	1.62	195.66	218
6702	W 33 x 118	G		1176	.068		195	3	1.59	199.59	221
6902	x 130	G		1134	.071		215	3.11	1.65	219.76	243
7102	x 141	G		1134	.071		233	3.11	1.65	237.76	263
7302	W 36 x 135	G		1170	.068		223	3.02	1.60	227.62	252
7502	x 150	G		1170	.068		248	3.02	1.60	252.62	279
7702	x 194	G		1125	.071		320	3.14	1.67	324.81	355
7902	x 231	G		1125	.071		380	3.14	1.67	384.81	425
8102	x 302	G		1035	.077		500	3.41	1.81	505.22	560
8490	For projects 75 to 99 tons, add						10%				
8492	50 to 74 tons, add						20%				
8494	25 to 49 tons, add						30%	10%			
8496	10 to 24 tons, add						50%	25%			
8498	2 to 9 tons, add						75%	50%			
8499	Less than 2 tons, add						100%	100%			
9000	Minimum labor/equipment charge		E-2	2	28	Job		1,225	870	2,095	3,125

05 12 23.77 Structural Steel Projects										
0010	**STRUCTURAL STEEL PROJECTS** R050516-30									
0015	Made from recycled materials									
0020	Shop fab'd for 100-ton, 1-2 story project, bolted connections									

05 12 23 – Structural Steel for Buildings

05 12 23.77 Structural Steel Projects			Crew	Daily Output	Labor-Hours	Unit	Material	Labor	Equipment	Total	Total Incl O&P
							2009 Bare Costs				
0200	Apartments, nursing homes, etc., 1 to 2 stories R050523-10	G	E-5	10.30	7.767	Ton	3,000	345	182	3,527	4,125
0300	3 to 6 stories	G	"	10.10	7.921		3,050	350	186	3,586	4,200
0400	7 to 15 stories R051223-10	G	E-6	14.20	9.014		3,125	395	144	3,664	4,300
0500	Over 15 stories	G	"	13.90	9.209		3,250	405	147	3,802	4,475
0700	Offices, hospitals, etc., steel bearing, 1 to 2 stories R051223-15	G	E-5	10.30	7.767		3,000	345	182	3,527	4,125
0800	3 to 6 stories	G	E-6	14.40	8.889		3,050	390	142	3,582	4,250
0900	7 to 15 stories R051223-20	G		14.20	9.014		3,125	395	144	3,664	4,300
1000	Over 15 stories	G	↓	13.90	9.209		3,250	405	147	3,802	4,475
1100	For multi-story masonry wall bearing construction, add R051223-25							30%			
1300	Industrial bldgs., 1 story, beams & girders, steel bearing	G	E-5	12.90	6.202		3,000	273	145	3,418	3,950
1400	Masonry bearing	G	"	10	8	↓	3,000	355	188	3,543	4,150
1500	Industrial bldgs., 1 story, under 10 tons,										
1510	steel from warehouse, trucked	G	E-2	7.50	7.467	Ton	3,600	325	232	4,157	4,775
1600	1 story with roof trusses, steel bearing	G	E-5	10.60	7.547		3,550	335	177	4,062	4,700
1700	Masonry bearing	G	"	8.30	9.639		3,550	425	226	4,201	4,925
1900	Monumental structures, banks, stores, etc., minimum	G	E-6	13	9.846		3,000	435	157	3,592	4,250
2000	Maximum	G	"	9	14.222		4,975	625	227	5,827	6,850
2200	Churches, minimum	G	E-5	11.60	6.897		2,800	305	162	3,267	3,800
2300	Maximum	G	"	5.20	15.385		3,725	680	360	4,765	5,725
2800	Power stations, fossil fuels, minimum	G	E-6	11	11.636		3,000	515	186	3,701	4,425
2900	Maximum	G		5.70	22.456		4,500	990	360	5,850	7,150
2950	Nuclear fuels, non-safety steel, minimum	G		7	18.286		3,000	805	292	4,097	5,075
3000	Maximum	G		5.50	23.273		4,500	1,025	370	5,895	7,200
3040	Safety steel, minimum	G		2.50	51.200		4,375	2,250	820	7,445	9,800
3070	Maximum	G	↓	1.50	85.333		5,750	3,750	1,375	10,875	14,600
3100	Roof trusses, minimum	G	E-5	13	6.154		4,200	271	144	4,615	5,275
3200	Maximum	G		8.30	9.639		5,100	425	226	5,751	6,625
3210	Schools, minimum	G		14.50	5.517		3,000	243	129	3,372	3,875
3220	Maximum	G	↓	8.30	9.639		4,375	425	226	5,026	5,850
3400	Welded construction, simple commercial bldgs., 1 to 2 stories	G	E-7	7.60	10.526		3,050	465	264	3,779	4,500
3500	7 to 15 stories	G	E-9	8.30	15.422		3,550	680	291	4,521	5,450
3700	Welded rigid frame, 1 story, minimum	G	E-7	15.80	5.063		3,125	223	127	3,475	3,975
3800	Maximum	G	"	5.50	14.545	↓	4,050	640	365	5,055	6,000
3810	Fabrication shop costs (included in project material cost, above)										
3820	Mini mill base price, A992	G				Ton	1,250			1,250	1,375
3830	Mill extra for delivery to shop						310			310	340
3840	Shop extra for shop drawings and detailing						255			255	281
3850	Shop fabricating and handling						955			955	1,050
3860	Shop sandblasting and primer coat of paint						130			130	143
3870	Shop delivery to the job site						100			100	110
3880	Total material cost, shop fabricated, primed, delivered					↓	3,000			3,000	3,300
3900	High strength steel mill spec extras:										
3950	A529, A572 (50 ksi) and A36: same as A992 steel (no extra)										
4000	Add to A992 price for A572 (60, 65 ksi)	G				Ton	100			100	110
4100	A242 and A588 Weathering	G				"	90			90	99
4200	Mill size extras for W-Shapes: 0 to 30 plf: no extra charge										
4210	Member sizes 31 to 65 plf, deduct	G				Ton	6			6	6.60
4220	Member sizes 66 to 100 plf, deduct	G					1			1	1.10
4230	Member sizes 101 to 387 plf, add	G				↓	34			34	37.50
4300	Column base plates, light, up to 150 lb	G	2 Sswk	2000	.008	Lb.	1.65	.36		2.01	2.48
4400	Heavy, over 150 lb	G	E-2	7500	.007	"	1.73	.33	.23	2.29	2.74
4600	Castellated beams, light sections, to 50#/L.F., minimum	G		10.70	5.234	Ton	3,150	228	163	3,541	4,050
4700	Maximum	G	↓	7	8	↓	3,450	350	249	4,049	4,700

05 12 Structural Steel Framing

05 12 23 – Structural Steel for Buildings

05 12 23.77 Structural Steel Projects			Crew	Daily Output	Labor-Hours	Unit	Material	2009 Bare Costs Labor	Equipment	Total	Total Incl O&P
4900	Heavy sections, over 50# per L.F., minimum	G	E-2	11.70	4.786	Ton	3,300	208	149	3,657	4,150
5000	Maximum	G		7.80	7.179		3,600	315	223	4,138	4,750
5390	For projects 75 to 99 tons, add						10%				
5392	50 to 74 tons, add						20%				
5394	25 to 49 tons, add						30%	10%			
5396	10 to 24 tons, add						50%	25%			
5398	2 to 9 tons, add						75%	50%			
5399	Less than 2 tons, add						100%	100%			

05 12 23.80 Subpurlins			Crew	Daily Output	Labor-Hours	Unit	Material	Labor	Equipment	Total	Total Incl O&P
0010	**SUBPURLINS** R051223-50										
0015	Made from recycled materials										
0020	Bulb tees, shop fabricated, painted, 32-5/8″ O.C., 40 psf L.L.										
0100	Type 178, max 8′-9″ span, 2.15 plf, 2″ high x 1-5/8″ wide	G	E-1	4200	.006	S.F.	1.83	.25	.03	2.11	2.49
0200	Type 218, max 10′-2″ span, 3.19 plf, 2-1/8″ high x 2-1/8″ wide	G	″	3100	.008		2.12	.34	.04	2.50	2.98
1420	For 24-5/8″ spacing, add						33%	33%			
1430	For 48-5/8″ spacing, deduct						50%	50%			

05 14 Structural Aluminum Framing

05 14 23 – Non-Exposed Structural Aluminum Framing

05 14 23.05 Aluminum Shapes			Crew	Daily Output	Labor-Hours	Unit	Material	Labor	Equipment	Total	Total Incl O&P
0010	**ALUMINUM SHAPES**										
0015	Made from recycled materials										
0020	Structural shapes, 1″ to 10″ members, under 1 ton	G	E-2	1050	.053	Lb.	3.39	2.32	1.66	7.37	9.70
0050	1 to 5 tons	G		1330	.042		3.15	1.83	1.31	6.29	8.15
0100	Over 5 tons	G		1330	.042		3.15	1.83	1.31	6.29	8.15
0300	Extrusions, over 5 tons, stock shapes	G		1330	.042		3.11	1.83	1.31	6.25	8.10
0400	Custom shapes	G		1330	.042		3.10	1.83	1.31	6.24	8.10

05 15 Wire Rope Assemblies

05 15 16 – Steel Wire Rope Assemblies

05 15 16.05 Accessories for Steel Wire Rope			Crew	Daily Output	Labor-Hours	Unit	Material	Labor	Equipment	Total	Total Incl O&P
0010	**ACCESSORIES FOR STEEL WIRE ROPE**										
0015	Made from recycled materials										
1500	Thimbles, heavy duty, 1/4″	G	E-17	160	.100	Ea.	.63	4.57		5.20	9.15
1510	1/2″	G		160	.100		2.77	4.57		7.34	11.50
1520	3/4″	G		105	.152		6.30	6.95		13.25	19.80
1530	1″	G		52	.308		12.60	14.05		26.65	40
1540	1-1/4″	G		38	.421		19.35	19.25		38.60	57
1550	1-1/2″	G		13	1.231		54.50	56.50		111	164
1560	1-3/4″	G		8	2		112	91.50		203.50	293
1570	2″	G		6	2.667		164	122		286	405
1580	2-1/4″	G		4	4		221	183		404	585
1600	Clips, 1/4″ diameter	G		160	.100		2.84	4.57		7.41	11.55
1610	3/8″ diameter	G		160	.100		3.11	4.57		7.68	11.85
1620	1/2″ diameter	G		160	.100		5	4.57		9.57	13.95
1630	3/4″ diameter	G		102	.157		8.10	7.15		15.25	22
1640	1″ diameter	G		64	.250		13.50	11.45		24.95	36
1650	1-1/4″ diameter	G		35	.457		22	21		43	63
1670	1-1/2″ diameter	G		26	.615		30	28		58	85

05 15 16 – Steel Wire Rope Assemblies

05 15 16.05 Accessories for Steel Wire Rope			Crew	Daily Output	Labor-Hours	Unit	2009 Bare Costs Material	Labor	Equipment	Total	Total Incl O&P
1680	1-3/4" diameter	G	E-17	16	1	Ea.	69.50	45.50		115	161
1690	2" diameter	G		12	1.333		77.50	61		138.50	198
1700	2-1/4" diameter	G		10	1.600		114	73		187	260
1800	Sockets, open swage, 1/4" diameter	G		160	.100		40.50	4.57		45.07	53.50
1810	1/2" diameter	G		77	.208		58.50	9.50		68	82
1820	3/4" diameter	G		19	.842		91.50	38.50		130	172
1830	1" diameter	G		9	1.778		163	81		244	330
1840	1-1/4" diameter	G		5	3.200		227	146		373	520
1850	1-1/2" diameter	G		3	5.333		495	244		739	995
1860	1-3/4" diameter	G		3	5.333		880	244		1,124	1,425
1870	2" diameter	G		1.50	10.667		1,325	485		1,810	2,375
1900	Closed swage, 1/4" diameter	G		160	.100		23.50	4.57		28.07	34.50
1910	1/2" diameter	G		104	.154		41	7.05		48.05	58
1920	3/4" diameter	G		32	.500		61.50	23		84.50	110
1930	1" diameter	G		15	1.067		108	49		157	210
1940	1-1/4" diameter	G		7	2.286		162	104		266	370
1950	1-1/2" diameter	G		4	4		293	183		476	665
1960	1-3/4" diameter	G		3	5.333		435	244		679	925
1970	2" diameter	G		2	8		840	365		1,205	1,600
2000	Open spelter, galv., 1/4" diameter	G		160	.100		49.50	4.57		54.07	62.50
2010	1/2" diameter	G		70	.229		51.50	10.45		61.95	76
2020	3/4" diameter	G		26	.615		77	28		105	137
2030	1" diameter	G		10	1.600		214	73		287	370
2040	1-1/4" diameter	G		5	3.200		305	146		451	605
2050	1-1/2" diameter	G		4	4		650	183		833	1,050
2060	1-3/4" diameter	G		2	8		1,125	365		1,490	1,925
2070	2" diameter	G		1.20	13.333		1,300	610		1,910	2,550
2080	2-1/2" diameter	G		1	16		2,400	730		3,130	4,000
2100	Closed spelter, galv., 1/4" diameter	G		160	.100		40	4.57		44.57	52.50
2110	1/2" diameter	G		88	.182		42.50	8.30		50.80	62.50
2120	3/4" diameter	G		30	.533		64.50	24.50		89	116
2130	1" diameter	G		13	1.231		138	56.50		194.50	255
2140	1-1/4" diameter	G		7	2.286		220	104		324	435
2150	1-1/2" diameter	G		6	2.667		475	122		597	745
2160	1-3/4" diameter	G		2.80	5.714		630	261		891	1,175
2170	2" diameter	G		2	8		780	365		1,145	1,525
2200	Jaw & jaw turnbuckles, 1/4" x 4"	G		160	.100		22.50	4.57		27.07	33.50
2250	1/2" x 6"	G		96	.167		28.50	7.60		36.10	45.50
2260	1/2" x 9"	G		77	.208		38	9.50		47.50	59.50
2270	1/2" x 12"	G		66	.242		42.50	11.10		53.60	67.50
2300	3/4" x 6"	G		38	.421		56	19.25		75.25	97
2310	3/4" x 9"	G		30	.533		61.50	24.50		86	113
2320	3/4" x 12"	G		28	.571		79.50	26		105.50	136
2330	3/4" x 18"	G		23	.696		95	32		127	163
2350	1" x 6"	G		17	.941		108	43		151	199
2360	1" x 12"	G		13	1.231		119	56.50		175.50	235
2370	1" x 18"	G		10	1.600		178	73		251	330
2380	1" x 24"	G		9	1.778		196	81		277	365
2400	1-1/4" x 12"	G		7	2.286		199	104		303	410
2410	1-1/4" x 18"	G		6.50	2.462		247	112		359	480
2420	1-1/4" x 24"	G		5.60	2.857		330	131		461	605
2450	1-1/2" x 12"	G		5.20	3.077		335	141		476	630
2460	1-1/2" x 18"	G		4	4		355	183		538	730

05 15 16.05 Accessories for Steel Wire Rope			Crew	Daily Output	Labor-Hours	Unit	Material	Labor	Equipment	Total	Total Incl O&P
							2009 Bare Costs				
2470	1-1/2" x 24"	G	E-17	3.20	5	Ea.	480	229		709	950
2500	1-3/4" x 18"	G		3.20	5		725	229		954	1,225
2510	1-3/4" x 24"	G		2.80	5.714		825	261		1,086	1,400
2550	2" x 24"	G		1.60	10		1,125	455		1,580	2,075

05 15 16.50 Steel Wire Rope

			Crew	Daily Output	Labor-Hours	Unit	Material	Labor	Equipment	Total	Total Incl O&P
0010	**STEEL WIRE ROPE**										
0015	Made from recycled materials										
0020	6 x 19, bright, fiber core, 5000' rolls, 1/2" diameter	G				L.F.	1.68			1.68	1.84
0050	Steel core	G					2.21			2.21	2.43
0100	Fiber core, 1" diameter	G					5.65			5.65	6.20
0150	Steel core	G					6.45			6.45	7.10
0300	6 x 19, galvanized, fiber core, 1/2" diameter	G					2.47			2.47	2.72
0350	Steel core	G					2.82			2.82	3.10
0400	Fiber core, 1" diameter	G					7.25			7.25	7.95
0450	Steel core	G					7.60			7.60	8.35
0500	6 x 7, bright, IPS, fiber core, <500 L.F. w/acc., 1/4" diameter	G	E-17	6400	.003		1.25	.11		1.36	1.58
0510	1/2" diameter	G		2100	.008		3.04	.35		3.39	3.99
0520	3/4" diameter	G		960	.017		5.50	.76		6.26	7.45
0550	6 x 19, bright, IPS, IWRC, <500 L.F. w/acc., 1/4" diameter	G		5760	.003		1.85	.13		1.98	2.28
0560	1/2" diameter	G		1730	.009		3	.42		3.42	4.08
0570	3/4" diameter	G		770	.021		5.20	.95		6.15	7.45
0580	1" diameter	G		420	.038		8.80	1.74		10.54	12.95
0590	1-1/4" diameter	G		290	.055		14.65	2.52		17.17	21
0600	1-1/2" diameter	G		192	.083		18	3.81		21.81	27
0610	1-3/4" diameter	G	E-18	240	.167		28.50	7.40	4.50	40.40	50
0620	2" diameter	G		160	.250		37	11.10	6.75	54.85	68
0630	2-1/4" diameter	G		160	.250		49	11.10	6.75	66.85	81.50
0650	6 x 37, bright, IPS, IWRC, <500 L.F. w/acc., 1/4" diameter	G	E-17	6400	.003		2.35	.11		2.46	2.79
0660	1/2" diameter	G		1730	.009		3.98	.42		4.40	5.15
0670	3/4" diameter	G		770	.021		6.45	.95		7.40	8.80
0680	1" diameter	G		430	.037		10.20	1.70		11.90	14.35
0690	1-1/4" diameter	G		290	.055		15.40	2.52		17.92	21.50
0700	1-1/2" diameter	G		190	.084		22	3.85		25.85	31
0710	1-3/4" diameter	G	E-18	260	.154		35	6.85	4.15	46	55.50
0720	2" diameter	G		200	.200		45.50	8.90	5.40	59.80	72
0730	2-1/4" diameter	G		160	.250		60	11.10	6.75	77.85	93.50
0800	6 x 19 & 6 x 37, swaged, 1/2" diameter	G	E-17	1220	.013		7.15	.60		7.75	8.95
0810	9/16" diameter	G		1120	.014		8.30	.65		8.95	10.30
0820	5/8" diameter	G		930	.017		9.85	.79		10.64	12.25
0830	3/4" diameter	G		640	.025		12.55	1.14		13.69	15.90
0840	7/8" diameter	G		480	.033		15.80	1.52		17.32	20
0850	1" diameter	G		350	.046		19.30	2.09		21.39	25
0860	1-1/8" diameter	G		288	.056		23.50	2.54		26.04	30.50
0870	1-1/4" diameter	G		230	.070		28.50	3.18		31.68	37.50
0880	1-3/8" diameter	G		192	.083		33	3.81		36.81	43.50
0890	1-1/2" diameter	G	E-18	300	.133		40.50	5.90	3.60	50	59

05 15 16.60 Galvanized Steel Wire Rope and Accessories

			Crew	Daily Output	Labor-Hours	Unit	Material	Labor	Equipment	Total	Total Incl O&P
0010	**GALVANIZED STEEL WIRE ROPE & ACCESSORIES**										
0015	Made from recycled materials										
3000	Aircraft cable, galvanized, 7 x 7 x 1/8"	G	E-17	5000	.003	L.F.	.23	.15		.38	.52
3100	Clamps, 1/8"	G	"	125	.128	Ea.	2.03	5.85		7.88	13.10

05 15 Wire Rope Assemblies

05 15 16 – Steel Wire Rope Assemblies

05 15 16.70	Temporary Cable Safety Railing		Crew	Daily Output	Labor-Hours	Unit	2009 Bare Costs Material	Labor	Equipment	Total	Total Incl O&P
0010	**TEMPORARY CABLE SAFETY RAILING**, Each 100′ strand incl.										
0020	2 eyebolts, 1 turnbuckle, 100′ cable, 2 thimbles, 6 clips										
0025	Made from recycled materials										
0100	One strand using 1/4″ cable & accessories	G	2 Sswk	4	4	C.L.F.	194	179		373	545
0200	1/2″ cable & accessories	G	″	2	8	″	390	360		750	1,100

05 21 Steel Joist Framing

05 21 13 – Deep Longspan Steel Joist Framing

05 21 13.50	Deep Longspan Joists		Crew	Daily Output	Labor-Hours	Unit	Material	Labor	Equipment	Total	Total Incl O&P
0010	**DEEP LONGSPAN JOISTS**										
3010	DLH series, 40-ton job lots, bolted cross bridging, shop primer										
3015	Made from recycled materials										
3020	Spans to 144′ (shipped in 2 pieces), minimum	G	E-7	16	5	Ton	2,225	221	126	2,572	2,975
3040	Average	G		13	6.154		2,400	271	155	2,826	3,300
3100	Maximum	G		11	7.273		2,900	320	183	3,403	3,975
3200	52DLH11, 26 Lb/LF	G		2000	.040	L.F.	30	1.76	1	32.76	37.50
3220	52DLH16, 45 Lb/LF	G		2000	.040		54	1.76	1	56.76	64
3240	56DLH11, 26 Lb/LF	G		2000	.040		31	1.76	1	33.76	39
3260	56DLH16, 46 Lb/LF	G		2000	.040		55.50	1.76	1	58.26	65.50
3280	60DLH12, 29 Lb/LF	G		2000	.040		35	1.76	1	37.76	43
3300	60DLH17, 52 Lb/LF	G		2000	.040		62.50	1.76	1	65.26	73
3320	64DLH12, 31 Lb/LF	G		2200	.036		37	1.60	.91	39.51	45
3340	64DLH17, 52 Lb/LF	G		2200	.036		62.50	1.60	.91	65.01	72.50
3360	68DLH13, 37 Lb/LF	G		2200	.036		44.50	1.60	.91	47.01	53
3380	68DLH18, 61 Lb/LF	G		2200	.036		73.50	1.60	.91	76.01	84.50
3400	72DLH14, 41 Lb/LF	G		2200	.036		49.50	1.60	.91	52.01	58
3420	72DLH19, 70 Lb/LF	G		2200	.036		84	1.60	.91	86.51	96.50
3500	For less than 40-ton job lots										
3502	For 30 to 39 tons, add						10%				
3504	20 to 29 tons, add						20%				
3506	10 to 19 tons, add						30%				
3507	5 to 9 tons, add						50%	25%			
3508	1 to 4 tons, add						75%	50%			
3509	Less than 1 ton, add						100%	100%			
4010	SLH series, 40-ton job lots, bolted cross bridging, shop primer										
4020	Spans to 200′ (shipped in 3 pieces), minimum	G	E-7	16	5	Ton	2,200	221	126	2,547	2,950
4040	Average	G		13	6.154		2,475	271	155	2,901	3,375
4060	Maximum	G		11	7.273		2,950	320	183	3,453	4,000
4200	80SLH15, 40 Lb/LF	G		1500	.053	L.F.	49.50	2.35	1.34	53.19	60
4220	80SLH20, 75 Lb/LF	G		1500	.053		92.50	2.35	1.34	96.19	108
4240	88SLH16, 46 Lb/LF	G		1500	.053		57	2.35	1.34	60.69	68
4260	88SLH21, 89 Lb/LF	G		1500	.053		110	2.35	1.34	113.69	127
4280	96SLH17, 52 Lb/LF	G		1500	.053		64	2.35	1.34	67.69	76
4300	96SLH22, 102 Lb/LF	G		1500	.053		126	2.35	1.34	129.69	145
4320	104SLH18, 59 Lb/LF	G		1800	.044		73	1.96	1.12	76.08	85
4340	104SLH23, 109 Lb/LF	G		1800	.044		135	1.96	1.12	138.08	153
4360	112SLH19, 67 Lb/LF	G		1800	.044		82.50	1.96	1.12	85.58	96
4380	112SLH24, 131 Lb/LF	G		1800	.044		162	1.96	1.12	165.08	183
4400	120SLH20, 77 Lb/LF	G		1800	.044		95	1.96	1.12	98.08	110
4420	120SLH25, 152 Lb/LF	G		1800	.044		188	1.96	1.12	191.08	211

05 21 Steel Joist Framing

05 21 13 – Deep Longspan Steel Joist Framing

05 21 13.50 Deep Longspan Joists		Crew	Daily Output	Labor-Hours	Unit	Material	Labor (2009 Bare Costs)	Equipment (2009 Bare Costs)	Total (2009 Bare Costs)	Total Incl O&P
6100	For less than 40-ton job lots									
6102	For 30 to 39 tons, add					10%				
6104	20 to 29 tons, add					20%				
6106	10 to 19 tons, add					30%				
6107	5 to 9 tons, add					50%	25%			
6108	1 to 4 tons, add					75%	50%			
6109	Less than 1 ton, add					100%	100%			

05 21 16 – Longspan Steel Joist Framing

05 21 16.50 Longspan Joists

			Crew	Daily Output	Labor-Hours	Unit	Material	Labor	Equipment	Total	Total Incl O&P
0010	**LONGSPAN JOISTS**										
2000	LH series, 40-ton job lots, bolted cross bridging, shop primer										
2015	Made from recycled materials										
2020	Spans to 96′, minimum	G	E-7	16	5	Ton	2,050	221	126	2,397	2,775
2040	Average	G		13	6.154		2,250	271	155	2,676	3,125
2080	Maximum	G		11	7.273		2,675	320	183	3,178	3,700
2200	18LH04, 12 Lb/LF	G		1400	.057	L.F.	13.55	2.52	1.44	17.51	21
2220	18LH08, 19 Lb/LF	G		1400	.057		21.50	2.52	1.44	25.46	29.50
2240	20LH04, 12 Lb/LF	G		1400	.057		13.55	2.52	1.44	17.51	21
2260	20LH08, 19 Lb/LF	G		1400	.057		21.50	2.52	1.44	25.46	29.50
2280	24LH05, 13 Lb/LF	G		1400	.057		14.70	2.52	1.44	18.66	22.50
2300	24LH10, 23 Lb/LF	G		1400	.057		26	2.52	1.44	29.96	34.50
2320	28LH06, 16 Lb/LF	G		1800	.044		18.05	1.96	1.12	21.13	24.50
2340	28LH11, 25 Lb/LF	G		1800	.044		28	1.96	1.12	31.08	36
2360	32LH08, 17 Lb/LF	G		1800	.044		19.20	1.96	1.12	22.28	26
2380	32LH13, 30 Lb/LF	G		1800	.044		34	1.96	1.12	37.08	42.50
2400	36LH09, 21 Lb/LF	G		1800	.044		23.50	1.96	1.12	26.58	31
2420	36LH14, 36 Lb/LF	G		1800	.044		40.50	1.96	1.12	43.58	49.50
2440	40LH10, 21 Lb/LF	G		2200	.036		23.50	1.60	.91	26.01	30
2460	40LH15, 36 Lb/LF	G		2200	.036		40.50	1.60	.91	43.01	48.50
2480	44LH11, 22 Lb/LF	G		2200	.036		25	1.60	.91	27.51	31.50
2500	44LH16, 42 Lb/LF	G		2200	.036		47.50	1.60	.91	50.01	56
2520	48LH11, 22 Lb/LF	G		2200	.036		25	1.60	.91	27.51	31.50
2540	48LH16, 42 Lb/LF	G		2200	.036		47.50	1.60	.91	50.01	56
2600	For less than 40-ton job lots										
2602	For 30 to 39 tons, add						10%				
2604	20 to 29 tons, add						20%				
2606	10 to 19 tons, add						30%				
2607	5 to 9 tons, add						50%	25%			
2608	1 to 4 tons, add						75%	50%			
2609	Less than 1 ton, add						100%	100%			
6000	For welded cross bridging, add							30%			

05 21 19 – Open Web Steel Joist Framing

05 21 19.10 Open Web Joists

			Crew	Daily Output	Labor-Hours	Unit	Material	Labor	Equipment	Total	Total Incl O&P
0010	**OPEN WEB JOISTS**										
0015	Made from recycled materials										
0020	K series, 40-ton lots, horiz. bridging, spans to 30′, shop primer, minimum	G	E-7	15	5.333	Ton	1,825	235	134	2,194	2,575
0050	Average	G		12	6.667		2,050	294	167	2,511	2,975
0080	Maximum	G		9	8.889		2,450	390	223	3,063	3,650
0130	8K1, 5.1 Lb/LF	G		1200	.067	L.F.	5.20	2.94	1.67	9.81	12.85
0140	10K1, 5.0 Lb/LF	G		1200	.067		5.10	2.94	1.67	9.71	12.75
0160	12K3, 5.7 Lb/LF	G		1500	.053		5.80	2.35	1.34	9.49	12.10
0180	14K3, 6.0 Lb/LF	G		1500	.053		6.10	2.35	1.34	9.79	12.45

05 21 19.10 Open Web Joists			Crew	Daily Output	Labor-Hours	Unit	Material	2009 Bare Costs Labor	Equipment	Total	Total Incl O&P
0200	16K3, 6.3 Lb/LF	G	E-7	1800	.044	L.F.	6.40	1.96	1.12	9.48	11.80
0220	16K6, 8.1 Lb/LF	G		1800	.044		8.25	1.96	1.12	11.33	13.85
0240	18K5, 7.7 Lb/LF	G		2000	.040		7.85	1.76	1	10.61	12.95
0260	18K9, 10.2 Lb/LF	G		2000	.040		10.40	1.76	1	13.16	15.75
0410	Span 30' to 50', minimum	G		17	4.706	Ton	1,775	208	118	2,101	2,450
0440	Average	G		17	4.706		2,000	208	118	2,326	2,700
0460	Maximum	G		10	8		2,125	355	201	2,681	3,175
0500	20K5, 8.2 Lb/LF	G		2000	.040	L.F.	8.20	1.76	1	10.96	13.30
0520	20K9, 10.8 Lb/LF	G		2000	.040		10.80	1.76	1	13.56	16.20
0540	22K5, 8.8 Lb/LF	G		2000	.040		8.80	1.76	1	11.56	14
0560	22K9, 11.3 Lb/LF	G		2000	.040		11.30	1.76	1	14.06	16.75
0580	24K6, 9.7 Lb/LF	G		2200	.036		9.70	1.60	.91	12.21	14.55
0600	24K10, 13.1 Lb/LF	G		2200	.036		13.10	1.60	.91	15.61	18.30
0620	26K6, 10.6 Lb/LF	G		2200	.036		10.60	1.60	.91	13.11	15.55
0640	26K10, 13.8 Lb/LF	G		2200	.036		13.80	1.60	.91	16.31	19.10
0660	28K8, 12.7 Lb/LF	G		2400	.033		12.70	1.47	.84	15.01	17.50
0680	28K12, 17.1 Lb/LF	G		2400	.033		17.10	1.47	.84	19.41	22.50
0700	30K8, 13.2 Lb/LF	G		2400	.033		13.20	1.47	.84	15.51	18.05
0720	30K12, 17.6 Lb/LF	G		2400	.033		17.60	1.47	.84	19.91	23
0800	For less than 40-ton job lots										
0802	For 30 to 39 tons, add						10%				
0804	20 to 29 tons, add						20%				
0806	10 to 19 tons, add						30%				
0807	5 to 9 tons, add						50%	25%			
0808	1 to 4 tons, add						75%	50%			
0809	Less than 1 ton, add						100%	100%			
1010	CS series, 40-ton job lots, horizontal bridging, shop primer										
1020	Spans to 30', minimum	G	E-7	15	5.333	Ton	1,875	235	134	2,244	2,650
1040	Average	G		12	6.667		2,100	294	167	2,561	3,025
1060	Maximum	G		9	8.889		2,475	390	223	3,088	3,675
1100	10CS2, 7.5 Lb/LF	G		1200	.067	L.F.	7.85	2.94	1.67	12.46	15.80
1120	12CS2, 8.0 Lb/LF	G		1500	.053		8.35	2.35	1.34	12.04	14.90
1140	14CS2, 8.0Lb/LF	G		1500	.053		8.35	2.35	1.34	12.04	14.90
1160	16CS2, 8.5 Lb/LF	G		1800	.044		8.90	1.96	1.12	11.98	14.55
1180	16CS4, 14.5 Lb/LF	G		1800	.044		15.15	1.96	1.12	18.23	21.50
1200	18CS2, 9.0 Lb/LF	G		2000	.040		9.40	1.76	1	12.16	14.65
1220	18CS4, 15.0 Lb/LF	G		2000	.040		15.70	1.76	1	18.46	21.50
1240	20CS2, 9.5 Lb/LF	G		2000	.040		9.95	1.76	1	12.71	15.25
1260	20CS4, 16.5 Lb/LF	G		2000	.040		17.25	1.76	1	20.01	23.50
1280	22CS2, 10.0 Lb/LF	G		2000	.040		10.45	1.76	1	13.21	15.80
1300	22CS4, 16.5 Lb/LF	G		2000	.040		17.25	1.76	1	20.01	23.50
1320	24CS2, 10.0 Lb/LF	G		2200	.036		10.45	1.60	.91	12.96	15.40
1340	24CS4, 16.5 Lb/LF	G		2200	.036		17.25	1.60	.91	19.76	23
1360	26CS2, 10.0 Lb/LF	G		2200	.036		10.45	1.60	.91	12.96	15.40
1380	26CS4, 16.5 Lb/LF	G		2200	.036		17.25	1.60	.91	19.76	23
1400	28CS2, 10.5 Lb/LF	G		2400	.033		11	1.47	.84	13.31	15.65
1420	28CS4, 16.5 Lb/LF	G		2400	.033		17.25	1.47	.84	19.56	22.50
1440	30CS2, 11.0 Lb/LF	G		2400	.033		11.50	1.47	.84	13.81	16.20
1460	30CS4, 16.5 Lb/LF	G		2400	.033		17.25	1.47	.84	19.56	22.50
1500	For less than 40-ton job lots										
1502	For 30 to 39 tons, add						10%				
1504	20 to 29 tons, add						20%				
1506	10 to 19 tons, add						30%				

05 21 Steel Joist Framing

05 21 19 – Open Web Steel Joist Framing

05 21 19.10 Open Web Joists		Crew	Daily Output	Labor-Hours	Unit	Material (2009 Bare Costs)	Labor (2009 Bare Costs)	Equipment (2009 Bare Costs)	Total (2009 Bare Costs)	Total Incl O&P
1507	5 to 9 tons, add					50%	25%			
1508	1 to 4 tons, add					75%	50%			
1509	Less than 1 ton, add					100%	100%			
6200	For shop prime paint other than mfrs. standard, add					20%				
6300	For bottom chord extensions, add per chord G				Ea.	44.50			44.50	48.50
6400	Individual steel bearing plate, 6" x 6" x 1/4" with J-hook G	1 Bric	160	.050	"	9	2.03		11.03	13.20
9000	Minimum labor and equipment, bar joists	F-6	1	40	Job		1,475	770	2,245	3,275

05 21 23 – Steel Joist Girder Framing

05 21 23.50 Joist Girders		Crew	Daily Output	Labor-Hours	Unit	Material	Labor	Equipment	Total	Total Incl O&P
0010	**JOIST GIRDERS**									
0015	Made from recycled materials									
7000	Joist girders, 40-ton job lots, shop primer, minimum G	E-5	15	5.333	Ton	1,850	235	125	2,210	2,600
7020	Average G		13	6.154		2,050	271	144	2,465	2,900
7040	Maximum G		11	7.273		2,150	320	170	2,640	3,125
7100	For less than 40-ton job lots									
7102	For 30 to 39 tons, add					10%				
7104	20 to 29 tons, add					20%				
7106	10 to 19 tons, add					30%				
7107	5 to 9 tons, add					50%	25%			
7108	1 to 4 tons, add					75%	50%			
7109	Less than 1 ton, add					100%	100%			
8000	Trusses, 40-ton job lots, shop fabricated WT chords, shop primer, average G	E-5	11	7.273	Ton	6,675	320	170	7,165	8,125
8100	For less than 40-ton job lots									
8102	For 30 to 39 tons, add					10%				
8104	20 to 29 tons, add					20%				
8106	10 to 19 tons, add					30%				
8107	5 to 9 tons, add					50%	25%			
8108	1 to 4 tons, add					75%	50%			
8109	Less than 1 ton, add					100%	100%			

05 31 Steel Decking

05 31 13 – Steel Floor Decking

05 31 13.50 Floor Decking		Crew	Daily Output	Labor-Hours	Unit	Material	Labor	Equipment	Total	Total Incl O&P
0010	**FLOOR DECKING** R053100-10									
0015	Made from recycled materials									
3200	Open decking, 3" deep, wide rib, 22 gauge, galvanized, under 50 squares G	E-4	3600	.009	S.F.	3.54	.40	.04	3.98	4.67
3300	20 gauge, under 50 squares G		3400	.009		4.12	.43	.04	4.59	5.35
3400	18 gauge, under 50 squares G		3200	.010		5.35	.45	.04	5.84	6.75
3500	16 gauge, under 50 squares G		3000	.011		7.05	.48	.04	7.57	8.70
3700	4-1/2" deep, long span roof, over 50 squares, 20 gauge G		2700	.012		6.60	.54	.05	7.19	8.30
3800	18 gauge G		2460	.013		8.50	.59	.05	9.14	10.50
3900	16 gauge G		2350	.014		6.35	.62	.06	7.03	8.20
4100	6" deep, long span, 18 gauge G		2000	.016		12.15	.72	.07	12.94	14.80
4200	16 gauge G		1930	.017		9.10	.75	.07	9.92	11.45
4300	14 gauge G		1860	.017		11.70	.78	.07	12.55	14.40
4500	7-1/2" deep, long span, 18 gauge G		1690	.019		13.35	.86	.08	14.29	16.40
4600	16 gauge G		1590	.020		10	.91	.08	10.99	12.80
4700	14 gauge G		1490	.021		12.90	.97	.09	13.96	16.05
4800	For painted instead of galvanized, deduct					2%				
5000	For acoustical perforated, with fiberglass, add				S.F.	1.74			1.74	1.91

05 31 Steel Decking

05 31 13 – Steel Floor Decking

05 31 13.50 Floor Decking			Crew	Daily Output	Labor-Hours	Unit	2009 Bare Costs Material	Labor	Equipment	Total	Total Incl O&P
5200	Non-cellular composite deck, galv., 2" deep, 22 gauge	G	E-4	3860	.008	S.F.	2.44	.37	.03	2.84	3.41
5300	20 gauge	G		3600	.009		2.71	.40	.04	3.15	3.76
5400	18 gauge	G		3380	.009		3.43	.43	.04	3.90	4.60
5500	16 gauge	G		3200	.010		4.30	.45	.04	4.79	5.60
5700	3" deep, galv., 22 gauge	G		3200	.010		2.66	.45	.04	3.15	3.82
5800	20 gauge	G		3000	.011		2.98	.48	.04	3.50	4.22
5900	18 gauge	G		2850	.011		3.65	.51	.05	4.21	5
6000	16 gauge	G	↓	2700	.012	↓	4.89	.54	.05	5.48	6.45
9000	Minimum labor/equipment charge		1 Sswk	1	8	Job		360		360	660

05 31 23 – Steel Roof Decking

05 31 23.50 Roof Decking			Crew	Daily Output	Labor-Hours	Unit	Material	Labor	Equipment	Total	Total Incl O&P
0010	**ROOF DECKING**										
0015	Made from recycled materials										
2100	Open type, galv., 1-1/2" deep wide rib, 22 gauge, under 50 squares	G	E-4	4500	.007	S.F.	2.58	.32	.03	2.93	3.46
2400	Over 500 squares	G		5100	.006		1.85	.28	.03	2.16	2.60
2600	20 gauge, under 50 squares	G		3865	.008		3.03	.37	.03	3.43	4.06
2700	Over 500 squares	G		4300	.007		2.18	.34	.03	2.55	3.05
2900	18 gauge, under 50 squares	G		3800	.008		3.91	.38	.04	4.33	5.05
3000	Over 500 squares	G		4300	.007		2.82	.34	.03	3.19	3.75
3050	16 gauge, under 50 squares	G		3700	.009		5.25	.39	.04	5.68	6.55
3100	Over 500 squares	G	↓	4200	.008	↓	3.79	.34	.03	4.16	4.85

05 31 33 – Steel Form Decking

05 31 33.50 Form Decking			Crew	Daily Output	Labor-Hours	Unit	Material	Labor	Equipment	Total	Total Incl O&P
0010	**FORM DECKING**										
0015	Made from recycled materials										
6100	Slab form, steel, 28 gauge, 9/16" deep, uncoated	G	E-4	4000	.008	S.F.	1.72	.36	.03	2.11	2.60
6200	Galvanized	G		4000	.008		1.52	.36	.03	1.91	2.38
6220	24 gauge, 1" deep, uncoated	G		3900	.008		1.87	.37	.03	2.27	2.79
6240	Galvanized	G		3900	.008		2.20	.37	.03	2.60	3.15
6300	24 gauge, 1-5/16" deep, uncoated	G		3800	.008		1.99	.38	.04	2.41	2.94
6400	Galvanized	G		3800	.008		2.34	.38	.04	2.76	3.32
6500	22 gauge, 1-5/16" deep, uncoated	G		3700	.009		2.50	.39	.04	2.93	3.51
6600	Galvanized	G		3700	.009		2.55	.39	.04	2.98	3.57
6700	22 gauge, 2" deep uncoated	G		3600	.009		3.28	.40	.04	3.72	4.39
6800	Galvanized	G	↓	3600	.009	↓	3.22	.40	.04	3.66	4.32
7000	Sheet metal edge closure form, 12" wide with 2 bends, galv										
7100	18 gauge	G	E-14	360	.022	L.F.	5.30	1.04	.37	6.71	8.20
7200	16 gauge	G	"	360	.022	"	7.20	1.04	.37	8.61	10.25

05 35 Raceway Decking Assemblies

05 35 13 – Steel Cellular Decking

05 35 13.50 Cellular Decking			Crew	Daily Output	Labor-Hours	Unit	Material	Labor	Equipment	Total	Total Incl O&P
0010	**CELLULAR DECKING**										
0015	Made from recycled materials										
0200	Cellular units, galv, 2" deep, 20-20 gauge, over 15 squares	G	E-4	1460	.022	S.F.	10.20	.99	.09	11.28	13.20
0250	18-20 gauge	G		1420	.023		11.60	1.02	.09	12.71	14.80
0300	18-18 gauge	G		1390	.023		11.95	1.04	.10	13.09	15.20
0320	16-18 gauge	G		1360	.024		14.20	1.06	.10	15.36	17.70
0340	16-16 gauge	G		1330	.024		15.85	1.09	.10	17.04	19.50
0400	3" deep, galvanized, 20-20 gauge	G	↓	1375	.023	↓	11.25	1.05	.10	12.40	14.40

05 35 Raceway Decking Assemblies

05 35 13 – Steel Cellular Decking

05 35 13.50 Cellular Decking			Crew	Daily Output	Labor-Hours	Unit	2009 Bare Costs Material	Labor	Equipment	Total	Total Incl O&P
0500	18-20 gauge	G	E-4	1350	.024	S.F.	13.60	1.07	.10	14.77	17.05
0600	18-18 gauge	G		1290	.025		13.55	1.12	.10	14.77	17.15
0700	16-18 gauge	G		1230	.026		15.25	1.18	.11	16.54	19.10
0800	16-16 gauge	G		1150	.028		16.65	1.26	.12	18.03	21
1000	4-1/2" deep, galvanized, 20-18 gauge	G		1100	.029		15.70	1.31	.12	17.13	19.80
1100	18-18 gauge	G		1040	.031		15.60	1.39	.13	17.12	19.85
1200	16-18 gauge	G		980	.033		17.55	1.48	.14	19.17	22
1300	16-16 gauge	G		935	.034		19.15	1.55	.14	20.84	24
1500	For acoustical deck, add						15%				
1700	For cells used for ventilation, add						15%				
1900	For multi-story or congested site, add							50%			
8000	Metal deck and trench, 2" thick, 20 gauge, combination										
8010	60% cellular, 40% non-cellular, inserts and trench	G	R-4	1100	.036	S.F.	20.50	1.66	.12	22.28	25.50

05 41 Structural Metal Stud Framing

05 41 13 – Load-Bearing Metal Stud Framing

05 41 13.05 Bracing

			Crew	Daily Output	Labor-Hours	Unit	Material	Labor	Equipment	Total	Total Incl O&P
0010	**BRACING**, shear wall X-bracing, per 10' x 10' bay, one face										
0015	Made of recycled materials										
0120	Metal strap, 20 ga x 4" wide	G	2 Carp	18	.889	Ea.	27	35.50		62.50	88
0130	6" wide	G		18	.889		42	35.50		77.50	105
0160	18 ga x 4" wide	G		16	1		39	40		79	109
0170	6" wide	G		16	1		57	40		97	129
0410	Continuous strap bracing, per horizontal row on both faces										
0420	Metal strap, 20 ga x 2" wide, studs 12" O.C.	G	1 Carp	7	1.143	C.L.F.	68	45.50		113.50	151
0430	16" O.C.	G		8	1		68	40		108	141
0440	24" O.C.	G		10	.800		68	32		100	128
0450	18 ga x 2" wide, studs 12" O.C.	G		6	1.333		94.50	53.50		148	192
0460	16" O.C.	G		7	1.143		94.50	45.50		140	180
0470	24" O.C.	G		8	1		94.50	40		134.50	170

05 41 13.10 Bridging

			Crew	Daily Output	Labor-Hours	Unit	Material	Labor	Equipment	Total	Total Incl O&P
0010	**BRIDGING**, solid between studs w/ 1-1/4" leg track, per stud bay										
0015	Made from recycled materials										
0200	Studs 12" O.C., 18 ga x 2-1/2" wide	G	1 Carp	125	.064	Ea.	1.16	2.56		3.72	5.50
0210	3-5/8" wide	G		120	.067		1.40	2.66		4.06	5.95
0220	4" wide	G		120	.067		1.48	2.66		4.14	6.05
0230	6" wide	G		115	.070		1.93	2.78		4.71	6.70
0240	8" wide	G		110	.073		2.45	2.91		5.36	7.50
0300	16 ga x 2-1/2" wide	G		115	.070		1.44	2.78		4.22	6.20
0310	3-5/8" wide	G		110	.073		1.76	2.91		4.67	6.75
0320	4" wide	G		110	.073		1.88	2.91		4.79	6.85
0330	6" wide	G		105	.076		2.42	3.04		5.46	7.65
0340	8" wide	G		100	.080		3.08	3.20		6.28	8.70
1200	Studs 16" O.C., 18 ga x 2-1/2" wide	G		125	.064		1.49	2.56		4.05	5.85
1210	3-5/8" wide	G		120	.067		1.80	2.66		4.46	6.40
1220	4" wide	G		120	.067		1.90	2.66		4.56	6.50
1230	6" wide	G		115	.070		2.48	2.78		5.26	7.30
1240	8" wide	G		110	.073		3.14	2.91		6.05	8.25
1300	16 ga x 2-1/2" wide	G		115	.070		1.85	2.78		4.63	6.60
1310	3-5/8" wide	G		110	.073		2.26	2.91		5.17	7.30
1320	4" wide	G		110	.073		2.41	2.91		5.32	7.45

05 41 Structural Metal Stud Framing

05 41 13 – Load-Bearing Metal Stud Framing

05 41 13.10 Bridging			Crew	Daily Output	Labor-Hours	Unit	Material	2009 Bare Costs Labor	Equipment	Total	Total Incl O&P
1330	6" wide	G	1 Carp	105	.076	Ea.	3.10	3.04		6.14	8.40
1340	8" wide	G		100	.080		3.94	3.20		7.14	9.65
2200	Studs 24" O.C., 18 ga x 2-1/2" wide	G		125	.064		2.15	2.56		4.71	6.60
2210	3-5/8" wide	G		120	.067		2.60	2.66		5.26	7.25
2220	4" wide	G		120	.067		2.75	2.66		5.41	7.40
2230	6" wide	G		115	.070		3.58	2.78		6.36	8.55
2240	8" wide	G		110	.073		4.54	2.91		7.45	9.80
2300	16 ga x 2-1/2" wide	G		115	.070		2.67	2.78		5.45	7.55
2310	3-5/8" wide	G		110	.073		3.27	2.91		6.18	8.40
2320	4" wide	G		110	.073		3.49	2.91		6.40	8.65
2330	6" wide	G		105	.076		4.49	3.04		7.53	9.95
2340	8" wide	G		100	.080		5.70	3.20		8.90	11.60
3000	Continuous bridging, per row										
3100	16 ga x 1-1/2" channel thru studs 12" O.C.	G	1 Carp	6	1.333	C.L.F.	65	53.50		118.50	160
3110	16" O.C.	G		7	1.143		65	45.50		110.50	147
3120	24" O.C.	G		8.80	.909		65	36.50		101.50	132
4100	2" x 2" angle x 18 ga, studs 12" O.C.	G		7	1.143		94.50	45.50		140	180
4110	16" O.C.	G		9	.889		94.50	35.50		130	163
4120	24" O.C.	G		12	.667		94.50	26.50		121	148
4200	16 ga, studs 12" O.C.	G		5	1.600		120	64		184	238
4210	16" O.C.	G		7	1.143		120	45.50		165.50	208
4220	24" O.C.	G		10	.800		120	32		152	185

05 41 13.25 Framing, Boxed Headers/Beams			Crew	Daily Output	Labor-Hours	Unit	Material	Labor	Equipment	Total	Total Incl O&P
0010	**FRAMING, BOXED HEADERS/BEAMS**										
0015	Made from recycled materials										
0200	Double, 18 ga x 6" deep	G	2 Carp	220	.073	L.F.	6.70	2.91		9.61	12.20
0210	8" deep	G		210	.076		7.45	3.04		10.49	13.20
0220	10" deep	G		200	.080		9	3.20		12.20	15.25
0230	12" deep	G		190	.084		9.90	3.36		13.26	16.45
0300	16 ga x 8" deep	G		180	.089		8.55	3.55		12.10	15.25
0310	10" deep	G		170	.094		10.30	3.76		14.06	17.50
0320	12" deep	G		160	.100		11.20	4		15.20	18.90
0400	14 ga x 10" deep	G		140	.114		11.95	4.57		16.52	20.50
0410	12" deep	G		130	.123		13.10	4.92		18.02	22.50
1210	Triple, 18 ga x 8" deep	G		170	.094		10.80	3.76		14.56	18.10
1220	10" deep	G		165	.097		13	3.87		16.87	20.50
1230	12" deep	G		160	.100		14.35	4		18.35	22.50
1300	16 ga x 8" deep	G		145	.110		12.45	4.41		16.86	21
1310	10" deep	G		140	.114		14.85	4.57		19.42	24
1320	12" deep	G		135	.119		16.20	4.73		20.93	25.50
1400	14 ga x 10" deep	G		115	.139		16.30	5.55		21.85	27
1410	12" deep	G		110	.145		18.10	5.80		23.90	29.50

05 41 13.30 Framing, Stud Walls			Crew	Daily Output	Labor-Hours	Unit	Material	Labor	Equipment	Total	Total Incl O&P
0010	**FRAMING, STUD WALLS** w/ top & bottom track, no openings,										
0020	Headers, beams, bridging or bracing										
0025	Made from recycled materials										
4100	8' high walls, 18 ga x 2-1/2" wide, studs 12" O.C.	G	2 Carp	54	.296	L.F.	11.25	11.85		23.10	32
4110	16" O.C.	G		77	.208		9	8.30		17.30	23.50
4120	24" O.C.	G		107	.150		6.75	5.95		12.70	17.25
4130	3-5/8" wide, studs 12" O.C.	G		53	.302		13.40	12.05		25.45	34.50
4140	16" O.C.	G		76	.211		10.70	8.40		19.10	25.50
4150	24" O.C.	G		105	.152		8.05	6.10		14.15	18.90

05 41 13 – Load-Bearing Metal Stud Framing

05 41 13.30 Framing, Stud Walls		Crew	Daily Output	Labor-Hours	Unit	Material	Labor	Equipment	Total	Total Incl O&P
						2009 Bare Costs				
4160	4" wide, studs 12" O.C. G	2 Carp	52	.308	L.F.	14	12.30		26.30	36
4170	16" O.C. G		74	.216		11.20	8.65		19.85	26.50
4180	24" O.C. G		103	.155		8.40	6.20		14.60	19.50
4190	6" wide, studs 12" O.C. G		51	.314		17.85	12.55		30.40	40
4200	16" O.C. G		73	.219		14.30	8.75		23.05	30
4210	24" O.C. G		101	.158		10.75	6.35		17.10	22.50
4220	8" wide, studs 12" O.C. G		50	.320		22	12.80		34.80	45
4230	16" O.C. G		72	.222		17.50	8.90		26.40	34
4240	24" O.C. G		100	.160		13.20	6.40		19.60	25
4300	16 ga x 2-1/2" wide, studs 12" O.C. G		47	.340		13.25	13.60		26.85	37
4310	16" O.C. G		68	.235		10.50	9.40		19.90	27
4320	24" O.C. G		94	.170		7.75	6.80		14.55	19.75
4330	3-5/8" wide, studs 12" O.C. G		46	.348		15.85	13.90		29.75	40.50
4340	16" O.C. G		66	.242		12.55	9.70		22.25	30
4350	24" O.C. G		92	.174		9.25	6.95		16.20	21.50
4360	4" wide, studs 12" O.C. G		45	.356		16.75	14.20		30.95	42
4370	16" O.C. G		65	.246		13.25	9.85		23.10	31
4380	24" O.C. G		90	.178		9.80	7.10		16.90	22.50
4390	6" wide, studs 12" O.C. G		44	.364		21	14.55		35.55	47
4400	16" O.C. G		64	.250		16.70	10		26.70	35
4410	24" O.C. G		88	.182		12.35	7.25		19.60	25.50
4420	8" wide, studs 12" O.C. G		43	.372		26	14.85		40.85	53
4430	16" O.C. G		63	.254		20.50	10.15		30.65	39.50
4440	24" O.C. G		86	.186		15.25	7.45		22.70	29
5100	10' high walls, 18 ga x 2-1/2" wide, studs 12" O.C. G		54	.296		13.50	11.85		25.35	34.50
5110	16" O.C. G		77	.208		10.70	8.30		19	25.50
5120	24" O.C. G		107	.150		7.85	5.95		13.80	18.50
5130	3-5/8" wide, studs 12" O.C. G		53	.302		16.05	12.05		28.10	37.50
5140	16" O.C. G		76	.211		12.70	8.40		21.10	28
5150	24" O.C. G		105	.152		9.35	6.10		15.45	20.50
5160	4" wide, studs 12" O.C. G		52	.308		16.80	12.30		29.10	39
5170	16" O.C. G		74	.216		13.30	8.65		21.95	29
5180	24" O.C. G		103	.155		9.80	6.20		16	21
5190	6" wide, studs 12" O.C. G		51	.314		21.50	12.55		34.05	44
5200	16" O.C. G		73	.219		16.95	8.75		25.70	33
5210	24" O.C. G		101	.158		12.55	6.35		18.90	24.50
5220	8" wide, studs 12" O.C. G		50	.320		26	12.80		38.80	49.50
5230	16" O.C. G		72	.222		20.50	8.90		29.40	37.50
5240	24" O.C. G		100	.160		15.35	6.40		21.75	27.50
5300	16 ga x 2-1/2" wide, studs 12" O.C. G		47	.340		16	13.60		29.60	40
5310	16" O.C. G		68	.235		12.55	9.40		21.95	29.50
5320	24" O.C. G		94	.170		9.10	6.80		15.90	21.50
5330	3-5/8" wide, studs 12" O.C. G		46	.348		19.15	13.90		33.05	44
5340	16" O.C. G		66	.242		15.05	9.70		24.75	32.50
5350	24" O.C. G		92	.174		10.90	6.95		17.85	23.50
5360	4" wide, studs 12" O.C. G		45	.356		20	14.20		34.20	45.50
5370	16" O.C. G		65	.246		15.85	9.85		25.70	33.50
5380	24" O.C. G		90	.178		11.50	7.10		18.60	24.50
5390	6" wide, studs 12" O.C. G		44	.364		25.50	14.55		40.05	52
5400	16" O.C. G		64	.250		19.95	10		29.95	38.50
5410	24" O.C. G		88	.182		14.55	7.25		21.80	28
5420	8" wide, studs 12" O.C. G		43	.372		31	14.85		45.85	59
5430	16" O.C. G	↓	63	.254	↓	24.50	10.15		34.65	44

05 41 13.30 Framing, Stud Walls		Crew	Daily Output	Labor-Hours	Unit	Material	Labor	Equipment	Total	Total Incl O&P
						2009 Bare Costs				
5440 24" O.C.	G	2 Carp	86	.186	L.F.	17.90	7.45		25.35	32
6190 12' high walls, 18 ga x 6" wide, studs 12" O.C.	G		41	.390		25	15.60		40.60	53
6200 16" O.C.	G		58	.276		19.60	11		30.60	39.50
6210 24" O.C.	G		81	.198		14.30	7.90		22.20	29
6220 8" wide, studs 12" O.C.	G		40	.400		30.50	16		46.50	60
6230 16" O.C.	G		57	.281		24	11.20		35.20	45
6240 24" O.C.	G		80	.200		17.50	8		25.50	32.50
6390 16 ga x 6" wide, studs 12" O.C.	G		35	.457		29.50	18.25		47.75	62.50
6400 16" O.C.	G		51	.314		23	12.55		35.55	46
6410 24" O.C.	G		70	.229		16.70	9.15		25.85	33.50
6420 8" wide, studs 12" O.C.	G		34	.471		36.50	18.80		55.30	71
6430 16" O.C.	G		50	.320		28.50	12.80		41.30	52.50
6440 24" O.C.	G		69	.232		20.50	9.25		29.75	38
6530 14 ga x 3-5/8" wide, studs 12" O.C.	G		34	.471		28	18.80		46.80	62
6540 16" O.C.	G		48	.333		22	13.30		35.30	46
6550 24" O.C.	G		65	.246		15.80	9.85		25.65	33.50
6560 4" wide, studs 12" O.C.	G		33	.485		29.50	19.35		48.85	64.50
6570 16" O.C.	G		47	.340		23	13.60		36.60	48
6580 24" O.C.	G		64	.250		16.65	10		26.65	35
6730 12 ga x 3-5/8" wide, studs 12" O.C.	G		31	.516		39	20.50		59.50	77
6740 16" O.C.	G		43	.372		30	14.85		44.85	57.50
6750 24" O.C.	G		59	.271		21	10.85		31.85	41.50
6760 4" wide, studs 12" O.C.	G		30	.533		41.50	21.50		63	81
6770 16" O.C.	G		42	.381		32	15.20		47.20	60.50
6780 24" O.C.	G		58	.276		22.50	11		33.50	43
7390 16' high walls, 16 ga x 6" wide, studs 12" O.C.	G		33	.485		38.50	19.35		57.85	74
7400 16" O.C.	G		48	.333		29.50	13.30		42.80	54.50
7410 24" O.C.	G		67	.239		21	9.55		30.55	39
7420 8" wide, studs 12" O.C.	G		32	.500		47	20		67	85
7430 16" O.C.	G		47	.340		36.50	13.60		50.10	62.50
7440 24" O.C.	G		66	.242		26	9.70		35.70	44.50
7560 14 ga x 4" wide, studs 12" O.C.	G		31	.516		38.50	20.50		59	76.50
7570 16" O.C.	G		45	.356		29.50	14.20		43.70	56
7580 24" O.C.	G		61	.262		21	10.50		31.50	40.50
7590 6" wide, studs 12" O.C.	G		30	.533		48.50	21.50		70	88.50
7600 16" O.C.	G		44	.364		37.50	14.55		52.05	65
7610 24" O.C.	G		60	.267		26.50	10.65		37.15	46.50
7760 12 ga x 4" wide, studs 12" O.C.	G		29	.552		54.50	22		76.50	96.50
7770 16" O.C.	G		40	.400		41.50	16		57.50	72.50
7780 24" O.C.	G		55	.291		29	11.60		40.60	51
7790 6" wide, studs 12" O.C.	G		28	.571		69	23		92	114
7800 16" O.C.	G		39	.410		53	16.40		69.40	85
7810 24" O.C.	G		54	.296		37	11.85		48.85	60
8590 20' high walls, 14 ga x 6" wide, studs 12" O.C.	G		29	.552		59.50	22		81.50	102
8600 16" O.C.	G		42	.381		45.50	15.20		60.70	75.50
8610 24" O.C.	G		57	.281		32	11.20		43.20	53.50
8620 8" wide, studs 12" O.C.	G		28	.571		72.50	23		95.50	118
8630 16" O.C.	G		41	.390		56	15.60		71.60	87
8640 24" O.C.	G		56	.286		39.50	11.40		50.90	62
8790 12 ga x 6" wide, studs 12" O.C.	G		27	.593		85	23.50		108.50	133
8800 16" O.C.	G		37	.432		65	17.30		82.30	100
8810 24" O.C.	G		51	.314		45	12.55		57.55	70
8820 8" wide, studs 12" O.C.	G	↓	26	.615	↓	103	24.50		127.50	155

05 41 Structural Metal Stud Framing

05 41 13 – Load-Bearing Metal Stud Framing

05 41 13.30 Framing, Stud Walls			Crew	Daily Output	Labor-Hours	Unit	Material	Labor	Equipment	Total	Total Incl O&P
8830	16″ O.C.	G	2 Carp	36	.444	L.F.	79	17.75		96.75	117
8840	24″ O.C.	G		50	.320		54.50	12.80		67.30	81
9000	Minimum labor/equipment charge			4	4	Job		160		160	264

Column group heading: 2009 Bare Costs (Material, Labor, Equipment, Total).

05 42 Cold-Formed Metal Joist Framing

05 42 13 – Cold-Formed Metal Floor Joist Framing

05 42 13.05 Bracing

Line	Description		Crew	Daily Output	Labor-Hours	Unit	Material	Labor	Equipment	Total	Total Incl O&P
0010	**BRACING**, continuous, per row, top & bottom										
0015	Made from recycled materials										
0120	Flat strap, 20 ga x 2″ wide, joists at 12″ O.C.	G	1 Carp	4.67	1.713	C.L.F.	71.50	68.50		140	192
0130	16″ O.C.	G		5.33	1.501		69	60		129	175
0140	24″ O.C.	G		6.66	1.201		66.50	48		114.50	152
0150	18 ga x 2″ wide, joists at 12″ O.C.	G		4	2		93.50	80		173.50	235
0160	16″ O.C.	G		4.67	1.713		92	68.50		160.50	214
0170	24″ O.C.	G		5.33	1.501		90.50	60		150.50	199

05 42 13.10 Bridging

Line	Description		Crew	Daily Output	Labor-Hours	Unit	Material	Labor	Equipment	Total	Total Incl O&P
0010	**BRIDGING**, solid between joists w/ 1-1/4″ leg track, per joist bay										
0015	Made from recycled materials										
0230	Joists 12″ O.C., 18 ga track x 6″ wide	G	1 Carp	80	.100	Ea.	1.93	4		5.93	8.70
0240	8″ wide	G		75	.107		2.45	4.26		6.71	9.75
0250	10″ wide	G		70	.114		3.02	4.57		7.59	10.90
0260	12″ wide	G		65	.123		3.50	4.92		8.42	11.95
0330	16 ga track x 6″ wide	G		70	.114		2.42	4.57		6.99	10.20
0340	8″ wide	G		65	.123		3.08	4.92		8	11.50
0350	10″ wide	G		60	.133		3.78	5.35		9.13	12.95
0360	12″ wide	G		55	.145		4.36	5.80		10.16	14.40
0440	14 ga track x 8″ wide	G		60	.133		3.87	5.35		9.22	13.05
0450	10″ wide	G		55	.145		4.76	5.80		10.56	14.85
0460	12″ wide	G		50	.160		5.50	6.40		11.90	16.60
0550	12 ga track x 10″ wide	G		45	.178		7	7.10		14.10	19.40
0560	12″ wide	G		40	.200		7.90	8		15.90	22
1230	16″ O.C., 18 ga track x 6″ wide	G		80	.100		2.48	4		6.48	9.30
1240	8″ wide	G		75	.107		3.14	4.26		7.40	10.50
1250	10″ wide	G		70	.114		3.88	4.57		8.45	11.80
1260	12″ wide	G		65	.123		4.49	4.92		9.41	13.05
1330	16 ga track x 6″ wide	G		70	.114		3.10	4.57		7.67	10.95
1340	8″ wide	G		65	.123		3.94	4.92		8.86	12.45
1350	10″ wide	G		60	.133		4.85	5.35		10.20	14.15
1360	12″ wide	G		55	.145		5.60	5.80		11.40	15.75
1440	14 ga track x 8″ wide	G		60	.133		4.97	5.35		10.32	14.25
1450	10″ wide	G		55	.145		6.10	5.80		11.90	16.30
1460	12″ wide	G		50	.160		7.05	6.40		13.45	18.30
1550	12 ga track x 10″ wide	G		45	.178		9	7.10		16.10	21.50
1560	12″ wide	G		40	.200		10.15	8		18.15	24.50
2230	24″ O.C., 18 ga track x 6″ wide	G		80	.100		3.58	4		7.58	10.55
2240	8″ wide	G		75	.107		4.54	4.26		8.80	12.05
2250	10″ wide	G		70	.114		5.60	4.57		10.17	13.70
2260	12″ wide	G		65	.123		6.50	4.92		11.42	15.25
2330	16 ga track x 6″ wide	G		70	.114		4.49	4.57		9.06	12.50
2340	8″ wide	G		65	.123		5.70	4.92		10.62	14.40
2350	10″ wide	G	2 Carp	60	.133		7	5.35		12.35	16.50

05 42 Cold-Formed Metal Joist Framing

05 42 13 – Cold-Formed Metal Floor Joist Framing

05 42 13.10 Bridging			Crew	Daily Output	Labor-Hours	Unit	2009 Bare Costs Material	Labor	Equipment	Total	Total Incl O&P
2360	12″ wide	G	1 Carp	55	.145	Ea.	8.10	5.80		13.90	18.50
2440	14 ga track x 8″ wide	G		60	.133		7.20	5.35		12.55	16.70
2450	10″ wide	G		55	.145		8.85	5.80		14.65	19.30
2460	12″ wide	G		50	.160		10.20	6.40		16.60	22
2550	12 ga track x 10″ wide	G		45	.178		13	7.10		20.10	26
2560	12″ wide	G	↓	40	.200	↓	14.70	8		22.70	29.50

05 42 13.25 Framing, Band Joist

			Crew	Daily Output	Labor-Hours	Unit	Material	Labor	Equipment	Total	Total Incl O&P
0010	**FRAMING, BAND JOIST** (track) fastened to bearing wall										
0015	Made from recycled materials										
0220	18 ga track x 6″ deep	G	2 Carp	1000	.016	L.F.	1.58	.64		2.22	2.79
0230	8″ deep	G		920	.017		2	.69		2.69	3.34
0240	10″ deep	G		860	.019		2.47	.74		3.21	3.94
0320	16 ga track x 6″ deep	G		900	.018		1.97	.71		2.68	3.34
0330	8″ deep	G		840	.019		2.51	.76		3.27	4.02
0340	10″ deep	G		780	.021		3.09	.82		3.91	4.75
0350	12″ deep	G		740	.022		3.56	.86		4.42	5.35
0430	14 ga track x 8″ deep	G		750	.021		3.16	.85		4.01	4.89
0440	10″ deep	G		720	.022		3.89	.89		4.78	5.75
0450	12″ deep	G		700	.023		4.49	.91		5.40	6.45
0540	12 ga track x 10″ deep	G		670	.024		5.70	.95		6.65	7.85
0550	12″ deep	G	↓	650	.025	↓	6.45	.98		7.43	8.70

05 42 13.30 Framing, Boxed Headers/Beams

			Crew	Daily Output	Labor-Hours	Unit	Material	Labor	Equipment	Total	Total Incl O&P
0010	**FRAMING, BOXED HEADERS/BEAMS**										
0015	Made from recycled materials										
0200	Double, 18 ga x 6″ deep	G	2 Carp	220	.073	L.F.	6.70	2.91		9.61	12.20
0210	8″ deep	G		210	.076		7.45	3.04		10.49	13.20
0220	10″ deep	G		200	.080		9	3.20		12.20	15.25
0230	12″ deep	G		190	.084		9.90	3.36		13.26	16.45
0300	16 ga x 8″ deep	G		180	.089		8.55	3.55		12.10	15.25
0310	10″ deep	G		170	.094		10.30	3.76		14.06	17.50
0320	12″ deep	G		160	.100		11.20	4		15.20	18.90
0400	14 ga x 10″ deep	G		140	.114		11.95	4.57		16.52	20.50
0410	12″ deep	G		130	.123		13.10	4.92		18.02	22.50
0500	12 ga x 10″ deep	G		110	.145		15.80	5.80		21.60	27
0510	12″ deep	G		100	.160		17.50	6.40		23.90	30
1210	Triple, 18 ga x 8″ deep	G		170	.094		10.80	3.76		14.56	18.10
1220	10″ deep	G		165	.097		13	3.87		16.87	20.50
1230	12″ deep	G		160	.100		14.35	4		18.35	22.50
1300	16 ga x 8″ deep	G		145	.110		12.45	4.41		16.86	21
1310	10″ deep	G		140	.114		14.85	4.57		19.42	24
1320	12″ deep	G		135	.119		16.20	4.73		20.93	25.50
1400	14 ga x 10″ deep	G		115	.139		17.35	5.55		22.90	28.50
1410	12″ deep	G		110	.145		19.15	5.80		24.95	30.50
1500	12 ga x 10″ deep	G		90	.178		23	7.10		30.10	37
1510	12″ deep	G	↓	85	.188	↓	25.50	7.50		33	41

05 42 13.40 Framing, Joists

			Crew	Daily Output	Labor-Hours	Unit	Material	Labor	Equipment	Total	Total Incl O&P
0010	**FRAMING, JOISTS**, no band joists (track), web stiffeners, headers,										
0020	Beams, bridging or bracing										
0025	Made from recycled materials										
0030	Joists (2″ flange) and fasteners, materials only										
0220	18 ga x 6″ deep	G				L.F.	2.08			2.08	2.29
0230	8″ deep	G				↓	2.47			2.47	2.71

05 42 Cold-Formed Metal Joist Framing

05 42 13 – Cold-Formed Metal Floor Joist Framing

05 42 13.40 Framing, Joists			Crew	Daily Output	Labor-Hours	Unit	2009 Bare Costs Material	Labor	Equipment	Total	Total Incl O&P
0240	10″ deep	G				L.F.	2.90			2.90	3.19
0320	16 ga x 6″ deep	G					2.54			2.54	2.80
0330	8″ deep	G					3.05			3.05	3.35
0340	10″ deep	G					3.56			3.56	3.92
0350	12″ deep	G					4.03			4.03	4.44
0430	14 ga x 8″ deep	G					3.84			3.84	4.23
0440	10″ deep	G					4.43			4.43	4.87
0450	12″ deep	G					5.05			5.05	5.55
0540	12 ga x 10″ deep	G					6.45			6.45	7.10
0550	12″ deep	G					7.35			7.35	8.10
1010	Installation of joists to band joists, beams & headers, labor only										
1220	18 ga x 6″ deep		2 Carp	110	.145	Ea.		5.80		5.80	9.60
1230	8″ deep			90	.178			7.10		7.10	11.70
1240	10″ deep			80	.200			8		8	13.20
1320	16 ga x 6″ deep			95	.168			6.75		6.75	11.10
1330	8″ deep			70	.229			9.15		9.15	15.05
1340	10″ deep			60	.267			10.65		10.65	17.60
1350	12″ deep			55	.291			11.60		11.60	19.20
1430	14 ga x 8″ deep			65	.246			9.85		9.85	16.25
1440	10″ deep			45	.356			14.20		14.20	23.50
1450	12″ deep			35	.457			18.25		18.25	30
1540	12 ga x 10″ deep			40	.400			16		16	26.50
1550	12″ deep			30	.533			21.50		21.50	35
9000	Minimum labor/equipment charge			4	4	Job		160		160	264

05 42 13.45 Framing, Web Stiffeners			Crew	Daily Output	Labor-Hours	Unit	Material	Labor	Equipment	Total	Total Incl O&P
0010	**FRAMING, WEB STIFFENERS** at joist bearing, fabricated from										
0020	Stud piece (1-5/8″ flange) to stiffen joist (2″ flange)										
0025	Made from recycled materials										
2120	For 6″ deep joist, with 18 ga x 2-1/2″ stud	G	1 Carp	120	.067	Ea.	2.49	2.66		5.15	7.15
2130	3-5/8″ stud	G		110	.073		2.75	2.91		5.66	7.80
2140	4″ stud	G		105	.076		2.66	3.04		5.70	7.95
2150	6″ stud	G		100	.080		2.92	3.20		6.12	8.50
2160	8″ stud	G		95	.084		3	3.36		6.36	8.85
2220	8″ deep joist, with 2-1/2″ stud	G		120	.067		2.73	2.66		5.39	7.40
2230	3-5/8″ stud	G		110	.073		2.96	2.91		5.87	8.05
2240	4″ stud	G		105	.076		2.91	3.04		5.95	8.20
2250	6″ stud	G		100	.080		3.20	3.20		6.40	8.80
2260	8″ stud	G		95	.084		3.44	3.36		6.80	9.35
2320	10″ deep joist, with 2-1/2″ stud	G		110	.073		3.85	2.91		6.76	9.05
2330	3-5/8″ stud	G		100	.080		4.23	3.20		7.43	9.95
2340	4″ stud	G		95	.084		4.18	3.36		7.54	10.15
2350	6″ stud	G		90	.089		4.55	3.55		8.10	10.85
2360	8″ stud	G		85	.094		4.62	3.76		8.38	11.30
2420	12″ deep joist, with 2-1/2″ stud	G		110	.073		4.07	2.91		6.98	9.25
2430	3-5/8″ stud	G		100	.080		4.42	3.20		7.62	10.15
2440	4″ stud	G		95	.084		4.34	3.36		7.70	10.30
2450	6″ stud	G		90	.089		4.78	3.55		8.33	11.10
2460	8″ stud	G		85	.094		5.15	3.76		8.91	11.85
3130	For 6″ deep joist, with 16 ga x 3-5/8″ stud	G		100	.080		2.89	3.20		6.09	8.50
3140	4″ stud	G		95	.084		2.87	3.36		6.23	8.70
3150	6″ stud	G		90	.089		3.15	3.55		6.70	9.30
3160	8″ stud	G		85	.094		3.31	3.76		7.07	9.85

05 42 Cold-Formed Metal Joist Framing

05 42 13 – Cold-Formed Metal Floor Joist Framing

05 42 13.45 Framing, Web Stiffeners			Crew	Daily Output	Labor-Hours	Unit	2009 Bare Costs Material	Labor	Equipment	Total	Total Incl O&P
3230	8″ deep joist, with 3-5/8″ stud	G	1 Carp	100	.080	Ea.	3.21	3.20		6.41	8.85
3240	4″ stud	G		95	.084		3.15	3.36		6.51	9
3250	6″ stud	G		90	.089		3.49	3.55		7.04	9.70
3260	8″ stud	G		85	.094		3.73	3.76		7.49	10.30
3330	10″ deep joist, with 3-5/8″ stud	G		85	.094		4.38	3.76		8.14	11
3340	4″ stud	G		80	.100		4.48	4		8.48	11.50
3350	6″ stud	G		75	.107		4.86	4.26		9.12	12.40
3360	8″ stud	G		70	.114		5.05	4.57		9.62	13.10
3430	12″ deep joist, with 3-5/8″ stud	G		85	.094		4.79	3.76		8.55	11.45
3440	4″ stud	G		80	.100		4.70	4		8.70	11.75
3450	6″ stud	G		75	.107		5.20	4.26		9.46	12.80
3460	8″ stud	G		70	.114		5.55	4.57		10.12	13.65
4230	For 8″ deep joist, with 14 ga x 3-5/8″ stud	G		90	.089		4.16	3.55		7.71	10.45
4240	4″ stud	G		85	.094		4.24	3.76		8	10.85
4250	6″ stud	G		80	.100		4.59	4		8.59	11.65
4260	8″ stud	G		75	.107		4.92	4.26		9.18	12.45
4330	10″ deep joist, with 3-5/8″ stud	G		75	.107		5.85	4.26		10.11	13.50
4340	4″ stud	G		70	.114		5.80	4.57		10.37	13.90
4350	6″ stud	G		65	.123		6.35	4.92		11.27	15.10
4360	8″ stud	G		60	.133		6.65	5.35		12	16.10
4430	12″ deep joist, with 3-5/8″ stud	G		75	.107		6.20	4.26		10.46	13.90
4440	4″ stud	G		70	.114		6.30	4.57		10.87	14.50
4450	6″ stud	G		65	.123		6.85	4.92		11.77	15.65
4460	8″ stud	G		60	.133		7.35	5.35		12.70	16.90
5330	For 10″ deep joist, with 12 ga x 3-5/8″ stud	G		65	.123		6.15	4.92		11.07	14.90
5340	4″ stud	G		60	.133		6.35	5.35		11.70	15.75
5350	6″ stud	G		55	.145		7	5.80		12.80	17.30
5360	8″ stud	G		50	.160		7.70	6.40		14.10	19
5430	12″ deep joist, with 3-5/8″ stud	G		65	.123		6.85	4.92		11.77	15.60
5440	4″ stud	G		60	.133		6.70	5.35		12.05	16.15
5450	6″ stud	G		55	.145		7.65	5.80		13.45	18
5460	8″ stud	G		50	.160		8.80	6.40		15.20	20

05 42 23 – Cold-Formed Metal Roof Joist Framing

05 42 23.05 Framing, Bracing			Crew	Daily Output	Labor-Hours	Unit	Material	Labor	Equipment	Total	Total Incl O&P
0010	**FRAMING, BRACING**										
0015	Made from recycled materials										
0020	Continuous bracing, per row										
0100	16 ga x 1-1/2″ channel thru rafters/trusses @ 16″ O.C.	G	1 Carp	4.50	1.778	C.L.F.	65	71		136	189
0120	24″ O.C.	G		6	1.333		65	53.50		118.50	160
0300	2″ x 2″ angle x 18 ga, rafters/trusses @ 16″ O.C.	G		6	1.333		94.50	53.50		148	192
0320	24″ O.C.	G		8	1		94.50	40		134.50	170
0400	16 ga, rafters/trusses @ 16″ O.C.	G		4.50	1.778		120	71		191	249
0420	24″ O.C.	G		6.50	1.231		120	49		169	213

05 42 23.10 Framing, Bridging			Crew	Daily Output	Labor-Hours	Unit	Material	Labor	Equipment	Total	Total Incl O&P
0010	**FRAMING, BRIDGING**										
0015	Made from recycled materials										
0020	Solid, between rafters w/ 1-1/4″ leg track, per rafter bay										
1200	Rafters 16″ O.C., 18 ga x 4″ deep	G	1 Carp	60	.133	Ea.	1.90	5.35		7.25	10.90
1210	6″ deep	G		57	.140		2.48	5.60		8.08	11.95
1220	8″ deep	G		55	.145		3.14	5.80		8.94	13.05
1230	10″ deep	G		52	.154		3.88	6.15		10.03	14.40
1240	12″ deep	G		50	.160		4.49	6.40		10.89	15.50

05 42 Cold-Formed Metal Joist Framing

05 42 23 – Cold-Formed Metal Roof Joist Framing

05 42 23.10 Framing, Bridging			Crew	Daily Output	Labor-Hours	Unit	2009 Bare Costs Material	Labor	Equipment	Total	Total Incl O&P
2200	24″ O.C., 18 ga x 4″ deep	G	1 Carp	60	.133	Ea.	2.75	5.35		8.10	11.80
2210	6″ deep	G		57	.140		3.58	5.60		9.18	13.20
2220	8″ deep	G		55	.145		4.54	5.80		10.34	14.60
2230	10″ deep	G		52	.154		5.60	6.15		11.75	16.30
2240	12″ deep	G		50	.160		6.50	6.40		12.90	17.70

05 42 23.50 Framing, Parapets			Crew	Daily Output	Labor-Hours	Unit	Material	Labor	Equipment	Total	Total Incl O&P
0010	**FRAMING, PARAPETS**										
0015	Made from recycled materials										
0100	3′ high installed on 1st story, 18 ga x 4″ wide studs, 12″ O.C.	G	2 Carp	100	.160	L.F.	7	6.40		13.40	18.25
0110	16″ O.C.	G		150	.107		5.95	4.26		10.21	13.60
0120	24″ O.C.	G		200	.080		4.92	3.20		8.12	10.70
0200	6″ wide studs, 12″ O.C.	G		100	.160		9	6.40		15.40	20.50
0210	16″ O.C.	G		150	.107		7.65	4.26		11.91	15.45
0220	24″ O.C.	G		200	.080		6.35	3.20		9.55	12.25
1100	Installed on 2nd story, 18 ga x 4″ wide studs, 12″ O.C.	G		95	.168		7	6.75		13.75	18.80
1110	16″ O.C.	G		145	.110		5.95	4.41		10.36	13.85
1120	24″ O.C.	G		190	.084		4.92	3.36		8.28	10.95
1200	6″ wide studs, 12″ O.C.	G		95	.168		9	6.75		15.75	21
1210	16″ O.C.	G		145	.110		7.65	4.41		12.06	15.70
1220	24″ O.C.	G		190	.084		6.35	3.36		9.71	12.50
2100	Installed on gable, 18 ga x 4″ wide studs, 12″ O.C.	G		85	.188		7	7.50		14.50	20
2110	16″ O.C.	G		130	.123		5.95	4.92		10.87	14.65
2120	24″ O.C.	G		170	.094		4.92	3.76		8.68	11.60
2200	6″ wide studs, 12″ O.C.	G		85	.188		9	7.50		16.50	22.50
2210	16″ O.C.	G		130	.123		7.65	4.92		12.57	16.50
2220	24″ O.C.	G		170	.094		6.35	3.76		10.11	13.15

05 42 23.60 Framing, Roof Rafters			Crew	Daily Output	Labor-Hours	Unit	Material	Labor	Equipment	Total	Total Incl O&P
0010	**FRAMING, ROOF RAFTERS**										
0015	Made from recycled materials										
0100	Boxed ridge beam, double, 18 ga x 6″ deep	G	2 Carp	160	.100	L.F.	6.70	4		10.70	14
0110	8″ deep	G		150	.107		7.45	4.26		11.71	15.25
0120	10″ deep	G		140	.114		9	4.57		13.57	17.50
0130	12″ deep	G		130	.123		9.90	4.92		14.82	19
0200	16 ga x 6″ deep	G		150	.107		7.60	4.26		11.86	15.40
0210	8″ deep	G		140	.114		8.55	4.57		13.12	16.95
0220	10″ deep	G		130	.123		10.30	4.92		15.22	19.40
0230	12″ deep	G		120	.133		11.20	5.35		16.55	21
1100	Rafters, 2″ flange, material only, 18 ga x 6″ deep	G					2.08			2.08	2.29
1110	8″ deep	G					2.47			2.47	2.71
1120	10″ deep	G					2.90			2.90	3.19
1130	12″ deep	G					3.37			3.37	3.71
1200	16 ga x 6″ deep	G					2.54			2.54	2.80
1210	8″ deep	G					3.05			3.05	3.35
1220	10″ deep	G					3.56			3.56	3.92
1230	12″ deep	G					4.03			4.03	4.44
2100	Installation only, ordinary rafter to 4:12 pitch, 18 ga x 6″ deep		2 Carp	35	.457	Ea.		18.25		18.25	30
2110	8″ deep			30	.533			21.50		21.50	35
2120	10″ deep			25	.640			25.50		25.50	42
2130	12″ deep			20	.800			32		32	53
2200	16 ga x 6″ deep			30	.533			21.50		21.50	35
2210	8″ deep			25	.640			25.50		25.50	42
2220	10″ deep			20	.800			32		32	53

05 42 Cold-Formed Metal Joist Framing

05 42 23 – Cold-Formed Metal Roof Joist Framing

05 42 23.60 Framing, Roof Rafters			Crew	Daily Output	Labor-Hours	Unit	2009 Bare Costs Material	Labor	Equipment	Total	Total Incl O&P
2230	12″ deep		2 Carp	15	1.067	Ea.		42.50		42.50	70.50
8100	Add to labor, ordinary rafters on steep roofs							25%			
8110	Dormers & complex roofs							50%			
8200	Hip & valley rafters to 4:12 pitch							25%			
8210	Steep roofs							50%			
8220	Dormers & complex roofs							75%			
8300	Hip & valley jack rafters to 4:12 pitch							50%			
8310	Steep roofs							75%			
8320	Dormers & complex roofs							100%			
9000	Minimum labor/equipment charge		2 Carp	4	4	Job		160		160	264

05 42 23.70 Framing, Soffits and Canopies

Code	Description		Crew	Daily Output	Labor-Hours	Unit	Material	Labor	Equipment	Total	Total Incl O&P
0010	**FRAMING, SOFFITS & CANOPIES**										
0015	Made from recycled materials										
0130	Continuous ledger track @ wall, studs @ 16″ O.C., 18 ga x 4″ wide	G	2 Carp	535	.030	L.F.	1.27	1.19		2.46	3.36
0140	6″ wide	G		500	.032		1.65	1.28		2.93	3.93
0150	8″ wide	G		465	.034		2.09	1.37		3.46	4.57
0160	10″ wide	G		430	.037		2.59	1.49		4.08	5.30
0230	Studs @ 24″ O.C., 18 ga x 4″ wide	G		800	.020		1.21	.80		2.01	2.65
0240	6″ wide	G		750	.021		1.58	.85		2.43	3.14
0250	8″ wide	G		700	.023		2	.91		2.91	3.70
0260	10″ wide	G		650	.025		2.47	.98		3.45	4.33
1000	Horizontal soffit and canopy members, material only										
1030	1-5/8″ flange studs, 18 ga x 4″ deep	G				L.F.	1.68			1.68	1.85
1040	6″ deep	G					2.12			2.12	2.34
1050	8″ deep	G					2.57			2.57	2.82
1140	2″ flange joists, 18 ga x 6″ deep	G					2.38			2.38	2.61
1150	8″ deep	G					2.82			2.82	3.10
1160	10″ deep	G					3.31			3.31	3.64
4030	Installation only, 18 ga, 1-5/8″ flange x 4″ deep		2 Carp	130	.123	Ea.		4.92		4.92	8.10
4040	6″ deep			110	.145			5.80		5.80	9.60
4050	8″ deep			90	.178			7.10		7.10	11.70
4140	2″ flange, 18 ga x 6″ deep			110	.145			5.80		5.80	9.60
4150	8″ deep			90	.178			7.10		7.10	11.70
4160	10″ deep			80	.200			8		8	13.20
6010	Clips to attach facia to rafter tails, 2″ x 2″ x 18 ga angle	G	1 Carp	120	.067		1.12	2.66		3.78	5.65
6020	16 ga angle	G	″	100	.080		1.42	3.20		4.62	6.85
9000	Minimum labor/equipment charge		2 Carp	4	4	Job		160		160	264

05 44 Cold-Formed Metal Trusses

05 44 13 – Cold-Formed Metal Roof Trusses

05 44 13.60 Framing, Roof Trusses

Code	Description		Crew	Daily Output	Labor-Hours	Unit	Material	Labor	Equipment	Total	Total Incl O&P
0010	**FRAMING, ROOF TRUSSES**										
0015	Made from recycled materials										
0020	Fabrication of trusses on ground, Fink (W) or King Post, to 4:12 pitch										
0120	18 ga x 4″ chords, 16′ span	G	2 Carp	12	1.333	Ea.	78.50	53.50		132	174
0130	20′ span	G		11	1.455		98	58		156	204
0140	24′ span	G		11	1.455		118	58		176	225
0150	28′ span	G		10	1.600		137	64		201	257
0160	32′ span	G		10	1.600		157	64		221	278
0250	6″ chords, 28′ span	G		9	1.778		173	71		244	310
0260	32′ span	G		9	1.778		198	71		269	335

05 44 13.60 Framing, Roof Trusses			Crew	Daily Output	Labor-Hours	Unit	2009 Bare Costs Material	Labor	Equipment	Total	Total Incl O&P
0270	36′ span	G	2 Carp	8	2	Ea.	223	80		303	375
0280	40′ span	G		8	2		248	80		328	405
1120	5:12 to 8:12 pitch, 18 ga x 4″ chords, 16′ span	G		10	1.600		89.50	64		153.50	205
1130	20′ span	G		9	1.778		112	71		183	240
1140	24′ span	G		9	1.778		134	71		205	265
1150	28′ span	G		8	2		157	80		237	305
1160	32′ span	G		8	2		179	80		259	330
1250	6″ chords, 28′ span	G		7	2.286		198	91.50		289.50	370
1260	32′ span	G		7	2.286		227	91.50		318.50	400
1270	36′ span	G		6	2.667		255	107		362	455
1280	40′ span	G		6	2.667		283	107		390	485
2120	9:12 to 12:12 pitch, 18 ga x 4″ chords, 16′ span	G		8	2		112	80		192	255
2130	20′ span	G		7	2.286		140	91.50		231.50	305
2140	24′ span	G		7	2.286		168	91.50		259.50	335
2150	28′ span	G		6	2.667		196	107		303	390
2160	32′ span	G		6	2.667		224	107		331	420
2250	6″ chords, 28′ span	G		5	3.200		248	128		376	485
2260	32′ span	G		5	3.200		283	128		411	520
2270	36′ span	G		4	4		320	160		480	615
2280	40′ span	G		4	4		355	160		515	655
4900	Minimum labor/equipment charge			4	4	Job		160		160	264
5120	Erection only of roof trusses, to 4:12 pitch, 16′ span		F-6	48	.833	Ea.		31	16	47	68
5130	20′ span			46	.870			32.50	16.70	49.20	71
5140	24′ span			44	.909			34	17.45	51.45	74
5150	28′ span			42	.952			35.50	18.30	53.80	77.50
5160	32′ span			40	1			37	19.20	56.20	81.50
5170	36′ span			38	1.053			39	20	59	86.50
5180	40′ span			36	1.111			41.50	21.50	63	91
5220	5:12 to 8:12 pitch, 16′ span			42	.952			35.50	18.30	53.80	77.50
5230	20′ span			40	1			37	19.20	56.20	81.50
5240	24′ span			38	1.053			39	20	59	86.50
5250	28′ span			36	1.111			41.50	21.50	63	91
5260	32′ span			34	1.176			43.50	22.50	66	96.50
5270	36′ span			32	1.250			46.50	24	70.50	103
5280	40′ span			30	1.333			49.50	25.50	75	109
5320	9:12 to 12:12 pitch, 16′ span			36	1.111			41.50	21.50	63	91
5330	20′ span			34	1.176			43.50	22.50	66	96.50
5340	24′ span			32	1.250			46.50	24	70.50	103
5350	28′ span			30	1.333			49.50	25.50	75	109
5360	32′ span			28	1.429			53	27.50	80.50	117
5370	36′ span			26	1.538			57	29.50	86.50	126
5380	40′ span			24	1.667			62	32	94	136
9000	Minimum labor/equipment charge			2	20	Job		745	385	1,130	1,650

05 51 Metal Stairs

05 51 13 – Metal Pan Stairs

05 51 13.50 Pan Stairs			Crew	Daily Output	Labor-Hours	Unit	Material	2009 Bare Costs Labor	Equipment	Total	Total Incl O&P
0010	**PAN STAIRS**, shop fabricated, steel stringers										
0015	Made from recycled materials										
0200	Cement fill metal pan, picket rail, 3′-6″ wide	G	E-4	35	.914	Riser	560	41.50	3.83	605.33	700
0300	4′-0″ wide	G		30	1.067		635	48	4.47	687.47	795
0350	Wall rail, both sides, 3′-6″ wide	G		53	.604		430	27.50	2.53	460.03	530
1500	Landing, steel pan, conventional	G		160	.200	S.F.	75	9.05	.84	84.89	100
1600	Pre-erected	G		255	.125	″	131	5.65	.53	137.18	155
1700	Pre-erected, steel pan tread, 3′-6″ wide, 2 line pipe rail	G	E-2	87	.644	Riser	620	28	20	668	750

05 51 16 – Metal Floor Plate Stairs

05 51 16.50 Floor Plate Stairs			Crew	Daily Output	Labor-Hours	Unit	Material	Labor	Equipment	Total	Total Incl O&P
0010	**FLOOR PLATE STAIRS**, shop fabricated, steel stringers										
0015	Made from recycled materials										
0400	Cast iron tread and pipe rail, 3′-6″ wide	G	E-4	35	.914	Riser	600	41.50	3.83	645.33	740
0500	Checkered plate tread, industrial, 3′-6″ wide	G		28	1.143		375	51.50	4.79	431.29	510
0550	Circular, for tanks, 3′-0″ wide	G		33	.970		410	44	4.06	458.06	540
0600	For isolated stairs, add							100%			
0800	Custom steel stairs, 3′-6″ wide, minimum	G	E-4	35	.914		560	41.50	3.83	605.33	700
0810	Average	G		30	1.067		750	48	4.47	802.47	920
0900	Maximum	G		20	1.600		935	72.50	6.70	1,014.20	1,175
1100	For 4′ wide stairs, add						5%	5%			
1300	For 5′ wide stairs, add						10%	10%			

05 51 19 – Metal Grating Stairs

05 51 19.50 Grating Stairs			Crew	Daily Output	Labor-Hours	Unit	Material	Labor	Equipment	Total	Total Incl O&P
0010	**GRATING STAIRS**, shop fabricated, steel stringers, safety nosing on treads										
0015	Made from recycled materials										
0020	Grating tread and pipe railing, 3′-6″ wide	G	E-4	35	.914	Riser	375	41.50	3.83	420.33	490
0100	4′-0″ wide	G		30	1.067	″	485	48	4.47	537.47	630
9000	Minimum labor/equipment charge			2	16	Job		725	67	792	1,425

05 51 23 – Metal Fire Escapes

05 51 23.25 Fire Escapes			Crew	Daily Output	Labor-Hours	Unit	Material	Labor	Equipment	Total	Total Incl O&P
0010	**FIRE ESCAPES**, shop fabricated										
0200	2′ wide balcony, 1″ x 1/4″ bars 1-1/2″ O.C.	G	1 Sswk	5	1.600	L.F.	65.50	71.50		137	204
0400	1st story cantilevered stair, standard	G	″	.09	88.889	Ea.	2,725	3,975		6,700	10,400
0500	Cable counterweight	G				″	2,550			2,550	2,800
0700	Platform & fixed stair, 36″ x 40″	G	1 Sswk	.17	47.059	Flight	1,200	2,100		3,300	5,225
0900	For 3′-6″ wide escapes, add to above						100%	150%			

05 51 23.50 Fire Escape Stairs			Crew	Daily Output	Labor-Hours	Unit	Material	Labor	Equipment	Total	Total Incl O&P
0010	**FIRE ESCAPE STAIRS**, shop fabricated										
0020	One story, disappearing, stainless steel	G	2 Sswk	20	.800	V.L.F.	223	36		259	310
0100	Portable ladder					Ea.	93			93	102
1100	Fire escape, galvanized steel, 8′-0″ to 10′-4″ ceiling	G	2 Carp	1	16		1,525	640		2,165	2,725
1110	10′-6″ to 13′-6″ ceiling	G	″	1	16		2,000	640		2,640	3,250

05 51 33 – Metal Ladders

05 51 33.13 Vertical Metal Ladders			Crew	Daily Output	Labor-Hours	Unit	Material	Labor	Equipment	Total	Total Incl O&P
0010	**VERTICAL METAL LADDERS**, shop fabricated										
0015	Made from recycled materials										
0020	Steel, 20″ wide, bolted to concrete, with cage	G	E-4	50	.640	V.L.F.	78.50	29	2.68	110.18	143
0100	Without cage	G		85	.376		41	17	1.58	59.58	78
0300	Aluminum, bolted to concrete, with cage	G		50	.640		111	29	2.68	142.68	178
0400	Without cage	G		85	.376		64.50	17	1.58	83.08	104

05 51 Metal Stairs

05 51 33 – Metal Ladders

05 51 33.13 Vertical Metal Ladders

	05 51 33.13 Vertical Metal Ladders		Crew	Daily Output	Labor-Hours	Unit	Material	2009 Bare Costs Labor	Equipment	Total	Total Incl O&P
9000	Minimum labor/equipment charge		E-4	2	16	Job		725	67	792	1,425

05 51 33.16 Inclined Metal Ladders

	05 51 33.16 Inclined Metal Ladders		Crew	Daily Output	Labor-Hours	Unit	Material	2009 Bare Costs Labor	Equipment	Total	Total Incl O&P
0010	**INCLINED METAL LADDERS**, shop fabricated										
3900	Industrial ships ladder, 3′ W, grating treads, 2 line pipe rail	G	E-4	30	1.067	Riser	203	48	4.47	255.47	315
4000	Aluminum	G	"	30	1.067	"	310	48	4.47	362.47	440

05 51 33.23 Alternating Tread Ladders

	05 51 33.23 Alternating Tread Ladders		Crew	Daily Output	Labor-Hours	Unit	Material	2009 Bare Costs Labor	Equipment	Total	Total Incl O&P
0010	**ALTERNATING TREAD LADDERS**, shop fabricated										
0015	Made from recycled materials										
1350	Alternating tread stair, 56/68°, steel, standard paint color	G	2 Sswk	50	.320	V.L.F.	168	14.30		182.30	212
1360	Non-standard paint color	G		50	.320		190	14.30		204.30	236
1370	Galvanized steel	G		50	.320		196	14.30		210.30	243
1380	Stainless steel	G		50	.320		282	14.30		296.30	335
1390	68°, aluminum	G		50	.320		206	14.30		220.30	254

05 52 Metal Railings

05 52 13 – Pipe and Tube Railings

05 52 13.50 Railings, Pipe

	05 52 13.50 Railings, Pipe		Crew	Daily Output	Labor-Hours	Unit	Material	2009 Bare Costs Labor	Equipment	Total	Total Incl O&P
0010	**RAILINGS, PIPE**, shop fab'd, 3′-6″ high, posts @ 5′ O.C.										
0015	Made from recycled materials										
0020	Aluminum, 2 rail, satin finish, 1-1/4″ diameter	G	E-4	160	.200	L.F.	34	9.05	.84	43.89	55
0030	Clear anodized	G		160	.200		42	9.05	.84	51.89	64
0040	Dark anodized	G		160	.200		47.50	9.05	.84	57.39	70
0080	1-1/2″ diameter, satin finish	G		160	.200		41	9.05	.84	50.89	62.50
0090	Clear anodized	G		160	.200		45.50	9.05	.84	55.39	67.50
0100	Dark anodized	G		160	.200		50.50	9.05	.84	60.39	73
0140	Aluminum, 3 rail, 1-1/4″ diam., satin finish	G		137	.234		52	10.55	.98	63.53	78
0150	Clear anodized	G		137	.234		65.50	10.55	.98	77.03	92.50
0160	Dark anodized	G		137	.234		72.50	10.55	.98	84.03	100
0200	1-1/2″ diameter, satin finish	G		137	.234		62.50	10.55	.98	74.03	89.50
0210	Clear anodized	G		137	.234		71	10.55	.98	82.53	98.50
0220	Dark anodized	G		137	.234		78	10.55	.98	89.53	106
0500	Steel, 2 rail, on stairs, primed, 1-1/4″ diameter	G		160	.200		28	9.05	.84	37.89	48.50
0520	1-1/2″ diameter	G		160	.200		30.50	9.05	.84	40.39	51.50
0540	Galvanized, 1-1/4″ diameter	G		160	.200		39	9.05	.84	48.89	60
0560	1-1/2″ diameter	G		160	.200		43.50	9.05	.84	53.39	65.50
0580	Steel, 3 rail, primed, 1-1/4″ diameter	G		137	.234		41.50	10.55	.98	53.03	66.50
0600	1-1/2″ diameter	G		137	.234		44.50	10.55	.98	56.03	69
0620	Galvanized, 1-1/4″ diameter	G		137	.234		58.50	10.55	.98	70.03	85
0640	1-1/2″ diameter	G		137	.234		69	10.55	.98	80.53	96.50
0700	Stainless steel, 2 rail, 1-1/4″ diam. #4 finish	G		137	.234		101	10.55	.98	112.53	132
0720	High polish	G		137	.234		162	10.55	.98	173.53	200
0740	Mirror polish	G		137	.234		203	10.55	.98	214.53	245
0760	Stainless steel, 3 rail, 1-1/2″ diam., #4 finish	G		120	.267		152	12.05	1.12	165.17	191
0770	High polish	G		120	.267		251	12.05	1.12	264.17	300
0780	Mirror finish	G		120	.267		305	12.05	1.12	318.17	360
0900	Wall rail, alum. pipe, 1-1/4″ diam., satin finish	G		213	.150		19.50	6.80	.63	26.93	35
0905	Clear anodized	G		213	.150		24	6.80	.63	31.43	39.50
0910	Dark anodized	G		213	.150		29	6.80	.63	36.43	45
0915	1-1/2″ diameter, satin finish	G		213	.150		21.50	6.80	.63	28.93	37
0920	Clear anodized	G		213	.150		27	6.80	.63	34.43	43.50

05 52 Metal Railings

05 52 13 – Pipe and Tube Railings

05 52 13.50 Railings, Pipe			Crew	Daily Output	Labor-Hours	Unit	2009 Bare Costs Material	Labor	Equipment	Total	Total Incl O&P
0925	Dark anodized	G	E-4	213	.150	L.F.	33.50	6.80	.63	40.93	50.50
0930	Steel pipe, 1-1/4" diameter, primed	G		213	.150		16.95	6.80	.63	24.38	32
0935	Galvanized	G		213	.150		24.50	6.80	.63	31.93	40.50
0940	1-1/2" diameter	G		176	.182		17.45	8.20	.76	26.41	35
0945	Galvanized	G		213	.150		24.50	6.80	.63	31.93	40.50
0955	Stainless steel pipe, 1-1/2" diam., #4 finish	G		107	.299		80.50	13.50	1.25	95.25	115
0960	High polish	G		107	.299		164	13.50	1.25	178.75	206
0965	Mirror polish	G		107	.299		193	13.50	1.25	207.75	238
2000	Aluminum pipe & picket railing, double top rail, pickets @ 4-1/2" OC										
2010	36" high, straight & level	G	2 Sswk	80	.200	L.F.	81	8.95		89.95	106
2020	Curved & level	G		60	.267		111	11.90		122.90	144
2030	Straight & sloped	G		40	.400		93	17.90		110.90	135
9000	Minimum labor/equipment charge		1 Sswk	2	4	Job		179		179	330

05 52 16 – Industrial Railings

05 52 16.50 Railings, Industrial			Crew	Daily Output	Labor-Hours	Unit	Material	Labor	Equipment	Total	Total Incl O&P
0010	**RAILINGS, INDUSTRIAL**, shop fab'd, 3'-6" high, posts @ 5' O.C.										
0020	2 rail, 3'-6" high, 1-1/2" pipe	G	E-4	255	.125	L.F.	31	5.65	.53	37.18	45
0100	2" angle rail	G	"	255	.125		28	5.65	.53	34.18	42
0200	For 4" high kick plate, 10 gauge, add	G					6.40			6.40	7.05
0300	1/4" thick, add	G					8.25			8.25	9.10
0500	For curved rails, add						30%	30%			
9000	Minimum labor/equipment charge		1 Sswk	2	4	Job		179		179	330

05 53 Metal Gratings

05 53 13 – Floor Grating Frame

05 53 13.50 Grating Frame			Crew	Daily Output	Labor-Hours	Unit	Material	Labor	Equipment	Total	Total Incl O&P
0010	**GRATING FRAME**, field fabricated										
0020	Aluminum, for gratings 1" to 1-1/2" deep	G	1 Sswk	70	.114	L.F.	2.88	5.10		7.98	12.60
0100	For each corner, add	G				Ea.	4.32			4.32	4.75

05 53 16 – Floor Grating Planks

05 53 16.50 Grating Planks			Crew	Daily Output	Labor-Hours	Unit	Material	Labor	Equipment	Total	Total Incl O&P
0010	**GRATING PLANKS**, field fabricated from planks										
0020	Aluminum, 9-1/2" wide, 14 ga., 2" rib	G	E-4	950	.034	L.F.	28	1.52	.14	29.66	34
0200	Galvanized steel, 9-1/2" wide, 14 ga., 2-1/2" rib	G		950	.034		19.50	1.52	.14	21.16	24.50
0300	4" rib	G		950	.034		23	1.52	.14	24.66	28
0500	12 gauge, 2-1/2" rib	G		950	.034		26	1.52	.14	27.66	31.50
0600	3" rib	G		950	.034		31	1.52	.14	32.66	37
0800	Stainless steel, type 304, 16 ga., 2" rib	G		950	.034		41	1.52	.14	42.66	48
0900	Type 316	G		950	.034		50.50	1.52	.14	52.16	58.50

05 53 19 – Aluminum Floor Grating

05 53 19.50 Floor Grating, Aluminum			Crew	Daily Output	Labor-Hours	Unit	Material	Labor	Equipment	Total	Total Incl O&P
0010	**FLOOR GRATING, ALUMINUM**, field fabricated from panels										
0015	Made from recycled materials										
0110	Bearing bars @ 1-3/16" O.C., cross bars @ 4" O.C.,										
0111	Up to 300 S.F., 1" x 1/8" bar	G	E-4	900	.036	S.F.	23.50	1.61	.15	25.26	29
0112	Over 300 S.F.	G		850	.038		21.50	1.70	.16	23.36	27
0113	1-1/4" x 1/8" bar, up to 300 S.F.	G		800	.040		28	1.81	.17	29.98	34
0114	Over 300 S.F.	G		1000	.032		25	1.45	.13	26.58	31
0122	1-1/4" x 3/16" bar, up to 300 S.F.	G		750	.043		36	1.93	.18	38.11	43.50

05 53 Metal Gratings

05 53 19 – Aluminum Floor Grating

05 53 19.50 Floor Grating, Aluminum			Crew	Daily Output	Labor-Hours	Unit	2009 Bare Costs Material	Labor	Equipment	Total	Total Incl O&P
0124	Over 300 S.F.	G	E-4	1000	.032	S.F.	32.50	1.45	.13	34.08	39
0132	1-1/2" x 3/16" bar, up to 300 S.F.	G		700	.046		41.50	2.07	.19	43.76	49.50
0134	Over 300 S.F.	G		1000	.032		37.50	1.45	.13	39.08	44.50
0136	1-3/4" x 3/16" bar, up to 300 S.F.	G		500	.064		50.50	2.89	.27	53.66	61
0138	Over 300 S.F.	G		1000	.032		46	1.45	.13	47.58	53.50
0146	2-1/4" x 3/16" bar, up to 300 S.F.	G		600	.053		69	2.41	.22	71.63	80.50
0148	Over 300 S.F.	G		1000	.032		62.50	1.45	.13	64.08	72
0162	Cross bars @ 2" O.C., 1" x 1/8", up to 300 S.F.	G		600	.053		30.50	2.41	.22	33.13	38
0164	Over 300 S.F.	G		1000	.032		27.50	1.45	.13	29.08	33.50
0172	1-1/4" x 3/16" bar, up to 300 S.F.	G		600	.053		41	2.41	.22	43.63	49.50
0174	Over 300 S.F.	G		1000	.032		37	1.45	.13	38.58	44
0182	1-1/2" x 3/16" bar, up to 300 S.F.	G		600	.053		46	2.41	.22	48.63	55
0184	Over 300 S.F.	G		1000	.032		42	1.45	.13	43.58	49
0186	1-3/4" x 3/16" bar, up to 300 S.F.	G		600	.053		69.50	2.41	.22	72.13	81
0188	Over 300 S.F.	G		1000	.032		63	1.45	.13	64.58	72.50
0200	For straight cuts, add					L.F.	3.60			3.60	3.96
0212	Close mesh, 1" x 1/8", up to 300 S.F.	G	E-4	520	.062	S.F.	28	2.78	.26	31.04	36.50
0214	Over 300 S.F.	G		920	.035		25.50	1.57	.15	27.22	31
0222	1-1/4" x 3/16", up to 300 S.F.	G		520	.062		42	2.78	.26	45.04	52
0224	Over 300 S.F.	G		920	.035		38.50	1.57	.15	40.22	45
0232	1-1/2" x 3/16", up to 300 S.F.	G		520	.062		49	2.78	.26	52.04	59
0234	Over 300 S.F.	G		920	.035		44.50	1.57	.15	46.22	52
0600	For aluminum checkered plate nosings, add	G				L.F.	6			6	6.60
0700	For straight toe plate, add	G					9.10			9.10	10
0800	For curved toe plate, add	G					10.60			10.60	11.65
1000	For cast aluminum abrasive nosings, add	G					8.85			8.85	9.75
1200	Expanded aluminum, .65# per S.F.	G	E-4	1050	.030	S.F.	11.75	1.38	.13	13.26	15.65
1600	Heavy duty, all extruded plank, 3/4" deep, 1.8 # per S.F.	G		1100	.029		30	1.31	.12	31.43	35.50
1700	1-1/4" deep, 2.9# per S.F.	G		1000	.032		35	1.45	.13	36.58	41.50
1800	1-3/4" deep, 4.2# per S.F.	G		925	.035		42	1.56	.14	43.70	49
1900	2-1/4" deep, 5.0# per S.F.	G		875	.037		50	1.65	.15	51.80	58
2100	For safety serrated surface, add						15%				

05 53 21 – Steel Floor Grating

05 53 21.50 Floor Grating, Steel			Crew	Daily Output	Labor-Hours	Unit	Material	Labor	Equipment	Total	Total Incl O&P
0010	**FLOOR GRATING, STEEL**, field fabricated from panels										
0015	Made from recycled materials										
0300	Platforms, to 12' high, rectangular	G	E-4	3150	.010	Lb.	2.40	.46	.04	2.90	3.54
0400	Circular	G	"	2300	.014	"	3	.63	.06	3.69	4.52
0410	Painted bearing bars @ 1-3/16"										
0412	Cross bars @ 4" O.C., 3/4" x 1/8" bar, up to 300 S.F.	G	E-2	500	.112	S.F.	8.10	4.88	3.48	16.46	21.50
0414	Over 300 S.F.	G		750	.075		7.35	3.25	2.32	12.92	16.45
0422	1-1/4" x 3/16", up to 300 S.F.	G		400	.140		12.50	6.10	4.35	22.95	29.50
0424	Over 300 S.F.	G		600	.093		11.35	4.06	2.90	18.31	23
0432	1-1/2" x 3/16", up to 300 S.F.	G		400	.140		14.40	6.10	4.35	24.85	31.50
0434	Over 300 S.F.	G		600	.093		13.10	4.06	2.90	20.06	25
0436	1-3/4" x 3/16", up to 300 S.F.	G		400	.140		19.90	6.10	4.35	30.35	37.50
0438	Over 300 S.F.	G		600	.093		18.10	4.06	2.90	25.06	30.50
0452	2-1/4" x 3/16", up to 300 S.F.	G		300	.187		23.50	8.15	5.80	37.45	46.50
0454	Over 300 S.F.	G		450	.124		21	5.40	3.87	30.27	37.50
0462	Cross bars @ 2" O.C., 3/4" x 1/8", up to 300 S.F.	G		500	.112		13.65	4.88	3.48	22.01	27.50
0464	Over 300 S.F.	G		750	.075		11.35	3.25	2.32	16.92	21
0472	1-1/4" x 3/16", up to 300 S.F.	G		400	.140		17.85	6.10	4.35	28.30	35

05 53 21 – Steel Floor Grating

05 53 21.50 Floor Grating, Steel		Crew	Daily Output	Labor-Hours	Unit	Material	2009 Bare Costs Labor	Equipment	Total	Total Incl O&P
0474	Over 300 S.F. G	E-2	600	.093	S.F.	14.85	4.06	2.90	21.81	27
0482	1-1/2" x 3/16", up to 300 S.F. G		400	.140		20	6.10	4.35	30.45	37.50
0484	Over 300 S.F. G		600	.093		16.75	4.06	2.90	23.71	29
0486	1-3/4" x 3/16", up to 300 S.F. G		400	.140		30.50	6.10	4.35	40.95	49
0488	Over 300 S.F. G		600	.093		25	4.06	2.90	31.96	38
0502	2-1/4" x 3/16", up to 300 S.F. G		300	.137		29.50	8.15	5.80	43.45	53.50
0504	Over 300 S.F. G	↓	450	.124	↓	24.50	5.40	3.87	33.77	41
0601	Painted bearing bars @ 15/16" O.C., cross bars @ 4" O.C.,									
0612	Up to 300 S.F., 3/4" x 3/16" bars G	E-4	850	.038	S.F.	11	1.70	.16	12.86	15.40
0622	1-1/4" x 3/16" bars G		600	.053		15.35	2.41	.22	17.98	21.50
0632	1-1/2" x 3/16" bars G		550	.058		17.75	2.63	.24	20.62	24.50
0636	1-3/4" x 3/16" bars G	↓	450	.071		22	3.21	.30	25.51	31
0652	2-1/4" x 3/16" bars G	E-2	300	.137		26.50	8.15	5.80	40.45	50
0662	Cross bars @ 2" O.C., up to 300 S.F., 3/4" x 3/16" G		500	.112		14.25	4.88	3.48	22.61	28
0672	1-1/4" x 3/16" bars G		400	.140		18.20	6.10	4.35	28.65	35.50
0682	1-1/2" x 3/16" bars G		400	.140		20.50	6.10	4.35	30.95	38
0686	1-3/4" x 3/16" bars G	↓	300	.137		26	8.15	5.80	39.95	49.50
0690	For galvanized grating, add				↓	25%				
0800	For straight cuts, add				L.F.	5.20			5.20	5.70
0900	For curved cuts, add					6.55			6.55	7.20
1000	For straight banding, add G					5.60			5.60	6.15
1100	For curved banding, add G					7.30			7.30	8.05
1200	For checkered plate nosings, add G					6.40			6.40	7.05
1300	For straight toe or kick plate, add G					11.50			11.50	12.65
1400	For curved toe or kick plate, add G					13			13	14.30
1500	For abrasive nosings, add G				↓	8.55			8.55	9.40
1600	For safety serrated surface, minimum, add					15%				
1700	Maximum, add					25%				
2000	Stainless steel gratings, close spaced, 1" x 1/8" bars, up to 300 S.F. G	E-4	450	.071	S.F.	105	3.21	.30	108.51	122
2100	Standard spacing, 3/4" x 1/8" bars G		500	.064		75.50	2.89	.27	78.66	88.50
2200	1-1/4" x 3/16" bars G		400	.080		116	3.62	.34	119.96	135
2400	Expanded steel grating, at ground, 3.0# per S.F. G		900	.036		7.30	1.61	.15	9.06	11.20
2500	3.14# per S.F. G		900	.036		7.85	1.61	.15	9.61	11.80
2600	4.0# per S.F. G		850	.038		9.70	1.70	.16	11.56	13.95
2650	4.27# per S.F. G		850	.038		10.40	1.70	.16	12.26	14.75
2700	5.0# per S.F. G		800	.040		12.35	1.81	.17	14.33	17.15
2800	6.25# per S.F. G		750	.043		16	1.93	.18	18.11	21.50
2900	7.0# per S.F. G	↓	700	.046		17.70	2.07	.19	19.96	23.50
3100	For flattened expanded steel grating, add					8%				
3300	For elevated installation above 15', add				↓		15%			

05 54 Metal Floor Plates

05 54 13 – Floor Plates

05 54 13.20 Checkered Plates			Crew	Daily Output	Labor-Hours	Unit	Material	2009 Bare Costs Labor	Equipment	Total	Total Incl O&P
0010	**CHECKERED PLATES**, steel, field fabricated										
0015	Made from recycled materials										
0020	1/4″ & 3/8″, 2000 to 5000 S.F., bolted	G	E-4	2900	.011	Lb.	.98	.50	.05	1.53	2.05
0100	Welded	G		4400	.007	″	.93	.33	.03	1.29	1.66
0300	Pit or trench cover and frame, 1/4″ plate, 2′ to 3′ wide	G		100	.320	S.F.	14.15	14.45	1.34	29.94	44
0400	For galvanizing, add	G				Lb.	.57			.57	.63
0500	Platforms, 1/4″ plate, no handrails included, rectangular	G	E-4	4200	.008		2.40	.34	.03	2.77	3.32
0600	Circular	G	″	2500	.013		3	.58	.05	3.63	4.43

05 54 13.70 Trench Covers			Crew	Daily Output	Labor-Hours	Unit	Material	Labor	Equipment	Total	Total Incl O&P
0010	**TRENCH COVERS**, field fabricated										
0020	Cast iron grating with bar stops and angle frame, to 18″ wide	G	1 Sswk	20	.400	L.F.	64.50	17.90		82.40	104
0100	Frame only (both sides of trench), 1″ grating	G		45	.178		1.97	7.95		9.92	16.85
0150	2″ grating	G		35	.229		5.40	10.20		15.60	25
0200	Aluminum, stock units, including frames and										
0210	3/8″ plain cover plate, 4″ opening	G	E-4	205	.156	L.F.	40.50	7.05	.65	48.20	59
0300	6″ opening	G		185	.173		50.50	7.80	.72	59.02	71
0400	10″ opening	G		170	.188		69	8.50	.79	78.29	92.50
0500	16″ opening	G		155	.206		95	9.35	.87	105.22	122
0700	Add per inch for additional widths to 24″	G					114			114	125
0900	For custom fabrication, add						50%				
1100	For 1/4″ plain cover plate, deduct						12%				
1500	For cover recessed for tile, 1/4″ thick, deduct						12%				
1600	3/8″ thick, add						5%				
1800	For checkered plate cover, 1/4″ thick, deduct						12%				
1900	3/8″ thick, add						2%				
2100	For slotted or round holes in cover, 1/4″ thick, add						3%				
2200	3/8″ thick, add						4%				
2300	For abrasive cover, add						12%				

05 55 Metal Stair Treads and Nosings

05 55 13 – Metal Stair Treads

05 55 13.50 Stair Treads			Crew	Daily Output	Labor-Hours	Unit	Material	Labor	Equipment	Total	Total Incl O&P
0010	**STAIR TREADS**										
0020	Aluminum grating, 3′ long, 1-1/2″ x 3/16″ rect. bars, 6″ wide	G	1 Sswk	24	.333	Ea.	65.50	14.90		80.40	99.50
0100	12″ wide	G		22	.364		96.50	16.25		112.75	136
0200	1-1/2″ x 3/16″ I-bars, 6″ wide	G		24	.333		55	14.90		69.90	88
0300	12″ wide	G		22	.364		88.50	16.25		104.75	128
0400	For abrasive nosings, add	G					17.15			17.15	18.85
0500	For narrow mesh, add						60%				
0600	Stair treads, not incl. stringers, see also Div. 03 15 05.85										
0700	Cast aluminum, abrasive, 3′ long x 12″ wide, 5/16″ thick	G	1 Sswk	15	.533	Ea.	145	24		169	204
0800	3/8″ thick	G		15	.533		154	24		178	214
0900	1/2″ thick	G		15	.533		169	24		193	230
1000	Cast bronze, abrasive, 3/8″ thick	G		8	1		660	44.50		704.50	810
1100	1/2″ thick	G		8	1		865	44.50		909.50	1,050
1200	Cast iron, abrasive, 3′ long x 12″ wide, 3/8″ thick	G		15	.533		128	24		152	184
1300	1/2″ thick	G		15	.533		147	24		171	206
1400	Fiberglass reinforced plastic with safety nosing,										
1500	1-1/2″ thick, 12″ wide, 24″ long		1 Sswk	22	.364	Ea.	144	16.25		160.25	189
1600	30″ long			22	.364		151	16.25		167.25	196
1700	36″ long			22	.364		165	16.25		181.25	212

05 55 Metal Stair Treads and Nosings

05 55 13 – Metal Stair Treads

05 55 13.50 Stair Treads			Crew	Daily Output	Labor-Hours	Unit	Material	2009 Bare Costs Labor	Equipment	Total	Total Incl O&P
2000	Steel grating, painted, 3′ long, 1-1/4″ x 3/16″ bars, 6″ wide	G	1 Sswk	20	.400	Ea.	55.50	17.90		73.40	94
2010	12″ wide	G		18	.444		64.50	19.85		84.35	108
2022	Plain, 1″ x 3/16″, 6″ wide	G		17	.471		50.50	21		71.50	94.50
2024	12″ wide	G		15	.533		61.50	24		85.50	112
2032	1-1/4″ x 3/16″, 6″ wide	G		17	.471		54.50	21		75.50	99
2034	12″ wide	G		15	.533		68	24		92	119
2062	Serrated, 1-1/4″ x 3/16″, 6″ wide	G		17	.471		38.50	21		59.50	81.50
2064	12″ wide	G		15	.533		64.50	24		88.50	115
2072	1-1/2″ x 3/16″, 6″ wide	G		17	.471		42	21		63	85.50
2074	12″ wide	G		15	.533		75.50	24		99.50	127
2092	For galvanizing, 1″ x 3/16″, add	G				S.F.	1.99			1.99	2.19
2094	1-1/4″ x 3/16″, add	G				″	4.42			4.42	4.86
2100	Add for abrasive nosing, 3′ long, painted	G				Ea.	17.35			17.35	19.05
2200	Galvanized	G					23.50			23.50	25.50
2500	Expanded steel, 2-1/2″ deep, 9″ x 3′ long, 18 gauge	G	1 Sswk	20	.400		30.50	17.90		48.40	66.50
2600	14 gauge	G	″	20	.400		38	17.90		55.90	74.50

05 56 Metal Castings

05 56 13 – Metal Construction Castings

05 56 13.50 Construction Castings			Crew	Daily Output	Labor-Hours	Unit	Material	Labor	Equipment	Total	Total Incl O&P
0010	**CONSTRUCTION CASTINGS**										
0020	Manhole covers and frames, see Div. 33 44 13.13										
0100	Column bases, cast iron, 16″ x 16″, approx. 65 lb.	G	E-4	46	.696	Ea.	142	31.50	2.91	176.41	218
0200	32″ x 32″, approx. 256 lb.	G	″	23	1.391		530	63	5.85	598.85	705
0400	Cast aluminum for wood columns, 8″ x 8″	G	1 Carp	32	.250		45.50	10		55.50	67
0500	12″ x 12″	G	″	32	.250		98.50	10		108.50	125
0600	Miscellaneous C.I. castings, light sections, less than 150 lb.	G	E-4	3200	.010	Lb.	2.20	.45	.04	2.69	3.31
1100	Heavy sections, more than 150 lb	G		4200	.008		1.07	.34	.03	1.44	1.86
1300	Special low volume items	G		3200	.010		3.82	.45	.04	4.31	5.10
1500	For ductile iron, add						100%				

05 58 Formed Metal Fabrications

05 58 21 – Formed Chain

05 58 21.05 Alloy Steel Chain			Crew	Daily Output	Labor-Hours	Unit	Material	Labor	Equipment	Total	Total Incl O&P
0010	**ALLOY STEEL CHAIN**, Grade 80										
0015	Self-colored, cut lengths, w/accessories, 1/4″	G	E-17	4	4	C.L.F.	455	183		638	840
0020	3/8″	G		2	8		615	365		980	1,350
0030	1/2″	G		1.20	13.333		1,100	610		1,710	2,350
0040	5/8″	G		.72	22.222		1,675	1,025		2,700	3,725
0050	3/4″	G	E-18	.48	83.333		2,675	3,700	2,250	8,625	12,100
0060	7/8″	G		.40	100		4,150	4,450	2,700	11,300	15,600
0070	1″	G		.35	114		5,700	5,075	3,075	13,850	18,800
0080	1-1/4″	G		.24	165		16,900	7,400	4,500	28,800	36,900
0110	Clevis slip hook, 1/4″	G				Ea.	13.10			13.10	14.40
0120	3/8″	G					18.60			18.60	20.50
0130	1/2″	G					33			33	36
0140	5/8″	G					47			47	52
0150	3/4″	G					102			102	112
0160	Eye/sling hook w/ hammerlock coupling, 7/8″	G					280			280	310

05 58 Formed Metal Fabrications

05 58 21 – Formed Chain

05 58 21.05 Alloy Steel Chain		Crew	Daily Output	Labor-Hours	Unit	Material	Labor	Equipment	Total	Total Incl O&P
0170 1"	G				Ea.	475			475	520
0180 1-1/4"	G				↓	700			700	770

05 58 23 – Formed Columns

05 58 23.10 Aluminum Columns		Crew	Daily Output	Labor-Hours	Unit	Material	Labor	Equipment	Total	Total Incl O&P
0010 **ALUMINUM COLUMNS**										
0015 Made from recycled materials										
0020 Aluminum, extruded, stock units, no cap or base, 6" diameter	G	E-4	240	.133	L.F.	11.65	6.05	.56	18.26	24.50
0100 8" diameter	G		170	.188		15	8.50	.79	24.29	33
0200 10" diameter	G		150	.213		20.50	9.65	.89	31.04	41.50
0300 12" diameter	G		140	.229		37	10.35	.96	48.31	60.50
0400 15" diameter	G	↓	120	.267	↓	44	12.05	1.12	57.17	72
0410 Caps and bases, plain, 6" diameter	G				Set	22			22	24
0420 8" diameter	G					29			29	31.50
0430 10" diameter	G					37			37	40.50
0440 12" diameter	G					50			50	55
0450 15" diameter	G					90			90	99
0460 Caps, ornamental, minimum	G					242			242	266
0470 Maximum	G				↓	1,075			1,075	1,200
0500 For square columns, add to column prices above					L.F.	50%				
0700 Residential, flat, 8' high, plain	G	E-4	20	1.600	Ea.	95	72.50	6.70	174.20	245
0720 Fancy	G		20	1.600		181	72.50	6.70	260.20	340
0740 Corner type, plain	G		20	1.600		165	72.50	6.70	244.20	325
0760 Fancy	G	↓	20	1.600	↓	320	72.50	6.70	399.20	490

05 58 25 – Formed Lamp Posts

05 58 25.40 Lamp Posts		Crew	Daily Output	Labor-Hours	Unit	Material	Labor	Equipment	Total	Total Incl O&P
0010 **LAMP POSTS**										
0020 Aluminum, 7' high, stock units, post only	G	1 Carp	16	.500	Ea.	28.50	20		48.50	64.50
0100 Mild steel, plain	G		16	.500	"	23	20		43	58.50
9000 Minimum labor/equipment charge		↓	4	2	Job		80		80	132

05 58 27 – Formed Guards

05 58 27.90 Window Guards		Crew	Daily Output	Labor-Hours	Unit	Material	Labor	Equipment	Total	Total Incl O&P
0010 **WINDOW GUARDS**, shop fabricated										
0015 Expanded metal, steel angle frame, permanent	G	E-4	350	.091	S.F.	26	4.13	.38	30.51	36.50
0025 Steel bars, 1/2" x 1/2", spaced 5" O.C.	G	"	290	.110	"	18	4.99	.46	23.45	29.50
0030 Hinge mounted, add	G				Opng.	52			52	57.50
0040 Removable type, add	G				"	33			33	36.50
0050 For galvanized guards, add					S.F.	35%				
0070 For pivoted or projected type, add						105%	40%			
0100 Mild steel, stock units, economy	G	E-4	405	.079		7.05	3.57	.33	10.95	14.70
0200 Deluxe	G		405	.079	↓	14.50	3.57	.33	18.40	23
0400 Woven wire, stock units, 3/8" channel frame, 3' x 5' opening	G		40	.800	Opng.	191	36	3.35	230.35	281
0500 4' x 6' opening	G	↓	38	.842		305	38	3.53	346.53	410
0800 Basket guards for above, add	G					261			261	287
1000 Swinging guards for above, add	G				↓	89.50			89.50	98.50
9000 Minimum labor/equipment charge		1 Sswk	2	4	Job		179		179	330

Column group "2009 Bare Costs" covers Material, Labor, Equipment, Total.

05 59 Metal Specialties

05 59 63 – Detention Enclosures

05 59 63.13 Enclosures, Detention		Crew	Daily Output	Labor-Hours	Unit	2009 Bare Costs Material	Labor	Equipment	Total	Total Incl O&P
0010	**ENCLOSURES, DETENTION**									
0500	Bar front, rolling, 7/8" bars, 4" O.C., 7' high, 5' wide, with hardware G	E-4	2	16	Ea.	5,850	725	67	6,642	7,850

05 71 Decorative Metal Stairs

05 71 13 – Fabricated Metal Spiral Stairs

05 71 13.50 Spiral Stairs

		Crew	Daily Output	Labor-Hours	Unit	Material	Labor	Equipment	Total	Total Incl O&P
0010	**SPIRAL STAIRS**, shop fabricated									
1810	Spiral aluminum, 5'-0" diameter, stock units G	E-4	45	.711	Riser	680	32	2.98	714.98	815
1820	Custom units G		45	.711		1,275	32	2.98	1,309.98	1,500
1900	Spiral, cast iron, 4'-0" diameter, ornamental, minimum G		45	.711		605	32	2.98	639.98	730
1920	Maximum G		25	1.280		825	58	5.35	888.35	1,025
2000	Spiral, steel, industrial checkered plate, 4' diameter G		45	.711		605	32	2.98	639.98	730
2200	Stock units, 6'-0" diameter G		40	.800		735	36	3.35	774.35	880
3110	Spiral steel, stock units, primed, flat metal tread, 3'-6" dia G	2 Carp	1.60	10	Flight	1,250	400		1,650	2,000
3120	4'-0" dia G		1.45	11.034		1,425	440		1,865	2,300
3130	4'-6" dia G		1.35	11.852		1,575	475		2,050	2,500
3140	5'-0" dia G		1.25	12.800		1,950	510		2,460	3,000
3210	Galvanized, 3'-6" dia G		1.60	10		2,100	400		2,500	2,975
3220	4'-0" dia G		1.45	11.034		2,350	440		2,790	3,300
3230	4'-6" dia G		1.35	11.852		2,550	475		3,025	3,575
3240	5'-0" dia G		1.25	12.800		3,050	510		3,560	4,200
3310	Checkered plate tread, 3'-6" dia G		1.45	11.034		1,450	440		1,890	2,325
3320	4'-0" dia G		1.35	11.852		1,650	475		2,125	2,575
3330	4'-6" dia G		1.25	12.800		1,800	510		2,310	2,825
3340	5'-0" dia G		1.15	13.913		2,200	555		2,755	3,350
3410	Galvanized, 3'-6" dia G		1.45	11.034		2,400	440		2,840	3,375
3420	4'-0" dia G		1.35	11.852		2,650	475		3,125	3,675
3430	4'-6" dia G		1.25	12.800		2,850	510		3,360	3,975
3440	5'-0" dia G		1.15	13.913		3,350	555		3,905	4,625
3510	Red oak tread on flat metal, 3'-6" dia		1.35	11.852		2,375	475		2,850	3,375
3520	4'-0" dia		1.25	12.800		2,575	510		3,085	3,675
3530	4'-6" dia		1.15	13.913		2,800	555		3,355	4,000
3540	5'-0" dia		1.05	15.238		3,250	610		3,860	4,575

05 73 Decorative Metal Railings

05 73 16 – Wire Rope Decorative Metal Railings

05 73 16.10 Cable Railings

		Crew	Daily Output	Labor-Hours	Unit	Material	Labor	Equipment	Total	Total Incl O&P
0010	**CABLE RAILINGS**, with 316 stainless steel 1 x 19 cable, 3/16" diameter									
0015	Made from recycled materials									
0100	1-3/4" diameter stainless steel posts x 42" high, cables 4" OC G	2 Sswk	25	.640	L.F.	43	28.50		71.50	100

05 73 23 – Ornamental Railings

05 73 23.50 Railings, Ornamental

		Crew	Daily Output	Labor-Hours	Unit	Material	Labor	Equipment	Total	Total Incl O&P
0010	**RAILINGS, ORNAMENTAL**, shop fab'd, 3'-6" high, posts @ 5' O.C.									
0020	Bronze or stainless, minimum G	1 Sswk	24	.333	L.F.	36.50	14.90		51.40	68
0100	Maximum G		9	.889		365	39.50		404.50	475
0200	Aluminum ornamental rail, minimum G		15	.533		36.50	24		60.50	84.50
0300	Maximum G		8	1		110	44.50		154.50	204
0400	Hand-forged wrought iron, minimum G		12	.667		115	30		145	181
0500	Maximum G		8	1		345	44.50		389.50	465

05 73 Decorative Metal Railings

05 73 23 – Ornamental Railings

05 73 23.50 Railings, Ornamental			Crew	Daily Output	Labor-Hours	Unit	2009 Bare Costs Material	Labor	Equipment	Total	Total Incl O&P
0600	Composite metal/wood/glass, minimum		1 Sswk	6	1.333	L.F.	220	59.50		279.50	350
0700	Maximum			5	1.600		440	71.50		511.50	615
9000	Minimum labor/equipment charge			2	4	Job		179		179	330

05 75 Decorative Formed Metal

05 75 13 – Columns

05 75 13.20 Columns, Ornamental			Crew	Daily Output	Labor-Hours	Unit	Material	Labor	Equipment	Total	Total Incl O&P
0010	**COLUMNS, ORNAMENTAL**, shop fabricated										
6400	Mild steel, flat, 9" wide, stock units, painted, plain	G	E-4	160	.200	V.L.F.	8.90	9.05	.84	18.79	27.50
6450	Fancy	G		160	.200		17	9.05	.84	26.89	36.50
6500	Corner columns, painted, plain	G		160	.200		15.50	9.05	.84	25.39	34.50
6550	Fancy	G		160	.200		30	9.05	.84	39.89	50.50

Estimating Tips

06 05 00 Common Work Results for Wood, Plastics, and Composites

- Common to any wood-framed structure are the accessory connector items such as screws, nails, adhesives, hangers, connector plates, straps, angles, and hold-downs. For typical wood-framed buildings, such as residential projects, the aggregate total for these items can be significant, especially in areas where seismic loading is a concern. For floor and wall framing, the material cost is based on 10 to 25 lbs. per MBF. Hold-downs, hangers, and other connectors should be taken off by the piece.

06 10 00 Carpentry

- Lumber is a traded commodity and therefore sensitive to supply and demand in the marketplace. Even in "budgetary" estimating of wood-framed projects, it is advisable to call local suppliers for the latest market pricing.
- Common quantity units for wood-framed projects are "thousand board feet" (MBF). A board foot is a volume of wood, 1" x 1' x 1', or 144 cubic inches. Board-foot quantities are generally calculated using nominal material dimensions—dressed sizes are ignored. Board foot per lineal foot of any stick of lumber can be calculated by dividing the nominal cross-sectional area by 12. As an example, 2,000 lineal feet of 2 x 12 equates to 4 MBF by dividing the nominal area, 2 x 12, by 12, which equals 2, and multiplying by 2,000 to give 4,000 board feet. This simple rule applies to all nominal dimensioned lumber.
- Waste is an issue of concern at the quantity takeoff for any area of construction. Framing lumber is sold in even foot lengths, i.e., 10', 12', 14', 16', and depending on spans, wall heights, and the grade of lumber, waste is inevitable. A rule of thumb for lumber waste is 5%–10% depending on material quality and the complexity of the framing.
- Wood in various forms and shapes is used in many projects, even where the main structural framing is steel, concrete, or masonry. Plywood as a back-up partition material and 2x boards used as blocking and cant strips around roof edges are two common examples. The estimator should ensure that the costs of all wood materials are included in the final estimate.

06 20 00 Finish Carpentry

- It is necessary to consider the grade of workmanship when estimating labor costs for erecting millwork and interior finish. In practice, there are three grades: premium, custom, and economy. The RSMeans daily output for base and case moldings is in the range of 200 to 250 L.F. per carpenter per day. This is appropriate for most average custom-grade projects. For premium projects, an adjustment to productivity of 25%–50% should be made, depending on the complexity of the job.

Reference Numbers

Reference numbers are shown in shaded boxes at the beginning of some major classifications. These numbers refer to related items in the Reference Section. The reference information may be an estimating procedure, an alternate pricing method, or technical information.

Note: Not all subdivisions listed here necessarily appear in this publication.

06 05 Common Work Results for Wood, Plastics and Composites

06 05 05 – Selective Wood and Plastics Demolition

06 05 05.10 Selective Demolition Wood Framing		Crew	Daily Output	Labor-Hours	Unit	Material	2009 Bare Costs Labor	Equipment	Total	Total Incl O&P
0010	**SELECTIVE DEMOLITION WOOD FRAMING** R024119-10									
0100	Timber connector, nailed, small	1 Clab	96	.083	Ea.		2.63		2.63	4.35
0110	Medium		60	.133			4.21		4.21	6.95
0120	Large		48	.167			5.25		5.25	8.70
0130	Bolted, small		48	.167			5.25		5.25	8.70
0140	Medium		32	.250			7.90		7.90	13.05
0150	Large		24	.333			10.55		10.55	17.40
2958	Beams, 2″ x 6″	2 Clab	1100	.015	L.F.		.46		.46	.76
2960	2″ x 8″		825	.019			.61		.61	1.01
2965	2″ x 10″		665	.024			.76		.76	1.25
2970	2″ x 12″		550	.029			.92		.92	1.52
2972	2″ x 14″		470	.034			1.08		1.08	1.78
2975	4″ x 8″	B-1	413	.058			1.88		1.88	3.09
2980	4″ x 10″		330	.073			2.35		2.35	3.87
2985	4″ x 12″		275	.087			2.82		2.82	4.65
3000	6″ x 8″		275	.087			2.82		2.82	4.65
3040	6″ x 10″		220	.109			3.52		3.52	5.80
3080	6″ x 12″		185	.130			4.19		4.19	6.90
3120	8″ x 12″		140	.171			5.55		5.55	9.15
3160	10″ x 12″		110	.218			7.05		7.05	11.60
3162	Alternate pricing method		1.10	21.818	M.B.F.		705		705	1,150
3170	Blocking, in 16″ OC wall framing, 2″ x 4″	1 Clab	600	.013	L.F.		.42		.42	.70
3172	2″ x 6″		400	.020			.63		.63	1.04
3174	In 24″ OC wall framing, 2″ x 4″		600	.013			.42		.42	.70
3176	2″ x 6″		400	.020			.63		.63	1.04
3178	Alt method, wood blocking removal from wood framing		.40	20	M.B.F.		630		630	1,050
3179	Wood blocking removal from steel framing		.36	22.222	″		700		700	1,150
3180	Bracing, let in, 1″ x 3″, studs 16″ OC		1050	.008	L.F.		.24		.24	.40
3181	Studs 24″ OC		1080	.007			.23		.23	.39
3182	1″ x 4″, studs 16″ OC		1050	.008			.24		.24	.40
3183	Studs 24″ OC		1080	.007			.23		.23	.39
3184	1″ x 6″, studs 16″ OC		1050	.008			.24		.24	.40
3185	Studs 24″ OC		1080	.007			.23		.23	.39
3186	2″ x 3″, studs 16″ OC		800	.010			.32		.32	.52
3187	Studs 24″ OC		830	.010			.30		.30	.50
3188	2″ x 4″, studs 16″ OC		800	.010			.32		.32	.52
3189	Studs 24″ OC		830	.010			.30		.30	.50
3190	2″ x 6″, studs 16″ OC		800	.010			.32		.32	.52
3191	Studs 24″ OC		830	.010			.30		.30	.50
3192	2″ x 8″, studs 16″ OC		800	.010			.32		.32	.52
3193	Studs 24″ OC		830	.010			.30		.30	.50
3194	″T″ shaped metal bracing, studs at 16″ OC		1060	.008			.24		.24	.39
3195	Studs at 24″ OC		1200	.007			.21		.21	.35
3196	Metal straps, studs at 16″ OC		1200	.007			.21		.21	.35
3197	Studs at 24″ OC		1240	.006			.20		.20	.34
3200	Columns, round, 8′ to 14′ tall		40	.200	Ea.		6.30		6.30	10.45
3202	Dimensional lumber sizes	2 Clab	1.10	14.545	M.B.F.		460		460	760
3250	Blocking, between joists	1 Clab	320	.025	Ea.		.79		.79	1.30
3252	Bridging, metal strap, between joists		320	.025	Pr.		.79		.79	1.30
3260	Door buck, studs, header & access., 8′ high 2″ x 4″ wall, 3′ wide		32	.250	Ea.		7.90		7.90	13.05
3261	4′ wide		32	.250			7.90		7.90	13.05
3262	5′ wide		32	.250			7.90		7.90	13.05
3263	6′ wide		32	.250			7.90		7.90	13.05

06 05 05 – Selective Wood and Plastics Demolition

06 05 05.10 Selective Demolition Wood Framing		Crew	Daily Output	Labor-Hours	Unit	Material	2009 Bare Costs Labor	Equipment	Total	Total Incl O&P
3264	8′ wide	1 Clab	30	.267	Ea.		8.45		8.45	13.90
3265	10′ wide		30	.267			8.45		8.45	13.90
3266	12′ wide		30	.267			8.45		8.45	13.90
3267	2″ x 6″ wall, 3′ wide		32	.250			7.90		7.90	13.05
3268	4′ wide		32	.250			7.90		7.90	13.05
3269	5′ wide		32	.250			7.90		7.90	13.05
3270	6′ wide		32	.250			7.90		7.90	13.05
3271	8′ wide		30	.267			8.45		8.45	13.90
3272	10′ wide		30	.267			8.45		8.45	13.90
3273	12′ wide		30	.267			8.45		8.45	13.90
3274	Window buck, studs, header & access, 8′ high 2″ x 4″ wall, 2′ wide		24	.333			10.55		10.55	17.40
3275	3′ wide		24	.333			10.55		10.55	17.40
3276	4′ wide		24	.333			10.55		10.55	17.40
3277	5′ wide		24	.333			10.55		10.55	17.40
3278	6′ wide		24	.333			10.55		10.55	17.40
3279	7′ wide		24	.333			10.55		10.55	17.40
3280	8′ wide		22	.364			11.50		11.50	18.95
3281	10′ wide		22	.364			11.50		11.50	18.95
3282	12′ wide		22	.364			11.50		11.50	18.95
3283	2″ x 6″ wall, 2′ wide		24	.333			10.55		10.55	17.40
3284	3′ wide		24	.333			10.55		10.55	17.40
3285	4′ wide		24	.333			10.55		10.55	17.40
3286	5′ wide		24	.333			10.55		10.55	17.40
3287	6′ wide		24	.333			10.55		10.55	17.40
3288	7′ wide		24	.333			10.55		10.55	17.40
3289	8′ wide		22	.364			11.50		11.50	18.95
3290	10′ wide		22	.364			11.50		11.50	18.95
3291	12′ wide		22	.364			11.50		11.50	18.95
3400	Fascia boards, 1″ x 6″		500	.016	L.F.		.51		.51	.83
3440	1″ x 8″		450	.018			.56		.56	.93
3480	1″ x 10″		400	.020			.63		.63	1.04
3490	2″ x 6″		450	.018			.56		.56	.93
3500	2″ x 8″		400	.020			.63		.63	1.04
3510	2″ x 10″		350	.023			.72		.72	1.19
3610	Furring, on wood walls or ceiling		4000	.002	S.F.		.06		.06	.10
3620	On masonry or concrete walls or ceiling		1200	.007	″		.21		.21	.35
3800	Headers over openings, 2 @ 2″ x 6″		110	.073	L.F.		2.30		2.30	3.79
3840	2 @ 2″ x 8″		100	.080			2.53		2.53	4.17
3880	2 @ 2″ x 10″		90	.089			2.81		2.81	4.64
3885	Alternate pricing method		.26	30.651	M.B.F.		970		970	1,600
3920	Joists, 1″ x 4″		1250	.006	L.F.		.20		.20	.33
3930	1″ x 6″		1135	.007			.22		.22	.37
3940	1″ x 8″		1000	.008			.25		.25	.42
3950	1″ x 10″		895	.009			.28		.28	.47
3960	1″ x 12″		765	.010			.33		.33	.55
4200	2″ x 4″	2 Clab	1000	.016			.51		.51	.83
4230	2″ x 6″		970	.016			.52		.52	.86
4240	2″ x 8″		940	.017			.54		.54	.89
4250	2″ x 10″		910	.018			.56		.56	.92
4280	2″ x 12″		880	.018			.57		.57	.95
4281	2″ x 14″		850	.019			.59		.59	.98
4282	Composite joists, 9-1/2″		960	.017			.53		.53	.87
4283	11-7/8″		930	.017			.54		.54	.90

06 05 Common Work Results for Wood, Plastics and Composites

06 05 05 – Selective Wood and Plastics Demolition

06 05 05.10 Selective Demolition Wood Framing		Crew	Daily Output	Labor-Hours	Unit	Material	2009 Bare Costs Labor	Equipment	Total	Total Incl O&P
4284	14"	2 Clab	897	.018	L.F.		.56		.56	.93
4285	16"		865	.019			.58		.58	.96
4290	Wood joists, alternate pricing method		1.50	10.667	M.B.F.		335		335	555
4500	Open web joist, 12" deep		500	.032	L.F.		1.01		1.01	1.67
4505	14" deep		475	.034			1.06		1.06	1.76
4510	16" deep		450	.036			1.12		1.12	1.85
4520	18" deep		425	.038			1.19		1.19	1.96
4560	1" x 3"	1 Clab	1200	.007			.21		.21	.35
4580	2" x 2"		1100	.007			.23		.23	.38
4590	2" x 4"		1000	.008			.25		.25	.42
4600	2" x 6"		1000	.008			.25		.25	.42
4601	2" x 8 or 2" x 10"		800	.010			.32		.32	.52
4602	4" x 6"		600	.013			.42		.42	.70
4604	4" x 8"		450	.018			.56		.56	.93
5400	Posts, 4" x 4"	2 Clab	800	.020			.63		.63	1.04
5405	4" x 6"		550	.029			.92		.92	1.52
5410	4" x 8"		440	.036			1.15		1.15	1.90
5425	4" x 10"		390	.041			1.30		1.30	2.14
5430	4" x 12"		350	.046			1.44		1.44	2.38
5440	6" x 6"		400	.040			1.26		1.26	2.09
5445	6" x 8"		350	.046			1.44		1.44	2.38
5450	6" x 10"		320	.050			1.58		1.58	2.61
5455	6" x 12"		290	.055			1.74		1.74	2.88
5480	8" x 8"		300	.053			1.69		1.69	2.78
5500	10" x 10"		240	.067			2.11		2.11	3.48
5660	Tongue and groove floor planks		2	8	M.B.F.		253		253	415
5750	Rafters, ordinary, 16" OC, 2" x 4"		880	.018	S.F.		.57		.57	.95
5755	2" x 6"		840	.019			.60		.60	.99
5760	2" x 8"		820	.020			.62		.62	1.02
5770	2" x 10"		820	.020			.62		.62	1.02
5780	2" x 12"		810	.020			.62		.62	1.03
5785	24" OC, 2" x 4"		1170	.014			.43		.43	.71
5786	2" x 6"		1117	.014			.45		.45	.75
5787	2" x 8"		1091	.015			.46		.46	.77
5788	2" x 10"		1091	.015			.46		.46	.77
5789	2" x 12"		1077	.015			.47		.47	.77
5795	Rafters, ordinary, 2" x 4" (alternate method)		862	.019	L.F.		.59		.59	.97
5800	2" x 6" (alternate method)		850	.019			.59		.59	.98
5840	2" x 8" (alternate method)		837	.019			.60		.60	1
5855	2" x 10" (alternate method)		825	.019			.61		.61	1.01
5865	2" x 12" (alternate method)		812	.020			.62		.62	1.03
5870	Sill plate, 2" x 4"	1 Clab	1170	.007			.22		.22	.36
5871	2" x 6"		780	.010			.32		.32	.54
5872	2" x 8"		586	.014			.43		.43	.71
5873	Alternate pricing method		.78	10.256	M.B.F.		325		325	535
5885	Ridge board, 1" x 4"	2 Clab	900	.018	L.F.		.56		.56	.93
5886	1" x 6"		875	.018			.58		.58	.95
5887	1" x 8"		850	.019			.59		.59	.98
5888	1" x 10"		825	.019			.61		.61	1.01
5889	1" x 12"		800	.020			.63		.63	1.04
5890	2" x 4"		900	.018			.56		.56	.93
5892	2" x 6"		875	.018			.58		.58	.95
5894	2" x 8"		850	.019			.59		.59	.98

06 05 Common Work Results for Wood, Plastics and Composites

06 05 05 – Selective Wood and Plastics Demolition

06 05 05.10 Selective Demolition Wood Framing		Crew	Daily Output	Labor-Hours	Unit	2009 Bare Costs Material	Labor	Equipment	Total	Total Incl O&P
5896	2″ x 10″	2 Clab	825	.019	L.F.		.61		.61	1.01
5898	2″ x 12″		800	.020			.63		.63	1.04
5900	Hip & valley rafters, 2″ x 6″		500	.032			1.01		1.01	1.67
5940	2″ x 8″		420	.038			1.20		1.20	1.99
6050	Rafter tie, 1″ x 4″		1250	.013			.40		.40	.67
6052	1″ x 6″		1135	.014			.45		.45	.74
6054	2″ x 4″		1000	.016			.51		.51	.83
6056	2″ x 6″		970	.016			.52		.52	.86
6070	Sleepers, on concrete, 1″ x 2″	1 Clab	4700	.002			.05		.05	.09
6075	1″ x 3″		4000	.002			.06		.06	.10
6080	2″ x 4″		3000	.003			.08		.08	.14
6085	2″ x 6″		2600	.003			.10		.10	.16
6086	Sheathing from roof, 5/16″	2 Clab	1600	.010	S.F.		.32		.32	.52
6088	3/8″		1525	.010			.33		.33	.55
6090	1/2″		1400	.011			.36		.36	.60
6092	5/8″		1300	.012			.39		.39	.64
6094	3/4″		1200	.013			.42		.42	.70
6096	Board sheathing from roof		1400	.011			.36		.36	.60
6100	Sheathing, from walls, 1/4″		1200	.013			.42		.42	.70
6110	5/16″		1175	.014			.43		.43	.71
6120	3/8″		1150	.014			.44		.44	.73
6130	1/2″		1125	.014			.45		.45	.74
6140	5/8″		1100	.015			.46		.46	.76
6150	3/4″		1075	.015			.47		.47	.78
6152	Board sheathing from walls		1500	.011			.34		.34	.56
6158	Subfloor, with boards		1050	.015			.48		.48	.79
6160	Plywood, 1/2″ thick		768	.021			.66		.66	1.09
6162	5/8″ thick		760	.021			.67		.67	1.10
6164	3/4″ thick		750	.021			.67		.67	1.11
6165	1-1/8″ thick		720	.022			.70		.70	1.16
6166	Underlayment, particle board, 3/8″ thick	1 Clab	780	.010			.32		.32	.54
6168	1/2″ thick		768	.010			.33		.33	.54
6170	5/8″ thick		760	.011			.33		.33	.55
6172	3/4″ thick		750	.011			.34		.34	.56
6200	Stairs and stringers, minimum	2 Clab	40	.400	Riser		12.65		12.65	21
6240	Maximum	″	26	.615	″		19.45		19.45	32
6300	Components, tread	1 Clab	110	.073	Ea.		2.30		2.30	3.79
6320	Riser		80	.100	″		3.16		3.16	5.20
6390	Stringer, 2″ x 10″		260	.031	L.F.		.97		.97	1.60
6400	2″ x 12″		260	.031			.97		.97	1.60
6410	3″ x 10″		250	.032			1.01		1.01	1.67
6420	3″ x 12″		250	.032			1.01		1.01	1.67
6590	Wood studs, 2″ x 3″	2 Clab	3076	.005			.16		.16	.27
6600	2″ x 4″		2000	.008			.25		.25	.42
6640	2″ x 6″		1600	.010			.32		.32	.52
6720	Wall framing, including studs, plates and blocking, 2″ x 4″	1 Clab	600	.013	S.F.		.42		.42	.70
6740	2″ x 6″		480	.017	″		.53		.53	.87
6750	Headers, 2″ x 4″		1125	.007	L.F.		.22		.22	.37
6755	2″ x 6″		1125	.007			.22		.22	.37
6760	2″ x 8″		1050	.008			.24		.24	.40
6765	2″ x 10″		1050	.008			.24		.24	.40
6770	2″ x 12″		1000	.008			.25		.25	.42
6780	4″ x 10″		525	.015			.48		.48	.79

06 05 Common Work Results for Wood, Plastics and Composites

06 05 05 – Selective Wood and Plastics Demolition

06 05 05.10 Selective Demolition Wood Framing		Crew	Daily Output	Labor-Hours	Unit	Material	Labor	Equipment	Total	Total Incl O&P
						2009 Bare Costs				
6785	4″ x 12″	1 Clab	500	.016	L.F.		.51		.51	.83
6790	6″ x 8″		560	.014			.45		.45	.75
6795	6″ x 10″		525	.015			.48		.48	.79
6797	6″ x 12″		500	.016			.51		.51	.83
8000	Soffit, T & G wood		520	.015	S.F.		.49		.49	.80
8010	Hardboard, vinyl or aluminum		640	.013			.40		.40	.65
8030	Plywood	2 Carp	315	.051			2.03		2.03	3.35
9000	Minimum labor/equipment charge	1 Clab	4	2	Job		63		63	104

06 05 05.20 Selective Demolition Millwork and Trim		Crew	Daily Output	Labor-Hours	Unit	Material	Labor	Equipment	Total	Total Incl O&P
0010	**SELECTIVE DEMOLITION MILLWORK AND TRIM** R024119-10									
1000	Cabinets, wood, base cabinets, per L.F.	2 Clab	80	.200	L.F.		6.30		6.30	10.45
1020	Wall cabinets, per L.F.	″	80	.200	″		6.30		6.30	10.45
1060	Remove and reset, base cabinets	2 Carp	18	.889	Ea.		35.50		35.50	58.50
1070	Wall cabinets	″	20	.800	″		32		32	53
1100	Steel, painted, base cabinets	2 Clab	60	.267	L.F.		8.45		8.45	13.90
1120	Wall cabinets		60	.267	″		8.45		8.45	13.90
1200	Casework, large area		320	.050	S.F.		1.58		1.58	2.61
1220	Selective		200	.080	″		2.53		2.53	4.17
1500	Counter top, minimum		200	.080	L.F.		2.53		2.53	4.17
1510	Maximum		120	.133			4.21		4.21	6.95
1550	Remove and reset, minimum	2 Carp	50	.320			12.80		12.80	21
1560	Maximum	″	40	.400			16		16	26.50
2000	Paneling, 4′ x 8′ sheets	2 Clab	2000	.008	S.F.		.25		.25	.42
2100	Boards, 1″ x 4″		700	.023			.72		.72	1.19
2120	1″ x 6″		750	.021			.67		.67	1.11
2140	1″ x 8″		800	.020			.63		.63	1.04
3000	Trim, baseboard, to 6″ wide		1200	.013	L.F.		.42		.42	.70
3040	Greater than 6″ and up to 12″ wide		1000	.016			.51		.51	.83
3080	Remove and reset, minimum	2 Carp	400	.040			1.60		1.60	2.64
3090	Maximum	″	300	.053			2.13		2.13	3.52
3100	Ceiling trim	2 Clab	1000	.016			.51		.51	.83
3120	Chair rail		1200	.013			.42		.42	.70
3140	Railings with balusters		240	.067			2.11		2.11	3.48
3160	Wainscoting		700	.023	S.F.		.72		.72	1.19
9000	Minimum labor/equipment charge	1 Clab	4	2	Job		63		63	104

06 05 23 – Wood, Plastic, and Composite Fastenings

06 05 23.10 Nails		Crew	Daily Output	Labor-Hours	Unit	Material	Labor	Equipment	Total	Total Incl O&P
0010	**NAILS**, material only, based upon 50# box purchase									
0020	Copper nails, plain				Lb.	10.20			10.20	11.20
0400	Stainless steel, plain					6.65			6.65	7.30
0500	Box, 3d to 20d, bright					.91			.91	1
0520	Galvanized					1.67			1.67	1.84
0600	Common, 3d to 60d, plain					1.11			1.11	1.22
0700	Galvanized					1.53			1.53	1.68
0800	Aluminum					4.45			4.45	4.90
1000	Annular or spiral thread, 4d to 60d, plain					1.82			1.82	2
1200	Galvanized					1.99			1.99	2.19
1400	Drywall nails, plain					.79			.79	.87
1600	Galvanized					1.59			1.59	1.75
1800	Finish nails, 4d to 10d, plain					1.11			1.11	1.22
2000	Galvanized					1.75			1.75	1.93
2100	Aluminum					4.10			4.10	4.51

06 05 23 – Wood, Plastic, and Composite Fastenings

06 05 23.10 Nails		Crew	Daily Output	Labor-Hours	Unit	Material	2009 Bare Costs Labor	Equipment	Total	Total Incl O&P
2300	Flooring nails, hardened steel, 2d to 10d, plain				Lb.	2.04			2.04	2.24
2400	Galvanized					3.88			3.88	4.27
2500	Gypsum lath nails, 1-1/8", 13 ga. flathead, blued					2.64			2.64	2.90
2600	Masonry nails, hardened steel, 3/4" to 3" long, plain					1.97			1.97	2.17
2700	Galvanized					2.78			2.78	3.06
2900	Roofing nails, threaded, galvanized					1.38			1.38	1.52
3100	Aluminum					4.80			4.80	5.30
3300	Compressed lead head, threaded, galvanized					2.50			2.50	2.75
3600	Siding nails, plain shank, galvanized					1.50			1.50	1.65
3800	Aluminum					4.11			4.11	4.52
5000	Add to prices above for cement coating					.10			.10	.11
5200	Zinc or tin plating					.13			.13	.14
5500	Vinyl coated sinkers, 8d to 16d				↓	.60			.60	.66
06 05 23.20 Pneumatic Nails										
0010	**PNEUMATIC NAILS**									
0020	Framing, per carton of 5000, 2"				Ea.	40			40	44
0100	2-3/8"					45			45	49.50
0200	Per carton of 4000, 3"					40			40	44
0300	3-1/4"					40			40	44
0400	Per carton of 5000, 2-3/8", galv.					60			60	66
0500	Per carton of 4000, 3", galv.					65			65	71.50
0600	3-1/4", galv.					82			82	90
0700	Roofing, per carton of 7200, 1"					35.50			35.50	39
0800	1-1/4"					35			35	38.50
0900	1-1/2"					38.50			38.50	42.50
1000	1-3/4"				↓	47.50			47.50	52.50
06 05 23.40 Sheet Metal Screws										
0010	**SHEET METAL SCREWS**									
0020	Steel, standard, #8 x 3/4", plain				C	3.30			3.30	3.63
0100	Galvanized					3.39			3.39	3.73
0300	#10 x 1", plain					4.25			4.25	4.68
0400	Galvanized					4.43			4.43	4.87
0600	With washers, #14 x 1", plain					11.50			11.50	12.65
0700	Galvanized					12.45			12.45	13.70
0900	#14 x 2", plain					15			15	16.50
1000	Galvanized					17.15			17.15	18.85
1500	Self-drilling, with washers, (pinch point) #8 x 3/4", plain					8			8	8.80
1600	Galvanized					8.70			8.70	9.55
1800	#10 x 3/4", plain					10			10	11
1900	Galvanized					10.95			10.95	12.05
3000	Stainless steel w/aluminum or neoprene washers, #14 x 1", plain					32			32	35
3100	#14 x 2", plain				↓	41			41	45
06 05 23.50 Wood Screws										
0010	**WOOD SCREWS**									
0020	#8 x 1" long, steel				C	5.10			5.10	5.60
0100	Brass					11.50			11.50	12.65
0200	#8, 2" long, steel					5.30			5.30	5.85
0300	Brass					19.50			19.50	21.50
0400	#10, 1" long, steel					3.71			3.71	4.08
0500	Brass					15.65			15.65	17.20
0600	#10, 2" long, steel					5.95			5.95	6.55
0700	Brass				↓	23.50			23.50	26

06 05 23 – Wood, Plastic, and Composite Fastenings

06 05 23.50 Wood Screws		Crew	Daily Output	Labor-Hours	Unit	2009 Bare Costs Material	Labor	Equipment	Total	Total Incl O&P
0800	#10, 3" long, steel				C	9.95			9.95	10.95
1000	#12, 2" long, steel					7.90			7.90	8.70
1100	Brass					31.50			31.50	34.50
1500	#12, 3" long, steel					11.45			11.45	12.60
2000	#12, 4" long, steel					25.50			25.50	28

06 05 23.60 Timber Connectors

		Crew	Daily Output	Labor-Hours	Unit	Material	Labor	Equipment	Total	Total Incl O&P
0010	**TIMBER CONNECTORS**									
0020	Add up cost of each part for total cost of connection									
0100	Connector plates, steel, with bolts, straight	2 Carp	75	.213	Ea.	25.50	8.50		34	42
0110	Tee, 7 gauge		50	.320		29.50	12.80		42.30	53
0120	T- Strap, 14 gauge, 12" x 8" x 2"		50	.320		29.50	12.80		42.30	53
0150	Anchor plates, 7 ga, 9" x 7"		75	.213		25.50	8.50		34	42
0200	Bolts, machine, sq. hd. with nut & washer, 1/2" diameter, 4" long	1 Carp	140	.057		.79	2.28		3.07	4.64
0300	7-1/2" long		130	.062		1.37	2.46		3.83	5.55
0500	3/4" diameter, 7-1/2" long		130	.062		4.12	2.46		6.58	8.60
0610	Machine bolts, w/ nut, washer, 3/4" diam., 15" L, HD's & beam hangers		95	.084		7.60	3.36		10.96	13.90
0800	Drilling bolt holes in timber, 1/2" diameter		450	.018	Inch		.71		.71	1.17
0900	1" diameter		350	.023	"		.91		.91	1.51
1100	Framing anchor, angle, 3" x 3" x 1-1/2", 12 ga		175	.046	Ea.	2.20	1.83		4.03	5.45
1150	Framing anchors, 18 gauge, 4-1/2" x 2-3/4"		175	.046		2.20	1.83		4.03	5.45
1160	Framing anchors, 18 gauge, 4-1/2" x 3"		175	.046		2.20	1.83		4.03	5.45
1170	Clip anchors plates, 18 gauge, 12" x 1-1/8"		175	.046		2.20	1.83		4.03	5.45
1250	Holdowns, 3 gauge base, 10 gauge body		8	1		21	40		61	89
1260	Holdowns, 7 gauge 11-1/16" x 3-1/4"		8	1		21	40		61	89
1270	Holdowns, 7 gauge 14-3/8" x 3-1/8"		8	1		21	40		61	89
1275	Holdowns, 12 gauge 8" x 2-1/2"		8	1		21	40		61	89
1300	Joist and beam hangers, 18 ga. galv., for 2" x 4" joist		175	.046		.61	1.83		2.44	3.68
1400	2" x 6" to 2" x 10" joist		165	.048		1.17	1.94		3.11	4.49
1600	16 ga. galv., 3" x 6" to 3" x 10" joist		160	.050		2.49	2		4.49	6.05
1700	3" x 10" to 3" x 14" joist		160	.050		4.14	2		6.14	7.85
1800	4" x 6" to 4" x 10" joist		155	.052		2.58	2.06		4.64	6.25
1900	4" x 10" to 4" x 14" joist		155	.052		4.23	2.06		6.29	8.05
2000	Two-2" x 6" to two-2" x 10" joists		150	.053		3.52	2.13		5.65	7.40
2100	Two-2" x 10" to two-2" x 14" joists		150	.053		3.94	2.13		6.07	7.85
2300	3/16" thick, 6" x 8" joist		145	.055		54.50	2.20		56.70	63.50
2400	6" x 10" joist		140	.057		57	2.28		59.28	66.50
2500	6" x 12" joist		135	.059		59.50	2.37		61.87	69
2700	1/4" thick, 6" x 14" joist		130	.062		61.50	2.46		63.96	72
2900	Plywood clips, extruded aluminum H clip, for 3/4" panels					.20			.20	.22
3000	Galvanized 18 ga. back-up clip					.16			.16	.18
3200	Post framing, 16 ga. galv. for 4" x 4" base, 2 piece	1 Carp	130	.062		13.80	2.46		16.26	19.20
3300	Cap		130	.062		19.15	2.46		21.61	25
3500	Rafter anchors, 18 ga. galv., 1-1/2" wide, 5-1/4" long		145	.055		.41	2.20		2.61	4.09
3600	10-3/4" long		145	.055		1.23	2.20		3.43	4.99
3800	Shear plates, 2-5/8" diameter		120	.067		2	2.66		4.66	6.60
3900	4" diameter		115	.070		4.75	2.78		7.53	9.85
4000	Sill anchors, embedded in concrete or block, 25-1/2" long		115	.070		10.70	2.78		13.48	16.35
4100	Spike grids, 3" x 6"		120	.067		.81	2.66		3.47	5.30
4400	Split rings, 2-1/2" diameter		120	.067		1.65	2.66		4.31	6.20
4500	4" diameter		110	.073		2.50	2.91		5.41	7.55
4550	Tie plate, 20 gauge, 7" x 3 1/8"		110	.073		2.50	2.91		5.41	7.55
4560	Tie plate, 20 gauge, 5" x 4 1/8"		110	.073		2.50	2.91		5.41	7.55

06 05 Common Work Results for Wood, Plastics and Composites

06 05 23 – Wood, Plastic, and Composite Fastenings

06 05 23.60 Timber Connectors		Crew	Daily Output	Labor-Hours	Unit	2009 Bare Costs Material	Labor	Equipment	Total	Total Incl O&P
4575	Twist straps, 18 gauge, 12" x 1 1/4"	1 Carp	110	.073	Ea.	2.50	2.91		5.41	7.55
4580	Twist straps, 18 gauge, 16" x 1 1/4"		110	.073		2.50	2.91		5.41	7.55
4600	Strap ties, 20 ga., 2 -1/16" wide, 12 13/16" long		180	.044		.80	1.78		2.58	3.81
4700	Strap ties, 16 ga., 1-3/8" wide, 12" long		180	.044		.80	1.78		2.58	3.81
4800	21-5/8" x 1-1/4"		160	.050		2.49	2		4.49	6.05
5000	Toothed rings, 2-5/8" or 4" diameter		90	.089		1.49	3.55		5.04	7.50
5200	Truss plates, nailed, 20 gauge, up to 32' span		17	.471	Truss	10.85	18.80		29.65	43
5400	Washers, 2" x 2" x 1/8"				Ea.	.34			.34	.37
5500	3" x 3" x 3/16"				"	.89			.89	.98
9000	Minimum labor/equipment charge	1 Carp	4	2	Job		30		80	132

06 05 23.70 Rough Hardware		Crew	Daily Output	Labor-Hours	Unit	Material	Labor	Equipment	Total	Total Incl O&P
0010	**ROUGH HARDWARE**, average percent of carpentry material									
0020	Minimum					.50%				
0200	Maximum					1.50%				
0210	In seismic or hurricane areas, up to					10%				

06 05 23.80 Metal Bracing		Crew	Daily Output	Labor-Hours	Unit	Material	Labor	Equipment	Total	Total Incl O&P
0010	**METAL BRACING**									
0302	Let-in, "T" shaped, 22 ga. galv. steel, studs at 16" O.C.	1 Carp	580	.014	L.F.	.75	.55		1.30	1.74
0402	Studs at 24" O.C.		600	.013		.75	.53		1.28	1.71
0502	16 ga. galv. steel straps, studs at 16" O.C.		600	.013		.97	.53		1.50	1.95
0602	Studs at 24" O.C.		620	.013		.97	.52		1.49	1.92

06 11 Wood Framing

06 11 10 – Framing with Dimensional, Engineered or Composite Lumber

06 11 10.02 Blocking		Crew	Daily Output	Labor-Hours	Unit	Material	Labor	Equipment	Total	Total Incl O&P
0010	**BLOCKING**									
2600	Miscellaneous, to wood construction									
2620	2" x 4"	1 Carp	.17	47.059	M.B.F.	435	1,875		2,310	3,575
2625	Pneumatic nailed		.21	38.095		435	1,525		1,960	2,975
2660	2" x 8"		.27	29.630		530	1,175		1,705	2,525
2665	Pneumatic nailed		.33	24.242		530	970		1,500	2,175
2720	To steel construction									
2740	2" x 4"	1 Carp	.14	57.143	M.B.F.	435	2,275		2,710	4,250
2780	2" x 8"		.21	38.095	"	530	1,525		2,055	3,075
9000	Minimum labor/equipment charge		4	2	Job		30		80	132

06 11 10.04 Wood Bracing		Crew	Daily Output	Labor-Hours	Unit	Material	Labor	Equipment	Total	Total Incl O&P
0010	**WOOD BRACING**									
0011	Let-in, with 1" x 6" boards, studs 16" O.C.	1 Carp	1.50	5.333	C.L.F.	64	213		277	420
0202	Studs @ 24" O.C.	"	230	.035	L.F.	.64	1.39		2.03	2.99

06 11 10.06 Bridging		Crew	Daily Output	Labor-Hours	Unit	Material	Labor	Equipment	Total	Total Incl O&P
0010	**BRIDGING**									
0011	Wood, for joists 16" O.C., 1" x 3"	1 Carp	1.30	6.154	C.Pr.	51	246		297	460
0017	Pneumatic nailed		170	.047	Pr.	.51	1.88		2.39	3.66
0102	2" x 3" bridging		130	.062		.47	2.46		2.93	4.58
0107	Pneumatic nailed		170	.047		.47	1.88		2.35	3.62
0302	Steel, galvanized, 18 ga., for 2" x 10" joists at 12" O.C.		130	.062		1.58	2.46		4.04	5.80
0402	24" O.C.		140	.057		1.69	2.28		3.97	5.65
0902	Compression type, 16" O.C., 2" x 8" joists		200	.040		1.64	1.60		3.24	4.44
1002	2" x 12" joists		200	.040		1.64	1.60		3.24	4.44

06 11 10.10 Beam and Girder Framing

Code	Description	Crew	Daily Output	Labor-Hours	Unit	Material	Labor (2009 Bare Costs)	Equipment (2009 Bare Costs)	Total (2009 Bare Costs)	Total Incl O&P
0010	**BEAM AND GIRDER FRAMING** R061110-30									
1000	Single, 2" x 6"	2 Carp	700	.023	L.F.	.52	.91		1.43	2.08
1005	Pneumatic nailed		812	.020		.52	.79		1.31	1.87
1020	2" x 8"		650	.025		.71	.98		1.69	2.40
1025	Pneumatic nailed		754	.021		.71	.85		1.56	2.18
1040	2" x 10"		600	.027		.99	1.07		2.06	2.85
1045	Pneumatic nailed		696	.023		.99	.92		1.91	2.61
1060	2" x 12"		550	.029		1.39	1.16		2.55	3.44
1065	Pneumatic nailed		638	.025		1.39	1		2.39	3.17
1080	2" x 14"		500	.032		2.14	1.28		3.42	4.46
1085	Pneumatic nailed		580	.028		2.14	1.10		3.24	4.17
1100	3" x 8"		550	.029		2.10	1.16		3.26	4.23
1120	3" x 10"		500	.032		2.94	1.28		4.22	5.35
1140	3" x 12"		450	.036		3.54	1.42		4.96	6.25
1160	3" x 14"	↓	400	.040		4.48	1.60		6.08	7.55
1170	4" x 6"	F-3	1100	.036		2.34	1.47	.70	4.51	5.75
1180	4" x 8"		1000	.040		2.79	1.62	.77	5.18	6.55
1200	4" x 10"		950	.042		3.68	1.70	.81	6.19	7.75
1220	4" x 12"		900	.044		4.74	1.80	.85	7.39	9.10
1240	4" x 14"	↓	850	.047		5.60	1.90	.90	8.40	10.25
2000	Double, 2" x 6"	2 Carp	625	.026		1.04	1.02		2.06	2.84
2005	Pneumatic nailed		725	.022		1.04	.88		1.92	2.61
2020	2" x 8"		575	.028		1.42	1.11		2.53	3.40
2025	Pneumatic nailed		667	.024		1.42	.96		2.38	3.14
2040	2" x 10"		550	.029		1.98	1.16		3.14	4.10
2045	Pneumatic nailed		638	.025		1.98	1		2.98	3.83
2060	2" x 12"		525	.030		2.77	1.22		3.99	5.05
2065	Pneumatic nailed		610	.026		2.77	1.05		3.82	4.78
2080	2" x 14"		475	.034		4.28	1.35		5.63	6.90
2085	Pneumatic nailed		551	.029		4.28	1.16		5.44	6.60
3000	Triple, 2" x 6"		550	.029		1.56	1.16		2.72	3.64
3005	Pneumatic nailed		638	.025		1.56	1		2.56	3.37
3020	2" x 8"		525	.030		2.12	1.22		3.34	4.34
3025	Pneumatic nailed		609	.026		2.12	1.05		3.17	4.06
3040	2" x 10"		500	.032		2.98	1.28		4.26	5.40
3045	Pneumatic nailed		580	.028		2.98	1.10		4.08	5.10
3060	2" x 12"		475	.034		4.16	1.35		5.51	6.80
3065	Pneumatic nailed		551	.029		4.16	1.16		5.32	6.50
3080	2" x 14"		450	.036		6.40	1.42		7.82	9.40
3085	Pneumatic nailed	↓	522	.031	↓	6.40	1.22		7.62	9.05
9000	Minimum labor/equipment charge	1 Carp	2	4	Job		160		160	264

06 11 10.12 Ceiling Framing

Code	Description	Crew	Daily Output	Labor-Hours	Unit	Material	Labor	Equipment	Total	Total Incl O&P
0010	**CEILING FRAMING**									
6000	Suspended, 2" x 3"	2 Carp	1000	.016	L.F.	.32	.64		.96	1.41
6050	2" x 4"		900	.018		.29	.71		1	1.49
6100	2" x 6"		800	.020		.52	.80		1.32	1.89
6150	2" x 8"	↓	650	.025	↓	.71	.98		1.69	2.40
9000	Minimum labor/equipment charge	1 Carp	4	2	Job		80		80	132

06 11 10.14 Posts and Columns

Code	Description	Crew	Daily Output	Labor-Hours	Unit	Material	Labor	Equipment	Total	Total Incl O&P
0010	**POSTS AND COLUMNS**									
0100	4" x 4"	2 Carp	390	.041	L.F.	1.56	1.64		3.20	4.43
0150	4" x 6"	↓	275	.058	↓	2.34	2.32		4.66	6.40

06 11 10 – Framing with Dimensional, Engineered or Composite Lumber

06 11 10.14 Posts and Columns

		Crew	Daily Output	Labor-Hours	Unit	2009 Bare Costs Material	Labor	Equipment	Total	Total Incl O&P
0200	4" x 8"	2 Carp	220	.073	L.F.	2.79	2.91		5.70	7.85
0250	6" x 6"		215	.074		4.87	2.97		7.84	10.25
0300	6" x 8"		175	.091		7.60	3.65		11.25	14.40
0350	6" x 10"		150	.107		10.90	4.26		15.16	19.05
9000	Minimum labor/equipment charge	1 Carp	2	4	Job		160		160	264

06 11 10.18 Joist Framing

		Crew	Daily Output	Labor-Hours	Unit	Material	Labor	Equipment	Total	Total Incl O&P
0010	JOIST FRAMING R061110-30									
2002	Joists, 2" x 4"	2 Carp	1250	.013	L.F.	.29	.51		.80	1.16
2007	Pneumatic nailed		1438	.011		.29	.44		.73	1.05
2100	2" x 6"		1250	.013		.52	.51		1.03	1.41
2105	Pneumatic nailed		1438	.011		.52	.44		.96	1.30
2152	2" x 8"		1100	.015		.71	.58		1.29	1.74
2157	Pneumatic nailed		1265	.013		.71	.51		1.22	1.61
2202	2" x 10"		900	.018		.99	.71		1.70	2.26
2207	Pneumatic nailed		1035	.015		.99	.62		1.61	2.11
2252	2" x 12"		875	.018		1.39	.73		2.12	2.73
2257	Pneumatic nailed		1006	.016		1.39	.64		2.03	2.57
2302	2" x 14"		770	.021		2.14	.83		2.97	3.72
2307	Pneumatic nailed		886	.018		2.14	.72		2.86	3.54
2352	3" x 6"		925	.017		1.52	.69		2.21	2.82
2402	3" x 10"		780	.021		2.94	.82		3.76	4.58
2452	3" x 12"		600	.027		3.54	1.07		4.61	5.65
2502	4" x 6"		800	.020		2.34	.80		3.14	3.89
2552	4" x 10"		600	.027		3.68	1.07		4.75	5.80
2602	4" x 12"		450	.036		4.74	1.42		6.16	7.55
2607	Sister joist, 2" x 6"		800	.020		.52	.80		1.32	1.89
2608	Pneumatic nailed		960	.017		.52	.67		1.19	1.67
2612	2" x 8"		640	.025		.71	1		1.71	2.43
2613	Pneumatic nailed		768	.021		.71	.83		1.54	2.15
2617	2" x 10"		535	.030		.99	1.19		2.18	3.06
2618	Pneumatic nailed		642	.025		.99	1		1.99	2.73
2622	2" x 12"		455	.035		1.39	1.40		2.79	3.84
2627	Pneumatic nailed		546	.029		1.39	1.17		2.56	3.45
3000	Composite wood joist 9-1/2" deep		.90	17.778	M.L.F.	2,000	710		2,710	3,375
3010	11-1/2" deep		.88	18.182		2,175	725		2,900	3,575
3020	14" deep		.82	19.512		2,800	780		3,580	4,350
3030	16" deep		.78	20.513		3,250	820		4,070	4,925
4000	Open web joist 12" deep		.88	18.182		3,050	725		3,775	4,550
4010	14" deep		.82	19.512		3,275	780		4,055	4,875
4020	16" deep		.78	20.513		3,250	820		4,070	4,925
4030	18" deep		.74	21.622		3,500	865		4,365	5,275
9000	Minimum labor/equipment charge	1 Carp	4	2	Job		80		80	132

06 11 10.20 Framing, Light

		Crew	Daily Output	Labor-Hours	Unit	Material	Labor	Equipment	Total	Total Incl O&P
0010	FRAMING, LIGHT									
0020	Average cost for all light framing	2 Carp	1.05	15.238	M.B.F.	555	610		1,165	1,600
6500	Sills, 4" x 4", treated		.78	20.513	"	1,125	820		1,945	2,575
6802	Sleepers on concrete, treated, 1" x 2"		2350	.007	L.F.	.38	.27		.65	.87

06 11 10.24 Miscellaneous Framing

		Crew	Daily Output	Labor-Hours	Unit	Material	Labor	Equipment	Total	Total Incl O&P
0010	MISCELLANEOUS FRAMING									
2002	Firestops, 2" x 4"	2 Carp	780	.021	L.F.	.29	.82		1.11	1.67
2007	Pneumatic nailed		952	.017		.29	.67		.96	1.43
2102	2" x 6"		600	.027		.52	1.07		1.59	2.33

06 11 10.24 Miscellaneous Framing		Crew	Daily Output	Labor-Hours	Unit	Material	2009 Bare Costs Labor	Equipment	Total	Total Incl O&P
2107	Pneumatic nailed	2 Carp	732	.022	L.F.	.52	.87		1.39	2.01
5002	Nailers, treated, wood construction, 2" x 4"		800	.020		.44	.80		1.24	1.81
5007	Pneumatic nailed		960	.017		.44	.67		1.11	1.59
5102	2" x 6"		750	.021		.69	.85		1.54	2.16
5107	Pneumatic nailed		900	.018		.69	.71		1.40	1.92
5122	2" x 8"		700	.023		1.05	.91		1.96	2.66
5127	Pneumatic nailed		840	.019		1.05	.76		1.81	2.41
5202	Steel construction, 2" x 4"		750	.021		.44	.85		1.29	1.90
5222	2" x 6"		700	.023		.69	.91		1.60	2.26
5242	2" x 8"		650	.025		1.05	.98		2.03	2.77
7002	Rough bucks, treated, for doors or windows, 2" x 6"		400	.040		.69	1.60		2.29	3.39
7007	Pneumatic nailed		480	.033		.69	1.33		2.02	2.95
7102	2" x 8"		380	.042		1.05	1.68		2.73	3.93
7107	Pneumatic nailed		456	.035		1.05	1.40		2.45	3.46
8000	Stair stringers, 2" x 10"		130	.123		.99	4.92		5.91	9.20
8100	2" x 12"		130	.123		1.39	4.92		6.31	9.60
8150	3" x 10"		125	.128		2.94	5.10		8.04	11.70
8200	3" x 12"		125	.128		3.54	5.10		8.64	12.35
9000	Minimum labor/equipment charge	1 Carp	4	2	Job		80		80	132

06 11 10.26 Partitions		Crew	Daily Output	Labor-Hours	Unit	Material	Labor	Equipment	Total	Total Incl O&P
0010	**PARTITIONS**									
0020	Single bottom and double top plate, no waste, std. & better lumber									
0182	2" x 4" studs, 8' high, studs 12" O.C.	2 Carp	80	.200	L.F.	3.51	8		11.51	17.05
0187	12" O.C., pneumatic nailed		96	.167		3.51	6.65		10.16	14.85
0202	16" O.C.		100	.160		2.87	6.40		9.27	13.70
0207	16" O.C., pneumatic nailed		120	.133		2.87	5.35		8.22	11.95
0302	24" O.C.		125	.128		2.23	5.10		7.33	10.90
0307	24" O.C., pneumatic nailed		150	.107		2.23	4.26		6.49	9.50
0382	10' high, studs 12" O.C.		80	.200		4.15	8		12.15	17.75
0387	12" O.C., pneumatic nailed		96	.167		4.15	6.65		10.80	15.55
0402	16" O.C.		100	.160		3.35	6.40		9.75	14.25
0407	16" O.C., pneumatic nailed		120	.133		3.35	5.35		8.70	12.50
0502	24" O.C.		125	.128		2.55	5.10		7.65	11.25
0507	24" O.C., pneumatic nailed		150	.107		2.55	4.26		6.81	9.85
0582	12' high, studs 12" O.C.		65	.246		4.78	9.85		14.63	21.50
0587	12" O.C., pneumatic nailed		78	.205		4.78	8.20		12.98	18.80
0602	16" O.C.		80	.200		3.83	8		11.83	17.40
0607	16" O.C., pneumatic nailed		96	.167		3.83	6.65		10.48	15.20
0701	24" O.C.		100	.160		2.87	6.40		9.27	13.70
0706	24" O.C., pneumatic nailed		120	.133		2.87	5.35		8.22	11.95
0782	2" x 6" studs, 8' high, studs 12" O.C.		70	.229		6.30	9.15		15.45	22
0787	12" O.C., pneumatic nailed		84	.190		6.30	7.60		13.90	19.50
0802	16" O.C.		90	.178		5.15	7.10		12.25	17.35
0807	16" O.C., pneumatic nailed		108	.148		5.15	5.90		11.05	15.40
0902	24" O.C.		115	.139		4.01	5.55		9.56	13.60
0907	24" O.C., pneumatic nailed		138	.116		4.01	4.63		8.64	12.05
0982	10' high, studs 12" O.C.		70	.229		7.45	9.15		16.60	23.50
0987	12" O.C., pneumatic nailed		84	.190		7.45	7.60		15.05	21
1002	16" O.C.		90	.178		6	7.10		13.10	18.30
1007	16" O.C., pneumatic nailed		108	.148		6	5.90		11.90	16.35
1102	24" O.C.		115	.139		4.58	5.55		10.13	14.25
1107	24" O.C., pneumatic nailed		138	.116		4.58	4.63		9.21	12.70

06 11 10 – Framing with Dimensional, Engineered or Composite Lumber

06 11 10.26 Partitions		Crew	Daily Output	Labor-Hours	Unit	2009 Bare Costs Material	Labor	Equipment	Total	Total Incl O&P
1182	12′ high, studs 12″ O.C.	2 Carp	55	.291	L.F.	8.60	11.60		20.20	28.50
1187	12″ O.C., pneumatic nailed		66	.242		8.60	9.70		18.30	25.50
1202	16″ O.C.		70	.229		6.90	9.15		16.05	22.50
1207	16″ O.C., pneumatic nailed		84	.190		6.90	7.60		14.50	20
1302	24″ O.C.		90	.178		5.15	7.10		12.25	17.35
1307	24″ O.C., pneumatic nailed		108	.148		5.15	5.90		11.05	15.40
1402	For horizontal blocking, 2″ x 4″, add		600	.027		.32	1.07		1.39	2.11
1502	2″ x 6″, add		600	.027		.57	1.07		1.64	2.39
1600	For openings, add		250	.064			2.56		2.56	4.22
1702	Headers for above openings, material only, add				B.F.	.58			.58	.64
9000	Minimum labor/equipment charge	1 Carp	4	2	Job		80		80	132

06 11 10.28 Porch or Deck Framing		Crew	Daily Output	Labor-Hours	Unit	Material	Labor	Equipment	Total	Total Incl O&P
0010	**PORCH OR DECK FRAMING**									
0100	Treated lumber, posts or columns, 4″ x 4″	2 Carp	390	.041	L.F.	1.52	1.64		3.16	4.38
0110	4″ x 6″		275	.058		2.24	2.32		4.56	6.30
0120	4″ x 8″		220	.073		3.21	2.91		6.12	8.35
0130	Girder, single, 4″ x 4″		675	.024		1.52	.95		2.47	3.23
0140	4″ x 6″		600	.027		2.24	1.07		3.31	4.22
0150	4″ x 8″		525	.030		3.21	1.22		4.43	5.55
0160	Double, 2″ x 4″		625	.026		.91	1.02		1.93	2.69
0170	2″ x 6″		600	.027		1.42	1.07		2.49	3.32
0180	2″ x 8″		575	.028		2.16	1.11		3.27	4.22
0190	2″ x 10″		550	.029		2.63	1.16		3.79	4.81
0200	2″ x 12″		525	.030		4.15	1.22		5.37	6.55
0210	Triple, 2″ x 4″		575	.028		1.37	1.11		2.48	3.35
0220	2″ x 6″		550	.029		2.13	1.16		3.29	4.26
0230	2″ x 8″		525	.030		3.24	1.22		4.46	5.55
0240	2″ x 10″		500	.032		3.95	1.28		5.23	6.45
0250	2″ x 12″		475	.034		6.20	1.35		7.55	9.05
0260	Ledger, bolted 4′ O.C., 2″ x 4″		400	.040		.55	1.60		2.15	3.25
0270	2″ x 6″		395	.041		.80	1.62		2.42	3.55
0280	2″ x 8″		390	.041		1.16	1.64		2.80	3.99
0300	2″ x 12″		380	.042		2.14	1.68		3.82	5.15
0310	Joists, 2″ x 4″		1250	.013		.46	.51		.97	1.34
0320	2″ x 6″		1250	.013		.71	.51		1.22	1.62
0330	2″ x 8″		1100	.015		1.08	.58		1.66	2.15
0340	2″ x 10″		900	.018		1.32	.71		2.03	2.62
0350	2″ x 12″		875	.018		1.58	.73		2.31	2.95
0440	Balusters, square, 2″ x 2″		660	.024		.34	.97		1.31	1.97
0450	Turned, 2″ x 2″		420	.038		.45	1.52		1.97	3
0460	Stair stringer, 2″ x 10″		130	.123		1.32	4.92		6.24	9.55
0470	2″ x 12″		130	.123		1.58	4.92		6.50	9.85
0480	Stair treads, 1″ x 4″		140	.114		.52	4.57		5.09	8.15
0490	2″ x 4″		140	.114		.46	4.57		5.03	8.05
0500	2″ x 6″		160	.100		.69	4		4.69	7.35
0510	5/4″ x 6″		160	.100		1.10	4		5.10	7.80
0520	Turned handrail post, 4″ x 4″		64	.250	Ea.	31.50	10		41.50	51.50
0530	Lattice panel, 4′ x 8′		1600	.010	S.F.	.68	.40		1.08	1.41
0540	Cedar, posts or columns, 4″ x 4″		390	.041	L.F.	3.21	1.64		4.85	6.25
0550	4″ x 6″		275	.058		4.86	2.32		7.18	9.20
0560	4″ x 8″		220	.073		7.30	2.91		10.21	12.80
0800	Decking, 1″ x 4″		550	.029		1.76	1.16		2.92	3.85

06 11 10 – Framing with Dimensional, Engineered or Composite Lumber

06 11 10.28 Porch or Deck Framing		Crew	Daily Output	Labor-Hours	Unit	2009 Bare Costs Material	Labor	Equipment	Total	Total Incl O&P
0810	2" x 4"	2 Carp	600	.027	L.F.	3.63	1.07		4.70	5.75
0820	2" x 6"		640	.025		6.95	1		7.95	9.30
0830	5/4" x 6"		640	.025		4.60	1		5.60	6.70
0840	Railings and trim, 1" x 4"		600	.027		1.76	1.07		2.83	3.69
0860	2" x 4"		600	.027		3.63	1.07		4.70	5.75
0870	2" x 6"		600	.027		6.95	1.07		8.02	9.40
0920	Stair treads, 1" x 4"		140	.114		1.76	4.57		6.33	9.50
0930	2" x 4"		140	.114		3.63	4.57		8.20	11.55
0940	2" x 6"		160	.100		6.95	4		10.95	14.25
0950	5/4" x 6"		160	.100		4.60	4		8.60	11.65
0980	Redwood, posts or columns, 4" x 4"		390	.041		6.45	1.64		8.09	9.80
0990	4" x 6"		275	.058		11.20	2.32		13.52	16.15
1000	4" x 8"		220	.073		20.50	2.91		23.41	27.50
1280	Railings and trim, 1" x 4"		600	.027		1.67	1.07		2.74	3.60
1310	2" x 6"		600	.027		4.93	1.07		6	7.15
1420	Alternative decking, wood / plastic composite, 5/4" x 6" G		640	.025		3.01	1		4.01	4.97
1430	Vinyl, 1-1/2" x 5-1/2"		640	.025		4.15	1		5.15	6.20
1440	1" x 4" square edge fir		550	.029		1.57	1.16		2.73	3.64
1450	1" x 4" tongue and groove fir		450	.036		1.41	1.42		2.83	3.90
1460	1" x 4" mahogany		550	.029		2.51	1.16		3.67	4.68
1462	5/4" x 6" PVC		550	.029		2.72	1.16		3.88	4.91
1470	Accessories, joist hangers, 2" x 4"	1 Carp	160	.050	Ea.	.61	2		2.61	3.97
1480	2" x 6" through 2" x 12"	"	150	.053		1.17	2.13		3.30	4.81
1530	Post footing, incl excav, backfill, tube form & concrete, 4' deep, 8" dia	F-7	12	2.667		10.60	95.50		106.10	169
1540	10" diameter		11	2.909		16	104		120	190
1550	12" diameter		10	3.200		21.50	115		136.50	213

06 11 10.30 Roof Framing		Crew	Daily Output	Labor-Hours	Unit	Material	Labor	Equipment	Total	Total Incl O&P
0010	**ROOF FRAMING**									
2000	Fascia boards, 2" x 8"	2 Carp	225	.071	L.F.	.71	2.84		3.55	5.45
2100	2" x 10"		180	.089		.99	3.55		4.54	6.95
5000	Rafters, to 4 in 12 pitch, 2" x 6", ordinary		1000	.016		.52	.64		1.16	1.63
5060	2" x 8", ordinary		950	.017		.71	.67		1.38	1.89
5300	Hip and valley rafters, 2" x 6", ordinary		760	.021		.52	.84		1.36	1.96
5360	2" x 8", ordinary		720	.022		.71	.89		1.60	2.25
5540	Hip and valley jacks, 2" x 6", ordinary		600	.027		.52	1.07		1.59	2.33
5600	2" x 8", ordinary		490	.033		.71	1.30		2.01	2.93
5761	For slopes steeper than 4 in 12, add						30%			
5770	For dormers or complex roofs, add						50%			
5780	Rafter tie, 1" x 4", #3	2 Carp	800	.020	L.F.	.44	.80		1.24	1.80
5800	Ridge board, #2 or better, 1" x 6"		600	.027		.64	1.07		1.71	2.46
5820	1" x 8"		550	.029		.84	1.16		2	2.84
5840	1" x 10"		500	.032		1.23	1.28		2.51	3.46
5860	2" x 6"		500	.032		.52	1.28		1.80	2.68
5880	2" x 8"		450	.036		.71	1.42		2.13	3.13
5900	2" x 10"		400	.040		.99	1.60		2.59	3.73
5920	Roof cants, split, 4" x 4"		650	.025		1.56	.98		2.54	3.34
5940	6" x 6"		600	.027		4.87	1.07		5.94	7.10
5960	Roof curbs, untreated, 2" x 6"		520	.031		.52	1.23		1.75	2.60
5980	2" x 12"		400	.040		1.39	1.60		2.99	4.16
6000	Sister rafters, 2" x 6"		800	.020		.52	.80		1.32	1.89
6020	2" x 8"		640	.025		.71	1		1.71	2.43
6040	2" x 10"		535	.030		.99	1.19		2.18	3.06

06 11 10.30 Roof Framing

Line	06 11 10.30 Roof Framing	Crew	Daily Output	Labor-Hours	Unit	2009 Bare Costs Material	Labor	Equipment	Total	Total Incl O&P
6060	2" x 12"	2 Carp	455	.035	L.F.	1.39	1.40		2.79	3.84
9000	Minimum labor/equipment charge	1 Carp	4	2	Job		80		80	132

06 11 10.32 Sill and Ledger Framing

Line	06 11 10.32 Sill and Ledger Framing	Crew	Daily Output	Labor-Hours	Unit	Material	Labor	Equipment	Total	Total Incl O&P
0010	**SILL AND LEDGER FRAMING**									
2002	Ledgers, nailed, 2" x 4"	2 Carp	755	.021	L.F.	.29	.85		1.14	1.72
2052	2" x 6"		600	.027		.52	1.07		1.59	2.33
2102	Bolted, not including bolts, 3" x 6"		325	.049		1.52	1.97		3.49	4.93
2152	3" x 12"		233	.069		3.54	2.74		6.28	8.40
2602	Mud sills, redwood, construction grade, 2" x 4"		895	.018		2.27	.71		2.98	3.67
2622	2" x 6"		780	.021		3.40	.82		4.22	5.10
4002	Sills, 2" x 4"		600	.027		.29	1.07		1.36	2.08
4052	2" x 6"		550	.029		.52	1.16		1.68	2.49
4082	2" x 8"		500	.032		.71	1.28		1.99	2.89
4202	Treated, 2" x 4"		550	.029		.44	1.16		1.60	2.41
4222	2" x 6"		500	.032		.69	1.28		1.97	2.86
4242	2" x 8"		450	.036		1.05	1.42		2.47	3.50
4402	4" x 4"		450	.036		1.49	1.42		2.91	3.99
4422	4" x 6"		350	.046		2.19	1.83		4.02	5.40
4462	4" x 8"		300	.053		3.14	2.13		5.27	7
4481	4" x 10"	↓	260	.062	↓	4.10	2.46		6.56	8.55
9000	Minimum labor/equipment charge	1 Carp	4	2	Job		80		80	132

06 11 10.34 Sleepers

Line	06 11 10.34 Sleepers	Crew	Daily Output	Labor-Hours	Unit	Material	Labor	Equipment	Total	Total Incl O&P
0010	**SLEEPERS**									
0100	On concrete, treated, 1" x 2"	2 Carp	2350	.007	L.F.	.38	.27		.65	.87
0150	1" x 3"		2000	.008		.52	.32		.84	1.10
0200	2" x 4"		1500	.011		.44	.43		.87	1.19
0250	2" x 6"	↓	1300	.012	↓	.69	.49		1.18	1.56
9000	Minimum labor/equipment charge	1 Carp	4	2	Job		80		80	132

06 11 10.36 Soffit and Canopy Framing

Line	06 11 10.36 Soffit and Canopy Framing	Crew	Daily Output	Labor-Hours	Unit	Material	Labor	Equipment	Total	Total Incl O&P
0010	**SOFFIT AND CANOPY FRAMING**									
1002	Canopy or soffit framing , 1" x 4"	2 Carp	900	.018	L.F.	.44	.71		1.15	1.65
1042	1" x 8"		750	.021		.84	.85		1.69	2.33
1102	2" x 4"		620	.026		.29	1.03		1.32	2.02
1142	2" x 8"		500	.032		.71	1.28		1.99	2.89
1202	3" x 4"		500	.032		.93	1.28		2.21	3.13
1242	3" x 10"	↓	300	.053	↓	2.94	2.13		5.07	6.75
9000	Minimum labor/equipment charge	1 Carp	4	2	Job		80		80	132

06 11 10.38 Treated Lumber Framing Material

Line	06 11 10.38 Treated Lumber Framing Material	Crew	Daily Output	Labor-Hours	Unit	Material	Labor	Equipment	Total	Total Incl O&P
0010	**TREATED LUMBER FRAMING MATERIAL**									
0100	2" x 4"				M.B.F.	660			660	730
0110	2" x 6"					685			685	755
0120	2" x 8"					785			785	865
0130	2" x 10"					765			765	845
0140	2" x 12"					1,025			1,025	1,125
0200	4" x 4"					1,125			1,125	1,225
0210	4" x 6"					1,100			1,100	1,200
0220	4" x 8"				↓	1,175			1,175	1,300

06 11 10.40 Wall Framing

Line	06 11 10.40 Wall Framing	Crew	Daily Output	Labor-Hours	Unit	Material	Labor	Equipment	Total	Total Incl O&P
0010	**WALL FRAMING** R061110-30									
0100	Door buck, studs, header, access, 8' H, 2"x4" wall, 3' W	1 Carp	32	.250	Ea.	12.65	10		22.65	30.50
0110	4' wide	↓	32	.250	↓	13.70	10		23.70	31.50

06 11 10.40 Wall Framing		Crew	Daily Output	Labor-Hours	Unit	Material	Labor (2009 Bare Costs)	Equipment	Total	Total Incl O&P
0120	5′ wide	1 Carp	32	.250	Ea.	16.70	10		26.70	35
0130	6′ wide		32	.250		18.10	10		28.10	36.50
0140	8′ wide		30	.267		25.50	10.65		36.15	45.50
0150	10′ wide		30	.267		37.50	10.65		48.15	59
0160	12′ wide		30	.267		61.50	10.65		72.15	85.50
0170	2″ x 6″ wall, 3′ wide		32	.250		20	10		30	38.50
0180	4′ wide		32	.250		21	10		31	39.50
0190	5′ wide		32	.250		24	10		34	43
0200	6′ wide		32	.250		25.50	10		35.50	44.50
0210	8′ wide		30	.267		33	10.65		43.65	54
0220	10′ wide		30	.267		45	10.65		55.65	67
0230	12′ wide		30	.267		69	10.65		79.65	93.50
0240	Window buck, studs, header & access, 8′ high 2″ x 4″ wall, 2′ wide		24	.333		12.90	13.30		26.20	36
0250	3′ wide		24	.333		15.55	13.30		28.85	39
0260	4′ wide		24	.333		17.20	13.30		30.50	41
0270	5′ wide		24	.333		20	13.30		33.30	44
0280	6′ wide		24	.333		22.50	13.30		35.80	46.50
0300	8′ wide		22	.364		31.50	14.55		46.05	59
0310	10′ wide		22	.364		44.50	14.55		59.05	73
0320	12′ wide		22	.364		70.50	14.55		85.05	102
0330	2″ x 6″ wall, 2′ wide		24	.333		22	13.30		35.30	46.50
0340	3′ wide		24	.333		25.50	13.30		38.80	50
0350	4′ wide		24	.333		27	13.30		40.30	52
0360	5′ wide		24	.333		30.50	13.30		43.80	55.50
0370	6′ wide		24	.333		33.50	13.30		46.80	58.50
0380	7′ wide		24	.333		40.50	13.30		53.80	66.50
0390	8′ wide		22	.364		44	14.55		58.55	72.50
0400	10′ wide		22	.364		57.50	14.55		72.05	87.50
0410	12′ wide		22	.364		84.50	14.55		99.05	117
2002	Headers over openings, 2″ x 6″	2 Carp	360	.044	L.F.	.52	1.78		2.30	3.50
2007	2″ x 6″, pneumatic nailed		432	.037		.52	1.48		2	3.01
2052	2″ x 8″		340	.047		.71	1.88		2.59	3.88
2057	2″ x 8″, pneumatic nailed		408	.039		.71	1.57		2.28	3.37
2100	2″ x 10″		320	.050		.99	2		2.99	4.39
2105	2″ x 10″, pneumatic nailed		384	.042		.99	1.66		2.65	3.84
2152	2″ x 12″		300	.053		1.39	2.13		3.52	5.05
2157	2″ x 12″, pneumatic nailed		360	.044		1.39	1.78		3.17	4.45
2202	4″ x 12″		190	.084		4.74	3.36		8.10	10.75
2207	4″ x 12″, pneumatic nailed		228	.070		4.74	2.80		7.54	9.85
2252	6″ x 12″		140	.114		9.15	4.57		13.72	17.65
5002	Plates, untreated, 2″ x 3″		850	.019		.32	.75		1.07	1.59
5007	2″ x 3″, pneumatic nailed		1020	.016		.32	.63		.95	1.38
5022	2″ x 4″		800	.020		.29	.80		1.09	1.64
5027	2″ x 4″, pneumatic nailed		960	.017		.29	.67		.96	1.42
5041	2″ x 6″		750	.021		.52	.85		1.37	1.98
5046	2″ x 6″, pneumatic nailed		900	.018		.52	.71		1.23	1.74
5122	Studs, 8′ high wall, 2″ x 3″		1200	.013		.32	.53		.85	1.23
5127	2″ x 3″, pneumatic nailed		1440	.011		.32	.44		.76	1.08
5142	2″ x 4″		1100	.015		.29	.58		.87	1.28
5147	2″ x 4″, pneumatic nailed		1320	.012		.29	.48		.77	1.12
5162	2″ x 6″		1000	.016		.52	.64		1.16	1.63
5167	2″ x 6″, pneumatic nailed		1200	.013		.52	.53		1.05	1.45
5182	3″ x 4″		800	.020		.93	.80		1.73	2.34

06 11 Wood Framing

06 11 10 – Framing with Dimensional, Engineered or Composite Lumber

06 11 10.40 Wall Framing		Crew	Daily Output	Labor-Hours	Unit	Material	2009 Bare Costs Labor	Equipment	Total	Total Incl O&P
5187	3" x 4", pneumatic nailed	2 Carp	960	.017	L.F.	.93	.67		1.60	2.12
8200	For 12' high walls, deduct						5%			
8220	For stub wall, 6' high, add						20%			
8240	3' high, add						40%			
8250	For second story & above, add						5%			
8300	For dormer & gable, add						15%			
9000	Minimum labor/equipment charge	1 Carp	4	2	Job		80		80	132

06 11 10.42 Furring

		Crew	Daily Output	Labor-Hours	Unit	Material	Labor	Equipment	Total	Total Incl O&P
0010	FURRING									
0012	Wood strips, 1" x 2", on walls, on wood	1 Carp	550	.015	L.F.	.25	.58		.83	1.24
0017	Pneumatic nailed		710	.011		.25	.45		.70	1.02
0300	On masonry		495	.016		.25	.65		.90	1.35
0400	On concrete		260	.031		.25	1.23		1.48	2.31
0600	1" x 3", on walls, on wood		550	.015		.34	.58		.92	1.34
0607	Pneumatic nailed		710	.011		.34	.45		.79	1.12
0700	On masonry		495	.016		.34	.65		.99	1.45
0800	On concrete		260	.031		.34	1.23		1.57	2.41
0850	On ceilings, on wood		350	.023		.34	.91		1.25	1.89
0857	Pneumatic nailed		450	.018		.34	.71		1.05	1.55
0900	On masonry		320	.025		.34	1		1.34	2.03
0950	On concrete		210	.038		.34	1.52		1.86	2.89
9000	Minimum labor/equipment charge		4	2	Job		80		80	132

06 11 10.44 Grounds

		Crew	Daily Output	Labor-Hours	Unit	Material	Labor	Equipment	Total	Total Incl O&P
0010	GROUNDS									
0020	For casework, 1" x 2" wood strips, on wood	1 Carp	330	.024	L.F.	.25	.97		1.22	1.88
0100	On masonry		285	.028		.25	1.12		1.37	2.13
0200	On concrete		250	.032		.25	1.28		1.53	2.39
0400	For plaster, 3/4" deep, on wood		450	.018		.25	.71		.96	1.45
0500	On masonry		225	.036		.25	1.42		1.67	2.63
0600	On concrete		175	.046		.25	1.83		2.08	3.29
0700	On metal lath		200	.040		.25	1.60		1.85	2.92
9000	Minimum labor/equipment charge		4	2	Job		80		80	132

06 12 Structural Panels

06 12 10 – Structural Insulated Panels

06 12 10.10 OSB Faced Panels

		Crew	Daily Output	Labor-Hours	Unit	Material	Labor	Equipment	Total	Total Incl O&P
0010	OSB FACED PANELS									
0100	Structural insul. panels, 7/16" OSB both faces, EPS insul, 3-5/8" T G	F-3	2075	.019	S.F.	2.84	.78	.37	3.99	4.81
0110	5-5/8" thick G		1725	.023		3.37	.94	.45	4.76	5.75
0120	7-3/8" thick G		1425	.028		3.84	1.14	.54	5.52	6.65
0130	9-3/8" thick G		1125	.036		4.40	1.44	.68	6.52	7.95
0140	7/16" OSB one face, EPS insul, 3-5/8" thick G		2175	.018		1.65	.74	.35	2.74	3.43
0150	5-5/8" thick G		1825	.022		2.06	.89	.42	3.37	4.18
0160	7-3/8" thick G		1525	.026		2.40	1.06	.50	3.96	4.93
0170	9-3/8" thick G		1225	.033		2.81	1.32	.63	4.76	5.95
0190	7/16" OSB - 1/2" GWB faces , EPS insul, 3-5/8" T G		2075	.019		3.03	.78	.37	4.18	5
0200	5-5/8" thick G		1725	.023		3.34	.94	.45	4.73	5.70
0210	7-3/8" thick G		1425	.028		3.65	1.14	.54	5.33	6.45
0220	9-3/8" thick G		1125	.036		4	1.44	.68	6.12	7.50
0240	7/16" OSB - 1/2" MRGWB faces , EPS insul, 3-5/8" T G		2075	.019		3.03	.78	.37	4.18	5

06 12 Structural Panels

06 12 10 – Structural Insulated Panels

06 12 10.10 OSB Faced Panels		Crew	Daily Output	Labor-Hours	Unit	Material	2009 Bare Costs Labor	Equipment	Total	Total Incl O&P
0250	5-5/8" thick	G	F-3	1725	.023	S.F.	3.34	.94	.45	4.73	5.70
0260	7-3/8" thick			1425	.028		3.65	1.14	.54	5.33	6.45
0270	9-3/8" thick		↓	1125	.036		4	1.44	.68	6.12	7.50
0300	For 1/2" GWB added to OSB skin, add	G					.60			.60	.66
0310	For 1/2" MRGWB added to OSB skin, add	G					.75			.75	.83
0320	For one T1-11 skin, add to OSB-OSB	G					.74			.74	.81
0330	For one 19/32" CDX skin, add to OSB-OSB	G				↓	.69			.69	.76
0500	Structural insulated panel, 7/16" OSB both sides, straw core										
0510	4-3/8" T, walls (w/sill, splines, plates)	G	F-6	2400	.017	S.F.	7.60	.62	.32	8.54	9.70
0520	Floors (w/ splines)	G		2400	.017		7.60	.62	.32	8.54	9.70
0530	Roof (w/ splines)	G		2400	.017		7.60	.62	.32	8.54	9.70
0550	7-7/8" T, walls (w/sill, splines, plates)	G		2400	.017		10.95	.62	.32	11.89	13.40
0560	Floors (w/ splines)	G		2400	.017		10.95	.62	.32	11.89	13.40
0570	Roof (w/ splines)	G	↓	2400	.017	↓	10.95	.62	.32	11.89	13.40

06 13 Heavy Timber

06 13 23 – Heavy Timber Construction

06 13 23.10 Heavy Framing		Crew	Daily Output	Labor-Hours	Unit	Material	Labor	Equipment	Total	Total Incl O&P
0010	**HEAVY FRAMING**									
0020	Beams, single 6" x 10"	2 Carp	1.10	14.545	M.B.F.	1,825	580		2,405	2,950
0100	Single 8" x 16"		1.20	13.333	"	2,900	535		3,435	4,075
0202	Built from 2" lumber, multiple 2" x 14"		900	.018	B.F.	.92	.71		1.63	2.18
0212	Built from 3" lumber, multiple 3" x 6"		700	.023		1.02	.91		1.93	2.63
0222	Multiple 3" x 8"		800	.020		1.05	.80		1.85	2.48
0232	Multiple 3" x 10"		900	.018		1.18	.71		1.89	2.46
0242	Multiple 3" x 12"		1000	.016		1.18	.64		1.82	2.36
0252	Built from 4" lumber, multiple 4" x 6"		800	.020		1.17	.80		1.97	2.61
0262	Multiple 4" x 8"		900	.018		1.05	.71		1.76	2.32
0272	Multiple 4" x 10"		1000	.016		1.10	.64		1.74	2.27
0282	Multiple 4" x 12"		1100	.015	↓	1.18	.58		1.76	2.26
0292	Columns, structural grade, 1500f, 4" x 4"		450	.036	L.F.	1.75	1.42		3.17	4.28
0302	6" x 6"		225	.071		4.85	2.84		7.69	10.05
0402	8" x 8"		240	.067		10.75	2.66		13.41	16.25
0502	10" x 10"		90	.178		19	7.10		26.10	32.50
0602	12" x 12"		70	.229	↓	28	9.15		37.15	45.50
0802	Floor planks, 2" thick, T & G, 2" x 6"		1050	.015	B.F.	1.54	.61		2.15	2.71
0902	2" x 10"		1100	.015		1.54	.58		2.12	2.66
1102	3" thick, 3" x 6"		1050	.015		1.30	.61		1.91	2.44
1202	3" x 10"		1100	.015		1.30	.58		1.88	2.39
1402	Girders, structural grade, 12" x 12"		800	.020		2.32	.80		3.12	3.88
1502	10" x 16"		1000	.016		2.28	.64		2.92	3.57
2302	Roof purlins, 4" thick, structural grade		1050	.015		1.46	.61		2.07	2.62
2502	Roof trusses, add timber connectors, Division 06 05 23.60	↓	450	.036	↓	1.34	1.42		2.76	3.83
9000	Minimum labor/equipment charge	1 Carp	2	4	Job		160		160	264

06 15 Wood Decking

06 15 16 – Wood Roof Decking

06 15 16.10 Solid Wood Roof Decking		Crew	Daily Output	Labor-Hours	Unit	Material	2009 Bare Costs Labor	Equipment	Total	Total Incl O&P
0010	SOLID WOOD ROOF DECKING									
0400	Cedar planks, 3" thick	2 Carp	320	.050	S.F.	6.20	2		8.20	10.15
0500	4" thick		250	.064		8.35	2.56		10.91	13.40
0702	Douglas fir, 3" thick		320	.050		2.33	2		4.33	5.85
0802	4" thick		250	.064		3.12	2.56		5.68	7.65
1002	Hemlock, 3" thick		320	.050		2.33	2		4.33	5.85
1102	4" thick		250	.064		3.11	2.56		5.67	7.65
1302	Western white spruce, 3" thick		320	.050		2.24	2		4.24	5.75
1402	4" thick		250	.064		2.99	2.56		5.55	7.50
9000	Minimum labor/equipment charge	1 Carp	2	4	Job		160		160	264

06 15 23 – Laminated Wood Decking

06 15 23.10 Laminated Roof Deck		Crew	Daily Output	Labor-Hours	Unit	Material	Labor	Equipment	Total	Total Incl O&P
0010	LAMINATED ROOF DECK									
0020	Pine or hemlock, 3" thick	2 Carp	425	.038	S.F.	3.36	1.50		4.86	6.20
0100	4" thick		325	.049		4.47	1.97		6.44	8.15
0300	Cedar, 3" thick		425	.038		4.11	1.50		5.61	7
0400	4" thick		325	.049		5.25	1.97		7.22	9
0600	Fir, 3" thick		425	.038		3.43	1.50		4.93	6.25
0700	4" thick		325	.049		4.29	1.97		6.26	7.95
9000	Minimum labor/equipment charge	1 Carp	3	2.667	Job		107		107	176

06 16 Sheathing

06 16 23 – Subflooring

06 16 23.10 Subfloor		Crew	Daily Output	Labor-Hours	Unit	Material	Labor	Equipment	Total	Total Incl O&P
0010	SUBFLOOR									
0011	Plywood, CDX, 1/2" thick	2 Carp	1500	.011	SF Flr.	.46	.43		.89	1.21
0017	Pneumatic nailed		1860	.009		.46	.34		.80	1.08
0102	5/8" thick		1350	.012		.67	.47		1.14	1.52
0107	Pneumatic nailed		1674	.010		.67	.38		1.05	1.37
0202	3/4" thick		1250	.013		.83	.51		1.34	1.75
0207	Pneumatic nailed		1550	.010		.83	.41		1.24	1.59
0302	1-1/8" thick, 2-4-1 including underlayment		1050	.015		1.63	.61		2.24	2.80
0440	With boards, 1" x 6", S4S, laid regular		900	.018		1.40	.71		2.11	2.70
0452	1" x 8", laid regular		1000	.016		1.34	.64		1.98	2.53
0462	Laid diagonal		850	.019		1.34	.75		2.09	2.71
0502	1" x 10", laid regular		1100	.015		1.54	.58		2.12	2.66
0602	Laid diagonal		900	.018		1.54	.71		2.25	2.87
8990	Subfloor adhesive, 3/8" bead	1 Carp	2300	.003	L.F.	.09	.14		.23	.33
9000	Minimum labor/equipment charge	"	4	2	Job		80		80	132

06 16 26 – Underlayment

06 16 26.10 Wood Product Underlayment		Crew	Daily Output	Labor-Hours	Unit	Material	Labor	Equipment	Total	Total Incl O&P
0010	WOOD PRODUCT UNDERLAYMENT R061636-20									
0030	Plywood, underlayment grade, 3/8" thick	2 Carp	1500	.011	SF Flr.	1.03	.43		1.46	1.83
0080	Pneumatic nailed		1860	.009		1.03	.34		1.37	1.70
0102	1/2" thick		1450	.011		1	.44		1.44	1.83
0107	Pneumatic nailed		1798	.009		1	.36		1.36	1.69
0202	5/8" thick		1400	.011		1.29	.46		1.75	2.17
0207	Pneumatic nailed		1736	.009		1.29	.37		1.66	2.03
0302	3/4" thick		1300	.012		1.33	.49		1.82	2.27
0306	Pneumatic nailed		1612	.010		1.33	.40		1.73	2.11

06 16 Sheathing

06 16 26 – Underlayment

06 16 26.10 Wood Product Underlayment		Crew	Daily Output	Labor-Hours	Unit	Material	Labor (2009 Bare Costs)	Equipment	Total	Total Incl O&P
0502	Particle board, 3/8" thick	2 Carp	1500	.011	SF Flr.	.41	.43		.84	1.15
0507	Pneumatic nailed		1860	.009		.41	.34		.75	1.02
0602	1/2" thick		1450	.011		.45	.44		.89	1.23
0607	Pneumatic nailed		1798	.009		.45	.36		.81	1.09
0802	5/8" thick		1400	.011		.54	.46		1	1.34
0807	Pneumatic nailed		1736	.009		.54	.37		.91	1.20
0902	3/4" thick		1300	.012		.65	.49		1.14	1.53
0907	Pneumatic nailed		1612	.010		.65	.40		1.05	1.37
0955	Particleboard, 100% recycled straw/wheat, 4' x 8' x 1/4" G		1450	.011	S.F.	.26	.44		.70	1.02
0960	4' x 8' x 3/8" G		1450	.011		.39	.44		.83	1.16
0965	4' x 8' x 1/2" G		1350	.012		.53	.47		1	1.36
0970	4' x 8' x 5/8" G		1300	.012		.63	.49		1.12	1.50
0975	4' x 8' x 3/4" G		1250	.013		.73	.51		1.24	1.64
0980	4' x 8' x 1" G		1150	.014		.98	.56		1.54	2
0985	4' x 8' x 1-1/4" G		1100	.015		1.15	.58		1.73	2.23
1102	Hardboard, underlayment grade, 4' x 4', .215" thick		1500	.011	SF Flr.	.50	.43		.93	1.25
9000	Minimum labor/equipment charge	1 Carp	4	2	Job		80		80	132

06 16 36 – Wood Panel Product Sheathing

06 16 36.10 Sheathing		Crew	Daily Output	Labor-Hours	Unit	Material	Labor	Equipment	Total	Total Incl O&P
0010	**SHEATHING** R061636-20									
0012	Plywood on roofs, CDX									
0032	5/16" thick	2 Carp	1600	.010	S.F.	.53	.40		.93	1.24
0037	Pneumatic nailed R061110-30		1952	.008		.53	.33		.86	1.12
0052	3/8" thick		1525	.010		.45	.42		.87	1.18
0057	Pneumatic nailed		1860	.009		.45	.34		.79	1.06
0102	1/2" thick		1400	.011		.46	.46		.92	1.26
0103	Pneumatic nailed		1708	.009		.46	.37		.83	1.13
0202	5/8" thick		1300	.012		.67	.49		1.16	1.55
0207	Pneumatic nailed		1586	.010		.67	.40		1.07	1.41
0302	3/4" thick		1200	.013		.83	.53		1.36	1.79
0307	Pneumatic nailed		1464	.011		.83	.44		1.27	1.63
0502	Plywood on walls with exterior CDX, 3/8" thick		1200	.013		.45	.53		.98	1.37
0507	Pneumatic nailed		1488	.011		.45	.43		.88	1.20
0602	1/2" thick		1125	.014		.46	.57		1.03	1.45
0607	Pneumatic nailed		1395	.011		.46	.46		.92	1.27
0702	5/8" thick		1050	.015		.67	.61		1.28	1.75
0707	Pneumatic nailed		1302	.012		.67	.49		1.16	1.55
0802	3/4" thick		975	.016		.83	.66		1.49	1.99
0807	Pneumatic nailed		1209	.013		.83	.53		1.36	1.78
0840	Oriented strand board, 7/16" thick G		1400	.011		.22	.46		.68	.99
0845	Pneumatic nailed G		1736	.009		.22	.37		.59	.85
0846	1/2" thick G		1325	.012		.22	.48		.70	1.04
0847	Pneumatic nailed G		1643	.010		.22	.39		.61	.88
0852	5/8" thick G		1250	.013		.35	.51		.86	1.23
0857	Pneumatic nailed G		1550	.010		.35	.41		.76	1.07
1000	For shear wall construction, add						20%			
1200	For structural 1 exterior plywood, add				S.F.	10%				
1402	With boards, on roof 1" x 6" boards, laid horizontal	2 Carp	725	.022		1.40	.88		2.28	2.99
1502	Laid diagonal		650	.025		1.40	.98		2.38	3.15
1702	1" x 8" boards, laid horizontal		875	.018		1.34	.73		2.07	2.68
1802	Laid diagonal		725	.022		1.34	.88		2.22	2.93
2000	For steep roofs, add						40%			

06 16 Sheathing

06 16 36 – Wood Panel Product Sheathing

06 16 36.10 Sheathing		Crew	Daily Output	Labor-Hours	Unit	2009 Bare Costs Material	Labor	Equipment	Total	Total Incl O&P
2200	For dormers, hips and valleys, add				S.F.	5%	50%			
2402	Boards on walls, 1" x 6" boards, laid regular	2 Carp	650	.025		1.40	.98		2.38	3.15
2502	Laid diagonal		585	.027		1.40	1.09		2.49	3.33
2702	1" x 8" boards, laid regular		765	.021		1.34	.84		2.18	2.85
2802	Laid diagonal		650	.025		1.34	.98		2.32	3.09
2852	Gypsum, weatherproof, 1/2" thick		1050	.015		.61	.61		1.22	1.68
2900	With embedded glass mats		1100	.015		.49	.58		1.07	1.50
3000	Wood fiber, regular, no vapor barrier, 1/2" thick		1200	.013		.56	.53		1.09	1.50
3100	5/8" thick		1200	.013		.71	.53		1.24	1.66
3300	No vapor barrier, in colors, 1/2" thick		1200	.013		.75	.53		1.28	1.71
3400	5/8" thick		1200	.013		.80	.53		1.33	1.76
3600	With vapor barrier one side, white, 1/2" thick		1200	.013		.55	.53		1.08	1.49
3700	Vapor barrier 2 sides, 1/2" thick		1200	.013		.77	.53		1.30	1.73
3800	Asphalt impregnated, 25/32" thick		1200	.013		.28	.53		.81	1.19
3850	Intermediate, 1/2" thick		1200	.013		.18	.53		.71	1.08
9000	Minimum labor/equipment charge	1 Carp	2	4	Job		160		160	264

06 17 Shop-Fabricated Structural Wood

06 17 33 – Wood I-Joists

06 17 33.10 Wood and Composite I-Joists

		Crew	Daily Output	Labor-Hours	Unit	Material	Labor	Equipment	Total	Total Incl O&P
0010	**WOOD AND COMPOSITE I-JOISTS**									
0100	Plywood webs, incl. bridging & blocking, panels 24" O.C.									
1200	15' to 24' span, 50 psf live load	F-5	2400	.013	SF Flr.	2.28	.54		2.82	3.40
1300	55 psf live load		2250	.014		2.47	.58		3.05	3.67
1400	24' to 30' span, 45 psf live load		2600	.012		3.18	.50		3.68	4.32
1500	55 psf live load		2400	.013		3.70	.54		4.24	4.96
1600	Tubular steel open webs, 45 psf, 24" O.C., 40' span	F-3	6250	.006		2.24	.26	.12	2.62	3.02
1700	55' span		7750	.005		2.17	.21	.10	2.48	2.84
1800	70' span		9250	.004		2.82	.17	.08	3.07	3.48
1900	85 psf live load, 26' span		2300	.017		2.63	.70	.33	3.66	4.41

06 17 53 – Shop-Fabricated Wood Trusses

06 17 53.10 Roof Trusses

		Crew	Daily Output	Labor-Hours	Unit	Material	Labor	Equipment	Total	Total Incl O&P
0010	**ROOF TRUSSES**									
0100	Fink (W) or King post type, 2'-0" O.C.									
0200	Metal plate connected, 4 in 12 slope									
0210	24' to 29' span	F-3	3000	.013	SF Flr.	2.06	.54	.26	2.86	3.43
0300	30' to 43' span		3000	.013		2.28	.54	.26	3.08	3.67
0400	44' to 60' span		3000	.013		2.52	.54	.26	3.32	3.93
0700	Glued and nailed, add					50%				
0800	Flat wood truss 16' to 29' span	F-3	3000	.013		2.46	.54	.26	3.26	3.87

06 18 Glued-Laminated Construction

06 18 13 – Glued-Laminated Beams

06 18 13.20 Laminated Framing		Crew	Daily Output	Labor-Hours	Unit	2009 Bare Costs Material	Labor	Equipment	Total	Total Incl O&P
0010	**LAMINATED FRAMING**									
0200	Straight roof beams, 20′ clear span, beams 8′ O.C.	F-3	2560	.016	SF Flr.	1.98	.63	.30	2.91	3.54
0300	Beams 16′ O.C.		3200	.013		1.43	.51	.24	2.18	2.66
0500	40′ clear span, beams 8′ O.C.		3200	.013		3.78	.51	.24	4.53	5.25
0600	Beams 16′ O.C.		3840	.010		3.09	.42	.20	3.71	4.31
0800	60′ clear span, beams 8′ O.C.	F-4	2880	.017		6.50	.66	.39	7.55	8.65
0900	Beams 16′ O.C.	″	3840	.013		4.84	.50	.29	5.63	6.45
1100	Tudor arches, 30′ to 40′ clear span, frames 8′ O.C.	F-3	1680	.024		8.45	.96	.46	9.87	11.40
1200	Frames 16′ O.C.	″	2240	.018		6.65	.72	.34	7.71	8.85
1400	50′ to 60′ clear span, frames 8′ O.C.	F-4	2200	.022		9.15	.87	.51	10.53	12
1500	Frames 16′ O.C.		2640	.018		7.80	.72	.43	8.95	10.20
1700	Radial arches, 60′ clear span, frames 8′ O.C.		1920	.025		8.55	1	.59	10.14	11.65
1800	Frames 16′ O.C.		2880	.017		6.55	.66	.39	7.60	8.75
2000	100′ clear span, frames 8′ O.C.		1600	.030		8.85	1.20	.70	10.75	12.45
2100	Frames 16′ O.C.		2400	.020		7.80	.80	.47	9.07	10.35
2300	120′ clear span, frames 8′ O.C.		1440	.033		11.75	1.33	.78	13.86	15.95
2400	Frames 16′ O.C.		1920	.025		10.75	1	.59	12.34	14.05
2600	Bowstring trusses, 20′ O.C., 40′ clear span	F-3	2400	.017		5.30	.67	.32	6.29	7.30
2700	60′ clear span	F-4	3600	.013		4.76	.53	.31	5.60	6.45
2800	100′ clear span		4000	.012		6.75	.48	.28	7.51	8.50
2900	120′ clear span		3600	.013		7.25	.53	.31	8.09	9.15
3100	For premium appearance, add to S.F. prices					5%				
3300	For industrial type, deduct					15%				
3500	For stain and varnish, add					5%				
3900	For 3/4″ laminations, add to straight					25%				
4100	Add to curved					15%				
4300	Alternate pricing method: (use nominal footage of									
4310	components). Straight beams, camber less than 6″	F-3	3.50	11.429	M.B.F.	2,925	465	220	3,610	4,225
4400	Columns, including hardware		2	20		3,150	810	385	4,345	5,225
4600	Curved members, radius over 32′		2.50	16		3,225	650	310	4,185	4,950
4700	Radius 10′ to 32′		3	13.333		3,200	540	256	3,996	4,675
4900	For complicated shapes, add maximum					100%				
5100	For pressure treating, add to straight					35%				
5200	Add to curved					45%				
6000	Laminated veneer members, southern pine or western species									
6050	1-3/4″ wide x 5-1/2″ deep	2 Carp	480	.033	L.F.	2.99	1.33		4.32	5.50
6100	9-1/2″ deep		480	.033		4.79	1.33		6.12	7.45
6150	14″ deep		450	.036		7	1.42		8.42	10.05
6200	18″ deep		450	.036		9.05	1.42		10.47	12.35
6300	Parallel strand members, southern pine or western species									
6350	1-3/4″ wide x 9-1/4″ deep	2 Carp	480	.033	L.F.	4.79	1.33		6.12	7.45
6400	11-1/4″ deep		450	.036		5.95	1.42		7.37	8.85
6450	14″ deep		400	.040		7	1.60		8.60	10.35
6500	3-1/2″ wide x 9-1/4″ deep		480	.033		15.85	1.33		17.18	19.65
6550	11-1/4″ deep		450	.036		19.15	1.42		20.57	23.50
6600	14″ deep		400	.040		23.50	1.60		25.10	28
6650	7″ wide x 9-1/4″ deep		450	.036		33	1.42		34.42	39
6700	11-1/4″ deep		420	.038		39.50	1.52		41.02	46
6750	14″ deep		400	.040		46.50	1.60		48.10	54
9000	Minimum labor/equipment charge	F-3	2.50	16	Job		650	310	960	1,400

06 22 Millwork

06 22 13 – Standard Pattern Wood Trim

06 22 13.15 Moldings, Base		Crew	Daily Output	Labor-Hours	Unit	2009 Bare Costs Material	Labor	Equipment	Total	Total Incl O&P
0010	**MOLDINGS, BASE**									
0500	Base, stock pine, 9/16" x 3-1/2"	1 Carp	240	.033	L.F.	2.22	1.33		3.55	4.64
0550	9/16" x 4-1/2"		200	.040	"	2.47	1.60		4.07	5.35
9000	Minimum labor/equipment charge		4	2	Job		80		80	132

06 22 13.30 Moldings, Casings		Crew	Daily Output	Labor-Hours	Unit	2009 Bare Costs Material	Labor	Equipment	Total	Total Incl O&P
0010	**MOLDINGS, CASINGS**									
0090	Apron, stock pine, 5/8" x 2"	1 Carp	250	.032	L.F.	1.24	1.28		2.52	3.47
0110	5/8" x 3-1/2"		220	.036		1.88	1.45		3.33	4.47
0300	Band, stock pine, 11/16" x 1-1/8"		270	.030		.82	1.18		2	2.85
0350	11/16" x 1-3/4"		250	.032		1.10	1.28		2.38	3.32
0700	Casing, stock pine, 11/16" x 2-1/2"		240	.033		1.33	1.33		2.66	3.66
0750	11/16" x 3-1/2"		215	.037		1.89	1.49		3.38	4.53
0760	Door & window casing, exterior, 1-1/4" x 2"		200	.040		2.14	1.60		3.74	4.99
0770	Finger jointed, 1-1/4" x 2"		200	.040		.99	1.60		2.59	3.73
9000	Minimum labor/equipment charge		4	2	Job		80		80	132

06 22 13.35 Moldings, Ceilings		Crew	Daily Output	Labor-Hours	Unit	2009 Bare Costs Material	Labor	Equipment	Total	Total Incl O&P
0010	**MOLDINGS, CEILINGS**									
0600	Bed, stock pine, 9/16" x 1-3/4"	1 Carp	270	.030	L.F.	.97	1.18		2.15	3.02
0650	9/16" x 2"		240	.033		1.23	1.33		2.56	3.55
1200	Cornice molding, stock pine, 9/16" x 1-3/4"		330	.024		1.02	.97		1.99	2.72
1300	9/16" x 2-1/4"		300	.027		1.24	1.07		2.31	3.12
2400	Cove scotia, stock pine, 9/16" x 1-3/4"		270	.030		.93	1.18		2.11	2.97
2500	11/16" x 2-3/4"		255	.031		1.35	1.25		2.60	3.56
2600	Crown, stock pine, 9/16" x 3-5/8"		250	.032		1.92	1.28		3.20	4.22
2700	11/16" x 4-5/8"		220	.036		3.22	1.45		4.67	5.95
9000	Minimum labor/equipment charge		4	2	Job		80		80	132

06 22 13.40 Moldings, Exterior		Crew	Daily Output	Labor-Hours	Unit	2009 Bare Costs Material	Labor	Equipment	Total	Total Incl O&P
0010	**MOLDINGS, EXTERIOR**									
1500	Cornice, boards, pine, 1" x 2"	1 Carp	330	.024	L.F.	.27	.97		1.24	1.90
1600	1" x 4"		250	.032		.73	1.28		2.01	2.91
1700	1" x 6"		250	.032		1.15	1.28		2.43	3.38
1800	1" x 8"		200	.040		1.34	1.60		2.94	4.11
1900	1" x 10"		180	.044		1.78	1.78		3.56	4.89
2000	1" x 12"		180	.044		2.31	1.78		4.09	5.45
2200	Three piece, built-up, pine, minimum		80	.100		2.15	4		6.15	8.95
2300	Maximum		65	.123		5.45	4.92		10.37	14.05
3000	Corner board, sterling pine, 1" x 4"		200	.040		.69	1.60		2.29	3.40
3100	1" x 6"		200	.040		1.05	1.60		2.65	3.80
3200	2" x 6"		165	.048		1.98	1.94		3.92	5.40
3300	2" x 8"		165	.048		2.63	1.94		4.57	6.10
3350	Fascia, sterling pine, 1" x 6"		250	.032		1.05	1.28		2.33	3.27
3370	1" x 8"		225	.036		1.43	1.42		2.85	3.92
3395	Grounds, 1" x 1" redwood		300	.027		.25	1.07		1.32	2.04
3400	Trim, back band, 11/16" x 1-1/16"		250	.032		.89	1.28		2.17	3.09
3500	Casing, 11/16" x 4-1/4"		250	.032		2.19	1.28		3.47	4.52
3600	Crown, 11/16" x 4-1/4"		250	.032		3.65	1.28		4.93	6.15
3700	1" x 4" and 1" x 6" railing w/ balusters 4" O.C.		22	.364		22.50	14.55		37.05	49
3800	Insect screen framing stock, 1-1/16" x 1-3/4"		395	.020		2.03	.81		2.84	3.57
4100	Verge board, sterling pine, 1" x 4"		200	.040		.69	1.60		2.29	3.40
4200	1" x 6"		200	.040		1.05	1.60		2.65	3.80
4300	2" x 6"		165	.048		1.98	1.94		3.92	5.40
4400	2" x 8"		165	.048		2.63	1.94		4.57	6.10

06 22 13 – Standard Pattern Wood Trim

06 22 13.40 Moldings, Exterior		Crew	Daily Output	Labor-Hours	Unit	Material	2009 Bare Costs Labor	Equipment	Total	Total Incl O&P
4700	For redwood trim, add					200%				
9000	Minimum labor/equipment charge	1 Carp	4	2	Job		80		80	132

06 22 13.45 Moldings, Trim

		Crew	Daily Output	Labor-Hours	Unit	Material	Labor	Equipment	Total	Total Incl O&P
0010	**MOLDINGS, TRIM**									
0200	Astragal, stock pine, 11/16" x 1-3/4"	1 Carp	255	.031	L.F.	1.19	1.25		2.44	3.38
0250	1-5/16" x 2-3/16"		240	.033		2.57	1.33		3.90	5.05
0800	Chair rail, stock pine, 5/8" x 2-1/2"		270	.030		1.45	1.18		2.63	3.55
0900	5/8" x 3-1/2"		240	.033		2.27	1.33		3.60	4.70
1000	Closet pole, stock pine, 1-1/8" diameter		200	.040		1.01	1.60		2.61	3.75
1100	Fir, 1-5/8" diameter		200	.040		1.67	1.60		3.27	4.48
3300	Half round, stock pine, 1/4" x 1/2"		270	.030		.24	1.18		1.42	2.21
3350	1/2" x 1"	↓	255	.031	↓	.61	1.25		1.86	2.74
3400	Handrail, fir, single piece, stock, hardware not included									
3450	1-1/2" x 1-3/4"	1 Carp	80	.100	L.F.	1.85	4		5.85	8.65
3470	Pine, 1-1/2" x 1-3/4"		80	.100		1.85	4		5.85	8.65
3500	1-1/2" x 2-1/2"		76	.105		2.30	4.21		6.51	9.50
3600	Lattice, stock pine, 1/4" x 1-1/8"		270	.030		.42	1.18		1.60	2.41
3700	1/4" x 1-3/4"		250	.032		.66	1.28		1.94	2.84
3800	Miscellaneous, custom, pine, 1" x 1"		270	.030		.34	1.18		1.52	2.32
3900	1" x 3"		240	.033		.69	1.33		2.02	2.96
4100	Birch or oak, nominal 1" x 1"		240	.033		.53	1.33		1.86	2.78
4200	Nominal 1" x 3"		215	.037		1.81	1.49		3.30	4.44
4400	Walnut, nominal 1" x 1"		215	.037		.87	1.49		2.36	3.41
4500	Nominal 1" x 3"		200	.040		2.61	1.60		4.21	5.50
4700	Teak, nominal 1" x 1"		215	.037		1.23	1.49		2.72	3.80
4800	Nominal 1" x 3"		200	.040		3.52	1.60		5.12	6.50
4900	Quarter round, stock pine, 1/4" x 1/4"		275	.029		.22	1.16		1.38	2.16
4950	3/4" x 3/4"		255	.031	↓	.59	1.25		1.84	2.72
5600	Wainscot moldings, 1-1/8" x 9/16", 2' high, minimum		76	.105	S.F.	11.05	4.21		15.26	19.10
5700	Maximum		65	.123	"	20	4.92		24.92	30.50
9000	Minimum labor/equipment charge	↓	4	2	Job		80		80	132

06 22 13.50 Moldings, Window and Door

		Crew	Daily Output	Labor-Hours	Unit	Material	Labor	Equipment	Total	Total Incl O&P
0010	**MOLDINGS, WINDOW AND DOOR**									
2800	Door moldings, stock, decorative, 1-1/8" wide, plain	1 Carp	17	.471	Set	45	18.80		63.80	80.50
2900	Detailed		17	.471	"	87.50	18.80		106.30	128
2960	Clear pine door jamb, no stops, 11/16" x 4-9/16"		240	.033	L.F.	5.05	1.33		6.38	7.75
3150	Door trim set, 1 head and 2 sides, pine, 2-1/2 wide		5.90	1.356	Opng.	22.50	54		76.50	115
3170	3-1/2" wide		5.30	1.509	"	32	60.50		92.50	135
3250	Glass beads, stock pine, 3/8" x 1/2"		275	.029	L.F.	.31	1.16		1.47	2.26
3270	3/8" x 7/8"		270	.030		.41	1.18		1.59	2.40
4850	Parting bead, stock pine, 3/8" x 3/4"		275	.029		.36	1.16		1.52	2.32
4870	1/2" x 3/4"		255	.031		.45	1.25		1.70	2.57
5000	Stool caps, stock pine, 11/16" x 3-1/2"		200	.040		2.27	1.60		3.87	5.15
5100	1-1/16" x 3-1/4"		150	.053	↓	3.59	2.13		5.72	7.45
5300	Threshold, oak, 3' long, inside, 5/8" x 3-5/8"		32	.250	Ea.	9	10		19	26.50
5400	Outside, 1-1/2" x 7-5/8"	↓	16	.500	"	37	20		57	73.50
5900	Window trim sets, including casings, header, stops,									
5910	stool and apron, 2-1/2" wide, minimum	1 Carp	13	.615	Opng.	32.50	24.50		57	76
5950	Average		10	.800		38	32		70	95
6000	Maximum		6	1.333	↓	62.50	53.50		116	157
9000	Minimum labor/equipment charge	↓	4	2	Job		80		80	132

06 22 Millwork

06 22 13 – Standard Pattern Wood Trim

06 22 13.60 Moldings, Soffits		Crew	Daily Output	Labor-Hours	Unit	Material	2009 Bare Costs Labor	Equipment	Total	Total Incl O&P
0010	**MOLDINGS, SOFFITS**									
0200	Soffits, pine, 1″ x 4″	2 Carp	420	.038	L.F.	.44	1.52		1.96	2.99
0210	1″ x 6″		420	.038		.64	1.52		2.16	3.21
0220	1″ x 8″		420	.038		.84	1.52		2.36	3.43
0230	1″ x 10″		400	.040		1.23	1.60		2.83	3.99
0240	1″ x 12″		400	.040		1.60	1.60		3.20	4.40
0250	STK cedar, 1″ x 4″		420	.038		.65	1.52		2.17	3.23
0260	1″ x 6″		420	.038		1.09	1.52		2.61	3.71
0270	1″ x 8″		420	.038		1.52	1.52		3.04	4.18
0280	1″ x 10″		400	.040		2.09	1.60		3.69	4.94
0290	1″ x 12″		400	.040		2.70	1.60		4.30	5.60
1000	Exterior AC plywood, 1/4″ thick		420	.038	S.F.	.85	1.52		2.37	3.45
1050	3/8″ thick		420	.038		1.03	1.52		2.55	3.64
1100	1/2″ thick		420	.038		1	1.52		2.52	3.61
1150	Polyvinyl chloride, white, solid	1 Carp	230	.035		1.09	1.39		2.48	3.49
1160	Perforated	″	230	.035		1.09	1.39		2.48	3.49
9000	Minimum labor/equipment charge	2 Carp	5	3.200	Job		128		128	211

06 25 Prefinished Paneling

06 25 13 – Prefinished Hardboard Paneling

06 25 13.10 Paneling, Hardboard		Crew	Daily Output	Labor-Hours	Unit	Material	Labor	Equipment	Total	Total Incl O&P
0010	**PANELING, HARDBOARD**									
0050	Not incl. furring or trim, hardboard, tempered, 1/8″ thick G	2 Carp	500	.032	S.F.	.38	1.28		1.66	2.53
0100	1/4″ thick G		500	.032		.53	1.28		1.81	2.69
0300	Tempered pegboard, 1/8″ thick G		500	.032		.47	1.28		1.75	2.63
0400	1/4″ thick G		500	.032		.65	1.28		1.93	2.83
0600	Untempered hardboard, natural finish, 1/8″ thick G		500	.032		.39	1.28		1.67	2.54
0700	1/4″ thick G		500	.032		.42	1.28		1.70	2.57
0900	Untempered pegboard, 1/8″ thick G		500	.032		.38	1.28		1.66	2.53
1000	1/4″ thick G		500	.032		.42	1.28		1.70	2.57
1200	Plastic faced hardboard, 1/8″ thick G		500	.032		.64	1.28		1.92	2.81
1300	1/4″ thick G		500	.032		.85	1.28		2.13	3.05
1500	Plastic faced pegboard, 1/8″ thick G		500	.032		.61	1.28		1.89	2.78
1600	1/4″ thick G		500	.032		.75	1.28		2.03	2.94
1800	Wood grained, plain or grooved, 1/4″ thick, minimum G		500	.032		.57	1.28		1.85	2.74
1900	Maximum G		425	.038		1.19	1.50		2.69	3.79
2100	Moldings for hardboard, wood or aluminum, minimum		500	.032	L.F.	.38	1.28		1.66	2.53
2200	Maximum		425	.038	″	1.08	1.50		2.58	3.67
9000	Minimum labor/equipment charge	1 Carp	2	4	Job		160		160	264

06 25 16 – Prefinished Plywood Paneling

06 25 16.10 Paneling, Plywood		Crew	Daily Output	Labor-Hours	Unit	Material	Labor	Equipment	Total	Total Incl O&P
0010	**PANELING, PLYWOOD**									
2400	Plywood, prefinished, 1/4″ thick, 4′ x 8′ sheets									
2410	with vertical grooves. Birch faced, minimum	2 Carp	500	.032	S.F.	.89	1.28		2.17	3.09
2420	Average		420	.038		1.35	1.52		2.87	4
2430	Maximum		350	.046		1.97	1.83		3.80	5.20
2600	Mahogany, African		400	.040		2.51	1.60		4.11	5.40
2700	Philippine (Lauan)		500	.032		1.08	1.28		2.36	3.30
2900	Oak or Cherry, minimum		500	.032		2.10	1.28		3.38	4.42
3000	Maximum		400	.040		3.22	1.60		4.82	6.20

06 25 Prefinished Paneling

06 25 16 – Prefinished Plywood Paneling

06 25 16.10 Paneling, Plywood		Crew	Daily Output	Labor-Hours	Unit	Material	2009 Bare Costs Labor	Equipment	Total	Total Incl O&P
3200	Rosewood	2 Carp	320	.050	S.F.	4.58	2		6.58	8.35
3400	Teak		400	.040		3.22	1.60		4.82	6.20
3600	Chestnut		375	.043		4.77	1.70		6.47	8.05
3800	Pecan		400	.040		2.06	1.60		3.66	4.91
3900	Walnut, minimum		500	.032		2.75	1.28		4.03	5.15
3950	Maximum		400	.040		5.20	1.60		6.80	8.40
4000	Plywood, prefinished, 3/4" thick, stock grades, minimum		320	.050		1.24	2		3.24	4.66
4100	Maximum		224	.071		5.40	2.85		8.25	10.60
4300	Architectural grade, minimum		224	.071		3.98	2.85		6.83	9.10
4400	Maximum		160	.100		6.10	4		10.10	13.30
4600	Plywood, "A" face, birch, V.C., 1/2" thick, natural		450	.036		1.89	1.42		3.31	4.43
4700	Select		450	.036		2.06	1.42		3.48	4.62
4900	Veneer core, 3/4" thick, natural		320	.050		1.99	2		3.99	5.50
5000	Select		320	.050		2.23	2		4.23	5.75
5200	Lumber core, 3/4" thick, natural		320	.050		2.99	2		4.99	6.60
5500	Plywood, knotty pine, 1/4" thick, A2 grade		450	.036		1.63	1.42		3.05	4.14
5600	A3 grade		450	.036		2.06	1.42		3.48	4.62
5800	3/4" thick, veneer core, A2 grade		320	.050		2.11	2		4.11	5.60
5900	A3 grade		320	.050		2.38	2		4.38	5.90
6100	Aromatic cedar, 1/4" thick, plywood		400	.040		2.08	1.60		3.68	4.93
6200	1/4" thick, particle board	↓	400	.040	↓	1.01	1.60		2.61	3.75
9000	Minimum labor/equipment charge	1 Carp	2	4	Job		160		160	264

06 25 26 – Panel System

06 25 26.10 Panel Systems		Crew	Daily Output	Labor-Hours	Unit	Material	Labor	Equipment	Total	Total Incl O&P
0010	PANEL SYSTEMS									
0100	Raised panel, eng. wood core w/ wood veneer, std., paint grade	2 Carp	300	.053	S.F.	11.90	2.13		14.03	16.60
0110	Oak veneer		300	.053		19.70	2.13		21.83	25
0120	Maple veneer		300	.053		25	2.13		27.13	31
0130	Cherry veneer		300	.053		32	2.13		34.13	38.50
0300	Class I fire rated, paint grade		300	.053		14.25	2.13		16.38	19.15
0310	Oak veneer		300	.053		23.50	2.13		25.63	29.50
0320	Maple veneer		300	.053		30	2.13		32.13	36.50
0330	Cherry veneer		300	.053		38.50	2.13		40.63	45.50
0510	Beadboard, 5/8" MDF, standard, primed		300	.053		8.40	2.13		10.53	12.75
0520	Oak veneer, unfinished		300	.053		14.60	2.13		16.73	19.55
0530	Maple veneer, unfinished		300	.053		17.25	2.13		19.38	22.50
0610	Rustic paneling, 5/8" MDF, standard, maple veneer, unfinished	↓	300	.053	↓	22	2.13		24.13	28

06 26 Board Paneling

06 26 13 – Profile Board Paneling

06 26 13.10 Paneling, Boards		Crew	Daily Output	Labor-Hours	Unit	Material	Labor	Equipment	Total	Total Incl O&P
0010	PANELING, BOARDS									
6400	Wood board paneling, 3/4" thick, knotty pine	2 Carp	300	.053	S.F.	1.44	2.13		3.57	5.10
6500	Rough sawn cedar		300	.053		1.85	2.13		3.98	5.55
6700	Redwood, clear, 1" x 4" boards		300	.053		4.30	2.13		6.43	8.25
6900	Aromatic cedar, closet lining, boards	↓	275	.058	↓	3.32	2.32		5.64	7.50
9000	Minimum labor/equipment charge	1 Carp	2	4	Job		160		160	264

06 43 13 – Wood Stairs

06 43 13.20 Prefabricated Wood Stairs		Crew	Daily Output	Labor-Hours	Unit	Material	2009 Bare Costs Labor	Equipment	Total	Total Incl O&P
0010	**PREFABRICATED WOOD STAIRS**									
0100	Box stairs, prefabricated, 3'-0" wide									
0110	Oak treads, up to 14 risers	2 Carp	39	.410	Riser	85	16.40		101.40	121
0600	With pine treads for carpet, up to 14 risers	"	39	.410	"	55	16.40		71.40	87.50
1100	For 4' wide stairs, add				Flight	25%				
1550	Stairs, prefabricated stair handrail with balusters	1 Carp	30	.267	L.F.	75	10.65		85.65	99.50
1700	Basement stairs, prefabricated, pine treads									
1710	Pine risers, 3' wide, up to 14 risers	2 Carp	52	.308	Riser	55	12.30		67.30	81
4000	Residential, wood, oak treads, prefabricated		1.50	10.667	Flight	1,100	425		1,525	1,900
4200	Built in place		.44	36.364	"	1,575	1,450		3,025	4,125
4400	Spiral, oak, 4'-6" diameter, unfinished, prefabricated,									
4500	incl. railing, 9' high	2 Carp	1.50	10.667	Flight	3,625	425		4,050	4,675
9000	Minimum labor/equipment charge	"	3	5.333	Job		213		213	350

06 43 13.40 Wood Stair Parts		Crew	Daily Output	Labor-Hours	Unit	Material	Labor	Equipment	Total	Total Incl O&P
0010	**WOOD STAIR PARTS**									
0020	Pin top balusters, 1-1/4", oak, 34"	1 Carp	96	.083	Ea.	9.15	3.33		12.48	15.55
0030	36"		96	.083		9.60	3.33		12.93	16.05
0040	42"		96	.083		11.65	3.33		14.98	18.30
0050	Poplar, 34"		96	.083		6.10	3.33		9.43	12.25
0060	36"		96	.083		6.50	3.33		9.83	12.65
0070	42"		96	.083		14.20	3.33		17.53	21
0080	Maple, 31"		96	.083		11.65	3.33		14.98	18.30
0090	34"		96	.083		12.95	3.33		16.28	19.75
0100	36"		96	.083		13.50	3.33		16.83	20.50
0110	39"		96	.083		15.45	3.33		18.78	22.50
0120	41"		96	.083		17.35	3.33		20.68	24.50
0130	Primed, 31"		96	.083		4.03	3.33		7.36	9.95
0140	34"		96	.083		4.33	3.33		7.66	10.25
0150	36"		96	.083		4.51	3.33		7.84	10.45
0160	39"		96	.083		5.05	3.33		8.38	11.05
0170	41"		96	.083		5.55	3.33		8.88	11.65
0180	Box top balusters, 1-1/4", oak, 34"		60	.133		9.40	5.35		14.75	19.15
0190	36"		60	.133		10.25	5.35		15.60	20
0200	42"		60	.133		11.45	5.35		16.80	21.50
0210	Poplar, 34"		60	.133		6	5.35		11.35	15.40
0220	36"		60	.133		6.75	5.35		12.10	16.20
0230	42"		60	.133		7.50	5.35		12.85	17.05
0240	Maple, 31"		60	.133		13.45	5.35		18.80	23.50
0250	34"		60	.133		14.15	5.35		19.50	24.50
0260	36"		60	.133		16.35	5.35		21.70	27
0270	39"		60	.133		16.60	5.35		21.95	27
0280	41"		60	.133		17.15	5.35		22.50	27.50
0290	Primed, 31"		60	.133		4.46	5.35		9.81	13.70
0300	34"		60	.133		4.79	5.35		10.14	14.05
0310	36"		60	.133		4.97	5.35		10.32	14.25
0320	39"		60	.133		5.60	5.35		10.95	15
0330	41"		60	.133		6.25	5.35		11.60	15.70
0340	Square balusters, cut from lineal stock, pine, 1-1/16" x 1-1/16"		180	.044	L.F.	1.32	1.78		3.10	4.38
0350	1-5/16" x 1-5/16"		180	.044		1.85	1.78		3.63	4.97
0360	1-5/8" x 1-5/8"		180	.044		3.03	1.78		4.81	6.25
0370	Turned newel, oak, 3-1/2" square, 50" high		8	[illegible]	Ea.	119	40		159	196
0380	65" high		8	[illegible]		154	40		194	236

06 43 13.40 Wood Stair Parts		Crew	Daily Output	Labor-Hours	Unit	Material	Labor	Equipment	Total	Total Incl O&P
						2009 Bare Costs				
0390	Poplar, 3-1/2" square, 50" high	1 Carp	8	1	Ea.	99	40		139	175
0400	65" high		8	1		128	40		168	206
0410	Maple, 3-1/2" square, 50" high		8	1		138	40		178	217
0420	65" high		8	1		160	40		200	241
0430	Square newel, oak, 3-1/2" square, 50" high		8	1		226	40		266	315
0440	65" high		8	1		242	40		282	330
0450	Poplar, 3-1/2" square, 50" high		8	1		215	40		255	300
0460	65" high		8	1		220	40		260	310
0470	Maple, 3" square, 50" high		8	1		243	40		283	335
0480	65" high		8	1		265	40		305	360
0490	Railings, oak, minimum		96	.083	L.F.	7.50	3.33		10.83	13.75
0500	Average		96	.083		9.20	3.33		12.53	15.60
0510	Maximum		96	.083		9.35	3.33		12.68	15.80
0520	Maple, minimum		96	.083		8.80	3.33		12.13	15.20
0530	Average		96	.083		11	3.33		14.33	17.60
0540	Maximum		96	.083		13.20	3.33		16.53	20
0550	Oak, for bending rail, minimum		48	.167		21	6.65		27.65	34
0560	Average		48	.167		30	6.65		36.65	44
0570	Maximum		48	.167		32	6.65		38.65	46.50
0580	Maple, for bending rail, minimum		48	.167		32	6.65		38.65	46.50
0590	Average		48	.167		35	6.65		41.65	49.50
0600	Maximum		48	.167		37.50	6.65		44.15	52
0610	Risers, oak, 3/4" x 8", 36" long		80	.100	Ea.	22	4		26	30.50
0620	42" long		80	.100		25.50	4		29.50	34.50
0630	48" long		80	.100		29.50	4		33.50	38.50
0640	54" long		80	.100		33	4		37	42.50
0650	60" long		80	.100		36.50	4		40.50	47
0660	72" long		80	.100		44	4		48	55
0670	Poplar, 3/4" x 8", 36" long		80	.100		14.35	4		18.35	22.50
0680	42" long		80	.100		16.75	4		20.75	25
0690	48" long		80	.100		19.15	4		23.15	27.50
0700	54" long		80	.100		21.50	4		25.50	30
0710	60" long		80	.100		24	4		28	33
0720	72" long		80	.100		28.50	4		32.50	38
0730	Pine, 1" x 8", 36" long		80	.100		2.51	4		6.51	9.35
0740	42" long		80	.100		2.93	4		6.93	9.80
0750	48" long		80	.100		3.34	4		7.34	10.30
0760	54" long		80	.100		3.76	4		7.76	10.75
0770	60" long		80	.100		4.18	4		8.18	11.20
0780	72" long		80	.100		5	4		9	12.10
0790	Treads, oak, no returns, 1-1/32" x 11-1/2" x 36" long		32	.250		43.50	10		53.50	64.50
0800	42" long		32	.250		50.50	10		60.50	72.50
0810	48" long		32	.250		58	10		68	80
0820	54" long		32	.250		65	10		75	88
0830	60" long		32	.250		72.50	10		82.50	96
0840	72" long		32	.250		87	10		97	112
0850	Mitred return one end, 1-1/32" x 11-1/2" x 36" long		24	.333		65	13.30		78.30	93.50
0860	42" long		24	.333		76	13.30		89.30	106
0870	48" long		24	.333		86.50	13.30		99.80	118
0880	54" long		24	.333		97.50	13.30		110.80	129
0890	60" long		24	.333		108	13.30		121.30	141
0900	72" long		24	.333		130	13.30		143.30	165
0910	Mitred return two ends, 1-1/32" x 11-1/2" x 36" long		12	.667		81	26.50		107.50	134

06 43 Wood Stairs and Railings

06 43 13 – Wood Stairs

06 43 13.40 Wood Stair Parts		Crew	Daily Output	Labor-Hours	Unit	2009 Bare Costs Material	Labor	Equipment	Total	Total Incl O&P
0920	42" long	1 Carp	12	.657	Ea.	95	26.50		121.50	148
0930	48" long		12	.657		108	26.50		134.50	163
0940	54" long		12	.657		122	26.50		148.50	178
0950	60" long		12	.657		135	26.50		161.50	193
0960	72" long		12	.657		162	26.50		188.50	223
0970	Starting step, oak, 48", bullnose		8	[illegible]		163	40		203	246
0980	Double end bullnose		8	[illegible]		193	40		233	278
0990	Half circle		8	[illegible]		216	40		256	305
1010	Double end half circle		8	[illegible]		286	40		326	380
1030	Skirt board, pine, 1" x 10"		55	.145	L.F.	1.23	5.80		7.03	10.95
1040	1" x 12"		52	.154	"	1.60	6.15		7.75	11.90
1050	Oak landing tread, 1-1/16" thick		54	.148	S.F.	8.25	5.90		14.15	18.85
1060	Oak cove molding		96	.083	L.F.	2.75	3.33		6.08	8.55
1070	Oak stringer molding		96	.083		2.75	3.33		6.08	8.55
1080	Oak shoe molding		96	.083		2.75	3.33		6.08	8.55
1090	Rail bolt, 5/16" x 3-1/2"		48	.167	Ea.	1.65	6.65		8.30	12.80
1100	5/16" x 4-1/2"		48	.167		2.20	6.65		8.85	13.40
1120	Newel post anchor		16	.500		27.50	20		47.50	63.50
1130	Tapered plug, 1/2"		240	.033		.28	1.33		1.61	2.51
1140	1"		240	.033		.39	1.33		1.72	2.63
1150	Rail bolt kit		24	.333		11	13.30		24.30	34
9000	Minimum labor/equipment charge		3	2.667	Job		107		107	176

06 43 16 – Wood Railings

06 43 16.10 Wood Handrails and Railings		Crew	Daily Output	Labor-Hours	Unit	Material	Labor	Equipment	Total	Total Incl O&P
0010	**WOOD HANDRAILS AND RAILINGS**									
0020	Custom design, architectural grade, hardwood, minimum	1 Carp	38	.211	L.F.	9.50	8.40		17.90	24.50
0100	Maximum		30	.267		51.50	10.65		62.15	74
0300	Stock interior railing with spindles 4" O.C., 4' long		40	.200		31.50	8		39.50	48
0400	8' long		48	.167		31.50	6.65		38.15	46
9000	Minimum labor/equipment charge		3	2.667	Job		107		107	176

06 44 Ornamental Woodwork

06 44 19 – Wood Grilles

06 44 19.10 Grilles		Crew	Daily Output	Labor-Hours	Unit	Material	Labor	Equipment	Total	Total Incl O&P
0010	**GRILLES** and panels, hardwood, sanded									
0020	2' x 4' to 4' x 8', custom designs, unfinished, minimum	1 Carp	38	.211	S.F.	14.65	8.40		23.05	30
0050	Average		30	.267		31.50	10.65		42.15	52.50
0100	Maximum		19	.421		49	16.80		65.80	81.50
0300	As above, but prefinished, minimum		38	.211		14.65	8.40		23.05	30
0400	Maximum		19	.421		55	16.80		71.80	88.50
9000	Minimum labor/equipment charge		2	4	Job		160		160	264

06 44 33 – Wood Mantels

06 44 33.10 Fireplace Mantels		Crew	Daily Output	Labor-Hours	Unit	Material	Labor	Equipment	Total	Total Incl O&P
0010	**FIREPLACE MANTELS**									
0015	6" molding, 6' x 3'-6" opening, minimum	1 Carp	5	1.600	Opng.	228	64		292	355
0100	Maximum		5	1.600		410	64		474	560
0300	Prefabricated pine, colonial type, stock, deluxe		2	4		1,250	160		1,410	1,675
0400	Economy		3	2.657		530	107		637	760
9000	Minimum labor/equipment charge		3	2.657	Job		107		107	176

06 44 Ornamental Woodwork

06 44 33 – Wood Mantels

06 44 33.20 Fireplace Mantel Beam		Crew	Daily Output	Labor-Hours	Unit	Material	2009 Bare Costs Labor	Equipment	Total	Total Incl O&P
0010	FIREPLACE MANTEL BEAM									
0020	Rough texture wood, 4" x 8"	1 Carp	36	.222	L.F.	6	8.90		14.90	21.50
0100	4" x 10"		35	.229	"	7	9.15		16.15	23
0300	Laminated hardwood, 2-1/4" x 10-1/2" wide, 6' long		5	1.600	Ea.	110	64		174	227
0400	8' long		5	1.600	"	150	64		214	271
0600	Brackets for above, rough sawn		12	.667	Pr.	10	26.50		36.50	55
0700	Laminated		12	.667	"	15	26.50		41.50	60.50
9000	Minimum labor/equipment charge		4	2	Job		80		80	132

06 44 39 – Wood Posts and Columns

06 44 39.10 Decorative Beams		Crew	Daily Output	Labor-Hours	Unit	Material	Labor	Equipment	Total	Total Incl O&P
0010	DECORATIVE BEAMS									
0020	Rough sawn cedar, non-load bearing, 4" x 4"	2 Carp	180	.089	L.F.	1.47	3.55		5.02	7.45
0100	4" x 6"		170	.094		2.21	3.76		5.97	8.65
0200	4" x 8"		160	.100		2.95	4		6.95	9.85
0300	4" x 10"		150	.107		3.68	4.26		7.94	11.10
0400	4" x 12"		140	.114		4.42	4.57		8.99	12.40
0500	8" x 8"		130	.123		7.50	4.92		12.42	16.35
9000	Minimum labor/equipment charge	1 Carp	3	2.667	Job		107		107	176

06 44 39.20 Columns		Crew	Daily Output	Labor-Hours	Unit	Material	Labor	Equipment	Total	Total Incl O&P
0010	COLUMNS									
0050	Aluminum, round colonial, 6" diameter	2 Carp	80	.200	V.L.F.	16	8		24	31
0100	8" diameter		62.25	.257		17	10.25		27.25	35.50
0200	10" diameter		55	.291		19.10	11.60		30.70	40
0250	Fir, stock units, hollow round, 6" diameter		80	.200		17.20	8		25.20	32
0300	8" diameter		80	.200		20.50	8		28.50	35.50
0350	10" diameter		70	.229		25.50	9.15		34.65	43.50
0400	Solid turned, to 8' high, 3-1/2" diameter		80	.200		8.40	8		16.40	22.50
0500	4-1/2" diameter		75	.213		12.80	8.50		21.30	28
0600	5-1/2" diameter		70	.229		16.40	9.15		25.55	33
0800	Square columns, built-up, 5" x 5"		65	.246		10.20	9.85		20.05	27.50
0900	Solid, 3-1/2" x 3-1/2"		130	.123		6.70	4.92		11.62	15.45
1600	Hemlock, tapered, T & G, 12" diam, 10' high		100	.160		31.50	6.40		37.90	45
1700	16' high		65	.246		46.50	9.85		56.35	67.50
1900	10' high, 14" diameter		100	.160		78.50	6.40		84.90	97
2000	18' high		65	.246		63.50	9.85		73.35	86.50
2200	18" diameter, 12' high		65	.246		114	9.85		123.85	142
2300	20' high		50	.320		80	12.80		92.80	109
2500	20" diameter, 14' high		40	.400		129	16		145	169
2600	20' high		35	.457		120	18.25		138.25	162
2800	For flat pilasters, deduct					33%				
3000	For splitting into halves, add				Ea.	116			116	128
4000	Rough sawn cedar posts, 4" x 4"	2 Carp	250	.064	V.L.F.	6.30	2.56		8.86	11.15
4100	4" x 6"		235	.068		15	2.72		17.72	21
4200	6" x 6"		220	.073		26.50	2.91		29.41	34
4300	8" x 8"		200	.080		60	3.20		63.20	71.50
9000	Minimum labor/equipment charge	1 Carp	3	2.667	Job		107		107	176

06 48 Wood Frames

06 48 13 – Exterior Wood Door Frames

06 48 13.10 Exterior Wood Door Frames and Accessories		Crew	Daily Output	Labor-Hours	Unit	2009 Bare Costs Material	Labor	Equipment	Total	Total Incl O&P
0010	**EXTERIOR WOOD DOOR FRAMES AND ACCESSORIES**									
0400	Exterior frame, incl. ext. trim, pine, 5/4 x 4-9/16" deep	2 Carp	375	.043	L.F.	5.55	1.70		7.25	8.90
0420	5-3/16" deep		375	.043		11.30	1.70		13	15.25
0440	6-9/16" deep		375	.043		8.50	1.70		10.20	12.15
0600	Oak, 5/4 x 4-9/16" deep		350	.046		12	1.83		13.83	16.20
0620	5-3/16" deep		350	.046		13.50	1.83		15.33	17.85
0640	6-9/16" deep		350	.046		15.30	1.83		17.13	19.85
0800	Walnut, 5/4 x 4-9/16" deep		350	.046		14	1.83		15.83	18.40
0820	5-3/16" deep		350	.046		18	1.83		19.83	23
0840	6-9/16" deep		350	.046		20	1.83		21.83	25
1000	Sills, 8/4 x 8" deep, oak, no horns		100	.160		19	6.40		25.40	31.50
1020	2" horns		100	.160		19	6.40		25.40	31.50
1040	3" horns		100	.160		19.95	6.40		26.35	32.50
1100	8/4 x 10" deep, oak, no horns		90	.178		21.50	7.10		28.60	35.50
1120	2" horns		90	.178		19.80	7.10		26.90	33.50
1140	3" horns		90	.178		20.50	7.10		27.60	34.50
2000	Exterior, colonial, frame & trim, 3' opng., in-swing, minimum		22	.727	Ea.	315	29		344	400
2010	Average		21	.762		470	30.50		500.50	565
2020	Maximum		20	.800		1,075	32		1,107	1,225
2100	5'-4" opening, in-swing, minimum		17	.941		360	37.50		397.50	455
2120	Maximum		15	1.067		1,075	42.50		1,117.50	1,250
2140	Out-swing, minimum		17	.941		370	37.50		407.50	465
2160	Maximum		15	1.067		1,100	42.50		1,142.50	1,300
2400	6'-0" opening, in-swing, minimum		16	1		345	40		385	445
2420	Maximum		10	1.600		1,100	64		1,164	1,325
2460	Out-swing, minimum		16	1		375	40		415	480
2480	Maximum		10	1.600		1,300	64		1,364	1,525
2600	For two sidelights, add, minimum		30	.533	Opng.	355	21.50		376.50	425
2620	Maximum		20	.800	"	1,125	32		1,157	1,300
2700	Custom birch frame, 3'-0" opening		16	1	Ea.	205	40		245	292
2750	6'-0" opening		16	1		305	40		345	400
2900	Exterior, modern, plain trim, 3' opng., in-swing, minimum		26	.615		35.50	24.50		60	79.50
2920	Average		24	.667		42	26.50		68.50	90
2940	Maximum		22	.727		50.50	29		79.50	104

06 48 16 – Interior Wood Door Frames

06 48 16.10 Interior Wood Door Jamb and Frames		Crew	Daily Output	Labor-Hours	Unit	Material	Labor	Equipment	Total	Total Incl O&P
0010	**INTERIOR WOOD DOOR JAMB AND FRAMES**									
3000	Interior frame, pine, 11/16" x 3-5/8" deep	2 Carp	375	.043	L.F.	5.50	1.70		7.20	8.85
3020	4-9/16" deep		375	.043		6.45	1.70		8.15	9.90
3200	Oak, 11/16" x 3-5/8" deep		350	.046		4.35	1.83		6.18	7.80
3220	4-9/16" deep		350	.046		4.50	1.83		6.33	7.95
3240	5-3/16" deep		350	.046		4.58	1.83		6.41	8.05
3400	Walnut, 11/16" x 3-5/8" deep		350	.046		7.30	1.83		9.13	11.05
3420	4-9/16" deep		350	.046		7.50	1.83		9.33	11.25
3440	5-3/16" deep		350	.046		7.90	1.83		9.73	11.65
3600	Pocket door frame		16	[illegible]	Ea.	67	40		107	140
3800	Threshold, oak, 5/8" x 3-5/8" deep		200	.030	L.F.	3.83	3.20		7.03	9.50
3820	4-5/8" deep		190	.034		4.58	3.36		7.94	10.60
3840	5-5/8" deep		180	.039		7.15	3.55		10.70	13.70
9000	Minimum labor/equipment charge	1 Carp	4	2	Job		80		80	132

06 49 Wood Screens and Exterior Wood Shutters

06 49 19 – Exterior Wood Shutters

06 49 19.10 Shutters, Exterior		Crew	Daily Output	Labor-Hours	Unit	2009 Bare Costs Material	Labor	Equipment	Total	Total Incl O&P
0010	**SHUTTERS, EXTERIOR**									
0012	Aluminum, louvered, 1'-4" wide, 3'-0" long	1 Carp	10	.800	Pr.	55.50	32		87.50	114
0200	4'-0" long		10	.800		66.50	32		98.50	126
0300	5'-4" long		10	.800		87.50	32		119.50	150
0400	6'-8" long		9	.889		112	35.50		147.50	182
1000	Pine, louvered, primed, each 1'-2" wide, 3'-3" long		10	.800		96.50	32		128.50	159
1100	4'-7" long		10	.800		130	32		162	196
1250	Each 1'-4" wide, 3'-0" long		10	.800		98.50	32		130.50	161
1350	5'-3" long		10	.800		148	32		180	216
1500	Each 1'-6" wide, 3'-3" long		10	.800		102	32		134	166
1600	4'-7" long		10	.800		144	32		176	211
1620	Hemlock, louvered, 1'-2" wide, 5'-7" long		10	.800		158	32		190	227
1630	Each 1'-4" wide, 2'-2" long		10	.800		98.50	32		130.50	161
1640	3'-0" long		10	.800		98.50	32		130.50	161
1650	3'-3" long		10	.800		106	32		138	170
1660	3'-11" long		10	.800		121	32		153	186
1670	4'-3" long		10	.800		118	32		150	183
1680	5'-3" long		10	.800		147	32		179	214
1690	5'-11" long		10	.800		166	32		198	236
1700	Door blinds, 6'-9" long, each 1'-3" wide		9	.889		167	35.50		202.50	243
1710	1'-6" wide		9	.889		180	35.50		215.50	257
1720	Hemlock, solid raised panel, each 1'-4" wide, 3'-3" long		10	.800		159	32		191	228
1730	3'-11" long		10	.800		186	32		218	258
1740	4'-3" long		10	.800		202	32		234	275
1750	4'-7" long		10	.800		217	32		249	292
1760	4'-11" long		10	.800		237	32		269	315
1770	5'-11" long		10	.800		269	32		301	350
1800	Door blinds, 6'-9" long, each 1'-3" wide		9	.889		305	35.50		340.50	395
1900	1'-6" wide		9	.889		330	35.50		365.50	425
2500	Polystyrene, solid raised panel, each 1'-4" wide, 3'-3" long		10	.800		40.50	32		72.50	97.50
2600	3'-11" long		10	.800		46.50	32		78.50	105
2700	4'-7" long		10	.800		51	32		83	109
2800	5'-3" long		10	.800		55	32		87	114
2900	6'-8" long		9	.889		66	35.50		101.50	131
4500	Polystyrene, louvered, each 1'-2" wide, 3'-3" long		10	.800		27.50	32		59.50	83.50
4600	4'-7" long		10	.800		34	32		66	90.50
4750	5'-3" long		10	.800		39	32		71	96
4850	6'-8" long		9	.889		46.50	35.50		82	110
6000	Vinyl, louvered, each 1'-2" x 4'-7" long		10	.800		34	32		66	90.50
6200	Each 1'-4" x 6'-8" long		9	.889		47.50	35.50		83	111
9000	Minimum labor/equipment charge		4	2	Job		80		80	132

06 51 Structural Plastic Shapes and Plates

06 51 13 – Plastic Lumber

06 51 13.10 Recycled Plastic Lumber

			Crew	Daily Output	Labor-Hours	Unit	2009 Bare Costs Material	Labor	Equipment	Total	Total Incl O&P
0010	RECYCLED PLASTIC LUMBER										
4000	Sheeting, recycled plastic, black or white, 4′ x 8′ x 1/8″	G	2 Carp	1100	.015	S.F.	1.39	.58		1.97	2.49
4010	4′ x 8′ x 3/16″	G		1100	.015		2.10	.58		2.68	3.27
4020	4′ x 8′ x 1/4″	G		950	.017		2.53	.67		3.20	3.89
4030	4′ x 8′ x 3/8″	G		950	.017		3.53	.67		4.20	4.99
4040	4′ x 8′ x 1/2″	G		900	.018		4.56	.71		5.27	6.15
4050	4′ x 8′ x 5/8″	G		900	.018		6	.71		6.71	7.75
4060	4′ x 8′ x 3/4″	G	↓	850	.019	↓	6.30	.75		7.05	8.20
4070	Add for colors					Ea.	5%			5%	6%
8500	100% recycled plastic, var colors, NLB, 2″ x 2″	G				L.F.	2.62			2.62	2.88
8510	2″ x 4″	G					2.80			2.80	3.08
8520	2″ x 6″	G					4.37			4.37	4.81
8530	2″ x 8″	G					6.05			6.05	6.65
8540	2″ x 10″	G					8.90			8.90	9.80
8550	5/4″ x 4″	G					2.20			2.20	2.42
8560	5/4″ x 6″	G					3.08			3.08	3.39
8570	1″ x 6″	G					2.08			2.08	2.29
8580	1/2″ x 8″	G					2.30			2.30	2.53
8590	2″ x 10″ T & G	G					6.35			6.35	7
8600	3″ x 10″ T & G	G					10.15			10.15	11.15
8610	Add for premium colors					↓	20%			20%	22%
8620	Mailbox post, recycled plastic, black/green , 4″x 4″x 6′ H, incl. box	G	1 Clab	3	2.667	Ea.	58	84.50		142.50	203

06 52 Plastic Structural Assemblies

06 52 10 – Fiberglass Structural Assemblies

06 52 10.40 Fiberglass Floor Grating

		Crew	Daily Output	Labor-Hours	Unit	Material	Labor	Equipment	Total	Total Incl O&P
0010	FIBERGLASS FLOOR GRATING									
0100	Reinforced polyester, fire retardant, 1″ x 4″ grid, 1″ thick	E-4	510	.063	S.F.	14.25	2.84	.26	17.35	21
0200	1-1/2″ x 6″ mesh, 1-1/2″ thick		500	.064		16.65	2.89	.27	19.81	24
0300	With grit surface, 1-1/2″ x 6″ grid, 1-1/2″ thick	↓	500	.064	↓	17	2.89	.27	20.16	24.50

06 63 Plastic Railings

06 63 10 – Plastic (PVC) Railings

06 63 10.10 Plastic Railings

		Crew	Daily Output	Labor-Hours	Unit	Material	Labor	Equipment	Total	Total Incl O&P
0010	PLASTIC RAILINGS									
0100	Horizontal PVC handrail with balusters, 3-1/2″ wide, 36″ high	1 Carp	96	.083	L.F.	23.50	3.33		26.83	31
0150	42″ high		96	.083		26.50	3.33		29.83	34.50
0200	Angled PVC handrail with balusters, 3-1/2″ wide, 36″ high		72	.111		27	4.44		31.44	37
0250	42″ high		72	.111		30	4.44		34.44	40.50
0300	Post sleeve for 4 x 4 post		96	.083	↓	12	3.33		15.33	18.70
0400	Post cap for 4 x 4 post, flat profile		48	.167	Ea.	11.35	6.65		18	23.50
0450	Newel post style profile		48	.167		21	6.65		27.65	34
0500	Raised corbeled profile		48	.167		22	6.65		28.65	35.50
0550	Post base trim for 4 x 4 post	↓	96	.083	↓	12.80	3.33		16.13	19.55

06 65 10 – PVC Trim

06 65 10.10 PVC Trim, Exterior		Crew	Daily Output	Labor-Hours	Unit	2009 Bare Costs Material	Labor	Equipment	Total	Total Incl O&P
0010	**PVC TRIM, EXTERIOR**									
0100	Cornerboards, 5/4" x 6" x 6"	1 Carp	240	.033	L.F.	5.70	1.33		7.03	8.50
0110	Door / window casing, 1" x 4"		200	.040		1.28	1.60		2.88	4.05
0120	1" x 6"		200	.040		2.02	1.60		3.62	4.86
0130	1" x 8"		195	.041		2.66	1.64		4.30	5.65
0140	1" x 10"		195	.041		3.39	1.64		5.03	6.45
0150	1" x 12"		190	.042		4.12	1.68		5.80	7.30
0160	5/4" x 4"		195	.041		1.63	1.64		3.27	4.50
0170	5/4" x 6"		195	.041		2.57	1.64		4.21	5.55
0180	5/4" x 8"		190	.042		3.38	1.68		5.06	6.50
0190	5/4" x 10"		190	.042		4.32	1.68		6	7.55
0200	5/4" x 12"		185	.043		5.25	1.73		6.98	8.65
0210	Fascia, 1" x 4"		250	.032		1.28	1.28		2.56	3.52
0220	1" x 6"		250	.032		2.02	1.28		3.30	4.33
0230	1" x 8"		225	.036		2.66	1.42		4.08	5.30
0240	1" x 10"		225	.036		3.39	1.42		4.81	6.10
0250	1" x 12"		200	.040		4.12	1.60		5.72	7.15
0260	5/4" x 4"		240	.033		1.63	1.33		2.96	3.99
0270	5/4" x 6"		240	.033		2.57	1.33		3.90	5.05
0280	5/4" x 8"		215	.037		3.38	1.49		4.87	6.15
0290	5/4" x 10"		215	.037		4.32	1.49		5.81	7.20
0300	5/4" x 12"		190	.042		5.25	1.68		6.93	8.60
0310	Frieze, 1" x 4"		250	.032		1.28	1.28		2.56	3.52
0320	1" x 6"		250	.032		2.02	1.28		3.30	4.33
0330	1" x 8"		225	.036		2.66	1.42		4.08	5.30
0340	1" x 10"		225	.036		3.39	1.42		4.81	6.10
0350	1" x 12"		200	.040		4.12	1.60		5.72	7.15
0360	5/4" x 4"		240	.033		1.63	1.33		2.96	3.99
0370	5/4" x 6"		240	.033		2.57	1.33		3.90	5.05
0380	5/4" x 8"		215	.037		3.38	1.49		4.87	6.15
0390	5/4" x 10"		215	.037		4.32	1.49		5.81	7.20
0400	5/4" x 12"		190	.042		5.25	1.68		6.93	8.60
0410	Rake, 1" x 4"		200	.040		1.28	1.60		2.88	4.05
0420	1" x 6"		200	.040		2.02	1.60		3.62	4.86
0430	1" x 8"		190	.042		2.66	1.68		4.34	5.70
0440	1" x 10"		190	.042		3.39	1.68		5.07	6.50
0450	1" x 12"		180	.044		4.12	1.78		5.90	7.45
0460	5/4" x 4"		195	.041		1.63	1.64		3.27	4.50
0470	5/4" x 6"		195	.041		2.57	1.64		4.21	5.55
0480	5/4" x 8"		185	.043		3.38	1.73		5.11	6.55
0490	5/4" x 10"		185	.043		4.32	1.73		6.05	7.60
0500	5/4" x 12"		175	.046		5.25	1.83		7.08	8.80
0510	Rake trim, 1" x 4"		225	.036		1.28	1.42		2.70	3.76
0520	1" x 6"		225	.036		2.02	1.42		3.44	4.57
0560	5/4" x 4"		220	.036		1.63	1.45		3.08	4.19
0570	5/4" x 6"		220	.036		2.57	1.45		4.02	5.25
0610	Soffit, 1" x 4"	2 Carp	420	.038		1.28	1.52		2.80	3.92
0620	1" x 6"		420	.038		2.02	1.52		3.54	4.73
0630	1" x 8"		420	.038		2.66	1.52		4.18	5.45
0640	1" x 10"		400	.040		3.39	1.60		4.99	6.35
0650	1" x 12"		400	.040		4.12	1.60		5.72	7.15
0660	5/4" x 4"		410	.039		1.63	1.56		3.19	4.36
0670	5/4" x 6"		410	.039		2.57	1.56		4.13	5.40

06 65 Plastic Simulated Wood Trim

06 65 10 – PVC Trim

06 65 10.10 PVC Trim, Exterior		Crew	Daily Output	Labor-Hours	Unit	2009 Bare Costs Material	Labor	Equipment	Total	Total Incl O&P
0680	5/4" x 8"	2 Carp	410	.039	L.F.	3.38	1.56		4.94	6.30
0690	5/4" x 10"		390	.041		4.32	1.64		5.96	7.45
0700	5/4" x 12"	↓	390	.041	↓	5.25	1.64		6.89	8.50

Division Notes

		CREW	DAILY OUTPUT	LABOR-HOURS	UNIT	2009 BARE COSTS MAT.	LABOR	EQUIP.	TOTAL	TOTAL INCL O&P

Estimating Tips

07 10 00 Dampproofing and Waterproofing

- Be sure of the job specifications before pricing this subdivision. The difference in cost between waterproofing and dampproofing can be great. Waterproofing will hold back standing water. Dampproofing prevents the transmission of water vapor. Also included in this section are vapor retarding membranes.

07 20 00 Thermal Protection

- Insulation and fireproofing products are measured by area, thickness, volume or R-value. Specifications may give only what the specific R-value should be in a certain situation. The estimator may need to choose the type of insulation to meet that R-value.

07 30 00 Steep Slope Roofing

07 40 00 Roofing and Siding Panels

- Many roofing and siding products are bought and sold by the square. One square is equal to an area that measures 100 square feet.

This simple change in unit of measure could create a large error if the estimator is not observant. Accessories necessary for a complete installation must be figured into any calculations for both material and labor.

07 50 00 Membrane Roofing

07 60 00 Flashing and Sheet Metal

07 70 00 Roofing and Wall Specialties and Accessories

- The items in these subdivisions compose a roofing system. No one component completes the installation, and all must be estimated. Built-up or single-ply membrane roofing systems are made up of many products and installation trades. Wood blocking at roof perimeters or penetrations, parapet coverings, reglets, roof drains, gutters, downspouts, sheet metal flashing, skylights, smoke vents, and roof hatches all need to be considered along with the roofing material. Several different installation trades will need to work together on the roofing system Inherent difficulties in the scheduling and coordination of various trades must be accounted for when estimating labor costs.

07 90 00 Joint Protection

- To complete the weather-tight shell, the sealants and caulkings must be estimated. Where different materials meet—at expansion joints, at flashing penetrations, and at hundreds of other locations throughout a construction project—they provide another line of defense against water penetration. Often, an entire system is based on the proper location and placement of caulking or sealants. The detailed drawings that are included as part of a set of architectural plans show typical locations for these materials. When caulking or sealants are shown at typical locations, this means the estimator must include them for all the locations where this detail is applicable. Be careful to keep different types of sealants separate, and remember to consider backer rods and primers if necessary.

Reference Numbers

Reference numbers are shown in shaded boxes at the beginning of some major classifications. These numbers refer to related items in the Reference Section. The reference information may be an estimating procedure, an alternate pricing method, or technical information.

Note: Not all subdivisions listed here necessarily appear in this publication.

07 01 Operation and Maint. of Thermal and Moisture Protection

07 01 50 – Maintenance of Membrane Roofing

07 01 50.10 Roof Coatings		Crew	Daily Output	Labor-Hours	Unit	Material	2009 Bare Costs Labor	Equipment	Total	Total Incl O&P
0010	**ROOF COATINGS**									
0012	Asphalt, brush grade, material only				Gal.	8.25			8.25	9.10
0200	Asphalt base, fibered aluminum coating [G]					12.20			12.20	13.45
0300	Asphalt primer, 5 gallon					5.95			5.95	6.55
0600	Coal tar pitch, 200 lb. barrels				Ton	840			840	925
0700	Tar roof cement, 5 gal. lots				Gal.	8.40			8.40	9.25
0800	Glass fibered roof & patching cement, 5 gallon				"	7.35			7.35	8.05
0900	Reinforcing glass membrane, 450 S.F./roll				Ea.	48.50			48.50	53
1000	Neoprene roof coating, 5 gal, 2 gal/sq				Gal.	24.50			24.50	26.50
1100	Roof patch & flashing cement, 5 gallon					11.70			11.70	12.85
1200	Roof resaturant, glass fibered, 3 gal/sq					7.75			7.75	8.55
1300	Mineral rubber, 3 gal/sq					4.99			4.99	5.50

07 05 Common Work Results for Thermal and Moisture Protection

07 05 05 – Selective Demolition

07 05 05.10 Selective Demo., Thermal and Moist. Protection		Crew	Daily Output	Labor-Hours	Unit	Material	Labor	Equipment	Total	Total Incl O&P
0010	**SELECTIVE DEMOLITION, THERMAL AND MOISTURE PROTECTION**									
0020	Caulking / sealant, to 1" x 1" joint R024119-10	1 Clab	600	.013	L.F.		.42		.42	.70
0120	Downspouts, including hangers		350	.023	"		.72		.72	1.19
0220	Flashing, sheet metal		290	.028	S.F.		.87		.87	1.44
0420	Gutters, aluminum or wood, edge hung		240	.033	L.F.		1.05		1.05	1.74
0520	Built-in		100	.080	"		2.53		2.53	4.17
0620	Insulation, air / vapor barrier		3500	.002	S.F.		.07		.07	.12
0670	Batts or blankets		1400	.006	C.F.		.18		.18	.30
0720	Foamed or sprayed in place	2 Clab	1000	.016	B.F.		.51		.51	.83
0770	Loose fitting	1 Clab	3000	.003	C.F.		.08		.08	.14
0870	Rigid board		3450	.002	B.F.		.07		.07	.12
1120	Roll roofing, cold adhesive		12	.667	Sq.		21		21	35
1170	Roof accessories, adjustable metal chimney flashing		9	.889	Ea.		28		28	46.50
1325	Plumbing vent flashing		32	.250	"		7.90		7.90	13.05
1375	Ridge vent strip, aluminum		310	.026	L.F.		.82		.82	1.35
1620	Skylight to 10 S.F.		8	1	Ea.		31.50		31.50	52
2120	Roof edge, aluminum soffit and fascia		570	.014	L.F.		.44		.44	.73
2170	Concrete coping, up to 12" wide	2 Clab	160	.100			3.16		3.16	5.20
2220	Drip edge	1 Clab	1650	.005			.15		.15	.25
2270	Gravel stop		1650	.005			.15		.15	.25
2370	Sheet metal coping, up to 12" wide		240	.033			1.05		1.05	1.74
2470	Roof insulation board, over 2" thick	B-2	7800	.005	B.F.		.16		.16	.27
2520	Up to 2" thick	"	3900	.010	S.F.		.33		.33	.54
2620	Roof ventilation, louvered gable vent	1 Clab	16	.500	Ea.		15.80		15.80	26
2675	Rafter vents		960	.008	"		.26		.26	.43
2720	Soffit vent and/or fascia vent		575	.014	L.F.		.44		.44	.73
2775	Soffit vent strip, aluminum, 3" to 4" wide		160	.050			1.58		1.58	2.61
2820	Roofing accessories, shingle moulding, to 1" x 4"		1600	.005			.16		.16	.26
2870	Cant strip	B-2	2000	.020			.64		.64	1.06
2920	Concrete block walkway	1 Clab	230	.035			1.10		1.10	1.81
3070	Roofing, felt paper, 15#		70	.114	Sq.		3.61		3.61	5.95
3125	#30 felt		30	.267	"		8.45		8.45	13.90
3170	Asphalt shingles, 1 layer	B-2	3500	.011	S.F.		.37		.37	.60
3370	Modified bitumen		26	1.538	Sq.		49		49	81.50
3420	Built-up, no gravel, 3 ply		25	1.600			51		51	84.50

07 05 Common Work Results for Thermal and Moisture Protection

07 05 05 – Selective Demolition

07 05 05.10 Selective Demo., Thermal and Moist. Protection		Crew	Daily Output	Labor-Hours	Unit	Material	2009 Bare Costs Labor	Equipment	Total	Total Incl O&P
3470	4 ply	B-2	21	1.905	Sq.		61		61	101
3620	5 ply		1600	.025	S.F.		.80		.80	1.32
3720	5 ply, with gravel		890	.045			1.44		1.44	2.37
3725	Gravel removal, minimum		5000	.008			.26		.26	.42
3730	Maximum		2000	.020			.64		.64	1.06
3870	Fiberglass sheet		1200	.033			1.07		1.07	1.76
4120	Slate shingles		1900	.021			.67		.67	1.11
4170	Ridge shingles, clay or slate		2000	.020	L.F.		.64		.64	1.06
4320	Single ply membrane, attached at seams		52	.769	Sq.		24.50		24.50	40.50
4370	Ballasted		75	.533			17.05		17.05	28
4420	Fully adhered		39	1.026			33		33	54
4550	Roof hatch, 2′-6″ x 3′-0″	1 Clab	10	.800	Ea.		25.50		25.50	41.50
4670	Wood shingles	B-2	2200	.018	S.F.		.58		.58	.96
4820	Sheet metal roofing	"	2150	.019			.60		.60	.98
4970	Siding, horizontal wood clapboards	1 Clab	380	.021			.67		.67	1.10
5025	Exterior insulation finish system	"	120	.067			2.11		2.11	3.48
5070	Tempered hardboard, remove and reset	1 Carp	380	.021			.84		.84	1.39
5120	Tempered hardboard sheet siding	"	375	.021			.85		.85	1.41
5170	Metal, corner strips	1 Clab	850	.009	L.F.		.30		.30	.49
5225	Horizontal strips		444	.018	S.F.		.57		.57	.94
5320	Vertical strips		400	.020			.63		.63	1.04
5520	Wood shingles		350	.023			.72		.72	1.19
5620	Stucco siding		360	.022			.70		.70	1.16
5670	Textured plywood		725	.011			.35		.35	.58
5720	Vinyl siding		510	.016			.50		.50	.82
5770	Corner strips		900	.009	L.F.		.28		.28	.46
5870	Wood, boards, vertical		400	.020	S.F.		.63		.63	1.04
5920	Waterproofing, protection / drain board	2 Clab	3900	.004	B.F.		.13		.13	.21
5970	Over 1/2″ thick		1750	.009	S.F.		.29		.29	.48
6020	To 1/2″ thick		2000	.008	"		.25		.25	.42
9000	Minimum labor/equipment charge	1 Clab	2	4	Job		126		126	209

07 11 Dampproofing

07 11 13 – Bituminous Dampproofing

07 11 13.10 Bituminous Asphalt Coating

		Crew	Daily Output	Labor-Hours	Unit	Material	Labor	Equipment	Total	Total Incl O&P
0010	**BITUMINOUS ASPHALT COATING**									
0030	Brushed on, below grade, 1 coat	1 Rofc	665	.012	S.F.	.16	.41		.57	.92
0100	2 coat		500	.016		.32	.55		.87	1.34
0300	Sprayed on, below grade, 1 coat, 25.6 S.F./gal.		830	.010		.16	.33		.49	.77
0400	2 coat, 20.5 S.F./gal.		500	.016		.32	.55		.87	1.33
0500	Asphalt coating, with fibers				Gal.	8.50			8.50	9.35
0600	Troweled on, asphalt with fibers, 1/16″ thick	1 Rofc	500	.016	S.F.	.37	.55		.92	1.39
0700	1/8″ thick		400	.020		.65	.69		1.34	1.94
1000	1/2″ thick		350	.023		2.12	.78		2.90	3.73
9000	Minimum labor/equipment charge		3	2.667	Job		91.50		91.50	163

07 11 16 – Cementitious Dampproofing

07 11 16.20 Cementitious Parging

		Crew	Daily Output	Labor-Hours	Unit	Material	Labor	Equipment	Total	Total Incl O&P
0010	**CEMENTITIOUS PARGING**									
0020	Portland cement, 2 coats, 1/2″ thick	D-1	250	.064	S.F.	.28	2.33		2.61	4.07
0100	Waterproofed Portland cement, 1/2″ thick, 2 coats	"	250	.064	"	2.62	2.33		4.95	6.65

07 12 Built-up Bituminous Waterproofing

07 12 13 – Built-Up Asphalt Waterproofing

07 12 13.20 Membrane Waterproofing		Crew	Daily Output	Labor-Hours	Unit	Material	Labor (2009 Bare Costs)	Equipment	Total	Total Incl O&P
0010	**MEMBRANE WATERPROOFING**									
0012	On slabs, 1 ply, felt, mopped	G-1	3000	.019	S.F.	.34	.60	.15	1.09	1.61
0100	On slabs, 1 ply, glass fiber fabric, mopped		2100	.027		.36	.85	.21	1.42	2.14
0300	On slabs, 2 ply, felt, mopped		2500	.022		.69	.72	.18	1.59	2.23
0400	On slabs, 2 ply, glass fiber fabric, mopped		1650	.034		.79	1.09	.27	2.15	3.10
0600	On slabs, 3 ply, felt, mopped		2100	.027		1.03	.85	.21	2.09	2.89
0700	On slabs, 3 ply, glass fiber fabric, mopped		1550	.036		1.07	1.16	.28	2.51	3.55
0710	Asphaltic hardboard protection board, 1/8" thick	2 Rofc	500	.032		.40	1.10		1.50	2.40
1000	1/4" EPS membrane protection board		3500	.005		.21	.16		.37	.51
1050	3/8" thick		3500	.005		.23	.16		.39	.53
1060	1/2" thick		3500	.005		.26	.16		.42	.57
1070	Fiberglass fabric, black, 20/10 mesh		116	.138	Sq.	18	4.72		22.72	28.50
1080	White, 20/10 mesh		116	.138	"	18	4.72		22.72	28.50
9000	Minimum labor/equipment charge		2	8	Job		274		274	490

07 13 Sheet Waterproofing

07 13 53 – Elastomeric Sheet Waterproofing

07 13 53.10 Elastomeric Sheet Waterproofing and Access.		Crew	Daily Output	Labor-Hours	Unit	Material	Labor (2009 Bare Costs)	Equipment	Total	Total Incl O&P
0010	**ELASTOMERIC SHEET WATERPROOFING AND ACCESS.**									
0090	EPDM, plain, 45 mils thick	2 Rofc	580	.028	S.F.	1.19	.95		2.14	3
0100	60 mils thick		570	.028		1.25	.96		2.21	3.09
0300	Nylon reinforced sheets, 45 mils thick		580	.028		1.16	.95		2.11	2.96
0400	60 mils thick		570	.028		1.47	.96		2.43	3.33
0600	Vulcanizing splicing tape for above, 2" wide				C.L.F.	43.50			43.50	47.50
0700	4" wide				"	99.50			99.50	109
0900	Adhesive, bonding, 60 SF per gal				Gal.	16.50			16.50	18.15
1000	Splicing, 75 SF per gal				"	32			32	35
1200	Neoprene sheets, plain, 45 mils thick	2 Rofc	580	.028	S.F.	1.68	.95		2.63	3.54
1300	60 mils thick		570	.028		2.86	.96		3.82	4.87
1500	Nylon reinforced, 45 mils thick		580	.028		1.75	.95		2.70	3.62
1600	60 mils thick		570	.028		2.46	.96		3.42	4.43
1800	120 mils thick		500	.032		4	1.10		5.10	6.35
1900	Adhesive, splicing, 150 S.F. per gal. per coat				Gal.	17.50			17.50	19.25
2100	Fiberglass reinforced, fluid applied, 1/8" thick	2 Rofc	500	.032	S.F.	1.88	1.10		2.98	4.03
2200	Polyethylene and rubberized asphalt sheets, 1/8" thick		550	.029		.73	1		1.73	2.58
2210	Asphaltic hardboard protection board, 1/8" thick		500	.032		.40	1.10		1.50	2.40
2220	1/4" thick		450	.036		.68	1.22		1.90	2.92
2400	Polyvinyl chloride sheets, plain, 10 mils thick		580	.028		.17	.95		1.12	1.88
2500	20 mils thick		570	.028		.28	.96		1.24	2.03
2700	30 mils thick		560	.029		.39	.98		1.37	2.18
3000	Adhesives, trowel grade, 40-100 SF per gal				Gal.	27			27	29.50
3100	Brush grade, 100-250 SF per gal.				"	27			27	29.50
3300	Bitumen modified polyurethane, fluid applied, 55 mils thick	2 Rofc	665	.024	S.F.	.68	.82		1.50	2.22
3600	Vinyl plastic, sprayed on, 25 to 40 mils thick		475	.034	"	1.20	1.15		2.35	3.38
9000	Minimum labor/equipment charge		2	8	Job		274		274	490

07 16 Cementitious and Reactive Waterproofing

07 16 16 – Crystalline Waterproofing

07 16 16.20 Cementitious Waterproofing

		Crew	Daily Output	Labor-Hours	Unit	Material	2009 Bare Costs Labor	Equipment	Total	Total Incl O&P
0010	**CEMENTITIOUS WATERPROOFING**									
0020	1/8" application, sprayed on	G-2A	1000	.024	S.F.	1.62	.73	.61	2.96	3.72
0030	2 coat, cementitious/metallic slurry, troweled, 1/4" thick	1 Cefi	2.48	3.226	C.S.F.	24	124		148	220
0040	3 coat, 3/8" thick		1.84	4.348		42.50	167		209.50	305
0050	4 coat, 1/2" thick	↓	1.20	6.667	↓	56.50	255		311.50	460

07 17 Bentonite Waterproofing

07 17 13 – Bentonite Panel Waterproofing

07 17 13.10 Bentonite

		Crew	Daily Output	Labor-Hours	Unit	Material	Labor	Equipment	Total	Total Incl O&P
0010	**BENTONITE**									
0020	Panels, 4' x 4', 3/16" thick	1 Rofc	625	.013	S.F.	1	.44		1.44	1.88
0100	Rolls, 3/8" thick, with geotextile fabric both sides	"	550	.015	"	1.17	.50		1.67	2.18
0300	Granular bentonite, 50 lb. bags (.625 C.F.)				Bag	26.50			26.50	29.50
0400	3/8" thick, troweled on	1 Rofc	475	.017	S.F.	1.34	.58		1.92	2.50
0500	Drain board, expanded polystyrene, binder encapsulated, 1-1/2" thick	1 Rohe	1600	.005		1.16	.13		1.29	1.51
0510	2" thick		1600	.005		1.56	.13		1.69	1.95
0520	3" thick		1600	.005		2.33	.13		2.46	2.79
0530	4" thick		1600	.005		3.09	.13		3.22	3.63
0600	With filter fabric, 1-1/2" thick		1600	.005		1.41	.13		1.54	1.78
0625	2" thick		1600	.005		1.85	.13		1.98	2.27
0650	3" thick		1600	.005		2.63	.13		2.76	3.12
0675	4" thick	↓	1600	.005	↓	3.41	.13		3.54	3.98

07 19 Water Repellents

07 19 19 – Silicone Water Repellents

07 19 19.10 Silicone Based Water Repellents

		Crew	Daily Output	Labor-Hours	Unit	Material	Labor	Equipment	Total	Total Incl O&P
0010	**SILICONE BASED WATER REPELLENTS**									
0020	Water base liquid, roller applied	2 Rofc	7000	.002	S.F.	.65	.08		.73	.86
0200	Silicone or stearate, sprayed on CMU, 1 coat	1 Rofc	4000	.002		.36	.07		.43	.52
0300	2 coats		3000	.003	↓	.73	.09		.82	.96
9000	Minimum labor/equipment charge	↓	3	2.667	Job		91.50		91.50	163

07 21 Thermal Insulation

07 21 13 – Board Insulation

07 21 13.10 Rigid Insulation

		Crew	Daily Output	Labor-Hours	Unit	Material	Labor	Equipment	Total	Total Incl O&P
0010	**RIGID INSULATION**, for walls									
0040	Fiberglass, 1.5#/CF, unfaced, 1" thick, R4.1 G	1 Carp	1000	.008	S.F.	.44	.32		.76	1.01
0060	1-1/2" thick, R6.2 G		1000	.008		.64	.32		.96	1.23
0080	2" thick, R8.3 G		1000	.008		.69	.32		1.01	1.29
0120	3" thick, R12.4 G		800	.010		.80	.40		1.20	1.54
0370	3#/CF, unfaced, 1" thick, R4.3 G		1000	.008		.49	.32		.81	1.07
0390	1-1/2" thick, R6.5 G		1000	.008		.95	.32		1.27	1.58
0400	2" thick, R8.7 G		890	.009		1.15	.36		1.51	1.86
0420	2-1/2" thick, R10.9 G		800	.010		1.01	.40		1.41	1.77
0440	3" thick, R13 G		800	.010		1.01	.40		1.41	1.77
0520	Foil faced, 1" thick, R4.3 G		1000	.008		.92	.32		1.24	1.54
0540	1-1/2" thick, R6.5 G	↓	1000	.008	↓	1.36	.32		1.68	2.03

07 21 Thermal Insulation

07 21 13 – Board Insulation

07 21 13.10 Rigid Insulation			Crew	Daily Output	Labor-Hours	Unit	2009 Bare Costs Material	Labor	Equipment	Total	Total Incl O&P
0560	2" thick, R8.7	G	1 Carp	890	.009	S.F.	1.71	.36		2.07	2.47
0580	2-1/2" thick, R10.9	G		800	.010		2.02	.40		2.42	2.88
0600	3" thick, R13	G		800	.010		2.19	.40		2.59	3.07
0670	6#/CF, unfaced, 1" thick, R4.3	G		1000	.008		.98	.32		1.30	1.61
0690	1-1/2" thick, R6.5	G		890	.009		1.50	.36		1.86	2.24
0700	2" thick, R8.7	G		800	.010		2.12	.40		2.52	2.99
0721	2-1/2" thick, R10.9	G		800	.010		2.32	.40		2.72	3.21
0741	3" thick, R13	G		730	.011		2.78	.44		3.22	3.78
0821	Foil faced, 1" thick, R4.3	G		1000	.008		1.38	.32		1.70	2.05
0840	1-1/2" thick, R6.5	G		890	.009		1.98	.36		2.34	2.77
0850	2" thick, R8.7	G		800	.010		2.59	.40		2.99	3.51
0880	2-1/2" thick, R10.9	G		800	.010		3.11	.40		3.51	4.08
0900	3" thick, R13	G		730	.011		3.72	.44		4.16	4.81
1500	Foamglass, 1-1/2" thick, R4.5	G		800	.010		1.37	.40		1.77	2.17
1550	3" thick, R9	G		730	.011		3.29	.44		3.73	4.34
1700	Perlite, 1" thick, R2.77	G		800	.010		.30	.40		.70	.99
1750	2" thick, R5.55	G		730	.011		.60	.44		1.04	1.38
1900	Extruded polystyrene, 25 PSI compressive strength, 1" thick, R5	G		800	.010		.51	.40		.91	1.22
1940	2" thick R10	G		730	.011		1.04	.44		1.48	1.86
1960	3" thick, R15	G		730	.011		1.40	.44		1.84	2.26
2100	Expanded polystyrene, 1" thick, R3.85	G		800	.010		.24	.40		.64	.92
2120	2" thick, R7.69	G		730	.011		.61	.44		1.05	1.39
2140	3" thick, R11.49	G		730	.011		.78	.44		1.22	1.58
9000	Minimum labor/equipment charge			4	2	Job		80		80	132

07 21 13.13 Foam Board Insulation			Crew	Daily Output	Labor-Hours	Unit	Material	Labor	Equipment	Total	Total Incl O&P
0010	**FOAM BOARD INSULATION**										
0600	Polystyrene, expanded, 1" thick, R4	G	1 Carp	680	.012	S.F.	.31	.47		.78	1.12
0700	2" thick, R8	G		675	.012	"	.62	.47		1.09	1.46
9000	Minimum labor/equipment charge			4	2	Job		80		80	132

07 21 16 – Blanket Insulation

07 21 16.20 Blanket Insulation for Walls			Crew	Daily Output	Labor-Hours	Unit	Material	Labor	Equipment	Total	Total Incl O&P
0010	**BLANKET INSULATION FOR WALLS**										
0020	Kraft faced fiberglass, 3-1/2" thick, R11, 15" wide	G	1 Carp	1350	.006	S.F.	.34	.24		.58	.76
0030	23" wide	G		1600	.005		.34	.20		.54	.70
0060	R13, 11" wide	G		1150	.007		.34	.28		.62	.83
0080	15" wide	G		1350	.006		.34	.24		.58	.76
0100	23" wide	G		1600	.005		.34	.20		.54	.70
0110	R-15, 11" wide	G		1150	.007		.41	.28		.69	.91
0120	15" wide	G		1350	.006		.41	.24		.65	.84
0130	23" wide	G		1600	.005		.41	.20		.61	.78
0140	6" thick, R19, 11" wide	G		1150	.007		.44	.28		.72	.94
0160	15" wide	G		1350	.006		.44	.24		.68	.87
0180	23" wide	G		1600	.005		.44	.20		.64	.81
0182	R21, 11" wide	G		1150	.007		.50	.28		.78	1.01
0184	15" wide	G		1350	.006		.50	.24		.74	.94
0186	23" wide	G		1600	.005		.50	.20		.70	.88
0188	9" thick, R-30, 11" wide	G		985	.008		.66	.32		.98	1.27
0200	15" wide	G		1150	.007		.66	.28		.94	1.19
0220	23" wide	G		1350	.006		.66	.24		.90	1.12
0230	12" thick, R38, 11" wide	G		985	.008		.95	.32		1.27	1.59
0240	15" wide	G		1150	.007		.95	.28		1.23	1.51
0260	23" wide	G		1350	.006		.95	.24		1.19	1.44

07 21 16 – Blanket Insulation

07 21 16.20 Blanket Insulation for Walls			Crew	Daily Output	Labor-Hours	Unit	Material	2009 Bare Costs Labor	Equipment	Total	Total Incl O&P
0410	Foil faced fiberglass, 3-1/2" thick, R13, 11" wide	G	1 Carp	1150	.007	S.F.	.57	.28		.85	1.09
0420	15" wide	G		1350	.006		.57	.24		.81	1.02
0440	23" wide	G		1600	.005		.57	.20		.77	.96
0442	R15, 11" wide	G		1150	.007		.56	.28		.84	1.08
0444	15" wide	G		1350	.006		.56	.24		.80	1.01
0446	23" wide	G		1600	.005		.56	.20		.76	.95
0448	6" thick, R19, 11" wide	G		1150	.007		.77	.28		1.05	1.31
0460	15" wide	G		1350	.006		.77	.24		1.01	1.24
0480	23" wide	G		1600	.005		.77	.20		.97	1.18
0482	R-21, 11" wide	G		1150	.007		.82	.28		1.10	1.36
0484	15" wide	G		1350	.006		.82	.24		1.06	1.29
0486	23" wide	G		1600	.005		.82	.20		1.02	1.23
0488	9" thick, R-30, 11" wide	G		985	.008		.89	.32		1.21	1.52
0500	9" thick, R30, 15" wide	G		1150	.007		.89	.28		1.17	1.44
0550	23" wide	G		1350	.006		.89	.24		1.13	1.37
0560	12" thick, R-38, 11" wide	G		985	.008		.90	.32		1.22	1.53
0570	15" wide	G		1150	.007		.90	.28		1.18	1.45
0580	23" wide	G		1350	.006		.90	.24		1.14	1.38
0620	Unfaced fiberglass, 3-1/2" thick, R-13, 11" wide	G		1150	.007		.29	.28		.57	.78
0820	15" wide	G		1350	.006		.29	.24		.53	.71
0830	23" wide	G		1600	.005		.29	.20		.49	.65
0832	R15, 11" wide	G		1150	.007		.37	.28		.65	.87
0836	23" wide	G		1600	.005		.37	.20		.57	.74
0838	6" thick, R19, 11" wide	G		1150	.007		.42	.28		.70	.92
0860	15" wide	G		1150	.007		.42	.28		.70	.92
0880	23" wide	G		1350	.006		.42	.24		.66	.85
0882	R-21, 11" wide	G		1150	.007		.46	.28		.74	.97
0886	15" wide	G		1350	.006		.46	.24		.70	.90
0888	23" wide	G		1600	.005		.46	.20		.66	.84
0890	9" thick, R30, 11" wide	G		985	.008		.75	.32		1.07	1.37
0900	15" wide	G		1150	.007		.75	.28		1.03	1.29
0920	23" wide	G		1350	.006		.75	.24		.99	1.22
0930	12" thick, R38, 11" wide	G		985	.008		1.28	.32		1.60	1.95
0940	15" wide	G		1000	.008		1.28	.32		1.60	1.94
0960	23" wide	G		1150	.007		1.28	.28		1.56	1.87
1300	Mineral fiber batts, kraft faced										
1320	3-1/2" thick, R12	G	1 Carp	1600	.005	S.F.	.38	.20		.58	.75
1340	6" thick, R19	G		1600	.005		.48	.20		.68	.86
1380	10" thick, R30	G		1350	.006		.73	.24		.97	1.19
1850	Friction fit wire insulation supports, 16" O.C.			960	.008	Ea.	.08	.33		.41	.64
9000	Minimum labor/equipment charge			4	2	Job		80		80	132

07 21 23 – Loose-Fill Insulation

07 21 23.10 Poured Loose-Fill Insulation			Crew	Daily Output	Labor-Hours	Unit	Material	Labor	Equipment	Total	Total Incl O&P
0010	**POURED LOOSE-FILL INSULATION**										
0020	Cellulose fiber, R3.8 per inch	G	1 Carp	200	.040	C.F.	.60	1.60		2.20	3.30
0040	Ceramic type (perlite), R3.2 per inch	G		200	.040		1.75	1.60		3.35	4.57
0080	Fiberglass wool, R4 per inch	G		200	.040		.46	1.60		2.06	3.15
0100	Mineral wool, R3 per inch	G		200	.040		.39	1.60		1.99	3.07
0300	Polystyrene, R4 per inch	G		200	.040		3.09	1.60		4.69	6.05
0400	Vermiculite or perlite, R2.7 per inch	G		200	.040		1.75	1.60		3.35	4.57
0700	Wood fiber, R3.85 per inch	G		200	.040		.70	1.60		2.30	3.41
9000	Minimum labor/equipment charge			4	2	Job		80		80	132

07 21 23 – Loose-Fill Insulation

07 21 23.20 Masonry Loose-Fill Insulation			Crew	Daily Output	Labor-Hours	Unit	2009 Bare Costs Material	Labor	Equipment	Total	Total Incl O&P
0010	**MASONRY LOOSE-FILL INSULATION**, vermiculite or perlite										
0100	In cores of concrete block, 4" thick wall, .115 CF/SF	G	D-1	4800	.003	S.F.	.20	.12		.32	.42
0200	6" thick wall, .175 CF/SF	G		3000	.005		.31	.19		.50	.65
0300	8" thick wall, .258 CF/SF	G		2400	.007		.45	.24		.69	.89
0400	10" thick wall, .340 CF/SF	G		1850	.009		.60	.31		.91	1.16
0500	12" thick wall, .422 CF/SF	G		1200	.013		.74	.48		1.22	1.59
0550	For sand fill, deduct from above	G					70%				
0600	Poured cavity wall, vermiculite or perlite, water repellant	G	D-1	250	.064	C.F.	1.75	2.33		4.08	5.70
0700	Foamed in place, urethane in 2-5/8" cavity	G	G-2A	1035	.023	S.F.	.42	.71	.58	1.71	2.33
0800	For each 1" added thickness, add	G	"	2372	.010	"	.12	.31	.26	.69	.94

07 21 26 – Blown Insulation

07 21 26.10 Blown Insulation			Crew	Daily Output	Labor-Hours	Unit	Material	Labor	Equipment	Total	Total Incl O&P
0010	**BLOWN INSULATION** Ceilings, with open access										
0020	Cellulose, 3-1/2" thick, R13	G	G-4	5000	.005	S.F.	.19	.15	.07	.41	.54
0030	5-3/16" thick, R19	G		3800	.006		.28	.20	.09	.57	.75
0050	6-1/2" thick, R22	G		3000	.008		.36	.26	.11	.73	.95
0100	8-11/16" thick, R30	G		2600	.009		.48	.30	.13	.91	1.16
0120	10-7/8" thick, R38	G		1800	.013		.62	.43	.19	1.24	1.59
1000	Fiberglass, 5.5" thick, R11	G		3800	.006		.19	.20	.09	.48	.65
1050	6" thick, R12	G		3000	.008		.23	.26	.11	.60	.80
1100	8.8" thick, R19	G		2200	.011		.33	.35	.15	.83	1.11
1200	10" thick, R22	G		1800	.013		.38	.43	.19	1	1.33
1300	11.5" thick, R26	G		1500	.016		.46	.52	.22	1.20	1.60
1350	13" thick, R30	G		1400	.017		.50	.55	.24	1.29	1.72
1450	16" thick, R38	G		1145	.021		.61	.68	.29	1.58	2.11
1500	20" thick, R49	G		920	.026		.77	.84	.36	1.97	2.63
2000	Mineral wool, 4" thick, R12	G		3500	.007		.23	.22	.10	.55	.72
2050	6" thick, R17	G		2500	.010		.26	.31	.13	.70	.95
2100	9" thick, R23	G		1750	.014		.34	.44	.19	.97	1.31
2500	Wall installation, incl. drilling & patching from outside, two 1"										
2510	diam. holes @ 16" O.C., top & mid-point of wall, add to above										
2700	For masonry	G	G-4	415	.058	S.F.	.06	1.87	.80	2.73	4.03
2800	For wood siding	G		840	.029		.06	.92	.40	1.38	2.03
2900	For stucco/plaster	G		665	.036		.06	1.16	.50	1.72	2.54
9000	Minimum labor/equipment charge			4	6	Job		194	83.50	277.50	410

07 21 27 – Reflective Insulation

07 21 27.10 Reflective Insulation Options			Crew	Daily Output	Labor-Hours	Unit	Material	Labor	Equipment	Total	Total Incl O&P
0010	**REFLECTIVE INSULATION OPTIONS**										
0020	Aluminum foil on reinforced scrim	G	1 Carp	19	.421	C.S.F.	14.20	16.80		31	43.50
0100	Reinforced with woven polyolefin	G		19	.421		17.20	16.80		34	47
0500	With single bubble air space, R8.8	G		15	.533		25.50	21.50		47	63
0600	With double bubble air space, R9.8	G		15	.533		26.50	21.50		48	64
9000	Minimum labor/equipment charge			4	2	Job		80		80	132

07 21 29 – Sprayed Insulation

07 21 29.10 Sprayed-On Insulation			Crew	Daily Output	Labor-Hours	Unit	Material	Labor	Equipment	Total	Total Incl O&P
0010	**SPRAYED-ON INSULATION**										
0020	Fibrous/cementitious, finished wall, 1" thick, R3.7	G	G-2	2050	.012	S.F.	.26	.39	.06	.71	.99
0100	Attic, 5.2" thick, R19	G	"	1550	.015	"	.42	.52	.08	1.02	1.39
0300	Closed cell, spray polyurethane foam, 2 pounds per cubic foot density										
0310	1" thick	G	G-2A	6000	.004	S.F.	.70	.12	.10	.92	1.09

07 21 Thermal Insulation

07 21 29 – Sprayed Insulation

07 21 29.10 Sprayed-On Insulation			Crew	Daily Output	Labor-Hours	Unit	Material	2009 Bare Costs Labor	Equipment	Total	Total Incl O&P
0320	2″ thick	G	G-2A	3000	.008	S.F.	1.40	.24	.20	1.84	2.18
0330	3″ thick	G		2000	.012		2.11	.36	.30	2.77	3.28
0335	3-1/2″ thick	G		1715	.014		2.45	.43	.35	3.23	3.82
0340	4″ thick	G		1500	.016		2.80	.49	.40	3.69	4.37
0350	5″ thick	G		1200	.020		3.50	.61	.50	4.61	5.45
0355	5-1/2″ thick	G		1090	.022		3.85	.67	.56	5.08	6
0360	6″ thick	G		1000	.024		4.22	.73	.61	5.56	6.60
9000	Minimum labor/equipment charge		G-2	2	12	Job		400	63.50	463.50	720

07 22 Roof and Deck Insulation

07 22 16 – Roof Board Insulation

07 22 16.10 Roof Deck Insulation			Crew	Daily Output	Labor-Hours	Unit	Material	Labor	Equipment	Total	Total Incl O&P
0010	**ROOF DECK INSULATION**										
0020	Fiberboard low density, 1/2″ thick R1.39	G	1 Rofc	1000	.008	S.F.	.24	.27		.51	.75
0030	1″ thick R2.78	G		800	.010		.42	.34		.76	1.07
0080	1 1/2″ thick R4.17	G		800	.010		.64	.34		.98	1.31
0100	2″ thick R5.56	G		800	.010		.85	.34		1.19	1.55
0110	Fiberboard high density, 1/2″ thick R1.3	G		1000	.008		.22	.27		.49	.73
0120	1″ thick R2.5	G		800	.010		.44	.34		.78	1.09
0130	1-1/2″ thick R3.8	G		800	.010		.66	.34		1	1.34
0200	Fiberglass, 3/4″ thick R2.78	G		1000	.008		.56	.27		.83	1.11
0400	15/16″ thick R3.70	G		1000	.008		.74	.27		1.01	1.30
0460	1-1/16″ thick R4.17	G		1000	.008		.93	.27		1.20	1.51
0600	1-5/16″ thick R5.26	G		1000	.008		1.26	.27		1.53	1.88
0650	2-1/16″ thick R8.33	G		800	.010		1.35	.34		1.69	2.10
0700	2-7/16″ thick R10	G		800	.010		1.55	.34		1.89	2.32
1500	Foamglass, 1-1/2″ thick R4.5	G		800	.010		1.35	.34		1.69	2.10
1530	3″ thick R9	G		700	.011		2.75	.39		3.14	3.73
1600	Tapered for drainage	G		600	.013	B.F.	1.16	.46		1.62	2.10
1650	Perlite, 1/2″ thick R1.32	G		1050	.008	S.F.	.34	.26		.60	.84
1655	3/4″ thick R2.08	G		800	.010		.37	.34		.71	1.02
1660	1″ thick R2.78	G		800	.010		.46	.34		.80	1.12
1670	1-1/2″ thick R4.17	G		800	.010		.48	.34		.82	1.14
1680	2″ thick R5.56	G		700	.011		.80	.39		1.19	1.58
1685	2-1/2″ thick R6.67	G		700	.011		.95	.39		1.34	1.75
1690	Tapered for drainage	G		800	.010	B.F.	.73	.34		1.07	1.41
1700	Polyisocyanurate, 2#/CF density, 3/4″ thick, R5.1	G		1500	.005	S.F.	.44	.18		.62	.81
1705	1″ thick R7.14	G		1400	.006		.50	.20		.70	.90
1715	1-1/2″ thick R10.87	G		1250	.006		.63	.22		.85	1.08
1725	2″ thick R14.29	G		1100	.007		.81	.25		1.06	1.33
1735	2-1/2″ thick R16.67	G		1050	.008		1.01	.26		1.27	1.58
1745	3″ thick R21.74	G		1000	.008		1.26	.27		1.53	1.88
1755	3-1/2″ thick R25	G		1000	.008		1.93	.27		2.20	2.61
1765	Tapered for drainage	G		1400	.006	B.F.	1.93	.20		2.13	2.47
1900	Extruded Polystyrene										
1910	15 PSI compressive strength, 1″ thick, R5	G	1 Rofc	1500	.005	S.F.	.49	.18		.67	.87
1920	2″ thick, R10	G		1250	.006		.62	.22		.84	1.07
1930	3″ thick R15	G		1000	.008		1.27	.27		1.54	1.89
1932	4″ thick R20	G		1000	.008		1.64	.27		1.91	2.29
1934	Tapered for drainage	G		1500	.005	B.F.	.53	.18		.71	.91
1940	25 PSI compressive strength, 1″ thick R5	G		1500	.005	S.F.	.67	.18		.85	1.07

07 22 Roof and Deck Insulation

07 22 16 – Roof Board Insulation

07 22 16.10 Roof Deck Insulation			Crew	Daily Output	Labor-Hours	Unit	Material	Labor (2009 Bare Costs)	Equipment	Total	Total Incl O&P
1942	2" thick R10	G	1 Rofc	1250	.006	S.F.	1.27	.22		1.49	1.79
1944	3" thick R15	G		1000	.008		1.93	.27		2.20	2.61
1946	4" thick R20	G		1000	.008		2.72	.27		2.99	3.48
1948	Tapered for drainage	G		1500	.005	B.F.	.56	.18		.74	.95
1950	40 psi compressive strength, 1" thick R5	G		1500	.005	S.F.	.51	.18		.69	.89
1952	2" thick R10	G		1250	.006		.96	.22		1.18	1.45
1954	3" thick R15	G		1000	.008		1.40	.27		1.67	2.03
1956	4" thick R20	G		1000	.008		1.88	.27		2.15	2.56
1958	Tapered for drainage	G		1400	.006	B.F.	.71	.20		.91	1.13
1960	60 PSI compressive strength, 1" thick R5	G		1450	.006	S.F.	.59	.19		.78	.99
1962	2" thick R10	G		1200	.007		1.05	.23		1.28	1.57
1964	3" thick R15	G		975	.008		1.55	.28		1.83	2.21
1966	4" thick R20	G		950	.008		2.17	.29		2.46	2.90
1968	Tapered for drainage	G		1400	.006	B.F.	.85	.20		1.05	1.29
2010	Expanded polystyrene, 1#/CF density, 3/4" thick R2.89	G		1500	.005	S.F.	.31	.18		.49	.67
2020	1" thick R3.85	G		1500	.005		.31	.18		.49	.67
2100	2" thick R7.69	G		1250	.006		.62	.22		.84	1.07
2110	3" thick R11.49	G		1250	.006		.91	.22		1.13	1.39
2120	4" thick R15.38	G		1200	.007		.98	.23		1.21	1.49
2130	5" thick R19.23	G		1150	.007		1.05	.24		1.29	1.59
2140	6" thick R23.26	G		1150	.007		1.23	.24		1.47	1.78
2150	Tapered for drainage	G		1500	.005	B.F.	.49	.18		.67	.87
2400	Composites with 2" EPS										
2410	1" fiberboard	G	1 Rofc	950	.008	S.F.	1.07	.29		1.36	1.69
2420	7/16" oriented strand board	G		800	.010		1.26	.34		1.60	2
2430	1/2" plywood	G		800	.010		1.37	.34		1.71	2.12
2440	1" perlite	G		800	.010		1.12	.34		1.46	1.84
2450	Composites with 1-1/2" polyisocyanurate										
2460	1" fiberboard	G	1 Rofc	800	.010	S.F.	1.46	.34		1.80	2.22
2470	1" perlite	G		850	.009		1.53	.32		1.85	2.26
2480	7/16" oriented strand board	G		800	.010		1.77	.34		2.11	2.56
9000	Minimum labor/equipment charge			3.25	2.462	Job		84.50		84.50	151

07 24 Exterior Insulation and Finish Systems

07 24 13 – Polymer Based Exterior Insulation and Finish Systems

07 24 13.10 Exterior Insulation and Finish Systems			Crew	Daily Output	Labor-Hours	Unit	Material	Labor (2009 Bare Costs)	Equipment	Total	Total Incl O&P
0010	EXTERIOR INSULATION AND FINISH SYSTEMS										
0095	Field applied, 1" EPS insulation	G	J-1	295	.136	S.F.	2.18	4.69	.43	7.30	10.40
0100	With 1/2" cement board sheathing	G		220	.182		2.90	6.30	.58	9.78	13.90
0105	2" EPS insulation	G		295	.136		2.55	4.69	.43	7.67	10.85
0110	With 1/2" cement board sheathing	G		220	.182		3.27	6.30	.58	10.15	14.35
0115	3" EPS insulation	G		295	.136		2.72	4.69	.43	7.84	11
0120	With 1/2" cement board sheathing	G		220	.182		3.44	6.30	.58	10.32	14.50
0125	4" EPS insulation	G		295	.136		3.16	4.69	.43	8.28	11.50
0130	With 1/2" cement board sheathing	G		220	.182		4.60	6.30	.58	11.48	15.80
0140	Premium finish add			1265	.032		.31	1.09	.10	1.50	2.21
0150	Heavy duty reinforcement add			914	.044		1.08	1.51	.14	2.73	3.78
0160	2.5#/S.Y. metal lath substrate add		1 Lath	75	.107	S.Y.	2.37	3.79		6.16	8.60
0170	3.4#/S.Y. metal lath substrate add		"	75	.107	"	2.58	3.79		6.37	8.85
0180	Color or texture change,		J-1	1265	.032	S.F.	.83	1.09	.10	2.02	2.78
0190	With substrate leveling base coat		1 Plas	530	.015		.83	.55		1.38	1.79

07 24 Exterior Insulation and Finish Systems

07 24 13 – Polymer Based Exterior Insulation and Finish Systems

07 24 13.10 Exterior Insulation and Finish Systems		Crew	Daily Output	Labor-Hours	Unit	2009 Bare Costs Material	Labor	Equipment	Total	Total Incl O&P
0210	With substrate sealing base coat	1 Pord	1224	.007	S.F.	.08	.23		.31	.46
0370	V groove shape in panel face				L.F.	.60			.60	.66
0380	U groove shape in panel face				″	.79			.79	.87
0433	Crack repair, acrylic rubber, fluid applied, 20 mils thick	1 Plas	350	.023	S.F.	1.43	.83		2.26	2.90
0437	50 mils thick, reinforced	″	200	.040	″	2.62	1.45		4.07	5.20
0440	For higher than one story, add						25%			

07 26 Vapor Retarders

07 26 10 – Vapor Retarders Made of Natural and Synthetic Materials

07 26 10.10 Vapor Retarders

Line	Description		Crew	Daily Output	Labor-Hours	Unit	Material	Labor	Equipment	Total	Total Incl O&P
0010	**VAPOR RETARDERS**										
0020	Aluminum and kraft laminated, foil 1 side	G	1 Carp	37	.216	Sq.	5.20	8.65		13.85	20
0100	Foil 2 sides	G		37	.216		8.70	8.65		17.35	24
0400	Asphalt felt sheathing paper, 15#			37	.216		4.77	8.65		13.42	19.50
0450	Housewrap, exterior, spun bonded polypropylene										
0470	Small roll	G	1 Carp	3800	.002	S.F.	.23	.08		.31	.39
0480	Large roll	G	″	4000	.002	″	.12	.08		.20	.26
0500	Material only, 3′ x 111.1′ roll	G				Ea.	75			75	82.50
0520	9′ x 111.1′ roll	G				″	120			120	132
0600	Polyethylene vapor barrier, standard, .002″ thick	G	1 Carp	37	.216	Sq.	1.09	8.65		9.74	15.45
0700	.004″ thick	G		37	.216		3.54	8.65		12.19	18.15
0900	.006″ thick	G		37	.216		5.55	8.65		14.20	20.50
1200	.010″ thick	G		37	.216		6.70	8.65		15.35	21.50
1300	Clear reinforced, fire retardant, .008″ thick	G		37	.216		10.15	8.65		18.80	25.50
1350	Cross laminated type, .003″ thick	G		37	.216		7.10	8.65		15.75	22
1400	.004″ thick	G		37	.216		7.80	8.65		16.45	23
1500	Red rosin paper, 5 sq rolls, 4 lbs. per square			37	.216		2.19	8.65		10.84	16.65
1600	5 lbs. per square			37	.216		2.85	8.65		11.50	17.40
1800	Reinf. waterproof, .002″ polyethylene backing, 1 side			37	.216		5.65	8.65		14.30	20.50
1900	2 sides			37	.216		7.45	8.65		16.10	22.50
2100	Asphalt felt roof deck vapor barrier, class 1 metal decks		1 Rofc	37	.216		18.65	7.40		26.05	33.50
2200	For all other decks		″	37	.216		13.85	7.40		21.25	28.50
2400	Waterproofed kraft with sisal or fiberglass fibers, minimum		1 Carp	37	.216		6.10	8.65		14.75	21
2500	Maximum			37	.216		15.10	8.65		23.75	31
9950	Minimum labor/equipment charge			4	2	Job		80		80	132

07 31 Shingles and Shakes

07 31 13 – Asphalt Shingles

07 31 13.10 Asphalt Roof Shingles

Line	Description	Crew	Daily Output	Labor-Hours	Unit	Material	Labor	Equipment	Total	Total Incl O&P
0010	**ASPHALT ROOF SHINGLES**									
0100	Standard strip shingles									
0150	Inorganic, class A, 210-235 lb/sq	1 Rofc	5.50	1.455	Sq.	50	50		100	144
0155	Pneumatic nailed		7	1.143		50	39		89	125
0200	Organic, class C, 235-240 lb/sq		5	1.600		51	55		106	154
0205	Pneumatic nailed		6.25	1.280		51	44		95	135
0250	Standard, laminated multi-layered shingles									
0300	Class A, 240-260 lb/sq	1 Rofc	4.50	1.778	Sq.	61.50	61		122.50	177
0305	Pneumatic nailed		5.63	1.422		61.50	48.50		110	155
0350	Class C, 260-300 lb/square, 4 bundles/square		4	2		67	68.50		135.50	196

07 31 Shingles and Shakes

07 31 13 – Asphalt Shingles

07 31 13.10 Asphalt Roof Shingles		Crew	Daily Output	Labor-Hours	Unit	Material	Labor	Equipment	Total	Total Incl O&P
						2009 Bare Costs				
0355	Pneumatic nailed	1 Rofc	5	1.600	Sq.	67	55		122	172
0400	Premium, laminated multi-layered shingles									
0450	Class A, 260-300 lb, 4 bundles/sq	1 Rofc	3.50	2.286	Sq.	80	78.50		158.50	228
0455	Pneumatic nailed		4.37	1.831		80	62.50		142.50	200
0500	Class C, 300-385 lb/square, 5 bundles/square		3	2.667		106	91.50		197.50	280
0505	Pneumatic nailed		3.75	2.133		106	73		179	247
0800	#15 felt underlayment		64	.125		4.77	4.28		9.05	12.90
0825	#30 felt underlayment		58	.138		9.55	4.72		14.27	18.95
0850	Self adhering polyethylene and rubberized asphalt underlayment		22	.364		55.50	12.45		67.95	83
0900	Ridge shingles		330	.024	L.F.	1.49	.83		2.32	3.12
0905	Pneumatic nailed		412.50	.019	"	1.49	.66		2.15	2.83
1000	For steep roofs (7 to 12 pitch or greater), add						50%			
9000	Minimum labor/equipment charge	1 Rofc	3	2.667	Job		91.50		91.50	163

07 31 16 – Metal Shingles

07 31 16.10 Aluminum Shingles		Crew	Daily Output	Labor-Hours	Unit	Material	Labor	Equipment	Total	Total Incl O&P
0010	**ALUMINUM SHINGLES**									
0020	Mill finish, .019 thick	1 Carp	5	1.600	Sq.	214	64		278	340
0100	.020" thick	"	5	1.600		208	64		272	335
0300	For colors, add					16			16	17.60
0600	Ridge cap, .024" thick	1 Carp	170	.047	L.F.	2.33	1.88		4.21	5.65
0700	End wall flashing, .024" thick		170	.047		1.80	1.88		3.68	5.10
0900	Valley section, .024" thick		170	.047		3.09	1.88		4.97	6.50
1000	Starter strip, .024" thick		400	.020		1.56	.80		2.36	3.04
1200	Side wall flashing, .024" thick		170	.047		1.81	1.88		3.69	5.10
9000	Minimum labor/equipment charge		3	2.667	Job		107		107	176

07 31 16.20 Steel Shingles		Crew	Daily Output	Labor-Hours	Unit	Material	Labor	Equipment	Total	Total Incl O&P
0010	**STEEL SHINGLES**									
0012	Galvanized, 26 gauge	1 Rots	2.20	3.636	Sq.	232	124		356	475
0200	24 gauge	"	2.20	3.636		244	124		368	490
0300	For colored galvanized shingles, add					55			55	60.50
0500	For 1" factory applied polystyrene insulation, add					40			40	44
9000	Minimum labor/equipment charge	1 Rots	3	2.667	Job		91		91	162

07 31 19 – Mineral-Fiber Cement Shingles

07 31 19.10 Fiber Cement Shingles		Crew	Daily Output	Labor-Hours	Unit	Material	Labor	Equipment	Total	Total Incl O&P
0010	**FIBER CEMENT SHINGLES**									
0012	Field shingles, 16" x 9.35", 500 lb per square	1 Rofc	2.20	3.636	Sq.	335	125		460	590
0110	Starters, 16" x 9.35"		3	2.667	C.L.F.	114	91.50		205.50	288
0120	Hip & ridge, 4.75" x 14"		1	8	"	825	274		1,099	1,400
0200	Shakes, 16" x 9.35", 550 lb per square		2.20	3.636	Sq.	305	125		430	555
0300	Hip & ridge, 4.75 x 14"		1	8	C.L.F.	825	274		1,099	1,400
0400	Hexagonal, 16" x 16"		3	2.667	Sq.	227	91.50		318.50	410
0500	Square, 16" x 16"		3	2.667		203	91.50		294.50	385
2000	For steep roofs (7/12 pitch or greater), add						50%			
9000	Minimum labor/equipment charge	1 Rofc	3	2.667	Job		91.50		91.50	163

07 31 26 – Slate Shingles

07 31 26.10 Slate Roof Shingles			Crew	Daily Output	Labor-Hours	Unit	Material	Labor	Equipment	Total	Total Incl O&P
0010	**SLATE ROOF SHINGLES**										
0100	Buckingham Virginia black, 3/16" - 1/4" thick	G	1 Rots	1.75	4.571	Sq.	475	156		631	805
0200	1/4" thick	G		1.75	4.571		475	156		631	805
0900	Pennsylvania black, Bangor, #1 clear	G		1.75	4.571		490	156		646	820
1200	Vermont, unfading, green, mottled green	G		1.75	4.571		480	156		636	805

07 31 Shingles and Shakes

07 31 26 – Slate Shingles

07 31 26.10 Slate Roof Shingles		Crew	Daily Output	Labor-Hours	Unit	Material	2009 Bare Costs Labor	Equipment	Total	Total Incl O&P
1300	Semi-weathering green & gray G	1 Rots	1.75	4.571	Sq.	350	156		506	660
1400	Purple G		1.75	4.571		425	156		581	745
1500	Black or gray G		1.75	4.571		460	156		616	785
2500	Slate roof repair, extensive replacement		1	8		590	273		863	1,125
2600	Repair individual pieces, scattered		19	.421	Ea.	5.90	14.35		20.25	32
9000	Minimum labor/equipment charge		3	2.667	Job		91		91	162

07 31 29 – Wood Shingles and Shakes

07 31 29.13 Wood Shingles		Crew	Daily Output	Labor-Hours	Unit	Material	Labor	Equipment	Total	Total Incl O&P
0010	**WOOD SHINGLES**									
0012	16″ No. 1 red cedar shingles, 5″ exposure, on roof	1 Carp	2.50	3.200	Sq.	330	128		458	575
0015	Pneumatic nailed		3.25	2.462		330	98.50		428.50	525
0200	7-1/2″ exposure, on walls		2.05	3.902		221	156		377	500
0205	Pneumatic nailed		2.67	2.996		221	120		341	440
0300	18″ No. 1 red cedar perfections, 5-1/2″ exposure, on roof		2.75	2.909		330	116		446	550
0305	Pneumatic nailed		3.57	2.241		330	89.50		419.50	510
0600	Resquared, and rebutted, 5-1/2″ exposure, on roof		3	2.667		365	107		472	580
0605	Pneumatic nailed		3.90	2.051		365	82		447	540
0900	7-1/2″ exposure, on walls		2.45	3.265		269	130		399	510
0905	Pneumatic nailed		3.18	2.516		269	101		370	460
1000	Add to above for fire retardant shingles, 16″ long					63			63	69.50
1050	18″ long					63			63	69.50
1060	Preformed ridge shingles	1 Carp	400	.020	L.F.	3.50	.80		4.30	5.15
2000	White cedar shingles, 16″ long, extras, 5″ exposure, on roof		2.40	3.333	Sq.	178	133		311	415
2005	Pneumatic nailed		3.12	2.564		178	102		280	365
2050	5″ exposure on walls		2	4		178	160		338	460
2055	Pneumatic nailed		2.60	3.077		178	123		301	400
2100	7-1/2″ exposure, on walls		2	4		127	160		287	405
2105	Pneumatic nailed		2.60	3.077		127	123		250	345
2150	"B" grade, 5″ exposure on walls		2	4		150	160		310	430
2155	Pneumatic nailed		2.60	3.077		150	123		273	370
2300	For 15# organic felt underlayment on roof, 1 layer, add		64	.125		4.77	4.99		9.76	13.50
2400	2 layers, add		32	.250		9.55	10		19.55	27
2600	For steep roofs (7/12 pitch or greater), add to above						50%			
2700	Panelized systems, No.1 cedar shingles on 5/16″ CDX plywood									
2800	On walls, 8′ strips, 7″ or 14″ exposure	2 Carp	700	.023	S.F.	5.55	.91		6.46	7.60
3500	On roofs, 8′ strips, 7″ or 14″ exposure	1 Carp	3	2.667	Sq.	555	107		662	785
3505	Pneumatic nailed		4	2	″	555	80		635	740
9000	Minimum labor/equipment charge		3	2.667	Job		107		107	176

07 31 29.16 Wood Shakes		Crew	Daily Output	Labor-Hours	Unit	Material	Labor	Equipment	Total	Total Incl O&P
0010	**WOOD SHAKES**									
1100	Hand-split red cedar shakes, 1/2″ thick x 24″ long, 10″ exp. on roof	1 Carp	2.50	3.200	Sq.	252	128		380	490
1105	Pneumatic nailed		3.25	2.462		252	98.50		350.50	440
1110	3/4″ thick x 24″ long, 10″ exp. on roof		2.25	3.556		252	142		394	510
1115	Pneumatic nailed		2.92	2.740		252	109		361	460
1200	1/2″ thick, 18″ long, 8-1/2″ exp. on roof		2	4		239	160		399	525
1205	Pneumatic nailed		2.60	3.077		239	123		362	465
1210	3/4″ thick x 18″ long, 8 1/2″ exp. on roof		1.80	4.444		239	178		417	555
1215	Pneumatic nailed		2.34	3.419		239	137		376	490
1255	10″ exp. on walls		2	4		231	160		391	520
1260	10″ exposure on walls, pneumatic nailed		2.60	3.077		231	123		354	455
1700	Add to above for fire retardant shakes, 24″ long					63			63	69.50
1800	18″ long					63			63	69.50

07 31 Shingles and Shakes

07 31 29 – Wood Shingles and Shakes

07 31 29.16 Wood Shakes		Crew	Daily Output	Labor-Hours	Unit	Material	Labor (2009 Bare Costs)	Equipment	Total	Total Incl O&P
1810	Ridge shakes	1 Carp	350	.023	L.F.	3.50	.91		4.41	5.35

07 32 Roof Tiles

07 32 13 – Clay Roof Tiles

07 32 13.10 Clay Tiles

		Crew	Daily Output	Labor-Hours	Unit	Material	Labor	Equipment	Total	Total Incl O&P
0010	CLAY TILES									
0200	Lanai tile or Classic tile, 158 pc per sq G	1 Rots	1.65	4.848	Sq.	500	165		665	845
0300	Americana, 158 pc per sq, most colors G		1.65	4.848		660	165		825	1,025
0350	Green, gray or brown G		1.65	4.848		625	165		790	980
0400	Blue G		1.65	4.848		625	165		790	980
0600	Spanish tile, 171 pc per sq, red G		1.80	4.444		325	152		477	625
0800	Buff, green, gray, brown G		1.80	4.444		600	152		752	930
0900	Glazed white		1.80	4.444		665	152		817	1,000
1100	Mission tile, 192 pc per sq, machine scored finish, red G		1.15	6.957		740	237		977	1,250
1700	French tile, 133 pc per sq, smooth finish, red G		1.35	5.926		675	202		877	1,100
1750	Blue or green G		1.35	5.926		870	202		1,072	1,325
1800	Norman black 317 pc per sq G		1	8		1,050	273		1,323	1,625
2200	Williamsburg tile, 158 pc per sq, aged cedar G		1.35	5.926		730	202		932	1,150
2250	Gray or green G		1.35	5.926		595	202		797	1,025
2510	One piece mission tile, natural red, 75 pc per square G		1.65	4.848		231	165		396	550
2530	Mission Tile, 134 pc per square G	↓	1.15	6.957		263	237		500	715
3000	For steep roofs (7/12 pitch or greater), add to above						50%			
3010	Clay tile, #15 felt underlayment	1 Rofc	64	.125		4.77	4.28		9.05	12.90
3020	Clay tile, #30 felt underlayment		58	.138		9.55	4.72		14.27	18.95
3040	Clay tile, polyethylene and rubberized asph. underlayment	↓	22	.364	↓	55.50	12.45		67.95	83
9000	Minimum labor/equipment charge	1 Rots	3	2.667	Job		91		91	162

07 32 16 – Concrete Roof Tiles

07 32 16.10 Concrete Tiles

		Crew	Daily Output	Labor-Hours	Unit	Material	Labor	Equipment	Total	Total Incl O&P
0010	CONCRETE TILES									
0020	Corrugated, 13" x 16-1/2", 90 per sq, 950 lb per sq									
0050	Earthtone colors, nailed to wood deck	1 Rots	1.35	5.926	Sq.	96.50	202		298.50	465
0150	Blues		1.35	5.926		97.50	202		299.50	465
0200	Greens		1.35	5.926		97.50	202		299.50	465
0250	Premium colors	↓	1.35	5.926	↓	214	202		416	595
0500	Shakes, 13" x 16-1/2", 90 per sq, 950 lb per sq									
0600	All colors, nailed to wood deck	1 Rots	1.50	5.333	Sq.	252	182		434	605
1500	Accessory pieces, ridge & hip, 10" x 16-1/2", 8 lbs. each				Ea.	3.40			3.40	3.74
1700	Rake, 6-1/2" x 16-3/4", 9 lbs. each					3.40			3.40	3.74
1800	Mansard hip, 10" x 16-1/2", 9.2 lbs. each					3.40			3.40	3.74
1900	Hip starter, 10" x 16-1/2", 10.5 lbs. each					10.70			10.70	11.80
2000	3 or 4 way apex, 10" each side, 11.5 lbs. each				↓	12.35			12.35	13.60
9000	Minimum labor/equipment charge	1 Rots	3	2.667	Job		91		91	162

07 32 19 – Metal Roof Tiles

07 32 19.10 Metal Roof Tiles

		Crew	Daily Output	Labor-Hours	Unit	Material	Labor	Equipment	Total	Total Incl O&P
0010	METAL ROOF TILES									
0020	Accessories included, .032" thick aluminum, mission tile	1 Carp	2.50	3.200	Sq.	815	128		943	1,100
0200	Spanish tiles		3	2.667	"	560	107		667	790
9000	Minimum labor/equipment charge	↓	3	2.667	Job		107		107	176

07 33 Natural Roof Coverings

07 33 63 – Vegetated Roofing

07 33 63.10 Green Roof Systems			Crew	Daily Output	Labor-Hours	Unit	Material	2009 Bare Costs Labor	Equipment	Total	Total Incl O&P
0010	**GREEN ROOF SYSTEMS**										
0020	Soil mixture for green roof 30% sand, 55% gravel, 15% soil										
0100	Hoist and spread soil mixture 4 inch depth up to five stories tall roof	G	B-13B	4000	.014	S.F.	.25	.48	.28	1.01	1.37
0150	6 inch depth	G		2667	.021		.38	.72	.42	1.52	2.05
0200	8 inch depth	G		2000	.028		.50	.96	.56	2.02	2.74
0250	10 inch depth	G		1600	.035		.63	1.20	.70	2.53	3.40
0300	12 inch depth	G		1335	.042		.76	1.43	.84	3.03	4.09
0350	Mobilization 55 ton crane to site	G	1 Eqhv	3.60	2.222	Ea.		94.50		94.50	149
0355	Hoisting cost to five stories per day (Avg. 28 picks per day)	G	B-13B	1	56	Day		1,925	1,125	3,050	4,350
0360	Mobilization or demobilization, 100 ton crane to site driver & escort	G	A-3E	2.50	6.400	Ea.		238	51.50	289.50	435
0365	Hoisting cost six to ten stories per day (Avg. 21 picks per day)	G	B-13C	1	56	Day		1,925	1,600	3,525	4,875
0370	Hoist and spread soil mixture 4 inch depth six to ten stories tall roof	G		4000	.014	S.F.	.25	.48	.40	1.13	1.50
0375	6 inch depth	G		2667	.021		.38	.72	.60	1.70	2.25
0380	8 inch depth	G		2000	.028		.50	.96	.81	2.27	3.01
0385	10 inch depth	G		1600	.035		.63	1.20	1.01	2.84	3.74
0390	12 inch depth	G		1335	.042		.76	1.43	1.21	3.40	4.49
0400	Green roof edging treated lumber 4″ x4″ no hoisting included	G	2 Carp	400	.040	L.F.	1.52	1.60		3.12	4.31
0410	4″ x 6″	G		400	.040		2.24	1.60		3.84	5.10
0420	4″ x 8″	G		360	.044		3.21	1.78		4.99	6.45
0430	4″ x 6″ double stacked	G		300	.053		4.47	2.13		6.60	8.45
0500	Green roof edging redwood lumber 4″ x4″ no hoisting included	G		400	.040		6.45	1.60		8.05	9.75
0510	4″ x 6″	G		400	.040		11.20	1.60		12.80	14.95
0520	4″ x 8″	G		360	.044		20.50	1.78		22.28	25.50
0530	4″ x 6″ double stacked	G		300	.053		22.50	2.13		24.63	28
0600	Planting sedum, light soil, potted, 2-1/4″ diameter, two per SF	G	1 Clab	420	.019	S.F.	7	.60		7.60	8.70
0610	one per SF	G	″	840	.010		3.50	.30		3.80	4.35
0630	Planting sedum mat per SF including shipping (4000 SF Minimum)	G	4 Clab	4000	.008		8.40	.25		8.65	9.65
0640	Installation sedum mat system (no soil required) per SF (4000 SF minimum)	G	″	4000	.008		11.70	.25		11.95	13.25
0645	Note: pricing of sedum mats shipped in full truck loads (4000-5000 SF)										

07 41 Roof Panels

07 41 13 – Metal Roof Panels

07 41 13.10 Aluminum Roof Panels		Crew	Daily Output	Labor-Hours	Unit	Material	Labor	Equipment	Total	Total Incl O&P
0010	**ALUMINUM ROOF PANELS**									
0020	Corrugated or ribbed, .0155″ thick, natural	G-3	1200	.027	S.F.	.87	1.05		1.92	2.66
0300	Painted		1200	.027		1.27	1.05		2.32	3.10
0400	Corrugated, .018″ thick, on steel frame, natural finish		1200	.027		1.15	1.05		2.20	2.97
0600	Painted		1200	.027		1.41	1.05		2.46	3.25
0700	Corrugated, on steel frame, natural, .024″ thick		1200	.027		1.64	1.05		2.69	3.50
0800	Painted		1200	.027		1.99	1.05		3.04	3.89
0900	.032″ thick, natural		1200	.027		2.11	1.05		3.16	4.02
1200	Painted		1200	.027		2.70	1.05		3.75	4.67
1300	V-Beam, on steel frame construction, .032″ thick, natural		1200	.027		2.23	1.05		3.28	4.15
1500	Painted		1200	.027		2.74	1.05		3.79	4.71
1600	.040″ thick, natural		1200	.027		2.72	1.05		3.77	4.69
1800	Painted		1200	.027		3.27	1.05		4.32	5.30
1900	.050″ thick, natural		1200	.027		3.26	1.05		4.31	5.30
2100	Painted		1200	.027		3.93	1.05		4.98	6
2200	For roofing on wood frame, deduct		4600	.007		.07	.27		.34	.52
2400	Ridge cap, .032″ thick, natural		800	.040	L.F.	2.58	1.58		4.16	5.40
9000	Minimum labor/equipment charge	1 Rofc	3	2.667	Job		91.50		91.50	163

07 41 Roof Panels

07 41 13 – Metal Roof Panels

07 41 13.20 Steel Roofing Panels		Crew	Daily Output	Labor-Hours	Unit	Material	2009 Bare Costs Labor	Equipment	Total	Total Incl O&P
0010	STEEL ROOFING PANELS									
0012	Corrugated or ribbed, on steel framing, 30 ga galv	G-3	1100	.029	S.F.	1.68	1.15		2.83	3.70
0100	28 ga		1050	.030		1.79	1.20		2.99	3.91
0300	26 ga		1000	.032		1.85	1.26		3.11	4.07
0400	24 ga		950	.034		2.91	1.33		4.24	5.35
0600	Colored, 28 ga		1050	.030		1.68	1.20		2.88	3.79
0700	26 ga		1000	.032		1.99	1.26		3.25	4.22
0710	Flat profile, 1-3/4" standing seams, 10" wide, standard finish, 26 ga		1000	.032		3.89	1.26		5.15	6.30
0715	24 ga		950	.034		4.51	1.33		5.84	7.10
0720	22 ga		900	.036		5.55	1.40		6.95	8.40
0725	Zinc aluminum alloy finish, 26 ga		1000	.032		3.05	1.26		4.31	5.40
0730	24 ga		950	.034		3.64	1.33		4.97	6.15
0735	22 ga		900	.036		4.17	1.40		5.57	6.85
0740	12" wide, standard finish, 26 ga		1000	.032		3.88	1.26		5.14	6.30
0745	24 ga		950	.034		5.10	1.33		6.43	7.75
0750	Zinc aluminum alloy finish, 26 ga		1000	.032		4.41	1.26		5.67	6.90
0755	24 ga		950	.034		3.63	1.33		4.96	6.15
0840	Flat profile, 1" x 3/8" batten, 12" wide, standard finish, 26 ga		1000	.032		3.42	1.26		4.68	5.80
0845	24 ga		950	.034		4.01	1.33		5.34	6.55
0850	22 ga		900	.036		4.80	1.40		6.20	7.55
0855	Zinc aluminum alloy finish, 26 ga		1000	.032		3.28	1.26		4.54	5.65
0860	24 ga		950	.034		3.66	1.33		4.99	6.15
0865	22 ga		900	.036		4.23	1.40		5.63	6.90
0870	16-1/2" wide, standard finish, 24 ga		950	.034		3.95	1.33		5.28	6.50
0875	22 ga		900	.036		4.42	1.40		5.82	7.10
0880	Zinc aluminum alloy finish, 24 ga		950	.034		3.45	1.33		4.78	5.95
0885	22 ga		900	.036		3.85	1.40		5.25	6.50
0890	Flat profile, 2" x 2" batten, 12" wide, standard finish, 26 ga		1000	.032		3.93	1.26		5.19	6.35
0895	24 ga		950	.034		4.69	1.33		6.02	7.30
0900	22 ga		900	.036		5.75	1.40		7.15	8.60
0905	Zinc aluminum alloy finish, 26 ga		1000	.032		3.66	1.26		4.92	6.05
0910	24 ga		950	.034		4.18	1.33		5.51	6.75
0915	22 ga		900	.036		4.87	1.40		6.27	7.60
0920	16-1/2" wide, standard finish, 24 ga		950	.034		4.32	1.33		5.65	6.90
0925	22 ga		900	.036		5.05	1.40		6.45	7.80
0930	Zinc aluminum alloy finish, 24 ga		950	.034		3.91	1.33		5.24	6.45
0935	22 ga		900	.036		4.45	1.40		5.85	7.15
1200	Ridge, galvanized, 10" wide		800	.040	L.F.	3.18	1.58		4.76	6.05
1210	20" wide		750	.043	"	4.93	1.68		6.61	8.10
9000	Minimum labor/equipment charge	1 Rofc	2	4	Job		137		137	245

07 41 33 – Plastic Roof Panels

07 41 33.10 Fiberglass Panels		Crew	Daily Output	Labor-Hours	Unit	Material	Labor	Equipment	Total	Total Incl O&P
0010	FIBERGLASS PANELS									
0012	Corrugated panels, roofing, 8 oz per S.F.	G-3	1000	.032	S.F.	1.58	1.26		2.84	3.77
0100	12 oz per S.F.		1000	.032		3.43	1.26		4.69	5.80
0300	Corrugated siding, 6 oz per S.F.		880	.036		1.58	1.43		3.01	4.05
0400	8 oz per S.F.		880	.036		1.58	1.43		3.01	4.05
0500	Fire retardant		880	.036		3.21	1.43		4.64	5.85
0600	12 oz. siding, textured		880	.036		3.32	1.43		4.75	5.95
0700	Fire retardant		880	.036		4.39	1.43		5.82	7.15
0900	Flat panels, 6 oz per S.F., clear or colors		880	.036		1.83	1.43		3.26	4.32
1100	Fire retardant, class A		880	.036		3.18	1.43		4.61	5.80

07 41 Roof Panels

07 41 33 – Plastic Roof Panels

07 41 33.10 Fiberglass Panels		Crew	Daily Output	Labor-Hours	Unit	2009 Bare Costs Material	Labor	Equipment	Total	Total Incl O&P
1300	8 oz per S.F., clear or colors	G-3	880	.036	S.F.	2.38	1.43		3.81	4.93
1700	Sandwich panels, fiberglass, 1-9/16" thick, panels to 20 S.F.		180	.178		23	7		30	37
1900	As above, but 2-3/4" thick, panels to 100 S.F.	↓	265	.121	↓	17.05	4.76		21.81	26.50
9000	Minimum labor/equipment charge	1 Rofc	2	4	Job		137		137	245

07 42 Wall Panels

07 42 13 – Metal Wall Panels

07 42 13.10 Mansard Panels		Crew	Daily Output	Labor-Hours	Unit	2009 Bare Costs Material	Labor	Equipment	Total	Total Incl O&P
0010	**MANSARD PANELS**									
0600	Aluminum, stock units, straight surfaces	1 Shee	115	.070	S.F.	3.04	3.28		6.32	8.55
0700	Concave or convex surfaces		75	.107	"	3.34	5.05		8.39	11.65
0800	For framing, to 5' high, add		115	.070	L.F.	3.34	3.28		6.62	8.85
0900	Soffits, to 1' wide		125	.064	S.F.	1.66	3.02		4.68	6.65
9000	Minimum labor/equipment charge	↓	2.50	3.200	Job		151		151	240

07 42 13.20 Aluminum Siding Panels		Crew	Daily Output	Labor-Hours	Unit	2009 Bare Costs Material	Labor	Equipment	Total	Total Incl O&P
0010	**ALUMINUM SIDING PANELS**									
0012	Corrugated, on steel framing, .019 thick, natural finish	G-3	775	.041	S.F.	1.35	1.63		2.98	4.12
0100	Painted		775	.041		1.45	1.63		3.08	4.23
0400	Farm type, .021" thick on steel frame, natural		775	.041		1.36	1.63		2.99	4.13
0600	Painted		775	.041		1.45	1.63		3.08	4.23
0700	Industrial type, corrugated, on steel, .024" thick, mill		775	.041		1.89	1.63		3.52	4.71
0900	Painted		775	.041		2.04	1.63		3.67	4.87
1000	.032" thick, mill		775	.041		2.17	1.63		3.80	5
1200	Painted		775	.041		2.65	1.63		4.28	5.55
1300	V-Beam, on steel frame, .032" thick, mill		775	.041		2.45	1.63		4.08	5.35
1500	Painted		775	.041		2.81	1.63		4.44	5.70
1600	.040" thick, mill		775	.041		2.98	1.63		4.61	5.90
1800	Painted		775	.041		3.49	1.63		5.12	6.45
1900	.050" thick, mill		775	.041		3.52	1.63		5.15	6.50
2100	Painted		775	.041		4.19	1.63		5.82	7.25
2200	Ribbed, 3" profile, on steel frame, .032" thick, natural		775	.041		2.12	1.63		3.75	4.96
2400	Painted		775	.041		2.69	1.63		4.32	5.60
2500	.040" thick, natural		775	.041		2.42	1.63		4.05	5.30
2700	Painted		775	.041		2.83	1.63		4.46	5.75
2750	.050" thick, natural		775	.041		2.79	1.63		4.42	5.70
2760	Painted		775	.041		3.20	1.63		4.83	6.15
3300	For siding on wood frame, deduct from above	↓	2800	.011	↓	.08	.45		.53	.82
3400	Screw fasteners, aluminum, self tapping, neoprene washer, 1"				M	183			183	202
3600	Stitch screws, self tapping, with neoprene washer, 5/8"				"	137			137	151
3630	Flashing, sidewall, .032" thick	G-3	800	.040	L.F.	2.54	1.58		4.12	5.35
3650	End wall, .040" thick		800	.040		2.97	1.58		4.55	5.80
3670	Closure strips, corrugated, .032" thick		800	.040		.75	1.58		2.33	3.37
3680	Ribbed, 4" or 8", .032" thick		800	.040		.75	1.58		2.33	3.37
3690	V-beam, .040" thick	↓	800	.040	↓	1.02	1.58		2.60	3.66
3800	Horizontal, colored clapboard, 8" wide, plain	2 Carp	515	.031	S.F.	1.57	1.24		2.81	3.78
3810	Insulated		515	.031		1.54	1.24		2.78	3.74
3830	8" embossed, painted		515	.031		1.71	1.24		2.95	3.93
3840	Insulated		515	.031		1.90	1.24		3.14	4.14
3860	12" painted, smooth		600	.027		1.59	1.07		2.66	3.51
3870	Insulated		600	.027		1.84	1.07		2.91	3.78
3890	12" embossed, painted	↓	600	.027	↓	1.77	1.07		2.84	3.71

07 42 Wall Panels

07 42 13 – Metal Wall Panels

07 42 13.20 Aluminum Siding Panels		Crew	Daily Output	Labor-Hours	Unit	2009 Bare Costs Material	Labor	Equipment	Total	Total Incl O&P
3900	Insulated	2 Carp	515	.031	S.F.	1.98	1.24		3.22	4.23
4000	Vertical board & batten, colored, non-insulated		515	.031		1.53	1.24		2.77	3.73
4200	For simulated wood design, add					.10			.10	.11
4300	Corners for above, outside	2 Carp	515	.031	V.L.F.	2.36	1.24		3.60	4.65
4500	Inside corners	"	515	.031	"	1.39	1.24		2.63	3.58
4600	Sandwich panels, 1" insulation, single story	G-3	395	.081	S.F.	6.85	3.19		10.04	12.70
4900	Multi-story	"	345	.093		9.65	3.65		13.30	16.50
5100	For baked enamel finish 1 side, add					.37			.37	.41
9000	Minimum labor/equipment charge	1 Carp	3	2.667	Job		107		107	176

07 42 13.30 Steel Siding

		Crew	Daily Output	Labor-Hours	Unit	Material	Labor	Equipment	Total	Total Incl O&P
0010	**STEEL SIDING**									
0020	Beveled, vinyl coated, 8" wide	1 Carp	265	.030	S.F.	1.81	1.21		3.02	3.98
0050	10" wide	"	275	.029		1.94	1.16		3.10	4.05
0080	Galv, corrugated or ribbed, on steel frame, 30 gauge	G-3	800	.040		1.21	1.58		2.79	3.87
0100	28 gauge		795	.040		1.27	1.59		2.86	3.96
0300	26 gauge		790	.041		1.78	1.60		3.38	4.54
0400	24 gauge		785	.041		1.79	1.61		3.40	4.56
0600	22 gauge		770	.042		2.06	1.64		3.70	4.91
0700	Colored, corrugated/ribbed, on steel frame, 10 yr fnsh, 28 ga.		800	.040		1.88	1.58		3.46	4.61
0900	26 gauge		795	.040		1.96	1.59		3.55	4.72
1000	24 gauge		790	.041		2.29	1.60		3.89	5.10
1020	20 gauge		785	.041		2.90	1.61		4.51	5.80
1200	Factory sandwich panel, 26 ga., 1" insulation, galvanized		380	.084		4.82	3.32		8.14	10.65
1300	Colored 1 side		380	.084		5.90	3.32		9.22	11.85
1500	Galvanized 2 sides		380	.084		7.30	3.32		10.62	13.40
1600	Colored 2 sides		380	.084		7.60	3.32		10.92	13.70
1800	Acrylic paint face, regular paint liner		380	.084		5.70	3.32		9.02	11.65
1900	For 2" thick polystyrene, add					.88			.88	.97
2000	22 ga, galv, 2" insulation, baked enamel exterior	G-3	360	.089		11.20	3.50		14.70	17.95
2100	Polyvinylidene exterior finish	"	360	.089		11.80	3.50		15.30	18.60
9000	Minimum labor/equipment charge	1 Carp	3	2.667	Job		107		107	176

07 44 Faced Panels

07 44 73 – Metal Faced Panels

07 44 73.10 Metal Faced Panels and Accessories

		Crew	Daily Output	Labor-Hours	Unit	Material	Labor	Equipment	Total	Total Incl O&P
0010	**METAL FACED PANELS AND ACCESSORIES**									
0400	Textured aluminum, 4' x 8' x 5/16" plywood backing, single face	2 Shee	375	.043	S.F.	3.73	2.01		5.74	7.30
0600	Double face		375	.043		4.44	2.01		6.45	8.10
0700	4' x 10' x 5/16" plywood backing, single face		375	.043		4.16	2.01		6.17	7.80
0900	Double face		375	.043		5.85	2.01		7.86	9.65
1000	4' x 12' x 5/16" plywood backing, single face		375	.043		5.90	2.01		7.91	9.70
1300	Smooth aluminum, 1/4" plywood panel, fluoropolymer finish, double face		375	.043		5.80	2.01		7.81	9.60
1350	Clear anodized finish, double face		375	.043		9.15	2.01		11.16	13.25
1400	Double face textured aluminum, structural panel, 1" EPS insulation		375	.043		5.05	2.01		7.06	8.75
1500	Accessories, outside corner	1 Shee	175	.046	L.F.	1.90	2.16		4.06	5.50
1600	Inside corner		175	.046		1.34	2.16		3.50	4.90
1800	Batten mounting clip		200	.040		.49	1.89		2.38	3.54
1900	Low profile batten		480	.017		.60	.79		1.39	1.91
2100	High profile batten		480	.017		1.41	.79		2.20	2.80
2200	Water table		200	.040		2.11	1.89		4	5.30
2400	Horizontal joint connector		200	.040		1.67	1.89		3.56	4.84

07 44 Faced Panels

07 44 73 – Metal Faced Panels

07 44 73.10 Metal Faced Panels and Accessories		Crew	Daily Output	Labor-Hours	Unit	Material	Labor	Equipment	Total	Total Incl O&P
						2009 Bare Costs				
2500	Corner cap	1 Shee	200	.040	L.F.	1.84	1.89		3.73	5
2700	H - moulding	↓	480	.017	↓	1.25	.79		2.04	2.63

07 46 Siding

07 46 23 – Wood Siding

07 46 23.10 Wood Board Siding

		Crew	Daily Output	Labor-Hours	Unit	Material	Labor	Equipment	Total	Total Incl O&P
0010	**WOOD BOARD SIDING**									
3200	Wood, cedar bevel, A grade, 1/2" x 6"	1 Carp	295	.027	S.F.	4.62	1.08		5.70	6.90
3300	1/2" x 8"		330	.024		4.97	.97		5.94	7.05
3500	3/4" x 10", clear grade		375	.021		6.70	.85		7.55	8.75
3600	"B" grade		375	.021		6.25	.85		7.10	8.30
3800	Cedar, rough sawn, 1" x 4", A grade, natural		220	.036		3.77	1.45		5.22	6.55
3900	Stained		220	.036		4.17	1.45		5.62	7
4100	1" x 12", board & batten, #3 & Btr., natural		420	.019		3.94	.76		4.70	5.60
4200	Stained		420	.019		4.20	.76		4.96	5.90
4400	1" x 8" channel siding, #3 & Btr., natural		330	.024		2.34	.97		3.31	4.17
4500	Stained		330	.024		2.50	.97		3.47	4.35
4700	Redwood, clear, beveled, vertical grain, 1/2" x 4"		220	.036		3.10	1.45		4.55	5.80
4750	1/2" x 6"		295	.027		3.10	1.08		4.18	5.20
4800	1/2" x 8"		330	.024		3.24	.97		4.21	5.15
5000	3/4" x 10"		375	.021		3.32	.85		4.17	5.05
5200	Channel siding, 1" x 10", B grade		375	.021		3.32	.85		4.17	5.05
5250	Redwood, T&G boards, B grade, 1" x 4"		220	.036		3.10	1.45		4.55	5.80
5270	1" x 8"		330	.024		3.24	.97		4.21	5.15
5400	White pine, rough sawn, 1" x 8", natural		330	.024		2.20	.97		3.17	4.02
5500	Stained		330	.024	↓	2.60	.97		3.57	4.46
9000	Minimum labor/equipment charge	↓	2	4	Job		160		160	264

07 46 29 – Plywood Siding

07 46 29.10 Plywood Siding Options

		Crew	Daily Output	Labor-Hours	Unit	Material	Labor	Equipment	Total	Total Incl O&P
0010	**PLYWOOD SIDING OPTIONS**									
0900	Plywood, medium density overlaid, 3/8" thick	2 Carp	750	.021	S.F.	.96	.85		1.81	2.47
1000	1/2" thick		700	.023		1.27	.91		2.18	2.91
1100	3/4" thick		650	.025		1.31	.98		2.29	3.06
1600	Texture 1-11, cedar, 5/8" thick, natural		675	.024		2.45	.95		3.40	4.26
1700	Factory stained		675	.024		1.96	.95		2.91	3.72
1900	Texture 1-11, fir, 5/8" thick, natural		675	.024		1.13	.95		2.08	2.80
2000	Factory stained		675	.024		1.73	.95		2.68	3.46
2050	Texture 1-11, S.Y.P., 5/8" thick, natural		675	.024		1.21	.95		2.16	2.89
2100	Factory stained		675	.024		1.25	.95		2.20	2.94
2200	Rough sawn cedar, 3/8" thick, natural		675	.024		1.21	.95		2.16	2.89
2300	Factory stained		675	.024		.97	.95		1.92	2.63
2500	Rough sawn fir, 3/8" thick, natural		675	.024		.77	.95		1.72	2.41
2600	Factory stained		675	.024		.97	.95		1.92	2.63
2800	Redwood, textured siding, 5/8" thick		675	.024		1.94	.95		2.89	3.69
3000	Polyvinyl chloride coated, 3/8" thick	↓	750	.021	↓	1.12	.85		1.97	2.64
9000	Minimum labor/equipment charge	1 Carp	2	4	Job		160		160	264

07 46 33 – Plastic Siding

07 46 33.10 Vinyl Siding

		Crew	Daily Output	Labor-Hours	Unit	Material	Labor	Equipment	Total	Total Incl O&P
0010	**VINYL SIDING**									
3995	Clapboard profile, woodgrain texture, .048 thick, double 4	2 Carp	495	.032	S.F.	.92	1.29		2.21	3.14

07 46 33 – Plastic Siding

07 46 33.10 Vinyl Siding

	07 46 33.10 Vinyl Siding	Crew	Daily Output	Labor-Hours	Unit	2009 Bare Costs Material	Labor	Equipment	Total	Total Incl O&P
4000	Double 5	2 Carp	550	.029	S.F.	.92	1.16		2.08	2.93
4005	Single 8		495	.032		.96	1.29		2.25	3.19
4010	Single 10		550	.029		1.15	1.16		2.31	3.19
4015	.044 thick, double 4		495	.032		.84	1.29		2.13	3.05
4020	Double 5		550	.029		.84	1.16		2	2.84
4025	.042 thick, double 4		495	.032		.84	1.29		2.13	3.05
4030	Double 5		550	.029		.84	1.16		2	2.84
4035	Cross sawn texture, .040 thick, double 4		495	.032		.66	1.29		1.95	2.86
4040	Double 5		550	.029		.66	1.16		1.82	2.65
4045	Smooth texture, .042 thick, double 4		495	.032		.75	1.29		2.04	2.96
4050	Double 5		550	.029		.75	1.16		1.91	2.75
4055	Single 8		495	.032		.96	1.29		2.25	3.19
4060	Cedar texture, .044 thick, double 4		495	.032		.92	1.29		2.21	3.14
4065	Double 6		600	.027		.92	1.07		1.99	2.77
4070	Dutch lap profile, woodgrain texture, .048 thick, double 5		550	.029		.92	1.16		2.08	2.93
4075	.044 thick, double 4.5		525	.030		.92	1.22		2.14	3.02
4080	.042 thick, double 4.5		525	.030		.75	1.22		1.97	2.84
4085	.040 thick, double 4.5		525	.030		.66	1.22		1.88	2.74
4090	Shingle profile, random grooves, double 7		400	.040		2.57	1.60		4.17	5.45
4095	Triple 5		400	.040		2.57	1.60		4.17	5.45
4100	Shake profile, 10″ wide		400	.040		2.19	1.60		3.79	5.05
4105	Vertical pattern, .046 thick, double 5		550	.029		1.22	1.16		2.38	3.26
4110	.044 thick, triple 3		550	.029		1.37	1.16		2.53	3.43
4115	.040 thick, triple 4		550	.029		1.22	1.16		2.38	3.26
4120	.040 thick, triple 2.66		550	.029		1.47	1.16		2.63	3.54
4125	Insulation, fan folded extruded polystyrene, 1/4″		2000	.008		.25	.32		.57	.81
4130	3/8″		2000	.008		.27	.32		.59	.83
4135	Accessories, J channel, 5/8″ pocket		700	.023	L.F.	.40	.91		1.31	1.94
4140	3/4″ pocket		695	.023		.44	.92		1.36	2
4145	1-1/4″ pocket		680	.024		.56	.94		1.50	2.16
4150	Flexible, 3/4″ pocket		600	.027		1.57	1.07		2.64	3.48
4155	Under sill finish trim		500	.032		.39	1.28		1.67	2.53
4160	Vinyl starter strip		700	.023		.31	.91		1.22	1.85
4165	Aluminum starter strip		700	.023		.22	.91		1.13	1.75
4170	Window casing, 2-1/2″ wide, 3/4″ pocket		510	.031		.99	1.25		2.24	3.15
4175	Outside corner, woodgrain finish, 4″ face, 3/4″ pocket		700	.023		1.65	.91		2.56	3.33
4180	5/8″ pocket		700	.023		1.67	.91		2.58	3.35
4185	Smooth finish, 4″ face, 3/4″ pocket		700	.023		1.73	.91		2.64	3.41
4190	7/8″ pocket		690	.023		1.78	.93		2.71	3.49
4195	1-1/4″ pocket		700	.023		1.16	.91		2.07	2.79
4200	Soffit and fascia, 1′ overhang, solid		120	.133		3.31	5.35		8.66	12.45
4205	Vented		120	.133		3.31	5.35		8.66	12.45
4210	2′ overhang, solid		110	.145		4.52	5.80		10.32	14.55
4215	Vented		110	.145		4.52	5.80		10.32	14.55
4220	Colors for siding and soffits, add				S.F.	.16			.16	.18
4225	Colors for accessories and trim, add				L.F.	.32			.32	.35
9000	Minimum labor/equipment charge	1 Carp	3	2.667	Job		107		107	176

07 46 46 – Mineral-Fiber Cement Siding

07 46 46.10 Fiber Cement Siding

	07 46 46.10 Fiber Cement Siding	Crew	Daily Output	Labor-Hours	Unit	Material	Labor	Equipment	Total	Total Incl O&P
0010	**FIBER CEMENT SIDING**									
0020	Lap siding, 5/16″ thick, 6″ wide, 4-3/4″ exposure, smooth texture	2 Carp	415	.039	S.F.	1.34	1.54		2.88	4.01
0025	Woodgrain texture		415	.039		1.34	1.54		2.88	4.01

07 46 Siding

07 46 46 – Mineral-Fiber Cement Siding

07 46 46.10 Fiber Cement Siding		Crew	Daily Output	Labor-Hours	Unit	Material	Labor	Equipment	Total	Total Incl O&P
						2009 Bare Costs				
0030	7-1/2" wide, 6-1/4" exposure, smooth texture	2 Carp	425	.033	S.F.	1.36	1.50		2.86	3.98
0035	Woodgrain texture		425	.033		1.36	1.50		2.86	3.98
0040	8" wide, 6-3/4" exposure, smooth texture		425	.033		1.51	1.50		3.01	4.14
0045	Roughsawn texture		425	.033		1.51	1.50		3.01	4.14
0050	9-1/2" wide, 8-1/4" exposure, smooth texture		440	.036		1.45	1.45		2.90	4
0055	Woodgrain texture		440	.036		1.45	1.45		2.90	4
0060	12" wide, 10-3/8" exposure, smooth texture		455	.035		1.34	1.40		2.74	3.79
0065	Woodgrain texture		455	.035		1.34	1.40		2.74	3.79
0070	Panel siding, 5/16" thick, smooth texture		750	.021		1.12	.85		1.97	2.65
0075	Stucco texture		750	.021		1.12	.85		1.97	2.65
0080	Grooved woodgrain texture		750	.021		1.12	.85		1.97	2.65
0085	V - grooved woodgrain texture		750	.021		1.12	.85		1.97	2.65
0090	Wood starter strip		400	.040	L.F.	.52	1.60		2.12	3.21

07 46 73 – Soffit

07 46 73.10 Soffit Options		Crew	Daily Output	Labor-Hours	Unit	Material	Labor	Equipment	Total	Total Incl O&P
0010	**SOFFIT OPTIONS**									
0012	Aluminum, residential, .020" thick	1 Carp	210	.038	S.F.	1.62	1.52		3.14	4.29
0100	Baked enamel on steel, 16 or 18 gauge		105	.076		5.35	3.04		8.39	10.90
0300	Polyvinyl chloride, white, solid		230	.035		1.09	1.39		2.48	3.49
0400	Perforated		230	.035		1.09	1.39		2.48	3.49
0500	For colors, add					.13			.13	.14
9000	Minimum labor/equipment charge	1 Carp	3	2.667	Job		107		107	176

07 51 Built-Up Bituminous Roofing

07 51 13 – Built-Up Asphalt Roofing

07 51 13.10 Built-Up Roofing Components		Crew	Daily Output	Labor-Hours	Unit	Material	Labor	Equipment	Total	Total Incl O&P
0010	**BUILT-UP ROOFING COMPONENTS**									
0012	Asphalt saturated felt, #30, 2 square per roll	1 Rofc	58	.138	Sq.	9.55	4.72		14.27	18.95
0200	#15, 4 sq per roll, plain or perforated, not mopped		58	.138		4.77	4.72		9.49	13.70
0300	Roll roofing, smooth, #65		15	.533		9.30	18.25		27.55	42.50
0500	#90		15	.533		32	18.25		50.25	67.50
0520	Mineralized		15	.533		33.50	18.25		51.75	69
0540	D.C. (Double coverage), 19" selvage edge		10	.800		49.50	27.50		77	104
0580	Adhesive (lap cement)				Gal.	4.50			4.50	4.95
0600	Steep, flat or dead level asphalt, 10 ton lots, bulk				Ton	445			445	490
0800	Packaged				"	755			755	835
9000	Minimum labor/equipment charge	1 Rofc	4	2	Job		68.50		68.50	122

07 51 13.13 Cold-Applied Built-Up Asphalt Roofing		Crew	Daily Output	Labor-Hours	Unit	Material	Labor	Equipment	Total	Total Incl O&P
0010	**COLD-APPLIED BUILT-UP ASPHALT ROOFING**									
0020	3 ply system, installation only (components listed below)	G-5	50	.800	Sq.		25	3.46	28.46	48.50
0100	Spunbond poly. fabric, 1.35 oz/SY, 36"W, 10.8 Sq/roll				Ea.	177			177	195
0200	49" wide, 14.6 Sq./roll					244			244	268
0300	2.10 oz./S.Y., 36" wide, 10.8 Sq./roll					266			266	293
0400	49" wide, 14.6 Sq./roll					360			360	395
0500	Base & finish coat, 3 gal./Sq., 5 gal./can				Gal.	4.70			4.70	5.15
0600	Coating, ceramic granules, 1/2 Sq./bag				Ea.	16.60			16.60	18.25
0700	Aluminum, 2 gal./Sq.				Gal.	13.35			13.35	14.70
0800	Emulsion, fibered or non-fibered, 4 gal./Sq.				"	6.15			6.15	6.75

07 51 13.20 Built-Up Roofing Systems		Crew	Daily Output	Labor-Hours	Unit	Material	Labor	Equipment	Total	Total Incl O&P
0010	**BUILT-UP ROOFING SYSTEMS**									

07 51 Built-Up Bituminous Roofing

07 51 13 – Built-Up Asphalt Roofing

07 51 13.20 Built-Up Roofing Systems		Crew	Daily Output	Labor-Hours	Unit	Material	2009 Bare Costs Labor	Equipment	Total	Total Incl O&P
0120	Asphalt flood coat with gravel/slag surfacing, not including									
0140	Insulation, flashing or wood nailers									
0200	Asphalt base sheet, 3 plies #15 asphalt felt, mopped	G-1	22	2.545	Sq.	84	81.50	20	185.50	260
0350	On nailable decks		21	2.667		88.50	85.50	21	195	273
0500	4 plies #15 asphalt felt, mopped		20	2.800		117	89.50	22	228.50	310
0550	On nailable decks		19	2.947		104	94.50	23	221.50	310
0700	Coated glass base sheet, 2 plies glass (type IV), mopped		22	2.545		85	81.50	20	186.50	261
0850	3 plies glass, mopped		20	2.800		102	89.50	22	213.50	296
0950	On nailable decks		19	2.947		95	94.50	23	212.50	299
1100	4 plies glass fiber felt (type IV), mopped		20	2.800		125	89.50	22	236.50	320
1150	On nailable decks		19	2.947		112	94.50	23	229.50	315
1200	Coated & saturated base sheet, 3 plies #15 asph. felt, mopped		20	2.800		93.50	89.50	22	205	287
1250	On nailable decks		19	2.947		87	94.50	23	204.50	290
1300	4 plies #15 asphalt felt, mopped		22	2.545		109	81.50	20	210.50	287
2000	Asphalt flood coat, smooth surface									
2200	Asphalt base sheet & 3 plies #15 asphalt felt, mopped	G-1	24	2.333	Sq.	89.50	74.50	18.35	182.35	252
2400	On nailable decks		23	2.435		83	78	19.15	180.15	252
2600	4 plies #15 asphalt felt, mopped		24	2.333		105	74.50	18.35	197.85	268
2700	On nailable decks		23	2.435		98.50	78	19.15	195.65	268
2900	Coated glass fiber base sheet, mopped, and 2 plies of									
2910	glass fiber felt (type IV)	G-1	25	2.240	Sq.	80	71.50	17.60	169.10	235
3100	On nailable decks		24	2.333		75	74.50	18.35	167.85	236
3200	3 plies, mopped		23	2.435		96.50	78	19.15	193.65	266
3300	On nailable decks		22	2.545		90	81.50	20	191.50	266
3800	4 plies glass fiber felt (type IV), mopped		23	2.435		113	78	19.15	210.15	284
3900	On nailable decks		22	2.545		106	81.50	20	207.50	284
4000	Coated & saturated base sheet, 3 plies #15 asph. felt, mopped		24	2.333		88.50	74.50	18.35	181.35	250
4200	On nailable decks		23	2.435		82	78	19.15	179.15	250
4300	4 plies #15 organic felt, mopped		22	2.545		104	81.50	20	205.50	281
4500	Coal tar pitch with gravel/slag surfacing									
4600	4 plies #15 tarred felt, mopped	G-1	21	2.667	Sq.	156	85.50	21	262.50	345
4800	3 plies glass fiber felt (type IV), mopped	"	19	2.947	"	127	94.50	23	244.50	335
5000	Coated glass fiber base sheet, and 2 plies of									
5010	glass fiber felt, (type IV), mopped	G-1	19	2.947	Sq.	127	94.50	23	244.50	335
5300	On nailable decks		18	3.111		113	99.50	24.50	237	330
5600	4 plies glass fiber felt (type IV), mopped		21	2.667		177	85.50	21	283.50	370
5800	On nailable decks		20	2.800		163	89.50	22	274.50	365

07 51 13.30 Cants		Crew	Daily Output	Labor-Hours	Unit	Material	Labor	Equipment	Total	Total Incl O&P
0010	**CANTS**									
0012	Lumber, treated, 4" x 4" cut diagonally	1 Rofc	325	.025	L.F.	1.52	.84		2.36	3.18
0100	Foamglass		325	.025		2.23	.84		3.07	3.96
0300	Mineral or fiber, trapezoidal, 1"x 4" x 48"		325	.025		.19	.84		1.03	1.72
0400	1-1/2" x 5-5/8" x 48"		325	.025		.31	.84		1.15	1.85
9000	Minimum labor/equipment charge		4	2	Job		68.50		68.50	122

07 51 13.40 Felts		Crew	Daily Output	Labor-Hours	Unit	Material	Labor	Equipment	Total	Total Incl O&P
0010	**FELTS**									
0012	Glass fibered roofing felt, #15, not mopped	1 Rofc	58	.138	Sq.	6	4.72		10.72	15.05
0300	Base sheet, #45, channel vented		58	.138		25.50	4.72		30.22	36.50
0400	#50, coated		58	.138		13.80	4.72		18.52	23.50
0500	Cap, mineral surfaced		58	.138		25.50	4.72		30.22	36.50
0600	Flashing membrane, #65		16	.500		39	17.15		56.15	73.50
0800	Coal tar fibered, #15, no mopping		58	.138		11.60	4.72		16.32	21

07 51 Built-Up Bituminous Roofing

07 51 13 – Built-Up Asphalt Roofing

	07 51 13.40 Felts	Crew	Daily Output	Labor-Hours	Unit	2009 Bare Costs Material	Labor	Equipment	Total	Total Incl O&P
0900	Asphalt felt, #15, 4 sq per roll, no mopping	1 Rofc	58	.138	Sq.	4.77	4.72		9.49	13.70
1100	#30, 2 sq per roll		58	.138		9.55	4.72		14.27	18.95
1200	Double coated, #33		58	.138		9.50	4.72		14.22	18.90
1400	#40, base sheet		58	.138		10.80	4.72		15.52	20.50
1450	Coated and saturated		58	.138		9.45	4.72		14.17	18.85
1500	Tarred felt, organic, #15, 4 sq rolls		58	.138		13.90	4.72		18.62	24
1550	#30, 2 sq roll		58	.138		24.50	4.72		29.22	35.50
1700	Add for mopping above felts, per ply, asphalt, 24 lb per sq	G-1	192	.292		9.10	9.35	2.29	20.74	29
1800	Coal tar mopping, 30 lb per sq		186	.301		12.60	9.65	2.37	24.62	33.50
1900	Flood coat, with asphalt, 60 lb per sq		60	.933		22.50	30	7.35	59.85	86.50
2000	With coal tar, 75 lb per sq		56	1		31.50	32	7.85	71.35	100
9000	Minimum labor/equipment charge	1 Rofc	4	2	Job		68.50		68.50	122

	07 51 13.50 Walkways for Built-Up Roofs	Crew	Daily Output	Labor-Hours	Unit	Material	Labor	Equipment	Total	Total Incl O&P
0010	**WALKWAYS FOR BUILT-UP ROOFS**									
0020	Asphalt impregnated, 3′ x 6′ x 1/2″ thick	1 Rofc	400	.020	S.F.	1.55	.69		2.24	2.93
0100	3′ x 3′ x 3/4″ thick	"	400	.020		2.93	.69		3.62	4.44
0300	Concrete patio blocks, 2″ thick, natural	1 Clab	115	.070		2.59	2.20		4.79	6.50
0400	Colors	"	115	.070		2.79	2.20		4.99	6.70
9000	Minimum labor/equipment charge	1 Rofc	2.75	2.909	Job		99.50		99.50	178

07 52 Modified Bituminous Membrane Roofing

07 52 13 – Atactic-Polypropylene-Modified Bituminous Membrane Roofing

	07 52 13.10 APP Modified Bituminous Membrane	Crew	Daily Output	Labor-Hours	Unit	Material	Labor	Equipment	Total	Total Incl O&P
0010	**APP MODIFIED BITUMINOUS MEMBRANE** R075213-30									
0020	Base sheet, #15 glass fiber felt, nailed to deck	1 Rofc	58	.138	Sq.	6.75	4.72		11.47	15.90
0030	Spot mopped to deck	G-1	295	.190		10.55	6.05	1.49	18.09	24
0040	Fully mopped to deck	"	192	.292		15.10	9.35	2.29	26.74	36
0050	#15 organic felt, nailed to deck	1 Rofc	58	.138		5.55	4.72		10.27	14.55
0060	Spot mopped to deck	G-1	295	.190		9.30	6.05	1.49	16.84	22.50
0070	Fully mopped to deck	"	192	.292		13.85	9.35	2.29	25.49	34.50
2100	APP mod., smooth surf. cap sheet, poly. reinf., torched, 160 mils	G-5	2100	.019	S.F.	.48	.59	.08	1.15	1.68
2150	170 mils		2100	.019		.55	.59	.08	1.22	1.76
2200	Granule surface cap sheet, poly. reinf., torched, 180 mils		2000	.020		.59	.62	.09	1.30	1.86
2250	Smooth surface flashing, torched, 160 mils		1260	.032		.48	.99	.14	1.61	2.44
2300	170 mils		1260	.032		.55	.99	.14	1.68	2.52
2350	Granule surface flashing, torched, 180 mils		1260	.032		.59	.99	.14	1.72	2.56
2400	Fibrated aluminum coating	1 Rofc	3800	.002		.12	.07		.19	.26

07 52 16 – Styrene-Butadiene-Styrene Modified Bituminous Membrane Roofing

	07 52 16.10 SBS Modified Bituminous Membrane	Crew	Daily Output	Labor-Hours	Unit	Material	Labor	Equipment	Total	Total Incl O&P
0010	**SBS MODIFIED BITUMINOUS MEMBRANE**									
0080	SBS modified, granule surf cap sheet, polyester rein., mopped									
0600	150 mils	G-1	2000	.028	S.F.	.53	.90	.22	1.65	2.42
1100	160 mils		2000	.028		.77	.90	.22	1.89	2.69
1500	Glass fiber reinforced, mopped, 160 mils		2000	.028		.49	.90	.22	1.61	2.38
1600	Smooth surface cap sheet, mopped, 145 mils		2100	.027		.49	.85	.21	1.55	2.29
1700	Smooth surface flashing, 145 mils		1260	.044		.49	1.42	.35	2.26	3.46
1800	150 mils		1260	.044		.48	1.42	.35	2.25	3.45
1900	Granular surface flashing, 150 mils		1260	.044		.53	1.42	.35	2.30	3.50
2000	160 mils		1260	.044		.77	1.42	.35	2.54	3.77

07 53 Elastomeric Membrane Roofing

07 53 16 – Chlorosulfonate Polyethylene Roofing

07 53 16.10 Chlorosulfonated Polyethylene Roofing		Crew	Daily Output	Labor-Hours	Unit	2009 Bare Costs Material	Labor	Equipment	Total	Total Incl O&P
0010	**CHLOROSULFONATED POLYETHYLENE ROOFING**									
0800	Chlorosulfonated polyethylene-hypalon (CSPE), 45 mils,									
0900	0.29 P.S.F., fully adhered	G-5	26	1.538	Sq.	185	48	6.65	239.65	297
1100	Loose-laid & ballasted with stone (10 P.S.F.)		51	.784		193	24.50	3.40	220.90	259
1200	Mechanically attached		35	1.143		186	35.50	4.95	226.45	274
1300	Plates with adhesive attachment		35	1.143		183	35.50	4.95	223.45	270

07 53 23 – Ethylene-Propylene-Diene-Monomer Roofing

07 53 23.20 Ethylene-Propylene-Diene-Monomer Roofing		Crew	Daily Output	Labor-Hours	Unit	Material	Labor	Equipment	Total	Total Incl O&P
0010	**ETHYLENE-PROPYLENE-DIENE-MONOMER ROOFING (E.P.D.M.)**									
3500	Ethylene propylene diene monomer (EPDM), 45 mils, 0.28 P.S.F.									
3600	Loose-laid & ballasted with stone (10 P.S.F.)	G-5	51	.784	Sq.	74	24.50	3.40	101.90	128
3700	Mechanically attached		35	1.143		64.50	35.50	4.95	104.95	140
3800	Fully adhered with adhesive		26	1.538		93	48	6.65	147.65	195
4500	60 mils, 0.40 P.S.F.									
4600	Loose-laid & ballasted with stone (10 P.S.F.)	G-5	51	.784	Sq.	92.50	24.50	3.40	120.40	149
4700	Mechanically attached		35	1.143		82	35.50	4.95	122.45	159
4800	Fully adhered with adhesive		26	1.538		111	48	6.65	165.65	215
4810	45 mil, .28 PSF, membrane only					37			37	40.50
4820	60 mil, .40 PSF, membrane only					53			53	58.50
4850	Seam tape for membrane, 3" x 100' roll				Ea.	43.50			43.50	47.50
4900	Batten strips, 10' sections					3.45			3.45	3.80
4910	Cover tape for batten strips, 6" x 100' roll					135			135	149
4930	Plate anchors				M	65			65	71.50
4970	Adhesive for fully adhered systems, 60 S.F./gal.				Gal.	17.65			17.65	19.40

07 53 29 – Polyisobutylene Roofing

07 53 29.10 Polyisobutylene Roofing		Crew	Daily Output	Labor-Hours	Unit	Material	Labor	Equipment	Total	Total Incl O&P
0010	**POLYISOBUTYLENE ROOFING**									
7500	Polyisobutylene (PIB), 100 mils, 0.57 P.S.F.									
7600	Loose-laid & ballasted with stone/gravel (10 P.S.F.)	G-5	51	.784	Sq.	181	24.50	3.40	208.90	246
7700	Partially adhered with adhesive		35	1.143		227	35.50	4.95	267.45	320
7800	Hot asphalt attachment		35	1.143		217	35.50	4.95	257.45	305
7900	Fully adhered with contact cement		26	1.538		234	48	6.65	288.65	350

07 54 Thermoplastic Membrane Roofing

07 54 19 – Polyvinyl-Chloride Roofing

07 54 19.10 Polyvinyl-Chloride Roofing (P.V.C.)		Crew	Daily Output	Labor-Hours	Unit	Material	Labor	Equipment	Total	Total Incl O&P
0010	**POLYVINYL-CHLORIDE ROOFING (P.V.C.)**									
8200	Heat welded seams									
8700	Reinforced, 48 mils, 0.33 P.S.F.									
8750	Loose-laid & ballasted with stone/gravel (12 P.S.F.)	G-5	51	.784	Sq.	116	24.50	3.40	143.90	175
8800	Mechanically attached		35	1.143		106	35.50	4.95	146.45	186
8850	Fully adhered with adhesive		26	1.538		155	48	6.65	209.65	264
8860	Reinforced, 60 mils, .40 P.S.F.									
8870	Loose-laid & ballasted with stone/gravel (12 P.S.F.)	G-5	51	.784	Sq.	118	24.50	3.40	145.90	177
8880	Mechanically attached		35	1.143		108	35.50	4.95	148.45	188
8890	Fully adhered with adhesive		26	1.538		157	48	6.65	211.65	266

07 54 23 – Thermoplastic Polyolefin Roofing

07 54 23.10 Thermoplastic Polyolefin Roofing (T.P.O)		Crew	Daily Output	Labor-Hours	Unit	Material	Labor	Equipment	Total	Total Incl O&P
0010	**THERMOPLASTIC POLYOLEFIN ROOFING (T.P.O.)**									

07 54 Thermoplastic Membrane Roofing

07 54 23 – Thermoplastic Polyolefin Roofing

07 54 23.10 Thermoplastic Polyolefin Roofing (T.P.O)		Crew	Daily Output	Labor-Hours	Unit	2009 Bare Costs Material	Labor	Equipment	Total	Total Incl O&P
0100	45 mils, loose laid & ballasted with stone(1/2 ton/sq.)	G-5	51	.784	Sq.	98.50	24.50	3.40	126.40	156
0120	Fully adhered		25	1.600		144	49.50	6.95	200.45	255
0140	Mechanically attached		34	1.176		72.50	36.50	5.10	114.10	151
0160	Self adhered		35	1.143		81.50	35.50	4.95	121.95	158
0180	60 mil membrane, heat welded seams, ballasted		50	.800		116	25	3.46	144.46	176
0200	Fully adhered		25	1.600		161	49.50	6.95	217.45	274
0220	Mechanically attached		34	1.176		106	36.50	5.10	147.60	188
0240	Self adhered	↓	35	1.143	↓	100	35.50	4.95	140.45	179

07 56 Fluid-Applied Roofing

07 56 10 – Fluid-Applied Roofing Elastomers

07 56 10.10 Elastomeric Roofing

Line	Description	Crew	Daily Output	Labor-Hours	Unit	Material	Labor	Equipment	Total	Total Incl O&P
0010	**ELASTOMERIC ROOFING**									
0110	Acrylic rubber, fluid applied, 20 mils thick	G-5	2000	.020	S.F.	2.27	.62	.09	2.98	3.71
0120	50 mils, reinforced		1200	.033		3.52	1.04	.14	4.70	5.90
0130	For walking surface, add	↓	900	.044		1.07	1.38	.19	2.64	3.86
0300	Hypalon neoprene, fluid applied, 20 mil thick, not-reinforced	G-1	1135	.049		2.57	1.58	.39	4.54	6.10
0600	Non-woven polyester, reinforced		960	.058		2.61	1.87	.46	4.94	6.70
0700	5 coat neoprene deck, 60 mil thick, under 10,000 SF		325	.172		5.75	5.50	1.35	12.60	17.65
0900	Over 10,000 SF		625	.090		5.35	2.87	.70	8.92	11.75
1300	Vinyl plastic traffic deck, sprayed, 2 to 4 mils thick		625	.090		1.68	2.87	.70	5.25	7.70
1500	Vinyl and neoprene membrane traffic deck	↓	1550	.036	↓	1.77	1.16	.28	3.21	4.32
9000	Minimum labor/equipment charge	1 Rofc	2	4	Job		137		137	245

07 57 Coated Foamed Roofing

07 57 13 – Sprayed Polyurethane Foam Roofing

07 57 13.10 Sprayed Polyurethane Foam Roofing (S.P.F.)

Line	Description	Crew	Daily Output	Labor-Hours	Unit	Material	Labor	Equipment	Total	Total Incl O&P
0010	**SPRAYED POLYURETHANE FOAM ROOFING (S.P.F.)**									
0100	Primer for metal substrate (when required)	G-2A	3000	.008	S.F.	.42	.24	.20	.86	1.10
0200	Primer for non-metal substrate (when required)		3000	.008		.16	.24	.20	.60	.82
0300	Closed cell spray, polyureathane foam, 3 lbs per CF density, 1", R6.7		15000	.002		.70	.05	.04	.79	.89
0400	2", R13.4		13125	.002		1.40	.06	.05	1.51	1.69
0500	3", R20.1		11485	.002		2.10	.06	.05	2.21	2.48
0700	Spray-on silicone coating	↓	2500	.010		.92	.29	.24	1.45	1.79
0800	Warranty 5-20 year manufacturer's									.15
0900	Warranty 20 year, no dollar limit				↓					.20

07 58 Roll Roofing

07 58 10 – Asphalt Roll Roofing

07 58 10.10 Roll Roofing

Line	Description	Crew	Daily Output	Labor-Hours	Unit	Material	Labor	Equipment	Total	Total Incl O&P
0010	**ROLL ROOFING**									
0100	Asphalt, mineral surface									
0200	1 ply #15 organic felt, 1 ply mineral surfaced									
0300	Selvage roofing, lap 19", nailed & mopped	G-1	27	2.074	Sq.	67.50	66.50	16.30	150.30	210
0400	3 plies glass fiber felt (type IV), 1 ply mineral surfaced									
0500	Selvage roofing, lapped 19", mopped	G-1	25	2.240	Sq.	104	71.50	17.60	193.10	261
0600	Coated glass fiber base sheet, 2 plies of glass fiber									

07 58 Roll Roofing

07 58 10 – Asphalt Roll Roofing

07 58 10.10 Roll Roofing		Crew	Daily Output	Labor-Hours	Unit	Material	Labor	Equipment	Total	Total Incl O&P
						2009 Bare Costs				
0700	Felt (type IV), 1 ply mineral surfaced selvage									
0800	Roofing, lapped 19", mopped	G-1	25	2.240	Sq.	112	71.50	17.60	201.10	270
0900	On nailable decks	"	24	2.333	"	102	74.50	18.35	194.85	266
1000	3 plies glass fiber felt (type III), 1 ply mineral surfaced									
1100	Selvage roofing, lapped 19", mopped	G-1	25	2.240	Sq.	104	71.50	17.60	193.10	261

07 61 Sheet Metal Roofing

07 61 13 – Standing Seam Sheet Metal Roofing

07 61 13.10 Standing Seam Sheet Metal Roofing, Field Fab.

		Crew	Daily Output	Labor-Hours	Unit	Material	Labor	Equipment	Total	Total Incl O&P
0010	**STANDING SEAM SHEET METAL ROOFING, FIELD FABRICATED**									
0400	Copper standing seam roofing, over 10 squares, 16 oz, 125 lb per sq	1 Shee	1.30	6.154	Sq.	1,075	290		1,365	1,625
0600	18 oz, 140 lb per sq	↓	1.20	6.667		1,200	315		1,515	1,800
0700	20 oz, 150 lb per sq	↓	1.10	7.273		1,300	345		1,645	2,000
1200	For abnormal conditions or small areas, add					25%	100%			
1300	For lead-coated copper, add				↓	25%				
9000	Minimum labor/equipment charge	1 Shee	2	4	Job		189		189	300

07 61 16 – Batten Seam Sheet Metal Roofing

07 61 16.10 Batten Seam Sheet Metal Roofing, Field Fabricated

		Crew	Daily Output	Labor-Hours	Unit	Material	Labor	Equipment	Total	Total Incl O&P
0010	**BATTEN SEAM SHEET METAL ROOFING, FIELD FABRICATED**									
0012	Copper batten seam roofing, over 10 sq, 16 oz, 130 lb per sq	1 Shee	1.10	7.273	Sq.	1,075	345		1,420	1,725
0020	Lead batten seam roofing, 5 lb. per SF		1.20	6.667		565	315		880	1,125
0100	Zinc/copper alloy batten seam roofing, .020 thick		1.20	6.667		1,050	315		1,365	1,675
0200	Copper roofing, batten seam, over 10 sq, 18 oz, 145 lb per sq		1	8		1,200	380		1,580	1,900
0300	20 oz, 160 lb per sq		1	8		1,300	380		1,680	2,050
0500	Stainless steel batten seam roofing, type 304, 28 gauge		1.20	6.667		685	315		1,000	1,250
0600	26 gauge		1.15	6.957		635	330		965	1,225
0800	Zinc, copper alloy roofing, batten seam, .027" thick		1.15	6.957		1,500	330		1,830	2,200
0900	.032" thick		1.10	7.273		1,800	345		2,145	2,525
1000	.040" thick		1.05	7.619	↓	2,150	360		2,510	2,950
9000	Minimum labor/equipment charge	↓	2	4	Job		189		189	300

07 61 19 – Flat Seam Sheet Metal Roofing

07 61 19.10 Flat Seam Sheet Metal Roofing, Field Fabricated

		Crew	Daily Output	Labor-Hours	Unit	Material	Labor	Equipment	Total	Total Incl O&P
0010	**FLAT SEAM SHEET METAL ROOFING, FIELD FABRICATED**									
0900	Copper flat seam roofing, over 10 squares, 16 oz, 115 lb per sq	1 Shee	1.20	6.667	Sq.	1,075	315		1,390	1,675
1000	20 oz, 145 lb per sq		1.10	7.273		1,300	345		1,645	2,000
1100	Lead flat seam roofing, 5 lb per SF		1.30	6.154	↓	565	290		855	1,075
9000	Minimum labor/equipment charge	↓	2.75	2.909	Job		137		137	218

07 65 Flexible Flashing

07 65 10 – Sheet Metal Flashing

07 65 10.10 Sheet Metal Flashing and Counter Flashing		Crew	Daily Output	Labor-Hours	Unit	2009 Bare Costs Material	Labor	Equipment	Total	Total Incl O&P
0010	SHEET METAL FLASHING AND COUNTER FLASHING									
0011	Including up to 4 bends									
0020	Aluminum, mill finish, .013" thick	1 Rofc	145	.055	S.F.	.67	1.89		2.56	4.11
0030	.016" thick		145	.055		.79	1.89		2.68	4.24
0060	.019" thick		145	.055		1.02	1.89		2.91	4.49
0100	.032" thick		145	.055		1.48	1.89		3.37	5
0200	.040" thick		145	.055		2.21	1.89		4.10	5.80
0300	.050" thick		145	.055		2.52	1.89		4.41	6.15
0325	Mill finish 5" x 7" step flashing, .016" thick		1920	.004	Ea.	.16	.14		.30	.44
0350	Mill finish 12" x 12" step flashing, .016" thick		1600	.005	"	.61	.17		.78	.98
0400	Painted finish, add				S.F.	.37			.37	.41
1000	Mastic-coated 2 sides, .005" thick	1 Rofc	330	.024		1.67	.83		2.50	3.32
1100	.016" thick		330	.024		1.84	.83		2.67	3.50
1600	Copper, 16 oz, sheets, under 1000 lbs.		115	.070		6.80	2.38		9.18	11.70
1700	Over 4000 lbs.		155	.052		8	1.77		9.77	11.95
1900	20 oz sheets, under 1000 lbs.		110	.073		7.90	2.49		10.39	13.15
2000	Over 4000 lbs.		145	.055		8	1.89		9.89	12.15
2200	24 oz sheets, under 1000 lbs.		105	.076		9.60	2.61		12.21	15.25
2300	Over 4000 lbs.		135	.059		9.75	2.03		11.78	14.30
2500	32 oz sheets, under 1000 lbs.		100	.080		12.80	2.74		15.54	19
2600	Over 4000 lbs.		130	.062		20	2.11		22.11	26
2700	W shape for valleys, 16 oz, 24" wide		100	.080	L.F.	14.65	2.74		17.39	21
5800	Lead, 2.5 lb. per SF, up to 12" wide		135	.059	S.F.	4.14	2.03		6.17	8.15
5900	Over 12" wide		135	.059		4.14	2.03		6.17	8.15
8650	Copper, 16 oz		100	.080		5.25	2.74		7.99	10.65
8900	Stainless steel sheets, 32 ga, .010" thick		155	.052		3.79	1.77		5.56	7.35
9000	28 ga, .015" thick		155	.052		4.70	1.77		6.47	8.30
9100	26 ga, .018" thick		155	.052		5.70	1.77		7.47	9.40
9200	24 ga, .025" thick		155	.052		7.40	1.77		9.17	11.30
9290	For mechanically keyed flashing, add					40%				
9320	Steel sheets, galvanized, 20 gauge	1 Rofc	130	.062	S.F.	1.19	2.11		3.30	5.05
9340	30 gauge		160	.050		.50	1.71		2.21	3.61
9400	Terne coated stainless steel, .015" thick, 28 ga		155	.052		7.35	1.77		9.12	11.25
9500	.018" thick, 26 ga		155	.052		8.30	1.77		10.07	12.25
9600	Zinc and copper alloy (brass), .020" thick		155	.052		5.15	1.77		6.92	8.80
9700	.027" thick		155	.052		6.90	1.77		8.67	10.75
9800	.032" thick		155	.052		8.05	1.77		9.82	12
9900	.040" thick		155	.052		9.85	1.77		11.62	13.95
9950	Minimum labor/equipment charge		3	2.667	Job		91.50		91.50	163

07 65 12 – Fabric and Mastic Flashings

07 65 12.10 Fabric and Mastic Flashing and Counter Flashing		Crew	Daily Output	Labor-Hours	Unit	Material	Labor	Equipment	Total	Total Incl O&P
0010	FABRIC AND MASTIC FLASHING AND COUNTER FLASHING									
1300	Asphalt flashing cement, 5 gallon				Gal.	5.80			5.80	6.40
4900	Fabric, asphalt-saturated cotton, specification grade	1 Rofc	35	.229	S.Y.	2.37	7.85		10.22	16.60
5000	Utility grade		35	.229		1.50	7.85		9.35	15.65
5200	Open-mesh fabric, saturated, 40 oz per S.Y.		35	.229		1.67	7.85		9.52	15.85
5300	Close-mesh fabric, saturated, 17 oz per S.Y.		35	.229		1.75	7.85		9.60	15.95
5500	Fiberglass, resin-coated		35	.229		1.43	7.85		9.28	15.55
5600	Asphalt-coated, 40 oz per S.Y.		35	.229		9.75	7.85		17.60	24.50
8500	Shower pan, bituminous membrane, 7 oz		155	.052	S.F.	1.83	1.77		3.60	5.15

07 65 Flexible Flashing

07 65 13 – Laminated Sheet Flashing

07 65 13.10 Laminated Sheet Flashing		Crew	Daily Output	Labor-Hours	Unit	Material	Labor	Equipment	Total	Total Incl O&P
						2009 Bare Costs				
0010	**LAMINATED SHEET FLASHING**, Including up to 4 bends									
0500	Fabric-backed 2 sides, .004" thick	1 Rofc	330	.024	S.F.	1.41	.83		2.24	3.03
0700	.005" thick		330	.024		1.67	.83		2.50	3.32
0750	Mastic-backed, self adhesive		460	.017		3.46	.60		4.06	4.87
0800	Mastic-coated 2 sides, .004" thick		330	.024		1.41	.83		2.24	3.03
2800	Copper, paperbacked 1 side, 2 oz		330	.024		1.42	.83		2.25	3.04
2900	3 oz		330	.024		1.85	.83		2.68	3.52
3100	Paperbacked 2 sides, 2 oz		330	.024		1.43	.83		2.26	3.05
3150	3 oz		330	.024		1.84	.83		2.67	3.50
3200	5 oz		330	.024		2.76	.83		3.59	4.52
3250	7 oz		330	.024		4.50	.83		5.33	6.45
3400	Mastic-backed 2 sides, copper, 2 oz		330	.024		1.69	.83		2.52	3.34
3500	3 oz		330	.024		2.08	.83		2.91	3.77
3700	5 oz		330	.024		3.05	.83		3.88	4.84
3800	Fabric-backed 2 sides, copper, 2 oz		330	.024		1.79	.83		2.62	3.45
4000	3 oz		330	.024		2.31	.83		3.14	4.02
4100	5 oz		330	.024		3.14	.83		3.97	4.93
4300	Copper-clad stainless steel, .015" thick, under 500 lbs.		115	.070		5.25	2.38		7.63	10
4400	Over 2000 lbs.		155	.052		5.05	1.77		6.82	8.70
4600	.018" thick, under 500 lbs.		100	.080		6.90	2.74		9.64	12.50
4700	Over 2000 lbs.		145	.055		5.05	1.89		6.94	8.90
6100	Lead-coated copper, fabric-backed, 2 oz		330	.024		2.42	.83		3.25	4.14
6200	5 oz		330	.024		2.78	.83		3.61	4.54
6400	Mastic-backed 2 sides, 2 oz		330	.024		1.89	.83		2.72	3.56
6500	5 oz		330	.024		2.34	.83		3.17	4.05
6700	Paperbacked 1 side, 2 oz		330	.024		1.63	.83		2.46	3.27
6800	3 oz		330	.024		1.92	.83		2.75	3.59
7000	Paperbacked 2 sides, 2 oz		330	.024		1.69	.83		2.52	3.34
7100	5 oz		330	.024		2.75	.83		3.58	4.51
8550	3 ply copper and fabric, 3 oz		155	.052		2.73	1.77		4.50	6.15
8600	7 oz		155	.052		5.70	1.77		7.47	9.40
8700	Lead on copper and fabric, 5 oz		155	.052		2.78	1.77		4.55	6.20
8800	7 oz		155	.052		4.94	1.77		6.71	8.60
9300	Stainless steel, paperbacked 2 sides, .005" thick		330	.024		3.34	.83		4.17	5.15

07 65 19 – Plastic Sheet Flashing

07 65 19.10 Plastic Sheet Flashing and Counter Flashing

		Crew	Daily Output	Labor-Hours	Unit	Material	Labor	Equipment	Total	Total Incl O&P
0010	**PLASTIC SHEET FLASHING AND COUNTER FLASHING**									
7300	Polyvinyl chloride, black, .010" thick	1 Rofc	285	.028	S.F.	.26	.96		1.22	2.01
7400	.020" thick		285	.028		.36	.96		1.32	2.12
7600	.030" thick		285	.028		.47	.96		1.43	2.24
7700	.056" thick		285	.028		1.11	.96		2.07	2.94
7900	Black or white for exposed roofs, .060" thick		285	.028		2.45	.96		3.41	4.42
8060	PVC tape, 5" x 45 mils, for joint covers, 100 L.F./roll				Ea.	120			120	132
8850	Polyvinyl chloride, .030" thick	1 Rofc	160	.050	S.F.	.43	1.71		2.14	3.53

07 65 23 – Rubber Sheet Flashing

07 65 23.10 Rubber Sheet Flashing and Counterflashing

		Crew	Daily Output	Labor-Hours	Unit	Material	Labor	Equipment	Total	Total Incl O&P
0010	**RUBBER SHEET FLASHING AND COUNTERFLASHING**									
8100	Rubber, butyl, 1/32" thick	1 Rofc	285	.028	S.F.	.96	.96		1.92	2.78
8200	1/16" thick		285	.028		1.43	.96		2.39	3.29
8300	Neoprene, cured, 1/16" thick		285	.028		2.03	.96		2.99	3.95
8400	1/8" thick		285	.028		4.09	.96		5.05	6.20

07 71 Roof Specialties

07 71 19 – Manufactured Gravel Stops and Fascias

07 71 19.10 Gravel Stop

07 71 19.10 Gravel Stop		Crew	Daily Output	Labor-Hours	Unit	Material	2009 Bare Costs Labor	Equipment	Total	Total Incl O&P
0010	**GRAVEL STOP**									
0020	Aluminum, .050" thick, 4" face height, mill finish	1 Shee	145	.055	L.F.	5.95	2.60		8.55	10.70
0080	Duranodic finish		145	.055		5.75	2.60		8.35	10.45
0100	Painted		145	.055		6.65	2.60		9.25	11.45
0300	6" face height		135	.059		6.40	2.80		9.20	11.45
0350	Duranodic finish		135	.059		6.75	2.80		9.55	11.90
0400	Painted		135	.059		7.85	2.80		10.65	13.10
0600	8" face height		125	.064		8	3.02		11.02	13.60
0650	Duranodic finish		125	.064		7.75	3.02		10.77	13.35
0700	Painted		125	.064		7.90	3.02		10.92	13.50
0900	12" face height, .080 thick, 2 piece		100	.080		10.60	3.78		14.38	17.70
0950	Duranodic finish		100	.080		9.75	3.78		13.53	16.75
1000	Painted		100	.080		11.55	3.78		15.33	18.70
1350	Galv steel, 24 ga., 4" leg, plain, with continuous cleat, 4" face		145	.055		2.34	2.60		4.94	6.70
1360	6" face height		145	.055		3.53	2.60		6.13	8
1500	Polyvinyl chloride, 6" face height		135	.059		4.32	2.80		7.12	9.20
1600	9" face height		125	.064		5.10	3.02		8.12	10.40
1800	Stainless steel, 24 ga., 6" face height		135	.059		11.15	2.80		13.95	16.70
1900	12" face height		100	.080		23.50	3.78		27.28	31.50
2100	20 ga., 6" face height		135	.059		12.60	2.80		15.40	18.30
2200	12" face height		100	.080		26.50	3.78		30.28	35.50
9000	Minimum labor/equipment charge		3.50	2.286	Job		108		108	171

07 71 19.30 Fascia

07 71 19.30 Fascia		Crew	Daily Output	Labor-Hours	Unit	Material	Labor	Equipment	Total	Total Incl O&P
0010	**FASCIA**									
0100	Aluminum, reverse board and batten, .032" thick, colored, no furring incl	1 Shee	145	.055	S.F.	5.75	2.60		8.35	10.45
0300	Steel, galv and enameled, stock, no furring, long panels		145	.055		3.76	2.60		6.36	8.30
0600	Short panels		115	.070		5	3.28		8.28	10.70
9000	Minimum labor/equipment charge		4	2	Job		94.50		94.50	150

07 71 23 – Manufactured Gutters and Downspouts

07 71 23.10 Downspouts

07 71 23.10 Downspouts		Crew	Daily Output	Labor-Hours	Unit	Material	Labor	Equipment	Total	Total Incl O&P
0010	**DOWNSPOUTS**									
0020	Aluminum 2" x 3", .020" thick, embossed	1 Shee	190	.042	L.F.	1.04	1.99		3.03	4.30
0100	Enameled		190	.042		1.59	1.99		3.58	4.91
0300	Enameled, .024" thick, 2" x 3"		180	.044		1.84	2.10		3.94	5.35
0400	3" x 4"		140	.057		2.55	2.70		5.25	7.10
0600	Round, corrugated aluminum, 3" diameter, .020" thick		190	.042		1.42	1.99		3.41	4.72
0700	4" diameter, .025" thick		140	.057		2.39	2.70		5.09	6.90
0900	Wire strainer, round, 2" diameter		155	.052	Ea.	3.32	2.44		5.76	7.50
1000	4" diameter		155	.052		6.10	2.44		8.54	10.60
1200	Rectangular, perforated, 2" x 3"		145	.055		2.30	2.60		4.90	6.65
1300	3" x 4"		145	.055		3.32	2.60		5.92	7.80
1500	Copper, round, 16 oz., stock, 2" diameter		190	.042	L.F.	9	1.99		10.99	13.05
1600	3" diameter		190	.042		8.60	1.99		10.59	12.60
1800	4" diameter		145	.055		9	2.60		11.60	14.05
1900	5" diameter		130	.062		14.75	2.90		17.65	21
2100	Rectangular, corrugated copper, stock, 2" x 3"		190	.042		7.15	1.99		9.14	11.05
2200	3" x 4"		145	.055		9.30	2.60		11.90	14.35
2400	Rectangular, plain copper, stock, 2" x 3"		190	.042		7	1.99		8.99	10.85
2500	3" x 4"		145	.055		9.10	2.60		11.70	14.15
2700	Wire strainers, rectangular, 2" x 3"		145	.055	Ea.	5.50	2.60		8.10	10.20
2800	3" x 4"		145	.055		6.60	2.60		9.20	11.40
3000	Round, 2" diameter		145	.055		4.40	2.60		7	9

07 71 23.10 Downspouts		Crew	Daily Output	Labor-Hours	Unit	Material	Labor	Equipment	Total	Total Incl O&P
						2009 Bare Costs				
3100	3" diameter	1 Shee	145	.055	Ea.	6.10	2.60		8.70	10.90
3300	4" diameter		145	.055		11.30	2.60		13.90	16.55
3400	5" diameter		115	.070		16.60	3.28		19.88	23.50
3600	Lead-coated copper, round, stock, 2" diameter		190	.042	L.F.	11.45	1.99		13.44	15.75
3700	3" diameter		190	.042		12.60	1.99		14.59	17
3900	4" diameter		145	.055		14.70	2.60		17.30	20.50
4000	5" diameter, corrugated		130	.062		23.50	2.90		26.40	30
4200	6" diameter, corrugated		105	.076		24	3.60		27.60	32
4300	Rectangular, corrugated, stock, 2" x 3"		190	.042		12.75	1.99		14.74	17.20
4500	Plain, stock, 2" x 3"		190	.042		9.75	1.99		11.74	13.90
4600	3" x 4"		145	.055		10	2.60		12.60	15.15
4800	Steel, galvanized, round, corrugated, 2" or 3" diameter, 28 gauge		190	.042		1.74	1.99		3.73	5.05
4900	4" diameter, 28 gauge		145	.055		2.48	2.60		5.08	6.85
5100	5" diameter, 28 gauge		130	.062		3.68	2.90		6.58	8.65
5200	26 gauge		130	.062		2.97	2.90		5.87	7.90
5400	6" diameter, 28 gauge		105	.076		4.83	3.60		8.43	11
5500	26 gauge		105	.076		6.10	3.60		9.70	12.45
5700	Rectangular, corrugated, 28 gauge, 2" x 3"		190	.042		1.60	1.99		3.59	4.92
5800	3" x 4"		145	.055		1.60	2.60		4.20	5.90
6000	Rectangular, plain, 28 gauge, galvanized, 2" x 3"		190	.042		1.74	1.99		3.73	5.05
6100	3" x 4"		145	.055		2.51	2.60		5.11	6.90
6300	Epoxy painted, 24 gauge, corrugated, 2" x 3"		190	.042		1.90	1.99		3.89	5.25
6400	3" x 4"		145	.055		2.12	2.60		4.72	6.45
6600	Wire strainers, rectangular, 2" x 3"		145	.055	Ea.	5.90	2.60		8.50	10.65
6700	3" x 4"		145	.055		11.75	2.60		14.35	17.10
6900	Round strainers, 2" or 3" diameter		145	.055		2.84	2.60		5.44	7.25
7000	4" diameter		145	.055		11.75	2.60		14.35	17.10
7800	Stainless steel tubing, schedule 5, 2" x 3" or 3" diameter		190	.042	L.F.	49.50	1.99		51.49	57
7900	3" x 4" or 4" diameter		145	.055		62.50	2.60		65.10	73
8100	4" x 5" or 5" diameter		135	.059		129	2.80		131.80	146
8200	Vinyl, rectangular, 2" x 3"		210	.038		.75	1.80		2.55	3.69
8300	Round, 2-1/2"		220	.036		.75	1.72		2.47	3.56
9000	Minimum labor/equipment charge		4	2	Job		94.50		94.50	150

07 71 23.20 Downspout Elbows

		Crew	Daily Output	Labor-Hours	Unit	Material	Labor	Equipment	Total	Total Incl O&P
0010	**DOWNSPOUT ELBOWS**									
0020	Aluminum, 2" x 3", embossed	1 Shee	100	.080	Ea.	1.17	3.78		4.95	7.30
0100	Enameled		100	.080		1.17	3.78		4.95	7.30
0200	3" x 4", .025" thick, embossed		100	.080		3.99	3.78		7.77	10.40
0300	Enameled		100	.080		5.15	3.78		8.93	11.65
0400	Round corrugated, 3", embossed, .020" thick		100	.080		2.38	3.78		6.16	8.60
0500	4", .025" thick		100	.080		6.50	3.78		10.28	13.15
0600	Copper, 16 oz. round, 2" diameter		100	.080		13.40	3.78		17.18	21
0700	3" diameter		100	.080		7.70	3.78		11.48	14.45
0800	4" diameter		100	.080		12.85	3.78		16.63	20
1000	2" x 3" corrugated		100	.080		7.10	3.78		10.88	13.80
1100	3" x 4" corrugated		100	.080		12	3.78		15.78	19.20
1300	Vinyl, 2-1/2" diameter, 45° or 75°		100	.080		3.50	3.78		7.28	9.85
1400	Tee Y junction		75	.107		12	5.05		17.05	21
9000	Minimum labor/equipment charge		4	2	Job		94.50		94.50	150

07 71 23.30 Gutters

		Crew	Daily Output	Labor-Hours	Unit	Material	Labor	Equipment	Total	Total Incl O&P
0010	**GUTTERS**									
0012	Aluminum, stock units, 5" K type, .027" thick, plain	1 Shee	120	.067	L.F.	2.32	3.15		5.47	7.55

07 71 Roof Specialties

07 71 23 – Manufactured Gutters and Downspouts

07 71 23.30 Gutters		Crew	Daily Output	Labor-Hours	Unit	Material	Labor	Equipment	Total	Total Incl O&P
						2009 Bare Costs				
0100	Enameled	1 Shee	120	.067	L.F.	2.45	3.15		5.60	7.70
0300	5" K type type, .032" thick, plain		120	.067		3.27	3.15		6.42	8.60
0400	Enameled		120	.067		2.92	3.15		6.07	8.20
0700	Copper, half round, 16 oz, stock units, 4" wide		120	.067		6.25	3.15		9.40	11.90
0900	5" wide		120	.067		6.30	3.15		9.45	11.95
1000	6" wide		115	.070		8.90	3.28		12.18	15
1200	K type, 16 oz, stock, 4" wide		120	.067		6.15	3.15		9.30	11.75
1300	5" wide		120	.067		7.30	3.15		10.45	13.05
1500	Lead coated copper, half round, stock, 4" wide		120	.067		10.50	3.15		13.65	16.55
1600	6" wide		115	.070		21	3.28		24.28	28
1800	K type, stock, 4" wide		120	.067		11.35	3.15		14.50	17.50
1900	5" wide		120	.067		14.25	3.15		17.40	20.50
2100	Stainless steel, half round or box, stock, 4" wide		120	.067		5.90	3.15		9.05	11.45
2200	5" wide		120	.067		6.35	3.15		9.50	11.95
2400	Steel, galv, half round or box, 28 ga, 5" wide, plain		120	.067		1.46	3.15		4.61	6.60
2500	Enameled		120	.067		1.51	3.15		4.66	6.65
2700	26 ga, stock, 5" wide		120	.067		1.59	3.15		4.74	6.75
2800	6" wide		120	.067		2.30	3.15		5.45	7.55
3000	Vinyl, O.G., 4" wide	1 Carp	110	.073		1.14	2.91		4.05	6.05
3100	5" wide		110	.073		1.35	2.91		4.26	6.30
3200	4" half round, stock units		110	.073		.91	2.91		3.82	5.80
3250	Joint connectors				Ea.	1.82			1.82	2
3300	Wood, clear treated cedar, fir or hemlock, 3" x 4"	1 Carp	100	.080	L.F.	7	3.20		10.20	13
3400	4" x 5"	"	100	.080	"	8	3.20		11.20	14.10
9000	Minimum labor/equipment charge	1 Shee	3.75	2.133	Job		101		101	160

07 71 23.35 Gutter Guard

Code	Description	Crew	Daily Output	Labor-Hours	Unit	Material	Labor	Equipment	Total	Total Incl O&P
0010	**GUTTER GUARD**									
0020	6" wide strip, aluminum mesh	1 Carp	500	.016	L.F.	2.18	.64		2.82	3.46
0100	Vinyl mesh		500	.016	"	1.45	.64		2.09	2.66
9000	Minimum labor/equipment charge		4	2	Job		80		80	132

07 71 26 – Reglets

07 71 26.10 Reglets and Accessories

Code	Description	Crew	Daily Output	Labor-Hours	Unit	Material	Labor	Equipment	Total	Total Incl O&P
0010	**REGLETS AND ACCESSORIES**									
0020	Aluminum, .025" thick, in concrete parapet	1 Carp	225	.036	L.F.	1.27	1.42		2.69	3.75
0100	Copper, 10 oz.		225	.036		2.41	1.42		3.83	5
0300	16 oz.		225	.036		4.71	1.42		6.13	7.55
0400	Galvanized steel, 24 gauge		225	.036		.86	1.42		2.28	3.30
0600	Stainless steel, .020" thick		225	.036		2.90	1.42		4.32	5.55
0700	Zinc and copper alloy, 20 oz.		225	.036		2.41	1.42		3.83	5
0900	Counter flashing for above, 12" wide, .032" aluminum	1 Shee	150	.053		1.50	2.52		4.02	5.65
1000	Copper, 10 oz.		150	.053		4.55	2.52		7.07	9
1200	16 oz.		150	.053		5.25	2.52		7.77	9.75
1300	Galvanized steel, .020" thick		150	.053		.85	2.52		3.37	4.94
1500	Stainless steel, .020" thick		150	.053		5	2.52		7.52	9.50
1600	Zinc and copper alloy, 20 oz.		150	.053		4.39	2.52		6.91	8.85
9000	Minimum labor/equipment charge	1 Carp	3	2.667	Job		107		107	176

07 71 29 – Manufactured Roof Expansion Joints

07 71 29.10 Expansion Joints

Code	Description	Crew	Daily Output	Labor-Hours	Unit	Material	Labor	Equipment	Total	Total Incl O&P
0010	**EXPANSION JOINTS**									
0300	Butyl or neoprene center with foam insulation, metal flanges									
0400	Aluminum, .032" thick for openings to 2-1/2"	1 Rofc	165	.048	L.F.	11.30	1.66		12.96	15.40

07 71 Roof Specialties

07 71 29 – Manufactured Roof Expansion Joints

07 71 29.10 Expansion Joints		Crew	Daily Output	Labor-Hours	Unit	2009 Bare Costs Material	Labor	Equipment	Total	Total Incl O&P
0600	For joint openings to 3-1/2"	1 Rofc	165	.048	L.F.	13.20	1.66		14.86	17.50
0610	For joint openings to 5"		165	.048		15.90	1.66		17.56	20.50
0620	For joint openings to 8"		165	.048		25	1.66		26.66	30.50
0700	Copper, 16 oz. for openings to 2-1/2"		165	.048		18.40	1.66		20.06	23
0900	For joint openings to 3-1/2"		165	.048		21	1.66		22.66	26
0910	For joint openings to 5"		165	.048		25	1.66		26.66	30
0920	For joint openings to 8"		165	.048		37	1.66		38.66	44
1000	Galvanized steel, 26 ga. for openings to 2-1/2"		165	.048		9.30	1.66		10.96	13.15
1200	For joint openings to 3-1/2"		165	.048		11	1.66		12.66	15.05
1210	For joint openings to 5"		165	.048		13.75	1.66		15.41	18.10
1220	For joint openings to 8"		165	.048		23.50	1.66		25.16	29
1300	Lead-coated copper, 16 oz. for openings to 2-1/2"		165	.048		32.50	1.66		34.16	39
1500	For joint openings to 3-1/2"		165	.048		38.50	1.66		40.16	45.50
1600	Stainless steel, .018", for openings to 2-1/2"		165	.048		14.15	1.66		15.81	18.50
1800	For joint openings to 3-1/2"		165	.048		16.10	1.66		17.76	20.50
1810	For joint openings to 5"		165	.048		19.95	1.66		21.61	25
1820	For joint openings to 8"		165	.048		30.50	1.66		32.16	36.50
1900	Neoprene, double-seal type with thick center, 4-1/2" wide		125	.064		13.20	2.19		15.39	18.45
1950	Polyethylene bellows, with galv steel flat flanges		100	.080		5.10	2.74		7.84	10.55
1960	With galvanized angle flanges		100	.080		5.65	2.74		8.39	11.10
2000	Roof joint with extruded aluminum cover, 2"	1 Shee	115	.070		35.50	3.28		38.78	44
2100	Roof joint, plastic curbs, foam center, standard	1 Rofc	100	.080		12.90	2.74		15.64	19.10
2200	Large		100	.080		17.25	2.74		19.99	24
2300	Transitions, regular, minimum		10	.800	Ea.	111	27.50		138.50	172
2350	Maximum		4	2		142	68.50		210.50	278
2400	Large, minimum		9	.889		163	30.50		193.50	235
2450	Maximum		3	2.667		171	91.50		262.50	350
2500	Roof to wall joint with extruded aluminum cover	1 Shee	115	.070	L.F.	30	3.28		33.28	38
2700	Wall joint, closed cell foam on PVC cover, 9" wide	1 Rofc	125	.064		4.60	2.19		6.79	8.95
2800	12" wide	"	115	.070		5.20	2.38		7.58	9.95
9000	Minimum labor/equipment charge	1 Shee	3	2.667	Job		126		126	200

07 71 43 – Drip Edge

07 71 43.10 Drip Edge, Rake Edge, Ice Belts		Crew	Daily Output	Labor-Hours	Unit	Material	Labor	Equipment	Total	Total Incl O&P
0010	**DRIP EDGE, RAKE EDGE, ICE BELTS**									
0020	Aluminum, .016" thick, 5" wide, mill finish	1 Carp	400	.020	L.F.	.42	.80		1.22	1.78
0100	White finish		400	.020		.46	.80		1.26	1.83
0200	8" wide, mill finish		400	.020		.56	.80		1.36	1.94
0300	Ice belt, 28" wide, mill finish		100	.080		3.80	3.20		7	9.50
0310	Vented, mill finish		400	.020		2.63	.80		3.43	4.21
0320	Painted finish		400	.020		2.72	.80		3.52	4.31
0400	Galvanized, 5" wide		400	.020		.54	.80		1.34	1.91
0500	8" wide, mill finish		400	.020		.78	.80		1.58	2.18
0510	Rake edge, aluminum, 1-1/2" x 1-1/2"		400	.020		.24	.80		1.04	1.58
0520	3-1/2" x 1-1/2"		400	.020		.46	.80		1.26	1.83
9000	Minimum labor/equipment charge		4	2	Job		80		80	132

07 72 Roof Accessories

07 72 23 – Relief Vents

07 72 23.10 Roof Vents		Crew	Daily Output	Labor-Hours	Unit	Material	Labor (2009 Bare Costs)	Equipment	Total	Total Incl O&P
0010	**ROOF VENTS**									
0020	Mushroom shape, for built-up roofs, aluminum	1 Rofc	30	.267	Ea.	68.50	9.15		77.65	92
0100	PVC, 6" high		30	.267	"	28.50	9.15		37.65	48
9000	Minimum labor/equipment charge		2.75	2.909	Job		99.50		99.50	178

07 72 26 – Ridge Vents

07 72 26.10 Ridge Vents and Accessories		Crew	Daily Output	Labor-Hours	Unit	Material	Labor	Equipment	Total	Total Incl O&P
0010	**RIDGE VENTS AND ACCESSORIES**									
0100	Aluminum strips, mill finish	1 Rofc	160	.050	L.F.	2.50	1.71		4.21	5.80
0150	Painted finish		160	.050	"	3.61	1.71		5.32	7.05
0200	Connectors		48	.167	Ea.	3.50	5.70		9.20	14.05
0300	End caps		48	.167	"	1.30	5.70		7	11.65
0400	Galvanized strips		160	.050	L.F.	2.14	1.71		3.85	5.40
0430	Molded polyethylene, shingles not included		160	.050	"	3.05	1.71		4.76	6.40
0440	End plugs		48	.167	Ea.	1.30	5.70		7	11.65
0450	Flexible roll, shingles not included		160	.050	L.F.	2.48	1.71		4.19	5.80
2300	Ridge vent strip, mill finish	1 Shee	155	.052	"	2.76	2.44		5.20	6.90

07 72 33 – Roof Hatches

07 72 33.10 Roof Hatch Options		Crew	Daily Output	Labor-Hours	Unit	Material	Labor	Equipment	Total	Total Incl O&P
0010	**ROOF HATCH OPTIONS**									
0500	2'-6" x 3', aluminum curb and cover	G-3	10	3.200	Ea.	985	126		1,111	1,275
0520	Galvanized steel curb and aluminum cover		10	3.200		595	126		721	860
0540	Galvanized steel curb and cover		10	3.200		500	126		626	755
0600	2'-6" x 4'-6", aluminum curb and cover		9	3.556		850	140		990	1,150
0800	Galvanized steel curb and aluminum cover		9	3.556		600	140		740	885
0900	Galvanized steel curb and cover		9	3.556		825	140		965	1,125
1100	4' x 4' aluminum curb and cover		8	4		1,675	158		1,833	2,075
1120	Galvanized steel curb and aluminum cover		8	4		1,675	158		1,833	2,075
1140	Galvanized steel curb and cover		8	4		1,600	158		1,758	2,025
1200	2'-6" x 8'-0", aluminum curb and cover		6.60	4.848		1,675	191		1,866	2,125
1400	Galvanized steel curb and aluminum cover		6.60	4.848		1,600	191		1,791	2,050
1500	Galvanized steel curb and cover		6.60	4.848		1,200	191		1,391	1,625
1800	For plexiglass panels, 2'-6" x 3'-0", add to above					400			400	440
9000	Minimum labor/equipment charge	2 Carp	2	8	Job		320		320	530

07 72 36 – Smoke Vents

07 72 36.10 Smoke Hatches		Crew	Daily Output	Labor-Hours	Unit	Material	Labor	Equipment	Total	Total Incl O&P
0010	**SMOKE HATCHES**									
0200	For 3'-0" long, add to roof hatches from Division 07 72 33.10				Ea.	25%	5%			
0250	For 4'-0" long, add to roof hatches from Division 07 72 33.10					20%	5%			
0300	For 8'-0" long, add to roof hatches from Division 07 72 33.10					10%	5%			

07 72 36.20 Smoke Vent Options		Crew	Daily Output	Labor-Hours	Unit	Material	Labor	Equipment	Total	Total Incl O&P
0010	**SMOKE VENT OPTIONS**									
0100	4' x 4' aluminum cover and frame	G-3	13	2.462	Ea.	1,900	97		1,997	2,250
0200	Galvanized steel cover and frame		13	2.462		1,675	97		1,772	1,975
0300	4' x 8' aluminum cover and frame		8	4		2,600	158		2,758	3,100
0400	Galvanized steel cover and frame		8	4		2,200	158		2,358	2,675
9000	Minimum labor/equipment charge	2 Carp	2	8	Job		320		320	530

07 72 53 – Snow Guards

07 72 53.10 Snow Guard Options		Crew	Daily Output	Labor-Hours	Unit	Material	Labor	Equipment	Total	Total Incl O&P
0010	**SNOW GUARD OPTIONS**									
0100	Slate & asphalt shingle roofs	1 Rofc	160	.050	Ea.	9.40	1.71		11.11	13.40

07 72 Roof Accessories

07 72 53 – Snow Guards

07 72 53.10 Snow Guard Options		Crew	Daily Output	Labor-Hours	Unit	2009 Bare Costs Material	Labor	Equipment	Total	Total Incl O&P
0200	Standing seam metal roofs	1 Rofc	48	.167	Ea.	14.60	5.70		20.30	26.50
0300	Surface mount for metal roofs		48	.167		7.70	5.70		13.40	18.65
0400	Double rail pipe type, including pipe		130	.062	L.F.	22.50	2.11		24.61	29

07 72 73 – Pitch Pockets

07 72 73.10 Pitch Pockets, Variable Sizes		Crew	Daily Output	Labor-Hours	Unit	Material	Labor	Equipment	Total	Total Incl O&P
0010	**PITCH POCKETS, VARIABLE SIZES**									
0100	Adjustable, 4" to 7", welded corners, 4" deep	1 Rofc	48	.167	Ea.	12.90	5.70		18.60	24.50
0200	Side extenders, 6"	"	240	.033	"	1.97	1.14		3.11	4.21

07 72 80 – Vents

07 72 80.30 Vent Options		Crew	Daily Output	Labor-Hours	Unit	Material	Labor	Equipment	Total	Total Incl O&P
0010	**VENT OPTIONS**									
0020	Plastic, for insulated decks, 1 per M.S.F., minimum	1 Rofc	40	.200	Ea.	13.50	6.85		20.35	27
0100	Maximum		20	.400		31	13.70		44.70	59
0300	Aluminum		30	.267		13.50	9.15		22.65	31
0800	Polystyrene baffles, 12" wide for 16" O.C. rafter spacing	1 Carp	90	.089		.40	3.55		3.95	6.30
0900	For 24" O.C. rafter spacing		110	.073		.60	2.91		3.51	5.45
9000	Minimum labor/equipment charge		3	2.667	Job		107		107	176

07 81 Applied Fireproofing

07 81 16 – Cementitious Fireproofing

07 81 16.10 Sprayed Cementitious Fireproofing		Crew	Daily Output	Labor-Hours	Unit	Material	Labor	Equipment	Total	Total Incl O&P
0010	**SPRAYED CEMENTITIOUS FIREPROOFING**									
0050	Not incl tamping or canvas protection									
0100	1" thick, on flat plate steel	G-2	3000	.008	S.F.	.53	.27	.04	.84	1.06
0200	Flat decking		2400	.010		.53	.33	.05	.91	1.18
0400	Beams		1500	.016		.53	.53	.08	1.14	1.54
0500	Corrugated or fluted decks		1250	.019		.79	.64	.10	1.53	2.02
0700	Columns, 1-1/8" thick		1100	.022		.59	.73	.12	1.44	1.96
0800	2-3/16" thick		700	.034		1.13	1.14	.18	2.45	3.29
0850	For tamping, add						10%			
0900	For canvas protection, add	G-2	5000	.005	S.F.	.07	.16	.03	.26	.37
1500	Intumescent epoxy fireproofing on wire mesh, 3/16" thick									
1550	1 hour rating, exterior use	G-2	136	.176	S.F.	6.90	5.90	.93	13.73	18.15
1600	Magnesium oxychloride, 35# to 40# density, 1/4" thick		3000	.008		1.44	.27	.04	1.75	2.06
1650	1/2" thick		2000	.012		2.90	.40	.06	3.36	3.91
1700	60# to 70# density, 1/4" thick		3000	.008		1.92	.27	.04	2.23	2.59
1750	1/2" thick		2000	.012		3.85	.40	.06	4.31	4.96
2000	Vermiculite cement, troweled or sprayed, 1/4" thick		3000	.008		1.31	.27	.04	1.62	1.92
2050	1/2" thick		2000	.012		2.60	.40	.06	3.06	3.58
9000	Minimum labor/equipment charge		3	8	Job		267	42	309	480

07 84 13 – Penetration Firestopping

07 84 13.10 Firestopping		Crew	Daily Output	Labor-Hours	Unit	Material	2009 Bare Costs Labor	Equipment	Total	Total Incl O&P
0010	**FIRESTOPPING** R078413-30									
0100	Metallic piping, non insulated									
0110	Through walls, 2" diameter	1 Carp	16	.500	Ea.	12.40	20		32.40	46.50
0120	4" diameter		14	.571		18.90	23		41.90	58.50
0130	6" diameter		12	.667		25.50	26.50		52	72
0140	12" diameter		10	.800		45	32		77	103
0150	Through floors, 2" diameter		32	.250		7.50	10		17.50	25
0160	4" diameter		28	.286		10.80	11.40		22.20	31
0170	6" diameter		24	.333		14.20	13.30		27.50	37.50
0180	12" diameter	↓	20	.400	↓	24	16		40	53
0190	Metallic piping, insulated									
0200	Through walls, 2" diameter	1 Carp	16	.500	Ea.	17.55	20		37.55	52.50
0210	4" diameter		14	.571		24	23		47	64
0220	6" diameter		12	.667		30.50	26.50		57	77.50
0230	12" diameter		10	.800		50	32		82	108
0240	Through floors, 2" diameter		32	.250		12.70	10		22.70	30.50
0250	4" diameter		28	.286		15.95	11.40		27.35	36.50
0260	6" diameter		24	.333		19.35	13.30		32.65	43.50
0270	12" diameter	↓	20	.400	↓	24	16		40	53
0280	Non metallic piping, non insulated									
0290	Through walls, 2" diameter	1 Carp	12	.667	Ea.	51	26.50		77.50	100
0300	4" diameter		10	.800		64	32		96	124
0310	6" diameter		8	1		89	40		129	164
0330	Through floors, 2" diameter		16	.500		40	20		60	77
0340	4" diameter		6	1.333		49.50	53.50		103	143
0350	6" diameter	↓	6	1.333	↓	59.50	53.50		113	154
0370	Ductwork, insulated & non insulated, round									
0380	Through walls, 6" diameter	1 Carp	12	.667	Ea.	26	26.50		52.50	72.50
0390	12" diameter		10	.800		51.50	32		83.50	110
0400	18" diameter		8	1		84	40		124	159
0410	Through floors, 6" diameter		16	.500		14.20	20		34.20	48.50
0420	12" diameter		14	.571		26	23		49	66
0430	18" diameter	↓	12	.667	↓	45	26.50		71.50	93.50
0440	Ductwork, insulated & non insulated, rectangular									
0450	With stiffener/closure angle, through walls, 6" x 12"	1 Carp	8	1	Ea.	21.50	40		61.50	89.50
0460	12" x 24"		6	1.333		28.50	53.50		82	120
0470	24" x 48"		4	2		81.50	80		161.50	222
0480	With stiffener/closure angle, through floors, 6" x 12"		10	.800		11.65	32		43.65	66
0490	12" x 24"		8	1		21	40		61	89
0500	24" x 48"	↓	6	1.333	↓	41	53.50		94.50	133
0510	Multi trade openings									
0520	Through walls, 6" x 12"	1 Carp	2	4	Ea.	45	160		205	315
0530	12" x 24"	"	1	8		182	320		502	730
0540	24" x 48"	2 Carp	1	16		730	640		1,370	1,850
0550	48" x 96"	"	.75	21.333		2,925	850		3,775	4,625
0560	Through floors, 6" x 12"	1 Carp	2	4		45	160		205	315
0570	12" x 24"	"	1	8		182	320		502	730
0580	24" x 48"	2 Carp	.75	21.333		730	850		1,580	2,200
0590	48" x 96"	"	.50	32	↓	2,925	1,275		4,200	5,325
0600	Structural penetrations, through walls									
0610	Steel beams, W8 x 10	1 Carp	8	1	Ea.	28.50	40		68.50	97.50
0620	W12 x 14		6	1.333		45	53.50		98.50	138
0630	W21 x 44	↓	5	1.600	↓	90.50	64		154.50	206

07 84 Firestopping

07 84 13 – Penetration Firestopping

07 84 13.10 Firestopping		Crew	Daily Output	Labor-Hours	Unit	Material	Labor	Equipment	Total	Total Incl O&P
0640	W36 x 135	1 Carp	3	2.667	Ea.	220	107		327	420
0650	Bar joists, 18" deep		6	1.333		41.50	53.50		95	134
0660	24" deep		6	1.333		51.50	53.50		105	145
0670	36" deep		5	1.600		77.50	64		141.50	192
0680	48" deep	↓	4	2	↓	90.50	80		170.50	232
0690	Construction joints, floor slab at exterior wall									
0700	Precast, brick, block or drywall exterior									
0710	2" wide joint	1 Carp	125	.064	L.F.	6.45	2.56		9.01	11.30
0720	4" wide joint	"	75	.107	"	12.90	4.26		17.16	21.50
0730	Metal panel, glass or curtain wall exterior									
0740	2" wide joint	1 Carp	40	.200	L.F.	15.30	8		23.30	30
0750	4" wide joint	"	25	.320	"	21	12.80		33.80	44
0760	Floor slab to drywall partition									
0770	Flat joint	1 Carp	100	.080	L.F.	6.35	3.20		9.55	12.25
0780	Fluted joint		50	.160		12.90	6.40		19.30	25
0790	Etched fluted joint	↓	75	.107	↓	8.40	4.26		12.66	16.30
0800	Floor slab to concrete/masonry partition									
0810	Flat joint	1 Carp	75	.107	L.F.	14.20	4.26		18.46	22.50
0820	Fluted joint	"	50	.160	"	16.80	6.40		23.20	29
0830	Concrete/CMU wall joints									
0840	1" wide	1 Carp	100	.080	L.F.	7.75	3.20		10.95	13.85
0850	2" wide		75	.107		14.20	4.26		18.46	22.50
0860	4" wide	↓	50	.160	↓	27	6.40		33.40	40.50
0870	Concrete/CMU floor joints									
0880	1" wide	1 Carp	200	.040	L.F.	3.88	1.60		5.48	6.90
0890	2" wide		150	.053		7.10	2.13		9.23	11.30
0900	4" wide	↓	100	.080	↓	13.55	3.20		16.75	20

07 92 Joint Sealants

07 92 10 – Caulking and Sealants

07 92 10.10 Caulking and Sealant Options

		Crew	Daily Output	Labor-Hours	Unit	Material	Labor	Equipment	Total	Total Incl O&P
0010	**CAULKING AND SEALANT OPTIONS**									
0020	Acoustical sealant, elastomeric, cartridges				Ea.	4.60			4.60	5.05
0032	Backer rod, polyethylene, 1/4" diameter	1 Bric	460	.017	L.F.	.04	.70		.74	1.19
0052	1/2" diameter		460	.017		.07	.70		.77	1.22
0072	3/4" diameter		460	.017		.09	.70		.79	1.24
0092	1" diameter	↓	460	.017	↓	.16	.70		.86	1.32
0100	Acrylic latex caulk, white									
0200	11 fl. oz cartridge				Ea.	2.14			2.14	2.35
0500	1/4" x 1/2"	1 Bric	248	.032	L.F.	.17	1.31		1.48	2.30
0600	1/2" x 1/2"		250	.032		.35	1.30		1.65	2.48
0800	3/4" x 3/4"		230	.035		.79	1.41		2.20	3.15
0900	3/4" x 1"		200	.040		1.05	1.62		2.67	3.77
1000	1" x 1"	↓	180	.044	↓	1.31	1.80		3.11	4.35
1400	Butyl based, bulk				Gal.	26			26	28.50
1500	Cartridges				"	31.50			31.50	34.50
1700	Bulk, in place 1/4" x 1/2", 154 L.F./gal.	1 Bric	230	.035	L.F.	.17	1.41		1.58	2.47
1800	1/2" x 1/2", 77 L.F./gal.	"	180	.044	"	.34	1.80		2.14	3.28
2000	Latex acrylic based, bulk				Gal.	27			27	30
2100	Cartridges					33.50			33.50	36.50
2300	Polysulfide compounds, 1 component, bulk				↓	51			51	56

07 92 Joint Sealants

07 92 10 – Caulking and Sealants

07 92 10.10 Caulking and Sealant Options		Crew	Daily Output	Labor-Hours	Unit	2009 Bare Costs Material	Labor	Equipment	Total	Total Incl O&P
2600	1 or 2 component, in place, 1/4" x 1/4", 308 L.F./gal.	1 Bric	145	.055	L.F.	.17	2.23		2.40	3.79
2700	1/2" x 1/4", 154 L.F./gal.		135	.059		.33	2.40		2.73	4.25
2900	3/4" x 3/8", 68 L.F./gal.		130	.062		.75	2.49		3.24	4.86
3000	1" x 1/2", 38 L.F./gal.		130	.062		1.35	2.49		3.84	5.50
3200	Polyurethane, 1 or 2 component				Gal.	54			54	59.50
3500	Bulk, in place, 1/4" x 1/4"	1 Bric	150	.053	L.F.	.18	2.16		2.34	3.68
3600	1/2" x 1/4"		145	.055		.35	2.23		2.58	4
3800	3/4" x 3/8", 68 L.F./gal.		130	.062		.79	2.49		3.28	4.90
3900	1" x 1/2"		110	.073		1.40	2.95		4.35	6.30
4100	Silicone rubber, bulk				Gal.	41			41	45
4200	Cartridges				"	41			41	45
4400	Neoprene gaskets, closed cell, adhesive, 1/8" x 3/8"	1 Bric	240	.033	L.F.	.26	1.35		1.61	2.47
4500	1/4" x 3/4"		215	.037		.58	1.51		2.09	3.08
4700	1/2" x 1"		200	.040		1.30	1.62		2.92	4.05
4800	3/4" x 1-1/2"		165	.048		1.44	1.96		3.40	4.76
5500	Resin epoxy coating, 2 component, heavy duty				Gal.	30.50			30.50	33.50
5802	Tapes, sealant, P.V.C. foam adhesive, 1/16" x 1/4"				L.F.	.06			.06	.06
5902	1/16" x 1/2"					.08			.08	.09
5952	1/16" x 1"					.14			.14	.15
6002	1/8" x 1/2"					.09			.09	.10
6200	Urethane foam, 2 component, handy pack, 1 C.F.				Ea.	34			34	37
6300	50.0 C.F. pack				C.F.	17.20			17.20	18.95
9000	Minimum labor/equipment charge	1 Bric	4	2	Job		81		81	131

07 95 Expansion Control

07 95 13 – Expansion Joint Cover Assemblies

07 95 13.50 Expansion Joint Assemblies		Crew	Daily Output	Labor-Hours	Unit	Material	Labor	Equipment	Total	Total Incl O&P
0010	**EXPANSION JOINT ASSEMBLIES**									
0200	Floor cover assemblies, 1" space, aluminum	1 Sswk	38	.211	L.F.	21.50	9.40		30.90	41
0300	Bronze		38	.211		43.50	9.40		52.90	65.50
0500	2" space, aluminum		38	.211		26	9.40		35.40	46
0600	Bronze		38	.211		47	9.40		56.40	69
0800	Wall and ceiling assemblies, 1" space, aluminum		38	.211		12.80	9.40		22.20	31.50
0900	Bronze		38	.211		40.50	9.40		49.90	62.50
1100	2" space, aluminum		38	.211		21.50	9.40		30.90	41.50
1200	Bronze		38	.211		50	9.40		59.40	72.50
1400	Floor to wall assemblies, 1" space, aluminum		38	.211		19.55	9.40		28.95	39
1500	Bronze or stainless		38	.211		49.50	9.40		58.90	72
1700	Gym floor angle covers, aluminum, 3" x 3" angle		46	.174		16.90	7.75		24.65	33
1800	3" x 4" angle		46	.174		19.95	7.75		27.70	36.50
2000	Roof closures, aluminum, flat roof, low profile, 1" space		57	.140		39	6.25		45.25	54.50
2100	High profile		57	.140		48	6.25		54.25	64
2300	Roof to wall, low profile, 1" space		57	.140		21.50	6.25		27.75	35.50
2400	High profile		57	.140		28	6.25		34.25	42
9000	Minimum labor/equipment charge		2	4	Job		179		179	330

Division Notes

	CREW	DAILY OUTPUT	LABOR-HOURS	UNIT	2009 BARE COSTS MAT.	LABOR	EQUIP.	TOTAL	TOTAL INCL O&P

Estimating Tips

08 10 00 Doors and Frames

All exterior doors should be addressed for their energy conservation.

- Most metal doors and frames look alike, but there may be significant differences among them. When estimating these items, be sure to choose the line item that most closely compares to the specification or door schedule requirements regarding:
  - — type of metal
  - — metal gauge
  - — door core material
  - — fire rating
  - — finish
- Wood and plastic doors vary considerably in price. The primary determinant is the veneer material. Lauan, birch, and oak are the most common veneers. Other variables include the following:
  - — hollow or solid core
  - — fire rating
  - — flush or raised panel
  - — finish

08 30 00 Specialty Doors and Frames

- There are many varieties of special doors, and they are usually priced per each. Add frames, hardware, or operators required for a complete installation.

08 40 00 Entrances, Storefronts, and Curtain Walls

- Glazed curtain walls consist of the metal tube framing and the glazing material. The cost data in this subdivision is presented for the metal tube framing alone or the composite wall. If your estimate requires a detailed takeoff of the framing, be sure to add the glazing cost.

08 50 00 Windows

- Most metal windows are delivered preglazed. However, some metal windows are priced without glass. Refer to 08 80 00 Glazing for glass pricing. The grade C indicates commercial grade windows, usually ASTM C-35.
- All wood windows are priced preglazed. The two glazing options priced are single pane float glass and insulating glass 1/2" thick. Add the cost of screens and grills if required.

08 70 00 Hardware

- Hardware costs add considerably to the cost of a door. The most efficient method to determine the hardware requirements for a project is to review the door schedule.
- Door hinges are priced by the pair, with most doors requiring 1-1/2 pairs per door. The hinge prices do not include installation labor because it is included in door installation. Hinges are classified according to the frequency of use.

08 80 00 Glazing

- Different openings require different types of glass. The three most common types are:
  - — float
  - — tempered
  - — insulating
- Most exterior windows are glazed with insulating glass. Entrance doors and window walls, where the glass is less than 18" from the floor, are generally glazed with tempered glass. Interior windows and some residential windows are glazed with float glass.
- Coastal communities are starting to address the use of impact-resistant glass.
- The insulation or 'u' value is a strong consideration.

Reference Numbers

Reference numbers are shown in shaded boxes at the beginning of some major classifications. These numbers refer to related items in the Reference Section. The reference information may be an estimating procedure, an alternate pricing method, or technical information.

Note: Not all subdivisions listed here necessarily appear in this publication.

08 01 53 – Operation and Maintenance of Plastic Windows

08 01 53.81 Solid Vinyl Replacement Windows			Crew	Daily Output	Labor-Hours	Unit	2009 Bare Costs Material	Labor	Equipment	Total	Total Incl O&P
0010	SOLID VINYL REPLACEMENT WINDOWS										
0020	Double hung, insulated glass, up to 83 united inches	G	2 Carp	8	2	Ea.	330	80		410	490
0040	84 to 93	G		8	2		360	80		440	525
0060	94 to 101	G		6	2.667		360	107		467	570
0080	102 to 111	G		6	2.667		380	107		487	590
0100	112 to 120	G		6	2.667		410	107		517	630
0120	For each united inch over 120 , add	G		800	.020	Inch	5.20	.80		6	7
0140	Casement windows, one operating sash , 42 to 60 united inches	G		8	2	Ea.	191	80		271	340
0160	61 to 70	G		8	2		217	80		297	370
0180	71 to 80	G		8	2		236	80		316	390
0200	81 to 96	G		8	2		252	80		332	410
0220	Two operating sash, 58 to 78 united inches	G		8	2		380	80		460	550
0240	79 to 88	G		8	2		405	80		485	580
0260	89 to 98	G		8	2		445	80		525	620
0280	99 to 108	G		6	2.667		465	107		572	690
0300	109 to 121	G		6	2.667		500	107		607	730
0320	Three operating sash, 73 to 108 united inches	G		8	2		600	80		680	790
0340	109 to 118	G		8	2		635	80		715	830
0360	119 to 128	G		6	2.667		655	107		762	895
0380	129 to 138	G		6	2.667		700	107		807	945
0400	139 to 156	G		6	2.667		740	107		847	990
0420	Four operating sash, 98 to 118 united inches	G		8	2		865	80		945	1,075
0440	119 to 128	G		8	2		930	80		1,010	1,150
0460	129 to 138	G		6	2.667		985	107		1,092	1,250
0480	139 to 148	G		6	2.667		1,050	107		1,157	1,325
0500	149 to 168	G		6	2.667		1,100	107		1,207	1,400
0520	169 to 178	G		6	2.667		1,200	107		1,307	1,500
0540	For venting unit to fixed unit, deduct	G					16			16	17.60
0560	Fixed picture window, up to 63 united inches	G	2 Carp	8	2		136	80		216	282
0580	64 to 83	G		8	2		160	80		240	310
0600	84 to 101	G		8	2		207	80		287	360
0620	For each united inch over 101, add	G		900	.018	Inch	2.29	.71		3	3.69
0640	Picture window opt., low E glazing, up to 101 united inches	G				Ea.	19.50			19.50	21.50
0660	102 to 124	G					26			26	28.50
0680	124 and over	G					38			38	42
0700	Options, low E glazing, up to w/gas 101 united inches	G					31			31	34
0720	102 to 124	G					31			31	34
0740	124 and over	G					31			31	34
0760	Muntins, between glazing, square, per lite						3			3	3.30
0780	Diamond shape, per full or partial diamond						4			4	4.40
0800	Cellulose fiber insulation, poured into sash balance cavity	G	1 Carp	36	.222	C.F.	.60	8.90		9.50	15.30
0820	Silicone caulking at perimeter	G	"	800	.010	L.F.	.13	.40		.53	.81

08 05 Common Work Results for Openings

08 05 05 – Selective Windows and Doors Demolition

08 05 05.10 Selective Demolition Doors		Crew	Daily Output	Labor-Hours	Unit	Material	2009 Bare Costs Labor	Equipment	Total	Total Incl O&P
0010	**SELECTIVE DEMOLITION DOORS** R024119-10									
0200	Doors, exterior, 1-3/4" thick, single, 3' x 7' high	1 Clab	16	.500	Ea.		15.80		15.80	26
0220	Double, 6' x 7' high		12	.667			21		21	35
0500	Interior, 1-3/8" thick, single, 3' x 7' high		20	.400			12.65		12.65	21
0520	Double, 6' x 7' high		16	.500			15.80		15.80	26
0700	Bi-folding, 3' x 6'-8" high		20	.400			12.65		12.65	21
0720	6' x 6'-8" high		18	.444			14.05		14.05	23
0900	Bi-passing, 3' x 6'-8" high		16	.500			15.80		15.80	26
0940	6' x 6'-8" high		14	.571			18.05		18.05	30
1500	Remove and reset, minimum	1 Carp	8	1			40		40	66
1520	Maximum		6	1.333			53.50		53.50	88
2000	Frames, including trim, metal		8	1			40		40	66
2200	Wood	2 Carp	32	.500			20		20	33
2201	Alternate pricing method	1 Carp	200	.040	L.F.		1.60		1.60	2.64
3000	Special doors, counter doors	2 Carp	6	2.667	Ea.		107		107	176
3100	Double acting		10	1.600			64		64	106
3200	Floor door (trap type), or access type		8	2			80		80	132
3300	Glass, sliding, including frames		12	1.333			53.50		53.50	88
3400	Overhead, commercial, 12' x 12' high		4	4			160		160	264
3440	up to 20' x 16' high		3	5.333			213		213	350
3445	up to 35' x 30' high		1	16			640		640	1,050
3500	Residential, 9' x 7' high		8	2			80		80	132
3540	16' x 7' high		7	2.286			91.50		91.50	151
3600	Remove and reset, minimum		4	4			160		160	264
3620	Maximum		2.50	6.400			256		256	420
3700	Roll-up grille		5	3.200			128		128	211
3800	Revolving door		2	8			320		320	530
3900	Storefront swing door		3	5.333			213		213	350
6600	Demo flexible transparent strip entrance	3 Shee	115	.209	SF Surf		9.85		9.85	15.65
7100	Remove double swing pneumatic doors, openers and sensors	2 Skwk	.50	32	Opng.		1,300		1,300	2,125
7110	Remove automatic operators, industrial, sliding doors, to 12' wide	"	.40	40	"		1,625		1,625	2,650
7550	Hangar door demo	2 Sswk	220	.073	S.F.		3.25		3.25	6
7570	Remove shock absorbing door	"	1.90	8.421	Opng.		375		375	695
9000	Minimum labor/equipment charge	1 Carp	4	2	Job		80		80	132

08 05 05.20 Selective Demolition of Windows		Crew	Daily Output	Labor-Hours	Unit	Material	2009 Bare Costs Labor	Equipment	Total	Total Incl O&P
0010	**SELECTIVE DEMOLITION OF WINDOWS** R024119-10									
0200	Aluminum, including trim, to 12 S.F.	1 Clab	16	.500	Ea.		15.80		15.80	26
0240	To 25 S.F.		11	.727			23		23	38
0280	To 50 S.F.		5	1.600			50.50		50.50	83.50
0320	Storm windows/screens, to 12 S.F.		27	.296			9.35		9.35	15.45
0360	To 25 S.F.		21	.381			12.05		12.05	19.85
0400	To 50 S.F.		16	.500			15.80		15.80	26
0600	Glass, minimum		200	.040	S.F.		1.26		1.26	2.09
0620	Maximum		150	.053	"		1.69		1.69	2.78
1000	Steel, including trim, to 12 S.F.		13	.615	Ea.		19.45		19.45	32
1020	To 25 S.F.		9	.889			28		28	46.50
1040	To 50 S.F.		4	2			63		63	104
2000	Wood, including trim, to 12 S.F.		22	.364			11.50		11.50	18.95
2020	To 25 S.F.		18	.444			14.05		14.05	23
2060	To 50 S.F.		13	.615			19.45		19.45	32
2065	To 180 S.F.		8	1			31.50		31.50	52
4300	Remove bay/bow window	2 Carp	6	2.667			107		107	176

08 05 Common Work Results for Openings

08 05 05 – Selective Windows and Doors Demolition

08 05 05.20 Selective Demolition of Windows		Crew	Daily Output	Labor-Hours	Unit	Material	2009 Bare Costs Labor	Equipment	Total	Total Incl O&P
4410	Remove skylight, plstc domes, flush/curb mtd	G-3	395	.081	S.F.		3.19		3.19	5.15
5020	Remove and reset window, minimum	1 Carp	6	1.333	Ea.		53.50		53.50	88
5040	Average		4	2			80		80	132
5080	Maximum		2	4			160		160	264
9000	Minimum labor/equipment charge	1 Clab	4	2	Job		63		63	104

08 11 Metal Doors and Frames

08 11 16 – Aluminum Doors and Frames

08 11 16.10 Entrance Doors and Frames

		Crew	Daily Output	Labor-Hours	Unit	Material	Labor	Equipment	Total	Total Incl O&P
0010	**ENTRANCE DOORS AND FRAMES** Aluminum, narrow stile									
0011	Including standard hardware, clear finish, no glass									
0020	3′-0″ x 7′-0″ opening	2 Sswk	2	8	Ea.	670	360		1,030	1,400
0030	3′-6″ x 7′-0″ opening		2	8		685	360		1,045	1,400
0100	3′-0″ x 10′-0″ opening, 3′ high transom		1.80	8.889		1,175	395		1,570	2,025
0200	3′-6″ x 10′-0″ opening, 3′ high transom		1.80	8.889		1,175	395		1,570	2,025
0280	5′-0″ x 7′-0″ opening		2	8		1,050	360		1,410	1,825
0300	6′-0″ x 7′-0″ opening		1.30	12.308		1,000	550		1,550	2,125
0400	6′-0″ x 10′-0″ opening, 3′ high transom		1.10	14.545	Pr.	1,250	650		1,900	2,575
0420	7′-0″ x 7′-0″ opening		1	16	″	1,075	715		1,790	2,500
0520	3′-0″ x 7′-0″ opening, wide stile		2	8	Ea.	885	360		1,245	1,625
0540	3′-6″ x 7′-0″ opening		2	8		960	360		1,320	1,700
0560	5′-0″ x 7′-0″ opening		2	8		1,400	360		1,760	2,175
0580	6′-0″ x 7′-0″ opening		1.30	12.308	Pr.	1,425	550		1,975	2,600
0600	7′-0″ x 7′-0″ opening		1	16	″	1,525	715		2,240	3,000
1100	For full vision doors, with 1/2″ glass, add				Leaf	55%				
1200	For non-standard size, add					67%				
1300	Light bronze finish, add					36%				
1400	Dark bronze finish, add					18%				
1500	For black finish, add					36%				
1600	Concealed panic device, add					1,050			1,050	1,175
1700	Electric striker release, add				Opng.	275			275	305
1800	Floor check, add				Leaf	810			810	890
1900	Concealed closer, add				″	540			540	595
9000	Minimum labor/equipment charge	2 Carp	4	4	Job		160		160	264

08 11 63 – Metal Screen and Storm Doors and Frames

08 11 63.23 Aluminum Screen and Storm Doors and Frames

		Crew	Daily Output	Labor-Hours	Unit	Material	Labor	Equipment	Total	Total Incl O&P
0010	**ALUMINUM SCREEN AND STORM DOORS AND FRAMES**									
0020	Combination storm and screen									
0400	Clear anodic coating, 6′-8″ x 2′-6″ wide	2 Carp	15	1.067	Ea.	172	42.50		214.50	260
0420	2′-8″ wide		14	1.143		156	45.50		201.50	247
0440	3′-0″ wide		14	1.143		156	45.50		201.50	247
0500	For 7′ door height, add					5%				
1000	Mill finish, 6′-8″ x 2′-6″ wide	2 Carp	15	1.067	Ea.	225	42.50		267.50	320
1020	2′-8″ wide		14	1.143		225	45.50		270.50	325
1040	3′-0″ wide		14	1.143		246	45.50		291.50	345
1100	For 7′-0″ door, add					5%				
1500	White painted, 6′-8″ x 2′-6″ wide	2 Carp	15	1.067		305	42.50		347.50	405
1520	2′-8″ wide		14	1.143		260	45.50		305.50	360
1540	3′-0″ wide		14	1.143		305	45.50		350.50	410
1600	For 7′-0″ door, add					5%				

08 11 Metal Doors and Frames

08 11 63 – Metal Screen and Storm Doors and Frames

08 11 63.23 Aluminum Screen and Storm Doors and Frames		Crew	Daily Output	Labor-Hours	Unit	2009 Bare Costs Material	Labor	Equipment	Total	Total Incl O&P
2000	Wood door & screen, see Div. 08 14 33.20									
9000	Minimum labor/equipment charge	1 Carp	4	2	Job		80		80	132

08 11 74 – Sliding Metal Grilles

08 11 74.10 Rolling Grille Supports		Crew	Daily Output	Labor-Hours	Unit	Material	Labor	Equipment	Total	Total Incl O&P
0010	ROLLING GRILLE SUPPORTS									
0020	Rolling grille supports, overhead framed	E-4	36	.889	L.F.	23	40	3.72	66.72	104

08 12 Metal Frames

08 12 13 – Hollow Metal Frames

08 12 13.13 Standard Hollow Metal Frames		Crew	Daily Output	Labor-Hours	Unit	Material	Labor	Equipment	Total	Total Incl O&P
0010	STANDARD HOLLOW METAL FRAMES									
0020	16 ga., up to 5-3/4" jamb depth									
0025	6'-8" high, 3'-0" wide, single	2 Carp	16	1	Ea.	123	40		163	202
0028	3'-6" wide, single		16	1		116	40		156	193
0030	4'-0" wide, single		16	1		116	40		156	193
0040	6'-0" wide, double		14	1.143		153	45.50		198.50	245
0045	8'-0" wide, double		14	1.143		159	45.50		204.50	251
0100	7'-0" high, 3'-0" wide, single		16	1		110	40		150	187
0110	3'-6" wide, single		16	1		119	40		159	197
0112	4'-0" wide, single		16	1		151	40		191	233
0140	6'-0" wide, double		14	1.143		160	45.50		205.50	252
0145	8'-0" wide, double		14	1.143		150	45.50		195.50	241
1000	16 ga., up to 4-7/8" deep, 7'-0" H, 3'-0" W, single		16	1		130	40		170	209
1140	6'-0" wide, double		14	1.143		158	45.50		203.50	249
2800	14 ga., up to 3-7/8" deep, 7'-0" high, 3'-0" wide, single		16	1		134	40		174	213
2840	6'-0" wide, double		14	1.143		161	45.50		206.50	253
3000	14 ga., up to 5-3/4" deep, 6'-8" high, 3'-0" wide, single		16	1		140	40		180	220
3002	3'-6" wide, single		16	1		140	40		180	220
3005	4'-0" wide, single		16	1		140	40		180	220
3600	up to 5-3/4" jamb depth, 7'-0" high, 4'-0" wide, single		15	1.067		111	42.50		153.50	193
3620	6'-0" wide, double		12	1.333		174	53.50		227.50	279
3640	8'-0" wide, double		12	1.333		164	53.50		217.50	268
3700	8'-0" high, 4'-0" wide, single		15	1.067		146	42.50		188.50	231
3740	8'-0" wide, double		12	1.333		181	53.50		234.50	287
4000	6-3/4" deep, 7'-0" high, 4'-0" wide, single		15	1.067		172	42.50		214.50	260
4020	6'-0" wide, double		12	1.333		204	53.50		257.50	310
4040	8'-0", wide double		12	1.333		222	53.50		275.50	330
4100	8'-0" high, 4'-0" wide, single		15	1.067		174	42.50		216.50	262
4140	8'-0" wide, double		12	1.333		202	53.50		255.50	310
4400	8-3/4" deep, 7'-0" high, 4'-0" wide, single		15	1.067		163	42.50		205.50	250
4440	8'-0" wide, double		12	1.333		231	53.50		284.50	345
4500	8'-0" high, 4'-0" wide, single		15	1.067		174	42.50		216.50	262
4540	8'-0" wide, double		12	1.333		203	53.50		256.50	310
4900	For welded frames, add					47.50			47.50	52.50
5400	14 ga., "B" label, up to 5-3/4" deep, 7'-0" high, 4'-0" wide, single	2 Carp	15	1.067		170	42.50		212.50	258
5440	8'-0" wide, double		12	1.333		208	53.50		261.50	315
5800	6-3/4" deep, 7'-0" high, 4'-0" wide, single		15	1.067		150	42.50		192.50	236
5840	8'-0" wide, double		12	1.333		249	53.50		302.50	360
6200	8-3/4" deep, 7'-0" high, 4'-0" wide, single		15	1.067		220	42.50		262.50	315
6240	8'-0" wide, double		12	1.333		245	53.50		298.50	360

08 12 Metal Frames

08 12 13 – Hollow Metal Frames

08 12 13.13 Standard Hollow Metal Frames		Crew	Daily Output	Labor-Hours	Unit	Material	2009 Bare Costs Labor	Equipment	Total	Total Incl O&P
6300	For "A" label use same price as "B" label									
6400	For baked enamel finish, add					30%	15%			
6500	For galvanizing, add					15%				
6600	For hospital stop, add				Ea.	282			282	310
7900	Transom lite frames, fixed, add	2 Carp	155	.103	S.F.	45.50	4.12		49.62	57
8000	Movable, add	"	130	.123	"	55	4.92		59.92	68.50
9000	Minimum labor/equipment charge	1 Carp	4	2	Job		80		80	132

08 12 13.20 Wrap Around Drywall Frames		Crew	Daily Output	Labor-Hours	Unit	Material	Labor	Equipment	Total	Total Incl O&P
0010	**WRAP AROUND DRYWALL FRAMES**									
0400	Wrap-around drywall frame, 16 ga., 6-1/4" x 7'-0" x 2'-6" wide	2 Carp	16	1	Ea.	139	40		179	219
0430	2'-8" wide		16	1		139	40		179	219
0460	3'-0" wide		15	1.067		139	42.50		181.50	224
0490	3'-6" wide		15	1.067		145	42.50		187.50	231
0520	5'-0" wide		14	1.143		155	45.50		200.50	247
0580	6'-0" wide		14	1.143		163	45.50		208.50	255
0610	7'-0" wide		13	1.231		174	49		223	272
2000	18 ga., 4-3/4" x 7'-0 x 2'-6" wide		16	1		123	40		163	201
2030	2'-8" wide		16	1		123	40		163	201
2060	3'-0" wide		15	1.067		123	42.50		165.50	206
2090	3'-6" wide		15	1.067		128	42.50		170.50	212
2120	5'-0" wide		14	1.143		139	45.50		184.50	229
2150	5'-4" wide		14	1.143		151	45.50		196.50	242
2200	6'-0" wide		14	1.143		151	45.50		196.50	242
2250	7'-0" wide		13	1.231		154	49		203	251
3700	Fire door frames, 6-1/4" x 7'-0" x 2'-6" wide		16	1		168	40		208	250
3730	2'-8" wide		16	1		168	40		208	250
3760	3'-0" wide		15	1.067		168	42.50		210.50	256
3820	5'-0" wide		14	1.143		188	45.50		233.50	283
3850	5'-4" wide		14	1.143		197	45.50		242.50	293
3880	6'-0" wide	↓	14	1.143	↓	197	45.50		242.50	293
9000	Minimum labor/equipment charge	1 Carp	2	4	Job		160		160	264

08 12 13.25 Channel Metal Frames		Crew	Daily Output	Labor-Hours	Unit	Material	Labor	Equipment	Total	Total Incl O&P
0010	**CHANNEL METAL FRAMES**									
0020	Steel channels with anchors and bar stops									
0100	6" channel @ 8.2#/L.F., 3' x 7' door, weighs 150#	E-4	13	2.462	Ea.	270	111	10.30	391.30	515
0200	8" channel @ 11.5#/L.F., 6' x 8' door, weighs 275#		9	3.556		495	161	14.90	670.90	860
0300	8' x 12' door, weighs 400#		6.50	4.923		720	223	20.50	963.50	1,225
0400	10" channel @ 15.3#/L.F., 10' x 10' door, weighs 500#		6	5.333		900	241	22.50	1,163.50	1,450
0500	12' x 12' door, weighs 600#		5.50	5.818		1,075	263	24.50	1,362.50	1,700
0600	12" channel @ 20.7#/L.F., 12' x 12' door, weighs 825#		4.50	7.111		1,475	320	30	1,825	2,250
0700	12' x 16' door, weighs 1000#	↓	4	8		1,800	360	33.50	2,193.50	2,675
0800	For frames without bar stops, light sections, deduct					15%				
0900	Heavy sections, deduct				↓	10%				
9000	Minimum labor/equipment charge	E-4	4	8	Job		360	33.50	393.50	705

08 13 13.13 Standard Hollow Metal Doors		Crew	Daily Output	Labor-Hours	Unit	Material	Labor	Equipment	Total	Total Incl O&P
						2009 Bare Costs				
0010	**STANDARD HOLLOW METAL DOORS** R081313-20									
0015	Flush, full panel, hollow core									
0020	1-3/8" thick, 20 ga., 2'-0"x 6'-8"	2 Carp	20	.800	Ea.	267	32		299	345
0040	2'-8" x 6'-8"		18	.889		274	35.50		309.50	360
0060	3'-0" x 6'-8"		17	.941		280	37.50		317.50	365
0100	3'-0" x 7'-0"		17	.941		295	37.50		332.50	385
0120	For vision lite, add					92.50			92.50	102
0140	For narrow lite, add					101			101	111
0320	Half glass, 20 ga., 2'-0" x 6'-8"	2 Carp	20	.800		395	32		427	490
0340	2'-8" x 6'-8"		18	.889		405	35.50		440.50	505
0360	3'-0" x 6'-8"		17	.941		410	37.50		447.50	515
0400	3'-0" x 7'-0"		17	.941		420	37.50		457.50	520
0410	1-3/8" thick, 18 ga., 2'-0"x 6'-8"		20	.800		320	32		352	405
0420	3'-0" x 6'-8"		17	.941		325	37.50		362.50	415
0425	3'-0" x 7'-0"		17	.941		330	37.50		367.50	425
0450	For vision lite, add					92.50			92.50	102
0452	For narrow lite, add					101			101	111
0460	Half glass, 18 ga., 2'-0" x 6'-8"	2 Carp	20	.800		450	32		482	550
0465	2'-8" x 6'-8"		18	.889		465	35.50		500.50	570
0470	3'-0" x 6'-8"		17	.941		455	37.50		492.50	560
0475	3'-0" x 7'-0"		17	.941		460	37.50		497.50	565
0500	Hollow core, 1-3/4" thick, full panel, 20 ga., 2'-8" x 6'-8"		18	.889		305	35.50		340.50	395
0520	3'-0" x 6'-8"		17	.941		280	37.50		317.50	365
0640	3'-0" x 7'-0"		17	.941		325	37.50		362.50	420
0680	4'-0" x 7'-0"		15	1.067		465	42.50		507.50	580
0700	4'-0" x 8'-0"		13	1.231		530	49		579	665
1000	18 ga., 2'-8" x 6'-8"		17	.941		310	37.50		347.50	400
1020	3'-0" x 6'-8"		16	1		299	40		339	395
1120	3'-0" x 7'-0"		17	.941		370	37.50		407.50	465
1180	4'-0" x 7'-0"		14	1.143		450	45.50		495.50	570
1200	4'-0" x 8'-0"		17	.941		535	37.50		572.50	645
1212	For vision lite, add					92.50			92.50	102
1214	For narrow lite, add					101			101	111
1230	Half glass, 20 ga., 2'-8" x 6'-8"	2 Carp	20	.800		430	32		462	530
1240	3'-0" x 6'-8"		18	.889		420	35.50		455.50	525
1260	3'-0" x 7'-0"		18	.889		425	35.50		460.50	525
1320	18 ga., 2'-8" x 6'-8"		18	.889		460	35.50		495.50	565
1340	3'-0" x 6'-8"		17	.941		455	37.50		492.50	560
1360	3'-0" x 7'-0"		17	.941		465	37.50		502.50	570
1380	4'-0" x 7'-0"		15	1.067		580	42.50		622.50	705
1400	4'-0" x 8'-0"		14	1.143		660	45.50		705.50	805
1720	Insulated, 1-3/4" thick, full panel, 18 ga., 3'-0" x 6'-8"		15	1.067		400	42.50		442.50	515
1740	2'-8" x 7'-0"		16	[illegible]		420	40		460	525
1760	3'-0" x 7'-0"		15	1.067		415	42.50		457.50	525
1800	4'-0" x 8'-0"		13	1.231		605	49		654	745
1805	For vision lite, add					92.50			92.50	102
1810	For narrow lite, add					101			101	111
1820	Half glass, 18 ga., 3'-0" x 6'-8"	2 Carp	16	[illegible]		535	40		575	655
1840	2'-8" x 7'-0"		17	.941		545	37.50		582.50	660
1860	3'-0" x 7'-0"		16	[illegible]		580	40		620	705
1900	4'-0" x 8'-0"		14	1.143		655	45.50		700.50	795
2000	For bottom louver, add					162			162	178
2020	For baked enamel finish, add					30%	15%			

08 13 13.13 Standard Hollow Metal Doors		Crew	Daily Output	Labor-Hours	Unit	Material	2009 Bare Costs Labor	Equipment	Total	Total Incl O&P
2040	For galvanizing, add					15%				
9000	Minimum labor/equipment charge	1 Carp	4	2	Job		80		80	132

08 13 13.15 Metal Fire Doors

		Crew	Daily Output	Labor-Hours	Unit	Material	Labor	Equipment	Total	Total Incl O&P
0010	**METAL FIRE DOORS** R081313-20									
0015	Steel, flush, "B" label, 90 minute									
0020	Full panel, 20 ga., 2'-0" x 6'-8"	2 Carp	20	.800	Ea.	340	32		372	430
0040	2'-8" x 6'-8"		18	.889		345	35.50		380.50	440
0060	3'-0" x 6'-8"		17	.941		345	37.50		382.50	440
0080	3'-0" x 7'-0"		17	.941		360	37.50		397.50	455
0140	18 ga., 3'-0" x 6'-8"		16	1		395	40		435	500
0160	2'-8" x 7'-0"		17	.941		415	37.50		452.50	515
0180	3'-0" x 7'-0"		16	1		395	40		435	500
0200	4'-0" x 7'-0"		15	1.067		520	42.50		562.50	640
0220	For "A" label, 3 hour, 18 ga., use same price as "B" label									
0240	For vision lite, add				Ea.	112			112	124
0520	Flush, "B" label 90 min., composite, 20 ga., 2'-0" x 6'-8"	2 Carp	18	.889		430	35.50		465.50	535
0540	2'-8" x 6'-8"		17	.941		435	37.50		472.50	540
0560	3'-0" x 6'-8"		16	1		440	40		480	545
0580	3'-0" x 7'-0"		16	1		450	40		490	560
0640	Flush, "A" label 3 hour, composite, 18 ga., 3'-0" x 6'-8"		15	1.067		495	42.50		537.50	615
0660	2'-8" x 7'-0"		16	1		515	40		555	635
0680	3'-0" x 7'-0"		15	1.067		505	42.50		547.50	625
0700	4'-0" x 7'-0"		14	1.143		610	45.50		655.50	750
9000	Minimum labor/equipment charge	1 Carp	4	2	Job		80		80	132

08 13 13.20 Residential Steel Doors

		Crew	Daily Output	Labor-Hours	Unit	Material	Labor	Equipment	Total	Total Incl O&P
0010	**RESIDENTIAL STEEL DOORS**									
0020	Prehung, insulated, exterior									
0030	Embossed, full panel, 2'-8" x 6'-8" G	2 Carp	17	.941	Ea.	277	37.50		314.50	365
0040	3'-0" x 6'-8" G		15	1.067		249	42.50		291.50	345
0060	3'-0" x 7'-0" G		15	1.067		291	42.50		333.50	390
0070	5'-4" x 6'-8", double G		8	2		575	80		655	765
0220	Half glass, 2'-8" x 6'-8"		17	.941		278	37.50		315.50	365
0240	3'-0" x 6'-8"		16	1		278	40		318	370
0260	3'-0" x 7'-0"		16	1		335	40		375	435
0270	5'-4" x 6'-8", double		8	2		570	80		650	760
0720	Raised plastic face, full panel, 2'-8" x 6'-8"		16	1		278	40		318	370
0740	3'-0" x 6'-8"		15	1.067		280	42.50		322.50	380
0760	3'-0" x 7'-0"		15	1.067		284	42.50		326.50	380
0780	5'-4" x 6'-8", double		8	2		525	80		605	710
0820	Half glass, 2'-8" x 6'-8"		17	.941		310	37.50		347.50	405
0840	3'-0" x 6'-8"		16	1		315	40		355	410
0860	3'-0" x 7'-0"		16	1		345	40		385	445
0880	5'-4" x 6'-8", double		8	2		670	80		750	865
1320	Flush face, full panel, 2'-8" x 6'-8"		16	1		230	40		270	320
1340	3'-0" x 6'-8"		15	1.067		230	42.50		272.50	325
1360	3'-0" x 7'-0"		15	1.067		290	42.50		332.50	390
1380	5'-4" x 6'-8", double		8	2		470	80		550	645
1420	Half glass, 2'-8" x 6'-8"		17	.941		289	37.50		326.50	380
1440	3'-0" x 6'-8"		16	1		289	40		329	385
1460	3'-0" x 7'-0"		16	1		335	40		375	435
1480	5'-4" x 6'-8", double		8	2		555	80		635	740
1500	Sidelight, full lite, 1'-0" x 6'-8"with grille					228			228	251

08 13 Metal Doors

08 13 13 – Hollow Metal Doors

08 13 13.20 Residential Steel Doors		Crew	Daily Output	Labor-Hours	Unit	Material	2009 Bare Costs Labor	Equipment	Total	Total Incl O&P
1510	1'-0" x 6'-8", low e				Ea.	248			248	273
1520	1'-0" x 6'-8", half lite					254			254	280
1530	1'-0" x 6'-8", half lite, low e					262			262	289
2300	Interior, residential, closet, bi-fold, 6'-8" x 2'-0" wide	2 Carp	16	1		148	40		188	229
2330	3'-0" wide		16	1		165	40		205	248
2360	4'-0" wide		15	1.067		251	42.50		293.50	345
2400	5'-0" wide		14	1.143		290	45.50		335.50	395
2420	6'-0" wide		13	1.231		325	49		374	440
9000	Minimum labor/equipment charge	1 Carp	4	2	Job		80		80	132

08 13 16 – Aluminum Doors

08 13 16.10 Commercial Aluminum Doors		Crew	Daily Output	Labor-Hours	Unit	Material	Labor	Equipment	Total	Total Incl O&P
0010	**COMMERCIAL ALUMINUM DOORS**, no glazing									
0020	Incl. hinges, push/pull, deadlock, cyl., threshold									
1000	Narrow stile, no glazing, standard hardware, 3'-0" x 7'-0", single	2 Carp	3	5.333	Ea.	590	213		803	995
1200	Pair of 3'-0" x 7'-0"		1.70	9.412	Pr.	1,175	375		1,550	1,900
1500	3'-6" x 7'-0", single		3	5.333	Ea.	850	213		1,063	1,275
2100	Medium stile, 3'-0" x 7'-0", single		3	5.333	"	720	213		933	1,150
2200	Pair of 3'-0" x 7'-0"		1.70	9.412	Pr.	1,400	375		1,775	2,175
2300	3'-6" x 7'-0", single		3	5.333	Ea.	930	213		1,143	1,375
5000	Flush panel doors, pair of 2'-6" x 7'-0"	2 Sswk	2	8	Pr.	1,275	360		1,635	2,050
5050	3'-0" x 7'-0", single		2.50	6.400	Ea.	640	286		926	1,225
5100	Pair of 3'-0" x 7'-0"		2	8	Pr.	1,275	360		1,635	2,050
5150	3'-6" x 7'-0", single		2.50	6.400	Ea.	760	286		1,046	1,375

08 13 73 – Sliding Metal Doors

08 13 73.10 Steel Sliding Doors		Crew	Daily Output	Labor-Hours	Unit	Material	Labor	Equipment	Total	Total Incl O&P
0010	**STEEL SLIDING DOORS**									
0020	Up to 50' x 18', electric, standard duty, minimum	L-5	360	.156	S.F.	26	6.95	2.19	35.14	43.50
0100	Maximum		340	.165		42	7.35	2.32	51.67	62
0500	Heavy duty, minimum		297	.189		34	8.40	2.66	45.06	55.50
0600	Maximum		277	.202		88	9.05	2.85	99.90	117

08 14 Wood Doors

08 14 13 – Carved Wood Doors

08 14 13.10 Types of Wood Doors, Carved		Crew	Daily Output	Labor-Hours	Unit	Material	Labor	Equipment	Total	Total Incl O&P
0010	**TYPES OF WOOD DOORS, CARVED**									
3000	Solid wood, 1-3/4" thick stile and rail									
3020	Mahogany, 3'-0" x 7'-0", minimum	2 Carp	14	1.143	Ea.	850	45.50		895.50	1,000
3030	Maximum		10	1.600		1,350	64		1,414	1,575
3040	3'-6" x 8'-0", minimum		10	1.600		980	64		1,044	1,175
3050	Maximum		8	2		1,800	80		1,880	2,100
3100	Pine, 3'-0" x 7'-0", minimum		14	1.143		415	45.50		460.50	530
3110	Maximum		10	1.600		705	64		769	880
3120	3'-6" x 8'-0", minimum		10	1.600		750	64		814	930
3130	Maximum		8	2		1,200	80		1,280	1,450
3200	Red oak, 3'-0" x 7'-0", minimum		14	1.143		1,600	45.50		1,645.50	1,850
3210	Maximum		10	1.600		1,925	64		1,989	2,225
3220	3'-6" x 8'-0", minimum		10	1.600		1,775	64		1,839	2,050
3230	Maximum		8	2		3,150	80		3,230	3,600
4000	Hand carved door, mahogany									
4020	3'-0" x 7'-0", minimum	2 Carp	14	1.143	Ea.	1,600	45.50		1,645.50	1,825

08 14 Wood Doors

08 14 13 – Carved Wood Doors

08 14 13.10 Types of Wood Doors, Carved		Crew	Daily Output	Labor-Hours	Unit	Material	Labor	Equipment	Total	Total Incl O&P
4030	Maximum	2 Carp	11	1.455	Ea.	3,125	58		3,183	3,550
4040	3'-6" x 8'-0", minimum		10	1.600		2,000	64		2,064	2,300
4050	Maximum		8	2		3,075	80		3,155	3,500
4200	Rose wood, 3'-0" x 7'-0", minimum		14	1.143		4,850	45.50		4,895.50	5,425
4210	Maximum		11	1.455		13,300	58		13,358	14,700
4220	3'-6" x 8'-0", minimum		10	1.600		5,475	64		5,539	6,125
4280	For 6'-8" high door, deduct from 7'-0" door					34			34	37.50
4400	For custom finish, add					370			370	405
4600	Side light, mahogany, 7'-0" x 1'-6" wide, minimum	2 Carp	18	.889		895	35.50		930.50	1,050
4610	Maximum		14	1.143		2,625	45.50		2,670.50	2,975
4620	8'-0" x 1'-6" wide, minimum		14	1.143		1,675	45.50		1,720.50	1,925
4630	Maximum		10	1.600		1,900	64		1,964	2,200
4640	Side light, oak, 7'-0" x 1'-6" wide, minimum		18	.889		1,025	35.50		1,060.50	1,175
4650	Maximum		14	1.143		1,825	45.50		1,870.50	2,100
4660	8'-0" x 1'-6" wide, minimum		14	1.143		970	45.50		1,015.50	1,150
4670	Maximum		10	1.600		1,825	64		1,889	2,125

08 14 16 – Flush Wood Doors

08 14 16.09 Smooth Wood Doors

08 14 16.09 Smooth Wood Doors		Crew	Daily Output	Labor-Hours	Unit	Material	Labor	Equipment	Total	Total Incl O&P
0010	**SMOOTH WOOD DOORS**									
0015	Flush, int., 1-3/8", 7 ply, hollow core,									
0020	Lauan face, 2'-0" x 6'-8"	2 Carp	17	.941	Ea.	32.50	37.50		70	97.50
0080	3'-0" x 6'-8"		17	.941		50	37.50		87.50	117
0100	4'-0" x 6'-8"		16	1		81	40		121	155
0120	Birch face, 2'-0" x 6'-8"		17	.941		44.50	37.50		82	111
0140	2'-6" x 6'-8"		17	.941		81	37.50		118.50	151
0180	3'-0" x 6'-8"		17	.941		90	37.50		127.50	161
0200	4'-0" x 6'-8"		16	1		110	40		150	187
0220	Oak face, 2'-0" x 6'-8"		17	.941		84.50	37.50		122	155
0240	2'-6" x 6'-8"		17	.941		90	37.50		127.50	161
0280	3'-0" x 6'-8"		17	.941		96	37.50		133.50	168
0300	4'-0" x 6'-8"		16	1		122	40		162	200
0320	Walnut face, 2'-0" x 6'-8"		17	.941		170	37.50		207.50	249
0340	2'-6" x 6'-8"		17	.941		174	37.50		211.50	253
0380	3'-0" x 6'-8"		17	.941		180	37.50		217.50	260
0400	4'-0" x 6'-8"		16	1		204	40		244	291
0430	For 7'-0" high, add					15.20			15.20	16.70
0440	For 8'-0" high, add					28			28	31
0480	For prefinishing, clear, add					34			34	37.50
0500	For prefinishing, stain, add					45			45	49.50
1320	M.D. overlay on hardboard, 2'-0" x 6'-8"	2 Carp	17	.941		94	37.50		131.50	165
1340	2'-6" x 6'-8"		17	.941		94	37.50		131.50	165
1380	3'-0" x 6'-8"		17	.941		112	37.50		149.50	185
1400	4'-0" x 6'-8"		16	1		154	40		194	235
1420	For 7'-0" high, add					8.50			8.50	9.35
1440	For 8'-0" high, add					23			23	25.50
1720	H.P. plastic laminate, 2'-0" x 6'-8"	2 Carp	16	1		224	40		264	310
1740	2'-6" x 6'-8"		16	1		224	40		264	310
1780	3'-0" x 6'-8"		15	1.067		268	42.50		310.50	365
1800	4'-0" x 6'-8"		14	1.143		355	45.50		400.50	470
1820	For 7'-0" high, add					9			9	9.90
1840	For 8'-0" high, add					23			23	25.50
2020	5 ply particle core, lauan face, 2'-6" x 6'-8"	2 Carp	15	1.067		82	42.50		124.50	161

08 14 16.09 Smooth Wood Doors		Crew	Daily Output	Labor-Hours	Unit	Material	2009 Bare Costs Labor	Equipment	Total	Total Incl O&P
2040	3'-0" x 6'-8"	2 Carp	14	1.143	Ea.	85.50	45.50		131	170
2080	3'-0" x 7'-0"		13	1.231		92.50	49		141.50	183
2100	4'-0" x 7'-0"		12	1.333		110	53.50		163.50	210
2120	Birch face, 2'-6" x 6'-8"		15	1.067		93.50	42.50		136	174
2140	3'-0" x 6'-8"		14	1.143		102	45.50		147.50	189
2180	3'-0" x 7'-0"		13	1.231		105	49		154	196
2200	4'-0" x 7'-0"		12	1.333		128	53.50		181.50	228
2220	Oak face, 2'-6" x 6'-8"		15	1.067		103	42.50		145.50	185
2240	3'-0" x 6'-8"		14	1.143		114	45.50		159.50	201
2280	3'-0" x 7'-0"		13	1.231		117	49		166	209
2300	4'-0" x 7'-0"		12	1.333		143	53.50		196.50	245
2320	Walnut face, 2'-0" x 6'-8"		15	1.067		114	42.50		156.50	197
2340	2'-6" x 6'-8"		14	1.143		131	45.50		176.50	220
2380	3'-0" x 6'-8"		13	1.231		147	49		196	243
2400	4'-0" x 6'-8"	↓	12	1.333		192	53.50		245.50	299
2440	For 8'-0" high, add					28			28	31
2460	For 8'-0" high walnut, add					15.35			15.35	16.90
2480	For solid wood core, add					34.50			34.50	38
2720	For prefinishing, clear, add					22.50			22.50	24.50
2740	For prefinishing, stain, add					50			50	55
3320	M.D. overlay on hardboard, 2'-6" x 6'-8"	2 Carp	14	1.143		104	45.50		149.50	190
3340	3'-0" x 6'-8"		13	1.231		109	49		158	200
3380	3'-0" x 7'-0"		12	1.333		111	53.50		164.50	210
3400	4'-0" x 7'-0"	↓	10	1.600		136	64		200	255
3440	For 8'-0" height, add					30.50			30.50	33.50
3460	For solid wood core, add					39.50			39.50	43.50
3720	H.P. plastic laminate, 2'-6" x 6'-8"	2 Carp	13	1.231		152	49		201	248
3740	3'-0" x 6'-8"		12	1.333		172	53.50		225.50	278
3780	3'-0" x 7'-0"		11	1.455		179	58		237	292
3800	4'-0" x 7'-0"	↓	8	2		217	80		297	370
3840	For 8'-0" height, add					30.50			30.50	33.50
3860	For solid wood core, add					37			37	40.50
4000	Exterior, flush, solid wood stave core, birch, 1-3/4" x 7'-0" x 2'-6"	2 Carp	15	1.067		162	42.50		204.50	249
4020	2'-8" wide		15	1.067		170	42.50		212.50	258
4040	3'-0" wide		14	1.143		185	45.50		230.50	280
4100	Oak faced 1-3/4" x 7'-0" x 2'-6" wide		15	1.067		179	42.50		221.50	268
4120	2'-8" wide		15	1.067		192	42.50		234.50	282
4140	3'-0" wide		14	1.143		204	45.50		249.50	300
4200	Walnut faced, 1-3/4" x 7'-0" x 2'-6" wide		15	1.067		263	42.50		305.50	360
4220	2'-8" wide		15	1.067		275	42.50		317.50	375
4240	3'-0" wide	↓	14	1.143		287	45.50		332.50	390
4300	For 6'-8" high door, deduct from 7'-0" door				↓	15			15	16.50
9000	Minimum labor/equipment charge	1 Carp	4	2	Job		80		80	132

08 14 16.10 Wood Doors Decorator		Crew	Daily Output	Labor-Hours	Unit	Material	Labor	Equipment	Total	Total Incl O&P
0010	**WOOD DOORS DECORATOR**									
0040	7 ply hollow core lauan face, 2'-6" x 6'-8"	2 Carp	17	.941	Ea.	36	37.50		73.50	102

08 14 16.20 Wood Fire Doors		Crew	Daily Output	Labor-Hours	Unit	Material	Labor	Equipment	Total	Total Incl O&P
0010	**WOOD FIRE DOORS**									
0020	Particle core, 7 face plys, "B" label,									
0040	1 hour, birch face, 1-3/4" x 2'-6" x 6'-8"	2 Carp	14	1.143	Ea.	320	45.50		365.50	425
0080	3'-0" x 6'-8"		13	1.231		330	49		379	445
0090	3'-0" x 7'-0"	↓	12	1.333	↓	345	53.50		398.50	470

08 14 Wood Doors

08 14 16 – Flush Wood Doors

08 14 16.20 Wood Fire Doors		Crew	Daily Output	Labor-Hours	Unit	Material	2009 Bare Costs Labor	Equipment	Total	Total Incl O&P
0100	4'-0" x 7'-0"	2 Carp	12	1.333	Ea.	500	53.50		553.50	640
0140	Oak face, 2'-6" x 6'-8"		14	1.143		310	45.50		355.50	420
0180	3'-0" x 6'-8"		13	1.231		325	49		374	435
0190	3'-0" x 7'-0"		12	1.333		340	53.50		393.50	460
0200	4'-0" x 7'-0"		12	1.333		440	53.50		493.50	575
0240	Walnut face, 2'-6" x 6'-8"		14	1.143		410	45.50		455.50	525
0280	3'-0" x 6'-8"		13	1.231		420	49		469	540
0290	3'-0" x 7'-0"		12	1.333		440	53.50		493.50	570
0300	4'-0" x 7'-0"		12	1.333		590	53.50		643.50	740
0440	M.D. overlay on hardboard, 2'-6" x 6'-8"		15	1.067		274	42.50		316.50	370
0480	3'-0" x 6'-8"		14	1.143		285	45.50		330.50	390
0490	3'-0" x 7'-0"		13	1.231		300	49		349	410
0500	4'-0" x 7'-0"		12	1.333		370	53.50		423.50	495
0740	90 minutes, birch face, 1-3/4" x 2'-6" x 6'-8"		14	1.143		310	45.50		355.50	415
0780	3'-0" x 6'-8"		13	1.231		305	49		354	415
0790	3'-0" x 7'-0"		12	1.333		340	53.50		393.50	465
0800	4'-0" x 7'-0"		12	1.333		415	53.50		468.50	545
0840	Oak face, 2'-6" x 6'-8"		14	1.143		282	45.50		327.50	385
0880	3'-0" x 6'-8"		13	1.231		292	49		341	400
0890	3'-0" x 7'-0"		12	1.333		305	53.50		358.50	425
0900	4'-0" x 7'-0"		12	1.333		425	53.50		478.50	555
0940	Walnut face, 2'-6" x 6'-8"		14	1.143		385	45.50		430.50	500
0980	3'-0" x 6'-8"		13	1.231		395	49		444	515
0990	3'-0" x 7'-0"		12	1.333		415	53.50		468.50	545
1000	4'-0" x 7'-0"		12	1.333		595	53.50		648.50	745
1140	M.D. overlay on hardboard, 2'-6" x 6'-8"		15	1.067		305	42.50		347.50	405
1180	3'-0" x 6'-8"		14	1.143		315	45.50		360.50	420
1190	3'-0" x 7'-0"		13	1.231		325	49		374	440
1200	4'-0" x 7'-0"		12	1.333		445	53.50		498.50	580
1240	For 8'-0" height, add					60			60	66
1260	For 8'-0" height walnut, add					75			75	82.50
2200	Custom architectural "B" label, flush, 1-3/4" thick, birch,									
2210	Solid core									
2220	2'-6" x 7'-0"	2 Carp	15	1.067	Ea.	289	42.50		331.50	390
2260	3'-0" x 7'-0"		14	1.143		300	45.50		345.50	405
2300	4'-0" x 7'-0"		13	1.231		395	49		444	515
2420	4'-0" x 8'-0"		11	1.455		375	58		433	510
2480	For oak veneer, add					50%				
2500	For walnut veneer, add					75%				
9000	Minimum labor/equipment charge	1 Carp	4	2	Job		80		80	132

08 14 23 – Clad Wood Doors

08 14 23.13 Metal-Faced Wood Doors		Crew	Daily Output	Labor-Hours	Unit	Material	Labor	Equipment	Total	Total Incl O&P
0010	**METAL-FACED WOOD DOORS**									
0020	Interior, flush type, 3' x 7'	2 Carp	4.30	3.721	Opng.	233	149		382	500
9000	Minimum labor/equipment charge	1 Carp	2	4	Job		160		160	264

08 14 23.20 Tin Clad Wood Doors		Crew	Daily Output	Labor-Hours	Unit	Material	Labor	Equipment	Total	Total Incl O&P
0010	**TIN CLAD WOOD DOORS**									
0020	3 ply, 6' x 7', double sliding, doors only	2 Carp	1	16	Opng.	1,825	640		2,465	3,075
1000	For electric operator, add	1 Elec	2	4	"	3,250	188		3,438	3,875
9000	Minimum labor/equipment charge	2 Carp	1	16	Job		640		640	1,050

08 14 33.10 Wood Doors Paneled		Crew	Daily Output	Labor-Hours	Unit	Material	Labor (2009 Bare Costs)	Equipment	Total	Total Incl O&P
0010	**WOOD DOORS PANELED**									
0020	Interior, six panel, hollow core, 1-3/8″ thick									
0040	Molded hardboard, 2′-0″ x 6′-8″	2 Carp	17	.941	Ea.	55	37.50		92.50	123
0060	2′-6″ x 6′-8″		17	.941		59.50	37.50		97	128
0070	2′-8″ x 6′-8″		17	.941		62	37.50		99.50	131
0080	3′-0″ x 6′-8″		17	.941		65.50	37.50		103	134
0140	Embossed print, molded hardboard, 2′-0″ x 6′-8″		17	.941		59.50	37.50		97	128
0160	2′-6″ x 6′-8″		17	.941		59.50	37.50		97	128
0180	3′-0″ x 6′-8″		17	.941		65.50	37.50		103	134
0540	Six panel, solid, 1-3/8″ thick, pine, 2′-0″ x 6′-8″		15	1.067		144	42.50		186.50	229
0560	2′-6″ x 6′-8″		14	1.143		162	45.50		207.50	254
0580	3′-0″ x 6′-8″		13	1.231		186	49		235	286
1020	Two panel, bored rail, solid, 1-3/8″ thick, pine, 1′-6″ x 6′-8″		16	1		263	40		303	355
1040	2′-0″ x 6′-8″		15	1.067		345	42.50		387.50	450
1060	2′-6″ x 6′-8″		14	1.143		395	45.50		440.50	510
1340	Two panel, solid, 1-3/8″ thick, fir, 2′-0″ x 6′-8″		15	1.067		144	42.50		186.50	229
1360	2′-6″ x 6′-8″		14	1.143		162	45.50		207.50	254
1380	3′-0″ x 6′-8″		13	1.231		395	49		444	515
1740	Five panel, solid, 1-3/8″ thick, fir, 2′-0″ x 6′-8″		15	1.067		258	42.50		300.50	355
1760	2′-6″ x 6′-8″		14	1.143		415	45.50		460.50	530
1780	3′-0″ x 6′-8″		13	1.231		415	49		464	535
9000	Minimum labor/equipment charge	1 Carp	4	2	Job		80		80	132

08 14 33.20 Wood Doors Residential		Crew	Daily Output	Labor-Hours	Unit	Material	Labor (2009 Bare Costs)	Equipment	Total	Total Incl O&P
0010	**WOOD DOORS RESIDENTIAL**									
0200	Exterior, combination storm & screen, pine									
0220	Cross buck, 6′-9″ x 2′-6″ wide	2 Carp	11	1.455	Ea.	325	58		383	455
0260	2′-8″ wide		10	1.600		276	64		340	410
0280	3′-0″ wide		9	1.778		282	71		353	425
0300	7′-1″ x 3′-0″ wide		9	1.778		325	71		396	470
0400	Full lite, 6′-9″ x 2′-6″ wide		11	1.455		296	58		354	420
0420	2′-8″ wide		10	1.600		296	64		360	430
0440	3′-0″ wide		9	1.778		300	71		371	445
0500	7′-1″ x 3′-0″ wide		9	1.778		325	71		396	470
0700	Dutch door, pine, 1-3/4″ x 6′-8″ x 2′-8″ wide, minimum		12	1.333		675	53.50		728.50	835
0720	Maximum		10	1.600		825	64		889	1,025
0800	3′-0″ wide, minimum		12	1.333		690	53.50		743.50	845
0820	Maximum		10	1.600		900	64		964	1,100
1000	Entrance door, colonial, 1-3/4″ x 6′-8″ x 2′-8″ wide		16	1		450	40		490	560
1020	6 panel pine, 3′-0″ wide		15	1.067		430	42.50		472.50	545
1100	8 panel pine, 2′-8″ wide		16	1		570	40		610	690
1120	3′-0″ wide		15	1.067		555	42.50		597.50	680
1200	For tempered safety glass lites, (min of 2) add					65.50			65.50	72
1300	Flush, birch, solid core, 1-3/4″ x 6′-8″ x 2′-8″ wide	2 Carp	16	1		102	40		142	179
1320	3′-0″ wide		15	1.067		114	42.50		156.50	196
1350	7′-0″ x 2′-8″ wide		16	1		112	40		152	189
1360	3′-0″ wide		15	1.067		119	42.50		161.50	202
1380	For tempered safety glass lites, add					97.50			97.50	107
2700	Interior, closet, bi-fold, w/hardware, no frame or trim incl.									
2720	Flush, birch, 6′-6″ or 6′-8″ x 2′-6″ wide	2 Carp	13	1.231	Ea.	51	49		100	137
2740	3′-0″ wide		13	1.231		55	49		104	141
2760	4′-0″ wide		12	1.333		97.50	53.50		151	195
2780	5′-0″ wide		11	1.455		100	58		158	206

08 14 33 – Stile and Rail Wood Doors

08 14 33.20 Wood Doors Residential		Crew	Daily Output	Labor-Hours	Unit	Material	Labor	Equipment	Total	Total Incl O&P
						2009 Bare Costs				
2800	6'-0" wide	2 Carp	10	1.600	Ea.	109	64		173	226
3000	Raised panel pine, 6'-6" or 6'-8" x 2'-6" wide		13	1.231		162	49		211	260
3020	3'-0" wide		13	1.231		254	49		303	360
3040	4'-0" wide		12	1.333		305	53.50		358.50	425
3060	5'-0" wide		11	1.455		365	58		423	495
3080	6'-0" wide		10	1.600		400	64		464	545
3200	Louvered, pine 6'-6" or 6'-8" x 2'-6" wide		13	1.231		106	49		155	197
3220	3'-0" wide		13	1.231		165	49		214	262
3240	4'-0" wide		12	1.333		198	53.50		251.50	305
3260	5'-0" wide		11	1.455		224	58		282	345
3280	6'-0" wide		10	1.600		248	64		312	380
4400	Bi-passing closet, incl. hardware and frame, no trim incl.									
4420	Flush, lauan, 6'-8" x 4'-0" wide	2 Carp	12	1.333	Opng.	169	53.50		222.50	273
4440	5'-0" wide		11	1.455		184	58		242	298
4460	6'-0" wide		10	1.600		198	64		262	325
4600	Flush, birch, 6'-8" x 4'-0" wide		12	1.333		213	53.50		266.50	320
4620	5'-0" wide		11	1.455		209	58		267	325
4640	6'-0" wide		10	1.600		256	64		320	385
4800	Louvered, pine, 6'-8" x 4'-0" wide		12	1.333		420	53.50		473.50	550
4820	5'-0" wide		11	1.455		390	58		448	520
4840	6'-0" wide		10	1.600		515	64		579	670
5000	Paneled, pine, 6'-8" x 4'-0" wide		12	1.333		495	53.50		548.50	635
5020	5'-0" wide		11	1.455		405	58		463	540
5040	6'-0" wide		10	1.600		590	64		654	755
6100	Folding accordion, closet, including track and frame									
6120	Vinyl, 2 layer, stock see Div. 10 22 26.13	2 Carp	10	1.600	Ea.	51.50	64		115.50	163
6140	Woven mahogany and vinyl, stock		10	1.600		43.50	64		107.50	154
6160	Wood slats with vinyl overlay, stock		10	1.600		139	64		203	259
6180	Economy vinyl, stock		10	1.600		28.50	64		92.50	138
6200	Rigid PVC		10	1.600		32	64		96	141
6220	For custom partition, add					25%	10%			
7310	Passage doors, flush, no frame included									
7320	Hardboard, hollow core, 1-3/8" x 6'-8" x 1'-6" wide	2 Carp	18	.889	Ea.	42.50	35.50		78	105
7330	2'-0" wide		18	.889		42	35.50		77.50	105
7340	2'-6" wide		18	.889		46.50	35.50		82	110
7350	2'-8" wide		18	.889		48	35.50		83.50	112
7360	3'-0" wide		17	.941		51	37.50		88.50	118
7420	Lauan, hollow core, 1-3/8" x 6'-8" x 1'-6" wide		18	.889		30.50	35.50		66	92
7440	2'-0" wide		18	.889		28	35.50		63.50	89
7450	2'-4" wide		18	.889		31.50	35.50		67	93
7460	2'-6" wide		18	.889		31.50	35.50		67	93
7480	2'-8" wide		18	.889		33	35.50		68.50	95
7500	3'-0" wide		17	.941		34.50	37.50		72	100
7700	Birch, hollow core, 1-3/8" x 6'-8" x 1'-6" wide		18	.889		37	35.50		72.50	99.50
7720	2'-0" wide		18	.889		41	35.50		76.50	104
7740	2'-6" wide		18	.889		49	35.50		84.50	112
7760	2'-8" wide		18	.889		49.50	35.50		85	113
7780	3'-0" wide		17	.941		51	37.50		88.50	118
8000	Pine louvered, 1-3/8" x 6'-8" x 1'-6" wide		19	.842		101	33.50		134.50	167
8020	2'-0" wide		18	.889		128	35.50		163.50	199
8040	2'-6" wide		18	.889		139	35.50		174.50	212
8060	2'-8" wide		18	.889		146	35.50		181.50	220
8080	3'-0" wide		17	.941		157	37.50		194.50	234

08 14 Wood Doors

08 14 33 – Stile and Rail Wood Doors

08 14 33.20 Wood Doors Residential		Crew	Daily Output	Labor-Hours	Unit	2009 Bare Costs Material	Labor	Equipment	Total	Total Incl O&P
8300	Pine paneled, 1-3/8" x 6'-8" x 1'-6" wide	2 Carp	19	.842	Ea.	111	33.50		144.50	178
8320	2'-0" wide		18	.839		128	35.50		163.50	199
8330	2'-4" wide		18	.839		150	35.50		185.50	224
8340	2'-6" wide		18	.839		158	35.50		193.50	232
8360	2'-8" wide		18	.839		161	35.50		196.50	236
8380	3'-0" wide	↓	17	.941	↓	165	37.50		202.50	244
8550	For over 20 doors, deduct					15%				
9900	Minimum labor/equipment charge	1 Carp	4	2	Job		80		80	132

08 14 40 – Interior Cafe Doors

08 14 40.10 Cafe Style Doors		Crew	Daily Output	Labor-Hours	Unit	Material	Labor	Equipment	Total	Total Incl O&P
0010	**CAFE STYLE DOORS**									
6520	Interior cafe doors, 2'-6" opening, stock, panel pine	2 Carp	16	1	Ea.	188	40		228	272
6540	3'-0" opening	"	16	1	"	195	40		235	281
6550	Louvered pine									
6560	2'-6" opening	2 Carp	16	1	Ea.	164	40		204	246
8000	3'-0" opening		16	1		165	40		205	248
8010	2'-6" opening, hardwood		16	1		283	40		323	375
8020	3'-0" opening	↓	16	1	↓	315	40		355	410
9000	Minimum labor/equipment charge	1 Carp	4	2	Job		80		80	132

08 16 Composite Doors

08 16 13 – Fiberglass Doors

08 16 13.10 Entrance Doors, Fiberous Glass		Crew	Daily Output	Labor-Hours	Unit	Material	Labor	Equipment	Total	Total Incl O&P
0010	**ENTRANCE DOORS, FIBEROUS GLASS**									
0020	Exterior, fiberglass, door, 2'-8" wide x 6'-8" high	2 Carp	15	1.067	Ea.	248	42.50		290.50	345
0040	3'-0" wide x 6'-8" high		15	1.067		248	42.50		290.50	345
0060	3'-0" wide x 7'-0" high		15	1.067		445	42.50		487.50	560
0080	3'-0" wide x 6'-8" high, with two lites		15	1.067		278	42.50		320.50	375
0100	3'-0" wide x 8'-0" high, with two lites		15	1.067		450	42.50		492.50	565
0110	Half glass, 3'-0" wide x 6'-8" high		15	1.067		287	42.50		329.50	385
0120	3'-0" wide x 6'-8" high, low e		15	1.067		315	42.50		357.50	420
0130	3'-0" wide x 8'-0" high		15	1.067		570	42.50		612.50	695
0140	3'-0" wide x 8'-0" high, low e	↓	15	1.067		630	42.50		672.50	760
0150	Side lights, 1'-0" wide x 6'-8" high,					232			232	255
0160	1'-0" wide x 6'-8" high, low e					246			246	271
0180	1'-0" wide x 6'-8" high, full glass					275			275	305
0190	1'-0" wide x 6'-8" high, low e				↓	284			284	315

08 17 Integrated Door Opening Assemblies

08 17 13 – Integrated Metal Door Opening Assemblies

08 17 13.20 Tubular Steel Swing Doors		Crew	Daily Output	Labor-Hours	Unit	Material	Labor	Equipment	Total	Total Incl O&P
0010	**TUBULAR STEEL SWING DOORS**									
0020	Tubular steel, 7' high, single, 3'-4" opening	2 Sswk	2.50	6.400	Ea.	830	286		1,116	1,450
0100	Double, 6'-0" opening	"	2	8	Pr.	1,050	360		1,410	1,800

08 17 23 – Integrated Wood Door Opening Assemblies

08 17 23.10 Pre-Hung Doors		Crew	Daily Output	Labor-Hours	Unit	Material	Labor	Equipment	Total	Total Incl O&P
0010	**PRE-HUNG DOORS**									
0300	Exterior, wood, comb. storm & screen, 6'-9" x 2'-6" wide	2 Carp	15	1.067	Ea.	289	42.50		331.50	390

08 17 Integrated Door Opening Assemblies

08 17 23 – Integrated Wood Door Opening Assemblies

08 17 23.10 Pre-Hung Doors		Crew	Daily Output	Labor-Hours	Unit	Material	Labor	Equipment	Total	Total Incl O&P
						2009 Bare Costs				
0320	2′-8″ wide	2 Carp	15	1.067	Ea.	289	42.50		331.50	390
0340	3′-0″ wide		15	1.067		296	42.50		338.50	395
0360	For 7′-0″ high door, add					22.50			22.50	25
1600	Entrance door, flush, birch, solid core									
1620	4-5/8″ solid jamb, 1-3/4″ x 6′-8″ x 2′-8″ wide	2 Carp	16	1	Ea.	285	40		325	380
1640	3′-0″ wide	″	16	1		350	40		390	450
1680	For 7′-0″ high door, add					20.50			20.50	22.50
2000	Entrance door, colonial, 6 panel pine									
2020	4-5/8″ solid jamb, 1-3/4″ x 6′-8″ x 2′-8″ wide	2 Carp	16	1	Ea.	530	40		570	645
2040	3′-0″ wide	″	16	1		530	40		570	645
2060	For 7′-0″ high door, add					52.50			52.50	58
2200	For 5-5/8″ solid jamb, add					40.50			40.50	44.50
4000	Interior, passage door, 4-5/8″ solid jamb									
4400	Lauan, flush, solid core, 1-3/8″ x 6′-8″ x 2′-6″ wide	2 Carp	17	.941	Ea.	182	37.50		219.50	262
4420	2′-8″ wide		17	.941		182	37.50		219.50	262
4440	3′-0″ wide		16	1		195	40		235	281
4600	Hollow core, 1-3/8″ x 6′-8″ x 2′-6″ wide		17	.941		123	37.50		160.50	197
4620	2′-8″ wide		17	.941		123	37.50		160.50	197
4640	3′-0″ wide		16	1		124	40		164	202
4700	For 7′-0″ high door, add					24			24	26.50
5000	Birch, flush, solid core, 1-3/8″ x 6′-8″ x 2′-6″ wide	2 Carp	17	.941		170	37.50		207.50	249
5020	2′-8″ wide		17	.941		195	37.50		232.50	277
5040	3′-0″ wide		16	1		205	40		245	292
5200	Hollow core, 1-3/8″ x 6′-8″ x 2′-6″ wide		17	.941		141	37.50		178.50	217
5220	2′-8″ wide		17	.941		147	37.50		184.50	223
5240	3′-0″ wide		16	1		148	40		188	228
5280	For 7′-0″ high door, add					21			21	23
5500	Hardboard paneled, 1-3/8″ x 6′-8″ x 2′-6″ wide	2 Carp	17	.941		142	37.50		179.50	218
5520	2′-8″ wide		17	.941		149	37.50		186.50	225
5540	3′-0″ wide		16	1		146	40		186	227
6000	Pine paneled, 1-3/8″ x 6′-8″ x 2′-6″ wide		17	.941		247	37.50		284.50	335
6020	2′-8″ wide		17	.941		267	37.50		304.50	355
6040	3′-0″ wide		16	1		271	40		311	365
6500	For 5-5/8″ solid jamb, add					14.60			14.60	16.05
6520	For split jamb, deduct					16.45			16.45	18.10
9000	Minimum labor/equipment charge	1 Carp	4	2	Job		80		80	132

08 31 Access Doors and Panels

08 31 13 – Access Doors and Frames

08 31 13.10 Types of Framed Access Doors

		Crew	Daily Output	Labor-Hours	Unit	Material	Labor	Equipment	Total	Total Incl O&P
0010	**TYPES OF FRAMED ACCESS DOORS**									
1000	Fire rated door with lock									
1100	Metal, 12″ x 12″	1 Carp	10	.800	Ea.	157	32		189	226
1150	18″ x 18″		9	.889		205	35.50		240.50	285
1200	24″ x 24″		9	.889		325	35.50		360.50	420
1250	24″ x 36″		8	1		335	40		375	430
1300	24″ x 48″		8	1		410	40		450	520
1350	36″ x 36″		7.50	1.067		495	42.50		537.50	615
1400	48″ x 48″		7.50	1.067		635	42.50		677.50	770
1600	Stainless steel, 12″ x 12″		10	.800		281	32		313	365
1650	18″ x 18″		9	.889		410	35.50		445.50	510

08 31 13.10 Types of Framed Access Doors		Crew	Daily Output	Labor-Hours	Unit	2009 Bare Costs Material	Labor	Equipment	Total	Total Incl O&P
1700	24" x 24"	1 Carp	9	.889	Ea.	500	35.50		535.50	610
1750	24" x 36"		8	[illegible]		640	40		680	770
2000	Flush door for finishing									
2100	Metal 8" x 8"	1 Carp	10	.800	Ea.	45.50	32		77.50	103
2150	12" x 12"	"	10	.800	"	50.50	32		82.50	109
3000	Recessed door for acoustic tile									
3100	Metal, 12" x 12"	1 Carp	4.50	1.778	Ea.	69.50	71		140.50	193
3150	12" x 24"		4.50	1.778		90.50	71		161.50	217
3200	24" x 24"		4	2		121	80		201	265
3250	24" x 36"		4	2		156	80		236	305
4000	Recessed door for drywall									
4100	Metal 12" x 12"	1 Carp	6	1.333	Ea.	77.50	53.50		131	173
4150	12" x 24"		5.50	1.455		115	58		173	222
4200	24" x 36"		5	1.600		183	64		247	305
6000	Standard door									
6100	Metal, 8" x 8"	1 Carp	10	.800	Ea.	40	32		72	97
6150	12" x 12"		10	.800		45.50	32		77.50	103
6200	18" x 18"		9	.889		62.50	35.50		98	128
6250	24" x 24"		9	.889		81.50	35.50		117	148
6300	24" x 36"		8	1		120	40		160	198
6350	36" x 36"		8	1		147	40		187	227
6500	Stainless steel, 8" x 8"		10	.800		80	32		112	141
6550	12" x 12"		10	.800		107	32		139	170
6600	18" x 18"		9	.889		197	35.50		232.50	276
6650	24" x 24"		9	.889		258	35.50		293.50	345
7010	Aluminum cover, Ceiling hatches, 2'-6" x 2'-6", single leaf, st fr	G-3	11	2.909		540	115		655	780
9000	Minimum labor/equipment charge	1 Carp	4	2	Job		80		80	132
08 31 13.20 Bulkhead/Cellar Doors										
0010	**BULKHEAD/CELLAR DOORS**									
0020	Steel, not incl. sides, 44" x 62"	1 Carp	5.50	1.455	Ea.	350	58		408	480
0100	52" x 73"		5.10	1.569		365	62.50		427.50	505
0500	With sides and foundation plates, 57" x 45" x 24"		4.70	1.702		650	68		718	825
0600	42" x 49" x 51"		4.30	1.860		755	74.50		829.50	955
9000	Minimum labor/equipment charge		2	4	Job		160		160	264
08 31 13.30 Commercial Floor Doors										
0010	**COMMERCIAL FLOOR DOORS**									
0020	Aluminum tile, steel frame, one leaf, 2' x 2' opng.	2 Sswk	3.50	4.571	Opng.	775	204		979	1,225
0050	3'-6" x 3'-6" opening		3.50	4.571		1,100	204		1,304	1,575
0500	Double leaf, 4' x 4' opening		3	5.333		1,575	238		1,813	2,175
0550	5' x 5' opening		3	5.333		2,800	238		3,038	3,525
9000	Minimum labor/equipment charge		2	8	Job		360		360	660
08 31 13.35 Industrial Floor Doors										
0010	**INDUSTRIAL FLOOR DOORS**									
0020	Steel 300 psf L.L., single leaf, 2' x 2', 175#	2 Sswk	6	2.667	Opng.	630	119		749	915
0050	3' x 3' opening, 300#		5.50	2.909		870	130		1,000	1,200
0300	Double leaf, 4' x 4' opening, 455#		5	3.200		1,575	143		1,718	2,000
0350	5' x 5' opening, 645#		4.50	3.556		2,400	159		2,559	2,925
1000	Aluminum, 300 psf L.L., single leaf, 2' x 2', 60#		6	2.667		625	119		744	905
1050	3' x 3' opening, 100#		5.50	2.909		970	130		1,100	1,325
1500	Double leaf, 4' x 4' opening, 160#		5	3.200		1,575	143		1,718	2,000
1550	5' x 5' opening, 235#		4.50	3.556		2,400	159		2,559	2,950
2000	Aluminum, 150 psf L.L., single leaf, 2' x 2', 60#		6	2.667		570	119		689	845

08 31 Access Doors and Panels

08 31 13 – Access Doors and Frames

08 31 13.35 Industrial Floor Doors

Code	Description	Crew	Daily Output	Labor-Hours	Unit	2009 Bare Costs Material	Labor	Equipment	Total	Total Incl O&P
2050	3′ x 3′ opening, 95#	2 Sswk	5.50	2.909	Opng.	850	130		980	1,175
2500	Double leaf, 4′ x 4′ opening, 150#		5	3.200		1,350	143		1,493	1,750
2550	5′ x 5′ opening, 230#		4.50	3.556		1,825	159		1,984	2,325
9000	Minimum labor/equipment charge		2	8	Job		360		360	660

08 31 13.40 Kennel Doors

Code	Description	Crew	Daily Output	Labor-Hours	Unit	Material	Labor	Equipment	Total	Total Incl O&P
0010	KENNEL DOORS									
0020	2 way, swinging type, 13″ x 19″ opening	2 Carp	11	1.455	Opng.	100	58		158	206
0100	17″ x 29″ opening	″	11	1.455	″	224	58		282	340

08 32 Sliding Glass Doors

08 32 19 – Sliding Wood-Framed Glass Doors

08 32 19.15 Sliding Glass Vinyl-Clad Wood Doors

Code	Description		Crew	Daily Output	Labor-Hours	Unit	Material	Labor	Equipment	Total	Total Incl O&P
0010	SLIDING GLASS VINYL-CLAD WOOD DOORS										
0012	Vinyl clad, 1″ insul. glass, 6′-0″ x 6′-10″ high	G	2 Carp	4	4	Opng.	1,350	160		1,510	1,750
0030	6′-0″ x 8′-0″ high	G		4	4	Ea.	1,950	160		2,110	2,425
0100	8′-0″ x 6′-10″ high	G		4	4	Opng.	2,000	160		2,160	2,475
0500	3 leaf, 9′-0″ x 6′-10″ high	G		3	5.333		2,400	213		2,613	2,975
0600	12′-0″ x 6′-10″ high	G		3	5.333		3,175	213		3,388	3,850
9000	Minimum labor/equipment charge		1 Carp	4	2	Job		80		80	132

08 33 Coiling Doors and Grilles

08 33 13 – Coiling Counter Doors

08 33 13.10 Counter Doors, Coiling Type

Code	Description	Crew	Daily Output	Labor-Hours	Unit	Material	Labor	Equipment	Total	Total Incl O&P
0010	COUNTER DOORS, COILING TYPE									
0020	Manual, incl. frm and hdwe, galv. stl., 4′ roll-up, 6′ long	2 Carp	2	8	Opng.	1,125	320		1,445	1,775
0300	Galvanized steel, UL label		1.80	8.889		1,200	355		1,555	1,900
0600	Stainless steel, 4′ high roll-up, 6′ long		2	8		2,050	320		2,370	2,775
0700	10′ long		1.80	8.889		2,450	355		2,805	3,275
2000	Aluminum, 4′ high, 4′ long		2.20	7.273		1,350	291		1,641	1,975
2020	6′ long		2	8		1,500	320		1,820	2,175
2040	8′ long		1.90	8.421		1,725	335		2,060	2,450
2060	10′ long		1.80	8.889		1,825	355		2,180	2,600
2080	14′ long		1.40	11.429		2,675	455		3,130	3,675
2100	6′ high, 4′ long		2	8		1,525	320		1,845	2,200
2120	6′ long		1.60	10		1,550	400		1,950	2,350
2140	10′ long		1.40	11.429		2,125	455		2,580	3,100
9000	Minimum labor/equipment charge	1 Carp	2	4	Job		160		160	264

08 33 16 – Coiling Counter Grilles

08 33 16.10 Coiling Grilles

Code	Description	Crew	Daily Output	Labor-Hours	Unit	Material	Labor	Equipment	Total	Total Incl O&P
0010	COILING GRILLES									
0015	Aluminum, manual, incl. frame, mill finish									
0020	Top coiling, 4′ high, 4′ long	2 Sswk	3.20	5	Opng.	1,500	224		1,724	2,075
0030	6′ long		3.20	5		1,750	224		1,974	2,350
0040	8′ long		2.40	6.667		1,925	298		2,223	2,675
0050	12′ long		2.40	6.667		2,350	298		2,648	3,150
0060	16′ long		1.60	10		2,525	445		2,970	3,600
0070	6′ high, 4′ long		3.20	5		1,750	224		1,974	2,350
0080	6′ long		3.20	5		1,650	224		1,874	2,225

08 33 Coiling Doors and Grilles

08 33 16 – Coiling Counter Grilles

08 33 16.10 Coiling Grilles		Crew	Daily Output	Labor-Hours	Unit	Material	Labor	Equipment	Total	Total Incl O&P
0090	8′ long	2 Sswk	2.40	6.667	Opng.	1,675	298		1,973	2,400
0100	12′ long		1.60	10		2,200	445		2,645	3,250
0110	16′ long		1.20	13.333		2,625	595		3,220	3,975
0200	Side coiling, 8′ high, 12′ long		.60	26.667		2,175	1,200		3,375	4,600
0220	18′ long		.50	32		3,450	1,425		4,875	6,450
0240	24′ long		.40	40		3,425	1,800		5,225	7,075
0260	12′ high, 12′ long		.50	32		3,175	1,425		4,600	6,150
0280	18′ long		.40	40		4,700	1,800		6,500	8,450
0300	24′ long		.28	57.143		6,300	2,550		8,850	11,700
9000	Minimum labor/equipment charge		1	16	Job		715		715	1,325

08 33 23 – Overhead Coiling Doors

08 33 23.10 Rolling Service Doors		Crew	Daily Output	Labor-Hours	Unit	Material	Labor	Equipment	Total	Total Incl O&P
0010	**ROLLING SERVICE DOORS** Steel, manual, 20 ga., incl. hardware									
0050	8′ x 8′ high	2 Sswk	1.60	10	Ea.	1,025	445		1,470	1,950
0100	10′ x 10′ high		1.40	11.429		1,750	510		2,260	2,875
0200	20′ x 10′ high		1	16		2,925	715		3,640	4,550
0300	12′ x 12′ high		1.20	13.333		1,850	595		2,445	3,125
0400	20′ x 12′ high		.90	17.778		1,975	795		2,770	3,650
0500	14′ x 14′ high		.80	20		2,825	895		3,720	4,750
0600	20′ x 16′ high		.60	26.667		3,225	1,200		4,425	5,750
0700	10′ x 20′ high		.50	32		2,300	1,425		3,725	5,175
1000	12′ x 12′, crank operated, crank on door side		.80	20		1,575	895		2,470	3,375
1100	Crank thru wall		.70	22.857		1,775	1,025		2,800	3,850
1300	For vision panel, add					305			305	340
1400	For 22 ga., deduct				S.F.	1.10			1.10	1.21
1600	3′ x 7′ pass door within rolling steel door, new construction				Ea.	1,625			1,625	1,800
1700	Existing construction	2 Sswk	2	8		1,750	360		2,110	2,575
2000	Class A fire doors, manual, 20 ga., 8′ x 8′ high		1.40	11.429		1,375	510		1,885	2,475
2100	10′ x 10′ high		1.10	14.545		1,900	650		2,550	3,275
2200	20′ x 10′ high		.80	20		3,900	895		4,795	5,950
2300	12′ x 12′ high		1	16		2,975	715		3,690	4,600
2400	20′ x 12′ high		.80	20		4,225	895		5,120	6,300
2500	14′ x 14′ high		.60	26.667		3,275	1,200		4,475	5,800
2600	20′ x 16′ high		.50	32		5,275	1,425		6,700	8,450
2700	10′ x 20′ high		.40	40		4,075	1,800		5,875	7,800
3000	For 18 ga. doors, add				S.F.	1.27			1.27	1.40
3300	For enamel finish, add				″	1.52			1.52	1.67
3600	For safety edge bottom bar, pneumatic, add				L.F.	19.80			19.80	22
3700	Electric, add					37.50			37.50	41
4000	For weatherstripping, extruded rubber, jambs, add					12.80			12.80	14.10
4100	Hood, add					9			9	9.90
4200	Sill, add					5.15			5.15	5.65
4500	Motor operators, to 14′ x 14′ opening	2 Sswk	5	3.200	Ea.	1,075	143		1,218	1,450
4600	Over 14′ x 14′, jack shaft type	″	5	3.200		1,050	143		1,193	1,425
4700	For fire door, additional fusible link, add					23.50			23.50	25.50
9000	Minimum labor/equipment charge	2 Sswk	1	16	Job		715		715	1,325

08 34 Special Function Doors

08 34 13 – Cold Storage Doors

08 34 13.10 Doors for Cold Area Storage		Crew	Daily Output	Labor-Hours	Unit	Material	2009 Bare Costs Labor	Equipment	Total	Total Incl O&P
0010	**DOORS FOR COLD AREA STORAGE**									
0020	Single, 20 ga. galvanized steel									
0300	Horizontal sliding, 5′ x 7′, manual operation, 3.5″ thick	2 Carp	2	8	Ea.	3,100	320		3,420	3,925
0400	4″ thick		2	8		3,800	320		4,120	4,700
0500	6″ thick		2	8		3,100	320		3,420	3,950
0800	5′ x 7′, power operation, 2″ thick		1.90	8.421		5,150	335		5,485	6,225
0900	4″ thick		1.90	8.421		5,250	335		5,585	6,325
1000	6″ thick		1.90	8.421		5,975	335		6,310	7,125
1300	9′ x 10′, manual operation, 2″ insulation		1.70	9.412		4,175	375		4,550	5,200
1400	4″ insulation		1.70	9.412		4,300	375		4,675	5,350
1500	6″ insulation		1.70	9.412		5,175	375		5,550	6,325
1800	Power operation, 2″ insulation		1.60	10		7,175	400		7,575	8,550
1900	4″ insulation		1.60	10		7,350	400		7,750	8,725
2000	6″ insulation		1.70	9.412		8,325	375		8,700	9,775
2300	For stainless steel face, add					22%				
3000	Hinged, lightweight, 3′ x 7′-0″, galvanized 1 face, 2″ thick	2 Carp	2	8	Ea.	1,350	320		1,670	2,000
3050	4″ thick		1.90	8.421		1,675	335		2,010	2,375
3300	Aluminum doors, 3′ x 7′-0″, 4″ thick		1.90	8.421		1,225	335		1,560	1,900
3350	6″ thick		1.40	11.429		2,200	455		2,655	3,150
3600	Stainless steel, 3′ x 7′-0″, 4″ thick		1.90	8.421		1,575	335		1,910	2,275
3650	6″ thick		1.40	11.429		2,650	455		3,105	3,675
3900	Painted, 3′ x 7′-0″, 4″ thick		1.90	8.421		1,125	335		1,460	1,800
3950	6″ thick		1.40	11.429		2,150	455		2,605	3,125
5000	Bi-parting, electric operated									
5010	6′ x 8′ opening, galv. faces, 4″ thick for cooler	2 Carp	.80	20	Opng.	6,750	800		7,550	8,750
5050	For freezer, 4″ thick		.80	20		7,425	800		8,225	9,500
5300	For door buck framing and door protection, add		2.50	6.400		550	256		806	1,025
6000	Galvanized batten door, galvanized hinges, 4′ x 7′		2	8		1,650	320		1,970	2,350
6050	6′ x 8′		1.80	8.889		2,275	355		2,630	3,075
6500	Fire door, 3 hr., 6′ x 8′, single slide		.80	20		7,800	800		8,600	9,900
6550	Double, bi-parting		.70	22.857		12,000	915		12,915	14,700
9000	Minimum labor/equipment charge	1 Carp	2	4	Job		160		160	264

08 34 16 – Hangar Doors

08 34 16.10 Aircraft Hangar Doors		Crew	Daily Output	Labor-Hours	Unit	Material	Labor	Equipment	Total	Total Incl O&P
0010	**AIRCRAFT HANGAR DOORS**									
0020	Bi-fold, ovhd., 20 PSF wind load, incl. elec. oper.									
0100	12′ high x 40′	2 Sswk	240	.067	S.F.	14.80	2.98		17.78	22
0200	16′ high x 60′		230	.070		17.95	3.11		21.06	25.50
0300	20′ high x 80′		220	.073		19.15	3.25		22.40	27

08 34 36 – Darkroom Doors

08 34 36.10 Various Types of Darkroom Doors		Crew	Daily Output	Labor-Hours	Unit	Material	Labor	Equipment	Total	Total Incl O&P
0010	**VARIOUS TYPES OF DARKROOM DOORS**									
0015	Revolving, standard, 2 way, 36″ diameter	2 Carp	3.10	5.161	Opng.	2,075	206		2,281	2,625
0020	41″ diameter		3.10	5.161		2,250	206		2,456	2,825
0050	3 way, 51″ diameter		1.40	11.429		2,800	455		3,255	3,825
1000	4 way, 49″ diameter		1.40	11.429		3,325	455		3,780	4,400
2000	Hinged safety, 2 way, 41″ diameter		2.30	6.957		2,625	278		2,903	3,350
2500	3 way, 51″ diameter		1.40	11.429		3,325	455		3,780	4,400
3000	Pop out safety, 2 way, 41″ diameter		3.10	5.161		3,275	206		3,481	3,950
4000	3 way, 51″ diameter		1.40	11.429		3,250	455		3,705	4,325
5000	Wheelchair-type, pop out, 51″ diameter		1.40	11.429		3,300	455		3,755	4,400
5020	72″ diameter		.90	17.778		6,775	710		7,485	8,625

08 34 Special Function Doors

08 34 36 – Darkroom Doors

08 34 36.10 Various Types of Darkroom Doors		Crew	Daily Output	Labor-Hours	Unit	Material	2009 Bare Costs Labor	Equipment	Total	Total Incl O&P
9300	For complete darkrooms, see Div. 13 21 53.50									

08 34 59 – Vault Doors and Day Gates

08 34 59.10 Secure Storage Doors

		Crew	Daily Output	Labor-Hours	Unit	Material	Labor	Equipment	Total	Total Incl O&P
0010	**SECURE STORAGE DOORS**									
0020	Door and frame, 32" x 78", clear opening									
0100	1 hour test, 32" door, weighs 750 lbs.	2 Sswk	1.50	10.667	Opng.	3,950	475		4,425	5,200
0200	2 hour test, 32" door, weighs 950 lbs.		1.30	12.308		4,200	550		4,750	5,650
0250	40" door, weighs 1130 lbs.		1	16		4,650	715		5,365	6,450
0300	4 hour test, 32" door, weighs 1025 lbs.		1.20	13.333		4,550	595		5,145	6,125
0350	40" door, weighs 1140 lbs.		.90	17.778		5,200	795		5,995	7,175
0500	For stainless steel front, including frame, add to above					1,900			1,900	2,100
0550	Back, add					1,900			1,900	2,100
0600	For time lock, two movement, add	1 Elec	2	4	Ea.	1,450	188		1,638	1,875
0650	Three movement, add	"	2	4		1,900	188		2,088	2,400
0800	Day gate, painted, steel, 32" wide	2 Sswk	1.50	10.667		1,850	475		2,325	2,925
0850	40" wide		1.40	11.429		2,050	510		2,560	3,200
0900	Aluminum, 32" wide		1.50	10.667		2,950	475		3,425	4,125
0950	40" wide		1.40	11.429		3,275	510		3,785	4,550
2050	Security vault door, class I, 3' wide, 3 1/2" thick	E-24	.19	165	Opng.	15,100	7,300	4,100	26,500	34,300
2100	Class II, 3' wide, 7" thick		.19	165		17,500	7,300	4,100	28,900	36,900
2150	Class III, 9R, 3' wide, 10" thick, minimum		.13	250		21,800	11,000	6,175	38,975	50,500
2160	Class V, type 1, 40" door		2.48	12.903	Ea.	6,300	565	320	7,185	8,275
2170	Class V, type 2, 40" door		2.48	12.903		6,250	565	320	7,135	8,225
2180	Day gate for class V vault	2 Sswk	2	8		1,300	360		1,660	2,075

08 34 63 – Detention Doors and Frames

08 34 63.13 Steel Detention Doors and Frames

		Crew	Daily Output	Labor-Hours	Unit	Material	Labor	Equipment	Total	Total Incl O&P
0010	**STEEL DETENTION DOORS AND FRAMES**									
1000	Doors & frames, 3' x 7', complete, with hardware, single plate	E-4	4	8	Ea.	4,200	360	33.50	4,593.50	5,325
1650	Double plate	"	4	8	"	5,175	360	33.50	5,568.50	6,400

08 34 73 – Sound Control Door Assemblies

08 34 73.10 Acoustical Doors

		Crew	Daily Output	Labor-Hours	Unit	Material	Labor	Equipment	Total	Total Incl O&P
0010	**ACOUSTICAL DOORS**									
0020	Including framed seals, 3' x 7', wood, 27 STC rating	2 Carp	1.50	10.667	Ea.	400	425		825	1,150
0100	Steel, 40 STC rating		1.50	10.667		2,800	425		3,225	3,775
0200	45 STC rating		1.50	10.667		3,300	425		3,725	4,325
0300	48 STC rating		1.50	10.667		3,800	425		4,225	4,875
0400	52 STC rating		1.50	10.667		4,000	425		4,425	5,100
9000	Minimum labor/equipment charge	1 Carp	4	2	Job		80		80	132

08 36 Panel Doors

08 36 13 – Sectional Doors

08 36 13.10 Overhead Commercial Doors

		Crew	Daily Output	Labor-Hours	Unit	Material	Labor	Equipment	Total	Total Incl O&P
0010	**OVERHEAD COMMERCIAL DOORS**									
1000	Stock, sectional, heavy duty, wood, 1-3/4" thick, 8' x 8' high	2 Carp	2	8	Ea.	755	320		1,075	1,350
1100	10' x 10' high		1.80	8.889		1,125	355		1,480	1,825
1200	12' x 12' high		1.50	10.667		1,625	425		2,050	2,500
1300	Chain hoist, 14' x 14' high		1.30	12.308		2,550	490		3,040	3,600
1400	12' x 16' high		1	16		2,500	640		3,140	3,800
1500	20' x 8' high		1.30	12.270		2,250	490		2,740	3,275

08 36 Panel Doors

08 36 13 – Sectional Doors

08 36 13.10 Overhead Commercial Doors		Crew	Daily Output	Labor-Hours	Unit	Material	2009 Bare Costs Labor	Equipment	Total	Total Incl O&P
1600	20' x 16' high	2 Carp	.65	24.615	Ea.	5,950	985		6,935	8,175
1800	Center mullion openings, 8' high		4	4		1,025	160		1,185	1,400
1900	20' high		2	8		1,825	320		2,145	2,525
2100	For medium duty custom door, deduct					5%	5%			
2150	For medium duty stock doors, deduct					10%	5%			
2300	Fiberglass and aluminum, heavy duty, sectional, 12' x 12' high	2 Carp	1.50	10.667	Ea.	2,550	425		2,975	3,500
2450	Chain hoist, 20' x 20' high		.50	32		6,450	1,275		7,725	9,200
2600	Steel, 24 ga. sectional, manual, 8' x 8' high		2	8		735	320		1,055	1,325
2650	10' x 10' high		1.80	8.889		940	355		1,295	1,600
2700	12' x 12' high		1.50	10.667		1,175	425		1,600	2,000
2800	Chain hoist, 20' x 14' high		.70	22.857		3,400	915		4,315	5,225
2850	For 1-1/4" rigid insulation and 26 ga. galv.									
2860	back panel, add				S.F.	3.58			3.58	3.94
2900	For electric trolley operator, 1/3 H.P., to 12' x 12', add	1 Carp	2	4	Ea.	925	160		1,085	1,300
2950	Over 12' x 12', 1/2 H.P., add	"	1	8	"	1,100	320		1,420	1,750
9000	Minimum labor/equipment charge	2 Carp	1.50	10.667	Job		425		425	705

08 36 13.20 Residential Garage Doors		Crew	Daily Output	Labor-Hours	Unit	Material	Labor	Equipment	Total	Total Incl O&P
0010	**RESIDENTIAL GARAGE DOORS**									
0050	Hinged, wood, custom, double door, 9' x 7'	2 Carp	4	4	Ea.	540	160		700	855
0070	16' x 7'		3	5.333		955	213		1,168	1,400
0200	Overhead, sectional, incl. hardware, fiberglass, 9' x 7', standard		5.28	3.030		715	121		836	985
0220	Deluxe		5.28	3.030		900	121		1,021	1,200
0300	16' x 7', standard		6	2.667		1,550	107		1,657	1,875
0320	Deluxe		6	2.667		1,925	107		2,032	2,275
0500	Hardboard, 9' x 7', standard		8	2		540	80		620	725
0520	Deluxe		8	2		765	80		845	970
0600	16' x 7', standard		6	2.667		1,125	107		1,232	1,400
0620	Deluxe		6	2.667		1,300	107		1,407	1,600
0700	Metal, 9' x 7', standard		5.28	3.030		555	121		676	815
0720	Deluxe		8	2		805	80		885	1,025
0800	16' x 7', standard		3	5.333		705	213		918	1,125
0820	Deluxe		6	2.667		1,375	107		1,482	1,675
0900	Wood, 9' x 7', standard		8	2		705	80		785	905
0920	Deluxe		8	2		2,025	80		2,105	2,350
1000	16' x 7', standard		6	2.667		1,425	107		1,532	1,725
1020	Deluxe		6	2.667		2,950	107		3,057	3,425
1800	Door hardware, sectional	1 Carp	4	2		310	80		390	470
1810	Door tracks only		4	2		139	80		219	285
1820	One side only		7	1.143		100	45.50		145.50	186
3000	Swing-up, including hardware, fiberglass, 9' x 7', standard	2 Carp	8	2		970	80		1,050	1,200
3020	Deluxe		8	2		790	80		870	1,000
3100	16' x 7', standard		6	2.667		1,225	107		1,332	1,525
3120	Deluxe		6	2.667		980	107		1,087	1,250
3200	Hardboard, 9' x 7', standard		8	2		345	80		425	510
3220	Deluxe		8	2		455	80		535	635
3300	16' x 7', standard		6	2.667		480	107		587	705
3320	Deluxe		6	2.667		715	107		822	960
3400	Metal, 9' x 7', standard		8	2		380	80		460	550
3420	Deluxe		8	2		800	80		880	1,000
3500	16' x 7', standard		6	2.667		600	107		707	835
3520	Deluxe		6	2.667		960	107		1,067	1,225
3600	Wood, 9' x 7', standard		8	2		450	80		530	625

08 36 Panel Doors

08 36 13 – Sectional Doors

08 36 13.20 Residential Garage Doors		Crew	Daily Output	Labor-Hours	Unit	2009 Bare Costs Material	Labor	Equipment	Total	Total Incl O&P
3620	Deluxe	2 Carp	8	2	Ea.	850	80		930	1,075
3700	16′ x 7′, standard		6	2.667		725	107		832	975
3720	Deluxe	↓	6	2.667		1,025	107		1,132	1,300
3900	Door hardware only, swing up	1 Carp	4	2		142	80		222	288
3920	One side only		7	1.143		69	45.50		114.50	152
4000	For electric operator, economy, add		8	1		405	40		445	510
4100	Deluxe, including remote control	↓	8	1	↓	595	40		635	715
4500	For transmitter/receiver control , add to operator				Total	97			97	107
4600	Transmitters, additional				"	45.50			45.50	50
9000	Minimum labor/equipment charge	1 Carp	2.50	3.200	Job		128		128	211

08 36 19 – Multi-Leaf Vertical Lift Doors

08 36 19.10 Sectional Vertical Lift Doors		Crew	Daily Output	Labor-Hours	Unit	Material	Labor	Equipment	Total	Total Incl O&P
0010	**SECTIONAL VERTICAL LIFT DOORS**									
0020	Motorized, 14 ga. stl, incl., frm and ctrl pnl									
0050	16′ x 16′ high	L-10	.50	48	Ea.	21,000	2,150	1,525	24,675	28,600
0100	10′ x 20′ high		1.30	18.462		29,000	825	590	30,415	34,000
0120	15′ x 20′ high		1.30	18.462		36,000	825	590	37,415	41,700
0140	20′ x 20′ high		1	24		41,600	1,075	770	43,445	48,400
0160	25′ x 20′ high		1	24		45,900	1,075	770	47,745	53,000
0170	32′ x 24′ high		.75	32		44,700	1,425	1,025	47,150	53,000
0180	20′ x 25′ high		1	24		47,600	1,075	770	49,445	55,000
0200	25′ x 25′ high		.70	34.286		53,500	1,525	1,100	56,125	63,000
0220	25′ x 30′ high		.70	34.286		62,000	1,525	1,100	64,625	72,500
0240	30′ x 30′ high		.70	34.286		71,500	1,525	1,100	74,125	82,500
0260	35′ x 30′ high	↓	.70	34.286	↓	78,000	1,525	1,100	80,625	89,500

08 36 23 – Telescoping Vertical Lift Doors

08 36 23.10 Telescoping Steel Doors		Crew	Daily Output	Labor-Hours	Unit	Material	Labor	Equipment	Total	Total Incl O&P
0010	**TELESCOPING STEEL DOORS**									
1000	Overhead, .03″ thick, electric operated, 10′ x 10′	E-3	.80	30	Ea.	12,500	1,350	168	14,018	16,500
2000	20′ x 10′		.60	40		16,900	1,825	224	18,949	22,200
3000	20′ x 16′	↓	.40	60	↓	18,600	2,725	335	21,660	25,900

08 38 Traffic Doors

08 38 13 – Flexible Strip Doors

08 38 13.10 Flexible Transparent Strip Doors		Crew	Daily Output	Labor-Hours	Unit	Material	Labor	Equipment	Total	Total Incl O&P
0010	**FLEXIBLE TRANSPARENT STRIP DOORS**									
0100	12″ strip width, 2/3 overlap	3 Shee	135	.178	SF Surf	7.50	8.40		15.90	21.50
0200	Full overlap		115	.209		6.30	9.85		16.15	22.50
0220	8″ strip width, 1/2 overlap		140	.171		5.75	8.10		13.85	19.20
0240	Full overlap	↓	120	.200	↓	7.60	9.45		17.05	23.50
0300	Add for suspension system, header mount				L.F.	8.90			8.90	9.80
0400	Wall mount				"	9			9	9.90

08 38 19 – Rigid Traffic Doors

08 38 19.20 Double Acting Swing Doors		Crew	Daily Output	Labor-Hours	Unit	Material	Labor	Equipment	Total	Total Incl O&P
0010	**DOUBLE ACTING SWING DOORS**									
0020	Including frame, closer, hardware and vision panel									
1000	.063″ aluminum, 7′-0″ high, 4′-0″ wide	2 Carp	4.20	3.810	Pr.	1,900	152		2,052	2,350
1025	6′-0″ wide		4	4		2,100	160		2,260	2,575
1050	6′-8″ wide	↓	4	4	↓	2,300	160		2,460	2,800

08 38 Traffic Doors

08 38 19 – Rigid Traffic Doors

08 38 19.20 Double Acting Swing Doors		Crew	Daily Output	Labor-Hours	Unit	Material	Labor	Equipment	Total	Total Incl O&P
						2009 Bare Costs				
2000	Solid core wood, 3/4″ thick, metal frame, stainless steel									
2010	base plate, 7′ high opening, 4′ wide	2 Carp	4	4	Pr.	2,200	160		2,360	2,700
2050	7′ wide		3.80	4.211	″	2,500	168		2,668	3,025
9000	Minimum labor/equipment charge		2	8	Job		320		320	530

08 38 19.30 Shock Absorbing Doors

Code	Description	Crew	Daily Output	Labor-Hours	Unit	Material	Labor	Equipment	Total	Total Incl O&P
0010	**SHOCK ABSORBING DOORS**									
0020	Rigid, no frame, 1-1/2″ thick, 5′ x 7′	2 Sswk	1.90	8.421	Opng.	1,400	375		1,775	2,250
0100	8′ x 8′		1.80	8.889		2,000	395		2,395	2,925
0500	Flexible, no frame, insulated, .16″ thick, economy, 5′ x 7′		2	8		1,750	360		2,110	2,575
0600	Deluxe		1.90	8.421		2,625	375		3,000	3,575
1000	8′ x 8′ opening, economy		2	8		2,750	360		3,110	3,675
1100	Deluxe		1.90	8.421		3,500	375		3,875	4,550
9000	Minimum labor/equipment charge		2	8	Job		360		360	660

08 41 Entrances and Storefronts

08 41 13 – Aluminum-Framed Entrances and Storefronts

08 41 13.10 Aluminum Swing Doors

Code	Description	Crew	Daily Output	Labor-Hours	Unit	Material	Labor	Equipment	Total	Total Incl O&P
0010	**ALUMINUM SWING DOORS**									
0015	Aluminum entrance 6′ x 7′	2 Sswk	.70	22.857	Opng.	6,500	1,025		7,525	9,050
9000	Minimum labor/equipment charge	″	1	16	Job		715		715	1,325

08 41 13.20 Tube Framing

Code	Description	Crew	Daily Output	Labor-Hours	Unit	Material	Labor	Equipment	Total	Total Incl O&P
0010	**TUBE FRAMING**, For window walls and store fronts, aluminum stock									
0050	Plain tube frame, mill finish, 1-3/4″ x 1-3/4″	2 Glaz	103	.155	L.F.	8.70	6		14.70	19.25
0150	1-3/4″ x 4″		98	.163		11.85	6.30		18.15	23
0200	1-3/4″ x 4-1/2″		95	.168		13.80	6.50		20.30	25.50
0250	2″ x 6″		89	.180		19.55	6.95		26.50	32.50
0350	4″ x 4″		87	.184		19.65	7.10		26.75	33
0400	4-1/2″ x 4-1/2″		85	.188		25	7.25		32.25	39
0450	Glass bead		240	.067		2.55	2.57		5.12	6.95
1000	Flush tube frame, mill finish, 1/4″ glass, 1-3/4″ x 4″, open header		80	.200		11.45	7.70		19.15	25
1050	Open sill		82	.195		9	7.55		16.55	22
1100	Closed back header		83	.193		16.05	7.45		23.50	29.50
1150	Closed back sill		85	.188		15.20	7.25		22.45	28.50
1160	Tube fmg, spandrel cover both sides, alum 1″ wide	1 Sswk	85	.094	S.F.	86	4.21		90.21	102
1170	Tube fmg, spandrel cover both sides, alum 2″ wide	″	85	.094	″	32.50	4.21		36.71	44
1200	Vertical mullion, one piece	2 Glaz	75	.213	L.F.	16.90	8.25		25.15	32
1250	Two piece		73	.219		18.10	8.45		26.55	33.50
1300	90° or 180° vertical corner post		75	.213		27.50	8.25		35.75	44
1400	1-3/4″ x 4-1/2″, open header		80	.200		13.90	7.70		21.60	28
1450	Open sill		82	.195		11.60	7.55		19.15	25
1500	Closed back header		83	.193		16.85	7.45		24.30	30.50
1550	Closed back sill		85	.188		16.35	7.25		23.60	29.50
1600	Vertical mullion, one piece		75	.213		18.30	8.25		26.55	33.50
1650	Two piece		73	.219		19.30	8.45		27.75	34.50
1700	90° or 180° vertical corner post		75	.213		19.70	8.25		27.95	35
2000	Flush tube frame, mil fin.,ins. glass w/thml brk, 2″ x 4-1/2″, open header		75	.213		14.50	8.25		22.75	29
2050	Open sill		77	.208		12.25	8		20.25	26.50
2100	Closed back header		78	.205		13.75	7.90		21.65	28
2150	Closed back sill		80	.200		14.70	7.70		22.40	28.50
2200	Vertical mullion, one piece		70	.229		15.25	8.80		24.05	31

08 41 Entrances and Storefronts

08 41 13 – Aluminum-Framed Entrances and Storefronts

08 41 13.20 Tube Framing		Crew	Daily Output	Labor-Hours	Unit	Material	2009 Bare Costs Labor	Equipment	Total	Total Incl O&P
2250	Two piece	2 Glaz	68	.235	L.F.	16.45	9.10		25.55	32.50
2300	90° or 180° vertical corner post		70	.229		15.70	8.80		24.50	31.50
5000	Flush tube frame, mill fin., thermal brk., 2-1/4"x 4-1/2", open header		74	.216		15.15	8.35		23.50	30
5050	Open sill		75	.213		13.45	8.25		21.70	28
5100	Vertical mullion, one piece		69	.232		16.50	8.95		25.45	32.50
5150	Two piece		67	.239		19.15	9.20		28.35	36
5200	90° or 180° vertical corner post		69	.232		16.85	8.95		25.80	33
6980	Door stop (snap in)	↓	380	.042	↓	2.85	1.63		4.48	5.75
7000	For joints, 90°, clip type, add				Ea.	23			23	25
7050	Screw spline joint, add					20.50			20.50	22.50
7100	For joint other than 90°, add				↓	43			43	47
8000	For bronze anodized aluminum, add					15%				
8020	For black finish, add					27%				
8050	For stainless steel materials, add					350%				
8100	For monumental grade, add					52%				
8150	For steel stiffener, add	2 Glaz	200	.080	L.F.	10.40	3.09		13.49	16.45
8200	For 2 to 5 stories, add per story				Story		6%			
9000	Minimum labor/equipment charge	2 Glaz	2	8	Job		310		310	500

08 41 19 – Stainless-Steel-Framed Entrances and Storefronts

08 41 19.10 Stainless-Steel and Glass Entrance Unit		Crew	Daily Output	Labor-Hours	Unit	Material	Labor	Equipment	Total	Total Incl O&P
0010	**STAINLESS-STEEL AND GLASS ENTRANCE UNIT**, narrow stiles									
0020	3′ x 7′ opening, including hardware, minimum	2 Sswk	1.60	10	Opng.	6,000	445		6,445	7,425
0050	Average		1.40	11.429		6,500	510		7,010	8,100
0100	Maximum	↓	1.20	13.333		6,900	595		7,495	8,700
1000	For solid bronze entrance units, statuary finish, add					62%				
1100	Without statuary finish, add				↓	45%				
2000	Balanced doors, 3′ x 7′, economy	2 Sswk	.90	17.778	Ea.	8,100	795		8,895	10,400
2100	Premium		.70	22.857	"	14,000	1,025		15,025	17,300
9000	Minimum labor/equipment charge	↓	2	8	Job		360		360	660

08 41 26 – All-Glass Entrances and Storefronts

08 41 26.10 Window Walls Aluminum, Stock		Crew	Daily Output	Labor-Hours	Unit	Material	Labor	Equipment	Total	Total Incl O&P
0010	**WINDOW WALLS ALUMINUM, STOCK**, including glazing									
0020	Minimum	H-2	160	.150	S.F.	39.50	5.45		44.95	52.50
0050	Average		140	.171		54	6.20		60.20	69.50
0100	Maximum	↓	110	.218	↓	152	7.90		159.90	180
0500	For translucent sandwich wall systems, see Div. 07 41 33.10									
0850	Cost of the above walls depends on material,									
0860	finish, repetition, and size of units.									
0870	The larger the opening, the lower the S.F. cost									
1200	Double glazed acoustical window wall for airports,									
1220	including 1" thick glass with 2" x 4-1/2" tube frame	H-2	40	.600	S.F.	99	22		121	145

08 42 Entrances

08 42 26 – All-Glass Entrances

08 42 26.10 Swinging Glass Doors		Crew	Daily Output	Labor-Hours	Unit	Material (2009 Bare Costs)	Labor (2009 Bare Costs)	Equipment (2009 Bare Costs)	Total (2009 Bare Costs)	Total Incl O&P
0010	**SWINGING GLASS DOORS**									
0020	Including hardware, 1/2" thick, tempered, 3' x 7' opening	2 Glaz	2	8	Opng.	1,950	310		2,260	2,650
0100	6' x 7' opening		1.40	11.429	"	3,800	440		4,240	4,875
9000	Minimum labor/equipment charge		2	8	Job		310		310	500

08 42 29 – Automatic Entrances

08 42 29.23 Sliding Automatic Entrances		Crew	Daily Output	Labor-Hours	Unit	Material	Labor	Equipment	Total	Total Incl O&P
0010	**SLIDING AUTOMATIC ENTRANCES** 12' x 7'-6" opng., 5' x 7' door, 2 way traffic									
0020	Mat or electronic activated, panic pushout, incl. operator & hardware,									
0030	not including glass or glazing	2 Glaz	.70	22.857	Opng.	7,700	880		8,580	9,900
9000	Minimum labor/equipment charge	"	.70	22.857	Job		880		880	1,425

08 42 33 – Revolving Door Entrances

08 42 33.10 Circular Rotating Entrance Doors		Crew	Daily Output	Labor-Hours	Unit	Material	Labor	Equipment	Total	Total Incl O&P
0010	**CIRCULAR ROTATING ENTRANCE DOORS**, Aluminum									
0020	6'-10" to 7' high, stock units, minimum	4 Sswk	.75	42.667	Opng.	18,800	1,900		20,700	24,100
0050	Average		.60	53.333		22,600	2,375		24,975	29,300
0100	Maximum		.45	71.111		39,100	3,175		42,275	48,900
1000	Stainless steel		.30	105		37,700	4,750		42,450	50,500
1100	Solid bronze		.15	213		44,000	9,525		53,525	66,000
1500	For automatic controls, add	2 Elec	2	8		14,300	375		14,675	16,300

08 42 36 – Balanced Door Entrances

08 42 36.10 Balanced Entrance Doors		Crew	Daily Output	Labor-Hours	Unit	Material	Labor	Equipment	Total	Total Incl O&P
0010	**BALANCED ENTRANCE DOORS**									
0020	Hardware & frame, alum. & glass, 3' x 7', econ.	2 Sswk	.90	17.778	Ea.	5,800	795		6,595	7,850
0150	Premium		.70	22.857	"	7,275	1,025		8,300	9,900
9000	Minimum labor/equipment charge		1	16	Job		715		715	1,325

08 43 Storefronts

08 43 13 – Aluminum-Framed Storefronts

08 43 13.10 Aluminum-Framed Entrance Doors		Crew	Daily Output	Labor-Hours	Unit	Material	Labor	Equipment	Total	Total Incl O&P
0010	**ALUMINUM-FRAMED ENTRANCE DOORS** (Frame Only)									
0020	Entrance, 3' x 7' opening, clear anodized finish	2 Sswk	7	2.286	Opng.	420	102		522	650
0040	Bronze finish		7	2.286		450	102		552	685
0500	6' x 7' opening, clear finish		6	2.667		340	119		459	595
0520	Bronze finish		6	2.667		480	119		599	750
1000	With 3' high transoms, 3' x 10' opening, clear finish		6.50	2.462		430	110		540	680
1050	Bronze finish		6.50	2.462		465	110		575	720
1100	Black finish		6.50	2.462		525	110		635	785
1500	With 3' high transoms, 6' x 10' opening, clear finish		5.50	2.909		515	130		645	805
1550	Bronze finish		5.50	2.909		555	130		685	850
1600	Black finish		5.50	2.909		635	130		765	940
9000	Minimum labor/equipment charge		4	4	Job		179		179	330

08 43 13.20 Storefront Systems		Crew	Daily Output	Labor-Hours	Unit	Material	Labor	Equipment	Total	Total Incl O&P
0010	**STOREFRONT SYSTEMS**, aluminum frame clear 3/8" plate glass									
0020	incl. 3' x 7' door with hardware (400 sq. ft. max. wall)									
0500	Wall height to 12' high, commercial grade	2 Glaz	150	.107	S.F.	18.70	4.12		22.82	27
0600	Institutional grade		130	.123		23	4.75		27.75	33
0700	Monumental grade		115	.139		35.50	5.35		40.85	47.50
1000	6' x 7' door with hardware, commercial grade		135	.119		18.90	4.57		23.47	28.50
1100	Institutional grade		115	.139		26	5.35		31.35	37

08 43 Storefronts

08 43 13 – Aluminum-Framed Storefronts

08 43 13.20 Storefront Systems		Crew	Daily Output	Labor-Hours	Unit	Material	Labor (2009 Bare Costs)	Equipment (2009 Bare Costs)	Total (2009 Bare Costs)	Total Incl O&P
1200	Monumental grade	2 Glaz	100	.160	S.F.	48	6.20		54.20	62.50
1500	For bronze anodized finish, add					15%				
1600	For black anodized finish, add					35%				
1700	For stainless steel framing, add to monumental					76%				
9000	Minimum labor/equipment charge	2 Glaz	1	16	Job		620		620	995

08 43 29 – Sliding Storefronts

08 43 29.10 Sliding Panels		Crew	Daily Output	Labor-Hours	Unit	Material	Labor	Equipment	Total	Total Incl O&P
0010	**SLIDING PANELS**									
0020	Mall fronts, aluminum & glass, 15′ x 9′ high	2 Glaz	1.30	12.308	Opng.	3,025	475		3,500	4,100
0100	24′ x 9′ high		.70	22.857		4,400	880		5,280	6,250
0200	48′ x 9′ high, with fixed panels		.90	17.778		8,200	685		8,885	10,100
0500	For bronze finish, add					17%				
9000	Minimum labor/equipment charge	2 Glaz	1	16	Job		620		620	995

08 44 Curtain Wall and Glazed Assemblies

08 44 13 – Glazed Aluminum Curtain Walls

08 44 13.10 Glazed Curtain Walls		Crew	Daily Output	Labor-Hours	Unit	Material	Labor	Equipment	Total	Total Incl O&P
0010	**GLAZED CURTAIN WALLS**, aluminum, stock, including glazing									
0020	Minimum	H-1	205	.156	S.F.	31	6.50		37.50	46
0050	Average, single glazed		195	.164		40.50	6.85		47.35	56.50
0150	Average, double glazed		180	.178		57.50	7.40		64.90	76
0200	Maximum		160	.200		157	8.35		165.35	187

08 45 Translucent Wall and Roof Assemblies

08 45 10 – Translucent Roof Assemblies

08 45 10.10 Skyroofs		Crew	Daily Output	Labor-Hours	Unit	Material	Labor	Equipment	Total	Total Incl O&P
0010	**SKYROOFS**, Translucent panels, 2-3/4″ thick									
0020	Under 500 S.F.	G-3	395	.081	SF Hor.	31.50	3.19		34.69	39.50
0100	Over 5000 S.F.		465	.069		29.50	2.71		32.21	36.50
0300	Continuous vaulted, semi-circular, to 8′ wide, double glazed G		145	.221		49	8.70		57.70	68
0400	Single glazed		160	.200		33.50	7.90		41.40	49
0600	To 20′ wide, single glazed		175	.183		37	7.20		44.20	52
0700	Over 20′ wide, single glazed		200	.160		42.50	6.30		48.80	57
0900	Motorized opening type, single glazed, 1/3 opening		145	.221		44	8.70		52.70	62
1000	Full opening		130	.246		50	9.70		59.70	70.50
1200	Pyramid type units, self-supporting, to 30′ clear opening,									
1300	square or circular, single glazed, minimum	G-3	200	.160	SF Hor.	25.50	6.30		31.80	38
1310	Average		165	.194		34.50	7.65		42.15	50.50
1400	Maximum		130	.246		50	9.70		59.70	70.50
1500	Grid type, 4′ to 10′ modules, single glass glazed, minimum		200	.160		33.50	6.30		39.80	47
1550	Maximum		128	.250		52.50	9.85		62.35	73.50
1600	Preformed acrylic, minimum		300	.107		40	4.20		44.20	51
1650	Maximum		175	.183		55	7.20		62.20	72
1800	Skyroofs,dome type, self-supporting, to 100′ clear opening, circular									
1900	Rise to span ratio = 0.20									
1920	Minimum	G-3	197	.162	SF Hor.	15	6.40		21.40	27
1950	Maximum		113	.283		50	11.15		61.15	73
2100	Rise to span ratio = 0.33, minimum		169	.189		30	7.45		37.45	45
2150	Maximum		101	.317		59	12.50		71.50	85

08 45 Translucent Wall and Roof Assemblies

08 45 10 – Translucent Roof Assemblies

08 45 10.10 Skyroofs		Crew	Daily Output	Labor-Hours	Unit	Material	2009 Bare Costs Labor	Equipment	Total	Total Incl O&P
2200	Rise to span ratio = 0.50, minimum	G-3	148	.216	SF Hor.	45.50	8.50		54	64
2250	Maximum		87	.368		66	14.50		80.50	96
2400	Ridge units, continuous, to 8′ wide, double G		130	.246		115	9.70		124.70	143
2500	Single G		200	.160		79	6.30		85.30	97
2700	Ridge and furrow units, over 4′ O.C., double, minimum G		200	.160		25	6.30		31.30	37.50
2750	Maximum G		120	.267		46	10.50		56.50	67.50
2800	Single, minimum		214	.150		21	5.90		26.90	32.50
2850	Maximum		153	.209		44.50	8.25		52.75	62.50
3000	Rolling roof, translucent panels, flat roof, minimum		253	.126	S.F.	19.10	4.98		24.08	29
3030	Maximum G		160	.200		38.50	7.90		46.40	54.50
3100	Lean-to skyroof, long span, double, minimum G		197	.162		25.50	6.40		31.90	38.50
3150	Maximum G		101	.317		51	12.50		63.50	76
3300	Single, minimum		321	.100		19.40	3.93		23.33	28
3350	Maximum		160	.200		32	7.90		39.90	47.50
9000	Minimum labor/equipment charge	2 Carp	8	2	Job		80		80	132

08 51 Metal Windows

08 51 13 – Aluminum Windows

08 51 13.10 Aluminum Sash

		Crew	Daily Output	Labor-Hours	Unit	Material	Labor	Equipment	Total	Total Incl O&P
0010	**ALUMINUM SASH**									
0020	Stock, grade C, glaze & trim not incl., casement	2 Sswk	200	.080	S.F.	36	3.58		39.58	46
0050	Double hung		200	.080		36.50	3.58		40.08	46.50
0100	Fixed casement		200	.080		15.65	3.58		19.23	24
0150	Picture window		200	.080		16.65	3.58		20.23	25
0200	Projected window		200	.080		33	3.58		36.58	42.50
0250	Single hung		200	.080		15.80	3.58		19.38	24
0300	Sliding		200	.080		20.50	3.58		24.08	29
1000	Mullions for above, tubular		240	.067	L.F.	5.45	2.98		8.43	11.45
2000	Custom aluminum sash, grade HC, glazing not included, minimum		200	.080	S.F.	38	3.58		41.58	48.50
2100	Maximum		85	.188	"	49.50	8.40		57.90	70
9000	Minimum labor/equipment charge	1 Sswk	2	4	Job		179		179	330

08 51 13.20 Aluminum Windows

		Crew	Daily Output	Labor-Hours	Unit	Material	Labor	Equipment	Total	Total Incl O&P
0010	**ALUMINUM WINDOWS**, incl. frame and glazing, commercial grade									
0020	See Div. 05 12 23.40									
1000	Stock units, casement, 3′-1″ x 3′-2″ opening	2 Sswk	10	1.600	Ea.	355	71.50		426.50	520
1050	Add for storms					115			115	126
1600	Projected, with screen, 3′-1″ x 3′-2″ opening	2 Sswk	10	1.600		335	71.50		406.50	500
1700	Add for storms					112			112	123
2000	4′-5″ x 5′-3″ opening	2 Sswk	8	2		380	89.50		469.50	585
2100	Add for storms					120			120	132
2500	Enamel finish windows, 3′-1″ x 3′-2″	2 Sswk	10	1.600		340	71.50		411.50	505
2600	4′-5″ x 5′-3″		8	2		385	89.50		474.50	590
3000	Single hung, 2′ x 3′ opening, enameled, standard glazed		10	1.600		198	71.50		269.50	350
3100	Insulating glass		10	1.600		240	71.50		311.50	395
3300	2′-8″ x 6′-8″ opening, standard glazed		8	2		350	89.50		439.50	550
3400	Insulating glass		8	2		450	89.50		539.50	660
3700	3′-4″ x 5′-0″ opening, standard glazed		9	1.778		286	79.50		365.50	460
3800	Insulating glass		9	1.778		315	79.50		394.50	495
3890	Awning type, 3′ x 3′ opening standard glass		14	1.143		405	51		456	540
3900	Insulating glass		14	1.143		430	51		481	570
3910	3′ x 4′ opening, standard glass		10	1.600		470	71.50		541.50	650

08 51 Metal Windows

08 51 13 – Aluminum Windows

08 51 13.20 Aluminum Windows		Crew	Daily Output	Labor-Hours	Unit	Material	2009 Bare Costs Labor	Equipment	Total	Total Incl O&P
3920	Insulating glass	2 Sswk	10	1.600	Ea.	545	71.50		616.50	725
3930	3′ x 5′-4″ opening, standard glass		10	1.600		570	71.50		641.50	755
3940	Insulating glass		10	1.600		670	71.50		741.50	865
3950	4′ x 5′-4″ opening, standard glass		9	1.778		625	79.50		704.50	830
3960	Insulating glass		9	1.778		750	79.50		829.50	970
4000	Sliding aluminum, 3′ x 2′ opening, standard glazed		10	1.600		209	71.50		280.50	360
4100	Insulating glass		10	1.600		224	71.50		295.50	380
4300	5′ x 3′ opening, standard glazed		9	1.778		320	79.50		399.50	495
4400	Insulating glass		9	1.778		370	79.50		449.50	555
4600	8′ x 4′ opening, standard glazed		6	2.667		335	119		454	590
4700	Insulating glass		6	2.667		545	119		664	815
5000	9′ x 5′ opening, standard glazed		4	4		510	179		689	890
5100	Insulating glass		4	4		820	179		999	1,225
5500	Sliding, with thermal barrier and screen, 6′ x 4′, 2 track		8	2		695	89.50		784.50	930
5700	4 track	↓	8	2		880	89.50		969.50	1,125
6000	For above units with bronze finish, add					12%				
6200	For installation in concrete openings, add				↓	6%				

08 51 23 – Steel Windows

08 51 23.10 Steel Sash		Crew	Daily Output	Labor-Hours	Unit	Material	Labor	Equipment	Total	Total Incl O&P
0010	**STEEL SASH** Custom units, glazing and trim not included									
0100	Casement, 100% vented	2 Sswk	200	.080	S.F.	48	3.58		51.58	59
0200	50% vented		200	.080		44	3.58		47.58	55
0300	Fixed		200	.080		30	3.58		33.58	39.50
1000	Projected, commercial, 40% vented		200	.080		48	3.58		51.58	59.50
1100	Intermediate, 50% vented		200	.080		54	3.58		57.58	65.50
1500	Industrial, horizontally pivoted		200	.080		49.50	3.58		53.08	61
1600	Fixed		200	.080		28.50	3.58		32.08	38
2000	Industrial security sash, 50% vented		200	.080		53.50	3.58		57.08	65.50
2100	Fixed		200	.080		43.50	3.58		47.08	54.50
2500	Picture window		200	.080		28	3.58		31.58	37.50
3000	Double hung		200	.080	↓	56	3.58		59.58	68
5000	Mullions for above, open interior face		240	.067	L.F.	9.80	2.98		12.78	16.25
5100	With interior cover	↓	240	.067	"	16.35	2.98		19.33	23.50
9000	Minimum labor/equipment charge	1 Sswk	2	4	Job		179		179	330

08 51 23.20 Steel Windows		Crew	Daily Output	Labor-Hours	Unit	Material	Labor	Equipment	Total	Total Incl O&P
0010	**STEEL WINDOWS** Stock, including frame, trim and insul. glass									
0020	See Div. 13 34 19.50									
1000	Custom units, double hung, 2′-8″ x 4′-6″ opening	2 Sswk	12	1.333	Ea.	680	59.50		739.50	860
1100	2′-4″ x 3′-9″ opening		12	1.333		565	59.50		624.50	730
1500	Commercial projected, 3′-9″ x 5′-5″ opening		10	1.600		1,200	71.50		1,271.50	1,425
1600	6′-9″ x 4′-1″ opening		7	2.286		1,575	102		1,677	1,925
2000	Intermediate projected, 2′-9″ x 4′-1″ opening		12	1.333		670	59.50		729.50	845
2100	4′-1″ x 5′-5″ opening	↓	10	1.600	↓	1,350	71.50		1,421.50	1,600
9000	Minimum labor/equipment charge	1 Sswk	3	2.667	Job		119		119	221

08 51 66 – Metal Window Screens

08 51 66.10 Screens		Crew	Daily Output	Labor-Hours	Unit	Material	Labor	Equipment	Total	Total Incl O&P
0010	**SCREENS**									
0020	For metal sash, aluminum or bronze mesh, flat screen	2 Sswk	1200	.013	S.F.	4.10	.60		4.70	5.60
0500	Wicket screen, inside window		1000	.016		6.30	.72		7.02	8.25
0800	Security screen, aluminum frame with stainless steel cloth		1200	.013		22.50	.60		23.10	26
0900	Steel grate, painted, on steel frame	↓	1600	.010	↓	12.40	.45		12.85	14.50

08 51 Metal Windows

08 51 66 – Metal Window Screens

08 51 66.10 Screens		Crew	Daily Output	Labor-Hours	Unit	Material	Labor	Equipment	Total	Total Incl O&P
1000	For solar louvers, add	2 Sswk	160	.100	S.F.	23.50	4.47		27.97	34
4000	See Div. 05 58 27.90									

08 52 Wood Windows

08 52 10 – Plain Wood Windows

08 52 10.20 Awning Window

		Crew	Daily Output	Labor-Hours	Unit	Material	Labor	Equipment	Total	Total Incl O&P
0010	**AWNING WINDOW**, Including frame, screens and grilles									
0100	Average quality, builders model, 34" x 22", double insulated glass	1 Carp	10	.800	Ea.	249	32		281	325
0200	Low E glass		10	.800		258	32		290	335
0300	40" x 28", double insulated glass		9	.889		310	35.50		345.50	400
0400	Low E Glass		9	.889		330	35.50		365.50	425
0500	48" x 36", double insulated glass		8	1		450	40		490	560
0600	Low E glass		8	1		475	40		515	590
1000	Vinyl clad, 34" x 22"		10	.800		259	32		291	340
1100	40" x 22"		10	.800		283	32		315	365
1200	36" x 28"		9	.889		300	35.50		335.50	390
1300	36" x 36"		9	.889		335	35.50		370.50	430
1400	48" x 28"		8	1		360	40		400	465
1500	60" x 36"		8	1		505	40		545	620
2200	Metal clad, 36" x 25"		9	.889		269	35.50		304.50	355
2300	40" x 30"		9	.889		335	35.50		370.50	430
2400	48" x 28"		8	1		345	40		385	445
2500	60" x 36"	↓	8	1		365	40		405	470
4000	Impact windows, minimum, add					60%				
4010	Impact windows, maximum, add				↓	160%				
9000	Minimum labor/equipment charge	1 Carp	4	2	Job		80		80	132

08 52 10.30 Wood Windows

		Crew	Daily Output	Labor-Hours	Unit	Material	Labor	Equipment	Total	Total Incl O&P
0010	**WOOD WINDOWS**, double hung									
3200	20 S.F. and over	1 Carp	106	.075	S.F.	16.65	3.02		19.67	23.50
3800	Triple glazing for above, add					8.10			8.10	8.90
6000	Replacement sash, double hung, double glazing, to 12 S.F.	1 Carp	64	.125		24	4.99		28.99	35
6100	12 S.F. to 20 S.F.		94	.085	↓	19.05	3.40		22.45	26.50
7000	Sash, single lite, 2'-0" x 2'-0" high		20	.400	Ea.	45	16		61	76
7050	2'-6" x 2'-0" high		19	.421		56.50	16.80		73.30	90
7100	2'-6" x 2'-6" high		18	.444		67.50	17.75		85.25	104
7150	3'-0" x 2'-0" high	↓	17	.471	↓	88	18.80		106.80	128

08 52 10.40 Casement Window

			Crew	Daily Output	Labor-Hours	Unit	Material	Labor	Equipment	Total	Total Incl O&P
0010	**CASEMENT WINDOW**, including frame, screen and grilles										
0100	Avg. quality, bldrs. model, 2'-0" x 3'-0" H, dbl. insulated glass	G	1 Carp	10	.800	Ea.	276	32		308	360
0150	Low E glass	G		10	.800		225	32		257	300
0200	2'-0" x 4'-6" high, double insulated glass	G		9	.889		345	35.50		380.50	440
0250	Low E glass	G		9	.889		263	35.50		298.50	350
0300	2'-4" x 6'-0" high, double insulated glass	G		8	1		380	40		420	485
0350	Low E glass	G		8	1		465	40		505	580
0522	Vinyl clad, premium, double insulated glass, 2'-0" x 3'-0"	G		10	.800		266	32		298	345
0524	2'-0" x 4'-0"	G		9	.889		310	35.50		345.50	400
0525	2'-0" x 5'-0"	G		8	1		355	40		395	455
0528	2'-0" x 6'-0"	G	↓	8	1	↓	355	40		395	455
2000	Bay, casement units, 8' x 5', w/screens, dbl Insul glass		2 Carp	2.50	6.400	Opng.	1,575	256		1,831	2,150
2100	Low E glass		"	2.50	6.400	"	1,650	256		1,906	2,225

08 52 Wood Windows

08 52 10 – Plain Wood Windows

08 52 10.40 Casement Window

	08 52 10.40 Casement Window		Crew	Daily Output	Labor-Hours	Unit	2009 Bare Costs Material	Labor	Equipment	Total	Total Incl O&P
8100	Metal clad, deluxe, dbl. insul. glass, 2'-0" x 3'-0" high	G	1 Carp	10	.800	Ea.	226	32		258	300
8120	2'-0" x 4'-0" high	G		9	.889		273	35.50		308.50	360
8140	2'-0" x 5'-0" high	G		8	1		310	40		350	405
8160	2'-0" x 6'-0" high	G		8	1		355	40		395	455
8190	For installation, add per leaf							15%			
8200	For multiple leaf units, deduct for stationary sash										
8220	2' high					Ea.	22			22	24
8240	4'-6" high						25			25	28
8260	6' high						33.50			33.50	37
8300	Impact windows, minimum, add						60%				
8310	Impact windows, maximum, add						160%				
9000	Minimum labor/equipment charge		1 Carp	3	2.667	Job		107		107	176

08 52 10.50 Double Hung

	08 52 10.50 Double Hung		Crew	Daily Output	Labor-Hours	Unit	Material	Labor	Equipment	Total	Total Incl O&P
0010	**DOUBLE HUNG**, Including frame, screens and grilles										
0100	Avg. quality, bldrs. model, 2'-0" x 3'-0" high, dbl insul. glass	G	1 Carp	10	.800	Ea.	184	32		216	255
0150	Low E glass	G		10	.800		208	32		240	282
0200	3'-0" x 4'-0" high, double insulated glass	G		9	.889		257	35.50		292.50	340
0250	Low E glass	G		9	.889		275	35.50		310.50	365
0300	4'-0" x 4'-6" high, double insulated glass	G		8	1		315	40		355	410
0350	Low E glass	G		8	1		340	40		380	440
1000	Vinyl clad, premium, double insulated glass, 2'-6" x 3'-0"	G		10	.800		225	32		257	300
1100	3'-0" x 3'-6"	G		10	.800		261	32		293	340
1200	3'-0" x 4'-0"	G		9	.889		370	35.50		405.50	465
1300	3'-0" x 4'-6"	G		9	.889		340	35.50		375.50	430
1400	3'-0" x 5'-0"	G		8	1		315	40		355	410
1500	3'-6" x 6'-0"	G		8	1		370	40		410	470
2000	Metal clad, deluxe, dbl. insul. glass, 2'-6" x 3'-0" high	G		10	.800		262	32		294	340
2100	3'-0" x 3'-6" high	G		10	.800		299	32		331	385
2200	3'-0" x 4'-0" high	G		9	.889		315	35.50		350.50	405
2300	3'-0" x 4'-6" high	G		9	.889		330	35.50		365.50	425
2400	3'-0" x 5'-0" high	G		8	1		360	40		400	460
2500	3'-6" x 6'-0" high	G		8	1		435	40		475	540
8000	Impact windows, minimum, add						60%				
8010	Impact windows, maximum, add						160%				
9000	Minimum labor/equipment charge		1 Carp	3	2.667	Job		107		107	176

08 52 10.55 Picture Window

	08 52 10.55 Picture Window		Crew	Daily Output	Labor-Hours	Unit	Material	Labor	Equipment	Total	Total Incl O&P
0010	**PICTURE WINDOW**, Including frame and grilles										
0100	Average quality, bldrs. model, 3'-6" x 4'-0" high, dbl insulated glass		2 Carp	12	1.333	Ea.	420	53.50		473.50	555
0150	Low E glass			12	1.333		465	53.50		518.50	600
0200	4'-0" x 4'-6" high, double insulated glass			11	1.455		475	58		533	620
0250	Low E glass			11	1.455		500	58		558	645
0300	5'-0" x 4'-0" high, double insulated glass			11	1.455		555	58		613	705
0350	Low E glass			11	1.455		575	58		633	730
0400	6'-0" x 4'-6" high, double insulated glass			10	1.600		605	64		669	770
0450	Low E glass			10	1.600		610	64		674	780

08 52 10.65 Wood Sash

	08 52 10.65 Wood Sash		Crew	Daily Output	Labor-Hours	Unit	Material	Labor	Equipment	Total	Total Incl O&P
0010	**WOOD SASH**, Including glazing but not trim										
0050	Custom, 5'-0" x 4'-0", 1" dbl. glazed, 3/16" thick lites		2 Carp	3.20	5	Ea.	167	200		367	515
0100	1/4" thick lites			5	3.200		172	128		300	400
0200	1" thick, triple glazed			5	3.200		395	128		523	640
0300	7'-0" x 4'-6" high, 1" double glazed, 3/16" thick lites			4.30	3.721		400	149		549	685
0400	1/4" thick lites			4.30	3.721		450	149		599	740

08 52 Wood Windows

08 52 10 – Plain Wood Windows

08 52 10.65 Wood Sash		Crew	Daily Output	Labor-Hours	Unit	Material	2009 Bare Costs Labor	Equipment	Total	Total Incl O&P
0500	1" thick, triple glazed	2 Carp	4.30	3.721	Ea.	515	149		664	810
0600	8'-6" x 5'-0" high, 1" double glazed, 3/16" thick lites		3.50	4.571		540	183		723	890
0700	1/4" thick lites		3.50	4.571		590	183		773	950
0800	1" thick, triple glazed		3.50	4.571		595	183		778	955
0900	Window frames only, based on perimeter length				L.F.	3.68			3.68	4.05
1200	Window sill, stock, per lineal foot					7.65			7.65	8.40
1250	Casing, stock					3.07			3.07	3.38

08 52 10.70 Sliding Windows			Crew	Daily Output	Labor-Hours	Unit	Material	Labor	Equipment	Total	Total Incl O&P
0010	**SLIDING WINDOWS**										
0100	Average quality, bldrs. model, 3'-0" x 3'-0" high, double insulated	G	1 Carp	10	.800	Ea.	244	32		276	320
0120	Low E glass	G		10	.800		264	32		296	345
0200	4'-0" x 3'-6" high, double insulated	G		9	.889		282	35.50		317.50	370
0220	Low E glass	G		9	.889		305	35.50		340.50	395
0300	6'-0" x 5'-0" high, double insulated	G		8	1		420	40		460	525
0320	Low E glass	G		8	1		460	40		500	570
9000	Minimum labor/equipment charge			3	2.667	Job		107		107	176

08 52 13 – Metal-Clad Wood Windows

08 52 13.10 Awning Windows, Metal-Clad		Crew	Daily Output	Labor-Hours	Unit	Material	Labor	Equipment	Total	Total Incl O&P
0010	**AWNING WINDOWS, METAL-CLAD**									
2000	Metal clad, awning deluxe, double insulated glass, 34" x 22"	1 Carp	10	.800	Ea.	242	32		274	320
2100	40" x 22"	"	10	.800	"	285	32		317	370

08 52 13.35 Picture and Sliding Windows Metal-Clad			Crew	Daily Output	Labor-Hours	Unit	Material	Labor	Equipment	Total	Total Incl O&P
0010	**PICTURE AND SLIDING WINDOWS METAL-CLAD**										
2000	Metal clad, dlx picture, dbl. insul. glass, 4'-0" x 4'-0" high		2 Carp	12	1.333	Ea.	360	53.50		413.50	490
2100	4'-0" x 6'-0" high			11	1.455		535	58		593	685
2200	5'-0" x 6'-0" high			10	1.600		590	64		654	750
2300	6'-0" x 6'-0" high			10	1.600		675	64		739	850
2400	Metal clad, dlx sliding, double insulated glass, 3'-0" x 3'-0" high	G	1 Carp	10	.800		320	32		352	410
2420	4'-0" x 3'-6" high	G		9	.889		395	35.50		430.50	495
2440	5'-0" x 4'-0" high	G		9	.889		475	35.50		510.50	580
2460	6'-0" x 5'-0" high	G		8	1		700	40		740	835
9000	Minimum labor/equipment charge		2 Carp	2.75	5.818	Job		232		232	385

08 52 16 – Plastic-Clad Wood Windows

08 52 16.10 Bow Window		Crew	Daily Output	Labor-Hours	Unit	Material	Labor	Equipment	Total	Total Incl O&P
0010	**BOW WINDOW** Including frames, screens, and grilles									
0020	End panels operable									
1000	Bow type, casement, wood, bldrs mdl, 8' x 5' dbl insltd glass, 4 panel	2 Carp	10	1.600	Ea.	1,500	64		1,564	1,750
1050	Low E glass		10	1.600		1,300	64		1,364	1,550
1100	10'-0" x 5'-0" , double insulated glass, 6 panels		6	2.667		1,350	107		1,457	1,675
1200	Low E glass, 6 panels		6	2.667		1,450	107		1,557	1,775
1300	Vinyl clad, bldrs model, double insulated glass, 6'-0" x 4'-0", 3 panel		10	1.600		1,475	64		1,539	1,725
1340	9'-0" x 4'-0", 4 panel		8	2		1,450	80		1,530	1,725
1380	10'-0" x 6'-0", 5 panels		7	2.286		2,250	91.50		2,341.50	2,625
1420	12'-0" x 6'-0", 6 panels		6	2.667		2,925	107		3,032	3,400
1600	Metal clad, casement, bldrs mdl, 6'-0" x 4'-0", dbl insltd gls, 3 panels		10	1.600		920	64		984	1,100
1640	9'-0" x 4'-0" , 4 panels		8	2		1,250	80		1,330	1,500
1680	10'-0" x 5'-0", 5 panels		7	2.286		1,725	91.50		1,816.50	2,050
1720	12'-0" x 6'-0", 6 panels		6	2.667		2,400	107		2,507	2,825
2000	Bay window, builders model, 8' x 5' dbl insul glass,		10	1.600		1,825	64		1,889	2,125
2050	Low E glass,		10	1.600		2,225	64		2,289	2,550
2100	12'-0" x 6'-0" , double insulated glass, 6 panels		6	2.667		2,300	107		2,407	2,700

08 52 16.10 Bow Window		Crew	Daily Output	Labor-Hours	Unit	Material	2009 Bare Costs Labor	Equipment	Total	Total Incl O&P
2200	Low E glass	2 Carp	6	2.667	Ea.	2,175	107		2,282	2,575
2280	6'-0" x 4'-0"		11	1.455		1,100	58		1,158	1,325
2300	Vinyl clad, premium, double insulated glass, 8'-0" x 5'-0"		10	1.600		1,300	64		1,364	1,525
2340	10'-0" x 5'-0"		8	2		1,850	80		1,930	2,150
2380	10'-0" x 6'-0"		7	2.286		1,925	91.50		2,016.50	2,275
2420	12'-0" x 6'-0"		6	2.667		2,300	107		2,407	2,700
2600	Metal clad, deluxe, dbl insul. glass, 8'-0" x 5'-0" high, 4 panels		10	1.600		1,625	64		1,689	1,875
2640	10'-0" x 5'-0" high, 5 panels		8	2		1,750	80		1,830	2,050
2680	10'-0" x 6'-0" high, 5 panels		7	2.286		2,050	91.50		2,141.50	2,425
2720	12'-0" x 6'-0" high, 6 panels		6	2.667		2,850	107		2,957	3,325
3000	Double hung, bldrs. model, bay, 8' x 4' high,dbl insulated glass		10	1.600		1,300	64		1,364	1,525
3050	Low E glass		10	1.600		1,375	64		1,439	1,625
3100	9'-0" x 5'-0" high, double insulated glass		6	2.667		1,400	107		1,507	1,700
3200	Low E glass		6	2.667		1,475	107		1,582	1,800
3300	Vinyl clad, premium, double insulated glass, 7'-0" x 4'-6"		10	1.600		1,325	64		1,389	1,575
3340	8'-0" x 4'-6"		8	2		1,375	80		1,455	1,625
3380	8'-0" x 5'-0"		7	2.286		1,425	91.50		1,516.50	1,725
3420	9'-0" x 5'-0"		6	2.667		1,475	107		1,582	1,800
3600	Metal clad, deluxe, dbl insul. glass, 7'-0" x 4'-0" high		10	1.600		1,225	64		1,289	1,450
3640	8'-0" x 4'-0" high		8	2		1,275	80		1,355	1,525
3680	8'-0" x 5'-0" high		7	2.286		1,325	91.50		1,416.50	1,600
3720	9'-0" x 5'-0" high		6	2.667		1,400	107		1,507	1,725
9000	Minimum labor/equipment charge		2.50	6.400	Job		256		256	420

08 52 16.30 Palladian Windows		Crew	Daily Output	Labor-Hours	Unit	Material	Labor	Equipment	Total	Total Incl O&P
0010	**PALLADIAN WINDOWS**									
0020	Vinyl clad, double insulated glass, including frame and grilles									
0040	3'-2" x 2'-6" high	2 Carp	11	1.455	Ea.	1,275	58		1,333	1,500
0060	3'-2" x 4'-10"		11	1.455		1,775	58		1,833	2,050
0080	3'-2" x 6'-4"		10	1.600		1,750	64		1,814	2,025
0100	4'-0" x 4'-0"		10	1.600		1,575	64		1,639	1,825
0120	4'-0" x 5'-4"	3 Carp	10	2.400		1,875	96		1,971	2,200
0140	4'-0" x 6'-0"		9	2.667		1,950	107		2,057	2,325
0160	4'-0" x 7'-4"		9	2.667		2,125	107		2,232	2,500
0180	5'-5" x 4'-10"		9	2.667		2,275	107		2,382	2,675
0200	5'-5" x 6'-10"		9	2.667		2,575	107		2,682	3,025
0220	5'-5" x 7'-9"		9	2.667		2,800	107		2,907	3,250
0240	6'-0" x 7'-11"		8	3		3,475	120		3,595	4,025
0260	8'-0" x 6'-0"		8	3		3,075	120		3,195	3,575

08 52 16.40 Transom Windows		Crew	Daily Output	Labor-Hours	Unit	Material	Labor	Equipment	Total	Total Incl O&P
0010	**TRANSOM WINDOWS**									
1000	Vinyl clad, premium, dbl. insul. glass, 4'-0" x 4'-0"	2 Carp	12	1.333	Ea.	485	53.50		538.50	620
1100	4'-0" x 6'-0"		11	1.455		910	58		968	1,100
1200	5'-0" x 6'-0"		10	1.600		1,000	64		1,064	1,225
1300	6'-0" x 6'-0"		10	1.600		1,025	64		1,089	1,225

08 52 16.70 Vinyl Clad, Premium, Dbl. Insulated Glass		Crew	Daily Output	Labor-Hours	Unit	Material	Labor	Equipment	Total	Total Incl O&P
0010	**VINYL CLAD, PREMIUM, DBL. INSULATED GLASS**									
1000	Sliding, 3'-0" x 3'-0" G	1 Carp	10	.800	Ea.	575	32		607	685
1050	4'-0" x 3'-6" G		9	.889		670	35.50		705.50	795
1100	5'-0" x 4'-0" G		9	.889		860	35.50		895.50	1,000
1150	6'-0" x 5'-0" G		8	1		1,075	40		1,115	1,275

08 52 Wood Windows

08 52 50 – Window Accessories

08 52 50.10 Window Grille or Muntin

		Crew	Daily Output	Labor-Hours	Unit	Material	2009 Bare Costs Labor	Equipment	Total	Total Incl O&P
0010	**WINDOW GRILLE OR MUNTIN**, snap in type									
0020	Standard pattern interior grilles									
2000	Wood, awning window, glass size 28" x 16" high	1 Carp	30	.267	Ea.	22.50	10.65		33.15	42.50
2060	44" x 24" high		32	.250		32.50	10		42.50	52
2100	Casement, glass size, 20" x 36" high		30	.267		27.50	10.65		38.15	48
2180	20" x 56" high		32	.250		39.50	10		49.50	60
2200	Double hung, glass size, 16" x 24" high		24	.333	Set	48	13.30		61.30	74.50
2280	32" x 32" high		34	.235	"	128	9.40		137.40	157
2500	Picture, glass size, 48" x 48" high		30	.267	Ea.	112	10.65		122.65	141
2580	60" x 68" high		28	.286	"	160	11.40		171.40	195
2600	Sliding, glass size, 14" x 36" high		24	.333	Set	25.50	13.30		38.80	50
2680	36" x 36" high		22	.364	"	39	14.55		53.55	67
9000	Minimum labor/equipment charge		5	1.600	Job		64		64	106

08 52 66 – Wood Window Screens

08 52 66.10 Wood Screens

		Crew	Daily Output	Labor-Hours	Unit	Material	Labor	Equipment	Total	Total Incl O&P
0010	**WOOD SCREENS**									
0020	Over 3 S.F., 3/4" frames	2 Carp	375	.043	S.F.	4.28	1.70		5.98	7.50
0100	1-1/8" frames	"	375	.043	"	7.60	1.70		9.30	11.15
9000	Minimum labor/equipment charge	1 Carp	4	2	Job		80		80	132

08 52 69 – Wood Storm Windows

08 52 69.10 Storm Windows

		Crew	Daily Output	Labor-Hours	Unit	Material	Labor	Equipment	Total	Total Incl O&P
0010	**STORM WINDOWS**, aluminum residential									
0300	Basement, mill finish, incl. fiberglass screen									
0320	1'-10" x 1'-0" high	2 Carp	30	.533	Ea.	32	21.50		53.50	70
0340	2'-9" x 1'-6" high	"	30	.533	"	34	21.50		55.50	72.50
1600	Double-hung, combination, storm & screen									
2000	Average quality, clear anodic coating, 2'-0" x 3'-5" high	2 Carp	30	.533	Ea.	82	21.50		103.50	125
2020	2'-6" x 5'-0" high		28	.571		102	23		125	150
2040	4'-0" x 6'-0" high		25	.640		120	25.50		145.50	174
2400	White painted, 2'-0" x 3'-5" high		30	.533		81	21.50		102.50	124
2420	2'-6" x 5'-0" high		28	.571		88	23		111	135
2440	4'-0" x 6'-0" high		25	.640		96	25.50		121.50	148
2600	Mill finish, 2'-0" x 3'-5" high		30	.533		73	21.50		94.50	116
2620	2'-6" x 5'-0" high		28	.571		82	23		105	128
2640	4'-0" x 6'-8" high		25	.640		92	25.50		117.50	143
9410	Minimum labor/equipment charge	1 Carp	4	2	Job		80		80	132

08 53 Plastic Windows

08 53 13 – Vinyl Windows

08 53 13.20 Vinyl Single Hung Windows

			Crew	Daily Output	Labor-Hours	Unit	Material	Labor	Equipment	Total	Total Incl O&P
0010	**VINYL SINGLE HUNG WINDOWS**										
0100	Grids, low E, J fin, ext. jambs, 21" x 53"	G	2 Carp	18	.889	Ea.	151	35.50		186.50	225
0110	21" x 57"	G		17	.941		155	37.50		192.50	232
0120	21" x 65"	G		16	1		161	40		201	243
0130	25" x 41"	G		20	.800		142	32		174	209
0140	25" x 49"	G		18	.889		157	35.50		192.50	232
0150	25" x 57"	G		17	.941		161	37.50		198.50	239
0160	25" x 65"	G		16	1		168	40		208	251
0170	29" x 41"	G		18	.889		151	35.50		186.50	226

08 53 Plastic Windows

08 53 13 – Vinyl Windows

08 53 13.20 Vinyl Single Hung Windows			Crew	Daily Output	Labor-Hours	Unit	Material	Labor	Equipment	Total	Total Incl O&P
							2009 Bare Costs				
0180	29" x 53"	G	2 Carp	18	.889	Ea.	162	35.50		197.50	238
0190	29" x 57"	G		17	.941		166	37.50		203.50	245
0200	29" x 65"	G		16	1		173	40		213	256
0210	33" x 41"	G		20	.800		157	32		189	225
0220	33" x 53"	G		18	.889		169	35.50		204.50	245
0230	33" x 57"	G		17	.941		173	37.50		210.50	252
0240	33" x 65"	G		16	1		180	40		220	264
0250	37" x 41"	G		20	.800		165	32		197	235
0260	37" x 53"	G		18	.889		178	35.50		213.50	254
0270	37" x 57"	G		17	.941		181	37.50		218.50	261
0280	37" x 65"	G		16	1		189	40		229	274

08 53 13.30 Vinyl Double Hung Windows			Crew	Daily Output	Labor-Hours	Unit	Material	Labor	Equipment	Total	Total Incl O&P
0010	**VINYL DOUBLE HUNG WINDOWS**										
0100	Grids, low E, J fin, ext. jambs, 21" x 53"	G	2 Carp	18	.889	Ea.	173	35.50		208.50	249
0102	21" x 37"	G		18	.889		154	35.50		189.50	228
0104	21" x 41"	G		18	.889		157	35.50		192.50	232
0106	21" x 49"	G		18	.889		165	35.50		200.50	241
0110	21" x 57"	G		17	.941		176	37.50		213.50	256
0120	21" x 65"	G		16	1		184	40		224	268
0128	25" x 37"	G		20	.800		162	32		194	232
0130	25" x 41"	G		20	.800		166	32		198	235
0140	25" x 49"	G		18	.889		171	35.50		206.50	247
0145	25" x 53"	G		18	.889		178	35.50		213.50	255
0150	25" x 57"	G		17	.941		179	37.50		216.50	259
0160	25" x 65"	G		16	1		190	40		230	275
0162	25" x 69"	G		16	1		198	40		238	284
0164	25" x 77"	G		16	1		209	40		249	296
0168	29" x 37"	G		18	.889		168	35.50		203.50	243
0170	29" x 41"	G		18	.889		171	35.50		206.50	247
0172	29" x 49"	G		18	.889		180	35.50		215.50	257
0180	29" x 53"	G		18	.889		184	35.50		219.50	261
0190	29" x 57"	G		17	.941		188	37.50		225.50	268
0200	29" x 65"	G		16	1		195	40		235	281
0202	29" x 69"	G		16	1		202	40		242	289
0205	29" x 77"	G		16	1		214	40		254	300
0208	33" x 37"	G		20	.800		172	32		204	242
0210	33" x 41"	G		20	.800		176	32		208	247
0215	33" x 49"	G		20	.800		186	32		218	257
0220	33" x 53"	G		18	.889		190	35.50		225.50	268
0230	33" x 57"	G		17	.941		194	37.50		231.50	275
0240	33" x 65"	G		16	1		199	40		239	285
0242	33" x 69"	G		16	1		211	40		251	298
0246	33" x 77"	G		16	1		222	40		262	310
0250	37" x 41"	G		20	.800		180	32		212	251
0255	37" x 49"	G		20	.800		190	32		222	262
0260	37" x 53"	G		18	.889		197	35.50		232.50	276
0270	37" x 57"	G		17	.941		202	37.50		239.50	284
0280	37" x 65"	G		16	1		207	40		247	294
0282	37" x 69"	G		16	1		212	40		252	299
0286	37" x 77"	G		16	1		231	40		271	320
0300	Solid vinyl, average quality, double insulated glass, 2'-0" x 3'-0"	G	1 Carp	10	.800		279	32		311	360
0310	3'-0" x 4'-0"	G		9	.889		175	35.50		210.50	252

08 53 13.30 Vinyl Double Hung Windows

			Crew	Daily Output	Labor-Hours	Unit	2009 Bare Costs Material	Labor	Equipment	Total	Total Incl O&P
0320	4'-0" x 4'-6"	G	1 Carp	8	1	Ea.	286	40		326	380
0330	Premium, double insulated glass, 2'-6" x 3'-0"	G		10	.800		194	32		226	266
0340	3'-0" x 3'-6"	G		9	.889		225	35.50		260.50	305
0350	3'-0" x 4'-0"	G		9	.889		238	35.50		273.50	320
0360	3'-0" x 4'-6"	G		9	.889		243	35.50		278.50	325
0370	3'-0" x 5'-0"	G		8	1		261	40		301	355
0380	3'-6" x 6'-0"	G		8	1		300	40		340	395

08 53 13.40 Vinyl Casement Windows

			Crew	Daily Output	Labor-Hours	Unit	Material	Labor	Equipment	Total	Total Incl O&P
0010	**VINYL CASEMENT WINDOWS**										
0100	Grids, low E, J fin, ext. jambs, 1 lt, 21" x 41"	G	2 Carp	20	.800	Ea.	225	32		257	300
0110	21" x 47"	G		20	.800		245	32		277	325
0120	21" x 53"	G		20	.800		265	32		297	345
0128	24" x 35"	G		19	.842		216	33.50		249.50	294
0130	24" x 41"	G		19	.842		235	33.50		268.50	315
0140	24" x 47"	G		19	.842		255	33.50		288.50	335
0150	24" x 53"	G		19	.842		275	33.50		308.50	355
0158	28" x 35"	G		19	.842		230	33.50		263.50	310
0160	28" x 41"	G		19	.842		249	33.50		282.50	330
0170	28" x 47"	G		19	.842		269	33.50		302.50	350
0180	28" x 53"	G		19	.842		297	33.50		330.50	380
0184	28" x 59"	G		19	.842		300	33.50		333.50	390
0188	Two lites, 33" x 35"	G		18	.889		370	35.50		405.50	465
0190	33" x 41"	G		18	.889		395	35.50		430.50	495
0200	33" x 47"	G		18	.889		425	35.50		460.50	530
0210	33" x 53"	G		18	.889		455	35.50		490.50	560
0212	33" x 59"	G		18	.889		485	35.50		520.50	590
0215	33" x 72"	G		18	.889		500	35.50		535.50	610
0220	41" x 41"	G		18	.889		435	35.50		470.50	535
0230	41" x 47"	G		18	.889		465	35.50		500.50	570
0240	41" x 53"	G		17	.941		490	37.50		527.50	600
0242	41" x 59"	G		17	.941		520	37.50		557.50	630
0246	41" x 72"	G		17	.941		545	37.50		582.50	655
0250	47" x 41"	G		17	.941		440	37.50		477.50	540
0260	47" x 47"	G		17	.941		465	37.50		502.50	575
0270	47" x 53"	G		17	.941		495	37.50		532.50	605
0272	47" x 59"	G		17	.941		540	37.50		577.50	655
0280	56" x 41"	G		15	1.067		470	42.50		512.50	585
0290	56" x 47"	G		15	1.067		495	42.50		537.50	615
0300	56" x 53"	G		15	1.067		540	42.50		582.50	665
0302	56" x 59"	G		15	1.067		565	42.50		607.50	690
0310	56" x 72"	G		15	1.067		610	42.50		652.50	745
0340	Solid vinyl, premium, double insulated glass, 2'-0" x 3'-0" high	G	1 Carp	10	.800		252	32		284	330
0360	2'-0" x 4'-0" high	G		9	.889		280	35.50		315.50	370
0380	2'-0" x 5'-0" high	G		8	1		276	40		316	370

08 53 13.50 Vinyl Picture Windows

			Crew	Daily Output	Labor-Hours	Unit	Material	Labor	Equipment	Total	Total Incl O&P
0010	**VINYL PICTURE WINDOWS**										
0100	Grids, low E, J fin, ext. jambs, 33" x 47"		2 Carp	12	1.333	Ea.	221	53.50		274.50	330
0110	35" x 71"			12	1.333		230	53.50		283.50	340
0120	41" x 47"			12	1.333		252	53.50		305.50	365
0130	41" x 71"			12	1.333		273	53.50		326.50	390
0140	47" x 47"			12	1.333		285	53.50		338.50	405
0150	47" x 71"			11	1.455		299	58		357	425

08 53 Plastic Windows

08 53 13 – Vinyl Windows

08 53 13.50 Vinyl Picture Windows		Crew	Daily Output	Labor-Hours	Unit	Material	Labor	Equipment	Total	Total Incl O&P
						2009 Bare Costs				
0160	53" x 47"	2 Carp	11	1.455	Ea.	280	58		338	405
0170	53" x 71"		11	1.455		293	58		351	420
0180	59" x 47"		11	1.455		320	58		378	445
0190	59" x 71"		11	1.455		340	58		398	470
0200	71" x 47"		10	1.600		350	64		414	495
0210	71" x 71"		10	1.600		370	64		434	515

08 54 Composite Windows

08 54 13 – Fiberglass Windows

08 54 13.10 Fiberglass Single Hung Windows		Crew	Daily Output	Labor-Hours	Unit	Material	Labor	Equipment	Total	Total Incl O&P
0010	**FIBERGLASS SINGLE HUNG WINDOWS**									
0100	Grids, low E, 18" x 24" G	2 Carp	18	.889	Ea.	255	35.50		290.50	340
0110	18" x 40" G		17	.941		275	37.50		312.50	365
0130	24" x 40" G		20	.800		295	32		327	380
0230	36" x 36" G		17	.941		290	37.50		327.50	380
0250	36" x 48" G		20	.800		330	32		362	420
0260	36" x 60" G		18	.889		350	35.50		385.50	445
0280	36" x 72" G		16	1		360	40		400	460
0290	48" x 40" G		16	1		360	40		400	460

08 56 Special Function Windows

08 56 63 – Detention Windows

08 56 63.13 Visitor Cubicle Windows		Crew	Daily Output	Labor-Hours	Unit	Material	Labor	Equipment	Total	Total Incl O&P
0010	**VISITOR CUBICLE WINDOWS**									
4000	Visitor cubicle, vision panel, no intercom	E-4	2	16	Ea.	2,925	725	67	3,717	4,650

08 62 Unit Skylights

08 62 13 – Domed Unit Skylights

08 62 13.20 Skylights		Crew	Daily Output	Labor-Hours	Unit	Material	Labor	Equipment	Total	Total Incl O&P
0010	**SKYLIGHTS**, Plastic domes, flush or curb mounted ten or									
0100	more units									
0300	Nominal size under 10 S.F., double G	G-3	130	.246	S.F.	26.50	9.70		36.20	44.50
0400	Single		160	.200		23.50	7.90		31.40	38
0600	10 S.F. to 20 S.F., double G		315	.102		22.50	4		26.50	31
0700	Single		395	.081		23	3.19		26.19	30
0900	20 S.F. to 30 S.F., double G		395	.081		21	3.19		24.19	28
1000	Single		465	.069		19.05	2.71		21.76	25.50
1200	30 S.F. to 65 S.F., double G		465	.069		21	2.71		23.71	27.50
1300	Single		610	.052		16.50	2.07		18.57	21.50
1500	For insulated 4" curbs, double, add					27%				
1600	Single, add					30%				
1800	For integral insulated 9" curbs, double, add					30%				
1900	Single, add					40%				
2120	Ventilating insulated plexiglass dome with									
2130	curb mounting, 36" x 36" G	G-3	12	2.667	Ea.	380	105		485	590
2150	52" x 52" G		12	2.667		570	105		675	800
2160	28" x 52" G		10	3.200		455	126		581	705

08 62 Unit Skylights

08 62 13 – Domed Unit Skylights

08 62 13.20 Skylights			Crew	Daily Output	Labor-Hours	Unit	Material	Labor	Equipment	Total	Total Incl O&P
							2009 Bare Costs				
2170	36″ x 52″	G	G-3	10	3.200	Ea.	495	126		621	750
2180	For electric opening system, add	G					292			292	320
2200	Field fabricated, factory type, aluminum and wire glass	G	G-3	120	.267	S.F.	15.10	10.50		25.60	33.50
2300	Insulated safety glass with aluminum frame	G		160	.200		88	7.90		95.90	110
2400	Sandwich panels, fiberglass, for walls, 1-9/16″ thick, to 250 SF	G		200	.160		16	6.30		22.30	28
2500	250 SF and up	G		265	.121		14.30	4.76		19.06	23.50
2700	As above, but for roofs, 2-3/4″ thick, to 250 SF	G		295	.108		23	4.27		27.27	32.50
2800	250 SF and up	G		330	.097		18.90	3.82		22.72	27

08 63 Metal-Framed Skylights

08 63 23 – Ridge Metal-Framed Skylights

08 63 23.10 Prefabricated

0010	**PREFABRICATED** glass block with metal frame										
0020	Minimum	G	G-3	265	.121	S.F.	50	4.76		54.76	62.50
0100	Maximum	G	″	160	.200	″	100	7.90		107.90	123

08 71 Door Hardware

08 71 13 – Automatic Door Operators

08 71 13.10 Automatic Openers Commercial

0010	**AUTOMATIC OPENERS COMMERCIAL**									
0020	Pneumatic, incl opener, motion sens, control box, tubing, compressor									
0050	For single swing door, per opening	2 Skwk	.80	20	Ea.	4,125	815		4,940	5,875
0100	Pair, per opening		.50	32	Opng.	6,875	1,300		8,175	9,700
1000	For single sliding door, per opening		.60	26.667		4,575	1,100		5,675	6,800
1300	Bi-parting pair		.50	32		6,900	1,300		8,200	9,725
1420	Electronic door opener incl motion sens, 12V control box, motor									
1450	For single swing door, per opening	2 Skwk	.80	20	Opng.	3,400	815		4,215	5,050
1500	Pair, per opening		.50	32		5,550	1,300		6,850	8,225
1600	For single sliding door, per opening		.60	26.667		3,675	1,100		4,775	5,825
1700	Bi-parting pair		.50	32		5,625	1,300		6,925	8,325
1750	Handicap actuator buttons, 2, including 12V DC wiring, add	1 Carp	1.50	5.333	Pr.	415	213		628	805

08 71 13.20 Automatic Openers Industrial

0010	**AUTOMATIC OPENERS INDUSTRIAL**									
0015	Sliding doors up to 6′ wide	2 Skwk	.60	26.667	Opng.	5,400	1,100		6,500	7,700
0200	To 12′ wide	″	.40	40	″	6,575	1,625		8,200	9,900
0400	Over 12′ wide, add per L.F. of excess				L.F.	720			720	790
1000	Swing doors, to 5′ wide	2 Skwk	.80	20	Ea.	3,150	815		3,965	4,775
1860	Add for controls, wall pushbutton, 3 button		4	4		192	163		355	475
1870	Ceiling pull cord		4.30	3.721		165	152		317	430

08 71 20 – Hardware

08 71 20.10 Bolts, Flush

0010	**BOLTS, FLUSH**									
0020	Standard, concealed	1 Carp	7	1.143	Ea.	21	45.50		66.50	99
0800	Automatic fire exit	″	5	1.600		287	64		351	420
1600	For electric release, add	1 Elec	3	2.667		116	125		241	320
3000	Barrel, brass, 2″ long	1 Carp	40	.200		4.90	8		12.90	18.60
3020	4″ long		40	.200		3.60	8		11.60	17.15
3060	6″ long		40	.200		10.10	8		18.10	24.50

08 71 20 – Hardware

08 71 20.15 Hardware

	08 71 20.15 Hardware	Crew	Daily Output	Labor-Hours	Unit	2009 Bare Costs Material	Labor	Equipment	Total	Total Incl O&P
0009	**HARDWARE**									
0010	Average percentage for hardware, total job cost									
0025	Minimum									1%
0050	Maximum									4%
0500	Total hardware for building, average distribution					85%	15%			
1000	Door hardware, apartment, interior				Door	140			140	154
1500	Hospital bedroom, minimum					310			310	340
2000	Maximum					700			700	770
2100	Pocket door				Ea.	134			134	147
2250	School, single exterior, incl. lever, not incl. panic device				Door	445			445	490
2500	Single interior, regular use, no lever included					315			315	345
2550	Including handicap lever					410			410	450
2600	Heavy use, incl. lever and closer					550			550	605
2850	Stairway, single interior					790			790	870
3100	Double exterior, with panic device				Pr.	1,075			1,075	1,200
3600	Toilet, public, single interior				Door	170			170	187
6020	Add for door holder, electro-magnetic	1 Elec	4	2	Ea.	86	94		180	240

08 71 20.20 Door Protectors

	08 71 20.20 Door Protectors	Crew	Daily Output	Labor-Hours	Unit	Material	Labor	Equipment	Total	Total Incl O&P
0010	**DOOR PROTECTORS**									
0020	1-3/4" x 3/4" U channel	2 Carp	80	.200	L.F.	31	8		39	47
0021	1-3/4" x 1-1/4" U channel		80	.200	"	15.30	8		23.30	30
1000	Tear drop, spring-stl, 8" high x 19" long		15	1.067	Ea.	87.50	42.50		130	167
1010	8" high x 32" long		15	1.067		108	42.50		150.50	190
1100	Tear drop, stainless stl, 8" high x 19" long		15	1.067		230	42.50		272.50	325
1200	8" high x 32" long		15	1.067		285	42.50		327.50	385

08 71 20.30 Door Closers

	08 71 20.30 Door Closers	Crew	Daily Output	Labor-Hours	Unit	Material	Labor	Equipment	Total	Total Incl O&P
0010	**DOOR CLOSERS**									
0020	Adjustable backcheck, 3 way mount, all sizes, regular arm	1 Carp	6	1.333	Ea.	156	53.50		209.50	260
0040	Hold open arm		6	1.333		168	53.50		221.50	273
0100	Fusible link		6.50	1.231		130	49		179	224
0200	Non sized, regular arm		6	1.333		147	53.50		200.50	250
0240	Hold open arm		6	1.333		191	53.50		244.50	299
0400	4 way mount, non sized, regular arm		6	1.333		208	53.50		261.50	315
0440	Hold open arm		6	1.333		214	53.50		267.50	325
2000	Backcheck and adjustable power, hinge face mount									
2010	All sizes, regular arm	1 Carp	6.50	1.231	Ea.	187	49		236	287
2040	Hold open arm		6.50	1.231		235	49		284	340
2400	Top jamb mount, all sizes, regular arm		6	1.333		184	53.50		237.50	290
2440	Hold open arm		6	1.333		203	53.50		256.50	310
2800	Top face mount, all sizes, regular arm		6.50	1.231		184	49		233	283
2840	Hold open arm		6.50	1.231		203	49		252	305
4000	Backcheck, overhead concealed, all sizes, regular arm		5.50	1.455		200	58		258	315
4040	Concealed arm		5	1.600		217	64		281	345
4400	Compact overhead, concealed, all sizes, regular arm		5.50	1.455		340	58		398	465
4440	Concealed arm		5	1.600		375	64		439	520
4800	Concealed in door, all sizes, regular arm		5.50	1.455		129	58		187	238
4840	Concealed arm		5	1.600		139	64		203	259
4900	Floor concealed, all sizes, single acting		2.20	3.636		164	145		309	420
4940	Double acting		2.20	3.636		210	145		355	470
5000	For cast aluminum cylinder, deduct					17.90			17.90	19.70
5040	For delayed action, add					31			31	34
5080	For fusible link arm, add					13.20			13.20	14.50

08 71 Door Hardware

08 71 20 – Hardware

08 71 20.30 Door Closers		Crew	Daily Output	Labor-Hours	Unit	Material	2009 Bare Costs Labor	Equipment	Total	Total Incl O&P
5120	For shock absorbing arm, add				Ea.	39			39	43
5160	For spring power adjustment, add					30			30	33
6000	Closer-holder, hinge face mount, all sizes, exposed arm	1 Carp	6.50	1.231		140	49		189	235
7000	Electronic closer-holder, hinge facemount, concealed arm		5	1.600		218	64		282	345
7400	With built-in detector		5	1.600		655	64		719	825
8000	Surface mounted, stand. duty, parallel arm, primed, traditional		6	1.333		176	53.50		229.50	282
8030	Light duty		6	1.333		114	53.50		167.50	214
8050	Heavy duty		6	1.333		203	53.50		256.50	310
8100	Standard duty, parallel arm, modern		6	1.333		193	53.50		246.50	300
8150	Heavy duty		6	1.333		213	53.50		266.50	320
9000	Minimum labor/equipment charge		4	2	Job		80		80	132

08 71 20.31 Door Closers		Crew	Daily Output	Labor-Hours	Unit	Material	Labor	Equipment	Total	Total Incl O&P
0010	**DOOR CLOSERS**									
0015	Door closer, rack and pinion	1 Carp	6.50	1.231	Ea.	148	49		197	243

08 71 20.35 Panic Devices		Crew	Daily Output	Labor-Hours	Unit	Material	Labor	Equipment	Total	Total Incl O&P
0010	**PANIC DEVICES**									
0015	For rim locks, single door exit only	1 Carp	6	1.333	Ea.	475	53.50		528.50	615
0020	Outside key and pull		5	1.600		520	64		584	675
0200	Bar and vertical rod, exit only		5	1.600		700	64		764	870
0210	Outside key and pull		4	2		775	80		855	985
0400	Bar and concealed rod		4	2		570	80		650	755
0600	Touch bar, exit only		6	1.333		485	53.50		538.50	625
0610	Outside key and pull		5	1.600		555	64		619	715
0700	Touch bar and vertical rod, exit only		5	1.600		630	64		694	795
0710	Outside key and pull		4	2		730	80		810	935
0800	Touch bar, low profile, exit only		6	1.333		355	53.50		408.50	480
0810	Outside key and pull		5	1.600		410	64		474	560
0900	Touch bar and vertical rod, low profile, exit only		5	1.600		515	64		579	670
0910	Outside key and pull		4	2		545	80		625	725
1000	Mortise, bar, exit only		4	2		515	80		595	695
1600	Touch bar, exit only		4	2		590	80		670	780
2000	Narrow stile, rim mounted, bar, exit only		6	1.333		620	53.50		673.50	770
2010	Outside key and pull		5	1.600		675	64		739	845
2200	Bar and vertical rod, exit only		5	1.600		635	64		699	805
2210	Outside key and pull		4	2		635	80		715	830
2400	Bar and concealed rod, exit only		3	2.667		745	107		852	990
3000	Mortise, bar, exit only		4	2		570	80		650	755
3600	Touch bar, exit only		4	2		825	80		905	1,050
6000	Trim, rim mounted, cylinder and pull		25	.320		111	12.80		123.80	143
6100	Cylinder, pull and thumb piece		20	.400		110	16		126	148
6200	Pull only		30	.267		64	10.65		74.65	87.50
6400	Mortise, cylinder and pull		25	.320		130	12.80		142.80	164
6500	Cylinder, pull and thumb piece		20	.400		125	16		141	165
6600	Pull only		30	.267		111	10.65		121.65	140
9000	Minimum labor/equipment charge		2.50	3.200	Job		128		128	211

08 71 20.40 Lockset		Crew	Daily Output	Labor-Hours	Unit	Material	Labor	Equipment	Total	Total Incl O&P
0010	**LOCKSET**, Standard duty									
0020	Non-keyed, passage	1 Carp	12	.667	Ea.	49	26.50		75.50	97.50
0100	Privacy		12	.667		61	26.50		87.50	111
0400	Keyed, single cylinder function		10	.800		83.50	32		115.50	145
0420	Hotel		8	1		120	40		160	198
0500	Lever handled, keyed, single cylinder function		10	.800		148	32		180	216

08 71 20.40 Lockset

Line	Description	Crew	Daily Output	Labor-Hours	Unit	Material	Labor	Equipment	Total	Total Incl O&P
						2009 Bare Costs				
1000	Heavy duty with sectional trim, non-keyed, passages	1 Carp	12	.667	Ea.	138	26.50		164.50	196
1100	Privacy		12	.667		173	26.50		199.50	234
1400	Keyed, single cylinder function		10	.800		210	32		242	284
1420	Hotel		8	1		310	40		350	405
1600	Communicating		10	.800		235	32		267	310
1690	For re-core cylinder, add					34			34	37.50
1700	Residential, interior door, minimum	1 Carp	16	.500		17	20		37	51.50
1720	Maximum		8	1		43.50	40		83.50	114
1800	Exterior, minimum		14	.571		36.50	23		59.50	77.50
1810	Average		8	1		69	40		109	142
1820	Maximum		8	1		156	40		196	238
9000	Minimum labor/equipment charge		6	1.333	Job		53.50		53.50	88

08 71 20.41 Dead Locks

Line	Description	Crew	Daily Output	Labor-Hours	Unit	Material	Labor	Equipment	Total	Total Incl O&P
0010	**DEAD LOCKS**									
0011	Mortise heavy duty outside key (security item)	1 Carp	9	.889	Ea.	150	35.50		185.50	224
0020	Double cylinder		9	.889		162	35.50		197.50	237
0100	Medium duty, outside key		10	.800		110	32		142	174
0110	Double cylinder		10	.800		142	32		174	209
1000	Tubular, standard duty, outside key		10	.800		61	32		93	120
1010	Double cylinder		10	.800		79	32		111	140
1200	Night latch, outside key		10	.800		78.50	32		110.50	140

08 71 20.42 Mortise Locksets

Line	Description	Crew	Daily Output	Labor-Hours	Unit	Material	Labor	Equipment	Total	Total Incl O&P
0010	**MORTISE LOCKSETS**, Comm., wrought knobs & full escutcheon trim									
0020	Non-keyed, passage, minimum	1 Carp	9	.889	Ea.	175	35.50		210.50	252
0030	Maximum		8	1		288	40		328	380
0040	Privacy, minimum		9	.889		190	35.50		225.50	268
0050	Maximum		8	1		310	40		350	405
0100	Keyed, office/entrance/apartment, minimum		8	1		220	40		260	310
0110	Maximum		7	1.143		370	45.50		415.50	480
0120	Single cylinder, typical, minimum		8	1		189	40		229	274
0130	Maximum		7	1.143		350	45.50		395.50	460
0200	Hotel, minimum		7	1.143		220	45.50		265.50	320
0210	Maximum		6	1.333		365	53.50		418.50	490
0300	Communication, double cylinder, minimum		8	1		215	40		255	305
0310	Maximum		7	1.143		286	45.50		331.50	390
1000	Wrought knobs and sectional trim, non-keyed, passage, minimum		10	.800		119	32		151	184
1010	Maximum		9	.889		230	35.50		265.50	310
1040	Privacy, minimum		10	.800		136	32		168	203
1050	Maximum		9	.889		250	35.50		285.50	335
1100	Keyed, entrance, office/apartment, minimum		9	.889		206	35.50		241.50	286
1110	Maximum		8	1		295	40		335	390
1120	Single cylinder, typical, minimum		9	.889		200	35.50		235.50	279
1130	Maximum		8	1		288	40		328	380
2000	Cast knobs and full escutcheon trim									
2010	Non-keyed, passage, minimum	1 Carp	9	.889	Ea.	249	35.50		284.50	335
2020	Maximum		8	1		400	40		440	505
2040	Privacy, minimum		9	.889		305	35.50		340.50	395
2050	Maximum		8	1		440	40		480	545
2120	Keyed, single cylinder, typical, minimum		8	1		310	40		350	405
2130	Maximum		7	1.143		480	45.50		525.50	605
2200	Hotel, minimum		7	1.143		320	45.50		365.50	425
2210	Maximum		6	1.333		570	53.50		623.50	715

08 71 Door Hardware

08 71 20 – Hardware

08 71 20.42 Mortise Locksets

08 71 20.42 Mortise Locksets		Crew	Daily Output	Labor-Hours	Unit	2009 Bare Costs Material	Labor	Equipment	Total	Total Incl O&P
3000	Cast knob and sectional trim, non-keyed, passage, minimum	1 Carp	10	.800	Ea.	196	32		228	269
3010	Maximum		10	.800		380	32		412	475
3040	Privacy, minimum		10	.800		220	32		252	295
3050	Maximum		10	.800		390	32		422	485
3100	Keyed, office/entrance/apartment, minimum		9	.889		250	35.50		285.50	335
3110	Maximum		9	.889		400	35.50		435.50	500
3120	Single cylinder, typical, minimum		9	.889		250	35.50		285.50	335
3130	Maximum		9	.889		490	35.50		525.50	600
3190	For re-core cylinder, add					33			33	36.50
3800	Cipher lockset w/key pad (security item)	1 Carp	13	.615		775	24.50		799.50	890
3900	Cipher lockset, with dial for swinging doors (security item)		13	.615		1,675	24.50		1,699.50	1,900
3920	with dial for swinging doors & drill resistant plate (security item)		12	.667		2,025	26.50		2,051.50	2,275
3950	Cipher lockset with dial for safe/vault door (security item)		12	.667		1,300	26.50		1,326.50	1,475
3980	Keyless, pushbutton type									
4000	Residential/light commercial, deadbolt, standard	1 Carp	9	.889	Ea.	114	35.50		149.50	184
4010	Heavy duty		9	.889		134	35.50		169.50	206
4020	Industrial, heavy duty, with deadbolt		9	.889		271	35.50		306.50	355
4030	Key override		9	.889		300	35.50		335.50	395
4040	Lever activated handle		9	.889		320	35.50		355.50	415
4050	Key override		9	.889		360	35.50		395.50	455
4060	Double sided pushbutton type		8	1		660	40		700	795
4070	Key override		8	1		655	40		695	785

08 71 20.45 Peepholes

08 71 20.45 Peepholes		Crew	Daily Output	Labor-Hours	Unit	Material	Labor	Equipment	Total	Total Incl O&P
0010	**PEEPHOLES**									
2010	Peephole	1 Carp	32	.250	Ea.	15.80	10		25.80	34

08 71 20.50 Door Stops

08 71 20.50 Door Stops		Crew	Daily Output	Labor-Hours	Unit	Material	Labor	Equipment	Total	Total Incl O&P
0010	**DOOR STOPS**									
0020	Holder & bumper, floor or wall	1 Carp	32	.250	Ea.	33.50	10		43.50	53.50
1300	Wall bumper, 4" diameter, with rubber pad, aluminum		32	.250		10.25	10		20.25	28
1600	Door bumper, floor type, aluminum		32	.250		5.85	10		15.85	23
1900	Plunger type, door mounted		32	.250		27	10		37	46
9000	Minimum labor/equipment charge		6	1.333	Job		53.50		53.50	88

08 71 20.55 Push-Pull Plates

08 71 20.55 Push-Pull Plates		Crew	Daily Output	Labor-Hours	Unit	Material	Labor	Equipment	Total	Total Incl O&P
0010	**PUSH-PULL PLATES**									
0100	Push plate, .050 thick, 4" x 16", aluminum	1 Carp	12	.667	Ea.	8.65	26.50		35.15	53.50
0500	Bronze		12	.667		20.50	26.50		47	66.50
1500	Pull handle and push bar, aluminum		11	.727		126	29		155	187
2000	Bronze		10	.800		163	32		195	233
3000	Push plate both sides, aluminum		14	.571		17.75	23		40.75	57
3500	Bronze		13	.615		41.50	24.50		66	86
4000	Door pull, designer style, cast aluminum, minimum		12	.667		67.50	26.50		94	119
5000	Maximum		8	1		365	40		405	465
6000	Cast bronze, minimum		12	.667		85.50	26.50		112	138
7000	Maximum		8	1		400	40		440	505
8000	Walnut, minimum		12	.667		65.50	26.50		92	116
9000	Maximum		8	1		365	40		405	465
9800	Minimum labor/equipment charge		5	1.600	Job		64		64	106

08 71 20.60 Entrance Locks

08 71 20.60 Entrance Locks		Crew	Daily Output	Labor-Hours	Unit	Material	Labor	Equipment	Total	Total Incl O&P
0010	**ENTRANCE LOCKS**									
0015	Cylinder, grip handle deadlocking latch	1 Carp	9	.889	Ea.	127	35.50		162.50	199
0020	Deadbolt		8	1		154	40		194	235
0100	Push and pull plate, dead bolt		8	1		146	40		186	227

08 71 20.60 Entrance Locks

Line	Description	Crew	Daily Output	Labor-Hours	Unit	Material	Labor	Equipment	Total	Total Incl O&P
0200	Push bar and pull, dead bolt, bronze	1 Carp	7	1.143	Ea.	265	45.50		310.50	370
0240	Push bar and pull bar, dead bolt, bronze		7	1.143		415	45.50		460.50	535
0900	For handicapped lever, add					161			161	177

08 71 20.65 Thresholds

Line	Description	Crew	Daily Output	Labor-Hours	Unit	Material	Labor	Equipment	Total	Total Incl O&P
0010	**THRESHOLDS**									
0011	Threshold 3′ long saddles aluminum	1 Carp	48	.167	L.F.	4.02	6.65		10.67	15.40
0100	Aluminum, 8″ wide, 1/2″ thick		12	.667	Ea.	38.50	26.50		65	86
0500	Bronze		60	.133	L.F.	38.50	5.35		43.85	51.50
0600	Bronze, panic threshold, 5″ wide, 1/2″ thick		12	.667	Ea.	65.50	26.50		92	116
0700	Rubber, 1/2″ thick, 5-1/2″ wide		20	.400		37.50	16		53.50	68
0800	2-3/4″ wide		20	.400		16	16		32	44
9000	Minimum labor/equipment charge		4	2	Job		80		80	132

08 71 20.70 Floor Checks

Line	Description	Crew	Daily Output	Labor-Hours	Unit	Material	Labor	Equipment	Total	Total Incl O&P
0010	**FLOOR CHECKS**									
0020	For over 3′ wide doors single acting	1 Carp	2.50	3.200	Ea.	510	128		638	770
0500	Double acting	″	2.50	3.200	″	660	128		788	935

08 71 20.75 Door Hardware Accessories

Line	Description	Crew	Daily Output	Labor-Hours	Unit	Material	Labor	Equipment	Total	Total Incl O&P
0010	**DOOR HARDWARE ACCESSORIES**									
0050	Door closing coordinator, 36″ (for paired openings up to 56″)	1 Carp	8	1	Ea.	89.50	40		129.50	165
0060	48″ (for paired openings up to 84″)		8	1		96	40		136	171
0070	56″ (for paired openings up to 96″)		8	1		105	40		145	182

08 71 20.80 Hasps

Line	Description	Crew	Daily Output	Labor-Hours	Unit	Material	Labor	Equipment	Total	Total Incl O&P
0010	**HASPS**, steel assembly									
0015	3″	1 Carp	26	.308	Ea.	3.93	12.30		16.23	25
0020	4-1/2″		13	.615		5.05	24.50		29.55	46
0040	6″		12.50	.640		8.40	25.50		33.90	51.50

08 71 20.90 Hinges

Line	Description	Crew	Daily Output	Labor-Hours	Unit	Material	Labor	Equipment	Total	Total Incl O&P
0010	**HINGES**									
0012	Full mortise, avg. freq., steel base, USP, 4-1/2″ x 4-1/2″				Pr.	24.50			24.50	27
0040	US26D					30.50			30.50	33.50
0080	US10A					35.50			35.50	39
0100	5″ x 5″, USP					41			41	45
0200	6″ x 6″, USP					87			87	95.50
0400	Brass base, 4-1/2″ x 4-1/2″, US10					50			50	55
0440	US26D					59.50			59.50	65.50
0480	US10B					66.50			66.50	73
0500	5″ x 5″, US10					73.50			73.50	81
0600	6″ x 6″, US10					125			125	137
0800	Stainless steel base, 4-1/2″ x 4-1/2″, US32					74.50			74.50	82
0900	For non removable pin, add(security item)				Ea.	4.24			4.24	4.66
0910	For floating pin, driven tips, add					3.09			3.09	3.40
0930	For hospital type tip on pin, add					13.35			13.35	14.70
0940	For steeple type tip on pin, add					11.70			11.70	12.85
0950	Full mortise, high frequency, steel base, 3-1/2″ x 3-1/2″, US26D				Pr.	25.50			25.50	28.50
1000	4-1/2″ x 4-1/2″, USP					58.50			58.50	64.50
1040	US26D					58.50			58.50	64.50
1080	US26					76.50			76.50	84
1100	5″ x 5″, USP					54.50			54.50	60
1200	6″ x 6″, USP					134			134	147
1300	8″ x 8″, USP					226			226	248
1400	Brass base, 3-1/2″ x 3-1/2″, US4					45			45	49.50

08 71 20.90 Hinges		Crew	Daily Output	Labor-Hours	Unit	Material	Labor	Equipment	Total	Total Incl O&P
						2009 Bare Costs				
1430	4-1/2" x 4-1/2", US10				Pr.	78.50			78.50	86.50
1440	US26D					96.50			96.50	106
1480	US10B					110			110	121
1500	5" x 5", US10					118			118	130
1600	6" x 6", US10					170			170	187
1700	8" x 8", US10					315			315	345
1800	Stainless steel base, 4-1/2" x 4-1/2", US32					125			125	137
1810	5" x 4-1/2", US32				▼	175			175	193
1930	For hospital type tip on pin, add				Ea.	11.65			11.65	12.80
1950	Full mortise, low frequency, steel base, 3-1/2" x 3-1/2", US26D				Pr.	10.55			10.55	11.60
2000	4-1/2" x 4-1/2", USP					12.05			12.05	13.25
2040	US26D					15.65			15.65	17.25
2080	US10A					19.65			19.65	21.50
2100	5" x 5", USP					30			30	33
2200	6" x 6", USP					59.50			59.50	65.50
2300	4-1/2" x 4-1/2", US3					17.40			17.40	19.15
2310	5" x 5", US3					43			43	47.50
2400	Brass bass, 4-1/2" x 4-1/2", US10					42			42	46
2440	US26D					49			49	54
2480	US10A					45			45	49.50
2500	5" x 5", US10					63.50			63.50	70
2800	Stainless steel base, 4-1/2" x 4-1/2", US32				▼	72			72	79.50
3000	Half surface, half mortise, or full surface, average frequency									
3010	Steel base, 4-1/2" x 4-1/2", USP				Pr.	38			38	41.50
3040	US26D					60.50			60.50	66.50
3080	US10A					81			81	89
3100	5" x 5", USP					75			75	82.50
3400	Brass base, 4-1/2" x 4-1/2", US10					143			143	157
3440	US26D					154			154	169
3480	US10B					169			169	185
3500	5" x 5", US10					182			182	200
3800	Stainless steel base, 4-1/2" x 4-1/2", US32				▼	197			197	216
4000	Half surface, half mortise or full surface, high frequency									
4010	Steel base, 4-1/2" x 4-1/2", USP				Pr.	75.50			75.50	83
4040	US26D					139			139	153
4080	US10A					168			168	185
4100	5" x 5", USP					148			148	163
4400	Brass base, 4-1/2" x 4-1/2", US10					172			172	189
4440	US26D					180			180	198
4480	US10B					205			205	226
4500	5" x 5", US10					204			204	224
4800	Stainless steel base, 4-1/2" x 4-1/2", US32				▼	226			226	249
5000	Half surface, half mortise, or full surface, low frequency									
5010	Steel base, 4-1/2" x 4-1/2", USP				Pr.	28.50			28.50	31
5040	US26D					35.50			35.50	39
5080	US10A				▼	57			57	62.50

08 71 20.91 Special Hinges		Crew	Daily Output	Labor-Hours	Unit	Material	Labor	Equipment	Total	Total Incl O&P
0010	**SPECIAL HINGES**									
0015	Paumelle, high frequency									
0020	Steel base, 6" x 4-1/2", US10				Pr.	139			139	153
0040	US26D					173			173	190
0080	US10A				▼	181			181	199

08 71 20.91 Special Hinges		Crew	Daily Output	Labor-Hours	Unit	2009 Bare Costs Material	Labor	Equipment	Total	Total Incl O&P
0100	Bronze base, 5″ x 4-1/2″, US10				Pr.	174			174	192
0140	US26D					192			192	211
0180	US3					202			202	223
0200	Paumelle, average frequency, steel base, 4-1/2″ x 3-1/2″, US10					94			94	104
0240	US26D					98.50			98.50	108
0280	US10A					101			101	111
0400	Olive knuckle, low frequency, brass base, 6″ x 4-1/2″, US10					159			159	175
0440	US26D					179			179	197
0480	US3					180			180	198
0800	Emergency door pivot, average frequency									
0810	Brass base, 4-7/8″ jamb plate, USP				Pr.	64.50			64.50	71
0840	US26D					71.50			71.50	78.50
0880	For emergency door stop and hold back, add					61			61	67
1000	Electric hinge with concealed conductor, average frequency									
1010	Steel base, 4-1/2″ x 4-1/2″, US26D				Pr.	310			310	340
1100	Bronze base, 4-1/2″ x 4-1/2″, US26D				″	315			315	345
1200	Electric hinge with concealed conductor, high frequency									
1210	Steel base, 4-1/2″ x 4-1/2″, US26D				Pr.	231			231	255
1400	Non template, full mortise, low frequency									
1410	Steel base, 4″ x 4″, USP				Pr.	16.45			16.45	18.10
1440	US26D					18.50			18.50	20.50
1480	US10A					22			22	24.50
1500	Non template, full mortise, average frequency									
1510	Steel base, 4″ x 4″, USP				Pr.	34.50			34.50	38
1540	US26D					33			33	36.50
1580	US10A					42.50			42.50	47
1600	Double weight, 800 lb., steel base, removable pin, 5″ x 6″, USP					128			128	141
1700	Steel base-welded pin, 5″ x 6″, USP					158			158	174
1800	Triple weight, 2000 lb., steel base, welded pin, 5″ x 6″, USP					147			147	162
2000	Pivot reinf., high frequency, steel base, 7-3/4″ door plate, USP					177			177	195
2040	US26D					156			156	171
2080	US10A					189			189	208
2200	Bronze base, 7-3/4″ door plate, US10					222			222	245
2240	US26D					232			232	255
2280	US10A					266			266	293
3000	Swing clear, full mortise, full or half surface, high frequency,									
3010	Steel base, 5″ high, USP				Pr.	154			154	169
3040	US26D					148			148	163
3080	US10A					187			187	206
3200	Swing clear, full mortise, average frequency									
3210	Steel base, 4-1/2″ high, USP				Pr.	122			122	135
3280	US10A				″	150			150	165
3400	Swing clear, half mortise, high frequency									
3410	Steel base, 4-1/2″ high, USP				Pr.	154			154	170
3440	US26D					159			159	175
3480	US10A					188			188	207
4000	Wide throw, average frequency, steel base, 4-1/2″ x 6″, USP					93.50			93.50	103
4040	US26D					99.50			99.50	110
4080	US10A					103			103	113
4100	5″ x 7″, USP					114			114	126
4200	High frequency, steel base, 4-1/2″ x 6″, USP					143			143	157
4240	US26D					166			166	183
4280	US10A					185			185	203

08 71 Door Hardware

08 71 20 – Hardware

08 71 20.91 Special Hinges		Crew	Daily Output	Labor-Hours	Unit	2009 Bare Costs Material	Labor	Equipment	Total	Total Incl O&P
4300	5″ x 7″, USP				Pr.	146			146	160
4400	Wide throw, low frequency, steel base, 4-1/2″ x 6″, USP					70			70	77
4440	US26D					86			86	94.50
4480	US10A					92			92	101
4500	5″ x 7″, USP					99.50			99.50	110
4600	Spring hinge, single acting, 6″ flange, steel				Ea.	52.50			52.50	57.50
4700	Brass					92			92	101
4900	Double acting, 6″ flange, steel					94			94	103
4950	Brass					153			153	168
9000	Continuous hinge, steel, full mortise, heavy duty	2 Carp	64	.250	L.F.	12.40	10		22.40	30

08 71 20.92 Mortised Hinges

		Crew	Daily Output	Labor-Hours	Unit	Material	Labor	Equipment	Total	Total Incl O&P
0010	**MORTISED HINGES**									
0200	Average frequency, steel plated, ball bearing, 3-1/2″ x 3-1/2″				Pr.	25			25	27.50
0300	Bronze, ball bearing					31.50			31.50	34.50
1100	Bronze, ball bearing					80.50			80.50	88.50
1300	Average frequency, steel plated, ball bearing, 4-1/2″ x 4-1/2″					31			31	34
1500	Bronze, ball bearing, to 36″ wide					33			33	36.50
1700	Low frequency, steel, plated, plain bearing					15.65			15.65	17.20
1900	Bronze, plain bearing					17.55			17.55	19.35

08 71 20.95 Kick Plates

		Crew	Daily Output	Labor-Hours	Unit	Material	Labor	Equipment	Total	Total Incl O&P
0010	**KICK PLATES**									
0020	Stainless steel, 6″ high, for 3′ door	1 Carp	15	.533	Ea.	31	21.50		52.50	69
0500	Bronze 6″ high, for 3′ door		15	.533		43.50	21.50		65	83
2000	Aluminum, .050, with 3 beveled edges, 10″ x 28″		15	.533		22	21.50		43.50	59
2010	10″ x 30″		15	.533		23.50	21.50		45	61
2020	10″ x 34″		15	.533		24.50	21.50		46	62
2040	10″ x 38″		15	.533		26.50	21.50		48	64.50
9000	Minimum labor/equipment charge		6	1.333	Job		53.50		53.50	88

08 71 21 – Astragals

08 71 21.10 Exterior Mouldings, Astragals

		Crew	Daily Output	Labor-Hours	Unit	Material	Labor	Equipment	Total	Total Incl O&P
0010	**EXTERIOR MOULDINGS, ASTRAGALS**									
0400	One piece, overlapping cadmium plated steel, flat, 3/16″ x 2″	1 Carp	90	.089	L.F.	3.40	3.55		6.95	9.60
0600	Prime coated steel, flat, 1/8″ x 3″		90	.089		4.65	3.55		8.20	10.95
0800	Stainless steel, flat, 3/32″ x 1-5/8″		90	.089		17.80	3.55		21.35	25.50
1000	Aluminum, flat, 1/8″ x 2″		90	.089		3.45	3.55		7	9.65
1200	Nail on, ″T″ extrusion		120	.067		.78	2.66		3.44	5.25
1300	Vinyl bulb insert		105	.076		1.25	3.04		4.29	6.40
1600	Screw on, ″T″ extrusion		90	.089		5.05	3.55		8.60	11.40
1700	Vinyl insert		75	.107		3.40	4.26		7.66	10.80
2000	″L″ extrusion, neoprene bulbs		75	.107		1.95	4.26		6.21	9.20
2100	Neoprene sponge insert		75	.107		5.95	4.26		10.21	13.60
2200	Magnetic		75	.107		9.80	4.26		14.06	17.85
2400	Spring hinged security seal, with cam		75	.107		6.30	4.26		10.56	14
2600	Spring loaded locking bolt, vinyl insert		45	.178		8.60	7.10		15.70	21
2800	Neoprene sponge strip, ″Z″ shaped, aluminum		60	.133		4.10	5.35		9.45	13.30
2900	Solid neoprene strip, nail on aluminum strip		90	.089		3.40	3.55		6.95	9.60
3000	One piece stile protection									
3020	Neoprene fabric loop, nail on aluminum strips	1 Carp	60	.133	L.F.	.65	5.35		6	9.50
3110	Flush mounted aluminum extrusion, 1/2″ x 1-1/4″		60	.133		3.26	5.35		8.61	12.40
3140	3/4″ x 1-3/8″		60	.133		4.15	5.35		9.50	13.35
3160	1-1/8″ x 1-3/4″		60	.133		6.60	5.35		11.95	16.05
3300	Mortise, 9/16″ x 3/4″		60	.133		3.50	5.35		8.85	12.65

08 71 Door Hardware

08 71 21 – Astragals

08 71 21.10 Exterior Mouldings, Astragals		Crew	Daily Output	Labor-Hours	Unit	2009 Bare Costs Material	Labor	Equipment	Total	Total Incl O&P
3320	13/16" x 1-3/8"	1 Carp	60	.133	L.F.	3.80	5.35		9.15	13
3600	Spring bronze strip, nail on type		105	.076		3	3.04		6.04	8.30
3620	Screw on, with retainer		75	.107		2.45	4.26		6.71	9.75
3800	Flexible stainless steel housing, pile insert, 1/2" door		105	.076		6.75	3.04		9.79	12.45
3820	3/4" door		105	.076		7.60	3.04		10.64	13.35
4000	Extruded aluminum retainer, flush mount, pile insert		105	.076		2.36	3.04		5.40	7.60
4080	Mortise, felt insert		90	.089		4.20	3.55		7.75	10.45
4160	Mortise with spring, pile insert		90	.089		3.30	3.55		6.85	9.50
4400	Rigid vinyl retainer, mortise, pile insert		105	.076		2.35	3.04		5.39	7.60
4600	Wool pile filler strip, aluminum backing		105	.076		2.36	3.04		5.40	7.60
5000	Two piece overlapping astragal, extruded aluminum retainer									
5010	Pile insert	1 Carp	60	.133	L.F.	3.06	5.35		8.41	12.15
5020	Vinyl bulb insert		60	.133		1.98	5.35		7.33	11
5040	Vinyl flap insert		60	.133		6.15	5.35		11.50	15.60
5060	Solid neoprene flap insert		60	.133		6.10	5.35		11.45	15.50
5080	Hypalon rubber flap insert		60	.133		6.25	5.35		11.60	15.65
5090	Snap on cover, pile insert		60	.133		7.10	5.35		12.45	16.60
5400	Magnetic aluminum, surface mounted		60	.133		24	5.35		29.35	35.50
5500	Interlocking aluminum, 5/8" x 1" neoprene bulb insert		45	.178		3.80	7.10		10.90	15.90
5600	Adjustable aluminum, 9/16" x 21/32", pile insert		45	.178		18.15	7.10		25.25	31.50
5790	For vinyl bulb, deduct					.50			.50	.55
5800	Magnetic, adjustable, 9/16" x 21/32"	1 Carp	45	.178		23.50	7.10		30.60	37
6000	Two piece stile protection									
6010	Cloth backed rubber loop, 1" gap, nail on aluminum strips	1 Carp	45	.178	L.F.	3.90	7.10		11	16
6040	Screw on aluminum strips		45	.178		6.10	7.10		13.20	18.40
6100	1-1/2" gap, screw on aluminum extrusion		45	.178		5.45	7.10		12.55	17.70
6240	Vinyl fabric loop, slotted aluminum extrusion, 1" gap		45	.178		1.90	7.10		9	13.80
6300	1-1/4" gap		45	.178		5.75	7.10		12.85	18.05

08 71 25 – Weatherstripping

08 71 25.10 Mechanical Seals, Weatherstripping		Crew	Daily Output	Labor-Hours	Unit	Material	Labor	Equipment	Total	Total Incl O&P
0010	**MECHANICAL SEALS, WEATHERSTRIPPING**									
1000	Doors, wood frame, interlocking, for 3' x 7' door, zinc	1 Carp	3	2.667	Opng.	15.50	107		122.50	193
1100	Bronze		3	2.667		24	107		131	203
1300	6' x 7' opening, zinc		2	4		16.90	160		176.90	283
1400	Bronze		2	4		32	160		192	299
1700	Wood frame, spring type, bronze									
1800	3' x 7' door	1 Carp	7.60	1.053	Opng.	21.50	42		63.50	93
1900	6' x 7' door	"	7	1.143	"	26.50	45.50		72	105
2200	Metal frame, spring type, bronze									
2300	3' x 7' door	1 Carp	3	2.667	Opng.	35.50	107		142.50	215
2400	6' x 7' door	"	2.50	3.200	"	46	128		174	262
2500	For stainless steel, spring type, add					133%				
2700	Metal frame, extruded sections, 3' x 7' door, aluminum	1 Carp	2	4	Opng.	45	160		205	315
2800	Bronze		2	4		114	160		274	390
3100	6' x 7' door, aluminum		1.20	6.667		57.50	266		323.50	505
3200	Bronze		1.20	6.667		134	266		400	585
3500	Threshold weatherstripping									
3650	Door sweep, flush mounted, aluminum	1 Carp	25	.320	Ea.	13.30	12.80		26.10	35.50
3700	Vinyl		25	.320		15.70	12.80		28.50	38.50
5000	Garage door bottom weatherstrip, 12' aluminum, clear		14	.571		21.50	23		44.50	61
5010	Bronze		14	.571		81.50	23		104.50	127
5050	Bottom protection, Rubber		14	.571		22	23		45	62

08 71 Door Hardware

08 71 25 – Weatherstripping

08 71 25.10 Mechanical Seals, Weatherstripping		Crew	Daily Output	Labor-Hours	Unit	Material	Labor	Equipment	Total	Total Incl O&P
5100	Threshold	1 Carp	14	.571	Ea.	88.50	23		111.50	135
9000	Minimum labor/equipment charge	↓	3	2.667	Job		107		107	176

08 74 Access Control Hardware

08 74 13 – Card Key Access Control Hardware

08 74 13.50 Card Key Access		Crew	Daily Output	Labor-Hours	Unit	Material	Labor	Equipment	Total	Total Incl O&P
0010	CARD KEY ACCESS									
0020	Card type, 1 time zone, minimum				Ea.	615			615	675
0040	Maximum					1,200			1,200	1,300
0060	3 time zones, minimum					1,400			1,400	1,550
0080	Maximum				↓	1,900			1,900	2,100
0100	System with printer, and control console, 3 zones				Total	9,150			9,150	10,100
0120	6 zones				"	12,000			12,000	13,200
0140	For each door, minimum, add				Ea.	1,400			1,400	1,550
0160	Maximum, add					2,000			2,000	2,200
0320	Card reader, int/ext, basic, excl striker/power/wiring	1 Elec	3	2.667		255	125		380	475
0330	Tag reader, int/ext, basic, excl striker/power/wiring	"	3	2.667	↓	176	125		301	385

08 74 16 – Keypad Access Control Hardware

08 74 16.50 Keypad Access		Crew	Daily Output	Labor-Hours	Unit	Material	Labor	Equipment	Total	Total Incl O&P
0010	KEYPAD ACCESS									
0340	Digital keypad, int/ext, basic, excl. striker/power/wiring	1 Elec	3	2.667	Ea.	63	125		188	263
0350	Lockset, mechanical push-button type, complete, incl hardware	1 Carp	4	2	"	229	80		309	385

08 74 19 – Biometric Identity Access Control Hardware

08 74 19.50 Biometric Identity Access		Crew	Daily Output	Labor-Hours	Unit	Material	Labor	Equipment	Total	Total Incl O&P
0010	BIOMETRIC IDENTITY ACCESS									
0200	Fingerprint scanner unit, excl striker/power supply	1 Elec	4	2	Ea.	1,225	94		1,319	1,500
0210	Fingerprint scanner unit, for computer keyboard access		8	1		1,000	47		1,047	1,175
0220	Hand geometry scanner, mem of 512 users, excl striker/powr		3	2.667		1,900	125		2,025	2,300
0230	Memory upgrade for, adds 9,700 user profiles		8	1		249	47		296	345
0240	Adds 32,500 user profiles		8	1		550	47		597	680
0250	Prison type, memory of 256 users, excl striker, power		3	2.667		2,450	125		2,575	2,900
0260	Memory upgrade for, adds 3,300 user profiles		8	1		199	47		246	292
0270	Adds 9,700 user profiles		8	1		400	47		447	515
0280	Adds 27,900 user profiles		8	1		550	47		597	680
0290	All weather, mem of 512 users, excl striker/pwr		3	2.667		3,675	125		3,800	4,225
0300	Facial & fingerprint scanner, combination unit, excl striker/power		3	2.667		4,200	125		4,325	4,825
0310	Access for, for initial setup, excl striker/power	↓	3	2.667	↓	1,000	125		1,125	1,300

08 74 23 – Access Control Accessories

08 74 23.50 Security Access Control Accessories		Crew	Daily Output	Labor-Hours	Unit	Material	Labor	Equipment	Total	Total Incl O&P
0010	SECURITY ACCESS CONTROL ACCESSORIES									
0360	Scanner/reader access, power supply/transf, 110V to 12/24V	1 Elec	4	2	Ea.	244	94		338	415
0370	Elec/mag strikers, 12/24V		4	2		69	94		163	221
0380	1 hour battery backup power supply	↓	4	2		29	94		123	177
0390	Deadbolt, digital, batt-operated, indoor/outdoor, complete, incl hardware	1 Carp	4	2	↓	139	80		219	285

08 75 Window Hardware

08 75 30 – Weatherstripping

08 75 30.10 Mechanical Weather Seals		Crew	Daily Output	Labor-Hours	Unit	Material	Labor	Equipment	Total	Total Incl O&P
						2009 Bare Costs				
0010	**MECHANICAL WEATHER SEALS**, Window, double hung, 3′ X 5′									
0020	Zinc	1 Carp	7.20	1.111	Opng.	12.30	44.50		56.80	87
0100	Bronze		7.20	1.111		25.50	44.50		70	102
0500	As above but heavy duty, zinc		4.60	1.739		17.30	69.50		86.80	134
0600	Bronze		4.60	1.739		29	69.50		98.50	147
9000	Minimum labor/equipment charge	1 Clab	4.60	1.739	Job		55		55	90.50

08 79 Hardware Accessories

08 79 13 – Key Storage Equipment

08 79 13.10 Key Cabinets		Crew	Daily Output	Labor-Hours	Unit	Material	Labor	Equipment	Total	Total Incl O&P
0010	**KEY CABINETS**									
0020	Wall mounted, 60 key capacity	1 Carp	20	.400	Ea.	81.50	16		97.50	116
0200	Drawer type, 600 key capacity	1 Clab	15	.533		725	16.85		741.85	830
0300	2,400 key capacity		20	.400		5,500	12.65		5,512.65	6,075
0400	Tray type, 20 key capacity		50	.160		45	5.05		50.05	58
0500	50 key capacity		40	.200		82.50	6.30		88.80	101

08 79 20 – Door Accessories

08 79 20.10 Door Hardware Accessories		Crew	Daily Output	Labor-Hours	Unit	Material	Labor	Equipment	Total	Total Incl O&P
0010	**DOOR HARDWARE ACCESSORIES**									
0140	Door bolt, surface, 4″	1 Carp	32	.250	Ea.	8.65	10		18.65	26
0160	Door latch	″	12	.667	″	8.05	26.50		34.55	53
0200	Sliding closet door									
0220	Track and hanger, single	1 Carp	10	.800	Ea.	49.50	32		81.50	108
0240	Double		8	1		72.50	40		112.50	146
0260	Door guide, single		48	.167		23.50	6.65		30.15	37
0280	Double		48	.167		32	6.65		38.65	46.50
0600	Deadbolt and lock cover plate, brass or stainless steel		30	.267		26.50	10.65		37.15	47
0620	Hole cover plate, brass or chrome		35	.229		6.95	9.15		16.10	22.50
2240	Mortise lockset, passage, lever handle		9	.889		194	35.50		229.50	272
4000	Security chain, standard		18	.444		7.40	17.75		25.15	37.50

08 81 Glass Glazing

08 81 10 – Float Glass

08 81 10.10 Various Types and Thickness of Float Glass		Crew	Daily Output	Labor-Hours	Unit	Material	Labor	Equipment	Total	Total Incl O&P
0010	**VARIOUS TYPES AND THICKNESS OF FLOAT GLASS**									
0020	3/16″ Plain	2 Glaz	130	.123	S.F.	4.48	4.75		9.23	12.60
0200	Tempered, clear		130	.123		6.05	4.75		10.80	14.30
0300	Tinted		130	.123		7.05	4.75		11.80	15.40
0600	1/4″ thick, clear, plain		120	.133		5.25	5.15		10.40	14.10
0700	Tinted		120	.133		7.40	5.15		12.55	16.45
0800	Tempered, clear		120	.133		7.45	5.15		12.60	16.50
0900	Tinted		120	.133		9.75	5.15		14.90	19
1600	3/8″ thick, clear, plain		75	.213		9.10	8.25		17.35	23.50
1700	Tinted		75	.213		14.80	8.25		23.05	29.50
1800	Tempered, clear		75	.213		15.10	8.25		23.35	30
1900	Tinted		75	.213		16.85	8.25		25.10	32
2200	1/2″ thick, clear, plain		55	.291		18.15	11.25		29.40	38
2300	Tinted		55	.291		26	11.25		37.25	46.50
2400	Tempered, clear		55	.291		22	11.25		33.25	42.50

08 81 Glass Glazing

08 81 10 – Float Glass

08 81 10.10 Various Types and Thickness of Float Glass		Crew	Daily Output	Labor-Hours	Unit	2009 Bare Costs Material	Labor	Equipment	Total	Total Incl O&P
2500	Tinted	2 Glaz	55	.291	S.F.	26	11.25		37.25	46.50
2800	5/8" thick, clear, plain		45	.356		26	13.70		39.70	50.50
2900	Tempered, clear		45	.356		29.50	13.70		43.20	54.50
3200	3/4" thick, clear, plain		35	.457		33.50	17.65		51.15	65
3300	Tempered, clear		35	.457		39	17.65		56.65	71.50
3600	1" thick, clear, plain		30	.533		55.50	20.50		76	94
8900	For low emissivity coating for 3/16" & 1/4" only, add to above					16%				
9000	Minimum labor/equipment charge	1 Glaz	2	4	Job		154		154	249

08 81 13 – Decorative Glass Glazing

08 81 13.10 Beveled Glass

		Crew	Daily Output	Labor-Hours	Unit	Material	Labor	Equipment	Total	Total Incl O&P
0010	**BEVELED GLASS**, with design patterns									
0020	Minimum	2 Glaz	150	.107	S.F.	55	4.12		59.12	67
0050	Average		125	.128		129	4.94		133.94	150
0100	Maximum		100	.160		229	6.20		235.20	261

08 81 13.20 Faceted Glass

		Crew	Daily Output	Labor-Hours	Unit	Material	Labor	Equipment	Total	Total Incl O&P
0010	**FACETED GLASS**, Color tinted 3/4" thick									
0020	Minimum	2 Glaz	95	.168	S.F.	46.50	6.50		53	61.50
0100	Maximum	"	75	.213	"	81.50	8.25		89.75	103

08 81 13.30 Sandblasted Glass

		Crew	Daily Output	Labor-Hours	Unit	Material	Labor	Equipment	Total	Total Incl O&P
0010	**SANDBLASTED GLASS**, float glass									
0020	1/8" thick	2 Glaz	160	.100	S.F.	10.25	3.86		14.11	17.50
0100	3/16" thick		130	.123		11.40	4.75		16.15	20
0500	1/4" thick		120	.133		11.80	5.15		16.95	21.50
0600	3/8" thick		75	.213		12.85	8.25		21.10	27.50

08 81 20 – Vision Panels

08 81 20.10 Full Vision

		Crew	Daily Output	Labor-Hours	Unit	Material	Labor	Equipment	Total	Total Incl O&P
0010	**FULL VISION**, window system with 3/4" glass mullions									
0020	Up to 10' high	H-2	130	.185	S.F.	57.50	6.70		64.20	74
0100	10' to 20' high, minimum		110	.218		61	7.90		68.90	80
0150	Average		100	.240		65.50	8.70		74.20	86.50
0200	Maximum		80	.300		74	10.90		84.90	98.50
9000	Minimum labor/equipment charge	1 Glaz	2	4	Job		154		154	249

08 81 25 – Glazing Variables

08 81 25.10 Applications of Glazing

		Crew	Daily Output	Labor-Hours	Unit	Material	Labor	Equipment	Total	Total Incl O&P
0010	**APPLICATIONS OF GLAZING**									
0500	For high rise glazing, exterior, add per S.F. per story				S.F.					.28
0600	For glass replacement, add				"		100%			
0700	For gasket settings, add				L.F.	5.55			5.55	6.10
0900	For sloped glazing, add				S.F.		25%			
2000	Fabrication, polished edges, 1/4" thick				Inch	.48			.48	.53
2100	1/2" thick					1.21			1.21	1.33
2500	Mitered edges, 1/4" thick					1.21			1.21	1.33
2600	1/2" thick					1.96			1.96	2.16

08 81 30 – Insulating Glass

08 81 30.10 Reduce Heat Transfer Glass

		Crew	Daily Output	Labor-Hours	Unit	Material	Labor	Equipment	Total	Total Incl O&P
0010	**REDUCE HEAT TRANSFER GLASS**, 2 lites 1/8" float, 1/2" thk under 15 S.F.									
0020	Clear G	2 Glaz	95	.168	S.F.	8.95	6.50		15.45	20.50
0100	Tinted G		95	.168		12.55	6.50		19.05	24.50
0200	2 lites 3/16" float, for 5/8" thk unit, 15 to 30 S.F., clear G		90	.178		12.95	6.85		19.80	25.50
0300	Tinted G		90	.178		13.30	6.85		20.15	25.50

08 81 Glass Glazing

08 81 30 – Insulating Glass

08 81 30.10 Reduce Heat Transfer Glass		Crew	Daily Output	Labor-Hours	Unit	2009 Bare Costs Material	Labor	Equipment	Total	Total Incl O&P
0400	1" thk, dbl. glazed, 1/4" float, 30-70 S.F., clear [G]	2 Glaz	75	.213	S.F.	15.20	8.25		23.45	30
0500	Tinted [G]		75	.213		22	8.25		30.25	38
0600	1" thick double glazed, 1/4" float, 1/4" wire		75	.213		21	8.25		29.25	36.50
0700	1/4" float, 1/4" tempered		75	.213		20.50	8.25		28.75	36.50
0800	1/4" wire, 1/4" tempered		75	.213		29	8.25		37.25	45.50
0900	Both lites, 1/4" wire		75	.213		35	8.25		43.25	52
2000	Both lites, light & heat reflective [G]		85	.188		29.50	7.25		36.75	44
2500	Heat reflective, film inside, 1" thick unit, clear [G]		85	.188		26	7.25		33.25	40
2600	Tinted [G]		85	.188		28	7.25		35.25	42.50
3000	Film on weatherside, clear, 1/2" thick unit [G]		95	.168		18.50	6.50		25	31
3100	5/8" thick unit [G]		90	.178		18.10	6.85		24.95	31
3200	1" thick unit [G]		85	.188		25.50	7.25		32.75	39.50
9000	Minimum labor/equipment charge	1 Glaz	2	4	Job		154		154	249

08 81 35 – Translucent Glass

08 81 35.10 Obscure Glass		Crew	Daily Output	Labor-Hours	Unit	Material	Labor	Equipment	Total	Total Incl O&P
0010	**OBSCURE GLASS**, 1/8" thick									
0020	Minimum	2 Glaz	140	.114	S.F.	10.45	4.41		14.86	18.60
0100	Maximum		125	.128		12.55	4.94		17.49	22
0300	7/32" thick, minimum		120	.133		11.60	5.15		16.75	21
0400	Maximum		105	.152		14.80	5.90		20.70	26

08 81 35.20 Patterned Glass		Crew	Daily Output	Labor-Hours	Unit	Material	Labor	Equipment	Total	Total Incl O&P
0010	**PATTERNED GLASS**, colored, 1/8" thick									
0020	Minimum	2 Glaz	140	.114	S.F.	8.60	4.41		13.01	16.55
0100	Maximum		125	.128		14.40	4.94		19.34	24
0300	7/32" thick, minimum		120	.133		10.70	5.15		15.85	20
0400	Maximum		105	.152		15.40	5.90		21.30	26.50

08 81 45 – Sheet Glass

08 81 45.10 Window Glass, Sheet		Crew	Daily Output	Labor-Hours	Unit	Material	Labor	Equipment	Total	Total Incl O&P
0010	**WINDOW GLASS, SHEET** gray									
0020	1/8" thick	2 Glaz	160	.100	S.F.	5.10	3.86		8.96	11.80
0200	1/4" thick	"	130	.123	"	6.35	4.75		11.10	14.65

08 81 50 – Spandrel Glass

08 81 50.10 Glass for Non Vision Areas		Crew	Daily Output	Labor-Hours	Unit	Material	Labor	Equipment	Total	Total Incl O&P
0010	**GLASS FOR NON VISION AREAS**, 1/4" thick standard colors									
0020	Up to 100 S.F.	2 Glaz	110	.145	S.F.	15	5.60		20.60	25.50
0200	1,000 to 2,000 S.F.	"	120	.133	"	13.95	5.15		19.10	23.50
0300	For custom colors, add				Total	10%				
0500	For 3/8" thick, add				S.F.	10.65			10.65	11.70
1000	For double coated, 1/4" thick, add					3.97			3.97	4.37
1200	For insulation on panels, add					6.55			6.55	7.20
2000	Panels, insulated, with aluminum backed fiberglass, 1" thick	2 Glaz	120	.133		15.85	5.15		21	25.50
2100	2" thick	"	120	.133		18.45	5.15		23.60	29
2500	With galvanized steel backing, add					5.50			5.50	6.05

08 81 55 – Window Glass

08 81 55.10 Sheet Glass		Crew	Daily Output	Labor-Hours	Unit	Material	Labor	Equipment	Total	Total Incl O&P
0010	**SHEET GLASS** (window), clear float, stops, putty bed									
0015	1/8" thick, clear float	2 Glaz	480	.033	S.F.	4.27	1.29		5.56	6.75
0500	3/16" thick, clear		480	.033		5.25	1.29		6.54	7.85
0600	Tinted		480	.033		7.15	1.29		8.44	9.90
0700	Tempered		480	.033		8.65	1.29		9.94	11.55

08 81 Glass Glazing

08 81 55 – Window Glass

08 81 55.10 Sheet Glass		Crew	Daily Output	Labor-Hours	Unit	Material	Labor	Equipment	Total	Total Incl O&P
						2009 Bare Costs				
9000	Minimum labor/equipment charge	2 Glaz	5	3.200	Job		124		124	199

08 81 65 – Wire Glass

08 81 65.10 Glass Reinforced With Wire		Crew	Daily Output	Labor-Hours	Unit	Material	Labor	Equipment	Total	Total Incl O&P
0010	GLASS REINFORCED WITH WIRE									
0012	1/4" thick rough obscure	2 Glaz	135	.119	S.F.	13.75	4.57		18.32	22.50
1000	Polished wire, 1/4" thick, diamond, clear	↓	135	.119	↓	18.75	4.57		23.32	28
1500	Pinstripe, obscure	↓	135	.119	↓	21	4.57		25.57	30.50

08 83 Mirrors

08 83 13 – Mirrored Glass Glazing

08 83 13.10 Mirrors		Crew	Daily Output	Labor-Hours	Unit	Material	Labor	Equipment	Total	Total Incl O&P
0010	MIRRORS, No frames, wall type, 1/4" plate glass, polished edge									
0100	Up to 5 S.F.	2 Glaz	125	.128	S.F.	9.10	4.94		14.04	17.95
0200	Over 5 S.F.		160	.100		8.80	3.86		12.66	15.90
0500	Door type, 1/4" plate glass, up to 12 S.F.		160	.100		7.85	3.86		11.71	14.85
1000	Float glass, up to 10 S.F., 1/8" thick		160	.100		5.05	3.86		8.91	11.80
1100	3/16" thick		150	.107		6.15	4.12		10.27	13.45
1500	12" x 12" wall tiles, square edge, clear		195	.082		1.78	3.17		4.95	7.05
1600	Veined		195	.082		4.69	3.17		7.86	10.25
2000	1/4" thick, stock sizes, one way transparent		125	.128		17.70	4.94		22.64	27.50
2010	Bathroom, unframed, laminated		160	.100		13.20	3.86		17.06	20.50
2500	Tempered	↓	160	.100	↓	16.65	3.86		20.51	24.50

08 83 13.15 Reflective Glass		Crew	Daily Output	Labor-Hours	Unit	Material	Labor	Equipment	Total	Total Incl O&P
0010	REFLECTIVE GLASS									
0100	1/4" float with fused metallic oxide fixed G	2 Glaz	115	.139	S.F.	15.40	5.35		20.75	25.50
0500	1/4" float glass with reflective applied coating G	"	115	.139	"	12.85	5.35		18.20	23

08 84 Plastic Glazing

08 84 10 – Plexiglass Glazing

08 84 10.10 Plexiglass Acrylic		Crew	Daily Output	Labor-Hours	Unit	Material	Labor	Equipment	Total	Total Incl O&P
0010	PLEXIGLASS ACRYLIC, clear, masked,									
0020	1/8" thick, cut sheets	2 Glaz	170	.094	S.F.	4.69	3.63		8.32	11
0200	Full sheets		195	.082		2.45	3.17		5.62	7.80
0500	1/4" thick, cut sheets		165	.097		8.30	3.74		12.04	15.15
0600	Full sheets		185	.086		4.50	3.34		7.84	10.35
0900	3/8" thick, cut sheets		155	.103		15.15	3.98		19.13	23
1000	Full sheets		180	.089		8.15	3.43		11.58	14.55
1300	1/2" thick, cut sheets		135	.119		17.45	4.57		22.02	26.50
1400	Full sheets		150	.107		16.90	4.12		21.02	25.50
1700	3/4" thick, cut sheets		115	.139		62	5.35		67.35	76.50
1800	Full sheets		130	.123		36	4.75		40.75	47
2100	1" thick, cut sheets		105	.152		70	5.90		75.90	86.50
2200	Full sheets		125	.128		43	4.94		47.94	55.50
3000	Colored, 1/8" thick, cut sheets		170	.094		14.40	3.63		18.03	21.50
3200	Full sheets		195	.082		9.30	3.17		12.47	15.35
3500	1/4" thick, cut sheets		165	.097		16.10	3.74		19.84	24
3600	Full sheets		185	.086		11	3.34		14.34	17.50
4000	Mirrors, untinted, cut sheets, 1/8" thick	↓	185	.086	↓	6.75	3.34		10.09	12.85

08 84 Plastic Glazing

08 84 10 – Plexiglass Glazing

08 84 10.10 Plexiglass Acrylic		Crew	Daily Output	Labor-Hours	Unit	2009 Bare Costs Material	Labor	Equipment	Total	Total Incl O&P
4200	1/4" thick	2 Glaz	180	.089	S.F.	10.90	3.43		14.33	17.55

08 84 20 – Polycarbonate

08 84 20.10 Thermoplastic		Crew	Daily Output	Labor-Hours	Unit	Material	Labor	Equipment	Total	Total Incl O&P
0010	**THERMOPLASTIC**, clear, masked, cut sheets									
0020	1/8" thick	2 Glaz	170	.094	S.F.	9.25	3.63		12.88	16
0500	3/16" thick		165	.097		10.55	3.74		14.29	17.65
1000	1/4" thick		155	.103		11.90	3.98		15.88	19.45
1500	3/8" thick	↓	150	.107	↓	21	4.12		25.12	29.50
9000	Minimum labor/equipment charge	1 Glaz	2	4	Job		154		154	249

08 84 30 – Vinyl Glass

08 84 30.10 Vinyl Sheets		Crew	Daily Output	Labor-Hours	Unit	Material	Labor	Equipment	Total	Total Incl O&P
0010	**VINYL SHEETS**, Steel mesh reinforced, stock sizes									
0020	.090" thick	2 Glaz	170	.094	S.F.	11.90	3.63		15.53	18.90
0500	.120" thick		170	.094		14.50	3.63		18.13	22
1000	.250" thick	↓	155	.103		19.50	3.98		23.48	28
1500	For non-standard sizes, add				↓	18%				

08 87 Glazing Surface Films

08 87 13 – Solar Control Films

08 87 13.10 Solar Films On Glass		Crew	Daily Output	Labor-Hours	Unit	Material	Labor	Equipment	Total	Total Incl O&P
0010	**SOLAR FILMS ON GLASS** (glass not included)									
2000	Minimum G	2 Glaz	180	.089	S.F.	6.20	3.43		9.63	12.40
2050	Maximum G	"	225	.071	"	14.40	2.74		17.14	20

08 87 16 – Safety Films

08 87 16.10 Window Protection Film		Crew	Daily Output	Labor-Hours	Unit	Material	Labor	Equipment	Total	Total Incl O&P
0010	**WINDOW PROTECTION FILM**									
2010	Window protection film, controls blast damage G	2 Glaz	80	.200	S.F.	6.20	7.70		13.90	19.30

08 87 53 – Security Films

08 87 53.10 Security Film		Crew	Daily Output	Labor-Hours	Unit	Material	Labor	Equipment	Total	Total Incl O&P
0010	**SECURITY FILM**, clear, 32000psi tensile strength, adhered to glass									
0100	.002" thick, daylight installation	H-2	950	.025	S.F.	1.45	.92		2.37	3.09
0150	.004" thick, daylight installation		800	.030		1.60	1.09		2.69	3.53
0200	.006" thick, daylight installation		700	.034		1.70	1.24		2.94	3.89
0210	Install for anchorage		600	.040		1.89	1.45		3.34	4.43
0400	.007" thick, daylight installation		600	.040		1.80	1.45		3.25	4.33
0410	Install for anchorage		500	.048		2	1.74		3.74	5
0500	.008" thick, daylight installation		500	.048		2.10	1.74		3.84	5.15
0510	Install for anchorage		500	.048		2.33	1.74		4.07	5.40
0600	.015" thick, daylight installation		400	.060		3.50	2.18		5.68	7.40
0610	Install for anchorage	↓	400	.060	↓	2.33	2.18		4.51	6.10
0900	Security film anchorage, mechanical attachment and cover plate	H-3	370	.043	L.F.	8.70	1.49		10.19	12
0950	Security film anchorage, wet glaze structural caulking	1 Glaz	225	.036	"	.83	1.37		2.20	3.12
1000	Adhered security film removal	1 Clab	275	.029	S.F.		.92		.92	1.52

08 88 Special Function Glazing

08 88 40 – Acoustical Glass Units

08 88 40.10 Sound Reduction Units		Crew	Daily Output	Labor-Hours	Unit	Material	Labor	Equipment	Total	Total Incl O&P
						2009 Bare Costs				
0010	**SOUND REDUCTION UNITS**, 1lite at 3/8", 1 lite at 3/16"									
0020	For 1" thick	2 Glaz	100	.160	S.F.	32	6.20		38.20	45.50
0100	For 4" thick	"	80	.200	"	55	7.70		62.70	73

08 88 56 – Ballistics-Resistant Glazing

08 88 56.10 Laminated Glass		Crew	Daily Output	Labor-Hours	Unit	Material	Labor	Equipment	Total	Total Incl O&P
0010	**LAMINATED GLASS**									
0020	Clear float .03" vinyl 1/4"	2 Glaz	90	.178	S.F.	11.10	6.85		17.95	23.50
0100	3/8" thick		78	.205		20	7.90		27.90	35.50
0200	.06" vinyl, 1/2" thick		65	.246		23.50	9.50		33	41.50
1000	5/8" thick		90	.178		28	6.85		34.85	41.50
2000	Bullet-resisting, 1-3/16" thick, to 15 S.F.		16	1		69.50	38.50		108	139
2100	Over 15 S.F.		16	1		88	38.50		126.50	159
2500	2-1/4" thick, to 15 S.F.		12	1.333		91	51.50		142.50	183
2600	Over 15 S.F.		12	1.333		81	51.50		132.50	172
2700	Level 2 (.357 magnum)		12	1.333		60	51.50		111.50	149
2750	Level 3 (.44 magnum)		12	1.333		65	51.50		116.50	155
2800	Level 4 (AK-47)		12	1.333		85	51.50		136.50	177
2850	Level 5 (M-16)		12	1.333		95	51.50		146.50	188
2900	Level (7.62 Armor Piercing)		12	1.333		112	51.50		163.50	206

08 91 Louvers

08 91 19 – Fixed Louvers

08 91 19.10 Aluminum Louvers		Crew	Daily Output	Labor-Hours	Unit	Material	Labor	Equipment	Total	Total Incl O&P
0010	**ALUMINUM LOUVERS**									
0020	Aluminum with screen, residential, 8" x 8"	1 Carp	38	.211	Ea.	10.50	8.40		18.90	25.50
0100	12" x 12"		38	.211		11.85	8.40		20.25	27
0200	12" x 18"		35	.229		13.50	9.15		22.65	30
0250	14" x 24"		30	.267		17	10.65		27.65	36.50
0300	18" x 24"		27	.296		24.50	11.85		36.35	46.50
0500	24" x 30"		24	.333		34	13.30		47.30	59.50
0700	Triangle, adjustable, small		20	.400		29	16		45	58.50
0800	Large		15	.533		50	21.50		71.50	90.50
1200	Extruded aluminum, see Div. 23 37 15.40									
2100	Midget, aluminum, 3/4" deep, 1" diameter	1 Carp	85	.094	Ea.	.77	3.76		4.53	7.05
2150	3" diameter		60	.133		1.62	5.35		6.97	10.60
2200	4" diameter		50	.160		3.07	6.40		9.47	13.95
2250	6" diameter		30	.267		3.63	10.65		14.28	21.50

08 95 Vents

08 95 13 – Soffit Vents

08 95 13.10 Wall Louvers		Crew	Daily Output	Labor-Hours	Unit	Material	Labor	Equipment	Total	Total Incl O&P
0010	**WALL LOUVERS**									
2330	Soffit vent, continuous, 3" wide, aluminum, mill finish	1 Carp	200	.040	L.F.	.53	1.60		2.13	3.22
2340	Baked enamel finish		200	.040	"	.48	1.60		2.08	3.17
2400	Under eaves vent, aluminum, mill finish, 16" x 4"		48	.167	Ea.	1.29	6.65		7.94	12.40
2500	16" x 8"		48	.167	"	1.49	6.65		8.14	12.65

08 95 Vents

08 95 16 – Wall Vents

08 95 16.10 Louvers		Crew	Daily Output	Labor-Hours	Unit	2009 Bare Costs Material	Labor	Equipment	Total	Total Incl O&P
0010	**LOUVERS**									
0020	Redwood, 2'-0" diameter, full circle	1 Carp	16	.500	Ea.	159	20		179	207
0100	Half circle		16	.500		152	20		172	200
0200	Octagonal		16	.500		121	20		141	166
0300	Triangular, 5/12 pitch, 5'-0" at base		16	.500		257	20		277	315
7000	Vinyl gable vent, 8" x 8"		38	.211		11.20	8.40		19.60	26.50
7020	12" x 12"		38	.211		23.50	8.40		31.90	40
7080	12" x 18"		35	.229		30	9.15		39.15	48
7200	18" x 24"		30	.267		36.50	10.65		47.15	58
9000	Minimum labor/equipment charge		3.50	2.286	Job		91.50		91.50	151

Division Notes

		CREW	DAILY OUTPUT	LABOR-HOURS	UNIT	2009 BARE COSTS MAT.	LABOR	EQUIP.	TOTAL	TOTAL INCL O&P

Estimating Tips

General

- Room Finish Schedule: A complete set of plans should contain a room finish schedule. If one is not available, it would be well worth the time and effort to obtain one.

09 20 00 Plaster and Gypsum Board

- Lath is estimated by the square yard plus a 5% allowance for waste. Furring, channels, and accessories are measured by the linear foot. An extra foot should be allowed for each accessory miter or stop.
- Plaster is also estimated by the square yard. Deductions for openings vary by preference, from zero deduction to 50% of all openings over 2 feet in width. The estimator should allow one extra square foot for each linear foot of horizontal interior or exterior angle located below the ceiling level. Also, double the areas of small radius work.
- Drywall accessories, studs, track, and acoustical caulking are all measured by the linear foot. Drywall taping is figured by the square foot. Gypsum wallboard is estimated by the square foot. No material deductions should be made for door or window openings under 32 S.F.

09 60 00 Flooring

- Tile and terrazzo areas are taken off on a square foot basis. Trim and base materials are measured by the linear foot. Accent tiles are listed per each. Two basic methods of installation are used. Mud set is approximately 30% more expensive than thin set. In terrazzo work, be sure to include the linear footage of embedded decorative strips, grounds, machine rubbing, and power cleanup.
- Wood flooring is available in strip, parquet, or block configuration. The latter two types are set in adhesives with quantities estimated by the square foot. The laying pattern will influence labor costs and material waste. In addition to the material and labor for laying wood floors, the estimator must make allowances for sanding and finishing these areas unless the flooring is prefinished.
- Sheet flooring is measured by the square yard. Roll widths vary, so consideration should be given to use the most economical width, as waste must be figured into the total quantity. Consider also the installation methods available, direct glue down or stretched.

09 70 00 Wall Finishes

- Wall coverings are estimated by the square foot. The area to be covered is measured, length by height of wall above baseboards, to calculate the square footage of each wall. This figure is divided by the number of square feet in the single roll which is being used. Deduct, in full, the areas of openings such as doors and windows. Where a pattern match is required allow 25%–30% waste.

09 80 00 Acoustic Treatment

- Acoustical systems fall into several categories. The takeoff of these materials should be by the square foot of area with a 5% allowance for waste. Do not forget about scaffolding, if applicable, when estimating these systems.

09 90 00 Painting and Coating

- A major portion of the work in painting involves surface preparation. Be sure to include cleaning, sanding, filling, and masking costs in the estimate.
- Protection of adjacent surfaces is not included in painting costs. When considering the method of paint application, an important factor is the amount of protection and masking required. These must be estimated separately and may be the determining factor in choosing the method of application.

Reference Numbers

Reference numbers are shown in shaded boxes at the beginning of some major classifications. These numbers refer to related items in the Reference Section. The reference information may be an estimating procedure, an alternate pricing method, or technical information.

Note: Not all subdivisions listed here necessarily appear in this publication.

09 01 Maintenance of Finishes

09 01 60 – Maintenance of Flooring

09 01 60.10 Carpet Maintenance		Crew	Daily Output	Labor-Hours	Unit	2009 Bare Costs Material	Labor	Equipment	Total	Total Incl O&P
0011	**CARPET MAINTENANCE** See Div. 01 93 13.09									

09 01 70 – Maintenance of Wall Finishes

09 01 70.10 Gypsum Wallboard Repairs

		Crew	Daily Output	Labor-Hours	Unit	Material	Labor	Equipment	Total	Total Incl O&P
0010	**GYPSUM WALLBOARD REPAIRS**									
0100	Fill and sand, pin / nail holes	1 Carp	960	.008	Ea.		.33		.33	.55
0110	Screw head pops		480	.017			.67		.67	1.10
0120	Dents, up to 2″ square		48	.167		.01	6.65		6.66	11
0130	2″ to 4″ square		24	.333		.03	13.30		13.33	22
0140	Cut square, patch, sand and finish, holes, up to 2″ square		12	.667		.04	26.50		26.54	44
0150	2″ to 4″ square		11	.727		.10	29		29.10	48
0160	4″ to 8″ square		10	.800		.25	32		32.25	53.50
0170	8″ to 12″ square		8	1		.48	40		40.48	66.50
0180	12″ to 32″ square		6	1.333		3.04	53.50		56.54	91.50
0210	16″ by 48″		5	1.600		2.31	64		66.31	109
0220	32″ by 48″		4	2		4.13	80		84.13	137
0230	48″ square		3.50	2.286		5.85	91.50		97.35	157
0240	60″ square		3.20	2.500		9.80	100		109.80	176
0500	Skim coat surface with joint compound		1600	.005	S.F.	.03	.20		.23	.36
9000	Minimum labor/equipment charge		2	4	Job		160		160	264

09 05 Common Work Results for Finishes

09 05 05 – Selective Finishes Demolition

09 05 05.10 Selective Demolition, Ceilings

		Crew	Daily Output	Labor-Hours	Unit	Material	Labor	Equipment	Total	Total Incl O&P
0010	**SELECTIVE DEMOLITION, CEILINGS** R024119-10									
0200	Ceiling, drywall, furred and nailed or screwed	2 Clab	800	.020	S.F.		.63		.63	1.04
0220	On metal frame		760	.021			.67		.67	1.10
0240	On suspension system, including system		720	.022			.70		.70	1.16
1000	Plaster, lime and horse hair, on wood lath, incl. lath		700	.023			.72		.72	1.19
1020	On metal lath		570	.028			.89		.89	1.46
1100	Gypsum, on gypsum lath		720	.022			.70		.70	1.16
1120	On metal lath		500	.032			1.01		1.01	1.67
1200	Suspended ceiling, mineral fiber, 2′ x 2′ or 2′ x 4′		1500	.011			.34		.34	.56
1250	On suspension system, incl. system		1200	.013			.42		.42	.70
1500	Tile, wood fiber, 12″ x 12″, glued		900	.018			.56		.56	.93
1540	Stapled		1500	.011			.34		.34	.56
1580	On suspension system, incl. system		760	.021			.67		.67	1.10
2000	Wood, tongue and groove, 1″ x 4″		1000	.016			.51		.51	.83
2040	1″ x 8″		1100	.015			.46		.46	.76
2400	Plywood or wood fiberboard, 4′ x 8′ sheets		1200	.013			.42		.42	.70
9000	Minimum labor/equipment charge	1 Clab	2	4	Job		126		126	209

09 05 05.20 Selective Demolition, Flooring

		Crew	Daily Output	Labor-Hours	Unit	Material	Labor	Equipment	Total	Total Incl O&P
0010	**SELECTIVE DEMOLITION, FLOORING** R024119-10									
0200	Brick with mortar	2 Clab	475	.034	S.F.		1.06		1.06	1.76
0400	Carpet, bonded, including surface scraping		2000	.008			.25		.25	.42
0480	Tackless		9000	.002			.06		.06	.09
0550	Carpet tile, releasable adhesive		5000	.003			.10		.10	.17
0560	Permanent adhesive		1850	.009			.27		.27	.45
0600	Composition, acrylic or epoxy		400	.040			1.26		1.26	2.09
0700	Concrete, scarify skin	A-1A	225	.036			1.45	.97	2.42	3.43
0800	Resilient, sheet goods	2 Clab	1400	.011			.36		.36	.60

09 05 05.20 Selective Demolition, Flooring

		Crew	Daily Output	Labor-Hours	Unit	Material	2009 Bare Costs Labor	Equipment	Total	Total Incl O&P
0820	For gym floors	2 Clab	900	.018	S.F.		.56		.56	.93
0850	Vinyl or rubber cove base	1 Clab	1000	.008	L.F.		.25		.25	.42
0860	Vinyl or rubber cove base, molded corner	″	1000	.008	Ea.		.25		.25	.42
0870	For glued and caulked installation, add to labor						50%			
0900	Vinyl composition tile, 12″ x 12″	2 Clab	1000	.016	S.F.		.51		.51	.83
2000	Tile, ceramic, thin set		675	.024			.75		.75	1.24
2020	Mud set		625	.026			.81		.81	1.34
2200	Marble, slate, thin set		675	.024			.75		.75	1.24
2220	Mud set		625	.026			.81		.81	1.34
2600	Terrazzo, thin set		450	.036			1.12		1.12	1.85
2620	Mud set		425	.038			1.19		1.19	1.96
2640	Terrazzo, cast in place		300	.053			1.69		1.69	2.78
3000	Wood, block, on end	1 Carp	400	.020			.80		.80	1.32
3200	Parquet		450	.018			.71		.71	1.17
3400	Strip flooring, interior, 2-1/4″ x 25/32″ thick		325	.025			.98		.98	1.62
3500	Exterior, porch flooring, 1″ x 4″		220	.036			1.45		1.45	2.40
3800	Subfloor, tongue and groove, 1″ x 6″		325	.025			.98		.98	1.62
3820	1″ x 8″		430	.019			.74		.74	1.23
3840	1″ x 10″		520	.015			.61		.61	1.01
4000	Plywood, nailed		600	.013			.53		.53	.88
4100	Glued and nailed		400	.020			.80		.80	1.32
4200	Hardboard, 1/4″ thick		760	.011			.42		.42	.69
8000	Remove flooring, bead blast, minimum	A-1A	1000	.008			.33	.22	.55	.77
8100	Maximum		400	.020			.82	.54	1.36	1.93
8150	Mastic only		1500	.005			.22	.15	.37	.51
9000	Minimum labor/equipment charge	1 Clab	4	2	Job		63		63	104

09 05 05.30 Selective Demolition, Walls and Partitions

		Crew	Daily Output	Labor-Hours	Unit	Material	Labor	Equipment	Total	Total Incl O&P
0010	**SELECTIVE DEMOLITION, WALLS AND PARTITIONS** R024119-10									
0020	Walls, concrete, reinforced	B-39	120	.400	C.F.		13.25	1.60	14.85	23.50
0025	Plain	″	160	.300			9.95	1.20	11.15	17.60
0100	Brick, 4″ to 12″ thick	B-9	220	.182			5.80	.87	6.67	10.55
0200	Concrete block, 4″ thick		1000	.040	S.F.		1.28	.19	1.47	2.32
0280	8″ thick		810	.049			1.58	.24	1.82	2.87
0300	Exterior stucco 1″ thick over mesh		3200	.013			.40	.06	.46	.73
1000	Drywall, nailed or screwed	1 Clab	1000	.008			.25		.25	.42
1010	2 layers		400	.020			.63		.63	1.04
1020	Glued and nailed		900	.009			.28		.28	.46
1500	Fiberboard, nailed		900	.009			.28		.28	.46
1520	Glued and nailed		800	.010			.32		.32	.52
1568	Plenum barrier, sheet lead		300	.027			.84		.84	1.39
2000	Movable walls, metal, 5′ high		300	.027			.84		.84	1.39
2020	8′ high		400	.020			.63		.63	1.04
2200	Metal or wood studs, finish 2 sides, fiberboard	B-1	520	.046			1.49		1.49	2.46
2250	Lath and plaster		260	.092			2.98		2.98	4.92
2300	Plasterboard (drywall)		520	.046			1.49		1.49	2.46
2350	Plywood		450	.053			1.72		1.72	2.84
2800	Paneling, 4′ x 8′ sheets	1 Clab	475	.017			.53		.53	.88
3000	Plaster, lime and horsehair, on wood lath		400	.020			.63		.63	1.04
3020	On metal lath		335	.024			.75		.75	1.25
3400	Gypsum or perlite, on gypsum lath		410	.020			.62		.62	1.02
3420	On metal lath		300	.027			.84		.84	1.39
3450	Plaster, interior gypsum, acoustic, or cement		60	.133	S.Y.		4.21		4.21	6.95

09 05 Common Work Results for Finishes

09 05 05 – Selective Finishes Demolition

09 05 05.30 Selective Demolition, Walls and Partitions		Crew	Daily Output	Labor-Hours	Unit	Material	2009 Bare Costs Labor	Equipment	Total	Total Incl O&P
3500	Stucco, on masonry	1 Clab	145	.055	S.Y.		1.74		1.74	2.88
3510	Commercial 3-coat		80	.100			3.16		3.16	5.20
3520	Interior stucco		25	.320			10.10		10.10	16.70
3750	Terra cotta block and plaster, to 6" thick	B-1	175	.137	S.F.		4.43		4.43	7.30
3760	Tile, ceramic, on walls, thin set	1 Clab	300	.027			.84		.84	1.39
3765	Mud set		250	.032			1.01		1.01	1.67
3800	Toilet partitions, slate or marble		5	1.600	Ea.		50.50		50.50	83.50
3820	Metal or plastic		8	1	"		31.50		31.50	52
5000	Wallcovering, vinyl	1 Pape	700	.011	S.F.		.41		.41	.65
5010	With release agent		1500	.005			.19		.19	.30
5025	Wallpaper, 2 layers or less, by hand		250	.032			1.13		1.13	1.81
5035	3 layers or more		165	.048			1.72		1.72	2.75
5040	Designer		480	.017			.59		.59	.94
9000	Minimum labor/equipment charge	1 Clab	4	2	Job		63		63	104

09 21 Plaster and Gypsum Board Assemblies

09 21 13 – Plaster Assemblies

09 21 13.10 Plaster Partition Wall		Crew	Daily Output	Labor-Hours	Unit	Material	Labor	Equipment	Total	Total Incl O&P
0010	**PLASTER PARTITION WALL**									
0400	Stud walls, 3.4 lb. metal lath, 3 coat gypsum plaster, 2 sides									
0600	2" x 4" wood studs, 16" O.C.	J-2	315	.152	S.F.	3.31	5.30	.40	9.01	12.60
0700	2-1/2" metal studs, 25 ga., 12" O.C.		325	.148		3.06	5.15	.39	8.60	12.05
0800	3-5/8" metal studs, 25 ga., 16" O.C.		320	.150		3.08	5.20	.40	8.68	12.20
0900	Gypsum lath, 2 coat vermiculite plaster, 2 sides									
1000	2" x 4" wood studs, 16" O.C.	J-2	355	.135	S.F.	3.69	4.70	.36	8.75	12
1200	2-1/2" metal studs, 25 ga., 12" O.C.		365	.132		3.27	4.57	.35	8.19	11.35
1300	3-5/8" metal studs, 25 ga., 16" O.C.		360	.133		3.36	4.64	.35	8.35	11.55

09 21 16 – Gypsum Board Assemblies

09 21 16.23 Gypsum Board Shaft Wall Assemblies		Crew	Daily Output	Labor-Hours	Unit	Material	Labor	Equipment	Total	Total Incl O&P
0010	**GYPSUM BOARD SHAFT WALL ASSEMBLIES**									
0030	1" thick coreboard wall liner on shaft side									
0040	2-hour assembly with double layer									
0060	5/8" fire rated gypsum board on room side	2 Carp	220	.073	S.F.	1.56	2.91		4.47	6.50
0100	3-hour assembly with triple layer									
0300	5/8" fire rated gypsum board on room side	2 Carp	180	.089	S.F.	1.90	3.55		5.45	7.95
0400	4-hour assembly, 1" coreboard, 5/8" fire rated gypsum board									
0600	and 3/4" galv. metal furring channels, 24" O.C., with									
0700	Double layer 5/8" fire rated gypsum board on room side	2 Carp	110	.145	S.F.	1.77	5.80		7.57	11.55
0900	For taping & finishing, add per side	1 Carp	1050	.008	"	.04	.30		.34	.55
1000	For insulation, see Div. 07 21									
5200	For work over 8' high, add	2 Carp	3060	.005	S.F.		.21		.21	.34
5300	For distribution cost over 3 stories high, add per story	"	6100	.003	"		.10		.10	.17

09 21 16.33 Partition Wall		Crew	Daily Output	Labor-Hours	Unit	Material	Labor	Equipment	Total	Total Incl O&P
0010	**PARTITION WALL** Stud wall, 8' to 12' high									
0050	1/2", interior, gypsum board, std, tape & finish 2 sides									
0500	Installed on and incl., 2" x 4" wood studs, 16" O.C.	2 Carp	310	.052	S.F.	1.03	2.06		3.09	4.53
1000	Metal studs, NLB, 25 ga., 16" O.C., 3-5/8" wide		350	.046		1.16	1.83		2.99	4.28
1200	6" wide		330	.048		1.39	1.94		3.33	4.72
1400	Water resistant, on 2" x 4" wood studs, 16" O.C.		310	.052		1.37	2.06		3.43	4.91
1600	Metal studs, NLB, 25 ga., 16" O.C., 3-5/8" wide		350	.046		1.50	1.83		3.33	4.65

09 21 16.33 Partition Wall		Crew	Daily Output	Labor-Hours	Unit	Material	Labor	Equipment	Total	Total Incl O&P
						2009 Bare Costs				
1800	6" wide	2 Carp	330	.048	S.F.	1.73	1.94		3.67	5.10
2000	Fire res., 2 layers, 1-1/2 hr., on 2" x 4" wood studs, 16" O.C.		210	.076		1.69	3.04		4.73	6.85
2200	Metal studs, NLB, 25 ga., 16" O.C., 3-5/8" wide		250	.064		1.82	2.56		4.38	6.20
2400	6" wide		230	.070		2.05	2.78		4.83	6.85
2600	Fire & water res., 2 layers, 1-1/2 hr., 2" x 4" studs, 16" O.C.		210	.076		1.69	3.04		4.73	6.85
2800	Metal studs, NLB, 25 ga., 16" O.C., 3-5/8" wide		250	.064		1.82	2.56		4.38	6.20
3000	6" wide	↓	230	.070	↓	2.05	2.78		4.83	6.85
3200	5/8", interior, gypsum board, std, tape & finish 2 sides									
3400	Installed on and including 2" x 4" wood studs, 16" O.C.	2 Carp	300	.053	S.F.	1.09	2.13		3.22	4.72
3600	24" O.C.		330	.048		1.02	1.94		2.96	4.32
3800	Metal studs, NLB, 25 ga., 16" O.C., 3-5/8" wide		340	.047		1.22	1.88		3.10	4.44
4000	6" wide		320	.050		1.45	2		3.45	4.89
4200	24" O.C., 3-5/8" wide		360	.044		1.10	1.78		2.88	4.14
4400	6" wide		340	.047		1.27	1.88		3.15	4.49
4800	Water resistant, on 2" x 4" wood studs, 16" O.C.		300	.053		1.35	2.13		3.48	5
5000	24" O.C.		330	.048		1.28	1.94		3.22	4.61
5200	Metal studs, NLB, 25 ga. 16" O.C., 3-5/8" wide		340	.047		1.48	1.88		3.36	4.72
5400	6" wide		320	.050		1.71	2		3.71	5.20
5600	24" O.C., 3-5/8" wide		360	.044		1.36	1.78		3.14	4.42
5800	6" wide		340	.047		1.53	1.88		3.41	4.78
6000	Fire resistant, 2 layers, 2 hr., on 2" x 4" wood studs, 16" O.C.		205	.078		1.78	3.12		4.90	7.10
6200	24" O.C.		235	.068		1.78	2.72		4.50	6.45
6400	Metal studs, NLB, 25 ga., 16" O.C., 3-5/8" wide		245	.065		2	2.61		4.61	6.50
6600	6" wide		225	.071		2.21	2.84		5.05	7.10
6800	24" O.C., 3-5/8" wide		265	.060		1.86	2.41		4.27	6
7000	6" wide		245	.065		2.03	2.61		4.64	6.55
7200	Fire & water resistant, 2 layers, 2 hr., 2" x 4" studs, 16" O.C.		205	.078		1.85	3.12		4.97	7.20
7400	24" O.C.		235	.068		1.78	2.72		4.50	6.45
7600	Metal studs, NLB, 25 ga., 16" O.C., 3-5/8" wide		245	.065		1.98	2.61		4.59	6.50
7800	6" wide		225	.071		2.21	2.84		5.05	7.10
8000	24" O.C., 3-5/8" wide		265	.060		1.86	2.41		4.27	6
8200	6" wide	↓	245	.065	↓	2.03	2.61		4.64	6.55
8600	1/2" blueboard, mesh tape both sides									
8620	Installed on and including 2" x 4" wood studs, 16" O.C.	2 Carp	300	.053	S.F.	1.09	2.13		3.22	4.72
8640	Metal studs, NLB, 25 ga., 16" O.C., 3-5/8" wide		340	.047		1.22	1.88		3.10	4.44
8660	6" wide	↓	320	.050	↓	1.45	2		3.45	4.89
9000	Exterior, 1/2" gypsum sheathing, 1/2" gypsum finished, interior,									
9100	including foil faced insulation, metal studs, 20 ga.									
9200	16" O.C., 3-5/8" wide	2 Carp	290	.055	S.F.	2.17	2.20		4.37	6.05
9400	6" wide	"	270	.059	"	2.42	2.37		4.79	6.55

09 22 Supports for Plaster and Gypsum Board

09 22 03 – Fastening Methods for Finishes

09 22 03.20 Drilling Plaster/Drywall		Crew	Daily Output	Labor-Hours	Unit	Material	Labor	Equipment	Total	Total Incl O&P
						2009 Bare Costs				
0010	**DRILLING PLASTER/DRYWALL**									
1100	Drilling & layout for drywall/plaster walls, up to 1" deep, no anchor									
1200	Holes, 1/4" diameter	1 Carp	150	.053	Ea.	.01	2.13		2.14	3.53
1300	3/8" diameter		140	.057		.01	2.28		2.29	3.78
1400	1/2" diameter		130	.062		.01	2.46		2.47	4.07
1500	3/4" diameter		120	.067		.01	2.66		2.67	4.41
1600	1" diameter		110	.073		.02	2.91		2.93	4.82
1700	1-1/4" diameter		100	.080		.03	3.20		3.23	5.35
1800	1-1/2" diameter	↓	90	.089		.04	3.55		3.59	5.90
1900	For ceiling installations, add				↓		40%			

09 22 13 – Metal Furring

09 22 13.13 Metal Channel Furring		Crew	Daily Output	Labor-Hours	Unit	Material	Labor	Equipment	Total	Total Incl O&P
0010	**METAL CHANNEL FURRING**									
0030	Beams and columns, 7/8" channels, galvanized, 12" O.C.	1 Lath	155	.052	S.F.	.38	1.83		2.21	3.31
0050	16" O.C.		170	.047		.31	1.67		1.98	2.98
0070	24" O.C.		185	.043		.21	1.54		1.75	2.66
0100	Ceilings, on steel, 7/8" channels, galvanized, 12" O.C.		210	.038		.34	1.35		1.69	2.52
0300	16" O.C.		290	.028		.31	.98		1.29	1.89
0400	24" O.C.		420	.019		.21	.68		.89	1.30
0600	1-5/8" channels, galvanized, 12" O.C.		190	.042		.46	1.50		1.96	2.87
0700	16" O.C.		260	.031		.41	1.09		1.50	2.19
0900	24" O.C.		390	.021		.28	.73		1.01	1.45
0930	7/8" channels with sound isolation clips, 12" O.C.		120	.067		1.53	2.37		3.90	5.40
0940	16" O.C.		100	.080		2.12	2.84		4.96	6.80
0950	24" O.C.		165	.048		1.39	1.72		3.11	4.25
0960	1-5/8" channels, galvanized, 12" O.C.		110	.073		1.64	2.59		4.23	5.90
0970	16" O.C.		100	.080		2.22	2.84		5.06	6.95
0980	24" O.C.		155	.052		1.46	1.83		3.29	4.51
1000	Walls, 7/8" channels, galvanized, 12" O.C.		235	.034		.34	1.21		1.55	2.29
1200	16" O.C.		265	.030		.31	1.07		1.38	2.04
1300	24" O.C.		350	.023		.21	.81		1.02	1.51
1500	1-5/8" channels, galvanized, 12" O.C.		210	.038		.46	1.35		1.81	2.65
1600	16" O.C.		240	.033		.41	1.18		1.59	2.33
1800	24" O.C.		305	.026		.28	.93		1.21	1.77
1920	7/8" channels with sound isolation clips, 12" O.C.		125	.064		1.53	2.28		3.81	5.25
1940	16" O.C.		100	.080		2.12	2.84		4.96	6.80
1950	24" O.C.		150	.053		1.39	1.90		3.29	4.52
1960	1-5/8" channels, galvanized, 12" O.C.		115	.070		1.64	2.47		4.11	5.70
1970	16" O.C.		95	.084		2.22	2.99		5.21	7.20
1980	24" O.C.		140	.057	↓	1.46	2.03		3.49	4.82
9000	Minimum labor/equipment charge	↓	4	2	Job		71		71	112

09 22 16 – Non-Structural Metal Framing

09 22 16.13 Metal Studs and Track		Crew	Daily Output	Labor-Hours	Unit	Material	Labor	Equipment	Total	Total Incl O&P
0010	**METAL STUDS AND TRACK**									
1600	Non-load bearing, galv, 8' high, 25 ga. 1-5/8" wide, 16" O.C.	1 Carp	619	.013	S.F.	.33	.52		.85	1.21
1610	24" O.C.		950	.008		.25	.34		.59	.83
1620	2-1/2" wide, 16" O.C.		613	.013		.42	.52		.94	1.32
1630	24" O.C.		938	.009		.31	.34		.65	.90
1640	3-5/8" wide, 16" O.C.		600	.013		.48	.53		1.01	1.40
1650	24" O.C.		925	.009		.36	.35		.71	.96
1660	4" wide, 16" O.C.		594	.013		.51	.54		1.05	1.45
1670	24" O.C.	↓	925	.009	↓	.38	.35		.73	.99

09 22 Supports for Plaster and Gypsum Board

09 22 16 – Non-Structural Metal Framing

09 22 16.13 Metal Studs and Track		Crew	Daily Output	Labor-Hours	Unit	Material	Labor (2009 Bare Costs)	Equipment (2009 Bare Costs)	Total (2009 Bare Costs)	Total Incl O&P
1680	6″ wide, 16″ O.C.	1 Carp	588	.014	S.F.	.72	.54		1.26	1.69
1690	24″ O.C.		906	.009		.54	.35		.89	1.17
1700	20 ga. studs, 1-5/8″ wide, 16″ O.C.		494	.016		.53	.65		1.18	1.65
1710	24″ O.C.		763	.010		.40	.42		.82	1.13
1720	2-1/2″ wide, 16″ O.C.		488	.016		.62	.65		1.27	1.76
1730	24″ O.C.		750	.011		.47	.43		.90	1.21
1740	3-5/8″ wide, 16″ O.C.		481	.017		.67	.66		1.33	1.84
1750	24″ O.C.		738	.011		.50	.43		.93	1.26
1760	4″ wide, 16″ O.C.		475	.017		.80	.67		1.47	1.98
1770	24″ O.C.		738	.011		.60	.43		1.03	1.37
1780	6″ wide, 16″ O.C.		469	.017		.94	.68		1.62	2.16
1790	24″ O.C.		725	.011		.70	.44		1.14	1.50
2000	Non-load bearing, galv, 10′ high, 25 ga. 1-5/8″ wide, 16″ O.C.		495	.016		.31	.65		.96	1.41
2100	24″ O.C.		760	.011		.23	.42		.65	.94
2200	2-1/2″ wide, 16″ O.C.		490	.016		.39	.65		1.04	1.51
2250	24″ O.C.		750	.011		.29	.43		.72	1.02
2300	3-5/8″ wide, 16″ O.C.		480	.017		.45	.67		1.12	1.60
2350	24″ O.C.		740	.011		.33	.43		.76	1.07
2400	4″ wide, 16″ O.C.		475	.017		.49	.67		1.16	1.64
2450	24″ O.C.		740	.011		.36	.43		.79	1.10
2500	6″ wide, 16″ O.C.		470	.017		.68	.68		1.36	1.87
2550	24″ O.C.		725	.011		.50	.44		.94	1.28
2600	20 ga. studs, 1-5/8″ wide, 16″ O.C.		395	.020		.50	.81		1.31	1.89
2650	24″ O.C.		610	.013		.37	.52		.89	1.27
2700	2-1/2″ wide, 16″ O.C.		390	.021		.59	.82		1.41	2
2750	24″ O.C.		600	.013		.43	.53		.96	1.36
2800	3-5/8″ wide, 16″ OC		385	.021		.64	.83		1.47	2.07
2850	24″ O.C.		590	.014		.47	.54		1.01	1.41
2900	4″ wide, 16″ O.C.		380	.021		.75	.84		1.59	2.22
2950	24″ O.C.		590	.014		.56	.54		1.10	1.50
3000	6″ wide, 16″ O.C.		375	.021		.89	.85		1.74	2.39
3050	24″ O.C.		580	.014		.65	.55		1.20	1.63
3060	Non-load bearing, galv, 12′ high, 25 ga. 1-5/8″ wide, 16″ O.C.		413	.019		.30	.77		1.07	1.61
3070	24″ O.C.		633	.013		.22	.51		.73	1.07
3080	2-1/2″ wide, 16″ O.C.		408	.020		.38	.78		1.16	1.71
3090	24″ O.C.		625	.013		.27	.51		.78	1.14
3100	3-5/8″ wide, 16″ O.C.		400	.020		.43	.80		1.23	1.79
3110	24″ O.C.		617	.013		.31	.52		.83	1.20
3120	4″ wide, 16″ O.C.		396	.020		.47	.81		1.28	1.84
3130	24″ O.C.		617	.013		.34	.52		.86	1.23
3140	6″ wide, 16″ O.C.		392	.020		.65	.82		1.47	2.07
3150	24″ O.C.		604	.013		.47	.53		1	1.39
3160	20 ga. studs, 1-5/8″ wide, 16″ O.C.		329	.024		.48	.97		1.45	2.13
3170	24″ O.C.		508	.016		.35	.63		.98	1.42
3180	2-1/2″ wide, 16″ O.C.		325	.025		.56	.98		1.54	2.24
3190	24″ O.C.		500	.016		.41	.64		1.05	1.51
3200	3-5/8″ wide, 16″ O.C.		321	.025		.61	1		1.61	2.31
3210	24″ O.C.		492	.016		.44	.65		1.09	1.56
3220	4″ wide, 16″ O.C.		317	.025		.72	1.01		1.73	2.45
3230	24″ O.C.		492	.016		.52	.65		1.17	1.65
3240	6″ wide, 16″ O.C.		313	.026		.85	1.02		1.87	2.63
3250	24″ O.C.	▼	483	.017	▼	.62	.66		1.28	1.77
5000	Load bearing studs, see Div. 05 41 13.30									

09 22 16 – Non-Structural Metal Framing

09 22 16.13 Metal Studs and Track		Crew	Daily Output	Labor-Hours	Unit	2009 Bare Costs Material	Labor	Equipment	Total	Total Incl O&P
9000	Minimum labor/equipment charge	1 Carp	4	2	Job		80		80	132

09 22 26 – Suspension Systems

09 22 26.13 Ceiling Suspension Systems		Crew	Daily Output	Labor-Hours	Unit	Material	Labor	Equipment	Total	Total Incl O&P
0010	**CEILING SUSPENSION SYSTEMS** For gypsum board or plaster									
8000	Suspended ceilings, including carriers									
8200	1-1/2" carriers, 24" O.C. with:									
8300	7/8" channels, 16" O.C.	1 Lath	215	.037	S.F.	.58	1.32		1.90	2.73
8320	24" O.C.		275	.029		.48	1.03		1.51	2.16
8400	1-5/8" channels, 16" O.C.		205	.039		.69	1.39		2.08	2.95
8420	24" O.C.		265	.030		.55	1.07		1.62	2.31
8600	2" carriers, 24" O.C. with:									
8700	7/8" channels, 16" O.C.	1 Lath	205	.039	S.F.	.56	1.39		1.95	2.80
8720	24" O.C.		265	.030		.46	1.07		1.53	2.20
8800	1-5/8" channels, 16" O.C.		195	.041		.66	1.46		2.12	3.03
8820	24" O.C.		255	.031		.53	1.12		1.65	2.34

09 22 36 – Lath

09 22 36.13 Gypsum Lath		Crew	Daily Output	Labor-Hours	Unit	Material	Labor	Equipment	Total	Total Incl O&P
0010	**GYPSUM LATH**									
0020	Plain or perforated, nailed, 3/8" thick	1 Lath	85	.094	S.Y.	5.50	3.35		8.85	11.35
0100	1/2" thick		80	.100		5.85	3.56		9.41	12.05
0300	Clipped to steel studs, 3/8" thick		75	.107		5.50	3.79		9.29	12.05
0400	1/2" thick		70	.114		5.85	4.06		9.91	12.85
1500	For ceiling installations, add		216	.037			1.32		1.32	2.08
1600	For columns and beams, add		170	.047			1.67		1.67	2.64
9000	Minimum labor/equipment charge		4.25	1.882	Job		67		67	106

09 22 36.23 Metal Lath		Crew	Daily Output	Labor-Hours	Unit	Material	Labor	Equipment	Total	Total Incl O&P
0010	**METAL LATH**									
0020	Diamond, expanded, 2.5 lb. per S.Y., painted				S.Y.	3.32			3.32	3.65
0100	Galvanized					3.47			3.47	3.82
0300	3.4 lb. per S.Y., painted					3.99			3.99	4.39
0400	Galvanized					4.09			4.09	4.50
0600	For 15# asphalt sheathing paper, add					.43			.43	.47
0900	Flat rib, 1/8" high, 2.75 lb., painted					3.62			3.62	3.98
1000	Foil backed					3.56			3.56	3.92
1200	3.4 lb. per S.Y., painted					4.61			4.61	5.05
1300	Galvanized					4.23			4.23	4.65
1500	For 15# asphalt sheathing paper, add					.43			.43	.47
1800	High rib, 3/8" high, 3.4 lb. per S.Y., painted					5.55			5.55	6.15
1900	Galvanized					5.55			5.55	6.15
2400	3/4" high, painted, .60 lb. per S.F.				S.F.	.58			.58	.64
2500	.75 lb. per S.F.				"	1.25			1.25	1.38
2800	Stucco mesh, painted, 3.6 lb.				S.Y.	4.30			4.30	4.73
3000	K-lath, perforated, absorbent paper, regular					4.43			4.43	4.87
3100	Heavy duty					5.25			5.25	5.75
3300	Waterproof, heavy duty, grade B backing					5.10			5.10	5.65
3400	Fire resistant backing					5.65			5.65	6.25
3600	2.5 lb. diamond painted, on wood framing, on walls	1 Lath	85	.094		3.32	3.35		6.67	8.95
3700	On ceilings		75	.107		3.32	3.79		7.11	9.65
3900	3.4 lb. diamond painted, on wood framing, on walls		80	.100		4.61	3.56		8.17	10.65
4000	On ceilings		70	.114		4.61	4.06		8.67	11.45
4200	3.4 lb. diamond painted, wired to steel framing		75	.107		4.61	3.79		8.40	11.05

09 22 36.23 Metal Lath

	09 22 36.23 Metal Lath	Crew	Daily Output	Labor-Hours	Unit	Material	Labor	Equipment	Total	Total Incl O&P
						2009 Bare Costs				
4300	On ceilings	1 Lath	60	.133	S.Y.	4.61	4.74		9.35	12.55
4500	Columns and beams, wired to steel		40	.200		4.61	7.10		11.71	16.30
4600	Cornices, wired to steel		35	.229		4.61	8.15		12.76	17.90
4800	Screwed to steel studs, 2.5 lb.		80	.100		3.32	3.56		6.88	9.25
4900	3.4 lb.		75	.107		3.99	3.79		7.78	10.40
5100	Rib lath, painted, wired to steel, on walls, 2.5 lb.		75	.107		3.62	3.79		7.41	10
5200	3.4 lb.		70	.114		5.55	4.06		9.61	12.55
5400	4.0 lb.		65	.123		5.70	4.38		10.08	13.20
5500	For self-furring lath, add					.10			.10	.11
5700	Suspended ceiling system, incl. 3.4 lb. diamond lath, painted	1 Lath	15	.533		4.23	18.95		23.18	34.50
5800	Galvanized	"	15	.533		7.50	18.95		26.45	38.50
6000	Hollow metal stud partitions, 3.4 lb. painted lath both sides									
6010	Non-load bearing, 25 ga., w/rib lath 2-1/2" studs, 12" O.C.	1 Lath	20.30	.394	S.Y.	15.65	14		29.65	39
6300	16" O.C.		21.10	.379		14.70	13.50		28.20	37.50
6350	24" O.C.		22.70	.352		13.75	12.55		26.30	35
6400	3-5/8" studs, 16" O.C.		19.50	.410		15.20	14.60		29.80	39.50
6600	24" O.C.		20.40	.392		14.10	13.95		28.05	37.50
6700	4" studs, 16" O.C.		20.40	.392		15.50	13.95		29.45	39
6900	24" O.C.		21.60	.370		14.35	13.15		27.50	37
7000	6" studs, 16" O.C.		19.50	.410		17.25	14.60		31.85	42
7100	24" O.C.		21.10	.379		15.65	13.50		29.15	38.50
7200	L.B. partitions, 16 ga., w/rib lath, 2-1/2" studs, 16" O.C.		20	.400		16.95	14.20		31.15	41
7300	3-5/8" studs, 16 ga.		19.70	.406		19.20	14.45		33.65	44
7500	4" studs, 16 ga.		19.50	.410		20	14.60		34.60	45
7600	6" studs, 16 ga.		18.70	.428		23.50	15.20		38.70	50
9000	Minimum labor/equipment charge		4.25	1.882	Job		67		67	106

09 22 36.43 Security Mesh

	09 22 36.43 Security Mesh	Crew	Daily Output	Labor-Hours	Unit	Material	Labor	Equipment	Total	Total Incl O&P
0010	**SECURITY MESH**, expanded metal, flat, screwed to framing									
0100	On walls, 3/4", 1.76 lb/S.F.	2 Carp	1500	.011	S.F.	1.60	.43		2.03	2.46
0110	1-1/2", 1.14 lb/S.F.		1600	.010		1.30	.40		1.70	2.09
0200	On ceilings, 3/4", 1.76 lb/S.F.		1350	.012		1.60	.47		2.07	2.54
0210	1-1/2", 1.14 lb/S.F.		1450	.011		1.30	.44		1.74	2.16

09 22 36.83 Accessories, Plaster

	09 22 36.83 Accessories, Plaster	Crew	Daily Output	Labor-Hours	Unit	Material	Labor	Equipment	Total	Total Incl O&P
0010	**ACCESSORIES, PLASTER**									
0020	Casing bead, expanded flange, galvanized	1 Lath	2.70	2.963	C.L.F.	46.50	105		151.50	218
0900	Channels, cold rolled, 16 ga., 3/4" deep, galvanized					34			34	37.50
1200	1-1/2" deep, 16 ga., galvanized					46			46	50.50
1620	Corner bead, expanded bullnose, 3/4" radius, #10, galvanized	1 Lath	2.60	3.077		34	109		143	211
1650	#1, galvanized		2.55	3.137		55	112		167	237
1670	Expanded wing, 2-3/4" wide, #1, galvanized		2.65	3.019		34.50	107		141.50	208
1700	Inside corner (corner rite), 3" x 3", painted		2.60	3.077		30	109		139	206
1750	Strip-ex, 4" wide, painted		2.55	3.137		30	112		142	209
1800	Expansion joint, 3/4" grounds, limited expansion, galv., 1 piece		2.70	2.963		109	105		214	286
2100	Extreme expansion, galvanized, 2 piece		2.60	3.077		173	109		282	365

09 23 Gypsum Plastering

09 23 13 – Acoustical Gypsum Plastering

09 23 13.10 Perlite or Vermiculite Plaster		Crew	Daily Output	Labor-Hours	Unit	Material	Labor	Equipment	Total	Total Incl O&P
						2009 Bare Costs				
0010	PERLITE OR VERMICULITE PLASTER									
0020	In 100 lb. bags, under 200 bags				Bag	14.40			14.40	15.85
0100	Over 200 bags				"	13.10			13.10	14.40
0300	2 coats, no lath included, on walls	J-1	92	.435	S.Y.	4.05	15.05	1.38	20.48	30
0400	On ceilings	"	79	.506		4.05	17.50	1.61	23.16	34
0600	On and incl. 3/8" gypsum lath, on metal studs	J-2	84	.571		9.60	19.85	1.51	30.96	44.50
0700	On ceilings	"	70	.686		9.60	24	1.81	35.41	50.50
0900	3 coats, no lath included, on walls	J-1	74	.541		6.30	18.70	1.71	26.71	39
1000	On ceilings	"	63	.635		6.30	22	2.01	30.31	44.50
1200	On and incl. painted metal lath, on metal studs	J-2	72	.667		11.85	23	1.76	36.61	52
1300	On ceilings		61	.787		11.85	27.50	2.08	41.43	59.50
1500	On and incl. suspended metal lath ceiling		37	1.297		10.55	45	3.42	58.97	88
1700	For irregular or curved surfaces, add to above						30%			
1800	For columns and beams, add to above						50%			
1900	For soffits, add to ceiling prices						40%			
9000	Minimum labor/equipment charge	1 Plas	1	8	Job		289		289	465

09 23 20 – Gypsum Plaster

09 23 20.10 Gypsum Plaster On Walls and Ceilings		Crew	Daily Output	Labor-Hours	Unit	Material	Labor	Equipment	Total	Total Incl O&P
0010	GYPSUM PLASTER ON WALLS AND CEILINGS									
0020	80# bag, less than 1 ton				Bag	15.20			15.20	16.70
0100	Over 1 ton				"	14.90			14.90	16.40
0300	2 coats, no lath included, on walls	J-1	105	.381	S.Y.	3.77	13.20	1.21	18.18	26.50
0400	On ceilings	"	92	.435		3.77	15.05	1.38	20.20	29.50
0600	On and incl. 3/8" gypsum lath on steel, on walls	J-2	97	.495		9.25	17.20	1.31	27.76	39
0700	On ceilings	"	83	.578		9.25	20	1.53	30.78	44
0900	3 coats, no lath included, on walls	J-1	87	.460		5.40	15.90	1.46	22.76	33
1000	On ceilings	"	78	.513		5.40	17.75	1.63	24.78	36
1200	On and including painted metal lath, on wood studs	J-2	86	.558		10.80	19.40	1.47	31.67	44.50
1300	On ceilings	"	76.50	.627		10.80	22	1.66	34.46	48.50
1600	For irregular or curved surfaces, add						30%			
1800	For columns & beams, add						50%			
9000	Minimum labor/equipment charge	1 Plas	1	8	Job		289		289	465

09 23 20.20 Gauging Plaster		Crew	Daily Output	Labor-Hours	Unit	Material	Labor	Equipment	Total	Total Incl O&P
0010	GAUGING PLASTER									
0020	100 lb. bags, less than 1 ton				Bag	19.30			19.30	21
0100	Over 1 ton				"	16.90			16.90	18.60

09 23 20.30 Keenes Cement		Crew	Daily Output	Labor-Hours	Unit	Material	Labor	Equipment	Total	Total Incl O&P
0010	KEENES CEMENT									
0020	In 100 lb. bags, less than 1 ton				Bag	27.50			27.50	30
0100	Over 1 ton				"	26			26	28.50
0300	Finish only, add to plaster prices, standard	J-1	215	.186	S.Y.	2.50	6.45	.59	9.54	13.75
0400	High quality	"	144	.278	"	2.53	9.60	.88	13.01	19.20

09 24 Portland Cement Plastering

09 24 23 – Portland Cement Stucco

09 24 23.40 Stucco		Crew	Daily Output	Labor-Hours	Unit	2009 Bare Costs Material	Labor	Equipment	Total	Total Incl O&P
0010	**STUCCO**									
0015	3 coats 1" thick, float finish, with mesh, on wood frame	J-2	63	.762	S.Y.	6.70	26.50	2.01	35.21	52
0100	On masonry construction, no mesh incl.	J-1	67	.597		2.41	20.50	1.89	24.80	37.50
0300	For trowel finish, add	1 Plas	170	.047			1.70		1.70	2.74
0400	For 3/4" thick, on masonry, deduct	J-1	880	.045		.62	1.57	.14	2.33	3.37
0600	For coloring and special finish, add, minimum		685	.058		.40	2.02	.19	2.61	3.89
0700	Maximum		200	.200		1.39	6.90	.63	8.92	13.40
0900	For soffits, add	J-2	155	.310		2.15	10.75	.82	13.72	20.50
1000	Exterior stucco, with bonding agent, 3 coats, on walls, no mesh incl.	J-1	200	.200		3.61	6.90	.63	11.14	15.80
1200	Ceilings		180	.222		3.61	7.70	.70	12.01	17.10
1300	Beams		80	.500		3.61	17.30	1.59	22.50	33.50
1500	Columns		100	.400		3.61	13.85	1.27	18.73	28
1550	Minimum labor/equipment charge	1 Plas	1	8	Job		289		289	465
1600	Mesh, painted, nailed to wood, 1.8 lb.	1 Lath	60	.133	S.Y.	5.75	4.74		10.49	13.85
1800	3.6 lb.		55	.145		4.30	5.15		9.45	12.90
1900	Wired to steel, painted, 1.8 lb.		53	.151		5.75	5.35		11.10	14.85
2100	3.6 lb.		50	.160		4.30	5.70		10	13.75
9000	Minimum labor/equipment charge		4	2	Job		71		71	112

09 26 Veneer Plastering

09 26 13 – Gypsum Veneer Plastering

09 26 13.20 Blueboard

		Crew	Daily Output	Labor-Hours	Unit	Material	Labor	Equipment	Total	Total Incl O&P
0010	**BLUEBOARD** For use with thin coat									
0100	plaster application (see Div. 09 26 13.80)									
1000	3/8" thick, on walls or ceilings, standard, no finish included	2 Carp	1900	.008	S.F.	.28	.34		.62	.87
1100	With thin coat plaster finish		875	.018		.36	.73		1.09	1.61
1400	On beams, columns, or soffits, standard, no finish included		675	.024		.32	.95		1.27	1.91
1450	With thin coat plaster finish		475	.034		.40	1.35		1.75	2.66
3000	1/2" thick, on walls or ceilings, standard, no finish included		1900	.008		.33	.34		.67	.92
3100	With thin coat plaster finish		875	.018		.41	.73		1.14	1.66
3300	Fire resistant, no finish included		1900	.008		.33	.34		.67	.92
3400	With thin coat plaster finish		875	.018		.41	.73		1.14	1.66
3450	On beams, columns, or soffits, standard, no finish included		675	.024		.38	.95		1.33	1.98
3500	With thin coat plaster finish		475	.034		.46	1.35		1.81	2.73
3700	Fire resistant, no finish included		675	.024		.38	.95		1.33	1.98
3800	With thin coat plaster finish		475	.034		.46	1.35		1.81	2.73
5000	5/8" thick, on walls or ceilings, fire resistant, no finish included		1900	.008		.34	.34		.68	.93
5100	With thin coat plaster finish		875	.018		.42	.73		1.15	1.67
5500	On beams, columns, or soffits, no finish included		675	.024		.39	.95		1.34	1.99
5600	With thin coat plaster finish		475	.034		.47	1.35		1.82	2.74
6000	For high ceilings, over 8' high, add		3060	.005			.21		.21	.34
6500	For over 3 stories high, add per story		6100	.003			.10		.10	.17
9000	Minimum labor/equipment charge	1 Carp	2	4	Job		160		160	264

09 26 13.80 Thin Coat Plaster

		Crew	Daily Output	Labor-Hours	Unit	Material	Labor	Equipment	Total	Total Incl O&P
0010	**THIN COAT PLASTER**									
0012	1 coat veneer, not incl. lath	J-1	3600	.011	S.F.	.08	.38	.04	.50	.75
1000	In 50 lb. bags				Bag	11.10			11.10	12.20

09 28 Backing Boards and Underlayments

09 28 13 – Cementitious Backing Boards

09 28 13.10 Cementitious Backerboard		Crew	Daily Output	Labor-Hours	Unit	Material	Labor	Equipment	Total	Total Incl O&P
0010	**CEMENTITIOUS BACKERBOARD**									
0070	Cementitious backerboard, on floor, 3' x 4'x 1/2" sheets	2 Carp	525	.030	S.F.	.79	1.22		2.01	2.87
0080	3' x 5' x 1/2" sheets		525	.030		.78	1.22		2	2.86
0090	3' x 6' x 1/2" sheets		525	.030		.72	1.22		1.94	2.80
0100	3' x 4'x 5/8" sheets		525	.030		1.04	1.22		2.26	3.15
0110	3' x 5' x 5/8" sheets		525	.030		1.03	1.22		2.25	3.14
0120	3' x 6' x 5/8" sheets		525	.030		.96	1.22		2.18	3.07
0150	On wall, 3' x 4'x 1/2" sheets		350	.046		.79	1.83		2.62	3.87
0160	3' x 5' x 1/2" sheets		350	.046		.78	1.83		2.61	3.86
0170	3' x 6' x 1/2" sheets		350	.046		.72	1.83		2.55	3.80
0180	3' x 4'x 5/8" sheets		350	.046		1.04	1.83		2.87	4.15
0190	3' x 5' x 5/8" sheets		350	.046		1.03	1.83		2.86	4.14
0200	3' x 6' x 5/8" sheets		350	.046		.96	1.83		2.79	4.07
0250	On counter, 3' x 4'x 1/2" sheets		180	.089		.79	3.55		4.34	6.70
0260	3' x 5' x 1/2" sheets		180	.089		.78	3.55		4.33	6.70
0270	3' x 6' x 1/2" sheets		180	.089		.72	3.55		4.27	6.65
0300	3' x 4'x 5/8" sheets		180	.089		1.04	3.55		4.59	7
0310	3' x 5' x 5/8" sheets		180	.089		1.03	3.55		4.58	7
0320	3' x 6' x 5/8" sheets	↓	180	.089	↓	.96	3.55		4.51	6.90

09 29 Gypsum Board

09 29 10 – Gypsum Board Panels

09 29 10.10 Gypsum Board Ceilings		Crew	Daily Output	Labor-Hours	Unit	Material	Labor	Equipment	Total	Total Incl O&P
0010	**GYPSUM BOARD CEILINGS**, fire rated, finished									
0100	Screwed to grid, channel or joists, 1/2" thick	2 Carp	765	.021	S.F.	.32	.84		1.16	1.73
0150	Mold resistant		765	.021		.54	.84		1.38	1.97
0200	5/8" thick		765	.021		.36	.84		1.20	1.78
0250	Mold resistant		765	.021		.66	.84		1.50	2.11
0300	Over 8' high, 1/2" thick		615	.026		.32	1.04		1.36	2.07
0350	Mold resistant		615	.026		.54	1.04		1.58	2.31
0400	5/8" thick		615	.026		.36	1.04		1.40	2.12
0450	Mold resistant	↓	615	.026	↓	.66	1.04		1.70	2.45
0600	Grid suspension system, direct hung									
0700	1-1/2" C.R.C., with 7/8" hi hat furring channel, 16" O.C.	2 Carp	1025	.016	S.F.	1.57	.62		2.19	2.76
0800	24" O.C.		1300	.012		1.43	.49		1.92	2.38
0900	3-5/8" C.R.C., with 7/8" hi hat furring channel, 16" O.C.		1025	.016		1.66	.62		2.28	2.86
1000	24" O.C.	↓	1300	.012	↓	1.44	.49		1.93	2.39

09 29 10.30 Gypsum Board		Crew	Daily Output	Labor-Hours	Unit	Material	Labor	Equipment	Total	Total Incl O&P
0010	**GYPSUM BOARD** on walls & ceilings R092910-10									
0100	Nailed or screwed to studs unless otherwise noted									
0150	3/8" thick, on walls, standard, no finish included	2 Carp	2000	.008	S.F.	.31	.32		.63	.87
0200	On ceilings, standard, no finish included		1800	.009		.31	.36		.67	.93
0250	On beams, columns, or soffits, no finish included		675	.024		.31	.95		1.26	1.90
0300	1/2" thick, on walls, standard, no finish included		2000	.008		.31	.32		.63	.87
0350	Taped and finished (level 4 finish)		965	.017		.35	.66		1.01	1.48
0390	With compound skim coat (level 5 finish)		775	.021		.40	.83		1.23	1.80
0400	Fire resistant, no finish included		2000	.008		.32	.32		.64	.88
0450	Taped and finished (level 4 finish)		965	.017		.36	.66		1.02	1.49
0490	With compound skim coat (level 5 finish)		775	.021		.41	.83		1.24	1.81
0500	Water resistant, no finish included		2000	.008		.48	.32		.80	1.06
0550	Taped and finished (level 4 finish)	↓	965	.017	↓	.52	.66		1.18	1.66

09 29 Gypsum Board

09 29 10 – Gypsum Board Panels

09 29 10.30 Gypsum Board		Crew	Daily Output	Labor-Hours	Unit	Material	Labor	Equipment	Total	Total Incl O&P
						2009 Bare Costs				
0590	With compound skim coat (level 5 finish)	2 Carp	775	.021	S.F.	.57	.83		1.40	1.98
0600	Prefinished, vinyl, clipped to studs		900	.018		.83	.71		1.54	2.08
0700	Mold resistant, no finish included		2000	.008		.49	.32		.81	1.07
0710	Taped and finished (level 4 finish)		965	.017		.53	.66		1.19	1.68
0720	With compound skim coat (level 5 finish)		775	.021		.58	.83		1.41	1.99
1000	On ceilings, standard, no finish included		1800	.009		.31	.36		.67	.93
1050	Taped and finished (level 4 finish)		765	.021		.35	.84		1.19	1.77
1090	With compound skim coat (level 5 finish)		610	.026		.40	1.05		1.45	2.17
1100	Fire resistant, no finish included		1800	.009		.32	.36		.68	.94
1150	Taped and finished (level 4 finish)		765	.021		.36	.84		1.20	1.78
1195	With compound skim coat (level 5 finish)		610	.026		.41	1.05		1.46	2.18
1200	Water resistant, no finish included		1800	.009		.48	.36		.84	1.12
1250	Taped and finished (level 4 finish)		765	.021		.52	.84		1.36	1.95
1290	With compound skim coat (level 5 finish)		610	.026		.57	1.05		1.62	2.35
1310	Mold resistant, no finish included		1800	.009		.49	.36		.85	1.13
1320	Taped and finished (level 4 finish)		765	.021		.53	.84		1.37	1.97
1330	With compound skim coat (level 5 finish)		610	.026		.58	1.05		1.63	2.36
1500	On beams, columns, or soffits, standard, no finish included		675	.024		.36	.95		1.31	1.95
1550	Taped and finished (level 4 finish)		475	.034		.35	1.35		1.70	2.61
1590	With compound skim coat (level 5 finish)		540	.030		.40	1.18		1.58	2.39
1600	Fire resistant, no finish included		675	.024		.32	.95		1.27	1.91
1650	Taped and finished (level 4 finish)		475	.034		.36	1.35		1.71	2.62
1690	With compound skim coat (level 5 finish)		540	.030		.41	1.18		1.59	2.40
1700	Water resistant, no finish included		675	.024		.55	.95		1.50	2.17
1750	Taped and finished (level 4 finish)		475	.034		.52	1.35		1.87	2.79
1790	With compound skim coat (level 5 finish)		540	.030		.57	1.18		1.75	2.57
1800	Mold resistant, no finish included		675	.024		.56	.95		1.51	2.18
1810	Taped and finished (level 4 finish)		475	.034		.53	1.35		1.88	2.81
1820	With compound skim coat (level 5 finish)		540	.030		.58	1.18		1.76	2.58
2000	5/8" thick, on walls, standard, no finish included		2000	.008		.34	.32		.66	.90
2050	Taped and finished (level 4 finish)		965	.017		.38	.66		1.04	1.51
2090	With compound skim coat (level 5 finish)		775	.021		.43	.83		1.26	1.83
2100	Fire resistant, no finish included		2000	.008		.36	.32		.68	.93
2150	Taped and finished (level 4 finish)		965	.017		.40	.66		1.06	1.53
2195	With compound skim coat (level 5 finish)		775	.021		.45	.83		1.28	1.85
2200	Water resistant, no finish included		2000	.008		.47	.32		.79	1.05
2250	Taped and finished (level 4 finish)		965	.017		.51	.66		1.17	1.65
2290	With compound skim coat (level 5 finish)		775	.021		.56	.83		1.39	1.97
2300	Prefinished, vinyl, clipped to studs		900	.018		1.02	.71		1.73	2.29
2510	Mold resistant, no finish included		2000	.008		.60	.32		.92	1.19
2520	Taped and finished (level 4 finish)		965	.017		.64	.66		1.30	1.80
2530	With compound skim coat (level 5 finish)		775	.021		.69	.83		1.52	2.12
3000	On ceilings, standard, no finish included		1800	.009		.34	.36		.70	.96
3050	Taped and finished (level 4 finish)		765	.021		.38	.84		1.22	1.80
3090	With compound skim coat (level 5 finish)		615	.026		.43	1.04		1.47	2.19
3100	Fire resistant, no finish included		1800	.009		.36	.36		.72	.99
3150	Taped and finished (level 4 finish)		765	.021		.40	.84		1.24	1.82
3190	With compound skim coat (level 5 finish)		615	.026		.45	1.04		1.49	2.21
3200	Water resistant, no finish included		1800	.009		.47	.36		.83	1.11
3250	Taped and finished (level 4 finish)		765	.021		.51	.84		1.35	1.94
3290	With compound skim coat (level 5 finish)		615	.026		.56	1.04		1.60	2.33
3300	Mold resistant, no finish included		1800	.009		.60	.36		.96	1.25
3310	Taped and finished (level 4 finish)	2 Carp	765	.021	S.F.	.64	.84		1.48	2.09

09 29 10.30 Gypsum Board

	09 29 10.30 Gypsum Board	Crew	Daily Output	Labor-Hours	Unit	Material	2009 Bare Costs Labor	Equipment	Total	Total Incl O&P
3320	With compound skim coat (level 5 finish)	2 Carp	615	.026	S.F.	.69	1.04		1.73	2.48
3500	On beams, columns, or soffits, no finish included		675	.024		.39	.95		1.34	1.99
3550	Taped and finished (level 4 finish)		475	.034		.44	1.35		1.79	2.70
3590	With compound skim coat (level 5 finish)		380	.042		.49	1.68		2.17	3.32
3600	Fire resistant, no finish included		675	.024		.41	.95		1.36	2.02
3650	Taped and finished (level 4 finish)		475	.034		.46	1.35		1.81	2.73
3690	With compound skim coat (level 5 finish)		380	.042		.45	1.68		2.13	3.27
3700	Water resistant, no finish included		675	.024		.54	.95		1.49	2.15
3750	Taped and finished (level 4 finish)		475	.034		.59	1.35		1.94	2.87
3790	With compound skim coat (level 5 finish)		380	.042		.56	1.68		2.24	3.39
3800	Mold resistant, no finish included		675	.024		.69	.95		1.64	2.32
3810	Taped and finished (level 4 finish)		475	.034		.74	1.35		2.09	3.03
3820	With compound skim coat (level 5 finish)		380	.042		.69	1.68		2.37	3.54
4000	Fireproofing, beams or columns, 2 layers, 1/2" thick, incl finish		330	.048		.68	1.94		2.62	3.95
4010	Mold resistant		330	.048		1.02	1.94		2.96	4.32
4050	5/8" thick		300	.053		.80	2.13		2.93	4.41
4060	Mold resistant		300	.053		1.28	2.13		3.41	4.93
4100	3 layers, 1/2" thick		225	.071		1	2.84		3.84	5.80
4110	Mold resistant		225	.071		1.51	2.84		4.35	6.35
4150	5/8" thick		210	.076		1.21	3.04		4.25	6.35
4160	Mold resistant		210	.076		1.93	3.04		4.97	7.10
5050	For 1" thick coreboard on columns		480	.033		.88	1.33		2.21	3.17
5100	For foil-backed board, add					.14			.14	.15
5200	For work over 8' high, add	2 Carp	3060	.005			.21		.21	.34
5270	For textured spray, add	2 Lath	1600	.010		.08	.36		.44	.65
5300	For distribution cost over 3 stories high, add per story	2 Carp	6100	.003			.10		.10	.17
5350	For finishing inner corners, add		950	.017	L.F.	.09	.67		.76	1.21
5355	For finishing outer corners, add		1250	.013		.21	.51		.72	1.07
5500	For acoustical sealant, add per bead	1 Carp	500	.016		.04	.64		.68	1.11
5550	Sealant, 1 quart tube				Ea.	6.85			6.85	7.55
5600	Sound deadening board, 1/4" gypsum	2 Carp	1800	.009	S.F.	.37	.36		.73	1
5650	1/2" wood fiber	"	1800	.009	"	.28	.36		.64	.90
9000	Minimum labor/equipment charge	1 Carp	2	4	Job		160		160	264

09 29 10.50 High Abuse Gypsum Board

	09 29 10.50 High Abuse Gypsum Board	Crew	Daily Output	Labor-Hours	Unit	Material	Labor	Equipment	Total	Total Incl O&P
0010	**HIGH ABUSE GYPSUM BOARD**, fiber reinforced, nailed or									
0100	screwed to studs unless otherwise noted									
0110	1/2" thick, on walls, no finish included	2 Carp	1800	.009	S.F.	.87	.36		1.23	1.55
0120	Taped and finished (level 4 finish)		870	.018		.91	.73		1.64	2.21
0130	With compound skim coat (level 5 finish)		700	.023		.96	.91		1.87	2.56
0150	On ceilings, no finish included		1620	.010		.87	.39		1.26	1.61
0160	Taped and finished (level 4 finish)		690	.023		.91	.93		1.84	2.53
0170	With compound skim coat (level 5 finish)		550	.029		.96	1.16		2.12	2.97
0210	5/8" thick, on walls, no finish included		1800	.009		.94	.36		1.30	1.62
0220	Taped and finished (level 4 finish)		870	.018		.98	.73		1.71	2.29
0230	With compound skim coat (level 5 finish)		700	.023		1.03	.91		1.94	2.64
0250	On ceilings, no finish included		1620	.010		.94	.39		1.33	1.68
0260	Taped and finished (level 4 finish)		690	.023		.98	.93		1.91	2.61
0270	With compound skim coat (level 5 finish)		550	.029		1.03	1.16		2.19	3.05
0310	5/8" thick, on walls, very high impact, no finish included		1800	.009		.96	.36		1.32	1.65
0320	Taped and finished (level 4 finish)		870	.018		1	.73		1.73	2.31
0330	With compound skim coat (level 5 finish)		700	.023		1.05	.91		1.96	2.66
0350	On ceilings, no finish included		1620	.010		.96	.39		1.35	1.71

09 29 Gypsum Board

09 29 10 – Gypsum Board Panels

09 29 10.50 High Abuse Gypsum Board		Crew	Daily Output	Labor-Hours	Unit	Material	2009 Bare Costs Labor	Equipment	Total	Total Incl O&P
0360	Taped and finished (level 4 finish)	2 Carp	690	.023	S.F.	1	.93		1.93	2.63
0370	With compound skim coat (level 5 finish)	↓	550	.029	↓	1.05	1.16		2.21	3.07
0400	High abuse, gypsum core, paper face									
0410	1/2" thick, on walls, no finish included	2 Carp	1800	.009	S.F.	.75	.36		1.11	1.42
0420	Taped and finished (level 4 finish)		870	.018		.79	.73		1.52	2.08
0430	With compound skim coat (level 5 finish)		700	.023		.84	.91		1.75	2.43
0450	On ceilings, no finish included		1620	.010		.75	.39		1.14	1.48
0460	Taped and finished (level 4 finish)		690	.023		.79	.93		1.72	2.40
0470	With compound skim coat (level 5 finish)		550	.029		.84	1.16		2	2.84
0510	5/8" thick, on walls, no finish included		1800	.009		.76	.36		1.12	1.43
0520	Taped and finished (level 4 finish)		870	.018		.80	.73		1.53	2.09
0530	With compound skim coat (level 5 finish)		700	.023		.85	.91		1.76	2.44
0550	On ceilings, no finish included		1620	.010		.76	.39		1.15	1.49
0560	Taped and finished (level 4 finish)		690	.023		.80	.93		1.73	2.41
0570	With compound skim coat (level 5 finish)		550	.029		.85	1.16		2.01	2.85
1000	For high ceilings, over 8' high, add		2750	.006			.23		.23	.38
1010	For distribution cost over 3 stories high, add per story	↓	5500	.003	↓		.12		.12	.19

09 29 15 – Gypsum Board Accessories

09 29 15.10 Accessories, Gypsum Board		Crew	Daily Output	Labor-Hours	Unit	Material	Labor	Equipment	Total	Total Incl O&P
0010	**ACCESSORIES, GYPSUM BOARD**									
0020	Casing bead, galvanized steel	1 Carp	2.90	2.759	C.L.F.	22	110		132	207
0100	Vinyl		3	2.667		22.50	107		129.50	201
0300	Corner bead, galvanized steel, 1" x 1"		4	2		13.95	80		93.95	147
0400	1-1/4" x 1-1/4"		3.50	2.286		18.15	91.50		109.65	171
0600	Vinyl		4	2		17.80	80		97.80	152
0900	Furring channel, galv. steel, 7/8" deep, standard		2.60	3.077		27.50	123		150.50	234
1000	Resilient		2.55	3.137		26.50	125		151.50	237
1100	J trim, galvanized steel, 1/2" wide		3	2.667		22.50	107		129.50	201
1120	5/8" wide		2.95	2.712		24.50	108		132.50	206
1140	L trim, galvanized		3	2.667		19.35	107		126.35	198
1150	U trim, galvanized	↓	2.95	2.712	↓	19.40	108		127.40	201
1160	Screws #6 x 1" A				M	7.20			7.20	7.90
1170	#6 x 1-5/8" A				"	10.55			10.55	11.60
1200	For stud partitions, see Div. 05 41 13.30 and 09 22 16.13									
1500	Z stud, galvanized steel, 1-1/2" wide	1 Carp	2.60	3.077	C.L.F.	37	123		160	244
1600	2" wide		2.55	3.137	"	55	125		180	268
9000	Minimum labor/equipment charge	↓	3	2.667	Job		107		107	176

09 30 Tiling

09 30 13 – Ceramic Tiling

09 30 13.10 Ceramic Tile		Crew	Daily Output	Labor-Hours	Unit	Material	Labor	Equipment	Total	Total Incl O&P
0010	**CERAMIC TILE**									
0050	Base, using 1' x 4" high pc. with 1" x 1" tiles, mud set	D-7	82	.195	L.F.	4.95	6.65		11.60	15.85
0100	Thin set	"	128	.125		4.53	4.26		8.79	11.65
0300	For 6" high base, 1" x 1" tile face, add					.72			.72	.79
0400	For 2" x 2" tile face, add to above					.38			.38	.42
0600	Cove base, 4-1/4" x 4-1/4" high, mud set	D-7	91	.176		3.93	6		9.93	13.70
0700	Thin set		128	.125		4.07	4.26		8.33	11.15
0900	6" x 4-1/4" high, mud set		100	.160		3.67	5.45		9.12	12.60
1000	Thin set	↓	137	.117	↓	3.47	3.98		7.45	10

09 30 13.10 Ceramic Tile		Crew	Daily Output	Labor-Hours	Unit	2009 Bare Costs Material	Labor	Equipment	Total	Total Incl O&P
1200	Sanitary cove base, 6" x 4-1/4" high, mud set	D-7	93	.172	L.F.	4	5.85		9.85	13.55
1300	Thin set		124	.129		3.84	4.40		8.24	11.10
1500	6" x 6" high, mud set		84	.190		4.94	6.50		11.44	15.60
1600	Thin set		117	.137		4.74	4.66		9.40	12.50
1800	Bathroom accessories, average		82	.195	Ea.	10.70	6.65		17.35	22
1900	Bathtub, 5', rec. 4-1/4" x 4-1/4" tile wainscot, adhesive set 6' high		2.90	5.517		156	188		344	465
2100	7' high wainscot		2.50	6.400		179	218		397	535
2200	8' high wainscot		2.20	7.273		190	248		438	600
2400	Bullnose trim, 4-1/4" x 4-1/4", mud set		82	.195	L.F.	3.68	6.65		10.33	14.45
2500	Thin set		128	.125		3.30	4.26		7.56	10.25
2700	6" x 4-1/4" bullnose trim, mud set		84	.190		2.88	6.50		9.38	13.30
2800	Thin set		124	.129		2.74	4.40		7.14	9.90
3000	Floors, natural clay, random or uniform, thin set, color group 1		183	.087	S.F.	4.34	2.98		7.32	9.45
3100	Color group 2		183	.087		4.66	2.98		7.64	9.80
3255	Floors, glazed, thin set, 6" x 6", color group 1		200	.080		3.73	2.73		6.46	8.35
3260	8" x 8" tile		250	.064		3.73	2.18		5.91	7.50
3270	12" x 12" tile		325	.049		4.62	1.68		6.30	7.70
3280	16" x 16" tile		550	.029		6.30	.99		7.29	8.50
3285	Border, 6" x 12" tile		275	.058		11.70	1.98		13.68	15.95
3290	3" x 12" tile		200	.080		34	2.73		36.73	42
3300	Porcelain type, 1 color, color group 2, 1" x 1"		183	.087		4.71	2.98		7.69	9.85
3310	2" x 2" or 2" x 1", thin set		190	.084		5.40	2.87		8.27	10.45
3350	For random blend, 2 colors, add					.88			.88	.97
3360	4 colors, add					1.24			1.24	1.36
3370	For color group 3, add					.50			.50	.55
3380	For abrasive non-slip tile, add					.49			.49	.54
4300	Specialty tile, 4-1/4" x 4-1/4" x 1/2", decorator finish	D-7	183	.087		10	2.98		12.98	15.65
4500	Add for epoxy grout, 1/16" joint, 1" x 1" tile		800	.020		.62	.68		1.30	1.75
4600	2" x 2" tile		820	.020		.59	.66		1.25	1.69
4800	Pregrouted sheets, walls, 4-1/4" x 4-1/4", 6" x 4-1/4"									
4810	and 8-1/2" x 4-1/4", 4 S.F. sheets, silicone grout	D-7	240	.067	S.F.	4.73	2.27		7	8.75
5100	Floors, unglazed, 2 S.F. sheets,									
5110	Urethane adhesive	D-7	180	.089	S.F.	4.71	3.03		7.74	9.95
5400	Walls, interior, thin set, 4-1/4" x 4-1/4" tile		190	.084		2.49	2.87		5.36	7.25
5500	6" x 4-1/4" tile		190	.084		2.86	2.87		5.73	7.65
5700	8-1/2" x 4-1/4" tile		190	.084		3.93	2.87		6.80	8.80
5800	6" x 6" tile		200	.080		3.34	2.73		6.07	7.95
5810	8" x 8" tile		225	.071		4.36	2.42		6.78	8.60
5820	12" x 12" tile		300	.053		3.62	1.82		5.44	6.80
5830	16" x 16" tile		500	.032		3.90	1.09		4.99	6
6000	Decorated wall tile, 4-1/4" x 4-1/4", minimum		270	.059		3.66	2.02		5.68	7.20
6100	Maximum		180	.089		42	3.03		45.03	50.50
6300	Exterior walls, frostproof, mud set, 4-1/4" x 4-1/4"		102	.157		7.30	5.35		12.65	16.35
6400	1-3/8" x 1-3/8"		93	.172		4.94	5.85		10.79	14.60
6600	Crystalline glazed, 4-1/4" x 4-1/4", mud set, plain		100	.160		4.43	5.45		9.88	13.40
6700	4-1/4" x 4-1/4", scored tile		100	.160		5.25	5.45		10.70	14.30
6900	6"x 6" plain		93	.172		5.65	5.85		11.50	15.35
7000	For epoxy grout, 1/16" joints, 4-1/4" tile, add		800	.020		.39	.68		1.07	1.50
7200	For tile set in dry mortar, add		1735	.009			.31		.31	.49
7300	For tile set in Portland cement mortar, add		290	.055		.27	1.88		2.15	3.24
9300	Ceramic tiles, recycled glass, standard colors, 2" x 2" thru 6" x 6" G		190	.084		16.70	2.87		19.57	23
9310	6" x 6" G		200	.080		16.70	2.73		19.43	22.50
9320	8" x 8" G		225	.071		18.10	2.42		20.52	23.50

09 30 Tiling

09 30 13 – Ceramic Tiling

09 30 13.10 Ceramic Tile			Crew	Daily Output	Labor-Hours	Unit	Material	2009 Bare Costs Labor	Equipment	Total	Total Incl O&P
9330	12"x12"	G	D-7	300	.053	S.F.	18.10	1.82		19.92	22.50
9340	Earthtones, 2" x 2" to 4" x 8"	G		190	.084		21	2.87		23.87	27.50
9350	6" x 6"	G		200	.080		21	2.73		23.73	27.50
9360	8" x 8"	G		225	.071		22.50	2.42		24.92	28.50
9370	12" x 12"	G		300	.053		22.50	1.82		24.32	27.50
9380	Deep colors, 2" x 2" to 4" x 8"	G		190	.084		30	2.87		32.87	37.50
9390	6" x 6"	G		200	.080		30	2.73		32.73	37.50
9400	8" x 8"	G		225	.071		31.50	2.42		33.92	39
9410	12" x1 2"	G		300	.053		31.50	1.82		33.32	38
9500	Minimum labor/equipment charge			3.25	4.923	Job		168		168	262

09 30 16 – Quarry Tiling

09 30 16.10 Quarry Tile		Crew	Daily Output	Labor-Hours	Unit	Material	Labor	Equipment	Total	Total Incl O&P
0010	**QUARRY TILE**									
0100	Base, cove or sanitary, mud set, to 5" high, 1/2" thick	D-7	110	.145	L.F.	5.60	4.96		10.56	13.90
0300	Bullnose trim, red, mud set, 6" x 6" x 1/2" thick		120	.133		4.36	4.54		8.90	11.90
0400	4" x 4" x 1/2" thick		110	.145		4.79	4.96		9.75	13
0600	4" x 8" x 1/2" thick, using 8" as edge		130	.123		4.25	4.19		8.44	11.20
0700	Floors, mud set, 1,000 S.F. lots, red, 4" x 4" x 1/2" thick		120	.133	S.F.	7.50	4.54		12.04	15.35
0900	6" x 6" x 1/2" thick		140	.114		7.05	3.90		10.95	13.85
1000	4" x 8" x 1/2" thick		130	.123		7.50	4.19		11.69	14.80
1300	For waxed coating, add					.67			.67	.74
1500	For non-standard colors, add					.40			.40	.44
1600	For abrasive surface, add					.47			.47	.52
1800	Brown tile, imported, 6" x 6" x 3/4"	D-7	120	.133		8.45	4.54		12.99	16.35
1900	8" x 8" x 1"		110	.145		9.10	4.96		14.06	17.75
2100	For thin set mortar application, deduct		700	.023			.78		.78	1.22
2200	For epoxy grout & mortar, 6" x 6" x 1/2", add		350	.046		1.91	1.56		3.47	4.54
2700	Stair tread, 6" x 6" x 3/4", plain		50	.320		5.25	10.90		16.15	23
2800	Abrasive		47	.340		5.30	11.60		16.90	24
3000	Wainscot, 6" x 6" x 1/2", thin set, red		105	.152		4.24	5.20		9.44	12.75
3100	Non-standard colors		105	.152		4.69	5.20		9.89	13.25
3300	Window sill, 6" wide, 3/4" thick		90	.173	L.F.	5.25	6.05		11.30	15.30
3400	Corners		80	.200	Ea.	5.85	6.80		12.65	17.10
9000	Minimum labor/equipment charge		3.25	4.923	Job		168		168	262

09 30 23 – Glass Mosaic Tiling

09 30 23.10 Glass Mosaics		Crew	Daily Output	Labor-Hours	Unit	Material	Labor	Equipment	Total	Total Incl O&P
0010	**GLASS MOSAICS** 3/4" tile on 12" sheets, standard grout									
0300	Color group 1 & 2	D-7	73	.219	S.F.	17.35	7.45		24.80	31
0350	Color group 3		73	.219		18.25	7.45		25.70	31.50
0400	Color group 4		73	.219		25	7.45		32.45	39
0450	Color group 5		73	.219		28	7.45		35.45	42.50
0500	Color group 6		73	.219		34	7.45		41.45	48.50
0600	Color group 7		73	.219		39	7.45		46.45	54.50
0700	Color group 8, golds, silvers & specialties		64	.250		56.50	8.50		65	76

09 30 29 – Metal Tiling

09 30 29.10 Metal Tile		Crew	Daily Output	Labor-Hours	Unit	Material	Labor	Equipment	Total	Total Incl O&P
0010	**METAL TILE** 4' x 4' sheet, 24 ga., tile pattern, nailed									
0200	Stainless steel	2 Carp	512	.031	S.F.	24.50	1.25		25.75	29
0400	Aluminized steel	"	512	.031	"	13.15	1.25		14.40	16.50
9000	Minimum labor/equipment charge	1 Carp	4	2	Job		80		80	132

09 51 23.10 Suspended Acoustic Ceiling Tiles		Crew	Daily Output	Labor-Hours	Unit	2009 Bare Costs Material	Labor	Equipment	Total	Total Incl O&P
0010	**SUSPENDED ACOUSTIC CEILING TILES**, not including									
0100	suspension system									
0300	Fiberglass boards, film faced, 2' x 2' or 2' x 4', 5/8" thick	1 Carp	625	.013	S.F.	.73	.51		1.24	1.64
0400	3/4" thick		600	.013		1.69	.53		2.22	2.74
0500	3" thick, thermal, R11		450	.018		1.53	.71		2.24	2.85
0600	Glass cloth faced fiberglass, 3/4" thick		500	.016		3.04	.64		3.68	4.40
0700	1" thick		485	.016		2.16	.66		2.82	3.47
0820	1-1/2" thick, nubby face		475	.017		2.66	.67		3.33	4.04
1110	Mineral fiber tile, lay-in, 2' x 2' or 2' x 4', 5/8" thick, fine texture		625	.013		.57	.51		1.08	1.47
1115	Rough textured		625	.013		1.26	.51		1.77	2.23
1125	3/4" thick, fine textured		600	.013		1.56	.53		2.09	2.60
1130	Rough textured		600	.013		1.88	.53		2.41	2.95
1135	Fissured		600	.013		2.45	.53		2.98	3.58
1150	Tegular, 5/8" thick, fine textured		470	.017		1.45	.68		2.13	2.72
1155	Rough textured		470	.017		1.90	.68		2.58	3.21
1165	3/4" thick, fine textured		450	.018		2.08	.71		2.79	3.46
1170	Rough textured		450	.018		2.35	.71		3.06	3.76
1175	Fissured		450	.018		3.63	.71		4.34	5.15
1180	For aluminum face, add					6.05			6.05	6.65
1185	For plastic film face, add					.94			.94	1.03
1190	For fire rating, add					.45			.45	.50
1300	Mirror faced panels, 15/16" thick, 2' x 2'	1 Carp	500	.016		10.45	.64		11.09	12.55
1900	Eggcrate, acrylic, 1/2" x 1/2" x 1/2" cubes		500	.016		1.82	.64		2.46	3.06
2100	Polystyrene eggcrate, 3/8" x 3/8" x 1/2" cubes		510	.016		1.53	.63		2.16	2.71
2200	1/2" x 1/2" x 1/2" cubes		500	.016		2.04	.64		2.68	3.30
2400	Luminous panels, prismatic, acrylic		400	.020		2.22	.80		3.02	3.76
2500	Polystyrene		400	.020		1.14	.80		1.94	2.57
2700	Flat white acrylic		400	.020		3.86	.80		4.66	5.55
2800	Polystyrene		400	.020		2.65	.80		3.45	4.24
3000	Drop pan, white, acrylic		400	.020		5.65	.80		6.45	7.55
3100	Polystyrene		400	.020		4.73	.80		5.53	6.50
3600	Perforated aluminum sheets, .024" thick, corrugated, painted		490	.016		2.25	.65		2.90	3.56
3700	Plain		500	.016		3.76	.64		4.40	5.20
3720	Mineral fiber, 24" x 24" or 48", reveal edge, painted, 5/8" thick		600	.013		1.38	.53		1.91	2.40
3740	3/4" thick		575	.014		2.24	.56		2.80	3.38
5020	66 – 78% recycled content, 3/4" thick G		600	.013		1.82	.53		2.35	2.88
5040	Mylar, 42% recycled content, 3/4" thick G		600	.013		4.28	.53		4.81	5.60
9000	Minimum labor/equipment charge		4	2	Job		80		80	132

09 51 23.30 Suspended Ceilings, Complete		Crew	Daily Output	Labor-Hours	Unit	Material	Labor	Equipment	Total	Total Incl O&P
0010	**SUSPENDED CEILINGS, COMPLETE** Including standard									
0100	suspension system but not incl. 1-1/2" carrier channels									
0600	Fiberglass ceiling board, 2' x 4' x 5/8", plain faced	1 Carp	500	.016	S.F.	1.36	.64		2	2.56
0700	Offices, 2' x 4' x 3/4"		380	.021		2.32	.84		3.16	3.94
0800	Mineral fiber, on 15/16" T bar susp. 2' x 2' x 3/4" lay-in board		345	.023		2.35	.93		3.28	4.11
0810	2' x 4' x 5/8" tile		380	.021		1.20	.84		2.04	2.71
0820	Tegular, 2' x 2' x 5/8" tile on 9/16" grid		250	.032		2.40	1.28		3.68	4.74
0830	2' x 4' x 3/4" tile		275	.029		2.87	1.16		4.03	5.10
0900	Luminous panels, prismatic, acrylic		255	.031		2.85	1.25		4.10	5.20
1200	Metal pan with acoustic pad, steel		75	.107		4.16	4.26		8.42	11.65
1300	Painted aluminum		75	.107		2.88	4.26		7.14	10.20
1500	Aluminum, degreased finish		75	.107		4.80	4.26		9.06	12.35
1600	Stainless steel		75	.107		9.05	4.26		13.31	17

09 51 Acoustical Ceilings

09 51 23 – Acoustical Tile Ceilings

09 51 23.30 Suspended Ceilings, Complete		Crew	Daily Output	Labor-Hours	Unit	2009 Bare Costs Material	Labor	Equipment	Total	Total Incl O&P
1800	Tile, Z bar suspension, 5/8" mineral fiber tile	1 Carp	150	.053	S.F.	2.16	2.13		4.29	5.90
1900	3/4" mineral fiber tile		150	.053		2.31	2.13		4.44	6.05
2402	For strip lighting, see Div. 26 51 13.50									
2500	For rooms under 500 S.F., add				S.F.		25%			
9000	Minimum labor/equipment charge	1 Carp	2	4	Job		160		160	264

09 51 53 – Direct-Applied Acoustical Ceilings

09 51 53.10 Ceiling Tile		Crew	Daily Output	Labor-Hours	Unit	Material	Labor	Equipment	Total	Total Incl O&P
0010	**CEILING TILE**, Stapled or cemented									
0100	12" x 12" or 12" x 24", not including furring									
0600	Mineral fiber, vinyl coated, 5/8" thick	1 Carp	300	.027	S.F.	1.85	1.07		2.92	3.80
0700	3/4" thick		300	.027		1.87	1.07		2.94	3.82
0900	Fire rated, 3/4" thick, plain faced		300	.027		1.40	1.07		2.47	3.30
1000	Plastic coated face		300	.027		1.18	1.07		2.25	3.06
1200	Aluminum faced, 5/8" thick, plain		300	.027		1.51	1.07		2.58	3.42
3700	Wall application of above, add		1000	.008			.32		.32	.53
3900	For ceiling primer, add					.13			.13	.14
4000	For ceiling cement, add					.37			.37	.41
9000	Minimum labor/equipment charge	1 Carp	4	2	Job		80		80	132

09 53 Acoustical Ceiling Suspension Assemblies

09 53 23 – Metal Acoustical Ceiling Suspension Assemblies

09 53 23.30 Ceiling Suspension Systems		Crew	Daily Output	Labor-Hours	Unit	Material	Labor	Equipment	Total	Total Incl O&P
0010	**CEILING SUSPENSION SYSTEMS** for boards and tile									
0050	Class A suspension system, 15/16" T bar, 2' x 4' grid	1 Carp	800	.010	S.F.	.63	.40		1.03	1.35
0300	2' x 2' grid		650	.012		.79	.49		1.28	1.67
0310	25% recycled steel, 2' x 4' grid G		800	.010		.73	.40		1.13	1.46
0320	2' x 2' grid G		650	.012		.92	.49		1.41	1.82
0350	For 9/16" grid, add					.16			.16	.18
0360	For fire rated grid, add					.09			.09	.10
0370	For colored grid, add					.21			.21	.23
0400	Concealed Z bar suspension system, 12" module	1 Carp	520	.015		.65	.61		1.26	1.73
0600	1-1/2" carrier channels, 4' O.C., add	"	470	.017		.15	.68		.83	1.29
0700	Carrier channels for ceilings with									
0900	recessed lighting fixtures, add	1 Carp	460	.017	S.F.	.28	.69		.97	1.45
1040	Hanging wire, 12 ga., 4' long		65	.123	C.S.F.	7.65	4.92		12.57	16.50
1080	8' long		65	.123	"	15.25	4.92		20.17	25
5010	Four way support for seismic compression strut # 12 wire		32	.250	Ea.	40	10		50	60.50
5015	# 9 wire		24	.333		68	13.30		81.30	97
5020	Seismic manual compression strut for suspended ceilings		32	.250		120	10		130	149
5030	Hydraulic		32	.250		90	10		100	116
5040	Seismic wall clips for suspended ceilings, 4' O.C.		48	.167		1	6.65		7.65	12.10
9000	Minimum labor/equipment charge		5	1.600	Job		64		64	106

09 63 Masonry Flooring

09 63 13 – Brick Flooring

09 63 13.10 Miscellaneous Brick Flooring		Crew	Daily Output	Labor-Hours	Unit	2009 Bare Costs Material	Labor	Equipment	Total	Total Incl O&P
0010	**MISCELLANEOUS BRICK FLOORING**									
0020	Acid proof shales, red, 8" x 3-3/4" x 1-1/4" thick	D-7	.43	37.209	M	820	1,275		2,095	2,875
0050	2-1/4" thick	D-1	.40	40		950	1,450		2,400	3,400
0200	Acid proof clay brick, 8" x 3-3/4" x 2-1/4" thick		.40	40		870	1,450		2,320	3,300
0250	9" x 4-1/2" x 3"		95	.168	S.F.	3.93	6.10		10.03	14.20
0260	Cast ceramic, pressed, 4" x 8" x 1/2", unglazed	D-7	100	.160		5.95	5.45		11.40	15.10
0270	Glazed		100	.160		7.95	5.45		13.40	17.25
0280	Hand molded flooring, 4" x 8" x 3/4", unglazed		95	.168		7.85	5.75		13.60	17.65
0290	Glazed		95	.168		9.85	5.75		15.60	19.85
0300	8" hexagonal, 3/4" thick, unglazed		85	.188		8.60	6.40		15	19.50
0310	Glazed		85	.188		15.55	6.40		21.95	27
0400	Heavy duty industrial, cement mortar bed, 2" thick, not incl. brick	D-1	80	.200		.79	7.25		8.04	12.60
0450	Acid proof joints, 1/4" wide	"	65	.246		1.37	8.95		10.32	15.95
0500	Pavers, 8" x 4", 1" to 1-1/4" thick, red	D-7	95	.168		3.47	5.75		9.22	12.80
0510	Ironspot	"	95	.168		4.89	5.75		10.64	14.40
0540	1-3/8" to 1-3/4" thick, red	D-1	95	.168		3.34	6.10		9.44	13.55
0560	Ironspot		95	.168		4.84	6.10		10.94	15.20
0580	2-1/4" thick, red		90	.178		3.40	6.45		9.85	14.20
0590	Ironspot		90	.178		5.30	6.45		11.75	16.25
0700	Paver, adobe brick, 6" x 12", 1/2" joint		42	.381		.90	13.85		14.75	23.50
0710	Mexican red, 12" x 12"	1 Tilf	48	.167		1.61	6.35		7.96	11.70
0720	Saltillo, 12" x 12"	"	48	.167		1.13	6.35		7.48	11.20
0800	For sidewalks and patios with pavers, see Div. 32 14 16.10									
0870	For epoxy joints, add	D-1	600	.027	S.F.	2.59	.97		3.56	4.42
0880	For Furan underlayment, add	"	600	.027		2.14	.97		3.11	3.92
0890	For waxed surface, steam cleaned, add	A-1H	1000	.008		.19	.25	.05	.49	.69
9000	Minimum labor/equipment charge	1 Bric	2	4	Job		162		162	262

09 63 40 – Stone Flooring

09 63 40.10 Marble		Crew	Daily Output	Labor-Hours	Unit	Material	Labor	Equipment	Total	Total Incl O&P
0010	**MARBLE**									
0020	Thin gauge tile, 12" x 6", 3/8", white Carara	D-7	60	.267	S.F.	11.15	9.10		20.25	26.50
0100	Travertine		60	.267		12.20	9.10		21.30	27.50
0200	12" x 12" x 3/8", thin set, floors		60	.267		6.05	9.10		15.15	21
0300	On walls		52	.308		9.35	10.50		19.85	26.50
1000	Marble threshold, 4" wide x 36" long x 5/8" thick, white		60	.267	Ea.	6.05	9.10		15.15	21
9000	Minimum labor/equipment charge		3	5.333	Job		182		182	284

09 63 40.20 Slate Tile		Crew	Daily Output	Labor-Hours	Unit	Material	Labor	Equipment	Total	Total Incl O&P
0010	**SLATE TILE**									
0020	Vermont, 6" x 6" x 1/4" thick, thin set	D-7	180	.089	S.F.	6.80	3.03		9.83	12.20
0200	See also Div. 32 14 40.10									
9000	Minimum labor/equipment charge	D-7	3	5.333	Job		182		182	284

09 63 40.95 Concrete Floors		Crew	Daily Output	Labor-Hours	Unit	Material	Labor	Equipment	Total	Total Incl O&P
0010	**CONCRETE FLOORS** And toppings, see Div. 03 35 29.30									

09 64 Wood Flooring

09 64 16 – Wood Block Flooring

09 64 16.10 End Grain Block Flooring		Crew	Daily Output	Labor-Hours	Unit	Material	2009 Bare Costs Labor	Equipment	Total	Total Incl O&P
0010	**END GRAIN BLOCK FLOORING**									
0020	End grain flooring, coated, 2" thick	1 Carp	295	.027	S.F.	3.29	1.08		4.37	5.40
0400	Natural finish, 1" thick, fir		125	.064		3.40	2.56		5.96	7.95
0600	1-1/2" thick, pine		125	.064		3.34	2.56		5.90	7.90
0700	2" thick, pine		125	.064		4.64	2.56		7.20	9.30
9000	Minimum labor/equipment charge		2	4	Job		160		160	264

09 64 19 – Wood Composition Flooring

09 64 19.10 Wood Composition

		Crew	Daily Output	Labor-Hours	Unit	Material	Labor	Equipment	Total	Total Incl O&P
0010	**WOOD COMPOSITION** Gym floors									
0100	2-1/4" x 6-7/8" x 3/8", on 2" grout setting bed	D-7	150	.107	S.F.	5.85	3.64		9.49	12.15
0200	Thin set, on concrete	"	250	.064		5.35	2.18		7.53	9.30
0300	Sanding and finishing, add	1 Carp	200	.040		.79	1.60		2.39	3.51

09 64 23 – Wood Parquet Flooring

09 64 23.10 Wood Parquet

		Crew	Daily Output	Labor-Hours	Unit	Material	Labor	Equipment	Total	Total Incl O&P
0010	**WOOD PARQUET** flooring									
5200	Parquetry, standard, 5/16" thick, not incl. finish, oak, minimum	1 Carp	160	.050	S.F.	4.41	2		6.41	8.15
5300	Maximum		100	.080		4.95	3.20		8.15	10.75
5500	Teak, minimum		160	.050		4.79	2		6.79	8.55
5600	Maximum		100	.080		8.35	3.20		11.55	14.50
5650	13/16" thick, select grade oak, minimum		160	.050		9.20	2		11.20	13.45
5700	Maximum		100	.080		14	3.20		17.20	20.50
5800	Custom parquetry, including finish, minimum		100	.080		15.45	3.20		18.65	22.50
5900	Maximum		50	.160		20.50	6.40		26.90	33
6700	Parquetry, prefinished white oak, 5/16" thick, minimum		160	.050		3.61	2		5.61	7.25
6800	Maximum		100	.080		7.30	3.20		10.50	13.35
7000	Walnut or teak, parquetry, minimum		160	.050		5.35	2		7.35	9.20
7100	Maximum		100	.080		9.35	3.20		12.55	15.60
7200	Acrylic wood parquet blocks, 12" x 12" x 5/16",									
7210	Irradiated, set in epoxy	1 Carp	160	.050	S.F.	7.80	2		9.80	11.90

09 64 29 – Wood Strip and Plank Flooring

09 64 29.10 Wood

		Crew	Daily Output	Labor-Hours	Unit	Material	Labor	Equipment	Total	Total Incl O&P
0010	**WOOD**									
0020	Fir, vertical grain, 1" x 4", not incl. finish, grade B & better	1 Carp	255	.031	S.F.	2.74	1.25		3.99	5.10
0100	C grade & better		255	.031		2.58	1.25		3.83	4.91
4000	Maple, strip, 25/32" x 2-1/4", not incl. finish, select		170	.047		5.25	1.88		7.13	8.90
4100	#2 & better		170	.047		3.32	1.88		5.20	6.75
4300	33/32" x 3-1/4", not incl. finish, #1 grade		170	.047		4.19	1.88		6.07	7.70
4400	#2 & better		170	.047		3.73	1.88		5.61	7.20
4600	Oak, white or red, 25/32" x 2-1/4", not incl. finish									
4700	#1 common	1 Carp	170	.047	S.F.	3.18	1.88		5.06	6.60
4900	Select quartered, 2-1/4" wide		170	.047		2.94	1.88		4.82	6.35
5000	Clear		170	.047		4.01	1.88		5.89	7.50
6100	Prefinished, white oak, prime grade, 2-1/4" wide		170	.047		6.40	1.88		8.28	10.15
6200	3-1/4" wide		185	.043		8.20	1.73		9.93	11.85
6400	Ranch plank		145	.055		7.90	2.20		10.10	12.35
6500	Hardwood blocks, 9" x 9", 25/32" thick		160	.050		5.75	2		7.75	9.65
7400	Yellow pine, 3/4" x 3-1/8", T & G, C & better, not incl. finish		200	.040		2.35	1.60		3.95	5.25
7550	Refinish old floors, see Div. 01 93 13.09									
7800	Sanding and finishing, 2 coats polyurethane	1 Clab	295	.027	S.F.	.78	.86		1.64	2.27
7900	Subfloor and underlayment, see Div. 06 16									
8015	Transition molding, 2 1/4" wide, 5' long	1 Carp	19.20	.417	Ea.	15	16.65		31.65	44

09 64 Wood Flooring

09 64 29 – Wood Strip and Plank Flooring

09 64 29.10 Wood			Crew	Daily Output	Labor-Hours	Unit	2009 Bare Costs Material	Labor	Equipment	Total	Total Incl O&P
8300	Floating floor, wood composition strip, complete		1 Clab	133	.060	S.F.	4.23	1.90		6.13	7.80
8310	Components, T & G wood composite strips						3.66			3.66	4.03
8320	Film						.15			.15	.16
8330	Foam						.30			.30	.33
8340	Adhesive						.17			.17	.18
8350	Installation kit					↓	.17			.17	.19
8360	Trim, 2" wide x 3' long					L.F.	2.49			2.49	2.74
8370	Reducer moulding					"	4.29			4.29	4.72
8600	Flooring, wood, bamboo strips, unfinished, 5/8" x 4" x 3'	G	1 Carp	255	.031	S.F.	4.44	1.25		5.69	6.95
8610	5/8" x 4" x 4'	G		275	.029		4.61	1.16		5.77	6.95
8620	5/8" x 4" x 6'	G		295	.027		5.05	1.08		6.13	7.35
8630	Finished, 5/8" x 4" x 3'	G		255	.031		4.89	1.25		6.14	7.45
8640	5/8" x 4" x 4'	G		275	.029		5.10	1.16		6.26	7.55
8650	5/8" x 4" x 6'	G		295	.027	↓	5.15	1.08		6.23	7.50
8660	Stair treads, unfinished, 1-1/16" x 11-1/2" x 4'	G		18	.444	Ea.	42.50	17.75		60.25	76.50
8670	Finished, 1-1/16" x 11-1/2" x 4'	G		18	.444		65	17.75		82.75	101
8680	Stair risers, unfinished, 5/8" x 7-1/2" x 4'	G		18	.444		15.75	17.75		33.50	47
8690	Finished, 5/8" x 7-1/2" x 4'	G		18	.444		30	17.75		47.75	62.50
8700	Stair nosing, unfinished, 6' long	G		16	.500		35	20		55	71.50
8710	Finished, 6' long	G		16	.500	↓	44	20		64	81
9000	Minimum labor/equipment charge		↓	2	4	Job		160		160	264

09 64 66 – Wood Athletic Flooring

09 64 66.10 Gymnasium Flooring		Crew	Daily Output	Labor-Hours	Unit	Material	Labor	Equipment	Total	Total Incl O&P
0010	GYMNASIUM FLOORING									
0600	Gym floor, in mastic, over 2 ply felt, #2 & better									
0700	25/32" thick maple	1 Carp	100	.080	S.F.	4.03	3.20		7.23	9.75
0900	33/32" thick maple		98	.082		4.65	3.26		7.91	10.50
1000	For 1/2" corkboard underlayment, add	↓	750	.011		.90	.43		1.33	1.69
1300	For #1 grade maple, add				↓	.47			.47	.52
1600	Maple flooring, over sleepers, #2 & better									
1700	25/32" thick	1 Carp	85	.094	S.F.	4.32	3.76		8.08	10.95
1900	33/32" thick	"	83	.096		5.05	3.85		8.90	11.90
2000	For #1 grade, add					.50			.50	.55
2200	For 3/4" subfloor, add	1 Carp	350	.023		1.12	.91		2.03	2.74
2300	With two 1/2" subfloors, 25/32" thick	"	69	.116	↓	5.40	4.63		10.03	13.60
2500	Maple, incl. finish, #2 & btr., 25/32" thick, on rubber									
2600	Sleepers, with two 1/2" subfloors	1 Carp	76	.105	S.F.	5.80	4.21		10.01	13.35
2800	With steel spline, double connection to channels	"	73	.110		6.20	4.38		10.58	14.05
2900	For 33/32" maple, add					.67			.67	.74
3100	For #1 grade maple, add					.50			.50	.55
3500	For termite proofing all of the above, add					.27			.27	.30
3700	Portable hardwood, prefinished panels	1 Carp	83	.096		7.75	3.85		11.60	14.85
3720	Insulated with polystyrene, 1" thick, add		165	.048		.67	1.94		2.61	3.94
3750	Running tracks, Sitka spruce surface, 25/32" x 2-1/4"		62	.129		14.15	5.15		19.30	24
3770	3/4" plywood surface, finished	↓	100	.080	↓	3.36	3.20		6.56	9

09 65 Resilient Flooring

09 65 10 – Resilient Tile Underlayment

09 65 10.10 Latex Underlayment		Crew	Daily Output	Labor-Hours	Unit	Material	2009 Bare Costs Labor	Equipment	Total	Total Incl O&P
0010	**LATEX UNDERLAYMENT**									
3600	Latex underlayment, 1/8" thk., cementitious for resilient flooring	1 Tilf	160	.050	S.F.	1.72	1.91		3.63	4.87
4000	Liquid, fortified				Gal.	38.50			38.50	42

09 65 13 – Resilient Base and Accessories

09 65 13.13 Resilient Base		Crew	Daily Output	Labor-Hours	Unit	Material	Labor	Equipment	Total	Total Incl O&P
0010	**RESILIENT BASE**									
0800	Base, cove, rubber or vinyl									
1100	Standard colors, 0.080" thick, 2-1/2" high	1 Tilf	315	.025	L.F.	.75	.97		1.72	2.34
1150	4" high		315	.025		.93	.97		1.90	2.53
1200	6" high		315	.025		1.25	.97		2.22	2.89
1450	1/8" thick, 2-1/2" high		315	.025		.75	.97		1.72	2.34
1500	4" high		315	.025		.71	.97		1.68	2.29
1550	6" high		315	.025		1.30	.97		2.27	2.94
1600	Corners, 2-1/2" high		315	.025	Ea.	1.45	.97		2.42	3.11
1630	4" high		315	.025		2.75	.97		3.72	4.54
1660	6" high		315	.025		2.85	.97		3.82	4.65

09 65 13.23 Resilient Stair Treads and Risers		Crew	Daily Output	Labor-Hours	Unit	Material	Labor	Equipment	Total	Total Incl O&P
0010	**RESILIENT STAIR TREADS AND RISERS**									
0300	Rubber, molded tread, 12" wide, 5/16" thick, black	1 Tilf	115	.070	L.F.	11.75	2.65		14.40	17.10
0400	Colors		115	.070		12.25	2.65		14.90	17.65
0600	1/4" thick, black		115	.070		11.50	2.65		14.15	16.80
0700	Colors		115	.070		12.25	2.65		14.90	17.65
0900	Grip strip safety tread, colors, 5/16" thick		115	.070		16	2.65		18.65	22
1000	3/16" thick		120	.067		11.15	2.54		13.69	16.25
1200	Landings, smooth sheet rubber, 1/8" thick		120	.067	S.F.	5.75	2.54		8.29	10.30
1300	3/16" thick		120	.067	"	7.25	2.54		9.79	11.95
1500	Nosings, 3" wide, 3/16" thick, black		140	.057	L.F.	3.15	2.18		5.33	6.90
1600	Colors		140	.057		3.11	2.18		5.29	6.85
1800	Risers, 7" high, 1/8" thick, flat		250	.032		6.15	1.22		7.37	8.70
1900	Coved		250	.032		5.05	1.22		6.27	7.50
2100	Vinyl, molded tread, 12" wide, colors, 1/8" thick		115	.070		5.65	2.65		8.30	10.35
2200	1/4" thick		115	.070		7.50	2.65		10.15	12.40
2300	Landing material, 1/8" thick		200	.040	S.F.	5.65	1.52		7.17	8.60
2400	Riser, 7" high, 1/8" thick, coved		175	.046	L.F.	3.25	1.74		4.99	6.30
2500	Tread and riser combined, 1/8" thick		80	.100	"	8.85	3.81		12.66	15.70
9000	Minimum labor/equipment charge		3	2.667	Job		102		102	159

09 65 16 – Resilient Sheet Flooring

09 65 16.10 Rubber and Vinyl Sheet Flooring		Crew	Daily Output	Labor-Hours	Unit	Material	Labor	Equipment	Total	Total Incl O&P
0010	**RUBBER AND VINYL SHEET FLOORING**									
5500	Linoleum, sheet goods	1 Tilf	360	.022	S.F.	4.55	.85		5.40	6.30
5900	Rubber, sheet goods, 36" wide, 1/8" thick		120	.067		6.25	2.54		8.79	10.85
5950	3/16" thick		100	.080		8.50	3.05		11.55	14.10
6000	1/4" thick		90	.089		10	3.39		13.39	16.30
8000	Vinyl sheet goods, backed, .065" thick, minimum		250	.032		3.50	1.22		4.72	5.75
8050	Maximum		200	.040		3.39	1.52		4.91	6.10
8100	.080" thick, minimum		230	.035		3.50	1.33		4.83	5.90
8150	Maximum		200	.040		4.95	1.52		6.47	7.85
8200	.125" thick, minimum		230	.035		4	1.33		5.33	6.45
8250	Maximum		200	.040		5.95	1.52		7.47	8.95
8700	Adhesive cement, 1 gallon per 200 to 300 S.F.				Gal.	24.50			24.50	26.50
8800	Asphalt primer, 1 gallon per 300 S.F.					11.65			11.65	12.80

09 65 Resilient Flooring

09 65 16 – Resilient Sheet Flooring

09 65 16.10 Rubber and Vinyl Sheet Flooring		Crew	Daily Output	Labor-Hours	Unit	Material	Labor	Equipment	Total	Total Incl O&P
8900	Emulsion, 1 gallon per 140 S.F.				Gal.	15			15	16.50

09 65 19 – Resilient Tile Flooring

09 65 19.10 Miscellaneous Resilient Tile Flooring		Crew	Daily Output	Labor-Hours	Unit	Material	Labor	Equipment	Total	Total Incl O&P
0010	**MISCELLANEOUS RESILIENT TILE FLOORING**									
2200	Cork tile, standard finish, 1/8" thick G	1 Tilf	315	.025	S.F.	5.75	.97		6.72	7.85
2250	3/16" thick G		315	.025		5.75	.97		6.72	7.85
2300	5/16" thick G		315	.025		7.15	.97		8.12	9.35
2350	1/2" thick G		315	.025		8.45	.97		9.42	10.80
2500	Urethane finish, 1/8" thick G		315	.025		6.85	.97		7.82	9.05
2550	3/16" thick G		315	.025		7.20	.97		8.17	9.40
2600	5/16" thick G		315	.025		9.20	.97		10.17	11.65
2650	1/2" thick G		315	.025		12.40	.97		13.37	15.15
6050	Rubber tile, marbleized colors, 12" x 12", 1/8" thick		400	.020		6.20	.76		6.96	8
6100	3/16" thick		400	.020		9.25	.76		10.01	11.40
6300	Special tile, plain colors, 1/8" thick		400	.020		9.25	.76		10.01	11.40
6350	3/16" thick		400	.020		9.50	.76		10.26	11.65
6410	Raised, radial or square, minimum		400	.020		11.50	.76		12.26	13.85
6430	Maximum		400	.020		11.50	.76		12.26	13.85
6450	For golf course, skating rink, etc., 1/4" thick		275	.029		11.50	1.11		12.61	14.40
6700	Synthetic turf, 3/8" thick	↓	90	.089	↓	5.65	3.39		9.04	11.50
6750	Interlocking 2' x 2' squares, 1/2" thick, not									
6810	cemented, for playgrounds, minimum	1 Tilf	210	.038	S.F.	4.50	1.45		5.95	7.20
6850	Maximum		190	.042		11.25	1.60		12.85	14.90
7000	Vinyl composition tile, 12" x 12", 1/16" thick		500	.016		.87	.61		1.48	1.91
7050	Embossed		500	.016		2.07	.61		2.68	3.23
7100	Marbleized		500	.016		2.07	.61		2.68	3.23
7150	Solid		500	.016		2.66	.61		3.27	3.88
7200	3/32" thick, embossed		500	.016		1.30	.61		1.91	2.38
7250	Marbleized		500	.016		2.37	.61		2.98	3.56
7300	Solid		500	.016		2.20	.61		2.81	3.37
7350	1/8" thick, marbleized		500	.016		1.70	.61		2.31	2.82
7400	Solid		500	.016		2.14	.61		2.75	3.30
7450	Conductive		500	.016		5.35	.61		5.96	6.85
7500	Vinyl tile, 12" x 12", .050" thick, minimum		500	.016		2.95	.61		3.56	4.20
7550	Maximum		500	.016		5.90	.61		6.51	7.45
7600	1/8" thick, minimum		500	.016		4.20	.61		4.81	5.55
7650	Solid colors		500	.016		5.85	.61		6.46	7.40
7700	Marbleized or Travertine pattern		500	.016		5.65	.61		6.26	7.15
7750	Florentine pattern		500	.016		5.70	.61		6.31	7.25
7800	Maximum		500	.016	↓	12.35	.61		12.96	14.55
9500	Minimum labor/equipment charge	↓	4	2	Job		76		76	119

09 65 33 – Conductive Resilient Flooring

09 65 33.10 Conductive Rubber and Vinyl Flooring		Crew	Daily Output	Labor-Hours	Unit	Material	Labor	Equipment	Total	Total Incl O&P
0010	**CONDUCTIVE RUBBER AND VINYL FLOORING**									
1700	Conductive flooring, rubber tile, 1/8" thick	1 Tilf	315	.025	S.F.	5	.97		5.97	7
1800	Homogeneous vinyl tile, 1/8" thick	"	315	.025	"	6.95	.97		7.92	9.15

Note: Material, Labor, Equipment and Total columns are 2009 Bare Costs.

09 66 Terrazzo Flooring

09 66 13 – Portland Cement Terrazzo Flooring

09 66 13.10 Portland Cement Terrazzo		Crew	Daily Output	Labor-Hours	Unit	2009 Bare Costs Material	Labor	Equipment	Total	Total Incl O&P
0010	**PORTLAND CEMENT TERRAZZO**, cast-in-place									
0020	Cove base, 6" high, 16ga. zinc	1 Mstz	20	.400	L.F.	3.32	15.10		18.42	27
0100	Curb, 6" high and 6" wide		6	1.333		5.25	50.50		55.75	84.50
0300	Divider strip for floors, 14 ga., 1-1/4" deep, zinc		375	.021		1.21	.80		2.01	2.59
0400	Brass		375	.021		2.23	.80		3.03	3.71
0600	Heavy top strip 1/4" thick, 1-1/4" deep, zinc		300	.027		1.76	1.01		2.77	3.51
0900	Galv. bottoms, brass		300	.027		3.04	1.01		4.05	4.91
1200	For thin set floors, 16 ga., 1/2" x 1/2", zinc		350	.023		.75	.86		1.61	2.18
1300	Brass		350	.023		1.64	.86		2.50	3.15
1500	Floor, bonded to concrete, 1-3/4" thick, gray cement	J-3	75	.213	S.F.	2.87	7.30	3.38	13.55	18.30
1600	White cement, mud set		75	.213		3.27	7.30	3.38	13.95	18.75
1800	Not bonded, 3" total thickness, gray cement		70	.229		3.59	7.85	3.62	15.06	20
1900	White cement, mud set		70	.229		3.92	7.85	3.62	15.39	20.50
2100	For Venetian terrazzo, 1" topping, add					50%	50%			
2200	For heavy duty abrasive terrazzo, add					50%	50%			
2700	Monolithic terrazzo, 1/2" thick									
2710	10' panels	J-3	125	.128	S.F.	2.61	4.39	2.03	9.03	11.95
3000	Stairs, cast in place, pan filled treads		30	.533	L.F.	2.61	18.30	8.45	29.36	40.50
3100	Treads and risers		14	1.143	"	5.40	39	18.10	62.50	87.50
3300	For stair landings, add to floor prices						50%			
3400	Stair stringers and fascia	J-3	30	.533	S.F.	4.37	18.30	8.45	31.12	42.50
3600	For abrasive metal nosings on stairs, add		150	.107	L.F.	7.70	3.66	1.69	13.05	16.05
3700	For abrasive surface finish, add		600	.027	S.F.	1.03	.91	.42	2.36	3.02
3900	For raised abrasive strips, add		150	.107	L.F.	.98	3.66	1.69	6.33	8.65
4000	Wainscot, bonded, 1-1/2" thick		30	.533	S.F.	3.32	18.30	8.45	30.07	41.50
4200	Epoxy terrazzo, 1/4" thick		40	.400	"	4.88	13.70	6.35	24.93	34
4300	Stone chips, onyx gemstone, per 50 lb. bag				Bag	16.30			16.30	17.95
9000	Minimum labor/equipment charge	1 Mstz	1	8	Job		300		300	470

09 66 13.30 Terrazzo, Precast		Crew	Daily Output	Labor-Hours	Unit	Material	Labor	Equipment	Total	Total Incl O&P
0010	**TERRAZZO, PRECAST**									
0020	Base, 6" high, straight	1 Mstz	70	.114	L.F.	10.40	4.31		14.71	18.20
0100	Cove		60	.133		12.15	5.05		17.20	21
0300	8" high, straight		60	.133		10.85	5.05		15.90	19.75
0400	Cove		50	.160		15.95	6.05		22	27
0600	For white cement, add					.43			.43	.47
0700	For 16 ga. zinc toe strip, add					1.60			1.60	1.76
0900	Curbs, 4" x 4" high	1 Mstz	40	.200		30.50	7.55		38.05	45.50
1000	8" x 8" high	"	30	.267		34.50	10.05		44.55	53.50
1200	Floor tiles, non-slip, 1" thick, 12" x 12"	D-1	60	.267	S.F.	17.70	9.70		27.40	35
1300	1-1/4" thick, 12" x 12"		60	.267		19.60	9.70		29.30	37
1500	16" x 16"		50	.320		21.50	11.65		33.15	42.50
1600	1-1/2" thick, 16" x 16"		45	.356		19.45	12.90		32.35	42.50
1800	For Venetian terrazzo, add					5.85			5.85	6.40
1900	For white cement, add					.55			.55	.61
2400	Stair treads, 1-1/2" thick, non-slip, three line pattern	2 Mstz	70	.229	L.F.	39.50	8.60		48.10	57
2500	Nosing and two lines		70	.229		39.50	8.60		48.10	57
2700	2" thick treads, straight		60	.267		43	10.05		53.05	62.50
2800	Curved		50	.320		56	12.05		68.05	80.50
3000	Stair risers, 1" thick, to 6" high, straight sections		60	.267		9.80	10.05		19.85	26.50
3100	Cove		50	.320		13.90	12.05		25.95	34
3300	Curved, 1" thick, to 6" high, vertical		48	.333		19.05	12.55		31.60	40.50
3400	Cove		38	.421		36	15.85		51.85	64.50

09 66 Terrazzo Flooring

09 66 13 – Portland Cement Terrazzo Flooring

09 66 13.30 Terrazzo, Precast		Crew	Daily Output	Labor-Hours	Unit	2009 Bare Costs Material	Labor	Equipment	Total	Total Incl O&P
3600	Stair tread and riser, single piece, straight, minimum	2 Mstz	60	.267	L.F.	50	10.05		60.05	70.50
3700	Maximum		40	.400		65	15.10		80.10	94.50
3900	Curved tread and riser, minimum		40	.400		70	15.10		85.10	101
4000	Maximum		32	.500		88.50	18.85		107.35	127
4200	Stair stringers, notched, 1″ thick		25	.640		28.50	24		52.50	69
4300	2″ thick		22	.727		34	27.50		61.50	80
4500	Stair landings, structural, non-slip, 1-1/2″ thick		85	.188	S.F.	31	7.10		38.10	45.50
4600	3″ thick		75	.213		44.50	8.05		52.55	61
4800	Wainscot, 12″ x 12″ x 1″ tiles	1 Mstz	12	.667		6.50	25		31.50	46.50
4900	16″ x 16″ x 1-1/2″ tiles	″	8	1		13.70	37.50		51.20	74
9500	Minimum labor/equipment charge	1 Tilf	2	4	Job		152		152	238

09 66 16 – Terrazzo Floor Tile

09 66 16.10 Tile or Terrazzo Base		Crew	Daily Output	Labor-Hours	Unit	Material	Labor	Equipment	Total	Total Incl O&P
0010	**TILE OR TERRAZZO BASE**									
0020	Scratch coat only	1 Mstz	150	.053	S.F.	.41	2.01		2.42	3.59
0500	Scratch and brown coat only		75	.107	″	.78	4.02		4.80	7.15
9000	Minimum labor/equipment charge		1	8	Job		300		300	470

09 66 33 – Conductive Terrazzo Flooring

09 66 33.10 Conductive Terrazzo		Crew	Daily Output	Labor-Hours	Unit	Material	Labor	Equipment	Total	Total Incl O&P
0010	**CONDUCTIVE TERRAZZO**									
2400	Bonded conductive floor for hospitals	J-3	90	.178	S.F.	3.92	6.10	2.81	12.83	16.95
2500	Epoxy terrazzo, 1/4″ thick, minimum		100	.160		4.43	5.50	2.53	12.46	16.25
2550	Average		75	.213		4.88	7.30	3.38	15.56	20.50
2600	Maximum		60	.267		5.35	9.15	4.22	18.72	25

09 67 Fluid-Applied Flooring

09 67 20 – Epoxy-Marble Chip Flooring

09 67 20.13 Elastomeric Liquid Flooring		Crew	Daily Output	Labor-Hours	Unit	Material	Labor	Equipment	Total	Total Incl O&P
0010	**ELASTOMERIC LIQUID FLOORING**									
0020	Cementitious acrylic, 1/4″ thick	C-6	520	.092	S.F.	1.45	3.05	.10	4.60	6.70
0100	3/8″ thick	″	450	.107		1.85	3.53	.12	5.50	7.90
0200	Methyl methachrylate, 1/4″ thick	C-8A	3000	.016		5.40	.55		5.95	6.85
0210	1/8″ thick	″	3000	.016		2.97	.55		3.52	4.15
0300	Cupric oxychloride, on bond coat, minimum	C-6	480	.100		3.15	3.31	.11	6.57	9
0400	Maximum		420	.114		5.25	3.78	.12	9.15	12.10
2400	Mastic, hot laid, 2 coat, 1-1/2″ thick, standard, minimum		690	.070		3.65	2.30	.08	6.03	7.85
2500	Maximum		520	.092		4.69	3.05	.10	7.84	10.25
2700	Acidproof, minimum		605	.079		4.69	2.62	.09	7.40	9.50
2800	Maximum		350	.137		6.50	4.53	.15	11.18	14.70
3000	Neoprene, trowelled on, 1/4″ thick, minimum		545	.088		3.59	2.91	.10	6.60	8.80
3100	Maximum		430	.112		4.94	3.69	.12	8.75	11.65
4300	Polyurethane, with suspended vinyl chips, minimum		1065	.045		7	1.49	.05	8.54	10.20
4500	Maximum		860	.056		10.05	1.84	.06	11.95	14.20

09 67 20.16 Epoxy Terrazzo		Crew	Daily Output	Labor-Hours	Unit	Material	Labor	Equipment	Total	Total Incl O&P
0010	**EPOXY TERRAZZO**									
1800	Epoxy terrazzo, 1/4″ thick, chemical resistant, minimum	J-3	200	.080	S.F.	5.25	2.74	1.27	9.26	11.50
1900	Maximum		150	.107		7.70	3.66	1.69	13.05	16.05
2100	Conductive, minimum		210	.076		6.90	2.61	1.21	10.72	13
2200	Maximum		160	.100		8.95	3.43	1.58	13.96	16.95

09 67 Fluid-Applied Flooring

09 67 20 – Epoxy-Marble Chip Flooring

09 67 20.19 Polyacrylate Terrazzo		Crew	Daily Output	Labor-Hours	Unit	Material	2009 Bare Costs Labor	Equipment	Total	Total Incl O&P
0010	**POLYACRYLATE TERRAZZO**									
3150	Polyacrylate, 1/4" thick, minimum	C-6	735	.065	S.F.	2.98	2.16	.07	5.21	6.90
3170	Maximum		480	.100		3.68	3.31	.11	7.10	9.55
3200	3/8" thick, minimum		620	.077		3.88	2.56	.08	6.52	8.55
3220	Maximum		480	.100		5.15	3.31	.11	8.57	11.15
3300	Conductive, 1/4" thick, minimum		450	.107		6.25	3.53	.12	9.90	12.75
3330	Maximum		305	.157		7.75	5.20	.17	13.12	17.20
3350	3/8" thick, minimum		365	.132		8.20	4.35	.14	12.69	16.25
3370	Maximum		255	.188		9.70	6.20	.20	16.10	21
3450	Granite, conductive, 1/4" thick, minimum		695	.069		7.90	2.28	.07	10.25	12.50
3470	Maximum		420	.114		10.20	3.78	.12	14.10	17.50
3500	3/8" thick, minimum		695	.069		11.55	2.28	.07	13.90	16.50
3520	Maximum	↓	380	.126	↓	13.95	4.17	.14	18.26	22.50

09 67 20.26 Quartz Flooring

		Crew	Daily Output	Labor-Hours	Unit	Material	Labor	Equipment	Total	Total Incl O&P
0010	**QUARTZ FLOORING**									
0600	Epoxy, with colored quartz chips, broadcast, minimum	C-6	675	.071	S.F.	2.49	2.35	.08	4.92	6.65
0700	Maximum		490	.098		3.02	3.24	.11	6.37	8.75
0900	Trowelled, minimum		560	.086		3.21	2.83	.09	6.13	8.25
1000	Maximum	↓	480	.100	↓	4.69	3.31	.11	8.11	10.65
1200	Heavy duty epoxy topping, 1/4" thick,									
1300	500 to 1,000 S.F.	C-6	420	.114	S.F.	4.67	3.78	.12	8.57	11.45
1500	1,000 to 2,000 S.F.		450	.107		3.87	3.53	.12	7.52	10.15
1600	Over 10,000 S.F.		480	.100		3.62	3.31	.11	7.04	9.50
3600	Polyester, with colored quartz chips, 1/16" thick, minimum		1065	.045		2.67	1.49	.05	4.21	5.40
3700	Maximum		560	.086		3.63	2.83	.09	6.55	8.70
3900	1/8" thick, minimum		810	.059		3.12	1.96	.06	5.14	6.70
4000	Maximum		675	.071		4.22	2.35	.08	6.65	8.55
4200	Polyester, heavy duty, compared to epoxy, add	↓	2590	.019	↓	1.27	.61	.02	1.90	2.42

09 68 Carpeting

09 68 05 – Carpet Accessories

09 68 05.11 Flooring Transition Strip

		Crew	Daily Output	Labor-Hours	Unit	Material	Labor	Equipment	Total	Total Incl O&P
0010	**FLOORING TRANSITION STRIP**									
0107	Clamp down brass divider, 12' strip, vinyl to carpet	1 Tilf	31.25	.256	Ea.	26.50	9.75		36.25	44.50
0117	Vinyl to hard surface	"	31.25	.256	"	26.50	9.75		36.25	44.50

09 68 10 – Carpet Pad

09 68 10.10 Commercial Grade Carpet Pad

		Crew	Daily Output	Labor-Hours	Unit	Material	Labor	Equipment	Total	Total Incl O&P
0010	**COMMERCIAL GRADE CARPET PAD**									
9000	Sponge rubber pad, minimum	1 Tilf	150	.053	S.Y.	4.20	2.03		6.23	7.80
9100	Maximum		150	.053		10	2.03		12.03	14.20
9200	Felt pad, minimum		150	.053		4.03	2.03		6.06	7.60
9300	Maximum		150	.053		7.05	2.03		9.08	10.95
9400	Bonded urethane pad, minimum		150	.053		4.57	2.03		6.60	8.25
9500	Maximum		150	.053		7.80	2.03		9.83	11.80
9600	Prime urethane pad, minimum		150	.053		2.62	2.03		4.65	6.05
9700	Maximum	↓	150	.053	↓	4.84	2.03		6.87	8.50

09 68 13 – Tile Carpeting

09 68 13.10 Carpet Tile

		Crew	Daily Output	Labor-Hours	Unit	Material	Labor	Equipment	Total	Total Incl O&P
0010	**CARPET TILE**									

09 68 Carpeting

09 68 13 – Tile Carpeting

09 68 13.10 Carpet Tile		Crew	Daily Output	Labor-Hours	Unit	Material	Labor	Equipment	Total	Total Incl O&P
						2009 Bare Costs				
0100	Tufted nylon, 18" x 18", hard back, 20 oz.	1 Tilf	150	.053	S.Y.	22	2.03		24.03	27.50
0110	26 oz.		150	.053		38	2.03		40.03	44.50
0200	Cushion back, 20 oz.		150	.053		28	2.03		30.03	33.50
0210	26 oz.		150	.053		43.50	2.03		45.53	51
1100	Tufted, 24" x 24", 24 oz. nylon		80	.100		29.50	3.81		33.31	38
1180	35 oz.		80	.100		34	3.81		37.81	43.50
5060	42 oz.	↓	80	.100	↓	45.50	3.81		49.31	56

09 68 16 – Sheet Carpeting

09 68 16.10 Sheet Carpet		Crew	Daily Output	Labor-Hours	Unit	Material	Labor	Equipment	Total	Total Incl O&P
0010	**SHEET CARPET**									
0700	Nylon, level loop, 26 oz., light to medium traffic	1 Tilf	75	.107	S.Y.	26	4.06		30.06	35
0720	28 oz., light to medium traffic		75	.107		27	4.06		31.06	36
0900	32 oz., medium traffic		75	.107		32.50	4.06		36.56	42
1100	40 oz., medium to heavy traffic		75	.107		48	4.06		52.06	59.50
2920	Nylon plush, 30 oz., medium traffic		57	.140		24	5.35		29.35	35
3000	36 oz., medium traffic		75	.107		32	4.06		36.06	41.50
3100	42 oz., medium to heavy traffic		70	.114		36	4.35		40.35	46.50
3200	46 oz., medium to heavy traffic		70	.114		42	4.35		46.35	53
3300	54 oz., heavy traffic		70	.114		46	4.35		50.35	57.50
3340	60 oz., heavy traffic		70	.114		64	4.35		68.35	77.50
3665	Olefin, 24 oz., light to medium traffic		75	.107		14.50	4.06		18.56	22.50
3670	26 oz., medium traffic		75	.107		15.50	4.06		19.56	23.50
3680	28 oz., medium to heavy traffic		75	.107		17	4.06		21.06	25
3700	32 oz., medium to heavy traffic		75	.107		22	4.06		26.06	30.50
3730	42 oz., heavy traffic		70	.114		28	4.35		32.35	37.50
4110	Wool, level loop, 40 oz., medium traffic		75	.107		107	4.06		111.06	124
4500	50 oz., medium to heavy traffic		75	.107		110	4.06		114.06	127
4700	Patterned, 32 oz., medium to heavy traffic		70	.114		109	4.35		113.35	127
4900	48 oz., heavy traffic	↓	70	.114	↓	112	4.35		116.35	130
5000	For less than full roll (approx. 1500 S.F.), add					25%				
5100	For small rooms, less than 12' wide, add						25%			
5200	For large open areas (no cuts), deduct						25%			
5600	For bound carpet baseboard, add	1 Tilf	300	.027	L.F.	2.25	1.02		3.27	4.07
5610	For stairs, not incl. price of carpet, add	"	30	.267	Riser		10.15		10.15	15.90
5620	For borders and patterns, add to labor						18%			
8950	For tackless, stretched installation, add padding to above									
9850	For "branded" fiber, add				S.Y.	25%				
9910	Minimum labor/equipment charge	1 Tilf	3	2.667	Job		102		102	159

09 68 20 – Athletic Carpet

09 68 20.10 Indoor Athletic Carpet		Crew	Daily Output	Labor-Hours	Unit	Material	Labor	Equipment	Total	Total Incl O&P
0010	**INDOOR ATHLETIC CARPET**									
3700	Polyethylene, in rolls, no base incl., landscape surfaces	1 Tilf	275	.029	S.F.	3.05	1.11		4.16	5.10
3800	Nylon action surface, 1/8" thick		275	.029		3.20	1.11		4.31	5.25
3900	1/4" thick		275	.029		4.61	1.11		5.72	6.80
4000	3/8" thick		275	.029		5.80	1.11		6.91	8.10
4100	Golf tee surface with foam back		235	.034		5.75	1.30		7.05	8.35
4200	Practice putting, knitted nylon surface	↓	235	.034	↓	4.86	1.30		6.16	7.40
4400	Polyurethane, thermoset, prefabricated in place, indoor									
4500	3/8" thick for basketball, gyms, etc.	1 Tilf	100	.080	S.F.	4.58	3.05		7.63	9.80
4600	1/2" thick for professional sports		95	.084		6.15	3.21		9.36	11.75
4700	Outdoor, 1/4" thick, smooth, for tennis		100	.080		4.76	3.05		7.81	10
4800	Rough, for track, 3/8" thick	↓	95	.084	↓	5.50	3.21		8.71	11.05

09 68 Carpeting

09 68 20 – Athletic Carpet

09 68 20.10 Indoor Athletic Carpet		Crew	Daily Output	Labor-Hours	Unit	Material	Labor	Equipment	Total	Total Incl O&P
						2009 Bare Costs				
5000	Poured in place, indoor, with finish, 1/4" thick	1 Tilf	80	.100	S.F.	3.38	3.81		7.19	9.65
5050	3/8" thick		65	.123		4.11	4.69		8.80	11.85
5100	1/2" thick		50	.160		5.50	6.10		11.60	15.60
5500	Polyvinyl chloride, sheet goods for gyms, 1/4" thick		80	.100		7.50	3.81		11.31	14.20
5600	3/8" thick		60	.133		8.50	5.10		13.60	17.30

09 69 Access Flooring

09 69 13 – Rigid-Grid Access Flooring

09 69 13.10 Access Floors		Crew	Daily Output	Labor-Hours	Unit	Material	Labor	Equipment	Total	Total Incl O&P
0010	**ACCESS FLOORS**									
0015	Access floor package including panel, pedestal, stringers & laminate cover									
0100	Computer room, greater than 6,000 S.F.	4 Carp	750	.043	S.F.	8.65	1.70		10.35	12.30
0110	Less than 6,000 S.F.	2 Carp	375	.043		9.95	1.70		11.65	13.75
0120	Office, greater than 6,000 S.F.	4 Carp	1050	.030		4.56	1.22		5.78	7
0250	Panels, particle board or steel, 1250# load, no covering, under 6,000 S.F.	2 Carp	600	.027		4.24	1.07		5.31	6.40
0300	Over 6,000 S.F.		640	.025		3.72	1		4.72	5.75
0400	Aluminum, 24" panels		500	.032		31.50	1.28		32.78	36.50
0600	For carpet covering, add					8.55			8.55	9.40
0700	For vinyl floor covering, add					6.70			6.70	7.35
0900	For high pressure laminate covering, add					5.45			5.45	6
0910	For snap on stringer system, add	2 Carp	1000	.016		1.44	.64		2.08	2.64
0950	Office applications, steel or concrete panels,									
0960	no covering, over 6,000 S.F.	2 Carp	960	.017	S.F.	10.10	.67		10.77	12.20
1000	Machine cutouts after initial installation	1 Carp	50	.160	Ea.	5.05	6.40		11.45	16.10
1050	Pedestals, 6" to 12"	2 Carp	85	.188		7.95	7.50		15.45	21
1100	Air conditioning grilles, 4" x 12"	1 Carp	17	.471		66	18.80		84.80	104
1150	4" x 18"	"	14	.571		90.50	23		113.50	137
1200	Approach ramps, minimum	2 Carp	60	.267	S.F.	25.50	10.65		36.15	45.50
1300	Maximum	"	40	.400	"	33.50	16		49.50	63
1500	Handrail, 2 rail, aluminum	1 Carp	15	.533	L.F.	99.50	21.50		121	144

09 72 Wall Coverings

09 72 23 – Wallpapering

09 72 23.10 Wallpaper		Crew	Daily Output	Labor-Hours	Unit	Material	Labor	Equipment	Total	Total Incl O&P
0010	**WALLPAPER** including sizing; add 10-30 percent waste @ takeoff									
0050	Aluminum foil	1 Pape	275	.029	S.F.	.97	1.03		2	2.72
0100	Copper sheets, .025" thick, vinyl backing		240	.033		5.20	1.18		6.38	7.60
0300	Phenolic backing		240	.033		6.75	1.18		7.93	9.30
0600	Cork tiles, light or dark, 12" x 12" x 3/16"		240	.033		4.25	1.18		5.43	6.55
0700	5/16" thick		235	.034		3.62	1.21		4.83	5.90
0900	1/4" basketweave		240	.033		5.55	1.18		6.73	8.05
1000	1/2" natural, non-directional pattern		240	.033		6.70	1.18		7.88	9.25
1100	3/4" natural, non-directional pattern		240	.033		10.90	1.18		12.08	13.90
1200	Granular surface, 12" x 36", 1/2" thick		385	.021		1.20	.74		1.94	2.50
1300	1" thick		370	.022		1.55	.77		2.32	2.93
1500	Polyurethane coated, 12" x 12" x 3/16" thick		240	.033		3.76	1.18		4.94	6.05
1600	5/16" thick		235	.034		5.35	1.21		6.56	7.85
1800	Cork wallpaper, paperbacked, natural		480	.017		1.90	.59		2.49	3.03
1900	Colors		480	.017		2.65	.59		3.24	3.86

09 72 Wall Coverings

09 72 23 – Wallpapering

09 72 23.10 Wallpaper		Crew	Daily Output	Labor-Hours	Unit	Material	2009 Bare Costs Labor	Equipment	Total	Total Incl O&P
2100	Flexible wood veneer, 1/32" thick, plain woods	1 Pape	100	.080	S.F.	2.25	2.84		5.09	7
2200	Exotic woods		95	.084		3.41	2.99		6.40	8.50
2400	Gypsum-based, fabric-backed, fire resistant									
2500	for masonry walls, minimum, 21 oz./S.Y.	1 Pape	800	.010	S.F.	.80	.35		1.15	1.45
2600	Average		720	.011		1.17	.39		1.56	1.92
2700	Maximum (small quantities)		640	.013		1.30	.44		1.74	2.14
2750	Acrylic, modified, semi-rigid PVC, .028" thick	2 Carp	330	.048		1.08	1.94		3.02	4.39
2800	.040" thick	"	320	.050		1.42	2		3.42	4.86
3000	Vinyl wall covering, fabric-backed, lightweight (12-15 oz./S.Y.)	1 Pape	640	.013		.65	.44		1.09	1.43
3300	Medium weight, type 2 (20-24 oz./S.Y.)		480	.017		.77	.59		1.36	1.79
3400	Heavy weight, type 3 (28 oz./S.Y.)		435	.018		1.59	.65		2.24	2.79
3600	Adhesive, 5 gal. lots (18SY/Gal.)				Gal.	10.15			10.15	11.15
3700	Wallpaper, average workmanship, solid pattern, low cost paper	1 Pape	640	.013	S.F.	.35	.44		.79	1.10
3900	basic patterns (matching required), avg. cost paper		535	.015		.63	.53		1.16	1.54
4000	Paper at $85 per double roll, quality workmanship		435	.018		2.35	.65		3	3.63
4100	Linen wall covering, paper backed									
4150	Flame treatment, minimum				S.F.	.85			.85	.94
4200	Grass cloths with lining paper, minimum	1 Pape	400	.020		.74	.71		1.45	1.94
4300	Maximum		350	.023		2.37	.81		3.18	3.91
5990	Wallpaper removal, 1 layer, minimum		800	.010		.06	.35		.41	.64
6000	Wallpaper removal, 3 layer, maximum		400	.020		.13	.71		.84	1.27
9000	Minimum labor/equipment charge		2	4	Job		142		142	227

09 77 Special Wall Surfacing

09 77 33 – Fiberglass Reinforced Panels

09 77 33.10 Fiberglass Reinforced Plastic Panels

		Crew	Daily Output	Labor-Hours	Unit	Material	Labor	Equipment	Total	Total Incl O&P
0010	**FIBERGLASS REINFORCED PLASTIC PANELS**, .090" thick									
0020	On walls, adhesive mounted, embossed surface	2 Carp	640	.025	S.F.	1.23	1		2.23	3
0030	Smooth surface		640	.025		1.51	1		2.51	3.31
0040	Fire rated, embossed surface		640	.025		2.19	1		3.19	4.06
0050	Nylon rivet mounted, on drywall, embossed surface		480	.033		1.15	1.33		2.48	3.47
0060	Smooth surface		480	.033		1.43	1.33		2.76	3.77
0070	Fire rated, embossed surface		480	.033		2.11	1.33		3.44	4.52
0080	On masonry, embossed surface		320	.050		1.15	2		3.15	4.57
0090	Smooth surface		320	.050		1.43	2		3.43	4.87
0100	Fire rated, embossed surface		320	.050		2.11	2		4.11	5.60
0110	Nylon rivet and adhesive mounted, on drywall, embossed surface		240	.067		1.30	2.66		3.96	5.85
0120	Smooth surface		240	.067		1.58	2.66		4.24	6.15
0130	Fire rated, embossed surface		240	.067		2.26	2.66		4.92	6.90
0140	On masonry, embossed surface		190	.084		1.30	3.36		4.66	7
0150	Smooth surface		190	.084		1.58	3.36		4.94	7.30
0160	Fire rated, embossed surface		190	.084		2.26	3.36		5.62	8.05
0170	For moldings add	1 Carp	250	.032	L.F.	.28	1.28		1.56	2.42
0180	On ceilings, for lay in grid system, embossed surface		400	.020	S.F.	1.23	.80		2.03	2.67
0190	Smooth surface		400	.020		1.51	.80		2.31	2.98
0200	Fire rated, embossed surface		400	.020		2.19	.80		2.99	3.73

09 77 43 – Panel Systems

09 77 43.20 Slatwall Panels and Accessories

		Crew	Daily Output	Labor-Hours	Unit	Material	Labor	Equipment	Total	Total Incl O&P
0010	**SLATWALL PANELS AND ACCESSORIES**									
0100	Slatwall panel, 4' x 8' x 3/4" T, MDF, paint grade	1 Carp	500	.016	S.F.	1.60	.64		2.24	2.82

09 77 Special Wall Surfacing

09 77 43 – Panel Systems

09 77 43.20 Slatwall Panels and Accessories		Crew	Daily Output	Labor-Hours	Unit	2009 Bare Costs Material	Labor	Equipment	Total	Total Incl O&P
0110	Melamine finish	1 Carp	500	.016	S.F.	2.28	.64		2.92	3.57
0120	High pressure plastic laminate finish		500	.016		3.70	.64		4.34	5.15
0130	Aluminum channel inserts, add					3.57			3.57	3.93
0200	Accessories, corner forms, 8′ L				L.F.	4.89			4.89	5.40
0210	T-connector, 8′ L					5.95			5.95	6.55
0220	J-mold, 8′ L					1.59			1.59	1.75
0230	Edge cap, 8′ L					1.16			1.16	1.28
0240	Finish end cap, 8′ L					3.66			3.66	4.03
0300	Display hook, metal, 4″ L				Ea.	1.09			1.09	1.20
0310	6″ L					1.19			1.19	1.31
0320	8″ L					1.27			1.27	1.40
0330	10″ L					1.43			1.43	1.57
0340	12″ L					1.58			1.58	1.74
0350	Acrylic, 4″ L					.91			.91	1
0360	6″ L					1.05			1.05	1.16
0370	8″ L					1.10			1.10	1.21
0380	10″ L					1.21			1.21	1.33
0400	Waterfall hanger, metal, 12″ - 16″					6.35			6.35	6.95
0410	Acrylic					10.60			10.60	11.65
0500	Shelf bracket, metal, 8″					6.70			6.70	7.35
0510	10″					6.90			6.90	7.60
0520	12″					7.15			7.15	7.90
0530	14″					7.60			7.60	8.35
0540	16″					9.05			9.05	10
0550	Acrylic, 8″					3.41			3.41	3.75
0560	10″					3.82			3.82	4.20
0570	12″					4.23			4.23	4.65
0580	14″					4.79			4.79	5.25
0600	Shelf, acrylic, 12″ x 16″ x 1/4″					27			27	29.50
0610	12″ x 24″ x 1/4″					45.50			45.50	50

09 81 Acoustic Insulation

09 81 16 – Acoustic Blanket Insulation

09 81 16.10 Sound Attenuation Blanket

		Crew	Daily Output	Labor-Hours	Unit	Material	Labor	Equipment	Total	Total Incl O&P
0010	**SOUND ATTENUATION BLANKET**									
0020	Blanket, 1″ thick	1 Carp	925	.009	S.F.	.25	.35		.60	.85
0500	1-1/2″ thick		920	.009		.25	.35		.60	.85
1000	2″ thick		915	.009		.34	.35		.69	.95
1500	3″ thick		910	.009		.50	.35		.85	1.13
3000	Thermal or acoustical batt above ceiling, 2″ thick		900	.009		.49	.36		.85	1.13
3100	3″ thick		900	.009		.74	.36		1.10	1.40
3200	4″ thick		900	.009		.92	.36		1.28	1.60
3400	Urethane plastic foam, open cell, on wall, 2″ thick	2 Carp	2050	.008		3.05	.31		3.36	3.87
3500	3″ thick		1550	.010		4.05	.41		4.46	5.15
3600	4″ thick		1050	.015		5.70	.61		6.31	7.25
3700	On ceiling, 2″ thick		1700	.009		3.04	.38		3.42	3.96
3800	3″ thick		1300	.012		4.05	.49		4.54	5.25
3900	4″ thick		900	.018		5.70	.71		6.41	7.40
4000	Nylon matting 0.4″ thick, with carbon black spinerette									
4010	plus polyester fabric, on floor	D-7	4000	.004	S.F.	2.26	.14		2.40	2.70
4200	Fiberglass reinf. backer board underlayment, 7/16″ thick, on floor	″	800	.020	″	1.90	.68		2.58	3.16

09 81 Acoustic Insulation

09 81 16 – Acoustic Blanket Insulation

09 81 16.10 Sound Attenuation Blanket		Crew	Daily Output	Labor-Hours	Unit	2009 Bare Costs Material	Labor	Equipment	Total	Total Incl O&P
9000	Minimum labor/equipment charge	1 Carp	5	1.600	Job		64		64	106

09 84 Acoustic Room Components

09 84 13 – Fixed Sound-Absorptive Panels

09 84 13.10 Fixed Panels

		Crew	Daily Output	Labor-Hours	Unit	Material	Labor	Equipment	Total	Total Incl O&P
0010	**FIXED PANELS** Perforated steel facing, painted with									
0100	Fiberglass or mineral filler, no backs, 2-1/4" thick, modular									
0200	space units, ceiling or wall hung, white or colored	1 Carp	100	.080	S.F.	10	3.20		13.20	16.30
0300	Fiberboard sound deadening panels, 1/2" thick	"	600	.013	"	.32	.53		.85	1.23
0500	Fiberglass panels, 4' x 8' x 1" thick, with									
0600	glass cloth face for walls, cemented	1 Carp	155	.052	S.F.	7	2.06		9.06	11.10
0700	1-1/2" thick, dacron covered, inner aluminum frame,									
0710	wall mounted	1 Carp	300	.027	S.F.	8.75	1.07		9.82	11.35
0900	Mineral fiberboard panels, fabric covered, 30"x 108",									
1000	3/4" thick, concealed spline, wall mounted	1 Carp	150	.053	S.F.	6.25	2.13		8.38	10.40
9000	Minimum labor/equipment charge	"	4	2	Job		80		80	132

09 84 36 – Sound-Absorbing Ceiling Units

09 84 36.10 Barriers

		Crew	Daily Output	Labor-Hours	Unit	Material	Labor	Equipment	Total	Total Incl O&P
0010	**BARRIERS** Plenum									
0600	Aluminum foil, fiberglass reinf., parallel with joists	1 Carp	275	.029	S.F.	.85	1.16		2.01	2.86
0700	Perpendicular to joists		180	.044		.85	1.78		2.63	3.87
0900	Aluminum mesh, kraft paperbacked		275	.029		.77	1.16		1.93	2.77
0970	Fiberglass batts, kraft faced, 3-1/2" thick		1400	.006		.36	.23		.59	.78
0980	6" thick		1300	.006		.65	.25		.90	1.13
1000	Sheet lead, 1 lb., 1/64" thick, perpendicular to joists		150	.053		8.40	2.13		10.53	12.70
1100	Vinyl foam reinforced, 1/8" thick, 1.0 lb. per S.F.		150	.053		9.80	2.13		11.93	14.25

09 91 Painting

09 91 03 – Paint Restoration

09 91 03.20 Sanding

		Crew	Daily Output	Labor-Hours	Unit	Material	Labor	Equipment	Total	Total Incl O&P
0010	**SANDING** and puttying interior trim, compared to R099100-10									
0100	Painting 1 coat, on quality work				L.F.		100%			
0300	Medium work						50%			
0400	Industrial grade						25%			
0500	Surface protection, placement and removal									
0510	Basic drop cloths	1 Pord	6400	.001	S.F.		.04		.04	.07
0520	Masking with paper		800	.010		.06	.35		.41	.63
0530	Volume cover up (using plastic sheathing, or building paper)		16000	.001			.02		.02	.03

09 91 03.30 Exterior Surface Preparation

		Crew	Daily Output	Labor-Hours	Unit	Material	Labor	Equipment	Total	Total Incl O&P
0010	**EXTERIOR SURFACE PREPARATION** R099100-10									
0015	Doors, per side, not incl. frames or trim									
0020	Scrape & sand									
0030	Wood, flush	1 Pord	616	.013	S.F.		.46		.46	.73
0040	Wood, detail		496	.016			.57		.57	.91
0050	Wood, louvered		280	.029			1.01		1.01	1.61
0060	Wood, overhead		616	.013			.46		.46	.73
0070	Wire brush									
0080	Metal, flush	1 Pord	640	.013	S.F.		.44		.44	.70

09 91 03.30 Exterior Surface Preparation		Crew	Daily Output	Labor-Hours	Unit	Material	2009 Bare Costs Labor	Equipment	Total	Total Incl O&P
0090	Metal, detail	1 Pord	520	.015	S.F.		.54		.54	.87
0100	Metal, louvered		360	.022			.78		.78	1.25
0110	Metal or fibr., overhead		640	.013			.44		.44	.70
0120	Metal, roll up		560	.014			.50		.50	.80
0130	Metal, bulkhead		640	.013			.44		.44	.70
0140	Power wash, based on 2500 lb. operating pressure									
0150	Metal, flush	A-1H	2240	.004	S.F.		.11	.02	.13	.22
0160	Metal, detail		2120	.004			.12	.02	.14	.23
0170	Metal, louvered		2000	.004			.13	.03	.16	.24
0180	Metal or fibr., overhead		2400	.003			.11	.02	.13	.19
0190	Metal, roll up		2400	.003			.11	.02	.13	.19
0200	Metal, bulkhead		2200	.004			.12	.02	.14	.22
0400	Windows, per side, not incl. trim									
0410	Scrape & sand									
0420	Wood, 1-2 lite	1 Pord	320	.025	S.F.		.88		.88	1.41
0430	Wood, 3-6 lite		280	.029			1.01		1.01	1.61
0440	Wood, 7-10 lite		240	.033			1.17		1.17	1.87
0450	Wood, 12 lite		200	.040			1.41		1.41	2.25
0460	Wood, Bay / Bow		320	.025			.88		.88	1.41
0470	Wire brush									
0480	Metal, 1-2 lite	1 Pord	480	.017	S.F.		.59		.59	.94
0490	Metal, 3-6 lite		400	.020			.70		.70	1.13
0500	Metal, Bay / Bow		480	.017			.59		.59	.94
0510	Power wash, based on 2500 lb. operating pressure									
0520	1-2 lite	A-1H	4400	.002	S.F.		.06	.01	.07	.10
0530	3-6 lite		4320	.002			.06	.01	.07	.11
0540	7-10 lite		4240	.002			.06	.01	.07	.11
0550	12 lite		4160	.002			.06	.01	.07	.11
0560	Bay / Bow		4400	.002			.06	.01	.07	.10
0600	Siding, scrape and sand, light=10-30%, med.=30-70%									
0610	Heavy=70-100% of surface to sand									
0650	Texture 1-11, light	1 Pord	480	.017	S.F.		.59		.59	.94
0660	Med.		440	.018			.64		.64	1.02
0670	Heavy		360	.022			.78		.78	1.25
0680	Wood shingles, shakes, light		440	.018			.64		.64	1.02
0690	Med.		360	.022			.78		.78	1.25
0700	Heavy		280	.029			1.01		1.01	1.61
0710	Clapboard, light		520	.015			.54		.54	.87
0720	Med.		480	.017			.59		.59	.94
0730	Heavy		400	.020			.70		.70	1.13
0740	Wire brush									
0750	Aluminum, light	1 Pord	600	.013	S.F.		.47		.47	.75
0760	Med.		520	.015			.54		.54	.87
0770	Heavy		440	.018			.64		.64	1.02
0780	Pressure wash, based on 2500 lb. operating pressure									
0790	Stucco	A-1H	3080	.003	S.F.		.08	.02	.10	.16
0800	Aluminum or vinyl		3200	.003			.08	.02	.10	.15
0810	Siding, masonry, brick & block		2400	.003			.11	.02	.13	.19
1300	Miscellaneous, wire brush									
1310	Metal, pedestrian gate	1 Pord	100	.080	S.F.		2.82		2.82	4.50
8000	For chemical washing, see Div. 04 01 30									
8010	For steam cleaning, see Div. 04 01 30.20									
8020	For sand blasting, see Div. 05 01 10.51 and 03 35 29.60									

09 91 03 – Paint Restoration

09 91 03.40 Interior Surface Preparation		Crew	Daily Output	Labor-Hours	Unit	Material	2009 Bare Costs Labor	Equipment	Total	Total Incl O&P
0010	**INTERIOR SURFACE PREPARATION**									
0020	Doors, per side, not incl. frames or trim									
0030	Scrape & sand									
0040	Wood, flush	1 Pord	616	.013	S.F.		.46		.46	.73
0050	Wood, detail		496	.016			.57		.57	.91
0060	Wood, louvered		280	.029			1.01		1.01	1.61
0070	Wire brush									
0080	Metal, flush	1 Pord	640	.013	S.F.		.44		.44	.70
0090	Metal, detail		520	.015			.54		.54	.87
0100	Metal, louvered		360	.022			.78		.78	1.25
0110	Hand wash									
0120	Wood, flush	1 Pord	2160	.004	S.F.		.13		.13	.21
0130	Wood, detailed		2000	.004			.14		.14	.23
0140	Wood, louvered		1360	.006			.21		.21	.33
0150	Metal, flush		2160	.004			.13		.13	.21
0160	Metal, detail		2000	.004			.14		.14	.23
0170	Metal, louvered		1360	.006			.21		.21	.33
0400	Windows, per side, not incl. trim									
0410	Scrape & sand									
0420	Wood, 1-2 lite	1 Pord	360	.022	S.F.		.78		.78	1.25
0430	Wood, 3-6 lite		320	.025			.88		.88	1.41
0440	Wood, 7-10 lite		280	.029			1.01		1.01	1.61
0450	Wood, 12 lite		240	.033			1.17		1.17	1.87
0460	Wood, Bay / Bow		360	.022			.78		.78	1.25
0470	Wire brush									
0480	Metal, 1-2 lite	1 Pord	520	.015	S.F.		.54		.54	.87
0490	Metal, 3-6 lite		440	.018			.64		.64	1.02
0500	Metal, Bay / Bow		520	.015			.54		.54	.87
0600	Walls, sanding, light=10-30%, medium - 30-70%,									
0610	heavy=70-100% of surface to sand									
0650	Walls, sand									
0660	Drywall, gypsum, plaster, light	1 Pord	3077	.003	S.F.		.09		.09	.15
0670	Drywall, gypsum, plaster, med.		2160	.004			.13		.13	.21
0680	Drywall, gypsum, plaster, heavy		923	.009			.31		.31	.49
0690	Wood, T&G, light		2400	.003			.12		.12	.19
0700	Wood, T&G, med.		1600	.005			.18		.18	.28
0710	Wood, T&G, heavy		800	.010			.35		.35	.56
0720	Walls, wash									
0730	Drywall, gypsum, plaster	1 Pord	3200	.003	S.F.		.09		.09	.14
0740	Wood, T&G		3200	.003			.09		.09	.14
0750	Masonry, brick & block, smooth		2800	.003			.10		.10	.16
0760	Masonry, brick & block, coarse		2000	.004			.14		.14	.23
8000	For chemical washing, see Div. 04 01 30									
8010	For steam cleaning, see Div. 04 01 30.20									
8020	For sand blasting, see Div. 03 35 29.60 and 05 01 10.51									

09 91 03.41 Scrape After Fire Damage		Crew	Daily Output	Labor-Hours	Unit	Material	Labor	Equipment	Total	Total Incl O&P
0010	**SCRAPE AFTER FIRE DAMAGE**									
0050	Boards, 1" x 4"	1 Pord	336	.024	L.F.		.84		.84	1.34
0060	1" x 6"		260	.031			1.08		1.08	1.73
0070	1" x 8"		207	.039			1.36		1.36	2.17
0080	1" x 10"		174	.046			1.62		1.62	2.59
0500	Framing, 2" x 4"		265	.030			1.06		1.06	1.70

09 91 Painting

09 91 03 – Paint Restoration

09 91 03.41 Scrape After Fire Damage		Crew	Daily Output	Labor-Hours	Unit	Material	2009 Bare Costs Labor	Equipment	Total	Total Incl O&P
0510	2" x 6"	1 Pord	221	.036	L.F.		1.27		1.27	2.04
0520	2" x 8"		190	.042			1.48		1.48	2.37
0530	2" x 10"		165	.048			1.71		1.71	2.73
0540	2" x 12"		144	.056			1.96		1.96	3.13
1000	Heavy framing, 3" x 4"		226	.035			1.25		1.25	1.99
1010	4" x 4"		210	.038			1.34		1.34	2.14
1020	4" x 6"		191	.042			1.47		1.47	2.36
1030	4" x 8"		165	.048			1.71		1.71	2.73
1040	4" x 10"		144	.056			1.96		1.96	3.13
1060	4" x 12"		131	.061			2.15		2.15	3.44
2900	For sealing, minimum		825	.010	S.F.	.14	.34		.48	.70
2920	Maximum		460	.017	"	.28	.61		.89	1.29
9000	Minimum labor/equipment charge		3	2.667	Job		94		94	150

09 91 13 – Exterior Painting

09 91 13.30 Fences		Crew	Daily Output	Labor-Hours	Unit	Material	Labor	Equipment	Total	Total Incl O&P
0010	**FENCES** R099100-20									
0100	Chain link or wire metal, one side, water base									
0110	Roll & brush, first coat	1 Pord	960	.008	S.F.	.05	.29		.34	.53
0120	Second coat		1280	.006		.05	.22		.27	.41
0130	Spray, first coat		2275	.004		.05	.12		.17	.26
0140	Second coat		2600	.003		.05	.11		.16	.23
0150	Picket, water base									
0160	Roll & brush, first coat	1 Pord	865	.009	S.F.	.06	.33		.39	.58
0170	Second coat		1050	.008		.06	.27		.33	.49
0180	Spray, first coat		2275	.004		.06	.12		.18	.26
0190	Second coat		2600	.003		.06	.11		.17	.23
0200	Stockade, water base									
0210	Roll & brush, first coat	1 Pord	1040	.008	S.F.	.06	.27		.33	.49
0220	Second coat		1200	.007		.06	.23		.29	.44
0230	Spray, first coat		2275	.004		.06	.12		.18	.26
0240	Second coat		2600	.003		.06	.11		.17	.23

09 91 13.42 Miscellaneous, Exterior		Crew	Daily Output	Labor-Hours	Unit	Material	Labor	Equipment	Total	Total Incl O&P
0010	**MISCELLANEOUS, EXTERIOR** R099100-20									
0015	For painting metals, see Div. 09 97 13.23									
0100	Railing, ext., decorative wood, incl. cap & baluster									
0110	Newels & spindles @ 12" O.C.									
0120	Brushwork, stain, sand, seal & varnish									
0130	First coat	1 Pord	90	.089	L.F.	.61	3.13		3.74	5.65
0140	Second coat	"	120	.067	"	.61	2.35		2.96	4.42
0150	Rough sawn wood, 42" high, 2" x 2" verticals, 6" O.C.									
0160	Brushwork, stain, each coat	1 Pord	90	.089	L.F.	.18	3.13		3.31	5.20
0170	Wrought iron, 1" rail, 1/2" sq. verticals									
0180	Brushwork, zinc chromate, 60" high, bars 6" O.C.									
0190	Primer	1 Pord	130	.062	L.F.	.60	2.17		2.77	4.12
0200	Finish coat		130	.062		.19	2.17		2.36	3.67
0210	Additional coat		190	.042		.22	1.48		1.70	2.61
0220	Shutters or blinds, single panel, 2' x 4', paint all sides									
0230	Brushwork, primer	1 Pord	20	.400	Ea.	.59	14.10		14.69	23
0240	Finish coat, exterior latex		20	.400		.44	14.10		14.54	23
0250	Primer & 1 coat, exterior latex		13	.615		.89	21.50		22.39	35.50
0260	Spray, primer		35	.229		.85	8.05		8.90	13.80
0270	Finish coat, exterior latex		35	.229		.93	8.05		8.98	13.85

09 91 Painting

09 91 13 – Exterior Painting

09 91 13.42 Miscellaneous, Exterior		Crew	Daily Output	Labor-Hours	Unit	Material	2009 Bare Costs Labor	Equipment	Total	Total Incl O&P
0280	Primer & 1 coat, exterior latex	1 Pord	20	.400	Ea.	.91	14.10		15.01	23.50
0290	For louvered shutters, add				S.F.	10%				
0300	Stair stringers, exterior, metal									
0310	Roll & brush, zinc chromate, to 14", each coat	1 Pord	320	.025	L.F.	.06	.88		.94	1.48
0320	Rough sawn wood, 4" x 12"									
0330	Roll & brush, exterior latex, each coat	1 Pord	215	.037	L.F.	.06	1.31		1.37	2.16
0340	Trellis/lattice, 2" x 2" @ 3" O.C. with 2" x 8" supports									
0350	Spray, latex, per side, each coat	1 Pord	475	.017	S.F.	.06	.59		.65	1.02
0450	Decking, ext., sealer, alkyd, brushwork, sealer coat		1140	.007		.09	.25		.34	.49
0460	1st coat		1140	.007		.07	.25		.32	.47
0470	2nd coat		1300	.006		.05	.22		.27	.41
0500	Paint, alkyd, brushwork, primer coat		1140	.007		.07	.25		.32	.47
0510	1st coat		1140	.007		.07	.25		.32	.47
0520	2nd coat		1300	.006		.05	.22		.27	.41
0600	Sand paint, alkyd, brushwork, 1 coat	↓	150	.053	↓	.11	1.88		1.99	3.12

09 91 13.60 Siding Exterior		Crew	Daily Output	Labor-Hours	Unit	Material	Labor	Equipment	Total	Total Incl O&P
0010	**SIDING EXTERIOR**, Alkyd (oil base) R099100-10									
0450	Steel siding, oil base, paint 1 coat, brushwork	2 Pord	2015	.008	S.F.	.06	.28		.34	.51
0500	Spray R099100-20		4550	.004		.09	.12		.21	.30
0800	Paint 2 coats, brushwork		1300	.012		.11	.43		.54	.82
1000	Spray		2750	.006		.17	.20		.37	.52
1200	Stucco, rough, oil base, paint 2 coats, brushwork		1300	.012		.11	.43		.54	.82
1400	Roller		1625	.010		.12	.35		.47	.68
1600	Spray		2925	.005		.13	.19		.32	.45
1800	Texture 1-11 or clapboard, oil base, primer coat, brushwork		1300	.012		.09	.43		.52	.79
2000	Spray		4550	.004		.09	.12		.21	.30
2100	Paint 1 coat, brushwork		1300	.012		.08	.43		.51	.78
2200	Spray		4550	.004		.08	.12		.20	.29
2400	Paint 2 coats, brushwork		810	.020		.17	.70		.87	1.29
2600	Spray		2600	.006		.19	.22		.41	.55
3000	Stain 1 coat, brushwork		1520	.011		.06	.37		.43	.66
3200	Spray		5320	.003		.07	.11		.18	.25
3400	Stain 2 coats, brushwork		950	.017		.12	.59		.71	1.09
4000	Spray		3050	.005		.14	.18		.32	.45
4200	Wood shingles, oil base primer coat, brushwork		1300	.012		.08	.43		.51	.78
4400	Spray		3900	.004		.08	.14		.22	.32
4600	Paint 1 coat, brushwork		1300	.012		.07	.43		.50	.77
4800	Spray		3900	.004		.09	.14		.23	.33
5000	Paint 2 coats, brushwork		810	.020		.14	.70		.84	1.26
5200	Spray		2275	.007		.13	.25		.38	.55
5800	Stain 1 coat, brushwork		1500	.011		.06	.38		.44	.67
6000	Spray		3900	.004		.06	.14		.20	.30
6500	Stain 2 coats, brushwork		950	.017		.12	.59		.71	1.09
7000	Spray	↓	2660	.006		.17	.21		.38	.53
8000	For latex paint, deduct					10%				
8100	For work over 12' H, from pipe scaffolding, add						15%			
8200	For work over 12' H, from extension ladder, add						25%			
8300	For work over 12' H, from swing staging, add				↓		35%			

09 91 13.62 Siding, Misc.		Crew	Daily Output	Labor-Hours	Unit	Material	Labor	Equipment	Total	Total Incl O&P
0010	**SIDING, MISC.**, latex paint R099100-10									
0100	Aluminum siding									
0110	Brushwork, primer R099100-20	2 Pord	2275	.007	S.F.	.05	.25		.30	.46

09 91 13 – Exterior Painting

09 91 13.62 Siding, Misc.		Crew	Daily Output	Labor-Hours	Unit	2009 Bare Costs Material	Labor	Equipment	Total	Total Incl O&P
0120	Finish coat, exterior latex	2 Pord	2275	.007	S.F.	.04	.25		.29	.45
0130	Primer & 1 coat exterior latex		1300	.012		.10	.43		.53	.80
0140	Primer & 2 coats exterior latex		975	.016		.15	.58		.73	1.08
0150	Mineral fiber shingles									
0160	Brushwork, primer	2 Pord	1495	.011	S.F.	.09	.38		.47	.70
0170	Finish coat, industrial enamel		1495	.011		.13	.38		.51	.74
0180	Primer & 1 coat enamel		810	.020		.22	.70		.92	1.35
0190	Primer & 2 coats enamel		540	.030		.35	1.04		1.39	2.06
0200	Roll, primer		1625	.010		.10	.35		.45	.66
0210	Finish coat, industrial enamel		1625	.010		.14	.35		.49	.71
0220	Primer & 1 coat enamel		975	.016		.24	.58		.82	1.19
0230	Primer & 2 coats enamel		650	.025		.38	.87		1.25	1.80
0240	Spray, primer		3900	.004		.08	.14		.22	.32
0250	Finish coat, industrial enamel		3900	.004		.12	.14		.26	.36
0260	Primer & 1 coat enamel		2275	.007		.20	.25		.45	.61
0270	Primer & 2 coats enamel		1625	.010		.31	.35		.66	.89
0280	Waterproof sealer, first coat		4485	.004		.08	.13		.21	.29
0290	Second coat		5235	.003		.07	.11		.18	.25
0300	Rough wood incl. shingles, shakes or rough sawn siding									
0310	Brushwork, primer	2 Pord	1280	.013	S.F.	.12	.44		.56	.83
0320	Finish coat, exterior latex		1280	.013		.07	.44		.51	.78
0330	Primer & 1 coat exterior latex		960	.017		.19	.59		.78	1.15
0340	Primer & 2 coats exterior latex		700	.023		.27	.80		1.07	1.58
0350	Roll, primer		2925	.005		.16	.19		.35	.48
0360	Finish coat, exterior latex		2925	.005		.09	.19		.28	.41
0370	Primer & 1 coat exterior latex		1790	.009		.25	.31		.56	.77
0380	Primer & 2 coats exterior latex		1300	.012		.34	.43		.77	1.06
0390	Spray, primer		3900	.004		.13	.14		.27	.38
0400	Finish coat, exterior latex		3900	.004		.07	.14		.21	.31
0410	Primer & 1 coat exterior latex		2600	.006		.20	.22		.42	.57
0420	Primer & 2 coats exterior latex		2080	.008		.27	.27		.54	.73
0430	Waterproof sealer, first coat		4485	.004		.14	.13		.27	.35
0440	Second coat		4485	.004		.08	.13		.21	.29
0450	Smooth wood incl. butt, T&G, beveled, drop or B&B siding									
0460	Brushwork, primer	2 Pord	2325	.007	S.F.	.09	.24		.33	.48
0470	Finish coat, exterior latex		1280	.013		.07	.44		.51	.78
0480	Primer & 1 coat exterior latex		800	.020		.16	.70		.86	1.31
0490	Primer & 2 coats exterior latex		630	.025		.23	.89		1.12	1.69
0500	Roll, primer		2275	.007		.09	.25		.34	.50
0510	Finish coat, exterior latex		2275	.007		.08	.25		.33	.49
0520	Primer & 1 coat exterior latex		1300	.012		.18	.43		.61	.88
0530	Primer & 2 coats exterior latex		975	.016		.26	.58		.84	1.20
0540	Spray, primer		4550	.004		.07	.12		.19	.28
0550	Finish coat, exterior latex		4550	.004		.07	.12		.19	.28
0560	Primer & 1 coat exterior latex		2600	.006		.14	.22		.36	.51
0570	Primer & 2 coats exterior latex		1950	.008		.21	.29		.50	.69
0580	Waterproof sealer, first coat		5230	.003		.08	.11		.19	.26
0590	Second coat		5980	.003		.08	.09		.17	.24
0600	For oil base paint, add					10%				

09 91 13.70 Doors and Windows, Exterior		Crew	Daily Output	Labor-Hours	Unit	Material	Labor	Equipment	Total	Total Incl O&P
0010	**DOORS AND WINDOWS, EXTERIOR** R099100-10									
0100	Door frames & trim, only									

09 91 13.70 Doors and Windows, Exterior		Crew	Daily Output	Labor-Hours	Unit	Material	2009 Bare Costs Labor	Equipment	Total	Total Incl O&P
0110	Brushwork, primer R099100-20	1 Pord	512	.016	L.F.	.05	.55		.60	.94
0120	Finish coat, exterior latex		512	.016		.05	.55		.60	.94
0130	Primer & 1 coat, exterior latex		300	.027		.11	.94		1.05	1.62
0135	2 coats, exterior latex, both sides		15	.533	Ea.	4.87	18.75		23.62	35.50
0140	Primer & 2 coats, exterior latex		265	.030	L.F.	.16	1.06		1.22	1.88
0150	Doors, flush, both sides, incl. frame & trim									
0160	Roll & brush, primer	1 Pord	10	.800	Ea.	4.05	28		32.05	49.50
0170	Finish coat, exterior latex		10	.800		4.15	28		32.15	49.50
0180	Primer & 1 coat, exterior latex		7	1.143		8.20	40		48.20	73.50
0190	Primer & 2 coats, exterior latex		5	1.600		12.35	56.50		68.85	104
0200	Brushwork, stain, sealer & 2 coats polyurethane		4	2		21	70.50		91.50	136
0210	Doors, French, both sides, 10-15 lite, incl. frame & trim									
0220	Brushwork, primer	1 Pord	6	1.333	Ea.	2.03	47		49.03	77
0230	Finish coat, exterior latex		6	1.333		2.07	47		49.07	77.50
0240	Primer & 1 coat, exterior latex		3	2.667		4.10	94		98.10	155
0250	Primer & 2 coats, exterior latex		2	4		6.05	141		147.05	232
0260	Brushwork, stain, sealer & 2 coats polyurethane		2.50	3.200		7.70	113		120.70	189
0270	Doors, louvered, both sides, incl. frame & trim									
0280	Brushwork, primer	1 Pord	7	1.143	Ea.	4.05	40		44.05	69
0290	Finish coat, exterior latex		7	1.143		4.15	40		44.15	69
0300	Primer & 1 coat, exterior latex		4	2		8.20	70.50		78.70	122
0310	Primer & 2 coats, exterior latex		3	2.667		12.10	94		106.10	163
0320	Brushwork, stain, sealer & 2 coats polyurethane		4.50	1.778		21	62.50		83.50	123
0330	Doors, panel, both sides, incl. frame & trim									
0340	Roll & brush, primer	1 Pord	6	1.333	Ea.	4.05	47		51.05	79.50
0350	Finish coat, exterior latex		6	1.333		4.15	47		51.15	79.50
0360	Primer & 1 coat, exterior latex		3	2.667		8.20	94		102.20	159
0370	Primer & 2 coats, exterior latex		2.50	3.200		12.10	113		125.10	193
0380	Brushwork, stain, sealer & 2 coats polyurethane		3	2.667		21	94		115	173
0400	Windows, per ext. side, based on 15 SF									
0410	1 to 6 lite									
0420	Brushwork, primer	1 Pord	13	.615	Ea.	.80	21.50		22.30	35.50
0430	Finish coat, exterior latex		13	.615		.82	21.50		22.32	35.50
0440	Primer & 1 coat, exterior latex		8	1		1.62	35		36.62	58.50
0450	Primer & 2 coats, exterior latex		6	1.333		2.39	47		49.39	77.50
0460	Stain, sealer & 1 coat varnish		7	1.143		3.04	40		43.04	68
0470	7 to 10 lite									
0480	Brushwork, primer	1 Pord	11	.727	Ea.	.80	25.50		26.30	42
0490	Finish coat, exterior latex		11	.727		.82	25.50		26.32	42
0500	Primer & 1 coat, exterior latex		7	1.143		1.62	40		41.62	66.50
0510	Primer & 2 coats, exterior latex		5	1.600		2.39	56.50		58.89	92.50
0520	Stain, sealer & 1 coat varnish		6	1.333		3.04	47		50.04	78.50
0530	12 lite									
0540	Brushwork, primer	1 Pord	10	.800	Ea.	.80	28		28.80	46
0550	Finish coat, exterior latex		10	.800		.82	28		28.82	46
0560	Primer & 1 coat, exterior latex		6	1.333		1.62	47		48.62	77
0570	Primer & 2 coats, exterior latex		5	1.600		2.39	56.50		58.89	92.50
0580	Stain, sealer & 1 coat varnish		6	1.333		3.06	47		50.06	78.50
0590	For oil base paint, add					10%				

09 91 13.80 Trim, Exterior		Crew	Daily Output	Labor-Hours	Unit	Material	Labor	Equipment	Total	Total Incl O&P
0010	**TRIM, EXTERIOR** R099100-10									
0100	Door frames & trim (see Doors, interior or exterior)									

09 91 Painting

09 91 13 – Exterior Painting

09 91 13.80 Trim, Exterior		Crew	Daily Output	Labor-Hours	Unit	Material	Labor	Equipment	Total	Total Incl O&P
						2009 Bare Costs				
0110	Fascia, latex paint, one coat coverage R099100-20									
0120	1" x 4", brushwork	1 Porc	640	.013	L.F.	.02	.44		.46	.72
0130	Roll		1280	.006		.02	.22		.24	.37
0140	Spray		2080	.004		.01	.14		.15	.24
0150	1" x 6" to 1" x 10", brushwork		640	.013		.06	.44		.50	.76
0160	Roll		1230	.007		.06	.23		.29	.44
0170	Spray		2100	.004		.05	.13		.18	.26
0180	1" x 12", brushwork		640	.013		.06	.44		.50	.76
0190	Roll		1050	.008		.06	.27		.33	.50
0200	Spray	↓	2200	.004	↓	.05	.13		.18	.25
0210	Gutters & downspouts, metal, zinc chromate paint									
0220	Brushwork, gutters, 5", first coat	1 Porc	640	.013	L.F.	.07	.44		.51	.77
0230	Second coat		960	.008		.06	.29		.35	.54
0240	Third coat		1280	.006		.05	.22		.27	.41
0250	Downspouts, 4", first coat		640	.013		.07	.44		.51	.77
0260	Second coat		960	.008		.06	.29		.35	.54
0270	Third coat	↓	1280	.006	↓	.05	.22		.27	.41
0280	Gutters & downspouts, wood									
0290	Brushwork, gutters, 5", primer	1 Porc	640	.013	L.F.	.05	.44		.49	.76
0300	Finish coat, exterior latex		640	.013		.05	.44		.49	.75
0310	Primer & 1 coat exterior latex		400	.020		.11	.70		.81	1.25
0320	Primer & 2 coats exterior latex		325	.025		.16	.87		1.03	1.56
0330	Downspouts, 4", primer		640	.013		.05	.44		.49	.76
0340	Finish coat, exterior latex		640	.013		.05	.44		.49	.75
0350	Primer & 1 coat exterior latex		400	.020		.11	.70		.81	1.25
0360	Primer & 2 coats exterior latex	↓	325	.025	↓	.08	.87		.95	1.47
0370	Molding, exterior, up to 14" wide									
0380	Brushwork, primer	1 Porc	640	.013	L.F.	.06	.44		.50	.77
0390	Finish coat, exterior latex		640	.013		.06	.44		.50	.76
0400	Primer & 1 coat exterior latex		400	.020		.13	.70		.83	1.27
0410	Primer & 2 coats exterior latex		315	.025		.13	.89		1.02	1.57
0420	Stain & fill		1050	.008		.07	.27		.34	.51
0430	Shellac		1850	.004		.08	.15		.23	.33
0440	Varnish	↓	1275	.006	↓	.09	.22		.31	.45

09 91 13.90 Walls, Masonry (CMU), Exterior		Crew	Daily Output	Labor-Hours	Unit	Material	Labor	Equipment	Total	Total Incl O&P
0350	**WALLS, MASONRY (CMU), EXTERIOR**									
0360	Concrete masonry units (CMU), smooth surface									
0370	Brushwork, latex, first coat	1 Porc	640	.013	S.F.	.04	.44		.48	.75
0380	Second coat		960	.008		.04	.29		.33	.51
0390	Waterproof sealer, first coat		736	.011		.23	.38		.61	.87
0400	Second coat		1104	.007		.23	.26		.49	.67
0410	Roll, latex, paint, first coat		1465	.005		.05	.19		.24	.37
0420	Second coat		1790	.004		.04	.16		.20	.29
0430	Waterproof sealer, first coat		1680	.005		.23	.17		.40	.53
0440	Second coat		2060	.004		.23	.14		.37	.48
0450	Spray, latex, paint, first coat		1950	.004		.04	.14		.18	.28
0460	Second coat		2600	.003		.03	.11		.14	.21
0470	Waterproof sealer, first coat		2245	.004		.23	.13		.36	.46
0480	Second coat	↓	2990	.003	↓	.23	.09		.32	.41
0490	Concrete masonry unit (CMU), porous									
0500	Brushwork, latex, first coat	1 Porc	640	.013	S.F.	.09	.44		.53	.80
0510	Second coat	↓	960	.008	↓	.04	.29		.33	.52

09 91 Painting

09 91 13 – Exterior Painting

09 91 13.90 Walls, Masonry (CMU), Exterior		Crew	Daily Output	Labor-Hours	Unit	Material	Labor	Equipment	Total	Total Incl O&P
						2009 Bare Costs				
0520	Waterproof sealer, first coat	1 Pord	736	.011	S.F.	.23	.38		.61	.87
0530	Second coat		1104	.007		.23	.26		.49	.67
0540	Roll latex, first coat		1465	.005		.07	.19		.26	.38
0550	Second coat		1790	.004		.04	.16		.20	.30
0560	Waterproof sealer, first coat		1680	.005		.23	.17		.40	.53
0570	Second coat		2060	.004		.23	.14		.37	.48
0580	Spray latex, first coat		1950	.004		.05	.14		.19	.28
0590	Second coat		2600	.003		.03	.11		.14	.21
0600	Waterproof sealer, first coat		2245	.004		.23	.13		.36	.46
0610	Second coat	↓	2990	.003	↓	.23	.09		.32	.41

09 91 23 – Interior Painting

09 91 23.20 Cabinets and Casework

		Crew	Daily Output	Labor-Hours	Unit	Material	Labor	Equipment	Total	Total Incl O&P
0010	**CABINETS AND CASEWORK** R099100-10									
1000	Primer coat, oil base, brushwork	1 Pord	650	.012	S.F.	.05	.43		.48	.75
2000	Paint, oil base, brushwork, 1 coat R099100-20		650	.012		.06	.43		.49	.76
2500	2 coats		400	.020		.12	.70		.82	1.26
3000	Stain, brushwork, wipe off		650	.012		.06	.43		.49	.76
4000	Shellac, 1 coat, brushwork		650	.012		.07	.43		.50	.77
4500	Varnish, 3 coats, brushwork, sand after 1st coat	↓	325	.025		.23	.87		1.10	1.63
5000	For latex paint, deduct				↓	10%				

09 91 23.33 Doors and Windows, Interior Alkyd (Oil Base)

		Crew	Daily Output	Labor-Hours	Unit	Material	Labor	Equipment	Total	Total Incl O&P
0010	**DOORS AND WINDOWS, INTERIOR ALKYD (OIL BASE)** R099100-10									
0500	Flush door & frame, 3′ x 7′, oil, primer, brushwork	1 Pord	10	.800	Ea.	2.34	28		30.34	47.50
1000	Paint, 1 coat R099100-20		10	.800		2.25	28		30.25	47.50
1200	2 coats		6	1.333		3.70	47		50.70	79
1220	3 coats		5	1.600		8.70	56.50		65.20	99.50
1400	Stain, brushwork, wipe off		18	.444		1.29	15.65		16.94	26.50
1600	Shellac, 1 coat, brushwork		25	.320		1.48	11.25		12.73	19.60
1800	Varnish, 3 coats, brushwork, sand after 1st coat		9	.889		4.77	31.50		36.27	55.50
2000	Panel door & frame, 3′ x 7′, oil, primer, brushwork		6	1.333		1.96	47		48.96	77
2200	Paint, 1 coat		6	1.333		2.25	47		49.25	77.50
2400	2 coats		3	2.667		6.45	94		100.45	157
2420	3 coats		2	4		8.45	141		149.45	234
2600	Stain, brushwork, panel door, 3′ x 7′, not incl. frame		16	.500		1.29	17.60		18.89	29.50
2800	Shellac, 1 coat, brushwork		22	.364		1.48	12.80		14.28	22
3000	Varnish, 3 coats, brushwork, sand after 1st coat	↓	7.50	1.067	↓	4.77	37.50		42.27	65.50
4400	Windows, including frame and trim, per side									
4600	Colonial type, 6/6 lites, 2′ x 3′, oil, primer, brushwork	1 Pord	14	.571	Ea.	.31	20		20.31	32.50
5800	Paint, 1 coat		14	.571		.36	20		20.36	32.50
6000	2 coats		9	.889		.69	31.50		32.19	51
6010	3 coats		7	1.143		1.03	40		41.03	65.50
6200	3′ x 5′ opening, 6/6 lites, primer coat, brushwork		12	.667		.78	23.50		24.28	38.50
6400	Paint, 1 coat		12	.667		.89	23.50		24.39	38.50
6600	2 coats		7	1.143		1.73	40		41.73	66.50
6610	3 coats		6	1.333		2.56	47		49.56	78
6800	4′ x 8′ opening, 6/6 lites, primer coat, brushwork		8	1		1.65	35		36.65	58.50
7000	Paint, 1 coat		8	1		1.90	35		36.90	58.50
7200	2 coats		5	1.600		3.68	56.50		60.18	94
7210	3 coats		4	2		5.45	70.50		75.95	119
7500	Standard, 6/6 lites, 2′ x 3′, primer		14	.571		.31	20		20.31	32.50
7520	Paint 1 coat		14	.571		.36	20		20.36	32.50
7540	2 coats	↓	9	.889	↓	.69	31.50		32.19	51

09 91 23.33 Doors and Windows, Interior Alkyd (Oil Base)		Crew	Daily Output	Labor-Hours	Unit	Material	2009 Bare Costs Labor	Equipment	Total	Total Incl O&P
7560	3 coats	1 Pord	7	1.143	Ea.	1.03	40		41.03	65.50
7580	3′ x 5′, 6/6 lites, primer		12	.667		.78	23.50		24.28	38.50
7600	Paint 1 coat		12	.667		.89	23.50		24.39	38.50
7620	2 coats		7	1.143		1.73	40		41.73	66.50
7640	3 coats		6	1.333		2.56	47		49.56	78
7660	4′ x 8′, primer		8	1		1.65	35		36.65	58.50
7680	Paint 1 coat		8	1		1.90	35		36.90	58.50
7700	2 coats		5	1.600		3.68	56.50		60.18	94
7720	3 coats		4	2		5.45	70.50		75.95	119
8000	Single lite type, 2′ x 3′, oil base, primer coat, brushwork		33	.242		.31	8.55		8.86	14
8200	Paint, 1 coat		33	.242		.36	8.55		8.91	14.05
8400	2 coats		20	.400		.69	14.10		14.79	23.50
8410	3 coats		16	.500		1.03	17.60		18.63	29
8600	3′ x 5′ opening, primer coat, brushwork		20	.400		.78	14.10		14.88	23.50
8800	Paint, 1 coat		20	.400		.89	14.10		14.99	23.50
8900	2 coats		13	.615		1.73	21.50		23.23	36.50
9010	3 coats		10	.800		2.56	28		30.56	48
9200	4′ x 8′ opening, primer coat, brushwork		14	.571		1.65	20		21.65	34
9400	Paint, 1 coat		14	.571		1.90	20		21.90	34
9600	2 coats		8	1		3.68	35		38.68	60.50
9610	3 coats	↓	7	1.143	↓	5.45	40		45.45	70.50

09 91 23.35 Doors and Windows, Interior Latex		Crew	Daily Output	Labor-Hours	Unit	Material	Labor	Equipment	Total	Total Incl O&P
0010	**DOORS & WINDOWS, INTERIOR LATEX** R099100-10									
0100	Doors, flush, both sides, incl. frame & trim									
0110	Roll & brush, primer	1 Pord	10	.800	Ea.	3.64	28		31.64	49
0120	Finish coat, latex		10	.800		4.21	28		32.21	49.50
0130	Primer & 1 coat latex		7	1.143		7.85	40		47.85	73
0140	Primer & 2 coats latex		5	1.600		11.80	56.50		68.30	103
0160	Spray, both sides, primer		20	.400		3.83	14.10		17.93	26.50
0170	Finish coat, latex		20	.400		4.42	14.10		18.52	27.50
0180	Primer & 1 coat latex		11	.727		8.30	25.50		33.80	50
0190	Primer & 2 coats latex	↓	8	1	↓	12.50	35		47.50	70.50
0200	Doors, French, both sides, 10-15 lite, incl. frame & trim									
0210	Roll & brush, primer	1 Pord	6	1.333	Ea.	1.82	47		48.82	77
0220	Finish coat, latex		6	1.333		2.10	47		49.10	77.50
0230	Primer & 1 coat latex		3	2.667		3.92	94		97.92	154
0240	Primer & 2 coats latex	↓	2	4	↓	5.90	141		146.90	232
0260	Doors, louvered, both sides, incl. frame & trim									
0270	Roll & brush, primer	1 Pord	7	1.143	Ea.	3.64	40		43.64	68.50
0280	Finish coat, latex		7	1.143		4.21	40		44.21	69
0290	Primer & 1 coat, latex		4	2		7.65	70.50		78.15	121
0300	Primer & 2 coats, latex		3	2.667		12.05	94		106.05	163
0320	Spray, both sides, primer		20	.400		3.83	14.10		17.93	26.50
0330	Finish coat, latex		20	.400		4.42	14.10		18.52	27.50
0340	Primer & 1 coat, latex		11	.727		8.30	25.50		33.80	50
0350	Primer & 2 coats, latex	↓	8	1	↓	12.75	35		47.75	70.50
0360	Doors, panel, both sides, incl. frame & trim									
0370	Roll & brush, primer	1 Pord	6	1.333	Ea.	3.83	47		50.83	79
0380	Finish coat, latex		6	1.333		4.21	47		51.21	79.50
0390	Primer & 1 coat, latex		3	2.667		7.85	94		101.85	159
0400	Primer & 2 coats, latex		2.50	3.200		12.05	113		125.05	193
0420	Spray, both sides, primer	↓	10	.800	↓	3.83	28		31.83	49

09 91 23.35 Doors and Windows, Interior Latex		Crew	Daily Output	Labor-Hours	Unit	Material	Labor	Equipment	Total	Total Incl O&P
0430	Finish coat, latex	1 Pord	10	.800	Ea.	4.42	28		32.42	50
0440	Primer & 1 coat, latex		5	1.600		8.30	56.50		64.80	99
0450	Primer & 2 coats, latex		4	2		12.75	70.50		83.25	127
0460	Windows, per interior side, based on 15 SF									
0470	1 to 6 lite									
0480	Brushwork, primer	1 Pord	13	.615	Ea.	.72	21.50		22.22	35.50
0490	Finish coat, enamel		13	.615		.83	21.50		22.33	35.50
0500	Primer & 1 coat enamel		8	1		1.55	35		36.55	58
0510	Primer & 2 coats enamel		6	1.333		2.38	47		49.38	77.50
0530	7 to 10 lite									
0540	Brushwork, primer	1 Pord	11	.727	Ea.	.72	25.50		26.22	42
0550	Finish coat, enamel		11	.727		.83	25.50		26.33	42
0560	Primer & 1 coat enamel		7	1.143		1.55	40		41.55	66
0570	Primer & 2 coats enamel		5	1.600		2.38	56.50		58.88	92.50
0590	12 lite									
0600	Brushwork, primer	1 Pord	10	.800	Ea.	.72	28		28.72	46
0610	Finish coat, enamel		10	.800		.83	28		28.83	46
0620	Primer & 1 coat enamel		6	1.333		1.55	47		48.55	76.50
0630	Primer & 2 coats enamel		5	1.600		2.38	56.50		58.88	92.50
0650	For oil base paint, add					10%				

09 91 23.39 Doors and Windows, Interior Latex, Zero Voc		Crew	Daily Output	Labor-Hours	Unit	Material	Labor	Equipment	Total	Total Incl O&P
0010	**DOORS & WINDOWS, INTERIOR LATEX, ZERO VOC**									
0100	Doors flush, both sides, incl. frame & trim									
0110	Roll & brush, primer G	1 Pord	10	.800	Ea.	4.75	28		32.75	50.50
0120	Finish coat, latex G		10	.800		4.82	28		32.82	50.50
0130	Primer & 1 coat latex G		7	1.143		9.55	40		49.55	75
0140	Primer & 2 coats latex G		5	1.600		14.10	56.50		70.60	106
0160	Spray, both sides, primer G		20	.400		5	14.10		19.10	28
0170	Finish coat, latex G		20	.400		5.05	14.10		19.15	28
0180	Primer & 1 coat latex G		11	.727		10.10	25.50		35.60	52
0190	Primer & 2 coats latex G		8	1		14.95	35		49.95	73
0200	Doors, French, both sides, 10-15 lite, incl. frame & trim									
0210	Roll & brush, primer G	1 Pord	6	1.333	Ea.	2.38	47		49.38	77.50
0220	Finish coat, latex G		6	1.333		2.41	47		49.41	77.50
0230	Primer & 1 coat latex G		3	2.667		4.79	94		98.79	155
0240	Primer & 2 coats latex G		2	4		7.05	141		148.05	233
0360	Doors, panel, both sides, incl. frame & trim									
0370	Roll & brush, primer G	1 Pord	6	1.333	Ea.	5	47		52	80.50
0380	Finish coat, latex G		6	1.333		4.82	47		51.82	80.50
0390	Primer & 1 coat, latex G		3	2.667		9.55	94		103.55	161
0400	Primer & 2 coats, latex G		2.50	3.200		14.40	113		127.40	196
0420	Spray, both sides, primer G		10	.800		5	28		33	50.50
0430	Finish coat, latex G		10	.800		5.05	28		33.05	50.50
0440	Primer & 1 coat, latex G		5	1.600		10.10	56.50		66.60	101
0450	Primer & 2 coats, latex G		4	2		15.25	70.50		85.75	130
0460	Windows, per interior side, based on 15 SF									
0470	1 to 6 lite									
0480	Brushwork, primer G	1 Pord	13	.615	Ea.	.94	21.50		22.44	35.50
0490	Finish coat, enamel G		13	.615		.95	21.50		22.45	35.50
0500	Primer & 1 coat enamel G		8	1		1.89	35		36.89	58.50
0510	Primer & 2 coats enamel G		6	1.333		2.84	47		49.84	78

09 91 23 – Interior Painting

09 91 23.40 Floors, Interior

	09 91 23.40 Floors, Interior	Crew	Daily Output	Labor-Hours	Unit	2009 Bare Costs Material	Labor	Equipment	Total	Total Incl O&P
0010	**FLOORS, INTERIOR** R099100-10									
0100	Concrete paint, latex									
0110	Brushwork									
0120	1st coat	1 Pord	975	.008	S.F.	.14	.29		.43	.62
0130	2nd coat		1150	.007		.10	.25		.35	.50
0140	3rd coat		1300	.006		.08	.22		.30	.43
0150	Roll									
0160	1st coat	1 Pord	2600	.003	S.F.	.19	.11		.30	.38
0170	2nd coat		3250	.002		.12	.09		.21	.27
0180	3rd coat		3900	.002		.09	.07		.16	.21
0190	Spray									
0200	1st coat	1 Pord	2600	.003	S.F.	.17	.11		.28	.35
0210	2nd coat		3250	.002		.09	.09		.18	.24
0220	3rd coat		3900	.002		.07	.07		.14	.20
0300	Acid stain and sealer									
0310	Stain, one coat	1 Pord	650	.012	S.F.	.11	.43		.54	.81
0320	Two coats		570	.014		.22	.49		.71	1.03
0330	Acrylic sealer, one coat		2600	.003		.15	.11		.26	.34
0340	Two coats		1400	.006		.30	.20		.50	.65

09 91 23.52 Miscellaneous, Interior

	09 91 23.52 Miscellaneous, Interior	Crew	Daily Output	Labor-Hours	Unit	2009 Bare Costs Material	Labor	Equipment	Total	Total Incl O&P
0010	**MISCELLANEOUS, INTERIOR** R099100-10									
2400	Floors, conc./wood, oil base, primer/sealer coat, brushwork	2 Pord	1950	.008	S.F.	.06	.29		.35	.52
2450	Roller		5200	.003		.06	.11		.17	.23
2600	Spray		6000	.003		.06	.09		.15	.21
2650	Paint 1 coat, brushwork		1950	.008		.08	.29		.37	.55
2800	Roller		5200	.003		.09	.11		.20	.27
2850	Spray		6000	.003		.09	.09		.18	.25
3000	Stain, wood floor, brushwork, 1 coat		4550	.004		.06	.12		.18	.27
3200	Roller		5200	.003		.06	.11		.17	.24
3250	Spray		6000	.003		.06	.09		.15	.22
3400	Varnish, wood floor, brushwork		4550	.004		.08	.12		.20	.28
3450	Roller		5200	.003		.08	.11		.19	.26
3600	Spray		6000	.003		.08	.09		.17	.24
3650	For dust proofing or anti skid, see Div. 03 35 29.30									
3800	Grilles, per side, oil base, primer coat, brushwork	1 Pord	520	.015	S.F.	.10	.54		.64	.98
3850	Spray		1140	.007		.11	.25		.36	.51
3880	Paint 1 coat, brushwork		520	.015		.12	.54		.66	1
3900	Spray		1140	.007		.13	.25		.38	.53
3920	Paint 2 coats, brushwork		325	.025		.23	.87		1.10	1.63
3940	Spray		650	.012		.26	.43		.69	.98
3950	Prime & paint 1 coat		325	.025		.22	.87		1.09	1.62
3960	Prime & paint 2 coats		270	.030		.22	1.04		1.26	1.91
4500	Louvers, one side, primer, brushwork		524	.015		.06	.54		.60	.92
4520	Paint one coat, brushwork		520	.015		.05	.54		.59	.93
4530	Spray		1140	.007		.06	.25		.31	.46
4540	Paint two coats, brushwork		325	.025		.11	.87		.98	1.50
4550	Spray		650	.012		.12	.43		.55	.82
4560	Paint three coats, brushwork		270	.030		.16	1.04		1.20	1.84
4570	Spray		500	.016		.18	.56		.74	1.09
5000	Pipe, 1 – 4" diameter, primer or sealer coat, oil base, brushwork	2 Pord	1250	.013	L.F.	.06	.45		.51	.78
5100	Spray		2165	.007		.06	.26		.32	.48
5200	Paint 1 coat, brushwork		1250	.013		.06	.45		.51	.79

09 91 23.52 Miscellaneous, Interior		Crew	Daily Output	Labor-Hours	Unit	2009 Bare Costs Material	Labor	Equipment	Total	Total Incl O&P
5300	Spray	2 Pord	2165	.007	L.F.	.06	.26		.32	.48
5350	Paint 2 coats, brushwork		775	.021		.11	.73		.84	1.28
5400	Spray		1240	.013		.12	.45		.57	.87
5420	Paint 3 coats, brushwork		775	.021		.17	.73		.90	1.34
5450	5" - 8" diameter, primer or sealer coat, brushwork		620	.026		.12	.91		1.03	1.58
5500	Spray		1085	.015		.19	.52		.71	1.04
5550	Paint 1 coat, brushwork		620	.026		.17	.91		1.08	1.64
5600	Spray		1085	.015		.19	.52		.71	1.04
5650	Paint 2 coats, brushwork		385	.042		.22	1.46		1.68	2.59
5700	Spray		620	.026		.25	.91		1.16	1.72
5720	Paint 3 coats, brushwork		385	.042		.33	1.46		1.79	2.71
5750	9" - 12" diameter, primer or sealer coat, brushwork		415	.039		.18	1.36		1.54	2.36
5800	Spray		725	.022		.20	.78		.98	1.46
5850	Paint 1 coat, brushwork		415	.039		.17	1.36		1.53	2.36
6000	Spray		725	.022		.19	.78		.97	1.45
6200	Paint 2 coats, brushwork		260	.062		.34	2.17		2.51	3.83
6250	Spray		415	.039		.37	1.36		1.73	2.58
6270	Paint 3 coats, brushwork		260	.062		.50	2.17		2.67	4.01
6300	13" - 16" diameter, primer or sealer coat, brushwork		310	.052		.24	1.82		2.06	3.16
6350	Spray		540	.030		.26	1.04		1.30	1.96
6400	Paint 1 coat, brushwork		310	.052		.23	1.82		2.05	3.15
6450	Spray		540	.030		.26	1.04		1.30	1.95
6500	Paint 2 coats, brushwork		195	.082		.45	2.89		3.34	5.10
6550	Spray		310	.052		.50	1.82		2.32	3.45
6600	Radiators, per side, primer, brushwork	1 Pord	520	.015	S.F.	.06	.54		.60	.93
6620	Paint, one coat		520	.015		.05	.54		.59	.93
6640	Two coats		340	.024		.11	.83		.94	1.44
6660	Three coats		283	.028		.16	1		1.16	1.76
7000	Trim, wood, incl. puttying, under 6" wide									
7200	Primer coat, oil base, brushwork	1 Pord	650	.012	L.F.	.03	.43		.46	.72
7250	Paint, 1 coat, brushwork		650	.012		.03	.43		.46	.72
7400	2 coats		400	.020		.06	.70		.76	1.19
7450	3 coats		325	.025		.09	.87		.96	1.47
7500	Over 6" wide, primer coat, brushwork		650	.012		.05	.43		.48	.75
7550	Paint, 1 coat, brushwork		650	.012		.06	.43		.49	.76
7600	2 coats		400	.020		.12	.70		.82	1.26
7650	3 coats		325	.025		.17	.87		1.04	1.57
8000	Cornice, simple design, primer coat, oil base, brushwork		650	.012	S.F.	.05	.43		.48	.75
8250	Paint, 1 coat		650	.012		.06	.43		.49	.76
8300	2 coats		400	.020		.12	.70		.82	1.26
8350	Ornate design, primer coat		350	.023		.05	.80		.85	1.35
8400	Paint, 1 coat		350	.023		.06	.80		.86	1.36
8450	2 coats		400	.020		.12	.70		.82	1.26
8600	Balustrades, primer coat, oil base, brushwork		520	.015		.05	.54		.59	.93
8650	Paint, 1 coat		520	.015		.06	.54		.60	.94
8700	2 coats		325	.025		.12	.87		.99	1.51
8900	Trusses and wood frames, primer coat, oil base, brushwork		800	.010		.05	.35		.40	.62
8950	Spray		1200	.007		.05	.23		.28	.44
9000	Paint 1 coat, brushwork		750	.011		.06	.38		.44	.67
9200	Spray		1200	.007		.07	.23		.30	.45
9220	Paint 2 coats, brushwork		500	.016		.12	.56		.68	1.03
9240	Spray		600	.013		.13	.47		.60	.89
9260	Stain, brushwork, wipe off		600	.013		.06	.47		.53	.82

09 91 Painting

09 91 23 – Interior Painting

09 91 23.52 Miscellaneous, Interior		Crew	Daily Output	Labor-Hours	Unit	2009 Bare Costs Material	Labor	Equipment	Total	Total Incl O&P
9280	Varnish, 3 coats, brushwork	1 Pord	275	.029	S.F.	.23	1.02		1.25	1.89
9350	For latex paint, deduct				↓	10%				

09 91 23.72 Walls and Ceilings, Interior		Crew	Daily Output	Labor-Hours	Unit	2009 Bare Costs Material	Labor	Equipment	Total	Total Incl O&P
0010	**WALLS AND CEILINGS, INTERIOR** R099100-10									
0100	Concrete, drywall or plaster, oil base, primer or sealer coat R099100-20									
0200	Smooth finish, brushwork	1 Pord	1150	.007	S.F.	.05	.25		.30	.45
0240	Roller		1350	.006		.05	.21		.26	.38
0280	Spray		2750	.003		.04	.10		.14	.20
0300	Sand finish, brushwork		975	.008		.05	.29		.34	.51
0340	Roller		1150	.007		.05	.25		.30	.45
0380	Spray		2275	.004		.04	.12		.16	.25
0400	Paint 1 coat, smooth finish, brushwork		1200	.007		.06	.23		.29	.44
0440	Roller		1300	.006		.06	.22		.28	.41
0480	Spray		2275	.004		.05	.12		.17	.26
0500	Sand finish, brushwork		1050	.008		.06	.27		.33	.49
0540	Roller		1600	.005		.06	.18		.24	.34
0580	Spray		2100	.004		.05	.13		.18	.27
0800	Paint 2 coats, smooth finish, brushwork		680	.012		.11	.41		.52	.79
0840	Roller		800	.010		.12	.35		.47	.69
0880	Spray		1625	.005		.10	.17		.27	.39
0900	Sand finish, brushwork		605	.013		.11	.47		.58	.86
0940	Roller		1020	.008		.12	.28		.40	.57
0980	Spray		1700	.005		.10	.17		.27	.37
1200	Paint 3 coats, smooth finish, brushwork		510	.016		.17	.55		.72	1.06
1240	Roller		650	.012		.18	.43		.61	.88
1280	Spray		1625	.005		.15	.17		.32	.45
1300	Sand finish, brushwork		454	.018		.17	.62		.79	1.17
1340	Roller		680	.012		.18	.41		.59	.85
1380	Spray		1133	.007		.15	.25		.40	.57
1600	Glaze coating, 2 coats, spray, clear		1200	.007		.49	.23		.72	.92
1640	Multicolor	↓	1200	.007	↓	.95	.23		1.18	1.43
1660	Painting walls, complete, including surface prep, primer &									
1670	2 coats finish, on drywall or plaster, with roller	1 Pord	325	.025	S.F.	.18	.87		1.05	1.58
1700	For latex paint, deduct					10%				
1800	For ceiling installations, add				↓		25%			
2000	Masonry or concrete block, oil base, primer or sealer coat									
2100	Smooth finish, brushwork	1 Pord	1224	.007	S.F.	.06	.23		.29	.44
2180	Spray		2400	.003		.09	.12		.21	.29
2200	Sand finish, brushwork		1089	.007		.10	.26		.36	.52
2280	Spray		2400	.003		.09	.12		.21	.29
2400	Paint 1 coat, smooth finish, brushwork		1100	.007		.10	.26		.36	.52
2480	Spray		2400	.003		.09	.12		.21	.29
2500	Sand finish, brushwork		979	.008		.10	.29		.39	.57
2580	Spray		2400	.003		.09	.12		.21	.29
2800	Paint 2 coats, smooth finish, brushwork		756	.011		.20	.37		.57	.82
2880	Spray		1360	.006		.18	.21		.39	.53
2900	Sand finish, brushwork		672	.012		.20	.42		.62	.89
2980	Spray		1360	.006		.18	.21		.39	.53
3200	Paint 3 coats, smooth finish, brushwork		560	.014		.30	.50		.80	1.13
3280	Spray		1088	.007		.28	.26		.54	.71
3300	Sand finish, brushwork		498	.016		.30	.57		.87	1.23
3380	Spray	↓	1088	.007	↓	.28	.26		.54	.71

09 91 23 – Interior Painting

09 91 23.72 Walls and Ceilings, Interior		Crew	Daily Output	Labor-Hours	Unit	2009 Bare Costs Material	Labor	Equipment	Total	Total Incl O&P
3600	Glaze coating, 3 coats, spray, clear	1 Pord	900	.009	S.F.	.70	.31		1.01	1.27
3620	Multicolor		900	.009		1.10	.31		1.41	1.71
4000	Block filler, 1 coat, brushwork		425	.019		.13	.66		.79	1.21
4100	Silicone, water repellent, 2 coats, spray		2000	.004		.30	.14		.44	.56
4120	For latex paint, deduct					10%				
8200	For work 8 – 15′ H, add						10%			
8300	For work over 15′ H, add						20%			
8400	For light textured surfaces, add						10%			
8410	Heavy textured, add						25%			

09 91 23.74 Walls and Ceilings, Interior, Zero VOC Latex		Crew	Daily Output	Labor-Hours	Unit	Material	Labor	Equipment	Total	Total Incl O&P
0010	**WALLS AND CEILINGS, INTERIOR, ZERO VOC LATEX**									
0100	Concrete, dry wall or plaster, latex, primer or sealer coat									
0200	Smooth finish, brushwork	1 Pord	1150	.007	S.F.	.06	.25		.31	.46
0240	Roller [G]		1350	.006		.06	.21		.27	.39
0280	Spray [G]		2750	.003		.05	.10		.15	.21
0300	Sand finish, brushwork [G]		975	.008		.06	.29		.35	.52
0340	Roller [G]		1150	.007		.07	.25		.32	.46
0380	Spray [G]		2275	.004		.05	.12		.17	.26
0400	Paint 1 coat, smooth finish, brushwork [G]		1200	.007		.06	.23		.29	.45
0440	Roller [G]		1300	.006		.06	.22		.28	.42
0480	Spray [G]		2275	.004		.05	.12		.17	.26
0500	Sand finish, brushwork [G]		1050	.008		.06	.27		.33	.49
0540	Roller [G]		1600	.005		.06	.18		.24	.35
0580	Spray [G]		2100	.004		.05	.13		.18	.27
0800	Paint 2 coats, smooth finish, brushwork [G]		680	.012		.12	.41		.53	.79
0840	Roller [G]		800	.010		.12	.35		.47	.69
0880	Spray [G]		1625	.005		.10	.17		.27	.39
0900	Sand finish, brushwork [G]		605	.013		.11	.47		.58	.87
0940	Roller [G]		1020	.008		.12	.28		.40	.57
0980	Spray [G]		1700	.005		.10	.17		.27	.37
1200	Paint 3 coats, smooth finish, brushwork [G]		510	.016		.17	.55		.72	1.07
1240	Roller [G]		650	.012		.18	.43		.61	.89
1280	Spray [G]		1625	.005		.16	.17		.33	.45
1800	For ceiling installations, add [G]						25%			
8200	For work 8 – 15′ H, add						10%			
8300	For work over 15′ H, add						20%			

09 91 23.75 Dry Fall Painting		Crew	Daily Output	Labor-Hours	Unit	Material	Labor	Equipment	Total	Total Incl O&P
0010	**DRY FALL PAINTING** R099100-10									
0100	Walls									
0200	Wallboard and smooth plaster, one coat, brush R099100-20	1 Pord	910	.009	S.F.	.05	.31		.36	.54
0210	Roll		1560	.005		.05	.18		.23	.34
0220	Spray		2600	.003		.05	.11		.16	.22
0230	Two coats, brush		520	.015		.10	.54		.64	.98
0240	Roll		877	.009		.10	.32		.42	.62
0250	Spray		1560	.005		.10	.18		.28	.40
0260	Concrete or textured plaster, one coat, brush		747	.011		.05	.38		.43	.65
0270	Roll		1300	.006		.05	.22		.27	.40
0280	Spray		1560	.005		.05	.18		.23	.34
0290	Two coats, brush		422	.019		.10	.67		.77	1.18
0300	Roll		747	.011		.10	.38		.48	.71
0310	Spray		1300	.006		.10	.22		.32	.46
0320	Concrete block, one coat, brush		747	.011		.05	.38		.43	.65

09 91 Painting

09 91 23 – Interior Painting

09 91 23.75 Dry Fall Painting		Crew	Daily Output	Labor-Hours	Unit	Material	2009 Bare Costs Labor	Equipment	Total	Total Incl O&P
0330	Roll	1 Pord	1300	.006	S.F.	.05	.22		.27	.40
0340	Spray		1560	.005		.05	.18		.23	.34
0350	Two coats, brush		422	.019		.10	.67		.77	1.18
0360	Roll		747	.011		.10	.38		.48	.71
0370	Spray		1300	.006		.10	.22		.32	.46
0380	Wood, one coat, brush		747	.011		.05	.38		.43	.65
0390	Roll		1300	.006		.05	.22		.27	.40
0400	Spray		877	.009		.05	.32		.37	.56
0410	Two coats, brush		487	.016		.10	.58		.68	1.03
0420	Roll		747	.011		.10	.38		.48	.71
0430	Spray	↓	650	.012	↓	.10	.43		.53	.80
0440	Ceilings									
0450	Wallboard and smooth plaster, one coat, brush	1 Pord	600	.013	S.F.	.05	.47		.52	.80
0460	Roll		1040	.008		.05	.27		.32	.48
0470	Spray		1560	.005		.05	.18		.23	.34
0480	Two coats, brush		341	.023		.10	.83		.93	1.43
0490	Roll		650	.012		.10	.43		.53	.80
0500	Spray		1300	.006		.10	.22		.32	.46
0510	Concrete or textured plaster, one coat, brush		487	.016		.05	.58		.63	.97
0520	Roll		877	.009		.05	.32		.37	.56
0530	Spray		1560	.005		.05	.18		.23	.34
0540	Two coats, brush		276	.029		.10	1.02		1.12	1.74
0550	Roll		520	.015		.10	.54		.64	.98
0560	Spray		1300	.006		.10	.22		.32	.46
0570	Structural steel, bar joists or metal deck, one coat, spray		1560	.005		.05	.18		.23	.34
0580	Two coats, spray	↓	1040	.008	↓	.10	.27		.37	.54

09 93 Staining and Transparent Finishing

09 93 23 – Interior Staining and Finishing

09 93 23.10 Varnish

		Crew	Daily Output	Labor-Hours	Unit	Material	Labor	Equipment	Total	Total Incl O&P
0010	**VARNISH**									
0012	1 coat + sealer, on wood trim, brush, no sanding included	1 Pord	400	.020	S.F.	.07	.70		.77	1.21
0100	Hardwood floors, 2 coats, no sanding included, roller		1890	.004	"	.15	.15		.30	.41
9000	Minimum labor/equipment charge	↓	4	2	Job		70.50		70.50	113

09 96 High-Performance Coatings

09 96 23 – Graffiti-Resistant Coatings

09 96 23.10 Graffiti Resistant Treatments

		Crew	Daily Output	Labor-Hours	Unit	Material	Labor	Equipment	Total	Total Incl O&P
0010	**GRAFFITI RESISTANT TREATMENTS**, sprayed on walls									
0100	Non-sacrificial, permanent non-stick coating, clear, on metals	1 Pord	2000	.004	S.F.	1.59	.14		1.73	1.98
0200	Concrete		2000	.004		1.81	.14		1.95	2.22
0300	Concrete block		2000	.004		2.34	.14		2.48	2.81
0400	Brick		2000	.004		2.66	.14		2.80	3.15
0500	Stone		2000	.004		2.66	.14		2.80	3.15
0600	Unpainted wood		2000	.004		3.06	.14		3.20	3.60
2000	Semi-permanent cross linking polymer primer, on metals		2000	.004		.40	.14		.54	.67
2100	Concrete		2000	.004		.48	.14		.62	.76
2200	Concrete block		2000	.004		.60	.14		.74	.89
2300	Brick	↓	2000	.004	↓	.48	.14		.62	.76

09 96 High-Performance Coatings

09 96 23 – Graffiti-Resistant Coatings

09 96 23.10 Graffiti Resistant Treatments		Crew	Daily Output	Labor-Hours	Unit	2009 Bare Costs Material	Labor	Equipment	Total	Total Incl O&P
2400	Stone	1 Pord	2000	.004	S.F.	.48	.14		.62	.76
2500	Unpainted wood		2000	.004		.67	.14		.81	.96
3000	Top coat, on metals		2000	.004		.43	.14		.57	.70
3100	Concrete		2000	.004		.49	.14		.63	.76
3200	Concrete block		2000	.004		.68	.14		.82	.98
3300	Brick		2000	.004		.57	.14		.71	.85
3400	Stone		2000	.004		.57	.14		.71	.85
3500	Unpainted wood		2000	.004		.68	.14		.82	.98
5000	Sacrificial, water based, on metal		2000	.004		.21	.14		.35	.46
5100	Concrete		2000	.004		.21	.14		.35	.46
5200	Concrete block		2000	.004		.21	.14		.35	.46
5300	Brick		2000	.004		.21	.14		.35	.46
5400	Stone		2000	.004		.21	.14		.35	.46
5500	Unpainted wood	↓	2000	.004	↓	.21	.14		.35	.46
8000	Cleaner for use after treatment									
8100	Towels or wipes, per package of 30				Ea.	.80			.80	.88
8200	Aerosol spray, 24 oz. can				"	19			19	21

09 96 56 – Epoxy Coatings

09 96 56.20 Wall Coatings		Crew	Daily Output	Labor-Hours	Unit	Material	Labor	Equipment	Total	Total Incl O&P
0010	**WALL COATINGS**									
0100	Acrylic glazed coatings, minimum	1 Pord	525	.015	S.F.	.28	.54		.82	1.17
0200	Maximum		305	.026		.60	.92		1.52	2.14
0300	Epoxy coatings, minimum		525	.015		.37	.54		.91	1.27
0400	Maximum		170	.047		1.12	1.66		2.78	3.88
0600	Exposed aggregate, troweled on, 1/16" to 1/4", minimum		235	.034		.56	1.20		1.76	2.53
0700	Maximum (epoxy or polyacrylate)		130	.062		1.20	2.17		3.37	4.78
0900	1/2" to 5/8" aggregate, minimum		130	.062		1.11	2.17		3.28	4.68
1000	Maximum		80	.100		1.90	3.52		5.42	7.75
1200	1" aggregate size, minimum		90	.089		1.93	3.13		5.06	7.10
1300	Maximum		55	.145		2.95	5.10		8.05	11.45
1500	Exposed aggregate, sprayed on, 1/8" aggregate, minimum		295	.027		.51	.95		1.46	2.09
1600	Maximum		145	.055		.96	1.94		2.90	4.16
1800	High build epoxy, 50 mil, minimum		390	.021		.62	.72		1.34	1.83
1900	Maximum		95	.084		1.06	2.96		4.02	5.90
2100	Laminated epoxy with fiberglass, minimum		295	.027		.67	.95		1.62	2.27
2200	Maximum		145	.055		1.19	1.94		3.13	4.41
2400	Sprayed perlite or vermiculite, 1/16" thick, minimum		2935	.003		.24	.10		.34	.41
2500	Maximum		640	.013		.68	.44		1.12	1.45
2700	Vinyl plastic wall coating, minimum		735	.011		.30	.38		.68	.94
2800	Maximum		240	.033		.75	1.17		1.92	2.70
3000	Urethane on smooth surface, 2 coats, minimum		1135	.007		.24	.25		.49	.66
3100	Maximum		665	.012		.53	.42		.95	1.26
3300	3 coat, minimum		840	.010		.32	.34		.66	.89
3400	Maximum		470	.017		.72	.60		1.32	1.75
3600	Ceramic-like glazed coating, cementitious, minimum		440	.018		.44	.64		1.08	1.50
3700	Maximum		345	.023		.74	.82		1.56	2.11
3900	Resin base, minimum		640	.013		.30	.44		.74	1.03
4000	Maximum	↓	330	.024	↓	.49	.85		1.34	1.90

09 97 10.10 Coatings and Paints		Crew	Daily Output	Labor-Hours	Unit	2009 Bare Costs Material	Labor	Equipment	Total	Total Incl O&P
0010	**COATINGS & PAINTS** in 5 gallon lots R099100-20									
0050	For 100 gallons or more, deduct					10%				10%
0100	Paint, Exterior alkyd (oil base)									
0200	Flat				Gal.	25			25	27.50
0300	Gloss					23.50			23.50	25.50
0400	Primer					22.50			22.50	24.50
0500	Latex (water base)									
0600	Acrylic stain				Gal.	25			25	27.50
0700	Gloss enamel					23			23	25.50
0800	Flat					20			20	22
0900	Primer					21.50			21.50	23.50
1000	Semi-gloss					25			25	27.50
1054	Low V.O.C.									
1055	Primer				Gal.	41			41	45
1060	Satin					39.50			39.50	43.50
1065	Door and trim					47.50			47.50	52
1100	Interior, alkyd (oil base)									
1200	Enamel undercoat				Gal.	24.50			24.50	27
1300	Flat					25			25	27.50
1400	Gloss					25.50			25.50	28
1500	Primer sealer					20.50			20.50	22.50
1600	Semi-gloss					25			25	27.50
1700	Latex (water base)									
1800	Enamel undercoat				Gal.	19.70			19.70	21.50
1900	Flat					19.65			19.65	21.50
2000	Floor and deck					23			23	25.50
2100	Gloss					24			24	26.50
2200	Primer sealer					19.15			19.15	21
2300	Semi-gloss					23.50			23.50	26
2320	Low V.O.C.									
2330	Wallboard primer				Gal.	29			29	32
2335	Flat					36.50			36.50	40
2340	Semi-gloss					36.50			36.50	40
2345	Eggshell					37.50			37.50	41.50
2350	Stain					41			41	45
2400	Masonry, Exterior									
2500	Alkali resistant primer				Gal.	20			20	22
2600	Block filler, epoxy					27			27	30
2700	Latex					13.45			13.45	14.80
2800	Latex, flat					17.55			17.55	19.35
2900	Semi-gloss					24			24	26.50
3000	Masonry, Interior									
3100	Alkali resistant primer				Gal.	21.50			21.50	23.50
3200	Block filler, epoxy					27			27	29.50
3300	Latex					15.60			15.60	17.20
3400	Floor, alkyd					35			35	38.50
3500	Latex					29			29	32
3600	Latex, flat acrylic					19.95			19.95	22
3700	Flat emulsion					16.75			16.75	18.45
3800	Sealer					16.20			16.20	17.85
3900	Semi-gloss					26.50			26.50	29.50
4000	Metal									
4100	Galvanizing paint				Gal.	28			28	30.50

09 97 Special Coatings

09 97 10 – Coatings and Paints

09 97 10.10 Coatings and Paints		Crew	Daily Output	Labor-Hours	Unit	Material	Labor	Equipment	Total	Total Incl O&P
4200	High heat				Gal.	54			54	59.50
4300	Heat resistant					32			32	35.50
4400	Machinery enamel, alkyd					35.50			35.50	39
4500	Metal pretreatment (polyvinyl butyral)					30			30	33
4600	Rust inhibitor, ferrous metal					31.50			31.50	35
4700	Zinc chromate					25			25	27.50
4800	Zinc rich primer				↓	119			119	131
4900	Varnish and Stain									
5000	Alkyd, clear				Gal.	30.50			30.50	33.50
5100	Polyurethane, clear					28.50			28.50	31
5200	Primer sealer					28.50			28.50	31
5300	Stain, semi-transparent					24.50			24.50	27
5400	Solid color				↓	25			25	27.50
5500	Coatings									
5600	Heavy duty									
5700	Acrylic urethane				Gal.	50			50	55
5800	Chlorinated rubber					58			58	63.50
5900	Coal tar epoxy					67			67	74
6000	Polyamide epoxy, finish					53.50			53.50	59
6100	Primer					47.50			47.50	52.50
6200	Silicone alkyd					41.50			41.50	45.50
6300	2 component solvent based acrylic epoxy					70.50			70.50	77.50
6400	Polyester epoxy					104			104	114
6500	Vinyl				↓	33			33	36.50
6600	Special/Miscellaneous									
6700	Aluminum				Gal.	36			36	39.50
6900	Dry fall out, flat					14.95			14.95	16.40
7000	Fire retardant, intumescent					45			45	49.50
7100	Linseed oil					16.50			16.50	18.15
7200	Shellac					28			28	31
7300	Swimming pool, epoxy or urethane base					46.50			46.50	51.50
7400	Rubber base					33.50			33.50	37
7500	Texture paint					13.90			13.90	15.30
7600	Turpentine					16.25			16.25	17.90
7700	Water repellent, 5% silicone					22.50			22.50	24.50
7800	Insulating additive				↓	14.95			14.95	16.45

(2009 Bare Costs: Material, Labor, Equipment, Total)

09 97 13 – Steel Coatings

09 97 13.23 Exterior Steel Coatings		Crew	Daily Output	Labor-Hours	Unit	Material	Labor	Equipment	Total	Total Incl O&P
0010	EXTERIOR STEEL COATINGS R050516-30									
6100	Cold galvanizing, brush in field	1 Psst	1100	.007	S.F.	.07	.26		.33	.58
6510	Paints & protective coatings, sprayed in field									
6520	Alkyds, primer	2 Psst	3600	.004	S.F.	.06	.16		.22	.37
6540	Gloss topcoats		3200	.005		.05	.18		.23	.39
6560	Silicone alkyd		3200	.005		.12	.18		.30	.47
6610	Epoxy, primer		3000	.005		.24	.19		.43	.63
6630	Intermediate or topcoat		2800	.006		.21	.21		.42	.63
6650	Enamel coat		2800	.006		.27	.21		.48	.68
6700	Epoxy ester, primer		2800	.006		.52	.21		.73	.96
6720	Topcoats		2800	.006		.18	.21		.39	.59
6810	Latex primer		3600	.004		.05	.16		.21	.37
6830	Topcoats		3200	.005		.06	.18		.24	.40
6910	Universal primers, one part, phenolic, modified alkyd	↓	2000	.008	↓	.10	.29		.39	.66

09 97 Special Coatings

09 97 13 – Steel Coatings

09 97 13.23 Exterior Steel Coatings		Crew	Daily Output	Labor-Hours	Unit	2009 Bare Costs Material	Labor	Equipment	Total	Total Incl O&P
6940	Two part, epoxy spray	2 Psst	2000	.008	S.F.	.28	.29		.57	.86
7000	Zinc rich primers, self cure, spray, inorganic		1800	.009		.60	.32		.92	1.27
7010	Epoxy, spray, organic	↓	1800	.009	↓	.21	.32		.53	.85
7020	Above one story, spray painting simple structures, add						25%			
7030	Intricate structures, add						50%			

Division Notes

		CREW	DAILY OUTPUT	LABOR-HOURS	UNIT	2009 BARE COSTS MAT.	LABOR	EQUIP.	TOTAL	TOTAL INCL O&P

Estimating Tips

General

- The items in this division are usually priced per square foot or each.
- Many items in Division 10 require some type of support system or special anchors that are not usually furnished with the item. The required anchors must be added to the estimate in the appropriate division.
- Some items in Division 10, such as lockers, may require assembly before installation. Verify the amount of assembly required. Assembly can often exceed installation time.

10 20 00 Interior Specialties

- Support angles and blocking are not included in the installation of toilet compartments, shower/dressing compartments, or cubicles. Appropriate line items from Divisions 5 or 6 may need to be added to support the installations.
- Toilet partitions are priced by the stall. A stall consists of a side wall, pilaster, and door with hardware. Toilet tissue holders and grab bars are extra.
- The required acoustical rating of a folding partition can have a significant impact on costs. Verify the sound transmission coefficient rating of the panel priced to the specification requirements.
- Grab bar installation does not include supplemental blocking or backing to support the required load. When grab bars are installed at an existing facility, provisions must be made to attach the grab bars to solid structure.

Reference Numbers

Reference numbers are shown in shaded boxes at the beginning of some major classifications. These numbers refer to related items in the Reference Section. The reference information may be an estimating procedure, an alternate pricing method, or technical information.

Note: Not all subdivisions listed here necessarily appear in this publication.

10 05 Common Work Results for Specialties

10 05 05 – Selective Specialties Demolition

10 05 05.10 Selective Demolition, Specialties

	10 05 05.10 Selective Demolition, Specialties	Crew	Daily Output	Labor-Hours	Unit	Material	2009 Bare Costs Labor	Equipment	Total	Total Incl O&P
0010	**SELECTIVE DEMOLITION, SPECIALTIES** R024119-10									
1100	Boards and panels, wall mounted	2 Clab	15	1.067	Ea.		33.50		33.50	55.50
1200	Cases, for directory and/or bulletin boards, incl doors		24	.667			21		21	35
1850	Shower partitions, cabinet or stall, incl base and door		8	2			63		63	104
1855	Shower receptor, terrazzo or concrete	1 Clab	14	.571			18.05		18.05	30
1900	Curtain track or rod, hospital type, ceiling mounted or suspended	"	220	.036	L.F.		1.15		1.15	1.90
1910	Toilet Cubicles, remove	2 Clab	8	2	Ea.		63		63	104
1920	Toilet Cubicles, remove	"	16	1	"		31.50		31.50	52
2650	Wall guard, misc. wall or corner protection	1 Clab	320	.025	L.F.		.79		.79	1.30
2750	Access floor, metal panel system, incl pedestals, covering	2 Clab	850	.019	S.F.		.59		.59	.98
3050	Fireplace, prefab, freestanding or wall hung, incl hood and screen	1 Clab	2	4	Ea.		126		126	209
3054	Chimney top, simulated brick, 4′ high	"	15	.533			16.85		16.85	28
3200	Stove, woodburning, cast iron	2 Clab	2	8			253		253	415
3440	Weathervane, residential	1 Clab	12	.667			21		21	35
3500	Flagpole, groundset, to 70′ high, excl base/fndn	K-1	1	16			565	243	808	1,200
3555	To 30′ high	"	2.50	6.400			227	97	324	480
4070	Remove traffic posts to 12′-0″ high	B-6	100	.240			8.20	2.94	11.14	16.50
4300	Letter, signs or plaques, exterior on wall	1 Clab	20	.400			12.65		12.65	21
4310	Signs, street, reflective aluminum, incl post and bracket		60	.133			4.21		4.21	6.95
4320	Door signs interior on door 6″ x 6″, selective demolition		20	.400			12.65		12.65	21
4550	Turnstiles, manual or electric	2 Clab	2	8			253		253	415
5050	Lockers	"	15	1.067	Opng.		33.50		33.50	55.50
5250	Cabinets, recessed	Q-12	12	1.333	Ea.		58		58	89.50
5260	Mail boxes, Horiz., Key Lock, front loading, Remove	1 Carp	34	.235	"		9.40		9.40	15.50
5350	Awning, fabric, including frame	2 Clab	100	.160	S.F.		5.05		5.05	8.35
6050	Partition, woven wire		1400	.011			.36		.36	.60
6100	Folding gate, security, door or window		500	.032			1.01		1.01	1.67
6580	Acoustic air wall		650	.025			.78		.78	1.28
7550	Telephone enclosure, exterior, post mounted		3	5.333	Ea.		169		169	278
8850	Scale, platform, excludes foundation or pit		.25	64	"		2,025		2,025	3,350

10 11 Visual Display Surfaces

10 11 13 – Chalkboards

10 11 13.13 Fixed Chalkboards

	10 11 13.13 Fixed Chalkboards	Crew	Daily Output	Labor-Hours	Unit	Material	Labor	Equipment	Total	Total Incl O&P
0010	**FIXED CHALKBOARDS** Porcelain enamel steel									
3900	Wall hung									
4000	Aluminum frame and chalktrough									
4200	3′ x 4′	2 Carp	16	1	Ea.	181	40		221	265
4300	3′ x 5′		15	1.067		242	42.50		284.50	335
4500	4′ x 8′		14	1.143		345	45.50		390.50	455
4600	4′ x 12′		13	1.231		505	49		554	635
4700	Wood frame and chalktrough									
4800	3′ x 4′	2 Carp	16	1	Ea.	169	40		209	252
5000	3′ x 5′		15	1.067		240	42.50		282.50	335
5100	4′ x 5′		14	1.143		248	45.50		293.50	350
5300	4′ x 8′		13	1.231		340	49		389	450
5400	Liquid chalk, white porcelain enamel, wall hung									
5420	Deluxe units, aluminum trim and chalktrough									
5450	4′ x 4′	2 Carp	16	1	Ea.	202	40		242	288
5500	4′ x 8′		14	1.143		310	45.50		355.50	420
5550	4′ x 12′		12	1.333		470	53.50		523.50	605

10 11 13.13 Fixed Chalkboards

	10 11 13.13 Fixed Chalkboards	Crew	Daily Output	Labor-Hours	Unit	2009 Bare Costs Material	Labor	Equipment	Total	Total Incl O&P
5700	Wood trim and chalktrough									
5900	4′ x 4′	2 Carp	16	1	Ea.	460	40		500	575
6000	4′ x 6′		15	1.067		550	42.50		592.50	675
6200	4′ x 8′		14	1.143		635	45.50		680.50	775
6250	Economy dry erase board, melamine, 36″ high		104	.154	S.F.	7.40	6.15		13.55	18.30
6300	Liquid chalk, felt tip markers				Ea.	1.62			1.62	1.78
6500	Erasers					1.24			1.24	1.36
6600	Board cleaner, 8 oz. bottle					4.28			4.28	4.71
9000	Minimum labor/equipment charge	2 Carp	3	5.333	Job		213		213	350

10 11 13.23 Modular-Support-Mounted Chalkboards

	Description	Crew	Daily Output	Labor-Hours	Unit	Material	Labor	Equipment	Total	Total Incl O&P
0010	**MODULAR-SUPPORT-MOUNTED CHALKBOARDS**									
0400	Sliding chalkboards									
0450	Vertical, one sliding board with back panel, wall mounted									
0500	8′ x 4′	2 Carp	8	2	Ea.	1,825	80		1,905	2,150
0520	8′ x 8′		7.50	2.133		2,575	85		2,660	2,975
0540	8′ x 12′		7	2.286		3,125	91.50		3,216.50	3,600
0600	Two sliding boards, with back panel									
0620	8′ x 4′	2 Carp	8	2	Ea.	2,750	80		2,830	3,150
0640	8′ x 8′		7.50	2.133		3,925	85		4,010	4,475
0660	8′ x 12′		7	2.286		4,775	91.50		4,866.50	5,400
0700	Horizontal, two track									
0800	4′ x 8′, 2 sliding panels	2 Carp	8	2	Ea.	1,700	80		1,780	2,000
0820	4′ x 12′, 2 sliding panels		7.50	2.133		2,175	85		2,260	2,550
0840	4′ x 16′, 4 sliding panels		7	2.286		2,925	91.50		3,016.50	3,375
0900	Four track, four sliding panels									
0920	4′ x 8′	2 Carp	8	2	Ea.	2,725	80		2,805	3,125
0940	4′ x 12′		7.50	2.133		3,500	85		3,585	4,000
0960	4′ x 16′		7	2.286		4,425	91.50		4,516.50	5,025
1200	Vertical, motor operated									
1400	One sliding panel with back panel									
1450	10′ x 4′	2 Carp	4	4	Ea.	6,125	160		6,285	7,025
1500	10′ x 10′		3.75	4.267		7,175	170		7,345	8,150
1550	10′ x 16′		3.50	4.571		8,225	183		8,408	9,350
1700	Two sliding panels with back panel									
1750	10′ x 4′	2 Carp	4	4	Ea.	9,825	160		9,985	11,100
1800	10′ x 10′		3.75	4.267		11,200	170		11,370	12,600
1850	10′ x 16′		3.50	4.571		12,700	183		12,883	14,200
2000	Three sliding panels with back panel									
2100	10′ x 4′	2 Carp	4	4	Ea.	12,800	160		12,960	14,400
2150	10′ x 10′		3.75	4.267		14,900	170		15,070	16,700
2200	10′ x 16′		3.50	4.571		17,000	183		17,183	19,000
2400	For projection screen, glass beaded, add				S.F.	10.20			10.20	11.20
2500	For remote control, 1 panel control, add				Ea.	480			480	525
2600	2 panel control, add				″	825			825	905
2800	For units without back panels, deduct				S.F.	6.45			6.45	7.10
2850	For liquid chalk porcelain panels, add				″	6.90			6.90	7.60
3000	Swing leaf, any comb. of chalkboard & cork, aluminum frame									
3100	Floor style, 6 panels									
3150	30″ x 40″ panels				Ea.	1,475			1,475	1,625
3200	48″ x 40″ panels				″	2,100			2,100	2,300
3300	Wall mounted, 6 panels									
3400	30″ x 40″ panels	2 Carp	16	1	Ea.	1,350	40		1,390	1,575

10 11 Visual Display Surfaces

10 11 13 – Chalkboards

10 11 13.23 Modular-Support-Mounted Chalkboards		Crew	Daily Output	Labor-Hours	Unit	Material	2009 Bare Costs Labor	Equipment	Total	Total Incl O&P
3450	48″ x 40″ panels	2 Carp	16	1	Ea.	1,750	40		1,790	2,000
3600	Extra panels for swing leaf units									
3700	30″ x 40″ panels				Ea.	282			282	310
3750	48″ x 40″ panels				″	335			335	370

10 11 13.43 Portable Chalkboards		Crew	Daily Output	Labor-Hours	Unit	Material	Labor	Equipment	Total	Total Incl O&P
0010	**PORTABLE CHALKBOARDS**									
0100	Freestanding, reversible									
0120	Economy, wood frame, 4′ x 6′									
0140	Chalkboard both sides				Ea.	565			565	620
0160	Chalkboard one side, cork other side				″	485			485	535
0200	Standard, lightweight satin finished aluminum, 4′ x 6′									
0220	Chalkboard both sides				Ea.	650			650	715
0240	Chalkboard one side, cork other side				″	580			580	640
0300	Deluxe, heavy duty extruded aluminum, 4′ x 6′									
0320	Chalkboard both sides				Ea.	1,050			1,050	1,150
0340	Chalkboard one side, cork other side				″	940			940	1,025

10 11 23 – Tackboards

10 11 23.10 Fixed Tackboards		Crew	Daily Output	Labor-Hours	Unit	Material	Labor	Equipment	Total	Total Incl O&P
0010	**FIXED TACKBOARDS**									
0020	Cork sheets, unbacked, no frame, 1/4″ thick	2 Carp	290	.055	S.F.	3.60	2.20		5.80	7.60
0100	1/2″ thick		290	.055		6.30	2.20		8.50	10.60
0300	Fabric-face, no frame, on 7/32″ cork underlay		290	.055		5.35	2.20		7.55	9.55
0400	On 1/4″ cork on 1/4″ hardboard		290	.055		7.50	2.20		9.70	11.90
0600	With edges wrapped		290	.055		7.90	2.20		10.10	12.35
0700	On 7/16″ fire retardant core		290	.055		7.35	2.20		9.55	11.75
0900	With edges wrapped		290	.055		11.90	2.20		14.10	16.75
1000	Designer fabric only, cut to size					2.02			2.02	2.22
1200	1/4″ vinyl cork, on 1/4″ hardboard, no frame	2 Carp	290	.055		8	2.20		10.20	12.45
1300	On 1/4″ coreboard		290	.055		7.95	2.20		10.15	12.40
2000	For map and display rail, economy, add		385	.042	L.F.	2.27	1.66		3.93	5.25
2100	Deluxe, add		350	.046	″	3.23	1.83		5.06	6.55
2120	Prefabricated, 1/4″ cork, 3′ x 5′ with aluminum frame		16	1	Ea.	109	40		149	186
2140	Wood frame		16	1		139	40		179	219
2160	4′ x 4′ with aluminum frame		16	1		92.50	40		132.50	167
2180	Wood frame		16	1		120	40		160	198
2200	4′ x 8′ with aluminum frame		14	1.143		162	45.50		207.50	254
2210	With wood frame		14	1.143		144	45.50		189.50	234
2220	4′ x 12′ with aluminum frame		12	1.333		240	53.50		293.50	350
2230	Bulletin board case, single glass door, with lock									
2240	36″ x 24″, economy	2 Carp	12	1.333	Ea.	258	53.50		311.50	370
2250	Deluxe		12	1.333		460	53.50		513.50	600
2260	42″ x 30″, economy		12	1.333		290	53.50		343.50	410
2270	Deluxe		12	1.333		430	53.50		483.50	565
2300	Glass enclosed cabinets, alum., cork panel, hinged doors									
2400	3′ x 3′, 1 door	2 Carp	12	1.333	Ea.	535	53.50		588.50	680
2500	4′ x 4′, 2 door		11	1.455		895	58		953	1,075
2600	4′ x 7′, 3 door		10	1.600		1,500	64		1,564	1,750
2800	4′ x 10′, 4 door		8	2		1,950	80		2,030	2,275
2900	For lights, add per door opening	1 Elec	13	.615		175	29		204	238
3100	Horizontal sliding units, 4 doors, 4′ x 8′, 8′ x 4′	2 Carp	9	1.778		1,700	71		1,771	2,000
3200	4′ x 12′		7	2.286		2,175	91.50		2,266.50	2,550
3400	8 doors, 4′ x 16′		5	3.200		2,900	128		3,028	3,400

10 11 Visual Display Surfaces

10 11 23 – Tackboards

10 11 23.10 Fixed Tackboards		Crew	Daily Output	Labor-Hours	Unit	Material	2009 Bare Costs Labor	Equipment	Total	Total Incl O&P
3500	4′ x 24′	2 Carp	4	4	Ea.	3,900	160		4,060	4,575
9000	Minimum labor/equipment charge	↓	4	4	Job		160		160	264

10 11 23.20 Control Boards		Crew	Daily Output	Labor-Hours	Unit	Material	Labor	Equipment	Total	Total Incl O&P
0010	**CONTROL BOARDS**									
0020	Magnetic, porcelain finish, 18″ x 24″, framed	2 Carp	8	2	Ea.	199	80		279	350
0100	24″ x 36″		7.50	2.133		255	85		340	420
0200	36″ x 48″		7	2.286		375	91.50		466.50	565
0300	48″ x 72″		6	2.667		720	107		827	965
0400	48″ x 96″	↓	5	3.200	↓	845	128		973	1,150

10 13 Directories

10 13 10 – Building Directories

10 13 10.10 Directory Boards		Crew	Daily Output	Labor-Hours	Unit	Material	Labor	Equipment	Total	Total Incl O&P
0010	**DIRECTORY BOARDS**									
0050	Plastic, glass covered, 30″ x 20″	2 Carp	3	5.333	Ea.	242	213		455	615
0100	36″ x 48″		2	8		875	320		1,195	1,500
0300	Grooved cork, 30″ x 20″		3	5.333		385	213		598	775
0400	36″ x 48″		2	8		520	320		840	1,100
0600	Black felt, 30″ x 20″		3	5.333		184	213		397	550
0700	36″ x 48″		2	8		365	320		685	930
0900	Outdoor, weatherproof, black plastic, 36″ x 24″		2	8		740	320		1,060	1,350
1000	36″ x 36″		1.50	10.667		850	425		1,275	1,650
1800	Indoor, economy, open face, 18″ x 24″		7	2.286		132	91.50		223.50	297
1900	24″ x 36″		7	2.286		144	91.50		235.50	310
2000	36″ x 24″		6	2.667		164	107		271	355
2100	36″ x 48″		6	2.667		246	107		353	445
2400	Building directory, alum., black felt panels, 1 door, 24″ x 18″		4	4		315	160		475	615
2500	36″ x 24″		3.50	4.571		355	183		538	690
2600	48″ x 32″		3	5.333		590	213		803	1,000
2700	2 door, 36″ x 48″		2.50	6.400		645	256		901	1,125
2800	36″ x 60″		2	8		865	320		1,185	1,475
2900	48″ x 60″	↓	1	16		940	640		1,580	2,075
3100	For bronze enamel finish, add					15%				
3200	For bronze anodized finish, add					25%				
3400	For illuminated directory, single door unit, add				↓	175			175	193
3500	For 6″ header panel, 6 letters per foot, add				L.F.	29			29	31.50
5000	Building directory, illuminated, 7″ sections, bronze finish									
5100	19″ x 36″, 2 sections, 80 name capacity				Ea.	2,500			2,500	2,775
5200	26-1/4″ x 36″, 3 sections, 120 name capacity					3,000			3,000	3,300
5300	33-5/8″ x 36″, 4 sections, 160 name capacity					3,425			3,425	3,775
5400	48-3/4″ x 36″, 6 sections, 240 name capacity					4,425			4,425	4,875
5500	63-1/2″ x 36″, 8 sections, 320 name capacity					5,375			5,375	5,925
5600	Engraving charge per namestrip				↓	13.40			13.40	14.75
9000	Minimum labor/equipment charge	1 Carp	1	8	Job		320		320	530

10 14 Signage

10 14 19 – Dimensional Letter Signage

10 14 19.10 Exterior Signs

	10 14 19.10 Exterior Signs	Crew	Daily Output	Labor-Hours	Unit	2009 Bare Costs Material	Labor	Equipment	Total	Total Incl O&P
0010	**EXTERIOR SIGNS**									
0020	Letters, 2″ high, 3/8″ deep, cast bronze	1 Carp	24	.333	Ea.	22	13.30		35.30	46.50
0140	1/2″ deep, cast aluminum		18	.444		22	17.75		39.75	54
0160	Cast bronze		32	.250		29	10		39	48.50
0300	6″ high, 5/8″ deep, cast aluminum		24	.333		27.50	13.30		40.80	52.50
0400	Cast bronze		24	.333		56.50	13.30		69.80	84
0600	8″ high, 3/4″ deep, cast aluminum		14	.571		33.50	23		56.50	74.50
0700	Cast bronze		20	.400		80.50	16		96.50	115
0900	10″ high, 1″ deep, cast aluminum		18	.444		49.50	17.75		67.25	84
1000	Bronze		18	.444		106	17.75		123.75	147
1200	12″ high, 1-1/4″ deep, cast aluminum		12	.667		52.50	26.50		79	102
1500	Cast bronze		18	.444		123	17.75		140.75	166
1600	14″ high, 2-5/16″ deep, cast aluminum		12	.667		75	26.50		101.50	127
1800	Fabricated stainless steel, 6″ high, 2″ deep		20	.400		53	16		69	84.50
1900	12″ high, 3″ deep		18	.444		118	17.75		135.75	159
2100	18″ high, 3″ deep		12	.667		228	26.50		254.50	294
2200	24″ high, 4″ deep		10	.800		385	32		417	475
2700	Acrylic, on high density foam, 12″ high, 2″ deep		20	.400		20.50	16		36.50	49
2800	18″ high, 2″ deep		18	.444		51	17.75		68.75	85.50
3900	Plaques, custom, 20″ x 30″, for up to 450 letters, cast aluminum	2 Carp	4	4		1,050	160		1,210	1,425
4000	Cast bronze		4	4		1,475	160		1,635	1,900
4200	30″ x 36″, up to 900 letters cast aluminum		3	5.333		1,875	213		2,088	2,400
4300	Cast bronze		3	5.333		3,475	213		3,688	4,175
4500	36″ x 48″, for up to 1300 letters, cast bronze		2	8		4,050	320		4,370	4,975
4800	Signs, reflective alum. directional signs, dbl. face, 2-way, w/bracket		30	.533		83.50	21.50		105	127
4900	4-way		30	.533		167	21.50		188.50	219
5100	Exit signs, 24 ga. alum., 14″ x 12″ surface mounted	1 Carp	30	.267		40.50	10.65		51.15	62
5200	10″ x 7″		20	.400		24	16		40	53
5400	Bracket mounted, double face, 12″ x 10″		30	.267		40.50	10.65		51.15	62
5500	Sticky back, stock decals, 14″ x 10″	1 Clab	50	.160		11.10	5.05		16.15	20.50
6000	Interior elec., wall mount, fiberglass panels, 2 lamps, 6″	1 Elec	8	1		106	47		153	190
6100	8″	″	8	1		85	47		132	166
6400	Replacement sign faces, 6″ or 8″	1 Clab	50	.160		47	5.05		52.05	60
9000	Minimum labor/equipment charge	1 Carp	4	2	Job		80		80	132

10 14 23 – Panel Signage

10 14 23.13 Engraved Interior Panel Signage

	10 14 23.13 Engraved Interior Panel Signage	Crew	Daily Output	Labor-Hours	Unit	Material	Labor	Equipment	Total	Total Incl O&P
0010	**ENGRAVED INTERIOR PANEL SIGNAGE**									
1010	Flexible door sign, adhesive back, w/Braille, 5/8″ letters, 4″ x 4″	1 Clab	32	.250	Ea.	26.50	7.90		34.40	42.50
1050	6″ x 6″		32	.250		39.50	7.90		47.40	56.50
1100	8″ x 2″		32	.250		27.50	7.90		35.40	43
1150	8″ x 4″		32	.250		35	7.90		42.90	51.50
1200	8″ x 8″		32	.250		48	7.90		55.90	66
1250	12″ x 2″		32	.250		35	7.90		42.90	51.50
1300	12″ x 6″		32	.250		50	7.90		57.90	68
1350	12″ x 12″		32	.250		99	7.90		106.90	122
1500	Graphic symbols, 2″ x 2″		32	.250		12	7.90		19.90	26.50
1550	6″ x 6″		32	.250		30	7.90		37.90	46
1600	8″ x 8″		32	.250		30	7.90		37.90	46

10 14 53 – Traffic Signage

10 14 53.20 Traffic Signs

	10 14 53.20 Traffic Signs	Crew	Daily Output	Labor-Hours	Unit	Material	Labor	Equipment	Total	Total Incl O&P
0010	**TRAFFIC SIGNS**									
0012	Stock, 24″ x 24″, no posts, .080″ alum. reflectorized	B-80	70	.457	Ea.	73.50	15.45	9.35	98.30	116

10 14 Signage

10 14 53 – Traffic Signage

10 14 53.20 Traffic Signs		Crew	Daily Output	Labor-Hours	Unit	Material	Labor (2009 Bare Costs)	Equipment	Total	Total Incl O&P
0100	High intensity	B-80	70	.457	Ea.	73.50	15.45	9.35	98.30	116
0300	30″ x 30″, reflectorized		70	.457		150	15.45	9.35	174.80	200
0400	High intensity		70	.457		150	15.45	9.35	174.80	200
0600	Guide and directional signs, 12″ x 18″, reflectorized		70	.457		48	15.45	9.35	72.80	88
0700	High intensity		70	.457		45.50	15.45	9.35	70.30	86
0900	18″ x 24″, stock signs, reflectorized		70	.457		53.50	15.45	9.35	78.30	94.50
1000	High intensity		70	.457		53.50	15.45	9.35	78.30	94.50
1200	24″ x 24″, stock signs, reflectorized		70	.457		66.50	15.45	9.35	91.30	109
1300	High intensity		70	.457		66.50	15.45	9.35	91.30	109
1500	Add to above for steel posts, galvanized, 10′-0″ upright, bolted		200	.160		24	5.40	3.27	32.67	39
1600	12′-0″ upright, bolted		140	.229		32	7.75	4.68	44.43	52.50
1800	Highway road signs, aluminum, over 20 S. F., reflectorized		350	.091	S.F.	30.50	3.09	1.87	35.46	40.50
2000	High intensity		350	.091		30.50	3.09	1.87	35.46	40.50
2200	Highway, suspended over road, 80 S.F. min., reflectorized		165	.194		30.50	6.55	3.97	41.02	48.50
2300	High intensity		165	.194		30.50	6.55	3.97	41.02	48.50

10 17 Telephone Specialties

10 17 16 – Telephone Enclosures

10 17 16.10 Commercial Telephone Enclosures		Crew	Daily Output	Labor-Hours	Unit	Material	Labor	Equipment	Total	Total Incl O&P
0010	**COMMERCIAL TELEPHONE ENCLOSURES**									
0300	Shelf type, wall hung, minimum	2 Carp	5	3.200	Ea.	1,300	128		1,428	1,625
0400	Maximum		5	3.200		2,725	128		2,853	3,175
0600	Booth type, painted steel, indoor or outdoor, minimum		1.50	10.667		3,475	425		3,900	4,525
0700	Maximum (stainless steel)		1.50	10.667		11,600	425		12,025	13,400
1300	Outdoor, acoustical, on post		3	5.333		1,525	213		1,738	2,025
1400	Phone carousel, pedestal mounted with dividers		.60	26.667		5,800	1,075		6,875	8,125
1900	Outdoor, drive-up type, wall mounted		4	4		930	160		1,090	1,300
2000	Post mounted, stainless steel posts		3	5.333		1,425	213		1,638	1,925
2200	Directory shelf, wall mounted, stainless steel									
2300	3 binders	2 Carp	8	2	Ea.	1,100	80		1,180	1,350
2500	4 binders		7	2.286		1,275	91.50		1,366.50	1,550
2600	5 binders		6	2.667		1,625	107		1,732	1,975
2800	Table type, stainless steel, 4 binders		8	2		1,300	80		1,380	1,575
2900	7 binders		7	2.286		1,625	91.50		1,716.50	1,950
9000	Minimum labor/equipment charge		4	4	Job		160		160	264

10 21 Compartments and Cubicles

10 21 13 – Toilet Compartments

10 21 13.13 Metal Toilet Compartments		Crew	Daily Output	Labor-Hours	Unit	Material	Labor	Equipment	Total	Total Incl O&P
0010	**METAL TOILET COMPARTMENTS**									
0110	Cubicles, ceiling hung									
0200	Powder coated steel	2 Carp	4	4	Ea.	465	160		625	780
0500	Stainless steel	″	4	4		1,775	160		1,935	2,225
0600	For handicap units, incl. 52″ grab bars, add					390			390	430
0900	Floor and ceiling anchored									
1000	Powder coated steel	2 Carp	5	3.200	Ea.	485	128		613	745
1300	Stainless steel	″	5	3.200		1,700	128		1,828	2,050
1400	For handicap units, incl. 52″ grab bars, add					276			276	305
1610	Floor anchored									

10 21 Compartments and Cubicles

10 21 13 – Toilet Compartments

10 21 13.13 Metal Toilet Compartments		Crew	Daily Output	Labor-Hours	Unit	Material	Labor	Equipment	Total	Total Incl O&P
						2009 Bare Costs				
1700	Powder coated steel	2 Carp	7	2.286	Ea.	545	91.50		636.50	745
2000	Stainless steel	"	7	2.286		1,525	91.50		1,616.50	1,825
2100	For handicap units, incl. 52" grab bars, add					273			273	300
2200	For juvenile units, deduct					39.50			39.50	43.50
2450	Floor anchored, headrail braced									
2500	Powder coated steel	2 Carp	6	2.667	Ea.	445	107		552	660
2804	Stainless steel	"	4.60	3.478		1,125	139		1,264	1,475
2900	For handicap units, incl. 52" grab bars, add					330			330	365
3000	Wall hung partitions, powder coated steel	2 Carp	7	2.286		595	91.50		686.50	805
3300	Stainless steel	"	7	2.286		1,575	91.50		1,666.50	1,875
3400	For handicap units, incl. 52" grab bars, add					330			330	365
4000	Screens, entrance, floor mounted, 58" high, 48" wide									
4200	Powder coated steel	2 Carp	15	1.067	Ea.	233	42.50		275.50	325
4500	Stainless steel	"	15	1.067	"	850	42.50		892.50	1,000
4650	Urinal screen, 18" wide									
4704	Powder coated steel	2 Carp	6.15	2.602	Ea.	184	104		288	375
5004	Stainless steel	"	6.15	2.602	"	580	104		684	810
5100	Floor mounted, head rail braced									
5300	Powder coated steel	2 Carp	8	2	Ea.	199	80		279	350
5600	Stainless steel	"	8	2	"	555	80		635	745
5750	Pilaster, flush									
5800	Powder coated steel	2 Carp	10	1.600	Ea.	259	64		323	390
6100	Stainless steel		10	1.600		655	64		719	825
6300	Post braced, powder coated steel		10	1.600		182	64		246	305
6600	Stainless steel		10	1.600		485	64		549	640
6700	Wall hung, bracket supported									
6800	Powder coated steel	2 Carp	10	1.600	Ea.	289	64		353	425
7100	Stainless steel		10	1.600		315	64		379	455
7400	Flange supported, powder coated steel		10	1.600		95	64		159	210
7700	Stainless steel		10	1.600		276	64		340	410
7800	Wedge type, powder coated steel		10	1.600		128	64		192	247
8100	Stainless steel		10	1.600		540	64		604	700
9000	Minimum labor/equipment charge	1 Carp	2.50	3.200	Job		128		128	211

10 21 13.14 Metal Toilet Compartment Components

		Crew	Daily Output	Labor-Hours	Unit	Material	Labor	Equipment	Total	Total Incl O&P
0010	**METAL TOILET COMPARTMENT COMPONENTS**									
0100	Pilasters									
0110	Overhead braced, powder coated steel, 7" wide x 82" high	2 Carp	22.20	.721	Ea.	76.50	29		105.50	132
0120	Stainless steel		22.20	.721		206	29		235	275
0130	Floor braced, powder coated steel, 7" wide x 70" high		23.30	.687		121	27.50		148.50	179
0140	Stainless steel		23.30	.687		310	27.50		337.50	385
0150	Ceiling hung, powder coated steel, 7" wide x 83" high		13.30	1.203		129	48		177	222
0160	Stainless steel		13.30	1.203		330	48		378	445
0170	Wall hung, powder coated steel, 3" wide x 58" high		18.90	.847		115	34		149	183
0180	Stainless steel		18.90	.847		201	34		235	277
0200	Panels									
0210	Powder coated steel, 31" wide x 58" high	2 Carp	18.90	.847	Ea.	134	34		168	203
0220	Stainless steel		18.90	.847		440	34		474	540
0230	Powder coated steel, 53" wide x 58" high		18.90	.847		171	34		205	244
0240	Stainless steel		18.90	.847		590	34		624	705
0250	Powder coated steel, 63" wide x 58" high		18.90	.847		206	34		240	282
0260	Stainless steel		18.90	.847		725	34		759	855
0300	Doors									

10 21 13.14 Metal Toilet Compartment Components		Crew	Daily Output	Labor-Hours	Unit	2009 Bare Costs Material	Labor	Equipment	Total	Total Incl O&P
0310	Powder coated steel, 24" wide x 58" high	2 Carp	14.10	1.135	Ea.	147	45.50		192.50	237
0320	Stainless steel		14.10	1.135		415	45.50		460.50	530
0330	Powder coated steel, 26" wide x 58" high		14.10	1.135		149	45.50		194.50	239
0340	Stainless steel		14.10	1.135		420	45.50		465.50	540
0350	Powder coated steel, 28" wide x 58" high		14.10	1.135		158	45.50		203.50	249
0360	Stainless steel		14.10	1.135		445	45.50		490.50	565
0370	Powder coated steel, 36" wide x 58" high		14.10	1.135		182	45.50		227.50	276
0380	Stainless steel		14.10	1.135		515	45.50		560.50	645
0400	Headrails									
0410	For powder coated steel, 62" long	2 Carp	65	.246	Ea.	27.50	9.85		37.35	46.50
0420	Stainless steel		65	.246		27.50	9.85		37.35	46.50
0430	For powder coated steel, 84" long		50	.320		35	12.80		47.80	59.50
0440	Stainless steel		50	.320		35	12.80		47.80	59.50
0450	For powder coated steel, 120" long		30	.533		50.50	21.50		72	90.50
0460	Stainless steel		30	.533		50.50	21.50		72	90.50
9000	Minimum labor/equipment charge	1 Carp	4	2	Job		80		80	132

10 21 13.16 Plastic-Laminate-Clad Toilet Compartments		Crew	Daily Output	Labor-Hours	Unit	Material	Labor	Equipment	Total	Total Incl O&P
0010	**PLASTIC-LAMINATE-CLAD TOILET COMPARTMENTS**									
0110	Cubicles, ceiling hung									
0300	Plastic laminate on particle board	2 Carp	4	4	Ea.	605	160		765	930
0600	For handicap units, incl. 52" grab bars, add				"	390			390	430
0900	Floor and ceiling anchored									
1100	Plastic laminate on particle board	2 Carp	5	3.200	Ea.	615	128		743	885
1400	For handicap units, incl. 52" grab bars, add				"	276			276	305
1610	Floor mounted									
1800	Plastic laminate on particle board	2 Carp	7	2.286	Ea.	545	91.50		636.50	750
2450	Floor mounted, headrail braced									
2600	Plastic laminate on particle board	2 Carp	6	2.667	Ea.	575	107		682	810
3400	For handicap units, incl. 52" grab bars, add					330			330	365
4300	Entrance screen, floor mtd., plas. lam., 58" high, 48" wide	2 Carp	15	1.067		480	42.50		522.50	600
4800	Urinal screen, 18" wide, ceiling braced, plastic laminate		8	2		180	80		260	330
5400	Floor mounted, headrail braced		8	2		365	80		445	530
5900	Pilaster, flush, plastic laminate		10	1.600		400	64		464	545
6400	Post braced, plastic laminate		10	1.600		288	64		352	420
6700	Wall hung, bracket supported									
6900	Plastic laminate on particle board	2 Carp	10	1.600	Ea.	98	64		162	214
7450	Flange supported									
7500	Plastic laminate on particle board	2 Carp	10	1.600	Ea.	203	64		267	330
9000	Minimum labor/equipment charge	1 Carp	2.50	3.200	Job		128		128	211

10 21 13.17 Plastic-Laminate Clad Toilet Compart. Comp.		Crew	Daily Output	Labor-Hours	Unit	Material	Labor	Equipment	Total	Total Incl O&P
0010	**PLASTIC-LAMINATE CLAD TOILET COMPARTMENT COMPONENTS**									
0100	Pilasters									
0110	Overhead braced, 7" wide x 82" high	2 Carp	22.20	.721	Ea.	104	29		133	163
0130	Floor anchored, 7" wide x 70" high		23.30	.687		106	27.50		133.50	163
0150	Ceiling hung, 7" wide x 83" high		13.30	1.203		105	48		153	195
0180	Wall hung, 3" wide x 58" high		18.90	.847		115	34		149	183
0200	Panels									
0210	31" wide x 58" high	2 Carp	18.90	.847	Ea.	144	34		178	214
0230	51" wide x 58" high		18.90	.847		203	34		237	280
0250	63" wide x 58" high		18.90	.847		241	34		275	320
0300	Doors									
0310	24" wide x 58" high	2 Carp	14.10	1.135	Ea.	142	45.50		187.50	231

10 21 13 – Toilet Compartments

10 21 13.17 Plastic-Laminate Clad Toilet Compart. Comp.		Crew	Daily Output	Labor-Hours	Unit	2009 Bare Costs Material	Labor	Equipment	Total	Total Incl O&P
0330	26" wide x 58" high	2 Carp	14.10	1.135	Ea.	147	45.50		192.50	236
0350	28" wide x 58" high		14.10	1.135		152	45.50		197.50	243
0370	36" wide x 58" high	↓	14.10	1.135	↓	181	45.50		226.50	274
0400	Headrails									
0410	62" long	2 Carp	65	.246	Ea.	31	9.85		40.85	50.50
0430	84" long		60	.267		35	10.65		45.65	56
0450	120" long	↓	30	.533	↓	50.50	21.50		72	90.50
9000	Minimum labor/equipment charge	1 Carp	4	2	Job		80		80	132

10 21 13.19 Plastic Toilet Compartments		Crew	Daily Output	Labor-Hours	Unit	Material	Labor	Equipment	Total	Total Incl O&P
0010	**PLASTIC TOILET COMPARTMENTS**									
0110	Cubicles, ceiling hung									
0250	Phenolic	2 Carp	4	4	Ea.	965	160		1,125	1,350
0600	For handicap units, incl. 52" grab bars, add				"	390			390	430
0900	Floor and ceiling anchored									
1050	Phenolic	2 Carp	5	3.200	Ea.	1,025	128		1,153	1,325
1400	For handicap units, incl. 52" grab bars, add				"	276			276	305
1610	Floor mounted									
1750	Phenolic	2 Carp	7	2.286	Ea.	1,275	91.50		1,366.50	1,550
2100	For handicap units, incl. 52" grab bars, add					273			273	300
2200	For juvenile units, deduct				↓	39.50			39.50	43.50
2450	Floor mounted, headrail braced									
2550	Phenolic	2 Carp	6	2.667	Ea.	815	107		922	1,075
9000	Minimum labor/equipment charge	1 Carp	2.50	3.200	Job		128		128	211

10 21 13.20 Plastic Toilet Compartment Components		Crew	Daily Output	Labor-Hours	Unit	Material	Labor	Equipment	Total	Total Incl O&P
0010	**PLASTIC TOILET COMPARTMENT COMPONENTS**									
0100	Pilasters									
0110	Overhead braced, polymer plastic, 7" wide x 82" high	2 Carp	22.20	.721	Ea.	167	29		196	232
0120	Phenolic		22.20	.721		179	29		208	245
0130	Floor braced, polymer plastic, 7" wide x 70" high		23.30	.687		157	27.50		184.50	218
0140	Phenolic		23.30	.687		166	27.50		193.50	229
0150	Ceiling hung, polymer plastic, 7" wide x 83" high		13.30	1.203		141	48		189	235
0160	Phenolic		13.30	1.203		188	48		236	286
0180	Wall hung, phenolic, 3" wide x 58" high	↓	18.90	.847	↓	122	34		156	190
0200	Panels									
0210	Polymer plastic, 31" high x 55" high	2 Carp	18.90	.847	Ea.	360	34		394	455
0220	Phenolic, 31" wide x 58" high		18.90	.847		279	34		313	360
0230	Polymer plastic, 51" wide x 55" high		18.90	.847		550	34		584	665
0240	Phenolic, 51" wide x 58" high		18.90	.847		420	34		454	520
0250	Polymer plastic, 63" wide x 55" high		18.90	.847		670	34		704	790
0260	Phenolic, 63" wide x 58" high	↓	18.90	.847	↓	505	34		539	610
0300	Doors									
0310	Polymer plastic, 24" wide x 55" high	2 Carp	14.10	1.135	Ea.	290	45.50		335.50	395
0320	Phenolic, 24" wide x 58" high		14.10	1.135		315	45.50		360.50	425
0330	polymer plastic, 26" high x 55" high		14.10	1.135		310	45.50		355.50	415
0340	Phenolic, 26" wide x 58" high		14.10	1.135		335	45.50		380.50	445
0350	Polymer plastic, 26" wide x 55" high		14.10	1.135		325	45.50		370.50	435
0360	Phenolic, 28" wide x 58" high		14.10	1.135		355	45.50		400.50	465
0370	Polymer plastic, 36" wide x 55" high		14.10	1.135		410	45.50		455.50	525
0380	Phenolic, 36" wide x 58" high	↓	14.10	1.135	↓	440	45.50		485.50	560
0400	Headrails									
0410	For polymer plastic, 62" long	2 Carp	65	.246	Ea.	40.50	9.85		50.35	61
0420	Phenolic	↓	65	.246	↓	27.50	9.85		37.35	47

10 21 Compartments and Cubicles

10 21 13 – Toilet Compartments

10 21 13.20 Plastic Toilet Compartment Components		Crew	Daily Output	Labor-Hours	Unit	Material	Labor	Equipment	Total	Total Incl O&P
						2009 Bare Costs				
0430	For polymer plastic, 84″ long	2 Carp	50	.320	Ea.	42.50	12.80		55.30	68
0440	Phenolic		50	.320		32.50	12.80		45.30	56.50
0450	For polymer plastic, 120″ long		30	.533		60	21.50		81.50	101
0460	Phenolic		30	.533		42.50	21.50		64	82
9000	Minimum labor/equipment charge	1 Carp	4	2	Job		80		80	132

10 21 13.40 Stone Toilet Compartments		Crew	Daily Output	Labor-Hours	Unit	Material	Labor	Equipment	Total	Total Incl O&P
0010	**STONE TOILET COMPARTMENTS**									
0100	Cubicles, ceiling hung, marble	2 Marb	2	8	Ea.	1,675	310		1,985	2,325
0600	For handicap units, incl. 52″ grab bars, add					390			390	430
0800	Floor & ceiling anchored, marble	2 Marb	2.50	6.400		1,825	247		2,072	2,400
1400	For handicap units, incl. 52″ grab bars, add					276			276	305
1600	Floor mounted, marble	2 Marb	3	5.333		1,075	206		1,281	1,500
2400	Floor mounted, headrail braced, marble	″	3	5.333		1,025	206		1,231	1,450
2900	For handicap units, incl. 52″ grab bars, add					330			330	365
4100	Entrance screen, floor mounted marble, 58″ high, 48″ wide	2 Marb	9	1.778		675	68.50		743.50	855
4600	Urinal screen, 18″ wide, ceiling braced, marble	D-1	6	2.667		675	97		772	900
5100	Floor mounted, head rail braced									
5200	Marble	D-1	6	2.667	Ea.	575	97		672	785
5700	Pilaster, flush, marble		9	1.778		750	64.50		814.50	930
6200	Post braced, marble		9	1.778		745	64.50		809.50	925
9000	Minimum labor/equipment charge	1 Carp	2.50	3.200	Job		128		128	211

10 21 16 – Shower and Dressing Compartments

10 21 16.10 Partitions, Shower		Crew	Daily Output	Labor-Hours	Unit	Material	Labor	Equipment	Total	Total Incl O&P
0010	**PARTITIONS, SHOWER** Floor mounted, no plumbing									
0100	Cabinet, incl. base, no door, painted steel, 1″ thick walls	2 Shee	5	3.200	Ea.	905	151		1,056	1,250
0300	With door, fiberglass		4.50	3.556		705	168		873	1,050
0600	Galvanized and painted steel, 1″ thick walls		5	3.200		955	151		1,106	1,300
0800	Stall, 1″ thick wall, no base, enameled steel		5	3.200		980	151		1,131	1,325
1200	Stainless steel		5	3.200		1,700	151		1,851	2,125
1400	For double entry type, no doors, deduct					10%				
1500	Circular fiberglass, cabinet 36″ diameter,	2 Shee	4	4		725	189		914	1,100
1700	One piece, 36″ diameter, less door		4	4		610	189		799	970
1800	With door		3.50	4.571		1,000	216		1,216	1,450
2000	Curved shell shower, no door needed		3	5.333		875	252		1,127	1,375
2300	For fiberglass seat, add to both above					124			124	136
2400	Glass stalls, with doors, no receptors, chrome on brass	2 Shee	3	5.333		1,475	252		1,727	2,025
2700	Anodized aluminum	″	4	4		1,025	189		1,214	1,425
2900	Marble shower stall, stock design, with shower door	2 Marb	1.20	13.333		2,200	515		2,715	3,250
3000	With curtain		1.30	12.308		1,950	475		2,425	2,925
3200	Receptors, precast terrazzo, 32″ x 32″		14	1.143		242	44		286	340
3300	48″ x 34″		9.50	1.684		330	65		395	465
3500	Plastic, simulated terrazzo receptor, 32″ x 32″		14	1.143		100	44		144	182
3600	32″ x 48″		12	1.333		190	51.50		241.50	293
3800	Precast concrete, colors, 32″ x 32″		14	1.143		187	44		231	278
3900	48″ x 48″		8	2		390	77.50		467.50	555
4100	Shower doors, economy plastic, 24″ wide	1 Shee	9	.889		128	42		170	207
4200	Tempered glass door, economy		8	1		169	47		216	261
4400	Folding, tempered glass, aluminum frame		6	1.333		385	63		448	525
4500	Sliding, tempered glass, 48″ opening		6	1.333		255	63		318	380
4700	Deluxe, tempered glass, chrome on brass frame, minimum		8	1		223	47		270	320
4800	Maximum		1	8		880	380		1,260	1,575
4850	On anodized aluminum frame, minimum		2	4		144	189		333	460

10 21 Compartments and Cubicles

10 21 16 – Shower and Dressing Compartments

10 21 16.10 Partitions, Shower		Crew	Daily Output	Labor-Hours	Unit	Material	2009 Bare Costs Labor	Equipment	Total	Total Incl O&P
4900	Maximum	1 Shee	1	8	Ea.	520	380		900	1,175
5100	Shower enclosure, tempered glass, anodized alum. frame									
5120	2 panel & door, corner unit, 32" x 32"	1 Shee	2	4	Ea.	505	189		694	860
5140	Neo-angle corner unit, 16" x 24" x 16"	"	2	4		930	189		1,119	1,325
5200	Shower surround, 3 wall, polypropylene, 32" x 32"	1 Carp	4	2		275	80		355	435
5220	PVC, 32" x 32"		4	2		340	80		420	505
5240	Fiberglass		4	2		370	80		450	540
5250	2 wall, polypropylene, 32" x 32"		4	2		282	80		362	440
5270	PVC		4	2		350	80		430	515
5290	Fiberglass		4	2		355	80		435	520
5300	Tub doors, tempered glass & frame, minimum	1 Shee	8	1		204	47		251	299
5400	Maximum		6	1.333		475	63		538	625
5600	Chrome plated, brass frame, minimum		8	1		270	47		317	370
5700	Maximum		6	1.333		660	63		723	825
5900	Tub/shower enclosure, temp. glass, alum. frame, minimum		2	4		365	189		554	700
6200	Maximum		1.50	5.333		760	252		1,012	1,225
6500	On chrome-plated brass frame, minimum		2	4		505	189		694	855
6600	Maximum		1.50	5.333		1,075	252		1,327	1,575
6800	Tub surround, 3 wall, polypropylene	1 Carp	4	2		228	80		308	385
6900	PVC		4	2		345	80		425	510
7000	Fiberglass, minimum		4	2		360	80		440	525
7100	Maximum		3	2.667		605	107		712	840
9990	Minimum labor/equipment charge	1 Shee	2.50	3.200	Job		151		151	240

10 21 23 – Cubicles

10 21 23.16 Cubicle Track and Hardware		Crew	Daily Output	Labor-Hours	Unit	Material	Labor	Equipment	Total	Total Incl O&P
0010	**CUBICLE TRACK AND HARDWARE**									
0020	Curtain track, box channel, ceiling mounted	1 Carp	135	.059	L.F.	5.70	2.37		8.07	10.20
0100	Suspended	"	100	.080	"	7	3.20		10.20	13
0300	Curtains, nylon mesh tops, fire resistant, 11 oz. per lineal yard									
0310	Polyester oxford cloth, 9' ceiling height	1 Carp	425	.019	L.F.	11.65	.75		12.40	14.05
0500	8' ceiling height		425	.019		9.85	.75		10.60	12.10
0700	Designer oxford cloth		425	.019		25	.75		25.75	28.50
0800	I.V. track systems									
0820	I.V. track, oval	1 Carp	135	.059	L.F.	5.10	2.37		7.47	9.50
0830	I.V. trolley		32	.250	Ea.	35	10		45	55
0840	I.V. pendent, (tree, 5 hook)		32	.250	"	125	10		135	155

10 22 Partitions

10 22 13 – Wire Mesh Partitions

10 22 13.10 Partitions, Woven Wire		Crew	Daily Output	Labor-Hours	Unit	Material	Labor	Equipment	Total	Total Incl O&P
0010	**PARTITIONS, WOVEN WIRE** For tool or stockroom enclosures									
0100	Channel frame, 1-1/2" diamond mesh, 10 ga. wire, painted									
0300	Wall panels, 4'-0" wide, 7' high	2 Carp	25	.640	Ea.	132	25.50		157.50	187
0400	8' high		23	.696		135	28		163	195
0600	10' high		18	.889		169	35.50		204.50	245
0700	For 5' wide panels, add					5%				
0900	Ceiling panels, 10' long, 2' wide	2 Carp	25	.640		118	25.50		143.50	172
1000	4' wide		15	1.067		171	42.50		213.50	259
1200	Panel with service window & shelf, 5' wide, 7' high		20	.800		315	32		347	405
1300	8' high		15	1.067		325	42.50		367.50	430

10 22 Partitions

10 22 13 – Wire Mesh Partitions

10 22 13.10 Partitions, Woven Wire		Crew	Daily Output	Labor-Hours	Unit	2009 Bare Costs Material	Labor	Equipment	Total	Total Incl O&P
1500	Sliding doors, full height, 3′ wide, 7′ high	2 Carp	6	2.667	Ea.	410	107		517	625
1600	10′ high		5	3.200		420	128		548	675
1800	6′ wide sliding door, 7′ full height		5	3.200		615	128		743	890
1900	10′ high		4	4		750	160		910	1,100
2100	Swinging doors, 3′ wide, 7′ high, no transom		6	2.667		291	107		398	495
2200	7′ high, 3′ transom		5	3.200		350	128		478	595

10 22 16 – Folding Gates

10 22 16.10 Security Gates

		Crew	Daily Output	Labor-Hours	Unit	Material	Labor	Equipment	Total	Total Incl O&P
0010	**SECURITY GATES** For roll up type, see Div. 08 33 13.10									
0300	Scissors type folding gate, ptd. steel, single, 6-1/2′ high, 5-1/2′wide	2 Sswk	4	4	Opng.	180	179		359	530
0350	6-1/2′ wide		4	4		187	179		366	535
0400	7-1/2′ wide		4	4		185	179		364	535
0600	Double gate, 8′ high, 8′ wide		2.50	6.400		300	286		586	860
0650	10′ wide		2.50	6.400		310	286		596	870
0700	12′ wide		2	8		455	360		815	1,150
0750	14′ wide		2	8		540	360		900	1,250
0900	Door gate, folding steel, 4′ wide, 61″ high		4	4		94.50	179		273.50	435
1000	71″ high		4	4		98	179		277	440
1200	81″ high		4	4		106	179		285	445
1300	Window gates, 2′ to 4′ wide, 31″ high		4	4		55	179		234	390
1500	55″ high		3.75	4.267		86	191		277	450
1600	79″ high		3.50	4.571		99.50	204		303.50	490

10 22 19 – Demountable Partitions

10 22 19.43 Demountable Composite Partitions

		Crew	Daily Output	Labor-Hours	Unit	Material	Labor	Equipment	Total	Total Incl O&P
0010	**DEMOUNTABLE COMPOSITE PARTITIONS**, add for doors									
0100	Do not deduct door openings from total L.F.									
0900	Demountable gypsum system on 2″ to 2-1/2″									
1000	steel studs, 9′ high, 3″ to 3-3/4″ thick									
1200	Vinyl clad gypsum	2 Carp	48	.333	L.F.	56.50	13.30		69.80	84.50
1300	Fabric clad gypsum		44	.364		141	14.55		155.55	179
1500	Steel clad gypsum		40	.400		153	16		169	195
1600	1.75 system, aluminum framing, vinyl clad hardboard,									
1800	paper honeycomb core panel, 1-3/4″ to 2-1/2″ thick									
1900	9′ high	2 Carp	48	.333	L.F.	95	13.30		108.30	127
2100	7′ high		60	.267		85.50	10.65		96.15	112
2200	5′ high		80	.200		72	8		80	92.50
2250	Unitized gypsum system									
2300	Unitized panel, 9′ high, 2″ to 2-1/2″ thick									
2350	Vinyl clad gypsum	2 Carp	48	.333	L.F.	123	13.30		136.30	157
2400	Fabric clad gypsum	"	44	.364	"	202	14.55		216.55	246
2500	Unitized mineral fiber system									
2510	Unitized panel, 9′ high, 2-1/4″ thick, aluminum frame									
2550	Vinyl clad mineral fiber	2 Carp	48	.333	L.F.	122	13.30		135.30	156
2600	Fabric clad mineral fiber	"	44	.364	"	181	14.55		195.55	224
2800	Movable steel walls, modular system									
2900	Unitized panels, 9′ high, 48″ wide									
3100	Baked enamel, pre-finished	2 Carp	60	.267	L.F.	138	10.65		148.65	170
3200	Fabric clad steel		56	.286	"	199	11.40		210.40	238
5310	Trackless wall, cork finish, semi-acoustic, 1-5/8″ thick, minimum		325	.049	S.F.	38.50	1.97		40.47	46
5320	Maximum		190	.084		36.50	3.36		39.86	45.50
5330	Acoustic, 2″ thick, minimum		305	.052		31	2.10		33.10	37.50
5340	Maximum		225	.071		53.50	2.84		56.34	63.50

10 22 Partitions

10 22 19 – Demountable Partitions

10 22 19.43 Demountable Composite Partitions		Crew	Daily Output	Labor-Hours	Unit	2009 Bare Costs Material	Labor	Equipment	Total	Total Incl O&P
5500	For acoustical partitions, add, minimum				S.F.	2.22			2.22	2.44
5550	Maximum					10.35			10.35	11.40
5700	For doors, see Div. 08 11 & 08 16									
5800	For door hardware, see Div. 08 71									
6100	In-plant modular office system, w/prehung hollow core door									
6200	3" thick polystyrene core panels									
6250	12' x 12', 2 wall	2 Clab	3.80	4.211	Ea.	3,125	133		3,258	3,675
6300	4 wall		1.90	8.421		4,400	266		4,666	5,275
6350	16' x 16', 2 wall		3.60	4.444		4,600	140		4,740	5,275
6400	4 wall		1.80	8.889		6,200	281		6,481	7,275
9000	Minimum labor/equipment charge	2 Carp	3	5.333	Job		213		213	350

10 22 23 – Portable Partitions, Screens, and Panels

10 22 23.13 Wall Screens

		Crew	Daily Output	Labor-Hours	Unit	Material	Labor	Equipment	Total	Total Incl O&P
0010	**WALL SCREENS**, divider panels, free standing, fiber core									
0020	Fabric face straight									
0100	3'-0" long, 4'-0" high	2 Carp	100	.160	L.F.	106	6.40		112.40	127
0200	5'-0" high		90	.178		111	7.10		118.10	134
0500	6'-0" high		75	.213		114	8.50		122.50	139
0900	5'-0" long, 4'-0" high		175	.091		76.50	3.65		80.15	90.50
1000	5'-0" high		150	.107		88	4.26		92.26	104
1500	6"-0" high		125	.128		91	5.10		96.10	108
1600	6'-0" long, 5'-0" high		162	.099		76	3.95		79.95	90
3200	Economical panels, fabric face, 4'-0" long, 5'-0" high		132	.121		42.50	4.84		47.34	54.50
3250	6'-0" high		112	.143		45	5.70		50.70	58.50
3300	5'-0" long, 5'-0" high		150	.107		36	4.26		40.26	46.50
3350	6'-0" high		125	.128		38	5.10		43.10	50
3450	Acoustical panels, 60 to 90 NRC, 3'-0" long, 5'-0" high		90	.178		79.50	7.10		86.60	99
3550	6'-0" high		75	.213		89.50	8.50		98	113
3600	5'-0" long, 5'-0" high		150	.107		63	4.26		67.26	76.50
3650	6'-0" high		125	.128		61	5.10		66.10	75.50
3700	6'-0" long, 5'-0" high		162	.099		56	3.95		59.95	68
3750	6'-0" high		138	.116		56	4.63		60.63	69
3800	Economy acoustical panels, 40 N.R.C., 4'-0" long, 5'-0" high		132	.121		42.50	4.84		47.34	54.50
3850	6'-0" high		112	.143		45	5.70		50.70	58.50
3900	5'-0" long, 6'-0" high		125	.128		38	5.10		43.10	50
3950	6'-0" long, 5'-0" high		162	.099		30	3.95		33.95	39.50
4000	Metal chalkboard, 6'-6" high, chalkboard, 1 side		125	.128		101	5.10		106.10	119
4100	Metal chalkboard, 2 sides		120	.133		115	5.35		120.35	136
4300	Tackboard, both sides		123	.130		91	5.20		96.20	109
9000	Minimum labor/equipment charge		3	5.333	Job		213		213	350

10 22 23.23 Movable Panel Systems

		Crew	Daily Output	Labor-Hours	Unit	Material	Labor	Equipment	Total	Total Incl O&P
0010	**MOVABLE PANEL SYSTEMS**									
0030	Fabric panel, class A fire rated									
0040	Minimum N.R.C. 0.95, S.T.C. 28, fabric edged									
0200	42" high, 24" wide	2 Clab	30	.533	Ea.	135	16.85		151.85	176
0220	36" wide		28	.571		186	18.05		204.05	235
0240	48" wide		26	.615		192	19.45		211.45	243
0260	60" wide		25	.640		299	20		319	365
0400	54" high, 24" wide		30	.533		141	16.85		157.85	183
0420	36" wide		28	.571		169	18.05		187.05	216
0440	48" wide		26	.615		195	19.45		214.45	247
0460	60" wide		25	.640		305	20		325	370

10 22 23.23 Movable Panel Systems		Crew	Daily Output	Labor-Hours	Unit	Material	2009 Bare Costs Labor	Equipment	Total	Total Incl O&P
0600	60" high,24" wide	2 Clab	29	.552	Ea.	186	17.45		203.45	234
0620	36" wide		27	.593		290	18.75		308.75	350
0640	48" wide		25	.640		299	20		319	365
0660	60" wide		24	.667		310	21		331	375
0800	72" high, 24" wide		29	.552		285	17.45		302.45	345
1000	36" wide		27	.593		296	18.75		314.75	355
1200	48" wide		25	.640		277	20		297	340
1400	60" wide		24	.667		455	21		476	535
2000	Hardwood edged, 42" high, 24" wide		30	.533		310	16.85		326.85	370
2100	36" wide		28	.571		335	18.05		353.05	395
2120	48" wide		26	.615		360	19.45		379.45	425
2180	60" wide		25	.640		395	20		415	470
2200	54" high, 24" wide		30	.533		320	16.85		336.85	380
2220	36" wide		28	.571		340	18.05		358.05	405
2240	48" wide		26	.615		395	19.45		414.45	465
2300	60" wide		25	.640		420	20		440	495
2400	60" high, 24" wide		29	.552		330	17.45		347.45	390
2420	36" wide		27	.593		375	18.75		393.75	445
2440	48" wide		25	.640		475	20		495	560
2460	60" wide		24	.667		520	21		541	605
2800	72" high, 24" wide		29	.552		345	17.45		362.45	410
2820	36" wide		27	.593		395	18.75		413.75	460
2840	48" wide		25	.640		505	20		525	590
2860	60" wide		24	.667		505	21		526	590
3000	Fabric panel, straight, N.R.C. <= .50									
3110	60" high, 24" wide	2 Clab	29	.552	Ea.	159	17.45		176.45	204
3120	30" wide		28	.571		162	18.05		180.05	208
3130	36" wide		27	.593		169	18.75		187.75	217
3140	48" wide		26	.615		170	19.45		189.45	218
3150	60" wide		25	.640		180	20		200	231
3160	72" wide		24	.667		179	21		200	232
3210	72" high, 24" wide		29	.552		165	17.45		182.45	211
3220	30" wide		28	.571		169	18.05		187.05	216
3230	36" wide		27	.593		169	18.75		187.75	217
3240	48" wide		26	.615		179	19.45		198.45	229
3250	60" wide		25	.640		189	20		209	242
3260	72" wide		24	.667		199	21		220	254
3310	N.R.C. => .75, 42" high, 24" wide		30	.533		179	16.85		195.85	225
3320	30" wide		29	.552		190	17.45		207.45	238
3330	36" wide		28	.571		195	18.05		213.05	245
3340	48" wide		27	.593		215	18.75		233.75	267
3350	60" wide		26	.615		240	19.45		259.45	295
3360	72" wide		25	.640		250	20		270	310
3410	60" high, 24" wide		29	.552		209	17.45		226.45	259
3420	30" wide		28	.571		219	18.05		237.05	271
3430	36" wide		27	.593		239	18.75		257.75	294
3440	48" wide		26	.615		259	19.45		278.45	315
3450	60" wide		25	.640		315	20		335	380
3460	72" wide		24	.667		335	21		356	405
3510	72" high, 24" wide		29	.552		239	17.45		256.45	292
3520	30" wide		28	.571		255	18.05		273.05	310
3530	36" wide		27	.593		269	18.75		287.75	325
3540	48" wide		26	.615		289	19.45		308.45	350

10 22 Partitions

10 22 23 – Portable Partitions, Screens, and Panels

10 22 23.23 Movable Panel Systems		Crew	Daily Output	Labor-Hours	Unit	Material	Labor	Equipment	Total	Total Incl O&P
3550	60" wide	2 Clab	25	.640	Ea.	305	20		325	370
3560	72" wide		24	.667		335	21		356	405
3710	Clear glass, straight, 60" high, 30" wide		28	.571		330	18.05		348.05	395
3720	36" wide		27	.593		340	18.75		358.75	405
3730	48" wide		26	.615		380	19.45		399.45	445
3740	72" high, 60" wide		25	.640		425	20		445	500
3810	Fabric and clear glass, straight, 60" high, 30" wide		28	.571		268	18.05		286.05	325
3820	36" wide		27	.593		281	18.75		299.75	340
3830	48" wide		26	.615		325	19.45		344.45	390
3840	72" high, 60" wide		25	.640		470	20		490	555
3910	Connector kit, straight					20			20	22
3920	Corner, 2-way					14.35			14.35	15.75
3930	3-way					11			11	12.10
3940	4-way					10.50			10.50	11.55
3950	T-leg					18.50			18.50	20.50

10 22 26 – Operable Partitions

10 22 26.13 Accordion Folding Partitions

		Crew	Daily Output	Labor-Hours	Unit	Material	Labor	Equipment	Total	Total Incl O&P
0010	**ACCORDION FOLDING PARTITIONS**									
0100	Vinyl covered, over 150 S.F., frame not included									
0300	Residential, 1.25 lb. per S.F., 8' maximum height	2 Carp	300	.053	S.F.	19.35	2.13		21.48	25
0400	Commercial, 1.75 lb. per S.F., 8' maximum height		225	.071		22	2.84		24.84	29
0600	2 lb. per S.F., 17' maximum height		150	.107		23	4.26		27.26	32
0700	Industrial, 4 lb. per S.F., 20' maximum height		75	.213		32.50	8.50		41	50
0900	Acoustical, 3 lb. per S.F., 17' maximum height		100	.160		25.50	6.40		31.90	38.50
1200	5 lb. per S.F., 20' maximum height		95	.168		35.50	6.75		42.25	50
1300	5.5 lb. per S.F., 17' maximum height		90	.178		41.50	7.10		48.60	57.50
1400	Fire rated, 4.5 psf, 20' maximum height		160	.100		41.50	4		45.50	52.50
1500	Vinyl clad wood or steel, electric operation, 5.0 psf		160	.100		48.50	4		52.50	59.50
1900	Wood, non-acoustic, birch or mahogany, to 10' high		300	.053		26	2.13		28.13	32
9000	Minimum labor/equipment charge		4	4	Job		160		160	264

10 22 26.33 Folding Panel Partitions

		Crew	Daily Output	Labor-Hours	Unit	Material	Labor	Equipment	Total	Total Incl O&P
0010	**FOLDING PANEL PARTITIONS**, acoustic, wood									
0100	Vinyl faced, to 18' high, 6 psf, minimum	2 Carp	60	.267	S.F.	49	10.65		59.65	71.50
0150	Average		45	.356		58.50	14.20		72.70	88
0200	Maximum		30	.533		75.50	21.50		97	119
0400	Formica or hardwood finish, minimum		60	.267		50.50	10.65		61.15	73
0500	Maximum		30	.533		54	21.50		75.50	94.50
0600	Wood, low acoustical type, 4.5 psf, to 14' high		50	.320		37	12.80		49.80	61.50
1100	Steel, acoustical, 9 to 12 lb. per S.F., vinyl faced, minimum		60	.267		52.50	10.65		63.15	75
1200	Maximum		30	.533		64	21.50		85.50	105
1700	Aluminum framed, acoustical, to 12' high, 5.5 psf, minimum		60	.267		35	10.65		45.65	56.50
1800	Maximum		30	.533		42.50	21.50		64	82
2000	6.5 lb. per S.F., minimum		60	.267		37	10.65		47.65	58
2100	Maximum		30	.533		45.50	21.50		67	85.50
9000	Minimum labor/equipment charge		4	4	Job		160		160	264

10 22 26.43 Sliding Partitions

		Crew	Daily Output	Labor-Hours	Unit	Material	Labor	Equipment	Total	Total Incl O&P
0010	**SLIDING PARTITIONS**									
0020	Acoustic air wall, 1-5/8" thick, minimum	2 Carp	375	.043	S.F.	30	1.70		31.70	36
0100	Maximum		365	.044		51.50	1.75		53.25	59.50
0300	2-1/4" thick, minimum		360	.044		33.50	1.78		35.28	40
0400	Maximum		330	.048		59	1.94		60.94	68
0600	For track type, add to above				L.F.	110			110	121

10 22 Partitions

10 22 26 – Operable Partitions

10 22 26.43 Sliding Partitions		Crew	Daily Output	Labor-Hours	Unit	Material	2009 Bare Costs Labor	Equipment	Total	Total Incl O&P
0700	Overhead track type, acoustical, 3" thick, 11 psf, minimum	2 Carp	350	.046	S.F.	75.50	1.83		77.33	86
0800	Maximum	"	300	.053	"	91	2.13		93.13	104

10 26 Wall and Door Protection

10 26 13 – Corner Guards

10 26 13.10 Metal Corner Guards

		Crew	Daily Output	Labor-Hours	Unit	Material	Labor	Equipment	Total	Total Incl O&P
0010	**METAL CORNER GUARDS**									
0020	Steel angle w/anchors, 1" x 1" x 1/4", 1.5#/L.F.	2 Carp	160	.100	L.F.	5.75	4		9.75	12.95
0100	2" x 2" x 1/4" angles, 3.2#/L.F.		150	.107		10.45	4.26		14.71	18.55
0200	3" x 3" x 5/16" angles, 6.1#/L.F.		140	.114		15.55	4.57		20.12	24.50
0300	4" x 4" x 5/16" angles, 8.2#/L.F.	↓	120	.133		18.95	5.35		24.30	30
0350	For angles drilled and anchored to masonry, add					15%	120%			
0370	Drilled and anchored to concrete, add					20%	170%			
0400	For galvanized angles, add					35%				
0450	For stainless steel angles, add				↓	100%				
0500	Steel door track/wheel guards, 4' - 0" high	E-4	22	1.455	Ea.	156	66	6.10	228.10	300
0800	Pipe bumper for truck doors, 8' long, 6" diameter, filled		20	1.600		435	72.50	6.70	514.20	615
0900	8" diameter	↓	20	1.600	↓	660	72.50	6.70	739.20	870
9000	Minimum labor/equipment charge	1 Carp	2	4	Job		160		160	264

10 26 13.20 Corner Protection

		Crew	Daily Output	Labor-Hours	Unit	Material	Labor	Equipment	Total	Total Incl O&P
0010	**CORNER PROTECTION**									
0100	Stainless steel, 16 ga., adhesive mount, 3-1/2" leg	1 Sswk	80	.100	L.F.	18.05	4.47		22.52	28
0200	12 ga. stainless, adhesive mount	"	80	.100		21	4.47		25.47	31.50
0300	For screw mount, add						10%			
0500	Vinyl acrylic, adhesive mount, 3" leg	1 Carp	128	.063		7.60	2.50		10.10	12.50
0550	1-1/2" leg		160	.050		4.67	2		6.67	8.45
0600	Screw mounted, 3" leg		80	.100		7.75	4		11.75	15.15
0650	1-1/2" leg		100	.080		4.12	3.20		7.32	9.85
0700	Clear plastic, screw mounted, 2-1/2"		60	.133		3.65	5.35		9	12.80
1000	Vinyl cover, alum. retainer, surface mount, 3" x 3"		48	.167		8.95	6.65		15.60	21
1050	2" x 2"		48	.167		9.30	6.65		15.95	21.50
1100	Flush mounted, 3" x 3"		32	.250		19.25	10		29.25	37.50
1150	2" x 2"	↓	32	.250	↓	15.65	10		25.65	34

10 26 16 – Bumper Guards

10 26 16.10 Wallguard

		Crew	Daily Output	Labor-Hours	Unit	Material	Labor	Equipment	Total	Total Incl O&P
0010	**WALLGUARD**									
0400	Rub rail, vinyl, adhesive mounted	1 Carp	185	.043	L.F.	7.70	1.73		9.43	11.30
0500	Neoprene, aluminum backing, 1-1/2" x 2"		110	.073		8.65	2.91		11.56	14.30
1000	Trolley rail, PVC, clipped to wall, 5" high		185	.043		7.65	1.73		9.38	11.25
1050	8" high		180	.044	↓	10.45	1.78		12.23	14.40
1200	Bed bumper, vinyl acrylic, alum. retainer, 21" long		10	.800	Ea.	39.50	32		71.50	96.50
1300	53" long with aligner		9	.889	"	100	35.50		135.50	169
1400	Bumper, vinyl cover, alum. retain., cush. mnt., 1-1/2" x 2-3/4"		80	.100	L.F.	13.95	4		17.95	22
1500	2" x 4-1/4"		80	.100		20.50	4		24.50	29
1600	Surface mounted, 1-3/4" x 3-5/8"		80	.100		11.70	4		15.70	19.50
2000	Crash rail, vinyl cover, alum. retainer, 1" x 4"		110	.073		10.80	2.91		13.71	16.70
2100	1" x 8"		90	.089		16.90	3.55		20.45	24.50
2150	Vinyl inserts, aluminum plate, 1" x 2-1/2"		110	.073		14.60	2.91		17.51	21
2200	1" x 5"	↓	90	.089	↓	22.50	3.55		26.05	31
3000	Handrail/bumper, vinyl cover, alum. retainer									

10 26 Wall and Door Protection

10 26 16 – Bumper Guards

10 26 16.10 Wallguard		Crew	Daily Output	Labor-Hours	Unit	2009 Bare Costs Material	Labor	Equipment	Total	Total Incl O&P
3010	Bracket mounted, flat rail, 5-1/2"	1 Carp	80	.100	L.F.	17.55	4		21.55	26
3100	6-1/2"		80	.100		22	4		26	30.50
3200	Bronze bracket, 1-3/4" diam. rail		80	.100		15.85	4		19.85	24

10 28 Toilet, Bath, and Laundry Accessories

10 28 13 – Toilet Accessories

10 28 13.13 Commercial Toilet Accessories

		Crew	Daily Output	Labor-Hours	Unit	Material	Labor	Equipment	Total	Total Incl O&P
0010	**COMMERCIAL TOILET ACCESSORIES**									
0200	Curtain rod, stainless steel, 5' long, 1" diameter	1 Carp	13	.615	Ea.	41	24.50		65.50	86
0300	1-1/4" diameter		13	.615		36.50	24.50		61	80.50
0400	Diaper changing station, horizontal, wall mounted, plastic		10	.800		280	32		312	360
0500	Dispenser units, combined soap & towel dispensers,									
0510	mirror and shelf, flush mounted	1 Carp	10	.800	Ea.	295	32		327	380
0600	Towel dispenser and waste receptacle,									
0610	18 gallon capacity	1 Carp	10	.800	Ea.	395	32		427	490
0800	Grab bar, straight, 1-1/4" diameter, stainless steel, 18" long		24	.333		29.50	13.30		42.80	54.50
0900	24" long		23	.348		32	13.90		45.90	58
1000	30" long		22	.364		34.50	14.55		49.05	62
1100	36" long		20	.400		36	16		52	66
1105	42" long		20	.400		42	16		58	72.50
1200	1-1/2" diameter, 24" long		23	.348		34.50	13.90		48.40	61
1300	36" long		20	.400		43.50	16		59.50	74.50
1310	42" long		18	.444		47	17.75		64.75	81
1500	Tub bar, 1-1/4" diameter, 24" x 36"		14	.571		109	23		132	157
1600	Plus vertical arm		12	.667		122	26.50		148.50	178
1900	End tub bar, 1" diameter, 90° angle, 16" x 32"		12	.667		153	26.50		179.50	212
2010	Tub/shower/toilet, 2-wall, 36" x 24"		12	.667		140	26.50		166.50	198
2300	Hand dryer, surface mounted, electric, 115 volt, 20 amp		4	2		380	80		460	545
2400	230 volt, 10 amp		4	2		815	80		895	1,025
2600	Hat and coat strip, stainless steel, 4 hook, 36" long		24	.333		54	13.30		67.30	81.50
2700	6 hook, 60" long		20	.400		59	16		75	91.50
3000	Mirror, with stainless steel 3/4" square frame, 18" x 24"		20	.400		54.50	16		70.50	86.50
3100	36" x 24"		15	.533		137	21.50		158.50	186
3200	48" x 24"		10	.800		173	32		205	243
3300	72" x 24"		6	1.333		330	53.50		383.50	455
3500	With 5" stainless steel shelf, 18" x 24"		20	.400		227	16		243	276
3600	36" x 24"		15	.533		300	21.50		321.50	365
3700	48" x 24"		10	.800		335	32		367	425
3800	72" x 24"		6	1.333		375	53.50		428.50	505
4100	Mop holder strip, stainless steel, 5 holders, 48" long		20	.400		80.50	16		96.50	115
4200	Napkin/tampon dispenser, recessed		15	.533		820	21.50		841.50	940
4300	Robe hook, single, regular		36	.222		15.50	8.90		24.40	31.50
4400	Heavy duty, concealed mounting		36	.222		15.55	8.90		24.45	32
4600	Soap dispenser, chrome, surface mounted, liquid		20	.400		47.50	16		63.50	78.50
4700	Powder		20	.400		46.50	16		62.50	78
5000	Recessed stainless steel, liquid		10	.800		140	32		172	207
5100	Powder		10	.800		213	32		245	287
5300	Soap tank, stainless steel, 1 gallon		10	.800		190	32		222	262
5400	5 gallon		5	1.600		262	64		326	395
5600	Shelf, stainless steel, 5" wide, 18 ga., 24" long		24	.333		63.50	13.30		76.80	91.50
5700	48" long		16	.500		100	20		120	143

10 28 Toilet, Bath, and Laundry Accessories

10 28 13 – Toilet Accessories

10 28 13.13 Commercial Toilet Accessories		Crew	Daily Output	Labor-Hours	Unit	Material	Labor	Equipment	Total	Total Incl O&P
						2009 Bare Costs				
5800	8″ wide shelf, 18 ga., 24″ long	1 Carp	22	.364	Ea.	77.50	14.55		92.05	109
5900	48″ long		14	.571		124	23		147	175
6000	Toilet seat cover dispenser, stainless steel, recessed		20	.400		143	16		159	184
6050	Surface mounted		15	.533		34.50	21.50		56	73
6100	Toilet tissue dispenser, surface mounted, SS, single roll		30	.267		19.45	10.65		30.10	39
6200	Double roll		24	.333		24	13.30		37.30	48
6400	Towel bar, stainless steel, 18″ long		23	.348		39.50	13.90		53.40	66.50
6500	30″ long		21	.381		101	15.20		116.20	136
6700	Towel dispenser, stainless steel, surface mounted		16	.500		42	20		62	79
6800	Flush mounted, recessed		10	.800		270	32		302	350
7000	Towel holder, hotel type, 2 guest size		20	.400		54.50	16		70.50	86.50
7200	Towel shelf, stainless steel, 24″ long, 8″ wide		20	.400		93	16		109	129
7400	Tumbler holder, tumbler only		30	.267		38.50	10.65		49.15	60
7500	Soap, tumbler & toothbrush		30	.267		25	10.65		35.65	45
7700	Wall urn ash receiver, surface mount, 11″ long		12	.667		124	26.50		150.50	180
7800	7-1/2″, long		18	.444		121	17.75		138.75	163
8000	Waste receptacles, stainless steel, with top, 13 gallon		10	.800		330	32		362	415
8100	36 gallon		8	1		515	40		555	630
9000	Minimum labor/equipment charge		5	1.600	Job		64		64	106

10 28 16 – Bath Accessories

10 28 16.20 Medicine Cabinets		Crew	Daily Output	Labor-Hours	Unit	Material	Labor	Equipment	Total	Total Incl O&P
0010	**MEDICINE CABINETS**									
0020	With mirror, st. st. frame, 16″ x 22″, unlighted	1 Carp	14	.571	Ea.	81.50	23		104.50	127
0100	Wood frame		14	.571		113	23		136	162
0300	Sliding mirror doors, 20″ x 16″ x 4-3/4″, unlighted		7	1.143		101	45.50		146.50	187
0400	24″ x 19″ x 8-1/2″, lighted		5	1.600		159	64		223	281
0600	Triple door, 30″ x 32″, unlighted, plywood body		7	1.143		239	45.50		284.50	340
0700	Steel body		7	1.143		315	45.50		360.50	420
0900	Oak door, wood body, beveled mirror, single door		7	1.143		154	45.50		199.50	245
1000	Double door		6	1.333		375	53.50		428.50	500
1200	Hotel cabinets, stainless, with lower shelf, unlighted		10	.800		202	32		234	275
1300	Lighted		5	1.600		300	64		364	435
9000	Minimum labor/equipment charge		4	2	Job		80		80	132

10 28 23 – Laundry Accessories

10 28 23.13 Built-In Ironing Boards		Crew	Daily Output	Labor-Hours	Unit	Material	Labor	Equipment	Total	Total Incl O&P
0010	**BUILT-IN IRONING BOARDS**									
0020	Including cabinet, board & light, minimum	1 Carp	2	4	Ea.	299	160		459	595
0100	Maximum, see also Div. 11 23 13.13	″	1.50	5.333	″	500	213		713	900

10 31 Manufactured Fireplaces

10 31 13 – Manufactured Fireplace Chimneys

10 31 13.10 Fireplace Chimneys		Crew	Daily Output	Labor-Hours	Unit	Material	Labor	Equipment	Total	Total Incl O&P
0010	**FIREPLACE CHIMNEYS**									
0500	Chimney dbl. wall, all stainless, over 8′-6″, 7″ diam., add to fireplace	1 Carp	33	.242	V.L.F.	70	9.70		79.70	93
0600	10″ diameter, add to fireplace		32	.250		93	10		103	119
0700	12″ diameter, add to fireplace		31	.258		111	10.30		121.30	139
0800	14″ diameter, add to fireplace		30	.267		180	10.65		190.65	216
1000	Simulated brick chimney top, 4′ high, 16″ x 16″		10	.800	Ea.	244	32		276	320
1100	24″ x 24″		7	1.143	″	455	45.50		500.50	575

10 31 Manufactured Fireplaces

10 31 13 – Manufactured Fireplace Chimneys

10 31 13.20 Chimney Accessories	Crew	Daily Output	Labor-Hours	Unit	Material	Labor	Equipment	Total	Total Incl O&P
0010 **CHIMNEY ACCESSORIES**									
0020 Chimney screens, galv., 13" x 13" flue	1 Bric	8	1	Ea.	52	40.50		92.50	123
0050 24" x 24" flue		5	1.600		121	65		186	238
0200 Stainless steel, 13" x 13" flue		8	1		320	40.50		360.50	415
0250 20" x 20" flue		5	1.600		435	65		500	585
2400 Squirrel and bird screens, galvanized, 8" x 8" flue		16	.500		45.50	20.50		66	83
2450 13" x 13" flue		12	.667		49	27		76	97.50
9000 Minimum labor/equipment charge		3.50	2.286	Job		92.50		92.50	150

10 31 16 – Manufactured Fireplace Forms

10 31 16.10 Fireplace Forms	Crew	Daily Output	Labor-Hours	Unit	Material	Labor	Equipment	Total	Total Incl O&P
0010 **FIREPLACE FORMS**									
1800 Fireplace forms, no accessories, 32" opening	1 Bric	3	2.667	Ea.	620	108		728	855
1900 36" opening		2.50	3.200		785	130		915	1,075
2000 40" opening		2	4		1,050	162		1,212	1,400
2100 78" opening		1.50	5.333		1,525	216		1,741	2,025

10 31 23 – Prefabricated Fireplaces

10 31 23.10 Fireplace, Prefabricated	Crew	Daily Output	Labor-Hours	Unit	Material	Labor	Equipment	Total	Total Incl O&P
0010 **FIREPLACE, PREFABRICATED**, free standing or wall hung									
0100 With hood & screen, minimum	1 Carp	1.30	6.154	Ea.	1,175	246		1,421	1,675
0150 Average		1	8		1,625	320		1,945	2,325
0200 Maximum		.90	8.889		3,125	355		3,480	4,000
1500 Simulated logs, gas fired, 40,000 BTU, 2' long, minimum		7	1.143	Set	400	45.50		445.50	515
1600 Maximum		6	1.333		940	53.50		993.50	1,125
1700 Electric, 1,500 BTU, 1'-6" long, minimum		7	1.143		166	45.50		211.50	259
1800 11,500 BTU, maximum		6	1.333		370	53.50		423.50	500

10 32 Fireplace Specialties

10 32 13 – Fireplace Dampers

10 32 13.10 Dampers	Crew	Daily Output	Labor-Hours	Unit	Material	Labor	Equipment	Total	Total Incl O&P
0010 **DAMPERS**									
0800 Damper, rotary control, steel, 30" opening	1 Bric	6	1.333	Ea.	77.50	54		131.50	173
0850 Cast iron, 30" opening		6	1.333		86	54		140	182
0880 36" opening		6	1.333		93.50	54		147.50	191
0900 48" opening		6	1.333		133	54		187	235
0920 60" opening		6	1.333		299	54		353	420
0950 72" opening		5	1.600		360	65		425	500
1000 84" opening, special order		5	1.600		770	65		835	950
1050 96" opening, special order		4	2		780	81		861	990
1200 Steel plate, poker control, 60" opening		8	1		273	40.50		313.50	365
1250 84" opening, special order		5	1.600		495	65		560	650
1400 "Universal" type, chain operated, 32" x 20" opening		8	1		212	40.50		252.50	299
1450 48" x 24" opening		5	1.600		315	65		380	455

10 32 23 – Fireplace Doors

10 32 23.10 Doors	Crew	Daily Output	Labor-Hours	Unit	Material	Labor	Equipment	Total	Total Incl O&P
0010 **DOORS**									
0400 Cleanout doors and frames, cast iron, 8" x 8"	1 Bric	12	.667	Ea.	36.50	27		63.50	83.50
0450 12" x 12"		10	.800		44	32.50		76.50	101
0500 18" x 24"		8	1		127	40.50		167.50	206
0550 Cast iron frame, steel door, 24" x 30"		5	1.600		275	65		340	405

10 32 Fireplace Specialties

10 32 23 – Fireplace Doors

10 32 23.10 Doors		Crew	Daily Output	Labor-Hours	Unit	Material	2009 Bare Costs Labor	Equipment	Total	Total Incl O&P
1600	Dutch Oven door and frame, cast iron, 12" x 15" opening	1 Bric	13	.615	Ea.	111	25		136	163
1650	Copper plated, 12" x 15" opening	↓	13	.615	↓	218	25		243	280

10 35 Stoves

10 35 13 – Heating Stoves

10 35 13.10 Woodburning Stoves

		Crew	Daily Output	Labor-Hours	Unit	Material	Labor	Equipment	Total	Total Incl O&P
0010	**WOODBURNING STOVES**									
0015	Cast iron, minimum	2 Carp	1.30	12.308	Ea.	1,125	490		1,615	2,050
0020	Average		1	16		1,525	640		2,165	2,725
0030	Maximum	↓	.80	20		2,750	800		3,550	4,350
0050	For gas log lighter, add				↓	37.50			37.50	41.50

10 44 Fire Protection Specialties

10 44 13 – Fire Extinguisher Cabinets

10 44 13.53 Fire Equipment Cabinets

		Crew	Daily Output	Labor-Hours	Unit	Material	Labor	Equipment	Total	Total Incl O&P
0010	**FIRE EQUIPMENT CABINETS**, not equipped, 20 ga. steel box									
0040	recessed, D.S. glass in door, box size given									
1000	Portable extinguisher, single, 8" x 12" x 27", alum. door & frame	Q-12	8	2	Ea.	124	86.50		210.50	271
1100	Steel door and frame		8	2		86.50	86.50		173	230
1200	Stainless steel door and frame		8	2		185	86.50		271.50	340
2000	Portable extinguisher, large, 8" x 12" x 36", alum. door & frame		8	2		212	86.50		298.50	370
2100	Steel door and frame		8	2		134	86.50		220.50	282
2200	Stainless steel door and frame		8	2		222	86.50		308.50	380
2500	8" x 16" x 38", aluminum door & frame		8	2		219	86.50		305.50	375
2600	Steel door and frame	↓	8	2	↓	157	86.50		243.50	305
3000	Hose rack assy., 1-1/2" valve & 100' hose, 24" x 40" x 5-1/2"									
3100	Aluminum door and frame	Q-12	6	2.667	Ea.	305	116		421	515
3200	Steel door and frame		6	2.667		272	116		388	480
3300	Stainless steel door and frame	↓	6	2.667	↓	530	116		646	765
4000	Hose rack assy., 2-1/2" x 1-1/2" valve, 100' hose, 24" x 40" x 8"									
4100	Aluminum door and frame	Q-12	6	2.667	Ea.	305	116		421	515
4200	Steel door and frame		6	2.667		208	116		324	410
4300	Stainless steel door and frame	↓	6	2.667	↓	410	116		526	630
5000	Hose rack assy., 2-1/2" x 1-1/2" valve, 100' hose									
5010	and extinguisher, 30" x 40" x 8"									
5100	Aluminum door and frame	Q-12	5	3.200	Ea.	390	139		529	645
5200	Steel door and frame		5	3.200		208	139		347	445
5300	Stainless steel door and frame	↓	5	3.200	↓	710	139		849	1,000
6000	Hose rack assy., 1-1/2" valve, 100' hose									
6010	and 2-1/2" FD valve, 24" x 44" x 8"									
6100	Aluminum door and frame	Q-12	5	3.200	Ea.	350	139		489	600
6200	Steel door and frame		5	3.200		223	139		362	460
6300	Stainless steel door and frame	↓	5	3.200	↓	460	139		599	725
7000	Hose rack assy., 1-1/2" valve & 100' hose, 2-1/2" FD valve									
7010	and extinguisher, 30" x 44" x 8"									
7100	Aluminum door and frame	Q-12	5	3.200	Ea.	410	139		549	665
7200	Steel door and frame		5	3.200		510	139		649	780
7300	Stainless steel door and frame	↓	5	3.200	↓	700	139		839	985
8000	Valve cabinet for 2-1/2" FD angle valve, 18" x 18" x 8"									

10 44 Fire Protection Specialties

10 44 13 – Fire Extinguisher Cabinets

10 44 13.53 Fire Equipment Cabinets		Crew	Daily Output	Labor-Hours	Unit	2009 Bare Costs Material	Labor	Equipment	Total	Total Incl O&P
8100	Aluminum door and frame	Q-12	12	1.333	Ea.	135	58		193	239
8200	Steel door and frame		12	1.333		111	58		169	213
8300	Stainless steel door and frame		12	1.333		182	58		240	290

10 44 16 – Fire Extinguishers

10 44 16.13 Portable Fire Extinguishers

		Crew	Daily Output	Labor-Hours	Unit	Material	Labor	Equipment	Total	Total Incl O&P
0010	**PORTABLE FIRE EXTINGUISHERS**									
0120	CO_2, portable with swivel horn, 5 lb.				Ea.	159			159	175
0140	With hose and "H" horn, 10 lb.					215			215	237
0160	15 lb.					278			278	305
0180	20 lb.					305			305	335
1000	Dry chemical, pressurized									
1040	Standard type, portable, painted, 2-1/2 lb.				Ea.	35			35	38.50
1060	5 lb.					61.50			61.50	67.50
1080	10 lb.					91			91	100
1100	20 lb.					135			135	149
1120	30 lb.					295			295	325
1300	Standard type, wheeled, 150 lb.					2,200			2,200	2,425
2000	ABC all purpose type, portable, 2-1/2 lb.					34			34	37
2060	5 lb.					49			49	54
2080	9-1/2 lb.					75			75	82.50
2100	20 lb.					108			108	119
3000	Dry chemical, outside cartridge to -65° F, painted, 9 lb.					220			220	242
3060	26 lb.					275			275	305
5000	Pressurized water, 2-1/2 gallon, stainless steel					107			107	118
5060	With anti-freeze					149			149	163
9400	Installation of extinguishers, 12 or more, on wood	1 Carp	30	.267			10.65		10.65	17.60
9420	On masonry or concrete	"	15	.533			21.50		21.50	35

10 44 16.16 Wheeled Fire Extinguisher Units

		Crew	Daily Output	Labor-Hours	Unit	Material	Labor	Equipment	Total	Total Incl O&P
0010	**WHEELED FIRE EXTINGUISHER UNITS**									
0350	CO_2, portable, with swivel horn									
0360	Wheeled type, cart mounted, 50 lb.				Ea.	1,325			1,325	1,450
0400	100 lb.				"	2,650			2,650	2,900
2200	ABC all purpose type									
2300	Wheeled, 45 lb.				Ea.	715			715	785
2360	150 lb.				"	1,750			1,750	1,925

10 51 Lockers

10 51 13 – Metal Lockers

10 51 13.10 Lockers

		Crew	Daily Output	Labor-Hours	Unit	Material	Labor	Equipment	Total	Total Incl O&P
0011	**LOCKERS** Steel, baked enamel									
0110	Single tier box locker, 12" x 15" x 72"	1 Shee	8	1	Ea.	183	47		230	276
0120	18" x 15" x 72"		8	1		205	47		252	300
0130	12" x 18" x 72"		8	1		198	47		245	293
0140	18" x 18" x 72"		8	1		212	47		259	310
0410	Double tier, 12" x 15" x 36"		21	.381		199	18		217	248
0420	18" x 15" x 36"		21	.381		251	18		269	305
0430	12" x 18" x 36"		21	.381		201	18		219	250
0440	18" x 18" x 36"		21	.381		263	18		281	320
0500	Two person, 18" x 15" x 72"		8	1		253	47		300	355
0510	18" x 18" x 72"		8	1		271	47		318	375

10 51 Lockers

10 51 13 – Metal Lockers

10 51 13.10 Lockers		Crew	Daily Output	Labor-Hours	Unit	Material	Labor	Equipment	Total	Total Incl O&P
						2009 Bare Costs				
0520	Duplex, 15" x 15" x 72"	1 Shee	8	1	Ea.	299	47		346	405
0530	15" x 21" x 72"		8	1		325	47		372	430
0600	5 tier box lockers, minimum		30	.267	Opng.	41.50	12.60		54.10	65.50
0700	Maximum		24	.333		54	15.75		69.75	84.50
0900	6 tier box lockers, minimum		36	.222		30.50	10.50		41	50
1000	Maximum		30	.267		44	12.60		56.60	68.50
1100	Wire meshed wardrobe, floor. mtd., open front varsity type		7.50	1.067	Ea.	217	50.50		267.50	320
2400	16-person locker unit with clothing rack									
2500	72 wide x 15" deep x 72" high	1 Shee	8	1	Ea.	450	47		497	570
2550	18" deep	"	8	1	"	470	47		517	590
3000	Wall mounted lockers, 4 person, with coat bar									
3100	48" wide x 18" deep x 12" high	1 Shee	8	1	Ea.	292	47		339	395
3250	Rack w/ 24 wire mesh baskets		1.50	5.333	Set	345	252		597	780
3260	30 baskets		1.25	6.400		267	300		567	775
3270	36 baskets		.95	8.421		350	395		745	1,025
3280	42 baskets		.80	10		385	470		855	1,175
3300	For built-in lock with 2 keys, add				Ea.	7.10			7.10	7.80
3600	For hanger rods, add				"	1.86			1.86	2.05
9000	Minimum labor/equipment charge	1 Shee	2.50	3.200	Job		151		151	240

10 51 26 – Plastic Lockers

10 51 26.13 Lockers		Crew	Daily Output	Labor-Hours	Unit	Material	Labor	Equipment	Total	Total Incl O&P
0011	**LOCKERS** Plastic, 30% recycled									
0110	Single tier box locker, 12" x 12" x 72" G	1 Shee	8	1	Ea.	400	47		447	515
0120	12" x 15" x 72" G		8	1		420	47		467	535
0130	12" x 18" x 72" G		8	1		430	47		477	550
0410	Double tier, 12" x 12" x 72" G		21	.331		400	18		418	470
0420	12" x 15" x 72" G		21	.331		400	18		418	470
0430	12" x 18" x 72" G		21	.331		420	18		438	490

10 51 53 – Locker Room Benches

10 51 53.10 Benches		Crew	Daily Output	Labor-Hours	Unit	Material	Labor	Equipment	Total	Total Incl O&P
0010	**BENCHES**									
2100	Locker bench, laminated maple, top only	1 Shee	100	.080	L.F.	13.90	3.78		17.68	21.50
2200	Pedestals, steel pipe	"	25	.320	Ea.	36	15.10		51.10	64

10 55 Postal Specialties

10 55 23 – Mail Boxes

10 55 23.10 Commercial Mail Boxes		Crew	Daily Output	Labor-Hours	Unit	Material	Labor	Equipment	Total	Total Incl O&P
0010	**COMMERCIAL MAIL BOXES**									
0020	Horiz., key lock, 5"H x 6"W x 15"D, alum., rear load	1 Carp	34	.235	Ea.	35	9.40		44.40	54
0100	Front loading		34	.235		36	9.40		45.40	55
0200	Double, 5"H x 12"W x 15"D, rear loading		26	.308		71.50	12.30		83.80	99.50
0300	Front loading		26	.308		66	12.30		78.30	93
0500	Quadruple, 10"H x 12"W x 15"D, rear loading		20	.400		91	16		107	127
0600	Front loading		20	.400		108	16		124	146
0800	Vertical, front load, 15"H x 5"W x 6"D, alum., per compartment		34	.235		30	9.40		39.40	48.50
0900	Bronze, duranodic finish		34	.235		40	9.40		49.40	59.50
1000	Steel, enameled		34	.235		30	9.40		39.40	48.50
1700	Alphabetical directories, 120 names		10	.800		126	32		158	191
1800	Letter collection box		6	1.333		705	53.50		758.50	865
1830	Lobby collection boxes, aluminum	2 Shee	5	3.200		2,025	151		2,176	2,475

10 55 Postal Specialties

10 55 23 – Mail Boxes

10 55 23.10 Commercial Mail Boxes		Crew	Daily Output	Labor-Hours	Unit	Material	2009 Bare Costs Labor	Equipment	Total	Total Incl O&P
1840	Bronze or stainless	2 Shee	4.50	3.556	Ea.	2,500	168		2,668	3,025
1900	Letter slot, residential	1 Carp	20	.400		86	16		102	121
2000	Post office type		8	1		110	40		150	187
2200	Post office counter window, with grille		2	4		550	160		710	870
2250	Key keeper, single key, aluminum		26	.308		38.50	12.30		50.80	63
2300	Steel, enameled		26	.308		107	12.30		119.30	139
9000	Minimum labor/equipment charge		5	1.600	Job		64		64	106

10 55 91 – Mail Chutes

10 55 91.10 Commercial Mail Chutes		Crew	Daily Output	Labor-Hours	Unit	Material	Labor	Equipment	Total	Total Incl O&P
0010	**COMMERCIAL MAIL CHUTES**									
0020	Aluminum & glass, 14-1/4" wide, 4-5/8" deep	2 Shee	4	4	Floor	705	189		894	1,075
0100	8-5/8" deep		3.80	4.211		855	199		1,054	1,250
0300	8-3/4" x 3-1/2", aluminum		5	3.200		650	151		801	955
0400	Bronze or stainless		4.50	3.556		1,000	168		1,168	1,375

10 56 Storage Assemblies

10 56 13 – Metal Storage Shelving

10 56 13.10 Shelving		Crew	Daily Output	Labor-Hours	Unit	Material	Labor	Equipment	Total	Total Incl O&P
0010	**SHELVING**									
0020	Metal, industrial, cross-braced, 3' wide, 12" deep	1 Sswk	175	.046	SF Shlf	6.55	2.04		8.59	11
0100	24" deep		330	.024		4.70	1.08		5.78	7.15
0300	4' wide, 12" deep		185	.043		6.05	1.93		7.98	10.25
0400	24" deep		380	.021		4.39	.94		5.33	6.55
1200	Enclosed sides, cross-braced back, 3' wide, 12" deep		175	.046		10.45	2.04		12.49	15.30
1300	24" deep		290	.028		9.20	1.23		10.43	12.45
1500	Fully enclosed, sides and back, 3' wide, 12" deep		150	.053		11.45	2.38		13.83	17
1600	24" deep		255	.031		7.65	1.40		9.05	11.05
1800	4' wide, 12" deep		150	.053		8.80	2.38		11.18	14.05
1900	24" deep		290	.028		6.45	1.23		7.68	9.40
2200	Wide span, 1600 lb. capacity per shelf, 6' wide, 24" deep		380	.021		7	.94		7.94	9.45
2400	36" deep		440	.018		8.30	.81		9.11	10.60
2600	8' wide, 24" deep		440	.018		8.65	.81		9.46	11
2800	36" deep		520	.015		7	.69		7.69	8.95
4000	Pallet racks, steel frame 5,000 lb. capacity, 8' long, 36" deep	2 Sswk	450	.036		8.60	1.59		10.19	12.40
4200	42" deep		500	.032		7.35	1.43		8.78	10.75
4400	48" deep		520	.031		6.95	1.38		8.33	10.20
9000	Minimum labor/equipment charge	1 Carp	4	2	Job		80		80	132

10 56 13.20 Parts Bins		Crew	Daily Output	Labor-Hours	Unit	Material	Labor	Equipment	Total	Total Incl O&P
0010	**PARTS BINS** Metal, gray baked enamel finish									
0100	6'-3" high, 3' wide									
0300	12 bins, 18" wide x 12" high, 12" deep	2 Clab	10	1.600	Ea.	281	50.50		331.50	395
0400	24" deep		10	1.600		355	50.50		405.50	475
0600	72 bins, 6" wide x 6" high, 12" deep		8	2		410	63		473	555
0700	18" deep		8	2		645	63		708	815
1000	7'-3" high, 3' wide									
1200	14 bins, 18" wide x 12" high, 12" deep	2 Clab	10	1.600	Ea.	274	50.50		324.50	385
1300	24" deep		10	1.600		360	50.50		410.50	480
1500	84 bins, 6" wide x 6" high, 12" deep		8	2		650	63		713	820
1600	24" deep		8	2		815	63		878	1,000

10 57 Wardrobe and Closet Specialties

10 57 13 – Hat and Coat Racks

10 57 13.10 Coat Racks and Wardrobes		Crew	Daily Output	Labor-Hours	Unit	Material	2009 Bare Costs Labor	Equipment	Total	Total Incl O&P
0010	**COAT RACKS AND WARDROBES**									
0020	Hat & coat rack, floor model, 6 hangers									
0050	Standing, beech wood, 21″ x 21″ x 72″, chrome				Ea.	230			230	253
0100	18 gauge tubular steel, 21″ x 21″ x 69″, wood walnut				″	266			266	293
0500	16 gauge steel frame, 22 gauge steel shelves									
0650	Single pedestal, 30″ x 18″ x 63″				Ea.	234			234	257
0800	Single face rack, 29″ x 18-1/2″ x 62″					221			221	243
0900	51″ x 18-1/2″ x 70″					365			365	400
0910	Double face rack, 39″ x 26″ x 70″					310			310	340
0920	63″ x 26″ x 70″					415			415	455
0940	For 2″ ball casters, add				Set	85.50			85.50	94
1400	Utility hook strips, 3/8″ x 2-1/2″ x 18″, 6 hooks	1 Carp	48	.167	Ea.	57.50	6.65		64.15	74.50
1500	34″ long, 12 hooks	″	48	.167	″	52.50	6.65		59.15	69
1650	Wall mounted racks, 16 gauge steel frame, 22 gauge steel shelves									
1850	12″ x 15″ x 26″, 6 hangers	1 Carp	32	.250	Ea.	153	10		163	185
2000	12″ x 15″ x 50″, 12 hangers	″	32	.250	″	179	10		189	214
2150	Wardrobe cabinet, steel, baked enamel finish									
2300	36″ x 21″ x 78″, incl. top shelf & hanger rod				Ea.	273			273	300

10 57 23 – Closet and Utility Shelving

10 57 23.19 Wood Closet and Utility Shelving		Crew	Daily Output	Labor-Hours	Unit	Material	Labor	Equipment	Total	Total Incl O&P
0010	**WOOD CLOSET AND UTILITY SHELVING**									
0020	Pine, clear grade, no edge band, 1″ x 8″	1 Carp	115	.070	L.F.	1.98	2.78		4.76	6.75
0100	1″ x 10″		110	.073		2.22	2.91		5.13	7.25
0200	1″ x 12″		105	.076		2.64	3.04		5.68	7.90
0600	Plywood, 3/4″ thick with lumber edge, 12″ wide		75	.107		1.67	4.26		5.93	8.90
0700	24″ wide		70	.114		3	4.57		7.57	10.85
0900	Bookcase, clear grade pine, shelves 12″ O.C., 8″ deep, /SF shelf		70	.114	S.F.	6.45	4.57		11.02	14.65
1000	12″ deep shelves		65	.123	″	8.60	4.92		13.52	17.55
1200	Adjustable closet rod and shelf, 12″ wide, 3′ long		20	.400	Ea.	8.50	16		24.50	36
1300	8′ long		15	.533	″	24	21.50		45.50	61.50
1500	Prefinished shelves with supports, stock, 8″ wide		75	.107	L.F.	3.80	4.26		8.06	11.25
1600	10″ wide		70	.114	″	4.21	4.57		8.78	12.20
9000	Minimum labor/equipment charge		4	2	Job		80		80	132

10 73 Protective Covers

10 73 13 – Awnings

10 73 13.10 Awnings, Fabric		Crew	Daily Output	Labor-Hours	Unit	Material	Labor	Equipment	Total	Total Incl O&P
0010	**AWNINGS, FABRIC**									
0020	Including acrylic canvas and frame, standard design									
0100	Door and window, slope, 3′ high, 4′ wide	1 Carp	4.50	1.778	Ea.	690	71		761	875
0110	6′ wide		3.50	2.286		890	91.50		981.50	1,125
0120	8′ wide		3	2.667		1,100	107		1,207	1,375
0200	Quarter round convex, 4′ wide		3	2.667		1,075	107		1,182	1,350
0210	6′ wide		2.25	3.556		1,400	142		1,542	1,750
0220	8′ wide		1.80	4.444		1,700	178		1,878	2,175
0300	Dome, 4′ wide		7.50	1.067		415	42.50		457.50	525
0310	6′ wide		3.50	2.286		935	91.50		1,026.50	1,175
0320	8′ wide		2	4		1,650	160		1,810	2,100
0350	Elongated dome, 4′ wide		1.33	6.015		1,550	240		1,790	2,100
0360	6′ wide		1.11	7.207		1,850	288		2,138	2,525

10 73 Protective Covers

10 73 13 – Awnings

10 73 13.10 Awnings, Fabric		Crew	Daily Output	Labor-Hours	Unit	Material	2009 Bare Costs Labor	Equipment	Total	Total Incl O&P
0370	8′ wide	1 Carp	1	8	Ea.	2,175	320		2,495	2,925
1000	Entry or walkway, peak, 12′ long, 4′ wide	2 Carp	.90	17.778		4,925	710		5,635	6,575
1010	6′ wide		.60	26.667		7,575	1,075		8,650	10,100
1020	8′ wide		.40	40		10,500	1,600		12,100	14,200
1100	Radius with dome end, 4′ wide		1.10	14.545		3,725	580		4,305	5,050
1110	6′ wide		.70	22.857		6,000	915		6,915	8,100
1120	8′ wide		.50	32		8,525	1,275		9,800	11,500
2000	Retractable lateral arm awning, manual									
2010	To 12′ wide, 8′ - 6″ projection	2 Carp	1.70	9.412	Ea.	1,100	375		1,475	1,850
2020	To 14′ wide, 8′ - 6″ projection		1.10	14.545		1,300	580		1,880	2,375
2030	To 19′ wide, 8′ - 6″ projection		.85	18.824		1,750	750		2,500	3,175
2040	To 24′ wide, 8′ - 6″ projection		.67	23.881		2,225	955		3,180	4,025
2050	Motor for above, add	1 Carp	2.67	3		955	120		1,075	1,250
3000	Patio/deck canopy with frame									
3010	12′ wide, 12′ projection	2 Carp	2	8	Ea.	1,575	320		1,895	2,250
3020	16′ wide, 14′ projection	"	1.20	13.333		2,450	535		2,985	3,550
9000	For fire retardant canvas, add					7%				
9010	For lettering or graphics, add					35%				
9020	For painted or coated acrylic canvas, deduct					8%				
9030	For translucent or opaque vinyl canvas, add					10%				
9040	For 6 or more units, deduct					20%	15%			

10 73 16 – Canopies

10 73 16.20 Metal Canopies		Crew	Daily Output	Labor-Hours	Unit	Material	Labor	Equipment	Total	Total Incl O&P
0010	**METAL CANOPIES**									
0020	Wall hung, .032″, aluminum, prefinished, 8′ x 10′	K-2	1.30	18.462	Ea.	2,150	755	187	3,092	3,900
0300	8′ x 20′		1.10	21.818		4,275	890	221	5,386	6,550
0500	10′ x 10′		1.30	18.462		2,400	755	187	3,342	4,175
0700	10′ x 20′		1.10	21.818		4,450	890	221	5,561	6,750
1000	12′ x 20′		1	24		5,475	980	243	6,698	8,050
1360	12′ x 30′		.80	30		8,050	1,225	305	9,580	11,400
1700	12′ x 40′		.60	40		9,800	1,625	405	11,830	14,200
1900	For free standing units, add					20%	10%			
2300	Aluminum entrance canopies, flat soffit, .032″									
2500	3′-6″ x 4′-0″, clear anodized	2 Carp	4	4	Ea.	900	160		1,060	1,250
2700	Bronze anodized		4	4		1,575	160		1,735	2,025
3000	Polyurethane painted		4	4		1,275	160		1,435	1,675
3300	4′-6″ x 10′-0″, clear anodized		2	8		2,475	320		2,795	3,250
3500	Bronze anodized		2	8		3,175	320		3,495	4,000
3700	Polyurethane painted		2	8		2,650	320		2,970	3,450
4000	Wall downspout, 10 L.F., clear anodized	1 Carp	7	1.143		154	45.50		199.50	246
4300	Bronze anodized		7	1.143		269	45.50		314.50	370
4500	Polyurethane painted		7	1.143		231	45.50		276.50	330
7000	Carport, baked vinyl finish, .032″, 20′ x 10′, no foundations, min.	K-2	4	6	Car	3,850	245	60.50	4,155.50	4,725
7250	Maximum		2	12	"	7,000	490	121	7,611	8,725
7500	Walkway cover, to 12′ wide, stl., vinyl finish, .032″,no fndtns., min.		250	.096	S.F.	22.50	3.91	.97	27.38	33
7750	Maximum		200	.120	"	24.50	4.89	1.21	30.60	37
9000	Minimum labor/equipment charge	2 Carp	2	8	Job		320		320	530

10 74 Manufactured Exterior Specialties

10 74 29 – Steeples

10 74 29.10 Prefabricated Steeples		Crew	Daily Output	Labor-Hours	Unit	2009 Bare Costs Material	Labor	Equipment	Total	Total Incl O&P
0010	**PREFABRICATED STEEPLES**									
4000	Steeples, translucent fiberglass, 30" square, 15' high	F-3	2	20	Ea.	5,450	810	385	6,645	7,750
4150	25' high		1.80	22.222		6,350	900	425	7,675	8,950
4350	Opaque fiberglass, 24" square, 14' high		2	20		4,525	810	385	5,720	6,750
4500	28' high		1.80	22.222		4,775	900	425	6,100	7,225
4600	Aluminum, baked finish, 16" square, 14' high					4,200			4,200	4,625
4620	20' high, 3'-6" base					7,075			7,075	7,800
4640	35' high, 8' base					27,400			27,400	30,200
4660	60' high, 14' base					59,000			59,000	65,000
4680	152' high, custom					495,000			495,000	544,500
4700	Porcelain enamel steeples, custom, 40' high	F-3	.50	80		14,100	3,250	1,550	18,900	22,600
4800	60' high	"	.30	133		24,500	5,400	2,575	32,475	38,700

10 74 33 – Weathervanes

10 74 33.10 Residential Weathervanes		Crew	Daily Output	Labor-Hours	Unit	Material	Labor	Equipment	Total	Total Incl O&P
0010	**RESIDENTIAL WEATHERVANES**									
0020	Residential types, minimum	1 Carp	8	1	Ea.	183	40		223	268
0100	Maximum		2	4	"	1,750	160		1,910	2,200
9000	Minimum labor/equipment charge		4	2	Job		80		80	132

10 74 46 – Window Wells

10 74 46.10 Area Window Wells		Crew	Daily Output	Labor-Hours	Unit	Material	Labor	Equipment	Total	Total Incl O&P
0010	**AREA WINDOW WELLS**, Galvanized steel									
0020	20 ga., 3'-2" wide, 1' deep	1 Sswk	29	.276	Ea.	14	12.35		26.35	38.50
0100	2' deep		23	.348		25	15.55		40.55	56.50
0300	16 ga., 3'-2" wide, 1' deep		29	.276		20	12.35		32.35	45
0400	3' deep		23	.348		39	15.55		54.55	72
0600	Welded grating for above, 15 lbs., painted		45	.178		46	7.95		53.95	65
0700	Galvanized		45	.178		89.50	7.95		97.45	113
0900	Translucent plastic cap for above		60	.133		15	5.95		20.95	27.50

10 75 Flagpoles

10 75 16 – Ground-Set Flagpoles

10 75 16.10 Flagpoles		Crew	Daily Output	Labor-Hours	Unit	Material	Labor	Equipment	Total	Total Incl O&P
0010	**FLAGPOLES**, Ground set									
0050	Not including base or foundation									
0100	Aluminum, tapered, ground set 20' high	K-1	2	8	Ea.	970	284	121	1,375	1,675
0200	25' high		1.70	9.412		1,150	335	143	1,628	1,950
0300	30' high		1.50	10.667		1,250	380	162	1,792	2,175
0400	35' high		1.40	11.429		1,700	405	173	2,278	2,700
0500	40' high		1.20	13.333		2,275	475	202	2,952	3,525
0600	50' high		1	16		3,125	565	243	3,933	4,625
0700	60' high		.90	17.778		5,225	630	270	6,125	7,075
0800	70' high		.80	20		7,900	710	305	8,915	10,200
1100	Counterbalanced, internal halyard, 20' high		1.80	8.889		2,225	315	135	2,675	3,125
1200	30' high		1.50	10.667		2,450	380	162	2,992	3,500
1300	40' high		1.30	12.308		5,350	435	187	5,972	6,825
1400	50' high		1	16		7,450	565	243	8,258	9,400
2820	Aluminum, electronically operated, 30' high		1.40	11.429		4,300	405	173	4,878	5,575
2840	35' high		1.30	12.308		4,625	435	187	5,247	6,025
2860	39' high		1.10	14.545		5,050	515	221	5,786	6,650
2880	45' high		1	16		6,450	565	243	7,258	8,300

10 75 Flagpoles

10 75 16 – Ground-Set Flagpoles

10 75 16.10 Flagpoles		Crew	Daily Output	Labor-Hours	Unit	2009 Bare Costs Material	Labor	Equipment	Total	Total Incl O&P
2900	50′ high	K-1	.90	17.778	Ea.	6,875	630	270	7,775	8,875
3000	Fiberglass, tapered, ground set, 23′ high		2	8		1,100	284	121	1,505	1,800
3100	29′-7″ high		1.50	10.667		1,400	380	162	1,942	2,325
3200	36′-1″ high		1.40	11.429		1,850	405	173	2,428	2,875
3300	39′-5″ high		1.20	13.333		2,150	475	202	2,827	3,375
3400	49′-2″ high		1	16		4,800	565	243	5,608	6,475
3500	59′ high		.90	17.778		6,300	630	270	7,200	8,275
4300	Steel, direct imbedded installation									
4400	Internal halyard, 20′ high	K-1	2.50	6.400	Ea.	1,550	227	97	1,874	2,200
4500	25′ high		2.50	6.400		1,750	227	97	2,074	2,375
4600	30′ high		2.30	6.957		2,050	247	105	2,402	2,775
4700	40′ high		2.10	7.619		3,050	270	116	3,436	3,925
4800	50′ high		1.90	8.421		4,500	299	128	4,927	5,575
5000	60′ high		1.80	8.889		6,800	315	135	7,250	8,150
5100	70′ high		1.60	10		8,300	355	152	8,807	9,875
5200	80′ high		1.40	11.429		10,900	405	173	11,478	12,900
5300	90′ high		1.20	13.333		15,000	475	202	15,677	17,500
5500	100′ high		1	16		15,500	565	243	16,308	18,300
6400	Wood poles, tapered, clear vertical grain fir with tilting									
6410	base, not incl. foundation, 4″ butt, 25′ high	K-1	1.90	8.421	Ea.	1,375	299	128	1,802	2,150
6800	6″ butt, 30′ high	″	1.30	12.308	″	1,925	435	187	2,547	3,050
7300	Foundations for flagpoles, including									
7400	excavation and concrete, to 35′ high poles	C-1	10	3.200	Ea.	540	121		661	795
7600	40′ to 50′ high		3.50	9.143		985	345		1,330	1,650
7700	Over 60′ high		2	16		1,100	605		1,705	2,200

10 75 23 – Wall-Mounted Flagpoles

10 75 23.10 Flagpoles		Crew	Daily Output	Labor-Hours	Unit	Material	Labor	Equipment	Total	Total Incl O&P
0010	**FLAGPOLES**, Structure mounted									
0100	Fiberglass, vertical wall set, 19′-8″ long	K-1	1.50	10.667	Ea.	1,250	380	162	1,792	2,175
0200	23′ long		1.40	11.429		1,325	405	173	1,903	2,325
0300	26′-3″ long		1.30	12.308		1,725	435	187	2,347	2,825
0800	19′-8″ long outrigger		1.30	12.308		1,275	435	187	1,897	2,325
1300	Aluminum, vertical wall set, tapered, with base, 20′ high		1.20	13.333		960	475	202	1,637	2,050
1400	29′-6″ high		1	16		2,125	565	243	2,933	3,525
2400	Outrigger poles with base, 12′ long		1.30	12.308		880	435	187	1,502	1,900
2500	14′ long		1	16		1,375	565	243	2,183	2,725

10 88 Scales

10 88 05 – Commercial Scales

10 88 05.10 Scales		Crew	Daily Output	Labor-Hours	Unit	Material	Labor	Equipment	Total	Total Incl O&P
0010	**SCALES** Built-in floor scale, not incl. foundations									
0100	Dial type, 5 ton capacity, 8′ x 6′ platform	3 Carp	.50	48	Ea.	7,275	1,925		9,200	11,200
0300	9′ x 7′ platform		.40	60		9,450	2,400		11,850	14,400
0400	10 ton capacity, steel, 8′ x 6′ platform		.40	60		11,400	2,400		13,800	16,600
0600	9′ x 7′ platform		.35	68.571		10,700	2,750		13,450	16,300
0700	Truck scales, incl. steel weigh bridge,									
0800	not including foundation, pits									
1550	Digital, electronic, 100 ton capacity, steel deck 12′ x 10′ platform	3 Carp	.20	120	Ea.	12,100	4,800		16,900	21,200
1600	40′ x 10′ platform		.14	171		24,600	6,850		31,450	38,400
1640	60′ x 10′ platform		.13	184		30,400	7,375		37,775	45,700

10 88 05 – Commercial Scales

10 88 05.10 Scales		Crew	Daily Output	Labor-Hours	Unit	2009 Bare Costs Material	Labor	Equipment	Total	Total Incl O&P
1680	70′ x 10′ platform	3 Carp	.12	200	Ea.	34,000	8,000		42,000	50,500
2000	For standard automatic printing device, add					2,450			2,450	2,700
2100	For remote reading electronic system, add					2,225			2,225	2,450
2300	Concrete foundation pits for above, 8′ x 6′, 5 C.Y. required	C-1	.50	64		925	2,425		3,350	5,025
2400	14′ x 6′ platform, 10 C.Y. required		.35	91.429		1,350	3,450		4,800	7,225
2600	50′ x 10′ platform, 30 C.Y. required		.25	128		1,825	4,850		6,675	10,000
2700	70′ x 10′ platform, 40 C.Y. required		.15	213		4,000	8,075		12,075	17,700
2750	Crane scales, dial, 1 ton capacity					970			970	1,075
2780	5 ton capacity					1,325			1,325	1,450
2800	Digital, 1 ton capacity					1,875			1,875	2,050
2850	10 ton capacity					4,375			4,375	4,825
2900	Low profile electronic warehouse scale,									
3000	not incl. printer, 4′ x 4′ platform, 10,000 lb. capacity	2 Carp	.30	53.333	Ea.	1,375	2,125		3,500	5,025
3300	5′ x 7′ platform, 10,000 lb. capacity		.25	64		4,000	2,550		6,550	8,625
3400	20,000 lb. capacity		.20	80		5,150	3,200		8,350	11,000
3500	For printers, incl. time, date & numbering, add					1,275			1,275	1,400
3800	Portable, beam type, capacity 1000#, platform 18″ x 24″					745			745	820
3900	Dial type, capacity 2000#, platform 24″ x 24″					1,275			1,275	1,425
4000	Digital type, capacity 1000#, platform 24″ x 30″					2,050			2,050	2,275
4100	Portable contractor truck scales, 50 ton cap., 40′ x 10′ platform					32,000			32,000	35,200
4200	60′ x 10′ platform					28,600			28,600	31,500

Division Notes

		CREW	DAILY OUTPUT	LABOR-HOURS	UNIT	2009 BARE COSTS MAT.	LABOR	EQUIP.	TOTAL	TOTAL INCL O&P

Estimating Tips

General

- The items in this division are usually priced per square foot or each. Many of these items are purchased by the owner for installation by the contractor. Check the specifications for responsibilities and include time for receiving, storage, installation, and mechanical and electrical hookups in the appropriate divisions.
- Many items in Division 11 require some type of support system that is not usually furnished with the item. Examples of these systems include blocking for the attachment of casework and support angles for ceiling-hung projection screens. The required blocking or supports must be added to the estimate in the appropriate division.
- Some items in Division 11 may require assembly or electrical hookups. Verify the amount of assembly required or the need for a hard electrical connection and add the appropriate costs.

Reference Numbers

Reference numbers are shown in shaded boxes at the beginning of some major classifications. These numbers refer to related items in the Reference Section. The reference information may be an estimating procedure, an alternate pricing method, or technical information.

Note: Not all subdivisions listed here necessarily appear in this publication.

11 05 05 – Selective Equipment Demolition

11 05 05.10 Selective Demolition		Crew	Daily Output	Labor-Hours	Unit	Material	2009 Bare Costs Labor	Equipment	Total	Total Incl O&P
0010	SELECTIVE DEMOLITION R024119-10									
0130	Central vacuum, motor unit, residential or commercial	1 Clab	2	4	Ea.		126		126	209
0210	Vault door and frame	2 Skwk	2	8			325		325	530
0215	Day gate, for vault	"	3	5.333			218		218	355
0380	Bank equipment, teller window, bullet resistant	1 Clab	1.20	6.667			211		211	350
0381	Counter	2 Clab	1.50	10.667	Station		335		335	555
0382	Drive-up window, including drawer and glass		1.50	10.667	"		335		335	555
0383	Thru-wall boxes and chests, selective demolition		2.50	6.400	Ea.		202		202	335
0384	Bullet resistant partitions		20	.800	L.F.		25.50		25.50	41.50
0385	Pneumatic tube system, 2 lane drive-up	L-3	.45	35.556	Ea.		1,550		1,550	2,475
0386	Safety deposit box	1 Clab	50	.160	Opng.		5.05		5.05	8.35
0387	Surveillance system, video, complete	2 Elec	2	8	Ea.		375		375	580
0410	Church equipment, misc moveable fixtures	2 Clab	1	16			505		505	835
0412	Steeple, to 28′ high	F-3	3	13.333			540	256	796	1,150
0414	40′ to 60′ high	"	.80	50			2,025	960	2,985	4,350
0510	Library equipment, bookshelves, wood, to 90″ high	1 Clab	20	.400	L.F.		12.65		12.65	21
0515	Carrels, hardwood, 36″ x 24″	"	9	.889	Ea.		28		28	46.50
0630	Stage equipment, light control panel	1 Elec	1	8	"		375		375	580
0632	Border lights		40	.200	L.F.		9.40		9.40	14.45
0634	Spotlights		8	1	Ea.		47		47	72.50
0636	Telescoping platforms and risers	2 Clab	175	.091	SF Stg.		2.89		2.89	4.77
1020	Barber equipment, hydraulic chair	1 Clab	40	.200	Ea.		6.30		6.30	10.45
1030	Checkout counter, supermarket or warehouse conveyor	2 Clab	18	.889			28		28	46.50
1040	Food cases, refrigerated or frozen	Q-5	6	2.667			118		118	184
1190	Laundry equipment, commercial	L-6	3	4			193		193	298
1360	Movie equipment, lamphouse, to 4000 watt, incl rectifier	1 Elec	4	2			94		94	145
1365	Sound system, incl amplifier	"	1.25	6.400			300		300	465
1410	Air compressor, to 5 H.P.	2 Clab	2.50	6.400			202		202	335
1412	Lubrication equipment, automotive, 3 reel type, incl pump, excl piping	L-4	1	24	Set		895		895	1,475
1414	Booth, spray paint, complete, to 26′ long	"	.80	30	"		1,125		1,125	1,825
1560	Parking equipment, cashier booth	B-22	2	15	Ea.		550	97	647	1,000
1600	Loading dock equipment, dock bumpers, rubber	1 Clab	50	.160	"		5.05		5.05	8.35
1610	Door seal for door perimeter	"	50	.160	L.F.		5.05		5.05	8.35
1620	Platform lifter, fixed, 6′ x 8′, 5000 lb capacity	E-16	1.50	10.667	Ea.		485	89.50	574.50	1,000
1630	Dock leveller	"	2	8			365	67	432	750
1640	Lights, single or double arm	1 Elec	8	1			47		47	72.50
1650	Shelter, fabric, truck or train	1 Clab	1.50	5.333			169		169	278
1790	Waste handling equipment, commercial compactor	L-4	2	12			450		450	730
1792	Commercial or municipal incinerator, gas	"	2	12			450		450	730
1795	Crematory, excluding building	Q-3	.25	128			5,950		5,950	9,225
1910	Detection equipment, cell bar front	E-4	4	8			360	33.50	393.50	705
1912	Cell door and frame		8	4			181	16.75	197.75	355
1914	Prefab cell, 4′ to 5′ wide, 7′ to 8′ high, 7′ deep		8	4			181	16.75	197.75	355
1916	Cot, bolted, single		40	.800			36	3.35	39.35	70.50
1918	Visitor cubicle		4	8			360	33.50	393.50	705
2850	Hydraulic gates, canal, flap, knife, slide or sluice, to 18″ diameter	L-5A	8	4			179	89.50	268.50	420
2852	19″ to 36″ diameter		6	5.333			238	119	357	555
2854	37″ to 48″ diameter		2	16			715	360	1,075	1,675
2856	49″ to 60″ diameter		1	32			1,425	715	2,140	3,350
2858	Over 60″ diameter		.30	106			4,775	2,400	7,175	11,100
3100	Sewage pumping system, prefabricated, to 1000 GPM	C-17D	.20	420			17,400	3,250	20,650	31,800
3110	Sewage treatment, holding tank for recirc chemical water closet	1 Plum	8	1			49		49	75.50
3900	Wastewater treatment system, to 1500 gallons	B-21	2	14			510	64.50	574.50	900

11 05 Common Work Results for Equipment

11 05 05 – Selective Equipment Demolition

11 05 05.10 Selective Demolition		Crew	Daily Output	Labor-Hours	Unit	2009 Bare Costs Material	Labor	Equipment	Total	Total Incl O&P
4050	Food storage equipment, walk-in refrigerator/freezer	2 Clab	64	.250	S.F.		7.90		7.90	13.05
4052	Shelving, stainless steel, 4 tier or dunnage rack	1 Clab	12	.667	Ea.		21		21	35
4100	Food preparation equipment, small countertop		18	.444			14.05		14.05	23
4150	Food delivery carts, heated cabinets		18	.444			14.05		14.05	23
4200	Cooking equipment, commercial range	Q-1	12	1.333			58.50		58.50	90.50
4250	Hood and ventilation equipment, kitchen exhaust hood, excl fire prot	1 Clab	3	2.667			84.50		84.50	139
4255	Fire protection system	Q-1	3	5.333			234		234	365
4300	Food dispensing equipment, countertop items	1 Clab	15	.533			16.85		16.85	28
4310	Serving counter	"	65	.123	L.F.		3.89		3.89	6.40
4350	Ice machine, ice cube maker, flakers and storage bins, to 2000 lb/day	Q-1	1.60	10	Ea.		440		440	680
4400	Cleaning and disposal, dishwasher, commercial, rack type	L-6	1	12			580		580	895
4410	Dishwasher hood	2 Clab	5	3.200			101		101	167
4420	Garbage disposal, commercial, to 5 H.P.	L-1	8	2			96		96	148
4540	Water heater, residential, to 80 gal/day	"	5	3.200			153		153	237
4542	Water softener, automatic	2 Plum	10	1.600			78		78	121
4544	Disappearing stairway, to 15′ floor height	2 Clab	6	2.667			84.50		84.50	139
4710	Darkroom equipment, light	L-7	10	2.800			108		108	176
4712	Heavy	"	1.50	18.667			720		720	1,175
4720	Doors	2 Clab	3.50	4.571	Opng.		144		144	238
4830	Bowling alley, complete, incl pinsetter, scorer, counters, misc supplies	4 Clab	.40	80	Lane		2,525		2,525	4,175
4840	Health club equipment, circuit training apparatus	2 Clab	2	8	Set		253		253	415
4842	Squat racks	"	10	1.600	Ea.		50.50		50.50	83.50
4860	School equipment, basketball backstop	L-2	2	8			281		281	465
4862	Table and benches, folding, in wall, 14′ long	L-4	4	6			224		224	365
4864	Bleachers, telescoping, to 30 tier	F-5	120	.267	Seat		10.80		10.80	17.80
4866	Boxing ring, elevated	L-4	.20	120	Ea.		4,475		4,475	7,325
4867	Boxing ring, floor level	"	2	12			450		450	730
4868	Exercise equipment	1 Clab	6	1.333			42		42	69.50
4870	Gym divider	L-4	1000	.024	S.F.		.90		.90	1.46
4875	Scoreboard	R-3	2	10	Ea.		465	64.50	529.50	785
4880	Shooting range, incl bullet traps, targets, excl structure	L-9	1	36	Point		1,300		1,300	2,200
5200	Vocational shop equipment	2 Clab	8	2	Ea.		63		63	104
6200	Fume hood, incl countertop, excl HVAC	"	6	2.667	L.F.		84.50		84.50	139
7100	Medical sterilizing, distiller, water, steam heated, 50 gal capacity	1 Plum	2.80	2.857	Ea.		139		139	216
7200	Medical equipment, surgery table, minor	1 Clab	1	8			253		253	415
7210	Surgical lights, doctors office, single or double arm	2 Elec	3	5.333			251		251	385
7300	Physical therapy, table	2 Clab	4	4			126		126	209
7310	Whirlpool bath, fixed, incl mixing valves	1 Plum	4	2			97.50		97.50	151
7400	Dental equipment, chair, electric or hydraulic	1 Clab	.75	10.667			335		335	555
7410	Central suction system	1 Plum	2	4			195		195	300
7420	Drill console with accessories	1 Clab	3.20	2.500			79		79	130
7430	X-ray unit	"	4	2			63		63	104
7440	X-ray developer	1 Plum	10	.800			39		39	60.50

11 05 10 – Equipment Installation

11 05 10.10 Industrial Equipment Installation		Crew	Daily Output	Labor-Hours	Unit	Material	Labor	Equipment	Total	Total Incl O&P
0010	**INDUSTRIAL EQUIPMENT INSTALLATION**									
0020	Industrial equipment, minimum	E-2	12	4.667	Ton		203	145	348	520
0200	Maximum	"	2	28	"		1,225	870	2,095	3,125

11 11 Vehicle Service Equipment

11 11 13 – Compressed-Air Vehicle Service Equipment

11 11 13.10 Compressed Air Equipment		Crew	Daily Output	Labor-Hours	Unit	Material	Labor	Equipment	Total	Total Incl O&P
0010	**COMPRESSED AIR EQUIPMENT**									
0030	Compressors, electric, 1-1/2 H.P., standard controls	L-4	1.50	16	Ea.	370	595		965	1,375
0550	Dual controls		1.50	16		695	595		1,290	1,750
0600	5 H.P., 115/230 volt, standard controls		1	24		1,900	895		2,795	3,550
0650	Dual controls		1	24		2,025	895		2,920	3,700

11 11 19 – Vehicle Lubrication Equipment

11 11 19.10 Lubrication Equipment		Crew	Daily Output	Labor-Hours	Unit	Material	Labor	Equipment	Total	Total Incl O&P
0010	**LUBRICATION EQUIPMENT**									
3000	Lube equipment, 3 reel type, with pumps, not including piping	L-4	.50	48	Set	7,900	1,800		9,700	11,600

11 11 33 – Vehicle Spray Painting Equipment

11 11 33.10 Spray Painting Equipment		Crew	Daily Output	Labor-Hours	Unit	Material	Labor	Equipment	Total	Total Incl O&P
0010	**SPRAY PAINTING EQUIPMENT**									
4000	Spray painting booth, 26′ long, complete	L-4	.40	60	Ea.	15,400	2,250		17,650	20,600

11 12 Parking Control Equipment

11 12 13 – Parking Key and Card Control Units

11 12 13.10 Parking Control Units		Crew	Daily Output	Labor-Hours	Unit	Material	Labor	Equipment	Total	Total Incl O&P
0010	**PARKING CONTROL UNITS**									
5100	Card reader	1 Elec	2	4	Ea.	1,825	188		2,013	2,300
5120	Proximity with customer display	2 Elec	1	16		5,500	750		6,250	7,200
6000	Parking control software, minimum	1 Elec	.50	16		21,900	750		22,650	25,300
6020	Maximum	"	.20	40		96,000	1,875		97,875	108,500

11 12 16 – Parking Ticket Dispensers

11 12 16.10 Ticket Dispensers		Crew	Daily Output	Labor-Hours	Unit	Material	Labor	Equipment	Total	Total Incl O&P
0010	**TICKET DISPENSERS**									
5900	Ticket spitter with time/date stamp, standard	2 Elec	2	8	Ea.	6,275	375		6,650	7,475
5920	Mag stripe encoding	"	2	8	"	18,600	375		18,975	21,000

11 12 26 – Parking Fee Collection Equipment

11 12 26.13 Parking Fee Coin Collection Equipment		Crew	Daily Output	Labor-Hours	Unit	Material	Labor	Equipment	Total	Total Incl O&P
0010	**PARKING FEE COIN COLLECTION EQUIPMENT**									
5200	Cashier booth, average	B-22	1	30	Ea.	9,575	1,100	194	10,869	12,500
5300	Collector station, pay on foot	2 Elec	.20	80		109,500	3,750		113,250	126,500
5320	Credit card only	"	.50	32		20,200	1,500		21,700	24,600

11 12 26.23 Fee Equipment		Crew	Daily Output	Labor-Hours	Unit	Material	Labor	Equipment	Total	Total Incl O&P
0010	**FEE EQUIPMENT**									
5600	Fee computer	1 Elec	1.50	5.333	Ea.	13,300	251		13,551	15,000

11 12 33 – Parking Gates

11 12 33.13 Parking Gate Equipment		Crew	Daily Output	Labor-Hours	Unit	Material	Labor	Equipment	Total	Total Incl O&P
0010	**PARKING GATE EQUIPMENT**									
5000	Barrier gate with programmable controller	2 Elec	3	5.333	Ea.	3,300	251		3,551	4,000
5020	Industrial		3	5.333		4,525	251		4,776	5,350
5500	Exit verifier		1	16		17,400	750		18,150	20,400
5700	Full sign, 4″ letters	1 Elec	2	4		1,225	188		1,413	1,625
5800	Inductive loop	2 Elec	4	4		169	188		357	475
5950	Vehicle detector, microprocessor based	1 Elec	3	2.667		385	125		510	620

2009 Bare Costs: Material, Labor, Equipment, Total

11 13 Loading Dock Equipment

11 13 13 – Loading Dock Bumpers

11 13 13.10 Dock Bumpers		Crew	Daily Output	Labor-Hours	Unit	2009 Bare Costs Material	Labor	Equipment	Total	Total Incl O&P
0010	**DOCK BUMPERS** Bolts not included									
0012	2″ x 6″ to 4″ x 8″, average	1 Carp	300	.027	B.F.	.63	1.07		1.70	2.45
0050	Bumpers, rubber blocks 4-1/2″ thick, 10″ high, 14″ long		26	.308	Ea.	45.50	12.30		57.80	70.50
0200	24″ long		22	.364		86.50	14.55		101.05	119
0300	36″ long		17	.471		82.50	18.80		101.30	122
0500	12″ high, 14″ long		25	.320		91	12.80		103.80	121
0550	24″ long		20	.400		101	16		117	138
0600	36″ long		15	.533		112	21.50		133.50	158
0800	Rubber blocks 6″ thick, 10″ high, 14″ long		22	.364		86.50	14.55		101.05	119
0850	24″ long		18	.444		110	17.75		127.75	151
0900	36″ long		13	.615		137	24.50		161.50	192
0910	20″ high, 11″ long		13	.615		133	24.50		157.50	188
0920	Extruded rubber bumpers, T section, 22″ x 22″ x 3″ thick		41	.195		51	7.80		58.80	69
0940	Molded rubber bumpers, 24″ x 12″ x 3″ thick		20	.400		50	16		66	81.50
1000	Welded installation of above bumpers	E-14	8	1		3.40	46.50	16.80	66.70	109
1100	For drilled anchors, add per anchor	1 Carp	36	.222		6.10	8.90		15	21.50
1300	Steel bumpers, see Div. 10 26 13.10									

11 13 16 – Loading Dock Seals and Shelters

11 13 16.10 Dock Seals and Shelters		Crew	Daily Output	Labor-Hours	Unit	Material	Labor	Equipment	Total	Total Incl O&P
0010	**DOCK SEALS AND SHELTERS**									
3600	Door seal for door perimeter, 12″ x 12″, vinyl covered	1 Carp	26	.308	L.F.	28.50	12.30		40.80	51.50
3900	Folding gates, see Div. 10 22 16.10									
6200	Shelters, fabric, for truck or train, scissor arms, minimum	1 Carp	1	8	Ea.	1,475	320		1,795	2,150
6300	Maximum	″	.50	16	″	1,925	640		2,565	3,150

11 13 19 – Stationary Loading Dock Equipment

11 13 19.10 Dock Equipment		Crew	Daily Output	Labor-Hours	Unit	Material	Labor	Equipment	Total	Total Incl O&P
0010	**DOCK EQUIPMENT**									
2200	Dock boards, heavy duty, 60″ x 60″, aluminum, 5,000 lb. capacity				Ea.	1,150			1,150	1,250
2700	9,000 lb. capacity					1,350			1,350	1,475
3200	15,000 lb. capacity					1,500			1,500	1,650
4200	Platform lifter, 6′ x 6′, portable, 3,000 lb. capacity					8,300			8,300	9,150
4250	4,000 lb. capacity					10,200			10,200	11,300
4400	Fixed, 6′ x 8′, 5,000 lb. capacity	E-16	.70	22.857		8,950	1,050	192	10,192	12,000
4500	Levelers, hinged for trucks, 10 ton capacity, 6′ x 8′		1.08	14.815		4,525	675	124	5,324	6,350
4650	7′ x 8′		1.08	14.815		4,975	675	124	5,774	6,850
4670	Air bag power operated, 10 ton cap., 6′x8′		1.08	14.815		5,250	675	124	6,049	7,150
4680	7′ x 8′		1.08	14.815		5,275	675	124	6,074	7,175
4700	Hydraulic, 10 ton capacity, 6′ x 8′		1.08	14.815		7,875	675	124	8,674	10,000
4800	7′ x 8′		1.08	14.815		8,475	675	124	9,274	10,700

11 13 26 – Loading Dock Lights

11 13 26.10 Dock Lights		Crew	Daily Output	Labor-Hours	Unit	Material	Labor	Equipment	Total	Total Incl O&P
0010	**DOCK LIGHTS**									
5000	Lights for loading docks, single arm, 24″ long	1 Elec	3.80	2.105	Ea.	143	99		242	310
5700	Double arm, 60″ long	″	3.80	2.105	″	173	99		272	340

11 14 Pedestrian Control Equipment

11 14 13 – Pedestrian Gates

11 14 13.13 Portable Posts and Railings		Crew	Daily Output	Labor-Hours	Unit	Material	Labor (2009 Bare Costs)	Equipment	Total	Total Incl O&P
0010	**PORTABLE POSTS AND RAILINGS**									
0020	Portable for pedestrian traffic control, standard, minimum				Ea.	104			104	114
0100	Maximum					178			178	196
0300	Deluxe posts, minimum					170			170	187
0400	Maximum				↓	310			310	340
0600	Ropes for above posts, plastic covered, 1-1/2" diameter				L.F.	10.15			10.15	11.15
0700	Chain core				"	10.35			10.35	11.40
1500	Portable security or safety barrier, black with 7' yellow strap				Ea.	235			235	258
1510	12' yellow strap					275			275	300
1550	Sign holder, standard design				↓	122			122	134

11 14 13.19 Turnstiles		Crew	Daily Output	Labor-Hours	Unit	Material	Labor	Equipment	Total	Total Incl O&P
0010	**TURNSTILES**									
0020	One way, 4 arm, 46" diameter, economy, manual	2 Carp	5	3.200	Ea.	780	128		908	1,075
0100	Electric		1.20	13.333		1,275	535		1,810	2,275
0300	High security, galv., 5'-5" diameter, 7' high, manual		1	16		3,850	640		4,490	5,300
0350	Electric		.60	26.667		4,350	1,075		5,425	6,550
0420	Three arm, 24" opening, light duty, manual		2	8		1,225	320		1,545	1,875
0450	Heavy duty		1.50	10.667		3,025	425		3,450	4,050
0460	Manual, with registering & controls, light duty		2	8		2,650	320		2,970	3,450
0470	Heavy duty		1.50	10.667		3,200	425		3,625	4,225
0480	Electric, heavy duty	↓	1.10	14.545		3,700	580		4,280	5,025
0500	For coin or token operating, add				↓	595			595	650
1200	One way gate with horizontal bars, 5'-5" diameter									
1300	7' high, recreation or transit type	2 Carp	.80	20	Ea.	4,275	800		5,075	6,050
1500	For electronic counter, add				"	204			204	225

11 16 Vault Equipment

11 16 16 – Safes

11 16 16.10 Commercial Safes		Crew	Daily Output	Labor-Hours	Unit	Material	Labor	Equipment	Total	Total Incl O&P
0010	**COMMERCIAL SAFES**									
0200	Office, 1 hr. rating, 30" x 18" x 18"				Ea.	2,175			2,175	2,400
0250	40" x 18" x 18"					4,025			4,025	4,425
0300	60" x 36" x 18", double door					7,950			7,950	8,750
0600	1 hr. rating, 27" x 19" x 16"					7,425			7,425	8,150
0700	63" x 34" x 16"					14,100			14,100	15,500
0750	Diskette, 1 hr., 14" x 12" x 11", inside					4,100			4,100	4,525
0800	Money, "B" label, 9" x 14" x 14"					525			525	580
0900	Tool resistive, 24" x 24" x 20"					3,575			3,575	3,925
1050	Tool and torch resistive, 24" x 24" x 20"					9,150			9,150	10,100
1150	Jewelers, 23" x 20" x 18"					11,700			11,700	12,800
1200	63" x 25" x 18"					22,400			22,400	24,700
1300	For handling into building, add, minimum	A-2	8.50	2.824			88.50	23	111.50	171
1400	Maximum	"	.78	30.769	↓		965	249	1,214	1,875

11 17 Teller and Service Equipment

11 17 13 – Teller Equipment Systems

11 17 13.10 Bank Equipment		Crew	Daily Output	Labor-Hours	Unit	Material	2009 Bare Costs Labor	Equipment	Total	Total Incl O&P
0010	**BANK EQUIPMENT**									
0020	Alarm system, police	2 Elec	1.60	10	Ea.	4,675	470		5,145	5,875
0100	With vault alarm	"	.40	40		18,600	1,875		20,475	23,300
0400	Bullet resistant teller window, 44" x 60"	1 Glaz	.60	13.333		3,225	515		3,740	4,375
0500	48" x 60"	"	.60	13.333		4,100	515		4,615	5,350
3000	Counters for banks, frontal only	2 Carp	1	16	Station	1,750	640		2,390	2,975
3100	Complete with steel undercounter	"	.50	32	"	3,400	1,275		4,675	5,850
4600	Door and frame, bullet-resistant, with vision panel, minimum	2 Sswk	1.10	14.545	Ea.	4,550	650		5,200	6,200
4700	Maximum		1.10	14.545		6,175	650		6,825	7,975
4800	Drive-up window, drawer & mike, not incl. glass, minimum		1	16		5,575	715		6,290	7,475
4900	Maximum		.50	32		8,875	1,425		10,300	12,400
5000	Night depository, with chest, minimum		1	16		7,250	715		7,965	9,300
5100	Maximum		.50	32		10,300	1,425		11,725	14,000
5200	Package receiver, painted		3.20	5		1,300	224		1,524	1,850
5300	Stainless steel		3.20	5		2,225	224		2,449	2,850
5400	Partitions, bullet-resistant, 1-3/16" glass, 8' high	2 Carp	10	1.600	L.F.	189	64		253	315
5450	Acrylic	"	10	1.600	"	360	64		424	500
5500	Pneumatic tube systems, 2 lane drive-up, complete	L-3	.25	64	Total	24,600	2,775		27,375	31,600
5550	With T.V. viewer	"	.20	80	"	47,600	3,475		51,075	58,000
5570	Safety deposit boxes, minimum	1 Sswk	44	.182	Opng.	54.50	8.15		62.65	75
5580	Maximum, 10" x 15" opening		19	.421		116	18.80		134.80	162
5590	Teller locker, average		15	.533		1,500	24		1,524	1,700
5600	Pass thru, bullet-res. window, painted steel, 24" x 36"	2 Sswk	1.60	10	Ea.	2,250	445		2,695	3,300
5700	48" x 48"		1.20	13.333		2,550	595		3,145	3,900
5800	72" x 40"		.80	20		3,450	895		4,345	5,425
5900	For stainless steel frames, add					20%				
6100	Surveillance system, video camera, complete	2 Elec	1	16	Ea.	10,400	750		11,150	12,700
6110	For each additional camera, add				"	945			945	1,050
6120	CCTV system, see Div. 27 41 19.10									
6200	Twenty-four hour teller, single unit,									
6300	automated deposit, cash and memo	L-3	.25	64	Ea.	42,800	2,775		45,575	51,500
7000	Vault front, see Div. 08 34 59.10									

11 19 Detention Equipment

11 19 30 – Detention Cell Equipment

11 19 30.10 Cell Equipment		Crew	Daily Output	Labor-Hours	Unit	Material	Labor	Equipment	Total	Total Incl O&P
0010	**CELL EQUIPMENT**									
3000	Toilet apparatus including wash basin, average	L-8	1.50	13.333	Ea.	3,350	555		3,905	4,600

11 21 Mercantile and Service Equipment

11 21 13 – Cash Registers and Checking Equipment

11 21 13.10 Checkout Counter		Crew	Daily Output	Labor-Hours	Unit	Material	Labor	Equipment	Total	Total Incl O&P
0010	**CHECKOUT COUNTER**									
0020	Supermarket conveyor, single belt	2 Clab	10	1.600	Ea.	2,475	50.50		2,525.50	2,800
0100	Double belt, power take-away		9	1.778		4,175	56		4,231	4,700
0400	Double belt, power take-away, incl. side scanning		7	2.286		5,200	72		5,272	5,825
0800	Warehouse or bulk type		6	2.667		6,075	84.50		6,159.50	6,825
1000	Scanning system, 2 lanes, w/registers, scan gun & memory				System	15,200			15,200	16,700
1100	10 lanes, single processor, full scan, with scales				"	144,500			144,500	158,500

11 21 Mercantile and Service Equipment

11 21 13 – Cash Registers and Checking Equipment

11 21 13.10 Checkout Counter		Crew	Daily Output	Labor-Hours	Unit	Material	Labor	Equipment	Total	Total Incl O&P
						2009 Bare Costs				
2000	Register, restaurant, minimum				Ea.	615			615	675
2100	Maximum					2,700			2,700	2,975
2150	Store, minimum					615			615	675
2200	Maximum				↓	2,700			2,700	2,975

11 21 53 – Barber and Beauty Shop Equipment

11 21 53.10 Barber Equipment		Crew	Daily Output	Labor-Hours	Unit	Material	Labor	Equipment	Total	Total Incl O&P
0010	**BARBER EQUIPMENT**									
0020	Chair, hydraulic, movable, minimum	1 Carp	24	.333	Ea.	405	13.30		418.30	465
0050	Maximum	"	16	.500		2,750	20		2,770	3,050
0200	Wall hung styling station with mirrors, minimum	L-2	8	2		355	70.50		425.50	505
0300	Maximum	"	4	4		1,825	141		1,966	2,250
0500	Sink, hair washing basin, rough plumbing not incl.	1 Plum	8	1		395	49		444	505
1000	Sterilizer, liquid solution for tools					128			128	140
1100	Total equipment, rule of thumb, per chair, minimum	L-8	1	20		1,525	835		2,360	3,025
1150	Maximum	"	1	20	↓	4,075	835		4,910	5,850

11 23 Commercial Laundry and Dry Cleaning Equipment

11 23 13 – Dry Cleaning Equipment

11 23 13.13 Dry Cleaners		Crew	Daily Output	Labor-Hours	Unit	Material	Labor	Equipment	Total	Total Incl O&P
0010	**DRY CLEANERS** Not incl. rough-in									
2000	Dry cleaners, electric, 20 lb. capacity	L-1	.20	80	Ea.	33,300	3,825		37,125	42,500
2050	25 lb. capacity		.17	94.118		45,100	4,500		49,600	56,500
2100	30 lb. capacity		.15	106		47,600	5,100		52,700	60,500
2150	60 lb. capacity	↓	.09	177	↓	73,000	8,500		81,500	93,000

11 23 16 – Drying and Conditioning Equipment

11 23 16.13 Dryers		Crew	Daily Output	Labor-Hours	Unit	Material	Labor	Equipment	Total	Total Incl O&P
0010	**DRYERS**, Not including rough-in									
1500	Industrial, 30 lb. capacity	1 Plum	2	4	Ea.	2,975	195		3,170	3,575
1600	50 lb. capacity	"	1.70	4.706		3,200	229		3,429	3,875
4700	Lint collector, ductwork not included, 8,000 to 10,000 C.F.M.	Q-10	.30	80	↓	8,575	3,525		12,100	15,000

11 23 19 – Finishing Equipment

11 23 19.13 Folders and Spreaders		Crew	Daily Output	Labor-Hours	Unit	Material	Labor	Equipment	Total	Total Incl O&P
0010	**FOLDERS AND SPREADERS**									
3500	Folders, blankets & sheets, minimum	1 Elec	.17	47.059	Ea.	30,600	2,200		32,800	37,100
3700	King size with automatic stacker		.10	80		55,500	3,750		59,250	67,000
3800	For conveyor delivery, add	↓	.45	17.778		6,000	835		6,835	7,875
4900	Spreader feeders, 240V, 2 station	L-6	.70	17.143		51,000	825		51,825	57,500
4920	4 station	"	.35	34.286	↓	61,500	1,650		63,150	70,000

11 23 23 – Commercial Ironing Equipment

11 23 23.13 Irons and Pressers		Crew	Daily Output	Labor-Hours	Unit	Material	Labor	Equipment	Total	Total Incl O&P
0010	**IRONS AND PRESSERS**									
4500	Ironers, institutional, 110", single roll	1 Elec	.20	40	Ea.	29,700	1,875		31,575	35,600
4800	Pressers, low capacity air operated	L-6	1.75	6.857		8,775	330		9,105	10,200
4820	Hand operated		1.75	6.857		8,125	330		8,455	9,450
4840	Ironer 48", 240V		3.50	3.429		103,000	165		103,165	114,000
6600	Hand operated presser		.70	17.143		5,525	825		6,350	7,350
6620	Mushroom press 115V	↓	.70	17.143	↓	6,925	825		7,750	8,875

11 23 Commercial Laundry and Dry Cleaning Equipment

11 23 26 – Commercial Washers and Extractors

11 23 26.13 Washers and Extractors		Crew	Daily Output	Labor-Hours	Unit	2009 Bare Costs Material	Labor	Equipment	Total	Total Incl O&P
0010	**WASHERS AND EXTRACTORS**, not including rough-in									
6000	Combination washer/extractor, 20 lb. capacity	L-6	1.50	8	Ea.	5,350	385		5,735	6,500
6100	30 lb. capacity		.80	15		8,500	725		9,225	10,500
6200	50 lb. capacity		.68	17.647		9,925	850		10,775	12,200
6300	75 lb. capacity		.30	40		18,600	1,925		20,525	23,500
6350	125 lb. capacity		.16	75		24,900	3,625		28,525	33,000
6380	Washer extractor/dryer, 110 lb., 240V		1	12		8,150	580		8,730	9,875
6400	Washer extractor, 135 lb, 240V		1	12		23,600	580		24,180	26,900
6450	Pass through		1	12		64,000	580		64,580	71,500
6500	200 lb. washer extractor		1	12		66,500	580		67,080	74,500
6550	Pass through		1	12		69,500	580		70,080	77,500
6600	Extractor, low capacity	↓	1.75	6.857	↓	6,625	330		6,955	7,775

11 23 33 – Coin-Operated Laundry Equipment

11 23 33.13 Coin Operated Washers and Dryers		Crew	Daily Output	Labor-Hours	Unit	Material	Labor	Equipment	Total	Total Incl O&P
0010	**COIN OPERATED WASHERS AND DRYERS**									
0990	Dryer, gas fired									
1000	Commercial, 30 lb. capacity, coin operated, single	1 Plum	3	2.667	Ea.	3,100	130		3,230	3,600
1100	Double stacked	"	2	4		6,275	195		6,470	7,200
4860	Coin dry cleaner 20 lb.	L-6	1.75	6.857	↓	25,900	330		26,230	29,000
5290	Clothes washer									
5300	Commercial, coin operated, average	1 Plum	3	2.667	Ea.	1,125	130		1,255	1,450

11 24 Maintenance Equipment

11 24 19 – Vacuum Cleaning Systems

11 24 19.10 Vacuum Cleaning		Crew	Daily Output	Labor-Hours	Unit	Material	Labor	Equipment	Total	Total Incl O&P
0010	**VACUUM CLEANING**									
0020	Central, 3 inlet, residential	1 Skwk	.90	8.889	Total	925	365		1,290	1,625
0200	Commercial		.70	11.429		1,050	465		1,515	1,925
0400	5 inlet system, residential		.50	16		1,225	655		1,880	2,425
0600	7 inlet system, commercial		.40	20		1,400	815		2,215	2,850
0800	9 inlet system, residential	↓	.30	26.667		3,000	1,100		4,100	5,075
4010	Rule of thumb: First 1200 S.F., installed				↓				1,150	1,275
4020	For each additional S.F., add				S.F.					.21

11 26 Unit Kitchens

11 26 13 – Metal Unit Kitchens

11 26 13.10 Commercial Unit Kitchens		Crew	Daily Output	Labor-Hours	Unit	Material	Labor	Equipment	Total	Total Incl O&P
0010	**COMMERCIAL UNIT KITCHENS**									
1500	Combination range, refrigerator and sink, 30" wide, minimum	L-1	2	8	Ea.	900	385		1,285	1,575
1550	Maximum		1	16		3,175	765		3,940	4,675
1570	60" wide, average		1.40	11.429		2,725	545		3,270	3,825
1590	72" wide, average		1.20	13.333		3,700	640		4,340	5,050
1600	Office model, 48" wide		2	8		2,750	385		3,135	3,625
1620	Refrigerator and sink only	↓	2.40	6.667	↓	2,225	320		2,545	2,925
1640	Combination range, refrigerator, sink, microwave									
1660	Oven and ice maker	L-1	.80	20	Ea.	3,950	960		4,910	5,825

11 27 Photographic Processing Equipment

11 27 13 – Darkroom Processing Equipment

11 27 13.10 Darkroom Equipment		Crew	Daily Output	Labor-Hours	Unit	2009 Bare Costs Material	Labor	Equipment	Total	Total Incl O&P
0010	DARKROOM EQUIPMENT									
0020	Developing sink, 5" deep, 24" x 48"	Q-1	2	8	Ea.	5,025	350		5,375	6,075
0050	48" x 52"		1.70	9.412		5,100	415		5,515	6,250
0200	10" deep, 24" x 48"		1.70	9.412		6,475	415		6,890	7,775
0250	24" x 108"		1.50	10.667		8,525	470		8,995	10,100
0500	Dryers, dehumidified filtered air, 36" x 25" x 68" high	L-7	6	4.667		4,800	180		4,980	5,575
0550	48" x 25" x 68" high		5	5.600		9,000	216		9,216	10,300
2000	Processors, automatic, color print, minimum		4	7		13,100	270		13,370	14,800
2050	Maximum		.60	46.667		20,600	1,800		22,400	25,600
2300	Black and white print, minimum		2	14		9,950	540		10,490	11,800
2350	Maximum		.80	35		59,500	1,350		60,850	67,500
2600	Manual processor, 16" x 20" maximum print size		2	14		9,175	540		9,715	11,000
2650	20" x 24" maximum print size		1	28		8,725	1,075		9,800	11,400
3000	Viewing lights, 20" x 24"		6	4.667		380	180		560	710
3100	20" x 24" with color correction		6	4.667		535	180		715	885
3500	Washers, round, minimum sheet 11" x 14"	Q-1	2	8		3,075	350		3,425	3,925
3550	Maximum sheet 20" x 24"		1	16		3,475	700		4,175	4,925
3800	Square, minimum sheet 20" x 24"		1	16		3,125	700		3,825	4,525
3900	Maximum sheet 50" x 56"		.80	20		4,850	880		5,730	6,675
4500	Combination tank sink, tray sink, washers, with									
4510	Dry side tables, average	Q-1	.45	35.556	Ea.	9,600	1,550		11,150	13,000

11 31 Residential Appliances

11 31 13 – Residential Kitchen Appliances

11 31 13.13 Cooking Equipment

		Crew	Daily Output	Labor-Hours	Unit	Material	Labor	Equipment	Total	Total Incl O&P
0010	COOKING EQUIPMENT									
0020	Cooking range, 30" free standing, 1 oven, minimum	2 Clab	10	1.600	Ea.	320	50.50		370.50	435
0050	Maximum		4	4		1,775	126		1,901	2,150
0150	2 oven, minimum		10	1.600		1,450	50.50		1,500.50	1,675
0200	Maximum		10	1.600		1,625	50.50		1,675.50	1,875
0350	Built-in, 30" wide, 1 oven, minimum	1 Elec	6	1.333		690	62.50		752.50	855
0400	Maximum	2 Carp	2	8		1,375	320		1,695	2,050
0500	2 oven, conventional, minimum		4	4		1,350	160		1,510	1,750
0550	1 conventional, 1 microwave, maximum		2	8		1,000	320		1,320	1,625
0700	Free-standing, 1 oven, 21" wide range, minimum	2 Clab	10	1.600		315	50.50		365.50	430
0750	21" wide, maximum	"	4	4		325	126		451	570
0900	Countertop cooktops, 4 burner, standard, minimum	1 Elec	6	1.333		244	62.50		306.50	365
0950	Maximum		3	2.667		585	125		710	840
1050	As above, but with grill and griddle attachment, minimum		6	1.333		1,675	62.50		1,737.50	1,950
1100	Maximum		3	2.667		3,375	125		3,500	3,900
1200	Induction cooktop, 30" wide		3	2.667		940	125		1,065	1,225
1250	Microwave oven, minimum		4	2		104	94		198	259
1300	Maximum		2	4		450	188		638	785

11 31 13.23 Refrigeration Equipment

		Crew	Daily Output	Labor-Hours	Unit	Material	Labor	Equipment	Total	Total Incl O&P
0010	REFRIGERATION EQUIPMENT									
2000	Deep freeze, 15 to 23 C.F., minimum	2 Clab	10	1.600	Ea.	440	50.50		490.50	570
2050	Maximum		5	3.200		615	101		716	840
2200	30 C.F., minimum		8	2		805	63		868	990
2250	Maximum		3	5.333		915	169		1,084	1,275
5200	Icemaker, automatic, 20 lb. per day	1 Plum	7	1.143		850	55.50		905.50	1,025
5350	51 lb. per day	"	2	4		1,175	195		1,370	1,600

11 31 Residential Appliances

11 31 13 – Residential Kitchen Appliances

11 31 13.23 Refrigeration Equipment			Crew	Daily Output	Labor-Hours	Unit	2009 Bare Costs Material	Labor	Equipment	Total	Total Incl O&P
5450	Refrigerator, no frost, 6 C.F.		2 Clab	15	1.067	Ea.	238	33.50		271.50	315
5500	Refrigerator, no frost, 10 C.F. to 12 C.F. minimum			10	1.600		445	50.50		495.50	575
5600	Maximum			6	2.667		520	84.50		604.50	715
5750	14 C.F. to 16 C.F., minimum			9	1.778		460	56		516	600
5800	Maximum			5	3.200		620	101		721	850
5950	18 C.F. to 20 C.F., minimum			8	2		575	63		638	740
6000	Maximum			4	4		1,000	126		1,126	1,300
6150	21 C.F. to 29 C.F., minimum			7	2.286		855	72		927	1,050
6200	Maximum			3	5.333		2,950	169		3,119	3,525
6790	Energy-star qualified, 18 CF, minimum	G	2 Carp	4	4		525	160		685	845
6795	Maximum	G		2	8		1,050	320		1,370	1,675
6797	21.7 CF, minimum	G		4	4		875	160		1,035	1,225
6799	Maximum	G		4	4		1,150	160		1,310	1,550

11 31 13.33 Kitchen Cleaning Equipment

			Crew	Daily Output	Labor-Hours	Unit	Material	Labor	Equipment	Total	Total Incl O&P
0010	**KITCHEN CLEANING EQUIPMENT**										
2750	Dishwasher, built-in, 2 cycles, minimum		L-1	4	4	Ea.	206	192		398	525
2800	Maximum			2	8		310	385		695	930
2950	4 or more cycles, minimum			4	4		310	192		502	635
2960	Average			4	4		410	192		602	745
3000	Maximum			2	8		1,000	385		1,385	1,725
3100	Energy-star qualified, minimum	G		4	4		315	192		507	640
3110	Maximum	G		2	8		1,000	385		1,385	1,700

11 31 13.43 Waste Disposal Equipment

			Crew	Daily Output	Labor-Hours	Unit	Material	Labor	Equipment	Total	Total Incl O&P
0010	**WASTE DISPOSAL EQUIPMENT**										
1750	Compactor, residential size, 4 to 1 compaction, minimum		1 Carp	5	1.600	Ea.	485	64		549	635
1800	Maximum		"	3	2.667		515	107		622	740
3300	Garbage disposal, sink type, minimum		L-1	10	1.600		66.50	76.50		143	192
3350	Maximum		"	10	1.600		192	76.50		268.50	330

11 31 13.53 Kitchen Ventilation Equipment

			Crew	Daily Output	Labor-Hours	Unit	Material	Labor	Equipment	Total	Total Incl O&P
0010	**KITCHEN VENTILATION EQUIPMENT**										
4150	Hood for range, 2 speed, vented, 30" wide, minimum		L-3	5	3.200	Ea.	52.50	139		191.50	281
4200	Maximum			3	5.333		810	232		1,042	1,275
4300	42" wide, minimum			5	3.200		250	139		389	500
4330	Custom			5	3.200		1,400	139		1,539	1,775
4350	Maximum			3	5.333		1,725	232		1,957	2,275
4500	For ventless hood, 2 speed, add						16.55			16.55	18.20
4650	For vented 1 speed, deduct from maximum						44			44	48.50

11 31 23 – Residential Laundry Appliances

11 31 23.13 Washers

			Crew	Daily Output	Labor-Hours	Unit	Material	Labor	Equipment	Total	Total Incl O&P
0010	**WASHERS**										
5000	Residential, 4 cycle, average		1 Plum	3	2.667	Ea.	800	130		930	1,075
6650	Washing machine, automatic, minimum			3	2.667		315	130		445	545
6700	Maximum			1	8		1,100	390		1,490	1,800
6750	Energy star qualified, front loading, minimum	G		3	2.667		510	130		640	760
6760	Maximum	G		1	8		1,600	390		1,990	2,350
6764	Top loading, minimum	G		3	2.667		485	130		615	735
6766	Maximum	G		3	2.667		990	130		1,120	1,300

11 31 23.23 Dryers

			Crew	Daily Output	Labor-Hours	Unit	Material	Labor	Equipment	Total	Total Incl O&P
0010	**DRYERS**										
0500	Gas fired residential, 16 lb. capacity, average		1 Plum	3	2.667	Ea.	625	130		755	890
6770	Electric, front loading, energy-star qualified, minimum	G	L-2	3	5.333		255	187		442	590

11 31 Residential Appliances

11 31 23 – Residential Laundry Appliances

11 31 23.23 Dryers		Crew	Daily Output	Labor-Hours	Unit	2009 Bare Costs Material	Labor	Equipment	Total	Total Incl O&P
6780	Maximum G	L-2	2	8	Ea.	1,425	281		1,706	2,025
7450	Vent kits for dryers	1 Carp	10	.800	↓	31.50	32		63.50	88

11 31 33 – Miscellaneous Residential Appliances

11 31 33.13 Sump Pumps

		Crew	Daily Output	Labor-Hours	Unit	Material	Labor	Equipment	Total	Total Incl O&P
0010	**SUMP PUMPS**									
6400	Cellar drainer, pedestal, 1/3 H.P., molded PVC base	1 Plum	3	2.667	Ea.	119	130		249	335
6450	Solid brass	"	2	4	"	263	195		458	590
6460	Sump pump, see also Div. 22 14 29.16									

11 31 33.23 Water Heaters

		Crew	Daily Output	Labor-Hours	Unit	Material	Labor	Equipment	Total	Total Incl O&P
0010	**WATER HEATERS**									
6900	Electric, glass lined, 30 gallon, minimum	L-1	5	3.200	Ea.	420	153		573	700
6950	Maximum		3	5.333		585	255		840	1,050
7100	80 gallon, minimum		2	8		760	385		1,145	1,425
7150	Maximum	↓	1	16		1,050	765		1,815	2,325
7180	Gas, glass lined, 30 gallon, minimum	2 Plum	5	3.200		565	156		721	860
7220	Maximum		3	5.333		785	260		1,045	1,275
7260	50 gallon, minimum		2.50	6.400		845	310		1,155	1,400
7300	Maximum	↓	1.50	10.667	↓	1,175	520		1,695	2,075
7310	Water heater, see also Div. 22 33 30.13									

11 31 33.43 Air Quality

		Crew	Daily Output	Labor-Hours	Unit	Material	Labor	Equipment	Total	Total Incl O&P
0010	**AIR QUALITY**									
2450	Dehumidifier, portable, automatic, 15 pint				Ea.	150			150	165
2550	40 pint					179			179	197
3550	Heater, electric, built-in, 1250 watt, ceiling type, minimum	1 Elec	4	2		96	94		190	250
3600	Maximum		3	2.667		156	125		281	365
3700	Wall type, minimum		4	2		143	94		237	300
3750	Maximum		3	2.667		154	125		279	360
3900	1500 watt wall type, with blower		4	2		143	94		237	300
3950	3000 watt	↓	3	2.667		292	125		417	515
4850	Humidifier, portable, 8 gallons per day					160			160	176
5000	15 gallons per day				↓	193			193	212

11 33 Retractable Stairs

11 33 10 – Disappearing Stairs

11 33 10.10 Disappearing Stairway

		Crew	Daily Output	Labor-Hours	Unit	Material	Labor	Equipment	Total	Total Incl O&P
0010	**DISAPPEARING STAIRWAY** No trim included									
0100	Custom grade, pine, 8′-6″ ceiling, minimum	1 Carp	4	2	Ea.	202	80		282	355
0150	Average		3.50	2.286		215	91.50		306.50	385
0200	Maximum		3	2.667		275	107		382	480
0500	Heavy duty, pivoted, from 7′-7″ to 12′-10″ floor to floor		3	2.667		650	107		757	890
0600	16′-0″ ceiling		2	4		1,325	160		1,485	1,725
0800	Economy folding, pine, 8′-6″ ceiling		4	2		195	80		275	345
0900	9′-6″ ceiling	↓	4	2		225	80		305	380
1100	Automatic electric, aluminum, floor to floor height, 8′ to 9′	2 Carp	1	16		8,250	640		8,890	10,100
1400	11′ to 12′		.90	17.778		8,875	710		9,585	11,000
1700	14′ to 15′	↓	.70	22.857	↓	9,475	915		10,390	11,900
9000	Minimum labor/equipment charge	1 Carp	2	4	Job		160		160	264

11 41 13 – Refrigerated Food Storage Cases

11 41 13.10 Refrigerated Food Cases		Crew	Daily Output	Labor-Hours	Unit	2009 Bare Costs Material	Labor	Equipment	Total	Total Incl O&P
0010	**REFRIGERATED FOOD CASES**									
0030	Dairy, multi-deck, 12′ long	Q-5	3	5.333	Ea.	9,925	237		10,162	11,300
0100	For rear sliding doors, add					1,425			1,425	1,550
0200	Delicatessen case, service deli, 12′ long, single deck	Q-5	3.90	4.103		6,675	182		6,857	7,625
0300	Multi-deck, 18 S.F. shelf display		3	5.333		7,425	237		7,662	8,550
0400	Freezer, self-contained, chest-type, 30 C.F.		3.90	4.103		6,700	182		6,882	7,650
0500	Glass door, upright, 78 C.F.		3.30	4.848		9,275	215		9,490	10,500
0600	Frozen food, chest type, 12′ long		3.30	4.848		6,700	215		6,915	7,700
0700	Glass door, reach-in, 5 door		3	5.333		12,900	237		13,137	14,500
0800	Island case, 12′ long, single deck		3.30	4.848		7,600	215		7,815	8,675
0900	Multi-deck		3	5.333		16,100	237		16,337	18,100
1000	Meat case, 12′ long, single deck		3.30	4.848		5,525	215		5,740	6,400
1050	Multi-deck		3.10	5.161		9,500	229		9,729	10,800
1100	Produce, 12′ long, single deck		3.30	4.848		7,300	215		7,515	8,350
1200	Multi-deck		3.10	5.161		7,975	229		8,204	9,125

11 41 13.20 Refrigerated Food Storage Equipment		Crew	Daily Output	Labor-Hours	Unit	Material	Labor	Equipment	Total	Total Incl O&P
0010	**REFRIGERATED FOOD STORAGE EQUIPMENT**									
2350	Cooler, reach-in, beverage, 6′ long	Q-1	6	2.667	Ea.	4,300	117		4,417	4,925
4300	Freezers, reach-in, 44 C.F.		4	4		3,150	176		3,326	3,725
4500	68 C.F.		3	5.333		4,350	234		4,584	5,150
4600	Freezer, pre-fab, 8′ x 8′ w/refrigeration	2 Carp	.45	35.556		10,200	1,425		11,625	13,600
4620	8′ x 12′		.35	45.714		12,100	1,825		13,925	16,300
4640	8′ x 16′		.25	64		15,000	2,550		17,550	20,700
4660	8′ x 20′		.17	94.118		18,100	3,750		21,850	26,100
4680	Reach-in, 1 compartment	Q-1	4	4		1,550	176		1,726	1,975
4700	2 compartment		3	5.333		4,100	234		4,334	4,875
8300	Refrigerators, reach-in type, 44 C.F.		5	3.200		6,775	140		6,915	7,675
8310	With glass doors, 68 C.F.		4	4		6,625	176		6,801	7,575
8320	Refrigerator, reach-in, 1 compartment	R-18	7.80	3.333		1,950	123		2,073	2,350
8330	2 compartment		6.20	4.194		2,900	154		3,054	3,425
8340	3 compartment		5.60	4.643		3,925	171		4,096	4,575
8350	Pre-fab, with refrigeration, 8′ x 8′	2 Carp	.45	35.556		5,375	1,425		6,800	8,250
8360	8′ x 12′		.35	45.714		7,125	1,825		8,950	10,900
8370	8′ x 16′		.25	64		10,400	2,550		12,950	15,600
8380	8′ x 20′		.17	94.118		13,100	3,750		16,850	20,600
8390	Pass-thru/roll-in, 1 compartment	R-18	7.80	3.333		3,550	123		3,673	4,100
8400	2 compartment		6.24	4.167		4,975	153		5,128	5,700
8410	3 compartment		5.60	4.643		5,950	171		6,121	6,825
8420	Walk-in, alum, door & floor only, no refrig, 6′ x 6′ x 7′-6″	2 Carp	1.40	11.429		8,150	455		8,605	9,700
8430	10′ x 6′ x 7′-6″		.55	29.091		11,700	1,150		12,850	14,700
8440	12′ x 14′ x 7′-6″		.25	64		16,300	2,550		18,850	22,100
8450	12′ x 20′ x 7′-6″		.17	94.118		20,000	3,750		23,750	28,200
8460	Refrigerated cabinets, mobile					3,425			3,425	3,775
8470	Refrigerator/freezer, reach-in, 1 compartment	R-18	5.60	4.643		4,425	171		4,596	5,150
8480	2 compartment	"	4.80	5.417		6,250	199		6,449	7,200

11 41 13.30 Wine Cellar		Crew	Daily Output	Labor-Hours	Unit	Material	Labor	Equipment	Total	Total Incl O&P
0010	**WINE CELLAR**, refrigerated, Redwood interior, carpeted, walk-in type									
0020	6′-8″ high, including racks									
0200	80″ W x 48″ D for 900 bottles	2 Carp	1.50	10.667	Ea.	3,350	425		3,775	4,375
0250	80″ W x 72″ D for 1300 bottles		1.33	12.030		4,425	480		4,905	5,650
0300	80″ W x 94″ D for 1900 bottles		1.17	13.675		5,725	545		6,270	7,200
0400	80″ W x 124″ D for 2500 bottles		1	16		6,925	640		7,565	8,675

11 41 Food Storage Equipment

11 41 13 – Refrigerated Food Storage Cases

11 41 13.30 Wine Cellar		Crew	Daily Output	Labor-Hours	Unit	2009 Bare Costs Material	Labor	Equipment	Total	Total Incl O&P
0600	Portable cabinets, red oak, reach-in temp.& humidity controlled									
0650	26-5/8"W x 26-1/2"D x 68"H for 235 bottles				Ea.	2,200			2,200	2,425
0660	32"W x 21-1/2"D x 73-1/2"H for 144 bottles					3,500			3,500	3,850
0670	32"W x 29-1/2"D x 73-1/2"H for 288 bottles					2,525			2,525	2,775
0680	39-1/2"W x 29-1/2"D x 86-1/2"H for 440 bottles					3,125			3,125	3,450
0690	52-1/2"W x 29-1/2"D x 73-1/2"H for 468 bottles					3,725			3,725	4,075
0700	52-1/2"W x 29-1/2"D x 86-1/2"H for 572 bottles				↓	3,825			3,825	4,225
0730	Portable, red oak, can be built-in with glass door									
0750	23-7/8"W x 24"D x 34-1/2"H for 50 bottles				Ea.	955			955	1,050

11 41 33 – Food Storage Shelving

11 41 33.20 Metal Food Storage Shelving		Crew	Daily Output	Labor-Hours	Unit	2009 Bare Costs Material	Labor	Equipment	Total	Total Incl O&P
0010	**METAL FOOD STORAGE SHELVING**									
8600	Stainless steel shelving, louvered 4-tier, 20" x 3'	1 Clab	6	1.333	Ea.	1,175	42		1,217	1,350
8605	20" x 4'		6	1.333		1,675	42		1,717	1,900
8610	20" x 6'		6	1.333		2,425	42		2,467	2,725
8615	24" x 3'		6	1.333		1,525	42		1,567	1,750
8620	24" x 4'		6	1.333		1,775	42		1,817	2,025
8625	24" x 6'		6	1.333		2,575	42		2,617	2,925
8630	Flat 4-tier, 20" x 3'		6	1.333		980	42		1,022	1,150
8635	20" x 4'		6	1.333		1,225	42		1,267	1,425
8640	20" x 5'		6	1.333		1,350	42		1,392	1,575
8645	24" x 3'		6	1.333		1,075	42		1,117	1,250
8650	24" x 4'		6	1.333		1,875	42		1,917	2,150
8655	24" x 6'		6	1.333		2,250	42		2,292	2,550
8700	Galvanized shelving, louvered 4-tier, 20" x 3'		6	1.333		595	42		637	720
8705	20" x 4'		6	1.333		660	42		702	800
8710	20" x 6'		6	1.333		700	42		742	840
8715	24" x 3'		6	1.333		525	42		567	645
8720	24" x 4'		6	1.333		710	42		752	850
8725	24" x 6'		6	1.333		1,025	42		1,067	1,200
8730	Flat 4-tier, 20" x 3'		6	1.333		425	42		467	540
8735	20" x 4'		6	1.333		480	42		522	600
8740	20" x 6'		6	1.333		835	42		877	985
8745	24" x 3'		6	1.333		430	42		472	545
8750	24" x 4'		6	1.333		615	42		657	750
8755	24" x 6'		6	1.333		865	42		907	1,025
8760	Stainless steel dunnage rack, 24" x 3'		8	1		395	31.50		426.50	485
8765	24" x 4'		8	1		775	31.50		806.50	905
8770	Galvanized dunnage rack, 24" x 3'		8	1		162	31.50		193.50	230
8775	24" x 4'	↓	8	1	↓	181	31.50		212.50	251

11 42 Food Preparation Equipment

11 42 10 – Commercial Food Preparation Equipment

11 42 10.10 Choppers, Mixers and Misc. Equipment		Crew	Daily Output	Labor-Hours	Unit	2009 Bare Costs Material	Labor	Equipment	Total	Total Incl O&P
0010	**CHOPPERS, MIXERS AND MISC. EQUIPMENT**									
1700	Choppers, 5 pounds	R-18	7	3.714	Ea.	1,625	137		1,762	2,000
1720	16 pounds		5	5.200		1,975	191		2,166	2,475
1740	35 to 40 pounds		4	6.500		2,975	239		3,214	3,650
1840	Coffee brewer, 5 burners	1 Plum	3	2.667		1,175	130		1,305	1,500
1850	Coffee urn, twin 6 gallon urns		2	4		2,525	195		2,720	3,075
1860	Single, 3 gallon		3	2.667		1,800	130		1,930	2,200
3000	Fast food equipment, total package, minimum	6 Skwk	.08	600		172,000	24,500		196,500	229,500
3100	Maximum	"	.07	685		234,500	28,000		262,500	303,500
3800	Food mixers, 20 quarts	L-7	7	4		2,700	154		2,854	3,200
3850	40 quarts		5.40	5.185		6,950	200		7,150	7,975
3900	60 quarts		5	5.600		8,925	216		9,141	10,200
4040	80 quarts		3.90	7.179		11,000	277		11,277	12,600
4080	130 quarts		2.20	12.727		16,600	490		17,090	19,000
4100	Floor type, 20 quarts		15	1.867		3,125	72		3,197	3,550
4120	60 quarts		14	2		7,725	77		7,802	8,625
4140	80 quarts		12	2.333		9,575	90		9,665	10,600
4160	140 quarts		8.60	3.256		18,200	126		18,326	20,200
6700	Peelers, small	R-18	8	3.250		1,450	119		1,569	1,800
6720	Large	"	6	4.333		4,375	159		4,534	5,075
6800	Pulper/extractor, close coupled, 5 HP	1 Plum	1.90	4.211		3,325	205		3,530	4,000
8580	Slicer with table	R-18	9	2.889		4,675	106		4,781	5,300

11 43 Food Delivery Carts and Conveyors

11 43 13 – Food Delivery Carts

11 43 13.10 Mobile Carts, Racks and Trays		Crew	Daily Output	Labor-Hours	Unit	Material	Labor	Equipment	Total	Total Incl O&P
0010	**MOBILE CARTS, RACKS AND TRAYS**									
1650	Cabinet, heated, 1 compartment, reach-in	R-18	5.60	4.643	Ea.	3,250	171		3,421	3,850
1655	Pass-thru roll-in		5.60	4.643		5,125	171		5,296	5,900
1660	2 compartment, reach-in		4.80	5.417		6,750	199		6,949	7,750
1670	Mobile					3,000			3,000	3,300
6850	Mobile rack w/pan slide					1,450			1,450	1,600
9180	Tray and silver dispenser, mobile	1 Clab	16	.500		775	15.80		790.80	875

11 44 Food Cooking Equipment

11 44 13 – Commercial Ranges

11 44 13.10 Cooking Equipment		Crew	Daily Output	Labor-Hours	Unit	Material	Labor	Equipment	Total	Total Incl O&P
0010	**COOKING EQUIPMENT**									
0020	Bake oven, gas, one section	Q-1	8	2	Ea.	4,300	88		4,388	4,875
0300	Two sections		7	2.286		10,100	100		10,200	11,400
0600	Three sections		6	2.667		13,200	117		13,317	14,700
0900	Electric convection, single deck	L-7	4	7		5,400	270		5,670	6,375
1300	Broiler, without oven, standard	Q-1	8	2		3,550	88		3,638	4,025
1550	Infrared	L-7	4	7		9,250	270		9,520	10,600
4750	Fryer, with twin baskets, modular model	Q-1	7	2.286		1,450	100		1,550	1,750
5000	Floor model, on 6" legs	"	5	3.200		2,400	140		2,540	2,850
5100	Extra single basket, large					99			99	109
5300	Griddle, SS, 24" plate, w/4" legs, elec, 208V, 3 phase, 3' long	Q-1	7	2.286		1,125	100		1,225	1,375
5550	4' long	"	6	2.667		1,450	117		1,567	1,775

11 44 Food Cooking Equipment

11 44 13 – Commercial Ranges

11 44 13.10 Cooking Equipment		Crew	Daily Output	Labor-Hours	Unit	Material	2009 Bare Costs Labor	Equipment	Total	Total Incl O&P
6200	Iced tea brewer	1 Plum	3.44	2.326	Ea.	670	113		783	910
6350	Kettle, w/steam jacket, tilting, w/positive lock, SS, 20 gallons	L-7	7	4		6,675	154		6,829	7,600
6600	60 gallons	"	6	4.667		8,750	180		8,930	9,900
6900	Range, restaurant type, 6 burners and 1 standard oven, 36" wide	Q-1	7	2.286		2,325	100		2,425	2,700
6950	Convection		7	2.286		4,700	100		4,800	5,325
7150	2 standard ovens, 24" griddle, 60" wide		6	2.667		7,450	117		7,567	8,375
7200	1 standard, 1 convection oven		6	2.667		6,700	117		6,817	7,550
7450	Heavy duty, single 34" standard oven, open top		5	3.200		4,850	140		4,990	5,550
7500	Convection oven		5	3.200		3,425	140		3,565	4,000
7700	Griddle top		6	2.667		2,625	117		2,742	3,075
7750	Convection oven		6	2.667		5,950	117		6,067	6,700
8850	Steamer, electric 27 KW	L-7	7	4		7,375	154		7,529	8,375
9100	Electric, 10 KW or gas 100,000 BTU	"	5	5.600		5,350	216		5,566	6,225
9150	Toaster, conveyor type, 16-22 slices per minute					1,100			1,100	1,200
9160	Pop-up, 2 slot					535			535	585
9200	For deluxe models of above equipment, add					75%				
9400	Rule of thumb: Equipment cost based									
9410	on kitchen work area									
9420	Office buildings, minimum	L-7	77	.364	S.F.	76.50	14.05		90.55	108
9450	Maximum		58	.483		129	18.60		147.60	173
9550	Public eating facilities, minimum		77	.364		100	14.05		114.05	134
9600	Maximum		46	.609		163	23.50		186.50	219
9750	Hospitals, minimum		58	.483		103	18.60		121.60	145
9800	Maximum		39	.718		190	27.50		217.50	254

11 46 Food Dispensing Equipment

11 46 16 – Service Line Equipment

11 46 16.10 Commercial Food Dispensing Equipment		Crew	Daily Output	Labor-Hours	Unit	Material	Labor	Equipment	Total	Total Incl O&P
0010	COMMERCIAL FOOD DISPENSING EQUIPMENT									
1050	Butter pat dispenser	1 Clab	13	.615	Ea.	865	19.45		884.45	985
1100	Bread dispenser, counter top		13	.615		825	19.45		844.45	940
1900	Cup and glass dispenser, drop in		4	2		1,200	63		1,263	1,425
1920	Disposable cup, drop in		16	.500		405	15.80		420.80	470
2650	Dish dispenser, drop in, 12"		11	.727		1,850	23		1,873	2,075
2660	Mobile		10	.800		2,225	25.50		2,250.50	2,500
3300	Food warmer, counter, 1.2 KW					670			670	735
3550	1.6 KW					1,850			1,850	2,025
3600	Well, hot food, built-in, rectangular, 12" x 20"	R-30	10	2.600		590	98		688	800
3610	Circular, 7 qt		10	2.600		330	98		428	515
3620	Refrigerated, 2 compartments		10	2.600		2,725	98		2,823	3,125
3630	3 compartments		9	2.889		3,000	109		3,109	3,475
3640	4 compartments		8	3.250		3,700	122		3,822	4,275
4720	Frost cold plate		9	2.889		15,600	109		15,709	17,300
5700	Hot chocolate dispenser	1 Plum	4	2		1,025	97.50		1,122.50	1,275
6250	Jet spray dispenser	R-18	4.50	5.778		3,000	212		3,212	3,650
6300	Juice dispenser, concentrate	"	4.50	5.778		2,100	212		2,312	2,650
6690	Milk dispenser, bulk, 2 flavor	R-30	8	3.250		1,425	122		1,547	1,775
6695	3 flavor	"	8	3.250		1,950	122		2,072	2,350
8800	Serving counter, straight	1 Carp	40	.200	L.F.	745	8		753	835
8820	Curved section	"	30	.267	"	935	10.65		945.65	1,050
8825	Solid surface, see Div. 12 36 61.16									

11 46 Food Dispensing Equipment

11 46 16 – Service Line Equipment

11 46 16.10 Commercial Food Dispensing Equipment		Crew	Daily Output	Labor-Hours	Unit	Material	2009 Bare Costs Labor	Equipment	Total	Total Incl O&P
8830	Soft serve ice cream machine, medium	R-18	11	2.364	Ea.	9,850	87		9,937	10,900
8840	Large	″	9	2.889	″	17,300	106		17,406	19,200

11 47 Ice Machines

11 47 10 – Commercial Ice Machines

11 47 10.10 Commercial Ice Equipment		Crew	Daily Output	Labor-Hours	Unit	Material	Labor	Equipment	Total	Total Incl O&P
0010	**COMMERCIAL ICE EQUIPMENT**									
5800	Ice cube maker, 50 pounds per day	Q-1	6	2.667	Ea.	1,425	117		1,542	1,750
5900	250 pounds per day		1.20	13.333		2,400	585		2,985	3,550
6050	500 pounds per day		4	4		3,000	176		3,176	3,575
6060	With bin		1.20	13.333		2,775	585		3,360	3,950
6090	1000 pounds per day, with bin		1	16		4,775	700		5,475	6,350
6100	Ice flakers, 300 pounds per day		1.60	10		3,050	440		3,490	4,025
6120	600 pounds per day		.95	16.842		3,875	740		4,615	5,400
6130	1000 pounds per day		.75	21.333		4,325	935		5,260	6,200
6140	2000 pounds per day		.65	24.615		18,200	1,075		19,275	21,800
6160	Ice storage bin, 500 pound capacity	Q-5	1	16		940	710		1,650	2,125
6180	1000 pound	″	.56	28.571		1,800	1,275		3,075	3,950

11 48 Cleaning and Disposal Equipment

11 48 13 – Commercial Dishwashers

11 48 13.10 Dishwashers		Crew	Daily Output	Labor-Hours	Unit	Material	Labor	Equipment	Total	Total Incl O&P
0010	**DISHWASHERS**									
2700	Dishwasher, commercial, rack type									
2720	10 to 12 racks per hour	Q-1	3.20	5	Ea.	4,200	219		4,419	4,975
2750	Semi-automatic 38 to 50 racks per hour	″	1.30	12.308		8,350	540		8,890	10,000
2800	Automatic, 190 to 230 racks per hour	L-6	.35	34.286		12,400	1,650		14,050	16,300
2820	235 to 275 racks per hour		.25	48		27,400	2,300		29,700	33,800
2840	8,750 to 12,500 dishes per hour		.10	120		48,000	5,775		53,775	62,000
2950	Dishwasher hood, canopy type	L-3A	10	1.200	L.F.	770	52.50		822.50	930
2960	Pant leg type	″	2.50	4.800	Ea.	7,375	210		7,585	8,450
5200	Garbage disposal 1.5 HP, 100 GPH	L-1	4.80	3.333		1,825	160		1,985	2,250
5210	3 HP, 120 GPH		4.60	3.478		2,300	167		2,467	2,775
5220	5 HP, 250 GPH		4.50	3.556		3,550	170		3,720	4,175
6750	Pot sink, 3 compartment	1 Plum	7.25	1.103	L.F.	845	54		899	1,025
6760	Pot washer, small		1.60	5	Ea.	8,900	244		9,144	10,200
6770	Large		1.20	6.667		14,600	325		14,925	16,600
9170	Trash compactor, small, up to 125 lb. compacted weight	L-4	4	6		20,800	224		21,024	23,300
9171	Trash Compactor for Furniture Row		12	2		16,800	74.50		16,874.50	18,600
9175	Large, up to 175 lb. compacted weight		3	8		25,400	299		25,699	28,400

11 52 Audio-Visual Equipment

11 52 13 – Projection Screens

11 52 13.10 Projection Screens, Wall or Ceiling Hung		Crew	Daily Output	Labor-Hours	Unit	2009 Bare Costs Material	Labor	Equipment	Total	Total Incl O&P
0010	**PROJECTION SCREENS, WALL OR CEILING HUNG**, matte white									
0100	Manually operated, economy	2 Carp	500	.032	S.F.	5.85	1.28		7.13	8.55
0300	Intermediate		450	.036		6.80	1.42		8.22	9.85
0400	Deluxe		400	.040		9.45	1.60		11.05	13
0600	Electric operated, matte white, 25 S.F., economy		5	3.200	Ea.	855	128		983	1,150
0700	Deluxe		4	4		1,775	160		1,935	2,225
0900	50 S.F., economy		3	5.333		975	213		1,188	1,425
1000	Deluxe		2	8		1,975	320		2,295	2,700
1200	Heavy duty, electric operated, 200 S.F.		1.50	10.667		3,900	425		4,325	5,000
1300	400 S.F.		1	16		4,800	640		5,440	6,350
1500	Rigid acrylic in wall, for rear projection, 1/4" thick	2 Glaz	30	.533	S.F.	47	20.50		67.50	85
1600	1/2" thick (maximum size 10' x 20')	"	25	.640	"	83.50	24.50		108	132
9000	Minimum labor/equipment charge	2 Carp	3	5.333	Job		213		213	350

11 52 16 – Projectors

11 52 16.10 Movie Equipment		Crew	Daily Output	Labor-Hours	Unit	Material	Labor	Equipment	Total	Total Incl O&P
0010	**MOVIE EQUIPMENT**									
0020	Changeover, minimum				Ea.	465			465	510
0100	Maximum					905			905	995
0400	Film transport, incl. platters and autowind, minimum					5,025			5,025	5,525
0500	Maximum					14,300			14,300	15,700
0800	Lamphouses, incl. rectifiers, xenon, 1,000 watt	1 Elec	2	4		6,650	188		6,838	7,600
0900	1,600 watt		2	4		7,100	188		7,288	8,100
1000	2,000 watt		1.50	5.333		7,625	251		7,876	8,750
1100	4,000 watt		1.50	5.333		9,425	251		9,676	10,800
1400	Lenses, anamorphic, minimum					1,275			1,275	1,400
1500	Maximum					2,875			2,875	3,150
1800	Flat 35 mm, minimum					1,100			1,100	1,225
1900	Maximum					1,725			1,725	1,900
2200	Pedestals, for projectors					1,475			1,475	1,625
2300	Console type					10,700			10,700	11,800
2600	Projector mechanisms, incl. soundhead, 35 mm, minimum					11,100			11,100	12,200
2700	Maximum					15,200			15,200	16,700
3000	Projection screens, rigid, in wall, acrylic, 1/4" thick	2 Glaz	195	.082	S.F.	42	3.17		45.17	51
3100	1/2" thick	"	130	.123	"	48.50	4.75		53.25	61
3300	Electric operated, heavy duty, 400 S.F.	2 Carp	1	16	Ea.	2,950	640		3,590	4,300
3320	Theater projection screens, matte white, including frames	"	200	.080	S.F.	6.65	3.20		9.85	12.60
3400	Also see Div. 11 52 13.10									
3700	Sound systems, incl. amplifier, mono, minimum	1 Elec	.90	8.889	Ea.	3,300	420		3,720	4,275
3800	Dolby/Super Sound, maximum		.40	20		18,100	940		19,040	21,400
4100	Dual system, 2 channel, front surround, minimum		.70	11.429		4,625	535		5,160	5,925
4200	Dolby/Super Sound, 4 channel, maximum		.40	20		16,500	940		17,440	19,700
4500	Sound heads, 35 mm					5,300			5,300	5,825
4900	Splicer, tape system, minimum					740			740	815
5000	Tape type, maximum					1,325			1,325	1,450
5300	Speakers, recessed behind screen, minimum	1 Elec	2	4		1,050	188		1,238	1,475
5400	Maximum	"	1	8		3,100	375		3,475	3,975
5700	Seating, painted steel, upholstered, minimum	2 Carp	35	.457		132	18.25		150.25	175
5800	Maximum	"	28	.571		420	23		443	505
6100	Rewind tables, minimum					2,650			2,650	2,900
6200	Maximum					4,725			4,725	5,175
7000	For automation, varying sophistication, minimum	1 Elec	1	8	System	2,375	375		2,750	3,200
7100	Maximum	2 Elec	.30	53.333	"	5,575	2,500		8,075	9,975

11 53 Laboratory Equipment

11 53 03 – Laboratory Test Equipment

11 53 03.13 Test Equipment		Crew	Daily Output	Labor-Hours	Unit	2009 Bare Costs Material	Labor	Equipment	Total	Total Incl O&P
0010	**TEST EQUIPMENT**									
1700	Thermometer, electric, portable				Ea.	340			340	370
1800	Titration unit, four 2000 ml reservoirs				"	5,500			5,500	6,050

11 53 13 – Laboratory Fume Hoods

11 53 13.13 Recirculating Laboratory Fume Hoods

Code	Description	Crew	Daily Output	Labor-Hours	Unit	Material	Labor	Equipment	Total	Total Incl O&P
0010	**RECIRCULATING LABORATORY FUME HOODS**									
0600	Fume hood, with countertop & base, not including HVAC									
0610	Simple, minimum	2 Carp	5.40	2.963	L.F.	520	118		638	765
0620	Complex, including fixtures		2.40	6.667		1,675	266		1,941	2,300
0630	Special, maximum	↓	1.70	9.412	↓	1,825	375		2,200	2,625
0670	Service fixtures, average				Ea.	273			273	300
0680	For sink assembly with hot and cold water, add	1 Plum	1.40	5.714		705	279		984	1,200
0750	Glove box, fiberglass, bacteriological					16,900			16,900	18,600
0760	Controlled atmosphere					17,500			17,500	19,200
0770	Radioisotope					16,900			16,900	18,600
0780	Carcinogenic				↓	16,900			16,900	18,600

11 53 13.23 Exhaust Hoods

Code	Description	Crew	Daily Output	Labor-Hours	Unit	Material	Labor	Equipment	Total	Total Incl O&P
0010	**EXHAUST HOODS**									
0650	Ductwork, minimum	2 Shee	1	16	Hood	3,575	755		4,330	5,125
0660	Maximum	"	.50	32	"	5,500	1,500		7,000	8,450

11 53 16 – Laboratory Incubators

11 53 16.13 Incubators

Code	Description	Crew	Daily Output	Labor-Hours	Unit	Material	Labor	Equipment	Total	Total Incl O&P
0010	**INCUBATORS**									
1000	Incubators, minimum				Ea.	3,250			3,250	3,575
1010	Maximum				"	19,200			19,200	21,100

11 53 19 – Laboratory Sterilizers

11 53 19.13 Sterilizers

Code	Description	Crew	Daily Output	Labor-Hours	Unit	Material	Labor	Equipment	Total	Total Incl O&P
0010	**STERILIZERS**									
0700	Glassware washer, undercounter, minimum	L-1	1.80	8.889	Ea.	5,450	425		5,875	6,650
0710	Maximum	"	1	16		12,100	765		12,865	14,600
1850	Utensil washer-sanitizer	1 Plum	2	4	↓	10,400	195		10,595	11,700

11 53 23 – Laboratory Refrigerators

11 53 23.13 Refrigerators

Code	Description	Crew	Daily Output	Labor-Hours	Unit	Material	Labor	Equipment	Total	Total Incl O&P
0010	**REFRIGERATORS**									
1200	Blood bank, 28.6 C.F. emergency signal				Ea.	7,350			7,350	8,075
1210	Reach-in, 16.9 C.F.				"	7,375			7,375	8,100

11 53 33 – Emergency Safety Appliances

11 53 33.13 Emergency Equipment

Code	Description	Crew	Daily Output	Labor-Hours	Unit	Material	Labor	Equipment	Total	Total Incl O&P
0010	**EMERGENCY EQUIPMENT**									
1400	Safety equipment, eye wash, hand held				Ea.	405			405	445
1450	Deluge shower				"	725			725	800

11 53 43 – Service Fittings and Accessories

11 53 43.13 Fittings

Code	Description	Crew	Daily Output	Labor-Hours	Unit	Material	Labor	Equipment	Total	Total Incl O&P
0010	**FITTINGS**									
1600	Sink, one piece plastic, flask wash, hose, free standing	1 Plum	1.60	5	Ea.	1,800	244		2,044	2,350
1610	Epoxy resin sink, 25" x 16" x 10"	"	2	4	"	198	195		393	520
1950	Utility table, acid resistant top with drawers	2 Carp	30	.533	L.F.	147	21.50		168.50	196
8000	Alternate pricing method: as percent of lab furniture									
8050	Installation, not incl. plumbing & duct work				% Furn.				20%	22%

11 53 Laboratory Equipment

11 53 43 – Service Fittings and Accessories

11 53 43.13 Fittings		Crew	Daily Output	Labor-Hours	Unit	2009 Bare Costs Material	Labor	Equipment	Total	Total Incl O&P
8100	Plumbing, final connections, simple system				% Furn.				9.09%	10%
8110	Moderately complex system								13.64%	15%
8120	Complex system								18.18%	20%
8150	Electrical, simple system								9.09%	10%
8160	Moderately complex system								18.18%	20%
8170	Complex system								31.90%	35%

11 57 Vocational Shop Equipment

11 57 10 – Shop Equipment

11 57 10.10 Vocational School Shop Equipment

		Crew	Daily Output	Labor-Hours	Unit	Material	Labor	Equipment	Total	Total Incl O&P
0010	**VOCATIONAL SCHOOL SHOP EQUIPMENT**									
0020	Benches, work, wood, average	2 Carp	5	3.200	Ea.	505	128		633	765
0100	Metal, average		5	3.200		370	128		498	615
0400	Combination belt & disc sander, 6"		4	4		1,275	160		1,435	1,675
0700	Drill press, floor mounted, 12", 1/2 H.P.		4	4		400	160		560	700
0800	Dust collector, not incl. ductwork, 6" diameter	1 Shee	1.10	7.273		3,000	345		3,345	3,850
1000	Grinders, double wheel, 1/2 H.P.	2 Carp	5	3.200		245	128		373	480
1300	Jointer, 4", 3/4 H.P.		4	4		1,250	160		1,410	1,650
1600	Kilns, 16 C.F., to 2000°		4	4		2,250	160		2,410	2,750
1900	Lathe, woodworking, 10", 1/2 H.P.		4	4		765	160		925	1,100
2200	Planer, 13" x 6"		4	4		1,250	160		1,410	1,650
2500	Potter's wheel, motorized		4	4		1,125	160		1,285	1,500
2800	Saws, band, 14", 3/4 H.P.		4	4		850	160		1,010	1,200
3100	Metal cutting band saw, 14"		4	4		2,000	160		2,160	2,475
3400	Radial arm saw, 10", 2 H.P.		4	4		975	160		1,135	1,350
3700	Scroll saw, 24"		4	4		575	160		735	900
4000	Table saw, 10", 3 H.P.		4	4		1,925	160		2,085	2,375
4300	Welder AC arc, 30 amp capacity		4	4		1,950	160		2,110	2,400

11 61 Theater and Stage Equipment

11 61 23 – Folding and Portable Stages

11 61 23.10 Portable Stages

		Crew	Daily Output	Labor-Hours	Unit	Material	Labor	Equipment	Total	Total Incl O&P
0010	**PORTABLE STAGES**									
1500	Flooring, portable oak parquet, 3' x 3' sections				S.F.	13.40			13.40	14.70
1600	Cart to carry 225 S.F. of flooring				Ea.	400			400	440
5000	Stages, portable with steps, folding legs, stock, 8" high				SF Stg.	25.50			25.50	28.50
5100	16" high					23.50			23.50	26
5200	32" high					35.50			35.50	39
5300	40" high					59.50			59.50	65
6000	Telescoping platforms, extruded alum., straight, minimum	4 Carp	157	.204		28	8.15		36.15	44
6100	Maximum		77	.416		37.50	16.60		54.10	69
6500	Pie-shaped, minimum		150	.213		60.50	8.50		69	80.50
6600	Maximum		70	.457		67.50	18.25		85.75	104
6800	For 3/4" plywood covered deck, deduct					3.85			3.85	4.24
7000	Band risers, steel frame, plywood deck, minimum	4 Carp	275	.116		30.50	4.65		35.15	41
7100	Maximum	"	138	.232		63.50	9.25		72.75	85
7500	Chairs for above, self-storing, minimum	2 Carp	43	.372	Ea.	99.50	14.85		114.35	134
7600	Maximum	"	40	.400	"	172	16		188	216

11 61 Theater and Stage Equipment

11 61 33 – Rigging Systems and Controls

11 61 33.10 Controls		Crew	Daily Output	Labor-Hours	Unit	Material	2009 Bare Costs Labor	Equipment	Total	Total Incl O&P
0010	**CONTROLS**									
0050	Control boards with dimmers and breakers, minimum	1 Elec	1	8	Ea.	11,300	375		11,675	13,100
0100	Average		.50	16		35,500	750		36,250	40,300
0150	Maximum		.20	40		119,000	1,875		120,875	134,000
8000	Rule of thumb: total stage equipment, minimum	4 Carp	100	.320	SF Stg.	97	12.80		109.80	128
8100	Maximum	"	25	1.280	"	545	51		596	685

11 61 43 – Stage Curtains

11 61 43.10 Curtains		Crew	Daily Output	Labor-Hours	Unit	Material	Labor	Equipment	Total	Total Incl O&P
0010	**CURTAINS**									
0500	Curtain track, straight, light duty	2 Carp	20	.800	L.F.	25.50	32		57.50	81
0600	Heavy duty		18	.889		59	35.50		94.50	124
0700	Curved sections		12	1.333		164	53.50		217.50	268
1000	Curtains, velour, medium weight		600	.027	S.F.	7.90	1.07		8.97	10.45
1150	Silica based yarn, inherently fire retardant		50	.320	"	15.20	12.80		28	37.50

11 62 Musical Equipment

11 62 16 – Carillons

11 62 16.10 Bell Tower Equipment		Crew	Daily Output	Labor-Hours	Unit	Material	Labor	Equipment	Total	Total Incl O&P
0010	**BELL TOWER EQUIPMENT**									
0300	Carillon, 4 octave (48 bells), with keyboard				System	892,500			892,500	982,000
0320	2 octave (24 bells)					420,000			420,000	462,000
0340	3 to 4 bell peal, minimum					105,000			105,000	115,500
0360	Maximum					630,000			630,000	693,000
0380	Cast bronze bell, average				Ea.	94,500			94,500	104,000
0400	Electronic, digital, minimum					15,800			15,800	17,300
0410	With keyboard, maximum					79,000			79,000	86,500

11 66 Athletic Equipment

11 66 13 – Exercise Equipment

11 66 13.10 Physical Training Equipment		Crew	Daily Output	Labor-Hours	Unit	Material	Labor	Equipment	Total	Total Incl O&P
0010	**PHYSICAL TRAINING EQUIPMENT**									
0020	Abdominal rack, 2 board capacity				Ea.	470			470	515
0050	Abdominal board, upholstered					525			525	575
0200	Bicycle trainer, minimum					815			815	895
0300	Deluxe, electric					4,150			4,150	4,575
0400	Barbell set, chrome plated steel, 25 lbs.					244			244	269
0420	100 lbs.					370			370	405
0450	200 lbs.					715			715	785
0500	Weight plates, cast iron, per lb.				Lb.	5			5	5.50
0520	Storage rack, 10 station				Ea.	790			790	870
0600	Circuit training apparatus, 12 machines minimum	2 Clab	1.25	12.800	Set	27,600	405		28,005	31,000
0700	Average		1	16		33,800	505		34,305	38,000
0800	Maximum		.75	21.333		40,100	675		40,775	45,200
0820	Dumbbell set, cast iron, with rack and 5 pair					570			570	630
0900	Squat racks	2 Clab	5	3.200	Ea.	765	101		866	1,000
1200	Multi-station gym machine, 5 station					7,950			7,950	8,750
1250	9 station					11,200			11,200	12,300
1280	Rowing machine, hydraulic					1,475			1,475	1,625

11 66 Athletic Equipment

11 66 13 – Exercise Equipment

11 66 13.10 Physical Training Equipment		Crew	Daily Output	Labor-Hours	Unit	2009 Bare Costs Material	Labor	Equipment	Total	Total Incl O&P
1300	Treadmill, manual				Ea.	1,450			1,450	1,575
1320	Motorized					3,500			3,500	3,850
1340	Electronic					5,450			5,450	6,000
1360	Cardio-testing					7,775			7,775	8,550
1400	Treatment/massage tables, minimum					535			535	590
1420	Deluxe, with accessories					685			685	755
4150	Exercise equipment, bicycle trainer					595			595	655
4180	Chinning bar, adjustable, wall mounted	1 Carp	5	1.600		279	64		343	410
4200	Exercise ladder, 16′ x 1′-7″, suspended	L-2	3	5.333		1,125	187		1,312	1,550
4210	High bar, floor plate attached	1 Carp	4	2		1,100	80		1,180	1,350
4240	Parallel bars, adjustable		4	2		2,475	80		2,555	2,825
4270	Uneven parallel bars, adjustable		4	2		2,675	80		2,755	3,050
4280	Wall mounted, adjustable	L-2	1.50	10.667	Set	835	375		1,210	1,525
4300	Rope, ceiling mounted, 18′ long	1 Carp	3.66	2.186	Ea.	220	87.50		307.50	385
4330	Side horse, vaulting		5	1.600		1,700	64		1,764	1,975
4360	Treadmill, motorized, deluxe, training type		5	1.600		3,500	64		3,564	3,950
4390	Weight lifting multi-station, minimum	2 Clab	1	16		395	505		900	1,275
4450	Maximum	"	.50	32		14,700	1,000		15,700	17,900

11 66 23 – Gymnasium Equipment

11 66 23.13 Basketball Equipment		Crew	Daily Output	Labor-Hours	Unit	Material	Labor	Equipment	Total	Total Incl O&P
0010	**BASKETBALL EQUIPMENT**									
1000	Backstops, wall mtd., 6′ extended, fixed, minimum	L-2	1	16	Ea.	1,150	560		1,710	2,175
1100	Maximum		1	16		1,675	560		2,235	2,775
1200	Swing up, minimum		1	16		1,275	560		1,835	2,325
1250	Maximum		1	16		5,350	560		5,910	6,825
1300	Portable, manual, heavy duty, spring operated		1.90	8.421		11,400	296		11,696	13,000
1400	Ceiling suspended, stationary, minimum		.78	20.513		2,850	720		3,570	4,350
1450	Fold up, with accessories, maximum		.40	40		6,100	1,400		7,500	9,050
1600	For electrically operated, add	1 Elec	1	8		2,000	375		2,375	2,775
5800	Wall pads, 1-1/2″ thick	2 Carp	640	.025	S.F.	6.80	1		7.80	9.15

11 66 23.19 Boxing Ring		Crew	Daily Output	Labor-Hours	Unit	Material	Labor	Equipment	Total	Total Incl O&P
0010	**BOXING RING**									
4100	Elevated, 22′ x 22′	L-4	.10	240	Ea.	6,700	8,950		15,650	22,000
4110	For cellular plastic foam padding, add		.10	240		850	8,950		9,800	15,500
4120	Floor level, including posts and ropes only, 20′ x 20′		.80	30		2,375	1,125		3,500	4,425
4130	Canvas, 30′ x 30′		5	4.800		1,200	179		1,379	1,600

11 66 23.47 Gym Mats		Crew	Daily Output	Labor-Hours	Unit	Material	Labor	Equipment	Total	Total Incl O&P
0010	**GYM MATS**									
5500	2″ thick, naugahyde covered				S.F.	3.59			3.59	3.95
5600	Vinyl/nylon covered					6.20			6.20	6.85
6000	Wrestling mats, 1″ thick, heavy duty					5.50			5.50	6.05

11 66 43 – Interior Scoreboards

11 66 43.10 Scoreboards		Crew	Daily Output	Labor-Hours	Unit	Material	Labor	Equipment	Total	Total Incl O&P
0010	**SCOREBOARDS**									
7000	Baseball, minimum	R-3	1.30	15.385	Ea.	3,225	710	99.50	4,034.50	4,750
7200	Maximum		.05	400		15,600	18,500	2,575	36,675	48,700
7300	Football, minimum		.86	23.256		3,975	1,075	150	5,200	6,225
7400	Maximum		.20	100		12,600	4,625	645	17,870	21,800
7500	Basketball (one side), minimum		2.07	9.662		2,225	445	62.50	2,732.50	3,200
7600	Maximum		.30	66.667		5,550	3,075	430	9,055	11,400
7700	Hockey-basketball (four sides), minimum		.25	80		5,675	3,700	515	9,890	12,500

11 66 Athletic Equipment

11 66 43 – Interior Scoreboards

11 66 43.10 Scoreboards		Crew	Daily Output	Labor-Hours	Unit	2009 Bare Costs Material	Labor	Equipment	Total	Total Incl O&P
7800	Maximum	R-3	.15	133	Ea.	5,600	6,175	860	12,635	16,600

11 66 53 – Gymnasium Dividers

11 66 53.10 Divider Curtains

		Crew	Daily Output	Labor-Hours	Unit	Material	Labor	Equipment	Total	Total Incl O&P
0010	**DIVIDER CURTAINS**									
4500	Gym divider curtain, mesh top, vinyl bottom, manual	L-4	500	.043	S.F.	8.05	1.79		9.84	11.80
4700	Electric roll up	L-7	400	.070	"	11.75	2.70		14.45	17.35

11 67 Recreational Equipment

11 67 13 – Bowling Alley Equipment

11 67 13.10 Bowling Alleys

		Crew	Daily Output	Labor-Hours	Unit	Material	Labor	Equipment	Total	Total Incl O&P
0010	**BOWLING ALLEYS** Including alley, pinsetter, scorer,									
0020	Counters and misc. supplies, minimum	4 Carp	.20	160	Lane	40,300	6,400		46,700	55,000
0150	Average		.19	163		44,200	6,725		50,925	59,500
0300	Maximum		.18	177		50,500	7,100		57,600	67,500
0400	Combo table ball rack, add					1,150			1,150	1,250
0600	For automatic scorer, add, minimum					8,225			8,225	9,050
0700	Maximum					9,875			9,875	10,900

11 67 23 – Shooting Range Equipment

11 67 23.10 Shooting Range

		Crew	Daily Output	Labor-Hours	Unit	Material	Labor	Equipment	Total	Total Incl O&P
0010	**SHOOTING RANGE** Incl. bullet traps, target provisions, controls,									
0100	Separators, ceiling system, etc. Not incl. structural shell									
0200	Commercial	L-9	.64	56.250	Point	18,000	2,050		20,050	23,300
0300	Law enforcement		.28	128		28,000	4,675		32,675	38,700
0400	National Guard armories		.71	50.704		15,000	1,850		16,850	19,600
0500	Reserve training centers		.71	50.704		12,000	1,850		13,850	16,300
0600	Schools and colleges		.32	112		12,000	4,075		16,075	20,100
0700	Major academies		.19	189		37,000	6,875		43,875	52,500
0800	For acoustical treatment, add					10%	10%			
0900	For lighting, add					28%	25%			
1000	For plumbing, add					5%	5%			
1100	For ventilating system, add, minimum					40%	40%			
1200	Add, average					25%	25%			
1300	Add, maximum					35%	35%			

11 68 Play Field Equipment and Structures

11 68 13 – Playground Equipment

11 68 13.10 Free-Standing Playground Equipment

		Crew	Daily Output	Labor-Hours	Unit	Material	Labor	Equipment	Total	Total Incl O&P
0010	**FREE-STANDING PLAYGROUND EQUIPMENT** See also individual items									
0200	Bike rack, 10′ long, permanent G	B-1	12	2	Ea.	620	64.50		684.50	785
0240	Climber, arch, 6′ high, 12′ long, 5′ wide		4	6		940	194		1,134	1,350
0260	Fitness trail, with signs, 9 to 10 stations, treated pine, minimum		.25	96		6,550	3,100		9,650	12,300
0270	Maximum		.17	141		9,300	4,550		13,850	17,700
0280	Metal, minimum		.25	96		4,175	3,100		7,275	9,675
0285	Maximum		.17	141		6,950	4,550		11,500	15,200
0300	Redwood, minimum		.25	96		13,200	3,100		16,300	19,600
0310	Maximum		.17	141		13,200	4,550		17,750	22,000
0320	16 to 20 station, treated pine, minimum		.17	141		7,400	4,550		11,950	15,700
0330	Maximum		.13	134		9,725	5,950		15,675	20,500

11 68 Play Field Equipment and Structures

11 68 13 – Playground Equipment

11 68 13.10 Free-Standing Playground Equipment		Crew	Daily Output	Labor-Hours	Unit	2009 Bare Costs Material	Labor	Equipment	Total	Total Incl O&P
0340	Metal, minimum	B-1	.17	141	Ea.	14,200	4,550		18,750	23,200
0350	Maximum		.13	184		16,800	5,950		22,750	28,300
0360	Redwood, minimum		.17	141		14,400	4,550		18,950	23,400
0370	Maximum		.13	184		19,900	5,950		25,850	31,700
0400	Horizontal monkey ladder, 14′ long, 6′ high		4	6		940	194		1,134	1,350
0590	Parallel bars, 10′ long		4	6		315	194		509	665
0600	Posts, tether ball set, 2-3/8″ O.D.		12	2		274	64.50		338.50	405
0800	Poles, multiple purpose, 10′-6″ long		12	2	Pr.	155	64.50		219.50	278
1000	Ground socket for movable posts, 2-3/8″ post		10	2.400		177	77.50		254.50	320
1100	3-1/2″ post		10	2.400		222	77.50		299.50	370
1300	See-saw, spring, steel, 2 units		6	4	Ea.	900	129		1,029	1,200
1400	4 units		4	6		1,350	194		1,544	1,825
1500	6 units		3	8		1,375	258		1,633	1,950
1700	Shelter, fiberglass golf tee, 3 person		4.60	5.217		3,300	168		3,468	3,900
1900	Slides, stainless steel bed, 12′ long, 6′ high		3	8		4,750	258		5,008	5,650
2000	20′ long, 10′ high		2	12		2,500	385		2,885	3,400
2200	Swings, plain seats, 8′ high, 4 seats		2	12		1,250	385		1,635	2,025
2300	8 seats		1.30	18.462		2,075	595		2,670	3,250
2500	12′ high, 4 seats		2	12		1,775	385		2,160	2,600
2600	8 seats		1.30	18.462		2,400	595		2,995	3,625
2800	Whirlers, 8′ diameter		3	8		3,925	258		4,183	4,725
2900	10′ diameter		3	8		4,300	258		4,558	5,150

11 68 13.20 Modular Playground		Crew	Daily Output	Labor-Hours	Unit	Material	Labor	Equipment	Total	Total Incl O&P
0010	**MODULAR PLAYGROUND** Basic components									
0100	Deck, square, steel, 48″ x 48″	B-1	1	24	Ea.	885	775		1,660	2,250
0110	Recycled polyurethane		1	24		880	775		1,655	2,250
0120	Triangular, steel, 48″ side		1	24		625	775		1,400	1,975
0130	Post, steel, 5″ square		18	1.333	L.F.	30	43		73	104
0140	Aluminum, 2-3/8″ square		20	1.200		25	38.50		63.50	91.50
0150	5″ square		18	1.333		33	43		76	108
0160	Roof, square poly, 54″ side		18	1.333	Ea.	1,250	43		1,293	1,450
0170	Wheelchair transfer module, for 3′ high deck		3	8	″	2,700	258		2,958	3,375
0180	Guardrail, pipe, 36″ high		60	.400	L.F.	125	12.90		137.90	160
0190	Steps, deck-to-deck, 3 – 8″ steps		8	3	Ea.	1,250	97		1,347	1,525
0200	Activity panel, crawl through panel		2	12		425	385		810	1,100
0210	Alphabet/spelling panel		2	12		620	385		1,005	1,325
0360	With guardrails		3	8		2,275	258		2,533	2,925
0370	Crawl tunnel, straight, 56″ long		4	6		1,250	194		1,444	1,700
0380	90°, 4′ long		4	6		1,250	194		1,444	1,700
1200	Slide, tunnel, for 56″ high deck		8	3		1,875	97		1,972	2,200
1210	Straight, poly		8	3		2,225	97		2,322	2,600
1220	Stainless steel, 54″ high deck		6	4		268	129		397	505
1230	Curved, poly, 40″ high deck		6	4		575	129		704	850
1240	Spiral slide, 56″ - 72″ high		5	4.800		2,750	155		2,905	3,275
1300	Ladder, vertical, for 24″ - 72″ high deck		5	4.800		620	155		775	935
1310	Horizontal, 8′ long		5	4.800		735	155		890	1,075
1320	Corkscrew climber, 6′ high		3	8		530	258		788	1,000
1330	Fire pole for 72″ high deck		6	4		575	129		704	850
1340	Bridge, ring climber, 8′ long		4	6		1,150	194		1,344	1,575
1350	Suspension		4	6	L.F.	530	194		724	905

11 68 Play Field Equipment and Structures

11 68 16 – Play Structures

11 68 16.10 Handball/Squash Court

	11 68 16.10 Handball/Squash Court	Crew	Daily Output	Labor-Hours	Unit	2009 Bare Costs Material	Labor	Equipment	Total	Total Incl O&P
0010	**HANDBALL/SQUASH COURT**, outdoor									
0900	Handball or squash court, outdoor, wood	2 Carp	.50	32	Ea.	4,425	1,275		5,700	6,975
1000	Masonry handball/squash court	D-1	.30	53.333	"	24,500	1,950		26,450	30,100

11 68 16.30 Platform/Paddle Tennis Court

	11 68 16.30 Platform/Paddle Tennis Court	Crew	Daily Output	Labor-Hours	Unit	Material	Labor	Equipment	Total	Total Incl O&P
0010	**PLATFORM/PADDLE TENNIS COURT** Complete with lighting, etc.									
0100	Aluminum slat deck with aluminum frame	B-1	.08	300	Court	51,000	9,675		60,675	72,000
0500	Aluminum slat deck with wood frame	C-1	.12	266		51,000	10,100		61,100	72,500
0800	Aluminum deck heater, add	B-1	1.18	20.339		5,050	655		5,705	6,625
0900	Douglas fir planking with wood frame 2" x 6" x 30'	C-1	.12	266		48,000	10,100		58,100	69,500
1000	Plywood deck with steel frame		.12	266		48,000	10,100		58,100	69,500
1100	Steel slat deck with wood frame	↓	.12	266	↓	34,000	10,100		44,100	54,000

11 68 33 – Athletic Field Equipment

11 68 33.13 Football Field Equipment

	11 68 33.13 Football Field Equipment	Crew	Daily Output	Labor-Hours	Unit	Material	Labor	Equipment	Total	Total Incl O&P
0010	**FOOTBALL FIELD EQUIPMENT**									
0020	Goal posts, steel, football, double post	B-1	1.50	16	Pr.	2,675	515		3,190	3,800
0100	Deluxe, single post		1.50	16		4,550	515		5,065	5,850
0300	Football, convertible to soccer		1.50	16		3,550	515		4,065	4,775
0500	Soccer, regulation	↓	2	12	↓	2,725	385		3,110	3,625

11 71 Medical Sterilizing Equipment

11 71 10 – Medical Sterilizers & Distillers

11 71 10.10 Sterilizers and Distillers

	11 71 10.10 Sterilizers and Distillers	Crew	Daily Output	Labor-Hours	Unit	Material	Labor	Equipment	Total	Total Incl O&P
0010	**STERILIZERS AND DISTILLERS**									
0700	Distiller, water, steam heated, 50 gal. capacity	1 Plum	1.40	5.714	Ea.	17,600	279		17,879	19,700
5600	Sterilizers, floor loading, 26" x 62" x 42", single door, steam					151,000			151,000	166,000
5650	Double door, steam					194,000			194,000	213,500
5800	General purpose, 20" x 20" x 38", single door					14,600			14,600	16,000
6000	Portable, counter top, steam, minimum					3,625			3,625	3,975
6020	Maximum					5,650			5,650	6,225
6050	Portable, counter top, gas, 17" x 15" x 32-1/2"					37,400			37,400	41,200
6150	Manual washer/sterilizer, 16" x 16" x 26"	1 Plum	2	4	↓	51,500	195		51,695	57,000
6200	Steam generators, electric 10 kW to 180 kW, freestanding									
6250	Minimum	1 Elec	3	2.667	Ea.	7,625	125		7,750	8,575
6300	Maximum	"	.70	11.429		27,400	535		27,935	31,000
8200	Bed pan washer-sanitizer	1 Plum	2	4	↓	6,800	195		6,995	7,775

11 72 Examination and Treatment Equipment

11 72 13 – Examination Equipment

11 72 13.13 Examination Equipment

	11 72 13.13 Examination Equipment	Crew	Daily Output	Labor-Hours	Unit	Material	Labor	Equipment	Total	Total Incl O&P
0010	**EXAMINATION EQUIPMENT**									
0300	Blood pressure unit, mercurial, wall				Ea.	189			189	208
0400	Diagnostic set, wall					620			620	680
4400	Scale, physician's, with height rod				↓	305			305	335

11 72 53 – Treatment Equipment

11 72 53.13 Medical Treatment Equipment

	11 72 53.13 Medical Treatment Equipment	Crew	Daily Output	Labor-Hours	Unit	Material	Labor	Equipment	Total	Total Incl O&P
0010	**MEDICAL TREATMENT EQUIPMENT**									
6500	Surgery table, minor minimum	1 Sswk	.70	11.429	Ea.	13,200	510		13,710	15,400

11 72 Examination and Treatment Equipment

11 72 53 – Treatment Equipment

11 72 53.13 Medical Treatment Equipment		Crew	Daily Output	Labor-Hours	Unit	Material	2009 Bare Costs Labor	Equipment	Total	Total Incl O&P
6520	Maximum	1 Sswk	.50	16	Ea.	24,800	715		25,515	28,600
6700	Surgical lights, doctor's office, single arm	2 Elec	2	8		1,225	375		1,600	1,925
6750	Dual arm	"	1	16	↓	5,525	750		6,275	7,225

11 73 Patient Care Equipment

11 73 10 – Patient Treatment Equipment

11 73 10.10 Treatment Equipment		Crew	Daily Output	Labor-Hours	Unit	Material	Labor	Equipment	Total	Total Incl O&P
0010	**TREATMENT EQUIPMENT**									
0750	Exam room furnishings, average per room				Ea.	6,150			6,150	6,750
1800	Heat therapy unit, humidified, 26" x 78" x 28"				"	3,225			3,225	3,550
2100	Hubbard tank with accessories, stainless steel,									
2110	125 GPM at 45 psi water pressure				Ea.	25,100			25,100	27,600
2150	For electric overhead hoist, add					2,725			2,725	3,000
2900	K-Module for heat therapy, 20 oz. capacity, 75° to 110° F					385			385	425
3600	Paraffin bath, 126° F, auto controlled					1,400			1,400	1,525
3900	Parallel bars for walking training, 12'-0"					1,150			1,150	1,275
4600	Station, dietary, medium, with ice					15,300			15,300	16,800
4700	Medicine					6,900			6,900	7,575
7000	Tables, physical therapy, walk off, electric	2 Carp	3	5.333		3,050	213		3,263	3,700
7150	Standard, vinyl top with base cabinets, minimum		3	5.333		1,250	213		1,463	1,725
7200	Maximum	↓	2	8		3,675	320		3,995	4,575
8400	Whirlpool bath, mobile, sst, 18" x 24" x 60"					4,275			4,275	4,725
8450	Fixed, incl. mixing valves	1 Plum	2	4	↓	8,775	195		8,970	9,950

11 74 Dental Equipment

11 74 10 – Dental Office Equipment

11 74 10.10 Diagnostic and Treatment Equipment		Crew	Daily Output	Labor-Hours	Unit	Material	Labor	Equipment	Total	Total Incl O&P
0010	**DIAGNOSTIC AND TREATMENT EQUIPMENT**									
0020	Central suction system, minimum	1 Plum	1.20	6.667	Ea.	1,425	325		1,750	2,050
0100	Maximum	"	.90	8.889		3,700	435		4,135	4,750
0300	Air compressor, minimum	1 Skwk	.80	10		2,975	410		3,385	3,950
0400	Maximum		.50	16		6,850	655		7,505	8,625
0600	Chair, electric or hydraulic, minimum		.50	16		2,675	655		3,330	4,000
0700	Maximum	↓	.25	32		13,700	1,300		15,000	17,200
0800	Doctor's/assistant's stool, minimum					139			139	153
0850	Maximum					770			770	850
1000	Drill console with accessories, minimum	1 Skwk	1.60	5		1,500	204		1,704	1,975
1100	Maximum		1.60	5		4,750	204		4,954	5,550
2000	Light, ceiling mounted, minimum		8	1		1,400	41		1,441	1,600
2100	Maximum	↓	8	1		1,900	41		1,941	2,150
2200	Unit light, minimum	2 Skwk	5.33	3.002		600	123		723	860
2210	Maximum		5.33	3.002		1,625	123		1,748	2,000
2220	Track light, minimum		3.20	5		2,050	204		2,254	2,600
2230	Maximum	↓	3.20	5		3,550	204		3,754	4,225
2300	Sterilizers, steam portable, minimum					1,225			1,225	1,350
2350	Maximum					9,900			9,900	10,900
2600	Steam, institutional					4,250			4,250	4,675
2650	Dry heat, electric, portable, 3 trays					1,150			1,150	1,275
2700	Ultra-sonic cleaner, portable, minimum				↓	244			244	268

11 74 Dental Equipment

11 74 10 – Dental Office Equipment

11 74 10.10 Diagnostic and Treatment Equipment		Crew	Daily Output	Labor-Hours	Unit	2009 Bare Costs Material	Labor	Equipment	Total	Total Incl O&P
2750	Maximum (institutional)				Ea.	1,450			1,450	1,600
3000	X-ray unit, wall, minimum	1 Skwk	4	2		2,200	81.50		2,281.50	2,525
3010	Maximum		4	2		4,900	81.50		4,981.50	5,500
3100	Panoramic unit		.60	13.333		13,900	545		14,445	16,200
3105	Deluxe, minimum	2 Skwk	1.60	10		16,000	410		16,410	18,300
3110	Maximum	"	1.60	10		51,500	410		51,910	57,000
3500	Developers, X-ray, average	1 Plum	5.33	1.501		4,725	73		4,798	5,325
3600	Maximum	"	5.33	1.501		7,700	73		7,773	8,600

11 76 Operating Room Equipment

11 76 10 – Operating Room Equipment

11 76 10.10 Surgical Equipment

		Crew	Daily Output	Labor-Hours	Unit	Material	Labor	Equipment	Total	Total Incl O&P
0010	**SURGICAL EQUIPMENT**									
5000	Scrub, surgical, stainless steel, single station, minimum	1 Plum	3	2.667	Ea.	3,925	130		4,055	4,525
5100	Maximum					7,900			7,900	8,700
6550	Major surgery table, minimum	1 Sswk	.50	16		29,900	715		30,615	34,200
6570	Maximum	"	.50	16		99,000	715		99,715	110,000
6800	Surgical lights, major operating room, dual head, minimum	2 Elec	1	16		16,600	750		17,350	19,500
6850	Maximum	"	1	16		27,700	750		28,450	31,700

11 77 Radiology Equipment

11 77 10 – Radiology Equipment

11 77 10.10 X-Ray Equipment

		Crew	Daily Output	Labor-Hours	Unit	Material	Labor	Equipment	Total	Total Incl O&P
0010	**X-RAY EQUIPMENT**									
8700	X-ray, mobile, minimum				Ea.	12,800			12,800	14,000
8750	Maximum					71,500			71,500	79,000
8900	Stationary, minimum					39,900			39,900	43,800
8950	Maximum					207,000			207,000	228,000
9150	Developing processors, minimum					11,100			11,100	12,200
9200	Maximum					29,100			29,100	32,000

11 78 Mortuary Equipment

11 78 13 – Mortuary Refrigerators

11 78 13.10 Mortuary and Autopsy Equipment

		Crew	Daily Output	Labor-Hours	Unit	Material	Labor	Equipment	Total	Total Incl O&P
0010	**MORTUARY AND AUTOPSY EQUIPMENT**									
0015	Autopsy table, standard	1 Plum	1	8	Ea.	8,450	390		8,840	9,875
0020	Deluxe	"	.60	13.333		16,800	650		17,450	19,400
3200	Mortuary refrigerator, end operated, 2 capacity					12,400			12,400	13,600
3300	6 capacity					22,200			22,200	24,500

11 78 16 – Crematorium Equipment

11 78 16.10 Crematory

		Crew	Daily Output	Labor-Hours	Unit	Material	Labor	Equipment	Total	Total Incl O&P
0010	**CREMATORY**									
1500	Crematory, not including building, 1 place	Q-3	.20	160	Ea.	59,500	7,425		66,925	77,000
1750	2 place	"	.10	320	"	85,000	14,900		99,900	116,500

11 82 Solid Waste Handling Equipment

11 82 19 – Packaged Incinerators

11 82 19.10 Packaged Gas Fired Incinerators		Crew	Daily Output	Labor-Hours	Unit	2009 Bare Costs Material	Labor	Equipment	Total	Total Incl O&P
0010	PACKAGED GAS FIRED INCINERATORS									
4400	Incinerator, gas, not incl. chimney, elec. or pipe, 50#/hr., minimum	Q-3	.80	40	Ea.	11,200	1,850		13,050	15,200
4420	Maximum		.70	45.714		35,000	2,125		37,125	41,800
4440	200 lb. per hr., minimum (batch type)		.60	53.333		62,000	2,475		64,475	72,000
4460	Maximum (with feeder)		.50	64		85,000	2,975		87,975	98,000
4480	400 lb. per hr., minimum (batch type)		.30	106		68,500	4,950		73,450	83,000
4500	Maximum (with feeder)		.25	128		90,000	5,950		95,950	108,000
4520	800 lb. per hr., with feeder, minimum		.20	160		112,000	7,425		119,425	134,500
4540	Maximum		.17	188		150,000	8,750		158,750	178,500
4560	1,200 lb. per hr., with feeder, minimum		.15	213		179,500	9,900		189,400	212,500
4580	Maximum		.11	290		146,000	13,500		159,500	181,500
4600	2,000 lb. per hr., with feeder, minimum		.10	320		504,000	14,900		518,900	577,500
4620	Maximum		.05	640		357,000	29,700		386,700	438,500
4700	For heat recovery system, add, minimum		.25	128		69,000	5,950		74,950	85,000
4710	Add, maximum		.11	290		220,500	13,500		234,000	263,500
4720	For automatic ash conveyer, add		.50	64		29,000	2,975		31,975	36,500
4750	Large municipal incinerators, incl. stack, minimum		.25	128	Ton/day	17,600	5,950		23,550	28,600
4850	Maximum		.10	320	"	46,800	14,900		61,700	74,500

11 82 26 – Waste Compactors and Destructors

11 82 26.10 Compactors		Crew	Daily Output	Labor-Hours	Unit	Material	Labor	Equipment	Total	Total Incl O&P
0010	COMPACTORS									
0020	Compactors, 115 volt, 250#/hr., chute fed	L-4	1	24	Ea.	10,900	895		11,795	13,500
0100	Hand fed		2.40	10		7,975	375		8,350	9,375
0300	Multi-bag, 230 volt, 600#/hr, chute fed		1	24		9,900	895		10,795	12,400
0400	Hand fed		1	24		8,500	895		9,395	10,800
0500	Containerized, hand fed, 2 to 6 C.Y. containers, 250#/hr.		1	24		10,500	895		11,395	13,100
0550	For chute fed, add per floor		1	24		1,200	895		2,095	2,800
1000	Heavy duty industrial compactor, 0.5 C.Y. capacity		1	24		7,100	895		7,995	9,275
1050	1.0 C.Y. capacity		1	24		10,800	895		11,695	13,400
1100	3 C.Y. capacity		.50	48		14,800	1,800		16,600	19,100
1150	5.0 C.Y. capacity		.50	48		18,200	1,800		20,000	23,000
1200	Combination shredder/compactor (5,000 lbs./hr.)		.50	48		36,100	1,800		37,900	42,600
1400	For handling hazardous waste materials, 55 gallon drum packer, std.					16,800			16,800	18,500
1410	55 gallon drum packer w/HEPA filter					20,900			20,900	23,000
1420	55 gallon drum packer w/charcoal & HEPA filter					28,000			28,000	30,800
1430	All of the above made explosion proof, add					12,800			12,800	14,000
5500	Shredder, municipal use, 35 ton per hour					262,000			262,000	288,500
5600	60 ton per hour					558,500			558,500	614,500
5750	Shredder & baler, 50 ton per day					523,500			523,500	576,000
5800	Shredder, industrial, minimum					20,600			20,600	22,700
5850	Maximum					110,500			110,500	121,500
5900	Baler, industrial, minimum					8,250			8,250	9,075
5950	Maximum					482,500			482,500	531,000
6000	Transfer station compactor, with power unit									
6050	and pedestal, not including pit, 50 ton per hour				Ea.	165,000			165,000	181,500

11 91 Religious Equipment

11 91 13 – Baptistries

11 91 13.10 Baptistry		Crew	Daily Output	Labor-Hours	Unit	Material	Labor	Equipment	Total	Total Incl O&P
						2009 Bare Costs				
0010	**BAPTISTRY**									
0150	Fiberglass, 3′-6″ deep, x 13′-7″ long,									
0160	steps at both ends, incl. plumbing, minimum	L-8	1	20	Ea.	3,825	835		4,660	5,550
0200	Maximum	″	.70	28.571		6,250	1,200		7,450	8,825
0250	Add for filter, heater and lights				↓	1,375			1,375	1,500

11 91 23 – Sanctuary Equipment

11 91 23.10 Sanctuary Furnishings		Crew	Daily Output	Labor-Hours	Unit	Material	Labor	Equipment	Total	Total Incl O&P
0010	**SANCTUARY FURNISHINGS**									
0020	Altar, wood, custom design, plain	1 Carp	1.40	5.714	Ea.	1,925	228		2,153	2,475
0050	Deluxe	″	.20	40		9,225	1,600		10,825	12,900
0070	Granite or marble, average	2 Marb	.50	32		10,700	1,225		11,925	13,700
0090	Deluxe	″	.20	80		30,400	3,100		33,500	38,400
0100	Arks, prefabricated, plain	2 Carp	.80	20		8,025	800		8,825	10,200
0130	Deluxe, maximum	″	.20	80		114,000	3,200		117,200	131,000
0500	Reconciliation room, wood, prefabricated, single, plain	1 Carp	.60	13.333		2,425	535		2,960	3,550
0550	Deluxe		.40	20		6,675	800		7,475	8,675
0650	Double, plain		.40	20		4,850	800		5,650	6,675
0700	Deluxe		.20	40		15,200	1,600		16,800	19,400
1000	Lecterns, wood, plain		5	1.600		750	64		814	930
1100	Deluxe		2	4		5,050	160		5,210	5,825
2000	Pulpits, hardwood, prefabricated, plain		2	4		1,225	160		1,385	1,625
2100	Deluxe		1.60	5	↓	8,325	200		8,525	9,475
2500	Railing, hardwood, average		25	.320	L.F.	166	12.80		178.80	204
3000	Seating, individual, oak, contour, laminated		21	.381	Person	147	15.20		162.20	186
3100	Cushion seat		21	.381		133	15.20		148.20	171
3200	Fully upholstered		21	.381		130	15.20		145.20	168
3300	Combination, self-rising	↓	21	.381		335	15.20		350.20	390
3500	For cherry, add				↓	30%				
5000	Wall cross, aluminum, extruded, 2″ x 2″ section	1 Carp	34	.235	L.F.	132	9.40		141.40	161
5150	4″ x 4″ section		29	.276		190	11		201	227
5300	Bronze, extruded, 1″ x 2″ section		31	.258		260	10.30		270.30	305
5350	2-1/2″ x 2-1/2″ section		34	.235		395	9.40		404.40	450
5450	Solid bar stock, 1/2″ x 3″ section		29	.276		515	11		526	590
5600	Fiberglass, stock		34	.235		107	9.40		116.40	133
5700	Stainless steel, 4″ deep, channel section		29	.276		420	11		431	480
5800	4″ deep box section	↓	29	.276	↓	570	11		581	645

Division Notes

	CREW	DAILY OUTPUT	LABOR-HOURS	UNIT	2009 BARE COSTS MAT.	LABOR	EQUIP.	TOTAL	TOTAL INCL O&P

Estimating Tips

General

- The items in this division are usually priced per square foot or each. Most of these items are purchased by the owner and placed by the supplier.
  Do not assume the items in Division 12 will be purchased and installed by the supplier. Check the specifications for responsibilities and include receiving, storage, installation, and mechanical and electrical hookups in the appropriate divisions.
- Some items in this division require some type of support system that is not usually furnished with the item. Examples of these systems include blocking for the attachment of casework and heavy drapery rods. The required blocking must be added to the estimate in the appropriate division.

Reference Numbers

Reference numbers are shown in shaded boxes at the beginning of some major classifications. These numbers refer to related items in the Reference Section. The reference information may be an estimating procedure, an alternate pricing method, or technical information.

Note: Not all subdivisions listed here necessarily appear in this publication.

12 05 Common Work Results for Furnishings

12 05 05 – SELECTIVE DEMOLITION

12 05 05.10 Selective Demolition, Interiors		Crew	Daily Output	Labor-Hours	Unit	Material (2009 Bare Costs)	Labor (2009 Bare Costs)	Equipment (2009 Bare Costs)	Total (2009 Bare Costs)	Total Incl O&P
0010	SELECTIVE DEMOLITION, INTERIORS R024119-10									
3100	Casework, metal base cabinets	2 Clab	20	.800	L.F.		25.50		25.50	41.50
3110	Cabinet base trim		400	.040			1.26		1.26	2.09
3120	Countertop, stainless steel		80	.200			6.30		6.30	10.45
3130	Wall cabinets, wood, 84" high		30	.533			16.85		16.85	28
3500	Laboratory casework, tall storage cabinets, 84" high		40	.400			12.65		12.65	21
3510	Wall cabinets, metal		40	.400			12.65		12.65	21
4830	Floor mats, recessed or link mats	1 Clab	300	.027	S.F.		.84		.84	1.39
4832	Skate lock tile		200	.040			1.26		1.26	2.09
4834	Duckboard		300	.027			.84		.84	1.39
4920	Blinds, interior, horizontal or vertical		150	.053	L.F.		1.69		1.69	2.78
4922	Wood folding panels		35	.229	Pr.		7.20		7.20	11.90
4924	Shades, interior		700	.011	S.F.		.36		.36	.60
4930	Drapery hardware, traverse rods		35	.229	Ea.		7.20		7.20	11.90
5200	Fixed seating, per seat	2 Carp	44	.364	"		14.55		14.55	24
6400	Booth, restaurant	2 Clab	80	.200	L.F.		6.30		6.30	10.45
7400	Office systems furniture, cubicle	"	3000	.005	S.F.		.17		.17	.28

12 05 13 – Fabrics

12 05 13.10 Upholstery Materials		Crew	Daily Output	Labor-Hours	Unit	Material	Labor	Equipment	Total	Total Incl O&P
0010	UPHOLSTERY MATERIALS									
1000	Fabrics, fabric blends, minimum				S.Y.	34			34	37.50
1020	Maximum					84			84	92.50
1200	Nylons, minimum					31.50			31.50	34.50
1220	Maximum					47			47	51.50
1400	Polyester, minimum					24.50			24.50	27
1420	Maximum					59.50			59.50	65.50
1600	Silk, minimum					76			76	83.50
1620	Maximum					131			131	144
1800	Wool, minimum					54.50			54.50	60
1820	Maximum					99			99	109
3000	Treatment, acoustic, 1/8" flame retardant					9.35			9.35	10.25
3100	Acrylic backing					4.18			4.18	4.60
3200	Flame proofing					8			8	8.80
3300	Paper backing					7.10			7.10	7.80
3400	Scotchguard or teflon					2.47			2.47	2.72
5000	Leather, minimum				S.F.	8.95			8.95	9.85
5020	Maximum				"	10.30			10.30	11.35
6000	Vinyl, minimum				S.Y.	22.50			22.50	24.50
6020	Maximum				"	36.50			36.50	40

12 12 Wall Decorations

12 12 19 – Framed Prints

12 12 19.10 Art Work		Crew	Daily Output	Labor-Hours	Unit	Material	Labor	Equipment	Total	Total Incl O&P
0010	ART WORK framed									
1000	Photography, minimum	1 Carp	36	.222	Ea.	74.50	8.90		83.40	96.50
1050	Maximum		30	.267		495	10.65		505.65	565
2000	Posters, minimum		36	.222		40	8.90		48.90	58.50
2050	Maximum		30	.267		565	10.65		575.65	645
3000	Reproductions, minimum		36	.222		75	8.90		83.90	97
3050	Maximum		30	.267		695	10.65		705.65	785

12 21 Window Blinds

12 21 13 – Horizontal Louver Blinds

12 21 13.13 Metal Horizontal Louver Blinds		Crew	Daily Output	Labor-Hours	Unit	2009 Bare Costs Material	Labor	Equipment	Total	Total Incl O&P
0010	**METAL HORIZONTAL LOUVER BLINDS**									
0020	Horizontal, 1" aluminum slats, solid color, stock	1 Carp	590	.014	S.F.	4.75	.54		5.29	6.15
0090	Custom, minimum		590	.014		5.25	.54		5.79	6.70
0100	Maximum		440	.018		7.15	.73		7.88	9.05
0250	2" aluminum slats, custom, minimum		590	.014		8.10	.54		8.64	9.85
0350	Maximum		440	.018		13.15	.73		13.88	15.70
0450	Stock, minimum		590	.014		4.86	.54		5.40	6.25
0500	Maximum		440	.018		7.90	.73		8.63	9.90
0600	2" steel slats, stock, minimum		590	.014		1.84	.54		2.38	2.91
0630	Maximum		440	.018		5.15	.73		5.88	6.90
0750	Custom, minimum		590	.014		1.67	.54		2.21	2.73
0850	Maximum		400	.020		8.25	.80		9.05	10.40
1000	Alternate method of figuring:									
1300	1" aluminum slats, 48" wide, 48" high	1 Carp	30	.267	Ea.	50.50	10.65		61.15	73
1320	72" high		29	.276		67	11		78	91.50
1340	96" high		28	.286		74	11.40		85.40	100
1400	72" wide, 72" high		25	.320		105	12.80		117.80	137
1420	96" high		23	.348		106	13.90		119.90	140
1480	96" wide, 96" high		20	.400		149	16		165	191
1490	For special colors, add					12%				

12 21 16 – Vertical Louver Blinds

12 21 16.13 Metal Vertical Louver Blinds		Crew	Daily Output	Labor-Hours	Unit	Material	Labor	Equipment	Total	Total Incl O&P
0010	**METAL VERTICAL LOUVER BLINDS**									
1500	Vertical, 3" to 5" PVC or cloth strips, minimum	1 Carp	460	.017	S.F.	5.40	.69		6.09	7.10
1600	Maximum		400	.020		11.75	.80		12.55	14.20
1800	4" aluminum slats, minimum		460	.017		7.20	.69		7.89	9.10
1900	Maximum		400	.020		9.10	.80		9.90	11.30
1950	Mylar mirror-finish strips, to 8" wide, minimum		460	.017		4.62	.69		5.31	6.25
1970	Maximum		400	.020		10.50	.80		11.30	12.85
1990	Alternate method of figuring:									
2000	2" aluminum slats, 48" wide, 48" high	1 Carp	30	.267	Ea.	86	10.65		96.65	112
2050	72" high		29	.275		110	11		121	139
2100	96" high		28	.285		134	11.40		145.40	167
2200	72" wide, 72" high		25	.320		133	12.80		145.80	168
2250	96" high		23	.343		161	13.90		174.90	200
2300	96" wide, 96" high		20	.400		217	16		233	266
2500	Mirror finish, 48" wide, 48" high		30	.267		87	10.65		97.65	113
2550	72" high		29	.275		105	11		116	133
2600	96" high		28	.286		124	11.40		135.40	156
2650	72" wide, 72" high		25	.320		147	12.80		159.80	183
2700	96" high		23	.348		176	13.90		189.90	217
2750	96" wide, 96" high		20	.400		234	16		250	284
2800	Decorative printed finish, 48" wide, 48" high		30	.267		85.50	10.65		96.15	112
2850	72" high		29	.276		104	11		115	132
2900	96" high		28	.286		123	11.40		134.40	154
2950	72" wide, 72" high		25	.320		144	12.80		156.80	179
2980	96" high		23	.348		179	13.90		192.90	220
2990	96" wide, 96" high		20	.400		234	16		250	284

12 22 Curtains and Drapes

12 22 13 – Draperies

12 22 13.10 Custom Draperies		Crew	Daily Output	Labor-Hours	Unit	Material	2009 Bare Costs Labor	Equipment	Total	Total Incl O&P
0010	**CUSTOM DRAPERIES** (Material only)									
0050	Lined, minimum				S.F.	5			5	5.50
0100	Maximum					14.25			14.25	15.70
0200	Unlined, minimum					2.77			2.77	3.05
0300	Maximum					9.35			9.35	10.25
0400	Lightproof type, add, minimum					1.39			1.39	1.53
0500	Maximum					8.30			8.30	9.15
0600	Fire resistant, add, minimum					1.06			1.06	1.17
0700	Maximum					3.59			3.59	3.95
0800	Valances, pleated, 10" to 18" depth, minimum				L.F.	1.55			1.55	1.71
0900	Maximum				"	8.95			8.95	9.85
2000	Alternate method, lined overlaps and returns, average									
2020	32" thru 48" wide, 26" to 39" long				Ea.	153			153	168
2040	40" to 63" long					155			155	170
2060	64" to 72" long					159			159	175
2080	73" to 81" long					180			180	198
2100	82" to 90" long					186			186	204
2200	91" to 99" long					199			199	219
2400	100" to 108" long					208			208	229
2600	109" to 120" long					219			219	241
2800	121" to 130" long					223			223	245
3000	48" thru 72" wide, 26" to 39" long					223			223	245
3050	40" to 63" long					238			238	262
3100	64" to 72" long					249			249	274
3150	73" to 81" long					261			261	287
3200	82" to 90" long					286			286	315
3250	91" to 99" long					296			296	325
3300	100" to 108" long					305			305	340
3350	109" to 120" long					330			330	365
3400	121" to 130" long					355			355	390
3450	64" thru 96" wide, 26" to 39" long					297			297	325
3500	40" to 63" long					320			320	350
3550	64" to 72" long					330			330	365
3600	73" to 81" long					355			355	390
3650	82" to 90" long					375			375	415
3700	91" to 99" long					400			400	440
3750	100" to 108" long					410			410	450
3800	109" to 120" long					435			435	480
3820	121" to 130" long					465			465	515
3840	80" thru 120" wide, 26" to 39" long					375			375	415
3860	40" to 63" long					400			400	440
3880	64"to 72" long					410			410	450
4000	73" to 81" long					445			445	490
4020	82" to 90" long					465			465	510
4080	91" to 99" long					490			490	540
4100	100" to 108" long					525			525	575
4150	109" to 120" long					540			540	595
4200	121" to 130" long					565			565	625
4250	96" thru 144" wide, 26" to 39" long					445			445	490
4300	40" to 63" long					475			475	525
4400	64" to 72" long					490			490	535
4500	73" to 81" long					520			520	570
4600	82" to 90" long					560			560	615

12 22 13 – Draperies

12 22 13.10 Custom Draperies		Crew	Daily Output	Labor-Hours	Unit	Material (2009 Bare Costs)	Labor (2009 Bare Costs)	Equipment (2009 Bare Costs)	Total (2009 Bare Costs)	Total Incl O&P
4700	91" to 99" long				Ea.	590			590	650
4800	100" to 108" long					625			625	685
4900	109" to 120" long					660			660	725
5000	121" to 130" long					705			705	775
5100	112" thru 168" wide, 26" to 39" long					525			525	575
5200	40" to 63" long					560			560	615
5300	64" to 72" long					580			580	640
5400	73" to 81" long					615			615	675
5500	82" to 90" long					650			650	715
5600	91" to 99" long					690			690	760
5700	100" to 108" long					730			730	805
5800	109" to 120" long					820			820	900
6000	121" to 130" long					875			875	965
6100	128" thru 192" wide, 26" to 39" long					590			590	650
6200	40" to 63" long					640			640	700
6300	64" to 72" long					660			660	730
6400	73" to 81" long					705			705	775
6450	82" to 90" long					740			740	815
6500	91" to 99" long					785			785	865
6550	100" to 108" long					830			830	915
6600	109" to 120" long					880			880	965
6650	121" to 130" long					935			935	1,025
6700	144" thru 216" wide, 26" to 39" long					675			675	740
6750	40" to 63" long					740			740	815
6800	64" to 72" long					790			790	865
6850	73" to 81" long					810			810	890
6880	82" to 90" long					845			845	930
6900	91" to 99" long					890			890	980
6920	100" to 108" long					935			935	1,025
6980	109" to 120" long					980			980	1,075
7000	121" to 130" long					1,050			1,050	1,175
7100	160" thru 240" wide, 26" to 39" long					740			740	815
7150	40" to 63" long					800			800	880
7200	64" to 72" long					820			820	900
7250	73" to 81" long					880			880	965
7300	82" to 90" long					935			935	1,025
7350	91" to 99" long					990			990	1,100
7400	100" to 108" long					1,025			1,025	1,150
7500	109" to 120" long					1,100			1,100	1,200
7600	121" to 130" long				↓	1,175			1,175	1,300
8800	Drapery installation, hardware & drapes,									
9000	Labor cost only, minimum	1 Clab	75	.107	L.F.		3.37		3.37	5.55
9100	Maximum	"	20	.400	"		12.65		12.65	21

12 22 16 – Drapery Track and Accessories

12 22 16.10 Drapery Hardware		Crew	Daily Output	Labor-Hours	Unit	Material	Labor	Equipment	Total	Total Incl O&P
0010	**DRAPERY HARDWARE**									
0030	Standard traverse, per foot, minimum	1 Carp	59	.136	L.F.	3.35	5.40		8.75	12.65
0100	Maximum		51	.157	"	7.90	6.25		14.15	19
0200	Decorative traverse, 28"-48", minimum		22	.364	Ea.	25	14.55		39.55	51.50
0220	Maximum		21	.381		43	15.20		58.20	72.50
0300	48"-84", minimum		20	.400		20.50	16		36.50	49
0320	Maximum	↓	19	.421	↓	58.50	16.80		75.30	92

12 22 Curtains and Drapes

12 22 16 – Drapery Track and Accessories

12 22 16.10 Drapery Hardware		Crew	Daily Output	Labor-Hours	Unit	Material	Labor	Equipment	Total	Total Incl O&P
						2009 Bare Costs				
0400	66"-120", minimum	1 Carp	18	.444	Ea.	62	17.75		79.75	97.50
0420	Maximum		17	.471		78	18.80		96.80	117
0500	84"-156", minimum		16	.500		65	20		85	105
0520	Maximum		15	.533		98.50	21.50		120	144
0600	130"-240", minimum		14	.571		31	23		54	71.50
0620	Maximum		13	.615		134	24.50		158.50	188
0700	Slide rings, each, minimum					1.41			1.41	1.55
0720	Maximum					2.32			2.32	2.55
3000	Ripplefold, snap-a-pleat system, 3' or less, minimum	1 Carp	15	.533		52	21.50		73.50	92.50
3020	Maximum	"	14	.571		70.50	23		93.50	116
3200	Each additional foot, add, minimum				L.F.	2.19			2.19	2.41
3220	Maximum				"	6.65			6.65	7.30
4000	Traverse rods, adjustable, 28" to 48"	1 Carp	22	.364	Ea.	20.50	14.55		35.05	46.50
4020	48" to 84"		20	.400		29	16		45	58.50
4040	66" to 120"		18	.444		40	17.75		57.75	73.50
4060	84" to 156"		16	.500		41	20		61	78
4080	100" to 180"		14	.571		42	23		65	83.50
4090	156" to 228"		13	.615		45	24.50		69.50	90.50
4100	228" to 312"		13	.615		58.50	24.50		83	105
4200	Double rods, adjustable, 30" to 48"		9	.889		37	35.50		72.50	99.50
4220	48" to 86"		9	.889		50	35.50		85.50	114
4240	86" to 150"		8	1		54.50	40		94.50	126
4260	100" to 180"		7	1.143		57	45.50		102.50	139
4300	Tray and curtain rod, adjustable, 30" to 48"		9	.889		22.50	35.50		58	83.50
4320	48" to 86"		9	.889		34	35.50		69.50	96
4340	86" to 150"		8	1		41	40		81	111
4360	100" to 180"		7	1.143		45.50	45.50		91	126
5000	Stationary rods, first 2'					7.45			7.45	8.20
5020	Each additional foot, add				L.F.	3.57			3.57	3.93

12 22 16.20 Blast Curtains

0010	**BLAST CURTAINS**, per LF horizontal opening width, off-white or gray fabric									
0100	Blast curtains, drapery system, complete, including hardware, minimum				L.F.				123	135
0120	Average								133	146
0140	Maximum								142	157

12 23 Interior Shutters

12 23 10 – Wood Interior Shutters

12 23 10.10 Wood Interior Shutters

0010	**WOOD INTERIOR SHUTTERS**, louvered									
0200	Two panel, 27" wide, 36" high	1 Carp	5	1.600	Set	114	64		178	231
0300	33" wide, 36" high		5	1.600		149	64		213	270
0500	47" wide, 36" high		5	1.600		199	64		263	325
1000	Four panel, 27" wide, 36" high		5	1.600		134	64		198	254
1100	33" wide, 36" high		5	1.600		165	64		229	287
1300	47" wide, 36" high		5	1.600		234	64		298	365

12 23 10.13 Wood Panels

0010	**WOOD PANELS**									
3000	Wood folding panels with movable louvers, 7" x 20" each	1 Carp	17	.471	Pr.	46	18.80		64.80	81.50
3300	8" x 28" each		17	.471		60	18.80		78.80	97
3450	9" x 36" each		17	.471		74	18.80		92.80	113

12 23 Interior Shutters

12 23 10 – Wood Interior Shutters

12 23 10.13 Wood Panels		Crew	Daily Output	Labor-Hours	Unit	Material	2009 Bare Costs Labor	Equipment	Total	Total Incl O&P
3600	10" x 40" each	1 Carp	17	.471	Pr.	80	13.80		98.80	119
4000	Fixed louver type, stock units, 8" x 20" each		17	.471		79.50	13.80		98.30	119
4150	10" x 28" each		17	.471		88	13.80		106.80	128
4300	12" x 36" each		17	.471		104	13.80		122.80	145
4450	18" x 40" each		17	.471		168	13.80		186.80	216
5000	Insert panel type, stock, 7" x 20" each		17	.471		19.25	13.80		38.05	52
5150	8" x 28" each		17	.471		35.50	13.80		54.30	70
5300	9" x 36" each		17	.471		45	13.80		63.80	80.50
5450	10" x 40" each		17	.471		48	13.80		66.80	84
5600	Raised panel type, stock, 10" x 24" each		17	.471		134	13.80		152.80	179
5650	12" x 26" each		17	.471		156	13.80		174.80	203
5700	14" x 30" each		17	.471		210	13.80		228.80	262
5750	16" x 36" each		17	.471		288	13.80		306.80	345
6000	For custom built pine, add					22%				
6500	For custom built hardwood blinds, add					42%				

12 24 Window Shades

12 24 13 – Roller Window Shades

12 24 13.10 Shades		Crew	Daily Output	Labor-Hours	Unit	Material	Labor	Equipment	Total	Total Incl O&P
0010	**SHADES**									
0020	Basswood, roll-up, stain finish, 3/8" slats	1 Carp	300	.027	S.F.	13.65	1.07		14.72	16.80
0200	7/8" slats		300	.027		12.90	1.07		13.97	15.95
0300	Vertical side slide, stain finish, 3/8" slats		300	.027		17.70	1.07		18.77	21
0400	7/8" slats		300	.027		17.70	1.07		18.77	21
0500	For fire retardant finishes, add					16%				
0600	For "B" rated finishes, add					20%				
0900	Mylar, single layer, non-heat reflective	1 Carp	685	.012		6.05	.47		6.52	7.45
1000	Double layered, heat reflective		685	.012		9.10	.47		9.57	10.75
1100	Triple layered, heat reflective		685	.012		10.55	.47		11.02	12.40
1200	For metal roller instead of wood, add per				Shade	4.12			4.12	4.53
1300	Vinyl coated cotton, standard	1 Carp	685	.012	S.F.	3.36	.47		3.83	4.47
1400	Lightproof decorator shades		685	.012		2.10	.47		2.57	3.08
1500	Vinyl, lightweight, 4 gauge		685	.012		.55	.47		1.02	1.38
1600	Heavyweight, 6 gauge		685	.012		1.67	.47		2.14	2.61
1700	Vinyl laminated fiberglass, 6 ga., translucent		685	.012		2.36	.47		2.83	3.37
1800	Lightproof		685	.012		3.83	.47		4.30	4.98
3000	Woven aluminum, 3/8" thick, lightproof and fireproof		350	.023		5.40	.91		6.31	7.40
5011	Insulative shades G		125	.064		10.75	2.56		13.31	16.05
6011	Solar screening, fiberglass G		85	.094		4.60	3.76		8.36	11.25
8011	Interior insulative shutter									
8111	Stock unit, 15" x 60" G	1 Carp	17	.471	Pr.	10.75	18.80		29.55	43

12 32 16 – Manufactured Plastic-Laminate-Clad Casework

12 32 16.20 Plastic Laminate Casework Doors

	12 32 16.20 Plastic Laminate Casework Doors	Crew	Daily Output	Labor-Hours	Unit	Material	Labor	Equipment	Total	Total Incl O&P
0010	**PLASTIC LAMINATE CASEWORK DOORS**									
1000	For casework frames, see Div. 12 32 23.15									
1100	For casework hardware, see Div. 12 32 23.35									
6000	Plastic laminate on particle board									
6100	12" wide, 18" high	1 Carp	25	.320	Ea.	10.50	12.80		23.30	32.50
6120	24" high		24	.333		14	13.30		27.30	37.50
6140	30" high		23	.348		17.50	13.90		31.40	42.50
6160	36" high		21	.381		21	15.20		36.20	48
6200	48" high		16	.500		28	20		48	64
6250	60" high		13	.615		35	24.50		59.50	79
6300	72" high		12	.667		42	26.50		68.50	90
6320	15" wide, 18" high		24.50	.327		13.10	13.05		26.15	36
6340	24" high		23.50	.340		17.50	13.60		31.10	42
6360	30" high		22.50	.356		22	14.20		36.20	47.50
6380	36" high		20.50	.390		26.50	15.60		42.10	54.50
6400	48" high		15.50	.516		35	20.50		55.50	72.50
6450	60" high		12.50	.640		44	25.50		69.50	90.50
6480	72" high		11.50	.696		52.50	28		80.50	104
6500	18" wide, 18" high		24	.333		15.75	13.30		29.05	39.50
6550	24" high		23	.348		21	13.90		34.90	46
6600	30" high		22	.364		26.50	14.55		41.05	53
6650	36" high		20	.400		31.50	16		47.50	61
6700	48" high		15	.533		42	21.50		63.50	81
6750	60" high		12	.667		52.50	26.50		79	102
6800	72" high	↓	11	.727	↓	63	29		92	118

(2009 Bare Costs: Material, Labor, Equipment, Total)

12 32 16.25 Plastic Laminate Drawer Fronts

	12 32 16.25 Plastic Laminate Drawer Fronts	Crew	Daily Output	Labor-Hours	Unit	Material	Labor	Equipment	Total	Total Incl O&P
0010	**PLASTIC LAMINATE DRAWER FRONTS**									
2800	Plastic laminate on particle board front									
3000	4" high, 12" wide	1 Carp	17	.471	Ea.	4.62	18.80		23.42	36
3200	18" wide		16	.500		6.95	20		26.95	40.50
3600	24" wide		15	.533		9.25	21.50		30.75	45
3800	6" high, 12" wide		16	.500		7	20		27	40.50
4000	18" wide		15	.533		10.50	21.50		32	46.50
4500	24" wide		14	.571		14	23		37	53
4800	9" high, 12" wide		15	.533		10.50	21.50		32	46.50
5000	18" wide		14	.571		15.75	23		38.75	55
5200	24" wide	↓	13	.615	↓	21	24.50		45.50	63.50

12 32 23 – Hardwood Casework

12 32 23.10 Manufactured Wood Casework, Stock Units

	12 32 23.10 Manufactured Wood Casework, Stock Units	Crew	Daily Output	Labor-Hours	Unit	Material	Labor	Equipment	Total	Total Incl O&P
0010	**MANUFACTURED WOOD CASEWORK, STOCK UNITS**									
0700	Kitchen base cabinets, hardwood, not incl. counter tops,									
0710	24" deep, 35" high, prefinished									
0800	One top drawer, one door below, 12" wide	2 Carp	24.80	.645	Ea.	227	26		253	293
0820	15" wide		24	.667		272	26.50		298.50	345
0840	18" wide		23.30	.687		297	27.50		324.50	370
0860	21" wide		22.70	.705		310	28		338	385
0880	24" wide		22.30	.717		360	28.50		388.50	450
1000	Four drawers, 12" wide		24.80	.645		335	26		361	410
1020	15" wide		24	.667		365	26.50		391.50	445
1040	18" wide		23.30	.687		385	27.50		412.50	465
1060	24" wide		22.30	.717		410	28.50		438.50	500
1200	Two top drawers, two doors below, 27" wide	↓	22	.727	↓	370	29		399	460

12 32 Manufactured Wood Casework

12 32 23 – Hardwood Casework

12 32 23.10 Manufactured Wood Casework, Stock Units		Crew	Daily Output	Labor-Hours	Unit	2009 Bare Costs Material	Labor	Equipment	Total	Total Incl O&P
1220	30" wide	2 Carp	21.40	.748	Ea.	370	30		400	460
1240	33" wide		20.90	.766		375	30.50		405.50	465
1260	36" wide		20.30	.788		390	31.50		421.50	480
1280	42" wide		19.80	.808		420	32.50		452.50	515
1300	48" wide		18.90	.847		450	34		484	550
1500	Range or sink base, two doors below, 30" wide		21.40	.748		300	30		330	380
1520	33" wide		20.90	.766		320	30.50		350.50	405
1540	36" wide		20.30	.788		340	31.50		371.50	420
1560	42" wide		19.80	.808		360	32.50		392.50	455
1580	48" wide		18.90	.847		380	34		414	470
1800	For sink front units, deduct					60			60	66
2000	Corner base cabinets, 36" wide, standard	2 Carp	18	.889		630	35.50		665.50	750
2100	Lazy Susan with revolving door	"	16.50	.970		540	38.50		578.50	660
4000	Kitchen wall cabinets, hardwood, 12" deep with two doors									
4050	12" high, 30" wide	2 Carp	24.80	.645	Ea.	197	26		223	260
4100	36" wide		24	.667		228	26.50		254.50	295
4400	15" high, 30" wide		24	.667		198	26.50		224.50	262
4420	33" wide		23.30	.687		223	27.50		250.50	291
4440	36" wide		22.70	.705		243	28		271	315
4450	42" wide		22.70	.705		284	28		312	355
4700	24" high, 30" wide		23.30	.687		310	27.50		337.50	385
4720	36" wide		22.70	.705		345	28		373	425
4740	42" wide		22.30	.717		335	28.50		363.50	415
5000	30" high, one door, 12" wide		22	.727		191	29		220	259
5020	15" wide		21.40	.748		210	30		240	281
5040	18" wide		20.90	.766		234	30.50		264.50	310
5060	24" wide		20.30	.788		271	31.50		302.50	350
5300	Two doors, 27" wide		19.80	.808		289	32.50		321.50	375
5320	30" wide		19.30	.829		320	33		353	405
5340	36" wide		18.80	.851		360	34		394	450
5360	42" wide		18.50	.865		385	34.50		419.50	480
5380	48" wide		18.40	.870		400	34.50		434.50	500
6000	Corner wall, 30" high, 24" wide		18	.889		209	35.50		244.50	289
6050	30" wide		17.20	.930		260	37		297	350
6100	36" wide		16.50	.970		290	38.50		328.50	385
6500	Revolving Lazy Susan		15.20	1.053		390	42		432	500
7000	Broom cabinet, 84" high, 24" deep, 18" wide		10	1.600		515	64		579	670
7500	Oven cabinets, 84" high, 24" deep, 27" wide		8	2		785	80		865	990
7750	Valance board trim		396	.040	L.F.	12.25	1.61		13.86	16.15
9000	For deluxe models of all cabinets, add					40%				
9500	For custom built in place, add					25%	10%			
9550	Rule of thumb, kitchen cabinets not including									
9560	appliances & counter top, minimum	2 Carp	30	.533	L.F.	130	21.50		151.50	178
9600	Maximum	"	25	.640	"	310	25.50		335.50	385
9610	For metal cabinets, see Div. 12 35 70.13									
9700	Minimum labor/equipment charge	1 Carp	3	2.667	Job		107		107	176

12 32 23.15 Manufactured Wood Casework Frames		Crew	Daily Output	Labor-Hours	Unit	Material	Labor	Equipment	Total	Total Incl O&P
0010	**MANUFACTURED WOOD CASEWORK FRAMES**									
0050	Base cabinets, counter storage, 36" high									
0100	One bay, 18" wide	1 Carp	2.70	2.963	Ea.	131	118		249	340
0200	24" wide		2.50	3.200		154	128		282	380
0300	36" wide		2.30	3.478		183	139		322	430

12 32 23 – Hardwood Casework

12 32 23.15 Manufactured Wood Casework Frames		Crew	Daily Output	Labor-Hours	Unit	Material	Labor	Equipment	Total	Total Incl O&P
						2009 Bare Costs				
0400	Two bay, 36" wide	1 Carp	2.20	3.636	Ea.	200	145		345	460
1000	48" wide		2	4		216	160		376	500
1050	72" wide		1.80	4.444		254	178		432	570
1100	Three bay, 54" wide		1.50	5.333		238	213		451	610
1200	72" wide		1.30	6.154		293	246		539	725
1800	108" wide		1.10	7.273		340	291		631	855
2000	Four bay, 72" wide		1.10	7.273		330	291		621	845
2100	96" wide		1	8		370	320		690	935
2500	144" wide		.90	8.889		445	355		800	1,075
2800	Bookcases, one bay, 7' high, 18" wide		2.40	3.333		154	133		287	390
3000	24" wide		2.30	3.478		177	139		316	425
3100	36" wide		2.20	3.636		216	145		361	480
3500	Two bay, 36" wide		1.60	5		224	200		424	575
3800	48" wide		1.50	5.333		278	213		491	655
4000	72" wide		1.40	5.714		340	228		568	750
4100	Three bay, 54" wide		1.20	6.667		370	266		636	845
4200	72" wide		1.10	7.273		410	291		701	930
4500	108" wide		1	8		525	320		845	1,100
4600	Four bay, 72" wide		.95	8.421		415	335		750	1,025
4700	96" wide		.90	8.889		510	355		865	1,150
5000	144" wide		.85	9.412		695	375		1,070	1,375
5100	Coat racks, one bay, 7' high, 24" wide		4.50	1.778		154	71		225	286
5200	36" wide		4.30	1.860		168	74.50		242.50	310
5300	Two bay, 48" wide		2.75	2.909		214	116		330	430
5600	72" wide		2.50	3.200		246	128		374	480
5800	Three bay, 72" wide		2.10	3.810		315	152		467	595
6000	108" wide		1.90	4.211		340	168		508	655
6100	Wall mounted cabinet, one bay, 24" high, 18" wide		3.60	2.222		85	89		174	241
6400	24" wide		3.50	2.286		92	91.50		183.50	252
6600	36" wide		3.40	2.353		116	94		210	283
6800	Two bay, 36" wide		2.20	3.636		124	145		269	375
7000	48" wide		2.10	3.810		139	152		291	405
7200	72" wide		2	4		177	160		337	460
7400	Three bay, 54" wide		1.70	4.706		154	188		342	480
7600	72" wide		1.60	5		185	200		385	535
7800	108" wide		1.50	5.333		231	213		444	605
8000	Four bay, 72" wide		1.40	5.714		184	228		412	580
8100	96" wide		1.30	6.154		216	246		462	645
8200	144" wide		1.20	6.667		290	266		556	760
8400	30" high, one bay, 18" wide		3.60	2.222		92	89		181	248
8600	24" wide		3.40	2.353		107	94		201	272
8800	36" wide		3.20	2.500		122	100		222	300
9000	Two bay, 36" wide		2.15	3.721		122	149		271	380
9100	48" wide		2	4		145	160		305	425
9200	72" wide		1.85	4.324		176	173		349	480
9400	Three bay, 54" wide		1.60	5		153	200		353	500
9600	72" wide		1.50	5.333		183	213		396	550
9650	108" wide		1.40	5.714		405	228		633	820
9700	Four bay, 72" wide		1.30	6.154		191	246		437	615
9750	96" wide		1.22	6.557		229	262		491	680
9780	144" wide		1.15	6.957		310	278		588	800
9800	Wardrobe, 7' high, single, 24" wide		2.70	2.963		170	118		288	380
9850	36" wide	↓	2.50	3.200	↓	255	128		383	490

12 32 23 – Hardwood Casework

12 32 23.15 Manufactured Wood Casework Frames		Crew	Daily Output	Labor-Hours	Unit	Material	2009 Bare Costs Labor	Equipment	Total	Total Incl O&P
9880	Partition & adjustable shelves, 48" wide	1 Carp	1.70	4.706	Ea.	216	188		404	550
9890	72" wide		1.55	5.161		345	206		551	720
9950	Partition, adjustable shelves & drawers, 48" wide		1.40	5.714		325	228		553	730
9960	72" wide		1.25	6.400		425	256		681	885
9970	Minimum labor/equipment charge		4	2	Job		80		80	132
12 32 23.20 Manufactured Hardwood Casework Doors										
0010	**MANUFACTURED HARDWOOD CASEWORK DOORS**									
2000	Glass panel, hardwood frame									
2200	12" wide, 18" high	1 Carp	34	.235	Ea.	18.75	9.40		28.15	36
2400	24" high		33	.242		25	9.70		34.70	43.50
2600	30" high		32	.250		31.50	10		41.50	51
2800	36" high		30	.267		37.50	10.65		48.15	59
3000	48" high		23	.343		50	13.90		63.90	78
3200	60" high		17	.471		62.50	18.80		81.30	100
3400	72" high		15	.533		75	21.50		96.50	118
3600	15" wide, 18" high		33	.242		23.50	9.70		33.20	42
3800	24" high		32	.250		31.50	10		41.50	51
4000	30" high		30	.267		39	10.65		49.65	60.50
4250	36" high		28	.286		47	11.40		58.40	70.50
4300	48" high		22	.364		62.50	14.55		77.05	93
4350	60" high		16	.500		78	20		98	119
4400	72" high		14	.571		94	23		117	141
4450	18" wide, 18" high		32	.250		28	10		38	47.50
4500	24" high		30	.267		37.50	10.65		48.15	59
4550	30" high		29	.276		47	11		58	69.50
4600	36" high		27	.296		56.50	11.85		68.35	81.50
4650	48" high		21	.381		75	15.20		90.20	108
4700	60" high		15	.533		94	21.50		115.50	138
4750	72" high		13	.615		113	24.50		137.50	165
5000	Hardwood, raised panel									
5100	12" wide, 18" high	1 Carp	16	.500	Ea.	27	20		47	62.50
5150	24" high		15.50	.516		36	20.50		56.50	73.50
5200	30" high		15	.533		45	21.50		66.50	84.50
5250	36" high		14	.571		54	23		77	97
5300	48" high		11	.727		72	29		101	127
5320	60" high		8	1		90	40		130	165
5340	72" high		7	1.143		108	45.50		153.50	195
5360	15" wide, 18" high		15.50	.516		34	20.50		54.50	71
5380	24" high		15	.533		45	21.50		66.50	84.50
5400	30" high		14.50	.552		56.50	22		78.50	98.50
5420	36" high		13.50	.593		67.50	23.50		91	114
5440	48" high		10.50	.762		90	30.50		120.50	150
5460	60" high		7.50	1.067		113	42.50		155.50	195
5480	72" high		6.50	1.231		135	49		184	230
5500	18" wide, 18" high		15	.533		40.50	21.50		62	79.50
5550	24" high		14.50	.552		54	22		76	96
5600	30" high		14	.571		67.50	23		90.50	112
5650	36" high		13	.615		81	24.50		105.50	130
5700	48" high		10	.800		108	32		140	172
5750	60" high		7	1.143		135	45.50		180.50	225
5800	72" high		6	1.333		162	53.50		215.50	266
9000	Minimum labor/equipment charge		4	2	Job		80		80	132

12 32 23.25 Manufactured Wood Casework Drawer Fronts		Crew	Daily Output	Labor-Hours	Unit	2009 Bare Costs Material	Labor	Equipment	Total	Total Incl O&P
0010	MANUFACTURED WOOD CASEWORK DRAWER FRONTS									
0100	Solid hardwood front									
1000	4" high, 12" wide	1 Carp	17	.471	Ea.	2.66	18.80		21.46	34
1200	18" wide		16	.500		4	20		24	37.50
1400	24" wide		15	.533		5.30	21.50		26.80	41
1600	6" high, 12" wide		16	.500		4	20		24	37.50
1800	18" wide		15	.533		6	21.50		27.50	41.50
2000	24" wide		14	.571		8	23		31	46.50
2200	9" high, 12" wide		15	.533		6	21.50		27.50	41.50
2400	18" wide		14	.571		9	23		32	47.50
2600	24" wide		13	.615		12	24.50		36.50	53.50
9000	Minimum labor/equipment charge		4	2	Job		80		80	132

12 32 23.30 Manufactured Wood Casework Vanities		Crew	Daily Output	Labor-Hours	Unit	Material	Labor	Equipment	Total	Total Incl O&P
0010	MANUFACTURED WOOD CASEWORK VANITIES									
8000	Vanity bases, 2 doors, 30" high, 21" deep, 24" wide	2 Carp	20	.800	Ea.	275	32		307	360
8050	30" wide		16	1		315	40		355	415
8100	36" wide		13.33	1.200		345	48		393	460
8150	48" wide		11.43	1.400		425	56		481	565
9000	For deluxe models of all vanities, add to above					40%				
9500	For custom built in place, add to above					25%	10%			

12 32 23.35 Manufactured Wood Casework Hardware		Crew	Daily Output	Labor-Hours	Unit	Material	Labor	Equipment	Total	Total Incl O&P
0010	MANUFACTURED WOOD CASEWORK HARDWARE									
1000	Catches, minimum	1 Carp	235	.034	Ea.	1	1.36		2.36	3.34
1020	Average		119.40	.067		3.29	2.68		5.97	8.05
1040	Maximum		80	.100		6.25	4		10.25	13.45
2000	Door/drawer pulls, handles									
2200	Handles and pulls, projecting, metal, minimum	1 Carp	160	.050	Ea.	4.43	2		6.43	8.15
2220	Average		95.24	.084		6.90	3.36		10.26	13.10
2240	Maximum		68	.118		9.40	4.70		14.10	18.05
2300	Wood, minimum		160	.050		4.67	2		6.67	8.45
2320	Average		95.24	.084		6.25	3.36		9.61	12.40
2340	Maximum		68	.118		8.55	4.70		13.25	17.20
2600	Flush, metal, minimum		160	.050		4.67	2		6.67	8.45
2620	Average		95.24	.084		6.25	3.36		9.61	12.40
2640	Maximum		68	.118		8.55	4.70		13.25	17.20
3000	Drawer tracks/glides, minimum		48	.167	Pr.	7.90	6.65		14.55	19.70
3020	Average		32	.250		13.40	10		23.40	31.50
3040	Maximum		24	.333		23	13.30		36.30	47.50
4000	Cabinet hinges, minimum		160	.050		2.69	2		4.69	6.25
4020	Average		95.24	.084		6	3.36		9.36	12.15
4040	Maximum		68	.118		10.15	4.70		14.85	18.95
5000	Cabinet locks, minimum		47.90	.167	Ea.	5.05	6.65		11.70	16.55
5020	Average		23.95	.334		7.90	13.35		21.25	30.50
5040	Maximum		16	.500		19.15	20		39.15	54

12 35 Specialty Casework

12 35 50 – Educational/Library Casework

12 35 50.13 Educational Casework		Crew	Daily Output	Labor-Hours	Unit	2009 Bare Costs Material	Labor	Equipment	Total	Total Incl O&P
0010	**EDUCATIONAL CASEWORK**									
5000	School, 24" deep, metal, 84" high units	2 Carp	15	1.067	L.F.	365	42.50		407.50	475
5150	Counter height units		20	.800		246	32		278	325
5450	Wood, custom fabricated, 32" high counter		20	.800		205	32		237	278
5600	Add for counter top		56	.286		22	11.40		33.40	43
5800	84" high wall units	↓	15	1.067	↓	400	42.50		442.50	510
6000	Laminated plastic finish is same price as wood									

12 35 53 – Laboratory Casework

12 35 53.13 Metal Laboratory Casework		Crew	Daily Output	Labor-Hours	Unit	Material	Labor	Equipment	Total	Total Incl O&P
0010	**METAL LABORATORY CASEWORK**									
0020	Cabinets, base, door units, metal	2 Carp	18	.889	L.F.	183	35.50		218.50	260
0300	Drawer units		18	.889		410	35.50		445.50	510
0700	Tall storage cabinets, open, 7' high		20	.800		390	32		422	485
0900	With glazed doors		20	.800		470	32		502	570
1300	Wall cabinets, metal, 12-1/2" deep, open		20	.800		129	32		161	195
1500	With doors	↓	20	.800	↓	268	32		300	345
6300	Rule of thumb: lab furniture including installation & connection									
6320	High school				S.F.				32.50	36
6340	College								48	53
6360	Clinical, health care								41.50	45.50
6380	Industrial				↓				67	73.50

12 35 59 – Display Casework

12 35 59.10 Display Cases		Crew	Daily Output	Labor-Hours	Unit	Material	Labor	Equipment	Total	Total Incl O&P
0010	**DISPLAY CASES** Free standing, all glass									
0020	Aluminum frame, 42" high x 36" x 12" deep	2 Carp	8	2	Ea.	1,200	80		1,280	1,450
0100	70" high x 48" x 18" deep	"	6	2.667		2,400	107		2,507	2,825
0500	For wood bases, add					9%				
0600	For hardwood frames, deduct					8%				
0700	For bronze, baked enamel finish, add				↓	10%				
2000	Wall mounted, glass front, aluminum frame									
2010	Non-illuminated, one section 3' x 4' x 1'-4"	2 Carp	5	3.200	Ea.	2,125	128		2,253	2,550
2100	5' x 4' x 1'-4"		5	3.200		2,300	128		2,428	2,750
2200	6' x 4' x 1'-4"		4	4		2,725	160		2,885	3,275
2500	Two sections, 8' x 4' x 1'-4"		2	8		1,775	320		2,095	2,475
2600	10' x 4' x 1'-4"		2	8		2,075	320		2,395	2,825
3000	Three sections, 16' x 4' x 1'-4"	↓	1.50	10.667	↓	3,600	425		4,025	4,675
3500	For fluorescent lights, add				Section	299			299	330
4000	Table exhibit cases, 2' wide, 3' high, 4' long, flat top	2 Carp	5	3.200	Ea.	1,050	128		1,178	1,350
4100	3' wide, 3' high, 4' long, sloping top	"	3	5.333	"	1,250	213		1,463	1,725

12 35 70 – Healthcare Casework

12 35 70.13 Hospital Casework		Crew	Daily Output	Labor-Hours	Unit	Material	Labor	Equipment	Total	Total Incl O&P
0010	**HOSPITAL CASEWORK**									
0500	Base cabinets, laminated plastic	2 Carp	10	1.600	L.F.	235	64		299	365
0700	Enameled steel		10	1.600		216	64		280	345
1000	Stainless steel	↓	10	1.600		430	64		494	580
1200	For all drawers, add					24.50			24.50	27
1300	Cabinet base trim, 4" high, enameled steel	2 Carp	200	.080		39.50	3.20		42.70	49
1400	Stainless steel		200	.080		78.50	3.20		81.70	92
1450	Countertop, laminated plastic, no backsplash		40	.400		40.50	16		56.50	71
1650	With backsplash		40	.400	↓	50.50	16		66.50	82
1800	For sink cutout, add	↓	12.20	1.311	Ea.		52.50		52.50	86.50

12 35 Specialty Casework

12 35 70 – Healthcare Casework

12 35 70.13 Hospital Casework		Crew	Daily Output	Labor-Hours	Unit	Material	Labor	Equipment	Total	Total Incl O&P
						2009 Bare Costs				
1900	Stainless steel counter top	2 Carp	40	.400	L.F.	131	16		147	171
2000	For drop-in stainless 43" x 21" sink, add				Ea.	860			860	950
2500	Wall cabinets, laminated plastic	2 Carp	15	1.067	L.F.	176	42.50		218.50	264
2600	Enameled steel		15	1.067		216	42.50		258.50	310
2700	Stainless steel	↓	15	1.067		430	42.50		472.50	545
2800	For glass doors, add				↓	29.50			29.50	32.50

12 35 70.16 Nurse Station Casework		Crew	Daily Output	Labor-Hours	Unit	Material	Labor	Equipment	Total	Total Incl O&P
0010	**NURSE STATION CASEWORK**									
2100	Door type, laminated plastic	2 Carp	10	1.600	L.F.	272	64		336	405
2200	Enameled steel		10	1.600		260	64		324	390
2300	Stainless steel	↓	10	1.600		520	64		584	675
2400	For drawer type, add				↓	221			221	243

12 35 80 – Commercial Kitchen Casework

12 35 80.13 Metal Kitchen Casework		Crew	Daily Output	Labor-Hours	Unit	Material	Labor	Equipment	Total	Total Incl O&P
0010	**METAL KITCHEN CASEWORK**									
3500	Base cabinets, metal, minimum	2 Carp	30	.533	L.F.	63.50	21.50		85	105
3600	Maximum		25	.640		161	25.50		186.50	219
3700	Wall cabinets, metal, minimum		30	.533		63.50	21.50		85	105
3800	Maximum	↓	25	.640	↓	145	25.50		170.50	202

12 36 Countertops

12 36 23 – Plastic Countertops

12 36 23.13 Plastic-Laminate-Clad Countertops		Crew	Daily Output	Labor-Hours	Unit	Material	Labor	Equipment	Total	Total Incl O&P
0010	**PLASTIC-LAMINATE-CLAD COUNTERTOPS**									
0020	Stock, 24" wide w/ backsplash, minimum	1 Carp	30	.267	L.F.	8.65	10.65		19.30	27
0100	Maximum	"	25	.320	"	18	12.80		30.80	41
1700	For end splash, add				Ea.	17			17	18.70
9000	Minimum labor/equipment charge	1 Carp	3.75	2.133	Job		85		85	141

12 36 23.30 Plastic Laminate Countertop Components		Crew	Daily Output	Labor-Hours	Unit	Material	Labor	Equipment	Total	Total Incl O&P
0010	**PLASTIC LAMINATE COUNTERTOP COMPONENTS**									
1000	Edging, 24" wide, 1-1/2" thick									
1500	Plastic laminate, minimum	1 Carp	25	.320	L.F.	3.40	12.80		16.20	24.50
1520	Average		24	.333		6.25	13.30		19.55	29
1540	Maximum		22	.364		9.35	14.55		23.90	34.50
2500	Hardwood, minimum		20	.400		4.20	16		20.20	31
2520	Average		18	.444		6.75	17.75		24.50	37
2540	Maximum		16	.500		9.65	20		29.65	43.50
2600	Backsplash, add to above, minimum		36	.222		1.36	8.90		10.26	16.15
2620	Average		35	.229		2.27	9.15		11.42	17.55
2640	Maximum	↓	34	.235		3.18	9.40		12.58	19
2700	For metal cove, add				↓	.80			.80	.88
2900	Postformed backsplash, add to above									
2920	Minimum	1 Carp	21	.381	L.F.	1.23	15.20		16.43	26.50
2940	Average		20	.400		2.44	16		18.44	29
2960	Maximum		19	.421	↓	3.68	16.80		20.48	32
3500	Well openings (for typewriters etc.)		2.50	3.200	Ea.		128		128	211
3900	Cutouts for sinks, lavatories	↓	12	.667	"		26.50		26.50	44

12 36 Countertops

12 36 53 – Laboratory Countertops

12 36 53.10 Laboratory Countertops and Sinks		Crew	Daily Output	Labor-Hours	Unit	2009 Bare Costs Material	Labor	Equipment	Total	Total Incl O&P
0010	**LABORATORY COUNTERTOPS AND SINKS**									
0020	Countertops, not incl. base cabinets, acidproof, minimum	2 Carp	82	.195	S.F.	31.50	7.80		39.30	47.50
0030	Maximum		70	.229		41	9.15		50.15	60.50
0040	Stainless steel		82	.195		95.50	7.80		103.30	118

12 36 61 – Simulated Stone Countertops

12 36 61.16 Solid Surface Countertops		Crew	Daily Output	Labor-Hours	Unit	Material	Labor	Equipment	Total	Total Incl O&P
0010	**SOLID SURFACE COUNTERTOPS**, Acrylic polymer									
0020	Pricing for orders of 100 L.F. or greater									
0100	25" wide, solid colors	2 Carp	28	.571	L.F.	44.50	23		67.50	86.50
0200	Patterned colors		28	.571		56.50	23		79.50	99.50
0300	Premium patterned colors		28	.571		70.50	23		93.50	116
0400	With silicone attached 4" backsplash, solid colors		27	.593		49	23.50		72.50	93
0500	Patterned colors		27	.593		62	23.50		85.50	107
0600	Premium patterned colors		27	.593		77	23.50		100.50	124
0700	With hard seam attached 4" backsplash, solid colors		23	.696		49	28		77	100
0800	Patterned colors		23	.696		62	28		90	114
0900	Premium patterned colors		23	.696		77	28		105	131
1000	Pricing for order of 51 – 99 L.F.									
1100	25" wide, solid colors	2 Carp	24	.667	L.F.	51.50	26.50		78	101
1200	Patterned colors		24	.667		65	26.50		91.50	116
1300	Premium patterned colors		24	.667		81.50	26.50		108	134
1400	With silicone attached 4" backsplash, solid colors		23	.696		56.50	28		84.50	108
1500	Patterned colors		23	.696		71.50	28		99.50	125
1600	Premium patterned colors		23	.696		89	28		117	144
1700	With hard seam attached 4" backsplash, solid colors		20	.800		56.50	32		88.50	115
1800	Patterned colors		20	.800		71.50	32		103.50	132
1900	Premium patterned colors		20	.800		89	32		121	151
2000	Pricing for order of 1 – 50 L.F.									
2100	25" wide, solid colors	2 Carp	20	.800	L.F.	60	32		92	119
2200	Patterned colors		20	.800		76.50	32		108.50	137
2300	Premium patterned colors		20	.800		95.50	32		127.50	158
2400	With silicone attached 4" backsplash, solid colors		19	.842		66	33.50		99.50	128
2500	Patterned colors		19	.842		83.50	33.50		117	148
2600	Premium patterned colors		19	.842		104	33.50		137.50	171
2700	With hard seam attached 4" backsplash, solid colors		15	1.067		66	42.50		108.50	143
2800	Patterned colors		15	1.067		83.50	42.50		126	163
2900	Premium patterned colors		15	1.067		104	42.50		146.50	186
3000	Sinks, pricing for order of 100 or greater units									
3100	Single bowl, hard seamed, solid colors, 13" x 17"	1 Carp	3	2.667	Ea.	300	107		407	505
3200	10" x 15"		7	1.143		139	45.50		184.50	229
3300	Cutouts for sinks		8	1			40		40	66
3400	Sinks, pricing for order of 51 – 99 units									
3500	Single bowl, hard seamed, solid colors, 13" x 17"	1 Carp	2.55	3.137	Ea.	345	125		470	585
3600	10" x 15"		6	1.333		160	53.50		213.50	264
3700	Cutouts for sinks		7	1.143			45.50		45.50	75.50
3800	Sinks, pricing for order of 1 – 50 units									
3900	Single bowl, hard seamed, solid colors, 13" x 17"	1 Carp	2	4	Ea.	405	160		565	710
4000	10" x 15"		4.55	1.758		188	70		258	325
4100	Cutouts for sinks		5.25	1.524			61		61	101
4200	Cooktop cutouts, pricing for 100 or greater units		4	2		22	80		102	157
4300	51 – 99 units		3.40	2.353		25.50	94		119.50	183
4400	1 – 50 units		3	2.667		30	107		137	209

12 36 Countertops

12 36 61 – Simulated Stone Countertops

12 36 61.17 Solid Surface Vanity Tops		Crew	Daily Output	Labor-Hours	Unit	Material	Labor	Equipment	Total	Total Incl O&P
0010	**SOLID SURFACE VANITY TOPS**									
0015	Solid surface, center bowl, 17" x 19"	1 Carp	12	.667	Ea.	189	26.50		215.50	252
0020	19" x 25"		12	.667		229	26.50		255.50	296
0030	19" x 31"		12	.667		278	26.50		304.50	350
0040	19" x 37"		12	.667		325	26.50		351.50	400
0050	22" x 25"		10	.800		202	32		234	275
0060	22" x 31"		10	.800		236	32		268	310
0070	22" x 37"		10	.800		274	32		306	355
0080	22" x 43"		10	.800		315	32		347	400
0090	22" x 49"		10	.800		345	32		377	435
0110	22" x 55"		8	1		395	40		435	500
0120	22" x 61"		8	1		450	40		490	560
0220	Double bowl, 22" x 61"		8	1		505	40		545	625
0230	Double bowl, 22" x 73"		8	1		705	40		745	840
0240	For aggregate colors, add					35%				
0250	For faucets and fittings, see Div. 22 41 39.10									

(Column headings Material, Labor, Equipment, Total are under "2009 Bare Costs".)

12 36 61.19 Engineered Stone Countertops		Crew	Daily Output	Labor-Hours	Unit	Material	Labor	Equipment	Total	Total Incl O&P
0010	**ENGINEERED STONE COUNTERTOPS**									
0100	25" wide, 4" backsplash, color group A, minimum	2 Carp	15	1.067	L.F.	17	42.50		59.50	89
0110	Maximum		15	1.067		42	42.50		84.50	117
0120	Color group B, minimum		15	1.067		22	42.50		64.50	94.50
0130	Maximum		15	1.067		49	42.50		91.50	125
0140	Color group C, minimum		15	1.067		29.50	42.50		72	103
0150	Maximum		15	1.067		60	42.50		102.50	137
0160	Color group D, minimum		15	1.067		37	42.50		79.50	111
0170	Maximum		15	1.067		70	42.50		112.50	148

12 43 Portable Lamps

12 43 13 – Lamps

12 43 13.23 Miscellaneous Lamps		Crew	Daily Output	Labor-Hours	Unit	Material	Labor	Equipment	Total	Total Incl O&P
0010	**MISCELLANEOUS LAMPS**									
1000	Ceramic, desk, minimum				Ea.	49			49	54
1020	Maximum					130			130	143
1200	End table, minimum					45			45	49.50
1220	Maximum					108			108	119
1300	Night stand, minimum					40			40	44
1320	Maximum					84			84	92.50
1400	Wall, minimum					54			54	59.50
1420	Maximum					200			200	220
2000	Glass, desk, minimum					191			191	210
2020	Maximum					555			555	610
2100	End table, minimum					185			185	204
2120	Maximum					730			730	805
2200	Floor, minimum					465			465	510
2250	Maximum					1,100			1,100	1,200
2280	Night stand, minimum					292			292	320
2300	Maximum					555			555	610
2350	Pendant, minimum					47.50			47.50	52
2400	Maximum					905			905	995
3000	Metal, desk, minimum					43			43	47.50

12 43 Portable Lamps

12 43 13 – Lamps

12 43 13.23 Miscellaneous Lamps		Crew	Daily Output	Labor-Hours	Unit	2009 Bare Costs Material	Labor	Equipment	Total	Total Incl O&P
3100	Maximum				Ea.	117			117	129
3200	End table, minimum					70			70	77
3250	Maximum					92			92	101
3300	Wall, minimum					67			67	73.50
3350	Maximum					123			123	135
3400	Floor, minimum					90			90	99
3450	Maximum					400			400	440
3500	Night stand, minimum					40			40	44
3550	Maximum					105			105	115
3600	Pendant, minimum					35			35	38.50
3650	Maximum					340			340	375
4000	Stone, desk, minimum					80			80	88
4100	Maximum					117			117	129
4200	End table, minimum					62			62	68
4300	Maximum					92			92	101
4400	Night stand, minimum					62			62	68
4500	Maximum					70			70	77
5000	Wall, minimum					58.50			58.50	64.50
6000	Maximum					92			92	101
8000	Replacement shades, lamp, desk, minimum					14			14	15.40
8020	Maximum					50			50	55
8040	End table, minimum					15.50			15.50	17.05
8060	Maximum					41			41	45
8100	Night stand minimum					13			13	14.30
8120	Maximum					36			36	39.50
8140	Wall, minimum					9			9	9.90
8160	Maximum				▼	14.50			14.50	15.95

12 45 Bedroom Furnishings

12 45 13 – Bed Linens

12 45 13.13 Blankets		Crew	Daily Output	Labor-Hours	Unit	Material	Labor	Equipment	Total	Total Incl O&P
0010	**BLANKETS**									
7700	Bedspreads, unquilted, twin, minimum				Ea.	33			33	36
7800	Maximum					286			286	315
7850	Full, minimum					39			39	43
7900	Maximum					410			410	450
8000	Queen, minimum					53.50			53.50	59
8100	Maximum					435			435	475
8200	Quilted, twin, minimum					77			77	85
8300	Maximum					365			365	400
8400	Full, minimum					96			96	106
8500	Maximum					490			490	540
8600	Queen, minimum					94.50			94.50	104
8700	Maximum				▼	525			525	575

12 46 Furnishing Accessories

12 46 13 – Ash Receptacles

12 46 13.10 Ash/Trash Receivers		Crew	Daily Output	Labor-Hours	Unit	Material	Labor	Equipment	Total	Total Incl O&P
						2009 Bare Costs				
0010	ASH/TRASH RECEIVERS									
1000	Ash urn, cylindrical metal									
1020	8" diameter, 20" high	1 Clab	60	.133	Ea.	125	4.21		129.21	144
1040	8" dia., 25" high		60	.133		190	4.21		194.21	216
1060	10" diameter, 26" high		60	.133		110	4.21		114.21	128
1080	12" dia., 30" high		60	.133		160	4.21		164.21	183
2000	Combination ash/trash urn, metal									
2020	8" diameter, 20" high	1 Clab	60	.133	Ea.	80	4.21		84.21	95
2040	8" dia., 25" high		60	.133		220	4.21		224.21	249
2050	10" diameter, 26" high		60	.133		95	4.21		99.21	111
2060	12" dia., 30" high		60	.133		240	4.21		244.21	271

12 46 19 – Clocks

12 46 19.50 Wall Clocks		Crew	Daily Output	Labor-Hours	Unit	Material	Labor	Equipment	Total	Total Incl O&P
0010	WALL CLOCKS									
0080	12" diameter, single face	1 Elec	8	1	Ea.	81	47		128	162
0100	Double face	"	6.20	1.290	"	154	60.50		214.50	264

12 46 33 – Waste Receptacles

12 46 33.13 Trash Receptacles		Crew	Daily Output	Labor-Hours	Unit	Material	Labor	Equipment	Total	Total Incl O&P
0010	TRASH RECEPTACLES									
4000	Trash receptacle, metal									
4020	8" diameter, 15" high	1 Clab	60	.133	Ea.	67	4.21		71.21	80.50
4040	10" diameter, 18" high		60	.133		125	4.21		129.21	144
4060	16" dia., 16" high		60	.133		150	4.21		154.21	172
4100	18" dia., 32" high		60	.133		270	4.21		274.21	305
5000	Plastic, fire resistant									
5020	Rectangular 11" x 8" x 12" high	1 Clab	60	.133	Ea.	13	4.21		17.21	21.50
5040	16" x 8" x 14" high	"	60	.133	"	19	4.21		23.21	28
5500	Plastic, with lid									
5520	35 gallon	1 Clab	60	.133	Ea.	211	4.21		215.21	239
5540	45 gallon		60	.133		280	4.21		284.21	315
5550	Plastic recycling barrel, w/lid & wheels, 32 gal G		60	.133		60	4.21		64.21	73
5560	65 gal G		60	.133		107	4.21		111.21	125
5570	95 gal G		60	.133		310	4.21		314.21	345

12 46 36 – Desk Accessories

12 46 36.10 Commercial Desk Accessories		Crew	Daily Output	Labor-Hours	Unit	Material	Labor	Equipment	Total	Total Incl O&P
0010	COMMERCIAL DESK ACCESSORIES									
0300	Bookends, minimum				Ea.	7.80			7.80	8.55
0320	Maximum					14			14	15.40
1000	Calendar with pad, minimum					8.90			8.90	9.80
1020	Maximum					35			35	38.50
1200	Carafe, tray, minimum					60.50			60.50	66.50
1220	Maximum					71			71	78
1300	Desk pad, minimum					22			22	24
1320	Maximum					75			75	82.50
1400	Double pen set with pens, minimum					50			50	55
1420	Maximum					85			85	93.50
1500	Letter tray, minimum					13			13	14.30
1520	Maximum					33			33	36.50
1600	Memo box, minimum					9			9	9.90
1620	Maximum					13			13	14.30

12 48 Rugs and Mats

12 48 13 – Entrance Floor Mats and Frames

12 48 13.13 Entrance Floor Mats		Crew	Daily Output	Labor-Hours	Unit	Material	Labor	Equipment	Total	Total Incl O&P
0010	**ENTRANCE FLOOR MATS**									
0020	Recessed, in-laid black rubber, 3/8" thick, solid	1 Clab	155	.052	S.F.	19.70	1.63		21.33	24
0050	Perforated		155	.052		13.40	1.63		15.03	17.45
0100	1/2" thick, solid		155	.052		16.10	1.63		17.73	20.50
0150	Perforated		155	.052		21.50	1.63		23.13	26
0200	In colors, 3/8" thick, solid		155	.052		17.35	1.63		18.98	22
0250	Perforated		155	.052		17.90	1.63		19.53	22.50
0300	1/2" thick, solid		155	.052		22.50	1.63		24.13	27
0350	Perforated		155	.052		23	1.63		24.63	28
2000	Recycled rubber tire tile, 12" x 12" x 3/8" thick G		125	.064		8.40	2.02		10.42	12.60
2510	Natural cocoa fiber, 1/2" thick G		125	.064		7.95	2.02		9.97	12.10
2520	3/4" thick G		125	.064		6.25	2.02		8.27	10.20
2530	1" thick G	↓	125	.064	↓	6.90	2.02		8.92	10.95

12 51 Office Furniture

12 51 16 – Case Goods

12 51 16.13 Metal Case Goods		Crew	Daily Output	Labor-Hours	Unit	Material	Labor	Equipment	Total	Total Incl O&P
0010	**METAL CASE GOODS**									
0020	Desks, 29" high, double pedestal, 30" x 60", metal, minimum				Ea.	450			450	495
0030	Maximum					1,125			1,125	1,225
0400	36" x 72", metal, minimum					645			645	710
0420	Maximum					1,400			1,400	1,550
0600	Desks, single pedestal, 30" x 60", metal, minimum					405			405	445
0620	Maximum					905			905	995
0720	Desks, secretarial, 30" x 60", metal, minimum					385			385	425
0730	Maximum					640			640	705
0740	Return, 20" x 42", minimum					263			263	290
0750	Maximum					405			405	445
0940	59" x 12" x 23" high, steel, minimum					219			219	241
0960	Maximum					280			280	305
0970	Keyboard shelf, standard					80.50			80.50	88.50
0980	Articulating					380			380	420
0990	Center drawer, 18" wide					105			105	115
1020	Credenza, 18" to 22" x 60" to 72" metal, minimum					380			380	420
1040	Maximum					1,050			1,050	1,150
1240	Bookcase, 36" x 12" x 29" high					120			120	131
1260	52" high					180			180	197
1320	Computer stand, mobile, 25" x 24" x 38" high					280			280	305
1370	Computer desk with hutch, 47" x 24" x 46" high					310			310	340
1400	Printer stand, 22" x 24" x 34" high				↓	202			202	222

12 51 16.16 Wood Case Goods		Crew	Daily Output	Labor-Hours	Unit	Material	Labor	Equipment	Total	Total Incl O&P
0010	**WOOD CASE GOODS**									
0150	Desk, 29" high, double pedestal, 30" x 60"									
0160	Wood, minimum				Ea.	395			395	435
0180	Maximum				"	2,175			2,175	2,375
0550	Desk, 29" high, double pedestal, 36" x 72"									
0560	Wood, minimum				Ea.	500			500	550
0580	Maximum				"	2,925			2,925	3,200
0630	Single pedestal, 30" x 60"									
0640	Wood, minimum				Ea.	525			525	575
0650	Maximum				↓	680			680	750

12 51 Office Furniture

12 51 16 – Case Goods

12 51 16.16 Wood Case Goods		Crew	Daily Output	Labor-Hours	Unit	2009 Bare Costs Material	Labor	Equipment	Total	Total Incl O&P
0670	Executive return, 24" x 42", with box, file, wood, minimum				Ea.	355			355	390
0680	Maximum					680			680	750
0790	Desk, 29" high, secretarial, 30" x 60"									
0800	Wood, minimum				Ea.	435			435	475
0810	Maximum					2,200			2,200	2,425
0820	Return, 20" x 42", minimum					266			266	292
0830	Maximum					900			900	990
0900	Desktop organizer, 72" x 14" x 36" high, wood, minimum					160			160	176
0920	Maximum					380			380	420
1110	Furniture, credenza, 29" high, 18" to 22" x 60" to 72"									
1120	Wood, minimum				Ea.	610			610	670
1140	Maximum					2,750			2,750	3,025
1150	Hutch/bookcase, minimum					345			345	380
1160	Maximum					1,700			1,700	1,875
1200	Bookcase, 36" x 12" x 30" high, wood					149			149	164
1220	48" high					252			252	277
1300	Computer stand, mobile, wood, 25" x 24" x 38" high					415			415	455
1360	Computer desk with hutch, wood, 47" x 24" x 46" high					420			420	465

12 51 16.26 Resinite Case Goods

0010	**RESINITE CASE GOODS**									
1340	Computer stand, resinite				Ea.	360			360	395
1380	Desk with hutch				"	395			395	435

12 51 19 – Filing Cabinets

12 51 19.13 Lateral Filing Cabinets

0010	**LATERAL FILING CABINETS**, metal, baked enamel finish									
1060	Lateral, 36" wide, minimum				Ea.	400			400	440
1080	Maximum					430			430	475
1160	Lateral, 36" wide, minimum					420			420	460
1180	Maximum					635			635	695
1200	Wood, 2 drawer, lateral, minimum					445			445	490
1210	Maximum					795			795	875

12 51 19.16 Vertical Filing Cabinets

0010	**VERTICAL FILING CABINETS**, metal, baked enamel finish									
1000	2 drawer, vertical, minimum				Ea.	145			145	159
1020	Maximum					315			315	350
1100	4 drawer, vertical, minimum					170			170	187
1120	Maximum					465			465	515

12 51 19.23 Flat Files

0010	**FLAT FILES**, metal, baked enamel finish									
2010	Steel, 5 drawer, 40" wide x 27" deep				Ea.	660			660	725
2020	46" wide x 33" deep					740			740	815
2030	8 drawer, 40" wide x 27" deep					870			870	955
2040	46" wide x 33" deep					1,250			1,250	1,400
2050	Base, 40" wide x 27" deep x 5" high					187			187	206
2060	46" wide x 33" deep					201			201	221

12 51 19.26 Hanging Files

0010	**HANGING FILES**, metal, baked enamel finish									
2100	File stand				Ea.	345			345	380
2110	Clamps, package of 6				"	139			139	152

12 51 23 – Office Tables

12 51 23.13 Wood Tables		Crew	Daily Output	Labor-Hours	Unit	2009 Bare Costs Material	Labor	Equipment	Total	Total Incl O&P
0010	**WOOD TABLES**									
5000	Sled base, laminate top, coffee, minimum				Ea.	187			187	205
5100	Maximum					455			455	500
5150	End, minimum					149			149	164
5200	Maximum					315			315	345
5250	Wood cube, coffee					530			530	585
5350	End					555			555	610
5700	Designer table, Mies Barcelona, 40" sq., glass top, s.s. legs					2,050			2,050	2,250
5750	Saarinen, 42" round, plastic top, metal pedestal base					1,175			1,175	1,300
5800	Aulenti coffee table, 45" square, marble top and leg base					7,400			7,400	8,150
5840	Bruer, Laccio, 21-1/2" x 19", plastic top, tubular legs					430			430	475
5860	F. knoll, 24" square, glass top, solid steel leg base, chrome					1,400			1,400	1,550

12 51 23.23 Metal Tables		Crew	Daily Output	Labor-Hours	Unit	Material	Labor	Equipment	Total	Total Incl O&P
0010	**METAL TABLES**									
5400	All metal drum, 14" diameter, 14" high				Ea.	385			385	420
5420	21" high					440			440	485
5480	18" diameter, 18" high					415			415	455
5500	22" diameter, 15" high					540			540	590
5550	20" high					590			590	650
5600	For 3/4" glass top, add				S.F.	52.50			52.50	58
5650	For 1" marble top, add				"	52.50			52.50	58
7500	Table, training, with modesty panel, 30" x 60"				Ea.	540			540	595
8000	Tables, modular									
8010	Rectangular, 24" x 48"				Ea.	585			585	640
8020	24" x 60"					650			650	715
8030	24" x 72"					370			370	410
8040	30" x 60"					815			815	895
8050	30" x 72"					855			855	945
8100	Corner triangle, 24" deep					259			259	285
8110	30" deep					291			291	320
8200	Half round, 30" x 60"					705			705	775
8300	Trapezoid, 30" x 60"					580			580	635
8400	Crescent, 30" x 60"					830			830	915

12 51 23.33 Conference Tables		Crew	Daily Output	Labor-Hours	Unit	Material	Labor	Equipment	Total	Total Incl O&P
0010	**CONFERENCE TABLES**									
6010	Segmented, oval, 72" x 144"				Ea.	3,500			3,500	3,850
6050	Boat, 96" x 42", minimum					740			740	810
6150	Maximum					2,800			2,800	3,075
6200	120" x 48", minimum					1,000			1,000	1,100
6250	Maximum					3,400			3,400	3,750
6300	144" x 48", minimum					1,175			1,175	1,300
6350	Maximum					4,350			4,350	4,775
6400	168" x 60", minimum					1,975			1,975	2,175
6450	Maximum					6,225			6,225	6,850
6500	192" x 60", minimum					2,300			2,300	2,525
6550	Maximum					6,925			6,925	7,600
6680	240" x 60", minimum					3,350			3,350	3,675
6700	Maximum					13,200			13,200	14,500
6720	Rectangle, 96" x 42", minimum					965			965	1,075
6740	Maximum					2,800			2,800	3,075
6760	120" x 48", minimum					2,100			2,100	2,300
6780	Maximum					3,600			3,600	3,950

12 51 Office Furniture

12 51 23 – Office Tables

12 51 23.33 Conference Tables		Crew	Daily Output	Labor-Hours	Unit	Material	Labor	Equipment	Total	Total Incl O&P
						2009 Bare Costs				
6800	144" x 60", minimum				Ea.	3,175			3,175	3,500
6820	Maximum					5,175			5,175	5,700
6840	168" x 60", minimum					3,875			3,875	4,275
6860	Maximum					6,225			6,225	6,850
6880	192" x 60", minimum					4,525			4,525	4,975
6900	Maximum					6,925			6,925	7,600
6960	240" x 60", minimum					6,000			6,000	6,600
6980	Maximum					10,600			10,600	11,700
7010	Powered, 52" x 108"					2,625			2,625	2,900
7020	52" x 120"					2,675			2,675	2,925
7030	52" x 132"				↓	2,850			2,850	3,125

12 52 Seating

12 52 13 – Chairs

12 52 13.10 Chairs, Folding and Stack

		Crew	Daily Output	Labor-Hours	Unit	Material	Labor	Equipment	Total	Total Incl O&P
0010	**CHAIRS, FOLDING & STACK**									
2000	Folding, all steel, baked enamel finish,									
2100	Form fitting seat and backrests				Ea.	24.50			24.50	27
2200	Upholstered seat and back					54			54	59
2300	Polypropylene seat and back w/frame					35.50			35.50	39.50
2500	Chair caddy for above				↓	254			254	279
5000	Stack chair									
5300	Hardwood frame, seat and back									
5320	Minimum				Ea.	46.50			46.50	51
5340	Maximum				"	223			223	246
5400	Upholstered seat and back									
5420	Minimum				Ea.	72			72	79.50
5440	Maximum				"	335			335	365
5600	Metal frame, upholstered seat and back									
5620	Minimum				Ea.	43			43	47.50
5640	Maximum				"	300			300	330
5700	Plastic shell, metal legs									
5720	Minimum				Ea.	56			56	61.50
5740	Maximum					127			127	140
5800	Chair caddy for above					235			235	259
5900	Tablet arms, minimum					149			149	164
5920	Maximum				↓	193			193	212

12 52 19 – Upholstered Seating

12 52 19.13 Upholstered Office Seating

		Crew	Daily Output	Labor-Hours	Unit	Material	Labor	Equipment	Total	Total Incl O&P
0010	**UPHOLSTERED OFFICE SEATING**									
3800	Lounge chair, upholstered, minimum				Ea.	208			208	229
3850	Maximum					1,750			1,750	1,900
4000	Sofa, two seat, upholstered, minimum					620			620	680
4100	Maximum					2,525			2,525	2,775
4650	Three seat, minimum					900			900	990
4680	Maximum					3,550			3,550	3,900
4700	Modular seating, lounge chair unit, upholstered, minimum					1,425			1,425	1,575
4750	Maximum					2,075			2,075	2,275
4780	Corner unit, minimum					980			980	1,075
4800	Maximum				↓	2,275			2,275	2,500

12 52 Seating

12 52 23 – Office Seating

12 52 23.13 Office Chairs		Crew	Daily Output	Labor-Hours	Unit	2009 Bare Costs Material	Labor	Equipment	Total	Total Incl O&P
0010	**OFFICE CHAIRS**									
2000	Standard office chair, executive, minimum				Ea.	255			255	280
2150	Maximum					1,750			1,750	1,950
2200	Management, minimum					183			183	201
2250	Maximum					2,000			2,000	2,200
2280	Task, minimum					117			117	128
2290	Maximum					550			550	605
2300	Arm kit, minimum					55.50			55.50	61
2320	Maximum					83			83	91.50
2340	Ergonomic, executive, minimum					405			405	445
2380	Maximum					850			850	935
2390	Management, minimum					340			340	375
2400	Maximum					1,175			1,175	1,300
2450	Task, minimum					152			152	167
2500	Maximum					570			570	625
2550	Arm kit, minimum					36			36	40
2600	Maximum					78.50			78.50	86
3000	Side/guest chairs, upholstered, sled base, metal, min.					145			145	159
3100	Maximum					490			490	540
3200	Wood, minimum					181			181	199
3250	Maximum					935			935	1,025
3400	Traditional, wood leg, minimum					144			144	158
3500	Maximum					840			840	925
3600	Conference chair, upholstered, metal leg, minimum					99			99	109
3700	Maximum				↓	685			685	755

12 52 23.23 Multiple Office Seating Units

		Crew	Daily Output	Labor-Hours	Unit	Material	Labor	Equipment	Total	Total Incl O&P
0010	**MULTIPLE OFFICE SEATING UNITS**									
1000	Area seating, full upholstered, 3 seat straight unit, minimum				Ea.	1,025			1,025	1,125
1020	Maximum					1,675			1,675	1,850
1100	Four seat with corner table, minimum					1,700			1,700	1,875
1120	Maximum					2,275			2,275	2,500
1200	2 seat with in-line table, minimum					1,025			1,025	1,125
1220	Maximum					1,175			1,175	1,300
2000	Individual seat with table, minimum					675			675	745
2020	Maximum				↓	830			830	915

12 54 Hospitality Furniture

12 54 13 – Hotel and Motel Furniture

12 54 13.10 Hotel Furniture

		Crew	Daily Output	Labor-Hours	Unit	Material	Labor	Equipment	Total	Total Incl O&P
0010	**HOTEL FURNITURE**									
0020	Standard quality set, minimum				Room	2,325			2,325	2,550
0200	Maximum				"	8,300			8,300	9,125
0300	Bed frame				Ea.	48			48	53
0400	Bench, upholstered, 42" x 18" x 18"					185			185	203
0420	18" x 18" x 23"					160			160	176
0500	Desk section, one drawer, 34" x 20" x 30"					190			190	209
0600	Free standing, 42" x 22" x 30"					250			250	275
0700	Desk chair, upholstered, minimum					80			80	88
0720	Maximum					140			140	154
1000	Dressers, uniplex, 2 drawer				↓	330			330	365

12 54 Hospitality Furniture

12 54 13 – Hotel and Motel Furniture

12 54 13.10 Hotel Furniture		Crew	Daily Output	Labor-Hours	Unit	2009 Bare Costs Material	Labor	Equipment	Total	Total Incl O&P
1100	3 drawer				Ea.	360			360	395
2000	Guest tables, 30" diameter					105			105	115
2100	34" diameter					135			135	148
3000	Headboards, free standing, twin					110			110	121
3050	Full					210			210	231
3100	Queen					220			220	242
3200	Wall mounted, twin					80			80	88
3250	Full					175			175	192
3300	Queen					195			195	214
4000	Lounge chair, full upholstered, minimum					310			310	340
4050	Maximum					435			435	480
4200	Open arms, minimum					85			85	93.50
4250	Maximum					390			390	430
5000	Mattress/box springs, twin					360			360	395
5050	Full					430			430	475
5100	Queen					490			490	540
5150	Mirror, framed, 29" x 45"					155			155	170
6000	Sleep sofas, twin, minimum					580			580	640
6050	Maximum					660			660	725
6100	Full, minimum					600			600	660
6150	Maximum					1,075			1,075	1,175
6200	Queen, minimum					610			610	670
6250	Maximum					1,100			1,100	1,225
7000	Table, wood top, cocktail, 54" x 24" x 16" high					295			295	325
7020	Corner, 30" x 30" x 21" high					255			255	280
7040	End, 22" x 28" x 21" high				↓	285			285	315

12 54 13.20 Mattress and Box Springs		Crew	Daily Output	Labor-Hours	Unit	Material	Labor	Equipment	Total	Total Incl O&P
0010	**MATTRESS & BOX SPRINGS** per set									
1000	Hospital, 34" x 84"				Ea.	440			440	480
2000	Hotel/motel, twin, minimum					360			360	395
2020	Maximum					560			560	615
2200	Full, minimum					430			430	475
2220	Maximum					680			680	750
2400	Queen, minimum					490			490	540
2420	Maximum					800			800	880
3000	Stow away bed with head board, twin size				↓	380			380	420

12 54 16 – Restaurant Furniture

12 54 16.10 Tables, Folding		Crew	Daily Output	Labor-Hours	Unit	Material	Labor	Equipment	Total	Total Incl O&P
0010	**TABLES, FOLDING** Laminated plastic tops									
1000	Tubular steel legs with glides									
1020	18" x 60", minimum				Ea.	210			210	231
1040	Maximum					1,175			1,175	1,300
1100	18" x 72", minimum					254			254	280
1120	Maximum					1,250			1,250	1,350
1200	18" x 96", minimum					282			282	310
1220	Maximum					1,500			1,500	1,650
1400	30" x 48", minimum					183			183	201
1420	Maximum					975			975	1,075
1500	30" x 60", minimum					179			179	197
1520	Maximum					1,400			1,400	1,525
1600	30" x 72", minimum					252			252	277
1620	Maximum				↓	1,600			1,600	1,775

12 54 16.10 Tables, Folding		Crew	Daily Output	Labor-Hours	Unit	Material	Labor	Equipment	Total	Total Incl O&P
						2009 Bare Costs				
1700	30" x 96", minimum				Ea.	218			218	239
1720	Maximum					2,025			2,025	2,250
1800	36" x 72", minimum					218			218	240
1820	Maximum					1,800			1,800	2,000
1840	36" x 96", minimum					257			257	283
1860	Maximum					2,375			2,375	2,600
2000	Round, wood stained, plywood top, 60" diameter, minimum					365			365	400
2020	Maximum					1,275			1,275	1,400
2200	72" diameter, minimum					325			325	360
2220	Maximum				↓	2,300			2,300	2,550
4000	Mobile storage carts									
4020	For 72" tables, flat, maximum 10 tables				Ea.	575			575	635
4040	96" tables, flat, maximum 10 tables					510			510	560
4060	Rounds, on edge, maximum 10 tables				↓	830			830	915

12 54 16.20 Furniture, Restaurant		Crew	Daily Output	Labor-Hours	Unit	Material	Labor	Equipment	Total	Total Incl O&P
0010	**FURNITURE, RESTAURANT**									
0020	Bars, built-in, front bar	1 Carp	5	1.600	L.F.	238	64		302	370
0200	Back bar	"	5	1.600	"	173	64		237	296
0300	Booth seating, see Div. 12 54 16.70									
2000	Chair, bentwood side chair, metal, minimum				Ea.	76			76	83.50
2020	Maximum					89			89	98
2100	Wood, minimum					159			159	175
2120	Maximum					185			185	204
2400	Bruer Cesca, cane seat & back, arms, minimum					440			440	480
2420	Maximum					515			515	565
2500	Side, minimum					440			440	480
2520	Maximum					515			515	565
2600	Upholstered seat & back, arms, minimum					124			124	136
2620	Maximum					350			350	385
2640	Side, minimum					340			340	375
2660	Maximum					400			400	440
2700	Corbusier, arm chair, cane seat and back, minimum					170			170	187
2720	Maximum					185			185	204
2740	Fledermaus, fabric seat, ash frame, minimum					157			157	173
2760	Maximum					193			193	212
2780	Hoffman arm, fabric seat, cane back, minimum					179			179	197
2800	Maximum					196			196	215
2820	Side, minimum					140			140	154
2840	Maximum					151			151	166
2860	Lombard, minimum					96.50			96.50	106
2880	Maximum					126			126	138
2900	Mies, side chair, leather seat and back, minimum					415			415	455
2920	Maximum					610			610	670
2940	Napoleon, upholstered seat, wood back, minimum					69			69	76
2960	Maximum					88			88	97
3000	Prague, arm, upholstered, minimum					128			128	141
3020	Maximum					276			276	305
3040	Side, upholstered, minimum					128			128	141
3060	Maximum					157			157	173
3080	Contemporary leather seat & back, chrome frame, min.					257			257	282
3100	Maximum					375			375	410
3120	Foam padded seat & back, s. steel barstock, min.				↓	685			685	755

12 54 16 – Restaurant Furniture

12 54 16.20 Furniture, Restaurant		Crew	Daily Output	Labor-Hours	Unit	2009 Bare Costs Material	Labor	Equipment	Total	Total Incl O&P
3140	Maximum				Ea.	730			730	800
3160	Chrome tube cantilever frame, padded arms, min.					106			106	116
3180	Maximum					271			271	298
4200	"Tub" style, minimum					181			181	199
4220	Maximum					410			410	450
4280	Bent wood arms and legs with back bow, minimum					50.50			50.50	55.50
4300	Maximum					335			335	365
4320	Sled base, minimum					106			106	116
4340	Maximum					335			335	365
4360	Armchair foam padded seat & back,carved wood frame min.					305			305	335
4380	Maximum					585			585	645
4400	Ornate wood frame and legs, minimum					330			330	360
4420	Maximum					385			385	425
4440	Heavy wood frame and legs, tailored back, minimum					520			520	575
4460	Maximum					605			605	665
4480	Button tufted back, minimum					340			340	370
4500	Maximum					780			780	860
4520	"Dining" wood frame and legs, minimum					252			252	277
4550	Maximum					294			294	325
4600	"Queen Ann" wood frame and legs, minimum					345			345	380
4620	Maximum					380			380	415
4660	Upholstered arms and wood legs, minimum					480			480	530
4680	Maximum					565			565	625
4700	Wicker, foam padded seat and back, barrel design, minimum					425			425	465
4720	Maximum					480			480	525
4740	Couch design, minimum					315			315	345
4760	Maximum					385			385	425
4780	Chrome tube round cantilever base, minimum					181			181	199
4800	Maximum					400			400	440
4820	Chrome tube square cantilever base, minimum					174			174	191
4840	Maximum					390			390	425
4860	Misc. chair, upholst. seat, wood arms, bow back & legs, min.					142			142	156
4880	Maximum					197			197	217
4900	Open curved wood back, minimum					320			320	350
4950	Maximum					355			355	390
5000	Upholstered seat and open back, wood frame arms, min.					293			293	320
5100	Maximum					340			340	375
5150	Wood frame sled base, minimum					164			164	181
5200	Maximum					272			272	299
5250	Wood saddle seat, "Windsor", wood frame & legs, min.					168			168	185
5300	Maximum					196			196	215
5350	Brass nail trim leather, minimum					197			197	217
5400	Maximum					225			225	247
5450	Wood frame, fabric, minimum					169			169	185
5500	Maximum					199			199	219
6000	Stools, upholst. seat & back, chrome tube cantilever frame, min.					264			264	291
6150	Swivel on chrome post mount with spread base, minimum					155			155	170
6250	"Vienna" wood back and legs, minimum					207			207	228
6350	"Napoleon" wood back and legs, minimum					71			71	78
6450	Wood swivel seat and curved spindle back, leg base					123			123	135
6600	Upholstered seat no back, bent wood legs					248			248	272
7000	With back & straight wood frame and legs					246			246	271
7200	Wood swivel seat, "Captain" wood back and legs				↓	157			157	172

12 54 16.20 Furniture, Restaurant		Crew	Daily Output	Labor-Hours	Unit	Material	Labor	Equipment	Total	Total Incl O&P
7400	Veneer seat and curved wood back with arms, leg base				Ea.	149			149	164
7600	Upholstered swivel seat no back, wood legs					59			59	65
8000	With back & wood legs					430			430	475
8200	Upholstered seat & wood back, wood legs					410			410	450
8400	Upholstered swivel seat and back, wood legs					167			167	184
8600	Square tube frame and legs					42.50			42.50	47
8800	Metal frame and legs					42.50			42.50	47

12 54 16.30 Table Bases

		Crew	Daily Output	Labor-Hours	Unit	Material	Labor	Equipment	Total	Total Incl O&P
0010	**TABLE BASES**									
0040	Dining height									
1000	Metal disk design, minimum				Ea.	78			78	85.50
1200	Maximum					207			207	227
1400	Wood, minimum					330			330	365
1600	Maximum					365			365	400
2000	Fluted wheel, minimum					90.50			90.50	99.50
2200	Maximum					192			192	212
2400	Heavy cast iron, minimum					79			79	87
2600	Maximum					192			192	211
2800	Hobnail design, minimum					93			93	102
3000	Maximum					283			283	310
3200	Manhole design, minimum					92.50			92.50	102
3400	Maximum					248			248	272
3600	Ring design, minimum					129			129	142
3800	Maximum					213			213	234
4000	Trumpet design, minimum					233			233	256
4200	Maximum					256			256	281
4400	Tubular, round or rectangular shape, minimum					92			92	101
4600	Maximum					193			193	212
5000	Cocktail height bases, minimum					107			107	117
5200	Maximum					118			118	129
5400	For foot ring, add, minimum					11.20			11.20	12.30
5600	Maximum					25.50			25.50	28

12 54 16.40 Table Tops

		Crew	Daily Output	Labor-Hours	Unit	Material	Labor	Equipment	Total	Total Incl O&P
0010	**TABLE TOPS** Laminated plastic top and edge									
0040	24" wide, 24" long, minimum				Ea.	60.50			60.50	66.50
1000	Maximum					235			235	258
1200	30" long, minimum					70			70	77
1400	Maximum					345			345	380
1600	36" long, minimum					88.50			88.50	97.50
1800	Maximum					365			365	405
2000	48" long, minimum					107			107	117
2200	Maximum					435			435	475
2400	60" long, minimum					173			173	190
2600	Maximum					675			675	745
2800	72" long, minimum					211			211	232
3000	Maximum					710			710	785
3050	30" wide, 30" long, minimum					73.50			73.50	81
3150	Maximum					435			435	475
3200	36" long, minimum					92.50			92.50	102
3250	Maximum					435			435	475
3400	42" long, minimum					95.50			95.50	105
3600	Maximum					250			250	275

12 54 16.40 Table Tops		Crew	Daily Output	Labor-Hours	Unit	2009 Bare Costs Material	Labor	Equipment	Total	Total Incl O&P
3800	48" long, minimum				Ea.	120			120	132
4000	Maximum					475			475	525
4200	60" long, minimum					170			170	187
4400	Maximum					720			720	795
4600	72" long, minimum					208			208	229
4800	Maximum					720			720	795
5000	36" wide, 36" long, minimum					98			98	108
5200	Maximum					700			700	770
5400	48" long, minimum					142			142	157
5600	Maximum					560			560	615
5800	60" long, minimum					202			202	223
6000	Maximum					720			720	795
6200	72" long, minimum					231			231	254
6400	Maximum					850			850	935
6500	42" wide, 42" long, minimum					176			176	194
6600	Maximum					700			700	770
6700	Round, 24" diameter, minimum					66.50			66.50	73
6800	Maximum					290			290	320
6900	30" diameter, minimum					87.50			87.50	96
7000	Maximum					435			435	475
7200	36" diameter, minimum					113			113	124
7400	Maximum					495			495	545
7600	42" diameter, minimum					180			180	198
8000	Maximum					710			710	785
8200	48" diameter, minimum					202			202	223
8400	Maximum					800			800	880
8800	54" diameter, minimum					325			325	355
9000	Maximum					1,050			1,050	1,150
9200	60" diameter, minimum					345			345	380
9400	Maximum				↓	1,300			1,300	1,425

12 54 16.50 Table Tops		Crew	Daily Output	Labor-Hours	Unit	Material	Labor	Equipment	Total	Total Incl O&P
0010	**TABLE TOPS** laminate top and hardwood edge									
0040	24" wide, 24" long, minimum				Ea.	192			192	211
1000	Maximum					610			610	675
1200	30" long, minimum					212			212	234
1400	Maximum					655			655	720
1600	36" long, minimum					199			199	219
1800	Maximum					690			690	755
2000	48" long, minimum					280			280	310
2100	Maximum					755			755	835
2200	60" long, minimum					400			400	440
2400	Maximum					830			830	910
2600	72" long, minimum					455			455	500
2800	Maximum					925			925	1,025
3000	30" wide, 30" long, minimum					232			232	255
3100	Maximum					695			695	765
3200	36" long, minimum					275			275	305
3400	Maximum					750			750	825
3600	42" long, minimum					273			273	300
3800	Maximum					925			925	1,025
4000	48" long, minimum					305			305	335
4100	Maximum				↓	805			805	890

12 54 16.50 Table Tops		Crew	Daily Output	Labor-Hours	Unit	Material	Labor	Equipment	Total	Total Incl O&P
4200	60" long, minimum				Ea.	410			410	450
4300	Maximum					865			865	950
4400	72" long, minimum					330			330	360
4600	Maximum					905			905	995
4800	36" wide, 36" long, minimum					282			282	310
5000	Maximum					765			765	845
5200	48" long, minimum					330			330	365
5400	Maximum					875			875	965
5600	60" long, minimum					470			470	515
5800	Maximum					970			970	1,075
6000	72" long, minimum					510			510	565
6200	Maximum					1,075			1,075	1,175
6400	42" wide, 42" long, minimum					340			340	375
6600	Maximum					930			930	1,025
6800	Round, 24" diameter, minimum					350			350	385
6900	Maximum					650			650	715
7000	30" diameter, minimum					385			385	425
7200	Maximum					730			730	805
7400	36" diameter, minimum					425			425	465
7600	Maximum					810			810	895
7800	42" diameter, minimum					490			490	540
7900	Maximum					995			995	1,100
8000	48" diameter, minimum					530			530	580
8150	Maximum					1,050			1,050	1,175
8200	54" diameter, minimum					790			790	870
8250	Maximum					1,400			1,400	1,525
8300	60" diameter, minimum					800			800	880
8350	Maximum				↓	1,400			1,400	1,525
12 54 16.60 Table Tops										
0010	**TABLE TOPS** Polyester resin top & edge									
0050	Add inlay top material to cost									
0100	24" wide, 24" long				Ea.	365			365	400
1200	30" long					365			365	400
1600	36" long					405			405	445
2000	48" long					415			415	460
2400	60" long					495			495	545
2800	72" long					580			580	635
3200	30" wide, 30" long					410			410	450
3600	36" long					485			485	535
4000	42" long					510			510	560
4400	48" long					535			535	590
4800	60" long					610			610	670
5200	72" long					740			740	815
5600	36" wide, 36" long					485			485	535
6000	48" long					585			585	645
6400	60" long					715			715	785
6800	72" long					855			855	945
7200	42" wide, 42" long					570			570	630
7600	Round, 24" diameter					545			545	600
8000	30" diameter					565			565	620
8400	36" diameter					670			670	735
8800	42" diameter				↓	900			900	990

Material, Labor, Equipment and Total columns: 2009 Bare Costs

12 54 16 – Restaurant Furniture

12 54 16.60 Table Tops		Crew	Daily Output	Labor-Hours	Unit	2009 Bare Costs Material	Labor	Equipment	Total	Total Incl O&P
9150	48" diameter				Ea.	775			775	855
9250	54" diameter				↓	1,350			1,350	1,475
9350	60" diameter				↓	1,400			1,400	1,550

12 54 16.70 Booths		Crew	Daily Output	Labor-Hours	Unit	2009 Bare Costs Material	Labor	Equipment	Total	Total Incl O&P
0010	**BOOTHS**									
1000	Banquet, upholstered seat and back, custom									
1500	Straight, minimum	2 Carp	40	.400	L.F.	161	16		177	204
1520	Maximum		36	.444		315	17.75		332.75	375
1600	"L" or "U" shape, minimum		35	.457		164	18.25		182.25	211
1620	Maximum	↓	30	.533	↓	292	21.50		313.50	355
1800	Upholstered outside finished backs for									
1810	single booths and custom banquets									
1820	Minimum	2 Carp	44	.364	L.F.	18.55	14.55		33.10	44.50
1840	Maximum	"	40	.400	"	55.50	16		71.50	87.50
3000	Fixed seating, one piece plastic chair and									
3010	plastic laminate table top									
3100	Two seat, 24" x 24" table, minimum	F-7	30	1.067	Ea.	650	38		688	780
3120	Maximum		26	1.231		925	44		969	1,100
3200	Four seat, 24" x 48" table, minimum		28	1.143		645	41		686	775
3220	Maximum		24	1.333		1,100	47.50		1,147.50	1,275
3300	Six seat, 24" x 76" table, minimum		26	1.231		1,175	44		1,219	1,350
3320	Maximum		22	1.455		1,600	52		1,652	1,850
3400	Eight seat, 24" x 102" table, minimum		20	1.600		1,550	57.50		1,607.50	1,800
3420	Maximum	↓	18	1.778	↓	2,025	63.50		2,088.50	2,325
4000	Free standing, wood fiber core with									
4010	plastic laminate face, single booth									
4100	24" wide	2 Carp	38	.421	Ea.	305	16.80		321.80	370
4150	48" wide		34	.471		370	18.80		388.80	440
4200	60" wide		30	.533		455	21.50		476.50	535
4300	Double booth, 24" wide		32	.500		455	20		475	535
4350	48" wide		28	.571		605	23		628	705
4400	60" wide	↓	26	.615	↓	765	24.50		789.50	880
4600	Upholstered seat and back									
4650	Foursome, single booth, minimum	2 Carp	38	.421	Ea.	345	16.80		361.80	410
4700	Maximum		30	.533		1,325	21.50		1,346.50	1,475
4800	Double booth, minimum		32	.500		765	20		785	875
4850	Maximum	↓	26	.615	↓	2,025	24.50		2,049.50	2,300
5000	Mount in floor, wood fiber core with									
5010	plastic laminate face, single booth									
5050	24" wide	F-7	30	1.067	Ea.	249	38		287	335
5100	48" wide		28	1.143		315	41		356	420
5150	60" wide		26	1.231		455	44		499	575
5200	Double booth, 24" wide		26	1.231		400	44		444	515
5250	48" wide		24	1.333		510	47.50		557.50	645
5300	60" wide	↓	22	1.455	↓	725	52		777	885

12 55 Detention Furniture

12 55 13 – Detention Bunks

12 55 13.13 Cots		Crew	Daily Output	Labor-Hours	Unit	2009 Bare Costs Material	Labor	Equipment	Total	Total Incl O&P
0010	**COTS**									
2500	Bolted, single, painted steel	E-4	20	1.600	Ea.	330	72.50	6.70	409.20	505
2700	Stainless steel	"	20	1.600	"	960	72.50	6.70	1,039.20	1,200

12 56 Institutional Furniture

12 56 33 – Classroom Furniture

12 56 33.10 Furniture, School

		Crew	Daily Output	Labor-Hours	Unit	Material	Labor	Equipment	Total	Total Incl O&P
0010	**FURNITURE, SCHOOL**									
0500	Classroom, movable chair & desk type, minimum				Set				73.50	81
0600	Maximum				"				134	148
1000	Chair, molded plastic									
1100	Integral tablet arm, minimum				Ea.	72.50			72.50	79.50
1150	Maximum					148			148	163
2000	Desk, single pedestal, top book compartment, minimum					53.50			53.50	59
2020	Maximum					67.50			67.50	74
2100	Side book compartment, minimum					77			77	84.50
2120	Maximum					97			97	107
2200	Flip top, minimum					95.50			95.50	105
2220	Maximum					122			122	134
3000	Preschool, moulded plastic chairs, minimum					28			28	30.50
3020	Maximum					46.50			46.50	51.50
3800	Tables, plastic laminate top, 24" wide, 36" long					73			73	80
3820	48" long					92			92	101
3840	60" long					104			104	115
3900	30" wide, 48" long					97.50			97.50	107
3920	60" long					140			140	155
3940	72" long					157			157	173
4000	36" wide, 36" long					85.50			85.50	94
4020	60" long					123			123	136
4040	72" long					217			217	239
4200	Round, 36" dia.					85			85	93
4220	42" dia.					97.50			97.50	107
4230	48" dia.					122			122	135
4240	60" dia.				↓	169			169	186

12 56 43 – Dormitory Furniture

12 56 43.10 Dormitory Furnishings

		Crew	Daily Output	Labor-Hours	Unit	Material	Labor	Equipment	Total	Total Incl O&P
0010	**DORMITORY FURNISHINGS**									
0200	Bookcase, two shelf				Ea.	145			145	159
0300	Bunkable bed, twin, minimum					270			270	297
0320	Maximum					650			650	715
1000	Chest, four drawer, minimum					420			420	460
1020	Maximum				↓	550			550	605
1050	Built-in, minimum	2 Carp	13	1.231	L.F.	135	49		184	230
1150	Maximum		10	1.600		250	64		314	380
1200	Desk top, built-in, laminated plastic, 24" deep, minimum		50	.320		37	12.80		49.80	61.50
1300	Maximum		40	.400		111	16		127	149
1450	30" deep, minimum		50	.320		47.50	12.80		60.30	73
1550	Maximum		40	.400		207	16		223	255
1750	Dressing unit, built-in, minimum		12	1.333		203	53.50		256.50	310
1850	Maximum	↓	8	2	↓	610	80		690	800

12 56 Institutional Furniture

12 56 43 – Dormitory Furniture

12 56 43.10 Dormitory Furnishings		Crew	Daily Output	Labor-Hours	Unit	2009 Bare Costs Material	Labor	Equipment	Total	Total Incl O&P
2000	Desk, single pedestal				Ea.	460			460	505
2100	Hutch/bookcase, with light					222			222	244
3000	Ladder					179			179	197
4000	Mirror, with frame					104			104	114
5000	Nightstand					220			220	243
6000	Wall unit, open shelving					530			530	580
7000	Wardrobe					550			550	605
8000	Rule of thumb: total cost for furniture, minimum				Student				2,125	2,350
8050	Maximum				″				4,100	4,450

12 56 51 – Library Furniture

12 56 51.10 Library Furnishings		Crew	Daily Output	Labor-Hours	Unit	Material	Labor	Equipment	Total	Total Incl O&P
0010	**LIBRARY FURNISHINGS**									
0100	Attendant desk, 36″ x 62″ x 29″ high	1 Carp	16	.500	Ea.	2,575	20		2,595	2,850
0200	Book display, "A" frame display, both sides, 42″ x 42″ x 60″ high		16	.500		1,475	20		1,495	1,650
0220	Table with bulletin board, 42″ x 24″ x 49″ high		16	.500		1,300	20		1,320	1,450
0300	Book trucks, descending platform,									
0320	Small, 14″ x 30″ x 35″ high	1 Carp	16	.500	Ea.	730	20		750	835
0340	Large, 14″ x 40″ x 42″ high		16	.500		765	20		785	875
0800	Card catalogue, 30 tray unit		16	.500		2,775	20		2,795	3,075
0840	60 tray unit		16	.500		5,075	20		5,095	5,600
0880	72 tray unit	2 Carp	16	1		6,225	40		6,265	6,900
0960	120 tray unit	″	16	1		9,475	40		9,515	10,500
1000	Carrels, single face, initial unit	1 Carp	16	.500		700	20		720	805
1050	Additional unit	″	16	.500		720	20		740	825
1500	Double face, initial unit	2 Carp	16	1		1,100	40		1,140	1,300
1550	Additional unit		16	1		935	40		975	1,100
1600	Cloverleaf		11	1.455		1,900	58		1,958	2,175
1710	Carrels, hardwood, 36″ x 24″, minimum	1 Carp	5	1.600		510	64		574	665
1720	Maximum		4	2		785	80		865	995
2000	Chairs, sled base, arms, minimum		24	.333		245	13.30		258.30	291
2050	Maximum		16	.500		273	20		293	335
2100	No arms, minimum		24	.333		200	13.30		213.30	242
2150	Maximum		16	.500		223	20		243	278
2500	Standard leg base, arms, minimum		24	.333		233	13.30		246.30	278
2520	Maximum		16	.500		375	20		395	450
2600	No arms, minimum		24	.333		163	13.30		176.30	201
2620	Maximum		16	.500		280	20		300	345
2700	Card catalog file, 60 trays, complete					5,050			5,050	5,550
2720	Alternate method: each tray					84			84	92.50
3000	Charge desk, modular unit, 35″ x 27″ x 39″ high									
3020	Wood front and edges, plastic laminate tops									
3100	Book return	1 Carp	16	.500	Ea.	865	20		885	990
3150	Book truck port		16	.500		1,000	20		1,020	1,125
3200	Card file drawer, 5 drawers		16	.500		2,025	20		2,045	2,250
3250	10 drawers		16	.500		2,425	20		2,445	2,700
3280	15 drawers		16	.500		2,675	20		2,695	2,975
3300	Card & legal file		16	.500		2,925	20		2,945	3,250
3350	Charging machine		16	.500		2,150	20		2,170	2,400
3400	Corner		16	.500		965	20		985	1,075
3450	Cupboard		16	.500		1,375	20		1,395	1,525
3500	Detachable end panel		16	.500		425	20		445	500
3550	Gate		16	.500		510	20		530	595

12 56 Institutional Furniture

12 56 51 – Library Furniture

12 56 51.10 Library Furnishings		Crew	Daily Output	Labor-Hours	Unit	Material	2009 Bare Costs Labor	Equipment	Total	Total Incl O&P
3600	Knee space	1 Carp	16	.500	Ea.	1,325	20		1,345	1,475
3650	Open storage		16	.500		1,275	20		1,295	1,425
3700	Station charge		16	.500		2,550	20		2,570	2,825
3750	Work station		16	.500		2,250	20		2,270	2,500
3800	Charging desk, built-in, with counter, plastic laminated top		7	1.143	L.F.	260	45.50		305.50	360
4000	Dictionary stand, stationary		16	.500	Ea.	835	20		855	955
4020	Revolving		16	.500		335	20		355	405
4200	Exhibit case, table style, 60″ x 28″ x 36″		11	.727		2,725	29		2,754	3,050
4500	Globe stand		16	.500		1,025	20		1,045	1,150
4800	Magazine rack		16	.500		1,425	20		1,445	1,600
5000	Newspaper rack		16	.500		930	20		950	1,050
6010	Bookshelf, mtl, 90″ high, 10″ shelf, dbl face		11.50	.696	L.F.	118	28		146	176
6020	Single face		12	.667	″	113	26.50		139.50	168
6050	For 8″ shelving, subtract from above					10%				
6060	For 12″ shelving, add to above					10%				
6070	For 42″ high with countertop, subtract from above					20%				
6100	Mobile compacted shelving, hand crank, 9′-0″ high									
6110	Double face, including track, 3′ section				Ea.	1,125			1,125	1,250
6150	For electrical operation, add					25%				
6200	Magazine shelving, 82″ high, 12″ deep, single face	1 Carp	11.50	.696	L.F.	109	28		137	166
6210	Double face		11.50	.696	″	105	28		133	161
7000	Tables, card catalog reference, 24″ x 60″ x 42″		16	.500	Ea.	945	20		965	1,075
7050	24″ x 60″ x 72″		16	.500		1,250	20		1,270	1,425
7100	Index, single tier, 48″ x 72″		16	.500		2,025	20		2,045	2,250
7150	Double tier, 48″ x 72″		16	.500		2,275	20		2,295	2,525
7200	Reading table, laminated top, 60″ x 36″					645			645	710
8000	Parsons table, 29″ high, plastic lam. top, wood legs & edges									
8010	36″ x 36″	2 Carp	16	1	Ea.	735	40		775	870
8020	36″ x 60″		16	1		740	40		780	880
8030	36″ x 72″		16	1		830	40		870	975
8040	36″ x 84″		16	1		1,175	40		1,215	1,375
8050	42″ x 90″		16	1		1,125	40		1,165	1,325
8060	48″ x 72″		16	1		985	40		1,025	1,150
8070	48″ x 120″		16	1		2,200	40		2,240	2,500
8100	Round, leg or pedestal base, 36″ diameter		16	1		615	40		655	740
8110	42″ diameter		16	1		850	40		890	1,000
8120	48″ diameter		16	1		935	40		975	1,100
8130	60″ diameter		16	1		1,200	40		1,240	1,375
8500	Study, panel ends, plastic laminate surfaces 29″ high, 36″ x 60″		16	1		915	40		955	1,075
8510	36″ x 72″		16	1		980	40		1,020	1,150
8515	36″ x 90″		16	1		1,025	40		1,065	1,225
8525	48″ x 72″		16	1		1,075	40		1,115	1,275

12 56 70 – Healthcare Furniture

12 56 70.10 Furniture, Hospital		Crew	Daily Output	Labor-Hours	Unit	Material	Labor	Equipment	Total	Total Incl O&P
0010	**FURNITURE, HOSPITAL**									
0020	Beds, manual, minimum				Ea.	730			730	800
0100	Maximum					2,500			2,500	2,750
0300	Manual and electric beds, minimum					925			925	1,025
0400	Maximum					2,825			2,825	3,125
0600	All electric hospital beds, minimum					1,325			1,325	1,450
0700	Maximum					4,100			4,100	4,525
0900	Manual, nursing home beds, minimum					810			810	890

12 56 Institutional Furniture

12 56 70 – Healthcare Furniture

12 56 70.10 Furniture, Hospital		Crew	Daily Output	Labor-Hours	Unit	2009 Bare Costs Material	Labor	Equipment	Total	Total Incl O&P
1000	Maximum				Ea.	1,675			1,675	1,850
1020	Overbed table, laminated top, minimum					345			345	380
1040	Maximum					860			860	945
1100	Patient wall systems, not incl. plumbing, minimum				Room	1,025			1,025	1,125
1200	Maximum				"	1,900			1,900	2,100
2000	Geriatric chairs, minimum				Ea.	470			470	515
2020	Maximum				"	705			705	780

12 59 Systems Furniture

12 59 13 – Panel-Hung Component System Furniture

12 59 13.10 Furniture, Office Systems		Crew	Daily Output	Labor-Hours	Unit	Material	Labor	Equipment	Total	Total Incl O&P
0010	**FURNITURE, OFFICE SYSTEMS**, Panel hung									
0100	Acoustic panel, 43" high, 24" wide, NRC .85				Ea.	230			230	253
0110	30" wide					257			257	282
0120	36" wide					284			284	310
0130	48" wide					340			340	375
0200	64" high, 30" wide					305			305	335
0210	36" wide					325			325	360
0220	42" wide					380			380	415
0230	48" wide					400			400	440
0240	60" wide					440			440	485
0400	Bookshelf, 36" wide					106			106	117
0410	42" wide					112			112	123
0420	48" wide					115			115	127
1000	Connectors, brackets and supports									
1200	Bracket, cantilever, 20" deep, 24" deep				Ea.	36			36	39.50
1220	Countertop, per pair					14.95			14.95	16.45
1240	Flat, 20" deep, 24" wide					18.45			18.45	20.50
1260	Worksurface kit, per pair					14.95			14.95	16.45
1300	Connector kit, ell, 43" high, 90" deep					64			64	70.50
1320	64" high					38			38	41.50
1340	Straight, 43" high					26.50			26.50	29
1360	64" high					26.50			26.50	29
1400	Support column for peninsula					92			92	101
2000	Countertop, 15" deep, 36" wide					121			121	133
2020	48" wide					135			135	149
3000	Duplex receptacle circuit 1					13.20			13.20	14.50
3010	Circuit 2					13.20			13.20	14.50
3100	Electric power harness, 36" wide					86			86	94.50
3120	42" wide					181			181	199
3140	60" wide					90.50			90.50	99.50
4000	End cover, panel, 43" high					26.50			26.50	29
4020	64" high					26.50			26.50	29
4100	Finish, variable height, 2-way					39.50			39.50	43.50
4600	Overhead cabinet with door, 42" wide					250			250	274
4620	60" wide					380			380	415
5000	Pedestal spacer, 22" deep, 15" wide					43			43	47.50
5100	28" deep, 15" wide					52			52	57
5200	Box, box, file, 22" deep, 26" high					300			300	330
5300	28" deep					315			315	345
5400	File, file, 22" deep, 26" high					280			280	310

12 59 Systems Furniture

12 59 13 – Panel-Hung Component System Furniture

12 59 13.10 Furniture, Office Systems		Crew	Daily Output	Labor-Hours	Unit	Material	Labor	Equipment	Total	Total Incl O&P
5600	Lateral file, 2 drawer, 30" wide				Ea.	395			395	435
6000	Power, base in-feed cable					96			96	105
7400	Task light, recessed, 30" - 36" wide					117			117	129
7420	42" - 48" wide					127			127	139
7600	60" wide					137			137	151
8000	Worksurface, radius edge, 24" deep, 36" wide					126			126	138
8020	42" wide					155			155	170
8040	48" wide					165			165	182
8060	60" wide					207			207	228
8080	72" wide					234			234	257
8200	30" deep, 42" wide					198			198	218
8220	60" wide					231			231	254
8240	72" wide					264			264	290
8400	Corner, 24" deep, 36" wide					305			305	335
8420	42" wide					345			345	380
8500	Peninsula, 36" wide, 66" long				↓	395			395	435
9000	For installation of systems furniture, add 5%									

(Material, Labor, Equipment and Total are under the heading "2009 Bare Costs".)

12 59 16 – Free-Standing Component System Furniture

12 59 16.10 Furniture, Office Systems		Crew	Daily Output	Labor-Hours	Unit	Material	Labor	Equipment	Total	Total Incl O&P
0010	**FURNITURE, OFFICE SYSTEMS**, Freestanding									
0100	Desk table, 24" deep x 48" wide				Ea.	540			540	595
0110	30" deep x 48" wide					550			550	605
0120	30" deep x 60" wide					560			560	615
0130	30" deep x 72" wide					620			620	680
0200	File, mobile, 2 drawer, 22" deep					375			375	415
0210	Suspended, 2 file					395			395	435
0220	Undercounter					430			430	470
0230	2 box, 1 file					460			460	505
0400	Keyboard platform with articulating arm					142			142	156
0500	End support leg					127			127	140
0800	Modesty panel, 48" wide					72.50			72.50	80
0810	Return					79.50			79.50	87
0820	Pencil/utility drawer					105			105	116
0830	Privacy screen, 48" wide x 17-20" high					211			211	233
0840	72" wide					335			335	370
0910	Printer stand, 30" deep x 36" wide					450			450	495
0920	Universal					460			460	505
0930	Paper basket					53			53	58
1010	Storage unit with doors, 60" wide					515			515	565
1020	72" wide					755			755	830
1210	Table, peninsula, 30" x 72"					580			580	640
1220	Return, 24" x 48"					400			400	440
1230	Round conference return, 42" diameter					395			395	435
1510	Work surface, 24" deep x 48" wide					251			251	276
1520	30" deep x 30" wide					231			231	254
1530	60" wide					310			310	345
1540	72" wide					325			325	360
1550	Corner bridge, 30" x 42"					174			174	191
1560	Peninsula, 30" x 60"					480			480	530
1570	Connector plate				↓	13.20			13.20	14.55

12 59 Systems Furniture

12 59 23 – Desk System Furniture

12 59 23.10 Work Stations		Crew	Daily Output	Labor-Hours	Unit	Material	Labor	Equipment	Total	Total Incl O&P
						2009 Bare Costs				
0010	**WORK STATIONS**									
1000	Secretarial work station, minimum				Ea.	2,975			2,975	3,250
1020	Maximum					10,100			10,100	11,100
1400	Management work station, minimum					3,375			3,375	3,725
1420	Maximum					10,900			10,900	12,000
1800	Executive work station, minimum					7,500			7,500	8,250
1820	Maximum				↓	19,300			19,300	21,200
1840	For installation of systems furniture, add 15%									

12 61 Fixed Audience Seating

12 61 13 – Upholstered Audience Seating

12 61 13.13 Auditorium Chairs		Crew	Daily Output	Labor-Hours	Unit	Material	Labor	Equipment	Total	Total Incl O&P
0010	**AUDITORIUM CHAIRS**									
2000	All veneer construction	2 Carp	22	.727	Ea.	175	29		204	241
2200	Veneer back, padded seat		22	.727		221	29		250	291
2350	Fully upholstered, spring seat	↓	22	.727		211	29		240	280
2450	For tablet arms, add					58			58	63.50
2500	For fire retardancy, CATB-133, add				↓	25			25	27.50

12 61 13.23 Lecture Hall Seating		Crew	Daily Output	Labor-Hours	Unit	Material	Labor	Equipment	Total	Total Incl O&P
0010	**LECTURE HALL SEATING**									
1000	Pedestal type, minimum	2 Carp	22	.727	Ea.	165	29		194	230
1200	Maximum	"	14.50	1.103	"	555	44		599	685

12 63 Stadium and Arena Seating

12 63 13 – Stadium and Arena Bench Seating

12 63 13.13 Bleachers		Crew	Daily Output	Labor-Hours	Unit	Material	Labor	Equipment	Total	Total Incl O&P
0010	**BLEACHERS**									
3000	Telescoping, manual to 15 tier, minimum	F-5	65	.492	Seat	76	19.90		95.90	117
3100	Maximum		60	.533		114	21.50		135.50	162
3300	16 to 20 tier, minimum		60	.533		183	21.50		204.50	237
3400	Maximum		55	.582		228	23.50		251.50	290
3600	21 to 30 tier, minimum		50	.640		190	26		216	252
3700	Maximum	↓	40	.800		228	32.50		260.50	305
3900	For integral power operation, add, minimum	2 Elec	300	.053		38	2.51		40.51	46
4000	Maximum	"	250	.064	↓	61	3.01		64.01	71.50
5000	Benches, folding, in wall, 14' table, 2 benches	L-4	2	12	Set	645	450		1,095	1,450

12 67 Pews and Benches

12 67 13 – Pews

12 67 13.13 Sanctuary Pews		Crew	Daily Output	Labor-Hours	Unit	Material	2009 Bare Costs Labor	Equipment	Total	Total Incl O&P
0010	**SANCTUARY PEWS**									
1500	Bench type, hardwood, minimum	1 Carp	20	.400	L.F.	74	16		90	108
1550	Maximum	"	15	.533		147	21.50		168.50	196
1570	For kneeler, add				↓	18.15			18.15	19.95

12 92 Interior Planters and Artificial Plants

12 92 13 – Interior Artificial Plants

12 92 13.10 Plants		Crew	Daily Output	Labor-Hours	Unit	Material	2009 Bare Costs Labor	Equipment	Total	Total Incl O&P
0010	**PLANTS** Permanent only, weighted									
0020	Preserved or polyester leaf, natural wood									
0030	Or molded trunk. For pots see Div. 12 92 33.10									
0040	For fill see Div. 31 23 16.00									
0100	Plants, acuba, 5′ high				Ea.	210			210	231
0200	Apidistra, 4′ high					198			198	217
0300	Beech, variegated, 5′ high					193			193	212
0400	Birds nest fern, 6′ high					345			345	380
0500	Croton, 4′ high					84.50			84.50	93
0600	Diffenbachia, 3′ high					78			78	86
0700	Ficus Benjamina, 3′ high					92.50			92.50	102
0720	6′ high					234			234	257
0760	Nitida, 3′ high					208			208	229
0780	6′ high					365			365	400
0800	Helicona					193			193	212
1200	Palm, green date fan, 5′ high					201			201	221
1220	7′ high					254			254	279
1280	Green chamdora date, 5′ high					182			182	200
1300	7′ high					278			278	305
1400	Green giant, 7′ high					430			430	475
1600	Rubber plant, 5′ high					147			147	161
1800	Schefflera, 3′ high					100			100	110
1820	4′ high					124			124	136
1840	5′ high					193			193	212
1860	6′ high					263			263	289
1880	7′ high					400			400	440
1900	Spathiphyllum, 3′ high				↓	139			139	153
4000	Trees, polyester or preserved, with									
4020	Natural trunks									
4100	Acuba, 10′ high				Ea.	1,100			1,100	1,200
4120	12′ high					1,525			1,525	1,700
4140	14′ high					2,150			2,150	2,375
4400	Bamboo, 10′ high					480			480	525
4420	12′ high					555			555	610
4800	Beech, 10′ high					1,475			1,475	1,625
4820	12′ high					1,700			1,700	1,875
4840	14′ high					2,325			2,325	2,550
5000	Birch, 10′ high					1,475			1,475	1,625
5020	12′ high					1,700			1,700	1,875
5040	14′ high					1,825			1,825	2,025
5060	16′ high					2,325			2,325	2,550
5080	18′ high					2,600			2,600	2,850
5500	Ficus Benjamina, 10′ high				↓	1,075			1,075	1,200

12 92 Interior Planters and Artificial Plants

12 92 13 – Interior Artificial Plants

12 92 13.10 Plants		Crew	Daily Output	Labor-Hours	Unit	Material	2009 Bare Costs Labor	Equipment	Total	Total Incl O&P
5520	12′ high				Ea.	1,400			1,400	1,525
5540	14′ high					1,550			1,550	1,700
5560	16′ high					2,150			2,150	2,375
5580	18′ high					3,250			3,250	3,575
6000	Magnolia, 10′ high					1,575			1,575	1,750
6020	12′ high					2,250			2,250	2,475
6040	14′ high					3,050			3,050	3,375
6500	Maple, 10′ high					169			169	186
6520	12′ high					2,000			2,000	2,200
6540	14′ high					2,475			2,475	2,725
7000	Palms, chamadora, 10′ high					310			310	340
7020	12′ high					395			395	435
7040	Date, 10′ high					425			425	465
7060	12′ high				▼	485			485	535

12 92 33 – Interior Planters

12 92 33.10 Planters		Crew	Daily Output	Labor-Hours	Unit	Material	Labor	Equipment	Total	Total Incl O&P
0010	**PLANTERS**									
1000	Fiberglass, hanging, 12″ diameter, 7″ high				Ea.	70			70	77
1100	15″ diameter, 7″ high					104			104	115
1200	36″ diameter, 8″ high					300			300	330
1500	Rectangular, 48″ long, 16″ high x 15″ wide					395			395	435
1550	16″ high x 24″ wide					410			410	450
1600	24″ high x 24″ wide					610			610	670
1650	60″ long, 30″ high, 28″ wide					535			535	585
1700	72″ long, 16″ high, 15″ wide					875			875	965
1750	21″ high, 24″ wide					965			965	1,075
1800	30″ high, 24″ wide					1,225			1,225	1,350
2000	Round, 12″ diameter, 13″ high					62.50			62.50	68.50
2050	25″ high					65.50			65.50	72
2150	14″ diameter, 15″ high					89			89	98
2200	16″ diameter, 16″ high					107			107	117
2250	18″ diameter, 19″ high					128			128	141
2300	23″ high					106			106	116
2350	20″ diameter, 16″ high					101			101	111
2400	18″ high					202			202	222
2450	21″ high					108			108	118
2500	22″ diameter, 10″ high					210			210	231
2550	24″ diameter, 16″ high					103			103	113
2600	19″ high					160			160	176
2650	25″ high					177			177	194
2700	36″ high					234			234	257
2750	48″ high					525			525	575
2800	30″ diameter, 16″ high					185			185	204
2850	18″ high					192			192	211
2900	21″ high					249			249	274
3000	24″ high					330			330	360
3350	27″ high					150			150	165
3400	36″ diameter, 16″ high					177			177	195
3450	18″ high					180			180	198
3500	21″ high					182			182	200
3550	24″ high					192			192	212
3600	27″ high				▼	206			206	226

12 92 33 – Interior Planters

12 92 33.10 Planters		Crew	Daily Output	Labor-Hours	Unit	2009 Bare Costs Material	Labor	Equipment	Total	Total Incl O&P
3650	30″ high				Ea.	226			226	249
3700	48″ diameter, 16″ high					395			395	435
3750	21″ high					236			236	260
3800	24″ high					258			258	284
3850	27″ high					279			279	305
3900	30″ high					420			420	460
3950	36″ high					1,000			1,000	1,100
4000	60″ diameter, 16″ high					910			910	1,000
4100	21″ high					1,025			1,025	1,150
4150	27″ high					1,225			1,225	1,350
4200	30″ high					1,475			1,475	1,600
4250	33″ high					1,400			1,400	1,550
4300	36″ high					1,400			1,400	1,550
4400	39″ high					1,525			1,525	1,675
5000	Square, 10″ side, 20″ high					150			150	165
5100	14″ side, 15″ high					178			178	196
5200	18″ side, 19″ high					199			199	218
5300	20″ side, 16″ high					246			246	271
5320	18″ high					299			299	330
5340	21″ high					325			325	355
5400	24″ side, 16″ high					390			390	425
5420	21″ high					420			420	465
5440	25″ high					460			460	505
5460	30″ side, 16″ high					520			520	575
5480	24″ high					590			590	650
5490	27″ high					740			740	815
5500	Round, 36″ diameter, 16″ high					720			720	790
5510	18″ high					735			735	810
5520	21″ high					750			750	825
5530	24″ high					765			765	840
5540	27″ high					775			775	850
5550	30″ high					890			890	980
5800	48″ diameter, 16″ high					905			905	995
5820	21″ high					940			940	1,025
5840	24″ high					1,125			1,125	1,225
5860	27″ high					1,175			1,175	1,300
5880	30″ high					1,300			1,300	1,425
5900	60″ diameter, 16″ high					1,225			1,225	1,350
5920	21″ high					1,350			1,350	1,475
5940	27″ high					1,475			1,475	1,625
5960	30″ high					1,675			1,675	1,850
5980	36″ high					1,975			1,975	2,175
6000	Metal bowl, 32″ diameter, 8″ high, minimum					445			445	490
6050	Maximum					590			590	650
6100	Rectangle, 30″ long x 12″ wide, 6″ high, minimum					273			273	300
6200	Maximum					535			535	590
6300	36″ long 12″ wide, 6″ high, minimum					540			540	590
6400	Maximum					560			560	620
6500	Square, 15″ side, minimum					540			540	590
6600	Maximum					590			590	650
6700	20″ side, minimum					850			850	935
6800	Maximum					925			925	1,025
6900	Round, 6″ diameter x 6″ high, minimum				▼	39.50			39.50	43.50

12 92 Interior Planters and Artificial Plants

12 92 33 – Interior Planters

12 92 33.10 Planters		Crew	Daily Output	Labor-Hours	Unit	Material	Labor	Equipment	Total	Total Incl O&P
7000	Maximum				Ea.	47			47	52
7100	8" diameter x 8" high, minimum					53			53	58
7200	Maximum					56.50			56.50	62
7300	10" diameter x 11" high, minimum					84.50			84.50	93
7400	Maximum					119			119	130
7420	12" diameter x 13" high, minimum					93.50			93.50	103
7440	Maximum					136			136	149
7500	14" diameter x 15" high, minimum					125			125	137
7550	Maximum					175			175	193
7580	16" diameter x 17" high, minimum					135			135	149
7600	Maximum					193			193	212
7620	18" diameter x 19" high, minimum					164			164	180
7640	Maximum					236			236	260
7680	22" diameter x 20" high, minimum					209			209	230
7700	Maximum					300			300	330
7750	24" diameter x 21" high, minimum					274			274	300
7800	Maximum					405			405	445
7850	31" diameter x 18" high, minimum					625			625	685
7900	Maximum					1,200			1,200	1,325
7950	38" diameter x 24" high, minimum					1,000			1,000	1,100
8000	Maximum					2,150			2,150	2,375
8050	48" diameter x 24" high, minimum					1,850			1,850	2,025
8150	Maximum				↓	2,075			2,075	2,275
8500	Plastic laminate faced, fiberglass liner, square									
8520	14" sq., 15" high				Ea.	400			400	440
8540	24" sq., 16" high					505			505	555
8580	36" sq., 21" high					710			710	780
8600	Rectangle 36" long, 12" wide, 10" high					430			430	475
8650	48" long, 12" wide, 10" high					480			480	530
8700	48" long, 12" wide, 24" high				↓	710			710	780
8750	Wood, fiberglass liner, square									
8780	14" square, 15" high, minimum				Ea.	288			288	315
8800	Maximum					355			355	390
8820	24" sq., 16" high, minimum					355			355	390
8840	Maximum					470			470	515
8860	36" sq., 21" high, minimum					455			455	500
8880	Maximum					685			685	750
9000	Rectangle, 36" long x 12" wide, 10" high, minimum					315			315	350
9050	Maximum					410			410	455
9100	48" long x 12" wide, 10" high, minimum					340			340	375
9120	Maximum					430			430	475
9200	48" long x 12" wide, 24" high, minimum					410			410	455
9300	Maximum					600			600	660
9400	Plastic cylinder, molded, 10" diameter, 10" high					11.65			11.65	12.85
9500	11" diameter, 11" high					37			37	40.50
9600	13" diameter, 12" high					48.50			48.50	53.50
9700	16" diameter, 14" high				↓	50.50			50.50	55.50

(Labor, Equipment and Total are printed under the spanning heading "2009 Bare Costs".)

12 93 Site Furnishings

12 93 23 – Trash and Litter Receptors

12 93 23.10 Trash Receptacles		Crew	Daily Output	Labor-Hours	Unit	Material	2009 Bare Costs Labor	Equipment	Total	Total Incl O&P
0010	**TRASH RECEPTACLES**									
0020	Fiberglass, 2′ square, 18″ high	2 Clab	30	.533	Ea.	320	16.85		336.85	385
0100	2′ square, 2′-6″ high		30	.533		450	16.85		466.85	525
0300	Circular, 2′ diameter, 18″ high		30	.533		289	16.85		305.85	350
0400	2′ diameter, 2′-6″ high		30	.533		385	16.85		401.85	455
0500	Recycled plastic, var colors, round, 32 Gal, 28″ x 38″ H G		5	3.200		290	101		391	485
0510	32 Gal, 31″ x 32″ H G		5	3.200		360	101		461	560
1000	Alum. frame, hardboard panels, steel drum base,									
1020	30 gal. capacity, silk screen on plastic finish	2 Clab	25	.640	Ea.	690	20		710	795
1040	Aggregate finish		25	.640		805	20		825	920
1100	50 gal. capacity, silk screen on plastic finish		20	.800		705	25.50		730.50	815
1140	Aggregate finish		20	.800		835	25.50		860.50	960
1200	Formed plastic liner, 14 gal., silk screen on plastic finish		40	.400		305	12.65		317.65	355
1240	Aggregate finish		40	.400		350	12.65		362.65	405
1300	30 gal. capacity, silk screen on plastic finish		35	.457		415	14.45		429.45	480
1340	Aggregate finish		35	.457		475	14.45		489.45	550
1400	Redwood slats, plastic liner, leg base, 14 gal. capacity,									
1420	Varnish w/routed message	2 Clab	40	.400	Ea.	365	12.65		377.65	420
2000	Concrete, precast, 2′ to 2-1/2′ wide, 3′ high, sandblasted	″	15	1.067	″	725	33.50		758.50	855
3000	Galv. steel frame and panels, leg base, poly bag retainer,									
3020	40 gal. capacity, silk screen on enamel finish	2 Clab	25	.640	Ea.	405	20		425	485
3040	Aggregate finish		25	.640		540	20		560	625
3200	Formed plastic liner, 50 gal., silk screen on enamel finish		20	.800		615	25.50		640.50	715
3240	Aggregate finish		20	.800		715	25.50		740.50	825
4000	Perforated steel, pole mounted, 12″ diam., 10 gal., painted		25	.640		127	20		147	174
4040	Redwood slats		25	.640		330	20		350	395
4100	22 gal. capacity, painted		25	.640		187	20		207	239
4140	Redwood slats		25	.640		395	20		415	470
4500	Galvanized steel street basket, 52 gal. capacity, unpainted		40	.400		269	12.65		281.65	315

12 93 23.20 Trash Closure		Crew	Daily Output	Labor-Hours	Unit	Material	Labor	Equipment	Total	Total Incl O&P
0010	**TRASH CLOSURE**									
0020	Steel with pullover cover, 2′-3″ wide, 4′-7″ high, 6′-2″ long	2 Clab	5	3.200	Ea.	1,175	101		1,276	1,475
0100	10′-1″ long		4	4		1,550	126		1,676	1,925
0300	Wood, 10′ wide, 6′ high, 10′ long		1.20	13.333		1,125	420		1,545	1,950

12 93 33 – Manufactured Planters

12 93 33.10 Planters		Crew	Daily Output	Labor-Hours	Unit	Material	Labor	Equipment	Total	Total Incl O&P
0010	**PLANTERS**									
0012	Concrete, sandblasted, precast, 48″ diameter, 24″ high	2 Clab	15	1.067	Ea.	635	33.50		668.50	755
0100	Fluted, precast, 7′ diameter, 36″ high		10	1.600		1,075	50.50		1,125.50	1,250
0300	Fiberglass, circular, 36″ diameter, 24″ high		15	1.067		470	33.50		503.50	575
0320	36″ diameter, 27″ high		12	1.333		500	42		542	620
0330	33″ high		15	1.067		515	33.50		548.50	620
0335	24″ diameter, 36″ high		15	1.067		420	33.50		453.50	520
0340	60″ diameter, 39″ high		8	2		1,125	63		1,188	1,325
0400	60″ diameter, 24″ high		10	1.600		800	50.50		850.50	960
0600	Square, 24″ side, 36″ high		15	1.067		455	33.50		488.50	560
0610	24″ side, 27″ high		12	1.333		300	42		342	400
0620	24″ side, 16″ high		20	.800		330	25.50		355.50	405
0700	48″ side, 36″ high		15	1.067		810	33.50		843.50	945
0900	Planter/bench, 72″ square, 36″ high		5	3.200		1,275	101		1,376	1,575
1000	96″ square, 27″ high		5	3.200		1,950	101		2,051	2,325
1200	Wood, square, 48″ side, 24″ high		15	1.067		1,000	33.50		1,033.50	1,150

12 93 Site Furnishings

12 93 33 – Manufactured Planters

12 93 33.10 Planters		Crew	Daily Output	Labor-Hours	Unit	2009 Bare Costs Material	Labor	Equipment	Total	Total Incl O&P
1300	Circular, 48" diameter, 30" high	2 Clab	10	1.600	Ea.	820	50.50		870.50	990
1500	72" diameter, 30" high		10	1.600		1,450	50.50		1,500.50	1,675
1600	Planter/bench, 72"		5	3.200		3,225	101		3,326	3,725
9000	Minimum labor/equipment charge	1 Clab	2	4	Job		126		126	209

12 93 43 – Site Seating and Tables

12 93 43.13 Site Seating		Crew	Daily Output	Labor-Hours	Unit	Material	Labor	Equipment	Total	Total Incl O&P
0010	**SITE SEATING**									
0012	Seating, benches, park, precast conc, w/backs, wood rails, 4' long	2 Clab	5	3.200	Ea.	430	101		531	640
0100	8' long		4	4		860	126		986	1,150
0300	Fiberglass, without back, one piece, 4' long		10	1.600		565	50.50		615.50	710
0400	8' long		7	2.286		1,175	72		1,247	1,400
0500	Steel barstock pedestals w/backs, 2" x 3" wood rails, 4' long		10	1.600		1,050	50.50		1,100.50	1,225
0510	8' long		7	2.286		1,250	72		1,322	1,500
0520	3" x 8" wood plank, 4' long		10	1.600		1,050	50.50		1,100.50	1,250
0530	8' long		7	2.286		1,100	72		1,172	1,350
0540	Backless, 4" x 4" wood plank, 4' square		10	1.600		1,025	50.50		1,075.50	1,200
0550	8' long		7	2.286		975	72		1,047	1,200
0600	Aluminum pedestals, with backs, aluminum slats, 8' long		8	2		335	63		398	475
0610	15' long		5	3.200		500	101		601	715
0620	Portable, aluminum slats, 8' long		8	2		335	63		398	475
0630	15' long		5	3.200		655	101		756	885
0800	Cast iron pedestals, back & arms, wood slats, 4' long		8	2		370	63		433	515
0820	8' long		5	3.200		1,075	101		1,176	1,375
0840	Backless, wood slats, 4' long		8	2		645	63		708	815
0860	8' long		5	3.200		815	101		916	1,075
1700	Steel frame, fir seat, 10' long		10	1.600		239	50.50		289.50	345
9000	Minimum labor/equipment charge		2	8	Job		253		253	415

Estimating Tips

General

- The items and systems in this division are usually estimated, purchased, supplied, and installed as a unit by one or more subcontractors. The estimator must ensure that all parties are operating from the same set of specifications and assumptions, and that all necessary items are estimated and will be provided. Many times the complex items and systems are covered, but the more common ones, such as excavation or a crane, are overlooked for the very reason that everyone assumes nobody could miss them. The estimator should be the central focus and be able to ensure that all systems are complete.
- Another area where problems can develop in this division is at the interface between systems. The estimator must ensure, for instance, that anchor bolts, nuts, and washers are estimated and included for the air-supported structures and pre-engineered buildings to be bolted to their foundations. Utility supply is a common area where essential items or pieces of equipment can be missed or overlooked due to the fact that each subcontractor may feel it is another's responsibility. The estimator should also be aware of certain items which may be supplied as part of a package but installed by others, and ensure that the installing contractor's estimate includes the cost of installation. Conversely, the estimator must also ensure that items are not costed by two different subcontractors, resulting in an inflated overall estimate.

13 30 00 Special Structures

- The foundations and floor slab, as well as rough mechanical and electrical, should be estimated, as this work is required for the assembly and erection of the structure. Generally, as noted in the book, the pre-engineered building comes as a shell and additional features, such as windows and doors, must be included by the estimator. Here again, the estimator must have a clear understanding of the scope of each portion of the work and all the necessary interfaces.

Reference Numbers

Reference numbers are shown in shaded boxes at the beginning of some major classifications. These numbers refer to related items in the Reference Section. The reference information may be an estimating procedure, an alternate pricing method, or technical information.

Note: Not all subdivisions listed here necessarily appear in this publication.

13 05 05 – Selective Special Construction Demolition

13 05 05.10 Selective Demolition, Air Supported Structures

	13 05 05.10 Selective Demolition, Air Supported Structures	Crew	Daily Output	Labor-Hours	Unit	2009 Bare Costs Material	Labor	Equipment	Total	Total Incl O&P
0010	SELECTIVE DEMOLITION, AIR SUPPORTED STRUCTURES									
0020	Tank covers, scrim, dbl layer, vinyl poly w/ hdw, blower & controls									
0050	Round and rectangular R024119-10	B-2	9000	.004	S.F.		.14		.14	.23
0100	Warehouse structures									
0120	Poly/vinyl fabric, 28 oz, incl tension cables & inflation system	4 Clab	9000	.004	SF Flr.		.11		.11	.19
0150	Reinforced vinyl, 12 oz., 3000 S.F.	"	5000	.006			.20		.20	.33
0200	12,000 to 24,000 S.F.	8 Clab	20000	.003			.10		.10	.17
0250	Tedlar vinyl fabric, 28 oz. w/liner, to 3000 S.F.	4 Clab	5000	.006			.20		.20	.33
0300	12,000 to 24,000 S.F.	8 Clab	20000	.003			.10		.10	.17
0350	Greenhouse/shelter, woven polyethylene with liner									
0400	3000 S.F.	4 Clab	5000	.006	SF Flr.		.20		.20	.33
0450	12,000 to 24,000 S.F.	8 Clab	20000	.003			.10		.10	.17
0500	Tennis/gymnasium, poly/vinyl fabric, 28 oz., incl thermal liner	4 Clab	9000	.004			.11		.11	.19
0600	Stadium/convention center, teflon coated fiberglass, incl thermal liner	9 Clab	40000	.002			.06		.06	.09
0700	Doors, air lock, 15′ long, 10′ x 10′	2 Carp	1.50	10.667	Ea.		425		425	705
0720	15′ x 15′		.80	20			800		800	1,325
0750	Revolving personnel door, 6′ dia. x 6′-6″ high		1.50	10.667			425		425	705

13 05 05.20 Selective Demolition, Garden Houses

	Description	Crew	Daily Output	Labor-Hours	Unit	Material	Labor	Equipment	Total	Total Incl O&P
0010	SELECTIVE DEMOLITION, GARDEN HOUSES R024119-10									
0020	Prefab, wood, excl foundation, average	2 Clab	400	.040	SF Flr.		1.26		1.26	2.09

13 05 05.25 Selective Demolition, Geodesic Domes

	Description	Crew	Daily Output	Labor-Hours	Unit	Material	Labor	Equipment	Total	Total Incl O&P
0010	SELECTIVE DEMOLITION, GEODESIC DOMES									
0050	Shell only, interlocking plywood panels, 30′ diameter	F-5	3.20	10	Ea.		405		405	670
0060	34′ diameter		2.30	13.913			565		565	930
0070	39′ diameter		2	16			645		645	1,075
0080	45′ diameter		2.20	14.545			590		590	970
0090	55′ diameter		2	16			645		645	1,075
0100	60′ diameter		2	16			645		645	1,075
0110	65′ diameter		1.60	20			810		810	1,325

13 05 05.30 Selective Demolition, Greenhouses

	Description	Crew	Daily Output	Labor-Hours	Unit	Material	Labor	Equipment	Total	Total Incl O&P
0010	SELECTIVE DEMOLITION, GREENHOUSES R024119-10									
0020	Resi-type, free standing, excl foundations, 9′ long x 8′ wide	2 Clab	160	.100	SF Flr.		3.16		3.16	5.20
0030	9′ long x 11′ wide		170	.094			2.97		2.97	4.91
0040	9′ long x 14′ wide		220	.073			2.30		2.30	3.79
0050	9′ long x 17′ wide		320	.050			1.58		1.58	2.61
0060	Lean-to type, 4′ wide		64	.250			7.90		7.90	13.05
0070	7′ wide		120	.133			4.21		4.21	6.95
0080	Geodesic hemisphere, 1/8″ plexiglass glazing, 8′ dia		4	4	Ea.		126		126	209
0090	24′ dia		.80	20			630		630	1,050
0100	48′ dia		.40	40			1,275		1,275	2,075

13 05 05.35 Selective Demolition, Hangars

	Description	Crew	Daily Output	Labor-Hours	Unit	Material	Labor	Equipment	Total	Total Incl O&P
0010	SELECTIVE DEMOLITION, HANGARS									
0020	T type hangars, prefab, steel , galv roof & walls, incl doors, excl fndtn	E-2	2550	.022	SF Flr.		.96	.68	1.64	2.45
0030	Circular type, prefab, steel frame, plastic skin, incl foundation, 80′ dia	"	.50	112	Total		4,875	3,475	8,350	12,500

13 05 05.45 Selective Demolition, Lightning Protection

	Description	Crew	Daily Output	Labor-Hours	Unit	Material	Labor	Equipment	Total	Total Incl O&P
0010	SELECTIVE DEMOLITION, LIGHTNING PROTECTION									
0020	Air terminal & base, copper, 3/8″ diam. x 10″, to 75′ h	1 Clab	16	.500	Ea.		15.80		15.80	26
0030	1/2″ diam. x 12″, over 75′ h		16	.500			15.80		15.80	26
0050	Aluminum, 1/2″ diam. x 12″, to 75′ h		16	.500			15.80		15.80	26
0060	5/8″ diam. x 12″, over 75′ h		16	.500			15.80		15.80	26
0070	Cable, copper, 220 lb per thousand feet, to 75′ high		640	.013	L.F.		.40		.40	.65

13 05 Common Work Results for Special Construction

13 05 05 – Selective Special Construction Demolition

13 05 05.45 Selective Demolition, Lightning Protection

		Crew	Daily Output	Labor-Hours	Unit	Material	2009 Bare Costs Labor	Equipment	Total	Total Incl O&P
0080	375 lb per thousand feet, over 75′ high	1 Clab	460	.017	L.F.		.55		.55	.91
0090	Aluminum, 101 lb per thousand feet, to 75′ high		560	.014			.45		.45	.75
0100	199 lb per thousand feet, over 75′ high		480	.017			.53		.53	.87
0110	Arrester, 175 V AC, to ground		16	.500	Ea.		15.80		15.80	26
0120	650 V AC, to ground		13	.615	"		19.45		19.45	32

13 05 05.50 Selective Demolition, Pre-Engineered Steel Buildings

		Crew	Daily Output	Labor-Hours	Unit	Material	Labor	Equipment	Total	Total Incl O&P
0010	**SELECTIVE DEMOLITION, PRE-ENGINEERED STEEL BUILDINGS**									
0500	Pre-engd steel bldgs, rigid frame, clear span & multi post, excl salvage									
0550	3,500 to 7,500 S.F	L-11	1350	.024	SF Flr.		.86	1.39	2.25	2.97
0600	7,501 to 12,500 S.F		2000	.016			.58	.94	1.52	2
0650	12,500 S.F or greater		2200	.015			.53	.85	1.38	1.82
0700	Pre-engd steel building components									
0710	Entrance canopy, including frame 4′ x 4′	E-24	8	4	Ea.		175	98.50	273.50	425
0720	4′ x 8′	"	7	4.571			201	113	314	485
0730	H.M doors, self framing, single leaf	2 Skwk	8	2			81.50		81.50	133
0740	Double leaf		5	3.200			131		131	213
0760	Gutter, eave type		600	.027	L.F.		1.09		1.09	1.77
0770	Sash, single slide, double slide or fixed		24	.667	Ea.		27		27	44.50
0780	Skylight, fiberglass, to 30 S.F.		16	1			41		41	66.50
0785	Roof vents, circular, 12″ to 24″ diameter		12	1.333			54.50		54.50	88.50
0790	Continuous, 10′ long		8	2			81.50		81.50	133
0900	Shelters, aluminum frame									
0910	Acrylic glazing, 3′ x 9′ x 8′ high	2 Skwk	2	8	Ea.		325		325	530
0920	9′ x 12′ x 8′ high	"	1.50	10.667	"		435		435	710

13 05 05.55 Selective Demolition, Residential Garages

		Crew	Daily Output	Labor-Hours	Unit	Material	Labor	Equipment	Total	Total Incl O&P
0010	**SELECTIVE DEMOLITION, RESIDENTIAL GARAGES** R024119-10									
0020	Garage, residential, prefab shell, stock, wood, single car	2 Clab	1.50	10.667	Ea.		335		335	555
0030	Two car	"	1.10	14.545	"		460		460	760

13 05 05.60 Selective Demolition, Silos

		Crew	Daily Output	Labor-Hours	Unit	Material	Labor	Equipment	Total	Total Incl O&P
0010	**SELECTIVE DEMOLITION, SILOS**									
0020	Conc stave, indstrl, conical/sloping bott, excl fndtn, 12′ dia, 35′ h	D-8	.22	181	Ea.		6,750		6,750	10,900
0030	16′ dia, 45′ h		.16	250			9,300		9,300	15,000
0040	25′ dia, 75′ h		.10	400			14,900		14,900	24,000
0050	Steel, factory fabricated, 30,000 gal cap, painted or epoxy lined	L-5	2	28			1,250	395	1,645	2,700

13 05 05.65 Selective Demolition, Sound Control

		Crew	Daily Output	Labor-Hours	Unit	Material	Labor	Equipment	Total	Total Incl O&P
0010	**SELECTIVE DEMOLITION, SOUND CONTROL** R024119-10									
0120	Acoustical enclosure, 4″ thick walls & ceiling panels, 8 lbs/SF	3 Carp	144	.167	SF Surf		6.65		6.65	11
0130	10.5 lbs/SF		128	.188			7.50		7.50	12.35
0140	Reverb chamber, parallel walls, 4″ thick		120	.200			8		8	13.20
0150	Skewed walls, parallel roof, 4″ thick		110	.218			8.70		8.70	14.40
0160	Skewed walls/roof, 4″ layer/air space		96	.250			10		10	16.50
0170	Sound-absorbing panels, painted metal, 2′-6″ x 8′, under 1,000 SF		430	.056			2.23		2.23	3.68
0180	Over 1,000 SF		480	.050			2		2	3.30
0190	Flexible transparent curtain, clear	3 Shee	430	.056			2.63		2.63	4.19
0192	50% clear, 50% foam		430	.056			2.63		2.63	4.19
0194	25% clear, 75% foam		430	.056			2.63		2.63	4.19
0196	100% foam		430	.056			2.63		2.63	4.19
0200	Audio-masking sys, incl speakers, amplfr, signal gnrtr, clng mntd, 5,000 SF									
0205	Ceiling mounted, 5,000 SF	2 Elec	4800	.003	S.F.		.16		.16	.24
0210	10,000 SF		5600	.003			.13		.13	.21
0220	Plenum mounted, 5,000 SF		7600	.002			.10		.10	.15
0230	10,000 SF		8800	.002			.09		.09	.13

13 05 05 – Selective Special Construction Demolition

13 05 05.70 Selective Demolition, Special Purpose Rooms		Crew	Daily Output	Labor-Hours	Unit	Material	2009 Bare Costs Labor	Equipment	Total	Total Incl O&P
0010	SELECTIVE DEMOLITION, SPECIAL PURPOSE ROOMS R024119-10									
0100	Audiometric rooms, under 500 S.F. surface	4 Carp	200	.160	SF Surf		6.40		6.40	10.55
0110	Over 500 S.F. surface	"	240	.133	"		5.35		5.35	8.80
0200	Clean rooms, 12' x 12' soft wall, class 100	1 Carp	.30	26.667	Ea.		1,075		1,075	1,750
0210	Class 1000		.30	26.667			1,075		1,075	1,750
0220	Class 10,000		.35	22.857			915		915	1,500
0230	Class 100,000		.35	22.857			915		915	1,500
0300	Darkrooms, shell complete, 8' high	2 Carp	220	.073	SF Flr.		2.91		2.91	4.80
0310	12' high		110	.145	"		5.80		5.80	9.60
0350	Darkrooms doors, mini-cylindrical, revolving		4	4	Ea.		160		160	264
0400	Music room, practice modular		140	.114	SF Surf		4.57		4.57	7.55
0500	Refrigeration structures and finishes									
0510	Wall finish, 2 coat portland cement plaster, 1/2" thick	1 Clab	200	.040	S.F.		1.26		1.26	2.09
0520	Fiberglass panels, 1/8" thick		400	.020			.63		.63	1.04
0530	Ceiling finish, polystyrene plastic, 1" to 2" thick		500	.016			.51		.51	.83
0540	4" thick		450	.018			.56		.56	.93
0550	Refrigerator, prefab aluminum walk-in, 7'-6" high, 6' x 6' OD	2 Carp	100	.160	SF Flr.		6.40		6.40	10.55
0560	10' x 10' OD		160	.100			4		4	6.60
0570	Over 150 S.F.		200	.080			3.20		3.20	5.30
0600	Sauna, prefabricated, including heater & controls, 7' high, to 30 S.F.		120	.133			5.35		5.35	8.80
0610	To 40 S.F.		140	.114			4.57		4.57	7.55
0620	To 60 S.F.		175	.091			3.65		3.65	6.05
0630	To 100 S.F.		220	.073			2.91		2.91	4.80
0640	To 130 S.F.		250	.064			2.56		2.56	4.22
0650	Steam bath, heater, timer, head, single, to 140 CF	1 Plum	2.20	3.636	Ea.		177		177	275
0660	To 300 CF		2.20	3.636			177		177	275
0670	Steam bath, comm. size, w/blow-down assembly, to 800 CF		1.80	4.444			217		217	335
0680	To 2500 CF		1.60	5			244		244	380
0690	Steam bath, comm. size, multiple, for motels, apts, 500 CF, 2 baths		2	4			195		195	300
0700	1,000 CF, 4 baths		1.40	5.714			279		279	430

13 05 05.75 Selective Demolition, Storage Tanks		Crew	Daily Output	Labor-Hours	Unit	Material	Labor	Equipment	Total	Total Incl O&P
0010	SELECTIVE DEMOLITION, STORAGE TANKS									
0500	Steel tank, single wall, above ground, not incl fdn, pumps or piping									
0510	Single wall, 275 gallon R024119-10	Q-1	3	5.333	Ea.		234		234	365
0520	550 thru 2,000 gallon	B-34P	2	12			485	268	753	1,075
0530	5,000 thru 10,000 gallon	B-34Q	2	12			490	610	1,100	1,450
0540	15,000 thru 30,000 gallon	B-34N	2	4			128	233	361	465
0600	Steel tank, double wall, above ground not incl fdn, pumps & piping									
0620	500 thru 2,000 gallon	B-34P	2	12	Ea.		485	268	753	1,075

13 05 05.85 Selective Demolition, Swimming Pool Equip		Crew	Daily Output	Labor-Hours	Unit	Material	Labor	Equipment	Total	Total Incl O&P
0010	SELECTIVE DEMOLITION, SWIMMING POOL EQUIP									
0020	Diving stand, stainless steel, 3 meter	2 Clab	3	5.333	Ea.		169		169	278
0030	1 meter		5	3.200			101		101	167
0040	Diving board, 16' long, aluminum		5.40	2.963			93.50		93.50	155
0050	Fiberglass		5.40	2.963			93.50		93.50	155
0070	Ladders, heavy duty, stainless steel, 2 tread		14	1.143			36		36	59.50
0080	4 tread		12	1.333			42		42	69.50
0090	Lifeguard chair, stainless steel, fixed		5	3.200			101		101	167
0100	Slide, tubular, fiberglass, aluminum handrails & ladder, 5', straight		4	4			126		126	209
0110	8', curved		6	2.667			84.50		84.50	139
0120	10', curved		3	5.333			169		169	278
0130	12' straight, with platform		2.50	6.400			202		202	335

13 05 Common Work Results for Special Construction

13 05 05 – Selective Special Construction Demolition

13 05 05.85 Selective Demolition, Swimming Pool Equip		Crew	Daily Output	Labor-Hours	Unit	Material	2009 Bare Costs Labor	Equipment	Total	Total Incl O&P
0140	Removable access ramp, stainless steel	2 Clab	4	4	Ea.		126		126	209
0150	Removable stairs, stainless steel, collapsible	↓	4	4	↓		126		126	209

13 05 05.90 Selective Demolition, Tension Structures		Crew	Daily Output	Labor-Hours	Unit	Material	Labor	Equipment	Total	Total Incl O&P
0010	**SELECTIVE DEMOLITION, TENSION STRUCTURES**									
0020	Tension structure, steel/alum frame, fabric shell, 60′ clear span, 6,000 SF	B-41	2000	.022	SF Flr.		.72	.13	.85	1.33
0030	12,000 SF		2200	.020			.65	.12	.77	1.20
0040	80′ clear span, 20,800 SF	↓	2440	.018			.59	.11	.70	1.09
0050	100′ clear span, 10,000 SF	L-5	4350	.013			.58	.18	.76	1.24
0060	26,000 SF		4600	.012			.54	.17	.71	1.18
0080	120′ clear span, 24,000 SF		6000	.009			.42	.13	.55	.90
0090	150′ clear span, 30,000 SF	↓	7000	.008			.36	.11	.47	.77
0100	200′ clear span, 40,000 SF	E-6	16000	.008	↓		.35	.13	.48	.78
0110	For roll-up door, 12′ x 14′	L-2	2	8	Ea.		281		281	465

13 05 05.95 Selective Demo, X-Ray/Radio Freq Protection		Crew	Daily Output	Labor-Hours	Unit	Material	Labor	Equipment	Total	Total Incl O&P
0010	**SELECTIVE DEMO, X-RAY/RADIO FREQ PROTECTION**									
0020	Shielding lead, lined door frame, excl hdwe, 1/16″ thick	1 Clab	4.80	1.667	Ea.		52.50		52.50	87
0030	Lead sheets, 1/16″ thick	2 Clab	270	.059	S.F.		1.87		1.87	3.09
0040	1/8″ thick		240	.067			2.11		2.11	3.48
0050	Lead shielding, 1/4″ thick		270	.059			1.87		1.87	3.09
0060	1/2″ thick	↓	240	.067	↓		2.11		2.11	3.48
0070	Lead glass, 1/4″ thick, 2.0 mm LE, 12″ x 16″	2 Glaz	26	.615	Ea.		24		24	38.50
0080	24″ x 36″		16	1			38.50		38.50	62
0090	36″ x 60″		4	4			154		154	249
0100	Lead glass window frame, with 1/16″ lead & voice passage, 36″ x 60″		4	4			154		154	249
0110	Lead glass window frame, 24″ x 36″	↓	16	1	↓		38.50		38.50	62
0120	Lead gypsum board, 5/8″ thick with 1/16″ lead	2 Clab	320	.050	S.F.		1.58		1.58	2.61
0130	1/8″ lead		280	.057			1.81		1.81	2.98
0140	1/32″ lead		400	.040	↓		1.26		1.26	2.09
0150	Butt joints, 1/8″ lead or thicker, 2″ x 7′ long batten strip		480	.033	Ea.		1.05		1.05	1.74
0160	X-ray protection, average radiography room, up to 300 SF, 1/16″ lead, min		.50	32	Total		1,000		1,000	1,675
0170	Maximum		.30	53.333			1,675		1,675	2,775
0180	Deep therapy X-ray room, 250 KV cap, up to 300 SF, 1/4″ lead, min		.20	80			2,525		2,525	4,175
0190	Maximum		.12	133	↓		4,225		4,225	6,950
0880	Radio frequency shielding, prefab or screen-type copper or steel, minimum		360	.044	SF Surf		1.40		1.40	2.32
0890	Average		310	.052			1.63		1.63	2.69
0895	Maximum	↓	290	.055	↓		1.74		1.74	2.88

13 11 Swimming Pools

13 11 13 – Below-Grade Swimming Pools

13 11 13.50 Swimming Pools		Crew	Daily Output	Labor-Hours	Unit	Material	Labor	Equipment	Total	Total Incl O&P
0010	**SWIMMING POOLS** Residential in-ground, vinyl lined, concrete									
0020	Sides including equipment, sand bottom	B-52	300	.187	SF Surf	12.60	6.90	1.65	21.15	27
0100	Metal or polystyrene sides	B-14	410	.117		10.55	3.88	.72	15.15	18.75
0200	Add for vermiculite bottom				↓	.81			.81	.89
0500	Gunite bottom and sides, white plaster finish									
0600	12′ x 30′ pool	B-52	145	.386	SF Surf	23.50	14.25	3.41	41.16	53.50
0720	16′ x 32′ pool		155	.361		21	13.30	3.19	37.49	49
0750	20′ x 40′ pool	↓	250	.224	↓	18.90	8.25	1.98	29.13	36.50
0810	Concrete bottom and sides, tile finish									
0820	12′ x 30′ pool	B-52	80	.700	SF Surf	23.50	26	6.20	55.70	75

13 11 Swimming Pools

13 11 13 – Below-Grade Swimming Pools

13 11 13.50 Swimming Pools		Crew	Daily Output	Labor-Hours	Unit	Material	Labor	Equipment	Total	Total Incl O&P
						2009 Bare Costs				
0830	16′ x 32′ pool	B-52	95	.589	SF Surf	19.60	21.50	5.20	46.30	63
0840	20′ x 40′ pool	↓	130	.431	↓	15.60	15.85	3.81	35.26	47.50
1100	Motel, gunite with plaster finish, incl. medium									
1150	capacity filtration & chlorination	B-52	115	.487	SF Surf	29	17.95	4.30	51.25	66
1200	Municipal, gunite with plaster finish, incl. high									
1250	capacity filtration & chlorination	B-52	100	.560	SF Surf	37.50	20.50	4.95	62.95	80.50
1350	Add for formed gutters				L.F.	55			55	60.50
1360	Add for stainless steel gutters				"	163			163	179
1600	For water heating system, see Div. 23 52 28.10									
1700	Filtration and deck equipment only, as % of total				Total				20%	20%
1800	Deck equipment, rule of thumb, 20′ x 40′ pool				SF Pool				1.18	1.30
1900	5000 S.F. pool				"				1.73	1.90
3000	Painting pools, preparation + 3 coats, 20′ x 40′ pool, epoxy	2 Pord	.33	48.485	Total	1,550	1,700		3,250	4,425
3100	Rubber base paint, 18 gallons	"	.33	48.485		1,000	1,700		2,700	3,825
3500	42′ x 82′ pool, 75 gallons, epoxy paint	3 Pord	.14	171		6,550	6,025		12,575	16,900
3600	Rubber base paint	"	.14	171	↓	4,125	6,025		10,150	14,200

13 11 46 – Swimming Pool Accessories

13 11 46.50 Swimming Pool Equipment		Crew	Daily Output	Labor-Hours	Unit	Material	Labor	Equipment	Total	Total Incl O&P
0010	**SWIMMING POOL EQUIPMENT**									
0020	Diving stand, stainless steel, 3 meter	2 Carp	.40	40	Ea.	9,375	1,600		10,975	13,000
0300	1 meter		2.70	5.926		5,475	237		5,712	6,425
0600	Diving boards, 16′ long, aluminum		2.70	5.926		3,075	237		3,312	3,775
0700	Fiberglass	↓	2.70	5.926	↓	2,425	237		2,662	3,075
1100	Gutter system, stainless steel, with grating, stock,									
1110	contains supply and drainage system	E-1	20	1.200	L.F.	205	52	6.70	263.70	325
1120	Integral gutter and 5′ high wall system, stainless steel	"	10	2.400	"	310	104	13.40	427.40	540
1200	Ladders, heavy duty, stainless steel, 2 tread	2 Carp	7	2.286	Ea.	490	91.50		581.50	690
1500	4 tread		6	2.667		655	107		762	895
1800	Lifeguard chair, stainless steel, fixed	↓	2.70	5.926		2,275	237		2,512	2,900
1900	Portable					2,550			2,550	2,800
2100	Lights, underwater, 12 volt, with transformer, 300 watt	1 Elec	1	8		243	375		618	850
2200	110 volt, 500 watt, standard		1	8		201	375		576	800
2400	Low water cutoff type	↓	1	8	↓	234	375		609	835
2800	Heaters, see Div. 23 52 28.10									
3000	Pool covers, reinforced vinyl	3 Clab	1800	.013	S.F.	.36	.42		.78	1.10
3050	Automatic, electric								8.77	9.65
3100	Vinyl water tube	3 Clab	3200	.008		.46	.24		.70	.90
3200	Maximum	"	3000	.008		.58	.25		.83	1.06
3250	Sealed air bubble polyethylene solar blanket, 16 mils				↓	.31			.31	.34
3300	Slides, tubular, fiberglass, aluminum handrails & ladder, 5′-0″, straight	2 Carp	1.60	10	Ea.	3,625	400		4,025	4,650
3320	8′-0″, curved		3	5.333		14,600	213		14,813	16,500
3400	10′-0″, curved		1	16		21,600	640		22,240	24,800
3420	12′-0″, straight with platform	↓	1.20	13.333	↓	5,100	535		5,635	6,500
4500	Hydraulic lift, movable pool bottom, single ram									
4520	Under 1,000 S.F. area	L-9	.03	1200	Ea.	195,000	43,600		238,600	288,000
4600	Four ram lift, over 1,000 S.F.	"	.02	1800		280,000	65,500		345,500	418,500
5000	Removable access ramp, stainless steel	2 Clab	2	8		8,475	253		8,728	9,750
5500	Removable stairs, stainless steel, collapsible	"	2	8	↓	4,450	253		4,703	5,300

13 17 Tubs and Pools

13 17 13 – Redwood Hot Tub System

13 17 13.10 Redwood Hot Tub System		Crew	Daily Output	Labor-Hours	Unit	2009 Bare Costs Material	Labor	Equipment	Total	Total Incl O&P
0010	REDWOOD HOT TUB SYSTEM									
7050	4′ diameter x 4′ deep	Q-1	1	16	Ea.	2,025	700		2,725	3,325
7100	5′ diameter x 4′ deep		1	16		2,625	700		3,325	3,975
7150	6′ diameter x 4′ deep		.80	20		3,125	880		4,005	4,800
7200	8′ diameter x 4′ deep	↓	.80	20	↓	4,350	880		5,230	6,125

13 17 33 – Whirlpool Tubs

13 17 33.10 Whirlpool Bath

		Crew	Daily Output	Labor-Hours	Unit	Material	Labor	Equipment	Total	Total Incl O&P
0010	WHIRLPOOL BATH									
6000	Whirlpool, bath with vented overflow, molded fiberglass									
6100	66″ x 48″ x 24″	Q-1	1	16	Ea.	3,050	700		3,750	4,475
6400	72″ x 36″ x 24″		1	16		3,000	700		3,700	4,400
6500	60″ x 30″ x 21″		1	16		2,575	700		3,275	3,925
6600	72″ x 42″ x 22″		1	16		4,025	700		4,725	5,550
6700	83″ x 65″	↓	.30	53.333	↓	5,250	2,350		7,600	9,400

13 18 Ice Rinks

13 18 13 – Ice Rink Floor Systems

13 18 13.50 Ice Skating

		Crew	Daily Output	Labor-Hours	Unit	Material	Labor	Equipment	Total	Total Incl O&P
0010	ICE SKATING Equipment incl. refrigeration, plumbing & cooling									
0020	coils & concrete slab, 85′ x 200′ rink									
0300	55° system, 5 mos., 100 ton				Total	596,000			596,000	655,500
0700	90° system, 12 mos., 135 ton				″	650,000			650,000	715,000
1200	Subsoil heating system (recycled from compressor), 85′ x 200′	Q-7	.27	118	Ea.	27,100	5,575		32,675	38,500
1300	Subsoil insulation, 2 lb. polystyrene with vapor barrier, 85′ x 200′	2 Carp	.14	114	″	32,600	4,575		37,175	43,300

13 18 16 – Ice Rink Dasher Boards

13 18 16.50 Ice Rink Dasher Boards

		Crew	Daily Output	Labor-Hours	Unit	Material	Labor	Equipment	Total	Total Incl O&P
0010	ICE RINK DASHER BOARDS									
1000	Dasher boards, 1/2″ H.D. polyethylene faced steel frame, 3′ acrylic									
1020	screen at sides, 5′ acrylic ends, 85′ x 200′	F-5	.06	533	Ea.	130,000	21,600		151,600	178,500
1100	Fiberglass & aluminum construction, same sides and ends	″	.06	533	″	140,000	21,600		161,600	189,500

13 21 Controlled Environment Rooms

13 21 13 – Clean Rooms

13 21 13.50 Clean Room Components

		Crew	Daily Output	Labor-Hours	Unit	Material	Labor	Equipment	Total	Total Incl O&P
0010	CLEAN ROOM COMPONENTS									
1100	Clean room, soft wall, 12′ x 12′, Class 100	1 Carp	.18	44.444	Ea.	15,200	1,775		16,975	19,600
1110	Class 1,000		.18	44.444		12,300	1,775		14,075	16,400
1120	Class 10,000		.21	38.095		10,800	1,525		12,325	14,300
1130	Class 100,000	↓	.21	38.095	↓	10,100	1,525		11,625	13,600
2800	Ceiling grid support, slotted channel struts 4′-0″ O.C., ea. way				S.F.				5.91	6.50
3000	Ceiling panel, vinyl coated foil on mineral substrate									
3020	Sealed, non-perforated				S.F.				1.27	1.40
4000	Ceiling panel seal, silicone sealant, 150 L.F./gal.	1 Carp	150	.053	L.F.	.28	2.13		2.41	3.83
4100	Two sided adhesive tape	″	240	.033	″	.13	1.33		1.46	2.34
4200	Clips, one per panel				Ea.	1.09			1.09	1.20
6000	HEPA filter, 2′x4′, 99.97% eff., 3″ dp beveled frame (silicone seal)					355			355	390
6040	6″ deep skirted frame (channel seal)					375			375	415
6100	99.99% efficient, 3″ deep beveled frame (silicone seal)				↓	370			370	405

13 21 Controlled Environment Rooms

13 21 13 – Clean Rooms

13 21 13.50 Clean Room Components		Crew	Daily Output	Labor-Hours	Unit	Material	Labor (2009 Bare Costs)	Equipment	Total	Total Incl O&P
6140	6″ deep skirted frame (channel seal)				Ea.	395			395	430
6200	99.999% efficient, 3″ deep beveled frame (silicone seal)					385			385	425
6240	6″ deep skirted frame (channel seal)					410			410	450
7000	Wall panel systems, including channel strut framing									
7020	Polyester coated aluminum, particle board				S.F.				18.18	20
7100	Porcelain coated aluminum, particle board								31.82	35
7400	Wall panel support, slotted channel struts, to 12′ high								16.36	18

13 21 26 – Cold Storage Rooms

13 21 26.50 Refrigeration		Crew	Daily Output	Labor-Hours	Unit	Material	Labor (2009 Bare Costs)	Equipment	Total	Total Incl O&P
0010	**REFRIGERATION**									
0020	Curbs, 12″ high, 4″ thick, concrete	2 Carp	58	.276	L.F.	4.14	11		15.14	23
1000	Doors, see Div. 08 34 13.10									
2400	Finishes, 2 coat portland cement plaster, 1/2″ thick	1 Plas	48	.167	S.F.	1.14	6.05		7.19	10.95
2500	For galvanized reinforcing mesh, add	1 Lath	335	.024		.76	.85		1.61	2.18
2700	3/16″ thick latex cement	1 Plas	88	.091		1.94	3.29		5.23	7.45
2900	For glass cloth reinforced ceilings, add	″	450	.018		.46	.64		1.10	1.54
3100	Fiberglass panels, 1/8″ thick	1 Carp	149.45	.054		2.57	2.14		4.71	6.35
3200	Polystyrene, plastic finish ceiling, 1″ thick		274	.029		2.37	1.17		3.54	4.54
3400	2″ thick		274	.029		2.70	1.17		3.87	4.90
3500	4″ thick		219	.037		3	1.46		4.46	5.70
3800	Floors, concrete, 4″ thick	1 Cefi	93	.086		1.13	3.29		4.42	6.40
3900	6″ thick	″	85	.094		1.68	3.60		5.28	7.50
4000	Insulation, 1″ to 6″ thick, cork				B.F.	1.10			1.10	1.21
4100	Urethane					1.08			1.08	1.19
4300	Polystyrene, regular					.73			.73	.80
4400	Bead board					.55			.55	.61
4600	Installation of above, add per layer	2 Carp	657.60	.024	S.F.	.36	.97		1.33	2
4700	Wall and ceiling juncture		298.90	.054	L.F.	1.77	2.14		3.91	5.50
4900	Partitions, galvanized sandwich panels, 4″ thick, stock		219.20	.073	S.F.	7.45	2.92		10.37	12.95
5000	Aluminum or fiberglass		219.20	.073	″	8.10	2.92		11.02	13.75
5200	Prefab walk-in, 7′-6″ high, aluminum, incl. door & floors,									
5210	not incl. partitions or refrigeration, 6′ x 6′ O.D. nominal	2 Carp	54.80	.292	SF Flr.	133	11.65		144.65	165
5500	10′ x 10′ O.D. nominal		82.20	.195		107	7.80		114.80	130
5700	12′ x 14′ O.D. nominal		109.60	.146		96	5.85		101.85	115
5800	12′ x 20′ O.D. nominal		109.60	.146		83.50	5.85		89.35	102
6100	For 8′-6″ high, add					5%				
6300	Rule of thumb for complete units, w/o doors & refrigeration, cooler	2 Carp	146	.110		120	4.38		124.38	139
6400	Freezer		109.60	.146		142	5.85		147.85	167
6600	Shelving, plated or galvanized, steel wire type		360	.044	SF Hor.	10.55	1.78		12.33	14.55
6700	Slat shelf type		375	.043		13	1.70		14.70	17.10
6900	For stainless steel shelving, add					300%				
7000	Vapor barrier, on wood walls	2 Carp	1644	.010	S.F.	.15	.39		.54	.81
7200	On masonry walls	″	1315	.012	″	.40	.49		.89	1.24
7500	For air curtain doors, see Div. 23 34 33.10									

13 21 48 – Sound-Conditioned Rooms

13 21 48.10 Anechoic Chambers		Crew	Daily Output	Labor-Hours	Unit	Material	Labor (2009 Bare Costs)	Equipment	Total	Total Incl O&P
0010	**ANECHOIC CHAMBERS** Standard units, 7′ ceiling heights									
0100	Area for pricing is net inside dimensions									
0300	200 cycles per second cutoff, 25 S.F. floor area				SF Flr.	1,600			1,600	1,750
0400	50 S.F.								1,045	1,150
0600	75 S.F.								1,000	1,100
0700	100 S.F.					1,200			1,200	1,325

13 21 Controlled Environment Rooms

13 21 48 – Sound-Conditioned Rooms

13 21 48.10 Anechoic Chambers		Crew	Daily Output	Labor-Hours	Unit	2009 Bare Costs Material	Labor	Equipment	Total	Total Incl O&P
0900	For 150 cycles per second cutoff, add to 100 S.F. room				SF Flr.					30%
1000	For 100 cycles per second cutoff, add to 100 S.F. room									45%
13 21 48.15 Audiometric Rooms										
0010	**AUDIOMETRIC ROOMS**									
0020	Under 500 S.F. surface	4 Carp	98	.327	SF Surf	51.50	13.05		64.55	78
0100	Over 500 S.F. surface	"	120	.267	"	49	10.65		59.65	71

13 21 53 – Darkrooms

13 21 53.50 Darkrooms		Crew	Daily Output	Labor-Hours	Unit	2009 Bare Costs Material	Labor	Equipment	Total	Total Incl O&P
0010	**DARKROOMS**									
0020	Shell, complete except for door, 64 S.F., 8′ high	2 Carp	128	.125	SF Flr.	53.50	4.99		58.49	67
0100	12′ high		64	.250		68.50	10		78.50	91.50
0500	120 S.F. floor, 8′ high		120	.133		37.50	5.35		42.85	50
0600	12′ high		60	.267		51.50	10.65		62.15	74
0800	240 S.F. floor, 8′ high		120	.133		27	5.35		32.35	38.50
0900	12′ high		60	.267		38	10.65		48.65	59
1200	Mini-cylindrical, revolving, unlined, 4′ diameter		3.50	4.571	Ea.	2,825	183		3,008	3,400
1400	5′-6″ diameter		2.50	6.400		3,850	256		4,106	4,650
1600	Add for lead lining, inner cylinder, 1/32″ thick					1,475			1,475	1,625
1700	1/16″ thick					3,975			3,975	4,375
1800	Add for lead lining, inner and outer cylinder, 1/32″ thick					2,725			2,725	3,000
1900	1/16″ thick					6,100			6,100	6,700
2000	For darkroom door, see Div. 08 34 36.10									

13 21 56 – Music Rooms

13 21 56.50 Music Rooms		Crew	Daily Output	Labor-Hours	Unit	2009 Bare Costs Material	Labor	Equipment	Total	Total Incl O&P
0010	**MUSIC ROOMS**									
0020	Practice room, modular, perforated steel, under 500 S.F.	2 Carp	70	.229	SF Surf	31.50	9.15		40.65	49.50
0100	Over 500 S.F.	"	80	.200	"	26.50	8		34.50	42.50

13 24 Special Activity Rooms

13 24 16 – Saunas

13 24 16.50 Saunas and Heaters		Crew	Daily Output	Labor-Hours	Unit	2009 Bare Costs Material	Labor	Equipment	Total	Total Incl O&P
0010	**SAUNAS AND HEATERS**									
0020	Prefabricated, incl. heater & controls, 7′ high, 6′ x 4′, C/C	L-7	2.20	12.727	Ea.	4,625	490		5,115	5,900
0050	6′ x 4′, C/P		2	14		4,275	540		4,815	5,575
0400	6′ x 5′, C/C		2	14		5,175	540		5,715	6,550
0450	6′ x 5′, C/P		2	14		4,800	540		5,340	6,175
0600	6′ x 6′, C/C		1.80	15.556		5,475	600		6,075	7,000
0650	6′ x 6′, C/P		1.80	15.556		5,125	600		5,725	6,600
0800	6′ x 9′, C/C		1.60	17.500		6,800	675		7,475	8,575
0850	6′ x 9′, C/P		1.60	17.500		6,300	675		6,975	8,050
1000	8′ x 12′, C/C		1.10	25.455		10,600	980		11,580	13,300
1050	8′ x 12′, C/P		1.10	25.455		9,725	980		10,705	12,300
1200	8′ x 8′, C/C		1.40	20		8,100	770		8,870	10,200
1250	8′ x 8′, C/P		1.40	20		7,600	770		8,370	9,625
1400	8′ x 10′, C/C		1.20	23.333		8,975	900		9,875	11,400
1450	8′ x 10′, C/P		1.20	23.333		8,350	900		9,250	10,700
1600	10′ x 12′, C/C		1	28		11,200	1,075		12,275	14,100
1650	10′ x 12′, C/P		1	28		10,100	1,075		11,175	12,900
1700	Door only, cedar, 2′x6′, with tempered insulated glass window	2 Carp	3.40	4.706		610	188		798	980
1800	Prehung, incl. jambs, pulls & hardware	"	12	1.333		615	53.50		668.50	765

13 24 Special Activity Rooms

13 24 16 – Saunas

13 24 16.50 Saunas and Heaters		Crew	Daily Output	Labor-Hours	Unit	Material	Labor (2009 Bare Costs)	Equipment	Total	Total Incl O&P
2500	Heaters only (incl. above), wall mounted, to 200 C.F.				Ea.	605			605	665
2750	To 300 C.F.					770			770	845
3000	Floor standing, to 720 C.F., 10,000 watts, w/controls	1 Elec	3	2.667		2,025	125		2,150	2,425
3250	To 1,000 C.F., 16,000 watts	"	3	2.667		2,525	125		2,650	2,975

13 24 26 – Steam Baths

13 24 26.50 Steam Baths and Components

		Crew	Daily Output	Labor-Hours	Unit	Material	Labor	Equipment	Total	Total Incl O&P
0010	**STEAM BATHS AND COMPONENTS**									
0020	Heater, timer & head, single, to 140 C.F.	1 Plum	1.20	6.667	Ea.	1,400	325		1,725	2,050
0500	To 300 C.F.		1.10	7.273		1,625	355		1,980	2,325
1000	Commercial size, with blow-down assembly, to 800 C.F.		.90	8.889		5,200	435		5,635	6,400
1500	To 2500 C.F.		.80	10		8,025	490		8,515	9,575
2000	Multiple, motels, apts., 2 baths, w/ blow-down assm., 500 C.F.	Q-1	1.30	12.308		5,900	540		6,440	7,350
2500	4 baths	"	.70	22.857		7,450	1,000		8,450	9,750
2700	Conversion unit for residential tub, including door					4,100			4,100	4,500

13 28 Athletic and Recreational Special Construction

13 28 33 – Athletic and Recreational Court Walls

13 28 33.50 Sport Court

		Crew	Daily Output	Labor-Hours	Unit	Material	Labor	Equipment	Total	Total Incl O&P
0010	**SPORT COURT**									
0020	Floors, No. 2 & better maple, 25/32" thick				SF Flr.				6.05	6.65
0100	Walls, laminated plastic bonded to galv. steel studs				SF Wall				7.68	8.45
0150	Laminated fiberglass surfacing, minimum								1.91	2.10
0180	Maximum								2.15	2.37
0300	Squash, regulation court in existing building, minimum				Court	35,000			35,000	38,500
0400	Maximum				"	39,000			39,000	42,900
0450	Rule of thumb for components:									
0470	Walls	3 Carp	.15	160	Court	10,500	6,400		16,900	22,200
0500	Floor	"	.25	96		8,300	3,825		12,125	15,500
0550	Lighting	2 Elec	.60	26.667		2,000	1,250		3,250	4,125
0600	Handball, racquetball court in existing building, minimum	C-1	.20	160		37,900	6,050		43,950	51,500
0800	Maximum	"	.10	320		41,000	12,100		53,100	65,000
0900	Rule of thumb for components: walls	3 Carp	.12	200		12,000	8,000		20,000	26,400
1000	Floor		.25	96		8,300	3,825		12,125	15,500
1100	Ceiling		.33	72.727		4,000	2,900		6,900	9,200
1200	Lighting	2 Elec	.60	26.667		2,100	1,250		3,350	4,225

13 31 Fabric Structures

13 31 13 – Air-Supported Fabric Structures

13 31 13.09 Air Supported Tank Covers

		Crew	Daily Output	Labor-Hours	Unit	Material	Labor	Equipment	Total	Total Incl O&P
0010	**AIR SUPPORTED TANK COVERS**, vinyl polyester									
0100	Scrim, double layer, with hardware, blower, standby & controls									
0200	Round, 75′ diameter	B-2	4500	.009	S.F.	9.75	.28		10.03	11.20
0300	100′ diameter		5000	.008		8.90	.26		9.16	10.20
0400	150′ diameter		5000	.008		7.05	.26		7.31	8.15
0500	Rectangular, 20′ x 20′		4500	.009		22	.28		22.28	25
0600	30′ x 40′		4500	.009		22	.28		22.28	25
0700	50′ x 60′		4500	.009		22	.28		22.28	25
0800	For single wall construction, deduct, minimum					.75			.75	.83
0900	Maximum					2.23			2.23	2.45

13 31 Fabric Structures

13 31 13 – Air-Supported Fabric Structures

Line	Description	Crew	Daily Output	Labor-Hours	Unit	Material (2009 Bare Costs)	Labor (2009 Bare Costs)	Equipment (2009 Bare Costs)	Total (2009 Bare Costs)	Total Incl O&P
	13 31 13.09 Air Supported Tank Covers									
1000	For maximum resistance to atmosphere or cold, add				S.F.	1.09			1.09	1.20
1100	For average shipping charges, add				Total	1,900			1,900	2,075
	13 31 13.13 Single-Walled Air-Supported Structures									
0010	**SINGLE-WALLED AIR-SUPPORTED STRUCTURES**									
0020	Site preparation, incl. anchor placement and utilities	B-11B	1000	.016	SF Flr.	1.11	.57	.25	1.93	2.40
0030	For concrete curb, see Div. 03 30 53.40									
0050	Warehouse, polyester/vinyl fabric, 28 oz., over 10 yr. life, welded									
0060	Seams, tension cables, primary & auxiliary inflation system,									
0070	airlock, personnel doors and liner									
0100	5,000 S.F.	4 Clab	5000	.006	SF Flr.	24	.20		24.20	27
0250	12,000 S.F.	"	6000	.005		17.20	.17		17.37	19.20
0400	24,000 S.F.	8 Clab	12000	.005		12.10	.17		12.27	13.60
0500	50,000 S.F.	"	12500	.005		11.20	.16		11.36	12.60
0700	12 oz. reinforced vinyl fabric, 5 yr. life, sewn seams,									
0710	accordian door, including liner									
0750	3000 S.F.	4 Clab	3000	.011	SF Flr.	11.45	.34		11.79	13.15
0800	12,000 S.F.	"	6000	.005		9.75	.17		9.92	11
0850	24,000 S.F.	8 Clab	12000	.005		8.25	.17		8.42	9.35
0950	Deduct for single layer					.98			.98	1.08
1000	Add for welded seams					1.01			1.01	1.11
1050	Add for double layer, welded seams included					2.08			2.08	2.29
1250	Tedlar/vinyl fabric, 28 oz., with liner, over 10 yr. life,									
1260	incl. overhead and personnel doors									
1300	3000 S.F.	4 Clab	3000	.011	SF Flr.	23.50	.34		23.84	26
1450	12,000 S.F.	"	6000	.005		16.45	.17		16.62	18.40
1550	24,000 S.F.	8 Clab	12000	.005		12.70	.17		12.87	14.30
1700	Deduct for single layer					1.44			1.44	1.58
2250	Greenhouse/shelter, woven polyethylene with liner, 2 yr. life,									
2260	sewn seams, including doors									
2300	3000 S.F.	4 Clab	3000	.011	SF Flr.	6.60	.34		6.94	7.80
2350	12,000 S.F.	"	6000	.005		6.70	.17		6.87	7.70
2450	24,000 S.F.	8 Clab	12000	.005		5.60	.17		5.77	6.45
2550	Deduct for single layer					.63			.63	.69
2600	Tennis/gymnasium, polyester/vinyl fabric, 28 oz., over 10 yr. life,									
2610	including thermal liner, heat and lights									
2650	7,200 S.F.	4 Clab	6000	.005	SF Flr.	22.50	.17		22.67	25
2750	13,000 S.F.	"	6500	.005		17.20	.16		17.36	19.15
2850	Over 24,000 S.F.	8 Clab	12000	.005		15.70	.17		15.87	17.60
2860	For low temperature conditions, add					1.09			1.09	1.20
2870	For average shipping charges, add				Total	5,325			5,325	5,875
2900	Thermal liner, translucent reinforced vinyl				SF Flr.	1.09			1.09	1.20
2950	Metalized mylar fabric and mesh, double liner				"	2.23			2.23	2.45
3050	Stadium/convention center, teflon coated fiberglass, heavy weight,									
3060	over 20 yr. life, incl. thermal liner and heating system									
3100	Minimum	9 Clab	26000	.003	SF Flr.	55	.09		55.09	60.50
3110	Maximum	"	19000	.004	"	65	.12		65.12	71.50
3400	Doors, air lock, 15′ long, 10′ x 10′	2 Carp	.80	20	Ea.	20,600	800		21,400	23,900
3600	15′ x 15′	"	.50	32		29,000	1,275		30,275	34,000
3700	For each added 5′ length, add					5,175			5,175	5,700
3900	Revolving personnel door, 6′ diameter, 6′-6″ high	2 Carp	.80	20		14,500	800		15,300	17,300
4200	Double wall, self supporting, shell only, minimum				SF Flr.				19.09	21
4300	Maximum				"				35.45	39

13 31 Fabric Structures

13 31 23 – Tensioned Fabric Structures

13 31 23.50 Tension Structures		Crew	Daily Output	Labor-Hours	Unit	Material	2009 Bare Costs Labor	Equipment	Total	Total Incl O&P
0010	**TENSION STRUCTURES** Rigid steel/alum. frame, vinyl coated poly									
0100	Fabric shell, 60′ clear span, not incl. foundations or floors									
0200	6,000 S.F.	B-41	1000	.044	SF Flr.	13.15	1.44	.26	14.85	17.10
0300	12,000 S.F.		1100	.040		11.55	1.31	.24	13.10	15.10
0400	80′ clear span, 20,800 S.F.		1220	.036		11.75	1.18	.22	13.15	15.10
0410	100′ clear span, 10,000 S.F.	L-5	2175	.026		13.10	1.15	.36	14.61	16.90
0430	26,000 S.F.		2300	.024		12	1.09	.34	13.43	15.55
0450	36,000 S.F.		2500	.022		11.75	1	.32	13.07	15.05
0460	120′ clear span, 24,000 S.F.		3000	.019		14.10	.83	.26	15.19	17.35
0470	150′ clear span, 30,000 S.F.		6000	.009		14.65	.42	.13	15.20	17
0480	200′ clear span, 40,000 S.F.	E-6	8000	.016		17.70	.71	.26	18.67	21
0500	For roll-up door, 12′ x 14′, add	L-2	1	16	Ea.	5,175	560		5,735	6,625
0600	For personnel doors, add, minimum				SF Flr.	5%				
0700	Add, maximum					15%				
0800	For site work, simple foundation, etc., add, minimum								1.25	1.95
0900	Add, maximum								2.75	3.05

13 34 Fabricated Engineered Structures

13 34 13 – Glazed Structures

13 34 13.13 Greenhouses		Crew	Daily Output	Labor-Hours	Unit	Material	Labor	Equipment	Total	Total Incl O&P
0010	**GREENHOUSES**, Shell only, stock units, not incl. 2′ stub walls,									
0020	foundation, floors, heat or compartments									
0300	Residential type, free standing, 8′-6″ long x 7′-6″ wide	2 Carp	59	.271	SF Flr.	47.50	10.85		58.35	70.50
0400	10′-6″ wide		85	.188		37	7.50		44.50	53.50
0600	13′-6″ wide		108	.148		35	5.90		40.90	48
0700	17′-0″ wide		160	.100		35	4		39	45
0900	Lean-to type, 3′-10″ wide		34	.471		38.50	18.80		57.30	73.50
1000	6′-10″ wide		58	.276		37.50	11		48.50	59
1500	Commercial, custom, truss frame, incl. equip., plumbing, elec.,									
1510	benches and controls, under 2,000 S.F., minimum				SF Flr.				27.27	30
1550	Maximum					31.50			31.50	34.50
1700	Over 5,000 S.F., minimum					21.50			21.50	23.50
1750	Maximum								27.27	30
2000	Institutional, custom, rigid frame, including compartments and									
2010	multi-controls, under 500 S.F., minimum				SF Flr.				70	77
2050	Maximum					76			76	84
2150	Over 2,000 S.F., minimum					30.50			30.50	33.50
2200	Maximum								59.09	65
2400	Concealed rigid frame, under 500 S.F., minimum								81.82	90
2450	Maximum								100	110
2550	Over 2,000 S.F., minimum								61.82	68
2600	Maximum								71.82	79
2800	Lean-to type, under 500 S.F., minimum								72.73	80
2850	Maximum								109.09	120
3000	Over 2,000 S.F., minimum								40	44
3050	Maximum								66.36	73
3600	For 1/4″ clear plate glass, add				SF Surf	1.86			1.86	2.05
3700	For 1/4″ tempered glass, add				″	4.21			4.21	4.63
3900	For cooling, add, minimum				SF Flr.	2.81			2.81	3.09
4000	Maximum					6.95			6.95	7.65

13 34 Fabricated Engineered Structures

13 34 13 – Glazed Structures

13 34 13.13 Greenhouses		Crew	Daily Output	Labor-Hours	Unit	2009 Bare Costs Material	Labor	Equipment	Total	Total Incl O&P
4200	For heaters, 13.6 MBH, add				SF Flr.	5.35			5.35	5.85
4300	60 MBH, add					2			2	2.20
4500	For benches, 2′ x 3′-6″, add				SF Hor.	24			24	26.50
4600	3′ x 10′, add				S.F.	12.95			12.95	14.25
4800	For controls, add, minimum				Total	2,400			2,400	2,650
4900	Maximum				"	14,300			14,300	15,700
5100	For humidification equipment, add				M.C.F.	6.15			6.15	6.75
5200	For vinyl shading, add				S.F.	1.29			1.29	1.42
6000	Geodesic hemisphere, 1/8″ plexiglass glazing									
6050	8′ diameter	2 Carp	2	8	Ea.	2,700	320		3,020	3,500
6150	24′ diameter		.35	45.714		13,900	1,825		15,725	18,300
6250	48′ diameter		.20	80		37,600	3,200		40,800	46,700

13 34 13.19 Swimming Pool Enclosures		Crew	Daily Output	Labor-Hours	Unit	Material	Labor	Equipment	Total	Total Incl O&P
0010	**SWIMMING POOL ENCLOSURES** Translucent, free standing									
0020	not including foundations, heat or light									
0200	Economy, minimum	2 Carp	200	.080	SF Hor.	14.80	3.20		18	21.50
0300	Maximum		100	.160		41.50	6.40		47.90	56
0400	Deluxe, minimum		100	.160		48.50	6.40		54.90	64
0600	Maximum		70	.229		225	9.15		234.15	262
0700	For motorized roof, 40% opening, solid roof, add					7.95			7.95	8.75
0800	Skylight type roof, add					8.50			8.50	9.35
0900	Air-inflated, including blowers and heaters, minimum								3.18	3.50
1000	Maximum								6.09	6.70

13 34 16 – Grandstands and Bleachers

13 34 16.13 Grandstands		Crew	Daily Output	Labor-Hours	Unit	Material	Labor	Equipment	Total	Total Incl O&P
0010	**GRANDSTANDS** Permanent, municipal, including foundation									
0050	Steel understructure w/aluminum closed deck, minimum				Seat				118.18	130
0100	Maximum								205	225
0300	Steel, minimum					55			55	60.50
0400	Maximum					110			110	121
0600	Aluminum, extruded, stock design, minimum								63.64	70
0700	Maximum								104.55	115
0900	Composite, steel, wood and plastic, stock design, minimum					130			130	143
1000	Maximum					260			260	286

13 34 16.53 Bleachers		Crew	Daily Output	Labor-Hours	Unit	Material	Labor	Equipment	Total	Total Incl O&P
0010	**BLEACHERS**									
0020	Bleachers, outdoor, portable, 3 to 5 tiers, to 300′ long, min	2 Sswk	120	.133	Seat	49.50	5.95		55.45	65.50
0100	Maximum, less than 15′ long, prefabricated		80	.200		71	8.95		79.95	94.50
0200	6 to 20 tiers, minimum, up to 300′ long		120	.133		88.50	5.95		94.45	109
0300	Max., under 15′, (highly prefabricated, on wheels)		80	.200		86	8.95		94.95	111
0500	Permanent grandstands, wood seat, steel frame, 24″ row									
0600	3 to 15 tiers, minimum	2 Sswk	60	.267	Seat	126	11.90		137.90	161
0700	Maximum		48	.333		141	14.90		155.90	183
0900	16 to 30 tiers, minimum		60	.267		192	11.90		203.90	234
0950	Average		55	.291		217	13		230	262
1000	Maximum		48	.333		233	14.90		247.90	284
1200	Seat backs only, 30″ row, fiberglass		160	.100		27.50	4.47		31.97	39
1300	Steel and wood		160	.100		32.50	4.47		36.97	44
1400	NOTE: average seating is 1.5′ in width									

13 34 19.50 Pre-Engineered Steel Buildings		Crew	Daily Output	Labor-Hours	Unit	2009 Bare Costs Material	Labor	Equipment	Total	Total Incl O&P
0010	**PRE-ENGINEERED STEEL BUILDINGS** R133419-10									
0100	Clear span rigid frame, 26 ga. colored roofing and siding									
0150	20 wide, 10′ eave height	E-2	425	.132	SF Flr.	10.70	5.75	4.10	20.55	26.50
0160	14′ eave height		350	.160		11.35	6.95	4.97	23.27	30.50
0170	16′ eave height		320	.175		11.95	7.60	5.45	25	32.50
0180	20′ eave height		275	.204		13.20	8.85	6.35	28.40	37.50
0190	24′ eave height		240	.233		15	10.15	7.25	32.40	42.50
0200	30′ to 40′ wide, 10′ eave height		535	.105		8.65	4.56	3.25	16.46	21
0300	14′ eave height		450	.124		9.15	5.40	3.87	18.42	24
0400	16′ eave height		415	.135		9.65	5.90	4.20	19.75	25.50
0500	20′ eave height		360	.156		10.55	6.75	4.84	22.14	29
0600	24′ eave height		320	.175		11.85	7.60	5.45	24.90	32.50
0700	50′ to 100′ wide, 10′ eave height		865	.065		7.45	2.82	2.01	12.28	15.40
0800	14′ eave height		770	.073		11.75	3.17	2.26	17.18	21
0900	16′ eave height		730	.077		8.40	3.34	2.38	14.12	17.80
1000	20′ eave height		660	.085		9.05	3.70	2.64	15.39	19.45
1100	24′ eave height		605	.093		9.95	4.03	2.88	16.86	21
1200	Clear span tapered beam frame, 26 ga. colored roofing/siding									
1300	30′ wide, 10′ eave height	E-2	535	.105	SF Flr.	9.60	4.56	3.25	17.41	22
1400	14′ eave height		450	.124		10.55	5.40	3.87	19.82	25.50
1500	16′ eave height		415	.135		11.30	5.90	4.20	21.40	27.50
1600	20′ eave height		360	.156		12.45	6.75	4.84	24.04	31
1700	40′ wide, 10′ eave height		600	.093		8.75	4.06	2.90	15.71	20
1800	14′ eave height		510	.110		9.55	4.78	3.41	17.74	23
1900	16′ eave height		475	.118		9.95	5.15	3.67	18.77	24
2000	20′ eave height		415	.135		10.95	5.90	4.20	21.05	27
2100	50′ to 80′ wide, 10′ eave height		770	.073		8.15	3.17	2.26	13.58	17.10
2200	14′ eave height		675	.083		8.85	3.61	2.58	15.04	19
2300	16′ eave height		635	.088		9.20	3.84	2.74	15.78	20
2400	20′ eave height		490	.114		13.85	4.98	3.55	22.38	28
2500	Single post 2-span frame, 26 ga. colored roofing and siding									
2600	80′ wide, 14′ eave height	E-2	740	.076	SF Flr.	7.15	3.30	2.35	12.80	16.35
2700	16′ eave height		695	.081		7.60	3.51	2.51	13.62	17.40
2800	20′ eave height		625	.090		8.30	3.90	2.79	14.99	19.15
2900	24′ eave height		570	.098		9	4.28	3.05	16.33	21
3000	100′ wide, 14′ eave height		835	.067		6.95	2.92	2.09	11.96	15.15
3100	16′ eave height		795	.070		6.80	3.07	2.19	12.06	15.35
3200	20′ eave height		730	.077		8	3.34	2.38	13.72	17.35
3300	24′ eave height		670	.084		8.65	3.64	2.60	14.89	18.90
3400	120′ wide, 14′ eave height		870	.064		6.90	2.80	2	11.70	14.80
3500	16′ eave height		830	.067		7.20	2.94	2.10	12.24	15.45
3600	20′ eave height		765	.073		7.75	3.19	2.28	13.22	16.70
3700	24′ eave height		705	.079		8.50	3.46	2.47	14.43	18.20
3800	Double post 3-span frame, 26 ga. colored roofing and siding									
3900	150′ wide, 14′ eave height	E-2	925	.061	SF Flr.	5.90	2.64	1.88	10.42	13.25
4000	16′ eave height		890	.063		6.10	2.74	1.96	10.80	13.75
4100	20′ eave height		820	.068		6.50	2.97	2.12	11.59	14.80
4200	24′ eave height		765	.073		7.60	3.19	2.28	13.07	16.55
4300	Triple post 4-span frame, 26 ga. colored roofing and siding									
4400	160′ wide, 14′ eave height	E-2	970	.058	SF Flr.	5.75	2.51	1.79	10.05	12.75
4500	16′ eave height		930	.060		6	2.62	1.87	10.49	13.35
4600	20′ eave height		870	.064		6	2.80	2	10.80	13.80

13 34 19.50 Pre-Engineered Steel Buildings		Crew	Daily Output	Labor-Hours	Unit	2009 Bare Costs Material	Labor	Equipment	Total	Total Incl O&P
4700	24′ eave height	E-2	815	.069	SF Flr.	7.10	2.99	2.14	12.23	15.50
4800	200′ wide, 14′ eave height		1030	.054		5.50	2.37	1.69	9.56	12.10
4900	16′ eave height		995	.056		5.65	2.45	1.75	9.85	12.50
5000	20′ eave height		935	.060		6.10	2.61	1.86	10.57	13.40
5100	24′ eave height	↓	885	.063	↓	7.10	2.76	1.97	11.83	14.90
5200	Accessory items: add to the basic building cost above									
5250	Eave overhang, 2′ wide, 26 ga., with soffit	E-2	360	.156	L.F.	32	6.75	4.84	43.59	53
5300	4′ wide, without soffit		300	.187		30.50	8.15	5.80	44.45	54.50
5350	With soffit		250	.224		43	9.75	6.95	59.70	72.50
5400	6′ wide, without soffit		250	.224		41.50	9.75	6.95	58.20	70.50
5450	With soffit		200	.280	↓	54	12.20	8.70	74.90	90.50
5500	Entrance canopy, incl. frame, 4′ x 4′		25	2.240	Ea.	420	97.50	69.50	587	715
5550	4′ x 8′		19	2.947	"	565	128	91.50	784.50	950
5600	End wall roof overhang, 4′ wide, without soffit		850	.066	L.F.	20	2.87	2.05	24.92	30
5650	With soffit	↓	500	.112	"	30.50	4.88	3.48	38.86	46
5700	Doors, H.M. self-framing, incl. butts, lockset and trim									
5750	Single leaf, 3070 (3′ x 7′), economy	2 Sswk	5	3.200	Opng.	435	143		578	745
5800	Deluxe		4	4		525	179		704	905
5825	Glazed		4	4		535	179		714	915
5850	3670 (3′-6″ x 7′)		4	4		575	179		754	965
5900	4070 (4′ x 7′)		3	5.333		610	238		848	1,125
5950	Double leaf, 6070 (6′ x 7′)		2	8		815	360		1,175	1,550
6000	Glazed	↓	2	8		1,000	360		1,360	1,750
6050	Framing only, for openings, 3′ x 7′	E-2	25	2.240		209	97.50	69.50	376	480
6100	10′ x 10′		21	2.667		690	116	83	889	1,050
6150	For windows below, 2020 (2′ x 2′)		25	2.240		221	97.50	69.50	388	495
6200	4030 (4′ x 3′)	↓	22	2.545	↓	271	111	79	461	580
6250	Flashings, 26 ga., corner or eave, painted	2 Sswk	240	.067	L.F.	5	2.98		7.98	11
6300	Galvanized		240	.067		4.12	2.98		7.10	10.05
6350	Rake flashing, painted		240	.067		5.40	2.98		8.38	11.40
6400	Galvanized		240	.067		4.50	2.98		7.48	10.45
6450	Ridge flashing, 18″ wide, painted		240	.067		6.75	2.98		9.73	12.90
6500	Galvanized		240	.067		6	2.98		8.98	12.10
6550	Gutter, eave type, 26 ga., painted		320	.050		7.05	2.24		9.29	11.95
6600	Galvanized		320	.050		3.54	2.24		5.78	8.05
6650	Valley type, between buildings, painted	↓	120	.133	↓	11.30	5.95		17.25	23.50
6710	Insulation, rated .6 lb density, unfaced 4″ thick, R13	2 Carp	2300	.007	S.F.	.31	.28		.59	.80
6720	6″ thick, R-19		2300	.007		.40	.28		.68	.90
6730	10″ thick, R-30	↓	2300	.007	↓	.74	.28		1.02	1.27
6750	Insulation, rated .6 lb density, poly/scrim/foil (PSF) faced									
6760	4″ thick R-13	2 Carp	2300	.007	S.F.	.43	.28		.71	.93
6770	6″ thick, R-19		2300	.007		.57	.28		.85	1.09
6780	10″ thick, R-30		2300	.007		1.06	.28		1.34	1.63
6800	Insulation, rated .6 lb density, vinyl faced 1-1/2″ thick, R5		2300	.007		.27	.28		.55	.76
6850	3″ thick, R10		2300	.007		.36	.28		.64	.86
6900	4″ thick, R13		2300	.007		.51	.28		.79	1.02
6920	6″ thick, R19		2300	.007		.71	.28		.99	1.24
6930	10″ thick, R30		2300	.007		.96	.28		1.24	1.52
6950	Foil/scrim/kraft (FSK) faced, 1-1/2″ thick, R5		2300	.007		.29	.28		.57	.78
7000	2″ thick, R6		2300	.007		.37	.28		.65	.87
7050	3″ thick, R10		2300	.007		.45	.28		.73	.96
7100	4″ thick, R13		2300	.007		.57	.28		.85	1.09
7110	6″ thick, R19	↓	2300	.007	↓	.61	.28		.89	1.13

13 34 Fabricated Engineered Structures

13 34 19 – Metal Building Systems

13 34 19.50 Pre-Engineered Steel Buildings		Crew	Daily Output	Labor-Hours	Unit	Material	Labor	Equipment	Total	Total Incl O&P
						2009 Bare Costs				
7120	10" thick, R-30	2 Carp	2300	.007	S.F.	1.03	.28		1.31	1.59
7150	Met polyester/scrim/kraft (PSK) facing,1-1/2" thk,R5		2300	.007		.47	.28		.75	.98
7200	2" thick, R6		2300	.007		.55	.28		.83	1.07
7250	3" thick, R11		2300	.007		.58	.28		.86	1.10
7300	4" thick, R13		2300	.007		.71	.28		.99	1.24
7310	6" thick, R19		2300	.007		.87	.28		1.15	1.42
7320	10" thick, R-30		2300	.007		1.05	.28		1.33	1.62
7350	Vinyl/scrim/foil (VSF), 1-1/2" thick, R5		2300	.007		.41	.28		.69	.91
7400	2" thick, R6		2300	.007		.52	.28		.80	1.03
7450	3" thick, R10		2300	.007		.67	.28		.95	1.20
7500	4" thick, R13		2300	.007		.78	.28		1.06	1.32
7510	Vinyl/scrim/vinyl (VSV) 4" thick, R-13		2300	.007		.50	.28		.78	1.01
7520	6" thick R-19		2300	.007		.63	.28		.91	1.15
7530	10" thick, R-30		2300	.007		1.01	.28		1.29	1.57
7580	10" thick, R-19		2300	.007		1.05	.28		1.33	1.62
7585	Vinyl/scrim/polyester (VSP), 4" thick, R-13		2300	.007		.44	.28		.72	.94
7590	6" thick, R-19		2300	.007		.60	.28		.88	1.12
7600	10" thick, R-30		2300	.007		1.08	.28		1.36	1.65
7635	Insulation installation, over the purlin, second layer, up to 4" thick, add						90%			
7640	Insulation installation, between the purlins, up to 4" thick, add						100%			
7650	Sash, single slide, glazed, with screens, 2020 (2' x 2')	E-1	22	1.091	Opng.	116	47.50	6.10	169.60	219
7700	3030 (3' x 3')		14	1.714		261	74.50	9.60	345.10	430
7750	4030 (4' x 3')		13	1.846		350	80.50	10.30	440.80	540
7800	6040 (6' x 4')		12	2		695	87	11.20	793.20	930
7850	Double slide sash, 3030 (3' x 3')		14	1.714		177	74.50	9.60	261.10	340
7900	6040 (6' x 4')		12	2		475	87	11.20	573.20	685
7950	Fixed glass, no screens, 3030 (3' x 3')		14	1.714		218	74.50	9.60	302.10	385
8000	6040 (6' x 4')		12	2		580	87	11.20	678.20	805
8050	Prefinished storm sash, 3030 (3' x 3')		70	.343		68	14.90	1.92	84.82	104
8100	Siding and roofing, see Div. 07 41 13.00 & 07 42 13.00									
8200	Skylight, fiberglass panels, to 30 S.F.	E-1	10	2.400	Ea.	103	104	13.40	220.40	315
8250	Larger sizes, add for excess over 30 S.F.	"	300	.080	S.F.	3.43	3.48	.45	7.36	10.40
8300	Roof vents, circular with damper, birdscreen									
8350	and operator hardware, painted									
8400	26 ga., 12" diameter	1 Sswk	4	2	Ea.	48.50	89.50		138	219
8450	20" diameter		3	2.667		99.50	119		218.50	330
8500	24 ga., 24" diameter		2	4		197	179		376	545
8550	Galvanized		2	4		165	179		344	510
8600	Continuous, 26 ga., 10' long, 9" wide	2 Sswk	4	4		21.50	179		200.50	355
8650	12" wide	"	4	4		25.50	179		204.50	360

13 34 23 – Fabricated Structures

13 34 23.10 Comfort Stations		Crew	Daily Output	Labor-Hours	Unit	Material	Labor	Equipment	Total	Total Incl O&P
0010	**COMFORT STATIONS** Prefab., stock, w/doors, windows & fixt.									
0100	Not incl. interior finish or electrical									
0300	Mobile, on steel frame, minimum				S.F.	46.50			46.50	51
0350	Maximum					75			75	82.50
0400	Permanent, including concrete slab, minimum	B-12J	50	.320		199	11.85	19.20	230.05	259
0500	Maximum	"	43	.372		288	13.80	22.50	324.30	360
0600	Alternate pricing method, mobile, minimum				Fixture	2,150			2,150	2,375
0650	Maximum					3,175			3,175	3,500
0700	Permanent, minimum	B-12J	.70	22.857		12,200	850	1,375	14,425	16,400
0750	Maximum	"	.50	32		20,400	1,175	1,925	23,500	26,400

13 34 23.15 Domes

	13 34 23.15 Domes	Crew	Daily Output	Labor-Hours	Unit	2009 Bare Costs Material	Labor	Equipment	Total	Total Incl O&P
0010	**DOMES**									
0020	Domes, rev alum, elec drive, for astronomy obsv shell only, stock units									
0600	10'-0" diameter, 800#, dome	2 Carp	.25	64	Ea.	11,000	2,550		13,550	16,300
0700	Base		.67	23.881		3,800	955		4,755	5,750
0900	18'-0" diameter, 2,500#, dome		.17	94.118		32,100	3,750		35,850	41,500
1000	Base		.33	48.485		10,600	1,925		12,525	14,900
1200	24'-0" diameter, 4,500#, dome		.08	200		59,500	8,000		67,500	78,500
1300	Base		.25	64		19,200	2,550		21,750	25,300
1500	Domes, bulk storage, shell only, dual radius hemisphere. arch, steel									
1600	framing, corrugated steel covering, 150' diameter	E-2	550	.102	SF Flr.	28	4.43	3.17	35.60	42
1700	400' diameter	"	720	.078		22.50	3.39	2.42	28.31	33.50
1800	Wood framing, wood decking, to 400' diameter	F-4	400	.120		20.50	4.78	2.81	28.09	33.50
1900	Radial framed wood (2" x 6"), 1/2" thick									
2000	plywood, asphalt shingles, 50' diameter	F-3	2000	.020	SF Flr.	38.50	.81	.38	39.69	44
2100	60' diameter		1900	.021		29.50	.85	.40	30.75	34.50
2200	72' diameter		1800	.022		26.50	.90	.43	27.83	31
2300	116' diameter		1730	.023		23	.94	.44	24.38	27.50
2400	150' diameter		1500	.027		20	1.08	.51	21.59	24.50

13 34 23.16 Fabricated Control Booths

	13 34 23.16 Fabricated Control Booths	Crew	Daily Output	Labor-Hours	Unit	Material	Labor	Equipment	Total	Total Incl O&P
0010	**FABRICATED CONTROL BOOTHS**									
0100	Guard House, prefab conc w/bullet resistant doors & windows, roof & wiring									
0110	8' x 8', Level III	L-10	1	24	Ea.	22,000	1,075	770	23,845	26,900
0120	8' x 8', Level IV	"	1	24	"	27,500	1,075	770	29,345	33,000

13 34 23.25 Garage Costs

	13 34 23.25 Garage Costs	Crew	Daily Output	Labor-Hours	Unit	Material	Labor	Equipment	Total	Total Incl O&P
0010	**GARAGE COSTS**									
0020	Public parking, average				Car				16,600	18,300
0100	See also Square Foot Costs in Reference Section									
0300	Residential, prefab shell, stock, wood, single car, minimum	2 Carp	1	16	Total	5,100	640		5,740	6,650
0350	Maximum		.67	23.881		6,700	955		7,655	8,950
0400	Two car, minimum		.67	23.881		7,625	955		8,580	9,950
0450	Maximum		.50	32		10,700	1,275		11,975	13,900

13 34 23.30 Garden House

	13 34 23.30 Garden House	Crew	Daily Output	Labor-Hours	Unit	Material	Labor	Equipment	Total	Total Incl O&P
0010	**GARDEN HOUSE** Prefab wood, no floors or foundations									
0100	32 to 200 S.F., minimum	2 Carp	200	.080	SF Flr.	32.50	3.20		35.70	41.50
0300	Maximum	"	48	.333	"	37	13.30		50.30	63

13 34 23.35 Geodesic Domes

	13 34 23.35 Geodesic Domes	Crew	Daily Output	Labor-Hours	Unit	Material	Labor	Equipment	Total	Total Incl O&P
0010	**GEODESIC DOMES** Shell only, interlocking plywood panels									
0400	30' diameter	F-5	1.60	20	Ea.	19,300	810		20,110	22,500
0500	34' diameter		1.14	28.070		22,800	1,125		23,925	26,900
0600	39' diameter		1	32		25,600	1,300		26,900	30,300
0700	45' diameter	F-3	1.13	35.556		31,500	1,450	685	33,635	37,800
0750	55' diameter		1	40		45,200	1,625	770	47,595	53,500
0800	60' diameter		1	40		58,500	1,625	770	60,895	68,000
0850	65' diameter		.80	50		71,500	2,025	960	74,485	83,500
1100	Aluminum panel, with 6" insulation									
1200	100' diameter				SF Flr.	41.50			41.50	45.50
1300	500' diameter				"	35.50			35.50	39
1600	Aluminum framed, plexiglass closure panels									
1700	40' diameter				SF Flr.	105			105	116
1800	200' diameter				"	84			84	92.50
2100	Aluminum framed, aluminum closure panels									

13 34 Fabricated Engineered Structures

13 34 23 – Fabricated Structures

13 34 23.35 Geodesic Domes		Crew	Daily Output	Labor-Hours	Unit	2009 Bare Costs Material	Labor	Equipment	Total	Total Incl O&P
2200	40′ diameter				SF Flr.	33.50			33.50	36.50
2300	100′ diameter					25			25	27.50
2400	200′ diameter					24			24	26
2500	For VRP faced bonded fiberglass insulation, add									10
2700	Aluminum framed, fiberglass sandwich panel closure									
2800	6′ diameter	2 Carp	150	.107	SF Flr.	37	4.26		41.26	47.50
2900	28′ diameter	"	350	.046	"	33.50	1.83		35.33	40

13 34 23.45 Kiosks

0010	**KIOSKS**									
0020	Round, 5′ diameter, 8′ high, 1/4″ fiberglass wall				Ea.	5,800			5,800	6,375
0100	1″ insulated double wall, fiberglass					6,600			6,600	7,250
0500	Rectangular, 5′ x 9′, 7′-6″ high, 1/4″ fiberglass wall					8,450			8,450	9,300
0600	1″ insulated double wall, fiberglass					10,000			10,000	11,000

13 34 23.60 Portable Booths

0010	**PORTABLE BOOTHS** Prefab aluminum with doors, windows, ext. roof									
0100	lights wiring & insulation, 15 S.F. building, O.D., painted, minimum				S.F.	310			310	340
0300	30 S.F. building, minimum					212			212	233
0400	50 S.F. building, minimum					157			157	173
0600	80 S.F. building, minimum					128			128	141
0700	100 S.F. building, minimum					120			120	132
0900	Acoustical booth, 27 Db @ 1,000 Hz, 15 S.F. floor				Ea.	3,625			3,625	3,975
1000	7′ x 7′-6″, including light & ventilation					7,425			7,425	8,175
1200	Ticket booth, galv. steel, not incl. foundations., 4′ x 4′					6,200			6,200	6,800
1300	4′ x 6′					7,300			7,300	8,025

13 34 23.70 Shelters

0010	**SHELTERS**									
0020	Aluminum frame, acrylic glazing, 3′ x 9′ x 8′ high	2 Sswk	1.14	14.035	Ea.	3,150	625		3,775	4,625
0100	9′ x 12′ x 8′ high	"	.73	21.918	"	7,325	980		8,305	9,875

13 34 43 – Aircraft Hangars

13 34 43.50 Hangars

0010	**HANGARS** Prefabricated steel T hangars, Galv. steel roof &									
0100	walls, incl. electric bi-folding doors, 4 or more units,									
0110	not including floors or foundations, minimum	E-2	1275	.044	SF Flr.	13.70	1.91	1.37	16.98	19.95
0130	Maximum		1063	.053		15.20	2.29	1.64	19.13	22.50
0900	With bottom rolling doors, minimum		1386	.040		13.30	1.76	1.26	16.32	19.15
1000	Maximum		966	.058		14.80	2.52	1.80	19.12	23
1200	Alternate pricing method:									
1300	Galv. roof and walls, electric bi-folding doors, minimum	E-2	1.06	52.830	Plane	17,100	2,300	1,650	21,050	24,700
1500	Maximum		.91	61.538		20,600	2,675	1,925	25,200	29,600
1600	With bottom rolling doors, minimum		1.25	44.800		12,900	1,950	1,400	16,250	19,200
1800	Maximum		.97	57.732		15,000	2,525	1,800	19,325	23,000
2000	Circular type, prefab., steel frame, plastic skin, electric									
2010	door, including foundations, 80′ diameter,									
2020	for up to 5 light planes, minimum	E-2	.50	112	Total	76,000	4,875	3,475	84,350	96,000
2200	Maximum	"	.25	224	"	85,500	9,750	6,975	102,225	119,000

13 34 53 – Agricultural Structures

13 34 53.50 Silos

0010	**SILOS** Concrete stave industrial, conical or sloping bottom									
0100	excluding foundation, 12′ diameter, 35′ high	D-8	.11	363	Ea.	42,400	13,500		55,900	68,500
0200	16′ diameter, 45′ high		.08	500		69,000	18,600		87,600	106,000
0400	25′ diameter, 75′ high		.05	800		159,000	29,700		188,700	223,000

13 34 Fabricated Engineered Structures

13 34 53 – Agricultural Structures

13 34 53.50 Silos		Crew	Daily Output	Labor-Hours	Unit	Material	2009 Bare Costs Labor	Equipment	Total	Total Incl O&P
0500	Steel, factory fab., 30,000 gallon cap., painted, minimum	L-5	1	56	Ea.	19,000	2,500	790	22,290	26,300
0700	Maximum		.50	112		30,200	5,000	1,575	36,775	44,100
0800	Epoxy lined, minimum		1	56		31,100	2,500	790	34,390	39,600
1000	Maximum		.50	112		39,400	5,000	1,575	45,975	54,000

13 34 63 – Natural Fiber Construction

13 34 63.50 Straw Bale Construction		Crew	Daily Output	Labor-Hours	Unit	Material	Labor	Equipment	Total	Total Incl O&P
0010	**STRAW BALE CONSTRUCTION**									
2000	Straw bales wall, incl labor & material complete, minimum				SF Flr.					130
2010	Maximum				"					160
2020	Straw bales in walls w/ modified post and beam frame G	2 Carp	320	.050	S.F.	4.28	2		6.28	8

13 36 Towers

13 36 13 – Metal Towers

13 36 13.50 Control Towers		Crew	Daily Output	Labor-Hours	Unit	Material	Labor	Equipment	Total	Total Incl O&P
0010	**CONTROL TOWERS**									
0020	Modular 12′ x 10′, incl. instruments, min.				Ea.	670,000			670,000	737,000
0100	Maximum								927,273	1,020,000
0500	With standard 40′ tower, average					935,000			935,000	1,028,500
1000	Temporary portable control towers, 8′ x 12′,									
1010	complete with one position communications, minimum				Ea.				266,000	293,000
2000	For fixed facilities, depending on height, minimum								57,200	63,000
2010	Maximum								115,000	127,000

13 42 Building Modules

13 42 63 – Detention Cell Modules

13 42 63.16 Steel Detention Cell Modules		Crew	Daily Output	Labor-Hours	Unit	Material	Labor	Equipment	Total	Total Incl O&P
0010	**STEEL DETENTION CELL MODULES**									
2000	Cells, prefab., 5′ to 6′ wide, 7′ to 8′ high, 7′ to 8′ deep,									
2010	bar front, cot, not incl. plumbing	E-4	1.50	21.333	Ea.	9,650	965	89.50	10,704.50	12,500

13 48 Sound, Vibration, and Seismic Control

13 48 13 – Manufactured Sound and Vibration Control Components

13 48 13.50 Audio Masking		Crew	Daily Output	Labor-Hours	Unit	Material	Labor	Equipment	Total	Total Incl O&P
0010	**AUDIO MASKING**, acoustical enclosure, 4″ thick wall and ceiling									
0020	8# per S.F., up to 12′ span	3 Carp	72	.333	SF Surf	32	13.30		45.30	57.50
0300	Better quality panels, 10.5# per S.F.		64	.375		36.50	15		51.50	64.50
0400	Reverb-chamber, 4″ thick, parallel walls		60	.400		45.50	16		61.50	76.50
0600	Skewed wall, parallel roof, 4″ thick panels		55	.436		52	17.45		69.45	86
0700	Skewed walls, skewed roof, 4″ layers, 4″ air space		48	.500		58.50	20		78.50	97.50
0900	Sound-absorbing panels, pntd mtl, 2′-6″ x 8′, under 1,000 S.F.		215	.112		12.10	4.46		16.56	20.50
1100	Over 1000 S.F.		240	.100		11.65	4		15.65	19.40
1200	Fabric faced		240	.100		9.45	4		13.45	16.95
1500	Flexible transparent curtain, clear	3 Shee	215	.112		7.30	5.25		12.55	16.35
1600	50% foam		215	.112		10.20	5.25		15.45	19.55
1700	75% foam		215	.112		10.20	5.25		15.45	19.55
1800	100% foam		215	.112		10.20	5.25		15.45	19.55
3100	Audio masking system, including speakers, amplification									

13 48 Sound, Vibration, and Seismic Control

13 48 13 – Manufactured Sound and Vibration Control Components

13 48 13.50 Audio Masking		Crew	Daily Output	Labor-Hours	Unit	2009 Bare Costs Material	Labor	Equipment	Total	Total Incl O&P
3110	and signal generator									
3200	Ceiling mounted, 5,000 S.F.	2 Elec	2400	.007	S.F.	1.25	.31		1.56	1.86
3300	10,000 S.F.		2800	.006		1.01	.27		1.28	1.52
3400	Plenum mounted, 5,000 S.F.		3800	.004		1.06	.20		1.26	1.47
3500	10,000 S.F.		4400	.004		.72	.17		.89	1.05

13 49 Radiation Protection

13 49 13 – Lead Sheet

13 49 13.50 Lead Sheets

		Crew	Daily Output	Labor-Hours	Unit	Material	Labor	Equipment	Total	Total Incl O&P
0010	**LEAD SHEETS**									
0300	Lead sheets, 1/16" thick	2 Lath	135	.119	S.F.	7.15	4.21		11.36	14.55
0400	1/8" thick		120	.133		14.45	4.74		19.19	23.50
0500	Lead shielding, 1/4" thick		135	.119		34	4.21		38.21	44
0550	1/2" thick		120	.133		59.50	4.74		64.24	73
0950	Lead headed nails (average 1 lb. per sheet)				Lb.	7.15			7.15	7.90
1000	Butt joints in 1/8" lead or thicker, 2" batten strip x 7' long	2 Lath	240	.067	Ea.	16.95	2.37		19.32	22.50
1200	X-ray protection, average radiography or fluoroscopy									
1210	room, up to 300 S.F. floor, 1/16" lead, minimum	2 Lath	.25	64	Total	8,075	2,275		10,350	12,500
1500	Maximum, 7'-0" walls	"	.15	106	"	9,700	3,800		13,500	16,700
1600	Deep therapy X-ray room, 250 kV capacity,									
1800	up to 300 S.F. floor, 1/4" lead, minimum	2 Lath	.08	200	Total	22,600	7,100		29,700	36,000
1900	Maximum, 7'-0" walls		.06	266	"	27,800	9,475		37,275	45,600
1999	Minimum labor/equipment charge		4.50	3.556	Job		126		126	200

13 49 19 – Lead-Lined Materials

13 49 19.50 Shielding Lead

		Crew	Daily Output	Labor-Hours	Unit	Material	Labor	Equipment	Total	Total Incl O&P
0010	**SHIELDING LEAD**									
0100	Laminated lead in wood doors, 1/16" thick, no hardware				S.F.	45			45	49.50
0200	Lead lined door frame, not incl. hardware,									
0210	1/16" thick lead, butt prepared for hardware	1 Lath	2.40	3.333	Ea.	575	119		694	815
0850	Window frame with 1/16" lead and voice passage, 36" x 60"	2 Glaz	2	8		1,125	310		1,435	1,725
0870	24" x 36" frame		8	2		900	77		977	1,125
0900	Lead gypsum board, 5/8" thick with 1/16" lead		160	.100	S.F.	7.25	3.86		11.11	14.20
0910	1/8" lead		140	.114		14.10	4.41		18.51	22.50
0930	1/32" lead	2 Lath	200	.080		4.43	2.84		7.27	9.35

13 49 21 – Lead Glazing

13 49 21.50 Lead Glazing

		Crew	Daily Output	Labor-Hours	Unit	Material	Labor	Equipment	Total	Total Incl O&P
0010	**LEAD GLAZING**									
0600	Lead glass, 1/4" thick, 2.0 mm LE, 12" x 16"	2 Glaz	13	1.231	Ea.	265	47.50		312.50	370
0700	24" x 36"		8	2		980	77		1,057	1,200
0800	36" x 60"		2	8		3,850	310		4,160	4,750
2000	X-ray viewing panels, clear lead plastic									
2010	7 mm thick, 0.3 mm LE, 2.3 lbs/S.F.	H-3	139	.115	S.F.	143	3.97		146.97	164
2020	12 mm thick, 0.5 mm LE, 3.9 lbs/S.F.		82	.195		194	6.70		200.70	224
2030	18 mm thick, 0.8 mm LE, 5.9 lbs/S.F.		54	.296		210	10.20		220.20	248
2040	22 mm thick, 1.0 mm LE, 7.2 lbs/S.F.		44	.364		215	12.55		227.55	257
2050	35 mm thick, 1.5 mm LE, 11.5 lbs/S.F.		28	.571		241	19.70		260.70	297
2060	46 mm thick, 2.0 mm LE, 15.0 lbs/S.F.		21	.762		315	26.50		341.50	395
2090	For panels 12 S.F. to 48 S.F., add crating charge				Ea.					50

13 49 Radiation Protection

13 49 23 – Modular Shielding Partitions

13 49 23.50 Modular Shielding Partitions		Crew	Daily Output	Labor-Hours	Unit	2009 Bare Costs Material	Labor	Equipment	Total	Total Incl O&P
0010	**MODULAR SHIELDING PARTITIONS**									
4000	X-ray barriers, modular, panels mounted within framework for									
4002	attaching to floor, wall or ceiling, upper portion is clear lead									
4005	plastic window panels 48″H, lower portion is opaque leaded									
4008	steel panels 36″H, structural supports not incl.									
4010	1-section barrier, 36″W x 84″H overall									
4020	0.5 mm LE panels	H-3	6.40	2.500	Ea.	4,525	86		4,611	5,125
4030	0.8 mm LE panels		6.40	2.500		4,825	86		4,911	5,450
4040	1.0 mm LE panels		5.33	3.002		4,900	103		5,003	5,575
4050	1.5 mm LE panels		5.33	3.002		5,225	103		5,328	5,925
4060	2-section barrier, 72″W x 84″H overall									
4070	0.5 mm LE panels	H-3	4	4	Ea.	9,250	138		9,388	10,400
4080	0.8 mm LE panels		4	4		9,775	138		9,913	11,000
4090	1.0 mm LE panels		3.56	4.494		9,975	155		10,130	11,300
5000	1.5 mm LE panels		3.20	5		11,300	172		11,472	12,800
5010	3-section barrier, 108″W x 84″H overall									
5020	0.5 mm LE panels	H-3	3.20	5	Ea.	13,900	172		14,072	15,600
5030	0.8 mm LE panels		3.20	5		14,700	172		14,872	16,400
5040	1.0 mm LE panels		2.67	5.993		14,900	206		15,106	16,700
5050	1.5 mm LE panels		2.46	6.504		16,200	224		16,424	18,200
7000	X-ray barriers, mobile, mounted within framework w/casters on									
7005	bottom, clear lead plastic window panels on upper portion,									
7010	opaque on lower, 30″W x 75″H overall, incl. framework									
7020	24″H upper w/0.5 mm LE, 48″H lower w/0.8 mm LE	1 Carp	16	.500	Ea.	2,400	20		2,420	2,675
7030	48″W x 75″H overall, incl. framework									
7040	36″H upper w/0.5 mm LE, 36″H lower w/0.8 mm LE	1 Carp	16	.500	Ea.	4,425	20		4,445	4,875
7050	36″H upper w/1.0 mm LE, 36″H lower w/1.5 mm LE	″	16	.500	″	5,250	20		5,270	5,800
7060	72″W x 75″H overall, incl. framework									
7070	36″H upper w/0.5 mm LE, 36″H lower w/0.8 mm LE	1 Carp	16	.500	Ea.	5,225	20		5,245	5,775
7080	36″H upper w/1.0 mm LE, 36″H lower w/1.5 mm LE	″	16	.500	″	6,550	20		6,570	7,225

13 49 33 – Radio Frequency Shielding

13 49 33.50 Shielding, Radio Frequency		Crew	Daily Output	Labor-Hours	Unit	Material	Labor	Equipment	Total	Total Incl O&P
0010	**SHIELDING, RADIO FREQUENCY**									
0020	Prefabricated or screen-type copper or steel, minimum	2 Carp	180	.089	SF Surf	31.50	3.55		35.05	40.50
0100	Average		155	.103		33.50	4.12		37.62	44
0150	Maximum		145	.110		35.50	4.41		39.91	46.50

13 53 Meteorological Instrumentation

13 53 09 – Weather Instrumentation

13 53 09.50 Weather Station		Crew	Daily Output	Labor-Hours	Unit	Material	Labor	Equipment	Total	Total Incl O&P
0010	**WEATHER STATION**									
0020	Remote recording, minimum				Ea.	3,000			3,000	3,300
0100	Maximum				″	27,000			27,000	29,700

Division Notes

	CREW	DAILY OUTPUT	LABOR-HOURS	UNIT	2009 BARE COSTS				TOTAL INCL O&P
					MAT.	LABOR	EQUIP.	TOTAL	

Estimating Tips

General

- Many products in Division 14 will require some type of support or blocking for installation not included with the item itself. Examples are supports for conveyors or tube systems, attachment points for lifts, and footings for hoists or cranes. Add these supports in the appropriate division.

14 10 00 Dumbwaiters
14 20 00 Elevators

- Dumbwaiters and elevators are estimated and purchased in a method similar to buying a car. The manufacturer has a base unit with standard features. Added to this base unit price will be whatever options the owner or specifications require. Increased load capacity, additional vertical travel, additional stops, higher speed, and cab finish options are items to be considered. When developing an estimate for dumbwaiters and elevators, remember that some items needed by the installers may have to be included as part of the general contract.

  Examples are:
  - — shaftway
  - — rail support brackets
  - — machine room
  - — electrical supply
  - — sill angles
  - — electrical connections
  - — pits
  - — roof penthouses
  - — pit ladders

Check the job specifications and drawings before pricing.

- Installation of elevators and handicapped lifts in historic structures can require significant additional costs. The associated structural requirements may involve cutting into and repairing finishes, mouldings, flooring, etc. The estimator must account for these special conditions.

14 30 00 Escalators and Moving Walks

- Escalators and moving walks are specialty items installed by specialty contractors. There are numerous options associated with these items. For specific options, contact a manufacturer or contractor. In a method similar to estimating dumbwaiters and elevators, you should verify the extent of general contract work and add items as necessary.

14 40 00 Lifts
14 90 00 Other Conveying Equipment

- Products such as correspondence lifts, chutes, and pneumatic tube systems, as well as other items specified in this subdivision, may require trained installers. The general contractor might not have any choice as to who will perform the installation or when it will be performed. Long lead times are often required for these products, making early decisions in scheduling necessary.

Reference Numbers

Reference numbers are shown in shaded boxes at the beginning of some major classifications. These numbers refer to related items in the Reference Section. The reference information may be an estimating procedure, an alternate pricing method, or technical information.

Note: Not all subdivisions listed here necessarily appear in this publication.

14 05 Common Work Results for Conveying Equipment

14 05 05 – Selective Conveying Equipment Demolition

14 05 05.10 Dumbwaiter Removal		Crew	Daily Output	Labor-Hours	Unit	Material	2009 Bare Costs Labor	Equipment	Total	Total Incl O&P
0010	**DUMBWAITER REMOVAL**									
0050	Dumbwaiter removal, cab, track and equipment	2 Elev	4	4	Stop		226		226	350
14 05 05.20 Elevator Removal										
0010	**ELEVATOR REMOVAL**									
0050	Elevator removal, cab, track and equipment	2 Elev	2	8	Stop		455		455	695
14 05 05.30 Escalator/Moving Walk Removal										
0010	**ESCALATOR / MOVING WALK REMOVAL**									
0050	Escalator removal, 10′ floor to floor	M-1	.14	228	Ea.		12,300	405	12,705	19,400
0100	Escalator removal, 15′ floor to floor		.12	266			14,300	475	14,775	22,600
0200	Escalator removal, 20′ floor to floor		.10	320			17,200	570	17,770	27,100
0300	Escalator removal, 25′ floor to floor		.08	400			21,500	710	22,210	33,900
0400	Moving ramp / walk removal		10	3.200	L.F.		172	5.70	177.70	271
14 05 05.40 Lift Removal										
0010	**LIFT REMOVAL**									
0050	Chair lift removal, minimum	2 Elev	2	8	Ea.		455		455	695
0060	maximum	"	1	16			905		905	1,400
0100	Single post lift removal	L-4	7.50	3.200			119		119	195
0200	Double post lift removal		5	4.800			179		179	293
0300	Four post lift removal		3	8			299		299	490
14 05 05.50 Distribution System Removal										
0010	**DISTRIBUTION SYSTEM REMOVAL**									
0050	Distribution system removal, motorized cart	4 Mill	.30	106	Station		4,425		4,425	6,950
0100	Distribution system removal, conveyor belt	2 Mill	80	.200	L.F.		8.30		8.30	13.05
0200	Distribution system removal, overhead monorail	1 Elev	50	.160			9.05		9.05	13.95
0300	Distribution system removal, pneumatic tube	2 Stpi	150	.107			5.25		5.25	8.15
0400	Remove linen / trash chute	2 Shee	8	2	Floor		94.50		94.50	150
14 05 05.60 Crane Removal										
0010	**CRANE REMOVAL**									
0050	Remove crane rail	E-4	300	.107	L.F.		4.82	.45	5.27	9.45
0100	Remove overhead bridge crane	M-4	1	36	Ea.		1,625	234	1,859	2,775

14 11 Manual Dumbwaiters

14 11 10 – Hand Operated Dumbwaiters

14 11 10.20 Manual Dumbwaiters		Crew	Daily Output	Labor-Hours	Unit	Material	Labor	Equipment	Total	Total Incl O&P
0010	**MANUAL DUMBWAITERS**									
0020	2 stop, minimum	2 Elev	.75	21.333	Ea.	2,925	1,200		4,125	5,075
0100	Maximum		.50	32	"	6,650	1,800		8,450	10,100
0300	For each additional stop, add		.75	21.333	Stop	1,050	1,200		2,250	3,025

14 12 Electric Dumbwaiters

14 12 10 – Dumbwaiters

14 12 10.10 Electric Dumbwaiters		Crew	Daily Output	Labor-Hours	Unit	2009 Bare Costs Material	Labor	Equipment	Total	Total Incl O&P
0010	**ELECTRIC DUMBWAITERS**									
0020	2 stop, minimum	2 Elev	.13	123	Ea.	7,375	6,975		14,350	18,800
0100	Maximum		.11	145	"	22,200	8,225		30,425	37,100
0600	For each additional stop, add	↓	.54	29.630	Stop	3,250	1,675		4,925	6,150
0750	Correspondence lift, 1 floor, 2 stop, 45 lb capacity	2 Elec	.20	80	Ea.	8,350	3,750		12,100	15,000

14 21 Electric Traction Elevators

14 21 13 – Electric Traction Freight Elevators

14 21 13.10 Electric Traction Freight Elevators and Options

		Crew	Daily Output	Labor-Hours	Unit	Material	Labor	Equipment	Total	Total Incl O&P
0010	**ELECTRIC TRACTION FREIGHT ELEVATORS AND OPTIONS** R142000-10									
0425	Electric freight, base unit, 4000 lb, 200 fpm, 4 stop, std. fin.	2 Elev	.05	320	Ea.	99,500	18,100		117,600	137,500
0450	For 5000 lb capacity, add R142000-40					5,600			5,600	6,150
0500	For 6000 lb capacity, add					17,400			17,400	19,200
0525	For 7000 lb capacity, add					21,500			21,500	23,600
0550	For 8000 lb capacity, add					29,600			29,600	32,600
0575	For 10000 lb capacity, add					32,500			32,500	35,800
0600	For 12000 lb capacity, add					39,400			39,400	43,300
0625	For 16000 lb capacity, add					50,000			50,000	55,000
0650	For 20000 lb capacity, add					59,500			59,500	65,500
0675	For increased speed, 250 fpm, add					22,500			22,500	24,700
0700	300 fpm, geared electric, add					28,400			28,400	31,300
0725	350 fpm, geared electric, add					34,300			34,300	37,700
0750	400 fpm, geared electric, add					39,900			39,900	43,800
0775	500 fpm, gearless electric, add					47,300			47,300	52,000
0800	600 fpm, gearless electric, add					57,500			57,500	63,000
0825	700 fpm, gearless electric, add					68,500			68,500	75,500
0850	800 fpm, gearless electric, add					79,000			79,000	87,000
0875	For class "B" loading, add					6,525			6,525	7,175
0900	For class "C-1" loading, add					9,075			9,075	9,975
0925	For class "C-2" loading, add					10,300			10,300	11,400
0950	For class "C-3" loading, add				↓	13,000			13,000	14,400
0975	For travel over 40 V.L.F., add	2 Elev	7.25	2.207	V.L.F.	540	125		665	785
1000	For number of stops over 4, add	"	.27	59.259	Stop	4,150	3,350		7,500	9,725

14 21 23 – Electric Traction Passenger Elevators

14 21 23.10 Electric Traction Passenger Elevators and Options

		Crew	Daily Output	Labor-Hours	Unit	Material	Labor	Equipment	Total	Total Incl O&P
0010	**ELECTRIC TRACTION PASSENGER ELEVATORS AND OPTIONS**									
1625	Electric pass., base unit, 2000 lb, 200 fpm, 4 stop, std. fin.	2 Elev	.05	320	Ea.	91,500	18,100		109,600	128,500
1650	For 2500 lb capacity, add					3,800			3,800	4,200
1675	For 3000 lb capacity, add					4,550			4,550	5,000
1700	For 3500 lb capacity, add					7,675			7,675	8,425
1725	For 4000 lb capacity, add					7,100			7,100	7,800
1750	For 4500 lb capacity, add					8,975			8,975	9,875
1775	For 5000 lb capacity, add					10,800			10,800	11,900
1800	For increased speed, 250 fpm, geared electric, add					7,000			7,000	7,700
1825	300 fpm, geared electric, add					10,600			10,600	11,700
1850	350 fpm, geared electric, add					12,100			12,100	13,300
1875	400 fpm, geared electric, add					15,400			15,400	17,000
1900	500 fpm, gearless electric, add					32,600			32,600	35,900
1925	600 fpm, gearless electric, add					36,100			36,100	39,700
1950	700 fpm, gearless electric, add				↓	41,000			41,000	45,100

14 21 Electric Traction Elevators

14 21 23 – Electric Traction Passenger Elevators

14 21 23.10 Electric Traction Passenger Elevators and Options		Crew	Daily Output	Labor-Hours	Unit	Material	2009 Bare Costs Labor	Equipment	Total	Total Incl O&P
1975	800 fpm, gearless electric, add				Ea.	46,800			46,800	51,500
2000	For travel over 40 V.L.F., add	2 Elev	7.25	2.207	V.L.F.	650	125		775	905
2025	For number of stops over 4, add		.27	59.259	Stop	7,825	3,350		11,175	13,800
2400	Electric hospital, base unit, 4000 lb, 200 fpm, 4 stop, std fin.		.05	320	Ea.	79,000	18,100		97,100	115,000
2425	For 4500 lb capacity, add					5,650			5,650	6,200
2450	For 5000 lb capacity, add					6,775			6,775	7,475
2475	For increased speed, 250 fpm, geared electric, add					7,125			7,125	7,850
2500	300 fpm, geared electric, add					10,600			10,600	11,700
2525	350 fpm, geared electric, add					12,200			12,200	13,400
2550	400 fpm, geared electric, add					15,400			15,400	17,000
2575	500 fpm, gearless electric, add					31,700			31,700	34,900
2600	600 fpm, gearless electric, add					36,100			36,100	39,700
2625	700 fpm, gearless electric, add					40,900			40,900	45,000
2650	800 fpm, gearless electric, add					46,800			46,800	51,500
2675	For travel over 40 V.L.F., add	2 Elev	7.25	2.207	V.L.F.	665	125		790	925
2700	For number of stops over 4, add	″	.27	59.259	Stop	4,150	3,350		7,500	9,725

14 21 33 – Electric Traction Residential Elevators

14 21 33.20 Residential Elevators		Crew	Daily Output	Labor-Hours	Unit	Material	Labor	Equipment	Total	Total Incl O&P
0010	**RESIDENTIAL ELEVATORS**									
7000	Residential, cab type, 1 floor, 2 stop, minimum	2 Elev	.20	80	Ea.	11,000	4,525		15,525	19,100
7100	Maximum		.10	160		18,600	9,050		27,650	34,400
7200	2 floor, 3 stop, minimum		.12	133		16,400	7,550		23,950	29,600
7300	Maximum		.06	266		26,700	15,100		41,800	52,500

14 24 Hydraulic Elevators

14 24 13 – Hydraulic Freight Elevators

14 24 13.10 Hydraulic Freight Elevators and Options		Crew	Daily Output	Labor-Hours	Unit	Material	Labor	Equipment	Total	Total Incl O&P
0010	**HYDRAULIC FREIGHT ELEVATORS AND OPTIONS**									
1025	Hydraulic freight, base unit, 2000 lb, 50 fpm, 2 stop, std. fin.	2 Elev	.10	160	Ea.	82,000	9,050		91,050	104,000
1050	For 2500 lb capacity, add					5,025			5,025	5,525
1075	For 3000 lb capacity, add					8,850			8,850	9,750
1100	For 3500 lb capacity, add					16,000			16,000	17,600
1125	For 4000 lb capacity, add					17,100			17,100	18,800
1150	For 4500 lb capacity, add					20,000			20,000	22,000
1175	For 5000 lb capacity, add					25,300			25,300	27,800
1200	For 6000 lb capacity, add					31,500			31,500	34,600
1225	For 7000 lb capacity, add					37,900			37,900	41,700
1250	For 8000 lb capacity, add					41,700			41,700	45,900
1275	For 10000 lb capacity, add					52,500			52,500	57,500
1300	For 12000 lb capacity, add					68,500			68,500	75,500
1325	For 16000 lb capacity, add					94,000			94,000	103,000
1350	For 20000 lb capacity, add					114,000			114,000	125,500
1375	For increased speed, 100 fpm, add					2,150			2,150	2,375
1400	125 fpm, add					4,650			4,650	5,100
1425	150 fpm, add					5,525			5,525	6,075
1450	175 fpm, add					9,400			9,400	10,400
1475	For class "B" loading, add					6,500			6,500	7,150
1500	For class "C-1" loading, add					9,025			9,025	9,925
1525	For class "C-2" loading, add					10,300			10,300	11,300
1550	For class "C-3" loading, add					13,000			13,000	14,300

14 24 Hydraulic Elevators

14 24 13 – Hydraulic Freight Elevators

14 24 13.10 Hydraulic Freight Elevators and Options		Crew	Daily Output	Labor-Hours	Unit	Material	2009 Bare Costs Labor	Equipment	Total	Total Incl O&P
1575	For travel over 20 V.L.F., add	2 Elev	7.25	2.207	V.L.F.	1,675	125		1,800	2,050
1600	For number of stops over 2, add	"	.27	59.259	Stop	4,575	3,350		7,925	10,200

14 24 23 – Hydraulic Passenger Elevators

14 24 23.10 Hydraulic Passenger Elevators and Options

		Crew	Daily Output	Labor-Hours	Unit	Material	Labor	Equipment	Total	Total Incl O&P
0010	**HYDRAULIC PASSENGER ELEVATORS AND OPTIONS**									
2050	Hyd. pass., base unit, 1500 lb, 100 fpm, 2 stop, std. fin.	2 Elev	.10	160	Ea.	39,900	9,050		48,950	57,500
2075	For 2000 lb capacity, add					2,025			2,025	2,250
2100	For 2500 lb. capacity, add					3,225			3,225	3,550
2125	For 3000 lb capacity, add					4,275			4,275	4,700
2150	For 3500 lb capacity, add					6,225			6,225	6,850
2175	For 4000 lb capacity, add					7,600			7,600	8,350
2200	For 4500 lb capacity, add					9,000			9,000	9,900
2225	For 5000 lb capacity, add					11,900			11,900	13,100
2250	For increased speed, 125 fpm, add					2,975			2,975	3,275
2275	150 fpm, add					3,775			3,775	4,150
2300	175 fpm, add					5,650			5,650	6,225
2325	200 fpm, add					7,850			7,850	8,625
2350	For travel over 12 V.L.F., add	2 Elev	7.25	2.207	V.L.F.	890	125		1,015	1,175
2375	For number of stops over 2, add		.27	59.259	Stop	2,575	3,350		5,925	7,975
2725	Hydraulic hospital, base unit, 4000 lb, 100 fpm, 2 stop, std. fin.		.10	160	Ea.	62,000	9,050		71,050	82,000
2750	For 4000 lb capacity, add					7,150			7,150	7,850
2775	For 4500 lb capacity, add					6,900			6,900	7,600
2800	For 5000 lb capacity, add					9,150			9,150	10,100
2825	For increased speed, 125 fpm, add					3,450			3,450	3,775
2850	150 fpm, add					4,150			4,150	4,550
2875	175 fpm, add					6,000			6,000	6,600
2900	200 fpm, add					7,275			7,275	8,000
2925	For travel over 12 V.L.F., add	2 Elev	7.25	2.207	V.L.F.	745	125		870	1,000
2950	For number of stops over 2, add	"	.27	59.259	Stop	4,200	3,350		7,550	9,775

14 27 Custom Elevator Cabs

14 27 13 – Custom Elevator Cab Finishes

14 27 13.10 Cab Finishes

		Crew	Daily Output	Labor-Hours	Unit	Material	Labor	Equipment	Total	Total Incl O&P
0010	**CAB FINISHES**									
3325	Passenger elevator cab finishes (based on 3500 lb cab size)									
3350	Acrylic panel ceiling				Ea.	780			780	860
3375	Aluminum eggcrate ceiling					1,225			1,225	1,350
3400	Stainless steel doors					2,975			2,975	3,250
3425	Carpet flooring					650			650	715
3450	Epoxy flooring					480			480	530
3475	Quarry tile flooring					825			825	910
3500	Slate flooring					1,525			1,525	1,675
3525	Textured rubber flooring					200			200	220
3550	Stainless steel walls					5,175			5,175	5,700
3575	Stainless steel returns at door					1,200			1,200	1,325
4450	Hospital elevator cab finishes (based on 3500 lb cab size)									
4475	Aluminum eggcrate ceiling				Ea.	1,200			1,200	1,325
4500	Stainless steel doors					2,675			2,675	2,950
4525	Epoxy flooring					375			375	415
4550	Quarry tile flooring					845			845	930

14 27 Custom Elevator Cabs

14 27 13 – Custom Elevator Cab Finishes

14 27 13.10 Cab Finishes		Crew	Daily Output	Labor-Hours	Unit	Material	2009 Bare Costs Labor	Equipment	Total	Total Incl O&P
4575	Textured rubber flooring				Ea.	720			720	790
4600	Stainless steel walls					5,175			5,175	5,700
4625	Stainless steel returns at door					1,200			1,200	1,325

14 28 Elevator Equipment and Controls

14 28 10 – Elevator Equipment and Control Options

14 28 10.10 Elevator Controls and Doors		Crew	Daily Output	Labor-Hours	Unit	Material	2009 Bare Costs Labor	Equipment	Total	Total Incl O&P
0010	ELEVATOR CONTROLS AND DOORS									
2975	Passenger elevator options									
3000	2 car group automatic controls	2 Elev	.66	24.242	Ea.	6,650	1,375		8,025	9,400
3025	3 car group automatic controls		.44	36.364		13,000	2,050		15,050	17,500
3050	4 car group automatic controls		.33	48.485		26,500	2,750		29,250	33,300
3075	5 car group automatic controls		.26	61.538		48,100	3,475		51,575	58,500
3100	6 car group automatic controls		.22	72.727		98,500	4,125		102,625	115,000
3125	Intercom service		3	5.333		1,375	300		1,675	2,000
3150	Duplex car selective collective		.66	24.242		12,200	1,375		13,575	15,600
3175	Center opening 1 speed doors		2	8		1,850	455		2,305	2,750
3200	Center opening 2 speed doors		2	8		2,550	455		3,005	3,500
3225	Rear opening doors (opposite front)		2	8		3,125	455		3,580	4,150
3250	Side opening 2 speed doors		2	8		4,900	455		5,355	6,075
3275	Automatic emergency power switching		.66	24.242		1,200	1,375		2,575	3,425
3300	Manual emergency power switching		8	2		500	113		613	725
3625	Hall finishes, stainless steel doors					2,000			2,000	2,200
3650	Stainless steel frames					1,450			1,450	1,600
3675	12 month maintenance contract								3,600	3,960
3700	Signal devices, hall lanterns	2 Elev	8	2		540	113		653	770
3725	Position indicators, up to 3		9.40	1.702		650	96.50		746.50	860
3750	Position indicators, per each over 3		32	.500		620	28.50		648.50	730
3775	High speed heavy duty door opener					4,200			4,200	4,625
3800	Variable voltage, O.H. gearless machine, min.	2 Elev	.16	100		70,500	5,650		76,150	86,000
3815	Maximum		.07	228		105,500	12,900		118,400	136,000
3825	Basement installed geared machine		.33	48.485		72,000	2,750		74,750	83,500
3850	Freight elevator options									
3875	Doors, bi-parting	2 Elev	.66	24.242	Ea.	9,975	1,375		11,350	13,100
3900	Power operated door and gate	"	.66	24.242		23,400	1,375		24,775	27,900
3925	Finishes, steel plate floor					2,475			2,475	2,725
3950	14 ga. 1/4" x 4' steel plate walls					1,900			1,900	2,100
3975	12 month maintenance contract								3,600	3,960
4000	Signal devices, hall lanterns	2 Elev	8	2		540	113		653	770
4025	Position indicators, up to 3		9.40	1.702		650	96.50		746.50	860
4050	Position indicators, per each over 3		32	.500		620	28.50		648.50	730
4075	Variable voltage basement installed geared machine		.66	24.242		19,600	1,375		20,975	23,700
4100	Hospital elevator options									
4125	2 car group automatic controls	2 Elev	.66	24.242	Ea.	6,650	1,375		8,025	9,400
4150	3 car group automatic controls		.44	36.364		13,000	2,050		15,050	17,500
4175	4 car group automatic controls		.33	48.485		26,500	2,750		29,250	33,300
4200	5 car group automatic controls		.26	61.538		48,100	3,475		51,575	58,500
4225	6 car group automatic controls		.22	72.727		98,500	4,125		102,625	115,000
4250	Intercom service		3	5.333		1,375	300		1,675	2,000
4275	Duplex car selective collective		.66	24.242		12,200	1,375		13,575	15,600
4300	Center opening 1 speed doors		2	8		1,850	455		2,305	2,750

14 28 Elevator Equipment and Controls

14 28 10 – Elevator Equipment and Control Options

14 28 10.10 Elevator Controls and Doors		Crew	Daily Output	Labor-Hours	Unit	2009 Bare Costs Material	Labor	Equipment	Total	Total Incl O&P
4325	Center opening 2 speed doors	2 Elev	2	8	Ea.	2,525	455		2,980	3,500
4350	Rear opening doors (opposite front)		2	8		3,125	455		3,580	4,150
4375	Side opening 2 speed doors		2	8		3,925	455		4,380	5,025
4400	Automatic emergency power switching		.66	24.242		1,200	1,375		2,575	3,425
4425	Manual emergency power switching		8	2		500	113		613	725
4675	Hall finishes, stainless steel doors					2,000			2,000	2,200
4700	Stainless steel frames					1,450			1,450	1,600
4725	12 month maintenance contract								3,600	3,960
4750	Signal devices, hall lanterns	2 Elev	8	2		540	113		653	770
4775	Position indicators, up to 3		9.40	1.702		650	96.50		746.50	860
4800	Position indicators, per each over 3		32	.500		620	28.50		648.50	730
4825	High speed heavy duty door opener					4,200			4,200	4,625
4850	Variable voltage, O.H. gearless machine, min.	2 Elev	.16	100		70,500	5,650		76,150	86,000
4865	Maximum		.07	228		105,500	12,900		118,400	136,000
4875	Basement installed geared machine		.33	48.485		26,900	2,750		29,650	33,700
5000	Drilling for piston, casing included, 18" diameter	B-48	80	.700	V.L.F.	45	24.50	36	105.50	129

14 31 Escalators

14 31 10 – Glass and Steel Escalators

14 31 10.10 Escalators		Crew	Daily Output	Labor-Hours	Unit	Material	Labor	Equipment	Total	Total Incl O&P
0010	**ESCALATORS** R143110-10									
1000	Glass, 32" wide x 10' floor to floor height	M-1	.07	457	Ea.	90,500	24,600	815	115,915	138,000
1010	48" wide x 10' floor to floor height		.07	457		98,000	24,600	815	123,415	146,500
1020	32" wide x 15' floor to floor height		.06	533		96,000	28,700	950	125,650	150,500
1030	48" wide x 15' floor to floor height		.06	533		102,000	28,700	950	131,650	157,000
1040	32" wide x 20' floor to floor height		.05	653		101,000	35,100	1,150	137,250	166,500
1050	48" wide x 20' floor to floor height		.05	653		110,000	35,100	1,150	146,250	176,500
1060	32" wide x 25' floor to floor height		.04	800		107,000	43,000	1,425	151,425	185,000
1070	48" wide x 25' floor to floor height		.04	800		124,000	43,000	1,425	168,425	204,000
1080	Enameled steel, 32" wide x 10' floor to floor height		.07	457		95,500	24,600	815	120,915	144,000
1090	48" wide x 10' floor to floor height		.07	457		104,000	24,600	815	129,415	153,000
1110	32" wide x 15' floor to floor height		.06	533		104,000	28,700	950	133,650	159,500
1120	48" wide x 15' floor to floor height		.06	533		104,000	28,700	950	133,650	159,500
1130	32" wide x 20' floor to floor height		.05	653		107,500	35,100	1,150	143,750	173,500
1140	48" wide x 20' floor to floor height		.05	653		115,000	35,100	1,150	151,250	182,000
1150	32" wide x 25' floor to floor height		.04	800		115,000	43,000	1,425	159,425	194,000
1160	48" wide x 25' floor to floor height		.04	800		115,000	43,000	1,425	159,425	194,000
1170	Stainless steel, 32" wide x 10' floor to floor height		.07	457		115,000	24,600	815	140,415	165,000
1180	48" wide x 10' floor to floor height		.07	457		115,000	24,600	815	140,415	165,000
1500	32" wide x 15' floor to floor height		.06	533		106,000	28,700	950	135,650	161,500
1700	48" wide x 15' floor to floor height		.06	533		107,500	28,700	950	137,150	163,500
2300	32" wide x 25' floor to floor height		.04	800		114,500	43,000	1,425	158,925	193,500
2500	48" wide x 25' floor to floor height		.04	800		138,500	43,000	1,425	182,925	219,500

14 32 Moving Walks

14 32 10 – Moving Walkways

14 32 10.10 Moving Walks		Crew	Daily Output	Labor-Hours	Unit	Material	2009 Bare Costs Labor	Equipment	Total	Total Incl O&P
0010	**MOVING WALKS** R143210-20									
0020	Walk, 27" tread width, minimum	M-1	6.50	4.923	L.F.	850	265	8.75	1,123.75	1,350
0100	300' to 500', maximum		4.43	7.223		1,175	390	12.85	1,577.85	1,925
0300	48" tread width walk, minimum		4.43	7.223		1,925	390	12.85	2,327.85	2,750
0400	100' to 350', maximum		3.82	8.377		2,250	450	14.90	2,714.90	3,175
0600	Ramp, 12° incline, 36" tread width, minimum		5.27	6.072		1,575	325	10.80	1,910.80	2,225
0700	70' to 90' maximum		3.82	8.377		2,250	450	14.90	2,714.90	3,175
0900	48" tread width, minimum		3.57	8.964		2,300	480	15.95	2,795.95	3,275
1000	40' to 70', maximum	↓	2.91	10.997	↓	2,900	590	19.55	3,509.55	4,125

14 42 Wheelchair Lifts

14 42 13 – Inclined Wheelchair Lifts

14 42 13.10 Inclined Wheelchair Lifts and Stairclimbers		Crew	Daily Output	Labor-Hours	Unit	Material	Labor	Equipment	Total	Total Incl O&P
0010	**INCLINED WHEELCHAIR LIFTS AND STAIRCLIMBERS**									
7700	Stair climber (chair lift), single seat, minimum	2 Elev	1	16	Ea.	5,325	905		6,230	7,275
7800	Maximum		.20	80		7,350	4,525		11,875	15,100
8700	Stair lift, minimum		1	16		14,500	905		15,405	17,300
8900	Maximum	↓	.20	80	↓	22,900	4,525		27,425	32,200

14 42 16 – Vertical Wheelchair Lifts

14 42 16.10 Wheelchair Lifts		Crew	Daily Output	Labor-Hours	Unit	Material	Labor	Equipment	Total	Total Incl O&P
0010	**WHEELCHAIR LIFTS**									
8000	Wheelchair lift, minimum	2 Elev	1	16	Ea.	7,325	905		8,230	9,450
8500	Maximum	"	.50	32	"	17,300	1,800		19,100	21,800

14 45 Vehicle Lifts

14 45 10 – Hydraulic Vehicle Lifts

14 45 10.10 Hydraulic Lifts		Crew	Daily Output	Labor-Hours	Unit	Material	Labor	Equipment	Total	Total Incl O&P
0010	**HYDRAULIC LIFTS**									
0200	Double post, 9000 lb frame	L-4	1.15	20.870	Ea.	6,000	780		6,780	7,875
2200	Hoists, single post, 8,000# capacity, swivel arms		.40	60		4,975	2,250		7,225	9,125
2400	Two posts, adjustable frames, 11,000# capacity		.25	96		6,400	3,575		9,975	12,900
2500	24,000# capacity		.15	160		8,550	5,975		14,525	19,200
2700	7,500# capacity, frame supports		.50	48		7,125	1,800		8,925	10,800
2800	Four post, roll on ramp		.50	48		6,400	1,800		8,200	9,975
2810	Hydraulic lifts, above ground, 2 post, clear floor, 6000 lb cap		2.67	8.989		7,025	335		7,360	8,275
2815	9000 lb capacity		2.29	10.480		16,700	390		17,090	18,900
2820	15,000 lb capacity		2	12		18,800	450		19,250	21,400
2825	30,000 lb capacity		1.60	15		41,600	560		42,160	46,700
2830	4 post, ramp style, 25,000 lb capacity		2	12		16,300	450		16,750	18,600
2835	35,000 lb capacity		1	24		76,000	895		76,895	85,000
2840	50,000 lb capacity		1	24		85,000	895		85,895	95,000
2845	75,000 lb capacity	↓	1	24		98,500	895		99,395	110,000
2850	For drive thru tracks, add, minimum					1,050			1,050	1,150
2855	Maximum					1,800			1,800	1,975
2860	Ramp extensions, 3' (set of 2)					860			860	945
2865	Rolling jack platform					3,000			3,000	3,300
2870	Elec/hyd jacking beam					8,000			8,000	8,800
2880	Scissor lift, portable, 6000 lb capacity				↓	7,850			7,850	8,625

14 51 Correspondence and Parcel Lifts

14 51 10 – Electric Correspondence and Parcel Lifts

14 51 10.10 Correspondence Lifts		Crew	Daily Output	Labor-Hours	Unit	2009 Bare Costs Material	Labor	Equipment	Total	Total Incl O&P
0010	**CORRESPONDENCE LIFTS**									
0020	1 floor, 2 stop, 25 lb capacity, electric	2 Elev	.20	80	Ea.	5,975	4,525		10,500	13,600
0100	Hand, 5 lb capacity	″	.20	80	″	2,250	4,525		6,775	9,475
14 51 10.20 Parcel Lifts										
0010	**PARCEL LIFTS**									
0020	20″ x 20″, 100 lb capacity, electric, per floor	2 Mill	.25	64	Ea.	9,350	2,650		12,000	14,500

14 91 Facility Chutes

14 91 33 – Laundry and Linen Chutes

14 91 33.10 Chutes		Crew	Daily Output	Labor-Hours	Unit	Material	Labor	Equipment	Total	Total Incl O&P
0011	**CHUTES**, linen, trash or refuse									
0050	Aluminized steel, 16 ga., 18″ diameter	2 Shee	3.50	4.571	Floor	1,100	216		1,316	1,575
0100	24″ diameter		3.20	5		1,175	236		1,411	1,650
0200	30″ diameter		3	5.333		1,325	252		1,577	1,850
0300	36″ diameter		2.80	5.714		1,500	270		1,770	2,075
0400	Galvanized steel, 16 ga., 18″ diameter		3.50	4.571		1,000	216		1,216	1,450
0500	24″ diameter		3.20	5		1,125	236		1,361	1,625
0600	30″ diameter		3	5.333		1,275	252		1,527	1,800
0700	36″ diameter		2.80	5.714		1,500	270		1,770	2,075
0800	Stainless steel, 18″ diameter		3.50	4.571		2,350	216		2,566	2,925
0900	24″ diameter		3.20	5		2,550	236		2,786	3,175
1000	30″ diameter		3	5.333		3,100	252		3,352	3,800
1005	36″ diameter		2.80	5.714		3,450	270		3,720	4,225
1200	Linen chute bottom collector, aluminized steel		4	4	Ea.	1,400	189		1,589	1,850
1300	Stainless steel		4	4		1,800	189		1,989	2,275
1500	Refuse, bottom hopper, aluminized steel, 18″ diameter		3	5.333		890	252		1,142	1,375
1600	24″ diameter		3	5.333		1,075	252		1,327	1,575
1800	36″ diameter		3	5.333		1,800	252		2,052	2,375

14 91 82 – Trash Chutes

14 91 82.10 Trash Chutes and Accessories		Crew	Daily Output	Labor-Hours	Unit	Material	Labor	Equipment	Total	Total Incl O&P
0010	**TRASH CHUTES AND ACCESSORIES**									
2900	Package chutes, spiral type, minimum	2 Shee	4.50	3.556	Floor	2,425	168		2,593	2,950
3000	Maximum	″	1.50	10.667	″	6,325	505		6,830	7,775
9000	Minimum labor/equipment charge	1 Shee	1	8	Job		380		380	600

14 92 Pneumatic Tube Systems

14 92 10 – Conventional, Automatic and Computer Controlled Pneumatic Tube Systems

14 92 10.10 Pneumatic Tube Systems		Crew	Daily Output	Labor-Hours	Unit	Material	Labor	Equipment	Total	Total Incl O&P
0010	**PNEUMATIC TUBE SYSTEMS**									
0020	100′ long, single tube, 2 stations, stock									
0100	3″ diameter	2 Stpi	.12	133	Total	6,600	6,575		13,175	17,500
0300	4″ diameter	″	.09	177	″	7,450	8,775		16,225	21,800
0400	Twin tube, two stations or more, conventional system									
0600	2-1/2″ round	2 Stpi	62.50	.256	L.F.	13.30	12.65		25.95	34.50
0700	3″ round		46	.348		15.25	17.15		32.40	43.50
0900	4″ round		49.60	.323		16.90	15.90		32.80	43
1000	4″ x 7″ oval		37.60	.426		24.50	21		45.50	59.50
1050	Add for blower		2	8	System	5,200	395		5,595	6,325

14 92 Pneumatic Tube Systems

14 92 10 – Conventional, Automatic and Computer Controlled Pneumatic Tube Systems

14 92 10.10 Pneumatic Tube Systems		Crew	Daily Output	Labor-Hours	Unit	2009 Bare Costs Material	Labor	Equipment	Total	Total Incl O&P
1110	Plus for each round station, add	2 Stpi	7.50	2.133	Ea.	585	105		690	810
1150	Plus for each oval station, add		7.50	2.133	"	585	105		690	810
1200	Alternate pricing method: base cost, minimum		.75	21.333	Total	6,375	1,050		7,425	8,650
1300	Maximum		.25	64	"	12,700	3,150		15,850	18,800
1500	Plus total system length, add, minimum		93.40	.171	L.F.	8.20	8.45		16.65	22
1600	Maximum		37.60	.426	"	24.50	21		45.50	59
1800	Completely automatic system, 4" round, 15 to 50 stations		.29	55.172	Station	20,100	2,725		22,825	26,300
2200	51 to 144 stations		.32	50		15,600	2,475		18,075	21,000
2400	6" round or 4" x 7" oval, 15 to 50 stations		.24	66.667		25,200	3,300		28,500	32,800
2800	51 to 144 stations	▼	.23	69.565	▼	21,100	3,425		24,525	28,500

Estimating Tips

Pipe for fire protection and all uses is located in Subdivision 22 11 13.

The labor adjustment factors listed in Subdivision 22 01 02.20 also apply to Division 21.

Many, but not all, areas require backflow protection in the fire system. It is advisable to check local building codes for specific requirements.

For your reference, the following is a list of the most applicable Fire Codes and Standards which may be purchased from the NFPA, 1 Batterymarch Park, Quincy, MA 02169-7471.

NFPA 1: Uniform Fire Code
NFPA 10: Portable Fire Extinguishers
NFPA 11: Low-, Medium-, and High-Expansion Foam
NFPA 12: Carbon Dioxide Extinguishing Systems (Also companion 12A)
NFPA 13: Installation of Sprinkler Systems (Also companion 13D, 13E, and 13R)
NFPA 14: Installation of Standpipe and Hose Systems
NFPA 15: Water Spray Fixed Systems for Fire Protection
NFPA 16: Installation of Foam-Water Sprinkler and Foam-Water Spray Systems
NFPA 17: Dry Chemical Extinguishing Systems (Also companion 17A)
NFPA 18: Wetting Agents
NFPA 20: Installation of Stationary Pumps for Fire Protection
NFPA 22: Water Tanks for Private Fire Protection
NFPA 24: Installation of Private Fire Service Mains and their Appurtenances
NFPA 25: Inspection, Testing and Maintenance of Water-Based Fire Protection

Reference Numbers

Reference numbers are shown in shaded boxes at the beginning of some major classifications. These numbers refer to related items in the Reference Section. The reference information may be an estimating procedure, an alternate pricing method, or technical information.

Note: Not all subdivisions listed here necessarily appear in this publication.

Note: **Trade Service,** *in part, has been used as a reference source for some of the material prices used in Division 21.*

21 05 Common Work Results for Fire Suppression

21 05 23 – General-Duty Valves for Water-Based Fire-Suppression Piping

21 05 23.50 General-Duty Valves for Water-Based Fire Supp.		Crew	Daily Output	Labor-Hours	Unit	Material	2009 Bare Costs Labor	Equipment	Total	Total Incl O&P
0010	GENERAL-DUTY VALVES FOR WATER-BASED FIRE SUPPRESSION PIPING									
6200	Valves and components									
6210	Alarm, includes									
6220	retard chamber, trim, gauges, alarm line strainer									
6260	3" size	Q-12	3	5.333	Ea.	1,225	231		1,456	1,700
6280	4" size	"	2	8		1,275	345		1,620	1,950
6300	6" size	Q-13	4	8		1,425	365		1,790	2,150
6320	8" size	"	3	10.667	↓	1,725	490		2,215	2,650
6500	Check, swing, C.I. body, brass fittings, auto. ball drip									
6520	4" size	Q-12	3	5.333	Ea.	255	231		486	640
6540	6" size	Q-13	4	8		470	365		835	1,100
6580	8" size	"	3	10.667	↓	890	490		1,380	1,750
6800	Check, wafer, butterfly type, C.I. body, bronze fittings									
6820	4" size	Q-12	4	4	Ea.	283	173		456	580
6840	6" size	Q-13	5.50	5.818		485	267		752	950
6860	8" size		5	6.400		875	294		1,169	1,425
6880	10" size	↓	4.50	7.111		1,325	325		1,650	1,975
8800	Flow control valve, includes trim and gauges, 2" size	Q-12	2	8		2,025	345		2,370	2,775
8820	3" size	"	1.50	10.667		2,375	460		2,835	3,325
8840	4" size	Q-13	2.80	11.429		2,825	525		3,350	3,925
8860	6" size	"	2	16		3,475	735		4,210	4,975
9200	Pressure operated relief valve, brass body	1 Spri	18	.444	↓	430	21.50		451.50	510
9600	Waterflow indicator, with recycling retard and									
9610	two single pole retard switches, 2" thru 6" pipe size	1 Spri	8	1	Ea.	219	48		267	315
9990	Minimum labor/equipment charge	"	3	2.667	Job		128		128	199

21 11 Facility Fire-Suppression Water-Service Piping

21 11 16 – Facility Fire Hydrants

21 11 16.50 Fire Hydrants for Buildings		Crew	Daily Output	Labor-Hours	Unit	Material	Labor	Equipment	Total	Total Incl O&P
0010	FIRE HYDRANTS FOR BUILDINGS									
3750	Hydrants, wall, w/caps, single, flush, polished brass									
3800	2-1/2" x 2-1/2"	Q-12	5	3.200	Ea.	175	139		314	410
3840	2-1/2" x 3"		5	3.200		355	139		494	605
3860	3" x 3"	↓	4.80	3.333		287	144		431	540
3900	For polished chrome, add				↓	20%				
3950	Double, flush, polished brass									
4000	2-1/2" x 2-1/2" x 4"	Q-12	5	3.200	Ea.	470	139		609	735
4040	2-1/2" x 2-1/2" x 6"		4.60	3.478		675	151		826	980
4080	3" x 3" x 4"		4.90	3.265		1,000	141		1,141	1,325
4120	3" x 3" x 6"	↓	4.50	3.556		1,025	154		1,179	1,400
4200	For polished chrome, add				↓	10%				
4350	Double, projecting, polished brass									
4400	2-1/2" x 2-1/2" x 4"	Q-12	5	3.200	Ea.	209	139		348	445
4450	2-1/2" x 2-1/2" x 6"	"	4.60	3.478	"	425	151		576	705
4460	Valve control, dbl. flush/projecting hydrant, cap &									
4470	chain, extension rod & cplg., escutcheon, polished brass	Q-12	8	2	Ea.	249	86.50		335.50	405
4480	Four-way square, flush, polished brass									
4540	2-1/2"(4) x 6"	Q-12	3.60	4.444	Ea.	2,175	193		2,368	2,700

21 11 19.50 Connections for the Fire-Department		Crew	Daily Output	Labor-Hours	Unit	Material	2009 Bare Costs Labor	Equipment	Total	Total Incl O&P
0010	**CONNECTIONS FOR THE FIRE-DEPARTMENT**									
6000	Roof manifold, horiz., brass, without valves & caps									
6040	2-1/2" x 2-1/2" x 4"	Q-12	4.80	3.333	Ea.	139	144		283	375
6060	2-1/2" x 2-1/2" x 6"		4.60	3.478		146	151		297	395
6080	2-1/2" x 2-1/2" x 2-1/2" x 4"		4.60	3.478		261	151		412	520
6090	2-1/2" x 2-1/2" x 2-1/2" x 6"		4.60	3.478		230	151		381	485
7000	Sprinkler line tester, cast brass					19.80			19.80	22
7140	Standpipe connections, wall, w/plugs & chains									
7160	Single, flush, brass, 2-1/2" x 2-1/2"	Q-12	5	3.200	Ea.	123	139		262	350
7180	2-1/2" x 3"	"	5	3.200	"	135	139		274	365
7240	For polished chrome, add					15%				
7280	Double, flush, polished brass									
7300	2-1/2" x 2-1/2" x 4"	Q-12	5	3.200	Ea.	425	139		564	685
7330	2-1/2" x 2-1/2" x 6"		4.60	3.478		595	151		746	890
7340	3" x 3" x 4"		4.90	3.265		775	141		916	1,075
7370	3" x 3" x 6"		4.50	3.556		765	154		919	1,075
7400	For polished chrome, add					15%				
7440	For sill cock combination, add				Ea.	50.50			50.50	56
7580	Double projecting, polished brass									
7600	2-1/2" x 2-1/2" x 4"	Q-12	5	3.200	Ea.	390	139		529	645
7630	2-1/2" x 2-1/2" x 6"	"	4.60	3.478	"	660	151		811	960
7680	For polished chrome, add					15%				
7900	Three way, flush, polished brass									
7920	2-1/2" (3) x 4"	Q-12	4.80	3.333	Ea.	1,175	144		1,319	1,500
7930	2-1/2" (3) x 6"	"	4.80	3.333		1,150	144		1,294	1,475
8000	For polished chrome, add					9%				
8020	Three way, projecting, polished brass									
8040	2-1/2"(3) x 4"	Q-12	4.80	3.333	Ea.	985	144		1,129	1,300
8070	2-1/2" (3) x 6"	"	4.60	3.478		990	151		1,141	1,325
8100	For polished chrome, add					12%				
8200	Four way, square, flush, polished brass,									
8240	2-1/2"(4) x 6"	Q-12	3.60	4.444	Ea.	2,475	193		2,668	3,025
8300	For polished chrome, add				"	10%				
8550	Wall, vertical, flush, cast brass									
8600	Two way, 2-1/2" x 2-1/2" x 4"	Q-12	5	3.200	Ea.	580	139		719	855
8660	Four way, 2-1/2"(4) x 6"		3.80	4.211		2,125	182		2,307	2,600
8680	Six way, 2-1/2"(6) x 6"		3.40	4.706		3,275	204		3,479	3,925
8700	For polished chrome, add					10%				
8800	Sidewalk siamese unit, polished brass, two way									
8820	2-1/2" x 2-1/2" x 4"	Q-12	2.50	6.400	Ea.	470	277		747	950
8850	2-1/2" x 2-1/2" x 6"		2	8		645	345		990	1,250
8860	3" x 3" x 4"		2.50	6.400		740	277		1,017	1,250
8890	3" x 3" x 6"		2	8		1,000	345		1,345	1,650
8940	For polished chrome, add					12%				
9100	Sidewalk siamese unit, polished brass, three way									
9120	2-1/2" x 2-1/2" x 2-1/2" x 6"	Q-12	2	8	Ea.	825	345		1,170	1,450
9160	For polished chrome, add				"	15%				
9990	Minimum labor/equipment charge	1 Spri	4	2	Job		96.50		96.50	149

21 12 13 – Fire-Suppression Hoses and Nozzles

21 12 13.50 Fire Hoses and Nozzles		Crew	Daily Output	Labor-Hours	Unit	Material (2009 Bare Costs)	Labor (2009 Bare Costs)	Equipment (2009 Bare Costs)	Total (2009 Bare Costs)	Total Incl O&P
0010	**FIRE HOSES AND NOZZLES** R211226-10									
0200	Adapters, rough brass, straight hose threads									
0220	One piece, female to male, rocker lugs R211226-20									
0240	1" x 1"				Ea.	39.50			39.50	43.50
0260	1-1/2" x 1"					39.50			39.50	43.50
0280	1-1/2" x 1-1/2"					12.50			12.50	13.75
0300	2" x 1-1/2"					54			54	59.50
0320	2" x 2"					40.50			40.50	44.50
0340	2-1/2" x 1-1/2"					11.75			11.75	12.95
0360	2-1/2" x 2"					31.50			31.50	34.50
0380	2-1/2" x 2-1/2"					18.40			18.40	20
0400	3" x 2-1/2"					103			103	113
0420	3" x 3"					81			81	89
0500	For polished brass, add					50%				
0520	For polished chrome, add				↓	75%				
0700	One piece, female to male, hexagon									
0740	1-1/2" x 3/4"				Ea.	33			33	36.50
0760	2" x 1-1/2"					69.50			69.50	76.50
0780	2-1/2" x 1"					133			133	146
0800	2-1/2" x 1-1/2"					45.50			45.50	50
0820	2-1/2" x 2"					31.50			31.50	34.50
0840	3" x 2-1/2"					57.50			57.50	63.50
0900	For polished chrome, add				↓	75%				
1100	Swivel, female to female, pin lugs									
1120	1-1/2" x 1-1/2"				Ea.	50.50			50.50	55.50
1200	2-1/2" x 2-1/2"				Ea.	98.50			98.50	108
1260	For polished brass, add					50%				
1280	For polished chrome, add				↓	75%				
1400	Couplings, sngl & dbl jacket, pin lug or rocker lug, cast brass									
1410	1-1/2"				Ea.	43.50			43.50	48
1420	2-1/2"				"	57.50			57.50	63
1500	For polished brass, add					20%				
1520	For polished chrome, add					40%				
1580	Reducing, F x M, interior installation, cast brass									
1590	2" x 1-1/2"				Ea.	54			54	59.50
1600	2-1/2" x 1-1/2"					11.75			11.75	12.95
1680	For polished brass, add					50%				
1720	For polished chrome, add				↓	75%				
2200	Hose, less couplings									
2260	Synthetic jacket, lined, 300 lb. test, 1-1/2" diameter	Q-12	2600	.006	L.F.	2.39	.27		2.66	3.04
2280	2-1/2" diameter		2200	.007		3.98	.32		4.30	4.87
2360	High strength, 500 lb. test, 1-1/2" diameter		2600	.006		2.47	.27		2.74	3.13
2380	2-1/2" diameter	↓	2200	.007	↓	4.29	.32		4.61	5.20
5000	Nipples, straight hose to tapered iron pipe, brass									
5060	Female to female, 1-1/2" x 1-1/2"				Ea.	17.40			17.40	19.15
5080	2" x 2"					18			18	19.80
5100	2-1/2" x 2-1/2"					28.50			28.50	31
5190	For polished chrome, add					75%				
5200	Double male or male to female, 1" x 1"					34.50			34.50	38
5220	1-1/2" x 1"					50.50			50.50	56
5230	1-1/2" x 1-1/2"					11.40			11.40	12.55
5260	2" x 1-1/2"					55			55	60.50
5270	2" x 2"				↓	69.50			69.50	76.50

21 12 13 – Fire-Suppression Hoses and Nozzles

21 12 13.50 Fire Hoses and Nozzles		Crew	Daily Output	Labor-Hours	Unit	2009 Bare Costs Material	Labor	Equipment	Total	Total Incl O&P
5280	2-1/2" x 1-1/2"				Ea.	45.50			45.50	50
5300	2-1/2" x 2"					37.50			37.50	41.50
5310	2-1/2" x 2-1/2"					21.50			21.50	23.50
5340	For polished chrome, add				↓	75%				
5600	Nozzles, brass									
5620	Adjustable fog, 3/4" booster line				Ea.	88.50			88.50	97.50
5630	1" booster line					104			104	115
5640	1-1/2" leader line					63.50			63.50	70
5660	2-1/2" direct connection					168			168	185
5680	2-1/2" playpipe nozzle				↓	176			176	193
5780	For chrome plated, add					8%				
5850	Electrical fire, adjustable fog, no shock									
5900	1-1/2"				Ea.	340			340	375
5920	2-1/2"					455			455	505
5980	For polished chrome, add				↓	6%				
6200	Heavy duty, comb. adj. fog and str. stream, with handle									
6210	1" booster line				Ea.	325			325	360
6240	1-1/2"					365			365	405
6260	2-1/2", for playpipe					655			655	720
6280	2-1/2" direct connection					455			455	505
6300	2-1/2" playpipe combination					635			635	700
6480	For polished chrome, add					7%				
6500	Plain fog, polished brass, 1-1/2"					86.50			86.50	95
6540	Chrome plated, 1-1/2"					83			83	91.50
6700	Plain stream, polished brass, 1-1/2" x 10"					39.50			39.50	43.50
6760	2-1/2" x 15" x 7/8" or 1-1/2"				↓	71.50			71.50	78.50
6860	For polished chrome, add					20%				
7000	Underwriters playpipe, 2-1/2" x 30" with 1-1/8" tip				Ea.	203			203	224
7040	Less tip					157			157	173
9200	Storage house, hose only, primed steel					705			705	775
9220	Aluminum					1,550			1,550	1,700
9280	Hose and hydrant house, primed steel					945			945	1,050
9300	Aluminum					1,375			1,375	1,500
9340	Tools, crowbar and brackets	1 Carp	12	.667		96	26.50		122.50	150
9360	Combination hydrant wrench and spanner				↓	25.50			25.50	28.50
9380	Fire axe and brackets									
9400	6 lb.	1 Carp	12	.667	Ea.	100	26.50		126.50	154
9900	Minimum labor/equipment charge	1 Plum	2	[illegible]	Job		195		195	300

21 12 16 – Fire-Suppression Hose Reels

21 12 16.50 Fire -Suppression Hose Reels		Crew	Daily Output	Labor-Hours	Unit	Material	Labor	Equipment	Total	Total Incl O&P
0010	**FIRE -SUPPRESSION HOSE REELS**									
2990	Hose reel, swinging, for 1-1/2" polyester neoprene lined hose									
3000	50' long	Q-12	14	1.143	Ea.	108	49.50		157.50	195
3020	100' long		14	1.143		151	49.50		200.50	243
3060	For 2-1/2" cotton rubber hose, 75' long		14	1.143		175	49.50		224.50	269
3100	150' long	↓	14	1.143	↓	202	49.50		251.50	299

21 12 19 – Fire-Suppression Hose Racks

21 12 19.50 Fire Hose Racks		Crew	Daily Output	Labor-Hours	Unit	Material	Labor	Equipment	Total	Total Incl O&P
0010	**FIRE HOSE RACKS**									
2600	Hose rack, swinging, for 1-1/2" diameter hose,									
2620	Enameled steel, 50' & 75' lengths of hose	Q-12	20	.800	Ea.	48.50	34.50		83	107
2640	100' and 125' lengths of hose	↓	20	.800	↓	49	34.50		83.50	108

21 12 Fire-Suppression Standpipes

21 12 19 – Fire-Suppression Hose Racks

21 12 19.50 Fire Hose Racks		Crew	Daily Output	Labor-Hours	Unit	Material	Labor	Equipment	Total	Total Incl O&P
						2009 Bare Costs				
2680	Chrome plated, 50′ and 75′ lengths of hose	Q-12	20	.800	Ea.	60.50	34.50		95	121
2700	100′ and 125′ lengths of hose		20	.800		60.50	34.50		95	121
2780	For hose rack nipple, 1-1/2″ polished brass, add					21.50			21.50	24
2820	2-1/2″ polished brass, add					42.50			42.50	47
2840	1-1/2″ polished chrome, add					32			32	35
2860	2-1/2″ polished chrome, add					45			45	49.50

21 12 23 – Fire-Suppression Hose Valves

21 12 23.70 Fire Hose Valves		Crew	Daily Output	Labor-Hours	Unit	Material	Labor	Equipment	Total	Total Incl O&P
0010	**FIRE HOSE VALVES**									
0020	Angle, combination pressure adjust/restricting, rough brass									
0030	1-1/2″ R211226-20	1 Spri	12	.667	Ea.	70	32		102	127
0040	2-1/2″	"	7	1.143	"	139	55		194	239
0042	Nonpressure adjustable/restricting, rough brass									
0044	1-1/2″	1 Spri	12	.667	Ea.	43.50	32		75.50	98
0046	2-1/2″	"	7	1.143	"	74	55		129	167
0050	For polished brass, add					30%				
0060	For polished chrome, add					40%				
0080	Wheel handle, 300 lb., 1-1/2″	1 Spri	12	.667	Ea.	58	32		90	114
0090	2-1/2″	"	7	1.143	"	60	55		115	152
0100	For polished brass, add					35%				
0110	For polished chrome, add					50%				
1000	Ball drip, automatic, rough brass, 1/2″	1 Spri	20	.400	Ea.	13.25	19.25		32.50	44.50
1010	3/4″	"	20	.400	"	15.05	19.25		34.30	46.50
1100	Ball, 175 lb., sprinkler system, FM/UL, threaded, bronze									
1120	Slow close									
1150	1″ size	1 Spri	19	.421	Ea.	165	20.50		185.50	213
1160	1-1/4″ size		15	.533		178	25.50		203.50	236
1170	1-1/2″ size		13	.615		234	29.50		263.50	305
1180	2″ size		11	.727		285	35		320	370
1190	2-1/2″ size	Q-12	15	1.067		385	46		431	490
1230	For supervisory switch kit, all sizes									
1240	One circuit, add	1 Spri	48	.167	Ea.	124	8.05		132.05	149
1280	Quarter turn for trim									
1300	1/2″ size	1 Spri	22	.364	Ea.	22	17.50		39.50	51
1310	3/4″ size		20	.400		23.50	19.25		42.75	56
1320	1″ size		19	.421		26	20.50		46.50	60
1330	1-1/4″ size		15	.533		42.50	25.50		68	87
1340	1-1/2″ size		13	.615		53.50	29.50		83	105
1350	2″ size		11	.727		64	35		99	125
1400	Caps, polished brass with chain, 3/4″					34.50			34.50	38
1420	1″					44			44	48.50
1440	1-1/2″					11.40			11.40	12.55
1460	2-1/2″					17.65			17.65	19.40
1480	3″					27.50			27.50	30.50
1900	Escutcheon plate, for angle valves, polished brass, 1-1/2″					11.05			11.05	12.15
1920	2-1/2″					26			26	29
1940	3″					24			24	26.50
1980	For polished chrome, add					15%				
2000	Foam, control valve, 3″	1 Spri	6	1.333		1,500	64		1,564	1,750
2020	Supply valve, 2-1/2″		7	1.143		173	55		228	276
2040	Proportioner, 8″		2	4		2,875	193		3,068	3,475
2060	Oscillating foam monitor with electric remote control	Q-12	5.33	3.002		9,300	130		9,430	10,400

21 12 Fire-Suppression Standpipes

21 12 23 – Fire-Suppression Hose Valves

21 12 23.70 Fire Hose Valves		Crew	Daily Output	Labor-Hours	Unit	2009 Bare Costs Material	Labor	Equipment	Total	Total Incl O&P
3000	Gate, hose, wheel handle, N.R.S., rough brass, 1-1/2"	1 Spri	12	.667	Ea.	111	32		143	172
3040	2-1/2", 300 lb.	"	7	1.143		156	55		211	257
3080	For polished brass, add					40%				
3090	For polished chrome, add					50%				
3800	Hydrant, screw type, crank handle, brass									
3840	2-1/2" size	Q-12	11	1.455	Ea.	244	63		307	365
3880	For chrome, same price									
4200	Hydrolator, vent and draining, rough brass, 1-1/2"	1 Spri	12	.667	Ea.	59	32		91	115
4280	For polished brass, add					50%				
4290	For polished chrome, add					90%				
5000	Pressure restricting, adjustable rough brass, 1-1/2"	1 Spri	12	.667		99	32		131	159
5020	2-1/2"	"	7	1.143		139	55		194	238
5080	For polished brass, add					30%				
5090	For polished chrome, add					45%				
8000	Wye, leader line, ball type, swivel female x male x male									
8040	2-1/2" x 1-1/2" x 1-1/2" polished brass				Ea.	214			214	236
8060	2-1/2" x 1-1/2" x 1-1/2" polished chrome				"	208			208	228

21 13 Fire-Suppression Sprinkler Systems

21 13 13 – Wet-Pipe Sprinkler Systems

21 13 13.50 Wet-Pipe Sprinkler System Components

		Crew	Daily Output	Labor-Hours	Unit	Material	Labor	Equipment	Total	Total Incl O&P
0010	**WET-PIPE SPRINKLER SYSTEM COMPONENTS**									
1100	Alarm, electric pressure switch (circuit closer)	1 Spri	26	.308	Ea.	241	14.80		255.80	288
1140	For explosion proof, max 20 PSI, contacts close or open		26	.308		480	14.80		494.80	555
1220	Water motor, complete with gong		4	2		305	96.50		401.50	485
1800	Firecycle system, controls, includes panel,									
1820	batteries, solenoid valves and pressure switches	Q-13	1	32	Ea.	10,800	1,475		12,275	14,200
1860	Detector	1 Spri	16	.500	"	525	24		549	615
1900	Flexible sprinkler head connectors									
1910	Braided stainless steel hose with mounting bracket									
1920	1/2" and 1" outlet size									
1930	24" length	1 Spri	32	.250	Ea.	24	12.05		36.05	45
1940	36" length		30	.267		28	12.85		40.85	50.50
1950	48" length		26	.308		31.50	14.80		46.30	57.50
1960	60" length		22	.364		35.50	17.50		53	66
1970	72" length		18	.444		39	21.50		60.50	76
1982	May replace hard-pipe armovers									
1984	For wet, pre-action, deluge or dry pipe systems									
2000	Release, emergency, manual, for hydraulic or pneumatic system	1 Spri	12	.667	Ea.	161	32		193	227
2060	Release, thermostatic, for hydraulic or pneumatic release line		20	.400		500	19.25		519.25	580
2200	Sprinkler cabinets, 6 head capacity		16	.500		59	24		83	103
2260	12 head capacity		16	.500		62.50	24		86.50	107
2340	Sprinkler head escutcheons, standard, brass tone, 1" size		40	.200		2.55	9.65		12.20	17.75
2360	Chrome, 1" size		40	.200		3.05	9.65		12.70	18.30
2400	Recessed type, brass tone		40	.200		2.67	9.65		12.32	17.90
2440	Chrome or white enamel		40	.200		2.68	9.65		12.33	17.90
2600	Sprinkler heads, not including supply piping									
3700	Standard spray, pendent or upright, brass, 135° to 286° F									
3720	1/2" NPT, 3/8" orifice	1 Spri	16	.500	Ea.	11.45	24		35.45	50
3730	1/2" NPT, 7/16" orifice		16	.500		11.35	24		35.35	50
3740	1/2" NPT, 1/2" orifice		16	.500		7.35	24		31.35	45.50

21 13 13 – Wet-Pipe Sprinkler Systems

21 13 13.50 Wet-Pipe Sprinkler System Components		Crew	Daily Output	Labor-Hours	Unit	Material	Labor	Equipment	Total	Total Incl O&P
						2009 Bare Costs				
3760	1/2" NPT, 17/32" orifice	1 Spri	16	.500	Ea.	9.60	24		33.60	48
3780	3/4" NPT, 17/32" orifice	↓	16	.500	↓	8.85	24		32.85	47.50
3800	For open sprinklers, deduct					15%				
3840	For chrome, add				Ea.	2.38			2.38	2.62
3860	For wax and lead coating, add					24.50			24.50	26.50
3880	For wax coating, add					14.50			14.50	15.95
3900	For lead coating, add				↓	15.65			15.65	17.20
3920	For 360° F, same cost									
3930	For 400° F, add				Ea.	48.50			48.50	53
3940	For 500° F, add				"	48.50			48.50	53
4200	Sidewall, vertical brass, 135 -286° F									
4240	1/2" NPT, 1/2" orifice	1 Spri	16	.500	Ea.	17.40	24		41.40	56.50
4280	3/4" NPT, 17/32" orifice	"	16	.500		52.50	24		76.50	95
4360	For satin chrome, add				↓	1.75			1.75	1.93
4400	For 360° F, same cost									
4410	For 400° F, add				Ea.	10.90			10.90	11.95
4420	For 500° F, add				"	10.90			10.90	11.95
4500	Sidewall, horizontal, brass, 135° to 286° F									
4520	1/2" NPT, 1/2" orifice	1 Spri	16	.500	Ea.	15	24		39	54
4540	For 360° F, same cost									
4800	Recessed pendent, brass, 135° to 286° F									
4820	1/2" NPT, 3/8" orifice	1 Spri	10	.800	Ea.	40	38.50		78.50	104
4830	1/2" NPT, 7/16" orifice		10	.800		18.75	38.50		57.25	80.50
4840	1/2" NPT, 1/2" orifice		10	.800		13.60	38.50		52.10	75
4860	1/2" NPT, 17/32" orifice	↓	10	.800		33.50	38.50		72	96.50
4900	For satin chrome, add				↓	2.65			2.65	2.92
5000	Recessed-vertical sidewall, brass, 135° to 286° F									
5020	1/2" NPT, 3/8" orifice	1 Spri	10	.800	Ea.	24	38.50		62.50	86
5030	1/2" NPT, 7/16" orifice		10	.800		24	38.50		62.50	86
5040	1/2" NPT, 1/2" orifice	↓	10	.800	↓	12.50	38.50		51	74
5100	For bright nickel, same cost									
5600	Concealed, complete with cover plate									
5620	1/2" NPT, 1/2" orifice, 135° F to 212° F	1 Spri	9	.889	Ea.	34	43		77	104
5800	Window, brass, 1/2" NPT, 1/4" orifice		16	.500		25.50	24		49.50	65.50
5810	1/2" NPT, 5/16" orifice		16	.500		25.50	24		49.50	65.50
5820	1/2" NPT, 3/8" orifice		16	.500		25.50	24		49.50	65.50
5830	1/2" NPT, 7/16" orifice		16	.500		25.50	24		49.50	65.50
5840	1/2" NPT, 1/2" orifice	↓	16	.500		27	24		51	67
5860	For polished chrome, add					2.69			2.69	2.96
5880	3/4" NPT, 5/8" orifice	1 Spri	16	.500		28	24		52	68.50
5890	3/4 NPT, 3/4" orifice	"	16	.500		28	24		52	68.50
6000	Sprinkler head guards, bright zinc, 1/2" NPT					4.52			4.52	4.97
6020	Bright zinc, 3/4" NPT					4.52			4.52	4.97
6100	Sprinkler head wrenches, standard head					26			26	28.50
6120	Recessed head					34			34	37.50
6160	Tamper switch, (valve supervisory switch)	1 Spri	16	.500	↓	81	24		105	127

21 13 16 – Dry-Pipe Sprinkler Systems

21 13 16.50 Dry-Pipe Sprinkler System Components		Crew	Daily Output	Labor-Hours	Unit	Material	Labor	Equipment	Total	Total Incl O&P
0010	**DRY-PIPE SPRINKLER SYSTEM COMPONENTS**									
0600	Accelerator	1 Spri	8	1	Ea.	560	48		608	690
0800	Air compressor for dry pipe system, automatic, complete R211313-20									
0820	280 gal. system capacity, 3/4 HP	1 Spri	1.30	6.154	Ea.	700	296		996	1,225

21 13 16 – Dry-Pipe Sprinkler Systems

21 13 16.50 Dry-Pipe Sprinkler System Components			Crew	Daily Output	Labor-Hours	Unit	2009 Bare Costs Material	Labor	Equipment	Total	Total Incl O&P
0860	520 gal. system capacity, 1 HP	R211313-30	1 Spri	1.30	6.154	Ea.	735	296		1,031	1,275
0910	650 gal. system capacity, 1-1/2 HP			1.30	6.154		745	296		1,041	1,275
0920	790 gal. system capacity, 2 HP			1.30	6.154		805	296		1,101	1,350
0960	Air pressure maintenance control			24	.333		256	16.05		272.05	305
1600	Dehydrator package, incl. valves and nipples			12	.667		525	32		557	630
2600	Sprinkler heads, not including supply piping										
2640	Dry, pendent, 1/2" orifice, 3/4" or 1" NPT										
2660	1/2" to 6" length		1 Spri	14	.571	Ea.	90.50	27.50		118	142
2670	6-1/4" to 8" length			14	.571		94.50	27.50		122	147
2680	8-1/4" to 12" length			14	.571		98.50	27.50		126	151
2690	12-1/4" to 15" length			14	.571		102	27.50		129.50	155
2700	15-1/4" to 18" length			14	.571		106	27.50		133.50	160
2710	18-1/4" to 21" length			13	.615		110	29.50		139.50	167
2720	21-1/4" to 24" length			13	.615		114	29.50		143.50	172
2730	24-1/4" to 27" length			13	.615		119	29.50		148.50	176
2740	27-1/4" to 30" length			13	.615		123	29.50		152.50	181
2750	30-1/4" to 33" length			13	.615		127	29.50		156.50	185
2760	33-1/4" to 36" length			13	.615		131	29.50		160.50	190
2780	36-1/4" to 39" length			12	.667		135	32		167	199
2790	39-1/4" to 42" length			12	.667		139	32		171	203
2800	For each inch or fraction, add						1.32			1.32	1.45
6330	Valves and components										
8000	Dry pipe air check valve, 3" size		Q-12	2	8	Ea.	1,325	345		1,670	2,000
8200	Dry pipe valve, incl. trim and gauges, 3" size			2	8		1,850	345		2,195	2,575
8220	4" size			1	16		2,000	695		2,695	3,300
8240	6" size		Q-13	2	16		2,375	735		3,110	3,775
8280	For accelerator trim with gauges, add		1 Spri	8	1		690	48		738	835

21 13 20 – Fire-Cycle Sprinkler Systems

21 13 20.50 Firecycle Fire-Suppression Sprinkler Systems		Crew	Daily Output	Labor-Hours	Unit	Material	Labor	Equipment	Total	Total Incl O&P
0010	**FIRECYCLE FIRE-SUPPRESSION SPRINKLER SYSTEMS**									
8400	Firecycle package, includes swing check									
8420	and flow control valves with required trim									
8440	2" size	Q-12	2	8	Ea.	4,025	345		4,370	4,975
8460	3" size		1.50	10.667		4,425	460		4,885	5,575
8480	4" size		1	16		4,950	695		5,645	6,525
8500	6" size	Q-13	1.40	22.857		6,300	1,050		7,350	8,575

21 13 26 – Deluge Fire-Suppression Sprinkler Systems

21 13 26.50 Deluge Fire-Suppression Sprinkler System Components		Crew	Daily Output	Labor-Hours	Unit	Material	Labor	Equipment	Total	Total Incl O&P
0010	**DELUGE FIRE-SUPPRESSION SPRINKLER SYSTEM COMPONENTS**									
1400	Deluge system, monitoring panel w/deluge valve & trim	1 Spri	18	.444	Ea.	8,300	21.50		8,321.50	9,175
6200	Valves and components									
7000	Deluge, assembly, incl. trim, pressure									
7020	operated relief, emergency release, gauges									
7040	2" size	Q-12	2	8	Ea.	2,000	345		2,345	2,750
7060	3" size		1.50	10.667		2,325	460		2,785	3,275
7080	4" size		1	16		2,750	695		3,445	4,100
7100	6" size	Q-13	1.80	17.778		3,375	815		4,190	5,000
7800	Pneumatic actuator, bronze, required on all									
7820	pneumatic release systems, any size deluge	1 Spri	18	.444	Ea.	305	21.50		326.50	370

21 13 Fire-Suppression Sprinkler Systems

21 13 39 – Foam-Water Systems

21 13 39.50 Foam-Water System Components		Crew	Daily Output	Labor-Hours	Unit	2009 Bare Costs Material	Labor	Equipment	Total	Total Incl O&P
0010	**FOAM-WATER SYSTEM COMPONENTS**									
2600	Sprinkler heads, not including supply piping									
3600	Foam-water, pendent or upright, 1/2" NPT	1 Spri	12	.667	Ea.	130	32		162	192

21 21 Carbon-Dioxide Fire-Extinguishing Systems

21 21 16 – Carbon-Dioxide Fire-Extinguishing Equipment

21 21 16.50 CO2 Fire Extinguishing System

		Crew	Daily Output	Labor-Hours	Unit	Material	Labor	Equipment	Total	Total Incl O&P
0010	**CO_2 FIRE EXTINGUISHING SYSTEM**									
0042	For detectors and control stations, see Div. 28 31 23.50									
0100	Control panel, single zone with batteries (2 zones det., 1 suppr.)	1 Elec	1	8	Ea.	1,350	375		1,725	2,050
0150	Multizone (4) with batteries (8 zones det., 4 suppr.)	"	.50	16		2,575	750		3,325	3,975
1000	Dispersion nozzle, CO_2, 3" x 5"	1 Plum	18	.444		52.50	21.50		74	91.50
2000	Extinguisher, CO_2 system, high pressure, 75 lb. cylinder	Q-1	6	2.667		1,000	117		1,117	1,275
2100	100 lb. cylinder	"	5	3.200		1,025	140		1,165	1,350
3000	Electro/mechanical release	L-1	4	4		131	192		323	440
3400	Manual pull station	1 Plum	6	1.333		47.50	65		112.50	153
4000	Pneumatic damper release	"	8	1		175	49		224	269

21 22 Clean-Agent Fire-Extinguishing Systems

21 22 16 – Clean-Agent Fire-Extinguishing Equipment

21 22 16.50 FM200 Fire Extinguishing System

		Crew	Daily Output	Labor-Hours	Unit	Material	Labor	Equipment	Total	Total Incl O&P
0010	**FM200 FIRE EXTINGUISHING SYSTEM**									
1100	Dispersion nozzle FM200, 1-1/2"	1 Plum	14	.571	Ea.	52.50	28		80.50	101
2400	Extinguisher, FM 200 system, filled, with mounting bracket									
2460	26 lb. container	Q-1	8	2	Ea.	1,800	88		1,888	2,100
2480	44 lb. container		7	2.286		2,400	100		2,500	2,800
2500	63 lb. container		6	2.667		2,800	117		2,917	3,250
2520	101 lb. container		5	3.200		3,750	140		3,890	4,350
2540	196 lb. container		4	4		6,100	176		6,276	6,975
6000	Average FM200 system, minimum				C.F.	1.38			1.38	1.52
6020	Maximum				"	2.75			2.75	3.03

21 31 Centrifugal Fire Pumps

21 31 13 – Electric-Drive, Centrifugal Fire Pumps

21 31 13.50 Electric-Drive Fire Pumps

		Crew	Daily Output	Labor-Hours	Unit	Material	Labor	Equipment	Total	Total Incl O&P
0010	**ELECTRIC-DRIVE FIRE PUMPS** Including controller, fittings and relief valve									
3100	250 GPM, 55 psi, 15 HP, 3,550 RPM, 2" pump	Q-13	.70	45.714	Ea.	16,000	2,100		18,100	20,800
3200	500 GPM, 50 psi, 27 HP, 1770 RPM, 4" pump		.68	47.059		19,800	2,150		21,950	25,200
3250	500 GPM, 100 psi, 47 HP, 3550 RPM, 3" pump		.66	48.485		24,000	2,225		26,225	29,900
3300	500 GPM, 125 psi, 64 HP, 3550 RPM, 3" pump		.62	51.613		26,300	2,375		28,675	32,600
3350	750 GPM, 50 psi, 44 HP, 1,770 RPM, 5" pump		.64	50		28,900	2,300		31,200	35,400
3400	750 GPM, 100 psi, 66 HP, 3550 RPM, 4" pump		.58	55.172		25,600	2,525		28,125	32,100
3450	750 GPM, 165 psi, 120 HP, 3550 RPM, 4" pump		.56	57.143		33,400	2,625		36,025	40,800
3500	1000 GPM, 50 psi, 48 HP 1770 RPM, 5" pump		.60	53.333		28,900	2,450		31,350	35,600
3550	1000 GPM, 100 psi, 86 HP, 3550 RPM, 5" pump		.54	59.259		32,400	2,725		35,125	39,900
3600	1000 GPM, 150 psi, 142 HP, 3550 RPM, 5" pump		.50	64		36,700	2,925		39,625	45,000
3650	1000 GPM, 200 psi, 245 HP, 1770 RPM, 6" pump		.36	88.889		79,500	4,075		83,575	94,000

21 31 Centrifugal Fire Pumps

21 31 13 – Electric-Drive, Centrifugal Fire Pumps

21 31 13.50 Electric-Drive Fire Pumps		Crew	Daily Output	Labor-Hours	Unit	2009 Bare Costs Material	Labor	Equipment	Total	Total Incl O&P
3660	1250 GPM, 75 psi, 75 HP, 1770 RPM, 5″ pump	Q-13	.55	58.182	Ea.	29,700	2,675		32,375	36,900
3700	1500 GPM, 50 psi, 66 HP, 1770 RPM, 6″ pump		.50	64		31,200	2,925		34,125	38,900
3750	1500 GPM, 100 psi, 139 HP, 1770 RPM, 6″ pump		.46	69.565		35,200	3,200		38,400	43,800
3800	1500 GPM, 150 psi, 200 HP, 1770 RPM, 6″ pump		.36	88.889		74,500	4,075		78,575	88,500
3850	1500 GPM, 200 psi, 279 HP, 1770 RPM, 6″ pump		.32	100		82,000	4,575		86,575	97,500
3900	2000 GPM, 100 psi, 167 HP, 1770 RPM, 6″ pump		.34	94.118		45,100	4,325		49,425	56,500
3950	2000 GPM, 150 psi, 292 HP, 1770 RPM, 6″ pump		.28	114		60,500	5,250		65,750	74,500
4000	2500 GPM, 100 psi, 213 HP, 1770 RPM, 8″ pump		.30	106		49,300	4,900		54,200	61,500
4040	2500 GPM, 135 psi, 339 HP, 1770 RPM, 8″ pump		.26	123		70,500	5,650		76,150	86,500
4100	3000 GPM, 100 psi, 250 HP, 1770 RPM, 8″ pump		.28	114		59,000	5,250		64,250	73,000
4150	3000 GPM, 140 psi, 428 HP, 1770 RPM, 10″ pump		.24	133		94,000	6,125		100,125	112,500
4200	3500 GPM, 100 psi, 300 HP, 1770 RPM, 10″ pump		.26	123		73,000	5,650		78,650	89,500
4250	3500 GPM, 140 psi, 450 HP, 1770 RPM, 10″ pump		.24	133		94,000	6,125		100,125	112,500
5000	For jockey pump 1″, 3 HP, with control, add	Q-12	2	8		3,550	345		3,895	4,450

21 31 16 – Diesel-Drive, Centrifugal Fire Pumps

21 31 16.50 Diesel-Drive Fire Pumps		Crew	Daily Output	Labor-Hours	Unit	Material	Labor	Equipment	Total	Total Incl O&P
0010	**DIESEL-DRIVE FIRE PUMPS** Including controller, fittings and relief valve									
0050	500 GPM, 50 psi, 27 HP, 4″ pump	Q-13	.64	50	Ea.	66,000	2,300		68,300	76,000
0100	500 GPM, 100 psi, 62 HP, 4″ pump		.60	53.333		70,000	2,450		72,450	81,000
0150	500 GPM, 125 psi, 78 HP, 4″ pump		.56	57.143		70,500	2,625		73,125	82,000
0200	750 GPM, 50 psi, 44 HP, 5″ pump		.60	53.333		70,500	2,450		72,950	81,500
0250	750 GPM, 100 psi, 80 HP, 4″ pump		.56	57.143		73,500	2,625		76,125	85,000
0300	750 GPM, 165 psi, 203 HP, 5″ pump		.52	61.538		76,000	2,825		78,825	88,000
0350	1000 GPM, 50 psi, 48 HP, 5″ pump		.58	55.172		71,000	2,525		73,525	82,000
0400	1000 GPM, 100 psi, 89 HP, 4″ pump		.56	57.143		75,500	2,625		78,125	87,000
0450	1000 GPM, 150 psi, 148 HP, 4″ pump		.48	66.667		74,500	3,050		77,550	86,500
0470	1000 GPM, 200 psi, 280 HP, 5″ pump		.40	80		93,500	3,675		97,175	108,500
0480	1250 GPM, 75 psi, 75 HP, 5″ pump		.54	59.259		71,500	2,725		74,225	83,000
0500	1500 GPM, 50 psi, 66 HP, 6″ pump		.50	64		76,500	2,925		79,425	88,500
0550	1500 GPM, 100 psi, 140 HP, 6″ pump		.46	69.565		78,000	3,200		81,200	91,000
0600	1500 GPM, 150 psi, 228 HP, 6″ pump		.42	76.190		88,000	3,500		91,500	102,500
0650	1500 GPM, 200 psi, 279 HP, 6″ pump		.38	84.211		138,000	3,850		141,850	158,000
0700	2,000 GPM, 100 psi, 167 HP, 6″ pump		.34	94.118		80,000	4,325		84,325	94,500
0750	2000 GPM, 150 psi, 284 HP, 6″ pump		.30	106		108,500	4,900		113,400	126,500
0800	2500 GPM, 100 psi, 213 HP, 8″ pump		.32	100		102,000	4,575		106,575	119,500
0820	2500 GPM, 150 psi, 365 HP, 8″ pump		.26	123		119,500	5,650		125,150	140,500
0850	3000 GPM, 100 psi, 250 HP, 8″ pump		.28	114		118,000	5,250		123,250	137,500
0900	3000 GPM, 150 psi, 384 HP, 10″ pump		.20	160		137,000	7,350		144,350	162,500
0950	3500 GPM, 100 psi, 300 HP, 10″ pump		.24	133		140,500	6,125		146,625	164,500
1000	3500 GPM, 150 psi, 518 HP, 10″ pump		.20	160		199,000	7,350		206,350	230,500

Division Notes

	CREW	DAILY OUTPUT	LABOR-HOURS	UNIT	2009 BARE COSTS MAT.	LABOR	EQUIP.	TOTAL	TOTAL INCL O&P

Estimating Tips

22 10 00 Plumbing Piping and Pumps

This subdivision is primarily basic pipe and related materials. The pipe may be used by any of the mechanical disciplines, i.e., plumbing, fire protection, heating, and air conditioning.

- The labor adjustment factors listed in Subdivision 22 01 02.20 apply throughout Divisions 21, 22, and 23. CAUTION: the correct percentage may vary for the same items. For example, the percentage add for the basic pipe installation should be based on the maximum height that the craftsman must install for that particular section. If the pipe is to be located 14′ above the floor but it is suspended on threaded rod from beams, the bottom flange of which is 18′ high (4′ rods), then the height is actually 18′ and the add is 20%. The pipe coverer, however, does not have to go above the 14′, and so his or her add should be 10%.
- Most pipe is priced first as straight pipe with a joint (coupling, weld, etc.) every 10′ and a hanger usually every 10′. There are exceptions with hanger spacing such as for cast iron pipe (5′) and plastic pipe (3 per 10′). Following each type of pipe there are several lines listing sizes and the amount to be subtracted to delete couplings and hangers. This is for pipe that is to be buried or supported together on trapeze hangers. The reason that the couplings are deleted is that these runs are usually long, and frequently longer lengths of pipe are used. By deleting the couplings, the estimator is expected to look up and add back the correct reduced number of couplings.
- When preparing an estimate, it may be necessary to approximate the fittings. Fittings usually run between 25% and 50% of the cost of the pipe. The lower percentage is for simpler runs, and the higher number is for complex areas, such as mechanical rooms.
- For historic restoration projects, the systems must be as invisible as possible, and pathways must be sought for pipes, conduit, and ductwork. While installations in accessible spaces (such as basements and attics) are relatively straightforward to estimate, labor costs may be more difficult to determine when delivery systems must be concealed.

22 40 00 Plumbing Fixtures

- Plumbing fixture costs usually require two lines: the fixture itself and its "rough-in, supply, and waste."
- In the Assemblies Section (Plumbing D2010) for the desired fixture, the System Components Group at the center of the page shows the fixture on the first line. The rest of the list (fittings, pipe, tubing, etc.) will total up to what we refer to in the Unit Price section as "Rough-in, supply, waste, and vent." Note that for most fixtures we allow a nominal 5′ of tubing to reach from the fixture to a main or riser.
- Remember that gas- and oil-fired units need venting.

Reference Numbers

Reference numbers are shown in shaded boxes at the beginning of some major classifications. These numbers refer to related items in the Reference Section. The reference information may be an estimating procedure, an alternate pricing method, or technical information.

Note: Not all subdivisions listed here necessarily appear in this publication.

Note: **Trade Service,** *in part, has been used as a reference source for some of the material prices used in Division 22.*

22 01 02 – Labor Adjustments

22 01 02.10 Boilers, General		Crew	Daily Output	Labor-Hours	Unit	Material	2009 Bare Costs Labor	Equipment	Total	Total Incl O&P
0010	**BOILERS, GENERAL**, Prices do not include flue piping, elec. wiring,									
0020	gas or oil piping, boiler base, pad, or tankless unless noted									
0100	Boiler H.P.: 10 KW = 34 lbs/steam/hr = 33,475 BTU/hr.									
0150	To convert SFR to BTU rating: Hot water, 150 x SFR;									
0160	Forced hot water, 180 x SFR; steam, 240 x SFR									

22 01 02.20 Labor Adjustment Factors		Crew	Daily Output	Labor-Hours	Unit	Material	Labor	Equipment	Total	Total Incl O&P
0010	**LABOR ADJUSTMENT FACTORS**, (For Div. 21,22 and 23)									
1000	Add to labor for elevated installation (Above floor level)									
1080	10′ to 14.5′ high						10%			
1100	15′ to 19.5′ high						20%			
1120	20′ to 24.5′ high						25%			
1140	25′ to 29.5′ high						35%			
1160	30′ to 34.5′ high						40%			
1180	35′ to 39.5′ high						50%			
1200	40′ and higher						55%			
2000	Add to labor for crawl space									
2100	3′ high						40%			
2140	4′ high						30%			
3000	Add to labor for multi-story building									
3100	Add per floor for floors 3 thru 19						2%			
3140	Add per floor for floors 20 and up						4%			
4000	Add to labor for working in existing occupied buildings									
4100	Hospital						35%			
4140	Office building						25%			
4180	School						20%			
4220	Factory or warehouse						15%			
4260	Multi dwelling						15%			
5000	Add to labor, miscellaneous									
5100	Cramped shaft						35%			
5140	Congested area						15%			
5180	Excessive heat or cold						30%			
9000	Labor factors, The above are reasonable suggestions, however									
9010	each project should be evaluated for its own peculiarities.									
9100	Other factors to be considered are:									
9140	Movement of material and equipment through finished areas									
9180	Equipment room									
9220	Attic space									
9260	No service road									
9300	Poor unloading/storage area									
9340	Congested site area/heavy traffic									

22 05 05.10 Plumbing Demolition

	22 05 05.10 Plumbing Demolition	Crew	Daily Output	Labor-Hours	Unit	Material	2009 Bare Costs Labor	Equipment	Total	Total Incl O&P
0010	**PLUMBING DEMOLITION** R220105-10									
0400	Air compressor, up thru 2 H.P.	Q-1	10	1.600	Ea.		70		70	109
0410	3 H.P. thru 7-1/2 H.P. R024119-10		5.60	2.857			125		125	194
0420	10 H.P. thru 15 H.P.		1.40	11.429			500		500	780
0430	20 H.P. thru 30 H.P.	Q-2	1.30	18.462			840		840	1,300
0500	Backflow preventer, up thru 2" diameter	1 Plum	17	.471			23		23	35.50
0510	2-1/2" thru 3" diameter	Q-1	10	1.600			70		70	109
0520	4" thru 6" diameter	"	5	3.200			140		140	218
0530	8" thru 10" diameter	Q-2	3	8			365		365	565
0700	Carriers and supports									
0710	Fountains, sinks, lavatories and urinals	1 Plum	14	.571	Ea.		28		28	43
0720	Water closets	"	12	.667	"		32.50		32.50	50.50
0730	Grinder pump or sewage ejector system									
0732	Simplex	Q-1	7	2.286	Ea.		100		100	156
0734	Duplex	"	2.80	5.714			251		251	390
0738	Hot water dispenser	1 Plum	36	.222			10.85		10.85	16.80
0740	Hydrant, wall		26	.308			15		15	23.50
0744	Ground		12	.667			32.50		32.50	50.50
0760	Cleanouts and drains, up thru 4" pipe diameter		10	.800			39		39	60.50
0764	5" thru 8" pipe diameter	Q-1	10	1.600			70		70	109
0780	Industrial safety fixtures	1 Plum	8	1			49		49	75.50
1020	Fixtures, including 10' piping									
1100	Bath tubs, cast iron	1 Plum	4	2	Ea.		97.50		97.50	151
1120	Fiberglass		6	1.333			65		65	101
1140	Steel		5	1.600			78		78	121
1150	Bidet	Q-1	7	2.286			100		100	156
1200	Lavatory, wall hung	1 Plum	10	.800			39		39	60.50
1220	Counter top		8	1			49		49	75.50
1300	Sink, single compartment		8	1			49		49	75.50
1320	Double compartment		7	1.143			55.50		55.50	86.50
1340	Shower, stall and receptor	Q-1	6	2.667			117		117	181
1350	Group	"	7	2.286			100		100	156
1400	Water closet, floor mounted	1 Plum	8	1			49		49	75.50
1420	Wall mounted	"	7	1.143			55.50		55.50	86.50
1440	Wash fountain, 36" diameter	Q-2	8	3			137		137	212
1442	54" diameter	"	7	3.429			156		156	242
1500	Urinal, floor mounted	1 Plum	4	2			97.50		97.50	151
1520	Wall mounted	"	7	1.143			55.50		55.50	86.50
1590	Whirl pool or hot tub	Q-1	2.60	6.154			270		270	420
1600	Water fountains, free standing	1 Plum	8	1			49		49	75.50
1620	Wall or deck mounted		6	1.333			65		65	101
1800	Medical gas specialties		8	1			49		49	75.50
1900	Piping fittings, single connection, up thru 1-1/2" diameter		30	.267			13		13	20
1910	2" thru 4" diameter		14	.571			28		28	43
1980	Pipe hanger/support removal		80	.100			4.88		4.88	7.55
1990	Glass pipe with fittings, 1" thru 3" diameter		200	.040	L.F.		1.95		1.95	3.02
1992	4" thru 6" diameter		150	.053			2.60		2.60	4.03
2000	Piping, metal, up thru 1-1/2" diameter		200	.040			1.95		1.95	3.02
2050	2" thru 3-1/2" diameter		150	.053			2.60		2.60	4.03
2100	4" thru 6" diameter	2 Plum	100	.160			7.80		7.80	12.10
2150	8" thru 14" diameter	"	60	.267			13		13	20
2153	16" thru 20" diameter	Q-18	70	.343			15.80	.80	16.60	25.50
2155	24" thru 26" diameter		55	.436			20	1.01	21.01	32

22 05 05.10 Plumbing Demolition		Crew	Daily Output	Labor-Hours	Unit	Material	2009 Bare Costs Labor	Equipment	Total	Total Incl O&P
2156	30″ thru 36″ diameter	Q-18	40	.600	L.F.		27.50	1.39	28.89	44.50
2160	Plastic pipe with fittings, up thru 1-1/2″ diameter	1 Plum	250	.032			1.56		1.56	2.42
2162	2″ thru 3″ diameter	″	200	.040			1.95		1.95	3.02
2164	4″ thru 6″ diameter	Q-1	200	.080			3.51		3.51	5.45
2166	8″ thru 14″ diameter		150	.107			4.68		4.68	7.25
2168	16″ diameter		100	.160			7		7	10.90
2170	Prison fixtures, lavatory or sink		18	.889	Ea.		39		39	60.50
2172	Shower		5.60	2.857			125		125	194
2174	Urinal or water closet		13	1.231			54		54	84
2180	Pumps, all fractional horse-power		12	1.333			58.50		58.50	90.50
2184	1 H.P. thru 5 H.P.		6	2.667			117		117	181
2186	7-1/2 H.P. thru 15 H.P.		2.50	6.400			281		281	435
2188	20 H.P. thru 25 H.P.	Q-2	4	6			273		273	425
2190	30 H.P. thru 60 H.P.		.80	30			1,375		1,375	2,125
2192	75 H.P. thru 100 H.P.		.60	40			1,825		1,825	2,825
2194	150 H.P.		.50	48			2,175		2,175	3,375
2198	Pump, sump or submersible	1 Plum	12	.667			32.50		32.50	50.50
2200	Receptors and interceptors, up thru 20 GPM	″	8	1			49		49	75.50
2204	25 thru 100 GPM	Q-1	6	2.667			117		117	181
2208	125 thru 300 GPM	″	2.40	6.667			293		293	455
2211	325 thru 500 GPM	Q-2	2.60	9.231			420		420	650
2212	Deduct for salvage, aluminum scrap				Ton				1,474	1,620
2214	Brass scrap								3,440	3,780
2216	Copper scrap								5,114	5,620
2218	Lead scrap								764	840
2220	Steel scrap								205	225
2230	Temperature maintenance cable	1 Plum	1200	.007	L.F.		.33		.33	.50
2250	Water heater, 40 gal.	″	6	1.333	Ea.		65		65	101
3100	Tanks, water heaters and liquid containers									
3110	Up thru 45 gallons	Q-1	22	.727	Ea.		32		32	49.50
3120	50 thru 120 gallons		14	1.143			50		50	78
3130	130 thru 240 gallons		7.60	2.105			92.50		92.50	143
3140	250 thru 500 gallons		5.40	2.963			130		130	202
3150	600 thru 1000 gallons	Q-2	1.60	15			685		685	1,050
3160	1100 thru 2000 gallons		.70	34.286			1,550		1,550	2,425
3170	2100 thru 4000 gallons		.50	48			2,175		2,175	3,375
6000	Remove and reset fixtures, minimum	1 Plum	6	1.333			65		65	101
6100	Maximum		4	2			97.50		97.50	151
9000	Minimum labor/equipment charge		2	4	Job		195		195	300
9100	Valve, metal valves, strainers and similar, up thru 1-1/2″ diameter	1 Stpi	28	.286	Ea.		14.10		14.10	22
9110	2″ thru 3″ diameter	Q-1	11	1.455			64		64	99
9120	4″ thru 6″ diameter	″	8	2			88		88	136
9130	8″ thru 14″ diameter	Q-2	8	3			137		137	212
9140	16″ thru 20″ diameter		2	12			545		545	845
9150	24″ diameter		1.20	20			910		910	1,400
9200	Valve, plastic, up thru 1-1/2″ diameter	1 Plum	42	.190			9.30		9.30	14.40
9210	2″ thru 3″ diameter		15	.533			26		26	40.50
9220	4″ thru 6″ diameter		12	.667			32.50		32.50	50.50
9300	Vent flashing and caps		55	.145			7.10		7.10	11
9350	Water filter, commercial, 1″ thru 1-1/2″	Q-1	2	8			350		350	545
9360	2″ thru 2-1/2″	″	1.60	10			440		440	680
9402	Water heaters									
9410	Up thru 245 GPH	Q-1	2.40	6.667	Ea.		293		293	455

22 05 Common Work Results for Plumbing

22 05 05 – Selective Plumbing Demolition

22 05 05.10 Plumbing Demolition		Crew	Daily Output	Labor-Hours	Unit	Material	2009 Bare Costs Labor	Equipment	Total	Total Incl O&P
9420	250 thru 756 GPH	Q-1	1.60	10	Ea.		440		440	680
9430	775 thru 1640 GPH	↓	.80	20			880		880	1,350
9440	1650 thru 4000 GPH	Q-2	.50	48			2,175		2,175	3,375
9470	Water softener	Q-1	2	8	↓		350		350	545

22 05 23 – General-Duty Valves for Plumbing Piping

22 05 23.10 Valves, Brass

		Crew	Daily Output	Labor-Hours	Unit	Material	Labor	Equipment	Total	Total Incl O&P
0010	**VALVES, BRASS**									
0032	For motorized valves, see Div. 23 09 53.10									
0500	Gas cocks, threaded									
0510	1/4"	1 Plum	26	.308	Ea.	8.65	15		23.65	33
0520	3/8"		24	.333		8.65	16.25		24.90	34.50
0530	1/2"		24	.333		8.65	16.25		24.90	34.50
0540	3/4"		22	.364		13.65	17.75		31.40	42.50
0550	1"		19	.421		17.95	20.50		38.45	51.50
0560	1-1/4"		15	.533		30	26		56	73.50
0570	1-1/2"		13	.615		38.50	30		68.50	89
0580	2"	↓	11	.727	↓	64.50	35.50		100	126
0672	For larger sizes use lubricated plug valve, Div. 23 05 23.70									

22 05 23.20 Valves, Bronze

		Crew	Daily Output	Labor-Hours	Unit	Material	Labor	Equipment	Total	Total Incl O&P
0010	**VALVES, BRONZE** R220523-90									
1020	Angle, 150 lb., rising stem, threaded									
1030	1/8"	1 Plum	24	.333	Ea.	87	16.25		103.25	121
1040	1/4"		24	.333		90	16.25		106.25	124
1050	3/8"		24	.333		90	16.25		106.25	124
1060	1/2"		22	.364		90	17.75		107.75	127
1070	3/4"		20	.400		123	19.50		142.50	165
1080	1"		19	.421		177	20.50		197.50	227
1090	1-1/4"		15	.533		227	26		253	291
1100	1-1/2"		13	.615		296	30		326	370
1110	2"	↓	11	.727	↓	480	35.50		515.50	580
1380	Ball, 150 psi, threaded									
1400	1/4"	1 Plum	24	.333	Ea.	10.30	16.25		26.55	36.50
1430	3/8"		24	.333		10.30	16.25		26.55	36.50
1450	1/2"		22	.364		10.30	17.75		28.05	39
1460	3/4"		20	.400		16.95	19.50		36.45	48.50
1470	1"		19	.421		21.50	20.50		42	55.50
1480	1-1/4"		15	.533		39.50	26		65.50	84
1490	1-1/2"		13	.615		51	30		81	103
1500	2"		11	.727		61.50	35.50		97	123
1510	2-1/2"		9	.889		207	43.50		250.50	295
1520	3"	↓	8	1	↓	315	49		364	420
1600	Butterfly, 175 psi, full port, solder or threaded ends									
1610	Stainless steel disc and stem									
1620	1/4"	1 Plum	24	.333	Ea.	11.60	16.25		27.85	38
1630	3/8"		24	.333		11.60	16.25		27.85	38
1640	1/2"		22	.364		12.20	17.75		29.95	41
1650	3/4"		20	.400		19.75	19.50		39.25	51.50
1660	1"		19	.421		24	20.50		44.50	58.50
1670	1-1/4"		15	.533		38	26		64	82.50
1680	1-1/2"		13	.615		50	30		80	102
1690	2"	↓	11	.727	↓	62.50	35.50		98	124
1750	Check, swing, class 150, regrinding disc, threaded									

22 05 23.20 Valves, Bronze		Crew	Daily Output	Labor-Hours	Unit	2009 Bare Costs Material	Labor	Equipment	Total	Total Incl O&P
1800	1/8"	1 Plum	24	.333	Ea.	42	16.25		58.25	71
1830	1/4"		24	.333		42	16.25		58.25	71
1840	3/8"		24	.333		44.50	16.25		60.75	73.50
1850	1/2"		24	.333		44.50	16.25		60.75	73.50
1860	3/4"		20	.400		64	19.50		83.50	100
1870	1"		19	.421		91.50	20.50		112	133
1880	1-1/4"		15	.533		132	26		158	186
1890	1-1/2"		13	.615		154	30		184	216
1900	2"		11	.727		226	35.50		261.50	305
1910	2-1/2"	Q-1	15	1.067		505	47		552	635
1920	3"	"	13	1.231		675	54		729	830
2000	For 200 lb, add					5%	10%			
2040	For 300 lb, add					15%	15%			
2060	Check swing, 300#, sweat, 3/8" size	1 Plum	24	.333	Ea.	29	16.25		45.25	57
2070	1/2"		24	.333		29	16.25		45.25	57
2080	3/4"		20	.400		36.50	19.50		56	70.50
2090	1"		19	.421		51	20.50		71.50	88
2100	1-1/4"		15	.533		66	26		92	114
2110	1-1/2"		13	.615		87.50	30		117.50	143
2120	2"		11	.727		132	35.50		167.50	201
2130	2-1/2"	Q-1	15	1.067		315	47		362	420
2140	3"	"	13	1.231		420	54		474	545
2350	Check, lift, class 150, horizontal composition disc, threaded									
2430	1/4"	1 Plum	24	.333	Ea.	126	16.25		142.25	163
2440	3/8"		24	.333		160	16.25		176.25	201
2450	1/2"		24	.333		139	16.25		155.25	178
2460	3/4"		20	.400		170	19.50		189.50	217
2470	1"		19	.421		241	20.50		261.50	297
2480	1-1/4"		15	.533		325	26		351	395
2490	1-1/2"		13	.615		385	30		415	470
2500	2"		11	.727		645	35.50		680.50	765
2850	Gate, N.R.S., soldered, 125 psi									
2900	3/8"	1 Plum	24	.333	Ea.	36.50	16.25		52.75	65.50
2920	1/2"		24	.333		31	16.25		47.25	59
2940	3/4"		20	.400		35.50	19.50		55	69
2950	1"		19	.421		50.50	20.50		71	87.50
2960	1-1/4"		15	.533		80.50	26		106.50	129
2970	1-1/2"		13	.615		85.50	30		115.50	141
2980	2"		11	.727		122	35.50		157.50	189
2990	2-1/2"	Q-1	15	1.067		305	47		352	410
3000	3"	"	13	1.231		400	54		454	525
3350	Threaded, class 150									
3410	1/4"	1 Plum	24	.333	Ea.	52.50	16.25		68.75	83
3420	3/8"		24	.333		52.50	16.25		68.75	83
3430	1/2"		24	.333		49	16.25		65.25	79
3440	3/4"		20	.400		56	19.50		75.50	92
3450	1"		19	.421		71.50	20.50		92	111
3460	1-1/4" size		15	.533		96	26		122	146
3470	1-1/2"		13	.615		138	30		168	198
3480	2"		11	.727		164	35.50		199.50	236
3490	2-1/2"	Q-1	15	1.067		385	47		432	495
3500	3" size	"	13	1.231		540	54		594	680
3850	Rising stem, soldered, 300 psi									

22 05 23.20 Valves, Bronze		Crew	Daily Output	Labor-Hours	Unit	2009 Bare Costs Material	Labor	Equipment	Total	Total Incl O&P
3900	3/8"	1 Plum	24	.333	Ea.	83.50	16.25		99.75	117
3920	1/2"		24	.333		83.50	16.25		99.75	117
3940	3/4"		20	.400		91	19.50		110.50	130
3950	1"		19	.421		132	20.50		152.50	177
3960	1-1/4"		15	.533		183	26		209	242
3970	1-1/2"		13	.615		215	30		245	284
3980	2"		11	.727		350	35.50		385.50	440
3990	2-1/2"	Q-1	15	1.067		780	47		827	935
4000	3"	"	13	1.231		1,100	54		1,154	1,300
4250	Threaded, class 150									
4310	1/4"	1 Plum	24	.333	Ea.	47	16.25		63.25	76.50
4320	3/8"		24	.333		47	16.25		63.25	76.50
4330	1/2"		24	.333		44	16.25		60.25	73.50
4340	3/4"		20	.400		51	19.50		70.50	86
4350	1"		19	.421		68.50	20.50		89	107
4360	1-1/4"		15	.533		92	26		118	142
4370	1-1/2"		13	.615		117	30		147	175
4380	2"		11	.727		158	35.50		193.50	228
4390	2-1/2"	Q-1	15	1.067		365	47		412	480
4400	3"	"	13	1.231		510	54		564	645
4500	For 300 psi, threaded, add					100%	15%			
4540	For chain operated type, add					15%				
4850	Globe, class 150, rising stem, threaded									
4920	1/4"	1 Plum	24	.333	Ea.	67	16.25		83.25	98.50
4940	3/8" size		24	.333		67	16.25		83.25	98.50
4950	1/2"		24	.333		67	16.25		83.25	98.50
4960	3/4"		20	.400		90	19.50		109.50	129
4970	1"		19	.421		141	20.50		161.50	187
4980	1-1/4"		15	.533		224	26		250	288
4990	1-1/2"		13	.615		273	30		303	345
5000	2"		11	.727		410	35.50		445.50	505
5010	2-1/2"	Q-1	15	1.067		825	47		872	980
5020	3"	"	13	1.231		1,175	54		1,229	1,375
5120	For 300 lb threaded, add					50%	15%			
5130	Globe, 300 lb., sweat, 3/8" size	1 Plum	24	.333	Ea.	83.50	16.25		99.75	117
5140	1/2"		24	.333		85.50	16.25		101.75	120
5150	3/4"		20	.400		117	19.50		136.50	158
5160	1"		19	.421		161	20.50		181.50	209
5170	1-1/4"		15	.533		256	26		282	320
5180	1-1/2"		13	.615		310	30		340	385
5190	2"		11	.727		455	35.50		490.50	555
5200	2-1/2"	Q-1	15	1.067		890	47		937	1,050
5210	3"	"	13	1.231		1,150	54		1,204	1,325
5600	Relief, pressure & temperature, self-closing, ASME, threaded									
5640	3/4"	1 Plum	28	.286	Ea.	118	13.95		131.95	152
5650	1"		24	.333		180	16.25		196.25	224
5660	1-1/4"		20	.400		345	19.50		364.50	410
5670	1-1/2"		18	.444		660	21.50		681.50	760
5680	2"		16	.500		720	24.50		744.50	830
5950	Pressure, poppet type, threaded									
6000	1/2"	1 Plum	30	.267	Ea.	33	13		46	56
6040	3/4"	"	28	.286	"	35	13.95		48.95	60
6400	Pressure, water, ASME, threaded									

22 05 23.20 Valves, Bronze		Crew	Daily Output	Labor-Hours	Unit	Material	Labor	Equipment	Total	Total Incl O&P
						2009 Bare Costs				
6440	3/4"	1 Plum	28	.286	Ea.	70	13.95		83.95	98.50
6450	1"		24	.333		149	16.25		165.25	189
6460	1-1/4"		20	.400		236	19.50		255.50	289
6470	1-1/2"		18	.444		325	21.50		346.50	395
6480	2"		16	.500		470	24.50		494.50	560
6490	2-1/2"	↓	15	.533	↓	2,250	26		2,276	2,525
6900	Reducing, water pressure									
6920	300 psi to 25-75 psi, threaded or sweat									
6940	1/2"	1 Plum	24	.333	Ea.	214	16.25		230.25	261
6950	3/4"		20	.400		214	19.50		233.50	266
6960	1"		19	.421		330	20.50		350.50	395
6970	1-1/4"		15	.533		595	26		621	695
6980	1-1/2"		13	.615		895	30		925	1,025
6990	2"	↓	11	.727		1,300	35.50		1,335.50	1,475
7100	For built-in by-pass or 10-35 psi, add				↓	10.75			10.75	11.80
7700	High capacity, 250 psi to 25-75 psi, threaded									
7740	1/2"	1 Plum	24	.333	Ea.	214	16.25		230.25	261
7780	3/4"		20	.400		214	19.50		233.50	266
7790	1"		19	.421		330	20.50		350.50	395
7800	1-1/4"		15	.533		595	26		621	695
7810	1-1/2"		13	.615		895	30		925	1,025
7820	2"		11	.727		1,300	35.50		1,335.50	1,475
7830	2-1/2"		9	.889		1,850	43.50		1,893.50	2,125
7840	3"	↓	8	1		2,025	49		2,074	2,300
7850	3" flanged (iron body)	Q-1	10	1.600		2,100	70		2,170	2,400
7860	4" flanged (iron body)	"	8	2	↓	3,450	88		3,538	3,925
7920	For higher pressure, add					25%				
8350	Tempering, water, sweat connections									
8400	1/2"	1 Plum	24	.333	Ea.	71	16.25		87.25	103
8440	3/4"	"	20	.400	"	87	19.50		106.50	126
8650	Threaded connections									
8700	1/2"	1 Plum	24	.333	Ea.	87	16.25		103.25	121
8740	3/4"		20	.400		370	19.50		389.50	440
8750	1"		19	.421		410	20.50		430.50	480
8760	1-1/4"		15	.533		650	26		676	755
8770	1-1/2"		13	.615		695	30		725	810
8780	2"		11	.727	↓	1,200	35.50		1,235.50	1,375
9000	Minimum labor/equipment charge	↓	4	2	Job		97.50		97.50	151

22 05 23.40 Valves, Lined, Corrosion Resistant/High Purity		Crew	Daily Output	Labor-Hours	Unit	Material	Labor	Equipment	Total	Total Incl O&P
0010	**VALVES, LINED, CORROSION RESISTANT/HIGH PURITY** R220523-90									
2000	Butterfly 150 lb. ductile iron,									
2010	Wafer type									
2030	TFE lined 3", lever handle	Q-1	8	2	Ea.	1,200	88		1,288	1,450
2050	4", lever handle	"	5	3.200		1,675	140		1,815	2,075
2070	6", gear operated	Q-2	4.50	5.333		2,375	243		2,618	2,975
2080	8", gear operated		4	6		3,325	273		3,598	4,100
2090	10", gear operated		3.50	6.857		4,325	310		4,635	5,250
2100	12", gear operated	↓	2.50	9.600	↓	5,175	435		5,610	6,375
3500	Check lift, 125 lb., cast iron flanged									
3510	Horizontal PPL or SL lined									
3530	1"	1 Plum	14	.571	Ea.	595	28		623	700
3540	1-1/2"	↓	11	.727	↓	720	35.50		755.50	845

22 05 23.40 Valves, Lined, Corrosion Resistant/High Purity		Crew	Daily Output	Labor-Hours	Unit	Material	2009 Bare Costs Labor	Equipment	Total	Total Incl O&P
3550	2"	1 Plum	8	1	Ea.	835	49		884	995
3560	2-1/2"	Q-1	5	3.200		1,100	140		1,240	1,425
3570	3"		4.50	3.556		1,375	156		1,531	1,750
3590	4"		3	5.333		1,800	234		2,034	2,350
3610	6"	Q-2	3	8		3,075	365		3,440	3,950
3620	8"	"	2.50	9.600		6,750	435		7,185	8,100
4250	Vertical PPL or SL lined									
4270	1"	1 Plum	14	.571	Ea.	525	28		553	620
4290	1-1/2"		11	.727		660	35.50		695.50	780
4300	2"		8	1		770	49		819	920
4310	2-1/2"	Q-1	5	3.200		1,050	140		1,190	1,375
4320	3"		4.50	3.556		1,150	156		1,306	1,525
4340	4"		3	5.333		1,525	234		1,759	2,050
4360	6"	Q-2	3	8		2,575	365		2,940	3,400
4370	8"	"	2.50	9.600		4,650	435		5,085	5,775
5000	Clamp type, ductile iron, 150 lb. flanged									
5010	TFE lined									
5030	1"size, lever handle	1 Plum	9	.889	Ea.	1,100	43.50		1,143.50	1,300
5050	1-1/2" size, lever handle		6	1.333		1,450	65		1,515	1,700
5060	2" size, lever handle		5	1.600		1,775	78		1,853	2,075
5080	3" size, lever handle	Q-1	4.50	3.556		2,250	156		2,406	2,725
5100	4" size, gear operated	"	3	5.333		3,175	234		3,409	3,875
5120	6" size, gear operated	Q-2	3	8		11,100	365		11,465	12,800
5130	8" size, gear operated	"	2.50	9.600		12,400	435		12,835	14,300
6000	Diaphragm type, cast iron, 125 lb. flanged									
6010	PTFE or VITON, lined									
6030	1" size, handwheel operated	1 Plum	9	.889	Ea.	266	43.50		309.50	360
6050	1-1/2" size, handwheel operated		6	1.333		330	65		395	465
6060	2" size, handwheel operated		5	1.600		400	78		478	565
6080	3" size, handwheel operated	Q-1	4.50	3.556		665	156		821	970
6100	4" size, handwheel operated	"	3	5.333		1,200	234		1,434	1,700
6120	6" size, handwheel operated	Q-2	3	8		2,050	365		2,415	2,825
6130	8" size, handwheel operated	"	2.50	9.600		4,150	435		4,585	5,250

22 05 23.60 Valves, Plastic

0010	**VALVES, PLASTIC** R220523-90									
1100	Angle, PVC, threaded									
1110	1/4"	1 Plum	26	.308	Ea.	61	15		76	90.50
1120	1/2"		26	.308		61	15		76	90.50
1130	3/4"		25	.320		72.50	15.60		88.10	104
1140	1"		23	.348		88	16.95		104.95	124
1150	Ball, PVC, socket or threaded, single union									
1230	1/2"	1 Plum	26	.308	Ea.	20.50	15		35.50	46
1240	3/4"		25	.320		23.50	15.60		39.10	50
1250	1"		23	.348		29.50	16.95		46.45	59
1260	1-1/4"		21	.381		39.50	18.55		58.05	72.50
1270	1-1/2"		20	.400		47	19.50		66.50	82
1280	2"		17	.471		68.50	23		91.50	111
1290	2-1/2"	Q-1	26	.615		288	27		315	355
1300	3"		24	.667		350	29.50		379.50	430
1310	4"		20	.800		590	35		625	705
1360	For PVC, flanged, add					100%	15%			
1450	Double union 1/2"	1 Plum	26	.308		30	15		45	56.50

22 05 23.60 Valves, Plastic		Crew	Daily Output	Labor-Hours	Unit	2009 Bare Costs Material	Labor	Equipment	Total	Total Incl O&P
1460	3/4"	1 Plum	25	.320	Ea.	35.50	15.60		51.10	63
1470	1"		23	.348		42	16.95		58.95	73
1480	1-1/4"		21	.381		58.50	18.55		77.05	93.50
1490	1-1/2"		20	.400		70.50	19.50		90	108
1500	2"		17	.471		93	23		116	138
1650	CPVC, socket or threaded, single union									
1700	1/2"	1 Plum	26	.308	Ea.	39	15		54	66
1720	3/4"		25	.320		49	15.60		64.60	77.50
1730	1"		23	.348		58	16.95		74.95	90.50
1750	1-1/4"		21	.381		93	18.55		111.55	131
1760	1-1/2"		20	.400		93	19.50		112.50	132
1770	2"		17	.471		128	23		151	177
1780	3"	Q-1	24	.667		380	29.50		409.50	465
1840	For CPVC, flanged, add					65%	15%			
1880	For true union, socket or threaded, add					50%	5%			
2050	Polypropylene, threaded									
2100	1/4"	1 Plum	26	.308	Ea.	39	15		54	66.50
2120	3/8"		26	.308		39	15		54	66.50
2130	1/2"		26	.308		39	15		54	66.50
2140	3/4"		25	.320		49	15.60		64.60	77.50
2150	1"		23	.348		57.50	16.95		74.45	90
2160	1-1/4"		21	.381		84	18.55		102.55	122
2170	1-1/2"		20	.400		97	19.50		116.50	136
2180	2"		17	.471		131	23		154	180
2190	3"	Q-1	24	.667		345	29.50		374.50	425
2200	4"	"	20	.800		575	35		610	690
2550	PVC, three way, socket or threaded									
2600	1/2"	1 Plum	26	.308	Ea.	48	15		63	76
2640	3/4"		25	.320		54.50	15.60		70.10	84
2650	1"		23	.348		58.50	16.95		75.45	91
2660	1-1/2"		20	.400		119	19.50		138.50	161
2670	2"		17	.471		159	23		182	210
2680	3"	Q-1	24	.667		345	29.50		374.50	425
2740	For flanged, add					60%	15%			
3150	Ball check, PVC, socket or threaded									
3200	1/4"	1 Plum	26	.308	Ea.	32	15		47	59
3220	3/8"		26	.308		32	15		47	59
3240	1/2"		26	.308		33.50	15		48.50	60.50
3250	3/4"		25	.320		36	15.60		51.60	63.50
3260	1"		23	.348		45	16.95		61.95	76
3270	1-1/4"		21	.381		75.50	18.55		94.05	112
3280	1-1/2"		20	.400		75.50	19.50		95	113
3290	2"		17	.471		103	23		126	149
3310	3"	Q-1	24	.667		285	29.50		314.50	360
3320	4"	"	20	.800		405	35		440	500
3360	For PVC, flanged, add					50%	15%			
3750	CPVC, socket or threaded									
3800	1/2"	1 Plum	26	.308	Ea.	47	15		62	75
3840	3/4"		25	.320		56	15.60		71.60	85.50
3850	1"		23	.348		66	16.95		82.95	99
3860	1-1/2"		20	.400		114	19.50		133.50	155
3870	2"		17	.471		153	23		176	205
3880	3"	Q-1	24	.667		415	29.50		444.50	500

22 05 23.60 Valves, Plastic		Crew	Daily Output	Labor-Hours	Unit	Material	Labor	Equipment	Total	Total Incl O&P
						2009 Bare Costs				
3920	4"	Q-1	20	.800	Ea.	555	35		590	665
3930	For CPVC, flanged, add					40%	15%			
4340	Polypropylene, threaded									
4360	1/2"	1 Plum	26	.308	Ea.	33.50	15		48.50	60.50
4400	3/4"		25	.320		39	15.60		54.60	67
4440	1"		23	.348		50.50	16.95		67.45	82
4450	1-1/2"		20	.400		97	19.50		116.50	136
4460	2"		17	.471		122	23		145	170
4500	For polypropylene flanged, add					200%	15%			
4850	Foot valve, PVC, socket or threaded									
4900	1/2"	1 Plum	34	.235	Ea.	57	11.45		68.45	80.50
4930	3/4"		32	.250		64	12.20		76.20	89.50
4940	1"		28	.286		83	13.95		96.95	113
4950	1-1/4"		27	.296		160	14.45		174.45	199
4960	1-1/2"		26	.308		160	15		175	200
4970	2"		24	.333		186	16.25		202.25	229
4980	3"		20	.400		445	19.50		464.50	520
4990	4"		18	.444		780	21.50		801.50	890
5000	For flanged, add					25%	10%			
5050	CPVC, socket or threaded									
5060	1/2"	1 Plum	34	.235	Ea.	59.50	11.45		70.95	83.50
5070	3/4"		32	.250		69.50	12.20		81.70	95.50
5080	1"		28	.286		85	13.95		98.95	115
5090	1-1/4"		27	.296		136	14.45		150.45	172
5100	1-1/2"		26	.308		136	15		151	173
5110	2"		24	.333		174	16.25		190.25	217
5120	3"		20	.400		355	19.50		374.50	420
5130	4"		18	.444		645	21.50		666.50	745
5140	For flanged, add					25%	10%			
5280	Needle valve, PVC, threaded									
5300	1/4"	1 Plum	26	.308	Ea.	41	15		56	68.50
5340	3/8"		26	.308		47.50	15		62.50	75.50
5360	1/2"		26	.308		47.50	15		62.50	75.50
5380	For polypropylene, add					10%				
5800	Y check, PVC, socket or threaded									
5820	1/2"	1 Plum	26	.308	Ea.	68	15		83	98
5840	3/4"		25	.320		78	15.60		93.60	110
5850	1"		23	.348		79.50	16.95		96.45	114
5860	1-1/4"		21	.381		125	18.55		143.55	167
5870	1-1/2"		20	.400		137	19.50		156.50	180
5880	2"		17	.471		170	23		193	223
5890	2-1/2"		15	.533		360	26		386	435
5900	3"	Q-1	24	.667		425	29.50		454.50	515
5910	4"	"	20	.800		585	35		620	700
5960	For PVC flanged, add					45%	15%			
6350	Y sediment strainer, PVC, socket or threaded									
6400	1/2"	1 Plum	26	.308	Ea.	38.50	15		53.50	65.50
6440	3/4"		24	.333		41.50	16.25		57.75	71
6450	1"		23	.348		50.50	16.95		67.45	82
6460	1-1/4"		21	.381		83	18.55		101.55	121
6470	1-1/2"		20	.400		83	19.50		102.50	122
6480	2"		17	.471		102	23		125	148
6490	2-1/2"		15	.533		249	26		275	315

22 05 23 – General-Duty Valves for Plumbing Piping

22 05 23.60 Valves, Plastic		Crew	Daily Output	Labor-Hours	Unit	Material (2009 Bare Costs)	Labor (2009 Bare Costs)	Equipment (2009 Bare Costs)	Total (2009 Bare Costs)	Total Incl O&P
6500	3"	Q-1	24	.667	Ea.	249	29.50		278.50	320
6510	4"	"	20	.800		415	35		450	510
6560	For PVC, flanged, add				↓	55%	15%			
9000	Minimum labor/equipment charge	1 Plum	3.50	2.286	Job		111		111	173

22 05 29 – Hangers and Supports for Plumbing Piping and Equipment

22 05 29.10 Hangers & Supp. for Plumb'g/HVAC Pipe/Equip.		Crew	Daily Output	Labor-Hours	Unit	Material	Labor	Equipment	Total	Total Incl O&P
0010	**HANGERS AND SUPPORTS FOR PLUMB'G/HVAC PIPE/EQUIP.**									
0011	TYPE numbers per MSS-SP58									
0050	Brackets									
0060	Beam side or wall, malleable iron, TYPE 34									
0070	3/8" threaded rod size	1 Plum	48	.167	Ea.	2.19	8.15		10.34	15
0080	1/2" threaded rod size		48	.167		4.22	8.15		12.37	17.25
0090	5/8" threaded rod size		48	.167		5.60	8.15		13.75	18.80
0100	3/4" threaded rod size		48	.167		6.75	8.15		14.90	20
0110	7/8" threaded rod size	↓	48	.167		7.85	8.15		16	21.50
0120	For concrete installation, add				↓		30%			
0150	Wall, welded steel, medium, TYPE 32									
0160	0 size, 12" wide, 18" deep	1 Plum	34	.235	Ea.	177	11.45		188.45	212
0170	1 size, 18" wide, 24" deep		34	.235		210	11.45		221.45	249
0180	2 size, 24" wide, 30" deep	↓	34	.235	↓	278	11.45		289.45	325
0300	Clamps									
0310	C-clamp, for mounting on steel beam flange, w/locknut, TYPE 23									
0320	3/8" threaded rod size	1 Plum	160	.050	Ea.	1.81	2.44		4.25	5.75
0330	1/2" threaded rod size		160	.050		2.30	2.44		4.74	6.30
0340	5/8" threaded rod size		160	.050		3.81	2.44		6.25	7.95
0350	3/4" threaded rod size	↓	160	.050	↓	5.80	2.44		8.24	10.20
0400	High temperature to 1050° F, alloy steel									
0410	4" pipe size	Q-1	106	.151	Ea.	26.50	6.60		33.10	39.50
0420	6" pipe size		106	.151		44.50	6.60		51.10	59.50
0430	8" pipe size		97	.165		50.50	7.25		57.75	66.50
0440	10" pipe size		84	.190		68	8.35		76.35	88
0450	12" pipe size		72	.222		79	9.75		88.75	102
0460	14" pipe size		64	.250		165	10.95		175.95	199
0470	16" pipe size	↓	56	.286	↓	188	12.55		200.55	226
0500	I-beam, for mounting on bottom flange, strap iron, TYPE 21									
0510	2" flange size	1 Plum	96	.083	Ea.	8.85	4.06		12.91	16
0520	3" flange size		95	.084		10.50	4.11		14.61	17.90
0530	4" flange size		93	.086		6.20	4.19		10.39	13.35
0540	5" flange size		92	.087		6.50	4.24		10.74	13.70
0550	6" flange size		90	.089		7.85	4.33		12.18	15.30
0560	7" flange size		88	.091		8.25	4.43		12.68	15.95
0570	8" flange size	↓	86	.093	↓	8.85	4.53		13.38	16.75
0600	One hole, vertical mounting, malleable iron									
0610	1/2" pipe size	1 Plum	160	.050	Ea.	.73	2.44		3.17	4.58
0620	3/4" pipe size		145	.055		.88	2.69		3.57	5.15
0630	1" pipe size		136	.059		1.30	2.87		4.17	5.90
0640	1-1/4" pipe size		128	.063		1.46	3.05		4.51	6.35
0650	1-1/2" pipe size		120	.067		1.64	3.25		4.89	6.85
0660	2" pipe size		112	.071		2.37	3.48		5.85	8
0670	2-1/2" pipe size		104	.077		4.62	3.75		8.37	10.90
0680	3" pipe size		96	.083		6.55	4.06		10.61	13.50
0690	3-1/2" pipe size	↓	90	.089	↓	10.65	4.33		14.98	18.45

22 05 29 – Hangers and Supports for Plumbing Piping and Equipment

	22 05 29.10 Hangers & Supp. for Plumb'g/HVAC Pipe/Equip.	Crew	Daily Output	Labor-Hours	Unit	2009 Bare Costs Material	Labor	Equipment	Total	Total Incl O&P
0700	4" pipe size	1 Plum	84	.095	Ea.	13.75	4.64		18.39	22.50
0750	Riser or extension pipe, carbon steel, TYPE 8									
0760	3/4" pipe size	1 Plum	48	.167	Ea.	3.05	8.15		11.20	15.95
0770	1" pipe size		47	.170		3.08	8.30		11.38	16.25
0780	1-1/4" pipe size		46	.174		3.84	8.50		12.34	17.35
0790	1-1/2" pipe size		45	.178		4.09	8.65		12.74	17.95
0800	2" pipe size		43	.186		4.25	9.05		13.30	18.75
0810	2-1/2" pipe size		41	.195		4.48	9.50		13.98	19.70
0820	3" pipe size		40	.200		4.74	9.75		14.49	20.50
0830	3-1/2" pipe size		39	.205		9.05	10		19.05	25.50
0840	4" pipe size		38	.211		6	10.25		16.25	22.50
0850	5" pipe size		37	.216		8.95	10.55		19.50	26
0860	6" pipe size		36	.222		10.25	10.85		21.10	28
0870	8" pipe size		34	.235		16.55	11.45		28	36
0880	10" pipe size		32	.250		23	12.20		35.20	44
0890	12" pipe size	↓	28	.286	↓	28	13.95		41.95	52.50
0900	For plastic coating 3/4" to 4", add					190%				
0910	For copper plating 3/4" to 4", add					58%				
0950	Two piece, complete, carbon steel, medium weight, TYPE 4									
0960	1/2" pipe size	Q-1	137	.117	Ea.	2.23	5.10		7.33	10.40
0970	3/4" pipe size		134	.119		2.23	5.25		7.48	10.60
0980	1" pipe size		132	.121		2.27	5.30		7.57	10.75
0990	1-1/4" pipe size		130	.123		2.94	5.40		8.34	11.65
1000	1-1/2" pipe size		126	.127		2.94	5.55		8.49	11.90
1010	2" pipe size		124	.129		3.42	5.65		9.07	12.55
1020	2-1/2" pipe size		120	.133		3.62	5.85		9.47	13.05
1030	3" pipe size		117	.137		4.02	6		10.02	13.70
1040	3-1/2" pipe size		114	.140		5.60	6.15		11.75	15.70
1050	4" pipe size		110	.145		5.60	6.40		12	16.05
1060	5" pipe size		106	.151		9.20	6.60		15.80	20.50
1070	6" pipe size		104	.154		14.05	6.75		20.80	26
1080	8" pipe size		100	.160		16.80	7		23.80	29.50
1090	10" pipe size		96	.167		30.50	7.30		37.80	45
1100	12" pipe size		89	.180		35	7.90		42.90	51
1110	14" pipe size		82	.195		67.50	8.55		76.05	88
1120	16" pipe size	↓	68	.235		73	10.30		83.30	96.50
1130	For galvanized, add				↓	45%				
1150	Insert, concrete									
1160	Wedge type, carbon steel body, malleable iron nut									
1170	1/4" threaded rod size	1 Plum	96	.083	Ea.	1.10	4.06		5.16	7.50
1180	3/8" threaded rod size		96	.083		1.15	4.06		5.21	7.55
1190	1/2" threaded rod size		96	.083		1.18	4.06		5.24	7.60
1200	5/8" threaded rod size		96	.083		1.20	4.06		5.26	7.60
1210	3/4" threaded rod size		96	.083		1.42	4.06		5.48	7.85
1220	7/8" threaded rod size	↓	96	.083		1.91	4.06		5.97	8.40
1230	For galvanized, add				↓	.60			.60	.66
1250	Pipe guide sized for insulation									
1260	No. 1, 1" pipe size, 1" thick insulation	1 Stpi	26	.308	Ea.	118	15.20		133.20	154
1270	No. 2, 1-1/4"-2" pipe size, 1" thick insulation		23	.348		138	17.15		155.15	178
1280	No. 3, 1-1/4"-2" pipe size, 1-1/2" thick insulation		21	.381		138	18.80		156.80	180
1290	No. 4, 2-1/2"-3-1/2" pipe size, 1-1/2" thick insulation		18	.444		138	22		160	185
1300	No. 5, 4"-5" pipe size, 1-1/2" thick insulation	↓	16	.500		155	24.50		179.50	210
1310	No. 6, 5"-6" pipe size, 2" thick insulation	Q-5	21	.762	↓	171	34		205	241

22 05 29.10 Hangers & Supp. for Plumb'g/HVAC Pipe/Equip.		Crew	Daily Output	Labor-Hours	Unit	2009 Bare Costs Material	Labor	Equipment	Total	Total Incl O&P
1320	No. 7, 8" pipe size, 2" thick insulation	Q-5	16	1	Ea.	300	44.50		344.50	400
1330	No. 8, 10" pipe size, 2" thick insulation		12	1.333		570	59		629	720
1340	No. 9, 12" pipe size, 2" thick insulation	Q-6	17	1.412		570	65		635	730
1350	No. 10, 12"-14" pipe size, 2-1/2" thick insulation		16	1.500		745	69		814	925
1360	No. 11, 16" pipe size, 2-1/2" thick insulation		10.50	2.286		745	105		850	985
1370	No. 12, 16"-18" pipe size, 3" thick insulation		9	2.667		785	123		908	1,050
1380	No. 13, 20" pipe size, 3" thick insulation		7.50	3.200		785	147		932	1,100
1390	No. 14, 24" pipe size, 3" thick insulation		7	3.429		1,000	158		1,158	1,350
1400	Bands									
1410	Adjustable band, carbon steel, for non-insulated pipe, TYPE 7									
1420	1/2" pipe size	Q-1	142	.113	Ea.	.49	4.94		5.43	8.20
1430	3/4" pipe size		140	.114		.49	5		5.49	8.35
1440	1" pipe size		137	.117		.49	5.10		5.59	8.50
1450	1-1/4" pipe size		134	.119		.55	5.25		5.80	8.75
1460	1-1/2" pipe size		131	.122		.57	5.35		5.92	8.95
1470	2" pipe size		129	.124		.57	5.45		6.02	9.10
1480	2-1/2" pipe size		125	.128		1.02	5.60		6.62	9.80
1490	3" pipe size		122	.131		1.12	5.75		6.87	10.15
1500	3-1/2" pipe size		119	.134		1.83	5.90		7.73	11.15
1510	4" pipe size		114	.140		1.83	6.15		7.98	11.55
1520	5" pipe size		110	.145		3.04	6.40		9.44	13.25
1530	6" pipe size		108	.148		3.17	6.50		9.67	13.60
1540	8" pipe size		104	.154		3.79	6.75		10.54	14.60
1550	For copper plated, add					50%				
1560	For galvanized, add					30%				
1570	For plastic coating, add					30%				
1600	Adjusting nut malleable iron, steel band, TYPE 9									
1610	1/2" pipe size, galvanized band	Q-1	137	.117	Ea.	.86	5.10		5.96	8.90
1620	3/4" pipe size, galvanized band		135	.119		.86	5.20		6.06	9
1630	1" pipe size, galvanized band		132	.121		.90	5.30		6.20	9.25
1640	1-1/4" pipe size, galvanized band		129	.124		.90	5.45		6.35	9.45
1650	1-1/2" pipe size, galvanized band		126	.127		.90	5.55		6.45	9.65
1660	2" pipe size, galvanized band		124	.129		.95	5.65		6.60	9.85
1670	2-1/2" pipe size, galvanized band		120	.133		1.67	5.85		7.52	10.90
1680	3" pipe size, galvanized band		117	.137		1.67	6		7.67	11.15
1690	3-1/2" pipe size, galvanized band		114	.140		2.79	6.15		8.94	12.60
1700	4" pipe size, cadmium plated band		110	.145		2.88	6.40		9.28	13.05
1740	For plastic coated band, add					35%				
1750	For completely copper coated, add					45%				
1800	Clevis, adjustable, carbon steel, for non-insulated pipe, TYPE 1									
1810	1/2" pipe size	Q-1	137	.117	Ea.	1.05	5.10		6.15	9.10
1820	3/4" pipe size		135	.119		1.05	5.20		6.25	9.20
1830	1" pipe size		132	.121		1.09	5.30		6.39	9.45
1840	1-1/4" pipe size		129	.124		1.17	5.45		6.62	9.75
1850	1-1/2" pipe size		126	.127		1.24	5.55		6.79	10
1860	2" pipe size		124	.129		1.48	5.65		7.13	10.45
1870	2-1/2" pipe size		120	.133		2.33	5.85		8.18	11.60
1880	3" pipe size		117	.137		2.90	6		8.90	12.50
1890	3-1/2" pipe size		114	.140		3.08	6.15		9.23	12.95
1900	4" pipe size		110	.145		3.54	6.40		9.94	13.80
1910	5" pipe size		106	.151		4.85	6.60		11.45	15.60
1920	6" pipe size		104	.154		5.65	6.75		12.40	16.65
1930	8" pipe size		100	.160		9.30	7		16.30	21

22 05 29.10 Hangers & Supp. for Plumb'g/HVAC Pipe/Equip.		Crew	Daily Output	Labor-Hours	Unit	2009 Bare Costs Material	Labor	Equipment	Total	Total Incl O&P
1940	10" pipe size	Q-1	96	.167	Ea.	16.75	7.30		24.05	30
1950	12" pipe size		89	.180		19.95	7.90		27.85	34.50
1960	14" pipe size		82	.195		32.50	8.55		41.05	49
1970	16" pipe size		68	.235		46	10.30		56.30	67
1980	For galvanized, add					66%				
1990	For copper plated 1/2" to 4", add					77%				
2000	For light weight 1/2" to 4", deduct					13%				
2010	Insulated pipe type, 3/4" to 12" pipe, add					180%				
2020	Insulated pipe type, chrome-moly U-strap, add					530%				
2250	Split ring, malleable iron, for non-insulated pipe, TYPE 11									
2260	1/2" pipe size	Q-1	137	.117	Ea.	5.60	5.10		10.70	14.10
2270	3/4" pipe size		135	.119		3.79	5.20		8.99	12.20
2280	1" pipe size		132	.121		4.01	5.30		9.31	12.65
2290	1-1/4" pipe size		129	.124		4.98	5.45		10.43	13.95
2300	1-1/2" pipe size		126	.127		6.05	5.55		11.60	15.30
2310	2" pipe size		124	.129		6.85	5.65		12.50	16.35
2320	2-1/2" pipe size		120	.133		9.95	5.85		15.80	19.95
2330	3" pipe size		117	.137		12.35	6		18.35	23
2340	3-1/2" pipe size		114	.140		13.35	6.15		19.50	24
2350	4" pipe size		110	.145		12.90	6.40		19.30	24
2360	5" pipe size		106	.151		17.15	6.60		23.75	29
2370	6" pipe size		104	.154		38.50	6.75		45.25	52.50
2380	8" pipe size		100	.160		62.50	7		69.50	79.50
2390	For copper plated, add					8%				
2532	Turnbuckle, TYPE 13									
2534	3/8"	1 Plum	80	.100	Ea.	14.45	4.88		19.33	23.50
2535	1/2"		72	.111		22	5.40		27.40	32.50
2536	5/8"		64	.125		25	6.10		31.10	37
2537	3/4"		56	.143		38	6.95		44.95	52.50
2538	7/8"		48	.167		47	8.15		55.15	64
2539	1"		40	.200		65	9.75		74.75	86.50
2540	1-1/4"		32	.250		75.50	12.20		87.70	102
2544	Eye rod, welded									
2546	3/8"	1 Plum	56	.143	Ea.	10.70	6.95		17.65	22.50
2547	1/2"		48	.167		15.85	8.15		24	30
2548	5/8"		40	.200		24	9.75		33.75	41.50
2549	3/4"		32	.250		28	12.20		40.20	50
2550	7/8"		24	.333		37	16.25		53.25	66
2551	1"		16	.500		47	24.50		71.50	89.50
2556	Eye rod, welded, linked									
2558	3/8"	1 Plum	56	.143	Ea.	61.50	6.95		68.45	78.50
2559	1/2"		48	.167		82.50	8.15		90.65	103
2560	5/8"		40	.200		118	9.75		127.75	145
2561	3/4"		32	.250		138	12.20		150.20	171
2562	7/8"		24	.333		190	16.25		206.25	234
2563	1"		16	.500		241	24.50		265.50	305
2650	Rods, carbon steel									
2660	Continuous thread									
2670	1/4" thread size	1 Plum	144	.056	L.F.	.14	2.71		2.85	4.35
2680	3/8" thread size		144	.056		.23	2.71		2.94	4.45
2690	1/2" thread size		144	.056		.46	2.71		3.17	4.71
2700	5/8" thread size		144	.056		.73	2.71		3.44	5
2710	3/4" thread size		144	.056		1.10	2.71		3.81	5.40

22 05 29.10	Hangers & Supp. for Plumb'g/HVAC Pipe/Equip.	Crew	Daily Output	Labor-Hours	Unit	2009 Bare Costs Material	Labor	Equipment	Total	Total Incl O&P
2720	7/8" thread size	1 Plum	144	.056	L.F.	1.78	2.71		4.49	6.15
2725	1/4" thread size, bright finish		144	.056		1.21	2.71		3.92	5.55
2726	1/2" thread size, bright finish		144	.056		2.07	2.71		4.78	6.50
2730	For galvanized, add					40%				
2750	Both ends machine threaded 18" length									
2760	3/8" thread size	1 Plum	240	.033	Ea.	3.94	1.62		5.56	6.85
2770	1/2" thread size		240	.033		7.55	1.62		9.17	10.80
2780	5/8" thread size		240	.033		10.85	1.62		12.47	14.45
2790	3/4" thread size		240	.033		16.70	1.62		18.32	21
2800	7/8" thread size		240	.033		20	1.62		21.62	25
2810	1" thread size		240	.033		28	1.62		29.62	33
2860	Pipe hanger assy, adj. clevis, saddle, rod, clamp, insul. allowance									
2864	1/2" pipe size	Q-5	35	.457	Ea.	29	20.50		49.50	63.50
2866	3/4" pipe size		34.80	.460		30	20.50		50.50	64.50
2868	1" pipe size		34.60	.462		27	20.50		47.50	62
2869	1-1/4" pipe size		34.30	.466		27.50	20.50		48	62
2870	1-1/2" pipe size		33.90	.472		28.50	21		49.50	63.50
2872	2" pipe size		33.30	.480		29.50	21.50		51	65.50
2874	2-1/2" pipe size		32.30	.495		36.50	22		58.50	74.50
2876	3" pipe size		31.20	.513		42.50	23		65.50	82
2880	4" pipe size		30.70	.521		45.50	23		68.50	86
2884	6" pipe size		29.80	.537		55.50	24		79.50	98
2888	8" pipe size		28	.571		78	25.50		103.50	125
2892	10" pipe size		25.20	.635		86	28		114	139
2896	12" pipe size		23.20	.690		138	30.50		168.50	200
2900	Rolls									
2910	Adjustable yoke, carbon steel with CI roll, TYPE 43									
2920	2-1/2" pipe size	Q-1	137	.117	Ea.	10.60	5.10		15.70	19.60
2930	3" pipe size		131	.122		11.85	5.35		17.20	21.50
2940	3-1/2" pipe size		124	.129		15.75	5.65		21.40	26
2950	4" pipe size		117	.137		15.75	6		21.75	26.50
2960	5" pipe size		110	.145		18.40	6.40		24.80	30
2970	6" pipe size		104	.154		23	6.75		29.75	35.50
2980	8" pipe size		96	.167		34	7.30		41.30	49
2990	10" pipe size		80	.200		40.50	8.80		49.30	58.50
3000	12" pipe size		68	.235		60	10.30		70.30	82
3010	14" pipe size		56	.286		166	12.55		178.55	201
3020	16" pipe size		48	.333		213	14.65		227.65	257
3050	Chair, carbon steel with CI roll									
3060	2" pipe size	1 Plum	68	.118	Ea.	13.95	5.75		19.70	24.50
3070	2-1/2" pipe size		65	.123		15.10	6		21.10	26
3080	3" pipe size		62	.129		15.45	6.30		21.75	27
3090	3-1/2" pipe size		60	.133		19.75	6.50		26.25	31.50
3100	4" pipe size		58	.138		22	6.70		28.70	35
3110	5" pipe size		56	.143		24	6.95		30.95	37.50
3120	6" pipe size		53	.151		33.50	7.35		40.85	48.50
3130	8" pipe size		50	.160		45	7.80		52.80	61.50
3140	10" pipe size		48	.167		56.50	8.15		64.65	75
3150	12" pipe size		46	.174		82.50	8.50		91	104
3170	Single pipe roll, (see line 2650 for rods), TYPE 41, 1" pipe size	Q-1	137	.117		11.20	5.10		16.30	20.50
3180	1-1/4" pipe size		131	.122		11.20	5.35		16.55	20.50
3190	1-1/2" pipe size		129	.124		11.20	5.45		16.65	21
3200	2" pipe size		124	.129		11.90	5.65		17.55	22

22 05 29.10 Hangers & Supp. for Plumb'g/HVAC Pipe/Equip.		Crew	Daily Output	Labor-Hours	Unit	2009 Bare Costs Material	Labor	Equipment	Total	Total Incl O&P
3210	2-1/2" pipe size	Q-1	118	.136	Ea.	12.80	5.95		18.75	23.50
3220	3" pipe size		115	.139		13.50	6.10		19.60	24.50
3230	3-1/2" pipe size		113	.142		15.70	6.20		21.90	27
3240	4" pipe size		112	.143		17.15	6.25		23.40	28.50
3250	5" pipe size		110	.145		19.05	6.40		25.45	31
3260	6" pipe size		101	.158		24.50	6.95		31.45	38
3270	8" pipe size		90	.178		38.50	7.80		46.30	54
3280	10" pipe size		80	.200		45.50	8.80		54.30	63.50
3290	12" pipe size		68	.235		65.50	10.30		75.80	88
3291	14" pipe size		56	.286		109	12.55		121.55	138
3292	16" pipe size		48	.333		147	14.65		161.65	185
3293	18" pipe size		40	.400		159	17.55		176.55	201
3294	20" pipe size		35	.457		217	20		237	270
3296	24" pipe size		30	.533		345	23.50		368.50	415
3297	30" pipe size		25	.640		680	28		708	795
3298	36" pipe size		20	.800		835	35		870	975
3300	Saddles (add vertical pipe riser, usually 3" diameter)									
3310	Pipe support, complete, adjust., CI saddle, TYPE 36									
3320	2-1/2" pipe size	1 Plum	96	.083	Ea.	118	4.06		122.06	135
3330	3" pipe size		88	.091		119	4.43		123.43	138
3340	3-1/2" pipe size		79	.101		120	4.94		124.94	140
3350	4" pipe size		68	.118		156	5.75		161.75	180
3360	5" pipe size		64	.125		158	6.10		164.10	183
3370	6" pipe size		59	.136		160	6.60		166.60	187
3380	8" pipe size		53	.151		170	7.35		177.35	198
3390	10" pipe size		50	.160		188	7.80		195.80	219
3400	12" pipe size		48	.167		195	8.15		203.15	228
3450	For standard pipe support, one piece, CI deduct					34%				
3460	For stanchion support, CI with steel yoke, deduct					60%				
3550	Insulation shield 1" thick, 1/2" pipe size, TYPE 40	1 Asbe	100	.080	Ea.	4.89	3.53		8.42	11.10
3560	3/4" pipe size		100	.080		6.10	3.53		9.63	12.40
3570	1" pipe size		98	.082		6.10	3.60		9.70	12.50
3580	1-1/4" pipe size		98	.082		6.50	3.60		10.10	12.95
3590	1-1/2" pipe size		96	.083		6.50	3.67		10.17	13.10
3600	2" pipe size		96	.083		6.75	3.67		10.42	13.35
3610	2-1/2" pipe size		94	.085		6.95	3.75		10.70	13.70
3620	3" pipe size		94	.085		7.45	3.75		11.20	14.20
3630	2" thick, 3-1/2" pipe size		92	.087		8.30	3.83		12.13	15.35
3640	4" pipe size		92	.087		9.10	3.83		12.93	16.20
3650	5" pipe size		90	.089		14.85	3.92		18.77	22.50
3660	6" pipe size		90	.089		16.95	3.92		20.87	25
3670	8" pipe size		88	.091		23.50	4.01		27.51	32
3680	10" pipe size		88	.091		34.50	4.01		38.51	44.50
3690	12" pipe size		86	.093		39.50	4.10		43.60	50
3700	14" pipe size		86	.093		79.50	4.10		83.60	94
3710	16" pipe size		84	.095		84.50	4.20		88.70	100
3720	18" pipe size		84	.095		93.50	4.20		97.70	110
3730	20" pipe size		82	.098		103	4.30		107.30	120
3732	24" pipe size		80	.100		125	4.41		129.41	144
3733	30" pipe size		78	.103		390	4.52		394.52	435
3735	36" pipe size		75	.107		590	4.70		594.70	660
3750	Covering protection saddle, TYPE 39									
3760	1" covering size									

22 05 29.10 Hangers & Supp. for Plumb'g/HVAC Pipe/Equip.		Crew	Daily Output	Labor-Hours	Unit	2009 Bare Costs Material	Labor	Equipment	Total	Total Incl O&P
3770	3/4" pipe size	1 Plum	68	.118	Ea.	8.55	5.75		14.30	18.30
3780	1" pipe size		68	.118		8.55	5.75		14.30	18.30
3790	1-1/4" pipe size		68	.118		8.55	5.75		14.30	18.30
3800	1-1/2" pipe size		66	.121		9.30	5.90		15.20	19.35
3810	2" pipe size		66	.121		9.30	5.90		15.20	19.35
3820	2-1/2" pipe size		64	.125		9.30	6.10		15.40	19.65
3830	3" pipe size		64	.125		13.35	6.10		19.45	24
3840	3-1/2" pipe size		62	.129		14.25	6.30		20.55	25.50
3850	4" pipe size		62	.129		14.25	6.30		20.55	25.50
3860	5" pipe size		60	.133		14.25	6.50		20.75	26
3870	6" pipe size	↓	60	.133	↓	17	6.50		23.50	29
3900	1-1/2" covering size									
3910	3/4" pipe size	1 Plum	68	.118	Ea.	10.70	5.75		16.45	20.50
3920	1" pipe size		68	.118		10.70	5.75		16.45	20.50
3930	1-1/4" pipe size		68	.118		10.70	5.75		16.45	20.50
3940	1-1/2" pipe size		66	.121		10.70	5.90		16.60	21
3950	2" pipe size		66	.121		11.60	5.90		17.50	22
3960	2-1/2" pipe size		64	.125		13.70	6.10		19.80	24.50
3970	3" pipe size		64	.125		13.70	6.10		19.80	24.50
3980	3-1/2" pipe size		62	.129		21.50	6.30		27.80	33.50
3990	4" pipe size		62	.129		13.90	6.30		20.20	25
4000	5" pipe size		60	.133		13.90	6.50		20.40	25.50
4010	6" pipe size		60	.133		19.90	6.50		26.40	32
4020	8" pipe size	↓	58	.138	↓	25	6.70		31.70	38
4028	2" covering size									
4029	2-1/2" pipe size	1 Plum	62	.129	Ea.	13.15	6.30		19.45	24
4080	10" pipe size		58	.138		25	6.70		31.70	38
4090	12" pipe size		56	.143		41	6.95		47.95	56
4100	14" pipe size		56	.143		41	6.95		47.95	56
4110	16" pipe size		54	.148		58	7.20		65.20	74.50
4120	18" pipe size		54	.148		58	7.20		65.20	74.50
4130	20" pipe size		52	.154		64	7.50		71.50	82
4150	24" pipe size		50	.160		75	7.80		82.80	94.50
4160	30" pipe size		48	.167		82	8.15		90.15	103
4180	36" pipe size	↓	45	.178	↓	90.50	8.65		99.15	113
4200	Sockets									
4210	Rod end, malleable iron, TYPE 16									
4220	1/4" thread size	1 Plum	240	.033	Ea.	1.18	1.62		2.80	3.82
4230	3/8" thread size		240	.033		1.27	1.62		2.89	3.92
4240	1/2" thread size		230	.035		1.52	1.70		3.22	4.30
4250	5/8" thread size		225	.036		3	1.73		4.73	6
4260	3/4" thread size		220	.036		4.89	1.77		6.66	8.15
4270	7/8" thread size	↓	210	.038		9.60	1.86		11.46	13.45
4290	Strap, 1/2" pipe size, TYPE 26	Q-1	142	.113		2.86	4.94		7.80	10.80
4300	3/4" pipe size		140	.114		2.92	5		7.92	11
4310	1" pipe size		137	.117		2.98	5.10		8.08	11.25
4320	1-1/4" pipe size		134	.119		3.35	5.25		8.60	11.85
4330	1-1/2" pipe size		131	.122		4.16	5.35		9.51	12.90
4340	2" pipe size		129	.124		4.31	5.45		9.76	13.20
4350	2-1/2" pipe size		125	.128		7.50	5.60		13.10	16.95
4360	3" pipe size		122	.131		8.25	5.75		14	17.95
4370	3-1/2" pipe size		119	.134		9.25	5.90		15.15	19.30
4380	4" pipe size	↓	114	.140	↓	10.30	6.15		16.45	21

22 05 29.10 Hangers & Supp. for Plumb'g/HVAC Pipe/Equip.		Crew	Daily Output	Labor-Hours	Unit	Material	2009 Bare Costs Labor	Equipment	Total	Total Incl O&P
4400	U-bolt, carbon steel									
4410	Standard, with nuts, TYPE 42									
4420	1/2" pipe size	1 Plum	160	.050	Ea.	1.28	2.44		3.72	5.20
4430	3/4" pipe size		158	.051		1.28	2.47		3.75	5.25
4450	1" pipe size		152	.053		1.43	2.57		4	5.55
4460	1-1/4" pipe size		148	.054		1.65	2.63		4.28	5.90
4470	1-1/2" pipe size		143	.056		1.81	2.73		4.54	6.20
4480	2" pipe size		139	.058		1.85	2.81		4.66	6.40
4490	2-1/2" pipe size		134	.060		3.06	2.91		5.97	7.90
4500	3" pipe size		128	.063		3.21	3.05		6.26	8.25
4510	3-1/2" pipe size		122	.066		3.36	3.20		6.56	8.65
4520	4" pipe size		117	.068		3.39	3.33		6.72	8.90
4530	5" pipe size		114	.070		3.70	3.42		7.12	9.35
4540	6" pipe size		111	.072		6.45	3.51		9.96	12.55
4550	8" pipe size		109	.073		7.25	3.58		10.83	13.50
4560	10" pipe size		107	.075		12.90	3.65		16.55	19.85
4570	12" pipe size		104	.077		17.25	3.75		21	25
4580	For plastic coating on 1/2" thru 6" size, add					150%				
4700	U-hook, carbon steel, requires mounting screws or bolts									
4710	3/4" thru 2" pipe size									
4720	6" long	1 Plum	96	.083	Ea.	.39	4.06		4.45	6.75
4730	8" long		96	.083		.51	4.06		4.57	6.85
4740	10" long		96	.083		.58	4.06		4.64	6.95
4750	12" long		96	.083		.66	4.06		4.72	7.05
4760	For copper plated, add					50%				
8000	Pipe clamp, plastic, 1/2" CTS	1 Plum	80	.100		.22	4.88		5.10	7.80
8010	3/4" CTS		73	.110		.23	5.35		5.58	8.55
8020	1" CTS		68	.118		.52	5.75		6.27	9.45
8080	Economy clamp, 1/4" CTS		175	.046		.05	2.23		2.28	3.52
8090	3/8" CTS		168	.048		.07	2.32		2.39	3.68
8100	1/2" CTS		160	.050		.07	2.44		2.51	3.86
8110	3/4" CTS		145	.055		.14	2.69		2.83	4.32
8200	Half clamp, 1/2" CTS		80	.100		.08	4.88		4.96	7.65
8210	3/4" CTS		73	.110		.12	5.35		5.47	8.45
8300	Suspension clamp, 1/2" CTS		80	.100		.22	4.88		5.10	7.80
8310	3/4" CTS		73	.110		.23	5.35		5.58	8.55
8320	1" CTS		68	.118		.52	5.75		6.27	9.45
8400	Insulator, 1/2" CTS		80	.100		.36	4.88		5.24	7.95
8410	3/4" CTS		73	.110		.37	5.35		5.72	8.70
8420	1" CTS		68	.118		.38	5.75		6.13	9.30
8800	Wire cable support system									
8810	Cable with hook terminal and locking device									
8830	2 mm, (.079 ") dia cable, (100 lb. cap.)									
8840	1 m, (3.3') length, with hook	1 Shee	96	.083	Ea.	2.36	3.93		6.29	8.85
8850	2 m, (6.6') length, with hook		84	.095		2.72	4.50		7.22	10.15
8860	3 m, (9.9') length, with hook		72	.111		3.09	5.25		8.34	11.75
8870	5 m, (16.4') length, with hook	Q-9	60	.267		3.87	11.35		15.22	22.50
8880	10 m, (32.8') length, with hook	″	30	.533		5.60	22.50		28.10	42
8900	3mm, (.118 ") dia cable, (200 lb. cap.)									
8910	1 m, (3.3') length, with hook	1 Shee	96	.083	Ea.	3.02	3.93		6.95	9.55
8920	2 m, (6.6') length, with hook		84	.095		3.40	4.50		7.90	10.90
8930	3 m, (9.9') length, with hook		72	.111		3.62	5.25		8.87	12.35
8940	5 m, (16.4') length, with hook	Q-9	60	.267		4.62	11.35		15.97	23

22 05 Common Work Results for Plumbing

22 05 29 – Hangers and Supports for Plumbing Piping and Equipment

22 05 29.10 Hangers & Supp. for Plumb'g/HVAC Pipe/Equip.		Crew	Daily Output	Labor-Hours	Unit	2009 Bare Costs Material	Labor	Equipment	Total	Total Incl O&P
8950	10 m, (32.8′) length, with hook	Q-9	30	.533	Ea.	6.65	22.50		29.15	43.50
9000	Cable system accessories									
9010	Anchor bolt, 3/8″, with nut	1 Shee	140	.057	Ea.	.90	2.70		3.60	5.30
9020	Air duct corner protector		160	.050		.41	2.36		2.77	4.20
9030	Air duct support attachment		140	.057		.15	2.70		2.85	4.46
9040	Flange clip, hammer-on style									
9044	For flange thickness 3/32″ - 9/64″, 160 lb. cap.	1 Shee	180	.044	Ea.	.21	2.10		2.31	3.56
9048	For flange thickness 1/8″ - 1/4″, 200 lb. cap.		160	.050		.21	2.36		2.57	3.98
9052	For flange thickness 5/16″ - 1/2″, 200 lb. cap.		150	.053		.44	2.52		2.96	4.48
9056	For flange thickness 9/16″ - 3/4″, 200 lb. cap.		140	.057		.58	2.70		3.28	4.93
9060	Wire insulation protection tube		180	.044	L.F.	.22	2.10		2.32	3.57
9070	Wire cutter				Ea.	25			25	27.50

22 05 33 – Heat Tracing for Plumbing Piping

22 05 33.20 Temperature Maintenance Cable

		Crew	Daily Output	Labor-Hours	Unit	Material	Labor	Equipment	Total	Total Incl O&P
0010	**TEMPERATURE MAINTENANCE CABLE**									
0040	Components									
0080	Heating cable									
0100	208 V									
0140	125° F	Q-1	1060.80	.015	L.F.	9.80	.66		10.46	11.80
0150	140° F	″	1060.80	.015	″	9.80	.66		10.46	11.80
0200	120 V									
0220	125° F	Q-1	1060.80	.015	L.F.	6.30	.66		6.96	8
0300	Power kit w/1 end seal	1 Elec	48.80	.164	Ea.	56	7.70		63.70	74
0310	Splice kit		35.50	.225		66.50	10.60		77.10	89.50
0320	End seal		160	.050		8.90	2.35		11.25	13.40
0330	Tee kit w/1 end seal		26.80	.299		70.50	14.05		84.55	99
0340	Powered splice w/2 end seals		20	.400		67	18.80		85.80	103
0350	Powered tee kit w/3 end seals		18	.444		86	21		107	127
0360	Cross kit w/2 end seals		18.60	.430		88.50	20		108.50	129
0500	Recommended thickness of fiberglass insulation									
0510	Pipe size									
0520	1/2″ - 1″ use 1″ insulation									
0530	1-1/4″ - 2″ use 1-1/2″ insulation									
0540	2-1/2″ - 6″ use 2″ insulation									
0560	NOTE: For pipe sizes 1-1/4″ and smaller use 1/4″ larger diameter									
0570	insulation to allow room for installation over cable.									

22 05 48 – Vibration and Seismic Controls for Plumbing Piping and Equipment

22 05 48.40 Vibration Absorbers

		Crew	Daily Output	Labor-Hours	Unit	Material	Labor	Equipment	Total	Total Incl O&P
0010	**VIBRATION ABSORBERS**									
0100	Hangers, neoprene flex									
0200	10 – 120 lb. capacity				Ea.	16.55			16.55	18.20
0220	75 – 550 lb. capacity					21.50			21.50	23.50
0240	250 – 1100 lb. capacity					44			44	48
0260	1000 – 4000 lb. capacity					69.50			69.50	76
0500	Spring flex, 60 lb. capacity					23			23	25.50
0520	450 lb. capacity					29			29	31.50
0540	900 lb. capacity					38.50			38.50	42
0560	1100 – 1300 lb. capacity					38.50			38.50	42
1000	Mounts, neoprene, 45 – 380 lb. capacity					11			11	12.10
1020	250 – 1100 lb. capacity					29			29	31.50
1040	1000 – 4000 lb. capacity					64.50			64.50	71
1100	Spring flex, 60 lb. capacity					45			45	49.50

22 05 48 – Vibration and Seismic Controls for Plumbing Piping and Equipment

22 05 48.40 Vibration Absorbers		Crew	Daily Output	Labor-Hours	Unit	Material	Labor	Equipment	Total	Total Incl O&P
						2009 Bare Costs				
1120	165 lb. capacity				Ea.	45			45	49.50
1140	260 lb. capacity					45			45	49.50
1160	450 lb. capacity					59.50			59.50	65.50
1180	600 lb. capacity					59.50			59.50	65.50
1200	750 lb. capacity					59.50			59.50	65.50
1220	900 lb. capacity					62			62	68
1240	1100 lb. capacity					65.50			65.50	72
1260	1300 lb. capacity					65.50			65.50	72
1280	1500 lb. capacity					96			96	105
1300	1800 lb. capacity					103			103	113
1320	2200 lb. capacity					101			101	111
1340	2600 lb. capacity					103			103	113
2000	Pads, cork rib, 18" x 18" x 1", 10-50 psi					104			104	114
2020	18" x 36" x 1", 10-50 psi					208			208	229
2100	Shear flexible pads, 18" x 18" x 3/8", 20-70 psi					49.50			49.50	54
2120	18" x 36" x 3/8", 20-70 psi					104			104	114
3000	Note overlap in capacities due to deflections									

22 05 53 – Identification for Plumbing Piping and Equipment

22 05 53.10 Piping System Identification Labels		Crew	Daily Output	Labor-Hours	Unit	Material	Labor	Equipment	Total	Total Incl O&P
0010	**PIPING SYSTEM IDENTIFICATION LABELS**									
0100	Indicate contents and flow direction									
0106	Pipe markers									
0110	Plastic snap around									
0114	1/2" pipe	1 Plum	80	.100	Ea.	6	4.88		10.88	14.15
0116	3/4" pipe		80	.100		6	4.88		10.88	14.15
0118	1" pipe		80	.100		6	4.88		10.88	14.15
0120	2" pipe		75	.107		7.10	5.20		12.30	15.85
0122	3" pipe		70	.114		11	5.55		16.55	21
0124	4" pipe		60	.133		11	6.50		17.50	22
0126	6" pipe		60	.133		11.60	6.50		18.10	23
0128	8" pipe		56	.143		15.40	6.95		22.35	28
0130	10" pipe		56	.143		15.40	6.95		22.35	28
0200	Over 10" pipe size		50	.160		18.70	7.80		26.50	32.50
1110	Self adhesive									
1114	1" pipe	1 Plum	80	.100	Ea.	3.86	4.88		8.74	11.80
1116	2" pipe		75	.107		5.10	5.20		10.30	13.65
1118	3" pipe		70	.114		9.60	5.55		15.15	19.20
1120	4" pipe		60	.133		9.60	6.50		16.10	20.50
1122	6" pipe		60	.133		10.70	6.50		17.20	22
1124	8" pipe		56	.143		14.60	6.95		21.55	27
1126	10" pipe		56	.143		14.60	6.95		21.55	27
1200	Over 10" pipe size		50	.160		17.80	7.80		25.60	31.50
2000	Valve tags									
2010	Numbered plus identifying legend									
2100	Brass, 2" diameter	1 Plum	40	.200	Ea.	2.70	9.75		12.45	18.05
2200	Plastic, 1-1/2" Dia.	"	40	.200	"	3.10	9.75		12.85	18.50

22 05 76 – Facility Drainage Piping Cleanouts

22 05 76.10 Cleanouts		Crew	Daily Output	Labor-Hours	Unit	Material	Labor	Equipment	Total	Total Incl O&P
0010	**CLEANOUTS**									
0060	Floor type									
0080	Round or square, scoriated nickel bronze top									
0100	2" pipe size	1 Plum	10	.800	Ea.	202	39		241	283

22 05 76 – Facility Drainage Piping Cleanouts

22 05 76.10 Cleanouts		Crew	Daily Output	Labor-Hours	Unit	2009 Bare Costs Material	Labor	Equipment	Total	Total Incl O&P
0120	3" pipe size	1 Plum	8	1	Ea.	213	49		262	310
0140	4" pipe size		6	1.333		285	65		350	415
0160	5" pipe size		4	2		465	97.50		562.50	660
0180	6" pipe size	Q-1	6	2.667		465	117		582	690
0200	8" pipe size	"	4	4		445	176		621	755
0340	Recessed for tile, same price									
0980	Round top, recessed for terrazzo									
1000	2" pipe size	1 Plum	9	.889	Ea.	202	43.50		245.50	289
1080	3" pipe size		6	1.333		213	65		278	335
1100	4" pipe size		4	2		285	97.50		382.50	465
1120	5" pipe size	Q-1	6	2.667		465	117		582	690
1140	6" pipe size		5	3.200		465	140		605	730
1160	8" pipe size		4	4		445	176		621	755
2000	Round scoriated nickel bronze top, extra heavy duty									
2060	2" pipe size	1 Plum	9	.889	Ea.	252	43.50		295.50	345
2080	3" pipe size		6	1.333		263	65		328	390
2100	4" pipe size		4	2		335	97.50		432.50	520
2120	5" pipe size	Q-1	6	2.667		515	117		632	745
2140	6" pipe size		5	3.200		515	140		655	785
2160	8" pipe size		4	4		495	176		671	815
4000	Wall type, square smooth cover, over wall frame									
4060	2" pipe size	1 Plum	14	.571	Ea.	226	28		254	292
4080	3" pipe size		12	.667		305	32.50		337.50	385
4100	4" pipe size		10	.800		261	39		300	350
4120	5" pipe size		9	.889		380	43.50		423.50	480
4140	6" pipe size		8	1		420	49		469	535
4160	8" pipe size	Q-1	11	1.455		560	64		624	715
5000	Extension, C.I.; bronze countersunk plug, 8" long									
5040	2" pipe size	1 Plum	16	.500	Ea.	118	24.50		142.50	168
5060	3" pipe size		14	.571		137	28		165	194
5080	4" pipe size		13	.615		140	30		170	201
5100	5" pipe size		12	.667		219	32.50		251.50	292
5120	6" pipe size		11	.727		266	35.50		301.50	350
9000	Minimum labor/equipment charge		3	2.667	Job		130		130	202

22 05 76.20 Cleanout Tees

Line	Description	Crew	Daily Output	Labor-Hours	Unit	Material	Labor	Equipment	Total	Total Incl O&P
0010	**CLEANOUT TEES**									
0100	Cast iron, B&S, with countersunk plug									
0200	2" pipe size	1 Plum	4	2	Ea.	188	97.50		285.50	360
0220	3" pipe size		3.60	2.222		205	108		313	395
0240	4" pipe size		3.30	2.424		256	118		374	465
0260	5" pipe size	Q-1	5.50	2.909		555	128		683	810
0280	6" pipe size	"	5	3.200		690	140		830	980
0300	8" pipe size	Q-3	5	6.400		790	297		1,087	1,325
0500	For round smooth access cover, same price									
0600	For round scoriated access cover, same price									
0700	For square smooth access cover, add				Ea.	60%				
4000	Plastic, tees and adapters. Add plugs									
4010	ABS, DWV									
4020	Cleanout tee, 1-1/2" pipe size	1 Plum	15	.533	Ea.	6.60	26		32.60	48
4030	2" pipe size	Q-1	27	.593		7.75	26		33.75	49
4040	3" pipe size		21	.762		15.55	33.50		49.05	69
4050	4" pipe size		16	1		27	44		71	98

22 05 76 – Facility Drainage Piping Cleanouts

22 05 76.20 Cleanout Tees		Crew	Daily Output	Labor-Hours	Unit	2009 Bare Costs Material	Labor	Equipment	Total	Total Incl O&P
4100	Cleanout plug, 1-1/2" pipe size	1 Plum	32	.250	Ea.	1	12.20		13.20	20
4110	2" pipe size	Q-1	56	.286		1.32	12.55		13.87	21
4120	3" pipe size		36	.444		2.13	19.50		21.63	32.50
4130	4" pipe size		30	.533		4.14	23.50		27.64	41
4180	Cleanout adapter fitting, 1-1/2" pipe size	1 Plum	32	.250		1.59	12.20		13.79	20.50
4190	2" pipe size	Q-1	56	.286		2.22	12.55		14.77	22
4200	3" pipe size		36	.444		6.25	19.50		25.75	37
4210	4" pipe size		30	.533		10.05	23.50		33.55	47.50
5000	PVC, DWV									
5010	Cleanout tee, 1-1/2" pipe size	1 Plum	15	.533	Ea.	4.31	26		30.31	45
5020	2" pipe size	Q-1	27	.593		5.30	26		31.30	46.50
5030	3" pipe size		21	.762		10.60	33.50		44.10	63.50
5040	4" pipe size		16	1		19.65	44		63.65	89.50
5090	Cleanout plug, 1-1/2" pipe size	1 Plum	32	.250		1.15	12.20		13.35	20
5100	2" pipe size	Q-1	56	.286		1.18	12.55		13.73	21
5110	3" pipe size		36	.444		2.30	19.50		21.80	32.50
5120	4" pipe size		30	.533		3.39	23.50		26.89	40
5130	6" pipe size		24	.667		11.25	29.50		40.75	58
5170	Cleanout adapter fitting, 1-1/2" pipe size	1 Plum	32	.250		1.41	12.20		13.61	20.50
5180	2" pipe size	Q-1	56	.286		1.98	12.55		14.53	21.50
5190	3" pipe size		36	.444		5.60	19.50		25.10	36
5200	4" pipe size		30	.533		9.20	23.50		32.70	46.50
5210	6" pipe size		24	.667		24.50	29.50		54	72.50
9000	Minimum labor/equipment charge	1 Plum	2.75	2.909	Job		142		142	220

22 07 Plumbing Insulation

22 07 16 – Plumbing Equipment Insulation

22 07 16.10 Insulation for Plumbing Equipment

		Crew	Daily Output	Labor-Hours	Unit	Material	Labor	Equipment	Total	Total Incl O&P
0010	INSULATION FOR PLUMBING EQUIPMENT									
2900	Domestic water heater wrap kit									
2920	1-1/2" with vinyl jacket, 20-60 gal. G	1 Plum	8	1	Ea.	16.05	49		65.05	93

22 07 19 – Plumbing Piping Insulation

22 07 19.10 Piping Insulation

		Crew	Daily Output	Labor-Hours	Unit	Material	Labor	Equipment	Total	Total Incl O&P
0010	PIPING INSULATION									
0100	Rule of thumb, as a percentage of total mechanical costs				Job				10%	
0110	Insulation req'd is based on the surface size/area to be covered									
2930	Insulated protectors, (ADA)									
2935	For exposed piping under sinks or lavatories									
2940	Vinyl coated foam, velcro tabs									
2945	P Trap, 1-1/4" or 1-1/2"	1 Plum	32	.250	Ea.	18.50	12.20		30.70	39.50
2960	Valve and supply cover									
2965	1/2", 3/8", and 7/16" pipe size	1 Plum	32	.250	Ea.	17.60	12.20		29.80	38.50
2970	Extension drain cover									
2975	1-1/4", or 1-1/2" pipe size	1 Plum	32	.250	Ea.	19.30	12.20		31.50	40.50
2980	Tailpiece offset (wheelchair)									
2985	1-1/4" pipe size	1 Plum	32	.250	Ea.	21.50	12.20		33.70	42.50
4000	Pipe covering (price copper tube one size less than IPS)									
4280	Cellular glass, closed cell foam, all service jacket, sealant,									
4281	working temp. (-450° F to +900° F), 0 water vapor transmission									
4284	1" wall,									

22 07 19 – Plumbing Piping Insulation

22 07 19.10 Piping Insulation		Crew	Daily Output	Labor-Hours	Unit	Material	Labor	Equipment	Total	Total Incl O&P
						2009 Bare Costs				
4286	1/2" iron pipe size	G	Q-14	120	.133	L.F.	1.82	5.30		7.12	10.55
4300	1-1/2" wall,										
4301	1" iron pipe size	G	Q-14	105	.152	L.F.	2.93	6.05		8.98	13
4304	2-1/2" iron pipe size	G		90	.178		4.91	7.05		11.96	16.80
4306	3" iron pipe size	G		85	.188		4.99	7.45		12.44	17.60
4308	4" iron pipe size	G		70	.229		6.60	9.05		15.65	22
4310	5" iron pipe size	G	↓	65	.246	↓	7.60	9.75		17.35	24
4320	2" wall,										
4322	1" iron pipe size	G	Q-14	100	.160	L.F.	4.35	6.35		10.70	15.05
4324	2-1/2" iron pipe size	G		85	.188		6.45	7.45		13.90	19.20
4326	3" iron pipe size	G		80	.200		6.60	7.95		14.55	20
4328	4" iron pipe size	G		65	.246		7.40	9.75		17.15	24
4330	5" iron pipe size	G		60	.267		8.90	10.60		19.50	27
4332	6" iron pipe size	G		50	.320		10.55	12.70		23.25	32
4336	8" iron pipe size	G		40	.400		13.60	15.90		29.50	40.50
4338	10" iron pipe size	G	↓	35	.457	↓	14.05	18.15		32.20	45
4350	2-1/2" wall,										
4360	12" iron pipe size	G	Q-14	32	.500	L.F.	22.50	19.85		42.35	56.50
4362	14" iron pipe size	G	"	28	.571	"	24	22.50		46.50	63
4370	3" wall,										
4378	6" iron pipe size	G	Q-14	48	.333	L.F.	14.05	13.25		27.30	37
4380	8" iron pipe size	G		38	.421		16.40	16.70		33.10	45
4382	10" iron pipe size	G		33	.485		21.50	19.25		40.75	55
4384	16" iron pipe size	G		25	.640		28	25.50		53.50	72
4386	18" iron pipe size	G		22	.727		30.50	29		59.50	80.50
4388	20" iron pipe size	G	↓	20	.800	↓	36	32		68	91
4400	3-1/2" wall,										
4412	12" iron pipe size	G	Q-14	27	.593	L.F.	31.50	23.50		55	73
4414	14" iron pipe size	G	"	25	.640	"	33.50	25.50		59	78
4430	4" wall,										
4446	16" iron pipe size	G	Q-14	22	.727	L.F.	34	29		63	84
4448	18" iron pipe size	G		20	.800		44	32		76	100
4450	20" iron pipe size	G	↓	18	.889	↓	45.50	35.50		81	108
4480	Fittings, average with fabric and mastic										
4484	1" wall,										
4486	1/2" iron pipe size	G	1 Asbe	40	.200	Ea.	3.23	8.80		12.03	17.80
4500	1-1/2" wall,										
4502	1" iron pipe size	G	1 Asbe	38	.211	Ea.	5.55	9.30		14.85	21
4504	2-1/2" iron pipe size	G		32	.250		6.90	11.05		17.95	25.50
4506	3" iron pipe size	G		30	.267		7.60	11.75		19.35	27.50
4508	4" iron pipe size	G		28	.286		11.70	12.60		24.30	33.50
4510	5" iron pipe size	G	↓	24	.333	↓	12.90	14.70		27.60	38
4520	2" wall,										
4522	1" iron pipe size	G	1 Asbe	36	.222	Ea.	6.45	9.80		16.25	23
4524	2-1/2" iron pipe size	G		30	.267		7.60	11.75		19.35	27.50
4526	3" iron pipe size	G		28	.286		9.15	12.60		21.75	30.50
4528	4" iron pipe size	G		24	.333		12.30	14.70		27	37.50
4530	5" iron pipe size	G		22	.364		15.50	16.05		31.55	43
4532	6" iron pipe size	G		20	.400		18.15	17.65		35.80	48.50
4536	8" iron pipe size	G		12	.667		27.50	29.50		57	78
4538	10" iron pipe size	G	↓	8	1	↓	34	44		78	109
4550	2-1/2" wall,										
4560	12" iron pipe size	G	1 Asbe	6	1.333	Ea.	58.50	59		117.50	160

22 07 19.10 Piping Insulation			Crew	Daily Output	Labor-Hours	Unit	Material	Labor	Equipment	Total	Total Incl O&P
							2009 Bare Costs				
4562	14″ iron pipe size	G	1 Asbe	4	2	Ea.	68	88		156	218
4570	3″ wall,										
4578	6″ iron pipe size	G	1 Asbe	16	.500	Ea.	21	22		43	58.50
4580	8″ iron pipe size	G		10	.800		31	35.50		66.50	91
4582	10″ iron pipe size	G		6	1.333		44	59		103	144
4584	16″ iron pipe size	G		2	4		105	176		281	400
4586	18″ iron pipe size	G		2	4		136	176		312	435
4588	20″ iron pipe size	G		2	4		160	176		336	460
4600	3-1/2″ wall,										
4612	12″ iron pipe size	G	1 Asbe	4	2	Ea.	76	88		164	227
4614	14″ iron pipe size	G	″	2	4	″	84.50	176		260.50	380
4630	4″ wall,										
4646	16″ iron pipe size	G	1 Asbe	2	4	Ea.	126	176		302	425
4648	18″ iron pipe size	G		2	4		155	176		331	455
4650	20″ iron pipe size	G		2	4		180	176		356	485
4900	Calcium silicate, with cover										
5100	1″ wall, 1/2″ iron pipe size	G	Q-14	170	.094	L.F.	2.56	3.74		6.30	8.85
5130	3/4″ iron pipe size	G		170	.094		2.58	3.74		6.32	8.90
5140	1″ iron pipe size	G		170	.094		2.50	3.74		6.24	8.80
5150	1-1/4″ iron pipe size	G		165	.097		2.56	3.85		6.41	9.05
5160	1-1/2″ iron pipe size	G		165	.097		2.60	3.85		6.45	9.10
5170	2″ iron pipe size	G		160	.100		2.98	3.97		6.95	9.70
5180	2-1/2″ iron pipe size	G		160	.100		3.17	3.97		7.14	9.90
5190	3″ iron pipe size	G		150	.107		3.43	4.23		7.66	10.60
5200	4″ iron pipe size	G		140	.114		4.25	4.54		8.79	12.05
5210	5″ iron pipe size	G		135	.119		4.48	4.71		9.19	12.55
5220	6″ iron pipe size	G		130	.123		4.76	4.89		9.65	13.15
5280	1-1/2″ wall, 1/2″ iron pipe size	G		150	.107		2.77	4.23		7	9.90
5310	3/4″ iron pipe size	G		150	.107		2.82	4.23		7.05	9.95
5320	1″ iron pipe size	G		150	.107		3.07	4.23		7.30	10.25
5330	1-1/4″ iron pipe size	G		145	.110		3.30	4.38		7.68	10.75
5340	1-1/2″ iron pipe size	G		145	.110		3.54	4.38		7.92	11
5350	2″ iron pipe size	G		140	.114		3.90	4.54		8.44	11.65
5360	2-1/2″ iron pipe size	G		140	.114		4.25	4.54		8.79	12.05
5370	3″ iron pipe size	G		135	.119		4.47	4.71		9.18	12.50
5380	4″ iron pipe size	G		125	.128		5.15	5.10		10.25	13.90
5390	5″ iron pipe size	G		120	.133		5.85	5.30		11.15	14.95
5400	6″ iron pipe size	G		110	.145		6	5.75		11.75	15.95
5460	2″ wall, 1/2″ iron pipe size	G		135	.119		4.27	4.71		8.98	12.30
5490	3/4″ iron pipe size	G		135	.119		4.49	4.71		9.20	12.55
5500	1″ iron pipe size	G		135	.119		4.73	4.71		9.44	12.80
5510	1-1/4″ iron pipe size	G		130	.123		5.05	4.89		9.94	13.50
5520	1-1/2″ iron pipe size	G		130	.123		5.30	4.89		10.19	13.70
5530	2″ iron pipe size	G		125	.128		5.55	5.10		10.65	14.30
5540	2-1/2″ iron pipe size	G		125	.128		6.65	5.10		11.75	15.50
5550	3″ iron pipe size	G		120	.133		6.70	5.30		12	15.90
5560	4″ iron pipe size	G		115	.139		7.75	5.50		13.25	17.45
5570	5″ iron pipe size	G		110	.145		8.80	5.75		14.55	19.05
5580	6″ iron pipe size	G		105	.152		9.65	6.05		15.70	20.50
5600	Calcium silicate, no cover										
5720	1″ wall, 1/2″ iron pipe size	G	Q-14	180	.089	L.F.	2.33	3.53		5.86	8.25
5740	3/4″ iron pipe size	G		180	.089		2.33	3.53		5.86	8.25
5750	1″ iron pipe size	G		180	.089		2.23	3.53		5.76	8.15

22 07 19 – Plumbing Piping Insulation

22 07 19.10 Piping Insulation		Crew	Daily Output	Labor-Hours	Unit	Material	Labor	Equipment	Total	Total Incl O&P
						2009 Bare Costs				
5760	1-1/4" iron pipe size	G	Q-14	175	.091	L.F.	2.27	3.63		5.90	8.35
5770	1-1/2" iron pipe size	G		175	.091		2.29	3.63		5.92	8.35
5780	2" iron pipe size	G		170	.094		2.64	3.74		6.38	8.95
5790	2-1/2" iron pipe size	G		170	.094		2.79	3.74		6.53	9.10
5800	3" iron pipe size	G		160	.100		3	3.97		6.97	9.70
5810	4" iron pipe size	G		150	.107		3.75	4.23		7.98	11
5820	5" iron pipe size	G		145	.110		3.90	4.38		8.28	11.40
5830	6" iron pipe size	G		140	.114		4.10	4.54		8.64	11.85
5900	1-1/2" wall, 1/2" iron pipe size	G		160	.100		2.46	3.97		6.43	9.10
5920	3/4" iron pipe size	G		160	.100		2.50	3.97		6.47	9.15
5930	1" iron pipe size	G		160	.100		2.73	3.97		6.70	9.40
5940	1-1/4" iron pipe size	G		155	.103		2.94	4.10		7.04	9.90
5950	1-1/2" iron pipe size	G		155	.103		3.16	4.10		7.26	10.15
5960	2" iron pipe size	G		150	.107		3.48	4.23		7.71	10.70
5970	2-1/2" iron pipe size	G		150	.107		3.80	4.23		8.03	11.05
5980	3" iron pipe size	G		145	.110		3.97	4.38		8.35	11.45
5990	4" iron pipe size	G		135	.119		4.60	4.71		9.31	12.65
6000	5" iron pipe size	G		130	.123		5.20	4.89		10.09	13.60
6010	6" iron pipe size	G		120	.133		5.30	5.30		10.60	14.35
6020	7" iron pipe size	G		115	.139		6.20	5.50		11.70	15.80
6030	8" iron pipe size	G		105	.152		7	6.05		13.05	17.50
6040	9" iron pipe size	G		100	.160		8.40	6.35		14.75	19.50
6050	10" iron pipe size	G		95	.168		9.30	6.70		16	21
6060	12" iron pipe size	G		90	.178		11	7.05		18.05	23.50
6070	14" iron pipe size	G		85	.188		12.45	7.45		19.90	26
6080	16" iron pipe size	G		80	.200		13.95	7.95		21.90	28
6090	18" iron pipe size	G		75	.213		15.40	8.45		23.85	30.50
6120	2" wall, 1/2" iron pipe size	G		145	.110		3.89	4.38		8.27	11.40
6140	3/4" iron pipe size	G		145	.110		4.10	4.38		8.48	11.60
6150	1" iron pipe size	G		145	.110		4.32	4.38		8.70	11.85
6160	1-1/4" iron pipe size	G		140	.114		4.63	4.54		9.17	12.45
6170	1-1/2" iron pipe size	G		140	.114		4.83	4.54		9.37	12.65
6180	2" iron pipe size	G		135	.119		5.05	4.71		9.76	13.20
6190	2-1/2" iron pipe size	G		135	.119		6.10	4.71		10.81	14.30
6200	3" iron pipe size	G		130	.123		6.10	4.89		10.99	14.65
6210	4" iron pipe size	G		125	.128		7.10	5.10		12.20	16
6220	5" iron pipe size	G		120	.133		8.10	5.30		13.40	17.45
6230	6" iron pipe size	G		115	.139		8.85	5.50		14.35	18.65
6240	7" iron pipe size	G		110	.145		9.55	5.75		15.30	19.90
6250	8" iron pipe size	G		105	.152		10.50	6.05		16.55	21.50
6260	9" iron pipe size	G		100	.160		11.90	6.35		18.25	23.50
6270	10" iron pipe size	G		95	.168		13.20	6.70		19.90	25.50
6280	12" iron pipe size	G		90	.178		14.70	7.05		21.75	27.50
6290	14" iron pipe size	G		85	.188		16.15	7.45		23.60	30
6300	16" iron pipe size	G		80	.200		17.80	7.95		25.75	32.50
6310	18" iron pipe size	G		75	.213		19.40	8.45		27.85	35
6320	20" iron pipe size	G		65	.246		24	9.75		33.75	42.50
6330	22" iron pipe size	G		60	.267		26.50	10.60		37.10	46
6340	24" iron pipe size	G		55	.291		27	11.55		38.55	48.50
6360	3" wall, 1/2" iron pipe size	G		115	.139		7.05	5.50		12.55	16.70
6380	3/4" iron pipe size	G		115	.139		7.10	5.50		12.60	16.75
6390	1" iron pipe size	G		115	.139		7.15	5.50		12.65	16.80
6400	1-1/4" iron pipe size	G	↓	110	.145	↓	7.25	5.75		13	17.35

22 07 19.10 Piping Insulation			Crew	Daily Output	Labor-Hours	Unit	Material	Labor	Equipment	Total	Total Incl O&P
							2009 Bare Costs				
6410	1-1/2" iron pipe size	G	Q-14	110	.145	L.F.	7.30	5.75		13.05	17.40
6420	2" iron pipe size	G		105	.152		7.55	6.05		13.60	18.10
6430	2-1/2" iron pipe size	G		105	.152		8.85	6.05		14.90	19.55
6440	3" iron pipe size	G		100	.160		8.95	6.35		15.30	20
6450	4" iron pipe size	G		95	.168		11.40	6.70		18.10	23.50
6460	5" iron pipe size	G		90	.178		12.40	7.05		19.45	25
6470	6" iron pipe size	G		90	.178		13.95	7.05		21	26.50
6480	7" iron pipe size	G		85	.188		15.50	7.45		22.95	29
6490	8" iron pipe size	G		85	.188		16.60	7.45		24.05	30.50
6500	9" iron pipe size	G		80	.200		18.80	7.95		26.75	33.50
6510	10" iron pipe size	G		75	.213		20	8.45		28.45	35.50
6520	12" iron pipe size	G		70	.229		22	9.05		31.05	39
6530	14" iron pipe size	G		65	.246		25	9.75		34.75	43.50
6540	16" iron pipe size	G		60	.267		27.50	10.60		38.10	47
6550	18" iron pipe size	G		55	.291		30	11.55		41.55	51.50
6560	20" iron pipe size	G		50	.320		36	12.70		48.70	60
6570	22" iron pipe size	G		45	.356		39	14.10		53.10	65.50
6580	24" iron pipe size	G	↓	40	.400	↓	42	15.90		57.90	71.50
6600	Fiberglass, with all service jacket										
6640	1/2" wall, 1/2" iron pipe size	G	Q-14	250	.064	L.F.	.73	2.54		3.27	4.91
6660	3/4" iron pipe size	G		240	.067		.83	2.65		3.48	5.20
6670	1" iron pipe size	G		230	.070		.86	2.76		3.62	5.40
6680	1-1/4" iron pipe size	G		220	.073		.91	2.89		3.80	5.65
6690	1-1/2" iron pipe size	G		220	.073		1.05	2.89		3.94	5.85
6700	2" iron pipe size	G		210	.076		1.12	3.02		4.14	6.10
6710	2-1/2" iron pipe size	G		200	.080		1.17	3.18		4.35	6.45
6840	1" wall, 1/2" iron pipe size	G		240	.067		.89	2.65		3.54	5.25
6860	3/4" iron pipe size	G		230	.070		.97	2.76		3.73	5.55
6870	1" iron pipe size	G		220	.073		1.04	2.89		3.93	5.80
6880	1-1/4" iron pipe size	G		210	.076		1.12	3.02		4.14	6.10
6890	1-1/2" iron pipe size	G		210	.076		1.21	3.02		4.23	6.20
6900	2" iron pipe size	G		200	.080		1.31	3.18		4.49	6.60
6910	2-1/2" iron pipe size	G		190	.084		1.49	3.34		4.83	7.05
6920	3" iron pipe size	G		180	.089		1.59	3.53		5.12	7.45
6930	3-1/2" iron pipe size	G		170	.094		1.73	3.74		5.47	7.95
6940	4" iron pipe size	G		150	.107		2.11	4.23		6.34	9.15
6950	5" iron pipe size	G		140	.114		2.38	4.54		6.92	9.95
6960	6" iron pipe size	G		120	.133		2.52	5.30		7.82	11.30
6970	7" iron pipe size	G		110	.145		3	5.75		8.75	12.65
6980	8" iron pipe size	G		100	.160		4.12	6.35		10.47	14.80
6990	9" iron pipe size	G		90	.178		4.34	7.05		11.39	16.15
7000	10" iron pipe size	G		90	.178		4.37	7.05		11.42	16.20
7010	12" iron pipe size	G		80	.200		4.76	7.95		12.71	18.10
7020	14" iron pipe size	G		80	.200		5.50	7.95		13.45	18.90
7030	16" iron pipe size	G		70	.229		7	9.05		16.05	22.50
7040	18" iron pipe size	G		70	.229		7.75	9.05		16.80	23
7050	20" iron pipe size	G		60	.267		8.65	10.60		19.25	26.50
7060	24" iron pipe size	G		60	.267		10.55	10.60		21.15	28.50
7080	1-1/2" wall, 1/2" iron pipe size	G		230	.070		1.67	2.76		4.43	6.30
7100	3/4" iron pipe size	G		220	.073		1.68	2.89		4.57	6.50
7110	1" iron pipe size	G		210	.076		1.80	3.02		4.82	6.85
7120	1-1/4" iron pipe size	G		200	.080		1.96	3.18		5.14	7.30
7130	1-1/2" iron pipe size	G	↓	200	.080	↓	2.05	3.18		5.23	7.40

22 07 19.10 Piping Insulation		Crew	Daily Output	Labor-Hours	Unit	Material	Labor	Equipment	Total	Total Incl O&P
						2009 Bare Costs				
7140 2" iron pipe size	G	Q-14	190	.084	L.F.	2.26	3.34		5.60	7.90
7150 2-1/2" iron pipe size	G		180	.089		2.42	3.53		5.95	8.35
7160 3" iron pipe size	G		170	.094		2.54	3.74		6.28	8.85
7170 3-1/2" iron pipe size	G		160	.100		2.78	3.97		6.75	9.45
7180 4" iron pipe size	G		140	.114		2.88	4.54		7.42	10.50
7190 5" iron pipe size	G		130	.123		3.23	4.89		8.12	11.45
7200 6" iron pipe size	G		110	.145		3.41	5.75		9.16	13.10
7210 7" iron pipe size	G		100	.160		3.79	6.35		10.14	14.40
7220 8" iron pipe size	G		90	.178		4.78	7.05		11.83	16.65
7230 9" iron pipe size	G		85	.188		4.98	7.45		12.43	17.60
7240 10" iron pipe size	G		80	.200		5.15	7.95		13.10	18.50
7250 12" iron pipe size	G		75	.213		5.85	8.45		14.30	20
7260 14" iron pipe size	G		70	.229		6.70	9.05		15.75	22
7270 16" iron pipe size	G		65	.246		8.75	9.75		18.50	25.50
7280 18" iron pipe size	G		60	.267		9.85	10.60		20.45	28
7290 20" iron pipe size	G		55	.291		10.05	11.55		21.60	30
7300 24" iron pipe size	G		50	.320		12.30	12.70		25	34
7320 2" wall, 1/2" iron pipe size	G		220	.073		2.60	2.89		5.49	7.55
7340 3/4" iron pipe size	G		210	.076		2.68	3.02		5.70	7.85
7350 1" iron pipe size	G		200	.080		2.86	3.18		6.04	8.30
7360 1-1/4" iron pipe size	G		190	.084		3.02	3.34		6.36	8.70
7370 1-1/2" iron pipe size	G		190	.084		3.14	3.34		6.48	8.85
7380 2" iron pipe size	G		180	.089		3.32	3.53		6.85	9.35
7390 2-1/2" iron pipe size	G		170	.094		3.57	3.74		7.31	10
7400 3" iron pipe size	G		160	.100		3.79	3.97		7.76	10.55
7410 3-1/2" iron pipe size	G		150	.107		4.10	4.23		8.33	11.35
7420 4" iron pipe size	G		130	.123		4.42	4.89		9.31	12.75
7430 5" iron pipe size	G		120	.133		5.05	5.30		10.35	14.10
7440 6" iron pipe size	G		100	.160		5.20	6.35		11.55	16
7450 7" iron pipe size	G		90	.178		5.95	7.05		13	17.90
7460 8" iron pipe size	G		80	.200		6.35	7.95		14.30	19.80
7470 9" iron pipe size	G		75	.213		6.95	8.45		15.40	21.50
7480 10" iron pipe size	G		70	.229		7.60	9.05		16.65	23
7490 12" iron pipe size	G		65	.246		8.50	9.75		18.25	25
7500 14" iron pipe size	G		60	.267		10.55	10.60		21.15	28.50
7510 16" iron pipe size	G		55	.291		11.60	11.55		23.15	31.50
7520 18" iron pipe size	G		50	.320		12.95	12.70		25.65	35
7530 20" iron pipe size	G		45	.356		14.65	14.10		28.75	39
7540 24" iron pipe size	G		40	.400		15.75	15.90		31.65	43
7560 2-1/2" wall, 1/2" iron pipe size	G		210	.076		3.08	3.02		6.10	8.30
7562 3/4" iron pipe size	G		200	.080		3.22	3.18		6.40	8.70
7564 1" iron pipe size	G		190	.084		3.35	3.34		6.69	9.10
7566 1-1/4" iron pipe size	G		185	.086		3.47	3.43		6.90	9.35
7568 1-1/2" iron pipe size	G		180	.089		3.65	3.53		7.18	9.70
7570 2" iron pipe size	G		170	.094		3.83	3.74		7.57	10.25
7572 2-1/2" iron pipe size	G		160	.100		4.41	3.97		8.38	11.25
7574 3" iron pipe size	G		150	.107		4.64	4.23		8.87	11.95
7576 3-1/2" iron pipe size	G		140	.114		5.05	4.54		9.59	12.90
7578 4" iron pipe size	G		120	.133		5.35	5.30		10.65	14.40
7580 5" iron pipe size	G		110	.145		7.10	5.75		12.85	17.15
7582 6" iron pipe size	G		90	.178		7.65	7.05		14.70	19.80
7584 7" iron pipe size	G		80	.200		7.70	7.95		15.65	21.50
7586 8" iron pipe size	G	↓	70	.229	↓	8.15	9.05		17.20	23.50

22 07 19.10 Piping Insulation			Crew	Daily Output	Labor-Hours	Unit	Material	2009 Bare Costs Labor	Equipment	Total	Total Incl O&P
7588	9" iron pipe size	G	Q-14	65	.246	L.F.	8.95	9.75		18.70	25.50
7590	10" iron pipe size	G		60	.267		9.75	10.60		20.35	28
7592	12" iron pipe size	G		55	.291		11.80	11.55		23.35	31.50
7594	14" iron pipe size	G		50	.320		13.30	12.70		26	35
7596	16" iron pipe size	G		45	.356		15.25	14.10		29.35	40
7598	18" iron pipe size	G		40	.400		16.55	15.90		32.45	43.50
7602	24" iron pipe size	G		30	.533		22	21		43	58
7620	3" wall, 1/2" iron pipe size	G		200	.080		3.94	3.18		7.12	9.50
7622	3/4" iron pipe size	G		190	.084		4.17	3.34		7.51	10
7624	1" iron pipe size	G		180	.089		4.41	3.53		7.94	10.55
7626	1-1/4" iron pipe size	G		175	.091		4.50	3.63		8.13	10.80
7628	1-1/2" iron pipe size	G		170	.094		4.70	3.74		8.44	11.20
7630	2" iron pipe size	G		160	.100		5.10	3.97		9.07	12
7632	2-1/2" iron pipe size	G		150	.107		5.30	4.23		9.53	12.70
7634	3" iron pipe size	G		140	.114		5.65	4.54		10.19	13.55
7636	3-1/2" iron pipe size	G		130	.123		6.20	4.89		11.09	14.75
7638	4" iron pipe size	G		110	.145		6.70	5.75		12.45	16.70
7640	5" iron pipe size	G		100	.160		7.55	6.35		13.90	18.55
7642	6" iron pipe size	G		80	.200		8.10	7.95		16.05	22
7644	7" iron pipe size	G		70	.229		9.25	9.05		18.30	25
7646	8" iron pipe size	G		60	.267		10.05	10.60		20.65	28
7648	9" iron pipe size	G		55	.291		10.85	11.55		22.40	30.50
7650	10" iron pipe size	G		50	.320		11.65	12.70		24.35	33.50
7652	12" iron pipe size	G		45	.356		14.50	14.10		28.60	39
7654	14" iron pipe size	G		40	.400		16.10	15.90		32	43
7656	16" iron pipe size	G		35	.457		18.25	18.15		36.40	49.50
7658	18" iron pipe size	G		32	.500		19.55	19.85		39.40	53.50
7660	20" iron pipe size	G		30	.533		21.50	21		42.50	57.50
7662	24" iron pipe size	G		28	.571		27	22.50		49.50	66
7664	26" iron pipe size	G		26	.615		29.50	24.50		54	72
7666	30" iron pipe size	G		24	.667		34	26.50		60.50	80.50
7800	For fiberglass with standard canvas jacket, deduct						5%				
7802	For fittings, add 3 L.F. for each fitting										
7804	plus 4 L.F. for each flange of the fitting										
7810	Finishes										
7814	For single layer of felt, add						10%	10%			
7816	For roofing paper, 45 lb to 55 lb, add						25%	10%			
7879	Rubber tubing, flexible closed cell foam										
7880	3/8" wall, 1/4" iron pipe size	G	1 Asbe	120	.067	L.F.	.55	2.94		3.49	5.35
7900	3/8" iron pipe size	G		120	.067		.68	2.94		3.62	5.50
7910	1/2" iron pipe size	G		115	.070		.70	3.07		3.77	5.75
7920	3/4" iron pipe size	G		115	.070		.85	3.07		3.92	5.90
7930	1" iron pipe size	G		110	.073		.98	3.21		4.19	6.30
7940	1-1/4" iron pipe size	G		110	.073		1.20	3.21		4.41	6.50
7950	1-1/2" iron pipe size	G		110	.073		1.30	3.21		4.51	6.65
8100	1/2" wall, 1/4" iron pipe size	G		90	.089		.44	3.92		4.36	6.85
8120	3/8" iron pipe size	G		90	.089		.49	3.92		4.41	6.90
8130	1/2" iron pipe size	G		89	.090		.54	3.96		4.50	7
8140	3/4" iron pipe size	G		89	.090		.60	3.96		4.56	7.05
8150	1" iron pipe size	G		88	.091		.67	4.01		4.68	7.25
8160	1-1/4" iron pipe size	G		87	.092		.77	4.06		4.83	7.40
8170	1-1/2" iron pipe size	G		87	.092		.94	4.06		5	7.60
8180	2" iron pipe size	G		86	.093		1.22	4.10		5.32	8

22 07 19.10 Piping Insulation			Crew	Daily Output	Labor-Hours	Unit	Material	Labor	Equipment	Total	Total Incl O&P
8190	2-1/2" iron pipe size	G	1 Asbe	86	.093	L.F.	1.54	4.10		5.64	8.35
8200	3" iron pipe size	G		85	.094		1.72	4.15		5.87	8.60
8210	3-1/2" iron pipe size	G		85	.094		2.39	4.15		6.54	9.35
8220	4" iron pipe size	G		80	.100		2.55	4.41		6.96	9.95
8230	5" iron pipe size	G		80	.100		3.51	4.41		7.92	11
8240	6" iron pipe size	G		75	.107		3.51	4.70		8.21	11.45
8300	3/4" wall, 1/4" iron pipe size	G		90	.089		.68	3.92		4.60	7.10
8320	3/8" iron pipe size	G		90	.089		.75	3.92		4.67	7.20
8330	1/2" iron pipe size	G		89	.090		.90	3.96		4.86	7.40
8340	3/4" iron pipe size	G		89	.090		1.09	3.96		5.05	7.60
8350	1" iron pipe size	G		88	.091		1.25	4.01		5.26	7.90
8360	1-1/4" iron pipe size	G		87	.092		1.68	4.06		5.74	8.40
8370	1-1/2" iron pipe size	G		87	.092		1.90	4.06		5.96	8.65
8380	2" iron pipe size	G		86	.093		2.23	4.10		6.33	9.10
8390	2-1/2" iron pipe size	G		86	.093		2.98	4.10		7.08	9.95
8400	3" iron pipe size	G		85	.094		3.40	4.15		7.55	10.45
8410	3-1/2" iron pipe size	G		85	.094		4.05	4.15		8.20	11.15
8420	4" iron pipe size	G		80	.100		4.31	4.41		8.72	11.90
8430	5" iron pipe size	G		80	.100		5.20	4.41		9.61	12.85
8440	6" iron pipe size	G		80	.100		6.30	4.41		10.71	14.05
8444	1" wall, 1/2" iron pipe size	G		86	.093		1.71	4.10		5.81	8.55
8445	3/4" iron pipe size	G		84	.095		2.07	4.20		6.27	9.10
8446	1" iron pipe size	G		84	.095		2.41	4.20		6.61	9.45
8447	1-1/4" iron pipe size	G		82	.098		2.72	4.30		7.02	9.95
8448	1-1/2" iron pipe size	G		82	.098		3.16	4.30		7.46	10.45
8449	2" iron pipe size	G		80	.100		4.23	4.41		8.64	11.80
8450	2-1/2" iron pipe size	G	↓	80	.100	↓	5.50	4.41		9.91	13.20
8456	Rubber insulation tape, 1/8" x 2" x 30'	G				Ea.	11.40			11.40	12.50
9600	Minimum labor/equipment charge		1 Plum	4	2	Job		97.50		97.50	151

22 07 19.30 Piping Insulation Protective Jacketing, PVC		Crew	Daily Output	Labor-Hours	Unit	Material	Labor	Equipment	Total	Total Incl O&P
0010	**PIPING INSULATION PROTECTIVE JACKETING, PVC**									
0100	PVC, white, 48" lengths cut from roll goods									
0120	20 mil thick									
0140	Size based on OD of insulation									
0150	1-1/2" ID	Q-14	270	.059	L.F.	.26	2.35		2.61	4.09
0152	2" ID		260	.062		.32	2.44		2.76	4.30
0154	2-1/2" ID		250	.064		.40	2.54		2.94	4.55
0156	3" ID		240	.067		.47	2.65		3.12	4.80
0158	3-1/2" ID		230	.070		.54	2.76		3.30	5.05
0160	4" ID		220	.073		.60	2.89		3.49	5.35
0162	4-1/2" ID		210	.076		.67	3.02		3.69	5.65
0164	5" ID		200	.080		.75	3.18		3.93	6
0166	5-1/2" ID		190	.084		.82	3.34		4.16	6.30
0168	6" ID		180	.089		.88	3.53		4.41	6.65
0170	6-1/2" ID		175	.091		.95	3.63		4.58	6.90
0172	7" ID		170	.094		1.03	3.74		4.77	7.20
0174	7-1/2" ID		164	.098		1.10	3.87		4.97	7.45
0176	8" ID		161	.099		1.18	3.95		5.13	7.70
0178	8-1/2" ID		158	.101		1.24	4.02		5.26	7.85
0180	9" ID		155	.103		1.31	4.10		5.41	8.10
0182	9-1/2" ID		152	.105		1.38	4.18		5.56	8.25
0184	10" ID	↓	149	.107	↓	1.46	4.26		5.72	8.50

Material, Labor, Equipment and Total columns: 2009 Bare Costs

22 07 19.30 Piping Insulation Protective Jacketing, PVC		Crew	Daily Output	Labor-Hours	Unit	2009 Bare Costs Material	Labor	Equipment	Total	Total Incl O&P
0186	10-1/2" ID	Q-14	146	.110	L.F.	1.52	4.35		5.87	8.70
0188	11" ID		143	.112		1.59	4.44		6.03	8.95
0190	11-1/2" ID		140	.114		1.66	4.54		6.20	9.20
0192	12" ID		137	.117		1.74	4.64		6.38	9.40
0194	12-1/2" ID		134	.119		1.80	4.74		6.54	9.65
0195	13" ID		132	.121		1.88	4.81		6.69	9.85
0196	13-1/2" ID		132	.121		1.95	4.81		6.76	9.95
0198	14" ID		130	.123		2.02	4.89		6.91	10.10
0200	15" ID		128	.125		2.15	4.96		7.11	10.40
0202	16" ID		126	.127		2.30	5.05		7.35	10.70
0204	17" ID		124	.129		2.43	5.10		7.53	10.95
0206	18" ID		122	.131		2.58	5.20		7.78	11.25
0208	19" ID		120	.133		2.72	5.30		8.02	11.55
0210	20" ID		118	.136		2.86	5.40		8.26	11.85
0212	21" ID		116	.138		3	5.50		8.50	12.15
0214	22" ID		114	.140		3.14	5.55		8.69	12.45
0216	23" ID		112	.143		3.28	5.65		8.93	12.75
0218	24" ID		110	.145		3.43	5.75		9.18	13.10
0220	25" ID		108	.148		3.56	5.90		9.46	13.40
0222	26" ID		106	.151		3.71	6		9.71	13.80
0224	27" ID		104	.154		3.84	6.10		9.94	14.10
0226	28" ID		102	.157		3.99	6.25		10.24	14.45
0228	29" ID		100	.160		4.13	6.35		10.48	14.80
0230	30" ID		98	.163		4.27	6.50		10.77	15.20
0300	For colors, add				Ea.	10%				
1000	30 mil thick									
1010	Size based on OD of insulation									
1020	2" ID	Q-14	260	.062	L.F.	.50	2.44		2.94	4.50
1022	2-1/2" ID		250	.064		.60	2.54		3.14	4.77
1024	3" ID		240	.067		.72	2.65		3.37	5.05
1026	3-1/2" ID		230	.070		.82	2.76		3.58	5.35
1028	4" ID		220	.073		.92	2.89		3.81	5.70
1030	4-1/2" ID		210	.076		1.03	3.02		4.05	6
1032	5" ID		200	.080		1.13	3.18		4.31	6.40
1034	5-1/2" ID		190	.084		1.24	3.34		4.58	6.75
1036	6" ID		180	.089		1.34	3.53		4.87	7.15
1038	6-1/2" ID		175	.091		1.46	3.63		5.09	7.45
1040	7" ID		170	.094		1.56	3.74		5.30	7.75
1042	7-1/2" ID		164	.098		1.68	3.87		5.55	8.10
1044	8" ID		161	.099		1.78	3.95		5.73	8.35
1046	8-1/2" ID		158	.101		1.89	4.02		5.91	8.60
1048	9" ID		155	.103		1.99	4.10		6.09	8.85
1050	9-1/2" ID		152	.105		2.11	4.18		6.29	9.05
1052	10" ID		149	.107		2.21	4.26		6.47	9.35
1054	10-1/2" ID		146	.110		2.32	4.35		6.67	9.60
1056	11" ID		143	.112		2.42	4.44		6.86	9.85
1058	11-1/2" ID		140	.114		2.53	4.54		7.07	10.15
1060	12" ID		137	.117		2.63	4.64		7.27	10.40
1062	12-1/2" ID		134	.119		2.74	4.74		7.48	10.65
1063	13" ID		132	.121		2.86	4.81		7.67	10.95
1064	13-1/2" ID		132	.121		2.96	4.81		7.77	11.05
1066	14" ID		130	.123		3.07	4.89		7.96	11.30
1068	15" ID		128	.125		3.28	4.96		8.24	11.65

22 07 19.30 Piping Insulation Protective Jacketing, PVC		Crew	Daily Output	Labor-Hours	Unit	Material	Labor	Equipment	Total	Total Incl O&P
						2009 Bare Costs				
1070	16″ ID	Q-14	126	.127	L.F.	3.49	5.05		8.54	12
1072	17″ ID		124	.129		3.71	5.10		8.81	12.40
1074	18″ ID		122	.131		3.92	5.20		9.12	12.70
1076	19″ ID		120	.133		4.13	5.30		9.43	13.10
1078	20″ ID		118	.136		4.35	5.40		9.75	13.50
1080	21″ ID		116	.138		4.56	5.50		10.06	13.85
1082	22″ ID		114	.140		4.77	5.55		10.32	14.25
1084	23″ ID		112	.143		4.98	5.65		10.63	14.65
1086	24″ ID		110	.145		5.20	5.75		10.95	15.10
1088	25″ ID		108	.148		5.40	5.90		11.30	15.45
1090	26″ ID		106	.151		5.65	6		11.65	15.90
1092	27″ ID		104	.154		5.85	6.10		11.95	16.35
1094	28″ ID		102	.157		6.05	6.25		12.30	16.70
1096	29″ ID		100	.160		6.30	6.35		12.65	17.15
1098	30″ ID	↓	98	.163	↓	6.50	6.50		13	17.65
1300	For colors, add				Ea.	10%				
2000	PVC, white, fitting covers									
2020	Fiberglass insulation inserts included with sizes 1-3/4″ thru 9-3/4″									
2030	Size is based on OD of insulation									
2040	90° Elbow fitting									
2060	1-3/4″	Q-14	135	.119	Ea.	.49	4.71		5.20	8.15
2062	2″		130	.123		.60	4.89		5.49	8.55
2064	2-1/4″		128	.125		.66	4.96		5.62	8.80
2068	2-1/2″		126	.127		.74	5.05		5.79	8.95
2070	2-3/4″		123	.130		.82	5.15		5.97	9.25
2072	3″		120	.133		.83	5.30		6.13	9.45
2074	3-3/8″		116	.138		.95	5.50		6.45	9.90
2076	3-3/4″		113	.142		1.08	5.60		6.68	10.30
2078	4-1/8″		110	.145		1.34	5.75		7.09	10.80
2080	4-3/4″		105	.152		1.61	6.05		7.66	11.55
2082	5-1/4″		100	.160		1.88	6.35		8.23	12.30
2084	5-3/4″		95	.168		2.52	6.70		9.22	13.55
2086	6-1/4″		90	.178		4.24	7.05		11.29	16.05
2088	6-3/4″		87	.184		4.49	7.30		11.79	16.75
2090	7-1/4″		85	.188		5.60	7.45		13.05	18.30
2092	7-3/4″		83	.193		5.80	7.65		13.45	18.80
2094	8-3/4″		80	.200		7.60	7.95		15.55	21
2096	9-3/4″		77	.208		10.15	8.25		18.40	24.50
2098	10-7/8″		74	.216		11.30	8.60		19.90	26.50
2100	11-7/8″		71	.225		12.95	8.95		21.90	28.50
2102	12-7/8″		68	.235		16.70	9.35		26.05	33.50
2104	14-1/8″		66	.242		18.55	9.60		28.15	36
2106	15-1/8″		64	.250		20	9.95		29.95	38
2108	16-1/8″		63	.254		22	10.10		32.10	40.50
2110	17-1/8″		62	.258		24.50	10.25		34.75	43.50
2112	18-1/8″		61	.262		30.50	10.40		40.90	50.50
2114	19-1/8″		60	.267		42.50	10.60		53.10	64
2116	20-1/8″	↓	59	.271	↓	55	10.75		65.75	78
2200	45° Elbow fitting									
2220	1-3/4″ thru 9-3/4″ same price as 90° Elbow fitting									
2320	10-7/8″	Q-14	74	.216	Ea.	11.30	8.60		19.90	26.50
2322	11-7/8″		71	.225		12.50	8.95		21.45	28
2324	12-7/8″	↓	68	.235	↓	13.90	9.35		23.25	30.50

22 07 Plumbing Insulation

22 07 19 – Plumbing Piping Insulation

22 07 19.30 Piping Insulation Protective Jacketing, PVC		Crew	Daily Output	Labor-Hours	Unit	Material	2009 Bare Costs Labor	Equipment	Total	Total Incl O&P
2326	14-1/8"	Q-14	66	.242	Ea.	15.85	9.60		25.45	33
2328	15-1/8"		64	.250		17.05	9.95		27	35
2330	16-1/8"		63	.254		19.50	10.10		29.60	38
2332	17-1/8"		62	.258		22	10.25		32.25	40.50
2334	18-1/8"		61	.262		27	10.40		37.40	46.50
2336	19-1/8"		60	.267		36.50	10.60		47.10	57
2338	20-1/8"	↓	59	.271	↓	41.50	10.75		52.25	63
2400	Tee fitting									
2410	1-3/4"	Q-14	96	.167	Ea.	.93	6.60		7.53	11.70
2412	2"		94	.170		1.03	6.75		7.78	12.10
2414	2-1/4"		91	.176		1.15	7		8.15	12.55
2416	2-1/2"		88	.182		1.25	7.20		8.45	13.05
2418	2-3/4"		85	.188		1.37	7.45		8.82	13.60
2420	3"		82	.195		1.49	7.75		9.24	14.20
2422	3-3/8"		79	.203		1.71	8.05		9.76	14.90
2424	3-3/4"		76	.211		1.96	8.35		10.31	15.65
2426	4-1/8"		73	.219		2.30	8.70		11	16.60
2428	4-3/4"		70	.229		2.88	9.05		11.93	17.80
2430	5-1/4"		67	.239		3.45	9.50		12.95	19.15
2432	5-3/4"		63	.254		4.57	10.10		14.67	21.50
2434	6-1/4"		60	.267		6	10.60		16.60	23.50
2436	6-3/4"		59	.271		7.45	10.75		18.20	25.50
2438	7-1/4"		57	.281		12.05	11.15		23.20	31.50
2440	7-3/4"		54	.296		13.20	11.75		24.95	33.50
2442	8-3/4"		52	.308		16.05	12.20		28.25	37.50
2444	9-3/4"		50	.320		18.95	12.70		31.65	41.50
2446	10-7/8"		48	.333		19.15	13.25		32.40	42.50
2448	11-7/8"		47	.340		21.50	13.50		35	45.50
2450	12-7/8"		46	.348		23.50	13.80		37.30	48.50
2452	14-1/8"		45	.356		25.50	14.10		39.60	51
2454	15-1/8"		44	.364		27.50	14.45		41.95	54
2456	16-1/8"		43	.372		30	14.75		44.75	57
2458	17-1/8"		42	.381		32	15.10		47.10	59.50
2460	18-1/8"		41	.390		35	15.50		50.50	63.50
2462	19-1/8"		40	.400		38.50	15.90		54.40	67.50
2464	20-1/8"	↓	39	.410	↓	41.50	16.30		57.80	72
4000	Mechanical grooved fitting cover, including insert									
4020	90° Elbow fitting									
4030	3/4" & 1"	Q-14	140	.114	Ea.	6.75	4.54		11.29	14.80
4040	1-1/4" & 1-1/2"		135	.119		7.05	4.71		11.76	15.35
4042	2"		130	.123		7.75	4.89		12.64	16.40
4044	2-1/2"		125	.128		8.60	5.10		13.70	17.65
4046	3"		120	.133		9.65	5.30		14.95	19.20
4048	3-1/2"		115	.139		11.20	5.50		16.70	21.50
4050	4"		110	.145		12.50	5.75		18.25	23
4052	5"		100	.160		15.45	6.35		21.80	27.50
4054	6"		90	.178		22	7.05		29.05	35.50
4056	8"		80	.200		25	7.95		32.95	40.50
4058	10"		75	.213		32	8.45		40.45	48.50
4060	12"		68	.235		46	9.35		55.35	65.50
4062	14"		65	.246		57	9.75		66.75	78.50
4064	16"		63	.254		77.50	10.10		87.60	101
4066	18"	↓	61	.262	↓	106	10.40		116.40	134

22 07 19.30 Piping Insulation Protective Jacketing, PVC		Crew	Daily Output	Labor-Hours	Unit	Material	2009 Bare Costs Labor	Equipment	Total	Total Incl O&P
4100	45° Elbow fitting									
4120	3/4" & 1"	Q-14	140	.114	Ea.	6.45	4.54		10.99	14.45
4130	1-1/4" & 1-1/2"		135	.119		6.75	4.71		11.46	15.05
4140	2"		130	.123		7.10	4.89		11.99	15.70
4142	2-1/2"		125	.128		7.85	5.10		12.95	16.85
4144	3"		120	.133		8.65	5.30		13.95	18.10
4146	3-1/2"		115	.139		10.10	5.50		15.60	20
4148	4"		110	.145		10.45	5.75		16.20	21
4150	5"		100	.160		13.95	6.35		20.30	25.50
4152	6"		90	.178		19.40	7.05		26.45	33
4154	8"		80	.200		22	7.95		29.95	37
4156	10"		75	.213		28	8.45		36.45	44
4158	12"		68	.235		39.50	9.35		48.85	58.50
4160	14"		65	.246		44.50	9.75		54.25	65
4162	16"		63	.254		65.50	10.10		75.60	89
4164	18"	↓	61	.262	↓	94	10.40		104.40	120
4200	Tee fitting									
4220	3/4" & 1"	Q-14	93	.172	Ea.	9.30	6.85		16.15	21.50
4230	1-1/4" & 1-1/2"		90	.178		9.90	7.05		16.95	22.50
4240	2"		87	.184		10.95	7.30		18.25	24
4242	2-1/2"		84	.190		12.20	7.55		19.75	25.50
4244	3"		80	.200		13.10	7.95		21.05	27.50
4246	3-1/2"		77	.208		15.25	8.25		23.50	30
4248	4"		73	.219		17.95	8.70		26.65	34
4250	5"		67	.239		21	9.50		30.50	38.50
4252	6"		60	.267		25	10.60		35.60	44.50
4254	8"		54	.296		31	11.75		42.75	53
4256	10"		50	.320		39.50	12.70		52.20	64
4258	12"		46	.348		52.50	13.80		66.30	80.50
4260	14"		43	.372		67	14.75		81.75	97.50
4262	16"		42	.381		89	15.10		104.10	123
4264	18"	↓	41	.390	↓	118	15.50		133.50	154

22 07 19.40 Pipe Insulation Protective Jacketing, Aluminum		Crew	Daily Output	Labor-Hours	Unit	Material	Labor	Equipment	Total	Total Incl O&P
0010	**PIPE INSULATION PROTECTIVE JACKETING, ALUMINUM**									
0100	Metal roll jacketing									
0120	Aluminum with polykraft moisture barrier									
0140	Smooth, based on OD of insulation, .010" thick									
0180	1/2" ID	Q-14	220	.073	L.F.	.26	2.89		3.15	4.96
0190	3/4" ID		215	.074		.35	2.95		3.30	5.15
0200	1" ID		210	.076		.43	3.02		3.45	5.35
0210	1-1/4" ID		205	.078		.51	3.10		3.61	5.55
0220	1-1/2" ID		202	.079		.59	3.14		3.73	5.75
0230	1-3/4" ID		199	.080		.67	3.19		3.86	5.90
0240	2" ID		195	.082		.75	3.26		4.01	6.10
0250	2-1/4" ID		191	.084		.83	3.33		4.16	6.30
0260	2-1/2" ID		187	.086		.92	3.40		4.32	6.50
0270	2-3/4" ID		184	.087		1	3.45		4.45	6.70
0280	3" ID		180	.089		1.07	3.53		4.60	6.90
0290	3-1/4" ID		176	.091		1.16	3.61		4.77	7.15
0300	3-1/2" ID		172	.093		1.24	3.69		4.93	7.30
0310	3-3/4" ID		169	.095		1.32	3.76		5.08	7.55
0320	4" ID	↓	165	.097	↓	1.40	3.85		5.25	7.80

22 07 19.40 Pipe Insulation Protective Jacketing, Aluminum		Crew	Daily Output	Labor-Hours	Unit	Material	2009 Bare Costs Labor	Equipment	Total	Total Incl O&P
0330	4-1/4" ID	Q-14	161	.099	L.F.	1.48	3.95		5.43	8.05
0340	4-1/2" ID		157	.102		1.56	4.05		5.61	8.25
0350	4-3/4" ID		154	.104		1.64	4.12		5.76	8.45
0360	5" ID		150	.107		1.73	4.23		5.96	8.75
0370	5-1/4" ID		146	.110		1.80	4.35		6.15	9.05
0380	5-1/2" ID		143	.112		1.88	4.44		6.32	9.25
0390	5-3/4" ID		139	.115		1.97	4.57		6.54	9.55
0400	6" ID		135	.119		2.05	4.71		6.76	9.85
0410	6-1/4" ID		133	.120		2.13	4.78		6.91	10.05
0420	6-1/2" ID		131	.122		2.21	4.85		7.06	10.30
0430	7" ID		128	.125		2.37	4.96		7.33	10.65
0440	7-1/4" ID		125	.128		2.45	5.10		7.55	10.90
0450	7-1/2" ID		123	.130		2.54	5.15		7.69	11.15
0460	8" ID		121	.132		2.69	5.25		7.94	11.45
0470	8-1/2" ID		119	.134		2.86	5.35		8.21	11.80
0480	9" ID		116	.138		3.02	5.50		8.52	12.15
0490	9-1/2" ID		114	.140		3.18	5.55		8.73	12.50
0500	10" ID		112	.143		3.35	5.65		9	12.85
0510	10-1/2" ID		110	.145		3.50	5.75		9.25	13.20
0520	11" ID		107	.150		3.67	5.95		9.62	13.65
0530	11-1/2" ID		105	.152		3.83	6.05		9.88	14
0540	12" ID		103	.155		3.99	6.15		10.14	14.35
0550	12-1/2" ID		100	.160		4.16	6.35		10.51	14.85
0560	13" ID		99	.162		4.31	6.40		10.71	15.15
0570	14" ID		98	.163		4.64	6.50		11.14	15.60
0580	15" ID		96	.167		4.97	6.60		11.57	16.15
0590	16" ID		95	.168		5.30	6.70		12	16.60
0600	17" ID		93	.172		5.60	6.85		12.45	17.20
0610	18" ID		92	.174		5.85	6.90		12.75	17.55
0620	19" ID		90	.178		6.25	7.05		13.30	18.30
0630	20" ID		89	.180		6.60	7.15		13.75	18.80
0640	21" ID		87	.184		6.90	7.30		14.20	19.40
0650	22" ID		86	.186		7.25	7.40		14.65	19.90
0660	23" ID		84	.190		7.55	7.55		15.10	20.50
0670	24" ID		83	.193		7.90	7.65		15.55	21
0700	For smooth .016" thick, add					45%	33%			
0710	For smooth .020" thick, add					27%	10%			
0720	For smooth .024" thick, add					52%	20%			
0730	For smooth .032" thick, add					104%	33%			
0800	For stucco embossed, add					1%				
0820	For corrugated, add					2.50%				
0900	White aluminum with polysurlyn moisture barrier									
0910	Smooth, % is an add to polykraft lines of same thickness									
0940	For smooth .016" thick, add				L.F.	35%				
0960	For smooth .024" thick, add				"	22%				
1000	Aluminum fitting covers									
1010	Size is based on OD of insulation									
1020	90 LR elbow, 2 piece									
1100	1-1/2"	Q-14	140	.114	Ea.	3.05	4.54		7.59	10.70
1110	1-3/4"		135	.119		3.05	4.71		7.76	10.95
1120	2"		130	.123		3.83	4.89		8.72	12.10
1130	2-1/4"		128	.125		3.83	4.96		8.79	12.25
1140	2-1/2"		126	.127		3.83	5.05		8.88	12.35

22 07 19 – Plumbing Piping Insulation

22 07 19.40 Pipe Insulation Protective Jacketing, Aluminum		Crew	Daily Output	Labor-Hours	Unit	Material	Labor	Equipment	Total	Total Incl O&P
						2009 Bare Costs				
1150	2-3/4"	Q-14	123	.130	Ea.	3.83	5.15		8.98	12.55
1160	3"		120	.133		4.16	5.30		9.46	13.15
1170	3-1/4"		117	.137		4.16	5.45		9.61	13.40
1180	3-1/2"		115	.139		4.91	5.50		10.41	14.35
1190	3-3/4"		113	.142		5.05	5.60		10.65	14.65
1200	4"		110	.145		5.25	5.75		11	15.15
1210	4-1/4"		108	.148		5.40	5.90		11.30	15.45
1220	4-1/2"		106	.151		5.40	6		11.40	15.65
1230	4-3/4"		104	.154		5.40	6.10		11.50	15.85
1240	5"		102	.157		6.10	6.25		12.35	16.75
1250	5-1/4"		100	.160		7.05	6.35		13.40	18
1260	5-1/2"		97	.165		8.80	6.55		15.35	20.50
1270	5-3/4"		95	.168		8.80	6.70		15.50	20.50
1280	6"		92	.174		7.50	6.90		14.40	19.40
1290	6-1/4"		90	.178		7.50	7.05		14.55	19.65
1300	6-1/2"		87	.184		9	7.30		16.30	21.50
1310	7"		85	.188		10.95	7.45		18.40	24
1320	7-1/4"		84	.190		10.95	7.55		18.50	24.50
1330	7-1/2"		83	.193		16.10	7.65		23.75	30
1340	8"		82	.195		12	7.75		19.75	26
1350	8-1/2"		80	.200		22	7.95		29.95	37.50
1360	9"		78	.205		22	8.15		30.15	37.50
1370	9-1/2"		77	.208		16.30	8.25		24.55	31.50
1380	10"		76	.211		16.30	8.35		24.65	31.50
1390	10-1/2"		75	.213		16.65	8.45		25.10	32
1400	11"		74	.216		16.65	8.60		25.25	32.50
1410	11-1/2"		72	.222		18.95	8.80		27.75	35.50
1420	12"		71	.225		18.95	8.95		27.90	35.50
1430	12-1/2"		69	.232		33.50	9.20		42.70	52
1440	13"		68	.235		33.50	9.35		42.85	52
1450	14"		66	.242		45	9.60		54.60	65.50
1460	15"		64	.250		47.50	9.95		57.45	68
1470	16"		63	.254		51.50	10.10		61.60	73.50
2000	45 Elbow, 2 piece									
2010	2-1/2"	Q-14	126	.127	Ea.	3.17	5.05		8.22	11.65
2020	2-3/4"		123	.130		3.17	5.15		8.32	11.85
2030	3"		120	.133		3.58	5.30		8.88	12.50
2040	3-1/4"		117	.137		3.58	5.45		9.03	12.75
2050	3-1/2"		115	.139		4.08	5.50		9.58	13.45
2060	3-3/4"		113	.142		4.08	5.60		9.68	13.60
2070	4"		110	.145		4.15	5.75		9.90	13.90
2080	4-1/4"		108	.148		4.73	5.90		10.63	14.70
2090	4-1/2"		106	.151		4.73	6		10.73	14.90
2100	4-3/4"		104	.154		4.73	6.10		10.83	15.10
2110	5"		102	.157		5.55	6.25		11.80	16.15
2120	5-1/4"		100	.160		5.55	6.35		11.90	16.35
2130	5-1/2"		97	.165		5.85	6.55		12.40	17
2140	6"		92	.174		5.85	6.90		12.75	17.55
2150	6-1/2"		87	.184		8.05	7.30		15.35	20.50
2160	7"		85	.188		8.05	7.45		15.50	21
2170	7-1/2"		83	.193		8.10	7.65		15.75	21.50
2180	8"		82	.195		8.10	7.75		15.85	21.50
2190	8-1/2"		80	.200		10.10	7.95		18.05	24

22 07 19.40 Pipe Insulation Protective Jacketing, Aluminum		Crew	Daily Output	Labor-Hours	Unit	2009 Bare Costs Material	Labor	Equipment	Total	Total Incl O&P
2200	9"	Q-14	78	.205	Ea.	10.10	8.15		18.25	24.50
2210	9-1/2"		77	.208		14.05	8.25		22.30	29
2220	10"		76	.211		14.05	8.35		22.40	29
2230	10-1/2"		75	.213		13.10	8.45		21.55	28
2240	11"		74	.216		13.10	8.60		21.70	28.50
2250	11-1/2"		72	.222		15.75	8.80		24.55	31.50
2260	12"		71	.225		15.75	8.95		24.70	32
2270	13"		68	.235		18.50	9.35		27.85	35.50
2280	14"		66	.242		22.50	9.60		32.10	40.50
2290	15"		64	.250		35	9.95		44.95	54.50
2300	16"		63	.254		38	10.10		48.10	58
2310	17"		62	.258		37	10.25		47.25	57
2320	18"		61	.262		42	10.40		52.40	63.50
2330	19"		60	.267		51.50	10.60		62.10	73.50
2340	20"		59	.271		50	10.75		60.75	72.50
2350	21"	↓	58	.276	↓	54	10.95		64.95	76.50
3000	Tee, 4 piece									
3010	2-1/2"	Q-14	88	.182	Ea.	24.50	7.20		31.70	38.50
3020	2-3/4"		86	.186		24.50	7.40		31.90	39
3030	3"		84	.190		28	7.55		35.55	43
3040	3-1/4"		82	.195		28	7.75		35.75	43
3050	3-1/2"		80	.200		29	7.95		36.95	45
3060	4"		78	.205		29.50	8.15		37.65	45.50
3070	4-1/4"		76	.211		31	8.35		39.35	47.50
3080	4-1/2"		74	.216		31	8.60		39.60	48
3090	4-3/4"		72	.222		31	8.80		39.80	48.50
3100	5"		70	.229		32	9.05		41.05	50
3110	5-1/4"		68	.235		32	9.35		41.35	50.50
3120	5-1/2"		66	.242		33.50	9.60		43.10	52.50
3130	6"		64	.250		33.50	9.95		43.45	53
3140	6-1/2"		60	.267		37	10.60		47.60	57.50
3150	7"		58	.276		37	10.95		47.95	58
3160	7-1/2"		56	.286		41	11.35		52.35	63.50
3170	8"		54	.296		41	11.75		52.75	64
3180	8-1/2"		52	.308		42	12.20		54.20	66
3190	9"		50	.320		42	12.70		54.70	66.50
3200	9-1/2"		49	.327		31	12.95		43.95	55.50
3210	10"		48	.333		31	13.25		44.25	56
3220	10-1/2"		47	.340		33	13.50		46.50	58.50
3230	11"		46	.348		33	13.80		46.80	59
3240	11-1/2"		45	.356		35	14.10		49.10	61.50
3250	12"		44	.364		35	14.45		49.45	62
3260	13"		43	.372		37	14.75		51.75	64.50
3270	14"		42	.381		38.50	15.10		53.60	67
3280	15"		41	.390		41.50	15.50		57	71
3290	16"		40	.400		43	15.90		58.90	72.50
3300	17"		39	.410		49	16.30		65.30	80.50
3310	18"		38	.421		51	16.70		67.70	83
3320	19"		37	.432		56.50	17.15		73.65	90
3330	20"		36	.444		57.50	17.65		75.15	92
3340	22"		35	.457		73.50	18.15		91.65	110
3350	23"		34	.471		75.50	18.70		94.20	113
3360	24"	↓	31	.516	↓	77.50	20.50		98	118

22 07 19.50 Pipe Insulation Protective Jacketing, St. Stl.		Crew	Daily Output	Labor-Hours	Unit	Material (2009 Bare Costs)	Labor (2009 Bare Costs)	Equipment (2009 Bare Costs)	Total (2009 Bare Costs)	Total Incl O&P
0010	**PIPE INSULATION PROTECTIVE JACKETING, STAINLESS STEEL**									
0100	Metal roll jacketing									
0120	Type 304 with moisture barrier									
0140	Smooth, based on OD of insulation, .010″ thick									
0260	2-1/2″ ID	Q-14	250	.064	L.F.	2.70	2.54		5.24	7.10
0270	2-3/4″ ID		245	.065		2.94	2.59		5.53	7.40
0280	3″ ID		240	.067		3.18	2.65		5.83	7.80
0290	3-1/4″ ID		235	.068		3.42	2.70		6.12	8.15
0300	3-1/2″ ID		230	.070		3.66	2.76		6.42	8.50
0310	3-3/4″ ID		225	.071		3.90	2.82		6.72	8.85
0320	4″ ID		220	.073		4.14	2.89		7.03	9.20
0330	4-1/4″ ID		215	.074		4.38	2.95		7.33	9.60
0340	4-1/2″ ID		210	.076		4.62	3.02		7.64	10
0350	5″ ID		200	.080		5.10	3.18		8.28	10.75
0360	5-1/2″ ID		190	.084		5.55	3.34		8.89	11.55
0370	6″ ID		180	.089		6.05	3.53		9.58	12.35
0380	6-1/2″ ID		175	.091		6.55	3.63		10.18	13.05
0390	7″ ID		170	.094		7	3.74		10.74	13.75
0400	7-1/2″ ID		164	.098		7.50	3.87		11.37	14.50
0410	8″ ID		161	.099		7.95	3.95		11.90	15.15
0420	8-1/2″ ID		158	.101		8.45	4.02		12.47	15.80
0430	9″ ID		155	.103		8.95	4.10		13.05	16.45
0440	9-1/2″ ID		152	.105		9.40	4.18		13.58	17.10
0450	10″ ID		149	.107		9.90	4.26		14.16	17.80
0460	10-1/2″ ID		146	.110		10.35	4.35		14.70	18.45
0470	11″ ID		143	.112		10.85	4.44		15.29	19.15
0480	12″ ID		137	.117		11.80	4.64		16.44	20.50
0490	13″ ID		132	.121		12.75	4.81		17.56	22
0500	14″ ID	↓	130	.123		13.70	4.89		18.59	23
0700	For smooth .016″ thick, add				↓	45%	33%			
1000	Stainless steel, Type 316, fitting covers									
1010	Size is based on OD of insulation									
1020	90 LR Elbow, 2 piece									
1100	1-1/2″	Q-14	126	.127	Ea.	19.55	5.05		24.60	29.50
1110	2-3/4″		123	.130		12.50	5.15		17.65	22
1120	3″		120	.133		13.05	5.30		18.35	23
1130	3-1/4″		117	.137		13.05	5.45		18.50	23
1140	3-1/2″		115	.139		13.75	5.50		19.25	24
1150	3-3/4″		113	.142		14.50	5.60		20.10	25
1160	4″		110	.145		15.95	5.75		21.70	27
1170	4-1/4″		108	.148		21	5.90		26.90	33
1180	4-1/2″		106	.151		21	6		27	33
1190	5″		102	.157		21.50	6.25		27.75	34
1200	5-1/2″		97	.165		32.50	6.55		39.05	46
1210	6″		92	.174		35.50	6.90		42.40	50
1220	6-1/2″		87	.184		49	7.30		56.30	65.50
1230	7″		85	.188		49	7.45		56.45	65.50
1240	7-1/2″		83	.193		57	7.65		64.65	75
1250	8″		80	.200		57	7.95		64.95	75.50
1260	8-1/2″		80	.200		59	7.95		66.95	78
1270	9″		78	.205		89.50	8.15		97.65	112
1280	9-1/2″	↓	77	.208	↓	88	8.25		96.25	110

22 07 Plumbing Insulation

22 07 19 – Plumbing Piping Insulation

22 07 19.50 Pipe Insulation Protective Jacketing, St. Stl.		Crew	Daily Output	Labor-Hours	Unit	2009 Bare Costs Material	Labor	Equipment	Total	Total Incl O&P
1290	10″	Q-14	76	.211	Ea.	88	8.35		96.35	111
1300	10-1/2″		75	.213		105	8.45		113.45	130
1310	11″		74	.216		101	8.60		109.60	125
1320	12″		71	.225		114	8.95		122.95	140
1330	13″		68	.235		158	9.35		167.35	189
1340	14″		66	.242		159	9.60		168.60	191
2000	45 Elbow, 2 piece									
2010	2-1/2″	Q-14	126	.127	Ea.	10.40	5.05		15.45	19.60
2020	2-3/4″		123	.130		10.40	5.15		15.55	19.80
2030	3″		120	.133		11.20	5.30		16.50	21
2040	3-1/4″		117	.137		11.20	5.45		16.65	21
2050	3-1/2″		115	.139		11.35	5.50		16.85	21.50
2060	3-3/4″		113	.142		11.35	5.60		16.95	21.50
2070	4″		110	.145		14.60	5.75		20.35	25.50
2080	4-1/4″		108	.148		20.50	5.90		26.40	32
2090	4-1/2″		106	.151		20.50	6		26.50	32
2100	4-3/4″		104	.154		20.50	6.10		26.60	32.50
2110	5″		102	.157		21	6.25		27.25	33
2120	5-1/2″		97	.165		21	6.55		27.55	33.50
2130	6″		92	.174		24.50	6.90		31.40	38
2140	6-1/2″		87	.184		24.50	7.30		31.80	39
2150	7″		85	.188		42	7.45		49.45	58
2160	7-1/2″		83	.193		42.50	7.65		50.15	59.50
2170	8″		82	.195		42.50	7.75		50.25	59.50
2180	8-1/2″		80	.200		50	7.95		57.95	68
2190	9″		78	.205		50	8.15		58.15	68
2200	9-1/2″		77	.208		58.50	8.25		66.75	78
2210	10″		76	.211		58.50	8.35		66.85	78
2220	10-1/2″		75	.213		70.50	8.45		78.95	91
2230	11″		74	.216		70.50	8.60		79.10	91.50
2240	12″		71	.225		76.50	8.95		85.45	98.50
2250	13″		68	.235		90	9.35		99.35	114

22 11 Facility Water Distribution

22 11 13 – Facility Water Distribution Piping

22 11 13.23 Pipe, Copper

		Crew	Daily Output	Labor-Hours	Unit	Material	Labor	Equipment	Total	Total Incl O&P
0010	**PIPE, COPPER**, Solder joints R221113-50									
1000	Type K tubing, couplings & clevis hanger assemblies 10′ O.C.									
1100	1/4″ diameter	1 Plum	84	.095	L.F.	3.43	4.64		8.07	10.95
1120	3/8″ diameter		82	.098		4.29	4.76		9.05	12.10
1140	1/2″ diameter		78	.103		5	5		10	13.25
1160	5/8″ diameter		77	.104		6.40	5.05		11.45	14.90
1180	3/4″ diameter		74	.108		9.15	5.25		14.40	18.25
1200	1″ diameter		66	.121		12.10	5.90		18	22.50
1220	1-1/4″ diameter		56	.143		15.10	6.95		22.05	27.50
1240	1-1/2″ diameter		50	.160		19.70	7.80		27.50	33.50
1260	2″ diameter		40	.200		30.50	9.75		40.25	48.50
1280	2-1/2″ diameter	Q-1	60	.267		45.50	11.70		57.20	68
1300	3″ diameter		54	.296		63.50	13		76.50	90
1320	3-1/2″ diameter		42	.381		85	16.70		101.70	120
1330	4″ diameter		38	.421		107	18.50		125.50	147

22 11 13.23 Pipe, Copper		Crew	Daily Output	Labor-Hours	Unit	Material	2009 Bare Costs Labor	Equipment	Total	Total Incl O&P
1340	5" diameter	Q-1	32	.500	L.F.	238	22		260	296
1360	6" diameter	Q-2	38	.632		345	28.50		373.50	425
1380	8" diameter	"	34	.706		615	32		647	725
1390	For other than full hard temper, add					13%				
1440	For silver solder, add						15%			
1800	For medical clean, (oxygen class), add					12%				
1950	To delete cplgs. & hngrs., 1/4"-1" pipe, subtract					27%	60%			
1960	1-1/4" -3" pipe, subtract					14%	52%			
1970	3-1/2"-5" pipe, subtract					10%	60%			
1980	6"-8" pipe, subtract					19%	53%			
2000	Type L tubing, couplings & clevis hanger assemblies 10' O.C.									
2100	1/4" diameter	1 Plum	88	.091	L.F.	2.23	4.43		6.66	9.30
2120	3/8" diameter		84	.095		3.26	4.64		7.90	10.80
2140	1/2" diameter		81	.099		3.83	4.82		8.65	11.65
2160	5/8" diameter		79	.101		5.45	4.94		10.39	13.65
2180	3/4" diameter		76	.105		6	5.15		11.15	14.55
2200	1" diameter		68	.118		9.05	5.75		14.80	18.85
2220	1-1/4" diameter		58	.138		12.50	6.70		19.20	24
2240	1-1/2" diameter		52	.154		16.10	7.50		23.60	29.50
2260	2" diameter		42	.190		25	9.30		34.30	42
2280	2-1/2" diameter	Q-1	62	.258		38	11.30		49.30	59.50
2300	3" diameter		56	.286		51.50	12.55		64.05	76.50
2320	3-1/2" diameter		43	.372		70	16.35		86.35	103
2340	4" diameter		39	.410		87	18		105	124
2360	5" diameter		34	.471		186	20.50		206.50	236
2380	6" diameter	Q-2	40	.600		256	27.50		283.50	325
2400	8" diameter	"	36	.667		450	30.50		480.50	540
2410	For other than full hard temper, add					21%				
2590	For silver solder, add						15%			
2900	For medical clean, (oxygen class), add					12%				
2940	To delete cplgs. & hngrs., 1/4"-1" pipe, subtract					37%	63%			
2960	1-1/4"-3" pipe, subtract					12%	53%			
2970	3-1/2"-5" pipe, subtract					12%	63%			
2980	6"-8" pipe, subtract					24%	55%			
3000	Type M tubing, couplings & clevis hanger assemblies 10' O.C.									
3100	1/4" diameter	1 Plum	90	.089	L.F.	2.62	4.33		6.95	9.60
3120	3/8" diameter		87	.092		2.65	4.48		7.13	9.85
3140	1/2" diameter		84	.095		2.79	4.64		7.43	10.25
3160	5/8" diameter		81	.099		4.45	4.82		9.27	12.35
3180	3/4" diameter		78	.103		4.41	5		9.41	12.60
3200	1" diameter		70	.114		6.95	5.55		12.50	16.25
3220	1-1/4" diameter		60	.133		10.25	6.50		16.75	21.50
3240	1-1/2" diameter		54	.148		14	7.20		21.20	26.50
3260	2" diameter		44	.182		22	8.85		30.85	38
3280	2-1/2" diameter	Q-1	64	.250		33	10.95		43.95	53.50
3300	3" diameter		58	.276		44.50	12.10		56.60	67.50
3320	3-1/2" diameter		45	.356		62.50	15.60		78.10	92.50
3340	4" diameter		40	.400		79.50	17.55		97.05	114
3360	5" diameter		36	.444		186	19.50		205.50	234
3370	6" diameter	Q-2	42	.571		256	26		282	325
3380	8" diameter	"	38	.632		450	28.50		478.50	540
3440	For silver solder, add						15%			
3960	To delete cplgs. & hngrs., 1/4"-1" pipe, subtract					35%	65%			

22 11 13.23 Pipe, Copper		Crew	Daily Output	Labor-Hours	Unit	2009 Bare Costs Material	Labor	Equipment	Total	Total Incl O&P
3970	1-1/4"-3" pipe, subtract					19%	56%			
3980	3-1/2"-5" pipe, subtract					13%	65%			
3990	6"-8" pipe, subtract					28%	58%			
4000	Type DWV tubing, couplings & clevis hanger assemblies 10' O.C.									
4100	1-1/4" diameter	1 Plum	60	.133	L.F.	10.90	6.50		17.40	22
4120	1-1/2" diameter		54	.148		13.70	7.20		20.90	26.50
4140	2" diameter		44	.182		18.45	8.85		27.30	34.50
4160	3" diameter	Q-1	58	.276		32.50	12.10		44.60	55
4180	4" diameter		40	.400		57.50	17.55		75.05	90.50
4200	5" diameter		36	.444		165	19.50		184.50	211
4220	6" diameter	Q-2	42	.571		235	26		261	300
4240	8" diameter	"	38	.632		575	28.50		603.50	675
4730	To delete cplgs. & hngrs., 1-1/4"-2" pipe, subtract					16%	53%			
4740	3"-4" pipe, subtract					13%	60%			
4750	5"-8" pipe, subtract					23%	58%			
5200	ACR tubing, type L, hard temper, cleaned and									
5220	capped, no couplings or hangers									
5240	3/8" OD				L.F.	2.06			2.06	2.27
5250	1/2" OD					3.26			3.26	3.59
5260	5/8" OD					3.97			3.97	4.37
5270	3/4" OD					5.60			5.60	6.15
5280	7/8" OD					6.20			6.20	6.85
5290	1-1/8" OD					9			9	9.90
5300	1-3/8" OD					12.05			12.05	13.25
5310	1-5/8" OD					15.45			15.45	17
5320	2-1/8" OD					24.50			24.50	27
5330	2-5/8" OD					36.50			36.50	40
5340	3-1/8" OD					49			49	53.50
5350	3-5/8" OD					64			64	70
5360	4-1/8" OD					81			81	89
5380	ACR tubing, type L, hard, cleaned and capped									
5381	No couplings or hangers									
5384	3/8"	1 Stpi	160	.050	L.F.	2.06	2.47		4.53	6.10
5385	1/2"		160	.050		3.26	2.47		5.73	7.40
5386	5/8"		160	.050		3.97	2.47		6.44	8.20
5387	3/4"		130	.062		5.60	3.04		8.64	10.85
5388	7/8"		130	.062		6.20	3.04		9.24	11.55
5389	1-1/8"		115	.070		9	3.43		12.43	15.25
5390	1-3/8"		100	.080		12.05	3.95		16	19.35
5391	1-5/8"		90	.089		15.45	4.39		19.84	24
5392	2-1/8"		80	.100		24.50	4.94		29.44	34.50
5393	2-5/8"	Q-5	125	.128		36.50	5.70		42.20	49
5394	3-1/8"		105	.152		49	6.75		55.75	64
5395	4-1/8"		95	.168		81	7.50		88.50	101
5800	Refrigeration tubing, dryseal, 60' coils									
5840	1/8" OD				Coil	41.50			41.50	45.50
5850	3/16" OD					48.50			48.50	53
5860	1/4" OD					57.50			57.50	63
5870	5/16" OD					74.50			74.50	82
5880	3/8" OD					85			85	93.50
5890	1/2" OD					112			112	123
5900	5/8" OD					151			151	166
5910	3/4" OD					182			182	200

22 11 13.23 Pipe, Copper		Crew	Daily Output	Labor-Hours	Unit	2009 Bare Costs Material	Labor	Equipment	Total	Total Incl O&P
5920	7/8″ OD				Coil	272			272	299
5930	1-1/8″ OD					390			390	430
5940	1-3/8″ OD					600			600	660
5950	1-5/8″ OD					760			760	835
9000	Minimum labor/equipment charge	1 Plum	4	2	Job		97.50		97.50	151
22 11 13.25 Pipe Fittings, Copper										
0010	**PIPE FITTINGS, COPPER**, Wrought unless otherwise noted									
0040	Solder joints, copper x copper									
0070	90° elbow, 1/4″	1 Plum	22	.364	Ea.	3.89	17.75		21.64	32
0090	3/8″		22	.364		3.69	17.75		21.44	31.50
0100	1/2″		20	.400		1.23	19.50		20.73	31.50
0110	5/8″		19	.421		4.34	20.50		24.84	37
0120	3/4″		19	.421		2.76	20.50		23.26	35
0130	1″		16	.500		6.80	24.50		31.30	45.50
0140	1-1/4″		15	.533		10.25	26		36.25	52
0150	1-1/2″		13	.615		16.10	30		46.10	64.50
0160	2″		11	.727		29	35.50		64.50	87
0170	2-1/2″	Q-1	13	1.231		58.50	54		112.50	148
0180	3″		11	1.455		78	64		142	185
0190	3-1/2″		10	1.600		272	70		342	410
0200	4″		9	1.778		200	78		278	340
0210	5″		6	2.667		835	117		952	1,100
0220	6″	Q-2	9	2.667		1,125	121		1,246	1,425
0230	8″	″	8	3		4,125	137		4,262	4,725
0250	45° elbow, 1/4″	1 Plum	22	.364		7.30	17.75		25.05	35.50
0270	3/8″		22	.364		5.90	17.75		23.65	34
0280	1/2″		20	.400		2.26	19.50		21.76	32.50
0290	5/8″		19	.421		11.30	20.50		31.80	44.50
0300	3/4″		19	.421		3.96	20.50		24.46	36.50
0310	1″		16	.500		9.95	24.50		34.45	49
0320	1-1/4″		15	.533		13.45	26		39.45	55.50
0330	1-1/2″		13	.615		16.20	30		46.20	64.50
0340	2″		11	.727		27	35.50		62.50	85
0350	2-1/2″	Q-1	13	1.231		57.50	54		111.50	148
0360	3″		13	1.231		85.50	54		139.50	178
0370	3-1/2″		10	1.600		144	70		214	267
0380	4″		9	1.778		182	78		260	320
0390	5″		6	2.667		705	117		822	955
0400	6″	Q-2	9	2.667		1,125	121		1,246	1,425
0410	8″	″	8	3		3,775	137		3,912	4,350
0450	Tee, 1/4″	1 Plum	14	.571		8.15	28		36.15	52
0470	3/8″		14	.571		6.25	28		34.25	50
0480	1/2″		13	.615		2.11	30		32.11	49
0490	5/8″		12	.667		13.55	32.50		46.05	65.50
0500	3/4″		12	.667		5.10	32.50		37.60	56
0510	1″		10	.800		15.70	39		54.70	78
0520	1-1/4″		9	.889		21.50	43.50		65	90.50
0530	1-1/2″		8	1		33	49		82	112
0540	2″		7	1.143		51.50	55.50		107	143
0550	2-1/2″	Q-1	8	2		104	88		192	250
0560	3″		7	2.286		159	100		259	330
0570	3-1/2″		6	2.667		460	117		577	685

22 11 13.25 Pipe Fittings, Copper		Crew	Daily Output	Labor-Hours	Unit	2009 Bare Costs Material	Labor	Equipment	Total	Total Incl O&P
0580	4"	Q-1	5	3.200	Ea.	385	140		525	640
0590	5"		4	4		1,250	176		1,426	1,650
0600	6"	Q-2	6	4		1,725	182		1,907	2,175
0610	8"	"	5	4.800		6,625	218		6,843	7,650
0612	Tee, reducing on the outlet, 1/4"	1 Plum	15	.533		13.15	26		39.15	55
0613	3/8"		15	.533		12.35	26		38.35	54
0614	1/2"		14	.571		10.85	28		38.85	55
0615	5/8"		13	.615		19.40	30		49.40	68
0616	3/4"		12	.667		7	32.50		39.50	58
0617	1"		11	.727		15.75	35.50		51.25	72.50
0618	1-1/4"		10	.800		29.50	39		68.50	93
0619	1-1/2"		9	.889		31.50	43.50		75	102
0620	2"		8	1		44.50	49		93.50	125
0621	2-1/2"	Q-1	9	1.778		119	78		197	251
0622	3"		8	2		133	88		221	283
0623	4"		6	2.667		245	117		362	450
0624	5"		5	3.200		1,250	140		1,390	1,600
0625	6"	Q-2	7	3.429		1,625	156		1,781	2,050
0626	8"	"	6	4		6,625	182		6,807	7,575
0630	Tee, reducing on the run, 1/4"	1 Plum	15	.533		16.45	26		42.45	58.50
0631	3/8"		15	.533		22	26		48	64.50
0632	1/2"		14	.571		14.90	28		42.90	59.50
0633	5/8"		13	.615		22	30		52	70.50
0634	3/4"		12	.667		6	32.50		38.50	57
0635	1"		11	.727		18.70	35.50		54.20	75.50
0636	1-1/4"		10	.800		29.50	39		68.50	93
0637	1-1/2"		9	.889		49.50	43.50		93	122
0638	2"		8	1		72	49		121	155
0639	2-1/2"	Q-1	9	1.778		157	78		235	293
0640	3"		8	2		231	88		319	390
0641	4"		6	2.667		510	117		627	745
0642	5"		5	3.200		1,200	140		1,340	1,525
0643	6"	Q-2	7	3.429		1,825	156		1,981	2,250
0644	8"	"	6	4		6,525	182		6,707	7,450
0650	Coupling, 1/4"	1 Plum	24	.333		.90	16.25		17.15	26
0670	3/8"		24	.333		1.19	16.25		17.44	26.50
0680	1/2"		22	.364		.94	17.75		18.69	28.50
0690	5/8"		21	.381		2.81	18.55		21.36	32
0700	3/4"		21	.381		1.87	18.55		20.42	31
0710	1"		18	.444		3.73	21.50		25.23	37.50
0715	1-1/4"		17	.471		6.65	23		29.65	43
0716	1-1/2"		15	.533		8.75	26		34.75	50
0718	2"		13	.615		14.65	30		44.65	62.50
0721	2-1/2"	Q-1	15	1.067		31	47		78	107
0722	3"		13	1.231		47	54		101	136
0724	3-1/2"		8	2		89.50	88		177.50	235
0726	4"		7	2.286		98.50	100		198.50	264
0728	5"		6	2.667		242	117		359	445
0731	6"	Q-2	8	3		400	137		537	650
0732	8"	"	7	3.429		1,300	156		1,456	1,675
0741	Coupling, reducing, concentric									
0743	1/2"	1 Plum	23	.348	Ea.	2.61	16.95		19.56	29.50
0745	3/4"		21.50	.372		4.19	18.15		22.34	32.50

22 11 Facility Water Distribution

22 11 13 – Facility Water Distribution Piping

22 11 13.25	Pipe Fittings, Copper	Crew	Daily Output	Labor-Hours	Unit	2009 Bare Costs Material	Labor	Equipment	Total	Total Incl O&P
0747	1"	1 Plum	19.50	.410	Ea.	7.50	20		27.50	39.50
0748	1-1/4"		18	.444		9	21.50		30.50	43.50
0749	1-1/2"		16	.500		15.20	24.50		39.70	54.50
0751	2"		14	.571		23.50	28		51.50	68.50
0752	2-1/2"		13	.615		57	30		87	109
0753	3"	Q-1	14	1.143		70.50	50		120.50	156
0755	4"	"	8	2		150	88		238	300
0757	5"	Q-2	7.50	3.200		550	146		696	830
0759	6"		7	3.429		855	156		1,011	1,175
0761	8"		6.50	3.692		1,400	168		1,568	1,800
0771	Cap, sweat									
0773	1/2"	1 Plum	40	.200	Ea.	.90	9.75		10.65	16.10
0775	3/4"		38	.211		1.67	10.25		11.92	17.75
0777	1"		32	.250		3.96	12.20		16.16	23.50
0778	1-1/4"		29	.276		5.30	13.45		18.75	27
0779	1-1/2"		26	.308		7.75	15		22.75	32
0781	2"		22	.364		14.30	17.75		32.05	43.50
0791	Flange, sweat									
0793	3"	Q-1	22	.727	Ea.	138	32		170	202
0795	4"		18	.889		148	39		187	224
0797	5"		12	1.333		360	58.50		418.50	485
0799	6"	Q-2	18	1.333		375	60.50		435.50	510
0801	8"	"	16	1.500		1,300	68.50		1,368.50	1,525
0850	Unions, 1/4"	1 Plum	21	.381		24	18.55		42.55	55.50
0870	3/8"		21	.381		24	18.55		42.55	55.50
0880	1/2"		19	.421		12.90	20.50		33.40	46
0890	5/8"		18	.444		55	21.50		76.50	94
0900	3/4"		18	.444		16.15	21.50		37.65	51.50
0910	1"		15	.533		28	26		54	71
0920	1-1/4"		14	.571		48	28		76	96
0930	1-1/2"		12	.667		63.50	32.50		96	121
0940	2"		10	.800		108	39		147	180
0950	2-1/2"	Q-1	12	1.333		230	58.50		288.50	345
0960	3"	"	10	1.600		615	70		685	785
0980	Adapter, copper x male IPS, 1/4"	1 Plum	20	.400		11.95	19.50		31.45	43
0990	3/8"		20	.400		6.35	19.50		25.85	37
1000	1/2"		18	.444		2.58	21.50		24.08	36.50
1010	3/4"		17	.471		4.33	23		27.33	40.50
1020	1"		15	.533		11.20	26		37.20	53
1030	1-1/4"		13	.615		16.55	30		46.55	64.50
1040	1-1/2"		12	.667		19	32.50		51.50	71.50
1050	2"		11	.727		32	35.50		67.50	90.50
1060	2-1/2"	Q-1	10.50	1.524		123	67		190	240
1070	3"		10	1.600		163	70		233	288
1080	3-1/2"		9	1.778		251	78		329	395
1090	4"		8	2		243	88		331	405
1200	5"		6	2.667		1,150	117		1,267	1,425
1210	6"	Q-2	8.50	2.824		1,400	128		1,528	1,750
1250	Cross, 1/2"	1 Plum	10	.800		17.45	39		56.45	79.50
1260	3/4"		9.50	.842		34	41		75	101
1270	1"		8	1		57.50	49		106.50	139
1280	1-1/4"		7.50	1.067		82.50	52		134.50	172
1290	1-1/2"		6.50	1.231		118	60		178	223

22 11 13.25 Pipe Fittings, Copper		Crew	Daily Output	Labor-Hours	Unit	2009 Bare Costs Material	Labor	Equipment	Total	Total Incl O&P
1300	2"	1 Plum	5.50	1.455	Ea.	223	71		294	355
1310	2-1/2"	Q-1	6.50	2.462		515	108		623	735
1320	3"	"	5.50	2.909	↓	645	128		773	910
1500	Tee fitting, mechanically formed, (Type 1)									
1520	1/2" run size, 3/8" to 1/2" branch size	1 Plum	80	.100	Ea.		4.88		4.88	7.55
1530	3/4" run size, 3/8" to 3/4" branch size		60	.133			6.50		6.50	10.10
1540	1" run size, 3/8" to 1" branch size		54	.148			7.20		7.20	11.20
1550	1-1/4" run size, 3/8" to 1-1/4" branch size		48	.167			8.15		8.15	12.60
1560	1-1/2" run size, 3/8" to 1-1/2" branch size		40	.200			9.75		9.75	15.10
1570	2" run size, 3/8" to 2" branch size		35	.229			11.15		11.15	17.30
1580	2-1/2" run size, 1/2" to 2" branch size		32	.250			12.20		12.20	18.90
1590	3" run size, 1" to 2" branch size		26	.308			15		15	23.50
1600	4" run size, 1" to 2" branch size	↓	24	.333	↓		16.25		16.25	25
1640	Tee fitting, mechanically formed, (Type 2)									
1650	2-1/2" run size, 2-1/2" branch size	1 Plum	12.50	.640	Ea.		31		31	48.50
1660	3" run size, 2-1/2" to 3" branch size		12	.667			32.50		32.50	50.50
1670	3-1/2" run size, 2-1/2" to 3-1/2" branch size		11	.727			35.50		35.50	55
1680	4" run size, 2-1/2" to 4" branch size		10.50	.762			37		37	57.50
1698	5" run size, 2" to 4" branch size		9.50	.842			41		41	63.50
1700	6" run size, 2" to 4" branch size		8.50	.941			46		46	71
1710	8" run size, 2" to 4" branch size	↓	7	1.143	↓		55.50		55.50	86.50
1800	ACR fittings, OD size									
1802	Tee, straight									
1808	5/8"	1 Stpi	12	.667	Ea.	2.11	33		35.11	53.50
1810	3/4"		12	.667		13.55	33		46.55	66
1812	7/8"		10	.800		5.10	39.50		44.60	66.50
1813	1"		10	.800		37	39.50		76.50	102
1814	1-1/8"		10	.800		15.70	39.50		55.20	78.50
1816	1-3/8"		9	.889		21.50	44		65.50	91.50
1818	1-5/8"		8	[illegible]		33	49.50		82.50	113
1820	2-1/8"	↓	7	1.143		51.50	56.50		108	144
1822	2-5/8"	Q-5	8	2		104	89		193	252
1824	3-1/8"		7	2.286		159	102		261	330
1826	4-1/8"	↓	5	3.200	↓	385	142		527	640
1830	90° elbow									
1836	5/8"	1 Stpi	19	.421	Ea.	4.13	21		25.13	36.50
1838	3/4"		19	.421		7.50	21		28.50	40.50
1840	7/8"		16	.500		7.45	24.50		31.95	46.50
1842	1-1/8"		16	.500		9.95	24.50		34.45	49.50
1844	1-3/8"		15	.533		10.25	26.50		36.75	52.50
1846	1-5/8"		13	.615		16	30.50		46.50	64.50
1848	2-1/8"	↓	11	.727		29	36		65	87.50
1850	2-5/8"	Q-5	13	1.231		58.50	54.50		113	149
1852	3-1/8"		11	1.455		78	64.50		142.50	186
1854	4-1/8"	↓	9	1.778	↓	200	79		279	340
1860	Coupling									
1866	5/8"	1 Stpi	21	.381	Ea.	.94	18.80		19.74	30
1868	3/4"		21	.381		2.83	18.80		21.63	32
1870	7/8"		18	.444		1.87	22		23.87	36
1871	1"		18	.444		6.35	22		28.35	41
1872	1-1/8"		18	.444		3.78	22		25.78	38
1874	1-3/8"		17	.471		6.65	23		29.65	43.50
1876	1-5/8"	↓	15	.533	↓	8.75	26.50		35.25	50.50

22 11 13.25 Pipe Fittings, Copper		Crew	Daily Output	Labor-Hours	Unit	Material	2009 Bare Costs Labor	Equipment	Total	Total Incl O&P
1878	2-1/8"	1 Stpi	13	.615	Ea.	14.65	30.50		45.15	63
1880	2-5/8"	Q-5	15	1.067		33	47.50		80.50	110
1882	3-1/8"		13	1.231		48	54.50		102.50	138
1884	4-1/8"		7	2.286		114	102		216	282
2000	DWV, solder joints, copper x copper									
2030	90° Elbow, 1-1/4"	1 Plum	13	.615	Ea.	11.90	30		41.90	59.50
2050	1-1/2"		12	.667		17.85	32.50		50.35	70
2070	2"		10	.800		23.50	39		62.50	86
2090	3"	Q-1	10	1.600		57.50	70		127.50	173
2100	4"	"	9	1.778		281	78		359	430
2150	45° Elbow, 1-1/4"	1 Plum	13	.615		10.95	30		40.95	58.50
2170	1-1/2"		12	.667		9.05	32.50		41.55	60.50
2180	2"		10	.800		21	39		60	83.50
2190	3"	Q-1	10	1.600		44.50	70		114.50	158
2200	4"	"	9	1.778		193	78		271	335
2250	Tee, Sanitary, 1-1/4"	1 Plum	9	.889		23.50	43.50		67	93
2270	1-1/2"		8	1		29.50	49		78.50	108
2290	2"		7	1.143		34	55.50		89.50	124
2310	3"	Q-1	7	2.286		126	100		226	295
2330	4"	"	6	2.667		320	117		437	535
2400	Coupling, 1-1/4"	1 Plum	14	.571		5.55	28		33.55	49
2420	1-1/2"		13	.615		6.90	30		36.90	54
2440	2"		11	.727		9.60	35.50		45.10	65.50
2460	3"	Q-1	11	1.455		18.65	64		82.65	120
2480	4"	"	10	1.600		59.50	70		129.50	175
2602	Traps, see Div. 22 13 16.60									
6992	Polybutylene/polyethylene pipe, See Div. 22 11 13.76 for plastic ftng.									
7000	Insert type Brass/copper, 100 psi @ 180° F, CTS									
7010	Adapter MPT 3/8" x 3/8" CTS	1 Plum	29	.276	Ea.	1.60	13.45		15.05	23
7020	1/2" x 1/2"		26	.308		1.60	15		16.60	25.50
7030	3/4" x 1/2"		26	.308		2.73	15		17.73	26.50
7040	3/4" x 3/4"		25	.320		2.36	15.60		17.96	26.50
7050	Adapter CTS 1/2" x 1/2" sweat		24	.333		.60	16.25		16.85	25.50
7060	3/4" x 3/4" sweat		22	.364		.75	17.75		18.50	28.50
7070	Coupler center set 3/8" CTS		25	.320		.78	15.60		16.38	25
7080	1/2" CTS		23	.348		.60	16.95		17.55	27
7090	3/4" CTS		22	.364		.75	17.75		18.50	28.50
7100	Elbow 90°, copper 3/8"		25	.320		1.71	15.60		17.31	26
7110	1/2" CTS		23	.348		1.50	16.95		18.45	28
7120	3/4" CTS		22	.364		1.33	17.75		19.08	29
7130	Tee copper 3/8" CTS		17	.471		2.03	23		25.03	37.50
7140	1/2" CTS		15	.533		1.39	26		27.39	42
7150	3/4" CTS		14	.571		2.14	28		30.14	45.50
7160	3/8" x 3/8" x 1/2"		16	.500		1.39	24.50		25.89	39.50
7170	1/2" x 3/8" x 1/2"		15	.533		1.39	26		27.39	42
7180	3/4" x 1/2" x 3/4"		14	.571		1.39	28		29.39	44.50
9000	Minimum labor/equipment charge		4	2	Job		97.50		97.50	151

22 11 13.44 Pipe, Steel										
0010	**PIPE, STEEL** R221113-50									
0020	All pipe sizes are to Spec. A-53 unless noted otherwise									
0032	Schedule 10, see Div. 22 11 13.48 0500									
0050	Schedule 40, threaded, with couplings, and clevis hanger									

22 11 13.44 Pipe, Steel		Crew	Daily Output	Labor-Hours	Unit	Material	2009 Bare Costs Labor	Equipment	Total	Total Incl O&P
0060	assemblies sized for covering, 10′ O.C.									
0540	Black, 1/4″ diameter	1 Plum	66	.121	L.F.	2.17	5.90		8.07	11.55
0550	3/8″ diameter		65	.123		2.46	6		8.46	12
0560	1/2″ diameter		63	.127		2.23	6.20		8.43	12.05
0570	3/4″ diameter		61	.131		2.64	6.40		9.04	12.80
0580	1″ diameter		53	.151		3.87	7.35		11.22	15.65
0590	1-1/4″ diameter	Q-1	89	.180		4.89	7.90		12.79	17.65
0600	1-1/2″ diameter		80	.200		5.75	8.80		14.55	19.95
0610	2″ diameter		64	.250		7.70	10.95		18.65	25.50
0620	2-1/2″ diameter		50	.320		12.05	14.05		26.10	35.50
0630	3″ diameter		43	.372		15.50	16.35		31.85	42.50
0640	3-1/2″ diameter		40	.400		25.50	17.55		43.05	55
0650	4″ diameter		36	.444		23	19.50		42.50	55
0809	A-106, gr. A/B, seamless w/cplgs. & clevis hanger assemblies									
0811	1/4″ diameter	1 Plum	66	.121	L.F.	7.10	5.90		13	16.95
0812	3/8″ diameter		65	.123		8.70	6		14.70	18.85
0813	1/2″ diameter		63	.127		6.70	6.20		12.90	16.95
0814	3/4″ diameter		61	.131		8.25	6.40		14.65	18.95
0815	1″ diameter		53	.151		9.15	7.35		16.50	21.50
0816	1-1/4″ diameter	Q-1	89	.180		11.20	7.90		19.10	24.50
0817	1-1/2″ diameter		80	.200		14.35	8.80		23.15	29.50
0819	2″ diameter		64	.250		14.75	10.95		25.70	33
0821	2-1/2″ diameter		50	.320		19.05	14.05		33.10	43
0822	3″ diameter		43	.372		25	16.35		41.35	53
0823	4″ diameter		36	.444		34	19.50		53.50	67
1220	To delete coupling & hanger, subtract									
1230	1/4″ diam. to 3/4″ diam.					31%	56%			
1240	1″ diam. to 1-1/2″ diam.					23%	51%			
1250	2″ diam. to 4″ diam.					23%	41%			
1280	All pipe sizes are to Spec. A-53 unless noted otherwise									
1281	Schedule 40, threaded, with couplings and clevis hanger									
1282	assemblies sized for covering, 10′ O. C.									
1290	Galvanized, 1/4″ diameter	1 Plum	66	.121	L.F.	2.36	5.90		8.26	11.75
1300	3/8″ diameter		65	.123		3.31	6		9.31	12.95
1310	1/2″ diameter		63	.127		3.11	6.20		9.31	13
1320	3/4″ diameter		61	.131		3.76	6.40		10.16	14.05
1330	1″ diameter		53	.151		5.10	7.35		12.45	17.05
1340	1-1/4″ diameter	Q-1	89	.180		6.55	7.90		14.45	19.45
1350	1-1/2″ diameter		80	.200		7.80	8.80		16.60	22
1360	2″ diameter		64	.250		10.35	10.95		21.30	28.50
1370	2-1/2″ diameter		50	.320		17.60	14.05		31.65	41.50
1380	3″ diameter		43	.372		22	16.35		38.35	49.50
1390	3-1/2″ diameter		40	.400		28	17.55		45.55	58
1400	4″ diameter		36	.444		32.50	19.50		52	65.50
1750	To delete coupling & hanger, subtract									
1760	1/4″ diam. to 3/4″ diam.					31%	56%			
1770	1″ diam. to 1-1/2″ diam.					23%	51%			
1780	2″ diam. to 4″ diam.					23%	41%			
2000	Welded, sch. 40, on yoke & roll hanger assys, sized for covering, 10′ O.C.									
2040	Black, 1″ diameter	Q-15	93	.172	L.F.	4.30	7.55	.60	12.45	17.10
2050	1-1/4″ diameter		84	.190		5.55	8.35	.66	14.56	19.80
2060	1-1/2″ diameter		76	.211		6.25	9.25	.73	16.23	22
2070	2″ diameter		61	.262		7.95	11.50	.91	20.36	27.50

22 11 13.44 Pipe, Steel		Crew	Daily Output	Labor-Hours	Unit	Material	Labor	Equipment	Total	Total Incl O&P
						2009 Bare Costs				
2080	2-1/2" diameter	Q-15	47	.340	L.F.	9.85	14.95	1.18	25.98	35
2090	3" diameter		43	.372		15.55	16.35	1.29	33.19	44
2100	3-1/2" diameter		39	.410		17.45	18	1.43	36.88	49
2110	4" diameter		37	.432		18.10	19	1.50	38.60	51
2120	5" diameter		32	.500		24.50	22	1.74	48.24	63
2130	6" diameter	Q-16	36	.667		31	30.50	1.55	63.05	82.50
2140	8" diameter		29	.828		46	37.50	1.92	85.42	112
2150	10" diameter		24	1		67	45.50	2.32	114.82	147
2160	12" diameter		19	1.263		80.50	57.50	2.93	140.93	181
2170	14" diameter, (two rod roll type hanger for 14" dia. and up)		15	1.600		91	73	3.71	167.71	217
2180	16" diameter, (two rod roll type hanger)		13	1.846		104	84	4.28	192.28	249
2190	18" diameter, (two rod roll type hanger)		11	2.182		121	99.50	5.05	225.55	293
2200	20" diameter, (two rod roll type hanger)		9	2.667		145	121	6.20	272.20	355
2220	24" diameter, (two rod roll type hanger)		8	3		201	137	6.95	344.95	440
2345	Sch. 40, A-53, gr. A/B, ERW, welded w/hngrs.									
2346	2" diameter	Q-15	61	.262	L.F.	6.50	11.50	.91	18.91	26
2347	2-1/2" diameter		47	.340		8.20	14.95	1.18	24.33	33.50
2348	3" diameter		43	.372		11.90	16.35	1.29	29.54	40
2349	4" diameter		38	.421		13.40	18.50	1.47	33.37	45
2350	6" diameter	Q-16	37	.649		45	29.50	1.50	76	97
2351	8" diameter		29	.828		58.50	37.50	1.92	97.92	125
2352	10" diameter		24	1		83	45.50	2.32	130.82	164
2363	.375" wall, A-53, gr. A/B, ERW, welded w/hngrs.									
2364	12" diameter	Q-16	20	1.200	L.F.	99	54.50	2.78	156.28	197
2365	14" diameter		17	1.412		109	64	3.28	176.28	223
2366	16" diameter		14	1.714		146	78	3.98	227.98	286
2560	To delete hanger, subtract									
2570	1" diam. to 1-1/2" diam.					15%	34%			
2580	2" diam. to 3-1/2" diam.					9%	21%			
2590	4" diam. to 12" diam.					5%	12%			
2596	14" diam. to 36" diam.					3%	10%			
3250	Flanged, 150 lb. weld neck, on yoke & roll hangers									
3260	sized for covering, 10' O.C.									
3290	Black, 1" diameter	Q-15	70	.229	L.F.	8.45	10.05	.80	19.30	25.50
3300	1-1/4" diameter		64	.250		9.70	10.95	.87	21.52	28.50
3310	1-1/2" diameter		58	.276		10.45	12.10	.96	23.51	31.50
3320	2" diameter		45	.356		12.60	15.60	1.24	29.44	39.50
3330	2-1/2" diameter		36	.444		14.95	19.50	1.55	36	48
3340	3" diameter		32	.500		21	22	1.74	44.74	59
3350	3-1/2" diameter		29	.552		25	24	1.92	50.92	67
3360	4" diameter		26	.615		26	27	2.14	55.14	73
3370	5" diameter		21	.762		36	33.50	2.65	72.15	94.50
3380	6" diameter	Q-16	25	.960		44	43.50	2.23	89.73	119
3390	8" diameter		19	1.263		66.50	57.50	2.93	126.93	166
3400	10" diameter		16	1.500		109	68.50	3.48	180.98	230
3410	12" diameter		14	1.714		130	78	3.98	211.98	267
3470	For 300 lb. flanges, add					63%				
3480	For 600 lb. flanges, add					310%				
3960	To delete flanges & hanger, subtract									
3970	1" diam. to 2" diam.					76%	65%			
3980	2-1/2" diam. to 4" diam.					62%	59%			
3990	5" diam. to 12" diam.					60%	46%			
4750	Schedule 80, threaded, with couplings, and clevis hanger assemblies									

22 11 13.44 Pipe, Steel		Crew	Daily Output	Labor-Hours	Unit	Material	2009 Bare Costs Labor	Equipment	Total	Total Incl O&P
4760	sized for covering, 10′ O.C.									
4790	Black, 1/4″ diameter	1 Plum	54	.148	L.F.	2.81	7.20		10.01	14.30
4800	3/8″ diameter		53	.151		3.60	7.35		10.95	15.35
4810	1/2″ diameter		52	.154		3.32	7.50		10.82	15.30
4820	3/4″ diameter		50	.160		4.05	7.80		11.85	16.55
4830	1″ diameter		45	.178		5.55	8.65		14.20	19.55
4840	1-1/4″ diameter	Q-1	75	.213		7.35	9.35		16.70	22.50
4850	1-1/2″ diameter		69	.232		8.60	10.15		18.75	25.50
4860	2″ diameter		56	.286		11.70	12.55		24.25	32.50
4870	2-1/2″ diameter		44	.364		18.10	15.95		34.05	45
4880	3″ diameter		38	.421		24	18.50		42.50	55
4890	3-1/2″ diameter		35	.457		28.50	20		48.50	62.50
4900	4″ diameter		32	.500		26	22		48	62.50
5061	A-106, gr. A/B seamless with cplgs. & clevis hanger assemblies, 1/4″ dia.	1 Plum	63	.127		6.60	6.20		12.80	16.85
5062	3/8″ diameter		62	.129		7.75	6.30		14.05	18.25
5063	1/2″ diameter		61	.131		6.75	6.40		13.15	17.30
5064	3/4″ diameter		57	.140		7.85	6.85		14.70	19.25
5065	1″ diameter		51	.157		9.40	7.65		17.05	22
5066	1-1/4″ diameter	Q-1	85	.188		12.50	8.25		20.75	26.50
5067	1-1/2″ diameter		77	.208		15.50	9.10		24.60	31
5071	2″ diameter		61	.262		18.10	11.50		29.60	38
5072	2-1/2″ diameter		48	.333		19.30	14.65		33.95	44
5073	3″ diameter		41	.390		25.50	17.10		42.60	54.50
5074	4″ diameter		35	.457		37.50	20		57.50	72
5430	To delete coupling & hanger, subtract									
5440	1/4″ diam. to 1/2″ diam.					31%	54%			
5450	3/4″ diam. to 1-1/2″ diam.					28%	49%			
5460	2″ diam. to 4″ diam.					21%	40%			
5510	Galvanized, 1/4″ diameter	1 Plum	54	.148	L.F.	4.20	7.20		11.40	15.80
5520	3/8″ diameter		53	.151		4.23	7.35		11.58	16.05
5530	1/2″ diameter		52	.154		4.34	7.50		11.84	16.40
5540	3/4″ diameter		50	.160		5.25	7.80		13.05	17.90
5550	1″ diameter		45	.178		7.35	8.65		16	21.50
5560	1-1/4″ diameter	Q-1	75	.213		9.85	9.35		19.20	25.50
5570	1-1/2″ diameter		69	.232		11.60	10.15		21.75	28.50
5580	2″ diameter		56	.286		16.05	12.55		28.60	37
5590	2-1/2″ diameter		44	.364		25	15.95		40.95	52.50
5600	3″ diameter		38	.421		33	18.50		51.50	64.50
5610	3-1/2″ diameter		35	.457		35.50	20		55.50	70
5620	4″ diameter		32	.500		46.50	22		68.50	85
5930	To delete coupling & hanger, subtract									
5940	1/4″ diam. to 1/2″ diam.					31%	54%			
5950	3/4″ diam. to 1-1/2″ diam.					28%	49%			
5960	2″ diam. to 4″ diam.					21%	40%			
6000	Welded, on yoke & roller hangers									
6010	sized for covering, 10′ O.C.									
6040	Black, 1″ diameter	Q-15	85	.188	L.F.	6	8.25	.66	14.91	20
6050	1-1/4″ diameter		79	.203		7.90	8.90	.70	17.50	23
6060	1-1/2″ diameter		72	.222		9.15	9.75	.77	19.67	26
6070	2″ diameter		57	.281		11.80	12.30	.98	25.08	33
6080	2-1/2″ diameter		44	.364		17.75	15.95	1.27	34.97	46
6090	3″ diameter		40	.400		23.50	17.55	1.39	42.44	54
6100	3-1/2″ diameter		34	.471		27.50	20.50	1.64	49.64	64.50

22 11 Facility Water Distribution

22 11 13 – Facility Water Distribution Piping

22 11 13.44 Pipe, Steel		Crew	Daily Output	Labor-Hours	Unit	2009 Bare Costs Material	Labor	Equipment	Total	Total Incl O&P
6110	4" diameter	Q-15	33	.485	L.F.	24.50	21.50	1.69	47.69	62
6120	5" diameter		26	.615		50.50	27	2.14	79.64	100
6130	6" diameter	Q-16	30	.800		48.50	36.50	1.86	86.86	112
6140	8" diameter		25	.960		84	43.50	2.23	129.73	162
6150	10" diameter		20	1.200		150	54.50	2.78	207.28	253
6160	12" diameter		15	1.600		231	73	3.71	307.71	370
6540	To delete hanger, subtract									
6550	1" diam. to 1-1/2" diam.					30%	14%			
6560	2" diam. to 3" diam.					23%	9%			
6570	3-1/2" diam. to 5" diam.					12%	6%			
6580	6" diam. to 12" diam.					10%	4%			
7250	Flanged, 300 lb. weld neck, on yoke & roll hangers									
7260	sized for covering, 10' O.C.									
7290	Black, 1" diameter	Q-15	66	.242	L.F.	11.10	10.65	.84	22.59	29.50
7300	1-1/4" diameter		61	.262		13	11.50	.91	25.41	33
7310	1-1/2" diameter		54	.296		14.30	13	1.03	28.33	37
7320	2" diameter		42	.381		17.65	16.70	1.33	35.68	47
7330	2-1/2" diameter		33	.485		24.50	21.50	1.69	47.69	62
7340	3" diameter		29	.552		31	24	1.92	56.92	73.50
7350	3-1/2" diameter		24	.667		40	29.50	2.32	71.82	92
7360	4" diameter		23	.696		37	30.50	2.42	69.92	90.50
7370	5" diameter		19	.842		69	37	2.93	108.93	137
7380	6" diameter	Q-16	23	1.043		66	47.50	2.42	115.92	149
7390	8" diameter		17	1.412		115	64	3.28	182.28	229
7400	10" diameter		14	1.714		205	78	3.98	286.98	350
7410	12" diameter		12	2		300	91	4.64	395.64	475
7470	For 600 lb. flanges, add					100%				
7940	To delete flanges & hanger, subtract									
7950	1" diam. to 1-1/2" diam.					75%	66%			
7960	2" diam. to 3" diam.					62%	60%			
7970	3-1/2" diam. to 5" diam.					54%	66%			
7980	6" diam. to 12" diam.					55%	62%			
8040	Galvanized, 1" diameter	Q-15	66	.242	L.F.	12.90	10.65	.84	24.39	31.50
8050	1-1/4" diameter		61	.262		15.50	11.50	.91	27.91	36
8060	1-1/2" diameter		54	.296		17.25	13	1.03	31.28	40
8070	2" diameter		42	.381		22	16.70	1.33	40.03	51.50
8080	2-1/2" diameter		33	.485		31.50	21.50	1.69	54.69	69.50
8090	3" diameter		29	.552		40	24	1.92	65.92	83.50
8100	3-1/2" diameter		24	.667		47	29.50	2.32	78.82	100
8110	4" diameter		23	.696		57.50	30.50	2.42	90.42	113
8120	5" diameter		19	.842		95.50	37	2.93	135.43	166
8130	6" diameter	Q-16	23	1.043		102	47.50	2.42	151.92	189
8140	8" diameter		17	1.412		158	64	3.28	225.28	277
8150	10" diameter		14	1.714		291	78	3.98	372.98	445
8160	12" diameter		12	2		410	91	4.64	505.64	595
8240	For 600 lb. flanges, add					100%				
8900	To delete flanges & hangers, subtract									
8910	1" diam. to 1-1/2" diam.					72%	66%			
8920	2" diam. to 3" diam.					59%	60%			
8930	3-1/2" diam. to 5" diam.					51%	66%			
8940	6" diam. to 12" diam.					49%	62%			
9000	Threading pipe labor, one end, all schedules through 80									
9010	1/4" through 3/4" pipe size	1 Plum	80	.100	Ea.		4.88		4.88	7.55

22 11 13.44 Pipe, Steel		Crew	Daily Output	Labor-Hours	Unit	2009 Bare Costs Material	Labor	Equipment	Total	Total Incl O&P
9020	1" through 2" pipe size	1 Plum	73	.110	Ea.		5.35		5.35	8.30
9030	2-1/2" pipe size		53	.151			7.35		7.35	11.40
9040	3" pipe size		50	.160			7.80		7.80	12.10
9050	3-1/2" pipe size	Q-1	89	.180			7.90		7.90	12.25
9060	4" pipe size		73	.219			9.60		9.60	14.90
9070	5" pipe size		53	.302			13.25		13.25	20.50
9080	6" pipe size		46	.348			15.25		15.25	23.50
9090	8" pipe size		29	.552			24		24	37.50
9100	10" pipe size		21	.762			33.50		33.50	52
9110	12" pipe size		13	1.231			54		54	84
9200	Welding labor per joint									
9210	Schedule 40,									
9230	1/2" pipe size	Q-15	32	.500	Ea.		22	1.74	23.74	36
9240	3/4" pipe size		27	.593			26	2.06	28.06	43
9250	1" pipe size		23	.696			30.50	2.42	32.92	50
9260	1-1/4" pipe size		20	.800			35	2.78	37.78	57.50
9270	1-1/2" pipe size		19	.842			37	2.93	39.93	60.50
9280	2" pipe size		16	1			44	3.48	47.48	72
9290	2-1/2" pipe size		13	1.231			54	4.28	58.28	88.50
9300	3" pipe size		12	1.333			58.50	4.64	63.14	95.50
9310	4" pipe size		10	1.600			70	5.55	75.55	115
9320	5" pipe size		9	1.778			78	6.20	84.20	128
9330	6" pipe size		8	2			88	6.95	94.95	144
9340	8" pipe size		5	3.200			140	11.15	151.15	230
9350	10" pipe size		4	4			176	13.90	189.90	287
9360	12" pipe size		3	5.333			234	18.55	252.55	385
9370	14" pipe size		2.60	6.154			270	21.50	291.50	445
9380	16" pipe size		2.20	7.273			320	25.50	345.50	525
9390	18" pipe size		2	8			350	28	378	575
9400	20" pipe size		1.80	8.889			390	31	421	640
9410	22" pipe size		1.70	9.412			415	33	448	675
9420	24" pipe size		1.50	10.667			470	37	507	765
9450	Schedule 80,									
9460	1/2" pipe size	Q-15	27	.593	Ea.		26	2.06	28.06	43
9470	3/4" pipe size		23	.696			30.50	2.42	32.92	50
9480	1" pipe size		20	.800			35	2.78	37.78	57.50
9490	1-1/4" pipe size		19	.842			37	2.93	39.93	60.50
9500	1-1/2" pipe size		18	.889			39	3.09	42.09	64
9510	2" pipe size		15	1.067			47	3.71	50.71	76.50
9520	2-1/2" pipe size		12	1.333			58.50	4.64	63.14	95.50
9530	3" pipe size		11	1.455			64	5.05	69.05	105
9540	4" pipe size		8	2			88	6.95	94.95	144
9550	5" pipe size		6	2.667			117	9.30	126.30	191
9560	6" pipe size		5	3.200			140	11.15	151.15	230
9570	8" pipe size		4	4			176	13.90	189.90	287
9580	10" pipe size		3	5.333			234	18.55	252.55	385
9590	12" pipe size		2	8			350	28	378	575
9600	14" pipe size	Q-16	2.60	9.231			420	21.50	441.50	675
9610	16" pipe size		2.30	10.435			475	24	499	760
9620	18" pipe size		2	12			545	28	573	875
9630	20" pipe size		1.80	13.333			605	31	636	975
9640	22" pipe size		1.60	15			685	35	720	1,100
9650	24" pipe size		1.50	16			730	37	767	1,175

22 11 13.44 Pipe, Steel		Crew	Daily Output	Labor-Hours	Unit	Material	Labor (2009 Bare Costs)	Equipment	Total	Total Incl O&P
9990	Minimum labor/equipment charge	1 Plum	3	2.667	Job		130		130	202

22 11 13.45 Pipe Fittings, Steel, Threaded

Code	Description	Crew	Daily Output	Labor-Hours	Unit	Material	Labor	Equipment	Total	Total Incl O&P
0010	PIPE FITTINGS, STEEL, THREADED R221113-50									
0020	Cast Iron									
0040	Standard weight, black									
0060	90° Elbow, straight									
0070	1/4"	1 Plum	16	.500	Ea.	5.80	24.50		30.30	44.50
0080	3/8"		16	.500		8.40	24.50		32.90	47.50
0090	1/2"		15	.533		3.69	26		29.69	44.50
0100	3/4"		14	.571		3.84	28		31.84	47
0110	1"		13	.615		4.54	30		34.54	51.50
0120	1-1/4"	Q-1	22	.727		6.45	32		38.45	56.50
0130	1-1/2"		20	.800		8.90	35		43.90	64.50
0140	2"		18	.889		13.95	39		52.95	76
0150	2-1/2"		14	1.143		33.50	50		83.50	115
0160	3"		10	1.600		54.50	70		124.50	169
0170	3-1/2"		8	2		149	88		237	299
0180	4"		6	2.667		101	117		218	293
0250	45° Elbow, straight									
0260	1/4"	1 Plum	16	.500	Ea.	6.95	24.50		31.45	45.50
0270	3/8"		16	.500		7.50	24.50		32	46.50
0280	1/2"		15	.533		5.65	26		31.65	46.50
0300	3/4"		14	.571		5.65	28		33.65	49.50
0320	1"		13	.615		6.65	30		36.65	54
0330	1-1/4"	Q-1	22	.727		8.95	32		40.95	59.50
0340	1-1/2"		20	.800		14.80	35		49.80	71
0350	2"		18	.889		17.10	39		56.10	79.50
0360	2-1/2"		14	1.143		44.50	50		94.50	127
0370	3"		10	1.600		70.50	70		140.50	187
0380	3-1/2"		8	2		156	88		244	310
0400	4"		6	2.667		146	117		263	340
0500	Tee, straight									
0510	1/4"	1 Plum	10	.800	Ea.	9.10	39		48.10	70.50
0520	3/8"		10	.800		8.85	39		47.85	70
0530	1/2"		9	.889		5.75	43.50		49.25	73.50
0540	3/4"		9	.889		6.70	43.50		50.20	74.50
0550	1"		8	1		5.95	49		54.95	82
0560	1-1/4"	Q-1	14	1.143		10.85	50		60.85	90
0570	1-1/2"		13	1.231		14.15	54		68.15	99.50
0580	2"		11	1.455		19.70	64		83.70	121
0590	2-1/2"		9	1.778		51	78		129	177
0600	3"		6	2.667		79	117		196	268
0610	3-1/2"		5	3.200		158	140		298	390
0620	4"		4	4		153	176		329	440
0700	Standard weight, galvanized cast iron									
0720	90° Elbow, straight									
0730	1/4"	1 Plum	16	.500	Ea.	9.10	24.50		33.60	48
0740	3/8"		16	.500		9.10	24.50		33.60	48
0750	1/2"		15	.533		6.90	26		32.90	48
0760	3/4"		14	.571		10.20	28		38.20	54
0770	1"		13	.615		11.75	30		41.75	59.50
0780	1-1/4"	Q-1	22	.727		18.30	32		50.30	69.50

22 11 13 – Facility Water Distribution Piping

22 11 13.45 Pipe Fittings, Steel, Threaded		Crew	Daily Output	Labor-Hours	Unit	Material	2009 Bare Costs Labor	Equipment	Total	Total Incl O&P
0790	1-1/2"	Q-1	20	.800	Ea.	25	35		60	82
0800	2"		18	.889		37	39		76	102
0810	2-1/2"		14	1.143		76	50		126	162
0820	3"		10	1.600		116	70		186	236
0830	3-1/2"		8	2		185	88		273	340
0840	4"		6	2.667		212	117		329	415
0900	45° Elbow, straight									
0910	1/4"	1 Plum	16	.500	Ea.	13	24.50		37.50	52.50
0920	3/8"		16	.500		14	24.50		38.50	53.50
0930	1/2"		15	.533		10.35	26		36.35	52
0940	3/4"		14	.571		11.90	28		39.90	56
0950	1"		13	.615		12.45	30		42.45	60
0960	1-1/4"	Q-1	22	.727		16.70	32		48.70	68
0970	1-1/2"		20	.800		31	35		66	88.50
0980	2"		18	.889		43.50	39		82.50	108
0990	2-1/2"		14	1.143		85	50		135	172
1000	3"		10	1.600		136	70		206	259
1010	3-1/2"		8	2		217	88		305	375
1020	4"		6	2.667		294	117		411	505
1100	Tee, straight									
1110	1/4"	1 Plum	10	.800	Ea.	16.85	39		55.85	79
1120	3/8"		10	.800		11.60	39		50.60	73.50
1130	1/2"		9	.889		10.70	43.50		54.20	79
1140	3/4"		9	.889		14.65	43.50		58.15	83
1150	1"		8	1		15.90	49		64.90	93
1160	1-1/4"	Q-1	14	1.143		28	50		78	109
1170	1-1/2"		13	1.231		37	54		91	125
1180	2"		11	1.455		46	64		110	150
1190	2-1/2"		9	1.778		99	78		177	230
1200	3"		6	2.667		256	117		373	460
1210	3-1/2"		5	3.200		263	140		403	505
1220	4"		4	4		294	176		470	595
1300	Extra heavy weight, black									
1310	Couplings, steel straight									
1320	1/4"	1 Plum	19	.421	Ea.	2.34	20.50		22.84	34.50
1330	3/8"		19	.421		2.56	20.50		23.06	35
1340	1/2"		19	.421		3.46	20.50		23.96	36
1350	3/4"		18	.444		3.70	21.50		25.20	37.50
1360	1"		15	.533		4.70	26		30.70	45.50
1370	1-1/4"	Q-1	26	.615		7.55	27		34.55	50.50
1380	1-1/2"		24	.667		7.55	29.50		37.05	54
1390	2"		21	.762		11.60	33.50		45.10	65
1400	2-1/2"		18	.889		17.05	39		56.05	79.50
1410	3"		14	1.143		20.50	50		70.50	101
1420	3-1/2"		12	1.333		27	58.50		85.50	121
1430	4"		10	1.600		32	70		102	144
1510	90° Elbow, straight									
1520	1/2"	1 Plum	15	.533	Ea.	18.25	26		44.25	60.50
1530	3/4"		14	.571		19.30	28		47.30	64
1540	1"		13	.615		23.50	30		53.50	72
1550	1-1/4"	Q-1	22	.727		35	32		67	87.50
1560	1-1/2"		20	.800		43	35		78	102
1580	2"		18	.889		53	39		92	119

22 11 13.45 Pipe Fittings, Steel, Threaded		Crew	Daily Output	Labor-Hours	Unit	2009 Bare Costs Material	Labor	Equipment	Total	Total Incl O&P
1590	2-1/2"	Q-1	14	1.143	Ea.	124	50		174	215
1600	3"		10	1.600		164	70		234	290
1610	4"		6	2.667		345	117		462	560
1650	45° Elbow, straight									
1660	1/2"	1 Plum	15	.533	Ea.	25.50	26		51.50	68.50
1670	3/4"		14	.571		25	28		53	70.50
1680	1"		13	.615		30	30		60	79.50
1690	1-1/4"	Q-1	22	.727		49.50	32		81.50	104
1700	1-1/2"		20	.800		54.50	35		89.50	115
1710	2"		18	.889		78	39		117	146
1720	2-1/2"		14	1.143		134	50		184	226
1800	Tee, straight									
1810	1/2"	1 Plum	9	.889	Ea.	29	43.50		72.50	98.50
1820	3/4"		9	.889		29	43.50		72.50	98.50
1830	1"		8	1		35	49		84	114
1840	1-1/4"	Q-1	14	1.143		52	50		102	135
1850	1-1/2"		13	1.231		66.50	54		120.50	158
1860	2"		11	1.455		83	64		147	190
1870	2-1/2"		9	1.778		177	78		255	315
1880	3"		6	2.667		241	117		358	445
1890	4"		4	4		465	176		641	780
4000	Standard weight, black									
4010	Couplings, steel straight, merchants									
4030	1/4"	1 Plum	19	.421	Ea.	.63	20.50		21.13	32.50
4040	3/8"		19	.421		.75	20.50		21.25	33
4050	1/2"		19	.421		.80	20.50		21.30	33
4060	3/4"		18	.444		1.03	21.50		22.53	34.50
4070	1"		15	.533		1.44	26		27.44	42
4080	1-1/4"	Q-1	26	.615		1.84	27		28.84	44
4090	1-1/2"		24	.667		2.33	29.50		31.83	48
4100	2"		21	.762		3.33	33.50		36.83	55.50
4110	2-1/2"		18	.889		9.50	39		48.50	71
4120	3"		14	1.143		13.30	50		63.30	92.50
4130	3-1/2"		12	1.333		23.50	58.50		82	117
4140	4"		10	1.600		23.50	70		93.50	135
4166	Plug, 1/4"	1 Plum	38	.211		1.84	10.25		12.09	17.90
4167	3/8"		38	.211		1.84	10.25		12.09	17.90
4168	1/2"		38	.211		1.63	10.25		11.88	17.70
4169	3/4"		32	.250		4.94	12.20		17.14	24.50
4170	1"		30	.267		5.25	13		18.25	26
4171	1-1/4"	Q-1	52	.308		6.05	13.50		19.55	27.50
4172	1-1/2"		48	.333		8.60	14.65		23.25	32
4173	2"		42	.381		11.10	16.70		27.80	38
4176	2-1/2"		36	.444		16.40	19.50		35.90	48
4180	4"		20	.800		28.50	35		63.50	85.50
5000	Malleable iron, 150 lb.									
5020	Black									
5040	90° elbow, straight									
5060	1/4"	1 Plum	16	.500	Ea.	2.94	24.50		27.44	41
5070	3/8"		16	.500		2.94	24.50		27.44	41
5080	1/2"		15	.533		2.94	26		28.94	43.50
5090	3/4"		14	.571		3.37	28		31.37	46.50
5100	1"		13	.615		4.27	30		34.27	51

22 11 13.45 Pipe Fittings, Steel, Threaded		Crew	Daily Output	Labor-Hours	Unit	2009 Bare Costs Material	Labor	Equipment	Total	Total Incl O&P
5110	1-1/4"	Q-1	22	.727	Ea.	7.05	32		39.05	57.50
5120	1-1/2"		20	.800		9.25	35		44.25	64.50
5130	2"		18	.889		16.70	39		55.70	79
5140	2-1/2"		14	1.143		35.50	50		85.50	117
5150	3"		10	1.600		52	70		122	167
5160	3-1/2"		8	2		139	88		227	289
5170	4"		6	2.667		112	117		229	305
5250	45° elbow, straight									
5270	1/4"	1 Plum	16	.500	Ea.	4.43	24.50		28.93	43
5280	3/8"		16	.500		4.43	24.50		28.93	43
5290	1/2"		15	.533		3.36	26		29.36	44
5300	3/4"		14	.571		4.16	28		32.16	47.50
5310	1"		13	.615		5.25	30		35.25	52.50
5320	1-1/4"	Q-1	22	.727		9.25	32		41.25	59.50
5330	1-1/2"		20	.800		11.45	35		46.45	67
5340	2"		18	.889		17.35	39		56.35	79.50
5350	2-1/2"		14	1.143		50	50		100	133
5360	3"		10	1.600		65	70		135	181
5370	3-1/2"		8	2		124	88		212	272
5380	4"		6	2.667		128	117		245	320
5450	Tee, straight									
5470	1/4"	1 Plum	10	.800	Ea.	4.27	39		43.27	65
5480	3/8"		10	.800		4.27	39		43.27	65
5490	1/2"		9	.889		2.72	43.50		46.22	70
5500	3/4"		9	.889		3.91	43.50		47.41	71.50
5510	1"		8	1		6.70	49		55.70	83
5520	1-1/4"	Q-1	14	1.143		10.85	50		60.85	90
5530	1-1/2"		13	1.231		13.50	54		67.50	99
5540	2"		11	1.455		23	64		87	125
5550	2-1/2"		9	1.778		49.50	78		127.50	176
5560	3"		6	2.667		73	117		190	262
5570	3-1/2"		5	3.200		164	140		304	400
5580	4"		4	4		176	176		352	465
5650	Coupling									
5670	1/4"	1 Plum	19	.421	Ea.	3.65	20.50		24.15	36
5680	3/8"		19	.421		3.65	20.50		24.15	36
5690	1/2"		19	.421		2.81	20.50		23.31	35
5700	3/4"		18	.444		3.30	21.50		24.80	37
5710	1"		15	.533		4.93	26		30.93	46
5720	1-1/4"	Q-1	26	.615		6.40	27		33.40	49
5730	1-1/2"		24	.667		8.65	29.50		38.15	55
5740	2"		21	.762		12.80	33.50		46.30	66
5750	2-1/2"		18	.889		35.50	39		74.50	99.50
5760	3"		14	1.143		48	50		98	131
5770	3-1/2"		12	1.333		96	58.50		154.50	197
5780	4"		10	1.600		96	70		166	215
5840	Reducer, concentric, 1/4"	1 Plum	19	.421		3.84	20.50		24.34	36
5850	3/8"		19	.421		4.93	20.50		25.43	37.50
5860	1/2"		19	.421		3.89	20.50		24.39	36.50
5870	3/4"		16	.500		4.60	24.50		29.10	43
5880	1"		15	.533		7.25	26		33.25	48.50
5890	1-1/4"	Q-1	26	.615		8.70	27		35.70	51.50
5900	1-1/2"		24	.667		12.45	29.50		41.95	59

22 11 13.45 Pipe Fittings, Steel, Threaded		Crew	Daily Output	Labor-Hours	Unit	Material	Labor	Equipment	Total	Total Incl O&P
						2009 Bare Costs				
5910	2"	Q-1	21	.762	Ea.	18	33.50		51.50	72
5911	2-1/2"		18	.889		40	39		79	105
5981	Bushing, 1/4"	1 Plum	19	.421		3.82	20.50		24.32	36
5982	3/8"		19	.421		4.20	20.50		24.70	36.50
5983	1/2"		19	.421		4.20	20.50		24.70	36.50
5984	3/4"		16	.500		4.04	24.50		28.54	42.50
5985	1"		15	.533		5.90	26		31.90	47
5986	1-1/4"	Q-1	26	.615		7.35	27		34.35	50
5987	1-1/2"		24	.667		8.30	29.50		37.80	54.50
5988	2"		21	.762		8.95	33.50		42.45	62
5989	Cap, 1/4"	1 Plum	38	.211		2.71	10.25		12.96	18.90
5991	3/8"		38	.211		2.71	10.25		12.96	18.90
5992	1/2"		38	.211		2.06	10.25		12.31	18.15
5993	3/4"		32	.250		2.79	12.20		14.99	22
5994	1"		30	.267		3.37	13		16.37	23.50
5995	1-1/4"	Q-1	52	.308		4.44	13.50		17.94	26
5996	1-1/2"		48	.333		6.10	14.65		20.75	29
5997	2"		42	.381		8.90	16.70		25.60	36
6000	For galvanized elbows, tees, and couplings add					20%				
6058	For galvanized reducers, caps and bushings add					20%				
7000	Union, with brass seat									
7010	1/4"	1 Plum	15	.533	Ea.	14.45	26		40.45	56.50
7020	3/8"		15	.533		9.95	26		35.95	51.50
7030	1/2"		14	.571		9.05	28		37.05	53
7040	3/4"		13	.615		10.35	30		40.35	58
7050	1"		12	.667		13.50	32.50		46	65.50
7060	1-1/4"	Q-1	21	.762		19.35	33.50		52.85	73.50
7070	1-1/2"		19	.842		24	37		61	84
7080	2"		17	.941		28	41.50		69.50	95
7090	2-1/2"		13	1.231		83.50	54		137.50	176
7100	3"		9	1.778		101	78		179	232
7250	For galvanized unions, add					15%				
9757	Forged steel, 3000 lb.									
9758	Black									
9760	90° Elbow, 1/4"	1 Plum	16	.500	Ea.	15.35	24.50		39.85	55
9761	3/8"		16	.500		15.35	24.50		39.85	55
9762	1/2"		15	.533		12.45	26		38.45	54
9763	3/4"		14	.571		14.50	28		42.50	59
9764	1"		13	.615		21	30		51	69.50
9765	1-1/4"	Q-1	22	.727		42.50	32		74.50	96
9766	1-1/2"		20	.800		54	35		89	114
9767	2"		18	.889		65	39		104	132
9780	45° Elbow 1/4"	1 Plum	16	.500		19.50	24.50		44	59.50
9781	3/8"		16	.500		19.50	24.50		44	59.50
9782	1/2"		15	.533		19.20	26		45.20	61.50
9783	3/4"		14	.571		22	28		50	67.50
9784	1"		13	.615		30	30		60	80
9785	1-1/4"	Q-1	22	.727		42	32		74	95.50
9786	1-1/2"		20	.800		60	35		95	121
9787	2"		18	.889		83	39		122	152
9800	Tee, 1/4"	1 Plum	10	.800		18.65	39		57.65	81
9801	3/8"		10	.800		18.65	39		57.65	81
9802	1/2"		9	.889		16.80	43.50		60.30	85.50

22 11 13.45 Pipe Fittings, Steel, Threaded

	22 11 13.45 Pipe Fittings, Steel, Threaded	Crew	Daily Output	Labor-Hours	Unit	Material	Labor	Equipment	Total	Total Incl O&P
9803	3/4"	1 Plum	9	.889	Ea.	22.50	43.50		66	91.50
9804	1"		8	1		29	49		78	108
9805	1-1/4"	Q-1	14	1.143		58	50		108	142
9806	1-1/2"		13	1.231		67.50	54		121.50	159
9807	2"		11	1.455		132	64		196	245
9820	Reducer, concentric, 1/4"	1 Plum	19	.421		10.35	20.50		30.85	43.50
9821	3/8"		19	.421		10.80	20.50		31.30	44
9822	1/2"		17	.471		10.80	23		33.80	47.50
9823	3/4"		16	.500		12.80	24.50		37.30	52
9824	1"		15	.533		16.65	26		42.65	59
9825	1-1/4"	Q-1	26	.615		27.50	27		54.50	72.50
9826	1-1/2"		24	.667		30	29.50		59.50	78.50
9827	2"		21	.762		43.50	33.50		77	100
9840	Cap, 1/4"	1 Plum	38	.211		5.75	10.25		16	22
9841	3/8"		38	.211		5.75	10.25		16	22
9842	1/2"		34	.235		5.60	11.45		17.05	24
9843	3/4"		32	.250		7.80	12.20		20	27.50
9844	1"		30	.267		11.95	13		24.95	33
9845	1-1/4"	Q-1	52	.308		18.90	13.50		32.40	42
9846	1-1/2"		48	.333		22	14.65		36.65	47
9847	2"		42	.381		32.50	16.70		49.20	61.50
9860	Plug, 1/4"	1 Plum	38	.211		2.54	10.25		12.79	18.70
9861	3/8"		38	.211		2.67	10.25		12.92	18.85
9862	1/2"		34	.235		2.80	11.45		14.25	21
9863	3/4"		32	.250		3.48	12.20		15.68	22.50
9864	1"		30	.267		5.35	13		18.35	26
9865	1-1/4"	Q-1	52	.308		10.80	13.50		24.30	33
9866	1-1/2"		48	.333		12.55	14.65		27.20	36.50
9867	2"		42	.381		19.85	16.70		36.55	48
9880	Union, bronze seat, 1/4"	1 Plum	15	.533		41.50	26		67.50	86.50
9881	3/8"		15	.533		41.50	26		67.50	86.50
9882	1/2"		14	.571		40	28		68	87
9883	3/4"		13	.615		53.50	30		83.50	105
9884	1"		12	.667		60.50	32.50		93	117
9885	1-1/4"	Q-1	21	.762		117	33.50		150.50	181
9886	1-1/2"		19	.842		121	37		158	191
9887	2"		17	.941		143	41.50		184.50	221
9900	Coupling, 1/4"	1 Plum	19	.421		5.55	20.50		26.05	38
9901	3/8"		19	.421		5.55	20.50		26.05	38
9902	1/2"		17	.471		4.81	23		27.81	41
9903	3/4"		16	.500		6.15	24.50		30.65	45
9904	1"		15	.533		10.90	26		36.90	52.50
9905	1-1/4"	Q-1	26	.615		18.15	27		45.15	62
9906	1-1/2"		24	.667		23.50	29.50		53	71.50
9907	2"		21	.762		28	33.50		61.50	83
9990	Minimum labor/equipment charge	1 Plum	4	2	Job		97.50		97.50	151

22 11 13.47 Pipe Fittings, Steel

	22 11 13.47 Pipe Fittings, Steel	Crew	Daily Output	Labor-Hours	Unit	Material	Labor	Equipment	Total	Total Incl O&P
0010	**PIPE FITTINGS, STEEL**, Flanged, Welded & Special									
0020	Flanged joints, C.I., standard weight, black. One gasket & bolt									
0040	set, mat'l only, required at each joint, not included (see line 0620)									
0060	90° Elbow, straight, 1-1/2" pipe size	Q-1	14	1.143	Ea.	320	50		370	430
0080	2" pipe size		13	1.231		194	54		248	297

22 11 13.47 Pipe Fittings, Steel		Crew	Daily Output	Labor-Hours	Unit	2009 Bare Costs Material	Labor	Equipment	Total	Total Incl O&P
0090	2-1/2" pipe size	Q-1	12	1.333	Ea.	210	58.50		268.50	320
0100	3" pipe size		11	1.455		175	64		239	291
0110	4" pipe size		8	2		216	88		304	375
0120	5" pipe size	↓	7	2.286		510	100		610	720
0130	6" pipe size	Q-2	9	2.667		340	121		461	560
0140	8" pipe size		8	3		585	137		722	855
0150	10" pipe size		7	3.429		1,275	156		1,431	1,650
0160	12" pipe size	↓	6	4		2,575	182		2,757	3,125
0200	45° Elbow, straight, 1-1/2" pipe size	Q-1	14	1.143		385	50		435	500
0220	2" pipe size		13	1.231		281	54		335	395
0230	2-1/2" pipe size		12	1.333		299	58.50		357.50	420
0240	3" pipe size		11	1.455		291	64		355	420
0250	4" pipe size		8	2		330	88		418	495
0260	5" pipe size	↓	7	2.286		785	100		885	1,025
0270	6" pipe size	Q-2	9	2.667		535	121		656	780
0280	8" pipe size		8	3		785	137		922	1,075
0290	10" pipe size		7	3.429		1,650	156		1,806	2,075
0300	12" pipe size	↓	6	4		2,525	182		2,707	3,050
0350	Tee, straight, 1-1/2" pipe size	Q-1	10	1.600		365	70		435	510
0370	2" pipe size		9	1.778		212	78		290	355
0380	2-1/2" pipe size		8	2		310	88		398	475
0390	3" pipe size		7	2.286		217	100		317	395
0400	4" pipe size		5	3.200		330	140		470	580
0410	5" pipe size	↓	4	4		880	176		1,056	1,250
0420	6" pipe size	Q-2	6	4		480	182		662	805
0430	8" pipe size		5	4.800		815	218		1,033	1,250
0440	10" pipe size		4	6		2,200	273		2,473	2,850
0450	12" pipe size	↓	3	8		3,450	365		3,815	4,375
0500	For galvanized elbows and tees, add					100%				
0520	For extra heavy weight elbows and tees, add					140%				
0620	Gasket and bolt set, 150#, 1/2" pipe size	1 Plum	20	.400		2.16	19.50		21.66	32.50
0622	3/4" pipe size		19	.421		2.82	20.50		23.32	35
0624	1" pipe size		18	.444		2.31	21.50		23.81	36
0626	1-1/4" pipe size		17	.471		2.43	23		25.43	38
0628	1-1/2" pipe size		15	.533		2.58	26		28.58	43.50
0630	2" pipe size		13	.615		4.69	30		34.69	51.50
0640	2-1/2" pipe size		12	.667		4.84	32.50		37.34	56
0650	3" pipe size		11	.727		5	35.50		40.50	60.50
0660	3-1/2" pipe size		9	.889		8.15	43.50		51.65	76
0670	4" pipe size		8	1		9	49		58	85.50
0680	5" pipe size		7	1.143		14.80	55.50		70.30	103
0690	6" pipe size		6	1.333		14.80	65		79.80	117
0700	8" pipe size		5	1.600		16.95	78		94.95	140
0710	10" pipe size		4.50	1.778		30.50	86.50		117	168
0720	12" pipe size		4.20	1.905		32.50	93		125.50	180
0730	14" pipe size		4	2		34.50	97.50		132	189
0740	16" pipe size		3	2.667		36	130		166	242
0750	18" pipe size		2.70	2.963		70	144		214	300
0760	20" pipe size		2.30	3.478		111	170		281	385
0780	24" pipe size		1.90	4.211		142	205		347	475
0790	26" pipe size		1.60	5		192	244		436	590
0810	30" pipe size		1.40	5.714		375	279		654	840
0830	36" pipe size	↓	1.10	7.273	↓	700	355		1,055	1,325

22 11 13.47 Pipe Fittings, Steel		Crew	Daily Output	Labor-Hours	Unit	Material	2009 Bare Costs Labor	Equipment	Total	Total Incl O&P
0850	For 300 lb gasket set, add					40%				
2000	Flanged unions, 125 lb., black, 1/2" pipe size	1 Plum	17	.471	Ea.	44	23		67	83.50
2040	3/4" pipe size		17	.471		57.50	23		80.50	99
2050	1" pipe size		16	.500		56	24.50		80.50	99.50
2060	1-1/4" pipe size	Q-1	28	.571		66	25		91	112
2070	1-1/2" pipe size		27	.593		61	26		87	108
2080	2" pipe size		26	.615		72	27		99	121
2090	2-1/2" pipe size		24	.667		97.50	29.50		127	153
2100	3" pipe size		22	.727		110	32		142	171
2110	3-1/2" pipe size		18	.889		187	39		226	266
2120	4" pipe size		16	1		150	44		194	233
2130	5" pipe size		14	1.143		350	50		400	465
2140	6" pipe size	Q-2	19	1.263		330	57.50		387.50	450
2150	8" pipe size	"	16	1.500		730	68.50		798.50	905
2200	For galvanized unions, add					150%				
2290	Threaded flange									
2300	Cast iron									
2310	Black, 125 lb., per flange									
2320	1" pipe size	1 Plum	27	.296	Ea.	27	14.45		41.45	52
2330	1-1/4" pipe size	Q-1	44	.364		32	15.95		47.95	60.50
2340	1-1/2" pipe size		40	.400		29.50	17.55		47.05	59
2350	2" pipe size		36	.444		30	19.50		49.50	63
2360	2-1/2" pipe size		28	.571		34.50	25		59.50	77
2370	3" pipe size		20	.800		44.50	35		79.50	104
2380	3-1/2" pipe size		16	1		62.50	44		106.50	137
2390	4" pipe size		12	1.333		60	58.50		118.50	157
2400	5" pipe size		10	1.600		85	70		155	203
2410	6" pipe size	Q-2	14	1.714		96	78		174	227
2420	8" pipe size		12	2		151	91		242	310
2430	10" pipe size		10	2.400		268	109		377	465
2440	12" pipe size		8	3		575	137		712	840
2460	For galvanized flanges, add					95%				
2490	Blind flange									
2492	Cast iron									
2494	Black, 125 lb., per flange									
2496	1" pipe size	1 Plum	27	.296	Ea.	42.50	14.45		56.95	69.50
2500	1-1/2" pipe size	Q-1	40	.400		48	17.55		65.55	79.50
2502	2" pipe size		36	.444		54	19.50		73.50	89.50
2504	2-1/2" pipe size		28	.571		59.50	25		84.50	105
2506	3" pipe size		20	.800		71.50	35		106.50	133
2508	4" pipe size		12	1.333		91	58.50		149.50	191
2510	5" pipe size		10	1.600		148	70		218	272
2512	6" pipe size	Q-2	14	1.714		161	78		239	298
2514	8" pipe size		12	2		253	91		344	420
2516	10" pipe size		10	2.400		375	109		484	580
2518	12" pipe size		8	3		680	137		817	960
2520	For galvanized flanges, add					80%				
2570	Threaded flange									
2580	Forged steel,									
2590	Black 150 lb., per flange									
2600	1/2" pipe size	1 Plum	30	.267	Ea.	21	13		34	43
2610	3/4" pipe size		28	.286		21	13.95		34.95	44.50
2620	1" pipe size		27	.296		21	14.45		35.45	45.50

22 11 13.47 Pipe Fittings, Steel		Crew	Daily Output	Labor-Hours	Unit	2009 Bare Costs Material	Labor	Equipment	Total	Total Incl O&P
2630	1-1/4" pipe size	Q-1	44	.364	Ea.	21	15.95		36.95	48
2640	1-1/2" pipe size		40	.400		21	17.55		38.55	50
2650	2" pipe size		36	.444		21	19.50		40.50	53
2660	2-1/2" pipe size		28	.571		26	25		51	67.50
2670	3" pipe size		20	.800		26	35		61	83
2690	4" pipe size		12	1.333		32	58.50		90.50	126
2700	5" pipe size		10	1.600		46.50	70		116.50	161
2710	6" pipe size	Q-2	14	1.714		54	78		132	181
2720	8" pipe size		12	2		88.50	91		179.50	239
2730	10" pipe size		10	2.400		166	109		275	350
2860	Black 300 lb., per flange									
2870	1/2" pipe size	1 Plum	30	.267	Ea.	23	13		36	45.50
2880	3/4" pipe size		28	.286		23	13.95		36.95	47
2890	1" pipe size		27	.296		23	14.45		37.45	48
2900	1-1/4" pipe size	Q-1	44	.364		23	15.95		38.95	50.50
2910	1-1/2" pipe size		40	.400		23	17.55		40.55	52.50
2920	2" pipe size		36	.444		24.50	19.50		44	57
2930	2-1/2" pipe size		28	.571		33.50	25		58.50	76
2940	3" pipe size		20	.800		37.50	35		72.50	95.50
2960	4" pipe size		12	1.333		53	58.50		111.50	149
2970	6" pipe size	Q-2	14	1.714		102	78		180	233
3000	Weld joint, butt, carbon steel, standard weight									
3040	90° elbow, long radius									
3050	1/2" pipe size	Q-15	16	1	Ea.	27.50	44	3.48	74.98	102
3060	3/4" pipe size		16	1		27.50	44	3.48	74.98	102
3070	1" pipe size		16	1		12.80	44	3.48	60.28	86
3080	1-1/4" pipe size		14	1.143		12.80	50	3.98	66.78	96.50
3090	1-1/2" pipe size		13	1.231		12.80	54	4.28	71.08	103
3100	2" pipe size		10	1.600		14.40	70	5.55	89.95	131
3110	2-1/2" pipe size		8	2		16.95	88	6.95	111.90	162
3120	3" pipe size		7	2.286		20.50	100	7.95	128.45	187
3130	4" pipe size		5	3.200		34	140	11.15	185.15	268
3136	5" pipe size		4	4		75	176	13.90	264.90	370
3140	6" pipe size	Q-16	5	4.800		74.50	218	11.15	303.65	435
3150	8" pipe size		3.75	6.400		141	291	14.85	446.85	620
3160	10" pipe size		3	8		283	365	18.55	666.55	895
3170	12" pipe size		2.50	9.600		420	435	22.50	877.50	1,150
3180	14" pipe size		2	12		625	545	28	1,198	1,550
3190	16" pipe size		1.50	16		840	730	37	1,607	2,100
3191	18" pipe size		1.25	19.200		1,125	875	44.50	2,044.50	2,625
3192	20" pipe size		1.15	20.870		1,525	950	48.50	2,523.50	3,200
3194	24" pipe size		1.02	23.529		2,225	1,075	54.50	3,354.50	4,150
3200	45° Elbow, long									
3210	1/2" pipe size	Q-15	16	1	Ea.	38.50	44	3.48	85.98	114
3220	3/4" pipe size		16	1		38.50	44	3.48	85.98	114
3230	1" pipe size		16	1		13.75	44	3.48	61.23	87
3240	1-1/4" pipe size		14	1.143		13.75	50	3.98	67.73	97.50
3250	1-1/2" pipe size		13	1.231		13.75	54	4.28	72.03	104
3260	2" pipe size		10	1.600		13.75	70	5.55	89.30	130
3270	2-1/2" pipe size		8	2		16	88	6.95	110.95	161
3280	3" pipe size		7	2.286		16.95	100	7.95	124.90	183
3290	4" pipe size		5	3.200		30.50	140	11.15	181.65	264
3296	5" pipe size		4	4		46.50	176	13.90	236.40	340

22 11 13.47 Pipe Fittings, Steel		Crew	Daily Output	Labor-Hours	Unit	2009 Bare Costs Material	Labor	Equipment	Total	Total Incl O&P
3300	6" pipe size	Q-16	5	4.800	Ea.	60	218	11.15	289.15	420
3310	8" pipe size		3.75	6.400		100	291	14.85	405.85	575
3320	10" pipe size		3	8		200	365	18.55	583.55	805
3330	12" pipe size		2.50	9.600		283	435	22.50	740.50	1,000
3340	14" pipe size		2	12		380	545	28	953	1,300
3341	16" pipe size		1.50	16		685	730	37	1,452	1,925
3342	18" pipe size		1.25	19.200		975	875	44.50	1,894.50	2,475
3343	20" pipe size		1.15	20.870		1,000	950	48.50	1,998.50	2,625
3345	24" pipe size		1.05	22.857		1,525	1,050	53	2,628	3,350
3346	26" pipe size		.85	28.235		1,700	1,275	65.50	3,040.50	3,925
3347	30" pipe size		.45	53.333		1,850	2,425	124	4,399	5,950
3349	36" pipe size		.38	63.158		2,050	2,875	147	5,072	6,850
3350	Tee, straight									
3360	1/2" pipe size	Q-15	10	1.600	Ea.	68.50	70	5.55	144.05	191
3370	3/4" pipe size		10	1.600		68.50	70	5.55	144.05	191
3380	1" pipe size		10	1.600		34	70	5.55	109.55	153
3390	1-1/4" pipe size		9	1.778		42.50	78	6.20	126.70	175
3400	1-1/2" pipe size		8	2		42.50	88	6.95	137.45	191
3410	2" pipe size		6	2.667		34	117	9.30	160.30	229
3420	2-1/2" pipe size		5	3.200		46.50	140	11.15	197.65	281
3430	3" pipe size		4	4		51.50	176	13.90	241.40	345
3440	4" pipe size		3	5.333		72.50	234	18.55	325.05	465
3446	5" pipe size		2.50	6.400		120	281	22.50	423.50	590
3450	6" pipe size	Q-16	3	8		130	365	18.55	513.55	730
3460	8" pipe size		2.50	9.600		228	435	22.50	685.50	950
3470	10" pipe size		2	12		450	545	28	1,023	1,375
3480	12" pipe size		1.60	15		630	685	35	1,350	1,775
3481	14" pipe size		1.30	18.462		1,125	840	43	2,008	2,600
3482	16" pipe size		1	24		1,250	1,100	55.50	2,405.50	3,125
3483	18" pipe size		.80	30		1,875	1,375	69.50	3,319.50	4,275
3484	20" pipe size		.75	32		3,025	1,450	74	4,549	5,675
3486	24" pipe size		.70	34.286		3,925	1,550	79.50	5,554.50	6,825
3487	26" pipe size		.55	43.636		4,325	1,975	101	6,401	7,925
3488	30" pipe size		.30	80		4,725	3,650	186	8,561	11,100
3490	36" pipe size		.25	96		5,225	4,375	223	9,823	12,800
3491	Eccentric reducer, 1-1/2" pipe size	Q-15	14	1.143		87.50	50	3.98	141.48	179
3492	2" pipe size		11	1.455		124	64	5.05	193.05	242
3493	2-1/2" pipe size		9	1.778		157	78	6.20	241.20	300
3494	3" pipe size		8	2		167	88	6.95	261.95	330
3495	4" pipe size		6	2.667		183	117	9.30	309.30	390
3496	6" pipe size	Q-16	5	4.800		320	218	11.15	549.15	700
3497	8" pipe size		4	6		365	273	13.90	651.90	840
3498	10" pipe size		3	8		425	365	18.55	808.55	1,050
3499	12" pipe size		2.50	9.600		605	435	22.50	1,062.50	1,375
3501	Cap, 1-1/2" pipe size	Q-15	28	.571		15.35	25	1.99	42.34	58
3502	2" pipe size		22	.727		15.05	32	2.53	49.58	69
3503	2-1/2" pipe size		18	.889		16.30	39	3.09	58.39	82
3504	3" pipe size		16	1		16.30	44	3.48	63.78	90
3505	4" pipe size		12	1.333		23.50	58.50	4.64	86.64	121
3506	6" pipe size	Q-16	10	2.400		40	109	5.55	154.55	219
3507	8" pipe size		8	3		60.50	137	6.95	204.45	286
3508	10" pipe size		6	4		109	182	9.30	300.30	410
3509	12" pipe size		5	4.800		165	218	11.15	394.15	535

22 11 13.47 Pipe Fittings, Steel		Crew	Daily Output	Labor-Hours	Unit	Material (2009 Bare Costs)	Labor (2009 Bare Costs)	Equipment (2009 Bare Costs)	Total (2009 Bare Costs)	Total Incl O&P
3511	14" pipe size	Q-16	4	6	Ea.	220	273	13.90	506.90	685
3512	16" pipe size		4	6		257	273	13.90	543.90	725
3513	18" pipe size		3	8		350	365	18.55	733.55	970
3517	Weld joint, butt, carbon steel, extra strong									
3519	90° elbow, long									
3520	1/2" pipe size	Q-15	13	1.231	Ea.	35	54	4.28	93.28	127
3530	3/4" pipe size		12	1.333		35	58.50	4.64	98.14	134
3540	1" pipe size		11	1.455		16.95	64	5.05	86	123
3550	1-1/4" pipe size		10	1.600		16.95	70	5.55	92.50	134
3560	1-1/2" pipe size		9	1.778		16.95	78	6.20	101.15	146
3570	2" pipe size		8	2		16.95	88	6.95	111.90	162
3580	2-1/2" pipe size		7	2.286		24.50	100	7.95	132.45	192
3590	3" pipe size		6	2.667		30.50	117	9.30	156.80	225
3600	4" pipe size		4	4		50.50	176	13.90	240.40	345
3606	5" pipe size		3.50	4.571		107	201	15.90	323.90	445
3610	6" pipe size	Q-16	4.50	5.333		127	243	12.35	382.35	530
3620	8" pipe size		3.50	6.857		244	310	15.90	569.90	770
3630	10" pipe size		2.50	9.600		545	435	22.50	1,002.50	1,300
3640	12" pipe size		2.25	10.667		630	485	25	1,140	1,475
3650	45° Elbow, long									
3660	1/2" pipe size	Q-15	13	1.231	Ea.	38.50	54	4.28	96.78	131
3670	3/4" pipe size		12	1.333		38.50	58.50	4.64	101.64	138
3680	1" pipe size		11	1.455		17.60	64	5.05	86.65	124
3690	1-1/4" pipe size		10	1.600		17.60	70	5.55	93.15	135
3700	1-1/2" pipe size		9	1.778		17.60	78	6.20	101.80	147
3710	2" pipe size		8	2		17.60	88	6.95	112.55	163
3720	2-1/2" pipe size		7	2.286		39	100	7.95	146.95	208
3730	3" pipe size		6	2.667		22.50	117	9.30	148.80	216
3740	4" pipe size		4	4		36	176	13.90	225.90	325
3746	5" pipe size		3.50	4.571		86	201	15.90	302.90	420
3750	6" pipe size	Q-16	4.50	5.333		99	243	12.35	354.35	500
3760	8" pipe size		3.50	6.857		172	310	15.90	497.90	695
3770	10" pipe size		2.50	9.600		330	435	22.50	787.50	1,075
3780	12" pipe size		2.25	10.667		495	485	25	1,005	1,325
3800	Tee, straight									
3810	1/2" pipe size	Q-15	9	1.778	Ea.	85.50	78	6.20	169.70	222
3820	3/4" pipe size		8.50	1.882		81.50	82.50	6.55	170.55	225
3830	1" pipe size		8	2		45	88	6.95	139.95	193
3840	1-1/4" pipe size		7	2.286		42	100	7.95	149.95	211
3850	1-1/2" pipe size		6	2.667		51	117	9.30	177.30	247
3860	2" pipe size		5	3.200		42	140	11.15	193.15	277
3870	2-1/2" pipe size		4	4		70	176	13.90	259.90	365
3880	3" pipe size		3.50	4.571		65.50	201	15.90	282.40	400
3890	4" pipe size		2.50	6.400		106	281	22.50	409.50	575
3896	5" pipe size		2.25	7.111		228	310	25	563	765
3900	6" pipe size	Q-16	2.25	10.667		186	485	25	696	985
3910	8" pipe size		2	12		350	545	28	923	1,250
3920	10" pipe size		1.75	13.714		600	625	32	1,257	1,675
3930	12" pipe size		1.50	16		910	730	37	1,677	2,175
4000	Eccentric reducer, 1-1/2" pipe size	Q-15	10	1.600		85	70	5.55	160.55	209
4010	2" pipe size		9	1.778		231	78	6.20	315.20	385
4020	2-1/2" pipe size		8	2		158	88	6.95	252.95	320
4030	3" pipe size		7	2.286		203	100	7.95	310.95	390

22 11 13.47 Pipe Fittings, Steel		Crew	Daily Output	Labor-Hours	Unit	2009 Bare Costs Material	Labor	Equipment	Total	Total Incl O&P
4040	4" pipe size	Q-15	5	3.200	Ea.	297	140	11.15	448.15	555
4046	5" pipe size		4.70	3.404		625	149	11.85	785.85	930
4050	6" pipe size	Q-16	4.50	5.333		415	243	12.35	670.35	850
4060	8" pipe size		3.50	6.857		460	310	15.90	785.90	1,000
4070	10" pipe size		2.50	9.600		585	435	22.50	1,042.50	1,350
4080	12" pipe size		2.25	10.667		640	485	25	1,150	1,475
4090	14" pipe size		2.10	11.429		1,250	520	26.50	1,796.50	2,200
4100	16" pipe size		1.90	12.632		1,250	575	29.50	1,854.50	2,300
4151	Cap, 1-1/2" pipe size	Q-15	24	.667		19.50	29.50	2.32	51.32	69.50
4152	2" pipe size		18	.889		15.05	39	3.09	57.14	80.50
4153	2-1/2" pipe size		16	1		21	44	3.48	68.48	95
4154	3" pipe size		14	1.143		23	50	3.98	76.98	108
4155	4" pipe size		10	1.600		32	70	5.55	107.55	150
4156	6" pipe size	Q-16	9	2.667		64.50	121	6.20	191.70	266
4157	8" pipe size		7	3.429		97	156	7.95	260.95	360
4158	10" pipe size		5	4.800		149	218	11.15	378.15	515
4159	12" pipe size		4	6		199	273	13.90	485.90	660
4200	Welding ring w/spacer pins, 2" pipe size					2.96			2.96	3.26
4210	2-1/2" pipe size					3.58			3.58	3.94
4220	3" pipe size					3.70			3.70	4.07
4230	4" pipe size					3.85			3.85	4.24
4236	5" pipe size					4.53			4.53	4.98
4240	6" pipe size					4.53			4.53	4.98
4250	8" pipe size					5.35			5.35	5.90
4260	10" pipe size					6.55			6.55	7.20
4270	12" pipe size					7.50			7.50	8.25
4280	14" pipe size					8.75			8.75	9.60
4290	16" pipe size					9.85			9.85	10.85
4300	18" pipe size					11.40			11.40	12.50
4310	20" pipe size					12.05			12.05	13.30
4330	24" pipe size					14.05			14.05	15.45
4340	26" pipe size					18.75			18.75	20.50
4350	30" pipe size					18.75			18.75	20.50
4370	36" pipe size					21			21	23
5000	Weld joint, socket, forged steel, 3000 lb., schedule 40 pipe									
5010	90° elbow, straight									
5020	1/4" pipe size	Q-15	22	.727	Ea.	18.75	32	2.53	53.28	73
5030	3/8" pipe size		22	.727		18.75	32	2.53	53.28	73
5040	1/2" pipe size		20	.800		10.45	35	2.78	48.23	69
5050	3/4" pipe size		20	.800		11.10	35	2.78	48.88	70
5060	1" pipe size		20	.800		14.10	35	2.78	51.88	73
5070	1-1/4" pipe size		18	.889		27	39	3.09	69.09	93.50
5080	1-1/2" pipe size		16	1		30	44	3.48	77.48	105
5090	2" pipe size		12	1.333		46	58.50	4.64	109.14	146
5100	2-1/2" pipe size		10	1.600		131	70	5.55	206.55	259
5110	3" pipe size		8	2		230	88	6.95	324.95	395
5120	4" pipe size		6	2.667		595	117	9.30	721.30	845
5130	45° Elbow, straight									
5134	1/4" pipe size	Q-15	22	.727	Ea.	18.85	32	2.53	53.38	73.50
5135	3/8" pipe size		22	.727		18.85	32	2.53	53.38	73.50
5136	1/2" pipe size		20	.800		13.95	35	2.78	51.73	73
5137	3/4" pipe size		20	.800		16.20	35	2.78	53.98	75.50
5140	1" pipe size		20	.800		21	35	2.78	58.78	81

22 11 13.47 Pipe Fittings, Steel		Crew	Daily Output	Labor-Hours	Unit	2009 Bare Costs Material	Labor	Equipment	Total	Total Incl O&P
5150	1-1/4" pipe size	Q-15	18	.889	Ea.	29	39	3.09	71.09	96
5160	1-1/2" pipe size		16	1		35.50	44	3.48	82.98	111
5170	2" pipe size		12	1.333		57.50	58.50	4.64	120.64	159
5180	2-1/2" pipe size		10	1.600		150	70	5.55	225.55	281
5190	3" pipe size		8	2		250	88	6.95	344.95	420
5200	4" pipe size		6	2.667		495	117	9.30	621.30	730
5250	Tee, straight									
5254	1/4" pipe size	Q-15	15	1.067	Ea.	20.50	47	3.71	71.21	99.50
5255	3/8" pipe size		15	1.067		20.50	47	3.71	71.21	99.50
5256	1/2" pipe size		13	1.231		12.95	54	4.28	71.23	103
5257	3/4" pipe size		13	1.231		15.70	54	4.28	73.98	106
5260	1" pipe size		13	1.231		21.50	54	4.28	79.78	112
5270	1-1/4" pipe size		12	1.333		33	58.50	4.64	96.14	132
5280	1-1/2" pipe size		11	1.455		44	64	5.05	113.05	153
5290	2" pipe size		8	2		63.50	88	6.95	158.45	213
5300	2-1/2" pipe size		6	2.667		182	117	9.30	308.30	390
5310	3" pipe size		5	3.200		435	140	11.15	586.15	710
5320	4" pipe size		4	4		700	176	13.90	889.90	1,050
5350	For reducing sizes, add					60%				
5450	Couplings									
5451	1/4" pipe size	Q-15	23	.696	Ea.	11.65	30.50	2.42	44.57	63
5452	3/8" pipe size		23	.696		11.65	30.50	2.42	44.57	63
5453	1/2" pipe size		21	.762		5.40	33.50	2.65	41.55	61
5454	3/4" pipe size		21	.762		6.95	33.50	2.65	43.10	62.50
5460	1" pipe size		20	.800		7.70	35	2.78	45.48	66
5470	1-1/4" pipe size		20	.800		13.60	35	2.78	51.38	72.50
5480	1-1/2" pipe size		18	.889		15.05	39	3.09	57.14	80.50
5490	2" pipe size		14	1.143		24	50	3.98	77.98	109
5500	2-1/2" pipe size		12	1.333		54	58.50	4.64	117.14	155
5510	3" pipe size		9	1.778		121	78	6.20	205.20	261
5520	4" pipe size		7	2.286		182	100	7.95	289.95	365
5570	Union, 1/4" pipe size		21	.762		25	33.50	2.65	61.15	82.50
5571	3/8" pipe size		21	.762		25	33.50	2.65	61.15	82.50
5572	1/2" pipe size		19	.842		21.50	37	2.93	61.43	84
5573	3/4" pipe size		19	.842		24.50	37	2.93	64.43	87.50
5574	1" pipe size		19	.842		31.50	37	2.93	71.43	95
5575	1-1/4" pipe size		17	.941		51.50	41.50	3.28	96.28	125
5576	1-1/2" pipe size		15	1.067		55	47	3.71	105.71	137
5577	2" pipe size		11	1.455		76	64	5.05	145.05	188
5600	Reducer, 1/4" pipe size		23	.696		27.50	30.50	2.42	60.42	80
5601	3/8" pipe size		23	.696		28	30.50	2.42	60.92	81
5602	1/2" pipe size		21	.762		19.10	33.50	2.65	55.25	76
5603	3/4" pipe size		21	.762		19.15	33.50	2.65	55.30	76
5604	1" pipe size		21	.762		21	33.50	2.65	57.15	78.50
5605	1-1/4" pipe size		19	.842		35	37	2.93	74.93	99
5607	1-1/2" pipe size		17	.941		41	41.50	3.28	85.78	113
5608	2" pipe size		13	1.231		45.50	54	4.28	103.78	139
5612	Cap, 1/4" pipe size		46	.348		11.15	15.25	1.21	27.61	37
5613	3/8" pipe size		46	.348		11.15	15.25	1.21	27.61	37
5614	1/2" pipe size		42	.381		6.90	16.70	1.33	24.93	35
5615	3/4" pipe size		42	.381		8.05	16.70	1.33	26.08	36.50
5616	1" pipe size		42	.381		12.20	16.70	1.33	30.23	41
5617	1-1/4" pipe size		38	.421		14.20	18.50	1.47	34.17	45.50

22 11 13.47 Pipe Fittings, Steel		Crew	Daily Output	Labor-Hours	Unit	Material	2009 Bare Costs Labor	Equipment	Total	Total Incl O&P
5618	1-1/2" pipe size	Q-15	34	.471	Ea.	19.65	20.50	1.64	41.79	55.50
5619	2" pipe size		26	.615		30.50	27	2.14	59.64	78
5630	T-O-L, 1/4" pipe size, nozzle		23	.696		6.10	30.50	2.42	39.02	57
5631	3/8" pipe size, nozzle		23	.696		6.10	30.50	2.42	39.02	57
5632	1/2" pipe size, nozzle		22	.727		6.10	32	2.53	40.63	59
5633	3/4" pipe size, nozzle		21	.762		7	33.50	2.65	43.15	62.50
5634	1" pipe size, nozzle		20	.800		8.15	35	2.78	45.93	66.50
5635	1-1/4" pipe size, nozzle		18	.889		12.95	39	3.09	55.04	78
5636	1-1/2" pipe size, nozzle		16	1		12.95	44	3.48	60.43	86
5637	2" pipe size, nozzle		12	1.333		14.75	58.50	4.64	77.89	112
5638	2-1/2" pipe size, nozzle		10	1.600		49.50	70	5.55	125.05	170
5639	4" pipe size, nozzle		6	2.667		112	117	9.30	238.30	315
5640	W-O-L, 1/4" pipe size, nozzle		23	.696		12.60	30.50	2.42	45.52	64
5641	3/8" pipe size, nozzle		23	.696		12.60	30.50	2.42	45.52	64
5642	1/2" pipe size, nozzle		22	.727		12.60	32	2.53	47.13	66
5643	3/4" pipe size, nozzle		21	.762		13.30	33.50	2.65	49.45	69.50
5644	1" pipe size, nozzle		20	.800		13.90	35	2.78	51.68	73
5645	1-1/4" pipe size, nozzle		18	.889		16.50	39	3.09	58.59	82
5646	1-1/2" pipe size, nozzle		16	1		16.50	44	3.48	63.98	90
5647	2" pipe size, nozzle		12	1.333		16.65	58.50	4.64	79.79	114
5648	2-1/2" pipe size, nozzle		10	1.600		38	70	5.55	113.55	157
5649	3" pipe size, nozzle		8	2		42	88	6.95	136.95	190
5650	4" pipe size, nozzle		6	2.667		53	117	9.30	179.30	249
5651	5" pipe size, nozzle		5	3.200		130	140	11.15	281.15	375
5652	6" pipe size, nozzle		4	4		149	176	13.90	338.90	450
5653	8" pipe size, nozzle		3	5.333		285	234	18.55	537.55	700
5654	10" pipe size, nozzle		2.60	6.154		520	270	21.50	811.50	1,025
5655	12" pipe size, nozzle		2.20	7.273		985	320	25.50	1,330.50	1,600
5656	14" pipe size, nozzle		2	8		2,350	350	28	2,728	3,150
5674	S-O-L, 1/4" pipe size, outlet		23	.696		7.55	30.50	2.42	40.47	58.50
5675	3/8" pipe size, outlet		23	.696		7.55	30.50	2.42	40.47	58.50
5676	1/2" pipe size, outlet		22	.727		7.55	32	2.53	42.08	60.50
5677	3/4" pipe size, outlet		21	.762		7.65	33.50	2.65	43.80	63.50
5678	1" pipe size, outlet		20	.800		8.50	35	2.78	46.28	67
5679	1-1/4" pipe size, outlet		18	.889		14.20	39	3.09	56.29	79.50
5680	1-1/2" pipe size, outlet		16	1		14.20	44	3.48	61.68	87.50
5681	2" pipe size, outlet		12	1.333		15.60	58.50	4.64	78.74	113
6000	Weld-on flange, forged steel									
6020	Slip-on, 150 lb. flange (welded front and back)									
6050	1/2" pipe size	Q-15	18	.889	Ea.	14	39	3.09	56.09	79.50
6060	3/4" pipe size		18	.889		14	39	3.09	56.09	79.50
6070	1" pipe size		17	.941		14	41.50	3.28	58.78	83
6080	1-1/4" pipe size		16	1		14	44	3.48	61.48	87
6090	1-1/2" pipe size		15	1.067		14	47	3.71	64.71	92
6100	2" pipe size		12	1.333		14	58.50	4.64	77.14	111
6110	2-1/2" pipe size		10	1.600		24.50	70	5.55	100.05	142
6120	3" pipe size		9	1.778		21.50	78	6.20	105.70	152
6130	3-1/2" pipe size		7	2.286		25.50	100	7.95	133.45	193
6140	4" pipe size		6	2.667		24.50	117	9.30	150.80	218
6150	5" pipe size		5	3.200		42	140	11.15	193.15	277
6160	6" pipe size	Q-16	6	4		39.50	182	9.30	230.80	335
6170	8" pipe size		5	4.800		63.50	218	11.15	292.65	420
6180	10" pipe size		4	6		104	273	13.90	390.90	555

22 11 13.47 Pipe Fittings, Steel		Crew	Daily Output	Labor-Hours	Unit	Material	Labor	Equipment	Total	Total Incl O&P
						2009 Bare Costs				
6190	12" pipe size	Q-16	3	8	Ea.	153	365	18.55	536.55	755
6191	14" pipe size		2.50	9.600		209	435	22.50	666.50	930
6192	16" pipe size		1.80	13.333		345	605	31	981	1,350
6200	300 lb. flange									
6210	1/2" pipe size	Q-15	17	.941	Ea.	18.50	41.50	3.28	63.28	88
6220	3/4" pipe size		17	.941		18.50	41.50	3.28	63.28	88
6230	1" pipe size		16	1		18.50	44	3.48	65.98	92.50
6240	1-1/4" pipe size		13	1.231		18.50	54	4.28	76.78	109
6250	1-1/2" pipe size		12	1.333		18.50	58.50	4.64	81.64	116
6260	2" pipe size		11	1.455		22	64	5.05	91.05	129
6270	2-1/2" pipe size		9	1.778		27.50	78	6.20	111.70	158
6280	3" pipe size		7	2.286		32	100	7.95	139.95	200
6290	4" pipe size		6	2.667		46.50	117	9.30	172.80	243
6300	5" pipe size		4	4		70	176	13.90	259.90	365
6310	6" pipe size	Q-16	5	4.800		74.50	218	11.15	303.65	435
6320	8" pipe size		4	6		128	273	13.90	414.90	580
6330	10" pipe size		3.40	7.059		202	320	16.40	538.40	740
6340	12" pipe size		2.80	8.571		266	390	19.90	675.90	920
6400	Welding neck, 150 lb. flange									
6410	1/2" pipe size	Q-15	40	.400	Ea.	19.90	17.55	1.39	38.84	50.50
6420	3/4" pipe size		36	.444		19.90	19.50	1.55	40.95	53.50
6430	1" pipe size		32	.500		19.90	22	1.74	43.64	58
6440	1-1/4" pipe size		29	.552		19.90	24	1.92	45.82	61.50
6450	1-1/2" pipe size		26	.615		19.90	27	2.14	49.04	66.50
6460	2" pipe size		20	.800		21.50	35	2.78	59.28	81
6470	2-1/2" pipe size		16	1		23.50	44	3.48	70.98	97.50
6480	3" pipe size		14	1.143		25	50	3.98	78.98	110
6500	4" pipe size		10	1.600		34.50	70	5.55	110.05	153
6510	5" pipe size		8	2		48.50	88	6.95	143.45	197
6520	6" pipe size	Q-16	10	2.400		54.50	109	5.55	169.05	235
6530	8" pipe size		7	3.429		85	156	7.95	248.95	345
6540	10" pipe size		6	4		143	182	9.30	334.30	450
6550	12" pipe size		5	4.800		206	218	11.15	435.15	580
6551	14" pipe size		4.50	5.333		278	243	12.35	533.35	695
6552	16" pipe size		3	8		425	365	18.55	808.55	1,050
6553	18" pipe size		2.50	9.600		620	435	22.50	1,077.50	1,375
6554	20" pipe size		2.30	10.435		720	475	24	1,219	1,550
6556	24" pipe size		2	12		1,025	545	28	1,598	2,000
6557	26" pipe size		1.70	14.118		1,100	640	33	1,773	2,250
6558	30" pipe size		.90	26.667		1,275	1,225	62	2,562	3,350
6559	36" pipe size		.75	32		1,450	1,450	74	2,974	3,925
6560	300 lb. flange									
6570	1/2" pipe size	Q-15	36	.444	Ea.	24.50	19.50	1.55	45.55	58.50
6580	3/4" pipe size		34	.471		24.50	20.50	1.64	46.64	61
6590	1" pipe size		30	.533		24.50	23.50	1.86	49.86	65.50
6600	1-1/4" pipe size		28	.571		24.50	25	1.99	51.49	68
6610	1-1/2" pipe size		24	.667		24.50	29.50	2.32	56.32	75
6620	2" pipe size		18	.889		27	39	3.09	69.09	93.50
6630	2-1/2" pipe size		14	1.143		31.50	50	3.98	85.48	117
6640	3" pipe size		12	1.333		36.50	58.50	4.64	99.64	136
6650	4" pipe size		8	2		58	88	6.95	152.95	208
6660	5" pipe size		7	2.286		85.50	100	7.95	193.45	259
6670	6" pipe size	Q-16	9	2.667		93	121	6.20	220.20	297

22 11 13.47 Pipe Fittings, Steel		Crew	Daily Output	Labor-Hours	Unit	2009 Bare Costs Material	Labor	Equipment	Total	Total Incl O&P
6680	8" pipe size	Q-16	6	4	Ea.	159	182	9.30	350.30	465
6690	10" pipe size		5	4.800		258	218	11.15	487.15	635
6700	12" pipe size		4	6		330	273	13.90	616.90	805
6710	14" pipe size		3.50	6.857		760	310	15.90	1,085.90	1,350
6720	16" pipe size		2	12		860	545	28	1,433	1,825
7740	Plain ends for plain end pipe, mechanically coupled									
7750	Cplg & labor required at joints not included, add 1 per									
7760	joint for installed price, see line 9180									
7770	Malleable iron, painted, unless noted otherwise									
7800	90° Elbow 1"				Ea.	68.50			68.50	75.50
7810	1-1/2"					81.50			81.50	89.50
7820	2"					122			122	134
7830	2-1/2"					142			142	157
7840	3"					148			148	163
7860	4"					168			168	184
7870	5" welded steel					193			193	212
7880	6"					244			244	268
7890	8" welded steel					460			460	505
7900	10" welded steel					605			605	665
7910	12" welded steel					675			675	740
7970	45° Elbow 1"					41			41	45
7980	1-1/2"					61.50			61.50	67.50
7990	2"					129			129	142
8000	2-1/2"					129			129	142
8010	3"					148			148	163
8030	4"					154			154	169
8040	5" welded steel					193			193	212
8050	6"					221			221	243
8060	8"					259			259	285
8070	10" welded steel					278			278	305
8080	12" welded steel					505			505	555
8140	Tee, straight 1"					77			77	84.50
8150	1-1/2"					99			99	109
8160	2"					99			99	109
8170	2-1/2"					129			129	142
8180	3"					199			199	219
8200	4"					286			286	315
8210	5" welded steel					390			390	430
8220	6"					340			340	375
8230	8" welded steel					490			490	540
8240	10" welded steel					760			760	835
8250	12" welded steel					935			935	1,025
8340	Segmentally welded steel, painted									
8390	Wye 2"				Ea.	124			124	136
8400	2-1/2"					124			124	136
8410	3"					139			139	152
8430	4"					207			207	228
8440	5"					251			251	277
8450	6"					345			345	380
8460	8"					450			450	495
8470	10"					660			660	730
8480	12"					1,025			1,025	1,125
8540	Wye, lateral 2"					134			134	148

22 11 13.47 Pipe Fittings, Steel		Crew	Daily Output	Labor-Hours	Unit	Material	Labor	Equipment	Total	Total Incl O&P
						2009 Bare Costs				
8550	2-1/2″				Ea.	155			155	170
8560	3″					183			183	202
8580	4″					253			253	278
8590	5″					430			430	475
8600	6″					445			445	485
8610	8″					745			745	820
8620	10″					770			770	850
8630	12″					1,425			1,425	1,550
8690	Cross, 2″					141			141	155
8700	2-1/2″					141			141	155
8710	3″					168			168	185
8730	4″					230			230	253
8740	5″					325			325	360
8750	6″					430			430	475
8760	8″					555			555	610
8770	10″					805			805	890
8780	12″					1,175			1,175	1,275
8800	Tees, reducing 2″ x 1″					84			84	92.50
8810	2″ x 1-1/2″					84			84	92.50
8820	3″ x 1″					84			84	92.50
8830	3″ x 1-1/2″					84			84	92.50
8840	3″ x 2″					84			84	92.50
8850	4″ x 1″					121			121	133
8860	4″ x 1-1/2″					121			121	133
8870	4″ x 2″					121			121	133
8880	4″ x 2-1/2″					125			125	138
8890	4″ x 3″					125			125	138
8900	6″ x 2″					199			199	218
8910	6″ x 3″					214			214	235
8920	6″ x 4″					260			260	286
8930	8″ x 2″					271			271	298
8940	8″ x 3″					277			277	305
8950	8″ x 4″					277			277	305
8960	8″ x 5″					286			286	315
8970	8″ x 6″					286			286	315
8980	10″ x 4″					410			410	450
8990	10″ x 6″					410			410	450
9000	10″ x 8″					435			435	480
9010	12″ x 6″					630			630	695
9020	12″ x 8″					635			635	695
9030	12″ x 10″				↓	655			655	725
9080	Adapter nipples 3″ long									
9090	1″				Ea.	11.40			11.40	12.50
9100	1-1/2″					11.40			11.40	12.50
9110	2″					11.40			11.40	12.50
9120	2-1/2″					13.25			13.25	14.60
9130	3″					16.25			16.25	17.85
9140	4″					27			27	29.50
9150	6″				↓	72			72	79.50
9180	Coupling, mechanical, plain end pipe to plain end pipe or fitting									
9190	1″	Q-1	29	.552	Ea.	41	24		65	83
9200	1-1/2″		28	.571		41	25		66	84.50
9210	2″	↓	27	.593	↓	41	26		67	86

22 11 13.47 Pipe Fittings, Steel		Crew	Daily Output	Labor-Hours	Unit	Material	2009 Bare Costs Labor	Equipment	Total	Total Incl O&P
9220	2-1/2″	Q-1	26	.615	Ea.	41	27		68	87.50
9230	3″		25	.640		60.50	28		88.50	110
9240	3-1/2″		24	.667		70.50	29.50		100	123
9250	4″		22	.727		70.50	32		102.50	127
9260	5″	Q-2	28	.857		100	39		139	171
9270	6″		24	1		122	45.50		167.50	205
9280	8″		19	1.263		216	57.50		273.50	325
9290	10″		16	1.500		281	68.50		349.50	415
9300	12″		12	2		350	91		441	530
9940	For galvanized fittings for plain end pipe, add					20%				
9990	Minimum labor/equipment charge	Q-15	3	5.333	Job		234	18.55	252.55	385

22 11 13.48 Pipe, Fittings and Valves, Steel, Grooved-Joint

		Crew	Daily Output	Labor-Hours	Unit	Material	Labor	Equipment	Total	Total Incl O&P
0010	**PIPE, FITTINGS AND VALVES, STEEL, GROOVED-JOINT**									
0012	Fittings are ductile iron. Steel fittings noted.									
0020	Pipe includes coupling & clevis type hanger assemblies, 10′ O.C.									
0500	Schedule 10, black									
0550	2″ diameter	1 Plum	43	.186	L.F.	5.30	9.05		14.35	19.90
0560	2-1/2″ diameter	Q-1	61	.262		6.90	11.50		18.40	25.50
0570	3″ diameter		55	.291		8.20	12.75		20.95	29
0580	3-1/2″ diameter		53	.302		10.60	13.25		23.85	32
0590	4″ diameter		49	.327		10.90	14.35		25.25	34
0600	5″ diameter		40	.400		14.20	17.55		31.75	42.50
0610	6″ diameter	Q-2	46	.522		18.40	23.50		41.90	57
0620	8″ diameter	″	41	.585		24.50	26.50		51	68.50
0700	To delete couplings & hangers, subtract									
0710	2″ diam. to 5″ diam.					25%	20%			
0720	6″ diam. to 8″ diam.					27%	15%			
1000	Schedule 40, black									
1040	3/4″ diameter	1 Plum	71	.113	L.F.	3.53	5.50		9.03	12.40
1050	1″ diameter		63	.127		4.23	6.20		10.43	14.25
1060	1-1/4″ diameter		58	.138		5.45	6.70		12.15	16.45
1070	1-1/2″ diameter		51	.157		6.30	7.65		13.95	18.80
1080	2″ diameter		40	.200		8.05	9.75		17.80	24
1090	2-1/2″ diameter	Q-1	57	.281		10.05	12.30		22.35	30
1100	3″ diameter		50	.320		16.10	14.05		30.15	39.50
1110	4″ diameter		45	.356		19.10	15.60		34.70	45
1120	5″ diameter		37	.432		26.50	19		45.50	58.50
1130	6″ diameter	Q-2	42	.571		33.50	26		59.50	77.50
1140	8″ diameter		37	.649		51.50	29.50		81	103
1150	10″ diameter		31	.774		77.50	35		112.50	140
1160	12″ diameter		27	.839		82.50	40.50		123	154
1170	14″ diameter		20	1.200		94.50	54.50		149	189
1180	16″ diameter		17	1.412		112	64		176	224
1190	18″ diameter		14	1.714		129	78		207	263
1200	20″ diameter		12	2		156	91		247	315
1210	24″ diameter		10	2.400		189	109		298	375
1740	To delete coupling & hanger, subtract									
1750	3/4″ diam. to 2″ diam.					65%	27%			
1760	2-1/2″ diam. to 5″ diam.					41%	18%			
1770	6″ diam. to 12″ diam.					31%	13%			
1780	14″ diam. to 24″ diam.					35%	10%			
1800	Galvanized									

22 11 13.48 Pipe, Fittings and Valves, Steel, Grooved-Joint		Crew	Daily Output	Labor-Hours	Unit	2009 Bare Costs Material	Labor	Equipment	Total	Total Incl O&P
1840	3/4" diameter	1 Plum	71	.113	L.F.	4.63	5.50		10.13	13.60
1850	1" diameter		63	.127		6.10	6.20		12.30	16.30
1860	1-1/4" diameter		58	.138		7.85	6.70		14.55	19.10
1870	1-1/2" diameter		51	.157		9.20	7.65		16.85	22
1880	2" diameter		40	.200		11.85	9.75		21.60	28
1890	2-1/2" diameter	Q-1	57	.281		17.95	12.30		30.25	39
1900	3" diameter		50	.320		23	14.05		37.05	47.50
1910	4" diameter		45	.356		30	15.60		45.60	57
1920	5" diameter		37	.432		34.50	19		53.50	67.50
1930	6" diameter	Q-2	42	.571		40.50	26		66.50	85
1940	8" diameter		37	.649		48.50	29.50		78	99
1950	10" diameter		31	.774		71.50	35		106.50	134
1960	12" diameter		27	.889		93	40.50		133.50	165
2540	To delete coupling & hanger, subtract									
2550	3/4" diam. to 2" diam.					36%	27%			
2560	2-1/2" diam. to 5" diam.					19%	18%			
2570	6" diam. to 12" diam.					14%	13%			
2600	Schedule 80, black									
2610	3/4" diameter	1 Plum	65	.123	L.F.	4.67	6		10.67	14.45
2650	1" diameter		61	.131		5.95	6.40		12.35	16.45
2660	1-1/4" diameter		55	.145		7.80	7.10		14.90	19.60
2670	1-1/2" diameter		49	.163		9.20	7.95		17.15	22.50
2680	2" diameter		38	.211		11.95	10.25		22.20	29
2690	2-1/2" diameter	Q-1	54	.296		18.05	13		31.05	40
2700	3" diameter		48	.333		24	14.65		38.65	48.50
2710	4" diameter		44	.364		25.50	15.95		41.45	53
2720	5" diameter		35	.457		52.50	20		72.50	88.50
2730	6" diameter	Q-2	40	.600		48	27.50		75.50	95.50
2740	8" diameter		35	.686		85.50	31		116.50	143
2750	10" diameter		29	.828		150	37.50		187.50	224
2760	12" diameter		24	1		229	45.50		274.50	320
3240	To delete coupling & hanger, subtract									
3250	3/4" diam. to 2" diam.					30%	25%			
3260	2-1/2" diam. to 5" diam.					14%	17%			
3270	6" diam. to 12" diam.					12%	12%			
3300	Galvanized									
3310	3/4" diameter	1 Plum	65	.123	L.F.	5.90	6		11.90	15.75
3350	1" diameter		61	.131		7.75	6.40		14.15	18.40
3360	1-1/4" diameter		55	.145		10.25	7.10		17.35	22.50
3370	1-1/2" diameter		46	.174		12.20	8.50		20.70	26.50
3380	2" diameter		38	.211		16.15	10.25		26.40	33.50
3390	2-1/2" diameter	Q-1	54	.296		24.50	13		37.50	47
3400	3" diameter		48	.333		32.50	14.65		47.15	58.50
3410	4" diameter		44	.364		46	15.95		61.95	75.50
3420	5" diameter		35	.457		77.50	20		97.50	117
3430	6" diameter	Q-2	40	.600		83.50	27.50		111	135
3440	8" diameter		35	.686		127	31		158	189
3450	10" diameter		29	.828		235	37.50		272.50	320
3460	12" diameter		24	1		335	45.50		380.50	440
3920	To delete coupling & hanger, subtract									
3930	3/4" diam. to 2" diam.					30%	25%			
3940	2-1/2" diam. to 5" diam.					15%	17%			
3950	6" diam. to 12" diam.					11%	12%			

22 11 13.48 Pipe, Fittings and Valves, Steel, Grooved-Joint		Crew	Daily Output	Labor-Hours	Unit	Material	2009 Bare Costs Labor	Equipment	Total	Total Incl O&P
3990	Fittings: coupling material required at joints not incl. in fitting price.									
3994	Add 1 selected coupling, material only, per joint for installed price.									
4000	Elbow, 90° or 45°, painted									
4030	3/4" diameter	1 Plum	50	.160	Ea.	36.50	7.80		44.30	52
4040	1" diameter		50	.160		19.35	7.80		27.15	33.50
4050	1-1/4" diameter		40	.200		19.35	9.75		29.10	36.50
4060	1-1/2" diameter		33	.242		19.35	11.80		31.15	40
4070	2" diameter		25	.320		19.35	15.60		34.95	45.50
4080	2-1/2" diameter	Q-1	40	.400		19.35	17.55		36.90	48.50
4090	3" diameter		33	.485		34.50	21.50		56	70.50
4100	4" diameter		25	.640		37.50	28		65.50	84.50
4110	5" diameter		20	.800		90	35		125	154
4120	6" diameter	Q-2	25	.960		106	43.50		149.50	184
4130	8" diameter		21	1.143		220	52		272	325
4140	10" diameter		18	1.333		400	60.50		460.50	535
4150	12" diameter		15	1.600		640	73		713	820
4170	14" diameter		12	2		835	91		926	1,050
4180	16" diameter		11	2.182		1,075	99.50		1,174.50	1,350
4190	18" diameter	Q-3	15	2.133		1,375	99		1,474	1,675
4200	20" diameter		13	2.462		1,825	114		1,939	2,175
4210	24" diameter		11	2.909		2,625	135		2,760	3,100
4250	For galvanized elbows, add					26%				
4690	Tee, painted									
4700	3/4" diameter	1 Plum	38	.211	Ea.	39	10.25		49.25	59
4740	1" diameter		33	.242		30	11.80		41.80	51.50
4750	1-1/4" diameter		27	.296		30	14.45		44.45	55.50
4760	1-1/2" diameter		22	.364		30	17.75		47.75	60.50
4770	2" diameter		17	.471		30	23		53	68.50
4780	2-1/2" diameter	Q-1	27	.593		30	26		56	73.50
4790	3" diameter		22	.727		41.50	32		73.50	95
4800	4" diameter		17	.941		63.50	41.50		105	134
4810	5" diameter		13	1.231		148	54		202	247
4820	6" diameter	Q-2	17	1.412		171	64		235	288
4830	8" diameter		14	1.714		375	78		453	535
4840	10" diameter		12	2		780	91		871	1,000
4850	12" diameter		10	2.400		1,100	109		1,209	1,375
4851	14" diameter		9	2.667		1,125	121		1,246	1,425
4852	16" diameter		8	3		1,275	137		1,412	1,600
4853	18" diameter	Q-3	11	2.909		1,575	135		1,710	1,950
4854	20" diameter		10	3.200		2,275	149		2,424	2,725
4855	24" diameter		8	4		3,450	186		3,636	4,100
4900	For galvanized tees, add					24%				
4939	Couplings									
4940	Flexible, standard, painted									
4950	3/4" diameter	1 Plum	100	.080	Ea.	10.40	3.90		14.30	17.50
4960	1" diameter		100	.080		10.40	3.90		14.30	17.50
4970	1-1/4" diameter		80	.100		13.80	4.88		18.68	23
4980	1-1/2" diameter		67	.119		15.15	5.80		20.95	25.50
4990	2" diameter		50	.160		16	7.80		23.80	29.50
5000	2-1/2" diameter	Q-1	80	.200		18.90	8.80		27.70	34.50
5010	3" diameter		67	.239		21	10.50		31.50	39.50
5020	3-1/2" diameter		57	.281		30.50	12.30		42.80	52.50
5030	4" diameter		50	.320		30.50	14.05		44.55	55.50

22 11 13.48 Pipe, Fittings and Valves, Steel, Grooved-Joint		Crew	Daily Output	Labor-Hours	Unit	2009 Bare Costs Material	Labor	Equipment	Total	Total Incl O&P
5040	5" diameter	Q-1	40	.400	Ea.	46	17.55		63.55	78
5050	6" diameter	Q-2	50	.480		54.50	22		76.50	94
5070	8" diameter		42	.571		89	26		115	139
5090	10" diameter		35	.686		147	31		178	211
5110	12" diameter		32	.750		168	34		202	238
5120	14" diameter		24	1		198	45.50		243.50	289
5130	16" diameter		20	1.200		260	54.50		314.50	370
5140	18" diameter		18	1.333		305	60.50		365.50	430
5150	20" diameter		16	1.500		405	68.50		473.50	550
5160	24" diameter		13	1.846		525	84		609	705
5200	For galvanized couplings, add					33%				
5750	Flange, w/groove gasket, black steel									
5754	See Div. 22 11 13.44 0620 for gasket & bolt set									
5760	ANSI class 125 and 150, painted									
5780	2" pipe size	1 Plum	23	.348	Ea.	65	16.95		81.95	98
5790	2-1/2" pipe size	Q-1	37	.432		80.50	19		99.50	118
5800	3" pipe size		31	.516		87	22.50		109.50	131
5820	4" pipe size		23	.696		116	30.50		146.50	176
5830	5" pipe size		19	.842		135	37		172	207
5840	6" pipe size	Q-2	23	1.043		147	47.50		194.50	236
5850	8" pipe size		17	1.412		167	64		231	283
5860	10" pipe size		14	1.714		264	78		342	410
5870	12" pipe size		12	2		345	91		436	520
5880	14" pipe size		10	2.400		770	109		879	1,025
5890	16" pipe size		9	2.667		895	121		1,016	1,175
5900	18" pipe size		6	4		1,100	182		1,282	1,475
5910	20" pipe size		5	4.800		1,325	218		1,543	1,800
5920	24" pipe size		4.50	5.333		1,700	243		1,943	2,225
8000	Butterfly valve, 2 position handle, with standard trim									
8010	1-1/2" pipe size	1 Plum	50	.160	Ea.	173	7.80		180.80	202
8020	2" pipe size	"	38	.211		173	10.25		183.25	206
8030	3" pipe size	Q-1	50	.320		249	14.05		263.05	295
8050	4" pipe size	"	38	.421		273	18.50		291.50	330
8070	6" pipe size	Q-2	38	.632		550	28.50		578.50	650
8080	8" pipe size		27	.889		790	40.50		830.50	935
8090	10" pipe size		20	1.200		1,400	54.50		1,454.50	1,625
8200	With stainless steel trim									
8240	1-1/2" pipe size	1 Plum	50	.160	Ea.	220	7.80		227.80	253
8250	2" pipe size	"	38	.211		220	10.25		230.25	257
8270	3" pipe size	Q-1	50	.320		295	14.05		309.05	345
8280	4" pipe size	"	38	.421		320	18.50		338.50	380
8300	6" pipe size	Q-2	38	.632		600	28.50		628.50	705
8310	8" pipe size		27	.889		1,575	40.50		1,615.50	1,800
8320	10" pipe size		20	1.200		2,850	54.50		2,904.50	3,200
9000	Cut one groove, labor									
9010	3/4" pipe size	Q-1	152	.105	Ea.		4.62		4.62	7.15
9020	1" pipe size		140	.114			5		5	7.80
9030	1-1/4" pipe size		124	.129			5.65		5.65	8.80
9040	1-1/2" pipe size		114	.140			6.15		6.15	9.55
9050	2" pipe size		104	.154			6.75		6.75	10.45
9060	2-1/2" pipe size		96	.167			7.30		7.30	11.35
9070	3" pipe size		88	.182			8		8	12.35
9080	3-1/2" pipe size		83	.193			8.45		8.45	13.10

22 11 13.48 Pipe, Fittings and Valves, Steel, Grooved-Joint		Crew	Daily Output	Labor-Hours	Unit	Material	2009 Bare Costs Labor	Equipment	Total	Total Incl O&P
9090	4" pipe size	Q-1	78	.205	Ea.		9		9	13.95
9100	5" pipe size		72	.222			9.75		9.75	15.10
9110	6" pipe size		70	.229			10.05		10.05	15.55
9120	8" pipe size		54	.296			13		13	20
9130	10" pipe size		38	.421			18.50		18.50	28.50
9140	12" pipe size		30	.533			23.50		23.50	36.50
9150	14" pipe size		20	.800			35		35	54.50
9160	16" pipe size		19	.842			37		37	57.50
9170	18" pipe size		18	.889			39		39	60.50
9180	20" pipe size		17	.941			41.50		41.50	64
9190	24" pipe size		15	1.067			47		47	72.50
9210	Roll one groove									
9220	3/4" pipe size	Q-1	266	.060	Ea.		2.64		2.64	4.09
9230	1" pipe size		228	.070			3.08		3.08	4.78
9240	1-1/4" pipe size		200	.080			3.51		3.51	5.45
9250	1-1/2" pipe size		178	.090			3.94		3.94	6.10
9260	2" pipe size		116	.133			6.05		6.05	9.40
9270	2-1/2" pipe size		110	.145			6.40		6.40	9.90
9280	3" pipe size		100	.160			7		7	10.90
9290	3-1/2" pipe size		94	.170			7.45		7.45	11.60
9300	4" pipe size		86	.186			8.15		8.15	12.65
9310	5" pipe size		84	.190			8.35		8.35	12.95
9320	6" pipe size		80	.200			8.80		8.80	13.60
9330	8" pipe size		66	.242			10.65		10.65	16.50
9340	10" pipe size		58	.276			12.10		12.10	18.75
9350	12" pipe size		46	.348			15.25		15.25	23.50
9360	14" pipe size		30	.533			23.50		23.50	36.50
9370	16" pipe size		28	.571			25		25	39
9380	18" pipe size		27	.593			26		26	40.50
9390	20" pipe size		25	.640			28		28	43.50
9400	24" pipe size		23	.696			30.50		30.50	47.50
9990	Minimum labor/equipment charge	1 Plum	4	2	Job		97.50		97.50	151

22 11 13.60 Tubing Fittings, Stainless Steel		Crew	Daily Output	Labor-Hours	Unit	Material	Labor	Equipment	Total	Total Incl O&P
0010	**TUBING FITTINGS, STAINLESS STEEL**									
5010	Type 304, no joints, no hangers									
5020	.035 wall									
5021	1/4"	1 Plum	160	.050	L.F.	3.20	2.44		5.64	7.30
5022	3/8"		160	.050		4.05	2.44		6.49	8.25
5023	1/2"		160	.050		5.20	2.44		7.64	9.55
5024	5/8"		160	.050		6.40	2.44		8.84	10.85
5025	3/4"		133	.060		7.70	2.93		10.63	13
5026	7/8"		133	.060		9.15	2.93		12.08	14.60
5027	1"		114	.070		10.95	3.42		14.37	17.35
5040	.049 wall									
5041	1/4"	1 Plum	160	.050	L.F.	3.77	2.44		6.21	7.95
5042	3/8"		160	.050		5.20	2.44		7.64	9.55
5043	1/2"		160	.050		5.80	2.44		8.24	10.15
5044	5/8"		160	.050		7.80	2.44		10.24	12.40
5045	3/4"		133	.060		8.30	2.93		11.23	13.70
5046	7/8"		133	.060		9.75	2.93		12.68	15.30
5047	1"		114	.070		10.35	3.42		13.77	16.70
5060	.065 wall									

22 11 13.60 Tubing Fittings, Stainless Steel

	22 11 13.60 Tubing Fittings, Stainless Steel	Crew	Daily Output	Labor-Hours	Unit	Material	Labor	Equipment	Total	Total Incl O&P
						2009 Bare Costs				
5061	1/4"	1 Plum	160	.050	L.F.	4.23	2.44		6.67	8.45
5062	3/8"		160	.050		5.85	2.44		8.29	10.20
5063	1/2"		160	.050		7.05	2.44		9.49	11.55
5064	5/8"		160	.050		8.60	2.44		11.04	13.25
5065	3/4"		133	.060		9.10	2.93		12.03	14.55
5066	7/8"		133	.060		11.75	2.93		14.68	17.45
5067	1"		114	.070		11.50	3.42		14.92	17.95
5210	Type 316									
5220	.035 wall									
5221	1/4"	1 Plum	160	.050	L.F.	3.99	2.44		6.43	8.15
5222	3/8"		160	.050		4.98	2.44		7.42	9.30
5223	1/2"		160	.050		8.20	2.44		10.64	12.85
5224	5/8"		160	.050		10.20	2.44		12.64	15
5225	3/4"		133	.060		13.05	2.93		15.98	18.90
5226	7/8"		133	.060		17.70	2.93		20.63	24
5227	1"		114	.070		19.50	3.42		22.92	27
5240	.049 wall									
5241	1/4"	1 Plum	160	.050	L.F.	5.05	2.44		7.49	9.35
5242	3/8"		160	.050		5.60	2.44		8.04	9.95
5243	1/2"		160	.050		7.50	2.44		9.94	12.05
5244	5/8"		160	.050		10.15	2.44		12.59	14.95
5245	3/4"		133	.060		10.30	2.93		13.23	15.90
5246	7/8"		133	.060		14.80	2.93		17.73	21
5247	1"		114	.070		14.90	3.42		18.32	21.50
5260	.065 wall									
5261	1/4"	1 Plum	160	.050	L.F.	5.05	2.44		7.49	9.35
5262	3/8"		160	.050		8.50	2.44		10.94	13.15
5263	1/2"		160	.050		9.30	2.44		11.74	14.05
5264	5/8"		160	.050		11.35	2.44		13.79	16.25
5265	3/4"		133	.060		12	2.93		14.93	17.75
5266	7/8"		133	.060		17.35	2.93		20.28	23.50
5267	1"		114	.070		18.45	3.42		21.87	26

22 11 13.61 Tubing Fittings, Stainless Steel

	22 11 13.61 Tubing Fittings, Stainless Steel	Crew	Daily Output	Labor-Hours	Unit	Material	Labor	Equipment	Total	Total Incl O&P
0010	**TUBING FITTINGS, STAINLESS STEEL**									
8200	Tube fittings, compression type									
8202	Type 316									
8204	90° elbow									
8206	1/4"	1 Plum	24	.333	Ea.	14.20	16.25		30.45	40.50
8207	3/8"		22	.364		17.35	17.75		35.10	46.50
8208	1/2"		22	.364		28.50	17.75		46.25	59
8209	5/8"		21	.381		32	18.55		50.55	64
8210	3/4"		21	.381		50.50	18.55		69.05	84.50
8211	7/8"		20	.400		77.50	19.50		97	115
8212	1"		20	.400		96.50	19.50		116	136
8220	Union tee									
8222	1/4"	1 Plum	15	.533	Ea.	19.95	26		45.95	62.50
8224	3/8"		15	.533		25.50	26		51.50	69
8225	1/2"		15	.533		39.50	26		65.50	84
8226	5/8"		14	.571		44.50	28		72.50	92
8227	3/4"		14	.571		59.50	28		87.50	109
8228	7/8"		13	.615		96	30		126	153
8229	1"		13	.615		127	30		157	187

22 11 13.61 Tubing Fittings, Stainless Steel

	22 11 13.61 Tubing Fittings, Stainless Steel	Crew	Daily Output	Labor-Hours	Unit	2009 Bare Costs Material	Labor	Equipment	Total	Total Incl O&P
8234	Union									
8236	1/4"	1 Plum	24	.333	Ea.	9.75	16.25		26	36
8237	3/8"		22	.364		12.90	17.75		30.65	41.50
8238	1/2"		22	.364		21	17.75		38.75	50.50
8239	5/8"		21	.381		27	18.55		45.55	59
8240	3/4"		21	.381		33.50	18.55		52.05	66
8241	7/8"		20	.400		55	19.50		74.50	90.50
8242	1"		20	.400		58.50	19.50		78	94
8250	Male connector									
8252	1/4" x 1/4"	1 Plum	24	.333	Ea.	6.30	16.25		22.55	32
8253	3/8" x 3/8"		22	.364		9.75	17.75		27.50	38.50
8254	1/2" x 1/2"		22	.364		14.50	17.75		32.25	43.50
8256	3/4" x 3/4"		21	.381		22.50	18.55		41.05	53.50
8258	1" x 1"		20	.400		39	19.50		58.50	72.50

22 11 13.64 Pipe, Stainless Steel

	22 11 13.64 Pipe, Stainless Steel	Crew	Daily Output	Labor-Hours	Unit	Material	Labor	Equipment	Total	Total Incl O&P
0010	**PIPE, STAINLESS STEEL**									
0020	Welded, with clevis type hanger assemblies, 10' O.C.									
0500	Schedule 5, type 304									
0540	1/2" diameter	Q-15	128	.125	L.F.	7.30	5.50	.44	13.24	17.05
0550	3/4" diameter		116	.138		9.35	6.05	.48	15.88	20
0560	1" diameter		103	.155		11.75	6.80	.54	19.09	24
0570	1-1/4" diameter		93	.172		16.50	7.55	.60	24.65	30.50
0580	1-1/2" diameter		85	.188		22.50	8.25	.66	31.41	38
0590	2" diameter		69	.232		37	10.15	.81	47.96	57.50
0600	2-1/2" diameter		53	.302		41	13.25	1.05	55.30	66.50
0610	3" diameter		48	.333		50	14.65	1.16	65.81	78.50
0620	4" diameter		44	.364		63.50	15.95	1.27	80.72	96
0630	5" diameter		36	.444		130	19.50	1.55	151.05	175
0640	6" diameter	Q-16	42	.571		120	26	1.33	147.33	173
0650	8" diameter		34	.706		185	32	1.64	218.64	255
0660	10" diameter		26	.923		256	42	2.14	300.14	350
0670	12" diameter		21	1.143		335	52	2.65	389.65	455
0700	To delete hangers, subtract									
0710	1/2" diam. to 1-1/2" diam.					8%	19%			
0720	2" diam. to 5" diam.					4%	9%			
0730	6" diam. to 12" diam.					3%	4%			
0750	For small quantities, add				L.F.	10%				
1250	Schedule 5, type 316									
1290	1/2" diameter	Q-15	128	.125	L.F.	9.55	5.50	.44	15.49	19.50
1300	3/4" diameter		116	.138		12.25	6.05	.48	18.78	23.50
1310	1" diameter		103	.155		16.75	6.80	.54	24.09	29.50
1320	1-1/4" diameter		93	.172		20.50	7.55	.60	28.65	35
1330	1-1/2" diameter		85	.188		29	8.25	.66	37.91	45.50
1340	2" diameter		69	.232		48	10.15	.81	58.96	69
1350	2-1/2" diameter		53	.302		56.50	13.25	1.05	70.80	83.50
1360	3" diameter		48	.333		71.50	14.65	1.16	87.31	103
1370	4" diameter		44	.364		89	15.95	1.27	106.22	124
1380	5" diameter		36	.444		171	19.50	1.55	192.05	220
1390	6" diameter	Q-16	42	.571		166	26	1.33	193.33	224
1400	8" diameter		34	.706		250	32	1.64	283.64	325
1410	10" diameter		26	.923		360	42	2.14	404.14	460
1420	12" diameter		21	1.143		460	52	2.65	514.65	595

22 11 13.64 Pipe, Stainless Steel		Crew	Daily Output	Labor-Hours	Unit	Material	2009 Bare Costs Labor	Equipment	Total	Total Incl O&P
1490	For small quantities, add				L.F.	10%				
1940	To delete hanger, subtract									
1950	1/2" diam. to 1-1/2" diam.					5%	19%			
1960	2" diam. to 5" diam.					3%	9%			
1970	6" diam. to 12" diam.					2%	4%			
2000	Schedule 10, type 304									
2040	1/4" diameter	Q-15	131	.122	L.F.	4.46	5.35	.43	10.24	13.65
2050	3/8" diameter		128	.125		4.69	5.50	.44	10.63	14.15
2060	1/2" diameter		125	.128		4.93	5.60	.45	10.98	14.65
2070	3/4" diameter		113	.142		7.35	6.20	.49	14.04	18.25
2080	1" diameter		100	.160		9.85	7	.56	17.41	22.50
2090	1-1/4" diameter		91	.176		9.90	7.70	.61	18.21	23.50
2100	1-1/2" diameter		83	.193		10.85	8.45	.67	19.97	26
2110	2" diameter		67	.239		13.25	10.50	.83	24.58	31.50
2120	2-1/2" diameter		51	.314		17.40	13.75	1.09	32.24	42
2130	3" diameter		46	.348		21.50	15.25	1.21	37.96	48.50
2140	4" diameter		42	.381		31.50	16.70	1.33	49.53	62.50
2150	5" diameter	↓	35	.457		42.50	20	1.59	64.09	80
2160	6" diameter	Q-16	40	.600		46	27.50	1.39	74.89	94.50
2170	8" diameter		33	.727		79.50	33	1.69	114.19	141
2180	10" diameter		25	.960		110	43.50	2.23	155.73	191
2190	12" diameter	↓	21	1.143		138	52	2.65	192.65	235
2250	For small quantities, add				↓	10%				
2650	To delete hanger, subtract									
2660	1/4" diam. to 3/4" diam.					9%	22%			
2670	1" diam. to 2" diam.					4%	15%			
2680	2-1/2" diam. to 5" diam.					3%	8%			
2690	6" diam. to 12" diam.					3%	4%			
2750	Schedule 10, type 316									
2790	1/4" diameter	Q-15	131	.122	L.F.	5.30	5.35	.43	11.08	14.55
2800	3/8" diameter		128	.125		6.25	5.50	.44	12.19	15.90
2810	1/2" diameter		125	.128		6.55	5.60	.45	12.60	16.45
2820	3/4" diameter		113	.142		8.60	6.20	.49	15.29	19.65
2830	1" diameter		100	.160		10.45	7	.56	18.01	23
2840	1-1/4" diameter		91	.176		15	7.70	.61	23.31	29
2850	1-1/2" diameter		83	.193		20.50	8.45	.67	29.62	36.50
2860	2" diameter		67	.239		18.60	10.50	.83	29.93	37.50
2870	2-1/2" diameter		51	.314		27.50	13.75	1.09	42.34	52.50
2880	3" diameter		46	.348		32	15.25	1.21	48.46	60
2890	4" diameter		42	.381		43	16.70	1.33	61.03	75
2900	5" diameter	↓	35	.457		60	20	1.59	81.59	99
2910	6" diameter	Q-16	40	.600		68.50	27.50	1.39	97.39	120
2920	8" diameter		33	.727		117	33	1.69	151.69	181
2930	10" diameter		25	.960		154	43.50	2.23	199.73	239
2940	12" diameter	↓	21	1.143		193	52	2.65	247.65	295
2990	For small quantities, add				↓	10%				
3430	To delete hanger, subtract									
3440	1/4" diam. to 3/4" diam.					6%	22%			
3450	1" diam. to 2" diam.					3%	15%			
3460	2-1/2" diam. to 5" diam.					2%	8%			
3470	6" diam. to 12" diam.					2%	4%			
3500	Threaded, couplings and clevis hanger assemblies, 10' O.C.									
3520	Schedule 40, type 304									

22 11 13 – Facility Water Distribution Piping

22 11 13.64 Pipe, Stainless Steel		Crew	Daily Output	Labor-Hours	Unit	Material	2009 Bare Costs Labor	Equipment	Total	Total Incl O&P
3540	1/4″ diameter	1 Plum	54	.148	L.F.	6.80	7.20		14	18.70
3550	3/8″ diameter		53	.151		7.10	7.35		14.45	19.20
3560	1/2″ diameter		52	.154		7.65	7.50		15.15	20
3570	3/4″ diameter		51	.157		9.60	7.65		17.25	22.50
3580	1″ diameter		45	.178		13.55	8.65		22.20	28.50
3590	1-1/4″ diameter	Q-1	76	.211		18.70	9.25		27.95	35
3600	1-1/2″ diameter		69	.232		22	10.15		32.15	40
3610	2″ diameter		57	.281		29.50	12.30		41.80	51.50
3620	2-1/2″ diameter		44	.364		54.50	15.95		70.45	85
3630	3″ diameter		38	.421		71.50	18.50		90	107
3640	4″ diameter	Q-2	51	.471		92.50	21.50		114	135
3740	For small quantities, add					10%				
4200	To delete couplings & hangers, subtract									
4210	1/4″ diam. to 3/4″ diam.					15%	56%			
4220	1″ diam. to 2″ diam.					18%	49%			
4230	2-1/2″ diam. to 4″ diam.					34%	40%			
4250	Schedule 40, type 316									
4290	1/4″ diameter	1 Plum	54	.148	L.F.	8.45	7.20		15.65	20.50
4300	3/8″ diameter		53	.151		8.90	7.35		16.25	21
4310	1/2″ diameter		52	.154		12.25	7.50		19.75	25
4320	3/4″ diameter		51	.157		15.95	7.65		23.60	29.50
4330	1″ diameter		45	.178		23	8.65		31.65	39
4340	1-1/4″ diameter	Q-1	76	.211		33	9.25		42.25	50.50
4350	1-1/2″ diameter		69	.232		36	10.15		46.15	55.50
4360	2″ diameter		57	.281		51.50	12.30		63.80	76
4370	2-1/2″ diameter		44	.364		86	15.95		101.95	120
4380	3″ diameter		38	.421		113	18.50		131.50	153
4390	4″ diameter	Q-2	51	.471		154	21.50		175.50	202
4490	For small quantities, add					10%				
4900	To delete couplings & hangers, subtract									
4910	1/4″ diam. to 3/4″ diam.					12%	56%			
4920	1″ diam. to 2″ diam.					14%	49%			
4930	2-1/2″ diam. to 4″ diam.					27%	40%			
5000	Schedule 80, type 304									
5040	1/4″ diameter	1 Plum	53	.151	L.F.	15.60	7.35		22.95	28.50
5050	3/8″ diameter		52	.154		16.85	7.50		24.35	30
5060	1/2″ diameter		51	.157		21.50	7.65		29.15	36
5070	3/4″ diameter		48	.167		25.50	8.15		33.65	40.50
5080	1″ diameter		43	.186		30.50	9.05		39.55	47.50
5090	1-1/4″ diameter	Q-1	73	.219		40	9.60		49.60	59
5100	1-1/2″ diameter		67	.239		50	10.50		60.50	72
5110	2″ diameter		54	.296		67.50	13		80.50	94
5190	For small quantities, add					10%				
5700	To delete couplings & hangers, subtract									
5710	1/4″ diam. to 3/4″ diam.					10%	53%			
5720	1″ diam. to 2″ diam.					14%	47%			
5750	Schedule 80, type 316									
5790	1/4″ diameter	1 Plum	53	.151	L.F.	18.15	7.35		25.50	31.50
5800	3/8″ diameter		52	.154		19.35	7.50		26.85	33
5810	1/2″ diameter		51	.157		25.50	7.65		33.15	40
5820	3/4″ diameter		48	.167		34.50	8.15		42.65	50.50
5830	1″ diameter		43	.186		41	9.05		50.05	59
5840	1-1/4″ diameter	Q-1	73	.219		58.50	9.60		68.10	79.50

	22 11 13.64 Pipe, Stainless Steel	Crew	Daily Output	Labor-Hours	Unit	2009 Bare Costs Material	Labor	Equipment	Total	Total Incl O&P
5850	1-1/2" diameter	Q-1	67	.239	L.F.	66	10.50		76.50	89
5860	2" diameter		54	.296		94	13		107	123
5950	For small quantities, add					10%				
7000	To delete couplings & hangers, subtract									
7010	1/4" diam. to 3/4" diam.					9%	53%			
7020	1" diam. to 2" diam.					14%	47%			
8000	Weld joints with clevis type hanger assemblies, 10' O.C.									
8010	Schedule 40, type 304									
8050	1/8" pipe size	Q-15	126	.127	L.F.	5.50	5.55	.44	11.49	15.20
8060	1/4" pipe size		125	.128		5.65	5.60	.45	11.70	15.45
8070	3/8" pipe size		122	.131		5.70	5.75	.46	11.91	15.70
8080	1/2" pipe size		118	.136		5.80	5.95	.47	12.22	16.15
8090	3/4" pipe size		109	.147		7.15	6.45	.51	14.11	18.45
8100	1" pipe size		95	.168		9.60	7.40	.59	17.59	22.50
8110	1-1/4" pipe size		86	.186		12.35	8.15	.65	21.15	27
8120	1-1/2" pipe size		78	.205		14.85	9	.71	24.56	31
8130	2" pipe size		62	.258		18.45	11.30	.90	30.65	39
8140	2-1/2" pipe size		49	.327		28.50	14.35	1.14	43.99	55
8150	3" pipe size		44	.364		36	15.95	1.27	53.22	66
8160	3-1/2" pipe size		44	.364		38.50	15.95	1.27	55.72	68.50
8170	4" pipe size		39	.410		40.50	18	1.43	59.93	74
8180	5" pipe size		32	.500		69	22	1.74	92.74	112
8190	6" pipe size	Q-16	37	.649		84	29.50	1.50	115	140
8200	8" pipe size		29	.828		118	37.50	1.92	157.42	191
8210	10" pipe size		24	1		199	45.50	2.32	246.82	292
8220	12" pipe size		20	1.200		226	54.50	2.78	283.28	335
8300	Schedule 40, type 316									
8310	1/8" pipe size	Q-15	126	.127	L.F.	7.45	5.55	.44	13.44	17.35
8320	1/4" pipe size		125	.128		7.05	5.60	.45	13.10	16.95
8330	3/8" pipe size		122	.131		7.25	5.75	.46	13.46	17.35
8340	1/2" pipe size		118	.136		10	5.95	.47	16.42	21
8350	3/4" pipe size		109	.147		13	6.45	.51	19.96	25
8360	1" pipe size		95	.168		18.40	7.40	.59	26.39	32.50
8370	1-1/4" pipe size		86	.186		25	8.15	.65	33.80	41.50
8380	1-1/2" pipe size		78	.205		27.50	9	.71	37.21	45
8390	2" pipe size		62	.258		38	11.30	.90	50.20	60.50
8400	2-1/2" pipe size		49	.327		55	14.35	1.14	70.49	84
8410	3" pipe size		44	.364		70.50	15.95	1.27	87.72	104
8420	3-1/2" pipe size		44	.364		83	15.95	1.27	100.22	118
8430	4" pipe size		39	.410		91	18	1.43	110.43	130
8440	5" pipe size		32	.500		123	22	1.74	146.74	172
8450	6" pipe size	Q-16	37	.649		144	29.50	1.50	175	207
8460	8" pipe size		29	.828		270	37.50	1.92	309.42	360
8470	10" pipe size		24	1		360	45.50	2.32	407.82	470
8480	12" pipe size		20	1.200		490	54.50	2.78	547.28	630
8500	Schedule 80, type 304									
8510	1/4" pipe size	Q-15	110	.145	L.F.	14.25	6.40	.51	21.16	26
8520	3/8" pipe size		109	.147		15.35	6.45	.51	22.31	27.50
8530	1/2" pipe size		106	.151		19.80	6.60	.53	26.93	33
8540	3/4" pipe size		96	.167		23	7.30	.58	30.88	37.50
8550	1" pipe size		87	.184		26.50	8.05	.64	35.19	42
8560	1-1/4" pipe size		81	.198		30	8.65	.69	39.34	47
8570	1-1/2" pipe size		74	.216		38.50	9.50	.75	48.75	57.50

22 11 13.64 Pipe, Stainless Steel		Crew	Daily Output	Labor-Hours	Unit	2009 Bare Costs Material	Labor	Equipment	Total	Total Incl O&P
8580	2" pipe size	Q-15	58	.276	L.F.	51.50	12.10	.96	64.56	77
8590	2-1/2" pipe size		46	.348		106	15.25	1.21	122.46	141
8600	3" pipe size		41	.390		142	17.10	1.36	160.46	184
8610	4" pipe size		33	.485		223	21.50	1.69	246.19	280
8630	6" pipe size	Q-16	30	.800		490	36.50	1.86	528.36	600
8640	Schedule 80, type 316									
8650	1/4" pipe size	Q-15	110	.145	L.F.	16.60	6.40	.51	23.51	28.50
8660	3/8" pipe size		109	.147		17.70	6.45	.51	24.66	30
8670	1/2" pipe size		106	.151		23	6.60	.53	30.13	36.50
8680	3/4" pipe size		96	.167		31	7.30	.58	38.88	46
8690	1" pipe size		87	.184		35	8.05	.64	43.69	51.50
8700	1-1/4" pipe size		81	.198		45.50	8.65	.69	54.84	64
8710	1-1/2" pipe size		74	.216		51	9.50	.75	61.25	71.50
8720	2" pipe size		58	.276		73	12.10	.96	86.06	100
8730	2-1/2" pipe size		46	.348		197	15.25	1.21	213.46	242
8740	3" pipe size		41	.390		230	17.10	1.36	248.46	281
8760	4" pipe size		33	.485		320	21.50	1.69	343.19	385
8770	6" pipe size	Q-16	30	.800		865	36.50	1.86	903.36	1,000
9100	Threading pipe labor, sst, one end, schedules 40 & 80									
9110	1/4" through 3/4" pipe size	1 Plum	61.50	.130	Ea.		6.35		6.35	9.85
9120	1" through 2" pipe size		55.90	.143			7		7	10.80
9130	2-1/2" pipe size		41.50	.193			9.40		9.40	14.55
9140	3" pipe size		38.50	.208			10.15		10.15	15.70
9150	3-1/2" pipe size	Q-1	68.40	.234			10.25		10.25	15.90
9160	4" pipe size		73	.219			9.60		9.60	14.90
9170	5" pipe size		40.70	.393			17.25		17.25	27
9180	6" pipe size		35.40	.452			19.85		19.85	31
9190	8" pipe size		22.30	.717			31.50		31.50	49
9200	10" pipe size		16.10	.994			43.50		43.50	67.50
9210	12" pipe size		12.30	1.301			57		57	88.50
9250	Welding labor per joint for stainless steel									
9260	Schedule 5 and 10									
9270	1/4" pipe size	Q-15	36	.444	Ea.		19.50	1.55	21.05	31.50
9280	3/8" pipe size		35	.457			20	1.59	21.59	33
9290	1/2" pipe size		35	.457			20	1.59	21.59	33
9300	3/4" pipe size		28	.571			25	1.99	26.99	41
9310	1" pipe size		25	.640			28	2.23	30.23	46
9320	1-1/4" pipe size		22	.727			32	2.53	34.53	52.50
9330	1-1/2" pipe size		21	.762			33.50	2.65	36.15	55
9340	2" pipe size		18	.889			39	3.09	42.09	64
9350	2-1/2" pipe size		12	1.333			58.50	4.64	63.14	95.50
9360	3" pipe size		9.73	1.644			72	5.70	77.70	118
9370	4" pipe size		7.37	2.171			95.50	7.55	103.05	156
9380	5" pipe size		6.15	2.602			114	9.05	123.05	187
9390	6" pipe size		5.71	2.802			123	9.75	132.75	202
9400	8" pipe size		3.69	4.336			190	15.10	205.10	310
9410	10" pipe size		2.91	5.498			241	19.15	260.15	395
9420	12" pipe size		2.31	6.926			305	24	329	495
9500	Schedule 40									
9510	1/4" pipe size	Q-15	28	.571	Ea.		25	1.99	26.99	41
9520	3/8" pipe size		27	.593			26	2.06	28.06	43
9530	1/2" pipe size		25.40	.630			27.50	2.19	29.69	45.50
9540	3/4" pipe size		22.22	.720			31.50	2.51	34.01	52

22 11 13.64 Pipe, Stainless Steel		Crew	Daily Output	Labor-Hours	Unit	Material	2009 Bare Costs Labor	Equipment	Total	Total Incl O&P
9550	1" pipe size	Q-15	20.25	.790	Ea.		34.50	2.75	37.25	57
9560	1-1/4" pipe size		18.82	.850			37.50	2.96	40.46	61.50
9570	1-1/2" pipe size		17.78	.900			39.50	3.13	42.63	64.50
9580	2" pipe size		15.09	1.060			46.50	3.69	50.19	76
9590	2-1/2" pipe size		7.96	2.010			88	7	95	145
9600	3" pipe size		6.43	2.488			109	8.65	117.65	179
9610	4" pipe size		4.88	3.279			144	11.40	155.40	236
9620	5" pipe size		4.26	3.756			165	13.05	178.05	270
9630	6" pipe size		3.77	4.244			186	14.75	200.75	305
9640	8" pipe size		2.44	6.557			288	23	311	470
9650	10" pipe size		1.92	8.333			365	29	394	595
9660	12" pipe size	↓	1.52	10.526	↓		460	36.50	496.50	755
9750	Schedule 80									
9760	1/4" pipe size	Q-15	21.55	.742	Ea.		32.50	2.58	35.08	53.50
9770	3/8" pipe size		20.75	.771			34	2.68	36.68	55.50
9780	1/2" pipe size		19.54	.819			36	2.85	38.85	58.50
9790	3/4" pipe size		17.09	.936			41	3.26	44.26	67
9800	1" pipe size		15.58	1.027			45	3.57	48.57	74
9810	1-1/4" pipe size		14.48	1.105			48.50	3.85	52.35	79
9820	1-1/2" pipe size		13.68	1.170			51.50	4.07	55.57	84
9830	2" pipe size		11.61	1.378			60.50	4.80	65.30	99.50
9840	2-1/2" pipe size		6.12	2.614			115	9.10	124.10	188
9850	3" pipe size		4.94	3.239			142	11.25	153.25	232
9860	4" pipe size		3.75	4.267			187	14.85	201.85	305
9870	5" pipe size		3.27	4.893			215	17.05	232.05	355
9880	6" pipe size		2.90	5.517			242	19.20	261.20	395
9890	8" pipe size		1.87	8.556			375	30	405	615
9900	10" pipe size		1.48	10.811			475	37.50	512.50	775
9910	12" pipe size		1.17	13.675			600	47.50	647.50	985
9920	Schedule 160, 1/2" pipe size		17	.941			41.50	3.28	44.78	67.50
9930	3/4" pipe size		14.81	1.080			47.50	3.76	51.26	77.50
9940	1" pipe size		13.50	1.185			52	4.12	56.12	85
9950	1-1/4" pipe size		12.55	1.275			56	4.44	60.44	92
9960	1-1/2" pipe size		11.85	1.350			59.50	4.70	64.20	97
9970	2" pipe size		10	1.600			70	5.55	75.55	115
9980	3" pipe size		4.28	3.738			164	13	177	268
9990	4" pipe size	↓	3.25	4.923	↓		216	17.15	233.15	355

22 11 13.66 Pipe Fittings, Stainless Steel		Crew	Daily Output	Labor-Hours	Unit	Material	Labor	Equipment	Total	Total Incl O&P
0010	**PIPE FITTINGS, STAINLESS STEEL**									
0100	Butt weld joint, schedule 5, type 304									
0120	90° Elbow, long									
0140	1/2"	Q-15	17.50	.914	Ea.	18	40	3.18	61.18	85.50
0150	3/4"		14	1.143		18	50	3.98	71.98	102
0160	1"		12.50	1.280		19	56	4.45	79.45	113
0170	1-1/4"		11	1.455		26	64	5.05	95.05	133
0180	1-1/2"		10.50	1.524		22	67	5.30	94.30	134
0190	2"		9	1.778		26	78	6.20	110.20	156
0200	2-1/2"		6	2.667		60	117	9.30	186.30	257
0210	3"		4.86	3.292		66	144	11.45	221.45	310
0220	3-1/2"		4.27	3.747		165	164	13.05	342.05	450
0230	4"		3.69	4.336		90	190	15.10	295.10	410
0240	5"	↓	3.08	5.195	↓	350	228	18.10	596.10	760

22 11 13.66 Pipe Fittings, Stainless Steel		Crew	Daily Output	Labor-Hours	Unit	2009 Bare Costs Material	Labor	Equipment	Total	Total Incl O&P
0250	6"	Q-16	4.29	5.594	Ea.	270	255	13	538	705
0260	8"		2.76	8.696		570	395	20	985	1,250
0270	10"		2.18	11.009		880	500	25.50	1,405.50	1,775
0280	12"		1.73	13.873		1,250	630	32	1,912	2,400
0320	For schedule 5, type 316, add					30%				
0600	45° Elbow, long									
0620	1/2"	Q-15	17.50	.914	Ea.	18	40	3.18	61.18	85.50
0630	3/4"		14	1.143		18	50	3.98	71.98	102
0640	1"		12.50	1.280		19	56	4.45	79.45	113
0650	1-1/4"		11	1.455		26	64	5.05	95.05	133
0660	1-1/2"		10.50	1.524		22	67	5.30	94.30	134
0670	2"		9	1.778		26	78	6.20	110.20	156
0680	2-1/2"		6	2.667		60	117	9.30	186.30	257
0690	3"		4.86	3.292		66	144	11.45	221.45	310
0700	3-1/2"		4.27	3.747		165	164	13.05	342.05	450
0710	4"		3.69	4.336		76	190	15.10	281.10	395
0720	5"		3.08	5.195		280	228	18.10	526.10	685
0730	6"	Q-16	4.29	5.594		190	255	13	458	620
0740	8"		2.76	8.696		400	395	20	815	1,075
0750	10"		2.18	11.009		700	500	25.50	1,225.50	1,575
0760	12"		1.73	13.873		875	630	32	1,537	1,975
0800	For schedule 5, type 316, add					25%				
1100	Tee, straight									
1130	1/2"	Q-15	11.66	1.372	Ea.	54	60	4.78	118.78	158
1140	3/4"		9.33	1.715		54	75.50	5.95	135.45	183
1150	1"		8.33	1.921		57	84.50	6.70	148.20	201
1160	1-1/4"		7.33	2.183		46	96	7.60	149.60	208
1170	1-1/2"		7	2.286		45	100	7.95	152.95	214
1180	2"		6	2.667		47	117	9.30	173.30	243
1190	2-1/2"		4	4		113	176	13.90	302.90	410
1200	3"		3.24	4.938		89	217	17.20	323.20	450
1210	3-1/2"		2.85	5.614		236	246	19.55	501.55	660
1220	4"		2.46	6.504		128	285	22.50	435.50	610
1230	5"		2	8		405	350	28	783	1,025
1240	6"	Q-16	2.85	8.421		320	385	19.55	724.55	965
1250	8"		1.84	13.043		680	595	30.50	1,305.50	1,700
1260	10"		1.45	16.552		1,100	755	38.50	1,893.50	2,425
1270	12"		1.15	20.870		1,525	950	48.50	2,523.50	3,200
1320	For schedule 5, type 316, add					25%				
2000	Butt weld joint, schedule 10, type 304									
2020	90° elbow, long									
2040	1/2"	Q-15	17	.941	Ea.	13.50	41.50	3.28	58.28	82.50
2050	3/4"		14	1.143		13.50	50	3.98	67.48	97
2060	1"		12.50	1.280		14.25	56	4.45	74.70	108
2070	1-1/4"		11	1.455		19.50	64	5.05	88.55	126
2080	1-1/2"		10.50	1.524		16.50	67	5.30	88.80	128
2090	2"		9	1.778		19.50	78	6.20	103.70	149
2100	2-1/2"		6	2.667		45	117	9.30	171.30	241
2110	3"		4.86	3.292		42	144	11.45	197.45	283
2120	3-1/2"		4.27	3.747		124	164	13.05	301.05	405
2130	4"		3.69	4.336		70.50	190	15.10	275.60	390
2140	5"		3.08	5.195		263	228	18.10	509.10	665
2150	6"	Q-16	4.29	5.594		203	255	13	471	630

22 11 13.66 Pipe Fittings, Stainless Steel		Crew	Daily Output	Labor-Hours	Unit	Material	Labor	Equipment	Total	Total Incl O&P
						2009 Bare Costs				
2160	8"	Q-16	2.76	8.696	Ea.	430	395	20	845	1,100
2170	10"		2.18	11.009		660	500	25.50	1,185.50	1,525
2180	12"	↓	1.73	13.873	↓	940	630	32	1,602	2,050
2500	45° elbow, long									
2520	1/2"	Q-15	17.50	.914	Ea.	13.50	40	3.18	56.68	80.50
2530	3/4"		14	1.143		13.50	50	3.98	67.48	97
2540	1"		12.50	1.280		14.25	56	4.45	74.70	108
2550	1-1/4"		11	1.455		19.50	64	5.05	88.55	126
2560	1-1/2"		10.50	1.524		16.50	67	5.30	88.80	128
2570	2"		9	1.778		19.50	78	6.20	103.70	149
2580	2-1/2"		6	2.667		45	117	9.30	171.30	241
2590	3"		4.86	3.292		34	144	11.45	189.45	274
2600	3-1/2"		4.27	3.747		124	164	13.05	301.05	405
2610	4"		3.69	4.336		57	190	15.10	262.10	375
2620	5"	↓	3.08	5.195		210	228	18.10	456.10	605
2630	6"	Q-16	4.29	5.594		143	255	13	411	565
2640	8"		2.76	8.696		300	395	20	715	965
2650	10"		2.18	11.009		525	500	25.50	1,050.50	1,375
2660	12"	↓	1.73	13.873	↓	655	630	32	1,317	1,725
3000	Tee, straight									
3030	1/2"	Q-15	11.66	1.372	Ea.	40.50	60	4.78	105.28	143
3040	3/4"		9.33	1.715		40.50	75.50	5.95	121.95	168
3050	1"		8.33	1.921		43	84.50	6.70	134.20	185
3060	1-1/4"		7.33	2.183		52.50	96	7.60	156.10	215
3070	1-1/2"		7	2.286		34	100	7.95	141.95	202
3080	2"		6	2.667		35.50	117	9.30	161.80	230
3090	2-1/2"		4	4		85	176	13.90	274.90	380
3100	3"		3.24	4.938		67	217	17.20	301.20	425
3110	3-1/2"		2.85	5.614		98.50	246	19.55	364.05	510
3120	4"		2.46	6.504		96	285	22.50	403.50	575
3130	5"	↓	2	8		305	350	28	683	910
3140	6"	Q-16	2.85	8.421		240	385	19.55	644.55	880
3150	8"		1.84	13.043		510	595	30.50	1,135.50	1,525
3151	10"		1.45	16.552		825	755	38.50	1,618.50	2,125
3152	12"	↓	1.15	20.870		1,150	950	48.50	2,148.50	2,775
3154	For schedule 10, type 316, add				↓	25%				
3281	Butt weld joint, schedule 40, type 304									
3284	90° Elbow, long, 1/2"	Q-15	12.70	1.260	Ea.	15.75	55.50	4.38	75.63	108
3288	3/4"		11.10	1.441		15.75	63.50	5	84.25	121
3289	1"		10.13	1.579		16.50	69.50	5.50	91.50	131
3290	1-1/4"		9.40	1.702		22	74.50	5.90	102.40	147
3300	1-1/2"		8.89	1.800		17.25	79	6.25	102.50	148
3310	2"		7.55	2.119		25	93	7.35	125.35	179
3320	2-1/2"		3.98	4.020		47.50	176	14	237.50	340
3330	3"		3.21	4.984		61	219	17.35	297.35	425
3340	3-1/2"		2.83	5.654		225	248	19.65	492.65	655
3350	4"		2.44	6.557		105	288	23	416	585
3360	5"	↓	2.13	7.512		355	330	26	711	930
3370	6"	Q-16	2.83	8.481		310	385	19.65	714.65	960
3380	8"		1.83	13.115		615	595	30.50	1,240.50	1,625
3390	10"		1.44	16.667		1,300	760	38.50	2,098.50	2,650
3400	12"	↓	1.14	21.053	↓	1,650	960	49	2,659	3,350
3410	For schedule 40, type 316, add					25%				

22 11 13.66 Pipe Fittings, Stainless Steel		Crew	Daily Output	Labor-Hours	Unit	Material	2009 Bare Costs Labor	Equipment	Total	Total Incl O&P
3460	45° Elbow, long, 1/2"	Q-15	12.70	1.260	Ea.	15.75	55.50	4.38	75.63	108
3470	3/4"		11.10	1.441		15.75	63.50	5	84.25	121
3480	1"		10.13	1.579		16.50	69.50	5.50	91.50	131
3490	1-1/4"		9.40	1.702		22	74.50	5.90	102.40	147
3500	1-1/2"		8.89	1.800		17.25	79	6.25	102.50	148
3510	2"		7.55	2.119		25	93	7.35	125.35	179
3520	2-1/2"		3.98	4.020		47.50	176	14	237.50	340
3530	3"		3.21	4.984		48	219	17.35	284.35	410
3540	3-1/2"		2.83	5.654		225	248	19.65	492.65	655
3550	4"		2.44	6.557		75	288	23	386	555
3560	5"		2.13	7.512		248	330	26	604	810
3570	6"	Q-16	2.83	8.481		218	385	19.65	622.65	860
3580	8"		1.83	13.115		430	595	30.50	1,055.50	1,425
3590	10"		1.44	16.667		910	760	38.50	1,708.50	2,225
3600	12"		1.14	21.053		1,150	960	49	2,159	2,800
3610	For schedule 40, type 316, add					25%				
3660	Tee, straight 1/2"	Q-15	8.46	1.891	Ea.	40.50	83	6.60	130.10	181
3670	3/4"		7.40	2.162		40.50	95	7.50	143	200
3680	1"		6.74	2.374		43	104	8.25	155.25	218
3690	1-1/4"		6.27	2.552		100	112	8.90	220.90	294
3700	1-1/2"		5.92	2.703		46	119	9.40	174.40	245
3710	2"		5.03	3.181		68.50	140	11.05	219.55	305
3720	2-1/2"		2.65	6.038		86.50	265	21	372.50	530
3730	3"		2.14	7.477		87	330	26	443	635
3740	3-1/2"		1.88	8.511		164	375	29.50	568.50	795
3750	4"		1.62	9.877		164	435	34.50	633.50	890
3760	5"		1.42	11.268		380	495	39	914	1,225
3770	6"	Q-16	1.88	12.766		335	580	29.50	944.50	1,300
3780	8"		1.22	19.672		735	895	45.50	1,675.50	2,250
3790	10"		.96	25		1,425	1,150	58	2,633	3,425
3800	12"		.76	31.579		1,900	1,425	73.50	3,398.50	4,400
3810	For schedule 40, type 316, add					25%				
3820	Tee, reducing on outlet, 3/4" x 1/2"	Q-15	7.73	2.070	Ea.	54	91	7.20	152.20	208
3822	1" x 1/2"		7.24	2.210		58.50	97	7.70	163.20	223
3824	1" x 3/4"		6.96	2.299		54	101	8	163	224
3826	1-1/4" x 1"		6.43	2.488		132	109	8.65	249.65	325
3828	1-1/2" x 1/2"		6.58	2.432		79.50	107	8.45	194.95	262
3830	1-1/2" x 3/4"		6.35	2.520		74.50	111	8.75	194.25	262
3832	1-1/2" x 1"		6.18	2.589		57	114	9	180	248
3834	2" x 1"		5.50	2.909		102	128	10.10	240.10	320
3836	2" x 1-1/2"		5.30	3.019		85.50	132	10.50	228	310
3838	2-1/2" x 2"		3.15	5.079		129	223	17.70	369.70	505
3840	3" x 1-1/2"		2.72	5.882		131	258	20.50	409.50	565
3842	3" x 2"		2.65	6.038		109	265	21	395	555
3844	4" x 2"		2.10	7.619		272	335	26.50	633.50	850
3846	4" x 3"		1.77	9.040		197	395	31.50	623.50	865
3848	5" x 4"		1.48	10.811		455	475	37.50	967.50	1,275
3850	6" x 3"	Q-16	2.19	10.959		505	500	25.50	1,030.50	1,350
3852	6" x 4"		2.04	11.765		435	535	27.50	997.50	1,350
3854	8" x 4"		1.46	16.438		1,025	750	38	1,813	2,325
3856	10" x 8"		.69	34.783		1,725	1,575	80.50	3,380.50	4,425
3858	12" x 10"		.55	43.636		2,275	1,975	101	4,351	5,700
3950	Reducer, concentric, 3/4" x 1/2"	Q-15	11.85	1.350		30	59.50	4.70	94.20	130

22 11 13.66 Pipe Fittings, Stainless Steel		Crew	Daily Output	Labor-Hours	Unit	2009 Bare Costs Material	Labor	Equipment	Total	Total Incl O&P
3952	1" x 3/4"	Q-15	10.60	1.509	Ea.	31.50	66	5.25	102.75	143
3954	1-1/4" x 3/4"		10.19	1.570		77.50	69	5.45	151.95	198
3956	1-1/4" x 1"		9.76	1.639		39	72	5.70	116.70	161
3958	1-1/2" x 3/4"		9.88	1.619		65.50	71	5.65	142.15	188
3960	1-1/2" x 1"		9.47	1.690		52.50	74	5.90	132.40	179
3962	2" x 1"		8.65	1.850		22.50	81	6.45	109.95	158
3964	2" x 1-1/2"		8.16	1.961		26.50	86	6.80	119.30	170
3966	2-1/2" x 1"		5.71	2.802		113	123	9.75	245.75	325
3968	2-1/2" x 2"		5.21	3.071		58	135	10.70	203.70	284
3970	3" x 1"		4.88	3.279		78	144	11.40	233.40	320
3972	3" x 1-1/2"		4.72	3.390		40	149	11.80	200.80	288
3974	3" x 2"		4.51	3.548		37.50	156	12.35	205.85	296
3976	4" x 2"		3.69	4.336		53.50	190	15.10	258.60	370
3978	4" x 3"		2.77	5.776		40	253	20	313	460
3980	5" x 3"		2.56	6.250		237	274	22	533	710
3982	5" x 4"	↓	2.27	7.048		191	310	24.50	525.50	715
3984	6" x 3"	Q-16	3.57	6.723		122	305	15.60	442.60	625
3986	6" x 4"		3.19	7.524		101	340	17.45	458.45	660
3988	8" x 4"		2.44	9.836		315	450	23	788	1,075
3990	8" x 6"		2.22	10.811		236	490	25	751	1,050
3992	10" x 6"		1.91	12.565		440	570	29	1,039	1,400
3994	10" x 8"		1.61	14.907		365	680	34.50	1,079.50	1,500
3995	12" x 6"		1.63	14.724		675	670	34	1,379	1,825
3996	12" x 8"		1.41	17.021		620	775	39.50	1,434.50	1,925
3997	12" x 10"	↓	1.27	18.898	↓	435	860	44	1,339	1,850
4000	Socket weld joint, 3000 lb., type 304									
4100	90° Elbow									
4140	1/4"	Q-15	13.47	1.188	Ea.	42	52	4.13	98.13	132
4150	3/8"		12.97	1.234		54.50	54	4.29	112.79	149
4160	1/2"		12.21	1.310		56.50	57.50	4.56	118.56	157
4170	3/4"		10.68	1.498		68.50	65.50	5.20	139.20	183
4180	1"		9.74	1.643		103	72	5.70	180.70	231
4190	1-1/4"		9.05	1.768		179	77.50	6.15	262.65	325
4200	1-1/2"		8.55	1.871		218	82	6.50	306.50	375
4210	2"	↓	7.26	2.204	↓	350	96.50	7.65	454.15	545
4300	45° Elbow									
4340	1/4"	Q-15	13.47	1.188	Ea.	78.50	52	4.13	134.63	172
4350	3/8"		12.97	1.234		78.50	54	4.29	136.79	175
4360	1/2"		12.21	1.310		78.50	57.50	4.56	140.56	181
4370	3/4"		10.68	1.498		89.50	65.50	5.20	160.20	206
4380	1"		9.74	1.643		129	72	5.70	206.70	260
4390	1-1/4"		9.05	1.768		213	77.50	6.15	296.65	360
4400	1-1/2"		8.55	1.871		213	82	6.50	301.50	370
4410	2"	↓	7.26	2.204	↓	390	96.50	7.65	494.15	590
4500	Tee									
4540	1/4"	Q-15	8.97	1.784	Ea.	66	78.50	6.20	150.70	200
4550	3/8"		8.64	1.852		66	81.50	6.45	153.95	206
4560	1/2"		8.13	1.968		81	86.50	6.85	174.35	231
4570	3/4"		7.12	2.247		93	98.50	7.80	199.30	265
4580	1"		6.48	2.469		125	108	8.60	241.60	315
4590	1-1/4"		6.03	2.653		223	116	9.25	348.25	435
4600	1-1/2"		5.69	2.812		320	123	9.80	452.80	550
4610	2"	↓	4.83	3.313	↓	485	145	11.55	641.55	770

22 11 Facility Water Distribution

22 11 13 – Facility Water Distribution Piping

22 11 13.66 Pipe Fittings, Stainless Steel		Crew	Daily Output	Labor-Hours	Unit	Material	2009 Bare Costs Labor	Equipment	Total	Total Incl O&P
5000	Socket weld joint, 3000 lb., type 316									
5100	90° Elbow									
5140	1/4"	Q-15	13.47	1.188	Ea.	51	52	4.13	107.13	142
5150	3/8"		12.97	1.234		60	54	4.29	118.29	155
5160	1/2"		12.21	1.310		72	57.50	4.56	134.06	173
5170	3/4"		10.68	1.498		95	65.50	5.20	165.70	212
5180	1"		9.74	1.643		135	72	5.70	212.70	266
5190	1-1/4"		9.05	1.768		240	77.50	6.15	323.65	390
5200	1-1/2"		8.55	1.871		271	82	6.50	359.50	435
5210	2"		7.26	2.204		460	96.50	7.65	564.15	670
5300	45° Elbow									
5340	1/4"	Q-15	13.47	1.138	Ea.	103	52	4.13	159.13	200
5350	3/8"		12.97	1.234		103	54	4.29	161.29	203
5360	1/2"		12.21	1.310		103	57.50	4.56	165.06	208
5370	3/4"		10.68	1.498		115	65.50	5.20	185.70	234
5380	1"		9.74	1.643		173	72	5.70	250.70	310
5390	1-1/4"		9.05	1.768		247	77.50	6.15	330.65	400
5400	1-1/2"		8.55	1.871		278	82	6.50	366.50	440
5410	2"		7.26	2.204		415	96.50	7.65	519.15	615
5500	Tee									
5540	1/4"	Q-15	8.97	1.784	Ea.	69.50	78.50	6.20	154.20	204
5550	3/8"		8.64	1.852		85	81.50	6.45	172.95	227
5560	1/2"		8.13	1.968		94.50	86.50	6.85	187.85	246
5570	3/4"		7.12	2.247		117	98.50	7.80	223.30	291
5580	1"		6.48	2.469		179	108	8.60	295.60	375
5590	1-1/4"		6.03	2.653		284	116	9.25	409.25	500
5600	1-1/2"		5.69	2.812		400	123	9.80	532.80	640
5610	2"		4.83	3.313		640	145	11.55	796.55	945
5700	For socket weld joint, 6000 lb., type 304 and 316, add					100%				
6000	Threaded companion flange									
6010	Stainless steel, 150 lb., type 304									
6020	1/2" diam.	1 Plum	30	.267	Ea.	35	13		48	58.50
6030	3/4" diam.		28	.286		39	13.95		52.95	64.50
6040	1" diam.		27	.296		43	14.45		57.45	70
6050	1-1/4" diam.	Q-1	44	.364		55	15.95		70.95	85.50
6060	1-1/2" diam.		40	.400		55	17.55		72.55	87.50
6070	2" diam.		36	.444		72	19.50		91.50	109
6080	2-1/2" diam.		28	.571		100	25		125	149
6090	3" diam.		20	.800		105	35		140	171
6110	4" diam.		12	1.333		144	58.50		202.50	249
6130	6" diam.	Q-2	14	1.714		259	78		337	405
6140	8" diam.	"	12	2		500	91		591	690
6150	For type 316 add					40%				
6260	Weld flanges, stainless steel, type 304									
6270	Slip on, 150 lb. (welded, front and back)									
6280	1/2" diam.	Q-15	12.70	1.260	Ea.	31	55.50	4.38	90.88	124
6290	3/4" diam.		11.11	1.440		32	63	5	100	139
6300	1" diam.		10.13	1.579		35	69.50	5.50	110	152
6310	1-1/4" diam.		9.41	1.700		48	74.50	5.90	128.40	176
6320	1-1/2" diam.		8.89	1.800		48	79	6.25	133.25	182
6330	2" diam.		7.55	2.119		61	93	7.35	161.35	219
6340	2-1/2" diam.		3.98	4.020		86	176	14	276	385
6350	3" diam.		3.21	4.984		92	219	17.35	328.35	460

22 11 13.66 Pipe Fittings, Stainless Steel		Crew	Daily Output	Labor-Hours	Unit	2009 Bare Costs Material	Labor	Equipment	Total	Total Incl O&P
6370	4″ diam.	Q-15	2.44	6.557	Ea.	125	288	23	436	610
6390	6″ diam.	Q-16	1.89	12.698		190	580	29.50	799.50	1,125
6400	8″ diam.	″	1.22	19.672		360	895	45.50	1,300.50	1,850
6410	For type 316, add					40%				
6530	Weld neck 150 lb.									
6540	1/2″ diam.	Q-15	25.40	.630	Ea.	16.85	27.50	2.19	46.54	64
6550	3/4″ diam.		22.22	.720		20.50	31.50	2.51	54.51	74.50
6560	1″ diam.		20.25	.790		25	34.50	2.75	62.25	84.50
6570	1-1/4″ diam.		18.82	.850		43	37.50	2.96	83.46	109
6580	1-1/2″ diam.		17.78	.900		38.50	39.50	3.13	81.13	107
6590	2″ diam.		15.09	1.060		40	46.50	3.69	90.19	120
6600	2-1/2″ diam.		7.96	2.010		64	88	7	159	215
6610	3″ diam.		6.43	2.488		69.50	109	8.65	187.15	255
6630	4″ diam.		4.88	3.279		107	144	11.40	262.40	355
6640	5″ diam.		4.26	3.756		148	165	13.05	326.05	435
6650	6″ diam.	Q-16	5.66	4.240		189	193	9.85	391.85	520
6652	8″ diam.		3.65	6.575		330	299	15.25	644.25	840
6654	10″ diam.		2.88	8.333		465	380	19.35	864.35	1,125
6656	12″ diam.		2.28	10.526		720	480	24.50	1,224.50	1,575
6670	For type 316 add					23%				
7000	Threaded joint, 150 lb., type 304									
7030	90° elbow									
7040	1/8″	1 Plum	13	.615	Ea.	24	30		54	73
7050	1/4″		13	.615		24	30		54	73
7070	3/8″		13	.615		26.50	30		56.50	75.50
7080	1/2″		12	.667		23.50	32.50		56	76.50
7090	3/4″		11	.727		27	35.50		62.50	84.50
7100	1″		10	.800		35.50	39		74.50	100
7110	1-1/4″	Q-1	17	.941		55.50	41.50		97	125
7120	1-1/2″		16	1		61	44		105	135
7130	2″		14	1.143		88	50		138	175
7140	2-1/2″		11	1.455		243	64		307	365
7150	3″		8	2		335	88		423	505
7160	4″	Q-2	11	2.182		450	99.50		549.50	650
7180	45° elbow									
7190	1/8″	1 Plum	13	.615	Ea.	35.50	30		65.50	85.50
7200	1/4″		13	.615		35.50	30		65.50	85.50
7210	3/8″		13	.615		33.50	30		63.50	83.50
7220	1/2″		12	.667		34	32.50		66.50	88
7230	3/4″		11	.727		26.50	35.50		62	84.50
7240	1″		10	.800		40.50	39		79.50	105
7250	1-1/4″	Q-1	17	.941		52	41.50		93.50	122
7260	1-1/2″		16	1		76.50	44		120.50	152
7270	2″		14	1.143		95.50	50		145.50	183
7280	2-1/2″		11	1.455		345	64		409	475
7290	3″		8	2		500	88		588	685
7300	4″	Q-2	11	2.182		905	99.50		1,004.50	1,150
7320	Tee, straight									
7330	1/8″	1 Plum	9	.889	Ea.	37	43.50		80.50	108
7340	1/4″		9	.889		37	43.50		80.50	108
7350	3/8″		9	.889		37.50	43.50		81	108
7360	1/2″		8	1		35.50	49		84.50	115
7370	3/4″		7	1.143		38	55.50		93.50	128

22 11 13.66 Pipe Fittings, Stainless Steel		Crew	Daily Output	Labor-Hours	Unit	2009 Bare Costs Material	Labor	Equipment	Total	Total Incl O&P
7380	1"	1 Plum	6.50	1.231	Ea.	44.50	60		104.50	142
7390	1-1/4"	Q-1	11	1.455		75	64		139	182
7400	1-1/2"		10	1.600		92	70		162	210
7410	2"		9	1.778		114	78		192	247
7420	2-1/2"		7	2.286		350	100		450	540
7430	3"		5	3.200		530	140		670	805
7440	4"	Q-2	7	3.429		1,300	156		1,456	1,700
7460	Coupling, straight									
7470	1/8"	1 Plum	19	.421	Ea.	9.70	20.50		30.20	42.50
7480	1/4"		19	.421		11.50	20.50		32	44.50
7490	3/8"		19	.421		13.80	20.50		34.30	47
7500	1/2"		19	.421		18.35	20.50		38.85	52
7510	3/4"		18	.444		24.50	21.50		46	60.50
7520	1"		15	.533		39.50	26		65.50	84
7530	1-1/4"	Q-1	26	.615		63.50	27		90.50	112
7540	1-1/2"		24	.667		71	29.50		100.50	124
7550	2"		21	.762		112	33.50		145.50	175
7560	2-1/2"		18	.889		260	39		299	345
7570	3"		14	1.143		355	50		405	470
7580	4"	Q-2	16	1.500		520	68.50		588.50	680
7600	Reducer, concentric, 1/2"	1 Plum	12	.667		18.35	32.50		50.85	70.50
7610	3/4"		11	.727		24.50	35.50		60	82
7612	1"		10	.800		42	39		81	107
7614	1-1/4"	Q-1	17	.941		89.50	41.50		131	163
7616	1-1/2"		16	1		98	44		142	176
7618	2"		14	1.143		115	50		165	204
7620	2-1/2"		11	1.455		380	64		444	515
7622	3"		8	2		420	88		508	600
7624	4"	Q-2	11	2.182		700	99.50		799.50	925
7710	Union									
7720	1/8"	1 Plum	12	.667	Ea.	45	32.50		77.50	100
7730	1/4"		12	.667		45	32.50		77.50	100
7740	3/8"		12	.667		53	32.50		85.50	109
7750	1/2"		11	.727		66	35.50		101.50	128
7760	3/4"		10	.800		90.50	39		129.50	160
7770	1"		9	.889		131	43.50		174.50	211
7780	1-1/4"	Q-1	16	1		305	44		349	405
7790	1-1/2"		15	1.067		330	47		377	440
7800	2"		13	1.231		420	54		474	545
7810	2-1/2"		10	1.600		790	70		860	980
7820	3"		7	2.286		1,050	100		1,150	1,300
7830	4"	Q-2	10	2.400		1,450	109		1,559	1,775
7850	For 150 lb., type 316, add					25%				
8750	Threaded joint, 2000 lb., type 304									
8770	90° Elbow									
8780	1/8"	1 Plum	13	.615	Ea.	31	30		61	80.50
8790	1/4"		13	.615		31	30		61	80.50
8800	3/8"		13	.615		38	30		68	88.50
8810	1/2"		12	.667		51.50	32.50		84	108
8820	3/4"		11	.727		60	35.50		95.50	121
8830	1"		10	.800		79.50	39		118.50	148
8840	1-1/4"	Q-1	17	.941		129	41.50		170.50	206
8850	1-1/2"		16	1		208	44		252	296

22 11 13.66 Pipe Fittings, Stainless Steel		Crew	Daily Output	Labor-Hours	Unit	Material (2009 Bare Costs)	Labor (2009 Bare Costs)	Equipment (2009 Bare Costs)	Total (2009 Bare Costs)	Total Incl O&P
8860	2"	Q-1	14	1.143	Ea.	292	50		342	400
8880	45° Elbow									
8890	1/8"	1 Plum	13	.615	Ea.	64	30		94	117
8900	1/4"		13	.615		64	30		94	117
8910	3/8"		13	.615		80.50	30		110.50	135
8920	1/2"		12	.667		80.50	32.50		113	139
8930	3/4"		11	.727		88.50	35.50		124	153
8940	1"		10	.800		106	39		145	177
8950	1-1/4"	Q-1	17	.941		184	41.50		225.50	267
8960	1-1/2"		16	1		247	44		291	340
8970	2"		14	1.143		285	50		335	395
8990	Tee, straight									
9000	1/8"	1 Plum	9	.889	Ea.	40	43.50		83.50	111
9010	1/4"		9	.889		40	43.50		83.50	111
9020	3/8"		9	.889		51	43.50		94.50	124
9030	1/2"		8	1		65.50	49		114.50	148
9040	3/4"		7	1.143		74.50	55.50		130	169
9050	1"		6.50	1.231		103	60		163	206
9060	1-1/4"	Q-1	11	1.455		173	64		237	290
9070	1-1/2"		10	1.600		289	70		359	425
9080	2"		9	1.778		400	78		478	560
9100	For couplings and unions use 3000 lb., type 304									
9120	2000 lb., type 316									
9130	90° Elbow									
9140	1/8"	1 Plum	13	.615	Ea.	38.50	30		68.50	88.50
9150	1/4"		13	.615		38.50	30		68.50	88.50
9160	3/8"		13	.615		45	30		75	96
9170	1/2"		12	.667		58	32.50		90.50	115
9180	3/4"		11	.727		71	35.50		106.50	134
9190	1"		10	.800		85.50	39		124.50	155
9200	1-1/4"	Q-1	17	.941		199	41.50		240.50	283
9210	1-1/2"		16	1		221	44		265	310
9220	2"		14	1.143		375	50		425	495
9240	45° Elbow									
9250	1/8"	1 Plum	13	.615	Ea.	81.50	30		111.50	136
9260	1/4"		13	.615		81.50	30		111.50	136
9270	3/8"		13	.615		81.50	30		111.50	136
9280	1/2"		12	.667		81.50	32.50		114	140
9300	3/4"		11	.727		91.50	35.50		127	156
9310	1"		10	.800		138	39		177	213
9320	1-1/4"	Q-1	17	.941		198	41.50		239.50	281
9330	1-1/2"		16	1		270	44		314	365
9340	2"		14	1.143		330	50		380	445
9360	Tee, straight									
9370	1/8"	1 Plum	9	.889	Ea.	49	43.50		92.50	121
9380	1/4"		9	.889		49	43.50		92.50	121
9390	3/8"		9	.889		59.50	43.50		103	133
9400	1/2"		8	1		74.50	49		123.50	157
9410	3/4"		7	1.143		92	55.50		147.50	188
9420	1"		6.50	1.231		140	60		200	247
9430	1-1/4"	Q-1	11	1.455		227	64		291	350
9440	1-1/2"		10	1.600		325	70		395	470
9450	2"		9	1.778		520	78		598	690

22 11 13.66 Pipe Fittings, Stainless Steel		Crew	Daily Output	Labor-Hours	Unit	2009 Bare Costs Material	Labor	Equipment	Total	Total Incl O&P
9470	For couplings and unions use 3000 lb., type 316									
9490	3000 lb., type 304									
9510	Coupling									
9520	1/8"	1 Plum	19	.421	Ea.	12.15	20.50		32.65	45.50
9530	1/4"		19	.421		13.20	20.50		33.70	46.50
9540	3/8"		19	.421		15.40	20.50		35.90	49
9550	1/2"		19	.421		18.55	20.50		39.05	52.50
9560	3/4"		18	.444		24.50	21.50		46	60.50
9570	1"		15	.533		42	26		68	86.50
9580	1-1/4"	Q-1	26	.615		102	27		129	154
9590	1-1/2"		24	.667		119	29.50		148.50	177
9600	2"		21	.762		157	33.50		190.50	224
9620	Union									
9630	1/8"	1 Plum	12	.657	Ea.	93	32.50		125.50	153
9640	1/4"		12	.657		93	32.50		125.50	153
9650	3/8"		12	.657		99.50	32.50		132	160
9660	1/2"		11	.727		99.50	35.50		135	164
9670	3/4"		10	.800		122	39		161	195
9680	1"		9	.889		188	43.50		231.50	274
9690	1-1/4"	Q-1	16	1		345	44		389	450
9700	1-1/2"		15	1.067		390	47		437	505
9710	2"		13	1.231		525	54		579	660
9730	3000 lb., type 316									
9750	Coupling									
9770	1/8"	1 Plum	19	.421	Ea.	13.70	20.50		34.20	47
9780	1/4"		19	.421		15.45	20.50		35.95	49
9790	3/8"		19	.421		16.55	20.50		37.05	50
9800	1/2"		19	.421		23.50	20.50		44	58
9810	3/4"		18	.444		34.50	21.50		56	71.50
9820	1"		15	.533		57.50	26		83.50	104
9830	1-1/4"	Q-1	26	.615		129	27		156	184
9840	1-1/2"		24	.667		151	29.50		180.50	212
9850	2"		21	.762		209	33.50		242.50	281
9870	Union									
9880	1/8"	1 Plum	12	.667	Ea.	102	32.50		134.50	163
9890	1/4"		12	.667		102	32.50		134.50	163
9900	3/8"		12	.667		118	32.50		150.50	181
9910	1/2"		11	.727		118	35.50		153.50	185
9920	3/4"		10	.300		155	39		194	231
9930	1"		9	.389		244	43.50		287.50	335
9940	1-1/4"	Q-1	16	1		425	44		469	540
9950	1-1/2"		15	1.067		525	47		572	655
9960	2"		13	1.231		670	54		724	820

22 11 13.74 Pipe, Plastic		Crew	Daily Output	Labor-Hours	Unit	Material	Labor	Equipment	Total	Total Incl O&P
0010	**PIPE, PLASTIC**									
0020	Fiberglass reinforced, couplings 10' O.C., clevis hanger assy's, 3 per 10'									
0080	General service									
0120	2" diameter	Q-1	59	.271	L.F.	5.90	11.90		17.80	25
0140	3" diameter		52	.308		8.05	13.50		21.55	30
0150	4" diameter		48	.333		10.85	14.65		25.50	34.50
0160	6" diameter		39	410		19.60	18		37.60	49.50
0170	8" diameter	Q-2	49	490		32.50	22.50		55	70

22 11 13.74 Pipe, Plastic		Crew	Daily Output	Labor-Hours	Unit	2009 Bare Costs Material	Labor	Equipment	Total	Total Incl O&P
0180	10" diameter	Q-2	41	.585	L.F.	49	26.50		75.50	95.50
0190	12" diameter		36	.667		63.50	30.50		94	117
0600	PVC, high impact/pressure, cplgs. 10' O.C., clevis hanger assy's, 3 per 10'									
1020	Schedule 80									
1070	1/2" diameter	1 Plum	50	.160	L.F.	3.03	7.80		10.83	15.45
1080	3/4" diameter		47	.170		3.84	8.30		12.14	17.05
1090	1" diameter		43	.186		5.15	9.05		14.20	19.70
1100	1-1/4" diameter		39	.205		6.90	10		16.90	23
1110	1-1/2" diameter		34	.235		8.15	11.45		19.60	27
1120	2" diameter	Q-1	55	.291		10.75	12.75		23.50	31.50
1140	3" diameter		50	.320		22	14.05		36.05	46
1150	4" diameter		46	.348		31.50	15.25		46.75	58.50
1170	6" diameter		38	.421		71.50	18.50		90	107
1730	To delete coupling & hangers, subtract									
1740	1/4" diam. to 1/2" diam.					62%	80%			
1750	3/4" diam. to 1-1/4" diam.					58%	73%			
1760	1-1/2" diam. to 6" diam.					40%	57%			
1770	8" diam. to 12" diam.					34%	50%			
1800	PVC, couplings 10' O.C., clevis hanger assy's, 3 per 10'									
1820	Schedule 40									
1860	1/2" diameter	1 Plum	54	.148	L.F.	1.43	7.20		8.63	12.75
1870	3/4" diameter		51	.157		1.67	7.65		9.32	13.70
1880	1" diameter		46	.174		1.96	8.50		10.46	15.30
1890	1-1/4" diameter		42	.190		2.49	9.30		11.79	17.15
1900	1-1/2" diameter		36	.222		2.73	10.85		13.58	19.80
1910	2" diameter	Q-1	59	.271		3.33	11.90		15.23	22
1920	2-1/2" diameter		56	.286		4.93	12.55		17.48	25
1930	3" diameter		53	.302		6.60	13.25		19.85	28
1940	4" diameter		48	.333		9.15	14.65		23.80	32.50
1950	5" diameter		43	.372		12.85	16.35		29.20	39.50
1960	6" diameter		39	.410		16.95	18		34.95	46.50
1970	8" diameter	Q-2	48	.500		24.50	23		47.50	62.50
1980	10" diameter		43	.558		55	25.50		80.50	100
1990	12" diameter		42	.571		66	26		92	114
2000	14" diameter		31	.774		116	35		151	182
2010	16" diameter		23	1.043		166	47.50		213.50	257
2340	To delete coupling & hangers, subtract									
2360	1/2" diam. to 1-1/4" diam.					65%	74%			
2370	1-1/2" diam. to 6" diam.					44%	57%			
2380	8" diam. to 12" diam.					41%	53%			
2390	14" diam. to 16" diam.					48%	45%			
2420	Schedule 80									
2440	1/4" diameter	1 Plum	58	.138	L.F.	1.09	6.70		7.79	11.65
2450	3/8" diameter		55	.145		1.09	7.10		8.19	12.20
2460	1/2" diameter		50	.160		1.52	7.80		9.32	13.75
2470	3/4" diameter		47	.170		1.79	8.30		10.09	14.80
2480	1" diameter		43	.186		2.14	9.05		11.19	16.40
2490	1-1/4" diameter		39	.205		2.76	10		12.76	18.55
2500	1-1/2" diameter		34	.235		3.07	11.45		14.52	21
2510	2" diameter	Q-1	55	.291		3.75	12.75		16.50	24
2520	2-1/2" diameter		52	.308		4.60	13.50		18.10	26
2530	3" diameter		50	.320		7.65	14.05		21.70	30.50
2540	4" diameter		46	.348		10.75	15.25		26	35.50

22 11 13 – Facility Water Distribution Piping

22 11 13.74 Pipe, Plastic		Crew	Daily Output	Labor-Hours	Unit	2009 Bare Costs Material	Labor	Equipment	Total	Total Incl O&P
2550	5" diameter	Q-1	42	.381	L.F.	15.25	16.70		31.95	43
2560	6" diameter		38	.421		24	18.50		42.50	55
2570	8" diameter	Q-2	47	.511		30.50	23		53.50	69.50
2580	10" diameter		42	.571		64.50	26		90.50	112
2590	12" diameter		38	.632		80	28.50		108.50	132
2830	To delete coupling & hangers, subtract									
2840	1/4" diam. to 1/2" diam.					66%	80%			
2850	3/4" diam. to 1-1/4" diam.					61%	73%			
2860	1-1/2" diam. to 6" diam.					41%	57%			
2870	8" diam. to 12" diam.					31%	50%			
2900	Schedule 120									
2910	1/2" diameter	1 Plum	50	.160	L.F.	1.65	7.80		9.45	13.90
2950	3/4" diameter		47	.170		1.95	8.30		10.25	15
2960	1" diameter		43	.186		2.38	9.05		11.43	16.65
2970	1-1/4" diameter		39	.205		3.10	10		13.10	18.90
2980	1-1/2" diameter		33	.242		3.49	11.80		15.29	22
2990	2" diameter	Q-1	54	.296		4.38	13		17.38	25
3000	2-1/2" diameter		52	.308		7.15	13.50		20.65	29
3010	3" diameter		49	.327		8.90	14.35		23.25	32
3020	4" diameter		45	.356		13.40	15.60		29	38.50
3030	6" diameter		37	.432		25.50	19		44.50	57.50
3240	To delete coupling & hangers, subtract									
3250	1/2" diam. to 1-1/4" diam.					52%	74%			
3260	1-1/2" diam. to 4" diam.					30%	57%			
3270	6" diam.					17%	50%			
3300	PVC, pressure, couplings 10' O.C., clevis hanger assy's, 3 per 10'									
3310	SDR 26, 160 psi									
3350	1-1/4" diameter	1 Plum	42	.190	L.F.	2.07	9.30		11.37	16.65
3360	1-1/2" diameter	"	36	.222		2.27	10.85		13.12	19.30
3370	2" diameter	Q-1	59	.271		2.71	11.90		14.61	21.50
3380	2-1/2" diameter		56	.286		3.97	12.55		16.52	24
3390	3" diameter		53	.302		5.60	13.25		18.85	26.50
3400	4" diameter		48	.333		8.10	14.65		22.75	31.50
3420	6" diameter		39	.410		16.30	18		34.30	46
3430	8" diameter	Q-2	48	.500		25	23		48	63
3660	To delete coupling & clevis hanger assy's, subtract									
3670	1-1/4" diam.					63%	68%			
3680	1-1/2" diam. to 4" diam.					48%	57%			
3690	6" diam. to 8" diam.					60%	54%			
3720	SDR 21, 200 psi, 1/2" diameter	1 Plum	54	.148	L.F.	1.21	7.20		8.41	12.55
3740	3/4" diameter		51	.157		1.48	7.65		9.13	13.50
3750	1" diameter		46	.174		1.64	8.50		10.14	14.95
3760	1-1/4" diameter		42	.190		2.15	9.30		11.45	16.75
3770	1-1/2" diameter		36	.222		2.38	10.85		13.23	19.40
3780	2" diameter	Q-1	59	.271		2.90	11.90		14.80	21.50
3790	2-1/2" diameter		56	.286		5.15	12.55		17.70	25
3800	3" diameter		53	.302		6	13.25		19.25	27
3810	4" diameter		48	.333		9.90	14.65		24.55	33.50
3830	6" diameter		39	.410		17.85	18		35.85	47.50
3840	8" diameter	Q-2	48	.500		27.50	23		50.50	66
4000	To delete coupling & hangers, subtract									
4010	1/2" diam. to 3/4" diam.					71%	77%			
4020	1" diam. to 1-1/4" diam.					63%	70%			

22 11 13.74 Pipe, Plastic		Crew	Daily Output	Labor-Hours	Unit	2009 Bare Costs Material	Labor	Equipment	Total	Total Incl O&P
4030	1-1/2″ diam. to 6″ diam.					44%	57%			
4040	8″ diam.					46%	54%			
4100	DWV type, schedule 40, couplings 10′ O.C., clevis hanger assy's, 3 per 10′									
4120	ABS									
4140	1-1/4″ diameter	1 Plum	42	.190	L.F.	1.58	9.30		10.88	16.15
4150	1-1/2″ diameter	″	36	.222		1.72	10.85		12.57	18.70
4160	2″ diameter	Q-1	59	.271		2.21	11.90		14.11	21
4170	3″ diameter		53	.302		4.94	13.25		18.19	26
4180	4″ diameter		48	.333		8	14.65		22.65	31.50
4190	6″ diameter		39	.410		16.30	18		34.30	46
4194	To delete coupling & hangers, subtract									
4196	1-1/4″ diam.					56%	62%			
4198	1-1/2″ diam. to 6″ diam.					50%	52%			
4210	ABS, schedule 40, foam core type									
4212	Plain end black									
4214	1-1/2″ diameter	1 Plum	39	.205	L.F.	1.76	10		11.76	17.45
4216	2″ diameter	Q-1	62	.258		2.17	11.30		13.47	19.95
4218	3″ diameter		56	.286		4.15	12.55		16.70	24
4220	4″ diameter		51	.314		5.75	13.75		19.50	28
4222	6″ diameter		42	.381		11.95	16.70		28.65	39
4240	To delete coupling & hangers, subtract									
4244	1-1/2″ diam. to 6″ diam.					43%	48%			
4400	PVC									
4410	1-1/4″ diameter	1 Plum	42	.190	L.F.	1.85	9.30		11.15	16.45
4420	1-1/2″ diameter	″	36	.222		1.82	10.85		12.67	18.80
4460	2″ diameter	Q-1	59	.271		2.20	11.90		14.10	21
4470	3″ diameter		53	.302		4.21	13.25		17.46	25
4480	4″ diameter		48	.333		5.75	14.65		20.40	29
4490	6″ diameter		39	.410		10.55	18		28.55	39.50
4500	8″ diameter	Q-2	48	.500		17.50	23		40.50	55
4510	To delete coupling & hangers, subtract									
4520	1-1/4″ diam. to 1-1/2″ diam.					48%	60%			
4530	2″ diam. to 8″ diam.					42%	54%			
4550	PVC, schedule 40, foam core type									
4552	Plain end, white									
4554	1-1/2″ diameter	1 Plum	39	.205	L.F.	1.56	10		11.56	17.20
4556	2″ diameter	Q-1	62	.258		1.88	11.30		13.18	19.60
4558	3″ diameter		56	.286		3.60	12.55		16.15	23.50
4560	4″ diameter		51	.314		4.83	13.75		18.58	27
4562	6″ diameter		42	.381		8.70	16.70		25.40	35.50
4564	8″ diameter	Q-2	51	.471		14.60	21.50		36.10	49
4568	10″ diameter		48	.500		19.25	23		42.25	56.50
4570	12″ diameter		46	.522		24	23.50		47.50	63.50
4580	To delete coupling & hangers, subtract									
4582	1-1/2″ dia to 2″ dia					58%	54%			
4584	3″ dia to 12″ dia					46%	42%			
4800	PVC, clear pipe, cplgs. 10′ O.C., clevis hanger assy's 3 per 10′, Sched. 40									
4840	1/4″ diameter	1 Plum	59	.136	L.F.	1.52	6.60		8.12	11.90
4850	3/8″ diameter		56	.143		1.79	6.95		8.74	12.75
4860	1/2″ diameter		54	.148		2.31	7.20		9.51	13.75
4870	3/4″ diameter		51	.157		2.85	7.65		10.50	15
4880	1″ diameter		46	.174		3.94	8.50		12.44	17.50
4890	1-1/4″ diameter		42	.190		5.10	9.30		14.40	20

22 11 13.74 Pipe, Plastic		Crew	Daily Output	Labor-Hours	Unit	2009 Bare Costs Material	Labor	Equipment	Total	Total Incl O&P
4900	1-1/2" diameter	1 Plum	36	.222	L.F.	5.95	10.85		16.80	23.50
4910	2" diameter	Q-1	59	.271		7.80	11.90		19.70	27
4920	2-1/2" diameter		56	.286		12.30	12.55		24.85	33
4930	3" diameter		53	.302		15.90	13.25		29.15	38
4940	3-1/2" diameter		50	.320		22.50	14.05		36.55	47
4950	4" diameter	↓	48	.333	↓	23.50	14.65		38.15	48.50
5250	To delete coupling & hangers, subtract									
5260	1/4" diam. to 3/8" diam.					60%	31%			
5270	1/2" diam. to 3/4" diam.					41%	77%			
5280	1" diam. to 1-1/2" diam.					26%	57%			
5290	2" diam. to 4" diam.					16%	58%			
5360	CPVC, couplings 10' O.C., clevis hanger assy's, 3 per 10'									
5380	Schedule 40									
5460	1/2" diameter	1 Plum	54	.148	L.F.	2.84	7.20		10.04	14.30
5470	3/4" diameter		51	.157		3.37	7.65		11.02	15.55
5480	1" diameter		46	.174		4.13	8.50		12.63	17.70
5490	1-1/4" diameter		42	.190		5.05	9.30		14.35	19.95
5500	1-1/2" diameter	↓	36	.222		5.70	10.85		16.55	23
5510	2" diameter	Q-1	59	.271		7.05	11.90		18.95	26
5520	2-1/2" diameter		56	.286		11.50	12.55		24.05	32
5530	3" diameter		53	.302		14.20	13.25		27.45	36
5540	4" diameter		48	.333		23.50	14.65		38.15	48
5550	6" diameter	↓	43	.372	↓	37	16.35		53.35	66
5730	To delete coupling & hangers, subtract									
5740	1/2" diam. to 3/4" diam.					37%	77%			
5750	1" diam. to 1-1/4" diam.					27%	70%			
5760	1-1/2" diam. to 3" diam.					21%	57%			
5770	4" diam. to 6" diam.					16%	57%			
5800	Schedule 80									
5860	1/2" diameter	1 Plum	50	.160	L.F.	2.99	7.80		10.79	15.40
5870	3/4" diameter		47	.170		3.60	8.30		11.90	16.80
5880	1" diameter		43	.186		4.45	9.05		13.50	18.95
5890	1-1/4" diameter		39	.205		5.55	10		15.55	21.50
5900	1-1/2" diameter	↓	34	.235		6.30	11.45		17.75	25
5910	2" diameter	Q-1	55	.291		8.05	12.75		20.80	28.50
5920	2-1/2" diameter		52	.308		12.85	13.50		26.35	35
5930	3" diameter		50	.320		16.15	14.05		30.20	40
5940	4" diameter		46	.348		26.50	15.25		41.75	52.50
5950	6" diameter	↓	38	.421		44	18.50		62.50	77
5960	8" diameter	Q-2	47	.511	↓	80.50	23		103.50	125
6060	To delete couplings & hangers, subtract									
6070	1/2" diam. to 3/4" diam.					44%	77%			
6080	1" diam. to 1-1/4" diam.					32%	71%			
6090	1-1/2" diam. to 4" diam.					25%	58%			
6100	6" diam. to 8" diam.					20%	53%			
6240	CTS, 1/2" diameter	1 Plum	54	.148	L.F.	1.35	7.20		8.55	12.70
6250	3/4" diameter		51	.157		1.93	7.65		9.58	14
6260	1" diameter		46	.174		3.21	8.50		11.71	16.70
6270	1 1/4"		42	.190		4.42	9.30		13.72	19.25
6280	1 1/2" diameter	↓	36	.222		5.65	10.85		16.50	23
6290	2" diameter	Q-1	59	.271	↓	9.10	11.90		21	28.50
6370	To delete coupling & hangers, subtract									
6380	1/2" diam.					51%	79%			

22 11 13.74 Pipe, Plastic		Crew	Daily Output	Labor-Hours	Unit	2009 Bare Costs Material	Labor	Equipment	Total	Total Incl O&P
6390	3/4" diam.					40%	76%			
6392	1" thru 2" diam.					72%	68%			
6500	Residential installation, plastic pipe									
6510	Couplings 10' O.C., strap hangers 3 per 10'									
6520	PVC, Schedule 40									
6530	1/2" diameter	1 Plum	138	.058	L.F.	1.23	2.83		4.06	5.75
6540	3/4" diameter		128	.063		1.36	3.05		4.41	6.25
6550	1" diameter		119	.067		1.62	3.28		4.90	6.90
6560	1-1/4" diameter		111	.072		1.98	3.51		5.49	7.65
6570	1-1/2" diameter		104	.077		2.39	3.75		6.14	8.45
6580	2" diameter	Q-1	197	.081		2.85	3.56		6.41	8.70
6590	2-1/2" diameter		162	.099		4.86	4.33		9.19	12.05
6600	4" diameter		123	.130		7.95	5.70		13.65	17.60
6700	PVC, DWV, Schedule 40									
6720	1-1/4" diameter	1 Plum	100	.080	L.F.	2.05	3.90		5.95	8.30
6730	1-1/2" diameter	"	94	.085		2.24	4.15		6.39	8.90
6740	2" diameter	Q-1	178	.090		2.60	3.94		6.54	8.95
6760	4" diameter	"	110	.145		7.10	6.40		13.50	17.70
7280	Polyethylene, flexible, no couplings or hangers									
7282	Note: For labor costs add 25% to the couplings and fittings labor total.									
7300	SDR 15, 100 psi									
7310	3/4" diameter				L.F.	.32			.32	.35
7350	1" diameter					.43			.43	.47
7360	1-1/4" diameter					.78			.78	.86
7370	1-1/2" diameter					.95			.95	1.05
7380	2" diameter					1.55			1.55	1.71
7700	SDR 9, 160 psi									
7710	1/2" diameter				L.F.	.43			.43	.47
7750	3/4" diameter					.43			.43	.47
7760	1" diameter					.65			.65	.72
7770	1-1/4" diameter					1.16			1.16	1.28
7780	1-1/2" diameter					1.62			1.62	1.78
7790	2" diameter					2.68			2.68	2.95
8120	SDR 9, 200 psi									
8150	3/4" diameter				L.F.	.38			.38	.42
8160	1" diameter					.63			.63	.69
8170	1-1/4" diameter					.95			.95	1.05
8420	SDR 7, 250 psi									
8440	3/4" diameter				L.F.	.55			.55	.61
8450	1" diameter					.88			.88	.97
8460	1-1/4" diameter					1.55			1.55	1.71
9900	Minimum labor/equipment charge	1 Plum	4	2	Job		97.50		97.50	151

22 11 13.76 Pipe Fittings, Plastic		Crew	Daily Output	Labor-Hours	Unit	Material	Labor	Equipment	Total	Total Incl O&P
0010	**PIPE FITTINGS, PLASTIC**									
0030	Epoxy resin, fiberglass reinforced, general service									
0090	Elbow, 90°, 2"	Q-1	33.10	.483	Ea.	79.50	21		100.50	121
0100	3"		20.80	.769		91.50	34		125.50	154
0110	4"		16.50	.970		126	42.50		168.50	204
0120	6"		10.10	1.584		182	69.50		251.50	310
0130	8"	Q-2	9.30	2.581		335	117		452	550
0140	10"		8.50	2.824		425	128		553	665
0150	12"		7.60	3.158		605	144		749	895

22 11 13.76 Pipe Fittings, Plastic		Crew	Daily Output	Labor-Hours	Unit	Material	2009 Bare Costs Labor	Equipment	Total	Total Incl O&P
0160	45° Elbow, same as 90°									
0290	Tee, 2"	Q-1	20	.800	Ea.	43.50	35		78.50	103
0300	3"		13.90	1.151		52.50	50.50		103	137
0310	4"		11	1.455		72	64		136	179
0320	6"		6.70	2.388		207	105		312	390
0330	8"	Q-2	6.20	3.871		1,300	176		1,476	1,700
0340	10"		5.70	4.211		1,475	192		1,667	1,925
0350	12"		5.10	4.706		1,800	214		2,014	2,300
0380	Couplings									
0410	2"	Q-1	33.10	.483	Ea.	14	21		35	48.50
0420	3"		20.80	.769		15	34		49	69
0430	4"		16.50	.970		20	42.50		62.50	88
0440	6"		10.10	1.584		48	69.50		117.50	161
0450	8"	Q-2	9.30	2.581		81.50	117		198.50	272
0460	10"		8.50	2.824		120	128		248	330
0470	12"		7.60	3.158		160	144		304	400
0473	High corrosion resistant couplings, add					30%				
2100	PVC schedule 80, socket joint									
2110	90° elbow, 1/2"	1 Plum	30.30	.264	Ea.	2.08	12.85		14.93	22
2130	3/4"		26	.308		2.66	15		17.66	26.50
2140	1"		22.70	.352		3.98	17.20		21.18	31
2150	1-1/4"		20.20	.396		5.75	19.30		25.05	36.50
2160	1-1/2"		18.20	.440		6.15	21.50		27.65	40
2170	2"	Q-1	33.10	.483		7.45	21		28.45	41
2180	3"		20.80	.769		19.50	34		53.50	74
2190	4"		16.50	.970		29.50	42.50		72	98.50
2200	6"		10.10	1.584		84.50	69.50		154	201
2210	8"	Q-2	9.30	2.581		233	117		350	440
2250	45° elbow, 1/2"	1 Plum	30.30	.264		3.92	12.85		16.77	24.50
2270	3/4"		26	.308		5.95	15		20.95	30
2280	1"		22.70	.352		9	17.20		26.20	36.50
2290	1-1/4"		20.20	.396		11.45	19.30		30.75	42.50
2300	1-1/2"		18.20	.440		13.50	21.50		35	48
2310	2"	Q-1	33.10	.483		17.50	21		38.50	52.50
2320	3"		20.80	.769		22.50	34		56.50	77
2330	4"		16.50	.970		80.50	42.50		123	155
2340	6"		10.10	1.584		102	69.50		171.50	220
2350	8"	Q-2	9.30	2.581		220	117		337	425
2400	Tee, 1/2"	1 Plum	20.20	.396		5.85	19.30		25.15	36.50
2420	3/4"		17.30	.462		6.15	22.50		28.65	42
2430	1"		15.20	.526		7.70	25.50		33.20	48.50
2440	1-1/4"		13.50	.593		21	29		50	68.50
2450	1-1/2"		12.10	.661		21	32		53	73.50
2460	2"	Q-1	20	.800		26.50	35		61.50	83.50
2470	3"		13.90	1.151		36	50.50		86.50	118
2480	4"		11	1.455		41.50	64		105.50	145
2490	6"		6.70	2.388		142	105		247	320
2500	8"	Q-2	6.20	3.871		296	176		472	600
2510	Flange, socket, 150 lb., 1/2"	1 Plum	55.60	.144		11.40	7		18.40	23.50
2514	3/4"		47.60	.168		12.20	8.20		20.40	26
2518	1"		41.70	.192		13.55	9.35		22.90	29.50
2522	1-1/2"		33.30	.240		14.30	11.70		26	34
2526	2"	Q-1	60.60	.264		19	11.60		30.60	39

22 11 13.76 Pipe Fittings, Plastic		Crew	Daily Output	Labor-Hours	Unit	2009 Bare Costs Material	Labor	Equipment	Total	Total Incl O&P
2530	4"	Q-1	30.30	.528	Ea.	41	23		64	81
2534	6"		18.50	.865		64.50	38		102.50	130
2538	8"	Q-2	17.10	1.404		115	64		179	226
2550	Coupling, 1/2"	1 Plum	30.30	.264		3.77	12.85		16.62	24
2570	3/4"		26	.308		5.10	15		20.10	29
2580	1"		22.70	.352		5.25	17.20		22.45	32.50
2590	1-1/4"		20.20	.396		8	19.30		27.30	39
2600	1-1/2"		18.20	.440		8.60	21.50		30.10	42.50
2610	2"	Q-1	33.10	.483		9.25	21		30.25	43
2620	3"		20.80	.769		22.50	34		56.50	77.50
2630	4"		16.50	.970		32.50	42.50		75	102
2640	6"		10.10	1.584		70.50	69.50		140	186
2650	8"	Q-2	9.30	2.581		96	117		213	287
2660	10"		8.50	2.824		330	128		458	560
2670	12"		7.60	3.158		380	144		524	640
2700	PVC (white), schedule 40, socket joints									
2760	90° elbow, 1/2"	1 Plum	33.30	.240	Ea.	.41	11.70		12.11	18.60
2770	3/4"		28.60	.280		.45	13.65		14.10	21.50
2780	1"		25	.320		.82	15.60		16.42	25
2790	1-1/4"		22.20	.360		1.44	17.55		18.99	28.50
2800	1-1/2"		20	.400		1.55	19.50		21.05	31.50
2810	2"	Q-1	36.40	.440		2.42	19.30		21.72	32.50
2820	2-1/2"		26.70	.599		7.40	26.50		33.90	49
2830	3"		22.90	.699		8.85	30.50		39.35	57
2840	4"		18.20	.879		15.80	38.50		54.30	77.50
2850	5"		12.10	1.322		41	58		99	135
2860	6"		11.10	1.441		50	63.50		113.50	154
2870	8"	Q-2	10.30	2.330		129	106		235	305
2980	45° elbow, 1/2"	1 Plum	33.30	.240		.67	11.70		12.37	18.90
2990	3/4"		28.60	.280		1.04	13.65		14.69	22
3000	1"		25	.320		1.25	15.60		16.85	25.50
3010	1-1/4"		22.20	.360		1.73	17.55		19.28	29
3020	1-1/2"		20	.400		2.18	19.50		21.68	32.50
3030	2"	Q-1	36.40	.440		2.84	19.30		22.14	33
3040	2-1/2"		26.70	.599		7.40	26.50		33.90	49
3050	3"		22.90	.699		11.50	30.50		42	60
3060	4"		18.20	.879		20.50	38.50		59	82.50
3070	5"		12.10	1.322		41	58		99	135
3080	6"		11.10	1.441		51	63.50		114.50	154
3090	8"	Q-2	10.30	2.330		122	106		228	299
3180	Tee, 1/2"	1 Plum	22.20	.360		.50	17.55		18.05	27.50
3190	3/4"		19	.421		.58	20.50		21.08	32.50
3200	1"		16.70	.479		1.08	23.50		24.58	37
3210	1-1/4"		14.80	.541		1.68	26.50		28.18	43
3220	1-1/2"		13.30	.602		2.06	29.50		31.56	48
3230	2"	Q-1	24.20	.661		2.99	29		31.99	48.50
3240	2-1/2"		17.80	.899		9.85	39.50		49.35	72
3250	3"		15.20	1.053		12.95	46		58.95	86
3260	4"		12.10	1.322		23.50	58		81.50	116
3270	5"		8.10	1.975		56.50	86.50		143	197
3280	6"		7.40	2.162		79	95		174	234
3290	8"	Q-2	6.80	3.529		183	161		344	450
3380	Coupling, 1/2"	1 Plum	33.30	.240		.27	11.70		11.97	18.45

22 11 13.76 Pipe Fittings, Plastic		Crew	Daily Output	Labor-Hours	Unit	2009 Bare Costs Material	Labor	Equipment	Total	Total Incl O&P
3390	3/4"	1 Plum	28.60	.280	Ea.	.37	13.65		14.02	21.50
3400	1"		25	.320		.66	15.60		16.26	24.50
3410	1-1/4"		22.20	.360		.89	17.55		18.44	28
3420	1-1/2"		20	.400		.95	19.50		20.45	31
3430	2"	Q-1	36.40	.440		1.45	19.30		20.75	31.50
3440	2-1/2"		26.70	.599		3.20	26.50		29.70	44.50
3450	3"		22.90	.699		5	30.50		35.50	53
3460	4"		18.20	.879		7.25	38.50		45.75	68
3470	5"		12.10	1.322		13.30	58		71.30	105
3480	6"		11.10	1.441		23	63.50		86.50	123
3490	8"	Q-2	10.30	2.330		43	106		149	211
4500	DWV, ABS, non pressure, socket joints									
4540	1/4 Bend, 1-1/4"	1 Plum	20.20	.396	Ea.	2.78	19.30		22.08	33
4560	1-1/2"	"	18.20	.440		1.96	21.50		23.46	35
4570	2"	Q-1	33.10	.483		3.04	21		24.04	36.50
4580	3"		20.80	.769		7.30	34		41.30	60.50
4590	4"		16.50	.970		14.05	42.50		56.55	81.50
4600	6"		10.10	1.584		66.50	69.50		136	182
4650	1/8 Bend, same as 1/4 Bend									
4800	Tee, sanitary									
4820	1-1/4"	1 Plum	13.50	.593	Ea.	3.31	29		32.31	48.50
4830	1-1/2"	"	12.10	.661		2.67	32		34.67	53
4840	2"	Q-1	20	.800		4.12	35		39.12	59
4850	3"		13.90	1.151		11.35	50.50		61.85	91
4860	4"		11	1.455		20.50	64		84.50	122
4862	Tee, sanitary, reducing, 2" x 1-1/2"		22	.727		3.94	32		35.94	54
4864	3" x 2"		15.30	1.046		8.25	46		54.25	80
4868	4" x 3"		12.10	1.322		19.20	58		77.20	111
4870	Combination Y and 1/8 bend									
4872	1-1/2"	1 Plum	12.10	.661	Ea.	6.35	32		38.35	57
4874	2"	Q-1	20	.800		7.55	35		42.55	63
4876	3"		13.90	1.151		16.35	50.50		66.85	96.50
4878	4"		11	1.455		33	64		97	135
4880	3" x 1-1/2"		15.50	1.032		15.55	45.50		61.05	87.50
4882	4" x 3"		12.10	1.322		26.50	58		84.50	119
4900	Wye, 1-1/4"	1 Plum	13.50	.593		3.82	29		32.82	49
4902	1-1/2"	"	12.10	.661		3.91	32		35.91	54.50
4904	2"	Q-1	20	.800		5.50	35		40.50	60.50
4906	3"		13.90	1.151		12.60	50.50		63.10	92.50
4908	4"		11	1.455		27	64		91	129
4910	6"		6.70	2.388		83	105		188	255
4918	3" x 1-1/2"		15.50	1.032		9.15	45.50		54.65	80.50
4920	4" x 3"		12.10	1.322		21	58		79	113
4922	6" x 4"		6.90	2.319		83	102		185	250
4930	Double Wye, 1-1/2"	1 Plum	9.10	.879		8.95	43		51.95	76.50
4932	2"	Q-1	16.60	.964		10.70	42.50		53.20	77.50
4934	3"		10.40	1.538		27.50	67.50		95	136
4936	4"		8.25	1.939		56	85		141	194
4940	2" x 1-1/2"		16.80	.952		10.70	42		52.70	77
4942	3" x 2"		10.60	1.509		20.50	66		86.50	126
4944	4" x 3"		8.45	1.893		44.50	83		127.50	178
4946	6" x 4"		7.25	2.207		87	97		184	246
4950	Reducer bushing, 2" x 1-1/2"		36.40	.440		1.48	19.30		20.78	31.50

22 11 13.76 Pipe Fittings, Plastic		Crew	Daily Output	Labor-Hours	Unit	2009 Bare Costs Material	Labor	Equipment	Total	Total Incl O&P
4952	3" x 1-1/2"	Q-1	27.30	.586	Ea.	5.55	25.50		31.05	46
4954	4" x 2"		18.20	.879		11.35	38.50		49.85	72.50
4956	6" x 4"		11.10	1.441		30.50	63.50		94	132
4960	Couplings, 1-1/2"	1 Plum	18.20	.440		.90	21.50		22.40	34
4962	2"	Q-1	33.10	.483		1.33	21		22.33	34.50
4963	3"		20.80	.769		3.54	34		37.54	56.50
4964	4"		16.50	.970		6.30	42.50		48.80	73
4966	6"		10.10	1.584		25	69.50		94.50	136
4970	2" x 1-1/2"		33.30	.480		2.46	21		23.46	35
4972	3" x 1-1/2"		21	.762		6.80	33.50		40.30	59.50
4974	4" x 3"		16.70	.958		12.05	42		54.05	78.50
4978	Closet flange, 4"	1 Plum	32	.250		5.75	12.20		17.95	25
4980	4" x 3"	"	34	.235		6.45	11.45		17.90	25
5000	DWV, PVC, schedule 40, socket joints									
5040	1/4 bend, 1-1/4"	1 Plum	20.20	.396	Ea.	5.70	19.30		25	36.50
5060	1-1/2"	"	18.20	.440		1.90	21.50		23.40	35
5070	2"	Q-1	33.10	.483		2.89	21		23.89	36
5080	3"		20.80	.769		7.30	34		41.30	60.50
5090	4"		16.50	.970		13.75	42.50		56.25	81
5100	6"		10.10	1.584		63.50	69.50		133	178
5105	8"	Q-2	9.30	2.581		92.50	117		209.50	284
5110	1/4 bend, long sweep, 1-1/2"	1 Plum	18.20	.440		3.41	21.50		24.91	37
5112	2"	Q-1	33.10	.483		3.68	21		24.68	37
5114	3"		20.80	.769		8.45	34		42.45	62
5116	4"		16.50	.970		16.15	42.50		58.65	84
5150	1/8 bend, 1-1/4"	1 Plum	20.20	.396		2.68	19.30		21.98	33
5170	1-1/2"	"	18.20	.440		1.36	21.50		22.86	34.50
5180	2"	Q-1	33.10	.483		2.10	21		23.10	35.50
5190	3"		20.80	.769		5.75	34		39.75	59
5200	4"		16.50	.970		9.90	42.50		52.40	77
5210	6"		10.10	1.584		41.50	69.50		111	154
5215	8"	Q-2	9.30	2.581		69.50	117		186.50	259
5250	Tee, sanitary 1-1/4"	1 Plum	13.50	.593		4.10	29		33.10	49.50
5254	1-1/2"	"	12.10	.661		2.57	32		34.57	53
5255	2"	Q-1	20	.800		3.78	35		38.78	58.50
5256	3"		13.90	1.151		9.70	50.50		60.20	89
5257	4"		11	1.455		16.85	64		80.85	118
5259	6"		6.70	2.388		73.50	105		178.50	244
5261	8"	Q-2	6.20	3.871		184	176		360	475
5264	2" x 1-1/2"	Q-1	22	.727		3.34	32		35.34	53
5266	3" x 1-1/2"		15.50	1.032		6.15	45.50		51.65	77.50
5268	4" x 3"		12.10	1.322		21.50	58		79.50	114
5271	6" x 4"		6.90	2.319		79	102		181	245
5314	Combination Y & 1/8 bend, 1-1/2"	1 Plum	12.10	.661		5.60	32		37.60	56
5315	2"	Q-1	20	.800		7.40	35		42.40	62.50
5317	3"		13.90	1.151		17.90	50.50		68.40	98
5318	4"		11	1.455		32	64		96	134
5324	Combination Y & 1/8 bend, reducing									
5325	2" x 2" x 1-1/2"	Q-1	22	.727	Ea.	8	32		40	58.50
5327	3" x 3" x 1-1/2"		15.50	1.032		13.85	45.50		59.35	86
5328	3" x 3" x 2"		15.30	1.046		10.65	46		56.65	82.50
5329	4" x 4" x 2"		12.20	1.311		17.75	57.50		75.25	109
5331	Wye, 1-1/4"	1 Plum	13.50	.593		5.25	29		34.25	51

22 11 13.76 Pipe Fittings, Plastic		Crew	Daily Output	Labor-Hours	Unit	2009 Bare Costs Material	Labor	Equipment	Total	Total Incl O&P
5332	1-1/2"	1 Plum	12.10	.661	Ea.	3.93	32		35.93	54.50
5333	2"	Q-1	20	.800		4.60	35		39.60	59.50
5334	3"		13.90	1.151		11.85	50.50		62.35	91.50
5335	4"		11	1.455		22	64		86	123
5336	6"		6.70	2.388		68	105		173	238
5337	8"	Q-2	6.20	3.871		185	176		361	475
5341	2" x 1-1/2"	Q-1	22	.727		5.65	32		37.65	56
5342	3" x 1-1/2"		15.50	1.032		7.95	45.50		53.45	79
5343	4" x 3"		12.10	1.322		17.50	58		75.50	109
5344	6" x 4"		6.90	2.319		50	102		152	213
5345	8" x 6"	Q-2	6.40	3.750		151	171		322	430
5347	Double wye, 1-1/2"	1 Plum	9.10	.879		8.90	43		51.90	76.50
5348	2"	Q-1	16.60	.964		9.55	42.50		52.05	76
5349	3"		10.40	1.538		24.50	67.50		92	132
5350	4"		8.25	1.939		50	85		135	187
5354	2" x 1-1/2"		16.80	.952		8.75	42		50.75	74.50
5355	3" x 2"		10.60	1.509		18.35	66		84.35	123
5356	4" x 3"		8.45	1.893		39.50	83		122.50	173
5357	6" x 4"		7.25	2.207		100	97		197	260
5374	Coupling, 1-1/4"	1 Plum	20.20	.396		2.48	19.30		21.78	32.50
5376	1-1/2"	"	18.20	.440		.66	21.50		22.16	33.50
5378	2"	Q-1	33.10	.483		.89	21		21.89	34
5380	3"		20.80	.769		3.25	34		37.25	56
5390	4"		16.50	.970		5.60	42.50		48.10	72
5400	6"		10.10	1.584		18.25	69.50		87.75	128
5402	8"	Q-2	9.30	2.581		45	117		162	231
5404	2" x 1-1/2"	Q-1	33.30	.480		1.97	21		22.97	34.50
5406	3" x 1-1/2"		21	.762		6.25	33.50		39.75	59
5408	4" x 3"		16.70	.958		10.10	42		52.10	76
5410	Reducer bushing, 2" x 1-1/4"		36.50	.438		2.51	19.25		21.76	33
5412	3" x 1-1/2"		27.30	.586		6.05	25.50		31.55	46.50
5414	4" x 2"		18.20	.879		11.25	38.50		49.75	72.50
5416	6" x 4"		11.10	1.441		30	63.50		93.50	131
5418	8" x 6"	Q-2	10.20	2.353		73	107		180	247
5425	Closet flange 4"	Q-1	32	.500		6.55	22		28.55	41.50
5426	4" x 3"	"	34	.471		5.80	20.50		26.30	38.50
5450	Solvent cement for PVC, industrial grade, per quart				Qt.	17.25			17.25	19
5500	CPVC, Schedule 80, threaded joints									
5540	90° Elbow, 1/4"	1 Plum	32	.250	Ea.	10.20	12.20		22.40	30
5560	1/2"		30.30	.264		5.90	12.85		18.75	26.50
5570	3/4"		26	.308		8.85	15		23.85	33.50
5580	1"		22.70	.352		12.40	17.20		29.60	40
5590	1-1/4"		20.20	.396		24	19.30		43.30	56.50
5600	1-1/2"		18.20	.440		26	21.50		47.50	61.50
5610	2"	Q-1	33.10	.483		34.50	21		55.50	71
5620	2-1/2"		24.20	.661		107	29		136	163
5630	3"		20.80	.769		115	34		149	179
5640	4"		16.50	.970		180	42.50		222.50	264
5650	6"		10.10	1.584		212	69.50		281.50	340
5700	45° Elbow same as 90° Elbow									
5850	Tee, 1/4"	1 Plum	22	.364	Ea.	19.80	17.75		37.55	49.50
5870	1/2"		20.20	.396		19.80	19.30		39.10	52
5880	3/4"		17.30	.462		28.50	22.50		51	66.50

22 11 Facility Water Distribution

22 11 13 – Facility Water Distribution Piping

22 11 13.76 Pipe Fittings, Plastic		Crew	Daily Output	Labor-Hours	Unit	2009 Bare Costs Material	Labor	Equipment	Total	Total Incl O&P
5890	1"	1 Plum	15.20	.526	Ea.	30.50	25.50		56	74
5900	1-1/4"		13.50	.593		31	29		60	79
5910	1-1/2"		12.10	.661		32.50	32		64.50	85.50
5920	2"	Q-1	20	.800		36	35		71	94
5930	2-1/2"		16.20	.988		177	43.50		220.50	262
5940	3"		13.90	1.151		206	50.50		256.50	305
5950	4"		11	1.455		485	64		549	635
5960	6"		6.70	2.388		580	105		685	805
6000	Coupling, 1/4"	1 Plum	32	.250		13	12.20		25.20	33
6020	1/2"		30.30	.264		13	12.85		25.85	34.50
6030	3/4"		26	.308		15.60	15		30.60	40.50
6040	1"		22.70	.352		17.25	17.20		34.45	45.50
6050	1-1/4"		20.20	.396		21	19.30		40.30	53
6060	1-1/2"		18.20	.440		22.50	21.50		44	57.50
6070	2"	Q-1	33.10	.483		26.50	21		47.50	62
6080	2-1/2"		24.20	.661		47	29		76	97
6090	3"		20.80	.769		55	34		89	113
6100	4"		16.50	.970		112	42.50		154.50	189
6110	6"		10.10	1.584		153	69.50		222.50	276
6120	8"	Q-2	9.30	2.581		325	117		442	540
6200	CTS, 100 psi at 180° F, hot and cold water									
6230	90° Elbow, 1/2"	1 Plum	20	.400	Ea.	.08	19.50		19.58	30
6250	3/4"		19	.421		.17	20.50		20.67	32
6251	1"		16	.500		.58	24.50		25.08	38.50
6252	1-1/4"		15	.533		.99	26		26.99	41.50
6253	1-1/2"		14	.571		1.59	28		29.59	45
6254	2"	Q-1	23	.696		3.43	30.50		33.93	51.50
6260	45° Elbow, 1/2"	1 Plum	20	.400		.27	19.50		19.77	30.50
6280	3/4"		19	.421		.27	20.50		20.77	32.50
6281	1"		16	.500		.53	24.50		25.03	38.50
6282	1-1/4"		15	.533		1.04	26		27.04	41.50
6283	1-1/2"		14	.571		1.65	28		29.65	45
6284	2"	Q-1	23	.696		3.48	30.50		33.98	51.50
6290	Tee, 1/2"	1 Plum	13	.615		.14	30		30.14	46.50
6310	3/4"		12	.667		.26	32.50		32.76	51
6311	1"		11	.727		1.20	35.50		36.70	56.50
6312	1-1/4"		10	.800		1.60	39		40.60	62.50
6313	1-1/2"		10	.800		2.60	39		41.60	63.50
6314	2"	Q-1	17	.941		4.22	41.50		45.72	68.50
6320	Coupling, 1/2"	1 Plum	22	.364		.10	17.75		17.85	27.50
6340	3/4"		21	.381		.14	18.55		18.69	29
6341	1"		18	.444		.50	21.50		22	34
6342	1-1/4"		17	.471		.62	23		23.62	36
6343	1-1/2"		16	.500		.88	24.50		25.38	39
6344	2"	Q-1	28	.571		1.75	25		26.75	41
6360	Solvent cement for CPVC, commercial grade, per quart				Qt.	17			17	18.70
7340	PVC flange, slip-on, Sch 80 std., 1/2"	1 Plum	22	.364	Ea.	11.40	17.75		29.15	40
7350	3/4"		21	.381		12.20	18.55		30.75	42.50
7360	1"		18	.444		13.55	21.50		35.05	48.50
7370	1-1/4"		17	.471		13.95	23		36.95	51
7380	1-1/2"		16	.500		14.30	24.50		38.80	53.50
7390	2"	Q-1	26	.615		19	27		46	63
7400	2-1/2"		24	.667		29.50	29.50		59	78

22 11 13.76 Pipe Fittings, Plastic		Crew	Daily Output	Labor-Hours	Unit	Material	2009 Bare Costs Labor	Equipment	Total	Total Incl O&P
7410	3"	Q-1	18	.889	Ea.	32.50	39		71.50	96
7420	4"		15	1.067		41	47		88	118
7430	6"		10	1.600		64.50	70		134.50	180
7440	8"	Q-2	11	2.182		115	99.50		214.50	281
7550	Union, schedule 40, socket joints, 1/2"	1 Plum	19	.421		3.97	20.50		24.47	36.50
7560	3/4"		18	.444		4.52	21.50		26.02	38.50
7570	1"		15	.533		4.65	26		30.65	45.50
7580	1-1/4"		14	.571		14	28		42	58.50
7590	1-1/2"		13	.615		15	30		45	63
7600	2"	Q-1	20	.800		21.50	35		56.50	78.50
7992	Polybutyl/polyethyl pipe, for copper fittings see Div. 22 11 13.25 7000									
8000	Compression type, PVC, 160 psi cold water									
8010	Coupling, 3/4" CTS	1 Plum	21	.381	Ea.	2.76	18.55		21.31	32
8020	1" CTS		18	.444		3.43	21.50		24.93	37.50
8030	1-1/4" CTS		17	.471		4.83	23		27.83	41
8040	1-1/2" CTS		16	.500		6.60	24.50		31.10	45.50
8050	2" CTS		15	.533		9.15	26		35.15	50.50
8060	Female adapter, 3/4" FPT x 3/4" CTS		23	.348		3.12	16.95		20.07	30
8070	3/4" FPT x 1" CTS		21	.381		3.80	18.55		22.35	33
8080	1" FPT x 1" CTS		20	.400		4.98	19.50		24.48	35.50
8090	1-1/4" FPT x 1-1/4" CTS		18	.444		6.50	21.50		28	40.50
8100	1-1/2" FPT x 1-1/2" CTS		16	.500		7.45	24.50		31.95	46
8110	2" FPT x 2" CTS		13	.615		9.05	30		39.05	56.50
8130	Male adapter, 3/4" MPT x 3/4" CTS		23	.348		3.02	16.95		19.97	30
8140	3/4" MPT x 1" CTS		21	.381		3.25	18.55		21.80	32.50
8150	1" MPT x 1" CTS		20	.400		4.22	19.50		23.72	34.50
8160	1-1/4" MPT x 1-1/4" CTS		18	.444		5.70	21.50		27.20	40
8170	1-1/2" MPT x 1-1/2" CTS		16	.500		6.85	24.50		31.35	45.50
8180	2" MPT x 2" CTS		13	.615		8.90	30		38.90	56.50
8200	Spigot adapter, 3/4" IPS x 3/4" CTS		23	.348		2.15	16.95		19.10	29
8210	3/4" IPS x 1" CTS		21	.381		2.76	18.55		21.31	32
8220	1" IPS x 1" CTS		20	.400		2.76	19.50		22.26	33
8230	1-1/4" IPS x 1-1/4" CTS		18	.444		4.30	21.50		25.80	38
8240	1-1/2" IPS x 1-1/2" CTS		16	.500		4.33	24.50		28.83	43
8250	2" IPS x 2" CTS		13	.615		5.40	30		35.40	52.50
8270	Price includes insert stiffeners									
8280	250 psi is same price as 160 psi									
8300	Insert type, nylon, 160 & 250 psi, cold water									
8310	Clamp ring stainless steel, 3/4" IPS	1 Plum	115	.070	Ea.	1.95	3.39		5.34	7.40
8320	1" IPS		107	.075		1.61	3.65		5.26	7.40
8330	1-1/4" IPS		101	.079		2	3.86		5.86	8.20
8340	1-1/2" IPS		95	.084		2.18	4.11		6.29	8.75
8350	2" IPS		85	.094		2.50	4.59		7.09	9.85
8370	Coupling, 3/4" IPS		22	.364		.74	17.75		18.49	28.50
8380	1" IPS		19	.421		.77	20.50		21.27	33
8390	1-1/4" IPS		18	.444		1.15	21.50		22.65	35
8400	1-1/2" IPS		17	.471		1.36	23		24.36	37
8410	2" IPS		16	.500		2.57	24.50		27.07	41
8430	Elbow, 90°, 3/4" IPS		22	.364		1.49	17.75		19.24	29
8440	1" IPS		19	.421		1.64	20.50		22.14	34
8450	1-1/4" IPS		18	.444		1.83	21.50		23.33	35.50
8460	1-1/2" IPS		17	.471		2.17	23		25.17	38
8470	2" IPS		16	.500		3.02	24.50		27.52	41.50

22 11 13 – Facility Water Distribution Piping

22 11 13.76 Pipe Fittings, Plastic		Crew	Daily Output	Labor-Hours	Unit	Material	Labor	Equipment	Total	Total Incl O&P
8490	Male adapter, 3/4" IPS x 3/4" MPT	1 Plum	25	.320	Ea.	.74	15.60		16.34	25
8500	1" IPS x 1" MPT		21	.381		.77	18.55		19.32	30
8510	1-1/4" IPS x 1-1/4" MPT		20	.400		1.22	19.50		20.72	31.50
8520	1-1/2" IPS x 1-1/2" MPT		18	.444		1.36	21.50		22.86	35
8530	2" IPS x 2" MPT		15	.533		2.60	26		28.60	43.50
8550	Tee, 3/4" IPS		14	.571		1.44	28		29.44	44.50
8560	1" IPS		13	.615		1.87	30		31.87	48.50
8570	1-1/4" IPS		12	.667		2.93	32.50		35.43	53.50
8580	1-1/2" IPS		11	.727		3.32	35.50		38.82	58.50
8590	2" IPS		10	.800		6.55	39		45.55	67.50
8610	Insert type, PVC, 100 psi @ 180° F, hot & cold water									
8620	Coupler, male, 3/8" CTS x 3/8" MPT	1 Plum	29	.276	Ea.	.61	13.45		14.06	21.50
8630	3/8" CTS x 1/2" MPT		28	.286		.61	13.95		14.56	22
8640	1/2" CTS x 1/2" MPT		27	.296		.61	14.45		15.06	23
8650	1/2" CTS x 3/4" MPT		26	.308		1.93	15		16.93	25.50
8660	3/4" CTS x 1/2" MPT		25	.320		1.73	15.60		17.33	26
8670	3/4" CTS x 3/4" MPT		25	.320		.74	15.60		16.34	25
8700	Coupling, 3/8" CTS x 1/2" CTS		25	.320		2.86	15.60		18.46	27
8710	1/2" CTS		23	.348		3.41	16.95		20.36	30.50
8730	3/4" CTS		22	.364		5.40	17.75		23.15	33.50
8750	Elbow 90°, 3/8" CTS		25	.320		3.90	15.60		19.50	28.50
8760	1/2" CTS		23	.348		5.60	16.95		22.55	32.50
8770	3/4" CTS		22	.364		6.25	17.75		24	34.50
8800	Rings, crimp, copper, 3/8" CTS		120	.067		.16	3.25		3.41	5.25
8810	1/2" CTS		117	.068		.14	3.33		3.47	5.30
8820	3/4" CTS		115	.070		.19	3.39		3.58	5.45
8850	Reducer tee, 3/8" x 3/8" x 1/2" CTS		17	.471		1.18	23		24.18	37
8860	1/2" x 3/8" x 1/2" CTS		15	.533		1.18	26		27.18	42
8870	3/4" x 1/2" x 1/2" CTS		14	.571		1.82	28		29.82	45
8890	3/4" x 3/4" x 1/2" CTS		14	.571		1.82	28		29.82	45
8900	3/4" x 1/2" x 3/8" CTS		14	.571		1.82	28		29.82	45
8930	Tee, 3/8" CTS		17	.471		1.40	23		24.40	37
8940	1/2" CTS		15	.533		1.58	26		27.58	42
8950	3/4" CTS		14	.571		1.71	28		29.71	45
8960	Copper rings included in fitting price									
9000	Flare type, assembled, acetal, hot & cold water									
9010	Coupling, 1/4" & 3/8" CTS	1 Plum	24	.333	Ea.	2.90	16.25		19.15	28
9020	1/2" CTS		22	.364		3.32	17.75		21.07	31
9030	3/4" CTS		21	.381		4.92	18.55		23.47	34.50
9040	1" CTS		18	.444		6.25	21.50		27.75	40.50
9050	Elbow 90°, 1/4" CTS		26	.308		3.23	15		18.23	27
9060	3/8" CTS		24	.333		3.46	16.25		19.71	29
9070	1/2" CTS		22	.364		4.10	17.75		21.85	32
9080	3/4" CTS		21	.381		6.25	18.55		24.80	36
9090	1" CTS		18	.444		7.90	21.50		29.40	42
9110	Tee ,1/4" & 3/8" CTS		15	.533		3.57	26		29.57	44.50
9120	1/2" CTS		14	.571		4.69	28		32.69	48
9130	3/4" CTS		13	.615		7.15	30		37.15	54.50
9140	1" CTS		12	.667		9.55	32.50		42.05	61
9552	For plastic hangers see Div. 22 05 29.10 8000									
9562	For copper/brass fittings see Div. 22 11 13.25 7000									
9900	Minimum labor/equipment charge	1 Plum	4	2	Job		97.50		97.50	151

22 11 19.10 Flexible Connectors

	22 11 19.10 Flexible Connectors	Crew	Daily Output	Labor-Hours	Unit	Material	2009 Bare Costs Labor	Equipment	Total	Total Incl O&P
0010	**FLEXIBLE CONNECTORS**, Corrugated, 7/8" O.D., 1/2" I.D.									
0050	Gas, seamless brass, steel fittings									
0200	12" long	1 Plum	36	.222	Ea.	15.10	10.85		25.95	33.50
0220	18" long		36	.222		18.35	10.85		29.20	37
0240	24" long		34	.235		22	11.45		33.45	42.50
0260	30" long		34	.235		24	11.45		35.45	44.50
0280	36" long		32	.250		26.50	12.20		38.70	48
0320	48" long		30	.267		33.50	13		46.50	57
0340	60" long		30	.267		40	13		53	64
0360	72" long		30	.267		46	13		59	70.50
2000	Water, copper tubing, dielectric separators									
2100	12" long	1 Plum	36	.222	Ea.	14	10.85		24.85	32
2220	15" long		36	.222		15.45	10.85		26.30	34
2240	18" long		36	.222		16.85	10.85		27.70	35.50
2260	24" long		34	.235		20.50	11.45		31.95	40.50
9000	Minimum labor/equipment charge		4	2	Job		97.50		97.50	151

22 11 19.14 Flexible Metal Hose

	22 11 19.14 Flexible Metal Hose	Crew	Daily Output	Labor-Hours	Unit	Material	Labor	Equipment	Total	Total Incl O&P
0010	**FLEXIBLE METAL HOSE**, Connectors, standard lengths									
0100	Bronze braided, bronze ends									
0120	3/8" diameter x 12"	1 Stpi	26	.308	Ea.	18.30	15.20		33.50	43.50
0140	1/2" diameter x 12"		24	.333		17.40	16.45		33.85	44.50
0160	3/4" diameter x 12"		20	.400		25.50	19.75		45.25	58.50
0180	1" diameter x 18"		19	.421		33	21		54	68
0200	1-1/2" diameter x 18"		13	.615		52	30.50		82.50	105
0220	2" diameter x 18"		11	.727		63	36		99	125
1000	Carbon steel ends									
1020	1/4" diameter x 12"	1 Stpi	28	.286	Ea.	13.60	14.10		27.70	37
1040	3/8" diameter x 12"		26	.308		13.95	15.20		29.15	39
1060	1/2" diameter x 12"		24	.333		15.40	16.45		31.85	42.50
1080	1/2" diameter x 24"		24	.333		34.50	16.45		50.95	63
1100	1/2" diameter x 36"		24	.333		40.50	16.45		56.95	70
1120	3/4" diameter x 12"		20	.400		25	19.75		44.75	58
1140	3/4" diameter x 24"		20	.400		24.50	19.75		44.25	57.50
1160	3/4" diameter x 36"		20	.400		51	19.75		70.75	86.50
1180	1" diameter x 18"		19	.421		31.50	21		52.50	66.50
1200	1" diameter x 30"		19	.421		51	21		72	88
1220	1" diameter x 36"		19	.421		60	21		81	98
1240	1-1/4" diameter x 18"		15	.533		43	26.50		69.50	88.50
1260	1-1/4" diameter x 36"		15	.533		79.50	26.50		106	128
1280	1-1/2" diameter x 18"		13	.615		61	30.50		91.50	114
1300	1-1/2" diameter x 36"		13	.615		87.50	30.50		118	143
1320	2" diameter x 24"		11	.727		88.50	36		124.50	153
1340	2" diameter x 36"		11	.727		109	36		145	176
1360	2-1/2" diameter x 24"		9	.889		202	44		246	291
1380	2-1/2" diameter x 36"		9	.889		263	44		307	355
1400	3" diameter x 24"		7	1.143		279	56.50		335.50	395
1420	3" diameter x 36"		7	1.143		375	56.50		431.50	500
2000	Carbon steel braid, carbon steel solid ends									
2100	1/2" diameter x 12"	1 Stpi	24	.333	Ea.	34	16.45		50.45	63
2120	3/4" diameter x 12"		20	.400		43	19.75		62.75	78
2140	1" diameter x 12"		19	.421		48.50	21		69.50	85
2160	1-1/4" diameter x 12"		15	.533		54	26.50		80.50	100

22 11 19.14 Flexible Metal Hose		Crew	Daily Output	Labor-Hours	Unit	Material	2009 Bare Costs Labor	Equipment	Total	Total Incl O&P
2180	1-1/2" diameter x 12"	1 Stpi	13	.615	Ea.	55.50	30.50		86	109
3000	Stainless steel braid, welded on carbon steel ends									
3100	1/2" diameter x 12"	1 Stpi	24	.333	Ea.	43.50	16.45		59.95	73
3120	3/4" diameter x 12"		20	.400		55	19.75		74.75	91
3140	3/4" diameter x 24"		20	.400		63.50	19.75		83.25	100
3160	3/4" diameter x 36"		20	.400		72	19.75		91.75	110
3180	1" diameter x 12"		19	.421		65.50	21		86.50	104
3200	1" diameter x 24"		19	.421		77.50	21		98.50	117
3220	1" diameter x 36"		19	.421		90.50	21		111.50	132
3240	1-1/4" diameter x 12"		15	.533		98.50	26.50		125	150
3260	1-1/4" diameter x 24"		15	.533		104	26.50		130.50	155
3280	1-1/4" diameter x 36"		15	.533		118	26.50		144.50	170
3300	1-1/2" diameter x 12"		13	.615		103	30.50		133.50	160
3320	1-1/2" diameter x 24"		13	.615		115	30.50		145.50	174
3340	1-1/2" diameter x 36"		13	.615		135	30.50		165.50	195
3400	Metal stainless steel braid, over corrugated stainless steel, flanged ends									
3410	150 PSI									
3420	1/2" diameter x 12"	1 Stpi	24	.333	Ea.	144	16.45		160.45	184
3430	1" diameter x 12"		20	.400		180	19.75		199.75	229
3440	1-1/2" diameter x 12"		15	.533		204	26.50		230.50	265
3450	2-1/2" diameter x 9"		12	.667		75	33		108	134
3460	3" diameter x 9"		9	.889		90	44		134	167
3470	4" diameter x 9"		7	1.143		110	56.50		166.50	209
3480	4" diameter x 30"		5	1.600		325	79		404	475
3490	4" diameter x 36"		4.80	1.667		350	82.50		432.50	515
3500	6" diameter x 11"		5	1.600		183	79		262	325
3510	6" diameter x 36"		3.80	2.105		505	104		609	720
3520	8" diameter x 12"		4	2		360	98.50		458.50	550
3530	10" diameter x 13"		3	2.667		490	132		622	745
3540	12" diameter x 14"	Q-5	4	4		670	178		848	1,000

22 11 19.18 Mixing Valve		Crew	Daily Output	Labor-Hours	Unit	Material	Labor	Equipment	Total	Total Incl O&P
0010	**MIXING VALVE**, Automatic, water tempering.									
0040	1/2" size	1 Stpi	19	.421	Ea.	435	21		456	510
0050	3/4" size		18	.444		435	22		457	515
0100	1" size		16	.500		625	24.50		649.50	725
0120	1-1/4" size		13	.615		905	30.50		935.50	1,050
0140	1-1/2" size		10	.800		1,075	39.50		1,114.50	1,225
0160	2" size		8	1		1,350	49.50		1,399.50	1,575
0170	2-1/2" size		6	1.333		1,350	66		1,416	1,600
0180	3" size		4	2		3,125	98.50		3,223.50	3,600
0190	4" size		3	2.667		3,125	132		3,257	3,650
9000	Minimum labor/equipment charge		5	1.600	Job		79		79	122

22 11 19.22 Pressure Reducing Valve		Crew	Daily Output	Labor-Hours	Unit	Material	Labor	Equipment	Total	Total Incl O&P
0010	**PRESSURE REDUCING VALVE**, Steam, pilot operated.									
0100	Threaded, iron body									
0200	1-1/2" size	1 Stpi	8	1	Ea.	1,000	49.50		1,049.50	1,175
0220	2" size	"	5	1.600	"	1,325	79		1,404	1,575
1000	Flanged, iron body, 125 lb. flanges									
1020	2" size	1 Stpi	8	1	Ea.	1,450	49.50		1,499.50	1,675
1040	2-1/2" size	"	4	2		1,750	98.50		1,848.50	2,075
1060	3" size	Q-5	4.50	3.556		1,775	158		1,933	2,200
1080	4" size	"	3	5.333		2,200	237		2,437	2,800

22 11 19.22 Pressure Reducing Valve		Crew	Daily Output	Labor-Hours	Unit	2009 Bare Costs Material	Labor	Equipment	Total	Total Incl O&P
1500	For 250 lb. flanges, add				Ea.	5%				

22 11 19.26 Pressure Regulators

		Crew	Daily Output	Labor-Hours	Unit	2009 Bare Costs Material	Labor	Equipment	Total	Total Incl O&P
0010	**PRESSURE REGULATORS**									
0200	Oil, light, hot water, ordinary steam, threaded									
0220	Bronze body, 1/4" size	1 Stpi	24	.333	Ea.	173	16.45		189.45	217
0230	3/8" size		24	.333		187	16.45		203.45	232
0240	1/2" size		24	.333		233	16.45		249.45	282
0250	3/4" size		20	.400		278	19.75		297.75	335
0260	1" size		19	.421		425	21		446	495
0270	1-1/4" size		15	.533		570	26.50		596.50	665
0280	1-1/2" size		13	.615		610	30.50		640.50	720
0290	2" size		11	.727		1,100	36		1,136	1,275
0320	Iron body, 1/4" size		24	.333		136	16.45		152.45	175
0330	3/8" size		24	.333		160	16.45		176.45	201
0340	1/2" size		24	.333		169	16.45		185.45	212
0350	3/4" size		20	.400		205	19.75		224.75	257
0360	1" size		19	.421		268	21		289	325
0370	1-1/4" size		15	.533		370	26.50		396.50	450
0380	1-1/2" size		13	.615		415	30.50		445.50	500
0390	2" size		11	.727		600	36		636	715
0500	Oil, heavy, viscous fluids, threaded									
0520	Bronze body, 3/8" size	1 Stpi	24	.333	Ea.	298	16.45		314.45	355
0530	1/2" size		24	.333		385	16.45		401.45	445
0540	3/4" size		20	.400		425	19.75		444.75	500
0550	1" size		19	.421		535	21		556	615
0560	1-1/4" size		15	.533		750	26.50		776.50	865
0570	1-1/2" size		13	.615		860	30.50		890.50	990
0600	Iron body, 3/8" size		24	.333		237	16.45		253.45	287
0620	1/2" size		24	.333		288	16.45		304.45	340
0630	3/4" size		20	.400		325	19.75		344.75	390
0640	1" size		19	.421		385	21		406	455
0650	1-1/4" size		15	.533		555	26.50		581.50	655
0660	1-1/2" size		13	.615		595	30.50		625.50	700
0800	Process steam, wet or super heated, monel trim, threaded									
0820	Bronze body, 1/4" size	1 Stpi	24	.333	Ea.	625	16.45		641.45	715
0830	3/8" size		24	.333		625	16.45		641.45	715
0840	1/2" size		24	.333		720	16.45		736.45	815
0850	3/4" size		20	.400		845	19.75		864.75	960
0860	1" size		19	.421		1,075	21		1,096	1,200
0870	1-1/4" size		15	.533		1,325	26.50		1,351.50	1,500
0880	1-1/2" size		13	.615		1,600	30.50		1,630.50	1,800
0890	2" size		11	.727		2,325	36		2,361	2,600
0920	Iron body, max 125 PSIG press out, 1/4" size		24	.333		136	16.45		152.45	175
0930	3/8" size		24	.333		120	16.45		136.45	158
0940	1/2" size		24	.333		127	16.45		143.45	165
0950	3/4" size		20	.400		154	19.75		173.75	200
0960	1" size		19	.421		201	21		222	253
0970	1-1/4" size		15	.533		279	26.50		305.50	345
0980	1-1/2" size		13	.615		310	30.50		340.50	385
0990	2" size		11	.727		450	36		486	550
3000	Steam, high capacity, bronze body, stainless steel trim									
3020	Threaded, 1/2" diameter	1 Stpi	24	.333	Ea.	1,300	16.45		1,316.45	1,450

22 11 19.26 Pressure Regulators		Crew	Daily Output	Labor-Hours	Unit	Material	Labor	Equipment	Total	Total Incl O&P
						2009 Bare Costs				
3030	3/4" diameter	1 Stpi	24	.333	Ea.	1,300	16.45		1,316.45	1,450
3040	1" diameter		19	.421		1,450	21		1,471	1,625
3060	1-1/4" diameter		15	.533		1,600	26.50		1,626.50	1,825
3080	1-1/2" diameter		13	.615		1,825	30.50		1,855.50	2,075
3100	2" diameter	↓	11	.727		2,250	36		2,286	2,525
3120	2-1/2" diameter	Q-5	12	1.333		2,825	59		2,884	3,200
3140	3" diameter	"	11	1.455	↓	3,200	64.50		3,264.50	3,625
3500	Flanged connection, iron body, 125 lb. W.S.P.									
3520	3" diameter	Q-5	11	1.455	Ea.	3,525	64.50		3,589.50	3,975
3540	4" diameter	"	5	3.200	"	4,425	142		4,567	5,100
9002	For water pressure regulators, see Div. 22 05 23.20									

22 11 19.34 Sleeves and Escutcheons		Crew	Daily Output	Labor-Hours	Unit	Material	Labor	Equipment	Total	Total Incl O&P
0010	**SLEEVES & ESCUTCHEONS**									
0100	Pipe sleeve									
0110	Steel, w/water stop, 12" long, with link seal									
0120	2" diam. for 1/2" carrier pipe	1 Plum	8.40	.952	Ea.	45	46.50		91.50	122
0130	2-1/2" diam. for 3/4" carrier pipe		8	1		51	49		100	132
0140	2-1/2" diam. for 1" carrier pipe		8	1		48	49		97	129
0150	3" diam. for 1-1/4" carrier pipe		7.20	1.111		62	54		116	152
0160	3-1/2" diam. for 1-1/2" carrier pipe		6.80	1.176		63	57.50		120.50	158
0170	4" diam. for 2" carrier pipe		6	1.333		68	65		133	176
0180	4" diam. for 2-1/2" carrier pipe		6	1.333		68	65		133	176
0190	5" diam. for 3" carrier pipe		5.40	1.481		81.50	72		153.50	202
0200	6" diam. for 4" carrier pipe	↓	4.80	1.667		92	81.50		173.50	227
0210	10" diam. for 6" carrier pipe	Q-1	8	2		159	88		247	310
0220	12" diam. for 8" carrier pipe		7.20	2.222		222	97.50		319.50	395
0230	14" diam. for 10" carrier pipe		6.40	2.500		256	110		366	450
0240	16" diam. for 12" carrier pipe		5.80	2.759		292	121		413	510
0250	18" diam. for 14" carrier pipe		5.20	3.077		395	135		530	645
0260	24" diam. for 18" carrier pipe		4	4		670	176		846	1,000
0270	24" diam. for 20" carrier pipe		4	4		585	176		761	910
0280	30" diam. for 24" carrier pipe	↓	3.20	5	↓	905	219		1,124	1,350
0500	Wall sleeve									
0510	Ductile iron with rubber gasket seal									
0520	3"	1 Plum	8.40	.952	Ea.	700	46.50		746.50	840
0530	4"		7.20	1.111		750	54		804	910
0540	6"		6	1.333		920	65		985	1,125
0550	8"		4	2		1,125	97.50		1,222.50	1,400
0560	10"		3	2.667		1,375	130		1,505	1,725
0570	12"	↓	2.40	3.333	↓	1,625	163		1,788	2,050
5000	Escutcheon									
5100	Split ring, pipe									
5110	Chrome plated									
5120	1/2"	1 Plum	160	.050	Ea.	1.05	2.44		3.49	4.94
5130	3/4"		160	.050		1.09	2.44		3.53	4.98
5140	1"		135	.059		1.17	2.89		4.06	5.75
5150	1-1/2"		115	.070		1.76	3.39		5.15	7.20
5160	2"		100	.080		1.85	3.90		5.75	8.10
5170	4"		80	.100		4.11	4.88		8.99	12.05
5180	6"	↓	68	.118	↓	6.85	5.75		12.60	16.45
5400	Shallow flange type									
5410	Chrome plated steel									

22 11 19.34 Sleeves and Escutcheons		Crew	Daily Output	Labor-Hours	Unit	Material	2009 Bare Costs Labor	Equipment	Total	Total Incl O&P
5420	1/2" CTS	1 Plum	180	.044	Ea.	.28	2.17		2.45	3.67
5430	3/4" CTS		180	.044		.28	2.17		2.45	3.67
5440	1/2" IPS		180	.044		.26	2.17		2.43	3.65
5450	3/4" IPS		180	.044		.28	2.17		2.45	3.67
5460	1" IPS		175	.046		.46	2.23		2.69	3.97
5470	1-1/2" IPS		170	.047		.72	2.29		3.01	4.35
5480	2" IPS		160	.050		.86	2.44		3.30	4.73

22 11 19.38 Water Supply Meters		Crew	Daily Output	Labor-Hours	Unit	Material	Labor	Equipment	Total	Total Incl O&P
0010	**WATER SUPPLY METERS**									
1000	Detector, serves dual systems such as fire and domestic or									
1020	process water, wide range cap., UL and FM approved									
1100	3" mainline x 2" by-pass, 400 GPM	Q-1	3.60	4.444	Ea.	6,250	195		6,445	7,175
1140	4" mainline x 2" by-pass, 700 GPM	"	2.50	6.400		6,250	281		6,531	7,300
1180	6" mainline x 3" by-pass, 1600 GPM	Q-2	2.60	9.231		9,550	420		9,970	11,200
1220	8" mainline x 4" by-pass, 2800 GPM		2.10	11.429		14,200	520		14,720	16,400
1260	10" mainline x 6" by-pass, 4400 GPM		2	12		20,300	545		20,845	23,100
1300	10"x12" mainlines x 6" by-pass,5400 GPM		1.70	14.118		27,500	640		28,140	31,200
2000	Domestic/commercial, bronze									
2020	Threaded									
2060	5/8" diameter, to 20 GPM	1 Plum	16	.500	Ea.	42	24.50		66.50	84
2080	3/4" diameter, to 30 GPM		14	.571		76.50	28		104.50	127
2100	1" diameter, to 50 GPM		12	.667		116	32.50		148.50	179
2300	Threaded/flanged									
2340	1-1/2" diameter, to 100 GPM	1 Plum	8	1	Ea.	284	49		333	385
2360	2" diameter, to 160 GPM	"	6	1.333	"	385	65		450	525
2600	Flanged, compound									
2640	3" diameter, 320 GPM	Q-1	3	5.333	Ea.	2,625	234		2,859	3,250
2660	4" diameter, to 500 GPM		1.50	10.667		4,200	470		4,670	5,325
2680	6" diameter, to 1,000 GPM		1	16		6,675	700		7,375	8,450
2700	8" diameter, to 1,800 GPM		.80	20		10,500	830		11,380	12,900
7000	Turbine									
7260	Flanged									
7300	2" diameter, to 160 GPM	1 Plum	7	1.143	Ea.	510	55.50		565.50	650
7320	3" diameter, to 450 GPM	Q-1	3.60	4.444		1,075	195		1,270	1,475
7340	4" diameter, to 650 GPM	"	2.50	6.400		1,775	281		2,056	2,375
7360	6" diameter, to 1800 GPM	Q-2	2.60	9.231		3,150	420		3,570	4,125
7380	8" diameter, to 2500 GPM		2.10	11.429		5,000	520		5,520	6,325
7400	10" diameter, to 5500 GPM		1.70	14.118		6,775	640		7,415	8,450
9000	Minimum labor/equipment charge	1 Plum	3.25	2.462	Job		120		120	186

22 11 19.42 Backflow Preventers		Crew	Daily Output	Labor-Hours	Unit	Material	Labor	Equipment	Total	Total Incl O&P
0010	**BACKFLOW PREVENTERS**, Includes valves									
0020	and four test cocks, corrosion resistant, automatic operation									
1000	Double check principle									
1010	Threaded, with ball valves									
1020	3/4" pipe size	1 Plum	16	.500	Ea.	223	24.50		247.50	283
1030	1" pipe size		14	.571		257	28		285	325
1040	1-1/2" pipe size		10	.800		670	39		709	800
1050	2" pipe size		7	1.143		800	55.50		855.50	965
1080	Threaded, with gate valves									
1100	3/4" pipe size	1 Plum	16	.500	Ea.	860	24.50		884.50	985
1120	1" pipe size		14	.571		865	28		893	1,000
1140	1-1/2" pipe size		10	.800		1,125	39		1,164	1,275

22 11 19.42 Backflow Preventers		Crew	Daily Output	Labor-Hours	Unit	Material	2009 Bare Costs Labor	Equipment	Total	Total Incl O&P
1160	2″ pipe size	1 Plum	7	1.143	Ea.	1,375	55.50		1,430.50	1,575
1300	Flanged, valves are OS&Y									
1380	3″ pipe size	Q-1	4.50	3.556	Ea.	3,525	156		3,681	4,125
1400	4″ pipe size	″	3	5.333		3,925	234		4,159	4,700
1420	6″ pipe size	Q-2	3	8		6,125	365		6,490	7,325
4000	Reduced pressure principle									
4100	Threaded, bronze, valves are ball									
4120	3/4″ pipe size	1 Plum	16	.500	Ea.	330	24.50		354.50	405
4140	1″ pipe size		14	.571		355	28		383	435
4150	1-1/4″ pipe size		12	.667		610	32.50		642.50	720
4160	1-1/2″ pipe size		10	.800		670	39		709	795
4180	2″ pipe size		7	1.143		750	55.50		805.50	910
5000	Flanged, bronze, valves are OS&Y									
5060	2-1/2″ pipe size	Q-1	5	3.200	Ea.	3,200	140		3,340	3,750
5080	3″ pipe size		4.50	3.556		3,350	156		3,506	3,950
5100	4″ pipe size		3	5.333		4,225	234		4,459	5,025
5120	6″ pipe size	Q-2	3	8		6,100	365		6,465	7,300
5600	Flanged, iron, valves are OS&Y									
5660	2-1/2″ pipe size	Q-1	5	3.200	Ea.	1,750	140		1,890	2,150
5680	3″ pipe size		4.50	3.556		2,700	156		2,856	3,200
5700	4″ pipe size		3	5.333		3,450	234		3,684	4,175
5720	6″ pipe size	Q-2	3	8		4,875	365		5,240	5,950
5740	8″ pipe size		2	12		8,600	545		9,145	10,300
5760	10″ pipe size		1	24		11,500	1,100		12,600	14,400
9010	Minimum labor/equipment charge	1 Plum	2	4	Job		195		195	300

22 11 19.50 Vacuum Breakers		Crew	Daily Output	Labor-Hours	Unit	Material	Labor	Equipment	Total	Total Incl O&P
0010	**VACUUM BREAKERS**									
0013	See also backflow preventers Div. 22 11 19.42									
1000	Anti-siphon continuous pressure type									
1010	Max. 150 PSI - 210° F									
1020	Bronze body									
1030	1/2″ size	1 Stpi	24	.333	Ea.	143	16.45		159.45	183
1040	3/4″ size		20	.400		143	19.75		162.75	188
1050	1″ size		19	.421		148	21		169	195
1060	1-1/4″ size		15	.533		291	26.50		317.50	360
1070	1-1/2″ size		13	.615		360	30.50		390.50	440
1080	2″ size		11	.727		370	36		406	460
1200	Max. 125 PSI with atmospheric vent									
1210	Brass, in-line construction									
1220	1/4″ size	1 Stpi	24	.333	Ea.	57.50	16.45		73.95	89
1230	3/8″ size	″	24	.333		57.50	16.45		73.95	89
1260	For polished chrome finish, add					13%				
2000	Anti-siphon, non-continuous pressure type									
2010	Hot or cold water 125 PSI - 210° F									
2020	Bronze body									
2030	1/4″ size	1 Stpi	24	.333	Ea.	36	16.45		52.45	65
2040	3/8″ size		24	.333		36	16.45		52.45	65
2050	1/2″ size		24	.333		41	16.45		57.45	70.50
2060	3/4″ size		20	.400		48.50	19.75		68.25	84
2070	1″ size		19	.421		75.50	21		96.50	115
2080	1-1/4″ size		15	.533		133	26.50		159.50	187
2090	1-1/2″ size		13	.615		137	30.50		167.50	198

22 11 19.50 Vacuum Breakers		Crew	Daily Output	Labor-Hours	Unit	Material	2009 Bare Costs Labor	Equipment	Total	Total Incl O&P
2100	2" size	1 Stpi	11	.727	Ea.	243	36		279	325
2110	2-1/2" size		8	1		695	49.50		744.50	840
2120	3" size		6	1.333		925	66		991	1,125
2150	For polished chrome finish, add					50%				

22 11 19.54 Water Hammer Arresters/Shock Absorbers

		Crew	Daily Output	Labor-Hours	Unit	Material	Labor	Equipment	Total	Total Incl O&P
0010	**WATER HAMMER ARRESTERS/SHOCK ABSORBERS**									
0490	Copper									
0500	3/4" male I.P.S. For 1 to 11 fixtures	1 Plum	12	.667	Ea.	17	32.50		49.50	69
0600	1" male I.P.S., For 12 to 32 fixtures		8	1		41.50	49		90.50	122
0700	1-1/4" male I.P.S. For 33 to 60 fixtures		8	1		44	49		93	124
0800	1-1/2" male I.P.S. For 61 to 113 fixtures		8	1		63	49		112	145
0900	2" male I.P.S.For 114 to 154 fixtures		8	1		92	49		141	177
1000	2-1/2" male I.P.S. For 155 to 330 fixtures		4	2		285	97.50		382.50	465
9000	Minimum labor/equipment charge		3.50	2.286	Job		111		111	173

22 11 19.64 Hydrants

		Crew	Daily Output	Labor-Hours	Unit	Material	Labor	Equipment	Total	Total Incl O&P
0010	**HYDRANTS**									
0050	Wall type, moderate climate, bronze, encased									
0200	3/4" IPS connection	1 Plum	16	.500	Ea.	535	24.50		559.50	630
0300	1" IPS connection	"	14	.571		615	28		643	720
0500	Anti-siphon type, 3/4" connection					465			465	510
1000	Non-freeze, bronze, exposed									
1100	3/4" IPS connection, 4" to 9" thick wall	1 Plum	14	.571	Ea.	365	28		393	445
1120	10" to 14" thick wall		12	.667		475	32.50		507.50	570
1140	15" to 19" thick wall		12	.667		475	32.50		507.50	570
1160	20" to 24" thick wall		10	.800		480	39		519	590
1200	For 1" IPS connection, add					15%	10%			
1240	For 3/4" adapter type vacuum breaker, add				Ea.	45.50			45.50	50
1280	For anti-siphon type, add				"	95			95	105
2000	Non-freeze bronze, encased, anti-siphon type									
2100	3/4" IPS connection, 5" to 9" thick wall	1 Plum	14	.571	Ea.	920	28		948	1,050
2120	10" to 14" thick wall		12	.667		950	32.50		982.50	1,100
2140	15" to 19" thick wall		12	.667		995	32.50		1,027.50	1,150
2160	20" to 24" thick wall		10	.800		1,025	39		1,064	1,175
2200	For 1" IPS connection, add					10%	10%			
3000	Ground box type, bronze frame, 3/4" IPS connection									
3080	Non-freeze, all bronze, polished face, set flush									
3100	2 feet depth of bury	1 Plum	8	1	Ea.	685	49		734	830
3120	3 feet depth of bury		8	1		735	49		784	885
3140	4 feet depth of bury		8	1		790	49		839	940
3160	5 feet depth of bury		7	1.143		845	55.50		900.50	1,025
3180	6 feet depth of bury		7	1.143		895	55.50		950.50	1,075
3200	7 feet depth of bury		6	1.333		950	65		1,015	1,150
3220	8 feet depth of bury		5	1.600		1,000	78		1,078	1,225
3240	9 feet depth of bury		4	2		1,050	97.50		1,147.50	1,300
3260	10 feet depth of bury		4	2		1,100	97.50		1,197.50	1,375
3400	For 1" IPS connection, add					15%	10%			
3450	For 1-1/4" IPS connection, add					325%	14%			
3500	For 1-1/2" connection, add					370%	18%			
3550	For 2" connection, add					445%	24%			
3600	For tapped drain port in box, add					61.50			61.50	68
4000	Non-freeze, CI body, bronze frame & scoriated cover									
4010	with hose storage									

22 11 19 – Domestic Water Piping Specialties

22 11 19.64 Hydrants		Crew	Daily Output	Labor-Hours	Unit	2009 Bare Costs Material	Labor	Equipment	Total	Total Incl O&P
4100	2 feet depth of bury	1 Plum	7	1.143	Ea.	1,250	55.50		1,305.50	1,475
4120	3 feet depth of bury		7	1.143		1,325	55.50		1,380.50	1,525
4140	4 feet depth of bury		7	1.143		1,350	55.50		1,405.50	1,575
4160	5 feet depth of bury		6.50	1.231		1,375	60		1,435	1,625
4180	6 feet depth of bury		6	1.333		1,400	65		1,465	1,650
4200	7 feet depth of bury		5.50	1.455		1,450	71		1,521	1,700
4220	8 feet depth of bury		5	1.600		1,500	78		1,578	1,775
4240	9 feet depth of bury		4.50	1.778		1,550	86.50		1,636.50	1,850
4260	10 feet depth of bury		4	2		1,600	97.50		1,697.50	1,925
4280	For 1" IPS connection, add					277			277	305
4300	For tapped drain port in box, add					61.50			61.50	68
5000	Moderate climate, all bronze, polished face									
5020	and scoriated cover, set flush									
5100	3/4" IPS connection	1 Plum	16	.500	Ea.	475	24.50		499.50	560
5120	1" IPS connection	"	14	.571		585	28		613	685
5200	For tapped drain port in box, add					61.50			61.50	68
6000	Ground post type, all non-freeze, all bronze, aluminum casing									
6010	guard, exposed head, 3/4" IPS connection									
6100	2 feet depth of bury	1 Plum	8	1	Ea.	670	49		719	810
6120	3 feet depth of bury		8	1		725	49		774	870
6140	4 feet depth of bury		8	1		795	49		844	950
6160	5 feet depth of bury		7	1.143		835	55.50		890.50	1,000
6180	6 feet depth of bury		7	1.143		895	55.50		950.50	1,075
6200	7 feet depth of bury		6	1.333		955	65		1,020	1,150
6220	8 feet depth of bury		5	1.600		1,025	78		1,103	1,250
6240	9 feet depth of bury		4	2		1,075	97.50		1,172.50	1,325
6260	10 feet depth of bury		4	2		1,150	97.50		1,247.50	1,400
6300	For 1" IPS connection, add					40%	10%			
6350	For 1-1/4" IPS connection, add					140%	14%			
6400	For 1-1/2" IPS connection, add					225%	18%			
6450	For 2" IPS connection, add					315%	24%			
9000	Minimum labor/equipment charge	1 Plum	3	2.667	Job		130		130	202

22 11 23 – Domestic Water Pumps

22 11 23.10 General Utility Pumps		Crew	Daily Output	Labor-Hours	Unit	Material	Labor	Equipment	Total	Total Incl O&P
0010	**GENERAL UTILITY PUMPS**									
2000	Single stage									
3000	Double suction,									
3190	75 HP, to 2500 GPM	Q-3	.28	114	Ea.	15,600	5,300		20,900	25,300
3220	100 HP, to 3000 GPM		.26	123		19,600	5,725		25,325	30,400
3240	150 HP, to 4000 GPM		.24	133		30,200	6,200		36,400	42,800

22 12 Facility Potable-Water Storage Tanks

22 12 23 – Facility Indoor Potable-Water Storage Tanks

22 12 23.13 Facility Steel, Indoor Pot.-Water Storage Tanks		Crew	Daily Output	Labor-Hours	Unit	Material	Labor	Equipment	Total	Total Incl O&P
						2009 Bare Costs				
0010	FACILITY STEEL, INDOOR POT.-WATER STORAGE TANKS									
2000	Galvanized steel, 15 gal., 14" diam., 26" LOA	1 Plum	12	.667	Ea.	990	32.50		1,022.50	1,125
2060	30 gal., 14" diam. x 49" LOA		11	.727		1,175	35.50		1,210.50	1,325
2080	80 gal., 20" diam. x 64" LOA		9	.889		1,825	43.50		1,868.50	2,075
2100	135 gal., 24" diam. x 75" LOA		6	1.333		2,650	65		2,715	3,025
2120	240 gal., 30" diam. x 86" LOA		4	2		5,200	97.50		5,297.50	5,875
2140	300 gal., 36" diam. x 76" LOA		3	2.667		7,375	130		7,505	8,300
2160	400 gal., 36" diam. x 100" LOA	Q-1	4	4		9,075	176		9,251	10,200
2180	500 gal., 36" diam., x 126" LOA	"	3	5.333		11,200	234		11,434	12,800
3000	Glass lined, P.E., 80 gal., 20" diam. x 60" LOA	1 Plum	9	.889		2,450	43.50		2,493.50	2,775
3060	140 gal., 24" diam. x 80" LOA		6	1.333		3,400	65		3,465	3,850
3080	225 gal., 30" diam. x 78" LOA		4	2		4,250	97.50		4,347.50	4,825
3100	325 gal., 36" diam. x 81" LOA		3	2.667		5,425	130		5,555	6,150
3120	460 gal., 42" diam. x 84" LOA	Q-1	4	4		5,700	176		5,876	6,550
3140	605 gal., 48" diam. x 87" LOA		3	5.333		10,900	234		11,134	12,400
3160	740 gal., 54" diam. x 91" LOA		3	5.333		12,300	234		12,534	13,900
3180	940 gal., 60" diam. x 93" LOA		2.50	6.400		14,200	281		14,481	16,000
3200	1330 gal., 66" diam. x 107" LOA		2	8		19,400	350		19,750	21,900
3220	1615 gal., 72" diam. x 110" LOA		1.50	10.667		22,100	470		22,570	25,000
3240	2285 gal., 84" diam. x 128" LOA		1	16		27,300	700		28,000	31,100
3260	3440 gal., 96" diam. x 157" LOA	Q-2	1.50	16		40,400	730		41,130	45,600

22 13 Facility Sanitary Sewerage

22 13 16 – Sanitary Waste and Vent Piping

22 13 16.20 Pipe, Cast Iron

		Crew	Daily Output	Labor-Hours	Unit	Material	Labor	Equipment	Total	Total Incl O&P
0010	**PIPE, CAST IRON**, Soil, on clevis hanger assemblies, 5' O.C. R221113-50									
0020	Single hub, service wt., lead & oakum joints 10' O.C.									
2120	2" diameter	Q-1	63	.254	L.F.	6.35	11.15		17.50	24.50
2140	3" diameter		60	.267		8.90	11.70		20.60	28
2160	4" diameter		55	.291		11.50	12.75		24.25	32.50
2180	5" diameter	Q-2	76	.316		15.85	14.35		30.20	40
2200	6" diameter	"	73	.329		19.50	14.95		34.45	44.50
2220	8" diameter	Q-3	59	.542		30.50	25		55.50	72.50
2240	10" diameter		54	.593		50.50	27.50		78	98
2260	12" diameter		48	.667		72	31		103	127
2261	15" diameter		40	.800		108	37		145	177
2320	For service weight, double hub, add					10%				
2340	For extra heavy, single hub, add					48%	4%			
2360	For extra heavy, double hub, add					71%	4%			
2400	Lead for caulking, (1#/dia in)	Q-1	160	.100	Lb.	1.04	4.39		5.43	7.95
2420	Oakum for caulking, (1/8#/dia in)	"	40	.400	"	3.95	17.55		21.50	31.50
2960	To delete hangers, subtract									
2970	2" diam. to 4" diam.					16%	19%			
2980	5" diam. to 8" diam.					14%	14%			
2990	10" diam. to 15" diam.					13%	19%			
3000	Single hub, service wt, push-on gasket joints 10' O.C.									
3010	2" diameter	Q-1	66	.242	L.F.	7	10.65		17.65	24
3020	3" diameter		63	.254		9.70	11.15		20.85	28
3030	4" diameter		57	.281		12.55	12.30		24.85	33
3040	5" diameter	Q-2	79	.304		17.45	13.80		31.25	40.50
3050	6" diameter	"	75	.320		21	14.55		35.55	46

22 13 16.20 Pipe, Cast Iron

	22 13 16.20 Pipe, Cast Iron	Crew	Daily Output	Labor-Hours	Unit	Material	2009 Bare Costs Labor	Equipment	Total	Total Incl O&P
3060	8" diameter	Q-3	62	.516	L.F.	34	24		58	74.50
3070	10" diameter		56	.571		56	26.50		82.50	103
3080	12" diameter		49	.653		78	30.50		108.50	133
3082	15" diameter		40	.800		116	37		153	185
3100	For service weight, double hub, add					65%				
3110	For extra heavy, single hub, add					48%	4%			
3120	For extra heavy, double hub, add					29%	4%			
3130	To delete hangers, subtract									
3140	2" diam. to 4" diam.					12%	21%			
3150	5" diam. to 8" diam.					10%	16%			
3160	10" diam. to 15" diam.					9%	21%			
4000	No hub, couplings 10' O.C.									
4100	1-1/2" diameter	Q-1	71	.225	L.F.	6.90	9.90		16.80	23
4120	2" diameter		67	.239		7.10	10.50		17.60	24
4140	3" diameter		64	.250		9.80	10.95		20.75	28
4160	4" diameter		58	.276		12.70	12.10		24.80	32.50
4180	5" diameter	Q-2	83	.289		18.05	13.15		31.20	40.50
4200	6" diameter	"	79	.304		21.50	13.80		35.30	45.50
4220	8" diameter	Q-3	69	.464		40	21.50		61.50	77.50
4240	10" diameter	"	61	.525		67	24.50		91.50	112
4280	To delete hangers, subtract									
4290	1-1/2" diam. to 6" diam.					22%	47%			
4300	8" diam. to 10" diam.					21%	44%			
9000	Minimum labor/equipment charge	1 Plum	4	2	Job		97.50		97.50	151

22 13 16.30 Pipe Fittings, Cast Iron

	22 13 16.30 Pipe Fittings, Cast Iron	Crew	Daily Output	Labor-Hours	Unit	Material	Labor	Equipment	Total	Total Incl O&P
0010	**PIPE FITTINGS, CAST IRON**, Soil									
0040	Hub and spigot, service weight, lead & oakum joints									
0080	1/4 bend, 2"	Q-1	16	1	Ea.	13.70	44		57.70	83
0120	3"		14	1.143		18.40	50		68.40	98.50
0140	4"		13	1.231		28.50	54		82.50	116
0160	5"	Q-2	18	1.333		40.50	60.50		101	139
0180	6"	"	17	1.412		50	64		114	155
0200	8"	Q-3	11	2.909		151	135		286	375
0220	10"		10	3.200		220	149		369	475
0224	12"		9	3.556		298	165		463	585
0266	Closet bend, 3" diameter with flange 10" x 16"	Q-1	14	1.143		78	50		128	164
0268	16"x16"		12	1.333		107	58.50		165.50	208
0270	Closet bend, 4" diameter, 1" x 4" ring, 6" x 16"		13	1.231		86.50	54		140.50	179
0280	8" x 16"		13	1.231		77.50	54		131.50	170
0290	10" x 12"		12	1.333		73.50	58.50		132	171
0300	10" x 18"		11	1.455		101	64		165	210
0310	12" x 16"		11	1.455		87	64		151	195
0330	16" x 16"		10	1.600		109	70		179	229
0340	1/8 bend, 2"		16	1		9.80	44		53.80	79
0350	3"		14	1.143		15.35	50		65.35	95
0360	4"		13	1.231		22.50	54		76.50	109
0380	5"	Q-2	18	1.333		31	60.50		91.50	128
0400	6"	"	17	1.412		38	64		102	142
0420	8"	Q-3	11	2.909		113	135		248	335
0440	10"		10	3.200		163	149		312	410
0460	12"		9	3.556		310	165		475	595
0500	Sanitary tee, 2"	Q-1	10	1.600		19.25	70		89.25	130

22 13 16.30 Pipe Fittings, Cast Iron		Crew	Daily Output	Labor-Hours	Unit	Material	2009 Bare Costs Labor	Equipment	Total	Total Incl O&P
0540	3"	Q-1	9	1.778	Ea.	31	78		109	155
0620	4"		8	2		38	88		126	178
0700	5"	Q-2	12	2		75.50	91		166.50	224
0800	6"	"	11	2.182		86	99.50		185.50	249
0880	8"	Q-3	7	4.571		227	212		439	580
1000	Tee, 2"	Q-1	10	1.600		25.50	70		95.50	137
1060	3"		9	1.778		41.50	78		119.50	167
1120	4"		8	2		53	88		141	195
1200	5"	Q-2	12	2		81.50	91		172.50	231
1300	6"	"	11	2.182		111	99.50		210.50	276
1380	8"	Q-3	7	4.571		345	212		557	710
1400	Combination Y and 1/8 bend									
1420	2"	Q-1	10	1.600	Ea.	24	70		94	136
1460	3"		9	1.778		36.50	78		114.50	162
1520	4"		8	2		50.50	88		138.50	192
1540	5"	Q-2	12	2		95.50	91		186.50	246
1560	6"		11	2.182		121	99.50		220.50	287
1580	8"		7	3.429		298	156		454	570
1582	12"	Q-3	6	5.333		665	248		913	1,125
1600	Double Y, 2"	Q-1	8	2		36.50	88		124.50	177
1610	3"		7	2.286		53	100		153	215
1620	4"		6.50	2.462		69.50	108		177.50	245
1630	5"	Q-2	9	2.667		119	121		240	320
1640	6"	"	8	3		179	137		316	410
1650	8"	Q-3	5.50	5.818		440	270		710	900
1660	10"		5	6.400		875	297		1,172	1,425
1670	12"		4.50	7.111		1,050	330		1,380	1,650
1740	Reducer, 3" x 2"	Q-1	15	1.067		13.45	47		60.45	87.50
1750	4" x 2"		14.50	1.103		14.15	48.50		62.65	90.50
1760	4" x 3"		14	1.143		17.50	50		67.50	97.50
1770	5" x 2"		14	1.143		34.50	50		84.50	116
1780	5" x 3"		13.50	1.185		34	52		86	118
1790	5" x 4"		13	1.231		22.50	54		76.50	109
1800	6" x 2"		13.50	1.185		33	52		85	117
1810	6" x 3"		13	1.231		34	54		88	122
1840	6" x 5"		11	1.455		38	64		102	141
1880	8" x 3"	Q-2	13.50	1.778		64.50	81		145.50	196
1900	8" x 4"		13	1.846		59	84		143	195
1920	8" x 5"		12	2		61	91		152	208
1940	8" x 6"		12	2		60.50	91		151.50	208
1960	Increaser, 2" x 3"	Q-1	15	1.067		32	47		79	108
1980	2" x 4"		14	1.143		32	50		82	114
2000	2" x 5"		13	1.231		39	54		93	127
2020	3" x 4"		13	1.231		35.50	54		89.50	123
2040	3" x 5"		13	1.231		39	54		93	127
2060	3" x 6"		12	1.333		47.50	58.50		106	143
2070	4" x 5"		13	1.231		38.50	54		92.50	127
2080	4" x 6"		12	1.333		48	58.50		106.50	144
2090	4" x 8"	Q-2	13	1.846		98	84		182	238
2100	5" x 6"	Q-1	11	1.455		71	64		135	177
2110	5" x 8"	Q-2	12	2		114	91		205	267
2120	6" x 8"		12	2		115	91		206	267
2130	6" x 10"		8	3		207	137		344	440

22 13 16.30 Pipe Fittings, Cast Iron		Crew	Daily Output	Labor-Hours	Unit	Material	2009 Bare Costs Labor	Equipment	Total	Total Incl O&P
2140	8" x 10"	Q-2	6.50	3.692	Ea.	208	168		376	490
2150	10" x 12"		5.50	4.364		370	199		569	720
2500	Y, 2"	Q-1	10	1.600		17.50	70		87.50	128
2510	3"		9	1.778		32.50	78		110.50	157
2520	4"		8	2		37	88		125	177
2530	5"	Q-2	12	2		73	91		164	222
2540	6"	"	11	2.182		100	99.50		199.50	264
2550	8"	Q-3	7	4.571		244	212		456	600
2560	10"		6	5.333		395	248		643	820
2570	12"		5	6.400		805	297		1,102	1,350
2580	15"		4	8		1,450	370		1,820	2,150
3000	For extra heavy, add					44%	4%			
3600	Hub and spigot, service weight gasket joint									
3605	Note: gaskets and joint labor have									
3606	been included with all listed fittings.									
3610	1/4 bend, 2"	Q-1	20	.800	Ea.	19.90	35		54.90	76.50
3620	3"		17	.941		26.50	41.50		68	93
3630	4"		15	1.067		39	47		86	116
3640	5"	Q-2	21	1.143		56	52		108	143
3650	6"	"	19	1.263		66.50	57.50		124	162
3660	8"	Q-3	12	2.667		187	124		311	400
3670	10"		11	2.909		276	135		411	515
3680	12"		10	3.200		360	149		509	630
3700	Closet bend, 3" diameter with ring 10" x 16"	Q-1	17	.941		86.50	41.50		128	159
3710	16" x 16"		15	1.067		115	47		162	199
3730	Closet bend, 4" diameter, 1" x 4" ring, 6" x 16"		15	1.067		97	47		144	179
3740	8" x 16"		15	1.067		88	47		135	170
3750	10" x 12"		14	1.143		83.50	50		133.50	170
3760	10" x 18"		13	1.231		111	54		165	206
3770	12" x 16"		13	1.231		97.50	54		151.50	191
3780	16" x 16"		12	1.333		119	58.50		177.50	222
3800	1/8 bend, 2"		20	.800		16	35		51	72
3810	3"		17	.941		23.50	41.50		65	90
3820	4"		15	1.067		32.50	47		79.50	109
3830	5"	Q-2	21	1.143		47	52		99	132
3840	6"	"	19	1.263		54.50	57.50		112	149
3850	8"	Q-3	12	2.667		150	124		274	355
3860	10"		11	2.909		219	135		354	450
3870	12"		10	3.200		375	149		524	640
3900	Sanitary Tee, 2"	Q-1	12	1.333		31.50	58.50		90	126
3910	3"		10	1.600		47	70		117	161
3920	4"		9	1.778		58.50	78		136.50	186
3930	5"	Q-2	13	1.846		107	84		191	248
3940	6"	"	11	2.182		119	99.50		218.50	285
3950	8"	Q-3	8.50	3.765		300	175		475	600
3980	Tee, 2"	Q-1	12	1.333		38	58.50		96.50	132
3990	3"		10	1.600		57.50	70		127.50	173
4000	4"		9	1.778		73.50	78		151.50	202
4010	5"	Q-2	13	1.846		113	84		197	255
4020	6"	"	11	2.182		144	99.50		243.50	310
4030	8"	Q-3	8	4		415	186		601	750
4060	Combination Y and 1/8 bend									
4070	2"	Q-1	12	1.333	Ea.	36.50	58.50		95	131

22 13 16.30 Pipe Fittings, Cast Iron		Crew	Daily Output	Labor-Hours	Unit	Material (2009 Bare Costs)	Labor (2009 Bare Costs)	Equipment (2009 Bare Costs)	Total (2009 Bare Costs)	Total Incl O&P
4080	3"	Q-1	10	1.600	Ea.	53	70		123	167
4090	4"		9	1.778		71	78		149	199
4100	5"	Q-2	13	1.846		127	84		211	270
4110	6"	"	11	2.182		154	99.50		253.50	325
4120	8"	Q-3	8	4		370	186		556	700
4121	12"	"	7	4.571		795	212		1,007	1,200
4160	Double Y, 2"	Q-1	10	1.600		55.50	70		125.50	170
4170	3"		8	2		77.50	88		165.50	221
4180	4"		7	2.286		100	100		200	266
4190	5"	Q-2	10	2.400		167	109		276	350
4200	6"	"	9	2.667		228	121		349	440
4210	8"	Q-3	6	5.333		550	248		798	985
4220	10"		5	6.400		1,050	297		1,347	1,600
4230	12"		4.50	7.111		1,250	330		1,580	1,875
4260	Reducer, 3" x 2"	Q-1	17	.941		27.50	41.50		69	94.50
4270	4" x 2"		16.50	.970		30.50	42.50		73	99.50
4280	4" x 3"		16	1		36	44		80	108
4290	5" x 2"		16	1		57	44		101	131
4300	5" x 3"		15.50	1.032		58	45.50		103.50	134
4310	5" x 4"		15	1.067		48.50	47		95.50	126
4320	6" x 2"		15.50	1.032		55.50	45.50		101	132
4330	6" x 3"		15	1.067		58.50	47		105.50	137
4340	6" x 5"		13	1.231		70.50	54		124.50	162
4360	8" x 3"	Q-2	15	1.600		109	73		182	233
4370	8" x 4"		15	1.600		106	73		179	229
4380	8" x 5"		14	1.714		113	78		191	246
4390	8" x 6"		14	1.714		113	78		191	246
4430	Increaser, 2" x 3"	Q-1	17	.941		40.50	41.50		82	109
4440	2" x 4"		16	1		42.50	44		86.50	115
4450	2" x 5"		15	1.067		55	47		102	133
4460	3" x 4"		15	1.067		45.50	47		92.50	123
4470	3" x 5"		15	1.067		55	47		102	133
4480	3" x 6"		14	1.143		64	50		114	149
4490	4" x 5"		15	1.067		54.50	47		101.50	133
4500	4" x 6"		14	1.143		64.50	50		114.50	149
4510	4" x 8"	Q-2	15	1.600		134	73		207	261
4520	5" x 6"	Q-1	13	1.231		87.50	54		141.50	181
4530	5" x 8"	Q-2	14	1.714		151	78		229	287
4540	6" x 8"		14	1.714		151	78		229	287
4550	6" x 10"		10	2.400		263	109		372	460
4560	8" x 10"		8.50	2.824		263	128		391	490
4570	10" x 12"		7.50	3.200		435	146		581	705
4600	Y, 2"	Q-1	12	1.333		30	58.50		88.50	124
4610	3"		10	1.600		48.50	70		118.50	163
4620	4"		9	1.778		57.50	78		135.50	185
4630	5"	Q-2	13	1.846		105	84		189	245
4640	6"	"	11	2.182		133	99.50		232.50	300
4650	8"	Q-3	8	4		315	186		501	640
4660	10"		7	4.571		505	212		717	885
4670	12"		6	5.333		935	248		1,183	1,400
4672	15"		5	6.400		1,600	297		1,897	2,200
4900	For extra heavy, add					44%	4%			
4940	Gasket and making push-on joint									

22 13 16.30 Pipe Fittings, Cast Iron		Crew	Daily Output	Labor-Hours	Unit	Material	Labor	Equipment	Total	Total Incl O&P
						2009 Bare Costs				
4950	2"	Q-1	40	.400	Ea.	6.20	17.55		23.75	34
4960	3"		35	.457		8.10	20		28.10	40
4970	4"		32	.500		10.25	22		32.25	45.50
4980	5"	Q-2	43	.558		15.90	25.50		41.40	57
4990	6"	"	40	.600		16.45	27.50		43.95	60.50
5000	8"	Q-3	32	1		36.50	46.50		83	112
5010	10"		29	1.103		55.50	51		106.50	141
5020	12"		25	1.280		63.50	59.50		123	162
5022	15"		21	1.524		76	71		147	194
5030	Note: gaskets and joint labor have									
5040	Been included with all listed fittings.									
5990	No hub									
6000	Cplg. & labor required at joints not incl. in fitting									
6010	price. Add 1 coupling per joint for installed price									
6020	1/4 Bend, 1-1/2"				Ea.	7.70			7.70	8.45
6060	2"					8.45			8.45	9.30
6080	3"					11.60			11.60	12.80
6120	4"					16.80			16.80	18.50
6140	5"					38.50			38.50	42.50
6160	6"					42.50			42.50	46.50
6180	8"					117			117	129
6184	1/4 Bend, long sweep, 1-1/2"					18.20			18.20	20
6186	2"					18.20			18.20	20
6188	3"					21.50			21.50	23.50
6189	4"					34.50			34.50	38
6190	5"					63.50			63.50	70
6191	6"					77.50			77.50	85
6192	8"					187			187	206
6193	10"					335			335	370
6200	1/8 Bend, 1-1/2"					6.45			6.45	7.10
6210	2"					6.45			6.45	7.10
6212	3"					9.75			9.75	10.70
6214	4"					12.30			12.30	13.55
6216	5"					26.50			26.50	29
6218	6"					28.50			28.50	31.50
6220	8"					82			82	90.50
6222	10"					153			153	168
6380	Sanitary Tee, tapped, 1-1/2"					14.20			14.20	15.65
6382	2" x 1-1/2"					12.80			12.80	14.10
6384	2"					14.35			14.35	15.80
6386	3" x 2"					20			20	22
6388	3"					37			37	40.50
6390	4" x 1-1/2"					17.70			17.70	19.50
6392	4" x 2"					20			20	22
6393	4"					20			20	22
6394	6" x 1-1/2"					44			44	48.50
6396	6" x 2"					45			45	49.50
6459	Sanitary Tee, 1-1/2"					10.70			10.70	11.80
6460	2"					11.60			11.60	12.80
6470	3"					14.20			14.20	15.65
6472	4"					22			22	24
6474	5"					62.50			62.50	69
6476	6"					64			64	70.50

22 13 16.30 Pipe Fittings, Cast Iron		Crew	Daily Output	Labor-Hours	Unit	Material	2009 Bare Costs Labor	Equipment	Total	Total Incl O&P
6478	8"				Ea.	259			259	285
6730	Y, 1-1/2"					10.85			10.85	11.95
6740	2"					10.85			10.85	11.95
6750	3"					15.45			15.45	17
6760	4"					25			25	27.50
6762	5"					58.50			58.50	64.50
6764	6"					67.50			67.50	74
6768	8"					159			159	174
6791	Y, reducing, 3" x 2"					11.60			11.60	12.80
6792	4" x 2"					16.15			16.15	17.80
6793	5" x 2"					36.50			36.50	40
6794	6" x 2"					40.50			40.50	44.50
6795	6" x 4"					52.50			52.50	58
6796	8" x 4"					90.50			90.50	99.50
6797	8" x 6"					111			111	122
6798	10" x 6"					250			250	275
6799	10" x 8"					300			300	330
6800	Double Y, 2"					16.80			16.80	18.50
6920	3"					31			31	34
7000	4"					63.50			63.50	69.50
7100	6"					112			112	123
7120	8"					320			320	350
7200	Combination Y and 1/8 Bend									
7220	1-1/2"				Ea.	11.55			11.55	12.70
7260	2"					12.30			12.30	13.55
7320	3"					18.70			18.70	20.50
7400	4"					34.50			34.50	37.50
7480	5"					75.50			75.50	83
7500	6"					103			103	113
7520	8"					236			236	260
7800	Reducer, 3" x 2"					5.90			5.90	6.45
7820	4" x 2"					9.10			9.10	10
7840	4" x 3"					9.10			9.10	10
8000	Coupling, standard (by CISPI Mfrs.)									
8020	1-1/2"	Q-1	48	.333	Ea.	14.45	14.65		29.10	38.50
8040	2"		44	.364		14.45	15.95		30.40	41
8080	3"		38	.421		17.25	18.50		35.75	47.50
8120	4"		33	.485		20.50	21.50		42	55.50
8160	5"	Q-2	44	.545		49.50	25		74.50	93
8180	6"	"	40	.600		51.50	27.50		79	99.50
8200	8"	Q-3	33	.970		97.50	45		142.50	177
8220	10"	"	26	1.231		129	57		186	231
8300	Coupling, cast iron clamp & neoprene gasket (by MG)									
8310	1-1/2"	Q-1	48	.333	Ea.	7.15	14.65		21.80	30.50
8320	2"		44	.364		7.50	15.95		23.45	33.50
8330	3"		38	.421		9	18.50		27.50	38.50
8340	4"		33	.485		12.30	21.50		33.80	46.50
8350	5"	Q-2	44	.545		20	25		45	60.50
8360	6"	"	40	.600		21	27.50		48.50	65.50
8380	8"	Q-3	33	.970		95.50	45		140.50	175
8400	10"	"	26	1.231		166	57		223	272
8600	Coupling, Stainless steel, heavy duty									
8620	1-1/2"	Q-1	48	.333	Ea.	10	14.65		24.65	33.50

22 13 16.30 Pipe Fittings, Cast Iron

	22 13 16.30 Pipe Fittings, Cast Iron	Crew	Daily Output	Labor-Hours	Unit	2009 Bare Costs Material	Labor	Equipment	Total	Total Incl O&P
8630	2″	Q-1	44	.364	Ea.	10.40	15.95		26.35	36.50
8640	2″ x 1-1/2″		44	.364		12.65	15.95		28.60	39
8650	3″		38	.421		11.25	18.50		29.75	41
8660	4″		33	.485		12.70	21.50		34.20	47
8670	4″ x 3″		33	.485		18.45	21.50		39.95	53.50
8680	5″	Q-2	44	.545		27.50	25		52.50	68.50
8690	6″	″	40	.600		30.50	27.50		58	76
8700	8″	Q-3	33	.970		52	45		97	127
8710	10″	″	26	1.231		65.50	57		122.50	161
9000	Minimum labor/equipment charge	1 Plum	4	2	Job		97.50		97.50	151

22 13 16.50 Shower Drains

	22 13 16.50 Shower Drains	Crew	Daily Output	Labor-Hours	Unit	Material	Labor	Equipment	Total	Total Incl O&P
0010	**SHOWER DRAINS**									
2780	Shower, with strainer, uniform diam. trap, bronze top									
2800	2″ and 3″ pipe size	Q-1	8	2	Ea.	298	88		386	465
2820	4″ pipe size	″	7	2.286		330	100		430	515
2840	For galvanized body, add					132			132	146
2860	With strainer, backwater valve, drum trap									
2880	1-1/2″, 2″ & 3″ pipe size	Q-1	8	2	Ea.	267	88		355	430
2890	4″ pipe size	″	7	2.286		370	100		470	565
2900	For galvanized body, add					132			132	146

22 13 16.60 Traps

	22 13 16.60 Traps	Crew	Daily Output	Labor-Hours	Unit	Material	Labor	Equipment	Total	Total Incl O&P
0010	**TRAPS**									
0030	Cast iron, service weight									
0050	Running P trap, without vent									
1100	2″	Q-1	16	1	Ea.	99	44		143	177
1140	3″		14	1.143		99	50		149	187
1150	4″		13	1.231		99	54		153	193
1160	6″	Q-2	17	1.412		430	64		494	575
1180	Running trap, single hub, with vent									
2080	3″ pipe size, 3″ vent	Q-1	14	1.143	Ea.	78.50	50		128.50	165
2120	4″ pipe size, 4″ vent	″	13	1.231		107	54		161	201
2140	5″ pipe size, 4″ vent	Q-2	11	2.182		170	99.50		269.50	340
2160	6″ pipe size, 4″ vent		10	2.400		460	109		569	675
2180	6″ pipe size, 6″ vent		8	3		500	137		637	760
2200	8″ pipe size, 4″ vent	Q-3	10	3.200		1,500	149		1,649	1,875
2220	8″ pipe size, 6″ vent	″	8	4		1,625	186		1,811	2,100
2300	For double hub, vent, add					10%	20%			
2800	S trap,									
2850	4″ pipe size	Q-1	13	1.231	Ea.	48	54		102	137
3000	P trap, B&S, 2″ pipe size		16	1		23.50	44		67.50	94
3040	3″ pipe size		14	1.143		35	50		85	117
3060	4″ pipe size		13	1.231		50.50	54		104.50	140
3080	5″ pipe size	Q-2	18	1.333		107	60.50		167.50	211
3100	6″ pipe size	″	17	1.412		156	64		220	271
3120	8″ pipe size	Q-3	11	2.909		470	135		605	725
3130	10″ pipe size	″	10	3.200		885	149		1,034	1,200
3150	P trap, no hub, 1-1/2″ pipe size	Q-1	17	.941		13.95	41.50		55.45	79.50
3160	2″ pipe size		16	1		13.20	44		57.20	82.50
3170	3″ pipe size		14	1.143		29	50		79	110
3180	4″ pipe size		13	1.231		50.50	54		104.50	140
3190	6″ pipe size	Q-2	17	1.412		123	64		187	235
3350	Deep seal trap, B&S									

22 13 16 – Sanitary Waste and Vent Piping

22 13 16.60 Traps		Crew	Daily Output	Labor-Hours	Unit	Material	2009 Bare Costs Labor	Equipment	Total	Total Incl O&P
3400	1-1/4" pipe size	Q-1	14	1.143	Ea.	40.50	50		90.50	123
3410	1-1/2" pipe size		14	1.143		40.50	50		90.50	123
3420	2" pipe size		14	1.143		31.50	50		81.50	113
3440	3" pipe size		12	1.333		43	58.50		101.50	138
3460	4" pipe size		11	1.455		68.50	64		132.50	175
3500	For trap primer connection, add					95.50			95.50	105
3540	For trap with floor cleanout, add					70%	5%			
3580	For trap with adjustable cleanout, add	Q-1	10	1.600	Ea.	215	70		285	345
4700	Copper, drainage, drum trap									
4800	3" x 5" solid, 1-1/2" pipe size	1 Plum	16	.500	Ea.	97.50	24.50		122	145
4840	3" x 6" swivel, 1-1/2" pipe size	"	16	.500	"	155	24.50		179.50	208
5100	P trap, standard pattern									
5200	1-1/4" pipe size	1 Plum	18	.444	Ea.	72	21.50		93.50	113
5240	1-1/2" pipe size		17	.471		69.50	23		92.50	112
5260	2" pipe size		15	.533		107	26		133	159
5280	3" pipe size		11	.727		258	35.50		293.50	340
5340	With cleanout and slip joint									
5360	1-1/4" pipe size	1 Plum	18	.444	Ea.	66.50	21.50		88	107
5400	1-1/2" pipe size		17	.471		106	23		129	153
5420	2" pipe size		15	.533		171	26		197	229
5460	For swivel, 1-1/2"					155			155	170
5750	Chromed brass, tubular, P trap, without cleanout, 20 Ga.									
5800	1-1/4" pipe size	1 Plum	18	.444	Ea.	16.95	21.50		38.45	52
5840	1-1/2" pipe size	"	17	.471	"	19.30	23		42.30	57
5900	With cleanout, 20 Ga.									
5940	1-1/4" pipe size	1 Plum	18	.444	Ea.	27.50	21.50		49	63.50
6000	1-1/2" pipe size	"	17	.471	"	29.50	23		52.50	68
6350	S trap, without cleanout, 20 Ga.									
6400	1-1/4" pipe size	1 Plum	18	.444	Ea.	30	21.50		51.50	66.50
6440	1-1/2" pipe size	"	17	.471	"	34	23		57	73
6550	With cleanout, 20 Ga.									
6600	1-1/4" pipe size	1 Plum	18	.444	Ea.	51.50	21.50		73	90.50
6640	1-1/2" pipe size	"	17	.471		56.50	23		79.50	97.50
6660	Corrosion resistant, glass, P trap, 1-1/2" pipe size	Q-1	17	.941		63	41.50		104.50	133
6670	2" pipe size		16	1		82.50	44		126.50	159
6680	3" pipe size		14	1.143		169	50		219	264
6690	4" pipe size		13	1.231		253	54		307	360
6700	6" pipe size	Q-2	17	1.412		970	64		1,034	1,175
6710	ABS DWV P trap, solvent weld joint									
6720	1-1/2" pipe size	1 Plum	18	.444	Ea.	6.75	21.50		28.25	41
6722	2" pipe size		17	.471		9.15	23		32.15	45.50
6724	3" pipe size		15	.533		35	26		61	79
6726	4" pipe size		14	.571		72	28		100	122
6760	PP DWV, dilution trap, 1-1/2" pipe size		16	.500		158	24.50		182.50	212
6770	P trap, 1-1/2" pipe size		17	.471		39	23		62	78.50
6780	2" pipe size		16	.500		58	24.50		82.50	102
6790	3" pipe size		14	.571		106	28		134	160
6800	4" pipe size		13	.615		179	30		209	244
6830	S trap, 1-1/2" pipe size		16	.500		32	24.50		56.50	73
6840	2" pipe size		15	.533		48	26		74	93.50
6850	Universal trap, 1-1/2" pipe size		14	.571		64	28		92	114
6860	PVC DWV hub x hub, basin trap, 1-1/4" pipe size		18	.444		8.95	21.50		30.45	43.50
6870	Sink P trap, 1-1/2" pipe size		18	.444		8.95	21.50		30.45	43.50

22 13 16.60 Traps		Crew	Daily Output	Labor-Hours	Unit	2009 Bare Costs Material	Labor	Equipment	Total	Total Incl O&P
6880	Tubular S trap, 1-1/2" pipe size	1 Plum	17	.471	Ea.	16.55	23		39.55	53.50
6890	PVC sch. 40 DWV, drum trap									
6900	1-1/2" pipe size	1 Plum	16	.500	Ea.	20.50	24.50		45	60.50
6910	P trap, 1-1/2" pipe size		18	.444		5.05	21.50		26.55	39
6920	2" pipe size		17	.471		6.65	23		29.65	43
6930	3" pipe size		15	.533		23	26		49	66
6940	4" pipe size		14	.571		53	28		81	101
6950	P trap w/clean out, 1-1/2" pipe size		18	.444		8.35	21.50		29.85	42.50
6960	2" pipe size		17	.471		14.20	23		37.20	51
6970	P trap adjustable, 1-1/2" pipe size		17	.471		6.45	23		29.45	42.50
6980	P trap adj. w/union & cleanout, 1-1/2" pipe size		16	.500		14.90	24.50		39.40	54.50
7000	Trap primer, flow through type, 1/2" diameter		24	.333		42.50	16.25		58.75	71.50
7100	With sediment strainer		22	.364		46.50	17.75		64.25	78.50
7450	Trap primer distribution unit									
7500	2 openings	1 Plum	18	.444	Ea.	28	21.50		49.50	64.50
7540	3 openings		17	.471		28	23		51	66.50
7560	4 openings		16	.500		28	24.50		52.50	69
7850	Trap primer manifold									
7900	2 outlet	1 Plum	18	.444	Ea.	55	21.50		76.50	94
7940	4 outlet		16	.500		88.50	24.50		113	136
7960	6 outlet		15	.533		123	26		149	176
7980	8 outlet		13	.615		156	30		186	218
9000	Minimum labor/equipment charge		3	2.667	Job		130		130	202

22 13 16.80 Vent Flashing and Caps		Crew	Daily Output	Labor-Hours	Unit	2009 Bare Costs Material	Labor	Equipment	Total	Total Incl O&P
0010	**VENT FLASHING AND CAPS**									
0120	Vent caps									
0140	Cast iron									
0180	2-1/2" - 3-5/8" pipe	1 Plum	21	.381	Ea.	40	18.55		58.55	73
0190	4" - 4-1/8" pipe	"	19	.421	"	48	20.50		68.50	85
0900	Vent flashing									
1000	Aluminum with lead ring									
1020	1-1/4" pipe	1 Plum	20	.400	Ea.	9.90	19.50		29.40	41
1030	1-1/2" pipe		20	.400		10.60	19.50		30.10	41.50
1040	2" pipe		18	.444		10.60	21.50		32.10	45
1050	3" pipe		17	.471		11.70	23		34.70	48.50
1060	4" pipe		16	.500		14.15	24.50		38.65	53.50
1350	Copper with neoprene ring									
1400	1-1/4" pipe	1 Plum	20	.400	Ea.	19.95	19.50		39.45	52
1430	1-1/2" pipe		20	.400		19.95	19.50		39.45	52
1440	2" pipe		18	.444		21	21.50		42.50	56.50
1450	3" pipe		17	.471		24.50	23		47.50	62.50
1460	4" pipe		16	.500		27	24.50		51.50	68
2000	Galvanized with neoprene ring									
2020	1-1/4" pipe	1 Plum	20	.400	Ea.	3.07	19.50		22.57	33.50
2030	1-1/2" pipe		20	.400		12.10	19.50		31.60	43.50
2040	2" pipe		18	.444		9.10	21.50		30.60	43.50
2050	3" pipe		17	.471		9.90	23		32.90	46.50
2060	4" pipe		16	.500		12.85	24.50		37.35	52
2980	Neoprene, one piece									
3000	1-1/4" pipe	1 Plum	24	.333	Ea.	6.30	16.25		22.55	32
3030	1-1/2" pipe		24	.333		6.90	16.25		23.15	32.50
3040	2" pipe		23	.348		7.25	16.95		24.20	34.50

22 13 16 – Sanitary Waste and Vent Piping

22 13 16.80 Vent Flashing and Caps		Crew	Daily Output	Labor-Hours	Unit	Material	2009 Bare Costs Labor	Equipment	Total	Total Incl O&P
3050	3" pipe	1 Plum	21	.381	Ea.	8	18.55		26.55	38
3060	4" pipe	↓	20	.400	↓	10.80	19.50		30.30	42
4000	Lead, 4#, 8" skirt, VTR									
4100	2" pipe	1 Plum	18	.444	Ea.	38	21.50		59.50	75
4110	3" pipe		17	.471		45	23		68	85
4120	4" pipe		16	.500		51.50	24.50		76	94.50
4130	6" pipe		14	.571	↓	73.50	28		101.50	124
9000	Minimum labor/equipment charge	↓	4	2	Job		97.50		97.50	151

22 13 19 – Sanitary Waste Piping Specialties

22 13 19.13 Sanitary Drains

		Crew	Daily Output	Labor-Hours	Unit	Material	Labor	Equipment	Total	Total Incl O&P
0010	**SANITARY DRAINS**									
0400	Deck, auto park, C.I., 13" top									
0440	3", 4", 5", and 6" pipe size	Q-1	8	2	Ea.	1,000	88		1,088	1,225
0480	For galvanized body, add				"	545			545	600
0800	Promenade, heelproof grate, C.I., 14" top									
0840	2", 3", and 4" pipe size	Q-1	10	1.600	Ea.	390	70		460	535
0860	5" and 6" pipe size		9	1.778		480	78		558	650
0880	8" pipe size	↓	8	2		570	88		658	760
0940	For galvanized body, add					286			286	315
0960	For polished bronze top, add				↓	263			263	290
1200	Promenade, heelproof grate, C.I., lateral, 14" top									
1240	2", 3" and 4" pipe size	Q-1	10	1.600	Ea.	505	70		575	665
1260	5" and 6" pipe size		9	1.778		600	78		678	780
1280	8" pipe size	↓	8	2		690	88		778	895
1340	For galvanized body, add					405			405	445
1360	For polished bronze top, add				↓	380			380	420
1500	Promenade, slotted grate, C.I., 11" top									
1540	2", 3", 4", 5", and 6" pipe size	Q-1	12	1.333	Ea.	350	58.50		408.50	475
1600	For galvanized body, add					161			161	177
1640	For polished bronze top, add				↓	188			188	207
2000	Floor, medium duty, C.I., deep flange, 7" dia top									
2040	2" and 3" pipe size	Q-1	12	1.333	Ea.	141	58.50		199.50	247
2080	For galvanized body, add					67			67	74
2120	For polished bronze top, add				↓	102			102	112
2160	Heavy duty, C.I., 12" dia anti-tilt grate									
2180	2", 3", 4", 5" and 6" pipe size	Q-1	10	1.600	Ea.	410	70		480	560
2220	For galvanized body, add					208			208	229
2240	For polished bronze top, add				↓	239			239	263
2300	X-Heavy duty, C.I., 15" antitilt grate									
2320	4", 5", 6", and 8" pipe size	Q-1	8	2	Ea.	880	88		968	1,100
2360	For galvanized body, add					340			340	370
2380	For polished bronze top, add				↓	350			350	385
2400	Heavy duty, with sediment bucket, C.I., 12" dia. loose grate									
2420	2", 3", 4", 5", and 6" pipe size	Q-1	9	1.778	Ea.	450	78		528	615
2440	For galvanized body, add					310			310	340
2460	For polished bronze top, add				↓	201			201	221
2500	Heavy duty, cleanout & trap w/bucket, C.I., 15" top									
2540	2", 3", and 4" pipe size	Q-1	6	2.667	Ea.	4,625	117		4,742	5,250
2560	For galvanized body, add					1,175			1,175	1,300
2580	For polished bronze top, add				↓	510			510	560
2600	Medium duty, with perforated SS basket, C.I., body,									
2610	18" top for refuse container washing area									

22 13 19 – Sanitary Waste Piping Specialties

22 13 19.13 Sanitary Drains		Crew	Daily Output	Labor-Hours	Unit	Material	Labor	Equipment	Total	Total Incl O&P
						2009 Bare Costs				
2620	2" thru 6" pipe size	Q-1	4	4	Ea.	2,800	176		2,976	3,350
2630	Acid resistant									
2638	PVC									
2640	2", 3" and 4" pipe size	Q-1	16	1	Ea.	232	44		276	325
2644	Cast iron, epoxy coated									
2646	2", 3" and 4" pipe size	Q-1	14	1.143	Ea.	232	50		282	335
2650	PVC or ABS thermoplastic									
2660	3" and 4" pipe size	Q-1	16	1	Ea.	258	44		302	350
2680	Extra heavy duty, oil intercepting, gas seal cone,									
2690	with cleanout, loose grate, C.I., body 16" top									
2700	3" and 4" diameter outlet, 4" slab depth	Q-1	4	4	Ea.	5,000	176		5,176	5,775
2720	3" and 4" diameter outlet, 8" slab depth		3	5.333		5,300	234		5,534	6,200
2740	3" &`4" dia. outlet, 10"-12" slab depth, 16"top		2	8		5,625	350		5,975	6,725
2910	Prison cell, vandal-proof, 1-1/2", and 2" diam. pipe		12	1.333		310	58.50		368.50	435
2920	3" pipe size		10	1.600		350	70		420	495
2930	Trap drain, light duty, backwater valve C.I. top									
2950	8" diameter top, 2" pipe size	Q-1	12	1.333	Ea.	272	58.50		330.50	390
2960	10" diameter top, 3" pipe size		10	1.600		375	70		445	525
2970	12" diameter top, 4" pipe size		8	2		525	88		613	710

22 13 19.14 Floor Receptors

		Crew	Daily Output	Labor-Hours	Unit	Material	Labor	Equipment	Total	Total Incl O&P
0010	**FLOOR RECEPTORS**, For connection to 2", 3" & 4" diameter pipe									
0200	12-1/2" square top, 25 sq in open area	Q-1	10	1.600	Ea.	680	70		750	860
0300	For grate with 4" diam. x 3-3/4" high funnel, add					137			137	150
0400	For grate with 6" diameter x 6" high funnel, add					202			202	222
0500	For full hinged grate with open center, add					113			113	125
0600	For aluminum bucket, add					102			102	112
0700	For acid-resisting bucket, add					192			192	211
0800	For galvanized bucket, add					125			125	138
0900	For stainless steel mesh bucket liner, add					152			152	167
1000	For bronze antisplash dome strainer, add					73.50			73.50	81
1100	For partial solid cover, add					36.50			36.50	40
1200	For trap primer connection, add					61.50			61.50	68
2000	12-5/8" diameter top, 40 sq. in. open area	Q-1	10	1.600		540	70		610	705
2100	For options, add same prices as square top									
3000	8" x 4" rectangular top, 7.5 sq. in. open area	Q-1	14	1.143	Ea.	525	50		575	655
3100	For trap primer connections, add					61.50			61.50	68
4000	24" x 16" rectangular top, 70 sq. in. open area	Q-1	4	4		3,050	176		3,226	3,650
4100	For trap primer connection, add					122			122	134
9000	Minimum labor/equipment charge	Q-1	3	5.333	Job		234		234	365

22 13 19.15 Sink Waste Treatment

		Crew	Daily Output	Labor-Hours	Unit	Material	Labor	Equipment	Total	Total Incl O&P
0010	**SINK WASTE TREATMENT**, System for commercial kitchens									
0100	includes clock timer, & fittings									
0200	System less chemical, wall mounted cabinet	1 Plum	16	.500	Ea.	490	24.50		514.50	580
2000	Chemical, 1 gallon, add					57			57	62.50
2100	6 gallons, add					256			256	281
2200	15 gallons, add					695			695	765
2300	30 gallons, add					1,300			1,300	1,425
2400	55 gallons, add					2,225			2,225	2,450

22 13 23 – Sanitary Waste Interceptors

22 13 23.10 Interceptors

		Crew	Daily Output	Labor-Hours	Unit	Material	Labor	Equipment	Total	Total Incl O&P
0010	**INTERCEPTORS**									
0150	Grease, cast iron, 4 GPM, 8 lb. fat capacity	1 Plum	4	2	Ea.	835	97.50		932.50	1,075

22 13 23 – Sanitary Waste Interceptors

22 13 23.10 Interceptors		Crew	Daily Output	Labor-Hours	Unit	Material	2009 Bare Costs Labor	Equipment	Total	Total Incl O&P
0200	7 GPM, 14 lb. fat capacity	1 Plum	4	2	Ea.	1,150	97.50		1,247.50	1,425
1000	10 GPM, 20 lb. fat capacity		4	2		1,375	97.50		1,472.50	1,650
1040	15 GPM, 30 lb. fat capacity		4	2		2,025	97.50		2,122.50	2,375
1060	20 GPM, 40 lb. fat capacity		3	2.667		2,475	130		2,605	2,925
1080	25 GPM, 50 lb. fat capacity	Q-1	3.50	4.571		2,775	201		2,976	3,350
1100	35 GPM, 70 lb. fat capacity		3	5.333		3,425	234		3,659	4,150
1110	50 GPM, 100 lb. fat capacity		2	8		4,550	350		4,900	5,550
1120	Fabricated steel, 50 GPM, 100 lb. fat capacity		2	8		9,575	350		9,925	11,000
1140	75 GPM, 150 lb. fat capacity		2	8		9,725	350		10,075	11,200
1160	100 GPM, 200 lb. fat capacity		2	8		10,900	350		11,250	12,500
1180	150 GPM, 300 lb. fat capacity		2	8		11,400	350		11,750	13,100
1200	200 GPM, 400 lb. fat capacity		1.50	10.667		16,600	470		17,070	19,000
1220	250 GPM, 500 lb. fat capacity		1.30	12.308		19,300	540		19,840	22,100
1240	300 GPM, 600 lb. fat capacity		1	16		22,800	700		23,500	26,100
1260	400 GPM, 800 lb. fat capacity	Q-2	1.20	20		27,900	910		28,810	32,000
1280	500 GPM, 1000 lb. fat capacity	"	1	24		33,000	1,100		34,100	38,000
1580	For seepage pan, add					7%				
3000	Hair, cast iron, 1-1/4" and 1-1/2" pipe connection	1 Plum	8	1	Ea.	315	49		364	420
3100	For chrome-plated cast iron, add					192			192	211
3200	For polished bronze, add					263			263	289
3400	Lint interceptor, fabricated steel									
3410	Size based on 10 GPM per machine									
3420	30 GPM, 2" pipe size	Q-1	3	5.333	Ea.	3,675	234		3,909	4,425
3430	70 GPM, 3" pipe size		2.50	6.400		4,275	281		4,556	5,125
3440	100 GPM, 4" pipe size		2	8		5,025	350		5,375	6,075
3450	200 GPM, 4" pipe size		1.50	10.667		6,100	470		6,570	7,425
3460	300 GPM, 6" pipe size		1	16		7,075	700		7,775	8,875
3470	400 GPM, 6" pipe size	Q-2	1.20	20		8,075	910		8,985	10,300
3480	500 GPM, 6" pipe size	"	1	24		8,750	1,100		9,850	11,300
4000	Oil, fabricated steel, 10 GPM, 2" pipe size	1 Plum	4	2		1,850	97.50		1,947.50	2,200
4100	15 GPM, 2" or 3" pipe size		4	2		2,550	97.50		2,647.50	2,950
4120	20 GPM, 2" or 3" pipe size		3	2.667		3,100	130		3,230	3,600
4140	25 GPM, 2" or 3" pipe size	Q-1	3.50	4.571		3,350	201		3,551	4,000
4160	35 GPM, 2", 3", or 4" pipe size		3	5.333		4,075	234		4,309	4,850
4180	50 GPM, 2", 3", or 4" pipe size		2	8		5,475	350		5,825	6,575
4200	75 GPM, 3" pipe size		2	8		9,500	350		9,850	11,000
4220	100 GPM, 3" pipe size		2	8		10,300	350		10,650	11,800
4240	150 GPM, 4" pipe size		2	8		12,800	350		13,150	14,500
4260	200 GPM, 4" pipe size		1.50	10.667		18,000	470		18,470	20,500
4280	250 GPM, 5" pipe size		1.30	12.308		20,900	540		21,440	23,800
4300	300 GPM, 5" pipe size		1	16		23,600	700		24,300	27,100
4320	400 GPM, 6" pipe size	Q-2	1.20	20		30,400	910		31,310	34,800
4340	500 GPM, 6" pipe size	"	1	24		38,300	1,100		39,400	43,800
5000	Sand interceptor, fabricated steel									
5020	20 GPM, 4" pipe size	Q-1	3	5.333	Ea.	4,600	234		4,834	5,425
5030	50 GPM, 4" pipe size		2.50	6.400		7,800	281		8,081	9,000
5040	150 GPM, 4" pipe size		2	8		23,500	350		23,850	26,400
5050	250 GPM, 6" pipe size		1.30	12.308		24,500	540		25,040	27,800
5060	500 GPM, 6" pipe size	Q-2	1	24		31,500	1,100		32,600	36,400
6000	Solids, precious metals recovery, C.I., 1-1/4" to 2" pipe	1 Plum	4	2		475	97.50		572.50	670
6100	Dental Lab., large, C.I., 1-1/2" to 2" pipe		3	2.667		1,650	130		1,780	2,025
9000	Minimum labor/equipment charge		3	2.667	Job		130		130	202

22 13 Facility Sanitary Sewerage

22 13 26 – Sanitary Waste Separators

22 13 26.10 Separators		Crew	Daily Output	Labor-Hours	Unit	Material	Labor (2009 Bare Costs)	Equipment	Total	Total Incl O&P
0010	**SEPARATORS**, Entrainment eliminator, steel body, 150 PSIG.									
0100	1/4" size	1 Stpi	24	.333	Ea.	197	16.45		213.45	243
0120	1/2" size		24	.333		207	16.45		223.45	254
0140	3/4" size		20	.400		221	19.75		240.75	274
0160	1" size		19	.421		291	21		312	350
0180	1-1/4" size		15	.533		330	26.50		356.50	405
0200	1-1/2" size		13	.615		365	30.50		395.50	445
0220	2" size		11	.727		405	36		441	500
0240	2-1/2" size	Q-5	15	1.067		1,650	47.50		1,697.50	1,875
0260	3" size		13	1.231		1,950	54.50		2,004.50	2,225
0280	4" size		10	1.600		2,200	71		2,271	2,525
0300	5" size		6	2.667		2,575	118		2,693	3,000
0320	6" size		3	5.333		2,850	237		3,087	3,500
0340	8" size	Q-6	4.40	5.455		3,525	251		3,776	4,275
0360	10" size	"	4	6		5,125	276		5,401	6,050
1000	For 300 PSIG, add					15%				

22 13 29 – Sanitary Sewerage Pumps

22 13 29.13 Wet-Pit-Mounted, Vertical Sewerage Pumps

		Crew	Daily Output	Labor-Hours	Unit	Material	Labor	Equipment	Total	Total Incl O&P
0010	**WET-PIT-MOUNTED, VERTICAL SEWERAGE PUMPS**									
0020	Controls incl. alarm/disconnect panel w/wire. Excavation not included									
0260	Simplex, 9 GPM at 60 PSIG, 91 gal. tank				Ea.	2,825			2,825	3,100
0300	Unit with manway, 26" I.D., 18" high					3,250			3,250	3,575
0340	26" I.D., 36" high					3,400			3,400	3,725
0380	43" I.D., 4' high					3,450			3,450	3,800
0600	Simplex, 9 GPM at 60 PSIG, 150 gal. tank, indoor					3,100			3,100	3,400
0700	Unit with manway, 26" I.D., 36" high					3,750			3,750	4,125
0740	26" I.D., 4' high					3,875			3,875	4,250
2000	Duplex, 18 GPM at 60 PSIG, 150 gal. tank, indoor					5,400			5,400	5,950
2060	Unit with manway, 43" I.D., 4' high					7,125			7,125	7,825
2400	For core only					1,525			1,525	1,675

22 13 29.14 Sewage Ejector Pumps

		Crew	Daily Output	Labor-Hours	Unit	Material	Labor	Equipment	Total	Total Incl O&P
0010	**SEWAGE EJECTOR PUMPS**, With operating and level controls									
0100	Simplex system incl. tank, cover, pump 15' head									
0500	37 gal PE tank, 12 GPM, 1/2 HP, 2" discharge	Q-1	3.20	5	Ea.	440	219		659	825
0510	3" discharge		3.10	5.161		480	226		706	875
0530	87 GPM, .7 HP, 2" discharge		3.20	5		670	219		889	1,075
0540	3" discharge		3.10	5.161		725	226		951	1,150
0600	45 gal. coated stl tank, 12 GPM, 1/2 HP, 2" discharge		3	5.333		780	234		1,014	1,225
0610	3" discharge		2.90	5.517		810	242		1,052	1,275
0630	87 GPM, .7 HP, 2" discharge		3	5.333		1,000	234		1,234	1,475
0640	3" discharge		2.90	5.517		1,050	242		1,292	1,525
0660	134 GPM, 1 HP, 2" discharge		2.80	5.714		1,075	251		1,326	1,600
0680	3" discharge		2.70	5.926		1,150	260		1,410	1,650
0700	70 gal. PE tank, 12 GPM, 1/2 HP, 2" discharge		2.60	6.154		850	270		1,120	1,350
0710	3" discharge		2.40	6.667		910	293		1,203	1,450
0730	87 GPM, 0.7 HP, 2" discharge		2.50	6.400		1,100	281		1,381	1,625
0740	3" discharge		2.30	6.957		1,150	305		1,455	1,750
0760	134 GPM, 1 HP, 2" discharge		2.20	7.273		1,200	320		1,520	1,800
0770	3" discharge		2	8		1,275	350		1,625	1,950
0800	75 gal. coated stl. tank, 12 GPM, 1/2 HP, 2 " discharge		2.40	6.667		935	293		1,228	1,475
0810	3" discharge		2.20	7.273		970	320		1,290	1,575

22 13 Facility Sanitary Sewerage

22 13 29 – Sanitary Sewerage Pumps

22 13 29.14 Sewage Ejector Pumps		Crew	Daily Output	Labor-Hours	Unit	Material	2009 Bare Costs Labor	Equipment	Total	Total Incl O&P
0830	87 GPM, .7 HP, 2" discharge	Q-1	2.30	6.957	Ea.	1,175	305		1,480	1,775
0840	3" discharge		2.10	7.619		1,225	335		1,560	1,875
0860	134 GPM, 1 HP, 2" discharge		2	8		1,250	350		1,600	1,925
0880	3" discharge		1.80	8.889		1,325	390		1,715	2,050
1040	Duplex system incl. tank, covers, pumps									
1060	110 gal. fiberglass tank, 24 GPM, 1/2 HP, 2" discharge	Q-1	1.60	10	Ea.	1,700	440		2,140	2,550
1080	3" discharge		1.40	11.429		1,800	500		2,300	2,750
1100	174 GPM, .7 HP, 2" discharge		1.50	10.667		2,200	470		2,670	3,125
1120	3" discharge		1.30	12.308		2,275	540		2,815	3,350
1140	268 GPM, 1 HP, 2" discharge		1.20	13.333		2,375	585		2,960	3,525
1160	3" discharge		1	16		2,475	700		3,175	3,800
1260	135 gal. coated stl. tank, 24 GPM, 1/2 HP, 2" discharge	Q-2	1.70	14.118		1,750	640		2,390	2,900
2000	3" discharge		1.60	15		1,850	685		2,535	3,075
2640	174 GPM, .7 HP, 2" discharge		1.60	15		2,275	685		2,960	3,550
2660	3" discharge		1.50	16		2,400	730		3,130	3,750
2700	268 GPM, 1 HP, 2" discharge		1.30	18.462		2,475	840		3,315	4,025
3040	3" discharge		1.10	21.818		2,625	995		3,620	4,450
3060	275 gal. coated stl. tank, 24 GPM, 1/2 HP, 2" discharge		1.50	16		2,175	730		2,905	3,525
3080	3" discharge		1.40	17.143		2,200	780		2,980	3,625
3100	174 GPM, .7 HP, 2" discharge		1.40	17.143		2,800	780		3,580	4,275
3120	3" discharge		1.30	18.462		2,975	840		3,815	4,575
3140	268 GPM, 1 HP, 2" discharge		1.10	21.818		3,075	995		4,070	4,925
3160	3" discharge		.90	26.667		3,250	1,225		4,475	5,450
3260	Pump system accessories, add									
3300	Alarm horn and lights, 115V mercury switch	Q-1	8	2	Ea.	93	88		181	238
3340	Switch, mag. contactor, alarm bell, light, 3 level control		5	3.200		455	140		595	720
3380	Alternator, mercury switch activated		4	4		825	176		1,001	1,175
9000	Minimum labor/equipment charge		2.50	6.400	Job		281		281	435

22 14 Facility Storm Drainage

22 14 23 – Storm Drainage Piping Specialties

22 14 23.33 Backwater Valves		Crew	Daily Output	Labor-Hours	Unit	Material	Labor	Equipment	Total	Total Incl O&P
0010	**BACKWATER VALVES**, C.I. Body.									
6980	Bronze gate and automatic flapper valves									
7000	3" and 4" pipe size	Q-1	13	1.231	Ea.	1,425	54		1,479	1,650
7100	5" and 6" pipe size	"	13	1.231	"	2,200	54		2,254	2,500
7240	Bronze flapper valve, bolted cover									
7260	2" pipe size	Q-1	16	1	Ea.	420	44		464	530
7280	3" pipe size		14.50	1.103		655	48.50		703.50	795
7300	4" pipe size		13	1.231		810	54		864	975
7320	5" pipe size	Q-2	18	1.333		970	60.50		1,030.50	1,175
7340	6" pipe size	"	17	1.412		1,150	64		1,214	1,375
7360	8" pipe size	Q-3	10	3.200		1,625	149		1,774	2,025
7380	10" pipe size	"	9	3.556		2,675	165		2,840	3,175
7500	For threaded cover, same cost									
7540	Revolving disk type, same cost as flapper type									

22 14 26 – Facility Storm Drains

22 14 26.13 Roof Drains		Crew	Daily Output	Labor-Hours	Unit	Material	Labor	Equipment	Total	Total Incl O&P
0010	**ROOF DRAINS**									
0140	Cornice, C.I., 45° or 90° outlet									

22 14 26.13 Roof Drains		Crew	Daily Output	Labor-Hours	Unit	2009 Bare Costs Material	Labor	Equipment	Total	Total Incl O&P
0200	3" and 4" pipe size	Q-1	12	1.333	Ea.	231	58.50		289.50	345
0260	For galvanized body, add					53			53	58.50
0280	For polished bronze dome, add					39			39	43
3860	Roof, flat metal deck, C.I. body, 12" C.I. dome									
3880	2" pipe size	Q-1	15	1.067	Ea.	202	47		249	295
3890	3" pipe size		14	1.143		277	50		327	385
3900	4" pipe size		13	1.231		277	54		331	390
3910	5" pipe size		12	1.333		480	58.50		538.50	620
3920	6" pipe size		10	1.600		480	70		550	640
4280	Integral expansion joint, C.I. body, 12" C.I. dome									
4300	2" pipe size	Q-1	8	2	Ea.	490	88		578	675
4320	3" pipe size		7	2.286		505	100		605	710
4340	4" pipe size		6	2.667		535	117		652	770
4360	5" pipe size		4	4		855	176		1,031	1,200
4380	6" pipe size		3	5.333		910	234		1,144	1,375
4400	8" pipe size		3	5.333		1,550	234		1,784	2,075
4440	For galvanized body, add					183			183	202
4480	Flexible neoprene bellows, no-hub connection									
4500	3" and 4" pipe size	Q-1	5	3.200	Ea.	118	140		258	350
4520	6" and 8" pipe size	"	4	4	"	264	176		440	560
4620	Main, all aluminum, 12" low profile dome									
4640	2", 3" and 4" pipe size	Q-1	14	1.143	Ea.	305	50		355	415
4660	5" and 6" pipe size		13	1.231		405	54		459	530
4680	8" pipe size		10	1.600		505	70		575	665
4690	Main, CI body, 12" poly. dome, 2", 3", & 4" pipe		8	2		228	88		316	385
4710	5" and 6" pipe size		6	2.667		325	117		442	540
4720	8" pipe size		4	4		425	176		601	735
4730	For underdeck clamp, add		22	.727		158	32		190	223
4740	For vandalproof dome, add					40			40	44
4750	For galvanized body, add					310			310	340
4760	Main, PVC body and dome, 2" pipe size	Q-1	14	1.143		115	50		165	205
4780	3" pipe size		14	1.143		115	50		165	205
4800	4" pipe size		14	1.143		115	50		165	205
4820	For underdeck clamp, add		24	.667		25	29.50		54.50	73
4900	Terrace planting area, with perforated overflow, C.I.									
4920	2", 3" and 4" pipe size	Q-1	8	2	Ea.	430	88		518	610
9000	Minimum labor/equipment charge	1 Plum	4	2	Job		97.50		97.50	151

22 14 26.16 Facility Area Drains		Crew	Daily Output	Labor-Hours	Unit	Material	Labor	Equipment	Total	Total Incl O&P
0010	**FACILITY AREA DRAINS**									
4980	Scupper floor, oblique strainer, C.I.									
5000	6" x 7" top, 2", 3" and 4" pipe size	Q-1	16	1	Ea.	199	44		243	287
5100	8" x 12" top, 5" and 6" pipe size	"	14	1.143		385	50		435	505
5160	For galvanized body, add					40%				
5200	For polished bronze strainer, add					85%				

22 14 26.19 Facility Trench Drains		Crew	Daily Output	Labor-Hours	Unit	Material	Labor	Equipment	Total	Total Incl O&P
0010	**FACILITY TRENCH DRAINS**									
5980	Trench, floor, hvy duty, modular, C.I., 12" x 12" top									
6000	2", 3", 4", 5", & 6" pipe size	Q-1	8	2	Ea.	630	88		718	825
6100	For unit with polished bronze top					940			940	1,025
6200	For 12" extension section, C.I. top					630			630	690
6240	For 12" extension section, polished bronze top					1,575			1,575	1,725
6600	Trench, floor, for cement concrete encasement									

22 14 Facility Storm Drainage

22 14 26 – Facility Storm Drains

22 14 26.19 Facility Trench Drains		Crew	Daily Output	Labor-Hours	Unit	Material	Labor	Equipment	Total	Total Incl O&P
						2009 Bare Costs				
6610	Not including trenching or concrete									
6640	Polyester polymer concrete									
6650	4″ internal width, with grate									
6660	Light duty steel grate	Q-1	120	.133	L.F.	26.50	5.85		32.35	38
6670	Medium duty steel grate		115	.139		33	6.10		39.10	46
6680	Heavy duty iron grate		110	.145		51	6.40		57.40	66
6700	12″ internal width, with grate									
6770	Heavy duty galvanized grate	Q-1	80	.200	L.F.	128	8.80		136.80	155
6800	Fiberglass									
6810	8″ internal width, with grate									
6820	Medium duty galvanized grate	Q-1	115	.139	L.F.	89	6.10		95.10	107
6830	Heavy duty iron grate	″	110	.145	″	86	6.40		92.40	104

22 14 29 – Sump Pumps

22 14 29.13 Wet-Pit-Mounted, Vertical Sump Pumps

		Crew	Daily Output	Labor-Hours	Unit	Material	Labor	Equipment	Total	Total Incl O&P
0010	**WET-PIT-MOUNTED, VERTICAL SUMP PUMPS**									
0400	Molded PVC base, 21 GPM at 15′ head, 1/3 HP	1 Plum	5	1.600	Ea.	119	78		197	252
0800	Iron base, 21 GPM at 15′ head, 1/3 HP		5	1.600		150	78		228	286
1200	Solid brass, 21 GPM at 15′ head, 1/3 HP		5	1.600		263	78		341	410

22 14 29.16 Submersible Sump Pumps

		Crew	Daily Output	Labor-Hours	Unit	Material	Labor	Equipment	Total	Total Incl O&P
0010	**SUBMERSIBLE SUMP PUMPS**									
7000	Sump pump, automatic									
7100	Plastic, 1-1/4″ discharge, 1/4 HP	1 Plum	6	1.333	Ea.	136	65		201	250
7140	1/3 HP		5	1.600		163	78		241	300
7160	1/2 HP		5	1.600		201	78		279	340
7180	1-1/2″ discharge, 1/2 HP		4	2		223	97.50		320.50	395
7500	Cast iron, 1-1/4″ discharge, 1/4 HP		6	1.333		156	65		221	272
7540	1/3 HP		6	1.333		184	65		249	305
7560	1/2 HP		5	1.600		222	78		300	365
9000	Minimum labor/equipment charge		4	2	Job		97.50		97.50	151

22 15 General Service Compressed-Air Systems

22 15 13 – General Service Compressed-Air Piping

22 15 13.10 Compressor Accessories

		Crew	Daily Output	Labor-Hours	Unit	Material	Labor	Equipment	Total	Total Incl O&P
0010	**COMPRESSOR ACCESSORIES**									
1700	Refrigerated air dryers with ambient air filters									
1710	10 CFM	Q-5	8	2	Ea.	700	89		789	910
1720	25 CFM		6.60	2.424		1,875	108		1,983	2,225
1730	50 CFM		6.20	2.581		2,200	115		2,315	2,600
1740	75 CFM		5.80	2.759		2,625	123		2,748	3,075
1750	100 CFM		5.60	2.857		2,875	127		3,002	3,350
4000	Couplers, air line, sleeve type									
4010	Female, connection size NPT									
4020	1/4″	1 Stpi	38	.211	Ea.	6.30	10.40		16.70	23
4030	3/8″		36	.222		7.10	10.95		18.05	25
4040	1/2″		35	.229		14.40	11.30		25.70	33.50
4050	3/4″		34	.235		14.40	11.60		26	34
4100	Male									
4110	1/4″	1 Stpi	38	.211	Ea.	6.35	10.40		16.75	23
4120	3/8″		36	.222		7.10	10.95		18.05	25
4130	1/2″		35	.229		14.40	11.30		25.70	33.50

22 15 General Service Compressed-Air Systems

22 15 13 – General Service Compressed-Air Piping

22 15 13.10 Compressor Accessories		Crew	Daily Output	Labor-Hours	Unit	Material	2009 Bare Costs Labor	Equipment	Total	Total Incl O&P
4140	3/4"	1 Stpi	34	.235	Ea.	14.40	11.60		26	34
4150	Coupler, combined male and female halves									
4160	1/2"	1 Stpi	17	.471	Ea.	29	23		52	67.50
4170	3/4"	"	15	.533	"	29	26.50		55.50	72.50

22 15 19 – General Service Packaged Air Compressors and Receivers

22 15 19.10 Air Compressors		Crew	Daily Output	Labor-Hours	Unit	Material	Labor	Equipment	Total	Total Incl O&P
0010	**AIR COMPRESSORS**									
5250	Air, reciprocating air cooled, splash lubricated, tank mounted									
5300	Single stage, 1 phase, 140 psi									
5303	1/2 HP, 30 gal tank	1 Stpi	3	2.667	Ea.	1,250	132		1,382	1,575
5305	3/4 HP, 30 gal tank		2.60	3.077		1,275	152		1,427	1,625
5307	1 HP, 30 gal tank		2.20	3.636		1,625	179		1,804	2,075
5309	2 HP, 30 gal tank	Q-5	4	4		2,075	178		2,253	2,575
5310	3 HP, 30 gal tank		3.60	4.444		2,500	197		2,697	3,050
5314	3 HP, 60 gal tank		3.50	4.571		3,200	203		3,403	3,825
5320	5 HP, 60 gal tank		3.20	5		3,275	222		3,497	3,950
5330	5 HP, 80 gal tank		3	5.333		3,550	237		3,787	4,300
5340	7.5 HP, 80 gal tank		2.60	6.154		5,725	273		5,998	6,700
5600	2 stage pkg., 3 phase									
5650	6 CFM at 125 psi 1-1/2 HP, 60 gal tank	Q-5	3	5.333	Ea.	2,750	237		2,987	3,400
5670	10.9 CFM at 125 psi, 3 HP, 80 gal tank		1.50	10.667		3,375	475		3,850	4,450
5680	38.7 CFM at 125 psi, 10 HP, 120 gal tank		.60	26.667		5,375	1,175		6,550	7,725
5690	105 CFM at 125 psi, 25 HP, 250 gal tank	Q-6	.60	40		11,000	1,850		12,850	15,100
5800	With single stage pump									
5850	8.3 CFM at 125 psi, 2 HP, 80 gal tank	Q-6	3.50	6.857	Ea.	3,225	315		3,540	4,025
5860	38.7 CFM at 125 psi, 10 HP, 120 gal tank	"	.90	26.667	"	5,800	1,225		7,025	8,275
6000	Reciprocating, 2 stage, tank mtd, 3 Ph., Cap. rated @175 PSIG									
6050	Pressure lubricated, hvy duty, 9.7 CFM, 3 HP, 120 Gal. tank	Q-5	1.30	12.308	Ea.	5,025	545		5,570	6,375
6054	5 CFM, 1-1/2 HP, 80 gal tank		2.80	5.714		2,925	254		3,179	3,600
6056	6.4 CFM, 2 HP, 80 gal tank		2	8		3,225	355		3,580	4,075
6058	8.1 CFM, 3 HP, 80 gal tank		1.70	9.412		3,375	420		3,795	4,375
6059	14.8 CFM, 5 HP, 80 gal tank		1	16		3,550	710		4,260	5,025
6060	16.5 CFM, 5 HP, 120 gal. tank		1	16		5,075	710		5,785	6,700
6063	13 CFM, 6 HP, 80 gal tank		.90	17.778		5,475	790		6,265	7,250
6066	19.8 CFM, 7.5 HP, 80 gal tank		.80	20		5,775	890		6,665	7,725
6070	25.8 CFM, 7-1/2 HP, 120 gal. tank		.80	20		7,575	890		8,465	9,700
6078	34.8 CFM, 10 HP, 80 gal. tank		.70	22.857		7,100	1,025		8,125	9,375
6080	34.8 CFM, 10 HP, 120 gal. tank		.60	26.667		7,450	1,175		8,625	10,000
6090	53.7 CFM, 15 HP, 120 gal. tank	Q-6	.80	30		9,350	1,375		10,725	12,500
6100	76.7 CFM, 20 HP, 120 gal. tank		.70	34.286		11,600	1,575		13,175	15,300
6104	76.7 CFM, 20 HP, 240 gal tank		.68	35.294		12,600	1,625		14,225	16,400
6110	90.1 CFM, 25 HP, 120 gal. tank		.63	38.095		12,100	1,750		13,850	16,000
6120	101 CFM, 30 HP, 120 gal. tank		.57	42.105		13,000	1,950		14,950	17,300
6130	101 CFM, 30 HP, 250 gal. tank		.52	46.154		14,100	2,125		16,225	18,800
6200	Oil-less, 13.6 CFM, 5 HP, 120 gal. tank	Q-5	.88	18.182		14,500	810		15,310	17,200
6210	13.6 CFM, 5 HP, 250 gal. tank		.80	20		15,900	890		16,790	18,900
6220	18.2 CFM, 7.5 HP, 120 gal. tank		.73	21.918		15,000	975		15,975	18,000
6230	18.2 CFM, 7.5 HP, 250 gal. tank		.67	23.881		16,100	1,050		17,150	19,400
6250	30.5 CFM, 10 HP, 120 gal. tank		.57	28.070		17,400	1,250		18,650	21,100
6260	30.5 CFM, 10 HP, 250 gal. tank		.53	30.189		18,500	1,350		19,850	22,500
6270	41.3 CFM, 15 HP, 120 gal. tank	Q-6	.70	34.286		19,000	1,575		20,575	23,400
6280	41.3 CFM, 15 HP, 250 gal. tank	"	.67	35.821		20,100	1,650		21,750	24,700

22 31 Domestic Water Softeners

22 31 13 – Residential Domestic Water Softeners

22 31 13.10 Residential Water Softeners		Crew	Daily Output	Labor-Hours	Unit	Material	2009 Bare Costs Labor	Equipment	Total	Total Incl O&P
0010	RESIDENTIAL WATER SOFTENERS									
7350	Water softener, automatic, to 30 grains per gallon	2 Plum	5	3.200	Ea.	410	156		566	690
7400	To 100 grains per gallon	"	4	4	"	680	195		875	1,050

22 31 16 – Commercial Domestic Water Softeners

22 31 16.10 Water Softeners		Crew	Daily Output	Labor-Hours	Unit	Material	Labor	Equipment	Total	Total Incl O&P
0010	WATER SOFTENERS									
5800	Softener systems, automatic, intermediate sizes									
5820	available, may be used in multiples.									
6000	Hardness capacity between regenerations and flow									
6100	150,000 grains, 37 GPM cont., 51 GPM peak	Q-1	1.20	13.333	Ea.	3,600	585		4,185	4,850
6200	300,000 grains, 81 GPM cont., 113 GPM peak		1	16		7,800	700		8,500	9,675
6300	750,000 grains, 160 GPM cont., 230 GPM peak		.80	20		10,200	880		11,080	12,600
6400	900,000 grains, 185 GPM cont., 270 GPM peak	↓	.70	22.857	↓	16,400	1,000		17,400	19,600

22 32 Domestic Water Filtration Equipment

22 32 19 – Domestic-Water Off-Floor Cartridge Filters

22 32 19.10 Water Filters		Crew	Daily Output	Labor-Hours	Unit	Material	Labor	Equipment	Total	Total Incl O&P
0010	**WATER FILTERS**, Purification and treatment.									
1000	Cartridge style, dirt and rust type	1 Plum	12	.667	Ea.	160	32.50		192.50	227
1200	Replacement cartridge		32	.250		15.20	12.20		27.40	35.50
1600	Taste and odor type		12	.667		196	32.50		228.50	266
1700	Replacement cartridge		32	.250		35.50	12.20		47.70	58
3000	Central unit, dirt/rust/odor/taste/scale		4	2		1,025	97.50		1,122.50	1,275
3100	Replacement cartridge, standard		20	.400		201	19.50		220.50	252
3600	Replacement cartridge, heavy duty	↓	20	.400	↓	226	19.50		245.50	278
8000	Commercial, fully automatic or push button automatic									
8200	Iron removal, 660 GPH, 1" pipe size	Q-1	1.50	10.667	Ea.	2,225	470		2,695	3,175
8240	1500 GPH, 1-1/4" pipe size		1	16		3,750	700		4,450	5,225
8280	2340 GPH, 1-1/2" pipe size		.80	20		4,125	880		5,005	5,875
8320	3420 GPH, 2" pipe size		.60	26.667		7,600	1,175		8,775	10,200
8360	4620 GPH, 2-1/2" pipe size		.50	32		12,100	1,400		13,500	15,500
8500	Neutralizer for acid water, 780 GPH, 1" pipe size		1.50	10.667		2,150	470		2,620	3,100
8540	1140 GPH, 1-1/4" pipe size		1	16		2,425	700		3,125	3,775
8580	1740 GPH, 1-1/2" pipe size		.80	20		3,500	880		4,380	5,225
8620	2520 GPH, 2" pipe size		.60	26.667		4,550	1,175		5,725	6,850
8660	3480 GPH, 2-1/2" pipe size		.50	32		7,600	1,400		9,000	10,500
8800	Sediment removal, 780 GPH, 1" pipe size		1.50	10.667		2,050	470		2,520	2,975
8840	1140 GPH, 1-1/4" pipe size		1	16		2,450	700		3,150	3,800
8880	1740 GPH, 1-1/2" pipe size		.80	20		3,250	880		4,130	4,925
8920	2520 GPH, 2" pipe size		.60	26.667		4,700	1,175		5,875	7,000
8960	3480 GPH, 2-1/2" pipe size		.50	32		7,400	1,400		8,800	10,300
9200	Taste and odor removal, 660 GPH, 1" pipe size		1.50	10.667		2,875	470		3,345	3,875
9240	1500 GPH, 1-1/4" pipe size		1	16		4,925	700		5,625	6,525
9280	2340 GPH, 1-1/2" pipe size		.80	20		5,600	880		6,480	7,525
9320	3420 GPH, 2" pipe size		.60	26.667		8,625	1,175		9,800	11,300
9360	4620 GPH, 2-1/2" pipe size	↓	.50	32	↓	13,300	1,400		14,700	16,800

22 33 Electric Domestic Water Heaters

22 33 13 – Instantaneous Electric Domestic Water Heaters

22 33 13.10 Hot Water Dispensers		Crew	Daily Output	Labor-Hours	Unit	2009 Bare Costs Material	Labor	Equipment	Total	Total Incl O&P
0010	HOT WATER DISPENSERS									
0160	Commercial, 100 cup, 11.3 amp	1 Plum	14	.571	Ea.	400	28		428	485
3180	Household, 60 cup	"	14	.571	"	190	28		218	251

22 33 13.20 Instantaneous Electric Point-Of-use Water Heaters		Crew	Daily Output	Labor-Hours	Unit	Material	Labor	Equipment	Total	Total Incl O&P
0010	INSTANTANEOUS ELECTRIC POINT-OF-USE WATER HEATERS									
8965	Point of use, electric, glass lined									
8969	Energy saver									
8970	2.5 gal. single element	1 Plum	2.80	2.857	Ea.	261	139		400	505
8971	4 gal. single element		2.80	2.857		276	139		415	520
8974	6 gal. single element		2.50	3.200		295	156		451	565
8975	10 gal. single element		2.50	3.200		315	156		471	585
8976	15 gal. single element		2.40	3.333		330	163		493	610
8977	20 gal. single element		2.40	3.333		360	163		523	645
8978	30 gal. single element		2.30	3.478		400	170		570	705
8979	40 gal. single element		2.20	3.636		510	177		687	840
8988	Commercial (ASHRAE energy std. 90)									
8989	6 gallon	1 Plum	2.50	3.200	Ea.	560	156		716	860
8990	10 gallon		2.50	3.200		600	156		756	900
8991	15 gallon		2.40	3.333		635	163		798	950
8992	20 gallon		2.40	3.333		680	163		843	1,000
8993	30 gallon		2.30	3.478		1,425	170		1,595	1,850
8995	Under the sink, copper, w/bracket									
8996	2.5 gallon	1 Plum	4	2	Ea.	249	97.50		346.50	425
9000	Minimum labor/equipment charge	"	1.75	4.571	Job		223		223	345

22 33 30 – Residential, Electric Domestic Water Heaters

22 33 30.13 Residential, Small-Capacity Elec. Water Heaters		Crew	Daily Output	Labor-Hours	Unit	Material	Labor	Equipment	Total	Total Incl O&P
0010	RESIDENTIAL, SMALL-CAPACITY ELECTRIC DOMESTIC WATER HEATERS									
1000	Residential, electric, glass lined tank, 5 yr, 10 gal., single element	1 Plum	2.30	3.478	Ea.	315	170		485	610
1040	20 gallon, single element		2.20	3.636		360	177		537	670
1060	30 gallon, double element		2.20	3.636		465	177		642	790
1080	40 gallon, double element		2	4		500	195		695	845
1100	52 gallon, double element		2	4		560	195		755	915
1120	66 gallon, double element		1.80	4.444		755	217		972	1,175
1140	80 gallon, double element		1.60	5		840	244		1,084	1,300
1180	120 gallon, double element		1.40	5.714		1,175	279		1,454	1,725

22 33 33 – Light-Commercial Electric Domestic Water Heaters

22 33 33.10 Commercial Electric Water Heaters		Crew	Daily Output	Labor-Hours	Unit	Material	Labor	Equipment	Total	Total Incl O&P
0010	COMMERCIAL ELECTRIC WATER HEATERS									
4000	Commercial, 100° rise. NOTE: for each size tank, a range of									
4010	heaters between the ones shown are available									
4020	Electric									
4100	5 gal., 3 kW, 12 GPH, 208V	1 Plum	2	4	Ea.	2,675	195		2,870	3,250
4120	10 gal., 6 kW, 25 GPH, 208V		2	4		2,975	195		3,170	3,575
4140	50 gal., 9 kW, 37 GPH, 208V		1.80	4.444		4,075	217		4,292	4,800
4160	50 gal., 36 kW, 148 GPH, 208V		1.80	4.444		6,325	217		6,542	7,275
4180	80 gal., 12 kW, 49 GPH, 208V		1.50	5.333		5,025	260		5,285	5,925
4200	80 gal., 36 kW, 148 GPH, 208V		1.50	5.333		7,050	260		7,310	8,150
4220	100 gal., 36 kW, 148 GPH, 208V		1.20	6.667		7,350	325		7,675	8,575
4240	120 gal., 36 kW, 148 GPH, 208V		1.20	6.667		7,650	325		7,975	8,900
4260	150 gal., 15 kW , 61 GPH, 480V		1	8		18,000	390		18,390	20,400
4280	150 gal., 120 kW, 490 GPH, 480V		1	8		25,300	390		25,690	28,400

22 33 Electric Domestic Water Heaters

22 33 33 – Light-Commercial Electric Domestic Water Heaters

22 33 33.10 Commercial Electric Water Heaters		Crew	Daily Output	Labor-Hours	Unit	2009 Bare Costs Material	Labor	Equipment	Total	Total Incl O&P
4300	200 gal., 15 kW, 61 GPH, 480V	Q-1	1.70	9.412	Ea.	19,700	415		20,115	22,200
4320	200 gal., 120 kW , 490 GPH, 480V		1.70	9.412		26,900	415		27,315	30,100
4340	250 gal., 15 kW, 61 GPH, 480V		1.50	10.667		20,200	470		20,670	23,000
4360	250 gal., 150 kW, 615 GPH, 480V		1.50	10.667		29,500	470		29,970	33,100
4380	300 gal., 30 kW, 123 GPH, 480V		1.30	12.308		21,900	540		22,440	24,900
4400	300 gal., 180 kW, 738 GPH, 480V		1.30	12.308		33,300	540		33,840	37,400
4420	350 gal., 30 kW, 123 GPH, 480V		1.10	14.545		24,000	640		24,640	27,400
4440	350 gal, 180 kW, 738 GPH, 480V		1.10	14.545		32,600	640		33,240	36,900
4460	400 gal., 30 kW, 123 GPH, 480V		1	16		27,100	700		27,800	30,900
4480	400 gal., 210 kW, 860 GPH, 480V		1	16		37,600	700		38,300	42,500
4500	500 gal., 30 kW, 123 GPH, 480V		.80	20		31,200	880		32,080	35,700
4520	500 gal., 240 kW, 984 GPH, 480V	↓	.80	20		47,300	880		48,180	53,500
4540	600 gal., 30 kW, 123 GPH, 480V	Q-2	1.20	20		36,100	910		37,010	41,100
4560	600 gal., 300 kW, 1230 GPH, 480V		1.20	20		54,000	910		54,910	61,000
4580	700 gal., 30 kW, 123 GPH, 480V		1	24		39,100	1,100		40,200	44,700
4600	700 gal., 300 kW, 1230 GPH, 480V		1	24		55,500	1,100		56,600	62,500
4620	800 gal., 60 kW, 245 GPH, 480V		.90	26.667		40,500	1,225		41,725	46,500
4640	800 gal., 300 kW, 1230 GPH, 480V		.90	26.667		57,500	1,225		58,725	65,500
4660	1000 gal., 60 kW, 245 GPH, 480V		.70	34.286		44,400	1,550		45,950	51,000
4680	1000 gal., 480 kW, 1970 GPH, 480V		.70	34.286		73,500	1,550		75,050	83,500
4700	1250 gal., 60 kW, 245 GPH, 480V		.60	40		50,000	1,825		51,825	58,000
4720	1250 gal., 480 kW, 1970 GPH, 480V		.60	40		76,000	1,825		77,825	87,000
4740	1500 gal., 60 kW, 245 GPH, 480V		.50	48		66,000	2,175		68,175	76,000
4760	1500 gal., 480 kW, 1970 GPH, 480V	↓	.50	48		106,500	2,175		108,675	120,500
5400	Modulating step control, 2-5 steps	1 Elec	5.30	1.509		1,975	71		2,046	2,275
5440	6-10 steps		3.20	2.500		2,550	118		2,668	3,000
5460	11-15 steps		2.70	2.963		3,100	139		3,239	3,625
5480	16-20 steps	↓	1.60	5	↓	3,600	235		3,835	4,300

22 34 Fuel-Fired Domestic Water Heaters

22 34 30 – Residential Gas Domestic Water Heaters

22 34 30.13 Residential, Atmos, Gas Domestic Wtr Heaters

		Crew	Daily Output	Labor-Hours	Unit	Material	Labor	Equipment	Total	Total Incl O&P
0010	RESIDENTIAL, ATMOSPHERIC, GAS DOMESTIC WATER HEATERS									
2000	Gas fired, foam lined tank, 10 yr, vent not incl.,									
2040	30 gallon	1 Plum	2	4	Ea.	630	195		825	990
2060	40 gallon		1.90	4.211		895	205		1,100	1,300
2080	50 gallon		1.80	4.444		935	217		1,152	1,350
2100	75 gallon		1.50	5.333		1,225	260		1,485	1,750
2120	100 gallon	↓	1.30	6.154	↓	1,525	300		1,825	2,150

22 34 36 – Commercial Gas Domestic Water Heaters

22 34 36.13 Commercial, Atmos., Gas Domestic Water Htrs.

		Crew	Daily Output	Labor-Hours	Unit	Material	Labor	Equipment	Total	Total Incl O&P
0010	COMMERCIAL, ATMOSPHERIC, GAS DOMESTIC WATER HEATERS									
6000	Gas fired, flush jacket, std. controls, vent not incl.									
6040	75 MBH input, 73 GPH	1 Plum	1.40	5.714	Ea.	1,725	279		2,004	2,325
6060	98 MBH input, 95 GPH		1.40	5.714		3,975	279		4,254	4,800
6080	120 MBH input, 110 GPH		1.20	6.667		4,575	325		4,900	5,525
6100	120 MBH input, 115 GPH		1.10	7.273		4,775	355		5,130	5,800
6120	140 MBH input, 130 GPH		1	8		6,050	390		6,440	7,275
6140	155 MBH input, 150 GPH		.80	10		5,900	490		6,390	7,225
6160	180 MBH input, 170 GPH	↓	.70	11.429	↓	6,250	555		6,805	7,750

22 34 Fuel-Fired Domestic Water Heaters

22 34 36 – Commercial Gas Domestic Water Heaters

22 34 36.13 Commercial, Atmos., Gas Domestic Water Htrs.		Crew	Daily Output	Labor-Hours	Unit	2009 Bare Costs Material	Labor	Equipment	Total	Total Incl O&P
6180	200 MBH input, 192 GPH	1 Plum	.60	13.333	Ea.	6,275	650		6,925	7,925
6200	250 MBH input, 245 GPH		.50	16		6,700	780		7,480	8,575
6220	260 MBH input, 250 GPH	Q-1	.80	20		7,125	880		8,005	9,200
6240	360 MBH input, 360 GPH		.80	20		8,400	880		9,280	10,600
6260	500 MBH input, 480 GPH		.70	22.857		11,800	1,000		12,800	14,600
6280	725 MBH input, 690 GPH		.60	26.667		17,700	1,175		18,875	21,200
6900	For low water cutoff, add	1 Plum	8	1		275	49		324	375
6960	For bronze body hot water circulator, add	"	4	2		1,575	97.50		1,672.50	1,875

22 34 36.45 Commercial Packaged Water Heater Systems		Crew	Daily Output	Labor-Hours	Unit	Material	Labor	Equipment	Total	Total Incl O&P
0010	**COMMERCIAL PACKAGED WATER HEATER SYSTEMS**									
1000	Car wash package, continuous duty, high recovery, gas fired									
1040	100° rise, 180 MBH input, 174 GPH	1 Plum	3	2.667	Ea.	4,575	130		4,705	5,250
1060	280 MBH input, 270 GPH		2.50	3.200		4,900	156		5,056	5,625
1080	400 MBH input, 386 GPH		2.50	3.200		5,625	156		5,781	6,450
1100	480 MBH input, 464 GPH		2	4		5,950	195		6,145	6,850
1120	605 MBH input, 584 GPH		1.50	5.333		6,775	260		7,035	7,875
1140	700 MBH input, 676 GPH		1	8		7,050	390		7,440	8,350
1160	1000 MBH input, 966 GPH	Q-1	1.60	10		7,850	440		8,290	9,325
1180	1200 MBH input, 1159 GPH		1.40	11.429		10,700	500		11,200	12,500
1200	1400 MBH input, 1353 GPH		1.20	13.333		11,600	585		12,185	13,700
3000	Combination dishwasher & general purpose, 2 temp. gas fired									
3040	124 GPH @ 140° rise; 434 GPH @ 40° rise	1 Plum	2	4	Ea.	4,675	195		4,870	5,450
3060	193 GPH @ 140° rise; 677 GPH @ 40° rise		1.60	5		5,350	244		5,594	6,250
3080	276 GPH @ 140° rise; 969 GPH @ 40° rise		1.40	5.714		6,075	279		6,354	7,125
3100	330 GPH @ 140° rise; 1160 GPH @ 40°rise		1.20	6.667		6,450	325		6,775	7,600
3120	417 GPH @ 140° rise; 1460 GPH @ 40°rise		1	8		7,275	390		7,665	8,625
3140	481 GPH @ 140° rise; 1685 GPH @ 40°rise	Q-1	1.40	11.429		8,400	500		8,900	10,000
3160	688 GPH @ 140° rise; 2410 GPH @ 40°rise		1.20	13.333		9,200	585		9,785	11,000
3180	828 GPH @ 140° rise; 2790 GPH @ 40°rise		1	16		12,100	700		12,800	14,400
3200	965 GPH @ 140° rise; 3370 GPH @ 40°rise		.80	20		13,000	880		13,880	15,700
3960	For unit base, add					360			360	395
5000	Coin laundry units, gas fired, 100° rise									
5020	Single heater,									
5040	280 MBH input, 270 GPH	1 Plum	1.60	5	Ea.	5,350	244		5,594	6,250
5060	400 MBH input, 386 GPH		1.30	6.154		7,050	300		7,350	8,225
5080	480 MBH input, 464 GPH		1	8		7,425	390		7,815	8,750
5100	605 MBH input, 584 GPH	Q-1	1.40	11.429		8,750	500		9,250	10,400
5120	700 MBH input, 676 GPH		1.20	13.333		9,200	585		9,785	11,000
5140	1000 MBH input, 966 GPH		1	16		10,000	700		10,700	12,100
5160	1200 MBH input, 1159 GPH		.90	17.778		13,000	780		13,780	15,500
5180	1400 MBH input, 1353 GPH		.70	22.857		14,400	1,000		15,400	17,500
6000	Multiple heater									
6040	560 MBH input, 540 GPH	1 Plum	1	8	Ea.	8,475	390		8,865	9,900
6060	800 MBH input, 772 GPH	Q-1	1.30	12.308		11,000	540		11,540	12,900
6080	960 MBH input, 928 GPH		1.20	13.333		12,400	585		12,985	14,500
6100	1210 MBH input, 1168 GPH		1	16		13,900	700		14,600	16,400
6120	1400 MBH input, 1352 GPH		1	16		14,600	700		15,300	17,100
6140	1700 MBH input, 1642 GPH		.80	20		16,600	880		17,480	19,700
6160	2000 MBH input, 1932 GPH		.70	22.857		20,100	1,000		21,100	23,700
6180	2400 MBH input, 2318 GPH	Q-2	.90	26.667		22,600	1,225		23,825	26,700
6200	2600 MBH input, 2512 GPH		.80	30		25,200	1,375		26,575	29,800
6220	2800 MBH input, 2706 GPH		.70	34.286		28,400	1,550		29,950	33,600

22 34 Fuel-Fired Domestic Water Heaters

22 34 36 – Commercial Gas Domestic Water Heaters

22 34 36.45 Commercial Packaged Water Heater Systems		Crew	Daily Output	Labor-Hours	Unit	Material	Labor	Equipment	Total	Total Incl O&P
						2009 Bare Costs				
6800	Standard system sizing is based on									
6820	80% of the washers operating at one time.									

22 34 46 – Oil-Fired Domestic Water Heaters

22 34 46.10 Residential Oil-Fired Water Heaters

		Crew	Daily Output	Labor-Hours	Unit	Material	Labor	Equipment	Total	Total Incl O&P
0010	**RESIDENTIAL OIL-FIRED WATER HEATERS**									
3000	Oil fired, glass lined tank, 5 yr, vent not included, 30 gallon	1 Plum	2	4	Ea.	665	195		860	1,025
3040	50 gallon		1.80	4.444		1,175	217		1,392	1,600
3060	70 gallon		1.50	5.333		1,725	260		1,985	2,300
3080	85 gallon		1.40	5.714		4,750	279		5,029	5,650

22 34 46.20 Commercial Oil-Fired Water Heaters

		Crew	Daily Output	Labor-Hours	Unit	Material	Labor	Equipment	Total	Total Incl O&P
0010	**COMMERCIAL OIL-FIRED WATER HEATERS**									
8000	Oil fired, glass lined, UL listed, std. controls, vent not incl.									
8060	140 gal., 140 MBH input, 134 GPH	Q-1	2.13	7.512	Ea.	14,200	330		14,530	16,100
8080	140 gal., 199 MBH input, 191 GPH		2	8		14,600	350		14,950	16,600
8100	140 gal., 255 MBH input, 247 GPH		1.60	10		15,000	440		15,440	17,200
8120	140 gal., 270 MBH input, 259 GPH		1.20	13.333		18,600	585		19,185	21,300
8140	140 gal., 400 MBH input, 384 GPH		1	16		19,000	700		19,700	22,000
8160	140 gal., 540 MBH input, 519 GPH		.96	16.667		19,900	730		20,630	23,000
8180	140 gal., 720 MBH input, 691 GPH		.92	17.391		20,300	765		21,065	23,500
8200	221 gal., 300 MBH input, 288 GPH		.88	18.182		26,800	800		27,600	30,700
8220	221 gal., 600 MBH input, 576 GPH		.86	18.605		29,900	815		30,715	34,200
8240	221 gal., 800 MBH input, 768 GPH		.82	19.512		30,100	855		30,955	34,400
8260	201 gal., 1000 MBH input, 960 GPH	Q-2	1.26	19.048		30,800	865		31,665	35,300
8280	201 gal., 1250 MBH input, 1200 GPH		1.22	19.672		31,100	895		31,995	35,600
8300	201 gal., 1500 MBH input, 1441 GPH		1.16	20.690		33,900	940		34,840	38,700
8320	411 gal., 600 MBH input, 576 GPH		1.12	21.429		34,200	975		35,175	39,100
8340	411 gal., 800 MBH input, 768 GPH		1.08	22.222		34,300	1,000		35,300	39,300
8360	411 gal., 1000 MBH input, 960 GPH		1.04	23.077		36,200	1,050		37,250	41,400
8380	411 gal., 1250 MBH input, 1200 GPH		.98	24.490		37,000	1,125		38,125	42,400
8400	397 gal., 1500 MBH input, 1441 GPH		.92	26.087		37,400	1,175		38,575	43,000
8420	397 gal., 1750 MBH input, 1681 GPH		.86	27.907		40,500	1,275		41,775	46,500
8430	397 gal., 2000 MBH input, 1921 GPH		.82	29.268		43,800	1,325		45,125	50,500
8440	375 gal., 2250 MBH input, 2161 GPH		.76	31.579		45,000	1,425		46,425	51,500
8450	375 gal., 2500 MBH input, 2401 GPH		.82	29.268		46,900	1,325		48,225	53,500
8900	For low water cutoff, add	1 Plum	8	1		275	49		324	375
8960	For bronze body hot water circulator, add	"	4	2		595	97.50		692.50	805

22 35 Domestic Water Heat Exchangers

22 35 30 – Water Heating by Steam

22 35 30.10 Water Heating Transfer Package

		Crew	Daily Output	Labor-Hours	Unit	Material	Labor	Equipment	Total	Total Incl O&P
0010	**WATER HEATING TRANSFER PACKAGE**, Complete controls,									
0020	expansion tank, converter, air separator									
1000	Hot water, 180° F enter, 200° F leaving, 15# steam									
1010	One pump system, 28 GPM	Q-6	.75	32	Ea.	15,600	1,475		17,075	19,500
1020	35 GPM		.70	34.286		17,200	1,575		18,775	21,400
1040	55 GPM		.65	36.923		19,500	1,700		21,200	24,200
1060	130 GPM		.55	43.636		24,200	2,000		26,200	29,700
1080	255 GPM		.40	60		32,900	2,775		35,675	40,500
1100	550 GPM		.30	80		41,600	3,675		45,275	51,500
1120	800 GPM		.25	96		49,400	4,425		53,825	61,500

22 35 Domestic Water Heat Exchangers

22 35 30 – Water Heating by Steam

22 35 30.10 Water Heating Transfer Package		Crew	Daily Output	Labor-Hours	Unit	2009 Bare Costs Material	Labor	Equipment	Total	Total Incl O&P
1220	Two pump system, 28 GPM	Q-6	.70	34.286	Ea.	21,300	1,575		22,875	25,900
1240	35 GPM		.65	36.923		25,300	1,700		27,000	30,500
1260	55 GPM		.60	40		25,800	1,850		27,650	31,300
1280	130 GPM		.50	48		33,100	2,200		35,300	39,800
1300	255 GPM		.35	68.571		44,600	3,150		47,750	54,000
1320	550 GPM		.25	96		56,000	4,425		60,425	68,500
1340	800 GPM		.20	120		65,000	5,525		70,525	80,000

22 41 Residential Plumbing Fixtures

22 41 06 – Plumbing Fixtures General

22 41 06.10 Plumbing Fixture Notes

		Crew	Daily Output	Labor-Hours	Unit	Material	Labor	Equipment	Total	Total Incl O&P
0010	**PLUMBING FIXTURE NOTES**, Incl. trim fittings unless otherwise noted									
0080	For rough-in, supply, waste, and vent, see add for each type									
0122	For electric water coolers, see Div. 22 47 16.10									
0160	For color, unless otherwise noted, add				Ea.	20%				

22 41 13 – Residential Water Closets, Urinals, and Bidets

22 41 13.10 Bidets

		Crew	Daily Output	Labor-Hours	Unit	Material	Labor	Equipment	Total	Total Incl O&P
0010	**BIDETS**									
0180	Vitreous china, with trim on fixture	Q-1	5	3.200	Ea.	650	140		790	935
0200	With trim for wall mounting		5	3.200		650	140		790	935
9600	For rough-in, supply, waste and vent, add		1.78	8.989		258	395		653	895

22 41 13.40 Water Closets

		Crew	Daily Output	Labor-Hours	Unit	Material	Labor	Equipment	Total	Total Incl O&P
0010	**WATER CLOSETS** R224000-30									
0022	For seats, see Div. 22 42 13.44									
0032	For automatic flush, see Div. 22 42 39.10 0972									
0150	Tank type, vitreous china, incl. seat, supply pipe w/stop									
0200	Wall hung									
0400	Two piece, close coupled	Q-1	5.30	3.019	Ea.	555	132		687	815
0960	For rough-in, supply, waste, vent and carrier		2.73	5.861		630	257		887	1,100
1000	Floor mounted, one piece		5.30	3.019		530	132		662	790
1020	One piece, low profile		5.30	3.019		400	132		532	650
1050	One piece combination		5.30	3.019		795	132		927	1,075
1100	Two piece, close coupled		5.30	3.019		184	132		316	410
1960	For color, add					30%				
1980	For rough-in, supply, waste and vent	Q-1	3.05	5.246	Ea.	275	230		505	655

22 41 13.44 Toilet Seats

		Crew	Daily Output	Labor-Hours	Unit	Material	Labor	Equipment	Total	Total Incl O&P
0010	**TOILET SEATS**									
0100	Molded composition, white									
0150	Industrial, w/o cover, open front, regular bowl	1 Plum	24	.333	Ea.	21	16.25		37.25	48
0200	With self-sustaining hinge		24	.333		22.50	16.25		38.75	49.50
0220	With self-sustaining check hinge		24	.333		22.50	16.25		38.75	49.50
0240	Extra heavy, with check hinge		24	.333		21	16.25		37.25	48
0260	Elongated bowl, same price									
0300	Junior size, w/o cover, open front	1 Plum	24	.333	Ea.	37	16.25		53.25	65.50
0320	Regular primary bowl, open front		24	.333		36	16.25		52.25	64.50
0340	Regular baby bowl, open front, check hinge		24	.333		37	16.25		53.25	65.50
0380	Open back & front, w/o cover, reg. or elongated bowl		24	.333		29	16.25		45.25	57
0400	Residential									
0420	Regular bowl, w/cover, closed front	1 Plum	24	.333	Ea.	29	16.25		45.25	56.50
0440	Open front	"	24	.333	"	26	16.25		42.25	54

22 41 Residential Plumbing Fixtures

22 41 13 – Residential Water Closets, Urinals, and Bidets

22 41 13.44 Toilet Seats		Crew	Daily Output	Labor-Hours	Unit	Material	2009 Bare Costs Labor	Equipment	Total	Total Incl O&P
0460	Elongated bowl, add					25%				
0500	Self-raising hinge, w/o cover, open front									
0520	Regular bowl	1 Plum	24	.333	Ea.	101	16.25		117.25	136
0540	Elongated bowl	"	24	.333	"	21	16.25		37.25	48
0700	Molded wood, white, with cover									
0720	Closed front, regular bowl, square back	1 Plum	24	.333	Ea.	9.90	16.25		26.15	36
0740	Extended back		24	.333		13.80	16.25		30.05	40
0780	Elongated bowl, square back		24	.333		13.95	16.25		30.20	40.50
0800	Open front		24	.333		16.45	16.25		32.70	43
0850	Decorator styles									
0890	Vinyl top, patterned	1 Plum	24	.333	Ea.	20.50	16.25		36.75	47.50
0900	Vinyl padded, plain colors, regular bowl		24	.333		20.50	16.25		36.75	47.50
0930	Elongated bowl		24	.333		21.50	16.25		37.75	48.50
1000	Solid plastic, white									
1030	Industrial, w/o cover, open front, regular bowl	1 Plum	24	.333	Ea.	26	16.25		42.25	54
1080	Extra heavy, concealed check hinge		24	.333		17.25	16.25		33.50	44
1100	Self-sustaining hinge		24	.333		21	16.25		37.25	48
1150	Elongated bowl		24	.333		29	16.25		45.25	57
1170	Concealed check		24	.333		17.25	16.25		33.50	44
1190	Self-sustaining hinge, concealed check		24	.333		47	16.25		63.25	77
1220	Residential, with cover, closed front, regular bowl		24	.333		41.50	16.25		57.75	70.50
1240	Elongated bowl		24	.333		50	16.25		66.25	80.50
1260	Open front, regular bowl		24	.333		36.50	16.25		52.75	65
1280	Elongated bowl		24	.333		44.50	16.25		60.75	74

22 41 16 – Residential Lavatories and Sinks

22 41 16.10 Lavatories

		Crew	Daily Output	Labor-Hours	Unit	Material	Labor	Equipment	Total	Total Incl O&P
0010	**LAVATORIES**, With trim, white unless noted otherwise R224000-30									
0500	Vanity top, porcelain enamel on cast iron									
0600	20″ x 18″	Q-1	6.40	2.500	Ea.	231	110		341	425
0640	33″ x 19″ oval		6.40	2.500		505	110		615	725
0680	20″ x 17″ oval		6.40	2.500		156	110		266	340
0720	19″ round		6.40	2.500		218	110		328	410
0760	20″ x 12″ triangular bowl		6.40	2.500		234	110		344	430
0860	For color, add					25%				
1000	Cultured marble, 19″ x 17″, single bowl	Q-1	6.40	2.500	Ea.	169	110		279	355
1040	25″ x 19″, single bowl		6.40	2.500		180	110		290	370
1080	31″ x 19″, single bowl		6.40	2.500		193	110		303	380
1120	25″ x 22″, single bowl		6.40	2.500		185	110		295	375
1160	37″ x 22″, single bowl		6.40	2.500		216	110		326	405
1200	49″ x 22″, single bowl		6.40	2.500		294	110		404	495
1580	For color, same price									
1900	Stainless steel, self-rimming, 25″ x 22″, single bowl, ledge	Q-1	6.40	2.500	Ea.	330	110		440	530
1960	17″ x 22″, single bowl		6.40	2.500		320	110		430	520
2040	18-3/4″ round		6.40	2.500		555	110		665	780
2600	Steel, enameled, 20″ x 17″, single bowl		5.80	2.759		159	121		280	365
2660	19″ round		5.80	2.759		156	121		277	360
2720	18″ round		5.80	2.759		137	121		258	340
2860	For color, add					10%				
2900	Vitreous china, 20″ x 16″, single bowl	Q-1	5.40	2.963	Ea.	260	130		390	485
2960	20″ x 17″, single bowl		5.40	2.963		176	130		306	395
3020	19″ round, single bowl		5.40	2.963		174	130		304	395
3080	19″ x 16″, single bowl		5.40	2.963		260	130		390	490

22 41 Residential Plumbing Fixtures

22 41 16 – Residential Lavatories and Sinks

22 41 16.10 Lavatories		Crew	Daily Output	Labor-Hours	Unit	Material	Labor	Equipment	Total	Total Incl O&P
						2009 Bare Costs				
3140	17" x 14", single bowl	Q-1	5.40	2.963	Ea.	214	130		344	435
3200	22" x 13", single bowl		5.40	2.963		267	130		397	495
3560	For color, add					50%				
3580	Rough-in, supply, waste and vent for all above lavatories	Q-1	2.30	6.957	Ea.	224	305		529	720
4000	Wall hung									
4040	Porcelain enamel on cast iron, 16" x 14", single bowl	Q-1	8	2	Ea.	410	88		498	585
4060	18" x 15" single bowl		8	2		330	88		418	495
4120	19" x 17", single bowl		8	2		345	88		433	510
4180	20" x 18", single bowl		8	2		297	88		385	460
4240	22" x 19", single bowl		8	2		545	88		633	735
4580	For color, add					30%				
6000	Vitreous china, 18" x 15", single bowl with backsplash	Q-1	7	2.286	Ea.	231	100		331	410
6060	19" x 17", single bowl		7	2.286		173	100		273	345
6120	24" x 20", single bowl		7	2.286		480	100		580	680
6210	27" x 20", wheelchair type		7	2.286		365	100		465	555
6500	For color, add					30%				
6960	Rough-in, supply, waste and vent for above lavatories	Q-1	1.66	9.639	Ea.	365	425		790	1,050
9000	Minimum labor/equipment charge	1 Plum	3	2.667	Job		130		130	202

22 41 16.30 Sinks

0010	**SINKS**, With faucets and drain R224000-30									
2000	Kitchen, counter top style, P.E. on C.I., 24" x 21" single bowl	Q-1	5.60	2.857	Ea.	272	125		397	495
2100	31" x 22" single bowl		5.60	2.857		278	125		403	500
2200	32" x 21" double bowl		4.80	3.333		370	146		516	635
2300	42" x 21" double bowl		4.80	3.333		500	146		646	775
3000	Stainless steel, self rimming, 19" x 18" single bowl		5.60	2.857		510	125		635	755
3100	25" x 22" single bowl		5.60	2.857		570	125		695	820
3200	33" x 22" double bowl		4.80	3.333		830	146		976	1,125
3300	43" x 22" double bowl		4.80	3.333		965	146		1,111	1,275
3400	22" x 43" triple bowl		4.40	3.636		975	160		1,135	1,325
3500	Corner double bowl each 14" x 16"		4.80	3.333		675	146		821	965
4000	Steel, enameled, with ledge, 24" x 21" single bowl		5.60	2.857		145	125		270	355
4100	32" x 21" double bowl		4.80	3.333		166	146		312	410
4960	For color sinks except stainless steel, add					10%				
4980	For rough-in, supply, waste and vent, counter top sinks	Q-1	2.14	7.477		266	330		596	805
5000	Kitchen, raised deck, P.E. on C.I.									
5100	32" x 21", dual level, double bowl	Q-1	2.60	6.154	Ea.	289	270		559	740
5200	42" x 21", double bowl & disposer well	"	2.20	7.273		805	320		1,125	1,375
5700	For color, add					20%				
5790	For rough-in, supply, waste & vent, sinks	Q-1	1.85	8.649		266	380		646	885

22 41 19 – Residential Bathtubs

22 41 19.10 Baths

0010	**BATHS** R224000-30									
0100	Tubs, recessed porcelain enamel on cast iron, with trim									
0180	48" x 42"	Q-1	4	4	Ea.	1,800	176		1,976	2,250
0220	72" x 36"	"	3	5.333	"	1,850	234		2,084	2,425
0300	Mat bottom									
0340	4'-6" long	Q-1	5	3.200	Ea.	920	140		1,060	1,225
0380	5' long		4.40	3.636		810	160		970	1,125
0420	5'-6" long		4	4		1,250	176		1,426	1,650
0480	Above floor drain, 5' long		4	4		800	176		976	1,150
0560	Corner 48" x 44"		4.40	3.636		1,800	160		1,960	2,225
0750	For color, add					30%				

22 41 19 – Residential Bathtubs

22 41 19.10 Baths		Crew	Daily Output	Labor-Hours	Unit	2009 Bare Costs Material	Labor	Equipment	Total	Total Incl O&P
2000	Enameled formed steel, 4'-6" long	Q-1	5.80	2.759	Ea.	365	121		486	590
2300	Above floor drain, 5' long	"	5.50	2.909	"	335	128		463	570
2350	For color, add					10%				
4000	Soaking, acrylic with pop-up drain, 60" x 32" x 21" deep	Q-1	5.50	2.909	Ea.	875	128		1,003	1,175
4100	60" x 48" x 18-1/2" deep		5	3.200		810	140		950	1,100
4200	72" x 45" x 21-1/2" deep		4.80	3.333		1,025	146		1,171	1,350
4600	Module tub & showerwall surround, molded fiberglass									
4610	5' long x 34" wide x 76" high	Q-1	4	4	Ea.	560	176		736	885
4750	Handicap with 1-1/2" OD grab bar, antiskid bottom									
4760	60" x 32-3/4" x 72" high	Q-1	4	4	Ea.	670	176		846	1,000
4770	60" x 30" x 71" high with molded seat		3.50	4.571		655	201		856	1,025
9600	Rough-in, supply, waste and vent, for all above tubs, add		2.07	7.729		300	340		640	855
9900	Minimum labor/equipment charge		3	5.333	Job		234		234	365

22 41 23 – Residential Shower Receptors and Basins

22 41 23.20 Showers		Crew	Daily Output	Labor-Hours	Unit	Material	Labor	Equipment	Total	Total Incl O&P
0010	**SHOWERS** R224000-30									
1500	Stall, with drain only. Add for valve and door/curtain									
1510	Baked enamel, molded stone receptor, 30" square	Q-1	5.20	3.077	Ea.	730	135		865	1,025
1520	32" square		5	3.200		365	140		505	620
1530	36" square		4.80	3.333		870	146		1,016	1,175
1540	Terrazzo receptor, 32" square		5	3.200		740	140		880	1,025
1560	36" square		4.80	3.333		1,150	146		1,296	1,500
1580	36" corner angle		4.80	3.333		1,275	146		1,421	1,625
1600	For color, add					10%				
3000	Fiberglass, one piece, with 3 walls, 32" x 32" square	Q-1	5.50	2.909	Ea.	460	128		588	710
3100	36" x 36" square	"	5.50	2.909	"	525	128		653	775
3200	Handicap, 1-1/2" O.D. grab bars, nonskid floor									
3210	48" x 34-1/2" x 72" corner seat	Q-1	5	3.200	Ea.	615	140		755	895
3220	60" x 34-1/2" x 72" corner seat		4	4		1,275	176		1,451	1,675
3230	48" x 34-1/2" x 72" fold up seat		5	3.200		2,125	140		2,265	2,575
3250	64" x 65-3/4" x 81-1/2" fold. seat, whlchr.		3.80	4.211		2,775	185		2,960	3,325
4000	Polypropylene, stall only, w/ molded-stone floor, 30" x 30"		2	8		500	350		850	1,100
4100	32" x 32"		2	8		515	350		865	1,100
4200	Rough-in, supply, waste and vent for above showers		2.05	7.805		395	340		735	965

22 41 23.40 Shower System Components		Crew	Daily Output	Labor-Hours	Unit	Material	Labor	Equipment	Total	Total Incl O&P
0010	**SHOWER SYSTEM COMPONENTS**									
4520	Showers, fiberglass receptor only	1 Plum	8	1	Ea.	187	49		236	281
5000	Built-in, head, arm, 2.5 GPM valve		4	2		208	97.50		305.50	380
5200	Head, arm, by-pass, integral stops, handles		3.60	2.222		198	108		306	385
5500	Head, water economizer [G]		24	.333		49.50	16.25		65.75	79.50
5800	Mixing valve, built-in		6	1.333		121	65		186	234
5900	Exposed		6	1.333		675	65		740	845

22 41 36 – Residential Laundry Trays

22 41 36.10 Laundry Sinks		Crew	Daily Output	Labor-Hours	Unit	Material	Labor	Equipment	Total	Total Incl O&P
0010	**LAUNDRY SINKS**, With trim									
0020	Porcelain enamel on cast iron, black iron frame									
0050	24" x 21", single compartment	Q-1	6	2.667	Ea.	415	117		532	635
0100	26" x 21", single compartment	"	6	2.667	"	405	117		522	625
2000	Molded stone, on wall hanger or legs									
2020	22" x 23", single compartment	Q-1	6	2.667	Ea.	143	117		260	340
2100	45" x 21", double compartment	"	5	3.200	"	213	140		353	455

22 41 36 – Residential Laundry Trays

22 41 36.10 Laundry Sinks		Crew	Daily Output	Labor-Hours	Unit	2009 Bare Costs Material	Labor	Equipment	Total	Total Incl O&P
3000	Plastic, on wall hanger or legs									
3020	18″ x 23″, single compartment	Q-1	6.50	2.462	Ea.	99.50	108		207.50	278
3100	20″ x 24″, single compartment		6.50	2.462		133	108		241	315
3200	36″ x 23″, double compartment		5.50	2.909		158	128		286	370
3300	40″ x 24″, double compartment		5.50	2.909		236	128		364	455
5000	Stainless steel, counter top, 22″ x 17″ single compartment		6	2.667		58	117		175	245
5200	33″ x 22″, double compartment		5	3.200		70	140		210	295
9600	Rough-in, supply, waste and vent, for all laundry sinks		2.14	7.477		266	330		596	805
9810	Minimum labor/equipment charge	1 Plum	3	2.667	Job		130		130	202

22 41 39 – Residential Faucets, Supplies and Trim

22 41 39.10 Faucets and Fittings		Crew	Daily Output	Labor-Hours	Unit	Material	Labor	Equipment	Total	Total Incl O&P
0010	**FAUCETS AND FITTINGS**									
0150	Bath, faucets, diverter spout combination, sweat	1 Plum	8	1	Ea.	104	49		153	190
0200	For integral stops, IPS unions, add					109			109	120
0300	Three valve combinations, spout, head, arm, flange, sweat	1 Plum	6	1.333		52.50	65		117.50	159
0400	For integral stops, IPS unions, add				Pr.	102			102	112
0420	Bath, press-bal mix valve w/diverter, spout, shower hd, arm/flange	1 Plum	8	1	Ea.	137	49		186	227
0500	Drain, central lift, 1-1/2″ IPS male		20	.400		40.50	19.50		60	74.50
0600	Trip lever, 1-1/2″ IPS male		20	.400		41	19.50		60.50	75
0700	Pop up, 1-1/2″ IPS male		18	.444		48.50	21.50		70	87
0800	Chain and stopper, 1-1/2″ IPS male		24	.333		32.50	16.25		48.75	60.50
0810	Bidet									
0812	Fitting, over the rim, swivel spray/pop-up drain	1 Plum	8	1	Ea.	175	49		224	268
1000	Kitchen sink faucets, top mount, cast spout		10	.800		55	39		94	121
1100	For spray, add		24	.333		16.15	16.25		32.40	43
1110	For basket strainer w/tail piece, add		24	.333		12.15	16.25		28.40	38.50
1200	Wall type, swing tube spout		10	.800		74.50	39		113.50	142
1240	For soap dish, add					3.60			3.60	3.96
1250	For basket strainer w/tail piece, add					40			40	44
1300	Single control lever handle									
1310	With pull out spray									
1320	Polished chrome	1 Plum	10	.800	Ea.	192	39		231	272
1330	Polished brass		10	.800		230	39		269	315
1340	White		10	.800		211	39		250	293
1348	With spray thru escutcheon									
1350	Polished chrome	1 Plum	10	.800	Ea.	152	39		191	228
1360	Polished brass		10	.800		183	39		222	262
1370	White		10	.800		164	39		203	241
2000	Laundry faucets, shelf type, IPS or copper unions		12	.667		45.50	32.50		78	101
2100	Lavatory faucet, centerset, without drain		10	.800		44.50	39		83.50	110
2120	With pop-up drain		6.66	1.201		61.50	58.50		120	159
2130	For acrylic handles, add					5.15			5.15	5.65
2150	Concealed, 12″ centers	1 Plum	10	.800		101	39		140	172
2160	With pop-up drain	"	6.66	1.201		118	58.50		176.50	221
2210	Porcelain cross handles and pop-up drain									
2220	Polished chrome	1 Plum	6.66	1.201	Ea.	126	58.50		184.50	230
2230	Polished brass	"	6.66	1.201	"	189	58.50		247.50	299
2260	Single lever handle and pop-up drain									
2270	Black nickel	1 Plum	6.66	1.201	Ea.	220	58.50		278.50	335
2280	Polished brass		6.66	1.201		220	58.50		278.50	335
2290	Polished chrome		6.66	1.201		169	58.50		227.50	277
2600	Shelfback, 4″ to 6″ centers, 17 Ga. tailpiece		10	.800		78	39		117	147

22 41 Residential Plumbing Fixtures

22 41 39 – Residential Faucets, Supplies and Trim

22 41 39.10 Faucets and Fittings		Crew	Daily Output	Labor-Hours	Unit	Material	Labor	Equipment	Total	Total Incl O&P
						2009 Bare Costs				
2650	With pop-up drain	1 Plum	6.66	1.201	Ea.	95	58.50		153.50	195
2700	Shampoo faucet with supply tube		24	.333		48.50	16.25		64.75	78.50
2800	Self-closing, center set		10	.800		131	39		170	205
2810	Automatic sensor and operator, with faucet head G		6.15	1.301		355	63.50		418.50	490
4000	Shower by-pass valve with union		18	.444		68.50	21.50		90	109
4100	Shower arm with flange and head		22	.364		69	17.75		86.75	104
4140	Shower, hand held, pin mount, massage action, chrome		22	.364		69	17.75		86.75	104
4142	Polished brass		22	.364		97	17.75		114.75	135
4144	Shower, hand held, wall mtd, adj. spray, 2 wall mounts, chrome		20	.400		119	19.50		138.50	160
4146	Polished brass		20	.400		242	19.50		261.50	296
4148	Shower, hand held head, bar mounted 24", adj. spray, chrome		20	.400		158	19.50		177.50	203
4150	Polished brass		20	.400		385	19.50		404.50	450
4200	Shower thermostatic mixing valve, concealed	↓	8	1		320	49		369	425
4204	For inlet strainer, check, and stops, add				↓	45			45	49.50
4220	Shower pressure balancing mixing valve,									
4230	With shower head, arm, flange and diverter tub spout									
4240	Chrome	1 Plum	6.14	1.303	Ea.	170	63.50		233.50	286
4250	Polished brass		6.14	1.303		238	63.50		301.50	360
4260	Satin		6.14	1.303		238	63.50		301.50	360
4270	Polished chrome/brass		6.14	1.303		196	63.50		259.50	315
5000	Sillcock, compact, brass, IPS or copper to hose	↓	24	.333	↓	8.05	16.25		24.30	34
6000	Stop and waste valves, bronze									
6100	Angle, solder end 1/2"	1 Plum	24	.333	Ea.	7.05	16.25		23.30	33
6110	3/4"		20	.400		7.75	19.50		27.25	38.50
6300	Straightway, solder end 3/8"		24	.333		6.15	16.25		22.40	32
6310	1/2"		24	.333		2.81	16.25		19.06	28
6320	3/4"		20	.400		3.48	19.50		22.98	34
6330	1"		19	.421		19.95	20.50		40.45	54
6400	Straightway, threaded 3/8"		24	.333		7.15	16.25		23.40	33
6410	1/2"		24	.333		3.82	16.25		20.07	29
6420	3/4"		20	.400		6.35	19.50		25.85	37
6430	1"		19	.421		12.60	20.50		33.10	46
7800	Water closet, wax gasket		96	.083		1.42	4.06		5.48	7.85
7820	Gasket toilet tank to bowl	↓	32	.250	↓	1.24	12.20		13.44	20.50
8000	Water supply stops, polished chrome plate									
8200	Angle, 3/8"	1 Plum	24	.333	Ea.	7.35	16.25		23.60	33
8300	1/2"		22	.364		7.35	17.75		25.10	35.50
8400	Straight, 3/8"		26	.308		7.75	15		22.75	32
8500	1/2"		24	.333		7.75	16.25		24	33.50
8600	Water closet, angle, w/flex riser, 3/8"		24	.333	↓	28	16.25		44.25	56
9000	Minimum labor/equipment charge	↓	4	2	Job		97.50		97.50	151

22 42 Commercial Plumbing Fixtures

22 42 13 – Commercial Water Closets, Urinals, and Bidets

22 42 13.30 Urinals

		Crew	Daily Output	Labor-Hours	Unit	Material	Labor (2009 Bare Costs)	Equipment	Total	Total Incl O&P
0010	**URINALS** R224000-30									
0102	For automatic flush see Div. 22 42 39.10 0972									
3000	Wall hung, vitreous china, with hanger & self-closing valve									
3100	Siphon jet type	Q-1	3	5.333	Ea.	248	234		482	640
3120	Blowout type		3	5.333		385	234		619	790
3300	Rough-in, supply, waste & vent		2.83	5.654		240	248		488	650
5000	Stall type, vitreous china, includes valve		2.50	6.400		575	281		856	1,075
6980	Rough-in, supply, waste and vent		1.99	8.040		350	355		705	930
8000	Waterless (no flush) urinal									
8010	Wall hung									
8014	Fiberglass reinforced polyester									
8020	Standard unit G	Q-1	21.30	.751	Ea.	385	33		418	470
8030	ADA compliant unit G	"	21.30	.751		400	33		433	490
8070	For solid color, add G					60			60	66
8080	For 2" brass flange, (new const.), add G	Q-1	96	.167		19.20	7.30		26.50	32.50
8090	Rough-in, supply, waste & vent G	"	3.86	4.145		202	182		384	505
8200	Vitreous china									
8220	Standard unit G	Q-1	21.30	.751	Ea.	280	33		313	360
8240	ADA compliant unit G	"	21.30	.751		320	33		353	400
8270	For solid color, add G					60			60	66
8290	Rough-in, supply, waste & vent G	Q-1	2.92	5.479		430	240		670	850
8400	Trap liquid									
8410	1 quart G				Ea.	14.10			14.10	15.50
8420	1 gallon G				"	55			55	60.50
9000	Minimum labor/equipment charge	Q-1	4	4	Job		176		176	272

22 42 13.40 Water Closets

		Crew	Daily Output	Labor-Hours	Unit	Material	Labor	Equipment	Total	Total Incl O&P
0010	**WATER CLOSETS**									
3000	Bowl only, with flush valve, seat									
3100	Wall hung	Q-1	5.80	2.759	Ea.	284	121		405	505
3200	For rough-in, supply, waste and vent, single WC		2.56	6.250		680	274		954	1,175
3300	Floor mounted		5.80	2.759		305	121		426	525
3350	With wall outlet		5.80	2.759		420	121		541	650
3400	For rough-in, supply, waste and vent, single WC		2.84	5.634		320	247		567	740
3500	Gang side by side carrier system, rough-in, supply, waste & vent									
3510	For single hook-up	Q-1	1.97	8.122	Ea.	840	355		1,195	1,475
3520	For each additional hook-up, add	"	2.14	7.477	"	790	330		1,120	1,375
3550	Gang back to back carrier system, rough-in, supply, waste & vent									
3560	For pair hook-up	Q-1	1.76	9.091	Pr.	1,300	400		1,700	2,075
3570	For each additional pair hook-up, add	"	1.81	8.840	"	1,250	390		1,640	1,975
4000	Water conserving systems									
4900	2 quart flush, round, residential G	Q-1	5.40	2.963	Ea.	1,275	130		1,405	1,600
5100	2 quart flush, elongated G		4.60	3.478		1,275	153		1,428	1,625
5140	For rough-in, supply, waste and vent G		1.94	8.247		300	360		660	890
5200	For remote valve, add G		24	.667		565	29.50		594.50	665
5300	For residential air compressor G		6	2.667		730	117		847	985
5500	For heavy duty industrial air compressor G		1	16		975	700		1,675	2,175
9000	Minimum labor/equipment charge		4	4	Job		176		176	272

22 42 16 – Commercial Lavatories and Sinks

22 42 16.10 Handwasher-Dryer Module

		Crew	Daily Output	Labor-Hours	Unit	Material	Labor	Equipment	Total	Total Incl O&P
0010	**HANDWASHER-DRYER MODULE**									
0030	Wall mounted									
0040	With electric dryer									

22 42 Commercial Plumbing Fixtures

22 42 16 – Commercial Lavatories and Sinks

22 42 16.10 Handwasher-Dryer Module		Crew	Daily Output	Labor-Hours	Unit	2009 Bare Costs Material	Labor	Equipment	Total	Total Incl O&P
0050	Sensor operated	Q-1	8	2	Ea.	3,650	88		3,738	4,125
0110	Sensor operated (ADA)		8	2		3,650	88		3,738	4,150
0140	Sensor operated (ADA), surface mounted	↓	8	2	↓	4,400	88		4,488	4,975
0150	With paper towels									
0180	Sensor operated	Q-1	8	2	Ea.	3,550	88		3,638	4,025
22 42 16.14 Lavatories										
0010	**LAVATORIES**, With trim, white unless noted otherwise									
0020	Commercial lavatories same as residential. See Div. 22 41 16.10									
22 42 16.34 Laboratory Countertops and Sinks										
0010	**LABORATORY COUNTERTOPS AND SINKS**									
0050	Laboratory sinks, corrosion resistant									
1000	Polyethylene, single sink, bench mounted, with									
1020	plug & waste fitting with 1-1/2" straight threads									
1030	2 drainboards, backnut & strainer									
1050	18-1/2" x 15-1/2" x 12-1/2" sink, 54" x 24" O.D.	Q-1	3	5.333	Ea.	1,025	234		1,259	1,525
1100	Single drainboard, backnut & strainer									
1130	18-1/2" x 15-1/2" x 12-1/2" sink, 47" x 24" O.D.	Q-1	3	5.333	Ea.	1,225	234		1,459	1,725
1150	18-1/2" x 15-1/2" x 12-1/2" sink, 70" x 24" O.D.	"	3	5.333	"	1,100	234		1,334	1,600
1290	Flanged 1-1/4" wide, rectangular with strainer									
1300	plug & waste fitting, 1-1/2" straight threads									
1320	12" x 12" x 8" sink, 14-1/2" x 14-1/2" O.D.	Q-1	4	4	Ea.	205	176		381	495
1340	16" x 16" x 8" sink, 18-1/2" x 18-1/2" O.D.		4	4		288	176		464	585
1360	21" x 18" x 10" sink, 23-1/2" x 20-1/2" O.D.		4	4		325	176		501	625
1490	For rough-in, supply, waste & vent, add	↓	2.02	7.921	↓	157	350		507	715
1600	Polypropylene									
1620	Cup sink, oval, integral strainers									
1640	6" x 3" I.D., 7" x 4" O.D.	Q-1	6	2.667	Ea.	105	117		222	296
1660	9" x 3" I.D., 10" x 4-1/2" O.D.	"	6	2.667		90	117		207	280
1740	1-1/2" diam. x 11" long					35			35	38.50
1980	For rough-in, supply, waste & vent, add	Q-1	1.70	9.412	↓	150	415		565	805
22 42 16.40 Service Sinks										
0010	**SERVICE SINKS**									
6650	Service, floor, corner, P.E. on C.I., 28" x 28"	Q-1	4.40	3.636	Ea.	625	160		785	935
6750	Vinyl coated rim guard, add					67.50			67.50	74
6760	Mop sink, molded stone, 24" x 36"	1 Plum	3.33	2.402		248	117		365	455
6770	Mop sink, molded stone, 24" x 36", w/rim 3 sides	"	3.33	2.402		233	117		350	440
6790	For rough-in, supply, waste & vent, floor service sinks	Q-1	1.64	9.756		705	430		1,135	1,450
7000	Service, wall, P.E. on C.I., roll rim, 22" x 18"		4	4		565	176		741	895
7100	24" x 20"	↓	4	4		625	176		801	955
7600	For stainless steel rim guard, add					105			105	116
7800	For stainless steel rim guard, front only, add					41			41	45
8600	Vitreous china, 22" x 20"	Q-1	4	4		465	176		641	785
8960	For stainless steel rim guard, front or side, add					32.50			32.50	35.50
8980	For rough-in, supply, waste & vent, wall service sinks	Q-1	1.30	12.308	↓	940	540		1,480	1,875
9000	Minimum labor/equipment charge	"	4	4	Job		176		176	272

22 42 23 – Commercial Shower Receptors and Basins

22 42 23.30 Group Showers		Crew	Daily Output	Labor-Hours	Unit	Material	Labor	Equipment	Total	Total Incl O&P
0010	**GROUP SHOWERS**									
6000	Group, w/pressure balancing valve, rough-in and rigging not included									
6800	Column, 6 heads, no receptors, less partitions	Q-1	3	5.333	Ea.	2,925	234		3,159	3,600
6900	With stainless steel partitions	↓	1	16	↓	8,975	700		9,675	11,000

22 42 Commercial Plumbing Fixtures

22 42 23 – Commercial Shower Receptors and Basins

22 42 23.30 Group Showers		Crew	Daily Output	Labor-Hours	Unit	Material	2009 Bare Costs Labor	Equipment	Total	Total Incl O&P
7600	5 heads, no receptors, less partitions	Q-1	3	5.333	Ea.	2,400	234		2,634	3,025
7620	4 heads (1 handicap) no receptors, less partitions		3	5.333		3,850	234		4,084	4,625
7700	With stainless steel partitions		1	16		7,375	700		8,075	9,225
8000	Wall, 2 heads, no receptors, less partitions		4	4		1,350	176		1,526	1,750
8100	With stainless steel partitions		2	8		3,125	350		3,475	4,000
9000	Minimum labor/equipment charge		4	4	Job		176		176	272

22 42 33 – Wash Fountains

22 42 33.20 Commercial Wash Fountains		Crew	Daily Output	Labor-Hours	Unit	Material	Labor	Equipment	Total	Total Incl O&P
0010	**COMMERCIAL WASH FOUNTAINS**									
1900	Group, foot control									
2000	Precast terrazzo, circular, 36″ diam., 5 or 6 persons	Q-2	3	8	Ea.	4,075	365		4,440	5,050
2100	54″ diameter for 8 or 10 persons		2.50	9.600		5,050	435		5,485	6,250
2400	Semi-circular, 36″ diam. for 3 persons		3	8		3,750	365		4,115	4,700
2500	54″ diam. for 4 or 5 persons		2.50	9.600		4,525	435		4,960	5,650
2700	Quarter circle (corner), 54″ for 3 persons		3.50	6.857		4,400	310		4,710	5,300
3000	Stainless steel, circular, 36″ diameter		3.50	6.857		3,975	310		4,285	4,850
3100	54″ diameter		2.80	8.571		5,150	390		5,540	6,275
3400	Semi-circular, 36″ diameter		3.50	6.857		3,475	310		3,785	4,300
3500	54″ diameter		2.80	8.571		4,450	390		4,840	5,500
5000	Thermoplastic, pre-assembled, circular, 36″ diameter		6	4		3,000	182		3,182	3,550
5100	54″ diameter		4	6		3,475	273		3,748	4,250
5400	Semi-circular, 36″ diameter		6	4		2,750	182		2,932	3,300
5600	54″ diameter		4	6		3,350	273		3,623	4,125
5610	Group, infrared control, barrier free									
5614	Precast terrazzo									
5620	Semi-circular 36″ diam. for 3 persons	Q-2	3	8	Ea.	5,725	365		6,090	6,850
5630	46″ diam. for 4 persons		2.80	8.571		6,225	390		6,615	7,450
5640	Circular, 54″ diam. for 8 persons, button control		2.50	9.600		7,475	435		7,910	8,900
5700	Rough-in, supply, waste and vent for above wash fountains	Q-1	1.82	8.791		410	385		795	1,050
6200	Duo for small washrooms, stainless steel		2	8		2,625	350		2,975	3,450
6400	Bowl with backsplash		2	8		2,175	350		2,525	2,950
6500	Rough-in, supply, waste & vent for duo fountains		2.02	7.921		227	350		577	790
9000	Minimum labor/equipment charge	Q-2	3	8	Job		365		365	565

22 42 39 – Commercial Faucets, Supplies, and Trim

22 42 39.10 Faucets and Fittings		Crew	Daily Output	Labor-Hours	Unit	Material	Labor	Equipment	Total	Total Incl O&P
0010	**FAUCETS AND FITTINGS**									
0840	Flush valves, with vacuum breaker									
0850	Water closet									
0860	Exposed, rear spud	1 Plum	8	1	Ea.	130	49		179	219
0870	Top spud		8	1		120	49		169	208
0880	Concealed, rear spud		8	1		171	49		220	265
0890	Top spud		8	1		141	49		190	231
0900	Wall hung		8	1		151	49		200	242
0920	Urinal									
0930	Exposed, stall	1 Plum	8	1	Ea.	120	49		169	208
0940	Wall, (washout)		8	1		120	49		169	208
0950	Pedestal, top spud		8	1		122	49		171	211
0960	Concealed, stall		8	1		138	49		187	227
0970	Wall (washout)		8	1		148	49		197	239
0971	Automatic flush sensor and operator for									
0972	urinals or water closets G	1 Plum	8	1	Ea.	405	49		454	520
3000	Service sink faucet, cast spout, pail hook, hose end	″	14	.571	″	80	28		108	131

22 42 39.30 Carriers and Supports		Crew	Daily Output	Labor-Hours	Unit	2009 Bare Costs Material	Labor	Equipment	Total	Total Incl O&P
0010	**CARRIERS AND SUPPORTS**, For plumbing fixtures									
0500	Drinking fountain, wall mounted									
0600	Plate type with studs, top back plate	1 Plum	7	1.143	Ea.	63	55.50		118.50	156
0700	Top front and back plate		7	1.143		77.50	55.50		133	172
0800	Top & bottom, front & back plates, w/bearing jacks		7	1.143		113	55.50		168.50	211
3000	Lavatory, concealed arm									
3050	Floor mounted, single									
3100	High back fixture	1 Plum	6	1.333	Ea.	350	65		415	485
3200	Flat slab fixture		6	1.333		296	65		361	425
3220	Paraplegic		6	1.333		355	65		420	490
3250	Floor mounted, back to back									
3300	High back fixtures	1 Plum	5	1.600	Ea.	500	78		578	670
3400	Flat slab fixtures		5	1.600		615	78		693	795
3430	Paraplegic		5	1.600		480	78		558	650
3500	Wall mounted, in stud or masonry									
3600	High back fixture	1 Plum	6	1.333	Ea.	215	65		280	335
3700	Flat slab fixture	"	6	1.333	"	186	65		251	305
4000	Exposed arm type, floor mounted									
4100	Single high back or flat slab fixture	1 Plum	6	1.333	Ea.	445	65		510	590
4200	Back to back, high back or flat slab fixtures		5	1.600		785	78		863	985
4300	Wall mounted, High back or flat slab lavatory		6	1.333		345	65		410	480
4600	Sink, floor mounted									
4650	Exposed arm system									
4700	Single heavy fixture	1 Plum	5	1.600	Ea.	585	78		663	765
4750	Single heavy sink with slab		5	1.600		750	78		828	945
4800	Back to back, standard fixtures		5	1.600		430	78		508	595
4850	Back to back, heavy fixtures		5	1.600		615	78		693	795
4900	Back to back, heavy sink with slab		5	1.600		615	78		693	795
4950	Exposed offset arm system									
5000	Single heavy deep fixture	1 Plum	5	1.600	Ea.	560	78		638	735
5100	Plate type system									
5200	With bearing jacks, single fixture	1 Plum	5	1.600	Ea.	640	78		718	825
5300	With exposed arms, single heavy fixture		5	1.600		825	78		903	1,025
5400	Wall mounted, exposed arms, single heavy fixture		5	1.600		300	78		378	450
6000	Urinal, floor mounted, 2" or 3" coupling, blowout type		6	1.333		375	65		440	510
6100	With fixture or hanger bolts, blowout or washout		6	1.333		266	65		331	395
6200	With bearing plate		6	1.333		266	65		331	395
6300	Wall mounted, plate type system		6	1.333		230	65		295	355
6980	Water closet, siphon jet									
7000	Horizontal, adjustable, caulk									
7040	Single, 4" pipe size	1 Plum	5.33	1.501	Ea.	500	73		573	665
7050	4" pipe size, paraplegic		5.33	1.501		500	73		573	665
7060	5" pipe size		5.33	1.501		855	73		928	1,050
7100	Double, 4" pipe size		5	1.600		930	78		1,008	1,150
7110	4" pipe size, paraplegic		5	1.600		930	78		1,008	1,150
7120	5" pipe size		5	1.600		1,125	78		1,203	1,375
7160	Horizontal, adjustable, extended, caulk									
7180	Single, 4" pipe size	1 Plum	5.33	1.501	Ea.	670	73		743	855
7200	5" pipe size		5.33	1.501		855	73		928	1,050
7240	Double, 4" pipe size		5	1.600		1,175	78		1,253	1,400
7260	5" pipe size		5	1.600		720	78		798	910
7400	Vertical, adjustable, caulk or thread									

22 42 39 – Commercial Faucets, Supplies, and Trim

22 42 39.30 Carriers and Supports		Crew	Daily Output	Labor-Hours	Unit	Material	Labor	Equipment	Total	Total Incl O&P
7440	Single, 4" pipe size	1 Plum	5.33	1.501	Ea.	685	73		758	870
7460	5" pipe size		5.33	1.501		845	73		918	1,050
7480	6" pipe size		5	1.600		965	78		1,043	1,200
7520	Double, 4" pipe size		5	1.600		1,025	78		1,103	1,250
7540	5" pipe size		5	1.600		1,175	78		1,253	1,425
7560	6" pipe size		4	2		1,300	97.50		1,397.50	1,575
7600	Vertical, adjustable, extended, caulk									
7620	Single, 4" pipe size	1 Plum	5.33	1.501	Ea.	685	73		758	870
7640	5" pipe size		5.33	1.501		845	73		918	1,050
7680	6" pipe size		5	1.600		965	78		1,043	1,200
7720	Double, 4" pipe size		5	1.600		1,025	78		1,103	1,250
7740	5" pipe size		5	1.600		1,175	78		1,253	1,425
7760	6" pipe size		4	2		1,300	97.50		1,397.50	1,575
7780	Water closet, blow out									
7800	Vertical offset, caulk or thread									
7820	Single, 4" pipe size	1 Plum	5.33	1.501	Ea.	505	73		578	670
7840	Double, 4" pipe size	"	5	1.600	"	865	78		943	1,075
7880	Vertical offset, extended, caulk									
7900	Single, 4" pipe size	1 Plum	5.33	1.501	Ea.	635	73		708	815
7920	Double, 4" pipe size	"	5	1.600	"	995	78		1,073	1,225
7960	Vertical, for floor mounted back-outlet									
7980	Single, 4" thread, 2" vent	1 Plum	5.33	1.501	Ea.	445	73		518	605
8000	Double, 4" thread, 2" vent	"	6	1.333	"	1,300	65		1,365	1,550
8040	Vertical, for floor mounted back-outlet, extended									
8060	Single, 4" caulk, 2" vent	1 Plum	6	1.333	Ea.	445	65		510	590
8080	Double, 4" caulk, 2" vent	"	6	1.333	"	1,275	65		1,340	1,500
8200	Water closet, residential									
8220	Vertical centerline, floor mount									
8240	Single, 3" caulk, 2" or 3" vent	1 Plum	6	1.333	Ea.	405	65		470	545
8260	4" caulk, 2" or 4" vent		6	1.333		520	65		585	675
8280	3" copper sweat, 3" vent		6	1.333		365	65		430	500
8300	4" copper sweat, 4" vent		6	1.333		440	65		505	585
8400	Vertical offset, floor mount									
8420	Single, 3" or 4" caulk, vent	1 Plum	4	2	Ea.	505	97.50		602.50	705
8440	3" or 4" copper sweat, vent		5	1.600		505	78		583	675
8460	Double, 3" or 4" caulk, vent		4	2		865	97.50		962.50	1,100
8480	3" or 4" copper sweat, vent		5	1.600		865	78		943	1,075
9000	Water cooler (electric), floor mounted									
9100	Plate type with bearing plate, single	1 Plum	6	1.333	Ea.	256	65		321	380
9120	Plate type with bearing plate, double		4	2		590	97.50		687.50	795
9140	Plate type with bearing plate, back to back		4	2		355	97.50		452.50	545
9990	Minimum labor/equipment charge		3.50	2.286	Job		111		111	173

22 43 Healthcare Plumbing Fixtures

22 43 13 – Healthcare Water Closets

22 43 13.40 Water Closets		Crew	Daily Output	Labor-Hours	Unit	2009 Bare Costs Material	Labor	Equipment	Total	Total Incl O&P
0010	WATER CLOSETS									
1000	Floor mounted, 1 piece, w/ seat and flush valve, ADA compliant 18″ high									
1150	With wall outlet	Q-1	5.30	3.019	Ea.	365	132		497	610
1180	For rough-in, supply, waste and vent		2.84	5.634		320	247		567	740
1200	With floor outlet		5.30	3.019		274	132		406	505
1800	For rough-in, supply, waste and vent		3.05	5.246		275	230		505	655
3000	Bowl only, with flush valve, seat									
3100	Wall hung	Q-1	5.80	2.759	Ea.	284	121		405	505
3150	Hospital type, slotted rim for bed pan									
3160	Elongated bowl	Q-1	5.80	2.759	Ea.	315	121		436	540
3200	For rough-in, supply, waste and vent, single WC	″	2.56	6.250	″	680	274		954	1,175
3300	Floor mounted									
3360	Hospital type, slotted rim for bed pan									
3370	Elongated bowl, top spud	Q-1	5	3.200	Ea.	305	140		445	560
3380	Elongated bowl, rear spud		5	3.200		320	140		460	575
3500	For rough-in, supply, waste and vent		3.05	5.246		275	230		505	655

22 43 16 – Healthcare Sinks

22 43 16.10 Sinks

		Crew	Daily Output	Labor-Hours	Unit	Material	Labor	Equipment	Total	Total Incl O&P
0010	SINKS									
0020	Vitreous china									
6702	Hospital type, without trim (see Div. 22 41 39.10)									
6710	20″ x 18″, contoured splash shield	Q-1	8	2	Ea.	130	88		218	279
6730	28″ x 20″, surgeon, side decks		8	2		515	88		603	705
6740	28″ x 22″, surgeon scrub-up, deep bowl		8	2		685	88		773	885
6750	20″ x 27″, patient, wheelchair		7	2.286		255	100		355	435
6760	30″ x 22″, all purpose		7	2.286		565	100		665	780
6770	30″ x 22″, plaster work		7	2.286		565	100		665	780
6820	20″ x 24″ clinic service, liquid/solid waste		6	2.667		755	117		872	1,000

22 43 19 – Healthcare Bathtubs and Showers

22 43 19.10 Bathtubs

		Crew	Daily Output	Labor-Hours	Unit	Material	Labor	Equipment	Total	Total Incl O&P
0010	BATHTUBS									
5002	Hospital type, with trim (see Div. 22 41 39.10)									
5050	Bathing pool, porcelain enamel on cast iron, grab bars									
5060	pop-up drain, 72″ x 36″	Q-1	3	5.333	Ea.	2,825	234		3,059	3,500
5100	Perineal (sitz), vitreous china		3	5.333		915	234		1,149	1,375
5120	For pedestal, vitreous china, add		8	2		135	88		223	285
5180	Pier tub, porcelain enamel on cast iron, 66-3/4″ x 30″		3	5.333		2,375	234		2,609	2,975
5200	Base, porcelain enamel on cast iron		8	2		1,100	88		1,188	1,350
5300	Whirlpool, porcelain enamel on cast iron, 72″ x 36″		1	16		2,950	700		3,650	4,350

22 43 19.20 Showers

		Crew	Daily Output	Labor-Hours	Unit	Material	Labor	Equipment	Total	Total Incl O&P
0010	SHOWERS									
5950	Module, handicap, SS panel, fixed & hand held head, control									
5960	valves, grab bar, curtain & rod, folding seat	1 Plum	4	2	Ea.	1,850	97.50		1,947.50	2,200

22 43 39 – Healthcare Faucets, Supplies, and Trim

22 43 39.10 Faucets and Fittings

		Crew	Daily Output	Labor-Hours	Unit	Material	Labor	Equipment	Total	Total Incl O&P
0010	FAUCETS AND FITTINGS									
2850	Medical, bedpan cleanser, with pedal valve,	1 Plum	12	.667	Ea.	600	32.50		632.50	710
2860	With screwdriver stop valve		12	.667		310	32.50		342.50	395
2870	With self-closing spray valve		12	.667		234	32.50		266.50	310
2900	Faucet, gooseneck spout, wrist handles, grid drain		10	.800		124	39		163	198
2940	Mixing valve, knee action, screwdriver stops		4	2		485	97.50		582.50	685

22 45 Emergency Plumbing Fixtures

22 45 13 – Emergency Showers

22 45 13.10 Emergency Showers		Crew	Daily Output	Labor-Hours	Unit	Material	2009 Bare Costs Labor	Equipment	Total	Total Incl O&P
0010	**EMERGENCY SHOWERS**, Rough-in not included									
5000	Shower, single head, drench, ball valve, pull, freestanding	Q-1	4	4	Ea.	320	176		496	625
5200	Horizontal or vertical supply		4	4		460	176		636	775
6000	Multi-nozzle, eye/face wash combination		4	4		590	176		766	920
6400	Multi-nozzle, 12 spray, shower only		4	4		1,600	176		1,776	2,025
6600	For freeze-proof, add		6	2.667		335	117		452	545
8000	Walk-thru decontamination with eye-face wash		2	8		2,675	350		3,025	3,500
8200	For freeze proof, add		4	4		540	176		716	865
9000	Minimum labor/equipment charge		3	5.333	Job		234		234	365

22 45 16 – Eyewash Equipment

22 45 16.10 Eyewash Safety Equipment		Crew	Daily Output	Labor-Hours	Unit	Material	Labor	Equipment	Total	Total Incl O&P
0010	**EYEWASH SAFETY EQUIPMENT**, Rough-in not included									
1000	Eye wash fountain									
1400	Plastic bowl, pedestal mounted	Q-1	4	4	Ea.	232	176		408	525
1600	Unmounted		4	4		160	176		336	450
1800	Wall mounted		4	4		166	176		342	455
2000	Stainless steel, pedestal mounted		4	4		360	176		536	665
2200	Unmounted		4	4		226	176		402	520
2400	Wall mounted		4	4		216	176		392	510

22 45 19 – Self-Contained Eyewash Equipment

22 45 19.10 Self-Contained Eyewash Safety Equipment		Crew	Daily Output	Labor-Hours	Unit	Material	Labor	Equipment	Total	Total Incl O&P
0010	**SELF-CONTAINED EYEWASH SAFETY EQUIPMENT**									
3000	Eye wash, portable, self-contained				Ea.	455			455	500

22 45 26 – Eye/Face Wash Equipment

22 45 26.10 Eye/Face Wash Safety Equipment		Crew	Daily Output	Labor-Hours	Unit	Material	Labor	Equipment	Total	Total Incl O&P
0010	**EYE/FACE WASH SAFETY EQUIPMENT**, Rough-in not included									
4000	Eye and face wash, combination fountain									
4200	Stainless steel, pedestal mounted	Q-1	4	4	Ea.	490	176		666	810
4400	Unmounted		4	4		345	176		521	650
4600	Wall mounted		4	4		370	176		546	675

22 46 Security Plumbing Fixtures

22 46 13 – Security Water Closets and Urinals

22 46 13.10 Security Water Closets and Urinals		Crew	Daily Output	Labor-Hours	Unit	Material	Labor	Equipment	Total	Total Incl O&P
0010	**SECURITY WATER CLOSETS AND URINALS,** Stainless steel									
2000	Urinal, back supply and flush									
2200	Wall hung	Q-1	4	4	Ea.	1,200	176		1,376	1,575
2240	Stall		2.50	6.400		1,350	281		1,631	1,925
2300	For urinal rough-in, supply, waste and vent		1.49	10.738		320	470		790	1,075
3000	Water closet, integral seat, back supply and flush									
3300	Wall hung, wall outlet	Q-1	5.80	2.759	Ea.	935	121		1,056	1,225
3400	Floor mount, wall outlet		5.80	2.759		1,075	121		1,196	1,375
3440	Floor mount, floor outlet		5.80	2.759		1,175	121		1,296	1,500
3480	For recessed tissue holder, add					90			90	99
3500	For water closet rough-in, supply, waste and vent	Q-1	1.19	13.445		295	590		885	1,250
5000	Water closet and lavatory units, push button filler valves,									
5010	soap & paper holders, seat									
5300	Wall hung	Q-1	5	3.200	Ea.	2,200	140		2,340	2,650
5400	Floor mount		5	3.200		2,100	140		2,240	2,525

22 46 Security Plumbing Fixtures

22 46 13 – Security Water Closets and Urinals

22 46 13.10 Security Water Closets and Urinals		Crew	Daily Output	Labor-Hours	Unit	Material	2009 Bare Costs Labor	Equipment	Total	Total Incl O&P
6300	For unit rough-in, supply, waste and vent	Q-1	1	16	Ea.	325	700		1,025	1,450

22 46 16 – Security Lavatories and Sinks

22 46 16.10 Security Lavatories

		Crew	Daily Output	Labor-Hours	Unit	Material	Labor	Equipment	Total	Total Incl O&P
0010	**SECURITY LAVATORIES,** Stainless steel									
1000	Lavatory, wall hung, push button filler valve									
1100	Rectangular bowl	Q-1	8	2	Ea.	990	88		1,078	1,225
1200	Oval bowl		8	2		955	88		1,043	1,175
1240	Oval bowl, corner mount		8	2		1,075	88		1,163	1,300
1300	For lavatory rough-in, supply, waste and vent	↓	1.50	10.667	↓	273	470		743	1,025

22 46 63 – Security Service Sink

22 46 63.10 Security Service Sink

		Crew	Daily Output	Labor-Hours	Unit	Material	Labor	Equipment	Total	Total Incl O&P
0010	**SECURITY SERVICE SINK,** Stainless steel									
1700	Service sink, with soap dish									
1740	24" x 19" size	Q-1	3	5.333	Ea.	1,650	234		1,884	2,175
1790	For sink rough-in, supply, waste and vent	"	.89	17.978	"	630	790		1,420	1,925

22 46 73 – Security Shower

22 46 73.10 Security Shower

		Crew	Daily Output	Labor-Hours	Unit	Material	Labor	Equipment	Total	Total Incl O&P
0010	**SECURITY SHOWER,** Stainless steel									
1800	Shower cabinet, unitized									
1840	36" x 36" x 88"	Q-1	2.20	7.273	Ea.	4,400	320		4,720	5,325
1900	Shower package for built-in									
1940	Hot & cold valves, recessed soap dish	Q-1	6	2.667	Ea.	555	117		672	790

22 47 Drinking Fountains and Water Coolers

22 47 13 – Drinking Fountains

22 47 13.10 Drinking Water Fountains

		Crew	Daily Output	Labor-Hours	Unit	Material	Labor	Equipment	Total	Total Incl O&P
0010	**DRINKING WATER FOUNTAINS,** For connection to cold water supply R224000-30									
0802	For remote water chiller, see Div. 22 47 23.10									
1000	Wall mounted, non-recessed									
1200	Aluminum,									
1280	Dual bubbler type	1 Plum	3.20	2.500	Ea.	1,950	122		2,072	2,350
1400	Bronze, with no back		4	2		935	97.50		1,032.50	1,175
1600	Cast iron, enameled, low back, single bubbler		4	2		580	97.50		677.50	790
1640	Dual bubbler type		3.20	2.500		1,125	122		1,247	1,425
1680	Triple bubbler type		3.20	2.500		1,475	122		1,597	1,825
1800	Cast aluminum, enameled, for correctional institutions		4	2		760	97.50		857.50	990
2000	Fiberglass, 12" back, single bubbler unit		4	2		1,150	97.50		1,247.50	1,425
2040	Dual bubbler		3.20	2.500		985	122		1,107	1,275
2080	Triple bubbler		3.20	2.500		1,275	122		1,397	1,600
2200	Polymarble, no back, single bubbler		4	2		610	97.50		707.50	820
2240	Dual bubbler		3.20	2.500		1,025	122		1,147	1,325
2280	Triple bubbler		3.20	2.500		1,400	122		1,522	1,725
2400	Precast stone, no back		4	2		615	97.50		712.50	830
2700	Stainless steel, single bubbler, no back		4	2		1,200	97.50		1,297.50	1,475
2740	With back		4	2		2,000	97.50		2,097.50	2,350
2780	Dual handle & wheelchair projection type		4	2		625	97.50		722.50	835
2820	Dual level for handicapped type		3.20	2.500		1,275	122		1,397	1,600
2840	Vandal resistant type	↓	4	2	↓	560	97.50		657.50	765
3300	Vitreous china									

22 47 13 – Drinking Fountains

22 47 13.10 Drinking Water Fountains		Crew	Daily Output	Labor-Hours	Unit	Material (2009 Bare Costs)	Labor (2009 Bare Costs)	Equipment (2009 Bare Costs)	Total (2009 Bare Costs)	Total Incl O&P
3340	7″ back	1 Plum	4	2	Ea.	510	97.50		607.50	710
3940	For vandal-resistant bottom plate, add					59.50			59.50	65.50
3960	For freeze-proof valve system, add	1 Plum	2	4		510	195		705	860
3980	For rough-in, supply and waste, add	″	2.21	3.620		171	176		347	465
4000	Wall mounted, semi-recessed									
4200	Poly-marble, single bubbler	1 Plum	4	2	Ea.	780	97.50		877.50	1,000
4600	Stainless steel, satin finish, single bubbler		4	2		905	97.50		1,002.50	1,150
4900	Vitreous china, single bubbler		4	2		610	97.50		707.50	820
5980	For rough-in, supply and waste, add		1.83	4.372		171	213		384	520
6000	Wall mounted, fully recessed									
6400	Poly-marble, single bubbler	1 Plum	4	2	Ea.	890	97.50		987.50	1,125
6440	For water glass filler, add					65.50			65.50	72
6800	Stainless steel, single bubbler	1 Plum	4	2		875	97.50		972.50	1,100
6900	Fountain and cuspidor combination		2	4		1,775	195		1,970	2,250
7560	For freeze-proof valve system, add		2	4		620	195		815	980
7580	For rough-in, supply and waste, add		1.83	4.372		171	213		384	520
7600	Floor mounted, pedestal type									
7700	Aluminum, architectural style, C.I. base	1 Plum	2	4	Ea.	530	195		725	880
7780	Wheelchair handicap unit		2	4		1,300	195		1,495	1,750
8000	Bronze, architectural style		2	4		1,375	195		1,570	1,800
8040	Enameled steel cylindrical column style		2	4		1,150	195		1,345	1,575
8200	Precast stone/concrete, cylindrical column		1	8		970	390		1,360	1,675
8240	Wheelchair handicap unit		1	8		1,275	390		1,665	2,000
8400	Stainless steel, architectural style		2	4		1,375	195		1,570	1,800
8600	Enameled iron, heavy duty service, 2 bubblers		2	4		1,275	195		1,470	1,700
8660	4 bubblers		2	4		1,350	195		1,545	1,775
8880	For freeze-proof valve system, add		2	4		385	195		580	725
8900	For rough-in, supply and waste, add		1.83	4.372		171	213		384	520
9000	Minimum labor/equipment charge		2	4	Job		195		195	300
9100	Deck mounted									
9500	Stainless steel, circular receptor	1 Plum	4	2	Ea.	350	97.50		447.50	535
9540	14″ x 9″ receptor		4	2		289	97.50		386.50	470
9580	25″ x 17″ deep receptor, with water glass filler		3	2.667		266	130		396	495
9760	White enameled steel, 14″ x 9″ receptor		4	2		300	97.50		397.50	480
9860	White enameled cast iron, 24″ x 16″ receptor		3	2.667		315	130		445	545
9980	For rough-in, supply and waste, add		1.83	4.372		171	213		384	520

22 47 16 – Pressure Water Coolers

22 47 16.10 Electric Water Coolers		Crew	Daily Output	Labor-Hours	Unit	Material	Labor	Equipment	Total	Total Incl O&P
0010	**ELECTRIC WATER COOLERS** R224000-30									
0100	Wall mounted, non-recessed									
0140	4 GPH	Q-1	4	4	Ea.	575	176		751	900
0160	8 GPH, barrier free, sensor operated		4	4		725	176		901	1,075
0180	8.2 GPH		4	4		645	176		821	980
0220	14.3 GPH		4	4		745	176		921	1,100
0600	For hot and cold water, add					165			165	182
0640	For stainless steel cabinet, add					84			84	92
1000	Dual height, 8.2 GPH	Q-1	3.80	4.211		930	185		1,115	1,300
1040	14.3 GPH	″	3.80	4.211		985	185		1,170	1,350
1240	For stainless steel cabinet, add					131			131	144
2600	Wheelchair type, 8 GPH	Q-1	4	4		1,650	176		1,826	2,075
3000	Simulated recessed, 8 GPH		4	4		845	176		1,021	1,200
3040	11.5 GPH		4	4		890	176		1,066	1,250

22 47 Drinking Fountains and Water Coolers

22 47 16 – Pressure Water Coolers

22 47 16.10 Electric Water Coolers		Crew	Daily Output	Labor-Hours	Unit	Material	2009 Bare Costs Labor	Equipment	Total	Total Incl O&P
3200	For glass filler, add				Ea.	84			84	92
3240	For stainless steel cabinet, add					58			58	64
3300	Semi-recessed, 8.1 GPH	Q-1	4	4		965	176		1,141	1,325
3320	12 GPH	"	4	4		1,025	176		1,201	1,400
3340	For glass filler, add					84			84	92
3360	For stainless steel cabinet, add					58			58	64
3400	Full recessed, stainless steel, 8 GPH	Q-1	3.50	4.571		1,525	201		1,726	1,975
3420	11.5 GPH	"	3.50	4.571		1,575	201		1,776	2,025
3460	For glass filler, add					154			154	169
3600	For mounting can only					168			168	185
4600	Floor mounted, flush-to-wall									
4640	4 GPH	1 Plum	3	2.667	Ea.	600	130		730	860
4680	8.2 GPH		3	2.667		695	130		825	965
4720	14.3 GPH		3	2.667		715	130		845	985
4960	14 GPH hot and cold water		3	2.667		1,000	130		1,130	1,300
4980	For stainless steel cabinet, add					128			128	141
5000	Dual height, 8.2 GPH	1 Plum	2	4		1,000	195		1,195	1,400
5040	14.3 GPH	"	2	4		1,050	195		1,245	1,450
5120	For stainless steel cabinet, add					168			168	185
5600	Explosion Proof, 16 GPH	1 Plum	3	2.667		1,475	130		1,605	1,825
5640	For stainless steel cabinet, add					110			110	121
6000	Refrigerator Compartment Type, 4.5 GPH	1 Plum	3	2.667		1,275	130		1,405	1,600
6600	Bottle Supply Type, 1.0 GPH		4	2		340	97.50		437.50	520
6640	Hot and cold, 1.0 GPH		4	2		460	97.50		557.50	655
9000	Minimum labor/equipment charge		2	4	Job		195		195	300
9800	For supply, waste & vent, all coolers		2.21	3.620	Ea.	171	176		347	465

22 47 23 – Remote Water Coolers

22 47 23.10 Remote Water Coolers		Crew	Daily Output	Labor-Hours	Unit	Material	Labor	Equipment	Total	Total Incl O&P
0010	**REMOTE WATER COOLERS,** 80° F inlet.									
0100	Air cooled, 50° F outlet, 115V, 4.1 GPH	1 Plum	6	1.333	Ea.	610	65		675	775
0200	5.7 GPH		5.50	1.455		860	71		931	1,050
0300	8.0 GPH		5	1.600		685	78		763	875
0400	10.0 GPH		4.50	1.778		815	86.50		901.50	1,025
0500	13.4 GPH		4	2		1,000	97.50		1,097.50	1,250
0700	29 GPH	Q-1	5	3.200		1,500	140		1,640	1,875
1000	230V, 32 GPH	"	5	3.200		1,675	140		1,815	2,050

22 51 Swimming Pool Plumbing Systems

22 51 19 – Swimming Pool Water Treatment Equipment

22 51 19.50 Swimming Pool Filtration Equipment		Crew	Daily Output	Labor-Hours	Unit	Material	Labor	Equipment	Total	Total Incl O&P
0010	**SWIMMING POOL FILTRATION EQUIPMENT**									
0900	Filter system, sand or diatomite type, incl. pump, 6,000 gal./hr.	2 Plum	1.80	8.889	Total	1,475	435		1,910	2,275
1020	Add for chlorination system, 800 S.F. pool		3	5.333	Ea.	78.50	260		338.50	490
1040	5,000 S.F. pool		3	5.333	"	1,950	260		2,210	2,525

22 52 Fountain Plumbing Systems

22 52 13 – Fountain Piping

22 52 13.10 Fountain Drains		Crew	Daily Output	Labor-Hours	Unit	Material	Labor	Equipment	Total	Total Incl O&P
						2009 Bare Costs				
0010	FOUNTAIN DRAINS									
2990	Pool main discharge, polished bronze top, C.I. body									
3000	8" diam. grate, 2", 3" & 4" pipe size	Q-1	8	2	Ea.	580	88		668	775
3100	12" diam. grate, 4", 5" & 6" pipe size	"	8	2		975	88		1,063	1,200
3160	For galvanized body, add				↓	233			233	257
3200	Gutter, bronze body, polished bronze top									
3220	9" grate, 1-1/2" & 2" pipe size	Q-1	8	2	Ea.	167	88		255	320
3240	10" grate, 2", 2-1/2" & 3" pipe size	"	7	2.286	"	174	100		274	345
3380	Overflow, 5" dome & 6" high pipe polished bronze,									
3400	cast iron body, 2", 3", & 4" pipe size	Q-1	10	1.600	Ea.	735	70		805	915
3580	Recirculating inlet, 6" sq face polished bronze,									
3600	C.I. body, 2", 2-1/2" & 3" pipe size	Q-1	10	1.600	Ea.	105	70		175	224
3620	For galvanized body, add				"	45			45	49.50

22 52 16 – Fountain Pumps

22 52 16.10 Fountain Water Pumps		Crew	Daily Output	Labor-Hours	Unit	Material	Labor	Equipment	Total	Total Incl O&P
0010	FOUNTAIN WATER PUMPS									
0100	Pump w/controls									
0200	Single phase, 100' cord, 1/2 H.P. pump	2 Skwk	4.40	3.636	Ea.	1,800	149		1,949	2,225
0300	3/4 H.P. pump		4.30	3.721		2,050	152		2,202	2,500
0400	1 H.P. pump		4.20	3.810		3,800	156		3,956	4,425
0500	1-1/2 H.P. pump		4.10	3.902		3,500	159		3,659	4,100
0600	2 H.P. pump		4	4		4,725	163		4,888	5,475
0700	Three phase, 200' cord, 5 H.P. pump		3.90	4.103		5,000	168		5,168	5,775
0800	7-1/2 H.P. pump		3.80	4.211		8,500	172		8,672	9,625
0900	10 H.P. pump		3.70	4.324		9,275	177		9,452	10,500
1000	15 H.P. pump	↓	3.60	4.444	↓	11,300	182		11,482	12,800

22 52 33 – Fountain Ancillary

22 52 33.10 Fountain Miscellaneous		Crew	Daily Output	Labor-Hours	Unit	Material	Labor	Equipment	Total	Total Incl O&P
0010	FOUNTAIN MISCELLANEOUS									
1100	Nozzles, minimum	2 Skwk	8	2	Ea.	150	81.50		231.50	298
1200	Maximum		8	2		300	81.50		381.50	465
1300	Lights w/mounting kits, 200 watt		18	.889		1,000	36.50		1,036.50	1,150
1400	300 watt		18	.889		1,200	36.50		1,236.50	1,375
1500	500 watt		18	.889		1,300	36.50		1,336.50	1,475
1600	Color blender	↓	12	1.333	↓	530	54.50		584.50	675

22 62 Vacuum Systems for Laboratory and Healthcare Facilities

22 62 19 – Vacuum Equipment for Laboratory and Healthcare Facilities

22 62 19.70 Healthcare Vacuum Equipment		Crew	Daily Output	Labor-Hours	Unit	Material	Labor	Equipment	Total	Total Incl O&P
0010	HEALTHCARE VACUUM EQUIPMENT									
1100	Vacuum system									
1110	Vacuum outlet alarm panel	1 Plum	3.20	2.500	Ea.	935	122		1,057	1,225

22 63 Gas Systems for Laboratory and Healthcare Facilities

22 63 13 – Gas Piping for Laboratory and Healthcare Facilities

22 63 13.70 Healthcare Gas Piping		Crew	Daily Output	Labor-Hours	Unit	2009 Bare Costs Material	Labor	Equipment	Total	Total Incl O&P
0010	**HEALTHCARE GAS PIPING**									
1000	Nitrogen or oxygen system									
1010	Cylinder manifold									
1020	5 cylinder	1 Plum	.80	10	Ea.	5,025	490		5,515	6,275
1026	10 cylinder	"	.40	20	"	6,100	975		7,075	8,200
3000	Outlets and valves									
3010	Recessed, wall mounted									
3012	Single outlet	1 Plum	3.20	2.500	Ea.	60	122		182	255
3100	Ceiling outlet									
3200	Zone valve with box									
3210	2"	1 Plum	3.20	2.500	Ea.	420	122		542	650
4000	Alarm panel, medical gases and vacuum									
4010	Alarm panel	1 Plum	3.20	2.500	Ea.	935	122		1,057	1,225

22 66 Chemical-Waste Systems for Lab. and Healthcare Facilities

22 66 53 – Laboratory Chemical-Waste and Vent Piping

22 66 53.30 Glass Pipe

		Crew	Daily Output	Labor-Hours	Unit	Material	Labor	Equipment	Total	Total Incl O&P
0010	**GLASS PIPE**, Borosilicate, couplings & clevis hanger assemblies, 10' O.C.									
0020	Drainage									
1100	1-1/2" diameter	Q-1	52	.308	L.F.	12.55	13.50		26.05	35
1120	2" diameter		44	.364		16.65	15.95		32.60	43.50
1140	3" diameter		39	.410		22.50	18		40.50	52.50
1160	4" diameter		30	.533		40.50	23.50		64	81
1180	6" diameter		26	.615		74	27		101	124
1870	To delete coupling & hanger, subtract									
1880	1-1/2" diam. to 2" diam.					19%	22%			
1890	3" diam. to 6" diam.					20%	17%			
2000	Process supply (pressure), beaded joints									
2040	1/2" diameter	1 Plum	36	.222	L.F.	4.27	10.85		15.12	21.50
2060	3/4" diameter		31	.258		4.84	12.60		17.44	25
2080	1" diameter		27	.296		17.90	14.45		32.35	42
2100	1-1/2" diameter	Q-1	47	.340		21.50	14.95		36.45	47
2120	2" diameter		39	.410		29	18		47	59.50
2140	3" diameter		34	.471		42	20.50		62.50	78
2160	4" diameter		25	.640		58.50	28		86.50	108
2180	6" diameter		21	.762		104	33.50		137.50	166
2860	To delete coupling & hanger, subtract									
2870	1/2" diam. to 1" diam.					25%	33%			
2880	1-1/2" diam. to 3" diam.					22%	21%			
2890	4" diam. to 6" diam.					23%	15%			
3800	Conical joint, transparent									
3980	6" diameter	Q-1	21	.762	L.F.	850	33.50		883.50	985
4500	To delete couplings & hangers, subtract									
4530	6" diam.					22%	26%			
9000	Minimum labor/equipment charge	1 Plum	4	2	Job		97.50		97.50	151

22 66 53.40 Pipe Fittings, Glass

		Crew	Daily Output	Labor-Hours	Unit	Material	Labor	Equipment	Total	Total Incl O&P
0010	**PIPE FITTINGS, GLASS**									
0020	Drainage, beaded ends									
0040	Coupling & labor required at joints not incl. in fitting									
0050	price. Add 1 per joint for installed price									
0070	90° Bend or sweep, 1-1/2"				Ea.	27.50			27.50	30.50

22 66 53 – Laboratory Chemical-Waste and Vent Piping

22 66 53.40 Pipe Fittings, Glass		Crew	Daily Output	Labor-Hours	Unit	2009 Bare Costs Material	Labor	Equipment	Total	Total Incl O&P
0090	2"				Ea.	35			35	38.50
0100	3"					57.50			57.50	63.50
0110	4"					92.50			92.50	102
0120	6" (sweep only)					281			281	310
0200	45° Bend or sweep same as 90°									
0350	Tee, single sanitary, 1-1/2"				Ea.	45			45	49
0370	2"					45			45	49
0380	3"					67			67	73.50
0390	4"					120			120	132
0400	6"					320			320	355
0410	Tee, straight, 1-1/2"					55.50			55.50	61
0430	2"					55.50			55.50	61
0440	3"					80			80	88
0450	4"					141			141	155
0460	6"					345			345	380
0500	Coupling, stainless steel, TFE seal ring									
0520	1-1/2"	Q-1	32	.500	Ea.	19.35	22		41.35	55.50
0530	2"		30	.533		24.50	23.50		48	63.50
0540	3"		25	.640		33	28		61	80
0550	4"		23	.696		56.50	30.50		87	110
0560	6"		20	.800		127	35		162	195
0600	Coupling, stainless steel, bead to plain end									
0610	1-1/2"	Q-1	36	.444	Ea.	25	19.50		44.50	57.50
0620	2"		34	.471		31	20.50		51.50	66
0630	3"		29	.552		53	24		77	95.50
0640	4"		27	.593		79	26		105	127
0650	6"		24	.667		266	29.50		295.50	340
2000	Process supply (pressure), beaded ends									
2050	90° Sweep elbow, 1/2"				Ea.	28			28	31
2070	3/4"					33			33	36.50
2080	1"					44			44	48
2090	1-1/2"					46			46	50.50
2100	2"					62			62	68
2120	3"					97.50			97.50	107
2130	4"					137			137	151
2140	6"					350			350	385
2150	45° Sweep elbow, same as 90°									
2250	Tee, straight, 1/2"				Ea.	47.50			47.50	52.50
2270	3/4"					47.50			47.50	52.50
2280	1"					53			53	58
2290	1-1/2"					78.50			78.50	86
2300	2"					93			93	102
2310	3"					113			113	124
2320	4"					190			190	209
2330	6"					360			360	400
2350	Coupling, Viton liner, for temperatures to 400° F									
2370	1/2"	Q-1	40	.400	Ea.	40.50	17.55		58.05	71.50
2380	3/4"		37	.432		46.50	19		65.50	80.50
2390	1"		35	.457		55	20		75	91.50
2400	1-1/2"		32	.500		62	22		84	103
2410	2"		30	.533		74.50	23.50		98	119
2420	3"		25	.640		137	28		165	195
2430	4"		23	.696		166	30.50		196.50	231

22 66 53.40 Pipe Fittings, Glass		Crew	Daily Output	Labor-Hours	Unit	2009 Bare Costs Material	Labor	Equipment	Total	Total Incl O&P
2440	6″	Q-1	20	.800	Ea.	390	35		425	485
2550	For beaded joint armored fittings, add					200%				
2600	Conical ends. Flange set, gasket & labor not incl. in fitting									
2620	price. Add 1 per joint for installed price.									
2650	90° Sweep elbow, 1″				Ea.	94.50			94.50	104
2670	1-1/2″					191			191	210
2680	2″					199			199	218
2690	3″					310			310	340
2700	4″					545			545	600
2710	6″					960			960	1,050
2750	Cross (straight), add					55%				
2850	Tee, add					20%				
9000	Minimum labor/equipment charge	1 Plum	4	2	Job		97.50		97.50	151

22 66 53.60 Corrosion Resistant Pipe		Crew	Daily Output	Labor-Hours	Unit	Material	Labor	Equipment	Total	Total Incl O&P
0010	**CORROSION RESISTANT PIPE**, No couplings or hangers									
0020	Iron alloy, drain, mechanical joint									
1000	1-1/2″ diameter	Q-1	70	.229	L.F.	34.50	10.05		44.55	53.50
1100	2″ diameter		66	.242		35.50	10.65		46.15	55.50
1120	3″ diameter		60	.267		45.50	11.70		57.20	68
1140	4″ diameter		52	.308		58.50	13.50		72	85.50
1980	Iron alloy, drain, B&S joint									
2000	2″ diameter	Q-1	54	.296	L.F.	42.50	13		55.50	66.50
2100	3″ diameter		52	.308		53.50	13.50		67	80
2120	4″ diameter		48	.333		74	14.65		88.65	104
2140	6″ diameter	Q-2	59	.407		119	18.50		137.50	160
2160	8″ diameter	"	54	.444		224	20		244	278
2980	Plastic, epoxy, fiberglass filament wound, B&S joint									
3000	2″ diameter	Q-1	62	.258	L.F.	9.40	11.30		20.70	28
3100	3″ diameter		51	.314		11.10	13.75		24.85	34
3120	4″ diameter		45	.356		16.10	15.60		31.70	41.50
3140	6″ diameter		32	.500		23	22		45	59
3160	8″ diameter	Q-2	38	.632		35.50	28.50		64	83.50
3180	10″ diameter		32	.750		49.50	34		83.50	108
3200	12″ diameter		28	.857		59	39		98	126
3980	Polyester, fiberglass filament wound, B&S joint									
4000	2″ diameter	Q-1	62	.258	L.F.	10.30	11.30		21.60	29
4100	3″ diameter		51	.314		14.05	13.75		27.80	37
4120	4″ diameter		45	.356		20	15.60		35.60	46
4140	6″ diameter		32	.500		29.50	22		51.50	66.50
4160	8″ diameter	Q-2	38	.632		69.50	28.50		98	121
4180	10″ diameter		32	.750		84.50	34		118.50	146
4200	12″ diameter		28	.857		85.50	39		124.50	155
4980	Polypropylene, acid resistant, fire retardant, schedule 40									
5000	1-1/2″ diameter	Q-1	68	.235	L.F.	4.98	10.30		15.28	21.50
5100	2″ diameter		62	.258		6.75	11.30		18.05	25
5120	3″ diameter		51	.314		13.80	13.75		27.55	36.50
5140	4″ diameter		45	.356		17.60	15.60		33.20	43.50
5160	6″ diameter		32	.500		35	22		57	72.50
5980	Proxylene, fire retardant, Schedule 40									
6000	1-1/2″ diameter	Q-1	68	.235	L.F.	7.70	10.30		18	24.50
6100	2″ diameter		62	.258		10.60	11.30		21.90	29
6120	3″ diameter		51	.314		19.15	13.75		32.90	42.50

22 66 53 – Laboratory Chemical-Waste and Vent Piping

22 66 53.60 Corrosion Resistant Pipe		Crew	Daily Output	Labor-Hours	Unit	Material (2009 Bare Costs)	Labor (2009 Bare Costs)	Equipment (2009 Bare Costs)	Total (2009 Bare Costs)	Total Incl O&P
6140	4" diameter	Q-1	45	.356	L.F.	27	15.60		42.60	54
6160	6" diameter	↓	32	.500	↓	46	22		68	84.50
6820	For Schedule 80, add				↓	35%	2%			
9800	Minimum labor/equipment charge	1 Plum	4	2	Job		97.50		97.50	151

22 66 53.70 Pipe Fittings, Corrosion Resistant		Crew	Daily Output	Labor-Hours	Unit	Material	Labor	Equipment	Total	Total Incl O&P
0010	**PIPE FITTINGS, CORROSION RESISTANT**									
0030	Iron alloy									
0050	Mechanical joint									
0060	1/4 Bend, 1-1/2"	Q-1	12	1.333	Ea.	47	58.50		105.50	142
0080	2"		10	1.600		76.50	70		146.50	193
0090	3"		9	1.778		92	78		170	222
0100	4"		8	2		105	88		193	252
0110	1/8 Bend, 1-1/2"		12	1.333		30	58.50		88.50	123
0130	2"		10	1.600		51	70		121	165
0140	3"		9	1.778		68	78		146	196
0150	4"	↓	8	2	↓	91	88		179	236
0160	Tee and Y, sanitary, straight									
0170	1-1/2"	Q-1	8	2	Ea.	51	88		139	192
0180	2"		7	2.286		68	100		168	231
0190	3"		6	2.667		105	117		222	297
0200	4"		5	3.200		194	140		334	430
0360	Coupling, 1-1/2"		14	1.143		35.50	50		85.50	118
0380	2"		12	1.333		41	58.50		99.50	136
0390	3"		11	1.455		42.50	64		106.50	146
0400	4"	↓	10	1.600	↓	48.50	70		118.50	163
0500	Bell & Spigot									
0510	1/4 and 1/16 bend, 2"	Q-1	16	1	Ea.	53.50	44		97.50	127
0520	3"		14	1.143		125	50		175	215
0530	4"	↓	13	1.231		128	54		182	224
0540	6"	Q-2	17	1.412		253	64		317	380
0550	8"	"	12	2		1,050	91		1,141	1,300
0620	1/8 bend, 2"	Q-1	16	1		60	44		104	134
0640	3"		14	1.143		111	50		161	200
0650	4"	↓	13	1.231		117	54		171	213
0660	6"	Q-2	17	1.412		212	64		276	335
0680	8"	"	12	2		825	91		916	1,050
0700	Tee, sanitary, 2"	Q-1	10	1.600		114	70		184	234
0710	3"		9	1.778		370	78		448	525
0720	4"	↓	8	2		305	88		393	470
0730	6"	Q-2	11	2.182		385	99.50		484.50	575
0740	8"	"	8	3		1,050	137		1,187	1,350
1800	Y, sanitary, 2"	Q-1	10	1.600		120	70		190	241
1820	3"		9	1.778		215	78		293	355
1830	4"	↓	8	2		193	88		281	350
1840	6"	Q-2	11	2.182		640	99.50		739.50	860
1850	8"	"	8	3	↓	1,775	137		1,912	2,150
3000	Epoxy, filament wound									
3030	Quick-lock joint									
3040	90° Elbow, 2"	Q-1	28	.571	Ea.	79.50	25		104.50	127
3060	3"		16	1		91.50	44		135.50	169
3070	4"		13	1.231		125	54		179	222
3080	6"	↓	8	2	↓	182	88		270	335

22 66 53.70 Pipe Fittings, Corrosion Resistant		Crew	Daily Output	Labor-Hours	Unit	2009 Bare Costs Material	Labor	Equipment	Total	Total Incl O&P
3090	8"	Q-2	9	2.667	Ea.	335	121		456	560
3100	10"		7	3.429		425	156		581	705
3110	12"		6	4		605	182		787	950
3120	45° Elbow, 2"	Q-1	28	.571		61.50	25		86.50	107
3130	3"		16	1		87.50	44		131.50	164
3140	4"		13	1.231		121	54		175	217
3150	6"		8	2		182	88		270	335
3160	8"	Q-2	9	2.667		335	121		456	560
3170	10"		7	3.429		425	156		581	705
3180	12"		6	4		605	182		787	950
3190	Tee, 2"	Q-1	19	.842		190	37		227	267
3200	3"		11	1.455		230	64		294	350
3210	4"		9	1.778		276	78		354	425
3220	6"		5	3.200		420	140		560	685
3230	8"	Q-2	6	4		525	182		707	860
3240	10"		5	4.800		735	218		953	1,150
3250	12"		4	6		1,025	273		1,298	1,550
4000	Polypropylene, acid resistant									
4020	Non-pressure, electrofusion joints									
4050	1/4 bend, 1-1/2"	1 Plum	16	.500	Ea.	9.60	24.50		34.10	48.50
4060	2"	Q-1	28	.571		18.60	25		43.60	59.50
4080	3"		17	.941		20.50	41.50		62	86.50
4090	4"		14	1.143		33.50	50		83.50	115
4110	6"		8	2		79	88		167	223
4150	1/4 Bend, long sweep									
4170	1-1/2"	1 Plum	16	.500	Ea.	10.80	24.50		35.30	50
4180	2"	Q-1	28	.571		18	25		43	59
4200	3"		17	.941		22	41.50		63.50	88.50
4210	4"		14	1.143		36	50		86	118
4250	1/8 bend, 1-1/2"	1 Plum	16	.500		9	24.50		33.50	48
4260	2"	Q-1	28	.571		11.40	25		36.40	51.50
4280	3"		17	.941		21.50	41.50		63	88
4290	4"		14	1.143		24	50		74	105
4310	6"		8	2		66.50	88		154.50	210
4400	Tee, sanitary									
4420	1-1/2"	1 Plum	10	.800	Ea.	12	39		51	73.50
4430	2"	Q-1	17	.941		14.40	41.50		55.90	80
4450	3"		11	1.455		29	64		93	131
4460	4"		9	1.778		42.50	78		120.50	168
4480	6"		5	3.200		168	140		308	405
4490	Tee, sanitary reducing, 2" x 2" x 1-1/2"		17	.941		14.40	41.50		55.90	80
4492	3" x 3" x 2"		11	1.455		29	64		93	131
4494	4" x 4" x 3"		9	1.778		41.50	78		119.50	167
4496	6" x 6" x 4"		5	3.200		139	140		279	370
4500	Tee/wye, long turn									
4520	1-1/2"	1 Plum	10	.800	Ea.	16.20	39		55.20	78.50
4530	2"	Q-1	17	.941		20.50	41.50		62	86.50
4550	3"		11	1.455		35	64		99	138
4570	4"		9	1.778		47.50	78		125.50	173
4650	Wye 45°, 1-1/2"	1 Plum	10	.800		13.20	39		52.20	75
4652	2"	Q-1	17	.941		18	41.50		59.50	84
4653	3"		11	1.455		30.50	64		94.50	133
4654	4"		9	1.778		44.50	78		122.50	170

22 66 53 – Laboratory Chemical-Waste and Vent Piping

22 66 53.70 Pipe Fittings, Corrosion Resistant		Crew	Daily Output	Labor-Hours	Unit	2009 Bare Costs Material	Labor	Equipment	Total	Total Incl O&P
4656	6″	Q-1	5	3.200	Ea.	114	140		254	345
9000	Minimum labor/equipment charge	1 Plum	4	2	Job		97.50		97.50	151

22 66 83 – Chemical-Waste Tanks

22 66 83.13 Chemical-Waste Dilution Tanks		Crew	Daily Output	Labor-Hours	Unit	Material	Labor	Equipment	Total	Total Incl O&P
0010	**CHEMICAL-WASTE DILUTION TANKS**									
7000	Tanks, covers included									
7800	Polyethylene									
7810	Continuous service to 220° F									
7830	2 gallon, 8″ x 8″ x 8″	Q-1	20	.800	Ea.	69	35		104	131
7850	7 gallon, 12″ x 12″ x 12″		20	.800		83.50	35		118.50	147
7870	15 gallon, 24″ x 12″ x 12″		17	.941		192	41.50		233.50	275
8010	30 gallon, 24″ x 18″ x 18″		12	1.333		254	58.50		312.50	370
8070	45 gallon, 24″ x 18″ x 24″		10	1.600		340	70		410	485
8080	90 gallon, 36″ x 24″ x 24″	↓	8	2	↓	580	88		668	775
8150	Polyethylene, heavy duty walls									
8160	Continuous service to 180° F									
8180	5 gallon, 11″ I.D. x 15″ deep	Q-1	20	.800	Ea.	47	35		82	106
8210	15 gallon, 14″ I.D. x 27″ deep		17	.941		81.50	41.50		123	154
8230	55 gallon, 22″ I.D. x 36″ deep		10	1.600		176	70		246	305
8250	100 gallon 28″ I.D. x 42″ deep		8	2		385	88		473	555
8270	200 gallon 36″ I.D. x 48″ deep		6	2.667		565	117		682	805
8290	360 gallon 48″ I.D. x 48″ deep	↓	5	3.200	↓	915	140		1,055	1,225

Estimating Tips

The labor adjustment factors listed in Subdivision 22 01 02.20 also apply to Division 23.

23 10 00 Facility Fuel Systems

- The prices in this subdivision for above- and below-ground storage tanks do not include foundations or hold-down slabs. The estimator should refer to Divisions 3 and 31 for foundation system pricing. In addition to the foundations, required tank accessories, such as tank gauges, leak detection devices, and additional manholes and piping, must be added to the tank prices.

23 50 00 Central Heating Equipment

- When estimating the cost of an HVAC system, check to see who is responsible for providing and installing the temperature control system. It is possible to overlook controls, assuming that they would be included in the electrical estimate.
- When looking up a boiler, be careful on specified capacity. Some manufacturers rate their products on output while others use input.
- Include HVAC insulation for pipe, boiler, and duct (wrap and liner).
- Be careful when looking up mechanical items to get the correct pressure rating and connection type (thread, weld, flange).

23 70 00 Central HVAC Equipment

- Combination heating and cooling units are sized by the air conditioning requirements. (See Reference No. R236000-20 for preliminary sizing guide.)
- A ton of air conditioning is nominally 400 CFM.
- Rectangular duct is taken off by the linear foot for each size, but its cost is usually estimated by the pound. Remember that SMACNA standards now base duct on internal pressure.
- Prefabricated duct is estimated and purchased like pipe: straight sections and fittings.
- Note that cranes or other lifting equipment are not included on any lines in Division 23. For example, if a crane is required to lift a heavy piece of pipe into place high above a gym floor, or to put a rooftop unit on the roof of a four-story building, etc., it must be added. Due to the potential for extreme variation—from nothing additional required to a major crane or helicopter—we feel that including a nominal amount for "lifting contingency" would be useless and detract from the accuracy of the estimate. When using equipment rental cost data from RSMeans, do not forget to include the cost of the operator(s).

Reference Numbers

Reference numbers are shown in shaded boxes at the beginning of some major classifications. These numbers refer to related items in the Reference Section. The reference information may be an estimating procedure, an alternate pricing method, or technical information.

Note: Not all subdivisions listed here necessarily appear in this publication.

Note: **Trade Service,** *in part, has been used as a reference source for some of the material prices used in Division 23.*

23 05 05 – Selective HVAC Demolition

23 05 05.10 HVAC Demolition		Crew	Daily Output	Labor-Hours	Unit	Material	2009 Bare Costs Labor	Equipment	Total	Total Incl O&P
0010	**HVAC DEMOLITION** R220105-10									
0100	Air conditioner, split unit, 3 ton	Q-5	2	8	Ea.		355		355	550
0150	Package unit, 3 ton R024119-10	Q-6	3	8			370		370	570
0190	Rooftop, self contained, up to 5 ton	1 Plum	1.20	6.667			325		325	505
0250	Air curtain	Q-9	20	.800	L.F.		34		34	54
0254	Air filters, up thru 16,000 CFM		20	.800	Ea.		34		34	54
0256	20,000 thru 60,000 CFM		16	1			42.50		42.50	67.50
0297	Boiler blowdown	Q-5	8	2			89		89	138
0298	Boilers									
0300	Electric, up thru 148 kW	Q-19	2	12	Ea.		545		545	840
0310	150 thru 518 kW	″	1	24			1,075		1,075	1,675
0320	550 thru 2000 kW	Q-21	.40	80			3,700		3,700	5,725
0330	2070 kW and up	″	.30	106			4,950		4,950	7,650
0340	Gas and/or oil, up thru 150 MBH	Q-7	2.20	14.545			685		685	1,050
0350	160 thru 2000 MBH		.80	40			1,875		1,875	2,925
0360	2100 thru 4500 MBH		.50	64			3,000		3,000	4,675
0370	4600 thru 7000 MBH		.30	106			5,025		5,025	7,775
0380	7100 thru 12,000 MBH		.16	200			9,400		9,400	14,600
0390	12,200 thru 25,000 MBH		.12	266			12,500		12,500	19,400
0400	Central station air handler unit, up thru 15 ton	Q-5	1.60	10			445		445	690
0410	17.5 thru 30 ton	″	.80	20			890		890	1,375
0430	Computer room unit									
0434	Air cooled split, up thru 10 ton	Q-5	.67	23.881	Ea.		1,050		1,050	1,650
0436	12 thru 23 ton		.53	30.189			1,350		1,350	2,075
0440	Chilled water, up thru 10 ton		1.30	12.308			545		545	850
0444	12 thru 23 ton		1	16			710		710	1,100
0450	Glycol system, up thru 10 ton		.53	30.189			1,350		1,350	2,075
0454	12 thru 23 ton		.40	40			1,775		1,775	2,750
0460	Water cooled, not including condenser, up thru 10 ton		.80	20			890		890	1,375
0464	12 thru 23 ton		.60	26.667			1,175		1,175	1,825
0600	Condenser, up thru 50 ton		1	16			710		710	1,100
0610	51 thru 100 ton	Q-6	.80	30			1,375		1,375	2,150
0620	101 thru 1000 ton	″	.16	150			6,900		6,900	10,700
0660	Condensing unit, up thru 10 ton	Q-5	1.25	12.800			570		570	880
0670	11 thru 50 ton	″	.40	40			1,775		1,775	2,750
0680	60 thru 100 ton	Q-6	.30	80			3,675		3,675	5,725
0700	Cooling tower, up thru 400 ton		.80	30			1,375		1,375	2,150
0710	450 thru 600 ton		.53	45.283			2,075		2,075	3,225
0720	700 thru 1300 ton		.40	60			2,775		2,775	4,275
0780	Dehumidifier, up thru 155 lb./hr.	Q-1	8	2			88		88	136
0790	240 lb./hr. and up	″	2	8			350		350	545
1560	Ductwork									
1570	Metal, steel, sst, fabricated	Q-9	1000	.016	Lb.		.68		.68	1.08
1580	Aluminum, fabricated		485	.033	″		1.40		1.40	2.23
1590	Spiral, prefabricated		400	.040	L.F.		1.70		1.70	2.70
1600	Fiberglass, prefabricated		400	.040			1.70		1.70	2.70
1610	Flex, prefabricated		500	.032			1.36		1.36	2.16
1620	Glass fiber reinforced plastic, prefabricated		280	.057			2.43		2.43	3.86
1630	Diffusers, registers or grills, up thru 20″ max dimension	1 Shee	50	.160	Ea.		7.55		7.55	12
1640	21 thru 36″ max dimension		36	.222			10.50		10.50	16.65
1650	Above 36″ max dimension		30	.267			12.60		12.60	20
1700	Evaporator, up thru 12,000 BTUH	Q-5	5.30	3.019			134		134	208
1710	12,500 thru 30,000 BTUH	″	2.70	5.926			263		263	410

23 05 05 – Selective HVAC Demolition

23 05 05.10 HVAC Demolition		Crew	Daily Output	Labor-Hours	Unit	Material	2009 Bare Costs Labor	Equipment	Total	Total Incl O&P
1720	31,000 BTUH and up	Q-6	1.50	16	Ea.		735		735	1,150
1730	Evaporative cooler, up thru 5 H.P.	Q-9	2.70	5.925			252		252	400
1740	10 thru 30 H.P.	″	.67	23.881			1,025		1,025	1,600
1750	Exhaust systems									
1760	Exhaust components	1 Shee	8	1	System		47		47	75
1770	Weld fume hoods	″	20	.400	Ea.		18.90		18.90	30
2120	Fans, up thru 1 H.P. or 2000 CFM	Q-9	8	2			85		85	135
2124	1-1/2 thru 10 H.P. or 20,000 CFM		5.30	3.019			128		128	204
2128	15 thru 30 H.P. or above 20,000 CFM		4	4			170		170	270
2150	Fan coil air conditioner, chilled water, up thru 7.5 ton	Q-5	14	1.143			51		51	78.50
2154	Direct expansion, up thru 10 ton		8	2			89		89	138
2158	11 thru 30 ton		2	8			355		355	550
2170	Flue shutter damper	Q-9	8	2			85		85	135
2200	Furnace, electric	Q-20	2	10			435		435	685
2300	Gas or oil, under 120 MBH	Q-9	4	4			170		170	270
2340	Over 120 MBH	″	3	5.333			227		227	360
2730	Heating and ventilating unit	Q-5	2.70	5.926			263		263	410
2740	Heater, electric, wall, baseboard and quartz	1 Elec	10	.800			37.50		37.50	58
2750	Heater, electric, unit, cabinet, fan and convector	″	8	1			47		47	72.50
2760	Heat exchanger, shell and tube type	Q-5	1.60	10			445		445	690
2770	Plate type	Q-6	.60	40			1,850		1,850	2,850
2810	Heat pump									
2820	Air source, split, up thru 10 ton	Q-5	.90	17.778	Ea.		790		790	1,225
2830	15 thru 25 ton	Q-6	.80	30			1,375		1,375	2,150
2850	Single package, up thru 12 ton	Q-5	1	16			710		710	1,100
2860	Water source, up thru 15 ton	″	.90	17.778			790		790	1,225
2870	20 thru 50 ton	Q-6	.80	30			1,375		1,375	2,150
2910	Heat recovery package, up thru 20,000 CFM	Q-5	2	8			355		355	550
2920	25,000 CFM and up		1.20	13.333			590		590	920
2930	Heat transfer package, up thru 130 GPM		.80	20			890		890	1,375
2934	255 thru 800 GPM		.42	38.095			1,700		1,700	2,625
2940	Humidifier		10.60	1.509			67		67	104
2961	Hydronic unit heaters, up thru 200 MBH		14	1.143			51		51	78.50
2962	Above 200 MBH		8	2			89		89	138
2964	Valance units		32	.500			22		22	34.50
2966	Radiant floor heating									
2967	System valves, controls, manifolds	Q-5	16	1	Ea.		44.50		44.50	69
2968	Per room distribution		8	2			89		89	138
2970	Hydronic heating, baseboard radiation		16	1			44.50		44.50	69
2976	Convectors and free standing radiators		18	.889			39.50		39.50	61
2980	Induced draft fan, up thru 1 H.P.	Q-9	4.60	3.478			148		148	235
2984	1-1/2 H.P. thru 7-1/2 H.P.	″	2.20	7.273			310		310	490
2988	Infra-red unit	Q-5	16	1			44.50		44.50	69
2992	Louvers	1 Shee	46	.174	S.F.		8.20		8.20	13.05
3000	Mechanical equipment, light items. Unit is weight, not cooling.	Q-5	.90	17.778	Ton		790		790	1,225
3600	Heavy items	″	1.10	14.545	″		645		645	1,000
3700	Deduct for salvage (when applicable), minimum				Job				73	80
3710	Maximum				″				455	500
3720	Make up air unit, up thru 6000 CFM	Q-5	3	5.333	Ea.		237		237	365
3730	6500 thru 30,000 CFM	″	1.60	10			445		445	690
3740	35,000 thru 75,000 CFM	Q-6	1	24			1,100		1,100	1,725
3800	Mixing boxes, constant and VAV	Q-9	18	.889			38		38	60
4000	Packaged terminal air conditioner, up thru 18,000 BTUH	Q-5	8	2			89		89	138

23 05 05 – Selective HVAC Demolition

23 05 05.10 HVAC Demolition		Crew	Daily Output	Labor-Hours	Unit	2009 Bare Costs Material	Labor	Equipment	Total	Total Incl O&P
4010	24,000 thru 48,000 BTUH	Q-5	2.80	5.714	Ea.		254		254	395
5000	Refrigerant compressor, reciprocating or scroll									
5010	Up thru 5 ton	1 Stpi	6	1.333	Ea.		66		66	102
5020	5.08 thru 10 ton	Q-5	6	2.667			118		118	184
5030	15 thru 50 ton	"	.40	40			1,775		1,775	2,750
5040	60 thru 130 ton	Q-6	.48	50			2,300		2,300	3,575
5100	Roof top air conditioner, up thru 10 ton	Q-5	1.40	11.429			510		510	785
5110	12 thru 40 ton	Q-6	1	24			1,100		1,100	1,725
5120	50 thru 140 ton		.50	48			2,200		2,200	3,425
5130	150 thru 300 ton		.30	80			3,675		3,675	5,725
6000	Self contained single package air conditioner, up thru 10 ton	Q-5	1.60	10			445		445	690
6010	15 thru 60 ton	Q-6	1.20	20			920		920	1,425
6100	Space heaters, up thru 200 MBH	Q-5	10	1.600			71		71	110
6110	Over 200 MBH		5	3.200			142		142	220
6200	Split ductless, both sections		8	2			89		89	138
6300	Steam condensate meter	1 Stpi	11	.727			36		36	55.50
6600	Thru-the-wall air conditioner	L-2	8	2			70.50		70.50	116
7000	Vent chimney, prefabricated, up thru 12" diameter	Q-9	94	.170	V.L.F.		7.25		7.25	11.50
7010	14" thru 36" diameter		40	.400			17		17	27
7020	38" thru 48" diameter		32	.500			21		21	34
7030	54" thru 60" diameter	Q-10	14	1.714			75.50		75.50	120
7400	Ventilators, up thru 14" neck diameter	Q-9	58	.276	Ea.		11.70		11.70	18.60
7410	16" thru 50" neck diameter		40	.400			17		17	27
7450	Relief vent, up thru 24" x 96"		22	.727			31		31	49
7460	48" x 60" thru 96" x 144"		10	1.600			68		68	108
8000	Water chiller up thru 10 ton	Q-5	2.50	6.400			284		284	440
8010	15 thru 100 ton	Q-6	.48	50			2,300		2,300	3,575
8020	110 thru 500 ton	Q-7	.29	110			5,175		5,175	8,050
8030	600 thru 1000 ton		.23	139			6,550		6,550	10,100
8040	1100 ton and up		.20	160			7,525		7,525	11,700
8400	Window air conditioner	1 Carp	16	.500			20		20	33
9000	Minimum labor/equipment charge	Q-6	3	8	Job		370		370	570

23 05 23 – General-Duty Valves for HVAC Piping

23 05 23.30 Valves, Iron Body		Crew	Daily Output	Labor-Hours	Unit	Material	Labor	Equipment	Total	Total Incl O&P
0010	**VALVES, IRON BODY** R220523-90									
0022	For grooved joint, see Div. 22 11 13.48									
0560	Butterfly, lug type, pneumatic operator, 2" size	1 Stpi	14	.571	Ea.	192	28		220	255
0570	3"	Q-1	8	2		205	88		293	360
0580	4"	"	5	3.200		230	140		370	470
0590	6"	Q-2	5	4.800		320	218		538	690
0600	8"		4.50	5.333		410	243		653	825
0610	10"		4	6		520	273		793	995
0620	12"		3	8		820	365		1,185	1,475
0630	14"		2.30	10.435		1,325	475		1,800	2,200
0640	18"		1.50	16		2,175	730		2,905	3,525
0650	20"		1	24		2,650	1,100		3,750	4,600
0790	Butterfly, lug type, gear operated, 2" size, 200 lb. except noted	1 Plum	14	.571		254	28		282	320
0800	2-1/2"	Q-1	9	1.778		255	78		333	400
0810	3"		8	2		268	88		356	430
0820	4"		5	3.200		305	140		445	555
0830	5"	Q-2	5	4.800		375	218		593	755
0840	6"		5	4.800		420	218		638	800

23 05 23 – General-Duty Valves for HVAC Piping

23 05 23.30 Valves, Iron Body		Crew	Daily Output	Labor-Hours	Unit	Material	2009 Bare Costs Labor	Equipment	Total	Total Incl O&P
0850	8"	Q-2	4.50	5.333	Ea.	540	243		783	970
0860	10"		4	6		715	273		988	1,200
0870	12"		3	8		1,050	365		1,415	1,725
0880	14", 150 lb.		2.30	10.435		1,975	475		2,450	2,900
0890	16", 150 lb.		1.75	13.714		2,525	625		3,150	3,775
0900	18", 150 lb.		1.50	16		3,625	730		4,355	5,100
0910	20", 150 lb.		1	24		4,575	1,100		5,675	6,725
0930	24", 150 lb.		.75	32		8,200	1,450		9,650	11,300
1020	Butterfly, wafer type, gear actuator, 200 lb.									
1030	2"	1 Plum	14	.571	Ea.	155	28		183	213
1040	2-1/2"	Q-1	9	1.778		159	78		237	296
1050	3"		8	2		165	88		253	320
1060	4"		5	3.200		188	140		328	425
1070	5"	Q-2	5	4.800		249	218		467	615
1080	6"		5	4.800		281	218		499	650
1090	8"		4.50	5.333		320	243		563	725
1100	10"		4	6		470	273		743	940
1110	12"		3	8		650	365		1,015	1,275
1200	Wafer type, lever actuator, 200 lb.									
1220	2"	1 Plum	14	.571	Ea.	113	28		141	167
1230	2-1/2"	Q-1	9	1.778		116	78		194	248
1240	3"		8	2		122	88		210	270
1250	4"		5	3.200		149	140		289	380
1260	5"	Q-2	5	4.800		207	218		425	570
1270	6"		5	4.800		250	218		468	615
1280	8"		4.50	5.333		370	243		613	780
1290	10"		4	6		515	273		788	990
1300	12"		3	8		930	365		1,295	1,600
1650	Gate, 125 lb., N.R.S.									
2150	Flanged									
2200	2"	1 Plum	5	1.600	Ea.	495	78		573	665
2240	2-1/2"	Q-1	5	3.200		510	140		650	780
2260	3"		4.50	3.556		570	156		726	865
2280	4"		3	5.333		815	234		1,049	1,250
2290	5"	Q-2	3.40	7.059		1,400	320		1,720	2,025
2300	6"		3	8		1,400	365		1,765	2,100
2320	8"		2.50	9.600		2,375	435		2,810	3,300
2340	10"		2.20	10.909		4,200	495		4,695	5,375
2360	12"		1.70	14.118		5,750	640		6,390	7,325
2370	14"		1.30	18.462		7,475	840		8,315	9,525
2380	16"		1	24		10,400	1,100		11,500	13,200
2420	For 250 lb flanged, add					200%	10%			
3550	OS&Y, 125 lb., flanged									
3600	2"	1 Plum	5	1.600	Ea.	310	78		388	460
3640	2-1/2"	Q-1	5	3.200		320	140		460	570
3660	3"		4.50	3.556		355	156		511	635
3670	3-1/2"		3	5.333		405	234		639	810
3680	4"		3	5.333		510	234		744	925
3690	5"	Q-2	3.40	7.059		840	320		1,160	1,425
3700	6"		3	8		840	365		1,205	1,500
3720	8"		2.50	9.600		1,500	435		1,935	2,325
3740	10"		2.20	10.909		2,725	495		3,220	3,775
3760	12"		1.70	14.118		3,650	640		4,290	5,025

23 05 23 – General-Duty Valves for HVAC Piping

23 05 23.30 Valves, Iron Body		Crew	Daily Output	Labor-Hours	Unit	2009 Bare Costs Material	Labor	Equipment	Total	Total Incl O&P
3770	14"	Q-2	1.30	18.462	Ea.	6,975	840		7,815	8,950
3780	16"		1	24		10,400	1,100		11,500	13,200
3790	18"		.80	30		13,500	1,375		14,875	16,900
3800	20"		.60	40		19,500	1,825		21,325	24,300
3830	24"		.50	48		27,800	2,175		29,975	34,000
3900	For 175 lb, flanged, add					200%	10%			
4350	Globe, OS&Y									
4540	Class 125, flanged									
4550	2"	1 Plum	5	1.600	Ea.	630	78		708	815
4560	2-1/2"	Q-1	5	3.200		635	140		775	920
4570	3"		4.50	3.556		775	156		931	1,100
4580	4"		3	5.333		1,100	234		1,334	1,600
4590	5"	Q-2	3.40	7.059		2,025	320		2,345	2,725
4600	6"		3	8		2,025	365		2,390	2,800
4610	8"		2.50	9.600		3,950	435		4,385	5,025
5040	Class 250, flanged									
5050	2"	1 Plum	4.50	1.778	Ea.	1,000	86.50		1,086.50	1,250
5060	2-1/2"	Q-1	4.50	3.556		1,325	156		1,481	1,700
5070	3"		4	4		1,350	176		1,526	1,775
5080	4"		2.70	5.926		2,000	260		2,260	2,600
5090	5"	Q-2	3	8		3,575	365		3,940	4,500
5100	6"		2.70	8.889		3,575	405		3,980	4,550
5110	8"		2.20	10.909		6,075	495		6,570	7,450
5120	10"		2	12		9,275	545		9,820	11,000
5130	12"		1.60	15		13,900	685		14,585	16,400
5450	Swing check, 125 lb., threaded									
5470	1"	1 Plum	13	.615	Ea.	560	30		590	660
5500	2"	"	11	.727		635	35.50		670.50	755
5540	2-1/2"	Q-1	15	1.067		645	47		692	780
5550	3"		13	1.231		715	54		769	870
5560	4"		10	1.600		1,225	70		1,295	1,450
5950	Flanged									
6000	2"	1 Plum	5	1.600	Ea.	230	78		308	375
6040	2-1/2"	Q-1	5	3.200		258	140		398	500
6050	3"		4.50	3.556		276	156		432	545
6060	4"		3	5.333		435	234		669	840
6070	6"	Q-2	3	8		740	365		1,105	1,375
6080	8"		2.50	9.600		1,400	435		1,835	2,200
6090	10"		2.20	10.909		2,375	495		2,870	3,400
6100	12"		1.70	14.118		3,850	640		4,490	5,250
6160	For 250 lb flanged, add					200%	20%			
6600	Silent check, bronze trim									
6610	Compact wafer type, for 125 or 150 lb. flanges									
6630	1-1/2"	1 Plum	11	.727	Ea.	107	35.50		142.50	172
6640	2"	"	9	.889		127	43.50		170.50	207
6650	2-1/2"	Q-1	9	1.778		139	78		217	274
6660	3"		8	2		150	88		238	300
6670	4"		5	3.200		191	140		331	430
6680	5"	Q-2	6	4		249	182		431	555
6690	6"		6	4		335	182		517	645
6700	8"		4.50	5.333		595	243		838	1,025
6710	10"		4	6		1,025	273		1,298	1,550
6720	12"		3	8		1,975	365		2,340	2,725

23 05 23 – General-Duty Valves for HVAC Piping

23 05 23.30 Valves, Iron Body		Crew	Daily Output	Labor-Hours	Unit	Material	2009 Bare Costs Labor	Equipment	Total	Total Incl O&P
9000	Minimum labor/equipment charge	1 Plum	3	2.667	Job		130		130	202

23 05 23.70 Valves, Semi-Steel		Crew	Daily Output	Labor-Hours	Unit	Material	2009 Bare Costs Labor	Equipment	Total	Total Incl O&P
0010	**VALVES, SEMI-STEEL**									
1020	Lubricated plug valve, threaded, 200 psi									
1030	1/2"	1 Plum	18	.444	Ea.	73	21.50		94.50	114
1040	3/4"		16	.500		73	24.50		97.50	118
1050	1"		14	.571		93	28		121	145
1060	1-1/4"		12	.667		110	32.50		142.50	172
1070	1-1/2"		11	.727		119	35.50		154.50	186
1080	2"	↓	8	1		142	49		191	232
1090	2-1/2"	Q-1	5	3.200		218	140		358	455
1100	3"	"	4.50	3.556	↓	268	156		424	535
6990	Flanged, 200 psi									
7000	2"	1 Plum	8	1	Ea.	176	49		225	270
7010	2-1/2"	Q-1	5	3.200		253	140		393	495
7020	3"		4.50	3.556		305	156		461	575
7030	4"		3	5.333		385	234		619	785
7036	5"	↓	2.50	6.400		840	281		1,121	1,350
7040	6"	Q-2	3	8		630	365		995	1,250
7050	8"		2.50	9.600		1,225	435		1,660	2,025
7060	10"		2.20	10.909		1,950	495		2,445	2,900
7070	12"	↓	1.70	14.118	↓	4,425	640		5,065	5,850

23 05 23.80 Valves, Steel		Crew	Daily Output	Labor-Hours	Unit	Material	2009 Bare Costs Labor	Equipment	Total	Total Incl O&P
0010	**VALVES, STEEL** R220523-90									
0800	Cast									
1350	Check valve, swing type, 150 lb., flanged									
1370	1"	1 Plum	10	.800	Ea.	510	39		549	620
1400	2"	"	8	1		640	49		689	780
1440	2-1/2"	Q-1	5	3.200		740	140		880	1,025
1450	3"		4.50	3.556		755	156		911	1,075
1460	4"	↓	3	5.333		1,075	234		1,309	1,575
1470	6"	Q-2	3	8		1,700	365		2,065	2,450
1480	8"		2.50	9.600		2,775	435		3,210	3,725
1490	10"		2.20	10.909		4,100	495		4,595	5,300
1500	12"		1.70	14.118		5,850	640		6,490	7,450
1510	14"		1.30	18.462		8,450	840		9,290	10,600
1520	16"	↓	1	24		9,600	1,100		10,700	12,300
1548	For 600 lb., flanged, add					110%	20%			
1571	300 lb., 2"	1 Plum	7.40	1.081		680	52.50		732.50	830
1572	2-1/2"	Q-1	4.20	3.810		1,025	167		1,192	1,375
1573	3"		4	4		1,025	176		1,201	1,400
1574	4"	↓	2.80	5.714		1,325	251		1,576	1,875
1575	6"	Q-2	2.90	8.276		2,575	375		2,950	3,425
1576	8"		2.40	10		3,725	455		4,180	4,800
1577	10"		2.10	11.429		5,525	520		6,045	6,875
1578	12"		1.60	15		7,700	685		8,385	9,525
1579	14"		1.20	20		11,700	910		12,610	14,200
1581	16"	↓	.90	26.667	↓	15,000	1,225		16,225	18,500
1950	Gate valve, 150 lb., flanged									
2000	2"	1 Plum	8	1	Ea.	685	49		734	830
2040	2-1/2"	Q-1	5	3.200		970	140		1,110	1,300
2050	3"	↓	4.50	3.556	↓	970	156		1,126	1,325

23 05 23.80 Valves, Steel		Crew	Daily Output	Labor-Hours	Unit	2009 Bare Costs Material	Labor	Equipment	Total	Total Incl O&P
2060	4"	Q-1	3	5.333	Ea.	1,200	234		1,434	1,700
2070	6"	Q-2	3	8		1,800	365		2,165	2,550
2080	8"		2.50	9.600		1,875	435		2,310	2,750
2090	10"		2.20	10.909		4,275	495		4,770	5,475
2100	12"		1.70	14.118		5,800	640		6,440	7,375
2110	14"		1.30	18.462		9,225	840		10,065	11,400
2120	16"		1	24		11,900	1,100		13,000	14,800
2130	18"		.80	30		14,500	1,375		15,875	18,000
2140	20"		.60	40		17,400	1,825		19,225	21,900
2650	300 lb., flanged									
2700	2"	1 Plum	7.40	1.081	Ea.	940	52.50		992.50	1,100
2740	2-1/2"	Q-1	4.20	3.810		1,275	167		1,442	1,650
2750	3"		4	4		1,275	176		1,451	1,675
2760	4"		2.80	5.714		1,775	251		2,026	2,350
2770	6"	Q-2	2.90	8.276		2,975	375		3,350	3,850
2780	8"		2.40	10		4,675	455		5,130	5,825
2790	10"		2.10	11.429		6,250	520		6,770	7,675
2800	12"		1.60	15		9,025	685		9,710	11,000
2810	14"		1.20	20		17,100	910		18,010	20,200
2820	16"		.90	26.667		21,500	1,225		22,725	25,500
2830	18"		.70	34.286		30,600	1,550		32,150	36,100
2840	20"		.50	48		34,100	2,175		36,275	40,900
3650	Globe valve, 150 lb., flanged									
3700	2"	1 Plum	8	1	Ea.	865	49		914	1,025
3740	2-1/2"	Q-1	5	3.200		1,100	140		1,240	1,425
3750	3"		4.50	3.556		1,100	156		1,256	1,450
3760	4"		3	5.333		1,600	234		1,834	2,125
3770	6"	Q-2	3	8		2,525	365		2,890	3,350
3780	8"		2.50	9.600		4,625	435		5,060	5,750
3790	10"		2.20	10.909		8,200	495		8,695	9,775
3800	12"		1.70	14.118		11,000	640		11,640	13,100
4080	300 lb., flanged									
4100	2"	1 Plum	7.40	1.081	Ea.	1,150	52.50		1,202.50	1,350
4140	2-1/2"	Q-1	4.20	3.810		1,575	167		1,742	1,975
4150	3"		4	4		1,575	176		1,751	2,000
4160	4"		2.80	5.714		2,175	251		2,426	2,800
4170	6"	Q-2	2.90	8.276		3,925	375		4,300	4,900
4180	8"		2.40	10		6,425	455		6,880	7,775
4190	10"		2.10	11.429		12,100	520		12,620	14,100
4200	12"		1.60	15		14,000	685		14,685	16,500
4680	600 lb., flanged									
4700	2"	1 Plum	7	1.143	Ea.	1,600	55.50		1,655.50	1,850
4740	2-1/2"	Q-1	4	4		2,400	176		2,576	2,925
4750	3"		3.60	4.444		2,400	195		2,595	2,950
4760	4"		2.50	6.400		3,875	281		4,156	4,700
4770	6"	Q-2	2.60	9.231		8,075	420		8,495	9,525
4780	8"	"	2.10	11.429		13,100	520		13,620	15,300
5150	Forged									
5340	Ball valve, 800 psi, threaded, 1/4" size	1 Plum	24	.333	Ea.	50.50	16.25		66.75	81
5350	3/8"		24	.333		50.50	16.25		66.75	80.50
5360	1/2"		24	.333		63	16.25		79.25	94
5370	3/4"		20	.400		83.50	19.50		103	122
5380	1"		19	.421		104	20.50		124.50	146

23 05 23 – General-Duty Valves for HVAC Piping

23 05 23.80 Valves, Steel		Crew	Daily Output	Labor-Hours	Unit	Material	2009 Bare Costs Labor	Equipment	Total	Total Incl O&P
5390	1-1/4"	1 Plum	15	.533	Ea.	138	26		164	192
5400	1-1/2"		13	.615		181	30		211	246
5410	2"		11	.727		239	35.50		274.50	320
5460	Ball valve, 800 lb., socket weld, 1/4" size	Q-15	19	.842		62.50	37	2.93	102.43	129
5470	3/8"		19	.842		62.50	37	2.93	102.43	129
5480	1/2"		19	.842		62.50	37	2.93	102.43	129
5490	3/4"		19	.842		83	37	2.93	122.93	152
5500	1"		15	1.067		103	47	3.71	153.71	191
5510	1-1/4"		13	1.231		138	54	4.28	196.28	241
5520	1-1/2"		11	1.455		180	64	5.05	249.05	305
5530	2"		8.50	1.882		255	82.50	6.55	344.05	415
5650	Check valve, class 800, horizontal, socket									
5698	Threaded									
5700	1/4"	1 Plum	24	.333	Ea.	76.50	16.25		92.75	109
5720	3/8"		24	.333		76.50	16.25		92.75	109
5730	1/2"		24	.333		76.50	16.25		92.75	109
5740	3/4"		20	.400		82	19.50		101.50	121
5750	1"		19	.421		96.50	20.50		117	138
5760	1-1/4"		15	.533		189	26		215	249
5770	1-1/2"		13	.615		189	30		219	255
5780	2"		11	.727		265	35.50		300.50	345
5840	For class 150, flanged, add					100%	15%			
5860	For class 300, flanged, add					120%	20%			
6100	Gate, class 800, OS&Y, socket									
6102	3/8"	Q-15	19	.842	Ea.	53.50	37	2.93	93.43	119
6103	1/2"		19	.842		53.50	37	2.93	93.43	119
6104	3/4"		19	.842		58.50	37	2.93	98.43	125
6105	1"		15	1.067		70.50	47	3.71	121.21	154
6106	1-1/4"		13	1.231		134	54	4.28	192.28	236
6107	1-1/2"		11	1.455		134	64	5.05	203.05	252
6108	2"		8.50	1.882		174	82.50	6.55	263.05	325
6118	Threaded									
6120	3/8"	1 Plum	24	.333	Ea.	53.50	16.25		69.75	83.50
6130	1/2"		24	.333		53.50	16.25		69.75	83.50
6140	3/4"		20	.400		58.50	19.50		78	94
6150	1"		19	.421		70.50	20.50		91	110
6160	1-1/4"		15	.533		134	26		160	188
6170	1-1/2"		13	.615		134	30		164	194
6180	2"		11	.727		174	35.50		209.50	247
6260	For OS&Y, flanged, add					100%	20%			
6700	Globe, OS&Y, class 800, socket									
6710	1/4"	Q-15	19	.842	Ea.	82	37	2.93	121.93	151
6720	3/8"		19	.842		82	37	2.93	121.93	151
6730	1/2"		19	.842		82	37	2.93	121.93	151
6740	3/4"		19	.842		93.50	37	2.93	133.43	164
6750	1"		15	1.067		122	47	3.71	172.71	212
6760	1-1/4"		13	1.231		240	54	4.28	298.28	355
6770	1-1/2"		11	1.455		240	64	5.05	309.05	370
6780	2"		8.50	1.882		305	82.50	6.55	394.05	475

23 05 23.90 Valves, Stainless Steel		Crew	Daily Output	Labor-Hours	Unit	Material	Labor	Equipment	Total	Total Incl O&P
0010	**VALVES, STAINLESS STEEL** R220523-90									
1610	Ball, threaded 1/4"	1 Stpi	24	.333	Ea.	40.50	16.45		56.95	70

23 05 23 – General-Duty Valves for HVAC Piping

23 05 23.90 Valves, Stainless Steel		Crew	Daily Output	Labor-Hours	Unit	Material	Labor	Equipment	Total	Total Incl O&P
						2009 Bare Costs				
1620	3/8"	1 Stpi	24	.333	Ea.	40.50	16.45		56.95	70
1630	1/2"		22	.364		40.50	17.95		58.45	72.50
1640	3/4"		20	.400		67.50	19.75		87.25	105
1650	1"		19	.421		82	21		103	123
1660	1-1/4		15	.533		156	26.50		182.50	213
1670	1-1/2"		13	.615		163	30.50		193.50	227
1680	2"		11	.727		212	36		248	289
1700	Check, 200 lb., threaded									
1710	1/4"	1 Plum	24	.333	Ea.	126	16.25		142.25	164
1720	1/2"		22	.364		126	17.75		143.75	167
1730	3/4"		20	.400		138	19.50		157.50	182
1750	1"		19	.421		176	20.50		196.50	226
1760	1-1/2"		13	.615		340	30		370	415
1770	2"		11	.727		570	35.50		605.50	680
1800	150 lb., flanged									
1810	2-1/2"	Q-1	5	3.200	Ea.	1,375	140		1,515	1,725
1820	3"		4.50	3.556		1,375	156		1,531	1,750
1830	4"		3	5.333		2,025	234		2,259	2,600
1840	6"	Q-2	3	8		3,625	365		3,990	4,575
1850	8"	"	2.50	9.600		6,050	435		6,485	7,350
2100	Gate, OS&Y, 150 lb., flanged									
2120	1/2"	1 Plum	18	.444	Ea.	400	21.50		421.50	475
2140	3/4"		16	.500		425	24.50		449.50	510
2150	1"		14	.571		525	28		553	625
2160	1-1/2"		11	.727		730	35.50		765.50	860
2170	2"		8	1		820	49		869	975
2180	2-1/2"	Q-1	5	3.200		1,350	140		1,490	1,725
2190	3"		4.50	3.556		2,000	156		2,156	2,450
2200	4"		3	5.333		2,000	234		2,234	2,575
2205	5"		2.80	5.714		3,450	251		3,701	4,200
2210	6"	Q-2	3	8		3,450	365		3,815	4,375
2220	8"		2.50	9.600		5,550	435		5,985	6,775
2230	10"		2.30	10.435		9,100	475		9,575	10,700
2240	12"		1.90	12.632		11,700	575		12,275	13,700
2260	For 300 lb., flanged, add					120%	15%			
2600	600 lb., flanged									
2620	1/2"	1 Plum	16	.500	Ea.	173	24.50		197.50	228
2640	3/4"		14	.571		187	28		215	249
2650	1"		12	.667		225	32.50		257.50	299
2660	1-1/2"		10	.800		360	39		399	455
2670	2"		7	1.143		495	55.50		550.50	630
2680	2-1/2"	Q-1	4	4		4,800	176		4,976	5,550
2690	3"	"	3.60	4.444		4,800	195		4,995	5,575
3100	Globe, OS&Y, 150 lb., flanged									
3120	1/2"	1 Plum	18	.444	Ea.	400	21.50		421.50	475
3140	3/4"		16	.500		445	24.50		469.50	530
3150	1"		14	.571		555	28		583	655
3160	1-1/2"		11	.727		825	35.50		860.50	960
3170	2"		8	1		1,000	49		1,049	1,175
3180	2-1/2"	Q-1	5	3.200		2,150	140		2,290	2,600
3190	3"		4.50	3.556		2,150	156		2,306	2,625
3200	4"		3	5.333		2,525	234		2,759	3,175
3210	6"	Q-2	3	8		4,450	365		4,815	5,475

23 05 93.10 Balancing, Air		Crew	Daily Output	Labor-Hours	Unit	Material (2009 Bare Costs)	Labor	Equipment	Total	Total Incl O&P
0010	**BALANCING, AIR**, (Subcontractor's quote incl. material and labor.)									
0900	Heating and ventilating equipment									
1000	Centrifugal fans, utility sets				Ea.				317.34	348.72
1100	Heating and ventilating unit								476	523.08
1200	In-line fan								476	523.08
1300	Propeller and wall fan								89.91	98.80
1400	Roof exhaust fan								211.56	232.48
2000	Air conditioning equipment, central station								687.56	755.56
2100	Built-up low pressure unit								634.67	697.44
2200	Built-up high pressure unit								740.45	813.68
2300	Built-up high pressure dual duct								1,163.56	1,278.64
2400	Built-up variable volume								1,375.12	1,511.12
2500	Multi-zone A.C. and heating unit								476	523.08
2600	For each zone over one, add								105.78	116.24
2700	Package A.C. unit								264.45	290.60
2800	Rooftop heating and cooling unit								370.22	406.84
3000	Supply, return, exhaust, registers & diffusers, avg. height ceiling								63.47	69.74
3100	High ceiling								95.20	104.62
3200	Floor height								52.89	58.12
3300	Off mixing box								42.31	46.50
3500	Induction unit								68.76	75.56
3600	Lab fume hood								317.34	348.72
3700	Linear supply								158.67	174.36
3800	Linear supply high								185.11	203.42
4000	Linear return								52.89	58.12
4100	Light troffers								63.47	69.74
4200	Moduline - master								63.47	69.74
4300	Moduline - slaves								31.73	34.87
4400	Regenerators								423.11	464.96
4500	Taps into ceiling plenums								79.33	87.18
4600	Variable volume boxes				↓				63.47	69.74

23 05 93.20 Balancing, Water		Crew	Daily Output	Labor-Hours	Unit	Material	Labor	Equipment	Total	Total Incl O&P
0010	**BALANCING, WATER**, (Subcontractor's quote incl. material and labor.)									
0050	Air cooled condenser				Ea.				188.68	207.34
0080	Boiler								379.52	417.05
0100	Cabinet unit heater								64.69	71.09
0200	Chiller								458.22	503.54
0300	Convector								53.91	59.24
0400	Converter								269.54	296.20
0500	Cooling tower								350.40	385.06
0600	Fan coil unit, unit ventilator								97.04	106.63
0700	Fin tube and radiant panels								107.82	118.48
0800	Main and duct re-heat coils								99.73	109.59
0810	Heat exchanger								99.73	109.59
0900	Main balancing cocks								80.86	88.86
1000	Pumps								237.20	260.66
1100	Unit heater				↓				75.47	82.94

23 05 93.50 Piping, Testing		Crew	Daily Output	Labor-Hours	Unit	Material	Labor	Equipment	Total	Total Incl O&P
0010	**PIPING, TESTING**									
0100	Nondestructive testing									
0110	Nondestructive hydraulic pressure test, isolate & 1 hr. hold									
0120	1" - 4" pipe									

23 05 Common Work Results for HVAC

23 05 93 – Testing, Adjusting, and Balancing for HVAC

23 05 93.50 Piping, Testing		Crew	Daily Output	Labor-Hours	Unit	2009 Bare Costs Material	Labor	Equipment	Total	Total Incl O&P
0140	0 – 250 L.F.	1 Stpi	1.33	6.015	Ea.		297		297	460
0160	250 – 500 L.F.	"	.80	10			495		495	765
0180	500 – 1000 L.F.	Q-5	1.14	14.035			625		625	965
0200	1000 – 2000 L.F.	"	.80	20			890		890	1,375
0300	6" - 10" pipe									
0320	0 – 250 L.F.	Q-5	1	16	Ea.		710		710	1,100
0340	250 – 500 L.F.		.73	21.918			975		975	1,500
0360	500 – 1000 L.F.		.53	30.189			1,350		1,350	2,075
0380	1000 – 2000 L.F.		.38	42.105			1,875		1,875	2,900
2000	X-Ray of welds									
2110	2" diam.	1 Stpi	8	1	Ea.	10.20	49.50		59.70	87.50
2120	3" diam.		8	1		10.20	49.50		59.70	87.50
2130	4" diam.		8	1		15.30	49.50		64.80	93.50
2140	6" diam.		8	1		15.30	49.50		64.80	93.50
2150	8" diam.		6.60	1.212		15.30	60		75.30	110
2160	10" diam.		6	1.333		20.50	66		86.50	125

23 07 HVAC Insulation

23 07 13 – Duct Insulation

23 07 13.10 Duct Thermal Insulation

		Crew	Daily Output	Labor-Hours	Unit	Material	Labor	Equipment	Total	Total Incl O&P
0010	**DUCT THERMAL INSULATION**									
0100	Rule of thumb, as a percentage of total mechanical costs				Job				10%	
0110	Insulation req'd is based on the surface size/area to be covered									
3000	Ductwork									
3020	Blanket type, fiberglass, flexible									
3140	FSK vapor barrier wrap, .75 lb. density									
3160	1" thick G	Q-14	350	.046	S.F.	.17	1.81		1.98	3.12
3170	1-1/2" thick G		320	.050		.20	1.99		2.19	3.43
3180	2" thick G		300	.053		.26	2.12		2.38	3.71
3190	3" thick G		260	.062		.39	2.44		2.83	4.38
3200	4" thick G		242	.066		.56	2.63		3.19	4.86
3212	Vinyl Jacket, .75 lb. density, 1-1/2" thick G		320	.050		.20	1.99		2.19	3.43
3280	Unfaced, 1 lb. density									
3310	1" thick G	Q-14	360	.044	S.F.	.16	1.76		1.92	3.03
3320	1-1/2" thick G		330	.048		.19	1.92		2.11	3.32
3330	2" thick G		310	.052		.26	2.05		2.31	3.60
3400	FSK facing, 1 lb. density									
3420	1-1/2" thick G	Q-14	310	.052	S.F.	.17	2.05		2.22	3.50
3430	2" thick G		300	.053		.19	2.12		2.31	3.63
3470	1-1/2" thick G		300	.053		.25	2.12		2.37	3.70
3480	2" thick G		290	.055		.32	2.19		2.51	3.89
3730	Sheet insulation									
3760	Polyethylene foam, closed cell, UV resistant									
3770	Standard temperature (-90° to +212° F)									
3771	1/4" thick G	Q-14	450	.036	S.F.	1.30	1.41		2.71	3.71
3772	3/8" thick G		440	.036		1.86	1.44		3.30	4.38
3773	1/2" thick G		420	.038		2.28	1.51		3.79	4.96
3774	3/4" thick G		400	.040		3.26	1.59		4.85	6.15
3775	1" thick G		380	.042		4.41	1.67		6.08	7.55
3776	1-1/2" thick G		360	.044		6.95	1.76		8.71	10.50
3777	2" thick G		340	.047		9.35	1.87		11.22	13.25

23 07 HVAC Insulation

23 07 13 – Duct Insulation

23 07 13.10 Duct Thermal Insulation			Crew	Daily Output	Labor-Hours	Unit	Material	2009 Bare Costs Labor	Equipment	Total	Total Incl O&P
3778	2-1/2" thick	G	Q-14	320	.050	S.F.	11.80	1.99		13.79	16.20
3779	Adhesive (see line 7878)										
3780	Foam, rubber										
3782	1" thick	G	1 Stpi	50	.160	S.F.	3.43	7.90		11.33	16
3795	Finishes										
3800	Stainless steel woven mesh		Q-14	100	.160	S.F.	.84	6.35		7.19	11.15
3820	18 oz. fiberglass cloth, pasted on			170	.094		.77	3.74		4.51	6.90
3900	8 oz. canvas, pasted on		↓	180	.089	↓	.49	3.53		4.02	6.25
7878	Contact cement, quart can					Ea.	9.20			9.20	10.10
9600	Minimum labor/equipment charge		1 Stpi	4	2	Job		98.50		98.50	153

23 07 16 – HVAC Equipment Insulation

23 07 16.10 HVAC Equipment Thermal Insulation			Crew	Daily Output	Labor-Hours	Unit	Material	2009 Bare Costs Labor	Equipment	Total	Total Incl O&P
0010	**HVAC EQUIPMENT THERMAL INSULATION**										
0100	Rule of thumb, as a percentage of total mechanical costs					Job				10%	
0110	Insulation req'd is based on the surface size/area to be covered										
1000	Boiler, 1-1/2" calcium silicate only	G	Q-14	110	.145	S.F.	2.65	5.75		8.40	12.25
1020	Plus 2" fiberglass	G	"	80	.200	"	3.51	7.95		11.46	16.70
2000	Breeching, 2" calcium silicate										
2020	Rectangular	G	Q-14	42	.381	S.F.	5.20	15.10		20.30	30
2040	Round	G	"	38.70	.413	"	5.45	16.40		21.85	32.50
2300	Calcium silicate block, + 200° F to + 1200° F										
2310	On irregular surfaces, valves and fittings										
2340	1" thick	G	Q-14	30	.533	S.F.	2.35	21		23.35	36.50
2360	1-1/2" thick	G		25	.640		2.65	25.50		28.15	44
2380	2" thick	G		22	.727		3.48	29		32.48	50.50
2400	3" thick	G	↓	18	.889	↓	5.35	35.50		40.85	63
2410	On plane surfaces										
2420	1" thick	G	Q-14	126	.127	S.F.	2.35	5.05		7.40	10.75
2430	1-1/2" thick	G		120	.133		2.65	5.30		7.95	11.45
2440	2" thick	G		100	.160		3.48	6.35		9.83	14.10
2450	3" thick	G	↓	70	.229	↓	5.35	9.05		14.40	20.50

23 09 Instrumentation and Control for HVAC

23 09 13 – Instrumentation and Control Devices for HVAC

23 09 13.60 Water Level Controls		Crew	Daily Output	Labor-Hours	Unit	Material	2009 Bare Costs Labor	Equipment	Total	Total Incl O&P
0010	**WATER LEVEL CONTROLS**									
1000	Electric water feeder	1 Stpi	12	.667	Ea.	229	33		262	305
2000	Feeder cut-off combination									
2100	Steam system up to 5000 sq. ft.	1 Stpi	12	.667	Ea.	490	33		523	590
2200	Steam system above 5000 sq. ft.		12	.667		605	33		638	715
2300	Steam and hot water, high pressure	↓	10	.800	↓	710	39.50		749.50	840
3000	Low water cut-off for hot water boiler, 50 psi maximum									
3100	1" top & bottom equalizing pipes, manual reset	1 Stpi	14	.571	Ea.	275	28		303	345
3200	1" top & bottom equalizing pipes		14	.571		275	28		303	345
3300	2-1/2" side connection for nipple-to-boiler	↓	14	.571	↓	248	28		276	315
4000	Low water cut-off for low pressure steam with quick hook-up ftgs.									
4100	For installation in gauge glass tappings	1 Stpi	16	.500	Ea.	198	24.50		222.50	257
4200	Built-in type, 2-1/2" tap - 3-1/8" insertion		16	.500		166	24.50		190.50	221
4300	Built-in type, 2-1/2" tap - 1-3/4" insertion		16	.500		176	24.50		200.50	233
4400	Side connection to 2-1/2" tapping	↓	16	.500	↓	189	24.50		213.50	247

23 09 13 – Instrumentation and Control Devices for HVAC

23 09 13.60 Water Level Controls		Crew	Daily Output	Labor-Hours	Unit	Material	Labor	Equipment	Total	Total Incl O&P
						2009 Bare Costs				
5000	Pump control, low water cut-off and alarm switch	1 Stpi	14	.571	Ea.	425	28		453	515

23 09 23 – Direct-Digital Control System for HVAC

23 09 23.10 Control Components/DDC Systems		Crew	Daily Output	Labor-Hours	Unit	Material	Labor	Equipment	Total	Total Incl O&P
0010	**CONTROL COMPONENTS/DDC SYSTEMS** (Sub's quote incl. M & L)									
0100	Analog inputs									
0110	Sensors (avg. 50' run in 1/2" EMT)									
0120	Duct temperature				Ea.				360	396
0130	Space temperature								577	634.64
0140	Duct humidity, +/- 3%								605.32	665.85
0150	Space humidity, +/- 2%								926.90	1,019.59
0160	Duct static pressure								491.82	541
0170	C.F.M./Transducer								662.07	728.28
0172	Water temp. (see Div. 23 21 20 for well tap add)								567.64	624.24
0174	Water flow (see Div. 23 21 20 for circuit sensor add)								2,080.80	2,288.88
0176	Water pressure differential (see Div. 23 21 20 for tap add)								851.24	936.36
0177	Steam flow (see Div. 23 21 20 for circuit sensor add)								2,080.80	2,288.88
0178	Steam pressure (see Div. 23 21 20 for tap add)								889.06	977.97
0180	K.W./Transducer								1,182.27	1,300.50
0182	K.W.H. totalization (not incl. elec. meter pulse xmtr.)								543.38	597.72
0190	Space static pressure				↓				927.27	1,020
1000	Analog outputs (avg. 50' run in 1/2" EMT)									
1010	P/I Transducer				Ea.				548.57	603.43
1020	Analog output, matl. in MUX								264.83	291.31
1030	Pneumatic (not incl. control device)								558	613.83
1040	Electric (not incl control device)				↓				331	364.14
2000	Status (Alarms)									
2100	Digital inputs (avg. 50' run in 1/2" EMT)									
2110	Freeze				Ea.				378.33	416.16
2120	Fire								340.49	374.54
2130	Differential pressure, (air)								520.20	572.22
2140	Differential pressure, (water)								850	935
2150	Current sensor								378.33	416.16
2160	Duct high temperature thermostat								496.55	546.21
2170	Duct smoke detector				↓				614.78	676.26
2200	Digital output (avg. 50' run in 1/2" EMT)									
2210	Start/stop				Ea.				296.73	326.40
2220	On/off (maintained contact)				"				510	561
3000	Controller M.U.X. panel, incl. function boards									
3100	48 point				Ea.				4,587.22	5,045.94
3110	128 point				"				6,290	6,918.66
3200	D.D.C. controller (avg. 50' run in conduit)									
3210	Mechanical room									
3214	16 point controller (incl. 120v/1ph power supply)				Ea.				1,937.45	3,121.20
3229	32 point controller (incl. 120v/1ph power supply)				"				4,729	5,202
3230	Includes software programming and checkout									
3260	Space									
3266	V.A.V. terminal box (incl. space temp. sensor)				Ea.				733	806.31
3280	Host computer (avg. 50' run in conduit)									
3281	Package complete with PC, keyboard,									
3282	printer, color CRT, modem, basic software				Ea.				8,512.36	9,363.60
4000	Front end costs									
4100	Computer (P.C.)/software program				Ea.				5,675	6,242.40

23 09 23 – Direct-Digital Control System for HVAC

23 09 23.10 Control Components/DDC Systems		Crew	Daily Output	Labor-Hours	Unit	Material	2009 Bare Costs Labor	Equipment	Total	Total Incl O&P
4200	Color graphics software				Ea.				3,405	3,745.44
4300	Color graphics slides								426	468.18
4350	Additional dot matrix printer				↓				851	936.36
4400	Communications trunk cable				L.F.				3.31	3.64
4500	Engineering labor, (not incl. dftg.)				Point				71.88	79.07
4600	Calibration labor								71.88	79.07
4700	Start-up, checkout labor				↓				108.76	119.64
4800	Drafting labor, as req'd									
5000	Communications bus (data transmission cable)									
5010	#18 twisted shielded pair in 1/2" EMT conduit				C.L.F.				331.04	364.14
8000	Applications software									
8050	Basic maintenance manager software (not incl. data base entry)				Ea.				1,702	1,872.72
8100	Time program				Point				5.95	6.55
8120	Duty cycle								11.86	13.05
8140	Optimum start/stop								35.94	39.53
8160	Demand limiting								17.78	19.56
8180	Enthalpy program				↓				35.94	39.53
8200	Boiler optimization				Ea.				1,064	1,170.45
8220	Chiller optimization				"				1,419	1,560.60
8240	Custom applications									
8260	Cost varies with complexity									

23 09 33 – Electric and Electronic Control System for HVAC

23 09 33.10 Electronic Control Systems		Crew	Daily Output	Labor-Hours	Unit	Material	Labor	Equipment	Total	Total Incl O&P
0010	**ELECTRONIC CONTROL SYSTEMS**									
0020	For electronic costs, add to Div. 23 09 43.10				Ea.					15%
9000	Minimum labor/equipment charge	1 Plum	8	1	Job		49		49	75.50

23 09 43 – Pneumatic Control System for HVAC

23 09 43.10 Pneumatic Control Systems		Crew	Daily Output	Labor-Hours	Unit	Material	Labor	Equipment	Total	Total Incl O&P
0010	**PNEUMATIC CONTROL SYSTEMS**									
0011	Including a nominal 50 Ft. of tubing. Add control panelboard if req'd.									
0100	Heating and ventilating, split system									
0200	Mixed air control, economizer cycle, panel readout, tubing									
0220	Up to 10 tons G	Q-19	.68	35.294	Ea.	3,575	1,600		5,175	6,400
0240	For 10 to 20 tons G		.63	37.915		3,825	1,725		5,550	6,850
0260	For over 20 tons G		.58	41.096		4,150	1,850		6,000	7,425
0270	Enthalpy cycle, up to 10 tons		.50	48.387		3,950	2,200		6,150	7,750
0280	For 10 to 20 tons		.46	52.174		4,250	2,350		6,600	8,325
0290	For over 20 tons	↓	.42	56.604	↓	4,625	2,575		7,200	9,050
0300	Heating coil, hot water, 3 way valve,									
0320	Freezestat, limit control on discharge, readout	Q-5	.69	23.088	Ea.	2,650	1,025		3,675	4,525
0500	Cooling coil, chilled water, room									
0520	Thermostat, 3 way valve	Q-5	2	8	Ea.	1,175	355		1,530	1,850
0600	Cooling tower, fan cycle, damper control,									
0620	Control system including water readout in/out at panel	Q-19	.67	35.821	Ea.	4,700	1,625		6,325	7,675
1000	Unit ventilator, day/night operation,									
1100	freezestat, ASHRAE, cycle 2	Q-19	.91	26.374	Ea.	2,600	1,200		3,800	4,700
2000	Compensated hot water from boiler, valve control,									
2100	readout and reset at panel, up to 60 GPM	Q-19	.55	43.956	Ea.	4,875	2,000		6,875	8,425
2120	For 120 GPM		.51	47.059		5,200	2,125		7,325	9,025
2140	For 240 GPM		.49	49.180		5,450	2,225		7,675	9,450
3000	Boiler room combustion air, damper to 5 SF, controls		1.37	17.582		2,350	795		3,145	3,800
3500	Fan coil, heating and cooling valves, 4 pipe control system	↓	3	8	↓	1,050	360		1,410	1,725

23 09 43 – Pneumatic Control System for HVAC

23 09 43.10 Pneumatic Control Systems		Crew	Daily Output	Labor-Hours	Unit	Material	Labor	Equipment	Total	Total Incl O&P
						2009 Bare Costs				
3600	Heat exchanger system controls	Q-19	.86	27.907	Ea.	2,275	1,275		3,550	4,450
3900	Multizone control (one per zone), includes thermostat, damper									
3910	motor and reset of discharge temperature	Q-5	.51	31.373	Ea.	2,350	1,400		3,750	4,725
4000	Pneumatic thermostat, including controlling room radiator valve	″	2.43	6.593		705	293		998	1,225
4040	Program energy saving optimizer G	Q-19	1.21	19.786		5,875	895		6,770	7,825
4060	Pump control system	″	3	8		1,075	360		1,435	1,750
4080	Reheat coil control system, not incl coil	Q-5	2.43	6.593		920	293		1,213	1,450
4500	Air supply for pneumatic control system									
4600	Tank mounted duplex compressor, starter, alternator,									
4620	piping, dryer, PRV station and filter									
4630	1/2 HP	Q-19	.68	35.139	Ea.	8,725	1,600		10,325	12,100
4640	3/4 HP		.64	37.383		9,150	1,700		10,850	12,700
4650	1 HP		.61	39.539		10,000	1,800		11,800	13,800
4660	1-1/2 HP		.58	41.739		10,700	1,900		12,600	14,600
4680	3 HP		.55	43.956		14,500	2,000		16,500	19,100
4690	5 HP		.42	57.143		25,300	2,575		27,875	31,900
4800	Main air supply, includes 3/8″ copper main and labor	Q-5	1.82	8.791	C.L.F.	293	390		683	930
4810	If poly tubing used, deduct									30%
7000	Static pressure control for air handling unit, includes pressure									
7010	sensor, receiver controller, readout and damper motors	Q-19	.64	37.383	Ea.	6,950	1,700		8,650	10,300
7020	If return air fan requires control, add									70%
8600	VAV boxes, incl. thermostat, damper motor, reheat coil & tubing	Q-5	1.46	10.989		1,100	490		1,590	1,950
8610	If no reheat coil, deduct									204

23 09 53 – Pneumatic and Electric Control System for HVAC

23 09 53.10 Control Components		Crew	Daily Output	Labor-Hours	Unit	Material	Labor	Equipment	Total	Total Incl O&P
0010	**CONTROL COMPONENTS**									
0700	Controller, receiver									
0730	Pneumatic, panel mount, single input	1 Plum	8	1	Ea.	410	49		459	525
0740	With conversion mounting bracket		8	1		410	49		459	525
0750	Dual input, with control point adjustment		7	1.143		565	55.50		620.50	705
0850	Electric, single snap switch	1 Elec	4	2		390	94		484	575
0860	Dual snap switches	″	3	2.667		525	125		650	775
1000	Enthalpy control, boiler water temperature control									
1010	governed by outdoor temperature, with timer	1 Elec	3	2.667	Ea.	330	125		455	560
2000	Gauges, pressure or vacuum									
2100	2″ diameter dial	1 Stpi	32	.250	Ea.	22	12.35		34.35	43
2200	2-1/2″ diameter dial		32	.250		24	12.35		36.35	45.50
2300	3-1/2″ diameter dial		32	.250		35.50	12.35		47.85	58.50
2400	4-1/2″ diameter dial		32	.250		52.50	12.35		64.85	77
2700	Flanged iron case, black ring									
2800	3-1/2″ diameter dial	1 Stpi	32	.250	Ea.	102	12.35		114.35	131
2900	4-1/2″ diameter dial		32	.250		105	12.35		117.35	135
3000	6″ diameter dial		32	.250		167	12.35		179.35	202
3300	For compound pressure-vacuum, add					18%				
3350	Humidistat									
3390	Electric operated	1 Shee	8	1	Ea.	69.50	47		116.50	151
3400	Relays									
3430	Pneumatic/electric	1 Plum	16	.500	Ea.	310	24.50		334.50	380
3440	Pneumatic proportioning		8	1		209	49		258	305
3450	Pneumatic switching		12	.667		142	32.50		174.50	207
3460	Selector, 3 point		6	1.333		101	65		166	212
3470	Pneumatic time delay		8	1		264	49		313	365

23 09 53.10 Control Components		Crew	Daily Output	Labor-Hours	Unit	2009 Bare Costs Material	Labor	Equipment	Total	Total Incl O&P
3500	Sensor, air operated									
3520	Humidity	1 Plum	16	.500	Ea.	340	24.50		364.50	410
3540	Pressure		16	.500		58	24.50		82.50	102
3560	Temperature		12	.667		152	32.50		184.50	218
3600	Electric operated									
3620	Humidity	1 Elec	8	1	Ea.	77	47		124	158
3650	Pressure		8	1		1,500	47		1,547	1,725
3680	Temperature		10	.800		93.50	37.50		131	161
4000	Thermometers									
4100	Dial type, 3-1/2" diameter, vapor type, union connection	1 Stpi	32	.250	Ea.	152	12.35		164.35	186
4120	Liquid type, union connection		32	.250		295	12.35		307.35	345
4500	Stem type, 6-1/2" case, 2" stem, 1/2" NPT		32	.250		40.50	12.35		52.85	63.50
4520	4" stem, 1/2" NPT		32	.250		50	12.35		62.35	73.50
4600	9" case, 3-1/2" stem, 3/4" NPT		28	.286		81.50	14.10		95.60	112
4620	6" stem, 3/4" NPT		28	.286		96	14.10		110.10	128
4640	8" stem, 3/4" NPT		28	.286		109	14.10		123.10	142
4660	12" stem, 1" NPT		26	.308		152	15.20		167.20	191
5000	Thermostats									
5030	Manual	1 Shee	8	[illegible]	Ea.	27	47		74	105
5040	1 set back, electric, timed G		8	[illegible]		95	47		142	179
5050	2 set back, electric, timed G		8	[illegible]		209	47		256	305
5100	Locking cover					21			21	23
5200	24 hour, automatic, clock G	1 Shee	8	[illegible]		115	47		162	202
5220	Electric, low voltage, 2 wire	1 Elec	13	.615		18	29		47	64.50
5230	3 wire	"	10	.800		22	37.50		59.50	82.50
5300	Transmitter, pneumatic									
5320	Temperature averaging element	Q-1	8	2	Ea.	116	88		204	263
5350	Pressure differential	1 Plum	7	1.143		975	55.50		1,030.50	1,150
5370	Humidity, duct		8	[illegible]		310	49		359	415
5380	Room		12	.667		310	32.50		342.50	390
5390	Temperature, with averaging element		6	1.333		164	65		229	281
6000	Valves, motorized zone									
6100	Sweat connections, 1/2" C x C	1 Stpi	20	.400	Ea.	140	19.75		159.75	185
6110	3/4" C x C		20	.400		141	19.75		160.75	186
6120	1" C x C		19	.421		183	21		204	234
6140	1/2" C x C, with end switch, 2 wire		20	.400		139	19.75		158.75	184
6150	3/4" C x C, with end switch, 2 wire		20	.400		148	19.75		167.75	194
6160	1" C x C, with end switch, 2 wire		19	.421		169	21		190	218
7090	Valves, motor controlled, including actuator									
7100	Electric motor actuated									
7200	Brass, two way, screwed									
7210	1/2" pipe size	L-6	36	.333	Ea.	202	16.05		218.05	248
7220	3/4" pipe size		30	.400		258	19.25		277.25	315
7230	1" pipe size		28	.429		300	20.50		320.50	360
7240	1-1/2" pipe size		19	.632		395	30.50		425.50	480
7250	2" pipe size		16	.750		650	36		686	770
7350	Brass, three way, screwed									
7360	1/2" pipe size	L-6	33	.364	Ea.	254	17.50		271.50	305
7370	3/4" pipe size		27	.444		296	21.50		317.50	360
7380	1" pipe size		25.50	.471		385	22.50		407.50	460
7384	1-1/4" pipe size		21	.571		490	27.50		517.50	585
7390	1-1/2" pipe size		17	.706		560	34		594	670
7400	2" pipe size		14	.857		745	41.50		786.50	885

23 09 Instrumentation and Control for HVAC

23 09 53 – Pneumatic and Electric Control System for HVAC

23 09 53.10 Control Components		Crew	Daily Output	Labor-Hours	Unit	Material	2009 Bare Costs Labor	Equipment	Total	Total Incl O&P
7550	Iron body, two way, flanged									
7560	2-1/2" pipe size	L-6	4	3	Ea.	1,075	145		1,220	1,400
7570	3" pipe size		3	4		1,150	193		1,343	1,575
7580	4" pipe size		2	6		2,025	289		2,314	2,675
7850	Iron body, three way, flanged									
7860	2-1/2" pipe size	L-6	3	4	Ea.	1,125	193		1,318	1,525
7870	3" pipe size		2.50	4.800		1,250	231		1,481	1,750
7880	4" pipe size		2	6		1,625	289		1,914	2,225
8000	Pneumatic, air operated									
8050	Brass, two way, screwed									
8060	1/2" pipe size, class 250	1 Plum	24	.333	Ea.	164	16.25		180.25	205
8070	3/4" pipe size, class 250		20	.400		196	19.50		215.50	246
8080	1" pipe size, class 250		19	.421		229	20.50		249.50	284
8090	1-1/4" pipe size, class 125		15	.533		285	26		311	355
8100	1-1/2" pipe size, class 125		13	.615		365	30		395	445
8110	2" pipe size, class 125		11	.727		420	35.50		455.50	520
8180	Brass, three way, screwed									
8190	1/2" pipe size, class 250	1 Plum	22	.364	Ea.	180	17.75		197.75	226
8200	3/4" pipe size, class 250		18	.444		223	21.50		244.50	279
8210	1" pipe size, class 250		17	.471		262	23		285	325
8214	1-1/4" pipe size, class 250		14	.571		365	28		393	450
8220	1-1/2" pipe size, class 125		11	.727		435	35.50		470.50	535
8230	2" pipe size, class 125		9	.889		525	43.50		568.50	640
8450	Iron body, two way, flanged									
8460	2-1/2" pipe size, 250 lb. flanges	Q-1	5	3.200	Ea.	1,700	140		1,840	2,100
8470	3" pipe size, 250 lb. flanges		4.50	3.556		1,800	156		1,956	2,225
8480	4" pipe size, 250 lb. flanges		3	5.333		2,100	234		2,334	2,675
8560	Iron body, three way, flanged									
8570	2-1/2" pipe size, class 125	Q-1	4.50	3.556	Ea.	960	156		1,116	1,300
8580	3" pipe size, class 125		4	4		1,100	176		1,276	1,500
8590	4" pipe size, class 125		2.50	6.400		2,225	281		2,506	2,875
8600	6" pipe size, class 125	Q-2	3	8		3,400	365		3,765	4,300
9000	Minimum labor/equipment charge	1 Plum	4	2	Job		97.50		97.50	151

23 11 Facility Fuel Piping

23 11 13 – Facility Fuel-Oil Piping

23 11 13.10 Fuel Oil Specialties

		Crew	Daily Output	Labor-Hours	Unit	Material	Labor	Equipment	Total	Total Incl O&P
0010	**FUEL OIL SPECIALTIES**									
0020	Foot valve, single poppet, metal to metal construction									
0040	Bevel seat, 1/2" diameter	1 Stpi	20	.400	Ea.	59	19.75		78.75	95.50
0060	3/4" diameter		18	.444		57	22		79	96.50
0080	1" diameter		16	.500		24	24.50		48.50	65
0100	1-1/4" diameter		15	.533		88	26.50		114.50	138
0120	1-1/2" diameter		13	.615		37	30.50		67.50	87.50
0140	2" diameter		11	.727		38.50	36		74.50	97.50
0400	Fuel fill box, flush type									
0408	Nonlocking, watertight									
0410	1-1/2" diameter	1 Stpi	12	.667	Ea.	19.40	33		52.40	72.50
0440	2" diameter		10	.800		16.75	39.50		56.25	79.50
0450	2-1/2" diameter		9	.889		56	44		100	130
0460	3" diameter		7	1.143		54	56.50		110.50	147

23 11 13 – Facility Fuel-Oil Piping

23 11 13.10 Fuel Oil Specialties		Crew	Daily Output	Labor-Hours	Unit	Material	2009 Bare Costs Labor	Equipment	Total	Total Incl O&P
0470	4" diameter	1 Stpi	5	1.600	Ea.	75	79		154	205
0500	Locking inner cover									
0510	2" diameter	1 Stpi	8	1	Ea.	82.50	49.50		132	167
0520	2-1/2" diameter		7	1.143		110	56.50		166.50	209
0530	3" diameter		5	1.600		118	79		197	252
0540	4" diameter		4	2		152	98.50		250.50	320
0600	Fuel system components									
0620	Spill container	1 Stpi	4	2	Ea.	585	98.50		683.50	795
0640	Fill adapter, 4", straight drop		8	1		47	49.50		96.50	128
0680	Fill cap, 4"		30	.267		31.50	13.15		44.65	55.50
0700	Extractor fitting, 4" x 1-1/2"		8	1		310	49.50		359.50	415
0740	Vapor hose adapter, 4"		8	1		90	49.50		139.50	176
0760	Wood gage stick, 10'					11			11	12.10
1000	Oil filters, 3/8" IPT., 20 gal. per hour	1 Stpi	20	.400		19.25	19.75		39	51.50
1020	32 gal. per hour		18	.444		34.50	22		56.50	72
1040	40 gal. per hour		16	.500		37	24.50		61.50	79.50
1060	50 gal. per hour		14	.571		38.50	28		66.50	85.50
2000	Remote tank gauging system, self contained									
2100	Single tank kit/8 sensors inputs w/printer	1 Stpi	2.50	3.200	Ea.	3,275	158		3,433	3,850
2120	Two tank kit/8 sensors inputs w/printer		2	4		4,425	197		4,622	5,150
3000	Valve, ball check, globe type, 3/8" diameter		24	.333		11.10	16.45		27.55	37.50
3500	Fusible, 3/8" diameter		24	.333		11.40	16.45		27.85	38
3600	1/2" diameter		24	.333		27.50	16.45		43.95	56
3610	3/4" diameter		20	.400		61	19.75		80.75	97.50
3620	1" diameter		19	.421		175	21		196	225
4000	Nonfusible, 3/8" diameter		24	.333		19.25	16.45		35.70	46.50
4500	Shutoff, gate type, lever handle, spring-fusible kit									
4520	1/4" diameter	1 Stpi	14	.571	Ea.	30	28		58	76.50
4540	3/8" diameter		12	.667		29	33		62	83
4560	1/2" diameter		10	.800		43	39.50		82.50	109
4570	3/4" diameter		8	1		74	49.50		123.50	158
4580	Lever handle, requires weight and fusible kit									
4600	1" diameter	1 Stpi	9	.889	Ea.	164	44		208	248
4620	1-1/4" diameter		8	1		189	49.50		238.50	285
4640	1-1/2" diameter		7	1.143		244	56.50		300.50	355
4660	2" diameter		6	1.333		305	66		371	435
4680	For fusible link, weight and braided wire, add					5%				
5000	Vent alarm, whistling signal					26			26	28.50
5500	Vent protector/breather, 1-1/4" diameter	1 Stpi	32	.250		10.40	12.35		22.75	30.50
5520	1-1/2" diameter		32	.250		12	12.35		24.35	32.50
5540	2" diameter		32	.250		21	12.35		33.35	42
5560	3" diameter		28	.286		33.50	14.10		47.60	59
5580	4" diameter		24	.333		37.50	16.45		53.95	67
5600	Dust cap, breather, 2"		40	.200		11.70	9.85		21.55	28
8000	Fuel oil and tank heaters									
8020	Electric, capacity rated at 230 volts									
8040	Immersion element in steel manifold									
8060	96 GPH at 50° F rise	Q-5	6.40	2.500	Ea.	870	111		981	1,125
8070	128 GPH at 50° F rise		6.20	2.581		900	115		1,015	1,175
8080	160 GPH at 50° F rise		5.90	2.712		1,000	120		1,120	1,275
8090	192 GPH at 50° F rise		5.50	2.909		1,100	129		1,229	1,400
8100	240 GPH at 50° F rise		5.10	3.137		1,250	139		1,389	1,600
8110	288 GPH at 50° F rise		4.60	3.478		1,425	155		1,580	1,800

23 11 Facility Fuel Piping

23 11 13 – Facility Fuel-Oil Piping

23 11 13.10 Fuel Oil Specialties		Crew	Daily Output	Labor-Hours	Unit	Material	Labor (2009 Bare Costs)	Equipment	Total	Total Incl O&P
8120	384 GPH at 50° F rise	Q-5	3.10	5.161	Ea.	1,650	229		1,879	2,175
8130	480 GPH at 50° F rise		2.30	6.957		2,075	310		2,385	2,750
8140	576 GPH at 50° F rise		2.10	7.619		2,075	340		2,415	2,800
8300	Suction stub, immersion type									
8320	75" long, 750 watts	1 Stpi	14	.571	Ea.	590	28		618	695
8330	99" long, 2000 watts		12	.667		830	33		863	965
8340	123" long, 3000 watts		10	.800		895	39.50		934.50	1,050
8660	Steam, cross flow, rated at 5 PSIG									
8680	42 GPH	Q-5	7	2.286	Ea.	1,375	102		1,477	1,650
8690	73 GPH		6.70	2.388		1,600	106		1,706	1,925
8700	112 GPH		6.20	2.581		1,650	115		1,765	2,000
8710	158 GPH		5.80	2.759		1,800	123		1,923	2,175
8720	187 GPH		4	4		2,475	178		2,653	3,000
8730	270 GPH		3.60	4.444		2,675	197		2,872	3,225
8740	365 GPH	Q-6	4.90	4.898		3,125	226		3,351	3,800
8750	635 GPH		3.70	6.486		3,950	299		4,249	4,825
8760	845 GPH		2.50	9.600		5,800	440		6,240	7,050
8770	1420 GPH		1.60	15		9,425	690		10,115	11,500
8780	2100 GPH		1.10	21.818		12,900	1,000		13,900	15,800

23 11 23 – Facility Natural-Gas Piping

23 11 23.10 Gas Meters		Crew	Daily Output	Labor-Hours	Unit	Material	Labor	Equipment	Total	Total Incl O&P
0010	**GAS METERS**									
4000	Residential									
4010	Gas meter, residential, 3/4" pipe size	1 Plum	14	.571	Ea.	131	28		159	187
4020	Gas meter, residential, 1" pipe size		12	.667		181	32.50		213.50	250
4030	Gas meter, residential, 1-1/4" pipe size		10	.800		198	39		237	279

23 12 Facility Fuel Pumps

23 12 13 – Facility Fuel-Oil Pumps

23 12 13.10 Pump and Motor Sets		Crew	Daily Output	Labor-Hours	Unit	Material	Labor	Equipment	Total	Total Incl O&P
0010	**PUMP AND MOTOR SETS**									
1810	Light fuel and diesel oils									
1820	20 GPH 1/3 HP	Q-5	6	2.667	Ea.	790	118		908	1,050
1830	27 GPH, 1/3 HP		6	2.667		790	118		908	1,050
1840	80 GPH, 1/3 HP		5	3.200		790	142		932	1,100
1850	145 GPH, 1/2 HP		4	4		835	178		1,013	1,200
1860	277 GPH, 1 HP		4	4		930	178		1,108	1,300
1870	700 GPH, 1-1/2 HP		3	5.333		1,925	237		2,162	2,500
1880	1000 GPH, 2 HP		3	5.333		2,325	237		2,562	2,925
1890	1800 GPH, 5 HP		1.80	8.889		3,200	395		3,595	4,100

23 12 16 – Facility Gasoline Dispensing Pumps

23 12 16.20 Fuel Dispensing Equipment		Crew	Daily Output	Labor-Hours	Unit	Material	Labor	Equipment	Total	Total Incl O&P
0010	**FUEL DISPENSING EQUIPMENT**									
1100	Product dispenser with vapor recovery for 6 nozzles, installed, not									
1110	including piping to storage tanks				Ea.	20,800			20,800	22,900

23 13 Facility Fuel-Storage Tanks

23 13 13 – Facility Underground Fuel-Oil, Storage Tanks

23 13 13.09 Single-Wall Steel Fuel-Oil Tanks		Crew	Daily Output	Labor-Hours	Unit	Material	Labor	Equipment	Total	Total Incl O&P
						2009 Bare Costs				
0010	**SINGLE-WALL STEEL FUEL-OIL TANKS**									
5000	Steel underground, sti-P3, set in place, not incl. hold-down bars.									
5500	Excavation, pad, pumps and piping not included									
5510	Single wall, 500 gallon capacity, 7 gauge shell	Q-5	2.70	5.926	Ea.	1,975	263		2,238	2,575
5520	1,000 gallon capacity, 7 gauge shell	"	2.50	6.400		3,150	284		3,434	3,925
5530	2,000 gallon capacity, 1/4" thick shell	Q-7	4.60	6.957		5,125	325		5,450	6,125
5535	2,500 gallon capacity, 7 gauge shell	Q-5	3	5.333		5,650	237		5,887	6,600
5540	5,000 gallon capacity, 1/4" thick shell	Q-7	3.20	10		8,250	470		8,720	9,800
5560	10,000 gallon capacity, 1/4" thick shell		2	16		10,200	750		10,950	12,400
5580	15,000 gallon capacity, 5/16" thick shell		1.70	18.824		12,700	885		13,585	15,300
5600	20,000 gallon capacity, 5/16" thick shell		1.50	21.333		21,100	1,000		22,100	24,800
5610	25,000 gallon capacity, 3/8" thick shell		1.30	24.615		25,700	1,150		26,850	30,100
5620	30,000 gallon capacity, 3/8" thick shell		1.10	29.091		31,700	1,375		33,075	37,000
5630	40,000 gallon capacity, 3/8" thick shell		.90	35.556		39,600	1,675		41,275	46,200
5640	50,000 gallon capacity, 3/8" thick shell		.80	40		49,500	1,875		51,375	57,500

23 13 13.13 Double-Wall Steel, Underground Fuel-Oil, Storage Tanks		Crew	Daily Output	Labor-Hours	Unit	Material	Labor	Equipment	Total	Total Incl O&P
0010	**DOUBLE-WALL STEEL, UNDERGROUND FUEL-OIL, STORAGE TANKS**									
6200	Steel, underground, 360°, double wall, U.L. listed,									
6210	with sti-P3 corrosion protection,									
6220	(dielectric coating, cathodic protection, electrical									
6230	isolation) 30 year warranty,									
6240	not incl. manholes or hold-downs.									
6250	500 gallon capacity	Q-5	2.40	6.667	Ea.	3,900	296		4,196	4,750
6260	1,000 gallon capactiy	"	2.25	7.111		5,700	315		6,015	6,775
6270	2,000 gallon capacity	Q-7	4.16	7.692		7,475	360		7,835	8,775
6280	3,000 gallon capacity		3.90	8.205		9,350	385		9,735	10,900
6290	4,000 gallon capacity		3.64	8.791		11,500	415		11,915	13,300
6300	5,000 gallon capacity		2.91	10.997		14,100	515		14,615	16,300
6310	6,000 gallon capacity		2.42	13.223		16,000	620		16,620	18,600
6320	8,000 gallon capacity		2.08	15.385		17,800	725		18,525	20,700
6330	10,000 gallon capacity		1.82	17.582		21,900	825		22,725	25,400
6340	12,000 gallon capacity		1.70	18.824		23,700	885		24,585	27,500
6350	15,000 gallon capacity		1.33	24.060		29,000	1,125		30,125	33,700
6360	20,000 gallon capacity		1.33	24.060		37,600	1,125		38,725	43,200
6370	25,000 gallon capacity		1.16	27.586		60,000	1,300		61,300	68,000
6380	30,000 gallon capacity		1.03	31.068		72,500	1,450		73,950	82,000
6390	40,000 gallon capacity		.80	40		93,000	1,875		94,875	105,000
6395	50,000 gallon capacity		.73	43.836		113,000	2,050		115,050	127,000
6400	For hold-downs 500-2000 gal, add		16	2	Set	151	94		245	310
6410	For hold-downs 3000-6000 gal, add		12	2.667		305	125		430	530
6420	For hold-downs 8000-12,000 gal, add		11	2.909		365	137		502	615
6430	For hold-downs 15,000 gal, add		9	3.556		530	167		697	840
6440	For hold-downs 20,000 gal, add		8	4		610	188		798	960
6450	For hold-downs 20,000 gal plus, add		6	5.333		790	251		1,041	1,250
6500	For manways, add				Ea.	1,500			1,500	1,650
6600	In place with hold-downs									
6652	550 gallon capacity	Q-5	1.84	8.696	Ea.	4,050	385		4,435	5,075

23 13 13.23 Glass-Fiber-Reinfcd-Plastic, Fuel-Oil, Storage		Crew	Daily Output	Labor-Hours	Unit	Material	Labor	Equipment	Total	Total Incl O&P
0010	**GLASS-FIBER-REINFCD-PLASTIC, UNDERGRND FUEL-OIL, STORAGE**									
0210	Fiberglass, underground, single wall, U.L. listed, not including									
0220	manway or hold-down strap									
0225	550 gallon capacity	Q-5	2.67	5.993	Ea.	3,000	266		3,266	3,725

23 13 Facility Fuel-Storage Tanks

23 13 13 – Facility Underground Fuel-Oil, Storage Tanks

23 13 13.23 Glass-Fiber-Reinfcd-Plastic, Fuel-Oil, Storage		Crew	Daily Output	Labor-Hours	Unit	2009 Bare Costs Material	Labor	Equipment	Total	Total Incl O&P
0230	1,000 gallon capacity	Q-5	2.46	6.504	Ea.	3,650	289		3,939	4,475
0240	2,000 gallon capacity	Q-7	4.57	7.002		5,750	330		6,080	6,825
0250	4,000 gallon capacity		3.55	9.014		8,150	425		8,575	9,625
0260	6,000 gallon capacity		2.67	11.985		9,050	565		9,615	10,900
0270	8,000 gallon capacity		2.29	13.974		10,800	655		11,455	12,800
0280	10,000 gallon capacity		2	16		12,200	750		12,950	14,600
0282	12,000 gallon capacity		1.88	17.021		20,500	800		21,300	23,900
0284	15,000 gallon capacity		1.68	19.048		22,600	895		23,495	26,200
0290	20,000 gallon capacity		1.45	22.069		24,400	1,025		25,425	28,400
0300	25,000 gallon capacity		1.28	25		36,100	1,175		37,275	41,500
0320	30,000 gallon capacity		1.14	28.070		43,500	1,325		44,825	49,900
0340	40,000 gallon capacity		.89	35.955		62,000	1,700		63,700	70,500
0360	48,000 gallon capacity		.81	39.506		73,000	1,850		74,850	83,000
0500	For manway, fittings and hold-downs, add					20%	15%			
0600	For manways, add					1,800			1,800	1,975
1000	For helical heating coil, add	Q-5	2.50	6.400		3,650	284		3,934	4,450
1020	Fiberglass, underground, double wall, U.L. listed									
1030	includes manways, not incl. hold-down straps									
1040	600 gallon capacity	Q-5	2.42	6.612	Ea.	6,650	294		6,944	7,750
1050	1,000 gallon capacity	″	2.25	7.111		8,350	315		8,665	9,700
1060	2,500 gallon capacity	Q-7	4.16	7.692		12,700	360		13,060	14,600
1070	3,000 gallon capacity		3.90	8.205		13,700	385		14,085	15,700
1080	4,000 gallon capacity		3.64	8.791		15,500	415		15,915	17,600
1090	6,000 gallon capacity		2.42	13.223		15,700	620		16,320	18,300
1100	8,000 gallon capacity		2.08	15.385		20,100	725		20,825	23,200
1110	10,000 gallon capacity		1.82	17.582		23,000	825		23,825	26,600
1120	12,000 gallon capacity		1.70	18.824		28,300	885		29,185	32,600
1122	15,000 gallon capacity		1.52	21.053		36,700	990		37,690	41,800
1124	20,000 gallon capacity		1.33	24.060		43,900	1,125		45,025	50,000
1126	25,000 gallon capacity		1.16	27.586		58,500	1,300		59,800	66,500
1128	30,000 gallon capacity		1.03	31.068		70,500	1,450		71,950	80,000
1140	For hold-down straps, add					2%	10%			
1150	For hold-downs 500-4000 gal., add	Q-7	16	2	Set	420	94		514	605
1160	For hold-downs 5000-15000 gal., add		8	4		835	188		1,023	1,200
1170	For hold-downs 20,000 gal., add		5.33	6.004		1,250	282		1,532	1,825
1180	For hold-downs 25,000 gal., add		4	8		1,675	375		2,050	2,425
1190	For hold-downs 30,000 gal., add		2.60	12.308		2,500	580		3,080	3,650
2210	Fiberglass, underground, single wall, U.L. listed, including									
2220	hold-down straps, no manways									
2225	550 gallon capacity	Q-5	2	8	Ea.	3,400	355		3,755	4,300
2230	1,000 gallon capacity	″	1.88	8.511		4,075	380		4,455	5,050
2240	2,000 gallon capacity	Q-7	3.55	9.014		6,150	425		6,575	7,425
2250	4,000 gallon capacity		2.90	11.034		8,575	520		9,095	10,300
2260	6,000 gallon capacity		2	16		9,900	750		10,650	12,100
2270	8,000 gallon capacity		1.78	17.978		11,600	845		12,445	14,100
2280	10,000 gallon capacity		1.60	20		13,000	940		13,940	15,800
2282	12,000 gallon capacity		1.52	21.053		21,400	990		22,390	25,000
2284	15,000 gallon capacity		1.39	23.022		23,400	1,075		24,475	27,500
2290	20,000 gallon capacity		1.14	28.070		25,600	1,325		26,925	30,300
2300	25,000 gallon capacity		.96	33.333		37,700	1,575		39,275	43,900
2320	30,000 gallon capacity		.80	40		46,000	1,875		47,875	53,500
3020	Fiberglass, underground, double wall, U.L. listed									
3030	includes manways and hold-down straps									

23 13 Facility Fuel-Storage Tanks

23 13 13 – Facility Underground Fuel-Oil, Storage Tanks

23 13 13.23 Glass-Fiber-Reinfcd-Plastic, Fuel-Oil, Storage		Crew	Daily Output	Labor-Hours	Unit	Material	Labor	Equipment	Total	Total Incl O&P
						2009 Bare Costs				
3040	600 gallon capacity	Q-5	1.86	8.602	Ea.	7,050	380		7,430	8,375
3050	1,000 gallon capacity	"	1.70	9.412		8,775	420		9,195	10,300
3060	2,500 gallon capacity	Q-7	3.29	9.726		13,100	455		13,555	15,100
3070	3,000 gallon capacity		3.13	10.224		14,100	480		14,580	16,300
3080	4,000 gallon capacity		2.93	10.922		15,900	515		16,415	18,300
3090	6,000 gallon capacity		1.86	17.204		16,500	810		17,310	19,500
3100	8,000 gallon capacity		1.65	19.394		21,000	910		21,910	24,500
3110	10,000 gallon capacity		1.48	21.622		23,900	1,025		24,925	27,800
3120	12,000 gallon capacity		1.40	22.857		29,200	1,075		30,275	33,800
3122	15,000 gallon capacity		1.28	25		37,500	1,175		38,675	43,100
3124	20,000 gallon capacity		1.06	30.189		45,200	1,425		46,625	52,000
3126	25,000 gallon capacity		.90	35.556		60,500	1,675		62,175	69,000
3128	30,000 gallon capacity		.74	43.243		73,000	2,025		75,025	83,000

23 13 23 – Facility Aboveground Fuel-Oil, Storage Tanks

23 13 23.13 Vertical, Steel, Aboveground Fuel-Oil, Storage Tanks

		Crew	Daily Output	Labor-Hours	Unit	Material	Labor	Equipment	Total	Total Incl O&P
0010	**VERTICAL, STEEL, ABOVEGROUND FUEL-OIL, STORAGE TANKS**									
4000	Fixed roof oil storage tanks, steel, (1 BBL=42 GAL w/ foundation 3'D x 1'W)									
4200	5,000 barrels				Ea.				157,273	173,000
4300	24,000 barrels								269,091	296,000
4500	56,000 barrels								535,000	588,500
4600	110,000 barrels								777,273	855,000
4800	143,000 barrels								996,818	1,096,500
4900	224,000 barrels								1,495,909	1,645,500
5100	Floating roof gasoline tanks, steel, 5,000 barrels								172,727	190,000
5200	25,000 barrels								295,455	325,000
5400	55,000 barrels								590,909	650,000
5500	100,000 barrels								854,545	940,000
5700	150,000 barrels								1,095,455	1,205,000
5800	225,000 barrels								1,645,455	1,810,000

23 13 23.16 Horizontal, Stl, Abvgrd Fuel-Oil, Storage Tanks

		Crew	Daily Output	Labor-Hours	Unit	Material	Labor	Equipment	Total	Total Incl O&P
0010	**HORIZONTAL, STEEL, ABOVEGROUND FUEL-OIL, STORAGE TANKS**									
3000	Steel, storage, above ground, including cradles, coating,									
3020	fittings, not including foundation, pumps or piping									
3040	Single wall, interior, 275 gallon	Q-5	5	3.200	Ea.	385	142		527	640
3060	550 gallon	"	2.70	5.926		2,275	263		2,538	2,925
3080	1,000 gallon	Q-7	5	6.400		2,875	300		3,175	3,625
3100	1,500 gallon		4.75	6.737		4,275	315		4,590	5,225
3120	2,000 gallon		4.60	6.957		4,925	325		5,250	5,900
3140	5,000 gallon		3.20	10		6,400	470		6,870	7,775
3150	10,000 gallon		2	16		18,400	750		19,150	21,500
3160	15,000 gallon		1.70	18.824		21,000	885		21,885	24,500
3170	20,000 gallon		1.45	22.069		26,300	1,025		27,325	30,500
3180	25,000 gallon		1.30	24.615		31,900	1,150		33,050	36,900
3190	30,000 gallon		1.10	29.091		36,100	1,375		37,475	41,800
3320	Double wall, 500 gallon capacity	Q-5	2.40	6.667		2,300	296		2,596	2,975
3330	2000 gallon capacity	Q-7	4.15	7.711		8,250	360		8,610	9,625
3340	4000 gallon capacity		3.60	8.889		14,700	420		15,120	16,900
3350	6000 gallon capacity		2.40	13.333		17,400	625		18,025	20,100
3360	8000 gallon capacity		2	16		22,300	750		23,050	25,700
3370	10000 gallon capacity		1.80	17.778		25,000	835		25,835	28,800
3380	15000 gallon capacity		1.50	21.333		37,900	1,000		38,900	43,300
3390	20000 gallon capacity		1.30	24.615		43,200	1,150		44,350	49,300

23 13 Facility Fuel-Storage Tanks

23 13 23 – Facility Aboveground Fuel-Oil, Storage Tanks

23 13 23.16 Horizontal, Stl, Abvgrd Fuel-Oil, Storage Tanks		Crew	Daily Output	Labor-Hours	Unit	2009 Bare Costs Material	Labor	Equipment	Total	Total Incl O&P
3400	25000 gallon capacity	Q-7	1.15	27.826	Ea.	52,500	1,300		53,800	59,500
3410	30000 gallon capacity		1	32		57,500	1,500		59,000	66,000

23 21 Hydronic Piping and Pumps

23 21 20 – Hydronic HVAC Piping Specialties

23 21 20.10 Air Control

		Crew	Daily Output	Labor-Hours	Unit	Material	Labor	Equipment	Total	Total Incl O&P
0010	**AIR CONTROL**									
0030	Air separator, with strainer									
0040	2″ diameter	Q-5	6	2.667	Ea.	750	118		868	1,000
0080	2-1/2″ diameter		5	3.200		850	142		992	1,150
0100	3″ diameter		4	4		1,300	178		1,478	1,700
0120	4″ diameter		3	5.333		1,875	237		2,112	2,450
0130	5″ diameter	Q-6	3.60	6.667		2,375	305		2,680	3,100
0140	6″ diameter		3.40	7.059		2,850	325		3,175	3,625
0160	8″ diameter		3	8		4,275	370		4,645	5,275
0180	10″ diameter		2.20	10.909		6,625	505		7,130	8,050
0200	12″ diameter		1.70	14.118		11,100	650		11,750	13,200

23 21 20.18 Automatic Air Vent

		Crew	Daily Output	Labor-Hours	Unit	Material	Labor	Equipment	Total	Total Incl O&P
0010	**AUTOMATIC AIR VENT**									
0020	Cast iron body, stainless steel internals, float type									
0060	1/2″ NPT inlet, 300 psi	1 Stpi	12	.667	Ea.	88	33		121	148
0140	3/4″ NPT inlet, 300 psi		12	.667		88	33		121	148
0180	1/2″ NPT inlet, 250 psi		10	.800		275	39.50		314.50	360
0220	3/4″ NPT inlet, 250 psi		10	.800		275	39.50		314.50	360
0260	1″ NPT inlet, 250 psi		10	.800		410	39.50		449.50	515
0340	1-1/2″ NPT inlet, 250 psi	Q-5	12	1.333		870	59		929	1,050
0380	2″ NPT inlet, 250 psi	″	12	1.333		870	59		929	1,050
0600	Forged steel body, stainless steel internals, float type									
0640	1/2″ NPT inlet, 750 psi	1 Stpi	12	.667	Ea.	920	33		953	1,050
0680	3/4″ NPT inlet, 750 psi		12	.667		920	33		953	1,050
0760	3/4″ NPT inlet, 1000 psi		10	.800		1,375	39.50		1,414.50	1,575
0800	1″ NPT inlet, 1000 psi	Q-5	12	1.333		1,375	59		1,434	1,625
0880	1-1/2″ NPT inlet, 1000 psi		10	1.600		3,875	71		3,946	4,375
0920	2″ NPT inlet, 1000 psi		10	1.600		3,875	71		3,946	4,375
1100	Formed steel body, noncorrosive									
1110	1/8″ NPT inlet 150 psi	1 Stpi	32	.250	Ea.	9.65	12.35		22	30
1120	1/4″ NPT inlet 150 psi		32	.250		32.50	12.35		44.85	55
1130	3/4″ NPT inlet 150 psi		32	.250		32.50	12.35		44.85	55
1300	Chrome plated brass, automatic/manual, for radiators									
1310	1/8″ NPT inlet, nickel plated brass	1 Stpi	32	.250	Ea.	5.65	12.35		18	25.50

23 21 20.22 Circuit Sensor

		Crew	Daily Output	Labor-Hours	Unit	Material	Labor	Equipment	Total	Total Incl O&P
0010	**CIRCUIT SENSOR**, Flow meter									
0020	Metering stations									
0040	Wafer orifice insert type									
0060	2-1/2″ pipe size	Q-5	12	1.333	Ea.	164	59		223	272
0100	3″ pipe size		11	1.455		180	64.50		244.50	298
0140	4″ pipe size		8	2		205	89		294	365
0180	5″ pipe size		7.30	2.192		262	97.50		359.50	440
0220	6″ pipe size		6.40	2.500		310	111		421	510
0260	8″ pipe size	Q-6	5.30	4.528		435	209		644	800

23 21 20 – Hydronic HVAC Piping Specialties

23 21 20.22 Circuit Sensor		Crew	Daily Output	Labor-Hours	Unit	Material	2009 Bare Costs Labor	Equipment	Total	Total Incl O&P
0280	10" pipe size	Q-6	4.60	5.217	Ea.	505	240		745	930
0360	12" pipe size		4.20	5.714		835	263		1,098	1,325
23 21 20.26 Circuit Setter										
0010	**CIRCUIT SETTER**, Balance valve									
0020	3/4" pipe size	1 Stpi	20	.400	Ea.	62	19.75		81.75	99
0040	1" pipe size		18	.444		80.50	22		102.50	123
0060	1-1/2" pipe size		12	.667		140	33		173	205
0080	2" pipe size		10	.800		200	39.50		239.50	281
0100	2-1/2" pipe size	Q-5	15	1.067		450	47.50		497.50	570
0120	3" pipe size		10	1.600		635	71		706	810
0136	3" pipe size, flanged		4	4		635	178		813	975
0140	4" pipe size		3	5.333		955	237		1,192	1,425
0200	For differential meter, accurate to 1%, add					735			735	810
23 21 20.34 Dielectric Unions										
0010	**DIELECTRIC UNIONS**, Standard gaskets for water and air.									
0020	250 psi maximum pressure									
0280	Female IPT to sweat, straight									
0300	1/2" pipe size	1 Plum	24	.333	Ea.	4.25	16.25		20.50	29.50
0340	3/4" pipe size		20	.400		4.79	19.50		24.29	35.50
0360	1" pipe size		19	.421		6.60	20.50		27.10	39.50
0380	1-1/4" pipe size		15	.533		10.25	26		36.25	52
0400	1-1/2" pipe size		13	.615		15.40	30		45.40	63.50
0420	2" pipe size		11	.727		21	35.50		56.50	78
0580	Female IPT to brass pipe thread, straight									
0600	1/2" pipe size	1 Plum	24	.333	Ea.	9.40	16.25		25.65	35.50
0640	3/4" pipe size		20	.400		10.35	19.50		29.85	41.50
0660	1" pipe size		19	.421		19.25	20.50		39.75	53
0680	1-1/4" pipe size		15	.533		23.50	26		49.50	66.50
0700	1-1/2" pipe size		13	.615		34.50	30		64.50	84
0720	2" pipe size		11	.727		67.50	35.50		103	129
0780	Female IPT to female IPT, straight									
0800	1/2" pipe size	1 Plum	24	.333	Ea.	8.80	16.25		25.05	34.50
0840	3/4" pipe size		20	.400		9.90	19.50		29.40	41
0860	1" pipe size		19	.421		13.25	20.50		33.75	46.50
0880	1-1/4" pipe size		15	.533		17.90	26		43.90	60
0900	1-1/2" pipe size		13	.615		27.50	30		57.50	76.50
0920	2" pipe size		11	.727		40	35.50		75.50	99
2000	175 psi maximum pressure									
2180	Female IPT to sweat									
2240	2" pipe size	1 Plum	9	.889	Ea.	116	43.50		159.50	195
2260	2-1/2" pipe size	Q-1	15	1.067		126	47		173	212
2280	3" pipe size		14	1.143		173	50		223	268
2300	4" pipe size		11	1.455		460	64		524	605
2480	Female IPT to brass pipe									
2500	1-1/2" pipe size	1 Plum	11	.727	Ea.	145	35.50		180.50	214
2540	2" pipe size	"	9	.889		173	43.50		216.50	257
2560	2-1/2" pipe size	Q-1	15	1.067		250	47		297	350
2580	3" pipe size		14	1.143		290	50		340	400
2600	4" pipe size		11	1.455		520	64		584	675
9000	Minimum labor/equipment charge	1 Plum	4	2	Job		97.50		97.50	151
23 21 20.38 Expansion Couplings										
0010	**EXPANSION COUPLINGS**, Hydronic									

23 21 20 – Hydronic HVAC Piping Specialties

23 21 20.38 Expansion Couplings

23 21 20.38 Expansion Couplings		Crew	Daily Output	Labor-Hours	Unit	2009 Bare Costs Material	Labor	Equipment	Total	Total Incl O&P
0100	Copper to copper, sweat									
0120	3/4" diameter	1 Stpi	20	.400	Ea.	32	19.75		51.75	66
0140	1" diameter		19	.421		41	21		62	77
0160	1-1/4" diameter	↓	15	.533	↓	55.50	26.50		82	102
1000	Baseboard riser fitting, 5" stub by coupling 12" long									
1020	1/2" diameter	1 Stpi	24	.333	Ea.	15.60	16.45		32.05	42.50
1040	3/4" diameter		20	.400		22	19.75		41.75	55
1060	1" diameter		19	.421		31	21		52	66
1080	1-1/4" diameter	↓	15	.533	↓	40.50	26.50		67	85.50
1180	9" Stub by tubing 8" long									
1200	1/2" diameter	1 Stpi	24	.333	Ea.	14	16.45		30.45	41
1220	3/4" diameter		20	.400		18.55	19.75		38.30	51
1240	1" diameter		19	.421		25.50	21		46.50	60.50
1260	1-1/4" diameter	↓	15	.533	↓	36	26.50		62.50	80.50

23 21 20.42 Expansion Joints

		Crew	Daily Output	Labor-Hours	Unit	Material	Labor	Equipment	Total	Total Incl O&P
0010	**EXPANSION JOINTS**									
0100	Bellows type, neoprene cover, flanged spool									
0140	6" face to face, 1-1/4" diameter	1 Stpi	11	.727	Ea.	248	36		284	330
0160	1-1/2" diameter	"	10.60	.755		248	37.50		285.50	330
0180	2" diameter	Q-5	13.30	1.203		251	53.50		304.50	360
0190	2-1/2" diameter		12.40	1.290		260	57.50		317.50	375
0200	3" diameter		11.40	1.404		291	62.50		353.50	415
0480	10" face to face, 2" diameter		13	1.231		360	54.50		414.50	480
0500	2-1/2" diameter		12	1.333		380	59		439	510
0520	3" diameter		11	1.455		390	64.50		454.50	525
0540	4" diameter		8	2		440	89		529	625
0560	5" diameter		7	2.286		525	102		627	735
0580	6" diameter		6	2.667		545	118		663	785
0600	8" diameter		5	3.200		650	142		792	935
0620	10" diameter		4.60	3.478		720	155		875	1,025
0640	12" diameter		4	4		890	178		1,068	1,250
0660	14" diameter	↓	3.80	4.211	↓	1,100	187		1,287	1,500

23 21 20.46 Expansion Tanks

		Crew	Daily Output	Labor-Hours	Unit	Material	Labor	Equipment	Total	Total Incl O&P
0010	**EXPANSION TANKS**									
1400	Plastic, corrosion resistant, see Plumbing Cost Data									
1507	Fiberglass and steel single / double wall storage, see Div. 23 13 13									
1512	Tank leak detection systems, see Div. 28 33 33.50									
2000	Steel, liquid expansion, ASME, painted, 15 gallon capacity	Q-5	17	.941	Ea.	440	42		482	550
2020	24 gallon capacity		14	1.143		470	51		521	600
2040	30 gallon capacity		12	1.333		495	59		554	635
2060	40 gallon capacity		10	1.600		580	71		651	745
2080	60 gallon capacity		8	2		695	89		784	905
2100	80 gallon capacity		7	2.286		745	102		847	975
2120	100 gallon capacity		6	2.667		1,000	118		1,118	1,275
2130	120 gallon capacity		5	3.200		1,075	142		1,217	1,400
2140	135 gallon capacity		4.50	3.556		1,125	158		1,283	1,475
2150	175 gallon capacity		4	4		1,750	178		1,928	2,200
2160	220 gallon capacity		3.60	4.444		2,000	197		2,197	2,500
2170	240 gallon capacity		3.30	4.848		2,075	215		2,290	2,600
2180	305 gallon capacity		3	5.333		2,925	237		3,162	3,600
2190	400 gallon capacity	↓	2.80	5.714	↓	3,600	254		3,854	4,375
2360	Galvanized									

23 21 20 – Hydronic HVAC Piping Specialties

23 21 20.46 Expansion Tanks		Crew	Daily Output	Labor-Hours	Unit	2009 Bare Costs Material	Labor	Equipment	Total	Total Incl O&P
2370	15 gallon capacity	Q-5	17	.941	Ea.	780	42		822	920
2380	24 gallon capacity		14	1.143		895	51		946	1,075
2390	30 gallon capacity		12	1.333		930	59		989	1,125
2400	40 gallon capacity		10	1.600		1,100	71		1,171	1,300
2410	60 gallon capacity		8	2		1,250	89		1,339	1,525
2420	80 gallon capacity		7	2.286		1,450	102		1,552	1,725
2430	100 gallon capacity		6	2.667		1,875	118		1,993	2,250
2440	120 gallon capacity		5	3.200		2,000	142		2,142	2,425
2450	135 gallon capacity		4.50	3.556		2,125	158		2,283	2,575
2460	175 gallon capacity		4	4		3,450	178		3,628	4,075
2470	220 gallon capacity		3.60	4.444		3,925	197		4,122	4,625
2480	240 gallon capacity		3.30	4.848		4,125	215		4,340	4,850
2490	305 gallon capacity		3	5.333		5,975	237		6,212	6,950
2500	400 gallon capacity		2.80	5.714		7,350	254		7,604	8,475
3000	Steel ASME expansion, rubber diaphragm, 19 gal. cap. accept.		12	1.333		1,925	59		1,984	2,225
3020	31 gallon capacity		8	2		2,150	89		2,239	2,525
3040	61 gallon capacity		6	2.667		3,025	118		3,143	3,500
3060	79 gallon capacity		5	3.200		3,075	142		3,217	3,600
3080	119 gallon capacity		4	4		3,250	178		3,428	3,850
3100	158 gallon capacity		3.80	4.211		4,525	187		4,712	5,275
3120	211 gallon capacity		3.30	4.848		5,225	215		5,440	6,075
3140	317 gallon capacity		2.80	5.714		6,825	254		7,079	7,900
3160	422 gallon capacity		2.60	6.154		10,100	273		10,373	11,500
3180	528 gallon capacity		2.40	6.667		11,100	296		11,396	12,700
9000	Minimum labor/equipment charge		4	4	Job		178		178	276

23 21 20.58 Hydronic Heating Control Valves		Crew	Daily Output	Labor-Hours	Unit	Material	Labor	Equipment	Total	Total Incl O&P
0010	**HYDRONIC HEATING CONTROL VALVES**									
0050	Hot water, nonelectric, thermostatic									
0100	Radiator supply, 1/2" diameter	1 Stpi	24	.333	Ea.	52.50	16.45		68.95	83
0120	3/4" diameter		20	.400		54	19.75		73.75	90
0140	1" diameter		19	.421		68.50	21		89.50	108
0160	1-1/4" diameter		15	.533		98	26.50		124.50	149
0500	For low pressure steam, add					25%				
1000	Manual, radiator supply									
1010	1/2" pipe size, angle union	1 Stpi	24	.333	Ea.	34.50	16.45		50.95	63.50
1020	3/4" pipe size, angle union		20	.400		44	19.75		63.75	79
1030	1" pipe size, angle union		19	.421		57	21		78	95
1100	Radiator, balancing, straight, sweat connections									
1110	1/2" pipe size	1 Stpi	24	.333	Ea.	13.35	16.45		29.80	40
1120	3/4" pipe size		20	.400		18.70	19.75		38.45	51
1130	1" pipe size		19	.421		30.50	21		51.50	65.50
1140	Balance and stop valve 1/2" size		22	.364		14.45	17.95		32.40	44
1150	3/4" size		20	.400		17.25	19.75		37	49.50
1160	1" size		19	.421		20.50	21		41.50	54.50
1170	1-1/4" size		15	.533		26	26.50		52.50	70
1200	Steam, radiator, supply									
1210	1/2" pipe size, angle union	1 Stpi	24	.333	Ea.	34.50	16.45		50.95	63
1220	3/4" pipe size, angle union		20	.400		37	19.75		56.75	71.50
1230	1" pipe size, angle union		19	.421		43	21		64	79.50
1240	1-1/4" pipe size, angle union		15	.533		55	26.50		81.50	102
8000	System balancing and shut-off									
8020	Butterfly, quarter turn, calibrated, threaded or solder									

23 21 20.58 Hydronic Heating Control Valves

	23 21 20.58 Hydronic Heating Control Valves	Crew	Daily Output	Labor-Hours	Unit	2009 Bare Costs Material	Labor	Equipment	Total	Total Incl O&P
8040	Bronze, -30° F to +350° F, pressure to 175 psi									
8060	1/2" size	1 Stpi	22	.364	Ea.	12.20	17.95		30.15	41.50
8070	3/4" size		20	.400		19.75	19.75		39.50	52
8080	1" size		19	.421		24	21		45	58.50
8090	1-1/4" size		15	.533		38	26.50		64.50	83
8100	1-1/2" size		13	.615		50	30.50		80.50	102
8110	2" size		11	.727		62.50	36		98.50	125

23 21 20.70 Steam Traps

	23 21 20.70 Steam Traps	Crew	Daily Output	Labor-Hours	Unit	Material	Labor	Equipment	Total	Total Incl O&P
0010	**STEAM TRAPS**									
0030	Cast iron body, threaded									
0040	Inverted bucket									
0050	1/2" pipe size	1 Stpi	12	.667	Ea.	128	33		161	191
0070	3/4" pipe size		10	.800		222	39.50		261.50	305
0100	1" pipe size		9	.889		340	44		384	445
0120	1-1/4" pipe size		8	1		515	49.50		564.50	645
1000	Float & thermostatic, 15 psi									
1010	3/4" pipe size	1 Stpi	16	.500	Ea.	103	24.50		127.50	152
1020	1" pipe size		15	.533		124	26.50		150.50	177
1030	1-1/4" pipe size		13	.615		150	30.50		180.50	212
1040	1-1/2" pipe size		9	.889		218	44		262	310
1060	2" pipe size		6	1.333		400	66		466	540
1290	Brass body, threaded									
1300	Thermostatic, angle union, 25 psi									
1310	1/2" pipe size	1 Stpi	24	.333	Ea.	49.50	16.45		65.95	80
1320	3/4" pipe size		20	.400		75	19.75		94.75	113
1330	1" pipe size		19	.421		111	21		132	154
9000	Minimum labor/equipment charge		4	2	Job		98.50		98.50	153

23 21 20.74 Strainers, Basket Type

	23 21 20.74 Strainers, Basket Type	Crew	Daily Output	Labor-Hours	Unit	Material	Labor	Equipment	Total	Total Incl O&P
0010	**STRAINERS, BASKET TYPE**, Perforated stainless steel basket									
0100	Brass or monel available									
2000	Simplex style									
2300	Bronze body									
2320	Screwed, 3/8" pipe size	1 Stpi	22	.364	Ea.	127	17.95		144.95	168
2340	1/2" pipe size		20	.400		129	19.75		148.75	173
2360	3/4" pipe size		17	.471		133	23		156	182
2380	1" pipe size		15	.533		173	26.50		199.50	231
2400	1-1/4" pipe size		13	.615		260	30.50		290.50	335
2420	1-1/2" pipe size		12	.667		268	33		301	345
2440	2" pipe size		10	.800		400	39.50		439.50	500
2460	2-1/2" pipe size	Q-5	15	1.067		570	47.50		617.50	700
2480	3" pipe size	"	14	1.143		825	51		876	985
2600	Flanged, 2" pipe size	1 Stpi	6	1.333		605	66		671	765
2620	2-1/2" pipe size	Q-5	4.50	3.556		950	158		1,108	1,300
2640	3" pipe size		3.50	4.571		1,075	203		1,278	1,500
2660	4" pipe size		3	5.333		1,800	237		2,037	2,350
2680	5" pipe size	Q-6	3.40	7.059		2,775	325		3,100	3,550
2700	6" pipe size		3	8		3,525	370		3,895	4,450
2710	8" pipe size		2.50	9.600		5,725	440		6,165	6,975
3600	Iron body									
3700	Screwed, 3/8" pipe size	1 Stpi	22	.364	Ea.	88.50	17.95		106.45	126
3720	1/2" pipe size		20	.400		88.50	19.75		108.25	128
3740	3/4" pipe size		17	.471		113	23		136	160

23 21 20 – Hydronic HVAC Piping Specialties

23 21 20.74 Strainers, Basket Type		Crew	Daily Output	Labor-Hours	Unit	Material	Labor (2009 Bare Costs)	Equipment	Total	Total Incl O&P
3760	1" pipe size	1 Stpi	15	.533	Ea.	115	26.50		141.50	167
3780	1-1/4" pipe size		13	.615		150	30.50		180.50	212
3800	1-1/2" pipe size		12	.667		166	33		199	233
3820	2" pipe size		10	.800		198	39.50		237.50	279
3840	2-1/2" pipe size	Q-5	15	1.067		265	47.50		312.50	365
3860	3" pipe size	"	14	1.143		320	51		371	435
4000	Flanged, 2" pipe size	1 Stpi	6	1.333		266	66		332	395
4020	2-1/2" pipe size	Q-5	4.50	3.556		360	158		518	640
4040	3" pipe size		3.50	4.571		375	203		578	730
4060	4" pipe size		3	5.333		575	237		812	1,000
4080	5" pipe size	Q-6	3.40	7.059		875	325		1,200	1,475
4100	6" pipe size		3	8		1,125	370		1,495	1,800
4120	8" pipe size		2.50	9.600		2,025	440		2,465	2,900
4140	10" pipe size		2.20	10.909		4,450	505		4,955	5,675
6000	Cast steel body									
6400	Screwed, 1" pipe size	1 Stpi	15	.533	Ea.	185	26.50		211.50	245
6410	1-1/4" pipe size		13	.615		278	30.50		308.50	350
6420	1-1/2" pipe size		12	.667		278	33		311	355
6440	2" pipe size		10	.800		380	39.50		419.50	480
6460	2-1/2" pipe size	Q-5	15	1.067		540	47.50		587.50	670
6480	3" pipe size	"	14	1.143		705	51		756	855
6560	Flanged, 2" pipe size	1 Stpi	6	1.333		560	66		626	715
6580	2-1/2" pipe size	Q-5	4.50	3.556		840	158		998	1,175
6600	3" pipe size		3.50	4.571		900	203		1,103	1,300
6620	4" pipe size		3	5.333		1,275	237		1,512	1,775
6640	6" pipe size	Q-6	3	8		2,350	370		2,720	3,150
6660	8" pipe size	"	2.50	9.600		3,775	440		4,215	4,825
7000	Stainless steel body									
7200	Screwed, 1" pipe size	1 Stpi	15	.533	Ea.	280	26.50		306.50	345
7210	1-1/4" pipe size		13	.615		430	30.50		460.50	520
7220	1-1/2" pipe size		12	.667		430	33		463	525
7240	2" pipe size		10	.800		645	39.50		684.50	765
7260	2-1/2" pipe size	Q-5	15	1.067		910	47.50		957.50	1,075
7280	3" pipe size	"	14	1.143		1,250	51		1,301	1,450
7400	Flanged, 2" pipe size	1 Stpi	6	1.333		920	66		986	1,125
7420	2-1/2" pipe size	Q-5	4.50	3.556		1,675	158		1,833	2,100
7440	3" pipe size		3.50	4.571		1,725	203		1,928	2,225
7460	4" pipe size		3	5.333		2,700	237		2,937	3,350
7480	6" pipe size	Q-6	3	8		4,700	370		5,070	5,750
7500	8" pipe size	"	2.50	9.600		6,850	440		7,290	8,200
8100	Duplex style									
8200	Bronze body									
8240	Screwed, 3/4" pipe size	1 Stpi	16	.500	Ea.	895	24.50		919.50	1,025
8260	1" pipe size		14	.571		895	28		923	1,025
8280	1-1/4" pipe size		12	.667		1,750	33		1,783	1,975
8300	1-1/2" pipe size		11	.727		1,750	36		1,786	1,975
8320	2" pipe size		9	.889		2,825	44		2,869	3,200
8340	2-1/2" pipe size	Q-5	14	1.143		3,650	51		3,701	4,100
8420	Flanged, 2" pipe size	1 Stpi	6	1.333		3,050	66		3,116	3,450
8440	2-1/2" pipe size	Q-5	4.50	3.556		4,200	158		4,358	4,875
8460	3" pipe size		3.50	4.571		4,575	203		4,778	5,375
8480	4" pipe size		3	5.333		6,575	237		6,812	7,600
8500	5" pipe size	Q-6	3.40	7.059		14,600	325		14,925	16,600

23 21 20 – Hydronic HVAC Piping Specialties

23 21 20.74 Strainers, Basket Type		Crew	Daily Output	Labor-Hours	Unit	2009 Bare Costs Material	Labor	Equipment	Total	Total Incl O&P
8520	6" pipe size	Q-6	3	8	Ea.	14,600	370		14,970	16,700
8700	Iron body									
8740	Screwed, 3/4" pipe size	1 Stpi	16	.500	Ea.	525	24.50		549.50	615
8760	1" pipe size		14	.571		525	28		553	620
8780	1-1/4" pipe size		12	.667		925	33		958	1,075
8800	1-1/2" pipe size		11	.727		925	36		961	1,075
8820	2" pipe size		9	.889		1,550	44		1,594	1,800
8840	2-1/2" pipe size	Q-5	14	1.143		1,725	51		1,776	1,975
9000	Flanged, 2" pipe size	1 Stpi	6	1.333		1,675	66		1,741	1,925
9020	2-1/2" pipe size	Q-5	4.50	3.556		1,775	158		1,933	2,200
9040	3" pipe size		3.50	4.571		1,950	203		2,153	2,450
9060	4" pipe size		3	5.333		3,275	237		3,512	3,975
9080	5" pipe size	Q-6	3.40	7.059		6,500	325		6,825	7,650
9100	6" pipe size		3	8		6,500	370		6,870	7,725
9120	8" pipe size		2.50	9.600		11,300	440		11,740	13,200
9140	10" pipe size		2.20	10.909		15,800	505		16,305	18,100
9160	12" pipe size		1.70	14.118		17,500	650		18,150	20,200
9170	14" pipe size		1.40	17.143		21,200	790		21,990	24,500
9180	16" pipe size		1	24		26,600	1,100		27,700	30,900
9300	Cast steel body									
9340	Screwed, 1" pipe size	1 Stpi	14	.571	Ea.	1,100	28		1,128	1,275
9360	1-1/2" pipe size		11	.727		1,800	36		1,836	2,025
9380	2" pipe size		9	.889		2,375	44		2,419	2,675
9460	Flanged, 2" pipe size		6	1.333		2,550	66		2,616	2,900
9480	2-1/2" pipe size	Q-5	4.50	3.556		4,100	158		4,258	4,750
9500	3" pipe size		3.50	4.571		4,475	203		4,678	5,250
9520	4" pipe size		3	5.333		5,525	237		5,762	6,450
9540	6" pipe size	Q-6	3	8		10,300	370		10,670	12,000
9560	8" pipe size	"	2.50	9.600		22,800	440		23,240	25,800
9700	Stainless steel body									
9740	Screwed, 1" pipe size	1 Stpi	14	.571	Ea.	1,425	28		1,453	1,625
9760	1-1/2" pipe size		11	.727		2,175	36		2,211	2,425
9780	2" pipe size		9	.889		3,075	44		3,119	3,450
9860	Flanged, 2" pipe size		6	1.333		3,300	66		3,366	3,725
9880	2-1/2" pipe size	Q-5	4.50	3.556		5,525	158		5,683	6,325
9900	3" pipe size		3.50	4.571		6,025	203		6,228	6,950
9920	4" pipe size		3	5.333		7,850	237		8,087	9,000
9940	6" pipe size	Q-6	3	8		12,000	370		12,370	13,800
9960	8" pipe size	"	2.50	9.600		33,500	440		33,940	37,500

23 21 20.76 Strainers, Y Type, Bronze Body		Crew	Daily Output	Labor-Hours	Unit	Material	Labor	Equipment	Total	Total Incl O&P
0010	**STRAINERS, Y TYPE, BRONZE BODY**									
0050	Screwed, 150 lb., 1/4" pipe size	1 Stpi	24	.333	Ea.	14.55	16.45		31	41.50
0070	3/8" pipe size		24	.333		19.20	16.45		35.65	46.50
0100	1/2" pipe size		20	.400		19.20	19.75		38.95	51.50
0120	3/4" pipe size		19	.421		21.50	21		42.50	56
0140	1" pipe size		17	.471		23	23		46	61
0150	1-1/4" pipe size		15	.533		51.50	26.50		78	98
0160	1-1/2" pipe size		14	.571		49.50	28		77.50	98
0180	2" pipe size		13	.615		66	30.50		96.50	120
0182	3" pipe size		12	.667		590	33		623	700
0200	300 lb., 2-1/2" pipe size	Q-5	17	.941		350	42		392	450
0220	3" pipe size		16	1		695	44.50		739.50	835

23 21 20 – Hydronic HVAC Piping Specialties

23 21 20.76 Strainers, Y Type, Bronze Body		Crew	Daily Output	Labor-Hours	Unit	2009 Bare Costs Material	Labor	Equipment	Total	Total Incl O&P
0240	4″ pipe size	Q-5	15	1.067	Ea.	1,575	47.50		1,622.50	1,825
0500	For 300 lb rating 1/4″ thru 2″, add					15%				
1000	Flanged, 150 lb., 1-1/2″ pipe size	1 Stpi	11	.727	Ea.	365	36		401	455
1020	2″ pipe size	″	8	1		440	49.50		489.50	560
1030	2-1/2″ pipe size	Q-5	5	3.200		650	142		792	935
1040	3″ pipe size		4.50	3.556		805	158		963	1,125
1060	4″ pipe size		3	5.333		1,225	237		1,462	1,725
1080	5″ pipe size	Q-6	3.40	7.059		1,225	325		1,550	1,850
1100	6″ pipe size		3	8		2,325	370		2,695	3,125
1106	8″ pipe size		2.60	9.231		2,550	425		2,975	3,475
1500	For 300 lb rating, add					40%				
9000	Minimum labor/equipment charge	1 Stpi	3.75	2.133	Job		105		105	163

23 21 20.78 Strainers, Y Type, Iron Body		Crew	Daily Output	Labor-Hours	Unit	Material	Labor	Equipment	Total	Total Incl O&P
0010	**STRAINERS, Y TYPE, IRON BODY**									
0050	Screwed, 250 lb., 1/4″ pipe size	1 Stpi	20	.400	Ea.	7.75	19.75		27.50	39
0070	3/8″ pipe size		20	.400		7.75	19.75		27.50	39
0100	1/2″ pipe size		20	.400		7.75	19.75		27.50	39
0120	3/4″ pipe size		18	.444		9.45	22		31.45	44.50
0140	1″ pipe size		16	.500		12.90	24.50		37.40	52.50
0150	1-1/4″ pipe size		15	.533		17.20	26.50		43.70	60
0160	1-1/2″ pipe size		12	.667		21	33		54	74
0180	2″ pipe size		8	1		32	49.50		81.50	112
0200	2-1/2″ pipe size	Q-5	12	1.333		174	59		233	283
0220	3″ pipe size		11	1.455		187	64.50		251.50	305
0240	4″ pipe size		5	3.200		315	142		457	570
0500	For galvanized body, add					50%				
1000	Flanged, 125 lb., 1-1/2″ pipe size	1 Stpi	11	.727	Ea.	101	36		137	167
1020	2″ pipe size	″	8	1		106	49.50		155.50	193
1030	2-1/2″ pipe size	Q-5	5	3.200		128	142		270	360
1040	3″ pipe size		4.50	3.556		142	158		300	400
1060	4″ pipe size		3	5.333		181	237		418	565
1080	5″ pipe size	Q-6	3.40	7.059		283	325		608	815
1100	6″ pipe size		3	8		283	370		653	880
1120	8″ pipe size		2.50	9.600		580	440		1,020	1,325
1140	10″ pipe size		2	12		1,100	555		1,655	2,075
1160	12″ pipe size		1.70	14.118		1,675	650		2,325	2,825
1170	14″ pipe size		1.30	18.462		3,075	850		3,925	4,700
1180	16″ pipe size		1	24		4,350	1,100		5,450	6,500
1500	For 250 lb rating, add					20%				
2000	For galvanized body, add					50%				
2500	For steel body, add					40%				

23 21 20.84 Thermoflo Indicator		Crew	Daily Output	Labor-Hours	Unit	Material	Labor	Equipment	Total	Total Incl O&P
0010	**THERMOFLO INDICATOR**, For balancing									
1000	Sweat connections, 1-1/4″ pipe size	1 Stpi	12	.667	Ea.	505	33		538	605
1020	1-1/2″ pipe size		10	.800		515	39.50		554.50	625
1040	2″ pipe size		8	1		540	49.50		589.50	670
1060	2-1/2″ pipe size		7	1.143		820	56.50		876.50	995
2000	Flange connections, 3″ pipe size	Q-5	5	3.200		980	142		1,122	1,300
2020	4″ pipe size		4	4		1,175	178		1,353	1,575
2030	5″ pipe size		3.50	4.571		1,475	203		1,678	1,950
2040	6″ pipe size		3	5.333		1,575	237		1,812	2,125
2060	8″ pipe size		2	8		1,925	355		2,280	2,675

23 21 20 – Hydronic HVAC Piping Specialties

23 21 20.88 Venturi Flow

	23 21 20.88 Venturi Flow	Crew	Daily Output	Labor-Hours	Unit	2009 Bare Costs Material	Labor	Equipment	Total	Total Incl O&P
0010	**VENTURI FLOW**, Measuring device									
0050	1/2" diameter	1 Stpi	24	.333	Ea.	228	16.45		244.45	277
0100	3/4" diameter		20	.400		211	19.75		230.75	263
0120	1" diameter		19	.421		225	21		246	280
0140	1-1/4" diameter		15	.533		280	26.50		306.50	350
0160	1-1/2" diameter		13	.615		293	30.50		323.50	365
0180	2" diameter		11	.727		299	36		335	385
0200	2-1/2" diameter	Q-5	16	1		410	44.50		454.50	520
0220	3" diameter		14	1.143		425	51		476	545
0240	4" diameter		11	1.455		635	64.50		699.50	795
0260	5" diameter	Q-6	4	6		835	276		1,111	1,350
0280	6" diameter		3.50	6.857		925	315		1,240	1,525
0300	8" diameter		3	8		1,175	370		1,545	1,875
0320	10" diameter		2	12		2,800	555		3,355	3,925
0500	For meter, add					1,675			1,675	1,850

23 21 20.94 Weld End Ball Joints

	23 21 20.94 Weld End Ball Joints	Crew	Daily Output	Labor-Hours	Unit	Material	Labor	Equipment	Total	Total Incl O&P
0010	**WELD END BALL JOINTS**, Steel									
0050	2-1/2" diameter	Q-17	13	1.231	Ea.	575	54.50	4.28	633.78	720
0100	3" diameter		12	1.333		675	59	4.64	738.64	840
0120	4" diameter		11	1.455		1,075	64.50	5.05	1,144.55	1,300
0140	5" diameter	Q-18	14	1.714		1,200	79	3.98	1,282.98	1,450
0160	6" diameter		12	2		1,950	92	4.64	2,046.64	2,300
0180	8" diameter		9	2.667		2,900	123	6.20	3,029.20	3,375
0200	10" diameter		8	3		3,775	138	6.95	3,919.95	4,375
0220	12" diameter		6	4		5,025	184	9.30	5,218.30	5,825

23 21 23 – Hydronic Pumps

23 21 23.13 In-Line Centrifugal Hydronic Pumps

	23 21 23.13 In-Line Centrifugal Hydronic Pumps	Crew	Daily Output	Labor-Hours	Unit	Material	Labor	Equipment	Total	Total Incl O&P
0010	**IN-LINE CENTRIFUGAL HYDRONIC PUMPS**									
0600	Bronze, sweat connections, 1/40 HP, in line									
0640	3/4" size	Q-1	16	1	Ea.	172	44		216	257
1000	Flange connection, 3/4" to 1-1/2" size									
1040	1/12 HP	Q-1	6	2.667	Ea.	445	117		562	670
1060	1/8 HP		6	2.667		750	117		867	1,000
1100	1/3 HP		6	2.667		855	117		972	1,125
1140	2" size, 1/6 HP		5	3.200		1,100	140		1,240	1,425
1180	2-1/2" size, 1/4 HP		5	3.200		1,375	140		1,515	1,725
1220	3" size, 1/4 HP		4	4		1,425	176		1,601	1,850
1260	1/3 HP		4	4		1,750	176		1,926	2,200
1300	1/2 HP		4	4		1,800	176		1,976	2,250
1340	3/4 HP		4	4		1,950	176		2,126	2,425
1380	1 HP		4	4		3,175	176		3,351	3,750
2000	Cast iron, flange connection									
2040	3/4" to 1-1/2" size, in line, 1/12 HP	Q-1	6	2.667	Ea.	287	117		404	495
2060	1/8 HP		6	2.667		480	117		597	705
2100	1/3 HP		6	2.667		535	117		652	770
2140	2" size, 1/6 HP		5	3.200		585	140		725	865
2180	2-1/2" size, 1/4 HP		5	3.200		740	140		880	1,025
2220	3" size, 1/4 HP		4	4		755	176		931	1,100
2260	1/3 HP		4	4		1,025	176		1,201	1,400
2300	1/2 HP		4	4		1,050	176		1,226	1,425
2340	3/4 HP		4	4		1,225	176		1,401	1,625

23 21 Hydronic Piping and Pumps

23 21 23 – Hydronic Pumps

23 21 23.13 In-Line Centrifugal Hydronic Pumps		Crew	Daily Output	Labor-Hours	Unit	2009 Bare Costs Material	Labor	Equipment	Total	Total Incl O&P
2380	1 HP	Q-1	4	4	Ea.	1,750	176		1,926	2,200
2600	For non-ferrous impeller, add					3%				
3000	High head, bronze impeller									
3030	1-1/2" size 1/2 HP	Q-1	5	3.200	Ea.	880	140		1,020	1,175
3040	1-1/2" size 3/4 HP		5	3.200		940	140		1,080	1,250
3050	2" size 1 HP		4	4		1,125	176		1,301	1,525
3090	2" size 1-1/2 HP		4	4		1,300	176		1,476	1,725
4000	Close coupled, end suction, bronze impeller									
4040	1-1/2" size, 1-1/2 HP, to 40 GPM	Q-1	3	5.333	Ea.	1,625	234		1,859	2,150
4090	2" size, 2 HP, to 50 GPM		3	5.333		1,925	234		2,159	2,475
4100	2" size, 3 HP, to 90 GPM		2.30	6.957		2,000	305		2,305	2,675
4190	2-1/2" size, 3 HP, to 150 GPM		2	8		2,175	350		2,525	2,925
4300	3" size, 5 HP, to 225 GPM		1.80	8.889		2,475	390		2,865	3,325
4410	3" size, 10 HP, to 350 GPM		1.60	10		3,325	440		3,765	4,325
4420	4" size, 7-1/2 HP, to 350 GPM		1.60	10		3,225	440		3,665	4,225
4520	4" size, 10 HP, to 600 GPM	Q-2	1.70	14.118		3,700	640		4,340	5,050
4530	5" size, 15 HP, to 1000 GPM		1.70	14.118		3,700	640		4,340	5,075
4610	5" size, 20 HP, to 1350 GPM		1.50	16		3,925	730		4,655	5,450
4620	5" size, 25 HP, to 1550 GPM		1.50	16		5,375	730		6,105	7,025
5000	Base mounted, bronze impeller, coupling guard									
5040	1-1/2" size, 1-1/2 HP, to 40 GPM	Q-1	2.30	6.957	Ea.	3,100	305		3,405	3,900
5090	2" size, 2 HP, to 50 GPM		2.30	6.957		3,400	305		3,705	4,200
5100	2" size, 3 HP, to 90 GPM		2	8		3,425	350		3,775	4,300
5190	2-1/2" size, 3 HP, to 150 GPM		1.80	8.889		3,700	390		4,090	4,675
5300	3" size, 5 HP, to 225 GPM		1.60	10		3,775	440		4,215	4,825
5410	4" size, 5 HP, to 350 GPM		1.50	10.667		4,025	470		4,495	5,150
5420	4" size, 7-1/2 HP, to 350 GPM		1.50	10.667		4,150	470		4,620	5,300
5520	5" size, 10 HP, to 600 GPM	Q-2	1.60	15		4,275	685		4,960	5,750
5530	5" size, 15 HP, to 1000 GPM		1.60	15		4,425	685		5,110	5,925
5610	6" size, 20 HP, to 1350 GPM		1.40	17.143		5,750	780		6,530	7,525
5620	6" size, 25 HP, to 1550 GPM		1.40	17.143		5,900	780		6,680	7,700
9000	Minimum labor/equipment charge	Q-1	3.25	4.923	Job		216		216	335

23 22 Steam and Condensate Piping and Pumps

23 22 13 – Steam and Condensate Heating Piping

23 22 13.23 Aboveground Steam and Condensate Piping

0010	**ABOVEGROUND STEAM AND CONDENSATE HEATING PIPING**									
0020	Condensate meter									
0100	500 lb. per hour	1 Stpi	14	.571	Ea.	2,300	28		2,328	2,575
0140	1500 lb. per hour		7	1.143		2,475	56.50		2,531.50	2,825
0160	3000 lb. per hour		5	1.600		3,325	79		3,404	3,775
0200	12,000 lb. per hour	Q-5	3.50	4.571		4,450	203		4,653	5,200

23 22 23 – Steam Condensate Pumps

23 22 23.10 Condensate Return System

0010	**CONDENSATE RETURN SYSTEM**									
0200	Simplex, 3/4 H.P. mtr, float switch, controls, 10 Gal. C.I. rcvr, 6-15GPM	Q-1	1	16	Ea.	2,500	700		3,200	3,850
1000	Duplex, 2 pumps, 3/4 H.P. motors, float switch,									
1060	alternator assembly, 15 Gal. C.I. receiver	Q-1	.50	32	Ea.	3,950	1,400		5,350	6,525

23 23 13 – Refrigerant Piping Valves

23 23 13.10 Valves		Crew	Daily Output	Labor-Hours	Unit	Material	2009 Bare Costs Labor	Equipment	Total	Total Incl O&P
0010	VALVES									
8100	Check valve, soldered									
8110	5/8"	1 Stpi	36	.222	Ea.	79.50	10.95		90.45	104
8114	7/8"		26	.308		108	15.20		123.20	143
8118	1-1/8"		18	.444		129	22		151	176
8122	1-3/8"		14	.571		179	28		207	240
8126	1-5/8"		13	.615		172	30.50		202.50	236
8130	2-1/8"		12	.667		256	33		289	335
8134	2-5/8"	Q-5	22	.727		1,625	32.50		1,657.50	1,850
8138	3-1/8"	"	20	.800		1,950	35.50		1,985.50	2,200
8500	Refrigeration valve, packless, soldered									
8510	1/2"	1 Stpi	38	.211	Ea.	29.50	10.40		39.90	48.50
8514	5/8"		36	.222		31	10.95		41.95	51
8518	7/8"		26	.308		273	15.20		288.20	325
8520	Packed, soldered									
8522	1-1/8"	1 Stpi	18	.444	Ea.	161	22		183	211
8526	1-3/8"		14	.571		293	28		321	365
8530	1-5/8"		13	.615		315	30.50		345.50	390
8534	2-1/8"		12	.667		500	33		533	600
8538	2-5/8"	Q-5	22	.727		680	32.50		712.50	795
8542	3-1/8"		20	.800		785	35.50		820.50	920
8546	4-1/8"		18	.889		1,375	39.50		1,414.50	1,575
8600	Solenoid valve, flange/solder									
8610	1/2"	1 Stpi	38	.211	Ea.	144	10.40		154.40	174
8614	5/8"		36	.222		178	10.95		188.95	212
8618	3/4"		30	.267		225	13.15		238.15	268
8622	7/8"		26	.308		283	15.20		298.20	335
8626	1-1/8"		18	.444		370	22		392	440
8630	1-3/8"		14	.571		445	28		473	535
8634	1-5/8"		13	.615		630	30.50		660.50	740
8638	2-1/8"		12	.667		690	33		723	810
8800	Thermostatic valve, flange/solder									
8810	1/2 – 3 ton, 3/8" x 5/8"	1 Stpi	9	.889	Ea.	131	44		175	212
8814	4 – 5 ton, 1/2" x 7/8"		7	1.143		131	56.50		187.50	232
8818	6 – 8 ton, 5/8" x 7/8"		5	1.600		131	79		210	266
8822	7 – 12 ton, 7/8" x 1-1/8"		4	2		153	98.50		251.50	320
8826	15 – 20 ton, 7/8" x 1-3/8"		3.20	2.500		153	123		276	360

23 23 16 – Refrigerant Piping Specialties

23 23 16.10 Refrigerant Piping Component Specialties		Crew	Daily Output	Labor-Hours	Unit	Material	Labor	Equipment	Total	Total Incl O&P
0010	REFRIGERANT PIPING COMPONENT SPECIALTIES									
0600	Accumulator									
0610	3/4"	1 Stpi	8.80	.909	Ea.	70.50	45		115.50	147
0614	7/8"		6.40	1.250		80.50	61.50		142	184
0618	1-1/8"		4.80	1.667		111	82.50		193.50	250
0622	1-3/8"		4	2		145	98.50		243.50	315
0626	1-5/8"		3.20	2.500		157	123		280	365
0630	2-1/8"		2.40	3.333		395	165		560	690
0700	Condensate drip/drain pan		24	.333		116	16.45		132.45	153
1000	Filter dryer									
1010	Replaceable core type, solder									
1020	1/2"	1 Stpi	20	.400	Ea.	108	19.75		127.75	150
1030	5/8"		19	.421		142	21		163	188

23 23 Refrigerant Piping

23 23 16 – Refrigerant Piping Specialties

23 23 16.10 Refrigerant Piping Component Specialties		Crew	Daily Output	Labor-Hours	Unit	Material	2009 Bare Costs Labor	Equipment	Total	Total Incl O&P
1040	7/8"	1 Stpi	18	.444	Ea.	142	22		164	190
1050	1-1/8"		15	.533		148	26.50		174.50	204
1060	1-3/8"		14	.571		148	28		176	207
1070	1-5/8"		12	.667		152	33		185	219
1080	2-1/8"		10	.800		155	39.50		194.50	231
1100	3-1/8"		8	1		630	49.50		679.50	765
1200	Sealed in-line, solder									
1210	1/4", 3 cubic inches	1 Stpi	22	.364	Ea.	8.10	17.95		26.05	37
1220	3/8", 5 cubic inches		21	.381		7.40	18.80		26.20	37
1230	1/2", 9 cubic inches		20	.400		12	19.75		31.75	43.50
1240	1/2", 16 cubic inches		20	.400		15.60	19.75		35.35	47.50
1250	5/8", 16 cubic inches		19	.421		16.35	21		37.35	50
1260	5/8", 30 cubic inches		19	.421		26.50	21		47.50	61
1270	7/8", 30 cubic inches		18	.444		28	22		50	65
1280	7/8", 41 cubic inches		17	.471		37.50	23		60.50	77.50
1290	1-1/8", 60 cubic inches		15	.533		52	26.50		78.50	98.50
4000	P-Trap, suction line, solder									
4010	5/8"	1 Stpi	19	.421	Ea.	52	21		73	89.50
4020	3/4"		19	.421		52	21		73	89.50
4030	7/8"		18	.444		56	22		78	95.50
4040	1-1/8"		15	.533		59	26.50		85.50	106
4050	1-3/8"		14	.571		107	28		135	162
4060	1-5/8"		12	.667		163	33		196	230
4070	2-1/8"		10	.800		335	39.50		374.50	425
5000	Sightglass									
5010	Moisture and liquid indicator, solder									
5020	1/4"	1 Stpi	22	.364	Ea.	19.30	17.95		37.25	49
5030	3/8"		21	.381		18.40	18.80		37.20	49.50
5040	1/2"		20	.400		22.50	19.75		42.25	55
5050	5/8"		19	.421		24	21		45	58
5060	7/8"		18	.444		42	22		64	80
5070	1-1/8"		15	.533		43	26.50		69.50	88.50
5080	1-3/8"		14	.571		32.50	28		60.50	79.50
5090	1-5/8"		12	.667		38	33		71	93
5100	2-1/8"		10	.800		47.50	39.50		87	113
7400	Vacuum pump set									
7410	Two stage, high vacuum continuous duty	1 Stpi	8	1	Ea.	3,000	49.50		3,049.50	3,375

23 23 23 – Refrigerants

23 23 23.10 Anti-Freeze

		Crew	Daily Output	Labor-Hours	Unit	Material	Labor	Equipment	Total	Total Incl O&P
0010	**ANTI-FREEZE**, Inhibited									
0900	Ethylene glycol concentrated									
1000	55 gallon drums, small quantities				Gal.	12.10			12.10	13.30
1200	Large quantities					12			12	13.20
2000	Propylene glycol, for solar heat, small quantities					13.85			13.85	15.25
2100	Large quantities					13.75			13.75	15.15

23 23 23.20 Refrigerant

		Crew	Daily Output	Labor-Hours	Unit	Material	Labor	Equipment	Total	Total Incl O&P
0010	**REFRIGERANT**									
4420	Refrigerant, R-22, 50 lb. disposable cylinder				Lb.	17.45			17.45	19.15
4440	Refrigerant, R-507, 25 lb. disposable cylinder				"	10.50			10.50	11.55

23 31 13.13 Rectangular Metal Ducts		Crew	Daily Output	Labor-Hours	Unit	Material	2009 Bare Costs Labor	Equipment	Total	Total Incl O&P
0010	**RECTANGULAR METAL DUCTS** R233100-40									
0020	Fabricated rectangular, includes fittings, joints, supports,									
0030	allowance for flexible connections, no insulation									
0031	NOTE: Fabrication and installation are combined									
0040	as LABOR cost. Approx. 25% fittings assumed.									
0042	Fabrication/Inst. is to commercial quality standards									
0043	(SMACNA or equiv.) for structure, sealing, leak testing, etc.									
0050	Add to labor for elevated installation									
0051	of fabricated ductwork									
0052	10′ to 15′ high						6%			
0053	15′ to 20′ high						12%			
0054	20′ to 25′ high						15%			
0055	25′ to 30′ high						21%			
0056	30′ to 35′ high						24%			
0057	35′ to 40′ high						30%			
0058	Over 40′ high						33%			
0072	For duct insulation see Div. 23 07 13.10 3000									
0100	Aluminum, alloy 3003-H14, under 100 lb.	Q-10	75	.320	Lb.	3.72	14.10		17.82	26.50
0110	100 to 500 lb.		80	.300		2.48	13.20		15.68	23.50
0120	500 to 1,000 lb.		95	.253		2.31	11.15		13.46	20
0140	1,000 to 2,000 lb.		120	.200		2.15	8.80		10.95	16.35
0150	2,000 to 5,000 lb.		130	.185		2.15	8.15		10.30	15.25
0160	Over 5,000 lb.		145	.166		2.15	7.30		9.45	13.95
0500	Galvanized steel, under 200 lb.		235	.102		1.12	4.50		5.62	8.40
0520	200 to 500 lb.		245	.098		.99	4.32		5.31	7.95
0540	500 to 1,000 lb.		255	.094		.97	4.15		5.12	7.65
0560	1,000 to 2,000 lb.		265	.091		.97	3.99		4.96	7.40
0570	2,000 to 5,000 lb.		275	.087		.95	3.84		4.79	7.15
0580	Over 5,000 lb.		285	.084		.93	3.71		4.64	6.90
0600	For large quantities special prices available from supplier									
1000	Stainless steel, type 304, under 100 lb.	Q-10	165	.145	Lb.	4.35	6.40		10.75	15
1020	100 to 500 lb.		175	.137		4.02	6.05		10.07	14
1030	500 to 1,000 lb.		190	.126		3.68	5.55		9.23	12.90
1040	1,000 to 2,000 lb.		200	.120		3.65	5.30		8.95	12.40
1050	2,000 to 5,000 lb.		225	.107		3.31	4.70		8.01	11.10
1060	Over 5,000 lb.		235	.102		3.24	4.50		7.74	10.70
1080	Note: Minimum order cost exceeds per lb. cost for min. wt.									
1100	For medium pressure ductwork, add				Lb.		15%			
1200	For high pressure ductwork, add						40%			
1210	For welded ductwork, add						85%			
1220	For 30% fittings, add						11%			
1224	For 40% fittings, add						34%			
1228	For 50% fittings, add						56%			
1232	For 60% fittings, add						79%			
1236	For 70% fittings, add						101%			
1240	For 80% fittings, add						124%			
1244	For 90% fittings, add						147%			
1248	For 100% fittings, add						169%			
1252	Note: Fittings add includes time for detailing and installation.									

23 31 13.16 Round and Flat-Oval Spiral Ducts

0010	**ROUND AND FLAT-OVAL SPIRAL DUCTS**									
1280	Add to labor for elevated installation									

23 31 13 – Metal Ducts

23 31 13.16 Round and Flat-Oval Spiral Ducts		Crew	Daily Output	Labor-Hours	Unit	Material	2009 Bare Costs Labor	Equipment	Total	Total Incl O&P
1282	of prefabricated (purchased) ductwork									
1283	10′ to 15′ high						10%			
1284	15′ to 20′ high						20%			
1285	20′ to 25′ high						25%			
1286	25′ to 30′ high						35%			
1287	30′ to 35′ high						40%			
1288	35′ to 40′ high						50%			
1289	Over 40′ high						55%			
5400	Spiral preformed, steel, galv., straight lengths, Max 10″ spwg.									
5410	4″ diameter, 26 ga.	Q-9	360	.044	L.F.	1.39	1.89		3.28	4.53
5416	5″ diameter, 26 ga.		320	.050		1.77	2.12		3.89	5.35
5420	6″ diameter, 26 ga.		280	.057		2.02	2.43		4.45	6.10
5425	7″ diameter, 26 ga.		240	.067		2.40	2.83		5.23	7.15
5430	8″ diameter, 26 ga.		200	.080		2.66	3.40		6.06	8.35
5440	10″ diameter, 26 ga.		160	.100		4.30	4.25		8.55	11.50
5450	12″ diameter, 26 ga.		120	.133		5.20	5.65		10.85	14.70
5460	14″ diameter, 26 ga.		80	.200		6.05	8.50		14.55	20
5480	16″ diameter, 24 ga.		60	.267		6.95	11.35		18.30	25.50
5490	18″ diameter, 24 ga.	↓	50	.320		9.50	13.60		23.10	32
5500	20″ diameter, 24 ga.	Q-10	65	.369		10.50	16.25		26.75	37.50
5510	22″ diameter, 24 ga.		60	.400		11.50	17.60		29.10	40.50
5520	24″ diameter, 24 ga.		55	.436		12.50	19.20		31.70	44.50
5540	30″ diameter, 22 ga.		45	.533		18.45	23.50		41.95	58
5600	36″ diameter, 22 ga.	↓	40	.600	↓	29.50	26.50		56	74.50
5800	Connector, 4″ diameter	Q-9	100	.160	Ea.	2.44	6.80		9.24	13.50
5810	5″ diameter		94	.170		2.61	7.25		9.86	14.35
5820	6″ diameter		88	.182		2.81	7.70		10.51	15.35
5840	8″ diameter		78	.205		3.63	8.70		12.33	17.85
5860	10″ diameter		70	.229		5.70	9.70		15.40	22
5880	12″ diameter		50	.320		6.90	13.60		20.50	29
5900	14″ diameter		44	.364		8	15.45		23.45	33.50
5920	16″ diameter		40	.400		9	17		26	37
5930	18″ diameter		37	.432		10.05	18.35		28.40	40
5940	20″ diameter		34	.471		10.90	20		30.90	44
5950	22″ diameter		31	.516		11.90	22		33.90	48
5960	24″ diameter		28	.571		12.75	24.50		37.25	52.50
5980	30″ diameter		22	.727		16.60	31		47.60	67.50
6000	36″ diameter		18	.889		25	38		63	87.50
6300	Elbow, 45°, 4″ diameter		60	.267		2.75	11.35		14.10	21
6310	5″ diameter		52	.308		3.54	13.05		16.59	25
6320	6″ diameter		44	.364		3.87	15.45		19.32	29
6340	8″ diameter		28	.571		5.10	24.50		29.60	44
6360	10″ diameter		18	.889		7.75	38		45.75	68.50
6380	12″ diameter		13	1.231		10.05	52.50		62.55	94
6400	14″ diameter		11	1.455		18.85	62		80.85	119
6420	16″ diameter		10	1.600		31	68		99	142
6430	18″ diameter	↓	9.60	1.667		38	71		109	155
6440	20″ diameter	Q-10	14	1.714		48	75.50		123.50	173
6450	22″ diameter		13	1.846		57.50	81.50		139	193
6460	24″ diameter		12	2		64.50	88		152.50	211
6480	30″ diameter		9	2.667		99	117		216	296
6500	36″ diameter	↓	7	3.429		120	151		271	370
6600	Elbow, 90°, 4″ diameter	Q-9	60	.267	↓	4.71	11.35		16.06	23

23 31 13.16 Round and Flat-Oval Spiral Ducts		Crew	Daily Output	Labor-Hours	Unit	2009 Bare Costs Material	Labor	Equipment	Total	Total Incl O&P
6610	5″ diameter	Q-9	52	.308	Ea.	5.40	13.05		18.45	27
6620	6″ diameter		44	.364		6.20	15.45		21.65	31.50
6625	7″ diameter		36	.444		6.55	18.90		25.45	37
6630	8″ diameter		28	.571		7.60	24.50		32.10	47
6640	10″ diameter		18	.889		10.85	38		48.85	72
6650	12″ diameter		13	1.231		14.35	52.50		66.85	99
6660	14″ diameter		11	1.455		27	62		89	128
6670	16″ diameter		10	1.600		42.50	68		110.50	155
6676	18″ diameter		9.60	1.667		55.50	71		126.50	174
6680	20″ diameter	Q-10	14	1.714		74	75.50		149.50	201
6684	22″ diameter		13	1.846		115	81.50		196.50	256
6690	24″ diameter		12	2		129	88		217	281
6700	30″ diameter		9	2.667		198	117		315	405
6710	36″ diameter		7	3.429		240	151		391	505
6800	Reducing coupling, 6″ x 4″	Q-9	46	.348		7	14.80		21.80	31
6820	8″ x 6″		40	.400		8.60	17		25.60	36.50
6840	10″ x 8″		32	.500		10.90	21		31.90	46
6860	12″ x 10″		24	.667		15.60	28.50		44.10	62
6880	14″ x 12″		20	.800		25.50	34		59.50	82.50
6900	16″ x 14″		18	.889		37.50	38		75.50	101
6920	18″ x 16″		16	1		46.50	42.50		89	119
6940	20″ x 18″	Q-10	24	1		56.50	44		100.50	132
6950	22″ x 20″		23	1.043		66	46		112	146
6960	24″ x 22″		22	1.091		76	48		124	160
6980	30″ x 28″		18	1.333		113	58.50		171.50	218
7000	36″ x 34″		16	1.500		152	66		218	272
7100	Tee, 90°, 4″ diameter	Q-9	40	.400		6.15	17		23.15	34
7110	5″ diameter		34.60	.462		8.20	19.65		27.85	40
7120	6″ diameter		30	.533		8.85	22.50		31.35	46
7130	8″ diameter		19	.842		11.55	36		47.55	69.50
7140	10″ diameter		12	1.333		16.45	56.50		72.95	108
7150	12″ diameter		9	1.778		21	75.50		96.50	143
7160	14″ diameter		7	2.286		40	97		137	198
7170	16″ diameter		6.70	2.388		56.50	101		157.50	223
7176	18″ diameter		6	2.667		78	113		191	266
7180	20″ diameter	Q-10	8.70	2.759		86	122		208	288
7190	24″ diameter		7	3.429		103	151		254	355
7200	30″ diameter		6	4		126	176		302	420
7210	36″ diameter		5	4.800		160	211		371	510
7400	Steel, PVC coated both sides, straight lengths									
7410	4″ diameter, 26 ga.	Q-9	360	.044	L.F.	2.52	1.89		4.41	5.75
7416	5″ diameter, 26 ga.		240	.067		3.14	2.83		5.97	7.95
7420	6″ diameter, 26 ga.		280	.057		3.49	2.43		5.92	7.70
7440	8″ diameter, 26 ga.		200	.080		4.82	3.40		8.22	10.70
7460	10″ diameter, 26 ga.		160	.100		6.50	4.25		10.75	13.90
7480	12″ diameter, 26 ga.		120	.133		8.05	5.65		13.70	17.85
7500	14″ diameter, 26 ga.		80	.200		11	8.50		19.50	25.50
7520	16″ diameter, 24 ga.		60	.267		15.55	11.35		26.90	35
7540	18″ diameter, 24 ga.		45	.356		16.75	15.10		31.85	42.50
7560	20″ diameter, 24 ga.	Q-10	65	.369		19.80	16.25		36.05	48
7580	24″ diameter, 24 ga.		55	.436		25	19.20		44.20	58
7600	30″ diameter, 22 ga.		45	.533		36.50	23.50		60	77.50
7620	36″ diameter, 22 ga.		40	.600		44.50	26.50		71	91

23 31 13.16 Round and Flat-Oval Spiral Ducts		Crew	Daily Output	Labor-Hours	Unit	2009 Bare Costs Material	Labor	Equipment	Total	Total Incl O&P
7890	Connector, 4″ diameter	Q-9	100	.160	Ea.	4.81	6.80		11.61	16.10
7896	5″ diameter		83	.193		5.15	8.20		13.35	18.65
7900	6″ diameter		88	.182		5.40	7.70		13.10	18.20
7920	8″ diameter		78	.205		6.85	8.70		15.55	21.50
7940	10″ diameter		70	.229		9.40	9.70		19.10	26
7960	12″ diameter		50	.320		12	13.60		25.60	34.50
7980	14″ diameter		44	.364		15.45	15.45		30.90	41.50
8000	16″ diameter		40	.400		17.10	17		34.10	46
8020	18″ diameter		37	.432		19.30	18.35		37.65	50
8040	20″ diameter		34	.471		21	20		41	55
8060	24″ diameter		28	.571		23	24.50		47.50	64
8080	30″ diameter		22	.727		30.50	31		61.50	82.50
8100	36″ diameter		18	.889		41	38		79	106
8390	Elbow, 45°, 4″ diameter		60	.267		6.05	11.35		17.40	24.50
8396	5″ diameter		36	.444		7.30	18.90		26.20	38
8400	6″ diameter		44	.364		7.95	15.45		23.40	33.50
8420	8″ diameter		28	.571		10.50	24.50		35	50
8440	10″ diameter		18	.889		15.25	38		53.25	77
8460	12″ diameter		13	1.231		22.50	52.50		75	108
8480	14″ diameter		11	1.455		40.50	62		102.50	143
8500	16″ diameter		10	1.600		58	68		126	172
8520	18″ diameter		9	1.778		67	75.50		142.50	194
8540	20″ diameter	Q-10	13	1.846		89.50	81.50		171	228
8560	24″ diameter		11	2.182		113	96		209	278
8580	30″ diameter		9	2.667		174	117		291	380
8600	36″ diameter		7	3.429		212	151		363	475
8800	Elbow, 90°, 4″ diameter	Q-9	60	.267		10.05	11.35		21.40	29
8804	5″ diameter		52	.308		11.45	13.05		24.50	33.50
8806	6″ diameter		44	.364		12.55	15.45		28	38.50
8808	8″ diameter		28	.571		15.65	24.50		40.15	55.50
8810	10″ diameter		18	.889		21	38		59	83
8812	12″ diameter		13	1.231		29	52.50		81.50	115
8814	14″ diameter		11	1.455		52	62		114	155
8816	16″ diameter		10	1.600		81	68		149	197
8818	18″ diameter		9.60	1.667		99	71		170	222
8820	20″ diameter	Q-10	14	1.714		126	75.50		201.50	259
8822	24″ diameter		12	2		227	88		315	390
8824	30″ diameter		9	2.667		350	117		467	570
8826	36″ diameter		7	3.429		425	151		576	705
8850	Tee, 90°, 4″ diameter	Q-9	40	.400		11.30	17		28.30	39.50
8854	5″ diameter		34.60	.462		14.90	19.65		34.55	47.50
8856	6″ diameter		30	.533		16.45	22.50		38.95	54
8858	8″ diameter		19	.842		21	36		57	80.50
8860	10″ diameter		12	1.333		27.50	56.50		84	121
8862	12″ diameter		9	1.778		39	75.50		114.50	163
8864	14″ diameter		7	2.286		68.50	97		165.50	229
8866	16″ diameter		6.70	2.388		99.50	101		200.50	270
8868	18″ diameter		6	2.667		144	113		257	340
8870	20″ diameter	Q-10	8.70	2.759		155	122		277	365
8872	24″ diameter		7	3.429		214	151		365	475
8874	30″ diameter		6	4		267	176		443	575
8876	36″ diameter		5	4.800		340	211		551	705
9000	Reducing coupling, tapered offset									

23 31 13 – Metal Ducts

23 31 13.16 Round and Flat-Oval Spiral Ducts		Crew	Daily Output	Labor-Hours	Unit	Material	2009 Bare Costs Labor	Equipment	Total	Total Incl O&P
9020	Diameter x any smaller diameter									
9040	6" x	Q-9	46	.348	Ea.	12.50	14.80		27.30	37.50
9060	8" x		40	.400		15.45	17		32.45	44
9080	10" x		32	.500		18.65	21		39.65	54.50
9100	12" x		24	.667		26	28.50		54.50	73.50
9120	14" x		20	.800		41.50	34		75.50	100
9140	16" x		18	.889		63.50	38		101.50	130
9160	18" x		16	1		77	42.50		119.50	152
9180	20" x	Q-10	24	1		94.50	44		138.50	174
9200	24" x		22	1.091		128	48		176	218
9220	30" x		18	1.333		181	58.50		239.50	293
9240	36" x		16	1.500		236	66		302	365
9400	Lateral 45°, straight barrel									
9410	4" size	Q-9	40	.400	Ea.	12.95	17		29.95	41.50
9416	5" size		34.60	.462		17.10	19.65		36.75	50
9420	6" size		30	.533		18.95	22.50		41.45	57
9440	8" size		19	.842		24.50	36		60.50	83.50
9460	10" size		12	1.333		35	56.50		91.50	129
9480	12" size		9	1.778		45	75.50		120.50	170
9500	14" size		7	2.286		78.50	97		175.50	241
9520	16" size		6.70	2.388		114	101		215	287
9540	18" size		6	2.667		166	113		279	360
9560	20" size	Q-10	8.70	2.759		179	122		301	390
9580	24" size		7	3.429		246	151		397	510
9600	30" size		6	4		305	176		481	620
9620	36" size		5	4.800		390	211		601	765
9640	For reducing lateral, add					40%				
9700	Register boot, elbow or perimeter									
9710	6" duct diameter	Q-9	60	.267	Ea.	26.50	11.35		37.85	47.50
9720	8" duct diameter		56	.286		32	12.15		44.15	54.50
9730	10" duct diameter		45	.356		43.50	15.10		58.60	72
9740	12" duct diameter		36	.444		43.50	18.90		62.40	78
9990	Minimum labor/equipment charge	1 Shee	3	2.667	Job		126		126	200

23 31 13.17 Round Grease Duct		Crew	Daily Output	Labor-Hours	Unit	Material	Labor	Equipment	Total	Total Incl O&P
0010	**ROUND GREASE DUCT**									
1280	Add to labor for elevated installation									
1282	of prefabricated (purchased) ductwork									
1283	10' to 15' high						10%			
1284	15' to 20' high						20%			
1285	20' to 25' high						25%			
1286	25' to 30' high						35%			
1287	30' to 35' high						40%			
1288	35' to 40' high						50%			
1289	Over 40' high						55%			
4020	Zero clearance, 2 hr. fire rated, UL listed, 3" double wall									
4030	304 interior, aluminized exterior, all sizes are ID									
4032	For 316 interior and aluminized exterior, add					6%				
4034	For 304 interior and 304 exterior, add					21%				
4036	For 316 interior and 316 exterior, add					40%				
4040	Straight									
4050	5"	Q-9	36.48	.439	L.F.	62.50	18.65		81.15	98
4052	6"		34.20	.468		79	19.85		98.85	118

23 31 13.17 Round Grease Duct		Crew	Daily Output	Labor-Hours	Unit	Material	2009 Bare Costs Labor	Equipment	Total	Total Incl O&P
4054	8″	Q-9	29.64	.540	L.F.	90.50	23		113.50	136
4056	10″		27.36	.585		102	25		127	152
4058	12″		25.10	.637		116	27		143	171
4060	14″		23.94	.668		134	28.50		162.50	192
4062	16″		22.80	.702		140	30		170	202
4064	18″		21.66	.739		172	31.50		203.50	240
4066	20″	Q-10	20.52	1.170		198	51.50		249.50	300
4068	22″		19.38	1.238		225	54.50		279.50	335
4070	24″		18.24	1.316		256	58		314	375
4072	26″		17.67	1.358		271	60		331	395
4074	28″		17.10	1.404		285	62		347	415
4076	30″		15.96	1.504		310	66		376	445
4078	32″		15.39	1.559		335	68.50		403.50	480
4080	36″		14.25	1.684		375	74		449	530
4100	Adjustable section, 30″ long									
4102	5″	Q-9	18.24	.877	Ea.	196	37.50		233.50	274
4104	6″		17.10	.936		245	40		285	330
4106	8″		14.82	1.080		250	46		296	350
4108	10″		13.68	1.170		320	49.50		369.50	430
4110	12″		12.54	1.276		365	54		419	485
4112	14″		11.97	1.337		420	57		477	550
4114	16″		11.40	1.404		435	59.50		494.50	575
4116	18″		10.83	1.477		535	63		598	690
4118	20″	Q-10	10.26	2.339		620	103		723	845
4120	22″		9.69	2.477		705	109		814	950
4122	24″		9.12	2.632		800	116		916	1,075
4124	26″		8.83	2.718		845	120		965	1,125
4126	28″		8.55	2.807		890	124		1,014	1,175
4128	30″		8.26	2.906		965	128		1,093	1,275
4130	32″		7.98	3.008		1,050	132		1,182	1,350
4132	36″		6.84	3.509		1,150	155		1,305	1,500
4140	Tee, 90°, grease									
4142	5″	Q-9	14.25	1.123	Ea.	520	47.50		567.50	645
4144	6″		13.68	1.170		610	49.50		659.50	750
4146	8″		12.54	1.276		700	54		754	855
4148	10″		11.97	1.337		820	57		877	990
4150	12″		11.40	1.404		955	59.50		1,014.50	1,150
4152	14″		10.26	1.559		1,125	66.50		1,191.50	1,350
4154	16″		9.12	1.754		1,275	74.50		1,349.50	1,525
4156	18″		7.98	2.005		1,500	85		1,585	1,775
4158	20″		9.69	1.651		1,675	70		1,745	1,950
4160	22″		8.26	1.937		1,875	82.50		1,957.50	2,200
4162	24″		6.84	2.339		2,075	99.50		2,174.50	2,425
4164	26″		6.55	2.443		2,350	104		2,454	2,750
4166	28″	Q-10	6.27	3.828		2,600	169		2,769	3,125
4168	30″		5.98	4.013		3,075	177		3,252	3,675
4170	32″		5.70	4.211		3,550	185		3,735	4,225
4172	36″		5.13	4.678		4,150	206		4,356	4,875
4180	Cleanout Tee Cap									
4182	5″	Q-9	22.80	.702	Ea.	152	30		182	216
4184	6″		21.09	.759		180	32		212	249
4186	8″		19.38	.826		183	35		218	257
4188	10″		18.24	.877		201	37.50		238.50	281

23 31 13.17 Round Grease Duct		Crew	Daily Output	Labor-Hours	Unit	Material	Labor	Equipment	Total	Total Incl O&P
						2009 Bare Costs				
4190	12"	Q-9	17.10	.936	Ea.	210	40		250	294
4192	14"		15.96	1.003		242	42.50		284.50	335
4194	16"		14.25	1.123		265	47.50		312.50	365
4196	18"		13.68	1.170		299	49.50		348.50	410
4198	20"	Q-10	15.39	1.559		325	68.50		393.50	465
4200	22"		12.54	1.914		365	84.50		449.50	535
4202	24"		11.97	2.005		410	88.50		498.50	590
4204	26"		11.40	2.105		455	92.50		547.50	650
4206	28"		10.83	2.216		505	97.50		602.50	710
4208	30"		10.26	2.339		560	103		663	785
4210	32"		9.69	2.477		620	109		729	855
4212	36"		8.55	2.807		770	124		894	1,050
4220	Elbow, 45°									
4222	5"	Q-9	18.24	.877	Ea.	284	37.50		321.50	370
4224	6"		17.10	.936		335	40		375	430
4226	8"		14.82	1.080		380	46		426	490
4228	10"		13.68	1.170		425	49.50		474.50	545
4230	12"		12.54	1.276		485	54		539	615
4232	14"		11.97	1.337		555	57		612	700
4234	16"		11.40	1.404		625	59.50		684.50	785
4236	18"		10.83	1.477		710	63		773	880
4238	20"	Q-10	10.26	2.339		810	103		913	1,050
4240	22"		9.69	2.477		915	109		1,024	1,175
4242	24"		9.12	2.632		1,050	116		1,166	1,325
4244	26"		8.83	2.718		1,100	120		1,220	1,425
4246	28"		8.55	2.807		1,175	124		1,299	1,500
4248	30"		8.26	2.906		1,375	128		1,503	1,725
4250	32"		7.98	3.008		1,575	132		1,707	1,950
4252	36"		6.84	3.509		2,225	155		2,380	2,700
4260	Elbow, 90°									
4262	5"	Q-9	18.24	.877	Ea.	450	37.50		487.50	555
4264	6"		17.10	.936		530	40		570	645
4266	8"		14.82	1.080		580	46		626	715
4268	10"		13.68	1.170		670	49.50		719.50	815
4270	12"		12.54	1.276		760	54		814	920
4272	14"		11.97	1.337		890	57		947	1,075
4274	16"		11.40	1.404		990	59.50		1,049.50	1,200
4276	18"		10.83	1.477		1,150	63		1,213	1,375
4278	20"	Q-10	10.26	2.339		1,275	103		1,378	1,600
4280	22"		9.69	2.477		1,450	109		1,559	1,750
4282	24"		9.12	2.632		1,600	116		1,716	1,925
4284	26"		8.83	2.718		1,800	120		1,920	2,175
4286	28"		8.55	2.807		2,000	124		2,124	2,400
4288	30"		8.26	2.906		2,325	128		2,453	2,750
4290	32"		7.98	3.008		2,650	132		2,782	3,125
4292	36"		6.84	3.509		3,250	155		3,405	3,825
4300	Support strap									
4302	5"	Q-9	27	.593	Ea.	22.50	25		47.50	64.50
4304	6"		25.50	.627		22.50	26.50		49	67
4306	8"		22	.727		23.50	31		54.50	74.50
4308	10"		20.50	.780		23.50	33		56.50	78
4310	12"		18.80	.851		23.50	36		59.50	83
4312	14"		17.90	.894		23.50	38		61.50	86

23 31 13 – Metal Ducts

23 31 13.17 Round Grease Duct		Crew	Daily Output	Labor-Hours	Unit	2009 Bare Costs Material	Labor	Equipment	Total	Total Incl O&P
4314	16"	Q-9	17	.941	Ea.	23.50	40		63.50	89
4316	18"	↓	16.20	.988		24.50	42		66.50	93.50
4318	20"	Q-10	15.40	1.558		25.50	68.50		94	137
4320	22"	"	14.50	1.655	↓	25.50	73		98.50	144
4340	Plate Support Assembly									
4342	5"	Q-9	17.10	.936	Ea.	106	40		146	179
4344	6"		14.82	1.030		124	46		170	209
4346	8"		12.54	1.276		139	54		193	239
4348	10"		11.40	1.404		152	59.50		211.50	263
4350	12"		10.26	1.559		158	66.50		224.50	279
4352	14"		9.69	1.651		193	70		263	325
4354	16"		9.12	1.754		201	74.50		275.50	340
4356	18"	↓	8.55	1.871		214	79.50		293.50	360
4358	20"	Q-10	9.12	2.632		225	116		341	430
4360	22"		8.55	2.807		232	124		356	450
4362	24"		7.98	3.008		238	132		370	475
4364	26"		7.69	3.121		292	137		429	540
4366	28"		7.41	3.239		330	143		473	590
4368	30"		7.12	3.371		335	148		483	605
4370	32"		6.84	3.509		380	155		535	660
4372	36"	↓	5.70	4.211	↓	475	185		660	815
4380	Wall Guide Assembly									
4382	5"	Q-9	20.20	.792	Ea.	124	33.50		157.50	190
4384	6"		19.10	.838		135	35.50		170.50	205
4386	8"		16.50	.970		141	41		182	221
4388	10"		15.30	1.046		148	44.50		192.50	234
4390	12"		14.10	1.135		155	48		203	247
4392	14"		13.40	1.194		167	50.50		217.50	265
4394	16"		12.75	1.255		169	53.50		222.50	271
4396	18"	↓	12.10	1.322		177	56		233	285
4398	20"	Q-10	11.50	2.087		186	92		278	350
4400	22"		10.90	2.202		191	97		288	365
4402	24"		10.20	2.353		193	104		297	375
4404	26"		9.90	2.424		201	107		308	390
4406	28"		9.60	2.500		216	110		326	415
4408	30"		9.30	2.581		225	114		339	430
4410	32"		9	2.667		235	117		352	445
4412	36"	↓	7.60	3.158	↓	256	139		395	505
4420	Hood Transition, Flanged or Unflanged									
4422	5"	Q-9	16.60	.964	Ea.	42.50	41		83.50	112
4424	6"		15.60	1.026		50	43.50		93.50	124
4426	8"		13.50	1.185		57	50.50		107.50	143
4428	10"		12.50	1.280		63.50	54.50		118	157
4430	12"		11.40	1.404		73.50	59.50		133	176
4432	14"		10.90	1.468		84.50	62.50		147	192
4434	16"		10.40	1.538		94.50	65.50		160	208
4436	18"	↓	9.80	1.633		109	69.50		178.50	230
4438	20"	Q-10	9.40	2.553		127	112		239	320
4440	22"		8.80	2.727		149	120		269	355
4442	24"		8.30	2.892		168	127		295	385
4444	26"		8.10	2.963		179	131		310	405
4446	28"		7.80	3.077		193	136		329	425
4448	30"	↓	7.50	3.200	↓	208	141		349	455

23 31 13.17 Round Grease Duct		Crew	Daily Output	Labor-Hours	Unit	Material	Labor	Equipment	Total	Total Incl O&P
						2009 Bare Costs				
4450	32″	Q-10	7.30	3.288	Ea.	218	145		363	470
4452	36″	↓	6.20	3.871	↓	264	171		435	560
4460	Fan Adapter									
4462	5″	Q-9	16	1	Ea.	240	42.50		282.50	330
4464	6″		15	1.067		290	45.50		335.50	390
4466	8″		13	1.231		325	52.50		377.50	440
4468	10″		12	1.333		360	56.50		416.50	490
4470	12″		11	1.455		415	62		477	560
4472	14″		10.50	1.524		480	64.50		544.50	635
4474	16″		10	1.600		550	68		618	710
4476	18″	↓	9.50	1.684		625	71.50		696.50	800
4478	20″	Q-10	9	2.667		730	117		847	985
4480	22″		8.50	2.824		850	124		974	1,125
4482	24″		8	3		875	132		1,007	1,175
4484	26″		7.70	3.117		905	137		1,042	1,225
4486	28″		7.50	3.200		925	141		1,066	1,250
4488	30″		7.20	3.333		955	147		1,102	1,275
4490	32″		7	3.429		985	151		1,136	1,325
4492	36″	↓	6	4	↓	1,025	176		1,201	1,400
4500	Tapered Increaser/Reducer									
4510	5″ x 10″ dia	Q-9	13.68	1.170	Ea.	465	49.50		514.50	590
4520	5″ x 20″ dia		10.30	1.553		725	66		791	905
4530	6″ x 10″ dia		13.40	1.194		465	50.50		515.50	590
4540	6″ x 20″ dia		10.10	1.584		765	67.50		832.50	945
4550	8″ x 10″ dia		13.20	1.212		440	51.50		491.50	560
4560	8″ x 20″ dia		9.90	1.616		765	68.50		833.50	955
4570	10″ x 20″ dia	↓	10.60	1.509		825	64		889	1,000
4580	10″ x 28″ dia	Q-10	8.60	2.791		1,125	123		1,248	1,450
4590	12″ x 20″ dia		8.80	2.727		870	120		990	1,150
4600	14″ x 28″ dia		8.40	2.857		1,300	126		1,426	1,625
4610	16″ x 20″ dia		8.60	2.791		870	123		993	1,150
4620	16″ x 30″ dia		8.30	2.892		1,425	127		1,552	1,775
4630	18″ x 24″ dia		8.80	2.727		1,050	120		1,170	1,350
4640	18″ x 30″ dia		8.10	2.963		1,425	131		1,556	1,775
4650	20″ x 24″ dia		8.30	2.892		995	127		1,122	1,300
4660	20″ x 32″ dia		8	3		1,525	132		1,657	1,875
4670	24″ mm x 26″ dia		8	3		995	132		1,127	1,300
4680	24″ mm x 36″ dia		7.30	3.288		1,750	145		1,895	2,150
4690	28″ x 30″ dia		7.90	3.038		1,200	134		1,334	1,550
4700	28″ x 36″ dia		7.10	3.380		1,750	149		1,899	2,150
4710	30″ x 36″ dia	↓	6.90	3.478	↓	1,800	153		1,953	2,225
4730	Many intermediate standard sizes of tapered fittings are available									

23 31 13.19 Metal Duct Fittings		Crew	Daily Output	Labor-Hours	Unit	Material	Labor	Equipment	Total	Total Incl O&P
0010	**METAL DUCT FITTINGS**									
0050	Air extractors, 12″ x 4″	1 Shee	24	.333	Ea.	15.65	15.75		31.40	42
0100	8″ x 6″		22	.364		15.65	17.15		32.80	44.50
0120	12″ x 6″		21	.381		21.50	18		39.50	52.50
0140	16″ x 6″		20	.400		27	18.90		45.90	60
0160	24″ x 6″		18	.444		45.50	21		66.50	83.50
0180	12″ x 8″		20	.400		24.50	18.90		43.40	56.50
0200	20″ x 8″		16	.500		35.50	23.50		59	76.50
0240	18″ x 10″	↓	14	.571	↓	34.50	27		61.50	81

23 31 13 – Metal Ducts

23 31 13.19 Metal Duct Fittings		Crew	Daily Output	Labor-Hours	Unit	Material	2009 Bare Costs Labor	Equipment	Total	Total Incl O&P
0260	24" x 10"	1 Shee	12	.667	Ea.	44.50	31.50		76	99
0280	24" x 12"		10	.800		48	38		86	113
0300	30" x 12"		8	1		60	47		107	141
2000	Fabrics for flexible connections, with metal edge		100	.080	L.F.	2.13	3.78		5.91	8.35
2100	Without metal edge		160	.050	"	1.53	2.36		3.89	5.45

23 31 16 – Nonmetal Ducts

23 31 16.13 Fibrous-Glass Ducts

		Crew	Daily Output	Labor-Hours	Unit	Material	Labor	Equipment	Total	Total Incl O&P
0010	**FIBROUS-GLASS DUCTS** R233100-40									
1280	Add to labor for elevated installation									
1282	of prefabricated (purchased) ductwork									
1283	10' to 15' high						10%			
1284	15' to 20' high						20%			
1285	20' to 25' high						25%			
1286	25' to 30' high						35%			
1287	30' to 35' high						40%			
1288	35' to 40' high						50%			
1289	Over 40' high						55%			
3490	Rigid fiberglass duct board, foil reinf. kraft facing									
3500	Rectangular, 1" thick, alum. faced, (FRK), std. weight	Q-10	350	.069	SF Surf	.73	3.02		3.75	5.60
9990	Minimum labor/equipment charge	1 Shee	3	2.667	Job		126		126	200

23 31 16.16 Thermoset Fiberglass-Reinforced Plastic Ducts

		Crew	Daily Output	Labor-Hours	Unit	Material	Labor	Equipment	Total	Total Incl O&P
0010	**THERMOSET FIBERGLASS-REINFORCED PLASTIC DUCTS** R233100-40									
1280	Add to labor for elevated installation									
1282	of prefabricated (purchased) ductwork									
1283	10' to 15' high						10%			
1284	15' to 20' high						20%			
1285	20' to 25' high						25%			
1286	25' to 30' high						35%			
1287	30' to 35' high						40%			
1288	35' to 40' high						50%			
1289	Over 40' high						55%			
3550	Rigid fiberglass reinforced plastic, FM approved									
3552	for acid fume and smoke exhaust system, nonflammable									
3554	Straight, 4" diameter	Q-9	200	.080	L.F.	10.60	3.40		14	17.05
3555	6" diameter		146	.110		13.90	4.66		18.56	22.50
3556	8" diameter		106	.151		16.85	6.40		23.25	28.50
3557	10" diameter		85	.188		20	8		28	34.50
3558	12" diameter		68	.235		23.50	10		33.50	42
3561	18" diameter		31	.516		43.50	22		65.50	82.50
3564	24" diameter	Q-10	37	.649		55.50	28.50		84	107
3584	Note: joints are cemented with									
3586	fiberglass resin, included in material cost.									
3590	Elbow, 90°, 4" diameter	Q-9	20.30	.788	Ea.	51	33.50		84.50	110
3591	6" diameter		13.80	1.159		71.50	49.50		121	157
3592	8" diameter		10	1.600		79	68		147	195
3593	10" diameter		7.80	2.051		103	87		190	252
3594	12" diameter		6.50	2.462		131	105		236	310
3597	18" diameter		4.40	3.636		274	154		428	545
3600	24" diameter	Q-10	4.90	4.898		410	216		626	795
3626	Elbow, 45°, 4" diameter	Q-9	22.30	.717		49	30.50		79.50	103
3627	6" diameter		15.20	1.053		68	44.50		112.50	146
3628	8" diameter		11	1.455		72.50	62		134.50	178

23 31 16.16 Thermoset Fiberglass-Reinforced Plastic Ducts		Crew	Daily Output	Labor-Hours	Unit	2009 Bare Costs Material	Labor	Equipment	Total	Total Incl O&P
3629	10″ diameter	Q-9	8.60	1.860	Ea.	76.50	79		155.50	210
3630	12″ diameter		7.20	2.222		83.50	94.50		178	242
3633	18″ diameter		4.84	3.306		143	140		283	380
3636	24″ diameter	Q-10	5.40	4.444		257	196		453	595
3660	Tee, 90°, 4″ diameter	Q-9	14.06	1.138		28	48.50		76.50	108
3661	6″ diameter		9.52	1.681		43	71.50		114.50	160
3662	8″ diameter		6.98	2.292		58.50	97.50		156	220
3663	10″ diameter		5.45	2.936		76.50	125		201.50	282
3664	12″ diameter		4.60	3.478		95.50	148		243.50	340
3667	18″ diameter		3.05	5.246		196	223		419	570
3670	24″ diameter	Q-10	3.46	6.936		295	305		600	810
3690	For Y @ 45°, add					22%				

23 31 16.19 PVC Ducts		Crew	Daily Output	Labor-Hours	Unit	Material	Labor	Equipment	Total	Total Incl O&P
0010	**PVC DUCTS**									
4000	Rigid plastic, corrosive fume resistant PVC									
4020	Straight, 6″ diameter	Q-9	220	.073	L.F.	7.50	3.09		10.59	13.15
4030	7″ diameter		160	.100		11.10	4.25		15.35	19
4040	8″ diameter		160	.100		11.10	4.25		15.35	19
4050	9″ diameter		140	.114		12.55	4.86		17.41	21.50
4060	10″ diameter		120	.133		14	5.65		19.65	24.50
4070	12″ diameter		100	.160		16.35	6.80		23.15	29
4080	14″ diameter		70	.229		21	9.70		30.70	39
4090	16″ diameter		62	.258		24.50	10.95		35.45	44.50
4100	18″ diameter		58	.276		38	11.70		49.70	60.50
4110	20″ diameter	Q-10	75	.320		49	14.10		63.10	76.50
4120	22″ diameter		60	.400		56	17.60		73.60	89.50
4130	24″ diameter		55	.436		63	19.20		82.20	100
4140	26″ diameter		42	.571		71.50	25		96.50	119
4150	28″ diameter		35	.686		77	30		107	133
4160	30″ diameter		28	.857		83	38		121	152
4170	36″ diameter		23	1.043		99.50	46		145.50	182
4180	42″ diameter		20	1.200		116	53		169	211
4200	48″ diameter		18	1.333		132	58.50		190.50	240
4250	Coupling, 6″ diameter	Q-9	88	.182	Ea.	9.45	7.70		17.15	22.50
4260	7″ diameter		82	.195		10.10	8.30		18.40	24.50
4270	8″ diameter		78	.205		10.55	8.70		19.25	25.50
4280	9″ diameter		74	.216		11.45	9.20		20.65	27
4290	10″ diameter		70	.229		11.45	9.70		21.15	28
4300	12″ diameter		55	.291		12.30	12.35		24.65	33
4310	14″ diameter		44	.364		14.95	15.45		30.40	41
4320	16″ diameter		40	.400		17.70	17		34.70	46.50
4330	18″ diameter		38	.421		20.50	17.90		38.40	51
4340	20″ diameter	Q-10	55	.436		35.50	19.20		54.70	69.50
4350	22″ diameter		50	.480		41	21		62	78.50
4360	24″ diameter		45	.533		46.50	23.50		70	88.50
4370	26″ diameter		39	.615		49	27		76	96.50
4380	28″ diameter		33	.727		51	32		83	108
4390	30″ diameter		27	.889		56	39		95	124
4400	36″ diameter		25	.960		88.50	42.50		131	164
4410	42″ diameter		23	1.043		99	46		145	182
4420	48″ diameter		21	1.143		115	50.50		165.50	207
4470	Elbow, 90°, 6″ diameter	Q-9	44	.364		74.50	15.45		89.95	107

23 31 HVAC Ducts and Casings

23 31 16 – Nonmetal Ducts

23 31 16.19 PVC Ducts		Crew	Daily Output	Labor-Hours	Unit	Material	2009 Bare Costs Labor	Equipment	Total	Total Incl O&P
4480	7″ diameter	Q-9	36	.444	Ea.	80	18.90		98.90	118
4490	8″ diameter		28	.571		80	24.50		104.50	127
4500	9″ diameter		22	.727		110	31		141	170
4510	10″ diameter		18	.889		130	38		168	203
4520	12″ diameter		15	1.067		177	45.50		222.50	266
4530	14″ diameter		11	1.455		242	62		304	365
4540	16″ diameter	↓	10	1.600		292	68		360	430
4550	18″ diameter	Q-10	15	1.600		315	70.50		385.50	460
4560	20″ diameter		14	1.714		335	75.50		410.50	490
4570	22″ diameter		13	1.846		425	81.50		506.50	600
4580	24″ diameter		12	2		480	88		568	670
4590	26″ diameter		11	2.182		550	96		646	760
4600	28″ diameter		10	2.400		650	106		756	885
4610	30″ diameter		9	2.667		770	117		887	1,025
4620	36″ diameter		7	3.429		1,050	151		1,201	1,400
4630	42″ diameter		5	4.800		1,450	211		1,661	1,925
4640	48″ diameter	↓	4	6	↓	1,900	264		2,164	2,525
4750	Elbow 45°, use 90° and deduct					45%				
9990	Minimum labor/equipment charge	1 Shee	3	2.667	Job		126		126	200

23 33 Air Duct Accessories

23 33 13 – Dampers

23 33 13.13 Volume-Control Dampers		Crew	Daily Output	Labor-Hours	Unit	Material	Labor	Equipment	Total	Total Incl O&P
0010	**VOLUME-CONTROL DAMPERS**									
5990	Multi-blade dampers, opposed blade, 8″ x 6″	1 Shee	24	.333	Ea.	20	15.75		35.75	47
5994	8″ x 8″		22	.364		21	17.15		38.15	50.50
5996	10″ x 10″		21	.381		24.50	18		42.50	55.50
6000	12″ x 12″		21	.381		27.50	18		45.50	58.50
6020	12″ x 18″		18	.444		36.50	21		57.50	74
6030	14″ x 10″		20	.400		26.50	18.90		45.40	59.50
6031	14″ x 14″		17	.471		32.50	22		54.50	71
6033	16″ x 12″		17	.471		32.50	22		54.50	71
6035	16″ x 16″		16	.500		40.50	23.50		64	82
6037	18″ x 16″		15	.533		44.50	25		69.50	89
6038	18″ x 18″		15	.533		48	25		73	93
6040	18″ x 24″		12	.667		62	31.50		93.50	118
6060	18″ x 28″		10	.800		72	38		110	139
6070	20″ x 16″		14	.571		48	27		75	96
6072	20″ x 20″		13	.615		57.50	29		86.50	110
6074	22″ x 18″		14	.571		57.50	27		84.50	107
6076	24″ x 16″		11	.727		57	34.50		91.50	117
6078	24″ x 20″		8	1		67	47		114	149
6080	24″ x 24″		8	1		78.50	47		125.50	162
6100	24″ x 28″		6	1.333		92	63		155	201
6110	26″ x 26″	↓	6	1.333		86.50	63		149.50	195
6120	28″ x 28″	Q-9	11	1.455		98	62		160	206
6130	30″ x 18″		10	1.600		83.50	68		151.50	200
6132	30″ x 24″		7	2.286		104	97		201	269
6133	30″ x 30″		6.60	2.424		124	103		227	300
6135	32″ x 32″		6.40	2.500		144	106		250	325
6151	36″ x 12″	↓	10	1.600	↓	65.50	68		133.50	180

23 33 13.13 Volume-Control Dampers		Crew	Daily Output	Labor-Hours	Unit	Material	Labor	Equipment	Total	Total Incl O&P
						2009 Bare Costs				
6152	36" x 16"	Q-9	8	2	Ea.	89.50	85		174.50	233
6158	36" x 36"		6	2.667		176	113		289	375
6160	44" x 28"		5.80	2.759		181	117		298	385
6180	48" x 36"		5.60	2.857		237	121		358	455
6200	56" x 36"		5.40	2.963		287	126		413	515
6220	60" x 36"		5.20	3.077		305	131		436	545
6240	60" x 44"		5	3.200		360	136		496	610
7500	Variable volume modulating motorized damper, incl. elect. mtr.									
7504	8" x 6"	1 Shee	15	.533	Ea.	109	25		134	160
7506	10" x 6"		14	.571		109	27		136	163
7510	10" x 10"		13	.615		112	29		141	170
7520	12" x 12"		12	.667		117	31.50		148.50	179
7522	12" x 16"		11	.727		118	34.50		152.50	185
7524	16" x 10"		12	.667		115	31.50		146.50	177
7526	16" x 14"		10	.800		120	38		158	191
7528	16" x 18"		9	.889		125	42		167	204
7540	18" x 12"		10	.800		120	38		158	192
7542	18" x 18"		8	1		127	47		174	215
7544	20" x 14"		8	1		130	47		177	218
7546	20" x 18"		7	1.143		136	54		190	236
7560	24" x 12"		8	1		132	47		179	220
7562	24" x 18"		7	1.143		143	54		197	243
7564	24" x 24"		6	1.333		148	63		211	262
7568	28" x 10"		7	1.143		132	54		186	231
7580	28" x 16"		6	1.333		149	63		212	264
7590	30" x 14"		5	1.600		140	75.50		215.50	274
7600	30" x 18"		4	2		180	94.50		274.50	350
7610	30" x 24"		3.80	2.105		239	99.50		338.50	420
7700	For thermostat, add		8	1		40.50	47		87.50	120
7800	For transformer 40 VA capacity, add		16	.500		18	23.50		41.50	57.50
8000	Multi-blade dampers, parallel blade									
8100	8" x 8"	1 Shee	24	.333	Ea.	71	15.75		86.75	103
8120	12" x 8"		22	.364		71	17.15		88.15	106
8140	16" x 10"		20	.400		91.50	18.90		110.40	130
8160	18" x 12"		18	.444		104	21		125	148
8180	22" x 12"		15	.533		106	25		131	156
8200	24" x 16"		11	.727		116	34.50		150.50	183
8220	28" x 16"		10	.800		130	38		168	203
8240	30" x 16"		8	1		134	47		181	223
8260	30" x 18"		7	1.143		159	54		213	261

23 33 13.16 Fire Dampers		Crew	Daily Output	Labor-Hours	Unit	Material	Labor	Equipment	Total	Total Incl O&P
0010	**FIRE DAMPERS**									
3000	Fire damper, curtain type, 1-1/2 hr rated, vertical, 6" x 6"	1 Shee	24	.333	Ea.	21	15.75		36.75	48
3020	8" x 6"		22	.364		21	17.15		38.15	50.50
3040	12" x 6"		22	.364		21	17.15		38.15	50.50
3060	20" x 6"		18	.444		24	21		45	60
3080	12" x 8"		22	.364		24	17.15		41.15	54
3100	24" x 8"		16	.500		29	23.50		52.50	69.50
3120	12" x 10"		21	.381		21	18		39	51.50
3140	24" x 10"		15	.533		31	25		56	74
3160	36" x 10"		12	.667		39.50	31.50		71	93.50
3180	16" x 12"		20	.400		23	18.90		41.90	55.50

23 33 13.16 Fire Dampers		Crew	Daily Output	Labor-Hours	Unit	Material	2009 Bare Costs Labor	Equipment	Total	Total Incl O&P
3200	24" x 12"	1 Shee	13	.615	Ea.	32.50	29		61.50	81.50
3220	48" x 12"		10	.800		49.50	38		87.50	115
3240	16" x 14"		18	.444		28.50	21		49.50	65
3260	24" x 14"		12	.667		35	31.50		66.50	88.50
3280	30" x 14"		11	.727		42	34.50		76.50	101
3300	18" x 16"		17	.471		34.50	22		56.50	73.50
3320	24" x 16"		11	.727		37.50	34.50		72	95.50
3340	36" x 16"		10	.800		50	38		88	115
3360	24" x 18"		10	.800		43.50	38		81.50	108
3380	48" x 18"		8	1		62.50	47		109.50	144
3400	24" x 20"		8	1		40.50	47		87.50	120
3420	36" x 20"		7	1.143		53.50	54		107.50	144
3440	24" x 22"		7	1.143		42	54		96	132
3460	30" x 22"		6	1.333		44.50	63		107.50	149
3480	26" x 24"		7	1.143		45.50	54		99.50	136
3500	48" x 24"	Q-9	12	1.333		70	56.50		126.50	167
3520	28" x 26"		13	1.231		51	52.50		103.50	140
3540	30" x 28"		12	1.333		56.50	56.50		113	153
3560	48" x 30"		11	1.455		82.50	62		144.50	189
3580	48" x 36"		10	1.600		90	68		158	207
3600	44" x 44"		10	1.600		94	68		162	211
3620	48" x 48"		8	2		124	85		209	271
3700	U.L. label included in above									
3800	For horizontal operation, add				Ea.	20%				
3900	For cap for blades out of air stream, add					20%				
4000	For 10" 22 ga., U.L. approved sleeve, add					35%				
4100	For 10" 22 ga., U.L. sleeve, 100% free area, add					65%				
4200	For oversize openings group dampers									
4500	Fire/smoke combination damper, louver type, UL									
4510	8" x 8"	1 Shee	22	.364	Ea.	126	17.15		143.15	167
4520	16" x 8"		20	.400		138	18.90		156.90	182
4540	18" x 8"		18	.444		138	21		159	186
4560	20" x 8"		16	.500		148	23.50		171.50	200
4580	10" x 10"		21	.381		143	18		161	186
4600	24" x 10"		15	.533		168	25		193	225
4620	30" x 10"		12	.667		173	31.50		204.50	241
4640	12" x 12"		20	.400		149	18.90		167.90	194
4660	18" x 12"		18	.444		152	21		173	201
4680	24" x 12"		13	.615		160	29		189	222
4700	30" x 12"		11	.727		166	34.50		200.50	238
4720	14" x 14"		17	.471		160	22		182	213
4740	16" x 14"		18	.444		165	21		186	215
4760	20" x 14"		14	.571		168	27		195	228
4780	24" x 14"		12	.667		172	31.50		203.50	239
4800	30" x 14"		10	.800		182	38		220	260
4820	16" x 16"		16	.500		178	23.50		201.50	233
4840	20" x 16"		14	.571		182	27		209	244
4860	24" x 16"		11	.727		188	34.50		222.50	262
4880	30" x 16"		8	1		196	47		243	291
4900	18" x 18"		15	.533		176	25		201	233
5000	24" x 18"		10	.800		188	38		226	267
5020	36" x 18"		7	1.143		206	54		260	310
5040	20" x 20"		13	.615		196	29		225	262

23 33 Air Duct Accessories

23 33 13 – Dampers

23 33 13.16 Fire Dampers		Crew	Daily Output	Labor-Hours	Unit	Material	Labor	Equipment	Total	Total Incl O&P
						2009 Bare Costs				
5060	24" x 20"	1 Shee	8	1	Ea.	201	47		248	296
5080	30" x 20"		8	1		211	47		258	310
5100	36" x 20"		7	1.143		222	54		276	330
5120	24" x 24"		8	1		217	47		264	315
5130	30" x 24"		7.60	1.053		206	49.50		255.50	305
5140	36" x 24"	Q-9	12	1.333		237	56.50		293.50	350
5141	36" x 30"		10	1.600		250	68		318	385
5142	40" x 36"		8	2		445	85		530	625
5143	48" x 48"		4	4		870	170		1,040	1,225
5150	Damper operator motor, 24 or 120 volt	1 Shee	16	.500		231	23.50		254.50	292

23 33 13.28 Splitter Damper Assembly		Crew	Daily Output	Labor-Hours	Unit	Material	Labor	Equipment	Total	Total Incl O&P
0009	**SPLITTER DAMPER ASSEMBLY**									
0010	Self locking, 1' rod	1 Shee	24	.333	Ea.	21.50	15.75		37.25	49
7020	3' rod		22	.364		28	17.15		45.15	58.50
7040	4' rod		20	.400		31.50	18.90		50.40	64.50
7060	6' rod		18	.444		38	21		59	75.50

23 33 19 – Duct Silencers

23 33 19.10 Duct Silencers		Crew	Daily Output	Labor-Hours	Unit	Material	Labor	Equipment	Total	Total Incl O&P
0009	**DUCT SILENCERS**									
0010	Silencers, noise control for air flow, duct				MCFM	54			54	59.50

23 33 23 – Turning Vanes

23 33 23.13 Air Turning Vanes		Crew	Daily Output	Labor-Hours	Unit	Material	Labor	Equipment	Total	Total Incl O&P
0010	**AIR TURNING VANES**									
9400	Turning vane components									
9410	Turning vane rail	1 Shee	160	.050	L.F.	.77	2.36		3.13	4.60
9420	Double thick, factory fab. vane		300	.027		1.12	1.26		2.38	3.23
9428	12" high set		170	.047		1.57	2.22		3.79	5.25
9432	14" high set		160	.050		1.82	2.36		4.18	5.75
9434	16" high set		150	.053		2.08	2.52		4.60	6.30
9436	18" high set		144	.056		2.34	2.62		4.96	6.75
9438	20" high set		138	.058		2.60	2.74		5.34	7.20
9440	22" high set		130	.062		2.86	2.90		5.76	7.75
9442	24" high set		124	.065		3.12	3.05		6.17	8.25
9444	26" high set		116	.069		3.38	3.26		6.64	8.85
9446	30" high set		112	.071		3.90	3.37		7.27	9.65
9900	Minimum labor/equipment charge		4	2	Job		94.50		94.50	150

23 33 33 – Duct-Mounting Access Doors

23 33 33.13 Duct Access Doors		Crew	Daily Output	Labor-Hours	Unit	Material	Labor	Equipment	Total	Total Incl O&P
0010	**DUCT ACCESS DOORS**									
1000	Duct access door, insulated, 6" x 6"	1 Shee	14	.571	Ea.	14.55	27		41.55	59
1020	10" x 10"		11	.727		17.60	34.50		52.10	74
1040	12" x 12"		10	.800		18.80	38		56.80	80.50
1050	12" x 18"		9	.889		33.50	42		75.50	104
1060	16" x 12"		9	.889		26.50	42		68.50	96
1070	18" x 18"		8	1		33.50	47		80.50	112
1074	24" x 18"		8	1		40	47		87	119
1080	24" x 24"		8	1		42.50	47		89.50	122

23 33 46 – Flexible Ducts

23 33 46.10 Flexible Air Ducts		Crew	Daily Output	Labor-Hours	Unit	Material	Labor	Equipment	Total	Total Incl O&P
0010	**FLEXIBLE AIR DUCTS** R233100-40									

23 33 Air Duct Accessories

23 33 46 – Flexible Ducts

23 33 46.10 Flexible Air Ducts		Crew	Daily Output	Labor-Hours	Unit	Material	Labor	Equipment	Total	Total Incl O&P
						2009 Bare Costs				
1280	Add to labor for elevated installation									
1282	of prefabricated (purchased) ductwork									
1283	10′ to 15′ high						10%			
1284	15′ to 20′ high						20%			
1285	20′ to 25′ high						25%			
1286	25′ to 30′ high						35%			
1287	30′ to 35′ high						40%			
1288	35′ to 40′ high						50%			
1289	Over 40′ high						55%			
1300	Flexible, coated fiberglass fabric on corr. resist. metal helix									
1400	pressure to 12″ (WG) UL-181									
1500	Non-insulated, 3″ diameter	Q-9	400	.040	L.F.	.99	1.70		2.69	3.79
1520	4″ diameter		360	.044		.99	1.89		2.88	4.09
1540	5″ diameter		320	.050		1.21	2.12		3.33	4.71
1560	6″ diameter		280	.057		1.49	2.43		3.92	5.50
1580	7″ diameter		240	.067		1.76	2.83		4.59	6.45
1600	8″ diameter		200	.080		2.04	3.40		5.44	7.65
1620	9″ diameter		180	.089		2.42	3.78		6.20	8.65
1640	10″ diameter		160	.100		2.53	4.25		6.78	9.55
1660	12″ diameter		120	.133		3.08	5.65		8.73	12.40
1680	14″ diameter		80	.200		3.74	8.50		12.24	17.60
1700	16″ diameter		60	.267		5.35	11.35		16.70	24
1800	For adjustable clamps, add				Ea.	.69			.69	.76
1900	Insulated, 1″ thick, PE jacket, 3″ diameter G	Q-9	380	.042	L.F.	2.09	1.79		3.88	5.15
1910	4″ diameter G		340	.047		2.09	2		4.09	5.50
1920	5″ diameter G		300	.053		2.09	2.27		4.36	5.90
1940	6″ diameter G		260	.062		2.31	2.61		4.92	6.70
1960	7″ diameter G		220	.073		2.70	3.09		5.79	7.90
1980	8″ diameter G		180	.089		2.92	3.78		6.70	9.20
2000	9″ diameter G		160	.100		3.30	4.25		7.55	10.40
2020	10″ diameter G		140	.114		3.41	4.86		8.27	11.45
2040	12″ diameter G		100	.160		4.24	6.80		11.04	15.45
2060	14″ diameter G		80	.200		5.10	8.50		13.60	19.15
2080	16″ diameter G		60	.267		6.40	11.35		17.75	25
2100	18″ diameter G		45	.356		8.10	15.10		23.20	33
2120	20″ diameter G	Q-10	65	.369		8.85	16.25		25.10	36
2500	Insulated, heavy duty, coated fiberglass fabric									
2520	4″ diameter G	Q-9	340	.047	L.F.	3.36	2		5.36	6.90
2540	5″ diameter G		300	.053		3.58	2.27		5.85	7.55
2560	6″ diameter G		260	.062		4.02	2.61		6.63	8.55
2580	7″ diameter G		220	.073		4.46	3.09		7.55	9.80
2600	8″ diameter G		180	.089		5.05	3.78		8.83	11.55
2620	9″ diameter G		160	.100		5.40	4.25		9.65	12.70
2640	10″ diameter G		140	.114		6.55	4.86		11.41	14.90
2660	12″ diameter G		100	.160		7	6.80		13.80	18.50
2680	14″ diameter G		80	.200		9.15	8.50		17.65	23.50
2700	16″ diameter G		60	.267		13.70	11.35		25.05	33
2720	18″ diameter G		45	.356		16.80	15.10		31.90	42.50
9990	Minimum labor/equipment charge	1 Shee	3	2.667	Job		125		126	200

23 33 53 – Duct Liners

23 33 53.10 Duct Liner Board		Crew	Daily Output	Labor-Hours	Unit	Material	Labor	Equipment	Total	Total Incl O&P
0010	**DUCT LINER BOARD**									

23 33 Air Duct Accessories

23 33 53 – Duct Liners

23 33 53.10 Duct Liner Board		Crew	Daily Output	Labor-Hours	Unit	2009 Bare Costs Material	Labor	Equipment	Total	Total Incl O&P
3490	Board type, fiberglass liner, 3 lb. density									
3500	Fire resistant, black pigmented, 1 side									
3520	1" thick G	Q-14	150	.107	S.F.	.73	4.23		4.96	7.65
3540	1-1/2" thick G		130	.123		.92	4.89		5.81	8.90
3560	2" thick G		120	.133		1.12	5.30		6.42	9.80
3600	FSK vapor barrier									
3620	1" thick G	Q-14	150	.107	S.F.	.73	4.23		4.96	7.65
3630	1-1/2" thick G		130	.123		.92	4.89		5.81	8.90
3640	2" thick G		120	.133		1.12	5.30		6.42	9.80
3680	No finish									
3700	1" thick G	Q-14	170	.094	S.F.	.38	3.74		4.12	6.45
3710	1-1/2" thick G		140	.114		.57	4.54		5.11	8
3720	2" thick G		130	.123		.77	4.89		5.66	8.75
3940	Board type, non-fibrous foam									
3950	Temperature, bacteria and fungi resistant									
3960	1" thick G	Q-14	150	.107	S.F.	2.25	4.23		6.48	9.35
3970	1-1/2" thick G		130	.123		3	4.89		7.89	11.20
3980	2" thick G		120	.133		3.60	5.30		8.90	12.50

23 34 HVAC Fans

23 34 13 – Axial HVAC Fans

23 34 13.10 Axial Flow HVAC Fans

		Crew	Daily Output	Labor-Hours	Unit	Material	Labor	Equipment	Total	Total Incl O&P
0010	**AXIAL FLOW HVAC FANS** R233400-10									
0020	Air conditioning and process air handling									
0030	Axial flow, compact, low sound, 2.5" S.P.									
0050	3,800 CFM, 5 HP	Q-20	3.40	5.882	Ea.	4,375	255		4,630	5,200
0080	6,400 CFM, 5 HP		2.80	7.143		4,875	310		5,185	5,875
0100	10,500 CFM, 7-1/2 HP		2.40	8.333		6,075	360		6,435	7,250
0120	15,600 CFM, 10 HP		1.60	12.500		7,650	540		8,190	9,275
0140	23,000 CFM, 15 HP		.70	28.571		11,700	1,250		12,950	14,900
0160	28,000 CFM, 20 HP		.40	50		13,100	2,175		15,275	17,800
1500	Vaneaxial, low pressure, 2000 CFM, 1/2 HP		3.60	5.556		1,825	241		2,066	2,375
1520	4,000 CFM, 1 HP		3.20	6.250		2,100	271		2,371	2,725
1540	8,000 CFM, 2 HP		2.80	7.143		2,675	310		2,985	3,425
1560	16,000 CFM, 5 HP		2.40	8.333		3,850	360		4,210	4,800

23 34 14 – Blower HVAC Fans

23 34 14.10 Blower Type HVAC Fans

		Crew	Daily Output	Labor-Hours	Unit	Material	Labor	Equipment	Total	Total Incl O&P
0010	**BLOWER TYPE HVAC FANS**									
2000	Blowers, direct drive with motor, complete									
2020	1045 CFM @ .5" S.P., 1/5 HP	Q-20	18	1.111	Ea.	208	48		256	305
2040	1385 CFM @ .5" S.P., 1/4 HP		18	1.111		202	48		250	298
2060	1640 CFM @ .5" S.P., 1/3 HP		18	1.111		232	48		280	330
2080	1760 CFM @ .5" S.P., 1/2 HP		18	1.111		233	48		281	330
2090	4 speed									
2100	1164 to 1739 CFM @ .5" S.P., 1/3 HP	Q-20	16	1.250	Ea.	264	54		318	375
2120	1467 to 2218 CFM @ 1.0" S.P., 3/4 HP	"	14	1.429	"	340	62		402	475
2500	Ceiling fan, right angle, extra quiet, 0.10" S.P.									
2520	95 CFM	Q-20	20	1	Ea.	228	43.50		271.50	320
2540	210 CFM		19	1.053		270	45.50		315.50	370
2560	385 CFM		18	1.111		345	48		393	450

23 34 14 – Blower HVAC Fans

23 34 14.10 Blower Type HVAC Fans		Crew	Daily Output	Labor-Hours	Unit	Material	Labor (2009 Bare Costs)	Equipment (2009 Bare Costs)	Total (2009 Bare Costs)	Total Incl O&P
2580	885 CFM	Q-20	16	1.250	Ea.	675	54		729	825
2600	1,650 CFM		13	1.538		935	66.50		1,001.50	1,125
2620	2,960 CFM		11	1.818		1,250	79		1,329	1,500
2640	For wall or roof cap, add	1 Shee	16	.500		228	23.50		251.50	289
2660	For straight thru fan, add					10%				
2680	For speed control switch, add	1 Elec	16	.500		125	23.50		148.50	173
7500	Utility set, steel construction, pedestal, 1/4" S.P.									
7520	Direct drive, 150 CFM, 1/8 HP	Q-20	6.40	3.125	Ea.	735	136		871	1,025
7540	485 CFM, 1/6 HP		5.80	3.448		925	150		1,075	1,250
7560	1950 CFM, 1/2 HP		4.80	4.167		1,075	181		1,256	1,475
7580	2410 CFM, 3/4 HP		4.40	4.545		2,000	197		2,197	2,500
7600	3328 CFM, 1-1/2 HP		3	6.667		2,225	289		2,514	2,900
7680	V-belt drive, drive cover, 3 phase									
7700	800 CFM, 1/4 HP	Q-20	6	3.333	Ea.	740	145		885	1,050
7720	1,300 CFM, 1/3 HP		5	4		775	174		949	1,125
7740	2,000 CFM, 1 HP		4.60	4.348		915	189		1,104	1,300
7760	2,900 CFM, 3/4 HP		4.20	4.762		1,250	207		1,457	1,675
7780	3,600 CFM, 3/4 HP		4	5		1,525	217		1,742	2,025
7800	4,800 CFM, 1 HP		3.50	5.714		1,800	248		2,048	2,375
7820	6,700 CFM, 1-1/2 HP		3	6.667		2,225	289		2,514	2,900
7830	7,500 CFM, 2 HP		2.50	8		3,000	345		3,345	3,875
7840	11,000 CFM, 3 HP		2	10		4,025	435		4,460	5,100
7860	13,000 CFM, 3 HP		1.60	12.500		4,075	540		4,615	5,350
7880	15,000 CFM, 5 HP		1	20		4,225	870		5,095	6,025
7900	17,000 CFM, 7-1/2 HP		.80	25		4,525	1,075		5,600	6,675
7920	20,000 CFM, 7-1/2 HP		.80	25		5,400	1,075		6,475	7,625

23 34 16 – Centrifugal HVAC Fans

23 34 16.10 Centrifugal Type HVAC Fans		Crew	Daily Output	Labor-Hours	Unit	Material	Labor	Equipment	Total	Total Incl O&P
0010	**CENTRIFUGAL TYPE HVAC FANS**									
0200	In-line centrifugal, supply/exhaust booster									
0220	aluminum wheel/hub, disconnect switch, 1/4" S.P.									
0240	500 CFM, 10" diameter connection	Q-20	3	6.667	Ea.	1,125	289		1,414	1,675
0260	1,380 CFM, 12" diameter connection		2	10		1,175	435		1,610	1,975
0280	1,520 CFM, 16" diameter connection		2	10		1,300	435		1,735	2,100
0300	2,560 CFM, 18" diameter connection		1	20		1,400	870		2,270	2,925
0320	3,480 CFM, 20" diameter connection		.80	25		1,675	1,075		2,750	3,525
0326	5,080 CFM, 20" diameter connection		.75	26.667		1,800	1,150		2,950	3,825
3500	Centrifugal, airfoil, motor and drive, complete									
3520	1000 CFM, 1/2 HP	Q-20	2.50	8	Ea.	1,525	345		1,870	2,250
3540	2,000 CFM, 1 HP		2	10		1,675	435		2,110	2,525
3560	4,000 CFM, 3 HP		1.80	11.111		2,350	480		2,830	3,325
3580	8,000 CFM, 7-1/2 HP		1.40	14.286		3,625	620		4,245	4,975
3600	12,000 CFM, 10 HP		1	20		4,375	870		5,245	6,200
4000	Single width, belt drive, not incl. motor, capacities									
4020	at 2000 FPM, 2.5" S.P. for indicated motor									
4040	6900 CFM, 5 HP	Q-9	2.40	6.667	Ea.	3,200	283		3,483	3,950
4060	10,340 CFM, 7-1/2 HP		2.20	7.273		4,450	310		4,760	5,375
4080	15,320 CFM, 10 HP		2	8		5,225	340		5,565	6,300
4100	22,780 CFM, 15 HP		1.80	8.889		7,775	380		8,155	9,150
4120	33,840 CFM, 20 HP		1.60	10		10,300	425		10,725	12,100
4140	41,400 CFM, 25 HP		1.40	11.429		13,700	485		14,185	15,900
4160	50,100 CFM, 30 HP		.80	20		17,400	850		18,250	20,500

23 34 16 – Centrifugal HVAC Fans

23 34 16.10 Centrifugal Type HVAC Fans		Crew	Daily Output	Labor-Hours	Unit	Material	Labor	Equipment	Total	Total Incl O&P
						2009 Bare Costs				
4200	Double width wheel, 12,420 CFM, 7.5 HP	Q-9	2.20	7.273	Ea.	4,550	310		4,860	5,525
4220	18,620 CFM, 15 HP		2	8		6,525	340		6,865	7,725
4240	27,580 CFM, 20 HP		1.80	8.889		7,975	380		8,355	9,375
4260	40,980 CFM, 25 HP		1.50	10.667		12,500	455		12,955	14,400
4280	60,920 CFM, 40 HP		1	16		16,800	680		17,480	19,600
4300	74,520 CFM, 50 HP		.80	20		19,900	850		20,750	23,300
4320	90,160 CFM, 50 HP		.70	22.857		27,300	970		28,270	31,600
4340	110,300 CFM, 60 HP		.50	32		36,900	1,350		38,250	42,800
4360	134,960 CFM, 75 HP	↓	.40	40	↓	49,100	1,700		50,800	56,500
4500	Corrosive fume resistant, plastic									
4600	roof ventilators, centrifugal, V belt drive, motor									
4620	1/4" S.P., 250 CFM, 1/4 HP	Q-20	6	3.333	Ea.	3,350	145		3,495	3,925
4640	895 CFM, 1/3 HP		5	4		3,650	174		3,824	4,300
4660	1630 CFM, 1/2 HP		4	5		4,325	217		4,542	5,100
4680	2240 CFM, 1 HP		3	6.667		4,500	289		4,789	5,400
4700	3810 CFM, 2 HP		2	10		5,000	435		5,435	6,175
4710	5000 CFM, 2 HP		1.80	11.111		10,500	480		10,980	12,400
4715	8000 CFM, 5 HP		1.40	14.286		11,700	620		12,320	13,900
4720	11760 CFM, 5 HP		1	20		11,900	870		12,770	14,500
4740	18810 CFM, 10 HP	↓	.70	28.571	↓	12,300	1,250		13,550	15,500
4800	For intermediate capacity, motors may be varied									
4810	For explosion proof motor, add				Ea.	15%				
5000	Utility set, centrifugal, V belt drive, motor									
5020	1/4" S.P., 1200 CFM, 1/4 HP	Q-20	6	3.333	Ea.	4,075	145		4,220	4,700
5040	1520 CFM, 1/3 HP		5	4		4,075	174		4,249	4,750
5060	1850 CFM, 1/2 HP		4	5		4,100	217		4,317	4,850
5080	2180 CFM, 3/4 HP		3	6.667		4,150	289		4,439	5,000
5100	1/2" S.P., 3600 CFM, 1 HP		2	10		6,000	435		6,435	7,275
5120	4250 CFM, 1-1/2 HP		1.60	12.500		6,075	540		6,615	7,550
5140	4800 CFM, 2 HP		1.40	14.286		6,150	620		6,770	7,750
5160	6920 CFM, 5 HP		1.30	15.385		6,350	665		7,015	8,025
5180	7700 CFM, 7-1/2 HP	↓	1.20	16.667	↓	6,550	725		7,275	8,350
5200	For explosion proof motor, add					15%				
5500	Fans, industrial exhauster, for air which may contain granular matl.									
5520	1000 CFM, 1-1/2 HP	Q-20	2.50	8	Ea.	2,475	345		2,820	3,275
5540	2000 CFM, 3 HP		2	10		3,050	435		3,485	4,050
5560	4000 CFM, 7-1/2 HP		1.80	11.111		4,350	480		4,830	5,525
5580	8000 CFM, 15 HP		1.40	14.286		5,625	620		6,245	7,175
5600	12,000 CFM, 30 HP	↓	1	20	↓	9,150	870		10,020	11,500
7000	Roof exhauster, centrifugal, aluminum housing, 12" galvanized									
7020	curb, bird screen, back draft damper, 1/4" S.P.									
7100	Direct drive, 320 CFM, 11" sq. damper	Q-20	7	2.857	Ea.	510	124		634	760
7120	600 CFM, 11" sq. damper		6	3.333		520	145		665	800
7140	815 CFM, 13" sq. damper		5	4		520	174		694	845
7160	1450 CFM, 13" sq. damper		4.20	4.762		665	207		872	1,050
7180	2050 CFM, 16" sq. damper		4	5		665	217		882	1,075
7200	V-belt drive, 1650 CFM, 12" sq. damper		6	3.333		915	145		1,060	1,225
7220	2750 CFM, 21" sq. damper		5	4		1,075	174		1,249	1,450
7230	3500 CFM, 21" sq. damper		4.50	4.444		1,175	193		1,368	1,600
7240	4910 CFM, 23" sq. damper		4	5		1,450	217		1,667	1,950
7260	8525 CFM, 28" sq. damper		3	6.667		1,825	289		2,114	2,450
7280	13,760 CFM, 35" sq. damper		2	10		2,500	435		2,935	3,425
7300	20,558 CFM, 43" sq. damper	↓	1	20	↓	5,300	870		6,170	7,225

23 34 HVAC Fans

23 34 16 – Centrifugal HVAC Fans

23 34 16.10 Centrifugal Type HVAC Fans		Crew	Daily Output	Labor-Hours	Unit	2009 Bare Costs Material	Labor	Equipment	Total	Total Incl O&P
7320	For 2 speed winding, add				Ea.	15%				
7340	For explosionproof motor, add					375			375	415
7360	For belt driven, top discharge, add					15%				
8500	Wall exhausters, centrifugal, auto damper, 1/8" S.P.									
8520	Direct drive, 610 CFM, 1/20 HP	Q-20	14	1.429	Ea.	330	62		392	465
8540	796 CFM, 1/12 HP		13	1.538		675	66.50		741.50	850
8560	822 CFM, 1/6 HP		12	1.667		840	72.50		912.50	1,050
8580	1,320 CFM, 1/4 HP		12	1.667		985	72.50		1,057.50	1,200
8600	1756 CFM, 1/4 HP		11	1.818		990	79		1,069	1,200
8620	1983 CFM, 1/4 HP		10	2		1,000	87		1,087	1,250
8640	2900 CFM, 1/2 HP		9	2.222		1,075	96.50		1,171.50	1,325
8660	3307 CFM, 3/4 HP		8	2.500		1,175	108		1,283	1,450
9500	V-belt drive, 3 phase									
9520	2,800 CFM, 1/4 HP	Q-20	9	2.222	Ea.	1,500	96.50		1,596.50	1,800
9540	3,740 CFM, 1/2 HP		8	2.500		1,550	108		1,658	1,900
9560	4400 CFM, 3/4 HP		7	2.857		1,575	124		1,699	1,950
9580	5700 CFM, 1-1/2 HP		6	3.333		1,650	145		1,795	2,025
9900	Minimum labor/equipment charge	1 Elec	4	2	Job		94		94	145

23 34 23 – HVAC Power Ventilators

23 34 23.10 HVAC Power Circulators and Ventilators		Crew	Daily Output	Labor-Hours	Unit	Material	Labor	Equipment	Total	Total Incl O&P
0010	**HVAC POWER CIRCULATORS AND VENTILATORS**									
3000	Paddle blade air circulator, 3 speed switch									
3020	42", 5,000 CFM high, 3000 CFM low G	1 Elec	2.40	3.333	Ea.	89	157		246	340
3040	52", 6,500 CFM high, 4000 CFM low G	"	2.20	3.636	"	96.50	171		267.50	370
3100	For antique white motor, same cost									
3200	For brass plated motor, same cost									
3300	For light adaptor kit, add G				Ea.	29			29	31.50
6000	Propeller exhaust, wall shutter, 1/4" S.P.									
6020	Direct drive, two speed									
6100	375 CFM, 1/10 HP	Q-20	10	2	Ea.	410	87		497	585
6120	730 CFM, 1/7 HP		9	2.222		455	96.50		551.50	650
6140	1000 CFM, 1/8 HP		8	2.500		635	108		743	870
6160	1890 CFM, 1/4 HP		7	2.857		645	124		769	905
6180	3275 CFM, 1/2 HP		6	3.333		660	145		805	955
6200	4720 CFM, 1 HP		5	4		1,025	174		1,199	1,400
6300	V-belt drive, 3 phase									
6320	6175 CFM, 3/4 HP	Q-20	5	4	Ea.	855	174		1,029	1,225
6340	7500 CFM, 3/4 HP		5	4		900	174		1,074	1,275
6360	10,100 CFM, 1 HP		4.50	4.444		1,100	193		1,293	1,500
6380	14,300 CFM, 1-1/2 HP		4	5		1,300	217		1,517	1,775
6400	19,800 CFM, 2 HP		3	6.667		1,475	289		1,764	2,075
6420	26,250 CFM, 3 HP		2.60	7.692		1,775	335		2,110	2,475
6440	38,500 CFM, 5 HP		2.20	9.091		2,100	395		2,495	2,925
6460	46,000 CFM, 7-1/2 HP		2	10		2,250	435		2,685	3,150
6480	51,500 CFM, 10 HP		1.80	11.111		2,325	480		2,805	3,325
6490	V-belt drive, 115V., residential, whole house									
6500	Ceiling-wall, 5200 CFM, 1/4 HP, 30" x 30"	1 Shee	6	1.333	Ea.	445	63		508	590
6530	13,200 CFM, 1/3 HP, 48" x 48"		4	2		540	94.50		634.50	745
6540	15,445 CFM, 1/2 HP, 48" x 48"		4	2		560	94.50		654.50	765
6550	17,025 CFM, 1/2 HP, 54" x 54"		4	2		1,075	94.50		1,169.50	1,350
6560	For two speed motor, add					20%				
6570	Shutter, automatic, ceiling/wall									

23 34 23 – HVAC Power Ventilators

23 34 23.10 HVAC Power Circulators and Ventilators		Crew	Daily Output	Labor-Hours	Unit	2009 Bare Costs Material	Labor	Equipment	Total	Total Incl O&P
6580	30" x 30"	1 Shee	8	1	Ea.	145	47		192	235
6590	36" x 36"		8	1		166	47		213	258
6600	42" x 42"		8	1		208	47		255	305
6610	48" x 48"		7	1.143		225	54		279	335
6620	54" x 54"		6	1.333		297	63		360	425
6630	Timer, shut off, to 12 Hr.		20	.400		48	18.90		66.90	82.50
6650	Residential, bath exhaust, grille, back draft damper									
6660	50 CFM	Q-20	24	.833	Ea.	40.50	36		76.50	102
6670	110 CFM		22	.909		66.50	39.50		106	135
6680	Light combination, squirrel cage, 100 watt, 70 CFM		24	.833		73.50	36		109.50	138
6700	Light/heater combination, ceiling mounted									
6710	70 CFM, 1450 watt	Q-20	24	.833	Ea.	89	36		125	155
6800	Heater combination, recessed, 70 CFM		24	.833		42	36		78	104
6820	With 2 infrared bulbs		23	.870		64	37.50		101.50	130
6900	Kitchen exhaust, grille, complete, 160 CFM		22	.909		76	39.50		115.50	146
6910	180 CFM		20	1		66.50	43.50		110	142
6920	270 CFM		18	1.111		117	48		165	204
6930	350 CFM		16	1.250		91	54		145	186
6940	Residential roof jacks and wall caps									
6944	Wall cap with back draft damper									
6946	3" & 4" dia. round duct	1 Shee	11	.727	Ea.	16.15	34.50		50.65	72.50
6948	6" dia. round duct	"	11	.727	"	39	34.50		73.50	97.50
6958	Roof jack with bird screen and back draft damper									
6960	3" & 4" dia. round duct	1 Shee	11	.727	Ea.	15	34.50		49.50	71
6962	3-1/4" x 10" rectangular duct	"	10	.800	"	28.50	38		66.50	91.50
6980	Transition									
6982	3-1/4" x 10" to 6" dia. round	1 Shee	20	.400	Ea.	18.55	18.90		37.45	50.50
8020	Attic, roof type									
8030	Aluminum dome, damper & curb									
8040	6" diameter, 300 CFM	1 Elec	16	.500	Ea.	380	23.50		403.50	455
8050	7" diameter, 450 CFM		15	.533		415	25		440	495
8060	9" diameter, 900 CFM		14	.571		455	27		482	540
8080	12" diameter, 1000 CFM (gravity)		10	.800		470	37.50		507.50	580
8090	16" diameter, 1500 CFM (gravity)		9	.889		570	42		612	690
8100	20" diameter, 2500 CFM (gravity)		8	1		695	47		742	840
8110	26" diameter, 4000 CFM (gravity)		7	1.143		845	53.50		898.50	1,025
8120	32" diameter, 6500 CFM (gravity)		6	1.333		1,150	62.50		1,212.50	1,375
8130	38" diameter, 8000 CFM (gravity)		5	1.600		1,725	75		1,800	2,025
8140	50" diameter, 13,000 CFM (gravity)		4	2		2,500	94		2,594	2,900
8160	Plastic, ABS dome									
8180	1050 CFM	1 Elec	14	.571	Ea.	139	27		166	194
8200	1600 CFM	"	12	.667	"	208	31.50		239.50	276
8240	Attic, wall type, with shutter, one speed									
8250	12" diameter, 1000 CFM	1 Elec	14	.571	Ea.	300	27		327	370
8260	14" diameter, 1500 CFM		12	.667		325	31.50		356.50	405
8270	16" diameter, 2000 CFM		9	.889		365	42		407	470
8290	Whole house, wall type, with shutter, one speed									
8300	30" diameter, 4800 CFM	1 Elec	7	1.143	Ea.	785	53.50		838.50	950
8310	36" diameter, 7000 CFM		6	1.333		855	62.50		917.50	1,025
8320	42" diameter, 10,000 CFM		5	1.600		960	75		1,035	1,175
8330	48" diameter, 16,000 CFM		4	2		1,200	94		1,294	1,450
8340	For two speed, add					71.50			71.50	79
8350	Whole house, lay-down type, with shutter, one speed									

23 34 23 – HVAC Power Ventilators

23 34 23.10 HVAC Power Circulators and Ventilators		Crew	Daily Output	Labor-Hours	Unit	Material	2009 Bare Costs Labor	Equipment	Total	Total Incl O&P
8360	30" diameter, 4500 CFM	1 Elec	8	1	Ea.	835	47		882	995
8370	36" diameter, 6500 CFM		7	1.143		900	53.50		953.50	1,075
8380	42" diameter, 9000 CFM		6	1.333		990	62.50		1,052.50	1,175
8390	48" diameter, 12,000 CFM	↓	5	1.600		1,125	75		1,200	1,350
8440	For two speed, add					54			54	59.50
8450	For 12 hour timer switch, add	1 Elec	32	.250	↓	54	11.75		65.75	77.50

23 34 33 – Air Curtains

23 34 33.10 Air Barrier Curtains

		Crew	Daily Output	Labor-Hours	Unit	Material	Labor	Equipment	Total	Total Incl O&P
0010	**AIR BARRIER CURTAINS**, Incl. motor starters, transformers,									
0050	door switches & temperature controls									
0100	Shipping and receiving doors, unheated, minimal wind stoppage									
0150	8' high, multiples of 3' wide	2 Shee	6	2.667	L.F.	410	126		536	655
0160	5' wide		10	1.600		350	75.50		425.50	505
0210	10' high, multiples of 4' wide		8	2		440	94.50		534.50	630
0250	12' high, 3'-6" wide		7	2.286		194	108		302	385
0260	12' wide		6	2.667		250	126		376	475
0350	16' high, 3'-6" wide		7	2.286		194	108		302	385
0360	12' wide	↓	6	2.667	↓	250	126		376	475
0500	Maximum wind stoppage									
0550	10' high, multiples of 4' wide	2 Shee	8	2	L.F.	320	94.50		414.50	500
0650	14' high, multiples of 4' wide		8	2		365	94.50		459.50	550
0750	20' high, multiples of 8' wide	↓	6	2.667	↓	950	126		1,076	1,250
1100	Heated, maximum wind stoppage, steam heat									
1150	10' high, multiples of 4' wide	2 Shee	6	2.667	L.F.	855	126		981	1,150
1250	14' high, multiples of 4' wide		6	2.667		365	126		491	600
1350	20' high, multiples of 8' wide	↓	4	4	↓	1,250	189		1,439	1,675
1500	Customer entrance doors, unheated, minimal wind stoppage									
1550	10' high, multiples of 3' wide	2 Shee	6	2.667	L.F.	305	126		431	535
1560	5' wide		10	1.600		350	75.50		425.50	505
1650	Maximum wind stoppage, 12' high, multiples of 4' wide	↓	8	2	↓	340	94.50		434.50	525
1700	Heated, minimal wind stoppage, electric heat									
1750	8' high, multiples of 3' wide	2 Shee	6	2.667	L.F.	380	126		506	620
1760	Multiples of 5' wide		10	1.600		310	75.50		385.50	465
1850	10' high, multiples of 3' wide		6	2.667		815	126		941	1,100
1860	Multiples of 5' wide	↓	10	1.600	↓	535	75.50		610.50	705
1950	Maximum wind stoppage, steam heat									
1960	12' high, multiples of 4' wide	2 Shee	8	2	L.F.	655	94.50		749.50	870
2000	Walk-in coolers and freezers, ambient air, minimal wind stoppage									
2050	8' high, multiples of 3' wide	2 Shee	6	2.667	L.F.	360	126		486	595
2060	Multiples of 5' wide		10	1.600		238	75.50		313.50	380
2250	Maximum wind stoppage, 12' high, multiples of 3' wide		6	2.667		222	126		348	445
2450	Conveyor openings or service windows, unheated, 5' high		5	3.200		228	151		379	490
2460	Heated, electric, 5' high, 2'-6" wide	↓	5	3.200	↓	146	151		297	400

23 35 16 – Engine Exhaust Systems

23 35 16.10 Engine Exhaust Removal Systems		Crew	Daily Output	Labor-Hours	Unit	Material	Labor	Equipment	Total	Total Incl O&P
						2009 Bare Costs				
0010	ENGINE EXHAUST REMOVAL SYSTEMS									
0500	Engine exhaust, garage, in-floor system									
0510	Single tube outlet assemblies									
0520	For transite pipe ducting, self-storing tube									
0530	3″ tubing adapter plate	1 Shee	16	.500	Ea.	217	23.50		240.50	277
0540	4″ tubing adapter plate		16	.500		217	23.50		240.50	277
0550	5″ tubing adapter plate		16	.500		217	23.50		240.50	277
0600	For vitrified tile ducting									
0610	3″ tubing adapter plate, self-storing tube	1 Shee	16	.500	Ea.	217	23.50		240.50	277
0620	4″ tubing adapter plate, self-storing tube		16	.500		217	23.50		240.50	277
0660	5″ tubing adapter plate, self-storing tube		16	.500		217	23.50		240.50	277
0800	Two tube outlet assemblies									
0810	For transite pipe ducting, self-storing tube									
0820	3″ tubing, dual exhaust adapter plate	1 Shee	16	.500	Ea.	217	23.50		240.50	277
0850	For vitrified tile ducting									
0860	3″ tubing, dual exhaust, self-storing tube	1 Shee	16	.500	Ea.	217	23.50		240.50	277
0870	3″ tubing, double outlet, non-storing tubes	″	16	.500	″	217	23.50		240.50	277
0900	Accessories for metal tubing, (overhead systems also)									
0910	Adapters, for metal tubing end									
0920	3″ tail pipe type				Ea.	48.50			48.50	53
0930	4″ tail pipe type					48.50			48.50	53
0940	5″ tail pipe type					48.50			48.50	53
0990	5″ diesel stack type					340			340	370
1000	6″ diesel stack type					340			340	370
1100	Bullnose (guide) required for in-floor assemblies									
1110	3″ tubing size				Ea.	24			24	26.50
1120	4″ tubing size					24			24	26.50
1130	5″ tubing size					24			24	26.50
1150	Plain rings, for tubing end									
1160	3″ tubing size				Ea.	11.50			11.50	12.65
1170	4″ tubing size				″	14.50			14.50	15.95
1200	Tubing, galvanized, flexible, (for overhead systems also)									
1210	3″ ID				L.F.	6.45			6.45	7.10
1220	4″ ID					7.80			7.80	8.55
1230	5″ ID					9.05			9.05	9.95
1240	6″ ID					10.40			10.40	11.40
1250	Stainless steel, flexible, (for overhead system, also)									
1260	3″ ID				L.F.	13.40			13.40	14.75
1270	4″ ID					17.45			17.45	19.20
1280	5″ ID					21			21	23
1290	6″ ID					25			25	27.50
1500	Engine exhaust, garage, overhead components, for neoprene tubing									
1510	Alternate metal tubing & accessories see above									
1550	Adapters, for neoprene tubing end									
1560	3″ tail pipe, adjustable, neoprene				Ea.	42.50			42.50	46.50
1570	3″ tail pipe, heavy wall neoprene					54.50			54.50	60
1580	4″ tail pipe, heavy wall neoprene					58			58	64
1590	5″ tail pipe, heavy wall neoprene					78.50			78.50	86.50
1650	Connectors, tubing									
1660	3″ interior, aluminum				Ea.	11.50			11.50	12.65
1670	4″ interior, aluminum					14.50			14.50	15.95
1710	5″ interior, neoprene					36			36	40
1750	3″ spiralock, neoprene					11.50			11.50	12.65

23 35 16 – Engine Exhaust Systems

23 35 16.10 Engine Exhaust Removal Systems		Crew	Daily Output	Labor-Hours	Unit	Material (2009 Bare Costs)	Labor (2009 Bare Costs)	Equipment (2009 Bare Costs)	Total (2009 Bare Costs)	Total Incl O&P
1760	4″ spiralock, neoprene				Ea.	14.50			14.50	15.95
1780	Y for 3″ ID tubing, neoprene, dual exhaust					30			30	33
1790	Y for 4″ ID tubing, aluminum, dual exhaust					54.50			54.50	60
1850	Elbows, aluminum, splice into tubing for strap									
1860	3″ neoprene tubing size				Ea.	36			36	40
1870	4″ neoprene tubing size					42.50			42.50	46.50
1900	Flange assemblies, connect tubing to overhead duct					40			40	44
2000	Hardware and accessories									
2020	Cable, galvanized, 1/8″ diameter				L.F.	.42			.42	.46
2040	Cleat, tie down cable or rope				Ea.	4.83			4.83	5.30
2060	Pulley					6.65			6.65	7.30
2080	Pulley hook, universal					4.83			4.83	5.30
2100	Rope, nylon, 1/4″ diameter				L.F.	.36			.36	.40
2120	Winch, 1″ diameter				Ea.	103			103	113
2150	Lifting strap, mounts on neoprene									
2160	3″ tubing size				Ea.	24			24	26.50
2170	4″ tubing size					24			24	26.50
2180	5″ tubing size					24			24	26.50
2190	6″ tubing size					24			24	26.50
2200	Tubing, neoprene, 11′ lengths									
2210	3″ ID				L.F.	7.25			7.25	8
2220	4″ ID					9.05			9.05	9.95
2230	5″ ID					16.30			16.30	17.95
2500	Engine exhaust, thru-door outlet									
2510	3″ tube size	1 Carp	16	.500	Ea.	42.50	20		62.50	79.50
2530	4″ tube size	″	16	.500	″	42.50	20		62.50	79.50

23 35 43 – Welding Fume Elimination Systems

23 35 43.10 Welding Fume Elimination System Components		Crew	Daily Output	Labor-Hours	Unit	Material	Labor	Equipment	Total	Total Incl O&P
0010	**WELDING FUME ELIMINATION SYSTEM COMPONENTS**									
7500	Welding fume elimination accessories for garage exhaust systems									
7600	Cut off (blast gate)									
7610	3″ tubing size, 3″ x 6″ opening	1 Shee	24	.333	Ea.	19.30	15.75		35.05	46.50
7620	4″ tubing size, 4″ x 8″ opening		24	.333		21.50	15.75		37.25	49
7630	5″ tubing size, 5″ x 10″ opening		24	.333		29	15.75		44.75	57
7640	6″ tubing size		24	.333		34	15.75		49.75	62
7650	8″ tubing size		24	.333		48.50	15.75		64.25	78
7700	Hoods, magnetic, with handle & screen									
7710	3″ tubing size, 3″ x 6″ opening	1 Shee	24	.333	Ea.	90.50	15.75		106.25	125
7720	4″ tubing size, 4″ x 8″ opening		24	.333		90.50	15.75		106.25	125
7730	5″ tubing size, 5″ x 10″ opening		24	.333		90.50	15.75		106.25	125

23 36 Air Terminal Units

23 36 13 – Constant-Air-Volume Units

23 36 13.10 Constant Volume Mixing Boxes		Crew	Daily Output	Labor-Hours	Unit	Material	2009 Bare Costs Labor	Equipment	Total	Total Incl O&P
0010	**CONSTANT VOLUME MIXING BOXES**									
5180	Mixing box, includes electric or pneumatic motor									
5192	Recommend use with silencer, see Div. 23 33 19.10 0010									
5200	Constant volume, 150 to 270 CFM	Q-9	12	1.333	Ea.	620	56.50		676.50	770
5210	270 to 600 CFM		11	1.455		640	62		702	800
5230	550 to 1000 CFM		9	1.778		640	75.50		715.50	820
5240	1000 to 1600 CFM		8	2		655	85		740	855
5250	1300 to 1900 CFM		6	2.667		670	113		783	915
5260	550 to 2640 CFM		5.60	2.857		735	121		856	1,000
5270	650 to 3120 CFM		5.20	3.077		785	131		916	1,075

23 36 16 – Variable-Air-Volume Units

23 36 16.10 Variable Volume Mixing Boxes		Crew	Daily Output	Labor-Hours	Unit	Material	Labor	Equipment	Total	Total Incl O&P
0010	**VARIABLE VOLUME MIXING BOXES**									
5180	Mixing box, includes electric or pneumatic motor									
5192	Recommend use with attenuator, see Div. 23 33 19.10 0010									
5500	VAV Cool only, pneumatic, pressure independent 300 to 600 CFM	Q-9	11	1.455	Ea.	400	62		462	540
5510	500 to 1000 CFM		9	1.778		410	75.50		485.50	570
5520	800 to 1600 CFM		9	1.778		425	75.50		500.50	585
5530	1100 to 2000 CFM		8	2		435	85		520	610
5540	1500 to 3000 CFM		7	2.286		460	97		557	665
5550	2000 to 4000 CFM		6	2.667		465	113		578	695
5560	For electric, w/thermostat, pressure dependent, add					23			23	25.50
5600	VAV Cool & HW coils, damper, actuator and t'stat									
5610	200 CFM	Q-9	11	1.455	Ea.	700	62		762	870
5620	400 CFM		10	1.600		705	68		773	885
5630	600 CFM		10	1.600		705	68		773	885
5640	800 CFM		8	2		725	85		810	935
5650	1000 CFM		8	2		725	85		810	935
5660	1250 CFM		6	2.667		795	113		908	1,050
5670	1500 CFM		6	2.667		795	113		908	1,050
5680	2000 CFM		4	4		865	170		1,035	1,225
5684	3000 CFM		3.80	4.211		940	179		1,119	1,300
5700	VAV Cool only, fan powered, damper, actuator, thermostat									
5710	200 CFM	Q-9	10	1.600	Ea.	955	68		1,023	1,150
5720	400 CFM		9	1.778		990	75.50		1,065.50	1,225
5730	600 CFM		9	1.778		990	75.50		1,065.50	1,225
5740	800 CFM		7	2.286		1,050	97		1,147	1,325
5750	1000 CFM		7	2.286		1,050	97		1,147	1,325
5760	1250 CFM		5	3.200		1,150	136		1,286	1,475
5770	1500 CFM		5	3.200		1,150	136		1,286	1,475
5780	2000 CFM		4	4		1,275	170		1,445	1,675
5800	VAV Fan powr'd, cooling, with HW coils, dampers, actuators, t'stat									
5810	200 CFM	Q-9	10	1.600	Ea.	1,150	68		1,218	1,375
5820	400 CFM		9	1.778		1,200	75.50		1,275.50	1,425
5830	600 CFM		9	1.778		1,200	75.50		1,275.50	1,425
5840	800 CFM		7	2.286		1,250	97		1,347	1,525
5850	1000 CFM		7	2.286		1,250	97		1,347	1,525
5860	1250 CFM		5	3.200		1,425	136		1,561	1,800
5870	1500 CFM		5	3.200		1,425	136		1,561	1,800
5880	2000 CFM		3.50	4.571		1,525	194		1,719	2,000

23 37 Air Outlets and Inlets

23 37 13 – Diffusers, Registers, and Grilles

23 37 13.10 Diffusers		Crew	Daily Output	Labor-Hours	Unit	Material	2009 Bare Costs Labor	Equipment	Total	Total Incl O&P
0010	**DIFFUSERS**, Aluminum, opposed blade damper unless noted									
0100	Ceiling, linear, also for sidewall									
0120	2″ wide	1 Shee	32	.250	L.F.	39.50	11.80		51.30	62.50
0140	3″ wide		30	.267		45	12.60		57.60	69.50
0160	4″ wide		26	.308		51.50	14.50		66	80
0180	6″ wide		24	.333		63.50	15.75		79.25	95
0200	8″ wide		22	.364		75	17.15		92.15	110
0220	10″ wide		20	.400		85	18.90		103.90	124
0240	12″ wide		18	.444		91	21		112	134
0260	For floor or sill application, add					15%				
0500	Perforated, 24″ x 24″ lay-in panel size, 6″ x 6″	1 Shee	16	.500	Ea.	56	23.50		79.50	99
0520	8″ x 8″		15	.533		57	25		82	103
0530	9″ x 9″		14	.571		61	27		88	111
0540	10″ x 10″		14	.571		63	27		90	113
0560	12″ x 12″		12	.667		64	31.50		95.50	121
0580	15″ x 15″		11	.727		127	34.50		161.50	195
0590	16″ x 16″		11	.727		106	34.50		140.50	172
0600	18″ x 18″		10	.800		112	38		150	183
0610	20″ x 20″		10	.800		129	38		167	202
0620	24″ x 24″		9	.889		161	42		203	244
1000	Rectangular, 1 to 4 way blow, 6″ x 6″		16	.500		45	23.50		68.50	87
1010	8″ x 8″		15	.533		54	25		79	99.50
1014	9″ x 9″		15	.533		59.50	25		84.50	106
1016	10″ x 10″		15	.533		64	25		89	111
1020	12″ x 6″		15	.533		65.50	25		90.50	112
1040	12″ x 9″		14	.571		71.50	27		98.50	122
1060	12″ x 12″		12	.667		72	31.50		103.50	129
1070	14″ x 6″		13	.615		71	29		100	124
1074	14″ x 14″		12	.667		101	31.50		132.50	161
1080	18″ x 12″		11	.727		126	34.50		160.50	194
1120	15″ x 15″		10	.800		103	38		141	173
1140	18″ x 15″		9	.889		148	42		190	230
1150	18″ x 18″		9	.889		126	42		168	206
1160	21″ x 21″		8	1		180	47		227	273
1170	24″ x 12″		10	.800		134	38		172	207
1180	24″ x 24″		7	1.143		256	54		310	365
1500	Round, butterfly damper, steel, diffuser size, 6″ diameter		18	.444		22	21		43	57.50
1520	8″ diameter		16	.500		23.50	23.50		47	63.50
1540	10″ diameter		14	.571		29	27		56	75
1560	12″ diameter		12	.667		38.50	31.50		70	92.50
1580	14″ diameter		10	.800		49	33		87	114
1600	18″ diameter		9	.889		117	42		159	196
1610	20″ diameter		9	.889		125	42		167	205
1620	22″ diameter		8	1		137	47		184	226
1630	24″ diameter		8	1		150	47		197	240
1640	28″ diameter		7	1.143		183	54		237	288
1650	32″ diameter		6	1.333		205	63		268	325
2000	T bar mounting, 24″ x 24″ lay-in frame, 6″ x 6″		16	.500		42.50	23.50		66	84.50
2020	8″ x 8″		14	.571		44.50	27		71.50	91.50
2040	12″ x 12″		12	.667		52	31.50		83.50	108
2060	16″ x 16″		11	.727		70	34.50		104.50	132
2080	18″ x 18″		10	.800		78	33		116	146
2500	Combination supply and return									

23 37 13.10 Diffusers		Crew	Daily Output	Labor-Hours	Unit	2009 Bare Costs Material	Labor	Equipment	Total	Total Incl O&P
2520	21" x 21" supply, 15" x 15" return	Q-9	10	1.600	Ea.	305	68		373	445
2540	24" x 24" supply, 18" x 18" return		9.50	1.684		385	71.50		456.50	540
2560	27" x 27" supply, 18" x 18" return		9	1.778		450	75.50		525.50	615
2580	30" x 30" supply, 21" x 21" return		8.50	1.882		570	80		650	750
2600	33" x 33" supply, 24" x 24" return		8	2		665	85		750	865
2620	36" x 36" supply, 24" x 24" return		7.50	2.133		780	90.50		870.50	1,000
3000	Baseboard, white enameled steel									
3100	18" long	1 Shee	20	.400	Ea.	11.75	18.90		30.65	43
3120	24" long		18	.444		23	21		44	58.50
3140	48" long		16	.500		42	23.50		65.50	83.50
3400	For matching return, deduct					10%				
4000	Floor, steel, adjustable pattern									
4100	2" x 10"	1 Shee	34	.235	Ea.	15.75	11.10		26.85	35
4120	2" x 12"		32	.250		17.30	11.80		29.10	38
4140	2" x 14"		30	.267		19.60	12.60		32.20	41.50
4200	4" x 10"		28	.286		18.20	13.50		31.70	41.50
4220	4" x 12"		26	.308		20	14.50		34.50	45
4240	4" x 14"		25	.320		22.50	15.10		37.60	49
4260	6" x 10"		26	.308		27.50	14.50		42	53
4280	6" x 12"		24	.333		28.50	15.75		44.25	56.50
4300	6" x 14"		22	.364		30	17.15		47.15	60.50
5000	Sidewall, aluminum, 3 way dispersion									
5100	8" x 4"	1 Shee	24	.333	Ea.	8.65	15.75		24.40	34.50
5110	8" x 6"		24	.333		9.25	15.75		25	35
5120	10" x 4"		22	.364		9.10	17.15		26.25	37.50
5130	10" x 6"		22	.364		9.50	17.15		26.65	38
5140	10" x 8"		20	.400		13.25	18.90		32.15	44.50
5160	12" x 4"		18	.444		9.75	21		30.75	44.50
5170	12" x 6"		17	.471		10.80	22		32.80	47.50
5180	12" x 8"		16	.500		14	23.50		37.50	53
5200	14" x 4"		15	.533		10.85	25		35.85	52
5220	14" x 6"		14	.571		12	27		39	56
5240	14" x 8"		13	.615		15.60	29		44.60	63
5260	16" x 6"		12.50	.640		17.15	30		47.15	67
6000	For steel diffusers instead of aluminum, deduct					10%				
9000	Minimum labor/equipment charge	1 Shee	4	2	Job		94.50		94.50	150

23 37 13.30 Grilles		Crew	Daily Output	Labor-Hours	Unit	Material	Labor	Equipment	Total	Total Incl O&P
0010	**GRILLES**									
0020	Aluminum									
0100	Air supply, single deflection, adjustable									
0120	8" x 4"	1 Shee	30	.267	Ea.	17.70	12.60		30.30	39.50
0140	8" x 8"		28	.286		21.50	13.50		35	45
0160	10" x 4"		24	.333		19.50	15.75		35.25	46.50
0180	10" x 10"		23	.348		25	16.40		41.40	53.50
0200	12" x 6"		23	.348		23	16.40		39.40	51.50
0220	12" x 12"		22	.364		30.50	17.15		47.65	61
0230	14" x 6"		23	.348		25.50	16.40		41.90	54.50
0240	14" x 8"		23	.348		27.50	16.40		43.90	56.50
0260	14" x 14"		22	.364		37	17.15		54.15	68.50
0270	18" x 8"		23	.348		31.50	16.40		47.90	61
0280	18" x 10"		23	.348		36.50	16.40		52.90	66
0300	18" x 18"		21	.381		56.50	18		74.50	91

23 37 13 – Diffusers, Registers, and Grilles

23 37 13.30 Grilles		Crew	Daily Output	Labor-Hours	Unit	Material	2009 Bare Costs Labor	Equipment	Total	Total Incl O&P
0320	20″ x 12″	1 Shee	22	.364	Ea.	41.50	17.15		58.65	73
0340	20″ x 20″		21	.381		68.50	18		86.50	104
0350	24″ x 8″		20	.400		43	18.90		61.90	77.50
0360	24″ x 14″		18	.444		56	21		77	95
0380	24″ x 24″		15	.533		97	25		122	147
0400	30″ x 8″		20	.400		51	18.90		69.90	86
0420	30″ x 10″		19	.421		67.50	19.85		87.35	106
0440	30″ x 12″		18	.444		62	21		83	102
0460	30″ x 16″		17	.471		55.50	22		77.50	96.50
0480	30″ x 18″		17	.471		90	22		112	135
0500	30″ x 30″		14	.571		160	27		187	219
0520	36″ x 12″		17	.471		77	22		99	120
0540	36″ x 16″		16	.500		98	23.50		121.50	146
0560	36″ x 18″		15	.533		109	25		134	159
0570	36″ x 20″		15	.533		119	25		144	171
0580	36″ x 24″		14	.571		154	27		181	213
0600	36″ x 28″		13	.615		189	29		218	254
0620	36″ x 30″		12	.667		208	31.50		239.50	279
0640	36″ x 32″		12	.667		228	31.50		259.50	300
0660	36″ x 34″		11	.727		248	34.50		282.50	330
0680	36″ x 36″		11	.727		270	34.50		304.50	350
0700	For double deflecting, add					70%				
1000	Air return, 6″ x 6″	1 Shee	26	.308		17.05	14.50		31.55	42
1020	10″ x 6″		24	.333		17.05	15.75		32.80	44
1040	14″ x 6″		23	.348		19.80	16.40		36.20	48
1060	10″ x 8″		23	.348		19.80	16.40		36.20	48
1080	16″ x 8″		22	.364		21.50	17.15		38.65	51
1100	12″ x 12″		22	.364		23	17.15		40.15	53
1120	24″ x 12″		18	.444		32	21		53	68.50
1140	30″ x 12″		16	.500		41.50	23.50		65	83
1160	14″ x 14″		22	.364		27.50	17.15		44.65	58
1180	16″ x 16″		22	.364		31	17.15		48.15	61.50
1200	18″ x 18″		21	.381		35	18		53	67
1220	24″ x 18″		16	.500		40	23.50		63.50	81.50
1240	36″ x 18″		15	.533		61.50	25		86.50	108
1260	24″ x 24″		15	.533		47.50	25		72.50	92
1280	36″ x 24″		14	.571		69	27		96	119
1300	48″ x 24″		12	.667		118	31.50		149.50	180
1320	48″ x 30″		11	.727		153	34.50		187.50	223
1340	36″ x 36″		13	.615		92	29		121	147
1360	48″ x 36″		11	.727		190	34.50		224.50	264
1380	48″ x 48″		8	1		234	47		281	330
2000	Door grilles, 12″ x 12″		22	.364		52.50	17.15		69.65	85
2020	18″ x 12″		22	.364		59.50	17.15		76.65	93
2040	24″ x 12″		18	.444		64.50	21		85.50	105
2060	18″ x 18″		18	.444		80	21		101	121
2080	24″ x 18″		16	.500		91	23.50		114.50	138
2100	24″ x 24″		15	.533		125	25		150	178
3000	Filter grille with filter, 12″ x 12″		24	.333		45.50	15.75		61.25	75
3020	18″ x 12″		20	.400		64	18.90		82.90	100
3040	24″ x 18″		18	.444		75	21		96	116
3060	24″ x 24″		16	.500		86.50	23.50		110	133
3080	30″ x 24″		14	.571		108	27		135	162

23 37 13.30 Grilles		Crew	Daily Output	Labor-Hours	Unit	Material	Labor	Equipment	Total	Total Incl O&P
3100	30" x 30"	1 Shee	13	.615	Ea.	127	29		156	186
3950	Eggcrate, framed, 6" x 6" opening		26	.308		18.70	14.50		33.20	43.50
3954	8" x 8" opening		24	.333		19.80	15.75		35.55	47
3960	10" x 10" opening		23	.348		21.50	16.40		37.90	49.50
3970	12" x 12" opening		22	.364		26	17.15		43.15	56
3980	14" x 14" opening		22	.364		27	17.15		44.15	57
3984	16" x 16" opening		21	.381		39.50	18		57.50	72
3990	18" x 18" opening		21	.381		45.50	18		63.50	78.50
4020	22" x 22" opening		17	.471		59.50	22		81.50	101
4040	24" x 24" opening		15	.533		75	25		100	123
4044	28" x 28" opening		14	.571		116	27		143	170
4048	36" x 36" opening		12	.667		132	31.50		163.50	195
4050	48" x 24" opening	↓	12	.667	↓	168	31.50		199.50	235
4060	Eggcrate, lay-in, T-bar system									
4070	48" x 24" sheet	1 Shee	40	.200	Ea.	90	9.45		99.45	114
5000	Transfer grille, vision proof, 8" x 4"		30	.267		32.50	12.60		45.10	56
5020	8" x 6"		28	.286		32.50	13.50		46	57.50
5040	8" x 8"		26	.308		36.50	14.50		51	63
5060	10" x 6"		24	.333		34	15.75		49.75	62.50
5080	10" x 10"		23	.348		43	16.40		59.40	73
5090	12" x 6"		23	.348		36.50	16.40		52.90	66
5100	12" x 10"		22	.364		47	17.15		64.15	79
5110	12" x 12"		22	.364		51	17.15		68.15	83.50
5120	14" x 10"		22	.364		51	17.15		68.15	83.50
5140	16" x 8"		21	.381		51	18		69	84.50
5160	18" x 12"		20	.400		66.50	18.90		85.40	103
5170	18" x 18"		21	.381		93.50	18		111.50	132
5180	20" x 12"		20	.400		71.50	18.90		90.40	109
5200	20" x 20"		19	.421		113	19.85		132.85	156
5220	24" x 12"		17	.471		87	22		109	132
5240	24" x 24"		16	.500		155	23.50		178.50	208
5260	30" x 6"		15	.533		71.50	25		96.50	119
5280	30" x 8"		15	.533		83.50	25		108.50	132
5300	30" x 12"		14	.571		101	27		128	154
5320	30" x 16"		13	.615		124	29		153	182
5340	30" x 20"		12	.667		156	31.50		187.50	222
5360	30" x 24"		11	.727		192	34.50		226.50	267
5380	30" x 30"	↓	10	.800		254	38		292	340
6000	For steel grilles instead of aluminum in above, deduct				↓	10%				
6200	Plastic, eggcrate, lay-in, T-bar system									
6210	48" x 24" sheet	1 Shee	50	.160	Ea.	22.50	7.55		30.05	37
6250	Steel door louver									
6270	With fire link, steel only									
6350	12" x 18"	1 Shee	22	.364	Ea.	187	17.15		204.15	234
6410	12" x 24"		21	.381		214	18		232	264
6470	18" x 18"		20	.400		209	18.90		227.90	260
6530	18" x 24"		19	.421		240	19.85		259.85	296
6600	18" x 36"		17	.471		279	22		301	340
6660	24" x 24"		15	.533		242	25		267	305
6720	24" x 30"		15	.533		279	25		304	345
6780	30" x 30"		13	.615		325	29		354	405
6840	30" x 36"		12	.667		380	31.50		411.50	465
6900	36" x 36"	↓	10	.800	↓	435	38		473	540

23 37 13.30 Grilles		Crew	Daily Output	Labor-Hours	Unit	Material	2009 Bare Costs Labor	Equipment	Total	Total Incl O&P
9000	Minimum labor/equipment charge	1 Shee	4	2	Job		94.50		94.50	150

23 37 13.60 Registers		Crew	Daily Output	Labor-Hours	Unit	Material	Labor	Equipment	Total	Total Incl O&P
0010	**REGISTERS**									
0980	Air supply									
1000	Ceiling/wall, O.B. damper, anodized aluminum									
1010	One or two way deflection, adj. curved face bars									
1014	6" x 6"	1 Shee	24	.333	Ea.	30.50	15.75		46.25	58.50
1020	8" x 4"		26	.308		30.50	14.50		45	56.50
1040	8" x 8"		24	.333		38	15.75		53.75	67
1060	10" x 6"		20	.400		35	18.90		53.90	68.50
1080	10" x 10"		19	.421		47	19.85		66.85	83.50
1100	12" x 6"		19	.421		35.50	19.85		55.35	70.50
1120	12" x 12"		18	.444		44.50	21		65.50	82.50
1140	14" x 8"		17	.471		48.50	22		70.50	89
1160	14" x 14"		18	.444		65	21		86	105
1170	16" x 16"		17	.471		72	22		94	115
1180	18" x 8"		18	.444		54.50	21		75.50	93.50
1200	18" x 18"		17	.471		99	22		121	145
1220	20" x 4"		19	.421		50	19.85		69.85	86.50
1240	20" x 6"		18	.444		50	21		71	88.50
1260	20" x 8"		18	.444		54	21		75	93
1280	20" x 20"		17	.471		120	22		142	169
1290	22" x 22"		15	.533		138	25		163	192
1300	24" x 4"		17	.471		61.50	22		83.50	103
1320	24" x 6"		16	.500		61.50	23.50		85	105
1340	24" x 8"		13	.615		67.50	29		96.50	120
1350	24" x 18"		12	.667		125	31.50		156.50	188
1360	24" x 24"		11	.727		170	34.50		204.50	242
1380	30" x 4"		16	.500		77	23.50		100.50	122
1400	30" x 6"		15	.533		77	25		102	125
1420	30" x 8"		14	.571		88	27		115	140
1440	30" x 24"		12	.667		219	31.50		250.50	291
1460	30" x 30"		10	.800		281	33		319	370
1504	4 way deflection, adjustable curved face bars									
1510	6" x 6"	1 Shee	26	.308	Ea.	36.50	14.50		51	63.50
1514	8" x 8"		24	.333		45.50	15.75		61.25	75.50
1518	10" x 10"		19	.421		56.50	19.85		76.35	94
1522	12" x 6"		19	.421		42.50	19.85		62.35	78
1526	12" x 12"		18	.444		53.50	21		74.50	92.50
1530	14" x 14"		18	.444		78	21		99	120
1534	16" x 16"		17	.471		86.50	22		108.50	131
1538	18" x 18"		17	.471		119	22		141	167
1542	22" x 22"		16	.500		166	23.50		189.50	221
1980	One way deflection, adj. vert. or horiz. face bars									
1990	6" x 6"	1 Shee	26	.308	Ea.	32	14.50		46.50	58
2000	8" x 4"		26	.308		31.50	14.50		46	57.50
2020	8" x 8"		24	.333		37	15.75		52.75	66
2040	10" x 6"		20	.400		37.50	18.90		56.40	71
2060	10" x 10"		19	.421		43.50	19.85		63.35	79.50
2080	12" x 6"		19	.421		40.50	19.85		60.35	76
2100	12" x 12"		18	.444		52.50	21		73.50	91.50
2120	14" x 8"		17	.471		48.50	22		70.50	88.50

23 37 13.60 Registers		Crew	Daily Output	Labor-Hours	Unit	Material	Labor	Equipment	Total	Total Incl O&P
						2009 Bare Costs				
2140	14" x 12"	1 Shee	18	.444	Ea.	58	21		79	97
2160	14" x 14"		18	.444		65.50	21		86.50	106
2180	16" x 6"		18	.444		44	21		65	81.50
2200	16" x 12"		18	.444		64	21		85	104
2220	16" x 16"		17	.471		81	22		103	125
2240	18" x 8"		18	.444		56	21		77	95.50
2260	18" x 12"		17	.471		68.50	22		90.50	111
2280	18" x 18"		16	.500		99.50	23.50		123	147
2300	20" x 10"		18	.444		67	21		88	107
2320	20" x 16"		16	.500		96.50	23.50		120	144
2340	20" x 20"		16	.500		122	23.50		145.50	172
2360	24" x 12"		15	.533		84	25		109	133
2380	24" x 16"		14	.571		112	27		139	166
2400	24" x 20"		12	.667		140	31.50		171.50	204
2420	24" x 24"		10	.800		171	38		209	248
2440	30" x 12"		13	.615		109	29		138	166
2460	30" x 16"		13	.615		132	29		161	192
2480	30" x 24"		11	.727		221	34.50		255.50	298
2500	30" x 30"		9	.889		283	42		325	375
2520	36" x 12"		11	.727		135	34.50		169.50	204
2540	36" x 24"		10	.800		272	38		310	360
2560	36" x 36"		8	1		455	47		502	575
2600	For 2 way deflect., adj. vert. or horiz. face bars, add					40%				
2700	Above registers in steel instead of aluminum, deduct					10%				
3000	Baseboard, hand adj. damper, enameled steel									
3012	8" x 6"	1 Shee	26	.308	Ea.	12.85	14.50		27.35	37
3020	10" x 6"		24	.333		14	15.75		29.75	40.50
3040	12" x 5"		23	.348		15.20	16.40		31.60	42.50
3060	12" x 6"		23	.348		15.20	16.40		31.60	42.50
3080	12" x 8"		22	.364		22	17.15		39.15	51.50
3100	14" x 6"		20	.400		16.45	18.90		35.35	48
4000	Floor, toe operated damper, enameled steel									
4020	4" x 8"	1 Shee	32	.250	Ea.	24	11.80		35.80	45
4040	4" x 12"		26	.308		28	14.50		42.50	53.50
4060	6" x 8"		28	.286		25.50	13.50		39	49.50
4080	6" x 14"		22	.364		32.50	17.15		49.65	63
4100	8" x 10"		22	.364		29	17.15		46.15	59.50
4120	8" x 16"		20	.400		48.50	18.90		67.40	83.50
4140	10" x 10"		20	.400		35	18.90		53.90	68.50
4160	10" x 16"		18	.444		53	21		74	91.50
4180	12" x 12"		18	.444		43	21		64	80.50
4200	12" x 24"		16	.500		148	23.50		171.50	201
4220	14" x 14"		16	.500		122	23.50		145.50	172
4240	14" x 20"		15	.533		163	25		188	219
4300	Spiral pipe supply register									
4310	Aluminum, double deflection, w/damper extractor									
4320	4" x 12", for 6" thru 10" diameter duct	1 Shee	25	.320	Ea.	57	15.10		72.10	87
4330	4" x 18", for 6" thru 10" diameter duct		18	.444		71	21		92	112
4340	6" x 12", for 8" thru 12" diameter duct		19	.421		62.50	19.85		82.35	100
4350	6" x 18", for 8" thru 12" diameter duct		18	.444		81	21		102	123
4360	6" x 24", for 8" thru 12" diameter duct		16	.500		100	23.50		123.50	148
4370	6" x 30", for 8" thru 12" diameter duct		15	.533		127	25		152	180
4380	8" x 18", for 10" thru 14" diameter duct		18	.444		86.50	21		107.50	129

23 37 13 – Diffusers, Registers, and Grilles

23 37 13.60 Registers		Crew	Daily Output	Labor-Hours	Unit	2009 Bare Costs Material	Labor	Equipment	Total	Total Incl O&P
4390	8" x 24", for 10" thru 14" diameter duct	1 Shee	15	.533	Ea.	108	25		133	159
4400	8" x 30", for 10" thru 14" diameter duct		14	.571		144	27		171	201
4410	10" x 24", for 12" thru 18" diameter duct		13	.615		119	29		148	177
4420	10" x 30", for 12" thru 18" diameter duct		12	.667		157	31.50		188.50	222
4430	10" x 36", for 12" thru 18" diameter duct		11	.727		195	34.50		229.50	269
4980	Air return									
5000	Ceiling or wall, fixed 45° face blades									
5010	Adjustable O.B. damper, anodized aluminum									
5020	4" x 8"	1 Shee	26	.308	Ea.	23	14.50		37.50	48.50
5040	6" x 8"		24	.333		26.50	15.75		42.25	54
5060	6" x 10"		19	.421		26.50	19.85		46.35	60.50
5080	6" x 16"		18	.444		34	21		55	71
5100	8" x 10"		19	.421		31	19.85		50.85	65.50
5120	8" x 12"		16	.500		32.50	23.50		56	73
5140	10" x 10"		18	.444		32.50	21		53.50	69
5160	10" x 16"		17	.471		40	22		62	79.50
5180	12" x 18"		18	.444		46	21		67	84.50
5200	12" x 30"		12	.667		70.50	31.50		102	128
5220	16" x 16"		17	.471		49	22		71	89.50
5240	18" x 18"		16	.500		58.50	23.50		82	102
5260	18" x 36"		10	.800		127	38		165	199
5280	24" x 24"		11	.727		89	34.50		123.50	153
5300	24" x 36"		8	1		140	47		187	229
5320	24" x 48"		6	1.333		209	63		272	330
6000	For steel construction instead of aluminum, deduct					10%				
9000	Minimum labor/equipment charge	1 Shee	4	2	Job		94.50		94.50	150

23 37 15 – Louvers

23 37 15.40 HVAC Louvers		Crew	Daily Output	Labor-Hours	Unit	Material	Labor	Equipment	Total	Total Incl O&P
0010	**HVAC LOUVERS**									
0100	Aluminum, extruded, with screen, mill finish									
1002	Brick vent, see also Div. 04 05 23.19									
1100	Standard, 4" deep, 8" wide, 5" high	1 Shee	24	.333	Ea.	34.50	15.75		50.25	62.50
1200	Modular, 4" deep, 7-3/4" wide, 5" high		24	.333		36	15.75		51.75	64.50
1300	Speed brick, 4" deep, 11-5/8" wide, 3-7/8" high		24	.333		36	15.75		51.75	64.50
1400	Fuel oil brick, 4" deep, 8" wide, 5" high		24	.333		61.50	15.75		77.25	93
2000	Cooling tower and mechanical equip., screens, light weight		40	.200	S.F.	15.40	9.45		24.85	32
2020	Standard weight		35	.229		41	10.80		51.80	62
2500	Dual combination, automatic, intake or exhaust		20	.400		56	18.90		74.90	91.50
2520	Manual operation		20	.400		41.50	18.90		60.40	76
2540	Electric or pneumatic operation		20	.400		41.50	18.90		60.40	76
2560	Motor, for electric or pneumatic		14	.571	Ea.	480	27		507	575
3000	Fixed blade, continuous line									
3100	Mullion type, stormproof	1 Shee	28	.286	S.F.	41.50	13.50		55	67.50
3200	Stormproof		28	.286		41.50	13.50		55	67.50
3300	Vertical line		28	.286		49.50	13.50		63	75.50
3500	For damper to use with above, add					50%	30%			
3520	Motor, for damper, electric or pneumatic	1 Shee	14	.571	Ea.	480	27		507	575
4000	Operating, 45°, manual, electric or pneumatic		24	.333	S.F.	41.50	15.75		57.25	71
4100	Motor, for electric or pneumatic		14	.571	Ea.	480	27		507	575
4200	Penthouse, roof		56	.143	S.F.	24.50	6.75		31.25	37.50
4300	Walls		40	.200		58	9.45		67.45	79
5000	Thinline, under 4" thick, fixed blade		40	.200		24	9.45		33.45	41.50

23 37 Air Outlets and Inlets

23 37 15 – Louvers

23 37 15.40 HVAC Louvers		Crew	Daily Output	Labor-Hours	Unit	2009 Bare Costs Material	Labor	Equipment	Total	Total Incl O&P
5010	Finishes, applied by mfr. at additional cost, available in colors									
5020	Prime coat only, add				S.F.	3.30			3.30	3.63
5040	Baked enamel finish coating, add					6.10			6.10	6.70
5060	Anodized finish, add					6.60			6.60	7.25
5080	Duranodic finish, add					12			12	13.20
5100	Fluoropolymer finish coating, add					18.90			18.90	21
9980	For small orders (under 10 pieces), add				↓	25%				

23 37 23 – HVAC Gravity Ventilators

23 37 23.10 HVAC Gravity Air Ventilators

		Crew	Daily Output	Labor-Hours	Unit	Material	Labor	Equipment	Total	Total Incl O&P
0010	**HVAC GRAVITY AIR VENTILATORS**, Incl. base and damper									
1280	Rotary ventilators, wind driven, galvanized									
1300	4" neck diameter	Q-9	20	.800	Ea.	35	34		69	92.50
1320	5" neck diameter		18	.889		39	38		77	103
1340	6" neck diameter		16	1		39	42.50		81.50	111
1360	8" neck diameter		14	1.143		40.50	48.50		89	122
1380	10" neck diameter		12	1.333		52.50	56.50		109	148
1400	12" neck diameter		10	1.600		56.50	68		124.50	171
1420	14" neck diameter		10	1.600		78.50	68		146.50	195
1440	16" neck diameter		9	1.778		95	75.50		170.50	225
1460	18" neck diameter, 1700 CFM		9	1.778		111	75.50		186.50	242
1480	20" neck diameter, 2100 CFM		8	2		123	85		208	270
1500	24" neck diameter, 3,100 CFM		8	2		165	85		250	315
1520	30" neck diameter, 4500 CFM		7	2.286		345	97		442	535
1540	36" neck diameter, 5,500 CFM	↓	6	2.667		485	113		598	715
1600	For aluminum, add					300%				
1620	For stainless steel, add				↓	600%				
2000	Stationary, gravity, syphon, galvanized									
2100	3" neck diameter, 40 CFM	Q-9	24	.667	Ea.	31.50	28.50		60	80
2120	4" neck diameter, 50 CFM		20	.800		34.50	34		68.50	92
2140	5" neck diameter, 58 CFM		18	.889		35	38		73	98.50
2160	6" neck diameter, 66 CFM		16	1		30	42.50		72.50	101
2180	7" neck diameter, 86 CFM		15	1.067		38	45.50		83.50	114
2200	8" neck diameter, 110 CFM		14	1.143		38	48.50		86.50	119
2220	10" neck diameter, 140 CFM		12	1.333		49.50	56.50		106	145
2240	12" neck diameter, 160 CFM		10	1.600		63.50	68		131.50	178
2260	14" neck diameter, 250 CFM		10	1.600		80	68		148	196
2280	16" neck diameter, 380 CFM		9	1.778		107	75.50		182.50	237
2300	18" neck diameter, 500 CFM		9	1.778		142	75.50		217.50	276
2320	20" neck diameter, 625 CFM		8	2		150	85		235	300
2340	24" neck diameter, 900 CFM		8	2		173	85		258	325
2360	30" neck diameter, 1375 CFM		7	2.286		395	97		492	590
2380	36" neck diameter, 2,000 CFM		6	2.667		540	113		653	775
2400	42" neck diameter, 3000 CFM	↓	4	4		765	170		935	1,100
2500	For aluminum, add					300%				
2520	For stainless steel, add					600%				
3000	Rotating chimney cap, galvanized, 4" neck diameter	Q-9	20	.800		27	34		61	84
3020	5" neck diameter		18	.889		28	38		66	90.50
3040	6" neck diameter		16	1		28.50	42.50		71	98.50
3060	7" neck diameter		15	1.067		30.50	45.50		76	106
3080	8" neck diameter		14	1.143		32	48.50		80.50	112
3100	10" neck diameter		12	1.333		41	56.50		97.50	135
3600	Stationary chimney rain cap, galvanized, 3" neck diameter	↓	24	.667	↓	24	28.50		52.50	71

23 37 Air Outlets and Inlets

23 37 23 – HVAC Gravity Ventilators

23 37 23.10 HVAC Gravity Air Ventilators		Crew	Daily Output	Labor-Hours	Unit	2009 Bare Costs Material	Labor	Equipment	Total	Total Incl O&P
3620	4" neck diameter	Q-9	20	.800	Ea.	24.50	34		58.50	80.50
3640	6" neck diameter		16	1		25.50	42.50		68	95.50
3680	8" neck diameter		14	1.143		27.50	48.50		76	107
3700	10" neck diameter		12	1.333		34.50	56.50		91	128
3720	12" neck diameter		10	1.600		57.50	68		125.50	171
3740	14" neck diameter		10	1.600		61.50	68		129.50	176
3760	16" neck diameter		9	1.778		75.50	75.50		151	203
4200	Stationary mushroom, aluminum, 16" orifice diameter		10	1.600		510	68		578	670
4220	26" orifice diameter		6.15	2.602		755	111		866	1,000
4230	30" orifice diameter		5.71	2.802		1,100	119		1,219	1,425
4240	38" orifice diameter		5	3.200		1,575	136		1,711	1,975
4250	42" orifice diameter		4.70	3.404		2,100	145		2,245	2,525
4260	50" orifice diameter	↓	4.44	3.604	↓	2,500	153		2,653	3,000
5000	Relief vent									
5500	Rectangular, aluminum, galvanized curb									
5510	intake/exhaust, 0.033" SP									
5580	500 CFM, 12" x 12"	Q-9	8.60	1.860	Ea.	590	79		669	775
5600	600 CFM, 12"x16"		8	2		665	85		750	865
5620	750 CFM, 12" x 20"		7.20	2.222		715	94.50		809.50	935
5640	1000 CFM, 12" x 24"		6.60	2.424		750	103		853	990
5660	1500 CFM, 12"x 36"		5.80	2.759		995	117		1,112	1,275
5680	3000 CFM, 20" x 42"		4	4		1,325	170		1,495	1,725
5700	5000 CFM, 20" x 72"		3	5.333		1,900	227		2,127	2,425
5720	6000 CFM, 24" x 72"		2.30	6.957		2,050	296		2,346	2,725
5740	10,000 CFM, 48" x 60"		1.80	8.889		2,600	380		2,980	3,450
5760	12,000 CFM, 48" x 72"		1.60	10		3,075	425		3,500	4,050
5780	13,750 CFM, 60" x 66"		1.40	11.429		3,400	485		3,885	4,500
5800	15,000 CFM, 60" x 72"		1.30	12.308		3,625	525		4,150	4,825
5820	18,000 CFM, 72" x 72"	↓	1.20	13.333	↓	4,125	565		4,690	5,450
5880	Size is throat area, volume is at 500 FPM									
7000	Note: sizes based on exhaust. Intake, with 0.125" SP									
7100	loss, approximately twice listed capacity.									
9000	Minimum labor/equipment charge	1 Plum	2	4	Job		195		195	300

23 38 Ventilation Hoods

23 38 13 – Commercial-Kitchen Hoods

23 38 13.10 Hood and Ventilation Equipment		Crew	Daily Output	Labor-Hours	Unit	Material	Labor	Equipment	Total	Total Incl O&P
0010	**HOOD AND VENTILATION EQUIPMENT**									
2970	Exhaust hood, sst, gutter on all sides, 4' x 4' x 2'	1 Carp	1.80	4.444	Ea.	4,050	178		4,228	4,750
2980	4' x 4' x 7'	"	1.60	5		6,350	200		6,550	7,325
7950	Hood fire protection system, minimum	Q-1	3	5.333		4,100	234		4,334	4,875
8050	Maximum	"	1	16	↓	31,600	700		32,300	35,900

23 41 Particulate Air Filtration

23 41 13 – Panel Air Filters

23 41 13.10 Panel Type Air Filters		Crew	Daily Output	Labor-Hours	Unit	Material	2009 Bare Costs Labor	Equipment	Total	Total Incl O&P
0010	**PANEL TYPE AIR FILTERS**									
2950	Mechanical media filtration units									
3000	High efficiency type, with frame, non-supported G				MCFM	45			45	49.50
3100	Supported type G				"	60			60	66
5500	Throwaway glass or paper media type				Ea.	3.35			3.35	3.69

23 41 16 – Renewable-Media Air Filters

23 41 16.10 Disposable Media Air Filters		Crew	Daily Output	Labor-Hours	Unit	Material	Labor	Equipment	Total	Total Incl O&P
0010	**DISPOSABLE MEDIA AIR FILTERS**									
5000	Renewable disposable roll				MCFM	250			250	275

23 41 19 – Washable Air Filters

23 41 19.10 Permanent Air Filters		Crew	Daily Output	Labor-Hours	Unit	Material	Labor	Equipment	Total	Total Incl O&P
0010	**PERMANENT AIR FILTERS**									
4500	Permanent washable G				MCFM	20			20	22

23 41 23 – Extended Surface Filters

23 41 23.10 Expanded Surface Filters		Crew	Daily Output	Labor-Hours	Unit	Material	Labor	Equipment	Total	Total Incl O&P
0010	**EXPANDED SURFACE FILTERS**									
4000	Medium efficiency, extended surface G				MCFM	5.50			5.50	6.05

23 42 Gas-Phase Air Filtration

23 42 13 – Activated-Carbon Air Filtration

23 42 13.10 Charcoal Type Air Filtration		Crew	Daily Output	Labor-Hours	Unit	Material	Labor	Equipment	Total	Total Incl O&P
0010	**CHARCOAL TYPE AIR FILTRATION**									
0050	Activated charcoal type, full flow				MCFM	600			600	660
0060	Full flow, impregnated media 12" deep					225			225	248
0070	HEPA filter & frame for field erection					300			300	330
0080	HEPA filter-diffuser, ceiling install.				↓	275			275	305

23 42 16 – Chemically-Impregnated Adsorption Air Filtration

23 42 16.10 Chemical Adsorption Air Filtration		Crew	Daily Output	Labor-Hours	Unit	Material	Labor	Equipment	Total	Total Incl O&P
0010	**CHEMICAL ADSORPTION AIR FILTRATION**									
0500	Chemical media filtration type									
1100	Industrial air fume & odor scrubber unit w/pump & motor									
1110	corrosion resistant PVC construction									
1120	Single pack filter, horizontal type									
1130	500 CFM	Q-9	14	1.143	Ea.	5,200	48.50		5,248.50	5,800
1140	1000 CFM		11	1.455		6,100	62		6,162	6,825
1150	2000 CFM		8	2		7,400	85		7,485	8,275
1160	3000 CFM		7	2.286		8,575	97		8,672	9,575
1170	5000 CFM		5	3.200		11,200	136		11,336	12,500
1180	8000 CFM		4	4		15,600	170		15,770	17,500
1190	12,000 CFM		3	5.333		20,700	227		20,927	23,200
1200	16,000 CFM		2.50	6.400		25,600	272		25,872	28,500
1210	20,000 CFM		2	8		32,100	340		32,440	35,800
1220	26,000 CFM	↓	1.50	10.667		36,900	455		37,355	41,300
1230	30,000 CFM	Q-10	2	12		42,000	530		42,530	46,900
1240	40,000 CFM	"	1.50	16		55,000	705		55,705	61,500
1250	50,000 CFM	Q-11	2	16		67,000	720		67,720	74,500
1260	55,000 CFM		1.80	17.778		73,000	800		73,800	81,500
1270	60,000 CFM	↓	1.50	21.333	↓	78,500	960		79,460	88,000
1300	Double pack filter, horizontal type									

23 42 16 – Chemically-Impregnated Adsorption Air Filtration

23 42 16.10 Chemical Adsorption Air Filtration		Crew	Daily Output	Labor-Hours	Unit	2009 Bare Costs Material	Labor	Equipment	Total	Total Incl O&P
1310	500 CFM	Q-9	10	1.600	Ea.	6,775	68		6,843	7,550
1320	1000 CFM		8	2		8,125	85		8,210	9,050
1330	2000 CFM		6	2.667		9,775	113		9,888	11,000
1340	3000 CFM		5	3.200		11,700	136		11,836	13,100
1350	5000 CFM		4	4		14,900	170		15,070	16,600
1360	8000 CFM		3	5.333		19,600	227		19,827	22,000
1370	12,000 CFM		2.50	6.400		26,600	272		26,872	29,700
1380	16,000 CFM		2	8		34,800	340		35,140	38,700
1390	20,000 CFM		1.50	10.667		42,100	455		42,555	47,000
1400	26,000 CFM		1	16		48,800	680		49,480	54,500
1410	30,000 CFM	Q-10	1.50	16		57,500	705		58,205	64,000
1420	40,000 CFM	"	1.30	18.462		73,500	815		74,315	82,000
1430	50,000 CFM	Q-11	1.50	21.333		91,500	960		92,460	102,500
1440	55,000 CFM		1.30	24.615		100,000	1,100		101,100	112,000
1450	60,000 CFM		1	32		108,500	1,450		109,950	122,000
1500	Single pack filter, vertical type									
1510	500 CFM	Q-9	24	.667	Ea.	4,975	28.50		5,003.50	5,525
1520	1000 CFM		18	.889		5,875	38		5,913	6,500
1530	2000 CFM		12	1.333		7,050	56.50		7,106.50	7,850
1540	3000 CFM		9	1.778		8,225	75.50		8,300.50	9,175
1550	5000 CFM		6	2.667		11,000	113		11,113	12,300
1560	8000 CFM		4	4		15,600	170		15,770	17,500
1570	12,000 CFM		3	5.333		21,200	227		21,427	23,800
1580	16,000 CFM		2	8		24,800	340		25,140	27,800
1590	20,000 CFM		1.80	8.889		31,300	380		31,680	35,100
1600	24,000 CFM		1.60	10		35,000	425		35,425	39,200
1650	Double pack filter, vertical type									
1660	500 CFM	Q-9	22	.727	Ea.	6,450	31		6,481	7,150
1670	1000 CFM		16	1		7,700	42.50		7,742.50	8,525
1680	2000 CFM		10	1.600		9,225	68		9,293	10,300
1690	3000 CFM		7	2.286		10,900	97		10,997	12,100
1700	5000 CFM		5	3.200		14,200	136		14,336	15,800
1710	8000 CFM		3	5.333		19,400	227		19,627	21,800
1720	12,000 CFM		2.50	6.400		26,300	272		26,572	29,300
1730	16,000 CFM		2	8		33,000	340		33,340	36,800
1740	20,000 CFM		1.50	10.667		41,200	455		41,655	46,100
1750	24,000 CFM		1	16		46,000	680		46,680	51,500
1800	Inlet or outlet transition, horizontal									
1810	Single pack to 12,000 CFM, add					4%				
1820	Single pack to 30,000 CFM, add					6%				
1830	Single pack to 60,000 CFM, add					8%				
1840	Double pack to 12,000 CFM, add					3%				
1850	Double pack to 30,000 CFM, add					5%				
1860	Double pack to 60,000 CFM, add					6%				
1870	Inlet or outlet transition, vertical									
1880	Single pack to 5000 CFM, add					2%				
1890	Single pack to 24,000 CFM, add					3%				
1900	Double pack to 24,000 CFM, add					2%				
9000	Minimum labor/equipment charge	1 Shee	2.25	3.556	Job		163		168	267

23 43 Electronic Air Cleaners

23 43 13 – Washable Electronic Air Cleaners

23 43 13.10 Electronic Air Cleaners		Crew	Daily Output	Labor-Hours	Unit	Material	2009 Bare Costs Labor	Equipment	Total	Total Incl O&P
0010	ELECTRONIC AIR CLEANERS									
2000	Electronic air cleaner, duct mounted									
2150	400 – 1000 CFM	1 Shee	2.30	3.478	Ea.	1,000	164		1,164	1,350
2200	1000 – 1400 CFM		2.20	3.636		1,225	172		1,397	1,625
2250	1400 – 2000 CFM		2.10	3.810		1,400	180		1,580	1,825

23 51 Breechings, Chimneys, and Stacks

23 51 13 – Draft Control Devices

23 51 13.13 Draft-Induction Fans		Crew	Daily Output	Labor-Hours	Unit	Material	Labor	Equipment	Total	Total Incl O&P
0010	DRAFT-INDUCTION FANS									
1000	Breeching installation									
1800	Hot gas, 600° F, variable pitch pulley and motor									
1840	6″ diam. inlet, 1/4 H.P., 1 phase, 400 CFM	Q-9	6	2.667	Ea.	1,500	113		1,613	1,825
1850	7″ diam. inlet, 1/4 H.P., 1 phase, 800 CFM		5	3.200		1,600	136		1,736	1,975
1860	8″ diam. inlet, 1/4 H.P., 1phase, 1120 CFM		4	4		2,325	170		2,495	2,825
1870	9″ diam. inlet, 3/4 H.P., 1 phase, 1440 CFM		3.60	4.444		2,750	189		2,939	3,325
1880	10″ diam. inlet, 3/4 H.P., 1 phase, 2000 CFM		3.30	4.848		3,100	206		3,306	3,725
1900	12″ diam. inlet, 3/4 H.P., 3 phase, 2960 CFM		3	5.333		3,175	227		3,402	3,850
1910	14″ diam. inlet, 1 H.P., 3 phase, 4160 CFM		2.60	6.154		3,750	261		4,011	4,550
1920	16″ diam. inlet, 2 H.P., 3 phase, 6720 CFM		2.30	6.957		4,100	296		4,396	4,975
1940	18″ diam. inlet, 3 H.P., 3 phase, 9120 CFM		2	8		4,925	340		5,265	5,975
1950	20″ diam. inlet, 3 H.P., 3 phase, 9760 CFM		1.50	10.667		5,375	455		5,830	6,625
1960	22″ diam. inlet, 5 H.P., 3 phase, 13,360 CFM		1	16		6,300	680		6,980	8,000
1980	24″ diam. inlet, 7-1/2 H.P., 3 phase, 17,760 CFM		.80	20		7,750	850		8,600	9,875
2300	For multi-blade damper at fan inlet, add					20%				
3600	Chimney-top installation									
3700	6″ size	1 Shee	8	1	Ea.	690	47		737	830
3740	8″ size		7	1.143		690	54		744	845
3750	10″ size		6.50	1.231		905	58		963	1,100
3780	13″ size		6	1.333		930	63		993	1,125
3880	For speed control switch, add					54.50			54.50	60
3920	For thermal fan control, add					54.50			54.50	60
5500	Flue shutter damper for draft control,									
5510	parallel blades									
5550	8″ size	Q-9	8	2	Ea.	520	85		605	710
5560	9″ size		7.50	2.133		565	90.50		655.50	765
5570	10″ size		7	2.286		585	97		682	795
5580	12″ size		6.50	2.462		640	105		745	870
5590	14″ size		6	2.667		725	113		838	980
5600	16″ size		5.50	2.909		790	124		914	1,075
5610	18″ size		5	3.200		865	136		1,001	1,175
5620	20″ size		4.50	3.556		960	151		1,111	1,300
5630	22″ size		4	4		1,050	170		1,220	1,425
5640	24″ size		3.50	4.571		1,150	194		1,344	1,575
5650	27″ size		3	5.333		1,175	227		1,402	1,650
5660	30″ size		2.50	6.400		1,225	272		1,497	1,775
5670	32″ size		2	8		1,275	340		1,615	1,975
5680	36″ size		1.50	10.667		1,375	455		1,830	2,250

23 51 13.16 Vent Dampers		Crew	Daily Output	Labor-Hours	Unit	Material	Labor	Equipment	Total	Total Incl O&P
0010	VENT DAMPERS									
5000	Vent damper, bi-metal, gas, 3″ diameter	Q-9	24	.667	Ea.	38	28.50		66.50	87

23 51 Breechings, Chimneys, and Stacks

23 51 13 – Draft Control Devices

23 51 13.16 Vent Dampers		Crew	Daily Output	Labor-Hours	Unit	Material	2009 Bare Costs Labor	Equipment	Total	Total Incl O&P
5010	4″ diameter	Q-9	24	.667	Ea.	38	28.50		66.50	87
5020	5″ diameter		23	.696		38	29.50		67.50	89
5030	6″ diameter		22	.727		38	31		69	91
5040	7″ diameter		21	.762		40.50	32.50		73	96
5050	8″ diameter	↓	20	.800	↓	44	34		78	103
9000	Minimum labor/equipment charge	1 Shee	4	2	Job		94.50		94.50	150

23 51 13.19 Barometric Dampers		Crew	Daily Output	Labor-Hours	Unit	Material	Labor	Equipment	Total	Total Incl O&P
0010	**BAROMETRIC DAMPERS**									
1000	Barometric, gas fired system only, 6″ size for 5″ and 6″ pipes	1 Shee	20	.400	Ea.	54	18.90		72.90	89.50
1020	7″ size, for 6″ and 7″ pipes		19	.421		59	19.85		78.85	96.50
1040	8″ size, for 7″ and 8″ pipes		18	.444		74.50	21		95.50	116
1060	9″ size, for 8″ and 9″ pipes	↓	16	.500	↓	87	23.50		110.50	134
2000	All fuel, oil, oil/gas, coal									
2020	10″ for 9″ and 10″ pipes	1 Shee	15	.533	Ea.	125	25		150	178
2040	12″ for 11″ and 12″ pipes		15	.533		163	25		188	220
2060	14″ for 13″ and 14″ pipes		14	.571		212	27		239	276
2080	16″ for 15″ and 16″ pipes		13	.615		297	29		326	370
2100	18″ for 17″ and 18″ pipes		12	.667		395	31.50		426.50	485
2120	20″ for 19″ and 21″ pipes	↓	10	.800		470	38		508	580
2140	24″ for 22″ and 25″ pipes	Q-9	12	1.333		580	56.50		636.50	725
2160	28″ for 26″ and 30″ pipes		10	1.600		715	68		783	900
2180	32″ for 31″ and 34″ pipes	↓	8	2		915	85		1,000	1,125
3260	For thermal switch for above, add	1 Shee	24	.333	↓	52.50	15.75		68.25	83

23 51 23 – Gas Vents

23 51 23.10 Gas Chimney Vents		Crew	Daily Output	Labor-Hours	Unit	Material	Labor	Equipment	Total	Total Incl O&P
0010	**GAS CHIMNEY VENTS,** Prefab metal, U.L. listed									
0020	Gas, double wall, galvanized steel									
0080	3″ diameter	Q-9	72	.222	V.L.F.	5.35	9.45		14.80	21
0100	4″ diameter		68	.235		6.70	10		16.70	23.50
0120	5″ diameter		64	.250		7.80	10.60		18.40	25.50
0140	6″ diameter		60	.267		9.10	11.35		20.45	28
0160	7″ diameter		56	.286		13.35	12.15		25.50	34
0180	8″ diameter		52	.308		14.85	13.05		27.90	37.50
0200	10″ diameter		48	.333		34	14.15		48.15	59.50
0220	12″ diameter		44	.364		45	15.45		60.45	74
0240	14″ diameter		42	.381		76	16.20		92.20	109
0260	16″ diameter		40	.400		103	17		120	140
0280	18″ diameter	↓	38	.421		133	17.90		150.90	175
0300	20″ diameter	Q-10	36	.667		139	29.50		168.50	200
0320	22″ diameter		34	.706		199	31		230	269
0340	24″ diameter		32	.750		245	33		278	320
0360	26″ diameter		31	.774		294	34		328	380
0380	28″ diameter		30	.800		310	35		345	395
0400	30″ diameter		28	.857		330	38		368	420
0420	32″ diameter		27	.889		380	39		419	480
0440	34″ diameter		26	.923		435	40.50		475.50	540
0460	36″ diameter		25	.960		460	42.50		502.50	570
0480	38″ diameter		24	1		500	44		544	620
0500	40″ diameter		23	1.043		560	46		606	690
0520	42″ diameter		22	1.091		585	48		633	720
0540	44″ diameter		21	1.143		645	50.50		695.50	790
0560	46″ diameter	↓	20	1.200	↓	710	53		763	870

23 51 23.10 Gas Chimney Vents		Crew	Daily Output	Labor-Hours	Unit	Material	Labor	Equipment	Total	Total Incl O&P
						2009 Bare Costs				
0580	48" diameter	Q-10	19	1.263	V.L.F.	790	55.50		845.50	955
0600	For 4", 5" and 6" oval, add					50%				
0650	Gas, double wall, galvanized steel, fittings									
0660	Elbow 45°, 3" diameter	Q-9	36	.444	Ea.	13.25	18.90		32.15	44.50
0670	4" diameter		34	.471		15.75	20		35.75	49.50
0680	5" diameter		32	.500		18.50	21		39.50	54.50
0690	6" diameter		30	.533		22	22.50		44.50	60.50
0700	7" diameter		28	.571		32	24.50		56.50	73.50
0710	8" diameter		26	.615		37	26		63	82
0720	10" diameter		24	.667		98	28.50		126.50	153
0730	12" diameter		22	.727		124	31		155	186
0740	14" diameter		21	.762		174	32.50		206.50	243
0750	16" diameter		20	.800		218	34		252	294
0760	18" diameter		19	.842		281	36		317	365
0770	20" diameter	Q-10	18	1.333		330	58.50		388.50	460
0780	22" diameter		17	1.412		510	62		572	660
0790	24" diameter		16	1.500		635	66		701	805
0800	26" diameter		16	1.500		1,250	66		1,316	1,475
0810	28" diameter		15	1.600		1,275	70.50		1,345.50	1,500
0820	30" diameter		14	1.714		1,350	75.50		1,425.50	1,625
0830	32" diameter		14	1.714		1,525	75.50		1,600.50	1,800
0840	34" diameter		13	1.846		1,675	81.50		1,756.50	1,975
0850	36" diameter		12	2		1,725	88		1,813	2,025
0860	38" diameter		12	2		1,850	88		1,938	2,175
0870	40" diameter		11	2.182		1,950	96		2,046	2,300
0880	42" diameter		11	2.182		2,050	96		2,146	2,400
0890	44" diameter		10	2.400		2,200	106		2,306	2,600
0900	46" diameter		10	2.400		2,350	106		2,456	2,750
0910	48" diameter		10	2.400		2,450	106		2,556	2,850
0950	Elbow 90°, adjustable, 3" diameter	Q-9	36	.444		22	18.90		40.90	54.50
0960	4" diameter		34	.471		27	20		47	61.50
0970	5" diameter		32	.500		31.50	21		52.50	68.50
0980	6" diameter		30	.533		37.50	22.50		60	77.50
0990	7" diameter		28	.571		59.50	24.50		84	104
1010	8" diameter		26	.615		63.50	26		89.50	111
1020	Wall thimble, 4 to 7" adjustable, 3" diameter		36	.444		14.05	18.90		32.95	45.50
1022	4" diameter		34	.471		15.65	20		35.65	49.50
1024	5" diameter		32	.500		14.35	21		35.35	50
1026	6" diameter		30	.533		21.50	22.50		44	59.50
1028	7" diameter		28	.571		25	24.50		49.50	66
1030	8" diameter		26	.615		26	26		52	70
1040	Roof flashing, 3" diameter		36	.444		24.50	18.90		43.40	57
1050	4" diameter		34	.471		28	20		48	63
1060	5" diameter		32	.500		29.50	21		50.50	66.50
1070	6" diameter		30	.533		31.50	22.50		54	70.50
1080	7" diameter		28	.571		53	24.50		77.50	96.50
1090	8" diameter		26	.615		54	26		80	101
1100	10" diameter		24	.667		83.50	28.50		112	137
1110	12" diameter		22	.727		106	31		137	166
1120	14" diameter		20	.800		124	34		158	191
1130	16" diameter		18	.889		154	38		192	229
1140	18" diameter		16	1		208	42.50		250.50	297
1150	20" diameter	Q-10	18	1.333		270	58.50		328.50	390

23 51 23.10 Gas Chimney Vents		Crew	Daily Output	Labor-Hours	Unit	Material	2009 Bare Costs Labor	Equipment	Total	Total Incl O&P
1160	22" diameter	Q-10	14	1.714	Ea.	340	75.50		415.50	495
1170	24" diameter		12	2		410	88		498	590
1200	Tee, 3" diameter	Q-9	27	.593		33	25		58	76
1210	4" diameter		26	.615		34	26		60	79
1220	5" diameter		25	.640		37	27		64	84
1230	6" diameter		24	.667		40	28.50		68.50	89
1240	7" diameter		23	.696		44	29.50		73.50	95.50
1250	8" diameter		22	.727		60.50	31		91.50	116
1260	10" diameter		21	.762		176	32.50		208.50	246
1270	12" diameter		20	.800		212	34		246	287
1280	14" diameter		18	.889		345	38		383	440
1290	16" diameter		16	1		490	42.50		532.50	610
1300	18" diameter		14	1.143		610	48.50		658.50	750
1310	20" diameter	Q-10	17	1.412		830	62		892	1,025
1320	22" diameter		13	1.846		1,025	81.50		1,106.50	1,250
1330	24" diameter		12	2		1,200	88		1,288	1,475
1340	26" diameter		12	2		1,725	88		1,813	2,050
1350	28" diameter		11	2.182		1,925	96		2,021	2,275
1360	30" diameter		10	2.400		2,100	106		2,206	2,475
1370	32" diameter		10	2.400		2,425	106		2,531	2,850
1380	34" diameter		9	2.667		2,450	117		2,567	2,875
1390	36" diameter		9	2.667		2,925	117		3,042	3,400
1400	38" diameter		8	3		2,950	132		3,082	3,425
1410	40" diameter		7	3.429		3,225	151		3,376	3,800
1420	42" diameter		6	4		3,500	176		3,676	4,125
1430	44" diameter		5	4.800		3,800	211		4,011	4,525
1440	46" diameter		5	4.800		4,175	211		4,386	4,925
1450	48" diameter		5	4.800		4,450	211		4,661	5,225
1460	Tee cap, 3" diameter	Q-9	45	.356		2.34	15.10		17.44	26.50
1470	4" diameter		42	.381		2.63	16.20		18.83	28.50
1490	6" diameter		37	.432		4.06	18.35		22.41	33.50
1510	8" diameter		34	.471		7.75	20		27.75	40.50
1530	12" diameter		30	.533		21.50	22.50		44	59.50
1550	16" diameter		25	.640		51	27		78	99
1570	20" diameter	Q-10	27	.889		61.50	39		100.50	130
1590	24" diameter		21	1.143		79.50	50.50		130	168
1610	28" diameter		19	1.263		325	55.50		380.50	445
1630	32" diameter		17	1.412		435	62		497	580
1650	36" diameter		15	1.600		450	70.50		520.50	605
1670	40" diameter		14	1.714		470	75.50		545.50	640
1690	44" diameter		12	2		490	88		578	680
1710	48" diameter		11	2.182		505	96		601	710
1750	Top, 3" diameter	Q-9	46	.348		12.30	14.80		27.10	37
1760	4" diameter		44	.364		16.20	15.45		31.65	42.50
1780	6" diameter		40	.400		22.50	17		39.50	51.50
1800	8" diameter		36	.444		43.50	18.90		62.40	77.50
1820	12" diameter		32	.500		165	21		186	215
1840	16" diameter		28	.571		340	24.50		364.50	415
1860	20" diameter	Q-10	28	.857		665	38		703	790
1880	24" diameter		20	1.200		1,125	53		1,178	1,325
1882	26" diameter		19	1.263		1,700	55.50		1,755.50	1,975
1884	28" diameter		18	1.333		1,950	58.50		2,008.50	2,225
1886	30" diameter		17	1.412		2,175	62		2,237	2,500

23 51 23 – Gas Vents

23 51 23.10 Gas Chimney Vents		Crew	Daily Output	Labor-Hours	Unit	2009 Bare Costs Material	Labor	Equipment	Total	Total Incl O&P
1888	32" diameter	Q-10	16	1.500	Ea.	2,300	66		2,366	2,625
1890	34" diameter		15	1.600		2,450	70.50		2,520.50	2,800
1892	36" diameter	↓	14	1.714	↓	2,625	75.50		2,700.50	3,025

23 51 26 – All-Fuel Vent Chimneys

23 51 26.10 All-Fuel Vent Chimneys, Press. Tight, Dbl. Wall		Crew	Daily Output	Labor-Hours	Unit	2009 Bare Costs Material	Labor	Equipment	Total	Total Incl O&P
0010	**ALL-FUEL VENT CHIMNEYS, PRESSURE TIGHT, DOUBLE WALL**									
3200	All fuel, pressure tight, double wall, U.L. listed, 1400° F.									
3210	304 stainless steel liner, aluminized steel outer jacket									
3220	6" diameter	Q-9	60	.267	L.F.	47	11.35		58.35	70
3221	8" diameter		52	.308		58	13.05		71.05	84.50
3222	10" diameter		48	.333		64.50	14.15		78.65	93.50
3223	12" diameter		44	.364		73.50	15.45		88.95	105
3224	14" diameter		42	.381		81	16.20		97.20	115
3225	16" diameter		40	.400		89.50	17		106.50	126
3226	18" diameter	↓	38	.421		101	17.90		118.90	140
3227	20" diameter	Q-10	36	.667		115	29.50		144.50	174
3228	24" diameter		32	.750		148	33		181	216
3229	28" diameter		30	.800		170	35		205	243
3230	32" diameter		27	.889		191	39		230	272
3231	36" diameter		25	.960		224	42.50		266.50	315
3232	42" diameter		22	1.091		258	48		306	360
3233	48" diameter	↓	19	1.263		300	55.50		355.50	420
3260	For 316 stainless steel liner add				↓	6%				
3280	All fuel, pressure tight, double wall fittings									
3284	304 stainless steel inner, aluminized steel jacket									
3288	Adjustable 18"/30" section									
3291	5" diameter	Q-9	32	.500	Ea.	121	21		142	167
3292	6" diameter		30	.533		153	22.50		175.50	204
3293	8" diameter		26	.615		176	26		202	236
3294	10" diameter		24	.667		198	28.50		226.50	263
3295	12" diameter		22	.727		225	31		256	297
3296	14" diameter		21	.762		260	32.50		292.50	340
3297	16" diameter		20	.800		270	34		304	350
3298	18" diameter	↓	19	.842		335	36		371	420
3299	20" diameter	Q-10	18	1.333		385	58.50		443.50	520
3300	24" diameter		16	1.500		495	66		561	650
3301	28" diameter		15	1.600		555	70.50		625.50	720
3302	32" diameter		14	1.714		650	75.50		725.50	835
3303	36" diameter		12	2		680	88		768	890
3304	42" diameter		11	2.182		745	96		841	975
3305	48" diameter	↓	10	2.400		875	106		981	1,125
3326	For 316 stainless steel liner, add				↓	15%				
3350	Elbow 90° fixed									
3353	5" diameter	Q-9	32	.500	Ea.	227	21		248	283
3354	6" diameter		30	.533		266	22.50		288.50	330
3355	8" diameter		26	.615		294	26		320	365
3356	10" diameter		24	.667		335	28.50		363.50	415
3357	12" diameter		22	.727		380	31		411	470
3358	14" diameter		21	.762		445	32.50		477.50	540
3359	16" diameter		20	.800		500	34		534	605
3360	18" diameter	↓	19	.842		580	36		616	695
3361	20" diameter	Q-10	18	1.333	↓	650	58.50		708.50	810

23 51 26.10 All-Fuel Vent Chimneys, Press. Tight, Dbl. Wall		Crew	Daily Output	Labor-Hours	Unit	2009 Bare Costs Material	Labor	Equipment	Total	Total Incl O&P
3362	24" diameter	Q-10	16	1.500	Ea.	805	66		871	990
3363	28" diameter		15	1.600		1,000	70.50		1,070.50	1,200
3364	32" diameter		14	1.714		1,350	75.50		1,425.50	1,600
3365	36" diameter		12	2		1,575	88		1,663	1,900
3366	42" diameter		11	2.182		2,400	96		2,496	2,800
3367	48" diameter		10	2.400		3,025	106		3,131	3,500
3380	For 316 stainless steel liner, add					20%				
3400	Elbow 45°									
3404	6" diameter	Q-9	30	.533	Ea.	168	22.50		190.50	220
3405	8" diameter		26	.615		190	26		216	251
3406	10" diameter		24	.667		213	28.50		241.50	280
3407	12" diameter		22	.727		243	31		274	315
3408	14" diameter		21	.762		280	32.50		312.50	360
3409	16" diameter		20	.800		315	34		349	400
3410	18" diameter		19	.842		355	36		391	445
3411	20" diameter	Q-10	18	1.333		410	58.50		468.50	545
3412	24" diameter		16	1.500		525	66		591	680
3413	28" diameter		15	1.600		595	70.50		665.50	765
3414	32" diameter		14	1.714		795	75.50		870.50	995
3415	36" diameter		12	2		1,125	88		1,213	1,375
3416	42" diameter		11	2.182		1,525	96		1,621	1,825
3417	48" diameter		10	2.400		2,050	106		2,156	2,425
3430	For 316 stainless steel liner, add					20%				
3450	Tee 90°									
3454	6" diameter	Q-9	24	.667	Ea.	175	28.50		203.50	238
3455	8" diameter		22	.727		202	31		233	271
3456	10" diameter		21	.762		236	32.50		268.50	310
3457	12" diameter		20	.800		275	34		309	355
3458	14" diameter		18	.889		325	38		363	420
3459	16" diameter		16	1		365	42.50		407.50	475
3460	18" diameter		14	1.143		430	48.50		478.50	550
3461	20" diameter	Q-10	17	1.412		485	62		547	635
3462	24" diameter		12	2		600	88		688	795
3463	28" diameter		11	2.182		745	96		841	975
3464	32" diameter		10	2.400		1,025	106		1,131	1,300
3465	36" diameter		9	2.667		1,200	117		1,317	1,500
3466	42" diameter		6	4		1,900	176		2,076	2,375
3467	48" diameter		5	4.800		2,425	211		2,636	3,000
3480	For Tee Cap, add					35%	20%			
3500	For 316 stainless steel liner, add					22%				
3520	Plate support, galvanized									
3524	6" diameter	Q-9	26	.615	Ea.	91	26		117	142
3525	8" diameter		22	.727		103	31		134	163
3526	10" diameter		20	.800		112	34		146	178
3527	12" diameter		18	.889		117	38		155	189
3528	14" diameter		17	.941		143	40		183	221
3529	16" diameter		16	1		150	42.50		192.50	233
3530	18" diameter		15	1.067		158	45.50		203.50	246
3531	20" diameter	Q-10	16	1.500		168	66		234	289
3532	24" diameter		14	1.714		176	75.50		251.50	315
3533	28" diameter		13	1.846		216	81.50		297.50	365
3534	32" diameter		12	2		281	88		369	450
3535	36" diameter		10	2.400		350	106		456	555

23 51 26.10 All-Fuel Vent Chimneys, Press. Tight, Dbl. Wall		Crew	Daily Output	Labor-Hours	Unit	Material	Labor	Equipment	Total	Total Incl O&P
						2009 Bare Costs				
3536	42" diameter	Q-10	9	2.667	Ea.	620	117		737	865
3537	48" diameter	↓	8	3	↓	745	132		877	1,025
3570	Bellows, lined, 316 stainless steel only									
3574	6" diameter	Q-9	30	.533	Ea.	440	22.50		462.50	515
3575	8" diameter		26	.615		475	26		501	565
3576	10" diameter		24	.667		525	28.50		553.50	620
3577	12" diameter		22	.727		580	31		611	690
3578	14" diameter		21	.762		655	32.50		687.50	770
3579	16" diameter		20	.800		725	34		759	850
3580	18" diameter	↓	19	.842		805	36		841	945
3581	20" diameter	Q-10	18	1.333		910	58.50		968.50	1,100
3582	24" diameter		16	1.500		1,150	66		1,216	1,350
3583	28" diameter		15	1.600		1,275	70.50		1,345.50	1,500
3584	32" diameter		14	1.714		1,475	75.50		1,550.50	1,750
3585	36" diameter		12	2		1,700	88		1,788	2,025
3586	42" diameter		11	2.182		2,525	96		2,621	2,950
3587	48" diameter	↓	10	2.400	↓	2,625	106		2,731	3,075
3600	Ventilated roof thimble, 304 stainless steel									
3620	6" diameter	Q-9	26	.615	Ea.	225	26		251	289
3624	8" diameter		22	.727		241	31		272	315
3625	10" diameter		20	.800		260	34		294	340
3626	12" diameter		18	.889		281	38		319	370
3627	14" diameter		17	.941		300	40		340	395
3628	16" diameter		16	1		320	42.50		362.50	425
3629	18" diameter	↓	15	1.067		345	45.50		390.50	450
3630	20" diameter	Q-10	16	1.500		370	66		436	515
3631	24" diameter		14	1.714		420	75.50		495.50	580
3632	28" diameter		13	1.846		480	81.50		561.50	660
3633	32" diameter		12	2		540	88		628	735
3634	36" diameter		10	2.400		620	106		726	850
3635	42" diameter		9	2.667		820	117		937	1,100
3636	48" diameter	↓	8	3	↓	1,025	132		1,157	1,350
3650	For 316 stainless steel, add					10%				
3670	Exit cone, 316 stainless steel only									
3674	6" diameter	Q-9	46	.348	Ea.	178	14.80		192.80	220
3675	8" diameter		42	.381		186	16.20		202.20	231
3676	10" diameter		40	.400		202	17		219	249
3677	12" diameter		38	.421		211	17.90		228.90	261
3678	14" diameter		37	.432		234	18.35		252.35	287
3679	16" diameter		36	.444		258	18.90		276.90	315
3680	18" diameter	↓	35	.457		315	19.40		334.40	380
3681	20" diameter	Q-10	28	.857		380	38		418	475
3682	24" diameter		26	.923		485	40.50		525.50	600
3683	28" diameter		25	.960		590	42.50		632.50	715
3684	32" diameter		24	1		735	44		779	875
3685	36" diameter		22	1.091		880	48		928	1,050
3686	42" diameter		21	1.143		1,125	50.50		1,175.50	1,325
3687	48" diameter	↓	20	1.200	↓	1,450	53		1,503	1,675
3720	Roof support assembly, incl. 30" pipe sect, 304 stainless steel									
3724	6" diameter	Q-9	25	.640	Ea.	259	27		286	330
3725	8" diameter		21	.762		282	32.50		314.50	360
3726	10" diameter		19	.842		310	36		346	395
3727	12" diameter	↓	17	.941	↓	330	40		370	425

23 51 26.10 All-Fuel Vent Chimneys, Press. Tight, Dbl. Wall		Crew	Daily Output	Labor-Hours	Unit	2009 Bare Costs Material	Labor	Equipment	Total	Total Incl O&P
3728	14" diameter	Q-9	16	1	Ea.	370	42.50		412.50	475
3729	16" diameter		15	1.067		390	45.50		435.50	500
3730	18" diameter		14	1.143		420	48.50		468.50	540
3731	20" diameter	Q-10	15	1.600		450	70.50		520.50	605
3732	24" diameter		13	1.846		500	81.50		581.50	680
3733	28" diameter		12	2		590	88		678	785
3734	32" diameter		11	2.182		705	96		801	930
3735	36" diameter		9	2.667		855	117		972	1,125
3736	42" diameter		8	3		1,350	132		1,482	1,675
3737	48" diameter		7	3.429		1,650	151		1,801	2,075
3750	For 316 stainless steel, add					20%				
3770	Stack cap, 316 stainless steel only									
3773	5" diameter	Q-9	48	.333	Ea.	169	14.15		183.15	209
3774	6" diameter		46	.348		198	14.80		212.80	242
3775	8" diameter		42	.381		229	16.20		245.20	277
3776	10" diameter		40	.400		270	17		287	325
3777	12" diameter		38	.421		315	17.90		332.90	375
3778	14" diameter		37	.432		370	18.35		388.35	435
3779	16" diameter		36	.444		420	18.90		438.90	495
3780	18" diameter		35	.457		475	19.40		494.40	550
3781	20" diameter	Q-10	28	.857		540	38		578	655
3782	24" diameter		26	.923		655	40.50		695.50	785
3783	28" diameter		25	.960		710	42.50		752.50	850
3784	32" diameter		24	1		820	44		864	970
3785	36" diameter		22	1.091		1,000	48		1,048	1,175
3786	42" diameter		21	1.143		1,225	50.50		1,275.50	1,425
3787	48" diameter		20	1.200		1,450	53		1,503	1,675
3800	Round storm collar									
3806	5" diameter	Q-9	32	.500	Ea.	26.50	21		47.50	63

23 51 26.30 All-Fuel Vent Chimneys, Double Wall, St. Stl.		Crew	Daily Output	Labor-Hours	Unit	Material	Labor	Equipment	Total	Total Incl O&P
0010	**ALL-FUEL VENT CHIMNEYS, DOUBLE WALL, STAINLESS STEEL**									
7800	All fuel, double wall, stainless steel, 6" diameter	Q-9	60	.267	V.L.F.	56.50	11.35		67.85	80
7802	7" diameter		56	.286		73	12.15		85.15	100
7804	8" diameter		52	.308		86.50	13.05		99.55	116
7806	10" diameter		48	.333		125	14.15		139.15	160
7808	12" diameter		44	.364		167	15.45		182.45	209
7810	14" diameter		42	.381		219	16.20		235.20	267
8000	All fuel, double wall, stainless steel fittings									
8010	Roof support 6" diameter	Q-9	30	.533	Ea.	103	22.50		125.50	150
8020	7" diameter		28	.571		117	24.50		141.50	167
8030	8" diameter		26	.615		130	26		156	185
8040	10" diameter		24	.667		150	28.50		178.50	210
8050	12" diameter		22	.727		181	31		212	248
8060	14" diameter		21	.762		229	32.50		261.50	305
8100	Elbow 15°, 6" diameter		30	.533		165	22.50		187.50	218
8120	7" diameter		28	.571		186	24.50		210.50	244
8140	8" diameter		26	.615		212	26		238	275
8160	10" diameter		24	.667		278	28.50		306.50	350
8180	12" diameter		22	.727		335	31		366	420
8200	14" diameter		21	.762		405	32.50		437.50	495
8300	Insulated tee with insulated tee cap, 6" diameter		30	.533		156	22.50		178.50	208
8340	7" diameter		28	.571		205	24.50		229.50	264

23 51 Breechings, Chimneys, and Stacks

23 51 26 – All-Fuel Vent Chimneys

23 51 26.30 All-Fuel Vent Chimneys, Double Wall, St. Stl.		Crew	Daily Output	Labor-Hours	Unit	Material	2009 Bare Costs Labor	Equipment	Total	Total Incl O&P
8360	8″ diameter	Q-9	26	.615	Ea.	231	26		257	296
8380	10″ diameter		24	.667		325	28.50		353.50	405
8400	12″ diameter		22	.727		455	31		486	550
8420	14″ diameter		21	.762		600	32.50		632.50	710
8500	Joist shield, 6″ diameter		30	.533		47.50	22.50		70	88.50
8510	7″ diameter		28	.571		51.50	24.50		76	95.50
8520	8″ diameter		26	.615		63.50	26		89.50	112
8530	10″ diameter		24	.667		86	28.50		114.50	140
8540	12″ diameter		22	.727		107	31		138	167
8550	14″ diameter		21	.762		133	32.50		165.50	198
8600	Round top, 6″ diameter		30	.533		53	22.50		75.50	94
8620	7″ diameter		28	.571		72	24.50		96.50	118
8640	8″ diameter		26	.615		97	26		123	149
8660	10″ diameter		24	.667		180	28.50		208.50	243
8680	12″ diameter		22	.727		249	31		280	320
8700	14″ diameter		21	.762		330	32.50		362.50	415
8800	Adjustable roof flashing, 6″ diameter		30	.533		63	22.50		85.50	105
8820	7″ diameter		28	.571		72	24.50		96.50	118
8840	8″ diameter		26	.615		78	26		104	128
8860	10″ diameter		24	.667		100	28.50		128.50	155
8880	12″ diameter		22	.727		130	31		161	191
8900	14″ diameter		21	.762		162	32.50		194.50	230

23 51 33 – Insulated Sectional Chimneys

23 51 33.10 Prefabricated Insulated Sectional Chimneys		Crew	Daily Output	Labor-Hours	Unit	Material	Labor	Equipment	Total	Total Incl O&P
0010	**PREFABRICATED INSULATED SECTIONAL CHIMNEYS**									
9000	High temp. (2000° F), steel jacket, acid resistant refractory lining									
9010	11 ga. galvanized jacket, U.L. listed									
9020	Straight section, 48″ long, 10″ diameter	Q-10	13.30	1.805	Ea.	400	79.50		479.50	565
9030	12″ diameter		11.20	2.143		440	94.50		534.50	630
9040	18″ diameter		7.40	3.243		615	143		758	900
9050	24″ diameter		4.60	5.217		840	230		1,070	1,300
9060	30″ diameter		3.70	6.486		1,075	286		1,361	1,625
9070	36″ diameter		2.70	8.889		1,325	390		1,715	2,075
9080	42″ diameter	Q-11	3.10	10.323		1,625	465		2,090	2,550
9090	48″ diameter		2.70	11.852		1,925	535		2,460	2,950
9100	54″ diameter		2.20	14.545		2,325	655		2,980	3,600
9110	60″ diameter		2	16		3,050	720		3,770	4,500
9120	Tee section, 10″ diameter	Q-10	4.40	5.455		855	240		1,095	1,325
9130	12″ diameter		3.70	6.486		900	286		1,186	1,450
9140	18″ diameter		2.40	10		1,225	440		1,665	2,050
9150	24″ diameter		1.50	16		1,675	705		2,380	2,975
9160	30″ diameter		1.20	20		2,350	880		3,230	4,000
9170	36″ diameter		.80	30		3,975	1,325		5,300	6,475
9180	42″ diameter	Q-11	1	32		5,225	1,450		6,675	8,025
9190	48″ diameter		.90	35.556		6,775	1,600		8,375	10,000
9200	54″ diameter		.70	45.714		9,900	2,050		11,950	14,200
9210	60″ diameter		.60	53.333		12,500	2,400		14,900	17,600
9220	Cleanout pier section, 10″ diameter	Q-10	3.50	6.857		655	300		955	1,200
9230	12″ diameter		2.50	9.600		690	425		1,115	1,425
9240	18″ diameter		1.90	12.632		885	555		1,440	1,850
9250	24″ diameter		1.30	18.462		1,125	815		1,940	2,525
9260	30″ diameter		1	24		1,375	1,050		2,425	3,175

23 51 Breechings, Chimneys, and Stacks

23 51 33 – Insulated Sectional Chimneys

23 51 33.10 Prefabricated Insulated Sectional Chimneys		Crew	Daily Output	Labor-Hours	Unit	2009 Bare Costs Material	Labor	Equipment	Total	Total Incl O&P
9270	36" diameter	Q-10	.75	32	Ea.	1,625	1,400		3,025	4,025
9280	42" diameter	Q-11	.90	35.556		2,050	1,600		3,650	4,825
9290	48" diameter		.75	42.667		2,475	1,925		4,400	5,775
9300	54" diameter		.60	53.333		2,950	2,400		5,350	7,050
9310	60" diameter		.50	64		3,775	2,875		6,650	8,725
9320	For drain, add					53%				
9330	Elbow, 30° and 45°, 10" diameter	Q-10	6.60	3.636		555	160		715	865
9340	12" diameter		5.60	4.286		600	189		789	955
9350	18" diameter		3.70	6.486		900	286		1,186	1,450
9360	24" diameter		2.30	10.435		1,350	460		1,810	2,200
9370	30" diameter		1.85	12.973		1,850	570		2,420	2,925
9380	36" diameter		1.30	18.462		2,225	815		3,040	3,750
9390	42" diameter	Q-11	1.60	20		3,050	900		3,950	4,800
9400	48" diameter		1.35	23.704		3,900	1,075		4,975	5,975
9410	54" diameter		1.10	29.091		5,050	1,300		6,350	7,625
9420	60" diameter		1	32		6,150	1,450		7,600	9,050
9430	For 60° and 90° elbow, add					112%				
9440	End cap, 10" diameter	Q-10	26	.923		955	40.50		995.50	1,125
9450	12" diameter		22	1.091		1,100	48		1,148	1,275
9460	18" diameter		15	1.600		1,350	70.50		1,420.50	1,575
9470	24" diameter		9	2.667		1,650	117		1,767	2,000
9480	30" diameter		7.30	3.288		2,050	145		2,195	2,475
9490	36" diameter		5	4.800		2,350	211		2,561	2,925
9500	42" diameter	Q-11	6.30	5.079		2,950	228		3,178	3,600
9510	48" diameter		5.30	6.038		3,475	271		3,746	4,225
9520	54" diameter		4.40	7.273		3,975	325		4,300	4,875
9530	60" diameter		3.90	8.205		4,675	370		5,045	5,725
9540	Increaser (1 diameter), 10" diameter	Q-10	6.60	3.636		475	160		635	775
9550	12" diameter		5.60	4.286		505	189		694	855
9560	18" diameter		3.70	6.486		740	286		1,026	1,275
9570	24" diameter		2.30	10.435		1,050	460		1,510	1,900
9580	30" diameter		1.85	12.973		1,325	570		1,895	2,350
9590	36" diameter		1.30	18.462		1,450	815		2,265	2,900
9592	42" diameter	Q-11	1.60	20		1,600	900		2,500	3,175
9594	48" diameter		1.35	23.704		1,925	1,075		3,000	3,800
9596	54" diameter		1.10	29.091		2,325	1,300		3,625	4,625
9600	For expansion joints, add to straight section					7%				
9610	For 1/4" hot rolled steel jacket, add					157%				
9620	For 2950° F very high temperature, add					89%				
9630	26 ga. aluminized jacket, straight section, 48" long									
9640	10" diameter	Q-10	15.30	1.569	V.L.F.	237	69		306	370
9650	12" diameter		12.90	1.860		252	82		334	410
9660	18" diameter		8.50	2.824		365	124		489	605
9670	24" diameter		5.30	4.528		520	199		719	885
9680	30" diameter		4.30	5.581		665	246		911	1,125
9690	36" diameter		3.10	7.742		845	340		1,185	1,475
9700	Accessories (all models)									
9710	Guy band, 10" diameter	Q-10	32	.750	Ea.	74	33		107	134
9720	12" diameter		30	.800		78	35		113	142
9730	18" diameter		26	.923		102	40.50		142.50	177
9740	24" diameter		24	1		149	44		193	233
9750	30" diameter		20	1.200		173	53		226	274
9760	36" diameter		18	1.333		191	58.50		249.50	305

23 51 Breechings, Chimneys, and Stacks

23 51 33 – Insulated Sectional Chimneys

23 51 33.10 Prefabricated Insulated Sectional Chimneys		Crew	Daily Output	Labor-Hours	Unit	2009 Bare Costs Material	Labor	Equipment	Total	Total Incl O&P
9770	42″ diameter	Q-11	22	1.455	Ea.	345	65.50		410.50	485
9780	48″ diameter		20	1.600		380	72		452	530
9790	54″ diameter		16	2		420	90		510	605
9800	60″ diameter		12	2.667		455	120		575	690
9810	Draw band, galv. stl., 11 gauge, 10″ diameter	Q-10	32	.750		65	33		98	124
9820	12″ diameter		30	.800		69	35		104	132
9830	18″ diameter		26	.923		86	40.50		126.50	159
9840	24″ diameter		24	1		110	44		154	191
9850	30″ diameter		20	1.200		127	53		180	223
9860	36″ diameter		18	1.333		146	58.50		204.50	254
9870	42″ diameter	Q-11	22	1.455		196	65.50		261.50	320
9880	48″ diameter		20	1.600		251	72		323	390
9890	54″ diameter		16	2		291	90		381	465
9900	60″ diameter		12	2.667		370	120		490	595
9910	Draw band, aluminized stl., 26 gauge, 10″ diameter	Q-10	32	.750		18.95	33		51.95	73.50
9920	12″ diameter		30	.800		18.95	35		53.95	77
9930	18″ diameter		26	.923		24	40.50		64.50	91
9940	24″ diameter		24	1		28	44		72	101
9950	30″ diameter		20	1.200		32	53		85	119
9960	36″ diameter		18	1.333		43	58.50		101.50	141
9962	42″ diameter	Q-11	22	1.455		43	65.50		108.50	151
9964	48″ diameter		20	1.600		51	72		123	170
9966	54″ diameter		16	2		51	90		141	199
9968	60″ diameter		12	2.667		56	120		176	253
9990	Minimum labor/equipment charge	Q-9	3	5.333	Job		227		227	360

23 52 Heating Boilers

23 52 13 – Electric Boilers

23 52 13.10 Electric Boilers, ASME

		Crew	Daily Output	Labor-Hours	Unit	Material	Labor	Equipment	Total	Total Incl O&P
0010	**ELECTRIC BOILERS, ASME**, Standard controls and trim. R235000-50									
1000	Steam, 6 KW, 20.5 MBH	Q-19	1.20	20	Ea.	3,375	905		4,280	5,125
1040	9 KW, 30.7 MBH		1.20	20		3,525	905		4,430	5,275
1080	24 KW, 81.8 MBH		1.10	21.818		4,200	990		5,190	6,150
1160	60 KW, 205 MBH		1	24		5,800	1,075		6,875	8,050
1240	148 KW, 505 MBH		.65	36.923		8,375	1,675		10,050	11,800
1280	222 KW, 758 MBH		.55	43.636		13,100	1,975		15,075	17,500
1320	300 KW, 1023 MBH		.40	60		17,000	2,725		19,725	22,900
1360	444 KW, 1515 MBH		.30	80		19,500	3,625		23,125	27,100
1400	592 KW, 2020 MBH	Q-21	.34	94.118		25,000	4,350		29,350	34,300
1480	814 KW, 2778 MBH		.25	128		31,000	5,925		36,925	43,300
1520	1036 KW, 3536 MBH		.20	160		33,400	7,400		40,800	48,200
1560	2070 KW, 7063 MBH		.18	177		49,000	8,225		57,225	66,500
1600	2,340 KW, 7984 MBH		.16	200		63,500	9,250		72,750	84,000
2000	Hot water, 7.5 KW, 25.6 MBH	Q-19	1.30	18.462		3,575	835		4,410	5,250
2040	30 KW, 102 MBH		1.20	20		3,825	905		4,730	5,600
2070	60 KW, 205 MBH		1.20	20		4,525	905		5,430	6,375
2100	90 KW, 307 MBH		1.10	21.818		5,225	990		6,215	7,275
2140	120 KW, 410 MBH		.90	26.667		5,875	1,200		7,075	8,325
2180	150 KW, 512 MBH		.65	36.923		6,625	1,675		8,300	9,850
2220	296 KW, 1010 MBH		.55	43.636		11,600	1,975		13,575	15,800
2300	444 KW, 1515 MBH		.35	68.571		15,500	3,100		18,600	21,900

23 52 Heating Boilers

23 52 13 – Electric Boilers

23 52 13.10 Electric Boilers, ASME			Crew	Daily Output	Labor-Hours	Unit	Material	2009 Bare Costs Labor	Equipment	Total	Total Incl O&P
2340	518 KW, 1768 MBH		Q-21	.44	72.727	Ea.	17,200	3,375		20,575	24,100
2420	740 KW, 2526 MBH			.39	82.051		20,800	3,800		24,600	28,700
2460	888 KW, 3031 MBH			.37	86.486		22,600	4,000		26,600	31,000
2500	1,036 KW, 3536 MBH			.34	94.118		26,900	4,350		31,250	36,400
2540	1440 KW, 4915 MBH			.32	100		34,200	4,625		38,825	44,800
2580	1680 KW, 5733 MBH			.30	106		41,000	4,950		45,950	53,000
2620	1980 KW, 6757 MBH			.28	114		46,100	5,300		51,400	58,500
2660	2220 KW, 7576 MBH			.26	123		53,000	5,700		58,700	67,500
2700	2610 KW, 8905 MBH			.24	133		60,500	6,175		66,675	76,000
2740	2970 KW, 10133 MBH			.21	152		64,500	7,050		71,550	82,000
2780	3240 KW, 11055 MBH			.18	177		72,000	8,225		80,225	91,500
2820	3600 KW, 12,283 MBH			.16	200		79,000	9,250		88,250	101,500
9000	Minimum labor/equipment charge		Q-20	1	20	Job		870		870	1,375

23 52 16 – Condensing Boilers

23 52 16.24 Condensing Boilers			Crew	Daily Output	Labor-Hours	Unit	Material	Labor	Equipment	Total	Total Incl O&P
0010	**CONDENSING BOILERS**, Cast iron										
0020	Packaged with standard controls, circulator and trim										
0030	Intermittent (spark) pilot, natural or LP gas										
0040	Hot water, DOE MBH output, (AFUE)										
0100	42 MBH, (84.0%)	G	Q-5	1.80	8.889	Ea.	1,300	395		1,695	2,050
0120	57 MBH, (84.3%)	G		1.60	10		1,450	445		1,895	2,300
0140	85 MBH, (84.0%)	G		1.40	11.429		1,600	510		2,110	2,525
0160	112 MBH, (83.7%)	G		1.20	13.333		1,800	590		2,390	2,900
0180	140 MBH, (83.3%)	G	Q-6	1.60	15		2,075	690		2,765	3,350
0200	167 MBH, (83.0%)	G		1.40	17.143		2,325	790		3,115	3,800
0220	194 MBH, (82.7%)	G		1.20	20		2,550	920		3,470	4,225

23 52 19 – Pulse Combustion Boilers

23 52 19.20 Pulse Type Combustion Boilers			Crew	Daily Output	Labor-Hours	Unit	Material	Labor	Equipment	Total	Total Incl O&P
0010	**PULSE TYPE COMBUSTION BOILERS**										
7990	Special feature gas fired boilers										
8000	Pulse combustion, standard controls / trim										
8050	88,000 BTU	G	Q-5	1.40	11.429	Ea.	4,775	510		5,285	6,025
8080	134,000 BTU	G	"	1.20	13.333	"	5,700	590		6,290	7,200

23 52 23 – Cast-Iron Boilers

23 52 23.20 Gas-Fired Boilers			Crew	Daily Output	Labor-Hours	Unit	Material	Labor	Equipment	Total	Total Incl O&P
0010	**GAS-FIRED BOILERS,** Natural or propane, standard controls, packaged.										
1000	Cast iron, with insulated jacket										
2000	Steam, gross output, 81 MBH		Q-7	1.40	22.857	Ea.	1,850	1,075		2,925	3,725
2020	102 MBH			1.30	24.615		2,100	1,150		3,250	4,100
2040	122 MBH			1	32		2,250	1,500		3,750	4,800
2060	163 MBH			.90	35.556		2,700	1,675		4,375	5,575
2080	203 MBH	R235000-50		.90	35.556		2,950	1,675		4,625	5,850
2100	240 MBH			.85	37.647		2,950	1,775		4,725	6,000
2120	280 MBH			.80	40		3,950	1,875		5,825	7,250
2140	320 MBH			.70	45.714		4,200	2,150		6,350	7,925
2160	360 MBH			.63	51.200		4,675	2,400		7,075	8,875
2180	400 MBH			.56	56.838		4,950	2,675		7,625	9,600
2200	440 MBH			.51	62.500		5,275	2,950		8,225	10,400
2220	544 MBH			.45	71.588		8,125	3,375		11,500	14,200
2240	765 MBH			.43	74.419		9,825	3,500		13,325	16,200
2260	892 MBH			.38	84.211		11,400	3,950		15,350	18,700

23 52 23.20 Gas-Fired Boilers		Crew	Daily Output	Labor-Hours	Unit	Material	Labor	Equipment	Total	Total Incl O&P
						2009 Bare Costs				
2280	1275 MBH	Q-7	.34	94.118	Ea.	13,900	4,425		18,325	22,200
2300	1530 MBH		.32	100		15,200	4,700		19,900	24,000
2320	1,875 MBH		.30	106		17,800	5,025		22,825	27,300
2340	2170 MBH		.26	122		19,000	5,750		24,750	29,800
2360	2675 MBH		.20	163		21,800	7,675		29,475	35,900
2380	3060 MBH		.19	172		23,100	8,100		31,200	38,000
2400	3570 MBH		.18	181		25,600	8,550		34,150	41,500
2420	4207 MBH		.16	205		28,000	9,650		37,650	45,700
2440	4,720 MBH		.15	207		29,600	9,775		39,375	47,700
2460	5660 MBH		.15	220		65,000	10,400		75,400	87,500
2480	6,100 MBH		.13	246		74,000	11,600		85,600	99,500
2500	6390 MBH		.12	266		78,000	12,500		90,500	105,500
2520	6680 MBH		.11	290		79,500	13,700		93,200	108,500
2540	6,970 MBH		.10	320		83,000	15,000		98,000	115,000
3000	Hot water, gross output, 80 MBH		1.46	21.918		1,700	1,025		2,725	3,475
3020	100 MBH		1.35	23.704		1,950	1,125		3,075	3,875
3040	122 MBH		1.10	29.091		2,100	1,375		3,475	4,450
3060	163 MBH		1	32		2,550	1,500		4,050	5,125
3080	203 MBH		1	32		2,800	1,500		4,300	5,425
3100	240 MBH		.95	33.684		2,950	1,575		4,525	5,700
3120	280 MBH		.90	35.556		3,700	1,675		5,375	6,675
3140	320 MBH		.80	40		4,000	1,875		5,875	7,325
3160	360 MBH		.71	45.070		4,475	2,125		6,600	8,200
3180	400 MBH		.64	50		4,775	2,350		7,125	8,900
3200	440 MBH		.58	54.983		5,100	2,575		7,675	9,600
3220	544 MBH		.51	62.992		7,875	2,950		10,825	13,300
3240	765 MBH		.46	70.022		9,525	3,300		12,825	15,600
3260	1,088 MBH		.40	80		11,900	3,750		15,650	18,900
3280	1,275 MBH		.36	89.888		13,600	4,225		17,825	21,500
3300	1,530 MBH		.31	104		14,900	4,925		19,825	24,100
3320	2,000 MBH		.26	125		21,000	5,875		26,875	32,200
3340	2,312 MBH		.22	148		21,000	6,975		27,975	33,900
3360	2,856 MBH		.20	160		23,100	7,525		30,625	37,200
3380	3,264 MBH		.18	179		24,300	8,450		32,750	39,900
3400	3,808 MBH		.16	195		26,800	9,175		35,975	43,700
3420	4,488 MBH		.15	210		29,100	9,900		39,000	47,300
3440	4,720 MBH		.15	220		62,500	10,400		72,900	84,500
3460	5,520 MBH		.14	228		80,000	10,700		90,700	104,500
3480	6,100 MBH		.13	250		82,000	11,800		93,800	108,000
3500	6,390 MBH		.11	285		86,500	13,400		99,900	116,500
3520	6,680 MBH		.10	310		86,500	14,600		101,100	118,000
3540	6,970 MBH		.09	359		89,500	16,900		106,400	124,500
7000	For tankless water heater, add					10%				
7050	For additional zone valves up to 312 MBH add					145			145	159
9900	Minimum labor/equipment charge	Q-6	1	24	Job		1,100		1,100	1,725

23 52 23.30 Gas/Oil Fired Boilers		Crew	Daily Output	Labor-Hours	Unit	Material	Labor	Equipment	Total	Total Incl O&P
0010	**GAS/OIL FIRED BOILERS,** Combination with burners and controls, packaged.									
1000	Cast iron with insulated jacket R235000-50									
2000	Steam, gross output, 720 MBH	Q-7	.43	74.074	Ea.	13,400	3,475		16,875	20,200
2020	810 MBH		.38	83.990		13,400	3,950		17,350	20,900
2040	1,084 MBH		.34	93.023		14,800	4,375		19,175	23,000
2060	1,360 MBH		.33	98.160		16,400	4,625		21,025	25,300

23 52 23.30 Gas/Oil Fired Boilers

23 52 23.30 Gas/Oil Fired Boilers		Crew	Daily Output	Labor-Hours	Unit	2009 Bare Costs Material	Labor	Equipment	Total	Total Incl O&P
2080	1,600 MBH	Q-7	.30	107	Ea.	17,800	5,025		22,825	27,400
2100	2,040 MBH		.25	130		23,100	6,125		29,225	34,900
2120	2,450 MBH		.21	156		25,300	7,350		32,650	39,300
2140	2,700 MBH		.19	165		26,600	7,800		34,400	41,400
2160	3,000 MBH		.18	175		28,100	8,275		36,375	43,700
2180	3,270 MBH		.17	183		29,500	8,650		38,150	45,800
2200	3,770 MBH		.17	191		60,500	9,000		69,500	80,500
2220	4,070 MBH		.16	200		63,500	9,400		72,900	84,500
2240	4,650 MBH		.15	210		67,000	9,900		76,900	89,500
2260	5,230 MBH		.14	223		74,500	10,500		85,000	98,500
2280	5,520 MBH		.14	235		81,000	11,100		92,100	106,000
2300	5,810 MBH		.13	248		82,000	11,700		93,700	108,000
2320	6,100 MBH		.12	260		82,500	12,200		94,700	110,000
2340	6,390 MBH		.11	296		87,000	13,900		100,900	117,000
2360	6,680 MBH		.10	320		91,500	15,000		106,500	124,000
2380	6,970 MBH		.09	372		92,000	17,500		109,500	128,500
2900	Hot water, gross output									
2910	200 MBH	Q-6	.62	39.024	Ea.	7,750	1,800		9,550	11,300
2920	300 MBH		.49	49.080		7,750	2,250		10,000	12,000
2930	400 MBH		.41	57.971		9,075	2,675		11,750	14,100
2940	500 MBH		.36	67.039		9,775	3,100		12,875	15,600
3000	584 MBH	Q-7	44	72.072		13,200	3,400		16,600	19,800
3020	876 MBH		41	79.012		14,300	3,725		18,025	21,500
3040	1,168 MBH		31	103		16,000	4,875		20,875	25,300
3060	1,460 MBH		28	113		18,700	5,325		24,025	28,800
3080	2,044 MBH		26	122		24,700	5,750		30,450	36,000
3100	2,628 MBH		21	150		28,900	7,075		35,975	42,800
3120	3,210 MBH		18	174		35,700	8,225		43,925	52,000
3140	3,796 MBH		17	186		39,900	8,750		48,650	57,500
3160	4,088 MBH		16	195		45,400	9,175		54,575	64,000
3180	4,672 MBH		16	203		49,500	9,575		59,075	69,500
3200	5,256 MBH		15	217		54,500	10,200		64,700	76,000
3220	6,000 MBH, 179 BHP		13	256		79,500	12,000		91,500	106,000
3240	7,130 MBH, 213 BHP		08	385		83,000	18,100		101,100	119,500
3260	9,800 MBH, 286 BHP		06	533		95,500	25,100		120,600	144,000
3280	10,900 MBH, 325.6 BHP		05	592		116,000	27,900		143,900	170,500
3290	12,200 MBH, 364.5 BHP		05	666		126,500	31,300		157,800	188,000
3300	13,500 MBH, 403.3 BHP		04	727		140,000	34,200		174,200	207,000

23 52 23.40 Oil-Fired Boilers

23 52 23.40 Oil-Fired Boilers		Crew	Daily Output	Labor-Hours	Unit	Material	Labor	Equipment	Total	Total Incl O&P
0010	**OIL-FIRED BOILERS,** Stand. cont., flame retention burner, packaged R235000-50									
1000	Cast iron, with insulated flush jacket									
2000	Steam, gross output, 109 MBH	Q-7	1.20	26.667	Ea.	1,875	1,250		3,125	4,025
2020	144 MBH		1.10	29.091		2,125	1,375		3,500	4,450
2040	173 MBH		1	32		2,375	1,500		3,875	4,925
2060	207 MBH		.90	35.556		2,575	1,675		4,250	5,425
2080	236 MBH		.85	37.647		3,050	1,775		4,825	6,125
2100	300 MBH		.70	45.714		3,650	2,150		5,800	7,350
2120	480 MBH		.50	64		4,675	3,000		7,675	9,825
2140	665 MBH		.45	71.111		6,400	3,350		9,750	12,200
2160	794 MBH		.41	78.049		7,575	3,675		11,250	14,000
2180	1,084 MBH		.38	85.106		8,725	4,000		12,725	15,800
2200	1,360 MBH		.33	98.160		10,100	4,625		14,725	18,300

23 52 23 – Cast-Iron Boilers

23 52 23.40 Oil-Fired Boilers		Crew	Daily Output	Labor-Hours	Unit	2009 Bare Costs Material	Labor	Equipment	Total	Total Incl O&P
2220	1,600 MBH	Q-7	.26	122	Ea.	11,500	5,750		17,250	21,500
2240	2,175 MBH		.24	133		14,700	6,300		21,000	25,900
2260	2,480 MBH		.21	156		17,700	7,350		25,050	30,900
2280	3,000 MBH		.19	170		20,500	8,000		28,500	34,900
2300	3,550 MBH		.17	187		22,700	8,800		31,500	38,600
2320	3,820 MBH		.16	200		24,000	9,400		33,400	41,000
2340	4,360 MBH		.15	214		26,300	10,100		36,400	44,600
2360	4,940 MBH		.14	225		55,500	10,600		66,100	77,500
2380	5,520 MBH		.14	235		65,500	11,100		76,600	89,000
2400	6,100 MBH		.13	256		75,000	12,000		87,000	101,000
2420	6,390 MBH		.11	290		79,000	13,700		92,700	107,500
2440	6,680 MBH		.10	313		81,000	14,700		95,700	112,000
2460	6,970 MBH		.09	363		84,000	17,100		101,100	119,000
3000	Hot water, same price as steam									
4000	For tankless coil in smaller sizes, add				Ea.	15%				

23 52 23.60 Solid-Fuel Boilers		Crew	Daily Output	Labor-Hours	Unit	Material	Labor	Equipment	Total	Total Incl O&P
0010	**SOLID-FUEL BOILERS** R235000-50									
3000	Stoker fired (coal) cast iron with flush jacket and									
3400	insulation, steam or water, gross output, 1280 MBH	Q-6	.36	66.667	Ea.	33,800	3,075		36,875	42,000
3420	1460 MBH		.30	80		37,100	3,675		40,775	46,500
3440	1640 MBH		.28	85.714		39,500	3,950		43,450	49,600
3460	1820 MBH		.26	92.308		42,300	4,250		46,550	53,000
3480	2000 MBH		.25	96		45,000	4,425		49,425	56,500
3500	2360 MBH		.23	104		51,000	4,800		55,800	63,500
3540	2725 MBH		.20	120		56,500	5,525		62,025	71,000
3800	2950 MBH	Q-7	.16	200		94,500	9,400		103,900	118,500
3820	3210 MBH		.15	213		99,500	10,000		109,500	125,000
3840	3480 MBH		.14	228		104,000	10,700		114,700	131,000
3860	3745 MBH		.14	228		109,000	10,700		119,700	136,500
3880	4000 MBH		.13	246		113,500	11,600		125,100	142,500
3900	4200 MBH		.13	246		118,500	11,600		130,100	148,000
3920	4400 MBH		.12	266		123,000	12,500		135,500	154,500
3940	4600 MBH		.12	266		127,500	12,500		140,000	160,000

23 52 26 – Steel Boilers

23 52 26.40 Oil-Fired Boilers		Crew	Daily Output	Labor-Hours	Unit	Material	Labor	Equipment	Total	Total Incl O&P
0010	**OIL-FIRED BOILERS**, Standard controls, flame retention burner									
5000	Steel, with insulated flush jacket									
7000	Hot water, gross output, 103 MBH	Q-6	1.60	15	Ea.	1,675	690		2,365	2,925
7020	122 MBH		1.45	16.506		1,775	760		2,535	3,125
7040	137 MBH		1.36	17.595		1,900	810		2,710	3,325
7060	168 MBH		1.30	18.405		2,000	850		2,850	3,500
7080	225 MBH		1.22	19.704		2,625	910		3,535	4,300
7100	315 MBH		.96	25.105		5,425	1,150		6,575	7,775
7120	420 MBH		.70	34.483		5,700	1,600		7,300	8,750
7140	525 MBH		.57	42.403		6,850	1,950		8,800	10,600
7180	735 MBH		.48	50.104		8,425	2,300		10,725	12,900
7220	1,050 MBH		.37	65.753		13,400	3,025		16,425	19,400
7280	2,310 MBH		.21	114		21,200	5,300		26,500	31,500
7320	3,150 MBH		.13	184		30,200	8,500		38,700	46,500
7340	For tankless coil in steam or hot water, add					7%				
9000	Minimum labor/equipment charge	Q-6	1.75	13.714	Job		630		630	980

23 52 Heating Boilers

23 52 26 – Steel Boilers

23 52 26.70 Packaged Water Tube Boilers		Crew	Daily Output	Labor-Hours	Unit	Material	2009 Bare Costs Labor	Equipment	Total	Total Incl O&P
0010	**PACKAGED WATER TUBE BOILERS**									
2000	Packaged water tube, #2 oil, steam or hot water, gross output									
2040	1200 MBH	Q-7	.50	64	Ea.	18,200	3,000		21,200	24,700
2060	1600 MBH		.40	80		22,900	3,750		26,650	31,000
2080	2400 MBH		.30	106		30,800	5,025		35,825	41,700
2100	3200 MBH		.25	128		35,200	6,025		41,225	48,000
2120	4800 MBH		.20	160		46,000	7,525		53,525	62,000
2140	For gas fired, add		.40	80		3,100	3,750		6,850	9,225

23 52 28 – Swimming Pool Boilers

23 52 28.10 Swimming Pool Heaters		Crew	Daily Output	Labor-Hours	Unit	Material	Labor	Equipment	Total	Total Incl O&P
0010	**SWIMMING POOL HEATERS**, Not including wiring, external									
0020	piping, base or pad,									
0160	Gas fired, input, 155 MBH	Q-6	1.50	16	Ea.	1,575	735		2,310	2,900
0200	199 MBH		1	24		1,700	1,100		2,800	3,600
0220	250 MBH		.70	34.286		1,850	1,575		3,425	4,475
0240	300 MBH		.60	40		1,925	1,850		3,775	4,975
0260	399 MBH		.50	48		2,175	2,200		4,375	5,800
0280	500 MBH		.40	60		7,225	2,775		10,000	12,200
0300	650 MBH		.35	68.571		7,675	3,150		10,825	13,400
0320	750 MBH		.33	72.727		8,400	3,350		11,750	14,400
0360	990 MBH		.22	109		11,300	5,025		16,325	20,200
0370	1,260 MBH		.21	114		13,200	5,275		18,475	22,700
0380	1,440 MBH		.19	126		14,100	5,825		19,925	24,500
0400	1,800 MBH		.14	171		15,700	7,900		23,600	29,500
0410	2,070 MBH		.13	184		18,500	8,500		27,000	33,600
1120	Oil fired, 315 MBH		.96	25		3,375	1,150		4,525	5,500
2000	Electric, 12 KW, 4,800 gallon pool	Q-19	3	8		1,950	360		2,310	2,700
2020	15 KW, 7,200 gallon pool		2.80	8.571		2,000	390		2,390	2,800
2040	24 KW, 9,600 gallon pool		2.40	10		2,275	455		2,730	3,200
2060	30 KW, 12,000 gallon pool		2	12		2,325	545		2,870	3,400
2080	36 KW, 14,400 gallon pool		1.60	15		2,675	680		3,355	4,000
2100	57 KW, 24,000 gallon pool		1.20	20		3,325	905		4,230	5,050
9000	To select pool heater: 12 BTUH x S.F. pool area									
9010	X temperature differential = required output									
9050	For electric, KW = gallons x 2.5 divided by 1000									
9100	For family home type pool, double the									
9110	Rated gallon capacity = 1/2° F rise per hour									

23 52 39 – Fire-Tube Boilers

23 52 39.13 Scotch Marine Boilers		Crew	Daily Output	Labor-Hours	Unit	Material	Labor	Equipment	Total	Total Incl O&P
0010	**SCOTCH MARINE BOILERS**									
1000	Packaged fire tube, #2 oil, gross output									
1020	3348 MBH, 100 HP	Q-7	.21	152	Ea.	52,500	7,175		59,675	69,000
1040	6696 MBH, 200 HP		.14	223		75,500	10,500		86,000	99,500
1060	10,044 MBH, 300 HP		.13	251		103,500	11,800		115,300	132,500
1080	16,740 MBH, 500 HP		.08	380		140,500	17,900		158,400	182,500
1100	23,435 MBH, 700 HP		.07	484		172,500	22,800		195,300	225,000
1120	To fire #6, add		.83	38.554		10,400	1,800		12,200	14,300
1140	To fire #6, and gas, add		.42	76.190		15,300	3,575		18,875	22,500
1160	For high pressure, add					17,000			17,000	18,700
1180	For duplex package feed system									
1200	To 3348 MBH boiler, add	Q-7	.54	59.259	Ea.	8,250	2,775		11,025	13,400

23 52 39 – Fire-Tube Boilers

23 52 39.13 Scotch Marine Boilers		Crew	Daily Output	Labor-Hours	Unit	2009 Bare Costs Material	Labor	Equipment	Total	Total Incl O&P
1220	To 6696 MBH boiler, add	Q-7	.41	78.049	Ea.	9,775	3,675		13,450	16,500
1240	To 10,044 MBH boiler, add		.38	84.211		11,200	3,950		15,150	18,500
1260	To 16,740 MBH boiler, add		.28	114		12,800	5,375		18,175	22,400
1280	To 23,435 MBH boiler, add		.25	128		15,100	6,025		21,125	25,900

23 52 84 – Boiler Blowdown

23 52 84.10 Boiler Blowdown Systems		Crew	Daily Output	Labor-Hours	Unit	Material	Labor	Equipment	Total	Total Incl O&P
0010	**BOILER BLOWDOWN SYSTEMS**									
1010	Boiler blowdown, auto/manual to 2000 MBH	Q-5	3.75	4.267	Ea.	3,125	190		3,315	3,725
1020	7300 MBH	"	3	5.333	"	4,350	237		4,587	5,175

23 52 88 – Burners

23 52 88.10 Replacement Type Burners		Crew	Daily Output	Labor-Hours	Unit	Material	Labor	Equipment	Total	Total Incl O&P
0010	**REPLACEMENT TYPE BURNERS**									
0990	Residential, conversion, gas fired, LP or natural									
1000	Gun type, atmospheric input 50 to 225 MBH	Q-1	2.50	6.400	Ea.	645	281		926	1,150
1020	100 to 400 MBH		2	8		1,075	350		1,425	1,725
1040	300 to 1000 MBH		1.70	9.412		3,300	415		3,715	4,275
2000	Commercial and industrial, gas/oil, input									
2050	400 MBH	Q-1	1.50	10.667	Ea.	3,400	470		3,870	4,475
2090	670 MBH		1.40	11.429		3,400	500		3,900	4,525
2140	1155 MBH		1.30	12.308		3,800	540		4,340	5,025
2200	1800 MBH		1.20	13.333		3,800	585		4,385	5,075
2260	3000 MBH		1.10	14.545		4,575	640		5,215	6,025
2320	4100 MBH		1	16		4,675	700		5,375	6,250
2500	Impinged jet, rectangular, burner only, 3-1/2" WC,									
2520	Gas fired, input, 420 to 640 MBH	Q-1	2	8	Ea.	825	350		1,175	1,450
2540	560 to 860 MBH		2	8		1,175	350		1,525	1,850
2560	630 to 950 MBH		1.90	8.421		1,225	370		1,595	1,925
2580	950 to 1200 MBH		1.80	8.889		1,750	390		2,140	2,525
2600	1120 to 1700 MBH		1.80	8.889		2,350	390		2,740	3,175
2620	1700 to 1900 MBH		1.70	9.412		2,650	415		3,065	3,550
2640	1400 to 2100 MBH		1.60	10		2,925	440		3,365	3,900
2660	1600 to 2400 MBH		1.60	10		3,075	440		3,515	4,075
2680	1700 to 2500 MBH		1.50	10.667		3,525	470		3,995	4,600
2700	1880 to 2880 MBH		1.40	11.429		3,700	500		4,200	4,850
2720	2260 to 3680 MBH		1.40	11.429		4,325	500		4,825	5,525
2740	2590 to 4320 MBH		1.30	12.308		5,275	540		5,815	6,650
2760	3020 to 5040 MBH		1.20	13.333		6,150	585		6,735	7,675
2780	3450 to 5760 MBH		1.20	13.333		7,050	585		7,635	8,650
2850	Burner pilot		50	.320		21	14.05		35.05	45
2860	Thermocouple		50	.320		21.50	14.05		35.55	46
2870	Thermocouple & pilot bracket		45	.356		84	15.60		99.60	117
2900	For 7" W.C. pressure, increase MBH 40%									
3000	Flame retention oil fired assembly, input									
3020	.50 to 2.25 GPH	Q-1	2.40	6.667	Ea.	289	293		582	775
3040	2.0 to 5.0 GPH		2	8		330	350		680	910
3060	3.0 to 7.0 GPH		1.80	8.889		590	390		980	1,250
3080	6.0 to 12.0 GPH		1.60	10		935	440		1,375	1,700
4600	Gas safety, shut off valve, 3/4" threaded	1 Stpi	20	.400		178	19.75		197.75	227
4610	1" threaded		19	.421		172	21		193	221
4620	1-1/4" threaded		15	.533		194	26.50		220.50	254
4630	1-1/2" threaded		13	.615		209	30.50		239.50	277
4640	2" threaded		11	.727		235	36		271	315

23 52 Heating Boilers

23 52 88 – Burners

23 52 88.10 Replacement Type Burners		Crew	Daily Output	Labor-Hours	Unit	Material	Labor	Equipment	Total	Total Incl O&P
4650	2-1/2" threaded	Q-1	15	1.067	Ea.	269	47		316	370
4660	3" threaded		13	1.231		370	54		424	490
4670	4" flanged		3	5.333		2,625	234		2,859	3,275
4680	6" flanged	Q-2	3	8		5,875	365		6,240	7,050

(Material, Labor, Equipment and Total are 2009 Bare Costs.)

23 54 Furnaces

23 54 13 – Electric-Resistance Furnaces

23 54 13.10 Electric Furnaces

		Crew	Daily Output	Labor-Hours	Unit	Material	Labor	Equipment	Total	Total Incl O&P
0010	**ELECTRIC FURNACES**, Hot air, blowers, std. controls R235000-50									
0011	not including gas, oil or flue piping									
1000	Electric, UL listed									
1020	10.2 MBH	Q-20	5	4	Ea.	360	174		534	670
1040	17.1 MBH		4.80	4.167		375	181		556	700
1060	27.3 MBH		4.60	4.348		455	189		644	800
1100	34.1 MBH		4.40	4.545		465	197		662	820
1120	51.6 MBH		4.20	4.762		575	207		782	960
1140	68.3 MBH		4	5		700	217		917	1,100
1160	85.3 MBH		3.80	5.263		765	228		993	1,200

23 54 16 – Fuel-Fired Furnaces

23 54 16.13 Gas-Fired Furnaces

		Crew	Daily Output	Labor-Hours	Unit	Material	Labor	Equipment	Total	Total Incl O&P
0010	**GAS-FIRED FURNACES**									
3000	Gas, AGA certified, upflow, direct drive models									
3020	45 MBH input	Q-9	4	4	Ea.	575	170		745	905
3040	60 MBH input		3.80	4.211		615	179		794	960
3060	75 MBH input		3.60	4.444		685	189		874	1,050
3100	100 MBH input		3.20	5		695	212		907	1,100
3120	125 MBH input		3	5.333		700	227		927	1,125
3130	150 MBH input		2.80	5.714		705	243		948	1,150
3140	200 MBH input		2.60	6.154		2,425	261		2,686	3,075
3160	300 MBH input		2.30	6.957		2,525	296		2,821	3,275
3180	400 MBH input		2	8		2,825	340		3,165	3,650

23 54 16.16 Oil-Fired Furnaces

		Crew	Daily Output	Labor-Hours	Unit	Material	Labor	Equipment	Total	Total Incl O&P
0010	**OIL-FIRED FURNACES**									
6000	Oil, UL listed, atomizing gun type burner									
6020	56 MBH output	Q-9	3.60	4.444	Ea.	1,850	189		2,039	2,325
6030	84 MBH output		3.50	4.571		1,850	194		2,044	2,350
6040	95 MBH output		3.40	4.706		1,875	200		2,075	2,400
6060	134 MBH output		3.20	5		1,950	212		2,162	2,500
6080	151 MBH output		3	5.333		2,275	227		2,502	2,850
6100	200 MBH input		2.60	6.154		2,325	261		2,586	2,975
6120	300 MBH input		2.30	6.957		2,825	296		3,121	3,575
6140	400 MBH input		2	8		3,275	340		3,615	4,150
9000	Minimum labor/equipment charge		2.75	5.818	Job		247		247	395

23 54 24 – Furnace Components for Cooling

23 54 24.10 Furnace Components and Combinations

		Crew	Daily Output	Labor-Hours	Unit	Material	Labor	Equipment	Total	Total Incl O&P
0010	**FURNACE COMPONENTS AND COMBINATIONS**									
0080	Coils, A/C evaporator, for gas or oil furnaces									
0090	Add-on, with holding charge									
0100	Upflow									
0120	1-1/2 ton cooling	Q-5	4	4	Ea.	158	178		336	450

23 54 Furnaces

23 54 24 – Furnace Components for Cooling

23 54 24.10 Furnace Components and Combinations		Crew	Daily Output	Labor-Hours	Unit	Material	Labor	Equipment	Total	Total Incl O&P
						2009 Bare Costs				
0130	2 ton cooling	Q-5	3.70	4.324	Ea.	188	192		380	505
0140	3 ton cooling		3.30	4.848		238	215		453	595
0150	4 ton cooling		3	5.333		325	237		562	725
0160	5 ton cooling		2.70	5.926		385	263		648	835
0300	Downflow									
0330	2-1/2 ton cooling	Q-5	3	5.333	Ea.	232	237		469	620
0340	3-1/2 ton cooling		2.60	6.154		276	273		549	730
0350	5 ton cooling		2.20	7.273		385	325		710	925
0600	Horizontal									
0630	2 ton cooling	Q-5	3.90	4.103	Ea.	156	182		338	455
0640	3 ton cooling		3.50	4.571		184	203		387	515
0650	4 ton cooling		3.20	5		224	222		446	590
0660	5 ton cooling		2.90	5.517		224	245		469	625
2000	Cased evaporator coils for air handlers									
2100	1-1/2 ton cooling	Q-5	4.40	3.636	Ea.	244	162		406	520
2110	2 ton cooling		4.10	3.902		248	173		421	540
2120	2-1/2 ton cooling		3.90	4.103		259	182		441	570
2130	3 ton cooling		3.70	4.324		286	192		478	615
2140	3-1/2 ton cooling		3.50	4.571		299	203		502	645
2150	4 ton cooling		3.20	5		360	222		582	740
2160	5 ton cooling		2.90	5.517		415	245		660	835
3010	Air handler, modular									
3100	With cased evaporator cooling coil									
3120	1-1/2 ton cooling	Q-5	3.80	4.211	Ea.	775	187		962	1,150
3130	2 ton cooling		3.50	4.571		865	203		1,068	1,275
3140	2-1/2 ton cooling		3.30	4.848		885	215		1,100	1,300
3150	3 ton cooling		3.10	5.161		945	229		1,174	1,375
3160	3-1/2 ton cooling		2.90	5.517		1,000	245		1,245	1,475
3170	4 ton cooling		2.50	6.400		1,125	284		1,409	1,700
3180	5 ton cooling		2.10	7.619		1,275	340		1,615	1,925
3500	With no cooling coil									
3520	1-1/2 ton coil size	Q-5	12	1.333	Ea.	575	59		634	720
3530	2 ton coil size		10	1.600		605	71		676	775
3540	2-1/2 ton coil size		10	1.600		605	71		676	775
3554	3 ton coil size		9	1.778		685	79		764	875
3560	3-1/2 ton coil size		9	1.778		745	79		824	940
3570	4 ton coil size		8.50	1.882		860	83.50		943.50	1,075
3580	5 ton coil size		8	2		1,000	89		1,089	1,250
4000	With heater									
4120	5 kW, 17.1 MBH	Q-5	16	1	Ea.	585	44.50		629.50	715
4130	7.5 kW, 25.6 MBH		15.60	1.026		590	45.50		635.50	720
4140	10 kW, 34.2 MBH		15.20	1.053		825	47		872	980
4150	12.5 KW, 42.7 MBH		14.80	1.081		910	48		958	1,075
4160	15 KW, 51.2 MBH		14.40	1.111		935	49.50		984.50	1,100
4170	25 KW, 85.4 MBH		14	1.143		1,050	51		1,101	1,225
4180	30 KW, 102 MBH		13	1.231		1,225	54.50		1,279.50	1,425

23 55 Fuel-Fired Heaters

23 55 13 – Fuel-Fired Duct Heaters

23 55 13.16 Gas-Fired Duct Heaters		Crew	Daily Output	Labor-Hours	Unit	Material	2009 Bare Costs Labor	Equipment	Total	Total Incl O&P
0010	**GAS-FIRED DUCT HEATERS**, Includes burner, controls, stainless steel R235000-50									
0020	heat exchanger. Gas fired, electric ignition									
0030	Indoor installation									
0080	100 MBH output	Q-5	5	3.200	Ea.	1,875	142		2,017	2,300
0100	120 MBH output		4	4		2,125	178		2,303	2,625
0130	200 MBH output		2.70	5.926		2,850	263		3,113	3,525
0140	240 MBH output		2.30	6.957		2,975	310		3,285	3,725
0160	280 MBH output		2	8		3,350	355		3,705	4,250
0180	320 MBH output		1.60	10		3,550	445		3,995	4,625
0300	For powered venter and adapter, add					460			460	510
0502	For required flue pipe, see Div. 23 51 23.10									
1000	Outdoor installation, with vent cap									
1020	75 MBH output	Q-5	4	4	Ea.	2,775	178		2,953	3,325
1040	94 MBH output		4	4		3,125	178		3,303	3,725
1060	120 MBH output		4	4		3,525	178		3,703	4,150
1080	157 MBH output		3.50	4.571		4,100	203		4,303	4,825
1100	187 MBH output		3	5.333		5,050	237		5,287	5,925
1120	225 MBH output		2.50	6.400		5,225	284		5,509	6,200
1140	300 MBH output		1.80	8.889		6,575	395		6,970	7,850
1160	375 MBH output		1.60	10		6,575	445		7,020	7,925
1180	450 MBH output		1.40	11.429		6,875	510		7,385	8,350
1200	600 MBH output		1	16		8,375	710		9,085	10,300
1300	Aluminized exchanger, subtract					15%				
1400	For powered venter, add					10%				
1500	For two stage gas valve, add					253			253	278

23 55 33 – Fuel-Fired Unit Heaters

23 55 33.13 Oil-Fired Unit Heaters		Crew	Daily Output	Labor-Hours	Unit	Material	Labor	Equipment	Total	Total Incl O&P
0010	**OIL-FIRED UNIT HEATERS**, Cabinet, grilles, fan, ctrl, burner, no piping									
6000	Oil fired, suspension mounted, 94 MBH output	Q-5	4	4	Ea.	3,400	178		3,578	4,000
6040	140 MBH output		3	5.333		3,625	237		3,862	4,350
6060	184 MBH output		3	5.333		3,900	237		4,137	4,675

23 55 33.16 Gas-Fired Unit Heaters		Crew	Daily Output	Labor-Hours	Unit	Material	Labor	Equipment	Total	Total Incl O&P
0010	**GAS-FIRED UNIT HEATERS**, Cabinet, grilles, fan, ctrls., burner, no piping									
0022	thermostat, no piping. For flue see Div. 23 51 23.10 R235000-50									
1000	Gas fired, floor mounted									
1100	60 MBH output	Q-5	10	1.600	Ea.	805	71		876	995
1120	80 MBH output		9	1.778		815	79		894	1,025
1140	100 MBH output		8	2		890	89		979	1,125
1160	120 MBH output		7	2.286		1,050	102		1,152	1,300
1180	180 MBH output		6	2.667		1,275	118		1,393	1,575
1500	Rooftop mounted, gravity vent, stainless steel exchanger									
1520	75 MBH output	Q-6	4	6	Ea.	5,225	276		5,501	6,175
1540	95 MBH output		3.60	6.667		5,325	305		5,630	6,325
1560	120 MBH output		3.30	7.273		5,700	335		6,035	6,800
1580	159 MBH output		3	8		5,875	370		6,245	7,025
1600	190 MBH output		2.60	9.231		6,625	425		7,050	7,925
1620	225 MBH output		2.30	10.435		6,725	480		7,205	8,150
1640	300 MBH output		1.90	12.632		8,025	580		8,605	9,725
1660	375 MBH output		1.40	17.143		9,100	790		9,890	11,200
1680	450 MBH output		1.20	20		9,275	920		10,195	11,600
1700	600 MBH output		1	24		11,200	1,100		12,300	14,000
1720	750 MBH output		.80	30		19,100	1,375		20,475	23,200

23 55 Fuel-Fired Heaters

23 55 33 – Fuel-Fired Unit Heaters

23 55 33.16 Gas-Fired Unit Heaters		Crew	Daily Output	Labor-Hours	Unit	2009 Bare Costs Material	Labor	Equipment	Total	Total Incl O&P
1740	900 MBH output	Q-6	.60	40	Ea.	21,400	1,850		23,250	26,500
1760	1200 MBH output		.30	80		26,100	3,675		29,775	34,400
1800	For power vent, add					10%				
1900	For aluminized steel exchanger, subtract					10%				
2000	Suspension mounted, propeller fan, 20 MBH output	Q-5	8.50	1.882		985	83.50		1,068.50	1,200
2020	40 MBH output		7.50	2.133		1,075	95		1,170	1,325
2040	60 MBH output		7	2.286		1,200	102		1,302	1,475
2060	80 MBH output		6	2.667		1,225	118		1,343	1,525
2080	100 MBH output		5.50	2.909		1,350	129		1,479	1,700
2100	130 MBH output		5	3.200		1,500	142		1,642	1,875
2120	140 MBH output		4.50	3.556		1,550	158		1,708	1,950
2140	160 MBH output		4	4		1,600	178		1,778	2,025
2160	180 MBH output		3.50	4.571		1,700	203		1,903	2,200
2180	200 MBH output		3	5.333		1,800	237		2,037	2,350
2200	240 MBH output		2.70	5.926		2,075	263		2,338	2,675
2220	280 MBH output		2.30	6.957		2,425	310		2,735	3,150
2240	320 MBH output		2	8		2,625	355		2,980	3,450
2500	For powered venter and adapter, add					355			355	390
3000	Suspension mounted, blower type, 40 MBH output	Q-5	6.80	2.353		605	105		710	825
3020	60 MBH output		6.60	2.424		775	108		883	1,025
3040	84 MBH output		5.80	2.759		810	123		933	1,075
3060	104 MBH output		5.20	3.077		920	137		1,057	1,200
3080	140 MBH output		4.30	3.721		1,025	165		1,190	1,375
3100	180 MBH output		3.30	4.848		1,150	215		1,365	1,575
3120	240 MBH output		2.50	6.400		1,500	284		1,784	2,100
3140	280 MBH output		2	8		1,875	355		2,230	2,625
4000	Suspension mounted, sealed combustion system,									
4020	Aluminized steel exchanger, powered vent									
4040	100 MBH output	Q-5	5	3.200	Ea.	1,800	142		1,942	2,225
4060	120 MBH output		4.70	3.404		2,025	151		2,176	2,450
4080	160 MBH output		3.70	4.324		2,200	192		2,392	2,725
4100	200 MBH output		2.90	5.517		2,650	245		2,895	3,300
4120	240 MBH output		2.50	6.400		2,850	284		3,134	3,575
4140	320 MBH output		1.70	9.412		3,400	420		3,820	4,375
5000	Wall furnace, 17.5 MBH output		6	2.667		615	118		733	865
5020	24 MBH output		5	3.200		640	142		782	925
5040	35 MBH output		4	4		770	178		948	1,125
9000	Minimum labor/equipment charge		3.50	4.571	Job		203		203	315

23 56 Solar Energy Heating Equipment

23 56 16 – Packaged Solar Heating Equipment

23 56 16.40 Solar Heating Systems		Crew	Daily Output	Labor-Hours	Unit	Material	Labor	Equipment	Total	Total Incl O&P
0010	**SOLAR HEATING SYSTEMS**									
0020	System/Package prices, not including connecting									
0030	pipe, insulation, or special heating/plumbing fixtures									
0152	For solar ultraviolet pipe insulation see Div. 22 07 19.10									
0500	Hot water, standard package, low temperature									
0540	1 collector, circulator, fittings, 65 gal. tank G	Q-1	.50	32	Ea.	1,125	1,400		2,525	3,400
0580	2 collectors, circulator, fittings, 120 gal. tank G		.40	40		1,825	1,750		3,575	4,750
0620	3 collectors, circulator, fittings, 120 gal. tank G		.34	47.059		2,500	2,075		4,575	5,950
0700	Medium temperature package									

23 56 Solar Energy Heating Equipment

23 56 16 – Packaged Solar Heating Equipment

23 56 16.40 Solar Heating Systems			Crew	Daily Output	Labor-Hours	Unit	Material	Labor (2009 Bare Costs)	Equipment (2009 Bare Costs)	Total (2009 Bare Costs)	Total Incl O&P
0720	1 collector, circulator, fittings, 80 gal. tank	G	Q-1	.50	32	Ea.	1,850	1,400		3,250	4,200
0740	2 collectors, circulator, fittings, 120 gal. tank	G		.40	40		2,625	1,750		4,375	5,625
0780	3 collectors, circulator, fittings, 120 gal. tank	G		.30	53.333		3,400	2,350		5,750	7,375
0980	For each additional 120 gal. tank, add	G					1,150			1,150	1,250

23 56 19 – Solar Heating Components

23 56 19.50 Solar Heating Ancillary			Crew	Daily Output	Labor-Hours	Unit	Material	Labor	Equipment	Total	Total Incl O&P
0010	**SOLAR HEATING ANCILLARY**										
2250	Controller, liquid temperature	G	1 Plum	5	1.600	Ea.	96	78		174	226
2300	Circulators, air										
2310	Blowers										
2330	100-300 S.F. system, 1/10 HP	G	Q-9	16	1	Ea.	229	42.50		271.50	320
2340	300-500 S.F. system, 1/5 HP	G		15	1.067		299	45.50		344.50	400
2350	Two speed, 100-300 S.F., 1/10 HP	G		14	1.143		136	48.50		184.50	227
2400	Reversible fan, 20" diameter, 2 speed	G		18	.889		108	38		146	179
2520	Space & DHW system, less duct work	G		.50	32		1,475	1,350		2,825	3,775
2550	Booster fan 6" diameter, 120 CFM	G		16	1		35	42.50		77.50	106
2570	6" diameter, 225 CFM	G		16	1		43	42.50		85.50	115
2580	8" diameter, 150 CFM	G		16	1		39.50	42.50		82	111
2590	8" diameter, 310 CFM	G		14	1.143		60.50	48.50		109	144
2600	8" diameter, 425 CFM	G		14	1.143		68	48.50		116.50	152
2650	Rheostat	G		32	.500		14.80	21		35.80	50.50
2660	Shutter/damper	G		12	1.333		49	56.50		105.50	144
2670	Shutter motor	G		16	1		119	42.50		161.50	199
2800	Circulators, liquid, 1/25 HP, 5.3 GPM	G	Q-1	14	1.143		227	50		277	330
2820	1/20 HP, 17 GPM	G		12	1.333		154	58.50		212.50	261
2850	1/20 HP, 17 GPM, stainless steel	G		12	1.333		243	58.50		301.50	360
2870	1/12 HP, 30 GPM	G		10	1.600		330	70		400	475
3000	Collector panels, air with aluminum absorber plate										
3010	Wall or roof mount										
3040	Flat black, plastic glazing										
3080	4' x 8'	G	Q-9	6	2.667	Ea.	635	113		748	875
3100	4' x 10'	G		5	3.200	"	780	136		916	1,075
3200	Flush roof mount, 10' to 16' x 22" wide	G		96	.167	L.F.	550	7.10		557.10	615
3210	Manifold, by L.F. width of collectors	G		160	.100	"	106	4.25		110.25	124
3300	Collector panels, liquid with copper absorber plate										
3320	Black chrome, tempered glass glazing										
3330	Alum. frame, 4' x 8', 5/32" single glazing	G	Q-1	9.50	1.684	Ea.	665	74		739	845
3390	Alum. frame, 4' x 10', 5/32" single glazing	G		6	2.667		805	117		922	1,075
3450	Flat black, alum. frame, 3.5' x 7.5'	G		9	1.778		560	78		638	740
3500	4' x 8'	G		5.50	2.909		690	128		818	955
3520	4' x 10'	G		10	1.600		830	70		900	1,025
3540	4' x 12.5'	G		5	3.200		1,025	140		1,165	1,350
3550	Liquid with fin tube absorber plate										
3560	Alum. frame 4' x 8' tempered glass	G	Q-1	10	1.600	Ea.	495	70		565	655
3580	Liquid with vacuum tubes, 4' x 6'-10"	G		9	1.778		850	78		928	1,050
3600	Liquid, full wetted, plastic, alum. frame, 3' x 10'	G		5	3.200		234	140		374	475
3650	Collector panel mounting, flat roof or ground rack	G		7	2.286		217	100		317	395
3670	Roof clamps	G		70	.229	Set	2.35	10.05		12.40	18.15
3700	Roof strap, teflon	G	1 Plum	205	.039	L.F.	19.95	1.90		21.85	25
3900	Differential controller with two sensors										
3930	Thermostat, hard wired	G	1 Plum	8	1	Ea.	79.50	49		128.50	163
3950	Line cord and receptacle	G		12	.667		345	32.50		377.50	430

23 56 Solar Energy Heating Equipment

23 56 19 – Solar Heating Components

23 56 19.50 Solar Heating Ancillary			Crew	Daily Output	Labor-Hours	Unit	2009 Bare Costs Material	Labor	Equipment	Total	Total Incl O&P
4050	Pool valve system	G	1 Plum	2.50	3.200	Ea.	264	156		420	535
4070	With 12 VAC actuator	G		2	4		276	195		471	605
4080	Pool pump system, 2" pipe size	G		6	1.333		172	65		237	291
4100	Five station with digital read-out	G		3	2.667		218	130		348	440
4150	Sensors										
4200	Brass plug, 1/2" MPT	G	1 Plum	32	.250	Ea.	17.95	12.20		30.15	38.50
4210	Brass plug, reversed	G		32	.250		24.50	12.20		36.70	46
4220	Freeze prevention	G		32	.250		22	12.20		34.20	43
4240	Screw attached	G		32	.250		8.35	12.20		20.55	28
4250	Brass, immersion	G		32	.250		26.50	12.20		38.70	48
4300	Heat exchanger										
4330	Fluid to air coil, up flow, 45 MBH	G	Q-1	4	4	Ea.	295	176		471	595
4380	70 MBH	G		3.50	4.571		335	201		536	675
4400	80 MBH	G		3	5.333		445	234		679	855
4580	Fluid to fluid package includes two circulating pumps										
4590	expansion tank, check valve, relief valve										
4600	controller, high temperature cutoff and sensors	G	Q-1	2.50	6.400	Ea.	695	281		976	1,200
4650	Heat transfer fluid										
4700	Propylene glycol, inhibited anti-freeze	G	1 Plum	28	.286	Gal.	13.85	13.95		27.80	37
4800	Solar storage tanks, knocked down										
4810	Air, galvanized steel clad, double wall, 4" fiberglass										
4820	insulation, 20 Mil PVC lining										
4870	4' high, 4' x 4', = 64 C.F./450 gallons	G	Q-9	2	8	Ea.	2,275	340		2,615	3,050
4880	4' x 8' = 128 C.F./900 gallons	G		1.50	10.667		3,225	455		3,680	4,275
4890	4' x 12' = 190 C.F./1300 gallons	G		1.30	12.308		4,300	525		4,825	5,550
4900	8' x 8' = 250 C.F./1700 gallons	G		1	16		4,800	680		5,480	6,350
5010	6'-3" high, 7' x 7' = 306 C.F./2000 gallons	G	Q-10	1.20	20		11,700	880		12,580	14,300
5020	7' x 10'-6" = 459 C.F./3000 gallons	G		.80	30		14,700	1,325		16,025	18,300
5030	7' x 14' = 613 C.F./4000 gallons	G		.60	40		17,200	1,750		18,950	21,800
5040	10'-6" x 10'-6" = 689 C.F./4500 gallons	G		.50	48		17,800	2,125		19,925	22,900
5050	10'-6" x 14' = 919 C.F./6000 gallons	G		.40	60		21,000	2,650		23,650	27,300
5060	14' x 14' = 1225 C.F./8000 gallons	G	Q-11	.40	80		24,500	3,600		28,100	32,600
5070	14' x 17'-6" = 1531 C.F./10,000 gallons	G		.30	106		27,600	4,800		32,400	37,900
5080	17'-6" x 17'-6" = 1914 C.F./12,500 gallons	G		.25	128		32,500	5,750		38,250	44,900
5090	17'-6" x 21' = 2297 C.F./15,000 gallons	G		.20	160		36,200	7,200		43,400	51,000
5100	21' x 21' = 2756 C.F./18,000 gallons	G		.18	177		41,500	8,000		49,500	58,500
5120	30 Mil reinforced Chemflex lining,										
5140	4' high, 4' x 4' = 64 C.F./450 gallons	G	Q-9	2	8	Ea.	2,525	340		2,865	3,325
5150	4' x 8' = 128 C.F./900 gallons	G		1.50	10.667		3,825	455		4,280	4,925
5160	4' x 12' = 190 C.F./1300 gallons	G		1.30	12.308		5,050	525		5,575	6,400
5170	8' x 8' = 250 C.F./1700 gallons	G		1	16		5,650	680		6,330	7,275
5190	6'-3" high, 7' x 7' = 306 C.F./2000 gallons	G	Q-10	1.20	20		12,500	880		13,380	15,100
5200	7' x 10'-6" = 459 C.F./3000 gallons	G		.80	30		15,200	1,325		16,525	18,900
5210	7' x 14' = 613 C.F./4000 gallons	G		.60	40		17,700	1,750		19,450	22,300
5220	10'-6" x 10'-6" = 689 C.F./4500 gallons	G		.50	48		18,300	2,125		20,425	23,500
5230	10'-6" x 14' = 919 C.F./6000 gallons	G		.40	60		21,500	2,650		24,150	27,800
5240	14' x 14' = 1225 C.F./8000 gallons	G	Q-11	.40	80		25,100	3,600		28,700	33,400
5250	14' x 17'-6" = 1531 C.F./10,000 gallons	G		.30	106		28,100	4,800		32,900	38,600
5260	17'-6" x 17'-6" = 1914 C.F./12,500 gallons	G		.25	128		33,100	5,750		38,850	45,600
5270	17'-6" x 21' = 2297 C.F./15,000 gallons	G		.20	160		37,300	7,200		44,500	52,500
5280	21' x 21' = 2756 C.F./18,000 gallons	G		.18	177		42,700	8,000		50,700	59,500
5290	30 Mil reinforced Hypalon lining, add						.02%				
7000	Solar control valves and vents										

23 56 Solar Energy Heating Equipment

23 56 19 – Solar Heating Components

23 56 19.50 Solar Heating Ancillary			Crew	Daily Output	Labor-Hours	Unit	Material	2009 Bare Costs Labor	Equipment	Total	Total Incl O&P
7050	Air purger, 1" pipe size	G	1 Plum	12	.667	Ea.	43.50	32.50		76	98
7070	Air eliminator, automatic 3/4" size	G		32	.250		28.50	12.20		40.70	50.50
7090	Air vent, automatic, 1/8" fitting	G		32	.250		12.65	12.20		24.85	33
7100	Manual, 1/8" NPT	G		32	.250		2.78	12.20		14.98	22
7120	Backflow preventer, 1/2" pipe size	G		16	.500		58.50	24.50		83	102
7130	3/4" pipe size	G		16	.500		101	24.50		125.50	149
7150	Balancing valve, 3/4" pipe size	G		20	.400		40.50	19.50		60	75
7180	Draindown valve, 1/2" copper tube	G		9	.889		202	43.50		245.50	289
7200	Flow control valve, 1/2" pipe size	G		22	.364		102	17.75		119.75	140
7220	Expansion tank, up to 5 gal.	G		32	.250		61.50	12.20		73.70	86.50
7250	Hydronic controller (aquastat)	G		8	1		144	49		193	235
7400	Pressure gauge, 2" dial	G		32	.250		22	12.20		34.20	43.50
7450	Relief valve, temp. and pressure 3/4" pipe size	G		30	.267		15.15	13		28.15	36.50
7500	Solenoid valve, normally closed										
7520	Brass, 3/4" NPT, 24V	G	1 Plum	9	.889	Ea.	103	43.50		146.50	180
7530	1" NPT, 24V	G		9	.889		815	43.50		858.50	960
7750	Vacuum relief valve, 3/4" pipe size	G		32	.250		26.50	12.20		38.70	48
7800	Thermometers										
7820	Digital temperature monitoring, 4 locations	G	1 Plum	2.50	3.200	Ea.	128	156		284	385
7900	Upright, 1/2" NPT	G		8	1		24	49		73	102
7970	Remote probe, 2" dial	G		8	1		30.50	49		79.50	109
7990	Stem, 2" dial, 9" stem	G		16	.500		20	24.50		44.50	60
8250	Water storage tank with heat exchanger and electric element										
8270	66 gal. with 2" x 2 lb. density insulation	G	1 Plum	1.60	5	Ea.	925	244		1,169	1,400
8300	80 gal. with 2" x 2 lb. density insulation	G		1.60	5		1,050	244		1,294	1,525
8380	120 gal. with 2" x 2 lb. density insulation	G		1.40	5.714		1,175	279		1,454	1,700
8400	120 gal. with 2" x 2 lb. density insul., 40 S.F. heat coil	G		1.40	5.714		1,500	279		1,779	2,075
8500	Water storage module, plastic										
8600	Tubular, 12" diameter, 4' high	G	1 Carp	48	.167	Ea.	103	6.65		109.65	124
8610	12" diameter, 8' high	G		40	.200		159	8		167	188
8620	18" diameter, 5' high	G		38	.211		174	8.40		182.40	205
8630	18" diameter, 10' high	G		32	.250		229	10		239	269
8640	58" diameter, 5' high	G	2 Carp	32	.500		490	20		510	570
8650	Cap, 12" diameter	G					19			19	21
8660	18" diameter	G					24			24	26.50
9000	Minimum labor/equipment charge	G	1 Plum	2	4	Job		195		195	300

23 57 Heat Exchangers for HVAC

23 57 16 – Steam-to-Water Heat Exchangers

23 57 16.10 Shell/Tube Type Steam-to-Water Heat Exch.		Crew	Daily Output	Labor-Hours	Unit	Material	Labor	Equipment	Total	Total Incl O&P
0010	**SHELL AND TUBE TYPE STEAM-TO-WATER HEAT EXCHANGERS**									
0016	Shell & tube type, 2 or 4 pass, 3/4" O.D. copper tubes,									
0020	C.I. heads, C.I. tube sheet, steel shell									
0100	Hot water 40° F to 180° F, by steam at 10 PSI									
0120	8 GPM	Q-5	6	2.667	Ea.	1,550	118		1,668	1,875
0140	10 GPM		5	3.200		2,350	142		2,492	2,800
0160	40 GPM		4	4		3,625	178		3,803	4,275
0180	64 GPM		2	8		5,550	355		5,905	6,650
0200	96 GPM		1	16		7,425	710		8,135	9,275
0220	120 GPM	Q-6	1.50	16		8,625	735		9,360	10,700
0240	168 GPM		1	24		12,000	1,100		13,100	14,900

23 57 Heat Exchangers for HVAC

23 57 16 – Steam-to-Water Heat Exchangers

23 57 16.10 Shell/Tube Type Steam-to-Water Heat Exch.		Crew	Daily Output	Labor-Hours	Unit	2009 Bare Costs Material	Labor	Equipment	Total	Total Incl O&P
0260	240 GPM	Q-6	.80	30	Ea.	18,700	1,375		20,075	22,800
0300	600 GPM		.70	34.286		49,000	1,575		50,575	56,500
0500	For bronze head and tube sheet, add					50%				

23 57 19 – Liquid-to-Liquid Heat Exchangers

23 57 19.13 Plate-Type, Liquid-to-Liquid Heat Exchangers

		Crew	Daily Output	Labor-Hours	Unit	Material	Labor	Equipment	Total	Total Incl O&P
0010	**PLATE-TYPE, LIQUID-TO-LIQUID HEAT EXCHANGERS**									
3000	Plate type,									
3100	400 GPM	Q-6	.80	30	Ea.	31,400	1,375		32,775	36,700
3120	800 GPM	"	.50	48		54,000	2,200		56,200	63,000
3140	1200 GPM	Q-7	.34	94.118		80,500	4,425		84,925	95,500
3160	1800 GPM	"	.24	133		106,500	6,275		112,775	126,500

23 57 19.16 Shell-Type, Liquid-to-Liquid Heat Exchangers

		Crew	Daily Output	Labor-Hours	Unit	Material	Labor	Equipment	Total	Total Incl O&P
0010	**SHELL-TYPE, LIQUID-TO-LIQUID HEAT EXCHANGERS**									
1000	Hot water 40° F to 140° F, by water at 200° F									
1020	7 GPM	Q-5	6	2.667	Ea.	1,900	118		2,018	2,275
1040	16 GPM		5	3.200		2,700	142		2,842	3,200
1060	34 GPM		4	4		4,100	178		4,278	4,775
1080	55 GPM		3	5.333		5,325	237		5,562	6,225
1100	74 GPM		1.50	10.667		7,400	475		7,875	8,875
1120	86 GPM		1.40	11.429		9,900	510		10,410	11,700
1140	112 GPM	Q-6	2	12		12,300	555		12,855	14,500
1160	126 GPM		1.80	13.333		15,400	615		16,015	17,900
1180	152 GPM		1	24		19,700	1,100		20,800	23,300

23 61 Refrigerant Compressors

23 61 15 – Rotary Refrigerant Compressors

23 61 15.10 Rotary Compressors

		Crew	Daily Output	Labor-Hours	Unit	Material	Labor	Equipment	Total	Total Incl O&P
0010	**ROTARY COMPRESSORS**									
0100	Refrigeration, hermetic, switches and protective devices									
0210	1.25 ton	1 Stpi	3	2.667	Ea.	221	132		353	445
0220	1.42 ton		3	2.667		223	132		355	450
0230	1.68 ton		2.80	2.857		243	141		384	485
0240	2.00 ton		2.60	3.077		250	152		402	510
0250	2.37 ton		2.50	3.200		282	158		440	555
0260	2.67 ton		2.40	3.333		296	165		461	580
0270	3.53 ton		2.30	3.478		325	172		497	620
0280	4.43 ton		2.20	3.636		415	179		594	735
0290	5.08 ton		2.10	3.810		455	188		643	790
0300	5.22 ton		2	4		640	197		837	1,000
0310	7.31 ton	Q-5	3	5.333		1,025	237		1,262	1,500
0320	9.95 ton		2.60	6.154		1,200	273		1,473	1,750
0330	11.6 ton		2.50	6.400		1,550	284		1,834	2,150
0340	15.25 ton		2.30	6.957		2,125	310		2,435	2,800
0350	17.7 ton		2.10	7.619		2,275	340		2,615	3,025

23 61 16 – Reciprocating Refrigerant Compressors

23 61 16.10 Reciprocating Compressors

		Crew	Daily Output	Labor-Hours	Unit	Material	Labor	Equipment	Total	Total Incl O&P
0010	**RECIPROCATING COMPRESSORS**									
0990	Refrigeration, recip. hermetic, switches & protective devices									
1000	10 ton	Q-5	1	16	Ea.	8,150	710		8,860	10,100
1100	20 ton	Q-6	.72	33.333		12,300	1,525		13,825	15,900

23 61 Refrigerant Compressors

23 61 16 – Reciprocating Refrigerant Compressors

23 61 16.10 Reciprocating Compressors		Crew	Daily Output	Labor-Hours	Unit	Material	2009 Bare Costs Labor	Equipment	Total	Total Incl O&P
1200	30 ton	Q-6	.64	37.500	Ea.	14,700	1,725		16,425	18,800
1300	40 ton		.44	54.545		16,000	2,525		18,525	21,400
1400	50 ton		.20	120		16,800	5,525		22,325	27,100
1500	75 ton	Q-7	.27	118		18,600	5,575		24,175	29,100
1600	130 ton	"	.21	152		21,200	7,175		28,375	34,400

23 61 19 – Scroll Refrigerant Compressors

23 61 19.10 Scroll Compressors		Crew	Daily Output	Labor-Hours	Unit	Material	Labor	Equipment	Total	Total Incl O&P
0010	**SCROLL COMPRESSORS**									
1800	Refrigeration, scroll type									
1810	1.9 ton	1 Stpi	2.70	2.963	Ea.	269	146		415	525
1820	2.35 ton		2.50	3.200		310	158		468	585
1830	2.82 ton		2.35	3.404		335	168		503	625
1840	3.3 ton		2.30	3.478		350	172		522	650
1850	3.83 ton		2.25	3.556		400	175		575	705
1860	4.1 ton		2.20	3.636		450	179		629	775
1870	4.8 ton		2.10	3.810		475	188		663	815
1880	5 ton		2.07	3.865		485	191		676	830

23 62 Packaged Compressor and Condenser Units

23 62 13 – Packaged Air-Cooled Refrigerant Compressor and Condenser Units

23 62 13.10 Packaged Air-Cooled Refrig. Condensing Units		Crew	Daily Output	Labor-Hours	Unit	Material	Labor	Equipment	Total	Total Incl O&P
0010	**PACKAGED AIR-COOLED REFRIGERANT CONDENSING UNITS**									
0020	Condensing unit									
0030	Air cooled, compressor, standard controls									
0050	1.5 ton	Q-5	2.50	6.400	Ea.	1,125	284		1,409	1,700
0100	2 ton		2.10	7.619		1,150	340		1,490	1,800
0200	2.5 ton		1.70	9.412		1,250	420		1,670	2,025
0300	3 ton		1.30	12.308		1,300	545		1,845	2,275
0350	3.5 ton		1.10	14.545		1,500	645		2,145	2,650
0400	4 ton		.90	17.778		1,700	790		2,490	3,075
0500	5 ton		.60	26.667		2,050	1,175		3,225	4,075
0550	7.5 ton		.55	29.091		3,350	1,300		4,650	5,700
0560	8.5 ton		.53	30.189		4,000	1,350		5,350	6,475
0600	10 ton		.50	32		4,350	1,425		5,775	6,975
0620	11 ton		.48	33.333		6,700	1,475		8,175	9,675
0650	15 ton		.40	40		7,150	1,775		8,925	10,600
0700	20 ton	Q-6	.40	60		9,950	2,775		12,725	15,300
0720	25 ton		.35	68.571		15,000	3,150		18,150	21,400
0750	30 ton		.30	80		17,100	3,675		20,775	24,500
0800	40 ton		.20	120		20,700	5,525		26,225	31,400
0840	50 ton		.18	133		24,700	6,150		30,850	36,600
0860	60 ton		.16	150		28,400	6,900		35,300	41,900
0900	75 ton		.14	171		32,500	7,900		40,400	47,900
1000	80 ton		.12	200		36,000	9,225		45,225	54,000
1100	100 ton		.09	266		50,500	12,300		62,800	74,500

23 62 23 – Packaged Water-Cooled Refrigerant Compressor and Condenser Units

23 62 23.10 Packaged Water-Cooled Refrig. Cond. Units		Crew	Daily Output	Labor-Hours	Unit	Material	Labor	Equipment	Total	Total Incl O&P
0010	**PACKAGED WATER-COOLED REFRIGERANT CONDENSING UNITS**									
2000	Water cooled, compressor, heat exchanger, controls									
2100	5 ton	Q-5	.70	22.857	Ea.	11,500	1,025		12,525	14,200

23 62 Packaged Compressor and Condenser Units

23 62 23 – Packaged Water-Cooled Refrigerant Compressor and Condenser Units

23 62 23.10 Packaged Water-Cooled Refrig. Cond. Units		Crew	Daily Output	Labor-Hours	Unit	2009 Bare Costs Material	Labor	Equipment	Total	Total Incl O&P
2200	15 ton	Q-5	.50	32	Ea.	18,100	1,425		19,525	22,100
2300	20 ton	Q-6	.40	60		20,200	2,775		22,975	26,500
2400	40 ton		.20	120		27,400	5,525		32,925	38,700
2500	100 ton		.11	218		54,500	10,100		64,600	75,000
9000	Minimum labor/equipment charge	Q-5	1.50	10.667	Job		475		475	735

23 63 Refrigerant Condensers

23 63 13 – Air-Cooled Refrigerant Condensers

23 63 13.10 Air-Cooled Refrig. Condensers		Crew	Daily Output	Labor-Hours	Unit	Material	Labor	Equipment	Total	Total Incl O&P
0010	**AIR-COOLED REFRIG. CONDENSERS**, Ratings for 30° F TD, R-22. R262213-27									
0080	Air cooled, belt drive, propeller fan									
0220	45 ton	Q-6	.70	34.286	Ea.	12,300	1,575		13,875	16,100
0240	50 ton		.69	34.985		12,700	1,600		14,300	16,500
0260	54 ton		.64	37.795		13,200	1,750		14,950	17,200
0280	59 ton		.58	41.308		14,200	1,900		16,100	18,600
0300	65 ton		.53	45.541		14,800	2,100		16,900	19,500
0320	73 ton		.47	51.173		16,700	2,350		19,050	22,100
0340	81 ton		.42	56.738		19,000	2,625		21,625	25,000
0360	86 ton		.40	60.302		19,800	2,775		22,575	26,100
0380	88 ton		.39	61.697		21,200	2,850		24,050	27,700
0400	101 ton	Q-7	.45	70.640		24,900	3,325		28,225	32,600
0500	159 ton		.31	102		37,200	4,825		42,025	48,400
0600	228 ton		.22	148		54,000	6,975		60,975	70,000
1500	May be specified single or multi-circuit									
1550	Air cooled, direct drive, propeller fan									
1590	1 ton	Q-5	3.80	4.211	Ea.	705	187		892	1,075
1600	1-1/2 ton		3.60	4.444		840	197		1,037	1,225
1620	2 ton		3.20	5		925	222		1,147	1,375
1630	3 ton		2.40	6.667		1,025	296		1,321	1,575
1640	5 ton		2	8		1,875	355		2,230	2,625
1650	8 ton		1.80	8.889		2,375	395		2,770	3,225
1660	10 ton		1.40	11.429		2,900	510		3,410	3,975
1670	12 ton		1.30	12.308		3,275	545		3,820	4,450
1680	14 ton		1.20	13.333		3,725	590		4,315	5,025
1690	16 ton		1.10	14.545		4,100	645		4,745	5,500
1700	21 ton		1	16		4,725	710		5,435	6,300
1720	26 ton		.84	19.002		5,300	845		6,145	7,125
1740	30 ton		.70	22.792		8,825	1,025		9,850	11,300
1760	41 ton	Q-6	.77	31.008		9,700	1,425		11,125	12,900
1780	52 ton		.66	36.419		13,800	1,675		15,475	17,800
1800	63 ton		.55	44.037		15,300	2,025		17,325	20,000
1820	76 ton		.45	52.980		17,300	2,450		19,750	22,900
1840	86 ton		.40	60		21,400	2,775		24,175	27,900
1860	97 ton		.35	67.989		24,400	3,125		27,525	31,700
1880	105 ton	Q-7	.44	73.563		28,100	3,450		31,550	36,300
1890	118 ton		.39	82.687		30,300	3,875		34,175	39,300
1900	126 ton		.36	88.154		35,200	4,150		39,350	45,100
1910	136 ton		.34	95.238		39,400	4,475		43,875	50,500
1920	142 ton		.32	99.379		40,400	4,675		45,075	51,500

23 63 Refrigerant Condensers

23 63 30 – Lubricants

23 63 30.10 Lubricant Oils		Crew	Daily Output	Labor-Hours	Unit	Material	Labor	Equipment	Total	Total Incl O&P
						2009 Bare Costs				
0010	**LUBRICANT OILS**									
8000	Oils									
8100	Lubricating									
8120	Oil, lubricating				Oz.	.49			.49	.54
8500	Refrigeration									
8520	Oil, refrigeration				Gal.	21			21	23.50
8525	Oil, refrigeration				Qt.	5.30			5.30	5.85

23 63 33 – Evaporative Refrigerant Condensers

23 63 33.10 Evaporative Condensers

		Crew	Daily Output	Labor-Hours	Unit	Material	Labor	Equipment	Total	Total Incl O&P
0010	**EVAPORATIVE CONDENSERS**									
3400	Evaporative, copper coil, pump, fan motor									
3440	10 ton	Q-5	.54	29.630	Ea.	5,325	1,325		6,650	7,925
3460	15 ton		.50	32		5,550	1,425		6,975	8,300
3480	20 ton		.47	34.043		5,775	1,525		7,300	8,700
3500	25 ton		.45	35.556		6,175	1,575		7,750	9,225
3520	30 ton		.42	38.095		6,575	1,700		8,275	9,850
3540	40 ton	Q-6	.49	48.980		8,700	2,250		10,950	13,100
3560	50 ton		.39	61.538		10,900	2,825		13,725	16,400
3580	65 ton		.35	68.571		12,100	3,150		15,250	18,200
3600	80 ton		.33	72.727		13,100	3,350		16,450	19,600
3620	90 ton		.29	82.759		14,400	3,825		18,225	21,800
3640	100 ton	Q-7	.36	88.889		15,700	4,175		19,875	23,800
3660	110 ton		.33	96.970		17,400	4,550		21,950	26,200
3680	125 ton		.30	106		18,900	5,025		23,925	28,600
3700	135 ton		.28	114		20,200	5,375		25,575	30,600
3720	150 ton		.25	128		22,300	6,025		28,325	33,800
3740	165 ton		.23	139		24,500	6,550		31,050	37,000
3760	185 ton		.22	145		25,800	6,850		32,650	39,000
3860	For fan damper control, add	Q-5	2	8		555	355		910	1,150

23 64 Packaged Water Chillers

23 64 13 – Absorption Water Chillers

23 64 13.13 Direct-Fired Absorption Water Chillers

		Crew	Daily Output	Labor-Hours	Unit	Material	Labor	Equipment	Total	Total Incl O&P
0010	**DIRECT-FIRED ABSORPTION WATER CHILLERS**									
3000	Gas fired, air cooled									
3220	5 ton	Q-5	.60	26.667	Ea.	9,525	1,175		10,700	12,300
3270	10 ton	"	.40	40	"	25,500	1,775		27,275	30,900
4000	Water cooled, duplex									
4130	100 ton	Q-7	.13	246	Ea.	136,000	11,600		147,600	167,500
4140	200 ton		.11	283		133,000	13,300		146,300	167,000
4150	300 ton		.11	299		219,500	14,100		233,600	263,500
4160	400 ton		.10	316		285,500	14,900		300,400	337,000
4170	500 ton		.10	329		359,500	15,500		375,000	419,000
4180	600 ton		.09	340		424,500	16,000		440,500	491,500
4190	700 ton		.09	359		470,500	16,900		487,400	543,500
4200	800 ton		.08	380		568,000	17,900		585,900	653,000
4210	900 ton		.08	400		658,000	18,800		676,800	753,000
4220	1000 ton		.08	421		731,500	19,800		751,300	835,000

23 64 13.16 Indirect-Fired Absorption Water Chillers

		Crew	Daily Output	Labor-Hours	Unit	Material	Labor	Equipment	Total	Total Incl O&P
0010	**INDIRECT-FIRED ABSORPTION WATER CHILLERS** R262213-27									

23 64 Packaged Water Chillers

23 64 13 – Absorption Water Chillers

23 64 13.16 Indirect-Fired Absorption Water Chillers		Crew	Daily Output	Labor-Hours	Unit	2009 Bare Costs Material	Labor	Equipment	Total	Total Incl O&P
0020	Steam or hot water, water cooled									
0050	100 ton	Q-7	.13	240	Ea.	125,000	11,300		136,300	155,000
0100	148 ton		.12	258		191,000	12,100		203,100	229,000
0200	200 ton		.12	275		212,500	13,000		225,500	253,500
0300	354 ton		.11	304		292,000	14,300		306,300	343,000
0400	420 ton		.10	323		319,000	15,200		334,200	374,500
0500	665 ton		.09	340		448,000	16,000		464,000	518,000
0600	750 ton		.09	363		523,500	17,100		540,600	602,000
0700	850 ton		.08	385		558,500	18,100		576,600	642,500
0800	955 ton		.08	410		597,500	19,300		616,800	687,000
0900	1125 ton		.08	421		689,500	19,800		709,300	789,000
1000	1250 ton		.07	444		741,500	20,900		762,400	848,500
1100	1465 ton		.07	463		879,500	21,800		901,300	1,001,000
1200	1660 ton	↓	.07	477	↓	1,034,500	22,500		1,057,000	1,173,000
2000	For two stage unit, add					80%	25%			
9000	Minimum labor/equipment charge	Q-5	1	16	Job		710		710	1,100

23 64 16 – Centrifugal Water Chillers

23 64 16.10 Centrifugal Type Water Chillers		Crew	Daily Output	Labor-Hours	Unit	Material	Labor	Equipment	Total	Total Incl O&P
0010	**CENTRIFUGAL TYPE WATER CHILLERS**, With standard controls									
0020	Centrifugal liquid chiller, water cooled									
0030	not including water tower R262213-27									
0100	2000 ton, (twin 1000 ton units)	Q-7	.07	477	Ea.	840,000	22,500		862,500	959,000
0274	Centrifugal, packaged unit, water cooled, not incl. tower									
0280	400 ton	Q-7	.11	283	Ea.	133,000	13,300		146,300	166,500
0320	1000 ton		.09	372		420,000	17,500		437,500	489,000
0340	1300 ton		.08	410		530,000	19,300		549,300	613,000
0360	1500 ton	↓	.08	426	↓	611,500	20,100		631,600	703,500

23 64 19 – Reciprocating Water Chillers

23 64 19.10 Reciprocating Type Water Chillers		Crew	Daily Output	Labor-Hours	Unit	Material	Labor	Equipment	Total	Total Incl O&P
0010	**RECIPROCATING TYPE WATER CHILLERS**, With standard controls.									
0494	Water chillers, integral air cooled condenser									
0546	70 ton cooling	Q-7	.27	119	Ea.	52,000	5,625		57,625	66,000
0554	90 ton cooling		.25	125		60,500	5,925		66,425	75,500
0600	100 ton cooling		.25	129		68,500	6,075		74,575	84,500
0620	110 ton cooling		.24	132		74,000	6,225		80,225	91,000
0630	130 ton cooling		.24	135		88,000	6,350		94,350	106,500
0640	150 ton cooling		.23	137		97,000	6,475		103,475	117,000
0650	175 ton cooling		.23	140		106,500	6,625		113,125	127,500
0654	190 ton cooling		.22	144		120,500	6,775		127,275	143,000
0660	210 ton cooling		.22	148		124,000	6,975		130,975	147,500
0662	240 ton cooling		.21	151		143,500	7,125		150,625	169,000
0664	270 ton cooling		.21	156		148,500	7,350		155,850	175,000
0666	300 ton cooling		.20	160		168,000	7,525		175,525	196,500
0668	330 ton cooling		.20	164		193,000	7,725		200,725	224,000
0670	360 ton cooling		.19	168		213,000	7,925		220,925	246,500
0672	390 ton cooling		.19	172		226,000	8,125		234,125	261,000
0674	420 ton cooling	↓	.18	177	↓	236,000	8,350		244,350	273,000
0980	Water cooled, multiple compressor, semi-hermetic, tower not incl.									
1000	15 ton cooling	Q-6	.36	65.934	Ea.	13,500	3,050		16,550	19,500
1020	20 ton cooling	Q-7	.41	78.049		14,800	3,675		18,475	22,000
1040	25 ton cooling		.36	89.888		16,600	4,225		20,825	24,900
1060	30 ton cooling	↓	.31	101	↓	17,500	4,800		22,300	26,600

23 64 Packaged Water Chillers

23 64 19 – Reciprocating Water Chillers

23 64 19.10 Reciprocating Type Water Chillers		Crew	Daily Output	Labor-Hours	Unit	2009 Bare Costs Material	Labor	Equipment	Total	Total Incl O&P
1080	40 ton cooling	Q-7	.30	108	Ea.	20,300	5,075		25,375	30,200
1100	50 ton cooling		.28	113		31,300	5,350		36,650	42,700
1120	60 ton cooling		.25	125		34,500	5,925		40,425	47,200
1130	70 ton cooling		.23	139		46,600	6,550		53,150	61,500
1140	80 ton cooling		.21	151		48,800	7,125		55,925	64,500
1150	90 ton cooling		.19	164		52,000	7,750		59,750	69,500
1160	100 ton cooling		.18	179		56,000	8,450		64,450	74,500
1170	110 ton cooling		.17	190		59,000	8,950		67,950	79,000
1180	120 ton cooling		.16	196		63,500	9,225		72,725	84,000
1200	140 ton cooling		.16	202		74,500	9,525		84,025	97,000
1210	160 ton cooling	↓	.15	210	↓	79,000	9,900		88,900	102,500
1300	Water cooled, single compressor, semi-hermetic, tower not incl.									
1320	40 ton cooling,	Q-7	.30	108	Ea.	16,000	5,075		21,075	25,500
1340	50 ton cooling		.28	113		19,800	5,350		25,150	30,100
1360	60 ton cooling	↓	.25	125	↓	22,000	5,925		27,925	33,400
1451	Water cooled, dual compressors, semi-hermetic, tower not incl.									
1500	80 ton cooling	Q-7	.14	222	Ea.	27,400	10,400		37,800	46,300
1520	100 ton cooling		.14	228		37,000	10,700		47,700	57,500
1540	120 ton cooling	↓	.14	231	↓	44,800	10,900		55,700	66,000
4000	Packaged chiller, remote air cooled condensers not incl.									
4020	15 ton cooling	Q-7	.30	108	Ea.	13,500	5,075		18,575	22,700
4030	20 ton cooling		.28	115		14,800	5,450		20,250	24,800
4040	25 ton cooling		.25	125		16,700	5,925		22,625	27,500
4050	30 ton cooling		.24	133		17,500	6,300		23,800	29,100
4060	40 ton cooling		.22	144		20,300	6,775		27,075	32,800
4070	50 ton cooling		.21	153		25,000	7,225		32,225	38,700
4080	60 ton cooling		.20	164		29,100	7,725		36,825	44,000
4090	70 ton cooling		.18	173		46,600	8,175		54,775	64,000
4100	80 ton cooling		.17	183		48,800	8,650		57,450	67,000
4110	90 ton cooling		.17	193		52,000	9,125		61,125	71,500
4120	100 ton cooling		.16	203		56,000	9,575		65,575	76,500
4130	110 ton cooling		.15	213		59,000	10,000		69,000	80,500
4140	120 ton cooling		.14	223		63,500	10,500		74,000	86,000
4150	140 ton cooling	↓	.14	233	↓	74,500	11,000		85,500	99,000

23 64 23 – Scroll Water Chillers

23 64 23.10 Scroll Water Chillers		Crew	Daily Output	Labor-Hours	Unit	Material	Labor	Equipment	Total	Total Incl O&P
0010	**SCROLL WATER CHILLERS,** With standard controls.									
0490	Packaged w/integral air cooled condenser, 15 ton cool	Q-7	.37	86.486	Ea.	16,500	4,075		20,575	24,400
0500	20 ton cooling		.34	94.118		21,800	4,425		26,225	30,800
0510	25 ton cooling		.34	94.118		22,800	4,425		27,225	31,900
0515	30 ton cooling		.31	101		24,400	4,800		29,200	34,200
0517	35 ton cooling		.31	104		27,300	4,925		32,225	37,700
0520	40 ton cooling		.30	108		29,500	5,075		34,575	40,400
0528	45 ton cooling		.29	109		31,400	5,175		36,575	42,500
0536	50 ton cooling	↓	.28	113	↓	34,300	5,325		39,625	46,000
0680	Scroll water cooled, single compressor, hermetic, tower not incl.									
0700	2 ton cooling	Q-5	.57	28.070	Ea.	3,225	1,250		4,475	5,475
0710	5 ton cooling		.57	28.070		3,850	1,250		5,100	6,175
0720	6 ton cooling		.42	38.005		4,850	1,700		6,550	7,975
0740	8 ton cooling	↓	.31	52.117		5,325	2,325		7,650	9,450
0760	10 ton cooling	Q-6	.36	67.039		6,125	3,100		9,225	11,500
0780	15 ton cooling	"	.33	72.727	↓	9,625	3,350		12,975	15,800

23 64 Packaged Water Chillers

23 64 23 – Scroll Water Chillers

23 64 23.10 Scroll Water Chillers		Crew	Daily Output	Labor-Hours	Unit	2009 Bare Costs Material	Labor	Equipment	Total	Total Incl O&P
0800	20 ton cooling	Q-7	.38	83.990	Ea.	10,900	3,950		14,850	18,100
0820	30 ton cooling	″	.33	96.096		12,200	4,525		16,725	20,500

23 64 26 – Rotary-Screw Water Chillers

23 64 26.10 Rotary-Screw Type Water Chillers

		Crew	Daily Output	Labor-Hours	Unit	Material	Labor	Equipment	Total	Total Incl O&P
0010	**ROTARY-SCREW TYPE WATER CHILLERS**, With standard controls									
0110	Screw, liquid chiller, air cooled, insulated evaporator									
0120	130 ton	Q-7	.14	228	Ea.	79,500	10,700		90,200	104,000
0124	160 ton		.13	246		98,000	11,600		109,600	125,500
0128	180 ton		.13	250		110,000	11,800		121,800	139,000
0132	210 ton		.12	258		121,500	12,100		133,600	152,500
0136	270 ton		.12	266		138,500	12,500		151,000	172,000
0140	320 ton		.12	275		174,000	13,000		187,000	211,500
0240	200 ton		.13	251		82,000	11,800		93,800	108,500
1450	Water cooled, tower not included									
1560	135 ton cooling, screw compressors	Q-7	.14	235	Ea.	55,000	11,100		66,100	77,500
1580	150 ton cooling, screw compressors		.13	240		63,500	11,300		74,800	87,500
1620	200 ton cooling, screw compressors		.13	250		87,500	11,800		99,300	114,500
1660	291 ton cooling, screw compressors		.12	260		91,500	12,200		103,700	120,000

23 64 33 – Direct Expansion Water Chillers

23 64 33.10 Direct Expansion Type Water Chillers

		Crew	Daily Output	Labor-Hours	Unit	Material	Labor	Equipment	Total	Total Incl O&P
0010	**DIRECT EXPANSION TYPE WATER CHILLERS**, With standard controls									
9000	Minimum labor/equipment charge	Q-6	1	24	Job		1,100		1,100	1,725

23 65 Cooling Towers

23 65 13 – Forced-Draft Cooling Towers

23 65 13.10 Forced-Draft Type Cooling Towers

		Crew	Daily Output	Labor-Hours	Unit	Material	Labor	Equipment	Total	Total Incl O&P
0010	**FORCED-DRAFT TYPE COOLING TOWERS**, Packaged units R262213-27									
0070	Galvanized steel									
0080	Induced draft, crossflow									
0100	Vertical, belt drive, 61 tons	Q-6	90	.267	TonAC	137	12.30		149.30	170
0150	100 ton		100	.240		150	11.05		161.05	182
0200	115 ton		109	.220		130	10.15		140.15	159
0250	131 ton		120	.200		169	9.20		178.20	200
0260	162 ton		132	.182		137	8.40		145.40	163
1000	For higher capacities, use multiples									
1500	Induced air, double flow									
1900	Vertical, gear drive, 167 ton	Q-6	126	.190	TonAC	133	8.80		141.80	160
2000	297 ton		129	.186		83	8.55		91.55	104
2100	582 ton		132	.182		67	8.40		75.40	86.50
2150	849 ton		142	.169		65	7.80		72.80	83.50
2200	1016 ton		150	.160		60	7.35		67.35	77.50
3000	For higher capacities, use multiples									
3500	For pumps and piping, add	Q-6	38	.632	TonAC	68.50	29		97.50	121
4000	For absorption systems, add				″	75%	75%			
5000	Fiberglass									
5010	Draw thru									
5100	60 ton	Q-6	1.50	16	Ea.	3,650	735		4,385	5,175
5120	125 ton		.99	24.242		7,475	1,125		8,600	9,925
5140	300 ton		.43	55.814		17,500	2,575		20,075	23,200
5160	600 ton		.22	109		31,400	5,025		36,425	42,300

23 65 Cooling Towers

23 65 13 – Forced-Draft Cooling Towers

23 65 13.10 Forced-Draft Type Cooling Towers		Crew	Daily Output	Labor-Hours	Unit	Material	Labor	Equipment	Total	Total Incl O&P
						2009 Bare Costs				
5180	1000 ton	Q-6	.15	160	Ea.	54,000	7,375		61,375	71,000
6000	Stainless steel									
6010	Induced draft, crossflow, horizontal, belt drive									
6100	57 ton	Q-6	1.50	16	Ea.	20,400	735		21,135	23,700
6120	91 ton		.99	24.242		24,700	1,125		25,825	28,800
6140	111 ton		.43	55.814		37,100	2,575		39,675	44,800
6160	126 ton		.22	109		37,100	5,025		42,125	48,600
6170	Induced draft, crossflow, vertical, gear drive									
6172	167 ton	Q-6	.75	32	Ea.	47,100	1,475		48,575	54,500
6174	297 ton		.43	55.814		52,500	2,575		55,075	61,500
6176	582 ton		.23	104		83,000	4,800		87,800	98,500
6178	849 ton		.17	141		117,000	6,500		123,500	138,500
6180	1016 ton		.15	160		130,000	7,375		137,375	154,500
9000	Minimum labor/equipment charge		1	24	Job		1,100		1,100	1,725

23 72 Air-to-Air Energy Recovery Equipment

23 72 13 – Heat-Wheel Air-to-Air Energy-Recovery Equipment

23 72 13.10 Heat-Wheel Air-to-air Energy Recovery Equip.

		Crew	Daily Output	Labor-Hours	Unit	Material	Labor	Equipment	Total	Total Incl O&P
0010	**HEAT-WHEEL AIR-TO-AIR ENERGY RECOVERY EQUIPMENT**									
0100	Air to air									
4000	Enthalpy recovery wheel									
4010	1000 max CFM G	Q-9	1.20	13.333	Ea.	6,050	565		6,615	7,550
4020	2000 max CFM G		1	16		7,075	680		7,755	8,850
4030	4000 max CFM G		.80	20		8,200	850		9,050	10,400
4040	6000 max CFM G		.70	22.857		9,575	970		10,545	12,100
4050	8000 max CFM G	Q-10	1	24		10,600	1,050		11,650	13,300
4060	10,000 max CFM G		.90	26.667		12,700	1,175		13,875	15,800
4070	20,000 max CFM G		.80	30		22,900	1,325		24,225	27,300
4080	25,000 max CFM G		.70	34.286		28,000	1,500		29,500	33,200
4090	30,000 max CFM G		.50	48		31,100	2,125		33,225	37,600
4100	40,000 max CFM G		.45	53.333		42,900	2,350		45,250	51,000
4110	50,000 max CFM G		.40	60		49,900	2,650		52,550	59,000
9000	Minimum labor/equipment charge	1 Shee	4	2	Job		94.50		94.50	150

23 73 Indoor Central-Station Air-Handling Units

23 73 13 – Modular Indoor Central-Station Air-Handling Units

23 73 13.20 Air Handling Units

		Crew	Daily Output	Labor-Hours	Unit	Material	Labor	Equipment	Total	Total Incl O&P
0010	**AIR HANDLING UNITS**, Chilled water R262213-27									
3000	Packaged, 3000 CFM, 7.5 ton	Q-6	1.70	14.118	Ea.	2,450	650		3,100	3,700
3050	3200 CFM, 8 ton		1.60	15		2,800	690		3,490	4,150
3100	4000 CFM, 10 ton		1.40	17.143		2,950	790		3,740	4,475
3150	4400 CFM, 11 ton		1.35	17.778		3,500	820		4,320	5,125
3200	6000 CFM, 15 ton		1	24		3,850	1,100		4,950	5,950
3250	7000 CFM, 17.5 ton		.90	26.667		4,600	1,225		5,825	6,950
3300	10,000 CFM, 25 ton		.65	36.923		6,950	1,700		8,650	10,300
3400	12,000 CFM, 30 ton		.50	48		7,550	2,200		9,750	11,700
3500	For hot water heat coil, add					30%	5%			

23 73 Indoor Central-Station Air-Handling Units

23 73 39 – Indoor, Direct Gas-Fired Heating and Ventilating Units

23 73 39.10 Make-Up Air Unit		Crew	Daily Output	Labor-Hours	Unit	2009 Bare Costs Material	Labor	Equipment	Total	Total Incl O&P
0010	MAKE-UP AIR UNIT R233400-10									
0020	Indoor suspension, natural/LP gas, direct fired,									
0032	standard control. For flue see Div. 23 51 23.10									
0040	70° F temperature rise, MBH is input									
0100	2000 CFM, 168 MBH	Q-6	3	8	Ea.	9,050	370		9,420	10,500
0120	3000 CFM, 252 MBH		2	12		9,300	555		9,855	11,100
0140	4000 CFM, 336 MBH		1.80	13.333		9,450	615		10,065	11,400
0160	6000 CFM, 502 MBH		1.50	16		11,400	735		12,135	13,700
0180	8000 CFM, 670 MBH		1.40	17.143		12,500	790		13,290	14,900
0200	10,000 CFM, 838 MBH		1.20	20		14,000	920		14,920	16,800
0220	12,000 CFM, 1005 MBH		1	24		14,200	1,100		15,300	17,300
0240	14,000 CFM, 1180 MBH		.94	25.532		14,400	1,175		15,575	17,600
0260	18,000 CFM, 1340 MBH		.86	27.907		14,500	1,275		15,775	17,900
0280	20,000 CFM, 1675 MBH		.76	31.579		17,800	1,450		19,250	21,900
0300	24,000 CFM, 2007 MBH	Q-7	1	32		18,500	1,500		20,000	22,600
0320	30,000 CFM, 2510 MBH		.96	33.333		18,900	1,575		20,475	23,200
0340	35,000 CFM, 2930 MBH		.92	34.783		23,900	1,625		25,525	28,800
0360	40,000 CFM, 3350 MBH		.88	36.364		24,500	1,700		26,200	29,600
0380	45,000 CFM, 3770 MBH		.84	38.095		25,100	1,800		26,900	30,400
0400	50,000 CFM, 4180 MBH		.80	40		25,400	1,875		27,275	30,800
0420	55,000 CFM, 4600 MBH		.75	42.667		32,400	2,000		34,400	38,700
0440	60,000 CFM, 5020 MBH		.70	45.714		33,600	2,150		35,750	40,300
0460	65,000 CFM, 5435 MBH		.60	53.333		33,800	2,500		36,300	41,100
0480	75,000 CFM, 6275 MBH		.50	64		35,600	3,000		38,600	43,900
0600	For discharge louver assembly, add					10%				
0700	For filters, add					20%				
0800	For air shut-off damper section, add					10%				
0900	For vertical unit, add					10%				

23 74 Packaged Outdoor HVAC Equipment

23 74 23 – Packaged, Outdoor, Heating-Only Makeup-Air Units

23 74 23.16 Pkgd.,Ind.-Fired,Outdoor, Heat-Only Makeup-Air		Crew	Daily Output	Labor-Hours	Unit	Material	Labor	Equipment	Total	Total Incl O&P
0010	PACKAGED, IND.-FIRED, OUTDOOR, HEAT-ONLY MAKEUP-AIR UNITS									
1000	Rooftop unit, natural gas, gravity vent, S.S. exchanger									
1010	70° F temperature rise, MBH is input									
1020	995 CFM, 75 MBH	Q-6	4	6	Ea.	6,075	276		6,351	7,100
1040	1245 CFM, 95 MBH		3.60	6.667		6,125	305		6,430	7,225
1060	1590 CFM, 120 MBH		3.30	7.273		6,600	335		6,935	7,800
1080	2090 CFM, 159 MBH		3	8		6,800	370		7,170	8,075
1100	2490 CFM, 190 MBH		2.60	9.231		7,800	425		8,225	9,225
1120	2985 CFM, 225 MBH		2.30	10.435		7,925	480		8,405	9,450
1140	3975 CFM, 300 MBH		1.90	12.632		9,375	580		9,955	11,200
1160	4970 CFM, 375 MBH		1.40	17.143		10,900	790		11,690	13,100
1180	5950 CFM, 450 MBH		1.20	20		11,400	920		12,320	14,000
1200	7940 CFM, 600 MBH		1	24		13,300	1,100		14,400	16,300
1220	9930 CFM, 750 MBH		.80	30		22,700	1,375		24,075	27,200
1240	11,900 CFM, 900 MBH		.60	40		25,400	1,850		27,250	30,800
1260	15,880 CFM, 1200 MBH		.30	80		30,500	3,675		34,175	39,200
1500	For power vent, add					8%				
1600	For cleanable filters, add				Ea.	176			176	194

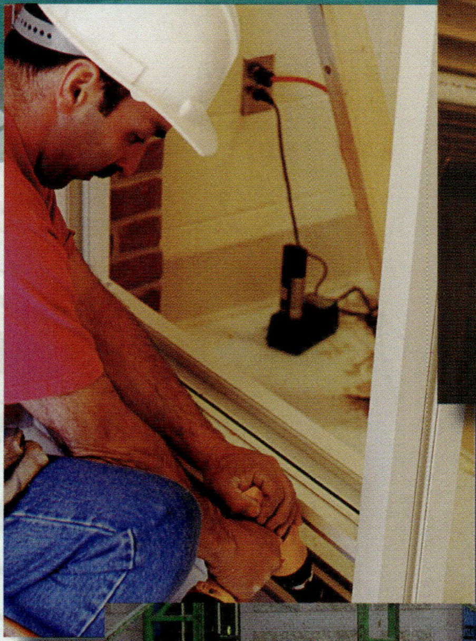

Means FacilitiesWorks™ Specialized Features...

- Technical evaluation of owner and contractor estimate.
- On-screen take-off capability.
- Powerful search engine tools to quickly find unit-price line items using the Means index and by customizing the index.
- Copy and paste estimates or estimate line items.
- Updated estimates to reflect price changes in seconds.
- Enhanced line-item estimating with the ability to include an unlimited amount of detailed notes as well as engineering calculations at the line-item level.
- Trackable pre-priced and non-priced line items and project percentages automatically done at the estimate level, project level, and contract level.
- Handles multiple coefficients.
- Means CCI can be applied uniquely by CSI division.

JOC Estimates by discipline, estimator, location, contractor or group

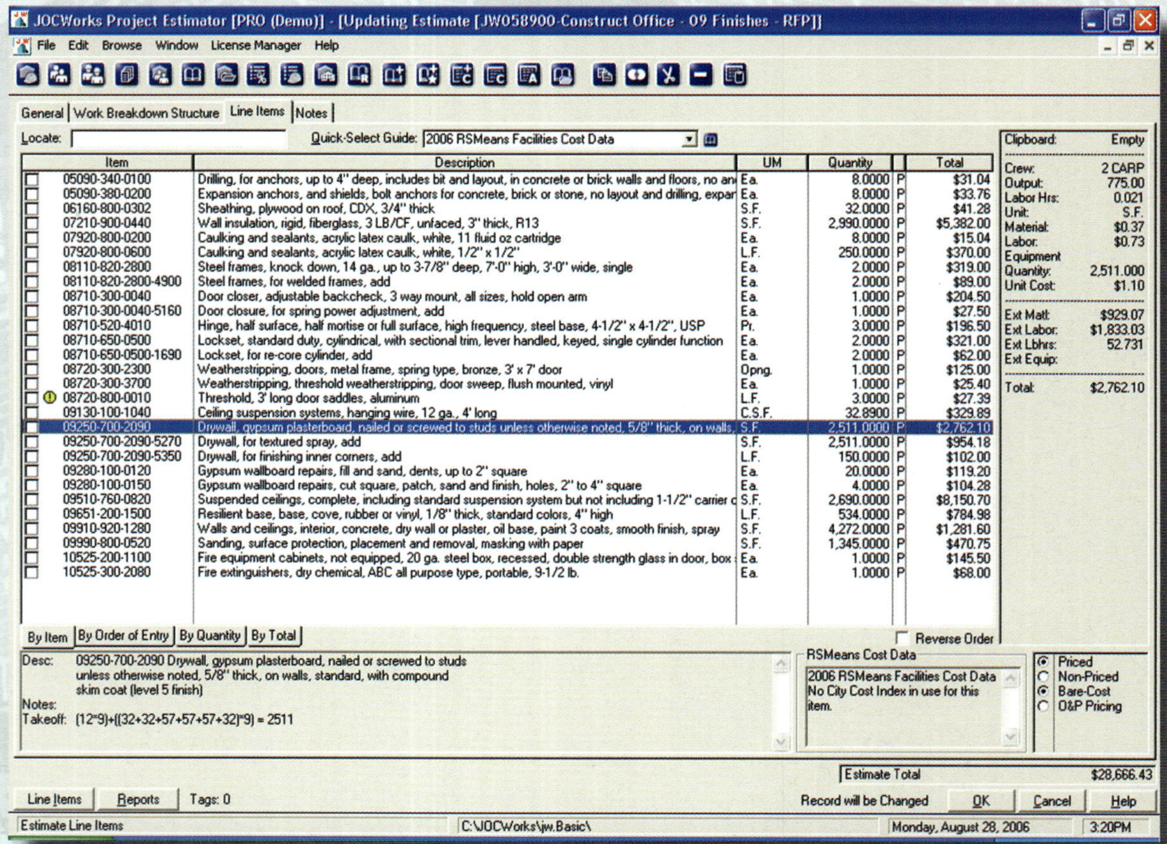

For more information, contact:
Robert Gair, Principal
RSMeans Business Solutions
202-285-7826
rgair@reedbusiness.com
www.rsmeans.com

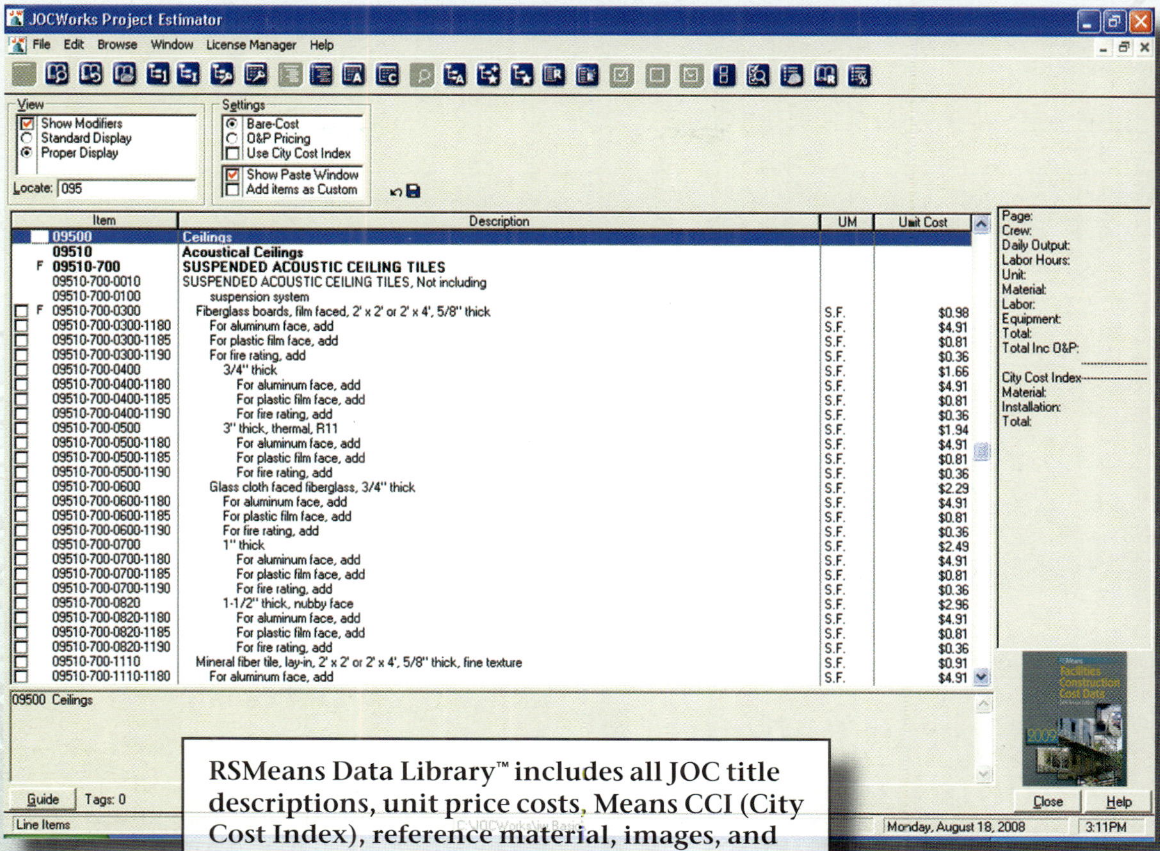

RSMeans Data Library™ includes all JOC title descriptions, unit price costs, Means CCI (City Cost Index), reference material, images, and graphs to find items quickly.

JOC is the preferred delivery method for renovation projects in many sectors and within states including Arizona, California, Florida, Minnesota, Missouri, Texas, and Washington.

Means FacilitiesWorks™

- Line item markups are linked to the material item
- Support of multiple- and dual-award contracts and pricing guides
- Quick modification tools that allow changes to one or more projects, estimates, or estimate line items
- Organization of projects by contract, contract performance period, contractor, location, estimator, and customer
- Milestones for tracking projects from design through warranty, including site visits, invoicing, change orders, and material submittals
- Complete description of line items: crew, material, labor, and equipment
- Provision of audit trail for estimate updates and changes
- Exportable reports to Microsoft Excel™, Microsoft Word™, and Adobe™ PDF

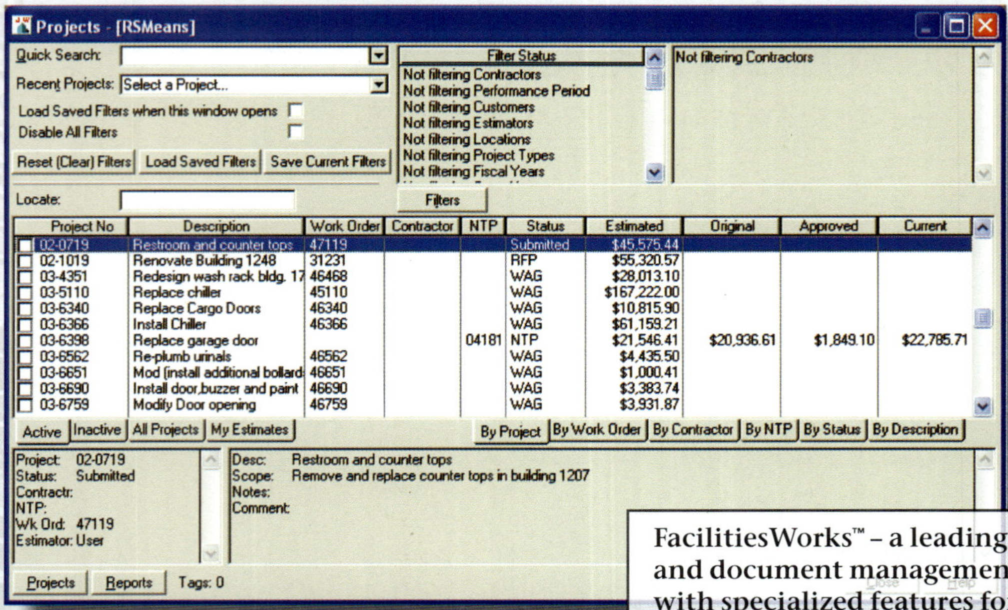

FacilitiesWorks™ – a leading project management and document management software program with specialized features for renovation and repair projects.

Facility-Owner Benefits of Adopting JOC

- Dramatically reduces the backlog of renovation projects
- Reduces administrative costs by accomplishing more projects with the same team
- Leverages purchasing and maintenance tasks
- Dramatically reduces change orders
- Decreases up-front design and administrative costs and time
- Eliminates multiple project solicitation
- Minimizes construction claims
- Dramatically increases facility productivity

RSMeans services for developing a JOC System

A JOC-Knowledgeable Staff: Facility Management staffs must have knowledge of the JOC process, how to interact with JOC contractors, and the concept of partnering that is integral to successful JOCs. This requires familiarity with the master bidding document (i.e., Unit Price Book) and the development of JOC project specifications and contracts.

The RSMeans Unit Price Book (UPB): This customized document is the line-item listing of JOC tasks and the components of those tasks. When both the JOC client and contractor agree to this document, negotiations are reduced to project-specific peculiarities.

Training Programs: JOC-specific training programs to understand how to organize a JOC program, how to effectively use RSMeans Unit Price Book for JOC, and how to use FacilitiesWorks™ software.

23 74 23 – Packaged, Outdoor, Heating-Only Makeup-Air Units

23 74 23.16 Pkgd.,Ind.-Fired,Outdoor, Heat-Only Makeup-Air		Crew	Daily Output	Labor-Hours	Unit	Material	2009 Bare Costs Labor	Equipment	Total	Total Incl O&P
1700	For electric modulating gas control, add				Ea.	1,900			1,900	2,075
1800	For aluminized steel exchanger, deduct					10%				
9000	Minimum labor/equipment charge	Q-6	2.75	8.727	Job		400		400	625

23 74 33 – Packaged, Outdoor, Heating and Cooling Makeup Air-Conditioners

23 74 33.10 Roof Top Air Conditioners

		Crew	Daily Output	Labor-Hours	Unit	Material	Labor	Equipment	Total	Total Incl O&P
0010	**ROOF TOP AIR CONDITIONERS**, Standard controls, curb, economizer.									
1000	Single zone, electric cool, gas heat									
1090	2 ton cooling, 55 MBH heating	Q-5	.93	17.204	Ea.	3,100	765		3,865	4,575
1100	3 ton cooling, 60 MBH heating		.70	22.857		3,200	1,025		4,225	5,100
1120	4 ton cooling, 95 MBH heating R262213-27		.61	26.403		3,750	1,175		4,925	5,950
1140	5 ton cooling, 112 MBH heating		.56	28.521		4,150	1,275		5,425	6,550
1145	6 ton cooling, 140 MBH heating		.52	30.769		4,900	1,375		6,275	7,525
1150	7.5 ton cooling, 170 MBH heating		.50	32.258		6,275	1,425		7,700	9,125
1160	10 ton cooling, 200 MBH heating	Q-6	.67	35.982		8,300	1,650		9,950	11,700
1170	12.5 ton cooling, 230 MBH heating		.63	37.975		9,400	1,750		11,150	13,000
1180	15 ton cooling, 270 MBH heating		.57	42.032		12,500	1,925		14,425	16,800
1190	18 ton cooling, 330 MBH heating		.52	45.889		15,400	2,125		17,525	20,200
1200	20 ton cooling, 360 MBH heating	Q-7	.67	47.976		20,100	2,250		22,350	25,600
1210	25 ton cooling, 450 MBH heating		.56	57.554		23,700	2,700		26,400	30,200
1220	30 ton cooling, 540 MBH heating		.47	68.376		25,800	3,225		29,025	33,300
1240	40 ton cooling, 675 MBH heating		.35	91.168		32,900	4,275		37,175	42,900
2000	Multizone, electric cool, gas heat, economizer									
2100	15 ton cooling, 360 MBH heating	Q-7	.61	52.545	Ea.	53,500	2,475		55,975	63,000
2120	20 ton cooling, 360 MBH heating		.53	60.038		57,500	2,825		60,325	68,000
2140	25 ton cooling, 450 MBH heating		.45	71.910		69,000	3,375		72,375	81,500
2160	28 ton cooling, 450 MBH heating		.41	79.012		78,500	3,725		82,225	92,500
2180	30 ton cooling, 540 MBH heating		.37	85.562		88,500	4,025		92,525	104,000
2200	40 ton cooling, 540 MBH heating		.28	113		101,000	5,350		106,350	119,500
2210	50 ton cooling, 540 MBH heating		.23	142		125,500	6,675		132,175	148,500
2220	70 ton cooling, 1500 MBH heating		.16	198		136,000	9,350		145,350	164,000
2240	80 ton cooling, 1500 MBH heating		.14	228		155,500	10,700		166,200	187,500
2260	90 ton cooling, 1500 MBH heating		.13	256		163,000	12,000		175,000	198,000
2280	105 ton cooling, 1500 MBH heating		.11	290		179,500	13,700		193,200	218,500
2400	For hot water heat coil, deduct					5%				
2500	For steam heat coil, deduct					2%				
2600	For electric heat, deduct					3%	5%			
5000	Single zone, electric cool only									
5050	3 ton cooling	Q-5	.88	18.203	Ea.	3,025	810		3,835	4,575
5060	4 ton cooling		.76	21.108		3,450	940		4,390	5,250
5070	5 ton cooling		.70	22.792		3,925	1,025		4,950	5,900
5080	6 ton cooling		.65	24.502		4,500	1,100		5,600	6,650
5090	7.5 ton cooling		.62	25.806		5,675	1,150		6,825	8,025
5100	8.5 ton cooling		.59	26.981		6,775	1,200		7,975	9,300
5110	10 ton cooling	Q-6	.83	28.812		7,075	1,325		8,400	9,825
5120	12 ton cooling		.80	30		8,900	1,375		10,275	12,000
5130	15 ton cooling		.71	33.613		12,200	1,550		13,750	15,800
5140	18 ton cooling		.65	36.697		14,500	1,700		16,200	18,500
5150	20 ton cooling	Q-7	.83	38.415		18,400	1,800		20,200	23,000
5160	25 ton cooling		.70	45.977		21,100	2,150		23,250	26,600
5170	30 ton cooling		.59	54.701		23,200	2,575		25,775	29,500
5180	40 ton cooling		.44	73.059		30,000	3,425		33,425	38,300
5400	For low heat, add					7%				

23 74 Packaged Outdoor HVAC Equipment

23 74 33 – Packaged, Outdoor, Heating and Cooling Makeup Air-Conditioners

23 74 33.10 Roof Top Air Conditioners		Crew	Daily Output	Labor-Hours	Unit	Material	2009 Bare Costs Labor	Equipment	Total	Total Incl O&P
5410	For high heat, add				Ea.	10%				
6000	Single zone electric cooling with variable volume distribution									
6020	20 ton cooling	Q-7	.72	44.199	Ea.	23,300	2,075		25,375	28,900
6030	25 ton cooling		.61	52.893		27,100	2,475		29,575	33,700
6040	30 ton cooling		.51	62.868		30,100	2,950		33,050	37,700
6050	40 ton cooling		.38	83.990		39,300	3,950		43,250	49,400
6060	50 ton cooling		.31	104		46,300	4,925		51,225	58,500
6070	60 ton cooling	↓	.25	125		50,500	5,925		56,425	64,500
6200	For low heat, add					7%				
6210	For high heat, add				↓	10%				
7000	Multizone, cool/heat, variable volume distribution									
7100	50 ton cooling	Q-7	.20	164	Ea.	93,500	7,725		101,225	114,500
7110	70 ton cooling		.14	228		130,500	10,700		141,200	160,000
7120	90 ton cooling		.11	296		150,000	13,900		163,900	186,500
7130	105 ton cooling		.10	333		168,000	15,700		183,700	209,000
7140	120 ton cooling		.08	380		192,000	17,900		209,900	239,000
7150	140 ton cooling	↓	.07	444	↓	205,500	20,900		226,400	258,500
7300	Penthouse unit, cool/heat, variable volume distribution									
7400	150 ton cooling	Q-7	.10	310	Ea.	336,000	14,600		350,600	392,000
7410	170 ton cooling		.10	320		359,000	15,000		374,000	418,500
7420	200 ton cooling		.10	326		422,500	15,400		437,900	488,500
7430	225 ton cooling		.09	340		462,000	16,000		478,000	533,500
7440	250 ton cooling		.09	359		476,500	16,900		493,400	550,500
7450	270 ton cooling		.09	363		515,000	17,100		532,100	593,000
7460	300 ton cooling	↓	.09	372	↓	572,000	17,500		589,500	656,500
9000	Minimum labor/equipment charge	Q-5	1.50	10.667	Job		475		475	735

23 76 Evaporative Air-Cooling Equipment

23 76 13 – Direct Evaporative Air Coolers

23 76 13.10 Evaporative Coolers			Crew	Daily Output	Labor-Hours	Unit	Material	Labor	Equipment	Total	Total Incl O&P
0010	**EVAPORATIVE COOLERS,** (Swamp Coolers) ducted, not incl. duct.										
0100	Side discharge style, capacities at .25" S.P.										
0120	1785 CFM, 1/3 HP, 115 V	G	Q-9	5	3.200	Ea.	465	136		601	730
0140	2740 CFM, 1/3 HP, 115 V	G		4.50	3.556		570	151		721	870
0160	3235 CFM, 1/2 HP, 115 V	G		4	4		585	170		755	915
0180	3615 CFM, 1/2 HP, 230 V	G		3.60	4.444		595	189		784	955
0200	4215 CFM, 3/4 HP, 230 V	G		3.20	5		760	212		972	1,175
0220	5255 CFM. 1 HP, 115/230 V	G		3	5.333		1,250	227		1,477	1,725
0240	6090 CFM, 1 HP, 230/460 V	G		2.80	5.714		1,750	243		1,993	2,300
0260	8300 CFM, 1-1/2 HP, 230/460 V	G		2.60	6.154		1,750	261		2,011	2,350
0280	8360 CFM, 1-1/2 HP, 230/460 V	G		2.20	7.273		2,400	310		2,710	3,125
0300	9725 CFM, 2 HP, 230/460 V	G		1.80	8.889		2,400	380		2,780	3,250
0320	11,715 CFM, 3 HP, 230/460 V	G		1.40	11.429		2,450	485		2,935	3,475
0340	14,410 CFM, 5 HP, 230/460 V	G	↓	1	16	↓	2,575	680		3,255	3,925
0400	For two-speed motor, add						5%				
0500	For down discharge style, add						10%				
1000	High capacities at .4" S.P.										
1040	15,000 CFM, 5 HP, 230/460 V	G	Q-9	.54	29.630	Ea.	13,300	1,250		14,550	16,600
1080	20,000 CFM, 10 HP, 230/460 V	G	"	.48	33.333		16,300	1,425		17,725	20,300
1100	25,000 CFM, 20 HP, 230/460 V	G	Q-10	.54	44.444		18,200	1,950		20,150	23,100
1140	30,000 CFM, 20 HP, 230/460 V	G	↓	.46	52.174	↓	20,900	2,300		23,200	26,700

23 76 Evaporative Air-Cooling Equipment

23 76 13 – Direct Evaporative Air Coolers

23 76 13.10 Evaporative Coolers		Crew	Daily Output	Labor-Hours	Unit	Material	2009 Bare Costs Labor	Equipment	Total	Total Incl O&P
1160	35,000 CFM, 20 HP, 230/460 V G	Q-10	44	54.545	Ea.	22,600	2,400		25,000	28,700
1180	40,000 CFM, 30 HP, 230/460 V G		42	57.143		23,800	2,525		26,325	30,100

23 76 16 – Indirect Evaporative Air Coolers

23 76 16.10 Evaporators

		Crew	Daily Output	Labor-Hours	Unit	Material	Labor	Equipment	Total	Total Incl O&P
0010	**EVAPORATORS**, DX coils, remote compressors not included.									
1000	Coolers, reach-in type, above freezing temperatures									
1300	Shallow depth, wall mount, 7 fins per inch, air defrost									
1310	600 BTUH, 8" fan, rust-proof core	Q-5	3.80	4.211	Ea.	380	187		567	710
1320	900 BTUH, 8" fan, rust-proof core		3.30	4.848		470	215		685	850
1330	1200 BTUH, 8" fan, rust-proof core		3	5.333		555	237		792	975
1340	1800 BTUH, 8" fan, rust-proof core		2.40	6.667		640	296		936	1,175
1350	2500 BTUH, 10" fan, rust-proof core		2.20	7.273		695	325		1,020	1,275
1360	3500 BTUH, 10" fan		2	8		815	355		1,170	1,450
1370	4500 BTUH, 10" fan		1.90	8.421		860	375		1,235	1,525
1390	For pan drain, add					45.50			45.50	50
1600	Undercounter refrigerators, ceiling or wall mount,									
1610	8 fins per inch, air defrost, rust-proof core									
1630	800 BTUH, one 6" fan	Q-5	5	3.200	Ea.	256	142		398	500
1640	1300 BTUH, two 6" fans		4	4		310	178		488	615
1650	1700 BTUH, two 6" fans		3.60	4.444		345	157		542	685
2000	Coolers, reach-in and walk-in types, above freezing temperatures									
2600	Two-way discharge, ceiling mount, 150-4100 CFM,									
2610	above 34° F applications, air defrost									
2630	900 BTUH, 7 fins per inch, 8" fan	Q-5	4	4	Ea.	271	178		449	575
2660	2500 BTUH, 7 fins per inch, 10" fan		2.60	6.154		535	273		808	1,000
2690	5500 BTUH, 7 fins per inch, 12" fan		1.70	9.412		1,050	420		1,470	1,800
2720	8500 BTUH, 8 fins per inch, 16" fan		1.20	13.333		1,300	590		1,890	2,375
2750	15,000 BTUH, 7 fins per inch, 18" fan		1.10	14.545		1,825	645		2,470	3,000
2770	24,000 BTUH, 7 fins per inch, two 16" fans		1	16		2,750	710		3,460	4,100
2790	30,000 BTUH, 7 fins per inch, two 18" fans		.90	17.778		3,400	790		4,190	4,950
2850	Two-way discharge, low profile, ceiling mount,									
2860	8 fins per inch, 200-570 CFM, air defrost									
2880	800 BTUH, one 6" fan	Q-5	4	4	Ea.	289	178		467	595
2890	1300 BTUH, two 6" fans		3.50	4.571		340	203		543	690
2900	1800 BTUH, two 6" fans		3.30	4.848		360	215		575	730
2910	2700 BTUH, three 6" fans		2.70	5.926		425	263		688	875
3000	Coolers, walk-in type, above freezing temperatures									
3300	General use, ceiling mount, 108-2080 CFM									
3320	600 BTUH, 7 fins per inch, 6" fan	Q-5	5.20	3.077	Ea.	225	137		362	460
3340	1200 BTUH, 7 fins per inch, 8" fan		3.70	4.324		310	192		502	640
3360	1800 BTUH, 7 fins per inch, 10" fan		2.70	5.926		435	263		698	890
3380	3500 BTUH, 7 fins per inch, 12" fan		2.40	6.667		640	296		936	1,175
3400	5500 BTUH, 8 fins per inch, 12" fan		2	8		715	355		1,070	1,350
3430	8500 BTUH, 8 fins per inch, 16" fan		1.40	11.429		1,075	510		1,585	1,950
3460	15,000 BTUH, 7 fins per inch, two 16" fans		1.10	14.545		1,700	645		2,345	2,875
3640	Low velocity, high latent load, ceiling mount,									
3650	1050-2420 CFM, air defrost, 6 fins per inch									
3670	6700 BTUH, two 10" fans	Q-5	1.30	12.308	Ea.	1,500	545		2,045	2,500
3680	10,000 BTUH, three 10" fans		1	16		1,750	710		2,460	3,025
3690	13,500 BTUH, three 10" fans		1	16		2,100	710		2,810	3,400
3700	18,000 BTUH, four 10" fans		1	16		2,800	710		3,510	4,175
3710	26,500 BTUH, four 10" fans		.80	20		3,450	890		4,340	5,150

23 76 16 – Indirect Evaporative Air Coolers

23 76 16.10 Evaporators		Crew	Daily Output	Labor-Hours	Unit	Material	2009 Bare Costs Labor	Equipment	Total	Total Incl O&P
3730	For electric defrost, add				Ea.	17%				
5000	Freezers and coolers, reach-in type, above 34° F									
5030	to sub-freezing temperature range, low latent load,									
5050	air defrost, 7 fins per inch									
5070	1200 BTUH, 8" fan, rustproof core	Q-5	4.40	3.636	Ea.	315	162		477	595
5080	1500 BTUH, 8" fan, rustproof core		4.20	3.810		325	169		494	620
5090	1800 BTUH, 8" fan, rustproof core		3.70	4.324		340	192		532	675
5100	2500 BTUH, 10" fan, rustproof core		2.90	5.517		390	245		635	805
5110	3500 BTUH, 10" fan, rustproof core		2.50	6.400		530	284		814	1,025
5120	4500 BTUH, 12" fan, rustproof core		2.10	7.619		705	340		1,045	1,300
6000	Freezers and coolers, walk-in type									
6050	1960-18,000 CFM, medium profile									
6060	Standard motor, 6 fins per inch, to -30° F, all aluminum									
6080	10,500 BTUH, air defrost, one 18" fan	Q-5	1.40	11.429	Ea.	1,275	510		1,785	2,200
6090	12,500 BTUH, air defrost, one 18" fan		1.40	11.429		1,400	510		1,910	2,300
6100	16,400 BTUH, air defrost, one 18" fan		1.30	12.308		1,500	545		2,045	2,500
6110	20,900 BTUH, air defrost, one 18" fans		1	16		1,750	710		2,460	3,025
6120	27,000 BTUH, air defrost, two 18" fans		1	16		2,600	710		3,310	3,950
6130	32,900 BTUH, air defrost, two 20" fans	Q-6	1.30	18.462		2,900	850		3,750	4,500
6140	39,000 BTUH, air defrost, two 20" fans		1.25	19.200		3,375	885		4,260	5,075
6150	44,100 BTUH, air defrost, three 20" fans		1	24		3,825	1,100		4,925	5,925
6155	53,000 BTUH, air defrost, three 20" fans		1	24		3,825	1,100		4,925	5,925
6160	66,200 BTUH, air defrost, four 20" fans		1	24		4,350	1,100		5,450	6,500
6170	78,000 BTUH, air defrost, four 20" fans		.80	30		5,400	1,375		6,775	8,100
6180	88,200 BTUH, air defrost, four 24" fans		.60	40		5,750	1,850		7,600	9,175
6190	110,000 BTUH, air defrost, four 24" fans		.50	48		6,525	2,200		8,725	10,600
6330	Hot gas defrost, standard motor units, add					35%				
6370	Electric defrost, 230V, standard motor, add					16%				
6410	460V, standard or high capacity motor, add					25%				
6800	Eight fins per inch, increases BTUH 75%									
6810	Standard capacity, add				Ea.	5%				
6830	Hot gas & electric defrost not recommended									
7000	800-4150 CFM, 12" fans									
7010	Four fins per inch									
7030	3400 BTUH, 1 fan	Q-5	2.80	5.714	Ea.	460	254		714	900
7040	4200 BTUH, 1 fan		2.40	6.667		470	296		766	975
7050	5300 BTUH, 1 fan		1.90	8.421		540	375		915	1,175
7060	6800 BTUH, 1 fan		1.70	9.412		595	420		1,015	1,300
7070	8400 BTUH, 2 fans		1.60	10		815	445		1,260	1,575
7080	10,500 BTUH, 2 fans		1.40	11.429		885	510		1,395	1,750
7090	13,000 BTUH, 3 fans		1.30	12.308		1,150	545		1,695	2,100
7100	17,000 BTUH, 4 fans		1.20	13.333		1,450	590		2,040	2,525
7110	21,500 BTUH, 5 fans		1	16		1,775	710		2,485	3,075
7160	Six fins per inch increases BTUH 35%, add					5%				
7180	Eight fins per inch increases BTUH 53%,									
7190	not recommended for below 35° F, add				Ea.	8%				
7210	For 12° F temperature differential, add					20%				
7230	For adjustable thermostat control, add					102			102	112
7240	For hot gas defrost, except 8 fin models									
7250	on applications below 35° F, add				Ea.	58%				
7260	For electric defrost, except 8 fin models									
7270	on applications below 35° F, add				Ea.	36%				
8000	Freezers, pass-thru door uprights, walk-in storage									

23 76 Evaporative Air-Cooling Equipment

23 76 16 – Indirect Evaporative Air Coolers

23 76 16.10 Evaporators		Crew	Daily Output	Labor-Hours	Unit	2009 Bare Costs Material	Labor	Equipment	Total	Total Incl O&P
8300	Low temperature, thin profile, electric defrost									
8310	138-1800 CFM, 5 fins per inch									
8320	900 BTUH, one 6" fan	Q-5	3.30	4.848	Ea.	360	215		575	730
8340	1500 BTUH, two 6" fans		2.20	7.273		535	325		860	1,100
8360	2600 BTUH, three 6" fans		1.90	8.421		770	375		1,145	1,425
8370	3300 BTUH, four 6" fans		1.70	9.412		815	420		1,235	1,550
8380	4400 BTUH, five 6" fans		1.40	11.429		1,200	510		1,710	2,100
8400	7000 BTUH, three 10" fans		1.30	12.308		1,450	545		1,995	2,450
8410	8700 BTUH, four 10" fans		1.10	14.545		1,650	645		2,295	2,800
8450	For air defrost, deduct					176			176	194

23 81 Decentralized Unitary HVAC Equipment

23 81 13 – Packaged Terminal Air-Conditioners

23 81 13.10 Packaged Cabinet Type Air-Conditioners		Crew	Daily Output	Labor-Hours	Unit	Material	Labor	Equipment	Total	Total Incl O&P
0010	**PACKAGED CABINET TYPE AIR-CONDITIONERS**, Cabinet, wall sleeve,									
0100	louver, electric heat, thermostat, manual changeover, 208 V									
0200	6,000 BTUH cooling, 8800 BTU heat R235000-50	Q-5	6	2.667	Ea.	625	118		743	875
0220	9,000 BTUH cooling, 13,900 BTU heat		5	3.200		720	142		862	1,000
0240	12,000 BTUH cooling, 13,900 BTU heat R262213-27		4	4		780	178		958	1,125
0260	15,000 BTUH cooling, 13,900 BTU heat		3	5.333		865	237		1,102	1,325
0280	18,000 BTUH cooling, 10 KW heat		2.40	6.667		1,575	296		1,871	2,200
0300	24,000 BTUH cooling, 10 KW heat		1.90	8.421		1,725	375		2,100	2,475
0320	30,000 BTUH cooling, 10 KW heat		1.40	11.429		1,875	510		2,385	2,850
0340	36,000 BTUH cooling, 10 KW heat		1.25	12.800		1,900	570		2,470	2,975
0360	42,000 BTUH cooling, 10 KW heat		1	16		2,400	710		3,110	3,725
0380	48,000 BTUH cooling, 10 KW heat		.90	17.778		2,600	790		3,390	4,075
0500	For hot water coil, increase heat by 10%, add					5%	10%			
1000	For steam, increase heat output by 30%, add					8%	10%			

23 81 19 – Self-Contained Air-Conditioners

23 81 19.10 Window Unit Air Conditioners		Crew	Daily Output	Labor-Hours	Unit	Material	Labor	Equipment	Total	Total Incl O&P
0010	**WINDOW UNIT AIR CONDITIONERS**									
4000	Portable/window, 15 amp 125V grounded receptacle required									
4060	5000 BTUH	1 Carp	8	1	Ea.	163	40		203	245
4340	6000 BTUH		8	1		265	40		305	360
4480	8000 BTUH		6	1.333		310	53.50		363.50	435
4500	10,000 BTUH		6	1.333		395	53.50		448.50	525
4520	12,000 BTUH	L-2	8	2		510	70.50		580.50	675
4600	Window/thru-the-wall, 15 amp 230V grounded receptacle required									
4780	17,000 BTUH	L-2	6	2.667	Ea.	695	93.50		788.50	920
4940	25,000 BTUH		4	4		880	141		1,021	1,200
4960	29,000 BTUH		4	4		950	141		1,091	1,275
9000	Minimum labor/equipment charge	1 Carp	2	4	Job		160		160	264

23 81 19.20 Self-Contained Single Package		Crew	Daily Output	Labor-Hours	Unit	Material	Labor	Equipment	Total	Total Incl O&P
0010	**SELF-CONTAINED SINGLE PACKAGE** R262213-27									
0100	Air cooled, for free blow or duct, not incl. remote condenser									
0200	3 ton cooling	Q-5	1	16	Ea.	3,200	710		3,910	4,625
0210	4 ton cooling	"	.80	20		3,500	890		4,390	5,225
0220	5 ton cooling	Q-6	1.20	20		3,825	920		4,745	5,625
0230	7.5 ton cooling	"	.90	26.667		4,975	1,225		6,200	7,375
0240	10 ton cooling	Q-7	1	32		6,600	1,500		8,100	9,575

23 81 19 – Self-Contained Air-Conditioners

23 81 19.20 Self-Contained Single Package		Crew	Daily Output	Labor-Hours	Unit	Material	2009 Bare Costs Labor	Equipment	Total	Total Incl O&P
0250	15 ton cooling	Q-7	.95	33.684	Ea.	9,200	1,575		10,775	12,600
0260	20 ton cooling		.90	35.556		13,300	1,675		14,975	17,200
0270	25 ton cooling		.85	37.647		15,400	1,775		17,175	19,700
0280	30 ton cooling		.80	40		19,100	1,875		20,975	23,900
0300	40 ton cooling		.60	53.333		35,200	2,500		37,700	42,600
0320	50 ton cooling		.50	64		44,100	3,000		47,100	53,000
0340	60 ton cooling	Q-8	.40	80		47,200	3,750	139	51,089	58,000
0490	For duct mounting no price change									
0500	For steam heating coils, add				Ea.	10%	10%			
1000	Water cooled for free blow or duct, not including tower									
1010	Constant volume									
1100	3 ton cooling	Q-6	1	24	Ea.	3,075	1,100		4,175	5,125
1120	5 ton cooling		1	24		4,025	1,100		5,125	6,150
1130	7.5 ton cooling		.80	30		5,725	1,375		7,100	8,450
1140	10 ton cooling	Q-7	.90	35.556		7,900	1,675		9,575	11,300
1150	15 ton cooling		.85	37.647		11,400	1,775		13,175	15,300
1160	20 ton cooling		.80	40		25,800	1,875		27,675	31,200
1170	25 ton cooling		.75	42.667		30,400	2,000		32,400	36,500
1180	30 ton cooling		.70	45.714		34,200	2,150		36,350	40,900
1200	40 ton cooling		.40	80		43,300	3,750		47,050	53,500
1220	50 ton cooling		.30	106		54,000	5,025		59,025	67,000
1240	60 ton cooling	Q-8	.30	106		63,000	5,025	186	68,211	77,000
1300	For hot water or steam heat coils, add					12%	10%			
9000	Minimum labor/equipment charge	Q-5	1	16	Job		710		710	1,100

23 81 23 – Computer-Room Air-Conditioners

23 81 23.10 Computer Room Units		Crew	Daily Output	Labor-Hours	Unit	Material	2009 Bare Costs Labor	Equipment	Total	Total Incl O&P
0010	**COMPUTER ROOM UNITS** R235000-50									
1000	Air cooled, includes remote condenser but not									
1020	interconnecting tubing or refrigerant R236000-30									
1080	3 ton	Q-5	.50	32	Ea.	15,600	1,425		17,025	19,400
1120	5 ton R262213-27		.45	35.556		16,700	1,575		18,275	20,900
1160	6 ton		.30	53.333		29,500	2,375		31,875	36,200
1200	8 ton		.27	59.259		31,400	2,625		34,025	38,600
1240	10 ton		.25	64		32,800	2,850		35,650	40,500
1280	15 ton		.22	72.727		36,100	3,225		39,325	44,700
1290	18 ton		.20	80		40,200	3,550		43,750	49,700
1300	20 ton	Q-6	.26	92.308		43,200	4,250		47,450	54,000
1320	22 ton		.24	100		43,800	4,600		48,400	55,500
1360	30 ton		.21	114		54,000	5,275		59,275	67,500
2200	Chilled water, for connection to									
2220	existing chiller system of adequate capacity									
2260	5 ton	Q-5	.74	21.622	Ea.	12,000	960		12,960	14,700
2280	6 ton		.52	30.769		12,000	1,375		13,375	15,300
2300	8 ton		.50	32		12,100	1,425		13,525	15,500
2320	10 ton		.49	32.653		12,200	1,450		13,650	15,700
2330	12 ton		.49	32.990		12,500	1,475		13,975	16,100
2360	15 ton		.48	33.333		12,900	1,475		14,375	16,500
2400	20 ton		.46	34.783		13,600	1,550		15,150	17,400
2440	25 ton	Q-6	.63	38.095		14,500	1,750		16,250	18,700
2460	30 ton		.57	42.105		15,700	1,950		17,650	20,300
2480	35 ton		.52	46.154		15,700	2,125		17,825	20,600
2500	45 ton		.46	52.174		16,800	2,400		19,200	22,200

23 81 Decentralized Unitary HVAC Equipment

23 81 23 – Computer-Room Air-Conditioners

23 81 23.10 Computer Room Units		Crew	Daily Output	Labor-Hours	Unit	2009 Bare Costs Material	Labor	Equipment	Total	Total Incl O&P
2520	50 ton	Q-6	.42	57.143	Ea.	19,400	2,625		22,025	25,400
2540	60 ton		.38	63.158		20,600	2,900		23,500	27,100
4000	Glycol system, complete except for interconnecting tubing									
4060	3 ton	Q-5	.40	40	Ea.	19,700	1,775		21,475	24,500
4100	5 ton		.38	42.105		21,600	1,875		23,475	26,600
4120	6 ton		.25	64		31,600	2,850		34,450	39,100
4140	8 ton		.23	69.565		34,700	3,100		37,800	42,900
4160	10 ton		.21	76.190		36,900	3,375		40,275	45,800
4200	15 ton	Q-6	.26	92.308		45,700	4,250		49,950	56,500
4240	20 ton		.24	100		50,500	4,600		55,100	62,500
4280	22 ton		.23	104		53,000	4,800		57,800	65,500
4430	30 ton		.22	109		65,500	5,025		70,525	80,000
8000	Water cooled system, not including condenser,									
8020	water supply or cooling tower									
8060	3 ton	Q-5	.62	25.806	Ea.	16,000	1,150		17,150	19,300
8100	5 ton		.54	29.630		17,400	1,325		18,725	21,200
8120	6 ton		.35	45.714		27,100	2,025		29,125	33,000
8140	8 ton		.33	48.485		28,500	2,150		30,650	34,700
8160	10 ton		.31	51.613		29,500	2,300		31,800	36,000
8200	15 ton		.27	59.259		34,500	2,625		37,125	42,100
8240	20 ton	Q-6	.33	63.158		37,200	2,900		40,100	45,400
8280	22 ton		.35	68.571		38,900	3,150		42,050	47,700
8300	30 ton		.30	80		48,500	3,675		52,175	59,000

23 81 26 – Split-System Air-Conditioners

23 81 26.10 Split Ductless Systems		Crew	Daily Output	Labor-Hours	Unit	Material	Labor	Equipment	Total	Total Incl O&P
0010	**SPLIT DUCTLESS SYSTEMS** R262213-27									
0100	Cooling only, single zone									
0110	Wall mount									
0120	3/4 ton cooling	Q-5	2	8	Ea.	1,175	355		1,530	1,825
0130	1 ton cooling		1.80	8.889		1,300	395		1,695	2,025
0140	1-1/2 ton cooling		1.60	10		1,800	445		2,245	2,675
0150	2 ton cooling		1.40	11.429		2,125	510		2,635	3,125
1000	Ceiling mount									
1020	2 ton cooling	Q-5	1.40	11.429	Ea.	1,450	510		1,960	2,350
1030	3 ton cooling	"	1.20	13.333	"	3,900	590		4,490	5,225
2000	T-Bar mount									
2010	2 ton cooling	Q-5	1.40	11.429	Ea.	3,400	510		3,910	4,525
2020	3 ton cooling		1.20	13.333		3,800	590		4,390	5,100
2030	3-1/2 ton cooling		1.10	14.545		4,200	645		4,845	5,625
3000	Multizone									
3010	Wall mount									
3020	2 @ 3/4 ton cooling	Q-5	1.80	8.889	Ea.	1,475	395		1,870	2,225
5000	Cooling / Heating									
5010	Wall mount									
5110	1 ton cooling	Q-5	1.70	9.412	Ea.	1,100	420		1,520	1,850
5120	1-1/2 ton cooling	"	1.50	10.667	"	1,750	475		2,225	2,650
5300	Ceiling mount									
5310	3 ton cooling	Q-5	1	16	Ea.	4,125	710		4,835	5,650
7000	Accessories for all split ductless systems									
7010	Add for ambient frost control	Q-5	8	2	Ea.	130	89		219	281
7020	Add for tube / wiring kit									
7030	15' kit	Q-5	32	.500	Ea.	38.50	22		60.50	77

23 81 26 – Split-System Air-Conditioners

23 81 26.10 Split Ductless Systems		Crew	Daily Output	Labor-Hours	Unit	2009 Bare Costs Material	Labor	Equipment	Total	Total Incl O&P
7040	35′ kit	Q-5	24	.667	Ea.	125	29.50		154.50	183

23 81 43 – Air-Source Unitary Heat Pumps

23 81 43.10 Air-Source Heat Pumps

		Crew	Daily Output	Labor-Hours	Unit	Material	Labor	Equipment	Total	Total Incl O&P
0010	**AIR-SOURCE HEAT PUMPS**, Not including interconnecting tubing.									
1000	Air to air, split system, not including curbs, pads, fan coil and ductwork									
1012	Outside condensing unit only, for fan coil see Div. 23 82 19.10									
1015	1.5 ton cooling, 7 MBH heat @ 0° F R235000-50	Q-5	2.40	6.667	Ea.	1,600	296		1,896	2,200
1020	2 ton cooling, 8.5 MBH heat @ 0° F		2	8		1,650	355		2,005	2,375
1030	2.5 ton cooling, 10 MBH heat @ 0° F R262213-27		1.60	10		1,850	445		2,295	2,725
1040	3 ton cooling, 13 MBH heat @ 0° F		1.20	13.333		2,100	590		2,690	3,225
1050	3.5 ton cooling, 18 MBH heat @ 0° F		1	16		2,250	710		2,960	3,575
1054	4 ton cooling, 24 MBH heat @ 0° F		.80	20		2,450	890		3,340	4,075
1060	5 ton cooling, 27 MBH heat @ 0° F		.50	32		2,675	1,425		4,100	5,150
1080	7.5 ton cooling, 33 MBH heat @ 0° F		.45	35.556		5,050	1,575		6,625	8,000
1100	10 ton cooling, 50 MBH heat @ 0° F	Q-6	.64	37.500		7,050	1,725		8,775	10,400
1120	15 ton cooling, 64 MBH heat @ 0° F		.50	48		10,200	2,200		12,400	14,600
1130	20 ton cooling, 85 MBH heat @ 0° F		.35	68.571		15,300	3,150		18,450	21,700
1140	25 ton cooling, 119 MBH heat @ 0° F		.25	96		18,400	4,425		22,825	27,200
1500	Single package, not including curbs, pads, or plenums									
1502	1/2 ton cooling, supplementary heat not incl.	Q-5	8	2	Ea.	1,675	89		1,764	2,000
1504	3/4 ton cooling, supplementary heat not incl.		6	2.667		1,850	118		1,968	2,200
1506	1 ton cooling, supplementary heat not incl.		4	4		2,025	178		2,203	2,500
1510	1.5 ton cooling, 5 MBH heat @ 0° F		1.55	10.323		2,400	460		2,860	3,350
1520	2 ton cooling, 6.5 MBH heat @ 0° F		1.50	10.667		2,575	475		3,050	3,550
1540	2.5 ton cooling, 8 MBH heat @ 0° F		1.40	11.429		2,800	510		3,310	3,875
1560	3 ton cooling, 10 MBH heat @ 0° F		1.20	13.333		3,050	590		3,640	4,275
1570	3.5 ton cooling, 11 MBH heat @ 0° F		1	16		3,375	710		4,085	4,825
1580	4 ton cooling, 13 MBH heat @ 0° F		.96	16.667		3,650	740		4,390	5,175
1620	5 ton cooling, 27 MBH heat @ 0° F		.65	24.615		3,900	1,100		5,000	6,000
1640	7.5 ton cooling, 35 MBH heat @ 0° F		.40	40		6,450	1,775		8,225	9,850
1648	10 ton cooling, 45 MBH heat @ 0° F	Q-6	.40	60		8,350	2,775		11,125	13,500
1652	12 ton cooling, 50 MBH heat @ 0° F	"	.36	66.667		9,625	3,075		12,700	15,400
1696	Supplementary electric heat coil incl., except as noted									

23 81 46 – Water-Source Unitary Heat Pumps

23 81 46.10 Water Source Heat Pumps

		Crew	Daily Output	Labor-Hours	Unit	Material	Labor	Equipment	Total	Total Incl O&P
0010	**WATER SOURCE HEAT PUMPS**, Not incl. connecting tubing or water source									
2000	Water source to air, single package									
2100	1 ton cooling, 13 MBH heat @ 75° F	Q-5	2	8	Ea.	1,300	355		1,655	1,975
2120	1.5 ton cooling, 17 MBH heat @ 75° F		1.80	8.889		1,525	395		1,920	2,275
2140	2 ton cooling, 19 MBH heat @ 75° F		1.70	9.412		1,550	420		1,970	2,350
2160	2.5 ton cooling, 25 MBH heat @ 75° F		1.60	10		1,675	445		2,120	2,550
2180	3 ton cooling, 27 MBH heat @ 75° F		1.40	11.429		1,800	510		2,310	2,750
2190	3.5 ton cooling, 29 MBH heat @ 75° F		1.30	12.308		1,875	545		2,420	2,925
2200	4 ton cooling, 31 MBH heat @ 75° F		1.20	13.333		2,100	590		2,690	3,225
2220	5 ton cooling, 29 MBH heat @ 75° F		.90	17.778		2,450	790		3,240	3,925
2240	7.5 ton cooling, 35 MBH heat @ 75° F		.60	26.667		6,100	1,175		7,275	8,525
2250	8.5 ton cooling, 40 MBH heat @ 75° F		.58	27.586		6,475	1,225		7,700	9,025
2260	10 ton cooling, 50 MBH heat @ 75° F		.53	30.189		6,700	1,350		8,050	9,450
2280	15 ton cooling, 64 MBH heat @ 75° F	Q-6	.47	51.064		14,000	2,350		16,350	19,100
2300	20 ton cooling, 100 MBH heat @ 75° F		.41	58.537		15,300	2,700		18,000	21,000
2310	25 ton cooling, 100 MBH heat @ 75° F		.32	75		20,800	3,450		24,250	28,300
2320	30 ton cooling, 128 MBH heat @ 75° F		.24	102		22,800	4,700		27,500	32,400

23 81 Decentralized Unitary HVAC Equipment

23 81 46 – Water-Source Unitary Heat Pumps

23 81 46.10 Water Source Heat Pumps		Crew	Daily Output	Labor-Hours	Unit	Material	2009 Bare Costs Labor	Equipment	Total	Total Incl O&P
2340	40 ton cooling, 200 MBH heat @ 75° F	Q-6	.21	117	Ea.	32,200	5,400		37,600	43,800
2360	50 ton cooling, 200 MBH heat @ 75° F		.15	160		36,400	7,375		43,775	51,500
3960	For supplementary heat coil, add					10%				
4000	For increase in capacity thru use									
4020	of solar collector, size boiler at 60%									
9000	Minimum labor/equipment charge	Q-5	1.75	9.143	Job		405		405	630

23 82 Convection Heating and Cooling Units

23 82 13 – Valance Heating and Cooling Units

23 82 13.16 Valance Units		Crew	Daily Output	Labor-Hours	Unit	Material	Labor	Equipment	Total	Total Incl O&P
0010	**VALANCE UNITS**									
6000	Valance units, complete with 1/2" cooling coil, enclosure									
6020	2 tube	Q-5	18	.889	L.F.	29.50	39.50		69	93.50
6040	3 tube		16	1		35	44.50		79.50	108
6060	4 tube		16	1		40	44.50		84.50	113
6080	5 tube		15	1.067		45.50	47.50		93	124
6100	6 tube		15	1.067		50.50	47.50		98	129
6120	8 tube		14	1.143		73	51		124	159
6200	For 3/4" cooling coil, add					10%				

23 82 16 – Air Coils

23 82 16.10 Flanged Coils		Crew	Daily Output	Labor-Hours	Unit	Material	Labor	Equipment	Total	Total Incl O&P
0010	**FLANGED COILS**									
0500	Chilled water cooling, 6 rows, 24" x 48"	Q-5	3.20	5	Ea.	3,575	222		3,797	4,275
1000	Direct expansion cooling, 6 rows, 24" x 48"		2.80	5.714		3,875	254		4,129	4,650
1500	Hot water heating, 1 row, 24" x 48"		4	4		1,425	178		1,603	1,850
2000	Steam heating, 1 row, 24" x 48"		3.06	5.229		2,025	232		2,257	2,575
9000	Minimum labor/equipment charge	1 Plum	1	8	Job		390		390	605

23 82 16.20 Duct Heaters		Crew	Daily Output	Labor-Hours	Unit	Material	Labor	Equipment	Total	Total Incl O&P
0010	**DUCT HEATERS**, Electric, 480 V, 3 Ph.									
0020	Finned tubular insert, 500 ° F									
0100	8" wide x 6" high, 4.0 kW	Q-20	16	1.250	Ea.	705	54		759	860
0120	12" high, 8.0 kW		15	1.333		1,175	58		1,233	1,375
0140	18" high, 12.0 kW		14	1.429		1,650	62		1,712	1,900
0160	24" high, 16.0 kW		13	1.538		2,100	66.50		2,166.50	2,425
0180	30" high, 20.0 kW		12	1.667		2,575	72.50		2,647.50	2,950
0300	12" wide x 6" high, 6.7 kW		15	1.333		750	58		808	915
0320	12" high, 13.3 kW		14	1.429		1,200	62		1,262	1,425
0340	18" high, 20.0 kW		13	1.538		1,700	66.50		1,766.50	1,975
0360	24" high, 26.7 kW		12	1.667		2,175	72.50		2,247.50	2,525
0380	30" high, 33.3 kW		11	1.818		2,650	79		2,729	3,050
0500	18" wide x 6" high, 13.3 kW		14	1.429		785	62		847	965
0520	12" high, 26.7 kW		13	1.538		1,375	66.50		1,441.50	1,625
0540	18" high, 40.0 kW		12	1.667		1,850	72.50		1,922.50	2,150
0560	24" high, 53.3 kW		11	1.818		2,450	79		2,529	2,825
0580	30" high, 66.7 kW		10	2		3,050	87		3,137	3,500
0700	24" wide x 6" high, 17.8 kW		13	1.538		865	66.50		931.50	1,050
0720	12" high, 35.6 kW		12	1.667		1,500	72.50		1,572.50	1,775
0740	18" high, 53.3 kW		11	1.818		2,050	79		2,129	2,375
0760	24" high, 71.1 kW		10	2		2,725	87		2,812	3,125
0780	30" high, 88.9 kW		9	2.222		3,325	96.50		3,421.50	3,825

23 82 16.20 Duct Heaters		Crew	Daily Output	Labor-Hours	Unit	2009 Bare Costs Material	Labor	Equipment	Total	Total Incl O&P
0900	30" wide x 6" high, 22.2 kW	Q-20	12	1.667	Ea.	920	72.50		992.50	1,125
0920	12" high, 44.4 kW		11	1.818		1,575	79		1,654	1,875
0940	18" high, 66.7 kW		10	2		2,175	87		2,262	2,525
0960	24" high, 88.9 kW		9	2.222		2,675	96.50		2,771.50	3,100
0980	30" high, 111.0 kW		8	2.500		3,525	108		3,633	4,050
1400	Note decreased kW available for									
1410	each duct size at same cost									
1420	See line 5000 for modifications and accessories									
2000	Finned tubular flange with insulated									
2020	terminal box, 500 ° F									
2100	12" wide x 36" high, 54 kW	Q-20	10	2	Ea.	3,475	87		3,562	3,950
2120	40" high, 60 kW		9	2.222		4,000	96.50		4,096.50	4,550
2200	24" wide x 36" high, 118.8 kW		9	2.222		4,150	96.50		4,246.50	4,700
2220	40" high, 132 kW		8	2.500		4,550	108		4,658	5,175
2400	36" wide x 8" high, 40 kW		11	1.818		1,750	79		1,829	2,050
2420	16" high, 80 kW		10	2		2,350	87		2,437	2,725
2440	24" high, 120 kW		9	2.222		3,075	96.50		3,171.50	3,550
2460	32" high, 160 kW		8	2.500		3,950	108		4,058	4,525
2480	36" high, 180 kW		7	2.857		4,875	124		4,999	5,550
2500	40" high, 200 kW		6	3.333		5,400	145		5,545	6,150
2600	40" wide x 8" high, 45 kW		11	1.818		1,850	79		1,929	2,150
2620	16" high, 90 kW		10	2		2,550	87		2,637	2,925
2640	24" high, 135 kW		9	2.222		3,225	96.50		3,321.50	3,700
2660	32" high, 180 kW		8	2.500		4,300	108		4,408	4,900
2680	36" high, 202.5 kW		7	2.857		4,950	124		5,074	5,650
2700	40" high, 225 kW		6	3.333		5,625	145		5,770	6,425
2800	48" wide x 8" high, 54.8 kW		10	2		1,925	87		2,012	2,250
2820	16" high, 109.8 kW		9	2.222		2,675	96.50		2,771.50	3,100
2840	24" high, 164.4 kW		8	2.500		3,400	108		3,508	3,925
2860	32" high, 219.2 kW		7	2.857		4,550	124		4,674	5,200
2880	36" high, 246.6 kW		6	3.333		5,225	145		5,370	5,975
2900	40" high, 274 kW		5	4		5,925	174		6,099	6,775
3000	56" wide x 8" high, 64 kW		9	2.222		2,200	96.50		2,296.50	2,575
3020	16" high, 128 kW		8	2.500		3,050	108		3,158	3,525
3040	24" high, 192 kW		7	2.857		3,700	124		3,824	4,275
3060	32" high, 256 kW		6	3.333		5,175	145		5,320	5,900
3080	36" high, 288kW		5	4		5,925	174		6,099	6,800
3100	40" high, 320kW		4	5		6,550	217		6,767	7,575
3200	64" wide x 8" high, 74kW		8	2.500		2,275	108		2,383	2,675
3220	16" high, 148kW		7	2.857		3,150	124		3,274	3,650
3240	24" high, 222kW		6	3.333		4,000	145		4,145	4,625
3260	32" high, 296kW		5	4		5,350	174		5,524	6,175
3280	36" high, 333kW		4	5		6,375	217		6,592	7,375
3300	40" high, 370kW		3	6.667		7,050	289		7,339	8,200
3800	Note decreased kW available for									
3820	each duct size at same cost									
5000	Duct heater modifications and accessories									
5120	T.C.O. limit auto or manual reset	Q-20	42	.476	Ea.	116	20.50		136.50	160
5140	Thermostat		28	.714		495	31		526	595
5160	Overheat thermocouple (removable)		7	2.857		700	124		824	965
5180	Fan interlock relay		18	1.111		168	48		216	261
5200	Air flow switch		20	1		146	43.50		189.50	229
5220	Split terminal box cover		100	.200		48	8.70		56.70	66.50

23 82 Convection Heating and Cooling Units

23 82 16 – Air Coils

23 82 16.20 Duct Heaters		Crew	Daily Output	Labor-Hours	Unit	Material	2009 Bare Costs Labor	Equipment	Total	Total Incl O&P
8000	To obtain BTU multiply kW by 3413									
9900	Minimum labor/equipment charge	1 Shee	4	2	Job		94.50		94.50	150

23 82 19 – Fan Coil Units

23 82 19.10 Fan Coil Air Conditioning		Crew	Daily Output	Labor-Hours	Unit	Material	2009 Bare Costs Labor	Equipment	Total	Total Incl O&P
0010	**FAN COIL AIR CONDITIONING** Cabinet mounted, filters, controls R262213-27									
0100	Chilled water, 1/2 ton cooling	Q-5	8	2	Ea.	675	89		764	885
0110	3/4 ton cooling		7	2.286		700	102		802	925
0120	1 ton cooling		6	2.667		775	118		893	1,050
0140	1.5 ton cooling		5.50	2.909		875	129		1,004	1,175
0150	2 ton cooling		5.25	3.048		1,100	135		1,235	1,400
0160	2.5 ton cooling		5	3.200		1,450	142		1,592	1,825
0180	3 ton cooling		4	4		1,650	178		1,828	2,100
0190	7.5 ton cooling	↓	2.70	5.926	↓	1,950	263		2,213	2,550
0262	For hot water coil, add					40%	10%			
0940	Direct expansion, for use w/air cooled condensing unit, 1.5 ton cooling	Q-5	5	3.200	Ea.	540	142		682	815
0950	2 ton cooling		4.80	3.333		575	148		723	865
0960	2.5 ton cooling		4.40	3.636		610	162		772	920
0970	3 ton cooling		3.30	4.211		750	187		937	1,125
0980	3.5 ton cooling		3.60	4.444		800	197		997	1,175
0990	4 ton cooling		3.40	4.706		900	209		1,109	1,325
1000	5 ton cooling		3	5.333		1,025	237		1,262	1,500
1020	7.5 ton cooling	↓	3	5.333		2,350	237		2,587	2,950
1040	10 ton cooling	Q-6	2.60	9.231		2,750	425		3,175	3,675
1042	11 ton cooling		2.30	10.435		3,500	480		3,980	4,600
1050	15 ton cooling		1.60	15		3,650	690		4,340	5,100
1060	20 ton cooling		.70	34.286		4,850	1,575		6,425	7,775
1070	25 ton cooling		.65	36.923		6,100	1,700		7,800	9,350
1080	30 ton cooling	↓	.60	40		6,850	1,850		8,700	10,400
1500	For hot water coil, add				↓	40%	10%			
1512	For condensing unit add see Div. 23 62									
9000	Minimum labor/equipment charge	Q-5	3.75	4.267	Job		190		190	294

23 82 19.20 Heating and Ventilating Units		Crew	Daily Output	Labor-Hours	Unit	Material	2009 Bare Costs Labor	Equipment	Total	Total Incl O&P
0010	**HEATING AND VENTILATING UNITS**, Classroom units R233400-10									
0020	Includes filter, heating/cooling coils, standard controls									
0080	750 CFM, 2 tons cooling R235000-50	Q-6	2	12	Ea.	3,550	555		4,105	4,750
0100	1000 CFM, 2-1/2 tons cooling		1.60	15		4,200	690		4,890	5,700
0120	1250 CFM, 3 tons cooling R262213-27		1.40	17.143		4,325	790		5,115	6,000
0140	1500 CFM, 4 tons cooling		.80	30		4,625	1,375		6,000	7,250
0160	2000 CFM 5 tons cooling	↓	.50	48		5,450	2,200		7,650	9,425
0500	For electric heat, add					35%				
1000	For no cooling, deduct				↓	25%	10%			

23 82 27 – Infrared Units

23 82 27.10 Infrared Type Heating Units		Crew	Daily Output	Labor-Hours	Unit	Material	2009 Bare Costs Labor	Equipment	Total	Total Incl O&P
0010	**INFRARED TYPE HEATING UNITS** R235000-50									
0020	Gas fired, unvented, electric ignition, 100% shutoff.									
0030	Piping and wiring not included									
0060	Input, 15 MBH	Q-5	7	2.286	Ea.	425	102		527	620
0100	30 MBH		6	2.667		430	118		548	660
0120	45 MBH		5	3.200		435	142		577	700
0140	50 MBH		4.50	3.556		450	158		608	740
0160	60 MBH	↓	4	4	↓	540	178		718	870

23 82 Convection Heating and Cooling Units

23 82 27 – Infrared Units

23 82 27.10 Infrared Type Heating Units		Crew	Daily Output	Labor-Hours	Unit	2009 Bare Costs Material	Labor	Equipment	Total	Total Incl O&P
0180	75 MBH	Q-5	3	5.333	Ea.	545	237		782	960
0200	90 MBH		2.50	6.400		640	284		924	1,150
0220	105 MBH		2	8		645	355		1,000	1,250
0240	120 MBH		2	8		900	355		1,255	1,550
2000	Electric, single or three phase									
2050	6 kW, 20,478 BTU	1 Elec	2.30	3.478	Ea.	400	163		563	690
2100	13.5 KW, 40,956 BTU		2.20	3.636		605	171		776	930
2150	24 KW, 81,912 BTU		2	4		1,200	188		1,388	1,600
3000	Oil fired, pump, controls, fusible valve, oil supply tank									
3050	91,000 BTU	Q-5	2.50	6.400	Ea.	6,000	284		6,284	7,050
3080	105,000 BTU		2.25	7.111		6,375	315		6,690	7,525
3110	119,000 BTU		2	8		6,975	355		7,330	8,225

23 82 29 – Radiators

23 82 29.10 Hydronic Heating

		Crew	Daily Output	Labor-Hours	Unit	Material	Labor	Equipment	Total	Total Incl O&P
0010	**HYDRONIC HEATING**, Terminal units, not incl. main supply pipe R235000-50									
1000	Radiation									
1100	Panel, baseboard, C.I., including supports, no covers	Q-5	46	.348	L.F.	32	15.45		47.45	59
3000	Radiators, cast iron									
3100	Free standing or wall hung, 6 tube, 25" high	Q-5	96	.167	Section	30	7.40		37.40	44.50
3150	4 tube 25" high		96	.167		27	7.40		34.40	41
3200	4 tube, 19" high		96	.167		23.50	7.40		30.90	37.50
3250	Adj. brackets, 2 per wall radiator up to 30 sections	1 Stpi	32	.250	Ea.	27	12.35		39.35	49
3500	Recessed, 20" high x 5" deep, without grille	Q-5	60	.267	Section	30.50	11.85		42.35	52
3600	For inlet grille, add				"	2.12			2.12	2.33
9000	Minimum labor/equipment charge	Q-5	3.20	5	Job		222		222	345
9500	To convert SFR to BTU rating: Hot water, 150 x SFR									
9510	Forced hot water, 180 x SFR; steam, 240 x SFR									

23 82 33 – Convectors

23 82 33.10 Convector Units

		Crew	Daily Output	Labor-Hours	Unit	Material	Labor	Equipment	Total	Total Incl O&P
0010	**CONVECTOR UNITS,** Terminal units, not incl. main supply pipe									
2204	Convector, multifin, 2 pipe w/ cabinet									
2210	17" H x 24" L	Q-5	10	1.600	Ea.	92	71		163	211
2214	17" H x 36" L		8.60	1.860		138	82.50		220.50	280
2218	17" H x 48" L		7.40	2.162		184	96		280	350
2222	21" H x 24" L		9	1.778		92	79		171	223
2226	21" H x 36" L		8.20	1.951		138	86.50		224.50	286
2228	21" H x 48" L		6.80	2.353		184	105		289	365
2240	For knob operated damper, add					140%				
2241	For metal trim strips, add	Q-5	64	.250	Ea.	18.30	11.10		29.40	37.50
2243	For snap-on inlet grille, add					10%	10%			
2245	For hinged access door, add	Q-5	64	.250	Ea.	34	11.10		45.10	54.50
2246	For air chamber, auto-venting, add	"	58	.276	"	5.65	12.25		17.90	25

23 82 36 – Finned-Tube Radiation Heaters

23 82 36.10 Finned Tube Radiation

		Crew	Daily Output	Labor-Hours	Unit	Material	Labor	Equipment	Total	Total Incl O&P
0010	**FINNED TUBE RADIATION**, Terminal units, not incl. main supply pipe									
1150	Fin tube, wall hung, 14" slope top cover, with damper									
1200	1-1/4" copper tube, 4-1/4" alum. fin	Q-5	38	.421	L.F.	39	18.70		57.70	71.50
1250	1-1/4" steel tube, 4-1/4" steel fin		36	.444		36.50	19.75		56.25	71
1255	2" steel tube, 4-1/4" steel fin		32	.500		39.50	22		61.50	78
1310	Baseboard, pkgd, 1/2" copper tube, alum. fin, 7" high		60	.267		7.20	11.85		19.05	26.50
1320	3/4" copper tube, alum. fin, 7" high		58	.276		7.65	12.25		19.90	27.50

23 82 Convection Heating and Cooling Units

23 82 36 – Finned-Tube Radiation Heaters

23 82 36.10 Finned Tube Radiation		Crew	Daily Output	Labor-Hours	Unit	2009 Bare Costs Material	Labor	Equipment	Total	Total Incl O&P
1340	1" copper tube, alum. fin, 8-7/8" high	Q-5	56	.286	L.F.	24.50	12.70		37.20	46
1360	1-1/4" copper tube, alum. fin, 8-7/8" high		54	.296		25	13.15		38.15	48
1380	1-1/4" IPS steel tube with steel fins		52	.308		24.50	13.65		38.15	48
1500	Note: fin tube may also require corners, caps, etc.									

23 82 39 – Unit Heaters

23 82 39.16 Propeller Unit Heaters

		Crew	Daily Output	Labor-Hours	Unit	Material	Labor	Equipment	Total	Total Incl O&P
0010	**PROPELLER UNIT HEATERS**									
3950	Unit heaters, propeller, 115 V 2 psi steam, 60° F entering air									
4000	Horizontal, 12 MBH	Q-5	12	1.333	Ea.	315	59		374	435
4020	28.2 MBH		10	1.600		405	71		476	555
4040	36.5 MBH		8	2		475	89		564	665
4060	43.9 MBH		8	2		475	89		564	665
4080	56.5 MBH		7.50	2.133		510	95		605	705
4100	65.6 MBH		7	2.286		520	102		622	725
4120	87.6 MBH		6.50	2.462		560	109		669	785
4140	96.8 MBH		6	2.667		650	118		768	905
4160	133.3 MBH		5	3.200		745	142		887	1,050
4180	157.6 MBH		4	4		915	178		1,093	1,275
4200	197.7 MBH		3	5.333		1,050	237		1,287	1,525
4220	257.2 MBH		2.50	6.400		1,275	284		1,559	1,850
4240	286.9 MBH		2	8		1,425	355		1,780	2,125
4250	326.0 MBH		1.90	8.421		1,825	375		2,200	2,575
4260	364 MBH		1.80	8.889		2,000	395		2,395	2,775
4270	404 MBH		1.60	10		2,500	445		2,945	3,450
4300	For vertical diffuser, add					174			174	191
4310	Vertical flow, 40 MBH	Q-5	11	1.455		465	64.50		529.50	615
4314	58.5 MBH		8	2		485	89		574	670
4318	92 MBH		7	2.286		595	102		697	810
4322	109.7 MBH		6	2.667		730	118		848	985
4326	131.0 MBH		4	4		730	178		908	1,075
4330	160.0 MBH		3	5.333		775	237		1,012	1,225
4334	194.0 MBH		2.20	7.273		895	325		1,220	1,475
4338	212.0 MBH		2.10	7.619		1,100	340		1,440	1,750
4342	247.0 MBH		1.96	8.163		1,250	365		1,615	1,925
4346	297.0 MBH		1.80	8.889		1,400	395		1,795	2,125
4350	333.0 MBH	Q-6	1.90	12.632		1,525	580		2,105	2,575
4354	420 MBH,(460 V)		1.80	13.333		1,875	615		2,490	3,000
4358	500 MBH, (460 V)		1.71	14.035		2,475	645		3,120	3,725
4362	570 MBH, (460 V)		1.40	17.143		3,725	790		4,515	5,325
4366	620 MBH, (460 V)		1.30	18.462		4,275	850		5,125	6,025
4370	960 MBH, (460 V)		1.10	21.818		7,200	1,000		8,200	9,450

23 83 16 – Radiant-Heating Hydronic Piping

23 83 16.10 Radiant Floor Heating		Crew	Daily Output	Labor-Hours	Unit	2009 Bare Costs Material	Labor	Equipment	Total	Total Incl O&P
0010	**RADIANT FLOOR HEATING**									
0100	Tubing, PEX (cross-linked polyethylene)									
0110	Oxygen barrier type for systems with ferrous materials									
0120	1/2"	Q-5	800	.020	L.F.	.94	.89		1.83	2.41
0130	3/4"		535	.030		1.15	1.33		2.48	3.33
0140	1"		400	.040		2.07	1.78		3.85	5.05
0200	Non barrier type for ferrous free systems									
0210	1/2"	Q-5	800	.020	L.F.	.51	.89		1.40	1.94
0220	3/4"		535	.030		.92	1.33		2.25	3.07
0230	1"		400	.040		1.60	1.78		3.38	4.52
1000	Manifolds									
1110	Brass									
1120	Valved									
1130	1", 2 circuit	Q-5	13	1.231	Ea.	47	54.50		101.50	137
1140	1", 3 circuit		11	1.455		63.50	64.50		128	170
1150	1", 4 circuit		9	1.778		87.50	79		166.50	219
1160	1-1/4", 3 circuit		10	1.600		73	71		144	190
1170	1-1/4", 4 circuit		8	2		88.50	89		177.50	236
1300	Valveless									
1310	1", 2 circuit	Q-5	14	1.143	Ea.	24.50	51		75.50	105
1320	1", 3 circuit		12	1.333		32	59		91	128
1330	1", 4 circuit		10	1.600		41.50	71		112.50	156
1340	1-1/4", 3 circuit		11	1.455		43.50	64.50		108	148
1350	1-1/4", 4 circuit		9	1.778		45	79		124	172
1400	Manifold connector kit									
1410	Mounting brackets, couplings, end cap, air vent, valve, plug									
1420	1"	Q-5	8	2	Ea.	81	89		170	227
1430	1-1/4"	"	7	2.286	"	95.50	102		197.50	262
1610	Copper, (cut to size)									
1620	1" x 1/2" x 72" Lg, 2" sweat drops, 24 circuit	Q-5	3.33	4.805	Ea.	84.50	213		297.50	425
1630	1-1/4" x 1/2" x 72" Lg, 2" sweat drops, 24 circuit		3.20	5		98	222		320	455
1640	1-1/2" x 1/2" x 72" Lg, 2" sweat drops, 24 circuit		3	5.333		120	237		357	495
1650	1-1/4" x 3/4" x 72" Lg, 2" sweat drops, 24 circuit		3.10	5.161		105	229		334	470
1660	1-1/2" x 3/4" x 72" Lg, 2" sweat drops, 24 circuit		2.90	5.517		127	245		372	520
1710	Copper, (modular)									
1720	1" x 1/2", 2" sweat drops, 2 circuit	Q-5	17.50	.914	Ea.	8.60	40.50		49.10	72.50
1730	1" x 1/2", 2" sweat drops, 3 circuit		14.50	1.103		12.05	49		61.05	89.50
1740	1" x 1/2", 2" sweat drops, 4 circuit		12.30	1.301		13.75	58		71.75	105
1750	1-1/4" x 1/2", 2" sweat drops, 2 circuit		17	.941		9.45	42		51.45	75.50
1760	1-1/4" x 1/2", 2" sweat drops, 3 circuit		14	1.143		13.75	51		64.75	93.50
1770	1-1/4" x 1/2", 2" sweat drops, 4 circuit		11.80	1.356		18.05	60		78.05	113
3000	Valves									
3110	Motorized zone valve operator, 24V	Q-5	30	.533	Ea.	49	23.50		72.50	91
3120	Motorized zone valve with operator complete, 24V									
3130	3/4"	Q-5	35	.457	Ea.	85	20.50		105.50	125
3140	1"		32	.500		96.50	22		118.50	141
3150	1-1/4"		29.60	.541		119	24		143	168
3200	Flow control/isolation valve, 1/2"		32	.500		27.50	22		49.50	65
3300	Thermostatic mixing valves									
3310	1/2"	Q-5	26.60	.602	Ea.	88.50	26.50		115	139
3320	3/4"		25	.640		107	28.50		135.50	162
3330	1"		21.40	.748		115	33		148	179
3400	3 Way mixing/diverting valve, manual, brass									

23 83 Radiant Heating Units

23 83 16 – Radiant-Heating Hydronic Piping

23 83 16.10 Radiant Floor Heating		Crew	Daily Output	Labor-Hours	Unit	2009 Bare Costs Material	Labor	Equipment	Total	Total Incl O&P
3410	1/2″	Q-5	21.30	.751	Ea.	117	33.50		150.50	181
3420	3/4″		20	.800		120	35.50		155.50	187
3430	1″		17.80	.899		139	40		179	215
3500	4 Way mixing / diverting valve, manual, brass									
3510	1/2″	Q-5	16	1	Ea.	122	44.50		166.50	203
3520	3/4″		15	1.067		136	47.50		183.50	223
3530	1″		13.30	1.203		152	53.50		205.50	250
3540	1-1/4″		11.40	1.404		171	62.50		233.50	285
3550	1-1/2″		10.60	1.509		224	67		291	350
3560	2″		10	1.600		242	71		313	375
3570	2-1/2″		8	2		665	89		754	875
3800	Motor control, 3 or 4 way, for valves up to 1-1/2″		30	.533		169	23.50		192.50	222
5000	Radiant floor heating, zone control panel									
5110	6 Zone control									
5120	For zone valves	Q-5	16	1	Ea.	158	44.50		202.50	243
5130	For circulators	″	16	1	″	225	44.50		269.50	315
7000	PEX tubing fittings, compression type									
7110	PEX tubing to brass valve conn., 1/2″	Q-5	54	.296	Ea.	2.72	13.15		15.87	23.50
7200	PEX x male NPT									
7210	1/2″ x 1/2″	Q-5	54	.296	Ea.	5.30	13.15		18.45	26.50
7220	3/4″ x 3/4″		46	.348		9.05	15.45		24.50	34
7230	1″ x 1″		40	.400		21.50	17.75		39.25	51
7300	PEX coupling									
7310	1/2″ x 1/2″	Q-5	50	.320	Ea.	6.45	14.20		20.65	29
7320	3/4″ x 3/4″		42	.381		18.50	16.95		35.45	47
7330	1″ x 1″		36	.444		25	19.75		44.75	58
7400	PEX x copper sweat									
7410	1/2″ x 1/2″	Q-5	38	.421	Ea.	4.73	18.70		23.43	34
7420	3/4″ x 3/4″		34	.471		9.90	21		30.90	43.50
7430	1″ x 1″		29	.552		21.50	24.50		46	61.50

23 83 33 – Electric Radiant Heaters

23 83 33.10 Electric Heating		Crew	Daily Output	Labor-Hours	Unit	Material	Labor	Equipment	Total	Total Incl O&P
0010	**ELECTRIC HEATING**, not incl. conduit or feed wiring. R235000-50									
1100	Rule of thumb: Baseboard units, including control	1 Elec	4.40	1.818	kW	92.50	85.50		178	233
1300	Baseboard heaters, 2′ long, 375 watt		8	1	Ea.	40	47		87	117
1400	3′ long, 500 watt		8	1		46	47		93	123
1600	4′ long, 750 watt		6.70	1.194		54	56		110	146
1800	5′ long, 935 watt		5.70	1.404		68	66		134	176
2000	6′ long, 1125 watt		5	1.600		74	75		149	198
2200	7′ long, 1310 watt		4.40	1.818		84	85.50		169.50	224
2400	8′ long, 1500 watt		4	2		92	94		186	246
2600	9′ long, 1680 watt		3.60	2.222		104	104		208	275
2800	10′ long, 1875 watt		3.30	2.424		193	114		307	385
2950	Wall heaters with fan, 120 to 277 volt									
3160	Recessed, residential, 750 watt	1 Elec	6	1.333	Ea.	98.50	62.50		161	205
3170	1000 watt		6	1.333		101	62.50		163.50	208
3180	1250 watt		5	1.600		111	75		186	238
3190	1500 watt		4	2		111	94		205	267
3210	2000 watt		4	2		117	94		211	274
3230	2500 watt		3.50	2.286		252	107		359	445
3240	3000 watt		3	2.667		380	125		505	615
3250	4000 watt		2.70	2.963		385	139		524	635

23 83 Radiant Heating Units

23 83 33 – Electric Radiant Heaters

23 83 33.10 Electric Heating		Crew	Daily Output	Labor-Hours	Unit	Material	2009 Bare Costs Labor	Equipment	Total	Total Incl O&P
3260	Commercial, 750 watt	1 Elec	6	1.333	Ea.	176	62.50		238.50	290
3270	1000 watt		6	1.333		176	62.50		238.50	290
3280	1250 watt		5	1.600		176	75		251	310
3290	1500 watt		4	2		176	94		270	340
3300	2000 watt		4	2		176	94		270	340
3310	2500 watt		3.50	2.286		176	107		283	360
3320	3000 watt		3	2.667		320	125		445	545
3330	4000 watt		2.70	2.963		320	139		459	565
3600	Thermostats, integral		16	.500		20.50	23.50		44	58.50
3800	Line voltage, 1 pole		8	1		31	47		78	107
3810	2 pole		8	1		33.50	47		80.50	110
3820	Low voltage, 1 pole		8	1		27.50	47		74.50	103
4000	Heat trace system, 400 degree									
4020	115V, 2.5 watts per L.F.	1 Elec	530	.015	L.F.	6.70	.71		7.41	8.45
4030	5 watts per L.F.		530	.015		6.70	.71		7.41	8.45
4050	10 watts per L.F.		530	.015		6.70	.71		7.41	8.45
4060	208V, 5 watts per L.F.		530	.015		6.70	.71		7.41	8.45
4080	480V, 8 watts per L.F.		530	.015		6.70	.71		7.41	8.45
4200	Heater raceway									
4260	Heat transfer cement									
4280	1 gallon				Ea.	60.50			60.50	67
4300	5 gallon				"	249			249	274
4320	Snap band, clamp									
4340	3/4" pipe size	1 Elec	470	.017	Ea.		.80		.80	1.23
4360	1" pipe size		444	.018			.85		.85	1.30
4380	1-1/4" pipe size		400	.020			.94		.94	1.45
4400	1-1/2" pipe size		355	.023			1.06		1.06	1.63
4420	2" pipe size		320	.025			1.18		1.18	1.81
4440	3" pipe size		160	.050			2.35		2.35	3.62
4460	4" pipe size		100	.080			3.76		3.76	5.80
4480	Thermostat NEMA 3R, 22 amp, 0-150 Deg, 10' cap.		8	1		214	47		261	310
4500	Thermostat NEMA 4X, 25 amp, 40 Deg, 5-1/2' cap.		7	1.143		214	53.50		267.50	320
4520	Thermostat NEMA 4X, 22 amp, 25-325 Deg, 10' cap.		7	1.143		575	53.50		628.50	720
4540	Thermostat NEMA 4X, 22 amp, 15-140 Deg,		6	1.333		485	62.50		547.50	630
4580	Thermostat NEMA 4,7,9, 22 amp, 25-325 Deg, 10' cap.		3.60	2.222		710	104		814	940
4600	Thermostat NEMA 4,7,9, 22 amp, 15-140 Deg,		3	2.667		705	125		830	970
4720	Fiberglass application tape, 36 yard roll		11	.727		66.50	34		100.50	126
5000	Radiant heating ceiling panels, 2' x 4', 500 watt		16	.500		253	23.50		276.50	315
5050	750 watt		16	.500		279	23.50		302.50	340
5200	For recessed plaster frame, add		32	.250		62	11.75		73.75	86.50
5300	Infrared quartz heaters, 120 volts, 1000 watts		6.70	1.194		164	56		220	267
5350	1500 watt		5	1.600		164	75		239	296
5400	240 volts, 1500 watt		5	1.600		164	75		239	296
5450	2000 watt		4	2		164	94		258	325
5500	3000 watt		3	2.667		189	125		314	400
5550	4000 watt		2.60	3.077		189	145		334	430
5570	Modulating control		.80	10		89	470		559	825
5600	Unit heaters, heavy duty, with fan & mounting bracket									
5650	Single phase, 208-240-277 volt, 3 kW	1 Elec	3.20	2.500	Ea.	405	118		523	625
5750	5 kW		2.40	3.333		420	157		577	700
5800	7 kW		1.90	4.211		645	198		843	1,025
5850	10 kW		1.30	6.154		735	289		1,024	1,250
5950	15 kW		.90	8.889		1,200	420		1,620	1,950

23 83 33.10 Electric Heating		Crew	Daily Output	Labor-Hours	Unit	Material	Labor	Equipment	Total	Total Incl O&P
						2009 Bare Costs				
6000	480 volt, 3 kW	1 Elec	3.30	2.424	Ea.	400	114		514	615
6020	4 kW		3	2.667		455	125		580	695
6040	5 kW		2.60	3.077		465	145		610	730
6060	7 kW		2	4		690	188		878	1,050
6080	10 kW		1.40	5.714		755	269		1,024	1,250
6100	13 kW		1.10	7.273		1,200	340		1,540	1,825
6120	15 kW		1	8		1,200	375		1,575	1,875
6140	20 kW		.90	8.889		1,550	420		1,970	2,350
6300	3 phase, 208-240 volt, 5 kW		2.40	3.333		395	157		552	675
6320	7 kW		1.90	4.211		610	198		808	975
6340	10 kW		1.30	6.154		660	289		949	1,175
6360	15 kW		.90	8.889		1,100	420		1,520	1,875
6380	20 kW		.70	11.429		1,575	535		2,110	2,550
6400	25 kW		.50	16		1,850	750		2,600	3,175
6500	480 volt, 5 kW		2.60	3.077		550	145		695	825
6520	7 kW		2	4		695	188		883	1,050
6540	10 kW		1.40	5.714		735	269		1,004	1,225
6560	13 kW		1.10	7.273		1,200	340		1,540	1,825
6580	15 kW		1	8		1,200	375		1,575	1,875
6600	20 kW		.90	8.889		1,550	420		1,970	2,350
6620	25 kW		.60	13.333		1,850	625		2,475	3,000
6630	30 kW		.70	11.429		2,150	535		2,685	3,175
6640	40 kW		.60	13.333		2,725	625		3,350	3,975
6650	50 kW		.50	16		3,300	750		4,050	4,775
6800	Vertical discharge heaters, with fan									
6820	Single phase, 208-240-277 volt, 10 kW	1 Elec	1.30	6.154	Ea.	680	289		969	1,200
6840	15 kW		.90	8.889		1,150	420		1,570	1,900
6900	3 phase, 208-240 volt, 10 kW		1.30	6.154		655	289		944	1,175
6920	15 kW		.90	8.889		1,100	420		1,520	1,875
6940	20 kW		.70	11.429		1,575	535		2,110	2,550
6960	25 kW		.50	16		1,850	750		2,600	3,175
6980	30 kW		.40	20		2,150	940		3,090	3,800
7000	40 kW		.36	22.222		2,750	1,050		3,800	4,625
7020	50 kW		.32	25		3,300	1,175		4,475	5,425
7100	480 volt, 10 kW		1.40	5.714		735	269		1,004	1,225
7120	15 kW		1	8		1,200	375		1,575	1,875
7140	20 kW		.90	8.889		1,550	420		1,970	2,350
7160	25 kW		.60	13.333		1,850	625		2,475	3,000
7180	30 kW		.50	16		2,150	750		2,900	3,500
7200	40 kW		.40	20		2,725	940		3,665	4,450
7220	50 kW		.35	22.857		3,300	1,075		4,375	5,275
7410	Sill height convector heaters, 5" high x 2' long, 500 watt		6.70	1.194		297	56		353	410
7420	3' long, 750 watt		6.50	1.231		350	58		408	475
7430	4' long, 1000 watt		6.20	1.290		405	60.50		465.50	540
7440	5' long, 1250 watt		5.50	1.455		460	68.50		528.50	610
7450	6' long, 1500 watt		4.80	1.667		520	78.50		598.50	695
7460	8' long, 2000 watt		3.60	2.222		710	104		814	940
7470	10' long, 2500 watt		3	2.667		880	125		1,005	1,175
7900	Cabinet convector heaters, 240 volt									
7920	3' long, 2000 watt	1 Elec	5.30	1.509	Ea.	1,775	71		1,846	2,075
7940	3000 watt		5.30	1.509		1,850	71		1,921	2,150
7960	4000 watt		5.30	1.509		1,925	71		1,996	2,200
7980	6000 watt		4.60	1.739		2,000	81.50		2,081.50	2,300

23 83 33 – Electric Radiant Heaters

23 83 33.10 Electric Heating		Crew	Daily Output	Labor-Hours	Unit	Material	Labor	Equipment	Total	Total Incl O&P
						2009 Bare Costs				
8000	8000 watt	1 Elec	4.60	1.739	Ea.	2,075	81.50		2,156.50	2,400
8020	4' long, 4000 watt		4.60	1.739		1,950	81.50		2,031.50	2,250
8040	6000 watt		4	2		2,025	94		2,119	2,375
8060	8000 watt		4	2		2,100	94		2,194	2,450
8080	10,000 watt		4	2		2,100	94		2,194	2,450
8100	Available also in 208 or 277 volt									
8200	Cabinet unit heaters, 120 to 277 volt, 1 pole,									
8220	wall mounted, 2 kW	1 Elec	4.60	1.739	Ea.	1,725	81.50		1,806.50	2,025
8230	3 kW		4.60	1.739		1,800	81.50		1,881.50	2,125
8240	4 kW		4.40	1.818		1,875	85.50		1,960.50	2,175
8250	5 kW		4.40	1.818		1,925	85.50		2,010.50	2,250
8260	6 kW		4.20	1.905		1,950	89.50		2,039.50	2,275
8270	8 kW		4	2		2,025	94		2,119	2,375
8280	10 kW		3.80	2.105		2,050	99		2,149	2,425
8290	12 kW		3.50	2.286		2,225	107		2,332	2,625
8300	13.5 kW		2.90	2.759		2,275	130		2,405	2,700
8310	16 kW		2.70	2.963		2,325	139		2,464	2,775
8320	20 kW		2.30	3.478		3,450	163		3,613	4,050
8330	24 kW		1.90	4.211		3,500	198		3,698	4,150
8350	Recessed, 2 kW		4.40	1.818		1,800	85.50		1,885.50	2,100
8370	3 kW		4.40	1.818		1,875	85.50		1,960.50	2,200
8380	4 kW		4.20	1.905		1,925	89.50		2,014.50	2,275
8390	5 kW		4.20	1.905		2,000	89.50		2,089.50	2,350
8400	6 kW		4	2		2,075	94		2,169	2,425
8410	8 kW		3.80	2.105		2,100	99		2,199	2,450
8420	10 kW		3.50	2.286		2,125	107		2,232	2,525
8430	12 kW		2.90	2.759		2,175	130		2,305	2,575
8440	13.5 kW		2.70	2.963		2,750	139		2,889	3,250
8450	16 kW		2.30	3.478		2,775	163		2,938	3,300
8460	20 kW		1.90	4.211		2,975	198		3,173	3,575
8470	24 kW		1.60	5		3,025	235		3,260	3,675
8490	Ceiling mounted, 2 kW		3.20	2.500		1,800	118		1,918	2,150
8510	3 kW		3.20	2.500		1,875	118		1,993	2,250
8520	4 kW		3	2.667		1,925	125		2,050	2,325
8530	5 kW		3	2.667		2,000	125		2,125	2,400
8540	6 kW		2.80	2.857		2,075	134		2,209	2,475
8550	8 kW		2.40	3.333		2,100	157		2,257	2,550
8560	10 kW		2.20	3.636		2,125	171		2,296	2,625
8570	12 kW		2	4		2,175	188		2,363	2,675
8580	13.5 kW		1.50	5.333		2,750	251		3,001	3,400
8590	16 kW		1.30	6.154		2,775	289		3,064	3,500
8600	20 kW		.90	8.889		2,975	420		3,395	3,925
8610	24 kW		.60	13.333		3,025	625		3,650	4,300
8630	208 to 480 V, 3 pole									
8650	Wall mounted, 2 kW	1 Elec	4.60	1.739	Ea.	1,925	81.50		2,006.50	2,250
8670	3 kW		4.60	1.739		2,025	81.50		2,106.50	2,350
8680	4 kW		4.40	1.818		2,075	85.50		2,160.50	2,400
8690	5 kW		4.40	1.818		2,125	85.50		2,210.50	2,475
8700	6 kW		4.20	1.905		2,150	89.50		2,239.50	2,500
8710	8 kW		4	2		2,225	94		2,319	2,600
8720	10 kW		3.80	2.105		2,250	99		2,349	2,625
8730	12 kW		3.50	2.286		2,300	107		2,407	2,700
8740	13.5 kW		2.90	2.759		2,350	130		2,480	2,775

23 83 Radiant Heating Units

23 83 33 – Electric Radiant Heaters

23 83 33.10 Electric Heating		Crew	Daily Output	Labor-Hours	Unit	2009 Bare Costs Material	Labor	Equipment	Total	Total Incl O&P
8750	16 kW	1 Elec	2.70	2.963	Ea.	2,375	139		2,514	2,850
8760	20 kW		2.30	3.478		3,275	163		3,438	3,850
8770	24 kW		1.90	4.211		3,325	198		3,523	3,950
8790	Recessed, 2 kW		4.40	1.818		2,000	85.50		2,085.50	2,325
8810	3 kW		4.40	1.818		2,075	85.50		2,160.50	2,425
8820	4 kW		4.20	1.905		2,125	89.50		2,214.50	2,475
8830	5 kW		4.20	1.905		2,200	89.50		2,289.50	2,575
8840	6 kW		4	2		2,225	94		2,319	2,575
8850	8 kW		3.80	2.105		2,275	99		2,374	2,675
8860	10 kW		3.50	2.286		2,325	107		2,432	2,750
8870	12 kW		2.90	2.759		2,375	130		2,505	2,800
8880	13.5 kW		2.70	2.963		2,400	139		2,539	2,875
8890	16 kW		2.30	3.478		2,450	163		2,613	2,950
8900	20 kW		1.90	4.211		3,350	198		3,548	3,975
8920	24 kW		1.60	5		3,400	235		3,635	4,075
8940	Ceiling mount, 2 kW		3.20	2.500		2,000	118		2,118	2,375
8950	3 kW		3.20	2.500		2,075	118		2,193	2,475
8960	4 kW		3	2.667		2,125	125		2,250	2,550
8970	5 kW		3	2.667		2,200	125		2,325	2,625
8980	6 kW		2.80	2.857		2,225	134		2,359	2,625
8990	8 kW		2.40	3.333		2,275	157		2,432	2,775
9000	10 kW		2.20	3.636		2,325	171		2,496	2,850
9020	13.5 kW		1.50	5.333		2,400	251		2,651	3,025
9030	16 kW		1.30	6.154		2,450	289		2,739	3,150
9040	20 kW		.90	8.889		3,350	420		3,770	4,325
9060	24 kW		.60	13.333		3,400	625		4,025	4,700
9990	Minimum labor/equipment charge		4	2	Job		94		94	145

23 84 Humidity Control Equipment

23 84 13 – Humidifiers

23 84 13.10 Humidifier Units

		Crew	Daily Output	Labor-Hours	Unit	Material	Labor	Equipment	Total	Total Incl O&P
0010	**HUMIDIFIER UNITS**									
0520	Steam, room or duct, filter, regulators, auto. controls, 220 V									
0540	11 lb. per hour R236000-30	Q-5	6	2.667	Ea.	2,275	118		2,393	2,675
0560	22 lb. per hour		5	3.200		2,525	142		2,667	3,000
0580	33 lb. per hour		4	4		2,575	178		2,753	3,125
0600	50 lb. per hour		4	4		3,175	178		3,353	3,750
0620	100 lb. per hour		3	5.333		3,800	237		4,037	4,550
0640	150 lb. per hour		2.50	6.400		5,025	284		5,309	5,975
0660	200 lb. per hour		2	8		6,200	355		6,555	7,375
0700	With blower									
0720	11 lb. per hour	Q-5	5.50	2.909	Ea.	3,175	129		3,304	3,700
0740	22 lb. per hour		4.75	3.368		3,425	150		3,575	3,975
0760	33 lb. per hour		3.75	4.267		3,475	190		3,665	4,125
0780	50 lb. per hour		3.50	4.571		4,200	203		4,403	4,925
0800	100 lb. per hour		2.75	5.818		4,725	259		4,984	5,600
0820	150 lb. per hour		2	8		7,000	355		7,355	8,250
0840	200 lb. per hour		1.50	10.667		8,225	475		8,700	9,750
5000	Furnace type, wheel bypass									
5020	10 GPD	1 Stpi	4	2	Ea.	192	98.50		290.50	365
5040	14 GPD		3.80	2.105		234	104		338	420

23 84 Humidity Control Equipment

23 84 13 – Humidifiers

23 84 13.10 Humidifier Units		Crew	Daily Output	Labor-Hours	Unit	Material	Labor	Equipment	Total	Total Incl O&P
5060	19 GPD	1 Stpi	3.60	2.222	Ea.	272	110		382	470
9000	Minimum labor/equipment charge	Q-5	3.50	4.571	Job		203		203	315

23 84 16 – Dehumidifiers

23 84 16.10 Dehumidifier Units

		Crew	Daily Output	Labor-Hours	Unit	Material	Labor	Equipment	Total	Total Incl O&P
0010	**DEHUMIDIFIER UNITS** R236000-30									
6000	Self contained with filters and standard controls									
6040	1.5 lb/hr, 50 cfm	1 Plum	8	1	Ea.	5,300	49		5,349	5,900
6060	3 lb/hr, 150 cfm	Q-1	12	1.333		5,525	58.50		5,583.50	6,175
6065	6 lb/hr, 150 cfm		9	1.778		9,000	78		9,078	10,000
6070	16 to 20 lb/hr, 600 cfm		5	3.200		18,300	140		18,440	20,400
6080	30 to 40 lb/hr, 1125 cfm		4	4		28,600	176		28,776	31,700
6090	60 to 75 lb/hr, 2250 cfm		3	5.333		37,300	234		37,534	41,400
6100	120 to 155 lb/hr, 4500 cfm		2	8		66,000	350		66,350	73,000
6110	240 to 310 lb/hr, 9000 cfm		1.50	10.667		94,500	470		94,970	104,500
6120	400 to 515 lb/hr, 15,000 cfm	Q-2	1.60	15		128,000	685		128,685	142,000
6130	530 to 690 lb/hr, 20,000 cfm		1.40	17.143		139,000	780		139,780	153,500
6140	800 to 1030 lb/hr, 30,000 cfm		1.20	20		158,000	910		158,910	175,500
6150	1060 to 1375 lb/hr, 40,000 cfm		1	24		220,000	1,100		221,100	243,500

Estimating Tips

26 05 00 Common Work Results for Electrical

- Conduit should be taken off in three main categories—power distribution, branch power, and branch lighting—so the estimator can concentrate on systems and components, therefore making it easier to ensure all items have been accounted for.
- For cost modifications for elevated conduit installation, add the percentages to labor according to the height of installation, and only to the quantities exceeding the different height levels, not to the total conduit quantities.
- Remember that aluminum wiring of equal ampacity is larger in diameter than copper and may require larger conduit.
- If more than three wires at a time are being pulled, deduct percentages from the labor hours of that grouping of wires.
- When taking off grounding systems, identify separately the type and size of wire, and list each unique type of ground connection.
- The estimator should take the weights of materials into consideration when completing a takeoff. Topics to consider include: How will the materials be supported? What methods of support are available? How high will the support structure have to reach? Will the final support structure be able to withstand the total burden? Is the support material included or separate from the fixture, equipment, and material specified?
- Do not overlook the costs for equipment used in the installation. If scaffolding or highlifts are available in the field, contractors may use them in lieu of the proposed ladders and rolling staging.

26 20 00 Low-Voltage Electrical Transmission

- Supports and concrete pads may be shown on drawings for the larger equipment, or the support system may be only a piece of plywood for the back of a panelboard. In either case, it must be included in the costs.

26 40 00 Electrical and Cathodic Protection

- When taking off cathodic protections systems, identify the type and size of cable, and list each unique type of anode connection.

26 50 00 Lighting

- Fixtures should be taken off room by room, using the fixture schedule, specifications, and the ceiling plan. For large concentrations of lighting fixtures in the same area, deduct the percentages from labor hours.

Reference Numbers

Reference numbers are shown in shaded boxes at the beginning of some major classifications. These numbers refer to related items in the Reference Section. The reference information may be an estimating procedure, an alternate pricing method, or technical information.

Note: Not all subdivisions listed here necessarily appear in this publication.

Note: **Trade Service,** *in part, has been used as a reference source for some of the material prices used in Division 26.*

26 01 Operation and Maintenance of Electrical Systems

26 01 40 – Operation and Maintenance of Electrical and Cathodic Protection Systems

26 01 40.51 Electrical Systems Repair and Replacement		Crew	Daily Output	Labor-Hours	Unit	Material	2009 Bare Costs Labor	Equipment	Total	Total Incl O&P
0010	**ELECTRICAL SYSTEMS REPAIR AND REPLACEMENT**									
3000	Remove and replace (reinstall), switch cover	1 Elec	60	.133	Ea.		6.25		6.25	9.65
3020	Outlet cover	"	60	.133	"		6.25		6.25	9.65

26 01 50 – Operation and Maintenance of Lighting

26 01 50.81 Luminaire Replacement		Crew	Daily Output	Labor-Hours	Unit	Material	Labor	Equipment	Total	Total Incl O&P
0010	**LUMINAIRE REPLACEMENT**									
3200	Remove and replace (reinstall), lighting fixture	1 Elec	4	2	Ea.		94		94	145

26 05 Common Work Results for Electrical

26 05 05 – Selective Electrical Demolition

26 05 05.10 Electrical Demolition		Crew	Daily Output	Labor-Hours	Unit	Material	Labor	Equipment	Total	Total Incl O&P
0010	**ELECTRICAL DEMOLITION** R260105-30									
0020	Conduit to 15' high, including fittings & hangers									
0100	Rigid galvanized steel, 1/2" to 1" diameter R024119-10	1 Elec	242	.033	L.F.		1.55		1.55	2.39
0120	1-1/4" to 2"	"	200	.040			1.88		1.88	2.89
0140	2-1/2" to 3-1/2"	2 Elec	302	.053			2.49		2.49	3.83
0160	4" to 6"	"	160	.100			4.70		4.70	7.25
0170	PVC #40, 1/2" to 1"	1 Elec	410	.020			.92		.92	1.41
0172	1-1/4" to 2"		350	.023			1.07		1.07	1.65
0174	2-1/2"		250	.032			1.50		1.50	2.31
0176	3" to 3-1/2"	2 Elec	340	.047			2.21		2.21	3.40
0178	4" to 6"	"	230	.070			3.27		3.27	5.05
0200	Electric metallic tubing (EMT), 1/2" to 1"	1 Elec	394	.020			.95		.95	1.47
0220	1-1/4" to 1-1/2"		326	.025			1.15		1.15	1.77
0240	2" to 3"		236	.034			1.59		1.59	2.45
0260	3-1/2" to 4"	2 Elec	310	.052			2.43		2.43	3.73
0270	Armored cable, (BX) avg. 50' runs									
0280	#14, 2 wire	1 Elec	690	.012	L.F.		.54		.54	.84
0290	#14, 3 wire		571	.014			.66		.66	1.01
0300	#12, 2 wire		605	.013			.62		.62	.96
0310	#12, 3 wire		514	.016			.73		.73	1.13
0320	#10, 2 wire		514	.016			.73		.73	1.13
0330	#10, 3 wire		425	.019			.88		.88	1.36
0340	#8, 3 wire		342	.023			1.10		1.10	1.69
0350	Non metallic sheathed cable (Romex)									
0360	#14, 2 wire	1 Elec	720	.011	L.F.		.52		.52	.80
0370	#14, 3 wire		657	.012			.57		.57	.88
0380	#12, 2 wire		629	.013			.60		.60	.92
0390	#10, 3 wire		450	.018			.84		.84	1.29
0400	Wiremold raceway, including fittings & hangers									
0420	No. 3000	1 Elec	250	.032	L.F.		1.50		1.50	2.31
0440	No. 4000		217	.037			1.73		1.73	2.67
0460	No. 6000		166	.048			2.26		2.26	3.48
0465	Telephone/power pole		12	.667	Ea.		31.50		31.50	48
0470	Non-metallic, straight section		480	.017	L.F.		.78		.78	1.21
0500	Channels, steel, including fittings & hangers									
0520	3/4" x 1-1/2"	1 Elec	308	.026	L.F.		1.22		1.22	1.88
0540	1-1/2" x 1-1/2"		269	.030			1.40		1.40	2.15
0560	1-1/2" x 1-7/8"		229	.035			1.64		1.64	2.53
0600	Copper bus duct, indoor, 3 phase									

26 05 05.10 Electrical Demolition		Crew	Daily Output	Labor-Hours	Unit	Material	2009 Bare Costs Labor	Equipment	Total	Total Incl O&P
0610	Including hangers & supports									
0620	225 amp	2 Elec	135	.119	L.F.		5.55		5.55	8.55
0640	400 amp		106	.151			7.10		7.10	10.90
0660	600 amp		86	.186			8.75		8.75	13.45
0680	1000 amp		60	.267			12.55		12.55	19.30
0700	1600 amp		40	.400			18.80		18.80	29
0720	3000 amp		10	1.600			75		75	116
0800	Plug-in switches, 600V 3 ph, incl. disconnecting									
0820	wire, conduit terminations, 30 amp	1 Elec	15.50	.516	Ea.		24.50		24.50	37.50
0840	60 amp		13.90	.576			27		27	41.50
0850	100 amp		10.40	.769			36		36	55.50
0860	200 amp		6.20	1.290			60.50		60.50	93.50
0880	400 amp	2 Elec	5.40	2.963			139		139	214
0900	600 amp		3.40	4.706			221		221	340
0920	800 amp		2.60	6.154			289		289	445
0940	1200 amp		2	8			375		375	580
0960	1600 amp		1.70	9.412			440		440	680
1010	Safety switches, 250 or 600V, incl. disconnection									
1050	of wire & conduit terminations									
1100	30 amp	1 Elec	12.30	.650	Ea.		30.50		30.50	47
1120	60 amp		8.80	.909			42.50		42.50	65.50
1140	100 amp		7.30	1.096			51.50		51.50	79
1160	200 amp		5	1.600			75		75	116
1180	400 amp	2 Elec	6.80	2.353			111		111	170
1200	600 amp	″	4.60	3.478			163		163	251
1210	Panel boards, incl. removal of all breakers,									
1220	conduit terminations & wire connections									
1230	3 wire, 120/240V, 100A, to 20 circuits	1 Elec	2.60	3.077	Ea.		145		145	222
1240	200 amps, to 42 circuits	2 Elec	2.60	6.154			289		289	445
1250	400 amps, to 42 circuits	″	2.20	7.273			340		340	525
1260	4 wire, 120/208V, 125A, to 20 circuits	1 Elec	2.40	3.333			157		157	241
1270	200 amps, to 42 circuits	2 Elec	2.40	6.667			315		315	480
1280	400 amps, to 42 circuits	″	1.92	8.333			390		390	605
1300	Transformer, dry type, 1 phase, incl. removal of									
1320	supports, wire & conduit terminations									
1340	1 kVA	1 Elec	7.70	1.039	Ea.		49		49	75
1360	5 kVA		4.70	1.702			80		80	123
1380	10 kVA		3.60	2.222			104		104	161
1400	37.5 kVA	2 Elec	3	5.333			251		251	385
1420	75 kVA	″	2.50	6.400			300		300	465
1440	3 phase to 600V, primary									
1460	3 kVA	1 Elec	3.85	2.078	Ea.		97.50		97.50	150
1480	15 kVA	2 Elec	4.20	3.810			179		179	275
1500	30 kVA		3.50	4.571			215		215	330
1510	45 kVA		3.10	5.161			243		243	375
1520	75 kVA		2.70	5.926			279		279	430
1530	112.5 kVA	R-3	2.90	6.897			320	44.50	364.50	545
1540	150 kVA		2.70	7.407			345	48	393	585
1560	500 kVA		1.40	14.286			660	92.50	752.50	1,125
1570	750 kVA		1.10	18.182			840	117	957	1,425
1600	Pull boxes & cabinets, sheet metal, incl. removal									
1620	of supports and conduit terminations									
1640	6″ x 6″ x 4″	1 Elec	31.10	.257	Ea.		12.10		12.10	18.60

26 05 05.10 Electrical Demolition		Crew	Daily Output	Labor-Hours	Unit	Material	2009 Bare Costs Labor	Equipment	Total	Total Incl O&P
1660	12" x 12" x 4"	1 Elec	23.30	.343	Ea.		16.15		16.15	25
1680	24" x 24" x 6"		12.30	.650			30.50		30.50	47
1700	36" x 36" x 8"		7.70	1.039			49		49	75
1720	Junction boxes, 4" sq. & oct.		80	.100			4.70		4.70	7.25
1740	Handy box		107	.075			3.51		3.51	5.40
1760	Switch box		107	.075			3.51		3.51	5.40
1780	Receptacle & switch plates		257	.031			1.46		1.46	2.25
1790	Receptacles & switches, 15 to 30 amp		135	.059			2.79		2.79	4.28
1800	Wire, THW-THWN-THHN, removed from									
1810	in place conduit, to 15' high									
1830	#14	1 Elec	65	.123	C.L.F.		5.80		5.80	8.90
1840	#12		55	.145			6.85		6.85	10.50
1850	#10		45.50	.176			8.25		8.25	12.70
1860	#8		40.40	.198			9.30		9.30	14.30
1870	#6		32.60	.245			11.55		11.55	17.75
1880	#4	2 Elec	53	.302			14.20		14.20	22
1890	#3		50	.320			15.05		15.05	23
1900	#2		44.60	.359			16.85		16.85	26
1910	1/0		33.20	.482			22.50		22.50	35
1920	2/0		29.20	.548			26		26	39.50
1930	3/0		25	.640			30		30	46.50
1940	4/0		22	.727			34		34	52.50
1950	250 kcmil		20	.800			37.50		37.50	58
1960	300 kcmil		19	.842			39.50		39.50	61
1970	350 kcmil		18	.889			42		42	64.50
1980	400 kcmil		17	.941			44		44	68
1990	500 kcmil		16.20	.988			46.50		46.50	71.50
2000	Interior fluorescent fixtures, incl. supports									
2010	& whips, to 15' high									
2100	Recessed drop-in 2' x 2', 2 lamp	2 Elec	35	.457	Ea.		21.50		21.50	33
2120	2' x 4', 2 lamp		33	.485			23		23	35
2140	2' x 4', 4 lamp		30	.533			25		25	38.50
2160	4' x 4', 4 lamp		20	.800			37.50		37.50	58
2180	Surface mount, acrylic lens & hinged frame									
2200	1' x 4', 2 lamp	2 Elec	44	.364	Ea.		17.10		17.10	26.50
2220	2' x 2', 2 lamp		44	.364			17.10		17.10	26.50
2260	2' x 4', 4 lamp		33	.485			23		23	35
2280	4' x 4', 4 lamp		23	.696			32.50		32.50	50.50
2300	Strip fixtures, surface mount									
2320	4' long, 1 lamp	2 Elec	53	.302	Ea.		14.20		14.20	22
2340	4' long, 2 lamp		50	.320			15.05		15.05	23
2360	8' long, 1 lamp		42	.381			17.90		17.90	27.50
2380	8' long, 2 lamp		40	.400			18.80		18.80	29
2400	Pendant mount, industrial, incl. removal									
2410	of chain or rod hangers, to 15' high									
2420	4' long, 2 lamp	2 Elec	35	.457	Ea.		21.50		21.50	33
2440	8' long, 2 lamp	"	27	.593	"		28		28	43
2460	Interior incandescent, surface, ceiling									
2470	or wall mount, to 12' high									
2480	Metal cylinder type, 75 Watt	2 Elec	62	.258	Ea.		12.15		12.15	18.65
2500	150 Watt	"	62	.258	"		12.15		12.15	18.65
2520	Metal halide, high bay									
2540	400 Watt	2 Elec	15	1.067	Ea.		50		50	77

26 05 05.10 Electrical Demolition		Crew	Daily Output	Labor-Hours	Unit	Material	2009 Bare Costs Labor	Equipment	Total	Total Incl O&P
2560	1000 Watt	2 Elec	12	1.333	Ea.		62.50		62.50	96.50
2580	150 Watt, low bay	↓	20	.800	↓		37.50		37.50	58
2600	Exterior fixtures, incandescent, wall mount									
2620	100 Watt	2 Elec	50	.320	Ea.		15.05		15.05	23
2640	Quartz, 500 Watt		33	.485			23		23	35
2660	1500 Watt	↓	27	.593	↓		28		28	43
2680	Wall pack, mercury vapor									
2700	175 Watt	2 Elec	25	.640	Ea.		30		30	46.50
2720	250 Watt	"	25	.640			30		30	46.50
7000	Weatherhead/mast, 2"	1 Elec	16	.500			23.50		23.50	36
7002	3"		10	.800			37.50		37.50	58
7004	3-1/2"		8.50	.941			44		44	68
7006	4"		8	1	↓		47		47	72.50
7100	Service entry cable, #6, +#6 neutral		420	.019	L.F.		.90		.90	1.38
7102	#4, +#4 neutral		360	.022			1.04		1.04	1.61
7104	#2, +#4 neutral		345	.023			1.09		1.09	1.68
7106	#2, +#2 neutral		330	.024	↓		1.14		1.14	1.75
9000	Minimum labor/equipment charge	↓	4	2	Job		94		94	145

26 05 05.15 Electrical Demolition, Grounding		Crew	Daily Output	Labor-Hours	Unit	Material	Labor	Equipment	Total	Total Incl O&P
0010	**ELECTRICAL DEMOLITION, GROUNDING** Addition									
0100	Ground clamp, bronze	1 Elec	64	.125	Ea.		5.90		5.90	9.05
0140	Water pipe ground clamp, bronze, heavy duty	"	24	.333	"		15.65		15.65	24
0200	Ground wire, bare armored	2 Elec	900	.018	L.F.		.84		.84	1.29
0240	Bare copper or aluminum	"	2800	.006	"		.27		.27	.41

26 05 05.20 Electrical Demolition, Wiring Methods		Crew	Daily Output	Labor-Hours	Unit	Material	Labor	Equipment	Total	Total Incl O&P
0010	**ELECTRICAL DEMOLITION, WIRING METHODS** Addition R024119-10									
0100	Armored cable, w/ PVC jacket, in cable tray									
0110	#6	1 Elec	9.30	.860	C.L.F.		40.50		40.50	62
0120	#4	2 Elec	16.20	.988			46.50		46.50	71.50
0130	#2		13.80	1.159			54.50		54.50	84
0140	#1		12	1.333			62.50		62.50	96.50
0150	1/0		10.80	1.481			69.50		69.50	107
0160	2/0		10.20	1.569			73.50		73.50	113
0170	3/0		9.60	1.667			78.50		78.50	121
0180	4/0	↓	9	1.778			83.50		83.50	129
0190	250 kcmil	3 Elec	10.80	2.222			104		104	161
0210	350 kcmil		9.90	2.424			114		114	175
0230	500 kcmil		9	2.667			125		125	193
0240	750 kcmil	↓	8.10	2.963	↓		139		139	214
1100	Control cable, 600 V or less									
1110	3 wires	1 Elec	24	.333	C.L.F.		15.65		15.65	24
1120	5 wires		20	.400			18.80		18.80	29
1130	7 wires		16.50	.485			23		23	35
1140	9 wires	↓	15	.533			25		25	38.50
1150	12 wires	2 Elec	26	.615			29		29	44.50
1160	15 wires		22	.727			34		34	52.50
1170	19 wires		19	.842			39.50		39.50	61
1180	25 wires	↓	16	1	↓		47		47	72.50
1500	Mineral insulated (MI) cable, 600 V									
1510	#10	1 Elec	4.80	1.667	C.L.F.		78.50		78.50	121
1520	#8		4.50	1.778			83.50		83.50	129
1530	#6	↓	4.20	1.905	↓		89.50		89.50	138

26 05 05 – Selective Electrical Demolition

26 05 05.20 Electrical Demolition, Wiring Methods		Crew	Daily Output	Labor-Hours	Unit	Material	2009 Bare Costs Labor	Equipment	Total	Total Incl O&P
1540	#4	2 Elec	7.20	2.222	C.L.F.		104		104	161
1550	#2		6.60	2.424			114		114	175
1560	#1		6.30	2.540			119		119	184
1570	1/0		6	2.667			125		125	193
1580	2/0		5.70	2.807			132		132	203
1590	3/0		5.40	2.963			139		139	214
1600	4/0		4.80	3.333			157		157	241
1610	250 kcmil	3 Elec	7.20	3.333			157		157	241
1620	500 kcmil	"	5.90	4.068			191		191	294
2000	Shielded cable, XLP shielding, to 35 kV									
2010	#4	2 Elec	13.20	1.212	C.L.F.		57		57	87.50
2020	#1		12	1.333			62.50		62.50	96.50
2030	1/0		11.40	1.404			66		66	101
2040	2/0		10.80	1.481			69.50		69.50	107
2050	4/0		9.60	1.667			78.50		78.50	121
2060	250 kcmil	3 Elec	13.50	1.778			83.50		83.50	129
2070	350 kcmil		11.70	2.051			96.50		96.50	148
2080	500 kcmil		11.20	2.143			101		101	155
2090	750 kcmil		10.80	2.222			104		104	161
3000	Modular flexible wiring									
3010	Cable set	1 Elec	120	.067	Ea.		3.13		3.13	4.82
3020	Conversion module		48	.167			7.85		7.85	12.05
3030	Switching assembly		96	.083			3.92		3.92	6
3200	Undercarpet									
3210	Power or telephone, flat cable	1 Elec	1320	.006	L.F.		.28		.28	.44
3220	Transition block assemblies w/ fitting		75	.107	Ea.		5		5	7.70
3230	Floor box with fitting		60	.133			6.25		6.25	9.65
3300	Data system, cable with connection		50	.160			7.50		7.50	11.55
4000	Cable tray, including fitting & support									
4010	Galvanized steel, 6" wide	2 Elec	310	.052	L.F.		2.43		2.43	3.73
4020	9" wide		295	.054			2.55		2.55	3.92
4030	12" wide		285	.056			2.64		2.64	4.06
4040	18" wide		270	.059			2.79		2.79	4.28
4050	24" wide		260	.062			2.89		2.89	4.45
4060	30" wide		240	.067			3.13		3.13	4.82
4070	36" wide		220	.073			3.42		3.42	5.25
4110	Aluminum, 6" wide		420	.038			1.79		1.79	2.75
4120	9" wide		415	.039			1.81		1.81	2.79
4130	12" wide		390	.041			1.93		1.93	2.97
4140	18" wide		370	.043			2.03		2.03	3.13
4150	24" wide		350	.046			2.15		2.15	3.30
4160	30" wide		325	.049			2.31		2.31	3.56
4170	36" wide		300	.053			2.51		2.51	3.86
4310	Cable channel, aluminum, 4" wide straight	1 Elec	240	.033			1.57		1.57	2.41
5000	Conduit nipples, with locknuts and bushings									
5020	1/2"	1 Elec	108	.074	Ea.		3.48		3.48	5.35
5040	3/4"		96	.083			3.92		3.92	6
5060	1"		81	.099			4.64		4.64	7.15
5080	1-1/4"		69	.116			5.45		5.45	8.40
5100	1-1/2"		60	.133			6.25		6.25	9.65
5120	2"		54	.148			6.95		6.95	10.70
5140	2-1/2"		45	.178			8.35		8.35	12.85
5160	3"		36	.222			10.45		10.45	16.05

26 05 05 – Selective Electrical Demolition

26 05 05.20 Electrical Demolition, Wiring Methods

	26 05 05.20 Electrical Demolition, Wiring Methods	Crew	Daily Output	Labor-Hours	Unit	Material	2009 Bare Costs Labor	Equipment	Total	Total Incl O&P
5180	3-1/2"	1 Elec	33	.242	Ea.		11.40		11.40	17.55
5200	4"		27	.296			13.95		13.95	21.50
5220	5"		21	.381			17.90		17.90	27.50
5240	6"		18	.444			21		21	32
5500	Electric nonmetallic tubing (ENT), flexible, 1/2" to 1" diameter		690	.012	L.F.		.54		.54	.84
5510	1-1/4" to 2" diameter		300	.027			1.25		1.25	1.93
5600	Flexible metallic tubing, steel, 3/8" to 3/4" diameter		600	.013			.63		.63	.96
5610	1" to 1-1/4" diameter		260	.031			1.45		1.45	2.22
5620	1-1/2" to 2" diameter		140	.057			2.69		2.69	4.13
5630	2-1/2" diameter		100	.080			3.76		3.76	5.80
5640	3" to 3-1/2" diameter	2 Elec	150	.107			5		5	7.70
5650	4" diameter	"	100	.160			7.50		7.50	11.55
5700	Sealtite flexible conduit, 3/8" to 3/4" diameter	1 Elec	420	.019			.90		.90	1.38
5710	1" to 1-1/4" diameter		180	.044			2.09		2.09	3.21
5720	1-1/2" to 2" diameter		120	.067			3.13		3.13	4.82
5730	2-1/2" diameter		80	.100			4.70		4.70	7.25
5740	3" diameter	2 Elec	150	.107			5		5	7.70
5750	4" diameter	"	100	.160			7.50		7.50	11.55
5800	Wiring duct, plastic, 1-1/2" to 2-1/2" wide	1 Elec	360	.022			1.04		1.04	1.61
5810	3" wide		330	.024			1.14		1.14	1.75
5820	4" wide		300	.027			1.25		1.25	1.93
6300	Wireway, with fittings and supports, to 15' high									
6310	2-1/2" x 2-1/2"	1 Elec	180	.044	L.F.		2.09		2.09	3.21
6320	4" x 4"	"	160	.050			2.35		2.35	3.62
6330	6" x 6"	2 Elec	240	.067			3.13		3.13	4.82
6340	8" x 8"		160	.100			4.70		4.70	7.25
6350	10" x 10"		120	.133			6.25		6.25	9.65
6360	12" x 12"		80	.200			9.40		9.40	14.45

26 05 05.25 Electrical Demolition, Electrical Power

	26 05 05.25 Electrical Demolition, Electrical Power	Crew	Daily Output	Labor-Hours	Unit	Material	2009 Bare Costs Labor	Equipment	Total	Total Incl O&P
0010	**ELECTRICAL DEMOLITION, ELECTRICAL POWER** R024119-10									
0100	Meter centers and sockets									
0120	Meter socket, 4 terminal R260105-30	1 Elec	8.40	.952	Ea.		45		45	69
0140	Trans-socket, 13 terminal, 400 A	"	3.60	2.222			104		104	161
0160	800 A	2 Elec	4.40	3.636			171		171	263
0200	Meter center, 400 A		5.80	2.759			130		130	199
0210	600 A		4	4			188		188	289
0220	800 A		3.30	4.848			228		228	350
0230	1200 A		2.80	5.714			269		269	415
0240	1600 A		2.50	6.400			300		300	465
0300	Base meter devices, 3 meter		3.60	4.444			209		209	320
0310	4 meter		3.30	4.848			228		228	350
0320	5 meter		2.90	5.517			259		259	400
0330	6 meter		2.20	7.273			340		340	525
0340	7 meter		2	8			375		375	580
0350	8 meter		1.90	8.421			395		395	610
0400	Branch meter devices									
0410	Socket w/ circuit breaker 200 A, 2 meter	2 Elec	3.30	4.848	Ea.		228		228	350
0420	3 meter		2.90	5.517			259		259	400
0430	4 meter		2.60	6.154			289		289	445
0450	Main circuit breaker, 400 A		5.80	2.759			130		130	199
0460	600 A		4	4			188		188	289
0470	800 A		3.30	4.848			228		228	350

26 05 05 – Selective Electrical Demolition

26 05 05.25 Electrical Demolition, Electrical Power		Crew	Daily Output	Labor-Hours	Unit	Material	2009 Bare Costs Labor	Equipment	Total	Total Incl O&P
0480	1200 A	2 Elec	2.80	5.714	Ea.		269		269	415
0490	1600 A		2.50	6.400			300		300	465
0500	Main lug terminal box, 800 A		3.40	4.706			221		221	340
0510	1200 A		2.60	6.154			289		289	445
1000	Motors, 230/460 V, 60 Hz, 3/4 HP	1 Elec	10.70	.748			35		35	54
1010	5 HP		9	.889			42		42	64.50
1020	10 HP		8	1			47		47	72.50
1030	15 HP		6.40	1.250			59		59	90.50
1040	20 HP	2 Elec	10.40	1.538			72.50		72.50	111
1050	50 HP		9.60	1.667			78.50		78.50	121
1060	75 HP		5.60	2.857			134		134	207
1070	100 HP	3 Elec	5.40	4.444			209		209	320
1080	150 HP		3.60	6.667			315		315	480
1090	200 HP		3	8			375		375	580
1200	Variable frequency drive, 460 V, for 5 HP motor size	1 Elec	3.20	2.500			118		118	181
1210	10 HP motor size	"	2.70	2.963			139		139	214
1220	20 HP motor size	2 Elec	3.60	4.444			209		209	320
1230	50 HP motor size	"	2.10	7.619			360		360	550
1240	75 HP motor size	R-3	2.20	9.091			420	58.50	478.50	715
1250	100 HP motor size		2	10			465	64.50	529.50	785
1260	150 HP motor size		2	10			465	64.50	529.50	785
1270	200 HP motor size		1.70	11.765			545	76	621	925
2000	Generator set w/ accessories, 3 phase 4 wire, 277/480 V									
2010	7.5 kW	3 Elec	2	12	Ea.		565		565	870
2020	20 kW		1.70	14.118			665		665	1,025
2040	30 kW		1.33	18.045			850		850	1,300
2060	50 kW		1	24			1,125		1,125	1,725
2080	100 kW		.75	32			1,500		1,500	2,325
2100	150 kW		.63	38.095			1,800		1,800	2,750
2120	250 kW		.59	40.678			1,900		1,900	2,950
2140	400 kW		.50	48			2,250		2,250	3,475
2160	500 kW		.44	54.545			2,575		2,575	3,950
2180	750 kW	4 Elec	.56	57.143			2,675		2,675	4,125
2200	1000 kW	"	.46	69.565			3,275		3,275	5,025
3000	Uninterruptible power supply system (UPS)									
3010	Single phase, 120 V, 1 kVA	1 Elec	3.20	2.500	Ea.		118		118	181
3020	2 kVA	2 Elec	3.60	4.444			209		209	320
3030	5 kVA	3 Elec	2.50	9.600			450		450	695
3040	10 kVA		2.30	10.435			490		490	755
3050	15 kVA		1.80	13.333			625		625	965
4000	Transformer, incl support, wire & conduit termination									
4010	Buck-boost, single phase, 120/240 V, 0.1 kVA	1 Elec	25	.320	Ea.		15.05		15.05	23
4020	0.5 kVA		12.50	.640			30		30	46.50
4030	1 kVA		6.30	1.270			59.50		59.50	92
4040	5 kVA		3.80	2.105			99		99	152
4800	5 kV or 15 kV primary, 277/480 V second, 112.5 kVA	R-3	3.50	5.714			265	37	302	450
4810	150 kVA		2.70	7.407			345	48	393	585
4820	225 kVA		2.30	8.696			405	56	461	680
4830	500 kVA		1.45	13.793			640	89	729	1,075
4840	750 kVA		1.33	15.038			695	97	792	1,175
4850	1000 kVA		1.25	16			740	103	843	1,275
4860	2000 kVA		1.05	19.048			880	123	1,003	1,475
4870	3000 kVA		.75	26.667			1,225	172	1,397	2,100

26 05 05 – Selective Electrical Demolition

26 05 05.25 Electrical Demolition, Electrical Power		Crew	Daily Output	Labor-Hours	Unit	Material	2009 Bare Costs Labor	Equipment	Total	Total Incl O&P
6010	Isolation panel, 3 kVA	1 Elec	2.30	3.478	Ea.		163		163	251
6020	5 kVA		2.20	3.636			171		171	263
6030	7.5 kVA		2.10	3.810			179		179	275
6040	10 kVA		1.80	4.444			209		209	320
6050	15 kVA		1.40	5.714			269		269	415
7000	Power filters & conditioners									
7100	Automatic voltage regulator	2 Elec	6	2.667	Ea.		125		125	193
7200	Capacitor, 1 kVAR	1 Elec	8.60	.930			43.50		43.50	67.50
7210	5 kVAR		5.80	1.379			65		65	99.50
7220	10 kVAR		4.80	1.667			78.50		78.50	121
7230	15 kVAR		4.20	1.905			89.50		89.50	138
7240	20 kVAR		3.50	2.286			107		107	165
7250	30 kVAR		3.40	2.353			111		111	170
7260	50 kVAR		3.20	2.500			118		118	181
7400	Computer isolator transformer									
7410	Single phase, 120/240 V, 0.5 kVAR	1 Elec	12.80	.625	Ea.		29.50		29.50	45
7420	1 kVAR		8.50	.941			44		44	68
7430	2.5 kVAR		6.40	1.250			59		59	90.50
7440	5 kVAR		3.70	2.162			102		102	156
7500	Computer regulator transformer									
7510	Single phase, 240 V, 0.5 kVAR	1 Elec	8.50	.941	Ea.		44		44	68
7520	1 kVAR		6.40	1.250			59		59	90.50
7530	2 kVAR		3.20	2.500			118		118	181
7540	Single phase, plug-in unit 120 V, 0.5 kVAR		26	.308			14.45		14.45	22.50
7550	1 kVAR		17	.471			22		22	34
7600	Power conditioner transformer									
7610	Single phase 115 V - 240 V, 3 kVA	2 Elec	5.10	3.137	Ea.		147		147	227
7620	5 kVA	"	3.70	4.324			203		203	315
7630	7.5 kVA	3 Elec	4.80	5			235		235	360
7640	10 kVA	"	4.30	5.581			262		262	405
7700	Transient suppressor/voltage regulator									
7710	Single phase 115 or 220 V, 1 kVA	1 Elec	8.50	.941	Ea.		44		44	68
7720	2 kVA		7.30	1.096			51.50		51.50	79
7730	4 kVA		6.80	1.176			55.50		55.50	85
7800	Transient voltage suppressor transformer									
7810	Single phase, 115 or 220 V, 3.6 kVA	1 Elec	12.80	.625	Ea.		29.50		29.50	45
7820	7.2 kVA		11.50	.696			32.50		32.50	50.50
7830	14.4 kVA		10.20	.784			37		37	56.50
7840	Single phase, plug-in, 120 V, 1.8 kVA		26	.308			14.45		14.45	22.50
8000	Power measurement & control									
8010	Switchboard instruments, 3 phase 4 wire, indicating unit	1 Elec	25	.320	Ea.		15.05		15.05	23
8020	Recording unit		12	.667			31.50		31.50	48
8100	3 current transformers, 3 phase 4 wire, 5 to 800 A		6.40	1.250			59		59	90.50
8110	1000 to 1500 A		4.20	1.905			89.50		89.50	138
8120	2000 to 4000 A		3.20	2.500			118		118	181

26 05 05.30 Electrical Demolition, Transmission and Distribution

		Crew	Daily Output	Labor-Hours	Unit	Material	Labor	Equipment	Total	Total Incl O&P
0010	**ELECTRICAL DEMOLITION, TRANSMISSION & DISTRIBUTION**									
0100	Load interrupter switch, 600 A, NEMA 1, 4.8 kV	R-3	1.33	15.038	Ea.		695	97	792	1,175
0120	13.8 kV (R024119-10)	"	1.27	15.748			730	102	832	1,225
0200	Lightning arrester, 4.8 kV	1 Elec	9	.889			42		42	64.50
0220	13.8 kV (R260105-30)		6.67	1.199			56.50		56.50	86.50
0300	Alarm or option items		3.33	2.402			113		113	174

26 05 Common Work Results for Electrical

26 05 05 – Selective Electrical Demolition

26 05 05.35 Electrical Demolition, L.V. Distribution		Crew	Daily Output	Labor-Hours	Unit	Material	2009 Bare Costs Labor	Equipment	Total	Total Incl O&P
0010	**ELECTRICAL DEMOLITION, L.V. DISTRIBUTION** Addition R024119-10									
0100	Circuit breakers in enclosure									
0120	Enclosed (NEMA 1), 600 V, 3 pole, 30 A R260105-30	1 Elec	12.30	.650	Ea.		30.50		30.50	47
0140	60 A		8.80	.909			42.50		42.50	65.50
0160	100 A		7.30	1.096			51.50		51.50	79
0180	225 A		5	1.600			75		75	116
0200	400 A	2 Elec	6.80	2.353			111		111	170
0220	600 A		4.60	3.478			163		163	251
0240	800 A		3.60	4.444			209		209	320
0260	1200 A		3.10	5.161			243		243	375
0280	1600 A		2.80	5.714			269		269	415
0300	2000 A		2.50	6.400			300		300	465
0400	Enclosed (NEMA 7), 600 V, 3 pole, 50 A	1 Elec	8.80	.909			42.50		42.50	65.50
0410	100 A		5.80	1.379			65		65	99.50
0420	150 A		3.80	2.105			99		99	152
0430	250 A	2 Elec	6.20	2.581			121		121	187
0440	400 A	″	4.60	3.478			163		163	251
0460	Manual motor starter, NEMA 1	1 Elec	24	.333			15.65		15.65	24
0480	NEMA 4 or NEMA 7		15	.533			25		25	38.50
0500	Time switches, single pole single throw		15.40	.519			24.50		24.50	37.50
0520	Photo cell		30	.267			12.55		12.55	19.30
0540	Load management device, 4 loads		7.70	1.039			49		49	75
0550	8 loads		3.80	2.105			99		99	152
0600	Transfer switches, enclosed, 30 A		12	.667			31.50		31.50	48
0610	60 A		9.50	.842			39.50		39.50	61
0620	100 A		6.50	1.231			58		58	89
0630	150 A	2 Elec	12	1.333			62.50		62.50	96.50
0640	260 A		10	1.600			75		75	116
0660	400 A		8	2			94		94	145
0670	600 A		5	3.200			150		150	231
0680	800 A		4	4			188		188	289
0690	1200 A		3.50	4.571			215		215	330
0700	1600 A		3	5.333			251		251	385
0710	2000 A		2.50	6.400			300		300	465
1000	Enclosed controller, NEMA 1, 30 A	1 Elec	13.80	.580			27.50		27.50	42
1020	60 A		11.50	.696			32.50		32.50	50.50
1040	100 A		9.60	.833			39		39	60.50
1060	150 A		7.70	1.039			49		49	75
1080	200 A		5.40	1.481			69.50		69.50	107
1100	400 A	2 Elec	6.90	2.319			109		109	168
1120	600 A		4.60	3.478			163		163	251
1140	800 A		3.80	4.211			198		198	305
1160	1200 A		3.10	5.161			243		243	375
1200	Control station, NEMA 1	1 Elec	30	.267			12.55		12.55	19.30
1220	NEMA 7		23	.348			16.35		16.35	25
1230	Control switches, push button		69	.116			5.45		5.45	8.40
1240	Indicating light unit		120	.067			3.13		3.13	4.82
1250	Relay		15	.533			25		25	38.50
2000	Motor control center components									
2020	Starter, NEMA 1, size 1	1 Elec	9	.889	Ea.		42		42	64.50
2040	Size 2	2 Elec	13.30	1.203			56.50		56.50	87
2060	Size 3		6.70	2.388			112		112	173

26 05 05.35 Electrical Demolition, L.V. Distribution		Crew	Daily Output	Labor-Hours	Unit	Material	2009 Bare Costs Labor	Equipment	Total	Total Incl O&P
2080	Size 4	2 Elec	5.30	3.019	Ea.		142		142	218
2100	Size 5	↓	3.30	4.848			228		228	350
2120	NEMA 7, size 1	1 Elec	8.70	.920			43		43	66.50
2140	Size 2	2 Elec	12.70	1.260			59		59	91
2160	Size 3		6.30	2.540			119		119	184
2180	Size 4		5	3.200			150		150	231
2200	Size 5	↓	3.20	5			235		235	360
2300	Fuse, light contactor, NEMA 1, 30 A	1 Elec	9	.889			42		42	64.50
2310	60 A		6.70	1.194			56		56	86.50
2320	100 A		3.30	2.424			114		114	175
2330	200 A	↓	2.70	2.963			139		139	214
2350	Motor control center, incoming section	2 Elec	4	4			188		188	289
2400	Starter & structure, 10 HP	1 Elec	9	.889			42		42	64.50
2410	25 HP	2 Elec	13.30	1.203			56.50		56.50	87
2420	50 HP		6.70	2.388			112		112	173
2430	75 HP		5.30	3.019			142		142	218
2440	100 HP		4.70	3.404			160		160	246
2450	200 HP		3.30	4.848			228		228	350
2460	400 HP	↓	2.70	5.926	↓		279		279	430
2500	Motor starter & control									
2510	Motor starter, NEMA 1, 5 HP	1 Elec	8.90	.899	Ea.		42.50		42.50	65
2520	10 HP	"	6.20	1.290			60.50		60.50	93.50
2530	25 HP	2 Elec	8.40	1.905			89.50		89.50	138
2540	50 HP		6.90	2.319			109		109	168
2550	100 HP		4.60	3.478			163		163	251
2560	200 HP		3.50	4.571			215		215	330
2570	400 HP	↓	3.10	5.161			243		243	375
2610	Motor starter, NEMA 7, 5 HP	1 Elec	6.20	1.290			60.50		60.50	93.50
2620	10 HP	"	4.20	1.905			89.50		89.50	138
2630	25 HP	2 Elec	6.90	2.319			109		109	168
2640	50 HP		4.60	3.478			163		163	251
2650	100 HP		3.50	4.571			215		215	330
2660	200 HP	↓	1.90	8.421			395		395	610
2710	Combination control unit, NEMA 1, 5 HP	1 Elec	6.90	1.159			54.50		54.50	84
2720	10 HP	"	5	1.600			75		75	116
2730	25 HP	2 Elec	7.70	2.078			97.50		97.50	150
2740	50 HP		5.10	3.137			147		147	227
2750	100 HP	↓	3.10	5.161			243		243	375
2810	NEMA 7, 5 HP	1 Elec	5	1.600			75		75	116
2820	10 HP	"	3.90	2.051			96.50		96.50	148
2830	25 HP	2 Elec	5.10	3.137			147		147	227
2840	50 HP		3.10	5.161			243		243	375
2850	100 HP		2.30	6.957			325		325	505
2860	200 HP	↓	1.50	10.667	↓		500		500	770
3000	Panelboard or load center circuit breaker									
3010	Bolt-on or plug in, 15 A to 50 A	1 Elec	20	.400	Ea.		18.80		18.80	29
3020	60 A to 70 A		16	.500			23.50		23.50	36
3030	Bolt-on, 80 A to 100 A		14	.571			27		27	41.50
3040	Up to 250 A		7.30	1.096			51.50		51.50	79
3050	Motor operated, 30 A		13	.615			29		29	44.50
3060	60 A		10	.800			37.50		37.50	58
3070	100 A	↓	8	1	↓		47		47	72.50
3200	Switchboard circuit breaker									

26 05 05.35 Electrical Demolition, L.V. Distribution		Crew	Daily Output	Labor-Hours	Unit	Material	2009 Bare Costs Labor	Equipment	Total	Total Incl O&P
3210	15 A - 60 A	1 Elec	19	.421	Ea.		19.80		19.80	30.50
3220	70 A - 100 A		14	.571			27		27	41.50
3230	125 A - 400 A		10	.800			37.50		37.50	58
3240	450 A - 600 A		5.30	1.509			71		71	109
3250	700 A - 800 A		4.30	1.860			87.50		87.50	135
3260	1000 A		3.30	2.424			114		114	175
3270	1200 A		2.70	2.963			139		139	214
3500	Switchboard, incoming section, 400 A	2 Elec	3.70	4.324			203		203	315
3510	600 A		3.30	4.848			228		228	350
3520	800 A		2.90	5.517			259		259	400
3530	1200 A		2.40	6.667			315		315	480
3540	1600 A		2.20	7.273			340		340	525
3550	2000 A		2.10	7.619			360		360	550
3560	3000 A		1.90	8.421			395		395	610
3570	4000 A		1.70	9.412			440		440	680
3610	Distribution section, 600 A		4	4			188		188	289
3620	800 A		3.60	4.444			209		209	320
3630	1200 A		3.10	5.161			243		243	375
3640	1600 A		2.90	5.517			259		259	400
3650	2000 A		2.70	5.926			279		279	430
3710	Transition section, 600 A		3.80	4.211			198		198	305
3720	800 A		3.30	4.848			228		228	350
3730	1200 A		2.70	5.926			279		279	430
3740	1600 A		2.40	6.667			315		315	480
3750	2000 A		2.20	7.273			340		340	525
3760	2500 A		2.10	7.619			360		360	550
3770	3000 A		1.90	8.421			395		395	610
4000	Bus duct, aluminum or copper, 30 A	1 Elec	200	.040	L.F.		1.88		1.88	2.89
4020	60 A		160	.050			2.35		2.35	3.62
4040	100 A		140	.057			2.69		2.69	4.13
5000	Feedrail, trolley busway, up to 60 A		160	.050			2.35		2.35	3.62
5020	100 A		120	.067			3.13		3.13	4.82
5040	Busway, 50 A		200	.040			1.88		1.88	2.89
6000	Fuse, 30 A		133	.060			2.83		2.83	4.35
6010	60 A		133	.060			2.83		2.83	4.35
6020	100 A		105	.076			3.58		3.58	5.50
6030	200 A		95	.084			3.96		3.96	6.10
6040	400 A		80	.100			4.70		4.70	7.25
6050	600 A		53	.151			7.10		7.10	10.90
6060	601 A - 1200 A		42	.190			8.95		8.95	13.75
6070	1500 A - 1600 A		34	.235			11.05		11.05	17
6080	Up to 2500 A		26	.308			14.45		14.45	22.50
6090	4000 A		21	.381			17.90		17.90	27.50
6100	4500 A - 5000 A		18	.444			21		21	32
6120	6000 A		15	.533			25		25	38.50
6200	Fuse plug or fustat		100	.080			3.76		3.76	5.80

26 05 05.50 Electrical Demolition, Lighting			Crew	Daily Output	Labor-Hours	Unit	Material	Labor	Equipment	Total	Total Incl O&P
0010	**ELECTRICAL DEMOLITION, LIGHTING** Addition	R024119-10									
0100	Fixture hanger, flexible, 1/2" diameter		1 Elec	36	.222	Ea.		10.45		10.45	16.05
0120	3/4" diameter	R260105-30	"	30	.267	"		12.55		12.55	19.30
3000	Light pole, anchor base, excl concrete bases										
3010	Metal light pole, 10'		2 Elec	24	.667	Ea.		31.50		31.50	48

26 05 05 – Selective Electrical Demolition

26 05 05.50 Electrical Demolition, Lighting		Crew	Daily Output	Labor-Hours	Unit	Material	2009 Bare Costs Labor	Equipment	Total	Total Incl O&P
3020	16′	2 Elec	18	.889	Ea.		42		42	64.50
3030	20′	R-3	8.70	2.299			106	14.85	120.85	180
3040	40′	"	6	3.333			154	21.50	175.50	262
3100	Wood light pole, 10′	2 Elec	36	.444			21		21	32
3120	20′	"	24	.667			31.50		31.50	48
3140	Bollard light, 42″	1 Elec	9	.889			42		42	64.50
3160	Walkway luminaire	"	8.10	.988			46.50		46.50	71.50
4000	Explosionproof									
4010	Metal halide, 175 W	1 Elec	8.70	.920	Ea.		43		43	66.50
4020	250 W	"	8.10	.988			46.50		46.50	71.50
4030	400 W	2 Elec	14.40	1.111			52		52	80.50
4050	High pressure sodium, 70 W	1 Elec	9	.889			42		42	64.50
4060	100 W		9	.889			42		42	64.50
4070	150 W		8.10	.988			46.50		46.50	71.50
4100	Incandescent		8.70	.920			43		43	66.50
4200	Flurorescent		8.10	.988			46.50		46.50	71.50
5000	Ballast, fluorescent fixture		24	.333			15.65		15.65	24
5040	High intensity discharge fixture		24	.333			15.65		15.65	24
5300	Exit and emergency lighting									
5310	Exit light	1 Elec	24	.333	Ea.		15.65		15.65	24
5320	Emergency battery pack lighting unit		12	.667			31.50		31.50	48
5330	Remote lamp only		80	.100			4.70		4.70	7.25
5340	Self-contained fluorescent lamp pack		30	.267			12.55		12.55	19.30
5800	Energy saving devices									
5810	Occupancy sensor	1 Elec	21	.381	Ea.		17.90		17.90	27.50
5820	Automatic wall switch		72	.111			5.20		5.20	8.05
5830	Remote power pack		30	.267			12.55		12.55	19.30
5840	Photoelectric control		24	.333			15.65		15.65	24
6000	Lamps									
6010	Fluorescent	1 Elec	200	.040	Ea.		1.88		1.88	2.89
6030	High intensity discharge lamp, up to 400 W		68	.118			5.55		5.55	8.50
6040	Up to 1000 W		45	.178			8.35		8.35	12.85
6050	Quartz		90	.089			4.18		4.18	6.45
6070	Incandescent		360	.022			1.04		1.04	1.61
6080	Exterior, PAR		250	.028			1.30		1.30	1.99
6090	Guards for fluorescent lamp		24	.333			15.65		15.65	24

26 05 13 – Medium-Voltage Cables

26 05 13.10 Cable Terminations

		Crew	Daily Output	Labor-Hours	Unit	Material	Labor	Equipment	Total	Total Incl O&P
0010	**CABLE TERMINATIONS,** 5 kV to 35 kV									
0100	Indoor, insulation diameter range .525″ to 1.025″									
0300	Padmount, 5 kV	1 Elec	8	1	Ea.	63.50	47		110.50	142
0400	15 kV		6.40	1.250		114	59		173	216
0500	25 kV		6	1.333		132	62.50		194.50	242
0600	35 kV		5.60	1.429		169	67		236	289
0700	insulation diameter range .975″ to 1.570″									
0800	Padmount, 5 kV	1 Elec	8	1	Ea.	77	47		124	158
0900	15 kV		6	1.333		152	62.50		214.50	264
1000	25 kV		5.60	1.429		175	67		242	295
1100	35 kV		5.30	1.509		209	71		280	340
1200	insulation diameter range 1.540″ to 1.900″									
1300	Padmount, 5 KV	1 Elec	7.40	1.081	Ea.	116	51		167	206
1400	15 kV		5.60	1.429		192	67		259	315

26 05 13.10 Cable Terminations

26 05 13.10 Cable Terminations		Crew	Daily Output	Labor-Hours	Unit	Material	2009 Bare Costs Labor	Equipment	Total	Total Incl O&P
1500	25 kV	1 Elec	5.30	1.509	Ea.	245	71		316	380
1600	35 kV		5	1.600		274	75		349	415
1700	Outdoor systems, #4 stranded to 1/0 stranded									
1800	5 kV	1 Elec	7.40	1.081	Ea.	125	51		176	216
1900	15 kV		5.30	1.509		155	71		226	280
2000	25 kV		5	1.600		175	75		250	310
2100	35 kV		4.80	1.667		204	78.50		282.50	345
2200	#1 solid to 4/0 stranded, 5 kV		6.90	1.159		140	54.50		194.50	238
2300	15 kV		5	1.600		155	75		230	287
2400	25 kV		4.80	1.667		209	78.50		287.50	350
2500	35 kV		4.60	1.739		209	81.50		290.50	355
2600	3/0 solid to 350 kcmil stranded, 5 kV		6.40	1.250		175	59		234	283
2700	15 kV		4.80	1.667		175	78.50		253.50	315
2800	25 kV		4.60	1.739		245	81.50		326.50	395
2900	35 kV		4.40	1.818		245	85.50		330.50	400
3000	400 kcmil compact to 750 kcmil stranded, 5 kV		6	1.333		227	62.50		289.50	345
3100	15 kV		4.60	1.739		227	81.50		308.50	375
3200	25 kV		4.40	1.818		274	85.50		359.50	430
3300	35 kV		4.20	1.905		274	89.50		363.50	440
3400	1000 kcmil, 5 kV		5.60	1.429		245	67		312	370
3500	15 kV		4.40	1.818		262	85.50		347.50	420
3600	25 kV		4.20	1.905		274	89.50		363.50	440
3700	35 kV		4	2		274	94		368	445

26 05 13.16 Medium-Voltage, Single Cable

26 05 13.16 Medium-Voltage, Single Cable		Crew	Daily Output	Labor-Hours	Unit	Material	Labor	Equipment	Total	Total Incl O&P
0010	**MEDIUM-VOLTAGE, SINGLE CABLE** Splicing & terminations not included									
0040	Copper, XLP shielding, 5 kV, #6	2 Elec	4.40	3.636	C.L.F.	168	171		339	450
0050	#4		4.40	3.636		218	171		389	505
0100	#2		4	4		258	188		446	575
0200	#1		4	4		296	188		484	615
0400	1/0		3.80	4.211		325	198		523	665
0600	2/0		3.60	4.444		410	209		619	770
0800	4/0		3.20	5		535	235		770	945
1000	250 kcmil	3 Elec	4.50	5.333		660	251		911	1,100
1200	350 kcmil		3.90	6.154		850	289		1,139	1,375
1400	500 kcmil		3.60	6.667		1,050	315		1,365	1,625
1600	15 kV, ungrounded neutral, #1	2 Elec	4	4		360	188		548	685
1800	1/0		3.80	4.211		430	198		628	780
2000	2/0		3.60	4.444		490	209		699	860
2200	4/0		3.20	5		650	235		885	1,075
2400	250 kcmil	3 Elec	4.50	5.333		720	251		971	1,175
2600	350 kcmil		3.90	6.154		910	289		1,199	1,450
2800	500 kcmil		3.60	6.667		1,125	315		1,440	1,700
3000	25 kV, grounded neutral, #1/0	2 Elec	3.60	4.444		595	209		804	975
3200	2/0		3.40	4.706		655	221		876	1,050
3400	4/0		3	5.333		820	251		1,071	1,275
3600	250 kcmil	3 Elec	4.20	5.714		1,025	269		1,294	1,550
3800	350 kcmil		3.60	6.667		1,200	315		1,515	1,800
3900	500 kcmil		3.30	7.273		1,400	340		1,740	2,075
4000	35 kV, grounded neutral, #1/0	2 Elec	3.40	4.706		630	221		851	1,025
4200	2/0		3.20	5		740	235		975	1,175
4400	4/0		2.80	5.714		930	269		1,199	1,450
4600	250 kcmil	3 Elec	3.90	6.154		1,100	289		1,389	1,650

26 05 13 – Medium-Voltage Cables

26 05 13.16 Medium-Voltage, Single Cable		Crew	Daily Output	Labor-Hours	Unit	Material	2009 Bare Costs Labor	Equipment	Total	Total Incl O&P
4800	350 kcmil	3 Elec	3.30	7.273	C.L.F.	1,300	340		1,640	1,975
5000	500 kcmil		3	8		1,550	375		1,925	2,275
5050	Aluminum, XLP shielding, 5 kV, #2	2 Elec	5	3.200		165	150		315	410
5070	#1		4.40	3.636		171	171		342	450
5090	1/0		4	4		197	188		385	505
5100	2/0		3.80	4.211		221	198		419	550
5150	4/0		3.60	4.444		262	209		471	610
5200	250 kcmil	3 Elec	4.80	5		315	235		550	710
5220	350 kcmil		4.50	5.333		375	251		626	795
5240	500 kcmil		3.90	6.154		480	289		769	970
5260	750 kcmil		3.60	6.667		635	315		950	1,175
5300	15 kV aluminum, XLP, #1	2 Elec	4.40	3.636		212	171		383	495
5320	1/0		4	4		219	188		407	530
5340	2/0		3.80	4.211		261	198		459	590
5360	4/0		3.60	4.444		289	209		498	640
5380	250 kcmil	3 Elec	4.80	5		345	235		580	740
5400	350 kcmil		4.50	5.333		390	251		641	810
5420	500 kcmil		3.90	6.154		535	289		824	1,025
5440	750 kcmil		3.60	6.667		710	315		1,025	1,275

26 05 19 – Low-Voltage Electrical Power Conductors and Cables

26 05 19.13 Undercarpet Electrical Power Cables

		Crew	Daily Output	Labor-Hours	Unit	Material	Labor	Equipment	Total	Total Incl O&P
0010	**UNDERCARPET ELECTRICAL POWER CABLES** R260519-80									
0020	Power System									
0100	Cable flat, 3 conductor, #12, w/attached bottom shield	1 Elec	982	.008	L.F.	4.94	.38		5.32	6.05
0200	Shield, top, steel		1768	.005	"	5.30	.21		5.51	6.15
0250	Splice, 3 conductor		48	.167	Ea.	16.10	7.35		23.95	30
0300	Top shield		96	.083		1.43	3.92		5.35	7.55
0350	Tap		40	.200		20.50	9.40		29.90	37.50
0400	Insulating patch, splice, tap, & end		48	.167		51	7.85		58.85	68
0450	Fold		230	.035			1.63		1.63	2.51
0500	Top shield, tap & fold		96	.083		1.43	3.92		5.35	7.55
0700	Transition, block assembly		77	.104		74.50	4.88		79.38	89
0750	Receptacle frame & base		32	.250		40.50	11.75		52.25	63
0800	Cover receptacle		120	.067		3.47	3.13		6.60	8.65
0850	Cover blank		160	.050		4.07	2.35		6.42	8.10
0860	Receptacle, direct connected, single		25	.320		87	15.05		102.05	119
0870	Dual		16	.500		143	23.50		166.50	193
0880	Combination Hi & Lo, tension		21	.381		105	17.90		122.90	144
0900	Box, floor with cover		20	.400		88.50	18.80		107.30	126
0920	Floor service w/barrier		4	2		250	94		344	420
1000	Wall, surface, with cover		20	.400		58	18.80		76.80	93
1100	Wall, flush, with cover		20	.400		40.50	18.80		59.30	74
1450	Cable flat, 5 conductor #12, w/attached bottom shield		800	.010	L.F.	8.10	.47		8.57	9.60
1550	Shield, top, steel		1768	.005	"	8.05	.21		8.26	9.20
1600	Splice, 5 conductor		48	.167	Ea.	26	7.85		33.85	40.50
1650	Top shield		96	.083		1.43	3.92		5.35	7.55
1700	Tap		48	.167		34	7.85		41.85	49.50
1750	Insulating patch, splice tap, & end		83	.096		51	4.53		55.53	63
1800	Transition, block assembly		77	.104		53.50	4.88		58.38	66
1850	Box, wall, flush with cover		20	.400		54	18.80		72.80	88.50
1900	Cable flat, 4 conductor, #12		933	.009	L.F.	6.55	.40		6.95	7.80
1950	3 conductor #10		982	.008		5.65	.38		6.03	6.85

26 05 19.13 Undercarpet Electrical Power Cables		Crew	Daily Output	Labor-Hours	Unit	Material	2009 Bare Costs Labor	Equipment	Total	Total Incl O&P
1960	4 conductor #10	1 Elec	933	.009	L.F.	7.45	.40		7.85	8.75
1970	5 conductor #10	↓	884	.009	↓	9.10	.43		9.53	10.65
2500	Telephone System									
2510	Transition fitting wall box, surface	1 Elec	24	.333	Ea.	43	15.65		58.65	71.50
2520	Flush		24	.333		43	15.65		58.65	71.50
2530	Flush, for PC board		24	.333		43	15.65		58.65	71.50
2540	Floor service box	↓	4	2		227	94		321	395
2550	Cover, surface					13.90			13.90	15.30
2560	Flush					13.90			13.90	15.30
2570	Flush for PC board					13.90			13.90	15.30
2700	Floor fitting w/duplex jack & cover	1 Elec	21	.381		43	17.90		60.90	74.50
2720	Low profile		53	.151		14.95	7.10		22.05	27.50
2740	Miniature w/duplex jack		53	.151		23	7.10		30.10	36.50
2760	25 pair kit		21	.381		45	17.90		62.90	77
2780	Low profile		53	.151		15.30	7.10		22.40	27.50
2800	Call director kit for 5 cable		19	.421		72	19.80		91.80	110
2820	4 pair kit		19	.421		85.50	19.80		105.30	125
2840	3 pair kit		19	.421		90.50	19.80		110.30	130
2860	Comb. 25 pair & 3 conductor power		21	.381		72	17.90		89.90	107
2880	5 conductor power		21	.381		82	17.90		99.90	118
2900	PC board, 8 per 3 pair		161	.050		62.50	2.34		64.84	72
2920	6 per 4 pair		161	.050		62.50	2.34		64.84	72
2940	3 pair adapter		161	.050		56.50	2.34		58.84	66
2950	Plug		77	.104		2.64	4.88		7.52	10.40
2960	Couplers		321	.025	↓	7.45	1.17		8.62	9.95
3000	Bottom shield for 25 pair cable		4420	.002	L.F.	.74	.09		.83	.94
3020	4 pair		4420	.002		.35	.09		.44	.52
3040	Top shield for 25 pair cable		4420	.002	↓	.74	.09		.83	.94
3100	Cable assembly, double-end, 50′, 25 pair		11.80	.678	Ea.	219	32		251	290
3110	3 pair		23.60	.339		64.50	15.95		80.45	95.50
3120	4 pair		23.60	.339	↓	72	15.95		87.95	104
3140	Bulk 3 pair		1473	.005	L.F.	1.10	.26		1.36	1.60
3160	4 pair	↓	1473	.005	"	1.38	.26		1.64	1.91
3500	Data System									
3520	Cable 25 conductor w/connection 40′, 75 ohm	1 Elec	14.50	.552	Ea.	65	26		91	112
3530	Single lead		22	.364		177	17.10		194.10	222
3540	Dual lead	↓	22	.364	↓	220	17.10		237.10	269
3560	Shields same for 25 conductor as 25 pair telephone									
3570	Single & dual, none required									
3590	BNC coax connectors, Plug	1 Elec	40	.200	Ea.	9.65	9.40		19.05	25
3600	TNC coax connectors, Plug	"	40	.200	"	12.40	9.40		21.80	28
3700	Cable-bulk									
3710	Single lead	1 Elec	1473	.005	L.F.	2.37	.26		2.63	3
3720	Dual lead	"	1473	.005	"	3.41	.26		3.67	4.14
3730	Hand tool crimp				Ea.	470			470	520
3740	Hand tool notch				"	16.90			16.90	18.60
3750	Boxes & floor fitting same as telephone									
3790	Data cable notching, 90°	1 Elec	97	.082	Ea.		3.88		3.88	5.95
3800	180°		60	.133			6.25		6.25	9.65
8100	Drill floor		160	.050	↓	1.87	2.35		4.22	5.70
8200	Marking floor		1600	.005	L.F.		.24		.24	.36
8300	Tape, hold down		6400	.001	"	.14	.06		.20	.24
8350	Tape primer, 500 ft. per can	↓	96	.083	Ea.	28.50	3.92		32.42	37.50

26 05 19.13 Undercarpet Electrical Power Cables		Crew	Daily Output	Labor-Hours	Unit	Material	2009 Bare Costs Labor	Equipment	Total	Total Incl O&P
8400	Tool, splicing				Ea.	228			228	250

26 05 19.20 Armored Cable		Crew	Daily Output	Labor-Hours	Unit	Material	2009 Bare Costs Labor	Equipment	Total	Total Incl O&P
0010	**ARMORED CABLE**									
0050	600 volt, copper (BX), #14, 2 conductor, solid	1 Elec	2.40	3.333	C.L.F.	77.50	157		234.50	325
0100	3 conductor, solid		2.20	3.636		121	171		292	395
0120	4 conductor, solid		2	4		162	188		350	465
0150	#12, 2 conductor, solid		2.30	3.478		79	163		242	340
0200	3 conductor, solid		2	4		128	188		316	430
0220	4 conductor, solid		1.80	4.444		169	209		378	505
0240	#12, 19 conductor, stranded		1.10	7.273		915	340		1,255	1,525
0250	#10, 2 conductor, solid		2	4		150	188		338	455
0300	3 conductor, solid		1.60	5		221	235		456	605
0320	4 conductor, solid		1.40	5.714		350	269		619	800
0350	#8, 3 conductor, solid		1.30	6.154		450	289		739	940
0370	4 conductor, stranded		1.10	7.273		635	340		975	1,225
0380	#6, 2 conductor, stranded		1.30	6.154		475	289		764	965
0400	3 conductor with PVC jacket, in cable tray, #6		3.10	2.581		415	121		536	640
0450	#4	2 Elec	5.40	2.963		535	139		674	805
0500	#2		4.60	3.478		710	163		873	1,025
0550	#1		4	4		965	188		1,153	1,350
0600	1/0		3.60	4.444		1,125	209		1,334	1,575
0650	2/0		3.40	4.706		1,375	221		1,596	1,875
0700	3/0		3.20	5		1,600	235		1,835	2,125
0750	4/0		3	5.333		1,850	251		2,101	2,400
0800	250 kcmil	3 Elec	3.60	6.667		2,075	315		2,390	2,750
0850	350 kcmil		3.30	7.273		2,775	340		3,115	3,575
0900	500 kcmil		3	8		3,600	375		3,975	4,550
0910	4 conductor with PVC jacket, in cable tray, #6	1 Elec	2.70	2.963		530	139		669	795
0920	#4	2 Elec	4.60	3.478		680	163		843	995
0930	#2		4	4		840	188		1,028	1,225
0940	#1		3.60	4.444		1,225	209		1,434	1,675
0950	1/0		3.40	4.706		1,400	221		1,621	1,900
0960	2/0		3.20	5		1,600	235		1,835	2,100
0970	3/0		3	5.333		1,900	251		2,151	2,475
0980	4/0		2.40	6.667		2,425	315		2,740	3,150
0990	250 kcmil	3 Elec	3.30	7.273		2,900	340		3,240	3,700
1000	350 kcmil		3	8		3,425	375		3,800	4,350
1010	500 kcmil		2.70	8.889		4,700	420		5,120	5,825
1050	5 kV, copper, 3 conductor with PVC jacket,									
1060	non-shielded, in cable tray, #4	2 Elec	380	.042	L.F.	8.20	1.98		10.18	12.05
1100	#2		360	.044		10.65	2.09		12.74	14.90
1200	#1		300	.053		13.55	2.51		16.06	18.75
1400	1/0		290	.055		15.65	2.59		18.24	21
1600	2/0		260	.062		18.10	2.89		20.99	24.50
2000	4/0		240	.067		24	3.13		27.13	31.50
2100	250 kcmil	3 Elec	330	.073		33	3.42		36.42	42
2150	350 kcmil		315	.076		41	3.58		44.58	50.50
2200	500 kcmil		270	.089		54	4.18		58.18	66
2400	15 kV, copper, 3 conductor with PVC jacket galv steel armored									
2500	grounded neutral, in cable tray, #2	2 Elec	300	.053	L.F.	17.20	2.51		19.71	23
2600	#1		280	.057		18.35	2.69		21.04	24
2800	1/0		260	.062		21	2.89		23.89	27.50

26 05 19 – Low-Voltage Electrical Power Conductors and Cables

26 05 19.20 Armored Cable		Crew	Daily Output	Labor-Hours	Unit	2009 Bare Costs Material	Labor	Equipment	Total	Total Incl O&P
2900	2/0	2 Elec	220	.073	L.F.	28	3.42		31.42	36
3000	4/0		190	.084		31.50	3.96		35.46	41
3100	250 kcmil	3 Elec	270	.089		35	4.18		39.18	45.50
3150	350 kcmil		240	.100		41.50	4.70		46.20	53
3200	500 kcmil		210	.114		55.50	5.35		60.85	70
3400	15 kV, copper, 3 conductor with PVC jacket,									
3450	ungrounded neutral, in cable tray, #2	2 Elec	260	.062	L.F.	18.55	2.89		21.44	25
3500	#1		230	.070		20.50	3.27		23.77	27.50
3600	1/0		200	.080		23.50	3.76		27.26	32
3700	2/0		190	.084		29	3.96		32.96	38
3800	4/0		160	.100		35	4.70		39.70	46
4000	250 kcmil	3 Elec	210	.114		40.50	5.35		45.85	53.50
4050	350 kcmil		195	.123		53.50	5.80		59.30	68
4100	500 kcmil		180	.133		65	6.25		71.25	81
4200	600 volt, aluminum, 3 conductor in cable tray with PVC jacket									
4300	#2	2 Elec	540	.030	L.F.	4.21	1.39		5.60	6.75
4400	#1		460	.035		4.67	1.63		6.30	7.65
4500	#1/0		400	.040		5.80	1.88		7.68	9.30
4600	#2/0		360	.044		5.90	2.09		7.99	9.70
4700	#3/0		340	.047		6.90	2.21		9.11	11
4800	#4/0		320	.050		8.35	2.35		10.70	12.75
4900	250 kcmil	3 Elec	450	.053		10	2.51		12.51	14.85
5000	350 kcmil		360	.067		11.95	3.13		15.08	17.90
5200	500 kcmil		330	.073		14.85	3.42		18.27	21.50
5300	750 kcmil		285	.084		19.15	3.96		23.11	27
5400	600 volt, aluminum, 4 conductor in cable tray with PVC jacket									
5410	#2	2 Elec	520	.031	L.F.	4.82	1.45		6.27	7.50
5430	#1		440	.036		5.90	1.71		7.61	9.15
5450	1/0		380	.042		6.90	1.98		8.88	10.65
5470	2/0		340	.047		7.05	2.21		9.26	11.15
5480	3/0		320	.050		8.25	2.35		10.60	12.70
5500	4/0		300	.053		9.70	2.51		12.21	14.55
5520	250 kcmil	3 Elec	420	.057		10.40	2.69		13.09	15.60
5540	350 kcmil		330	.073		13.45	3.42		16.87	20
5560	500 kcmil		300	.080		16.85	3.76		20.61	24.50
5580	750 kcmil		270	.089		23	4.18		27.18	32
5600	5 kV, aluminum, unshielded in cable tray, #2 with PVC jacket	2 Elec	380	.042		5.90	1.98		7.88	9.55
5700	#1 with PVC jacket		360	.044		6.60	2.09		8.69	10.45
5800	1/0 with PVC jacket		300	.053		6.75	2.51		9.26	11.25
6000	2/0 with PVC jacket		290	.055		6.90	2.59		9.49	11.60
6200	3/0 with PVC jacket		260	.062		8.25	2.89		11.14	13.55
6300	4/0 with PVC jacket		240	.067		9.70	3.13		12.83	15.50
6400	250 kcmil with PVC jacket	3 Elec	330	.073		10.65	3.42		14.07	16.95
6500	350 kcmil with PVC jacket		315	.076		12.45	3.58		16.03	19.20
6600	500 kcmil with PVC jacket		300	.080		14.75	3.76		18.51	22
6800	750 kcmil with PVC jacket		270	.089		18.20	4.18		22.38	26.50
6900	15 kV, aluminum, shielded-grounded, #2 with PVC jacket	2 Elec	320	.050		13.40	2.35		15.75	18.35
7000	#1 with PVC jacket		300	.053		13.75	2.51		16.26	19
7200	1/0 with PVC jacket		280	.057		14.85	2.69		17.54	20.50
7300	2/0 with PVC jacket		260	.062		15.05	2.89		17.94	21
7400	3/0 with PVC jacket		240	.067		16.90	3.13		20.03	23.50
7500	4/0 with PVC jacket		220	.073		17.35	3.42		20.77	24.50
7600	250 kcmil with PVC jacket	3 Elec	300	.080		19.05	3.76		22.81	27

26 05 19.20 Armored Cable		Crew	Daily Output	Labor-Hours	Unit	2009 Bare Costs Material	Labor	Equipment	Total	Total Incl O&P
7700	350 kcmil with PVC jacket	3 Elec	270	.089	L.F.	22.50	4.18		26.68	31
7800	500 kcmil with PVC jacket		240	.100		27	4.70		31.70	37
8000	750 kcmil with PVC jacket		204	.118		32	5.55		37.55	43.50
8200	15 kV, aluminum, shielded-ungrounded, #1 with PVC jacket	2 Elec	250	.064		16.75	3.01		19.76	23
8300	1/0 with PVC jacket		230	.070		17.35	3.27		20.62	24
8400	2/0 with PVC jacket		210	.076		19.05	3.58		22.63	26.50
8500	3/0 with PVC jacket		200	.080		19.35	3.76		23.11	27.50
8600	4/0 with PVC jacket		190	.084		21	3.96		24.96	29
8700	250 kcmil with PVC jacket	3 Elec	270	.089		22.50	4.18		26.68	31.50
8800	350 kcmil with PVC jacket		240	.100		26	4.70		30.70	36
8900	500 kcmil with PVC jacket		210	.114		31	5.35		36.35	43
8950	750 kcmil with PVC jacket		174	.138		38.50	6.50		45	52.50
9010	600 volt, copper (MC) steel clad, #14, 2 wire	1 Elec	2.40	3.333	C.L.F.	77.50	157		234.50	325
9020	3 wire		2.20	3.636		120	171		291	395
9030	4 wire		2	4		162	188		350	465
9040	#12, 2 wire		2.30	3.478		79	163		242	340
9050	3 wire		2	4		128	188		316	430
9060	4 wire		1.80	4.444		169	209		378	505
9070	#10, 2 wire		2	4		150	188		338	455
9080	3 wire		1.60	5		221	235		456	605
9090	4 wire		1.40	5.714		350	269		619	800
9091	For health care facilities cable, add, minimum					5%				
9092	Maximum					20%				
9100	#8, 2 wire, stranded	1 Elec	1.80	4.444	C.L.F.	305	209		514	655
9110	3 wire, stranded		1.30	6.154		455	289		744	945
9120	4 wire, stranded		1.10	7.273		640	340		980	1,225
9130	#6, 2 wire, stranded		1.30	6.154		475	289		764	965
9200	600 volt, copper (MC) aluminum clad, #14, 2 wire		2.65	3.019		77.50	142		219.50	305
9210	3 wire		2.45	3.265		120	153		273	370
9220	4 wire		2.20	3.636		162	171		333	440
9230	#12, 2 wire		2.55	3.137		79	147		226	315
9240	3 wire		2.20	3.636		128	171		299	405
9250	4 wire		2	4		168	188		356	475
9260	#10, 2 wire		2.20	3.636		150	171		321	430
9270	3 wire		1.80	4.444		221	209		430	565
9280	4 wire		1.55	5.161		350	243		593	760
9600	Alum (MC) aluminum clad, #6, 3 conductor w/#6 grnd		1.67	4.790		490	225		715	885
9610	4 conductor w/#6 grnd		1.64	4.878		535	229		764	945
9620	#4, 3 conductor w/#6 grnd	2 Elec	2.86	5.594		510	263		773	965
9630	4 conductor w/#6 grnd		2.82	5.674		555	267		822	1,025
9640	#2, 3 conductor w/#4 grnd		2.50	6.400		590	300		890	1,125
9650	4 conductor w/#4 grnd		2.47	6.478		660	305		965	1,200
9660	#1, 3 conductor w/#4 grnd		2	8		760	375		1,135	1,425
9670	4 conductor w/#4 grnd		1.98	8.081		900	380		1,280	1,575
9680	1/0, 3 conductor w/#4 grnd		1.82	8.791		865	415		1,280	1,575
9690	4 conductor w/#4 grnd		1.79	8.939		1,050	420		1,470	1,800
9700	2/0, 3 conductor w/#4 grnd		1.75	9.143		910	430		1,340	1,650
9710	4 conductor w/#4 grnd		1.71	9.357		1,175	440		1,615	1,975
9720	3/0, 3 conductor w/#4 grnd		1.71	9.357		1,175	440		1,615	1,950
9730	4 conductor w/#4 grnd		1.67	9.581		1,450	450		1,900	2,300
9740	4/0, 3 conductor w/#2 grnd		1.67	9.581		1,300	450		1,750	2,125
9750	4 conductor w/#2 grnd		1.61	9.938		1,625	465		2,090	2,525
9760	250 kcmil, 3 conductor w/#1 grnd	3 Elec	2.42	9.917		1,550	465		2,015	2,425

26 05 19 – Low-Voltage Electrical Power Conductors and Cables

26 05 19.20 Armored Cable		Crew	Daily Output	Labor-Hours	Unit	Material	2009 Bare Costs Labor	Equipment	Total	Total Incl O&P
9770	4 conductor w/#1 grnd	3 Elec	2.36	10.169	C.L.F.	2,100	480		2,580	3,025
9775	300 kcmil, 4 conductor w/ #1 grnd		2.22	10.811		2,275	510		2,785	3,300
9780	350 kcmil, 3 conductor w/ 1/0 grnd		2.19	10.959		2,150	515		2,665	3,175
9790	4 conductor w/ 1/0 grnd		2.10	11.429		2,775	535		3,310	3,875
9800	500 kcmil, 3 conductor w/#1 grnd		2.10	11.429		2,750	535		3,285	3,850
9810	4 conductor w/ 2/0 grnd		2	12		3,375	565		3,940	4,600
9840	750 kcmil, 3 conductor w/ 1/0 grnd		1.95	12.308		4,000	580		4,580	5,300
9850	4 conductor w/ 3/0 grnd		1.89	12.698		4,725	595		5,320	6,100
9900	Minimum labor/equipment charge	1 Elec	4	2	Job		94		94	145

26 05 19.25 Cable Connectors		Crew	Daily Output	Labor-Hours	Unit	Material	Labor	Equipment	Total	Total Incl O&P
0010	**CABLE CONNECTORS**									
0100	600 volt, nonmetallic, #14-2 wire	1 Elec	160	.050	Ea.	1.09	2.35		3.44	4.82
0200	#14-3 wire to #12-2 wire		133	.060		1.09	2.83		3.92	5.55
0300	#12-3 wire to #10-2 wire		114	.070		1.09	3.30		4.39	6.25
0400	#10-3 wire to #14-4 and #12-4 wire		100	.080		1.09	3.76		4.85	7
0500	#8-3 wire to #10-4 wire		80	.100		2.07	4.70		6.77	9.55
0600	#6-3 wire		40	.200		2.79	9.40		12.19	17.50
0800	SER aluminum, 3 #8 insulated + 1 #8 ground		32	.250		3.21	11.75		14.96	21.50
0900	3 #6 + 1 #6 ground		24	.333		3.21	15.65		18.86	27.50
1000	3 #4 + 1 #6 ground		22	.364		5	17.10		22.10	32
1100	3 #2 + 1 #4 ground		20	.400		9.10	18.80		27.90	39
1200	3 1/0 + 1 #2 ground		18	.444		23.50	21		44.50	58
1400	3 2/0 + 1 #1 ground		16	.500		23.50	23.50		47	62
1600	3 4/0 + 1 #2/0 ground		14	.571		28.50	27		55.50	72.50
1800	600 volt, armored, #14-2 wire		80	.100		.66	4.70		5.36	8
2200	#14-4, #12-3 and #10-2 wire		40	.200		.66	9.40		10.06	15.20
2400	#12-4, #10-3 and #8-2 wire		32	.250		1.56	11.75		13.31	19.80
2600	#8-3 and #10-4 wire		26	.308		2.72	14.45		17.17	25.50
2650	#8-4 wire		22	.364		4.76	17.10		21.86	32
2700	PVC jacket connector, #6-3 wire, #6-4 wire		16	.500		12.15	23.50		35.65	49.50
2800	#4-3 wire, #4-4 wire		16	.500		12.15	23.50		35.65	49.50
2900	#2-3 wire		12	.667		12.15	31.50		43.65	61.50
3000	#1-3 wire, #2-4 wire		12	.667		17.10	31.50		48.60	67
3200	1/0-3 wire		11	.727		17.10	34		51.10	71.50
3400	2/0-3 wire, 1/0-4 wire		10	.800		17.10	37.50		54.60	77
3500	3/0-3 wire, 2/0-4 wire		9	.889		35	42		77	103
3600	4/0-3 wire, 3/0-4 wire		7	1.143		35	53.50		88.50	121
3800	250 kcmil-3 wire, 4/0-4 wire		6	1.333		48.50	62.50		111	150
4000	350 kcmil-3 wire, 250 kcmil-4 wire		5	1.600		48.50	75		123.50	170
4100	350 kcmil-4 wire		4	2		189	94		283	355
4200	500 kcmil-3 wire		4	2		189	94		283	355
4250	500 kcmil-4 wire, 750 kcmil-3 wire		3.50	2.286		247	107		354	435
4300	750 kcmil-4 wire		3	2.667		247	125		372	465
4400	5 kV, armored, #4		8	1		66.50	47		113.50	146
4600	#2		8	1		66.50	47		113.50	146
4800	#1		8	1		66.50	47		113.50	146
5000	1/0		6.40	1.250		82	59		141	181
5200	2/0		5.30	1.509		82	71		153	199
5500	4/0		4	2		108	94		202	264
5600	250 kcmil		3.60	2.222		133	104		237	305
5650	350 kcmil		3.20	2.500		156	118		274	355
5700	500 kcmil		2.50	3.200		156	150		306	405

26 05 19.25 Cable Connectors

26 05 19.25 Cable Connectors		Crew	Daily Output	Labor-Hours	Unit	Material	2009 Bare Costs Labor	Equipment	Total	Total Incl O&P
5720	750 kcmil	1 Elec	2.20	3.636	Ea.	228	171		399	515
5750	1000 kcmil		2	4		305	188		493	625
5800	15 kV, armored, #1		4	2		104	94		198	259
5900	1/0		4	2		104	94		198	259
6000	3/0		3.60	2.222		134	104		238	310
6100	4/0		3.40	2.353		134	111		245	315
6200	250 kcmil		3.20	2.500		148	118		266	345
6300	350 kcmil		2.70	2.963		172	139		311	405
6400	500 kcmil		2	4		172	188		360	480

26 05 19.30 Cable Splicing

26 05 19.30 Cable Splicing		Crew	Daily Output	Labor-Hours	Unit	Material	Labor	Equipment	Total	Total Incl O&P
0010	**CABLE SPLICING** URD or similar, ideal conditions									
0100	#6 stranded to #1 stranded, 5 kV	1 Elec	4	2	Ea.	124	94		218	281
0120	15 kV		3.60	2.222		124	104		228	297
0140	25 kV		3.20	2.500		124	118		242	315
0200	#1 stranded to 4/0 stranded, 5 kV		3.60	2.222		129	104		233	305
0210	15 kV		3.20	2.500		129	118		247	325
0220	25 kV		2.80	2.857		129	134		263	350
0300	4/0 stranded to 500 kcmil stranded, 5 kV		3.30	2.424		290	114		404	495
0310	15 kV		2.90	2.759		290	130		420	520
0320	25 kV		2.50	3.200		290	150		440	550
0400	500 kcmil, 5 kV		3.20	2.500		290	118		408	500
0410	15 kV		2.80	2.857		290	134		424	525
0420	25 kV		2.30	3.478		290	163		453	570
0500	600 kcmil, 5 kV		2.90	2.759		290	130		420	520
0510	15 kV		2.40	3.333		290	157		447	560
0520	25 kV		2	4		290	188		478	610
0600	750 kcmil, 5 kV		2.60	3.077		340	145		485	595
0610	15 kV		2.20	3.636		340	171		511	640
0620	25 kV		1.90	4.211		340	198		538	680
0700	1000 kcmil, 5 kV		2.30	3.478		370	163		533	655
0710	15 kV		1.90	4.211		370	198		568	710
0720	25 kV		1.60	5		370	235		605	765

26 05 19.35 Cable Terminations

26 05 19.35 Cable Terminations		Crew	Daily Output	Labor-Hours	Unit	Material	Labor	Equipment	Total	Total Incl O&P
0010	**CABLE TERMINATIONS**									
0015	Wire connectors, screw type, #22 to #14	1 Elec	260	.031	Ea.	.08	1.45		1.53	2.31
0020	#18 to #12		240	.033		.10	1.57		1.67	2.52
0025	#18 to #10		240	.033		.15	1.57		1.72	2.58
0030	Screw-on connectors, insulated, #18 to #12		240	.033		.11	1.57		1.68	2.53
0035	#16 to #10		230	.035		.14	1.63		1.77	2.66
0040	#14 to #8		210	.038		.30	1.79		2.09	3.08
0045	#12 to #6		180	.044		.50	2.09		2.59	3.76
0050	Terminal lugs, solderless, #16 to #10		50	.160		.56	7.50		8.06	12.15
0100	#8 to #4		30	.267		.87	12.55		13.42	20.50
0150	#2 to #1		22	.364		1.17	17.10		18.27	28
0200	1/0 to 2/0		16	.500		2.22	23.50		25.72	38.50
0250	3/0		12	.667		4	31.50		35.50	52.50
0300	4/0		11	.727		4	34		38	57
0350	250 kcmil		9	.889		4	42		46	69
0400	350 kcmil		7	1.143		5.15	53.50		58.65	88
0450	500 kcmil		6	1.333		10.05	62.50		72.55	108
0500	600 kcmil		5.80	1.379		10.75	65		75.75	111
0550	750 kcmil		5.20	1.538		11.80	72.50		84.30	124

26 05 19.35 Cable Terminations		Crew	Daily Output	Labor-Hours	Unit	Material	2009 Bare Costs Labor	Equipment	Total	Total Incl O&P
0600	Split bolt connectors, tapped, #6	1 Elec	16	.500	Ea.	4.75	23.50		28.25	41.50
0650	#4		14	.571		5.70	27		32.70	48
0700	#2		12	.667		8.65	31.50		40.15	57.50
0750	#1		11	.727		10.90	34		44.90	64.50
0800	1/0		10	.800		10.90	37.50		48.40	70
0850	2/0		9	.889		16.20	42		58.20	82.50
0900	3/0		7.20	1.111		25	52		77	108
1000	4/0		6.40	1.250		28	59		87	121
1100	250 kcmil		5.70	1.404		28	66		94	132
1200	300 kcmil		5.30	1.509		51	71		122	165
1400	350 kcmil		4.60	1.739		51	81.50		132.50	182
1500	500 kcmil		4	2		66	94		160	218
1600	Crimp 1 hole lugs, copper or aluminum, 600 volt									
1620	#14	1 Elec	60	.133	Ea.	.43	6.25		6.68	10.10
1630	#12		50	.160		.58	7.50		8.08	12.20
1640	#10		45	.178		.58	8.35		8.93	13.50
1780	#8		36	.222		2.30	10.45		12.75	18.60
1800	#6		30	.267		2.92	12.55		15.47	22.50
2000	#4		27	.296		3.50	13.95		17.45	25.50
2200	#2		24	.333		6.45	15.65		22.10	31
2400	#1		20	.400		6.60	18.80		25.40	36.50
2500	1/0		17.50	.457		7.20	21.50		28.70	41
2600	2/0		15	.533		8.25	25		33.25	47.50
2800	3/0		12	.667		9.70	31.50		41.20	58.50
3000	4/0		11	.727		10.95	34		44.95	64.50
3200	250 kcmil		9	.889		12.70	42		54.70	78.50
3400	300 kcmil		8	1		15.30	47		62.30	89.50
3500	350 kcmil		7	1.143		15.05	53.50		68.55	99
3600	400 kcmil		6.50	1.231		19.10	58		77.10	110
3800	500 kcmil		6	1.333		21	62.50		83.50	120
4000	600 kcmil		5.80	1.379		39.50	65		104.50	143
4200	700 kcmil		5.50	1.455		47	68.50		115.50	157
4400	750 kcmil		5.20	1.538		47	72.50		119.50	163
4500	Crimp 2-way connectors, copper or alum., 600 volt,									
4510	#14	1 Elec	60	.133	Ea.	.66	6.25		6.91	10.40
4520	#12		50	.160		1.04	7.50		8.54	12.70
4530	#10		45	.178		1.04	8.35		9.39	14
4540	#8		27	.296		3.20	13.95		17.15	25
4600	#6		25	.320		5.45	15.05		20.50	29
4800	#4		23	.348		6.20	16.35		22.55	32
5000	#2		20	.400		7.90	18.80		26.70	37.50
5200	#1		16	.500		9.15	23.50		32.65	46
5400	1/0		13	.615		10.10	29		39.10	55.50
5420	2/0		12	.667		10.40	31.50		41.90	59.50
5440	3/0		11	.727		12.40	34		46.40	66
5460	4/0		10	.800		13.40	37.50		50.90	73
5480	250 kcmil		9	.889		16.10	42		58.10	82
5500	300 kcmil		8.50	.941		17.80	44		61.80	87.50
5520	350 kcmil		8	1		17.90	47		64.90	92
5540	400 kcmil		7.30	1.096		22	51.50		73.50	103
5560	500 kcmil		6.20	1.290		27	60.50		87.50	124
5580	600 kcmil		5.50	1.455		42.50	68.50		111	152
5600	700 kcmil		4.50	1.778		49.50	83.50		133	184

26 05 19.35 Cable Terminations		Crew	Daily Output	Labor-Hours	Unit	2009 Bare Costs Material	Labor	Equipment	Total	Total Incl O&P
5620	750 kcmil	1 Elec	4	2	Ea.	49.50	94		143.50	200
7000	Compression equipment adapter, aluminum wire, #6		30	.267		10.30	12.55		22.85	30.50
7020	#4		27	.296		10.80	13.95		24.75	33.50
7040	#2		24	.333		11.40	15.65		27.05	36.50
7060	#1		20	.400		13.10	18.80		31.90	43.50
7080	1/0		18	.444		13.60	21		34.60	47
7100	2/0		15	.533		26	25		51	67
7140	4/0		11	.727		26	34		60	81
7160	250 kcmil		9	.889		26	42		68	93
7180	300 kcmil		8	1		31	47		78	107
7200	350 kcmil		7	1.143		31	53.50		84.50	117
7220	400 kcmil		6.50	1.231		40	58		98	133
7240	500 kcmil		6	1.333		40	62.50		102.50	141
7260	600 kcmil		5.80	1.379		54.50	65		119.50	160
7280	750 kcmil		5.20	1.538		54.50	72.50		127	171
8000	Compression tool, hand					1,225			1,225	1,325
8100	Hydraulic					2,525			2,525	2,775
8500	Hydraulic dies					380			380	420

26 05 19.50 Mineral Insulated Cable

		Crew	Daily Output	Labor-Hours	Unit	Material	Labor	Equipment	Total	Total Incl O&P
0010	**MINERAL INSULATED CABLE** 600 volt									
0100	1 conductor, #12	1 Elec	1.60	5	C.L.F.	310	235		545	700
0200	#10		1.60	5		405	235		640	805
0400	#8		1.50	5.333		475	251		726	905
0500	#6		1.40	5.714		530	269		799	1,000
0600	#4	2 Elec	2.40	6.667		735	315		1,050	1,300
0800	#2		2.20	7.273		985	340		1,325	1,600
0900	#1		2.10	7.619		1,125	360		1,485	1,775
1000	1/0		2	8		1,300	375		1,675	2,000
1100	2/0		1.90	8.421		1,550	395		1,945	2,300
1200	3/0		1.80	8.889		1,825	420		2,245	2,650
1400	4/0		1.60	10		2,100	470		2,570	3,025
1410	250 kcmil	3 Elec	2.40	10		2,375	470		2,845	3,350
1420	350 kcmil		1.95	12.308		2,650	580		3,230	3,825
1430	500 kcmil		1.95	12.308		3,325	580		3,905	4,550
1500	2 conductor, #12	1 Elec	1.40	5.714		640	269		909	1,125
1600	#10		1.20	6.667		775	315		1,090	1,325
1800	#8		1.10	7.273		965	340		1,305	1,575
2000	#6		1.05	7.619		1,225	360		1,585	1,900
2100	#4	2 Elec	2	8		1,625	375		2,000	2,350
2200	3 conductor, #12	1 Elec	1.20	6.667		805	315		1,120	1,375
2400	#10		1.10	7.273		930	340		1,270	1,550
2600	#8		1.05	7.619		1,125	360		1,485	1,775
2800	#6		1	8		1,450	375		1,825	2,175
3000	#4	2 Elec	1.80	8.889		1,775	420		2,195	2,625
3100	4 conductor, #12	1 Elec	1.20	6.667		855	315		1,170	1,425
3200	#10		1.10	7.273		1,025	340		1,365	1,650
3400	#8		1	8		1,325	375		1,700	2,025
3600	#6		.90	8.889		1,650	420		2,070	2,475
3620	7 conductor, #12		1.10	7.273		1,075	340		1,415	1,725
3640	#10		1	8		1,350	375		1,725	2,050
3800	Terminations, 600 volt, 1 conductor, #12		8	1	Ea.	13.55	47		60.55	87.50
4000	#10		7.60	1.053		13.55	49.50		63.05	91

26 05 19.50 Mineral Insulated Cable		Crew	Daily Output	Labor-Hours	Unit	Material	2009 Bare Costs Labor	Equipment	Total	Total Incl O&P
4100	#8	1 Elec	7.30	1.096	Ea.	13.55	51.50		65.05	94
4200	#6		6.70	1.194		13.55	56		69.55	101
4400	#4		6.20	1.290		13.55	60.50		74.05	108
4600	#2		5.70	1.404		20.50	66		86.50	124
4800	#1		5.30	1.509		20.50	71		91.50	132
5000	1/0		5	1.600		20.50	75		95.50	139
5100	2/0		4.70	1.702		20.50	80		100.50	146
5200	3/0		4.30	1.860		20.50	87.50		108	158
5400	4/0		4	2		45.50	94		139.50	195
5410	250 kcmil		4	2		45.50	94		139.50	195
5420	350 kcmil		4	2		81	94		175	234
5430	500 kcmil		4	2		81	94		175	234
5500	2 conductor, #12		6.70	1.194		13.55	56		69.55	101
5600	#10		6.40	1.250		20.50	59		79.50	113
5800	#8		6.20	1.290		20.50	60.50		81	116
6000	#6		5.70	1.404		20.50	66		86.50	124
6200	#4		5.30	1.509		45.50	71		116.50	159
6400	3 conductor, #12		5.70	1.404		20.50	66		86.50	124
6500	#10		5.50	1.455		20.50	68.50		89	128
6600	#8		5.20	1.538		20.50	72.50		93	134
6800	#6		4.80	1.667		20.50	78.50		99	144
7200	#4		4.60	1.739		45.50	81.50		127	176
7400	4 conductor, #12		4.60	1.739		22.50	81.50		104	151
7500	#10		4.40	1.818		22.50	85.50		108	156
7600	#8		4.20	1.905		22.50	89.50		112	163
8400	#6		4	2		47.50	94		141.50	198
8500	7 conductor, #12		3.50	2.286		22.50	107		129.50	190
8600	#10		3	2.667		49	125		174	247
8800	Crimping tool, plier type					49.50			49.50	54.50
9000	Stripping tool					187			187	205
9200	Hand vise					55.50			55.50	61.50
9500	Minimum labor/equipment charge	1 Elec	4	2	Job		94		94	145

26 05 19.55 Non-Metallic Sheathed Cable		Crew	Daily Output	Labor-Hours	Unit	Material	Labor	Equipment	Total	Total Incl O&P
0010	**NON-METALLIC SHEATHED CABLE** 600 volt									
0100	Copper with ground wire, (Romex)									
0150	#14, 2 conductor	1 Elec	2.70	2.963	C.L.F.	32.50	139		171.50	250
0200	3 conductor		2.40	3.333		50	157		207	296
0220	4 conductor		2.20	3.636		76	171		247	345
0250	#12, 2 conductor		2.50	3.200		50	150		200	286
0300	3 conductor		2.20	3.636		70	171		241	340
0320	4 conductor		2	4		118	188		306	420
0350	#10, 2 conductor		2.20	3.636		78	171		249	350
0400	3 conductor		1.80	4.444		111	209		320	440
0420	4 conductor		1.60	5		163	235		398	540
0450	#8, 3 conductor		1.50	5.333		185	251		436	590
0480	4 conductor		1.40	5.714		280	269		549	725
0500	#6, 3 conductor		1.40	5.714		300	269		569	745
0520	#4, 3 conductor	2 Elec	2.40	6.667		530	315		845	1,050
0540	#2, 3 conductor	"	2.20	7.273		790	340		1,130	1,400
0550	SE type SER aluminum cable, 3 RHW and									
0600	1 bare neutral, 3 #8 & 1 #8	1 Elec	1.60	5	C.L.F.	161	235		396	535
0650	3 #6 & 1 #6	"	1.40	5.714		182	269		451	615

26 05 19 – Low-Voltage Electrical Power Conductors and Cables

26 05 19.55 Non-Metallic Sheathed Cable		Crew	Daily Output	Labor-Hours	Unit	2009 Bare Costs Material	Labor	Equipment	Total	Total Incl O&P
0700	3 #4 & 1 #6	2 Elec	2.40	6.667	C.L.F.	204	315		519	705
0750	3 #2 & 1 #4		2.20	7.273		300	340		640	855
0800	3 #1/0 & 1 #2		2	8		455	375		830	1,075
0850	3 #2/0 & 1 #1		1.80	8.889		535	420		955	1,225
0900	3 #4/0 & 1 #2/0		1.60	10		765	470		1,235	1,575
1450	UF underground feeder cable, copper with ground, #14, 2 conductor	1 Elec	4	2		35	94		129	184
1500	#12, 2 conductor		3.50	2.286		53	107		160	223
1550	#10, 2 conductor		3	2.667		82	125		207	283
1600	#14, 3 conductor		3.50	2.286		49.50	107		156.50	220
1650	#12, 3 conductor		3	2.667		73.50	125		198.50	274
1700	#10, 3 conductor		2.50	3.200		114	150		264	355
2400	SEU service entrance cable, copper 2 conductors, #8 + #8 neutral		1.50	5.333		152	251		403	550
2600	#6 + #8 neutral		1.30	6.154		227	289		516	695
2800	#6 + #6 neutral		1.30	6.154		258	289		547	730
3000	#4 + #6 neutral	2 Elec	2.20	7.273		355	340		695	915
3200	#4 + #4 neutral		2.20	7.273		400	340		740	965
3400	#3 + #5 neutral		2.10	7.619		465	360		825	1,075
3600	#3 + #3 neutral		2.10	7.619		500	360		860	1,100
3800	#2 + #4 neutral		2	8		550	375		925	1,175
4000	#1 + #1 neutral		1.90	8.421		805	395		1,200	1,500
4200	1/0 + 1/0 neutral		1.80	8.889		1,000	420		1,420	1,750
4400	2/0 + 2/0 neutral		1.70	9.412		1,250	440		1,690	2,050
4600	3/0 + 3/0 neutral		1.60	10		1,575	470		2,045	2,450
4620	4/0 + 4/0 neutral		1.45	11.034		1,750	520		2,270	2,700
4800	Aluminum 2 conductors, #8 + #8 neutral	1 Elec	1.60	5		133	235		368	505
5000	#6 + #6 neutral	"	1.40	5.714		134	269		403	560
5100	#4 + #6 neutral	2 Elec	2.50	6.400		171	300		471	655
5200	#4 + #4 neutral		2.40	6.667		171	315		486	670
5300	#2 + #4 neutral		2.30	6.957		228	325		553	755
5400	#2 + #2 neutral		2.20	7.273		228	340		568	775
5450	1/0 + #2 neutral		2.10	7.619		350	360		710	930
5500	1/0 + 1/0 neutral		2	8		350	375		725	960
5550	2/0 + #1 neutral		1.90	8.421		400	395		795	1,050
5600	2/0 + 2/0 neutral		1.80	8.889		400	420		820	1,075
5800	3/0 + 1/0 neutral		1.70	9.412		555	440		995	1,300
6000	3/0 + 3/0 neutral		1.70	9.412		555	440		995	1,300
6200	4/0 + 2/0 neutral		1.60	10		560	470		1,030	1,350
6400	4/0 + 4/0 neutral		1.60	10		560	470		1,030	1,350
6500	Service entrance cap for copper SEU									
6600	100 amp	1 Elec	12	.667	Ea.	12.15	31.50		43.65	61.50
6700	150 amp		10	.800		22	37.50		59.50	82.50
6800	200 amp		8	1		33.50	47		80.50	110
9000	Minimum labor/equipment charge		4	2	Job		94		94	145

26 05 19.75 Modular Flexible Wiring System		Crew	Daily Output	Labor-Hours	Unit	Material	Labor	Equipment	Total	Total Incl O&P
0010	**MODULAR FLEXIBLE WIRING SYSTEM**									
0050	Standard selector cable, 3 conductor, 15′ long	1 Elec	40	.200	Ea.	47.50	9.40		56.90	66.50
0100	Conversion Module		16	.500		17.20	23.50		40.70	55
0150	Cable set, 3 conductor, 15′ long		24	.333		18.45	15.65		34.10	44.50
0200	Switching assembly, 3 conductor, 10′ long		32	.250		36.50	11.75		48.25	58

26 05 19.90 Wire		Crew	Daily Output	Labor-Hours	Unit	Material	Labor	Equipment	Total	Total Incl O&P
0010	**WIRE** R260519-90									
0020	600 volt type THW, copper, solid, #14	1 Elec	13	.615	C.L.F.	10.30	29		39.30	56

26 05 19.90 Wire		Crew	Daily Output	Labor-Hours	Unit	Material	Labor	Equipment	Total	Total Incl O&P
						2009 Bare Costs				
0030	#12	1 Elec	11	.727	C.L.F.	15.90	34		49.90	70
0040	#10		10	.800		25	37.50		62.50	85.50
0050	Stranded, #14 R260519-92		13	.615		10.40	29		39.40	56
0100	#12		11	.727		16.25	34		50.25	70.50
0120	#10 R260533-22		10	.800		25	37.50		62.50	85.50
0140	#8		8	1		43.50	47		90.50	121
0160	#6		6.50	1.231		67.50	58		125.50	164
0180	#4	2 Elec	10.60	1.509		106	71		177	226
0200	#3		10	1.600		134	75		209	264
0220	#2		9	1.778		168	83.50		251.50	315
0240	#1		8	2		213	94		307	380
0260	1/0		6.60	2.424		259	114		373	460
0280	2/0		5.80	2.759		325	130		455	560
0300	3/0		5	3.200		410	150		560	680
0350	4/0		4.40	3.636		515	171		686	830
0400	250 kcmil	3 Elec	6	4		610	188		798	960
0420	300 kcmil		5.70	4.211		725	198		923	1,100
0450	350 kcmil		5.40	4.444		850	209		1,059	1,250
0480	400 kcmil		5.10	4.706		970	221		1,191	1,425
0490	500 kcmil		4.80	5		1,175	235		1,410	1,650
0500	600 kcmil		3.90	6.154		1,400	289		1,689	2,000
0510	750 kcmil		3.30	7.273		2,600	340		2,940	3,400
0520	1000 kcmil		2.70	8.889		3,450	420		3,870	4,450
0540	Aluminum, stranded, #6	1 Elec	8	1		31.50	47		78.50	108
0560	#4	2 Elec	13	1.231		39.50	58		97.50	133
0580	#2		10.60	1.509		53.50	71		124.50	168
0600	#1		9	1.778		78.50	83.50		162	215
0620	1/0		8	2		94	94		188	248
0640	2/0		7.20	2.222		112	104		216	284
0680	3/0		6.60	2.424		138	114		252	325
0700	4/0		6.20	2.581		154	121		275	355
0720	250 kcmil	3 Elec	8.70	2.759		188	130		318	405
0740	300 kcmil		8.10	2.963		259	139		398	500
0760	350 kcmil		7.50	3.200		264	150		414	520
0780	400 kcmil		6.90	3.478		310	163		473	590
0800	500 kcmil		6	4		340	188		528	665
0850	600 kcmil		5.70	4.211		430	198		628	780
0880	700 kcmil		5.10	4.706		500	221		721	885
0900	750 kcmil		4.80	5		505	235		740	915
0920	Type THWN-THHN, copper, solid, #14	1 Elec	13	.615		10.30	29		39.30	56
0940	#12		11	.727		15.90	34		49.90	70
0960	#10		10	.800		25	37.50		62.50	85.50
1000	Stranded, #14		13	.615		10.40	29		39.40	56
1200	#12		11	.727		16.25	34		50.25	70.50
1250	#10		10	.800		25	37.50		62.50	85.50
1300	#8		8	1		43.50	47		90.50	121
1350	#6		6.50	1.231		67.50	58		125.50	164
1400	#4	2 Elec	10.60	1.509		106	71		177	226
1450	#3		10	1.600		134	75		209	264
1500	#2		9	1.778		168	83.50		251.50	315
1550	#1		8	2		213	94		307	380
1600	1/0		6.60	2.424		259	114		373	460
1650	2/0		5.80	2.759		325	130		455	560

26 05 19.90 Wire		Crew	Daily Output	Labor-Hours	Unit	Material	2009 Bare Costs Labor	Equipment	Total	Total Incl O&P
1700	3/0	2 Elec	5	3.200	C.L.F.	410	150		560	680
2000	4/0		4.40	3.636		515	171		686	830
2200	250 kcmil	3 Elec	6	4		610	188		798	960
2400	300 kcmil		5.70	4.211		725	198		923	1,100
2600	350 kcmil		5.40	4.444		850	209		1,059	1,250
2700	400 kcmil		5.10	4.706		970	221		1,191	1,425
2800	500 kcmil		4.80	5		1,175	235		1,410	1,650
2900	600 volt, copper type XHHW, solid, #14	1 Elec	13	.615		20.50	29		49.50	67
2920	#12		11	.727		30.50	34		64.50	86
2940	#10		10	.800		45	37.50		82.50	108
3000	Stranded, #14		13	.615		21.50	29		50.50	68
3020	#12		11	.727		32	34		66	88
3040	#10		10	.800		49	37.50		86.50	112
3060	#8		8	1		75	47		122	155
3080	#6		6.50	1.231		111	58		169	212
3100	#4	2 Elec	10.60	1.509		172	71		243	298
3120	#2		9	1.778		268	83.50		351.50	425
3140	#1		8	2		345	94		439	525
3160	1/0		6.60	2.424		425	114		539	645
3180	2/0		5.80	2.759		535	130		665	785
3200	3/0		5	3.200		670	150		820	965
3220	4/0		4.40	3.636		840	171		1,011	1,175
3240	250 kcmil	3 Elec	6	4		880	188		1,068	1,250
3260	300 kcmil		5.70	4.211		910	198		1,108	1,300
3280	350 kcmil		5.40	4.444		1,050	209		1,259	1,500
3300	400 kcmil		5.10	4.706		1,200	221		1,421	1,675
3320	500 kcmil		4.80	5		1,450	235		1,685	1,950
3340	600 kcmil		3.90	6.154		1,750	289		2,039	2,375
3360	750 kcmil		3.30	7.273		2,900	340		3,240	3,725
3380	1000 kcmil		2.40	10		3,875	470		4,345	4,975
5020	600 volt, aluminum type XHHW, stranded, #6	1 Elec	8	1		32	47		79	108
5040	#4	2 Elec	13	1.231		39.50	58		97.50	133
5060	#2		10.60	1.509		53.50	71		124.50	168
5080	#1		9	1.778		78.50	83.50		162	215
5100	1/0		8	2		94	94		188	248
5120	2/0		7.20	2.222		112	104		216	284
5140	3/0		6.60	2.424		138	114		252	325
5160	4/0		6.20	2.581		154	121		275	355
5180	250 kcmil	3 Elec	8.70	2.759		188	130		318	405
5200	300 kcmil		8.10	2.963		259	139		398	500
5220	350 kcmil		7.50	3.200		264	150		414	520
5240	400 kcmil		6.90	3.478		310	163		473	590
5260	500 kcmil		6	4		340	188		528	665
5280	600 kcmil		5.70	4.211		430	198		628	780
5300	700 kcmil		5.40	4.444		500	209		709	865
5320	750 kcmil		5.10	4.706		505	221		726	895
5340	1000 kcmil		3.60	6.667		750	315		1,065	1,300
5400	600 volt, copper type XLPE-USE(RHW), solid, #12	1 Elec	11	.727		32.50	34		66.50	88
5420	#10		10	.800		47	37.50		84.50	110
5440	Stranded, #14		13	.615		32	29		61	79.50
5460	#12		11	.727		38	34		72	94.50
5480	#10		10	.800		57	37.50		94.50	121
5500	#8		8	1		78	47		125	159

26 05 19 – Low-Voltage Electrical Power Conductors and Cables

26 05 19.90 Wire		Crew	Daily Output	Labor-Hours	Unit	Material	2009 Bare Costs Labor	Equipment	Total	Total Incl O&P
5520	#6	1 Elec	6.50	1.231	C.L.F.	118	58		176	218
5540	#4	2 Elec	10.60	1.509		185	71		256	310
5560	#2		9	1.778		290	83.50		373.50	450
5580	#1		8	2		365	94		459	550
5600	1/0		6.60	2.424		460	114		574	680
5620	2/0		5.80	2.759		575	130		705	835
5640	3/0		5	3.200		720	150		870	1,025
5660	4/0		4.40	3.636		820	171		991	1,175
5680	250 kcmil	3 Elec	6	4		865	188		1,053	1,250
5700	300 kcmil		5.70	4.211		1,025	198		1,223	1,450
5720	350 kcmil		5.40	4.444		1,200	209		1,409	1,650
5740	400 kcmil		5.10	4.706		1,375	221		1,596	1,850
5760	500 kcmil		4.80	5		1,700	235		1,935	2,225
5780	600 kcmil		3.90	6.154		1,925	289		2,214	2,575
5800	750 kcmil		3.30	7.273		3,200	340		3,540	4,050
5820	1000 kcmil		2.70	8.889		4,250	420		4,670	5,325
5840	600 volt, aluminum type XLPE-USE (RHW), stranded, #6	1 Elec	8	1		39	47		86	116
5860	#4	2 Elec	13	1.231		45	58		103	139
5880	#2		10.60	1.509		62	71		133	177
5900	#1		9	1.778		86.50	83.50		170	224
5920	1/0		8	2		106	94		200	262
5940	2/0		7.20	2.222		125	104		229	298
5960	3/0		6.60	2.424		147	114		261	335
5980	4/0		6.20	2.581		163	121		284	365
6000	250 kcmil	3 Elec	8.70	2.759		222	130		352	445
6020	300 kcmil		8.10	2.963		288	139		427	530
6040	350 kcmil		7.50	3.200		294	150		444	555
6060	400 kcmil		6.90	3.478		360	163		523	645
6080	500 kcmil		6	4		395	188		583	725
6100	600 kcmil		5.70	4.211		510	198		708	865
6110	700 kcmil		5.40	4.444		585	209		794	960
6120	750 kcmil		5.10	4.706		590	221		811	990
9000	Minimum labor/equipment charge	1 Elec	4	2	Job		94		94	145

26 05 23 – Control-Voltage Electrical Power Cables

26 05 23.10 Control Cable

		Crew	Daily Output	Labor-Hours	Unit	Material	Labor	Equipment	Total	Total Incl O&P
0010	**CONTROL CABLE**									
0020	600 volt, copper, #14 THWN wire with PVC jacket, 2 wires	1 Elec	9	.889	C.L.F.	27.50	42		69.50	94.50
0030	3 wires		8	1		38.50	47		85.50	115
0100	4 wires		7	1.143		47.50	53.50		101	135
0150	5 wires		6.50	1.231		57.50	58		115.50	153
0200	6 wires		6	1.333		82.50	62.50		145	188
0300	8 wires		5.30	1.509		103	71		174	222
0400	10 wires		4.80	1.667		121	78.50		199.50	254
0500	12 wires		4.30	1.860		140	87.50		227.50	288
0600	14 wires		3.80	2.105		167	99		266	335
0700	16 wires		3.50	2.286		170	107		277	350
0800	18 wires		3.30	2.424		192	114		306	385
0810	19 wires		3.10	2.581		212	121		333	420
0900	20 wires		3	2.667		222	125		347	440
1000	22 wires		2.80	2.857		230	134		364	460

26 05 23.20 Special Wires and Fittings

0010	**SPECIAL WIRES & FITTINGS**									

26 05 23.20 Special Wires and Fittings		Crew	Daily Output	Labor-Hours	Unit	2009 Bare Costs Material	Labor	Equipment	Total	Total Incl O&P
0100	Fixture TFFN 600 volt 90°C stranded, #18	1 Elec	13	.615	C.L.F.	9.70	29		38.70	55
0150	#16		13	.615		13.10	29		42.10	59
0500	Thermostat, no jacket, twisted, #18-2 conductor		8	1		8.70	47		55.70	82
0550	#18-3 conductor		7	1.143		14.30	53.50		67.80	98.50
0600	#18-4 conductor		6.50	1.231		20.50	58		78.50	112
0650	#18-5 conductor		6	1.333		24	62.50		86.50	123
0700	#18-6 conductor		5.50	1.455		26	68.50		94.50	134
0750	#18-7 conductor		5	1.600		34	75		109	154
0800	#18-8 conductor		4.80	1.667		42	78.50		120.50	167
1500	Fire alarm FEP teflon 150 volt to 200°C									
1550	#22, 1 pair	1 Elec	10	.800	C.L.F.	76.50	37.50		114	142
1600	2 pair		8	1		127	47		174	212
1650	4 pair		7	1.143		194	53.50		247.50	297
1700	6 pair		6	1.333		252	62.50		314.50	375
1750	8 pair		5.50	1.455		320	68.50		388.50	455
1800	10 pair		5	1.600		385	75		460	535
1850	#18, 1 pair		8	1		87	47		134	169
1900	2 pair		6.50	1.231		167	58		225	272
1950	4 pair		4.80	1.667		252	78.50		330.50	400
2000	6 pair		4	2		330	94		424	510
2050	8 pair		3.50	2.286		435	107		542	640
2100	10 pair		3	2.667		495	125		620	740
2460	Tray cable, type TC, copper, #16-2 conductor		9.40	.851		20.50	40		60.50	84.50
2464	#16-3 conductor		8.40	.952		27.50	45		72.50	99.50
2468	#16-4 conductor		7.30	1.096		34	51.50		85.50	116
2472	#16-5 conductor		6.70	1.194		41	56		97	132
2476	#16-7 conductor		5.50	1.455		54.50	68.50		123	165
2480	#16-9 conductor		5	1.600		64	75		139	186
2484	#16-12 conductor	2 Elec	8.80	1.818		88	85.50		173.50	228
2488	#16-15 conductor		7.20	2.222		115	104		219	287
2492	#16-19 conductor		6.60	2.424		139	114		253	330
2496	#16-25 conductor		6.40	2.500		203	118		321	405
2500	#14-2 conductor	1 Elec	9	.889		27.50	42		69.50	94.50
2520	#14-3 conductor		8	1		39	47		86	116
2540	#14-4 conductor		7	1.143		47.50	53.50		101	135
2560	#14-5 conductor		6.50	1.231		57.50	58		115.50	153
2564	#14-7 conductor		5.30	1.509		82.50	71		153.50	200
2568	#14-9 conductor	2 Elec	9.60	1.667		103	78.50		181.50	234
2572	#14-12 conductor		8.60	1.860		140	87.50		227.50	288
2576	#14-15 conductor		7.20	2.222		171	104		275	350
2578	#14-19 conductor		6.20	2.581		212	121		333	420
2582	#14-25 conductor		5.30	3.019		279	142		421	525
2590	#12-2 conductor	1 Elec	8.40	.952		39.50	45		84.50	113
2592	#12-3 conductor		7.60	1.053		57.50	49.50		107	140
2594	#12-4 conductor		6.60	1.212		74.50	57		131.50	170
2596	#12-5 conductor		6.20	1.290		87.50	60.50		148	190
2598	#12-7 conductor	2 Elec	10.40	1.538		124	72.50		196.50	248
2602	#12-9 conductor		8.80	1.818		162	85.50		247.50	310
2604	#12-12 conductor		7.80	2.051		203	96.50		299.50	370
2606	#12-15 conductor		6.80	2.353		243	111		354	435
2608	#12-19 conductor		6	2.667		315	125		440	540
2610	#12-25 conductor	3 Elec	7.70	3.117		395	146		541	660
2618	#10-2 conductor	1 Elec	8	1		57.50	47		104.50	136

26 05 23 – Control-Voltage Electrical Power Cables

26 05 23.20 Special Wires and Fittings		Crew	Daily Output	Labor-Hours	Unit	Material	2009 Bare Costs Labor	Equipment	Total	Total Incl O&P
2622	#10-3 conductor	1 Elec	7.30	1.096	C.L.F.	87.50	51.50		139	175
2624	#10-4 conductor	2 Elec	12.80	1.250		113	59		172	215
2626	#10-5 conductor		11.80	1.356		138	63.50		201.50	249
2628	#10-7 conductor		9.40	1.702		190	80		270	330
2630	#10-9 conductor		8.40	1.905		243	89.50		332.50	405
2632	#10-12 conductor		7.20	2.222		315	104		419	505
2640	300 V, copper braided shield, PVC jacket									
2650	2 conductor #18 stranded	1 Elec	7	1.143	C.L.F.	58.50	53.50		112	147
2660	3 conductor #18 stranded	"	6	1.333	"	82.50	62.50		145	188
3000	Strain relief grip for cable									
3050	Cord, top, #12-3	1 Elec	40	.200	Ea.	15.05	9.40		24.45	31
3060	#12-4		40	.200		15.05	9.40		24.45	31
3070	#12-5		39	.205		17.50	9.65		27.15	34
3100	#10-3		39	.205		17.50	9.65		27.15	34
3110	#10-4		38	.211		17.50	9.90		27.40	34.50
3120	#10-5		38	.211		19.10	9.90		29	36
3200	Bottom, #12-3		40	.200		38.50	9.40		47.90	57
3210	#12-4		40	.200		38.50	9.40		47.90	57
3220	#12-5		39	.205		39	9.65		48.65	58
3230	#10-3		39	.205		41.50	9.65		51.15	61
3300	#10-4		38	.211		45	9.90		54.90	64.50
3310	#10-5		38	.211		49	9.90		58.90	69
3400	Cable ties, standard, 4" length		190	.042		.16	1.98		2.14	3.22
3410	7" length		160	.050		.25	2.35		2.60	3.90
3420	14.5" length		90	.089		.55	4.18		4.73	7.05
3430	Heavy, 14.5" length		80	.100		.80	4.70		5.50	8.15
9000	Minimum labor/equipment charge		4	2	Job		94		94	145

26 05 26 – Grounding and Bonding for Electrical Systems

26 05 26.80 Grounding		Crew	Daily Output	Labor-Hours	Unit	Material	Labor	Equipment	Total	Total Incl O&P
0010	**GROUNDING**									
0030	Rod, copper clad, 8' long, 1/2" diameter	1 Elec	5.50	1.455	Ea.	14.80	68.50		83.30	121
0040	5/8" diameter		5.50	1.455		16.40	68.50		84.90	123
0050	3/4" diameter		5.30	1.509		27.50	71		98.50	140
0080	10' long, 1/2" diameter		4.80	1.667		17.35	78.50		95.85	140
0090	5/8" diameter		4.60	1.739		19.35	81.50		100.85	148
0100	3/4" diameter		4.40	1.818		33	85.50		118.50	168
0130	15' long, 3/4" diameter		4	2		92	94		186	246
0150	Coupling, bronze, 1/2" diameter					4.20			4.20	4.62
0160	5/8" diameter					6.20			6.20	6.80
0170	3/4" diameter					14.05			14.05	15.45
0190	Drive studs, 1/2" diameter					7.90			7.90	8.65
0210	5/8" diameter					9.60			9.60	10.55
0220	3/4" diameter					10.80			10.80	11.90
0230	Clamp, bronze, 1/2" diameter	1 Elec	32	.250		4.68	11.75		16.43	23.50
0240	5/8" diameter		32	.250		5.10	11.75		16.85	23.50
0250	3/4" diameter		32	.250		6.70	11.75		18.45	25.50
0260	Wire ground bare armored, #8-1 conductor		2	4	C.L.F.	109	188		297	410
0270	#6-1 conductor		1.80	4.444		131	209		340	465
0280	#4-1 conductor		1.60	5		175	235		410	555
0320	Bare copper wire, #14 solid		14	.571		9.65	27		36.65	52
0330	#12		13	.615		15	29		44	61
0340	#10		12	.667		24	31.50		55.50	74

26 05 26 – Grounding and Bonding for Electrical Systems

26 05 26.80 Grounding		Crew	Daily Output	Labor-Hours	Unit	2009 Bare Costs Material	Labor	Equipment	Total	Total Incl O&P
0350	#8	1 Elec	11	.727	C.L.F.	38.50	34		72.50	94.50
0360	#6		10	.800		60.50	37.50		98	125
0370	#4		8	1		102	47		149	185
0380	#2		5	1.600		163	75		238	295
0390	Bare copper wire, stranded, #8		11	.727		42	34		76	98.50
0400	#6		10	.800		66	37.50		103.50	131
0450	#4	2 Elec	16	1		105	47		152	188
0600	#2		10	1.600		167	75		242	299
0650	#1		9	1.778		210	83.50		293.50	360
0700	1/0		8	2		239	94		333	410
0750	2/0		7.20	2.222		300	104		404	490
0800	3/0		6.60	2.424		380	114		494	590
1000	4/0		5.70	2.807		480	132		612	730
1200	250 kcmil	3 Elec	7.20	3.333		565	157		722	860
1210	300 kcmil		6.60	3.636		710	171		881	1,050
1220	350 kcmil		6	4		790	188		978	1,150
1230	400 kcmil		5.70	4.211		905	198		1,103	1,300
1240	500 kcmil		5.10	4.706		1,125	221		1,346	1,600
1260	750 kcmil		3.60	6.667		1,950	315		2,265	2,625
1270	1000 kcmil		3	8		2,400	375		2,775	3,225
1360	Bare aluminum, stranded, #6	1 Elec	9	.889		21	42		63	87.50
1370	#4	2 Elec	16	1		24	47		71	99
1380	#2		13	1.231		35.50	58		93.50	128
1390	#1		10.60	1.509		44	71		115	158
1400	1/0		9	1.778		50	83.50		133.50	184
1410	2/0		8	2		63	94		157	215
1420	3/0		7.20	2.222		77	104		181	246
1430	4/0		6.60	2.424		96	114		210	281
1440	250 kcmil	3 Elec	9.30	2.581		118	121		239	315
1450	300 kcmil		8.70	2.759		155	130		285	370
1460	400 kcmil		7.50	3.200		182	150		332	430
1470	500 kcmil		6.90	3.478		218	163		381	490
1480	600 kcmil		6	4		258	188		446	575
1490	700 kcmil		5.70	4.211		298	198		496	635
1500	750 kcmil		5.10	4.706		335	221		556	710
1510	1000 kcmil		4.80	5		465	235		700	875
1800	Water pipe ground clamps, heavy duty									
2000	Bronze, 1/2" to 1" diameter	1 Elec	8	1	Ea.	26	47		73	101
2100	1-1/4" to 2" diameter		8	1		34	47		81	110
2200	2-1/2" to 3" diameter		6	1.333		57	62.50		119.50	160
2730	Exothermic weld, 4/0 wire to 1" ground rod		7	1.143		11.45	53.50		64.95	95
2740	4/0 wire to building steel		7	1.143		9.10	53.50		62.60	92.50
2750	4/0 wire to motor frame		7	1.143		9.10	53.50		62.60	92.50
2760	4/0 wire to 4/0 wire		7	1.143		7.95	53.50		61.45	91.50
2770	4/0 wire to #4 wire		7	1.143		7.95	53.50		61.45	91.50
2780	4/0 wire to #8 wire		7	1.143		7.95	53.50		61.45	91.50
2790	Mold, reusable, for above					131			131	144
2800	Brazed connections, #6 wire	1 Elec	12	.667		13.45	31.50		44.95	63
3000	#2 wire		10	.800		18	37.50		55.50	78
3100	3/0 wire		8	1		27	47		74	102
3200	4/0 wire		7	1.143		31	53.50		84.50	117
3400	250 kcmil wire		5	1.600		36	75		111	156
3600	500 kcmil wire		4	2		44.50	94		138.50	194

26 05 26.80 Grounding		Crew	Daily Output	Labor-Hours	Unit	2009 Bare Costs Material	Labor	Equipment	Total	Total Incl O&P
3700	Insulated ground wire, copper #14	1 Elec	13	.615	C.L.F.	10.40	29		39.40	56
3710	#12		11	.727		16.25	34		50.25	70.50
3720	#10		10	.800		25	37.50		62.50	85.50
3730	#8		8	1		43.50	47		90.50	121
3740	#6		6.50	1.231		67.50	58		125.50	164
3750	#4	2 Elec	10.60	1.509		106	71		177	226
3770	#2		9	1.778		168	83.50		251.50	315
3780	#1		8	2		213	94		307	380
3790	1/0		6.60	2.424		259	114		373	460
3800	2/0		5.80	2.759		325	130		455	560
3810	3/0		5	3.200		410	150		560	680
3820	4/0		4.40	3.636		515	171		686	830
3830	250 kcmil	3 Elec	6	4		610	188		798	960
3840	300 kcmil		5.70	4.211		725	198		923	1,100
3850	350 kcmil		5.40	4.444		850	209		1,059	1,250
3860	400 kcmil		5.10	4.706		970	221		1,191	1,425
3870	500 kcmil		4.80	5		1,175	235		1,410	1,650
3880	600 kcmil		3.90	6.154		1,400	289		1,689	2,000
3890	750 kcmil		3.30	7.273		2,600	340		2,940	3,400
3900	1000 kcmil		2.70	8.889		3,450	420		3,870	4,450
3960	Insulated ground wire, aluminum, #6	1 Elec	8	1		31.50	47		78.50	108
3970	#4	2 Elec	13	1.231		39.50	58		97.50	133
3980	#2		10.60	1.509		53.50	71		124.50	168
3990	#1		9	1.778		78.50	83.50		162	215
4000	1/0		8	2		94	94		188	248
4010	2/0		7.20	2.222		112	104		216	284
4020	3/0		6.60	2.424		138	114		252	325
4030	4/0		6.20	2.581		154	121		275	355
4040	250 kcmil	3 Elec	8.70	2.759		188	130		318	405
4050	300 kcmil		8.10	2.963		259	139		398	500
4060	350 kcmil		7.50	3.200		264	150		414	520
4070	400 kcmil		6.90	3.478		310	163		473	590
4080	500 kcmil		6	4		340	188		528	665
4090	600 kcmil		5.70	4.211		430	198		628	780
4100	700 kcmil		5.10	4.706		500	221		721	885
4110	750 kcmil		4.80	5		505	235		740	915
5000	Copper Electrolytic ground rod system									
5010	Includes augering hole, mixing bentonite clay,									
5020	Installing rod, and terminating ground wire									
5100	Straight Vertical type, 2″ Dia.									
5120	8.5′ long, Clamp Connection	1 Elec	2.67	2.996	Ea.	685	141		826	965
5130	With exothermic weld Connection		1.95	4.103		685	193		878	1,050
5140	10′ long		2.35	3.404		760	160		920	1,075
5150	With exothermic weld Connection		1.78	4.494		760	211		971	1,150
5160	12′ long		2.16	3.704		920	174		1,094	1,275
5170	With exothermic weld Connection		1.67	4.790		920	225		1,145	1,350
5180	20′ long		1.74	4.598		1,400	216		1,616	1,875
5190	With exothermic weld Connection		1.40	5.714		1,450	269		1,719	2,025
5195	40′ long with exothermic weld connection	2 Elec	2	8		2,700	375		3,075	3,525
5200	L-Shaped, 2″ Dia.									
5220	4′ Vert. x 10′ Horz., Clamp Connection	1 Elec	5.33	1.501	Ea.	1,100	70.50		1,170.50	1,300
5230	With exothermic weld Connection	″	3.08	2.597	″	1,100	122		1,222	1,400
5300	Protective Box at grade level, with breather slots									

26 05 26 – Grounding and Bonding for Electrical Systems

26 05 26.80 Grounding		Crew	Daily Output	Labor-Hours	Unit	Material	Labor	Equipment	Total	Total Incl O&P
						2009 Bare Costs				
5320	Round 12" long, Plastic	1 Elec	32	.250	Ea.	45	11.75		56.75	67.50
5330	Concrete	"	16	.500		75	23.50		98.50	119
5400	Bentonite Clay, 50# bag, 1 per 10' of rod					38			38	42
9000	Minimum labor/equipment charge	1 Elec	4	2	Job		94		94	145

26 05 29 – Hangers and Supports for Electrical Systems

26 05 29.20 Hangers

		Crew	Daily Output	Labor-Hours	Unit	Material	Labor	Equipment	Total	Total Incl O&P
0010	**HANGERS** R260533-21									
0030	Conduit supports									
0050	Strap w/2 holes, rigid steel conduit									
0100	1/2" diameter	1 Elec	470	.017	Ea.	.23	.80		1.03	1.48
0150	3/4" diameter		440	.018		.27	.85		1.12	1.61
0200	1" diameter		400	.020		.44	.94		1.38	1.93
0300	1-1/4" diameter		355	.023		.69	1.06		1.75	2.39
0350	1-1/2" diameter		320	.025		.81	1.18		1.99	2.70
0400	2" diameter		266	.030		.95	1.41		2.36	3.22
0500	2-1/2" diameter		160	.050		1.96	2.35		4.31	5.80
0550	3" diameter		133	.060		2.43	2.83		5.26	7
0600	3-1/2" diameter		100	.080		3.18	3.76		6.94	9.30
0650	4" diameter		80	.100		3.43	4.70		8.13	11
0700	EMT, 1/2" diameter		470	.017		.21	.80		1.01	1.46
0800	3/4" diameter		440	.018		.25	.85		1.10	1.59
0850	1" diameter		400	.020		.47	.94		1.41	1.97
0900	1-1/4" diameter		355	.023		.75	1.06		1.81	2.46
0950	1-1/2" diameter		320	.025		.79	1.18		1.97	2.68
1000	2" diameter		266	.030		1.13	1.41		2.54	3.41
1100	2-1/2" diameter		160	.050		1.94	2.35		4.29	5.75
1150	3" diameter		133	.060		2.60	2.83		5.43	7.20
1200	3-1/2" diameter		100	.080		2.93	3.76		6.69	9
1250	4" diameter		80	.100		3.31	4.70		8.01	10.90
1400	Hanger, with bolt, 1/2" diameter		200	.040		.61	1.88		2.49	3.56
1450	3/4" diameter		190	.042		.63	1.98		2.61	3.73
1500	1" diameter		176	.045		1.03	2.14		3.17	4.42
1550	1-1/4" diameter		160	.050		1.47	2.35		3.82	5.25
1600	1-1/2" diameter		140	.057		1.81	2.69		4.50	6.10
1650	2" diameter		130	.062		1.94	2.89		4.83	6.60
1700	2-1/2" diameter		100	.080		2.53	3.76		6.29	8.60
1750	3" diameter		64	.125		2.97	5.90		8.87	12.30
1800	3-1/2" diameter		50	.160		4.37	7.50		11.87	16.35
1850	4" diameter		40	.200		9.95	9.40		19.35	25.50
1900	Riser clamps, conduit, 1/2" diameter		40	.200		11.25	9.40		20.65	27
1950	3/4" diameter		36	.222		11.25	10.45		21.70	28.50
2000	1" diameter		30	.267		11.35	12.55		23.90	32
2100	1-1/4" diameter		27	.296		15.10	13.95		29.05	38
2150	1-1/2" diameter		27	.296		15.95	13.95		29.90	39
2200	2" diameter		20	.400		16.70	18.80		35.50	47.50
2250	2-1/2" diameter		20	.400		17.45	18.80		36.25	48
2300	3" diameter		18	.444		18.85	21		39.85	52.50
2350	3-1/2" diameter		18	.444		24	21		45	58.50
2400	4" diameter		14	.571		25	27		52	69
2500	Threaded rod, painted, 1/4" diameter		260	.031	L.F.	2.21	1.45		3.66	4.65
2600	3/8" diameter		200	.040		2.91	1.88		4.79	6.10
2700	1/2" diameter		140	.057		5.15	2.69		7.84	9.85

26 05 29 – Hangers and Supports for Electrical Systems

26 05 29.20 Hangers		Crew	Daily Output	Labor-Hours	Unit	2009 Bare Costs Material	Labor	Equipment	Total	Total Incl O&P
2800	5/8" diameter	1 Elec	100	.080	L.F.	6.40	3.76		10.16	12.80
2900	3/4" diameter	↓	60	.133	↓	8	6.25		14.25	18.45
2940	Couplings painted, 1/4" diameter				C	320			320	355
2960	3/8" diameter					520			520	570
2970	1/2" diameter					755			755	830
2980	5/8" diameter					1,175			1,175	1,275
2990	3/4" diameter					1,725			1,725	1,900
3000	Nuts, galvanized, 1/4" diameter					14			14	15.40
3050	3/8" diameter					21			21	23
3100	1/2" diameter					47			47	51.50
3150	5/8" diameter					71			71	78
3200	3/4" diameter					105			105	116
3250	Washers, galvanized, 1/4" diameter					14.50			14.50	15.95
3300	3/8" diameter					20.50			20.50	22.50
3350	1/2" diameter					32			32	35
3400	5/8" diameter					58			58	64
3450	3/4" diameter					85			85	93.50
3500	Lock washers, galvanized, 1/4" diameter					12.30			12.30	13.55
3550	3/8" diameter					17.25			17.25	19
3600	1/2" diameter					19.90			19.90	22
3650	5/8" diameter					34.50			34.50	38
3700	3/4" diameter				↓	54			54	59.50
3800	Channels, steel, 3/4" x 1-1/2"	1 Elec	80	.100	L.F.	6.75	4.70		11.45	14.70
3900	1-1/2" x 1-1/2"		70	.114		9.05	5.35		14.40	18.20
4000	1-7/8" x 1-1/2"		60	.133		22	6.25		28.25	33.50
4100	3" x 1-1/2"		50	.160	↓	25.50	7.50		33	39.50
4200	Spring nuts, long, 1/4"		120	.067	Ea.	1.96	3.13		5.09	7
4250	3/8"		100	.080		2.72	3.76		6.48	8.80
4300	1/2"		80	.100		2.90	4.70		7.60	10.45
4350	Spring nuts, short, 1/4"		120	.067		2.42	3.13		5.55	7.50
4400	3/8"		100	.080		2.64	3.76		6.40	8.70
4450	1/2"		80	.100	↓	3.14	4.70		7.84	10.70
4500	Closure strip		200	.040	L.F.	3.63	1.88		5.51	6.90
4550	End cap		60	.133	Ea.	1.49	6.25		7.74	11.30
4600	End connector 3/4" conduit		40	.200		5.20	9.40		14.60	20
4650	Junction box, 1 channel		16	.500		29	23.50		52.50	68
4700	2 channel		14	.571		35	27		62	80
4750	3 channel		12	.667		40	31.50		71.50	92
4800	4 channel		10	.800		44	37.50		81.50	107
4850	Spliceplate		40	.200		10.90	9.40		20.30	26.50
4900	Continuous concrete insert, 1-1/2" deep, 1' long		16	.500		16.50	23.50		40	54
4950	2' long		14	.571		20.50	27		47.50	64
5000	3' long		12	.667		22.50	31.50		54	73
5050	4' long		10	.800		28.50	37.50		66	89
5100	6' long		8	1		42.50	47		89.50	119
5150	3/4" deep, 1' long		16	.500		13.20	23.50		36.70	50.50
5200	2' long		14	.571		16.40	27		43.40	59.50
5250	3' long		12	.667		18.10	31.50		49.60	68
5300	4' long		10	.800		22.50	37.50		60	83
5350	6' long		8	1		34	47		81	110
5400	90° angle fitting 2-1/8" x 2-1/8"		60	.133		3.15	6.25		9.40	13.10
5450	Supports, suspension rod type, small		60	.133		13.10	6.25		19.35	24
5500	Large	↓	40	.200	↓	14.45	9.40		23.85	30.50

26 05 29.20 Hangers		Crew	Daily Output	Labor-Hours	Unit	2009 Bare Costs Material	Labor	Equipment	Total	Total Incl O&P
5550	Beam clamp, small	1 Elec	60	.133	Ea.	6.05	6.25		12.30	16.30
5600	Large		40	.200		7.35	9.40		16.75	22.50
5650	U-support, small		60	.133		4	6.25		10.25	14.05
5700	Large		40	.200		10.40	9.40		19.80	26
5750	Concrete insert, cast, for up to 1/2" threaded rod		16	.500		3.19	23.50		26.69	39.50
5800	Beam clamp, 1/4" clamp, for 1/4" threaded drop rod		32	.250		2.70	11.75		14.45	21
5900	3/8" clamp, for 3/8" threaded drop rod		32	.250		7.95	11.75		19.70	27
6000	Strap, rigid conduit, 1/2" diameter		540	.015		1.44	.70		2.14	2.65
6050	3/4" diameter		440	.018		1.60	.85		2.45	3.07
6100	1" diameter		420	.019		1.75	.90		2.65	3.31
6150	1-1/4" diameter		400	.020		2	.94		2.94	3.65
6200	1-1/2" diameter		400	.020		2.40	.94		3.34	4.09
6250	2" diameter		267	.030		2.60	1.41		4.01	5.05
6300	2-1/2" diameter		267	.030		3	1.41		4.41	5.45
6350	3" diameter		160	.050		3.30	2.35		5.65	7.25
6400	3-1/2" diameter		133	.060		4.40	2.83		7.23	9.20
6450	4" diameter		100	.080		4.60	3.76		8.36	10.85
6500	5" diameter		80	.100		10.30	4.70		15	18.60
6550	6" diameter		60	.133		11.70	6.25		17.95	22.50
6600	EMT, 1/2" diameter		540	.015		1.45	.70		2.15	2.67
6650	3/4" diameter		440	.018		1.55	.85		2.40	3.02
6700	1" diameter		420	.019		1.73	.90		2.63	3.28
6750	1-1/4" diameter		400	.020		1.93	.94		2.87	3.57
6800	1-1/2" diameter		400	.020		2.40	.94		3.34	4.09
6850	2" diameter		267	.030		2.55	1.41		3.96	4.98
6900	2-1/2" diameter		267	.030		2.90	1.41		4.31	5.35
6950	3" diameter		160	.050		3.10	2.35		5.45	7.05
6970	3-1/2" diameter		133	.060		3.50	2.83		6.33	8.20
6990	4" diameter		100	.080		4	3.76		7.76	10.20
7000	Clip, 1 hole for rigid conduit, 1/2" diameter		500	.016		.70	.75		1.45	1.93
7050	3/4" diameter		470	.017		1.02	.80		1.82	2.35
7100	1" diameter		440	.018		1.27	.85		2.12	2.71
7150	1-1/4" diameter		400	.020		2.75	.94		3.69	4.48
7200	1-1/2" diameter		355	.023		3.15	1.06		4.21	5.10
7250	2" diameter		320	.025		6.15	1.18		7.33	8.55
7300	2-1/2" diameter		266	.030		13.70	1.41		15.11	17.20
7350	3" diameter		160	.050		19.60	2.35		21.95	25
7400	3-1/2" diameter		133	.060		28	2.83		30.83	35
7450	4" diameter		100	.080		62.50	3.76		66.26	75
7500	5" diameter		80	.100		196	4.70		200.70	223
7550	6" diameter		60	.133		211	6.25		217.25	242
7820	Conduit hangers, with bolt & 12" rod, 1/2" diameter		150	.053		3.52	2.51		6.03	7.75
7830	3/4" diameter		145	.055		3.54	2.59		6.13	7.90
7840	1" diameter		135	.059		3.94	2.79		6.73	8.60
7850	1-1/4" diameter		120	.067		4.38	3.13		7.51	9.65
7860	1-1/2" diameter		110	.073		4.72	3.42		8.14	10.45
7870	2" diameter		100	.080		7.10	3.76		10.86	13.60
7880	2-1/2" diameter		80	.100		7.70	4.70		12.40	15.70
7890	3" diameter		60	.133		8.15	6.25		14.40	18.60
7900	3-1/2" diameter		45	.178		9.55	8.35		17.90	23.50
7910	4" diameter		35	.229		15.15	10.75		25.90	33
7920	5" diameter		30	.267		16.35	12.55		28.90	37.50
7930	6" diameter	↓	25	.320	↓	30	15.05		45.05	56

26 05 29.20 Hangers		Crew	Daily Output	Labor-Hours	Unit	Material	Labor	Equipment	Total	Total Incl O&P
						2009 Bare Costs				
7950	Jay clamp, 1/2" diameter	1 Elec	32	.250	Ea.	5.20	11.75		16.95	24
7960	3/4" diameter		32	.250		5.20	11.75		16.95	24
7970	1" diameter		32	.250		5.20	11.75		16.95	24
7980	1-1/4" diameter		30	.267		7.55	12.55		20.10	27.50
7990	1-1/2" diameter		30	.267		7.55	12.55		20.10	27.50
8000	2" diameter		30	.267		9.40	12.55		21.95	29.50
8010	2-1/2" diameter		28	.286		13.40	13.45		26.85	35.50
8020	3" diameter		28	.286		18.20	13.45		31.65	40.50
8030	3-1/2" diameter		25	.320		23	15.05		38.05	48.50
8040	4" diameter		25	.320		40.50	15.05		55.55	68
8050	5" diameter		20	.400		103	18.80		121.80	142
8060	6" diameter		16	.500		194	23.50		217.50	249
8070	Channels, 3/4" x 1-1/2" w/12" rods for 1/2" to 1" conduit		30	.267		13.05	12.55		25.60	33.50
8080	1-1/2" x 1-1/2" w/12" rods for 1-1/4" to 2" conduit		28	.286		14.20	13.45		27.65	36
8090	1-1/2" x 1-1/2" w/12" rods for 2-1/2" to 4" conduit		26	.308		17.65	14.45		32.10	42
8100	1-1/2" x 1-7/8" w/12" rods for 5" to 6" conduit		24	.333		52	15.65		67.65	81
8110	Beam clamp, conduit, plastic coated steel, 1/2" diam.		30	.267		20	12.55		32.55	41.50
8120	3/4" diameter		30	.267		21	12.55		33.55	42.50
8130	1" diameter		30	.267		21.50	12.55		34.05	43
8140	1-1/4" diameter		28	.286		29	13.45		42.45	52.50
8150	1-1/2" diameter		28	.286		35.50	13.45		48.95	59.50
8160	2" diameter		28	.286		46	13.45		59.45	71.50
8170	2-1/2" diameter		26	.308		50.50	14.45		64.95	78.50
8180	3" diameter		26	.308		57	14.45		71.45	85
8190	3-1/2" diameter		23	.348		58.50	16.35		74.85	89.50
8200	4" diameter		23	.348		64	16.35		80.35	95.50
8210	5" diameter		18	.444		188	21		209	239
8220	Channels, plastic coated									
8250	3/4" x 1-1/2", w/12" rods for 1/2" to 1" conduit	1 Elec	28	.286	Ea.	30	13.45		43.45	53.50
8260	1-1/2" x 1-1/2", w/12" rods for 1-1/4" to 2" conduit		26	.308		34	14.45		48.45	60
8270	1-1/2" x 1-1/2", w/12" rods for 2-1/2" to 3-1/2" conduit		24	.333		37	15.65		52.65	64.50
8280	1-1/2" x 1-7/8", w/12" rods for 4" to 5" conduit		22	.364		61.50	17.10		78.60	94
8290	1-1/2" x 1-7/8", w/12" rods for 6" conduit		20	.400		67.50	18.80		86.30	103
8320	Conduit hangers, plastic coated steel, with bolt & 12" rod, 1/2" diam.		140	.057		17.40	2.69		20.09	23.50
8330	3/4" diameter		135	.059		17.85	2.79		20.64	24
8340	1" diameter		125	.064		18.25	3.01		21.26	24.50
8350	1-1/4" diameter		110	.073		19.35	3.42		22.77	27
8360	1-1/2" diameter		100	.080		21.50	3.76		25.26	30
8370	2" diameter		90	.089		24	4.18		28.18	33
8380	2-1/2" diameter		70	.114		28.50	5.35		33.85	40
8390	3" diameter		50	.160		34.50	7.50		42	49.50
8400	3-1/2" diameter		35	.229		36.50	10.75		47.25	56.50
8410	4" diameter		25	.320		51	15.05		66.05	79.50
8420	5" diameter		20	.400		56	18.80		74.80	90.50
9000	Parallel type, conduit beam clamp, 1/2"		32	.250		4.50	11.75		16.25	23
9010	3/4"		32	.250		4.95	11.75		16.70	23.50
9020	1"		32	.250		5.30	11.75		17.05	24
9030	1-1/4"		30	.267		7.25	12.55		19.80	27.50
9040	1-1/2"		30	.267		8.25	12.55		20.80	28.50
9050	2"		30	.267		10.70	12.55		23.25	31
9060	2-1/2"		28	.286		14.20	13.45		27.65	36
9070	3"		28	.286		18.10	13.45		31.55	40.50
9090	4"		25	.320		23	15.05		38.05	48.50

26 05 Common Work Results for Electrical

26 05 29 – Hangers and Supports for Electrical Systems

	26 05 29.20 Hangers	Crew	Daily Output	Labor-Hours	Unit	2009 Bare Costs Material	Labor	Equipment	Total	Total Incl O&P
9110	Right angle, conduit beam clamp, 1/2"	1 Elec	32	.250	Ea.	3	11.75		14.75	21.50
9120	3/4"		32	.250		3.10	11.75		14.85	21.50
9130	1"		32	.250		3.40	11.75		15.15	22
9140	1-1/4"		30	.267		4.20	12.55		16.75	24
9150	1-1/2"		30	.267		4.70	12.55		17.25	24.50
9160	2"		30	.267		6.85	12.55		19.40	27
9170	2-1/2"		28	.286		8.50	13.45		21.95	30
9180	3"		28	.286		9.80	13.45		23.25	31.50
9190	3-1/2"		25	.320		12.25	15.05		27.30	36.50
9200	4"		25	.320		13.30	15.05		28.35	37.50
9230	Adjustable, conduit hanger, 1/2"		32	.250		3.70	11.75		15.45	22
9240	3/4"		32	.250		3.45	11.75		15.20	22
9250	1"		32	.250		4.25	11.75		16	23
9260	1-1/4"		30	.267		5.10	12.55		17.65	25
9270	1-1/2"		30	.267		5.85	12.55		18.40	26
9280	2"		30	.267		6.30	12.55		18.85	26.50
9290	2-1/2"		28	.286		7.85	13.45		21.30	29
9300	3"		28	.286		8.30	13.45		21.75	29.50
9310	3-1/2"		25	.320		10.70	15.05		25.75	35
9320	4"		25	.320		18.75	15.05		33.80	43.50
9330	5"		20	.400		24.50	18.80		43.30	56
9340	6"		16	.500		30.50	23.50		54	69.50
9350	Combination conduit hanger, 3/8"		32	.250		8.55	11.75		20.30	27.50
9360	Adjustable flange 3/8"		32	.250		10.05	11.75		21.80	29

26 05 33 – Raceway and Boxes for Electrical Systems

	26 05 33.05 Conduit	Crew	Daily Output	Labor-Hours	Unit	Material	Labor	Equipment	Total	Total Incl O&P
0010	**CONDUIT** To 15' high, includes 2 terminations, 2 elbows, R260533-20									
0020	11 beam clamps, and 11 couplings per 100 L.F.									
0300	Aluminum, 1/2" diameter	1 Elec	100	.080	L.F.	2.30	3.76		6.06	8.35
0500	3/4" diameter		90	.089		3.13	4.18		7.31	9.90
0700	1" diameter		80	.100		4.13	4.70		8.83	11.80
1000	1-1/4" diameter		70	.114		5.55	5.35		10.90	14.35
1030	1-1/2" diameter		65	.123		6.70	5.80		12.50	16.25
1050	2" diameter		60	.133		9.10	6.25		15.35	19.70
1070	2-1/2" diameter		50	.160		14.95	7.50		22.45	28
1100	3" diameter	2 Elec	90	.178		20.50	8.35		28.85	35.50
1130	3-1/2" diameter		80	.200		27	9.40		36.40	44.50
1140	4" diameter		70	.229		33	10.75		43.75	52.50
1150	5" diameter		50	.320		71.50	15.05		86.55	102
1160	6" diameter		40	.400		106	18.80		124.80	145
1161	Field bends, 45° to 90°, 1/2" diameter	1 Elec	53	.151	Ea.		7.10		7.10	10.90
1162	3/4" diameter		47	.170			8		8	12.30
1163	1" diameter		44	.182			8.55		8.55	13.15
1164	1-1/4" diameter		23	.348			16.35		16.35	25
1165	1-1/2" diameter		21	.381			17.90		17.90	27.50
1166	2" diameter		16	.500			23.50		23.50	36
1170	Elbows, 1/2" diameter		40	.200		13.15	9.40		22.55	29
1200	3/4" diameter		32	.250		17.90	11.75		29.65	38
1230	1" diameter		28	.286		25	13.45		38.45	48
1250	1-1/4" diameter		24	.333		39.50	15.65		55.15	67.50
1270	1-1/2" diameter		20	.400		53	18.80		71.80	87
1300	2" diameter		16	.500		77.50	23.50		101	122

26 05 33.05 Conduit		Crew	Daily Output	Labor-Hours	Unit	2009 Bare Costs Material	Labor	Equipment	Total	Total Incl O&P
1330	2-1/2" diameter	1 Elec	12	.667	Ea.	131	31.50		162.50	192
1350	3" diameter		8	1		202	47		249	296
1370	3-1/2" diameter		6	1.333		315	62.50		377.50	440
1400	4" diameter		5	1.600		375	75		450	525
1410	5" diameter		4	2		1,100	94		1,194	1,375
1420	6" diameter		2.50	3.200		1,525	150		1,675	1,900
1430	Couplings, 1/2" diameter					4.34			4.34	4.77
1450	3/4" diameter					6.60			6.60	7.25
1470	1" diameter					8.70			8.70	9.55
1500	1-1/4" diameter					10.55			10.55	11.65
1530	1-1/2" diameter					12.25			12.25	13.50
1550	2" diameter					17.35			17.35	19.10
1570	2-1/2" diameter					39			39	43
1600	3" diameter					51			51	56
1630	3-1/2" diameter					70			70	77
1650	4" diameter					84.50			84.50	93
1670	5" diameter					248			248	273
1690	6" diameter					390			390	425
1750	Rigid galvanized steel, 1/2" diameter	1 Elec	90	.089	L.F.	2.96	4.18		7.14	9.70
1770	3/4" diameter		80	.100		3.23	4.70		7.93	10.80
1800	1" diameter		65	.123		4.44	5.80		10.24	13.80
1830	1-1/4" diameter		60	.133		6.15	6.25		12.40	16.40
1850	1-1/2" diameter		55	.145		7.15	6.85		14	18.40
1870	2" diameter		45	.178		9.25	8.35		17.60	23
1900	2-1/2" diameter		35	.229		17.35	10.75		28.10	35.50
1930	3" diameter	2 Elec	50	.320		21.50	15.05		36.55	46.50
1950	3-1/2" diameter		44	.364		26.50	17.10		43.60	55.50
1970	4" diameter		40	.400		31	18.80		49.80	63
1980	5" diameter		30	.533		61	25		86	106
1990	6" diameter		20	.800		91.50	37.50		129	159
1991	Field bends, 45° to 90°, 1/2" diameter	1 Elec	44	.182	Ea.		8.55		8.55	13.15
1992	3/4" diameter		40	.200			9.40		9.40	14.45
1993	1" diameter		36	.222			10.45		10.45	16.05
1994	1-1/4" diameter		19	.421			19.80		19.80	30.50
1995	1-1/2" diameter		18	.444			21		21	32
1996	2" diameter		13	.615			29		29	44.50
2000	Elbows, 1/2" diameter		32	.250		12.80	11.75		24.55	32
2030	3/4" diameter		28	.286		13.30	13.45		26.75	35
2050	1" diameter		24	.333		20.50	15.65		36.15	46.50
2070	1-1/4" diameter		18	.444		28	21		49	63
2100	1-1/2" diameter		16	.500		34.50	23.50		58	74
2130	2" diameter		12	.667		50	31.50		81.50	103
2150	2-1/2" diameter		8	1		93.50	47		140.50	176
2170	3" diameter		6	1.333		130	62.50		192.50	240
2200	3-1/2" diameter		4.20	1.905		205	89.50		294.50	365
2220	4" diameter		4	2		232	94		326	400
2230	5" diameter		3.50	2.286		645	107		752	875
2240	6" diameter		2	4		970	188		1,158	1,375
2250	Couplings, 1/2" diameter					2.95			2.95	3.25
2270	3/4" diameter					3.62			3.62	3.98
2300	1" diameter					5.35			5.35	5.90
2330	1-1/4" diameter					6.70			6.70	7.40
2350	1-1/2" diameter					8.50			8.50	9.35

26 05 33.05 Conduit		Crew	Daily Output	Labor-Hours	Unit	Material	2009 Bare Costs Labor	Equipment	Total	Total Incl O&P
2370	2" diameter				Ea.	11.20			11.20	12.35
2400	2-1/2" diameter					27.50			27.50	30.50
2430	3" diameter					35.50			35.50	39.50
2450	3-1/2" diameter					48			48	53
2470	4" diameter					48			48	53
2480	5" diameter					111			111	122
2490	6" diameter					155			155	170
2500	Steel, intermediate conduit (IMC), 1/2" diameter	1 Elec	100	.080	L.F.	2.04	3.76		5.80	8.05
2530	3/4" diameter		90	.089		2.44	4.18		6.62	9.15
2550	1" diameter		70	.114		3.41	5.35		8.76	12
2570	1-1/4" diameter		65	.123		4.47	5.80		10.27	13.80
2600	1-1/2" diameter		60	.133		5.40	6.25		11.65	15.60
2630	2" diameter		50	.160		6.90	7.50		14.40	19.15
2650	2-1/2" diameter		40	.200		13.40	9.40		22.80	29
2670	3" diameter	2 Elec	60	.267		17.85	12.55		30.40	39
2700	3-1/2" diameter		54	.296		23	13.95		36.95	46.50
2730	4" diameter		50	.320		25	15.05		40.05	50.50
2731	Field bends, 45° to 90°, 1/2" diameter	1 Elec	44	.182	Ea.		8.55		8.55	13.15
2732	3/4" diameter		40	.200			9.40		9.40	14.45
2733	1" diameter		36	.222			10.45		10.45	16.05
2734	1-1/4" diameter		19	.421			19.80		19.80	30.50
2735	1-1/2" diameter		18	.444			21		21	32
2736	2" diameter		13	.615			29		29	44.50
2750	Elbows, 1/2" diameter		32	.250		12.05	11.75		23.80	31.50
2770	3/4" diameter		28	.286		12.60	13.45		26.05	34.50
2800	1" diameter		24	.333		19.30	15.65		34.95	45
2830	1-1/4" diameter		18	.444		27	21		48	61.50
2850	1-1/2" diameter		16	.500		33	23.50		56.50	72
2870	2" diameter		12	.667		47.50	31.50		79	100
2900	2-1/2" diameter		8	1		88.50	47		135.50	170
2930	3" diameter		6	1.333		122	62.50		184.50	232
2950	3-1/2" diameter		4.20	1.905		194	89.50		283.50	350
2970	4" diameter		4	2		219	94		313	385
3000	Couplings, 1/2" diameter					2.95			2.95	3.25
3030	3/4" diameter					3.62			3.62	3.98
3050	1" diameter					5.35			5.35	5.90
3070	1-1/4" diameter					6.70			6.70	7.40
3100	1-1/2" diameter					8.50			8.50	9.35
3130	2" diameter					11.20			11.20	12.35
3150	2-1/2" diameter					27.50			27.50	30.50
3170	3" diameter					35.50			35.50	39.50
3200	3-1/2" diameter					48			48	53
3230	4" diameter					48			48	53
4100	Rigid steel, plastic coated, 40 mil. thick									
4130	1/2" diameter	1 Elec	80	.100	L.F.	7.65	4.70		12.35	15.65
4150	3/4" diameter		70	.114		8.50	5.35		13.85	17.60
4170	1" diameter		55	.145		10.35	6.85		17.20	22
4200	1-1/4" diameter		50	.160		13.20	7.50		20.70	26
4230	1-1/2" diameter		45	.178		16.10	8.35		24.45	30.50
4250	2" diameter		35	.229		21	10.75		31.75	40
4270	2-1/2" diameter		25	.320		32	15.05		47.05	58.50
4300	3" diameter	2 Elec	44	.364		40	17.10		57.10	70.50
4330	3-1/2" diameter		40	.400		48.50	18.80		67.30	82.50

26 05 33.05 Conduit		Crew	Daily Output	Labor-Hours	Unit	2009 Bare Costs Material	Labor	Equipment	Total	Total Incl O&P
4350	4" diameter	2 Elec	36	.444	L.F.	56.50	21		77.50	94
4370	5" diameter		30	.533		121	25		146	172
4400	Elbows, 1/2" diameter	1 Elec	28	.286	Ea.	18.40	13.45		31.85	41
4430	3/4" diameter		24	.333		19.10	15.65		34.75	45
4450	1" diameter		18	.444		22	21		43	56
4470	1-1/4" diameter		16	.500		27	23.50		50.50	65.50
4500	1-1/2" diameter		12	.667		33	31.50		64.50	84.50
4530	2" diameter		8	1		46	47		93	124
4550	2-1/2" diameter		6	1.333		87	62.50		149.50	193
4570	3" diameter		4.20	1.905		139	89.50		228.50	291
4600	3-1/2" diameter		4	2		183	94		277	345
4630	4" diameter		3.80	2.105		199	99		298	370
4650	5" diameter		3.50	2.286		480	107		587	690
4680	Couplings, 1/2" diameter					5.25			5.25	5.80
4700	3/4" diameter					5.50			5.50	6.05
4730	1" diameter					7.15			7.15	7.85
4750	1-1/4" diameter					8.35			8.35	9.15
4770	1-1/2" diameter					9.90			9.90	10.90
4800	2" diameter					14.50			14.50	15.95
4830	2-1/2" diameter					36			36	39.50
4850	3" diameter					43.50			43.50	48
4870	3-1/2" diameter					56			56	61.50
4900	4" diameter					65.50			65.50	72
4950	5" diameter					213			213	234
5000	Electric metallic tubing (EMT), 1/2" diameter	1 Elec	170	.047	L.F.	.66	2.21		2.87	4.12
5020	3/4" diameter		130	.062		1.05	2.89		3.94	5.60
5040	1" diameter		115	.070		1.84	3.27		5.11	7.05
5060	1-1/4" diameter		100	.080		2.95	3.76		6.71	9.05
5080	1-1/2" diameter		90	.089		3.78	4.18		7.96	10.60
5100	2" diameter		80	.100		4.88	4.70		9.58	12.60
5120	2-1/2" diameter		60	.133		11.70	6.25		17.95	22.50
5140	3" diameter	2 Elec	100	.160		13.75	7.50		21.25	26.50
5160	3-1/2" diameter		90	.178		17.40	8.35		25.75	32
5180	4" diameter		80	.200		18.80	9.40		28.20	35
5200	Field bends, 45° to 90°, 1/2" diameter	1 Elec	89	.090	Ea.		4.22		4.22	6.50
5220	3/4" diameter		80	.100			4.70		4.70	7.25
5240	1" diameter		73	.110			5.15		5.15	7.90
5260	1-1/4" diameter		38	.211			9.90		9.90	15.20
5280	1-1/2" diameter		36	.222			10.45		10.45	16.05
5300	2" diameter		26	.308			14.45		14.45	22.50
5320	Offsets, 1/2" diameter		65	.123			5.80		5.80	8.90
5340	3/4" diameter		62	.129			6.05		6.05	9.35
5360	1" diameter		53	.151			7.10		7.10	10.90
5380	1-1/4" diameter		30	.267			12.55		12.55	19.30
5400	1-1/2" diameter		28	.286			13.45		13.45	20.50
5420	2" diameter		20	.400			18.80		18.80	29
5700	Elbows, 1" diameter		40	.200		8.20	9.40		17.60	23.50
5720	1-1/4" diameter		32	.250		10.15	11.75		21.90	29.50
5740	1-1/2" diameter		24	.333		11.85	15.65		27.50	37
5760	2" diameter		20	.400		17.40	18.80		36.20	48
5780	2-1/2" diameter		12	.667		42.50	31.50		74	94.50
5800	3" diameter		9	.889		63	42		105	134
5820	3-1/2" diameter		7	1.143		84.50	53.50		138	176

26 05 33.05 Conduit		Crew	Daily Output	Labor-Hours	Unit	Material	2009 Bare Costs Labor	Equipment	Total	Total Incl O&P
5840	4" diameter	1 Elec	6	1.333	Ea.	99.50	62.50		162	206
6200	Couplings, set screw, steel, 1/2" diameter					1.70			1.70	1.87
6220	3/4" diameter					2.56			2.56	2.82
6240	1" diameter					4.28			4.28	4.71
6260	1-1/4" diameter					9			9	9.90
6280	1-1/2" diameter					12.50			12.50	13.75
6300	2" diameter					16.75			16.75	18.45
6320	2-1/2" diameter					53.50			53.50	59
6340	3" diameter					57.50			57.50	63
6360	3-1/2" diameter					63.50			63.50	70
6380	4" diameter					68.50			68.50	75.50
6500	Box connectors, set screw, steel, 1/2" diameter	1 Elec	120	.067		1.10	3.13		4.23	6.05
6520	3/4" diameter		110	.073		1.79	3.42		5.21	7.20
6540	1" diameter		90	.089		3.36	4.18		7.54	10.15
6560	1-1/4" diameter		70	.114		6.60	5.35		11.95	15.50
6580	1-1/2" diameter		60	.133		9.50	6.25		15.75	20
6600	2" diameter		50	.160		13.15	7.50		20.65	26
6620	2-1/2" diameter		36	.222		46	10.45		56.45	66.50
6640	3" diameter		27	.296		55.50	13.95		69.45	82.50
6680	3-1/2" diameter		21	.381		77	17.90		94.90	112
6700	4" diameter		16	.500		86	23.50		109.50	131
6740	Insulated box connectors, set screw, steel, 1/2" diameter		120	.067		1.51	3.13		4.64	6.50
6760	3/4" diameter		110	.073		2.38	3.42		5.80	7.85
6780	1" diameter		90	.089		4.38	4.18		8.56	11.25
6800	1-1/4" diameter		70	.114		8	5.35		13.35	17.05
6820	1-1/2" diameter		60	.133		11.25	6.25		17.50	22
6840	2" diameter		50	.160		16.35	7.50		23.85	29.50
6860	2-1/2" diameter		36	.222		78	10.45		88.45	102
6880	3" diameter		27	.296		91.50	13.95		105.45	123
6900	3-1/2" diameter		21	.381		128	17.90		145.90	169
6920	4" diameter		16	.500		140	23.50		163.50	190
7000	EMT to conduit adapters, 1/2" diameter (compression)		70	.114		4.68	5.35		10.03	13.40
7020	3/4" diameter		60	.133		6.65	6.25		12.90	17
7040	1" diameter		50	.160		10.10	7.50		17.60	22.50
7060	1-1/4" diameter		40	.200		17.55	9.40		26.95	34
7080	1-1/2" diameter		30	.267		21.50	12.55		34.05	43.50
7100	2" diameter		25	.320		31	15.05		46.05	57.50
7200	EMT to Greenfield adapters, 1/2" to 3/8" diameter (compression)		90	.089		3.15	4.18		7.33	9.90
7220	1/2" diameter		90	.089		5.95	4.18		10.13	13
7240	3/4" diameter		80	.100		7.60	4.70		12.30	15.60
7260	1" diameter		70	.114		20	5.35		25.35	30.50
7270	1-1/4" diameter		60	.133		25.50	6.25		31.75	37.50
7280	1-1/2" diameter		50	.160		28.50	7.50		36	43
7290	2" diameter		40	.200		42.50	9.40		51.90	61.50
7400	EMT,LB, LR or LL fittings with covers, 1/2" dia, set screw		24	.333		11.20	15.65		26.85	36.50
7420	3/4" diameter		20	.400		13.80	18.80		32.60	44
7440	1" diameter		16	.500		21	23.50		44.50	59.50
7450	1-1/4" diameter		13	.615		31.50	29		60.50	79
7460	1-1/2" diameter		11	.727		37.50	34		71.50	93.50
7470	2" diameter		9	.889		59	42		101	130
7600	EMT, "T" fittings with covers, 1/2" diameter, set screw		16	.500		14.45	23.50		37.95	52
7620	3/4" diameter		15	.533		18.30	25		43.30	58.50
7640	1" diameter		12	.667		26.50	31.50		58	77.50

26 05 33 – Raceway and Boxes for Electrical Systems

26 05 33.05 Conduit		Crew	Daily Output	Labor-Hours	Unit	Material	Labor	Equipment	Total	Total Incl O&P
						2009 Bare Costs				
7650	1-1/4" diameter	1 Elec	11	.727	Ea.	38	34		72	94
7660	1-1/2" diameter		10	.800		47	37.50		84.50	110
7670	2" diameter		8	1		62	47		109	141
8000	EMT, expansion fittings, no jumper, 1/2" diameter		24	.333		46.50	15.65		62.15	75
8020	3/4" diameter		20	.400		54	18.80		72.80	88.50
8040	1" diameter		16	.500		63.50	23.50		87	106
8060	1-1/4" diameter		13	.615		86	29		115	139
8080	1-1/2" diameter		11	.727		118	34		152	183
8100	2" diameter		9	.889		176	42		218	259
8110	2-1/2" diameter		7	1.143		278	53.50		331.50	390
8120	3" diameter		6	1.333		340	62.50		402.50	470
8140	4" diameter		5	1.600		595	75		670	770
8200	Split adapter, 1/2" diameter		110	.073		3.30	3.42		6.72	8.90
8210	3/4" diameter		90	.089		2.95	4.18		7.13	9.70
8220	1" diameter		70	.114		4.10	5.35		9.45	12.75
8230	1-1/4" diameter		60	.133		6.55	6.25		12.80	16.85
8240	1-1/2" diameter		50	.160		10.10	7.50		17.60	22.50
8250	2" diameter		36	.222		29.50	10.45		39.95	48.50
8300	1 hole clips, 1/2" diameter		500	.016		.43	.75		1.18	1.63
8320	3/4" diameter		470	.017		.57	.80		1.37	1.86
8340	1" diameter		444	.018		1.04	.85		1.89	2.44
8360	1-1/4" diameter		400	.020		1.39	.94		2.33	2.98
8380	1-1/2" diameter		355	.023		2.13	1.06		3.19	3.97
8400	2" diameter		320	.025		3.34	1.18		4.52	5.50
8420	2-1/2" diameter		266	.030		6.20	1.41		7.61	8.95
8440	3" diameter		160	.050		7.80	2.35		10.15	12.20
8460	3-1/2" diameter		133	.060		12.30	2.83		15.13	17.90
8480	4" diameter		100	.080		15.90	3.76		19.66	23.50
8500	Clamp back spacers, 1/2" diameter		500	.016		1.30	.75		2.05	2.59
8510	3/4" diameter		470	.017		1.55	.80		2.35	2.94
8520	1" diameter		444	.018		2.70	.85		3.55	4.27
8530	1-1/4" diameter		400	.020		4.80	.94		5.74	6.75
8540	1-1/2" diameter		355	.023		5.40	1.06		6.46	7.60
8550	2" diameter		320	.025		9.15	1.18		10.33	11.85
8560	2-1/2" diameter		266	.030		17.60	1.41		19.01	21.50
8570	3" diameter		160	.050		25	2.35		27.35	31
8580	3-1/2" diameter		133	.060		34	2.83		36.83	42
8590	4" diameter		100	.080		74	3.76		77.76	87.50
8600	Offset connectors, 1/2" diameter		40	.200		5.90	9.40		15.30	21
8610	3/4" diameter		32	.250		8.50	11.75		20.25	27.50
8620	1" diameter		24	.333		9.30	15.65		24.95	34.50
8650	90° pulling elbows, female, 1/2" diameter, with gasket		24	.333		5.30	15.65		20.95	30
8660	3/4" diameter		20	.400		8.20	18.80		27	38
8700	Couplings, compression, 1/2" diameter, steel					3.45			3.45	3.80
8710	3/4" diameter					4.80			4.80	5.30
8720	1" diameter					8.30			8.30	9.15
8730	1-1/4" diameter					15.60			15.60	17.15
8740	1-1/2" diameter					23			23	25
8750	2" diameter					31			31	34
8760	2-1/2" diameter					139			139	153
8770	3" diameter					174			174	191
8780	3-1/2" diameter					282			282	310
8790	4" diameter					289			289	320

26 05 33.05 Conduit		Crew	Daily Output	Labor-Hours	Unit	2009 Bare Costs Material	Labor	Equipment	Total	Total Incl O&P
8800	Box connectors, compression, 1/2″ diam., steel	1 Elec	120	.067	Ea.	2.70	3.13		5.83	7.80
8810	3/4″ diameter		110	.073		3.95	3.42		7.37	9.60
8820	1″ diameter		90	.089		6.45	4.18		10.63	13.55
8830	1-1/4″ diameter		70	.114		12.75	5.35		18.10	22.50
8840	1-1/2″ diameter		60	.133		18.60	6.25		24.85	30
8850	2″ diameter		50	.160		26.50	7.50		34	40.50
8860	2-1/2″ diameter		36	.222		83.50	10.45		93.95	108
8870	3″ diameter		27	.296		105	13.95		118.95	138
8880	3-1/2″ diameter		21	.381		159	17.90		176.90	203
8890	4″ diameter		16	.500		176	23.50		199.50	229
8900	Box connectors, insulated compression, 1/2″ diam., steel		120	.067		3.48	3.13		6.61	8.65
8910	3/4″ diameter		110	.073		4.78	3.42		8.20	10.50
8920	1″ diameter		90	.089		7.95	4.18		12.13	15.20
8930	1-1/4″ diameter		70	.114		16.55	5.35		21.90	26.50
8940	1-1/2″ diameter		60	.133		25	6.25		31.25	37
8950	2″ diameter		50	.160		34.50	7.50		42	49.50
8960	2-1/2″ diameter		36	.222		103	10.45		113.45	130
8970	3″ diameter		27	.296		141	13.95		154.95	177
8980	3-1/2″ diameter		21	.381		232	17.90		249.90	284
8990	4″ diameter		16	.500		209	23.50		232.50	265
9100	PVC, schedule 40, 1/2″ diameter		190	.042	L.F.	1.19	1.98		3.17	4.35
9110	3/4″ diameter		145	.055		1.46	2.59		4.05	5.60
9120	1″ diameter		125	.064		2.42	3.01		5.43	7.30
9130	1-1/4″ diameter		110	.073		3.21	3.42		6.63	8.80
9140	1-1/2″ diameter		100	.080		3.68	3.76		7.44	9.85
9150	2″ diameter		90	.089		4.67	4.18		8.85	11.60
9160	2-1/2″ diameter		65	.123		8.15	5.80		13.95	17.90
9170	3″ diameter	2 Elec	110	.145		9.15	6.85		16	20.50
9180	3-1/2″ diameter		100	.160		12.35	7.50		19.85	25
9190	4″ diameter		90	.178		13.20	8.35		21.55	27.50
9200	5″ diameter		70	.229		18.65	10.75		29.40	37
9210	6″ diameter		60	.267		25.50	12.55		38.05	47.50
9220	Elbows, 1/2″ diameter	1 Elec	50	.160	Ea.	1.58	7.50		9.08	13.30
9230	3/4″ diameter		42	.190		1.59	8.95		10.54	15.50
9240	1″ diameter		35	.229		2.69	10.75		13.44	19.50
9250	1-1/4″ diameter		28	.286		3.83	13.45		17.28	24.50
9260	1-1/2″ diameter		20	.400		5.20	18.80		24	34.50
9270	2″ diameter		16	.500		7.55	23.50		31.05	44.50
9280	2-1/2″ diameter		11	.727		13.15	34		47.15	67
9290	3″ diameter		9	.889		23	42		65	90
9300	3-1/2″ diameter		7	1.143		32	53.50		85.50	118
9310	4″ diameter		6	1.333		40	62.50		102.50	141
9320	5″ diameter		4	2		70	94		164	222
9330	6″ diameter		3	2.667		119	125		244	325
9340	Field bends, 45° & 90°, 1/2″ diameter		45	.178			8.35		8.35	12.85
9350	3/4″ diameter		40	.200			9.40		9.40	14.45
9360	1″ diameter		35	.229			10.75		10.75	16.55
9370	1-1/4″ diameter		32	.250			11.75		11.75	18.10
9380	1-1/2″ diameter		27	.296			13.95		13.95	21.50
9390	2″ diameter		20	.400			18.80		18.80	29
9400	2-1/2″ diameter		16	.500			23.50		23.50	36
9410	3″ diameter		13	.615			29		29	44.50
9420	3-1/2″ diameter		12	.667			31.50		31.50	48

26 05 33 – Raceway and Boxes for Electrical Systems

26 05 33.05 Conduit		Crew	Daily Output	Labor-Hours	Unit	Material	2009 Bare Costs Labor	Equipment	Total	Total Incl O&P
9430	4″ diameter	1 Elec	10	.800	Ea.		37.50		37.50	58
9440	5″ diameter		9	.889			42		42	64.50
9450	6″ diameter		8	1			47		47	72.50
9460	PVC adapters, 1/2″ diameter		50	.160		.51	7.50		8.01	12.10
9470	3/4″ diameter		42	.190		.94	8.95		9.89	14.80
9480	1″ diameter		38	.211		1.19	9.90		11.09	16.50
9490	1-1/4″ diameter		35	.229		1.59	10.75		12.34	18.30
9500	1-1/2″ diameter		32	.250		1.85	11.75		13.60	20
9510	2″ diameter		27	.296		2.66	13.95		16.61	24.50
9520	2-1/2″ diameter		23	.348		4.57	16.35		20.92	30
9530	3″ diameter		18	.444		6.75	21		27.75	39.50
9540	3-1/2″ diameter		13	.615		8.70	29		37.70	54
9550	4″ diameter		11	.727		11.80	34		45.80	65.50
9560	5″ diameter		8	1		23	47		70	98
9570	6″ diameter		6	1.333		28	62.50		90.50	128
9580	PVC-LB, LR or LL fittings & covers									
9590	1/2″ diameter	1 Elec	20	.400	Ea.	7.10	18.80		25.90	37
9600	3/4″ diameter		16	.500		9.15	23.50		32.65	46
9610	1″ diameter		12	.667		10.05	31.50		41.55	59
9620	1-1/4″ diameter		9	.889		15.30	42		57.30	81.50
9630	1-1/2″ diameter		7	1.143		18.35	53.50		71.85	103
9640	2″ diameter		6	1.333		35	62.50		97.50	135
9650	2-1/2″ diameter		6	1.333		127	62.50		189.50	237
9660	3″ diameter		5	1.600		131	75		206	261
9670	3-1/2″ diameter		4	2		140	94		234	299
9680	4″ diameter		3	2.667		145	125		270	350
9690	PVC-tee fitting & cover									
9700	1/2″	1 Elec	14	.571	Ea.	9.90	27		36.90	52.50
9710	3/4″		13	.615		11.25	29		40.25	57
9720	1″		10	.800		12.60	37.50		50.10	72
9730	1-1/4″		9	.889		20.50	42		62.50	87
9740	1-1/2″		8	1		25	47		72	100
9750	2″		7	1.143		38.50	53.50		92	125
9760	PVC-reducers, 3/4″ x 1/2″ diameter					1.67			1.67	1.84
9770	1″ x 1/2″ diameter					4.05			4.05	4.46
9780	1″ x 3/4″ diameter					4.32			4.32	4.75
9790	1-1/4″ x 3/4″ diameter					5.65			5.65	6.25
9800	1-1/4″ x 1″ diameter					5.65			5.65	6.20
9810	1-1/2″ x 1-1/4″ diameter					5.85			5.85	6.45
9820	2″ x 1-1/4″ diameter					7			7	7.70
9830	2-1/2″ x 2″ diameter					23			23	25
9840	3″ x 2″ diameter					24.50			24.50	27
9850	4″ x 3″ diameter					30.50			30.50	33.50
9860	Cement, quart					26.50			26.50	29.50
9870	Gallon					114			114	125
9880	Heat bender, to 6″ diameter					1,350			1,350	1,475
9900	Add to labor for higher elevated installation									
9910	15′ to 20′ high, add						10%			
9920	20′ to 25′ high, add						20%			
9930	25′ to 30′ high, add						25%			
9940	30′ to 35′ high, add						30%			
9950	35′ to 40′ high, add						35%			
9960	Over 40′ high, add						40%			

26 05 33 – Raceway and Boxes for Electrical Systems

26 05 33.05 Conduit

	26 05 33.05 Conduit	Crew	Daily Output	Labor-Hours	Unit	2009 Bare Costs Material	Labor	Equipment	Total	Total Incl O&P
9990	Minimum labor/equipment charge	1 Elec	4	2	Job		94		94	145

26 05 33.10 Conduit

	26 05 33.10 Conduit	Crew	Daily Output	Labor-Hours	Unit	Material	Labor	Equipment	Total	Total Incl O&P
0010	**CONDUIT** To 15' high, includes couplings only									
0200	Electric metallic tubing, 1/2"diameter	1 Elec	435	.018	L.F.	.49	.86		1.35	1.87
0220	3/4" diameter		253	.032		.86	1.49		2.35	3.24
0240	1" diameter		207	.039		1.44	1.82		3.26	4.37
0260	1-1/4" diameter		173	.046		2.45	2.17		4.62	6.05
0280	1-1/2" diameter		153	.052		3.18	2.46		5.64	7.30
0300	2" diameter		130	.062		4.05	2.89		6.94	8.90
0320	2-1/2" diameter		92	.087		9.50	4.09		13.59	16.75
0340	3" diameter	2 Elec	148	.108		10.90	5.10		16	19.80
0360	3-1/2" diameter		134	.119		13.65	5.60		19.25	23.50
0380	4" diameter		114	.140		14.25	6.60		20.85	26
0500	Steel rigid galvanized, 1/2" diameter	1 Elec	146	.055		2.11	2.58		4.69	6.30
0520	3/4" diameter		125	.064		2.37	3.01		5.38	7.25
0540	1" diameter		93	.086		3.42	4.04		7.46	9.95
0560	1-1/4" diameter		88	.091		4.69	4.27		8.96	11.70
0580	1-1/2" diameter		80	.100		5.55	4.70		10.25	13.35
0600	2" diameter		65	.123		7.10	5.80		12.90	16.70
0620	2-1/2" diameter		48	.167		13.40	7.85		21.25	27
0640	3" diameter	2 Elec	64	.250		16.60	11.75		28.35	36.50
0660	3-1/2" diameter		60	.267		20.50	12.55		33.05	42
0680	4" diameter		52	.308		23	14.45		37.45	48
0700	5" diameter		50	.320		44	15.05		59.05	71
0720	6" diameter		48	.333		63	15.65		78.65	93
1000	Steel intermediate conduit (IMC), 1/2 diameter	1 Elec	155	.052		1.32	2.43		3.75	5.20
1010	3/4" diameter		130	.062		1.63	2.89		4.52	6.25
1020	1" diameter		100	.080		2.45	3.76		6.21	8.50
1030	1-1/4" diameter		93	.086		3.11	4.04		7.15	9.60
1040	1-1/2" diameter		85	.094		3.91	4.42		8.33	11.10
1050	2" diameter		70	.114		5.10	5.35		10.45	13.85
1060	2-1/2" diameter		53	.151		10.70	7.10		17.80	22.50
1070	3" diameter	2 Elec	80	.200		13.75	9.40		23.15	29.50
1080	3-1/2" diameter		70	.229		16.85	10.75		27.60	35
1090	4" diameter		60	.267		18	12.55		30.55	39

26 05 33.20 Conduit Nipples

	26 05 33.20 Conduit Nipples	Crew	Daily Output	Labor-Hours	Unit	Material	Labor	Equipment	Total	Total Incl O&P
0010	**CONDUIT NIPPLES** With locknuts and bushings									
0100	Aluminum, 1/2" diameter, close	1 Elec	36	.222	Ea.	6.50	10.45		16.95	23
0120	1-1/2" long		36	.222		6.65	10.45		17.10	23.50
0140	2" long		36	.222		7	10.45		17.45	24
0160	2-1/2" long		36	.222		7.85	10.45		18.30	24.50
0180	3" long		36	.222		8.15	10.45		18.60	25
0200	3-1/2" long		36	.222		8.65	10.45		19.10	25.50
0220	4" long		36	.222		9.15	10.45		19.60	26
0240	5" long		36	.222		10.15	10.45		20.60	27
0260	6" long		36	.222		10.60	10.45		21.05	27.50
0280	8" long		36	.222		13.45	10.45		23.90	31
0300	10" long		36	.222		15.90	10.45		26.35	33.50
0320	12" long		36	.222		18	10.45		28.45	36
0340	3/4" diameter, close		32	.250		9.15	11.75		20.90	28
0360	1-1/2" long		32	.250		9.35	11.75		21.10	28.50
0380	2" long		32	.250		9.60	11.75		21.35	28.50

26 05 33.20 Conduit Nipples		Crew	Daily Output	Labor-Hours	Unit	2009 Bare Costs Material	Labor	Equipment	Total	Total Incl O&P
0400	2-1/2" long	1 Elec	32	.250	Ea.	10	11.75		21.75	29
0420	3" long		32	.250		10.55	11.75		22.30	29.50
0440	3-1/2" long		32	.250		10.85	11.75		22.60	30
0460	4" long		32	.250		11.25	11.75		23	30.50
0480	5" long		32	.250		12.80	11.75		24.55	32
0500	6" long		32	.250		14.10	11.75		25.85	33.50
0520	8" long		32	.250		17.50	11.75		29.25	37.50
0540	10" long		32	.250		19.95	11.75		31.70	40
0560	12" long		32	.250		23.50	11.75		35.25	44
0580	1" diameter, close		27	.296		12.90	13.95		26.85	35.50
0600	2" long		27	.296		13.70	13.95		27.65	36.50
0620	2-1/2" long		27	.296		14.45	13.95		28.40	37.50
0640	3" long		27	.296		15.10	13.95		29.05	38
0660	3-1/2" long		27	.296		16.35	13.95		30.30	39.50
0680	4" long		27	.296		17.30	13.95		31.25	40.50
0700	5" long		27	.296		19.40	13.95		33.35	43
0720	6" long		27	.296		22	13.95		35.95	45.50
0740	8" long		27	.296		25.50	13.95		39.45	50
0760	10" long		27	.296		31	13.95		44.95	55.50
0780	12" long		27	.296		35.50	13.95		49.45	60.50
0800	1-1/4" diameter, close		23	.348		17.75	16.35		34.10	44.50
0820	2" long		23	.348		17.95	16.35		34.30	45
0840	2-1/2" long		23	.348		18.95	16.35		35.30	46
0860	3" long		23	.348		20.50	16.35		36.85	47.50
0880	3-1/2" long		23	.348		22	16.35		38.35	49.50
0900	4" long		23	.348		23.50	16.35		39.85	50.50
0920	5" long		23	.348		26	16.35		42.35	53.50
0940	6" long		23	.348		29	16.35		45.35	57
0960	8" long		23	.348		34.50	16.35		50.85	63
0980	10" long		23	.348		40.50	16.35		56.85	69.50
1000	12" long		23	.348		46	16.35		62.35	75.50
1020	1-1/2" diameter, close		20	.400		24	18.80		42.80	55.50
1040	2" long		20	.400		24.50	18.80		43.30	56
1060	2-1/2" long		20	.400		25.50	18.80		44.30	57
1080	3" long		20	.400		27	18.80		45.80	59
1100	3-1/2" long		20	.400		31	18.80		49.80	63
1120	4" long		20	.400		31	18.80		49.80	63
1140	5" long		20	.400		34	18.80		52.80	66
1160	6" long		20	.400		37	18.80		55.80	70
1180	8" long		20	.400		44	18.80		62.80	77.50
1200	10" long		20	.400		51	18.80		69.80	85.50
1220	12" long		20	.400		58	18.80		76.80	92.50
1240	2" diameter, close		18	.444		33.50	21		54.50	69
1260	2-1/2" long		18	.444		34.50	21		55.50	70
1280	3" long		18	.444		36.50	21		57.50	72
1300	3-1/2" long		18	.444		39.50	21		60.50	75.50
1320	4" long		18	.444		41	21		62	77
1340	5" long		18	.444		45	21		66	81.50
1360	6" long		18	.444		49.50	21		70.50	86
1380	8" long		18	.444		58	21		79	95.50
1400	10" long		18	.444		66.50	21		87.50	105
1420	12" long		18	.444		75.50	21		96.50	115
1440	2-1/2" diameter, close		15	.533		71	25		96	117

26 05 33.20 Conduit Nipples		Crew	Daily Output	Labor-Hours	Unit	Material	Labor	Equipment	Total	Total Incl O&P
						2009 Bare Costs				
1460	3″ long	1 Elec	15	.533	Ea.	72	25		97	118
1480	3-1/2″ long		15	.533		75.50	25		100.50	122
1500	4″ long		15	.533		77.50	25		102.50	124
1520	5″ long		15	.533		82.50	25		107.50	130
1540	6″ long		15	.533		87	25		112	134
1560	8″ long		15	.533		99.50	25		124.50	149
1580	10″ long		15	.533		113	25		138	163
1600	12″ long		15	.533		125	25		150	176
1620	3″ diameter, close		12	.667		81.50	31.50		113	138
1640	3″ long		12	.667		85.50	31.50		117	142
1660	3-1/2″ long		12	.667		93	31.50		124.50	150
1680	4″ long		12	.667		94.50	31.50		126	152
1700	5″ long		12	.667		101	31.50		132.50	160
1720	6″ long		12	.667		109	31.50		140.50	168
1740	8″ long		12	.667		126	31.50		157.50	187
1760	10″ long		12	.667		143	31.50		174.50	205
1780	12″ long		12	.667		160	31.50		191.50	224
1800	3-1/2″ diameter, close		11	.727		147	34		181	214
1820	4″ long		11	.727		161	34		195	231
1840	5″ long		11	.727		169	34		203	239
1860	6″ long		11	.727		179	34		213	250
1880	8″ long		11	.727		197	34		231	270
1900	10″ long		11	.727		219	34		253	294
1920	12″ long		11	.727		239	34		273	315
1940	4″ diameter, close		9	.889		176	42		218	259
1960	4″ long		9	.889		190	42		232	273
1980	5″ long		9	.889		199	42		241	284
2000	6″ long		9	.889		212	42		254	298
2020	8″ long		9	.889		236	42		278	325
2040	10″ long		9	.889		260	42		302	350
2060	12″ long		9	.889		284	42		326	375
2080	5″ diameter, close		7	1.143		385	53.50		438.50	510
2100	5″ long		7	1.143		410	53.50		463.50	535
2120	6″ long		7	1.143		415	53.50		468.50	540
2140	8″ long		7	1.143		460	53.50		513.50	590
2160	10″ long		7	1.143		495	53.50		548.50	630
2180	12″ long		7	1.143		535	53.50		588.50	670
2200	6″ diameter, close		6	1.333		630	62.50		692.50	790
2220	5″ long		6	1.333		670	62.50		732.50	835
2240	6″ long		6	1.333		685	62.50		747.50	850
2260	8″ long		6	1.333		745	62.50		807.50	915
2280	10″ long		6	1.333		795	62.50		857.50	970
2300	12″ long		6	1.333		830	62.50		892.50	1,000
2320	Rigid galvanized steel, 1/2″ diameter, close		32	.250		3.77	11.75		15.52	22.50
2340	1-1/2″ long		32	.250		4.20	11.75		15.95	22.50
2360	2″ long		32	.250		4.56	11.75		16.31	23
2380	2-1/2″ long		32	.250		4.79	11.75		16.54	23.50
2400	3″ long		32	.250		5.05	11.75		16.80	23.50
2420	3-1/2″ long		32	.250		5.40	11.75		17.15	24
2440	4″ long		32	.250		5.70	11.75		17.45	24.50
2460	5″ long		32	.250		6.20	11.75		17.95	25
2480	6″ long		32	.250		7.10	11.75		18.85	26
2500	8″ long		32	.250		11.20	11.75		22.95	30.50

26 05 33 – Raceway and Boxes for Electrical Systems

26 05 33.20 Conduit Nipples		Crew	Daily Output	Labor-Hours	Unit	2009 Bare Costs Material	Labor	Equipment	Total	Total Incl O&P
2520	10" long	1 Elec	32	.250	Ea.	12.65	11.75		24.40	32
2540	12" long		32	.250		14.30	11.75		26.05	34
2560	3/4" diameter, close		27	.296		5.10	13.95		19.05	27
2580	2" long		27	.296		5.65	13.95		19.60	27.50
2600	2-1/2" long		27	.296		6.05	13.95		20	28
2620	3" long		27	.296		6.35	13.95		20.30	28.50
2640	3-1/2" long		27	.296		6.55	13.95		20.50	28.50
2660	4" long		27	.296		7.10	13.95		21.05	29.50
2680	5" long		27	.296		7.85	13.95		21.80	30
2700	6" long		27	.296		8.75	13.95		22.70	31
2720	8" long		27	.296		13	13.95		26.95	36
2740	10" long		27	.296		15.15	13.95		29.10	38
2760	12" long		27	.296		16.75	13.95		30.70	40
2780	1" diameter, close		23	.348		7.80	16.35		24.15	33.50
2800	2" long		23	.348		8.20	16.35		24.55	34
2820	2-1/2" long		23	.348		8.60	16.35		24.95	34.50
2840	3" long		23	.348		9.20	16.35		25.55	35
2860	3-1/2" long		23	.348		9.95	16.35		26.30	36
2880	4" long		23	.348		10.45	16.35		26.80	36.50
2900	5" long		23	.348		11.35	16.35		27.70	37.50
2920	6" long		23	.348		12.10	16.35		28.45	38.50
2940	8" long		23	.348		17.05	16.35		33.40	44
2960	10" long		23	.348		21	16.35		37.35	48
2980	12" long		23	.348		23.50	16.35		39.85	50.50
3000	1-1/4" diameter, close		20	.400		10.35	18.80		29.15	40.50
3020	2" long		20	.400		10.75	18.80		29.55	41
3040	3" long		20	.400		11.85	18.80		30.65	42
3060	3-1/2" long		20	.400		12.80	18.80		31.60	43
3080	4" long		20	.400		13.30	18.80		32.10	43.50
3100	5" long		20	.400		14.65	18.80		33.45	45
3120	6" long		20	.400		15.85	18.80		34.65	46.50
3140	8" long		20	.400		23	18.80		41.80	54
3160	10" long		20	.400		27.50	18.80		46.30	59.50
3180	12" long		20	.400		31.50	18.80		50.30	63.50
3200	1-1/2" diameter, close		18	.444		14	21		35	47.50
3220	2" long		18	.444		14.40	21		35.40	48
3240	2-1/2" long		18	.444		15.30	21		36.30	49
3260	3" long		18	.444		15.90	21		36.90	49.50
3280	3-1/2" long		18	.444		17.20	21		38.20	51
3300	4" long		18	.444		18.05	21		39.05	52
3320	5" long		18	.444		19.35	21		40.35	53.50
3340	6" long		18	.444		22	21		43	56.50
3360	8" long		18	.444		30	21		51	65
3380	10" long		18	.444		34.50	21		55.50	70
3400	12" long		18	.444		37	21		58	72.50
3420	2" diameter, close		16	.500		19.25	23.50		42.75	57
3440	2-1/2" long		16	.500		20.50	23.50		44	58.50
3460	3" long		16	.500		22	23.50		45.50	60.50
3480	3-1/2" long		16	.500		23.50	23.50		47	62
3500	4" long		16	.500		25	23.50		48.50	63.50
3520	5" long		16	.500		27.50	23.50		51	66
3540	6" long		16	.500		30	23.50		53.50	68.50
3560	8" long		16	.500		38	23.50		61.50	78

26 05 33.20 Conduit Nipples		Crew	Daily Output	Labor-Hours	Unit	Material	2009 Bare Costs Labor	Equipment	Total	Total Incl O&P
3580	10" long	1 Elec	16	.500	Ea.	43.50	23.50		67	83.50
3600	12" long		16	.500		48	23.50		71.50	88.50
3620	2-1/2" diameter, close		13	.615		48.50	29		77.50	97.50
3640	3" long		13	.615		48.50	29		77.50	98
3660	3-1/2" long		13	.615		53	29		82	103
3680	4" long		13	.615		54	29		83	104
3700	5" long		13	.615		59.50	29		88.50	110
3720	6" long		13	.615		64	29		93	115
3740	8" long		13	.615		76	29		105	128
3760	10" long		13	.615		84	29		113	137
3780	12" long		13	.615		93.50	29		122.50	148
3800	3" diameter, close		12	.667		59	31.50		90.50	113
3820	3" long		12	.667		64	31.50		95.50	119
3900	3-1/2" long		12	.667		64	31.50		95.50	119
3920	4" long		12	.667		66.50	31.50		98	122
3940	5" long		12	.667		72.50	31.50		104	128
3960	6" long		12	.667		78	31.50		109.50	134
3980	8" long		12	.667		92	31.50		123.50	149
4000	10" long		12	.667		103	31.50		134.50	161
4020	12" long		12	.667		118	31.50		149.50	178
4040	3-1/2" diameter, close		10	.800		93.50	37.50		131	161
4060	4" long		10	.800		103	37.50		140.50	171
4080	5" long		10	.800		108	37.50		145.50	177
4100	6" long		10	.800		115	37.50		152.50	185
4120	8" long		10	.800		129	37.50		166.50	200
4140	10" long		10	.800		143	37.50		180.50	216
4160	12" long		10	.800		157	37.50		194.50	231
4180	4" diameter, close		8	1		113	47		160	198
4200	4" long		8	1		122	47		169	207
4220	5" long		8	1		130	47		177	216
4240	6" long		8	1		137	47		184	223
4260	8" long		8	1		152	47		199	240
4280	10" long		8	1		170	47		217	260
4300	12" long		8	1		188	47		235	280
4320	5" diameter, close		6	1.333		227	62.50		289.50	345
4340	5" long		6	1.333		261	62.50		323.50	385
4360	6" long		6	1.333		270	62.50		332.50	395
4380	8" long		6	1.333		293	62.50		355.50	415
4400	10" long		6	1.333		315	62.50		377.50	445
4420	12" long		6	1.333		350	62.50		412.50	480
4440	6" diameter, close		5	1.600		330	75		405	480
4460	5" long		5	1.600		365	75		440	520
4480	6" long		5	1.600		380	75		455	535
4500	8" long		5	1.600		405	75		480	560
4520	10" long		5	1.600		450	75		525	610
4540	12" long		5	1.600		475	75		550	635
4560	Plastic coated, 40 mil thick, 1/2" diameter, 2" long		32	.250		15.55	11.75		27.30	35
4580	2-1/2" long		32	.250		17.30	11.75		29.05	37
4600	3" long		32	.250		17.55	11.75		29.30	37.50
4680	3-1/2" long		32	.250		19.30	11.75		31.05	39
4700	4" long		32	.250		19.45	11.75		31.20	39.50
4720	5" long		32	.250		19.65	11.75		31.40	39.50
4740	6" long		32	.250		20	11.75		31.75	40

26 05 33.20 Conduit Nipples		Crew	Daily Output	Labor-Hours	Unit	2009 Bare Costs Material	Labor	Equipment	Total	Total Incl O&P
4760	8" long	1 Elec	32	.250	Ea.	21	11.75		32.75	41
4780	10" long		32	.250		22	11.75		33.75	42
4800	12" long		32	.250		22.50	11.75		34.25	43
4820	3/4" diameter, 2" long		26	.308		16.65	14.45		31.10	41
4840	2-1/2" long		26	.308		18.50	14.45		32.95	43
4860	3" long		26	.308		18.60	14.45		33.05	43
4880	3-1/2" long		26	.308		20.50	14.45		34.95	45
4900	4" long		26	.308		20.50	14.45		34.95	45.50
4920	5" long		26	.308		21	14.45		35.45	45.50
4940	6" long		26	.308		21.50	14.45		35.95	46
4960	8" long		26	.308		22	14.45		36.45	47
4980	10" long		26	.308		23	14.45		37.45	47.50
5000	12" long		26	.308		24	14.45		38.45	48.50
5020	1" diameter, 2" long		22	.364		18.65	17.10		35.75	47
5040	2-1/2" long		22	.364		20.50	17.10		37.60	49
5060	3" long		22	.364		20.50	17.10		37.60	49.50
5080	3-1/2" long		22	.364		22.50	17.10		39.60	51.50
5100	4" long		22	.364		23	17.10		40.10	51.50
5120	5" long		22	.364		23.50	17.10		40.60	52
5140	6" long		22	.364		24	17.10		41.10	52.50
5160	8" long		22	.364		25	17.10		42.10	54
5180	10" long		22	.364		26	17.10		43.10	55
5200	12" long		22	.364		29	17.10		46.10	58.50
5220	1-1/4" diameter, 2" long		18	.444		23.50	21		44.50	58
5240	2-1/2" long		18	.444		25.50	21		46.50	60
5260	3" long		18	.444		25.50	21		46.50	60
5280	3-1/2" long		18	.444		25.50	21		46.50	60
5300	4" long		18	.444		28	21		49	62.50
5320	5" long		18	.444		28.50	21		49.50	63
5340	6" long		18	.444		29.50	21		50.50	64
5360	8" long		18	.444		29.50	21		50.50	64.50
5380	10" long		18	.444		33	21		54	68.50
5400	12" long		18	.444		38.50	21		59.50	74.50
5420	1-1/2" diameter, 2" long		16	.500		26.50	23.50		50	65
5440	2-1/2" long		16	.500		28.50	23.50		52	67.50
5460	3" long		16	.500		29	23.50		52.50	68
5480	3-1/2" long		16	.500		30	23.50		53.50	69
5500	4" long		16	.500		31.50	23.50		55	70.50
5520	5" long		16	.500		32	23.50		55.50	71
5540	6" long		16	.500		34.50	23.50		58	74
5560	8" long		16	.500		37	23.50		60.50	76.50
5580	10" long		16	.500		44	23.50		67.50	84.50
5600	12" long		16	.500		50.50	23.50		74	91.50
5620	2" diameter, 2-1/2" long		14	.571		35	27		62	80
5640	3" long		14	.571		35.50	27		62.50	80.50
5660	3-1/2" long		14	.571		38	27		65	83.50
5680	4" long		14	.571		41	27		68	86.50
5700	5" long		14	.571		42	27		69	87.50
5720	6" long		14	.571		43.50	27		70.50	89.50
5740	8" long		14	.571		48.50	27		75.50	95
5760	10" long		14	.571		58.50	27		85.50	106
5780	12" long		14	.571		65.50	27		92.50	114
5800	2-1/2" diameter, 3-1/2" long		12	.667		68	31.50		99.50	123

26 05 33.20 Conduit Nipples		Crew	Daily Output	Labor-Hours	Unit	Material	Labor	Equipment	Total	Total Incl O&P
						2009 Bare Costs				
5820	4″ long	1 Elec	12	.667	Ea.	69.50	31.50		101	125
5840	5″ long		12	.667		80.50	31.50		112	137
5860	6″ long		12	.667		82.50	31.50		114	139
5880	8″ long		12	.667		93	31.50		124.50	151
5900	10″ long		12	.667		102	31.50		133.50	160
5920	12″ long		12	.667		115	31.50		146.50	175
5940	3″ diameter, 3-1/2″ long		11	.727		80.50	34		114.50	141
5960	4″ long		11	.727		85	34		119	146
5980	5″ long		11	.727		91	34		125	153
6000	6″ long		11	.727		99	34		133	162
6020	8″ long		11	.727		110	34		144	174
6040	10″ long		11	.727		126	34		160	192
6060	12″ long		11	.727		146	34		180	213
6080	3-1/2″ diameter, 4″ long		9	.889		126	42		168	204
6100	5″ long		9	.889		128	42		170	205
6120	6″ long		9	.889		135	42		177	213
6140	8″ long		9	.889		153	42		195	233
6160	10″ long		9	.889		172	42		214	254
6180	12″ long		9	.889		184	42		226	268
6200	4″ diameter, 4″ long		7.50	1.067		145	50		195	237
6220	5″ long		7.50	1.067		149	50		199	240
6240	6″ long		7.50	1.067		161	50		211	254
6260	8″ long		7.50	1.067		179	50		229	274
6280	10″ long		7.50	1.067		212	50		262	310
6300	12″ long		7.50	1.067		222	50		272	320
6320	5″ diameter, 5″ long		5.50	1.455		268	68.50		336.50	400
6340	6″ long		5.50	1.455		286	68.50		354.50	420
6360	8″ long		5.50	1.455		299	68.50		367.50	435
6380	10″ long		5.50	1.455		325	68.50		393.50	460
6400	12″ long		5.50	1.455		330	68.50		398.50	465
6420	6″ diameter, 5″ long		4.50	1.778		380	83.50		463.50	550
6440	6″ long		4.50	1.778		400	83.50		483.50	565
6460	8″ long		4.50	1.778		415	83.50		498.50	585
6480	10″ long		4.50	1.778		440	83.50		523.50	615
6500	12″ long		4.50	1.778		470	83.50		553.50	645

26 05 33.25 Conduit Fittings for Rigid Galvanized Steel		Crew	Daily Output	Labor-Hours	Unit	Material	Labor	Equipment	Total	Total Incl O&P
0010	**CONDUIT FITTINGS FOR RIGID GALVANIZED STEEL**									
0050	Standard, locknuts, 1/2″ diameter				Ea.	.23			.23	.25
0100	3/4″ diameter					.41			.41	.45
0300	1″ diameter					.65			.65	.72
0500	1-1/4″ diameter					.89			.89	.98
0700	1-1/2″ diameter					1.47			1.47	1.62
1000	2″ diameter					2.14			2.14	2.35
1030	2-1/2″ diameter					5.95			5.95	6.55
1050	3″ diameter					7.65			7.65	8.40
1070	3-1/2″ diameter					12.95			12.95	14.25
1100	4″ diameter					16.15			16.15	17.75
1110	5″ diameter					34.50			34.50	38
1120	6″ diameter					59			59	65
1130	Bushings, plastic, 1/2″ diameter	1 Elec	40	.200		.23	9.40		9.63	14.70
1150	3/4″ diameter		32	.250		.38	11.75		12.13	18.50
1170	1″ diameter		28	.286		.63	13.45		14.08	21

26 05 33 – Raceway and Boxes for Electrical Systems

26 05 33.25 Conduit Fittings for Rigid Galvanized Steel		Crew	Daily Output	Labor-Hours	Unit	Material	Labor	Equipment	Total	Total Incl O&P
						2009 Bare Costs				
1200	1-1/4" diameter	1 Elec	24	.333	Ea.	.96	15.65		16.61	25
1230	1-1/2" diameter		18	.444		1.29	21		22.29	33.50
1250	2" diameter		15	.533		2.41	25		27.41	41
1270	2-1/2" diameter		13	.615		5.60	29		34.60	50.50
1300	3" diameter		12	.667		5.80	31.50		37.30	54.50
1330	3-1/2" diameter		11	.727		7.60	34		41.60	61
1350	4" diameter		9	.889		9.35	42		51.35	75
1360	5" diameter		7	1.143		19.90	53.50		73.40	105
1370	6" diameter		5	1.600		38	75		113	158
1390	Steel, 1/2" diameter		40	.200		.55	9.40		9.95	15.05
1400	3/4" diameter		32	.250		.73	11.75		12.48	18.90
1430	1" diameter		28	.286		1.10	13.45		14.55	21.50
1450	Steel insulated, 1-1/4" diameter		24	.333		6.75	15.65		22.40	31.50
1470	1-1/2" diameter		18	.444		8.55	21		29.55	41.50
1500	2" diameter		15	.533		12.30	25		37.30	52
1530	2-1/2" diameter		13	.615		22.50	29		51.50	69.50
1550	3" diameter		12	.667		30.50	31.50		62	82
1570	3-1/2" diameter		11	.727		39	34		73	95.50
1600	4" diameter		9	.889		49.50	42		91.50	119
1610	5" diameter		7	1.143		108	53.50		161.50	202
1620	6" diameter		5	1.600		166	75		241	298
1630	Sealing locknuts, 1/2" diameter		40	.200		1.70	9.40		11.10	16.30
1650	3/4" diameter		32	.250		2	11.75		13.75	20.50
1670	1" diameter		28	.286		2.91	13.45		16.36	23.50
1700	1-1/4" diameter		24	.333		4.95	15.65		20.60	29.50
1730	1-1/2" diameter		18	.444		6.30	21		27.30	39
1750	2" diameter		15	.533		8	25		33	47.50
1760	Grounding bushing, insulated, 1/2" diameter		32	.250		5.10	11.75		16.85	23.50
1770	3/4" diameter		28	.286		6.45	13.45		19.90	27.50
1780	1" diameter		20	.400		7.30	18.80		26.10	37
1800	1-1/4" diameter		18	.444		9.20	21		30.20	42
1830	1-1/2" diameter		16	.500		9.90	23.50		33.40	47
1850	2" diameter		13	.615		12.05	29		41.05	58
1870	2-1/2" diameter		12	.667		22	31.50		53.50	72
1900	3" diameter		11	.727		28	34		62	83
1930	3-1/2" diameter		9	.889		37	42		79	105
1950	4" diameter		8	1		39	47		86	116
1960	5" diameter		6	1.333		99.50	62.50		162	207
1970	6" diameter		4	2		152	94		246	310
1990	Coupling, with set screw, 1/2" diameter		50	.160		5.20	7.50		12.70	17.25
2000	3/4" diameter		40	.200		6.55	9.40		15.95	21.50
2030	1" diameter		35	.229		10.80	10.75		21.55	28.50
2050	1-1/4" diameter		28	.286		18.40	13.45		31.85	40.50
2070	1-1/2" diameter		23	.348		23.50	16.35		39.85	51
2090	2" diameter		20	.400		52.50	18.80		71.30	87
2100	2-1/2" diameter		18	.444		113	21		134	156
2110	3" diameter		15	.533		134	25		159	187
2120	3-1/2" diameter		12	.667		192	31.50		223.50	259
2130	4" diameter		10	.800		251	37.50		288.50	335
2140	5" diameter		9	.889		405	42		447	510
2150	6" diameter		8	1		525	47		572	655
2160	Box connector with set screw, plain, 1/2" diameter		70	.114		3.40	5.35		8.75	12
2170	3/4" diameter	↓	60	.133	↓	4.80	6.25		11.05	14.95

26 05 33.25 Conduit Fittings for Rigid Galvanized Steel		Crew	Daily Output	Labor-Hours	Unit	2009 Bare Costs Material	Labor	Equipment	Total	Total Incl O&P
2180	1" diameter	1 Elec	50	.160	Ea.	7.60	7.50		15.10	19.90
2190	Insulated, 1-1/4" diameter		40	.200		15.80	9.40		25.20	32
2200	1-1/2" diameter		30	.267		22.50	12.55		35.05	44.50
2210	2" diameter		20	.400		45	18.80		63.80	78.50
2220	2-1/2" diameter		18	.444		109	21		130	152
2230	3" diameter		15	.533		145	25		170	198
2240	3-1/2" diameter		12	.667		203	31.50		234.50	271
2250	4" diameter		10	.800		268	37.50		305.50	355
2260	5" diameter		9	.889		325	42		367	420
2270	6" diameter		8	1		415	47		462	530
2280	LB, LR or LL fittings & covers, 1/2" diameter		16	.500		11.95	23.50		35.45	49
2290	3/4" diameter		13	.615		13.95	29		42.95	60
2300	1" diameter		11	.727		20.50	34		54.50	75.50
2330	1-1/4" diameter		8	1		31.50	47		78.50	107
2350	1-1/2" diameter		6	1.333		39	62.50		101.50	140
2370	2" diameter		5	1.600		65	75		140	188
2380	2-1/2" diameter		4	2		129	94		223	287
2390	3" diameter		3.50	2.286		168	107		275	350
2400	3-1/2" diameter		3	2.667		273	125		398	495
2410	4" diameter		2.50	3.200		305	150		455	565
2420	T fittings, with cover, 1/2" diameter		12	.667		14.40	31.50		45.90	64
2430	3/4" diameter		11	.727		16.75	34		50.75	71
2440	1" diameter		9	.889		26	42		68	93.50
2450	1-1/4" diameter		6	1.333		35.50	62.50		98	136
2470	1-1/2" diameter		5	1.600		45.50	75		120.50	166
2500	2" diameter		4	2		69.50	94		163.50	222
2510	2-1/2" diameter		3.50	2.286		142	107		249	320
2520	3" diameter		3	2.667		190	125		315	400
2530	3-1/2" diameter		2.50	3.200		262	150		412	520
2540	4" diameter		2	4		355	188		543	680
2550	Nipples chase, plain, 1/2" diameter		40	.200		1.02	9.40		10.42	15.55
2560	3/4" diameter		32	.250		1.27	11.75		13.02	19.50
2570	1" diameter		28	.286		2.59	13.45		16.04	23.50
2600	Insulated, 1-1/4" diameter		24	.333		9.65	15.65		25.30	34.50
2630	1-1/2" diameter		18	.444		12.70	21		33.70	46
2650	2" diameter		15	.533		19.20	25		44.20	59.50
2660	2-1/2" diameter		12	.667		45.50	31.50		77	98.50
2670	3" diameter		10	.800		49.50	37.50		87	112
2680	3-1/2" diameter		9	.889		69.50	42		111.50	141
2690	4" diameter		8	1		105	47		152	189
2700	5" diameter		7	1.143		310	53.50		363.50	425
2710	6" diameter		6	1.333		480	62.50		542.50	625
2720	Nipples offset, plain, 1/2" diameter		40	.200		6.35	9.40		15.75	21.50
2730	3/4" diameter		32	.250		6.60	11.75		18.35	25.50
2740	1" diameter		24	.333		8.65	15.65		24.30	33.50
2750	Insulated, 1-1/4" diameter		20	.400		36	18.80		54.80	68.50
2760	1-1/2" diameter		18	.444		45.50	21		66.50	82
2770	2" diameter		16	.500		70.50	23.50		94	114
2780	3" diameter		14	.571		143	27		170	199
2850	Coupling, expansion, 1/2" diameter		12	.667		46.50	31.50		78	99
2880	3/4" diameter		10	.800		54	37.50		91.50	118
2900	1" diameter		8	1		63.50	47		110.50	143
2920	1-1/4" diameter		6.40	1.250		86	59		145	185

26 05 33.25 Conduit Fittings for Rigid Galvanized Steel		Crew	Daily Output	Labor-Hours	Unit	Material	2009 Bare Costs Labor	Equipment	Total	Total Incl O&P
2940	1-1/2" diameter	1 Elec	5.30	1.509	Ea.	118	71		189	239
2960	2" diameter		4.60	1.739		176	81.50		257.50	320
2980	2-1/2" diameter		3.60	2.222		278	104		382	465
3000	3" diameter		3	2.667		340	125		465	570
3020	3-1/2" diameter		2.80	2.857		505	134		639	760
3040	4" diameter		2.40	3.333		595	157		752	895
3060	5" diameter		2	4		1,250	188		1,438	1,675
3080	6" diameter		1.80	4.444		2,025	209		2,234	2,550
3100	Expansion deflection, 1/2" diameter		12	.667		165	31.50		196.50	230
3120	3/4" diameter		12	.667		176	31.50		207.50	242
3140	1" diameter		10	.800		222	37.50		259.50	300
3160	1-1/4" diameter		6.40	1.250		277	59		336	395
3180	1-1/2" diameter		5.30	1.509		294	71		365	435
3200	2" diameter		4.60	1.739		380	81.50		461.50	540
3220	2-1/2" diameter		3.60	2.222		535	104		639	750
3240	3" diameter		3	2.667		675	125		800	940
3260	3-1/2" diameter		2.80	2.857		805	134		939	1,100
3280	4" diameter		2.40	3.333		895	157		1,052	1,225
3300	5" diameter		2	4		1,425	188		1,613	1,875
3320	6" diameter		1.80	4.444		2,450	209		2,659	3,025
3340	Ericson, 1/2" diameter		16	.500		5.50	23.50		29	42
3360	3/4" diameter		14	.571		7	27		34	49
3380	1" diameter		11	.727		13	34		47	67
3400	1-1/4" diameter		8	1		27.50	47		74.50	103
3420	1-1/2" diameter		7	1.143		33	53.50		86.50	119
3440	2" diameter		5	1.600		66.50	75		141.50	189
3460	2-1/2" diameter		4	2		162	94		256	325
3480	3" diameter		3.50	2.286		245	107		352	435
3500	3-1/2" diameter		3	2.667		410	125		535	645
3520	4" diameter		2.70	2.963		480	139		619	745
3540	5" diameter		2.50	3.200		885	150		1,035	1,200
3560	6" diameter		2.30	3.478		1,175	163		1,338	1,550
3580	Split, 1/2" diameter		32	.250		4.85	11.75		16.60	23.50
3600	3/4" diameter		27	.296		6.05	13.95		20	28
3620	1" diameter		20	.400		11.20	18.80		30	41.50
3640	1-1/4" diameter		16	.500		16.20	23.50		39.70	54
3660	1-1/2" diameter		14	.571		20.50	27		47.50	64
3680	2" diameter		12	.667		40.50	31.50		72	92.50
3700	2-1/2" diameter		10	.800		94	37.50		131.50	161
3720	3" diameter		9	.889		141	42		183	220
3740	3-1/2" diameter		8	1		227	47		274	325
3760	4" diameter		7	1.143		268	53.50		321.50	380
3780	5" diameter		6	1.333		465	62.50		527.50	605
3800	6" diameter		5	1.600		620	75		695	800
4600	Reducing bushings, 3/4" to 1/2" diameter		54	.148		1.61	6.95		8.56	12.45
4620	1" to 3/4" diameter		46	.174		2.50	8.15		10.65	15.30
4640	1-1/4" to 1" diameter		40	.200		5.80	9.40		15.20	21
4660	1-1/2" to 1-1/4" diameter		36	.222		8.30	10.45		18.75	25
4680	2" to 1-1/2" diameter		32	.250		16.85	11.75		28.60	36.50
4740	2-1/2" to 2" diameter		30	.267		18.70	12.55		31.25	40
4760	3" to 2-1/2" diameter		28	.286		24	13.45		37.45	47
4800	Through-wall seal, 1/2" diameter		8	1		176	47		223	267
4820	3/4" diameter	↓	7.50	1.067	↓	176	50		226	271

26 05 33.25 Conduit Fittings for Rigid Galvanized Steel		Crew	Daily Output	Labor-Hours	Unit	2009 Bare Costs Material	Labor	Equipment	Total	Total Incl O&P
4840	1" diameter	1 Elec	6.50	1.231	Ea.	176	58		234	283
4860	1-1/4" diameter		5.50	1.455		243	68.50		311.50	370
4880	1-1/2" diameter		5	1.600		243	75		318	385
4900	2" diameter		4.20	1.905		264	89.50		353.50	430
4920	2-1/2" diameter		3.50	2.286		282	107		389	475
4940	3" diameter		3	2.667		282	125		407	505
4960	3-1/2" diameter		2.50	3.200		440	150		590	715
4980	4" diameter		2	4		505	188		693	845
5000	5" diameter		1.50	5.333		645	251		896	1,100
5020	6" diameter		1	8		645	375		1,020	1,300
5100	Cable supports, 2 or more wires									
5120	1-1/2" diameter	1 Elec	8	1	Ea.	69	47		116	148
5140	2" diameter		6	1.333		103	62.50		165.50	210
5160	2-1/2" diameter		4	2		115	94		209	271
5180	3" diameter		3.50	2.286		155	107		262	335
5200	3-1/2" diameter		2.60	3.077		205	145		350	445
5220	4" diameter		2	4		255	188		443	570
5240	5" diameter		1.50	5.333		475	251		726	910
5260	6" diameter		1	8		800	375		1,175	1,450
5280	Service entrance cap, 1/2" diameter		16	.500		9.05	23.50		32.55	46
5300	3/4" diameter		13	.615		10.50	29		39.50	56
5320	1" diameter		10	.800		12.15	37.50		49.65	71.50
5340	1-1/4" diameter		8	1		18.35	47		65.35	92.50
5360	1-1/2" diameter		6.50	1.231		33	58		91	126
5380	2" diameter		5.50	1.455		50.50	68.50		119	161
5400	2-1/2" diameter		4	2		157	94		251	320
5420	3" diameter		3.40	2.353		239	111		350	435
5440	3-1/2" diameter		3	2.667		293	125		418	515
5460	4" diameter		2.70	2.963		445	139		584	705
5600	Fire stop fittings, to 3/4" diameter		24	.333		122	15.65		137.65	158
5610	1" diameter		22	.364		149	17.10		166.10	191
5620	1-1/2" diameter		20	.400		186	18.30		204.80	234
5640	2" diameter		16	.500		289	23.50		312.50	355
5660	3" diameter		12	.667		365	31.50		396.50	450
5680	4" diameter		10	.800		480	37.50		517.50	590
5700	6" diameter		8	1		830	47		877	990
5750	90° pull elbows steel, female, 1/2" diameter		16	.500		9.65	23.50		33.15	46.50
5760	3/4" diameter		13	.615		11.25	29		40.25	57
5780	1" diameter		11	.727		19	34		53	73.50
5800	1-1/4" diameter		8	1		28.50	47		75.50	104
5820	1-1/2" diameter		6	1.333		40	62.50		102.50	141
5840	2" diameter		5	1.600		65	75		140	188
6000	Explosion proof, flexible coupling									
6010	1/2" diameter, 4" long	1 Elec	12	.667	Ea.	113	31.50		144.50	172
6020	6" long		12	.667		126	31.50		157.50	187
6050	12" long		12	.667		161	31.50		192.50	225
6070	18" long		12	.667		197	31.50		228.50	265
6090	24" long		12	.667		232	31.50		263.50	305
6110	30" long		12	.667		268	31.50		299.50	345
6130	36" long		12	.667		305	31.50		336.50	385
6140	3/4" diameter, 4" long		10	.800		136	37.50		173.50	208
6150	6" long		10	.800		154	37.50		191.50	227
6180	12" long		10	.800		196	37.50		233.50	274

26 05 33 – Raceway and Boxes for Electrical Systems

26 05 33.25 Conduit Fittings for Rigid Galvanized Steel		Crew	Daily Output	Labor-Hours	Unit	Material	2009 Bare Costs Labor	Equipment	Total	Total Incl O&P
6200	18" long	1 Elec	10	.800	Ea.	252	37.50		289.50	335
6220	24" long		10	.800		305	37.50		342.50	395
6240	30" long		10	.800		355	37.50		392.50	450
6260	36" long		10	.800		405	37.50		442.50	505
6270	1" diameter, 6" long		8	1		274	47		321	375
6300	12" long		8	1		345	47		392	455
6320	18" long		8	1		415	47		462	535
6340	24" long		8	1		490	47		537	615
6360	30" long		8	1		665	47		712	810
6380	36" long		8	1		755	47		802	905
6390	1-1/4" diameter, 12" long		6.40	1.250		530	59		589	670
6410	18" long		6.40	1.250		720	59		779	885
6430	24" long		6.40	1.250		860	59		919	1,025
6450	30" long		6.40	1.250		970	59		1,029	1,175
6470	36" long		6.40	1.250		1,225	59		1,284	1,450
6480	1-1/2" diameter, 12" long		5.30	1.509		705	71		776	885
6500	18" long		5.30	1.509		820	71		891	1,000
6520	24" long		5.30	1.509		930	71		1,001	1,125
6540	30" long		5.30	1.509		1,250	71		1,321	1,475
6560	36" long		5.30	1.509		1,425	71		1,496	1,675
6570	2" diameter, 12" long		4.60	1.739		1,075	81.50		1,156.50	1,300
6590	18" long		4.60	1.739		1,325	81.50		1,406.50	1,600
6610	24" long		4.60	1.739		1,450	81.50		1,531.50	1,725
6630	30" long		4.60	1.739		1,600	81.50		1,681.50	1,875
6650	36" long		4.60	1.739		1,950	81.50		2,031.50	2,275
7000	Close up plug, 1/2" diameter, explosion proof		40	.200		2.40	9.40		11.80	17.10
7010	3/4" diameter		32	.250		2.80	11.75		14.55	21
7020	1" diameter		28	.286		3.35	13.45		16.80	24
7030	1-1/4" diameter		24	.333		3.63	15.65		19.28	28
7040	1-1/2" diameter		18	.444		5.15	21		26.15	37.50
7050	2" diameter		15	.533		8.75	25		33.75	48
7060	2-1/2" diameter		13	.615		13.55	29		42.55	59.50
7070	3" diameter		12	.667		19.80	31.50		51.30	70
7080	3-1/2" diameter		11	.727		25.50	34		59.50	81
7090	4" diameter		9	.889		34.50	42		76.50	102
7091	Elbow female, 45°, 1/2"		16	.500		10.20	23.50		33.70	47
7092	3/4"		13	.615		11	29		40	56.50
7093	1"		11	.727		14.85	34		48.85	69
7094	1-1/4"		8	1		21	47		68	95.50
7095	1-1/2"		6	1.333		21.50	62.50		84	121
7096	2"		5	1.600		26.50	75		101.50	146
7097	2-1/2"		4.50	1.778		72	83.50		155.50	208
7098	3"		4.20	1.905		78	89.50		167.50	224
7099	3-1/2"		4	2		118	94		212	275
7100	4"		3.80	2.105		145	99		244	310
7101	90°, 1/2"		16	.500		9.60	23.50		33.10	46.50
7102	3/4"		13	.615		10.50	29		39.50	56
7103	1"		11	.727		14.15	34		48.15	68
7104	1-1/4"		8	1		22	47		69	97
7105	1-1/2"		6	1.333		32	62.50		94.50	132
7106	2"		5	1.600		51	75		126	173
7107	2-1/2"		4.50	1.778		92.50	83.50		176	231
7110	Elbows 90°, long male & female, 1/2" diameter, explosion proof		16	.500		14.15	23.50		37.65	51.50

26 05 33.25 Conduit Fittings for Rigid Galvanized Steel		Crew	Daily Output	Labor-Hours	Unit	2009 Bare Costs Material	Labor	Equipment	Total	Total Incl O&P
7120	3/4" diameter	1 Elec	13	.615	Ea.	15.90	29		44.90	62
7130	1" diameter		11	.727		22	34		56	77
7140	1-1/4" diameter		8	1		25.50	47		72.50	101
7150	1-1/2" diameter		6	1.333		38.50	62.50		101	139
7160	2" diameter		5	1.600		57.50	75		132.50	180
7170	Capped elbow, 1/2" diameter, explosion proof		11	.727		13.30	34		47.30	67
7180	3/4" diameter		8	1		14.90	47		61.90	89
7190	1" diameter		6	1.333		18.40	62.50		80.90	117
7200	1-1/4" diameter		5	1.600		44	75		119	165
7210	Pulling elbow, 1/2" diameter, explosion proof		11	.727		64	34		98	123
7220	3/4" diameter		8	1		66.50	47		113.50	146
7230	1" diameter		6	1.333		174	62.50		236.50	288
7240	1-1/4" diameter		5	1.600		180	75		255	315
7250	1-1/2" diameter		5	1.600		266	75		341	410
7260	2" diameter		4	2		276	94		370	450
7270	2-1/2" diameter		3.50	2.286		605	107		712	835
7280	3" diameter		3	2.667		585	125		710	840
7290	3-1/2" diameter		2.50	3.200		1,100	150		1,250	1,425
7300	4" diameter		2.20	3.636		1,075	171		1,246	1,450
7310	LB conduit body, 1/2" diameter		11	.727		41	34		75	97.50
7320	3/4" diameter		8	1		47	47		94	124
7330	T conduit body, 1/2" diameter		9	.889		43	42		85	112
7340	3/4" diameter		6	1.333		50.50	62.50		113	152
7350	Explosion proof, round box w/cover, 3 threaded hubs, 1/2" diameter		8	1		46.50	47		93.50	124
7351	3/4" diameter		8	1		50.50	47		97.50	128
7352	1" diameter		7.50	1.067		62.50	50		112.50	146
7353	1-1/4" diameter		7	1.143		115	53.50		168.50	210
7354	1-1/2" diameter		7	1.143		209	53.50		262.50	315
7355	2" diameter		6	1.333		216	62.50		278.50	335
7356	Round box w/cover & mtng flange, 3 threaded hubs, 1/2" diameter		8	1		59	47		106	138
7357	3/4" diameter		8	1		62	47		109	141
7358	4 threaded hubs, 1" diameter		7	1.143		60	53.50		113.50	149
7400	Unions, 1/2" diameter		20	.400		10.60	18.80		29.40	40.50
7410	3/4" - 1/2" diameter		16	.500		15.40	23.50		38.90	53
7420	3/4" diameter		16	.500		14.60	23.50		38.10	52
7430	1" diameter		14	.571		26.50	27		53.50	70.50
7440	1-1/4" diameter		12	.667		39.50	31.50		71	91.50
7450	1-1/2" diameter		10	.800		51	37.50		88.50	114
7460	2" diameter		8.50	.941		65.50	44		109.50	140
7480	2-1/2" diameter		8	1		100	47		147	183
7490	3" diameter		7	1.143		136	53.50		189.50	233
7500	3-1/2" diameter		6	1.333		213	62.50		275.50	330
7510	4" diameter		5	1.600		237	75		312	375
7680	Reducer, 3/4" to 1/2"		54	.148		2.30	6.95		9.25	13.25
7690	1" to 1/2"		46	.174		3.24	8.15		11.39	16.10
7700	1" to 3/4"		46	.174		3.24	8.15		11.39	16.10
7710	1-1/4" to 3/4"		40	.200		4.68	9.40		14.08	19.60
7720	1-1/4" to 1"		40	.200		4.68	9.40		14.08	19.60
7730	1-1/2" to 1"		36	.222		7.65	10.45		18.10	24.50
7740	1-1/2" to 1-1/4"		36	.222		7.65	10.45		18.10	24.50
7750	2" to 3/4"		32	.250		14.85	11.75		26.60	34.50
7760	2" to 1-1/4"		32	.250		10.35	11.75		22.10	29.50
7770	2" to 1-1/2"		32	.250		11	11.75		22.75	30

26 05 33.25 Conduit Fittings for Rigid Galvanized Steel		Crew	Daily Output	Labor-Hours	Unit	Material	2009 Bare Costs Labor	Equipment	Total	Total Incl O&P
7780	2-1/2" to 1-1/2"	1 Elec	30	.267	Ea.	17.75	12.55		30.30	39
7790	3" to 2"		30	.267		21	12.55		33.55	42.50
7800	3-1/2" to 2-1/2"		28	.286		41	13.45		54.45	65.50
7810	4" to 3"		28	.286		47.50	13.45		60.95	73
7820	Sealing fitting, vertical/horizontal, 1/2" diameter		14.50	.552		15.85	26		41.85	57.50
7830	3/4" diameter		13.30	.602		18.60	28.50		47.10	64
7840	1" diameter		11.40	.702		24	33		57	77
7850	1-1/4" diameter		10	.800		29	37.50		66.50	90
7860	1-1/2" diameter		8.80	.909		44	42.50		86.50	114
7870	2" diameter		8	1		57	47		104	135
7880	2-1/2" diameter		6.70	1.194		80	56		136	175
7890	3" diameter		5.70	1.404		98.50	66		164.50	209
7900	3-1/2" diameter		4.70	1.702		263	80		343	410
7910	4" diameter		4	2		415	94		509	600
7920	Sealing hubs, 1" by 1-1/2"		12	.667		28.50	31.50		60	79.50
7930	1-1/4" by 2"		10	.800		32	37.50		69.50	93.50
7940	1-1/2" by 2"		9	.889		61.50	42		103.50	132
7950	2" by 2-1/2"		8	1		81	47		128	162
7960	3" by 4"		7	1.143		139	53.50		192.50	236
7970	4" by 5"		6	1.333		290	62.50		352.50	415
7980	Drain, 1/2"		32	.250		47	11.75		58.75	70
7990	Breather, 1/2"		32	.250		47	11.75		58.75	70
8000	Plastic coated 40 mil thick									
8010	LB, LR or LL conduit body w/cover, 1/2" diameter	1 Elec	13	.615	Ea.	51	29		80	101
8020	3/4" diameter		11	.727		57	34		91	115
8030	1" diameter		8	1		76	47		123	156
8040	1-1/4" diameter		6	1.333		110	62.50		172.50	218
8050	1-1/2" diameter		5	1.600		135	75		210	264
8060	2" diameter		4.50	1.778		199	83.50		282.50	345
8070	2-1/2" diameter		4	2		365	94		459	545
8080	3" diameter		3.50	2.286		460	107		567	670
8090	3-1/2" diameter		3	2.667		675	125		800	940
8100	4" diameter		2.50	3.200		755	150		905	1,075
8150	T conduit body with cover, 1/2" diameter		11	.727		59	34		93	118
8160	3/4" diameter		9	.889		74	42		116	146
8170	1" diameter		6	1.333		88.50	62.50		151	194
8180	1-1/4" diameter		5	1.600		124	75		199	253
8190	1-1/2" diameter		4.50	1.778		155	83.50		238.50	300
8200	2" diameter		4	2		226	94		320	395
8210	2-1/2" diameter		3.50	2.286		380	107		487	585
8220	3" diameter		3	2.667		510	125		635	755
8230	3-1/2" diameter		2.50	3.200		675	150		825	975
8240	4" diameter		2	4		735	188		923	1,100
8300	FS conduit body, 1 gang, 3/4" diameter		11	.727		58.50	34		92.50	117
8310	1" diameter		10	.800		63.50	37.50		101	128
8350	2 gang, 3/4" diameter		9	.889		101	42		143	176
8360	1" diameter		8	1		120	47		167	205
8400	Duplex receptacle cover		64	.125		43	5.90		48.90	56
8410	Switch cover		64	.125		58.50	5.90		64.40	73
8420	Switch, vaportight cover		53	.151		134	7.10		141.10	158
8430	Blank, cover		64	.125		39	5.90		44.90	52
8520	FSC conduit body, 1 gang, 3/4" diameter		10	.800		69	37.50		106.50	134
8530	1" diameter		9	.889		77.50	42		119.50	150

26 05 33.25 Conduit Fittings for Rigid Galvanized Steel		Crew	Daily Output	Labor-Hours	Unit	Material	2009 Bare Costs Labor	Equipment	Total	Total Incl O&P
8550	2 gang, 3/4" diameter	1 Elec	8	1	Ea.	116	47		163	201
8560	1" diameter		7	1.143		129	53.50		182.50	224
8590	Conduit hubs, 1/2" diameter		18	.444		38.50	21		59.50	74.50
8600	3/4" diameter		16	.500		44	23.50		67.50	84
8610	1" diameter		14	.571		55	27		82	102
8620	1-1/4" diameter		12	.667		63	31.50		94.50	118
8630	1-1/2" diameter		10	.800		72	37.50		109.50	137
8640	2" diameter		8.80	.909		104	42.50		146.50	180
8650	2-1/2" diameter		8.50	.941		162	44		206	246
8660	3" diameter		8	1		223	47		270	320
8670	3-1/2" diameter		7.50	1.067		289	50		339	395
8680	4" diameter		7	1.143		355	53.50		408.50	475
8690	5" diameter		6	1.333		425	62.50		487.50	560
8700	Plastic coated 40 mil thick									
8710	Pipe strap, stamped 1 hole, 1/2" diameter	1 Elec	470	.017	Ea.	10.85	.80		11.65	13.15
8720	3/4" diameter		440	.018		10.90	.85		11.75	13.25
8730	1" diameter		400	.020		12.45	.94		13.39	15.15
8740	1-1/4" diameter		355	.023		15.45	1.06		16.51	18.60
8750	1-1/2" diameter		320	.025		17.35	1.18		18.53	21
8760	2" diameter		266	.030		19.55	1.41		20.96	23.50
8770	2-1/2" diameter		200	.040		23.50	1.88		25.38	29
8780	3" diameter		133	.060		25	2.83		27.83	32
8790	3-1/2" diameter		110	.073		26.50	3.42		29.92	35
8800	4" diameter		90	.089		33	4.18		37.18	42.50
8810	5" diameter		70	.114		134	5.35		139.35	155
8840	Clamp back spacers, 3/4" diameter		440	.018		16.45	.85		17.30	19.40
8850	1" diameter		400	.020		21.50	.94		22.44	25
8860	1-1/4" diameter		355	.023		22	1.06		23.06	25.50
8870	1-1/2" diameter		320	.025		30	1.18		31.18	35
8880	2" diameter		266	.030		47	1.41		48.41	53.50
8900	3" diameter		133	.060		106	2.83		108.83	121
8920	4" diameter		90	.089		154	4.18		158.18	176
8950	Touch-up plastic coating, spray, 12 oz.					43.50			43.50	48
8960	Sealing fittings, 1/2" diameter	1 Elec	11	.727		63.50	34		97.50	122
8970	3/4" diameter		9	.889		65	42		107	136
8980	1" diameter		7.50	1.067		78	50		128	163
8990	1-1/4" diameter		6.50	1.231		94	58		152	192
9000	1-1/2" diameter		5.50	1.455		129	68.50		197.50	247
9010	2" diameter		4.80	1.667		156	78.50		234.50	293
9020	2-1/2" diameter		4	2		227	94		321	395
9030	3" diameter		3.50	2.286		285	107		392	480
9040	3-1/2" diameter		3	2.667		715	125		840	980
9050	4" diameter		2.50	3.200		1,150	150		1,300	1,500
9060	5" diameter		1.70	4.706		1,525	221		1,746	2,025
9070	Unions, 1/2" diameter		18	.444		53	21		74	90.50
9080	3/4" diameter		15	.533		54	25		79	98
9090	1" diameter		13	.615		72.50	29		101.50	124
9100	1-1/4" diameter		11	.727		117	34		151	181
9110	1-1/2" diameter		9.50	.842		141	39.50		180.50	216
9120	2" diameter		8	1		187	47		234	279
9130	2-1/2" diameter		7.50	1.067		284	50		334	385
9140	3" diameter		6.80	1.176		390	55.50		445.50	515
9150	3-1/2" diameter		5.80	1.379		485	65		550	635

26 05 33 – Raceway and Boxes for Electrical Systems

26 05 33.25 Conduit Fittings for Rigid Galvanized Steel		Crew	Daily Output	Labor-Hours	Unit	2009 Bare Costs Material	Labor	Equipment	Total	Total Incl O&P
9160	4" diameter	1 Elec	4.80	1.667	Ea.	625	78.50		703.50	805
9170	5" diameter	↓	4	2	↓	875	94		969	1,100

26 05 33.30 Electrical Nonmetallic Tubing (ENT)		Crew	Daily Output	Labor-Hours	Unit	Material	Labor	Equipment	Total	Total Incl O&P
0010	**ELECTRICAL NONMETALLIC TUBING (ENT)**									
0050	Flexible, 1/2" diameter	1 Elec	270	.030	L.F.	.66	1.39		2.05	2.87
0100	3/4" diameter		230	.035		.91	1.63		2.54	3.51
0200	1" diameter		145	.055		1.46	2.59		4.05	5.60
0210	1-1/4" diameter		125	.064		2.03	3.01		5.04	6.85
0220	1-1/2" diameter		100	.080		2.70	3.76		6.46	8.75
0230	2" diameter		75	.107	↓	3.15	5		8.15	11.15
0300	Connectors, to outlet box, 1/2" diameter		230	.035	Ea.	1.30	1.63		2.93	3.94
0310	3/4" diameter		210	.038		2.40	1.79		4.19	5.40
0320	1" diameter		200	.040		3.80	1.88		5.68	7.05
0400	Couplings, to conduit, 1/2" diameter		145	.055		1.42	2.59		4.01	5.55
0410	3/4" diameter		130	.062		2.40	2.89		5.29	7.10
0420	1" diameter	↓	125	.064	↓	3.65	3.01		6.66	8.65

26 05 33.35 Flexible Metallic Conduit		Crew	Daily Output	Labor-Hours	Unit	Material	Labor	Equipment	Total	Total Incl O&P
0010	**FLEXIBLE METALLIC CONDUIT**									
0050	Steel, 3/8" diameter	1 Elec	200	.040	L.F.	.43	1.88		2.31	3.36
0100	1/2" diameter		200	.040		.49	1.88		2.37	3.43
0200	3/4" diameter		160	.050		.66	2.35		3.01	4.35
0250	1" diameter		100	.080		1.30	3.76		5.06	7.25
0300	1-1/4" diameter		70	.114		1.59	5.35		6.94	10
0350	1-1/2" diameter		50	.160		2.64	7.50		10.14	14.45
0370	2" diameter		40	.200		3.30	9.40		12.70	18.10
0380	2-1/2" diameter	↓	30	.267		4	12.55		16.55	23.50
0390	3" diameter	2 Elec	50	.320		7	15.05		22.05	30.50
0400	3-1/2" diameter		40	.400		7.95	18.80		26.75	38
0410	4" diameter	↓	30	.533	↓	9.05	25		34.05	48.50
0420	Connectors, plain, 3/8" diameter	1 Elec	100	.080	Ea.	2.88	3.76		6.64	8.95
0430	1/2" diameter		80	.100		3.36	4.70		8.06	10.95
0440	3/4" diameter		70	.114		3.82	5.35		9.17	12.45
0450	1" diameter		50	.160		8.05	7.50		15.55	20.50
0452	1-1/4" diameter		45	.178		9.90	8.35		18.25	24
0454	1-1/2" diameter		40	.200		14.65	9.40		24.05	30.50
0456	2" diameter		28	.286		22	13.45		35.45	44.50
0458	2-1/2" diameter		25	.320		37.50	15.05		52.55	64
0460	3" diameter		20	.400		54	18.80		72.80	88
0462	3-1/2" diameter		16	.500		156	23.50		179.50	208
0464	4" diameter		13	.615		200	29		229	265
0490	Insulated, 1" diameter		40	.200		8.80	9.40		18.20	24
0500	1-1/4" diameter		40	.200		15.75	9.40		25.15	32
0550	1-1/2" diameter		32	.250		22.50	11.75		34.25	43
0600	2" diameter		23	.348		40.50	16.35		56.85	69.50
0610	2-1/2" diameter		20	.400		106	18.80		124.80	146
0620	3" diameter		17	.471		140	22		162	188
0630	3-1/2" diameter		13	.615		239	29		268	310
0640	4" diameter		10	.800		315	37.50		352.50	405
0650	Connectors 90°, plain, 3/8" diameter		80	.100		2.10	4.70		6.80	9.55
0660	1/2" diameter		60	.133		3.50	6.25		9.75	13.50
0700	3/4" diameter		50	.160		6.35	7.50		13.85	18.55
0750	1" diameter	↓	40	.200	↓	11.10	9.40		20.50	26.50

26 05 33.35 Flexible Metallic Conduit		Crew	Daily Output	Labor-Hours	Unit	Material	2009 Bare Costs Labor	Equipment	Total	Total Incl O&P
0790	Insulated, 1" diameter	1 Elec	40	.200	Ea.	16.50	9.40		25.90	32.50
0800	1-1/4" diameter		30	.267		35.50	12.55		48.05	58.50
0850	1-1/2" diameter		23	.348		62	16.35		78.35	93
0900	2" diameter		18	.444		82	21		103	122
0910	2-1/2" diameter		16	.500		175	23.50		198.50	229
0920	3" diameter		14	.571		219	27		246	283
0930	3-1/2" diameter		11	.727		640	34		674	755
0940	4" diameter		8	1		960	47		1,007	1,125
0960	Couplings, to flexible conduit, 1/2" diameter		50	.160		2	7.50		9.50	13.75
0970	3/4" diameter		40	.200		3.30	9.40		12.70	18.10
0980	1" diameter		35	.229		5.70	10.75		16.45	23
0990	1-1/4" diameter		28	.286		12.60	13.45		26.05	34.50
1000	1-1/2" diameter		23	.348		16.30	16.35		32.65	43
1010	2" diameter		20	.400		33	18.80		51.80	65.50
1020	2-1/2" diameter		18	.444		45	21		66	81.50
1030	3" diameter		15	.533		113	25		138	163
1070	Sealtite, 3/8" diameter		140	.057	L.F.	1.38	2.69		4.07	5.65
1080	1/2" diameter		140	.057		1.64	2.69		4.33	5.95
1090	3/4" diameter		100	.080		2.30	3.76		6.06	8.35
1100	1" diameter		70	.114		3.46	5.35		8.81	12.05
1200	1-1/4" diameter		50	.160		4.75	7.50		12.25	16.80
1300	1-1/2" diameter		40	.200		5.50	9.40		14.90	20.50
1400	2" diameter		30	.267		6.85	12.55		19.40	27
1410	2-1/2" diameter		27	.296		11.75	13.95		25.70	34.50
1420	3" diameter	2 Elec	50	.320		16.25	15.05		31.30	41
1440	4" diameter	"	30	.533		24.50	25		49.50	65.50
1490	Connectors, plain, 3/8" diameter	1 Elec	70	.114	Ea.	5.10	5.35		10.45	13.85
1500	1/2" diameter		70	.114		5	5.35		10.35	13.75
1700	3/4" diameter		50	.160		7.30	7.50		14.80	19.60
1900	1" diameter		40	.200		12.60	9.40		22	28.50
1910	Insulated, 1" diameter		40	.200		16.20	9.40		25.60	32.50
2000	1-1/4" diameter		32	.250		25.50	11.75		37.25	46
2100	1-1/2" diameter		27	.296		33	13.95		46.95	58
2200	2" diameter		20	.400		59	18.80		77.80	94
2210	2-1/2" diameter		15	.533		350	25		375	425
2220	3" diameter		12	.667		390	31.50		421.50	480
2240	4" diameter		8	1		485	47		532	605
2290	Connectors, 90°, 3/8" diameter		70	.114		6.25	5.35		11.60	15.15
2300	1/2" diameter		70	.114		6.05	5.35		11.40	14.90
2400	3/4" diameter		50	.160		9.35	7.50		16.85	22
2600	1" diameter		40	.200		19.20	9.40		28.60	35.50
2790	Insulated, 1" diameter		40	.200		25.50	9.40		34.90	42.50
2800	1-1/4" diameter		32	.250		41.50	11.75		53.25	64
3000	1-1/2" diameter		27	.296		47	13.95		60.95	73.50
3100	2" diameter		20	.400		65.50	18.80		84.30	102
3110	2-1/2" diameter		14	.571		310	27		337	380
3120	3" diameter		11	.727		355	34		389	445
3140	4" diameter		7	1.143		460	53.50		513.50	590
4300	Coupling sealtite to rigid, 1/2" diameter		20	.400		5.35	18.80		24.15	35
4500	3/4" diameter		18	.444		7.60	21		28.60	40.50
4800	1" diameter		14	.571		10.40	27		37.40	53
4900	1-1/4" diameter		12	.667		17.35	31.50		48.85	67
5000	1-1/2" diameter		11	.727		31.50	34		65.50	87

26 05 33 – Raceway and Boxes for Electrical Systems

26 05 33.35 Flexible Metallic Conduit		Crew	Daily Output	Labor-Hours	Unit	Material	Labor	Equipment	Total	Total Incl O&P
						2009 Bare Costs				
5100	2" diameter	1 Elec	10	.800	Ea.	53.50	37.50		91	117
5110	2-1/2" diameter		9.50	.842		207	39.50		246.50	289
5120	3" diameter		9	.889		220	42		262	305
5130	3-1/2" diameter		9	.889		272	42		314	365
5140	4" diameter		8.50	.941		272	44		316	365

26 05 33.45 Wireway		Crew	Daily Output	Labor-Hours	Unit	Material	Labor	Equipment	Total	Total Incl O&P
0010	**WIREWAY** to 15' high									
0100	Screw cover, NEMA 1 w/ fittings and supports, 2-1/2" x 2-1/2"	1 Elec	45	.178	L.F.	11.45	8.35		19.80	25.50
0200	4" x 4"	"	40	.200		12.65	9.40		22.05	28.50
0400	6" x 6"	2 Elec	60	.267		19.70	12.55		32.25	41
0600	8" x 8"		40	.400		33	18.80		51.80	65.50
0620	10" x 10"		30	.533		42	25		67	84.50
0640	12" x 12"		20	.800		60	37.50		97.50	124
0800	Elbows, 90°, 2-1/2"	1 Elec	24	.333	Ea.	35.50	15.65		51.15	63
1000	4"		20	.400		40.50	18.80		59.30	73.50
1200	6"		18	.444		46	21		67	82.50
1400	8"		16	.500		73.50	23.50		97	117
1420	10"		12	.667		81.50	31.50		113	138
1440	12"		10	.800		132	37.50		169.50	203
1500	Elbows, 45°, 2-1/2"		24	.333		33.50	15.65		49.15	61
1510	4"		20	.400		43.50	18.80		62.30	77
1520	6"		18	.444		45.50	21		66.50	82
1530	8"		16	.500		73.50	23.50		97	117
1540	10"		12	.667		87	31.50		118.50	144
1550	12"		10	.800		185	37.50		222.50	262
1600	"T" box, 2-1/2"		18	.444		41	21		62	77
1800	4"		16	.500		49	23.50		72.50	90
2000	6"		14	.571		55	27		82	102
2200	8"		12	.667		102	31.50		133.50	161
2220	10"		10	.800		118	37.50		155.50	188
2240	12"		8	1		196	47		243	289
2300	Cross, 2-1/2"		16	.500		45.50	23.50		69	86
2310	4"		14	.571		55.50	27		82.50	103
2320	6"		12	.667		68.50	31.50		100	124
2400	Panel adapter, 2-1/2"		24	.333		15.20	15.65		30.85	40.50
2600	4"		20	.400		17.20	18.80		36	48
2800	6"		18	.444		22.50	21		43.50	56.50
3000	8"		16	.500		32	23.50		55.50	71
3020	10"		14	.571		40.50	27		67.50	86
3040	12"		12	.667		51.50	31.50		83	105
3200	Reducer, 4" to 2-1/2"		24	.333		24	15.65		39.65	50
3400	6" to 4"		20	.400		45.50	18.80		64.30	79
3600	8" to 6"		18	.444		53	21		74	90
3620	10" to 8"		16	.500		60.50	23.50		84	103
3640	12" to 10"		14	.571		75.50	27		102.50	125
3780	End cap, 2-1/2"		24	.333		5.10	15.65		20.75	29.50
3800	4"		20	.400		6.25	18.80		25.05	36
4000	6"		18	.444		7.50	21		28.50	40.50
4200	8"		16	.500		10.20	23.50		33.70	47
4220	10"		14	.571		14.50	27		41.50	57.50
4240	12"		12	.667		23	31.50		54.50	73.50
4300	U-connector, 2-1/2"		200	.040		5.10	1.88		6.98	8.50

26 05 33.45 Wireway		Crew	Daily Output	Labor-Hours	Unit	2009 Bare Costs Material	Labor	Equipment	Total	Total Incl O&P
4320	4″	1 Elec	200	.040	Ea.	6.25	1.88		8.13	9.80
4340	6″		180	.044		7.50	2.09		9.59	11.45
4360	8″		170	.047		12.20	2.21		14.41	16.80
4380	10″		150	.053		16.35	2.51		18.86	22
4400	12″		130	.062		30.50	2.89		33.39	38
4420	Hanger, 2-1/2″		100	.080		8.90	3.76		12.66	15.60
4430	4″		100	.080		11.30	3.76		15.06	18.25
4440	6″		80	.100		20.50	4.70		25.20	30
4450	8″		65	.123		27	5.80		32.80	38.50
4460	10″		50	.160		38.50	7.50		46	53.50
4470	12″		40	.200		71.50	9.40		80.90	93.50
4475	Screw cover, NEMA 3R w/ fittings and supports, 4″ x 4″		36	.222	L.F.	27	10.45		37.45	45.50
4480	6″ x 6″	2 Elec	55	.291		34	13.65		47.65	58.50
4485	8″ x 8″		36	.444		55	21		76	92.50
4490	12″ x 12″		18	.889		77	42		119	149
4500	Hinged cover, with fittings and supports, 2-1/2″ x 2-1/2″	1 Elec	60	.133		11.45	6.25		17.70	22.50
4520	4″ x 4″	″	45	.178		12.65	8.35		21	27
4540	6″ x 6″	2 Elec	80	.200		20	9.40		29.40	36.50
4560	8″ x 8″		60	.267		33	12.55		45.55	56
4580	10″ x 10″		50	.320		42	15.05		57.05	69
4600	12″ x 12″		24	.667		60	31.50		91.50	114
4700	Elbows 90°, hinged cover, 2-1/2″ x 2-1/2″	1 Elec	32	.250	Ea.	35.50	11.75		47.25	57
4720	4″		27	.296		40.50	13.95		54.45	66
4730	6″		23	.348		46	16.35		62.35	75.50
4740	8″		18	.444		73.50	21		94.50	113
4750	10″		14	.571		81.50	27		108.50	132
4760	12″		12	.667		132	31.50		163.50	193
4800	Tee box, hinged cover, 2-1/2″ x 2-1/2″		23	.348		41	16.35		57.35	70
4810	4″		20	.400		49	18.80		67.80	83
4820	6″		18	.444		55	21		76	92.50
4830	8″		16	.500		102	23.50		125.50	149
4840	10″		12	.667		118	31.50		149.50	178
4860	12″		10	.800		196	37.50		233.50	274
4880	Cross box, hinged cover, 2-1/2″ x 2-1/2″		18	.444		45.50	21		66.50	82
4900	4″		16	.500		55.50	23.50		79	97
4920	6″		13	.615		68.50	29		97.50	120
4940	8″		11	.727		102	34		136	165
4960	10″		10	.800		175	37.50		212.50	251
4980	12″		9	.889		191	42		233	275
5000	Flanged, oil tite w/screw cover, 2-1/2″ x 2-1/2″		40	.200	L.F.	29	9.40		38.40	46.50
5020	4″ x 4″		35	.229		34	10.75		44.75	54
5040	6″ x 6″	2 Elec	60	.267		49	12.55		61.55	73
5060	8″ x 8″	″	50	.320		67	15.05		82.05	96.50
5120	Elbows 90°, flanged, 2-1/2″ x 2-1/2″	1 Elec	23	.348	Ea.	65	16.35		81.35	96.50
5140	4″		20	.400		82.50	18.80		101.30	120
5160	6″		18	.444		102	21		123	144
5180	8″		15	.533		151	25		176	205
5240	Tee box, flanged, 2-1/2″ x 2-1/2″		18	.444		89.50	21		110.50	131
5260	4″		16	.500		102	23.50		125.50	148
5280	6″		15	.533		138	25		163	191
5300	8″		13	.615		200	29		229	265
5360	Cross box, flanged, 2-1/2″ x 2-1/2″		15	.533		110	25		135	160
5380	4″		13	.615		144	29		173	204

26 05 33 – Raceway and Boxes for Electrical Systems

26 05 33.45 Wireway		Crew	Daily Output	Labor-Hours	Unit	Material	Labor	Equipment	Total	Total Incl O&P
						2009 Bare Costs				
5400	6"	1 Elec	12	.667	Ea.	186	31.50		217.50	253
5420	8"		10	.800		220	37.50		257.50	300
5480	Flange gasket, 2-1/2"		160	.050		3.24	2.35		5.59	7.20
5500	4"		80	.100		4.41	4.70		9.11	12.10
5520	6"		53	.151		5.90	7.10		13	17.40
5530	8"		40	.200		7.85	9.40		17.25	23

26 05 33.50 Outlet Boxes		Crew	Daily Output	Labor-Hours	Unit	Material	Labor	Equipment	Total	Total Incl O&P
0010	**OUTLET BOXES**									
0020	Pressed steel, octagon, 4"	1 Elec	20	.400	Ea.	3.09	18.80		21.89	32.50
0040	For Romex or BX		20	.400		4.65	18.80		23.45	34
0050	For Romex or BX, with bracket		20	.400		7.30	18.80		26.10	37
0060	Covers, blank		64	.125		1.29	5.90		7.19	10.45
0100	Extension rings		40	.200		5.15	9.40		14.55	20
0150	Square, 4"		20	.400		2.38	18.80		21.18	31.50
0160	For Romex or BX		20	.400		8.25	18.80		27.05	38
0170	For Romex or BX, with bracket		20	.400		8.95	18.80		27.75	39
0200	Extension rings		40	.200		5.20	9.40		14.60	20
0220	2-1/8" deep, 1" KO		20	.400		6.90	18.80		25.70	36.50
0250	Covers, blank		64	.125		1.40	5.90		7.30	10.60
0260	Raised device		64	.125		2.65	5.90		8.55	11.95
0300	Plaster rings		64	.125		2.84	5.90		8.74	12.15
0350	Square, 4-11/16"		20	.400		7.25	18.80		26.05	37
0370	2-1/8" deep, 3/4" to 1-1/4" KO		20	.400		8.75	18.80		27.55	38.50
0400	Extension rings		40	.200		11.65	9.40		21.05	27.50
0450	Covers, blank		53	.151		2.63	7.10		9.73	13.80
0460	Raised device		53	.151		7.85	7.10		14.95	19.50
0500	Plaster rings		53	.151		7.75	7.10		14.85	19.45
0550	Handy box		27	.296		3.42	13.95		17.37	25.50
0560	Covers, device		64	.125		1.20	5.90		7.10	10.35
0600	Extension rings		54	.148		4.10	6.95		11.05	15.20
0652	Switchbox		24	.333		4.97	15.65		20.62	29.50
0660	Romex or BX		27	.296		7.10	13.95		21.05	29.50
0670	with bracket		27	.296		8.20	13.95		22.15	30.50
0680	Partition, metal		27	.296		3.12	13.95		17.07	25
0700	Masonry, 1 gang, 2-1/2" deep		27	.296		9.45	13.95		23.40	32
0710	3-1/2" deep		27	.296		9.50	13.95		23.45	32
0750	2 gang, 2-1/2" deep		20	.400		17.75	18.80		36.55	48.50
0760	3-1/2" deep		20	.400		14.30	18.80		33.10	45
0800	3 gang, 2-1/2" deep		13	.615		21.50	29		50.50	68
0850	4 gang, 2-1/2" deep		10	.800		24.50	37.50		62	85
0860	5 gang, 2-1/2" deep		9	.889		26	42		68	93
0870	6 gang, 2-1/2" deep		8	1		54	47		101	132
0880	Masonry thru-the-wall, 1 gang, 4" block		16	.500		21	23.50		44.50	59
0890	6" block		16	.500		24	23.50		47.50	62.50
0900	8" block		16	.500		27	23.50		50.50	66
0920	2 gang, 6" block		16	.500		25	23.50		48.50	63.50
0940	Bar hanger with 3/8" stud, for wood and masonry boxes		53	.151		4.45	7.10		11.55	15.80
0950	Concrete, set flush, 4" deep		20	.400		14.80	18.80		33.60	45.50
1000	Plate with 3/8" stud		80	.100		6	4.70		10.70	13.85
1100	Concrete, floor, 1 gang		5.30	1.509		85	71		156	203
1150	2 gang		4	2		134	94		228	292
1200	3 gang		2.70	2.963		190	139		329	425

26 05 33.50 Outlet Boxes

	26 05 33.50 Outlet Boxes	Crew	Daily Output	Labor-Hours	Unit	Material	2009 Bare Costs Labor	Equipment	Total	Total Incl O&P
1250	For duplex receptacle, pedestal mounted, add	1 Elec	24	.333	Ea.	84.50	15.65		100.15	117
1270	Flush mounted, add		27	.296		23	13.95		36.95	47
1300	For telephone, pedestal mounted, add		30	.267		90	12.55		102.55	118
1350	Carpet flange, 1 gang		53	.151		45.50	7.10		52.60	61.50
1400	Cast, 1 gang, FS (2" deep), 1/2" hub		12	.667		18.85	31.50		50.35	68.50
1410	3/4" hub		12	.667		19.85	31.50		51.35	70
1420	FD (2-11/16" deep), 1/2" hub		12	.667		18.90	31.50		50.40	69
1430	3/4" hub		12	.667		20.50	31.50		52	70.50
1450	2 gang, FS, 1/2" hub		10	.800		33.50	37.50		71	95
1460	3/4" hub		10	.800		34.50	37.50		72	96
1470	FD, 1/2" hub		10	.800		39.50	37.50		77	102
1480	3/4" hub		10	.800		40.50	37.50		78	103
1500	3 gang, FS, 3/4" hub		9	.889		55	42		97	125
1510	Switch cover, 1 gang, FS		64	.125		4.80	5.90		10.70	14.35
1520	2 gang		53	.151		7.50	7.10		14.60	19.15
1530	Duplex receptacle cover, 1 gang, FS		64	.125		4.80	5.90		10.70	14.35
1540	2 gang, FS		53	.151		7.50	7.10		14.60	19.15
1550	Weatherproof switch cover		64	.125		7.85	5.90		13.75	17.70
1600	Weatherproof receptacle cover		64	.125		7.85	5.90		13.75	17.70
1750	FSC, 1 gang, 1/2" hub		11	.727		21	34		55	75.50
1760	3/4" hub		11	.727		23	34		57	77.50
1770	2 gang, 1/2" hub		9	.889		35.50	42		77.50	104
1780	3/4" hub		9	.889		39.50	42		81.50	108
1790	FDC, 1 gang, 1/2" hub		11	.727		25	34		59	80
1800	3/4" hub		11	.727		27	34		61	82
1810	2 gang, 1/2" hub		9	.889		44	42		86	113
1820	3/4" hub		9	.889		42.50	42		84.50	112
2000	Poke-thru fitting, fire rated, for 3-3/4" floor		6.80	1.176		111	55.50		166.50	208
2040	For 7" floor		6.80	1.176		111	55.50		166.50	208
2100	Pedestal, 15 amp, duplex receptacle & blank plate		5.25	1.524		122	71.50		193.50	244
2120	Duplex receptacle and telephone plate		5.25	1.524		122	71.50		193.50	244
2140	Pedestal, 20 amp, duplex recept. & phone plate		5	1.600		123	75		198	251
2160	Telephone plate, both sides		5.25	1.524		117	71.50		188.50	239
2200	Abandonment plate		32	.250		35.50	11.75		47.25	57
9000	Minimum labor/equipment charge		4	2	Job		94		94	145

26 05 33.60 Outlet Boxes, Plastic

	26 05 33.60 Outlet Boxes, Plastic	Crew	Daily Output	Labor-Hours	Unit	Material	Labor	Equipment	Total	Total Incl O&P
0010	**OUTLET BOXES, PLASTIC**									
0050	4" diameter, round with 2 mounting nails	1 Elec	25	.320	Ea.	2.96	15.05		18.01	26.50
0100	Bar hanger mounted		25	.320		5.35	15.05		20.40	29
0200	4", square with 2 mounting nails		25	.320		4.44	15.05		19.49	28
0300	Plaster ring		64	.125		1.84	5.90		7.74	11.05
0400	Switch box with 2 mounting nails, 1 gang		30	.267		2.01	12.55		14.56	21.50
0500	2 gang		25	.320		4.12	15.05		19.17	27.50
0600	3 gang		20	.400		6.50	18.80		25.30	36
0700	Old work box		30	.267		3.96	12.55		16.51	23.50
1400	PVC, FSS, 1 gang, 1/2" hub		14	.571		21.50	27		48.50	65
1410	3/4" hub		14	.571		23	27		50	67
1420	FD, 1 gang for variable terminations		14	.571		20	27		47	63.50
1450	FS, 2 gang for variable terminations		12	.667		27	31.50		58.50	77.50
1480	Blank cover, FS, 1 gang		64	.125		5.70	5.90		11.60	15.30
1500	2 gang		53	.151		6.45	7.10		13.55	18
1510	Switch cover, FS, 1 gang		64	.125		13.60	5.90		19.50	24

26 05 33.60 Outlet Boxes, Plastic		Crew	Daily Output	Labor-Hours	Unit	Material	2009 Bare Costs Labor	Equipment	Total	Total Incl O&P
1520	2 gang	1 Elec	53	.151	Ea.	21.50	7.10		28.60	35
1530	Duplex receptacle cover, FS, 1 gang		64	.125		19.40	5.90		25.30	30.50
1540	2 gang		53	.151		21	7.10		28.10	34.50
1750	FSC, 1 gang, 1/2″ hub		13	.615		21	29		50	67.50
1760	3/4″ hub		13	.615		23	29		52	70
1770	FSC, 2 gang, 1/2″ hub		11	.727		27.50	34		61.50	82.50
1780	3/4″ hub		11	.727		27.50	34		61.50	82.50
1790	FDC, 1 gang, 1/2″ hub		13	.615		20	29		49	66.50
1800	3/4″ hub		13	.615		22	29		51	68.50
1810	Weatherproof, T box w/ 3 holes		14	.571		20.50	27		47.50	64.50
1820	4″ diameter round w/ 5 holes		14	.571		30	27		57	74.50
9000	Minimum labor/equipment charge		4	2	Job		94		94	145

26 05 33.65 Pull Boxes		Crew	Daily Output	Labor-Hours	Unit	Material	Labor	Equipment	Total	Total Incl O&P
0010	**PULL BOXES**									
0100	Sheet metal, pull box, NEMA 1, type SC, 6″ W x 6″ H x 4″ D	1 Elec	8	1	Ea.	12.55	47		59.55	86.50
0180	6″ W x 8″ H x 4″ D		8	1		14.70	47		61.70	88.50
0200	8″ W x 8″ H x 4″ D		8	1		17.20	47		64.20	91.50
0210	10″ W x 10″ H x 4″ D		7	1.143		22.50	53.50		76	108
0220	12″ W x 12″ H x 4″ D		6.50	1.231		29	58		87	121
0230	15″ W x 15″ H x 4″ D		5.20	1.538		39.50	72.50		112	154
0240	18″ W x 18″ H x 4″ D		4.40	1.818		54.50	85.50		140	191
0250	6″ W x 6″ H x 6″ D		8	1		15.10	47		62.10	89
0260	8″ W x 8″ H x 6″ D		7.50	1.067		20.50	50		70.50	99.50
0270	10″ W x 10″ H x 6″ D		5.50	1.455		27	68.50		95.50	135
0300	10″ W x 12″ H x 6″ D		5.30	1.509		30.50	71		101.50	143
0310	12″ W x 12″ H x 6″ D		5.20	1.538		34	72.50		106.50	149
0320	15″ W x 15″ H x 6″ D		4.60	1.739		51	81.50		132.50	182
0330	18″ W x 18″ H x 6″ D		4.20	1.905		60.50	89.50		150	205
0340	24″ W x 24″ H x 6″ D		3.20	2.500		123	118		241	315
0350	12″ W x 12″ H x 8″ D		5	1.600		38	75		113	158
0360	15″ W x 15″ H x 8″ D		4.50	1.778		78.50	83.50		162	215
0370	18″ W x 18″ H x 8″ D		4	2		120	94		214	277
0380	24″ W x 18″ H x 6″ D		3.70	2.162		95.50	102		197.50	261
0400	16″ W x 20″ H x 8″ D		4	2		114	94		208	270
0500	20″ W x 24″ H x 8″ D		3.20	2.500		134	118		252	330
0510	24″ W x 24″ H x 8″ D		3	2.667		136	125		261	345
0600	24″ W x 36″ H x 8″ D		2.70	2.963		190	139		329	425
0610	30″ W x 30″ H x 8″ D		2.70	2.963		252	139		391	490
0620	36″ W x 36″ H x 8″ D		2	4		310	188		498	630
0630	24″ W x 24″ H x 10″ D		2.50	3.200		156	150		306	405
0650	Pull box, hinged , NEMA 1, 6″ W x 6″ H x 4″ D		8	1		12.90	47		59.90	86.50
0660	8″ W x 8″ H x 4″ D		8	1		17.65	47		64.65	92
0670	10″ W x 10″ H x 4″ D		7	1.143		23.50	53.50		77	108
0680	12″ W x 12″ H x 4″ D		6	1.333		30	62.50		92.50	130
0690	15″ W x 15″ H x 4″ D		5.20	1.538		46	72.50		118.50	162
0700	18″ W x 18″ H x 4″ D		4.40	1.818		53	85.50		138.50	190
0710	6″ W x 6″ H x 6″ D		8	1		15.60	47		62.60	89.50
0720	8″ W x 8″ H x 6″ D		7.50	1.067		21	50		71	101
0730	10″ W x 10″ H x 6″ D		5.50	1.455		27.50	68.50		96	136
0740	12″ W x 12″ H x 6″ D		5.20	1.538		35	72.50		107.50	150
0800	12″ W x 16″ H x 6″ D		4.70	1.702		42	80		122	170
0810	15″ W x 15″ H x 6″ D		4.60	1.739		49	81.50		130.50	180

26 05 33.65 Pull Boxes		Crew	Daily Output	Labor-Hours	Unit	2009 Bare Costs Material	Labor	Equipment	Total	Total Incl O&P
0820	18" W x 18" H x 6" D	1 Elec	4.20	1.905	Ea.	62.50	89.50		152	207
1000	20" W x 20" H x 6" D		3.60	2.222		85	104		189	255
1010	24" W x 24" H x 6" D		3.20	2.500		127	118		245	320
1020	12" W x 12" H x 8" D		5	1.600		72.50	75		147.50	196
1030	15" W x 15" H x 8" D		4.50	1.778		103	83.50		186.50	242
1040	18" W x 18" H x 8" D		4	2		145	94		239	305
1200	20" W x 20" H x 8" D		3.20	2.500		170	118		288	370
1210	24" W x 24" H x 8" D		3	2.667		194	125		319	405
1220	30" W x 30" H x 8" D		2.70	2.963		276	139		415	520
1400	24" W x 36" H x 8" D		2.70	2.963		272	139		411	515
1600	24" W x 42" H x 8" D		2	4		410	188		598	745
1610	36" W x 36" H x 8" D		2	4		380	188		568	710
2100	Pull box, NEMA 3R, type SC, raintight & weatherproof									
2150	6" L x 6" W x 6" D	1 Elec	10	.800	Ea.	21.50	37.50		59	82
2200	8" L x 6" W x 6" D		8	1		27	47		74	103
2250	10" L x 6" W x 6" D		7	1.143		36	53.50		89.50	122
2300	12" L x 12" W x 6" D		5	1.600		51	75		126	172
2350	16" L x 16" W x 6" D		4.50	1.778		101	83.50		184.50	240
2400	20" L x 20" W x 6" D		4	2		138	94		232	297
2450	24" L x 18" W x 8" D		3	2.667		152	125		277	360
2500	24" L x 24" W x 10" D		2.50	3.200		204	150		354	455
2550	30" L x 24" W x 12" D		2	4		370	188		558	695
2600	36" L x 36" W x 12" D		1.50	5.333		485	251		736	920
2800	Cast iron, pull boxes for surface mounting									
3000	NEMA 4, watertight & dust tight									
3050	6" L x 6" W x 6" D	1 Elec	4	2	Ea.	214	94		308	380
3100	8" L x 6" W x 6" D		3.20	2.500		232	118		350	435
3150	10" L x 6" W x 6" D		2.50	3.200		289	150		439	550
3200	12" L x 12" W x 6" D		2.30	3.478		515	163		678	815
3250	16" L x 16" W x 6" D		1.30	6.154		1,200	289		1,489	1,775
3300	20" L x 20" W x 6" D		.80	10		2,250	470		2,720	3,200
3350	24" L x 18" W x 8" D		.70	11.429		2,375	535		2,910	3,425
3400	24" L x 24" W x 10" D		.50	16		3,625	750		4,375	5,150
3450	30" L x 24" W x 12" D		.40	20		6,075	940		7,015	8,125
3500	36" L x 36" W x 12" D		.20	40		7,500	1,875		9,375	11,200
3510	NEMA 4 clamp cover, 6" L x 6" W x 4" D		4	2		231	94		325	400
3520	8" L x 6" W x 4" D		4	2		270	94		364	440
4000	NEMA 7, explosionproof									
4050	6" L x 6" W x 6" D	1 Elec	2	4	Ea.	730	188		918	1,100
4100	8" L x 6" W x 6" D		1.80	4.444		835	209		1,044	1,250
4150	10" L x 6" W x 6" D		1.60	5		1,025	235		1,260	1,475
4200	12" L x 12" W x 6" D		1	8		1,675	375		2,050	2,425
4250	16" L x 14" W x 6" D		.60	13.333		2,325	625		2,950	3,550
4300	18" L x 18" W x 8" D		.50	16		4,125	750		4,875	5,675
4350	24" L x 18" W x 8" D		.40	20		5,525	940		6,465	7,525
4400	24" L x 24" W x 10" D		.30	26.667		7,150	1,250		8,400	9,800
4450	30" L x 24" W x 12" D		.20	40		10,900	1,875		12,775	14,900
5000	NEMA 9, dust tight 6" L x 6" W x 6" D		3.20	2.500		209	118		327	410
5050	8" L x 6" W x 6" D		2.70	2.963		250	139		389	490
5100	10" L x 6" W x 6" D		2	4		335	188		523	660
5150	12" L x 12" W x 6" D		1.60	5		665	235		900	1,100
5200	16" L x 16" W x 6" D		1	8		1,150	375		1,525	1,825
5250	18" L x 18" W x 8" D		.70	11.429		1,800	535		2,335	2,800

26 05 33.65 Pull Boxes		Crew	Daily Output	Labor-Hours	Unit	2009 Bare Costs Material	Labor	Equipment	Total	Total Incl O&P
5300	24" L x 18" W x 8" D	1 Elec	.60	13.333	Ea.	2,525	625		3,150	3,750
5350	24" L x 24" W x 10" D		.40	20		3,250	940		4,190	5,025
5400	30" L x 24" W x 12" D		.30	26.667		5,125	1,250		6,375	7,575
6000	J.I.C. wiring boxes, NEMA 12, dust tight & drip tight									
6050	6" L x 8" W x 4" D	1 Elec	10	.800	Ea.	52	37.50		89.50	115
6100	8" L x 10" W x 4" D		8	1		65	47		112	144
6150	12" L x 14" W x 6" D		5.30	1.509		105	71		176	225
6200	14" L x 16" W x 6" D		4.70	1.702		126	80		206	262
6250	16" L x 20" W x 6" D		4.40	1.818		267	85.50		352.50	425
6300	24" L x 30" W x 6" D		3.20	2.500		395	118		513	615
6350	24" L x 30" W x 8" D		2.90	2.759		420	130		550	660
6400	24" L x 36" W x 8" D		2.70	2.963		465	139		604	730
6450	24" L x 42" W x 8" D		2.30	3.478		510	163		673	810
6500	24" L x 48" W x 8" D		2	4		555	188		743	900

26 05 33.95 Cutting and Drilling		Crew	Daily Output	Labor-Hours	Unit	Material	Labor	Equipment	Total	Total Incl O&P
0010	**CUTTING AND DRILLING**									
0100	Hole drilling to 10' high, concrete wall									
0110	8" thick, 1/2" pipe size	R-31	12	.667	Ea.	4.71	31.50	4.41	40.62	58
0120	3/4" pipe size		12	.667		4.71	31.50	4.41	40.62	58
0130	1" pipe size		9.50	.842		9.50	39.50	5.55	54.55	77.50
0140	1-1/4" pipe size		9.50	.842		9.50	39.50	5.55	54.55	77.50
0150	1-1/2" pipe size		9.50	.842		9.50	39.50	5.55	54.55	77.50
0160	2" pipe size		4.40	1.818		10.15	85.50	12.05	107.70	155
0170	2-1/2" pipe size		4.40	1.818		10.15	85.50	12.05	107.70	155
0180	3" pipe size		4.40	1.818		10.15	85.50	12.05	107.70	155
0190	3-1/2" pipe size		3.30	2.424		11	114	16.05	141.05	205
0200	4" pipe size		3.30	2.424		11	114	16.05	141.05	205
0500	12" thick, 1/2" pipe size		9.40	.851		7.25	40	5.65	52.90	75.50
0520	3/4" pipe size		9.40	.851		7.25	40	5.65	52.90	75.50
0540	1" pipe size		7.30	1.096		13.75	51.50	7.25	72.50	102
0560	1-1/4" pipe size		7.30	1.096		13.75	51.50	7.25	72.50	102
0570	1-1/2" pipe size		7.30	1.096		13.75	51.50	7.25	72.50	102
0580	2" pipe size		3.60	2.222		15.70	104	14.70	134.40	194
0590	2-1/2" pipe size		3.60	2.222		15.70	104	14.70	134.40	194
0600	3" pipe size		3.60	2.222		15.70	104	14.70	134.40	194
0610	3-1/2" pipe size		2.80	2.857		18.25	134	18.90	171.15	248
0630	4" pipe size		2.50	3.200		18.25	150	21	189.25	275
0650	16" thick, 1/2" pipe size		7.60	1.053		9.75	49.50	6.95	66.20	94.50
0670	3/4" pipe size		7	1.143		9.75	53.50	7.55	70.80	102
0690	1" pipe size		6	1.333		18	62.50	8.85	89.35	126
0710	1-1/4" pipe size		5.50	1.455		18	68.50	9.65	96.15	135
0730	1-1/2" pipe size		5.50	1.455		18	68.50	9.65	96.15	135
0750	2" pipe size		3	2.667		21	125	17.65	163.65	236
0770	2-1/2" pipe size		2.70	2.963		21	139	19.60	179.60	259
0790	3" pipe size		2.50	3.200		21	150	21	192	278
0810	3-1/2" pipe size		2.30	3.478		25.50	163	23	211.50	305
0830	4" pipe size		2	4		25.50	188	26.50	240	345
0850	20" thick, 1/2" pipe size		6.40	1.250		12.25	59	8.30	79.55	113
0870	3/4" pipe size		6	1.333		12.25	62.50	8.85	83.60	120
0890	1" pipe size		5	1.600		22	75	10.60	107.60	152
0910	1-1/4" pipe size		4.80	1.667		22	78.50	11.05	111.55	158
0930	1-1/2" pipe size		4.60	1.739		22	81.50	11.50	115	163

26 05 33.95 Cutting and Drilling		Crew	Daily Output	Labor-Hours	Unit	Material	Labor	Equipment	Total	Total Incl O&P
						2009 Bare Costs				
0950	2" pipe size	R-31	2.70	2.963	Ea.	26.50	139	19.60	185.10	265
0970	2-1/2" pipe size		2.40	3.333		26.50	157	22	205.50	295
0990	3" pipe size		2.20	3.636		26.50	171	24	221.50	320
1010	3-1/2" pipe size		2	4		32.50	188	26.50	247	355
1030	4" pipe size		1.70	4.706		32.50	221	31	284.50	410
1050	24" thick, 1/2" pipe size		5.50	1.455		14.80	68.50	9.65	92.95	132
1070	3/4" pipe size		5.10	1.569		14.80	73.50	10.40	98.70	141
1090	1" pipe size		4.30	1.860		26.50	87.50	12.30	126.30	178
1110	1-1/4" pipe size		4	2		26.50	94	13.25	133.75	189
1130	1-1/2" pipe size		4	2		26.50	94	13.25	133.75	189
1150	2" pipe size		2.40	3.333		32	157	22	211	300
1170	2-1/2" pipe size		2.20	3.636		32	171	24	227	325
1190	3" pipe size		2	4		32	188	26.50	246.50	355
1210	3-1/2" pipe size		1.80	4.444		40	209	29.50	278.50	395
1230	4" pipe size		1.50	5.333		40	251	35.50	326.50	470
1500	Brick wall, 8" thick, 1/2" pipe size		18	.444		4.71	21	2.94	28.65	40.50
1520	3/4" pipe size		18	.444		4.71	21	2.94	28.65	40.50
1540	1" pipe size		13.30	.602		9.50	28.50	3.98	41.98	58.50
1560	1-1/4" pipe size		13.30	.602		9.50	28.50	3.98	41.98	58.50
1580	1-1/2" pipe size		13.30	.602		9.50	28.50	3.98	41.98	58.50
1600	2" pipe size		5.70	1.404		10.15	66	9.30	85.45	122
1620	2-1/2" pipe size		5.70	1.404		10.15	66	9.30	85.45	122
1640	3" pipe size		5.70	1.404		10.15	66	9.30	85.45	122
1660	3-1/2" pipe size		4.40	1.818		11	85.50	12.05	108.55	156
1680	4" pipe size		4	2		11	94	13.25	118.25	172
1700	12" thick, 1/2" pipe size		14.50	.552		7.25	26	3.65	36.90	52
1720	3/4" pipe size		14.50	.552		7.25	26	3.65	36.90	52
1740	1" pipe size		11	.727		13.75	34	4.81	52.56	73
1760	1-1/4" pipe size		11	.727		13.75	34	4.81	52.56	73
1780	1-1/2" pipe size		11	.727		13.75	34	4.81	52.56	73
1800	2" pipe size		5	1.600		15.70	75	10.60	101.30	145
1820	2-1/2" pipe size		5	1.600		15.70	75	10.60	101.30	145
1840	3" pipe size		5	1.600		15.70	75	10.60	101.30	145
1860	3-1/2" pipe size		3.80	2.105		18.25	99	13.95	131.20	187
1880	4" pipe size		3.30	2.424		18.25	114	16.05	148.30	213
1900	16" thick, 1/2" pipe size		12.30	.650		9.75	30.50	4.31	44.56	62.50
1920	3/4" pipe size		12.30	.650		9.75	30.50	4.31	44.56	62.50
1940	1" pipe size		9.30	.860		18	40.50	5.70	64.20	88
1960	1-1/4" pipe size		9.30	.860		18	40.50	5.70	64.20	88
1980	1-1/2" pipe size		9.30	.860		18	40.50	5.70	64.20	88
2000	2" pipe size		4.40	1.818		21	85.50	12.05	118.55	168
2010	2-1/2" pipe size		4.40	1.818		21	85.50	12.05	118.55	168
2030	3" pipe size		4.40	1.818		21	85.50	12.05	118.55	168
2050	3-1/2" pipe size		3.30	2.424		25.50	114	16.05	155.55	221
2070	4" pipe size		3	2.667		25.50	125	17.65	168.15	240
2090	20" thick, 1/2" pipe size		10.70	.748		12.25	35	4.95	52.20	73
2110	3/4" pipe size		10.70	.748		12.25	35	4.95	52.20	73
2130	1" pipe size		8	1		22	47	6.60	75.60	104
2150	1-1/4" pipe size		8	1		22	47	6.60	75.60	104
2170	1-1/2" pipe size		8	1		22	47	6.60	75.60	104
2190	2" pipe size		4	2		26.50	94	13.25	133.75	189
2210	2-1/2" pipe size		4	2		26.50	94	13.25	133.75	189
2230	3" pipe size		4	2		26.50	94	13.25	133.75	189

26 05 33 – Raceway and Boxes for Electrical Systems

26 05 33.95 Cutting and Drilling		Crew	Daily Output	Labor-Hours	Unit	Material	2009 Bare Costs Labor	Equipment	Total	Total Incl O&P
2250	3-1/2" pipe size	R-31	3	2.667	Ea.	32.50	125	17.65	175.15	248
2270	4" pipe size		2.70	2.963		32.50	139	19.60	191.10	272
2290	24" thick, 1/2" pipe size		9.40	.851		14.80	40	5.65	60.45	84
2310	3/4" pipe size		9.40	.851		14.80	40	5.65	60.45	84
2330	1" pipe size		7.10	1.127		26.50	53	7.45	86.95	119
2350	1-1/4" pipe size		7.10	1.127		26.50	53	7.45	86.95	119
2370	1-1/2" pipe size		7.10	1.127		26.50	53	7.45	86.95	119
2390	2" pipe size		3.60	2.222		32	104	14.70	150.70	213
2410	2-1/2" pipe size		3.60	2.222		32	104	14.70	150.70	213
2430	3" pipe size		3.60	2.222		32	104	14.70	150.70	213
2450	3-1/2" pipe size		2.80	2.857		40	134	18.90	192.90	272
2470	4" pipe size		2.50	3.200		40	150	21	211	299
3000	Knockouts to 8' high, metal boxes & enclosures									
3020	With hole saw, 1/2" pipe size	1 Elec	53	.151	Ea.		7.10		7.10	10.90
3040	3/4" pipe size		47	.170			8		8	12.30
3050	1" pipe size		40	.200			9.40		9.40	14.45
3060	1-1/4" pipe size		36	.222			10.45		10.45	16.05
3070	1-1/2" pipe size		32	.250			11.75		11.75	18.10
3080	2" pipe size		27	.296			13.95		13.95	21.50
3090	2-1/2" pipe size		20	.400			18.80		18.80	29
4010	3" pipe size		16	.500			23.50		23.50	36
4030	3-1/2" pipe size		13	.615			29		29	44.50
4050	4" pipe size		11	.727			34		34	52.50
4070	With hand punch set, 1/2" pipe size		40	.200			9.40		9.40	14.45
4090	3/4" pipe size		32	.250			11.75		11.75	18.10
4110	1" pipe size		30	.267			12.55		12.55	19.30
4130	1-1/4" pipe size		28	.286			13.45		13.45	20.50
4150	1-1/2" pipe size		26	.308			14.45		14.45	22.50
4170	2" pipe size		20	.400			18.80		18.80	29
4190	2-1/2" pipe size		17	.471			22		22	34
4200	3" pipe size		15	.533			25		25	38.50
4220	3-1/2" pipe size		12	.667			31.50		31.50	48
4240	4" pipe size		10	.800			37.50		37.50	58
4260	With hydraulic punch, 1/2" pipe size		44	.182			8.55		8.55	13.15
4280	3/4" pipe size		38	.211			9.90		9.90	15.20
4300	1" pipe size		38	.211			9.90		9.90	15.20
4320	1-1/4" pipe size		38	.211			9.90		9.90	15.20
4340	1-1/2" pipe size		38	.211			9.90		9.90	15.20
4360	2" pipe size		32	.250			11.75		11.75	18.10
4380	2-1/2" pipe size		27	.296			13.95		13.95	21.50
4400	3" pipe size		23	.348			16.35		16.35	25
4420	3-1/2" pipe size		20	.400			18.80		18.80	29
4440	4" pipe size		18	.444			21		21	32

26 05 36 – Cable Trays for Electrical Systems

26 05 36.10 Cable Tray Ladder Type		Crew	Daily Output	Labor-Hours	Unit	Material	Labor	Equipment	Total	Total Incl O&P
0010	**CABLE TRAY LADDER TYPE** w/ ftngs. & supports, 4" dp., to 15' elev.									
0160	Galvanized steel tray									
0170	4" rung spacing, 6" wide	2 Elec	98	.163	L.F.	13.25	7.65		20.90	26.50
0180	9" wide		92	.174		14.40	8.15		22.55	28.50
0200	12" wide		86	.186		15.90	8.75		24.65	31
0400	18" wide		82	.195		18.45	9.15		27.60	34.50
0600	24" wide		78	.205		21.50	9.65		31.15	38.50

26 05 36.10 Cable Tray Ladder Type		Crew	Daily Output	Labor-Hours	Unit	2009 Bare Costs Material	Labor	Equipment	Total	Total Incl O&P
0650	30″ wide	2 Elec	68	.235	L.F.	27	11.05		38.05	46.50
0700	36″ wide		60	.267		30	12.55		42.55	52
0800	6″ rung spacing, 6″ wide		100	.160		12.05	7.50		19.55	25
0850	9″ wide		94	.170		13.25	8		21.25	27
0860	12″ wide		88	.182		13.90	8.55		22.45	28.50
0870	18″ wide		84	.190		15.85	8.95		24.80	31
0880	24″ wide		80	.200		17.70	9.40		27.10	34
0890	30″ wide		70	.229		21	10.75		31.75	39.50
0900	36″ wide		64	.250		23	11.75		34.75	43.50
0910	9″ rung spacing, 6″ wide		102	.157		11.30	7.35		18.65	24
0920	9″ wide		98	.163		11.60	7.65		19.25	24.50
0930	12″ wide		94	.170		13.35	8		21.35	27
0940	18″ wide		90	.178		15	8.35		23.35	29.50
0950	24″ wide		86	.186		17.15	8.75		25.90	32.50
0960	30″ wide		80	.200		18.75	9.40		28.15	35
0970	36″ wide		74	.216		20	10.15		30.15	37.50
0980	12″ rung spacing, 6″ wide		106	.151		11.15	7.10		18.25	23
0990	9″ wide		104	.154		11.40	7.25		18.65	23.50
1000	12″ wide		100	.160		12.05	7.50		19.55	25
1010	18″ wide		96	.167		12.95	7.85		20.80	26.50
1020	24″ wide		94	.170		13.90	8		21.90	27.50
1030	30″ wide		88	.182		15.80	8.55		24.35	30.50
1040	36″ wide		84	.190		16.65	8.95		25.60	32
1041	18″ rung spacing, 6″ wide		108	.148		11.05	6.95		18	23
1042	9″ wide		106	.151		11.20	7.10		18.30	23.50
1043	12″ wide		102	.157		11.95	7.35		19.30	24.50
1044	18″ wide		98	.163		12.55	7.65		20.20	25.50
1045	24″ wide		96	.167		12.95	7.85		20.80	26.50
1046	30″ wide		90	.178		13.65	8.35		22	28
1047	36″ wide		86	.186		14	8.75		22.75	29
1050	Elbows horiz. 9″ rung spacing, 90°, 12″ radius, 6″ wide		9.60	1.667	Ea.	51.50	78.50		130	178
1060	9″ wide		8.40	1.905		54.50	89.50		144	198
1070	12″ wide		7.60	2.105		59.50	99		158.50	217
1080	18″ wide		6.20	2.581		78	121		199	273
1090	24″ wide		5.40	2.963		87.50	139		226.50	310
1100	30″ wide		4.80	3.333		115	157		272	370
1110	36″ wide		4.20	3.810		134	179		313	425
1120	90° 24″ radius, 6″ wide		9.20	1.739		106	81.50		187.50	243
1130	9″ wide		8	2		109	94		203	265
1140	12″ wide		7.20	2.222		112	104		216	285
1150	18″ wide		5.80	2.759		122	130		252	335
1160	24″ wide		5	3.200		131	150		281	375
1170	30″ wide		4.40	3.636		142	171		313	420
1180	36″ wide		3.80	4.211		168	198		366	490
1190	90° 36″ radius, 6″ wide		8.80	1.818		117	85.50		202.50	260
1200	9″ wide		7.60	2.105		123	99		222	288
1210	12″ wide		6.80	2.353		129	111		240	310
1220	18″ wide		5.40	2.963		148	139		287	375
1230	24″ wide		4.60	3.478		162	163		325	430
1240	30″ wide		4	4		195	188		383	505
1250	36″ wide		3.40	4.706		212	221		433	575
1260	45° 12″ radius, 6″ wide		13.20	1.212		40	57		97	132
1270	9″ wide		11	1.455		41.50	68.50		110	151

26 05 36 – Cable Trays for Electrical Systems

26 05 36.10 Cable Tray Ladder Type		Crew	Daily Output	Labor-Hours	Unit	Material	2009 Bare Costs Labor	Equipment	Total	Total Incl O&P
1280	12" wide	2 Elec	9.60	1.667	Ea.	43	78.50		121.50	169
1290	18" wide		7.60	2.105		45.50	99		144.50	202
1300	24" wide		6.20	2.581		58.50	121		179.50	252
1310	30" wide		5.40	2.963		74.50	139		213.50	296
1320	36" wide		4.60	3.478		78	163		241	335
1330	45° 24" radius, 6" wide		12.80	1.250		50.50	59		109.50	146
1340	9" wide		10.60	1.509		53.50	71		124.50	168
1350	12" wide		9.20	1.739		60	81.50		141.50	192
1360	18" wide		7.20	2.222		61.50	104		165.50	229
1370	24" wide		5.80	2.759		74.50	130		204.50	281
1380	30" wide		5	3.200		89	150		239	330
1390	36" wide		4.20	3.810		102	179		281	385
1400	45° 36" radius, 6" wide		12.40	1.290		72	60.50		132.50	173
1410	9" wide		10.20	1.569		76.50	73.50		150	197
1420	12" wide		8.80	1.818		78	85.50		163.50	217
1430	18" wide		6.80	2.353		84	111		195	263
1440	24" wide		5.40	2.963		89	139		228	310
1450	30" wide		4.60	3.478		115	163		278	380
1460	36" wide	↓	3.80	4.211	↓	128	198		326	445
1470	Elbows horizontal, 4" rung spacing, use 9" rung x 1.50									
1480	6" rung spacing use 9" rung x 1.20									
1490	12" rung spacing use 9" rung x .93									
1500	Elbows vert. 9" rung spacing, 90°, 12" radius, 6" wide	2 Elec	9.60	1.667	Ea.	70.50	78.50		149	199
1510	9" wide		8.40	1.905		72	89.50		161.50	218
1520	12" wide		7.60	2.105		74.50	99		173.50	234
1530	18" wide		6.20	2.581		78	121		199	273
1540	24" wide		5.40	2.963		81	139		220	305
1550	30" wide		4.80	3.333		87.50	157		244.50	335
1560	36" wide		4.20	3.810		90.50	179		269.50	375
1570	24" radius, 6" wide		9.20	1.739		100	81.50		181.50	236
1580	9" wide		8	2		103	94		197	258
1590	12" wide		7.20	2.222		106	104		210	278
1600	18" wide		5.80	2.759		112	130		242	325
1610	24" wide		5	3.200		119	150		269	360
1620	30" wide		4.40	3.636		123	171		294	400
1630	36" wide		3.80	4.211		139	198		337	460
1640	36" radius, 6" wide		8.80	1.818		134	85.50		219.50	279
1650	9" wide		7.60	2.105		139	99		238	305
1660	12" wide		6.80	2.353		140	111		251	325
1670	18" wide		5.40	2.963		153	139		292	380
1680	24" wide		4.60	3.478		154	163		317	420
1690	30" wide		4	4		175	188		363	480
1700	36" wide	↓	3.40	4.706	↓	187	221		408	545
1710	Elbows vertical, 4" rung spacing, use 9" rung x 1.25									
1720	6" rung spacing, use 9" rung x 1.15									
1730	12" rung spacing, use 9" rung x .90									
1740	Tee horizontal, 9" rung spacing, 12" radius, 6" wide	2 Elec	5	3.200	Ea.	109	150		259	350
1750	9" wide		4.60	3.478		117	163		280	380
1760	12" wide		4.40	3.636		123	171		294	400
1770	18" wide		4	4		140	188		328	445
1780	24" wide		3.60	4.444		165	209		374	500
1790	30" wide		3.40	4.706		198	221		419	560
1800	36" wide	↓	3	5.333	↓	225	251		476	630

26 05 36 – Cable Trays for Electrical Systems

26 05 36.10 Cable Tray Ladder Type		Crew	Daily Output	Labor-Hours	Unit	Material (2009 Bare Costs)	Labor (2009 Bare Costs)	Equipment (2009 Bare Costs)	Total (2009 Bare Costs)	Total Incl O&P
1810	24" radius, 6" wide	2 Elec	4.60	3.478	Ea.	170	163		333	440
1820	9" wide		4.20	3.810		178	179		357	470
1830	12" wide		4	4		184	188		372	490
1840	18" wide		3.60	4.444		200	209		409	540
1850	24" wide		3.20	5		217	235		452	600
1860	30" wide		3	5.333		236	251		487	645
1870	36" wide		2.60	6.154		305	289		594	780
1880	36" radius, 6" wide		4.20	3.810		254	179		433	555
1890	9" wide		3.80	4.211		265	198		463	595
1900	12" wide		3.60	4.444		282	209		491	630
1910	18" wide		3.20	5		305	235		540	695
1920	24" wide		2.80	5.714		340	269		609	790
1930	30" wide		2.60	6.154		375	289		664	855
1940	36" wide		2.20	7.273		400	340		740	965
1980	Tee vertical, 9" rung spacing, 12" radius, 6" wide		5.40	2.963		186	139		325	420
1990	9" wide		5.20	3.077		187	145		332	430
2000	12" wide		5	3.200		189	150		339	440
2010	18" wide		4.60	3.478		190	163		353	460
2020	24" wide		4.40	3.636		195	171		366	480
2030	30" wide		4	4		207	188		395	515
2040	36" wide		3.60	4.444		215	209		424	555
2050	24" radius, 6" wide		5	3.200		335	150		485	595
2060	9" wide		4.80	3.333		340	157		497	610
2070	12" wide		4.60	3.478		345	163		508	630
2080	18" wide		4.20	3.810		355	179		534	665
2090	24" wide		4	4		365	188		553	690
2100	30" wide		3.60	4.444		375	209		584	730
2110	36" wide		3.20	5		390	235		625	790
2120	36" radius, 6" wide		4.60	3.478		625	163		788	935
2130	9" wide		4.40	3.636		635	171		806	960
2140	12" wide		4.20	3.810		640	179		819	980
2150	18" wide		3.80	4.211		650	198		848	1,025
2160	24" wide		3.60	4.444		655	209		864	1,050
2170	30" wide		3.20	5		680	235		915	1,100
2180	36" wide		2.80	5.714		690	269		959	1,175
2190	Tee, 4" rung spacing, use 9" rung x 1.30									
2200	6" rung spacing, use 9" rung x 1.20									
2210	12" rung spacing, use 9" rung x .90									
2220	Cross horizontal, 9" rung spacing, 12" radius, 6" wide	2 Elec	4	4	Ea.	135	188		323	435
2230	9" wide		3.80	4.211		141	198		339	460
2240	12" wide		3.60	4.444		154	209		363	490
2250	18" wide		3.40	4.706		168	221		389	525
2260	24" wide		3	5.333		187	251		438	590
2270	30" wide		2.80	5.714		226	269		495	665
2280	36" wide		2.60	6.154		251	289		540	720
2290	24" radius, 6" wide		3.60	4.444		262	209		471	610
2300	9" wide		3.40	4.706		270	221		491	635
2310	12" wide		3.20	5		278	235		513	665
2320	18" wide		3	5.333		296	251		547	710
2330	24" wide		2.60	6.154		315	289		604	790
2340	30" wide		2.40	6.667		335	315		650	850
2350	36" wide		2.20	7.273		390	340		730	955
2360	36" radius, 6" wide		3.20	5		365	235		600	760

26 05 36 – Cable Trays for Electrical Systems

26 05 36.10 Cable Tray Ladder Type		Crew	Daily Output	Labor-Hours	Unit	Material	Labor	Equipment	Total	Total Incl O&P
						2009 Bare Costs				
2370	9" wide	2 Elec	3	5.333	Ea.	380	251		631	805
2380	12" wide		2.80	5.714		400	269		669	855
2390	18" wide		2.60	6.154		440	289		729	925
2400	24" wide		2.20	7.273		485	340		825	1,050
2410	30" wide		2	8		525	375		900	1,150
2420	36" wide		1.80	8.889		570	420		990	1,275
2430	Cross horizontal, 4" rung spacing, use 9" rung x 1.30									
2440	6" rung spacing, use 9" rung x 1.20									
2450	12" rung spacing, use 9" rung x .90									
2460	Reducer, 9" to 6" wide tray	2 Elec	13	1.231	Ea.	75	58		133	172
2470	12" to 9" wide tray		12	1.333		76.50	62.50		139	181
2480	18" to 12" wide tray		10.40	1.538		76.50	72.50		149	195
2490	24" to 18" wide tray		9	1.778		78	83.50		161.50	215
2500	30" to 24" wide tray		8	2		79.50	94		173.50	233
2510	36" to 30" wide tray		7	2.286		89	107		196	263
2511	Reducer, 18" to 6" wide tray		10.40	1.538		79.50	72.50		152	199
2512	24" to 12" wide tray		9	1.778		81	83.50		164.50	218
2513	30" to 18" wide tray		8	2		89	94		183	243
2514	30" to 12" wide tray		8	2		86	94		180	240
2515	36" to 24" wide tray		7	2.286		90.50	107		197.50	265
2516	36" to 18" wide tray		7	2.286		90.50	107		197.50	265
2517	36" to 12" wide tray		7	2.286		90.50	107		197.50	265
2520	Dropout or end plate, 6" wide		32	.500		8.35	23.50		31.85	45
2530	9" wide		28	.571		9.85	27		36.85	52.50
2540	12" wide		26	.615		10.75	29		39.75	56.50
2550	18" wide		22	.727		12.65	34		46.65	66.50
2560	24" wide		20	.800		15.30	37.50		52.80	75
2570	30" wide		18	.889		16.70	42		58.70	83
2580	36" wide		16	1		19.20	47		66.20	93.50
2590	Tray connector		48	.333		14.80	15.65		30.45	40.50
3200	Aluminum tray, 4" deep, 6" rung spacing, 6" wide		134	.119	L.F.	16.60	5.60		22.20	27
3210	9" wide		128	.125		17.60	5.90		23.50	28.50
3220	12" wide		124	.129		18.60	6.05		24.65	30
3230	18" wide		114	.140		20.50	6.60		27.10	33
3240	24" wide		106	.151		24	7.10		31.10	37.50
3250	30" wide		100	.160		26.50	7.50		34	40.50
3260	36" wide		94	.170		29	8		37	44
3270	9" rung spacing, 6" wide		140	.114		13.50	5.35		18.85	23
3280	9" wide		134	.119		14.15	5.60		19.75	24
3290	12" wide		130	.123		14.40	5.80		20.20	25
3300	18" wide		122	.131		16.85	6.15		23	28
3310	24" wide		116	.138		19.55	6.50		26.05	31.50
3320	30" wide		108	.148		21.50	6.95		28.45	34.50
3330	36" wide		100	.160		23	7.50		30.50	36.50
3340	12" rung spacing, 6" wide		146	.110		14.35	5.15		19.50	23.50
3350	9" wide		140	.114		14.50	5.35		19.85	24
3360	12" wide		134	.119		14.95	5.60		20.55	25
3370	18" wide		128	.125		15.85	5.90		21.75	26.50
3380	24" wide		124	.129		17.60	6.05		23.65	28.50
3390	30" wide		114	.140		18.85	6.60		25.45	30.50
3400	36" wide		106	.151		19.80	7.10		26.90	33
3401	18" rung spacing, 6" wide		150	.107		14.25	5		19.25	23.50
3402	9" wide tray		144	.111		14.40	5.20		19.60	24

26 05 36 – Cable Trays for Electrical Systems

26 05 36.10 Cable Tray Ladder Type		Crew	Daily Output	Labor-Hours	Unit	2009 Bare Costs Material	Labor	Equipment	Total	Total Incl O&P
3403	12" wide tray	2 Elec	140	.114	L.F.	14.95	5.35		20.30	24.50
3404	18" wide tray		134	.119		15.75	5.60		21.35	26
3405	24" wide tray		130	.123		17	5.80		22.80	27.50
3406	30" wide tray		120	.133		17.60	6.25		23.85	29
3407	36" wide tray		110	.145		18.35	6.85		25.20	30.50
3410	Elbows horiz. 9" rung spacing, 90° 12" radius, 6" wide		9.60	1.667	Ea.	64	78.50		142.50	192
3420	9" wide		8.40	1.905		65.50	89.50		155	211
3430	12" wide		7.50	2.105		78	99		177	238
3440	18" wide		6.20	2.581		87.50	121		208.50	284
3450	24" wide		5.40	2.963		109	139		248	335
3460	30" wide		4.80	3.333		115	157		272	370
3470	36" wide		4.20	3.810		139	179		318	430
3480	24" radius, 6" wide		9.20	1.739		114	81.50		195.50	251
3490	9" wide		8	2		118	94		212	275
3500	12" wide		7.20	2.222		124	104		228	297
3510	18" wide		5.80	2.759		135	130		265	345
3520	24" wide		5	3.200		148	150		298	395
3530	30" wide		4.40	3.636		161	171		332	440
3540	36" wide		3.80	4.211		183	198		381	505
3550	90°, 36" radius, 6" wide		8.80	1.818		125	85.50		210.50	268
3560	9" wide		7.60	2.105		128	99		227	293
3570	12" wide		6.80	2.353		144	111		255	330
3580	18" wide		5.40	2.963		161	139		300	390
3590	24" wide		4.60	3.478		181	163		344	450
3600	30" wide		4	4		200	188		388	510
3610	36" wide		3.40	4.706		229	221		450	590
3620	45° 12" radius, 6" wide		13.20	1.212		42	57		99	134
3630	9" wide		11	1.455		43.50	68.50		112	153
3640	12" wide		9.60	1.667		45	78.50		123.50	171
3650	18" wide		7.60	2.105		53.50	99		152.50	211
3660	24" wide		6.20	2.581		66.50	121		187.50	260
3670	30" wide		5.40	2.963		76.50	139		215.50	298
3680	36" wide		4.60	3.478		79.50	163		242.50	340
3690	45° 24" radius, 6" wide		12.80	1.250		56	59		115	153
3700	9" wide		10.60	1.509		64	71		135	180
3710	12" wide		9.20	1.739		65.50	81.50		147	198
3720	18" wide		7.20	2.222		70	104		174	238
3730	24" wide		5.80	2.759		81	130		211	288
3740	30" wide		5	3.200		96.50	150		246.50	335
3750	36" wide		4.20	3.810		103	179		282	390
3760	45° 36" radius, 6" wide		12.40	1.290		76.50	60.50		137	178
3770	9" wide		10.20	1.569		78	73.50		151.50	199
3780	12" wide		8.80	1.818		79.50	85.50		165	219
3790	18" wide		6.80	2.353		87.50	111		198.50	266
3800	24" wide		5.40	2.963		108	139		247	330
3810	30" wide		4.60	3.478		114	163		277	375
3820	36" wide		3.80	4.211		125	198		323	440
3830	Elbows horizontal, 4" rung spacing, use 9" rung x 1.50									
3840	6" rung spacing, use 9" rung x 1.20									
3850	12" rung spacing, use 9" rung x .93									
3860	Elbows vertical 9" rung spacing, 90° 12" radius, 6" wide	2 Elec	9.60	1.667	Ea.	89	78.50		167.50	219
3870	9" wide		8.40	1.905		93.50	89.50		183	241
3880	12" wide		7.60	2.105		95	99		194	257

26 05 36 – Cable Trays for Electrical Systems

26 05 36.10 Cable Tray Ladder Type		Crew	Daily Output	Labor-Hours	Unit	Material	Labor	Equipment	Total	Total Incl O&P
						2009 Bare Costs				
3890	18" wide	2 Elec	6.20	2.581	Ea.	98.50	121		219.50	295
3900	24" wide		5.40	2.963		105	139		244	330
3910	30" wide		4.80	3.333		109	157		266	360
3920	36" wide		4.20	3.810		112	179		291	400
3930	24" radius, 6" wide		9.20	1.739		96.50	81.50		178	232
3940	9" wide		8	2		100	94		194	255
3950	12" wide		7.20	2.222		103	104		207	274
3960	18" wide		5.80	2.759		109	130		239	320
3970	24" wide		5	3.200		115	150		265	360
3980	30" wide		4.40	3.636		122	171		293	395
3990	36" wide		3.80	4.211		129	198		327	445
4000	36" radius, 6" wide		8.80	1.818		150	85.50		235.50	296
4010	9" wide		7.60	2.105		153	99		252	320
4020	12" wide		6.80	2.353		162	111		273	350
4030	18" wide		5.40	2.963		168	139		307	400
4040	24" wide		4.60	3.478		175	163		338	445
4050	30" wide		4	4		189	188		377	495
4060	36" wide		3.40	4.706		197	221		418	555
4070	Elbows vertical, 4" rung spacing, use 9" rung x 1.25									
4080	6" rung spacing, use 9" rung x 1.15									
4090	12" rung spacing, use 9" rung x .90									
4100	Tee horizontal 9" rung spacing, 12" radius, 6" wide	2 Elec	5	3.200	Ea.	99	150		249	340
4110	9" wide		4.60	3.478		106	163		269	370
4120	12" wide		4.40	3.636		108	171		279	380
4130	18" wide		4.20	3.810		128	179		307	415
4140	24" wide		4	4		140	188		328	445
4150	30" wide		3.60	4.444		164	209		373	500
4160	36" wide		3.40	4.706		197	221		418	555
4170	24" radius, 6" wide		4.60	3.478		168	163		331	435
4180	9" wide		4.20	3.810		178	179		357	470
4190	12" wide		4	4		183	188		371	490
4200	18" wide		3.80	4.211		200	198		398	525
4210	24" wide		3.60	4.444		218	209		427	560
4220	30" wide		3.20	5		237	235		472	620
4230	36" wide		3	5.333		365	251		616	785
4240	36" radius, 6" wide		4.20	3.810		239	179		418	540
4250	9" wide		3.80	4.211		245	198		443	575
4260	12" wide		3.60	4.444		268	209		477	615
4270	18" wide		3.40	4.706		282	221		503	650
4280	24" wide		3.20	5		310	235		545	705
4290	30" wide		2.80	5.714		395	269		664	850
4300	36" wide		2.60	6.154		435	289		724	925
4310	Tee vertical 9" rung spacing, 12" radius, 6" wide		5.40	2.963		222	139		361	460
4320	9" wide		5.20	3.077		229	145		374	475
4330	12" wide		5	3.200		231	150		381	485
4340	18" wide		4.60	3.478		236	163		399	510
4350	24" wide		4.40	3.636		254	171		425	545
4360	30" wide		4.20	3.810		259	179		438	560
4370	36" wide		4	4		268	188		456	585
4380	24" radius, 6" wide		5	3.200		375	150		525	645
4390	9" wide		4.80	3.333		380	157		537	660
4400	12" wide		4.60	3.478		385	163		548	675
4410	18" wide		4.20	3.810		395	179		574	710

26 05 36.10 Cable Tray Ladder Type		Crew	Daily Output	Labor-Hours	Unit	2009 Bare Costs Material	Labor	Equipment	Total	Total Incl O&P
4420	24" wide	2 Elec	4	4	Ea.	405	188		593	740
4430	30" wide		3.80	4.211		420	198		618	765
4440	36" wide		3.60	4.444		455	209		664	820
4450	36" radius, 6" wide		4.60	3.478		825	163		988	1,150
4460	9" wide		4.40	3.636		845	171		1,016	1,200
4470	12" wide		4.20	3.810		875	179		1,054	1,250
4480	18" wide		3.80	4.211		885	198		1,083	1,275
4490	24" wide		3.60	4.444		910	209		1,119	1,325
4500	30" wide		3.40	4.706		915	221		1,136	1,350
4510	36" wide		3.20	5		935	235		1,170	1,375
4520	Tees, 4" rung spacing, use 9" rung x 1.30									
4530	6" rung spacing, use 9" rung x 1.20									
4540	12" rung spacing, use 9" rung x .90									
4550	Cross horizontal 9" rung spacing, 12" radius, 6" wide	2 Elec	4.40	3.636	Ea.	137	171		308	415
4560	9" wide		4.20	3.810		153	179		332	445
4570	12" wide		4	4		154	188		342	460
4580	18" wide		3.60	4.444		187	209		396	525
4590	24" wide		3.40	4.706		193	221		414	555
4600	30" wide		3	5.333		228	251		479	635
4610	36" wide		2.80	5.714		284	269		553	725
4620	24" radius, 6" wide		4	4		264	188		452	580
4630	9" wide		3.80	4.211		273	198		471	605
4640	12" wide		3.60	4.444		281	209		490	630
4650	18" wide		3.20	5		295	235		530	685
4660	24" wide		3	5.333		320	251		571	740
4670	30" wide		2.60	6.154		345	289		634	825
4680	36" wide		2.40	6.667		390	315		705	905
4690	36" radius, 6" wide		3.60	4.444		310	209		519	665
4700	9" wide		3.40	4.706		325	221		546	695
4710	12" wide		3.20	5		345	235		580	740
4720	18" wide		2.80	5.714		365	269		634	820
4730	24" wide		2.60	6.154		455	289		744	945
4740	30" wide		2.20	7.273		500	340		840	1,075
4750	36" wide		2	8		560	375		935	1,200
4760	Cross horizontal, 4" rung spacing, use 9" rung x 1.30									
4770	6" rung spacing, use 9" rung x 1.20									
4780	12" rung spacing, use 9" rung x .90									
4790	Reducer, 9" to 6" wide tray	2 Elec	16	1	Ea.	73.50	47		120.50	153
4800	12" to 9" wide tray		14	1.143		76.50	53.50		130	167
4810	18" to 12" wide tray		12.40	1.290		76.50	60.50		137	178
4820	24" to 18" wide tray		10.60	1.509		78	71		149	195
4830	30" to 24" wide tray		9.20	1.739		79.50	81.50		161	214
4840	36" to 30" wide tray		8	2		84	94		178	238
4841	Reducer, 18" to 6" wide tray		12.40	1.290		76.50	60.50		137	178
4842	24" to 12" wide tray		10.60	1.509		78	71		149	195
4843	30" to 18" wide tray		9.20	1.739		79.50	81.50		161	214
4844	30" to 12" wide tray		9.20	1.739		79.50	81.50		161	214
4845	36" to 24" wide tray		8	2		84	94		178	238
4846	36" to 18" wide tray		8	2		84	94		178	238
4847	36" to 12" wide tray		8	2		84	94		178	238
4850	Dropout or end plate, 6" wide		32	.500		9.65	23.50		33.15	46.50
4860	9" wide tray		28	.571		10.30	27		37.30	53
4870	12" wide tray		26	.615		11.25	29		40.25	57

26 05 36.10 Cable Tray Ladder Type

26 05 36.10 Cable Tray Ladder Type		Crew	Daily Output	Labor-Hours	Unit	2009 Bare Costs Material	Labor	Equipment	Total	Total Incl O&P
4880	18" wide tray	2 Elec	22	.727	Ea.	15	34		49	69
4890	24" wide tray		20	.800		17.65	37.50		55.15	77.50
4900	30" wide tray		18	.889		21	42		63	87.50
4910	36" wide tray		16	1		23	47		70	98
4920	Tray connector	↓	48	.333	↓	14.05	15.65		29.70	39.50
9200	Splice plate	1 Elec	48	.167	Pr.	14.65	7.85		22.50	28
9210	Expansion joint		48	.167		17.65	7.85		25.50	31.50
9220	Horizontal hinged		48	.167		14.65	7.85		22.50	28
9230	Vertical hinged		48	.167	↓	17.65	7.85		25.50	31.50
9240	Ladder hanger, vertical		28	.286	Ea.	4.37	13.45		17.82	25.50
9250	Ladder to channel connector		24	.333		51	15.65		66.65	80.50
9260	Ladder to box connector 30" wide		19	.421		51	19.80		70.80	87
9270	24" wide		20	.400		51	18.80		69.80	85.50
9280	18" wide		21	.381		49.50	17.90		67.40	82
9290	12" wide		22	.364		46	17.10		63.10	77.50
9300	9" wide		23	.348		41.50	16.35		57.85	70.50
9310	6" wide		24	.333		41.50	15.65		57.15	69.50
9320	Ladder floor flange		24	.333		33	15.65		48.65	60.50
9330	Cable roller for tray 30" wide		10	.800		272	37.50		309.50	355
9340	24" wide		11	.727		231	34		265	305
9350	18" wide		12	.667		220	31.50		251.50	290
9360	12" wide		13	.615		173	29		202	236
9370	9" wide		14	.571		154	27		181	211
9380	6" wide		15	.533		125	25		150	176
9390	Pulley, single wheel		12	.667		284	31.50		315.50	365
9400	Triple wheel		10	.800		565	37.50		602.50	685
9440	Nylon cable tie, 14" long		80	.100		.44	4.70		5.14	7.75
9450	Ladder, hold down clamp		60	.133		9.35	6.25		15.60	19.95
9460	Cable clamp		60	.133		9.50	6.25		15.75	20
9470	Wall bracket, 30" wide tray		19	.421		50	19.80		69.80	85.50
9480	24" wide tray		20	.400		39	18.80		57.80	72
9490	18" wide tray		21	.381		35	17.90		52.90	66
9500	12" wide tray		22	.364		32.50	17.10		49.60	62.50
9510	9" wide tray		23	.348		28.50	16.35		44.85	56.50
9520	6" wide tray	↓	24	.333	↓	28	15.65		43.65	55

26 05 36.20 Cable Tray Solid Bottom

26 05 36.20 Cable Tray Solid Bottom		Crew	Daily Output	Labor-Hours	Unit	Material	Labor	Equipment	Total	Total Incl O&P
0010	**CABLE TRAY SOLID BOTTOM** w/ ftngs. & supports, 3" dp, to 15' high									
0220	Galvanized steel, tray, 6" wide	2 Elec	120	.133	L.F.	10.70	6.25		16.95	21.50
0240	12" wide		100	.160		13.70	7.50		21.20	26.50
0260	18" wide		70	.229		16.65	10.75		27.40	35
0280	24" wide		60	.267		19.55	12.55		32.10	41
0300	30" wide		50	.320		23	15.05		38.05	48.50
0320	36" wide		44	.364	↓	33	17.10		50.10	63
0340	Elbow horizontal 90°, 12" radius, 6" wide		9.60	1.667	Ea.	81	78.50		159.50	210
0360	12" wide		6.80	2.353		91	111		202	270
0370	18" wide		5.40	2.963		104	139		243	330
0380	24" wide		4.40	3.636		126	171		297	400
0390	30" wide		3.80	4.211		151	198		349	470
0400	36" wide		3.40	4.706		172	221		393	530
0420	24" radius, 6" wide		9.20	1.739		115	81.50		196.50	253
0440	12" wide		6.40	2.500		134	118		252	330
0450	18" wide	↓	5	3.200	↓	156	150		306	405

26 05 36.20 Cable Tray Solid Bottom		Crew	Daily Output	Labor-Hours	Unit	Material	2009 Bare Costs Labor	Equipment	Total	Total Incl O&P
0460	24" wide	2 Elec	4	4	Ea.	181	188		369	490
0470	30" wide		3.40	4.706		212	221		433	575
0480	36" wide		3	5.333		239	251		490	650
0500	36" radius, 6" wide		8.80	1.818		168	85.50		253.50	315
0520	12" wide		6	2.667		190	125		315	400
0530	18" wide		4.60	3.478		232	163		395	505
0540	24" wide		3.60	4.444		251	209		460	595
0550	30" wide		3	5.333		305	251		556	720
0560	36" wide		2.60	6.154		340	289		629	820
0580	Elbow vertical 90°, 12" radius, 6" wide		9.60	1.667		98.50	78.50		177	229
0600	12" wide		6.80	2.353		105	111		216	285
0610	18" wide		5.40	2.963		106	139		245	330
0620	24" wide		4.40	3.636		123	171		294	400
0630	30" wide		3.80	4.211		131	198		329	450
0640	36" wide		3.40	4.706		136	221		357	490
0670	24" radius, 6" wide		9.20	1.739		139	81.50		220.50	279
0690	12" wide		6.40	2.500		153	118		271	350
0700	18" wide		5	3.200		165	150		315	415
0710	24" wide		4	4		179	188		367	485
0720	30" wide		3.40	4.706		190	221		411	550
0730	36" wide		3	5.333		207	251		458	615
0750	36" radius, 6" wide		8.80	1.818		190	85.50		275.50	340
0770	12" wide		6.60	2.424		212	114		326	410
0780	18" wide		4.60	3.478		232	163		395	505
0790	24" wide		3.60	4.444		254	209		463	600
0800	30" wide		3	5.333		278	251		529	690
0810	36" wide		2.60	6.154		300	289		589	775
0840	Tee horizontal, 12" radius, 6" wide		5	3.200		114	150		264	355
0860	12" wide		4	4		123	188		311	425
0870	18" wide		3.40	4.706		147	221		368	500
0880	24" wide		2.80	5.714		164	269		433	595
0890	30" wide		2.60	6.154		187	289		476	650
0900	36" wide		2.20	7.273		214	340		554	760
0940	24" radius, 6" wide		4.60	3.478		176	163		339	445
0960	12" wide		3.60	4.444		206	209		415	545
0970	18" wide		3	5.333		229	251		480	635
0980	24" wide		2.40	6.667		310	315		625	820
0990	30" wide		2.20	7.273		335	340		675	895
1000	36" wide		1.80	8.889		365	420		785	1,050
1020	36" radius, 6" wide		4.20	3.810		278	179		457	580
1040	12" wide		3.20	5		310	235		545	700
1050	18" wide		2.60	6.154		335	289		624	815
1060	24" wide		2.20	7.273		435	340		775	1,000
1070	30" wide		2	8		470	375		845	1,100
1080	36" wide		1.60	10		485	470		955	1,250
1100	Tee vertical, 12" radius, 6" wide		5	3.200		187	150		337	435
1120	12" wide		4	4		195	188		383	505
1130	18" wide		3.60	4.444		198	209		407	540
1140	24" wide		3.40	4.706		217	221		438	580
1150	30" wide		3	5.333		232	251		483	640
1160	36" wide		2.60	6.154		240	289		529	710
1180	24" radius, 6" wide		4.60	3.478		276	163		439	555
1200	12" wide		3.60	4.444		289	209		498	635

26 05 36.20 Cable Tray Solid Bottom		Crew	Daily Output	Labor-Hours	Unit	2009 Bare Costs Material	Labor	Equipment	Total	Total Incl O&P
1210	18" wide	2 Elec	3.20	5	Ea.	305	235		540	695
1220	24" wide		3	5.333		320	251		571	735
1230	30" wide		2.60	6.154		350	289		639	830
1240	36" wide		2.20	7.273		365	340		705	925
1260	36" radius, 6" wide		4.20	3.810		430	179		609	750
1280	12" wide		3.20	5		440	235		675	840
1290	18" wide		2.80	5.714		470	269		739	935
1300	24" wide		2.60	6.154		485	289		774	980
1310	30" wide		2.20	7.273		570	340		910	1,150
1320	36" wide		2	8		600	375		975	1,250
1340	Cross horizontal, 12" radius, 6" wide		4	4		136	188		324	440
1360	12" wide		3.40	4.706		148	221		369	505
1370	18" wide		2.80	5.714		173	269		442	605
1380	24" wide		2.40	6.667		195	315		510	695
1390	30" wide		2	8		217	375		592	820
1400	36" wide		1.80	8.889		239	420		659	910
1420	24" radius, 6" wide		3.60	4.444		248	209		457	595
1440	12" wide		3	5.333		278	251		529	690
1450	18" wide		2.40	6.667		310	315		625	820
1460	24" wide		2	8		400	375		775	1,025
1470	30" wide		1.80	8.889		430	420		850	1,125
1480	36" wide		1.60	10		455	470		925	1,225
1500	36" radius, 6" wide		3.20	5		405	235		640	805
1520	12" wide		2.60	6.154		445	289		734	935
1530	18" wide		2	8		485	375		860	1,125
1540	24" wide		1.80	8.889		575	420		995	1,275
1550	30" wide		1.60	10		625	470		1,095	1,400
1560	36" wide		1.40	11.429		665	535		1,200	1,550
1580	Drop out or end plate, 6" wide		32	.500		16.55	23.50		40.05	54
1600	12" wide		26	.615		20.50	29		49.50	67
1610	18" wide		22	.727		22	34		56	77
1620	24" wide		20	.800		26	37.50		63.50	86.50
1630	30" wide		18	.889		29	42		71	96.50
1640	36" wide		16	1		31	47		78	107
1660	Reducer, 12" to 6" wide		12	1.333		79.50	62.50		142	184
1680	18" to 12" wide		10.60	1.509		81	71		152	198
1700	18" to 6" wide		10.60	1.509		81	71		152	198
1720	24" to 18" wide		9.20	1.739		84	81.50		165.50	219
1740	24" to 12" wide		9.20	1.739		84	81.50		165.50	219
1760	30" to 24" wide		8	2		90.50	94		184.50	245
1780	30" to 18" wide		8	2		90.50	94		184.50	245
1800	30" to 12" wide		8	2		90.50	94		184.50	245
1820	36" to 30" wide		7.20	2.222		93.50	104		197.50	264
1840	36" to 24" wide		7.20	2.222		93.50	104		197.50	264
1860	36" to 18" wide		7.20	2.222		93.50	104		197.50	264
1880	36" to 12" wide		7.20	2.222		93.50	104		197.50	264
2000	Aluminum tray, 6" wide		150	.107	L.F.	12.95	5		17.95	22
2020	12" wide		130	.123		17	5.80		22.80	27.50
2030	18" wide		100	.160		21.50	7.50		29	35
2040	24" wide		90	.178		27	8.35		35.35	42.50
2050	30" wide		70	.229		31.50	10.75		42.25	51
2060	36" wide		64	.250		39.50	11.75		51.25	61.50
2080	Elbow horizontal 90°, 12" radius, 6" wide		9.60	1.667	Ea.	104	78.50		182.50	235

26 05 36 – Cable Trays for Electrical Systems

26 05 36.20 Cable Tray Solid Bottom		Crew	Daily Output	Labor-Hours	Unit	Material	Labor	Equipment	Total	Total Incl O&P
						2009 Bare Costs				
2100	12″ wide	2 Elec	7.60	2.105	Ea.	117	99		216	281
2110	18″ wide		6.80	2.353		144	111		255	330
2120	24″ wide		5.80	2.759		164	130		294	380
2130	30″ wide		5	3.200		200	150		350	450
2140	36″ wide		4.40	3.636		225	171		396	510
2160	24″ radius, 6″ wide		9.20	1.739		148	81.50		229.50	289
2180	12″ wide		7.20	2.222		175	104		279	355
2190	18″ wide		6.40	2.500		200	118		318	400
2200	24″ wide		5.40	2.963		246	139		385	485
2210	30″ wide		4.60	3.478		282	163		445	560
2220	36″ wide		4	4		325	188		513	645
2240	36″ radius, 6″ wide		8.80	1.818		218	85.50		303.50	370
2260	12″ wide		6.80	2.353		262	111		373	460
2270	18″ wide		6	2.667		310	125		435	540
2280	24″ wide		5	3.200		335	150		485	595
2290	30″ wide		4.20	3.810		380	179		559	690
2300	36″ wide		3.60	4.444		445	209		654	805
2320	Elbow vertical 90°, 12″ radius, 6″ wide		9.60	1.667		123	78.50		201.50	257
2340	12″ wide		7.60	2.105		128	99		227	293
2350	18″ wide		6.80	2.353		145	111		256	330
2360	24″ wide		5.80	2.759		153	130		283	365
2370	30″ wide		5	3.200		162	150		312	410
2380	36″ wide		4.40	3.636		167	171		338	445
2400	24″ radius, 6″ wide		9.20	1.739		175	81.50		256.50	320
2420	12″ wide		7.20	2.222		190	104		294	370
2430	18″ wide		6.40	2.500		204	118		322	405
2440	24″ wide		5.40	2.963		222	139		361	460
2450	30″ wide		4.60	3.478		236	163		399	510
2460	36″ wide		4	4		259	188		447	575
2480	36″ radius, 6″ wide		8.80	1.818		231	85.50		316.50	385
2500	12″ wide		6.80	2.353		259	111		370	455
2510	18″ wide		6	2.667		282	125		407	505
2520	24″ wide		5	3.200		293	150		443	555
2530	30″ wide		4.20	3.810		325	179		504	630
2540	36″ wide		3.60	4.444		335	209		544	690
2560	Tee horizontal, 12″ radius, 6″ wide		5	3.200		161	150		311	410
2580	12″ wide		4.40	3.636		195	171		366	480
2590	18″ wide		4	4		218	188		406	530
2600	24″ wide		3.60	4.444		245	209		454	590
2610	30″ wide		3	5.333		282	251		533	695
2620	36″ wide		2.40	6.667		325	315		640	835
2640	24″ radius, 6″ wide		4.60	3.478		265	163		428	545
2660	12″ wide		4	4		298	188		486	620
2670	18″ wide		3.60	4.444		335	209		544	690
2680	24″ wide		3	5.333		415	251		666	840
2690	30″ wide		2.40	6.667		470	315		785	1,000
2700	36″ wide		2.20	7.273		525	340		865	1,100
2720	36″ radius, 6″ wide		4.20	3.810		430	179		609	750
2740	12″ wide		3.60	4.444		470	209		679	840
2750	18″ wide		3.20	5		545	235		780	955
2760	24″ wide		2.60	6.154		635	289		924	1,150
2770	30″ wide		2	8		695	375		1,070	1,350
2780	36″ wide		1.80	8.889		780	420		1,200	1,500

26 05 36 – Cable Trays for Electrical Systems

26 05 36.20 Cable Tray Solid Bottom		Crew	Daily Output	Labor-Hours	Unit	2009 Bare Costs Material	Labor	Equipment	Total	Total Incl O&P
2800	Tee vertical, 12" radius, 6" wide	2 Elec	5	3.200	Ea.	229	150		379	485
2820	12" wide		4.40	3.636		234	171		405	520
2830	18" wide		4.20	3.810		243	179		422	545
2840	24" wide		4	4		259	188		447	575
2850	30" wide		3.60	4.444		282	209		491	630
2860	36" wide		3	5.333		298	251		549	715
2880	24" radius, 6" wide		4.60	3.478		335	163		498	620
2900	12" wide		4	4		355	188		543	685
2910	18" wide		3.80	4.211		380	198		578	725
2920	24" wide		3.60	4.444		415	209		624	775
2930	30" wide		3.20	5		455	235		690	860
2940	36" wide		2.60	6.154		480	289		769	970
2960	36" radius, 6" wide		4.20	3.810		525	179		704	855
2980	12" wide		3.40	4.706		550	221		771	945
2990	18" wide		3.40	4.706		585	221		806	980
3000	24" wide		3.20	5		600	235		835	1,025
3010	30" wide		2.80	5.714		650	269		919	1,125
3020	36" wide		2.20	7.273		715	340		1,055	1,325
3040	Cross horizontal, 12" radius, 6" wide		4.40	3.636		204	171		375	490
3060	12" wide		4	4		231	188		419	545
3070	18" wide		3.40	4.706		262	221		483	630
3080	24" wide		2.80	5.714		298	269		567	745
3090	30" wide		2.60	6.154		335	289		624	815
3100	36" wide		2.20	7.273		375	340		715	935
3120	24" radius, 6" wide		4	4		355	188		543	685
3140	12" wide		3.60	4.444		415	209		624	775
3150	18" wide		3	5.333		450	251		701	880
3160	24" wide		2.40	6.667		545	315		860	1,075
3170	30" wide		2.20	7.273		600	340		940	1,175
3180	36" wide		1.80	8.889		625	420		1,045	1,325
3200	36" radius, 6" wide		3.60	4.444		625	209		834	1,000
3220	12" wide		3.20	5		655	235		890	1,075
3230	18" wide		2.60	6.154		730	289		1,019	1,250
3240	24" wide		2	8		795	375		1,170	1,450
3250	30" wide		1.80	8.889		860	420		1,280	1,600
3260	36" wide		1.60	10		1,000	470		1,470	1,825
3280	Dropout, or end plate, 6" wide		32	.500		20.50	23.50		44	58.50
3300	12" wide		26	.615		23	29		52	70
3310	18" wide		22	.727		28	34		62	83
3320	24" wide		20	.800		30.50	37.50		68	91.50
3330	30" wide		18	.889		35.50	42		77.50	104
3340	36" wide		16	1		41.50	47		88.50	118
3380	Reducer, 12" to 6" wide		14	1.143		101	53.50		154.50	194
3400	18" to 12" wide		12	1.333		104	62.50		166.50	211
3420	18" to 6" wide		12	1.333		104	62.50		166.50	211
3440	24" to 18" wide		10.60	1.509		110	71		181	230
3460	24" to 12" wide		10.60	1.509		110	71		181	230
3480	30" to 24" wide		9.20	1.739		120	81.50		201.50	258
3500	30" to 18" wide		9.20	1.739		120	81.50		201.50	258
3520	30" to 12" wide		9.20	1.739		120	81.50		201.50	258
3540	36" to 30" wide		8	2		126	94		220	284
3560	36" to 24" wide		8	2		128	94		222	286
3580	36" to 18" wide		8	2		128	94		222	286

26 05 36.20 Cable Tray Solid Bottom		Crew	Daily Output	Labor-Hours	Unit	Material	2009 Bare Costs Labor	Equipment	Total	Total Incl O&P
3600	36" to 12" wide	2 Elec	8	2	Ea.	131	94		225	289

26 05 36.30 Cable Tray Trough

		Crew	Daily Output	Labor-Hours	Unit	Material	Labor	Equipment	Total	Total Incl O&P
0010	**CABLE TRAY TROUGH** vented, w/ ftngs & supports, 6" dp, to 15' high									
0200	Galvanized steel, tray, 6" wide	2 Elec	90	.178	L.F.	14.05	8.35		22.40	28.50
0240	12" wide		80	.200		17.15	9.40		26.55	33.50
0260	18" wide		70	.229		20.50	10.75		31.25	39.50
0280	24" wide		60	.267		23.50	12.55		36.05	45.50
0300	30" wide		50	.320		34	15.05		49.05	60.50
0320	36" wide		40	.400		38	18.80		56.80	70.50
0340	Elbow horizontal 90°, 12" radius, 6" wide		7.60	2.105	Ea.	86.50	99		185.50	248
0360	12" wide		5.60	2.857		103	134		237	320
0370	18" wide		4.40	3.636		119	171		290	395
0380	24" wide		3.60	4.444		147	209		356	480
0390	30" wide		3.20	5		165	235		400	540
0400	36" wide		2.80	5.714		192	269		461	625
0420	24" radius, 6" wide		7.20	2.222		128	104		232	300
0440	12" wide		5.20	3.077		148	145		293	385
0450	18" wide		4	4		178	188		366	485
0460	24" wide		3.20	5		201	235		436	580
0470	30" wide		2.80	5.714		232	269		501	670
0480	36" wide		2.40	6.667		270	315		585	775
0500	36" radius, 6" wide		6.80	2.353		192	111		303	380
0520	12" wide		4.80	3.333		218	157		375	480
0530	18" wide		3.60	4.444		262	209		471	610
0540	24" wide		2.80	5.714		278	269		547	720
0550	30" wide		2.40	6.667		310	315		625	825
0560	36" wide		2	8		350	375		725	965
0580	Elbow vertical 90°, 12" radius, 6" wide		7.60	2.105		112	99		211	276
0600	12" wide		5.60	2.857		125	134		259	345
0610	18" wide		4.40	3.636		126	171		297	400
0620	24" wide		3.60	4.444		145	209		354	480
0630	30" wide		3.20	5		148	235		383	525
0640	36" wide		2.80	5.714		156	269		425	585
0660	24" radius, 6" wide		7.20	2.222		161	104		265	340
0680	12" wide		5.20	3.077		170	145		315	410
0690	18" wide		4	4		189	188		377	495
0700	24" wide		3.20	5		195	235		430	575
0710	30" wide		2.80	5.714		218	269		487	655
0720	36" wide		2.40	6.667		229	315		544	730
0740	36" radius, 6" wide		6.80	2.353		217	111		328	410
0760	12" wide		4.80	3.333		231	157		388	495
0770	18" wide		3.60	4.444		256	209		465	600
0780	24" wide		2.80	5.714		275	269		544	715
0790	30" wide		2.40	6.667		281	315		596	790
0800	36" wide		2	8		310	375		685	925
0820	Tee horizontal, 12" radius, 6" wide		4	4		131	188		319	435
0840	12" wide		3.20	5		147	235		382	520
0850	18" wide		2.80	5.714		165	269		434	595
0860	24" wide		2.40	6.667		192	315		507	690
0870	30" wide		2.20	7.273		218	340		558	765
0880	36" wide		2	8		240	375		615	845
0900	24" radius, 6" wide		3.60	4.444		217	209		426	560

26 05 36.30 Cable Tray Trough		Crew	Daily Output	Labor-Hours	Unit	Material	Labor	Equipment	Total	Total Incl O&P
						2009 Bare Costs				
0920	12" wide	2 Elec	2.80	5.714	Ea.	237	269		506	675
0930	18" wide		2.40	6.667		265	315		580	770
0940	24" wide		2	8		345	375		720	960
0950	30" wide		1.80	8.889		375	420		795	1,050
0960	36" wide		1.60	10		410	470		880	1,175
0980	36" radius, 6" wide		3.20	5		310	235		545	705
1000	12" wide		2.40	6.667		355	315		670	870
1010	18" wide		2	8		390	375		765	1,000
1020	24" wide		1.60	10		470	470		940	1,250
1030	30" wide		1.40	11.429		495	535		1,030	1,375
1040	36" wide		1.20	13.333		585	625		1,210	1,600
1060	Tee vertical, 12" radius, 6" wide		4	4		218	188		406	530
1080	12" wide		3.20	5		222	235		457	605
1090	18" wide		3	5.333		228	251		479	635
1100	24" wide		2.80	5.714		240	269		509	680
1110	30" wide		2.60	6.154		254	289		543	725
1120	36" wide		2.20	7.273		259	340		599	810
1140	24" radius, 6" wide		3.60	4.444		293	209		502	645
1160	12" wide		2.80	5.714		310	269		579	755
1170	18" wide		2.60	6.154		325	289		614	800
1180	24" wide		2.40	6.667		350	315		665	865
1190	30" wide		2.20	7.273		365	340		705	925
1200	36" wide		1.80	8.889		395	420		815	1,075
1220	36" radius, 6" wide		3.20	5		470	235		705	875
1240	12" wide		2.40	6.667		480	315		795	1,000
1250	18" wide		2.20	7.273		495	340		835	1,075
1260	24" wide		2	8		525	375		900	1,150
1270	30" wide		1.80	8.889		610	420		1,030	1,325
1280	36" wide		1.40	11.429		615	535		1,150	1,500
1300	Cross horizontal, 12" radius, 6" wide		3.20	5		168	235		403	545
1320	12" wide		2.80	5.714		170	269		439	600
1330	18" wide		2.40	6.667		190	315		505	690
1340	24" wide		2	8		200	375		575	800
1350	30" wide		1.80	8.889		225	420		645	890
1360	36" wide		1.60	10		251	470		721	1,000
1380	24" radius, 6" wide		2.80	5.714		265	269		534	705
1400	12" wide		2.40	6.667		278	315		593	785
1410	18" wide		2	8		305	375		680	915
1420	24" wide		1.60	10		390	470		860	1,150
1430	30" wide		1.40	11.429		430	535		965	1,300
1440	36" wide		1.20	13.333		455	625		1,080	1,475
1460	36" radius, 6" wide		2.40	6.667		455	315		770	980
1480	12" wide		2	8		470	375		845	1,100
1490	18" wide		1.60	10		480	470		950	1,250
1500	24" wide		1.20	13.333		575	625		1,200	1,600
1510	30" wide		1	16		635	750		1,385	1,850
1520	36" wide		.80	20		680	940		1,620	2,200
1540	Dropout or end plate, 6" wide		26	.615		19.50	29		48.50	66
1560	12" wide		22	.727		24	34		58	79
1580	18" wide		20	.800		26	37.50		63.50	87
1600	24" wide		18	.889		30	42		72	97.50
1620	30" wide		16	1		32.50	47		79.50	108
1640	36" wide		13.40	1.194		35.50	56		91.50	126

26 05 36 – Cable Trays for Electrical Systems

26 05 36.30 Cable Tray Trough		Crew	Daily Output	Labor-Hours	Unit	Material	2009 Bare Costs Labor	Equipment	Total	Total Incl O&P
1660	Reducer, 12" to 6" wide	2 Elec	9.40	1.702	Ea.	82	80		162	213
1680	18" to 12" wide		8.40	1.905		85	89.50		174.50	232
1700	18" to 6" wide		8.40	1.905		85	89.50		174.50	232
1720	24" to 18" wide		7.20	2.222		91	104		195	261
1740	24" to 12" wide		7.20	2.222		91	104		195	261
1760	30" to 24" wide		6.40	2.500		94	118		212	284
1780	30" to 18" wide		6.40	2.500		94	118		212	284
1800	30" to 12" wide		6.40	2.500		94	118		212	284
1820	36" to 30" wide		5.80	2.759		101	130		231	310
1840	36" to 24" wide		5.80	2.759		101	130		231	310
1860	36" to 18" wide		5.80	2.759		102	130		232	310
1880	36" to 12" wide		5.80	2.759		102	130		232	310
2000	Aluminum, tray, vented, 6" wide		120	.133	L.F.	17	6.25		23.25	28.50
2010	9" wide		110	.145		19.15	6.85		26	31.50
2020	12" wide		100	.160		21.50	7.50		29	35
2030	18" wide		90	.178		25.50	8.35		33.85	41
2040	24" wide		80	.200		30	9.40		39.40	47.50
2050	30" wide		70	.229		40	10.75		50.75	60.50
2060	36" wide		60	.267		44	12.55		56.55	68
2080	Elbow horiz. 90°, 12" radius, 6" wide		7.60	2.105	Ea.	112	99		211	275
2090	9" wide		7	2.286		122	107		229	299
2100	12" wide		6.20	2.581		129	121		250	330
2110	18" wide		5.60	2.857		151	134		285	375
2120	24" wide		4.60	3.478		175	163		338	445
2130	30" wide		4	4		215	188		403	525
2140	36" wide		3.50	4.444		236	209		445	580
2160	24" radius, 6" wide		7.20	2.222		162	104		266	340
2180	12" wide		5.80	2.759		190	130		320	410
2190	18" wide		5.20	3.077		218	145		363	460
2200	24" wide		4.20	3.810		250	179		429	550
2210	30" wide		3.60	4.444		285	209		494	635
2220	36" wide		3.20	5		310	235		545	700
2240	36" radius, 6" wide		6.80	2.353		229	111		340	420
2260	12" wide		5.40	2.963		261	139		400	500
2270	18" wide		4.80	3.333		296	157		453	565
2280	24" wide		3.80	4.211		325	198		523	660
2290	30" wide		3.40	4.706		380	221		601	760
2300	36" wide		2.80	5.714		415	269		684	870
2320	Elbow vertical 90°, 12" radius, 6" wide		7.60	2.105		136	99		235	300
2330	9" wide		7	2.286		145	107		252	325
2340	12" wide		6.20	2.581		147	121		268	350
2350	18" wide		5.60	2.857		151	134		285	375
2360	24" wide		4.60	3.478		167	163		330	435
2370	30" wide		4	4		175	188		363	480
2380	36" wide		3.60	4.444		176	209		385	515
2400	24" radius, 6" wide		7.20	2.222		184	104		288	365
2420	12" wide		5.80	2.759		200	130		330	420
2430	18" wide		5.20	3.077		217	145		362	460
2440	24" wide		4.20	3.810		220	179		399	515
2450	30" wide		3.60	4.444		232	209		441	575
2460	36" wide		3.20	5		243	235		478	630
2480	36" radius, 6" wide		6.80	2.353		228	111		339	420
2500	12" wide		5.40	2.963		243	139		382	480

26 05 36.30 Cable Tray Trough		Crew	Daily Output	Labor-Hours	Unit	2009 Bare Costs Material	Labor	Equipment	Total	Total Incl O&P
2510	18" wide	2 Elec	4.80	3.333	Ea.	273	157		430	540
2520	24" wide		3.80	4.211		295	198		493	630
2530	30" wide		3.40	4.706		325	221		546	695
2540	36" wide		2.80	5.714		335	269		604	780
2560	Tee horizontal, 12" radius, 6" wide		4	4		170	188		358	475
2570	9" wide		3.80	4.211		175	198		373	495
2580	12" wide		3.60	4.444		190	209		399	530
2590	18" wide		3.20	5		220	235		455	600
2600	24" wide		2.80	5.714		253	269		522	695
2610	30" wide		2.40	6.667		271	315		586	780
2620	36" wide		2.20	7.273		310	340		650	865
2640	24" radius, 6" wide		3.60	4.444		259	209		468	605
2660	12" wide		3.20	5		295	235		530	685
2670	18" wide		2.80	5.714		325	269		594	770
2680	24" wide		2.40	6.667		405	315		720	925
2690	30" wide		2	8		435	375		810	1,050
2700	36" wide		1.80	8.889		510	420		930	1,200
2720	36" radius, 6" wide		3.20	5		415	235		650	820
2740	12" wide		2.80	5.714		480	269		749	940
2750	18" wide		2.40	6.667		545	315		860	1,075
2760	24" wide		2	8		635	375		1,010	1,275
2770	30" wide		1.60	10		675	470		1,145	1,475
2780	36" wide		1.40	11.429		780	535		1,315	1,675
2800	Tee vertical, 12" radius, 6" wide		4	4		234	188		422	545
2810	9" wide		3.80	4.211		236	198		434	565
2820	12" wide		3.60	4.444		253	209		462	600
2830	18" wide		3.40	4.706		254	221		475	620
2840	24" wide		3.20	5		261	235		496	645
2850	30" wide		3	5.333		273	251		524	685
2860	36" wide		2.60	6.154		284	289		573	755
2880	24" radius, 6" wide		3.60	4.444		330	209		539	685
2900	12" wide		3.20	5		355	235		590	755
2910	18" wide		3	5.333		380	251		631	805
2920	24" wide		2.80	5.714		405	269		674	860
2930	30" wide		2.60	6.154		440	289		729	925
2940	36" wide		2.20	7.273		445	340		785	1,025
2960	36" radius, 6" wide		3.20	5		510	235		745	925
2980	12" wide		2.80	5.714		545	269		814	1,000
2990	18" wide		2.60	6.154		560	289		849	1,050
3000	24" wide		2.40	6.667		600	315		915	1,150
3010	30" wide		2.20	7.273		650	340		990	1,250
3020	36" wide		1.80	8.889		680	420		1,100	1,400
3040	Cross horizontal, 12" radius, 6" wide		3.60	4.444		212	209		421	555
3050	9" wide		3.40	4.706		228	221		449	590
3060	12" wide		3.20	5		234	235		469	615
3070	18" wide		2.80	5.714		245	269		514	685
3080	24" wide		2.40	6.667		273	315		588	780
3090	30" wide		2.20	7.273		330	340		670	890
3100	36" wide		1.80	8.889		400	420		820	1,075
3120	24" radius, 6" wide		3.20	5		380	235		615	780
3140	12" wide		2.80	5.714		415	269		684	875
3150	18" wide		2.40	6.667		450	315		765	975
3160	24" wide		2	8		520	375		895	1,150

26 05 36.30 Cable Tray Trough		Crew	Daily Output	Labor-Hours	Unit	2009 Bare Costs Material	Labor	Equipment	Total	Total Incl O&P
3170	30" wide	2 Elec	1.80	8.889	Ea.	560	420		980	1,250
3180	36" wide		1.40	11.429		635	535		1,170	1,525
3200	36" radius, 6" wide		2.80	5.714		635	269		904	1,100
3220	12" wide		2.40	6.667		665	315		980	1,200
3230	18" wide		2	8		720	375		1,095	1,375
3240	24" wide		1.60	10		835	470		1,305	1,650
3250	30" wide		1.40	11.429		910	535		1,445	1,825
3260	36" wide		1.20	13.333		1,050	625		1,675	2,125
3280	Dropout, or end plate, 6" wide		26	.615		21	29		50	68
3300	12" wide		22	.727		25	34		59	80
3310	18" wide		20	.800		30.50	37.50		68	91.50
3320	24" wide		18	.889		36.50	42		78.50	105
3330	30" wide		16	1		39	47		86	116
3340	36" wide		14	1.143		44.50	53.50		98	132
3370	Reducer, 9" to 6" wide		12	1.333		104	62.50		166.50	211
3380	12" to 6" wide		11.40	1.404		109	66		175	221
3390	12" to 9" wide		11.40	1.404		109	66		175	221
3400	18" to 12" wide		9.60	1.667		117	78.50		195.50	250
3420	18" to 6" wide		9.60	1.667		117	78.50		195.50	250
3430	18" to 9" wide		9.60	1.667		117	78.50		195.50	250
3440	24" to 18" wide		8.40	1.905		125	89.50		214.50	275
3460	24" to 12" wide		8.40	1.905		125	89.50		214.50	275
3470	24" to 9" wide		8.40	1.905		126	89.50		215.50	277
3475	24" to 6" wide		8.40	1.905		126	89.50		215.50	277
3480	30" to 24" wide		7.20	2.222		128	104		232	300
3500	30" to 18" wide		7.20	2.222		128	104		232	300
3520	30" to 12" wide		7.20	2.222		131	104		235	305
3540	36" to 30" wide		6.40	2.500		133	118		251	325
3560	36" to 24" wide		6.40	2.500		133	118		251	325
3580	36" to 18" wide		6.40	2.500		133	118		251	325
3600	36" to 12" wide		6.40	2.500		133	118		251	325
3610	Elbow horizontal 60°, 12" radius, 6" wide		7.80	2.051		90.50	96.50		187	248
3620	9" wide		7.20	2.222		99	104		203	270
3630	12" wide		6.40	2.500		109	118		227	300
3640	18" wide		5.80	2.759		117	130		247	330
3650	24" wide		4.80	3.333		144	157		301	400
3680	Elbow horizontal 45°, 12" radius, 6" wide		8	2		78	94		172	231
3690	9" wide		7.40	2.162		82.50	102		184.50	247
3700	12" wide		6.60	2.424		87.50	114		201.50	271
3710	18" wide		6	2.667		98.50	125		223.50	300
3720	24" wide		5	3.200		114	150		264	355
3750	Elbow horizontal, 30° 12" radius, 6" wide		8.20	1.951		68	91.50		159.50	216
3760	9" wide		7.60	2.105		71.50	99		170.50	231
3770	12" wide		6.80	2.353		76.50	111		187.50	254
3780	18" wide		6.20	2.581		82.50	121		203.50	278
3790	24" wide		5.20	3.077		90.50	145		235.50	320
3820	Elbow vertical 60° in/outside, 12" radius, 6" wide		7.80	2.051		112	96.50		208.50	272
3830	9" wide		7.20	2.222		114	104		218	286
3840	12" wide		6.40	2.500		115	118		233	310
3850	18" wide		5.80	2.759		120	130		250	330
3860	24" wide		4.80	3.333		125	157		282	380
3890	Elbow vertical 45° in/outside, 12" radius, 6" wide		8	2		90.50	94		184.50	245
3900	9" wide		7.40	2.162		96	102		198	261

26 05 36 – Cable Trays for Electrical Systems

26 05 36.30 Cable Tray Trough		Crew	Daily Output	Labor-Hours	Unit	2009 Bare Costs Material	Labor	Equipment	Total	Total Incl O&P
3910	12″ wide	2 Elec	6.60	2.424	Ea.	99	114		213	284
3920	18″ wide		6	2.667		102	125		227	305
3930	24″ wide		5	3.200		112	150		262	355
3960	Elbow vertical 30° in/outside, 12″ radius, 6″ wide		8.20	1.951		78	91.50		169.50	227
3970	9″ wide		7.60	2.105		82.50	99		181.50	243
3980	12″ wide		6.80	2.353		84	111		195	263
3990	18″ wide		6.20	2.581		87.50	121		208.50	284
4000	24″ wide		5.20	3.077		89	145		234	320
4250	Reducer, left or right hand, 24″ to 18″ wide		8.40	1.905		119	89.50		208.50	268
4260	24″ to 12″ wide		8.40	1.905		119	89.50		208.50	268
4270	24″ to 9″ wide		8.40	1.905		119	89.50		208.50	268
4280	24″ to 6″ wide		8.40	1.905		120	89.50		209.50	270
4290	18″ to 12″ wide		9.60	1.667		111	78.50		189.50	243
4300	18″ to 9″ wide		9.60	1.667		111	78.50		189.50	243
4310	18″ to 6″ wide		9.60	1.667		111	78.50		189.50	243
4320	12″ to 9″ wide		11.40	1.404		106	66		172	218
4330	12″ to 6″ wide		11.40	1.404		106	66		172	218
4340	9″ to 6″ wide		12	1.333		103	62.50		165.50	210
4350	Splice plate	1 Elec	48	.167		9.85	7.85		17.70	23
4360	Splice plate, expansion joint		48	.167		10.05	7.85		17.90	23
4370	Splice plate, hinged, horizontal		48	.167		8.10	7.85		15.95	21
4380	Vertical		48	.167		11.55	7.85		19.40	25
4390	Trough, hanger, vertical		28	.286		32.50	13.45		45.95	56
4400	Box connector, 24″ wide		20	.400		44	18.80		62.80	77
4410	18″ wide		21	.381		36.50	17.90		54.40	67.50
4420	12″ wide		22	.364		35	17.10		52.10	65
4430	9″ wide		23	.348		33	16.35		49.35	61.50
4440	6″ wide		24	.333		31.50	15.65		47.15	59
4450	Floor flange		24	.333		33	15.65		48.65	60.50
4460	Hold down clamp		60	.133		3.82	6.25		10.07	13.85
4520	Wall bracket, 24″ wide tray		20	.400		32.50	18.80		51.30	65
4530	18″ wide tray		21	.381		32	17.90		49.90	62.50
4540	12″ wide tray		22	.364		17	17.10		34.10	45
4550	9″ wide tray		23	.348		15.20	16.35		31.55	41.50
4560	6″ wide tray		24	.333		13.85	15.65		29.50	39
5000	Cable channel aluminum, vented, 1-1/4″ deep, 4″ wide, straight		80	.100	L.F.	12.35	4.70		17.05	21
5010	Elbow horizontal, 36″ radius, 90°		5	1.600	Ea.	222	75		297	360
5020	60°		5.50	1.455		170	68.50		238.50	292
5030	45°		6	1.333		138	62.50		200.50	249
5040	30°		6.50	1.231		119	58		177	220
5050	Adjustable		6	1.333		114	62.50		176.50	222
5060	Elbow vertical, 36″ radius, 90°		5	1.600		238	75		313	380
5070	60°		5.50	1.455		186	68.50		254.50	310
5080	45°		6	1.333		153	62.50		215.50	266
5090	30°		6.50	1.231		136	58		194	238
5100	Adjustable		6	1.333		114	62.50		176.50	222
5110	Splice plate, hinged, horizontal		48	.167		8.50	7.85		16.35	21.50
5120	Splice plate, hinged, vertical		48	.167		12.05	7.85		19.90	25.50
5130	Hanger, vertical		28	.286		13.30	13.45		26.75	35
5140	Single		28	.286		20	13.45		33.45	42.50
5150	Double		20	.400		20.50	18.80		39.30	52
5160	Channel to box connector		24	.333		27	15.65		42.65	54
5170	Hold down clip		80	.100		3.84	4.70		8.54	11.45

26 05 36.30 Cable Tray Trough		Crew	Daily Output	Labor-Hours	Unit	2009 Bare Costs Material	Labor	Equipment	Total	Total Incl O&P
5180	Wall bracket, single	1 Elec	28	.286	Ea.	11	13.45		24.45	32.50
5190	Double		20	.400		14.15	18.80		32.95	44.50
5200	Cable roller		16	.500		173	23.50		196.50	227
5210	Splice plate		48	.167		5.30	7.85		13.15	17.90

26 05 36.40 Cable Tray, Covers and Dividers		Crew	Daily Output	Labor-Hours	Unit	Material	Labor	Equipment	Total	Total Incl O&P
0010	**CABLE TRAY, COVERS AND DIVIDERS** To 15′ high									
0100	Covers, ventilated galv. steel, straight, 6″ wide tray size	2 Elec	520	.031	L.F.	5.15	1.45		6.60	7.85
0200	9″ wide tray size		460	.035		6.30	1.63		7.93	9.45
0300	12″ wide tray size		400	.040		7.55	1.88		9.43	11.25
0400	18″ wide tray size		300	.053		10.30	2.51		12.81	15.20
0500	24″ wide tray size		220	.073		12.65	3.42		16.07	19.15
0600	30″ wide tray size		180	.089		15.30	4.18		19.48	23.50
0700	36″ wide tray size		160	.100		17.45	4.70		22.15	26.50
1000	Elbow horizontal 90°, 12″ radius, 6″ wide tray size		150	.107	Ea.	38	5		43	49.50
1020	9″ wide tray size		128	.125		42	5.90		47.90	55.50
1040	12″ wide tray size		108	.148		44.50	6.95		51.45	59.50
1060	18″ wide tray size		84	.190		61.50	8.95		70.45	82
1080	24″ wide tray size		66	.242		74.50	11.40		85.90	99.50
1100	30″ wide tray size		60	.267		92.50	12.55		105.05	121
1120	36″ wide tray size		50	.320		112	15.05		127.05	146
1160	24″ radius, 6″ wide tray size		136	.118		63.50	5.55		69.05	78
1180	9″ wide tray size		116	.138		65	6.50		71.50	81.50
1200	12″ wide tray size		96	.167		72	7.85		79.85	91.50
1220	18″ wide tray size		76	.211		89	9.90		98.90	113
1240	24″ wide tray size		60	.267		109	12.55		121.55	139
1260	30″ wide tray size		52	.308		143	14.45		157.45	180
1280	36″ wide tray size		44	.364		165	17.10		182.10	209
1320	36″ radius, 6″ wide tray size		120	.133		92.50	6.25		98.75	112
1340	9″ wide tray size		104	.154		102	7.25		109.25	123
1360	12″ wide tray size		84	.190		109	8.95		117.95	134
1380	18″ wide tray size		72	.222		139	10.45		149.45	169
1400	24″ wide tray size		52	.308		165	14.45		179.45	205
1420	30″ wide tray size		46	.348		197	16.35		213.35	241
1440	36″ wide tray size		40	.400		231	18.80		249.80	283
1480	Elbow horizontal 45°, 12″ radius, 6″ wide tray size		150	.107		27	5		32	37.50
1500	9″ wide tray size		128	.125		32.50	5.90		38.40	44.50
1520	12″ wide tray size		108	.148		35.50	6.95		42.45	49.50
1540	18″ wide tray size		88	.182		43	8.55		51.55	60.50
1560	24″ wide tray size		76	.211		50	9.90		59.90	70.50
1580	30″ wide tray size		66	.242		59.50	11.40		70.90	82.50
1600	36″ wide tray size		60	.267		66.50	12.55		79.05	92.50
1640	24″ radius, 6″ wide tray size		136	.118		40	5.55		45.55	52.50
1660	9″ wide tray size		116	.138		44.50	6.50		51	59
1680	12″ wide tray size		96	.167		50	7.85		57.85	67.50
1700	18″ wide tray size		80	.200		57.50	9.40		66.90	78
1720	24″ wide tray size		70	.229		66.50	10.75		77.25	89.50
1740	30″ wide tray size		60	.267		84.50	12.55		97.05	112
1760	36″ wide tray size		52	.308		92.50	14.45		106.95	125
1800	36″ radius, 6″ wide tray size		120	.133		57.50	6.25		63.75	73
1820	9″ wide tray size		104	.154		65	7.25		72.25	82.50
1840	12″ wide tray size		84	.190		71.50	8.95		80.45	92.50
1860	18″ wide tray size		76	.211		82	9.90		91.90	105

26 05 36.40 Cable Tray, Covers and Dividers		Crew	Daily Output	Labor-Hours	Unit	Material	Labor	Equipment	Total	Total Incl O&P
						2009 Bare Costs				
1880	24" wide tray size	2 Elec	62	.258	Ea.	100	12.15		112.15	129
1900	30" wide tray size		52	.308		109	14.45		123.45	143
1920	36" wide tray size		48	.333		131	15.65		146.65	168
1960	Elbow vertical 90°, 12" radius, 6" wide tray size		150	.107		32	5		37	42.50
1980	9" wide tray size		128	.125		32.50	5.90		38.40	44.50
2000	12" wide tray size		108	.148		35	6.95		41.95	49
2020	18" wide tray size		88	.182		40	8.55		48.55	57
2040	24" wide tray size		68	.235		40.50	11.05		51.55	61.50
2060	30" wide tray size		60	.267		44.50	12.55		57.05	68.50
2080	36" wide tray size		50	.320		57.50	15.05		72.55	86.50
2120	24" radius, 6" wide tray size		136	.118		39	5.55		44.55	51.50
2140	9" wide tray size		116	.138		43	6.50		49.50	57.50
2160	12" wide tray size		96	.167		44.50	7.85		52.35	61
2180	18" wide tray size		80	.200		59.50	9.40		68.90	79.50
2200	24" wide tray size		62	.258		66.50	12.15		78.65	91.50
2220	30" wide tray size		52	.308		75.50	14.45		89.95	106
2240	36" wide tray size		44	.364		86	17.10		103.10	121
2280	36" radius, 6" wide tray size		120	.133		45.50	6.25		51.75	59.50
2300	9" wide tray size		104	.154		57.50	7.25		64.75	74.50
2320	12" wide tray size		84	.190		65	8.95		73.95	85.50
2340	18" wide tray size		76	.211		82	9.90		91.90	105
2350	24" wide tray size		54	.296		91	13.95		104.95	122
2360	30" wide tray size		46	.348		109	16.35		125.35	145
2370	36" wide tray size		40	.400		128	18.80		146.80	170
2400	Tee horizontal, 12" radius, 6" wide tray size		92	.174		57.50	8.15		65.65	76
2410	9" wide tray size		80	.200		59.50	9.40		68.90	79.50
2420	12" wide tray size		68	.235		67	11.05		78.05	91
2430	18" wide tray size		60	.267		81	12.55		93.55	108
2440	24" wide tray size		52	.308		102	14.45		116.45	135
2460	30" wide tray size		36	.444		120	21		141	164
2470	36" wide tray size		30	.533		145	25		170	199
2500	24" radius, 6" wide tray size		88	.182		90.50	8.55		99.05	113
2510	9" wide tray size		76	.211		104	9.90		113.90	129
2520	12" wide tray size		64	.250		107	11.75		118.75	136
2530	18" wide tray size		56	.286		137	13.45		150.45	172
2540	24" wide tray size		48	.333		211	15.65		226.65	256
2560	30" wide tray size		32	.500		240	23.50		263.50	300
2570	36" wide tray size		26	.615		265	29		294	335
2600	36" radius, 6" wide tray size		84	.190		164	8.95		172.95	194
2610	9" wide tray size		72	.222		167	10.45		177.45	200
2620	12" wide tray size		60	.267		186	12.55		198.55	223
2630	18" wide tray size		52	.308		217	14.45		231.45	262
2640	24" wide tray size		44	.364		285	17.10		302.10	340
2660	30" wide tray size		28	.571		310	27		337	380
2670	36" wide tray size		22	.727		355	34		389	445
2700	Cross horizontal, 12" radius, 6" wide tray size		68	.235		86	11.05		97.05	112
2710	9" wide tray size		64	.250		91	11.75		102.75	118
2720	12" wide tray size		60	.267		101	12.55		113.55	131
2730	18" wide tray size		52	.308		120	14.45		134.45	155
2740	24" wide tray size		36	.444		147	21		168	193
2760	30" wide tray size		30	.533		167	25		192	223
2770	36" wide tray size		28	.571		195	27		222	257
2800	24" radius, 6" wide tray size	↓	64	.250	↓	164	11.75		175.75	198

26 05 36.40 Cable Tray, Covers and Dividers		Crew	Daily Output	Labor-Hours	Unit	Material	2009 Bare Costs Labor	Equipment	Total	Total Incl O&P
2810	9" wide tray size	2 Elec	60	.267	Ea.	178	12.55		190.55	215
2820	12" wide tray size		56	.286		192	13.45		205.45	232
2830	18" wide tray size		48	.333		229	15.65		244.65	276
2840	24" wide tray size		32	.500		279	23.50		302.50	340
2860	30" wide tray size		26	.615		315	29		344	390
2870	36" wide tray size		24	.667		355	31.50		386.50	440
2900	36" radius, 6" wide tray size		60	.267		275	12.55		287.55	320
2910	9" wide tray size		56	.286		284	13.45		297.45	330
2920	12" wide tray size		52	.308		298	14.45		312.45	355
2930	18" wide tray size		44	.364		345	17.10		362.10	405
2940	24" wide tray size		28	.571		440	27		467	520
2960	30" wide tray size		22	.727		470	34		504	575
2970	36" wide tray size		20	.800		500	37.50		537.50	615
3000	Reducer, 9" to 6" wide tray size		128	.125		36.50	5.90		42.40	49
3010	12" to 6" wide tray size		108	.148		38	6.95		44.95	52.50
3020	12" to 9" wide tray size		108	.148		38	6.95		44.95	52.50
3030	18" to 12" wide tray size		88	.182		40.50	8.55		49.05	57.50
3050	18" to 6" wide tray size		88	.182		40.50	8.55		49.05	57.50
3060	24" to 18" wide tray size		80	.200		57	9.40		66.40	77
3070	24" to 12" wide tray size		80	.200		52	9.40		61.40	71.50
3090	30" to 24" wide tray size		70	.229		60	10.75		70.75	82.50
3100	30" to 18" wide tray size		70	.229		60	10.75		70.75	82.50
3110	30" to 12" wide tray size		70	.229		52	10.75		62.75	73.50
3140	36" to 30" wide tray size		64	.250		65.50	11.75		77.25	90
3150	36" to 24" wide tray size		64	.250		65.50	11.75		77.25	90
3160	36" to 18" wide tray size		64	.250		65.50	11.75		77.25	90
3170	36" to 12" wide tray size		64	.250		65.50	11.75		77.25	90
3250	Covers, aluminum, straight, 6" wide tray size		520	.031	L.F.	4.99	1.45		6.44	7.70
3270	9" wide tray size		460	.035		6	1.63		7.63	9.10
3290	12" wide tray size		400	.040		7.20	1.88		9.08	10.80
3310	18" wide tray size		320	.050		9.50	2.35		11.85	14.05
3330	24" wide tray size		260	.062		11.85	2.89		14.74	17.50
3350	30" wide tray size		200	.080		13.10	3.76		16.86	20
3370	36" wide tray size		180	.089		13.90	4.18		18.08	21.50
3400	Elbow horizontal 90°, 12" radius, 6" wide tray size		150	.107	Ea.	37.50	5		42.50	48.50
3410	9" wide tray size		128	.125		39	5.90		44.90	52
3420	12" wide tray size		108	.148		42	6.95		48.95	57
3430	18" wide tray size		88	.182		55	8.55		63.55	73.50
3440	24" wide tray size		70	.229		69.50	10.75		80.25	93
3460	30" wide tray size		64	.250		83.50	11.75		95.25	110
3470	36" wide tray size		54	.296		102	13.95		115.95	134
3500	24" radius, 6" wide tray size		136	.118		51	5.55		56.55	65
3510	9" wide tray size		116	.138		63.50	6.50		70	79.50
3520	12" wide tray size		96	.167		69.50	7.85		77.35	88.50
3530	18" wide tray size		80	.200		82.50	9.40		91.90	105
3540	24" wide tray size		64	.250		102	11.75		113.75	130
3560	30" wide tray size		56	.286		125	13.45		138.45	158
3570	36" wide tray size		48	.333		154	15.65		169.65	194
3600	36" radius, 6" wide tray size		120	.133		87.50	6.25		93.75	106
3610	9" wide tray size		104	.154		96	7.25		103.25	116
3620	12" wide tray size		84	.190		109	8.95		117.95	134
3630	18" wide tray size		76	.211		129	9.90		138.90	157
3640	24" wide tray size		56	.286		158	13.45		171.45	194

26 05 36 – Cable Trays for Electrical Systems

26 05 36.40 Cable Tray, Covers and Dividers		Crew	Daily Output	Labor-Hours	Unit	Material	Labor	Equipment	Total	Total Incl O&P
						2009 Bare Costs				
3660	30″ wide tray size	2 Elec	50	.320	Ea.	181	15.05		196.05	222
3670	36″ wide tray size		44	.364		212	17.10		229.10	260
3700	Elbow horizontal 45°, 12″ radius, 6″ wide tray size		150	.107		27.50	5		32.50	38
3710	9″ wide tray size		128	.125		28.50	5.90		34.40	40
3720	12″ wide tray size		108	.148		32	6.95		38.95	45.50
3730	18″ wide tray size		88	.182		36.50	8.55		45.05	53
3740	24″ wide tray size		80	.200		40.50	9.40		49.90	59
3760	30″ wide tray size		70	.229		51	10.75		61.75	73
3770	36″ wide tray size		64	.250		61.50	11.75		73.25	86
3800	24″ radius, 6″ wide tray size		136	.118		32	5.55		37.55	43.50
3810	9″ wide tray size		116	.138		40.50	6.50		47	54.50
3820	12″ wide tray size		96	.167		42	7.85		49.85	58.50
3830	18″ wide tray size		80	.200		51	9.40		60.40	71
3840	24″ wide tray size		72	.222		63.50	10.45		73.95	85.50
3860	30″ wide tray size		64	.250		73	11.75		84.75	98.50
3870	36″ wide tray size		56	.286		84.50	13.45		97.95	114
3900	36″ radius, 6″ wide tray size		120	.133		55	6.25		61.25	70
3910	9″ wide tray size		104	.154		60	7.25		67.25	77
3920	12″ wide tray size		84	.190		63.50	8.95		72.45	83.50
3930	18″ wide tray size		76	.211		76.50	9.90		86.40	99
3940	24″ wide tray size		64	.250		92.50	11.75		104.25	120
3960	30″ wide tray size		56	.286		105	13.45		118.45	137
3970	36″ wide tray size		50	.320		120	15.05		135.05	155
4000	Elbow vertical 90°, 12″ radius, 6″ wide tray size		150	.107		32	5		37	42.50
4010	9″ wide tray size		128	.125		32	5.90		37.90	44
4020	12″ wide tray size		108	.148		35	6.95		41.95	49
4030	18″ wide tray size		88	.182		40.50	8.55		49.05	57.50
4040	24″ wide tray size		70	.229		42	10.75		52.75	63
4060	30″ wide tray size		64	.250		43	11.75		54.75	65.50
4070	36″ wide tray size		54	.296		52	13.95		65.95	78.50
4100	24″ radius, 6″ wide tray size		136	.118		36.50	5.55		42.05	48.50
4110	9″ wide tray size		116	.138		40	6.50		46.50	54
4120	12″ wide tray size		96	.167		43	7.85		50.85	59.50
4130	18″ wide tray size		80	.200		52	9.40		61.40	71.50
4140	24″ wide tray size		64	.250		57	11.75		68.75	80.50
4160	30″ wide tray size		56	.286		73	13.45		86.45	101
4170	36″ wide tray size		48	.333		79.50	15.65		95.15	112
4200	36″ radius, 6″ wide tray size		120	.133		42	6.25		48.25	56
4210	9″ wide tray size		104	.154		52	7.25		59.25	68
4220	12″ wide tray size		84	.190		60	8.95		68.95	80
4230	18″ wide tray size		76	.211		73	9.90		82.90	95.50
4240	24″ wide tray size		56	.286		86	13.45		99.45	115
4260	30″ wide tray size		50	.320		109	15.05		124.05	143
4270	36″ wide tray size		44	.364		117	17.10		134.10	156
4300	Tee horizontal, 12″ radius, 6″ wide tray size		108	.148		52	6.95		58.95	67.50
4310	9″ wide tray size		88	.182		55	8.55		63.55	73.50
4320	12″ wide tray size		80	.200		61.50	9.40		70.90	82.50
4330	18″ wide tray size		68	.235		73	11.05		84.05	97.50
4340	24″ wide tray size		56	.286		92.50	13.45		105.95	123
4360	30″ wide tray size		44	.364		109	17.10		126.10	147
4370	36″ wide tray size		36	.444		131	21		152	176
4400	24″ radius, 6″ wide tray size		96	.167		84.50	7.85		92.35	105
4410	9″ wide tray size		80	.200		94	9.40		103.40	117

26 05 36.40 Cable Tray, Covers and Dividers		Crew	Daily Output	Labor-Hours	Unit	Material	Labor	Equipment	Total	Total Incl O&P
						2009 Bare Costs				
4420	12" wide tray size	2 Elec	72	.222	Ea.	105	10.45		115.45	132
4430	18" wide tray size		60	.267		123	12.55		135.55	155
4440	24" wide tray size		48	.333		197	15.65		212.65	240
4460	30" wide tray size		40	.400		217	18.80		235.80	268
4470	36" wide tray size		32	.500		242	23.50		265.50	300
4500	36" radius, 6" wide tray size		88	.182		154	8.55		162.55	183
4510	9" wide tray size		72	.222		158	10.45		168.45	189
4520	12" wide tray size		64	.250		173	11.75		184.75	208
4530	18" wide tray size		56	.286		197	13.45		210.45	237
4540	24" wide tray size		44	.364		250	17.10		267.10	300
4560	30" wide tray size		36	.444		281	21		302	340
4570	36" wide tray size		28	.571		315	27		342	390
4600	Cross horizontal, 12" radius, 6" wide tray size		80	.200		79.50	9.40		88.90	102
4610	9" wide tray size		72	.222		84	10.45		94.45	109
4620	12" wide tray size		64	.250		93.50	11.75		105.25	121
4630	18" wide tray size		56	.286		114	13.45		127.45	146
4640	24" wide tray size		48	.333		131	15.65		146.65	168
4660	30" wide tray size		40	.400		156	18.80		174.80	201
4670	36" wide tray size		32	.500		179	23.50		202.50	233
4700	24" radius, 6" wide tray size		72	.222		156	10.45		166.45	188
4710	9" wide tray size		64	.250		167	11.75		178.75	202
4720	12" wide tray size		56	.286		179	13.45		192.45	218
4730	18" wide tray size		48	.333		209	15.65		224.65	254
4740	24" wide tray size		40	.400		250	18.80		268.80	305
4760	30" wide tray size		32	.500		292	23.50		315.50	355
4770	36" wide tray size		24	.667		315	31.50		346.50	400
4800	36" radius, 6" wide tray size		64	.250		250	11.75		261.75	293
4810	9" wide tray size		56	.286		261	13.45		274.45	310
4820	12" wide tray size		50	.320		281	15.05		296.05	335
4830	18" wide tray size		44	.364		310	17.10		327.10	365
4840	24" wide tray size		36	.444		380	21		401	450
4860	30" wide tray size		28	.571		440	27		467	520
4870	36" wide tray size		22	.727		480	34		514	580
4900	Reducer, 9" to 6" wide tray size		128	.125		40	5.90		45.90	53
4910	12" to 6" wide tray size		108	.148		41.50	6.95		48.45	56
4920	12" to 9" wide tray size		108	.148		41.50	6.95		48.45	56
4930	18" to 12" wide tray size		88	.182		45.50	8.55		54.05	63
4950	18" to 6" wide tray size		88	.182		45.50	8.55		54.05	63
4960	24" to 18" wide tray size		80	.200		58.50	9.40		67.90	78.50
4970	24" to 12" wide tray size		80	.200		50	9.40		59.40	70
4990	30" to 24" wide tray size		70	.229		61	10.75		71.75	83.50
5000	30" to 18" wide tray size		70	.229		61	10.75		71.75	83.50
5010	30" to 12" wide tray size		70	.229		61	10.75		71.75	83.50
5040	36" to 30" wide tray size		64	.250		67.50	11.75		79.25	92
5050	36" to 24" wide tray size		64	.250		67.50	11.75		79.25	92
5060	36" to 18" wide tray size		64	.250		67.50	11.75		79.25	92
5070	36" to 12" wide tray size		64	.250		67.50	11.75		79.25	92
5710	Tray cover hold down clamp	1 Elec	60	.133		10.75	6.25		17	21.50
8000	Divider strip, straight, galvanized, 3" deep		200	.040	L.F.	4.52	1.88		6.40	7.85
8020	4" deep		180	.044		5.55	2.09		7.64	9.30
8040	6" deep		160	.050		7.25	2.35		9.60	11.60
8060	Aluminum, straight, 3" deep		210	.038		4.52	1.79		6.31	7.70
8080	4" deep		190	.042		5.60	1.98		7.58	9.25

26 05 36 – Cable Trays for Electrical Systems

26 05 36.40 Cable Tray, Covers and Dividers		Crew	Daily Output	Labor-Hours	Unit	Material	2009 Bare Costs Labor	Equipment	Total	Total Incl O&P
8100	6" deep	1 Elec	170	.047	L.F.	7.20	2.21		9.41	11.30
8110	Divider strip vertical fitting 3" deep									
8120	12" radius, galvanized, 30°	1 Elec	28	.286	Ea.	22	13.45		35.45	44.50
8140	45°		27	.296		27.50	13.95		41.45	51.50
8160	60°		26	.308		29	14.45		43.45	54.50
8180	90°		25	.320		36.50	15.05		51.55	63
8200	Aluminum,30°		29	.276		18.55	12.95		31.50	40.50
8220	45°		28	.286		22	13.45		35.45	44.50
8240	60°		27	.296		25.50	13.95		39.45	49.50
8260	90°		26	.308		32.50	14.45		46.95	58
8280	24" radius, galvanized, 30°		25	.320		33	15.05		48.05	59.50
8300	45°		24	.333		37.50	15.65		53.15	65
8320	60°		23	.348		47	16.35		63.35	77
8340	90°		22	.364		64	17.10		81.10	97
8360	Aluminum, 30°		26	.308		31	14.45		45.45	56.50
8380	45°		25	.320		35.50	15.05		50.55	62.50
8400	60°		24	.333		44.50	15.65		60.15	73
8420	90°		23	.348		61	16.35		77.35	92
8440	36" radius, galvanized,30°		22	.364		44.50	17.10		61.60	75.50
8460	45°		21	.381		52	17.90		69.90	84.50
8480	60°		20	.400		61	18.80		79.80	96
8500	90°		19	.421		84.50	19.80		104.30	124
8520	Aluminum, 30°		23	.348		47	16.35		63.35	77
8540	45°		22	.364		61	17.10		78.10	93.50
8560	60°		21	.381		78.50	17.90		96.40	114
8570	90°	↓	20	.400	↓	103	18.80		121.80	142
8590	Divider strip vertical fitting 4" deep									
8600	12" radius, galvanized, 30°	1 Elec	27	.296	Ea.	28	13.95		41.95	52
8610	45°		26	.308		32.50	14.45		46.95	58
8620	60°		25	.320		36.50	15.05		51.55	63
8630	90°		24	.333		44.50	15.65		60.15	73
8640	Aluminum, 30°		28	.286		26	13.45		39.45	49.50
8650	45°		27	.296		30.50	13.95		44.45	55
8660	60°		26	.308		35	14.45		49.45	61
8670	90°		25	.320		41.50	15.05		56.55	68.50
8680	24" radius, galvanized, 30°		24	.333		44.50	15.65		60.15	73
8690	45°		23	.348		56	16.35		72.35	86.50
8700	60°		22	.364		63.50	17.10		80.60	96
8710	90°		21	.381		84.50	17.90		102.40	121
8720	Aluminum,30°		25	.320		41.50	15.05		56.55	68.50
8730	45°		24	.333		50	15.65		65.65	79.50
8740	60°		23	.348		59.50	16.35		75.85	90
8750	90°		22	.364		81	17.10		98.10	116
8760	36" radius, galvanized,30°		23	.348		52	16.35		68.35	82
8770	45°		22	.364		59.50	17.10		76.60	91.50
8780	60°		21	.381		74.50	17.90		92.40	110
8790	90°		20	.400		101	18.80		119.80	141
8800	Aluminum,30°		24	.333		63.50	15.65		79.15	93.50
8810	45°		23	.348		81	16.35		97.35	114
8820	60°		22	.364		99	17.10		116.10	136
8830	90°	↓	21	.381	↓	125	17.90		142.90	165
8840	Divider strip vertical fitting 6" deep									
8850	12" radius, galvanized, 30°	1 Elec	24	.333	Ea.	30.50	15.65		46.15	57.50

26 05 Common Work Results for Electrical

26 05 36 – Cable Trays for Electrical Systems

26 05 36.40 Cable Tray, Covers and Dividers		Crew	Daily Output	Labor-Hours	Unit	Material	2009 Bare Costs Labor	Equipment	Total	Total Incl O&P
8860	45°	1 Elec	23	.348	Ea.	34	15.35		50.35	62.50
8870	60°		22	.364		39	17.10		56.10	69.50
8880	90°		21	.381		48.50	17.90		66.40	81
8890	Aluminum, 30°		25	.320		28.50	15.05		43.55	54.50
8900	45°		24	.333		34	15.65		49.65	61.50
8910	60°		23	.348		35.50	16.35		51.85	64.50
8920	90°		22	.364		42	17.10		59.10	73
8930	24" radius, galvanized, 30°		23	.348		44.50	16.35		60.85	74
8940	45°		22	.364		56	17.10		73.10	88
8950	60°		21	.381		64	17.90		81.90	98
8960	90°		20	.400		84.50	18.80		103.30	122
8970	Aluminum, 30°		24	.333		42	15.65		57.65	70.50
8980	45°		23	.348		57.50	16.35		73.85	88.50
8990	60°		22	.364		63.50	17.10		80.60	96
9000	90°		21	.381		86	17.90		103.90	122
9010	36" radius, galvanized, 30°		22	.364		52	17.10		69.10	83.50
9020	45°		21	.381		64	17.90		81.90	98
9030	60°		20	.400		84.50	18.80		103.30	122
9040	90°		19	.421		109	19.80		128.80	151
9050	Aluminum, 30°		23	.348		64	16.35		80.35	95.50
9060	45°		22	.364		86	17.10		103.10	121
9070	60°		21	.381		101	17.90		118.90	140
9080	90°		20	.400		122	18.80		140.80	163
9120	Divider strip, horizontal fitting, galvanized, 3" deep		33	.242		31	11.40		42.40	52
9130	4" deep		30	.267		35	12.55		47.55	58
9140	6" deep		27	.296		44.50	13.95		58.45	70.50
9150	Aluminum, 3" deep		35	.229		29.50	10.75		40.25	49
9160	4" deep		32	.250		32.50	11.75		44.25	53.50
9170	6" deep		29	.276		43	12.95		55.95	67.50
9300	Divider strip protector		300	.027	L.F.	3.20	1.25		4.45	5.45
9310	Fastener, ladder tray				Ea.	.58			.58	.64
9320	Trough or solid bottom tray				"	.39			.39	.43
9900	Add to labor for higher elevated installation									
9910	15' to 20' high add						10%			
9920	20' to 25' high add						20%			
9930	25' to 30' high add						25%			
9940	30' to 35' high add						30%			
9960	Over 40' high add						40%			

26 05 39 – Underfloor Raceways for Electrical Systems

26 05 39.30 Conduit In Concrete Slab		Crew	Daily Output	Labor-Hours	Unit	Material	Labor	Equipment	Total	Total Incl O&P
0010	**CONDUIT IN CONCRETE SLAB** Including terminations,									
0020	fittings and supports									
3230	PVC, schedule 40, 1/2" diameter	1 Elec	270	.030	L.F.	.93	1.39		2.32	3.16
3250	3/4" diameter		230	.035		1.14	1.63		2.77	3.76
3270	1" diameter		200	.040		1.55	1.88		3.43	4.60
3300	1-1/4" diameter		170	.047		2.18	2.21		4.39	5.80
3330	1-1/2" diameter		140	.057		2.60	2.69		5.29	7
3350	2" diameter		120	.067		3.35	3.13		6.48	8.50
3370	2-1/2" diameter		90	.089		5.60	4.18		9.78	12.60
3400	3" diameter	2 Elec	160	.100		6.95	4.70		11.65	14.85
3430	3-1/2" diameter		120	.133		9.05	6.25		15.30	19.60
3440	4" diameter		100	.160		10	7.50		17.50	22.50

26 05 39.30 Conduit In Concrete Slab

	26 05 39.30 Conduit In Concrete Slab	Crew	Daily Output	Labor-Hours	Unit	2009 Bare Costs Material	Labor	Equipment	Total	Total Incl O&P
3450	5" diameter	2 Elec	80	.200	L.F.	14.90	9.40		24.30	31
3460	6" diameter	↓	60	.267	↓	20.50	12.55		33.05	42
3530	Sweeps, 1" diameter, 30" radius	1 Elec	32	.250	Ea.	25	11.75		36.75	45.50
3550	1-1/4" diameter		24	.333		31	15.65		46.65	58.50
3570	1-1/2" diameter		21	.381		32.50	17.90		50.40	63.50
3600	2" diameter		18	.444		33.50	21		54.50	69
3630	2-1/2" diameter		14	.571		58	27		85	106
3650	3" diameter		10	.800		66	37.50		103.50	131
3670	3-1/2" diameter		8	1		99	47		146	182
3700	4" diameter		7	1.143		94.50	53.50		148	187
3710	5" diameter	↓	6	1.333		145	62.50		207.50	256
3730	Couplings, 1/2" diameter					.41			.41	.45
3750	3/4" diameter					.50			.50	.55
3770	1" diameter					.77			.77	.85
3800	1-1/4" diameter					1.02			1.02	1.12
3830	1-1/2" diameter					1.40			1.40	1.54
3850	2" diameter					1.86			1.86	2.05
3870	2-1/2" diameter					3.30			3.30	3.63
3900	3" diameter					5.40			5.40	5.95
3930	3-1/2" diameter					6			6	6.60
3950	4" diameter					8.35			8.35	9.20
3960	5" diameter					21			21	23.50
3970	6" diameter					27			27	30
4030	End bells 1" diameter, PVC	1 Elec	60	.133		5.70	6.25		11.95	15.90
4050	1-1/4" diameter		53	.151		7.10	7.10		14.20	18.70
4100	1-1/2" diameter		48	.167		7.10	7.85		14.95	19.85
4150	2" diameter		34	.235		10.50	11.05		21.55	28.50
4170	2-1/2" diameter		27	.296		11.60	13.95		25.55	34.50
4200	3" diameter		20	.400		12.25	18.80		31.05	42.50
4250	3-1/2" diameter		16	.500		14.50	23.50		38	52
4300	4" diameter		14	.571		15.80	27		42.80	59
4310	5" diameter		12	.667		25	31.50		56.50	75.50
4320	6" diameter		9	.889	↓	27	42		69	94.50
4350	Rigid galvanized steel, 1/2" diameter		200	.040	L.F.	2.73	1.88		4.61	5.90
4400	3/4" diameter		170	.047		3.01	2.21		5.22	6.70
4450	1" diameter		130	.062		4.21	2.89		7.10	9.10
4500	1-1/4" diameter		110	.073		5.65	3.42		9.07	11.45
4600	1-1/2" diameter		100	.080		6.70	3.76		10.46	13.15
4800	2" diameter		90	.089	↓	8.55	4.18		12.73	15.90
9000	Minimum labor/equipment charge	↓	4	2	Job		94		94	145

26 05 39.40 Conduit In Trench

	26 05 39.40 Conduit In Trench	Crew	Daily Output	Labor-Hours	Unit	Material	Labor	Equipment	Total	Total Incl O&P
0010	**CONDUIT IN TRENCH** Includes terminations and fittings									
0020	Does not include excavation or backfill, see Div. 31 23 16.00									
0200	Rigid galvanized steel, 2" diameter	1 Elec	150	.053	L.F.	8.20	2.51		10.71	12.90
0400	2-1/2" diameter	"	100	.080		15.60	3.76		19.36	23
0600	3" diameter	2 Elec	160	.100		19.60	4.70		24.30	29
0800	3-1/2" diameter		140	.114		25.50	5.35		30.85	36.50
1000	4" diameter		100	.160		28.50	7.50		36	43
1200	5" diameter		80	.200		58.50	9.40		67.90	79
1400	6" diameter	↓	60	.267	↓	85.50	12.55		98.05	113
9000	Minimum labor/equipment charge	1 Elec	4	2	Job		94		94	145

Base 30 psf Roof Load
20 psf Wind load

floors
foundations
anchor bolts
interior finishes
utilities

Skylights
gutters / down spouts

Electric
conduit
wire type/condition
panel/breakers
fixtures/outlets

(?) Site Visit
- contracted Shop

Site visit (undrground + fuel)
#2 from pole to panel
100 Amp Panel (# cirbreaker)
#12 THN from panel to light/fixtures
4 (? # GFIC outlets)
4+4=8 light exterior lighting

26 05 43 – Underground Ducts and Raceways for Electrical Systems

26 05 43.10 Trench Duct		Crew	Daily Output	Labor-Hours	Unit	Material	2009 Bare Costs Labor	Equipment	Total	Total Incl O&P
0010	**TRENCH DUCT** Steel with cover									
0020	Standard adjustable, depths to 4″									
0100	Straight, single compartment, 9″ wide	2 Elec	40	.400	L.F.	104	18.80		122.80	144
0200	12″ wide		32	.500		118	23.50		141.50	166
0400	18″ wide		26	.615		157	29		186	218
0600	24″ wide		22	.727		202	34		236	275
0700	27″ wide		21	.762		217	36		253	294
0800	30″ wide		20	.800		244	37.50		281.50	325
1000	36″ wide		16	1		293	47		340	400
1020	Two compartment, 9″ wide		38	.421		118	19.80		137.80	161
1030	12″ wide		30	.533		135	25		160	187
1040	18″ wide		24	.667		165	31.50		196.50	230
1050	24″ wide		20	.800		216	37.50		253.50	295
1060	30″ wide		18	.889		263	42		305	355
1070	36″ wide		14	1.143		305	53.50		358.50	420
1090	Three compartment, 9″ wide		36	.444		135	21		156	180
1100	12″ wide		28	.571		149	27		176	205
1110	18″ wide		22	.727		187	34		221	258
1120	24″ wide		18	.889		235	42		277	325
1130	30″ wide		16	1		283	47		330	385
1140	36″ wide		12	1.333		330	62.50		392.50	460
1200	Horizontal elbow, 9″ wide		5.40	2.963	Ea.	385	139		524	640
1400	12″ wide		4.60	3.478		445	163		608	740
1600	18″ wide		4	4		570	188		758	915
1800	24″ wide		3.20	5		800	235		1,035	1,250
1900	27″ wide		3	5.333		910	251		1,161	1,375
2000	30″ wide		2.60	6.154		1,075	289		1,364	1,625
2200	36″ wide		2.40	6.667		1,400	315		1,715	2,025
2220	Two compartment, 9″ wide		3.80	4.211		625	198		823	990
2230	12″ wide		3	5.333		690	251		941	1,150
2240	18″ wide		2.40	6.667		830	315		1,145	1,400
2250	24″ wide		2	8		1,050	375		1,425	1,725
2260	30″ wide		1.80	8.889		1,400	420		1,820	2,200
2270	36″ wide		1.60	10		1,700	470		2,170	2,600
2290	Three compartment, 9″ wide		3.60	4.444		650	209		859	1,025
2300	12″ wide		2.80	5.714		730	269		999	1,225
2310	18″ wide		2.20	7.273		890	340		1,230	1,500
2320	24″ wide		1.80	8.889		1,125	420		1,545	1,875
2330	30″ wide		1.60	10		1,450	470		1,920	2,325
2350	36″ wide		1.40	11.429		1,775	535		2,310	2,775
2400	Vertical elbow, 9″ wide		5.40	2.963		135	139		274	360
2600	12″ wide		4.60	3.478		146	163		309	410
2800	18″ wide		4	4		168	188		356	475
3000	24″ wide		3.20	5		208	235		443	590
3100	27″ wide		3	5.333		217	251		468	625
3200	30″ wide		2.60	6.154		230	289		519	700
3400	36″ wide		2.40	6.667		253	315		568	760
3600	Cross, 9″ wide		4	4		630	188		818	985
3800	12″ wide		3.20	5		670	235		905	1,100
4000	18″ wide		2.60	6.154		800	289		1,089	1,325
4200	24″ wide		2.20	7.273		1,025	340		1,365	1,650
4300	27″ wide		2.20	7.273		1,200	340		1,540	1,825

26 05 43.10 Trench Duct		Crew	Daily Output	Labor-Hours	Unit	Material	Labor	Equipment	Total	Total Incl O&P
						2009 Bare Costs				
4400	30" wide	2 Elec	2	8	Ea.	1,325	375		1,700	2,025
4600	36" wide		1.80	8.889		1,650	420		2,070	2,450
4620	Two compartment, 9" wide		3.80	4.211		660	198		858	1,025
4630	12" wide		3	5.333		695	251		946	1,150
4640	18" wide		2.40	6.667		835	315		1,150	1,400
4650	24" wide		2	8		1,075	375		1,450	1,750
4660	30" wide		1.80	8.889		1,400	420		1,820	2,200
4670	36" wide		1.60	10		1,700	470		2,170	2,600
4690	Three compartment, 9" wide		3.60	4.444		670	209		879	1,050
4700	12" wide		2.80	5.714		765	269		1,034	1,250
4710	18" wide		2.20	7.273		905	340		1,245	1,525
4720	24" wide		1.80	8.889		1,125	420		1,545	1,900
4730	30" wide		1.60	10		1,475	470		1,945	2,350
4740	36" wide		1.40	11.429		1,825	535		2,360	2,825
4800	End closure, 9" wide		14.40	1.111		39.50	52		91.50	124
5000	12" wide		12	1.333		45	62.50		107.50	146
5200	18" wide		10	1.600		69	75		144	192
5400	24" wide		8	2		91	94		185	245
5500	27" wide		7	2.286		105	107		212	281
5600	30" wide		6.60	2.424		114	114		228	300
5800	36" wide		5.80	2.759		136	130		266	350
6000	Tees, 9" wide		4	4		385	188		573	710
6200	12" wide		3.60	4.444		445	209		654	810
6400	18" wide		3.20	5		570	235		805	985
6600	24" wide		3	5.333		820	251		1,071	1,275
6700	27" wide		2.80	5.714		905	269		1,174	1,400
6800	30" wide		2.60	6.154		1,075	289		1,364	1,625
7000	36" wide		2	8		1,400	375		1,775	2,125
7020	Two compartment, 9" wide		3.80	4.211		445	198		643	795
7030	12" wide		3.40	4.706		480	221		701	865
7040	18" wide		3	5.333		640	251		891	1,100
7050	24" wide		2.80	5.714		865	269		1,134	1,375
7060	30" wide		2.40	6.667		1,175	315		1,490	1,775
7070	36" wide		1.90	8.421		1,475	395		1,870	2,225
7090	Three compartment, 9" wide		3.60	4.444		510	209		719	880
7100	12" wide		3.20	5		535	235		770	945
7110	18" wide		2.80	5.714		675	269		944	1,150
7120	24" wide		2.60	6.154		935	289		1,224	1,475
7130	30" wide		2.20	7.273		1,225	340		1,565	1,875
7140	36" wide		1.80	8.889		1,550	420		1,970	2,375
7200	Riser, and cabinet connector, 9" wide		5.40	2.963		168	139		307	400
7400	12" wide		4.60	3.478		196	163		359	465
7600	18" wide		4	4		240	188		428	555
7800	24" wide		3.20	5		291	235		526	680
7900	27" wide		3	5.333		300	251		551	715
8000	30" wide		2.60	6.154		335	289		624	810
8200	36" wide		2	8		390	375		765	1,000
8400	Insert assembly, cell to conduit adapter, 1-1/4"	1 Elec	16	.500		66.50	23.50		90	109
8500	Adjustable partition	"	320	.025	L.F.	23.50	1.18		24.68	28
8600	Depth of duct over 4", per 1", add					10.30			10.30	11.30
8700	Support post	1 Elec	240	.033		23.50	1.57		25.07	28.50
8800	Cover double tile trim, 2 sides					36.50			36.50	40
8900	4 sides					108			108	119

26 05 43 – Underground Ducts and Raceways for Electrical Systems

26 05 43.10 Trench Duct		Crew	Daily Output	Labor-Hours	Unit	Material	2009 Bare Costs Labor	Equipment	Total	Total Incl O&P
9160	Trench duct 3-1/2" x 4-1/2", add				L.F.	9.80			9.80	10.75
9170	Trench duct 4" x 5", add					9.80			9.80	10.75
9200	For carpet trim, add					33			33	36.50
9210	For double carpet trim, add				↓	101			101	111

26 05 43.20 Underfloor Duct		Crew	Daily Output	Labor-Hours	Unit	Material	Labor	Equipment	Total	Total Incl O&P
0010	**UNDERFLOOR DUCT**									
0100	Duct, 1-3/8" x 3-1/8" blank, standard	2 Elec	160	.100	L.F.	12.70	4.70		17.40	21
0200	1-3/8" x 7-1/4" blank, super duct		120	.133		25.50	6.25		31.75	37.50
0400	7/8" or 1-3/8" insert type, 24" O.C., 1-3/8" x 3-1/8", std.		140	.114		17	5.35		22.35	27
0600	1-3/8" x 7-1/4", super duct	↓	100	.160	↓	29.50	7.50		37	44
0800	Junction box, single duct, 1 level, 3-1/8"	1 Elec	4	2	Ea.	380	94		474	565
0820	3-1/8" x 7-1/4"		4	2		445	94		539	635
0840	2 level, 3-1/8" upper & lower		3.20	2.500		445	118		563	670
0860	3-1/8" upper, 7-1/4" lower		2.70	2.963		445	139		584	705
0880	Carpet pan for above		80	.100		315	4.70		319.70	355
0900	Terrazzo pan for above		67	.119		740	5.60		745.60	825
1000	Junction box, single duct, 1 level, 7-1/4"		2.70	2.963		445	139		584	705
1020	2 level, 7-1/4" upper & lower		2.70	2.963		510	139		649	775
1040	2 duct, two 3-1/8" upper & lower		3.20	2.500		675	118		793	925
1200	1 level, 2 duct, 3-1/8"		3.20	2.500		510	118		628	740
1220	Carpet pan for above boxes		80	.100		315	4.70		319.70	355
1240	Terrazzo pan for above boxes		67	.119		740	5.60		745.60	825
1260	Junction box, 1 level, two 3-1/8" x one 3-1/8" + one 7-1/4"		2.30	3.478		865	163		1,028	1,200
1280	2 level, two 3-1/8" upper, one 3-1/8" + one 7-1/4" lower		2	4		950	188		1,138	1,350
1300	Carpet pan for above boxes		80	.100		315	4.70		319.70	355
1320	Terrazzo pan for above boxes		67	.119		740	5.60		745.60	825
1400	Junction box, 1 level, 2 duct, 7-1/4"		2.30	3.478		1,300	163		1,463	1,700
1420	Two 3-1/8" + one 7-1/4"		2	4		1,300	188		1,488	1,750
1440	Carpet pan for above		80	.100		315	4.70		319.70	355
1460	Terrazzo pan for above		67	.119		740	5.60		745.60	825
1580	Junction box, 1 level, one 3-1/8" + one 7-1/4" x same		2.30	3.478		865	163		1,028	1,200
1600	Triple duct, 3-1/8"		2.30	3.478		865	163		1,028	1,200
1700	Junction box, 1 level, one 3-1/8" + two 7-1/4"		2	4		1,450	188		1,638	1,900
1720	Carpet pan for above		80	.100		315	4.70		319.70	355
1740	Terrazzo pan for above		67	.119		740	5.60		745.60	825
1800	Insert to conduit adapter, 3/4" & 1"		32	.250		31	11.75		42.75	52.50
2000	Support, single cell		27	.296		47	13.95		60.95	73
2200	Super duct		16	.500		47	23.50		70.50	87.50
2400	Double cell		16	.500		47	23.50		70.50	87.50
2600	Triple cell		11	.727		47	34		81	104
2800	Vertical elbow, standard duct		10	.800		84	37.50		121.50	151
3000	Super duct		8	1		84	47		131	165
3200	Cabinet connector, standard duct		32	.250		63.50	11.75		75.25	88
3400	Super duct		27	.296		63.50	13.95		77.45	91.50
3600	Conduit adapter, 1" to 1-1/4"		32	.250		63.50	11.75		75.25	88
3800	2" to 1-1/4"		27	.296		76	13.95		89.95	105
4000	Outlet, low tension (tele, computer, etc.)		8	1		89	47		136	171
4200	High tension, receptacle (120 volt)		8	1		89	47		136	171
4300	End closure, standard duct		160	.050		3.56	2.35		5.91	7.55
4310	Super duct		160	.050		6.75	2.35		9.10	11
4350	Elbow, horiz., standard duct		26	.308		212	14.45		226.45	256
4360	Super duct	↓	26	.308	↓	212	14.45		226.45	256

26 05 43 – Underground Ducts and Raceways for Electrical Systems

26 05 43.20 Underfloor Duct		Crew	Daily Output	Labor-Hours	Unit	2009 Bare Costs Material	Labor	Equipment	Total	Total Incl O&P
4380	Elbow, offset, standard duct	1 Elec	26	.308	Ea.	84	14.45		98.45	115
4390	Super duct		26	.308		84	14.45		98.45	115
4400	Marker screw assembly for inserts		50	.160		15.05	7.50		22.55	28
4410	Y take off, standard duct		26	.308		127	14.45		141.45	163
4420	Super duct		26	.308		127	14.45		141.45	163
4430	Box opening plug, standard duct		160	.050		11.85	2.35		14.20	16.65
4440	Super duct		160	.050		11.85	2.35		14.20	16.65
4450	Sleeve coupling, standard duct		160	.050		32	2.35		34.35	38.50
4460	Super duct		160	.050		39	2.35		41.35	46.50
4470	Conduit adapter, standard duct, 3/4"		32	.250		63.50	11.75		75.25	88
4480	1" or 1-1/4"		32	.250		63.50	11.75		75.25	88
4500	1-1/2"	↓	32	.250	↓	63.50	11.75		75.25	88

26 05 80 – Wiring Connections

26 05 80.10 Motor Connections

		Crew	Daily Output	Labor-Hours	Unit	Material	Labor	Equipment	Total	Total Incl O&P
0010	**MOTOR CONNECTIONS**									
0020	Flexible conduit and fittings, 115 volt, 1 phase, up to 1 HP motor	1 Elec	8	1	Ea.	9.10	47		56.10	82.50
0050	2 HP motor		6.50	1.231		9.35	58		67.35	99.50
0100	3 HP motor		5.50	1.455		14.55	68.50		83.05	121
0120	230 volt, 10 HP motor, 3 phase		4.20	1.905		15.60	89.50		105.10	155
0150	15 HP motor		3.30	2.424		27	114		141	205
0200	25 HP motor		2.70	2.963		31	139		170	248
0400	50 HP motor		2.20	3.636		79	171		250	350
0600	100 HP motor		1.50	5.333		194	251		445	600
1500	460 volt, 5 HP motor, 3 phase		8	1		9.95	47		56.95	83.50
1520	10 HP motor		8	1		9.95	47		56.95	83.50
1530	25 HP motor		6	1.333		17.50	62.50		80	116
1540	30 HP motor		6	1.333		17.50	62.50		80	116
1550	40 HP motor		5	1.600		27	75		102	146
1560	50 HP motor		5	1.600		29	75		104	148
1570	60 HP motor		3.80	2.105		32.50	99		131.50	188
1580	75 HP motor		3.50	2.286		45.50	107		152.50	216
1590	100 HP motor		2.50	3.200		68.50	150		218.50	305
1600	125 HP motor		2	4		90	188		278	390
1610	150 HP motor		1.80	4.444		95	209		304	425
1620	200 HP motor		1.50	5.333		150	251		401	550
2005	460 Volt, 5 HP motor, 3 Phase, w/sealtite		8	1		14.95	47		61.95	89
2010	10 HP motor		8	1		14.95	47		61.95	89
2015	25 HP motor		6	1.333		27	62.50		89.50	126
2020	30 HP motor		6	1.333		27	62.50		89.50	126
2025	40 HP motor		5	1.600		49.50	75		124.50	171
2030	50 HP motor		5	1.600		48.50	75		123.50	170
2035	60 HP motor		3.80	2.105		78	99		177	238
2040	75 HP motor		3.50	2.286		81.50	107		188.50	255
2045	100 HP motor		2.50	3.200		110	150		260	350
2055	150 HP motor		1.80	4.444		116	209		325	450
2060	200 HP motor		1.50	5.333	↓	785	251		1,036	1,250
9000	Minimum labor/equipment charge	↓	4	2	Job		94		94	145

26 05 90 – Residential Applications

26 05 90.10 Residential Wiring

		Crew	Daily Output	Labor-Hours	Unit	Material	Labor	Equipment	Total	Total Incl O&P
0010	**RESIDENTIAL WIRING**									
0020	20' avg. runs and #14/2 wiring incl. unless otherwise noted									
1000	Service & panel, includes 24' SE-AL cable, service eye, meter,									

26 05 90.10 Residential Wiring		Crew	Daily Output	Labor-Hours	Unit	Material	2009 Bare Costs Labor	Equipment	Total	Total Incl O&P
1010	Socket, panel board, main bkr., ground rod, 15 or 20 amp									
1020	1-pole circuit breakers, and misc. hardware									
1100	100 amp, with 10 branch breakers	1 Elec	1.19	6.723	Ea.	585	315		900	1,125
1110	With PVC conduit and wire		.92	8.696		660	410		1,070	1,350
1120	With RGS conduit and wire		.73	10.959		840	515		1,355	1,725
1150	150 amp, with 14 branch breakers		1.03	7.767		895	365		1,260	1,550
1170	With PVC conduit and wire		.82	9.756		1,050	460		1,510	1,850
1180	With RGS conduit and wire		.67	11.940		1,400	560		1,960	2,400
1200	200 amp, with 18 branch breakers	2 Elec	1.80	8.889		1,175	420		1,595	1,950
1220	With PVC conduit and wire		1.46	10.959		1,325	515		1,840	2,275
1230	With RGS conduit and wire		1.24	12.903		1,800	605		2,405	2,900
1800	Lightning surge suppressor for above services, add	1 Elec	32	.250		50.50	11.75		62.25	73.50
2000	Switch devices									
2100	Single pole, 15 amp, Ivory, with a 1-gang box, cover plate,									
2110	Type NM (Romex) cable	1 Elec	17.10	.468	Ea.	11.70	22		33.70	47
2120	Type MC (BX) cable		14.30	.559		28	26.50		54.50	71.50
2130	EMT & wire		5.71	1.401		34.50	66		100.50	139
2150	3-way, #14/3, type NM cable		14.55	.550		17.25	26		43.25	59
2170	Type MC cable		12.31	.650		39	30.50		69.50	90
2180	EMT & wire		5	1.600		38.50	75		113.50	159
2200	4-way, #14/3, type NM cable		14.55	.550		33.50	26		59.50	77
2220	Type MC cable		12.31	.650		55.50	30.50		86	108
2230	EMT & wire		5	1.600		55	75		130	177
2250	S.P., 20 amp, #12/2, type NM cable		13.33	.600		19.70	28		47.70	65
2270	Type MC cable		11.43	.700		33	33		66	87
2280	EMT & wire		4.85	1.649		44.50	77.50		122	168
2290	S.P. rotary dimmer, 600W, no wiring		17	.471		19.90	22		41.90	56
2300	S.P. rotary dimmer, 600W, type NM cable		14.55	.550		26.50	26		52.50	69
2320	Type MC cable		12.31	.650		43	30.50		73.50	94
2330	EMT & wire		5	1.600		51.50	75		126.50	173
2350	3-way rotary dimmer, type NM cable		13.33	.600		22.50	28		50.50	68
2370	Type MC cable		11.43	.700		39	33		72	93.50
2380	EMT & wire		4.85	1.649		47.50	77.50		125	171
2400	Interval timer wall switch, 20 amp, 1-30 min., #12/2									
2410	Type NM cable	1 Elec	14.55	.550	Ea.	53	26		79	98
2420	Type MC cable		12.31	.650		60	30.50		90.50	113
2430	EMT & wire		5	1.600		77.50	75		152.50	202
2500	Decorator style									
2510	S.P., 15 amp, type NM cable	1 Elec	17.10	.468	Ea.	16.15	22		38.15	52
2520	Type MC cable		14.30	.559		32.50	26.50		59	76.50
2530	EMT & wire		5.71	1.401		39	66		105	144
2550	3-way, #14/3, type NM cable		14.55	.550		21.50	26		47.50	64
2570	Type MC cable		12.31	.650		43.50	30.50		74	94.50
2580	EMT & wire		5	1.600		43	75		118	164
2600	4-way, #14/3, type NM cable		14.55	.550		38	26		64	82
2620	Type MC cable		12.31	.650		60	30.50		90.50	113
2630	EMT & wire		5	1.600		59.50	75		134.50	182
2650	S.P., 20 amp, #12/2, type NM cable		13.33	.600		24	28		52	70
2670	Type MC cable		11.43	.700		37.50	33		70.50	92
2680	EMT & wire		4.85	1.649		49	77.50		126.50	173
2700	S.P., slide dimmer, type NM cable		17.10	.468		30	22		52	67.50
2720	Type MC cable		14.30	.559		46.50	26.50		73	92
2730	EMT & wire		5.71	1.401		55	66		121	162

26 05 Common Work Results for Electrical

26 05 90 – Residential Applications

26 05 90.10 Residential Wiring		Crew	Daily Output	Labor-Hours	Unit	Material	Labor (2009 Bare Costs)	Equipment	Total	Total Incl O&P
2750	S.P., touch dimmer, type NM cable	1 Elec	17.10	.468	Ea.	27	22		49	64
2770	Type MC cable		14.30	.559		43.50	26.50		70	88.50
2780	EMT & wire		5.71	1.401		52	66		118	159
2800	3-way touch dimmer, type NM cable		13.33	.600		47	28		75	95.50
2820	Type MC cable		11.43	.700		63.50	33		96.50	121
2830	EMT & wire		4.85	1.649		72	77.50		149.50	198
3000	Combination devices									
3100	S.P. switch/15 amp recpt., Ivory, 1-gang box, plate									
3110	Type NM cable	1 Elec	11.43	.700	Ea.	22.50	33		55.50	75.50
3120	Type MC cable		10	.800		39	37.50		76.50	101
3130	EMT & wire		4.40	1.818		47.50	85.50		133	183
3150	S.P. switch/pilot light, type NM cable		11.43	.700		23.50	33		56.50	76
3170	Type MC cable		10	.800		40	37.50		77.50	102
3180	EMT & wire		4.43	1.806		48	85		133	184
3190	2-S.P. switches, 2-#14/2, no wiring		14	.571		7.80	27		34.80	50
3200	2-S.P. switches, 2-#14/2, type NM cables		10	.800		28	37.50		65.50	89
3220	Type MC cable		8.89	.900		53.50	42.50		96	124
3230	EMT & wire		4.10	1.951		52.50	91.50		144	199
3250	3-way switch/15 amp recpt., #14/3, type NM cable		10	.800		31.50	37.50		69	93
3270	Type MC cable		8.89	.900		53.50	42.50		96	124
3280	EMT & wire		4.10	1.951		53	91.50		144.50	200
3300	2-3 way switches, 2-#14/3, type NM cables		8.89	.900		44.50	42.50		87	114
3320	Type MC cable		8	1		80	47		127	161
3330	EMT & wire		4	2		62	94		156	213
3350	S.P. switch/20 amp recpt., #12/2, type NM cable		10	.800		36	37.50		73.50	97.50
3370	Type MC cable		8.89	.900		43	42.50		85.50	113
3380	EMT & wire		4.10	1.951		60.50	91.50		152	208
3400	Decorator style									
3410	S.P. switch/15 amp recpt., type NM cable	1 Elec	11.43	.700	Ea.	27	33		60	80
3420	Type MC cable		10	.800		43.50	37.50		81	106
3430	EMT & wire		4.40	1.818		52	85.50		137.50	188
3450	S.P. switch/pilot light, type NM cable		11.43	.700		28	33		61	81
3470	Type MC cable		10	.800		44.50	37.50		82	107
3480	EMT & wire		4.40	1.818		52.50	85.50		138	189
3500	2-S.P. switches, 2-#14/2, type NM cables		10	.800		32.50	37.50		70	94
3520	Type MC cable		8.89	.900		58	42.50		100.50	129
3530	EMT & wire		4.10	1.951		57	91.50		148.50	204
3550	3-way/15 amp recpt., #14/3, type NM cable		10	.800		36	37.50		73.50	97.50
3570	Type MC cable		8.89	.900		58	42.50		100.50	129
3580	EMT & wire		4.10	1.951		57.50	91.50		149	205
3650	2-3 way switches, 2-#14/3, type NM cables		8.89	.900		48.50	42.50		91	119
3670	Type MC cable		8	1		84.50	47		131.50	166
3680	EMT & wire		4	2		66.50	94		160.50	218
3700	S.P. switch/20 amp recpt., #12/2, type NM cable		10	.800		40.50	37.50		78	103
3720	Type MC cable		8.89	.900		47.50	42.50		90	118
3730	EMT & wire		4.10	1.951		65	91.50		156.50	213
4000	Receptacle devices									
4010	Duplex outlet, 15 amp recpt., Ivory, 1-gang box, plate									
4015	Type NM cable	1 Elec	14.55	.550	Ea.	10.20	26		36.20	51
4020	Type MC cable		12.31	.650		26.50	30.50		57	76.50
4030	EMT & wire		5.33	1.501		33	70.50		103.50	146
4050	With #12/2, type NM cable		12.31	.650		13.65	30.50		44.15	62
4070	Type MC cable		10.67	.750		27	35		62	83.50

26 05 90.10 Residential Wiring		Crew	Daily Output	Labor-Hours	Unit	Material	2009 Bare Costs Labor	Equipment	Total	Total Incl O&P
4080	EMT & wire	1 Elec	4.71	1.699	Ea.	38.50	80		118.50	166
4100	20 amp recpt., #12/2, type NM cable		12.31	.650		22.50	30.50		53	71.50
4120	Type MC cable		10.67	.750		35.50	35		70.50	93
4130	EMT & wire		4.71	1.699		47	80		127	175
4140	For GFI see Div. 26 05 90.10 line 4300 below									
4150	Decorator style, 15 amp recpt., type NM cable	1 Elec	14.55	.550	Ea.	14.65	26		40.65	56
4170	Type MC cable		12.31	.650		31	30.50		61.50	81
4180	EMT & wire		5.33	1.501		37.50	70.50		108	150
4200	With #12/2, type NM cable		12.31	.650		18.10	30.50		48.60	67
4220	Type MC cable		10.67	.750		31.50	35		66.50	88.50
4230	EMT & wire		4.71	1.699		43	80		123	170
4250	20 amp recpt. #12/2, type NM cable		12.31	.650		26.50	30.50		57	76.50
4270	Type MC cable		10.67	.750		40	35		75	98
4280	EMT & wire		4.71	1.699		51.50	80		131.50	180
4300	GFI, 15 amp recpt., type NM cable		12.31	.650		41.50	30.50		72	92.50
4320	Type MC cable		10.67	.750		58	35		93	118
4330	EMT & wire		4.71	1.699		64.50	80		144.50	194
4350	GFI with #12/2, type NM cable		10.67	.750		45	35		80	104
4370	Type MC cable		9.20	.870		58.50	41		99.50	127
4380	EMT & wire		4.21	1.900		69.50	89.50		159	214
4400	20 amp recpt., #12/2 type NM cable		10.67	.750		47	35		82	106
4420	Type MC cable		9.20	.870		60	41		101	129
4430	EMT & wire		4.21	1.900		71.50	89.50		161	216
4500	Weather-proof cover for above receptacles, add		32	.250		5.25	11.75		17	24
4550	Air conditioner outlet, 20 amp-240 volt recpt.									
4560	30′ of #12/2, 2 pole circuit breaker									
4570	Type NM cable	1 Elec	10	.800	Ea.	63	37.50		100.50	127
4580	Type MC cable		9	.889		79	42		121	152
4590	EMT & wire		4	2		87.50	94		181.50	241
4600	Decorator style, type NM cable		10	.800		67.50	37.50		105	132
4620	Type MC cable		9	.889		84	42		126	157
4630	EMT & wire		4	2		92	94		186	246
4650	Dryer outlet, 30 amp-240 volt recpt., 20′ of #10/3									
4660	2 pole circuit breaker									
4670	Type NM cable	1 Elec	6.41	1.248	Ea.	75	58.50		133.50	173
4680	Type MC cable		5.71	1.401		83	66		149	193
4690	EMT & wire		3.48	2.299		89	108		197	264
4700	Range outlet, 50 amp-240 volt recpt., 30′ of #8/3									
4710	Type NM cable	1 Elec	4.21	1.900	Ea.	110	89.50		199.50	257
4720	Type MC cable		4	2		195	94		289	360
4730	EMT & wire		2.96	2.703		121	127		248	330
4750	Central vacuum outlet, Type NM cable		6.40	1.250		69.50	59		128.50	167
4770	Type MC cable		5.71	1.401		94	66		160	204
4780	EMT & wire		3.48	2.299		96.50	108		204.50	272
4800	30 amp-110 volt locking recpt., #10/2 circ. bkr.									
4810	Type NM cable	1 Elec	6.20	1.290	Ea.	80.50	60.50		141	182
4820	Type MC cable		5.40	1.481		111	69.50		180.50	229
4830	EMT & wire		3.20	2.500		110	118		228	300
4900	Low voltage outlets									
4910	Telephone recpt., 20′ of 4/C phone wire	1 Elec	26	.308	Ea.	10.30	14.45		24.75	34
4920	TV recpt., 20′ of RG59U coax wire, F type connector	″	16	.500	″	18.35	23.50		41.85	56
4950	Door bell chime, transformer, 2 buttons, 60′ of bellwire									
4970	Economy model	1 Elec	11.50	.696	Ea.	75	32.50		107.50	133

26 05 90.10 Residential Wiring		Crew	Daily Output	Labor-Hours	Unit	Material	Labor	Equipment	Total	Total Incl O&P
						2009 Bare Costs				
4980	Custom model	1 Elec	11.50	.696	Ea.	116	32.50		148.50	179
4990	Luxury model, 3 buttons	↓	9.50	.842	↓	310	39.50		349.50	400
6000	Lighting outlets									
6050	Wire only (for fixture), type NM cable	1 Elec	32	.250	Ea.	8.45	11.75		20.20	27.50
6070	Type MC cable		24	.333		18.55	15.65		34.20	44.50
6080	EMT & wire		10	.800		23.50	37.50		61	83.50
6100	Box (4"), and wire (for fixture), type NM cable		25	.320		17.45	15.05		32.50	42
6120	Type MC cable		20	.400		27.50	18.80		46.30	59.50
6130	EMT & wire	↓	11	.727	↓	32.50	34		66.50	88
6200	Fixtures (use with lines 6050 or 6100 above)									
6210	Canopy style, economy grade	1 Elec	40	.200	Ea.	29	9.40		38.40	46.50
6220	Custom grade		40	.200		51	9.40		60.40	70.50
6250	Dining room chandelier, economy grade		19	.421		82.50	19.80		102.30	122
6260	Custom grade		19	.421		243	19.80		262.80	298
6270	Luxury grade		15	.533		540	25		565	635
6310	Kitchen fixture (fluorescent), economy grade		30	.267		61.50	12.55		74.05	87
6320	Custom grade		25	.320		168	15.05		183.05	208
6350	Outdoor, wall mounted, economy grade		30	.267		29.50	12.55		42.05	52
6360	Custom grade		30	.267		108	12.55		120.55	138
6370	Luxury grade		25	.320		244	15.05		259.05	291
6410	Outdoor PAR floodlights, 1 lamp, 150 watt		20	.400		32	18.80		50.80	64
6420	2 lamp, 150 watt each		20	.400		52	18.80		70.80	86
6425	Motion sensing, 2 lamp, 150 watt each		20	.400		70	18.80		88.80	106
6430	For infrared security sensor, add		32	.250		111	11.75		122.75	140
6450	Outdoor, quartz-halogen, 300 watt flood		20	.400		41	18.80		59.80	74
6600	Recessed downlight, round, pre-wired, 50 or 75 watt trim		30	.267		42	12.55		54.55	65.50
6610	With shower light trim		30	.267		52	12.55		64.55	76.50
6620	With wall washer trim		28	.286		62.50	13.45		75.95	89.50
6630	With eye-ball trim	↓	28	.286		62.50	13.45		75.95	89.50
6640	For direct contact with insulation, add					2.10			2.10	2.31
6700	Porcelain lamp holder	1 Elec	40	.200		4.10	9.40		13.50	18.95
6710	With pull switch		40	.200		4.59	9.40		13.99	19.50
6750	Fluorescent strip, 1-20 watt tube, wrap around diffuser, 24"		24	.333		54	15.65		69.65	83.50
6760	1-40 watt tube, 48"		24	.333		66	15.65		81.65	96.50
6770	2-40 watt tubes, 48"		20	.400		80	18.80		98.80	117
6780	With residential ballast		20	.400		90	18.80		108.80	128
6800	Bathroom heat lamp, 1-250 watt		28	.286		46	13.45		59.45	71
6810	2-250 watt lamps	↓	28	.286	↓	71.50	13.45		84.95	99
6820	For timer switch, see Div. 26 05 90.10 line 2400									
6900	Outdoor post lamp, incl. post, fixture, 35' of #14/2									
6910	Type NMC cable	1 Elec	3.50	2.286	Ea.	203	107		310	390
6920	Photo-eye, add		27	.296		32.50	13.95		46.45	57.50
6950	Clock dial time switch, 24 hr., w/enclosure, type NM cable		11.43	.700		63	33		96	120
6970	Type MC cable		11	.727		79.50	34		113.50	140
6980	EMT & wire	↓	4.85	1.649	↓	85.50	77.50		163	213
7000	Alarm systems									
7050	Smoke detectors, box, #14/3, type NM cable	1 Elec	14.55	.550	Ea.	37.50	26		63.50	81
7070	Type MC cable		12.31	.650		53	30.50		83.50	105
7080	EMT & wire	↓	5	1.600		52.50	75		127.50	174
7090	For relay output to security system, add				↓	12.90			12.90	14.20
8000	Residential equipment									
8050	Disposal hook-up, incl. switch, outlet box, 3' of flex									
8060	20 amp-1 pole circ. bkr., and 25' of #12/2									

26 05 90.10 Residential Wiring		Crew	Daily Output	Labor-Hours	Unit	Material	Labor	Equipment	Total	Total Incl O&P
8070	Type NM cable	1 Elec	10	.800	Ea.	33.50	37.50		71	95
8080	Type MC cable		8	1		48.50	47		95.50	126
8090	EMT & wire		5	1.600		61	75		136	184
8100	Trash compactor or dishwasher hook-up, incl. outlet box,									
8110	3′ of flex, 15 amp-1 pole circ. bkr., and 25′ of #14/2									
8120	Type NM cable	1 Elec	10	.800	Ea.	24.50	37.50		62	85
8130	Type MC cable		8	1		43.50	47		90.50	121
8140	EMT & wire		5	1.600		52.50	75		127.50	174
8150	Hot water sink dispensor hook-up, use line 8100									
8200	Vent/exhaust fan hook-up, type NM cable	1 Elec	32	.250	Ea.	8.45	11.75		20.20	27.50
8220	Type MC cable		24	.333		18.55	15.65		34.20	44.50
8230	EMT & wire		10	.800		23.50	37.50		61	83.50
8250	Bathroom vent fan, 50 CFM (use with above hook-up)									
8260	Economy model	1 Elec	15	.533	Ea.	24.50	25		49.50	65.50
8270	Low noise model		15	.533		32	25		57	73.50
8280	Custom model		12	.667		117	31.50		148.50	177
8300	Bathroom or kitchen vent fan, 110 CFM									
8310	Economy model	1 Elec	15	.533	Ea.	66.50	25		91.50	112
8320	Low noise model	"	15	.533	"	86	25		111	133
8350	Paddle fan, variable speed (w/o lights)									
8360	Economy model (AC motor)	1 Elec	10	.800	Ea.	105	37.50		142.50	174
8370	Custom model (AC motor)		10	.800		182	37.50		219.50	258
8380	Luxury model (DC motor)		8	1		360	47		407	470
8390	Remote speed switch for above, add		12	.667		28.50	31.50		60	79.50
8500	Whole house exhaust fan, ceiling mount, 36″, variable speed									
8510	Remote switch, incl. shutters, 20 amp-1 pole circ. bkr.									
8520	30′ of #12/2, type NM cable	1 Elec	4	2	Ea.	985	94		1,079	1,225
8530	Type MC cable		3.50	2.286		1,000	107		1,107	1,275
8540	EMT & wire		3	2.667		1,025	125		1,150	1,325
8600	Whirlpool tub hook-up, incl. timer switch, outlet box									
8610	3′ of flex, 20 amp-1 pole GFI circ. bkr.									
8620	30′ of #12/2, type NM cable	1 Elec	5	1.600	Ea.	127	75		202	255
8630	Type MC cable		4.20	1.905		135	89.50		224.50	287
8640	EMT & wire		3.40	2.353		146	111		257	330
8650	Hot water heater hook-up, incl. 1-2 pole circ. bkr., box;									
8660	3′ of flex, 20′ of #10/2, type NM cable	1 Elec	5	1.600	Ea.	36	75		111	156
8670	Type MC cable		4.20	1.905		58	89.50		147.50	202
8680	EMT & wire		3.40	2.353		55.50	111		166.50	231
9000	Heating/air conditioning									
9050	Furnace/boiler hook-up, incl. firestat, local on-off switch									
9060	Emergency switch, and 40′ of type NM cable	1 Elec	4	2	Ea.	57	94		151	208
9070	Type MC cable		3.50	2.286		82.50	107		189.50	256
9080	EMT & wire		1.50	5.333		94	251		345	490
9100	Air conditioner hook-up, incl. local 60 amp disc. switch									
9110	3′ sealtite, 40 amp, 2 pole circuit breaker									
9130	40′ of #8/2, type NM cable	1 Elec	3.50	2.286	Ea.	229	107		336	415
9140	Type MC cable		3	2.667		360	125		485	590
9150	EMT & wire		1.30	6.154		268	239		557	740
9200	Heat pump hook-up, 1-40 & 1-100 amp 2 pole circ. bkr.									
9210	Local disconnect switch, 3′ sealtite									
9220	40′ of #8/2 & 30′ of #3/2									
9230	Type NM cable	1 Elec	1.30	6.154	Ea.	595	239		884	1,100
9240	Type MC cable		1.08	7.407		865	350		1,215	1,475

26 05 Common Work Results for Electrical

26 05 90 – Residential Applications

26 05 90.10 Residential Wiring		Crew	Daily Output	Labor-Hours	Unit	2009 Bare Costs Material	Labor	Equipment	Total	Total Incl O&P
9250	EMT & wire	1 Elec	.94	8.511	Ea.	675	400		1,075	1,350
9500	Thermostat hook-up, using low voltage wire									
9520	Heating only, 25′ of #18-3	1 Elec	24	.333	Ea.	7.70	15.65		23.35	32.50
9530	Heating/cooling, 25′ of #18-4	"	20	.400	"	9.15	18.80		27.95	39

26 09 Instrumentation and Control for Electrical Systems

26 09 13 – Electrical Power Monitoring and Control

26 09 13.10 Switchboard Instruments

		Crew	Daily Output	Labor-Hours	Unit	Material	Labor	Equipment	Total	Total Incl O&P
0010	**SWITCHBOARD INSTRUMENTS** 3 phase, 4 wire									
0100	AC indicating, ammeter & switch	1 Elec	8	1	Ea.	2,050	47		2,097	2,325
0200	Voltmeter & switch		8	1		2,050	47		2,097	2,325
0300	Wattmeter		8	1		4,050	47		4,097	4,525
0400	AC recording, ammeter		4	2		7,225	94		7,319	8,100
0500	Voltmeter		4	2		7,225	94		7,319	8,100
0600	Ground fault protection, zero sequence		2.70	2.963		6,375	139		6,514	7,250
0700	Ground return path		2.70	2.963		6,375	139		6,514	7,250
0800	3 current transformers, 5 to 800 amp		2	4		2,975	188		3,163	3,575
0900	1000 to 1500 amp		1.30	6.154		4,275	289		4,564	5,150
1200	2000 to 4000 amp		1	8		5,050	375		5,425	6,125
1300	Fused potential transformer, maximum 600 volt		8	1		1,125	47		1,172	1,300

26 09 13.20 Voltage Monitor Systems

		Crew	Daily Output	Labor-Hours	Unit	Material	Labor	Equipment	Total	Total Incl O&P
0010	**VOLTAGE MONITOR SYSTEMS** (test equipment)									
0100	AC voltage monitor system, 120/240 V, one-channel				Ea.	3,450			3,450	3,800
0110	Modem adapter					430			430	475
0120	Add-on detector only					1,800			1,800	2,000
0150	AC voltage remote monitor sys., 3 channel, 120, 230, or 480 V					6,275			6,275	6,900
0160	With internal modem					6,625			6,625	7,275
0170	Combination temperature and humidity probe					970			970	1,075
0180	Add-on detector only					4,550			4,550	5,000
0190	With internal modem					4,950			4,950	5,450

26 12 Medium-Voltage Transformers

26 12 19 – Pad-Mounted, Liquid-Filled, Medium-Voltage Transformers

26 12 19.10 Transformer, Oil-Filled

		Crew	Daily Output	Labor-Hours	Unit	Material	Labor	Equipment	Total	Total Incl O&P
0010	**TRANSFORMER, OIL-FILLED** primary delta or Y,									
0050	Pad mounted 5 kV or 15 kV, with taps, 277/480 V secondary, 3 phase									
0100	150 kVA	R-3	.65	30.769	Ea.	9,400	1,425	199	11,024	12,700
0110	225 kVA		.55	36.364		9,625	1,675	235	11,535	13,500
0200	300 kVA		.45	44.444		12,000	2,050	287	14,337	16,700
0300	500 kVA		.40	50		18,800	2,325	325	21,450	24,600
0400	750 kVA		.38	52.632		21,700	2,425	340	24,465	28,000
0500	1000 kVA		.26	76.923		25,700	3,550	495	29,745	34,300
0600	1500 kVA		.23	86.957		30,500	4,025	560	35,085	40,400
0700	2000 kVA		.20	100		38,500	4,625	645	43,770	50,500
0710	2500 kVA		.19	105		46,600	4,875	680	52,155	59,500
0720	3000 kVA		.17	117		56,000	5,450	760	62,210	71,500
0800	3750 kVA		.16	125		72,000	5,800	810	78,610	89,500

26 12 19.20 Transformer, Liquid-Filled

		Crew	Daily Output	Labor-Hours	Unit	Material	Labor	Equipment	Total	Total Incl O&P
0010	**TRANSFORMER, LIQUID-FILLED** Pad mounted									

26 12 Medium-Voltage Transformers

26 12 19 – Pad-Mounted, Liquid-Filled, Medium-Voltage Transformers

26 12 19.20 Transformer, Liquid-Filled		Crew	Daily Output	Labor-Hours	Unit	Material	2009 Bare Costs Labor	Equipment	Total	Total Incl O&P
0020	5 kV or 15 kV primary, 277/480 volt secondary, 3 phase									
0050	225 kVA	R-3	.55	36.364	Ea.	13,300	1,675	235	15,210	17,500
0100	300 kVA		.45	44.444		15,800	2,050	287	18,137	20,900
0200	500 kVA		.40	50		20,000	2,325	325	22,650	25,800
0250	750 kVA		.38	52.632		25,800	2,425	340	28,565	32,500
0300	1000 kVA		.26	76.923		29,900	3,550	495	33,945	38,900
0350	1500 kVA		.23	86.957		34,900	4,025	560	39,485	45,200
0400	2000 kVA		.20	100		43,200	4,625	645	48,470	55,500
0450	2500 kVA		.19	105		51,500	4,875	680	57,055	65,000

26 13 Medium-Voltage Switchgear

26 13 16 – Medium-Voltage Fusible Interrupter Switchgear

26 13 16.10 Switchgear

		Crew	Daily Output	Labor-Hours	Unit	Material	Labor	Equipment	Total	Total Incl O&P
0010	**SWITCHGEAR**, Incorporate switch with cable connections, transformer,									
0100	& Low Voltage section									
0200	Load interrupter switch, 600 amp, 2 position									
0300	NEMA 1, 4.8 KV, 300 kVA & below w/CLF fuses	R-3	.40	50	Ea.	19,600	2,325	325	22,250	25,400
0400	400 kVA & above w/CLF fuses		.38	52.632		21,800	2,425	340	24,565	28,200
0500	Non fusible		.41	48.780		15,900	2,250	315	18,465	21,200
0600	13.8 kV, 300 kVA & below w/CLF fuses		.38	52.632		24,600	2,425	340	27,365	31,200
0700	400 kVA & above w/CLF fuses		.36	55.556		24,600	2,575	360	27,535	31,400
0800	Non fusible		.40	50		18,600	2,325	325	21,250	24,400
0900	Cable lugs for 2 feeders 4.8 kV or 13.8 kV	1 Elec	8	1		595	47		642	730
1000	Pothead, one 3 conductor or three 1 conductor		4	2		2,850	94		2,944	3,300
1100	Two 3 conductor or six 1 conductor		2	4		5,650	188		5,838	6,500
1200	Key interlocks		8	1		660	47		707	800
1300	Lightning arresters, Distribution class (no charge)									
1400	Intermediate class or line type 4.8 kV	1 Elec	2.70	2.963	Ea.	3,200	139		3,339	3,750
1500	13.8 kV		2	4		4,250	188		4,438	4,975
1600	Station class, 4.8 kV		2.70	2.963		5,500	139		5,639	6,275
1700	13.8 kV		2	4		9,450	188		9,638	10,700
1800	Transformers, 4800 volts to 480/277 volts, 75 kVA	R-3	.68	29.412		16,700	1,350	190	18,240	20,700
1900	112.5 kVA		.65	30.769		20,400	1,425	199	22,024	24,900
2000	150 kVA		.57	35.088		23,200	1,625	227	25,052	28,300
2100	225 kVA		.48	41.667		26,700	1,925	269	28,894	32,700
2200	300 kVA		.41	48.780		29,800	2,250	315	32,365	36,600
2300	500 kVA		.36	55.556		39,300	2,575	360	42,235	47,600
2400	750 kVA		.29	68.966		44,600	3,200	445	48,245	54,500
2500	13,800 volts to 480/277 volts, 75 kVA		.61	32.787		23,600	1,525	212	25,337	28,500
2600	112.5 kVA		.55	36.364		31,300	1,675	235	33,210	37,300
2700	150 kVA		.49	40.816		31,600	1,900	264	33,764	37,900
2800	225 kVA		.41	48.780		36,400	2,250	315	38,965	43,900
2900	300 kVA		.37	54.054		37,200	2,500	350	40,050	45,200
3000	500 kVA		.31	64.516		41,200	3,000	415	44,615	50,500
3100	750 kVA		.42	47.733		45,300	2,200	310	47,810	53,500
3200	Forced air cooling & temperature alarm	1 Elec	1	8		3,650	375		4,025	4,575
3300	Low voltage components									
3400	Maximum panel height 49-1/2", single or twin row									
3500	Breaker heights, type FA or FH, 6"									
3600	type KA or KH, 8"									
3700	type LA, 11"									

26 13 Medium-Voltage Switchgear

26 13 16 – Medium-Voltage Fusible Interrupter Switchgear

26 13 16.10 Switchgear		Crew	Daily Output	Labor-Hours	Unit	2009 Bare Costs Material	Labor	Equipment	Total	Total Incl O&P
3800	type MA, 14"									
3900	Breakers, 2 pole, 15 to 60 amp, type FA	1 Elec	5.60	1.429	Ea.	385	67		452	530
4000	70 to 100 amp, type FA		4.20	1.905		490	89.50		579.50	675
4100	15 to 60 amp, type FH		5.60	1.429		665	67		732	835
4200	70 to 100 amp, type FH		4.20	1.905		775	89.50		864.50	990
4300	125 to 225 amp, type KA		3.40	2.353		1,125	111		1,236	1,425
4400	125 to 225 amp, type KH		3.40	2.353		2,775	111		2,886	3,225
4500	125 to 400 amp, type LA		2.50	3.200		2,100	150		2,250	2,525
4600	125 to 600 amp, type MA		1.80	4.444		3,275	209		3,484	3,925
4700	700 & 800 amp, type MA		1.50	5.333		4,225	251		4,476	5,025
4800	3 pole, 15 to 60 amp, type FA		5.30	1.509		495	71		566	655
4900	70 to 100 amp, type FA		4	2		610	94		704	815
5000	15 to 60 amp, type FH		5.30	1.509		775	71		846	965
5100	70 to 100 amp, type FH		4	2		880	94		974	1,125
5200	125 to 225 amp, type KA		3.20	2.500		1,425	118		1,543	1,750
5300	125 to 225 amp, type KH		3.20	2.500		3,350	118		3,468	3,850
5400	125 to 400 amp, type LA		2.30	3.478		2,525	163		2,688	3,025
5500	125 to 600 amp, type MA		1.60	5		4,150	235		4,385	4,925
5600	700 & 800 amp, type MA		1.30	6.154		5,450	289		5,739	6,450

26 22 Low-Voltage Transformers

26 22 13 – Low-Voltage Distribution Transformers

26 22 13.10 Transformer, Dry-Type

		Crew	Daily Output	Labor-Hours	Unit	Material	Labor	Equipment	Total	Total Incl O&P
0010	**TRANSFORMER, DRY-TYPE**									
0050	Single phase, 240/480 volt primary, 120/240 volt secondary									
0100	1 kVA	1 Elec	2	4	Ea.	266	188		454	580
0300	2 kVA		1.60	5		400	235		635	800
0500	3 kVA		1.40	5.714		495	269		764	960
0700	5 kVA		1.20	6.667		680	315		995	1,225
0900	7.5 kVA	2 Elec	2.20	7.273		945	340		1,285	1,575
1100	10 kVA		1.60	10		1,200	470		1,670	2,050
1300	15 kVA		1.20	13.333		1,650	625		2,275	2,800
1500	25 kVA		1	16		2,250	750		3,000	3,625
1700	37.5 kVA		.80	20		2,925	940		3,865	4,675
1900	50 kVA		.70	22.857		3,475	1,075		4,550	5,475
2100	75 kVA		.65	24.615		4,600	1,150		5,750	6,825
2110	100 kVA	R-3	.90	22.222		5,975	1,025	144	7,144	8,325
2120	167 kVA	"	.80	25		9,925	1,150	162	11,237	12,900
2190	480V primary 120/240V secondary, nonvent., 15 kVA	2 Elec	1.20	13.333		1,675	625		2,300	2,825
2200	25 kVA		.90	17.778		2,450	835		3,285	3,975
2210	37 kVA		.75	21.333		2,925	1,000		3,925	4,775
2220	50 kVA		.65	24.615		3,475	1,150		4,625	5,600
2230	75 kVA		.60	26.667		4,600	1,250		5,850	6,975
2240	100 kVA		.50	32		5,975	1,500		7,475	8,900
2250	Low operating temperature(80°C), 25 kVA		1	16		4,475	750		5,225	6,075
2260	37 kVA		.80	20		4,825	940		5,765	6,750
2270	50 kVA		.70	22.857		6,275	1,075		7,350	8,550
2280	75 kVA		.65	24.615		10,000	1,150		11,150	12,800
2290	100 kVA		.55	29.091		10,400	1,375		11,775	13,600
2300	3 phase, 480 volt primary 120/208 volt secondary									
2310	Ventilated, 3 kVA	1 Elec	1	8	Ea.	875	375		1,250	1,550

26 22 Low-Voltage Transformers

26 22 13 – Low-Voltage Distribution Transformers

26 22 13.10 Transformer, Dry-Type		Crew	Daily Output	Labor-Hours	Unit	Material	2009 Bare Costs Labor	Equipment	Total	Total Incl O&P
2700	6 kVA	1 Elec	.80	10	Ea.	1,200	470		1,670	2,050
2900	9 kVA		.70	11.429		1,375	535		1,910	2,325
3100	15 kVA	2 Elec	1.10	14.545		1,825	685		2,510	3,050
3300	30 kVA		.90	17.778		2,125	835		2,960	3,625
3500	45 kVA		.80	20		2,550	940		3,490	4,275
3700	75 kVA		.70	22.857		3,850	1,075		4,925	5,875
3900	112.5 kVA	R-3	.90	22.222		5,125	1,025	144	6,294	7,375
4100	150 kVA		.85	23.529		6,675	1,100	152	7,927	9,175
4300	225 kVA		.65	30.769		9,050	1,425	199	10,674	12,400
4500	300 kVA		.55	36.364		11,400	1,675	235	13,310	15,500
4700	500 kVA		.45	44.444		18,900	2,050	287	21,237	24,300
4800	750 kVA		.35	57.143		33,200	2,650	370	36,220	41,000
4820	1000 kVA		.32	62.500		38,000	2,900	405	41,305	46,700
4850	K-4 rated, 15 kVA	2 Elec	1.10	14.545		2,200	685		2,885	3,450
4855	30 kVA		.90	17.778		3,350	835		4,185	4,975
4860	45 kVA		.80	20		3,875	940		4,815	5,725
4865	75 kVA		.70	22.857		5,300	1,075		6,375	7,475
4870	112.5 kVA	R-3	.90	22.222		8,825	1,025	144	9,994	11,500
4875	150 kVA		.85	23.529		11,800	1,100	152	13,052	14,800
4880	225 kVA		.65	30.769		18,000	1,425	199	19,624	22,200
4885	300 kVA		.55	36.364		24,900	1,675	235	26,810	30,300
4890	500 kVA		.45	44.444		34,400	2,050	287	36,737	41,300
4900	K-13 rated, 15 kVA	2 Elec	1.10	14.545		2,875	685		3,560	4,200
4905	30 kVA		.90	17.778		3,950	835		4,785	5,625
4910	45 kVA		.80	20		4,650	940		5,590	6,550
4915	75 kVA		.70	22.857		6,575	1,075		7,650	8,875
4920	112.5 kVA	R-3	.90	22.222		13,500	1,025	144	14,669	16,700
4925	150 kVA		.85	23.529		16,500	1,100	152	17,752	19,900
4930	225 kVA		.65	30.769		24,000	1,425	199	25,624	28,800
4935	300 kVA		.55	36.364		33,700	1,675	235	35,610	40,000
4940	500 kVA		.45	44.444		51,500	2,050	287	53,837	60,000
5020	480 volt primary 120/208 volt secondary									
5030	Nonventilated, 15 kVA	2 Elec	1.10	14.545	Ea.	3,875	685		4,560	5,325
5040	30 kVA		.80	20		4,350	940		5,290	6,250
5050	45 kVA		.70	22.857		5,950	1,075		7,025	8,200
5060	75 kVA		.65	24.615		8,650	1,150		9,800	11,300
5070	112.5 kVA	R-3	.85	23.529		11,800	1,100	152	13,052	14,700
5081	150 kVA		.85	23.529		14,200	1,100	152	15,452	17,500
5090	225 kVA		.60	33.333		17,400	1,550	215	19,165	21,800
5100	300 kVA		.50	40		20,800	1,850	258	22,908	26,000
5200	Low operating temperature (80°C), 30 kVA	2 Elec	.90	17.778		3,550	835		4,385	5,175
5210	45 kVA		.80	20		5,350	940		6,290	7,325
5220	75 kVA		.70	22.857		7,125	1,075		8,200	9,475
5230	112.5 kVA	R-3	.90	22.222		12,600	1,025	144	13,769	15,600
5240	150 kVA		.85	23.529		15,900	1,100	152	17,152	19,300
5250	225 kVA		.65	30.769		22,000	1,425	199	23,624	26,600
5260	300 kVA		.55	36.364		26,300	1,675	235	28,210	31,800
5270	500 kVA		.45	44.444		39,600	2,050	287	41,937	47,000
5380	3 phase, 5 kV primary 277/480 volt secondary									
5400	High voltage, 112.5 kVA	R-3	.85	23.529	Ea.	15,400	1,100	152	16,652	18,700
5410	150 kVA		.65	30.769		16,700	1,425	199	18,324	20,800
5420	225 kVA		.55	36.364		19,700	1,675	235	21,610	24,600
5430	300 kVA		.45	44.444		25,200	2,050	287	27,537	31,200

26 22 13 – Low-Voltage Distribution Transformers

26 22 13.10 Transformer, Dry-Type		Crew	Daily Output	Labor-Hours	Unit	2009 Bare Costs Material	Labor	Equipment	Total	Total Incl O&P
5440	500 kVA	R-3	.35	57.143	Ea.	32,500	2,650	370	35,520	40,200
5450	750 kVA		.32	62.500		48,900	2,900	405	52,205	59,000
5460	1000 kVA		.30	66.667		57,500	3,075	430	61,005	68,500
5470	1500 kVA		.27	74.074		66,500	3,425	480	70,405	79,500
5480	2000 kVA		.25	80		78,000	3,700	515	82,215	92,500
5490	2500 kVA		.20	100		88,000	4,625	645	93,270	105,000
5500	3000 kVA		.18	111		115,500	5,150	720	121,370	135,500
5590	15 kV primary 277/480 volt secondary									
5600	High voltage, 112.5 kVA	R-3	.85	23.529	Ea.	23,400	1,100	152	24,652	27,600
5610	150 kVA		.65	30.769		27,300	1,425	199	28,924	32,500
5620	225 kVA		.55	36.364		30,200	1,675	235	32,110	36,100
5630	300 kVA		.45	44.444		35,600	2,050	287	37,937	42,600
5640	500 kVA		.35	57.143		45,000	2,650	370	48,020	54,000
5650	750 kVA		.32	62.500		56,500	2,900	405	59,805	67,500
5660	1000 kVA		.30	66.667		64,500	3,075	430	68,005	76,500
5670	1500 kVA		.27	74.074		74,500	3,425	480	78,405	88,000
5680	2000 kVA		.25	80		83,000	3,700	515	87,215	97,500
5690	2500 kVA		.20	100		96,000	4,625	645	101,270	113,500
5700	3000 kVA		.18	111		114,000	5,150	720	119,870	134,000
6000	2400V primary, 480V secondary, 300 KVA		.45	44.444		25,200	2,050	287	27,537	31,200
6010	500 KVA		.35	57.143		32,500	2,650	370	35,520	40,200
6020	750 KVA		.32	62.500		45,200	2,900	405	48,505	54,500
9000	Minimum labor/equipment charge	1 Elec	1	8	Job		375		375	580

26 22 13.20 Isolating Panels

		Crew	Daily Output	Labor-Hours	Unit	Material	Labor	Equipment	Total	Total Incl O&P
0010	**ISOLATING PANELS** used with isolating transformers									
0100	Critical care area, 8 circuit, 3 kVA	1 Elec	.58	13.793	Ea.	7,725	650		8,375	9,500
0200	5 kVA		.54	14.815		7,950	695		8,645	9,825
0400	7.5 kVA		.52	15.385		8,100	725		8,825	10,000
0600	10 kVA		.44	18.182		8,375	855		9,230	10,600
0800	Operating room power & lighting, 8 circuit, 3 kVA		.58	13.793		6,475	650		7,125	8,125
1000	5 kVA		.54	14.815		6,825	695		7,520	8,575
1200	7.5 kVA		.52	15.385		7,450	725		8,175	9,300
1400	10 kVA		.44	18.182		7,950	855		8,805	10,100
1600	X-ray systems, 15 kVA, 90 amp		.44	18.182		16,500	855		17,355	19,500
1800	25 kVA, 125 amp		.36	22.222		16,900	1,050		17,950	20,200

26 22 13.30 Isolating Transformer

		Crew	Daily Output	Labor-Hours	Unit	Material	Labor	Equipment	Total	Total Incl O&P
0010	**ISOLATING TRANSFORMER**									
0100	Single phase, 120/240 volt primary, 120/240 volt secondary									
0200	0.50 kVA	1 Elec	4	2	Ea.	355	94		449	535
0400	1 kVA		2	4		445	188		633	775
0600	2 kVA		1.60	5		665	235		900	1,100
0800	3 kVA		1.40	5.714		760	269		1,029	1,250
1000	5 kVA		1.20	6.667		1,000	315		1,315	1,575
1200	7.5 kVA		1.10	7.273		1,275	340		1,615	1,925
1400	10 kVA		.80	10		1,625	470		2,095	2,500
1600	15 kVA		.60	13.333		2,075	625		2,700	3,250
1800	25 kVA		.50	16		2,975	750		3,725	4,425
1810	37.5 kVA	2 Elec	.80	20		4,900	940		5,840	6,850
1820	75 kVA	"	.65	24.615		7,375	1,150		8,525	9,900
1830	3 phase, 120/240 primary, 120/208V secondary, 112.5 kVA	R-3	.90	22.222		9,325	1,025	144	10,494	12,100
1840	150 kVA		.85	23.529		11,800	1,100	152	13,052	14,800
1850	225 kVA		.65	30.769		16,500	1,425	199	18,124	20,500

26 22 13 – Low-Voltage Distribution Transformers

26 22 13.30 Isolating Transformer		Crew	Daily Output	Labor-Hours	Unit	Material	2009 Bare Costs Labor	Equipment	Total	Total Incl O&P
1860	300 kVA	R-3	.55	36.364	Ea.	21,900	1,675	235	23,810	27,000
1870	500 kVA		.45	44.444		36,500	2,050	287	38,837	43,700
1880	750 kVA		.35	57.143		36,900	2,650	370	39,920	45,100

26 22 13.90 Transformer Handling		Crew	Daily Output	Labor-Hours	Unit	Material	Labor	Equipment	Total	Total Incl O&P
0010	**TRANSFORMER HANDLING** Add to normal labor cost in restricted areas									
5000	Transformers									
5150	15 kVA, approximately 200 pounds	2 Elec	2.70	5.926	Ea.		279		279	430
5160	25 kVA, approximately 300 pounds		2.50	6.400			300		300	465
5170	37.5 kVA, approximately 400 pounds		2.30	6.957			325		325	505
5180	50 kVA, approximately 500 pounds		2	8			375		375	580
5190	75 kVA, approximately 600 pounds		1.80	8.889			420		420	645
5200	100 kVA, approximately 700 pounds		1.60	10			470		470	725
5210	112.5 kVA, approximately 800 pounds	3 Elec	2.20	10.909			515		515	790
5220	125 kVA, approximately 900 pounds		2	12			565		565	870
5230	150 kVA, approximately 1000 pounds		1.80	13.333			625		625	965
5240	167 kVA, approximately 1200 pounds		1.60	15			705		705	1,075
5250	200 kVA, approximately 1400 pounds		1.40	17.143			805		805	1,250
5260	225 kVA, approximately 1600 pounds		1.30	18.462			870		870	1,325
5270	250 kVA, approximately 1800 pounds		1.10	21.818			1,025		1,025	1,575
5280	300 kVA, approximately 2000 pounds		1	24			1,125		1,125	1,725
5290	500 kVA, approximately 3000 pounds		.75	32			1,500		1,500	2,325
5300	600 kVA, approximately 3500 pounds		.67	35.821			1,675		1,675	2,600
5310	750 kVA, approximately 4000 pounds		.60	40			1,875		1,875	2,900
5320	1000 kVA, approximately 5000 pounds		.50	48			2,250		2,250	3,475

26 22 16 – Low-Voltage Buck-Boost Transformers

26 22 16.10 Buck-Boost Transformer		Crew	Daily Output	Labor-Hours	Unit	Material	Labor	Equipment	Total	Total Incl O&P
0010	**BUCK-BOOST TRANSFORMER**									
0100	Single phase, 120/240 V primary, 12/24 V secondary									
0200	0.10 kVA	1 Elec	8	1	Ea.	88.50	47		135.50	170
0400	0.25 kVA		5.70	1.404		130	66		196	244
0600	0.50 kVA		4	2		178	94		272	340
0800	0.75 kVA		3.10	2.581		229	121		350	440
1000	1.0 kVA		2	4		284	188		472	600
1200	1.5 kVA		1.80	4.444		350	209		559	705
1400	2.0 kVA		1.60	5		420	235		655	825
1600	3.0 kVA		1.40	5.714		565	269		834	1,050
1800	5.0 kVA		1.20	6.667		745	315		1,060	1,300
2000	3 phase, 240 V primary, 208/120 V secondary, 15 kVA	2 Elec	2.40	6.667		2,275	315		2,590	3,000
2200	30 kVA		1.60	10		2,550	470		3,020	3,525
2400	45 kVA		1.40	11.429		3,075	535		3,610	4,200
2600	75 kVA		1.20	13.333		3,725	625		4,350	5,050
2800	112.5 kVA	R-3	1.40	14.286		4,625	660	92.50	5,377.50	6,200
3000	150 kVA		1.10	18.182		6,150	840	117	7,107	8,175
3200	225 kVA		1	20		8,000	925	129	9,054	10,400
3400	300 kVA		.90	22.222		10,900	1,025	144	12,069	13,700

26 24 13 – Switchboards

26 24 13.10 Incoming Switchboards		Crew	Daily Output	Labor-Hours	Unit	Material	Labor	Equipment	Total	Total Incl O&P
						2009 Bare Costs				
0010	**INCOMING SWITCHBOARDS** main service section									
0100	Aluminum bus bars, not including CT's or PT's									
0200	No main disconnect, includes CT compartment									
0300	120/208 volt, 4 wire, 600 amp	2 Elec	1	16	Ea.	4,250	750		5,000	5,825
0400	800 amp		.88	18.182		4,250	855		5,105	6,000
0500	1000 amp		.80	20		5,100	940		6,040	7,050
0600	1200 amp		.72	22.222		5,100	1,050		6,150	7,200
0700	1600 amp		.66	24.242		5,100	1,150		6,250	7,350
0800	2000 amp		.62	25.806		5,500	1,225		6,725	7,925
1000	3000 amp		.56	28.571		7,250	1,350		8,600	10,100
1200	277/480 volt, 4 wire, 600 amp		1	16		4,250	750		5,000	5,825
1300	800 amp		.88	18.182		4,250	855		5,105	6,000
1400	1000 amp		.80	20		5,375	940		6,315	7,375
1500	1200 amp		.72	22.222		5,375	1,050		6,425	7,525
1600	1600 amp		.66	24.242		5,375	1,150		6,525	7,675
1700	2000 amp		.62	25.806		5,500	1,225		6,725	7,925
1800	3000 amp		.56	28.571		7,250	1,350		8,600	10,100
1900	4000 amp		.52	30.769		8,975	1,450		10,425	12,100
2000	Fused switch & CT compartment									
2100	120/208 volt, 4 wire, 400 amp	2 Elec	1.12	14.286	Ea.	4,075	670		4,745	5,500
2200	600 amp		.94	17.021		4,825	800		5,625	6,525
2300	800 amp		.84	19.048		11,500	895		12,395	14,000
2400	1200 amp		.68	23.529		14,900	1,100		16,000	18,100
2500	277/480 volt, 4 wire, 400 amp		1.14	14.035		4,275	660		4,935	5,725
2600	600 amp		.94	17.021		4,875	800		5,675	6,600
2700	800 amp		.84	19.048		11,500	895		12,395	14,000
2800	1200 amp		.68	23.529		14,900	1,100		16,000	18,100
2900	Pressure switch & CT compartment									
3000	120/208 volt, 4 wire, 800 amp	2 Elec	.80	20	Ea.	10,300	940		11,240	12,800
3100	1200 amp		.66	24.242		19,900	1,150		21,050	23,700
3200	1600 amp		.62	25.806		21,200	1,225		22,425	25,200
3300	2000 amp		.56	28.571		22,500	1,350		23,850	26,900
3310	2500 amp		.50	32		27,700	1,500		29,200	32,800
3320	3000 amp		.44	36.364		37,400	1,700		39,100	43,700
3330	4000 amp		.40	40		48,000	1,875		49,875	56,000
3400	277/480 volt, 4 wire, 800 amp, with ground fault		.80	20		17,000	940		17,940	20,200
3600	1200 amp, with ground fault		.66	24.242		22,000	1,150		23,150	26,000
4000	1600 amp, with ground fault		.62	25.806		23,900	1,225		25,125	28,200
4200	2000 amp, with ground fault		.56	28.571		25,800	1,350		27,150	30,500
4400	Circuit breaker, molded case & CT compartment									
4600	3 pole, 4 wire, 600 amp	2 Elec	.94	17.021	Ea.	8,875	800		9,675	11,000
4800	800 amp		.84	19.048		10,600	895		11,495	13,100
5000	1200 amp		.68	23.529		14,500	1,100		15,600	17,600
5100	Copper bus bars, not incl. CT's or PT's, add, minimum					15%				

26 24 13.20 In Plant Distribution Switchboards		Crew	Daily Output	Labor-Hours	Unit	Material	Labor	Equipment	Total	Total Incl O&P
0010	**IN PLANT DISTRIBUTION SWITCHBOARDS**									
0100	Main lugs only, to 600 volt, 3 pole, 3 wire, 200 amp	2 Elec	1.20	13.333	Ea.	1,525	625		2,150	2,650
0110	400 amp		1.20	13.333		1,525	625		2,150	2,650
0120	600 amp		1.20	13.333		1,575	625		2,200	2,700
0130	800 amp		1.08	14.815		1,700	695		2,395	2,950
0140	1200 amp		.92	17.391		2,075	815		2,890	3,525
0150	1600 amp		.86	18.605		2,725	875		3,600	4,350

26 24 13.20 In Plant Distribution Switchboards		Crew	Daily Output	Labor-Hours	Unit	2009 Bare Costs Material	Labor	Equipment	Total	Total Incl O&P
0160	2000 amp	2 Elec	.82	19.512	Ea.	3,025	915		3,940	4,725
0250	To 480 volt, 3 pole, 4 wire, 200 amp		1.20	13.333		1,325	625		1,950	2,425
0260	400 amp		1.20	13.333		1,525	625		2,150	2,650
0270	600 amp		1.20	13.333		1,675	625		2,300	2,800
0280	800 amp		1.08	14.815		1,825	695		2,520	3,075
0290	1200 amp		.92	17.391		2,250	815		3,065	3,725
0300	1600 amp		.86	18.605		2,575	875		3,450	4,175
0310	2000 amp		.82	19.512		3,000	915		3,915	4,700
0400	Main circuit breaker, to 600 volt, 3 pole, 3 wire, 200 amp		1.20	13.333		3,250	625		3,875	4,550
0410	400 amp		1.14	14.035		3,250	660		3,910	4,600
0420	600 amp		1.10	14.545		4,100	685		4,785	5,575
0430	800 amp		1.04	15.385		6,875	725		7,600	8,675
0440	1200 amp		.88	18.182		9,000	855		9,855	11,200
0450	1600 amp		.84	19.048		14,400	895		15,295	17,300
0460	2000 amp		.80	20		15,400	940		16,340	18,400
0550	277/480 volt, 3 pole, 4 wire, 200 amp		1.20	13.333		3,400	625		4,025	4,725
0560	400 amp		1.14	14.035		3,400	660		4,060	4,775
0570	600 amp		1.10	14.545		4,275	685		4,960	5,750
0580	800 amp		1.04	15.385		7,200	725		7,925	9,025
0590	1200 amp		.88	18.182		9,300	855		10,155	11,500
0600	1600 amp		.84	19.048		14,400	895		15,295	17,300
0610	2000 amp		.80	20		15,400	940		16,340	18,400
0700	Main fusible switch w/fuse, 208/240 volt, 3 pole, 3 wire, 200 amp		1.20	13.333		3,575	625		4,200	4,900
0710	400 amp		1.14	14.035		3,575	660		4,235	4,950
0720	600 amp		1.10	14.545		4,350	685		5,035	5,825
0730	800 amp		1.04	15.385		8,900	725		9,625	10,900
0740	1200 amp		.88	18.182		10,400	855		11,255	12,700
0800	120/208, 120/240 volt, 3 pole, 4 wire, 200 amp		1.20	13.333		3,175	625		3,800	4,475
0810	400 amp		1.14	14.035		3,175	660		3,835	4,525
0820	600 amp		1.10	14.545		4,150	685		4,835	5,625
0830	800 amp		1.04	15.385		6,600	725		7,325	8,375
0840	1200 amp		.88	18.182		7,675	855		8,530	9,775
0900	480 or 600 volt, 3 pole, 3 wire, 200 amp		1.20	13.333		3,450	625		4,075	4,750
0910	400 amp		1.14	14.035		3,450	660		4,110	4,800
0920	600 amp		1.10	14.545		4,125	685		4,810	5,575
0930	800 amp		1.04	15.385		6,350	725		7,075	8,100
0940	1200 amp		.88	18.182		7,375	855		8,230	9,425
1000	277 or 480 volt, 3 pole, 4 wire, 200 amp		1.20	13.333		3,575	625		4,200	4,900
1010	400 amp		1.14	14.035		3,575	660		4,235	4,950
1020	600 amp		1.10	14.545		4,300	685		4,985	5,800
1030	800 amp		1.04	15.385		6,600	725		7,325	8,375
1040	1200 amp		.88	18.182		7,675	855		8,530	9,775
1120	1600 amp		.76	21.053		14,000	990		14,990	16,900
1130	2000 amp		.68	23.529		18,400	1,100		19,500	22,000
1150	Pressure switch, bolted, 3 pole, 208/240 volt, 3 wire, 800 amp		.96	16.667		10,400	785		11,185	12,600
1160	1200 amp		.80	20		13,200	940		14,140	16,100
1170	1600 amp		.76	21.053		15,100	990		16,090	18,100
1180	2000 amp		.68	23.529		17,200	1,100		18,300	20,700
1200	120/208 or 120/240 volt, 3 pole, 4 wire, 800 amp		.96	16.667		8,200	785		8,985	10,200
1210	1200 amp		.80	20		9,575	940		10,515	12,000
1220	1600 amp		.76	21.053		15,100	990		16,090	18,100
1230	2000 amp		.68	23.529		17,200	1,100		18,300	20,700
1300	480 or 600 volt, 3 wire, 800 amp		.96	16.667		10,400	785		11,185	12,600

26 24 13 – Switchboards

26 24 13.20 In Plant Distribution Switchboards		Crew	Daily Output	Labor-Hours	Unit	Material	Labor	Equipment	Total	Total Incl O&P
						2009 Bare Costs				
1310	1200 amp	2 Elec	.80	20	Ea.	14,500	940		15,440	17,500
1320	1600 amp		.76	21.053		16,400	990		17,390	19,500
1330	2000 amp		.68	23.529		18,400	1,100		19,500	22,000
1400	277-480 volt, 4 wire, 800 amp		.96	16.667		10,400	785		11,185	12,600
1410	1200 amp		.80	20		14,500	940		15,440	17,500
1420	1600 amp		.76	21.053		16,400	990		17,390	19,500
1430	2000 amp		.68	23.529		18,400	1,100		19,500	22,000
1500	Main ground fault protector, 1200-2000 amp		5.40	2.963		5,200	139		5,339	5,925
1600	Busway connection, 200 amp		5.40	2.963		460	139		599	720
1610	400 amp		4.60	3.478		460	163		623	755
1620	600 amp		4	4		460	188		648	795
1630	800 amp		3.20	5		460	235		695	865
1640	1200 amp		2.60	6.154		460	289		749	950
1650	1600 amp		2.40	6.667		950	315		1,265	1,525
1660	2000 amp		2	8		950	375		1,325	1,625
1700	Shunt trip for remote operation 200 amp		8	2		1,025	94		1,119	1,275
1710	400 amp		8	2		1,650	94		1,744	1,975
1720	600 amp		8	2		1,950	94		2,044	2,275
1730	800 amp		8	2		2,500	94		2,594	2,900
1740	1200-2000 amp		8	2		5,300	94		5,394	6,000
1800	Motor operated main breaker 200 amp		8	2		3,425	94		3,519	3,900
1810	400 amp		8	2		3,425	94		3,519	3,900
1820	600 amp		8	2		2,725	94		2,819	3,150
1830	800 amp		8	2		3,225	94		3,319	3,700
1840	1200-2000 amp		8	2		5,400	94		5,494	6,075
1900	Current/potential transformer metering compartment 200-800 amp		5.40	2.963		2,225	139		2,364	2,675
1940	1200 amp		5.40	2.963		3,075	139		3,214	3,625
1950	1600-2000 amp		5.40	2.963		3,075	139		3,214	3,625
2000	With watt meter 200-800 amp		4	4		8,775	188		8,963	9,950
2040	1200 amp		4	4		10,800	188		10,988	12,200
2050	1600-2000 amp		4	4		10,800	188		10,988	12,200
2100	Split bus 60-200 amp	1 Elec	5.30	1.509		310	71		381	455
2130	400 amp	2 Elec	4.60	3.478		540	163		703	845
2140	600 amp		3.60	4.444		660	209		869	1,050
2150	800 amp		2.60	6.154		840	289		1,129	1,375
2170	1200 amp		2	8		960	375		1,335	1,625
2250	Contactor control 60 amp	1 Elec	2	4		2,050	188		2,238	2,550
2260	100 amp		1.50	5.333		2,300	251		2,551	2,900
2270	200 amp		1	8		3,450	375		3,825	4,375
2280	400 amp	2 Elec	1	16		10,500	750		11,250	12,800
2290	600 amp		.84	19.048		11,800	895		12,695	14,400
2300	800 amp		.72	22.222		14,000	1,050		15,050	17,000
2500	Modifier, two distribution sections, add		.80	20		3,300	940		4,240	5,075
2520	Three distribution sections, add		.40	40		7,375	1,875		9,250	11,000
2560	Auxiliary pull section, 20", add		2	8		1,750	375		2,125	2,500
2580	24", add		1.80	8.889		1,750	420		2,170	2,575
2600	30", add		1.60	10		1,750	470		2,220	2,650
2620	36", add		1.40	11.429		2,100	535		2,635	3,125
2640	Dog house, 12", add		2.40	6.667		360	315		675	875
2660	18", add		2	8		720	375		1,095	1,375
3000	Transition section between switchboard and transformer									
3050	or motor control center, 4 wire alum. bus, 600 amp	2 Elec	1.14	14.035	Ea.	2,550	660		3,210	3,825
3100	800 amp		1	16		2,850	750		3,600	4,300

26 24 13.20 In Plant Distribution Switchboards		Crew	Daily Output	Labor-Hours	Unit	2009 Bare Costs Material	Labor	Equipment	Total	Total Incl O&P
3150	1000 amp	2 Elec	.38	18.182	Ea.	3,225	855		4,080	4,850
3200	1200 amp		.80	20		3,525	940		4,465	5,350
3250	1600 amp		.72	22.222		4,175	1,050		5,225	6,200
3300	2000 amp		.66	24.242		4,800	1,150		5,950	7,025
3350	2500 amp		.62	25.806		5,600	1,225		6,825	8,025
3400	3000 amp		.56	28.571		6,450	1,350		7,800	9,175
4000	Weatherproof construction, per vertical section		1.76	9.091		3,675	425		4,100	4,700

26 24 13.30 Distribution Switchboards Section		Crew	Daily Output	Labor-Hours	Unit	Material	Labor	Equipment	Total	Total Incl O&P
0010	**DISTRIBUTION SWITCHBOARDS SECTION**									
0100	Aluminum bus bars, not including breakers									
0160	Subfeed lug-rated at 60 amp	2 Elec	1.30	12.308	Ea.	1,725	580		2,305	2,800
0170	100 amp		1.26	12.698		1,725	595		2,320	2,825
0180	200 amp		1.20	13.333		1,725	625		2,350	2,875
0190	400 amp		1.10	14.545		1,750	685		2,435	2,975
0200	120/208 or 277/480 volt, 4 wire, 600 amp		1	16		2,175	750		2,925	3,550
0300	800 amp		.88	18.182		2,875	855		3,730	4,475
0400	1000 amp		.80	20		2,875	940		3,815	4,600
0500	1200 amp		.72	22.222		3,550	1,050		4,600	5,500
0600	1600 amp		.66	24.242		4,075	1,150		5,225	6,225
0700	2000 amp		.62	25.806		4,775	1,225		6,000	7,125
0800	2500 amp		.60	26.667		5,375	1,250		6,625	7,850
0900	3000 amp		.56	28.571		6,525	1,350		7,875	9,250
0950	4000 amp		.52	30.769		9,525	1,450		10,975	12,700

26 24 13.40 Switchboards Feeder Section		Crew	Daily Output	Labor-Hours	Unit	Material	Labor	Equipment	Total	Total Incl O&P
0010	**SWITCHBOARDS FEEDER SECTION** group mounted devices									
0030	Circuit breakers									
0160	FA frame, 15 to 60 amp, 240 volt, 1 pole	1 Elec	8	1	Ea.	108	47		155	192
0170	2 pole		7	1.143		217	53.50		270.50	320
0180	3 pole		5.30	1.509		310	71		381	455
0210	480 volt, 1 pole		8	1		137	47		184	224
0220	2 pole		7	1.143		355	53.50		408.50	475
0230	3 pole		5.30	1.509		455	71		526	610
0260	600 volt, 2 pole		7	1.143		425	53.50		478.50	555
0270	3 pole		5.30	1.509		530	71		601	695
0280	FA frame, 70 to 100 amp, 240 volt, 1 pole		7	1.143		143	53.50		196.50	240
0310	2 pole		5	1.600		335	75		410	485
0320	3 pole		4	2		385	94		479	570
0330	480 volt, 1 pole		7	1.143		171	53.50		224.50	271
0360	2 pole		5	1.600		455	75		530	615
0370	3 pole		4	2		545	94		639	745
0380	600 volt, 2 pole		5	1.600		515	75		590	685
0410	3 pole		4	2		635	94		729	845
0420	KA frame, 70 to 225 amp		3.20	2.500		1,200	118		1,318	1,500
0430	LA frame, 125 to 400 amp		2.30	3.478		2,700	163		2,863	3,225
0460	MA frame, 450 to 600 amp		1.60	5		4,475	235		4,710	5,250
0470	700 to 800 amp		1.30	6.154		5,800	289		6,089	6,825
0480	MAL frame, 1000 amp		1	8		6,025	375		6,400	7,200
0490	PA frame, 1200 amp		.80	10		12,200	470		12,670	14,200
0500	Branch circuit, fusible switch, 600 volt, double 30/30 amp		4	2		1,200	94		1,294	1,450
0550	60/60 amp		3.20	2.500		1,200	118		1,318	1,500
0600	100/100 amp		2.70	2.963		1,675	139		1,814	2,075
0650	Single, 30 amp		5.30	1.509		780	71		851	965

26 24 13.40 Switchboards Feeder Section		Crew	Daily Output	Labor-Hours	Unit	Material	2009 Bare Costs Labor	Equipment	Total	Total Incl O&P
0700	60 amp	1 Elec	4.70	1.702	Ea.	780	80		860	980
0750	100 amp		4	2		1,125	94		1,219	1,400
0800	200 amp		2.70	2.963		1,675	139		1,814	2,075
0850	400 amp		2.30	3.478		3,175	163		3,338	3,725
0900	600 amp		1.80	4.444		3,775	209		3,984	4,475
0950	800 amp		1.30	6.154		6,150	289		6,439	7,225
1000	1200 amp		.80	10		7,350	470		7,820	8,800
1080	Branch circuit, circuit breakers, high interrupting capacity									
1100	60 amp, 240, 480 or 600 volt, 1 pole	1 Elec	8	1	Ea.	277	47		324	380
1120	2 pole		7	1.143		665	53.50		718.50	815
1140	3 pole		5.30	1.509		790	71		861	980
1150	100 amp, 240, 480 or 600 volt, 1 pole		7	1.143		310	53.50		363.50	425
1160	2 pole		5	1.600		795	75		870	990
1180	3 pole		4	2		890	94		984	1,125
1200	225 amp, 240, 480 or 600 volt, 2 pole		3.50	2.286		2,775	107		2,882	3,225
1220	3 pole		3.20	2.500		3,350	118		3,468	3,875
1240	400 amp, 240, 480 or 600 volt, 2 pole		2.50	3.200		3,700	150		3,850	4,300
1260	3 pole		2.30	3.478		4,450	163		4,613	5,150
1280	600 amp, 240, 480 or 600 volt, 2 pole		1.80	4.444		4,500	209		4,709	5,300
1300	3 pole		1.60	5		5,425	235		5,660	6,325
1320	800 amp, 240, 480 or 600 volt, 2 pole		1.50	5.333		5,525	251		5,776	6,450
1340	3 pole		1.30	6.154		6,900	289		7,189	8,050
1360	1000 amp, 240, 480 or 600 volt, 2 pole		1.10	7.273		10,000	340		10,340	11,500
1380	3 pole		1	8		10,900	375		11,275	12,600
1400	1200 amp, 240, 480 or 600 volt, 2 pole		.90	8.889		7,325	420		7,745	8,700
1420	3 pole		.80	10		11,700	470		12,170	13,600
1700	Fusible switch, 240 V, 60 amp, 2 pole		3.20	2.500		480	118		598	705
1720	3 pole		3	2.667		655	125		780	915
1740	100 amp, 2 pole		2.70	2.963		485	139		624	750
1760	3 pole		2.50	3.200		735	150		885	1,050
1780	200 amp, 2 pole		2	4		1,025	188		1,213	1,425
1820	400 amp, 2 pole		1.50	5.333		1,900	251		2,151	2,475
1840	3 pole		1.30	6.154		2,375	289		2,664	3,075
1860	600 amp, 2 pole		1	8		2,700	375		3,075	3,550
1880	3 pole		.90	8.889		3,325	420		3,745	4,300
1900	240-600 V, 800 amp, 2 pole		.70	11.429		6,025	535		6,560	7,450
1920	3 pole		.60	13.333		7,375	625		8,000	9,100
2000	600 V, 60 amp, 2 pole		3.20	2.500		725	118		843	975
2040	100 amp, 2 pole		2.70	2.963		745	139		884	1,025
2080	200 amp, 2 pole		2	4		1,300	188		1,488	1,725
2120	400 amp, 2 pole		1.50	5.333		2,575	251		2,826	3,200
2160	600 amp, 2 pole		1	8		3,125	375		3,500	4,000
2500	Branch circuit, circuit breakers, 60 amp, 600 volt, 3 pole		5.30	1.509		690	71		761	870
2520	240, 480 or 600 volt, 1 pole		8	1		112	47		159	196
2540	240 volt, 2 pole		7	1.143		192	53.50		245.50	294
2560	480 or 600 volt, 2 pole		7	1.143		405	53.50		458.50	535
2580	240 volt, 3 pole		5.30	1.509		310	71		381	455
2600	480 volt, 3 pole		5.30	1.509		540	71		611	705
2620	100 amp, 600 volt, 2 pole		5	1.600		520	75		595	685
2640	3 pole		4	2		655	94		749	865
2660	480 volt, 2 pole		5	1.600		450	75		525	610
2680	240 volt, 2 pole		5	1.600		256	75		331	400
2700	3 pole		4	2		410	94		504	595

26 24 Switchboards and Panelboards

26 24 13 – Switchboards

26 24 13.40 Switchboards Feeder Section		Crew	Daily Output	Labor-Hours	Unit	Material	2009 Bare Costs Labor	Equipment	Total	Total Incl O&P
2720	480 volt, 3 pole	1 Elec	4	2	Ea.	600	94		694	810
2740	225 amp, 240, 480 or 600 volt, 2 pole		3.50	2.286		680	107		787	915
2760	3 pole		3.20	2.500		785	118		903	1,050
2780	400 amp, 240, 480 or 600 volt, 2 pole		2.50	3.200		1,575	150		1,725	1,950
2800	3 pole		2.30	3.478		1,825	163		1,988	2,250
2820	600 amp, 240 or 480 volt, 2 pole		1.80	4.444		2,550	209		2,759	3,150
2840	3 pole		1.60	5		3,150	235		3,385	3,800
2860	800 amp, 240, 480 volt or 600 volt, 2 pole		1.50	5.333		3,850	251		4,101	4,600
2880	3 pole		1.30	6.154		4,500	289		4,789	5,400
2900	1000 amp, 240, 480 or 600 volt, 2 pole		1.10	7.273		4,725	340		5,065	5,725
2920	480 volt, 600 volt, 3 pole		1	8		5,425	375		5,800	6,525
2940	1200 amp, 240, 480 or 600 volt, 2 pole		.90	8.889		6,650	420		7,070	7,975
2960	3 pole		.80	10		7,275	470		7,745	8,725
2980	600 volt, 3 pole	↓	.80	10	↓	7,275	470		7,745	8,725

26 24 16 – Panelboards

26 24 16.10 Load Centers

		Crew	Daily Output	Labor-Hours	Unit	Material	Labor	Equipment	Total	Total Incl O&P
0010	**LOAD CENTERS** (residential type) R262416-50									
0100	3 wire, 120/240V, 1 phase, including 1 pole plug-in breakers									
0200	100 amp main lugs, indoor, 8 circuits	1 Elec	1.40	5.714	Ea.	176	269		445	610
0300	12 circuits		1.20	6.667		246	315		561	750
0400	Rainproof, 8 circuits		1.40	5.714		211	269		480	645
0500	12 circuits	↓	1.20	6.667		298	315		613	805
0600	200 amp main lugs, indoor, 16 circuits	R-1A	1.80	8.889		340	345		685	920
0700	20 circuits		1.50	10.667		425	410		835	1,125
0800	24 circuits		1.30	12.308		560	475		1,035	1,375
0900	30 circuits		1.20	13.333		645	515		1,160	1,525
1000	40 circuits		.80	20		895	775		1,670	2,200
1200	Rainproof, 16 circuits		1.80	8.889		405	345		750	990
1300	20 circuits		1.50	10.667		485	410		895	1,175
1400	24 circuits		1.30	12.308		725	475		1,200	1,550
1500	30 circuits		1.20	13.333		815	515		1,330	1,700
1600	40 circuits		.80	20		1,075	775		1,850	2,400
1800	400 amp main lugs, indoor, 42 circuits		.72	22.222		1,250	860		2,110	2,725
1900	Rainproof, 42 circuit	↓	.72	22.222	↓	1,475	860		2,335	2,975
2200	Plug in breakers, 20 amp, 1 pole, 4 wire, 120/208 volts									
2210	125 amp main lugs, indoor, 12 circuits	1 Elec	1.20	6.667	Ea.	320	315		635	835
2300	18 circuits		.80	10		455	470		925	1,225
2400	Rainproof, 12 circuits		1.20	6.667		370	315		685	885
2500	18 circuits	↓	.80	10		495	470		965	1,275
2600	200 amp main lugs, indoor, 24 circuits	R-1A	1.30	12.308		610	475		1,085	1,425
2700	30 circuits		1.20	13.333		695	515		1,210	1,575
2800	36 circuits		1	16		840	620		1,460	1,900
2900	42 circuits		.80	20		925	775		1,700	2,250
3000	Rainproof, 24 circuits		1.30	12.308		680	475		1,155	1,500
3100	30 circuits		1.20	13.333		765	515		1,280	1,650
3200	36 circuits		1	16		1,050	620		1,670	2,150
3300	42 circuits		.80	20		1,150	775		1,925	2,475
3500	400 amp main lugs, indoor, 42 circuits		.72	22.222		1,400	860		2,260	2,875
3600	Rainproof, 42 circuits	↓	.72	22.222	↓	1,675	860		2,535	3,175
3700	Plug-in breakers, 20 amp, 1 pole, 3 wire, 120/240 volts									
3800	100 amp main breaker, indoor, 12 circuits	1 Elec	1.20	6.667	Ea.	345	315		660	855
3900	18 circuits	"	.80	10	↓	460	470		930	1,225

26 24 16.10 Load Centers		Crew	Daily Output	Labor-Hours	Unit	Material	Labor	Equipment	Total	Total Incl O&P
						2009 Bare Costs				
4000	200 amp main breaker, indoor, 20 circuits	R-1A	1.50	10.667	Ea.	620	410		1,030	1,325
4200	24 circuits		1.30	12.308		750	475		1,225	1,575
4300	30 circuits		1.20	13.333		835	515		1,350	1,725
4400	40 circuits		.90	17.778		1,050	685		1,735	2,225
4500	Rainproof, 20 circuits		1.50	10.667		675	410		1,085	1,400
4600	24 circuits		1.30	12.308		840	475		1,315	1,675
4700	30 circuits		1.20	13.333		925	515		1,440	1,850
4800	40 circuits		.90	17.778		1,150	685		1,835	2,325
5000	400 amp main breaker, indoor, 42 circuits		.72	22.222		3,000	860		3,860	4,625
5100	Rainproof, 42 circuits	↓	.72	22.222	↓	3,350	860		4,210	5,025
5300	Plug in breakers, 20 amp, 1 pole, 4 wire, 120/208 volts									
5400	200 amp main breaker, indoor, 30 circuits	R-1A	1.20	13.333	Ea.	1,275	515		1,790	2,225
5500	42 circuits		.80	20		1,550	775		2,325	2,925
5600	Rainproof, 30 circuits		1.20	13.333		1,350	515		1,865	2,300
5700	42 circuits	↓	.80	20	↓	1,650	775		2,425	3,025

26 24 16.20 Panelboard and Load Center Circuit Breakers		Crew	Daily Output	Labor-Hours	Unit	Material	Labor	Equipment	Total	Total Incl O&P
0010	**PANELBOARD AND LOAD CENTER CIRCUIT BREAKERS**									
0050	Bolt-on, 10,000 amp IC, 120 volt, 1 pole									
0100	15 to 50 amp	1 Elec	10	.800	Ea.	15.80	37.50		53.30	75.50
0200	60 amp		8	1		15.80	47		62.80	90
0300	70 amp	↓	8	1	↓	30	47		77	106
0350	240 volt, 2 pole									
0400	15 to 50 amp	1 Elec	8	1	Ea.	35	47		82	111
0500	60 amp		7.50	1.067		35	50		85	115
0600	80 to 100 amp		5	1.600		89.50	75		164.50	214
0700	3 pole, 15 to 60 amp		6.20	1.290		110	60.50		170.50	215
0800	70 amp		5	1.600		138	75		213	268
0900	80 to 100 amp		3.60	2.222		157	104		261	335
1000	22,000 amp I.C., 240 volt, 2 pole, 70 – 225 amp		2.70	2.963		675	139		814	960
1100	3 pole, 70 – 225 amp		2.30	3.478		750	163		913	1,075
1200	14,000 amp I.C., 277 volts, 1 pole, 15 – 30 amp		8	1		42	47		89	119
1300	22,000 amp I.C., 480 volts, 2 pole, 70 – 225 amp		2.70	2.963		675	139		814	960
1400	3 pole, 70 – 225 amp		2.30	3.478		835	163		998	1,175
2000	Plug-in panel or load center, 120/240 volt, to 60 amp, 1 pole		12	.667		12.25	31.50		43.75	61.50
2010	2 pole		9	.889		27.50	42		69.50	95
2020	3 pole		7.50	1.067		95.50	50		145.50	182
2030	100 amp, 2 pole		6	1.333		128	62.50		190.50	238
2040	3 pole		4.50	1.778		142	83.50		225.50	285
2050	150 to 200 amp, 2 pole		3	2.667		234	125		359	450
2060	Plug-in tandem, 120/240 V, 2-15 A, 1 pole		11	.727		31.50	34		65.50	87.50
2070	1-15 A & 1-20 A		11	.727		31.50	34		65.50	87.50
2080	2-20 A		11	.727		31.50	34		65.50	87.50
2082	Arc fault circuit interrupter, 120/240 V, 1-15 A & 1-20 A, 1 pole		11	.727		75	34		109	135
2100	High interrupting capacity, 120/240 volt, plug-in, 30 amp, 1 pole		12	.667		24	31.50		55.50	74.50
2110	60 amp, 2 pole		9	.889		56	42		98	126
2120	3 pole		7.50	1.067		145	50		195	237
2130	100 amp, 2 pole		6	1.333		161	62.50		223.50	274
2140	3 pole		4.50	1.778		209	83.50		292.50	360
2150	125 amp, 2 pole		3	2.667		430	125		555	670
2200	Bolt-on, 30 amp, 1 pole		10	.800		31	37.50		68.50	92
2210	60 amp, 2 pole		7.50	1.067		65	50		115	149
2220	3 pole	↓	6.20	1.290	↓	167	60.50		227.50	278

26 24 16.20 Panelboard and Load Center Circuit Breakers		Crew	Daily Output	Labor-Hours	Unit	2009 Bare Costs Material	Labor	Equipment	Total	Total Incl O&P
2230	100 amp, 2 pole	1 Elec	5	1.600	Ea.	186	75		261	320
2240	3 pole		3.60	2.222		243	104		347	430
2300	Ground fault, 240 volt, 30 amp, 1 pole		7	1.143		95	53.50		148.50	187
2310	2 pole		6	1.333		170	62.50		232.50	284
2350	Key operated, 240 volt, 1 pole, 30 amp		7	1.143		89.50	53.50		143	181
2360	Switched neutral, 240 volt, 30 amp, 2 pole		6	1.333		39.50	62.50		102	140
2370	3 pole		5.50	1.455		59.50	68.50		128	170
2400	Shunt trip, for 240 volt breaker, 60 amp, 1 pole		4	2		77.50	94		171.50	230
2410	2 pole		3.50	2.286		77.50	107		184.50	250
2420	3 pole		3	2.667		77.50	125		202.50	278
2430	100 amp, 2 pole		3	2.667		77.50	125		202.50	278
2440	3 pole		2.50	3.200		77.50	150		227.50	315
2450	150 amp, 2 pole		2	4		184	188		372	490
2500	Auxiliary switch, for 240 volt breaker, 60 amp, 1 pole		4	2		70.50	94		164.50	223
2510	2 pole		3.50	2.286		70.50	107		177.50	243
2520	3 pole		3	2.667		70.50	125		195.50	271
2530	100 amp, 2 pole		3	2.667		70.50	125		195.50	271
2540	3 pole		2.50	3.200		70.50	150		220.50	310
2550	150 amp, 2 pole		2	4		117	188		305	420
2600	Panel or load center, 277/480 volt, plug-in, 30 amp, 1 pole		12	.667		57	31.50		88.50	111
2610	60 amp, 2 pole		9	.889		171	42		213	253
2620	3 pole		7.50	1.067		250	50		300	350
2650	Bolt-on, 60 amp, 2 pole		7.50	1.067		171	50		221	265
2660	3 pole		6.20	1.290		250	60.50		310.50	370
2700	I-line, 277/480 volt, 30 amp, 1 pole		8	1		65	47		112	144
2710	60 amp, 2 pole		7.50	1.067		229	50		279	330
2720	3 pole		6.20	1.290		298	60.50		358.50	425
2730	100 amp, 1 pole		7.50	1.067		126	50		176	215
2740	2 pole		5	1.600		298	75		373	445
2750	3 pole		3.50	2.286		350	107		457	550
2800	High interrupting capacity, 277/480 volt, plug-in, 30 amp, 1 pole		12	.667		330	31.50		361.50	410
2810	60 amp, 2 pole		9	.889		505	42		547	625
2820	3 pole		7	1.143		590	53.50		643.50	735
2830	Bolt-on, 30 amp, 1 pole		8	1		330	47		377	435
2840	60 amp, 2 pole		7.50	1.067		505	50		555	635
2850	3 pole		6.20	1.290		590	60.50		650.50	745
2900	I-line, 30 amp, 1 pole		8	1		330	47		377	435
2910	60 amp, 2 pole		7.50	1.067		470	50		520	590
2920	3 pole		6.20	1.290		520	60.50		580.50	665
2930	100 amp, 1 pole		7.50	1.067		530	50		580	660
2940	2 pole		5	1.600		530	75		605	700
2950	3 pole		3.60	2.222		585	104		689	805
2960	Shunt trip, 277/480V breaker, remote oper., 30 amp, 1 pole		4	2		350	94		444	530
2970	60 amp, 2 pole		3.50	2.286		515	107		622	730
2980	3 pole		3	2.667		585	125		710	840
2990	100 amp, 1 pole		3.50	2.286		410	107		517	620
3000	2 pole		3	2.667		585	125		710	840
3010	3 pole		2.50	3.200		640	150		790	930
3050	Under voltage trip, 277/480 volt breaker, 30 amp, 1 pole		4	2		350	94		444	530
3060	60 amp, 2 pole		3.50	2.286		515	107		622	730
3070	3 pole		3	2.667		585	125		710	840
3080	100 amp, 1 pole		3.50	2.286		410	107		517	620
3090	2 pole		3	2.667		585	125		710	840

26 24 16.20 Panelboard and Load Center Circuit Breakers		Crew	Daily Output	Labor-Hours	Unit	Material	2009 Bare Costs Labor	Equipment	Total	Total Incl O&P
3100	3 pole	1 Elec	2.50	3.200	Ea.	640	150		790	930
3150	Motor operated, 277/480 volt breaker, 30 amp, 1 pole		4	2		625	94		719	835
3160	60 amp, 2 pole		3.50	2.286		790	107		897	1,025
3170	3 pole		3	2.667		860	125		985	1,150
3180	100 amp, 1 pole		3.50	2.286		685	107		792	920
3190	2 pole		3	2.667		860	125		985	1,150
3200	3 pole		2.50	3.200		910	150		1,060	1,225
3250	Panelboard spacers, per pole		40	.200		4.74	9.40		14.14	19.65
9000	Minimum labor/equipment charge		3	2.667	Job		125		125	193

26 24 16.30 Panelboards Commercial Applications		Crew	Daily Output	Labor-Hours	Unit	Material	Labor	Equipment	Total	Total Incl O&P
0010	**PANELBOARDS COMMERCIAL APPLICATIONS** R262416-50									
0050	NQOD, w/20 amp 1 pole bolt-on circuit breakers									
0100	3 wire, 120/240 volts, 100 amp main lugs									
0150	10 circuits	1 Elec	1	8	Ea.	495	375		870	1,125
0200	14 circuits		.88	9.091		580	425		1,005	1,300
0250	18 circuits		.75	10.667		635	500		1,135	1,475
0300	20 circuits		.65	12.308		715	580		1,295	1,675
0350	225 amp main lugs, 24 circuits	2 Elec	1.20	13.333		810	625		1,435	1,850
0400	30 circuits		.90	17.778		945	835		1,780	2,300
0450	36 circuits		.80	20		1,075	940		2,015	2,625
0500	38 circuits		.72	22.222		1,150	1,050		2,200	2,875
0550	42 circuits		.66	24.242		1,200	1,150		2,350	3,075
0600	4 wire, 120/208 volts, 100 amp main lugs, 12 circuits	1 Elec	1	8		560	375		935	1,200
0650	16 circuits		.75	10.667		645	500		1,145	1,475
0700	20 circuits		.65	12.308		750	580		1,330	1,725
0750	24 circuits		.60	13.333		815	625		1,440	1,850
0800	30 circuits		.53	15.094		940	710		1,650	2,125
0850	225 amp main lugs, 32 circuits	2 Elec	.90	17.778		1,050	835		1,885	2,425
0900	34 circuits		.84	19.048		1,075	895		1,970	2,575
0950	36 circuits		.80	20		1,100	940		2,040	2,675
1000	42 circuits		.68	23.529		1,250	1,100		2,350	3,075
1040	225 amp main lugs, NEMA 7, 12 circuits		1	16		3,275	750		4,025	4,750
1100	24 circuits		.40	40		3,925	1,875		5,800	7,225
1200	NEHB,w/20 amp, 1 pole bolt-on circuit breakers									
1250	4 wire, 277/480 volts, 100 amp main lugs, 12 circuits	1 Elec	.88	9.091	Ea.	1,075	425		1,500	1,825
1300	20 circuits	"	.60	13.333		1,575	625		2,200	2,700
1350	225 amp main lugs, 24 circuits	2 Elec	.90	17.778		1,825	835		2,660	3,275
1400	30 circuits		.80	20		2,200	940		3,140	3,850
1450	36 circuits		.72	22.222		2,550	1,050		3,600	4,400
1500	42 circuits		.60	26.667		2,925	1,250		4,175	5,125
1510	225 amp main lugs, NEMA 7, 12 circuits		.90	17.778		4,575	835		5,410	6,300
1590	24 circuits		.30	53.333		5,575	2,500		8,075	10,000
1600	NQOD panel, w/20 amp, 1 pole, circuit breakers									
1650	3 wire, 120/240 volt with main circuit breaker									
1700	100 amp main, 12 circuits	1 Elec	.80	10	Ea.	685	470		1,155	1,475
1750	20 circuits	"	.60	13.333		875	625		1,500	1,925
1800	225 amp main, 30 circuits	2 Elec	.68	23.529		1,675	1,100		2,775	3,550
1850	42 circuits		.52	30.769		1,950	1,450		3,400	4,375
1900	400 amp main, 30 circuits		.54	29.630		2,325	1,400		3,725	4,700
1950	42 circuits		.50	32		2,600	1,500		4,100	5,175
2000	4 wire, 120/208 volts with main circuit breaker									
2050	100 amp main, 24 circuits	1 Elec	.47	17.021	Ea.	1,025	800		1,825	2,350

26 24 16.30 Panelboards Commercial Applications		Crew	Daily Output	Labor-Hours	Unit	2009 Bare Costs Material	Labor	Equipment	Total	Total Incl O&P
2100	30 circuits	1 Elec	.40	20	Ea.	1,150	940		2,090	2,725
2200	225 amp main, 32 circuits	2 Elec	.72	22.222		1,950	1,050		3,000	3,750
2250	42 circuits		.56	28.571		2,150	1,350		3,500	4,425
2300	400 amp main, 42 circuits		.48	33.333		2,900	1,575		4,475	5,575
2350	600 amp main, 42 circuits		.40	40		4,300	1,875		6,175	7,625
2400	NEHB, with 20 amp, 1 pole circuit breaker									
2450	4 wire, 277/480 volts with main circuit breaker									
2500	100 amp main, 24 circuits	1 Elec	.42	19.048	Ea.	2,100	895		2,995	3,700
2550	30 circuits	"	.38	21.053		2,475	990		3,465	4,250
2600	225 amp main, 30 circuits	2 Elec	.72	22.222		3,100	1,050		4,150	5,025
2650	42 circuits		.56	28.571		3,850	1,350		5,200	6,300
2700	400 amp main, 42 circuits		.46	34.783		4,625	1,625		6,250	7,600
2750	600 amp main, 42 circuits		.38	42.105		6,350	1,975		8,325	10,000
2900	Note: the following line items don't include branch circuit breakers									
2910	For branch circuit breakers information, refer to									
2920	Division 26 24 16.20									
3010	Main lug, no main breaker, 240 volt, 1 pole, 3 wire, 100 amp	1 Elec	2.30	3.478	Ea.	445	163		608	740
3020	225 amp	2 Elec	2.40	6.667		540	315		855	1,075
3030	400 amp	"	1.80	8.889		810	420		1,230	1,525
3060	3 pole, 3 wire, 100 amp	1 Elec	2.30	3.478		485	163		648	780
3070	225 amp	2 Elec	2.40	6.667		575	315		890	1,100
3080	400 amp		1.80	8.889		880	420		1,300	1,625
3090	600 amp		1.60	10		1,025	470		1,495	1,850
3110	3 pole, 4 wire, 100 amp	1 Elec	2.30	3.478		540	163		703	840
3120	225 amp	2 Elec	2.40	6.667		680	315		995	1,225
3130	400 amp		1.80	8.889		880	420		1,300	1,625
3140	600 amp		1.60	10		1,025	470		1,495	1,850
3160	480 volt, 3 pole, 3 wire, 100 amp	1 Elec	2.30	3.478		650	163		813	965
3170	225 amp	2 Elec	2.40	6.667		790	315		1,105	1,350
3180	400 amp		1.80	8.889		1,125	420		1,545	1,875
3190	600 amp		1.60	10		1,250	470		1,720	2,100
3210	277/480 volt, 3 pole, 4 wire, 100 amp	1 Elec	2.30	3.478		630	163		793	945
3220	225 amp	2 Elec	2.40	6.667		775	315		1,090	1,325
3230	400 amp		1.80	8.889		1,100	420		1,520	1,850
3240	600 amp		1.60	10		1,225	470		1,695	2,075
3260	Main circuit breaker, 240 volt, 1 pole, 3 wire, 100 amp	1 Elec	2	4		610	188		798	960
3270	225 amp	2 Elec	2	8		1,275	375		1,650	1,975
3280	400 amp	"	1.60	10		2,025	470		2,495	2,975
3310	3 pole, 3 wire, 100 amp	1 Elec	2	4		705	188		893	1,075
3320	225 amp	2 Elec	2	8		1,475	375		1,850	2,200
3330	400 amp		1.60	10		2,325	470		2,795	3,300
3360	120/208 volt, 3 pole, 4 wire, 100 amp		4	4		705	188		893	1,075
3370	225 amp		2	8		1,475	375		1,850	2,200
3380	400 amp		1.60	10		2,325	470		2,795	3,300
3410	480 volt, 3 pole, 3 wire, 100 amp	1 Elec	2	4		1,000	188		1,188	1,400
3420	225 amp	2 Elec	2	8		1,750	375		2,125	2,500
3430	400 amp		1.60	10		2,625	470		3,095	3,600
3460	277/480 volt, 3 pole, 4 wire, 100 amp		4	4		975	188		1,163	1,375
3470	225 amp		2	8		1,700	375		2,075	2,450
3480	400 amp		1.60	10		2,650	470		3,120	3,625
3510	Main circuit breaker, HIC, 240 volt, 1 pole, 3 wire, 100 amp	1 Elec	2	4		950	188		1,138	1,350
3520	225 amp	2 Elec	2	8		2,550	375		2,925	3,375
3530	400 amp	"	1.60	10		3,425	470		3,895	4,475

26 24 16 – Panelboards

26 24 16.30 Panelboards Commercial Applications		Crew	Daily Output	Labor-Hours	Unit	Material	2009 Bare Costs Labor	Equipment	Total	Total Incl O&P
3560	3 pole, 3 wire, 100 amp	1 Elec	2	4	Ea.	1,050	188		1,238	1,475
3570	225 amp	2 Elec	2	8		2,875	375		3,250	3,725
3580	400 amp	"	1.60	10		3,825	470		4,295	4,925
3610	120/208 volt, 3 pole, 4 wire, 100 amp	1 Elec	2	4		1,050	188		1,238	1,475
3620	225 amp	2 Elec	2	8		2,875	375		3,250	3,725
3630	400 amp	"	1.60	10		3,825	470		4,295	4,925
3660	480 volt, 3 pole, 3 wire, 100 amp	1 Elec	2	4		1,600	188		1,788	2,050
3670	225 amp	2 Elec	2	8		3,200	375		3,575	4,100
3680	400 amp	"	1.60	10		4,075	470		4,545	5,225
3710	277/480 volt, 3 pole, 4 wire, 100 amp	1 Elec	2	4		1,525	188		1,713	1,975
3720	225 amp	2 Elec	2	8		3,050	375		3,425	3,925
3730	400 amp	"	1.60	10		4,050	470		4,520	5,175
3760	Main circuit breaker, shunt trip, 100 amp	1 Elec	1.20	6.667		790	315		1,105	1,350
3770	225 amp	2 Elec	1.60	10		1,875	470		2,345	2,775
3780	400 amp	"	1.40	11.429	↓	2,725	535		3,260	3,825
9000	Minimum labor/equipment charge	1 Elec	1	8	Job		375		375	580

26 24 19 – Motor-Control Centers

26 24 19.20 Motor Control Center Components		Crew	Daily Output	Labor-Hours	Unit	Material	Labor	Equipment	Total	Total Incl O&P
0010	**MOTOR CONTROL CENTER COMPONENTS**									
0100	Starter, size 1, FVNR, NEMA 1, type A, fusible	1 Elec	2.70	2.963	Ea.	1,625	139		1,764	2,025
0120	Circuit breaker		2.70	2.963		1,775	139		1,914	2,175
0140	Type B, fusible		2.70	2.963		1,800	139		1,939	2,200
0160	Circuit breaker		2.70	2.963		1,950	139		2,089	2,375
0180	NEMA 12, type A, fusible		2.60	3.077		1,650	145		1,795	2,050
0200	Circuit breaker		2.60	3.077		1,800	145		1,945	2,200
0220	Type B, fusible		2.60	3.077		1,825	145		1,970	2,225
0240	Circuit breaker		2.60	3.077		1,975	145		2,120	2,400
0300	Starter, size 1, FVR, NEMA 1, type A, fusible		2	4		2,350	188		2,538	2,875
0320	Circuit breaker		2	4		2,350	188		2,538	2,875
0340	Type B, fusible		2	4		2,575	188		2,763	3,125
0360	Circuit breaker		2	4		2,575	188		2,763	3,125
0380	NEMA 12, type A, fusible		1.90	4.211		2,375	198		2,573	2,900
0400	Circuit breaker		1.90	4.211		2,375	198		2,573	2,900
0420	Type B, fusible		1.90	4.211		2,600	198		2,798	3,175
0440	Circuit breaker	↓	1.90	4.211	↓	2,600	198		2,798	3,175
0490	Starter size 1, 2 speed, separate winding									
0500	NEMA 1, type A, fusible	1 Elec	2.60	3.077	Ea.	3,075	145		3,220	3,625
0520	Circuit breaker		2.60	3.077		3,075	145		3,220	3,625
0540	Type B, fusible		2.60	3.077		3,400	145		3,545	3,950
0560	Circuit breaker		2.60	3.077		3,400	145		3,545	3,950
0580	NEMA 12, type A, fusible		2.50	3.200		3,150	150		3,300	3,675
0600	Circuit breaker		2.50	3.200		3,150	150		3,300	3,675
0620	Type B, fusible		2.50	3.200		3,450	150		3,600	4,025
0640	Circuit breaker	↓	2.50	3.200	↓	3,450	150		3,600	4,025
0650	Starter size 1, 2 speed, consequent pole									
0660	NEMA 1, type A, fusible	1 Elec	2.60	3.077	Ea.	3,075	145		3,220	3,625
0680	Circuit breaker		2.60	3.077		3,075	145		3,220	3,625
0700	Type B, fusible		2.60	3.077		3,400	145		3,545	3,950
0720	Circuit breaker		2.60	3.077		3,400	145		3,545	3,950
0740	NEMA 12, type A, fusible		2.50	3.200		3,150	150		3,300	3,675
0760	Circuit breaker		2.50	3.200		3,150	150		3,300	3,675
0780	Type B, fusible	↓	2.50	3.200	↓	3,425	150		3,575	4,000

26 24 19.20 Motor Control Center Components		Crew	Daily Output	Labor-Hours	Unit	2009 Bare Costs Material	Labor	Equipment	Total	Total Incl O&P
0800	Circuit breaker	1 Elec	2.50	3.200	Ea.	3,450	150		3,600	4,025
0810	Starter size 1, 2 speed, space only									
0820	NEMA 1, type A, fusible	1 Elec	16	.500	Ea.	690	23.50		713.50	795
0840	Circuit breaker		16	.500		690	23.50		713.50	795
0860	Type B, fusible		16	.500		690	23.50		713.50	795
0880	Circuit breaker		16	.500		690	23.50		713.50	795
0900	NEMA 12, type A, fusible		15	.533		720	25		745	835
0920	Circuit breaker		15	.533		720	25		745	835
0940	Type B, fusible		15	.533		720	25		745	835
0960	Circuit breaker		15	.533		720	25		745	835
1100	Starter size 2, FVNR, NEMA 1, type A, fusible	2 Elec	4	4		1,850	188		2,038	2,325
1120	Circuit breaker		4	4		2,025	188		2,213	2,525
1140	Type B, fusible		4	4		2,025	188		2,213	2,525
1160	Circuit breaker		4	4		2,200	188		2,388	2,725
1180	NEMA 12, type A, fusible		3.80	4.211		1,875	198		2,073	2,350
1200	Circuit breaker		3.80	4.211		2,050	198		2,248	2,550
1220	Type B, fusible		3.80	4.211		2,050	198		2,248	2,550
1240	Circuit breaker		3.80	4.211		2,250	198		2,448	2,775
1300	FVR, NEMA 1, type A, fusible		3.20	5		3,125	235		3,360	3,775
1320	Circuit breaker		3.20	5		3,125	235		3,360	3,775
1340	Type B, fusible		3.20	5		3,425	235		3,660	4,125
1360	Circuit breaker		3.20	5		3,425	235		3,660	4,125
1380	NEMA type 12, type A, fusible		3	5.333		3,175	251		3,426	3,875
1400	Circuit breaker		3	5.333		3,175	251		3,426	3,875
1420	Type B, fusible		3	5.333		3,500	251		3,751	4,225
1440	Circuit breaker		3	5.333		3,500	251		3,751	4,225
1490	Starter size 2, 2 speed, separate winding									
1500	NEMA 1, type A, fusible	2 Elec	3.80	4.211	Ea.	3,500	198		3,698	4,150
1520	Circuit breaker		3.80	4.211		3,500	198		3,698	4,150
1540	Type B, fusible		3.80	4.211		3,825	198		4,023	4,525
1560	Circuit breaker		3.80	4.211		3,825	198		4,023	4,525
1570	NEMA 12, type A, fusible		3.60	4.444		3,550	209		3,759	4,225
1580	Circuit breaker		3.60	4.444		3,550	209		3,759	4,225
1600	Type B, fusible		3.60	4.444		3,900	209		4,109	4,600
1620	Circuit breaker		3.60	4.444		3,900	209		4,109	4,600
1630	Starter size 2, 2 speed, consequent pole									
1640	NEMA 1, type A, fusible	2 Elec	3.80	4.211	Ea.	3,950	198		4,148	4,650
1660	Circuit breaker		3.80	4.211		3,950	198		4,148	4,650
1680	Type B, fusible		3.80	4.211		4,200	198		4,398	4,900
1700	Circuit breaker		3.80	4.211		4,200	198		4,398	4,900
1720	NEMA 12, type A, fusible		3.80	4.211		3,975	198		4,173	4,675
1740	Circuit breaker		3.60	4.444		4,025	209		4,234	4,750
1760	Type B, fusible		3.60	4.444		4,250	209		4,459	5,000
1780	Circuit breaker		3.60	4.444		4,250	209		4,459	5,000
1830	Starter size 2, autotransformer									
1840	NEMA 1, type A, fusible	2 Elec	3.40	4.706	Ea.	6,175	221		6,396	7,150
1860	Circuit breaker		3.40	4.706		6,325	221		6,546	7,300
1880	Type B, fusible		3.40	4.706		6,750	221		6,971	7,775
1900	Circuit breaker		3.40	4.706		6,750	221		6,971	7,775
1920	NEMA 12, type A, fusible		3.20	5		6,300	235		6,535	7,275
1940	Circuit breaker		3.20	5		6,300	235		6,535	7,275
1960	Type B, fusible		3.20	5		6,850	235		7,085	7,900
1980	Circuit breaker		3.20	5		6,850	235		7,085	7,900

26 24 19.20 Motor Control Center Components		Crew	Daily Output	Labor-Hours	Unit	Material	Labor	Equipment	Total	Total Incl O&P
						2009 Bare Costs				
2030	Starter size 2, space only									
2040	NEMA 1, type A, fusible	1 Elec	16	.500	Ea.	690	23.50		713.50	795
2060	Circuit breaker		16	.500		690	23.50		713.50	795
2080	Type B, fusible		16	.500		690	23.50		713.50	795
2100	Circuit breaker		16	.500		690	23.50		713.50	795
2120	NEMA 12, type A, fusible		15	.533		720	25		745	835
2140	Circuit breaker		15	.533		720	25		745	835
2160	Type B, fusible		15	.533		720	25		745	835
2180	Circuit breaker		15	.533		720	25		745	835
2300	Starter size 3, FVNR, NEMA 1, type A, fusible	2 Elec	2	8		3,525	375		3,900	4,450
2320	Circuit breaker		2	8		3,125	375		3,500	4,000
2340	Type B, fusible		2	8		3,875	375		4,250	4,825
2360	Circuit breaker		2	8		3,450	375		3,825	4,375
2380	NEMA 12, type A, fusible		1.90	8.421		3,575	395		3,970	4,550
2400	Circuit breaker		1.90	8.421		3,175	395		3,570	4,100
2420	Type B, fusible		1.90	8.421		4,000	395		4,395	5,000
2440	Circuit breaker		1.90	8.421		3,500	395		3,895	4,450
2500	Starter size 3, FVR, NEMA 1, type A, fusible		1.60	10		4,800	470		5,270	6,000
2520	Circuit breaker		1.60	10		4,575	470		5,045	5,750
2540	Type B, fusible		1.60	10		5,225	470		5,695	6,475
2560	Circuit breaker		1.60	10		5,000	470		5,470	6,225
2580	NEMA 12, type A, fusible		1.50	10.667		4,900	500		5,400	6,150
2600	Circuit breaker		1.50	10.667		4,675	500		5,175	5,900
2620	Type B, fusible		1.50	10.667		5,325	500		5,825	6,625
2640	Circuit breaker		1.50	10.667		5,525	500		6,025	6,850
2690	Starter size 3, 2 speed, separate winding									
2700	NEMA 1, type A, fusible	2 Elec	2	8	Ea.	5,525	375		5,900	6,650
2720	Circuit breaker		2	8		4,875	375		5,250	5,925
2740	Type B, fusible		2	8		6,025	375		6,400	7,225
2760	Circuit breaker		2	8		5,325	375		5,700	6,425
2780	NEMA 12, type A, fusible		1.90	8.421		5,625	395		6,020	6,800
2800	Circuit breaker		1.90	8.421		4,975	395		5,370	6,075
2820	Type B, fusible		1.90	8.421		6,150	395		6,545	7,375
2840	Circuit breaker		1.90	8.421		5,425	395		5,820	6,575
2850	Starter size 3, 2 speed, consequent pole									
2860	NEMA 1, type A, fusible	2 Elec	2	8	Ea.	6,150	375		6,525	7,350
2880	Circuit breaker		2	8		5,500	375		5,875	6,625
2900	Type B, fusible		2	8		6,675	375		7,050	7,925
2920	Circuit breaker		2	8		5,975	375		6,350	7,125
2940	NEMA 12, type A, fusible		1.90	8.421		6,275	395		6,670	7,500
2960	Circuit breaker		1.90	8.421		5,600	395		5,995	6,775
2980	Type B, fusible		1.90	8.421		6,800	395		7,195	8,075
3000	Circuit breaker		1.90	8.421		6,075	395		6,470	7,275
3100	Starter size 3, autotransformer, NEMA 1, type A, fusible		1.60	10		7,650	470		8,120	9,150
3120	Circuit breaker		1.60	10		7,650	470		8,120	9,150
3140	Type B, fusible		1.60	10		8,375	470		8,845	9,925
3160	Circuit breaker		1.60	10		8,375	470		8,845	9,925
3180	NEMA 12, type A, fusible		1.50	10.667		7,800	500		8,300	9,350
3200	Circuit breaker		1.50	10.667		7,800	500		8,300	9,350
3220	Type B, fusible		1.50	10.667		8,500	500		9,000	10,100
3240	Circuit breaker		1.50	10.667		8,500	500		9,000	10,100
3260	Starter size 3, space only, NEMA 1, type A, fusible	1 Elec	15	.533		1,200	25		1,225	1,350
3280	Circuit breaker		15	.533		925	25		950	1,075

26 24 19.20 Motor Control Center Components		Crew	Daily Output	Labor-Hours	Unit	Material	2009 Bare Costs Labor	Equipment	Total	Total Incl O&P
3300	Type B, fusible	1 Elec	15	.533	Ea.	1,200	25		1,225	1,350
3320	Circuit breaker		15	.533		925	25		950	1,075
3340	NEMA 12, type A, fusible		14	.571		1,250	27		1,277	1,425
3360	Circuit breaker		14	.571		980	27		1,007	1,125
3380	Type B, fusible		14	.571		1,250	27		1,277	1,425
3400	Circuit breaker		14	.571		980	27		1,007	1,125
3500	Starter size 4, FVNR, NEMA 1, type A, fusible	2 Elec	1.60	10		4,575	470		5,045	5,750
3520	Circuit breaker		1.60	10		4,150	470		4,620	5,300
3540	Type B, fusible		1.60	10		5,000	470		5,470	6,225
3560	Circuit breaker		1.60	10		4,575	470		5,045	5,750
3580	NEMA 12, type A, fusible		1.50	10.667		4,700	500		5,200	5,925
3600	Circuit breaker		1.50	10.667		4,225	500		4,725	5,425
3620	Type B, fusible		1.50	10.667		5,125	500		5,625	6,400
3640	Circuit breaker		1.50	10.667		4,650	500		5,150	5,900
3700	Starter size 4, FVR, NEMA 1, type A, fusible		1.20	13.333		6,250	625		6,875	7,875
3720	Circuit breaker		1.20	13.333		5,650	625		6,275	7,175
3740	Type B, fusible		1.20	13.333		6,825	625		7,450	8,500
3760	Circuit breaker		1.20	13.333		6,225	625		6,850	7,800
3780	NEMA 12, type A, fusible		1.16	13.793		6,400	650		7,050	8,050
3800	Circuit breaker		1.16	13.793		5,725	650		6,375	7,300
3820	Type B, fusible		1.16	13.793		5,775	650		6,425	7,375
3840	Circuit breaker		1.16	13.793		6,300	650		6,950	7,925
3890	Starter size 4, 2 speed, separate windings									
3900	NEMA 1, type A, fusible	2 Elec	1.60	10	Ea.	7,900	470		8,370	9,425
3920	Circuit breaker		1.60	10		5,925	470		6,395	7,250
3940	Type B, fusible		1.60	10		8,675	470		9,145	10,300
3960	Circuit breaker		1.60	10		6,500	470		6,970	7,875
3980	NEMA 12, type A, fusible		1.50	10.667		8,050	500		8,550	9,625
4000	Circuit breaker		1.50	10.667		6,025	500		6,525	7,425
4020	Type B, fusible		1.50	10.667		8,825	500		9,325	10,500
4040	Circuit breaker		1.50	10.667		6,600	500		7,100	8,050
4050	Starter size 4, 2 speed, consequent pole									
4060	NEMA 1, type A, fusible	2 Elec	1.60	10	Ea.	9,225	470		9,695	10,900
4080	Circuit breaker		1.60	10		6,875	470		7,345	8,275
4100	Type B, fusible		1.60	10		10,200	470		10,670	11,900
4120	Circuit breaker		1.60	10		7,525	470		7,995	9,000
4140	NEMA 12, type A, fusible		1.50	10.667		9,375	500		9,875	11,100
4160	Circuit breaker		1.50	10.667		6,975	500		7,475	8,425
4180	Type B, fusible		1.50	10.667		10,300	500		10,800	12,100
4200	Circuit breaker		1.50	10.667		7,625	500		8,125	9,175
4300	Starter size 4, autotransformer, NEMA 1, type A, fusible		1.30	12.308		9,100	580		9,680	10,900
4320	Circuit breaker		1.30	12.308		9,175	580		9,755	11,000
4340	Type B, fusible		1.30	12.308		10,000	580		10,580	11,900
4360	Circuit breaker		1.30	12.308		10,000	580		10,580	11,900
4380	NEMA 12, type A, fusible		1.24	12.903		9,250	605		9,855	11,100
4400	Circuit breaker		1.24	12.903		9,300	605		9,905	11,100
4420	Type B, fusible		1.24	12.903		10,200	605		10,805	12,100
4440	Circuit breaker		1.24	12.903		10,100	605		10,705	12,100
4500	Starter size 4, space only, NEMA 1, type A, fusible	1 Elec	14	.571		1,650	27		1,677	1,875
4520	Circuit breaker		14	.571		1,200	27		1,227	1,350
4540	Type B, fusible		14	.571		1,650	27		1,677	1,875
4560	Circuit breaker		14	.571		1,200	27		1,227	1,350
4580	NEMA 12, type A, fusible		13	.615		1,750	29		1,779	1,975

26 24 19.20 Motor Control Center Components		Crew	Daily Output	Labor-Hours	Unit	Material	Labor	Equipment	Total	Total Incl O&P
						2009 Bare Costs				
4600	Circuit breaker	1 Elec	13	.615	Ea.	1,250	29		1,279	1,425
4620	Type B, fusible		13	.615		1,750	29		1,779	1,975
4640	Circuit breaker		13	.615		1,250	29		1,279	1,425
4800	Starter size 5, FVNR, NEMA 1, type A, fusible	2 Elec	1	16		9,450	750		10,200	11,600
4820	Circuit breaker		1	16		6,700	750		7,450	8,525
4840	Type B, fusible		1	16		10,400	750		11,150	12,600
4860	Circuit breaker		1	16		7,375	750		8,125	9,275
4880	NEMA 12, type A, fusible		.96	16.667		9,575	785		10,360	11,700
4900	Circuit breaker		.96	16.667		6,825	785		7,610	8,725
4920	Type B, fusible		.96	16.667		10,500	785		11,285	12,800
4940	Circuit breaker		.96	16.667		7,500	785		8,285	9,475
5000	Starter size 5, FVR, NEMA 1, type A, fusible		.80	20		13,900	940		14,840	16,800
5020	Circuit breaker		.80	20		10,900	940		11,840	13,500
5040	Type B, fusible		.80	20		15,300	940		16,240	18,300
5060	Circuit breaker		.80	20		12,000	940		12,940	14,700
5080	NEMA 12, type A, fusible		.76	21.053		14,200	990		15,190	17,100
5100	Circuit breaker		.76	21.053		11,100	990		12,090	13,700
5120	Type B, fusible		.76	21.053		15,500	990		16,490	18,600
5140	Circuit breaker		.76	21.053		12,200	990		13,190	14,900
5190	Starter size 5, 2 speed, separate windings									
5200	NEMA 1, type A, fusible	2 Elec	1	16	Ea.	19,200	750		19,950	22,300
5220	Circuit breaker		1	16		14,300	750		15,050	17,000
5240	Type B, fusible		1	16		21,100	750		21,850	24,400
5260	Circuit breaker		1	16		15,700	750		16,450	18,500
5280	NEMA 12, type A, fusible		.96	16.667		19,500	785		20,285	22,600
5300	Circuit breaker		.96	16.667		14,500	785		15,285	17,100
5320	Type B, fusible		.96	16.667		21,400	785		22,185	24,700
5340	Circuit breaker		.96	16.667		15,900	785		16,685	18,700
5400	Starter size 5, autotransformer, NEMA 1, type A, fusible		.70	22.857		15,700	1,075		16,775	19,000
5420	Circuit breaker		.70	22.857		13,100	1,075		14,175	16,100
5440	Type B, fusible		.70	22.857		17,300	1,075		18,375	20,700
5460	Circuit breaker		.70	22.857		14,300	1,075		15,375	17,500
5480	NEMA 12, type A, fusible		.68	23.529		15,900	1,100		17,000	19,200
5500	Circuit breaker		.68	23.529		13,200	1,100		14,300	16,200
5520	Type B, fusible		.68	23.529		17,400	1,100		18,500	20,900
5540	Circuit breakers		.68	23.529		14,500	1,100		15,600	17,600
5600	Starter size 5, space only, NEMA 1, type A, fusible	1 Elec	12	.667		2,100	31.50		2,131.50	2,375
5620	Circuit breaker		12	.667		2,100	31.50		2,131.50	2,375
5640	Type B, fusible		12	.667		2,100	31.50		2,131.50	2,375
5660	Circuit breaker		12	.667		1,375	31.50		1,406.50	1,575
5680	NEMA 12, type A, fusible		11	.727		2,250	34		2,284	2,525
5700	Circuit breaker		11	.727		1,475	34		1,509	1,675
5720	Type B, fusible		11	.727		2,250	34		2,284	2,525
5740	Circuit breaker		11	.727		1,475	34		1,509	1,675
5800	Fuse, light contactor NEMA 1, type A, 30 amp		2.70	2.963		1,625	139		1,764	2,025
5820	60 amp		2	4		1,850	188		2,038	2,325
5840	100 amp		1	8		3,525	375		3,900	4,450
5860	200 amp		.80	10		8,175	470		8,645	9,700
5880	Type B, 30 amp		2.70	2.963		1,775	139		1,914	2,175
5900	60 amp		2	4		2,000	188		2,188	2,475
5920	100 amp		1	8		3,875	375		4,250	4,825
5940	200 amp	2 Elec	1.60	10		8,950	470		9,420	10,600
5960	NEMA 12, type A, 30 amp	1 Elec	2.60	3.077		1,650	145		1,795	2,050

26 24 19.20 Motor Control Center Components		Crew	Daily Output	Labor-Hours	Unit	2009 Bare Costs Material	Labor	Equipment	Total	Total Incl O&P
5980	60 amp	1 Elec	1.90	4.211	Ea.	1,875	198		2,073	2,350
6000	100 amp		.95	8.421		3,575	395		3,970	4,550
6020	200 amp	2 Elec	1.50	10.667		8,300	500		8,800	9,900
6040	Type B, 30 amp	1 Elec	2.60	3.077		1,800	145		1,945	2,200
6060	60 amp		1.90	4.211		2,025	198		2,223	2,525
6080	100 amp		.95	8.421		3,950	395		4,345	4,925
6100	200 amp	2 Elec	1.50	10.667		9,075	500		9,575	10,700
6200	Circuit breaker, light contactor NEMA 1, type A, 30 amp	1 Elec	2.70	2.963		1,775	139		1,914	2,175
6220	60 amp		2	4		2,025	188		2,213	2,525
6240	100 amp		1	8		3,125	375		3,500	4,000
6260	200 amp	2 Elec	1.60	10		6,775	470		7,245	8,175
6280	Type B, 30 amp	1 Elec	2.70	2.963		1,950	139		2,089	2,375
6300	60 amp		2	4		2,225	188		2,413	2,750
6320	100 amp		1	8		3,400	375		3,775	4,325
6340	200 amp	2 Elec	1.60	10		7,375	470		7,845	8,850
6360	NEMA 12, type A, 30 amp	1 Elec	2.60	3.077		1,800	145		1,945	2,200
6380	60 amp		1.90	4.211		2,050	198		2,248	2,550
6400	100 amp		.95	8.421		3,175	395		3,570	4,100
6420	200 amp	2 Elec	1.50	10.667		6,800	500		7,300	8,250
6440	Type B, 30 amp	1 Elec	2.60	3.077		2,125	145		2,270	2,550
6460	60 amp		1.90	4.211		2,275	198		2,473	2,800
6480	100 amp		.95	8.421		3,475	395		3,870	4,400
6500	200 amp	2 Elec	1.50	10.667		7,475	500		7,975	9,000
6600	Fusible switch, NEMA 1, type A, 30 amp	1 Elec	5.30	1.509		1,075	71		1,146	1,300
6620	60 amp		5	1.600		1,150	75		1,225	1,400
6640	100 amp		4	2		1,275	94		1,369	1,550
6660	200 amp		3.20	2.500		2,200	118		2,318	2,600
6680	400 amp	2 Elec	4.60	3.478		5,175	163		5,338	5,950
6700	600 amp		3.20	5		5,600	235		5,835	6,525
6720	800 amp		2.60	6.154		15,400	289		15,689	17,300
6740	NEMA 12, type A, 30 amp	1 Elec	5.20	1.538		1,100	72.50		1,172.50	1,325
6760	60 amp		4.90	1.633		1,175	76.50		1,251.50	1,425
6780	100 amp		3.90	2.051		1,300	96.50		1,396.50	1,575
6800	200 amp		3.10	2.581		2,250	121		2,371	2,650
6820	400 amp	2 Elec	4.40	3.636		5,300	171		5,471	6,100
6840	600 amp		3	5.333		5,775	251		6,026	6,725
6860	800 amp		2.40	6.667		15,600	315		15,915	17,600
6900	Circuit breaker, NEMA 1, type A, 30 amp	1 Elec	5.30	1.509		995	71		1,066	1,200
6920	60 amp		5	1.600		995	75		1,070	1,225
6940	100 amp		4	2		995	94		1,089	1,250
6960	225 amp		3.20	2.500		1,775	118		1,893	2,125
6980	400 amp	2 Elec	4.60	3.478		3,375	163		3,538	3,950
7000	600 amp		3.20	5		3,950	235		4,185	4,675
7020	800 amp		2.60	6.154		8,225	289		8,514	9,500
7040	NEMA 12, type A, 30 amp	1 Elec	5.20	1.538		1,025	72.50		1,097.50	1,225
7060	60 amp		4.90	1.633		1,025	76.50		1,101.50	1,250
7080	100 amp		3.90	2.051		1,025	96.50		1,121.50	1,275
7100	225 amp		3.10	2.581		1,800	121		1,921	2,150
7120	400 amp	2 Elec	4.40	3.636		3,425	171		3,596	4,050
7140	600 amp		3	5.333		4,000	251		4,251	4,775
7160	800 amp		2.40	6.667		8,400	315		8,715	9,700
7300	Incoming line, main lug only, 600 amp, alum, NEMA 1		1.60	10		1,250	470		1,720	2,100
7320	NEMA 12		1.50	10.667		1,275	500		1,775	2,200

	26 24 19.20 Motor Control Center Components	Crew	Daily Output	Labor-Hours	Unit	Material	2009 Bare Costs Labor	Equipment	Total	Total Incl O&P
7340	Copper, NEMA 1	2 Elec	1.60	10	Ea.	1,325	470		1,795	2,175
7360	800 amp, alum., NEMA 1		1.50	10.667		3,150	500		3,650	4,250
7380	NEMA 12		1.40	11.429		3,200	535		3,735	4,350
7400	Copper, NEMA 1		1.50	10.667		3,275	500		3,775	4,375
7420	1200 amp, copper, NEMA 1		1.40	11.429		3,400	535		3,935	4,550
7440	Incoming line, fusible switch, 400 amp, alum., NEMA 1		1.20	13.333		4,325	625		4,950	5,750
7460	NEMA 12		1.10	14.545		4,425	685		5,110	5,900
7480	Copper, NEMA 1		1.20	13.333		4,400	625		5,025	5,825
7500	600 amp, alum., NEMA 1		1.10	14.545		5,300	685		5,985	6,875
7520	NEMA 12		1	16		5,375	750		6,125	7,075
7540	Copper, NEMA 1		1.10	14.545		5,350	685		6,035	6,950
7560	Incoming line, circuit breaker, 225 amp, alum., NEMA 1		1.20	13.333		2,250	625		2,875	3,450
7580	NEMA 12		1.10	14.545		2,300	685		2,985	3,575
7600	Copper, NEMA 1		1.20	13.333		2,325	625		2,950	3,525
7620	400 amp, alum., NEMA 1		1.20	13.333		3,375	625		4,000	4,675
7640	NEMA 12		1.10	14.545		3,425	685		4,110	4,825
7660	Copper, NEMA 1		1.20	13.333		3,450	625		4,075	4,750
7680	600 amp, alum., NEMA 1		1.10	14.545		3,950	685		4,635	5,375
7700	NEMA 12		1	16		4,000	750		4,750	5,550
7720	Copper, NEMA 1		1.10	14.545		4,000	685		4,685	5,450
7740	800 amp, copper, NEMA 1		.90	17.778		8,225	835		9,060	10,300
7760	Incoming line, for copper bus, add					124			124	137
7780	For 65000 amp bus bracing, add					185			185	203
7800	For NEMA 3R enclosure, add					5,800			5,800	6,375
7820	For NEMA 12 enclosure, add					160			160	176
7840	For 1/4" x 1" ground bus, add	1 Elec	16	.500		103	23.50		126.50	149
7860	For 1/4" x 2" ground bus, add	"	12	.667		103	31.50		134.50	161
7880	Main rating basic section, alum., NEMA 1, 600 amp	2 Elec	1.60	10			470		470	725
7900	800 amp		1.40	11.429		335	535		870	1,200
7920	1200 amp		1.20	13.333		660	625		1,285	1,700
7940	For copper bus, add					495			495	545
7960	For 65000 amp bus bracing, add					335			335	365
7980	For NEMA 3R enclosure, add					5,800			5,800	6,375
8000	For NEMA 12, enclosure, add					160			160	176
8020	For 1/4" x 1" ground bus, add	1 Elec	16	.500		103	23.50		126.50	149
8040	For 1/4" x 2" ground bus, add		12	.667		103	31.50		134.50	161
8060	Unit devices, pilot light, standard		16	.500		89	23.50		112.50	134
8080	Pilot light, push to test		16	.500		124	23.50		147.50	173
8100	Pilot light, standard, and push button		12	.667		213	31.50		244.50	282
8120	Pilot light, push to test, and push button		12	.667		249	31.50		280.50	320
8140	Pilot light, standard, and select switch		12	.667		213	31.50		244.50	282
8160	Pilot light, push to test, and select switch		12	.667		249	31.50		280.50	320

26 24 19.30 Motor Control Center

		Crew	Daily Output	Labor-Hours	Unit	Material	Labor	Equipment	Total	Total Incl O&P
0010	**MOTOR CONTROL CENTER** Consists of starters & structures									
0050	Starters, class 1, type B, comb. MCP, FVNR, with									
0100	control transformer, 10 HP, size 1, 12" high	1 Elec	2.70	2.963	Ea.	1,775	139		1,914	2,175
0200	25 HP, size 2, 18" high	2 Elec	4	4		2,025	188		2,213	2,525
0300	50 HP, size 3, 24" high		2	8		3,125	375		3,500	4,000
0350	75 HP, size 4, 24" high		1.60	10		4,150	470		4,620	5,300
0400	100 HP, size 4, 30" high		1.40	11.429		5,475	535		6,010	6,850
0500	200 HP, size 5, 48" high		1	16		8,175	750		8,925	10,200
0600	400 HP, size 6, 72" high		.80	20		16,800	940		17,740	20,000

26 24 19 – Motor-Control Centers

26 24 19.30 Motor Control Center		Crew	Daily Output	Labor-Hours	Unit	Material	Labor	Equipment	Total	Total Incl O&P
						2009 Bare Costs				
0800	Structures, 600 amp, 22,000 rms, takes any									
0900	combination of starters up to 72" high	2 Elec	1.60	10	Ea.	2,475	470		2,945	3,450
1000	Back to back, 72" front & 66" back	"	1.20	13.333		3,125	625		3,750	4,425
1100	For copper bus add per structure					277			277	305
1200	For NEMA 12, add per structure					160			160	176
1300	For 42,000 rms, add per structure					216			216	237
1400	For 100,000 rms, size 1 & 2, add					710			710	780
1500	Size 3, add					1,150			1,150	1,250
1600	Size 4, add					925			925	1,025
1700	For pilot lights, add per starter	1 Elec	16	.500		124	23.50		147.50	173
1800	For push button, add per starter		16	.500		124	23.50		147.50	173
1900	For auxiliary contacts, add per starter		16	.500		259	23.50		282.50	320

26 24 19.40 Motor Starters and Controls		Crew	Daily Output	Labor-Hours	Unit	Material	Labor	Equipment	Total	Total Incl O&P
0010	**MOTOR STARTERS AND CONTROLS**									
0050	Magnetic, FVNR, with enclosure and heaters, 480 volt									
0080	2 HP, size 00	1 Elec	3.50	2.286	Ea.	205	107		312	390
0100	5 HP, size 0		2.30	3.478		248	163		411	525
0200	10 HP, size 1		1.60	5		278	235		513	665
0300	25 HP, size 2	2 Elec	2.20	7.273		525	340		865	1,100
0400	50 HP, size 3		1.80	8.889		855	420		1,275	1,575
0500	100 HP, size 4		1.20	13.333		1,900	625		2,525	3,050
0600	200 HP, size 5		.90	17.778		4,425	835		5,260	6,150
0610	400 HP, size 6		.80	20		17,300	940		18,240	20,600
0620	NEMA 7, 5 HP, size 0	1 Elec	1.60	5		1,300	235		1,535	1,775
0630	10 HP, size 1	"	1.10	7.273		1,350	340		1,690	2,000
0640	25 HP, size 2	2 Elec	1.80	8.889		2,175	420		2,595	3,050
0650	50 HP, size 3		1.20	13.333		3,275	625		3,900	4,575
0660	100 HP, size 4		.90	17.778		5,275	835		6,110	7,075
0670	200 HP, size 5		.50	32		12,600	1,500		14,100	16,100
0700	Combination, with motor circuit protectors, 5 HP, size 0	1 Elec	1.80	4.444		805	209		1,014	1,200
0800	10 HP, size 1	"	1.30	6.154		835	289		1,124	1,375
0900	25 HP, size 2	2 Elec	2	8		1,175	375		1,550	1,875
1000	50 HP, size 3		1.32	12.121		1,700	570		2,270	2,725
1200	100 HP, size 4		.80	20		3,675	940		4,615	5,500
1220	NEMA 7, 5 HP, size 0	1 Elec	1.30	6.154		2,300	289		2,589	2,975
1230	10 HP, size 1	"	1	8		2,350	375		2,725	3,175
1240	25 HP, size 2	2 Elec	1.32	12.121		3,150	570		3,720	4,350
1250	50 HP, size 3		.80	20		5,200	940		6,140	7,175
1260	100 HP, size 4		.60	26.667		8,125	1,250		9,375	10,900
1270	200 HP, size 5		.40	40		17,600	1,875		19,475	22,300
1400	Combination, with fused switch, 5 HP, size 0	1 Elec	1.80	4.444		615	209		824	995
1600	10 HP, size 1	"	1.30	6.154		660	289		949	1,175
1800	25 HP, size 2	2 Elec	2	8		1,075	375		1,450	1,750
2000	50 HP, size 3		1.32	12.121		1,800	570		2,370	2,850
2200	100 HP, size 4		.80	20		3,150	940		4,090	4,925
2610	NEMA 4, with start-stop pushbutton size 1	1 Elec	1.30	6.154		1,650	289		1,939	2,275
2620	Size 2	2 Elec	2	8		2,200	375		2,575	3,000
2630	Size 3		1.32	12.121		3,500	570		4,070	4,725
2640	Size 4		.80	20		5,375	940		6,315	7,375
2650	NEMA 4, FVNR, including control transformer									
2660	Size 1	2 Elec	2.60	6.154	Ea.	1,525	289		1,814	2,125
2670	Size 2		2	8		2,175	375		2,550	2,950

26 24 19.40 Motor Starters and Controls		Crew	Daily Output	Labor-Hours	Unit	Material	Labor	Equipment	Total	Total Incl O&P
						2009 Bare Costs				
2680	Size 3	2 Elec	1.32	12.121	Ea.	3,450	570		4,020	4,675
2690	Size 4	↓	.80	20		5,325	940		6,265	7,325
2710	Magnetic, FVR, control circuit transformer, NEMA 1, size 1	1 Elec	1.30	6.154		895	289		1,184	1,425
2720	Size 2	2 Elec	2	8		1,425	375		1,800	2,150
2730	Size 3		1.32	12.121		2,175	570		2,745	3,275
2740	Size 4	↓	.80	20		4,750	940		5,690	6,675
2760	NEMA 4, size 1	1 Elec	1.10	7.273		1,300	340		1,640	1,950
2770	Size 2	2 Elec	1.60	10		2,075	470		2,545	3,000
2780	Size 3		1.20	13.333		3,125	625		3,750	4,400
2790	Size 4	↓	.70	22.857		6,400	1,075		7,475	8,700
2820	NEMA 12, size 1	1 Elec	1.10	7.273		1,050	340		1,390	1,675
2830	Size 2	2 Elec	1.60	10		1,675	470		2,145	2,550
2840	Size 3		1.20	13.333		2,625	625		3,250	3,875
2850	Size 4	↓	.70	22.857		5,425	1,075		6,500	7,600
2870	Combination FVR, fused, w/control XFMR & PB, NEMA 1, size 1	1 Elec	1	8		1,550	375		1,925	2,275
2880	Size 2	2 Elec	1.50	10.667		2,200	500		2,700	3,200
2890	Size 3		1.10	14.545		3,275	685		3,960	4,650
2900	Size 4	↓	.70	22.857		6,625	1,075		7,700	8,950
2910	NEMA 4, size 1	1 Elec	.90	8.889		2,300	420		2,720	3,175
2920	Size 2	2 Elec	1.40	11.429		3,350	535		3,885	4,500
2930	Size 3		1	16		5,275	750		6,025	6,950
2940	Size 4	↓	.60	26.667		8,750	1,250		10,000	11,600
2950	NEMA 12, size 1	1 Elec	1	8		1,750	375		2,125	2,500
2960	Size 2	2 Elec	1.40	11.429		2,475	535		3,010	3,550
2970	Size 3		1	16		3,625	750		4,375	5,150
2980	Size 4	↓	.60	26.667		7,200	1,250		8,450	9,850
3010	Manual, single phase, w/pilot, 1 pole 120V NEMA 1	1 Elec	6.40	1.250		64.50	59		123.50	161
3020	NEMA 4		4	2		360	94		454	540
3030	2 pole, 120/240 V, NEMA 1		6.40	1.250		70.50	59		129.50	168
3040	NEMA 4		4	2		365	94		459	550
3041	3 phase, 3 pole 600V, NEMA 1		5.50	1.455		251	68.50		319.50	380
3042	NEMA 4		3.50	2.286		490	107		597	705
3043	NEMA 12	↓	3.50	2.286		294	107		401	490
3070	Auxiliary contact, normally open				↓	73.50			73.50	81
3500	Magnetic FVNR with NEMA 12, enclosure & heaters, 480 volt									
3600	5 HP, size 0	1 Elec	2.20	3.636	Ea.	234	171		405	520
3700	10 HP, size 1	"	1.50	5.333		350	251		601	770
3800	25 HP, size 2	2 Elec	2	8		660	375		1,035	1,300
3900	50 HP, size 3		1.60	10		1,025	470		1,495	1,850
4000	100 HP, size 4		1	16		2,425	750		3,175	3,825
4100	200 HP, size 5	↓	.80	20		5,825	940		6,765	7,850
4200	Combination, with motor circuit protectors, 5 HP, size 0	1 Elec	1.70	4.706		775	221		996	1,200
4300	10 HP, size 1	"	1.20	6.667		805	315		1,120	1,375
4400	25 HP, size 2	2 Elec	1.80	8.889		1,200	420		1,620	1,975
4500	50 HP, size 3		1.20	13.333		1,950	625		2,575	3,125
4600	100 HP, size 4	↓	.74	21.622		4,425	1,025		5,450	6,450
4700	Combination, with fused switch, 5 HP, size 0	1 Elec	1.70	4.706		750	221		971	1,175
4800	10 HP, size 1	"	1.20	6.667		780	315		1,095	1,350
4900	25 HP, size 2	2 Elec	1.80	8.889		1,200	420		1,620	1,950
5000	50 HP, size 3		1.20	13.333		1,900	625		2,525	3,075
5100	100 HP, size 4	↓	.74	21.622	↓	3,875	1,025		4,900	5,825
5200	Factory installed controls, adders to size 0 thru 5									
5300	Start-stop push button	1 Elec	32	.250	Ea.	49	11.75		60.75	72

26 24 19.40 Motor Starters and Controls		Crew	Daily Output	Labor-Hours	Unit	2009 Bare Costs Material	Labor	Equipment	Total	Total Incl O&P
5400	Hand-off-auto-selector switch	1 Elec	32	.250	Ea.	49	11.75		60.75	72
5500	Pilot light		32	.250		92	11.75		103.75	119
5600	Start-stop-pilot		32	.250		141	11.75		152.75	173
5700	Auxiliary contact, NO or NC		32	.250		67.50	11.75		79.25	92
5800	NO-NC		32	.250		135	11.75		146.75	166
5810	Magnetic FVR, NEMA 7, w/heaters, size 1		.66	12.121		2,375	570		2,945	3,500
5830	Size 2	2 Elec	1.10	14.545		3,975	685		4,660	5,425
5840	Size 3		.70	22.857		6,400	1,075		7,475	8,700
5850	Size 4		.60	26.667		7,600	1,250		8,850	10,300
5860	Combination w/circuit breakers, heaters, control XFMR PB, size 1	1 Elec	.60	13.333		1,725	625		2,350	2,850
5870	Size 2	2 Elec	.80	20		2,200	940		3,140	3,875
5880	Size 3		.50	32		2,950	1,500		4,450	5,550
5890	Size 4		.40	40		5,825	1,875		7,700	9,300
5900	Manual, 240 volt, .75 HP motor	1 Elec	4	2		52	94		146	202
5910	2 HP motor		4	2		141	94		235	300
6000	Magnetic, 240 volt, 1 or 2 pole, .75 HP motor		4	2		205	94		299	370
6020	2 HP motor		4	2		226	94		320	395
6040	5 HP motor		3	2.667		325	125		450	550
6060	10 HP motor		2.30	3.478		805	163		968	1,125
6100	3 pole, .75 HP motor		3	2.667		205	125		330	420
6120	5 HP motor		2.30	3.478		278	163		441	555
6140	10 HP motor		1.60	5		525	235		760	935
6160	15 HP motor		1.60	5		525	235		760	935
6180	20 HP motor		1.10	7.273		855	340		1,195	1,475
6200	25 HP motor	2 Elec	2.20	7.273		855	340		1,195	1,475
6210	30 HP motor		1.80	8.889		855	420		1,275	1,575
6220	40 HP motor		1.80	8.889		1,900	420		2,320	2,725
6230	50 HP motor		1.80	8.889		1,900	420		2,320	2,725
6240	60 HP motor		1.20	13.333		4,425	625		5,050	5,850
6250	75 HP motor		1.20	13.333		4,425	625		5,050	5,850
6260	100 HP motor		1.20	13.333		4,425	625		5,050	5,850
6270	125 HP motor		.90	17.778		12,400	835		13,235	15,000
6280	150 HP motor		.90	17.778		12,400	835		13,235	15,000
6290	200 HP motor		.90	17.778		12,400	835		13,235	15,000
6400	Starter & nonfused disconnect, 240 volt, 1-2 pole, .75 HP motor	1 Elec	2	4		297	188		485	615
6410	2 HP motor		2	4		320	188		508	640
6420	5 HP motor		1.80	4.444		415	209		624	780
6430	10 HP motor		1.40	5.714		925	269		1,194	1,450
6440	3 pole, .75 HP motor		1.60	5		297	235		532	685
6450	5 HP motor		1.40	5.714		370	269		639	820
6460	10 HP motor		1.10	7.273		645	340		985	1,225
6470	15 HP motor		1	8		645	375		1,020	1,300
6480	20 HP motor	2 Elec	1.50	10.667		1,125	500		1,625	2,025
6490	25 HP motor		1.50	10.667		1,125	500		1,625	2,025
6500	30 HP motor		1.30	12.308		1,125	580		1,705	2,150
6510	40 HP motor		1.24	12.903		2,425	605		3,030	3,575
6520	50 HP motor		1.12	14.286		2,425	670		3,095	3,675
6530	60 HP motor		.90	17.778		4,950	835		5,785	6,725
6540	75 HP motor		.76	21.053		4,950	990		5,940	6,975
6550	100 HP motor		.70	22.857		4,950	1,075		6,025	7,100
6560	125 HP motor		.60	26.667		13,700	1,250		14,950	17,000
6570	150 HP motor		.52	30.769		13,700	1,450		15,150	17,300
6580	200 HP motor		.50	32		13,700	1,500		15,200	17,400

26 24 19 – Motor-Control Centers

26 24 19.40 Motor Starters and Controls		Crew	Daily Output	Labor-Hours	Unit	Material	2009 Bare Costs Labor	Equipment	Total	Total Incl O&P
6600	Starter & fused disconnect, 240 volt, 1-2 pole, .75 HP motor	1 Elec	2	4	Ea.	320	188		508	640
6610	2 HP motor		2	4		340	188		528	665
6620	5 HP motor		1.80	4.444		440	209		649	800
6630	10 HP motor		1.40	5.714		995	269		1,264	1,525
6640	3 pole, .75 HP motor		1.60	5		320	235		555	710
6650	5 HP motor		1.40	5.714		390	269		659	845
6660	10 HP motor		1.10	7.273		715	340		1,055	1,300
6690	15 HP motor		1	8		715	375		1,090	1,375
6700	20 HP motor	2 Elec	1.60	10		1,175	470		1,645	2,025
6710	25 HP motor		1.60	10		1,175	470		1,645	2,025
6720	30 HP motor		1.40	11.429		1,175	535		1,710	2,125
6730	40 HP motor		1.20	13.333		2,600	625		3,225	3,850
6740	50 HP motor		1.20	13.333		2,600	625		3,225	3,850
6750	60 HP motor		.90	17.778		5,150	835		5,985	6,925
6760	75 HP motor		.90	17.778		5,150	835		5,985	6,925
6770	100 HP motor		.70	22.857		5,150	1,075		6,225	7,300
6780	125 HP motor		.54	29.630		14,200	1,400		15,600	17,800
6790	Combination starter & nonfusible disconnect									
6800	240 volt, 1-2 pole, .75 HP motor	1 Elec	2	4	Ea.	585	188		773	935
6810	2 HP motor		2	4		585	188		773	935
6820	5 HP motor		1.50	5.333		615	251		866	1,050
6830	10 HP motor		1.20	6.667		960	315		1,275	1,525
6840	3 pole, .75 HP motor		1.80	4.444		585	209		794	965
6850	5 HP motor		1.30	6.154		615	289		904	1,125
6860	10 HP motor		1	8		960	375		1,335	1,625
6870	15 HP motor		1	8		960	375		1,335	1,625
6880	20 HP motor	2 Elec	1.32	12.121		1,575	570		2,145	2,600
6890	25 HP motor		1.32	12.121		1,575	570		2,145	2,600
6900	30 HP motor		1.32	12.121		1,575	570		2,145	2,600
6910	40 HP motor		.80	20		3,025	940		3,965	4,775
6920	50 HP motor		.80	20		3,025	940		3,965	4,775
6930	60 HP motor		.70	22.857		6,725	1,075		7,800	9,050
6940	75 HP motor		.70	22.857		6,725	1,075		7,800	9,050
6950	100 HP motor		.70	22.857		6,725	1,075		7,800	9,050
6960	125 HP motor		.60	26.667		17,700	1,250		18,950	21,400
6970	150 HP motor		.60	26.667		17,700	1,250		18,950	21,400
6980	200 HP motor		.60	26.667		17,700	1,250		18,950	21,400
6990	Combination starter and fused disconnect									
7000	240 volt, 1-2 pole, .75 HP motor	1 Elec	2	4	Ea.	605	188		793	955
7010	2 HP motor		2	4		605	188		793	955
7020	5 HP motor		1.50	5.333		635	251		886	1,075
7030	10 HP motor		1.20	6.667		980	315		1,295	1,550
7040	3 pole, .75 HP motor		1.80	4.444		605	209		814	985
7050	5 HP motor		1.30	6.154		635	289		924	1,150
7060	10 HP motor		1	8		980	375		1,355	1,650
7070	15 HP motor		1	8		980	375		1,355	1,650
7080	20 HP motor	2 Elec	1.32	12.121		1,625	570		2,195	2,675
7090	25 HP motor		1.32	12.121		1,625	570		2,195	2,675
7100	30 HP motor		1.32	12.121		1,625	570		2,195	2,675
7110	40 HP motor		.80	20		3,125	940		4,065	4,875
7120	50 HP motor		.80	20		3,125	940		4,065	4,875
7130	60 HP motor		.80	20		6,950	940		7,890	9,100
7140	75 HP motor		.70	22.857		6,950	1,075		8,025	9,300

26 24 19.40 Motor Starters and Controls		Crew	Daily Output	Labor-Hours	Unit	Material	2009 Bare Costs Labor	Equipment	Total	Total Incl O&P
7150	100 HP motor	2 Elec	.70	22.857	Ea.	6,950	1,075		8,025	9,300
7160	125 HP motor		.70	22.857		18,400	1,075		19,475	21,900
7170	150 HP motor		.60	26.667		18,400	1,250		19,650	22,100
7180	200 HP motor		.60	26.667		18,400	1,250		19,650	22,100
7190	Combination starter & circuit breaker disconnect									
7200	240 volt, 1-2 pole, .75 HP motor	1 Elec	2	4	Ea.	610	188		798	960
7210	2 HP motor		2	4		610	188		798	960
7220	5 HP motor		1.50	5.333		640	251		891	1,100
7230	10 HP motor		1.20	6.667		975	315		1,290	1,550
7240	3 pole, .75 HP motor		1.80	4.444		625	209		834	1,000
7250	5 HP motor		1.30	6.154		660	289		949	1,175
7260	10 HP motor		1	8		995	375		1,370	1,675
7270	15 HP motor		1	8		995	375		1,370	1,675
7280	20 HP motor	2 Elec	1.32	12.121		1,700	570		2,270	2,725
7290	25 HP motor		1.32	12.121		1,700	570		2,270	2,725
7300	30 HP motor		1.32	12.121		1,700	570		2,270	2,725
7310	40 HP motor		.80	20		3,675	940		4,615	5,500
7320	50 HP motor		.80	20		3,675	940		4,615	5,500
7330	60 HP motor		.80	20		8,500	940		9,440	10,800
7340	75 HP motor		.70	22.857		8,500	1,075		9,575	11,000
7350	100 HP motor		.70	22.857		8,500	1,075		9,575	11,000
7360	125 HP motor		.70	22.857		18,400	1,075		19,475	22,000
7370	150 HP motor		.60	26.667		18,400	1,250		19,650	22,200
7380	200 HP motor		.60	26.667		18,400	1,250		19,650	22,200
7400	Magnetic FVNR with enclosure & heaters, 2 pole,									
7410	230 volt, 1 HP size 00	1 Elec	4	2	Ea.	187	94		281	350
7420	2 HP, size 0		4	2		208	94		302	375
7430	3 HP, size 1		3	2.667		239	125		364	455
7440	5 HP, size 1p		3	2.667		305	125		430	530
7450	115 volt, 1/3 HP, size 00		4	2		187	94		281	350
7460	1 HP, size 0		4	2		208	94		302	375
7470	2 HP, size 1		3	2.667		239	125		364	455
7480	3 HP, size 1P		3	2.667		297	125		422	520
7500	3 pole, 480 volt, 600 HP, size 7	2 Elec	.70	22.857		16,500	1,075		17,575	19,900
7590	Magnetic FVNR with heater, NEMA 1									
7600	600 volt, 3 pole, 5 HP motor	1 Elec	2.30	3.478	Ea.	248	163		411	525
7610	10 HP motor	"	1.60	5		278	235		513	665
7620	25 HP motor	2 Elec	2.20	7.273		525	340		865	1,100
7630	30 HP motor		1.80	8.889		855	420		1,275	1,575
7640	40 HP motor		1.80	8.889		855	420		1,275	1,575
7650	50 HP motor		1.80	8.889		855	420		1,275	1,575
7660	60 HP motor		1.20	13.333		1,900	625		2,525	3,050
7670	75 HP motor		1.20	13.333		1,900	625		2,525	3,050
7680	100 HP motor		1.20	13.333		1,900	625		2,525	3,050
7690	125 HP motor		.90	17.778		4,425	835		5,260	6,150
7700	150 HP motor		.90	17.778		4,425	835		5,260	6,150
7710	200 HP motor		.90	17.778		4,425	835		5,260	6,150
7750	Starter & nonfused disconnect, 600 volt, 3 pole, 5 HP motor	1 Elec	1.40	5.714		410	269		679	865
7760	10 HP motor	"	1.10	7.273		440	340		780	1,000
7770	25 HP motor	2 Elec	1.50	10.667		685	500		1,185	1,525
7780	30 HP motor		1.30	12.308		1,150	580		1,730	2,150
7790	40 HP motor		1.30	12.308		1,150	580		1,730	2,150
7800	50 HP motor		1.30	12.308		1,150	580		1,730	2,150

26 24 19.40 Motor Starters and Controls		Crew	Daily Output	Labor-Hours	Unit	2009 Bare Costs Material	Labor	Equipment	Total	Total Incl O&P
7810	60 HP motor	2 Elec	.92	17.391	Ea.	2,350	815		3,165	3,850
7820	75 HP motor		.92	17.391		2,350	815		3,165	3,850
7830	100 HP motor		.84	19.048		2,350	895		3,245	3,975
7840	125 HP motor		.70	22.857		5,125	1,075		6,200	7,300
7850	150 HP motor		.70	22.857		5,125	1,075		6,200	7,300
7860	200 HP motor		.60	26.667		5,125	1,250		6,375	7,575
7870	Starter & fused disconnect, 600 volt, 3 pole, 5 HP motor	1 Elec	1.40	5.714		555	269		824	1,025
7880	10 HP motor	″	1.10	7.273		590	340		930	1,175
7890	25 HP motor	2 Elec	1.50	10.667		835	500		1,335	1,675
7900	30 HP motor		1.30	12.308		1,225	580		1,805	2,250
7910	40 HP motor		1.30	12.308		1,225	580		1,805	2,250
7920	50 HP motor		1.30	12.308		1,225	580		1,805	2,250
7930	60 HP motor		.92	17.391		2,575	815		3,390	4,100
7940	75 HP motor		.92	17.391		2,575	815		3,390	4,100
7950	100 HP motor		.84	19.048		2,575	895		3,470	4,225
7960	125 HP motor		.70	22.857		5,425	1,075		6,500	7,625
7970	150 HP motor		.70	22.857		5,425	1,075		6,500	7,625
7980	200 HP motor		.60	26.667		5,425	1,250		6,675	7,900
7990	Combination starter and nonfusible disconnect									
8000	600 volt, 3 pole, 5 HP motor	1 Elec	1.80	4.444	Ea.	585	209		794	965
8010	10 HP motor	″	1.30	6.154		615	289		904	1,125
8020	25 HP motor	2 Elec	2	8		960	375		1,335	1,625
8030	30 HP motor		1.32	12.121		1,625	570		2,195	2,675
8040	40 HP motor		1.32	12.121		1,625	570		2,195	2,675
8050	50 HP motor		1.32	12.121		1,575	570		2,145	2,600
8060	60 HP motor		.80	20		3,125	940		4,065	4,875
8070	75 HP motor		.80	20		3,125	940		4,065	4,875
8080	100 HP motor		.80	20		3,025	940		3,965	4,775
8090	125 HP motor		.70	22.857		6,950	1,075		8,025	9,300
8100	150 HP motor		.70	22.857		6,950	1,075		8,025	9,300
8110	200 HP motor		.70	22.857		6,950	1,075		8,025	9,300
8140	Combination starter and fused disconnect									
8150	600 volt, 3 pole, 5 HP motor	1 Elec	1.80	4.444	Ea.	605	209		814	985
8160	10 HP motor	″	1.30	6.154		660	289		949	1,175
8170	25 HP motor	2 Elec	2	8		1,075	375		1,450	1,750
8180	30 HP motor		1.32	12.121		1,675	570		2,245	2,700
8190	40 HP motor		1.32	12.121		1,800	570		2,370	2,850
8200	50 HP motor		1.32	12.121		1,800	570		2,370	2,850
8210	60 HP motor		.80	20		3,150	940		4,090	4,900
8220	75 HP motor		.80	20		3,150	940		4,090	4,900
8230	100 HP motor		.80	20		3,150	940		4,090	4,900
8240	125 HP motor		.70	22.857		6,950	1,075		8,025	9,300
8250	150 HP motor		.70	22.857		6,950	1,075		8,025	9,300
8260	200 HP motor		.70	22.857		6,950	1,075		8,025	9,300
8290	Combination starter & circuit breaker disconnect									
8300	600 volt, 3 pole, 5 HP motor	1 Elec	1.80	4.444	Ea.	805	209		1,014	1,200
8310	10 HP motor	″	1.30	6.154		835	289		1,124	1,375
8320	25 HP motor	2 Elec	2	8		1,175	375		1,550	1,875
8330	30 HP motor		1.32	12.121		1,700	570		2,270	2,725
8340	40 HP motor		1.32	12.121		1,700	570		2,270	2,725
8350	50 HP motor		1.32	12.121		1,700	570		2,270	2,725
8360	60 HP motor		.80	20		3,675	940		4,615	5,500
8370	75 HP motor		.80	20		3,675	940		4,615	5,500

26 24 Switchboards and Panelboards

26 24 19 – Motor-Control Centers

26 24 19.40 Motor Starters and Controls		Crew	Daily Output	Labor-Hours	Unit	2009 Bare Costs Material	Labor	Equipment	Total	Total Incl O&P
8380	100 HP motor	2 Elec	.80	20	Ea.	3,675	940		4,615	5,500
8390	125 HP motor		.70	22.857		8,500	1,075		9,575	11,000
8400	150 HP motor		.70	22.857		8,500	1,075		9,575	11,000
8410	200 HP motor		.70	22.857		8,500	1,075		9,575	11,000
8430	Starter & circuit breaker disconnect									
8440	600 volt, 3 pole, 5 HP motor	1 Elec	1.40	5.714	Ea.	860	269		1,129	1,350
8450	10 HP motor	"	1.10	7.273		890	340		1,230	1,500
8460	25 HP motor	2 Elec	1.50	10.667		1,125	500		1,625	2,025
8470	30 HP motor		1.30	12.308		1,475	580		2,055	2,500
8480	40 HP motor		1.30	12.308		1,475	580		2,055	2,500
8490	50 HP motor		1.30	12.308		1,475	580		2,055	2,500
8500	60 HP motor		.92	17.391		3,500	815		4,315	5,100
8510	75 HP motor		.92	17.391		3,500	815		4,315	5,100
8520	100 HP motor		.84	19.048		3,500	895		4,395	5,225
8530	125 HP motor		.70	22.857		6,050	1,075		7,125	8,300
8540	150 HP motor		.70	22.857		6,050	1,075		7,125	8,300
8550	200 HP motor		.60	26.667		7,200	1,250		8,450	9,825
8900	240 volt, 1-2 pole, .75 HP motor	1 Elec	2	4		815	188		1,003	1,175
8910	2 HP motor		2	4		835	188		1,023	1,200
8920	5 HP motor		1.80	4.444		935	209		1,144	1,350
8930	10 HP motor		1.40	5.714		1,425	269		1,694	1,975
8950	3 pole, .75 HP motor		1.60	5		815	235		1,050	1,250
8970	5 HP motor		1.40	5.714		890	269		1,159	1,400
8980	10 HP motor		1.10	7.273		1,125	340		1,465	1,775
8990	15 HP motor		1	8		1,125	375		1,500	1,825
9100	20 HP motor	2 Elec	1.50	10.667		1,550	500		2,050	2,475
9110	25 HP motor		1.50	10.667		1,550	500		2,050	2,475
9120	30 HP motor		1.30	12.308		1,550	580		2,130	2,600
9130	40 HP motor		1.24	12.903		3,500	605		4,105	4,775
9140	50 HP motor		1.12	14.286		3,500	670		4,170	4,875
9150	60 HP motor		.90	17.778		7,200	835		8,035	9,175
9160	75 HP motor		.76	21.053		7,200	990		8,190	9,425
9170	100 HP motor		.70	22.857		7,200	1,075		8,275	9,550
9180	125 HP motor		.60	26.667		16,400	1,250		17,650	20,000
9190	150 HP motor		.52	30.769		16,400	1,450		17,850	20,300
9200	200 HP motor		.50	32		16,400	1,500		17,900	20,400

26 25 Enclosed Bus Assemblies

26 25 13 – Bus Duct/Busway and Fittings

26 25 13.10 Aluminum Bus Duct		Crew	Daily Output	Labor-Hours	Unit	Material	Labor	Equipment	Total	Total Incl O&P
0010	**ALUMINUM BUS DUCT** 10 ft. long									
0050	Indoor 3 pole 4 wire, plug-in, straight section, 225 amp	2 Elec	44	.364	L.F.	166	17.10		183.10	209
0100	400 amp		36	.444		166	21		187	214
0150	600 amp		32	.500		166	23.50		189.50	218
0200	800 amp		26	.615		191	29		220	256
0250	1000 amp		24	.667		217	31.50		248.50	287
0300	1350 amp		22	.727		330	34		364	420
0310	1600 amp		18	.889		385	42		427	485
0320	2000 amp		16	1		445	47		492	565
0330	2500 amp		14	1.143		575	53.50		628.50	715
0340	3000 amp		12	1.333		650	62.50		712.50	810

26 25 Enclosed Bus Assemblies

26 25 13 – Bus Duct/Busway and Fittings

26 25 13.10 Aluminum Bus Duct		Crew	Daily Output	Labor-Hours	Unit	2009 Bare Costs Material	Labor	Equipment	Total	Total Incl O&P
0350	Feeder, 600 amp	2 Elec	34	.471	L.F.	153	22		175	202
0400	800 amp		28	.571		179	27		206	239
0450	1000 amp		26	.615		204	29		233	270
0455	1200 amp		25	.640		204	30		234	272
0500	1350 amp		24	.667		320	31.50		351.50	400
0550	1600 amp		20	.800		370	37.50		407.50	465
0600	2000 amp		18	.889		435	42		477	540
0620	2500 amp		14	1.143		560	53.50		613.50	705
0630	3000 amp		12	1.333		640	62.50		702.50	795
0640	4000 amp		10	1.600		920	75		995	1,125
0650	Elbow, 225 amp		4.40	3.636	Ea.	900	171		1,071	1,250
0700	400 amp		3.80	4.211		900	198		1,098	1,300
0750	600 amp		3.40	4.706		900	221		1,121	1,325
0800	800 amp		3	5.333		940	251		1,191	1,400
0850	1000 amp		2.80	5.714		1,225	269		1,494	1,775
0870	1200 amp		2.70	5.926		1,500	279		1,779	2,075
0900	1350 amp		2.60	6.154		1,625	289		1,914	2,225
0950	1600 amp		2.40	6.667		1,925	315		2,240	2,600
1000	2000 amp		2	8		2,125	375		2,500	2,900
1020	2500 amp		1.80	8.889		2,500	420		2,920	3,400
1030	3000 amp		1.60	10		2,875	470		3,345	3,900
1040	4000 amp		1.40	11.429		4,650	535		5,185	5,950
1100	Cable tap box end, 225 amp		3.60	4.444		1,275	209		1,484	1,725
1150	400 amp		3.20	5		1,275	235		1,510	1,750
1200	600 amp		2.60	6.154		1,275	289		1,564	1,850
1250	800 amp		2.20	7.273		1,375	340		1,715	2,050
1300	1000 amp		2	8		1,500	375		1,875	2,225
1320	1200 amp		2	8		1,800	375		2,175	2,550
1350	1350 amp		1.60	10		1,950	470		2,420	2,875
1400	1600 amp		1.40	11.429		2,200	535		2,735	3,250
1450	2000 amp		1.20	13.333		2,525	625		3,150	3,750
1460	2500 amp		1	16		3,075	750		3,825	4,525
1470	3000 amp		.80	20		3,400	940		4,340	5,175
1480	4000 amp		.60	26.667		4,525	1,250		5,775	6,900
1500	Switchboard stub, 225 amp		5.80	2.759		1,125	130		1,255	1,425
1550	400 amp		5.40	2.963		1,125	139		1,264	1,450
1600	600 amp		4.60	3.478		1,125	163		1,288	1,475
1650	800 amp		4	4		1,275	188		1,463	1,700
1700	1000 amp		3.20	5		1,425	235		1,660	1,900
1720	1200 amp		3.10	5.161		1,550	243		1,793	2,100
1750	1350 amp		3	5.333		1,750	251		2,001	2,300
1800	1600 amp		2.60	6.154		1,925	289		2,214	2,575
1850	2000 amp		2.40	6.667		2,225	315		2,540	2,925
1860	2500 amp		2.20	7.273		2,650	340		2,990	3,425
1870	3000 amp		2	8		3,075	375		3,450	3,975
1880	4000 amp		1.80	8.889		3,850	420		4,270	4,900
1890	Tee fittings, 225 amp		3.20	5		1,200	235		1,435	1,675
1900	400 amp		2.80	5.714		1,200	269		1,469	1,750
1950	600 amp		2.60	6.154		1,200	289		1,489	1,775
2000	800 amp		2.40	6.667		1,275	315		1,590	1,900
2050	1000 amp		2.20	7.273		1,375	340		1,715	2,025
2070	1200 amp		2.10	7.619		1,600	360		1,960	2,300
2100	1350 amp		2	8		2,200	375		2,575	3,000

26 25 13 – Bus Duct/Busway and Fittings

26 25 13.10 Aluminum Bus Duct		Crew	Daily Output	Labor-Hours	Unit	Material	Labor	Equipment	Total	Total Incl O&P
						2009 Bare Costs				
2150	1600 amp	2 Elec	1.60	10	Ea.	2,650	470		3,120	3,650
2200	2000 amp		1.20	13.333		2,950	625		3,575	4,200
2220	2500 amp		1	16		3,525	750		4,275	5,025
2230	3000 amp		.80	20		4,000	940		4,940	5,850
2240	4000 amp		.60	26.667		6,675	1,250		7,925	9,250
2300	Wall flange, 600 amp		20	.800		150	37.50		187.50	223
2310	800 amp		16	1		150	47		197	238
2320	1000 amp		13	1.231		150	58		208	254
2325	1200 amp		12	1.333		150	62.50		212.50	262
2330	1350 amp		10.80	1.481		150	69.50		219.50	272
2340	1600 amp		9	1.778		150	83.50		233.50	294
2350	2000 amp		8	2		150	94		244	310
2360	2500 amp		6.60	2.424		150	114		264	340
2370	3000 amp		5.40	2.963		150	139		289	380
2380	4000 amp		4	4		150	188		338	455
2390	5000 amp		3	5.333		150	251		401	550
2400	Vapor barrier		8	2		335	94		429	515
2420	Roof flange kit		4	4		740	188		928	1,100
2600	Expansion fitting, 225 amp		10	1.600		1,475	75		1,550	1,750
2610	400 amp		8	2		1,475	94		1,569	1,775
2620	600 amp		6	2.667		1,475	125		1,600	1,825
2630	800 amp		4.60	3.478		1,725	163		1,888	2,150
2640	1000 amp		4	4		1,925	188		2,113	2,425
2650	1350 amp		3.60	4.444		2,825	209		3,034	3,425
2660	1600 amp		3.20	5		3,375	235		3,610	4,050
2670	2000 amp		2.80	5.714		3,750	269		4,019	4,550
2680	2500 amp		2.40	6.667		4,550	315		4,865	5,475
2690	3000 amp		2	8		5,225	375		5,600	6,325
2700	4000 amp		1.60	10		6,875	470		7,345	8,300
2800	Reducer nonfused, 400 amp		8	2		1,175	94		1,269	1,450
2810	600 amp		6	2.667		1,175	125		1,300	1,500
2820	800 amp		4.60	3.478		1,425	163		1,588	1,825
2830	1000 amp		4	4		1,650	188		1,838	2,125
2840	1350 amp		3.60	4.444		2,175	209		2,384	2,725
2850	1600 amp		3.20	5		2,950	235		3,185	3,600
2860	2000 amp		2.80	5.714		3,375	269		3,644	4,150
2870	2500 amp		2.40	6.667		4,250	315		4,565	5,150
2880	3000 amp		2	8		4,925	375		5,300	6,000
2890	4000 amp		1.60	10		6,575	470		7,045	7,950
2950	Reducer fuse included, 225 amp		4.40	3.636		3,475	171		3,646	4,100
2960	400 amp		4.20	3.810		3,550	179		3,729	4,175
2970	600 amp		3.60	4.444		4,175	209		4,384	4,925
2980	800 amp		3.20	5		6,650	235		6,885	7,650
2990	1000 amp		3	5.333		7,600	251		7,851	8,725
3000	1200 amp		2.80	5.714		7,600	269		7,869	8,775
3010	1600 amp		2.20	7.273		17,300	340		17,640	19,600
3020	2000 amp		1.80	8.889		19,200	420		19,620	21,800
3100	Reducer circuit breaker, 225 amp		4.40	3.636		3,425	171		3,596	4,050
3110	400 amp		4.20	3.810		4,175	179		4,354	4,875
3120	600 amp		3.60	4.444		5,925	209		6,134	6,850
3130	800 amp		3.20	5		6,950	235		7,185	8,000
3140	1000 amp		3	5.333		7,875	251		8,126	9,050
3150	1200 amp		2.80	5.714		9,475	269		9,744	10,800

26 25 13 – Bus Duct/Busway and Fittings

26 25 13.10 Aluminum Bus Duct		Crew	Daily Output	Labor-Hours	Unit	Material	Labor	Equipment	Total	Total Incl O&P
						2009 Bare Costs				
3160	1600 amp	2 Elec	2.20	7.273	Ea.	14,000	340		14,340	15,900
3170	2000 amp		1.80	8.889		15,300	420		15,720	17,500
3250	Reducer circuit breaker, 75,000 AIC, 225 amp		4.40	3.636		5,350	171		5,521	6,175
3260	400 amp		4.20	3.810		5,350	179		5,529	6,175
3270	600 amp		3.60	4.444		7,175	209		7,384	8,200
3280	800 amp		3.20	5		7,875	235		8,110	9,000
3290	1000 amp		3	5.333		12,700	251		12,951	14,300
3300	1200 amp		2.80	5.714		12,700	269		12,969	14,300
3310	1600 amp		2.20	7.273		14,000	340		14,340	15,900
3320	2000 amp		1.80	8.889		15,300	420		15,720	17,500
3400	Reducer circuit breaker CLF 225 amp		4.40	3.636		5,500	171		5,671	6,325
3410	400 amp		4.20	3.810		6,550	179		6,729	7,475
3420	600 amp		3.60	4.444		9,800	209		10,009	11,100
3430	800 amp		3.20	5		10,200	235		10,435	11,600
3440	1000 amp		3	5.333		10,700	251		10,951	12,100
3450	1200 amp		2.80	5.714		13,900	269		14,169	15,700
3460	1600 amp		2.20	7.273		14,000	340		14,340	15,900
3470	2000 amp		1.80	8.889		15,300	420		15,720	17,500
3550	Ground bus added to bus duct, 225 amp		320	.050	L.F.	45	2.35		47.35	53.50
3560	400 amp		320	.050		45	2.35		47.35	53.50
3570	600 amp		280	.057		45	2.69		47.69	54
3580	800 amp		240	.067		45	3.13		48.13	55
3590	1000 amp		200	.080		45	3.76		48.76	56
3600	1350 amp		180	.089		45	4.18		49.18	56.50
3610	1600 amp		160	.100		45	4.70		49.70	57.50
3620	2000 amp		160	.100		45	4.70		49.70	57.50
3630	2500 amp		140	.114		45	5.35		50.35	58.50
3640	3000 amp		120	.133		45	6.25		51.25	59.50
3650	4000 amp		100	.160		45	7.50		52.50	61.50
3810	High short circuit, 400 amp		36	.444		153	21		174	200
3820	600 amp		32	.500		153	23.50		176.50	204
3830	800 amp		26	.615		179	29		208	242
3840	1000 amp		24	.667		191	31.50		222.50	259
3850	1350 amp		22	.727		255	34		289	335
3860	1600 amp		18	.889		293	42		335	390
3870	2000 amp		16	1		345	47		392	455
3880	2500 amp		14	1.143		575	53.50		628.50	715
3890	3000 amp		12	1.333		650	62.50		712.50	810
3920	Cross, 225 amp		5.60	2.857	Ea.	1,825	134		1,959	2,200
3930	400 amp		4.60	3.478		1,825	163		1,988	2,250
3940	600 amp		4	4		1,825	188		2,013	2,300
3950	800 amp		3.40	4.706		1,900	221		2,121	2,450
3960	1000 amp		3	5.333		2,000	251		2,251	2,575
3970	1350 amp		2.80	5.714		3,250	269		3,519	4,000
3980	1600 amp		2.20	7.273		3,850	340		4,190	4,775
3990	2000 amp		1.80	8.889		4,225	420		4,645	5,300
4000	2500 amp		1.60	10		5,025	470		5,495	6,250
4010	3000 amp		1.20	13.333		5,775	625		6,400	7,325
4020	4000 amp		1	16		8,825	750		9,575	10,900
4040	Cable tap box center, 225 amp		3.60	4.444		1,275	209		1,484	1,725
4050	400 amp		3.20	5		1,275	235		1,510	1,750
4060	600 amp		2.60	6.154		1,275	289		1,564	1,850
4070	800 amp		2.20	7.273		1,375	340		1,715	2,050

26 25 Enclosed Bus Assemblies

26 25 13 – Bus Duct/Busway and Fittings

26 25 13.10 Aluminum Bus Duct		Crew	Daily Output	Labor-Hours	Unit	Material	2009 Bare Costs Labor	Equipment	Total	Total Incl O&P
4080	1000 amp	2 Elec	2	8	Ea.	1,500	375		1,875	2,225
4090	1350 amp		1.60	10		1,950	470		2,420	2,875
4100	1600 amp		1.40	11.429		2,200	535		2,735	3,250
4110	2000 amp		1.20	13.333		2,525	625		3,150	3,750
4120	2500 amp		1	16		3,075	750		3,825	4,525
4130	3000 amp		.80	20		3,400	940		4,340	5,175
4140	4000 amp		.60	26.667		4,525	1,250		5,775	6,900
4500	Weatherproof 3 pole 4 wire, feeder, 600 amp		30	.533	L.F.	183	25		208	241
4520	800 amp		24	.667		215	31.50		246.50	284
4540	1000 amp		22	.727		245	34		279	320
4550	1200 amp		21	.762		340	36		376	430
4560	1350 amp		20	.800		385	37.50		422.50	480
4580	1600 amp		17	.941		445	44		489	560
4600	2000 amp		16	1		520	47		567	650
4620	2500 amp		12	1.333		675	62.50		737.50	835
4640	3000 amp		10	1.600		765	75		840	955
4660	4000 amp		8	2		1,100	94		1,194	1,350
5000	Indoor 3 pole, 3 wire, feeder, 600 amp		40	.400		140	18.80		158.80	183
5010	800 amp		32	.500		166	23.50		189.50	218
5020	1000 amp		30	.533		179	25		204	236
5025	1200 amp		29	.552		179	26		205	237
5030	1350 amp		28	.571		242	27		269	310
5040	1600 amp		24	.667		281	31.50		312.50	360
5050	2000 amp		20	.800		330	37.50		367.50	425
5060	2500 amp		16	1		460	47		507	580
5070	3000 amp		14	1.143		550	53.50		603.50	690
5080	4000 amp		12	1.333		665	62.50		727.50	825
5200	Plug-in type, 225 amp		50	.320		153	15.05		168.05	191
5210	400 amp		42	.381		153	17.90		170.90	196
5220	600 amp		36	.444		153	21		174	200
5230	800 amp		30	.533		179	25		204	236
5240	1000 amp		28	.571		191	27		218	253
5245	1200 amp		27	.593		217	28		245	282
5250	1350 amp		26	.615		255	29		284	325
5260	1600 amp		20	.800		293	37.50		330.50	385
5270	2000 amp		18	.889		345	42		387	445
5280	2500 amp		16	1		470	47		517	595
5290	3000 amp		14	1.143		560	53.50		613.50	705
5300	4000 amp		12	1.333		675	62.50		737.50	840
5330	High short circuit, 400 amp		42	.381		153	17.90		170.90	196
5340	600 amp		36	.444		153	21		174	200
5350	800 amp		30	.533		179	25		204	236
5360	1000 amp		28	.571		191	27		218	253
5370	1350 amp		26	.615		255	29		284	325
5380	1600 amp		20	.800		293	37.50		330.50	385
5390	2000 amp		18	.889		345	42		387	445
5400	2500 amp		16	1		540	47		587	670
5410	3000 amp		14	1.143		560	53.50		613.50	705
5440	Elbow, 225 amp		5	3.200	Ea.	780	150		930	1,075
5450	400 amp		4.40	3.636		780	171		951	1,125
5460	600 amp		4	4		780	188		968	1,150
5470	800 amp		3.40	4.706		830	221		1,051	1,250
5480	1000 amp		3.20	5		855	235		1,090	1,300

26 25 13.10 Aluminum Bus Duct		Crew	Daily Output	Labor-Hours	Unit	Material	2009 Bare Costs Labor	Equipment	Total	Total Incl O&P
5485	1200 amp	2 Elec	3.10	5.161	Ea.	905	243		1,148	1,375
5490	1350 amp		3	5.333		985	251		1,236	1,450
5500	1600 amp		2.80	5.714		1,500	269		1,769	2,075
5510	2000 amp		2.40	6.667		1,650	315		1,965	2,300
5520	2500 amp		2	8		2,050	375		2,425	2,825
5530	3000 amp		1.80	8.889		2,425	420		2,845	3,325
5540	4000 amp		1.60	10		3,225	470		3,695	4,275
5560	Tee fittings, 225 amp		3.60	4.444		1,075	209		1,284	1,525
5570	400 amp		3.20	5		1,075	235		1,310	1,550
5580	600 amp		3	5.333		1,075	251		1,326	1,575
5590	800 amp		2.80	5.714		1,150	269		1,419	1,700
5600	1000 amp		2.60	6.154		1,200	289		1,489	1,775
5605	1200 amp		2.50	6.400		1,275	300		1,575	1,875
5610	1350 amp		2.40	6.667		1,775	315		2,090	2,425
5620	1600 amp		1.80	8.889		2,100	420		2,520	2,950
5630	2000 amp		1.40	11.429		2,325	535		2,860	3,400
5640	2500 amp		1.20	13.333		2,900	625		3,525	4,175
5650	3000 amp		1	16		3,400	750		4,150	4,900
5660	4000 amp		.70	22.857		5,000	1,075		6,075	7,125
5680	Cross, 225 amp		6.40	2.500		1,750	118		1,868	2,100
5690	400 amp		5.40	2.963		1,750	139		1,889	2,150
5700	600 amp		4.60	3.478		1,750	163		1,913	2,175
5710	800 amp		4	4		1,850	188		2,038	2,350
5720	1000 amp		3.60	4.444		1,900	209		2,109	2,425
5730	1350 amp		3.20	5		2,850	235		3,085	3,475
5740	1600 amp		2.60	6.154		3,325	289		3,614	4,100
5750	2000 amp		2.20	7.273		3,625	340		3,965	4,525
5760	2500 amp		1.80	8.889		4,375	420		4,795	5,475
5770	3000 amp		1.40	11.429		5,225	535		5,760	6,575
5780	4000 amp		1.20	13.333		7,275	625		7,900	8,975
5800	Expansion fitting, 225 amp		11.60	1.379		1,275	65		1,340	1,500
5810	400 amp		9.20	1.739		1,275	81.50		1,356.50	1,525
5820	600 amp		7	2.286		1,275	107		1,382	1,575
5830	800 amp		5.20	3.077		1,500	145		1,645	1,875
5840	1000 amp		4.60	3.478		1,650	163		1,813	2,050
5850	1350 amp		4.20	3.810		2,000	179		2,179	2,475
5860	1600 amp		3.60	4.444		2,475	209		2,684	3,050
5870	2000 amp		3.20	5		2,925	235		3,160	3,575
5880	2500 amp		2.80	5.714		3,325	269		3,594	4,075
5890	3000 amp		2.40	6.667		3,875	315		4,190	4,750
5900	4000 amp		1.80	8.889		5,325	420		5,745	6,500
5940	Reducer, nonfused, 400 amp		9.20	1.739		1,025	81.50		1,106.50	1,250
5950	600 amp		7	2.286		1,025	107		1,132	1,300
5960	800 amp		5.20	3.077		1,075	145		1,220	1,400
5970	1000 amp		4.60	3.478		1,325	163		1,488	1,700
5980	1350 amp		4.20	3.810		1,950	179		2,129	2,425
5990	1600 amp		3.60	4.444		2,225	209		2,434	2,775
6000	2000 amp		3.20	5		2,600	235		2,835	3,200
6010	2500 amp		2.80	5.714		3,325	269		3,594	4,075
6020	3000 amp		2.20	7.273		3,875	340		4,215	4,800
6030	4000 amp		1.80	8.889		5,325	420		5,745	6,500
6050	Reducer, fuse included, 225 amp		5	3.200		2,675	150		2,825	3,150
6060	400 amp	↓	4.80	3.333	↓	3,550	157		3,707	4,150

26 25 13 – Bus Duct/Busway and Fittings

26 25 13.10 Aluminum Bus Duct		Crew	Daily Output	Labor-Hours	Unit	Material	2009 Bare Costs Labor	Equipment	Total	Total Incl O&P
6070	600 amp	2 Elec	4.20	3.810	Ea.	4,525	179		4,704	5,250
6080	800 amp		3.60	4.444		7,025	209		7,234	8,050
6090	1000 amp		3.40	4.706		7,650	221		7,871	8,775
6100	1350 amp		3.20	5		14,000	235		14,235	15,800
6110	1600 amp		2.60	6.154		16,600	289		16,889	18,600
6120	2000 amp		2	8		19,200	375		19,575	21,800
6160	Reducer, circuit breaker, 225 amp		5	3.200		3,275	150		3,425	3,825
6170	400 amp		4.80	3.333		4,025	157		4,182	4,675
6180	600 amp		4.20	3.810		5,775	179		5,954	6,625
6190	800 amp		3.60	4.444		6,725	209		6,934	7,725
6200	1000 amp		3.40	4.706		7,650	221		7,871	8,775
6210	1350 amp		3.20	5		9,250	235		9,485	10,600
6220	1600 amp		2.60	6.154		13,800	289		14,089	15,600
6230	2000 amp		2	8		15,100	375		15,475	17,200
6270	Cable tap box center, 225 amp		4.20	3.810		1,225	179		1,404	1,625
6280	400 amp		3.60	4.444		1,225	209		1,434	1,675
6290	600 amp		3	5.333		1,225	251		1,476	1,725
6300	800 amp		2.60	6.154		1,325	289		1,614	1,925
6310	1000 amp		2.40	6.667		1,425	315		1,740	2,025
6320	1350 amp		1.80	8.889		1,700	420		2,120	2,525
6330	1600 amp		1.60	10		1,900	470		2,370	2,825
6340	2000 amp		1.40	11.429		2,200	535		2,735	3,225
6350	2500 amp		1.20	13.333		2,725	625		3,350	3,975
6360	3000 amp		1	16		3,100	750		3,850	4,550
6370	4000 amp		.70	22.857		3,675	1,075		4,750	5,675
6390	Cable tap box end, 225 amp		4.20	3.810		745	179		924	1,100
6400	400 amp		3.60	4.444		745	209		954	1,150
6410	600 amp		3	5.333		745	251		996	1,200
6420	800 amp		2.60	6.154		815	289		1,104	1,350
6430	1000 amp		2.40	6.667		880	315		1,195	1,450
6435	1200 amp		2.10	7.619		955	360		1,315	1,600
6440	1350 amp		1.80	8.889		1,075	420		1,495	1,825
6450	1600 amp		1.60	10		1,200	470		1,670	2,050
6460	2000 amp		1.40	11.429		1,350	535		1,885	2,325
6470	2500 amp		1.20	13.333		1,575	625		2,200	2,700
6480	3000 amp		1	16		1,875	750		2,625	3,200
6490	4000 amp		.70	22.857		2,225	1,075		3,300	4,100
7000	Weatherproof 3 pole 3 wire, feeder, 600 amp		34	.471	L.F.	181	22		203	233
7020	800 amp		28	.571		200	27		227	261
7040	1000 amp		26	.615		215	29		244	281
7050	1200 amp		25	.640		251	30		281	325
7060	1350 amp		24	.667		291	31.50		322.50	370
7080	1600 amp		20	.800		335	37.50		372.50	430
7100	2000 amp		18	.889		400	42		442	505
7120	2500 amp		14	1.143		550	53.50		603.50	690
7140	3000 amp		12	1.333		660	62.50		722.50	820
7160	4000 amp		10	1.600		795	75		870	990

26 25 13.20 Bus Duct		Crew	Daily Output	Labor-Hours	Unit	Material	Labor	Equipment	Total	Total Incl O&P
0010	**BUS DUCT** 100 amp and less, aluminum or copper, plug-in									
0080	Bus duct, 3 pole 3 wire, 100 amp	1 Elec	42	.190	L.F.	28	8.95		36.95	44.50
0110	Elbow		4	2	Ea.	72	94		166	224
0120	Tee		2	4		104	188		292	405

26 25 13 – Bus Duct/Busway and Fittings

26 25 13.20 Bus Duct		Crew	Daily Output	Labor-Hours	Unit	2009 Bare Costs Material	Labor	Equipment	Total	Total Incl O&P
0130	Wall flange	1 Elec	8	1	Ea.	9.30	47		56.30	82.50
0140	Ground kit		16	.500		19.70	23.50		43.20	57.50
0180	3 pole 4 wire, 100 amp		40	.200	L.F.	32	9.40		41.40	49.50
0200	Cable tap box		3.10	2.581	Ea.	87	121		208	283
0300	End closure		16	.500		11.60	23.50		35.10	49
0400	Elbow		4	2		72	94		166	224
0500	Tee		2	4		104	188		292	405
0600	Hangers		10	.800		6.95	37.50		44.45	65.50
0700	Circuit breakers, 15 to 50 amp, 1 pole		8	1		186	47		233	277
0800	15 to 60 amp, 2 pole		6.70	1.194		725	56		781	885
0900	3 pole		5.30	1.509		725	71		796	910
1000	60 to 100 amp, 1 pole		6.70	1.194		725	56		781	885
1100	70 to 100 amp, 2 pole		5.30	1.509		725	71		796	910
1200	3 pole		4.50	1.778		725	83.50		808.50	930
1220	Switch, nonfused, 3 pole, 4 wire		8	1		98.50	47		145.50	181
1240	Fused, 3 fuses, 4 wire, 30 amp		8	1		355	47		402	470
1260	60 amp		5.30	1.509		375	71		446	520
1280	100 amp		4.50	1.778		625	83.50		708.50	815
1300	Plug, fusible, 3 pole 250 volt, 30 amp		5.30	1.509		355	71		426	505
1310	60 amp		5.30	1.509		425	71		496	575
1320	100 amp		4.50	1.778		625	83.50		708.50	815
1330	3 pole 480 volt, 30 amp		5.30	1.509		410	71		481	560
1340	60 amp		5.30	1.509		440	71		511	595
1350	100 amp		4.50	1.778		640	83.50		723.50	830
1360	Circuit breaker, 3 pole 250 volt, 60 amp		5.30	1.509		725	71		796	910
1370	3 pole 480 volt, 100 amp		4.50	1.778		725	83.50		808.50	930
2000	Bus duct, 2 wire, 250 volt, 30 amp		60	.133	L.F.	3.29	6.25		9.54	13.25
2100	60 amp		50	.160		3.29	7.50		10.79	15.15
2200	300 volt, 30 amp		60	.133		3.29	6.25		9.54	13.25
2300	60 amp		50	.160		3.29	7.50		10.79	15.15
2400	3 wire, 250 volt, 30 amp		60	.133		3.29	6.25		9.54	13.25
2500	60 amp		50	.160		3.29	7.50		10.79	15.15
2600	480/277 volt, 30 amp		60	.133		3.29	6.25		9.54	13.25
2700	60 amp		50	.160		3.29	7.50		10.79	15.15
2750	End feed, 300 volt 2 wire max. 30 amp		6	1.333	Ea.	60.50	62.50		123	163
2800	60 amp		5.50	1.455		60.50	68.50		129	172
2850	30 amp miniature		6	1.333		60.50	62.50		123	163
2900	3 wire, 30 amp		6	1.333		76	62.50		138.50	180
2950	60 amp		5.50	1.455		76	68.50		144.50	189
3000	30 amp miniature		6	1.333		76	62.50		138.50	180
3050	Center feed, 300 volt 2 wire, 30 amp		6	1.333		83.50	62.50		146	189
3100	60 amp		5.50	1.455		83.50	68.50		152	197
3150	3 wire, 30 amp		6	1.333		95	62.50		157.50	202
3200	60 amp		5.50	1.455		95	68.50		163.50	210
3220	Elbow, 30 amp		6	1.333		58	62.50		120.50	161
3240	60 amp		5.50	1.455		58	68.50		126.50	169
3260	End cap		40	.200		11.60	9.40		21	27
3280	Strength beam, 10 ft.		15	.533		32.50	25		57.50	74
3300	Hanger		24	.333		6.95	15.65		22.60	31.50
3320	Tap box, nonfusible		6.30	1.270		98.50	59.50		158	200
3340	Fusible switch 30 amp, 1 fuse		6	1.333		430	62.50		492.50	565
3360	2 fuse		6	1.333		435	62.50		497.50	575
3380	3 fuse		6	1.333		440	62.50		502.50	580

26 25 Enclosed Bus Assemblies

26 25 13 – Bus Duct/Busway and Fittings

26 25 13.20 Bus Duct

	26 25 13.20 Bus Duct	Crew	Daily Output	Labor-Hours	Unit	Material	2009 Bare Costs Labor	Equipment	Total	Total Incl O&P
3400	Circuit breaker handle on cover, 1 pole	1 Elec	6	1.333	Ea.	69.50	62.50		132	173
3420	2 pole		6	1.333		75.50	62.50		138	180
3440	3 pole		6	1.333		98.50	62.50		161	205
3460	Circuit breaker external operhandle, 1 pole		6	1.333		69.50	62.50		132	173
3480	2 pole		6	1.333		75.50	62.50		138	180
3500	3 pole		6	1.333		98.50	62.50		161	205
3520	Terminal plug only		16	.500		12.75	23.50		36.25	50
3540	Terminal with receptacle		16	.500		16.55	23.50		40.05	54
3560	Fixture plug		16	.500		11.35	23.50		34.85	48.50
4000	Copper bus duct, lighting, 2 wire 300 volt, 20 amp		70	.114	L.F.	3.60	5.35		8.95	12.20
4020	35 amp		60	.133		3.60	6.25		9.85	13.60
4040	50 amp		55	.145		3.60	6.85		10.45	14.45
4060	60 amp		50	.160		3.60	7.50		11.10	15.50
4080	3 wire 300 volt, 20 amp		70	.114		3.93	5.35		9.28	12.55
4100	35 amp		60	.133		3.93	6.25		10.18	13.95
4120	50 amp		55	.145		3.93	6.85		10.78	14.80
4140	60 amp		50	.160		3.93	7.50		11.43	15.85
4160	Feeder in box, end, 1 circuit		6	1.333	Ea.	69.50	62.50		132	173
4180	2 circuit		5.50	1.455		72	68.50		140.50	184
4200	Center, 1 circuit		6	1.333		95	62.50		157.50	202
4220	2 circuit		5.50	1.455		98.50	68.50		167	213
4240	End cap		40	.200		11.60	9.40		21	27
4260	Hanger, surface mount		24	.333		6.95	15.65		22.60	31.50
4280	Coupling		40	.200		8.80	9.40		18.20	24

26 25 13.30 Copper Bus Duct

	26 25 13.30 Copper Bus Duct	Crew	Daily Output	Labor-Hours	Unit	Material	Labor	Equipment	Total	Total Incl O&P
0010	**COPPER BUS DUCT**									
0100	Weatherproof 3 pole 4 wire, feeder duct, 600 amp	2 Elec	24	.667	L.F.	195	31.50		226.50	262
0110	800 amp		18	.889		237	42		279	325
0120	1000 amp		17	.941		264	44		308	360
0125	1200 amp		16.50	.970		320	45.50		365.50	425
0130	1350 amp		16	1		390	47		437	505
0140	1600 amp		12	1.333		445	62.50		507.50	585
0150	2000 amp		10	1.600		570	75		645	745
0160	2500 amp		7	2.286		710	107		817	945
0170	3000 amp		5	3.200		865	150		1,015	1,175
0180	4000 amp		3.60	4.444		1,150	209		1,359	1,575
0200	Indoor 3 pole 4 wire, plug-in, bus duct high short circuit, 400 amp		32	.500		186	23.50		209.50	240
0210	600 amp		26	.615		186	29		215	249
0220	800 amp		20	.800		220	37.50		257.50	300
0230	1000 amp		18	.889		244	42		286	335
0240	1350 amp		16	1		350	47		397	460
0250	1600 amp		12	1.333		395	62.50		457.50	530
0260	2000 amp		10	1.600		500	75		575	665
0270	2500 amp		8	2		615	94		709	820
0280	3000 amp		6	2.667		740	125		865	1,000
0310	Cross, 225 amp		3	5.333	Ea.	2,200	251		2,451	2,800
0320	400 amp		2.80	5.714		2,200	269		2,469	2,850
0330	600 amp		2.60	6.154		2,200	289		2,489	2,875
0340	800 amp		2.20	7.273		2,375	340		2,715	3,150
0350	1000 amp		2	8		2,675	375		3,050	3,500
0360	1350 amp		1.80	8.889		3,000	420		3,420	3,975
0370	1600 amp		1.70	9.412		3,350	440		3,790	4,350

26 25 13.30 Copper Bus Duct		Crew	Daily Output	Labor-Hours	Unit	Material	Labor (2009 Bare Costs)	Equipment	Total	Total Incl O&P
0380	2000 amp	2 Elec	1.60	10	Ea.	5,500	470		5,970	6,775
0390	2500 amp		1.40	11.429		6,650	535		7,185	8,150
0400	3000 amp		1.20	13.333		7,675	625		8,300	9,425
0410	4000 amp		1	16		10,000	750		10,750	12,200
0430	Expansion fitting, 225 amp		5.40	2.963		1,150	139		1,289	1,500
0440	400 amp		4.60	3.478		1,300	163		1,463	1,675
0450	600 amp		4	4		1,625	188		1,813	2,075
0460	800 amp		3.40	4.706		1,900	221		2,121	2,450
0470	1000 amp		3	5.333		2,175	251		2,426	2,775
0480	1350 amp		2.80	5.714		2,750	269		3,019	3,450
0490	1600 amp		2.60	6.154		3,825	289		4,114	4,650
0500	2000 amp		2.20	7.273		4,400	340		4,740	5,375
0510	2500 amp		1.80	8.889		5,350	420		5,770	6,550
0520	3000 amp		1.60	10		6,325	470		6,795	7,700
0530	4000 amp		1.20	13.333		8,150	625		8,775	9,950
0550	Reducer nonfused, 225 amp		5.40	2.963		1,325	139		1,464	1,675
0560	400 amp		4.60	3.478		1,325	163		1,488	1,700
0570	600 amp		4	4		1,325	188		1,513	1,750
0580	800 amp		3.40	4.706		1,600	221		1,821	2,100
0590	1000 amp		3	5.333		1,925	251		2,176	2,500
0600	1350 amp		2.80	5.714		2,900	269		3,169	3,625
0610	1600 amp		2.60	6.154		3,350	289		3,639	4,125
0620	2000 amp		2.20	7.273		4,025	340		4,365	4,950
0630	2500 amp		1.80	8.889		5,075	420		5,495	6,225
0640	3000 amp		1.60	10		6,025	470		6,495	7,375
0650	4000 amp		1.20	13.333		7,850	625		8,475	9,625
0670	Reducer fuse included, 225 amp		4.40	3.636		2,825	171		2,996	3,375
0680	400 amp		4.20	3.810		3,525	179		3,704	4,175
0690	600 amp		3.60	4.444		4,350	209		4,559	5,125
0700	800 amp		3.20	5		6,150	235		6,385	7,125
0710	1000 amp		3	5.333		7,700	251		7,951	8,850
0720	1350 amp		2.80	5.714		7,825	269		8,094	9,025
0730	1600 amp		2.20	7.273		17,100	340		17,440	19,300
0740	2000 amp		1.80	8.889		19,200	420		19,620	21,800
0790	Reducer, circuit breaker, 225 amp		4.40	3.636		3,575	171		3,746	4,200
0800	400 amp		4.20	3.810		4,175	179		4,354	4,875
0810	600 amp		3.60	4.444		5,925	209		6,134	6,850
0820	800 amp		3.20	5		6,950	235		7,185	8,000
0830	1000 amp		3	5.333		7,875	251		8,126	9,050
0840	1350 amp		2.80	5.714		9,475	269		9,744	10,800
0850	1600 amp		2.20	7.273		14,000	340		14,340	15,900
0860	2000 amp		1.80	8.889		15,300	420		15,720	17,500
0910	Cable tap box, center, 225 amp		3.20	5		1,325	235		1,560	1,825
0920	400 amp		2.60	6.154		1,325	289		1,614	1,925
0930	600 amp		2.20	7.273		1,325	340		1,665	2,000
0940	800 amp		2	8		1,475	375		1,850	2,200
0950	1000 amp		1.60	10		1,600	470		2,070	2,475
0960	1350 amp		1.40	11.429		2,025	535		2,560	3,050
0970	1600 amp		1.20	13.333		2,250	625		2,875	3,450
0980	2000 amp		1	16		2,700	750		3,450	4,125
1040	2500 amp		.80	20		3,200	940		4,140	4,975
1060	3000 amp		.60	26.667		3,700	1,250		4,950	6,000
1080	4000 amp		.40	40		4,650	1,875		6,525	8,025

26 25 13.30 Copper Bus Duct		Crew	Daily Output	Labor-Hours	Unit	2009 Bare Costs Material	Labor	Equipment	Total	Total Incl O&P
1800	Weatherproof 3 pole 3 wire, feeder duct, 600 amp	2 Elec	28	.571	L.F.	195	27		222	256
1820	800 amp		22	.727		237	34		271	315
1840	1000 amp		20	.800		264	37.50		301.50	350
1850	1200 amp		19	.842		274	39.50		313.50	360
1860	1350 amp		18	.889		390	42		432	495
1880	1600 amp		14	1.143		445	53.50		498.50	575
1900	2000 amp		12	1.333		570	62.50		632.50	725
1920	2500 amp		8	2		710	94		804	925
1940	3000 amp		6	2.667		865	125		990	1,150
1960	4000 amp		4	4		1,150	188		1,338	1,550
2000	Indoor 3 pole 3 wire, feeder duct, 600 amp		32	.500		162	23.50		185.50	215
2010	800 amp		26	.615		197	29		226	262
2020	1000 amp		24	.667		220	31.50		251.50	290
2025	1200 amp		22	.727		290	34		324	375
2030	1350 amp		20	.800		325	37.50		362.50	415
2040	1600 amp		16	1		370	47		417	485
2050	2000 amp		14	1.143		475	53.50		528.50	610
2060	2500 amp		10	1.600		590	75		665	765
2070	3000 amp		8	2		720	94		814	935
2080	4000 amp		6	2.667		950	125		1,075	1,250
2090	5000 amp		5	3.200		1,150	150		1,300	1,500
2200	Indoor 3 pole 3 wire, bus duct plug-in, 225 amp		46	.348		186	16.35		202.35	229
2210	400 amp		36	.444		186	21		207	236
2220	600 amp		30	.533		186	25		211	243
2230	800 amp		24	.667		220	31.50		251.50	290
2240	1000 amp		20	.800		244	37.50		281.50	325
2250	1350 amp		18	.889		350	42		392	450
2260	1600 amp		14	1.143		395	53.50		448.50	520
2270	2000 amp		12	1.333		500	62.50		562.50	645
2280	2500 amp		10	1.600		615	75		690	790
2290	3000 amp		8	2		740	94		834	960
2330	High short circuit, 400 amp		36	.444		186	21		207	236
2340	600 amp		30	.533		186	25		211	243
2350	800 amp		24	.667		220	31.50		251.50	290
2360	1000 amp		20	.800		244	37.50		281.50	325
2370	1350 amp		18	.889		350	42		392	450
2380	1600 amp		14	1.143		395	53.50		448.50	520
2390	2000 amp		12	1.333		500	62.50		562.50	645
2400	2500 amp		10	1.600		615	75		690	790
2410	3000 amp		8	2		740	94		834	960
2440	Elbows, 225 amp		4.60	3.478	Ea.	930	163		1,093	1,275
2450	400 amp		4.20	3.810		930	179		1,109	1,300
2460	600 amp		3.60	4.444		930	209		1,139	1,350
2470	800 amp		3.20	5		1,000	235		1,235	1,450
2480	1000 amp		3	5.333		1,050	251		1,301	1,525
2485	1200 amp		2.90	5.517		1,125	259		1,384	1,625
2490	1350 amp		2.80	5.714		1,250	269		1,519	1,800
2500	1600 amp		2.60	6.154		1,350	289		1,639	1,925
2510	2000 amp		2	8		1,625	375		2,000	2,375
2520	2500 amp		1.80	8.889		2,475	420		2,895	3,375
2530	3000 amp		1.60	10		2,875	470		3,345	3,875
2540	4000 amp		1.40	11.429		3,650	535		4,185	4,850
2560	Tee fittings, 225 amp		2.80	5.714		1,100	269		1,369	1,625

26 25 13.30 Copper Bus Duct		Crew	Daily Output	Labor-Hours	Unit	Material	Labor	Equipment	Total	Total Incl O&P
						2009 Bare Costs				
2570	400 amp	2 Elec	2.40	6.667	Ea.	1,100	315		1,415	1,675
2580	600 amp		2	8		1,100	375		1,475	1,775
2590	800 amp		1.80	8.889		1,200	420		1,620	1,950
2600	1000 amp		1.60	10		1,325	470		1,795	2,200
2605	1200 amp		1.50	10.667		1,425	500		1,925	2,350
2610	1350 amp		1.40	11.429		1,575	535		2,110	2,550
2620	1600 amp		1.20	13.333		1,825	625		2,450	2,975
2630	2000 amp		1	16		3,025	750		3,775	4,475
2640	2500 amp		.70	22.857		3,550	1,075		4,625	5,550
2650	3000 amp		.60	26.667		4,125	1,250		5,375	6,450
2660	4000 amp		.50	32		5,325	1,500		6,825	8,175
2680	Cross, 225 amp		3.60	4.444		1,675	209		1,884	2,150
2690	400 amp		3.20	5		1,675	235		1,910	2,175
2700	600 amp		3	5.333		1,675	251		1,926	2,200
2710	800 amp		2.60	6.154		2,000	289		2,289	2,650
2720	1000 amp		2.40	6.667		2,100	315		2,415	2,775
2730	1350 amp		2.20	7.273		2,500	340		2,840	3,275
2740	1600 amp		2	8		2,700	375		3,075	3,525
2750	2000 amp		1.80	8.889		4,275	420		4,695	5,350
2760	2500 amp		1.60	10		4,950	470		5,420	6,175
2770	3000 amp		1.40	11.429		5,725	535		6,260	7,125
2780	4000 amp		1	16		7,325	750		8,075	9,200
2800	Expansion fitting, 225 amp		6.40	2.500		1,325	118		1,443	1,650
2810	400 amp		5.40	2.963		1,325	139		1,464	1,700
2820	600 amp		4.60	3.478		1,325	163		1,488	1,725
2830	800 amp		4	4		1,600	188		1,788	2,050
2840	1000 amp		3.60	4.444		1,750	209		1,959	2,250
2850	1350 amp		3.20	5		2,175	235		2,410	2,750
2860	1600 amp		3	5.333		2,400	251		2,651	3,000
2870	2000 amp		2.60	6.154		2,900	289		3,189	3,625
2880	2500 amp		2.20	7.273		4,050	340		4,390	4,975
2890	3000 amp		1.80	8.889		4,750	420		5,170	5,875
2900	4000 amp		1.40	11.429		6,075	535		6,610	7,500
2920	Reducer nonfused, 225 amp		6.40	2.500		1,075	118		1,193	1,375
2930	400 amp		5.40	2.963		1,075	139		1,214	1,425
2940	600 amp		4.60	3.478		1,075	163		1,238	1,450
2950	800 amp		4	4		1,275	188		1,463	1,700
2960	1000 amp		3.60	4.444		1,425	209		1,634	1,875
2970	1350 amp		3.20	5		1,875	235		2,110	2,425
2980	1600 amp		3	5.333		2,125	251		2,376	2,725
2990	2000 amp		2.60	6.154		2,550	289		2,839	3,275
3000	2500 amp		2.20	7.273		3,750	340		4,090	4,650
3010	3000 amp		1.80	8.889		4,425	420		4,845	5,525
3020	4000 amp		1.40	11.429		5,750	535		6,285	7,150
3040	Reducer fuse included, 225 amp		5	3.200		2,575	150		2,725	3,050
3050	400 amp		4.80	3.333		3,425	157		3,582	4,000
3060	600 amp		4.20	3.810		4,075	179		4,254	4,750
3070	800 amp		3.60	4.444		5,800	209		6,009	6,700
3080	1000 amp		3.40	4.706		6,875	221		7,096	7,900
3090	1350 amp		3.20	5		7,000	235		7,235	8,050
3100	1600 amp		2.60	6.154		16,300	289		16,589	18,300
3110	2000 amp		2	8		18,400	375		18,775	20,900
3160	Reducer circuit breaker, 225 amp		5	3.200		3,275	150		3,425	3,825

26 25 13.30 Copper Bus Duct		Crew	Daily Output	Labor-Hours	Unit	2009 Bare Costs Material	Labor	Equipment	Total	Total Incl O&P
3170	400 amp	2 Elec	4.80	3.333	Ea.	4,025	157		4,182	4,675
3180	600 amp		4.20	3.810		5,775	179		5,954	6,625
3190	800 amp		3.60	4.444		6,725	209		6,934	7,725
3200	1000 amp		3.40	4.706		7,650	221		7,871	8,775
3210	1350 amp		3.20	5		9,250	235		9,485	10,600
3220	1600 amp		2.60	6.154		13,800	289		14,089	15,600
3230	2000 amp		2	8		15,100	375		15,475	17,200
3280	3 pole, 3 wire, cable tap box center, 225 amp		3.60	4.444		1,525	209		1,734	2,000
3290	400 amp		3	5.333		1,525	251		1,776	2,050
3300	600 amp		2.60	6.154		1,525	289		1,814	2,125
3310	800 amp		2.40	6.667		1,700	315		2,015	2,350
3320	1000 amp		1.80	8.889		1,825	420		2,245	2,675
3330	1350 amp		1.60	10		2,375	470		2,845	3,325
3340	1600 amp		1.40	11.429		2,650	535		3,185	3,725
3350	2000 amp		1.20	13.333		3,200	625		3,825	4,500
3360	2500 amp		1	16		3,825	750		4,575	5,350
3370	3000 amp		.70	22.857		4,450	1,075		5,525	6,525
3380	4000 amp		.50	32		5,625	1,500		7,125	8,525
3400	Cable tap box end, 225 amp		3.60	4.444		820	209		1,029	1,225
3410	400 amp		3	5.333		820	251		1,071	1,300
3420	600 amp		2.60	6.154		920	289		1,209	1,450
3430	800 amp		2.40	6.667		920	315		1,235	1,475
3440	1000 amp		1.80	8.889		1,025	420		1,445	1,775
3445	1200 amp		1.70	9.412		1,150	440		1,590	1,925
3450	1350 amp		1.60	10		1,275	470		1,745	2,125
3460	1600 amp		1.40	11.429		1,450	535		1,985	2,425
3470	2000 amp		1.20	13.333		1,650	625		2,275	2,800
3480	2500 amp		1	16		1,975	750		2,725	3,325
3490	3000 amp		.70	22.857		2,300	1,075		3,375	4,175
3500	4000 amp		.50	32		2,900	1,500		4,400	5,525
4600	Plug-in, fusible switch w/3 fuses, 3 pole, 250 volt, 30 amp	1 Elec	4	2		320	94		414	500
4610	60 amp		3.60	2.222		420	104		524	620
4620	100 amp		2.70	2.963		600	139		739	875
4630	200 amp	2 Elec	3.20	5		1,000	235		1,235	1,450
4640	400 amp		1.40	11.429		2,625	535		3,160	3,725
4650	600 amp		.90	17.778		3,650	835		4,485	5,275
4700	4 pole, 120/208 volt, 30 amp	1 Elec	3.90	2.051		440	96.50		536.50	630
4710	60 amp		3.50	2.286		475	107		582	685
4720	100 amp		2.60	3.077		660	145		805	945
4730	200 amp	2 Elec	3	5.333		1,100	251		1,351	1,600
4740	400 amp		1.30	12.308		2,600	580		3,180	3,750
4750	600 amp		.80	20		3,650	940		4,590	5,450
4800	3 pole, 480 volt, 30 amp	1 Elec	4	2		330	94		424	505
4810	60 amp		3.60	2.222		350	104		454	545
4820	100 amp		2.70	2.963		585	139		724	860
4830	200 amp	2 Elec	3.20	5		1,000	235		1,235	1,450
4840	400 amp		1.40	11.429		2,350	535		2,885	3,400
4850	600 amp		.90	17.778		3,350	835		4,185	4,950
4860	800 amp		.66	24.242		11,000	1,150		12,150	14,000
4870	1000 amp		.60	26.667		12,300	1,250		13,550	15,400
4880	1200 amp		.50	32		12,300	1,500		13,800	15,900
4890	1600 amp		.44	36.364		14,300	1,700		16,000	18,300
4900	4 pole, 277/480 volt, 30 amp	1 Elec	3.90	2.051		475	96.50		571.50	670

26 25 13.30 Copper Bus Duct		Crew	Daily Output	Labor-Hours	Unit	Material	2009 Bare Costs Labor	Equipment	Total	Total Incl O&P
4910	60 amp	1 Elec	3.50	2.286	Ea.	505	107		612	725
4920	100 amp	↓	2.60	3.077		740	145		885	1,025
4930	200 amp	2 Elec	3	5.333		1,475	251		1,726	2,000
4940	400 amp		1.30	12.308		2,775	580		3,355	3,950
4950	600 amp		.80	20		3,800	940		4,740	5,625
5050	800 amp		.60	26.667		12,500	1,250		13,750	15,600
5060	1000 amp		.56	28.571		14,400	1,350		15,750	17,900
5070	1200 amp		.48	33.333		14,500	1,575		16,075	18,400
5080	1600 amp	↓	.42	38.095		16,900	1,800		18,700	21,300
5150	Fusible with starter, 3 pole 250 volt, 30 amp	1 Elec	3.50	2.286		1,675	107		1,782	2,000
5160	60 amp		3.20	2.500		1,775	118		1,893	2,125
5170	100 amp	↓	2.50	3.200		2,000	150		2,150	2,425
5180	200 amp	2 Elec	2.80	5.714		3,300	269		3,569	4,050
5200	3 pole 480 volt, 30 amp	1 Elec	3.50	2.286		1,675	107		1,782	2,000
5210	60 amp		3.20	2.500		1,775	118		1,893	2,125
5220	100 amp	↓	2.50	3.200		2,000	150		2,150	2,425
5230	200 amp	2 Elec	2.80	5.714		3,300	269		3,569	4,050
5300	Fusible with contactor, 3 pole 250 volt, 30 amp	1 Elec	3.50	2.286		1,625	107		1,732	1,975
5310	60 amp		3.20	2.500		2,075	118		2,193	2,450
5320	100 amp	↓	2.50	3.200		2,900	150		3,050	3,400
5330	200 amp	2 Elec	2.80	5.714		3,300	269		3,569	4,075
5400	3 pole 480 volt, 30 amp	1 Elec	3.50	2.286		1,750	107		1,857	2,100
5410	60 amp		3.20	2.500		2,475	118		2,593	2,900
5420	100 amp	↓	2.50	3.200		3,375	150		3,525	3,950
5430	200 amp	2 Elec	2.80	5.714		3,475	269		3,744	4,250
5450	Fusible with capacitor, 3 pole 250 volt, 30 amp	1 Elec	3	2.667		4,225	125		4,350	4,850
5460	60 amp		2	4		4,925	188		5,113	5,700
5500	3 pole 480 volt, 30 amp		3	2.667		3,525	125		3,650	4,100
5510	60 amp		2	4		4,425	188		4,613	5,150
5600	Circuit breaker, 3 pole, 250 volt, 60 amp		4.50	1.778		450	83.50		533.50	625
5610	100 amp		3.20	2.500		555	118		673	790
5650	4 pole, 120/208 volt, 60 amp		4.40	1.818		510	85.50		595.50	690
5660	100 amp		3.10	2.581		605	121		726	850
5700	3 pole, 4 wire 277/480 volt, 60 amp		4.30	1.860		685	87.50		772.50	890
5710	100 amp	↓	3	2.667		765	125		890	1,025
5720	225 amp	2 Elec	3.20	5		1,700	235		1,935	2,200
5730	400 amp		1.20	13.333		3,500	625		4,125	4,850
5740	600 amp		.96	16.667		4,725	785		5,510	6,400
5750	700 amp		.60	26.667		5,975	1,250		7,225	8,500
5760	800 amp		.60	26.667		5,975	1,250		7,225	8,500
5770	900 amp		.54	29.630		7,875	1,400		9,275	10,800
5780	1000 amp		.54	29.630		7,875	1,400		9,275	10,800
5790	1200 amp	↓	.42	38.095		9,475	1,800		11,275	13,200
5810	Circuit breaker w/HIC fuses, 3 pole 480 volt, 60 amp	1 Elec	4.40	1.818		865	85.50		950.50	1,075
5820	100 amp	"	3.10	2.581		950	121		1,071	1,225
5830	225 amp	2 Elec	3.40	4.706		3,025	221		3,246	3,675
5840	400 amp		1.40	11.429		4,800	535		5,335	6,100
5850	600 amp		1	16		4,850	750		5,600	6,500
5860	700 amp		.64	25		6,450	1,175		7,625	8,875
5870	800 amp		.64	25		6,450	1,175		7,625	8,875
5880	900 amp		.56	28.571		13,900	1,350		15,250	17,400
5890	1000 amp	↓	.56	28.571		13,900	1,350		15,250	17,400
5950	3 pole 4 wire 277/480 volt, 60 amp	1 Elec	4.30	1.860	↓	865	87.50		952.50	1,075

26 25 13 – Bus Duct/Busway and Fittings

26 25 13.30 Copper Bus Duct		Crew	Daily Output	Labor-Hours	Unit	2009 Bare Costs Material	Labor	Equipment	Total	Total Incl O&P
5960	100 amp	1 Elec	3	2.667	Ea.	950	125		1,075	1,250
5970	225 amp	2 Elec	3	5.333		3,025	251		3,276	3,700
5980	400 amp		1.10	14.545		4,800	685		5,485	6,325
5990	600 amp		.94	17.021		4,850	800		5,650	6,575
6000	700 amp		.58	27.586		6,450	1,300		7,750	9,075
6010	800 amp		.58	27.586		6,450	1,300		7,750	9,075
6020	900 amp		.52	30.769		13,900	1,450		15,350	17,500
6030	1000 amp		.52	30.769		13,900	1,450		15,350	17,500
6040	1200 amp		.40	40		13,900	1,875		15,775	18,200
6100	Circuit breaker with starter, 3 pole 250 volt, 60 amp	1 Elec	3.20	2.500		1,250	118		1,368	1,550
6110	100 amp	"	2.50	3.200		1,750	150		1,900	2,150
6120	225 amp	2 Elec	3	5.333		2,025	251		2,276	2,600
6130	3 pole 480 volt, 60 amp	1 Elec	3.20	2.500		1,250	118		1,368	1,550
6140	100 amp	"	2.50	3.200		1,750	150		1,900	2,150
6150	225 amp	2 Elec	3	5.333		2,025	251		2,276	2,600
6200	Circuit breaker with contactor, 3 pole 250 volt, 60 amp	1 Elec	3.20	2.500		1,200	118		1,318	1,475
6210	100 amp	"	2.50	3.200		1,625	150		1,775	2,000
6220	225 amp	2 Elec	3	5.333		1,875	251		2,126	2,450
6250	3 pole 480 volt, 60 amp	1 Elec	3.20	2.500		1,200	118		1,318	1,475
6260	100 amp	"	2.50	3.200		1,625	150		1,775	2,000
6270	225 amp	2 Elec	3	5.333		1,875	251		2,126	2,450
6300	Circuit breaker with capacitor, 3 pole 250 volt, 60 amp	1 Elec	2	4		5,475	188		5,663	6,325
6310	3 pole 480 volt, 60 amp		2	4		5,075	188		5,263	5,900
6400	Add control transformer with pilot light to above starter		16	.500		370	23.50		393.50	445
6410	Switch, fusible, mechanically held contactor optional		16	.500		980	23.50		1,003.50	1,100
6430	Circuit breaker, mechanically held contactor optional		16	.500		980	23.50		1,003.50	1,100
6450	Ground neutralizer, 3 pole		16	.500		45	23.50		68.50	86

26 25 13.40 Copper Bus Duct		Crew	Daily Output	Labor-Hours	Unit	Material	Labor	Equipment	Total	Total Incl O&P
0010	**COPPER BUS DUCT** 10 ft. long									
0050	Indoor 3 pole 4 wire, plug-in, straight section, 225 amp	2 Elec	40	.400	L.F.	186	18.80		204.80	233
1000	400 amp		32	.500		186	23.50		209.50	240
1500	600 amp		26	.615		186	29		215	249
2400	800 amp		20	.800		220	37.50		257.50	300
2450	1000 amp		18	.889		244	42		286	335
2470	1200 amp		17	.941		315	44		359	415
2500	1350 amp		16	1		350	47		397	460
2510	1600 amp		12	1.333		395	62.50		457.50	530
2520	2000 amp		10	1.600		500	75		575	665
2530	2500 amp		8	2		615	94		709	820
2540	3000 amp		6	2.667		740	125		865	1,000
2550	Feeder, 600 amp		28	.571		162	27		189	221
2600	800 amp		22	.727		197	34		231	270
2700	1000 amp		20	.800		220	37.50		257.50	300
2750	1200 amp		19	.842		290	39.50		329.50	380
2800	1350 amp		18	.889		325	42		367	420
2900	1600 amp		14	1.143		370	53.50		423.50	495
3000	2000 amp		12	1.333		475	62.50		537.50	620
3010	2500 amp		8	2		590	94		684	795
3020	3000 amp		6	2.667		720	125		845	985
3030	4000 amp		4	4		950	188		1,138	1,350
3040	5000 amp		2	8		1,150	375		1,525	1,850
3100	Elbows, 225 amp		4	4	Ea.	1,100	188		1,288	1,500

26 25 13.40 Copper Bus Duct		Crew	Daily Output	Labor-Hours	Unit	Material	Labor (2009 Bare Costs)	Equipment	Total	Total Incl O&P
3200	400 amp	2 Elec	3.60	4.444	Ea.	1,100	209		1,309	1,525
3300	600 amp		3.20	5		1,100	235		1,335	1,550
3400	800 amp		2.80	5.714		1,200	269		1,469	1,725
3500	1000 amp		2.60	6.154		1,325	289		1,614	1,925
3550	1200 amp		2.50	6.400		1,475	300		1,775	2,100
3600	1350 amp		2.40	6.667		1,575	315		1,890	2,200
3700	1600 amp		2.20	7.273		1,700	340		2,040	2,400
3800	2000 amp		1.80	8.889		2,100	420		2,520	2,950
3810	2500 amp		1.60	10		3,325	470		3,795	4,375
3820	3000 amp		1.40	11.429		3,850	535		4,385	5,050
3830	4000 amp		1.20	13.333		4,975	625		5,600	6,450
3840	5000 amp		1	16		8,025	750		8,775	9,975
4000	End box, 225 amp		34	.471		148	22		170	197
4100	400 amp		32	.500		148	23.50		171.50	199
4200	600 amp		28	.571		148	27		175	205
4300	800 amp		26	.615		148	29		177	208
4400	1000 amp		24	.667		148	31.50		179.50	211
4410	1200 amp		23	.696		148	32.50		180.50	214
4500	1350 amp		22	.727		148	34		182	216
4600	1600 amp		20	.800		148	37.50		185.50	221
4700	2000 amp		18	.889		182	42		224	265
4710	2500 amp		16	1		182	47		229	273
4720	3000 amp		14	1.143		182	53.50		235.50	283
4730	4000 amp		12	1.333		220	62.50		282.50	340
4740	5000 amp		10	1.600		220	75		295	360
4800	Cable tap box end, 225 amp		3.20	5		1,100	235		1,335	1,575
5000	400 amp		2.60	6.154		1,100	289		1,389	1,675
5100	600 amp		2.20	7.273		1,100	340		1,440	1,750
5200	800 amp		2	8		1,250	375		1,625	1,950
5300	1000 amp		1.60	10		1,375	470		1,845	2,225
5350	1200 amp		1.50	10.667		1,525	500		2,025	2,475
5400	1350 amp		1.40	11.429		1,625	535		2,160	2,625
5500	1600 amp		1.20	13.333		1,850	625		2,475	3,000
5600	2000 amp		1	16		2,100	750		2,850	3,450
5610	2500 amp		.80	20		2,500	940		3,440	4,200
5620	3000 amp		.60	26.667		2,825	1,250		4,075	5,050
5630	4000 amp		.40	40		3,600	1,875		5,475	6,850
5640	5000 amp		.20	80		4,200	3,750		7,950	10,400
5700	Switchboard stub, 225 amp		5.40	2.963		1,125	139		1,264	1,450
5800	400 amp		4.60	3.478		1,125	163		1,288	1,475
5900	600 amp		4	4		1,125	188		1,313	1,525
6000	800 amp		3.20	5		1,350	235		1,585	1,825
6100	1000 amp		3	5.333		1,550	251		1,801	2,100
6150	1200 amp		2.80	5.714		1,825	269		2,094	2,425
6200	1350 amp		2.60	6.154		2,000	289		2,289	2,650
6300	1600 amp		2.40	6.667		2,275	315		2,590	2,975
6400	2000 amp		2	8		2,750	375		3,125	3,600
6410	2500 amp		1.80	8.889		3,325	420		3,745	4,325
6420	3000 amp		1.60	10		3,925	470		4,395	5,025
6430	4000 amp		1.40	11.429		5,100	535		5,635	6,425
6440	5000 amp		1.20	13.333		6,250	625		6,875	7,850
6490	Tee fittings, 225 amp		2.40	6.667		1,525	315		1,840	2,150
6500	400 amp		2	8		1,525	375		1,900	2,250

26 25 13.40 Copper Bus Duct		Crew	Daily Output	Labor-Hours	Unit	Material	Labor	Equipment	Total	Total Incl O&P
						2009 Bare Costs				
6600	600 amp	2 Elec	1.80	8.889	Ea.	1,525	420		1,945	2,325
6700	800 amp		1.60	10		1,750	470		2,220	2,650
6750	1000 amp		1.40	11.429		2,025	535		2,560	3,050
6770	1200 amp		1.30	12.308		2,250	580		2,830	3,400
6800	1350 amp		1.20	13.333		2,500	625		3,125	3,750
7000	1600 amp		1	16		2,850	750		3,600	4,300
7100	2000 amp		.80	20		3,400	940		4,340	5,175
7110	2500 amp		.60	26.667		4,175	1,250		5,425	6,500
7120	3000 amp		.50	32		4,850	1,500		6,350	7,650
7130	4000 amp		.40	40		6,275	1,875		8,150	9,800
7140	5000 amp		.20	80		7,425	3,750		11,175	13,900
7200	Plug-in fusible switches w/3 fuses, 600 volt, 3 pole, 30 amp	1 Elec	4	2		660	94		754	870
7300	60 amp		3.60	2.222		740	104		844	975
7400	100 amp		2.70	2.963		1,125	139		1,264	1,475
7500	200 amp	2 Elec	3.20	5		2,025	235		2,260	2,575
7600	400 amp		1.40	11.429		5,925	535		6,460	7,350
7700	600 amp		.90	17.778		6,725	835		7,560	8,675
7800	800 amp		.66	24.242		9,300	1,150		10,450	12,000
7900	1200 amp		.50	32		17,400	1,500		18,900	21,500
7910	1600 amp		.44	36.364		17,400	1,700		19,100	21,800
8000	Plug-in circuit breakers, molded case, 15 to 50 amp	1 Elec	4.40	1.818		625	85.50		710.50	820
8100	70 to 100 amp	"	3.10	2.581		695	121		816	950
8200	150 to 225 amp	2 Elec	3.40	4.706		1,875	221		2,096	2,425
8300	250 to 400 amp		1.40	11.429		3,300	535		3,835	4,450
8400	500 to 600 amp		1	16		4,475	750		5,225	6,075
8500	700 to 800 amp		.64	25		5,500	1,175		6,675	7,850
8600	900 to 1000 amp		.56	28.571		7,875	1,350		9,225	10,800
8700	1200 amp		.44	36.364		9,475	1,700		11,175	13,000
8720	1400 amp		.40	40		14,000	1,875		15,875	18,300
8730	1600 amp		.40	40		15,300	1,875		17,175	19,800
8750	Circuit breakers, with current limiting fuse, 15 to 50 amp	1 Elec	4.40	1.818		1,250	85.50		1,335.50	1,500
8760	70 to 100 amp	"	3.10	2.581		1,475	121		1,596	1,800
8770	150 to 225 amp	2 Elec	3.40	4.706		3,200	221		3,421	3,850
8780	250 to 400 amp		1.40	11.429		4,925	535		5,460	6,250
8790	500 to 600 amp		1	16		5,675	750		6,425	7,400
8800	700 to 800 amp		.64	25		9,350	1,175		10,525	12,100
8810	900 to 1000 amp		.56	28.571		10,700	1,350		12,050	13,800
8850	Combination starter FVNR, fusible switch, NEMA size 0, 30 amp	1 Elec	2	4		1,675	188		1,863	2,125
8860	NEMA size 1, 60 amp		1.80	4.444		1,775	209		1,984	2,275
8870	NEMA size 2, 100 amp		1.30	6.154		2,225	289		2,514	2,900
8880	NEMA size 3, 200 amp	2 Elec	2	8		3,600	375		3,975	4,525
8900	Circuit breaker, NEMA size 0, 30 amp	1 Elec	2	4		1,700	188		1,888	2,175
8910	NEMA size 1, 60 amp		1.80	4.444		1,775	209		1,984	2,275
8920	NEMA size 2, 100 amp		1.30	6.154		2,550	289		2,839	3,250
8930	NEMA size 3, 200 amp	2 Elec	2	8		3,250	375		3,625	4,150
8950	Combination contactor, fusible switch, NEMA size 0, 30 amp	1 Elec	2	4		985	188		1,173	1,375
8960	NEMA size 1, 60 amp		1.80	4.444		1,025	209		1,234	1,450
8970	NEMA size 2, 100 amp		1.30	6.154		1,500	289		1,789	2,100
8980	NEMA size 3, 200 amp	2 Elec	2	8		1,750	375		2,125	2,500
9000	Circuit breaker, NEMA size 0, 30 amp	1 Elec	2	4		1,125	188		1,313	1,550
9010	NEMA size 1, 60 amp		1.80	4.444		1,150	209		1,359	1,600
9020	NEMA size 2, 100 amp		1.30	6.154		1,800	289		2,089	2,425
9030	NEMA size 3, 200 amp	2 Elec	2	8		2,175	375		2,550	2,975

26 25 13.40 Copper Bus Duct		Crew	Daily Output	Labor-Hours	Unit	2009 Bare Costs Material	Labor	Equipment	Total	Total Incl O&P
9050	Control transformer for above, NEMA size 0, 30 amp	1 Elec	8	1	Ea.	191	47		238	284
9060	NEMA size 1, 60 amp		8	1		191	47		238	284
9070	NEMA size 2, 100 amp		7	1.143		267	53.50		320.50	375
9080	NEMA size 3, 200 amp	2 Elec	14	1.143		370	53.50		423.50	495
9100	Comb. fusible switch & lighting control, electrically held, 30 amp	1 Elec	2	4		785	188		973	1,150
9110	60 amp		1.80	4.444		1,125	209		1,334	1,575
9120	100 amp		1.30	6.154		1,450	289		1,739	2,050
9130	200 amp	2 Elec	2	8		3,575	375		3,950	4,500
9150	Mechanically held, 30 amp	1 Elec	2	4		980	188		1,168	1,375
9160	60 amp		1.80	4.444		1,450	209		1,659	1,925
9170	100 amp		1.30	6.154		1,875	289		2,164	2,525
9180	200 amp	2 Elec	2	8		3,825	375		4,200	4,775
9200	Ground bus added to bus duct, 225 amp		320	.050	L.F.	35	2.35		37.35	42
9210	400 amp		240	.067		35	3.13		38.13	43.50
9220	600 amp		240	.067		35	3.13		38.13	43.50
9230	800 amp		160	.100		40.50	4.70		45.20	52
9240	1000 amp		160	.100		46.50	4.70		51.20	58.50
9250	1350 amp		140	.114		69.50	5.35		74.85	85
9260	1600 amp		120	.133		75.50	6.25		81.75	92.50
9270	2000 amp		110	.145		98.50	6.85		105.35	119
9280	2500 amp		100	.160		122	7.50		129.50	146
9290	3000 amp		90	.178		145	8.35		153.35	173
9300	4000 amp		80	.200		191	9.40		200.40	225
9310	5000 amp		70	.229		232	10.75		242.75	272
9320	High short circuit bracing, add					15.65			15.65	17.25

26 25 13.60 Copper or Aluminum Bus Duct Fittings		Crew	Daily Output	Labor-Hours	Unit	Material	Labor	Equipment	Total	Total Incl O&P
0010	**COPPER OR ALUMINUM BUS DUCT FITTINGS**									
0100	Flange, wall, with vapor barrier, 225 amp	2 Elec	6.20	2.581	Ea.	685	121		806	940
0110	400 amp		6	2.667		685	125		810	950
0120	600 amp		5.80	2.759		685	130		815	955
0130	800 amp		5.40	2.963		685	139		824	970
0140	1000 amp		5	3.200		685	150		835	985
0145	1200 amp		4.80	3.333		685	157		842	995
0150	1350 amp		4.60	3.478		685	163		848	1,000
0160	1600 amp		4.20	3.810		685	179		864	1,025
0170	2000 amp		4	4		685	188		873	1,050
0180	2500 amp		3.60	4.444		685	209		894	1,075
0190	3000 amp		3.20	5		685	235		920	1,125
0200	4000 amp		2.60	6.154		795	289		1,084	1,325
0300	Roof, 225 amp		6.20	2.581		740	121		861	1,000
0310	400 amp		6	2.667		740	125		865	1,000
0320	600 amp		5.80	2.759		740	130		870	1,025
0330	800 amp		5.40	2.963		740	139		879	1,025
0340	1000 amp		5	3.200		740	150		890	1,050
0345	1200 amp		4.80	3.333		740	157		897	1,050
0350	1350 amp		4.60	3.478		740	163		903	1,075
0360	1600 amp		4.20	3.810		740	179		919	1,100
0370	2000 amp		4	4		740	188		928	1,100
0380	2500 amp		3.60	4.444		740	209		949	1,125
0390	3000 amp		3.20	5		740	235		975	1,175
0400	4000 amp		2.60	6.154		740	289		1,029	1,250
0420	Support, floor mounted, 225 amp		20	.800		150	37.50		187.50	223

26 25 Enclosed Bus Assemblies

26 25 13 – Bus Duct/Busway and Fittings

26 25 13.60 Copper or Aluminum Bus Duct Fittings		Crew	Daily Output	Labor-Hours	Unit	Material	2009 Bare Costs Labor	Equipment	Total	Total Incl O&P
0430	400 amp	2 Elec	20	.800	Ea.	150	37.50		187.50	223
0440	600 amp		18	.889		150	42		192	230
0450	800 amp		16	1		150	47		197	238
0460	1000 amp		13	1.231		150	58		208	254
0465	1200 amp		11.80	1.356		150	63.50		213.50	263
0470	1350 amp		10.60	1.509		150	71		221	274
0480	1600 amp		9.20	1.739		150	81.50		231.50	291
0490	2000 amp		8	2		150	94		244	310
0500	2500 amp		6.40	2.500		150	118		268	345
0510	3000 amp		5.40	2.963		150	139		289	380
0520	4000 amp		4	4		150	188		338	455
0540	Weather stop, 225 amp		12	1.333		445	62.50		507.50	580
0550	400 amp		10	1.600		445	75		520	600
0560	600 amp		9	1.778		445	83.50		528.50	615
0570	800 amp		8	2		445	94		539	630
0580	1000 amp		6.40	2.500		445	118		563	665
0585	1200 amp		5.90	2.712		445	127		572	680
0590	1350 amp		5.40	2.963		445	139		584	700
0600	1600 amp		4.60	3.478		445	163		608	735
0610	2000 amp		4	4		445	188		633	775
0620	2500 amp		3.20	5		445	235		680	845
0630	3000 amp		2.60	6.154		445	289		734	930
0640	4000 amp		2	8		445	375		820	1,075
0660	End closure, 225 amp		34	.471		148	22		170	197
0670	400 amp		32	.500		148	23.50		171.50	199
0680	600 amp		28	.571		148	27		175	205
0690	800 amp		26	.615		148	29		177	208
0700	1000 amp		24	.667		148	31.50		179.50	211
0705	1200 amp		23	.696		148	32.50		180.50	214
0710	1350 amp		22	.727		148	34		182	216
0720	1600 amp		20	.800		148	37.50		185.50	221
0730	2000 amp		18	.889		182	42		224	265
0740	2500 amp		16	1		182	47		229	273
0750	3000 amp		14	1.143		182	53.50		235.50	283
0760	4000 amp		12	1.333		220	62.50		282.50	340
0780	Switchboard stub, 3 pole 3 wire, 225 amp		6	2.667		895	125		1,020	1,175
0790	400 amp		5.20	3.077		895	145		1,040	1,200
0800	600 amp		4.60	3.478		895	163		1,058	1,225
0810	800 amp		3.60	4.444		1,075	209		1,284	1,525
0820	1000 amp		3.40	4.706		1,275	221		1,496	1,750
0825	1200 amp		3.20	5		1,275	235		1,510	1,750
0830	1350 amp		3	5.333		1,525	251		1,776	2,050
0840	1600 amp		2.80	5.714		1,775	269		2,044	2,375
0850	2000 amp		2.40	6.667		2,075	315		2,390	2,775
0860	2500 amp		2	8		2,525	375		2,900	3,350
0870	3000 amp		1.80	8.889		2,900	420		3,320	3,825
0880	4000 amp		1.60	10		3,750	470		4,220	4,850
0890	5000 amp		1.40	11.429		4,600	535		5,135	5,875
0900	3 pole 4 wire, 225 amp		5.40	2.963		1,125	139		1,264	1,450
0910	400 amp		4.60	3.478		1,125	163		1,288	1,475
0920	600 amp		4	4		1,125	188		1,313	1,525
0930	800 amp		3.20	5		1,350	235		1,585	1,825
0940	1000 amp		3	5.333		1,550	251		1,801	2,100

26 25 13.60 Copper or Aluminum Bus Duct Fittings		Crew	Daily Output	Labor-Hours	Unit	Material	2009 Bare Costs Labor	Equipment	Total	Total Incl O&P
0950	1350 amp	2 Elec	2.60	6.154	Ea.	2,000	289		2,289	2,650
0960	1600 amp		2.40	6.667		2,275	315		2,590	2,975
0970	2000 amp		2	8		2,750	375		3,125	3,600
0980	2500 amp		1.80	8.889		3,350	420		3,770	4,325
0990	3000 amp		1.60	10		3,925	470		4,395	5,025
1000	4000 amp		1.40	11.429		5,075	535		5,610	6,425
1050	Service head, weatherproof, 3 pole 3 wire, 225 amp		3	5.333		1,525	251		1,776	2,050
1060	400 amp		2.80	5.714		1,525	269		1,794	2,100
1070	600 amp		2.60	6.154		1,525	289		1,814	2,125
1080	800 amp		2.40	6.667		1,700	315		2,015	2,350
1090	1000 amp		2	8		1,850	375		2,225	2,600
1100	1350 amp		1.80	8.889		2,400	420		2,820	3,275
1110	1600 amp		1.60	10		2,675	470		3,145	3,675
1120	2000 amp		1.40	11.429		3,250	535		3,785	4,400
1130	2500 amp		1.20	13.333		3,875	625		4,500	5,250
1140	3000 amp		.90	17.778		4,525	835		5,360	6,250
1150	4000 amp		.70	22.857		5,750	1,075		6,825	7,975
1200	3 pole 4 wire, 225 amp		2.60	6.154		1,700	289		1,989	2,325
1210	400 amp		2.40	6.667		1,700	315		2,015	2,350
1220	600 amp		2.20	7.273		1,700	340		2,040	2,400
1230	800 amp		2	8		1,975	375		2,350	2,750
1240	1000 amp		1.70	9.412		2,275	440		2,715	3,175
1250	1350 amp		1.50	10.667		2,700	500		3,200	3,750
1260	1600 amp		1.40	11.429		2,950	535		3,485	4,050
1270	2000 amp		1.20	13.333		3,925	625		4,550	5,300
1280	2500 amp		1	16		4,850	750		5,600	6,475
1290	3000 amp		.80	20		5,675	940		6,615	7,700
1300	4000 amp		.60	26.667		7,350	1,250		8,600	10,000
1350	Flanged end, 3 pole 3 wire, 225 amp		6	2.667		825	125		950	1,100
1360	400 amp		5.20	3.077		825	145		970	1,125
1370	600 amp		4.60	3.478		825	163		988	1,150
1380	800 amp		3.60	4.444		915	209		1,124	1,325
1390	1000 amp		3.40	4.706		1,025	221		1,246	1,475
1395	1200 amp		3.20	5		1,150	235		1,385	1,625
1400	1350 amp		3	5.333		1,225	251		1,476	1,725
1410	1600 amp		2.80	5.714		1,400	269		1,669	1,975
1420	2000 amp		2.40	6.667		1,650	315		1,965	2,275
1430	2500 amp		2	8		1,925	375		2,300	2,700
1440	3000 amp		1.80	8.889		2,200	420		2,620	3,075
1450	4000 amp		1.60	10		2,725	470		3,195	3,725
1500	3 pole 4 wire, 225 amp		5.40	2.963		950	139		1,089	1,275
1510	400 amp		4.60	3.478		950	163		1,113	1,300
1520	600 amp		4	4		950	188		1,138	1,350
1530	800 amp		3.20	5		1,125	235		1,360	1,575
1540	1000 amp		3	5.333		1,250	251		1,501	1,750
1545	1200 amp		2.80	5.714		1,425	269		1,694	1,975
1550	1350 amp		2.60	6.154		1,525	289		1,814	2,125
1560	1600 amp		2.40	6.667		1,750	315		2,065	2,400
1570	2000 amp		2	8		2,050	375		2,425	2,825
1580	2500 amp		1.80	8.889		2,450	420		2,870	3,350
1590	3000 amp		1.60	10		2,800	470		3,270	3,800
1600	4000 amp		1.40	11.429		3,625	535		4,160	4,800
1650	Hanger, standard, 225 amp		64	.250		17.40	11.75		29.15	37.50

26 25 13 – Bus Duct/Busway and Fittings

26 25 13.60 Copper or Aluminum Bus Duct Fittings		Crew	Daily Output	Labor-Hours	Unit	2009 Bare Costs Material	Labor	Equipment	Total	Total Incl O&P
1660	400 amp	2 Elec	48	.333	Ea.	17.40	15.65		33.05	43
1670	600 amp		40	.400		17.40	18.80		36.20	48
1680	800 amp		32	.500		17.40	23.50		40.90	55
1690	1000 amp		24	.667		17.40	31.50		48.90	67
1695	1200 amp		22	.727		17.40	34		51.40	71.50
1700	1350 amp		20	.800		17.40	37.50		54.90	77
1710	1600 amp		20	.800		17.40	37.50		54.90	77
1720	2000 amp		18	.889		17.40	42		59.40	83.50
1730	2500 amp		16	1		17.40	47		64.40	91.50
1740	3000 amp		16	1		17.40	47		64.40	91.50
1750	4000 amp		16	1		17.40	47		64.40	91.50
1800	Spring type, 225 amp		16	1		77.50	47		124.50	158
1810	400 amp		14	1.143		77.50	53.50		131	168
1820	600 amp		14	1.143		77.50	53.50		131	168
1830	800 amp		14	1.143		77.50	53.50		131	168
1840	1000 amp		14	1.143		77.50	53.50		131	168
1845	1200 amp		14	1.143		77.50	53.50		131	168
1850	1350 amp		14	1.143		77.50	53.50		131	168
1860	1600 amp		12	1.333		77.50	62.50		140	182
1870	2000 amp		12	1.333		77.50	62.50		140	182
1880	2500 amp		12	1.333		77.50	62.50		140	182
1890	3000 amp		10	1.600		77.50	75		152.50	202
1900	4000 amp		10	1.600		77.50	75		152.50	202

26 25 13.70 Feedrail		Crew	Daily Output	Labor-Hours	Unit	Material	Labor	Equipment	Total	Total Incl O&P
0010	**FEEDRAIL**, 12 foot mounting									
0050	Trolley busway, 3 pole									
0100	300 volt 60 amp, plain, 10 ft. lengths	1 Elec	50	.160	L.F.	25	7.50		32.50	39
0300	Door track		50	.160		27.50	7.50		35	42
0500	Curved track		30	.267		365	12.55		377.55	420
0700	Coupling				Ea.	16.25			16.25	17.90
0900	Center feed	1 Elec	5.30	1.509		44	71		115	157
1100	End feed		5.30	1.509		52.50	71		123.50	167
1300	Hanger set		24	.333		6.25	15.65		21.90	31
3000	600 volt 100 amp, plain, 10 ft. lengths		35	.229	L.F.	61.50	10.75		72.25	84
3300	Door track		35	.229	"	71.50	10.75		82.25	95
3700	Coupling				Ea.	51.50			51.50	56.50
4000	End cap	1 Elec	40	.200		44	9.40		53.40	62.50
4200	End feed		4	2		268	94		362	440
4500	Trolley, 600 volt, 20 amp		5.30	1.509		345	71		416	490
4700	30 amp		5.30	1.509		345	71		416	490
4900	Duplex, 40 amp		4	2		730	94		824	945
5000	60 amp		4	2		730	94		824	945
5300	Fusible, 20 amp		4	2		880	94		974	1,125
5500	30 amp		4	2		880	94		974	1,125
5900	300 volt, 20 amp		5.30	1.509		269	71		340	405
6000	30 amp		5.30	1.509		299	71		370	440
6300	Fusible, 20 amp		4.70	1.702		365	80		445	525
6500	30 amp		4.70	1.702		520	80		600	695
7300	Busway, 250 volt 50 amp, 2 wire		70	.114	L.F.	18	5.35		23.35	28
7330	Coupling				Ea.	35			35	38.50
7340	Center feed	1 Elec	6	1.333		400	62.50		462.50	535
7350	End feed		6	1.333		72.50	62.50		135	177

26 25 Enclosed Bus Assemblies

26 25 13 – Bus Duct/Busway and Fittings

26 25 13.70 Feedrail		Crew	Daily Output	Labor-Hours	Unit	2009 Bare Costs Material	Labor	Equipment	Total	Total Incl O&P
7360	End cap	1 Elec	40	.200	Ea.	26.50	9.40		35.90	43.50
7370	Hanger set		24	.333		6.25	15.65		21.90	31
7400	125/250 volt 50 amp, 3 wire		60	.133	L.F.	18.90	6.25		25.15	30.50
7430	Coupling		6	1.333	Ea.	42.50	62.50		105	144
7440	Center feed		6	1.333		425	62.50		487.50	565
7450	End feed		6	1.333		77.50	62.50		140	182
7460	End cap		40	.200		29	9.40		38.40	46
7470	Hanger set		24	.333		6.25	15.65		21.90	31
7480	Trolley, 250 volt, 2 pole, 20 amp		6	1.333		44	62.50		106.50	145
7490	30 amp		6	1.333		44	62.50		106.50	145
7500	125/250 volt, 3 pole, 20 amp		6	1.333		46.50	62.50		109	148
7510	30 amp		6	1.333		46.50	62.50		109	148
8000	Cleaning tools, 300 volt, dust remover					81.50			81.50	89.50
8100	Bus bar cleaner					143			143	157
8300	600 volt, dust remover, 60 amp					300			300	330
8400	100 amp					415			415	455
8600	Bus bar cleaner, 60 amp					365			365	405
8700	100 amp					510			510	560

26 27 Low-Voltage Distribution Equipment

26 27 13 – Electricity Metering

26 27 13.10 Meter Centers and Sockets		Crew	Daily Output	Labor-Hours	Unit	Material	Labor	Equipment	Total	Total Incl O&P
0010	**METER CENTERS AND SOCKETS**									
0100	Sockets, single position, 4 terminal, 100 amp	1 Elec	3.20	2.500	Ea.	43	118		161	229
0200	150 amp		2.30	3.478		50.50	163		213.50	305
0300	200 amp		1.90	4.211		72.50	198		270.50	385
0400	Transformer rated, 20 amp		3.20	2.500		111	118		229	305
0500	Double position, 4 terminal, 100 amp		2.80	2.857		177	134		311	400
0600	150 amp		2.10	3.810		208	179		387	505
0700	200 amp		1.70	4.706		430	221		651	810
0800	Trans-socket, 13 terminal, 3 CT mounts, 400 amp		1	8		1,075	375		1,450	1,750
0900	800 amp	2 Elec	1.20	13.333		1,175	625		1,800	2,275
2000	Meter center, main fusible switch, 1P 3W 120/240 volt									
2030	400 amp	2 Elec	1.60	10	Ea.	1,775	470		2,245	2,675
2040	600 amp		1.10	14.545		3,100	685		3,785	4,450
2050	800 amp		.90	17.778		4,850	835		5,685	6,600
2060	Rainproof 1P 3W 120/240 volt, 400 amp		1.60	10		1,775	470		2,245	2,675
2070	600 amp		1.10	14.545		3,100	685		3,785	4,450
2080	800 amp		.90	17.778		4,850	835		5,685	6,600
2100	3P 4W 120/208V, 400 amp		1.60	10		2,025	470		2,495	2,975
2110	600 amp		1.10	14.545		3,800	685		4,485	5,225
2120	800 amp		.90	17.778		7,050	835		7,885	9,050
2130	Rainproof 3P 4W 120/208V, 400 amp		1.60	10		2,025	470		2,495	2,975
2140	600 amp		1.10	14.545		3,800	685		4,485	5,225
2150	800 amp		.90	17.778		7,050	835		7,885	9,050
2170	Main circuit breaker, 1P 3W 120/240V									
2180	400 amp	2 Elec	1.60	10	Ea.	3,175	470		3,645	4,225
2190	600 amp		1.10	14.545		4,275	685		4,960	5,750
2200	800 amp		.90	17.778		5,000	835		5,835	6,775
2210	1000 amp		.80	20		6,875	940		7,815	9,025
2220	1200 amp		.76	21.053		9,300	990		10,290	11,700

26 27 13.10 Meter Centers and Sockets		Crew	Daily Output	Labor-Hours	Unit	Material	Labor	Equipment	Total	Total Incl O&P
2230	1600 amp	2 Elec	.68	23.529	Ea.	20,000	1,100		21,100	23,700
2240	Rainproof 1P 3W 120/240V, 400 amp		1.60	10		3,175	470		3,645	4,225
2250	600 amp		1.10	14.545		4,275	685		4,960	5,750
2260	800 amp		.90	17.778		5,000	835		5,835	6,775
2270	1000 amp		.80	20		6,875	940		7,815	9,025
2280	1200 amp		.76	21.053		9,300	990		10,290	11,700
2300	3P 4W 120/208V, 400 amp		1.60	10		3,625	470		4,095	4,700
2310	600 amp		1.10	14.545		5,100	685		5,785	6,650
2320	800 amp		.90	17.778		6,075	835		6,910	7,950
2330	1000 amp		.80	20		7,975	940		8,915	10,200
2340	1200 amp		.76	21.053		10,200	990		11,190	12,700
2350	1600 amp		.68	23.529		20,000	1,100		21,100	23,700
2360	Rainproof 3P 4W 120/208V, 400 amp		1.60	10		3,625	470		4,095	4,700
2370	600 amp		1.10	14.545		5,100	685		5,785	6,650
2380	800 amp		.90	17.778		6,075	835		6,910	7,950
2390	1000 amp		.76	21.053		7,975	990		8,965	10,300
2400	1200 amp	↓	.68	23.529	↓	10,200	1,100		11,300	12,900
2420	Main lugs terminal box, 1P 3W 120/240V									
2430	800 amp	2 Elec	.94	17.021	Ea.	595	800		1,395	1,875
2440	1200 amp		.72	22.222		1,200	1,050		2,250	2,925
2450	Rainproof 1P 3W 120/240V, 225 amp		2.40	6.667		405	315		720	925
2460	800 amp		.94	17.021		595	800		1,395	1,875
2470	1200 amp		.72	22.222		1,200	1,050		2,250	2,925
2500	3P 4W 120/208V, 800 amp		.94	17.021		665	800		1,465	1,950
2510	1200 amp		.72	22.222		1,325	1,050		2,375	3,075
2520	Rainproof 3P 4W 120/208V, 225 amp		2.40	6.667		405	315		720	925
2530	800 amp		.94	17.021		665	800		1,465	1,950
2540	1200 amp	↓	.72	22.222	↓	1,325	1,050		2,375	3,075
2590	Basic meter device									
2600	1P 3W 120/240V 4 jaw 125A sockets, 3 meter	2 Elec	1	16	Ea.	685	750		1,435	1,900
2610	4 meter		.90	17.778		820	835		1,655	2,175
2620	5 meter		.80	20		1,025	940		1,965	2,575
2630	6 meter		.60	26.667		1,175	1,250		2,425	3,225
2640	7 meter		.56	28.571		1,500	1,350		2,850	3,725
2650	8 meter		.52	30.769		1,650	1,450		3,100	4,025
2660	10 meter	↓	.48	33.333	↓	2,050	1,575		3,625	4,650
2680	Rainproof 1P 3W 120/240V 4 jaw 125A sockets									
2690	3 meter	2 Elec	1	16	Ea.	685	750		1,435	1,900
2700	4 meter		.90	17.778		820	835		1,655	2,175
2710	6 meter		.60	26.667		1,175	1,250		2,425	3,225
2720	7 meter		.56	28.571		1,500	1,350		2,850	3,725
2730	8 meter	↓	.52	30.769	↓	1,650	1,450		3,100	4,025
2750	1P 3W 120/240V 4 jaw sockets									
2760	with 125A circuit breaker, 3 meter	2 Elec	1	16	Ea.	1,275	750		2,025	2,550
2770	4 meter		.90	17.778		1,625	835		2,460	3,050
2780	5 meter		.80	20		2,025	940		2,965	3,675
2790	6 meter		.60	26.667		2,375	1,250		3,625	4,525
2800	7 meter		.56	28.571		2,900	1,350		4,250	5,250
2810	8 meter		.52	30.769		3,225	1,450		4,675	5,775
2820	10 meter	↓	.48	33.333	↓	4,050	1,575		5,625	6,850
2830	Rainproof 1P 3W 120/240V 4 jaw sockets									
2840	with 125A circuit breaker, 3 meter	2 Elec	1	16	Ea.	1,275	750		2,025	2,550
2850	4 meter	↓	.90	17.778	↓	1,625	835		2,460	3,050

(Material, Labor, Equipment and Total are 2009 Bare Costs.)

26 27 13 – Electricity Metering

26 27 13.10 Meter Centers and Sockets		Crew	Daily Output	Labor-Hours	Unit	Material (2009 Bare Costs)	Labor (2009 Bare Costs)	Equipment (2009 Bare Costs)	Total (2009 Bare Costs)	Total Incl O&P
2870	6 meter	2 Elec	.60	26.667	Ea.	2,375	1,250		3,625	4,525
2880	7 meter		.56	28.571		2,900	1,350		4,250	5,250
2890	8 meter		.52	30.769		3,225	1,450		4,675	5,775
2920	1P 3W on 3P 4W 120/208V system 5 jaw									
2930	125A sockets, 3 meter	2 Elec	1	16	Ea.	685	750		1,435	1,900
2940	4 meter		.90	17.778		820	835		1,655	2,175
2950	5 meter		.80	20		1,025	940		1,965	2,575
2960	6 meter		.60	26.667		1,175	1,250		2,425	3,225
2970	7 meter		.56	28.571		1,500	1,350		2,850	3,725
2980	8 meter		.52	30.769		1,650	1,450		3,100	4,025
2990	10 meter		.48	33.333		2,050	1,575		3,625	4,650
3000	Rainproof 1P 3W on 3P 4W 120/208V system									
3020	5 jaw 125A sockets, 3 meter	2 Elec	1	16	Ea.	685	750		1,435	1,900
3030	4 meter		.90	17.778		820	835		1,655	2,175
3050	6 meter		.60	26.667		1,175	1,250		2,425	3,225
3060	7 meter		.56	28.571		1,500	1,350		2,850	3,725
3070	8 meter		.52	30.769		1,650	1,450		3,100	4,025
3090	1P 3W on 3P 4W 120/208V system 5 jaw sockets									
3100	With 125A circuit breaker, 3 meter	2 Elec	1	16	Ea.	1,275	750		2,025	2,550
3110	4 meter		.90	17.778		1,625	835		2,460	3,050
3120	5 meter		.80	20		2,025	940		2,965	3,675
3130	6 meter		.60	26.667		2,375	1,250		3,625	4,525
3140	7 meter		.56	28.571		2,900	1,350		4,250	5,250
3150	8 meter		.52	30.769		3,225	1,450		4,675	5,775
3160	10 meter		.48	33.333		4,050	1,575		5,625	6,850
3170	Rainproof 1P 3W on 3P 4W 120/208V system									
3180	5 jaw sockets w/125A circuit breaker, 3 meter	2 Elec	1	16	Ea.	1,275	750		2,025	2,550
3190	4 meter		.90	17.778		1,625	835		2,460	3,050
3210	6 meter		.60	26.667		2,375	1,250		3,625	4,525
3220	7 meter		.56	28.571		2,900	1,350		4,250	5,250
3230	8 meter		.52	30.769		3,225	1,450		4,675	5,775
3250	1P 3W 120/240V 4 jaw sockets									
3260	with 200A circuit breaker, 3 meter	2 Elec	1	16	Ea.	1,925	750		2,675	3,250
3270	4 meter		.90	17.778		2,600	835		3,435	4,125
3290	6 meter		.60	26.667		3,850	1,250		5,100	6,150
3300	7 meter		.56	28.571		4,525	1,350		5,875	7,050
3310	8 meter		.56	28.571		5,200	1,350		6,550	7,775
3330	Rainproof 1P 3W 120/240V 4 jaw sockets									
3350	with 200A circuit breaker, 3 meter	2 Elec	1	16	Ea.	1,925	750		2,675	3,250
3360	4 meter		.90	17.778		2,600	835		3,435	4,125
3380	6 meter		.60	26.667		3,850	1,250		5,100	6,150
3390	7 meter		.56	28.571		4,525	1,350		5,875	7,050
3400	8 meter		.52	30.769		5,200	1,450		6,650	7,925
3420	1P 3W on 3P 4W 120/208V 5 jaw sockets									
3430	with 200A circuit breaker, 3 meter	2 Elec	1	16	Ea.	1,925	750		2,675	3,250
3440	4 meter		.90	17.778		2,600	835		3,435	4,125
3460	6 meter		.60	26.667		3,850	1,250		5,100	6,150
3470	7 meter		.56	28.571		4,525	1,350		5,875	7,050
3480	8 meter		.52	30.769		5,200	1,450		6,650	7,925
3500	Rainproof 1P 3W on 3P 4W 120/208V 5 jaw socket									
3510	with 200A circuit breaker, 3 meter	2 Elec	1	16	Ea.	1,925	750		2,675	3,250
3520	4 meter		.90	17.778		2,600	835		3,435	4,125
3540	6 meter		.60	26.667		3,850	1,250		5,100	6,150

26 27 13 – Electricity Metering

26 27 13.10 Meter Centers and Sockets		Crew	Daily Output	Labor-Hours	Unit	Material	2009 Bare Costs Labor	Equipment	Total	Total Incl O&P
3550	7 meter	2 Elec	.56	28.571	Ea.	4,525	1,350		5,875	7,050
3560	8 meter		.52	30.769		5,200	1,450		6,650	7,925
3600	Automatic circuit closing, add					74			74	81.50
3610	Manual circuit closing, add					84.50			84.50	93
3650	Branch meter device									
3660	3P 4W 208/120 or 240/120 V 7 jaw sockets									
3670	with 200A circuit breaker, 2 meter	2 Elec	.90	17.778	Ea.	3,200	835		4,035	4,800
3680	3 meter		.80	20		4,800	940		5,740	6,750
3690	4 meter		.70	22.857		6,425	1,075		7,500	8,700
3700	Main circuit breaker 42,000 rms, 400 amp		1.60	10		2,800	470		3,270	3,800
3710	600 amp		1.10	14.545		4,525	685		5,210	6,025
3720	800 amp		.90	17.778		6,075	835		6,910	7,950
3730	Rainproof main circ. breaker 42,000 rms, 400 amp		1.60	10		2,800	470		3,270	3,800
3740	600 amp		1.10	14.545		4,525	685		5,210	6,025
3750	800 amp		.90	17.778		6,075	835		6,910	7,950
3760	Main circuit breaker 65,000 rms, 400 amp		1.60	10		3,800	470		4,270	4,900
3770	600 amp		1.10	14.545		5,350	685		6,035	6,925
3780	800 amp		.90	17.778		6,075	835		6,910	7,950
3790	1000 amp		.80	20		7,975	940		8,915	10,200
3800	1200 amp		.76	21.053		10,200	990		11,190	12,700
3810	1600 amp		.68	23.529		20,000	1,100		21,100	23,700
3820	Rainproof main circ. breaker 65,000 rms, 400 amp		1.60	10		3,800	470		4,270	4,900
3830	600 amp		1.10	14.545		5,350	685		6,035	6,925
3840	800 amp		.90	17.778		6,050	835		6,885	7,950
3850	1000 amp		.80	20		7,975	940		8,915	10,200
3860	1200 amp		.76	21.053		10,200	990		11,190	12,700
3880	Main circuit breaker 100,000 rms, 400 amp		1.60	10		3,800	470		4,270	4,900
3890	600 amp		1.10	14.545		5,350	685		6,035	6,925
3900	800 amp		.90	17.778		6,275	835		7,110	8,200
3910	Rainproof main circ. breaker 100,000 rms, 400 amp		1.60	10		3,800	470		4,270	4,900
3920	600 amp		1.10	14.545		5,350	685		6,035	6,925
3930	800 amp		.90	17.778		6,275	835		7,110	8,200
3940	Main lugs terminal box, 800 amp		.94	17.021		665	800		1,465	1,950
3950	1600 amp		.72	22.222		1,850	1,050		2,900	3,650
3960	Rainproof, 800 amp		.94	17.021		665	800		1,465	1,950
3970	1600 amp		.72	22.222		1,850	1,050		2,900	3,650
9000	Minimum labor/equipment charge	1 Elec	3	2.667	Job		125		125	193

26 27 16 – Electrical Cabinets and Enclosures

26 27 16.10 Cabinets		Crew	Daily Output	Labor-Hours	Unit	Material	Labor	Equipment	Total	Total Incl O&P
0010	**CABINETS**									
7000	Cabinets, current transformer									
7050	Single door, 24" H x 24" W x 10" D	1 Elec	1.60	5	Ea.	206	235		441	585
7100	30" H x 24" W x 10" D		1.30	6.154		210	289		499	675
7150	36" H x 24" W x 10" D		1.10	7.273		215	340		555	760
7200	30" H x 30" W x 10" D		1	8		225	375		600	830
7250	36" H x 30" W x 10" D		.90	8.889		305	420		725	980
7300	36" H x 36" W x 10" D		.80	10		325	470		795	1,075
7500	Double door, 48" H x 36" W x 10" D		.60	13.333		625	625		1,250	1,650
7550	24" H x 24" W x 12" D		1	8		274	375		649	880
7600	Telephone with wood backboard									
7620	Single door, 12" H x 12" W x 4" D	1 Elec	5.30	1.509	Ea.	117	71		188	238
7650	18" H x 12" W x 4" D		4.70	1.702		161	80		241	300

26 27 16 – Electrical Cabinets and Enclosures

26 27 16.10 Cabinets		Crew	Daily Output	Labor-Hours	Unit	Material	2009 Bare Costs Labor	Equipment	Total	Total Incl O&P
7700	24" H x 12" W x 4" D	1 Elec	4.20	1.905	Ea.	183	89.50		272.50	340
7720	18" H x 18" W x 4" D		4.20	1.905		198	89.50		287.50	355
7750	24" H x 18" W x 4" D		4	2		270	94		364	440
7780	36" H x 36" W x 4" D		3.60	2.222		370	104		474	570
7800	24" H x 24" W x 6" D		3.60	2.222		355	104		459	550
7820	30" H x 24" W x 6" D		3.20	2.500		460	118		578	685
7850	30" H x 30" W x 6" D		2.70	2.963		530	139		669	800
7880	36" H x 30" W x 6" D		2.50	3.200		565	150		715	850
7900	48" H x 36" W x 6" D		2.20	3.636		1,000	171		1,171	1,375
7920	Double door, 48" H x 36" W x 6" D		2	4		1,025	188		1,213	1,450
8000	NEMA 12, double door, floor mounted									
8020	54" H x 42" W x 8" D	2 Elec	6	2.667	Ea.	1,075	125		1,200	1,375
8040	60" H x 48" W x 8" D		5.40	2.963		1,475	139		1,614	1,825
8060	60" H x 48" W x 10" D		5.40	2.963		1,475	139		1,614	1,850
8080	60" H x 60" W x 10" D		5	3.200		1,700	150		1,850	2,075
8100	72" H x 60" W x 10" D		4	4		1,925	188		2,113	2,425
8120	72" H x 72" W x 10" D		3.40	4.706		2,150	221		2,371	2,725
8140	60" H x 48" W x 12" D		3.40	4.706		1,525	221		1,746	2,025
8160	60" H x 60" W x 12" D		3.20	5		1,700	235		1,935	2,225
8180	72" H x 60" W x 12" D		3	5.333		1,725	251		1,976	2,275
8200	72" H x 72" W x 12" D		3	5.333		1,950	251		2,201	2,500
8220	60" H x 48" W x 16" D		3.20	5		1,600	235		1,835	2,125
8240	72" H x 72" W x 16" D		2.60	6.154		2,275	289		2,564	2,950
8260	60" H x 48" W x 20" D		3	5.333		1,750	251		2,001	2,300
8280	72" H x 72" W x 20" D		2.20	7.273		2,475	340		2,815	3,250
8300	60" H x 48" W x 24" D		2.60	6.154		1,850	289		2,139	2,475
8320	72" H x 72" W x 24" D		2	8		2,600	375		2,975	3,450
8340	Pushbutton enclosure, oiltight									
8360	3-1/2" H x 3-1/4" W x 2-3/4" D, for 1 P.B.	1 Elec	12	.667	Ea.	49	31.50		80.50	102
8380	5-3/4" H x 3-1/4" W x 2-3/4" D, for 2 P.B.		11	.727		49	34		83	107
8400	8" H x 3-1/4" W x 2-3/4" D, for 3 P.B.		10.50	.762		67	36		103	129
8420	10-1/4" H x 3-1/4" W x 2-3/4" D, for 4 P.B.		10.50	.762		69	36		105	131
8440	7-1/4" H x 6-1/4" W x 3" D, for 4 P.B.		10	.800		73	37.50		110.50	139
8460	12-1/2" H x 3-1/4" W x 3" D, for 5 P.B.		9	.889		77	42		119	150
8480	9-1/2" H x 6-1/4" W x 3" D, for 6 P.B.		8.50	.941		88	44		132	165
8500	9-1/2" H x 8-1/2" W x 3" D, for 9 P.B.		8	1		95	47		142	178
8510	11-3/4" H x 8-1/2" W x 3" D, for 12 P.B.		7	1.143		105	53.50		158.50	199
8520	11-3/4" H x 10-3/4" W x 3" D, for 16 P.B.		6.50	1.231		136	58		194	239
8540	14" H x 10-3/4" W x 3" D, for 20 P.B.		5	1.600		154	75		229	285
8560	14" H x 13" W x 3" D, for 25 P.B.		4.50	1.778		175	83.50		258.50	320
8580	Sloping front pushbutton enclosures									
8600	3-1/2" H x 7-3/4" W x 4-7/8" D, for 3 P.B	1 Elec	10	.800	Ea.	81	37.50		118.50	147
8620	7-1/4" H x 8-1/2" W x 6-3/4" D, for 6 P.B.		8	1		119	47		166	204
8640	9-1/2" H x 8-1/2" W x 7-7/8" D, for 9 P.B.		7	1.143		131	53.50		184.50	227
8660	11-1/4" H x 8-1/2" W x 9" D, for 12 P.B.		5	1.600		154	75		229	285
8680	11-3/4" H x 10" W x 9" D, for 16 P.B.		5	1.600		172	75		247	305
8700	11-3/4" H x 13" W x 9" D, for 20 P.B.		5	1.600		201	75		276	335
8720	14" H x 13" W x 10-1/8" D, for 25 P.B.		4.50	1.778		226	83.50		309.50	380
8740	Pedestals, not including P.B. enclosure or base									
8760	Straight column 4" x 4"	1 Elec	4.50	1.778	Ea.	205	83.50		288.50	355
8780	6" x 6"		4	2		305	94		399	485
8800	Angled column 4" x 4"		4.50	1.778		240	83.50		323.50	395
8820	6" x 6"		4	2		365	94		459	545

26 27 16.10 Cabinets		Crew	Daily Output	Labor-Hours	Unit	Material	2009 Bare Costs Labor	Equipment	Total	Total Incl O&P
8840	Pedestal, base 18" x 18"	1 Elec	10	.800	Ea.	125	37.50		162.50	196
8860	24" x 24"		9	.889		282	42		324	375
8900	Electronic rack enclosures									
8920	72" H x 25" W x 24" D	1 Elec	1.50	5.333	Ea.	2,100	251		2,351	2,700
8940	72" H x 30" W x 24" D		1.50	5.333		2,250	251		2,501	2,850
8960	72" H x 25" W x 30" D		1.30	6.154		2,275	289		2,564	2,950
8980	72" H x 25" W x 36" D		1.20	6.667		2,425	315		2,740	3,150
9000	72" H x 30" W x 36" D		1.20	6.667		2,575	315		2,890	3,325
9020	NEMA 12 & 4 enclosure panels									
9040	12" x 24"	1 Elec	20	.400	Ea.	29	18.80		47.80	61
9060	16" x 12"		20	.400		20.50	18.80		39.30	52
9080	20" x 16"		20	.400		30.50	18.80		49.30	62.50
9100	20" x 20"		19	.421		36.50	19.80		56.30	71
9120	24" x 20"		18	.444		47.50	21		68.50	84.50
9140	24" x 24"		17	.471		54.50	22		76.50	94
9160	30" x 20"		16	.500		57	23.50		80.50	98.50
9180	30" x 24"		16	.500		66	23.50		89.50	109
9200	36" x 24"		15	.533		79	25		104	126
9220	36" x 30"		15	.533		105	25		130	155
9240	42" x 24"		15	.533		91	25		116	139
9260	42" x 30"		14	.571		121	27		148	175
9280	42" x 36"		14	.571		142	27		169	198
9300	48" x 24"		14	.571		103	27		130	155
9320	48" x 30"		14	.571		135	27		162	191
9340	48" x 36"		13	.615		160	29		189	221
9360	60" x 36"		12	.667		197	31.50		228.50	265
9400	Wiring trough steel JIC, clamp cover									
9490	4" x 4", 12" long	1 Elec	12	.667	Ea.	73	31.50		104.50	129
9510	24" long		10	.800		93.50	37.50		131	161
9530	36" long		8	1		114	47		161	198
9540	48" long		7	1.143		134	53.50		187.50	230
9550	60" long		6	1.333		154	62.50		216.50	266
9560	6" x 6", 12" long		11	.727		96	34		130	159
9580	24" long		9	.889		125	42		167	203
9600	36" long		7	1.143		157	53.50		210.50	256
9610	48" long		6	1.333		189	62.50		251.50	305
9620	60" long		5	1.600		222	75		297	360
9990	Minimum labor/equipment charge		2	4	Job		188		188	289

26 27 16.20 Cabinets and Enclosures		Crew	Daily Output	Labor-Hours	Unit	Material	Labor	Equipment	Total	Total Incl O&P
0010	**CABINETS AND ENCLOSURES** Nonmetallic									
0080	Enclosures fiberglass NEMA 4X									
0100	Wall mount, quick release latch door, 20"H x 16"W x 6"D	1 Elec	4.80	1.667	Ea.	500	78.50		578.50	670
0110	20"H x 20"W x 6"D		4.50	1.778		595	83.50		678.50	785
0120	24"H x 20"W x 6"D		4.20	1.905		640	89.50		729.50	845
0130	20"H x 16"W x 8"D		4.50	1.778		570	83.50		653.50	755
0140	20"H x 20"W x 8"D		4.20	1.905		630	89.50		719.50	835
0150	24"H x 24"W x 8"D		3.80	2.105		710	99		809	930
0160	30"H x 24"W x 8"D		3.20	2.500		790	118		908	1,050
0170	36"H x 30"W x 8"D		3	2.667		1,100	125		1,225	1,400
0180	20"H x 16"W x 10"D		3.50	2.286		650	107		757	880
0190	20"H x 20"W x 10"D		3.20	2.500		700	118		818	950
0200	24"H x 20"W x 10"D		3	2.667		750	125		875	1,025

26 27 16 – Electrical Cabinets and Enclosures

26 27 16.20 Cabinets and Enclosures		Crew	Daily Output	Labor-Hours	Unit	Material	Labor	Equipment	Total	Total Incl O&P
0210	30"H x 24"W x 10"D	1 Elec	2.80	2.857	Ea.	855	134		989	1,150
0220	20"H x 16"W x 12"D		3	2.667		720	125		845	985
0230	20"H x 20"W x 12"D		2.80	2.857		755	134		889	1,025
0240	24"H x 24"W x 12"D		2.60	3.077		825	145		970	1,125
0250	30"H x 24"W x 12"D		2.40	3.333		910	157		1,067	1,250
0260	36"H x 30"W x 12"D		2.20	3.636		1,275	171		1,446	1,675
0270	36"H x 36"W x 12"D		2.10	3.810		1,400	179		1,579	1,800
0280	48"H x 36"W x 12"D		2	4		1,625	188		1,813	2,100
0290	60"H x 36"W x 12"D		1.80	4.444		1,825	209		2,034	2,325
0300	30"H x 24"W x 16"D		1.40	5.714		1,075	269		1,344	1,625
0310	48"H x 36"W x 16"D		1.20	6.667		1,825	315		2,140	2,475
0320	60"H x 36"W x 16"D		1	8		2,050	375		2,425	2,825
0480	Freestanding, one door, 72"H x 25"W x 25"D		.80	10		3,525	470		3,995	4,600
0490	Two doors with two panels, 72"H x 49"W x 24"D		.50	16		8,325	750		9,075	10,300
0500	Floor stand kits, for NEMA 4 & 12, 20"W or more		24	.333		103	15.65		118.65	137
0510	6"H x 10"D		24	.333		113	15.65		128.65	148
0520	6"H x 12"D		24	.333		129	15.65		144.65	166
0530	6"H x 18"D		24	.333		152	15.65		167.65	191
0540	12"H x 8"D		22	.364		132	17.10		149.10	172
0550	12"H x 10"D		22	.364		144	17.10		161.10	185
0560	12"H x 12"D		22	.364		153	17.10		170.10	195
0570	12"H x 16"D		22	.364		173	17.10		190.10	217
0580	12"H x 18"D		22	.364		182	17.10		199.10	227
0590	12"H x 20"D		22	.364		191	17.10		208.10	237
0600	18"H x 8"D		20	.400		162	18.80		180.80	207
0610	18"H x 10"D		20	.400		173	18.80		191.80	219
0620	18"H x 12"D		20	.400		182	18.80		200.80	229
0630	18"H x 16"D		20	.400		204	18.80		222.80	253
0640	24"H x 8"D		16	.500		191	23.50		214.50	246
0650	24"H x 10"D		16	.500		204	23.50		227.50	260
0660	24"H x 12"D		16	.500		213	23.50		236.50	270
0670	24"H x 16"D		16	.500		234	23.50		257.50	293
0680	Small, screw cover, 5-1/2"H x 4"W x 4-15/16"D		12	.667		62	31.50		93.50	116
0690	7-1/2"H x 4"W x 4-15/16"D		12	.667		65	31.50		96.50	120
0700	7-1/2"H x 6"W x 5-3/16"D		10	.800		75	37.50		112.50	141
0710	9-1/2"H x 6"W x 5-11/16"D		10	.800		78	37.50		115.50	144
0720	11-1/2"H x 8"W x 6-11/16"D		8	1		114	47		161	198
0730	13-1/2"H x 10"W x 7-3/16"D		7	1.143		140	53.50		193.50	237
0740	15-1/2"H x 12"W x 8-3/16"D		6	1.333		185	62.50		247.50	300
0750	17-1/2"H x 14"W x 8-11/16"D		5	1.600		213	75		288	350
0760	Screw cover with window, 6"H x 4"W x 5"D		12	.667		101	31.50		132.50	159
0770	8"H x 4"W x 5"D		11	.727		107	34		141	171
0780	8"H x 6"W x 5"D		11	.727		152	34		186	220
0790	10"H x 6"W x 6"D		10	.800		159	37.50		196.50	233
0800	12"H x 8"W x 7"D		8	1		207	47		254	300
0810	14"H x 10"W x 7"D		7	1.143		246	53.50		299.50	355
0820	16"H x 12"W x 8"D		6	1.333		295	62.50		357.50	420
0830	18"H x 14"W x 9"D		5	1.600		355	75		430	505
0840	Quick-release latch cover, 5-1/2"H x 4"W x 5"D		12	.667		80.50	31.50		112	137
0850	7-1/2"H x 4"W x 5"D		12	.667		83.50	31.50		115	140
0860	7-1/2"H x 6"W x 5-1/4"D		10	.800		96.50	37.50		134	164
0870	9-1/2"H x 6"W x 5-3/4"D		10	.800		100	37.50		137.50	168
0880	11-1/2"H x 8"W x 6-3/4"D		8	1		147	47		194	235

26 27 16.20 Cabinets and Enclosures		Crew	Daily Output	Labor-Hours	Unit	2009 Bare Costs Material	Labor	Equipment	Total	Total Incl O&P
0890	13-1/2"H x 10"W x 7-1/4"D	1 Elec	7	1.143	Ea.	181	53.50		234.50	282
0900	15-1/2"H x 12"W x 8-1/4"D		6	1.333		233	62.50		295.50	355
0910	17-1/2"H x 14"W x 8-3/4"D		5	1.600		276	75		351	420
0920	Pushbutton, 1 hole 5-1/2"H x 4"W x 4-15/16"D		12	.667		59.50	31.50		91	113
0930	2 hole 7-1/2"H x 4"W x 4-15/16"D		11	.727		66.50	34		100.50	126
0940	4 hole 7-1/2"H x 6"W x 5-3/16"D		10.50	.762		79.50	36		115.50	143
0950	6 hole 9-1/2"H x 6"W x 5-11/16"D		9	.889		97	42		139	172
0960	8 hole 11-1/2"H x 8"W x 6-11/16"D		8.50	.941		123	44		167	203
0970	12 hole 13-1/2"H x 10"W x 7-3/16"D		8	1		159	47		206	248
0980	20 hole 15-1/2"H x 12"W x 8-3/16"D		5	1.600		217	75		292	355
0990	30 hole 17-1/2"H x 14"W x 8-11/16"D		4.50	1.778		241	83.50		324.50	395
1450	Enclosures polyester NEMA 4X									
1460	Small, screw cover,									
1500	3-15/16"H x 3-15/16"W x 3-1/16"D	1 Elec	12	.667	Ea.	49.50	31.50		81	103
1510	5-3/16"H x 3-5/16"W x 3-1/16"D		12	.667		48.50	31.50		80	101
1520	5-7/8"H x 3-7/8"W x 4-3/16"D		12	.667		51.50	31.50		83	105
1530	5-7/8"H x 5-7/8"W x 4-3/16"D		12	.667		58.50	31.50		90	112
1540	7-5/8"H x 3-5/16"W x 3-1/16"D		12	.667		53	31.50		84.50	107
1550	10-3/16"H x 3-5/16"W x 3-1/16"D		10	.800		63.50	37.50		101	128
1560	Clear cover, 3-15/16"H x 3-15/16"W x 2-7/8"D		12	.667		56	31.50		87.50	110
1570	5-3/16"H x 3-5/16"W x 2-7/8"D		12	.667		60	31.50		91.50	114
1580	5-7/8"H x 3-7/8"W x 4"D		12	.667		74.50	31.50		106	130
1590	5-7/8"H x 5-7/8"W x 4"D		12	.667		92	31.50		123.50	149
1600	7-5/8"H x 3-5/16"W x 2-7/8"D		12	.667		67	31.50		98.50	122
1610	10-3/16"H x 3-5/16"W x 2-7/8"D		10	.800		84	37.50		121.50	150
1620	Pushbutton, 1 hole, 5-5/16"H x 3-5/16"W x 3-1/16"D		12	.667		44	31.50		75.50	96.50
1630	2 hole, 7-5/8"H x 3-5/16"W x 3-1/8"D		11	.727		49	34		83	107
1640	3 hole, 10-3/16"H x 3-5/16"W x 3-1/16"D		10.50	.762		58.50	36		94.50	120
8000	Wireway fiberglass, straight sect. screwcover, 12" L, 4" W x 4" D		40	.200		150	9.40		159.40	179
8010	6" W x 6" D		30	.267		202	12.55		214.55	241
8020	24" L, 4" W x 4" D		20	.400		190	18.80		208.80	238
8030	6" W x 6" D		15	.533		293	25		318	360
8040	36" L, 4" W x 4" D		13.30	.602		233	28.50		261.50	300
8050	6" W x 6" D		10	.800		370	37.50		407.50	465
8060	48" L, 4" W x 4" D		10	.800		274	37.50		311.50	360
8070	6" W x 6" D		7.50	1.067		445	50		495	565
8080	60" L, 4" W x 4" D		8	1		315	47		362	420
8090	6" W x 6" D		6	1.333		520	62.50		582.50	665
8100	Elbow, 90°, 4" W x 4" D		20	.400		149	18.80		167.80	193
8110	6" W x 6" D		18	.444		298	21		319	360
8120	Elbow, 45°, 4" W x 4" D		20	.400		149	18.80		167.80	193
8130	6" W x 6" D		18	.444		292	21		313	350
8140	Tee, 4" W x 4" D		16	.500		190	23.50		213.50	245
8150	6" W x 6" D		14	.571		345	27		372	420
8160	Cross, 4" W x 4" D		14	.571		267	27		294	335
8170	6" W x 6" D		12	.667		500	31.50		531.50	600
8180	Cut-off fitting, w/flange & adhesive, 4" W x 4" D		18	.444		115	21		136	159
8190	6" W x 6" D		16	.500		233	23.50		256.50	292
8200	Flexible ftng., hvy neoprene coated nylon, 4" W x 4" D		20	.400		190	18.80		208.80	238
8210	6" W x 6" D		18	.444		284	21		305	340
8220	Closure plate, fiberglass, 4" W x 4" D		20	.400		36.50	18.80		55.30	69.50
8230	6" W x 6" D		18	.444		41.50	21		62.50	77.50
8240	Box connector, stainless steel type 304, 4" W x 4" D		20	.400		64.50	18.80		83.30	99.50

26 27 Low-Voltage Distribution Equipment

26 27 16 – Electrical Cabinets and Enclosures

26 27 16.20 Cabinets and Enclosures		Crew	Daily Output	Labor-Hours	Unit	2009 Bare Costs Material	Labor	Equipment	Total	Total Incl O&P
8250	6″ W x 6″ D	1 Elec	18	.444	Ea.	70.50	21		91.50	110
8260	Hanger, 4″ W x 4″ D		100	.080		16.10	3.76		19.86	23.50
8270	6″ W x 6″ D		80	.100		19.20	4.70		23.90	28.50
8280	Straight tube section fiberglass, 4″W x 4″D, 12″ long		40	.200		130	9.40		139.40	157
8290	24″ long		20	.400		151	18.80		169.80	195
8300	36″ long		13.30	.602		187	28.50		215.50	250
8310	48″ long		10	.800		208	37.50		245.50	287
8320	60″ long		8	1		235	47		282	330
8330	120″ long		4	2		360	94		454	540

26 27 19 – Multi-Outlet Assemblies

26 27 19.10 Wiring Duct

Code	Description	Crew	Daily Output	Labor-Hours	Unit	Material	Labor	Equipment	Total	Total Incl O&P
0010	**WIRING DUCT** Plastic									
1250	PVC, snap-in slots, adhesive backed									
1270	1-1/2″W x 2″H	2 Elec	120	.133	L.F.	5.65	6.25		11.90	15.85
1280	1-1/2″W x 3″H		120	.133		6.90	6.25		13.15	17.25
1290	1-1/2″W x 4″H		120	.133		8.10	6.25		14.35	18.55
1300	2″W x 1″H		120	.133		5.35	6.25		11.60	15.55
1310	2″W x 1-1/2″H		120	.133		5.65	6.25		11.90	15.85
1320	2″W x 2″H		120	.133		5.75	6.25		12	16
1330	2″W x 2-1/2″H		120	.133		7.05	6.25		13.30	17.40
1340	2″W x 3″H		120	.133		7.30	6.25		13.55	17.70
1350	2″W x 4″H		120	.133		8.60	6.25		14.85	19.10
1360	2-1/2″W x 3″H		120	.133		7.65	6.25		13.90	18.05
1370	3″W x 1″H		110	.145		6	6.85		12.85	17.10
1380	3″W x 1-1/4″H		110	.145		7	6.85		13.85	18.20
1390	3″W x 2″H		110	.145		7.05	6.85		13.90	18.25
1400	3″W x 3″H		110	.145		8.30	6.85		15.15	19.65
1410	3″W x 4″H		110	.145		10.20	6.85		17.05	21.50
1420	3″W x 5″H		110	.145		13.10	6.85		19.95	25
1430	4″W x 1-1/2″H		100	.160		7.05	7.50		14.55	19.30
1440	4″W x 2″H		100	.160		7.90	7.50		15.40	20.50
1450	4″W x 3″H		100	.160		9.10	7.50		16.60	21.50
1460	4″W x 4″H		100	.160		11.10	7.50		18.60	24
1470	4″W x 5″H		100	.160		15.35	7.50		22.85	28.50
1550	Cover, 1-1/2″W		200	.080		1.05	3.76		4.81	6.95
1560	2″W		200	.080		1.28	3.76		5.04	7.20
1570	2-1/2″W		200	.080		1.62	3.76		5.38	7.60
1580	3″W		200	.080		1.94	3.76		5.70	7.95
1590	4″W		200	.080		2.36	3.76		6.12	8.40

26 27 23 – Indoor Service Poles

26 27 23.40 Surface Raceway

Code	Description	Crew	Daily Output	Labor-Hours	Unit	Material	Labor	Equipment	Total	Total Incl O&P
0010	**SURFACE RACEWAY**									
0090	Metal, straight section									
0100	No. 500	1 Elec	100	.080	L.F.	.98	3.76		4.74	6.90
0110	No. 700		100	.080		1.10	3.76		4.86	7
0400	No. 1500, small pancake		90	.089		1.99	4.18		6.17	8.65
0600	No. 2000, base & cover, blank		90	.089		2.03	4.18		6.21	8.70
0610	Receptacle, 6″ O.C.		40	.200		11.35	9.40		20.75	27
0620	12″ O.C.		44	.182		7.40	8.55		15.95	21.50
0630	18″ O.C.		46	.174		7.80	8.15		15.95	21
0650	30″ O.C.		50	.160		5.40	7.50		12.90	17.50
0660	60″ O.C.		50	.160		5	7.50		12.50	17.05

26 27 23.40 Surface Raceway		Crew	Daily Output	Labor-Hours	Unit	Material	2009 Bare Costs Labor	Equipment	Total	Total Incl O&P
0670	No. 2200, base & cover, blank	1 Elec	80	.100	L.F.	3	4.70		7.70	10.55
0700	Receptacle, 18" O.C.		36	.222		7.40	10.45		17.85	24
0720	30" O.C.		40	.200		7	9.40		16.40	22
0730	60" O.C.		40	.200		6.20	9.40		15.60	21.50
0800	No. 3000, base & cover, blank		75	.107		3.91	5		8.91	12
0810	Receptacle, 6" O.C.		45	.178		35	8.35		43.35	51.50
0820	12" O.C.		62	.129		19.80	6.05		25.85	31.50
0830	18" O.C.		64	.125		16	5.90		21.90	26.50
0840	24" O.C.		66	.121		12	5.70		17.70	22
0850	30" O.C.		68	.118		10.80	5.55		16.35	20.50
0860	60" O.C.		70	.114		7.90	5.35		13.25	16.95
1000	No. 4000, base & cover, blank		65	.123		6.40	5.80		12.20	15.95
1010	Receptacle, 6" O.C.		41	.195		48.50	9.15		57.65	67
1020	12" O.C.		52	.154		29.50	7.25		36.75	43
1030	18" O.C.		54	.148		24	6.95		30.95	37
1040	24" O.C.		56	.143		19.60	6.70		26.30	32
1050	30" O.C.		58	.138		17.50	6.50		24	29
1060	60" O.C.		60	.133		13.60	6.25		19.85	24.50
1200	No. 6000, base & cover, blank		50	.160		10.60	7.50		18.10	23
1210	Receptacle, 6" O.C.		30	.267		59	12.55		71.55	84.50
1220	12" O.C.		37	.216		38	10.15		48.15	57.50
1230	18" O.C.		39	.205		32	9.65		41.65	50.50
1240	24" O.C.		41	.195		26.50	9.15		35.65	43
1250	30" O.C.		43	.186		25	8.75		33.75	41
1260	60" O.C.		45	.178		19.40	8.35		27.75	34.50
2400	Fittings, elbows, No. 500		40	.200	Ea.	1.76	9.40		11.16	16.40
2800	Elbow cover, No. 2000		40	.200		3.35	9.40		12.75	18.15
2880	Tee, No. 500		42	.190		3.40	8.95		12.35	17.50
2900	No. 2000		27	.296		11.15	13.95		25.10	34
3000	Switch box, No. 500		16	.500		11.55	23.50		35.05	48.50
3400	Telephone outlet, No. 1500		16	.500		13.80	23.50		37.30	51
3600	Junction box, No. 1500		16	.500		9.40	23.50		32.90	46.50
3800	Plugmold wired sections, No. 2000									
4000	1 circuit, 6 outlets, 3 ft. long	1 Elec	8	1	Ea.	35.50	47		82.50	112
4100	2 circuits, 8 outlets, 6 ft. long		5.30	1.509		57	71		128	172
4110	Tele-power pole, alum, w/ 2 recept, 10'		4	2		192	94		286	355
4120	12'		3.85	2.078		230	97.50		327.50	405
4130	15'		3.70	2.162		300	102		402	485
4140	Steel, w/ 2 recept, 10'		4	2		123	94		217	280
4150	One phone fitting, 10'		4	2		174	94		268	335
4160	Alum, 4 outlets, 10'		3.70	2.162		246	102		348	425
4300	Overhead distribution systems, 125 volt									
4800	No. 2000, entrance end fitting	1 Elec	20	.400	Ea.	4.75	18.80		23.55	34.50
5000	Blank end fitting		40	.200		2.10	9.40		11.50	16.75
5200	Supporting clip		40	.200		1.20	9.40		10.60	15.75
5800	No. 3000, entrance end fitting		20	.400		8.60	18.80		27.40	38.50
6000	Blank end fitting		40	.200		2.45	9.40		11.85	17.15
6020	Internal elbow		20	.400		11.20	18.80		30	41.50
6030	External elbow		20	.400		15.90	18.80		34.70	46.50
6040	Device bracket		53	.151		3.80	7.10		10.90	15.10
6400	Hanger clamp		32	.250		5.80	11.75		17.55	24.50
7000	No. 4000 Base		90	.089	L.F.	4.10	4.18		8.28	10.95
7200	Divider		100	.080	"	.80	3.76		4.56	6.70

26 27 23 – Indoor Service Poles

26 27 23.40 Surface Raceway		Crew	Daily Output	Labor-Hours	Unit	2009 Bare Costs Material	Labor	Equipment	Total	Total Incl O&P
7400	Entrance end fitting	1 Elec	16	.500	Ea.	22	23.50		45.50	60
7600	Blank end fitting		40	.200		5.80	9.40		15.20	21
7610	Recp. & tele. cover		53	.151		9.80	7.10		16.90	21.50
7620	External elbow		16	.500		32	23.50		55.50	71.50
7630	Coupling		53	.151		4.90	7.10		12	16.30
7640	Divider clip & coupling		80	.100		.97	4.70		5.67	8.30
7650	Panel connector		16	.500		20.50	23.50		44	58.50
7800	Take off connector		16	.500		67.50	23.50		91	111
8000	No. 6000, take off connector		16	.500		80.50	23.50		104	125
8100	Take off fitting		16	.500		59.50	23.50		83	102
8200	Hanger clamp		32	.250		14.45	11.75		26.20	34
8230	Coupling					7.10			7.10	7.80
8240	One gang device plate	1 Elec	53	.151		7.70	7.10		14.80	19.35
8250	Two gang device plate		40	.200		9.45	9.40		18.85	25
8260	Blank end fitting		40	.200		8.15	9.40		17.55	23.50
8270	Combination elbow		14	.571		33.50	27		60.50	78
8300	Panel connector		16	.500		14.65	23.50		38.15	52
8500	Chan-L-Wire system installed in 1-5/8" x 1-5/8" strut. Strut									
8600	not incl., 30 amp, 4 wire, 3 phase	1 Elec	200	.040	L.F.	4.60	1.88		6.48	7.95
8700	Junction box		8	1	Ea.	31	47		78	107
8800	Insulating end cap		40	.200		8.20	9.40		17.60	23.50
8900	Strut splice plate		40	.200		10.85	9.40		20.25	26.50
9000	Tap		40	.200		21.50	9.40		30.90	38
9100	Fixture hanger		60	.133		9.30	6.25		15.55	19.90
9200	Pulling tool					83			83	91.50
9300	Non-metallic, straight section									
9310	7/16" x 7/8", base & cover, blank	1 Elec	160	.050	L.F.	1.60	2.35		3.95	5.40
9320	Base & cover w/ adhesive		160	.050		1.75	2.35		4.10	5.55
9340	7/16" x 1-5/16", base & cover, blank		145	.055		1.75	2.59		4.34	5.90
9350	Base & cover w/ adhesive		145	.055		2.05	2.59		4.64	6.25
9370	11/16" x 2-1/4", base & cover, blank		130	.062		2.50	2.89		5.39	7.20
9380	Base & cover w/ adhesive		130	.062		2.85	2.89		5.74	7.60
9400	Fittings, elbows, 7/16" x 7/8"		50	.160	Ea.	1.75	7.50		9.25	13.50
9410	7/16" x 1-5/16"		45	.178		1.82	8.35		10.17	14.85
9420	11/16" x 2-1/4"		40	.200		1.97	9.40		11.37	16.60
9430	Tees, 7/16" x 7/8"		35	.229		2.30	10.75		13.05	19.10
9440	7/16" x 1-5/16"		32	.250		2.37	11.75		14.12	20.50
9450	11/16" x 2-1/4"		30	.267		2.40	12.55		14.95	22
9460	Cover clip, 7/16" x 7/8"		80	.100		.46	4.70		5.16	7.75
9470	7/16" x 1-5/16"		72	.111		.41	5.20		5.61	8.50
9480	11/16" x 2-1/4"		64	.125		.70	5.90		6.60	9.80
9490	Blank end, 7/16" x 7/8"		50	.160		.66	7.50		8.16	12.30
9500	7/16" x 1-5/16"		45	.178		.74	8.35		9.09	13.65
9510	11/16" x 2-1/4"		40	.200		1.12	9.40		10.52	15.70
9520	Round fixture box 5.5" dia x 1"		25	.320		10.55	15.05		25.60	34.50
9530	Device box, 1 gang		30	.267		4.70	12.55		17.25	24.50
9540	2 gang		25	.320		6.95	15.05		22	30.50
9990	Minimum labor/equipment charge		5	1.600	Job		75		75	116

26 27 26 – Wiring Devices

26 27 26.10 Low Voltage Switching		Crew	Daily Output	Labor-Hours	Unit	Material	Labor	Equipment	Total	Total Incl O&P
0010	LOW VOLTAGE SWITCHING									
3600	Relays, 120 V or 277 V standard	1 Elec	12	.667	Ea.	31	31.50		62.50	82

26 27 26 – Wiring Devices

26 27 26.10 Low Voltage Switching		Crew	Daily Output	Labor-Hours	Unit	Material	Labor	Equipment	Total	Total Incl O&P
3800	Flush switch, standard	1 Elec	40	.200	Ea.	10.75	9.40		20.15	26.50
4000	Interchangeable		40	.200		14.05	9.40		23.45	30
4100	Surface switch, standard		40	.200		7.90	9.40		17.30	23
4200	Transformer 115 V to 25 V		12	.667		110	31.50		141.50	169
4400	Master control, 12 circuit, manual		4	2		112	94		206	269
4500	25 circuit, motorized		4	2		122	94		216	279
4600	Rectifier, silicon		12	.667		36.50	31.50		68	88
4800	Switchplates, 1 gang, 1, 2 or 3 switch, plastic		80	.100		3.59	4.70		8.29	11.20
5000	Stainless steel		80	.100		9.65	4.70		14.35	17.90
5400	2 gang, 3 switch, stainless steel		53	.151		18.60	7.10		25.70	31.50
5500	4 switch, plastic		53	.151		8	7.10		15.10	19.70
5600	2 gang, 4 switch, stainless steel		53	.151		19.45	7.10		26.55	32.50
5700	6 switch, stainless steel		53	.151		43	7.10		50.10	58
5800	3 gang, 9 switch, stainless steel		32	.250		59.50	11.75		71.25	83.50
5900	Receptacle, triple, 1 return, 1 feed		26	.308		38	14.45		52.45	64.50
6000	2 feed		20	.400		38	18.30		56.80	71
6100	Relay gang boxes, flush or surface, 6 gang		5.30	1.509		65.50	71		136.50	181
6200	12 gang		4.70	1.702		69	80		149	199
6400	18 gang		4	2		85.50	94		179.50	240
6500	Frame, to hold up to 6 relays		12	.667		57	31.50		88.50	111
7200	Control wire, 2 conductor		6.30	1.270	C.L.F.	27.50	59.50		87	122
7400	3 conductor		5	1.600		39	75		114	159
7600	19 conductor		2.50	3.200		315	150		465	575
7800	26 conductor		2	4		395	188		583	725
8000	Weatherproof, 3 conductor		5	1.600		82	75		157	206

26 27 26.20 Wiring Devices Elements

		Crew	Daily Output	Labor-Hours	Unit	Material	Labor	Equipment	Total	Total Incl O&P
0010	**WIRING DEVICES ELEMENTS** R262726-90									
0200	Toggle switch, quiet type, single pole, 15 amp	1 Elec	40	.200	Ea.	5.35	9.40		14.75	20.50
0500	20 amp		27	.296		7.40	13.95		21.35	29.50
0510	30 amp		23	.348		24.50	16.35		40.85	52
0530	Lock handle, 20 amp		27	.296		25	13.95		38.95	49
0540	Security key, 20 amp		26	.308		70.50	14.45		84.95	100
0550	Rocker, 15 amp		40	.200		6	9.40		15.40	21
0560	20 amp		27	.296		13.70	13.95		27.65	36.50
0600	3 way, 15 amp		23	.348		7.65	16.35		24	33.50
0800	20 amp		18	.444		8.90	21		29.90	42
0810	30 amp		9	.889		32	42		74	99.50
0830	Lock handle, 20 amp		18	.444		28	21		49	63
0840	Security key, 20 amp		17	.471		76.50	22		98.50	118
0850	Rocker, 15 amp		23	.348		8.45	16.35		24.80	34.50
0860	20 amp		18	.444		19.85	21		40.85	54
0900	4 way, 15 amp		15	.533		25	25		50	66
1000	20 amp		11	.727		39.50	34		73.50	96
1020	Lock handle, 20 amp		11	.727		62	34		96	121
1030	Rocker, 15 amp		15	.533		32.50	25		57.50	74.50
1040	20 amp		11	.727		49	34		83	106
1100	Toggle switch, quiet type, double pole, 15 amp		15	.533		16.25	25		41.25	56.50
1200	20 amp		11	.727		15.90	34		49.90	70
1210	30 amp		9	.889		32.50	42		74.50	100
1230	Lock handle, 20 amp		11	.727		28.50	34		62.50	84
1250	Security key, 20 amp		10	.800		76.50	37.50		114	143
1420	Toggle switch quiet type, 1 pole, 2 throw center off, 15 amp		23	.348		54	16.35		70.35	84.50

26 27 26.20 Wiring Devices Elements		Crew	Daily Output	Labor-Hours	Unit	2009 Bare Costs Material	Labor	Equipment	Total	Total Incl O&P
1440	20 amp	1 Elec	18	.444	Ea.	58.50	21		79.50	96.50
1460	Lock handle, 20 amp		18	.444		60	21		81	98
1480	Momentary contact, 15 amp		23	.348		23.50	16.35		39.85	51
1500	20 amp		18	.444		35.50	21		56.50	71
1520	Momentary contact, lock handle, 20 amp		18	.444		41	21		62	77
1650	Dimmer switch, 120 volt, incandescent, 600 watt, 1 pole (G)		16	.500		13.10	23.50		36.60	50.50
1700	600 watt, 3 way (G)		12	.667		9.10	31.50		40.60	58
1750	1000 watt, 1 pole (G)		16	.500		46	23.50		69.50	86.50
1800	1000 watt, 3 way (G)		12	.667		64	31.50		95.50	119
2000	1500 watt, 1 pole (G)		11	.727		87	34		121	148
2100	2000 watt, 1 pole (G)		8	1		130	47		177	216
2110	Fluorescent, 600 watt (G)		15	.533		71.50	25		96.50	118
2120	1000 watt (G)		15	.533		89.50	25		114.50	137
2130	1500 watt (G)		10	.800		168	37.50		205.50	243
2160	Explosionproof, toggle switch, wall, single pole 20 amp		5.30	1.509		181	71		252	310
2180	Receptacle, single outlet, 20 amp		5.30	1.509		310	71		381	450
2190	30 amp		4	2		580	94		674	780
2290	60 amp		2.50	3.200		660	150		810	955
2360	Plug, 20 amp		16	.500		134	23.50		157.50	183
2370	30 amp		12	.667		238	31.50		269.50	310
2380	60 amp		8	1		335	47		382	440
2410	Furnace, thermal cutoff switch with plate		26	.308		13.60	14.45		28.05	37.50
2460	Receptacle, duplex, 120 volt, grounded, 15 amp		40	.200		1.35	9.40		10.75	15.95
2470	20 amp		27	.296		9.95	13.95		23.90	32.50
2480	Ground fault interrupting, 15 amp		27	.296		32.50	13.95		46.45	57.50
2482	20 amp		27	.296		34.50	13.95		48.45	59.50
2490	Dryer, 30 amp		15	.533		6.05	25		31.05	45
2500	Range, 50 amp		11	.727		13.10	34		47.10	67
2540	Isolated ground receptacle, duplex, 20 amp		27	.296		24	13.95		37.95	47.50
2542	Quad, 20 amp		20	.400		43.50	18.80		62.30	77
2550	Simplex, 20 amp		27	.296		26.50	13.95		40.45	51
2560	Simplex, 30 amp		15	.533		31	25		56	72.50
2570	Cable reel w/ receptacle 50′ w/3#12, 120 V, 20 A	2 Elec	2.67	5.999		720	282		1,002	1,225
2600	Wall plates, stainless steel, 1 gang	1 Elec	80	.100		2.30	4.70		7	9.80
2800	2 gang		53	.151		4.20	7.10		11.30	15.50
3000	3 gang		32	.250		6.50	11.75		18.25	25.50
3100	4 gang		27	.296		10.80	13.95		24.75	33.50
3110	Brown plastic, 1 gang		80	.100		.34	4.70		5.04	7.60
3120	2 gang		53	.151		.83	7.10		7.93	11.80
3130	3 gang		32	.250		1.24	11.75		12.99	19.45
3140	4 gang		27	.296		3.27	13.95		17.22	25
3150	Brushed brass, 1 gang		80	.100		4.80	4.70		9.50	12.55
3160	Anodized aluminum, 1 gang		80	.100		2.70	4.70		7.40	10.20
3170	Switch cover, weatherproof, 1 gang		60	.133		8.95	6.25		15.20	19.45
3180	Vandal proof lock, 1 gang		60	.133		10.95	6.25		17.20	21.50
3200	Lampholder, keyless		26	.308		12.30	14.45		26.75	36
3400	Pullchain with receptacle		22	.364		13.35	17.10		30.45	41
3500	Pilot light, neon with jewel		27	.296		13.55	13.95		27.50	36.50
3600	Receptacle, 20 amp, 250 volt, NEMA 6		27	.296		17.60	13.95		31.55	41
3620	277 volt NEMA 7		27	.296		20	13.95		33.95	43.50
3640	125/250 volt NEMA 10		27	.296		21	13.95		34.95	44.50
3680	125/250 volt NEMA 14		25	.320		26.50	15.05		41.55	52.50
3700	3 pole, 250 volt NEMA 15		25	.320		27.50	15.05		42.55	53.50

26 27 26.20 Wiring Devices Elements		Crew	Daily Output	Labor-Hours	Unit	Material	2009 Bare Costs Labor	Equipment	Total	Total Incl O&P
3720	120/208 volt NEMA 18	1 Elec	25	.320	Ea.	29	15.05		44.05	55
3740	30 amp, 125 volt NEMA 5		15	.533		18.80	25		43.80	59
3760	250 volt NEMA 6		15	.533		21.50	25		46.50	62
3780	277 volt NEMA 7		15	.533		27.50	25		52.50	68.50
3820	125/250 volt NEMA 14		14	.571		59	27		86	106
3840	3 pole, 250 volt NEMA 15		14	.571		55	27		82	102
3880	50 amp, 125 volt NEMA 5		11	.727		24.50	34		58.50	79.50
3900	250 volt NEMA 6		11	.727		24.50	34		58.50	79.50
3920	277 volt NEMA 7		11	.727		32	34		66	87.50
3960	125/250 volt NEMA 14		10	.800		76	37.50		113.50	142
3980	3 pole 250 volt NEMA 15		10	.800		75	37.50		112.50	141
4020	60 amp, 125/250 volt, NEMA 14		8	1		84.50	47		131.50	165
4040	3 pole, 250 volt NEMA 15		8	1		88.50	47		135.50	170
4060	120/208 volt NEMA 18		8	1		85.50	47		132.50	167
4100	Receptacle locking, 20 amp, 125 volt NEMA L5		27	.296		18.95	13.95		32.90	42.50
4120	250 volt NEMA L6		27	.296		18.95	13.95		32.90	42.50
4140	277 volt NEMA L7		27	.296		18.95	13.95		32.90	42.50
4150	3 pole, 250 volt, NEMA L11		27	.296		29.50	13.95		43.45	54
4160	20 amp, 480 volt NEMA L8		27	.296		24	13.95		37.95	48
4180	600 volt NEMA L9		27	.296		36	13.95		49.95	61
4200	125/250 volt NEMA L10		27	.296		29.50	13.95		43.45	54
4230	125/250 volt NEMA L14		25	.320		26.50	15.05		41.55	52
4280	250 volt NEMA L15		25	.320		26.50	15.05		41.55	52
4300	480 volt NEMA L16		25	.320		31.50	15.05		46.55	57.50
4320	3 phase, 120/208 volt NEMA L18		25	.320		40.50	15.05		55.55	67.50
4340	277/480 volt NEMA L19		25	.320		41	15.05		56.05	68
4360	347/600 volt NEMA L20		25	.320		41	15.05		56.05	68
4380	120/208 volt NEMA L21		23	.348		32	16.35		48.35	60.50
4400	277/480 volt NEMA L22		23	.348		38.50	16.35		54.85	67.50
4420	347/600 volt NEMA L23		23	.348		38.50	16.35		54.85	67.50
4440	30 amp, 125 volt NEMA L5		15	.533		26.50	25		51.50	68
4460	250 volt NEMA L6		15	.533		29	25		54	70.50
4480	277 volt NEMA L7		15	.533		29	25		54	70.50
4500	480 volt NEMA L8		15	.533		34.50	25		59.50	76.50
4520	600 volt NEMA L9		15	.533		34.50	25		59.50	76.50
4540	125/250 volt NEMA L10		15	.533		33	25		58	75
4560	3 phase, 250 volt NEMA L11		15	.533		33	25		58	75
4620	125/250 volt NEMA L14		14	.571		42	27		69	87.50
4640	250 volt NEMA L15		14	.571		42	27		69	88
4660	480 volt NEMA L16		14	.571		49	27		76	95.50
4680	600 volt NEMA L17		14	.571		48	27		75	94
4700	120/208 volt NEMA L18		14	.571		60	27		87	108
4720	277/480 volt NEMA L19		14	.571		60	27		87	108
4740	347/600 volt NEMA L20		14	.571		60	27		87	108
4760	120/208 volt NEMA L21		13	.615		44.50	29		73.50	93.50
4780	277/480 volt NEMA L22		13	.615		53.50	29		82.50	103
4800	347/600 volt NEMA L23		13	.615		66.50	29		95.50	118
4840	Receptacle, corrosion resistant, 15 or 20 amp, 125 volt NEMA L5		27	.296		27.50	13.95		41.45	52
4860	250 volt NEMA L6		27	.296		27.50	13.95		41.45	52
4900	Receptacle, cover plate, phenolic plastic, NEMA 5 & 6		80	.100		.60	4.70		5.30	7.90
4910	NEMA 7-23		80	.100		.70	4.70		5.40	8
4920	Stainless steel, NEMA 5 & 6		80	.100		2.10	4.70		6.80	9.55
4930	NEMA 7-23		80	.100		3.20	4.70		7.90	10.75

26 27 26 – Wiring Devices

26 27 26.20 Wiring Devices Elements		Crew	Daily Output	Labor-Hours	Unit	Material	Labor	Equipment	Total	Total Incl O&P
4940	Brushed brass NEMA 5 & 6	1 Elec	80	.100	Ea.	5.25	4.70		9.95	13.05
4950	NEMA 7-23		80	.100		5.70	4.70		10.40	13.50
4960	Anodized aluminum, NEMA 5 & 6		80	.100		2.90	4.70		7.60	10.45
4970	NEMA 7-23		80	.100		4.75	4.70		9.45	12.50
4980	Weatherproof NEMA 7-23		60	.133		30.50	6.25		36.75	43.50
5100	Plug, 20 amp, 250 volt, NEMA 6		30	.267		16.10	12.55		28.65	37
5110	277 volt NEMA 7		30	.267		17.40	12.55		29.95	38.50
5120	3 pole, 120/250 volt, NEMA 10		26	.308		22.50	14.45		36.95	47
5130	125/250 volt NEMA 14		26	.308		44.50	14.45		58.95	71.50
5140	250 volt NEMA 15		26	.308		46.50	14.45		60.95	74
5150	120/208 volt NEMA 8		26	.308		48	14.45		62.45	75
5160	30 amp, 125 volt NEMA 5		13	.615		45.50	29		74.50	94.50
5170	250 volt NEMA 6		13	.615		47	29		76	96.50
5180	277 volt NEMA 7		13	.615		58	29		87	109
5190	125/250 volt NEMA 14		13	.615		52	29		81	102
5200	3 pole, 250 volt NEMA 15		12	.667		54.50	31.50		86	108
5210	50 amp, 125 volt NEMA 5		9	.889		56.50	42		98.50	127
5220	250 volt NEMA 6		9	.889		59	42		101	130
5230	277 volt NEMA 7		9	.889		63	42		105	134
5240	125/250 volt NEMA 14		9	.889		62.50	42		104.50	134
5250	3 pole, 250 volt NEMA 15		8	1		65.50	47		112.50	145
5260	60 amp, 125/250 volt NEMA 14		7	1.143		73	53.50		126.50	163
5270	3 pole, 250 volt NEMA 15		7	1.143		78	53.50		131.50	169
5280	120/208 volt NEMA 18		7	1.143		75.50	53.50		129	166
5300	Plug angle, 20 amp, 250 volt NEMA 6		30	.267		22	12.55		34.55	43.50
5310	30 amp, 125 volt NEMA 5		13	.615		45.50	29		74.50	94.50
5320	250 volt NEMA 6		13	.615		47	29		76	96.50
5330	277 volt NEMA 7		13	.615		58	29		87	109
5340	125/250 volt NEMA 14		13	.615		57.50	29		86.50	108
5350	3 pole, 250 volt NEMA 15		12	.667		60.50	31.50		92	115
5360	50 amp, 125 volt NEMA 5		9	.889		50	42		92	120
5370	250 volt NEMA 6		9	.889		49	42		91	119
5380	277 volt NEMA 7		9	.889		63.50	42		105.50	135
5390	125/250 volt NEMA 14		9	.889		70	42		112	141
5400	3 pole, 250 volt NEMA 15		8	1		73	47		120	153
5410	60 amp, 125/250 volt NEMA 14		7	1.143		81.50	53.50		135	173
5420	3 pole, 250 volt NEMA 15		7	1.143		85	53.50		138.50	176
5430	120/208 volt NEMA 18		7	1.143		85	53.50		138.50	176
5500	Plug, locking, 20 amp, 125 volt NEMA L5		30	.267		14.85	12.55		27.40	35.50
5510	250 volt NEMA L6		30	.267		14.85	12.55		27.40	35.50
5520	277 volt NEMA L7		30	.267		14.85	12.55		27.40	35.50
5530	480 volt NEMA L8		30	.267		18.65	12.55		31.20	40
5540	600 volt NEMA L9		30	.267		22	12.55		34.55	44
5550	3 pole, 125/250 volt NEMA L10		26	.308		20	14.45		34.45	44.50
5560	250 volt NEMA L11		26	.308		20	14.45		34.45	44.50
5570	480 volt NEMA L12		26	.308		23.50	14.45		37.95	48
5580	125/250 volt NEMA L14		26	.308		22.50	14.45		36.95	47.50
5590	250 volt NEMA L15		26	.308		24	14.45		38.45	49
5600	480 volt NEMA L16		26	.308		28.50	14.45		42.95	53.50
5610	4 pole, 120/208 volt NEMA L18		24	.333		36	15.65		51.65	63.50
5620	277/480 volt NEMA L19		24	.333		36	15.65		51.65	63.50
5630	347/600 volt NEMA L20		24	.333		36	15.65		51.65	63.50
5640	120/208 volt NEMA L21		24	.333		29	15.65		44.65	56

26 27 26.20 Wiring Devices Elements		Crew	Daily Output	Labor-Hours	Unit	Material	Labor	Equipment	Total	Total Incl O&P
						2009 Bare Costs				
5650	277/480 volt NEMA L22	1 Elec	24	.333	Ea.	34.50	15.65		50.15	62
5660	347/600 volt NEMA L23		24	.333		34.50	15.65		50.15	62
5670	30 amp, 125 volt NEMA L5		13	.615		23	29		52	70
5680	250 volt NEMA L6		13	.615		23	29		52	70
5690	277 volt NEMA L7		13	.615		22.50	29		51.50	69.50
5700	480 volt NEMA L8		13	.615		27.50	29		56.50	74.50
5710	600 volt NEMA L9		13	.615		26.50	29		55.50	74
5720	3 pole, 125/250 volt NEMA L10		11	.727		21.50	34		55.50	76.50
5730	250 volt NEMA L11		11	.727		21.50	34		55.50	76.50
5760	125/250 volt NEMA L14		11	.727		30.50	34		64.50	86.50
5770	250 volt NEMA L15		11	.727		31	34		65	86.50
5780	480 volt NEMA L16		11	.727		36	34		70	92.50
5790	600 volt NEMA L17		11	.727		37	34		71	93
5800	4 pole, 120/208 volt NEMA L18		10	.800		45	37.50		82.50	108
5810	120/208 volt NEMA L19		10	.800		45	37.50		82.50	108
5820	347/600 volt NEMA L20		10	.800		45	37.50		82.50	108
5830	120/208 volt NEMA L21		10	.800		33	37.50		70.50	94.50
5840	277/480 volt NEMA L22		10	.800		41	37.50		78.50	103
5850	347/600 volt NEMA L23		10	.800		51	37.50		88.50	114
6000	Connector, 20 amp, 250 volt NEMA 6		30	.267		25.50	12.55		38.05	47.50
6010	277 volt NEMA 7		30	.267		28.50	12.55		41.05	51
6020	3 pole, 120/250 volt NEMA 10		26	.308		32	14.45		46.45	58
6030	125/250 volt NEMA 14		26	.308		32	14.45		46.45	58
6040	250 volt NEMA 15		26	.308		32.50	14.45		46.95	58
6050	120/208 volt NEMA 18		26	.308		35.50	14.45		49.95	61.50
6060	30 amp, 125 volt NEMA 5		13	.615		56	29		85	106
6070	250 volt NEMA 6		13	.615		56	29		85	106
6080	277 volt NEMA 7		13	.615		56	29		85	106
6110	50 amp, 125 volt NEMA 5		9	.889		77	42		119	149
6120	250 volt NEMA 6		9	.889		77	42		119	149
6130	277 volt NEMA 7		9	.889		77	42		119	149
6200	Connector, locking, 20 amp, 125 volt NEMA L5		30	.267		23	12.55		35.55	45
6210	250 volt NEMA L6		30	.267		23	12.55		35.55	45
6220	277 volt NEMA L7		30	.267		23	12.55		35.55	45
6230	480 volt NEMA L8		30	.267		34.50	12.55		47.05	57
6240	600 volt NEMA L9		30	.267		34.50	12.55		47.05	57
6250	3 pole, 125/250 volt NEMA L10		26	.308		31.50	14.45		45.95	57
6260	250 volt NEMA L11		26	.308		31.50	14.45		45.95	57
6280	125/250 volt NEMA L14		26	.308		31.50	14.45		45.95	57
6290	250 volt NEMA L15		26	.308		31.50	14.45		45.95	57
6300	480 volt NEMA L16		26	.308		38	14.45		52.45	64
6310	4 pole, 120/208 volt NEMA L18		24	.333		50	15.65		65.65	79
6320	277/480 volt NEMA L19		24	.333		50	15.65		65.65	79
6330	347/600 volt NEMA L20		24	.333		50	15.65		65.65	79
6340	120/208 volt NEMA L21		24	.333		47	15.65		62.65	76
6350	277/480 volt NEMA L22		24	.333		59.50	15.65		75.15	89.50
6360	347/600 volt NEMA L23		24	.333		60	15.65		75.65	90
6370	30 amp, 125 volt NEMA L5		13	.615		46	29		75	95
6380	250 volt NEMA L6		13	.615		46	29		75	95
6390	277 volt NEMA L7		13	.615		46	29		75	95
6400	480 volt NEMA L8		13	.615		54.50	29		83.50	105
6410	600 volt NEMA L9		13	.615		55	29		84	105
6420	3 pole, 125/250 volt NEMA L10		11	.727		61	34		95	120

26 27 Low-Voltage Distribution Equipment

26 27 26 – Wiring Devices

26 27 26.20 Wiring Devices Elements		Crew	Daily Output	Labor-Hours	Unit	Material	Labor	Equipment	Total	Total Incl O&P
						2009 Bare Costs				
6430	250 volt NEMA L11	1 Elec	11	.727	Ea.	61	34		95	120
6460	125/250 volt NEMA L14		11	.727		63.50	34		97.50	122
6470	250 volt NEMA L15		11	.727		63.50	34		97.50	123
6480	480 volt NEMA L16		11	.727		76.50	34		110.50	137
6490	600 volt NEMA L17		11	.727		76.50	34		110.50	137
6500	4 pole, 120/208 volt NEMA L18		10	.800		95	37.50		132.50	163
6510	120/208 volt NEMA L19		10	.800		95.50	37.50		133	163
6520	347/600 volt NEMA L20		10	.800		95.50	37.50		133	163
6530	120/208 volt NEMA L21		10	.800		63	37.50		100.50	128
6540	277/480 volt NEMA L22		10	.800		75.50	37.50		113	141
6550	347/600 volt NEMA L23		10	.800		93.50	37.50		131	161
7000	Receptacle computer, 250 volt, 15 amp, 3 pole 4 wire		8	1		72	47		119	152
7010	20 amp, 2 pole 3 wire		8	1		72	47		119	152
7020	30 amp, 2 pole 3 wire		6.50	1.231		113	58		171	214
7030	30 amp, 3 pole 4 wire		6.50	1.231		121	58		179	222
7040	60 amp, 3 pole 4 wire		4.50	1.778		203	83.50		286.50	355
7050	100 amp, 3 pole 4 wire		3	2.667		257	125		382	475
7100	Connector computer, 250 volt, 15 amp, 3 pole 4 wire		27	.296		100	13.95		113.95	132
7110	20 amp, 2 pole 3 wire		27	.296		100	13.95		113.95	132
7120	30 amp, 2 pole 3 wire		15	.533		158	25		183	213
7130	30 amp, 3 pole 4 wire		15	.533		149	25		174	203
7140	60 amp, 3 pole 4 wire		8	1		256	47		303	355
7150	100 amp, 3 pole 4 wire		4	2		355	94		449	540
7200	Plug, computer, 250 volt, 15 amp, 3 pole 4 wire		27	.296		91	13.95		104.95	122
7210	20 amp, 2 pole, 3 wire		27	.296		91	13.95		104.95	122
7220	30 amp, 2 pole, 3 wire		15	.533		133	25		158	186
7230	30 amp, 3 pole, 4 wire		15	.533		144	25		169	197
7240	60 amp, 3 pole, 4 wire		8	1		214	47		261	310
7250	100 amp, 3 pole, 4 wire		4	2		278	94		372	450
7300	Connector adapter to flexible conduit, 1/2"		60	.133		4.44	6.25		10.69	14.55
7310	3/4"		50	.160		6.50	7.50		14	18.70
7320	1-1/4"		30	.267		16.25	12.55		28.80	37
7330	1-1/2"		23	.348		23.50	16.35		39.85	50.50
9000	Minimum labor/equipment charge		4	2	Job		94		94	145

26 27 73 – Door Chimes

26 27 73.10 Doorbell System		Crew	Daily Output	Labor-Hours	Unit	Material	Labor	Equipment	Total	Total Incl O&P
0010	**DOORBELL SYSTEM**, incl. transformer, button & signal									
0100	6" bell	1 Elec	4	2	Ea.	107	94		201	262
0200	Buzzer		4	2		84.50	94		178.50	238
1000	Door chimes, 2 notes, minimum		16	.500		29.50	23.50		53	68
1020	Maximum		12	.667		150	31.50		181.50	213
1100	Tube type, 3 tube system		12	.667		222	31.50		253.50	293
1180	4 tube system		10	.800		355	37.50		392.50	450
1900	For transformer & button, minimum add		5	1.600		16.15	75		91.15	134
1960	Maximum, add		4.50	1.778		48.50	83.50		132	183
3000	For push button only, minimum		24	.333		3.27	15.65		18.92	27.50
3100	Maximum		20	.400		25.50	18.80		44.30	57
3200	Bell transformer		16	.500		23	23.50		46.50	61
9000	Minimum labor/equipment charge		4	2	Job		94		94	145

26 28 13.10 Fuse Elements		Crew	Daily Output	Labor-Hours	Unit	Material	2009 Bare Costs Labor	Equipment	Total	Total Incl O&P
0010	**FUSE ELEMENTS**									
0020	Cartridge, nonrenewable									
0050	250 volt, 30 amp	1 Elec	50	.160	Ea.	1.98	7.50		9.48	13.75
0100	60 amp		50	.160		3.34	7.50		10.84	15.20
0150	100 amp		40	.200		14.05	9.40		23.45	30
0200	200 amp		36	.222		34	10.45		44.45	53.50
0250	400 amp		30	.267		76.50	12.55		89.05	104
0300	600 amp		24	.333		131	15.65		146.65	168
0400	600 volt, 30 amp		40	.200		9.95	9.40		19.35	25.50
0450	60 amp		40	.200		15.75	9.40		25.15	32
0500	100 amp		36	.222		29.50	10.45		39.95	48.50
0550	200 amp		30	.267		79	12.55		91.55	106
0600	400 amp		24	.333		166	15.65		181.65	207
0650	600 amp		20	.400		250	18.80		268.80	305
0800	Dual element, time delay, 250 volt, 30 amp		50	.160		4.93	7.50		12.43	16.95
0850	60 amp		50	.160		9.05	7.50		16.55	21.50
0900	100 amp		40	.200		20.50	9.40		29.90	37
0950	200 amp		36	.222		44.50	10.45		54.95	65
1000	400 amp		30	.267		80.50	12.55		93.05	108
1050	600 amp		24	.333		132	15.65		147.65	169
1300	600 volt, 15 to 30 amp		40	.200		10.95	9.40		20.35	26.50
1350	35 to 60 amp		40	.200		18.85	9.40		28.25	35
1400	70 to 100 amp		36	.222		39	10.45		49.45	59
1450	110 to 200 amp		30	.267		78	12.55		90.55	105
1500	225 to 400 amp		24	.333		156	15.65		171.65	195
1550	600 amp		20	.400		226	18.80		244.80	277
1800	Class RK1, high capacity, 250 volt, 30 amp		50	.160		7	7.50		14.50	19.25
1850	60 amp		50	.160		12.80	7.50		20.30	25.50
1900	100 amp		40	.200		28.50	9.40		37.90	46
1950	200 amp		36	.222		63	10.45		73.45	85.50
2000	400 amp		30	.267		114	12.55		126.55	144
2050	600 amp		24	.333		186	15.65		201.65	229
2200	600 volt, 30 amp		40	.200		15.90	9.40		25.30	32
2250	60 amp		40	.200		27.50	9.40		36.90	44.50
2300	100 amp		36	.222		56.50	10.45		66.95	78
2350	200 amp		30	.267		113	12.55		125.55	143
2400	400 amp		24	.333		226	15.65		241.65	273
2450	600 amp		20	.400		330	18.80		348.80	390
2700	Class J, current limiting, 250 or 600 volt, 30 amp		40	.200		19.60	9.40		29	36
2750	60 amp		40	.200		32	9.40		41.40	49.50
2800	100 amp		36	.222		48.50	10.45		58.95	69.50
2850	200 amp		30	.267		91	12.55		103.55	119
2900	400 amp		24	.333		201	15.65		216.65	245
2950	600 amp		20	.400		284	18.80		302.80	340
3100	Class L, current limiting, 250 or 600 volt, 601 to 1200 amp		16	.500		495	23.50		518.50	580
3150	1500-1600 amp		13	.615		635	29		664	745
3200	1800-2000 amp		10	.800		845	37.50		882.50	990
3250	2500 amp		10	.800		1,125	37.50		1,162.50	1,300
3300	3000 amp		8	1		1,300	47		1,347	1,500
3350	3500-4000 amp		8	1		1,775	47		1,822	2,050
3400	4500-5000 amp		6.70	1.194		3,450	56		3,506	3,875
3450	6000 amp		5.70	1.404		4,625	66		4,691	5,200
3600	Plug, 120 volt, 1 to 10 amp		50	.160		1.69	7.50		9.19	13.40

26 28 Low-Voltage Circuit Protective Devices

26 28 13 – Fuses

26 28 13.10 Fuse Elements		Crew	Daily Output	Labor-Hours	Unit	Material	Labor	Equipment	Total	Total Incl O&P
						2009 Bare Costs				
3650	15 to 30 amp	1 Elec	50	.160	Ea.	2.40	7.50		9.90	14.20
3700	Dual element 0.3 to 14 amp		50	.160		9.30	7.50		16.80	22
3750	15 to 30 amp		50	.160		2.70	7.50		10.20	14.50
3800	Fustat, 120 volt, 15 to 30 amp		50	.160		3.90	7.50		11.40	15.85
3850	0.3 to 14 amp		50	.160		10.25	7.50		17.75	23
3900	Adapters 0.3 to 10 amp		50	.160		4.05	7.50		11.55	16
3950	15 to 30 amp	↓	50	.160	↓	4.05	7.50		11.55	16

26 28 16 – Enclosed Switches and Circuit Breakers

26 28 16.10 Circuit Breakers

		Crew	Daily Output	Labor-Hours	Unit	Material	Labor	Equipment	Total	Total Incl O&P
0010	**CIRCUIT BREAKERS** (in enclosure)									
0100	Enclosed (NEMA 1), 600 volt, 3 pole, 30 amp	1 Elec	3.20	2.500	Ea.	610	118		728	850
0200	60 amp		2.80	2.857		610	134		744	875
0400	100 amp		2.30	3.478		700	163		863	1,025
0600	225 amp	↓	1.50	5.333		1,625	251		1,876	2,150
0700	400 amp	2 Elec	1.60	10		2,750	470		3,220	3,750
0800	600 amp		1.20	13.333		4,000	625		4,625	5,375
1000	800 amp		.94	17.021		5,200	800		6,000	6,950
1200	1000 amp		.84	19.048		6,575	895		7,470	8,600
1220	1200 amp		.80	20		8,400	940		9,340	10,700
1240	1600 amp		.72	22.222		16,100	1,050		17,150	19,300
1260	2000 amp		.64	25		17,500	1,175		18,675	21,000
1400	1200 amp with ground fault		.80	20		17,100	940		18,040	20,300
1600	1600 amp with ground fault		.72	22.222		18,300	1,050		19,350	21,700
1800	2000 amp with ground fault	↓	.64	25		19,600	1,175		20,775	23,300
2000	Disconnect, 240 volt 3 pole, 5 HP motor	1 Elec	3.20	2.500		430	118		548	650
2020	10 HP motor		3.20	2.500		430	118		548	650
2040	15 HP motor		2.80	2.857		430	134		564	675
2060	20 HP motor		2.30	3.478		520	163		683	820
2080	25 HP motor		2.30	3.478		520	163		683	820
2100	30 HP motor		2.30	3.478		520	163		683	820
2120	40 HP motor		2	4		890	188		1,078	1,275
2140	50 HP motor		1.50	5.333		890	251		1,141	1,375
2160	60 HP motor	↓	1.50	5.333		2,050	251		2,301	2,625
2180	75 HP motor	2 Elec	2	8		2,050	375		2,425	2,825
2200	100 HP motor		1.60	10		2,050	470		2,520	2,975
2220	125 HP motor		1.60	10		2,050	470		2,520	2,975
2240	150 HP motor		1.20	13.333		4,000	625		4,625	5,375
2260	200 HP motor	↓	1.20	13.333		5,200	625		5,825	6,700
2300	Enclosed (NEMA 7), explosion proof, 600 volt 3 pole, 50 amp	1 Elec	2.30	3.478		1,300	163		1,463	1,675
2350	100 amp		1.50	5.333		1,350	251		1,601	1,875
2400	150 amp	↓	1	8		3,250	375		3,625	4,150
2450	250 amp	2 Elec	1.60	10		4,075	470		4,545	5,200
2500	400 amp	"	1.20	13.333	↓	4,575	625		5,200	6,025
9000	Minimum labor/equipment charge	1 Elec	4	2	Job		94		94	145

26 28 16.20 Safety Switches

		Crew	Daily Output	Labor-Hours	Unit	Material	Labor	Equipment	Total	Total Incl O&P
0010	**SAFETY SWITCHES**									
0100	General duty 240 volt, 3 pole NEMA 1, fusible, 30 amp	1 Elec	3.20	2.500	Ea.	113	118		231	305
0200	60 amp		2.30	3.478		192	163		355	460
0300	100 amp		1.90	4.211		330	198		528	665
0400	200 amp	↓	1.30	6.154		710	289		999	1,225
0500	400 amp	2 Elec	1.80	8.889		1,775	420		2,195	2,625
0600	600 amp	"	1.20	13.333	↓	3,325	625		3,950	4,650

26 28 16.20 Safety Switches		Crew	Daily Output	Labor-Hours	Unit	Material	Labor	Equipment	Total	Total Incl O&P
						2009 Bare Costs				
0610	Nonfusible, 30 amp	1 Elec	3.20	2.500	Ea.	92	118		210	282
0650	60 amp		2.30	3.478		122	163		285	385
0700	100 amp		1.90	4.211		282	198		480	615
0750	200 amp		1.30	6.154		520	289		809	1,025
0800	400 amp	2 Elec	1.80	8.889		1,250	420		1,670	2,025
0850	600 amp	"	1.20	13.333		2,400	625		3,025	3,625
1100	Heavy duty, 600 volt, 3 pole NEMA 1 nonfused									
1110	30 amp	1 Elec	3.20	2.500	Ea.	161	118		279	360
1500	60 amp		2.30	3.478		292	163		455	570
1700	100 amp		1.90	4.211		460	198		658	810
1900	200 amp		1.30	6.154		700	289		989	1,200
2100	400 amp	2 Elec	1.80	8.889		1,550	420		1,970	2,375
2300	600 amp		1.20	13.333		2,775	625		3,400	4,025
2500	800 amp		.94	17.021		5,675	800		6,475	7,475
2700	1200 amp		.80	20		7,600	940		8,540	9,825
2900	Heavy duty, 240 volt, 3 pole NEMA 1 fusible									
2910	30 amp	1 Elec	3.20	2.500	Ea.	183	118		301	380
3000	60 amp		2.30	3.478		310	163		473	590
3300	100 amp		1.90	4.211		490	198		688	840
3500	200 amp		1.30	6.154		840	289		1,129	1,375
3700	400 amp	2 Elec	1.80	8.889		2,150	420		2,570	3,025
3900	600 amp		1.20	13.333		3,700	625		4,325	5,050
4100	800 amp		.94	17.021		8,700	800		9,500	10,800
4300	1200 amp		.80	20		11,400	940		12,340	14,100
4340	2 pole fusible, 30 amp	1 Elec	3.50	2.286		138	107		245	315
4350	600 volt, 3 pole, fusible, 30 amp		3.20	2.500		310	118		428	520
4380	60 amp		2.30	3.478		375	163		538	665
4400	100 amp		1.90	4.211		685	198		883	1,050
4420	200 amp		1.30	6.154		995	289		1,284	1,550
4440	400 amp	2 Elec	1.80	8.889		2,575	420		2,995	3,500
4450	600 amp		1.20	13.333		4,350	625		4,975	5,750
4460	800 amp		.94	17.021		8,700	800		9,500	10,800
4480	1200 amp		.80	20		11,400	940		12,340	14,100
4500	240 volt 3 pole NEMA 3R (no hubs), fusible									
4510	30 amp	1 Elec	3.10	2.581	Ea.	330	121		451	545
4700	60 amp		2.20	3.636		515	171		686	830
4900	100 amp		1.80	4.444		740	209		949	1,125
5100	200 amp		1.20	6.667		1,025	315		1,340	1,600
5300	400 amp	2 Elec	1.60	10		2,250	470		2,720	3,200
5500	600 amp	"	1	16		4,700	750		5,450	6,325
5510	Heavy duty, 600 volt, 3 pole 3ph. NEMA 3R fusible, 30 amp	1 Elec	3.10	2.581		520	121		641	760
5520	60 amp		2.20	3.636		610	171		781	935
5530	100 amp		1.80	4.444		950	209		1,159	1,375
5540	200 amp		1.20	6.667		1,325	315		1,640	1,925
5550	400 amp	2 Elec	1.60	10		3,075	470		3,545	4,100
5700	600 volt, 3 pole NEMA 3R nonfused									
5710	30 amp	1 Elec	3.10	2.581	Ea.	290	121		411	505
5900	60 amp		2.20	3.636		510	171		681	825
6100	100 amp		1.80	4.444		710	209		919	1,100
6300	200 amp		1.20	6.667		860	315		1,175	1,425
6500	400 amp	2 Elec	1.60	10		2,150	470		2,620	3,075
6700	600 amp	"	1	16		4,275	750		5,025	5,850
6900	600 volt, 6 pole NEMA 3R nonfused, 30 amp	1 Elec	2.70	2.963		1,850	139		1,989	2,250

26 28 16 – Enclosed Switches and Circuit Breakers

26 28 16.20 Safety Switches		Crew	Daily Output	Labor-Hours	Unit	Material	2009 Bare Costs Labor	Equipment	Total	Total Incl O&P
7100	60 amp	1 Elec	2	4	Ea.	2,125	188		2,313	2,650
7300	100 amp		1.50	5.333		2,625	251		2,876	3,250
7500	200 amp		1.20	6.667		5,900	315		6,215	6,975
7600	600 volt, 3 pole NEMA 7 explosion proof nonfused									
7610	30 amp	1 Elec	2.20	3.636	Ea.	1,275	171		1,446	1,675
7620	60 amp		1.80	4.444		1,550	209		1,759	2,025
7630	100 amp		1.20	6.667		1,875	315		2,190	2,525
7640	200 amp		.80	10		3,875	470		4,345	5,000
7710	600 volt 6 pole, NEMA 3R fusible, 30 amp		2.70	2.963		2,075	139		2,214	2,500
7900	60 amp		2	4		2,475	188		2,663	3,025
8100	100 amp		1.50	5.333		3,025	251		3,276	3,700
8110	240 volt 3 pole, NEMA 12 fusible, 30 amp		3.10	2.581		400	121		521	625
8120	60 amp		2.20	3.636		540	171		711	860
8130	100 amp		1.80	4.444		795	209		1,004	1,200
8140	200 amp		1.20	6.667		1,125	315		1,440	1,725
8150	400 amp	2 Elec	1.60	10		2,600	470		3,070	3,600
8160	600 amp	"	1	16		4,225	750		4,975	5,800
8180	600 volt 3 pole, NEMA 12 fusible, 30 amp	1 Elec	3.10	2.581		525	121		646	765
8190	60 amp		2.20	3.636		540	171		711	860
8200	100 amp		1.80	4.444		870	209		1,079	1,275
8210	200 amp		1.20	6.667		1,350	315		1,665	1,975
8220	400 amp	2 Elec	1.60	10		2,975	470		3,445	4,000
8230	600 amp	"	1	16		5,025	750		5,775	6,675
8240	600 volt 3 pole, NEMA 12 nonfused, 30 amp	1 Elec	3.10	2.581		370	121		491	595
8250	60 amp		2.20	3.636		480	171		651	795
8260	100 amp		1.80	4.444		680	209		889	1,075
8270	200 amp		1.20	6.667		905	315		1,220	1,475
8280	400 amp	2 Elec	1.60	10		2,250	470		2,720	3,200
8290	600 amp	"	1	16		3,650	750		4,400	5,175
8310	600 volt, 3 pole NEMA 4 fusible, 30 amp	1 Elec	3	2.667		1,450	125		1,575	1,775
8320	60 amp		2.20	3.636		1,600	171		1,771	2,025
8330	100 amp		1.80	4.444		3,550	209		3,759	4,225
8340	200 amp		1.20	6.667		4,500	315		4,815	5,425
8350	400 amp	2 Elec	1.60	10		8,875	470		9,345	10,500
8360	600 volt 3 pole NEMA 4 nonfused, 30 amp	1 Elec	3	2.667		1,200	125		1,325	1,525
8370	60 amp		2.20	3.636		1,425	171		1,596	1,825
8380	100 amp		1.80	4.444		2,900	209		3,109	3,500
8390	200 amp		1.20	6.667		3,950	315		4,265	4,825
8400	400 amp	2 Elec	1.60	10		8,000	470		8,470	9,525
8490	Motor starters, manual, single phase, NEMA 1	1 Elec	6.40	1.250		68.50	59		127.50	166
8500	NEMA 4		4	2		192	94		286	355
8700	NEMA 7		4	2		210	94		304	375
8900	NEMA 1 with pilot		6.40	1.250		88	59		147	188
8920	3 pole, NEMA 1, 230/460 volt, 5 HP, size 0		3.50	2.286		203	107		310	390
8940	10 HP, size 1		2	4		240	188		428	555
9010	Disc. switch, 600V 3 pole fusible, 30 amp, to 10 HP motor		3.20	2.500		330	118		448	545
9050	60 amp, to 30 HP motor		2.30	3.478		760	163		923	1,075
9070	100 amp, to 60 HP motor		1.90	4.211		760	198		958	1,150
9100	200 amp, to 125 HP motor		1.30	6.154		1,150	289		1,439	1,700
9110	400 amp, to 200 HP motor	2 Elec	1.80	8.889		2,875	420		3,295	3,825
9990	Minimum labor/equipment charge	1 Elec	3	2.667	Job		125		125	193

26 28 Low-Voltage Circuit Protective Devices

26 28 16 – Enclosed Switches and Circuit Breakers

26 28 16.40 Time Switches		Crew	Daily Output	Labor-Hours	Unit	Material	2009 Bare Costs Labor	Equipment	Total	Total Incl O&P
0010	**TIME SWITCHES**									
0100	Single pole, single throw, 24 hour dial	1 Elec	4	2	Ea.	107	94		201	263
0200	24 hour dial with reserve power		3.60	2.222		460	104		564	665
0300	Astronomic dial		3.60	2.222		184	104		288	365
0400	Astronomic dial with reserve power		3.30	2.424		595	114		709	830
0500	7 day calendar dial		3.30	2.424		146	114		260	335
0600	7 day calendar dial with reserve power		3.20	2.500		655	118		773	900
0700	Photo cell 2000 watt		8	1		19	47		66	93.50
1080	Load management device, 4 loads		2	4		1,100	188		1,288	1,500
1100	8 loads		1	8		1,800	375		2,175	2,550
9000	Minimum labor/equipment charge		3.50	2.286	Job		107		107	165

26 29 Low-Voltage Controllers

26 29 13 – Enclosed Controllers

26 29 13.10 Contactors, AC		Crew	Daily Output	Labor-Hours	Unit	Material	Labor	Equipment	Total	Total Incl O&P
0010	**CONTACTORS, AC** Enclosed (NEMA 1)									
0050	Lighting, 600 volt 3 pole, electrically held									
0100	20 amp	1 Elec	4	2	Ea.	292	94		386	465
0200	30 amp		3.60	2.222		310	104		414	500
0300	60 amp		3	2.667		615	125		740	875
0400	100 amp		2.50	3.200		1,025	150		1,175	1,350
0500	200 amp		1.40	5.714		2,400	269		2,669	3,075
0600	300 amp	2 Elec	1.60	10		5,125	470		5,595	6,350
0800	600 volt 3 pole, mechanically held, 30 amp	1 Elec	3.60	2.222		465	104		569	675
0900	60 amp		3	2.667		925	125		1,050	1,225
1000	75 amp		2.80	2.857		1,325	134		1,459	1,650
1100	100 amp		2.50	3.200		1,325	150		1,475	1,675
1200	150 amp		2	4		3,600	188		3,788	4,250
1300	200 amp		1.40	5.714		3,600	269		3,869	4,375
1500	Magnetic with auxiliary contact, size 00, 9 amp		4	2		190	94		284	355
1600	Size 0, 18 amp		4	2		226	94		320	395
1700	Size 1, 27 amp		3.60	2.222		257	104		361	445
1800	Size 2, 45 amp		3	2.667		475	125		600	720
1900	Size 3, 90 amp		2.50	3.200		770	150		920	1,075
2000	Size 4, 135 amp		2.30	3.478		1,775	163		1,938	2,200
2100	Size 5, 270 amp	2 Elec	1.80	8.889		3,700	420		4,120	4,725
2200	Size 6, 540 amp		1.20	13.333		10,900	625		11,525	12,900
2300	Size 7, 810 amp		1	16		14,600	750		15,350	17,200
2310	Size 8, 1215 amp		.80	20		22,700	940		23,640	26,500
2500	Magnetic, 240 volt, 1-2 pole, .75 HP motor	1 Elec	4	2		153	94		247	315
2520	2 HP motor		3.60	2.222		171	104		275	350
2540	5 HP motor		2.50	3.200		415	150		565	690
2560	10 HP motor		1.40	5.714		685	269		954	1,175
2600	240 volt or less, 3 pole, .75 HP motor		4	2		153	94		247	315
2620	5 HP motor		3.60	2.222		190	104		294	370
2640	10 HP motor		3.60	2.222		220	104		324	405
2660	15 HP motor		2.50	3.200		440	150		590	715
2700	25 HP motor		2.50	3.200		440	150		590	715
2720	30 HP motor	2 Elec	2.80	5.714		735	269		1,004	1,225
2740	40 HP motor		2.80	5.714		735	269		1,004	1,225

26 29 13 – Enclosed Controllers

26 29 13.10 Contactors, AC		Crew	Daily Output	Labor-Hours	Unit	Material	2009 Bare Costs Labor	Equipment	Total	Total Incl O&P
2760	50 HP motor	2 Elec	1.60	10	Ea.	735	470		1,205	1,525
2800	75 HP motor		1.60	10		1,725	470		2,195	2,625
2820	100 HP motor		1	16		1,725	750		2,475	3,050
2860	150 HP motor		1	16		3,675	750		4,425	5,200
2880	200 HP motor	↓	1	16		3,675	750		4,425	5,200
3000	600 volt, 3 pole, 5 HP motor	1 Elec	4	2		190	94		284	355
3020	10 HP motor		3.60	2.222		220	104		324	405
3040	25 HP motor		3	2.667		440	125		565	680
3100	50 HP motor	↓	2.50	3.200		735	150		885	1,050
3160	100 HP motor	2 Elec	2.80	5.714		1,725	269		1,994	2,325
3220	200 HP motor	"	1.60	10	↓	3,675	470		4,145	4,775

26 29 13.20 Control Stations		Crew	Daily Output	Labor-Hours	Unit	Material	Labor	Equipment	Total	Total Incl O&P
0010	**CONTROL STATIONS**									
0050	NEMA 1, heavy duty, stop/start	1 Elec	8	1	Ea.	141	47		188	228
0100	Stop/start, pilot light		6.20	1.290		192	60.50		252.50	305
0200	Hand/off/automatic		6.20	1.290		104	60.50		164.50	208
0400	Stop/start/reverse		5.30	1.509		190	71		261	320
0500	NEMA 7, heavy duty, stop/start		6	1.333		320	62.50		382.50	450
0600	Stop/start, pilot light		4	2		390	94		484	575
0700	NEMA 7 or 9, 1 element		6	1.333		261	62.50		323.50	385
0800	2 element		6	1.333		340	62.50		402.50	470
0900	3 element		4	2		625	94		719	835
0910	Selector switch, 2 position		6	1.333		261	62.50		323.50	385
0920	3 position		4	2		261	94		355	435
0930	Oiltight, 1 element		8	1		98	47		145	181
0940	2 element		6.20	1.290		141	60.50		201.50	249
0950	3 element		5.30	1.509		190	71		261	320
0960	Selector switch, 2 position		6.20	1.290		104	60.50		164.50	208
0970	3 position	↓	5.30	1.509	↓	104	71		175	223

26 29 13.30 Control Switches		Crew	Daily Output	Labor-Hours	Unit	Material	Labor	Equipment	Total	Total Incl O&P
0010	**CONTROL SWITCHES** Field installed									
6000	Push button 600V 10A, momentary contact									
6150	Standard operator with colored button	1 Elec	34	.235	Ea.	16.50	11.05		27.55	35
6160	With single block 1NO 1NC		18	.444		35	21		56	70.50
6170	With double block 2NO 2NC		15	.533		53	25		78	97
6180	Stnd operator w/mushroom button 1-9/16" diam.	↓	34	.235	↓	35	11.05		46.05	55.50
6190	Stnd operator w/mushroom button 2-1/4" diam.									
6200	With single block 1NO 1NC	1 Elec	18	.444	Ea.	53	21		74	90.50
6210	With double block 2NO 2NC		15	.533		71.50	25		96.50	118
6500	Maintained contact, selector operator		34	.235		53	11.05		64.05	75.50
6510	With single block 1NO 1NC		18	.444		71.50	21		92.50	111
6520	With double block 2NO 2NC		15	.533		90	25		115	138
6560	Spring-return selector operator		34	.235		53	11.05		64.05	75.50
6570	With single block 1NO 1NC		18	.444		71.50	21		92.50	111
6580	With double block 2NO 2NC	↓	15	.533	↓	90	25		115	138
6620	Transformer operator w/illuminated									
6630	button 6V #12 lamp	1 Elec	32	.250	Ea.	90	11.75		101.75	117
6640	With single block 1NO 1NC w/guard		16	.500		108	23.50		131.50	155
6650	With double block 2NO 2NC w/guard		13	.615		127	29		156	184
6690	Combination operator		34	.235		53	11.05		64.05	75.50
6700	With single block 1NO 1NC		18	.444		71.50	21		92.50	111
6710	With double block 2NO 2NC	↓	15	.533	↓	90	25		115	138

26 29 Low-Voltage Controllers

26 29 13 – Enclosed Controllers

26 29 13.30 Control Switches		Crew	Daily Output	Labor-Hours	Unit	2009 Bare Costs Material	Labor	Equipment	Total	Total Incl O&P
9000	Indicating light unit, full voltage									
9010	110-125V front mount	1 Elec	32	.250	Ea.	71.50	11.75		83.25	97
9020	130V resistor type		32	.250		53	11.75		64.75	76.50
9030	6V transformer type		32	.250		65.50	11.75		77.25	90

26 29 13.40 Relays

		Crew	Daily Output	Labor-Hours	Unit	Material	Labor	Equipment	Total	Total Incl O&P
0010	**RELAYS** Enclosed (NEMA 1)									
0050	600 volt AC, 1 pole, 12 amp	1 Elec	5.30	1.509	Ea.	92	71		163	210
0100	2 pole, 12 amp		5	1.600		92	75		167	217
0200	4 pole, 10 amp		4.50	1.778		122	83.50		205.50	264
0500	250 volt DC, 1 pole, 15 amp		5.30	1.509		120	71		191	241
0600	2 pole, 10 amp		5	1.600		115	75		190	243
0700	4 pole, 4 amp		4.50	1.778		152	83.50		235.50	296

26 29 23 – Variable-Frequency Motor Controllers

26 29 23.10 Variable Frequency Drives/Adj. Frequency Drives

			Crew	Daily Output	Labor-Hours	Unit	Material	Labor	Equipment	Total	Total Incl O&P
0010	**VARIABLE FREQUENCY DRIVES/ADJ. FREQUENCY DRIVES**										
0100	Enclosed (NEMA 1), 460 volt, for 3 HP motor size	G	1 Elec	.80	10	Ea.	1,575	470		2,045	2,450
0110	5 HP motor size	G		.80	10		1,700	470		2,170	2,600
0120	7.5 HP motor size	G		.67	11.940		2,025	560		2,585	3,100
0130	10 HP motor size	G		.67	11.940		2,025	560		2,585	3,100
0140	15 HP motor size	G	2 Elec	.89	17.978		2,325	845		3,170	3,875
0150	20 HP motor size	G		.89	17.978		3,450	845		4,295	5,100
0160	25 HP motor size	G		.67	23.881		4,025	1,125		5,150	6,150
0170	30 HP motor size	G		.67	23.881		4,900	1,125		6,025	7,100
0180	40 HP motor size	G		.67	23.881		7,175	1,125		8,300	9,625
0190	50 HP motor size	G		.53	30.189		7,825	1,425		9,250	10,800
0200	60 HP motor size	G	R-3	.56	35.714		8,575	1,650	231	10,456	12,200
0210	75 HP motor size	G		.56	35.714		11,700	1,650	231	13,581	15,700
0220	100 HP motor size	G		.50	40		12,000	1,850	258	14,108	16,200
0230	125 HP motor size	G		.50	40		13,200	1,850	258	15,308	17,700
0240	150 HP motor size	G		.50	40		16,900	1,850	258	19,008	21,700
0250	200 HP motor size	G		.42	47.619		21,000	2,200	310	23,510	26,800
1100	Custom-engineered, 460 volt, for 3 HP motor size	G	1 Elec	.56	14.286		2,600	670		3,270	3,875
1110	5 HP motor size	G		.56	14.286		2,725	670		3,395	4,025
1120	7.5 HP motor size	G		.47	17.021		3,275	800		4,075	4,825
1130	10 HP motor size	G		.47	17.021		3,275	800		4,075	4,825
1140	15 HP motor size	G	2 Elec	.62	25.806		3,575	1,225		4,800	5,825
1150	20 HP motor size	G		.62	25.806		4,900	1,225		6,125	7,250
1160	25 HP motor size	G		.47	34.043		5,325	1,600		6,925	8,300
1170	30 HP motor size	G		.47	34.043		7,250	1,600		8,850	10,400
1180	40 HP motor size	G		.47	34.043		8,075	1,600		9,675	11,400
1190	50 HP motor size	G		.37	43.243		8,075	2,025		10,100	12,000
1200	60 HP motor size	G	R-3	.39	51.282		12,100	2,375	330	14,805	17,300
1210	75 HP motor size	G		.39	51.282		14,000	2,375	330	16,705	19,300
1220	100 HP motor size	G		.35	57.143		14,500	2,650	370	17,520	20,500
1230	125 HP motor size	G		.35	57.143		15,300	2,650	370	18,320	21,400
1240	150 HP motor size	G		.35	57.143		17,200	2,650	370	20,220	23,400
1250	200 HP motor size	G		.29	68.966		23,800	3,200	445	27,445	31,600
2000	For complex & special design systems to meet specific										
2010	requirements, obtain quote from vendor.										

26 31 Photovoltaic Collectors

26 31 13 – Photovoltaics

26 31 13.50 Solar Energy - Photovoltaics			Crew	Daily Output	Labor-Hours	Unit	2009 Bare Costs Material	Labor	Equipment	Total	Total Incl O&P
0010	**SOLAR ENERGY - PHOTOVOLTAICS**										
0220	Alt energy source, photovoltaic module, 6 watt, 15 volts	G	1 Elec	8	1	Ea.	83	47		130	164
0230	10 watt, 16.3 volts	G		8	1		110	47		157	193
0240	20 watt, 14.5 volts	G		8	1		181	47		228	272
0250	36 watt, 17 volts	G		8	1		246	47		293	345
0260	55 watt, 17 volts	G		8	1		335	47		382	440
0270	75 watt, 17 volts	G		8	1		455	47		502	575
0280	130 watt, 33 volts	G		8	1		620	47		667	755
0290	140 watt, 33 volts	G		8	1		875	47		922	1,050
0300	150 watt, 33 volts	G		8	1		1,000	47		1,047	1,175
0310	DC to AC inverter for, 12 V 2,000 watt	G		4	2		1,075	94		1,169	1,325
0320	12 V, 2,500 watt	G		4	2		2,225	94		2,319	2,600
0330	24 V, 2,500 watt	G		4	2		2,325	94		2,419	2,725
0340	12 V, 3,000 watt	G		3	2.667		2,400	125		2,525	2,850
0350	24 V, 3,000 watt	G		3	2.667		2,675	125		2,800	3,150
0360	24 V, 4,000 watt	G		2	4		2,200	188		2,388	2,725
0370	48 V, 4,000 watt	G		2	4		2,200	188		2,388	2,725
0380	48 V, 5,500 watt	G		2	4		2,750	188		2,938	3,325
0390	PV components, combiner box, 10 lug, NEMA 3R enclosure	G		4	2		161	94		255	320
0400	Fuse, 15 A for combiner box	G		40	.200		10.95	9.40		20.35	26.50
0410	Battery charger controller w/temperature sensor	G		4	2		455	94		549	645
0420	Digital readout panel, displays hours, volts, amps, etc	G		4	2		80.50	94		174.50	234
0430	Deep cycle solar battery, 6V, 180 Ah (C/20)	G		8	1		122	47		169	207
0440	Battery interconn, 15" AWG #2/0, sealed w/copper ring lugs	G		16	.500		7.80	23.50		31.30	44.50
0442	Battery interconn, 24" AWG #2/0, sealed w/copper ring lugs	G		16	.500		9.60	23.50		33.10	46.50
0444	Battery interconn, 60" AWG #2/0, sealed w/copper ring lugs	G		16	.500		12.60	23.50		36.10	50
0446	Batt temp computer probe, RJ11 jack, 15' cord	G		16	.500		21	23.50		44.50	59
0450	System disconnect, DC 175 Amp circuit breaker	G		8	1		202	47		249	296
0460	Conduit box for inverter	G		8	1		66	47		113	145
0470	Low voltage disconnect	G		8	1		33	47		80	109
0480	Vented battery enclosure, wood	G	1 Carp	2	4		140	160		300	420
0490	Roof mounting frame, metal, for 6 modules	G	1 Rofc	4	2		1,075	68.50		1,143.50	1,325

26 32 Packaged Generator Assemblies

26 32 13 – Engine Generators

26 32 13.13 Diesel-Engine-Driven Generator Sets		Crew	Daily Output	Labor-Hours	Unit	Material	Labor	Equipment	Total	Total Incl O&P
0010	**DIESEL-ENGINE-DRIVEN GENERATOR SETS**									
2000	Diesel engine, including battery, charger,									
2010	muffler, automatic transfer switch & day tank, 30 kW	R-3	.55	36.364	Ea.	19,600	1,675	235	21,510	24,500
2100	50 kW		.42	47.619		24,200	2,200	310	26,710	30,300
2110	60 kW		.39	51.282		26,100	2,375	330	28,805	32,700
2200	75 kW		.35	57.143		31,400	2,650	370	34,420	39,000
2300	100 kW		.31	64.516		34,900	3,000	415	38,315	43,500
2400	125 kW		.29	68.966		36,800	3,200	445	40,445	45,900
2500	150 kW		.26	76.923		42,200	3,550	495	46,245	52,500
2600	175 kW		.25	80		45,900	3,700	515	50,115	57,000
2700	200 kW		.24	83.333		47,400	3,850	540	51,790	58,500
2800	250 kW		.23	86.957		56,000	4,025	560	60,585	68,500
2850	275 kW		.22	90.909		56,000	4,200	585	60,785	68,500
2900	300 kW		.22	90.909		60,500	4,200	585	65,285	73,500
3000	350 kW		.20	100		68,000	4,625	645	73,270	83,000

26 32 Packaged Generator Assemblies

26 32 13 – Engine Generators

26 32 13.13 Diesel-Engine-Driven Generator Sets		Crew	Daily Output	Labor-Hours	Unit	2009 Bare Costs Material	Labor	Equipment	Total	Total Incl O&P
3100	400 kW	R-3	.19	105	Ea.	84,000	4,875	680	89,555	101,000
3200	500 kW		.18	111		106,000	5,150	720	111,870	125,000
3220	600 kW		.17	117		139,000	5,450	760	145,210	162,500
3230	650 kW	R-13	.38	110		165,000	4,975	540	170,515	190,000
3240	750 kW		.38	110		174,500	4,975	540	180,015	200,500
3250	800 kW		.36	116		181,000	5,250	570	186,820	207,500
3260	900 kW		.31	135		209,500	6,100	660	216,260	240,000
3270	1000 kW		.27	155		215,500	7,000	760	223,260	248,500

26 32 13.16 Gas-Engine-Driven Generator Sets

		Crew	Daily Output	Labor-Hours	Unit	Material	Labor	Equipment	Total	Total Incl O&P
0010	**GAS-ENGINE-DRIVEN GENERATOR SETS**									
0020	Gas or gasoline operated, includes battery,									
0050	charger, muffler & transfer switch									
0200	3 phase 4 wire, 277/480 volt, 7.5 kW	R-3	.83	24.096	Ea.	7,700	1,125	156	8,981	10,400
0300	11.5 kW		.71	28.169		10,900	1,300	182	12,382	14,200
0400	20 kW		.63	31.746		12,900	1,475	205	14,580	16,600
0500	35 kW		.55	36.364		15,300	1,675	235	17,210	19,700
0520	60 kW		.50	40		20,200	1,850	258	22,308	25,300
0600	80 kW		.40	50		25,100	2,325	325	27,750	31,500
0700	100 kW		.33	60.606		27,500	2,800	390	30,690	35,100
0800	125 kW		.28	71.429		56,500	3,300	460	60,260	67,500
0900	185 kW		.25	80		74,500	3,700	515	78,715	88,500

26 33 Battery Equipment

26 33 53 – Static Uninterruptible Power Supply

26 33 53.10 Uninterruptible Power Supply/Conditioner Transformers

		Crew	Daily Output	Labor-Hours	Unit	Material	Labor	Equipment	Total	Total Incl O&P
0010	**UNINTERRUPTIBLE POWER SUPPLY/CONDITIONER TRANSFORMERS**									
0100	Volt. regulating, isolating trans., w/invert. & 10 min. battery pack									
0110	Single-phase, 120 V, 0.35 kVA	1 Elec	2.29	3.493	Ea.	1,000	164		1,164	1,350
0120	0.5 kVA		2	4		1,050	188		1,238	1,450
0130	For additional 55 min. battery, add to .35 kVA		2.29	3.493		590	164		754	905
0140	Add to 0.5 kVA		1.14	7.018		620	330		950	1,175
0150	Single-phase, 120 V, 0.75 kVA		.80	10		1,325	470		1,795	2,175
0160	1.0 kVA		.80	10		1,900	470		2,370	2,825
0170	1.5 kVA	2 Elec	1.14	14.035		3,275	660		3,935	4,625
0180	2 kVA	"	.89	17.978		3,550	845		4,395	5,200
0190	3 kVA	R-3	.63	31.746		4,300	1,475	205	5,980	7,225
0200	5 kVA		.42	47.619		6,225	2,200	310	8,735	10,600
0210	7.5 kVA		.33	60.606		7,950	2,800	390	11,140	13,500
0220	10 kVA		.28	71.429		8,450	3,300	460	12,210	14,900
0230	15 kVA		.22	90.909		11,700	4,200	585	16,485	20,000
0500	For options & accessories add to above, minimum									10%
0520	Maximum									35%
0600	For complex & special design systems to meet specific									
0610	requirements, obtain quote from vendor									

26 35 Power Filters and Conditioners

26 35 13 – Capacitors

26 35 13.10 Capacitors Indoor		Crew	Daily Output	Labor-Hours	Unit	Material	Labor	Equipment	Total	Total Incl O&P
						2009 Bare Costs				
0010	CAPACITORS INDOOR									
0020	240 volts, single & 3 phase, 0.5 kVAR	1 Elec	2.70	2.963	Ea.	385	139		524	635
0100	1.0 kVAR		2.70	2.963		460	139		599	725
0150	2.5 kVAR		2	4		520	188		708	860
0200	5.0 kVAR		1.80	4.444		620	209		829	1,000
0250	7.5 kVAR		1.60	5		725	235		960	1,150
0300	10 kVAR		1.50	5.333		870	251		1,121	1,350
0350	15 kVAR		1.30	6.154		1,200	289		1,489	1,775
0400	20 kVAR		1.10	7.273		1,500	340		1,840	2,175
0450	25 kVAR		1	8		1,750	375		2,125	2,500
1000	480 volts, single & 3 phase, 1 kVAR		2.70	2.963		350	139		489	600
1050	2 kVAR		2.70	2.963		400	139		539	655
1100	5 kVAR		2	4		505	188		693	845
1150	7.5 kVAR		2	4		545	188		733	890
1200	10 kVAR		2	4		610	188		798	960
1250	15 kVAR		2	4		755	188		943	1,125
1300	20 kVAR		1.60	5		840	235		1,075	1,275
1350	30 kVAR		1.50	5.333		1,050	251		1,301	1,525
1400	40 kVAR		1.20	6.667		1,325	315		1,640	1,950
1450	50 kVAR		1.10	7.273		1,525	340		1,865	2,200
2000	600 volts, single & 3 phase, 1 kVAR		2.70	2.963		360	139		499	610
2050	2 kVAR		2.70	2.963		415	139		554	670
2100	5 kVAR		2	4		505	188		693	845
2150	7.5 kVAR		2	4		545	188		733	890
2200	10 kVAR		2	4		610	188		798	960
2250	15 kVAR		1.60	5		755	235		990	1,200
2300	20 kVAR		1.60	5		840	235		1,075	1,275
2350	25 kVAR		1.50	5.333		940	251		1,191	1,400
2400	35 kVAR		1.40	5.714		1,200	269		1,469	1,725
2450	50 kVAR		1.30	6.154		1,525	289		1,814	2,125

26 35 26 – Harmonic Filters

26 35 26.10 Computer Isolation Transformer

		Crew	Daily Output	Labor-Hours	Unit	Material	Labor	Equipment	Total	Total Incl O&P
0010	COMPUTER ISOLATION TRANSFORMER									
0100	Computer grade									
0110	Single-phase, 120/240 V, 0.5 kVA	1 Elec	4	2	Ea.	435	94		529	625
0120	1.0 kVA		2.67	2.996		615	141		756	895
0130	2.5 kVA		2	4		940	188		1,128	1,325
0140	5 kVA		1.14	7.018		1,075	330		1,405	1,675

26 35 26.20 Computer Regulator Transformer

		Crew	Daily Output	Labor-Hours	Unit	Material	Labor	Equipment	Total	Total Incl O&P
0010	COMPUTER REGULATOR TRANSFORMER									
0100	Ferro-resonant, constant voltage, variable transformer									
0110	Single-phase, 240 V, 0.5 kVA	1 Elec	2.67	2.996	Ea.	590	141		731	865
0120	1.0 kVA		2	4		810	188		998	1,175
0130	2.0 kVA		1	8		1,375	375		1,750	2,100
0210	Plug-in unit 120 V, 0.14 kVA		8	1		340	47		387	450
0220	0.25 kVA		8	1		400	47		447	515
0230	0.5 kVA		8	1		590	47		637	725
0240	1.0 kVA		5.33	1.501		810	70.50		880.50	1,000
0250	2.0 kVA		4	2		1,375	94		1,469	1,675

26 35 26.30 Power Conditioner Transformer

		Crew	Daily Output	Labor-Hours	Unit	Material	Labor	Equipment	Total	Total Incl O&P
0010	POWER CONDITIONER TRANSFORMER									
0100	Electronic solid state, buck-boost, transformer, w/tap switch									

26 35 Power Filters and Conditioners

26 35 26 – Harmonic Filters

26 35 26.30 Power Conditioner Transformer		Crew	Daily Output	Labor-Hours	Unit	Material	2009 Bare Costs Labor	Equipment	Total	Total Incl O&P
0110	Single-phase, 115 V, 3.0 kVA, + or - 3% accuracy	2 Elec	1.60	10	Ea.	3,175	470		3,645	4,225
0120	208, 220, 230, or 240 V, 5.0 kVA, + or - 1.5% accuracy	3 Elec	1.60	15		4,125	705		4,830	5,600
0130	5.0 kVA, + or - 6% accuracy	2 Elec	1.14	14.035		3,675	660		4,335	5,075
0140	7.5 kVA, + or - 1.5% accuracy	3 Elec	1.50	16		5,225	750		5,975	6,900
0150	7.5 kVA, + or - 6% accuracy		1.60	15		4,350	705		5,055	5,875
0160	10.0 kVA, + or - 1.5% accuracy		1.33	18.045		6,975	850		7,825	8,975
0170	10.0 kVA, + or - 6% accuracy		1.41	17.021		5,925	800		6,725	7,750

26 35 26.40 Transient Voltage Suppressor Transformer		Crew	Daily Output	Labor-Hours	Unit	Material	Labor	Equipment	Total	Total Incl O&P
0010	**TRANSIENT VOLTAGE SUPPRESSOR TRANSFORMER**									
0110	Single-phase, 120 V, 1.8 kVA	1 Elec	4	2	Ea.	620	94		714	830
0120	3.6 kVA		4	2		1,400	94		1,494	1,700
0130	7.2 kVA		3.20	2.500		1,825	118		1,943	2,175
0150	240 V, 3.6 kVA		4	2		1,400	94		1,494	1,700
0160	7.2 kVA		4	2		1,825	94		1,919	2,150
0170	14.4 kVA		3.20	2.500		2,350	118		2,468	2,750
0210	Plug-in unit, 120 V, 1.8 kVA		8	1		565	47		612	695

26 35 53 – Voltage Regulators

26 35 53.10 Automatic Voltage Regulators		Crew	Daily Output	Labor-Hours	Unit	Material	Labor	Equipment	Total	Total Incl O&P
0010	**AUTOMATIC VOLTAGE REGULATORS**									
0100	Computer grade, solid state, variable trans. volt. regulator									
0110	Single-phase, 120 V, 8.6 kVA	2 Elec	1.33	12.030	Ea.	6,425	565		6,990	7,950
0120	17.3 kVA		1.14	14.035		7,600	660		8,260	9,400
0130	208/240 V, 7.5/8.6 kVA		1.33	12.030		6,425	565		6,990	7,950
0140	13.5/15.6 kVA		1.33	12.030		7,600	565		8,165	9,250
0150	27.0/31.2 kVA		1.14	14.035		9,600	660		10,260	11,600
0210	Two-phase, single control, 208/240 V, 15.0/17.3 kVA		1.14	14.035		7,600	660		8,260	9,400
0220	Individual phase control, 15.0/17.3 kVA		1.14	14.035		7,600	660		8,260	9,400
0230	30.0/34.6 kVA	3 Elec	1.33	18.045		9,600	850		10,450	11,900
0310	Three-phase single control, 208/240 V, 26/30 kVA	2 Elec	1	16		7,600	750		8,350	9,525
0320	380/480 V, 24/30 kVA	"	1	16		7,600	750		8,350	9,525
0330	43/54 kVA	3 Elec	1.33	18.045		13,800	850		14,650	16,500
0340	Individual phase control, 208 V, 26 kVA	"	1.33	18.045		7,600	850		8,450	9,675
0350	52 kVA	R-3	.91	21.978		9,600	1,025	142	10,767	12,300
0360	340/480 V, 24/30 kVA	2 Elec	1	16		7,600	750		8,350	9,525
0370	43/54 kVA	"	1	16		9,600	750		10,350	11,800
0380	48/60 kVA	3 Elec	1.33	18.045		13,900	850		14,750	16,600
0390	86/108 kVA	R-3	.91	21.978		15,400	1,025	142	16,567	18,700
0500	Standard grade, solid state, variable transformer volt. regulator									
0510	Single-phase, 115 V, 2.3 kVA	1 Elec	2.29	3.493	Ea.	1,875	164		2,039	2,300
0520	4.2 kVA		2	4		3,000	188		3,188	3,600
0530	6.6 kVA		1.14	7.018		3,675	330		4,005	4,525
0540	13.0 kVA		1.14	7.018		6,325	330		6,655	7,450
0550	16.6 kVA	2 Elec	1.23	13.008		7,475	610		8,085	9,175
0610	230 V, 8.3 kVA		1.33	12.030		6,325	565		6,890	7,825
0620	21.4 kVA		1.23	13.008		7,475	610		8,085	9,175
0630	29.9 kVA		1.23	13.008		7,475	610		8,085	9,175
0710	460 V, 9.2 kVA		1.33	12.030		6,325	565		6,890	7,825
0720	20.7 kVA		1.23	13.008		7,475	610		8,085	9,175
0810	Three-phase, 230 V, 13.1 kVA	3 Elec	1.41	17.021		6,325	800		7,125	8,175
0820	19.1 kVA		1.41	17.021		7,475	800		8,275	9,450
0830	25.1 kVA		1.23	19.512		7,475	915		8,390	9,625
0840	57.8 kVA	R-3	.95	21.053		13,600	975	136	14,711	16,500

26 35 Power Filters and Conditioners

26 35 53 – Voltage Regulators

26 35 53.10 Automatic Voltage Regulators		Crew	Daily Output	Labor-Hours	Unit	Material	2009 Bare Costs Labor	Equipment	Total	Total Incl O&P
0850	74.9 kVA	R-3	.91	21.978	Ea.	13,600	1,025	142	14,767	16,600
0910	460 V, 14.3 kVA	3 Elec	1.41	17.021		6,325	800		7,125	8,175
0920	19.1 kVA		1.41	17.021		7,475	800		8,275	9,450
0930	27.9 kVA		1.23	19.512		7,475	915		8,390	9,625
0940	59.8 kVA	R-3	1	20		13,600	925	129	14,654	16,500
0950	79.7 kVA		.95	21.053		15,200	975	136	16,311	18,300
0960	118 kVA		.95	21.053		16,000	975	136	17,111	19,200
1000	Laboratory grade, precision, electronic voltage regulator									
1110	Single-phase, 115 V, 0.5 kVA	1 Elec	2.29	3.493	Ea.	1,575	164		1,739	1,975
1120	1.0 kVA		2	4		1,675	188		1,863	2,150
1130	3.0 kVA		.80	10		2,350	470		2,820	3,300
1140	6.0 kVA	2 Elec	1.46	10.959		4,350	515		4,865	5,600
1150	10.0 kVA	3 Elec	1	24		5,650	1,125		6,775	7,950
1160	15.0 kVA	"	1.50	16		6,450	750		7,200	8,250
1210	230 V, 3.0 kVA	1 Elec	.80	10		2,650	470		3,120	3,650
1220	6.0 kVA	2 Elec	1.46	10.959		4,450	515		4,965	5,700
1230	10.0 kVA	3 Elec	1.71	14.035		5,900	660		6,560	7,525
1240	15.0 kVA	"	1.60	15		6,675	705		7,380	8,425

26 35 53.30 Transient Suppressor/Voltage Regulator		Crew	Daily Output	Labor-Hours	Unit	Material	Labor	Equipment	Total	Total Incl O&P
0010	**TRANSIENT SUPPRESSOR/VOLTAGE REGULATOR** (without isolation)									
0110	Single-phase, 115 V, 1.0 kVA	1 Elec	2.67	2.996	Ea.	1,100	141		1,241	1,425
0120	2.0 kVA		2.29	3.493		1,500	164		1,664	1,875
0130	4.0 kVA		2.13	3.756		1,825	177		2,002	2,300
0140	220 V, 1.0 kVA		2.67	2.996		1,100	141		1,241	1,425
0150	2.0 kVA		2.29	3.493		1,500	164		1,664	1,875
0160	4.0 kVA		2.13	3.756		1,925	177		2,102	2,400
0210	Plug-in unit, 120 V, 1.0 kVA		8	1		1,050	47		1,097	1,225
0220	2.0 kVA		8	1		1,475	47		1,522	1,700

26 36 Transfer Switches

26 36 13 – Manual Transfer Switches

26 36 13.10 Non-Automatic Transfer Switches		Crew	Daily Output	Labor-Hours	Unit	Material	Labor	Equipment	Total	Total Incl O&P
0010	**NON-AUTOMATIC TRANSFER SWITCHES** enclosed									
0100	Manual operated, 480 volt 3 pole, 30 amp	1 Elec	2.30	3.478	Ea.	1,550	163		1,713	1,950
0150	60 amp		1.90	4.211		1,550	198		1,748	2,000
0200	100 amp		1.30	6.154		1,550	289		1,839	2,150
0250	200 amp	2 Elec	2	8		3,075	375		3,450	3,950
0300	400 amp		1.60	10		4,475	470		4,945	5,650
0350	600 amp		1	16		6,775	750		7,525	8,600
1000	250 volt 3 pole, 30 amp	1 Elec	2.30	3.478		1,550	163		1,713	1,950
1100	60 amp		1.90	4.211		1,550	198		1,748	2,000
1150	100 amp		1.30	6.154		1,550	289		1,839	2,150
1200	200 amp	2 Elec	2	8		3,075	375		3,450	3,950
1300	600 amp	"	1	16		6,775	750		7,525	8,600
1500	Electrically operated, 480 volt 3 pole, 60 amp	1 Elec	1.90	4.211		2,950	198		3,148	3,550
1600	100 amp	"	1.30	6.154		2,950	289		3,239	3,700
1650	200 amp	2 Elec	2	8		4,850	375		5,225	5,925
1700	400 amp		1.60	10		6,800	470		7,270	8,200
1750	600 amp		1	16		9,800	750		10,550	12,000
2000	250 volt 3 pole, 30 amp	1 Elec	2.30	3.478		2,950	163		3,113	3,500
2050	60 amp	"	1.90	4.211		2,950	198		3,148	3,550

26 36 Transfer Switches

26 36 13 – Manual Transfer Switches

26 36 13.10 Non-Automatic Transfer Switches		Crew	Daily Output	Labor-Hours	Unit	Material	Labor	Equipment	Total	Total Incl O&P
2150	200 amp	2 Elec	2	8	Ea.	4,850	375		5,225	5,925
2200	400 amp		1.60	10		6,800	470		7,270	8,200
2250	600 amp		1	16		9,800	750		10,550	12,000
2500	NEMA 3R, 480 volt 3 pole, 60 amp	1 Elec	1.80	4.444		3,475	209		3,684	4,150
2550	100 amp	"	1.20	6.667		3,475	315		3,790	4,300
2600	200 amp	2 Elec	1.80	8.889		5,500	420		5,920	6,700
2650	400 amp	"	1.40	11.429		7,450	535		7,985	9,000
2800	NEMA 3R, 250 volt 3 pole solid state, 100 amp	1 Elec	1.20	6.667		3,475	315		3,790	4,300
2850	150 amp	2 Elec	1.80	8.889		4,625	420		5,045	5,725
2900	250 volt 2 pole solid state, 100 amp	1 Elec	1.30	6.154		3,375	289		3,664	4,175
2950	150 amp	2 Elec	2	8		4,475	375		4,850	5,500

26 36 23 – Automatic Transfer Switches

26 36 23.10 Automatic Transfer Switch Devices		Crew	Daily Output	Labor-Hours	Unit	Material	Labor	Equipment	Total	Total Incl O&P
0010	**AUTOMATIC TRANSFER SWITCH DEVICES**									
0015	Switches, enclosed 120/240 volt, 2 pole, 30 amp	1 Elec	2.40	3.333	Ea.	2,925	157		3,082	3,475
0020	70 amp		2	4		2,925	188		3,113	3,525
0030	100 amp		1.35	5.926		2,925	279		3,204	3,650
0040	225 amp	2 Elec	2.10	7.619		4,325	360		4,685	5,300
0050	400 amp		1.70	9.412		6,575	440		7,015	7,925
0060	600 amp		1.06	15.094		9,675	710		10,385	11,700
0070	800 amp		.84	19.048		11,400	895		12,295	13,900
0100	Switches, enclosed 480 volt, 3 pole, 30 amp	1 Elec	2.30	3.478		3,275	163		3,438	3,850
0200	60 amp		1.90	4.211		3,275	198		3,473	3,900
0300	100 amp		1.30	6.154		3,275	289		3,564	4,050
0400	150 amp	2 Elec	2.40	6.667		4,025	315		4,340	4,900
0500	225 amp		2	8		5,150	375		5,525	6,225
0600	260 amp		2	8		5,950	375		6,325	7,125
0700	400 amp		1.60	10		6,975	470		7,445	8,400
0800	600 amp		1	16		10,700	750		11,450	13,000
0900	800 amp		.80	20		12,600	940		13,540	15,400
1000	1000 amp		.76	21.053		16,600	990		17,590	19,800
1100	1200 amp		.70	22.857		22,800	1,075		23,875	26,800
1200	1600 amp		.60	26.667		26,000	1,250		27,250	30,500
1300	2000 amp		.50	32		29,000	1,500		30,500	34,300
1600	Accessories, time delay on engine starting					255			255	280
1700	Adjustable time delay on retransfer					255			255	280
1800	Shunt trips for customer connections					455			455	500
1900	Maintenance select switch					104			104	114
2000	Auxiliary contact when normal fails					120			120	132
2100	Pilot light-emergency					104			104	114
2200	Pilot light-normal					104			104	114
2300	Auxiliary contact-closed on normal					120			120	132
2400	Auxiliary contact-closed on emergency					120			120	132
2500	Emergency source sensing, frequency relay					525			525	580

26 41 Facility Lightning Protection

26 41 13 – Lightning Protection for Structures

26 41 13.13 Lightning Protection for Buildings		Crew	Daily Output	Labor-Hours	Unit	2009 Bare Costs Material	Labor	Equipment	Total	Total Incl O&P
0010	**LIGHTNING PROTECTION FOR BUILDINGS**									
0200	Air terminals & base, copper									
0400	3/8" diameter x 10" (to 75' high)	1 Elec	8	1	Ea.	23	47		70	98
0500	1/2" diameter x 12" (over 75' high)		8	1		28.50	47		75.50	104
1000	Aluminum, 1/2" diameter x 12" (to 75' high)		8	1		3.53	47		50.53	76.50
1100	5/8" diameter x 12" (over 75' high)		8	1		5	47		52	78
2000	Cable, copper, 220 lb. per thousand ft. (to 75' high)		320	.025	L.F.	2.73	1.18		3.91	4.81
2100	375 lb. per thousand ft. (over 75' high)		230	.035		4.94	1.63		6.57	7.95
2500	Aluminum, 101 lb. per thousand ft. (to 75' high)		280	.029		.82	1.34		2.16	2.97
2600	199 lb. per thousand ft. (over 75' high)		240	.033		1.26	1.57		2.83	3.80
3000	Arrester, 175 volt AC to ground		8	1	Ea.	59	47		106	138
3100	650 volt AC to ground		6.70	1.194	"	68	56		124	162

26 42 Cathodic Protection

26 42 16 – Passive Cathodic Protection for Underground Storage Tank

26 42 16.50 Cathodic Protection Wiring Methods

		Crew	Daily Output	Labor-Hours	Unit	Material	Labor	Equipment	Total	Total Incl O&P
0010	**CATHODIC PROTECTION WIRING METHODS**									
1000	Anodes, magnesium type, 9 #	R-15	18.50	2.595	Ea.	47	119	17.75	183.75	255
1010	17 #		13	3.692		84	169	25.50	278.50	380
1020	32 #		10	4.800		153	220	33	406	545
1030	48 #		7.20	6.667		230	305	45.50	580.50	775
1100	Graphite type w/ epoxy cap, 3" x 60" (32 #)	R-22	8.40	4.438		82.50	177		259.50	370
1110	4" x 80" (68 #)		6	6.213		157	248		405	565
1120	6" x 72" (80 #)		5.20	7.169		1,275	286		1,561	1,850
1130	6" x 36" (45 #)		9.60	3.883		645	155		800	955
2000	Rectifiers, silicon type, air cooled, 28 V/10 A	R-19	3.50	5.714		2,025	269		2,294	2,650
2010	20 V/20 A		3.50	5.714		2,050	269		2,319	2,700
2100	Oil immersed, 28 V/10 A		3	6.667		3,425	315		3,740	4,250
2110	20 V/20 A		3	6.667		3,525	315		3,840	4,350
3000	Anode backfill, coke breeze	R-22	3850	.010	Lb.	.14	.39		.53	.76
4000	Cable, HMWPE, No. 8		2.40	15.533	M.L.F.	320	620		940	1,325
4010	No. 6		2.40	15.533		480	620		1,100	1,500
4020	No. 4		2.40	15.533		695	620		1,315	1,725
4030	No. 2		2.40	15.533		1,100	620		1,720	2,175
4040	No. 1		2.20	16.945		1,450	675		2,125	2,675
4050	No. 1/0		2.20	16.945		1,775	675		2,450	3,025
4060	No. 2/0		2.20	16.945		2,175	675		2,850	3,450
4070	No. 4/0		2	18.640		4,600	745		5,345	6,250
5000	Test station, 7 terminal box, flush curb type w/lockable cover	R-19	12	1.667	Ea.	70.50	78.50		149	199
5010	Reference cell, 2" dia PVC conduit, cplg, plug, set flush	"	4.80	4.167	"	148	196		344	465

169 25.50 194.50

26 51 13.10 Fixture Hangers

26 51 13.10 Fixture Hangers		Crew	Daily Output	Labor-Hours	Unit	2009 Bare Costs Material	Labor	Equipment	Total	Total Incl O&P
0010	**FIXTURE HANGERS**									
0220	Box hub cover	1 Elec	32	.250	Ea.	3.90	11.75		15.65	22.50
0240	Canopy		12	.667		8.80	31.50		40.30	57.50
0260	Connecting block		40	.200		4.10	9.40		13.50	18.95
0280	Cushion hanger		16	.500		19.50	23.50		43	57.50
0300	Box hanger, with mounting strap		8	1		7.80	47		54.80	81
0320	Connecting block		40	.200		1.90	9.40		11.30	16.55
0340	Flexible, 1/2" diameter, 4" long		12	.667		10.80	31.50		42.30	60
0360	6" long		12	.667		11.55	31.50		43.05	60.50
0380	8" long		12	.667		12.75	31.50		44.25	62
0400	10" long		12	.667		13.45	31.50		44.95	63
0420	12" long		12	.667		14.50	31.50		46	64
0440	15" long		12	.667		15.25	31.50		46.75	65
0460	18" long		12	.667		17.30	31.50		48.80	67
0480	3/4" diameter, 4" long		10	.800		12.85	37.50		50.35	72
0500	6" long		10	.800		14.30	37.50		51.80	74
0520	8" long		10	.800		15.55	37.50		53.05	75
0540	10" long		10	.800		16.40	37.50		53.90	76
0560	12" long		10	.800		17.80	37.50		55.30	77.50
0580	15" long		10	.800		19.50	37.50		57	79.50
0600	18" long		10	.800		21.50	37.50		59	82

26 51 13.40 Interior HID Fixtures

26 51 13.40 Interior HID Fixtures		Crew	Daily Output	Labor-Hours	Unit	Material	Labor	Equipment	Total	Total Incl O&P
0010	**INTERIOR HID FIXTURES** Incl. lamps, and mounting hardware									
0700	High pressure sodium, recessed, round, 70 watt	1 Elec	3.50	2.286	Ea.	420	107		527	625
0720	100 watt		3.50	2.286		445	107		552	655
0740	150 watt		3.20	2.500		495	118		613	725
0760	Square, 70 watt		3.60	2.222		420	104		524	620
0780	100 watt		3.60	2.222		445	104		549	650
0820	250 watt		3	2.667		605	125		730	860
0840	1000 watt	2 Elec	4.80	3.333		1,200	157		1,357	1,575
0860	Surface, round, 70 watt	1 Elec	3	2.667		595	125		720	850
0880	100 watt		3	2.667		610	125		735	865
0900	150 watt		2.70	2.963		640	139		779	920
0920	Square, 70 watt		3	2.667		580	125		705	830
0940	100 watt		3	2.667		610	125		735	865
0980	250 watt		2.50	3.200		645	150		795	940
1040	Pendent, round, 70 watt		3	2.667		635	125		760	895
1060	100 watt		3	2.667		610	125		735	865
1080	150 watt		2.70	2.963		620	139		759	895
1100	Square, 70 watt		3	2.667		650	125		775	910
1120	100 watt		3	2.667		660	125		785	920
1140	150 watt		2.70	2.963		675	139		814	960
1160	250 watt		2.50	3.200		915	150		1,065	1,225
1180	400 watt		2.40	3.333		960	157		1,117	1,300
1220	Wall, round, 70 watt		3	2.667		545	125		670	795
1240	100 watt		3	2.667		555	125		680	805
1260	150 watt		2.70	2.963		570	139		709	845
1300	Square, 70 watt		3	2.667		580	125		705	830
1320	100 watt		3	2.667		610	125		735	865
1340	150 watt		2.70	2.963		645	139		784	925
1360	250 watt		2.50	3.200		690	150		840	990
1380	400 watt	2 Elec	4.80	3.333		850	157		1,007	1,175

26 51 13 – Interior Lighting Fixtures, Lamps, and Ballasts

26 51 13.40 Interior HID Fixtures		Crew	Daily Output	Labor-Hours	Unit	2009 Bare Costs Material	Labor	Equipment	Total	Total Incl O&P
1400	1000 watt	2 Elec	3.60	4.444	Ea.	1,200	209		1,409	1,650
1500	Metal halide, recessed, round, 175 watt	1 Elec	3.40	2.353		420	111		531	635
1520	250 watt	"	3.20	2.500		465	118		583	695
1540	400 watt	2 Elec	5.80	2.759		665	130		795	930
1580	Square, 175 watt	1 Elec	3.40	2.353		385	111		496	590
1640	Surface, round, 175 watt		2.90	2.759		515	130		645	770
1660	250 watt		2.70	2.963		815	139		954	1,125
1680	400 watt	2 Elec	4.80	3.333		915	157		1,072	1,250
1720	Square, 175 watt	1 Elec	2.90	2.759		560	130		690	820
1800	Pendent, round, 175 watt		2.90	2.759		610	130		740	870
1820	250 watt		2.70	2.963		885	139		1,024	1,175
1840	400 watt	2 Elec	4.80	3.333		980	157		1,137	1,325
1880	Square, 175 watt	1 Elec	2.90	2.759		595	130		725	855
1900	250 watt	"	2.70	2.963		600	139		739	875
1920	400 watt	2 Elec	4.80	3.333		970	157		1,127	1,325
1980	Wall, round, 175 watt	1 Elec	2.90	2.759		515	130		645	770
2000	250 watt	"	2.70	2.963		815	139		954	1,100
2020	400 watt	2 Elec	4.80	3.333		835	157		992	1,150
2060	Square, 175 watt	1 Elec	2.90	2.759		545	130		675	800
2080	250 watt	"	2.70	2.963		590	139		729	865
2100	400 watt	2 Elec	4.80	3.333		830	157		987	1,150
2800	High pressure sodium, vaporproof, recessed, 70 watt	1 Elec	3.50	2.286		560	107		667	780
2820	100 watt		3.50	2.286		570	107		677	795
2840	150 watt		3.20	2.500		585	118		703	825
2900	Surface, 70 watt		3	2.667		630	125		755	890
2920	100 watt		3	2.667		660	125		785	920
2940	150 watt		2.70	2.963		685	139		824	970
3000	Pendent, 70 watt		3	2.667		625	125		750	880
3020	100 watt		3	2.667		645	125		770	905
3040	150 watt		2.70	2.963		680	139		819	965
3100	Wall, 70 watt		3	2.667		670	125		795	935
3120	100 watt		3	2.667		700	125		825	965
3140	150 watt		2.70	2.963		730	139		869	1,025
3200	Metal halide, vaporproof, recessed, 175 watt		3.40	2.353		540	111		651	765
3220	250 watt		3.20	2.500		575	118		693	815
3240	400 watt	2 Elec	5.80	2.759		720	130		850	990
3260	1000 watt	"	4.80	3.333		1,325	157		1,482	1,700
3280	Surface, 175 watt	1 Elec	2.90	2.759		525	130		655	775
3300	250 watt	"	2.70	2.963		820	139		959	1,125
3320	400 watt	2 Elec	4.80	3.333		1,025	157		1,182	1,375
3340	1000 watt	"	3.60	4.444		1,475	209		1,684	1,950
3360	Pendent, 175 watt	1 Elec	2.90	2.759		545	130		675	800
3380	250 watt	"	2.70	2.963		845	139		984	1,150
3400	400 watt	2 Elec	4.80	3.333		1,000	157		1,157	1,350
3420	1000 watt	"	3.60	4.444		1,625	209		1,834	2,125
3440	Wall, 175 watt	1 Elec	2.90	2.759		595	130		725	855
3460	250 watt	"	2.70	2.963		890	139		1,029	1,200
3480	400 watt	2 Elec	4.80	3.333		1,050	157		1,207	1,425
3500	1000 watt	"	3.60	4.444		1,700	209		1,909	2,175

26 51 13.50 Interior Lighting Fixtures

0010	**INTERIOR LIGHTING FIXTURES** Including lamps, mounting R265113-40									
0030	hardware and connections									

26 51 13 – Interior Lighting Fixtures, Lamps, and Ballasts

26 51 13.50 Interior Lighting Fixtures		Crew	Daily Output	Labor-Hours	Unit	2009 Bare Costs Material	Labor	Equipment	Total	Total Incl O&P
0100	Fluorescent, C.W. lamps, troffer, recess mounted in grid, RS R265723-05									
0200	Acrylic lens, 1'W x 4'L, two 40 watt	1 Elec	5.70	1.404	Ea.	53	66		119	159
0210	1'W x 4'L, three 40 watt R265723-20		5.40	1.481		61.50	69.50		131	175
0300	2'W x 2'L, two U40 watt		5.70	1.404		56	66		122	163
0400	2'W x 4'L, two 40 watt		5.30	1.509		56	71		127	171
0500	2'W x 4'L, three 40 watt		5	1.600		60.50	75		135.50	183
0600	2'W x 4'L, four 40 watt		4.70	1.702		64	80		144	194
0700	4'W x 4'L, four 40 watt	2 Elec	6.40	2.500		335	118		453	550
0800	4'W x 4'L, six 40 watt		6.20	2.581		345	121		466	565
0900	4'W x 4'L, eight 40 watt		5.80	2.759		360	130		490	595
0910	Acrylic lens, 1'W x 4'L, two 32 watt G	1 Elec	5.70	1.404		66	66		132	174
0930	2'W x 2'L, two U32 watt G		5.70	1.404		80	66		146	189
0940	2'W x 4'L, two 32 watt G		5.30	1.509		72	71		143	189
0950	2'W x 4'L, three 32 watt G		5	1.600		76.50	75		151.50	200
0960	2'W x 4'L, four 32 watt G		4.70	1.702		78.50	80		158.50	210
1000	Surface mounted, RS									
1030	Acrylic lens with hinged & latched door frame									
1100	1'W x 4'L, two 40 watt	1 Elec	7	1.143	Ea.	74	53.50		127.50	164
1110	1'W x 4'L, three 40 watt		6.70	1.194		76.50	56		132.50	171
1200	2'W x 2'L, two U40 watt		7	1.143		79.50	53.50		133	170
1300	2'W x 4'L, two 40 watt		6.20	1.290		91	60.50		151.50	194
1400	2'W x 4'L, three 40 watt		5.70	1.404		92	66		158	202
1500	2'W x 4'L, four 40 watt		5.30	1.509		94	71		165	212
1600	4'W x 4'L, four 40 watt	2 Elec	7.20	2.222		465	104		569	670
1700	4'W x 4'L, six 40 watt		6.60	2.424		505	114		619	730
1800	4'W x 4'L, eight 40 watt		6.20	2.581		525	121		646	760
1900	2'W x 8'L, four 40 watt		6.40	2.500		182	118		300	380
2000	2'W x 8'L, eight 40 watt		6.20	2.581		196	121		317	405
2010	Acrylic wrap around lens									
2020	6"W x 4'L, one 40 watt	1 Elec	8	1	Ea.	74.50	47		121.50	155
2030	6"W x 8'L, two 40 watt	2 Elec	8	2		81.50	94		175.50	235
2040	11"W x 4'L, two 40 watt	1 Elec	7	1.143		49.50	53.50		103	137
2050	11"W x 8'L, four 40 watt	2 Elec	6.60	2.424		82	114		196	266
2060	16"W x 4'L, four 40 watt	1 Elec	5.30	1.509		82	71		153	200
2070	16"W x 8'L, eight 40 watt	2 Elec	6.40	2.500		179	118		297	380
2080	2'W x 2'L, two U40 watt	1 Elec	7	1.143		100	53.50		153.50	193
2100	Strip fixture									
2130	Surface mounted									
2200	4' long, one 40 watt, RS	1 Elec	8.50	.941	Ea.	31	44		75	102
2300	4' long, two 40 watt, RS		8	1		33.50	47		80.50	110
2310	4' long, two 32 watt T8, RS G		8	1		62.50	47		109.50	142
2400	4' long, one 40 watt, SL		8	1		45	47		92	122
2500	4' long, two 40 watt, SL		7	1.143		61.50	53.50		115	150
2580	8' long, one 60 watt T8, SL G	2 Elec	13.40	1.194		77.50	56		133.50	172
2590	8' long, two 60 watt T8, SL G		12.40	1.290		81	60.50		141.50	183
2600	8' long, one 75 watt, SL		13.40	1.194		46.50	56		102.50	138
2700	8' long, two 75 watt, SL		12.40	1.290		56	60.50		116.50	155
2800	4' long, two 60 watt, HO	1 Elec	6.70	1.194		90	56		146	186
2810	4' long, two 54 watt, T5HO G	"	6.70	1.194		157	56		213	260
2900	8' long, two 110 watt, HO	2 Elec	10.60	1.509		95	71		166	214
2910	4' long, two 115 watt, VHO	1 Elec	6.50	1.231		125	58		183	227
2920	8' long, two 215 watt, VHO	2 Elec	10.40	1.538		135	72.50		207.50	260
2950	High bay, 16" W x 4' L, four 54 watt, T5HO G		8.90	1.798		198	84.50		282.50	350

26 51 13 – Interior Lighting Fixtures, Lamps, and Ballasts

26 51 13.50 Interior Lighting Fixtures			Crew	Daily Output	Labor-Hours	Unit	2009 Bare Costs Material	Labor	Equipment	Total	Total Incl O&P
2952	2′ W x 4′ L, six 54 watt, T5HO	G	2 Elec	8.50	1.882	Ea.	285	88.50		373.50	450
2954	2′ W x 4′ L, six 32 watt, T8	G		8.50	1.882		169	88.50		257.50	320
3000	Strip, pendent mounted, industrial, white porcelain enamel										
3100	4′ long, two 40 watt, RS		1 Elec	5.70	1.404	Ea.	51	66		117	157
3110	4′ long, two 32 watt T8, RS	G		5.70	1.404		72	66		138	181
3200	4′ long, two 60 watt, HO			5	1.600		81	75		156	205
3290	8′ long, two 60 watt T8, SL	G	2 Elec	8.80	1.818		111	85.50		196.50	253
3300	8′ long, two 75 watt, SL			8.80	1.818		96	85.50		181.50	237
3400	8′ long, two 110 watt, HO			8	2		123	94		217	280
3410	Acrylic finish, 4′ long, two 40 watt, RS		1 Elec	5.70	1.404		84	66		150	194
3420	4′ long, two 60 watt, HO			5	1.600		155	75		230	287
3430	4′ long, two 115 watt, VHO			4.80	1.667		200	78.50		278.50	340
3440	8′ long, two 75 watt, SL		2 Elec	8.80	1.818		164	85.50		249.50	310
3450	8′ long, two 110 watt, HO			8	2		185	94		279	350
3460	8′ long, two 215 watt, VHO			7.60	2.105		262	99		361	440
3470	Troffer, air handling, 2′W x 4′L with four 40 watt, RS		1 Elec	4	2		97	94		191	252
3480	2′W x 2′L with two U40 watt RS			5.50	1.455		84.50	68.50		153	198
3490	Air connector insulated, 5″ diameter			20	.400		66	18.80		84.80	102
3500	6″ diameter			20	.400		67.50	18.80		86.30	103
3510	Troffer parabolic lay-in, 1′W x 4′L with one 32 W	G		5.70	1.404		112	66		178	224
3520	1′W x 4′L with two 32 W	G		5.30	1.509		115	71		186	236
3525	2′W x 2′L with two U32 W	G		5.70	1.404		113	66		179	225
3530	2′W x 4′L with three 32 W	G		5	1.600		127	75		202	256
3535	Downlight, recess mounted	G		8	1		108	47		155	192
3540	Wall washer, recess mounted	G		8	1		108	47		155	192
3550	Direct/indirect, 4′ long, stl., pendent mtd.	G		5	1.600		159	75		234	291
3560	4′ long, alum., pendent mtd.	G		5	1.600		330	75		405	475
3565	Prefabricated cove, 4′ long, stl. continuous row	G		5	1.600		199	75		274	335
3570	4′ long, alum. continuous row	G		5	1.600		340	75		415	490
4220	Metal halide, integral ballast, ceiling, recess mounted										
4230	prismatic glass lens, floating door										
4240	2′W x 2′L, 250 watt		1 Elec	3.20	2.500	Ea.	310	118		428	520
4250	2′W x 2′L, 400 watt		2 Elec	5.80	2.759		355	130		485	590
4260	Surface mounted, 2′W x 2′L, 250 watt		1 Elec	2.70	2.963		335	139		474	585
4270	400 watt		2 Elec	4.80	3.333		395	157		552	675
4280	High bay, aluminum reflector,										
4290	Single unit, 400 watt		2 Elec	4.60	3.478	Ea.	405	163		568	695
4300	Single unit, 1000 watt			4	4		580	188		768	930
4310	Twin unit, 400 watt			3.20	5		805	235		1,040	1,250
4320	Low bay, aluminum reflector, 250W DX lamp		1 Elec	3.20	2.500		355	118		473	570
4330	400 watt lamp		2 Elec	5	3.200		515	150		665	795
4340	High pressure sodium integral ballast ceiling, recess mounted										
4350	prismatic glass lens, floating door										
4360	2′W x 2′L, 150 watt lamp		1 Elec	3.20	2.500	Ea.	380	118		498	595
4370	2′W x 2′L, 400 watt lamp		2 Elec	5.80	2.759		450	130		580	695
4380	Surface mounted, 2′W x 2′L, 150 watt lamp		1 Elec	2.70	2.963		460	139		599	725
4390	400 watt lamp		2 Elec	4.80	3.333		515	157		672	810
4400	High bay, aluminum reflector,										
4410	Single unit, 400 watt lamp		2 Elec	4.60	3.478	Ea.	370	163		533	660
4430	Single unit, 1000 watt lamp		″	4	4		535	188		723	880
4440	Low bay, aluminum reflector, 150 watt lamp		1 Elec	3.20	2.500		320	118		438	535
4445	High bay H.I.D. quartz restrike		″	16	.500		160	23.50		183.50	212
4450	Incandescent, high hat can, round alzak reflector, prewired										

26 51 13.50 Interior Lighting Fixtures		Crew	Daily Output	Labor-Hours	Unit	2009 Bare Costs Material	Labor	Equipment	Total	Total Incl O&P
4470	100 watt	1 Elec	8	1	Ea.	69	47		116	149
4480	150 watt		8	1		102	47		149	185
4500	300 watt		6.70	1.194		236	56		292	345
4520	Round with reflector and baffles, 150 watt		8	1		49	47		96	127
4540	Round with concentric louver, 150 watt PAR		8	1		77	47		124	157
4600	Square glass lens with metal trim, prewired									
4630	100 watt	1 Elec	6.70	1.194	Ea.	52.50	56		108.50	145
4700	200 watt		6.70	1.194		95	56		151	192
4800	300 watt		5.70	1.404		140	66		206	255
4810	500 watt		5	1.600		275	75		350	420
4900	Ceiling/wall, surface mounted, metal cylinder, 75 watt		10	.800		53.50	37.50		91	117
4920	150 watt		10	.800		77	37.50		114.50	143
4930	300 watt		8	1		164	47		211	253
5000	500 watt		6.70	1.194		350	56		406	470
5010	Square, 100 watt		8	1		107	47		154	191
5020	150 watt		8	1		117	47		164	202
5030	300 watt		7	1.143		325	53.50		378.50	445
5040	500 watt		6	1.333		330	62.50		392.50	460
5200	Ceiling, surface mounted, opal glass drum									
5300	8", one 60 watt lamp	1 Elec	10	.800	Ea.	43.50	37.50		81	106
5400	10", two 60 watt lamps		8	1		49	47		96	127
5500	12", four 60 watt lamps		6.70	1.194		67	56		123	160
5510	Pendent, round, 100 watt		8	1		107	47		154	191
5520	150 watt		8	1		118	47		165	203
5530	300 watt		6.70	1.194		165	56		221	269
5540	500 watt		5.50	1.455		315	68.50		383.50	450
5550	Square, 100 watt		6.70	1.194		146	56		202	248
5560	150 watt		6.70	1.194		152	56		208	254
5570	300 watt		5.70	1.404		225	66		291	350
5580	500 watt		5	1.600		305	75		380	450
5600	Wall, round, 100 watt		8	1		62	47		109	141
5620	300 watt		8	1		110	47		157	194
5630	500 watt		6.70	1.194		370	56		426	490
5640	Square, 100 watt		8	1		101	47		148	184
5650	150 watt		8	1		103	47		150	186
5660	300 watt		7	1.143		160	53.50		213.50	259
5670	500 watt		6	1.333		285	62.50		347.50	410
6010	Vapor tight, incandescent, ceiling mounted, 200 watt		6.20	1.290		64.50	60.50		125	164
6020	Recessed, 200 watt		6.70	1.194		104	56		160	201
6030	Pendent, 200 watt		6.70	1.194		64.50	56		120.50	157
6040	Wall, 200 watt		8	1		73	47		120	153
6100	Fluorescent, surface mounted, 2 lamps, 4'L, RS, 40 watt		3.20	2.500		113	118		231	305
6110	Industrial, 2 lamps 4' long in tandem, 430 MA		2.20	3.636		208	171		379	490
6130	2 lamps 4' long, 800 MA		1.90	4.211		178	198		376	500
6160	Pendent, indust, 2 lamps 4'L in tandem, 430 MA		1.90	4.211		243	198		441	575
6170	2 lamps 4' long, 430 MA		2.30	3.478		163	163		326	430
6180	2 lamps 4' long, 800 MA		1.70	4.706		205	221		426	565
6850	Vandalproof, surface mounted, fluorescent, two 40 watt		3.20	2.500		239	118		357	445
6860	Incandescent, one 150 watt		8	1		73	47		120	153
6900	Mirror light, fluorescent, RS, acrylic enclosure, two 40 watt		8	1		116	47		163	201
6910	One 40 watt		8	1		94	47		141	176
6920	One 20 watt		12	.667		74	31.50		105.50	130
7000	Low bay, aluminum reflector. 70 watt, high pressure sodium		4	2		265	94		359	435

26 51 13.50 Interior Lighting Fixtures		Crew	Daily Output	Labor-Hours	Unit	Material	2009 Bare Costs Labor	Equipment	Total	Total Incl O&P
7010	250 watt	1 Elec	3.20	2.500	Ea.	355	118		473	570
7020	400 watt	2 Elec	5	3.200	↓	360	150		510	625
7500	Ballast replacement, by weight of ballast, to 15′ high									
7520	Indoor fluorescent, less than 2 lbs.	1 Elec	10	.800	Ea.	22	37.50		59.50	82.50
7540	Two 40W, watt reducer, 2 to 5 lbs.		9.40	.851		39.50	40		79.50	105
7560	Two F96 slimline, over 5 lbs.		8	1		73	47		120	153
7580	Vaportite ballast, less than 2 lbs.		9.40	.851		22	40		62	86
7600	2 lbs. to 5 lbs.		8.90	.899		39.50	42.50		82	109
7620	Over 5 lbs.		7.60	1.053		73	49.50		122.50	157
7630	Electronic ballast for two tubes		8	1		35.50	47		82.50	112
7640	Dimmable ballast one lamp G		8	1		81.50	47		128.50	162
7650	Dimmable ballast two-lamp G	↓	7.60	1.053		133	49.50		182.50	222
7690	Emergency ballast (factory installed in fixture)				↓	145			145	160
7990	Decorator									
8000	Pendent RLM in colors, shallow dome, 12″ diam. 100 W	1 Elec	8	1	Ea.	79	47		126	160
8010	Regular dome, 12″ diam., 100 watt		8	1		81.50	47		128.50	162
8020	16″ diam., 200 watt		7	1.143		83	53.50		136.50	174
8030	18″ diam., 500 watt		6	1.333		100	62.50		162.50	207
8100	Picture framing light, minimum		16	.500		76	23.50		99.50	120
8110	Maximum		16	.500		100	23.50		123.50	146
8150	Miniature low voltage, recessed, pinhole		8	1		159	47		206	248
8160	Star		8	1		131	47		178	217
8170	Adjustable cone		8	1		178	47		225	268
8180	Eyeball		8	1		145	47		192	232
8190	Cone		8	1		127	47		174	213
8200	Coilex baffle		8	1		125	47		172	210
8210	Surface mounted, adjustable cylinder	↓	8	1	↓	129	47		176	214
8250	Chandeliers, incandescent									
8260	24″ diam. x 42″ high, 6 light candle	1 Elec	6	1.333	Ea.	390	62.50		452.50	525
8270	24″ diam. x 42″ high, 6 light candle w/glass shade		6	1.333		360	62.50		422.50	490
8280	17″ diam. x 12″ high, 8 light w/glass panels		8	1		277	47		324	380
8290	32″ diam. x 48″H, 10 light bohemian lead crystal		4	2		655	94		749	865
8300	27″ diam. x 29″H, 10 light bohemian lead crystal		4	2		570	94		664	775
8310	21″ diam. x 9″ high 6 light sculptured ice crystal	↓	8	1	↓	445	47		492	565
8500	Accent lights, on floor or edge, 0.5W low volt incandescent									
8520	incl. transformer & fastenings, based on 100′ lengths									
8550	Lights in clear tubing, 12″ on center	1 Elec	230	.035	L.F.	8.45	1.63		10.08	11.80
8560	6″ on center		160	.050		11	2.35		13.35	15.70
8570	4″ on center		130	.062		16.85	2.89		19.74	23
8580	3″ on center		125	.064		18.70	3.01		21.71	25
8590	2″ on center		100	.080		27	3.76		30.76	36
8600	Carpet, lights both sides 6″ OC, in alum. extrusion		270	.030		25.50	1.39		26.89	30
8610	In bronze extrusion		270	.030		29	1.39		30.39	33.50
8620	Carpet-bare floor, lights 18″ OC, in alum. extrusion		270	.030		20.50	1.39		21.89	24.50
8630	In bronze extrusion		270	.030		24	1.39		25.39	28
8640	Carpet edge-wall, lights 6″ OC in alum. extrusion		270	.030		25.50	1.39		26.89	30
8650	In bronze extrusion		270	.030		29	1.39		30.39	33.50
8660	Bare floor, lights 18″ OC, in aluminum extrusion		300	.027		20.50	1.25		21.75	24.50
8670	In bronze extrusion		300	.027		24	1.25		25.25	28
8680	Bare floor conduit, aluminum extrusion		300	.027		6.75	1.25		8	9.35
8690	In bronze extrusion		300	.027	↓	13.45	1.25		14.70	16.75
8700	Step edge to 36″, lights 6″ OC, in alum. extrusion		100	.080	Ea.	68	3.76		71.76	80.50
8710	In bronze extrusion	↓	100	.080	↓	70.50	3.76		74.26	83.50

26 51 Interior Lighting

26 51 13 – Interior Lighting Fixtures, Lamps, and Ballasts

26 51 13.50 Interior Lighting Fixtures		Crew	Daily Output	Labor-Hours	Unit	Material	2009 Bare Costs Labor	Equipment	Total	Total Incl O&P
8720	Step edge to 54", lights 6" OC, in alum. extrusion	1 Elec	100	.080	Ea.	102	3.76		105.76	118
8730	In bronze extrusion		100	.080		107	3.76		110.76	124
8740	Step edge to 72", lights 6" OC, in alum. extrusion		100	.080		136	3.76		139.76	155
8750	In bronze extrusion		100	.080		148	3.76		151.76	169
8760	Connector, male		32	.250		2.51	11.75		14.26	21
8770	Female with pigtail		32	.250		5.30	11.75		17.05	24
8780	Clamps		400	.020		.50	.94		1.44	2
8790	Transformers, 50 watt		8	1		69.50	47		116.50	149
8800	250 watt		4	2		229	94		323	395
8810	1000 watt		2.70	2.963		440	139		579	700
9000	Minimum labor/equipment charge		3	2.667	Job		125		125	193

26 51 13.70 Residential Fixtures		Crew	Daily Output	Labor-Hours	Unit	Material	Labor	Equipment	Total	Total Incl O&P
0010	**RESIDENTIAL FIXTURES**									
0400	Fluorescent, interior, surface, circline, 32 watt & 40 watt	1 Elec	20	.400	Ea.	91	18.80		109.80	129
0500	2' x 2', two U 40 watt		8	1		110	47		157	194
0700	Shallow under cabinet, two 20 watt		16	.500		47.50	23.50		71	88.50
0900	Wall mounted, 4'L, one 40 watt, with baffle		10	.800		135	37.50		172.50	207
2000	Incandescent, exterior lantern, wall mounted, 60 watt		16	.500		37	23.50		60.50	76.50
2100	Post light, 150W, with 7' post		4	2		117	94		211	274
2500	Lamp holder, weatherproof with 150W PAR		16	.500		23	23.50		46.50	61
2550	With reflector and guard		12	.667		58.50	31.50		90	113
2600	Interior pendent, globe with shade, 150 watt		20	.400		152	18.80		170.80	196
9000	Minimum labor/equipment charge		4	2	Job		94		94	145

26 51 13.90 Ballast, Replacement HID		Crew	Daily Output	Labor-Hours	Unit	Material	Labor	Equipment	Total	Total Incl O&P
0010	**BALLAST, REPLACEMENT HID**									
7510	Multi-tap 120/208/240/277 volt									
7550	High pressure sodium, 70 watt	1 Elec	10	.800	Ea.	196	37.50		233.50	273
7560	100 watt		9.40	.851		205	40		245	288
7570	150 watt		9	.889		222	42		264	310
7580	250 watt		8.50	.941		330	44		374	435
7590	400 watt		7	1.143		370	53.50		423.50	495
7600	1000 watt		6	1.333		515	62.50		577.50	660
7610	Metal halide, 175 watt		8	1		121	47		168	206
7620	250 watt		8	1		155	47		202	244
7630	400 watt		7	1.143		194	53.50		247.50	297
7640	1000 watt		6	1.333		335	62.50		397.50	460
7650	1500 watt		5	1.600		415	75		490	570

26 52 Emergency Lighting

26 52 13 – Emergency Lighting Equipments

26 52 13.10 Emergency Lighting and Battery Units		Crew	Daily Output	Labor-Hours	Unit	Material	Labor	Equipment	Total	Total Incl O&P
0010	**EMERGENCY LIGHTING AND BATTERY UNITS**									
0300	Emergency light units, battery operated									
0350	Twin sealed beam light, 25 watt, 6 volt each									
0500	Lead battery operated	1 Elec	4	2	Ea.	129	94		223	287
0700	Nickel cadmium battery operated		4	2		605	94		699	810
0780	Additional remote mount, sealed beam, 25W 6V		26.70	.300		27.50	14.10		41.60	52
0790	Twin sealed beam light, 25W 6V each		26.70	.300		52	14.10		66.10	78.50
0900	Self-contained fluorescent lamp pack		10	.800		146	37.50		183.50	219
9000	Minimum labor/equipment charge		4	2	Job		94		94	145

26 53 Exit Signs

26 53 13 – Exit Lighting

26 53 13.10 Exit Lighting Fixtures		Crew	Daily Output	Labor-Hours	Unit	2009 Bare Costs Material	Labor	Equipment	Total	Total Incl O&P
0010	EXIT LIGHTING FIXTURES									
0080	Exit light ceiling or wall mount, incandescent, single face	1 Elec	8	1	Ea.	42.50	47		89.50	120
0100	Double face		6.70	1.194		49	56		105	141
0120	Explosion proof		3.80	2.105		480	99		579	680
0150	Fluorescent, single face		8	1		65	47		112	144
0160	Double face		6.70	1.194		73	56		129	167
0200	L.E.D. standard, single face G		8	1		65.50	47		112.50	145
0220	Double face G		6.70	1.194		73	56		129	167
0230	L.E.D. vandal-resistant, single face G		7.27	1.100		210	51.50		261.50	310
0240	L.E.D. w/battery unit, single face G		4.40	1.818		138	85.50		223.50	282
0260	Double face G		4	2		140	94		234	299
0262	L.E.D. w/battery unit, vandal-resistant, single face G		4.40	1.818		345	85.50		430.50	510
0270	Combination emergency light units and exit sign	↓	4	2		188	94		282	350
1780	With emergency battery, explosion proof	R-19	7.70	2.597	↓	2,500	122		2,622	2,975
9000	Minimum labor/equipment charge	1 Elec	4	2	Job		94		94	145

26 54 Classified Location Lighting

26 54 13 – Classified Lighting

26 54 13.20 Explosionproof		Crew	Daily Output	Labor-Hours	Unit	Material	Labor	Equipment	Total	Total Incl O&P
0010	**EXPLOSIONPROOF**, incl lamps, mounting hardware and connections									
6310	Metal halide with ballast, ceiling, surface mounted, 175 watt	1 Elec	2.90	2.759	Ea.	865	130		995	1,150
6320	250 watt	"	2.70	2.963		1,050	139		1,189	1,375
6330	400 watt	2 Elec	4.80	3.333		1,100	157		1,257	1,475
6340	Ceiling, pendent mounted, 175 watt	1 Elec	2.60	3.077		820	145		965	1,125
6350	250 watt	"	2.40	3.333		995	157		1,152	1,350
6360	400 watt	2 Elec	4.20	3.810		1,075	179		1,254	1,450
6370	Wall, surface mounted, 175 watt	1 Elec	2.90	2.759		930	130		1,060	1,225
6380	250 watt	"	2.70	2.963		1,100	139		1,239	1,450
6390	400 watt	2 Elec	4.80	3.333		1,175	157		1,332	1,550
6400	High pressure sodium, ceiling surface mounted, 70 watt	1 Elec	3	2.667		935	125		1,060	1,225
6410	100 watt		3	2.667		970	125		1,095	1,275
6420	150 watt		2.70	2.963		1,000	139		1,139	1,325
6430	Pendent mounted, 70 watt		2.70	2.963		895	139		1,034	1,200
6440	100 watt		2.70	2.963		925	139		1,064	1,250
6450	150 watt		2.40	3.333		955	157		1,112	1,300
6460	Wall mounted, 70 watt		3	2.667		1,000	125		1,125	1,300
6470	100 watt		3	2.667		1,025	125		1,150	1,325
6480	150 watt		2.70	2.963		1,075	139		1,214	1,400
6510	Incandescent, ceiling mounted, 200 watt		4	2		835	94		929	1,075
6520	Pendent mounted, 200 watt		3.50	2.286		715	107		822	955
6530	Wall mounted, 200 watt		4	2		830	94		924	1,050
6600	Fluorescent, RS, 4' long, ceiling mounted, two 40 watt		2.70	2.963		2,000	139		2,139	2,425
6610	Three 40 watt		2.20	3.636		2,900	171		3,071	3,450
6620	Four 40 watt		1.90	4.211		3,725	198		3,923	4,400
6630	Pendent mounted, two 40 watt		2.30	3.478		2,325	163		2,488	2,825
6640	Three 40 watt		1.90	4.211		3,325	198		3,523	3,950
6650	Four 40 watt	↓	1.70	4.706	↓	4,375	221		4,596	5,150

26 55 Special Purpose Lighting

26 55 59 – Display Lighting

26 55 59.10 Track Lighting		Crew	Daily Output	Labor-Hours	Unit	Material	Labor	Equipment	Total	Total Incl O&P
						2009 Bare Costs				
0010	**TRACK LIGHTING**									
0080	Track, 1 circuit, 4′ section	1 Elec	6.70	1.194	Ea.	49.50	56		105.50	141
0100	8′ section		5.30	1.509		81	71		152	199
0200	12′ section		4.40	1.818		131	85.50		216.50	275
0300	3 circuits, 4′ section		6.70	1.194		68	56		124	162
0400	8′ section		5.30	1.509		106	71		177	226
0500	12′ section		4.40	1.818		176	85.50		261.50	325
1000	Feed kit, surface mounting		16	.500		12.20	23.50		35.70	49.50
1100	End cover		24	.333		4.75	15.65		20.40	29.50
1200	Feed kit, stem mounting, 1 circuit		16	.500		36.50	23.50		60	76.50
1300	3 circuit		16	.500		36.50	23.50		60	76.50
2000	Electrical joiner, for continuous runs, 1 circuit		32	.250		17.60	11.75		29.35	37.50
2100	3 circuit		32	.250		42.50	11.75		54.25	64.50
2200	Fixtures, spotlight, 75W PAR halogen		16	.500		105	23.50		128.50	152
2210	50W MR16 halogen		16	.500		128	23.50		151.50	177
3000	Wall washer, 250 watt tungsten halogen		16	.500		122	23.50		145.50	170
3100	Low voltage, 25/50 watt, 1 circuit		16	.500		123	23.50		146.50	171
3120	3 circuit		16	.500		127	23.50		150.50	176
9000	Minimum labor/equipment charge		3	2.667	Job		125		125	193

26 55 61 – Theatrical Lighting

26 55 61.10 Lights		Crew	Daily Output	Labor-Hours	Unit	Material	Labor	Equipment	Total	Total Incl O&P
0010	**LIGHTS**									
2000	Lights, border, quartz, reflector, vented,									
2100	colored or white	1 Elec	20	.400	L.F.	173	18.80		191.80	219
2500	Spotlight, follow spot, with transformer, 2,100 watt	"	4	2	Ea.	3,075	94		3,169	3,525
2600	For no transformer, deduct					910			910	1,000
3000	Stationary spot, fresnel quartz, 6″ lens	1 Elec	4	2		131	94		225	289
3100	8″ lens		4	2		231	94		325	400
3500	Ellipsoidal quartz, 1,000W, 6″ lens		4	2		335	94		429	515
3600	12″ lens		4	2		600	94		694	805
4000	Strobe light, 1 to 15 flashes per second, quartz		3	2.667		750	125		875	1,025
4500	Color wheel, portable, five hole, motorized		4	2		194	94		288	360

26 56 Exterior Lighting

26 56 13 – Lighting Poles and Standards

26 56 13.10 Lighting Poles		Crew	Daily Output	Labor-Hours	Unit	Material	Labor	Equipment	Total	Total Incl O&P
0010	**LIGHTING POLES**									
2800	Light poles, anchor base									
2820	not including concrete bases									
2840	Aluminum pole, 8′ high	1 Elec	4	2	Ea.	665	94		759	875
2850	10′ high		4	2		695	94		789	910
2860	12′ high		3.80	2.105		725	99		824	945
2870	14′ high		3.40	2.353		755	111		866	1,000
2880	16′ high		3	2.667		830	125		955	1,100
3000	20′ high	R-3	2.90	6.897		865	320	44.50	1,229.50	1,500
3200	30′ high		2.60	7.692		1,700	355	49.50	2,104.50	2,450
3400	35′ high		2.30	8.696		1,825	405	56	2,286	2,675
3600	40′ high		2	10		2,075	465	64.50	2,604.50	3,075
3800	Bracket arms, 1 arm	1 Elec	8	1		114	47		161	198
4000	2 arms		8	1		228	47		275	325

26 56 Exterior Lighting

26 56 13 – Lighting Poles and Standards

26 56 13.10 Lighting Poles		Crew	Daily Output	Labor-Hours	Unit	Material	Labor	Equipment	Total	Total Incl O&P
						2009 Bare Costs				
4200	3 arms	1 Elec	5.30	1.509	Ea.	345	71		416	485
4400	4 arms		5.30	1.509		455	71		526	615
4500	Steel pole, galvanized, 8′ high		3.80	2.105		580	99		679	790
4510	10′ high		3.70	2.162		605	102		707	825
4520	12′ high		3.40	2.353		655	111		766	890
4530	14′ high		3.10	2.581		695	121		816	950
4540	16′ high		2.90	2.759		740	130		870	1,000
4550	18′ high		2.70	2.963		780	139		919	1,075
4600	20′ high	R-3	2.60	7.692		1,025	355	49.50	1,429.50	1,725
4800	30′ high		2.30	8.696		1,200	405	56	1,661	2,000
5000	35′ high		2.20	9.091		1,325	420	58.50	1,803.50	2,175
5200	40′ high		1.70	11.765		1,625	545	76	2,246	2,725
5400	Bracket arms, 1 arm	1 Elec	8	1		168	47		215	258
5600	2 arms		8	1		261	47		308	360
5800	3 arms		5.30	1.509		283	71		354	420
6000	4 arms		5.30	1.509		395	71		466	545
6100	Fiberglass pole, 1 or 2 fixtures, 20′ high	R-3	4	5		615	232	32.50	879.50	1,075
6200	30′ high		3.60	5.556		765	257	36	1,058	1,275
6300	35′ high		3.20	6.250		1,200	289	40.50	1,529.50	1,800
6400	40′ high		2.80	7.143		1,400	330	46	1,776	2,075
6420	Wood pole, 4-1/2″ x 5-1/8″, 8′ high	1 Elec	6	1.333		305	62.50		367.50	430
6430	10′ high		6	1.333		345	62.50		407.50	475
6440	12′ high		5.70	1.404		435	66		501	580
6450	15′ high		5	1.600		510	75		585	675
6460	20′ high		4	2		615	94		709	820
7300	Transformer bases, not including concrete bases									
7320	Maximum pole size, steel, 40′ high	1 Elec	2	4	Ea.	1,300	188		1,488	1,725
7340	Cast aluminum, 30′ high		3	2.667		700	125		825	965
7350	40′ high		2.50	3.200		1,050	150		1,200	1,400

26 56 19 – Roadway Lighting

26 56 19.20 Roadway Area Luminaire

		Crew	Daily Output	Labor-Hours	Unit	Material	Labor	Equipment	Total	Total Incl O&P
0010	**ROADWAY AREA LUMINAIRE**									
2650	Roadway area luminaire, low pressure sodium, 135 watt	1 Elec	2	4	Ea.	620	188		808	975
2700	180 watt	″	2	4		655	188		843	1,000
2750	Metal halide, 400 watt	2 Elec	4.40	3.636		510	171		681	825
2760	1000 watt		4	4		575	188		763	925
2780	High pressure sodium, 400 watt		4.40	3.636		530	171		701	845
2790	1000 watt		4	4		605	188		793	955

26 56 23 – Area Lighting

26 56 23.10 Exterior Fixtures

		Crew	Daily Output	Labor-Hours	Unit	Material	Labor	Equipment	Total	Total Incl O&P
0010	**EXTERIOR FIXTURES** With lamps									
0200	Wall mounted, incandescent, 100 watt	1 Elec	8	1	Ea.	31.50	47		78.50	107
0400	Quartz, 500 watt		5.30	1.509		54	71		125	169
0420	1500 watt		4.20	1.905		109	89.50		198.50	258
1100	Wall pack, low pressure sodium, 35 watt		4	2		234	94		328	400
1150	55 watt		4	2		278	94		372	450
1160	High pressure sodium, 70 watt		4	2		220	94		314	385
1170	150 watt		4	2		251	94		345	420
1180	Metal Halide, 175 watt		4	2		255	94		349	425
1190	250 watt		4	2		290	94		384	465
1195	400 watt		4	2		335	94		429	510

26 56 Exterior Lighting

26 56 26 – Landscape Lighting

26 56 26.20 Landscape Fixtures		Crew	Daily Output	Labor-Hours	Unit	2009 Bare Costs Material	Labor	Equipment	Total	Total Incl O&P
0010	**LANDSCAPE FIXTURES**									
7380	Landscape recessed uplight, incl. housing, ballast, transformer									
7390	& reflector, not incl. conduit, wire, trench									
7420	Incandescent, 250 watt	1 Elec	5	1.600	Ea.	550	75		625	720
7440	Quartz, 250 watt		5	1.600		525	75		600	690
7460	500 watt		4	2		535	94		629	735

26 56 33 – Walkway Lighting

26 56 33.10 Walkway Luminaire

		Crew	Daily Output	Labor-Hours	Unit	Material	Labor	Equipment	Total	Total Incl O&P
0010	**WALKWAY LUMINAIRE**									
6500	Bollard light, lamp & ballast, 42" high with polycarbonate lens									
6800	Metal halide, 175 watt	1 Elec	3	2.667	Ea.	750	125		875	1,025
6900	High pressure sodium, 70 watt		3	2.667		770	125		895	1,050
7000	100 watt		3	2.667		770	125		895	1,050
7100	150 watt		3	2.667		750	125		875	1,025
7200	Incandescent, 150 watt		3	2.667		540	125		665	790
7810	Walkway luminaire, square 16", metal halide 250 watt		2.70	2.963		605	139		744	880
7820	High pressure sodium, 70 watt		3	2.667		695	125		820	955
7830	100 watt		3	2.667		705	125		830	970
7840	150 watt		3	2.667		705	125		830	970
7850	200 watt		3	2.667		710	125		835	975
7910	Round 19", metal halide, 250 watt		2.70	2.963		880	139		1,019	1,175
7920	High pressure sodium, 70 watt		3	2.667		970	125		1,095	1,275
7930	100 watt		3	2.667		970	125		1,095	1,275
7940	150 watt		3	2.667		975	125		1,100	1,275
7950	250 watt		2.70	2.963		1,025	139		1,164	1,350
8000	Sphere 14" opal, incandescent, 200 watt		4	2		285	94		379	460
8020	Sphere 18" opal, incandescent, 300 watt		3.50	2.286		345	107		452	545
8040	Sphere 16" clear, high pressure sodium, 70 watt		3	2.667		595	125		720	850
8050	100 watt		3	2.667		635	125		760	895
8100	Cube 16" opal, incandescent, 300 watt		3.50	2.286		380	107		487	585
8120	High pressure sodium, 70 watt		3	2.667		555	125		680	805
8130	100 watt		3	2.667		570	125		695	820
8230	Lantern, high pressure sodium, 70 watt		3	2.667		490	125		615	735
8240	100 watt		3	2.667		530	125		655	775
8250	150 watt		3	2.667		500	125		625	740
8260	250 watt		2.70	2.963		695	139		834	975
8270	Incandescent, 300 watt		3.50	2.286		365	107		472	565
8330	Reflector 22" w/globe, high pressure sodium, 70 watt		3	2.667		465	125		590	705
8340	100 watt		3	2.667		470	125		595	710
8350	150 watt		3	2.667		475	125		600	720
8360	250 watt		2.70	2.963		610	139		749	885
9000	Minimum labor/equipment charge		3.75	2.133	Job		100		100	154

26 56 36 – Flood Lighting

26 56 36.20 Floodlights

		Crew	Daily Output	Labor-Hours	Unit	Material	Labor	Equipment	Total	Total Incl O&P
0010	**FLOODLIGHTS** with ballast and lamp,									
1400	pole mounted, pole not included									
1950	Metal halide, 175 watt	1 Elec	2.70	2.963	Ea.	325	139		464	575
2000	400 watt	2 Elec	4.40	3.636		405	171		576	710
2200	1000 watt		4	4		555	188		743	900
2210	1500 watt		3.70	4.324		580	203		783	955
2250	Low pressure sodium, 55 watt	1 Elec	2.70	2.963		565	139		704	835

26 56 Exterior Lighting

26 56 36 – Flood Lighting

26 56 36.20 Floodlights		Crew	Daily Output	Labor-Hours	Unit	Material	2009 Bare Costs Labor	Equipment	Total	Total Incl O&P
2270	90 watt	1 Elec	2	4	Ea.	620	188		808	975
2290	180 watt		2	4		790	188		978	1,150
2340	High pressure sodium, 70 watt		2.70	2.963		235	139		374	475
2360	100 watt		2.70	2.963		242	139		381	480
2380	150 watt		2.70	2.963		282	139		421	525
2400	400 watt	2 Elec	4.40	3.636		365	171		536	665
2600	1000 watt	"	4	4		625	188		813	980
2610	Incandescent, 300 watt	1 Elec	4	2		100	94		194	255
2620	500 watt	"	4	2		160	94		254	320
2630	1000 watt	2 Elec	6	2.667		172	125		297	380
2640	1500 watt	"	6	2.667		189	125		314	400

26 61 Lighting Systems and Accessories

26 61 13 – Lighting Accessories

26 61 13.10 Energy Saving Lighting Devices

			Crew	Daily Output	Labor-Hours	Unit	Material	Labor	Equipment	Total	Total Incl O&P
0010	**ENERGY SAVING LIGHTING DEVICES**										
0100	Occupancy sensors infrared, ceiling mounted	G	1 Elec	7	1.143	Ea.	104	53.50		157.50	197
0150	Automatic wall switches	G		24	.333		64	15.65		79.65	94.50
0200	Remote power pack	G		10	.800		33.50	37.50		71	95
0250	Photoelectric control, S.P.S.T. 120 V	G		8	1		14.10	47		61.10	88
0300	S.P.S.T. 208 V/277 V	G		8	1		18.95	47		65.95	93.50
0350	D.P.S.T. 120 V	G		6	1.333		151	62.50		213.50	263
0400	D.P.S.T. 208 V/277 V	G		6	1.333		156	62.50		218.50	269
0450	S.P.D.T. 208 V/277 V	G		6	1.333		186	62.50		248.50	300

26 61 13.30 Fixture Whips

		Crew	Daily Output	Labor-Hours	Unit	Material	Labor	Equipment	Total	Total Incl O&P
0010	**FIXTURE WHIPS**									
0080	3/8" Greenfield, 2 connectors, 6' long									
0100	TFFN wire, three #18	1 Elec	32	.250	Ea.	8.35	11.75		20.10	27.50
0150	Four #18		28	.286		8.40	13.45		21.85	30
0200	Three #16		32	.250		8.45	11.75		20.20	27.50
0250	Four #16		28	.286		8.60	13.45		22.05	30
0300	THHN wire, three #14		32	.250		10.40	11.75		22.15	29.50
0350	Four #14		28	.286		11.25	13.45		24.70	33
0360	Three #12		32	.250		13	11.75		24.75	32.50

26 61 23 – Lamps Applications

26 61 23.10 Lamps

			Crew	Daily Output	Labor-Hours	Unit	Material	Labor	Equipment	Total	Total Incl O&P
0010	**LAMPS**										
0080	Fluorescent, rapid start, cool white, 2' long, 20 watt		1 Elec	1	8	C	335	375		710	945
0100	4' long, 40 watt			.90	8.889		365	420		785	1,050
0120	3' long, 30 watt			.90	8.889		415	420		835	1,100
0125	3' long, 25 watt energy saver	G		.90	8.889		780	420		1,200	1,500
0150	U-40 watt			.80	10		1,100	470		1,570	1,950
0155	U-34 watt energy saver	G		.80	10		1,100	470		1,570	1,950
0170	4' long, 34 watt energy saver	G		.90	8.889		345	420		765	1,025
0176	2' long, T8, 17 W energy saver	G		1	8		490	375		865	1,125
0178	3' long, T8, 25 W energy saver	G		.90	8.889		490	420		910	1,175
0180	4' long, T8, 32 watt energy saver	G		.90	8.889		315	420		735	990
0200	Slimline, 4' long, 40 watt			.90	8.889		825	420		1,245	1,550
0210	4' long, 30 watt energy saver	G		.90	8.889		825	420		1,245	1,550
0300	8' long, 75 watt			.80	10		950	470		1,420	1,775

26 61 23 – Lamps Applications

26 61 23.10 Lamps		Crew	Daily Output	Labor-Hours	Unit	Material	Labor	Equipment	Total	Total Incl O&P
0350	8′ long, 60 watt energy saver G	1 Elec	.80	10	C	560	470		1,030	1,350
0400	High output, 4′ long, 60 watt		.90	8.889		945	420		1,365	1,675
0410	8′ long, 95 watt energy saver G		.80	10		960	470		1,430	1,775
0500	8′ long, 110 watt		.80	10		960	470		1,430	1,775
0512	2′ long, T5, 14 watt energy saver G		1	8		1,400	375		1,775	2,125
0514	3′ long, T5, 21 watt energy saver G		.90	8.889		1,400	420		1,820	2,200
0516	4′ long, T5, 28 watt energy saver G		.90	8.889		1,225	420		1,645	1,975
0517	4′ long, T5, 54 watt energy saver G		.90	8.889		1,500	420		1,920	2,300
0520	Very high output, 4′ long, 110 watt		.90	8.889		2,450	420		2,870	3,350
0525	8′ long, 195 watt energy saver G		.70	11.429		2,575	535		3,110	3,650
0550	8′ long, 215 watt		.70	11.429		2,525	535		3,060	3,600
0554	Full spectrum, 4′ long, 60 watt		.90	8.889		1,775	420		2,195	2,625
0556	6′ long, 85 watt		.90	8.889		1,950	420		2,370	2,775
0558	8′ long, 110 watt		.80	10		1,650	470		2,120	2,550
0560	Twin tube compact lamp G		.90	8.889		575	420		995	1,275
0570	Double twin tube compact lamp G		.80	10		1,550	470		2,020	2,450
0600	Mercury vapor, mogul base, deluxe white, 100 watt		.30	26.667		3,100	1,250		4,350	5,350
0650	175 watt		.30	26.667		2,250	1,250		3,500	4,375
0700	250 watt		.30	26.667		4,400	1,250		5,650	6,775
0800	400 watt		.30	26.667		3,550	1,250		4,800	5,825
0900	1000 watt		.20	40		8,275	1,875		10,150	12,000
1000	Metal halide, mogul base, 175 watt		.30	26.667		3,900	1,250		5,150	6,200
1100	250 watt		.30	26.667		4,400	1,250		5,650	6,775
1200	400 watt		.30	26.667		4,175	1,250		5,425	6,525
1300	1000 watt		.20	40		10,100	1,875		11,975	14,000
1320	1000 watt, 125,000 initial lumens		.20	40		19,200	1,875		21,075	24,100
1330	1500 watt		.20	40		18,000	1,875		19,875	22,800
1350	High pressure sodium, 70 watt		.30	26.667		4,525	1,250		5,775	6,900
1360	100 watt		.30	26.667		4,750	1,250		6,000	7,150
1370	150 watt		.30	26.667		4,850	1,250		6,100	7,275
1380	250 watt		.30	26.667		5,175	1,250		6,425	7,600
1400	400 watt		.30	26.667		5,300	1,250		6,550	7,750
1450	1000 watt		.20	40		14,600	1,875		16,475	19,000
1500	Low pressure sodium, 35 watt		.30	26.667		9,650	1,250		10,900	12,500
1550	55 watt		.30	26.667		9,725	1,250		10,975	12,600
1600	90 watt		.30	26.667		11,300	1,250		12,550	14,300
1650	135 watt		.20	40		15,700	1,875		17,575	20,200
1700	180 watt		.20	40		17,100	1,875		18,975	21,700
1750	Quartz line, clear, 500 watt		1.10	7.273		1,000	340		1,340	1,625
1760	1500 watt		.20	40		4,250	1,875		6,125	7,575
1762	Spot, MR 16, 50 watt		1.30	6.154		900	289		1,189	1,425
1770	Tungsten halogen, T4, 400 watt		1.10	7.273		3,800	340		4,140	4,700
1775	T3, 1200 watt		.30	26.667		4,675	1,250		5,925	7,075
1778	PAR 30, 50 watt		1.30	6.154		1,075	289		1,364	1,650
1780	PAR 38, 90 watt		1.30	6.154		860	289		1,149	1,400
1800	Incandescent, interior, A21, 100 watt		1.60	5		180	235		415	560
1900	A21, 150 watt		1.60	5		176	235		411	555
2000	A23, 200 watt		1.60	5		246	235		481	630
2200	PS 35, 300 watt		1.60	5		680	235		915	1,100
2210	PS 35, 500 watt		1.60	5		1,150	235		1,385	1,625
2230	PS 52, 1000 watt		1.30	6.154		2,250	289		2,539	2,925
2240	PS 52, 1500 watt		1.30	6.154		6,050	289		6,339	7,100
2300	R30, 75 watt	↓	1.30	6.154	↓	595	289		884	1,100

26 61 Lighting Systems and Accessories

26 61 23 – Lamps Applications

26 61 23.10 Lamps		Crew	Daily Output	Labor-Hours	Unit	2009 Bare Costs Material	Labor	Equipment	Total	Total Incl O&P
2400	R40, 100 watt	1 Elec	1.30	6.154	C	605	289		894	1,100
2500	Exterior, PAR 38, 75 watt		1.30	6.154		1,350	289		1,639	1,925
2600	PAR 38, 150 watt		1.30	6.154		1,550	289		1,839	2,150
2700	PAR 46, 200 watt		1.10	7.273		2,950	340		3,290	3,775
2800	PAR 56, 300 watt		1.10	7.273		3,850	340		4,190	4,750
3000	Guards, fluorescent lamp, 4′ long		1	8		1,025	375		1,400	1,700
3200	8′ long		.90	8.889		2,050	420		2,470	2,900
9000	Minimum labor/equipment charge		4	2	Job		94		94	145

26 71 Electrical Machines

26 71 13 – Motors Applications

26 71 13.10 Handling

		Crew	Daily Output	Labor-Hours	Unit	Material	Labor	Equipment	Total	Total Incl O&P
0010	**HANDLING** Add to normal labor cost for restricted areas									
5000	Motors									
5100	1/2 HP, 23 pounds	1 Elec	4	2	Ea.		94		94	145
5110	3/4 HP, 28 pounds		4	2			94		94	145
5120	1 HP, 33 pounds		4	2			94		94	145
5130	1-1/2 HP, 44 pounds		3.20	2.500			118		118	181
5140	2 HP, 56 pounds		3	2.667			125		125	193
5150	3 HP, 71 pounds		2.30	3.478			163		163	251
5160	5 HP, 82 pounds		1.90	4.211			198		198	305
5170	7-1/2 HP, 124 pounds		1.50	5.333			251		251	385
5180	10 HP, 144 pounds		1.20	6.667			315		315	480
5190	15 HP, 185 pounds		1	8			375		375	580
5200	20 HP, 214 pounds	2 Elec	1.50	10.667			500		500	770
5210	25 HP, 266 pounds		1.40	11.429			535		535	825
5220	30 HP, 310 pounds		1.20	13.333			625		625	965
5230	40 HP, 400 pounds		1	16			750		750	1,150
5240	50 HP, 450 pounds		.90	17.778			835		835	1,275
5250	75 HP, 680 pounds		.80	20			940		940	1,450
5260	100 HP, 870 pounds	3 Elec	1	24			1,125		1,125	1,725
5270	125 HP, 940 pounds		.80	30			1,400		1,400	2,175
5280	150 HP, 1200 pounds		.70	34.286			1,600		1,600	2,475
5290	175 HP, 1300 pounds		.60	40			1,875		1,875	2,900
5300	200 HP, 1400 pounds		.50	48			2,250		2,250	3,475

26 71 13.20 Motors

		Crew	Daily Output	Labor-Hours	Unit	Material	Labor	Equipment	Total	Total Incl O&P
0010	**MOTORS** 230/460 volts, 60 HZ									
0050	Dripproof, premium efficiency, 1.15 service factor									
0060	1800 RPM, 1/4 HP	1 Elec	5.33	1.501	Ea.	164	70.50		234.50	289
0070	1/3 HP		5.33	1.501		173	70.50		243.50	299
0080	1/2 HP		5.33	1.501		225	70.50		295.50	355
0090	3/4 HP		5.33	1.501		252	70.50		322.50	385
0100	1 HP		4.50	1.778		264	83.50		347.50	420
0150	2 HP		4.50	1.778		267	83.50		350.50	425
0200	3 HP		4.50	1.778		269	83.50		352.50	425
0250	5 HP		4.50	1.778		370	83.50		453.50	540
0300	7.5 HP		4.20	1.905		510	89.50		599.50	700
0350	10 HP		4	2		635	94		729	845
0400	15 HP		3.20	2.500		785	118		903	1,050
0450	20 HP	2 Elec	5.20	3.077		1,050	145		1,195	1,400
0500	25 HP		5	3.200		1,225	150		1,375	1,575

26 71 13.20 Motors		Crew	Daily Output	Labor-Hours	Unit	2009 Bare Costs Material	Labor	Equipment	Total	Total Incl O&P
0550	30 HP	2 Elec	4.80	3.333	Ea.	1,425	157		1,582	1,825
0600	40 HP		4	4		1,775	188		1,963	2,250
0650	50 HP		3.20	5		2,075	235		2,310	2,625
0700	60 HP		2.80	5.714		2,525	269		2,794	3,225
0750	75 HP		2.40	6.667		3,075	315		3,390	3,875
0800	100 HP	3 Elec	2.70	8.889		3,925	420		4,345	4,975
0850	125 HP		2.10	11.429		4,700	535		5,235	6,000
0900	150 HP		1.80	13.333		6,800	625		7,425	8,450
0950	200 HP		1.50	16		8,600	750		9,350	10,600
1000	1200 RPM, 1 HP	1 Elec	4.50	1.778		264	83.50		347.50	420
1050	2 HP		4.50	1.778		365	83.50		448.50	530
1100	3 HP		4.50	1.778		515	83.50		598.50	695
1150	5 HP		4.50	1.778		580	83.50		663.50	770
1200	3600 RPM, 2 HP		4.50	1.778		305	83.50		388.50	470
1250	3 HP		4.50	1.778		335	83.50		418.50	500
1300	5 HP		4.50	1.778		365	83.50		448.50	530
1350	Totally enclosed, premium efficiency 1.15 service factor									
1360	1800 RPM, 1/4 HP	1 Elec	5.33	1.501	Ea.	178	70.50		248.50	305
1370	1/3 HP		5.33	1.501		198	70.50		268.50	325
1380	1/2 HP		5.33	1.501		234	70.50		304.50	365
1390	3/4 HP		5.33	1.501		252	70.50		322.50	385
1400	1 HP		4.50	1.778		282	83.50		365.50	440
1450	2 HP		4.50	1.778		340	83.50		423.50	505
1500	3 HP		4.50	1.778		410	83.50		493.50	580
1550	5 HP		4.50	1.778		465	83.50		548.50	645
1600	7.5 HP		4.20	1.905		675	89.50		764.50	880
1650	10 HP		4	2		810	94		904	1,025
1700	15 HP		3.20	2.500		1,150	118		1,268	1,425
1750	20 HP	2 Elec	5.20	3.077		1,325	145		1,470	1,675
1800	25 HP		5	3.200		1,525	150		1,675	1,900
1850	30 HP		4.80	3.333		1,800	157		1,957	2,250
1900	40 HP		4	4		2,325	188		2,513	2,850
1950	50 HP		3.20	5		2,850	235		3,085	3,500
2000	60 HP		2.80	5.714		4,250	269		4,519	5,100
2050	75 HP		2.40	6.667		5,475	315		5,790	6,500
2100	100 HP	3 Elec	2.70	8.889		8,400	420		8,820	9,875
2150	125 HP		2.10	11.429		11,000	535		11,535	12,900
2200	150 HP		1.80	13.333		12,800	625		13,425	15,100
2250	200 HP		1.50	16		15,600	750		16,350	18,400
2300	1200 RPM, 1 HP	1 Elec	4.50	1.778		281	83.50		364.50	440
2350	2 HP		4.50	1.778		320	83.50		403.50	485
2400	3 HP		4.50	1.778		435	83.50		518.50	610
2450	5 HP		4.50	1.778		605	83.50		688.50	795
2500	3600 RPM, 2 HP		4.50	1.778		255	83.50		338.50	410
2550	3 HP		4.50	1.778		310	83.50		393.50	470
2600	5 HP		4.50	1.778		390	83.50		473.50	560

Division Notes

		CREW	DAILY OUTPUT	LABOR-HOURS	UNIT	2009 BARE COSTS MAT.	LABOR	EQUIP.	TOTAL	TOTAL INCL O&P

Estimating Tips

27 20 00 Data Communications
27 30 00 Voice Communications
27 40 00 Audio-Video Communications

When estimating material costs for special systems, it is always prudent to obtain manufacturers' quotations for equipment prices and special installation requirements which will affect the total costs.

Reference Numbers

Reference numbers are shown in shaded boxes at the beginning of some major classifications. These numbers refer to related items in the Reference Section. The reference information may be an estimating procedure, an alternate pricing method, or technical information.

Note: Not all subdivisions listed here necessarily appear in this publication.

Note: **Trade Service,** *in part, has been used as a reference source for some of the material prices used in Division 27.*

27 01 Operation and Maintenance of Communications Systems

27 01 30 – Operation and Maintenance of Voice Communications

27 01 30.51 Operation and Maintenance of Voice Equipment		Crew	Daily Output	Labor-Hours	Unit	Material	2009 Bare Costs Labor	Equipment	Total	Total Incl O&P
0010	OPERATION AND MAINTENANCE OF VOICE EQUIPMENT									
3400	Remove and replace (reinstall), speaker	1 Elec	6	1.333	Ea.		62.50		62.50	96.50

27 05 Common Work Results for Communications

27 05 05 – Selective Communications Demolition

27 05 05.20 Electrical Demolition, Communications

		Crew	Daily Output	Labor-Hours	Unit	Material	Labor	Equipment	Total	Total Incl O&P
0010	ELECTRICAL DEMOLITION, COMMUNICATIONS R024119-10									
0100	Fiber optics									
0120	Cable R260105-30	1 Elec	2400	.003	L.F.		.16		.16	.24
0160	Multi-channel rack enclosure		6	1.333	Ea.		62.50		62.50	96.50
0180	Patch panel	↓	18	.444	"		21		21	32
0200	Communication cables & fittings									
0220	Voice/data outlet	1 Elec	140	.057	Ea.		2.69		2.69	4.13
0240	Telephone cable		2800	.003	L.F.		.13		.13	.21
0260	Phone jack		135	.059	Ea.		2.79		2.79	4.28
0300	High performance cable, 2 pair		3000	.003	L.F.		.13		.13	.19
0320	4 pair		2100	.004			.18		.18	.28
0340	25 pair	↓	900	.009	↓		.42		.42	.64
1000	Nurse call system									
1020	Station	1 Elec	24	.333	Ea.		15.65		15.65	24
1040	Standard call button		24	.333			15.65		15.65	24
1060	Corridor dome light or zone indictor	↓	24	.333			15.65		15.65	24
1080	Master control station	2 Elec	2	8	↓		375		375	580

27 05 05.30 Electrical Demolition, Sound and Video

		Crew	Daily Output	Labor-Hours	Unit	Material	Labor	Equipment	Total	Total Incl O&P
0010	ELECTRICAL DEMOLITION, SOUND & VIDEO R024119-10									
0100	Cables									
0120	TV antenna lead-in cable R260105-30	1 Elec	2100	.004	L.F.		.18		.18	.28
0140	Sound cable		2400	.003			.16		.16	.24
0160	Microphone cable		2400	.003			.16		.16	.24
0180	Coaxial cable		2400	.003	↓		.16		.16	.24
0200	Doorbell system, not including wires, cables, and conduit		16	.500	Ea.		23.50		23.50	36
0220	Door chime or devices	↓	36	.222	"		10.45		10.45	16.05
0300	Public address system, not including wires, cables, and conduit									
0320	Conventional office	1 Elec	16	.500	Speaker		23.50		23.50	36
0340	Conventional industrial	"	8	1	"		47		47	72.50
0400	Sound system, not including wires, cables, and conduit									
0410	Components	1 Elec	24	.333	Ea.		15.65		15.65	24
0420	Intercom, master station		6	1.333			62.50		62.50	96.50
0440	Remote station		24	.333			15.65		15.65	24
0460	Intercom outlets		24	.333			15.65		15.65	24
0480	Handset	↓	12	.667	↓		31.50		31.50	48
0500	Emergency call system, not including wires, cables, and conduit									
0520	Annunciator	1 Elec	4	2	Ea.		94		94	145
0540	Devices	"	16	.500			23.50		23.50	36
0600	Master door, buzzer type unit	2 Elec	1.60	10	↓		470		470	725
0800	TV System, not including wires, cables, and conduit									
0820	Master TV antenna system, per outlet	1 Elec	39	.205	Outlet		9.65		9.65	14.85
0840	School application, per outlet		16	.500	"		23.50		23.50	36
0860	Amplifier		12	.667	Ea.		31.50		31.50	48
0880	Antenna	↓	6	1.333	"		62.50		62.50	96.50
0900	One camera & one monitor	2 Elec	7.80	2.051	Total		96.50		96.50	148

27 05 Common Work Results for Communications

27 05 05 – Selective Communications Demolition

27 05 05.30 Electrical Demolition, Sound and Video		Crew	Daily Output	Labor-Hours	Unit	Material	2009 Bare Costs Labor	Equipment	Total	Total Incl O&P
0920	One camera	1 Elec	8	1	Ea.		47		47	72.50

27 13 Communications Backbone Cabling

27 13 23 – Communications Optical Fiber Backbone Cabling

27 13 23.13 Communications Optical Fiber

		Crew	Daily Output	Labor-Hours	Unit	Material	Labor	Equipment	Total	Total Incl O&P
0010	**COMMUNICATIONS OPTICAL FIBER**									
0040	Specialized tools & techniques cause installation costs to vary.									
0070	Fiber optic, cable, bulk simplex, minimum R271323-40	1 Elec	8	1	C.L.F.	23	47		70	98
0080	Bulk plenum quad, maximum	″	2.29	3.493	″	60	164		224	320
0150	Jumper				Ea.	55.50			55.50	61
0200	Pigtail					30			30	33
0300	Connector	1 Elec	24	.333		18.45	15.65		34.10	44.50
0350	Finger splice		32	.250		32	11.75		43.75	53
0400	Transceiver (low cost bi-directional)		8	1		300	47		347	405
0450	Rack housing, 4 rack spaces, 12 panels (144 fibers)		2	4		375	188		563	705
0500	Patch panel, 12 ports		6	1.333		242	62.50		304.50	365

27 15 Communications Horizontal Cabling

27 15 10 – Special Communications Cabling

27 15 10.23 Sound and Video Cables and Fittings

		Crew	Daily Output	Labor-Hours	Unit	Material	Labor	Equipment	Total	Total Incl O&P
0010	**SOUND AND VIDEO CABLES & FITTINGS**									
0900	TV antenna lead-in, 300 ohm, #20-2 conductor	1 Elec	7	1.143	C.L.F.	26	53.50		79.50	111
0950	Coaxial, feeder outlet		7	1.143		32	53.50		85.50	118
1000	Coaxial, main riser		6	1.333		49	62.50		111.50	151
1100	Sound, shielded with drain, #22-2 conductor		8	1		27.50	47		74.50	103
1150	#22-3 conductor		7.50	1.067		35	50		85	116
1200	#22-4 conductor		6.50	1.231		42	58		100	135
1250	Nonshielded, #22-2 conductor		10	.800		15.35	37.50		52.85	75
1300	#22-3 conductor		9	.889		22	42		64	88.50
1350	#22-4 conductor		8	1		28.50	47		75.50	104
1400	Microphone cable		8	1		96	47		143	179

27 15 13 – Communications Copper Horizontal Cabling

27 15 13.13 Communication Cables

		Crew	Daily Output	Labor-Hours	Unit	Material	Labor	Equipment	Total	Total Incl O&P
0010	**COMMUNICATION CABLES**									
2200	Telephone twisted, PVC insulation, #22-2 conductor	1 Elec	10	.800	C.L.F.	10.50	37.50		48	69.50
2250	#22-3 conductor		9	.889		13.80	42		55.80	79.50
2300	#22-4 conductor		8	1		16	47		63	90
2350	#18-2 conductor		9	.889		12.80	42		54.80	78.50
2370	Telephone jack, eight pins		32	.250	Ea.	7.50	11.75		19.25	26.50
5000	High performance unshielded twisted pair (UTP)									
5100	Cable, category 3, #24, 2 pair solid, PVC jacket R271513-75	1 Elec	10	.800	C.L.F.	8	37.50		45.50	67
5200	4 pair solid, PVC jacket		7	1.143		12	53.50		65.50	95.50
5300	25 pair solid, PVC jacket		3	2.667		64	125		189	264
5400	2 pair solid, plenum		10	.800		11	37.50		48.50	70
5500	4 pair solid, plenum		7	1.143		16	53.50		69.50	100
5600	25 pair solid, plenum		3	2.667		96	125		221	299
5700	4 pair stranded, PVC jacket		7	1.143		28	53.50		81.50	114
7000	Category 5, #24, 4 pair solid, PVC jacket		7	1.143		14.50	53.50		68	98.50
7100	4 pair solid, plenum		7	1.143		44.50	53.50		98	132

27 15 Communications Horizontal Cabling

27 15 13 – Communications Copper Horizontal Cabling

27 15 13.13 Communication Cables		Crew	Daily Output	Labor-Hours	Unit	2009 Bare Costs Material	Labor	Equipment	Total	Total Incl O&P
7200	4 pair stranded, PVC jacket	1 Elec	7	1.143	C.L.F.	21.50	53.50		75	107
7210	Category 5e, #24, 4 pair solid, PVC jacket		7	1.143		12.85	53.50		66.35	96.50
7212	4 pair solid, plenum		7	1.143		37	53.50		90.50	123
7214	4 pair stranded, PVC jacket		7	1.143		26	53.50		79.50	111
7240	Category 6, #24, 4 pair solid, PVC jacket		7	1.143		22	53.50		75.50	107
7242	4 pair solid, plenum		7	1.143		62	53.50		115.50	151
7244	4 pair stranded, PVC jacket		7	1.143		25	53.50		78.50	110
7300	Connector, RJ-45, category 5		80	.100	Ea.	1.21	4.70		5.91	8.60
7302	Shielded RJ-45, category 5		72	.111		3.45	5.20		8.65	11.85
7310	Jack, UTP RJ-45, category 3		72	.111		3.51	5.20		8.71	11.90
7312	Category 5		65	.123		4.54	5.80		10.34	13.90
7314	Category 5e		65	.123		4.54	5.80		10.34	13.90
7316	Category 6		65	.123		4.54	5.80		10.34	13.90
7322	Jack, shielded RJ-45, category 5		60	.133		5.70	6.25		11.95	15.90
7324	Category 5e		60	.133		5.70	6.25		11.95	15.90
7326	Category 6		60	.133		5.70	6.25		11.95	15.90
7400	Voice/data expansion module, category 5e		8	1		42.50	47		89.50	120

27 15 33 – Communications Coaxial Horizontal Cabling

27 15 33.10 Coaxial Cable and Fittings		Crew	Daily Output	Labor-Hours	Unit	Material	Labor	Equipment	Total	Total Incl O&P
0010	**COAXIAL CABLE & FITTINGS**									
3500	Coaxial connectors, 50 ohm impedance quick disconnect									
3540	BNC plug, for RG A/U #58 cable	1 Elec	42	.190	Ea.	4.58	8.95		13.53	18.80
3550	RG A/U #59 cable		42	.190		4.58	8.95		13.53	18.80
3560	RG A/U #62 cable		42	.190		4.58	8.95		13.53	18.80
3600	BNC jack, for RG A/U #58 cable		42	.190		4.73	8.95		13.68	18.95
3610	RG A/U #59 cable		42	.190		4.73	8.95		13.68	18.95
3620	RG A/U #62 cable		42	.190		4.73	8.95		13.68	18.95
3660	BNC panel jack, for RG A/U #58 cable		40	.200		7.35	9.40		16.75	22.50
3670	RG A/U #59 cable		40	.200		7.35	9.40		16.75	22.50
3680	RG A/U #62 cable		40	.200		7.35	9.40		16.75	22.50
3720	BNC bulkhead jack, for RG A/U #58 cable		40	.200		8.55	9.40		17.95	24
3730	RG A/U #59 cable		40	.200		8.55	9.40		17.95	24
3740	RG A/U #62 cable		40	.200		8.55	9.40		17.95	24
3850	Coaxial cable, RG A/U 58, 50 ohm		8	1	C.L.F.	44	47		91	121
3860	RG A/U 59, 75 ohm		8	1		36	47		83	112
3870	RG A/U 62, 93 ohm		8	1		41	47		88	118
3875	RG 6/U, 75 ohm		8	1		32	47		79	108
3950	Fire rated, RG A/U 58, 50 ohm		8	1		91	47		138	173
3960	RG A/U 59, 75 ohm		8	1		121	47		168	206
3970	RG A/U 62, 93 ohm		8	1		107	47		154	191

27 15 43 – Communications Faceplates and Connectors

27 15 43.13 Communication Outlets		Crew	Daily Output	Labor-Hours	Unit	Material	Labor	Equipment	Total	Total Incl O&P
0010	**COMMUNICATION OUTLETS**									
0100	Voice/data devices not included									
0120	Voice/Data outlets, single opening	1 Elec	48	.167	Ea.	7.20	7.85		15.05	19.95
0140	Two jack openings		48	.167		2.30	7.85		10.15	14.60
0160	One jack & one 3/4" round opening		48	.167		7.20	7.85		15.05	19.95
0180	One jack & one twinaxial opening		48	.167		7.20	7.85		15.05	19.95
0200	One jack & one connector cabling opening		48	.167		7.20	7.85		15.05	19.95
0220	Two 3/8" coaxial openings		48	.167		7.20	7.85		15.05	19.95
0300	Data outlets, single opening		48	.167		7.20	7.85		15.05	19.95
0320	One 25-pin subminiature opening		48	.167		7.20	7.85		15.05	19.95

27 32 Voice Communications Telephone Sets, Faxes and Modems

27 32 36 – TTY Equipment

27 32 36.10 TTY Telephone Equipment		Crew	Daily Output	Labor-Hours	Unit	2009 Bare Costs Material	Labor	Equipment	Total	Total Incl O&P
0010	**TTY TELEPHONE EQUIPMENT**									
1620	Telephone, TTY, compact, pocket type				Ea.	299			299	330
1630	Advanced, desk type	2 Elec	20	.800		490	37.50		527.50	600
1640	Full-featured public, wall type	"	4	4		995	188		1,183	1,400

27 41 Audio-Video Systems

27 41 19 – Portable Audio-Video Equipment

27 41 19.10 T.V. Systems		Crew	Daily Output	Labor-Hours	Unit	Material	Labor	Equipment	Total	Total Incl O&P
0010	**T.V. SYSTEMS**, not including rough-in wires, cables & conduits									
0100	Master TV antenna system									
0200	VHF reception & distribution, 12 outlets	1 Elec	6	1.333	Outlet	203	62.50		265.50	320
0400	30 outlets		10	.800		134	37.50		171.50	205
0600	100 outlets		13	.615		137	29		166	196
0800	VHF & UHF reception & distribution, 12 outlets		6	1.333		202	62.50		264.50	320
1000	30 outlets		10	.800		134	37.50		171.50	205
1200	100 outlets		13	.615		136	29		165	195
1400	School and deluxe systems, 12 outlets		2.40	3.333		266	157		423	535
1600	30 outlets		4	2		233	94		327	400
1800	80 outlets		5.30	1.509		224	71		295	355
1900	Amplifier		4	2	Ea.	670	94		764	880
1910	Antenna		2	4	"	320	188		508	645

27 51 Distributed Audio-Video Communications Systems

27 51 16 – Public Address and Mass Notification Systems

27 51 16.10 Public Address System		Crew	Daily Output	Labor-Hours	Unit	Material	Labor	Equipment	Total	Total Incl O&P
0010	**PUBLIC ADDRESS SYSTEM**									
0100	Conventional, office	1 Elec	5.33	1.501	Speaker	127	70.50		197.50	248
0200	Industrial		2.70	2.963	"	244	139		383	485
9000	Minimum labor/equipment charge		3.50	2.286	Job		107		107	165

27 51 19 – Sound Masking Systems

27 51 19.10 Sound System		Crew	Daily Output	Labor-Hours	Unit	Material	Labor	Equipment	Total	Total Incl O&P
0010	**SOUND SYSTEM**, not including rough-in wires, cables & conduits									
0100	Components, projector outlet	1 Elec	8	1	Ea.	58.50	47		105.50	137
0200	Microphone		4	2		65	94		159	217
0400	Speakers, ceiling or wall		8	1		110	47		157	194
0600	Trumpets		4	2		205	94		299	370
0800	Privacy switch		8	1		82	47		129	163
1000	Monitor panel		4	2		365	94		459	545
1200	Antenna, AM/FM		4	2		204	94		298	370
1400	Volume control		8	1		82	47		129	163
1600	Amplifier, 250 watts		1	8		1,625	375		2,000	2,375
1800	Cabinets		1	8		795	375		1,170	1,450
2000	Intercom, 25 station capacity, master station	2 Elec	2	8		1,925	375		2,300	2,675
2020	11 station capacity	"	4	4		900	188		1,088	1,275
2200	Remote station	1 Elec	8	1		153	47		200	242
2400	Intercom outlets		8	1		90	47		137	172
2600	Handset		4	2		298	94		392	475
2800	Emergency call system, 12 zones, annunciator		1.30	6.154		900	289		1,189	1,425
3000	Bell		5.30	1.509		93	71		164	211

27 51 Distributed Audio-Video Communications Systems

27 51 19 – Sound Masking Systems

27 51 19.10 Sound System		Crew	Daily Output	Labor-Hours	Unit	2009 Bare Costs Material	Labor	Equipment	Total	Total Incl O&P
3200	Light or relay	1 Elec	8	1	Ea.	46.50	47		93.50	124
3400	Transformer		4	2		204	94		298	370
3600	House telephone, talking station		1.60	5		440	235		675	840
3800	Press to talk, release to listen		5.30	1.509		102	71		173	221
4000	System-on button					61			61	67
4200	Door release	1 Elec	4	2		109	94		203	265
4400	Combination speaker and microphone		8	1		186	47		233	277
4600	Termination box		3.20	2.500		58.50	118		176.50	246
4800	Amplifier or power supply		5.30	1.509		670	71		741	850
5000	Vestibule door unit		16	.500	Name	124	23.50		147.50	172
5200	Strip cabinet		27	.296	Ea.	233	13.95		246.95	279
5400	Directory		16	.500		110	23.50		133.50	157
6000	Master door, button buzzer type, 100 unit	2 Elec	.54	29.630		1,125	1,400		2,525	3,375
6020	200 unit		.30	53.333		2,100	2,500		4,600	6,175
6040	300 unit		.20	80		3,200	3,750		6,950	9,300
6060	Transformer	1 Elec	8	1		28.50	47		75.50	104
6080	Door opener		5.30	1.509		41	71		112	154
6100	Buzzer with door release and plate		4	2		41	94		135	190
6200	Intercom type, 100 unit	2 Elec	.54	29.630		1,400	1,400		2,800	3,675
6220	200 unit		.30	53.333		2,725	2,500		5,225	6,825
6240	300 unit		.20	80		4,100	3,750		7,850	10,300
6260	Amplifier	1 Elec	2	4		204	188		392	515
6280	Speaker with door release		4	2		61	94		155	212
9000	Minimum labor/equipment charge		3.50	2.286	Job		107		107	165

27 52 Healthcare Communications and Monitoring Systems

27 52 23 – Nurse Call/Code Blue Systems

27 52 23.10 Nurse Call Systems		Crew	Daily Output	Labor-Hours	Unit	Material	Labor	Equipment	Total	Total Incl O&P
0010	**NURSE CALL SYSTEMS**									
0100	Single bedside call station	1 Elec	8	1	Ea.	220	47		267	315
0200	Ceiling speaker station		8	1		66	47		113	145
0400	Emergency call station		8	1		111	47		158	195
0600	Pillow speaker		8	1		205	47		252	299
0800	Double bedside call station		4	2		221	94		315	390
1000	Duty station		4	2		170	94		264	330
1200	Standard call button		8	1		90	47		137	172
1400	Lights, corridor, dome or zone indicator		8	1		61	47		108	140
1600	Master control station for 20 stations	2 Elec	.65	24.615	Total	3,900	1,150		5,050	6,075

27 53 Distributed Systems

27 53 13 – Clock Systems

27 53 13.50 Clock Equipments		Crew	Daily Output	Labor-Hours	Unit	Material	Labor	Equipment	Total	Total Incl O&P
0010	**CLOCK EQUIPMENTS**, not including wires & conduits									
0100	Time system components, master controller	1 Elec	.33	24.242	Ea.	1,750	1,150		2,900	3,675
0200	Program bell		8	1		57.50	47		104.50	136
0400	Combination clock & speaker		3.20	2.500		202	118		320	405
0600	Frequency generator		2	4		6,800	188		6,988	7,775
0800	Job time automatic stamp recorder, minimum		4	2		475	94		569	670
1000	Maximum		4	2		730	94		824	945

27 53 13 – Clock Systems

27 53 13.50 Clock Equipments		Crew	Daily Output	Labor-Hours	Unit	Material	2009 Bare Costs Labor	Equipment	Total	Total Incl O&P
1200	Time stamp for correspondence, hand operated				Ea.	365			365	400
1400	Fully automatic					530			530	585
1600	Master time clock system, clocks & bells, 20 room	4 Elec	.20	160		4,375	7,525		11,900	16,400
1800	50 room	"	.08	400		10,100	18,800		28,900	40,000
2000	Time clock, 100 cards in & out, 1 color	1 Elec	3.20	2.500		1,025	113		1,143	1,300
2200	2 colors		3.20	2.500		1,100	113		1,218	1,375
2400	With 3 circuit program device, minimum		2	4		400	183		588	730
2600	Maximum		2	4		585	183		773	930
2800	Metal rack for 25 cards		7	1.143		64.50	53.50		118	153
3000	Watchman's tour station		8	1		67.50	47		114.50	147
3200	Annunciator with zone indication		1	8		246	375		621	850
3400	Time clock with tape		1	8		655	375		1,030	1,300

Division Notes

		CREW	DAILY OUTPUT	LABOR-HOURS	UNIT	2009 BARE COSTS MAT.	LABOR	EQUIP.	TOTAL	TOTAL INCL O&P

Division 28 - Electronic Safety and Security

Estimating Tips

- When estimating material costs for electronic safety and security systems, it is always prudent to obtain manufacturers' quotations for equipment prices and special installation requirements that affect the total cost.
- Fire alarm systems consist of control panels, annunciator panels, battery with rack, charger, and fire alarm actuating and indicating devices. Some fire alarm systems include speakers, telephone lines, door closer controls, and other components. Be careful not to overlook the costs related to installation for these items. Also be aware of costs for integrated automation instrumentation and terminal devices, control equipment, control wiring, and programming.
- Security equipment includes items such as CCTV, access control, and other detection and identification systems to perform alert and alarm functions. Be sure to consider the costs related to installation for this security equipment, such as for integrated automation instrumentation and terminal devices, control equipment, control wiring, and programming.

Reference Numbers

Reference numbers are shown in shaded boxes at the beginning of some major classifications. These numbers refer to related items in the Reference Section. The reference information may be an estimating procedure, an alternate pricing method, or technical information.

Note: Not all subdivisions listed here necessarily appear in this publication.

28 01 Operation and Maint. of Electronic Safety and Security

28 01 30 – Operation and Maint. of Electronic Detection and Alarm

28 01 30.51 Maint. and Admin. of Elec. Detection and Alarm		Crew	Daily Output	Labor-Hours	Unit	2009 Bare Costs Material	Labor	Equipment	Total	Total Incl O&P
0010	**MAINT. AND ADMIN. OF ELEC. DETECTION AND ALARM**									
3300	Remove and replace (reinstall), fire alarm device	1 Elec	5.33	1.501	Ea.		70.50		70.50	109

28 13 Access Control

28 13 53 – Security Access Detection

28 13 53.13 Security Access Metal Detectors

		Crew	Daily Output	Labor-Hours	Unit	Material	Labor	Equipment	Total	Total Incl O&P
0010	**SECURITY ACCESS METAL DETECTORS**									
0240	Metal detector, hand-held, wand type, unit only				Ea.	81.50			81.50	90
0250	Metal detector, walk through portal type, single zone	1 Elec	2	4		2,750	188		2,938	3,325
0260	Multi-zone	″	2	4		3,500	188		3,688	4,150

28 13 53.16 Security Access X-Ray Equipment

		Crew	Daily Output	Labor-Hours	Unit	Material	Labor	Equipment	Total	Total Incl O&P
0010	**SECURITY ACCESS X-RAY EQUIPMENT**									
0290	X-ray machine, desk top, for mail/small packages/letters	1 Elec	4	2	Ea.	3,000	94		3,094	3,450
0300	Conveyor type, incl monitor, minimum		2	4		14,000	188		14,188	15,700
0310	Maximum		2	4		25,000	188		25,188	27,800
0320	X-ray machine, large unit, for airports, incl monitor, min	2 Elec	1	16		35,000	750		35,750	39,700
0330	Maximum	″	.50	32		60,000	1,500		61,500	68,500

28 13 53.23 Security Access Explosive Detection Equipment

		Crew	Daily Output	Labor-Hours	Unit	Material	Labor	Equipment	Total	Total Incl O&P
0010	**SECURITY ACCESS EXPLOSIVE DETECTION EQUIPMENT**									
0270	Explosives detector, walk through portal type	1 Elec	2	4	Ea.	3,500	188		3,688	4,150
0280	Hand-held, battery operated				″				25,500	28,050

28 16 Intrusion Detection

28 16 16 – Intrusion Detection Systems Infrastructure

28 16 16.50 Intrusion Detection

		Crew	Daily Output	Labor-Hours	Unit	Material	Labor	Equipment	Total	Total Incl O&P
0010	**INTRUSION DETECTION**, not including wires & conduits									
0100	Burglar alarm, battery operated, mechanical trigger	1 Elec	4	2	Ea.	272	94		366	445
0200	Electrical trigger		4	2		325	94		419	500
0400	For outside key control, add		8	1		77	47		124	157
0800	Card reader, flush type, standard		2.70	2.963		910	139		1,049	1,225
1000	Multi-code		2.70	2.963		1,175	139		1,314	1,525
1010	Card reader, proximity type		2.70	2.963		355	139		494	605
1200	Door switches, hinge switch		5.30	1.509		57.50	71		128.50	172
1400	Magnetic switch		5.30	1.509		67.50	71		138.50	184
1600	Exit control locks, horn alarm		4	2		340	94		434	520
1800	Flashing light alarm		4	2		380	94		474	565
2000	Indicating panels, 1 channel		2.70	2.963		360	139		499	610
2200	10 channel	2 Elec	3.20	5		1,225	235		1,460	1,700
2400	20 channel		2	8		2,400	375		2,775	3,225
2600	40 channel		1.14	14.035		4,375	660		5,035	5,825
2800	Ultrasonic motion detector, 12 volt	1 Elec	2.30	3.478		225	163		388	500
3000	Infrared photoelectric detector		2.30	3.478		186	163		349	455
3200	Passive infrared detector		2.30	3.478		278	163		441	555
3400	Glass break alarm switch		8	1		46.50	47		93.50	124
3420	Switchmats, 30″ x 5′		5.30	1.509		83	71		154	201
3440	30″ x 25′		4	2		199	94		293	365
3460	Police connect panel		4	2		239	94		333	410
3480	Telephone dialer		5.30	1.509		375	71		446	525
3500	Alarm bell		4	2		76	94		170	229

28 16 Intrusion Detection

28 16 16 – Intrusion Detection Systems Infrastructure

28 16 16.50 Intrusion Detection		Crew	Daily Output	Labor-Hours	Unit	2009 Bare Costs Material	Labor	Equipment	Total	Total Incl O&P
3520	Siren	1 Elec	4	2	Ea.	143	94		237	300
3540	Microwave detector, 10′ to 200′		2	4		655	188		843	1,000
3560	10′ to 350′		2	4		1,900	188		2,088	2,400

28 23 Video Surveillance

28 23 13 – Video Surveillance Control and Management Systems

28 23 13.10 Closed Circuit Television System

		Crew	Daily Output	Labor-Hours	Unit	Material	Labor	Equipment	Total	Total Incl O&P
0010	**CLOSED CIRCUIT TELEVISION SYSTEM**									
2000	Surveillance, one station (camera & monitor)	2 Elec	2.60	6.154	Total	1,300	289		1,589	1,875
2200	For additional camera stations, add	1 Elec	2.70	2.963	Ea.	720	139		859	1,000
2400	Industrial quality, one station (camera & monitor)	2 Elec	2.60	6.154	Total	2,675	289		2,964	3,400
2600	For additional camera stations, add	1 Elec	2.70	2.963	Ea.	1,625	139		1,764	2,025
2610	For low light, add		2.70	2.963		1,325	139		1,464	1,675
2620	For very low light, add		2.70	2.963		9,675	139		9,814	10,900
2800	For weatherproof camera station, add		1.30	6.154		1,000	289		1,289	1,550
3000	For pan and tilt, add		1.30	6.154		2,600	289		2,889	3,325
3200	For zoom lens - remote control, add, minimum		2	4		2,400	188		2,588	2,950
3400	Maximum		2	4		8,775	188		8,963	9,975
3410	For automatic iris for low light, add		2	4		2,100	188		2,288	2,625
3600	Educational T.V. studio, basic 3 camera system, black & white,									
3800	electrical & electronic equip. only, minimum	4 Elec	.80	40	Total	12,500	1,875		14,375	16,600
4000	Maximum (full console)		.28	114		53,000	5,375		58,375	67,000
4100	As above, but color system, minimum		.28	114		70,500	5,375		75,875	86,000
4120	Maximum		.12	266		306,000	12,500		318,500	356,000
4200	For film chain, black & white, add	1 Elec	1	8	Ea.	14,300	375		14,675	16,300
4250	Color, add		.25	32		17,400	1,500		18,900	21,400
4400	For video tape recorders, add, minimum		1	8		3,000	375		3,375	3,875
4600	Maximum	4 Elec	.40	80		25,000	3,750		28,750	33,300

28 23 23 – Video Surveillance Systems Infrastructure

28 23 23.50 Video Surveillance Equipments

		Crew	Daily Output	Labor-Hours	Unit	Material	Labor	Equipment	Total	Total Incl O&P
0010	**VIDEO SURVEILLANCE EQUIPMENTS**									
0200	Video cameras, wireless, hidden in exit signs, clocks, etc, incl receiver	1 Elec	3	2.667	Ea.	149	125		274	355
0210	Accessories for VCR, single camera		3	2.667		570	125		695	820
0220	For multiple cameras		3	2.667		1,450	125		1,575	1,800
0230	Video cameras, wireless, for under vehicle searching, complete		2	4		9,975	188		10,163	11,300

28 31 Fire Detection and Alarm

28 31 23 – Fire Detection and Alarm Annunciation Panels and Fire Stations

28 31 23.50 Alarm Panels and Devices

		Crew	Daily Output	Labor-Hours	Unit	Material	Labor	Equipment	Total	Total Incl O&P
0010	**ALARM PANELS AND DEVICES**, not including wires & conduits									
3594	Fire, alarm control panel									
3600	4 zone	2 Elec	2	8	Ea.	615	375		990	1,250
3800	8 zone		1	16		675	750		1,425	1,900
4000	12 zone		.67	23.988		2,300	1,125		3,425	4,250
4020	Alarm device	1 Elec	8	1		135	47		182	222
4050	Actuating device	"	8	1		325	47		372	430
4160	Alarm control panel, addressable w/o voice, up to 200 points	2 Elec	1.14	13.998		2,350	660		3,010	3,600
4170	addressable w/ voice, up to 400 points	"	.73	22.008		9,000	1,025		10,025	11,500
4175	Addressable interface device	1 Elec	7.25	1.103		227	52		279	330

28 31 Fire Detection and Alarm

28 31 23 – Fire Detection and Alarm Annunciation Panels and Fire Stations

28 31 23.50 Alarm Panels and Devices		Crew	Daily Output	Labor-Hours	Unit	2009 Bare Costs Material	Labor	Equipment	Total	Total Incl O&P
4200	Battery and rack	1 Elec	4	2	Ea.	805	94		899	1,025
4400	Automatic charger		8	1		495	47		542	620
4600	Signal bell		8	1		57.50	47		104.50	136
4800	Trouble buzzer or manual station		8	1		41	47		88	118
5600	Strobe and horn		5.30	1.509		105	71		176	225
5610	Strobe and horn (ADA type)		5.30	1.509		105	71		176	225
5620	Visual alarm (ADA type)		6.70	1.194		52.50	56		108.50	145
5800	Fire alarm horn		6.70	1.194		40.50	56		96.50	131
6000	Door holder, electro-magnetic		4	2		86	94		180	240
6200	Combination holder and closer		3.20	2.500		475	118		593	705
6400	Code transmitter		4	2		765	94		859	985
6600	Drill switch		8	1		96	47		143	178
6800	Master box		2.70	2.963		3,425	139		3,564	4,000
7000	Break glass station		8	1		55.50	47		102.50	134
7010	Break glass station, addressable		7.25	1.103		75.50	52		127.50	163
7800	Remote annunciator, 8 zone lamp		1.80	4.444		201	209		410	540
8000	12 zone lamp	2 Elec	2.60	6.154		345	289		634	825
8200	16 zone lamp	"	2.20	7.273		345	340		685	905

28 31 43 – Fire Detection Sensors

28 31 43.50 Fire and Heat Detectors		Crew	Daily Output	Labor-Hours	Unit	Material	Labor	Equipment	Total	Total Incl O&P
0010	**FIRE & HEAT DETECTORS**									
5000	Detector, rate of rise	1 Elec	8	1	Ea.	39	47		86	115
5010	Detector, heat (addressable type)		7.25	1.103		166	52		218	263
5100	Fixed temperature		8	1		32.50	47		79.50	109

28 31 46 – Smoke Detection Sensors

28 31 46.50 Smoke Detectors		Crew	Daily Output	Labor-Hours	Unit	Material	Labor	Equipment	Total	Total Incl O&P
0010	**SMOKE DETECTORS**									
5200	Smoke detector, ceiling type	1 Elec	6.20	1.290	Ea.	87.50	60.50		148	190
5240	Smoke detector (addressable type)		6	1.333		166	62.50		228.50	280
5400	Duct type		3.20	2.500		277	118		395	485
5420	Duct (addressable type)		3.20	2.500		315	118		433	530

28 32 Radiation Detection and Alarm

28 32 33 – Radiation and Alarm Detection Sensors

28 32 33.50 Radiation Detection Systems		Crew	Daily Output	Labor-Hours	Unit	Material	Labor	Equipment	Total	Total Incl O&P
0010	**RADIATION DETECTION SYSTEMS**									
9410	Minimum labor/equipment charge	1 Elec	4	2	Job		94		94	145

28 33 33 – Fuel-Gas Detection Sensors

28 33 33.50 Tank Leak Detection Systems		Crew	Daily Output	Labor-Hours	Unit	Material	2009 Bare Costs Labor	Equipment	Total	Total Incl O&P
0010	**TANK LEAK DETECTION SYSTEMS** Liquid and vapor									
0100	For hydrocarbons and hazardous liquids/vapors									
0120	Controller, data acquisition, incl. printer, modem, RS232 port									
0140	24 channel, for use with all probes				Ea.	9,500			9,500	10,500
0160	9 channel, for external monitoring				"	6,000			6,000	6,600
0200	Probes									
0210	Well monitoring									
0220	Liquid phase detection				Ea.	760			760	835
0230	Hydrocarbon vapor, fixed position					250			250	275
0240	Hydrocarbon vapor, float mounted					250			250	275
0250	Both liquid and vapor hydrocarbon				↓	250			250	275
0300	Secondary containment, liquid phase									
0310	Pipe trench/manway sump				Ea.	250			250	275
0320	Double wall pipe and manual sump					250			250	275
0330	Double wall fiberglass annular space					250			250	275
0340	Double wall steel tank annular space				↓	250			250	275
0500	Accessories									
0510	Modem, non-dedicated phone line				Ea.	273			273	300
0600	Monitoring, internal									
0610	Automatic tank gauge, incl. overfill				Ea.	1,200			1,200	1,325
0620	Product line				"	1,175			1,175	1,300
0700	Monitoring, special									
0710	Cathodic protection				Ea.	650			650	715
0720	Annular space chemical monitor				"	885			885	970

		CREW	DAILY OUTPUT	LABOR-HOURS	UNIT	2009 BARE COSTS MAT.	LABOR	EQUIP.	TOTAL	TOTAL INCL O&P

Estimating Tips

31 05 00 Common Work Results for Earthwork

- Estimating the actual cost of performing earthwork requires careful consideration of the variables involved. This includes items such as type of soil, whether water will be encountered, dewatering, whether banks need bracing, disposal of excavated earth, and length of haul to fill or spoil sites, etc. If the project has large quantities of cut or fill, consider raising or lowering the site to reduce costs, while paying close attention to the effect on site drainage and utilities.
- If the project has large quantities of fill, creating a borrow pit on the site can significantly lower the costs.
- It is very important to consider what time of year the project is scheduled for completion. Bad weather can create large cost overruns from dewatering, site repair, and lost productivity from cold weather.

Reference Numbers

Reference numbers are shown in shaded boxes at the beginning of some major classifications. These numbers refer to related items in the Reference Section. The reference information may be an estimating procedure, an alternate pricing method, or technical information.

Note: Not all subdivisions listed here necessarily appear in this publication.

31 05 Common Work Results for Earthwork

31 05 23 – Cement and Concrete for Earthwork

31 05 23.30 Plant Mixed Bituminous Concrete		Crew	Daily Output	Labor-Hours	Unit	Material	Labor (2009 Bare Costs)	Equipment (2009 Bare Costs)	Total (2009 Bare Costs)	Total Incl O&P
0010	**PLANT MIXED BITUMINOUS CONCRETE**									
0020	Asphaltic concrete plant mix (145 LB per C.F.)				Ton	57			57	62.50
0040	Asphaltic concrete less than 300 tons add trucking costs									
0050	See Div. 31 23 23.20 for hauling costs									
0200	All weather patching mix, hot				Ton	62			62	68
0250	Cold patch					62			62	68
0300	Berm mix					62			62	68
0400	Base mix					57			57	62.50
0500	Binder mix					57			57	62.50
0600	Sand or sheet mix					62			62	68

31 05 23.40 Recycled Plant Mixed Bituminous Concrete		Crew	Daily Output	Labor-Hours	Unit	Material	Labor	Equipment	Total	Total Incl O&P
0010	**RECYCLED PLANT MIXED BITUMINOUS CONCRETE**									
0200	Reclaimed pavement in stockpile G				Ton	26			26	28.50
0400	Recycled pavement, at plant, ratio old: new, 70:30 G					37			37	40.50
0600	Ratio old: new, 30:70 G					51			51	56.50

31 06 Schedules for Earthwork

31 06 60 – Schedules for Special Foundations and Load Bearing Elements

31 06 60.14 Piling Special Costs		Crew	Daily Output	Labor-Hours	Unit	Material	Labor	Equipment	Total	Total Incl O&P
0010	**PILING SPECIAL COSTS**									
0011	Piling special costs, pile caps, see Div. 03 30 53.40									
0500	Cutoffs, concrete piles, plain	1 Pile	5.50	1.455	Ea.		56		56	94
0600	With steel thin shell, add		38	.211			8.10		8.10	13.55
0700	Steel pile or "H" piles		19	.421			16.20		16.20	27
0800	Wood piles		38	.211			8.10		8.10	13.55
0900	Pre-augering up to 30′ deep, average soil, 24″ diameter	B-43	180	.267	L.F.		9.25	14.05	23.30	30.50
0920	36″ diameter		115	.417			14.45	22	36.45	47.50
0960	48″ diameter		70	.686			24	36	60	78.50
0980	60″ diameter		50	.960			33	50.50	83.50	110
1000	Testing, any type piles, test load is twice the design load									
1050	50 ton design load, 100 ton test				Ea.				17,000	18,700
1100	100 ton design load, 200 ton test								22,000	24,200
1150	150 ton design load, 300 ton test								28,000	30,800
1200	200 ton design load, 400 ton test								30,000	33,000
1250	400 ton design load, 800 ton test								34,996	38,923
1500	Wet conditions, soft damp ground									
1600	Requiring mats for crane, add								40%	40%
1700	Barge mounted driving rig, add								30%	30%

31 06 60.15 Mobilization		Crew	Daily Output	Labor-Hours	Unit	Material	Labor	Equipment	Total	Total Incl O&P
0010	**MOBILIZATION**									
0020	Set up & remove, air compressor, 600 C.F.M.	A-5	3.30	5.455	Ea.		172	14.75	186.75	300
0100	1200 C.F.M.	"	2.20	8.182			258	22	280	450
0200	Crane, with pile leads and pile hammer, 75 ton	B-19	.60	106			4,225	2,950	7,175	10,100
0300	150 ton	"	.36	177			7,025	4,900	11,925	16,900
0500	Drill rig, for caissons, to 36″, minimum	B-43	2	24			830	1,275	2,105	2,750
0520	Maximum		.50	96			3,325	5,075	8,400	11,000
0600	Up to 84″		1	48			1,650	2,525	4,175	5,475
0800	Auxiliary boiler, for steam small	A-5	1.66	10.843			340	29.50	369.50	595
0900	Large	"	.83	21.687			685	58.50	743.50	1,200
1100	Rule of thumb: complete pile driving set up, small	B-19	.45	142			5,625	3,925	9,550	13,500

31 06 Schedules for Earthwork

31 06 60 – Schedules for Special Foundations and Load Bearing Elements

31 06 60.15 Mobilization		Crew	Daily Output	Labor-Hours	Unit	Material	2009 Bare Costs Labor	Equipment	Total	Total Incl O&P
1200	Large	B-19	.27	237	Ea.		9,375	6,525	15,900	22,500

31 11 Clearing and Grubbing Land

31 11 10 – Clearing and Grubbing Land

31 11 10.10 Clear and Grub Site

		Crew	Daily Output	Labor-Hours	Unit	Material	Labor	Equipment	Total	Total Incl O&P
0010	**CLEAR AND GRUB SITE**									
0020	Cut & chip light trees to 6" diam.	B-7	1	48	Acre		1,600	1,300	2,900	4,050
0150	Grub stumps and remove	B-30	2	12			420	965	1,385	1,725
0200	Cut & chip medium, trees to 12" diam.	B-7	.70	68.571			2,300	1,850	4,150	5,775
0250	Grub stumps and remove	B-30	1	24			840	1,925	2,765	3,475
0300	Cut & chip heavy, trees to 24" diam.	B-7	.30	160			5,375	4,300	9,675	13,500
0350	Grub stumps and remove	B-30	.50	48			1,675	3,875	5,550	6,975
0400	If burning is allowed, reduce cut & chip									40%
3000	Chipping stumps, to 18" deep, 12" diam.	B-86	20	.400	Ea.		16.55	10.05	26.60	37
3040	18" diameter		16	.500			20.50	12.55	33.05	46.50
3080	24" diameter		14	.571			23.50	14.35	37.85	53
3100	30" diameter		12	.667			27.50	16.75	44.25	62
3120	36" diameter		10	.800			33	20	53	74
3160	48" diameter		8	1			41.50	25	66.50	92.50
5000	Tree thinning, feller buncher, conifer									
5080	Up to 8" diameter	B-93	240	.033	Ea.		1.38	2.34	3.72	4.74
5120	12" diameter		160	.050			2.07	3.51	5.58	7.10
5240	Hardwood, up to 4" diameter		240	.033			1.38	2.34	3.72	4.74
5280	8" diameter		180	.044			1.84	3.12	4.96	6.30
5320	12" diameter		120	.067			2.76	4.68	7.44	9.50
7000	Tree removal, congested area, aerial lift truck									
7040	8" diameter	B-85	7	5.714	Ea.		192	125	317	450
7080	12" diameter		6	6.667			224	145	369	525
7120	18" diameter		5	8			269	174	443	630
7160	24" diameter		4	10			335	218	553	790
7240	36" diameter		3	13.333			450	291	741	1,050
7280	48" diameter		2	20			670	435	1,105	1,575

31 13 Selective Tree and Shrub Removal and Trimming

31 13 13 – Selective Tree and Shrub Removal

31 13 13.20 Selective Tree Removal

		Crew	Daily Output	Labor-Hours	Unit	Material	Labor	Equipment	Total	Total Incl O&P
0010	**SELECTIVE TREE REMOVAL**									
0011	With tractor, large tract, firm									
0020	level terrain, no boulders, less than 12" diam. trees									
0300	300 HP dozer, up to 400 trees/acre, 0 to 25% hardwoods	B-10M	.75	16	Acre		610	1,900	2,510	3,050
0340	25% to 50% hardwoods		.60	20			760	2,375	3,135	3,825
0370	75% to 100% hardwoods		.45	26.667			1,025	3,150	4,175	5,100
0400	500 trees/acre, 0% to 25% hardwoods		.60	20			760	2,375	3,135	3,825
0440	25% to 50% hardwoods		.48	25			955	2,975	3,930	4,775
0470	75% to 100% hardwoods		.36	33.333			1,275	3,950	5,225	6,375
0500	More than 600 trees/acre, 0 to 25% hardwoods		.52	23.077			880	2,725	3,605	4,400
0540	25% to 50% hardwoods		.42	28.571			1,100	3,400	4,500	5,450
0570	75% to 100% hardwoods		.31	38.710			1,475	4,600	6,075	7,400
0900	Large tract clearing per tree									

31 13 Selective Tree and Shrub Removal and Trimming

31 13 13 – Selective Tree and Shrub Removal

31 13 13.20 Selective Tree Removal		Crew	Daily Output	Labor-Hours	Unit	Material	2009 Bare Costs Labor	Equipment	Total	Total Incl O&P
1500	300 HP dozer, to 12" diameter, softwood	B-10M	320	.038	Ea.		1.43	4.45	5.88	7.15
1550	Hardwood		100	.120			4.57	14.25	18.82	23
1600	12" to 24" diameter, softwood		200	.060			2.29	7.10	9.39	11.50
1650	Hardwood		80	.150			5.70	17.80	23.50	28.50
1700	24" to 36" diameter, softwood		100	.120			4.57	14.25	18.82	23
1750	Hardwood		50	.240			9.15	28.50	37.65	46
1800	36" to 48" diameter, softwood		70	.171			6.55	20.50	27.05	33
1850	Hardwood		35	.343			13.05	40.50	53.55	65.50
2000	Stump removal on site by hydraulic backhoe, 1-1/2 C.Y.									
2040	4" to 6" diameter	B-17	60	.533	Ea.		17.90	10.35	28.25	40.50
2050	8" to 12" diameter	B-30	33	.727			25.50	58.50	84	106
2100	14" to 24" diameter		25	.960			33.50	77.50	111	140
2150	26" to 36" diameter		16	1.500			52.50	121	173.50	218
3000	Remove selective trees, on site using chain saws and chipper,									
3050	not incl. stumps, up to 6" diameter	B-7	18	2.667	Ea.		89.50	71.50	161	225
3100	8" to 12" diameter		12	4			134	108	242	335
3150	14" to 24" diameter		10	4.800			161	129	290	405
3200	26" to 36" diameter		8	6			201	161	362	510
3300	Machine load, 2 mile haul to dump, 12" diam. tree, add								200	300

31 14 Earth Stripping and Stockpiling

31 14 13 – Soil Stripping and Stockpiling

31 14 13.23 Topsoil Stripping and Stockpiling

		Crew	Daily Output	Labor-Hours	Unit	Material	Labor	Equipment	Total	Total Incl O&P
0010	**TOPSOIL STRIPPING AND STOCKPILING**									
0020	200 H.P. dozer, ideal conditions	B-10B	2300	.005	C.Y.		.20	.47	.67	.84
0100	Adverse conditions	"	1150	.010			.40	.94	1.34	1.66
0200	300 H.P. dozer, ideal conditions	B-10M	3000	.004			.15	.47	.62	.76
0300	Adverse conditions	"	1650	.007			.28	.86	1.14	1.39
0400	400 H.P. dozer, ideal conditions	B-10X	3900	.003			.12	.47	.59	.71
0500	Adverse conditions	"	2000	.006			.23	.92	1.15	1.38
0600	Clay, dry and soft, 200 H.P. dozer, ideal conditions	B-10B	1600	.008			.29	.68	.97	1.20
0700	Adverse conditions	"	800	.015			.57	1.35	1.92	2.40
1000	Medium hard, 300 H.P. dozer, ideal conditions	B-10M	2000	.006			.23	.71	.94	1.14
1100	Adverse conditions	"	1100	.011			.42	1.29	1.71	2.08
1200	Very hard, 400 H.P. dozer, ideal conditions	B-10X	2600	.005			.18	.71	.89	1.06
1300	Adverse conditions	"	1340	.009			.34	1.38	1.72	2.06

31 22 Grading

31 22 16 – Fine Grading

31 22 16.10 Finish Grading

		Crew	Daily Output	Labor-Hours	Unit	Material	Labor	Equipment	Total	Total Incl O&P
0010	**FINISH GRADING**									
0012	Finish grading area to be paved with grader, small area	B-11L	400	.040	S.Y.		1.46	1.38	2.84	3.85
0016	Finish grading lagoon bottoms		4	4	M.S.F.		146	138	284	385
0100	Large area		2000	.008	S.Y.		.29	.28	.57	.77
0200	Grade subgrade for base course, roadways		3500	.005			.17	.16	.33	.44
1020	For large parking lots	B-32C	5000	.010			.35	.37	.72	.98
1050	For small irregular areas	"	2000	.024			.88	.93	1.81	2.45
1100	Fine grade for slab on grade, machine	B-11L	1040	.015			.56	.53	1.09	1.48
1150	Hand grading	B-18	700	.034			1.11	.06	1.17	1.89

31 22 Grading

31 22 16 – Fine Grading

31 22 16.10 Finish Grading		Crew	Daily Output	Labor-Hours	Unit	Material	2009 Bare Costs Labor	Equipment	Total	Total Incl O&P
1200	Fine grade granular base for sidewalks and bikeways	B-62	1200	.020	S.Y.		.68	.12	.80	1.25
2550	Hand grade select gravel	2 Clab	60	.267	C.S.F.		8.45		8.45	13.90
3000	Hand grade select gravel, including compaction, 4" deep	B-18	555	.043	S.Y.		1.40	.07	1.47	2.38
3100	6" deep		400	.060			1.94	.10	2.04	3.31
3120	8" deep		300	.080			2.58	.13	2.71	4.40
3300	Finishing grading slopes, gentle	B-11L	8900	.002			.07	.06	.13	.18
3310	Steep slopes	"	7100	.002			.08	.08	.16	.22
9000	Minimum labor/equipment charge, hand grading	1 Clab	2	4	Job		126		126	209
9100	Minimum labor/equipment charge, machine grading	B-11L	2	8	"		292	275	567	775

31 23 Excavation and Fill

31 23 16 – Excavation

31 23 16.13 Excavating, Trench		Crew	Daily Output	Labor-Hours	Unit	Material	Labor	Equipment	Total	Total Incl O&P
0010	**EXCAVATING, TRENCH**									
0011	Or continuous footing									
0020	Common earth with no sheeting or dewatering included									
0050	1' to 4' deep, 3/8 C.Y. excavator	B-11C	150	.107	B.C.Y.		3.89	1.96	5.85	8.40
0060	1/2 C.Y. excavator	B-11M	200	.080			2.92	1.72	4.64	6.60
0090	4' to 6' deep, 1/2 C.Y. excavator	"	200	.080			2.92	1.72	4.64	6.60
0100	5/8 C.Y. excavator	B-12Q	250	.064			2.37	2.07	4.44	6.10
0300	1/2 C.Y. excavator, truck mounted	B-12J	200	.080			2.97	4.79	7.76	10
0500	6' to 10' deep, 3/4 C.Y. excavator	B-12F	225	.071			2.64	2.66	5.30	7.15
0600	1 C.Y. excavator, truck mounted	B-12K	400	.040			1.48	2.83	4.31	5.50
0900	10' to 14' deep, 3/4 C.Y. excavator	B-12F	200	.080			2.97	2.99	5.96	8.05
1000	1-1/2 C.Y. excavator	B-12B	540	.030			1.10	1.60	2.70	3.52
1300	14' to 20' deep, 1 C.Y. excavator	B-12A	320	.050			1.85	2.09	3.94	5.30
1340	20' to 24' deep, 1 C.Y. excavator	"	288	.056			2.06	2.32	4.38	5.85
1352	4' to 6' deep, 1/2 C.Y. excavator w/ trench box	B-13H	188	.085			3.16	5.70	8.86	11.30
1354	5/8 C.Y. excavator	"	235	.068			2.52	4.56	7.08	9.05
1362	6' to 10' deep, 3/4 C.Y. excavator w/trench box	B-13G	212	.075			2.80	3.36	6.16	8.20
1374	10' to 14' deep, 3/4 C.Y. excavator w/trench box	"	188	.085			3.16	3.79	6.95	9.20
1376	1-1/2 C.Y. excavator/ trench box	B-13E	508	.032			1.17	1.93	3.10	4
1381	14' to 20' deep, 1 C.Y. excavator w/trench box	B-13D	301	.053			1.97	2.60	4.57	6.05
1386	20' to 24' deep, 1 C.Y. excavator w/trench box	"	271	.059			2.19	2.89	5.08	6.70
1391	Shoring by SF/day trench wall protected loose mat., 4' W	B-6	3200	.008	SF Wall	.43	.26	.09	.78	.98
1392	Rent shoring per week per SF wall protected, loose mat., 4' W					1.18			1.18	1.30
1395	Hydraulic shoring, SF trench wall protected stable mat., 4' W	2 Clab	2700	.006		.19	.19		.38	.52
1397	semi-stable material, 4' W	"	2400	.007		.26	.21		.47	.63
1398	Rent hydraulic shoring per day/SF wall, stable mat., 4' W					.19			.19	.21
1399	semi-stable material					.26			.26	.28
1400	By hand with pick and shovel 2' to 6' deep, light soil	1 Clab	8	1	B.C.Y.		31.50		31.50	52
1500	Heavy soil	"	4	2	"		63		63	104
1700	For tamping backfilled trenches, air tamp, add	A-1G	100	.080	E.C.Y.		2.53	.49	3.02	4.71
1900	Vibrating plate, add	B-18	180	.133	"		4.30	.22	4.52	7.35
2100	Trim sides and bottom for concrete pours, common earth		1500	.016	S.F.		.52	.03	.55	.88
2300	Hardpan		600	.040	"		1.29	.07	1.36	2.20
2400	Pier and spread footing excavation, add to above				B.C.Y.				30%	30%
3000	Backfill trench, F.E. loader, wheel mtd., 1 C.Y. bucket									
3020	Minimal haul	B-10R	400	.030	L.C.Y.		1.14	.60	1.74	2.48
3040	100' haul		200	.060			2.29	1.20	3.49	4.96
3060	200' haul		100	.120			4.57	2.39	6.96	9.95

31 23 16 – Excavation

31 23 16.13 Excavating, Trench		Crew	Daily Output	Labor-Hours	Unit	Material	2009 Bare Costs Labor	Equipment	Total	Total Incl O&P
5020	Loam & Sandy clay with no sheeting or dewatering included									
5050	1' to 4' deep, 3/8 C.Y. tractor loader/backhoe	B-11C	162	.099	B.C.Y.		3.60	1.81	5.41	7.80
5060	1/2 C.Y. excavator	B-11M	216	.074			2.70	1.60	4.30	6.10
5080	4' to 6' deep, 1/2 C.Y. excavator	"	216	.074			2.70	1.60	4.30	6.10
5090	5/8 C.Y. excavator	B-12Q	276	.058			2.15	1.87	4.02	5.50
5130	1/2 C.Y. excavator, truck mounted	B-12J	216	.074			2.75	4.44	7.19	9.30
5140	6' to 10' deep, 3/4 C.Y. excavator	B-12F	243	.066			2.44	2.46	4.90	6.65
5160	1 C.Y. excavator, truck mounted	B-12K	432	.037			1.37	2.62	3.99	5.10
5190	10' to 14' deep, 3/4 C.Y. excavator	B-12F	216	.074			2.75	2.77	5.52	7.45
5210	1-1/2 C.Y. excavator	B-12B	583	.027			1.02	1.48	2.50	3.26
5250	14' to 20' deep, 1 C.Y. excavator	B-12A	346	.046			1.71	1.93	3.64	4.88
5300	20' to 24' deep, 1 C.Y. excavator	"	311	.051			1.91	2.15	4.06	5.45
5352	4' to 6' deep, 1/2 C.Y. excavator w/trench box	B-13H	205	.078			2.89	5.25	8.14	10.40
5354	5/8 C.Y. excavator w/trench box	"	257	.062			2.31	4.17	6.48	8.30
5362	6' to 10' deep, 3/4 C.Y. excavator w/trench box	B-13G	231	.069			2.57	3.08	5.65	7.50
5370	10' to 14' deep, 3/4 C.Y. excavator w/trench box	"	205	.078			2.89	3.47	6.36	8.45
5374	1-1/2 C.Y. excavator w/trench box	B-13E	554	.029			1.07	1.77	2.84	3.66
5382	14' to 20' deep, 1 C.Y. excavator w/ trench box	B-13D	329	.049			1.80	2.38	4.18	5.50
5392	20' to 24' deep, 1 C.Y. excavator w/ trench box	"	295	.054	↓		2.01	2.65	4.66	6.15
6020	Sand & gravel with no sheeting or dewatering included									
6050	1' to 4' deep, 3/8 C.Y. excavator	B-11C	165	.097	B.C.Y.		3.54	1.78	5.32	7.65
6060	1/2 C.Y. excavator	B-11M	220	.073			2.65	1.57	4.22	6
6080	4' to 6' deep, 1/2 C.Y. excavator	"	220	.073			2.65	1.57	4.22	6
6090	5/8 C.Y. excavator	B-12Q	275	.058			2.16	1.88	4.04	5.55
6130	1/2 C.Y. excavator, truck mounted	B-12J	220	.073			2.70	4.36	7.06	9.10
6140	6' to 10' deep, 3/4 C.Y. excavator	B-12F	248	.065			2.39	2.41	4.80	6.50
6160	1 C.Y. excavator, truck mounted	B-12K	440	.036			1.35	2.57	3.92	4.99
6190	10' to 14' deep, 3/4 C.Y. excavator	B-12F	220	.073			2.70	2.72	5.42	7.30
6210	1-1/2 C.Y. excavator	B-12B	594	.027			1	1.46	2.46	3.20
6250	14' to 20' deep, 1 C.Y. excavator	B-12A	352	.045			1.69	1.90	3.59	4.80
6300	20' to 24' deep, 1 C.Y. excavator	"	317	.050			1.87	2.11	3.98	5.35
6352	4' to 6' deep, 1/2 C.Y. excavator w/ trench box	B-13H	209	.077			2.84	5.15	7.99	10.20
6354	5/8 C.Y. excavator w/ trench box	"	261	.061			2.27	4.11	6.38	8.15
6362	6' to 10' deep, 3/4 C.Y. excavator w/ trench box	B-13G	236	.068			2.51	3.02	5.53	7.35
6370	10' to 14' deep, 3/4 C.Y. excavator w/trench box	"	209	.077			2.84	3.41	6.25	8.30
6374	1-1/2 C.Y. excavator w/trench box	B-13E	564	.028			1.05	1.74	2.79	3.60
6382	14' to 20' deep, 1 C.Y. excavator w/trench box	B-13D	334	.048			1.78	2.34	4.12	5.45
6392	20' to 24' deep, 1 C.Y. excavator w/ trench box	"	301	.053	↓		1.97	2.60	4.57	6.05
7020	Dense hard clay with no sheeting or dewatering included									
7050	1' to 4' deep, 3/8 C.Y. excavator	B-11C	132	.121	B.C.Y.		4.42	2.23	6.65	9.55
7060	1/2 C.Y. excavator	B-11M	176	.091			3.32	1.96	5.28	7.50
7080	4' to 6' deep, 1/2 C.Y. excavator	"	176	.091			3.32	1.96	5.28	7.50
7090	5/8 C.Y. excavator	B-12Q	220	.073			2.70	2.35	5.05	6.90
7130	1/2 C.Y. excavator, truck mounted	B-12J	176	.091			3.37	5.45	8.82	11.40
7140	6' to 10' deep, 3/4 C.Y. excavator	B-12F	198	.081			3	3.02	6.02	8.15
7160	1 C.Y. excavator, truck mounted	B-12K	352	.045			1.69	3.21	4.90	6.25
7190	10' to 14' deep, 3/4 C.Y. excavator	B-12F	176	.091			3.37	3.40	6.77	9.15
7210	1-1/2 C.Y. excavator	B-12B	475	.034			1.25	1.82	3.07	4.01
7250	14' to 20' deep, 1 C.Y. excavator	B-12A	282	.057			2.10	2.37	4.47	6
7300	20' to 24' deep, 1 C.Y. excavator	"	254	.063	↓		2.34	2.63	4.97	6.65
9000	Minimum labor/equipment charge	1 Clab	4	2	Job		63		63	104

31 23 16.14 Excavating, Utility Trench		Crew	Daily Output	Labor-Hours	Unit	Material	2009 Bare Costs Labor	Equipment	Total	Total Incl O&P
0010	**EXCAVATING, UTILITY TRENCH**									
0011	Common earth									
0050	Trenching with chain trencher, 12 H.P., operator walking									
0100	4" wide trench, 12" deep	B-53	800	.010	L.F.		.39	.07	.46	.69
0150	18" deep		750	.011			.42	.08	.50	.75
0200	24" deep		700	.011			.45	.09	.54	.79
0300	6" wide trench, 12" deep		650	.012			.48	.09	.57	.86
0350	18" deep		600	.013			.52	.10	.62	.93
0400	24" deep		550	.015			.57	.11	.68	1.01
0450	36" deep		450	.018			.69	.13	.82	1.24
0600	8" wide trench, 12" deep		475	.017			.66	.13	.79	1.17
0650	18" deep		400	.020			.78	.15	.93	1.39
0700	24" deep		350	.023			.89	.17	1.06	1.59
0750	36" deep		300	.027			1.04	.20	1.24	1.86
0900	Minimum labor/equipment charge		2	4	Job		156	30	186	279
1000	Backfill by hand including compaction, add									
1050	4" wide trench, 12" deep	A-1G	800	.010	L.F.		.32	.06	.38	.59
1100	18" deep		530	.015			.48	.09	.57	.89
1150	24" deep		400	.020			.63	.12	.75	1.18
1300	6" wide trench, 12" deep		540	.015			.47	.09	.56	.87
1350	18" deep		405	.020			.62	.12	.74	1.16
1400	24" deep		270	.030			.94	.18	1.12	1.75
1450	36" deep		180	.044			1.40	.27	1.67	2.62
1600	8" wide trench, 12" deep		400	.020			.63	.12	.75	1.18
1650	18" deep		265	.030			.95	.19	1.14	1.78
1700	24" deep		200	.040			1.26	.25	1.51	2.36
1750	36" deep		135	.059			1.87	.37	2.24	3.49
2000	Chain trencher, 40 H.P. operator riding									
2050	6" wide trench and backfill, 12" deep	B-54	1200	.007	L.F.		.26	.25	.51	.69
2100	18" deep		1000	.008			.31	.30	.61	.82
2150	24" deep		975	.008			.32	.31	.63	.84
2200	36" deep		900	.009			.35	.33	.68	.92
2250	48" deep		750	.011			.42	.40	.82	1.10
2300	60" deep		650	.012			.48	.46	.94	1.27
2400	8" wide trench and backfill, 12" deep		1000	.008			.31	.30	.61	.82
2450	18" deep		950	.008			.33	.32	.65	.87
2500	24" deep		900	.009			.35	.33	.68	.92
2550	36" deep		800	.010			.39	.38	.77	1.02
2600	48" deep		650	.012			.48	.46	.94	1.27
2700	12" wide trench and backfill, 12" deep		975	.008			.32	.31	.63	.84
2750	18" deep		860	.009			.36	.35	.71	.95
2800	24" deep		800	.010			.39	.38	.77	1.02
2850	36" deep		725	.011			.43	.42	.85	1.14
3000	16" wide trench and backfill, 12" deep		835	.010			.37	.36	.73	.99
3050	18" deep		750	.011			.42	.40	.82	1.10
3100	24" deep		700	.011			.45	.43	.88	1.17
3200	Compaction with vibratory plate, add								50%	50%
5100	Hand excavate and trim for pipe bells after trench excavation									
5200	8" pipe	1 Clab	155	.052	L.F.		1.63		1.63	2.69
5300	18" pipe	"	130	.062	"		1.94		1.94	3.21
9000	Minimum labor/equipment charge	A-1G	4	2	Job		63	12.35	75.35	118

31 23 16.16 Structural Excavation for Minor Structures

	31 23 16.16 Structural Excavation for Minor Structures	Crew	Daily Output	Labor-Hours	Unit	Material	2009 Bare Costs Labor	Equipment	Total	Total Incl O&P
0010	**STRUCTURAL EXCAVATION FOR MINOR STRUCTURES**									
0015	Hand, pits to 6′ deep, sandy soil	1 Clab	8	1	B.C.Y.		31.50		31.50	52
0100	Heavy soil or clay		4	2			63		63	104
0300	Pits 6′ to 12′ deep, sandy soil		5	1.600			50.50		50.50	83.50
0500	Heavy soil or clay		3	2.667			84.50		84.50	139
0700	Pits 12′ to 18′ deep, sandy soil		4	2			63		63	104
0900	Heavy soil or clay		2	4			126		126	209
1100	Hand loading trucks from stock pile, sandy soil		12	.667			21		21	35
1300	Heavy soil or clay		8	1			31.50		31.50	52
1500	For wet or muck hand excavation, add to above				%				50%	50%
1550	Excavation rock by hand/air tool	B-9	3.40	11.765	B.C.Y.		375	56.50	431.50	680
9000	Minimum labor/equipment charge	1 Clab	4	2	Job		63		63	104

31 23 16.26 Rock Removal

	31 23 16.26 Rock Removal	Crew	Daily Output	Labor-Hours	Unit	Material	Labor	Equipment	Total	Total Incl O&P
0010	**ROCK REMOVAL**									
0015	Drilling only rock, 2″ hole for rock bolts	B-47	316	.076	L.F.		2.64	4.17	6.81	8.85
0800	2-1/2″ hole for pre-splitting		600	.040			1.39	2.20	3.59	4.67
4600	Quarry operations, 2-1/2″ to 3-1/2″ diameter		715	.034			1.17	1.84	3.01	3.92
4610	6″ diameter drill holes	B-47A	1350	.018			.67	.72	1.39	1.86

31 23 16.30 Drilling and Blasting Rock

	31 23 16.30 Drilling and Blasting Rock	Crew	Daily Output	Labor-Hours	Unit	Material	Labor	Equipment	Total	Total Incl O&P
0010	**DRILLING AND BLASTING ROCK**									
0020	Rock, open face, under 1500 C.Y.	B-47	225	.107	B.C.Y.	2.75	3.71	5.85	12.31	15.50
0100	Over 1500 C.Y.		300	.080		2.75	2.78	4.39	9.92	12.35
0200	Areas where blasting mats are required, under 1500 C.Y.		175	.137		2.75	4.77	7.55	15.07	19.10
0250	Over 1500 C.Y.		250	.096		2.75	3.34	5.25	11.34	14.25
0300	Bulk drilling and blasting, can vary greatly, average								10	11
0500	Pits, average								40	44
1300	Deep hole method, up to 1500 C.Y.	B-47	50	.480		2.75	16.70	26.50	45.95	59
1400	Over 1500 C.Y.		66	.364		2.75	12.65	19.95	35.35	45.50
1900	Restricted areas, up to 1500 C.Y.		13	1.846		2.75	64	101	167.75	218
2000	Over 1500 C.Y.		20	1.200		2.75	41.50	66	110.25	143
2200	Trenches, up to 1500 C.Y.		22	1.091		8	38	60	106	136
2300	Over 1500 C.Y.		26	.923		8	32	50.50	90.50	116
2500	Pier holes, up to 1500 C.Y.		22	1.091		2.75	38	60	100.75	131
2600	Over 1500 C.Y.		31	.774		2.75	27	42.50	72.25	93.50
2800	Boulders under 1/2 C.Y., loaded on truck, no hauling	B-100	80	.150			5.70	9.60	15.30	19.65
2900	Boulders, drilled, blasted	B-47	100	.240		2.75	8.35	13.15	24.25	31
3100	Jackhammer operators with foreman compressor, air tools	B-9	1	40	Day		1,275	192	1,467	2,300
3300	Track drill, compressor, operator and foreman	B-47	1	24	″		835	1,325	2,160	2,800
3500	Blasting caps				Ea.	5.10			5.10	5.65
3700	Explosives					.38			.38	.41
3900	Blasting mats, rent, for first day					120			120	132
4000	Per added day					40			40	44
4200	Preblast survey for 6 room house, individual lot, minimum	A-6	2.40	6.667			265	29	294	460
4300	Maximum	″	1.35	11.852			470	52	522	820
4500	City block within zone of influence, minimum	A-8	25200	.001	S.F.		.05		.05	.09
4600	Maximum	″	15100	.002	″		.09		.09	.15
5000	Excavate and load boulders, less than 0.5 C.Y.	B-10T	80	.150	B.C.Y.		5.70	4.95	10.65	14.55
5020	0.5 C.Y. to 1 C.Y.	B-10U	100	.120			4.57	8.55	13.12	16.70
5200	Excavate and load blasted rock, 3 C.Y. power shovel	B-12T	1530	.010			.39	1.04	1.43	1.77
5400	Haul boulders, 25 Ton off-highway dump, 1 mile round trip	B-34E	330	.024			.77	3.88	4.65	5.55
5420	2 mile round trip		275	.029			.93	4.65	5.58	6.60
5440	3 mile round trip		225	.036			1.14	5.70	6.84	8.10

31 23 16.30 Drilling and Blasting Rock

	31 23 16.30 Drilling and Blasting Rock	Crew	Daily Output	Labor-Hours	Unit	Material	2009 Bare Costs Labor	Equipment	Total	Total Incl O&P
5460	4 mile round trip	B-34E	200	.040	B.C.Y.		1.28	6.40	7.68	9.15
5600	Bury boulders on site, less than 0.5 C.Y., 300 H.P. dozer									
5620	150' haul	B-10M	310	.039	B.C.Y.		1.47	4.59	6.06	7.40
5640	300' haul		210	.057			2.18	6.80	8.98	10.90
5800	0.5 to 1 C.Y., 300 H.P. dozer, 150' haul		300	.040			1.52	4.74	6.26	7.65
5820	300' haul		200	.060			2.29	7.10	9.39	11.50

31 23 16.32 Ripping

	31 23 16.32 Ripping	Crew	Daily Output	Labor-Hours	Unit	Material	Labor	Equipment	Total	Total Incl O&P
0010	**RIPPING**									
0020	Ripping, trap rock, soft, 300 HP dozer, ideal conditions	B-11S	700	.017	B.C.Y.		.65	2.15	2.80	3.40
1400	Ideal conditions		710	.017			.64	2.12	2.76	3.36
1500	Adverse conditions		660	.018			.69	2.28	2.97	3.60
1600	Medium hard, 300 HP dozer, ideal conditons		600	.020			.76	2.50	3.26	3.97
1700	Adverse conditions		540	.022			.85	2.78	3.63	4.41
2000	Very hard, 410 HP dozer, ideal conditions	B-11T	350	.034			1.31	5.55	6.86	8.20
2100	Adverse conditions	"	310	.039			1.47	6.25	7.72	9.25
2200	Shale, soft, 300 HP dozer, ideal conditons	B-11S	1500	.008			.30	1	1.30	1.59
2300	Adverse conditions		1350	.009			.34	1.11	1.45	1.76
2400	Medium hard, 300 HP dozer, ideal conditons		1200	.010			.38	1.25	1.63	1.99
2500	Adverse conditions		1030	.011			.42	1.39	1.81	2.20
2600	Very hard, 410 HP dozer, ideal conditons	B-11T	800	.015			.57	2.42	2.99	3.58
2700	Adverse conditions	"	720	.017			.64	2.69	3.33	3.97
3000	Dozing ripped material, 200 HP, 100' haul	B-10B	700	.017			.65	1.55	2.20	2.74
3050	300' haul	"	250	.048			1.83	4.33	6.16	7.70
3200	300 HP, 100' haul	B-10M	1150	.010			.40	1.24	1.64	1.99
3250	300' haul	"	400	.030			1.14	3.56	4.70	5.75
3400	410 HP, 100' haul	B-10X	1680	.007			.27	1.10	1.37	1.64
3450	300' haul	"	600	.020			.76	3.08	3.84	4.61

31 23 16.42 Excavating, Bulk Bank Measure

	31 23 16.42 Excavating, Bulk Bank Measure	Crew	Daily Output	Labor-Hours	Unit	Material	Labor	Equipment	Total	Total Incl O&P
0010	**EXCAVATING, BULK BANK MEASURE**									
0011	Common earth piled									
0020	For loading onto trucks, add								15%	15%
0050	For mobilization and demobilization, see Div. 01 54 36.50									
0100	For hauling, see Div. 31 23 23.20									
0200	Excavator, hydraulic, crawler mtd., 1 C.Y. cap. = 100 C.Y./hr.	B-12A	800	.020	B.C.Y.		.74	.84	1.58	2.11
0250	1-1/2 C.Y. cap. = 125 C.Y./hr.	B-12B	1000	.016			.59	.87	1.46	1.90
0260	2 C.Y. cap. = 165 C.Y./hr.	B-12C	1320	.012			.45	.90	1.35	1.70
0300	3 C.Y. cap. = 260 C.Y./hr.	B-12D	2080	.008			.29	1.04	1.33	1.60
0305	3.5 C.Y. cap. = 300 C.Y./hr.	"	2400	.007			.25	.90	1.15	1.39
0310	Wheel mounted, 1/2 C.Y. cap. = 40 C.Y./hr.	B-12E	320	.050			1.85	1.23	3.08	4.34
0360	3/4 C.Y. cap. = 60 C.Y./hr.	B-12F	480	.033			1.24	1.25	2.49	3.35
0500	Clamshell, 1/2 C.Y. cap. = 20 C.Y./hr.	B-12G	160	.100			3.71	4.07	7.78	10.45
0550	1 C.Y. cap. = 35 C.Y./hr.	B-12H	280	.057			2.12	3.92	6.04	7.70
0950	Dragline, 1/2 C.Y. cap. = 30 C.Y./hr.	B-12I	240	.067			2.47	3.37	5.84	7.70
1000	3/4 C.Y. cap. = 35 C.Y./hr.	"	280	.057			2.12	2.89	5.01	6.60
1050	1-1/2 C.Y. cap. = 65 C.Y./hr.	B-12P	520	.031			1.14	2.09	3.23	4.13
1100	3 C.Y. cap. = 112 C.Y./hr.	B-12V	900	.018			.66	1.71	2.37	2.95
1200	Front end loader, track mtd., 1-1/2 C.Y. cap. = 70 C.Y./hr.	B-10N	560	.021			.82	.66	1.48	2.02
1250	2-1/2 C.Y. cap. = 95 C.Y./hr.	B-10O	760	.016			.60	1.01	1.61	2.07
1300	3 C.Y. cap. = 130 C.Y./hr.	B-10P	1040	.012			.44	.95	1.39	1.75
1350	5 C.Y. cap. = 160 C.Y./hr.	B-10Q	1280	.009			.36	1.13	1.49	1.82
1500	Wheel mounted, 3/4 C.Y. cap. = 45 C.Y./hr.	B-10R	360	.033			1.27	.66	1.93	2.75
1550	1-1/2 C.Y. cap. = 80 C.Y./hr.	B-10S	640	.019			.71	.51	1.22	1.70

31 23 16.42 Excavating, Bulk Bank Measure

	31 23 16.42 Excavating, Bulk Bank Measure	Crew	Daily Output	Labor-Hours	Unit	Material	2009 Bare Costs Labor	Equipment	Total	Total Incl O&P
1601	3 C.Y. cap. = 140 C.Y./hr.	B-10T	1120	.011	B.C.Y.		.41	.35	.76	1.04
1650	5 C.Y. cap. = 185 C.Y./hr.	B-10U	1480	.008			.31	.58	.89	1.13
1800	Hydraulic excavator, truck mtd. 1/2 C.Y. = 30 C.Y./hr.	B-12J	240	.067			2.47	4	6.47	8.35
1850	48 inch bucket, 1 C.Y. = 45 C.Y./hr.	B-12K	360	.044			1.65	3.14	4.79	6.10
3700	Shovel, 1/2 C.Y. capacity = 55 C.Y./hr.	B-12L	440	.036			1.35	1.52	2.87	3.84
3750	3/4 C.Y. capacity = 85 C.Y./hr.	B-12M	680	.024			.87	1.25	2.12	2.77
3800	1 C.Y. capacity = 120 C.Y./hr.	B-12N	960	.017			.62	1.16	1.78	2.27
3850	1-1/2 C.Y. capacity = 160 C.Y./hr.	B-12O	1280	.013			.46	.88	1.34	1.71
3900	3 C.Y. cap. = 250 C.Y./hr.	B-12T	2000	.008			.30	.80	1.10	1.36
4000	For soft soil or sand, deduct								15%	15%
4100	For heavy soil or stiff clay, add								60%	60%
4200	For wet excavation with clamshell or dragline, add								100%	100%
4250	All other equipment, add								50%	50%
4400	Clamshell in sheeting or cofferdam, minimum	B-12H	160	.100			3.71	6.85	10.56	13.50
4450	Maximum	"	60	.267			9.90	18.30	28.20	36
5000	Excavating, bulk bank measure, sandy clay & loam piled									
5020	For loading onto trucks, add								15%	15%
5100	Excavator, hydraulic, crawler mtd., 1 C.Y. cap. = 120 C.Y./hr.	B-12A	960	.017	B.C.Y.		.62	.70	1.32	1.76
5150	1-1/2 C.Y. cap. = 150 C.Y./hr.	B-12B	1200	.013			.49	.72	1.21	1.58
5300	2 C.Y. cap. = 195 C.Y./hr.	B-12C	1560	.010			.38	.76	1.14	1.44
5400	3 C.Y. cap. = 300 C.Y./hr.	B-12D	2400	.007			.25	.90	1.15	1.39
5500	3.5 C.Y. cap. = 350 C.Y./hr.	"	2800	.006			.21	.77	.98	1.19
5610	Wheel mounted, 1/2 C.Y. cap. = 44 C.Y./hr.	B-12E	352	.045			1.69	1.12	2.81	3.94
5660	3/4 C.Y. cap. = 66 C.Y./hr.	B-12F	528	.030			1.12	1.13	2.25	3.05
8000	For hauling excavated material, see Div. 31 23 23.20									
9000	Minimum labor/equipment charge	B-10L	2	6	Job		229	200	429	585

31 23 16.46 Excavating, Bulk, Dozer

	31 23 16.46 Excavating, Bulk, Dozer	Crew	Daily Output	Labor-Hours	Unit	Material	Labor	Equipment	Total	Total Incl O&P
0010	**EXCAVATING, BULK, DOZER**									
0011	Open site									
2000	80 H.P., 50' haul, sand & gravel	B-10L	460	.026	B.C.Y.		.99	.87	1.86	2.54
2010	Sandy clay & loam		440	.027			1.04	.91	1.95	2.66
2020	Common earth		400	.030			1.14	1	2.14	2.92
2040	Clay		250	.048			1.83	1.60	3.43	4.68
2200	150' haul, sand & gravel		230	.052			1.99	1.74	3.73	5.10
2210	Sandy clay & loam		220	.055			2.08	1.81	3.89	5.30
2220	Common earth		200	.060			2.29	2	4.29	5.85
2240	Clay		125	.096			3.66	3.19	6.85	9.35
2400	300' haul, sand & gravel		120	.100			3.81	3.33	7.14	9.75
2410	Sandy clay & loam		115	.104			3.98	3.47	7.45	10.15
2420	Common earth		100	.120			4.57	3.99	8.56	11.70
2440	Clay		65	.185			7.05	6.15	13.20	17.95
3000	105 H.P., 50' haul, sand & gravel	B-10W	700	.017			.65	.86	1.51	1.98
3010	Sandy clay & loam		680	.018			.67	.88	1.55	2.04
3020	Common earth		610	.020			.75	.98	1.73	2.28
3040	Clay		385	.031			1.19	1.56	2.75	3.61
3200	150' haul, sand & gravel		310	.039			1.47	1.94	3.41	4.48
3210	Sandy clay & loam		300	.040			1.52	2	3.52	4.63
3220	Common earth		270	.044			1.69	2.22	3.91	5.15
3240	Clay		170	.071			2.69	3.53	6.22	8.20
3300	300' haul, sand & gravel		140	.086			3.27	4.29	7.56	9.90
3310	Sandy clay & loam		135	.089			3.39	4.45	7.84	10.30
3320	Common earth		120	.100			3.81	5	8.81	11.60

31 23 16.46 Excavating, Bulk, Dozer

31 23 16.46 Excavating, Bulk, Dozer		Crew	Daily Output	Labor-Hours	Unit	Material	2009 Bare Costs Labor	Equipment	Total	Total Incl O&P
3340	Clay	B-10W	100	.120	B.C.Y.		4.57	6	10.57	13.90
4000	200 H.P., 50′ haul, sand & gravel	B-10B	1400	.009			.33	.77	1.10	1.37
4010	Sandy clay & loam		1360	.009			.34	.80	1.14	1.41
4020	Common earth		1230	.010			.37	.88	1.25	1.56
4040	Clay		770	.016			.59	1.40	1.99	2.50
4200	150′ haul, sand & gravel		595	.020			.77	1.82	2.59	3.23
4210	Sandy clay & loam		580	.021			.79	1.87	2.66	3.31
4220	Common earth		516	.023			.89	2.10	2.99	3.72
4240	Clay		325	.037			1.41	3.33	4.74	5.90
4400	300′ haul, sand & gravel		310	.039			1.47	3.49	4.96	6.20
4410	Sandy clay & loam		300	.040			1.52	3.61	5.13	6.40
4420	Common earth		270	.044			1.69	4.01	5.70	7.10
4440	Clay		170	.071			2.69	6.35	9.04	11.30
5000	300 H.P., 50′ haul, sand & gravel	B-10M	1900	.006			.24	.75	.99	1.20
5010	Sandy clay & loam		1850	.006			.25	.77	1.02	1.24
5020	Common earth		1650	.007			.28	.86	1.14	1.39
5040	Clay		1025	.012			.45	1.39	1.84	2.24
5200	150′ haul, sand & gravel		920	.013			.50	1.55	2.05	2.49
5210	Sandy clay & loam		895	.013			.51	1.59	2.10	2.56
5220	Common earth		800	.015			.57	1.78	2.35	2.87
5240	Clay		500	.024			.91	2.85	3.76	4.59
5400	300′ haul, sand & gravel		470	.026			.97	3.03	4	4.88
5410	Sandy clay & loam		455	.026			1	3.13	4.13	5.05
5420	Common earth		410	.029			1.12	3.47	4.59	5.60
5440	Clay		250	.048			1.83	5.70	7.53	9.15

31 23 16.50 Excavation, Bulk, Scrapers

31 23 16.50 Excavation, Bulk, Scrapers		Crew	Daily Output	Labor-Hours	Unit	Material	Labor	Equipment	Total	Total Incl O&P
0010	**EXCAVATION, BULK, SCRAPERS**									
0100	Elev. scraper 11 C.Y., sand & gravel 1500′ haul, 1/4 dozer	B-33F	690	.020	B.C.Y.		.78	2.10	2.88	3.56
0150	3000′ haul		610	.023			.89	2.38	3.27	4.03
0200	5000′ haul		505	.028			1.07	2.87	3.94	4.86
0300	Common earth, 1500′ haul		600	.023			.90	2.42	3.32	4.09
0350	3000′ haul		530	.026			1.02	2.74	3.76	4.63
0400	5000′ haul		440	.032			1.23	3.30	4.53	5.60
0410	Sandy clay & loam, 1500′ haul		648	.022			.83	2.24	3.07	3.79
0420	3000′ haul		572	.024			.94	2.54	3.48	4.29
0430	5000′ haul		475	.029			1.14	3.05	4.19	5.15
0500	Clay, 1500′ haul		375	.037			1.44	3.87	5.31	6.55
0550	3000′ haul		330	.042			1.64	4.40	6.04	7.45
0600	5000′ haul		275	.051			1.96	5.30	7.26	8.90
1000	Self propelled scraper, 14 C.Y. 1/4 push dozer, sand									
1050	Sand and gravel, 1500′ haul	B-33D	920	.015	B.C.Y.		.59	2.03	2.62	3.16
1100	3000′ haul		805	.017			.67	2.32	2.99	3.62
1200	5000′ haul		645	.022			.84	2.89	3.73	4.51
1300	Common earth, 1500′ haul		800	.018			.67	2.33	3	3.63
1350	3000′ haul		700	.020			.77	2.66	3.43	4.16
1400	5000′ haul		560	.025			.96	3.33	4.29	5.20
1420	Sandy clay & loam, 1500′ haul		864	.016			.62	2.16	2.78	3.36
1430	3000′ haul		786	.018			.69	2.37	3.06	3.70
1440	5000′ haul		605	.023			.89	3.08	3.97	4.81
1500	Clay, 1500′ haul		500	.028			1.08	3.73	4.81	5.80
1550	3000′ haul		440	.032			1.23	4.24	5.47	6.60
1600	5000′ haul		350	.040			1.54	5.35	6.89	8.30

31 23 16 – Excavation

31 23 16.50 Excavation, Bulk, Scrapers		Crew	Daily Output	Labor-Hours	Unit	Material	2009 Bare Costs Labor	Equipment	Total	Total Incl O&P
2000	21 C.Y., 1/4 push dozer, sand & gravel, 1500′ haul	B-33E	1180	.012	B.C.Y.		.46	1.93	2.39	2.85
2100	3000′ haul		910	.015			.59	2.50	3.09	3.69
2200	5000′ haul		750	.019			.72	3.04	3.76	4.49
2300	Common earth, 1500′ haul		1030	.014			.52	2.21	2.73	3.26
2350	3000′ haul		790	.018			.68	2.89	3.57	4.26
2400	5000′ haul		650	.022			.83	3.51	4.34	5.20
2420	Sandy clay & loam, 1500′ haul		1112	.013			.49	2.05	2.54	3.03
2430	3000′ haul		854	.016			.63	2.67	3.30	3.95
2440	5000′ haul		702	.020			.77	3.25	4.02	4.79
2500	Clay, 1500′ haul		645	.022			.84	3.54	4.38	5.20
2550	3000′ haul		495	.028			1.09	4.61	5.70	6.80
2600	5000′ haul	↓	405	.035			1.33	5.65	6.98	8.30
2700	Towed, 10 C.Y., 1/4 push dozer, sand & gravel, 1500′ haul	B-33B	560	.025			.96	3.45	4.41	5.35
2720	3000′ haul		450	.031			1.20	4.29	5.49	6.65
2730	5000′ haul		365	.038			1.48	5.30	6.78	8.15
2750	Common earth, 1500′ haul		420	.033			1.29	4.60	5.89	7.10
2770	3000′ haul		400	.035			1.35	4.83	6.18	7.45
2780	5000′ haul		310	.045			1.74	6.25	7.99	9.60
2785	Sandy clay & Loam, 1500′ haul		454	.031			1.19	4.26	5.45	6.55
2790	3000′ haul		432	.032			1.25	4.47	5.72	6.90
2795	5000′ haul		340	.041			1.59	5.70	7.29	8.80
2800	Clay, 1500′ haul		315	.044			1.71	6.15	7.86	9.50
2820	3000′ haul		300	.047			1.80	6.45	8.25	9.95
2840	5000′ haul	↓	225	.062			2.40	8.60	11	13.25
2900	15 C.Y., 1/4 push dozer, sand & gravel, 1500′ haul	B-33C	800	.018			.67	2.43	3.10	3.75
2920	3000′ haul		640	.022			.84	3.04	3.88	4.69
2940	5000′ haul		520	.027			1.04	3.74	4.78	5.75
2960	Common earth, 1500′ haul		600	.023			.90	3.25	4.15	5
2980	3000′ haul		560	.025			.96	3.48	4.44	5.35
3000	5000′ haul		440	.032			1.23	4.43	5.66	6.80
3005	Sandy clay & Loam, 1500′ haul		648	.022			.83	3	3.83	4.64
3010	3000′ haul		605	.023			.89	3.22	4.11	4.96
3015	5000′ haul		475	.029			1.14	4.10	5.24	6.30
3020	Clay, 1500′ haul		450	.031			1.20	4.33	5.53	6.65
3040	3000′ haul		420	.033			1.29	4.64	5.93	7.15
3060	5000′ haul	↓	320	.044	↓		1.69	6.10	7.79	9.40

31 23 19 – Dewatering

31 23 19.20 Dewatering Systems		Crew	Daily Output	Labor-Hours	Unit	Material	Labor	Equipment	Total	Total Incl O&P
0010	**DEWATERING SYSTEMS**									
0020	Excavate drainage trench, 2′ wide, 2′ deep	B-11C	90	.178	C.Y.		6.50	3.26	9.76	14
0100	2′ wide, 3′ deep, with backhoe loader	″	135	.119			4.32	2.18	6.50	9.35
0200	Excavate sump pits by hand, light soil	1 Clab	7.10	1.127			35.50		35.50	59
0300	Heavy soil	″	3.50	2.286	↓		72		72	119
0500	Pumping 8 hr., attended 2 hrs. per day, including 20 L.F.									
0550	of suction hose & 100 L.F. discharge hose									
0600	2″ diaphragm pump used for 8 hours	B-10H	4	3	Day		114	16.85	130.85	201
0650	4″ diaphragm pump used for 8 hours	B-10I	4	3			114	23.50	137.50	208
0800	8 hrs. attended, 2″ diaphragm pump	B-10H	1	12			455	67.50	522.50	805
0900	3″ centrifugal pump	B-10J	1	12			455	76	531	815
1000	4″ diaphragm pump	B-10I	1	12			455	94.50	549.50	835
1100	6″ centrifugal pump	B-10K	1	12	↓		455	335	790	1,100
1300	CMP, incl. excavation 3′ deep, 12″ diameter	B-6	115	.209	L.F.	13.65	7.10	2.55	23.30	29.50

31 23 19 – Dewatering

31 23 19.20 Dewatering Systems

	31 23 19.20 Dewatering Systems	Crew	Daily Output	Labor-Hours	Unit	2009 Bare Costs Material	Labor	Equipment	Total	Total Incl O&P
1400	18″ diameter	B-6	100	.240	L.F.	17	8.20	2.94	28.14	35
1600	Sump hole construction, incl. excavation and gravel, pit		1250	.019	C.F.	.96	.65	.24	1.85	2.37
1700	With 12″ gravel collar, 12″ pipe, corrugated, 16 ga.		70	.343	L.F.	22.50	11.70	4.20	38.40	48.50
1800	15″ pipe, corrugated, 16 ga.		55	.436		29	14.85	5.35	49.20	61.50
1900	18″ pipe, corrugated, 16 ga.		50	.480		33.50	16.35	5.90	55.75	70
2000	24″ pipe, corrugated, 14 ga.		40	.600		40.50	20.50	7.35	68.35	85.50
2200	Wood lining, up to 4′ x 4′, add		300	.080	SFCA	16.65	2.73	.98	20.36	24
9950	See Div. 31 23 19.40 for wellpoints									
9960	See Div. 31 23 19.30 for deep well systems									

31 23 19.30 Wells

	31 23 19.30 Wells	Crew	Daily Output	Labor-Hours	Unit	Material	Labor	Equipment	Total	Total Incl O&P
0010	**WELLS**									
0011	For dewatering 10′ to 20′ deep, 2′ diameter									
0020	with steel casing, minimum	B-6	165	.145	V.L.F.	47	4.96	1.78	53.74	61.50
0050	Average		98	.245		47	8.35	3	58.35	68.50
0100	Maximum		49	.490		47	16.70	6	69.70	85

31 23 19.40 Wellpoints

	31 23 19.40 Wellpoints	Crew	Daily Output	Labor-Hours	Unit	Material	Labor	Equipment	Total	Total Incl O&P
0010	**WELLPOINTS**									
0011	For equipment rental, see 01 54 33 in Reference Section									
0100	Installation and removal of single stage system									
0110	Labor only, .75 labor-hours per L.F., minimum	1 Clab	10.70	.748	LF Hdr		23.50		23.50	39
0200	2.0 labor-hours per L.F., maximum	″	4	2	″		63		63	104
0400	Pump operation, 4 @ 6 hr. shifts									
0410	Per 24 hour day	4 Eqlt	1.27	25.197	Day		985		985	1,550
0500	Per 168 hour week, 160 hr. straight, 8 hr. double time		.18	177	Week		6,950		6,950	10,900
0550	Per 4.3 week month		.04	800	Month		31,200		31,200	49,200
0600	Complete installation, operation, equipment rental, fuel &									
0610	removal of system with 2″ wellpoints 5′ O.C.									
0700	100′ long header, 6″ diameter, first month	4 Eqlt	3.23	9.907	LF Hdr	157	385		542	785
0800	Thereafter, per month		4.13	7.748		125	305		430	615
1000	200′ long header, 8″ diameter, first month		6	5.333		142	208		350	485
1100	Thereafter, per month		8.39	3.814		70.50	149		219.50	310
1300	500′ long header, 8″ diameter, first month		10.63	3.010		55	118		173	246
1400	Thereafter, per month		20.91	1.530		39	60		99	137
1600	1,000′ long header, 10″ diameter, first month		11.62	2.754		47	108		155	221
1700	Thereafter, per month		41.81	.765		23.50	30		53.50	73
1900	Note: above figures include pumping 168 hrs. per week									
1910	and include the pump operator and one stand-by pump.									

31 23 23 – Fill

31 23 23.13 Backfill

	31 23 23.13 Backfill	Crew	Daily Output	Labor-Hours	Unit	Material	Labor	Equipment	Total	Total Incl O&P
0010	**BACKFILL**									
0015	By hand, no compaction, light soil	1 Clab	14	.571	L.C.Y.		18.05		18.05	30
0100	Heavy soil		11	.727	″		23		23	38
0300	Compaction in 6″ layers, hand tamp, add to above		20.60	.388	E.C.Y.		12.25		12.25	20.50
0400	Roller compaction operator walking, add	B-10A	100	.120			4.57	1.45	6.02	8.90
0500	Air tamp, add	B-9D	190	.211			6.75	1.17	7.92	12.40
0600	Vibrating plate, add	A-1D	60	.133			4.21	.53	4.74	7.55
0800	Compaction in 12″ layers, hand tamp, add to above	1 Clab	34	.235			7.45		7.45	12.25
0900	Roller compaction operator walking, add	B-10A	150	.080			3.05	.97	4.02	5.90
1000	Air tamp, add	B-9	285	.140			4.49	.67	5.16	8.15
1100	Vibrating plate, add	A-1E	90	.089			2.81	.44	3.25	5.10

31 23 23.14 Backfill, Structural		Crew	Daily Output	Labor-Hours	Unit	Material	2009 Bare Costs Labor	Equipment	Total	Total Incl O&P
0010	**BACKFILL, STRUCTURAL**									
0011	Dozer or F.E. loader									
0020	From existing stockpile, no compaction									
2000	80 H.P., 50' haul, sand & gravel	B-10L	1100	.011	L.C.Y.		.42	.36	.78	1.06
2010	Sandy clay & loam		1070	.011			.43	.37	.80	1.09
2020	Common earth		975	.012			.47	.41	.88	1.20
2040	Clay		850	.014			.54	.47	1.01	1.38
2200	150' haul, sand & gravel		550	.022			.83	.73	1.56	2.13
2210	Sandy clay & loam		535	.022			.85	.75	1.60	2.18
2220	Common earth		490	.024			.93	.81	1.74	2.39
2240	Clay		425	.028			1.08	.94	2.02	2.75
2400	300' haul, sand & gravel		370	.032			1.24	1.08	2.32	3.16
2410	Sandy clay & loam		360	.033			1.27	1.11	2.38	3.24
2420	Common earth		330	.036			1.39	1.21	2.60	3.54
2440	Clay		290	.041			1.58	1.38	2.96	4.02
3000	105 H.P., 50' haul, sand & gravel	B-10W	1350	.009			.34	.44	.78	1.03
3010	Sandy clay & loam		1325	.009			.35	.45	.80	1.05
3020	Common earth		1225	.010			.37	.49	.86	1.14
3040	Clay		1100	.011			.42	.55	.97	1.26
3200	150' haul, sand & gravel		670	.018			.68	.90	1.58	2.08
3210	Sandy clay & loam		655	.018			.70	.92	1.62	2.12
3220	Common earth		610	.020			.75	.98	1.73	2.28
3240	Clay		550	.022			.83	1.09	1.92	2.53
3300	300' haul, sand & gravel		465	.026			.98	1.29	2.27	2.99
3310	Sandy clay & loam		455	.026			1	1.32	2.32	3.05
3320	Common earth		415	.029			1.10	1.45	2.55	3.35
3340	Clay		370	.032			1.24	1.62	2.86	3.76
4000	200 H.P., 50' haul, sand & gravel	B-10B	2500	.005			.18	.43	.61	.77
4010	Sandy clay & loam		2435	.005			.19	.44	.63	.79
4020	Common earth		2200	.005			.21	.49	.70	.87
4040	Clay		1950	.006			.23	.55	.78	.98
4200	150' haul, sand & gravel		1225	.010			.37	.88	1.25	1.57
4210	Sandy clay & loam		1200	.010			.38	.90	1.28	1.60
4220	Common earth		1100	.011			.42	.98	1.40	1.74
4240	Clay		975	.012			.47	1.11	1.58	1.97
4400	300' haul, sand & gravel		805	.015			.57	1.34	1.91	2.39
4410	Sandy clay & loam		790	.015			.58	1.37	1.95	2.43
4420	Common earth		735	.016			.62	1.47	2.09	2.61
4440	Clay		660	.018			.69	1.64	2.33	2.90
5000	300 H.P., 50' haul, sand & gravel	B-10M	3170	.004			.14	.45	.59	.72
5010	Sandy clay & loam		3110	.004			.15	.46	.61	.73
5020	Common earth		2900	.004			.16	.49	.65	.79
5040	Clay		2700	.004			.17	.53	.70	.85
5200	150' haul, sand & gravel		2200	.005			.21	.65	.86	1.04
5210	Sandy clay & loam		2150	.006			.21	.66	.87	1.07
5220	Common earth		1950	.006			.23	.73	.96	1.17
5240	Clay		1700	.007			.27	.84	1.11	1.35
5400	300' haul, sand & gravel		1500	.008			.30	.95	1.25	1.53
5410	Sandy clay & loam		1470	.008			.31	.97	1.28	1.56
5420	Common earth		1350	.009			.34	1.05	1.39	1.70
5440	Clay		1225	.010			.37	1.16	1.53	1.88

31 23 23 – Fill

31 23 23.15 Borrow, Loading And/Or Spreading		Crew	Daily Output	Labor-Hours	Unit	Material	2009 Bare Costs Labor	Equipment	Total	Total Incl O&P
0010	**BORROW, LOADING AND/OR SPREADING**									
0020	Material only, bank run gravel				Ton	19.55			19.55	21.50
0100	Crushed stone, 1-1/2" to 3/4" size					19.45			19.45	21.50
0150	3/8" size					25.50			25.50	28
0200	Dead or bank run sand					15.25			15.25	16.75
0500	Haul 2 mi. spread, 200HP dozer, bank run gravel	B-15	1100	.025			.88	1.95	2.83	3.57
0540	Crushed stone, 1-1/2" to 3/4" size		1150	.024			.84	1.87	2.71	3.41
0560	3/8" size		1150	.024			.84	1.87	2.71	3.41
0600	Bank run or dead sand		1000	.028			.97	2.15	3.12	3.93
1000	Hand spread, bank run gravel	A-5	33	.545			17.20	1.47	18.67	30
1040	Crushed stone, 1-1/2" to 3/4" size		38	.474			14.95	1.28	16.23	26
1060	3/8" size		40	.450			14.20	1.22	15.42	25
1100	Bank run or dead sand		33	.545			17.20	1.47	18.67	30
1800	Delivery charge, minimum 20 tons, 1 hr round trip, add	B-34B	130	.062			1.97	4.10	6.07	7.75
1820	1-1/2 hr. round trip, add		93	.086			2.75	5.75	8.50	10.80
1840	2 hr. round trip, add		65	.123			3.93	8.20	12.13	15.45
7000	Topsoil or loam from stockpile, shovel, 1 C.Y. bucket	B-12N	840	.019	B.C.Y.	22.50	.71	1.33	24.54	27.50
7010	1-1/2 C.Y. bucket	B-12O	1135	.014		22.50	.52	1	24.02	27
7020	3 C.Y. bucket	B-12T	1800	.009		22.50	.33	.89	23.72	26.50
7030	Front end loader, wheel mounted									
7050	3/4 C.Y. bucket	B-10R	550	.022	B.C.Y.	22.50	.83	.43	23.76	27
7060	1-1/2 C.Y. bucket	B-10S	970	.012		22.50	.47	.34	23.31	26
7070	3 C.Y. bucket	B-10T	1575	.008		22.50	.29	.25	23.04	25.50
7080	5 C.Y. bucket	B-10U	2600	.005		22.50	.18	.33	23.01	25.50
31 23 23.16 Fill By Borrow and Utility Bedding										
0010	**FILL BY BORROW AND UTILITY BEDDING**									
0015	Fill by borrow, load, 1 mile haul, spread with dozer									
0020	for embankments	B-15	1200	.023	L.C.Y.	7.35	.81	1.79	9.95	11.40
0035	Select fill for shoulders & embankments	"	1200	.023	"	12.60	.81	1.79	15.20	17.15
0040	Fill,for hauling over 1 mile,add to above per C.Y., see Div. 31 23 23.20				Mile				1.60	1.85
0049	Utility bedding, for pipe & conduit, not incl. compaction									
0050	Crushed or screened bank run gravel	B-6	150	.160	L.C.Y.	31	5.45	1.96	38.41	45
0100	Crushed stone 3/4" to 1/2"		150	.160		43.50	5.45	1.96	50.91	59
0200	Sand, dead or bank		150	.160		10.50	5.45	1.96	17.91	22.50
0500	Compacting bedding in trench	A-1D	90	.089	E.C.Y.		2.81	.35	3.16	5.05
0600	If material source exceeds 2 miles, add for extra mileage.									
0610	See Div. 31 23 23 .20 for hauling mileage add.									
31 23 23.17 General Fill										
0010	**GENERAL FILL**									
0011	Spread dumped material, no compaction									
0020	By dozer, no compaction	B-10B	1000	.012	L.C.Y.		.46	1.08	1.54	1.92
0100	By hand	1 Clab	12	.667	"		21		21	35
0150	Spread fill, from stockpile with 2-1/2 C.Y. F.E. loader									
0170	130 H.P., 300' haul	B-10P	600	.020	L.C.Y.		.76	1.65	2.41	3.04
0190	With dozer 300 H.P., 300' haul	B-10M	600	.020	"		.76	2.37	3.13	3.83
0500	Gravel fill, compacted, under floor slabs, 4" deep	B-37	10000	.005	S.F.	.25	.16	.01	.42	.56
0600	6" deep		8600	.006		.38	.19	.02	.59	.74
0700	9" deep		7200	.007		.63	.22	.02	.87	1.07
0800	12" deep		6000	.008		.88	.27	.02	1.17	1.43
1000	Alternate pricing method, 4" deep		120	.400	E.C.Y.	18.90	13.25	1.15	33.30	44
1100	6" deep		160	.300		18.90	9.95	.86	29.71	38.50
1200	9" deep		200	.240		18.90	7.95	.69	27.54	35

31 23 23.17 General Fill	Crew	Daily Output	Labor-Hours	Unit	Material	2009 Bare Costs Labor	Equipment	Total	Total Incl O&P
1300 12" deep	B-37	220	.218	E.C.Y.	18.90	7.25	.63	26.78	33.50
1500 For fill under exterior paving, see Div. 32 11 23.23									
1600 For flowable fill, see Div. 03 31 05.35									
9000 Minimum labor/equipment charge	1 Clab	4	2	Job		63		63	104

31 23 23.20 Hauling	Crew	Daily Output	Labor-Hours	Unit	Material	Labor	Equipment	Total	Total Incl O&P
0010 HAULING									
0011 Excavated or borrow, loose cubic yards									
0012 no loading equipment, including hauling,waiting, loading/dumping									
0013 time per cycle (wait, load, travel, unload or dump & return)									
0014 8 CY truck, 15 MPH ave, cycle 0.5 miles, 10 min. wait/Ld./Uld.	B-34A	320	.025	L.C.Y.		.80	1.02	1.82	2.43
0016 cycle 1 mile		272	.029			.94	1.20	2.14	2.86
0018 cycle 2 miles		208	.038			1.23	1.57	2.80	3.74
0020 cycle 4 miles		144	.056			1.78	2.27	4.05	5.40
0022 cycle 6 miles		112	.071			2.28	2.91	5.19	6.95
0024 cycle 8 miles		88	.091			2.90	3.71	6.61	8.85
0026 20 MPH ave,cycle 0.5 mile		336	.024			.76	.97	1.73	2.32
0028 cycle 1 mile		296	.027			.86	1.10	1.96	2.63
0030 cycle 2 miles		240	.033			1.06	1.36	2.42	3.24
0032 cycle 4 miles		176	.045			1.45	1.85	3.30	4.42
0034 cycle 6 miles		136	.059			1.88	2.40	4.28	5.70
0036 cycle 8 miles		112	.071			2.28	2.91	5.19	6.95
0044 25 MPH ave, cycle 4 miles		192	.042			1.33	1.70	3.03	4.05
0046 cycle 6 miles		160	.050			1.60	2.04	3.64	4.86
0048 cycle 8 miles		128	.063			2	2.55	4.55	6.10
0050 30 MPH ave, cycle 4 miles		216	.037			1.18	1.51	2.69	3.60
0052 cycle 6 miles		176	.045			1.45	1.85	3.30	4.42
0054 cycle 8 miles		144	.056			1.78	2.27	4.05	5.40
0114 15 MPH ave, cycle 0.5 mile, 15 min. wait/Ld./Uld.		224	.036			1.14	1.46	2.60	3.47
0116 cycle 1 mile		200	.040			1.28	1.63	2.91	3.89
0118 cycle 2 miles		168	.048			1.52	1.94	3.46	4.63
0120 cycle 4 miles		120	.067			2.13	2.72	4.85	6.50
0122 cycle 6 miles		96	.083			2.66	3.40	6.06	8.10
0124 cycle 8 miles		80	.100			3.20	4.08	7.28	9.75
0126 20 MPH ave, cycle 0.5 mile		232	.034			1.10	1.41	2.51	3.36
0128 cycle 1 mile		208	.038			1.23	1.57	2.80	3.74
0130 cycle 2 miles		184	.043			1.39	1.77	3.16	4.23
0132 cycle 4 miles		144	.056			1.78	2.27	4.05	5.40
0134 cycle 6 miles		112	.071			2.28	2.91	5.19	6.95
0136 cycle 8 miles		96	.083			2.66	3.40	6.06	8.10
0144 25 MPH ave, cycle 4 miles		152	.053			1.68	2.15	3.83	5.10
0146 cycle 6 miles		128	.063			2	2.55	4.55	6.10
0148 cycle 8 miles		112	.071			2.28	2.91	5.19	6.95
0150 30 MPH ave, cycle 4 miles		168	.048			1.52	1.94	3.46	4.63
0152 cycle 6 miles		144	.056			1.78	2.27	4.05	5.40
0154 cycle 8 miles		120	.067			2.13	2.72	4.85	6.50
0214 15 MPH ave, cycle 0.5 mile, 20 min wait/Ld./Uld.		176	.045			1.45	1.85	3.30	4.42
0216 cycle 1 mile		160	.050			1.60	2.04	3.64	4.86
0218 cycle 2 miles		136	.059			1.88	2.40	4.28	5.70
0220 cycle 4 miles		104	.077			2.46	3.14	5.60	7.50
0222 cycle 6 miles		88	.091			2.90	3.71	6.61	8.85
0224 cycle 8 miles		72	.111			3.55	4.53	8.08	10.80
0226 20 MPH ave, cycle 0.5 mile	↓	176	.045	↓		1.45	1.85	3.30	4.42

31 23 23.20 Hauling		Crew	Daily Output	Labor-Hours	Unit	Material	2009 Bare Costs Labor	Equipment	Total	Total Incl O&P
0228	cycle 1 mile	B-34A	168	.048	L.C.Y.		1.52	1.94	3.46	4.63
0230	cycle 2 miles		144	.056			1.78	2.27	4.05	5.40
0232	cycle 4 miles		120	.067			2.13	2.72	4.85	6.50
0234	cycle 6 miles		96	.083			2.66	3.40	6.06	8.10
0236	cycle 8 miles		88	.091			2.90	3.71	6.61	8.85
0244	25 MPH ave, cycle 4 miles		128	.063			2	2.55	4.55	6.10
0246	cycle 6 miles		112	.071			2.28	2.91	5.19	6.95
0248	cycle 8 miles		96	.083			2.66	3.40	6.06	8.10
0250	30 MPH ave, cycle 4 miles		136	.059			1.88	2.40	4.28	5.70
0252	cycle 6 miles		120	.067			2.13	2.72	4.85	6.50
0254	cycle 8 miles		104	.077			2.46	3.14	5.60	7.50
0314	15 MPH ave, cycle 0.5 mile, 25 min wait/Ld./Uld.		144	.056			1.78	2.27	4.05	5.40
0316	cycle 1 mile		128	.063			2	2.55	4.55	6.10
0318	cycle 2 miles		112	.071			2.28	2.91	5.19	6.95
0320	cycle 4 miles		96	.083			2.66	3.40	6.06	8.10
0322	cycle 6 miles		80	.100			3.20	4.08	7.28	9.75
0324	cycle 8 miles		64	.125			3.99	5.10	9.09	12.15
0326	20 MPH ave, cycle 0.5 mile		144	.056			1.78	2.27	4.05	5.40
0328	cycle 1 mile		136	.059			1.88	2.40	4.28	5.70
0330	cycle 2 miles		120	.067			2.13	2.72	4.85	6.50
0332	cycle 4 miles		104	.077			2.46	3.14	5.60	7.50
0334	cycle 6 miles		88	.091			2.90	3.71	6.61	8.85
0336	cycle 8 miles		80	.100			3.20	4.08	7.28	9.75
0344	25 MPH ave, cycle 4 miles		112	.071			2.28	2.91	5.19	6.95
0346	cycle 6 miles		96	.083			2.66	3.40	6.06	8.10
0348	cycle 8 miles		88	.091			2.90	3.71	6.61	8.85
0350	30 MPH ave, cycle 4 miles		112	.071			2.28	2.91	5.19	6.95
0352	cycle 6 miles		104	.077			2.46	3.14	5.60	7.50
0354	cycle 8 miles		96	.083			2.66	3.40	6.06	8.10
0414	15 MPH ave, cycle 0.5 mile, 30 min wait/Ld./Uld.		120	.067			2.13	2.72	4.85	6.50
0416	cycle 1 mile		112	.071			2.28	2.91	5.19	6.95
0418	cycle 2 miles		96	.083			2.66	3.40	6.06	8.10
0420	cycle 4 miles		80	.100			3.20	4.08	7.28	9.75
0422	cycle 6 miles		72	.111			3.55	4.53	8.08	10.80
0424	cycle 8 miles		64	.125			3.99	5.10	9.09	12.15
0426	20 MPH ave, cycle 0.5 mile		120	.067			2.13	2.72	4.85	6.50
0428	cycle 1 mile		112	.071			2.28	2.91	5.19	6.95
0430	cycle 2 miles		104	.077			2.46	3.14	5.60	7.50
0432	cycle 4 miles		88	.091			2.90	3.71	6.61	8.85
0434	cycle 6 miles		80	.100			3.20	4.08	7.28	9.75
0436	cycle 8 miles		72	.111			3.55	4.53	8.08	10.80
0444	25 MPH ave, cycle 4 miles		96	.083			2.66	3.40	6.06	8.10
0446	cycle 6 miles		88	.091			2.90	3.71	6.61	8.85
0448	cycle 8 miles		80	.100			3.20	4.08	7.28	9.75
0450	30 MPH ave, cycle 4 miles		96	.083			2.66	3.40	6.06	8.10
0452	cycle 6 miles		88	.091			2.90	3.71	6.61	8.85
0454	cycle 8 miles		80	.100			3.20	4.08	7.28	9.75
0514	15 MPH ave, cycle 0.5 mile, 35 min wait/Ld./Uld.		104	.077			2.46	3.14	5.60	7.50
0516	cycle 1 mile		96	.083			2.66	3.40	6.06	8.10
0518	cycle 2 miles		88	.091			2.90	3.71	6.61	8.85
0520	cycle 4 miles		72	.111			3.55	4.53	8.08	10.80
0522	cycle 6 miles		64	.125			3.99	5.10	9.09	12.15
0524	cycle 8 miles		56	.143			4.56	5.85	10.41	13.90

31 23 23 – Fill

31 23 23.20 Hauling		Crew	Daily Output	Labor-Hours	Unit	Material	2009 Bare Costs Labor	Equipment	Total	Total Incl O&P
0526	20 MPH ave, cycle 0.5 mile	B-34A	104	.077	L.C.Y.		2.46	3.14	5.60	7.50
0528	cycle 1 mile		96	.083			2.66	3.40	6.06	8.10
0530	cycle 2 miles		96	.083			2.66	3.40	6.06	8.10
0532	cycle 4 miles		80	.100			3.20	4.08	7.28	9.75
0534	cycle 6 miles		72	.111			3.55	4.53	8.08	10.80
0536	cycle 8 miles		64	.125			3.99	5.10	9.09	12.15
0544	25 MPH ave, cycle 4 miles		88	.091			2.90	3.71	6.61	8.85
0546	cycle 6 miles		80	.100			3.20	4.08	7.28	9.75
0548	cycle 8 miles		72	.111			3.55	4.53	8.08	10.80
0550	30 MPH ave, cycle 4 miles		88	.091			2.90	3.71	6.61	8.85
0552	cycle 6 miles		80	.100			3.20	4.08	7.28	9.75
0554	cycle 8 miles		72	.111			3.55	4.53	8.08	10.80
1014	12 CY truck, cycle 0.5 mile, 15 MPH ave, 15 min. wait/Ld./Uld.	B-34B	336	.024			.76	1.59	2.35	3
1016	cycle 1 mile		300	.027			.85	1.78	2.63	3.35
1018	cycle 2 miles		252	.032			1.01	2.12	3.13	3.99
1020	cycle 4 miles		180	.044			1.42	2.96	4.38	5.60
1022	cycle 6 miles		144	.056			1.78	3.70	5.48	7
1024	cycle 8 miles		120	.067			2.13	4.44	6.57	8.40
1025	cycle 10 miles		96	.083			2.66	5.55	8.21	10.45
1026	20 MPH ave, cycle 0.5 mile		348	.023			.73	1.53	2.26	2.88
1028	cycle 1 mile		312	.026			.82	1.71	2.53	3.22
1030	cycle 2 miles		276	.029			.93	1.93	2.86	3.64
1032	cycle 4 miles		216	.037			1.18	2.47	3.65	4.65
1034	cycle 6 miles		168	.048			1.52	3.17	4.69	6
1036	cycle 8 miles		144	.056			1.78	3.70	5.48	7
1038	cycle 10 miles		120	.067			2.13	4.44	6.57	8.40
1040	25 MPH ave, cycle 4 miles		228	.035			1.12	2.34	3.46	4.41
1042	cycle 6 miles		192	.042			1.33	2.78	4.11	5.25
1044	cycle 8 miles		168	.048			1.52	3.17	4.69	6
1046	cycle 10 miles		144	.056			1.78	3.70	5.48	7
1050	30 MPH ave, cycle 4 miles		252	.032			1.01	2.12	3.13	3.99
1052	cycle 6 miles		216	.037			1.18	2.47	3.65	4.65
1054	cycle 8 miles		180	.044			1.42	2.96	4.38	5.60
1056	cycle 10 miles		156	.051			1.64	3.42	5.06	6.45
1060	35 MPH ave, cycle 4 miles		264	.030			.97	2.02	2.99	3.81
1062	cycle 6 miles		228	.035			1.12	2.34	3.46	4.41
1064	cycle 8 miles		204	.039			1.25	2.61	3.86	4.92
1066	cycle 10 miles		180	.044			1.42	2.96	4.38	5.60
1068	cycle 20 miles		120	.067			2.13	4.44	6.57	8.40
1069	cycle 30 miles		84	.095			3.04	6.35	9.39	12
1070	cycle 40 miles		72	.111			3.55	7.40	10.95	13.95
1072	40 MPH ave, cycle 6 miles		240	.033			1.06	2.22	3.28	4.18
1074	cycle 8 miles		216	.037			1.18	2.47	3.65	4.65
1076	cycle 10 miles		192	.042			1.33	2.78	4.11	5.25
1078	cycle 20 miles		120	.067			2.13	4.44	6.57	8.40
1080	cycle 30 miles		96	.083			2.66	5.55	8.21	10.45
1082	cycle 40 miles		72	.111			3.55	7.40	10.95	13.95
1084	cycle 50 miles		60	.133			4.26	8.90	13.16	16.75
1094	45 MPH ave, cycle 8 miles		216	.037			1.18	2.47	3.65	4.65
1096	cycle 10 miles		204	.039			1.25	2.61	3.86	4.92
1098	cycle 20 miles		132	.061			1.94	4.04	5.98	7.60
1100	cycle 30 miles		108	.074			2.37	4.94	7.31	9.35
1102	cycle 40 miles		84	.095			3.04	6.35	9.39	12

31 23 23.20 Hauling		Crew	Daily Output	Labor-Hours	Unit	Material	2009 Bare Costs Labor	Equipment	Total	Total Incl O&P
1104	cycle 50 miles	B-34B	72	.111	L.C.Y.		3.55	7.40	10.95	13.95
1106	50 MPH ave, cycle 10 miles		216	.037			1.18	2.47	3.65	4.65
1108	cycle 20 miles		144	.056			1.78	3.70	5.48	7
1110	cycle 30 miles		108	.074			2.37	4.94	7.31	9.35
1112	cycle 40 miles		84	.095			3.04	6.35	9.39	12
1114	cycle 50 miles		72	.111			3.55	7.40	10.95	13.95
1214	15 MPH ave, cycle 0.5 mile, 20 min. wait/Ld./Uld.		264	.030			.97	2.02	2.99	3.81
1216	cycle 1 mile		240	.033			1.06	2.22	3.28	4.18
1218	cycle 2 miles		204	.039			1.25	2.61	3.86	4.92
1220	cycle 4 miles		156	.051			1.64	3.42	5.06	6.45
1222	cycle 6 miles		132	.061			1.94	4.04	5.98	7.60
1224	cycle 8 miles		108	.074			2.37	4.94	7.31	9.35
1225	cycle 10 miles		96	.083			2.66	5.55	8.21	10.45
1226	20 MPH ave, cycle 0.5 mile		264	.030			.97	2.02	2.99	3.81
1228	cycle 1 mile		252	.032			1.01	2.12	3.13	3.99
1230	cycle 2 miles		216	.037			1.18	2.47	3.65	4.65
1232	cycle 4 miles		180	.044			1.42	2.96	4.38	5.60
1234	cycle 6 miles		144	.056			1.78	3.70	5.48	7
1236	cycle 8 miles		132	.061			1.94	4.04	5.98	7.60
1238	cycle 10 miles		108	.074			2.37	4.94	7.31	9.35
1240	25 MPH ave, cycle 4 miles		192	.042			1.33	2.78	4.11	5.25
1242	cycle 6 miles		168	.048			1.52	3.17	4.69	6
1244	cycle 8 miles		144	.056			1.78	3.70	5.48	7
1246	cycle 10 miles		132	.061			1.94	4.04	5.98	7.60
1250	30 MPH ave, cycle 4 miles		204	.039			1.25	2.61	3.86	4.92
1252	cycle 6 miles		180	.044			1.42	2.96	4.38	5.60
1254	cycle 8 miles		156	.051			1.64	3.42	5.06	6.45
1256	cycle 10 miles		144	.056			1.78	3.70	5.48	7
1260	35 MPH ave, cycle 4 miles		216	.037			1.18	2.47	3.65	4.65
1262	cycle 6 miles		192	.042			1.33	2.78	4.11	5.25
1264	cycle 8 miles		168	.048			1.52	3.17	4.69	6
1266	cycle 10 miles		156	.051			1.64	3.42	5.06	6.45
1268	cycle 20 miles		108	.074			2.37	4.94	7.31	9.35
1269	cycle 30 miles		72	.111			3.55	7.40	10.95	13.95
1270	cycle 40 miles		60	.133			4.26	8.90	13.16	16.75
1272	40 MPH ave, cycle 6 miles		192	.042			1.33	2.78	4.11	5.25
1274	cycle 8 miles		180	.044			1.42	2.96	4.38	5.60
1276	cycle 10 miles		156	.051			1.64	3.42	5.06	6.45
1278	cycle 20 miles		108	.074			2.37	4.94	7.31	9.35
1280	cycle 30 miles		84	.095			3.04	6.35	9.39	12
1282	cycle 40 miles		72	.111			3.55	7.40	10.95	13.95
1284	cycle 50 miles		60	.133			4.26	8.90	13.16	16.75
1294	45 MPH ave, cycle 8 miles		180	.044			1.42	2.96	4.38	5.60
1296	cycle 10 miles		168	.048			1.52	3.17	4.69	6
1298	cycle 20 miles		120	.067			2.13	4.44	6.57	8.40
1300	cycle 30 miles		96	.083			2.66	5.55	8.21	10.45
1302	cycle 40 miles		72	.111			3.55	7.40	10.95	13.95
1304	cycle 50 miles		60	.133			4.26	8.90	13.16	16.75
1306	50 MPH ave, cycle 10 miles		180	.044			1.42	2.96	4.38	5.60
1308	cycle 20 miles		132	.061			1.94	4.04	5.98	7.60
1310	cycle 30 miles		96	.083			2.66	5.55	8.21	10.45
1312	cycle 40 miles	B-34B	84	.095	L.C.Y.		3.04	6.35	9.39	12
1314	cycle 50 miles		72	.111			3.55	7.40	10.95	13.95

31 23 23.20 Hauling		Crew	Daily Output	Labor-Hours	Unit	Material	2009 Bare Costs Labor	Equipment	Total	Total Incl O&P
1414	15 MPH ave, cycle 0.5 mile, 25 min. wait/Ld./Uld.	B-34B	204	.039	L.C.Y.		1.25	2.61	3.86	4.92
1416	cycle 1 mile		192	.042			1.33	2.78	4.11	5.25
1418	cycle 2 miles		168	.048			1.52	3.17	4.69	6
1420	cycle 4 miles		132	.061			1.94	4.04	5.98	7.60
1422	cycle 6 miles		120	.067			2.13	4.44	6.57	8.40
1424	cycle 8 miles		96	.083			2.66	5.55	8.21	10.45
1425	cycle 10 miles		84	.095			3.04	6.35	9.39	12
1426	20 MPH ave, cycle 0.5 mile		216	.037			1.18	2.47	3.65	4.65
1428	cycle 1 mile		204	.039			1.25	2.61	3.86	4.92
1430	cycle 2 miles		180	.044			1.42	2.96	4.38	5.60
1432	cycle 4 miles		156	.051			1.64	3.42	5.06	6.45
1434	cycle 6 miles		132	.061			1.94	4.04	5.98	7.60
1436	cycle 8 miles		120	.067			2.13	4.44	6.57	8.40
1438	cycle 10 miles		96	.083			2.66	5.55	8.21	10.45
1440	25 MPH ave, cycle 4 miles		168	.048			1.52	3.17	4.69	6
1442	cycle 6 miles		144	.056			1.78	3.70	5.48	7
1444	cycle 8 miles		132	.061			1.94	4.04	5.98	7.60
1446	cycle 10 miles		108	.074			2.37	4.94	7.31	9.35
1450	30 MPH ave, cycle 4 miles		168	.048			1.52	3.17	4.69	6
1452	cycle 6 miles		156	.051			1.64	3.42	5.06	6.45
1454	cycle 8 miles		132	.061			1.94	4.04	5.98	7.60
1456	cycle 10 miles		120	.067			2.13	4.44	6.57	8.40
1460	35 MPH ave, cycle 4 miles		180	.044			1.42	2.96	4.38	5.60
1462	cycle 6 miles		156	.051			1.64	3.42	5.06	6.45
1464	cycle 8 miles		144	.056			1.78	3.70	5.48	7
1466	cycle 10 miles		132	.061			1.94	4.04	5.98	7.60
1468	cycle 20 miles		96	.083			2.66	5.55	8.21	10.45
1469	cycle 30 miles		72	.111			3.55	7.40	10.95	13.95
1470	cycle 40 miles		60	.133			4.26	8.90	13.16	16.75
1472	40 MPH ave, cycle 6 miles		168	.048			1.52	3.17	4.69	6
1474	cycle 8 miles		156	.051			1.64	3.42	5.06	6.45
1476	cycle 10 miles		144	.056			1.78	3.70	5.48	7
1478	cycle 20 miles		96	.083			2.66	5.55	8.21	10.45
1480	cycle 30 miles		84	.095			3.04	6.35	9.39	12
1482	cycle 40 miles		60	.133			4.26	8.90	13.16	16.75
1484	cycle 50 miles		60	.133			4.26	8.90	13.16	16.75
1494	45 MPH ave, cycle 8 miles		156	.051			1.64	3.42	5.06	6.45
1496	cycle 10 miles		144	.056			1.78	3.70	5.48	7
1498	cycle 20 miles		108	.074			2.37	4.94	7.31	9.35
1500	cycle 30 miles		84	.095			3.04	6.35	9.39	12
1502	cycle 40 miles		72	.111			3.55	7.40	10.95	13.95
1504	cycle 50 miles		60	.133			4.26	8.90	13.16	16.75
1506	50 MPH ave, cycle 10 miles		156	.051			1.64	3.42	5.06	6.45
1508	cycle 20 miles		120	.067			2.13	4.44	6.57	8.40
1510	cycle 30 miles		96	.083			2.66	5.55	8.21	10.45
1512	cycle 40 miles		72	.111			3.55	7.40	10.95	13.95
1514	cycle 50 miles		60	.133			4.26	8.90	13.16	16.75
1614	15 MPH, cycle 0.5 mile, 30 min. wait/Ld./Uld.		180	.044			1.42	2.96	4.38	5.60
1616	cycle 1 mile		168	.048			1.52	3.17	4.69	6
1618	cycle 2 miles		144	.056			1.78	3.70	5.48	7
1620	cycle 4 miles		120	.067			2.13	4.44	6.57	8.40
1622	cycle 6 miles		108	.074			2.37	4.94	7.31	9.35
1624	cycle 8 miles		84	.095			3.04	6.35	9.39	12

31 23 23.20 Hauling		Crew	Daily Output	Labor-Hours	Unit	Material	2009 Bare Costs Labor	Equipment	Total	Total Incl O&P
1625	cycle 10 miles	B-34B	84	.095	L.C.Y.		3.04	6.35	9.39	12
1626	20 MPH ave, cycle 0.5 mile		180	.044			1.42	2.96	4.38	5.60
1628	cycle 1 mile		168	.048			1.52	3.17	4.69	6
1630	cycle 2 miles		156	.051			1.64	3.42	5.06	6.45
1632	cycle 4 miles		132	.061			1.94	4.04	5.98	7.60
1634	cycle 6 miles		120	.067			2.13	4.44	6.57	8.40
1636	cycle 8 miles		108	.074			2.37	4.94	7.31	9.35
1638	cycle 10 miles		96	.083			2.66	5.55	8.21	10.45
1640	25 MPH ave, cycle 4 miles		144	.056			1.78	3.70	5.48	7
1642	cycle 6 miles		132	.061			1.94	4.04	5.98	7.60
1644	cycle 8 miles		108	.074			2.37	4.94	7.31	9.35
1646	cycle 10 miles		108	.074			2.37	4.94	7.31	9.35
1650	30 MPH ave, cycle 4 miles		144	.056			1.78	3.70	5.48	7
1652	cycle 6 miles		132	.061			1.94	4.04	5.98	7.60
1654	cycle 8 miles		120	.067			2.13	4.44	6.57	8.40
1656	cycle 10 miles		108	.074			2.37	4.94	7.31	9.35
1660	35 MPH ave, cycle 4 miles		156	.051			1.64	3.42	5.06	6.45
1662	cycle 6 miles		144	.056			1.78	3.70	5.48	7
1664	cycle 8 miles		132	.061			1.94	4.04	5.98	7.60
1666	cycle 10 miles		120	.067			2.13	4.44	6.57	8.40
1668	cycle 20 miles		84	.095			3.04	6.35	9.39	12
1669	cycle 30 miles		72	.111			3.55	7.40	10.95	13.95
1670	cycle 40 miles		60	.133			4.26	8.90	13.16	16.75
1672	40 MPH, cycle 6 miles		144	.056			1.78	3.70	5.48	7
1674	cycle 8 miles		132	.061			1.94	4.04	5.98	7.60
1676	cycle 10 miles		120	.067			2.13	4.44	6.57	8.40
1678	cycle 20 miles		96	.083			2.66	5.55	8.21	10.45
1680	cycle 30 miles		72	.111			3.55	7.40	10.95	13.95
1682	cycle 40 miles		60	.133			4.26	8.90	13.16	16.75
1684	cycle 50 miles		48	.167			5.35	11.10	16.45	21
1694	45 MPH ave, cycle 8 miles		144	.056			1.78	3.70	5.48	7
1696	cycle 10 miles		132	.061			1.94	4.04	5.98	7.60
1698	cycle 20 miles		96	.083			2.66	5.55	8.21	10.45
1700	cycle 30 miles		84	.095			3.04	6.35	9.39	12
1702	cycle 40 miles		60	.133			4.26	8.90	13.16	16.75
1704	cycle 50 miles		60	.133			4.26	8.90	13.16	16.75
1706	50 MPH ave, cycle 10 miles		132	.061			1.94	4.04	5.98	7.60
1708	cycle 20 miles		108	.074			2.37	4.94	7.31	9.35
1710	cycle 30 miles		84	.095			3.04	6.35	9.39	12
1712	cycle 40 miles		72	.111			3.55	7.40	10.95	13.95
1714	cycle 50 miles		60	.133			4.26	8.90	13.16	16.75
2000	Hauling, 8 CY truck, small project cost per hour	B-34A	8	1	Hr.		32	41	73	97.50
2100	12 CY Truck	B-34B	8	1			32	66.50	98.50	126
2150	16.5 CY Truck	B-34C	8	1			32	74.50	106.50	134
2200	20 CY Truck	B-34D	8	1			32	76	108	136
2300	Grading at dump, or embankment if required, by dozer	B-10B	1000	.012	L.C.Y.		.46	1.08	1.54	1.92
2310	Spotter at fill or cut, if required	1 Clab	8	1	Hr.		31.50		31.50	52
3014	16.5 CY truck, 15 min. wait/Ld./Uld., 15 MPH, cycle 0.5 mile	B-34C	462	.017	L.C.Y.		.55	1.29	1.84	2.33
3016	cycle 1 mile		413	.019			.62	1.44	2.06	2.59
3018	cycle 2 miles		347	.023			.74	1.71	2.45	3.09
3020	cycle 4 miles		248	.032			1.03	2.40	3.43	4.33
3022	cycle 6 miles		198	.040			1.29	3	4.29	5.40
3024	cycle 8 miles		165	.048			1.55	3.60	5.15	6.50

31 23 23.20 Hauling		Crew	Daily Output	Labor-Hours	Unit	Material	2009 Bare Costs Labor	Equipment	Total	Total Incl O&P
3025	cycle 10 miles	B-34C	132	.061	L.C.Y.		1.94	4.50	6.44	8.10
3026	20 MPH ave, cycle 0.5 mile		479	.017			.53	1.24	1.77	2.23
3028	cycle 1 mile		429	.019			.60	1.39	1.99	2.50
3030	cycle 2 miles		380	.021			.67	1.56	2.23	2.82
3032	cycle 4 miles		281	.028			.91	2.12	3.03	3.82
3034	cycle 6 miles		231	.035			1.11	2.57	3.68	4.64
3036	cycle 8 miles		198	.040			1.29	3	4.29	5.40
3038	cycle 10 miles		165	.048			1.55	3.60	5.15	6.50
3040	25 MPH ave, cycle 4 miles		314	.025			.81	1.89	2.70	3.41
3042	cycle 6 miles		264	.030			.97	2.25	3.22	4.07
3044	cycle 8 miles		231	.035			1.11	2.57	3.68	4.64
3046	cycle 10 miles		198	.040			1.29	3	4.29	5.40
3050	30 MPH ave, cycle 4 miles		347	.023			.74	1.71	2.45	3.09
3052	cycle 6 miles		281	.028			.91	2.12	3.03	3.82
3054	cycle 8 miles		248	.032			1.03	2.40	3.43	4.33
3056	cycle 10 miles		215	.037			1.19	2.76	3.95	4.99
3060	35 MPH ave, cycle 4 miles		363	.022			.70	1.64	2.34	2.95
3062	cycle 6 miles		314	.025			.81	1.89	2.70	3.41
3064	cycle 8 miles		264	.030			.97	2.25	3.22	4.07
3066	cycle 10 miles		248	.032			1.03	2.40	3.43	4.33
3068	cycle 20 miles		149	.054			1.72	3.99	5.71	7.20
3070	cycle 30 miles		116	.069			2.20	5.10	7.30	9.25
3072	cycle 40 miles		83	.096			3.08	7.15	10.23	12.95
3074	40 MPH ave, cycle 6 miles		330	.024			.77	1.80	2.57	3.25
3076	cycle 8 miles		281	.028			.91	2.12	3.03	3.82
3078	cycle 10 miles		264	.030			.97	2.25	3.22	4.07
3080	cycle 20 miles		165	.048			1.55	3.60	5.15	6.50
3082	cycle 30 miles		132	.061			1.94	4.50	6.44	8.10
3084	cycle 40 miles		99	.081			2.58	6	8.58	10.85
3086	cycle 50 miles		83	.096			3.08	7.15	10.23	12.95
3094	45 MPH ave, cycle 8 miles		297	.027			.86	2	2.86	3.61
3096	cycle 10 miles		281	.028			.91	2.12	3.03	3.82
3098	cycle 20 miles		182	.044			1.40	3.27	4.67	5.90
3100	cycle 30 miles		132	.061			1.94	4.50	6.44	8.10
3102	cycle 40 miles		116	.069			2.20	5.10	7.30	9.25
3104	cycle 50 miles		99	.081			2.58	6	8.58	10.85
3106	50 MPH ave, cycle 10 miles		281	.028			.91	2.12	3.03	3.82
3108	cycle 20 miles		198	.040			1.29	3	4.29	5.40
3110	cycle 30 miles		149	.054			1.72	3.99	5.71	7.20
3112	cycle 40 miles		116	.069			2.20	5.10	7.30	9.25
3114	cycle 50 miles		99	.081			2.58	6	8.58	10.85
3214	20 min. wait/Ld./Uld., 15 MPH, cycle 0.5 mile		363	.022			.70	1.64	2.34	2.95
3216	cycle 1 mile		330	.024			.77	1.80	2.57	3.25
3218	cycle 2 miles		281	.028			.91	2.12	3.03	3.82
3220	cycle 4 miles		215	.037			1.19	2.76	3.95	4.99
3222	cycle 6 miles		182	.044			1.40	3.27	4.67	5.90
3224	cycle 8 miles		149	.054			1.72	3.99	5.71	7.20
3225	cycle 10 miles		132	.061			1.94	4.50	6.44	8.10
3226	20 MPH ave, cycle 0.5 mile		363	.022			.70	1.64	2.34	2.95
3228	cycle 1 mile		347	.023			.74	1.71	2.45	3.09
3230	cycle 2 miles		297	.027			.86	2	2.86	3.61
3232	cycle 4 miles		248	.032			1.03	2.40	3.43	4.33
3234	cycle 6 miles	↓	198	.040	↓		1.29	3	4.29	5.40

31 23 23.20 Hauling		Crew	Daily Output	Labor-Hours	Unit	Material	2009 Bare Costs Labor	Equipment	Total	Total Incl O&P
3236	cycle 8 miles	B-34C	182	.044	L.C.Y.		1.40	3.27	4.67	5.90
3238	cycle 10 miles		149	.054			1.72	3.99	5.71	7.20
3240	25 MPH ave, cycle 4 miles		264	.030			.97	2.25	3.22	4.07
3242	cycle 6 miles		231	.035			1.11	2.57	3.68	4.64
3244	cycle 8 miles		198	.040			1.29	3	4.29	5.40
3246	cycle 10 miles		182	.044			1.40	3.27	4.67	5.90
3250	30 MPH ave, cycle 4 miles		281	.028			.91	2.12	3.03	3.82
3252	cycle 6 miles		248	.032			1.03	2.40	3.43	4.33
3254	cycle 8 miles		215	.037			1.19	2.76	3.95	4.99
3256	cycle 10 miles		198	.040			1.29	3	4.29	5.40
3260	35 MPH ave, cycle 4 miles		297	.027			.86	2	2.86	3.61
3262	cycle 6 miles		264	.030			.97	2.25	3.22	4.07
3264	cycle 8 miles		231	.035			1.11	2.57	3.68	4.64
3266	cycle 10 miles		215	.037			1.19	2.76	3.95	4.99
3268	cycle 20 miles		149	.054			1.72	3.99	5.71	7.20
3270	cycle 30 miles		99	.081			2.58	6	8.58	10.85
3272	cycle 40 miles		83	.096			3.08	7.15	10.23	12.95
3274	40 MPH ave, cycle 6 miles		264	.030			.97	2.25	3.22	4.07
3276	cycle 8 miles		248	.032			1.03	2.40	3.43	4.33
3278	cycle 10 miles		215	.037			1.19	2.76	3.95	4.99
3280	cycle 20 miles		149	.054			1.72	3.99	5.71	7.20
3282	cycle 30 miles		116	.069			2.20	5.10	7.30	9.25
3284	cycle 40 miles		99	.081			2.58	6	8.58	10.85
3286	cycle 50 miles		83	.096			3.08	7.15	10.23	12.95
3294	45 MPH ave, cycle 8 miles		248	.032			1.03	2.40	3.43	4.33
3296	cycle 10 miles		231	.035			1.11	2.57	3.68	4.64
3298	cycle 20 miles		165	.048			1.55	3.60	5.15	6.50
3300	cycle 30 miles		132	.061			1.94	4.50	6.44	8.10
3302	cycle 40 miles		99	.081			2.58	6	8.58	10.85
3304	cycle 50 miles		83	.096			3.08	7.15	10.23	12.95
3306	50 MPH ave, cycle 10 miles		248	.032			1.03	2.40	3.43	4.33
3308	cycle 20 miles		182	.044			1.40	3.27	4.67	5.90
3310	cycle 30 miles		132	.061			1.94	4.50	6.44	8.10
3312	cycle 40 miles		116	.069			2.20	5.10	7.30	9.25
3314	cycle 50 miles		99	.081			2.58	6	8.58	10.85
3414	25 min. wait/Ld./Uld., 15 MPH, cycle 0.5 mile		281	.028			.91	2.12	3.03	3.82
3416	cycle 1 mile		264	.030			.97	2.25	3.22	4.07
3418	cycle 2 miles		231	.035			1.11	2.57	3.68	4.64
3420	cycle 4 miles		182	.044			1.40	3.27	4.67	5.90
3422	cycle 6 miles		165	.048			1.55	3.60	5.15	6.50
3424	cycle 8 miles		132	.061			1.94	4.50	6.44	8.10
3425	cycle 10 miles		116	.069			2.20	5.10	7.30	9.25
3426	20 MPH ave, cycle 0.5 mile		297	.027			.86	2	2.86	3.61
3428	cycle 1 mile		281	.028			.91	2.12	3.03	3.82
3430	cycle 2 miles		248	.032			1.03	2.40	3.43	4.33
3432	cycle 4 miles		215	.037			1.19	2.76	3.95	4.99
3434	cycle 6 miles		182	.044			1.40	3.27	4.67	5.90
3436	cycle 8 miles		165	.048			1.55	3.60	5.15	6.50
3438	cycle 10 miles		132	.061			1.94	4.50	6.44	8.10
3440	25 MPH ave, cycle 4 miles		231	.035			1.11	2.57	3.68	4.64
3442	cycle 6 miles		198	.040			1.29	3	4.29	5.40
3444	cycle 8 miles		182	.044			1.40	3.27	4.67	5.90
3446	cycle 10 miles		165	.048			1.55	3.60	5.15	6.50

31 23 23.20 Hauling		Crew	Daily Output	Labor-Hours	Unit	Material	2009 Bare Costs Labor	Equipment	Total	Total Incl O&P
3450	30 MPH ave, cycle 4 miles	B-34C	231	.035	L.C.Y.		1.11	2.57	3.68	4.64
3452	cycle 6 miles		215	.037			1.19	2.76	3.95	4.99
3454	cycle 8 miles		182	.044			1.40	3.27	4.67	5.90
3456	cycle 10 miles		165	.048			1.55	3.60	5.15	6.50
3460	35 MPH ave, cycle 4 miles		248	.032			1.03	2.40	3.43	4.33
3462	cycle 6 miles		215	.037			1.19	2.76	3.95	4.99
3464	cycle 8 miles		198	.040			1.29	3	4.29	5.40
3466	cycle 10 miles		182	.044			1.40	3.27	4.67	5.90
3468	cycle 20 miles		132	.061			1.94	4.50	6.44	8.10
3470	cycle 30 miles		99	.081			2.58	6	8.58	10.85
3472	cycle 40 miles		83	.096			3.08	7.15	10.23	12.95
3474	40 MPH, cycle 6 miles		231	.035			1.11	2.57	3.68	4.64
3476	cycle 8 miles		215	.037			1.19	2.76	3.95	4.99
3478	cycle 10 miles		198	.040			1.29	3	4.29	5.40
3480	cycle 20 miles		132	.061			1.94	4.50	6.44	8.10
3482	cycle 30 miles		116	.069			2.20	5.10	7.30	9.25
3484	cycle 40 miles		83	.096			3.08	7.15	10.23	12.95
3486	cycle 50 miles		83	.096			3.08	7.15	10.23	12.95
3494	45 MPH ave, cycle 8 miles		215	.037			1.19	2.76	3.95	4.99
3496	cycle 10 miles		198	.040			1.29	3	4.29	5.40
3498	cycle 20 miles		149	.054			1.72	3.99	5.71	7.20
3500	cycle 30 miles		116	.069			2.20	5.10	7.30	9.25
3502	cycle 40 miles		99	.081			2.58	6	8.58	10.85
3504	cycle 50 miles		83	.096			3.08	7.15	10.23	12.95
3506	50 MPH ave, cycle 10 miles		215	.037			1.19	2.76	3.95	4.99
3508	cycle 20 miles		165	.048			1.55	3.60	5.15	6.50
3510	cycle 30 miles		132	.061			1.94	4.50	6.44	8.10
3512	cycle 40 miles		99	.081			2.58	6	8.58	10.85
3514	cycle 50 miles		83	.096			3.08	7.15	10.23	12.95
3614	30 min. wait/Ld./Uld., 15 MPH, cycle 0.5 mile		248	.032			1.03	2.40	3.43	4.33
3616	cycle 1 mile		231	.035			1.11	2.57	3.68	4.64
3618	cycle 2 miles		198	.040			1.29	3	4.29	5.40
3620	cycle 4 miles		165	.048			1.55	3.60	5.15	6.50
3622	cycle 6 miles		149	.054			1.72	3.99	5.71	7.20
3624	cycle 8 miles		116	.069			2.20	5.10	7.30	9.25
3625	cycle 10 miles		116	.069			2.20	5.10	7.30	9.25
3626	20 MPH ave, cycle 0.5 mile		248	.032			1.03	2.40	3.43	4.33
3628	cycle 1 mile		231	.035			1.11	2.57	3.68	4.64
3630	cycle 2 miles		215	.037			1.19	2.76	3.95	4.99
3632	cycle 4 miles		182	.044			1.40	3.27	4.67	5.90
3634	cycle 6 miles		165	.048			1.55	3.60	5.15	6.50
3636	cycle 8 miles		149	.054			1.72	3.99	5.71	7.20
3638	cycle 10 miles		132	.061			1.94	4.50	6.44	8.10
3640	25 MPH ave, cycle 4 miles		198	.040			1.29	3	4.29	5.40
3642	cycle 6 miles		182	.044			1.40	3.27	4.67	5.90
3644	cycle 8 miles		149	.054			1.72	3.99	5.71	7.20
3646	cycle 10 miles		149	.054			1.72	3.99	5.71	7.20
3650	30 MPH ave, cycle 4 miles		198	.040			1.29	3	4.29	5.40
3652	cycle 6 miles		182	.044			1.40	3.27	4.67	5.90
3654	cycle 8 miles		165	.048			1.55	3.60	5.15	6.50
3656	cycle 10 miles		149	.054			1.72	3.99	5.71	7.20
3660	35 MPH ave, cycle 4 miles		215	.037			1.19	2.76	3.95	4.99
3662	cycle 6 miles	↓	198	.040	↓		1.29	3	4.29	5.40

31 23 23.20 Hauling		Crew	Daily Output	Labor-Hours	Unit	Material	2009 Bare Costs Labor	Equipment	Total	Total Incl O&P
3664	cycle 8 miles	B-34C	182	.044	L.C.Y.		1.40	3.27	4.67	5.90
3666	cycle 10 miles		165	.048			1.55	3.60	5.15	6.50
3668	cycle 20 miles		116	.069			2.20	5.10	7.30	9.25
3670	cycle 30 miles		99	.081			2.58	6	8.58	10.85
3672	cycle 40 miles		83	.096			3.08	7.15	10.23	12.95
3674	40 MPH, cycle 6 miles		198	.040			1.29	3	4.29	5.40
3676	cycle 8 miles		182	.044			1.40	3.27	4.67	5.90
3678	cycle 10 miles		165	.048			1.55	3.60	5.15	6.50
3680	cycle 20 miles		132	.061			1.94	4.50	6.44	8.10
3682	cycle 30 miles		99	.081			2.58	6	8.58	10.85
3684	cycle 40 miles		83	.096			3.08	7.15	10.23	12.95
3686	cycle 50 miles		66	.121			3.87	9	12.87	16.25
3694	45 MPH ave, cycle 8 miles		198	.040			1.29	3	4.29	5.40
3696	cycle 10 miles		182	.044			1.40	3.27	4.67	5.90
3698	cycle 20 miles		132	.061			1.94	4.50	6.44	8.10
3700	cycle 30 miles		116	.069			2.20	5.10	7.30	9.25
3702	cycle 40 miles		83	.096			3.08	7.15	10.23	12.95
3704	cycle 50 miles		83	.096			3.08	7.15	10.23	12.95
3706	50 MPH ave, cycle 10 miles		182	.044			1.40	3.27	4.67	5.90
3708	cycle 20 miles		149	.054			1.72	3.99	5.71	7.20
3710	cycle 30 miles		116	.069			2.20	5.10	7.30	9.25
3712	cycle 40 miles		99	.081			2.58	6	8.58	10.85
3714	cycle 50 miles		83	.096			3.08	7.15	10.23	12.95
4014	20 CY truck, 15 min. wait/Ld./Uld., 15 MPH, cycle 0.5 mile	B-34D	560	.014			.46	1.09	1.55	1.95
4016	cycle 1 mile		500	.016			.51	1.22	1.73	2.18
4018	cycle 2 miles		420	.019			.61	1.45	2.06	2.60
4020	cycle 4 miles		300	.027			.85	2.03	2.88	3.63
4022	cycle 6 miles		240	.033			1.06	2.54	3.60	4.53
4024	cycle 8 miles		200	.040			1.28	3.05	4.33	5.45
4025	cycle 10 miles		160	.050			1.60	3.81	5.41	6.80
4026	20 MPH ave, cycle 0.5 mile		580	.014			.44	1.05	1.49	1.87
4028	cycle 1 mile		520	.015			.49	1.17	1.66	2.10
4030	cycle 2 miles		450	.017			.56	1.32	1.88	2.37
4032	cycle 4 miles		340	.024			.75	1.79	2.54	3.20
4034	cycle 6 miles		280	.029			.91	2.18	3.09	3.89
4036	cycle 8 miles		240	.033			1.06	2.54	3.60	4.53
4038	cycle 10 miles		200	.040			1.28	3.05	4.33	5.45
4040	25 MPH ave, cycle 4 miles		380	.021			.67	1.60	2.27	2.86
4042	cycle 6 miles		320	.025			.80	1.90	2.70	3.40
4044	cycle 8 miles		260	.029			.91	2.18	3.09	3.89
4046	cycle 10 miles		240	.033			1.06	2.54	3.60	4.53
4050	30 MPH ave, cycle 4 miles		420	.019			.61	1.45	2.06	2.60
4052	cycle 6 miles		340	.024			.75	1.79	2.54	3.20
4054	cycle 8 miles		300	.027			.85	2.03	2.88	3.63
4056	cycle 10 miles		260	.031			.98	2.34	3.32	4.19
4060	35 MPH ave, cycle 4 miles		440	.018			.58	1.38	1.96	2.47
4062	cycle 6 miles		380	.021			.67	1.60	2.27	2.86
4064	cycle 8 miles		320	.025			.80	1.90	2.70	3.40
4066	cycle 10 miles		300	.027			.85	2.03	2.88	3.63
4068	cycle 20 miles		180	.044			1.42	3.38	4.80	6.05
4070	cycle 30 miles		140	.057			1.83	4.35	6.18	7.75
4072	cycle 40 miles		100	.080			2.56	6.10	8.66	10.90
4074	40 MPH ave, cycle 6 miles		400	.020			.64	1.52	2.16	2.72

31 23 23.20 Hauling		Crew	Daily Output	Labor-Hours	Unit	Material	2009 Bare Costs Labor	Equipment	Total	Total Incl O&P
4076	cycle 8 miles	B-34D	340	.024	L.C.Y.		.75	1.79	2.54	3.20
4078	cycle 10 miles		320	.025			.80	1.90	2.70	3.40
4080	cycle 20 miles		200	.040			1.28	3.05	4.33	5.45
4082	cycle 30 miles		160	.050			1.60	3.81	5.41	6.80
4084	cycle 40 miles		120	.067			2.13	5.10	7.23	9.10
4086	cycle 50 miles		100	.080			2.56	6.10	8.66	10.90
4094	45 MPH ave, cycle 8 miles		360	.022			.71	1.69	2.40	3.02
4096	cycle 10 miles		340	.024			.75	1.79	2.54	3.20
4098	cycle 20 miles		220	.036			1.16	2.77	3.93	4.94
4100	cycle 30 miles		160	.050			1.60	3.81	5.41	6.80
4102	cycle 40 miles		140	.057			1.83	4.35	6.18	7.75
4104	cycle 50 miles		120	.067			2.13	5.10	7.23	9.10
4106	50 MPH ave, cycle 10 miles		340	.024			.75	1.79	2.54	3.20
4108	cycle 20 miles		240	.033			1.06	2.54	3.60	4.53
4110	cycle 30 miles		180	.044			1.42	3.38	4.80	6.05
4112	cycle 40 miles		140	.057			1.83	4.35	6.18	7.75
4114	cycle 50 miles		120	.067			2.13	5.10	7.23	9.10
4214	20 min. wait/Ld./Uld., 15 MPH, cycle 0.5 mile		440	.018			.58	1.38	1.96	2.47
4216	cycle 1 mile		400	.020			.64	1.52	2.16	2.72
4218	cycle 2 miles		340	.024			.75	1.79	2.54	3.20
4220	cycle 4 miles		260	.031			.98	2.34	3.32	4.19
4222	cycle 6 miles		220	.036			1.16	2.77	3.93	4.94
4224	cycle 8 miles		180	.044			1.42	3.38	4.80	6.05
4225	cycle 10 miles		160	.050			1.60	3.81	5.41	6.80
4226	20 MPH ave, cycle 0.5 mile		440	.018			.58	1.38	1.96	2.47
4228	cycle 1 mile		420	.019			.61	1.45	2.06	2.60
4230	cycle 2 miles		360	.022			.71	1.69	2.40	3.02
4232	cycle 4 miles		300	.027			.85	2.03	2.88	3.63
4234	cycle 6 miles		240	.033			1.06	2.54	3.60	4.53
4236	cycle 8 miles		220	.036			1.16	2.77	3.93	4.94
4238	cycle 10 miles		180	.044			1.42	3.38	4.80	6.05
4240	25 MPH ave, cycle 4 miles		320	.025			.80	1.90	2.70	3.40
4242	cycle 6 miles		280	.029			.91	2.18	3.09	3.89
4244	cycle 8 miles		240	.033			1.06	2.54	3.60	4.53
4246	cycle 10 miles		220	.036			1.16	2.77	3.93	4.94
4250	30 MPH ave, cycle 4 miles		340	.024			.75	1.79	2.54	3.20
4252	cycle 6 miles		300	.027			.85	2.03	2.88	3.63
4254	cycle 8 miles		260	.031			.98	2.34	3.32	4.19
4256	cycle 10 miles		240	.033			1.06	2.54	3.60	4.53
4260	35 MPH ave, cycle 4 miles		360	.022			.71	1.69	2.40	3.02
4262	cycle 6 miles		320	.025			.80	1.90	2.70	3.40
4264	cycle 8 miles		280	.029			.91	2.18	3.09	3.89
4266	cycle 10 miles		260	.031			.98	2.34	3.32	4.19
4268	cycle 20 miles		180	.044			1.42	3.38	4.80	6.05
4270	cycle 30 miles		120	.067			2.13	5.10	7.23	9.10
4272	cycle 40 miles		100	.080			2.56	6.10	8.66	10.90
4274	40 MPH ave, cycle 6 miles		320	.025			.80	1.90	2.70	3.40
4276	cycle 8 miles		300	.027			.85	2.03	2.88	3.63
4278	cycle 10 miles		260	.031			.98	2.34	3.32	4.19
4280	cycle 20 miles		180	.044			1.42	3.38	4.80	6.05
4282	cycle 30 miles		140	.057			1.83	4.35	6.18	7.75
4284	cycle 40 miles		120	.067			2.13	5.10	7.23	9.10
4286	cycle 50 miles		100	.080			2.56	6.10	8.66	10.90

31 23 23.20 Hauling		Crew	Daily Output	Labor-Hours	Unit	Material	2009 Bare Costs Labor	Equipment	Total	Total Incl O&P
4294	45 MPH ave, cycle 8 miles	B-34D	300	.027	L.C.Y.		.85	2.03	2.88	3.63
4296	cycle 10 miles		280	.029			.91	2.18	3.09	3.89
4298	cycle 20 miles		200	.040			1.28	3.05	4.33	5.45
4300	cycle 30 miles		160	.050			1.60	3.81	5.41	6.80
4302	cycle 40 miles		120	.067			2.13	5.10	7.23	9.10
4304	cycle 50 miles		100	.080			2.56	6.10	8.66	10.90
4306	50 MPH ave, cycle 10 miles		300	.027			.85	2.03	2.88	3.63
4308	cycle 20 miles		220	.036			1.16	2.77	3.93	4.94
4310	cycle 30 miles		180	.044			1.42	3.38	4.80	6.05
4312	cycle 40 miles		140	.057			1.83	4.35	6.18	7.75
4314	cycle 50 miles		120	.067			2.13	5.10	7.23	9.10
4414	25 min. wait/Ld./Uld., 15 MPH, cycle 0.5 mile		340	.024			.75	1.79	2.54	3.20
4416	cycle 1 mile		320	.025			.80	1.90	2.70	3.40
4418	cycle 2 miles		280	.029			.91	2.18	3.09	3.89
4420	cycle 4 miles		220	.036			1.16	2.77	3.93	4.94
4422	cycle 6 miles		200	.040			1.28	3.05	4.33	5.45
4424	cycle 8 miles		160	.050			1.60	3.81	5.41	6.80
4425	cycle 10 miles		140	.057			1.83	4.35	6.18	7.75
4426	20 MPH ave, cycle 0.5 mile		360	.022			.71	1.69	2.40	3.02
4428	cycle 1 mile		340	.024			.75	1.79	2.54	3.20
4430	cycle 2 miles		300	.027			.85	2.03	2.88	3.63
4432	cycle 4 miles		260	.031			.98	2.34	3.32	4.19
4434	cycle 6 miles		220	.036			1.16	2.77	3.93	4.94
4436	cycle 8 miles		200	.040			1.28	3.05	4.33	5.45
4438	cycle 10 miles		160	.050			1.60	3.81	5.41	6.80
4440	25 MPH ave, cycle 4 miles		280	.029			.91	2.18	3.09	3.89
4442	cycle 6 miles		240	.033			1.06	2.54	3.60	4.53
4444	cycle 8 miles		220	.036			1.16	2.77	3.93	4.94
4446	cycle 10 miles		200	.040			1.28	3.05	4.33	5.45
4450	30 MPH ave, cycle 4 miles		280	.029			91	2.18	3.09	3.89
4452	cycle 6 miles		260	.031			98	2.34	3.32	4.19
4454	cycle 8 miles		220	.036			1.16	2.77	3.93	4.94
4456	cycle 10 miles		200	.040			1.28	3.05	4.33	5.45
4460	35 MPH ave, cycle 4 miles		300	.027			.85	2.03	2.88	3.63
4462	cycle 6 miles		260	.031			.98	2.34	3.32	4.19
4464	cycle 8 miles		240	.033			1.06	2.54	3.60	4.53
4466	cycle 10 miles		220	.036			1.16	2.77	3.93	4.94
4468	cycle 20 miles		160	.050			1.60	3.81	5.41	6.80
4470	cycle 30 miles		120	.067			2.13	5.10	7.23	9.10
4472	cycle 40 miles		100	.080			2.56	6.10	8.66	10.90
4474	40 MPH, cycle 6 miles		280	.029			.91	2.18	3.09	3.89
4476	cycle 8 miles		260	.031			.98	2.34	3.32	4.19
4478	cycle 10 miles		240	.033			1.06	2.54	3.60	4.53
4480	cycle 20 miles		160	.050			1.60	3.81	5.41	6.80
4482	cycle 30 miles		140	.057			1.83	4.35	6.18	7.75
4484	cycle 40 miles		100	.080			2.56	6.10	8.66	10.90
4486	cycle 50 miles		100	.080			2.56	6.10	8.66	10.90
4494	45 MPH ave, cycle 8 miles		260	.031			.98	2.34	3.32	4.19
4496	cycle 10 miles		240	.033			1.06	2.54	3.60	4.53
4498	cycle 20 miles		180	.044			1.42	3.38	4.80	6.05
4500	cycle 30 miles		140	.057			1.83	4.35	6.18	7.75
4502	cycle 40 miles		120	.067			2.13	5.10	7.23	9.10
4504	cycle 50 miles		100	.080			2.56	6.10	8.66	10.90

31 23 23.20 Hauling		Crew	Daily Output	Labor-Hours	Unit	Material	2009 Bare Costs Labor	Equipment	Total	Total Incl O&P
4506	50 MPH ave, cycle 10 miles	B-34D	260	.031	L.C.Y.		.98	2.34	3.32	4.19
4508	cycle 20 miles		200	.040			1.28	3.05	4.33	5.45
4510	cycle 30 miles		160	.050			1.60	3.81	5.41	6.80
4512	cycle 40 miles		120	.067			2.13	5.10	7.23	9.10
4514	cycle 50 miles		100	.080			2.56	6.10	8.66	10.90
4614	30 min. wait/Ld./Uld., 15 MPH, cycle 0.5 mile		300	.027			.85	2.03	2.88	3.63
4616	cycle 1 mile		280	.029			.91	2.18	3.09	3.89
4618	cycle 2 miles		240	.033			1.06	2.54	3.60	4.53
4620	cycle 4 miles		200	.040			1.28	3.05	4.33	5.45
4622	cycle 6 miles		180	.044			1.42	3.38	4.80	6.05
4624	cycle 8 miles		140	.057			1.83	4.35	6.18	7.75
4625	cycle 10 miles		140	.057			1.83	4.35	6.18	7.75
4626	20 MPH ave, cycle 0.5 mile		300	.027			.85	2.03	2.88	3.63
4628	cycle 1 mile		280	.029			.91	2.18	3.09	3.89
4630	cycle 2 miles		260	.031			.98	2.34	3.32	4.19
4632	cycle 4 miles		220	.036			1.16	2.77	3.93	4.94
4634	cycle 6 miles		200	.040			1.28	3.05	4.33	5.45
4636	cycle 8 miles		180	.044			1.42	3.38	4.80	6.05
4638	cycle 10 miles		160	.050			1.60	3.81	5.41	6.80
4640	25 MPH ave, cycle 4 miles		240	.033			1.06	2.54	3.60	4.53
4642	cycle 6 miles		220	.036			1.16	2.77	3.93	4.94
4644	cycle 8 miles		180	.044			1.42	3.38	4.80	6.05
4646	cycle 10 miles		180	.044			1.42	3.38	4.80	6.05
4650	30 MPH ave, cycle 4 miles		240	.033			1.06	2.54	3.60	4.53
4652	cycle 6 miles		220	.036			1.16	2.77	3.93	4.94
4654	cycle 8 miles		200	.040			1.28	3.05	4.33	5.45
4656	cycle 10 miles		180	.044			1.42	3.38	4.80	6.05
4660	35 MPH ave, cycle 4 miles		260	.031			.98	2.34	3.32	4.19
4662	cycle 6 miles		240	.033			1.06	2.54	3.60	4.53
4664	cycle 8 miles		220	.036			1.16	2.77	3.93	4.94
4666	cycle 10 miles		200	.040			1.28	3.05	4.33	5.45
4668	cycle 20 miles		140	.057			1.83	4.35	6.18	7.75
4670	cycle 30 miles		120	.067			2.13	5.10	7.23	9.10
4672	cycle 40 miles		100	.080			2.56	6.10	8.66	10.90
4674	40 MPH, cycle 6 miles		240	.033			1.06	2.54	3.60	4.53
4676	cycle 8 miles		220	.036			1.16	2.77	3.93	4.94
4678	cycle 10 miles		200	.040			1.28	3.05	4.33	5.45
4680	cycle 20 miles		160	.050			1.60	3.81	5.41	6.80
4682	cycle 30 miles		120	.067			2.13	5.10	7.23	9.10
4684	cycle 40 miles		100	.080			2.56	6.10	8.66	10.90
4686	cycle 50 miles		80	.100			3.20	7.60	10.80	13.60
4694	45 MPH ave, cycle 8 miles		220	.036			1.16	2.77	3.93	4.94
4696	cycle 10 miles		220	.036			1.16	2.77	3.93	4.94
4698	cycle 20 miles		160	.050			1.60	3.81	5.41	6.80
4700	cycle 30 miles		140	.057			1.83	4.35	6.18	7.75
4702	cycle 40 miles		100	.080			2.56	6.10	8.66	10.90
4704	cycle 50 miles		100	.080			2.56	6.10	8.66	10.90
4706	50 MPH ave, cycle 10 miles		220	.036			1.16	2.77	3.93	4.94
4708	cycle 20 miles		180	.044			1.42	3.38	4.80	6.05
4710	cycle 30 miles		140	.057			1.83	4.35	6.18	7.75
4712	cycle 40 miles		120	.067			2.13	5.10	7.23	9.10
4714	cycle 50 miles		100	.080			2.56	6.10	8.66	10.90
5000	22 CY off-road, 15 min. wait/Ld./Uld., 5 MPH, cycle 2000 ft	B-34F	528	.015			.48	2.17	2.65	3.17

31 23 23.20 Hauling		Crew	Daily Output	Labor-Hours	Unit	Material	2009 Bare Costs Labor	Equipment	Total	Total Incl O&P
5010	cycle 3000 ft	B-34F	484	.017	L.C.Y.		.53	2.36	2.89	3.47
5020	cycle 4000 ft		440	.018			.58	2.60	3.18	3.81
5030	cycle 0.5 mile		506	.016			.51	2.26	2.77	3.32
5040	cycle 1 mile		374	.021			.68	3.06	3.74	4.48
5050	cycle 2 miles		264	.030			.97	4.33	5.30	6.35
5060	10 MPH, cycle 2000 ft		594	.013			.43	1.93	2.36	2.83
5070	cycle 3000 ft		572	.014			.45	2	2.45	2.93
5080	cycle 4000 ft		528	.015			.48	2.17	2.65	3.17
5090	cycle 0.5 mile		572	.014			.45	2	2.45	2.93
5100	cycle 1 mile		506	.016			.51	2.26	2.77	3.32
5110	cycle 2 miles		374	.021			.68	3.06	3.74	4.48
5120	cycle 4 miles		264	.030			.97	4.33	5.30	6.35
5130	15 MPH, cycle 2000 ft		638	.013			.40	1.79	2.19	2.63
5140	cycle 3000 ft		594	.013			.43	1.93	2.36	2.83
5150	cycle 4000 ft		572	.014			.45	2	2.45	2.93
5160	cycle 0.5 mile		616	.013			.42	1.86	2.28	2.72
5170	cycle 1 mile		550	.015			.46	2.08	2.54	3.05
5180	cycle 2 miles		462	.017			.55	2.48	3.03	3.63
5190	cycle 4 miles		330	.024			.77	3.47	4.24	5.10
5200	20 MPH, cycle 2 miles		506	.016			.51	2.26	2.77	3.32
5210	cycle 4 miles		374	.021			.68	3.06	3.74	4.48
5220	25 MPH, cycle 2 miles		528	.015			.48	2.17	2.65	3.17
5230	cycle 4 miles		418	.019			.61	2.74	3.35	4.01
5300	20 min. wait/Ld./Uld., 5 MPH, cycle 2000 ft		418	.019			.61	2.74	3.35	4.01
5310	cycle 3000 ft		396	.020			.65	2.89	3.54	4.24
5320	cycle 4000 ft		352	.023			.73	3.25	3.98	4.77
5330	cycle 0.5 mile		396	.020			.65	2.89	3.54	4.24
5340	cycle 1 mile		330	.024			.77	3.47	4.24	5.10
5350	cycle 2 miles		242	.033			1.06	4.73	5.79	6.95
5360	10 MPH, cycle 2000 ft		462	.017			.55	2.48	3.03	3.63
5370	cycle 3000 ft		440	.018			.58	2.60	3.18	3.81
5380	cycle 4000 ft		418	.019			.61	2.74	3.35	4.01
5390	cycle 0.5 mile		462	.017			.55	2.48	3.03	3.63
5400	cycle 1 mile		396	.020			.65	2.89	3.54	4.24
5410	cycle 2 miles		330	.024			.77	3.47	4.24	5.10
5420	cycle 4 miles		242	.033			1.06	4.73	5.79	6.95
5430	15 MPH, cycle 2000 ft		484	.017			.53	2.36	2.89	3.47
5440	cycle 3000 ft		462	.017			.55	2.48	3.03	3.63
5450	cycle 4000 ft		462	.017			.55	2.48	3.03	3.63
5460	cycle 0.5 mile		484	.017			.53	2.36	2.89	3.47
5470	cycle 1 mile		440	.018			.58	2.60	3.18	3.81
5480	cycle 2 miles		374	.021			.68	3.06	3.74	4.48
5490	cycle 4 miles		286	.028			.89	4	4.89	5.85
5500	20 MPH, cycle 2 miles		396	.020			.65	2.89	3.54	4.24
5510	cycle 4 miles		330	.024			.77	3.47	4.24	5.10
5520	25 MPH, cycle 2 miles		418	.019			.61	2.74	3.35	4.01
5530	cycle 4 miles		352	.023			.73	3.25	3.98	4.77
5600	25 min. wait/Ld./Uld., 5 MPH, cycle 2000 ft		352	.023			.73	3.25	3.98	4.77
5610	cycle 3000 ft		330	.024			.77	3.47	4.24	5.10
5620	cycle 4000 ft		308	.026			.83	3.71	4.54	5.45
5630	cycle 0.5 mile		330	.024			.77	3.47	4.24	5.10
5640	cycle 1 mile		286	.028			.89	4	4.89	5.85
5650	cycle 2 miles		220	.036			1.16	5.20	6.36	7.60

31 23 23.20 Hauling		Crew	Daily Output	Labor-Hours	Unit	Material	2009 Bare Costs Labor	Equipment	Total	Total Incl O&P
5660	10 MPH, cycle 2000 ft	B-34F	374	.021	L.C.Y.		.68	3.06	3.74	4.48
5670	cycle 3000 ft		374	.021			.68	3.06	3.74	4.48
5680	cycle 4000 ft		352	.023			.73	3.25	3.98	4.77
5690	cycle 0.5 mile		374	.021			.68	3.06	3.74	4.48
5700	cycle 1 mile		330	.024			.77	3.47	4.24	5.10
5710	cycle 2 miles		286	.028			.89	4	4.89	5.85
5720	cycle 4 miles		220	.036			1.16	5.20	6.36	7.60
5730	15 MPH, cycle 2000 ft		396	.020			.65	2.89	3.54	4.24
5740	cycle 3000 ft		374	.021			.68	3.06	3.74	4.48
5750	cycle 4000 ft		374	.021			.68	3.06	3.74	4.48
5760	cycle 0.5 mile		374	.021			.68	3.06	3.74	4.48
5770	cycle 1 mile		352	.023			.73	3.25	3.98	4.77
5780	cycle 2 miles		308	.026			.83	3.71	4.54	5.45
5790	cycle 4 miles		242	.033			1.06	4.73	5.79	6.95
5800	20 MPH, cycle 2 miles		330	.024			.77	3.47	4.24	5.10
5810	cycle 4 miles		286	.028			.89	4	4.89	5.85
5820	25 MPH, cycle 2 miles		352	.023			.73	3.25	3.98	4.77
5830	cycle 4 miles		308	.026			.83	3.71	4.54	5.45
6000	34 CY off-road, 15 min. wait/Ld./Uld., 5 MPH, cycle 2000 ft	B-34G	816	.010			.31	1.94	2.25	2.64
6010	cycle 3000 ft		748	.011			.34	2.12	2.46	2.89
6020	cycle 4000 ft		680	.012			.38	2.33	2.71	3.18
6030	cycle 0.5 mile		782	.010			.33	2.02	2.35	2.77
6040	cycle 1 mile		578	.014			.44	2.74	3.18	3.73
6050	cycle 2 miles		408	.020			.63	3.88	4.51	5.30
6060	10 MPH, cycle 2000 ft		918	.009			.28	1.72	2	2.35
6070	cycle 3000 ft		884	.009			.29	1.79	2.08	2.44
6080	cycle 4000 ft		816	.010			.31	1.94	2.25	2.64
6090	cycle 0.5 mile		884	.009			.29	1.79	2.08	2.44
6100	cycle 1 mile		782	.010			.33	2.02	2.35	2.77
6110	cycle 2 miles		578	.014			.44	2.74	3.18	3.73
6120	cycle 4 miles		408	.020			.63	3.88	4.51	5.30
6130	15 MPH, cycle 2000 ft		986	.008			.26	1.60	1.86	2.18
6140	cycle 3000 ft		918	.009			.28	1.72	2	2.35
6150	cycle 4000 ft		884	.009			.29	1.79	2.08	2.44
6160	cycle 0.5 mile		952	.008			.27	1.66	1.93	2.27
6170	cycle 1 mile		850	.009			.30	1.86	2.16	2.54
6180	cycle 2 miles		714	.011			.36	2.21	2.57	3.03
6190	cycle 4 miles		510	.016			.50	3.10	3.60	4.23
6200	20 MPH, cycle 2 miles		782	.010			.33	2.02	2.35	2.77
6210	cycle 4 miles		578	.014			.44	2.74	3.18	3.73
6220	25 MPH, cycle 2 miles		816	.010			.31	1.94	2.25	2.64
6230	cycle 4 miles		646	.012			.40	2.45	2.85	3.34
6300	20 min. wait/Ld./Uld., 5 MPH, cycle 2000 ft		646	.012			.40	2.45	2.85	3.34
6310	cycle 3000 ft		612	.013			.42	2.58	3	3.52
6320	cycle 4000 ft		544	.015			.47	2.91	3.38	3.97
6330	cycle 0.5 mile		612	.013			.42	2.58	3	3.52
6340	cycle 1 mile		510	.016			.50	3.10	3.60	4.23
6350	cycle 2 miles		374	.021			.68	4.23	4.91	5.75
6360	10 MPH, cycle 2000 ft		714	.011			.36	2.21	2.57	3.03
6370	cycle 3000 ft		680	.012			.38	2.33	2.71	3.18
6380	cycle 4000 ft		646	.012			.40	2.45	2.85	3.34
6390	cycle 0.5 mile		714	.011			.36	2.21	2.57	3.03
6400	cycle 1 mile		612	.013			.42	2.58	3	3.52

31 23 23.20 Hauling		Crew	Daily Output	Labor-Hours	Unit	Material	2009 Bare Costs Labor	Equipment	Total	Total Incl O&P
6410	cycle 2 miles	B-34G	510	.016	L.C.Y.		.50	3.10	3.60	4.23
6420	cycle 4 miles		374	.021			.68	4.23	4.91	5.75
6430	15 MPH, cycle 2000 ft		748	.011			.34	2.12	2.46	2.89
6440	cycle 3000 ft		714	.011			.36	2.21	2.57	3.03
6450	cycle 4000 ft		714	.011			.36	2.21	2.57	3.03
6460	cycle 0.5 mile		748	.011			.34	2.12	2.46	2.89
6470	cycle 1 mile		680	.012			.38	2.33	2.71	3.18
6480	cycle 2 miles		578	.014			.44	2.74	3.18	3.73
6490	cycle 4 miles		442	.018			.58	3.58	4.16	4.89
6500	20 MPH, cycle 2 miles		612	.013			.42	2.58	3	3.52
6510	cycle 4 miles		510	.016			.50	3.10	3.60	4.23
6520	25 MPH, cycle 2 miles		646	.012			.40	2.45	2.85	3.34
6530	cycle 4 miles		544	.015			.47	2.91	3.38	3.97
6600	25 min. wait/Ld./Uld., 5 MPH, cycle 2000 ft		544	.015			.47	2.91	3.38	3.97
6610	cycle 3000 ft		510	.016			.50	3.10	3.60	4.23
6620	cycle 4000 ft		476	.017			.54	3.32	3.86	4.54
6630	cycle 0.5 mile		510	.016			.50	3.10	3.60	4.23
6640	cycle 1 mile		442	.018			.58	3.58	4.16	4.89
6650	cycle 2 miles		340	.024			.75	4.65	5.40	6.35
6660	10 MPH, cycle 2000 ft		578	.014			.44	2.74	3.18	3.73
6670	cycle 3000 ft		578	.014			.44	2.74	3.18	3.73
6680	cycle 4000 ft		544	.015			.47	2.91	3.38	3.97
6690	cycle 0.5 mile		578	.014			.44	2.74	3.18	3.73
6700	cycle 1 mile		510	.016			.50	3.10	3.60	4.23
6710	cycle 2 miles		442	.018			.58	3.58	4.16	4.89
6720	cycle 4 miles		340	.024			.75	4.65	5.40	6.35
6730	15 MPH, cycle 2000 ft		612	.013			.42	2.58	3	3.52
6740	cycle 3000 ft		578	.014			.44	2.74	3.18	3.73
6750	cycle 4000 ft		578	.014			.44	2.74	3.18	3.73
6760	cycle 0.5 mile		612	.013			.42	2.58	3	3.52
6770	cycle 1 mile		544	.015			.47	2.91	3.38	3.97
6780	cycle 2 miles		476	.017			.54	3.32	3.86	4.54
6790	cycle 4 miles		374	.021			.68	4.23	4.91	5.75
6800	20 MPH, cycle 2 miles		510	.016			.50	3.10	3.60	4.23
6810	cycle 4 miles		442	.018			.58	3.58	4.16	4.89
6820	25 MPH, cycle 2 miles		544	.015			.47	2.91	3.38	3.97
6830	cycle 4 miles		476	.017			.54	3.32	3.86	4.54
7000	42 CY off-road, 20 min. wait/Ld./Uld., 5 MPH, cycle 2000 ft	B-34H	798	.010			.32	2	2.32	2.73
7010	cycle 3000 ft		756	.011			.34	2.11	2.45	2.87
7020	cycle 4000 ft		672	.012			.38	2.37	2.75	3.23
7030	cycle 0.5 mile		756	.011			.34	2.11	2.45	2.87
7040	cycle 1 mile		630	.013			.41	2.53	2.94	3.44
7050	cycle 2 miles		462	.017			.55	3.45	4	4.71
7060	10 MPH, cycle 2000 ft		882	.009			.29	1.81	2.10	2.46
7070	cycle 3000 ft		840	.010			.30	1.90	2.20	2.59
7080	cycle 4000 ft		798	.010			.32	2	2.32	2.73
7090	cycle 0.5 mile		882	.009			.29	1.81	2.10	2.46
7100	cycle 1 mile		798	.010			.32	2	2.32	2.73
7110	cycle 2 miles		630	.013			.41	2.53	2.94	3.44
7120	cycle 4 miles		462	.017			.55	3.45	4	4.71
7130	15 MPH, cycle 2000 ft		924	.009			.28	1.73	2.01	2.35
7140	cycle 3000 ft		882	.009			.29	1.81	2.10	2.46
7150	cycle 4000 ft		882	.009			.29	1.81	2.10	2.46

31 23 23.20 Hauling		Crew	Daily Output	Labor-Hours	Unit	Material	2009 Bare Costs Labor	Equipment	Total	Total Incl O&P
7160	cycle 0.5 mile	B-34H	882	.009	L.C.Y.		.29	1.81	2.10	2.46
7170	cycle 1 mile		840	.010			.30	1.90	2.20	2.59
7180	cycle 2 miles		714	.011			.36	2.23	2.59	3.04
7190	cycle 4 miles		546	.015			.47	2.92	3.39	3.98
7200	20 MPH, cycle 2 miles		756	.011			.34	2.11	2.45	2.87
7210	cycle 4 miles		630	.013			.41	2.53	2.94	3.44
7220	25 MPH, cycle 2 miles		798	.010			.32	2	2.32	2.73
7230	cycle 4 miles		672	.012			.38	2.37	2.75	3.23
7300	25 min. wait/Ld./Uld., 5 MPH, cycle 2000 ft		672	.012			.38	2.37	2.75	3.23
7310	cycle 3000 ft		630	.013			.41	2.53	2.94	3.44
7320	cycle 4000 ft		588	.014			.43	2.71	3.14	3.69
7330	cycle 0.5 mile		630	.013			.41	2.53	2.94	3.44
7340	cycle 1 mile		546	.015			.47	2.92	3.39	3.98
7350	cycle 2 miles		378	.021			.68	4.22	4.90	5.75
7360	10 MPH, cycle 2000 ft		714	.011			.36	2.23	2.59	3.04
7370	cycle 3000 ft		714	.011			.36	2.23	2.59	3.04
7380	cycle 4000 ft		672	.012			.38	2.37	2.75	3.23
7390	cycle 0.5 mile		714	.011			.36	2.23	2.59	3.04
7400	cycle 1 mile		630	.013			.41	2.53	2.94	3.44
7410	cycle 2 miles		546	.015			.47	2.92	3.39	3.98
7420	cycle 4 miles		378	.021			.68	4.22	4.90	5.75
7430	15 MPH, cycle 2000 ft		756	.011			.34	2.11	2.45	2.87
7440	cycle 3000 ft		714	.011			.36	2.23	2.59	3.04
7450	cycle 4000 ft		714	.011			.36	2.23	2.59	3.04
7460	cycle 0.5 mile		714	.011			.36	2.23	2.59	3.04
7470	cycle 1 mile		672	.012			.38	2.37	2.75	3.23
7480	cycle 2 miles		588	.014			.43	2.71	3.14	3.69
7490	cycle 4 miles		462	.017			.55	3.45	4	4.71
7500	20 MPH, cycle 2 miles		630	.013			.41	2.53	2.94	3.44
7510	cycle 4 miles		546	.015			.47	2.92	3.39	3.98
7520	25 MPH, cycle 2 miles		672	.012			.38	2.37	2.75	3.23
7530	cycle 4 miles		588	.014			.43	2.71	3.14	3.69
8000	60 CY off-road, 20 min. wait/Ld./Uld., 5 MPH, cycle 2000 ft	B-34J	1140	.007			.22	1.80	2.02	2.35
8010	cycle 3000 ft		1080	.007			.24	1.90	2.14	2.48
8020	cycle 4000 ft		960	.008			.27	2.13	2.40	2.78
8030	cycle 0.5 mile		1080	.007			.24	1.90	2.14	2.48
8040	cycle 1 mile		900	.009			.28	2.27	2.55	2.97
8050	cycle 2 miles		660	.012			.39	3.10	3.49	4.04
8060	10 MPH, cycle 2000 ft		1260	.006			.20	1.62	1.82	2.12
8070	cycle 3000 ft		1200	.007			.21	1.71	1.92	2.23
8080	cycle 4000 ft		1140	.007			.22	1.80	2.02	2.35
8090	cycle 0.5 mile		1260	.006			.20	1.62	1.82	2.12
8100	cycle 1 mile		1080	.007			.24	1.90	2.14	2.48
8110	cycle 2 miles		900	.009			.28	2.27	2.55	2.97
8120	cycle 4 miles		660	.012			.39	3.10	3.49	4.04
8130	15 MPH, cycle 2000 ft		1320	.006			.19	1.55	1.74	2.03
8140	cycle 3000 ft		1260	.006			.20	1.62	1.82	2.12
8150	cycle 4000 ft		1260	.006			.20	1.62	1.82	2.12
8160	cycle 0.5 mile		1320	.006			.19	1.55	1.74	2.03
8170	cycle 1 mile		1200	.007			.21	1.71	1.92	2.23
8180	cycle 2 miles		1020	.008			.25	2.01	2.26	2.62
8190	cycle 4 miles		780	.010			.33	2.63	2.96	3.43
8200	20 MPH, cycle 2 miles		1080	.007			.24	1.90	2.14	2.48

31 23 23.20 Hauling		Crew	Daily Output	Labor-Hours	Unit	Material	2009 Bare Costs Labor	Equipment	Total	Total Incl O&P
8210	cycle 4 miles	B-34J	900	.009	L.C.Y.		.28	2.27	2.55	2.97
8220	25 MPH, cycle 2 miles		1140	.007			.22	1.80	2.02	2.35
8230	cycle 4 miles		960	.008			.27	2.13	2.40	2.78
8300	25 min. wait/Ld./Uld., 5 MPH, cycle 2000 ft		960	.008			.27	2.13	2.40	2.78
8310	cycle 3000 ft		900	.009			.28	2.27	2.55	2.97
8320	cycle 4000 ft		840	.010			.30	2.44	2.74	3.18
8330	cycle 0.5 mile		900	.009			.28	2.27	2.55	2.97
8340	cycle 1 mile		780	.010			.33	2.63	2.96	3.43
8350	cycle 2 miles		600	.013			.43	3.41	3.84	4.45
8360	10 MPH, cycle 2000 ft		1020	.008			.25	2.01	2.26	2.62
8370	cycle 3000 ft		1020	.008			.25	2.01	2.26	2.62
8380	cycle 4000 ft		960	.008			.27	2.13	2.40	2.78
8390	cycle 0.5 mile		1020	.008			.25	2.01	2.26	2.62
8400	cycle 1 mile		900	.009			.28	2.27	2.55	2.97
8410	cycle 2 miles		780	.010			.33	2.63	2.96	3.43
8420	cycle 4 miles		600	.013			.43	3.41	3.84	4.45
8430	15 MPH, cycle 2000 ft		1080	.007			.24	1.90	2.14	2.48
8440	cycle 3000 ft		1020	.008			.25	2.01	2.26	2.62
8450	cycle 4000 ft		1020	.008			.25	2.01	2.26	2.62
8460	cycle 0.5 mile		1080	.007			.24	1.90	2.14	2.48
8470	cycle 1 mile		960	.008			.27	2.13	2.40	2.78
8480	cycle 2 miles		840	.010			.30	2.44	2.74	3.18
8490	cycle 4 miles		660	.012			.39	3.10	3.49	4.04
8500	20 MPH, cycle 2 miles		900	.009			.28	2.27	2.55	2.97
8510	cycle 4 miles		780	.010			.33	2.63	2.96	3.43
8520	25 MPH, cycle 2 miles		960	.008			.27	2.13	2.40	2.78
8530	cycle 4 miles		840	.010			.30	2.44	2.74	3.18

31 23 23.23 Compaction		Crew	Daily Output	Labor-Hours	Unit	Material	Labor	Equipment	Total	Total Incl O&P
0010	**COMPACTION**									
5000	Riding, vibrating roller, 6" lifts, 2 passes	B-10Y	3000	.004	E.C.Y.		.15	.16	.31	.41
5020	3 passes		2300	.005			.20	.20	.40	.54
5040	4 passes		1900	.006			.24	.25	.49	.65
5060	12" lifts, 2 passes		5200	.002			.09	.09	.18	.24
5080	3 passes		3500	.003			.13	.13	.26	.36
5100	4 passes		2600	.005			.18	.18	.36	.48
5600	Sheepsfoot or wobbly wheel roller, 6" lifts, 2 passes	B-10G	2400	.005			.19	.48	.67	.82
5620	3 passes		1735	.007			.26	.66	.92	1.14
5640	4 passes		1300	.009			.35	.88	1.23	1.53
5680	12" lifts, 2 passes		5200	.002			.09	.22	.31	.38
5700	3 passes		3500	.003			.13	.33	.46	.57
5720	4 passes		2600	.005			.18	.44	.62	.76
7000	Walk behind, vibrating plate 18" wide, 6" lifts, 2 passes	A-1D	200	.040			1.26	.16	1.42	2.26
7020	3 passes		185	.043			1.37	.17	1.54	2.45
7040	4 passes		140	.057			1.81	.23	2.04	3.23
7200	12" lifts, 2 passes	A-1E	560	.014			.45	.07	.52	.83
7220	3 passes		375	.021			.67	.10	.77	1.23
7240	4 passes		280	.029			.90	.14	1.04	1.64
7500	Vibrating roller 24" wide, 6" lifts, 2 passes	B-10A	420	.029			1.09	.35	1.44	2.12
7520	3 passes		280	.043			1.63	.52	2.15	3.17
7540	4 passes		210	.057			2.18	.69	2.87	4.23
7600	12" lifts, 2 passes		840	.014			.54	.17	.71	1.06
7620	3 passes		560	.021			.82	.26	1.08	1.59

31 23 Excavation and Fill

31 23 23 – Fill

31 23 23.23 Compaction		Crew	Daily Output	Labor-Hours	Unit	Material	2009 Bare Costs Labor	Equipment	Total	Total Incl O&P
7640	4 passes	B-10A	420	.029	E.C.Y.		1.09	.35	1.44	2.12
8000	Rammer tamper, 6″ to 11″, 4″ lifts, 2 passes	A-1F	130	.062			1.94	.33	2.27	3.58
8050	3 passes		97	.082			2.61	.45	3.06	4.79
8100	4 passes		65	.123			3.89	.67	4.56	7.15
8200	8″ lifts, 2 passes		260	.031			.97	.17	1.14	1.78
8250	3 passes		195	.041			1.30	.22	1.52	2.38
8300	4 passes		130	.062			1.94	.33	2.27	3.58
8400	13″ to 18″, 4″ lifts, 2 passes	A-1G	390	.021			.65	.13	.78	1.21
8450	3 passes		290	.028			.87	.17	1.04	1.63
8500	4 passes		195	.041			1.30	.25	1.55	2.42
8600	8″ lifts, 2 passes		780	.010			.32	.06	.38	.61
8650	3 passes		585	.014			.43	.08	.51	.80
8700	4 passes		390	.021			.65	.13	.78	1.21
9000	Water, 3000 gal. truck, 3 mile haul	B-45	1888	.008		1.14	.31	.41	1.86	2.22
9010	6 mile haul		1444	.011		1.14	.41	.54	2.09	2.51
9020	12 mile haul		1000	.016		1.14	.59	.78	2.51	3.06
9030	6000 gal. wagon, 3 mile haul	B-59	2000	.004		1.14	.13	.21	1.48	1.70
9040	6 mile haul	″	1600	.005		1.14	.16	.27	1.57	1.81
9900	Minimum labor/equipment charge	1 Clab	4	2	Job		63		63	104

31 23 23.24 Compaction, Structural		Crew	Daily Output	Labor-Hours	Unit	Material	Labor	Equipment	Total	Total Incl O&P
0010	**COMPACTION, STRUCTURAL**									
0020	Steel wheel tandem roller, 5 tons	B-10E	8	1.500	Hr.		57	17.20	74.20	110

31 25 Erosion and Sedimentation Controls

31 25 13 – Erosion Controls

31 25 13.10 Synthetic Erosion Control			Crew	Daily Output	Labor-Hours	Unit	Material	Labor	Equipment	Total	Total Incl O&P
0010	**SYNTHETIC EROSION CONTROL**										
0020	Jute mesh, 100 SY per roll, 4′ wide, stapled	G	B-80A	2400	.010	S.Y.	1.06	.32	.10	1.48	1.80
0060	Nylon, 3 dimensional geomatrix, 9 mil thick	G		700	.034		4.95	1.08	.35	6.38	7.60
0062	12 mil thick	G		515	.047		6.70	1.47	.47	8.64	10.35
0064	18 mil thick	G		460	.052		7.10	1.65	.53	9.28	11.10
0100	Plastic netting, stapled, 2″ x 1″ mesh, 20 mil	G	B-1	2500	.010		.81	.31		1.12	1.40
0120	Revegetation mat, webbed	G	2 Clab	1000	.016		6.50	.51		7.01	8
0200	Polypropylene mesh, stapled, 6.5 oz./S.Y.	G	B-1	2500	.010		1.67	.31		1.98	2.35
0300	Tobacco netting, or jute mesh #2, stapled	G	″	2500	.010		.09	.31		.40	.61
1000	Silt fence, polypropylene, 3′ high, ideal conditions	G	2 Clab	1600	.010	L.F.	.39	.32		.71	.95
1100	Adverse conditions	G	″	950	.017	″	.39	.53		.92	1.31
1130	Cellular confinement, poly, 3-dimen, 8′ x 20′ panels, 4″ deep cell	G	B-6	1600	.015	S.F.	2.07	.51	.18	2.76	3.31
1140	8″ deep cells	G	″	1200	.020	″	2.84	.68	.24	3.76	4.50
1200	Place and remove hay bales	G	A-2	3	8	Ton	171	251	65	487	675
1250	Hay bales, staked	G	″	2500	.010	L.F.	6.85	.30	.08	7.23	8.15

31 31 Soil Treatment

31 31 16 – Termite Control

31 31 16.13 Chemical Termite Control		Crew	Daily Output	Labor-Hours	Unit	2009 Bare Costs Material	Labor	Equipment	Total	Total Incl O&P
0010	**CHEMICAL TERMITE CONTROL**									
0020	Slab and walls, residential	1 Skwk	1200	.007	SF Flr.	.32	.27		.59	.79
0100	Commercial, minimum		2496	.003		.34	.13		.47	.58
0200	Maximum		1645	.005		.51	.20		.71	.88
0390	Minimum labor/equipment charge		4	2	Job		81.50		81.50	133
0400	Insecticides for termite control, minimum		14.20	.563	Gal.	13.05	23		36.05	52
0500	Maximum		11	.727	"	22.50	29.50		52	73
3000	Soil poisoning (sterilization)	1 Clab	4496	.002	S.F.	.67	.06		.73	.83
3100	Herbicide application from truck	B-59	19000	.001	S.Y.		.01	.02	.03	.04

31 32 Soil Stabilization

31 32 13 – Soil Mixing Stabilization

31 32 13.13 Asphalt Soil Stabilization

		Crew	Daily Output	Labor-Hours	Unit	Material	Labor	Equipment	Total	Total Incl O&P
0010	**ASPHALT SOIL STABILIZATION**									
0011	Including scarifying and compaction									
0020	Asphalt, 1-1/2″ deep, 1/2 gal/S.Y.	B-75	4000	.014	S.Y.	1.09	53	1.17	2.79	3.32
0040	3/4 gal/S.Y.		4000	.014		1.63	.53	1.17	3.33	3.92
0100	3″ deep, 1 gal/S.Y.		3500	.016		2.17	.60	1.34	4.11	4.83
0140	1-1/2 gal/S.Y.		3500	.016		3.26	.60	1.34	5.20	6
0200	6″ deep, 2 gal/S.Y.		3000	.019		4.34	.70	1.57	6.61	7.60
0240	3 gal/S.Y.		3000	.019		6.50	.70	1.57	8.77	10
0300	8″ deep, 2-2/3 gal/S.Y.		2800	.020		5.80	.75	1.68	8.23	9.40
0340	4 gal/S.Y.		2800	.020		8.70	.75	1.68	11.13	12.60
0500	12″ deep, 4 gal/S.Y.		5000	.011		8.70	.42	.94	10.06	11.25
0540	6 gal/S.Y.		2600	.022		13	.81	1.81	15.62	17.60

31 32 13.16 Cement Soil Stabilization

		Crew	Daily Output	Labor-Hours	Unit	Material	Labor	Equipment	Total	Total Incl O&P
0010	**CEMENT SOIL STABILIZATION**									
0011	Including scarifying and compaction									
1020	Cement, 4% mix, by volume, 6″ deep	B-74	1100	.058	S.Y.	1.59	2.14	4.21	7.94	9.80
1030	8″ deep		1050	.061		2.08	2.24	4.41	8.73	10.75
1060	12″ deep		960	.067		3.12	2.45	4.82	10.39	12.65
1100	6% mix, 6″ deep		1100	.058		2.29	2.14	4.21	8.64	10.60
1120	8″ deep		1050	.061		2.98	2.24	4.41	9.63	11.75
1160	12″ deep		960	.067		4.50	2.45	4.82	11.77	14.20
1200	9% mix, 6″ deep		1100	.058		3.47	2.14	4.21	9.82	11.90
1220	8″ deep		1050	.061		4.50	2.24	4.41	11.15	13.40
1260	12″ deep		960	.067		6.80	2.45	4.82	14.07	16.70
1300	12% mix, 6″ deep		1100	.058		4.50	2.14	4.21	10.85	13
1320	8″ deep		1050	.061		6.05	2.24	4.41	12.70	15.10
1360	12″ deep		960	.067		9	2.45	4.82	16.27	19.15

31 32 13.19 Lime Soil Stabilization

		Crew	Daily Output	Labor-Hours	Unit	Material	Labor	Equipment	Total	Total Incl O&P
0010	**LIME SOIL STABILIZATION**									
0011	Including scarifying and compaction									
2020	Hydrated lime, for base, 2% mix by weight, 6″ deep	B-74	1800	.036	S.Y.	1.96	1.31	2.57	5.84	7.10
2030	8″ deep		1700	.038		2.81	1.39	2.72	6.92	8.30
2060	12″ deep		1550	.041		3.97	1.52	2.99	8.48	10.10
2100	4% mix, 6″ deep		1800	.036		2.93	1.31	2.57	6.81	8.15
2120	8″ deep		1700	.038		3.96	1.39	2.72	8.07	9.55
2160	12″ deep		1550	.041		5.90	1.52	2.99	10.41	12.20
2200	6% mix, 6″ deep		1800	.036		4.44	1.31	2.57	8.32	9.80

31 32 Soil Stabilization

31 32 13 – Soil Mixing Stabilization

31 32 13.19 Lime Soil Stabilization		Crew	Daily Output	Labor-Hours	Unit	Material	2009 Bare Costs Labor	Equipment	Total	Total Incl O&P
2220	8″ deep	B-74	1700	.038	S.Y.	5.90	1.39	2.72	10.01	11.70
2260	12″ deep	↓	1550	.041	↓	8.80	1.52	2.99	13.31	15.40

31 32 13.30 Calcium Chloride

		Crew	Daily Output	Labor-Hours	Unit	Material	Labor	Equipment	Total	Total Incl O&P
0010	**CALCIUM CHLORIDE**									
0020	Calcium chloride delivered, 100 LB bags, truckload lots				Ton	560			560	620
0030	Solution, 4 lb. flake per gallon, tank truck delivery				Gal.	1.20			1.20	1.32

31 32 19 – Geosynthetic Soil Stabilization and Layer Separation

31 32 19.16 Geotextile Soil Stabilization

		Crew	Daily Output	Labor-Hours	Unit	Material	Labor	Equipment	Total	Total Incl O&P
0010	**GEOTEXTILE SOIL STABILIZATION**									
1500	Geotextile fabric, woven, 200 lb. tensile strength	2 Clab	2500	.006	S.Y.	2.17	.20		2.37	2.72
1510	Heavy Duty, 600 lb. tensile strength		2400	.007		1.94	.21		2.15	2.48
1550	Non-woven, 120 lb. tensile strength	↓	2500	.006	↓	1.10	.20		1.30	1.54

31 33 Rock Stabilization

31 33 13 – Rock Bolting and Grouting

31 33 13.10 Rock Bolting

		Crew	Daily Output	Labor-Hours	Unit	Material	Labor	Equipment	Total	Total Incl O&P
0010	**ROCK BOLTING**									
2020	Hollow core, prestressable anchor, 1″ diameter, 5′ long	2 Skwk	32	.500	Ea.	98	20.50		118.50	142
2025	10′ long		24	.667		196	27		223	261
2060	2″ diameter, 5′ long		32	.500		365	20.50		385.50	440
2065	10′ long		24	.667		735	27		762	850
2100	Super high-tensile, 3/4″ diameter, 5′ long		32	.500		22.50	20.50		43	58
2105	10′ long		24	.667		44.50	27		71.50	93.50
2160	2″ diameter, 5′ long		32	.500		198	20.50		218.50	252
2165	10′ long	↓	24	.667		395	27		422	480
4400	Drill hole for rock bolt, 1-3/4″ diam., 5′ long (for 3/4″ bolt)	B-56	17	.941			33.50	76.50	110	138
4405	10′ long		9	1.778			63	145	208	260
4420	2″ diameter, 5′ long (for 1″ bolt)		13	1.231			43.50	100	143.50	180
4425	10′ long		7	2.286			81	186	267	335
4460	3-1/2″ diameter, 5′ long (for 2″ bolt)		10	1.600			56.50	130	186.50	234
4465	10′ long	↓	5	3.200	↓		113	260	373	470

31 36 Gabions

31 36 13 – Gabion Boxes

31 36 13.10 Gabion Box Systems

		Crew	Daily Output	Labor-Hours	Unit	Material	Labor	Equipment	Total	Total Incl O&P
0010	**GABION BOX SYSTEMS**									
0400	Gabions, galvanized steel mesh mats or boxes, stone filled, 6″ deep	B-13	200	.280	S.Y.	24	9.55	3.95	37.50	46.50
0500	9″ deep		163	.344		37	11.75	4.84	53.59	65
0600	12″ deep		153	.366		40.50	12.50	5.15	58.15	71
0700	18″ deep		102	.549		57.50	18.75	7.75	84	103
0800	36″ deep	↓	60	.933	↓	82	32	13.15	127.15	157

31 37 Riprap

31 37 13 – Machined Riprap

31 37 13.10 Rip-Rap and Rock Lining		Crew	Daily Output	Labor-Hours	Unit	Material	2009 Bare Costs Labor	Equipment	Total	Total Incl O&P
0010	**RIP-RAP AND ROCK LINING**									
0011	Random, broken stone									
0100	Machine placed for slope protection	B-12G	62	.258	L.C.Y.	31	9.55	10.50	51.05	61
0110	3/8 to 1/4 C.Y. pieces, grouted	B-13	80	.700	S.Y.	69.50	24	9.85	103.35	126
0200	18" minimum thickness, not grouted	"	53	1.057	"	19.25	36	14.90	70.15	96
0300	Dumped, 50 lb. average	B-11A	800	.020	Ton	28	.73	1.35	30.08	33
0350	100 lb. average		700	.023		40	.83	1.55	42.38	47
0370	300 lb. average		600	.027		46.50	.97	1.80	49.27	54.50

31 41 Shoring

31 41 13 – Timber Shoring

31 41 13.10 Building Shoring		Crew	Daily Output	Labor-Hours	Unit	Material	Labor	Equipment	Total	Total Incl O&P
0010	**BUILDING SHORING**									
0020	Shoring, existing building, with timber, no salvage allowance	B-51	2.20	21.818	M.B.F.	695	695	88.50	1,478.50	2,000
1000	On cribbing with 35 ton screw jacks, per box and jack		3.60	13.333	Jack	63.50	425	54	542.50	830
1090	Minimum labor/equipment charge		2	24	Ea.		765	97	862	1,350

31 41 16 – Sheet Piling

31 41 16.10 Sheet Piling Systems		Crew	Daily Output	Labor-Hours	Unit	Material	Labor	Equipment	Total	Total Incl O&P
0010	**SHEET PILING SYSTEMS**									
0020	Sheet piling steel, not incl. wales, 22 psf, 15' excav., left in place	B-40	10.81	5.920	Ton	1,225	234	305	1,764	2,075
0100	Drive, extract & salvage		6	10.667		505	420	550	1,475	1,850
0300	20' deep excavation, 27 psf, left in place		12.95	4.942		1,225	195	256	1,676	1,950
0400	Drive, extract & salvage		6.55	9.771		505	385	505	1,395	1,750
0600	25' deep excavation, 38 psf, left in place		19	3.368		1,225	133	174	1,532	1,750
0700	Drive, extract & salvage		10.50	6.095		505	241	315	1,061	1,300
0900	40' deep excavation, 38 psf, left in place		21.20	3.019		1,225	119	156	1,500	1,725
1000	Drive, extract & salvage		12.25	5.224		505	207	270	982	1,200
1200	15' deep excavation, 22 psf, left in place		983	.065	S.F.	14.35	2.58	3.37	20.30	23.50
1300	Drive, extract & salvage		545	.117		5.65	4.64	6.10	16.39	20.50
1500	20' deep excavation, 27 psf, left in place		960	.067		18.05	2.64	3.45	24.14	28
1600	Drive, extract & salvage		485	.132		7.35	5.20	6.85	19.40	24
1800	25' deep excavation, 38 psf, left in place		1000	.064		26.50	2.53	3.31	32.34	37
1900	Drive, extract & salvage		553	.116		10.05	4.53	6	20.63	25
2100	Rent steel sheet piling and wales, first month				Ton	263			263	289
2200	Per added month					26.50			26.50	29
2300	Rental piling left in place, add to rental					980			980	1,075
2500	Wales, connections & struts, 2/3 salvage					275			275	300
2700	High strength piling, 50,000 psi, add					66.50			66.50	73
2800	55,000 psi, add					70.50			70.50	77.50
3000	Tie rod, not upset, 1-1/2" to 4" diameter with turnbuckle					2,200			2,200	2,400
3100	No turnbuckle					1,675			1,675	1,850
3300	Upset, 1-3/4" to 4" diameter with turnbuckle					2,450			2,450	2,700
3400	No turnbuckle					2,125			2,125	2,350
3600	Lightweight, 18" to 28" wide, 7 ga., 9.22 psf, and									
3610	9 ga., 8.6 psf, minimum				Lb.	.90			.90	.99
3700	Average					.98			.98	1.08
3750	Maximum					1.12			1.12	1.23
3900	Wood, solid sheeting, incl. wales, braces and spacers,									
3910	drive, extract & salvage, 8' deep excavation	B-31	330	.121	S.F.	1.47	4.08	.55	6.10	8.95
4000	10' deep, 50 S.F./hr. in & 150 S.F./hr. out		300	.133		1.52	4.49	.60	6.61	9.75

31 41 Shoring

31 41 16 – Sheet Piling

31 41 16.10 Sheet Piling Systems		Crew	Daily Output	Labor-Hours	Unit	2009 Bare Costs Material	Labor	Equipment	Total	Total Incl O&P
4100	12′ deep, 45 S.F./hr. in & 135 S.F./hr. out	B-31	270	.148	S.F.	1.56	4.99	.67	7.22	10.70
4200	14′ deep, 42 S.F./hr. in & 126 S.F./hr. out		250	.160		1.61	5.40	.72	7.73	11.45
4300	16′ deep, 40 S.F./hr. in & 120 S.F./hr. out		240	.167		1.66	5.60	.75	8.01	11.90
4400	18′ deep, 38 S.F./hr. in & 114 S.F./hr. out		230	.174		1.71	5.85	.79	8.35	12.40
4500	20′ deep, 35 S.F./hr. in & 105 S.F./hr. out		210	.190		1.77	6.40	.86	9.03	13.50
4520	Left in place, 8′ deep, 55 S.F./hr.		440	.091		2.65	3.06	.41	6.12	8.40
4540	10′ deep, 50 S.F./hr.		400	.100		2.79	3.37	.45	6.61	9.10
4560	12′ deep, 45 S.F./hr.		360	.111		2.95	3.74	.50	7.19	9.95
4565	14′ deep, 42 S.F./hr.		335	.119		3.12	4.02	.54	7.68	10.65
4570	16′ deep, 40 S.F./hr.		320	.125		3.32	4.21	.57	8.10	11.20
4580	18′ deep, 38 S.F./hr.		305	.131		3.54	4.42	.59	8.55	11.85
4590	20′ deep, 35 S.F./hr.		280	.143		3.79	4.81	.65	9.25	12.85
4700	Alternate pricing, left in place, 8′ deep		1.76	22.727	M.B.F.	595	765	103	1,463	2,050
4800	Drive, extract and salvage, 8′ deep		1.32	30.303	"	530	1,025	137	1,692	2,400
4990	Minimum labor/equipment charge		2	20	Job		675	90.50	765.50	1,200
5000	For treated lumber add cost of treatment to lumber									

31 43 Concrete Raising

31 43 13 – Pressure Grouting

31 43 13.13 Concrete Pressure Grouting		Crew	Daily Output	Labor-Hours	Unit	Material	Labor	Equipment	Total	Total Incl O&P
0010	**CONCRETE PRESSURE GROUTING**									
0020	Grouting, pressure, cement & sand, 1:1 mix, minimum	B-61	124	.323	Bag	12.40	10.80	2.64	25.84	34
0100	Maximum		51	.784	"	12.40	26.50	6.40	45.30	63.50
0200	Cement and sand, 1:1 mix, minimum		250	.160	C.F.	25	5.35	1.31	31.66	37.50
0300	Maximum		100	.400		37	13.40	3.27	53.67	66.50
0400	Epoxy cement grout, minimum		137	.292		109	9.80	2.39	121.19	139
0500	Maximum		57	.702		109	23.50	5.75	138.25	165
0600	Structural epoxy grout					82			82	90
0700	Alternate pricing method: (Add for materials)									
0710	5 person crew and equipment	B-61	1	40	Day		1,350	325	1,675	2,525

31 45 Vibroflotation and Densification

31 45 13 – Vibroflotation

31 45 13.10 Vibroflotation Densification		Crew	Daily Output	Labor-Hours	Unit	Material	Labor	Equipment	Total	Total Incl O&P
0010	**VIBROFLOTATION DENSIFICATION**									
0900	Vibroflotation compacted sand cylinder, minimum	B-60	750	.075	V.L.F.		2.71	2.66	5.37	7.30
0950	Maximum		325	.172			6.25	6.15	12.40	16.80
1100	Vibro replacement compacted stone cylinder, minimum		500	.112			4.07	3.99	8.06	10.90
1150	Maximum		250	.224			8.15	8	16.15	22
1300	Mobilization and demobilization, minimum		.47	119	Total		4,325	4,250	8,575	11,600
1400	Maximum		.14	400	"		14,500	14,300	28,800	39,000

31 46 Needle Beams

31 46 13 – Cantilever Needle Beams

31 46 13.10 Needle Beams		Crew	Daily Output	Labor-Hours	Unit	2009 Bare Costs Material	Labor	Equipment	Total	Total Incl O&P
0010	**NEEDLE BEAMS**									
0011	Incl. wood shoring 10′ x 10′ opening									
0400	Block, concrete, 8″ thick	B-9	7.10	5.634	Ea.	45.50	180	27	252.50	380
0420	12″ thick		6.70	5.970		52.50	191	28.50	272	405
0800	Brick, 4″ thick with 8″ backup block		5.70	7.018		52.50	225	33.50	311	465
1000	Brick, solid, 8″ thick		6.20	6.452		45.50	206	31	282.50	425
1040	12″ thick		4.90	8.163		52.50	261	39	352.50	530
1080	16″ thick		4.50	8.889		67	284	42.50	393.50	590
2000	Add for additional floors of shoring	B-1	6	4		45.50	129		174.50	263
9000	Minimum labor/equipment charge	"	2	12	Job		385		385	640

31 48 Underpinning

31 48 13 – Underpinning Piers

31 48 13.10 Underpinning Foundations

Line	Description	Crew	Daily Output	Labor-Hours	Unit	Material	Labor	Equipment	Total	Total Incl O&P
0010	**UNDERPINNING FOUNDATIONS**									
0011	Including excavation,									
0020	forming, reinforcing, concrete and equipment									
0100	5′ to 16′ below grade, 100 to 500 C.Y.	B-52	2.30	24.348	C.Y.	278	895	215	1,388	2,025
0200	Over 500 C.Y.		2.50	22.400		250	825	198	1,273	1,850
0400	16′ to 25′ below grade, 100 to 500 C.Y.		2	28		305	1,025	248	1,578	2,300
0500	Over 500 C.Y.		2.10	26.667		289	985	236	1,510	2,175
0700	26′ to 40′ below grade, 100 to 500 C.Y.		1.50	35		335	1,300	310	1,945	2,800
0800	Over 500 C.Y.		1.30	31.111		305	1,150	275	1,730	2,500
0900	For under 50 C.Y., add					10%	40%			
1000	For 50 C.Y. to 100 C.Y., add					5%	20%			

31 52 Cofferdams

31 52 16 – Timber Cofferdams

31 52 16.10 Cofferdams

Line	Description	Crew	Daily Output	Labor-Hours	Unit	Material	Labor	Equipment	Total	Total Incl O&P
0010	**COFFERDAMS**									
0011	Incl. mobilization and temporary sheeting									
0080	Soldier beams & lagging H piles with 3″ wood sheeting									
0090	horizontal between piles, including removal of wales & braces									
0100	No hydrostatic head, 15′ deep, 1 line of braces, minimum	B-50	545	.206	S.F.	6.95	7.75	4.05	18.75	25
0200	Maximum		495	.226		7.70	8.55	4.46	20.71	27.50
0400	15′ to 22′ deep with 2 lines of braces, 10″ H, minimum		360	.311		8.15	11.75	6.15	26.05	35
0500	Maximum		330	.339		9.25	12.80	6.70	28.75	38.50
0700	23′ to 35′ deep with 3 lines of braces, 12″ H, minimum		325	.345		10.65	13	6.80	30.45	40.50
0800	Maximum		295	.380		11.55	14.35	7.50	33.40	44.50
1000	36′ to 45′ deep with 4 lines of braces, 14″ H, minimum		290	.386		11.95	14.60	7.60	34.15	45.50
1100	Maximum		265	.423		12.60	15.95	8.35	36.90	49.50
1300	No hydrostatic head, left in place, 15′ dp., 1 line of braces, min.		635	.176		9.25	6.65	3.48	19.38	25
1400	Maximum		575	.195		9.90	7.35	3.84	21.09	27
1600	15′ to 22′ deep with 2 lines of braces, minimum		455	.246		13.85	9.30	4.85	28	36
1700	Maximum		415	.270		15.40	10.20	5.30	30.90	39.50
1900	23′ to 35′ deep with 3 lines of braces, minimum		420	.267		16.50	10.05	5.25	31.80	40.50
2000	Maximum		380	.295		18.25	11.15	5.80	35.20	45
2200	36′ to 45′ deep with 4 lines of braces, minimum		385	.291		19.80	11	5.75	36.55	46.50
2300	Maximum		350	.320		23	12.10	6.30	41.40	52.50

31 52 Cofferdams

31 52 16 – Timber Cofferdams

31 52 16.10 Cofferdams		Crew	Daily Output	Labor-Hours	Unit	Material	Labor	Equipment	Total	Total Incl O&P
						2009 Bare Costs				
2350	Lagging only, 3″ thick wood between piles 8′ O.C., minimum	B-46	400	.120	S.F.	1.54	4.25	.09	5.88	8.85
2370	Maximum		250	.192		2.31	6.80	.15	9.26	14
2400	Open sheeting no bracing, for trenches to 10′ deep, min.		1736	.028		.69	.98	.02	1.69	2.41
2450	Maximum		1510	.032		.77	1.12	.03	1.92	2.75
2500	Tie-back method, add to open sheeting, add, minimum								20%	20%
2550	Maximum								60%	60%
2700	Tie-backs only, based on tie-backs total length, minimum	B-46	86.80	.553	L.F.	13.30	19.55	.44	33.29	47.50
2750	Maximum		38.50	1.247	″	23.50	44	.98	68.48	100
3500	Tie-backs only, typical average, 25′ long		2	24	Ea.	585	850	18.95	1,453.95	2,100
3600	35′ long		1.58	30.380	″	780	1,075	24	1,879	2,675
4500	Trench box, 7′ deep, 16′ x 8′, see 01 54 33 in Reference Section				Day				165.60	182
4600	20′ x 10′, see 01 54 33 in Reference Section				″				298.68	329
5200	Wood sheeting, in trench, jacks at 4′ O.C., 8′ deep	B-1	800	.030	S.F.	.53	.97		1.50	2.19
5250	12′ deep		700	.034		.63	1.11		1.74	2.52
5300	15′ deep		600	.040		.87	1.29		2.16	3.08

31 56 Slurry Walls

31 56 23 – Lean Concrete Slurry Walls

31 56 23.20 Slurry Trench

0010	**SLURRY TRENCH**									
0011	Excavated slurry trench in wet soils									
0020	backfilled with 3000 PSI concrete, no reinforcing steel									
0050	Minimum	C-7	333	.216	C.F.	7.50	7.40	3.60	18.50	24
0100	Maximum		200	.360	″	12.55	12.30	6	30.85	40.50
0200	Alternate pricing method, minimum		150	.480	S.F.	14.95	16.45	8	39.40	52
0300	Maximum		120	.600		22.50	20.50	10	53	69
0500	Reinforced slurry trench, minimum	B-48	177	.316		11.20	11.15	16.15	38.50	48
0600	Maximum	″	69	.812		37.50	28.50	41.50	107.50	133
0800	Haul for disposal, 2 mile haul, excavated material, add	B-34B	99	.081	C.Y.		2.58	5.40	7.98	10.15
0900	Haul bentonite castings for disposal, add	″	40	.200	″		6.40	13.35	19.75	25

31 62 Driven Piles

31 62 13 – Concrete Piles

31 62 13.23 Prestressed Concrete Piles

0010	**PRESTRESSED CONCRETE PILES**, 200 piles									
0020	Unless specified otherwise, not incl. pile caps or mobilization									
2200	Precast, prestressed, 50′ long, 12″ diam., 2-3/8″ wall	B-19	720	.089	V.L.F.	19.60	3.52	2.45	25.57	30
2300	14″ diameter, 2-1/2″ wall		680	.094		26.50	3.72	2.59	32.81	38
2500	16″ diameter, 3″ wall		640	.100		37.50	3.96	2.76	44.22	51
2600	18″ diameter, 3″ wall	B-19A	600	.107		47.50	4.22	3.82	55.54	63.50
2800	20″ diameter, 3-1/2″ wall		560	.114		55.50	4.52	4.09	64.11	73
2900	24″ diameter, 3-1/2″ wall		520	.123		68.50	4.87	4.41	77.78	88.50
2920	36″ diameter, 4-1/2″ wall	B-19	400	.160		91	6.35	4.41	101.76	115
2940	54″ diameter, 5″ wall		340	.188		121	7.45	5.20	133.65	152
2960	66″ diameter, 6″ wall		220	.291		183	11.50	8	202.50	229
3100	Precast, prestressed, 40′ long, 10″ thick, square		700	.091		14.40	3.62	2.52	20.54	24.50
3200	12″ thick, square		680	.094		18.15	3.72	2.59	24.46	29
3400	14″ thick, square		600	.107		21.50	4.22	2.94	28.66	33.50
3500	Octagonal		640	.100		29.50	3.96	2.76	36.22	42

31 62 13 – Concrete Piles

31 62 13.23 Prestressed Concrete Piles		Crew	Daily Output	Labor-Hours	Unit	Material	2009 Bare Costs Labor	Equipment	Total	Total Incl O&P
3700	16" thick, square	B-19	560	.114	V.L.F.	34	4.52	3.15	41.67	48.50
3800	Octagonal		600	.107		35.50	4.22	2.94	42.66	49
4000	18" thick, square	B-19A	520	.123		41	4.87	4.41	50.28	58
4100	Octagonal	B-19	560	.114		42	4.52	3.15	49.67	57.50
4300	20" thick, square	B-19A	480	.133		50	5.25	4.78	60.03	69.50
4400	Octagonal	B-19	520	.123		46.50	4.87	3.39	54.76	62.50
4600	24" thick, square	B-19A	440	.145		72	5.75	5.20	82.95	94
4700	Octagonal	B-19	480	.133		67	5.25	3.67	75.92	86
4730	Precast, prestressed, 60' long, 10" thick, square		700	.091		15.15	3.62	2.52	21.29	25.50
4740	12" thick, square (60' long)		680	.094		18.85	3.72	2.59	25.16	29.50
4750	Mobilization for 10,000 L.F. pile job, add		3300	.019			.77	.53	1.30	1.84
4800	25,000 L.F. pile job, add		8500	.008			.30	.21	.51	.72

31 62 16 – Steel Piles

31 62 16.13 Sheet Steel Piles

		Crew	Daily Output	Labor-Hours	Unit	Material	Labor	Equipment	Total	Total Incl O&P
0010	**SHEET STEEL PILES**									
0100	Step tapered, round, concrete filled									
0110	8" tip, 60 ton capacity, 30' depth	B-19	760	.084	V.L.F.	10.15	3.33	2.32	15.80	19.15
0120	60' depth		740	.086		11.45	3.42	2.38	17.25	21
0130	80' depth		700	.091		11.85	3.62	2.52	17.99	21.50
0150	10" tip, 90 ton capacity, 30' depth		700	.091		12.45	3.62	2.52	18.59	22.50
0160	60' depth		690	.093		12.85	3.67	2.56	19.08	23
0170	80' depth		670	.096		13.85	3.78	2.63	20.26	24.50
0190	12" tip, 120 ton capacity, 30' depth		660	.097		16.60	3.84	2.67	23.11	27.50
0200	60' depth, 12" diameter		630	.102		16.65	4.02	2.80	23.47	28
0210	80' depth		590	.108		15.20	4.29	2.99	22.48	27
0250	"H" Sections, 50' long, HP8 x 36		640	.100		17.40	3.96	2.76	24.12	28.50
0400	HP10 X 42		610	.105		20.50	4.15	2.89	27.54	32.50
0500	HP10 X 57		610	.105		27.50	4.15	2.89	34.54	40.50
0700	HP12 X 53		590	.108		26	4.29	2.99	33.28	39
0800	HP12 X 74	B-19A	590	.108		36.50	4.29	3.89	44.68	51.50
1000	HP14 X 73		540	.119		36	4.69	4.25	44.94	52
1100	HP14 X 89		540	.119		44	4.69	4.25	52.94	61
1300	HP14 X 102		510	.125		50	4.96	4.50	59.46	68.50
1400	HP14 X 117		510	.125		57.50	4.96	4.50	66.96	76.50
1600	Splice on standard points, not in leads, 8" or 10"	1 Sswl	5	1.600	Ea.	114	71.50		185.50	257
1700	12" or 14"		4	2		166	89.50		255.50	350
1900	Heavy duty points, not in leads, 10" wide		4	2		176	89.50		265.50	360
2100	14" wide		3.50	2.286		228	102		330	440

31 62 19 – Timber Piles

31 62 19.10 Wood Piles

		Crew	Daily Output	Labor-Hours	Unit	Material	Labor	Equipment	Total	Total Incl O&P
0010	**WOOD PILES**									
0011	Friction or end bearing, not including									
0100	Untreated piles, up to 30' long, 12" butts, 8" points	B-19	625	.102	V.L.F.	7.25	4.05	2.82	14.12	17.65
0200	30' to 39' long, 12" butts, 8" points		700	.091		7.25	3.62	2.52	13.39	16.60
0300	40' to 49' long, 12" butts, 7" points		720	.089		7.25	3.52	2.45	13.22	16.40
0400	50' to 59' long, 13"butts, 7" points		800	.080		7.30	3.16	2.20	12.66	15.65
0500	60' to 69' long, 13" butts, 7" points		840	.076		8.25	3.01	2.10	13.36	16.30
0600	70' to 80' long, 13" butts, 6" points		840	.076		9.15	3.01	2.10	14.26	17.30
0800	Treated piles, 12 lb. per C.F.,									
0810	friction or end bearing, ASTM class B									
1000	Up to 30' long, 12" butts, 8" points	B-19	625	.102	V.L.F.	10.70	4.05	2.82	17.57	21.50
1100	30' to 39' long, 12" butts, 8" points		700	.091		10.45	3.62	2.52	16.59	20

31 62 Driven Piles

31 62 19 – Timber Piles

31 62 19.10 Wood Piles		Crew	Daily Output	Labor-Hours	Unit	2009 Bare Costs Material	Labor	Equipment	Total	Total Incl O&P
1200	40′ to 49′ long, 12″ butts, 7″ points	B-19	720	.089	V.L.F.	11.15	3.52	2.45	17.12	21
1300	50′ to 59′ long, 13″ butts, 7″ points		800	.080		12.35	3.16	2.20	17.71	21
1400	60′ to 69′ long, 13″ butts, 6″ points	B-19A	840	.076		16.85	3.01	2.73	22.59	26.50
1500	70′ to 80′ long, 13″ butts, 6″ points	″	840	.076		21	3.01	2.73	26.74	31.50
1600	Treated piles, C.C.A., 2.5# per C.F.									
1610	8″ butts, 10′ long	B-19	400	.160	V.L.F.	7.75	6.35	4.41	18.51	24
1620	11′ to 16′ long		500	.128		7.75	5.05	3.53	16.33	20.50
1630	17′ to 20′ long		575	.111		7.75	4.40	3.07	15.22	19.10
1640	10″ butts, 10′ to 16′ long		500	.128		8.85	5.05	3.53	17.43	22
1650	17′ to 20′ long		575	.111		8.85	4.40	3.07	16.32	20.50
1660	21′ to 40′ long		700	.091		8.85	3.62	2.52	14.99	18.40
1670	12″ butts, 10′ to 20′ long		575	.111		9.60	4.40	3.07	17.07	21
1680	21′ to 35′ long		650	.098		9.60	3.89	2.71	16.20	19.90
1690	36′ to 40′ long		700	.091		9.60	3.62	2.52	15.74	19.20
1695	14″ butts. to 40′ long		700	.091		13.70	3.62	2.52	19.84	24
1700	Boot for pile tip, minimum	1 Pile	27	.296	Ea.	22	11.40		33.40	43
1800	Maximum		21	.381		65.50	14.65		80.15	96.50
2000	Point for pile tip, minimum		20	.400		22	15.40		37.40	50
2100	Maximum		15	.533		78.50	20.50		99	121
2300	Splice for piles over 50′ long, minimum	B-46	35	1.371		54.50	48.50	1.08	104.08	142
2400	Maximum		20	2.400		65.50	85	1.90	152.40	215
2600	Concrete encasement with wire mesh and tube		331	.145	V.L.F.	10.25	5.15	.11	15.51	20
2700	Mobilization for 10,000 L.F. pile job, add	B-19	3300	.019			.77	.53	1.30	1.84
2800	25,000 L.F. pile job, add	″	8500	.008			.30	.21	.51	.72

31 62 23 – Composite Piles

31 62 23.13 Concrete Filled Steel Piles

		Crew	Daily Output	Labor-Hours	Unit	Material	Labor	Equipment	Total	Total Incl O&P
0010	**CONCRETE FILLED STEEL PILES** no mobilization or demobilization									
2600	Pipe piles, 50′ lg. 8″ diam., 29 lb. per L.F., no concrete	B-19	500	.128	V.L.F.	18.80	5.05	3.53	27.38	32.50
2700	Concrete filled		460	.139		19.80	5.50	3.83	29.13	35
2900	10″ diameter, 34 lb. per L.F., no concrete		500	.128		23.50	5.05	3.53	32.08	38
3000	Concrete filled		450	.142		25.50	5.60	3.92	35.02	41.50
3200	12″ diameter, 44 lb. per L.F., no concrete		475	.135		29	5.35	3.71	38.06	45
3300	Concrete filled		415	.154		29.50	6.10	4.25	39.85	47
3500	14″ diameter, 46 lb. per L.F., no concrete		430	.149		31	5.90	4.10	41	48
3600	Concrete filled		355	.180		32.50	7.15	4.97	44.62	53
3800	16″ diameter, 52 lb. per L.F., no concrete		385	.166		34.50	6.55	4.58	45.63	54
3900	Concrete filled		335	.191		37.50	7.55	5.25	50.30	59
4100	18″ diameter, 59 lb. per L.F., no concrete		355	.180		45.50	7.15	4.97	57.62	67
4200	Concrete filled		310	.206		44	8.15	5.70	57.85	68
4400	Splices for pipe piles, not in leads, 8″ diameter	1 Sswl	4.67	1.713	Ea.	84.50	76.50		161	235
4500	14″ diameter		3.79	2.111		111	94.50		205.50	297
4600	16″ diameter		3.03	2.640		137	118		255	370
4650	18″ diameter		4.50	1.778		181	79.50		260.50	345
4800	Points, standard, 8″ diameter		4.61	1.735		94.50	77.50		172	248
4900	14″ diameter		4.05	1.975		131	88.50		219.50	310
5000	16″ diameter		3.37	2.374		160	106		266	375
5050	18″ diameter		5	1.600		250	71.50		321.50	405
5200	Points, heavy duty, 10″ diameter		2.89	2.768		65.50	124		189.50	300
5300	14″ or 16″ diameter		2.02	3.960		105	177		282	445
5500	For reinforcing steel, add		1150	.007	Lb.	.78	.31		1.09	1.44
5700	For thick wall sections, add				″	.89			.89	.98
6020	Steel pipe pile end plates, 8″ diameter	1 Sswl	14	.571	Ea.	72	25.50		97.50	127

31 62 Driven Piles

31 62 23 – Composite Piles

31 62 23.13 Concrete Filled Steel Piles		Crew	Daily Output	Labor-Hours	Unit	2009 Bare Costs Material	Labor	Equipment	Total	Total Incl O&P
6050	10″ diameter	1 Sswl	14	.571	Ea.	75.50	25.50		101	131
6100	12″ diameter		12	.667		80	30		110	143
6150	14″ diameter		10	.800		83	36		119	158
6200	16″ diameter		9	.889		88.50	39.50		128	171
6250	18″ diameter		8	1		99.50	44.50		144	192
6300	Steel pipe pile shoes, 8″ diameter		12	.667		153	30		183	224
6350	10″ diameter		12	.667		159	30		189	230
6400	12″ diameter		10	.800		170	36		206	253
6450	14″ diameter		9	.889		196	39.50		235.50	289
6500	16″ diameter		8	1		217	44.50		261.50	320
6550	18″ diameter	↓	6	1.333	↓	243	59.50		302.50	380

31 62 33 – Drilled Micropiles

31 62 33.10 Drilled Micropiles Metal Pipe		Crew	Daily Output	Labor-Hours	Unit	Material	Labor	Equipment	Total	Total Incl O&P
0010	**DRILLED MICROPILES METAL PIPE**									
0011	No mobilization or demobilization									
5000	Pressure grouted pin pile, 5″ diam., cased, up to 50 ton,									
5040	End bearing, less than 20′	B-48	90	.622	V.L.F.	44	22	32	98	119
5080	More than 40′		135	.415		26	14.65	21	61.65	75.50
5120	Friction, loose sand and gravel		107	.523		44	18.45	26.50	88.95	108
5160	Dense sand and gravel		135	.415		26	14.65	21	61.65	75.50
5200	Uncased, up to 10 ton capacity, 20′	↓	135	.415	↓	26	14.65	21	61.65	75.50

31 63 Bored Piles

31 63 26 – Drilled Caissons

31 63 26.13 Fixed End Cassion Piles		Crew	Daily Output	Labor-Hours	Unit	Material	Labor	Equipment	Total	Total Incl O&P
0010	**FIXED END CASSION PILES**									
0015	Including excavation, concrete, 50 lbs reinforcing									
0020	per C.Y., not incl. mobilization, boulder removal, disposal									
0100	Open style, machine drilled, to 50′ deep, in stable ground, no									
0110	casings or ground water, 18″ diam., 0.065 C.Y./L.F.	B-43	200	.240	V.L.F.	9.05	8.30	12.65	30	37.50
0200	24″ diameter, 0.116 C.Y./L.F.		190	.253		16.20	8.75	13.35	38.30	46.50
0300	30″ diameter, 0.182 C.Y./L.F.		150	.320		25.50	11.10	16.90	53.50	64.50
0400	36″ diameter, 0.262 C.Y./L.F.		125	.384		36.50	13.30	20.50	70.30	84
0500	48″ diameter, 0.465 C.Y./L.F.		100	.480		65	16.60	25.50	107.10	127
0600	60″ diameter, 0.727 C.Y./L.F.		90	.533		102	18.45	28	148.45	173
0700	72″ diameter, 1.05 C.Y./L.F.		80	.600		147	21	31.50	199.50	231
0800	84″ diameter, 1.43 C.Y./L.F.	↓	75	.640	↓	200	22	34	256	293
1000	For bell excavation and concrete, add									
1020	4′ bell diameter, 24″ shaft, 0.444 C.Y.	B-43	20	2.400	Ea.	45	83	127	255	325
1040	6′ bell diameter, 30″ shaft, 1.57 C.Y.		5.70	8.421		159	292	445	896	1,150
1060	8′ bell diameter, 36″ shaft, 3.72 C.Y.		2.40	20		375	695	1,050	2,120	2,700
1080	9′ bell diameter, 48″ shaft, 4.48 C.Y.		2	24		450	830	1,275	2,555	3,250
1100	10′ bell diameter, 60″ shaft, 5.24 C.Y.		1.70	28.235		530	980	1,500	3,010	3,800
1120	12′ bell diameter, 72″ shaft, 8.74 C.Y.		1	48		885	1,650	2,525	5,060	6,450
1140	14′ bell diameter, 84″ shaft, 13.6 C.Y.	↓	.70	68.571	↓	1,375	2,375	3,625	7,375	9,325
1200	Open style, machine drilled, to 50′ deep, in wet ground, pulled									
1300	casing and pumping, 18″ diameter, 0.065 C.Y./L.F.	B-48	160	.350	V.L.F.	9.05	12.35	17.90	39.30	49.50
1400	24″ diameter, 0.116 C.Y./L.F.		125	.448		16.20	15.80	23	55	68.50
1500	30″ diameter, 0.182 C.Y./L.F.		85	.659		25.50	23	33.50	82	103
1600	36″ diameter, 0.262 C.Y./L.F.	↓	60	.933	↓	36.50	33	47.50	117	146

31 63 26 – Drilled Caissons

31 63 26.13 Fixed End Cassion Piles		Crew	Daily Output	Labor-Hours	Unit	Material	2009 Bare Costs Labor	Equipment	Total	Total Incl O&P
1700	48" diameter, 0.465 C.Y./L.F.	B-49	55	1.600	V.L.F.	65	58.50	66.50	190	240
1800	60" diameter, 0.727 C.Y./L.F.		35	2.514		102	92	104	298	375
1900	72" diameter, 1.05 C.Y./L.F.		30	2.933		147	108	122	377	470
2000	84" diameter, 1.43 C.Y./L.F.		25	3.520		200	129	146	475	590
2100	For bell excavation and concrete, add									
2120	4' bell diameter, 24" shaft, 0.444 C.Y.	B-48	19.80	2.828	Ea.	45	99.50	144	288.50	370
2140	6' bell diameter, 30" shaft, 1.57 C.Y.		5.70	9.825		159	345	500	1,004	1,275
2160	8' bell diameter, 36" shaft, 3.72 C.Y.		2.40	23.333		375	825	1,200	2,400	3,050
2180	9' bell diameter, 48" shaft, 4.48 C.Y.	B-49	3.30	26.667		450	975	1,100	2,525	3,300
2200	10' bell diameter, 60" shaft, 5.24 C.Y.		2.80	31.429		530	1,150	1,300	2,980	3,875
2220	12' bell diameter, 72" shaft, 8.74 C.Y.		1.60	55		885	2,025	2,275	5,185	6,725
2240	14' bell diameter, 84" shaft, 13.6 C.Y.		1	88		1,375	3,225	3,650	8,250	10,800
2300	Open style, machine drilled, to 50' deep, in soft rocks and									
2400	medium hard shales, 18" diameter, 0.065 C.Y./L.F.	B-49	50	1.760	V.L.F.	9.05	64.50	73	146.55	194
2500	24" diameter, 0.116 C.Y./L.F.		30	2.933		16.20	108	122	246.20	325
2600	30" diameter, 0.182 C.Y./L.F.		20	4.400		25.50	161	183	369.50	490
2700	36" diameter, 0.262 C.Y./L.F.		15	5.867		36.50	215	243	494.50	660
2800	48" diameter, 0.465 C.Y./L.F.		10	8.800		65	325	365	755	990
2900	60" diameter, 0.727 C.Y./L.F.		7	12.571		102	460	520	1,082	1,425
3000	72" diameter, 1.05 C.Y./L.F.		6	14.667		147	540	610	1,297	1,700
3100	84" diameter, 1.43 C.Y./L.F.		5	17.600		200	645	730	1,575	2,075
3200	For bell excavation and concrete, add									
3220	4' bell diameter, 24" shaft, 0.444 C.Y.	B-49	10.90	8.073	Ea.	45	296	335	676	900
3240	6' bell diameter, 30" shaft, 1.57 C.Y.		3.10	28.387		159	1,050	1,175	2,384	3,150
3260	8' bell diameter, 36" shaft, 3.72 C.Y.		1.30	67.692		375	2,475	2,800	5,650	7,525
3280	9' bell diameter, 48" shaft, 4.48 C.Y.		1.10	80		450	2,925	3,325	6,700	8,900
3300	10' bell diameter, 60" shaft, 5.24 C.Y.		.90	97.778		530	3,575	4,050	8,155	10,800
3320	12' bell diameter, 72" shaft, 8.74 C.Y.		.60	146		885	5,375	6,075	12,335	16,400
3340	14' bell diameter, 84" shaft, 13.6 C.Y.		.40	220		1,375	8,075	9,125	18,575	24,500
3600	For rock excavation, sockets, add, minimum		120	.733	C.F.		27	30.50	57.50	77
3650	Average		95	.926			34	38.50	72.50	97.50
3700	Maximum		48	1.833			67	76	143	193
3900	For 50' to 100' deep, add				V.L.F.				7%	7%
4000	For 100' to 150' deep, add								25%	25%
4100	For 150' to 200' deep, add								30%	30%
4200	For casings left in place, add				Lb.	.95			.95	1.05
4300	For other than 50 lb. reinf. per C.Y., add or deduct				"	1.01			1.01	1.11
4400	For steel "I" beam cores, add	B-49	8.30	10.602	Ton	2,400	390	440	3,230	3,775
4500	Load and haul excess excavation, 2 miles	B-34B	178	.045	L.C.Y.		1.44	2.99	4.43	5.65

31 63 26.16 Concrete Caissons for Marine Construction		Crew	Daily Output	Labor-Hours	Unit	Material	Labor	Equipment	Total	Total Incl O&P
0010	**CONCRETE CAISSONS FOR MARINE CONSTRUCTION**									
0100	Caissons, incl. mobilization and demobilization, up to 50 miles									
0200	Uncased shafts, 30 to 80 tons cap., 17" diam., 10' depth	B-44	88	.727	V.L.F.	20	28.50	19.35	67.85	90
0300	25' depth		165	.388		14.45	15.10	10.30	39.85	52
0400	80-150 ton capacity, 22" diameter, 10' depth		80	.800		25.50	31	21.50	78	103
0500	20' depth		130	.492		20	19.15	13.10	52.25	68
0700	Cased shafts, 10 to 30 ton capacity, 10-5/8" diam., 20' depth		175	.366		14.45	14.25	9.75	38.45	50
0800	30' depth		240	.267		13.45	10.35	7.10	30.90	39.50
0850	30 to 60 ton capacity, 12" diameter, 20' depth		160	.400		20	15.55	10.65	46.20	59
0900	40' depth		230	.278		15.55	10.80	7.40	33.75	43
1000	80 to 100 ton capacity, 16" diameter, 20' depth		160	.400		29	15.55	10.65	55.20	68.50
1100	40' depth		230	.278		27	10.80	7.40	45.20	55.50

31 63 Bored Piles

31 63 26 – Drilled Caissons

31 63 26.16 Concrete Caissons for Marine Construction		Crew	Daily Output	Labor-Hours	Unit	2009 Bare Costs Material	Labor	Equipment	Total	Total Incl O&P
1200	110 to 140 ton capacity, 17-5/8" diameter, 20' depth	B-44	160	.400	V.L.F.	31	15.55	10.65	57.20	71
1300	40' depth		230	.278		29	10.80	7.40	47.20	57.50
1400	140 to 175 ton capacity, 19" diameter, 20' depth		130	.492		33.50	19.15	13.10	65.75	83
1500	40' depth	↓	210	.305	↓	31	11.85	8.10	50.95	62.50
1700	Over 30' long, L.F. cost tends to be lower									
1900	Maximum depth is about 90'									

31 63 29 – Drilled Concrete Piers and Shafts

31 63 29.13 Uncased Drilled Concrete Piers

		Crew	Daily Output	Labor-Hours	Unit	Material	Labor	Equipment	Total	Total Incl O&P
0010	**UNCASED DRILLED CONCRETE PIERS**									
0020	Unless specified otherwise, not incl. pile caps or mobilization									
0100	Cast in place, thin wall shell pile, straight sided,									
0110	not incl. reinforcing, 8" diam., 16 ga., 5.8 lb./L.F.	B-19	700	.091	V.L.F.	9.25	3.62	2.52	15.39	18.85
0200	10" diameter, 16 ga. corrugated, 7.3 lb./L.F.		650	.098		12.15	3.89	2.71	18.75	22.50
0300	12" diameter, 16 ga. corrugated, 8.7 lb./L.F.		600	.107		15.75	4.22	2.94	22.91	27.50
0400	14" diameter, 16 ga. corrugated, 10.0 lb./L.F.		550	.116		18.55	4.60	3.21	26.36	31.50
0500	16" diameter, 16 ga. corrugated, 11.6 lb./L.F.	↓	500	.128	↓	22.50	5.05	3.53	31.08	37
0800	Cast in place friction pile, 50' long, fluted,									
0810	tapered steel, 4000 psi concrete, no reinforcing									
0900	12" diameter, 7 ga.	B-19	600	.107	V.L.F.	28	4.22	2.94	35.16	40.50
1000	14" diameter, 7 ga.		560	.114		30.50	4.52	3.15	38.17	44.50
1100	16" diameter, 7 ga.		520	.123		35.50	4.87	3.39	43.76	50.50
1200	18" diameter, 7 ga.	↓	480	.133	↓	41.50	5.25	3.67	50.42	58.50
1300	End bearing, fluted, constant diameter,									
1320	4000 psi concrete, no reinforcing									
1340	12" diameter, 7 ga.	B-19	600	.107	V.L.F.	29	4.22	2.94	36.16	42
1360	14" diameter, 7 ga.		560	.114		36.50	4.52	3.15	44.17	51
1380	16" diameter, 7 ga.		520	.123		42	4.87	3.39	50.26	58
1400	18" diameter, 7 ga.	↓	480	.133	↓	46.50	5.25	3.67	55.42	63.50

31 63 29.20 Cast In Place Piles, Adds

		Crew	Daily Output	Labor-Hours	Unit	Material	Labor	Equipment	Total	Total Incl O&P
0010	**CAST IN PLACE PILES, ADDS**									
1500	For reinforcing steel, add				Lb.	1.55			1.55	1.71
1700	For ball or pedestal end, add	B-19	11	5.818	C.Y.	149	230	160	539	715
1900	For lengths above 60', concrete, add	"	11	5.818	"	155	230	160	545	720
2000	For steel thin shell, pipe only				Lb.	1.05			1.05	1.16

Division Notes

		CREW	DAILY OUTPUT	LABOR-HOURS	UNIT	2009 BARE COSTS MAT.	LABOR	EQUIP.	TOTAL	TOTAL INCL O&P

Estimating Tips

32 01 00 Operations and Maintenance of Exterior Improvements

- Recycling of asphalt pavement is becoming very popular and is an alternative to removal and replacement of asphalt pavement. It can be a good value engineering proposal if removed pavement can be recycled either at the site or another site that is reasonably close to the project site.

32 10 00 Bases, Ballasts, and Paving

- When estimating paving, keep in mind the project schedule. If an asphaltic paving project is in a colder climate and runs through to the spring, consider placing the base course in the autumn and then topping it in the spring, just prior to completion. This could save considerable costs in spring repair. Keep in mind that prices for asphalt and concrete are generally higher in the cold seasons.

32 90 00 Planting

- The timing of planting and guarantee specifications often dictate the costs for establishing tree and shrub growth and a stand of grass or ground cover. Establish the work performance schedule to coincide with the local planting season. Maintenance and growth guarantees can add from 20%–100% to the total landscaping cost. The cost to replace trees and shrubs can be as high as 5% of the total cost, depending on the planting zone, soil conditions, and time of year.

Reference Numbers

Reference numbers are shown in shaded boxes at the beginning of some major classifications. These numbers refer to related items in the Reference Section. The reference information may be an estimating procedure, an alternate pricing method, or technical information.

Note: Not all subdivisions listed here necessarily appear in this publication.

32 01 13 – Flexible Paving Surface Treatment

32 01 13.61 Slurry Seal (Latex Modified)

	32 01 13.61 Slurry Seal (Latex Modified)	Crew	Daily Output	Labor-Hours	Unit	2009 Bare Costs Material	Labor	Equipment	Total	Total Incl O&P
0010	**SLURRY SEAL (LATEX MODIFIED)**									
3600	Waterproofing, membrane, tar and fabric, small area	B-63	233	.172	S.Y.	11.60	5.70	.64	17.94	22.50
3640	Large area		1435	.028		10.65	.92	.10	11.67	13.35
3680	Preformed rubberized asphalt, small area		100	.400		15.70	13.25	1.49	30.44	40.50
3720	Large area		367	.109		14.30	3.61	.41	18.32	22
3780	Rubberized asphalt (latex) seal	B-45	5000	.003		2.26	.12	.16	2.54	2.85
9000	Minimum labor/equipment charge	1 Clab	2	4	Job		126		126	209

32 01 13.62 Asphalt Surface Treatment

	32 01 13.62 Asphalt Surface Treatment	Crew	Daily Output	Labor-Hours	Unit	Material	Labor	Equipment	Total	Total Incl O&P
0010	**ASPHALT SURFACE TREATMENT**									
3000	Pavement overlay, polypropylene									
3040	6 oz. per S.Y., ideal conditions	B-63	10000	.004	S.Y.	1.82	.13	.01	1.96	2.24
3080	Adverse conditions		1000	.040		2.44	1.32	.15	3.91	5
3120	4 oz. per S.Y., ideal conditions		10000	.004		1.25	.13	.01	1.39	1.62
3160	Adverse conditions		1000	.040		1.61	1.32	.15	3.08	4.09
3200	Tack coat, emulsion, .05 gal per S.Y., 1000 S.Y	B-45	2500	.006		.46	.23	.31	1	1.23
3240	10,000 S.Y.		10000	.002		.37	.06	.08	.51	.58
3270	.10 gal per S.Y., 1000 S.Y.		2500	.006		.86	.23	.31	1.40	1.67
3275	10,000 S.Y.		10000	.002		.86	.06	.08	1	1.13
3280	.15 gal per S.Y., 1000 S.Y.		2500	.006		1.27	.23	.31	1.81	2.11
3320	10,000 S.Y.		10000	.002		1.01	.06	.08	1.15	1.29

32 01 13.64 Sand Seal

	32 01 13.64 Sand Seal	Crew	Daily Output	Labor-Hours	Unit	Material	Labor	Equipment	Total	Total Incl O&P
0010	**SAND SEAL**									
2080	Sand sealing, sharp sand, asphalt emulsion, small area	B-91	10000	.006	S.Y.	1.24	.24	.22	1.70	1.98
2120	Roadway or large area	"	18000	.004	"	1.07	.13	.12	1.32	1.51
3000	Sealing random cracks, min 1/2" wide, to 1-1/2", 1,000 L.F.	B-77	2800	.014	L.F.	.98	.46	.18	1.62	2.02
3040	10,000 L.F.		4000	.010	"	.75	.32	.12	1.19	1.50
3080	Alternate method, 1,000 L.F.		200	.200	Gal.	14.65	6.35	2.46	23.46	29.50
3120	10,000 L.F.		325	.123	"	11.80	3.92	1.51	17.23	21
3200	Multi-cracks (flooding), 1 coat, small area	B-92	460	.070	S.Y.	4.41	2.23	1.09	7.73	9.75
3240	Large area		2850	.011		3.92	.36	.18	4.46	5.10
3280	2 coat, small area		230	.139		14.50	4.47	2.19	21.16	25.50
3320	Large area		1425	.022		13.20	.72	.35	14.27	16.10
3360	Alternate method, small area		115	.278	Gal.	15.15	8.95	4.37	28.47	36
3400	Large area		715	.045	"	14.25	1.44	.70	16.39	18.80

32 01 13.66 Fog Seal

	32 01 13.66 Fog Seal	Crew	Daily Output	Labor-Hours	Unit	Material	Labor	Equipment	Total	Total Incl O&P
0010	**FOG SEAL**									
0012	Sealcoating, 2 coat coal tar pitch emulsion over 10,000 SY	B-45	5000	.003	S.Y.	1	.12	.16	1.28	1.46
0030	1000 to 10,000 S.Y.	"	3000	.005		1	.20	.26	1.46	1.70
0100	Under 1000 S.Y.	B-1	1050	.023		1	.74		1.74	2.32
0300	Petroleum resistant, over 10,000 S.Y.	B-45	5000	.003		1.24	.12	.16	1.52	1.72
0320	1000 to 10,000 S.Y.	"	3000	.005		1.24	.20	.26	1.70	1.96
0400	Under 1000 S.Y.	B-1	1050	.023		1.24	.74		1.98	2.58
0600	Non-skid pavement renewal, over 10,000 S.Y.	B-45	5000	.003		1.43	.12	.16	1.71	1.93
0620	1000 to 10,000 S.Y.	"	3000	.005		1.43	.20	.26	1.89	2.17
0700	Under 1000 S.Y.	B-1	1050	.023		1.43	.74		2.17	2.79
0800	Prepare and clean surface for above	A-2	8545	.003			.09	.02	.11	.18
1000	Hand seal asphalt curbing	B-1	4420	.005	L.F.	.73	.18		.91	1.09
1900	Asphalt surface treatment, single course, small area									
1901	0.30 gal/S.Y. asphalt material, 20#/S.Y. aggregate	B-91	5000	.013	S.Y.	1.70	.47	.44	2.61	3.11
1910	Roadway or large area		10000	.006		1.56	.24	.22	2.02	2.34
1950	Asphalt surface treatment, dbl. course for small area		3000	.021		3.15	.78	.73	4.66	5.55
1960	Roadway or large area		6000	.011		2.84	.39	.36	3.59	4.15

32 01 13 – Flexible Paving Surface Treatment

32 01 13.66 Fog Seal		Crew	Daily Output	Labor-Hours	Unit	2009 Bare Costs Material	Labor	Equipment	Total	Total Incl O&P
1980	Asphalt surface treatment,single course, for shoulders	B-91	7500	.009	S.Y.	1.81	.31	.29	2.41	2.81

32 01 13.68 Slurry Seal

		Crew	Daily Output	Labor-Hours	Unit	Material	Labor	Equipment	Total	Total Incl O&P
0010	**SLURRY SEAL**									
0100	Slurry seal, type I, 8 lbs agg./S.Y., 1 coat, small or irregular area	B-90	2800	.023	S.Y.	1.48	.77	.78	3.03	3.74
0150	Roadway or large area		10000	.006		1.48	.22	.22	1.92	2.22
0200	Type II, 12 lbs aggregate/S.Y., 2 coats,small or irregular area		2000	.032		2.96	1.08	1.09	5.13	6.20
0250	Roadway or large area		8000	.008		2.96	.27	.27	3.50	4
0300	Type III, 20 lbs aggregate/S.Y., 2 coats, small or irregular area		1800	.036		3.49	1.20	1.21	5.90	7.10
0350	Roadway or large area		6000	.011		3.49	.36	.36	4.21	4.83
0400	Slurry seal, thermoplastic coal-tar, type I, small or irregular area		2400	.027		3.08	.90	.91	4.89	5.85
0450	Roadway or large area		8000	.008		3.08	.27	.27	3.62	4.13
0500	Type II, small or irregular area		2400	.027		3.94	.90	.91	5.75	6.80
0550	Roadway or large area		7800	.008		3.94	.28	.28	4.50	5.10

32 01 16 – Flexible Paving Rehabilitation

32 01 16.71 Cold Milling Asphalt Paving

		Crew	Daily Output	Labor-Hours	Unit	Material	Labor	Equipment	Total	Total Incl O&P
0010	**COLD MILLING ASPHALT PAVING**									
5200	Cold planing & cleaning, 1" to 3" asphalt pavmt., over 25,000 S.Y.	B-71	6000	.009	S.Y.		.34	1.07	1.41	1.72
5280	5,000 S.Y. to 10,000 S.Y.	"	4000	.014	"		.50	1.61	2.11	2.58
5300	Asphalt pavement removal from conc. base, no haul									
5320	Rip, load & sweep 1" to 3"	B-70	8000	.007	S.Y.		.25	.20	.45	.62
5330	3" to 6" deep	"	5000	.011			.40	.31	.71	.99
5340	Profile grooving, asphalt pavement load & sweep, 1" deep	B-71	12500	.004			.16	.51	.67	.83
5350	3" deep		9000	.006			.22	.71	.93	1.15
5360	6" deep		5000	.011			.40	1.29	1.69	2.07

32 01 16.73 In Place Cold Reused Asphalt Paving

		Crew	Daily Output	Labor-Hours	Unit	Material	Labor	Equipment	Total	Total Incl O&P
0010	**IN PLACE COLD REUSED ASPHALT PAVING**									
5000	Reclamation, pulverizing and blending with existing base									
5040	Aggregate base, 4" thick pavement, over 15,000 S.Y.	B-73	2400	.027	S.Y.		1.01	2.04	3.05	3.87
5080	5,000 S.Y. to 15,000 S.Y.		2200	.029			1.10	2.23	3.33	4.21
5120	8" thick pavement, over 15,000 S.Y.		2200	.029			1.10	2.23	3.33	4.21
5160	5,000 S.Y. to 15,000 S.Y.		2000	.032			1.21	2.45	3.66	4.64

32 01 16.74 In Place Hot Reused Asphalt Paving

		Crew	Daily Output	Labor-Hours	Unit	Material	Labor	Equipment	Total	Total Incl O&P
0010	**IN PLACE HOT REUSED ASPHALT PAVING**									
5500	Recycle asphalt pavement at site									
5520	Remove, rejuvenate and spread 4" deep G	B-72	2500	.026	S.Y.	3.85	.94	4.34	9.13	10.50
5521	6" deep G	"	2000	.032	"	5.65	1.18	5.45	12.28	14.05

32 01 30 – Operation and Maintenance of Site Improvements

32 01 30.10 Site Maintenance

		Crew	Daily Output	Labor-Hours	Unit	Material	Labor	Equipment	Total	Total Incl O&P
0010	**SITE MAINTENANCE** R019313-10									
0800	Flower bed maintenance									
0810	Cultivate bed, no mulch	1 Clab	14	.571	M.S.F.		18.05		18.05	30
0830	Fall clean-up of flower bed, including pick-up mulch for re-use		1	8			253		253	415
0840	Fertilize flower bed, dry granular 3 lb./c.s.f.		85	.094		11.60	2.97		14.57	17.65
1000	Mulch - see Div. 32 91 13.16									
1100	Plant bed preparation see Div. 32 91 13.16									
1110	Plant from flats	1 Clab	800	.010	S.F.		.32		.32	.52
1120	Planting, ground cover see Div. 32 93 13.20									
1130	Police, hand pickup	1 Clab	30	.267	M.S.F.		8.45		8.45	13.90
1140	Vacuum (outside)		48	.167			5.25		5.25	8.70
1200	Spring prepare		2	4			126		126	209
1300	Weed mulched bed		20	.400			12.65		12.65	21

32 01 30 – Operation and Maintenance of Site Improvements

32 01 30.10 Site Maintenance		Crew	Daily Output	Labor-Hours	Unit	Material	2009 Bare Costs Labor	Equipment	Total	Total Incl O&P
1310	Unmulched bed	1 Clab	8	1	M.S.F.		31.50		31.50	52
1550	General site work maintenance									
1560	Clearing brush with brush saw & rake	1 Clab	565	.014	S.Y.		.45		.45	.74
1570	By hand	"	280	.029			.90		.90	1.49
1580	With dozer, ball and chain, light clearing	B-11A	3675	.004			.16	.29	.45	.57
1590	Medium clearing	"	3110	.005			.19	.35	.54	.68
1600	Ground cover planting see Div. 32 93 13.20									
1610	Grounds, policing, incl. picking up trash & other debris	1 Clab	300	.027	M.S.F.		.84		.84	1.39
1750	Site improvements, (bleachers, playground equipment,									
3000	Lawn maintenance									
3010	Aerate lawn, 18" cultivating width, walk behind	A-1K	95	.084	M.S.F.		2.66	.96	3.62	5.45
3040	48" cultivating width	B-66	750	.011			.42	.31	.73	1
3060	72" cultivating width	"	1100	.007			.28	.21	.49	.68
3100	Edge lawn, by hand at walks	1 Clab	16	.500	C.L.F.		15.80		15.80	26
3150	At planting beds		7	1.143			36		36	59.50
3200	Using gas powered edger at walks		88	.091			2.87		2.87	4.74
3250	At planting beds		24	.333			10.55		10.55	17.40
3260	Vacuum, 30" gas, outdoors with hose		96	.083	M.L.F.		2.63		2.63	4.35
3350	Edging material for lawns & planting beds see Div. 32 94 13.20									
3400	Weed lawn, by hand	1 Clab	3	2.667	M.S.F.		84.50		84.50	139
3800	Lawn bed preparation see Div. 32 91 13.23									
4500	Rake leaves or lawn, by hand	1 Clab	7.50	1.067	M.S.F.		33.50		33.50	55.50
4510	Power rake	"	45	.178	"		5.60		5.60	9.25
4700	Seeding lawn, see Div. 32 92 19.14									
4750	Sodding, see Div. 32 92 23.10									
5900	Road & walk maintenance									
5910	Asphaltic concrete paving, cold patch, 2" thick	B-37	350	.137	S.Y.	6.60	4.55	.39	11.54	15.15
5913	3" thick	"	260	.185	"	9.45	6.15	.53	16.13	21
5915	De-icing roads and walks									
5920	Calcium Chloride in truckload lots see Div. 31 32 13.30									
6000	Ice melting comp., 90% Calc. Chlor., effec. to -30° F									
6010	50-80 lb. poly bags, med. applic. 19 lbs./M.S.F., by hand	1 Clab	60	.133	M.S.F.	18.60	4.21		22.81	27.50
6050	With hand operated rotary spreader		110	.073		18.60	2.30		20.90	24.50
6100	Rock salt, med. applic. on road & walkway, by hand		60	.133		4.66	4.21		8.87	12.10
6110	With hand operated rotary spreader		110	.073		4.66	2.30		6.96	8.95
6130	Hosing, sidewalks & other paved areas		30	.267			8.45		8.45	13.90
6150	Painting lines on pavement see Div. 32 17 23.13									
6250	Seal coating bituminous surfaces see Div. 32 01 13.61									
6260	Sidewalk, brick pavers, steam cleaning	A-1H	950	.008	S.F.	.05	.27	.05	.37	.56
6400	Sweep walk by hand	1 Clab	15	.533	M.S.F.		16.85		16.85	28
6410	Power vacuum		100	.080			2.53		2.53	4.17
6420	Drives & parking areas with power vacuum		120	.067			2.11		2.11	3.48
6600	Shrub maintenance									
6640	Shrub bed fertilize dry granular 3 lbs./M.S.F.	1 Clab	85	.094	M.S.F.	1.11	2.97		4.08	6.15
6800	Weed, by handhoe		8	1			31.50		31.50	52
6810	Spray out		32	.250			7.90		7.90	13.05
6820	Spray after mulch		48	.167			5.25		5.25	8.70
7100	Tree maintenance									
7140	Clear and grub trees, see Div. 31 11 10.10									
7160	Cutting and piling trees, see Div. 31 13 13.20									
7200	Fertilize, tablets, slow release, 30 gram/tree	1 Clab	100	.080	Ea.	.43	2.53		2.96	4.64
7280	Guying, including stakes, guy wire & wrap, see Div. 32 94 50.10									
7300	Planting, trees, Deciduous, in prep. beds, see Div. 32 93 43.20									

32 01 30 – Operation and Maintenance of Site Improvements

32 01 30.10 Site Maintenance

	32 01 30.10 Site Maintenance	Crew	Daily Output	Labor-Hours	Unit	2009 Bare Costs Material	Labor	Equipment	Total	Total Incl O&P
7400	Removal, trees see Div. 32 96 43.20									
7420	Pest control, spray	1 Clab	24	.333	Ea.	19.80	10.55		30.35	39.50
7430	Systemic	"	48	.167	"	19.95	5.25		25.20	30.50

32 01 30.20 Snow Removal

	32 01 30.20 Snow Removal	Crew	Daily Output	Labor-Hours	Unit	Material	Labor	Equipment	Total	Total Incl O&P
0010	**SNOW REMOVAL**									
0020	Plowing, 12 ton truck, 2"-4" deep	B-34A	250	.032	M.S.F.		1.02	1.31	2.33	3.12
0040	4"-10" deep		200	.040			1.28	1.63	2.91	3.89
0060	10"-15" deep		150	.053			1.70	2.18	3.88	5.20
0080	Pickup truck, 2"-4" deep	A-3A	175	.046			1.41	.73	2.14	3.13
0100	4"-10" deep		130	.062			1.90	.99	2.89	4.21
0120	10"-15" deep		75	.107			3.30	1.71	5.01	7.30
0140	Load and haul snow, 1 mile round trip	A-3B	230	.070	C.Y.		2.55	4.04	6.59	8.50
0160	2 mile round trip		175	.091			3.35	5.30	8.65	11.20
0180	3 mile round trip		150	.107			3.91	6.20	10.11	13.05
0200	4 mile round trip		130	.123			4.51	7.15	11.66	15.05
0220	5 mile round trip		120	.133			4.89	7.75	12.64	16.35
0240	Clearing with wheeled skid steer loader, 1 C.Y.	A-3C	80	.100			3.91	3.25	7.16	9.70
0260	Spread sand and salt mix	B-34A	375	.021	M.S.F.	5.55	.68	.87	7.10	8.20
0280	Sidewalks and drives, by hand	1 Clab	780	.010	C.F.		.32		.32	.54
0300	Power, 24" blower	A-1M	1200	.007	"		.21	.04	.25	.40
0320	2"-4" deep, single driveway (10' x 50')		8	1	Ea.		31.50	6.60	38.10	59.50
0340	Double driveway (20' x 50')		4.50	1.778			56	11.70	67.70	105
0360	4"-10" deep, single driveway		6	1.333			42	8.80	50.80	79
0380	Double driveway		3.25	2.462			78	16.20	94.20	146
0400	10"-15" deep, single driveway		4	2			63	13.20	76.20	119
0420	Double driveway		2.25	3.556			112	23.50	135.50	211
0440	For heavy wet snow, add								20%	
9000	Minimum labor and equipment charge	A-3A	2	4	Job		124	64	188	274

32 01 90 – Operation and Maintenance of Planting

32 01 90.13 Fertilizing

	32 01 90.13 Fertilizing	Crew	Daily Output	Labor-Hours	Unit	Material	Labor	Equipment	Total	Total Incl O&P
0010	**FERTILIZING**									
0100	Dry granular, 4#/M.S.F., hand spread	1 Clab	24	.333	M.S.F.	2.22	10.55		12.77	19.85
0110	Push rotary		140	.057	"	2.22	1.81		4.03	5.40
0112	Push rotary, per 1076 feet squared		130	.062	Ea.	2.22	1.94		4.16	5.65
0120	Tractor towed spreader, 8'	B-66	500	.016	M.S.F.	2.22	.62	.46	3.30	3.93
0130	12' spread		800	.010		2.22	.39	.29	2.90	3.37
0140	Truck whirlwind spreader		1200	.007		2.22	.26	.19	2.67	3.06
0180	Water soluable, hydro spread, 1.5 # /MSF	B-64	600	.027		2.25	.83	.55	3.63	4.46
0190	Add for weed control					.38			.38	.42

32 01 90.19 Mowing

	32 01 90.19 Mowing	Crew	Daily Output	Labor-Hours	Unit	Material	Labor	Equipment	Total	Total Incl O&P
0010	**MOWING**									
1650	Mowing brush, tractor with rotary mower									
1660	Light density	B-84	22	.364	M.S.F.		15.05	12.45	27.50	37
1670	Medium density		13	.615			25.50	21	46.50	63
1680	Heavy density		9	.889			37	30.50	67.50	91.50
2000	Mowing, brush/grass, tractor,rotary mower, highway/airport median		13	.615			25.50	21	46.50	63
2010	Traffic safety flashing truck for highway/airport median mowing	A-2B	1	8	Day		248	194	442	620
4000	Lawn mowing, improved areas, 16" hand push	1 Clab	48	.167	M.S.F.		5.25		5.25	8.70
4050	Power mower, 18" - 22"		65	.123			3.89		3.89	6.40
4100	22" - 30"		110	.073			2.30		2.30	3.79
4150	30" - 32"		140	.057			1.81		1.81	2.98
4160	Riding mower, 36" - 44"	B-66	300	.027			1.04	.77	1.81	2.49

32 01 90 – Operation and Maintenance of Planting

32 01 90.19 Mowing		Crew	Daily Output	Labor-Hours	Unit	Material	2009 Bare Costs Labor	Equipment	Total	Total Incl O&P
4170	48" - 58"	B-66	480	.017	M.S.F.		.65	.48	1.13	1.55
4175	Mowing with tractor & attachments									
4180	3 gang reel, 7'	B-66	930	.009	M.S.F.		.34	.25	.59	.80
4190	5 gang reel, 12'		1200	.007			.26	.19	.45	.62
4200	Cutter or sickle-bar, 5', rough terrain		210	.038			1.49	1.10	2.59	3.55
4210	Cutter or sickle-bar, 5', smooth terrain		340	.024			.92	.68	1.60	2.20
4220	Drainage channel, 5' sickle bar		5	1.600	Mile		62.50	46	108.50	150
4250	Lawnmower, rotary type, sharpen (all sizes)	1 Clab	10	.800	Ea.		25.50		25.50	41.50
4260	Repair or replace part		7	1.143	"		36		36	59.50
5000	Edge trimming with weed whacker		5760	.001	L.F.		.04		.04	.07

32 01 90.23 Pruning		Crew	Daily Output	Labor-Hours	Unit	Material	Labor	Equipment	Total	Total Incl O&P
0010	**PRUNING**									
0020	1-1/2" caliper	1 Clab	84	.095	Ea.		3.01		3.01	4.97
0030	2" caliper		70	.114			3.61		3.61	5.95
0040	2-1/2" caliper		50	.160			5.05		5.05	8.35
0050	3" caliper		30	.267			8.45		8.45	13.90
0060	4" caliper, by hand	2 Clab	21	.762			24		24	39.50
0070	Aerial lift equipment	B-85	38	1.053			35.50	23	58.50	83
0100	6" caliper, by hand	2 Clab	12	1.333			42		42	69.50
0110	Aerial lift equipment	B-85	20	2			67	43.50	110.50	158
0200	9" caliper, by hand	2 Clab	7.50	2.133			67.50		67.50	111
0210	Aerial lift equipment	B-85	12.50	3.200			108	70	178	252
0300	12" caliper, by hand	2 Clab	6.50	2.462			78		78	128
0310	Aerial lift equipment	B-85	10.80	3.704			125	81	206	292
0400	18" caliper by hand	2 Clab	5.60	2.857			90.50		90.50	149
0410	Aerial lift equipment	B-85	9.30	4.301			145	94	239	340
0500	24" caliper, by hand	2 Clab	4.60	3.478			110		110	181
0510	Aerial lift equipment	B-85	7.70	5.195			175	113	288	410
0600	30" caliper, by hand	2 Clab	3.70	4.324			137		137	226
0610	Aerial lift equipment	B-85	6.20	6.452			217	141	358	510
0700	36" caliper, by hand	2 Clab	2.70	5.926			187		187	310
0710	Aerial lift equipment	B-85	4.50	8.889			299	194	493	700
0800	48" caliper, by hand	2 Clab	1.70	9.412			297		297	490
0810	Aerial lift equipment	B-85	2.80	14.286			480	310	790	1,125

32 01 90.24 Shrub Pruning		Crew	Daily Output	Labor-Hours	Unit	Material	Labor	Equipment	Total	Total Incl O&P
0010	**SHRUB PRUNING**									
6700	Prune, shrub bed	1 Clab	7	1.143	M.S.F.		36		36	59.50
6710	Shrub under 3' height		190	.042	Ea.		1.33		1.33	2.20
6720	4' height		90	.089			2.81		2.81	4.64
6730	Over 6'		50	.160			5.05		5.05	8.35
7350	Prune trees from ground		20	.400			12.65		12.65	21
7360	High work		8	1			31.50		31.50	52

32 01 90.26 Watering		Crew	Daily Output	Labor-Hours	Unit	Material	Labor	Equipment	Total	Total Incl O&P
0010	**WATERING**									
4900	Water lawn or planting bed with hose, 1" of water	1 Clab	16	.500	M.S.F.		15.80		15.80	26
4910	50' soaker hoses, in place		82	.098			3.08		3.08	5.10
4920	60' soaker hoses, in place		89	.090			2.84		2.84	4.69
7500	Water trees or shrubs, under 1" caliper		32	.250	Ea.		7.90		7.90	13.05
7550	1" - 3" caliper		17	.471			14.85		14.85	24.50
7600	3" - 4" caliper		12	.667			21		21	35
7650	Over 4" caliper		10	.800			25.50		25.50	41.50
9000	For sprinkler irrigation systems, see Div. 32 84 23.10									

32 01 Operation and Maintenance of Exterior Improvements

32 01 90 – Operation and Maintenance of Planting

32 01 90.29 Topsoil Preservation		Crew	Daily Output	Labor-Hours	Unit	2009 Bare Costs Material	Labor	Equipment	Total	Total Incl O&P
0010	**TOPSOIL PRESERVATION**									
0100	Weed planting bed	1 Clab	800	.010	S.Y.		.32		.32	.52

32 06 Schedules for Exterior Improvements

32 06 10 – Schedules for Bases, Ballasts, and Paving

32 06 10.10 Sidewalks, Driveways and Patios

		Crew	Daily Output	Labor-Hours	Unit	Material	Labor	Equipment	Total	Total Incl O&P
0010	**SIDEWALKS, DRIVEWAYS AND PATIOS** No base									
0020	Asphaltic concrete, 2" thick	B-37	720	.067	S.Y.	6.05	2.21	.19	8.45	10.50
0100	2-1/2" thick	"	650	.073	"	7.65	2.41	.21	10.27	12.60
0110	Bedding for brick or stone, mortar, 1" thick	D-1	300	.053	S.F.	.77	1.94		2.71	3.98
0120	2" thick	"	200	.080		1.92	2.91		4.83	6.80
0130	Sand, 2" thick	B-18	8000	.003		.25	.10		.35	.44
0140	4" thick	"	4000	.006	↓	.49	.19	.01	.69	.87
0300	Concrete, 3000 psi, CIP, 6 x 6 - W1.4 x W1.4 mesh,									
0310	broomed finish, no base, 4" thick	B-24	600	.040	S.F.	1.68	1.46		3.14	4.22
0350	5" thick		545	.044		2.24	1.61		3.85	5.10
0400	6" thick	↓	510	.047		2.62	1.72		4.34	5.65
0450	For bank run gravel base, 4" thick, add	B-18	2500	.010		.64	.31	.02	.97	1.23
0520	8" thick, add	"	1600	.015		1.28	.48	.02	1.78	2.24
0550	Exposed aggregate finish, add to above, minimum	B-24	1875	.013		.15	.47		.62	.92
0600	Maximum	"	455	.053	↓	.49	1.93		2.42	3.66
0850	Splash block, precast concrete	1 Clab	150	.053	Ea.	9.85	1.69		11.54	13.65
0950	Concrete tree grate, 5' square	B-6	25	.960		420	32.50	11.75	464.25	530
0960	Cast iron tree grate with frame, 2 piece, round, 5' diameter		25	.960		1,175	32.50	11.75	1,219.25	1,350
0980	Square, 5' side	↓	25	.960	↓	1,175	32.50	11.75	1,219.25	1,350
1000	Crushed stone, 1" thick, white marble	2 Clab	1700	.009	S.F.	.22	.30		.52	.73
1050	Bluestone		1700	.009		.24	.30		.54	.76
1070	Granite chips	↓	1700	.009		.23	.30		.53	.74
1660	Limestone pavers, 3" thick	D-1	72	.222		9.30	8.05		17.35	23.50
1670	4" thick		70	.229		12.35	8.30		20.65	27
1680	5" thick	↓	68	.235		15.45	8.55		24	31
1700	Redwood, prefabricated, 4' x 4' sections	2 Carp	316	.051		8.25	2.02		10.27	12.45
1750	Redwood planks, 1" thick, on sleepers	"	240	.067		5.80	2.65		8.46	10.75
1830	1-1/2" thick	B-28	167	.144		4.33	5.35		9.68	13.55
1840	2" thick		167	.144		5.60	5.35		10.95	14.95
1850	3" thick		150	.160		8.25	5.95		14.20	18.85
1860	4" thick		150	.160		10.80	5.95		16.75	21.50
1870	5" thick	↓	150	.160	↓	13.75	5.95		19.70	25
2100	River or beach stone, stock	B-1	18	1.333	Ton	30	43		73	104
2150	Quarried	"	18	1.333	"	52.50	43		95.50	129
2160	Load, dump, and spread stone with skid steer, 100' haul	B-62	24	1	C.Y.		34	6.20	40.20	62.50
2165	200' haul		18	1.333			45.50	8.30	53.80	82.50
2168	300' haul	↓	12	2	↓		68	12.45	80.45	125
2170	Shale paver, 2-1/4" thick	D-1	200	.080	S.F.	3.10	2.91		6.01	8.10
2200	Coarse washed sand bed, 1"	B-62	1350	.018	S.Y.	1.36	.61	.11	2.08	2.60
2250	Stone dust, 4" thick	"	900	.027	"	3.33	.91	.17	4.41	5.30
2300	Tile thinset pavers, 3/8" thick	D-1	300	.053	S.F.	3.40	1.94		5.34	6.85
2350	3/4" thick	"	280	.057	"	5.40	2.08		7.48	9.30
2400	Wood rounds, cypress	B-1	175	.137	Ea.	9.75	4.43		14.18	18.05
9000	Minimum labor/equipment charge	D-1	2	8	Job		291		291	470

32 06 Schedules for Exterior Improvements

32 06 10 – Schedules for Bases, Ballasts, and Paving

32 06 10.20 Steps		Crew	Daily Output	Labor-Hours	Unit	2009 Bare Costs Material	Labor	Equipment	Total	Total Incl O&P
0010	**STEPS**									
0011	Incl. excav., borrow & concrete base as required									
0100	Brick steps	B-24	35	.686	LF Riser	11.40	25		36.40	53
0200	Railroad ties	2 Clab	25	.640		3.30	20		23.30	37
0300	Bluestone treads, 12" x 2" or 12" x 1-1/2"	B-24	30	.800	↓	27	29.50		56.50	77
0490	Minimum labor/equipment charge	D-1	2	8	Job		291		291	470
0600	Precast concrete, see Div. 03 41 23.50									
4000	Edging, redwood, 2" x 4"	2 Carp	330	.048	L.F.	2.48	1.94		4.42	5.90
4025	Steel edge strips, incl. stakes, 1/4" x 5"	B-1	390	.062		4.25	1.99		6.24	7.95
4050	Edging, landscape timber or railroad ties, 6" x 8"	2 Carp	170	.094	↓	3.10	3.76		6.86	9.60

32 11 Base Courses

32 11 23 – Aggregate Base Courses

32 11 23.23 Base Course Drainage Layers		Crew	Daily Output	Labor-Hours	Unit	Material	Labor	Equipment	Total	Total Incl O&P
0010	**BASE COURSE DRAINAGE LAYERS**									
0011	For roadways and large areas									
0050	Crushed 3/4" stone base, compacted, 3" deep	B-36C	5200	.008	S.Y.	4.23	.29	.63	5.15	5.80
0100	6" deep		5000	.008		8.45	.30	.66	9.41	10.50
0200	9" deep		4600	.009		12.70	.33	.71	13.74	15.30
0300	12" deep	↓	4200	.010		16.95	.36	.78	18.09	20
0301	Crushed 1-1/2" stone base, compacted to 4" deep	B-36B	6000	.011		5.85	.39	.62	6.86	7.70
0302	6" deep		5400	.012		8.75	.44	.69	9.88	11.05
0303	8" deep		4500	.014		11.65	.52	.83	13	14.55
0304	12" deep	↓	3800	.017	↓	17.50	.62	.99	19.11	21.50
0310	Minimum labor/equipment charge	B-36	1	40	Job		1,425	1,350	2,775	3,800
0350	Bank run gravel, spread and compacted									
0370	6" deep	B-32	6000	.005	S.Y.	5.35	.21	.31	5.87	6.50
0390	9" deep		4900	.007		8	.25	.38	8.63	9.60
0400	12" deep	↓	4200	.008	↓	10.65	.30	.44	11.39	12.65
1500	Alternate method to figure base course									
1510	Crushed stone, 3/4", compacted, 3" deep	B-36C	435	.092	E.C.Y.	43.50	3.49	7.55	54.54	62
1511	6" deep	B-36B	835	.077		43.50	2.82	4.49	50.81	57.50
1512	9" deep		1150	.056		43.50	2.05	3.26	48.81	55
1513	12" deep		1400	.046		43.50	1.68	2.68	47.86	53.50
1520	Crushed stone, 1-1/2", compacted 4" deep		665	.096		43.50	3.54	5.65	52.69	60
1521	6" deep		900	.071		43.50	2.61	4.16	50.27	57
1522	8" deep		1000	.064		43.50	2.35	3.75	49.60	56
1523	12" deep	↓	1265	.051		43.50	1.86	2.96	48.32	54.50
1530	Gravel, bank run, compacted, 6" deep	B-36C	835	.048		27.50	1.82	3.93	33.25	37
1531	9" deep		1150	.035		27.50	1.32	2.86	31.68	35.50
1532	12" deep	↓	1400	.029	↓	27.50	1.08	2.35	30.93	34.50
2010	Crushed stone, 3/4" maximum size, 3" deep	B-36	540	.074	Ton	26.50	2.66	2.50	31.66	36
2011	6" deep		1625	.025		26.50	.88	.83	28.21	31.50
2012	9" deep		1785	.022		26.50	.80	.76	28.06	31
2013	12" deep		1950	.021		26.50	.74	.69	27.93	31
2020	Crushed stone, 1-1/2" maximum size, 4" deep		720	.056		26.50	1.99	1.88	30.37	34.50
2021	6" deep		815	.049		26.50	1.76	1.66	29.92	33.50
2022	8" deep		835	.048		26.50	1.72	1.62	29.84	33.50
2023	12" deep	↓	975	.041		26.50	1.47	1.38	29.35	33
2030	Bank run gravel, 6" deep	B-32A	875	.027	↓	18.40	1.05	1.31	20.76	23

32 11 Base Courses

32 11 23 – Aggregate Base Courses

32 11 23.23 Base Course Drainage Layers		Crew	Daily Output	Labor-Hours	Unit	2009 Bare Costs Material	Labor	Equipment	Total	Total Incl O&P
2031	9″ deep	B-32A	970	.025	Ton	18.40	.94	1.18	20.52	23
2032	12″ deep	↓	1060	.023	↓	18.40	.86	1.08	20.34	22.50
6000	Stabilization fabric, polypropylene, 6 oz./S.Y.	B-6	10000	.002	S.Y.	1.46	.08	.03	1.57	1.77
6900	For small and irregular areas, add						50%	50%		
6990	Minimum labor/equipment charge	1 Clab	3	2.667	Job		84.50		84.50	139
7000	Prepare and roll sub-base, small areas to 2500 S.Y.	B-32A	1500	.016	S.Y.		.61	.76	1.37	1.81
8000	Large areas over 2500 S.Y.	B-32	3700	.009	″		.34	.50	.84	1.08
9000	Minimum labor/equipment charge	1 Clab	4	2	Job		63		63	104

32 11 26 – Asphaltic Base Courses

32 11 26.13 Plant Mix Asphaltic Base Courses		Crew	Daily Output	Labor-Hours	Unit	Material	Labor	Equipment	Total	Total Incl O&P
0010	**PLANT MIX ASPHALTIC BASE COURSES**									
0011	Roadways and large paved areas									
0500	Bituminous concrete, 4″ thick	B-25	4545	.019	S.Y.	12.85	.67	.55	14.07	15.80
0550	6″ thick		3700	.024		18.80	.82	.67	20.29	22.50
0560	8″ thick		3000	.029		25	1.01	.83	26.84	30
0570	10″ thick	↓	2545	.035	↓	31	1.19	.97	33.16	37
2000	Alternate method to figure base course									
2005	Bituminous concrete, 4″ thick	B-25	1000	.088	Ton	57	3.03	2.48	62.51	70
2006	6″ thick		1220	.072		57	2.48	2.03	61.51	69
2007	8″ thick		1320	.067		57	2.30	1.88	61.18	68.50
2008	10″ thick	↓	1400	.063	↓	57	2.16	1.77	60.93	68
8900	For small and irregular areas, add						50%	50%		

32 11 26.19 Bituminous-Stabilized Base Courses		Crew	Daily Output	Labor-Hours	Unit	Material	Labor	Equipment	Total	Total Incl O&P
0010	**BITUMINOUS-STABILIZED BASE COURSES**									
0020	And large paved areas									
0700	Liquid application to gravel base, asphalt emulsion	B-45	6000	.003	Gal.	5.85	.10	.13	6.08	6.70
0800	Prime and seal, cut back asphalt		6000	.003	″	6.90	.10	.13	7.13	7.90
1000	Macadam penetration crushed stone, 2 gal. per S.Y., 4″ thick		6000	.003	S.Y.	11.70	.10	.13	11.93	13.15
1100	6″ thick, 3 gal. per S.Y.		4000	.004		17.50	.15	.20	17.85	19.70
1200	8″ thick, 4 gal. per S.Y.	↓	3000	.005	↓	23.50	.20	.26	23.96	26
8900	For small and irregular areas, add						50%	50%		

32 12 Flexible Paving

32 12 16 – Asphalt Paving

32 12 16.13 Plant-Mix Asphalt Paving		Crew	Daily Output	Labor-Hours	Unit	Material	Labor	Equipment	Total	Total Incl O&P
0010	**PLANT-MIX ASPHALT PAVING**									
0020	And large paved areas with no hauling included									
0025	See Div. 31 23 23.20 for hauling costs									
0080	Binder course, 1-1/2″ thick	B-25	7725	.011	S.Y.	4.66	.39	.32	5.37	6.15
0120	2″ thick		6345	.014		6.20	.48	.39	7.07	8
0130	2-1/2″ thick		5620	.016		7.75	.54	.44	8.73	9.85
0160	3″ thick		4905	.018		9.30	.62	.51	10.43	11.80
0170	3-1/2″ thick		4520	.019		10.85	.67	.55	12.07	13.65
0200	4″ thick	↓	4140	.021		12.40	.73	.60	13.73	15.50
0300	Wearing course, 1″ thick	B-25B	10575	.009		3.08	.32	.26	3.66	4.18
0340	1-1/2″ thick		7725	.012		5.15	.44	.35	5.94	6.80
0380	2″ thick		6345	.015		6.95	.53	.43	7.91	9
0420	2-1/2″ thick		5480	.018		8.55	.61	.49	9.65	10.95
0460	3″ thick		4900	.020		10.20	.69	.55	11.44	12.95
0470	3-1/2″ thick	↓	4520	.021	↓	12	.74	.60	13.34	15

32 12 Flexible Paving

32 12 16 – Asphalt Paving

32 12 16.13 Plant-Mix Asphalt Paving		Crew	Daily Output	Labor-Hours	Unit	2009 Bare Costs Material	Labor	Equipment	Total	Total Incl O&P
0480	4 " thick	B-25B	4140	.023	S.Y.	13.70	.81	.65	15.16	17.10
0500	Open graded friction course	B-25C	5000	.010		3.29	.34	.43	4.06	4.64
0800	Alternate method of figuring paving costs									
0810	Binder course, 1-1/2" thick	B-25	630	.140	Ton	57	4.81	3.93	65.74	74.50
0811	2" thick		690	.128		57	4.39	3.59	64.98	73.50
0812	3" thick		800	.110		57	3.79	3.10	63.89	72
0813	4" thick		900	.098		57	3.37	2.75	63.12	71
0850	Wearing course, 1" thick	B-25B	575	.167		62	5.85	4.71	72.56	82.50
0851	1-1/2" thick		630	.152		62	5.35	4.30	71.65	81.50
0852	2" thick		690	.139		62	4.87	3.93	70.80	80
0853	2-1/2" thick		765	.125		62	4.39	3.54	69.93	79
0854	3" thick		800	.120		62	4.20	3.39	69.59	78.50
1000	Pavement replacement over trench, 2" thick	B-37	90	.533	S.Y.	6.40	17.70	1.53	25.63	37.50
1050	4" thick		70	.686		12.65	23	1.97	37.62	53
1080	6" thick		55	.873		20	29	2.50	51.50	72.50

32 12 16.14 Asphaltic Concrete Paving, Parking Lots and Driveways

		Crew	Daily Output	Labor-Hours	Unit	Material	Labor	Equipment	Total	Total Incl O&P
0011	**ASPHALTIC CONCRETE PAVING, PARKING LOTS & DRIVEWAYS**									
0020	6" stone base, 2" binder course, 1" topping	B-25C	9000	.005	S.F.	2.07	.19	.24	2.50	2.84
0300	Binder course, 1-1/2" thick		35000	.001		.52	.05	.06	.63	.72
0400	2" thick		25000	.002		.67	.07	.09	.83	.95
0500	3" thick		15000	.003		1.04	.11	.14	1.29	1.48
0600	4" thick		10800	.004		1.36	.16	.20	1.72	1.96
0800	Sand finish course, 3/4" thick		41000	.001		.30	.04	.05	.39	.45
0900	1" thick		34000	.001		.36	.05	.06	.47	.55
1000	Fill pot holes, hot mix, 2" thick	B-16	4200	.008		.77	.25	.13	1.15	1.38
1100	4" thick		3500	.009		1.12	.29	.15	1.56	1.88
1120	6" thick		3100	.010		1.50	.33	.17	2	2.39
1140	Cold patch, 2" thick	B-51	3000	.016		.78	.51	.06	1.35	1.76
1160	4" thick		2700	.018		1.48	.57	.07	2.12	2.63
1180	6" thick		1900	.025		2.30	.80	.10	3.20	3.97

32 13 Rigid Paving

32 13 13 – Concrete Paving

32 13 13.23 Concrete Paving Surface Treatment

		Crew	Daily Output	Labor-Hours	Unit	Material	Labor	Equipment	Total	Total Incl O&P
0010	**CONCRETE PAVING SURFACE TREATMENT**									
0015	Including joints, finishing and curing									
0020	Fixed form, 12' pass, unreinforced, 6" thick	B-26	3000	.029	S.Y.	21.50	1.04	1.02	23.56	27
0030	7" thick		2850	.031		29.50	1.09	1.07	31.66	35.50
0100	8" thick		2750	.032		32.50	1.13	1.11	34.74	39
0110	8" thick, small area		1375	.064		32.50	2.26	2.22	36.98	42
0200	9" thick		2500	.035		36.50	1.24	1.22	38.96	44
0300	10" thick		2100	.042		40	1.48	1.45	42.93	48
0310	10" thick, small area		1050	.084		40	2.96	2.90	45.86	52
0400	12" thick		1800	.049		45.50	1.73	1.69	48.92	54.50
0410	Conc pavement, w/jt,fnsh&curing,fix form,24' pass,unreinforced,6"T		6000	.015		20	.52	.51	21.03	23.50
0420	7" thick		5700	.015		26	.55	.54	27.09	30
0430	8" thick		5500	.016		29	.57	.55	30.12	33.50
0440	9" thick		5000	.018		33.50	.62	.61	34.73	38
0450	10" thick		4200	.021		36.50	.74	.73	37.97	42
0460	12" thick		3600	.024		42	.86	.85	43.71	48.50
0470	15" thick		3000	.029		55	1.04	1.02	57.06	63.50

32 13 Rigid Paving

32 13 13 – Concrete Paving

32 13 13.23 Concrete Paving Surface Treatment		Crew	Daily Output	Labor-Hours	Unit	2009 Bare Costs Material	Labor	Equipment	Total	Total Incl O&P
0500	Fixed form 12′ pass 15″ thick	B-26	1500	.059	S.Y.	58.50	2.07	2.03	62.60	70
0505	Minimum labor/equipment charge	1 Cefi	1	8	Job		305		305	480
0510	For small irregular areas, add				%		100%			
0520	Welded wire fabric, sheets for rigid paving 2.33 lbs/SY	2 Rodm	389	.041	S.Y.	1.62	1.83		3.45	4.91
0530	Reinforcing steel for rigid paving 12 lbs/SY		666	.024		9.80	1.07		10.87	12.55
0540	Reinforcing steel for rigid paving 18 lbs/SY		444	.036		14.65	1.61		16.26	18.90
0610	For under 10′ pass, add								7.30	8.55
0620	Slip form, 12′ pass, unreinforced, 6″ thick	B-26A	5600	.016		21	.56	.57	22.13	24.50
0622	7″ thick		5600	.016		28.50	.56	.57	29.63	32.50
0624	8″ thick		5300	.017		31.50	.59	.60	32.69	36
0626	9″ thick		4820	.018		35.50	.65	.66	36.81	41
0628	10″ thick		4050	.022		39	.77	.79	40.56	44.50
0630	12″ thick		3470	.025		44.50	.90	.92	46.32	51
0632	15″ thick		2890	.030		57	1.08	1.10	59.18	65.50
0640	Slip form, 24′ pass, unreinforced, 6″ thick		11200	.008		20	.28	.29	20.57	23
0642	7″ thick		11200	.008		25.50	.28	.29	26.07	29
0644	8″ thick		10600	.008		28.50	.29	.30	29.09	32.50
0646	9″ thick		9640	.009		32.50	.32	.33	33.15	37
0648	10″ thick		8100	.011		36	.38	.39	36.77	40.50
0650	12″ thick		6940	.013		41.50	.45	.46	42.41	46.50
0652	15″ thick		5780	.015		52.50	.54	.55	53.59	59.50
0700	Finishing, broom finish small areas	2 Cefi	120	.133			5.10		5.10	8
0710	Transverse joint support dowels	C-1	350	.091	Ea.	4.55	3.46		8.01	10.70
0720	Transverse contraction joints, saw cut & grind	A-1B	120	.067	L.F.		2.11	1.18	3.29	4.77
0730	Transverse expansion joints, incl. premolded bit. jt. filler	C-1	150	.213		2.10	8.10		10.20	15.65
0740	Transverse construction joint using bulkhead	″	73	.438		3.05	16.60		19.65	31
0750	Longitudinal joint tie bars, grouted	B-23	70	.571	Ea.	4.73	18.30	39.50	62.53	78.50
1000	Curing, with sprayed membrane by hand	2 Clab	1500	.011	S.Y.	.47	.34		.81	1.08
1650	For integral coloring, see Div. 03 05 13.20									

32 14 Unit Paving

32 14 13 – Precast Concrete Unit Paving

32 14 13.13 Interlocking Precast Concrete Unit Paving		Crew	Daily Output	Labor-Hours	Unit	Material	Labor	Equipment	Total	Total Incl O&P
0010	**INTERLOCKING PRECAST CONCRETE UNIT PAVING**									
0020	″V″ blocks for retaining soil	D-1	205	.078	S.F.	8.60	2.84		11.44	14.05

32 14 13.16 Precast Concrete Unit Paving Slabs		Crew	Daily Output	Labor-Hours	Unit	Material	Labor	Equipment	Total	Total Incl O&P
0010	**PRECAST CONCRETE UNIT PAVING SLABS**									
0710	Precast concrete patio blocks, 2-3/8″ thick, colors, 8″ x 16″	D-1	265	.060	S.F.	1.50	2.19		3.69	5.20
0715	12″ x 12″		300	.053		1.95	1.94		3.89	5.30
0720	16″ x 16″		335	.048		2.15	1.74		3.89	5.15
0730	24″ x 24″		510	.031		2.67	1.14		3.81	4.78
0740	Green, 8″ x 16″		265	.060		1.94	2.19		4.13	5.70
0750	Exposed local aggregate, natural	2 Bric	250	.064		6.45	2.59		9.04	11.30
0800	Colors		250	.064		7.05	2.59		9.64	11.95
0850	Exposed granite or limestone aggregate		250	.064		7.80	2.59		10.39	12.75
0900	Exposed white tumblestone aggregate		250	.064		4.61	2.59		7.20	9.25

32 14 16 – Brick Unit Paving

32 14 16.10 Brick Paving		Crew	Daily Output	Labor-Hours	Unit	Material	Labor	Equipment	Total	Total Incl O&P
0010	**BRICK PAVING**									
0012	4″ x 8″ x 1-1/2″, without joints (4.5 brick/S.F.)	D-1	110	.145	S.F.	2.78	5.30		8.08	11.60

32 14 Unit Paving

32 14 16 – Brick Unit Paving

32 14 16.10 Brick Paving		Crew	Daily Output	Labor-Hours	Unit	2009 Bare Costs Material	Labor	Equipment	Total	Total Incl O&P
0100	Grouted, 3/8" joint (3.9 brick/S.F.)	D-1	90	.178	S.F.	3.38	6.45		9.83	14.15
0200	4" x 8" x 2-1/4", without joints (4.5 bricks/S.F.)		110	.145		3.74	5.30		9.04	12.65
0300	Grouted, 3/8" joint (3.9 brick/S.F.)		90	.178		3.45	6.45		9.90	14.25
0500	Bedding, asphalt, 3/4" thick	B-25	5130	.017		.57	.59	.48	1.64	2.12
0540	Course washed sand bed, 1" thick	B-18	5000	.005		.28	.15	.01	.44	.57
0580	Mortar, 1" thick	D-1	300	.053		.64	1.94		2.58	3.83
0620	2" thick		200	.080		1.28	2.91		4.19	6.10
2000	Brick pavers, laid on edge, 7.2 per S.F.		70	.229		2.74	8.30		11.04	16.45
2500	For 4" thick concrete bed and joints, add		595	.027		1.19	.98		2.17	2.89
2800	For steam cleaning, add	A-1H	950	.008		.06	.27	.05	.38	.57
9000	Minimum labor/equipment charge	1 Bric	2	4	Job		162		162	262

32 14 23 – Asphalt Unit Paving

32 14 23.10 Asphalt Blocks

		Crew	Daily Output	Labor-Hours	Unit	Material	Labor	Equipment	Total	Total Incl O&P
0010	**ASPHALT BLOCKS**									
0020	Rectangular, 6" x 12" x 1-1/4", w/bed & neopr. adhesive	D-1	135	.119	S.F.	7.65	4.31		11.96	15.40
0100	3" thick		130	.123		10.75	4.47		15.22	19.05
0300	Hexagonal tile, 8" wide, 1-1/4" thick		135	.119		7.65	4.31		11.96	15.40
0400	2" thick		130	.123		10.75	4.47		15.22	19.05
0500	Square, 8" x 8", 1-1/4" thick		135	.119		7.65	4.31		11.96	15.40
0600	2" thick		130	.123		10.75	4.47		15.22	19.05
0900	For exposed aggregate (ground finish) add					.53			.53	.58
0910	For colors, add					.40			.40	.44
9000	Minimum labor/equipment charge	1 Bric	2	4	Job		162		162	262

32 14 40 – Stone Paving

32 14 40.10 Stone Pavers

		Crew	Daily Output	Labor-Hours	Unit	Material	Labor	Equipment	Total	Total Incl O&P
0010	**STONE PAVERS**									
1100	Flagging, bluestone, irregular, 1" thick,	D-1	81	.198	S.F.	5.95	7.20		13.15	18.15
1110	1-1/2" thick		90	.178		7	6.45		13.45	18.15
1120	Pavers, 1/2" thick		110	.145		9.90	5.30		15.20	19.45
1130	3/4" thick		95	.168		12.60	6.10		18.70	24
1140	1" thick		81	.198		13.50	7.20		20.70	26.50
1150	Snapped random rectangular, 1" thick		92	.174		9	6.30		15.30	20
1200	1-1/2" thick		85	.188		10.80	6.85		17.65	23
1250	2" thick		83	.193		12.60	7		19.60	25
1300	Slate, natural cleft, irregular, 3/4" thick		92	.174		6.95	6.30		13.25	17.85
1310	1" thick		85	.188		8.10	6.85		14.95	19.95
1350	Random rectangular, gauged, 1/2" thick		105	.152		15.05	5.55		20.60	25.50
1400	Random rectangular, butt joint, gauged, 1/4" thick		150	.107		16.20	3.88		20.08	24
1450	For sand rubbed finish, add					7.55			7.55	8.30
1500	For interior setting, add								25%	25%
1550	Granite blocks, 3-1/2" x 3-1/2" x 3-1/2"	D-1	92	.174	S.F.	10	6.30		16.30	21
1560	4" x 4" x 4"		95	.168		10.55	6.10		16.65	21.50
1600	4" to 12" long, 3" to 5" wide, 3" to 5" thick		98	.163		8.35	5.95		14.30	18.75
1650	6" to 15" long, 3" to 6" wide, 3" to 5" thick		105	.152		4.45	5.55		10	13.85

32 16 13 – Concrete Curbs and Gutters

32 16 13.13 Cast-in-Place Concrete Curbs and Gutters		Crew	Daily Output	Labor-Hours	Unit	2009 Bare Costs Material	Labor	Equipment	Total	Total Incl O&P
0010	**CAST-IN-PLACE CONCRETE CURBS AND GUTTERS**									
0300	Concrete, wood forms, 6″ x 18″, straight	C-2A	500	.096	L.F.	4.84	3.71		8.55	11.35
0400	6″ x 18″, radius	″	200	.240	↓	4.92	9.25		14.17	20.50
0415	Machine formed, 6″ x 18″, straight	B-69A	2000	.024		3.89	.83	.38	5.10	6.05
0416	6″ x 18″, radius	″	900	.053	↓	4.06	1.85	.84	6.75	8.40
0421	Curb and gutter, straight									
0422	with 6″ high curb and 6″ thick gutter, wood forms									
0430	24″ wide, .055 C.Y. per L.F.	C-2A	375	.128	L.F.	17.30	4.94		22.24	27
0435	30″ wide, .066 C.Y. per L.F.	″	340	.141	″	19.10	5.45		24.55	30

32 16 13.26 Precast Concrete Curbs		Crew	Daily Output	Labor-Hours	Unit	Material	Labor	Equipment	Total	Total Incl O&P
0010	**PRECAST CONCRETE CURBS**									
0550	Precast, 6″ x 18″, straight	B-29	700	.080	L.F.	10.50	2.74	1.37	14.61	17.50
0600	6″ x 18″, radius	″	325	.172	″	12.10	5.90	2.95	20.95	26

32 16 19 – Asphalt Curbs

32 16 19.10 Bituminous Concrete Curbs		Crew	Daily Output	Labor-Hours	Unit	Material	Labor	Equipment	Total	Total Incl O&P
0010	**BITUMINOUS CONCRETE CURBS**									
0012	Curbs, asphaltic, machine formed, 8″ wide, 6″ high, 40 L.F./ton	B-27	1000	.032	L.F.	1.24	1.03	.27	2.54	3.36
0100	8″ wide, 8″ high, 30 L.F. per ton		900	.036		1.43	1.14	.30	2.87	3.78
0150	Asphaltic berm, 12″ W, 3″-6″ H, 35 L.F./ton, before pavement	↓	700	.046		1.59	1.47	.39	3.45	4.60
0200	12″ W, 1-1/2″ to 4″ H, 60 L.F. per ton, laid with pavement	B-2	1050	.038	↓	.97	1.22		2.19	3.07

32 16 40 – Stone Curbs

32 16 40.13 Cut Stone Curbs		Crew	Daily Output	Labor-Hours	Unit	Material	Labor	Equipment	Total	Total Incl O&P
0010	**CUT STONE CURBS**									
1000	Granite, split face, straight, 5″ x 16″	D-13	500	.096	L.F.	12.05	3.68	1.20	16.93	20.50
1300	Radius curbing, 6″ x 18″, over 10′ radius	B-29	260	.215	″	19.40	7.35	3.69	30.44	37.50
1400	Corners, 2′ radius		80	.700	Ea.	65	24	12	101	124
1600	Edging, 4-1/2″ x 12″, straight		300	.187	L.F.	6.05	6.40	3.20	15.65	20.50
1800	Curb inlets, (guttermouth) straight	↓	41	1.366	Ea.	145	46.50	23.50	215	261
2000	Indian granite (belgian block)									
2100	Jumbo, 10-1/2″ x 7-1/2″ x 4″, grey	D-1	150	.107	L.F.	2.21	3.88		6.09	8.70
2150	Pink		150	.107		2.88	3.88		6.76	9.40
2200	Regular, 9″ x 4-1/2″ x 4-1/2″, grey		160	.100		2	3.63		5.63	8.10
2250	Pink		160	.100		2.76	3.63		6.39	8.95
2300	Cubes, 4″ x 4″ x 4″, grey		175	.091		1.91	3.32		5.23	7.45
2350	Pink		175	.091		2	3.32		5.32	7.55
2400	6″ x 6″ x 6″, pink	↓	155	.103	↓	4.96	3.75		8.71	11.50
2500	Alternate pricing method for indian granite									
2550	Jumbo, 10-1/2″ x 7-1/2″ x 4″ (30 lb), grey				Ton	126			126	139
2600	Pink					167			167	184
2650	Regular, 9″ x 4-1/2″ x 4-1/2″ (20 lb), grey					141			141	156
2700	Pink					193			193	212
2750	Cubes, 4″ x 4″ x 4″ (5 lb), grey					231			231	255
2800	Pink					257			257	283
2850	6″ x 6″ x 6″ (25 lb), pink					193			193	212
2900	For pallets, add				↓	20			20	22

32 17 Paving Specialties

32 17 13 – Parking Bumpers

32 17 13.13 Metal Parking Bumpers		Crew	Daily Output	Labor-Hours	Unit	2009 Bare Costs Material	Labor	Equipment	Total	Total Incl O&P
0010	METAL PARKING BUMPERS									
0015	Bumper rails for garages, 12 Ga. rail, 6" wide, with steel									
0020	posts 12'-6" O.C., minimum	E-4	190	.168	L.F.	17.90	7.60	.71	26.21	34.50
0030	Average		165	.194		22.50	8.75	.81	32.06	41.50
0100	Maximum		140	.229		27	10.35	.96	38.31	49.50
0300	12" channel rail, minimum		160	.200		22.50	9.05	.84	32.39	42
0400	Maximum	↓	120	.267	↓	33.50	12.05	1.12	46.67	60.50
1300	Pipe bollards, conc filled/paint, 8' L x 4' D hole, 6" diam.	B-6	20	1.200	Ea.	410	41	14.70	465.70	535
1400	8" diam.		15	1.600		620	54.50	19.60	694.10	790
1500	12" diam.		12	2		805	68	24.50	897.50	1,025
1592	Bollards, steel, 3' H, retractable, incl hydraulic controls, min		4	6		40,700	204	73.50	40,977.50	45,100
1594	Max	↓	2	12		45,300	410	147	45,857	50,500
2030	Folding with individual padlocks	B-2	50	.800	↓	990	25.50		1,015.50	1,150
8000	Parking lot control, see Div. 11 12 13.10									
9000	Minimum labor/equipment charge	E-4	2	16	Job		725	67	792	1,425

32 17 13.16 Plastic Parking Bumpers		Crew	Daily Output	Labor-Hours	Unit	Material	Labor	Equipment	Total	Total Incl O&P
0010	PLASTIC PARKING BUMPERS									
1200	Thermoplastic, 6" x 10" x 6'-0"	B-2	120	.333	Ea.	108	10.65		118.65	137

32 17 13.19 Precast Concrete Parking Bumpers		Crew	Daily Output	Labor-Hours	Unit	Material	Labor	Equipment	Total	Total Incl O&P
0010	PRECAST CONCRETE PARKING BUMPERS									
1000	Wheel stops, precast concrete incl. dowels, 6" x 10" x 6'-0"	B-2	120	.333	Ea.	58.50	10.65		69.15	82
1100	8" x 13" x 6'-0"	"	120	.333	"	68	10.65		78.65	92

32 17 13.26 Wood Parking Bumpers		Crew	Daily Output	Labor-Hours	Unit	Material	Labor	Equipment	Total	Total Incl O&P
0010	WOOD PARKING BUMPERS									
0020	Parking barriers, timber w/saddles, treated type									
0100	4" x 4" for cars	B-2	520	.077	L.F.	2.98	2.46		5.44	7.35
0200	6" x 6" for trucks		520	.077	"	5.95	2.46		8.41	10.60
0600	Flexible fixed stanchion, 2' high, 3" diameter	↓	100	.400	Ea.	34.50	12.80		47.30	59

32 17 23 – Pavement Markings

32 17 23.13 Painted Pavement Markings		Crew	Daily Output	Labor-Hours	Unit	Material	Labor	Equipment	Total	Total Incl O&P
0010	PAINTED PAVEMENT MARKINGS									
0020	Acrylic waterborne, white or yellow, 4" wide	B-78	20000	.002	L.F.	.25	.08	.03	.36	.44
0200	6" wide		11000	.004		.19	.14	.05	.38	.49
0500	8" wide		10000	.005		.33	.15	.05	.53	.67
0600	12" wide		4000	.012	↓	.60	.38	.13	1.11	1.43
0620	Arrows or gore lines		2300	.021	S.F.	.68	.66	.22	1.56	2.09
0640	Temporary paint, white or yellow	↓	15000	.003	L.F.	.20	.10	.03	.33	.43
0660	Removal	1 Clab	300	.027			.84		.84	1.39
0680	Temporary tape	2 Clab	1500	.011		1.74	.34		2.08	2.47
0710	Thermoplastic, white or yellow, 4" wide	B-79	15000	.003		.87	.09	.08	1.04	1.19
0730	6" wide		14000	.003		1.26	.09	.08	1.43	1.63
0740	8" wide		12000	.003		1.70	.11	.10	1.91	2.15
0750	12" wide		6000	.007	↓	2.52	.21	.20	2.93	3.35
0760	Arrows		660	.061	S.F.	1.82	1.93	1.79	5.54	7.15
0770	Gore lines		2500	.016		1.22	.51	.47	2.20	2.70
0780	Letters	↓	660	.061	↓	1.51	1.93	1.79	5.23	6.80
0782	Thermoplastic material				Ton	1,075			1,075	1,200
0784	Glass beads, add				M.L.F.	9.20			9.20	10.10
1000	Airport painted markings									
1050	Traffic safety flashing truck for airport painting	A-2B	1	8	Day		248	194	442	620
1100	Painting, white or yellow, taxiway markings	B-79	4000	.010	S.F.	.25	.32	.29	.86	1.13

32 17 Paving Specialties

32 17 23 – Pavement Markings

32 17 23.13 Painted Pavement Markings		Crew	Daily Output	Labor-Hours	Unit	Material	2009 Bare Costs Labor	Equipment	Total	Total Incl O&P
1200	Runway markings	B-79	3500	.011	S.F.	.25	.36	.34	.95	1.25
1300	Pavement location or direction signs		2500	.016		.25	.51	.47	1.23	1.64
1350	Mobilization airport pavement painting		4	10	Ea.		320	295	615	850
1400	Paint markings or pavement signs removal daytime	B-78B	400	.045	S.F.		1.46	.80	2.26	3.27
1500	Removal nighttime		335	.054	"		1.74	.96	2.70	3.91
1600	Mobilization pavement paint removal		4	4.500	Ea.		145	80.50	226.50	330

32 17 23.14 Pavement Parking Markings		Crew	Daily Output	Labor-Hours	Unit	Material	Labor	Equipment	Total	Total Incl O&P
0010	**PAVEMENT PARKING MARKINGS**									
0790	Layout of pavement marking	A-2	25000	.001	L.F.		.03	.01	.04	.06
0800	Lines on pvmt, parking stall, paint, white, 4" wide	B-78	440	.109	Stall	6	3.47	1.15	10.62	13.55
1000	Street letters and numbers	"	1600	.030	S.F.	.63	.95	.32	1.90	2.61
1100	Pavement marking letter, 6"	2 Pord	400	.040	Ea.	7.40	1.41		8.81	10.40
1110	12" letter		272	.059		7.40	2.07		9.47	11.45
1120	24" letter		160	.100		7.40	3.52		10.92	13.80
1130	36" letter		84	.190		6.45	6.70		13.15	17.80
1140	42" letter		84	.190		7.40	6.70		14.10	18.85
1150	72" letter		40	.400		12.25	14.10		26.35	36
1200	Handicap symbol		40	.400		15	14.10		29.10	39

32 18 Athletic and Recreational Surfacing

32 18 13 – Synthetic Grass Surfacing

32 18 13.10 Artificial Grass Surfacing		Crew	Daily Output	Labor-Hours	Unit	Material	Labor	Equipment	Total	Total Incl O&P
0010	**ARTIFICIAL GRASS SURFACING**									
0015	Not including asphalt base or drainage,									
0020	but including cushion pad, over 50,000 S.F.									
0200	1/2" pile and 5/16" cushion pad, standard	C-17	3200	.025	S.F.	10.30	1.03		11.33	13
0300	Deluxe		2560	.031		12.20	1.29		13.49	15.50
0500	1/2" pile and 5/8" cushion pad, standard		2844	.028		15	1.16		16.16	18.40
0600	Deluxe		2327	.034		16.40	1.42		17.82	20.50
0800	For asphaltic concrete base, 2-1/2" thick,									
0900	with 6" crushed stone sub-base, add	B-25	12000	.007	S.F.	1.86	.25	.21	2.32	2.68

32 18 23 – Athletic Surfacing

32 18 23.33 Running Track Surfacing		Crew	Daily Output	Labor-Hours	Unit	Material	Labor	Equipment	Total	Total Incl O&P
0010	**RUNNING TRACK SURFACING**									
0020	Running track, asphalt, incl base, 3" thick	B-37	300	.160	S.Y.	20.50	5.30	.46	26.26	31.50
0100	Surface, latex rubber system, 3/8" thick, black	B-20	125	.192		9.30	6.80		16.10	21.50
0150	Colors		125	.192		16.50	6.80		23.30	29.50
0300	Urethane rubber system, 3/8" thick, black		120	.200		25	7.05		32.05	38.50
0400	Color coating		115	.209		30.50	7.40		37.90	45.50

32 18 23.53 Tennis Court Surfacing		Crew	Daily Output	Labor-Hours	Unit	Material	Labor	Equipment	Total	Total Incl O&P
0010	**TENNIS COURT SURFACING**									
0020	Tennis court, asphalt, incl. base, 2-1/2" thick, one court	B-37	450	.107	S.Y.	28	3.54	.31	31.85	36.50
0200	Two courts		675	.071		12.55	2.36	.20	15.11	17.90
0300	Clay courts		360	.133		37	4.42	.38	41.80	48.50
0400	Pulverized natural greenstone with 4" base, fast dry		250	.192		35.50	6.35	.55	42.40	50.50
0800	Rubber-acrylic base resilient pavement		600	.080		52.50	2.65	.23	55.38	62.50
1000	Colored sealer, acrylic emulsion, 3 coats	2 Clab	800	.020		6.35	.63		6.98	8
1100	3 coat, 2 colors	"	900	.018		8.80	.56		9.36	10.60
1200	For preparing old courts, add	1 Clab	825	.010			.31		.31	.51
1400	Posts for nets, 3-1/2" diameter with eye bolts	B-1	3.40	7.059	Pr.	210	228		438	605

32 18 Athletic and Recreational Surfacing

32 18 23 – Athletic Surfacing

32 18 23.53 Tennis Court Surfacing		Crew	Daily Output	Labor-Hours	Unit	2009 Bare Costs Material	Labor	Equipment	Total	Total Incl O&P
1500	With pulley & reel	B-1	3.40	7.059	Pr.	299	228		527	705
1700	Net, 42′ long, nylon thread with binder		50	.480	Ea.	277	15.50		292.50	330
1800	All metal		6.50	3.692	″	490	119		609	735
2000	Paint markings on asphalt, 2 coats	1 Pord	1.78	4.494	Court	97	158		255	360
2200	Complete court with fence, etc., asphaltic conc., minimum	B-37	.20	240		21,300	7,975	690	29,965	37,300
2300	Maximum		.16	300		26,500	9,950	860	37,310	46,400
2800	Clay courts, minimum		.20	240		23,000	7,975	690	31,665	39,200
2900	Maximum		.16	300		27,700	9,950	860	38,510	47,700

32 31 Fences and Gates

32 31 13 – Chain Link Fences and Gates

32 31 13.10 Chain Link Gates and Fences		Crew	Daily Output	Labor-Hours	Unit	Material	Labor	Equipment	Total	Total Incl O&P
0010	**CHAIN LINK GATES AND FENCES**									
4750	Gate, transom for 10′ fence, galv. steel, single, 3′ x 7′ x 3′	B-80A	52	.462	Ea.	350	14.60	4.67	369.27	415
4752	4′ x 7′ x 3′		10	2.400		370	76	24.50	470.50	555
4754	3′ x 10′		8	3		390	95	30.50	515.50	620
4756	4′ x 10′		10	2.400		410	76	24.50	510.50	600
4758	Double transom, 10′ x 7′ x 3′	B-80B	10	3.200		510	107	22.50	639.50	760
4760	12′ x 7′ x 3′		6	5.333		560	178	37.50	775.50	945
4762	14′ x 7′ x 3′		5	6.400		600	214	45	859	1,050
4764	10′ x 10′		4	8		850	268	56	1,174	1,425
4766	12′ x 10′		7	4.571		905	153	32	1,090	1,275
4768	14′ x 10′		7	4.571		955	153	32	1,140	1,325
4780	Vinyl clad, single transom, 3′ x 7′ x 3′		6	5.333		525	178	37.50	740.50	910
4782	4′ x 7′ x 3′		10	3.200		565	107	22.50	694.50	820
4784	3′ x 10′		8	4		580	134	28	742	890
4786	4′ x 10′		10	3.200		605	107	22.50	734.50	865
4788	Double transom, 10′ x 7′ x 3′		10	3.200		790	107	22.50	919.50	1,075
4790	12′ x 7′ x 3′		6	5.333		845	178	37.50	1,060.50	1,250
4792	14′ x 7′ x 3′		5	6.400		905	214	45	1,164	1,400
4794	10′ x 10′		4	8		1,300	268	56	1,624	1,925
4798	12′ x 12′		7	4.571		1,350	153	32	1,535	1,775
4799	14′ x 14′		7	4.571		1,400	153	32	1,585	1,825

32 31 13.20 Fence, Chain Link Industrial		Crew	Daily Output	Labor-Hours	Unit	Material	Labor	Equipment	Total	Total Incl O&P
0010	**FENCE, CHAIN LINK INDUSTRIAL**									
0011	Schedule 40, including concrete									
0020	3 strands barb wire, 2″ post @ 10′ O.C., set in concrete, 6′ H									
0200	9 ga. wire, galv. steel, in concrete	B-80C	240	.100	L.F.	15.25	3.14	.84	19.23	23
0300	Aluminized steel		240	.100		19.55	3.14	.84	23.53	27.50
0500	6 ga. wire, galv. steel		240	.100		24	3.14	.84	27.98	32.50
0600	Aluminized steel		240	.100		27.50	3.14	.84	31.48	36
0800	6 ga. wire, 6′ high but omit barbed wire, galv. steel		250	.096		23	3.01	.81	26.82	31.50
0900	Aluminized steel, in concrete		250	.096		32.50	3.01	.81	36.32	42
0920	8′ H, 6 ga. wire, 2-1/2″ line post, galv. steel, in concrete		180	.133		37	4.18	1.12	42.30	49
0940	Aluminized steel, in concrete		180	.133		45.50	4.18	1.12	50.80	58
1400	Gate for 6′ high fence, 1-5/8″ frame, 3′ wide, galv. steel		10	2.400	Ea.	177	75.50	20	272.50	340
1500	Aluminized steel, in concrete		10	2.400	″	218	75.50	20	313.50	385
2000	5′-0″ high fence, 9 ga., no barbed wire, 2″ line post, in concrete									
2010	10′ O.C., 1-5/8″ top rail, in concrete									
2100	Galvanized steel, in concrete	B-80C	300	.080	L.F.	13	2.51	.67	16.18	19.15
2200	Aluminized steel, in concrete		300	.080	″	15.65	2.51	.67	18.83	22

32 31 13 – Chain Link Fences and Gates

32 31 13.20 Fence, Chain Link Industrial		Crew	Daily Output	Labor-Hours	Unit	2009 Bare Costs Material	Labor	Equipment	Total	Total Incl O&P
2400	Gate, 4' wide, 5' high, 2" frame, galv. steel, in concrete	B-80C	10	2.400	Ea.	191	75.50	20	286.50	355
2500	Aluminized steel, in concrete		10	2.400	"	214	75.50	20	309.50	380
3100	Overhead slide gate, chain link, 6' high, to 18' wide, in concrete		38	.632	L.F.	167	19.80	5.30	192.10	221
3110	Cantilever type, in concrete	B-80	48	.667		77.50	22.50	13.65	113.65	137
3120	8' high, in concrete		24	1.333		111	45	27.50	183.50	226
3130	10' high, in concrete		18	1.778		131	60	36.50	227.50	282
5000	Double swing gates, incl. posts & hardware, in concrete									
5010	5' high, 12' opening, in concrete	B-80C	3.40	7.059	Opng.	495	222	59.50	776.50	975
5020	20' opening, in concrete		2.80	8.571		675	269	72	1,016	1,275
5060	6' high, 12' opening, in concrete		3.20	7.500		835	235	63	1,133	1,375
5070	20' opening, in concrete		2.60	9.231		1,150	290	77.50	1,517.50	1,825
5080	8' high, 12' opening, in concrete	B-80	2.13	15.002		1,300	505	305	2,110	2,600
5090	20' opening, in concrete		1.45	22.069		1,700	745	450	2,895	3,600
5100	10' high, 12' opening, in concrete		1.31	24.427		1,625	825	500	2,950	3,700
5110	20' opening, in concrete		1.03	31.068		2,450	1,050	635	4,135	5,075
5120	12' high, 12' opening, in concrete		1.05	30.476		2,375	1,025	625	4,025	4,975
5130	20' opening, in concrete		.85	37.647		3,050	1,275	770	5,095	6,275
5190	For aluminized steel add					20%				
7055	Braces, galv. steel	B-80A	960	.025	L.F.	2.15	.79	.25	3.19	3.95
7056	Aluminized steel	"	960	.025	"	2.58	.79	.25	3.62	4.42
7071	Privacy slats, vertical, vinyl	1 Clab	500	.016	S.F.	1.70	.51		2.21	2.70
7072	Redwood		450	.018		3.05	.56		3.61	4.29
7073	Diagonal, aluminum		300	.027		3.95	.34		4.79	5.75
9000	Minimum labor/equipment charge	B-80	2	16	Job		540	325	865	1,250

32 31 13.25 Fence, Chain Link Residential		Crew	Daily Output	Labor-Hours	Unit	Material	Labor	Equipment	Total	Total Incl O&P
0010	**FENCE, CHAIN LINK RESIDENTIAL**									
0011	Schedule 20, 11 gauge wire, 1-5/8" post									
0020	10' O.C., 1-3/8" top rail, 2" corner post, galv. stl. 3' high	B-80C	500	.048	L.F.	4.42	1.51	.40	6.33	7.80
0050	4' high		400	.060		5.10	1.88	.50	7.48	9.25
0100	6' high		200	.120		6.15	3.77	1.01	10.93	14.10
0150	Add for gate 3' wide, 1-3/8" frame, 3' high		12	2	Ea.	63.50	63	16.80	143.30	191
0170	4' high		10	2.400		71.50	75.50	20	167	225
0190	6' high		10	2.400		99.50	75.50	20	195	256
0200	Add for gate 4' wide, 1-3/8" frame, 3' high		9	2.667		75	83.50	22.50	181	245
0220	4' high		9	2.667		80	83.50	22.50	186	251
0240	6' high		8	3		95	94	25	214	287
0350	Aluminized steel, 11 ga. wire, 3' high		500	.048	L.F.	6.05	1.51	.40	7.96	9.55
0380	4' high		400	.060		7.75	1.88	.50	10.13	12.20
0400	6' high		200	.120		10.90	3.77	1.01	15.68	19.30
0450	Add for gate 3' wide, 1-3/8" frame, 3' high		12	2	Ea.	83	63	16.80	162.80	213
0470	4' high		10	2.400		136	75.50	20	231.50	296
0490	6' high		10	2.400		171	75.50	20	266.50	335
0500	Add for gate 4' wide, 1-3/8" frame, 3' high		10	2.400		94	75.50	20	189.50	250
0520	4' high		9	2.667		105	83.50	22.50	211	278
0540	6' high		8	3		121	94	25	240	315
0620	Vinyl covered, 9 ga. wire, 3' high		500	.048	L.F.	4.78	1.51	.40	6.69	8.15
0640	4' high		400	.060		5.55	1.88	.50	7.93	9.75
0660	6' high		200	.120		7.05	3.77	1.01	11.83	15.10
0720	Add for gate 3' wide, 1-3/8" frame, 3' high		12	2	Ea.	92	63	16.80	171.80	223
0740	4' high		10	2.400		104	75.50	20	199.50	260
0760	6' high		10	2.400		163	75.50	20	258.50	325
0780	Add for gate 4' wide, 1-3/8" frame, 3' high		10	2.400		106	75.50	20	201.50	262

32 31 Fences and Gates

32 31 13 – Chain Link Fences and Gates

32 31 13.25 Fence, Chain Link Residential		Crew	Daily Output	Labor-Hours	Unit	Material	2009 Bare Costs Labor	Equipment	Total	Total Incl O&P
0800	4′ high	B-80C	9	2.667	Ea.	121	83.50	22.50	227	296
0820	6′ high	↓	8	3	↓	131	94	25	250	325
9000	Minimum labor/equipment charge	B-1	2	12	Job		385		385	640

32 31 13.26 Tennis Court Fences and Gates

		Crew	Daily Output	Labor-Hours	Unit	Material	Labor	Equipment	Total	Total Incl O&P
0010	**TENNIS COURT FENCES AND GATES**									
0860	Tennis courts, 11 ga. wire, 2-1/2″ post set									
0870	in concrete, 10′ O.C., 1-5/8″ top rail									
0900	10′ high	B-80	190	.168	L.F.	14.15	5.70	3.45	23.30	28.50
0920	12′ high		170	.188	″	16.75	6.35	3.85	26.95	33
1000	Add for gate 4′ wide, 1-5/8″ frame 7′ high		10	3.200	Ea.	135	108	65.50	308.50	395
1040	Aluminized steel, 11 ga. wire 10′ high		190	.168	L.F.	17.60	5.70	3.45	26.75	32.50
1100	12′ high		170	.188	″	22	6.35	3.85	32.20	38.50
1140	Add for gate 4′ wide, 1-5/8″ frame, 7′ high		10	3.200	Ea.	158	108	65.50	331.50	420
1250	Vinyl covered, 9 ga. wire, 10′ high		190	.168	L.F.	16.75	5.70	3.45	25.90	31.50
1300	12′ high	↓	170	.188	″	28	6.35	3.85	38.20	45
1310	Fence, CL, tennis ct, transom gate, single, galv, 4′ x 7′ x 3′	B-80A	8.72	2.752	Ea.	305	87	28	420	510
1400	Add for gate 4′ wide, 1-5/8″ frame, 7′ high	B-80	10	3.200	″	158	108	65.50	331.50	420

32 31 13.30 Fence, Chain Link, Gates and Posts

		Crew	Daily Output	Labor-Hours	Unit	Material	Labor	Equipment	Total	Total Incl O&P
0010	**FENCE, CHAIN LINK, GATES & POSTS**									
0011	(1/3 post length in ground)									
6580	Line posts, galvanized, 2-1/2″ OD, set in conc., 4′	B-80	80	.400	Ea.	27.50	13.50	8.20	49.20	61
6585	5′		76	.421		34	14.25	8.60	56.85	70
6590	6′		74	.432		32.50	14.60	8.85	55.95	70
6595	7′		72	.444		36.50	15	9.10	60.60	74.50
6600	8′		69	.464		39	15.70	9.50	64.20	78.50
6610	H-beam, 1-7/8″, 4′		83	.386		30	13.05	7.90	50.95	62.50
6615	5′		81	.395		34.50	13.35	8.10	55.95	68.50
6620	6′		78	.410		40	13.85	8.40	62.25	76
6625	7′		75	.427		43.50	14.40	8.75	66.65	80.50
6630	8′		73	.438		47.50	14.80	8.95	71.25	86
6635	Vinyl coated, 2-1/2″ OD, set in conc., 4′		79	.405		51	13.70	8.30	73	88
6640	5′		77	.416		58.50	14.05	8.50	81.05	97
6645	6′		74	.432		69	14.60	8.85	92.45	110
6650	7′		72	.444		77.50	15	9.10	101.60	120
6655	8′		69	.464		83.50	15.70	9.50	108.70	128
6660	End gate post, steel, 3″ OD, set in conc.,4′		68	.471		64	15.90	9.65	89.55	107
6665	5′		65	.492		76.50	16.65	10.05	103.20	122
6670	6′		63	.508		77.50	17.15	10.40	105.05	125
6675	7′		61	.525		99	17.75	10.75	127.50	150
6680	8′		59	.542		110	18.35	11.10	139.45	163
6685	Vinyl, 4′		68	.471		97.50	15.90	9.65	123.05	144
6690	5′		65	.492		108	16.65	10.05	134.70	157
6695	6′		63	.508		132	17.15	10.40	159.55	184
6700	7′		61	.525		129	17.75	10.75	157.50	183
6705	8′		59	.542		157	18.35	11.10	186.45	215
6710	Corner post, galv. steel, 4″ OD, set in conc., 4′		65	.492		98.50	16.65	10.05	125.20	147
6715	6′		63	.508		138	17.15	10.40	165.55	191
6720	7′		61	.525		157	17.75	10.75	185.50	213
6725	8′		65	.492		178	16.65	10.05	204.70	234
6730	Vinyl, 5′		65	.492		173	16.65	10.05	199.70	228
6735	6′		63	.508		226	17.15	10.40	253.55	287
6740	7′	↓	61	.525	↓	209	17.75	10.75	237.50	270

32 31 13 – Chain Link Fences and Gates

32 31 13.30 Fence, Chain Link, Gates and Posts		Crew	Daily Output	Labor-Hours	Unit	2009 Bare Costs Material	Labor	Equipment	Total	Total Incl O&P
6745	8′	B-80	59	.542	Ea.	269	13.35	11.10	298.45	340
7031	For corner, end, & pull post bracing, add					20%	15%			
7770	Gates, sliding w/overhead support, 4′ high	B-80B	35	.914	L.F.	52.50	30.50	6.45	89.45	115
7775	5′ high		32	1		56.50	33.50	7.05	97.05	124
7780	6′ high		28	1.143		61	38	8.05	107.05	138
7785	7′ high		25	1.280		80	43	9	132	167
7790	8′ high		23	1.391		99.50	46.50	9.80	155.80	196
7795	Cantilever, manual, exp. roller, (pr), 40′ wide x 8′ high	B-22	1	30	Ea.	4,100	1,100	194	5,394	6,525
7800	30′ wide x 8′ high		1	30		3,450	1,100	194	4,744	5,800
7805	24′ wide x 8′ high		1	30		3,050	1,100	194	4,344	5,375
7810	Motor operators for gates, (no elec wiring), 3′ wide swing	2 Skwk	.50	32		1,550	1,300		2,850	3,825
7815	Up to 20′ wide swing		.50	32		5,150	1,300		6,450	7,775
7820	Up to 45′ sliding		.50	32		5,175	1,300		6,475	7,825
7825	Overhead gate, 6′ to 18′ wide, sliding/cantilever		45	.356	L.F.	207	14.50		221.50	252
7830	Gate operators, digital receiver		7	2.286	Ea.	545	93.50		638.50	750
7835	Two button transmitter		24	.667		70	27		97	122
7840	3 button station		14	1.143		87.50	46.50		134	173
7845	Master slave system		4	4		320	163		483	620
7900	Auger fence post hole, 3′ deep, medium soil, by hand	1 Clab	30	.267			8.45		8.45	13.90
7925	By machine	B-80	175	.183			6.20	3.74	9.94	14.15
7950	Rock, with jackhammer	B-9	32	1.250			40	6	46	72.50
7975	With rock drill	B-47C	65	.246			8.70	20.50	29.20	36.50

32 31 13.33 Chain Link Backstops		Crew	Daily Output	Labor-Hours	Unit	Material	Labor	Equipment	Total	Total Incl O&P
0010	**CHAIN LINK BACKSTOPS**									
0015	Backstops, baseball, prefabricated, 30′ wide, 12′ high & 1 overhang	B-1	1	24	Ea.	3,125	775		3,900	4,725
0100	40′ wide, 12′ high & 2 overhangs	"	.75	32		5,400	1,025		6,425	7,625
0110	Regulation, galvanized	B-13	2.40	23.333		7,775	800	330	8,905	10,200
0120	Vinyl coated	"	2.40	23.333		15,900	800	330	17,030	19,200
0180	Softball, prefabricated, no overhang	B-1	1.60	15		2,825	485		3,310	3,900
0200	Softball, regulation, galv., 3″ posts @8′, 6 & 9 ga. mesh, 14′ H	B-13	8.80	6.364		9,225	218	89.50	9,532.50	10,600
0205	18′ high		.75	74.667		11,000	2,550	1,050	14,600	17,400
0210	20′ high		.72	77.778		11,600	2,650	1,100	15,350	18,200
0215	22′ high		.70	80		13,700	2,725	1,125	17,550	20,700
0220	24′ high		.60	92.838		12,300	3,175	1,300	16,775	20,100
0250	Vinyl coated, 14′ high		1.20	46.667		13,900	1,600	660	16,160	18,600
0255	18′ high		.75	74.667		16,100	2,550	1,050	19,700	23,000
0260	20′ high		.72	77.778		17,000	2,650	1,100	20,750	24,200
0265	22′ high		.70	80		19,700	2,725	1,125	23,550	27,400
0270	24′ high		.60	93.333		20,700	3,200	1,325	25,225	29,400
0300	Basketball, steel, single goal		3.04	18.421		1,250	630	260	2,140	2,650
0400	Double goal		1.92	29.167		810	995	410	2,215	2,975
0600	Tennis, wire mesh with pair of ends	B-1	2.48	9.677	Set	1,900	310		2,210	2,600
0700	Enclosed court	"	1.30	18.462	Ea.	5,400	595		5,995	6,925

32 31 13.40 Fence, Fabric and Accessories		Crew	Daily Output	Labor-Hours	Unit	Material	Labor	Equipment	Total	Total Incl O&P
0010	**FENCE, FABRIC & ACCESSORIES**									
1000	Fabric, 9 ga., galv., 1.2 oz. coat, 2″ chain link, 4′	B-80A	304	.079	L.F.	3.98	2.49	.80	7.27	9.40
1150	5′		285	.084		4.80	2.66	.85	8.31	10.65
1200	6′		266	.090		8.10	2.85	.91	11.86	14.60
1250	7′		247	.097		11.05	3.07	.98	15.10	18.30
1300	8′		228	.105		13.90	3.33	1.06	18.29	22
1400	9 ga., fused, 4′		304	.079		6.10	2.49	.80	9.39	11.70
1450	5′		285	.084		7.25	2.66	.85	10.76	13.30

32 31 Fences and Gates

32 31 13 – Chain Link Fences and Gates

32 31 13.40 Fence, Fabric and Accessories		Crew	Daily Output	Labor-Hours	Unit	2009 Bare Costs Material	Labor	Equipment	Total	Total Incl O&P
1500	6′	B-80A	266	.090	L.F.	8.60	2.85	.91	12.36	15.20
1550	7′		247	.097		10	3.07	.98	14.05	17.15
1600	8′		228	.105		14.40	3.33	1.06	18.79	22.50
1650	Barbed wire, galv., cost per strand		2280	.011		.15	.33	.11	.59	.84
1700	Vinyl coated		2280	.011		.27	.33	.11	.71	.97
1750	Extension arms, 3 strands		143	.168	Ea.	9.10	5.30	1.70	16.10	20.50
1800	6 strands, 2-3/8″		119	.202		11.65	6.35	2.04	20.04	25.50
1850	Eye tops, 2-3/8″		143	.168		3.92	5.30	1.70	10.92	14.95
1900	Top rail, incl. tie wires, 1-5/8″, galv.		912	.026	L.F.	3.43	.83	.27	4.53	5.45
1950	Vinyl coated		912	.026		5.60	.83	.27	6.70	7.80
2100	Rail, middle/bottom, w/tie wire, 1-5/8″, galv.		912	.026		3.02	.83	.27	4.12	4.98
2150	Vinyl coated		912	.026		4.33	.83	.27	5.43	6.40
2200	Reinforcing wire, coiled spring, 7 ga. galv.		2279	.011		.19	.33	.11	.63	.88
2250	9 ga., vinyl coated		2282	.011		.28	.33	.11	.72	.98

32 31 13.53 High-Security Chain Link Fences, Gates and Systems		Crew	Daily Output	Labor-Hours	Unit	Material	Labor	Equipment	Total	Total Incl O&P
0010	**HIGH-SECURITY CHAIN LINK FENCES, GATES AND SYSTEMS**									
0100	Fence, chain link, security, 7′ H, standard FE-7, incl excavation & posts	B-80C	480	.050	L.F.	53	1.57	.42	54.99	61.50
0200	Fence, barbed wire, security, 7′ high, with 3 wire barbed wire arm	″	400	.060	″	7.25	1.88	.50	9.63	11.65
0300	Complete systems, including material and installation									
0310	Taunt wire fence detection system				M.L.F.				23,000	25,300
0410	Microwave fence detection system								38,000	41,800
0510	Passive magnetic fence detection system								18,000	19,800
0610	Infrared fence detection system								12,000	13,200
0710	Strain relief fence detection system								23,000	25,300
0810	Electro-shock fence detection system								33,000	36,300
0910	Photo-electric fence detection system								15,000	16,500

32 31 19 – Decorative Metal Fences and Gates

32 31 19.10 Decorative Fence		Crew	Daily Output	Labor-Hours	Unit	Material	Labor	Equipment	Total	Total Incl O&P
0010	**DECORATIVE FENCE**									
5300	Tubular picket, steel, 6′ sections, 1-9/16″ posts, 4′ high	B-80C	300	.080	L.F.	31	2.51	.67	34.18	39.50
5400	2″ posts, 5′ high		240	.100		43	3.14	.84	46.98	53.50
5600	2″ posts, 6′ high		200	.120		49	3.77	1.01	53.78	61
5700	Staggered picket 1-9/16″ posts, 4′ high		300	.080		28	2.51	.67	31.18	36
5800	2″ posts, 5′ high		240	.100		46	3.14	.84	49.98	57
5900	2″ posts, 6′ high		200	.120		48	3.77	1.01	52.78	60
6200	Gates, 4′ high, 3′ wide	B-1	10	2.400	Ea.	271	77.50		348.50	425
6300	5′ high, 3′ wide		10	2.400		350	77.50		427.50	515
6400	6′ high, 3′ wide		10	2.400		360	77.50		437.50	530
6500	4′ wide		10	2.400		420	77.50		497.50	595

32 31 23 – Plastic Fences and Gates

32 31 23.10 Fence, Vinyl		Crew	Daily Output	Labor-Hours	Unit	Material	Labor	Equipment	Total	Total Incl O&P
0010	**FENCE, VINYL**									
0011	White, steel reinforced, stainless steel fasteners									
0020	Picket, 4″ x 4″ posts @ 6′ - 0″ OC, 3′ high	B-1	140	.171	L.F.	19.85	5.55		25.40	31
0030	4′ high		130	.185		23	5.95		28.95	35
0040	5′ high		120	.200		26	6.45		32.45	39.50
0100	Board (semi-privacy), 5″ x 5″ posts @ 7′ - 6″ OC, 5′ high		130	.185		27.50	5.95		33.45	40
0120	6′ high		125	.192		31	6.20		37.20	44
0200	Basketweave, 5″ x 5″ posts @ 7′ - 6″ OC, 5′ high		160	.150		24	4.84		28.84	34.50
0220	6′ high		150	.160		28.50	5.15		33.65	40
0300	Privacy, 5″ x 5″ posts @ 7′ - 6″ OC, 5′ high		130	.185		29.50	5.95		35.45	42.50

32 31 Fences and Gates

32 31 23 – Plastic Fences and Gates

32 31 23.10 Fence, Vinyl		Crew	Daily Output	Labor-Hours	Unit	Material	Labor	Equipment	Total	Total Incl O&P
						2009 Bare Costs				
0320	6′ high	B-1	150	.160	L.F.	33.50	5.15		38.65	45.50
0350	Gate, 5′ high		9	2.667	Ea.	415	86		501	595
0360	6′ high		9	2.667		425	86		511	605
0400	For posts set in concrete, add		25	.960		8.45	31		39.45	60.50

32 31 26 – Wire Fences and Gates

32 31 26.10 Fences, Misc. Metal

		Crew	Daily Output	Labor-Hours	Unit	Material	Labor	Equipment	Total	Total Incl O&P
0010	**FENCES, MISC. METAL**									
0012	Chicken wire, posts @ 4′, 1″ mesh, 4′ high	B-80C	410	.059	L.F.	2.16	1.84	.49	4.49	5.95
0100	2″ mesh, 6′ high		350	.069		1.95	2.15	.58	4.68	6.30
0200	Galv. steel, 12 ga., 2″ x 4″ mesh, posts 5′ O.C., 3′ high		300	.080		3.12	2.51	.67	6.30	8.30
0300	5′ high		300	.080		4.16	2.51	.67	7.34	9.45
0400	14 ga., 1″ x 2″ mesh, 3′ high		300	.080		3.31	2.51	.67	6.49	8.50
0500	5′ high		300	.080		4.58	2.51	.67	7.76	9.90
1000	Kennel fencing, 1-1/2″ mesh, 6′ long, 3′-6″ wide, 6′-2″ high	2 Clab	4	4	Ea.	520	126		646	780
1050	12′ long		4	4		625	126		751	895
1200	Top covers, 1-1/2″ mesh, 6′ long		15	1.067		106	33.50		139.50	172
1250	12′ long		12	1.333		169	42		211	256
1300	For kennel doors, see Div. 08 31 13.40									
4500	Security fence, prison grade, set in concrete, 12′ high	B-80	25	1.280	L.F.	46	43.50	26	115.50	150
4600	16′ high	″	20	1.600	″	55	54	32.50	141.50	185

32 31 26.20 Wire Fencing, General

		Crew	Daily Output	Labor-Hours	Unit	Material	Labor	Equipment	Total	Total Incl O&P
0010	**WIRE FENCING, GENERAL**									
0015	Barbed wire, galvanized, domestic steel, hi-tensile 15-1/2 ga.				M.L.F.	35			35	38.50
0020	Standard, 12-3/4 ga.					46.50			46.50	51.50
0210	Barbless wire, 2-strand galvanized, 12-1/2 ga.					46.50			46.50	51.50
0500	Helical razor ribbon, stainless steel, 18″ dia x 18″ spacing				C.L.F.	136			136	149
0600	Hardware cloth galv., 1/4″ mesh, 23 ga., 2′ wide				C.S.F.	61			61	67
0700	3′ wide					59			59	65
0900	1/2″ mesh, 19 ga., 2′ wide					53.50			53.50	59
1000	4′ wide					52.50			52.50	58
1200	Chain link fabric, steel, 2″ mesh, 6 ga, galvanized					201			201	221
1300	9 ga, galvanized					99			99	109
1350	Vinyl coated					83			83	91.50
1360	Aluminized					129			129	142
1400	2-1/4″ mesh, 11.5 ga, galvanized					67			67	73.50
1600	1-3/4″ mesh (tennis courts), 11.5 ga (core), vinyl coated					95			95	104
1700	9 ga, galvanized					85.50			85.50	94
2100	Welded wire fabric, galvanized, 1″ x 2″, 14 ga.					57			57	62.50
2200	2″ x 4″, 12-1/2 ga.					19.45			19.45	21.50

32 31 29 – Wood Fences and Gates

32 31 29.10 Fence, Wood

		Crew	Daily Output	Labor-Hours	Unit	Material	Labor	Equipment	Total	Total Incl O&P
0010	**FENCE, WOOD**									
0011	Basket weave, 3/8″ x 4″ boards, 2″ x 4″									
0020	stringers on spreaders, 4″ x 4″ posts									
0050	No. 1 cedar, 6′ high	B-80C	160	.150	L.F.	9.10	4.71	1.26	15.07	19.15
0070	Treated pine, 6′ high	″	150	.160	″	10.95	5	1.35	17.30	22
0200	Board fence, 1″ x 4″ boards, 2″ x 4″ rails, 4″ x 4″ post									
0220	Preservative treated, 2 rail, 3′ high	B-80C	145	.166	L.F.	6.75	5.20	1.39	13.34	17.55
0240	4′ high		135	.178		7.40	5.60	1.50	14.50	19
0260	3 rail, 5′ high		130	.185		8.35	5.80	1.55	15.70	20.50
0300	6′ high		125	.192		9.70	6.05	1.61	17.36	22.50

32 31 29 – Wood Fences and Gates

32 31 29.10 Fence, Wood		Crew	Daily Output	Labor-Hours	Unit	Material	Labor	Equipment	Total	Total Incl O&P
						2009 Bare Costs				
0320	No. 2 grade western cedar, 2 rail, 3′ high	B-80C	145	.166	L.F.	7.35	5.20	1.39	13.94	18.20
0340	4′ high		135	.178		8.80	5.60	1.50	15.90	20.50
0360	3 rail, 5′ high		130	.185		10.15	5.80	1.55	17.50	22.50
0400	6′ high		125	.192		11.10	6.05	1.61	18.76	24
0420	No. 1 grade cedar, 2 rail, 3′ high		145	.166		11.15	5.20	1.39	17.74	22.50
0440	4′ high		135	.178		12.60	5.60	1.50	19.70	24.50
0460	3 rail, 5′ high		130	.185		14.85	5.80	1.55	22.20	27.50
0500	6′ high		125	.192		16.60	6.05	1.61	24.26	30
0540	Shadow box, 1″ x 6″ board, 2″ x 4″ rail, 4″ x 4″post									
0560	Pine, pressure treated, 3 rail, 6′ high	B-80C	150	.160	L.F.	12.45	5	1.35	18.80	23.50
0600	Gate, 3′-6″ wide		8	3	Ea.	75.50	94	25	194.50	266
0620	No. 1 cedar, 3 rail, 4′ high		130	.185	L.F.	15.25	5.80	1.55	22.60	28
0640	6′ high		125	.192		18.85	6.05	1.61	26.51	32.50
0860	Open rail fence, split rails, 2 rail 3′ high, no. 1 cedar		160	.150		6.15	4.71	1.26	12.12	15.95
0870	No. 2 cedar		160	.150		4.86	4.71	1.26	10.83	14.50
0880	3 rail, 4′ high, no. 1 cedar		150	.160		8.25	5	1.35	14.60	18.85
0890	No. 2 cedar		150	.160		5.80	5	1.35	12.15	16.15
0920	Rustic rails, 2 rail 3′ high, no. 1 cedar		160	.150		3.98	4.71	1.26	9.95	13.50
0930	No. 2 cedar		160	.150		3.87	4.71	1.26	9.84	13.40
0940	3 rail, 4′ high		150	.160		5.40	5	1.35	11.75	15.65
0950	No. 2 cedar		150	.160		4.10	5	1.35	10.45	14.25
0960	Picket fence, gothic, pressure treated pine									
1000	2 rail, 3′ high	B-80C	140	.171	L.F.	4.60	5.40	1.44	11.44	15.50
1020	3 rail, 4′ high		130	.185	″	5.45	5.80	1.55	12.80	17.25
1040	Gate, 3′-6″ wide		9	2.667	Ea.	48.50	83.50	22.50	154.50	216
1060	No. 2 cedar, 2 rail, 3′ high		140	.171	L.F.	5.75	5.40	1.44	12.59	16.80
1100	3 rail, 4′ high		130	.185	″	5.90	5.80	1.55	13.25	17.75
1120	Gate, 3′-6″ wide		9	2.667	Ea.	57.50	83.50	22.50	163.50	226
1140	No. 1 cedar, 2 rail 3′ high		140	.171	L.F.	11.50	5.40	1.44	18.34	23
1160	3 rail, 4′ high		130	.185	″	13.40	5.80	1.55	20.75	26
1170	Gate, 3′-6″ wide		9	2.667	Ea.	127	83.50	22.50	233	305
1200	Rustic picket, molded pine, 2 rail, 3′ high		140	.171	L.F.	5.50	5.40	1.44	12.34	16.50
1220	No. 1 cedar, 2 rail, 3′ high		140	.171		7.35	5.40	1.44	14.19	18.55
1240	Stockade fence, no. 1 cedar, 3-1/4″ rails, 6′ high		160	.150		10.85	4.71	1.26	16.82	21
1260	8′ high		155	.155		14	4.86	1.30	20.16	25
1300	No. 2 cedar, treated wood rails, 6′ high		160	.150		10.85	4.71	1.26	16.82	21
1320	Gate, 3′-6″ wide		8	3	Ea.	63.50	94	25	182.50	253
1360	Treated pine, treated rails, 6′ high		160	.150	L.F.	10.60	4.71	1.26	16.57	21
1400	8′ high		150	.160	″	15.90	5	1.35	22.25	27

32 31 29.20 Fence, Wood Rail

		Crew	Daily Output	Labor-Hours	Unit	Material	Labor	Equipment	Total	Total Incl O&P
0010	**FENCE, WOOD RAIL**									
0012	Picket, No. 2 cedar, Gothic, 2 rail, 3′ high	B-1	160	.150	L.F.	5.85	4.84		10.69	14.45
0050	Gate, 3′-6″ wide	B-80C	9	2.667	Ea.	50.50	83.50	22.50	156.50	218
0400	3 rail, 4′ high		150	.160	L.F.	6.75	5	1.35	13.10	17.15
0500	Gate, 3′-6″ wide		9	2.667	Ea.	60.50	83.50	22.50	166.50	229
5000	Fence rail, redwood, 2″ x 4″, merch grade 8′	B-1	2400	.010	L.F.	1.15	.32		1.47	1.80
5050	Select grade, 8′		2400	.010	″	3.45	.32		3.77	4.33
6000	Fence post, select redwood, earthpacked & treated, 4″ x 4″ x 6′		96	.250	Ea.	11.05	8.05		19.10	25.50
6010	4″ x 4″ x 8′		96	.250		13.10	8.05		21.15	27.50
6020	Set in concrete, 4″ x 4″ x 6′		50	.480		14.20	15.50		29.70	41
6030	4″ x 4″ x 8′		50	.480		16.95	15.50		32.45	44
6040	Wood post, 4′ high, set in concrete, incl. concrete		50	.480		8.30	15.50		23.80	34.50

32 31 Fences and Gates

32 31 29 – Wood Fences and Gates

32 31 29.20	Fence, Wood Rail	Crew	Daily Output	Labor-Hours	Unit	Material	Labor	Equipment	Total	Total Incl O&P
						2009 Bare Costs				
6050	Earth packed	B-1	96	.250	Ea.	5.70	8.05		13.75	19.55
6060	6′ high, set in concrete, incl. concrete		50	.480		10.65	15.50		26.15	37
6070	Earth packed		96	.250		7.75	8.05		15.80	22
9000	Minimum labor/equipment charge	1 Clab	2	4	Job		126		126	209

32 32 Retaining Walls

32 32 13 – Cast-in-Place Concrete Retaining Walls

32 32 13.10 Retaining Walls, Cast Concrete

		Crew	Daily Output	Labor-Hours	Unit	Material	Labor	Equipment	Total	Total Incl O&P
0010	**RETAINING WALLS, CAST CONCRETE**									
1800	Concrete gravity wall with vertical face including excavation & backfill									
1850	No reinforcing									
1900	6′ high, level embankment	C-17C	36	2.306	L.F.	74	95	13.50	182.50	251
2000	33° slope embankment		32	2.594		67.50	107	15.20	189.70	265
2200	8′ high, no surcharge		27	3.074		92	127	18	237	325
2300	33° slope embankment		24	3.458		111	143	20.50	274.50	375
2500	10′ high, level embankment		19	4.368		131	180	25.50	336.50	465
2600	33° slope embankment		18	4.611		182	190	27	399	540
2800	Reinforced concrete cantilever, incl. excavation, backfill & reinf.									
2900	6′ high, 33° slope embankment	C-17C	35	2.371	L.F.	67.50	98	13.90	179.40	248
3000	8′ high, 33° slope embankment		29	2.862		77.50	118	16.75	212.25	296
3100	10′ high, 33° slope embankment		20	4.150		101	171	24.50	296.50	415
3200	20′ high, 500 lb. per L.F. surcharge		7.50	11.067		305	455	65	825	1,150
3500	Concrete cribbing, incl. excavation and backfill									
3700	12′ high, open face	B-13	210	.267	S.F.	33.50	9.10	3.76	46.36	56
3900	Closed face	"	210	.267	"	31.50	9.10	3.76	44.36	53.50
4100	Concrete filled slurry trench, see Div. 31 56 23.20									

32 32 23 – Segmental Retaining Walls

32 32 23.13 Segmental Conc. Unit Masonry Retaining Walls

		Crew	Daily Output	Labor-Hours	Unit	Material	Labor	Equipment	Total	Total Incl O&P
0010	**SEGMENTAL CONC. UNIT MASONRY RETAINING WALLS**									
7100	Segmental Retaining Wall system, incl pins, and void fill									
7120	base not included									
7140	Large unit, 8″ high x 18″ wide x 20″ deep, 3 plane split	B-62	300	.080	S.F.	10.30	2.73	.50	13.53	16.25
7150	Straight split		300	.080		10.30	2.73	.50	13.53	16.25
7160	Medium, ltwt, 8″ high x 18″ wide x 12″ deep, 3 plane split		400	.060		11.50	2.04	.37	13.91	16.40
7170	Straight split		400	.060		8.65	2.04	.37	11.06	13.25
7180	Small unit, 4″ x 18″ x 10″ deep, 3 plane split		400	.060		12.25	2.04	.37	14.66	17.25
7190	Straight split		400	.060		12.95	2.04	.37	15.36	18
7200	Cap unit, 3 plane split		300	.080		14.50	2.73	.50	17.73	21
7210	Cap unit, straight split		300	.080		14.50	2.73	.50	17.73	21
7260	For reinforcing, add								4.25	4.68
8000	For higher walls, add components as necessary									

32 32 26 – Metal Crib Retaining Walls

32 32 26.10 Metal Bin Retaining Walls

		Crew	Daily Output	Labor-Hours	Unit	Material	Labor	Equipment	Total	Total Incl O&P
0010	**METAL BIN RETAINING WALLS**									
0011	Aluminized steel bin, excavation									
0020	and backfill not included, 10′ wide									
0100	4′ high, 5.5′ deep	B-13	650	.086	S.F.	23.50	2.95	1.21	27.66	32
0200	8′ high, 5.5′ deep		615	.091		27	3.11	1.28	31.39	36
0300	10′ high, 7.7′ deep		580	.097		28.50	3.30	1.36	33.16	38
0400	12′ high, 7.7′ deep		530	.106		30.50	3.61	1.49	35.60	41

32 32 Retaining Walls

32 32 26 – Metal Crib Retaining Walls

32 32 26.10 Metal Bin Retaining Walls		Crew	Daily Output	Labor-Hours	Unit	Material	Labor	Equipment	Total	Total Incl O&P
						2009 Bare Costs				
0500	16′ high, 7.7′ deep	B-13	515	.109	S.F.	32.50	3.72	1.53	37.75	43
0600	16′ high, 9.9′ deep		500	.112		34	3.83	1.58	39.41	45.50
0700	20′ high, 9.9′ deep		470	.119		38.50	4.07	1.68	44.25	50.50
0800	20′ high, 12.1′ deep		460	.122		41	4.16	1.72	46.88	53.50
0900	24′ high, 12.1′ deep		455	.123		43.50	4.21	1.74	49.45	57
1000	24′ high, 14.3′ deep		450	.124		45.50	4.25	1.75	51.50	59
1100	28′ high, 14.3′ deep		440	.127		47	4.35	1.79	53.14	61
1300	For plain galvanized bin type walls, deduct					10%				

32 32 36 – Gabion Retaining Walls

32 32 36.10 Stone Gabion Retaining Walls

		Crew	Daily Output	Labor-Hours	Unit	Material	Labor	Equipment	Total	Total Incl O&P
0010	**STONE GABION RETAINING WALLS**									
4300	Stone filled gabions, not incl. excavation,									
4310	Stone, delivered, 3′ wide									
4340	Galvanized, 6′ long, 1′ high	B-13	113	.496	Ea.	75.50	16.95	7	99.45	119
4400	1′-6″ high		50	1.120		95	38.50	15.80	149.30	183
4490	3′-0″ high		13	4.308		167	147	60.50	374.50	490
4590	9′ long, 1′ high		50	1.120		113	38.50	15.80	167.30	204
4650	1′-6″ high		22	2.545		143	87	36	266	340
4690	3′-0″ high		6	9.333		245	320	132	697	935
4890	12′ long, 1′ high		28	2		147	68.50	28	243.50	305
4950	1′-6″ high		13	4.308		189	147	60.50	396.50	515
4990	3′-0″ high		3	18.667		320	640	263	1,223	1,675
5200	PVC coated, 6′ long, 1′ high		113	.496		92.50	16.95	7	116.45	137
5250	1′-6″ high		50	1.120		113	38.50	15.80	167.30	203
5300	3′ high		13	4.308		184	147	60.50	391.50	510
5500	9′ long, 1′ high		50	1.120		123	38.50	15.80	177.30	214
5550	1′-6″ high		22	2.545		162	87	36	285	360
5600	3′ high		6	9.333		270	320	132	722	960
5800	12′ long, 1′ high		28	2		167	68.50	28	263.50	325
5850	1′-6″ high		13	4.308		224	147	60.50	431.50	555
5900	3′ high		3	18.667		355	640	263	1,258	1,700
6000	Galvanized, 6′ long, 1′ high		75	.747	C.Y.	74.50	25.50	10.55	110.55	135
6010	1′-6″ high		50	1.120		95	38.50	15.80	149.30	183
6020	3′-0″ high		25	2.240		83.50	76.50	31.50	191.50	250
6030	9′ long, 1′ high		50	1.120		113	38.50	15.80	167.30	204
6040	1′-6″ high		33.30	1.682		95.50	57.50	23.50	176.50	225
6050	3′-0″ high		16.70	3.353		81.50	115	47.50	244	330
6060	12′ long, 1′ high		37.50	1.493		110	51	21	182	227
6070	1′-6″ high		25	2.240		94.50	76.50	31.50	202.50	263
6080	3′-0″ high		12.50	4.480		80.50	153	63	296.50	405
6100	PVC coated, 6′ long, 1′ high		75	.747		139	25.50	10.55	175.05	205
6110	1′-6″ high		50	1.120		113	38.50	15.80	167.30	203
6120	3′ high		25	2.240		92	76.50	31.50	200	260
6130	9′ long, 1′ high		50	1.120		123	38.50	15.80	177.30	214
6140	1′-6″ high		33.30	1.682		108	57.50	23.50	189	239
6150	3′ high		16.67	3.359		90	115	47.50	252.50	340
6160	12′ long, 1′ high		37.50	1.493		125	51	21	197	244
6170	1′-6″ high		25	2.240		101	76.50	31.50	209	271
6180	3′ high		12.50	4.480		88	153	63	304	415

32 32 60 – Stone Retaining Walls

32 32 60.10 Retaining Walls, Stone

0010	**RETAINING WALLS, STONE**									

32 32 Retaining Walls

32 32 60 – Stone Retaining Walls

32 32 60.10 Retaining Walls, Stone		Crew	Daily Output	Labor-Hours	Unit	Material	2009 Bare Costs Labor	Equipment	Total	Total Incl O&P
0015	Including excavation, concrete footing and									
0020	stone 3′ below grade. Price is exposed face area.									
0200	Decorative random stone, to 6′ high, 1′-6″ thick, dry set	D-1	35	.457	S.F.	47	16.60		63.60	79
0300	Mortar set		40	.400		47	14.55		61.55	75.50
0500	Cut stone, to 6′ high, 1′-6″ thick, dry set		35	.457		47	16.60		63.60	79
0600	Mortar set		40	.400		47	14.55		61.55	75.50
0800	Random stone, 6′ to 10′ high, 2′ thick, dry set		45	.356		47	12.90		59.90	73
0900	Mortar set		50	.320		47	11.65		58.65	71
1100	Cut stone, 6′ to 10′ high, 2′ thick, dry set		45	.356		47	12.90		59.90	73
1200	Mortar set		50	.320		47	11.65		58.65	71
5100	Setting stone, dry		100	.160	C.F.		5.80		5.80	9.40
5600	With mortar		120	.133	"		4.84		4.84	7.85
9000	Minimum labor/equipment charge		2	8	Job		291		291	470

32 34 Fabricated Bridges

32 34 20 – Fabricated Pedestrian Bridges

32 34 20.10 Bridges, Pedestrian		Crew	Daily Output	Labor-Hours	Unit	Material	2009 Bare Costs Labor	Equipment	Total	Total Incl O&P
0010	**BRIDGES, PEDESTRIAN**									
0011	Spans over streams, roadways, etc.									
0020	including erection, not including foundations									
0050	Precast concrete, complete in place, 8′ wide, 60′ span	E-2	215	.260	S.F.	67.50	11.35	8.10	86.95	103
0100	100′ span		135	.303		74	13.20	9.40	96.60	115
0150	120′ span		160	.350		80.50	15.25	10.90	106.65	127
0200	150′ span		145	.386		83.50	16.80	12	112.30	135
0300	Steel, trussed or arch spans, compl. in place, 8′ wide, 40′ span		320	.175		88	7.60	5.45	101.05	116
0400	50′ span		395	.142		79	6.15	4.41	89.56	102
0500	60′ span		465	.120		79	5.25	3.74	87.99	100
0600	80′ span		570	.098		94	4.28	3.05	101.33	114
0700	100′ span		465	.120		132	5.25	3.74	140.99	158
0800	120′ span		365	.153		167	6.70	4.77	178.47	200
0900	150′ span		310	.181		177	7.85	5.60	190.45	215
1000	160′ span		255	.220		177	9.55	6.85	193.40	220
1100	10′ wide, 80′ span		640	.088		97.50	3.81	2.72	104.03	117
1200	120′ span		415	.135		126	5.90	4.20	136.10	154
1300	150′ span		445	.126		142	5.50	3.91	151.41	170
1400	200′ span		205	.273		151	11.90	8.50	171.40	197
1600	Wood, laminated type, complete in place, 80′ span	C-12	203	.236		58	9.30	3.79	71.09	83.50
1700	130′ span	"	153	.314		60.50	12.35	5.05	77.90	92

32 84 Planting Irrigation

32 84 23 – Underground Sprinklers

32 84 23.10 Sprinkler Irrigation System		Crew	Daily Output	Labor-Hours	Unit	Material	2009 Bare Costs Labor	Equipment	Total	Total Incl O&P
0010	**SPRINKLER IRRIGATION SYSTEM**									
0011	For lawns									
0100	Golf course with fully automatic system	C-17	.05	1600	9 holes	97,500	66,000		163,500	215,000
0200	24′ diam. head at 15′ O.C incl. piping, auto oper., minimum	B-20	70	.343	Head	22	12.10		34.10	44
0300	Maximum		40	.600		50.50	21		71.50	90.50
0500	60′ diam. head at 40′ O.C. incl. piping, auto oper., minimum		28	.857		66.50	30.50		97	123
0600	Maximum		23	1.043		186	37		223	265

32 84 23 – Underground Sprinklers

32 84 23.10 Sprinkler Irrigation System		Crew	Daily Output	Labor-Hours	Unit	2009 Bare Costs Material	Labor	Equipment	Total	Total Incl O&P
0800	Residential system, custom, 1″ supply	B-20	2000	.012	S.F.	.32	.42		.74	1.05
0900	1-1/2″ supply		1800	.013	″	.39	.47		.86	1.20
1020	Pop up spray head w/risers, hi-pop, full circle pattern, 4″	2 Skwk	76	.211	Ea.	4.26	8.60		12.86	18.70
1030	1/2 circle pattern, 4″		76	.211		4.52	8.60		13.12	18.95
1040	6″, full circle pattern		76	.211		10.20	8.60		18.80	25.50
1050	1/2 circle pattern, 6″		76	.211		10.35	8.60		18.95	25.50
1060	12″, full circle pattern		76	.211		15.10	8.60		23.70	30.50
1070	1/2 circle pattern, 12″		76	.211		14.90	8.60		23.50	30.50
1080	Pop up bubbler head w/risers, hi-pop bubbler head, 4″		76	.211		4.62	8.60		13.22	19.10
1090	6″		76	.211		10.95	8.60		19.55	26
1100	12″		76	.211		15.15	8.60		23.75	30.50
1110	Impact full/part circle sprinklers, 28′-54′ 25-60 PSI		37	.432		14.75	17.65		32.40	45
1120	Spaced 37′-49′ @ 25-50 PSI		37	.432		31.50	17.65		49.15	63.50
1130	Spaced 43′-61′ @ 30-60 PSI		37	.432		62	17.65		79.65	97
1140	Spaced 54′-78′ @ 40-80 PSI		37	.432		128	17.65		145.65	170
1145	Impact rotor pop-up full/part commercial circle sprinklers									
1150	Spaced 42′-65′ 35-80 PSI	2 Skwk	25	.640	Ea.	24.50	26		50.50	69.50
1160	Spaced 48′-76′ 45-85 PSI	″	25	.640	″	12.10	26		38.10	56
1165	Impact rotor pop-up part. circle comm., 53′-75′, 55-100 PSI, w/ accessories									
1170	Plastic case, metal cover	2 Skwk	25	.640	Ea.	173	26		199	233
1180	Rubber cover		25	.640		131	26		157	187
1190	Iron case, metal cover		22	.727		162	29.50		191.50	227
1200	Rubber cover		22	.727		181	29.50		210.50	248
1250	Plastic case, 2 nozzle, metal cover		25	.640		152	26		178	211
1260	Rubber cover		25	.640		162	26		188	221
1270	Iron case, 2 nozzle, metal cover		22	.727		218	29.50		247.50	289
1280	Rubber cover		22	.727		220	29.50		249.50	291
1282	Impact rotor pop-up full circle commercial, 39′-99′, 30-100 PSI									
1284	Plastic case, metal cover	2 Skwk	25	.640	Ea.	153	26		179	211
1286	Rubber cover		25	.640		170	26		196	230
1288	Iron case, metal cover		22	.727		222	29.50		251.50	293
1290	Rubber cover		22	.727		232	29.50		261.50	305
1292	Plastic case, 2 nozzle, metal cover		22	.727		165	29.50		194.50	230
1294	Rubber cover		22	.727		171	29.50		200.50	237
1296	Iron case, 2 nozzle, metal cover		20	.800		212	32.50		244.50	286
1298	Rubber cover		20	.800		227	32.50		259.50	305
1305	Electric remote control valve, plastic, 3/4″		18	.889		18.55	36.50		55.05	79.50
1310	1″		18	.889		21.50	36.50		58	82.50
1320	1-1/2″		18	.889		111	36.50		147.50	181
1330	2″		18	.889		130	36.50		166.50	201
1335	Quick coupling valves, brass, locking cover									
1340	Inlet coupling valve, 3/4″	2 Skwk	18.75	.853	Ea.	49.50	35		84.50	112
1350	1″		18.75	.853		57	35		92	120
1360	Controller valve boxes, 6″ round boxes		18.75	.853		5.05	35		40.05	62.50
1370	10″ round boxes		14.25	1.123		12.50	46		58.50	88.50
1380	12″ square box		9.75	1.641		27.50	67		94.50	140
1388	Electromech. control, 14 day 3-60 min, auto start to 23/day									
1390	4 station	2 Skwk	1.04	15.385	Ea.	203	630		833	1,250
1400	7 station		.64	25		320	1,025		1,345	2,025
1410	12 station		.40	40		375	1,625		2,000	3,075
1420	Dual programs, 18 station		.24	66.667		1,225	2,725		3,950	5,775
1430	23 station		.16	100		1,975	4,075		6,050	8,825
1435	Backflow preventer, bronze, 0-175 PSI, w/valves, test cocks									

32 84 Planting Irrigation

32 84 23 – Underground Sprinklers

32 84 23.10 Sprinkler Irrigation System		Crew	Daily Output	Labor-Hours	Unit	2009 Bare Costs Material	Labor	Equipment	Total	Total Incl O&P
1440	3/4"	2 Skwk	2	8	Ea.	118	325		443	660
1450	1"		2	8		137	325		462	680
1460	1-1/2"		2	8		273	325		598	830
1470	2"	↓	2	8	↓	310	325		635	870
1475	Pressure vacuum breaker, brass, 15-150 PSI									
1480	3/4"	2 Skwk	2	8	Ea.	73.50	325		398.50	610
1490	1"		2	8		105	325		430	645
1500	1-1/2"		2	8		200	325		525	750
1510	2"	↓	2	8	↓	257	325		582	810

32 91 Planting Preparation

32 91 13 – Soil Preparation

32 91 13.16 Mulching		Crew	Daily Output	Labor-Hours	Unit	Material	Labor	Equipment	Total	Total Incl O&P
0010	**MULCHING**									
0100	Aged barks, 3" deep, hand spread	1 Clab	100	.080	S.Y.	2.88	2.53		5.41	7.35
0150	Skid steer loader	B-63	13.50	2.963	M.S.F.	320	98	11.05	429.05	520
0200	Hay, 1" deep, hand spread	1 Clab	475	.017	S.Y.	.47	.53		1	1.40
0250	Power mulcher, small	B-64	180	.089	M.S.F.	52	2.78	1.84	56.62	64
0350	Large	B-65	530	.030	"	52	.94	.88	53.82	60
0400	Humus peat, 1" deep, hand spread	1 Clab	700	.011	S.Y.	4.50	.36		4.86	5.55
0450	Push spreader	"	2500	.003	"	4.50	.10		4.60	5.10
0550	Tractor spreader	B-66	700	.011	M.S.F.	500	.45	.33	500.78	550
0600	Oat straw, 1" deep, hand spread	1 Clab	475	.017	S.Y.	.46	.53		.99	1.39
0650	Power mulcher, small	B-64	180	.089	M.S.F.	51	2.78	1.84	55.62	62.50
0700	Large	B-65	530	.030	"	51	.94	.88	52.82	58.50
0750	Add for asphaltic emulsion	B-45	1770	.009	Gal.	2.24	.33	.44	3.01	3.48
0800	Peat moss, 1" deep, hand spread	1 Clab	900	.009	S.Y.	1.85	.28		2.13	2.50
0850	Push spreader	"	2500	.003	"	1.85	.10		1.95	2.21
0950	Tractor spreader	B-66	700	.011	M.S.F.	206	45	.33	206.78	227
1000	Polyethylene film, 6 mil.	2 Clab	2000	.008	S.Y.	.19	.25		.44	.63
1100	Redwood nuggets, 3" deep, hand spread	1 Clab	150	.053	"	4.65	1.69		6.34	7.90
1150	Skid steer loader	B-63	13.50	2.963	M.S.F.	515	98	11.05	624.05	740
1200	Stone mulch, hand spread, ceramic chips, economy	1 Clab	125	.064	S.Y.	6.90	2.02		8.92	10.95
1250	Deluxe	"	95	.084	"	10.55	2.66		13.21	16.05
1300	Granite chips	B-1	10	2.400	C.Y.	34.50	77.50		112	166
1400	Marble chips		10	2.400		129	77.50		206.50	270
1600	Pea gravel		28	.857		68.50	27.50		96	121
1700	Quartz	↓	10	2.400	↓	166	77.50		243.50	310
1800	Tar paper, 15 Lb. felt	1 Clab	800	.010	S.Y.	.43	.32		.75	.99
1900	Wood chips, 2" deep, hand spread	"	220	.036	"	1.88	1.15		3.03	3.97
1950	Skid steer loader	B-63	20.30	1.970	M.S.F.	209	65	7.35	281.35	345

32 91 13.23 Structural Soil Mixing		Crew	Daily Output	Labor-Hours	Unit	Material	Labor	Equipment	Total	Total Incl O&P
0010	**STRUCTURAL SOIL MIXING**									
0100	Rake topsoil, site material, harley rock rake, ideal	B-6	33	.727	M.S.F.		25	8.90	33.90	50
0200	Adverse	"	7	3.429			117	42	159	235
0300	Screened loam, york rake and finish, ideal	B-62	24	1			34	6.20	40.20	62.50
0400	Adverse	"	20	1.200			41	7.45	48.45	74.50
1000	Remove topsoil & stock pile on site, 75 HP dozer, 6" deep, 50' haul	B-10L	30	.400			15.25	13.30	28.55	39
1050	300' haul		6.10	1.967			75	65.50	140.50	192
1100	12" deep, 50' haul		15.50	.774			29.50	26	55.50	75.50
1150	300' haul	↓	3.10	3.871	↓		147	129	276	375

32 91 Planting Preparation

32 91 13 – Soil Preparation

32 91 13.23 Structural Soil Mixing		Crew	Daily Output	Labor-Hours	Unit	Material	2009 Bare Costs Labor	Equipment	Total	Total Incl O&P
1200	200 HP dozer, 6″ deep, 50′ haul	B-10B	125	.096	M.S.F.		3.66	8.65	12.31	15.35
1250	300′ haul		30.70	.391			14.90	35.50	50.40	63
1300	12″ deep, 50′ haul		62	.194			7.35	17.45	24.80	31
1350	300′ haul		15.40	.779			29.50	70.50	100	125
1400	Alternate method, 75 HP dozer, 50′ haul	B-10L	860	.014	C.Y.		.53	.46	.99	1.36
1450	300′ haul	″	114	.105			4.01	3.50	7.51	10.25
1500	200 HP dozer, 50′ haul	B-10B	2660	.005			.17	.41	.58	.72
1600	300′ haul	″	570	.021			.80	1.90	2.70	3.37
1800	Rolling topsoil, hand push roller	1 Clab	3200	.003	S.F.		.08		.08	.13
1850	Tractor drawn roller	B-66	10666	.001	″		.03	.02	.05	.07
2000	Root raking and loading, residential, no boulders	B-6	53.30	.450	M.S.F.		15.35	5.50	20.85	31
2100	With boulders		32	.750			25.50	9.20	34.70	51.50
2200	Municipal, no boulders		200	.120			4.09	1.47	5.56	8.25
2300	With boulders		120	.200			6.80	2.45	9.25	13.75
2400	Large commercial, no boulders	B-10B	400	.030			1.14	2.71	3.85	4.80
2500	With boulders	″	240	.050			1.91	4.51	6.42	8
3000	Scarify subsoil, residential, skid steer loader w/scarifiers, 50 HP	B-66	32	.250			9.75	7.25	17	23.50
3050	Municipal, skid steer loader w/scarifiers, 50 HP	″	120	.067			2.60	1.93	4.53	6.20
3100	Large commercial, 75 HP, dozer w/scarifier	B-10L	240	.050			1.91	1.66	3.57	4.87
3500	Screen topsoil from stockpile, vibrating screen, wet material (organic)	B-10P	200	.060	C.Y.		2.29	4.95	7.24	9.10
3550	Dry material	″	300	.040			1.52	3.30	4.82	6.05
3600	Mixing with conditioners, manure and peat	B-10R	550	.022			.83	.43	1.26	1.81
3650	Mobilization add for 2 days or less operation	B-34K	3	2.667	Job		85	261	346	430
3800	Spread conditioned topsoil, 6″ deep, by hand	B-1	360	.067	S.Y.	5	2.15		7.15	9.05
3850	300 HP dozer	B-10M	27	.444	M.S.F.	540	16.95	52.50	609.45	680
4000	Spread soil conditioners, alum. sulfate, 1#/S.Y., hand push spreader	1 Clab	17500	.001	S.Y.	16.70	.01		16.71	18.35
4050	Tractor spreader	B-66	700	.011	M.S.F.	1,850	.45	.33	1,850.78	2,050
4100	Fertilizer, 0.2#/S.Y., push spreader	1 Clab	17500	.001	S.Y.	.08	.01		.09	.11
4150	Tractor spreader	B-66	700	.011	M.S.F.	8.90	.45	.33	9.68	10.85
4200	Ground limestone, 1#/S.Y., push spreader	1 Clab	17500	.001	S.Y.	.09	.01		.10	.12
4250	Tractor spreader	B-66	700	.011	M.S.F.	10	.45	.33	10.78	12.05
4300	Lusoil, 3#/S.Y., push spreader	1 Clab	17500	.001	S.Y.	.52	.01		.53	.59
4350	Tractor spreader	B-66	700	.011	M.S.F.	58	.45	.33	58.78	64.50
4400	Manure, 18#/S.Y., push spreader	1 Clab	2500	.003	S.Y.	3.05	.10		3.15	3.53
4450	Tractor spreader	B-66	280	.029	M.S.F.	340	1.12	.83	341.95	380
4500	Perlite, 1″ deep, push spreader	1 Clab	17500	.001	S.Y.	9.35	.01		9.36	10.30
4550	Tractor spreader	B-66	700	.011	M.S.F.	1,050	.45	.33	1,050.78	1,150
4600	Vermiculite, push spreader	1 Clab	17500	.001	S.Y.	2.90	.01		2.91	3.21
4650	Tractor spreader	B-66	700	.011	M.S.F.	320	.45	.33	320.78	355
5000	Spread topsoil, skid steer loader and hand dress	B-62	270	.089	C.Y.	22.50	3.03	.55	26.08	30.50
5100	Articulated loader and hand dress	B-100	320	.038		22.50	1.43	2.40	26.33	30
5200	Articulated loader and 75HP dozer	B-10M	500	.024		22.50	.91	2.85	26.26	29.50
5300	Road grader and hand dress	B-11L	1000	.016		22.50	.58	.55	23.63	26.50
6000	Tilling topsoil, 20 HP tractor, disk harrow, 2″ deep	B-66	450	.018	M.S.F.		.69	.51	1.20	1.66
6050	4″ deep		360	.022			.87	.64	1.51	2.08
6100	6″ deep		270	.030			1.16	.86	2.02	2.76
6150	26″ rototiller, 2″ deep	A-1J	1250	.006	S.Y.		.20	.05	.25	.38
6200	4″ deep		1000	.008			.25	.06	.31	.48
6250	6″ deep		750	.011			.34	.08	.42	.64
7000	Lawn maintenance, see Div. 32 01 30.10									

32 91 13.26 Planting Beds										
0010	**PLANTING BEDS**									

32 91 Planting Preparation

32 91 13 – Soil Preparation

32 91 13.26 Planting Beds		Crew	Daily Output	Labor-Hours	Unit	Material	2009 Bare Costs Labor	Equipment	Total	Total Incl O&P
0100	Backfill planting pit, by hand, on site topsoil	2 Clab	18	.889	C.Y.		28		28	46.50
0200	Prepared planting mix, by hand	"	24	.667			21		21	35
0300	Skid steer loader, on site topsoil	B-62	340	.071			2.41	.44	2.85	4.38
0400	Prepared planting mix	"	410	.059			2	.36	2.36	3.63
1000	Excavate planting pit, by hand, sandy soil	2 Clab	16	1			31.50		31.50	52
1100	Heavy soil or clay	"	8	2			63		63	104
1200	1/2 C.Y. backhoe, sandy soil	B-11C	150	.107			3.89	1.96	5.85	8.40
1300	Heavy soil or clay	"	115	.139			5.10	2.55	7.65	10.95
2000	Mix planting soil, incl. loam, manure, peat, by hand	2 Clab	60	.267		39.50	8.45		47.95	57.50
2100	Skid steer loader	B-62	150	.160	▼	39.50	5.45	1	45.95	53.50
3000	Pile sod, skid steer loader	"	2800	.009	S.Y.		.29	.05	.34	.53
3100	By hand	2 Clab	400	.040			1.26		1.26	2.09
4000	Remove sod, F.E. loader	B-10S	2000	.006			.23	.16	.39	.54
4100	Sod cutter	B-12K	3200	.005			.19	.35	.54	.69
4200	By hand	2 Clab	240	.067	▼		2.11		2.11	3.48

32 92 Turf and Grasses

32 92 19 – Seeding

32 92 19.14 Seeding, Athletic Fields		Crew	Daily Output	Labor-Hours	Unit	Material	2009 Bare Costs Labor	Equipment	Total	Total Incl O&P
0010	**SEEDING, ATHLETIC FIELDS** R329219-50									
0020	Seeding, athletic fields, athletic field mix, 8#/MSFpush spreader	1 Clab	8	1	M.S.F.	15.35	31.50		46.85	69
0100	Tractor spreader	B-66	52	.154		15.35	6	4.45	25.80	31
0200	Hydro or air seeding, with mulch & fertil.	B-81	80	.300		16.90	10.50	7.25	34.65	43.50
0400	Birdsfoot trefoil, .45#/M.S.F., push spreader	1 Clab	8	1		7.80	31.50		39.30	60.50
0500	Tractor spreader	B-66	52	.154		7.80	6	4.45	18.25	23
0600	Hydro or air seeding,with mulch & fertil.	B-81	80	.300		15	10.50	7.25	32.75	41.50
0800	Bluegrass, 4#/M.S.F., common, push spreader	1 Clab	8	1		16.35	31.50		47.85	70
0900	Tractor spreader	B-66	52	.154		16.35	6	4.45	26.80	32.50
1000	Hydro or air seeding, with mulch & fertil.	B-81	80	.300		27	10.50	7.25	44.75	54.50
1100	Baron, push spreader	1 Clab	8	1		22	31.50		53.50	76
1200	Tractor spreader	B-66	52	.154		22	6	4.45	32.45	38.50
1300	Hydro or air seeding, with mulch & fertil.	B-81	80	.300		30	10.50	7.25	47.75	58
1500	Clover, 0.67#/M.S.F., white, push spreader	1 Clab	8	1		1.43	31.50		32.93	53.50
1600	Tractor spreader	B-66	52	.154		1.43	6	4.45	11.88	15.90
1700	Hydro or air seeding, with mulch and fertil.	B-81	80	.300		7.85	10.50	7.25	25.60	33.50
1800	Ladino, push spreader	1 Clab	8	1		5.60	31.50		37.10	58
1900	Tractor spreader	B-66	52	.154		5.60	6	4.45	16.05	20.50
2000	Hydro or air seeding, with mulch and fertil.	B-81	80	.300		24.50	10.50	7.25	42.25	52
2200	Fescue 5.5#/M.S.F., tall, push spreader	1 Clab	8	1		11.10	31.50		42.60	64
2300	Tractor spreader	B-66	52	.154		11.10	6	4.45	21.55	26.50
2400	Hydro or air seeding, with mulch and fertilizer	B-81	80	.300		36.50	10.50	7.25	54.25	65.50
2500	Chewing, push spreader	1 Clab	8	1		11.10	31.50		42.60	64
2600	Tractor spreader	B-66	52	.154		11.10	6	4.45	21.55	26.50
2700	Hydro or air seeding, with mulch and fertil.	B-81	80	.300		36.50	10.50	7.25	54.25	65.50
2900	Crown vetch, 4#/M.S.F., push spreader	1 Clab	8	1		41	31.50		72.50	97
3000	Tractor spreader	B-66	52	.154		41	6	4.45	51.45	59.50
3100	Hydro or air seeding, with mulch and fertilizer	B-81	80	.300		56	10.50	7.25	73.75	86.50
3300	Rye, 10#/M.S.F., annual, push spreader	1 Clab	8	1		23	31.50		54.50	77
3400	Tractor spreader	B-66	52	.154		23	6	4.45	33.45	39.50
3500	Hydro or air seeding, with mulch and fertilizer	B-81	80	.300		50	10.50	7.25	67.75	80
3600	Fine textured, push spreader	1 Clab	8	1	▼	23	31.50		54.50	77

32 92 Turf and Grasses

32 92 19 – Seeding

32 92 19.14 Seeding, Athletic Fields		Crew	Daily Output	Labor-Hours	Unit	Material	Labor	Equipment	Total	Total Incl O&P
						2009 Bare Costs				
3700	Tractor spreader	B-66	52	.154	M.S.F.	23	6	4.45	33.45	39.50
3800	Hydro or air seeding, with mulch and fertilizer	B-81	80	.300		50	10.50	7.25	67.75	80
4000	Shade mix, 6#/M.S.F., push spreader	1 Clab	8	1		10.70	31.50		42.20	64
4100	Tractor spreader	B-66	52	.154		10.70	6	4.45	21.15	26
4200	Hydro or air seeding, with mulch and fertilizer	B-81	80	.300		23.50	10.50	7.25	41.25	51
4400	Slope mix, 6#/M.S.F., push spreader	1 Clab	8	1		10.70	31.50		42.20	64
4500	Tractor spreader	B-66	52	.154		10.70	6	4.45	21.15	26
4600	Hydro or air seeding, with mulch and fertilizer	B-81	80	.300		27	10.50	7.25	44.75	54.50
4800	Turf mix, 4#/M.S.F., push spreader	1 Clab	8	1		7.15	31.50		38.65	60
4900	Tractor spreader	B-66	52	.154		7.15	6	4.45	17.60	22
5000	Hydro or air seeding, with mulch and fertilizer	B-81	80	.300		17.90	10.50	7.25	35.65	44.50
5200	Utility mix, 7#/M.S.F., push spreader	1 Clab	8	1		12.45	31.50		43.95	65.50
5300	Tractor spreader	B-66	52	.154		12.45	6	4.45	22.90	28
5400	Hydro or air seeding, with mulch and fertilizer	B-81	80	.300		46.50	10.50	7.25	64.25	76.50
5600	Wildflower, .10#/M.S.F., push spreader	1 Clab	8	1		4.40	31.50		35.90	57
5700	Tractor spreader	B-66	52	.154		4.40	6	4.45	14.85	19.20
5800	Hydro or air seeding, with mulch and fertilizer	B-81	80	.300		24	10.50	7.25	41.75	51.50
7025	Fertilizer, mechanical spread	1 Clab	1.75	4.571	Acre	3.97	144		147.97	242
7060	Limestone, mechanical spread		1.74	4.598	"	3.97	145		148.97	244
9000	Minimum labor/equipment charge		4	2	Job		63		63	104

32 92 23 – Sodding

32 92 23.10 Sodding Systems		Crew	Daily Output	Labor-Hours	Unit	Material	Labor	Equipment	Total	Total Incl O&P
0010	**SODDING SYSTEMS**									
0020	Sodding, 1" deep, bluegrass sod, on level ground, over 8 MSF	B-63	22	1.818	M.S.F.	245	60	6.80	311.80	375
0200	4 M.S.F.		17	2.353		270	78	8.80	356.80	435
0300	1000 S.F.		13.50	2.963		295	98	11.05	404.05	495
0500	Sloped ground, over 8 M.S.F.		6	6.667		245	221	25	491	660
0600	4 M.S.F.		5	8		270	265	30	565	760
0700	1000 S.F.		4	10		295	330	37.50	662.50	905
1000	Bent grass sod, on level ground, over 6 M.S.F.		20	2		545	66	7.45	618.45	710
1100	3 M.S.F.		18	2.222		605	73.50	8.30	686.80	795
1200	Sodding 1000 S.F. or less		14	2.857		695	94.50	10.65	800.15	925
1500	Sloped ground, over 6 M.S.F.		15	2.667		545	88	9.95	642.95	750
1600	3 M.S.F.		13.50	2.963		605	98	11.05	714.05	835
1700	1000 S.F.		12	3.333		695	110	12.45	817.45	955

32 93 Plants

32 93 10 – General Planting Costs

32 93 10.12 Travel		Crew	Daily Output	Labor-Hours	Unit	Material	Labor	Equipment	Total	Total Incl O&P
0010	**TRAVEL** to all nursery items, for 10 to 20 miles, add									
0015	10 to 20 miles one way, add				All					5%
0100	30 to 50 miles one way, add				"					10%

32 93 13 – Ground Covers

32 93 13.20 Ground Cover and Vines		Crew	Daily Output	Labor-Hours	Unit	Material	Labor	Equipment	Total	Total Incl O&P
0010	**GROUND COVER AND VINES** Planting only, no preparation									
0100	Ajuga, 1 year, bare root	B-1	9	2.667	C	70	86		156	219
0150	Potted, 2 year		6	4		255	129		384	495
0200	Berberis, potted, 2 year		6	4		270	129		399	510
0250	Cotoneaster, 15" to 18", shady areas, B & B		.60	40		765	1,300		2,065	2,975
0300	Boston ivy, on bank, 1 year, bare root		6	4		122	129		251	345

32 93 Plants

32 93 13 – Ground Covers

32 93 13.20 Ground Cover and Vines		Crew	Daily Output	Labor-Hours	Unit	2009 Bare Costs Material	Labor	Equipment	Total	Total Incl O&P
0350	Potted, 2 year	B-1	6	4	C	305	129		434	550
0400	English ivy, 1 year, bare root		9	2.667		71	86		157	220
0450	Potted, 2 year		6	4		118	129		247	345
0500	Halls honeysuckle, 1 year, bare root		5	4.800		138	155		293	410
0550	Potted, 2 year		4	6		390	194		584	745
0600	Memorial rose, 9" to 12", 1 gallon container		3	8		95	258		353	530
0650	Potted, 2 gallon container		2	12		165	385		550	820
0700	Pachysandra, 1 year, bare root		10	2.400		14.20	77.50		91.70	144
0750	Potted, 2 year		6	4		100	129		229	325
0800	Vinca minor, 1 year, bare root		10	2.400		108	77.50		185.50	247
0850	Potted, 2 year		6	4		95	129		224	320
0900	Woodbine, on bank, 1/2 year, bare root		6	4		115	129		244	340
0950	Potted, 2 year		4	6		290	194		484	640
2000	Alternate method of figuring									
2100	Ajuga, field division, 4000/M.S.F.	B-1	.23	104	M.S.F.	2,225	3,375		5,600	8,000
2300	Boston ivy, 1 year, 60/M.S.F.		10	2.400		73	77.50		150.50	209
2400	English ivy, 1 yr., 500/M.S.F.		1.80	13.333		310	430		740	1,050
2500	Halls honeysuckle, 1 yr., 333/M.S.F.		1.50	16		175	515		690	1,050
2600	Memorial rose, 9"-12", 1 gal., 333/M.S.F.		.90	26.667		575	860		1,435	2,050
2700	Pachysandra, 1 yr., 4000/M.S.F.		.25	96		600	3,100		3,700	5,750
2800	Vinca minor, rooted cutting, 2000/M.S.F.		1	24		1,250	775		2,025	2,650
2900	Woodbine, 1 yr., 60/M.S.F.		10	2.400		73	77.50		150.50	209

32 93 33 – Shrubs

32 93 33.10 Shrubs and Trees

		Crew	Daily Output	Labor-Hours	Unit	Material	Labor	Equipment	Total	Total Incl O&P
0010	**SHRUBS AND TREES**									
0011	Evergreen, in prepared beds, B & B									
0100	Arborvitae pyramidal, 4'-5'	B-17	30	1.067	Ea.	47	36	20.50	103.50	133
0150	Globe, 12"-15"	B-1	96	.250		21	8.05		29.05	36.50
0200	Balsam, fraser, 6' - 7'	B-17	30	1.067		122	36	20.50	178.50	215
0300	Cedar, blue, 8'-10'		18	1.778		233	59.50	34.50	327	390
0350	Japanese, 4' - 5'		55	.582		99	19.50	11.30	129.80	153
0400	Cypress, hinoki, 15" - 18"	B-1	80	.300		61	9.70		70.70	83
0500	Hemlock, canadian, 2-1/2'-3'		36	.667		33.50	21.50		55	72.50
0550	Holly, Savannah, 8' - 10' H		9.68	2.479		475	80		555	655
0600	Juniper, andorra, 18"-24"		80	.300		16.75	9.70		26.45	34.50
0620	Wiltoni, 15"-18"		80	.300		18.10	9.70		27.80	36
0640	Skyrocket, 4-1/2'-5'	B-17	55	.582		55	19.50	11.30	85.80	104
0660	Blue pfitzer, 2'-2-1/2'	B-1	44	.545		25.50	17.60		43.10	57
0680	Ketleerie, 2-1/2'-3'		50	.480		33.50	15.50		49	62.50
0700	Pine, black, 2-1/2'-3'		50	.480		45.50	15.50		61	75.50
0720	Mugo, 18"-24"		60	.400		36	12.90		48.90	61
0740	White, 4'-5'	B-17	75	.427		56.50	14.30	8.25	79.05	95
0800	Spruce, blue, 18"-24"	B-1	60	.400		38	12.90		50.90	63.50
0820	Dwarf alberta, 18" - 24"	"	60	.400		25	12.90		37.90	49
0840	Norway, 4'-5'	B-17	75	.427		65	14.30	8.25	87.55	104
0900	Yew, denisforma, 12"-15"	B-1	60	.400		25.50	12.90		38.40	50
1000	Capitata, 18"-24"		30	.800		24	26		50	68.50
1100	Hicksi, 2'-2-1/2'		30	.800		32	26		58	77.50

32 93 33.20 Shrubs

		Crew	Daily Output	Labor-Hours	Unit	Material	Labor	Equipment	Total	Total Incl O&P
0010	**SHRUBS**									
0011	Broadleaf Evergreen, planted in prepared beds									
0100	Andromeda, 15"-18", container	B-1	96	.250	Ea.	26.50	8.05		34.55	42.50

32 93 Plants

32 93 33 – Shrubs

32 93 33.20 Shrubs		Crew	Daily Output	Labor-Hours	Unit	Material	Labor	Equipment	Total	Total Incl O&P
						2009 Bare Costs				
0200	Azalea, 15" - 18", container	B-1	96	.250	Ea.	32	8.05		40.05	48.50
0300	Barberry, 9"-12", container		130	.185		11.75	5.95		17.70	23
0400	Boxwood, 15"-18", B&B		96	.250		29	8.05		37.05	45
0500	Euonymus, emerald gaiety, 12" to 15", container		115	.209		21	6.75		27.75	34
0600	Holly, 15"-18", B & B		96	.250		18.60	8.05		26.65	34
0700	Leucothoe, 15" - 18", container		96	.250		15.50	8.05		23.55	30.50
0800	Mahonia, 18" - 24", container		80	.300		32	9.70		41.70	51
0900	Mount laurel, 18" - 24", B & B		80	.300		56	9.70		65.70	77.50
1000	Paxistema, 9 – 12" high		130	.185		18.25	5.95		24.20	30
1100	Rhododendron, 18"-24", container		48	.500		31.50	16.15		47.65	61
1200	Rosemary, 1 gal container		600	.040		62.50	1.29		63.79	71
2000	Deciduous, planted in prepared beds, amelanchier, 2'-3', B & B		57	.421		90	13.60		103.60	122
2100	Azalea, 15"-18", B & B		96	.250		24.50	8.05		32.55	40.50
2200	Barberry, 2' - 3', B & B		57	.421		23.50	13.60		37.10	48.50
2300	Bayberry, 2'-3', B & B		57	.421		28	13.60		41.60	53
2400	Boston ivy, 2 year, container		600	.040		17.65	1.29		18.94	21.50
2500	Corylus, 3' - 4', B & B	B-17	75	.427		40	14.30	8.25	62.55	76.50
2600	Cotoneaster, 15"-18", B & B	B-1	80	.300		16.50	9.70		26.20	34
2700	Deutzia, 12" - 15", B & B	"	96	.250		11.50	8.05		19.55	26
2800	Dogwood, 3'-4', B & B	B-17	40	.800		26	27	15.50	68.50	89
2900	Euonymus, alatus compacta, 15" to 18", container	B-1	80	.300		22	9.70		31.70	40
3000	Flowering almond, 2' - 3', container	"	36	.667		18	21.50		39.50	55.50
3100	Flowering currant, 3' - 4', container	B-17	75	.427		24	14.30	8.25	46.55	59
3200	Forsythia, 2'-3', container	B-1	60	.400		20	12.90		32.90	43.50
3300	Hibiscus, 3'-4', B & B	B-17	75	.427		15	14.30	8.25	37.55	49
3400	Honeysuckle, 3'-4', B & B	B-1	60	.400		23	12.90		35.90	47
3500	Hydrangea, 2'-3', B & B	"	57	.421		28.50	13.60		42.10	54
3600	Lilac, 3'-4', B & B	B-17	40	.800		27	27	15.50	69.50	90
3700	Mockorange, 3' - 4', B & B	B-1	36	.667		24	21.50		45.50	62
3800	Osier willow, 2' - 3', B & B		57	.421		20.50	13.60		34.10	45
3900	Privet, bare root, 18"-24"		80	.300		13.15	9.70		22.85	30.50
4000	Pyracantha, 2' - 3', container		80	.300		57	9.70		66.70	78.50
4100	Quince, 2'-3', B & B		57	.421		24	13.60		37.60	49
4200	Russian olive, 3'-4', B & B	B-17	75	.427		24	14.30	8.25	46.55	58.50
4300	Snowberry, 2' - 3', B & B	B-1	57	.421		12.70	13.60		26.30	36.50
4400	Spirea, 3'-4', B & B	"	70	.343		17.15	11.05		28.20	37
4500	Viburnum, 3'-4', B & B	B-17	40	.800		29.50	27	15.50	72	93
4600	Weigela, 3' - 4', B & B	B-1	70	.343		13.05	11.05		24.10	32.50

32 93 43 – Trees

32 93 43.10 Planting		Crew	Daily Output	Labor-Hours	Unit	Material	Labor	Equipment	Total	Total Incl O&P
0010	**PLANTING**									
0011	Trees, shrubs and ground cover									
0100	Light soil									
0110	Bare root seedlings, 3" to 5" height	1 Clab	960	.008	Ea.		.26		.26	.43
0120	6" to 10"		520	.015			.49		.49	.80
0130	11" to 16"		370	.022			.68		.68	1.13
0140	17" to 24"		210	.038			1.20		1.20	1.99
0200	Potted, 2-1/4" diameter		840	.010			.30		.30	.50
0210	3" diameter		700	.011			.36		.36	.60
0220	4" diameter		620	.013			.41		.41	.67
0300	Container, 1 gallon	2 Clab	84	.190			6		6	9.95
0310	2 gallon		52	.308			9.70		9.70	16.05

32 93 43.10 Planting			Crew	Daily Output	Labor-Hours	Unit	2009 Bare Costs Material	Labor	Equipment	Total	Total Incl O&P
0320	3 gallon		2 Clab	40	.400	Ea.		12.65		12.65	21
0330	5 gallon		↓	29	.552	↓		17.45		17.45	29

32 93 43.20 Trees

			Crew	Daily Output	Labor-Hours	Unit	Material	Labor	Equipment	Total	Total Incl O&P
0010	**TREES**										
0011	Deciduous, in prep. beds, balled & burlapped (B&B)										
0100	Ash, 2" caliper	G	B-17	8	4	Ea.	115	134	77.50	326.50	430
0200	Beech, 5'-6'	G		50	.640		223	21.50	12.40	256.90	294
0300	Birch, 6'-8', 3 stems	G		20	1.600		123	53.50	31	207.50	257
0400	Cherry, 6' - 8', 1" caliper	G		24	1.333		81.50	44.50	26	152	191
0500	Crabapple, 6'-8'	G		20	1.600		160	53.50	31	244.50	298
0600	Dogwood, 4'-5'	G		40	.800		68.50	27	15.50	111	136
0700	Eastern redbud 4'-5'	G		40	.800		133	27	15.50	175.50	208
0800	Elm, 8'-10'	G		20	1.600		119	53.50	31	203.50	253
0900	Ginkgo, 6'-7'	G		24	1.333		163	44.50	26	233.50	280
1000	Hawthorn, 8'-10', 1" caliper	G		20	1.600		136	53.50	31	220.50	272
1100	Honeylocust, 10'-12', 1-1/2" caliper	G		10	3.200		149	107	62	318	405
1200	Laburnum, 6' - 8', 1" caliper	G		24	1.333		55	44.50	26	125.50	162
1300	Larch, 8'	G		32	1		95	33.50	19.40	147.90	181
1400	Linden, 8'-10', 1" caliper	G		20	1.600		107	53.50	31	191.50	240
1500	Magnolia, 4'-5'	G		20	1.600		79.50	53.50	31	164	209
1600	Maple, red, 8'-10', 1-1/2" caliper	G		10	3.200		174	107	62	343	435
1700	Mountain ash, 8'-10', 1" caliper	G		16	2		181	67	39	287	350
1800	Oak, 2-1/2"-3" caliper	G		6	5.333		258	179	103	540	690
1900	Pagoda, 6' - 8'	G		20	1.600		90	53.50	31	174.50	221
2000	Pear, 6' - 8', 1" caliper	G		20	1.600		133	53.50	31	217.50	268
2100	Planetree, 9'-11', 1-1/4" caliper	G		10	3.200		105	107	62	274	360
2200	Plum, 6'-8', 1" caliper	G		20	1.600		91	53.50	31	175.50	222
2300	Poplar, 9'-11', 1-1/4" caliper	G		10	3.200		50	107	62	219	298
2400	Shadbush, 4' - 5'	G		60	.533		67.50	17.90	10.35	95.75	115
2500	Sumac, 2'-3'	G		75	.427		23	14.30	8.25	45.55	57.50
2600	Tupelo, 5' - 6'	G		40	.800		50	27	15.50	92.50	116
2700	Tulip, 5'-6'	G		40	.800		49	27	15.50	91.50	114
2800	Willow, 6'-8', 1" caliper	G	↓	20	1.600	↓	64	53.50	31	148.50	192
9000	Minimum labor/equipment charge		1 Clab	4	2	Job		63		63	104

32 94 Planting Accessories

32 94 13 – Landscape Edging

32 94 13.20 Edging

		Crew	Daily Output	Labor-Hours	Unit	Material	Labor	Equipment	Total	Total Incl O&P
0010	**EDGING**									
0050	Aluminum alloy, including stakes, 1/8" x 4", mill finish	B-1	390	.062	L.F.	3.10	1.99		5.09	6.70
0051	Black paint		390	.062		3.60	1.99		5.59	7.25
0052	Black anodized		390	.062		4.15	1.99		6.14	7.85
0060	3/16" x 4", mill finish		380	.063		4.46	2.04		6.50	8.25
0061	Black paint		380	.063		5.15	2.04		7.19	9
0062	Black anodized		380	.063		6	2.04		8.04	9.95
0070	1/8" x 5-1/2" mill finish		370	.065		4.50	2.09		6.59	8.40
0071	Black paint		370	.065		5.35	2.09		7.44	9.30
0072	Black anodized		370	.065		6.10	2.09		8.19	10.15
0080	3/16" x 5-1/2" mill finish		360	.067		6	2.15		8.15	10.15
0081	Black paint		360	.067		6.80	2.15		8.95	11
0082	Black anodized	↓	360	.067	↓	7.85	2.15		10	12.20

32 94 Planting Accessories

32 94 13 – Landscape Edging

32 94 13.20 Edging		Crew	Daily Output	Labor-Hours	Unit	Material	Labor	Equipment	Total	Total Incl O&P
						2009 Bare Costs				
0100	Brick, set horizontally, 1-1/2 bricks per L.F.	D-1	370	.043	L.F.	1.67	1.57		3.24	4.38
0150	Set vertically, 3 bricks per L.F.	"	135	.119		4.20	4.31		8.51	11.55
0200	Corrugated aluminum, roll, 4" wide	1 Carp	650	.012		.70	.49		1.19	1.58
0250	6" wide	"	550	.015	↓	.88	.58		1.46	1.92
0300	Concrete, cast in place, see Div. 03 30 53.40									
0350	Granite, 5" x 16", straight	B-29	300	.187	L.F.	12.05	6.40	3.20	21.65	27
0400	Polyethylene grass barrier, 5" x 1/8"	D-1	400	.040		.80	1.45		2.25	3.23
0500	Precast scallops, green, 2" x 8" x 16"		400	.040		1.10	1.45		2.55	3.56
0550	2" x 8" x 16" other than green	↓	400	.040		.88	1.45		2.33	3.32
0600	Railroad ties, 6" x 8"	2 Carp	170	.094		3.10	3.76		6.86	9.60
0650	7" x 9"		136	.118		3.44	4.70		8.14	11.55
0750	Redwood 2" x 4"	↓	330	.048		2.27	1.94		4.21	5.70
0800	Steel edge strips, incl. stakes, 1/4" x 5"	B-1	390	.062		4.25	1.99		6.24	7.95
0850	3/16" x 4"	"	390	.062		3.36	1.99		5.35	6.95
0900	Hardwood, pressure treated, 4" x 6"	2 Carp	250	.064		2.50	2.56		5.06	6.95
0940	6" x 6"		200	.080		3.58	3.20		6.78	9.25
0980	6" x 8"		170	.094		4.08	3.76		7.84	10.70
1000	Pine, pressure treated, 1" x 4"		500	.032		.55	1.28		1.83	2.72
1040	2" x 4"		330	.048		.92	1.94		2.86	4.22
1080	4" x 6"		250	.064		3.08	2.56		5.64	7.60
1100	6" x 6"		200	.080		4.62	3.20		7.82	10.40
1140	6" x 8"	↓	170	.094	↓	6.05	3.76		9.81	12.85
9000	Minimum labor/equipment charge	1 Carp	4	2	Job		80		80	132

32 94 50 – Tree Guying

32 94 50.10 Tree Guying Systems		Crew	Daily Output	Labor-Hours	Unit	Material	Labor	Equipment	Total	Total Incl O&P
0010	**TREE GUYING SYSTEMS**									
0015	Tree guying Including stakes, guy wire and wrap									
0100	Less than 3" caliper, 2 stakes	2 Clab	35	.457	Ea.	17.50	14.45		31.95	43.50
0200	3" to 4" caliper, 3 stakes	"	21	.762	"	20	24		44	61.50
1000	Including arrowhead anchor, cable, turnbuckles and wrap									
1100	Less than 3" caliper, 3" anchors	2 Clab	20	.800	Ea.	51	25.50		76.50	97.50
1200	3" to 6" caliper, 4" anchors		15	1.067		51	33.50		84.50	112
1300	6" caliper, 6" anchors		12	1.333		95	42		137	175
1400	8" caliper, 8" anchors	↓	9	1.778	↓	107	56		163	210

32 96 Transplanting

32 96 23 – Plant and Bulb Transplanting

32 96 23.23 Planting		Crew	Daily Output	Labor-Hours	Unit	Material	Labor	Equipment	Total	Total Incl O&P
0010	**PLANTING**									
0012	Moving shrubs on site, 12" ball	B-62	28	.857	Ea.		29	5.35	34.35	53.50
0100	24" ball	"	22	1.091	"		37	6.80	43.80	68

32 96 23.43 Moving Trees		Crew	Daily Output	Labor-Hours	Unit	Material	Labor	Equipment	Total	Total Incl O&P
0290	**MOVING TREES**, On site									
0300	Moving trees on site, 36" ball	B-6	3.75	6.400	Ea.		218	78.50	296.50	440
0400	60" ball	"	1	24	"		820	294	1,114	1,650

32 96 43 – Tree Transplanting

32 96 43.20 Tree Removal		Crew	Daily Output	Labor-Hours	Unit	Material	Labor	Equipment	Total	Total Incl O&P
0010	**TREE REMOVAL**									
0100	Dig & lace, shrubs, broadleaf evergreen, 18"-24" high	B-1	55	.436	Ea.		14.10		14.10	23
0200	2'-3'	"	35	.686	↓		22		22	36.50

32 96 43.20 Tree Removal		Crew	Daily Output	Labor-Hours	Unit	Material	2009 Bare Costs Labor	Equipment	Total	Total Incl O&P
0300	3′-4′	B-6	30	.800	Ea.		27.50	9.80	37.30	55
0400	4′-5′	"	20	1.200			41	14.70	55.70	82.50
1000	Deciduous, 12″-15″	B-1	110	.218			7.05		7.05	11.60
1100	18″-24″		65	.369			11.90		11.90	19.65
1200	2′-3′		55	.436			14.10		14.10	23
1300	3′-4′	B-6	50	.480			16.35	5.90	22.25	33
2000	Evergreen, 18″-24″	B-1	55	.436			14.10		14.10	23
2100	2′-0″ to 2′-6″		50	.480			15.50		15.50	25.50
2200	2′-6″ to 3′-0″		35	.686			22		22	36.50
2300	3′-0″ to 3′-6″		20	1.200			38.50		38.50	64
3000	Trees, deciduous, small, 2′-3′		55	.436			14.10		14.10	23
3100	3′-4′	B-6	50	.480			16.35	5.90	22.25	33
3200	4′-5′		35	.686			23.50	8.40	31.90	47.50
3300	5′-6′		30	.800			27.50	9.80	37.30	55
4000	Shade, 5′-6′		50	.480			16.35	5.90	22.25	33
4100	6′-8′		35	.686			23.50	8.40	31.90	47.50
4200	8′-10′		25	.960			32.50	11.75	44.25	66
4300	2″ caliper		12	2			68	24.50	92.50	138
5000	Evergreen, 4′-5′		35	.686			23.50	8.40	31.90	47.50
5100	5′-6′		25	.960			32.50	11.75	44.25	66
5200	6′-7′		19	1.263			43	15.45	58.45	87
5300	7′-8′		15	1.600			54.50	19.60	74.10	110
5400	8′-10′		11	2.182			74.50	26.50	101	151

Division Notes

		DAILY	LABOR-		2009 BARE COSTS				TOTAL
	CREW	OUTPUT	HOURS	UNIT	MAT.	LABOR	EQUIP.	TOTAL	INCL O&P

Estimating Tips

33 10 00 Water Utilities
33 30 00 Sanitary Sewerage Utilities
33 40 00 Storm Drainage Utilities

- Never assume that the water, sewer, and drainage lines will go in at the early stages of the project. Consider the site access needs before dividing the site in half with open trenches, loose pipe, and machinery obstructions. Always inspect the site to establish that the site drawings are complete. Check off all existing utilities on your drawings as you locate them. If you find any discrepancies, mark up the site plan for further research. Differing site conditions can be very costly if discovered later in the project.
- See also Section 33 01 00 for restoration of pipe where removal/replacement may be undesirable. Use of new types of piping materials can reduce the overall project cost. Owners/design engineers should consider the installing contractor as a valuable source of current information on utility products and local conditions that could lead to significant cost savings.

Reference Numbers

Reference numbers are shown in shaded boxes at the beginning of some major classifications. These numbers refer to related items in the Reference Section. The reference information may be an estimating procedure, an alternate pricing method, or technical information.

Note: Not all subdivisions listed here necessarily appear in this publication.

Note: **Trade Service,** *in part, has been used as a reference source for some of the material prices used in Division 33.*

33 01 10 – Operation and Maintenance of Water Utilities

33 01 10.10 Corrosion Resistance		Crew	Daily Output	Labor-Hours	Unit	Material	Labor	Equipment	Total	Total Incl O&P
						2009 Bare Costs				
0010	**CORROSION RESISTANCE**									
0012	Wrap & coat, add to pipe, 4" diameter				L.F.	2			2	2.20
0020	5" diameter					2.08			2.08	2.29
0040	6" diameter					2.08			2.08	2.29
0060	8" diameter					3.22			3.22	3.54
0080	10" diameter					4.81			4.81	5.30
0100	12" diameter					4.81			4.81	5.30
0120	14" diameter					7.05			7.05	7.75
0140	16" diameter					7.05			7.05	7.75
0160	18" diameter					7.15			7.15	7.85
0180	20" diameter					7.45			7.45	8.20
0200	24" diameter					9.10			9.10	10.05
0220	Small diameter pipe, 1" diameter, add					1.50			1.50	1.65
0240	2" diameter					1.65			1.65	1.82
0260	2-1/2" diameter					1.82			1.82	2
0280	3" diameter					1.82			1.82	2
0300	Fittings, field covered, add				S.F.	9.50			9.50	10.45
0500	Coating, bituminous, per diameter inch, 1 coat, add				L.F.	.37			.37	.41
0540	3 coat					.57			.57	.63
0560	Coal tar epoxy, per diameter inch, 1 coat, add					.21			.21	.23
0600	3 coat					.46			.46	.51
1000	Polyethylene H.D. extruded, .025" thk., 1/2" diameter add					.33			.33	.36
1020	3/4" diameter					.35			.35	.39
1040	1" diameter					.39			.39	.43
1060	1-1/4" diameter					.48			.48	.53
1080	1-1/2" diameter					.53			.53	.58
1100	.030" thk., 2" diameter					.57			.57	.63
1120	2-1/2" diameter					.67			.67	.74
1140	.035" thk., 3" diameter					.84			.84	.92
1160	3-1/2" diameter					.95			.95	1.05
1180	4" diameter					1.06			1.06	1.17
1200	.040" thk, 5" diameter					1.46			1.46	1.61
1220	6" diameter					1.52			1.52	1.67
1240	8" diameter					1.98			1.98	2.18
1260	10" diameter					2.43			2.43	2.67
1280	12" diameter					2.91			2.91	3.20
1300	.060" thk., 14" diameter					4			4	4.40
1320	16" diameter					4.68			4.68	5.15
1340	18" diameter					5.55			5.55	6.10
1360	20" diameter					6.25			6.25	6.85
1380	Fittings, field wrapped, add				S.F.	6.45			6.45	7.10

33 01 10.20 Pipe Repair

		Crew	Daily Output	Labor-Hours	Unit	Material	Labor	Equipment	Total	Total Incl O&P
0010	**PIPE REPAIR**									
0020	Not including excavation or backfill									
0100	Clamp, stainless steel, lightweight, for steel pipe									
0110	3" long, 1/2" diameter pipe	1 Plum	34	.235	Ea.	13.80	11.45		25.25	33
0120	3/4" diameter pipe		32	.250		14.20	12.20		26.40	34.50
0130	1" diameter pipe		30	.267		15.25	13		28.25	37
0140	1-1/4" diameter pipe		28	.286		16.15	13.95		30.10	39.50
0150	1-1/2" diameter pipe		26	.308		16.70	15		31.70	42
0160	2" diameter pipe		24	.333		18.25	16.25		34.50	45
0170	2-1/2" diameter pipe		23	.348		19.85	16.95		36.80	48.50

33 01 Operation and Maintenance of Utilities

33 01 10 – Operation and Maintenance of Water Utilities

33 01 10.20 Pipe Repair		Crew	Daily Output	Labor-Hours	Unit	2009 Bare Costs Material	Labor	Equipment	Total	Total Incl O&P
0180	3" diameter pipe	1 Plum	22	.364	Ea.	22	17.75		39.75	51.50
0190	3-1/2" diameter pipe		21	.381		23.50	18.55		42.05	55
0200	4" diameter pipe	B-20	56	.429		24.50	15.15		39.65	52
0210	5" diameter pipe		53	.453		28	16		44	57.50
0220	6" diameter pipe		48	.500		31.50	17.70		49.20	63.50
0230	8" diameter pipe		30	.800		36.50	28.50		65	86.50
0240	10" diameter pipe		28	.857		94	30.50		124.50	153
0250	12" diameter pipe		24	1		100	35.50		135.50	168
0260	14" diameter pipe		22	1.091		104	38.50		142.50	179
0270	16" diameter pipe		20	1.200		113	42.50		155.50	194
0280	18" diameter pipe		18	1.333		119	47		166	209
0290	20" diameter pipe		16	1.500		131	53		184	232
0300	24" diameter pipe		14	1.714		146	60.50		206.50	261
0360	For 6" long, add					100%	40%			
0370	For 9" long, add					200%	100%			
0380	For 12" long, add					300%	150%			
0390	For 18" long, add					500%	200%			
0400	Pipe Freezing for live repairs of systems 3/8 inch to 6 inch									
0410	Note: Pipe Freezing can also be used to install a valve into a live system									
0420	Pipe Freezing each side 3/8 inch	2 Skwk	8	2	Ea.	495	81.50		576.50	680
0425	Pipe Freezing each side 3/8 inch, second location same kit		8	2		19.95	81.50		101.45	155
0430	Pipe Freezing each side 3/4 inch		8	2		495	81.50		576.50	680
0435	Pipe Freezing each side 3/4 inch, second location same kit		8	2		19.95	81.50		101.45	155
0440	Pipe Freezing each side 1 1/2 inch		6	2.667		495	109		604	720
0445	Pipe Freezing each side 1 1/2 inch, second location same kit		6	2.667		19.95	109		128.95	199
0450	Pipe Freezing each side 2 inch		6	2.667		735	109		844	980
0455	Pipe Freezing each side 2 inch, second location same kit		6	2.667		19.95	109		128.95	199
0460	Pipe Freezing each side 2 1/2 inch-3 inch		6	2.667		1,025	109		1,134	1,300
0465	Pipe Freeze each side 2 1/2-3 inch, second location same kit		6	2.667		40	109		149	221
0470	Pipe Freezing each side 4 inch		4	4		1,675	163		1,838	2,100
0475	Pipe Freezing each side 4 inch, second location same kit		4	4		65	163		228	340
0480	Pipe Freezing each side 5-6 inch		4	4		4,025	163		4,188	4,700
0485	Pipe Freezing each side 5-6 inch, second location same kit		4	4		195	163		358	480
0490	Pipe Frz extra 20 lb CO_2 cylinders (3/8 to 2 " - 1 ea, 3 " -2 ea)					218			218	240
0500	Pipe Freezing extra 50 lb CO_2 cylinders (4 " - 2 ea, 5-6 " -6 ea)					490			490	535
1000	Clamp, stainless steel, with threaded service tap									
1040	Full seal for iron, steel, PVC pipe									
1100	6" long, 2" diameter pipe	1 Plum	17	.471	Ea.	177	23		200	231
1110	2-1/2" diameter pipe		16	.500		188	24.50		212.50	245
1120	3" diameter pipe		15.60	.513		200	25		225	259
1130	3-1/2" diameter pipe		15	.533		209	26		235	271
1140	4" diameter pipe	B-20	40	.600		230	21		251	287
1150	6" diameter pipe		34	.706		265	25		290	330
1160	8" diameter pipe		21	1.143		330	40.50		370.50	430
1170	10" diameter pipe		20	1.200		365	42.50		407.50	470
1180	12" diameter pipe		17	1.412		415	50		465	540
1205	For 9" long, add					20%	45%			
1210	For 12" long, add					40%	80%			
1220	For 18" long, add					70%	110%			
1600	Clamp, stainless steel, single section									
1640	Full seal for iron, steel, PVC pipe									
1700	6" long, 2" diameter pipe	1 Plum	17	.471	Ea.	92	23		115	137
1710	2-1/2" diameter pipe		16	.500		100	24.50		124.50	148

33 01 Operation and Maintenance of Utilities

33 01 10 – Operation and Maintenance of Water Utilities

	33 01 10.20 Pipe Repair	Crew	Daily Output	Labor-Hours	Unit	2009 Bare Costs Material	Labor	Equipment	Total	Total Incl O&P
1720	3″ diameter pipe	1 Plum	15.60	.513	Ea.	104	25		129	154
1730	3-1/2″ diameter pipe		15	.533		115	26		141	167
1740	4″ diameter pipe	B-20	40	.600		125	21		146	173
1750	6″ diameter pipe		34	.706		156	25		181	213
1760	8″ diameter pipe		21	1.143		209	40.50		249.50	297
1770	10″ diameter pipe		20	1.200		240	42.50		282.50	335
1780	12″ diameter pipe		17	1.412		261	50		311	370
1800	For 9″ long, add					40%	45%			
1810	For 12″ long, add					60%	80%			
1820	For 18″ long, add					120%	110%			
2000	Clamp, stainless steel, two section									
2040	Full seal, for iron, steel, PVC pipe									
2100	6″ long, 4″ diameter pipe	B-20	24	1	Ea.	177	35.50		212.50	253
2110	6″ diameter pipe		20	1.200		209	42.50		251.50	300
2120	8″ diameter pipe		13	1.846		230	65.50		295.50	360
2130	10″ diameter pipe		12	2		335	70.50		405.50	480
2140	12″ diameter pipe		10	2.400		365	85		450	540
2200	9″ long, 4″ diameter pipe		16	1.500		230	53		283	340
2210	6″ diameter pipe		13	1.846		261	65.50		326.50	395
2220	8″ diameter pipe		9	2.667		271	94.50		365.50	455
2230	10″ diameter pipe		8	3		375	106		481	590
2240	12″ diameter pipe		7	3.429		415	121		536	660
2250	14″ diameter pipe		6.40	3.750		625	133		758	910
2260	16″ diameter pipe		6	4		655	141		796	955
2270	18″ diameter pipe		5	4.800		680	170		850	1,025
2280	20″ diameter pipe		4.60	5.217		730	184		914	1,100
2290	24″ diameter pipe		4	6		1,250	212		1,462	1,725
2320	For 12″ long, add to 9″					15%	25%			
2330	For 18″ long, add to 9″					70%	55%			
8000	For internal cleaning and inspection, see Div. 33 01 30.16									
8100	For pipe testing, see Div. 23 05 93.50									

33 01 30 – Operation and Maintenance of Sewer Utilities

	33 01 30.16 TV Inspection of Sewer Pipelines	Crew	Daily Output	Labor-Hours	Unit	Material	Labor	Equipment	Total	Total Incl O&P
0010	**TV INSPECTION OF SEWER PIPELINES**									
0100	Pipe internal cleaning & inspection, cleaning, pressure pipe systems									
0120	Pig method, lengths 1000′ to 10,000′									
0140	4″ diameter thru 24″ diameter, minimum				L.F.				3.50	4.05
0160	Maximum				″				8.50	9.80
6000	Sewage/sanitary systems									
6100	Power rodder with header & cutters									
6110	Mobilization charge, minimum				Total				425	490
6120	Mobilization charge, maximum				″				1,000	1,150
6140	Cleaning 4-12 ″ diameter				L.F.				3.50	4.05
6190	14-24″ diameter								4	4.60
6240	30″ diameter								6	6.90
6250	36″ diameter								7	8.05
6260	48″ diameter								8	9.20
6270	60″ diameter								9	10.35
6280	72″ diameter								10	11.50
9000	Inspection, television camera with video									
9060	up to 500 linear feet				Total				750	865

33 01 30.71 Pipebursting

	33 01 30.71 Pipebursting	Crew	Daily Output	Labor-Hours	Unit	2009 Bare Costs Material	Labor	Equipment	Total	Total Incl O&P
0010	**PIPEBURSTING**									
0011	300′ runs, replace with HDPE pipe									
0020	Not including excavation, backfill, shoring, or dewatering									
0100	6″ to 15″ diameter, minimum				L.F.				100	110
0200	Maximum								200	220
0300	18″ to 36″ diameter, minimum								225	248
0400	Maximum								400	440
0500	Mobilize and demobilize, minimum				Job				3,000	3,300
0600	Maximum				″				30,000	33,000

33 01 30.72 Relining Sewers

	33 01 30.72 Relining Sewers	Crew	Daily Output	Labor-Hours	Unit	Material	Labor	Equipment	Total	Total Incl O&P
0010	**RELINING SEWERS**									
0011	With cement inc bypass & cleaning									
0020	Less than 10,000 L.F., urban, 6″ to 10″	C-17E	130	.615	L.F.	8.50	25.50	.66	34.66	51.50
0050	10″ to 12″		125	.640		10.45	26.50	.68	37.63	55
0070	12″ to 16″		115	.696		10.70	28.50	.74	39.94	59
0100	16″ to 20″		95	.842		12.55	34.50	.90	47.95	71.50
0200	24″ to 36″		90	.889		13.55	36.50	.95	51	75.50
0300	48″ to 72″		80	1		21.50	41.50	1.07	64.07	91.50
0500	Rural, 6″ to 10″		180	.444		8.50	18.35	.48	27.33	40
0550	10″ to 12″		175	.457		10.45	18.85	.49	29.79	42.50
0570	12″ to 16″		160	.500		10.90	20.50	.54	31.94	46
0600	16″ to 20″		135	.593		11.50	24.50	.63	36.63	53.50
0700	24″ to 36″		125	.640		13.75	26.50	.68	40.93	59
0800	48″ to 72″		100	.800		21.50	33	.86	55.36	78
1000	Greater than 10,000 L.F., urban, 6″ to 10″		160	.500		8.50	20.50	.54	29.54	43.50
1050	10″ to 12″		155	.516		10.30	21.50	.55	32.35	46.50
1070	12″ to 16″		140	.571		10.70	23.50	.61	34.81	51
1100	16″ to 20″		120	.667		11.50	27.50	.71	39.71	58.50
1200	24″ to 36″		115	.696		13.75	28.50	.74	42.99	62.50
1300	48″ to 72″		95	.842		21.50	34.50	.90	56.90	81
1500	Rural, 6″ to 10″		215	.372		8.50	15.35	.40	24.25	35
1550	10″ to 12″		210	.381		10.45	15.70	.41	26.56	37.50
1570	12″ to 16″		185	.432		10.70	17.85	.46	29.01	41.50
1600	16″ to 20″		150	.533		11.50	22	.57	34.07	49.50
1700	24″ to 36″		140	.571		13.75	23.50	.61	37.86	54.50
1800	48″ to 72″		120	.667		21.50	27.50	.71	49.71	70

33 01 30.74 HDPE Pipe Lining

	33 01 30.74 HDPE Pipe Lining	Crew	Daily Output	Labor-Hours	Unit	Material	Labor	Equipment	Total	Total Incl O&P
0010	**HDPE PIPE LINING**, excludes cleaning and video inspection									
0020	Pipe relined with one pipe size smaller than original (4″ for 6 ″)									
0100	6″ diameter, original size	B-6B	600	.080	L.F.	3.61	2.58	1.09	7.28	9.45
0150	8″ diameter, original size		600	.080		10.60	2.58	1.09	14.27	17.10
0200	10″ diameter, original size		600	.080		12.20	2.58	1.09	15.87	18.90
0250	12″ diameter, original size		400	.120		18.90	3.87	1.63	24.40	29
0300	14″ diameter, original size		400	.120		26.50	3.87	1.63	32	37.50

33 05 Common Work Results for Utilities

33 05 16 – Utility Structures

33 05 16.13 Utility Boxes

	33 05 16.13 Utility Boxes	Crew	Daily Output	Labor-Hours	Unit	2009 Bare Costs Material	Labor	Equipment	Total	Total Incl O&P
0010	**UTILITY BOXES** Precast concrete, 6" thick									
0040	4' x 6' x 6' high, I.D.	B-13	2	28	Ea.	1,800	955	395	3,150	3,950
0050	5' x 10' x 6' high, I.D.		2	28		2,225	955	395	3,575	4,425
0100	6' x 10' x 6' high, I.D.		2	28		2,300	955	395	3,650	4,500
0150	5' x 12' x 6' high, I.D.		2	28		2,450	955	395	3,800	4,675
0200	6' x 12' x 6' high, I.D.		1.80	31.111		2,725	1,075	440	4,240	5,200
0250	6' x 13' x 6' high, I.D.		1.50	37.333		3,600	1,275	525	5,400	6,600
0300	8' x 14' x 7' high, I.D.		1	56		3,875	1,925	790	6,590	8,250
0350	Hand hole, precast concrete, 1-1/2" thick									
0400	1'-0" x 2'-0" x 1'-9", I.D., light duty	B-1	4	6	Ea.	380	194		574	740
0450	4'-6" x 3'-2" x 2'-0", O.D., heavy duty	B-6	3	8		1,175	273	98	1,546	1,825
0460	Meter pit, 4' x 4', 4' deep		2	12		995	410	147	1,552	1,925
0470	6' deep		1.60	15		1,425	510	184	2,119	2,575
0480	8' deep		1.40	17.143		1,875	585	210	2,670	3,225
0490	10' deep		1.20	20		2,375	680	245	3,300	3,975
0500	15' deep		1	24		3,475	820	294	4,589	5,475
0510	6' x 6', 4' deep		1.40	17.143		1,475	585	210	2,270	2,800
0520	6' deep		1.20	20		2,200	680	245	3,125	3,800
0530	8' deep		1	24		2,950	820	294	4,064	4,875
0540	10' deep		.80	30		3,675	1,025	365	5,065	6,100
0550	15' deep		.60	40		5,575	1,375	490	7,440	8,900

33 05 23 – Trenchless Utility Installation

33 05 23.19 Microtunneling

	33 05 23.19 Microtunneling	Crew	Daily Output	Labor-Hours	Unit	Material	Labor	Equipment	Total	Total Incl O&P
0010	**MICROTUNNELING**									
0011	Not including excavation, backfill, shoring,									
0020	or dewatering, average 50'/day, slurry method									
0100	24" to 48" outside diameter, minimum				L.F.				800	880
0110	Adverse conditions, add				%				50%	50%
1000	Rent microtunneling machine, average monthly lease				Month				90,000	99,000
1010	Operating technician				Day				560	620
1100	Mobilization and demobilization, minimum				Job				40,000	44,000
1110	Maximum				"				400,000	440,000

33 05 23.20 Horizontal Boring

	33 05 23.20 Horizontal Boring	Crew	Daily Output	Labor-Hours	Unit	Material	Labor	Equipment	Total	Total Incl O&P
0010	**HORIZONTAL BORING**									
0011	Casing only, 100' minimum,									
0020	not incl. jacking pits or dewatering									
0100	Roadwork, 1/2" thick wall, 24" diameter casing	B-42	20	3.200	L.F.	119	114	68.50	301.50	395
0200	36" diameter		16	4		189	142	85.50	416.50	540
0300	48" diameter		15	4.267		278	152	91	521	655
0500	Railroad work, 24" diameter		15	4.267		119	152	91	362	485
0600	36" diameter		14	4.571		189	162	97.50	448.50	585
0700	48" diameter		12	5.333		278	189	114	581	745
0900	For ledge, add									20%
1000	Small diameter boring, 3", sandy soil	B-82	900	.018		29	.63	.09	29.72	32.50
1040	Rocky soil	"	500	.032		29	1.13	.16	30.29	33.50

33 05 26 – Utility Line Signs, Markers, and Flags

33 05 26.10 Utility Accessories

	33 05 26.10 Utility Accessories	Crew	Daily Output	Labor-Hours	Unit	Material	Labor	Equipment	Total	Total Incl O&P
0010	**UTILITY ACCESSORIES**									
0400	Underground tape, detectable, reinforced, alum. foil core, 2"	1 Clab	150	.053	C.L.F.	2.10	1.69		3.79	5.10
0500	6"		140	.057	"	5.25	1.81		7.06	8.80
9000	Minimum labor/equipment charge		4	2	Job		63		63	104

33 11 13.10 Water Supply, Concrete Pipe

33 11 13.10 Water Supply, Concrete Pipe		Crew	Daily Output	Labor-Hours	Unit	2009 Bare Costs Material	Labor	Equipment	Total	Total Incl O&P
0010	**WATER SUPPLY, CONCRETE PIPE**									
0020	Not including excavation or backfill, without gaskets									
3000	Conc. cylinder pipe (CCP), 150 PSI, 12" diam.	B-13	192	.292	L.F.	87.50	9.95	4.11	101.56	117
3010	24" diameter	"	128	.438		109	14.95	6.15	130.10	151
3040	36" diameter	B-13B	96	.583		135	19.95	11.70	166.65	193
3050	48" diameter		64	.875		184	30	17.60	231.60	271
3060	Prestressed (PCCP), 150 PSI, 60" diameter		60	.933		305	32	18.75	355.75	410
3070	72" diameter		60	.933		380	32	18.75	430.75	490
3080	84" diameter		40	1.400		495	48	28	571	655
3090	96" diameter	B-13C	40	1.400		730	48	40.50	818.50	930
3100	108" diameter		32	1.750		1,000	60	50.50	1,110.50	1,250
3102	120" diameter		16	3.500		1,475	120	101	1,696	1,925
3104	144" diameter		16	3.500		1,825	120	101	2,046	2,300
3110	Conc. cylinder pipe (CCP), 150 PSI, elbow, 90°, 12" diameter	B-13	32	1.750	Ea.	1,175	60	24.50	1,259.50	1,425
3140	24" diameter	"	6	9.333		2,275	320	132	2,727	3,175
3150	36" diameter	B-13B	15	3.733		4,475	128	75	4,678	5,225
3160	48" diameter		12	4.667		8,850	160	94	9,104	10,100
3170	Prestressed (PCCP), 150 PSI, elbow, 90°, 60" diameter		10	5.600		16,100	191	113	16,404	18,100
3180	72" diameter		4	14		25,200	480	281	25,961	28,800
3190	84" diameter		3.20	17.500		30,700	600	350	31,650	35,100
3200	96" diameter		2	28		41,100	955	565	42,620	47,400
3210	108" diameter	B-13C	1.20	46.667		61,500	1,600	1,350	64,450	71,500
3220	120" diameter		.40	140		52,500	4,775	4,025	61,300	70,000
3225	144" diameter		.30	184		98,000	6,300	5,300	109,600	123,500
3230	Concrete cylinder pipe (CCP), 150 PSI, elbow, 45°, 12" diameter	B-13	24	2.333		730	80	33	843	970
3250	24" diameter	"	6	9.333		920	320	132	1,372	1,700
3260	36" diameter	B-13B	4	14		2,500	480	281	3,261	3,850
3270	48" diameter		3	18.667		4,550	640	375	5,565	6,475
3280	Prestressed, (PCCP), 150 PSI, elbow, 45°, 60" diameter		2	28		10,000	955	565	11,520	13,200
3290	72" diameter		1.60	35		15,600	1,200	705	17,505	19,900
3300	84" diameter		1.30	42.945		23,900	1,475	865	26,240	29,600
3310	96" diameter		1	56		29,600	1,925	1,125	32,650	37,000
3320	108" diameter	B-13C	.65	84.337		37,300	2,875	2,425	42,600	48,400
3330	120" diameter		.40	140		37,700	4,775	4,025	46,500	53,500
3340	144" diameter		.30	184		60,500	6,300	5,300	72,100	82,500

33 11 13.15 Water Supply, Ductile Iron Pipe

33 11 13.15 Water Supply, Ductile Iron Pipe		Crew	Daily Output	Labor-Hours	Unit	Material	Labor	Equipment	Total	Total Incl O&P
0010	**WATER SUPPLY, DUCTILE IRON PIPE** R331113-80									
0011	Cement lined									
0020	Not including excavation or backfill									
2000	Pipe, class 50 water piping, 18' lengths									
2020	Mechanical joint, 4" diameter	B-21A	200	.200	L.F.	15.35	7.80	3.01	26.16	32.50
2040	6" diameter		160	.250		17.80	9.80	3.76	31.36	39
2060	8" diameter		133.33	.300		19.70	11.75	4.52	35.97	45
2080	10" diameter		114.29	.350		26.50	13.70	5.25	45.45	57
2100	12" diameter		105.26	.380		33	14.85	5.70	53.55	66
2120	14" diameter		100	.400		42	15.65	6	63.65	77.50
2140	16" diameter		72.73	.550		45.50	21.50	8.30	75.30	93
2160	18" diameter		68.97	.580		57.50	22.50	8.75	88.75	109
2170	20" diameter		57.14	.700		67	27.50	10.55	105.05	129
2180	24" diameter		47.06	.850		86	33	12.80	131.80	162
3000	Tyton, push-on joint, 4" diameter		400	.100		11.50	3.91	1.51	16.92	20.50
3020	6" diameter		333.33	.120		14.70	4.69	1.81	21.20	25.50

33 11 13.15 Water Supply, Ductile Iron Pipe		Crew	Daily Output	Labor-Hours	Unit	2009 Bare Costs Material	Labor	Equipment	Total	Total Incl O&P
3040	8″ diameter	B-21A	200	.200	L.F.	20.50	7.80	3.01	31.31	38.50
3060	10″ diameter		181.82	.220		29	8.60	3.31	40.91	49
3080	12″ diameter		160	.250		30.50	9.80	3.76	44.06	53
3100	14″ diameter		133.33	.300		33.50	11.75	4.52	49.77	60
3120	16″ diameter		114.29	.350		54	13.70	5.25	72.95	87
3140	18″ diameter		100	.400		60	15.65	6	81.65	97.50
3160	20″ diameter		88.89	.450		66	17.60	6.75	90.35	108
3180	24″ diameter		76.92	.520		78	20.50	7.85	106.35	127
8000	Fittings, mechanical joint									
8006	90° bend, 4″ diameter	B-20A	16	2	Ea.	213	76.50		289.50	355
8020	6″ diameter		12.80	2.500		292	95.50		387.50	470
8040	8″ diameter		10.67	2.999		425	115		540	650
8060	10″ diameter	B-21A	11.43	3.500		465	137	52.50	654.50	785
8080	12″ diameter		10.53	3.799		840	149	57	1,046	1,225
8100	14″ diameter		10	4		1,275	156	60	1,491	1,725
8120	16″ diameter		7.27	5.502		1,900	215	83	2,198	2,525
8140	18″ diameter		6.90	5.797		3,150	227	87.50	3,464.50	3,925
8160	20″ diameter		5.71	7.005		4,250	274	105	4,629	5,225
8180	24″ diameter		4.70	8.511		4,675	335	128	5,138	5,825
8200	Wye or tee, 4″ diameter	B-20A	10.67	2.999		325	115		440	540
8220	6″ diameter		8.53	3.751		435	143		578	705
8240	8″ diameter		7.11	4.501		620	172		792	955
8260	10″ diameter	B-21A	7.62	5.249		1,275	205	79	1,559	1,800
8280	12″ diameter		7.02	5.698		1,625	223	86	1,934	2,225
8300	14″ diameter		6.67	5.997		1,800	234	90.50	2,124.50	2,475
8320	16″ diameter		4.85	8.247		1,875	320	124	2,319	2,725
8340	18″ diameter		4.60	8.696		4,900	340	131	5,371	6,075
8360	20″ diameter		3.81	10.499		7,675	410	158	8,243	9,275
8380	24″ diameter		3.14	12.739		9,100	500	192	9,792	11,000
8398	45° bends, 4″ diameter	B-20A	16	2		167	76.50		243.50	305
8400	6″ diameter	″	12.80	2.500		231	95.50		326.50	405
8410	12″ diameter	B-21A	10.53	3.799		505	149	57	711	855
8420	16″ diameter		7.27	5.502		1,200	215	83	1,498	1,750
8430	20″ diameter		5.71	7.005		2,650	274	105	3,029	3,450
8440	24″ diameter		4.70	8.511		3,650	335	128	4,113	4,675
8450	Decreaser, 6″ x 4″ diameter	B-20A	14.22	2.250		179	86		265	335
8460	8″ x 6″ diameter	″	11.64	2.749		229	105		334	420
8470	10″ x 6 ″ diameter	B-21A	13.33	3.001		290	117	45	452	555
8480	12″ x 6″ diameter		12.70	3.150		355	123	47.50	525.50	640
8490	16″ x 6″ diameter		10	4		820	156	60	1,036	1,225
8500	20″ x 6″ diameter		8.42	4.751		2,025	186	71.50	2,282.50	2,600
8552	For water utility valves see Div. 33 12 16									
8600	12″ diameter	B-21	3	9.333	Ea.	2,700	340	43	3,083	3,575
8610	14″ diameter		2	14		3,325	510	64.50	3,899.50	4,550
8620	16″ diameter		2	14		5,125	510	64.50	5,699.50	6,525
8700	Joint restraint, ductile iron mechanical joints									
8710	4″ diameter	B-20A	32	1	Ea.	33	38		71	97
8720	6″ diameter		25.60	1.250		39.50	48		87.50	120
8730	8″ diameter		21.33	1.500		57	57.50		114.50	154
8740	10″ diameter		18.28	1.751		83	67		150	199
8750	12″ diameter		16.84	1.900		116	72.50		188.50	244
8760	14″ diameter		16	2		159	76.50		235.50	297
8770	16″ diameter		11.64	2.749		210	105		315	400

33 11 13.15 Water Supply, Ductile Iron Pipe		Crew	Daily Output	Labor-Hours	Unit	Material	2009 Bare Costs Labor	Equipment	Total	Total Incl O&P
8780	18" diameter	B-20A	11.03	2.901	Ea.	292	111		403	495
8785	20" diameter		9.14	3.501		360	134		494	610
8790	24" diameter		7.53	4.250		435	163		598	740
9600	Steel sleeve and tap, 4" diameter	B-20	3	8		610	283		893	1,125
9620	6" diameter		2	12		650	425		1,075	1,400
9630	8" diameter		2	12		890	425		1,315	1,675

33 11 13.20 Water Supply, Polyethylene Pipe, C901

		Crew	Daily Output	Labor-Hours	Unit	Material	Labor	Equipment	Total	Total Incl O&P
0010	**WATER SUPPLY, POLYETHYLENE PIPE, C901** R331113-80									
0020	Not including excavation or backfill									
1000	Piping, 160 P.S.I., 3/4" diameter	Q-1A	525	.019	L.F.	.47	.94		1.41	1.97
1120	1" diameter		485	.021		.74	1.01		1.75	2.38
1140	1-1/2" diameter		450	.022		1.60	1.09		2.69	3.45
1160	2" diameter		365	.027		2.76	1.35		4.11	5.15

33 11 13.25 Water Supply, Polyvinyl Chloride Pipe

		Crew	Daily Output	Labor-Hours	Unit	Material	Labor	Equipment	Total	Total Incl O&P
0010	**WATER SUPPLY, POLYVINYL CHLORIDE PIPE** R331113-80									
0020	Not including excavation or backfill, unless specified									
2100	AWWA Class 150, S.D.R. 18, 1-1/2" diameter	Q-1A	750	.013	L.F.	.34	.66		1	1.39
2120	2" diameter		686	.015		.51	.72		1.23	1.67
2140	2-1/2" diameter		500	.020		.87	.98		1.85	2.48
2160	3" diameter	B-20	430	.056		1.09	1.97		3.06	4.44
2180	4" diameter		375	.064		1.65	2.26		3.91	5.55
2200	6" diameter		316	.076		3.30	2.68		5.98	8.05
2210	8" diameter		260	.092		5.70	3.26		8.96	11.60
3010	AWWA C905, PR 100, DR 25									
3030	14" diameter	B-20A	213	.150	L.F.	19	5.75		24.75	30
3040	16" diameter		200	.160		31.50	6.10		37.60	44.50
3050	18" diameter		160	.200		38.50	7.65		46.15	54.50
3060	20" diameter		133	.241		53	9.20		62.20	73
3070	24" diameter		107	.299		68	11.45		79.45	93
3080	30" diameter		80	.400		110	15.30		125.30	146
3090	36" diameter		80	.400		170	15.30		185.30	212
3100	42" diameter		60	.533		230	20.50		250.50	286
3200	48" diameter		60	.533		300	20.50		320.50	365
4520	Pressure pipe Class 150, SDR 18, AWWA C900, 4" diameter		380	.084		2.32	3.22		5.54	7.70
4530	6" diameter		316	.101		4.59	3.87		8.46	11.20
4540	8" diameter		264	.121		8.05	4.64		12.69	16.25
4550	10" diameter		220	.145		12.20	5.55		17.75	22.50
4560	12" diameter		186	.172		17.25	6.60		23.85	29.50
8000	Fittings with rubber gasket									
8003	Class 150, D.R. 18									
8006	90° Bend , 4" diameter	B-20	100	.240	Ea.	51	8.50		59.50	70
8020	6" diameter		90	.267		92	9.45		101.45	116
8040	8" diameter		80	.300		174	10.60		184.60	208
8060	10" diameter		50	.480		265	16.95		281.95	320
8080	12" diameter		30	.800		405	28.50		433.50	490
8100	Tee, 4" diameter		90	.267		54	9.45		63.45	75
8120	6" diameter		80	.300		145	10.60		155.60	177
8140	8" diameter		70	.343		203	12.10		215.10	243
8160	10" diameter		40	.600		430	21		451	505
8180	12" diameter		20	1.200		600	42.50		642.50	730
8200	45° Bend, 4" diameter		100	.240		50.50	8.50		59	69.50
8220	6" diameter		90	.267		91	9.45		100.45	115

33 11 13.25 Water Supply, Polyvinyl Chloride Pipe		Crew	Daily Output	Labor-Hours	Unit	2009 Bare Costs Material	Labor	Equipment	Total	Total Incl O&P
8240	8″ diameter	B-20	50	.480	Ea.	171	16.95		187.95	216
8260	10″ diameter		50	.480		270	16.95		286.95	325
8280	12″ diameter		30	.800		360	28.50		388.50	440
8300	Reducing tee 6″x4″		100	.240		119	8.50		127.50	145
8320	8″ x 6″		90	.267		287	9.45		296.45	330
8330	10″ x 6″		90	.267		330	9.45		339.45	380
8340	10″x8″		90	.267		355	9.45		364.45	405
8350	12″ x 6″		90	.267		445	9.45		454.45	505
8360	12″ x 8″		90	.267		495	9.45		504.45	560
8400	Tapped service tee (threaded type) 6″ x 6″ x 3/4″		100	.240		77.50	8.50		86	99
8430	6″ x 6″ x 1″		90	.267		77.50	9.45		86.95	100
8440	6″ x 6″ x 1 1/2″		90	.267		77.50	9.45		86.95	100
8450	6″ x 6″ x 2″		90	.267		77.50	9.45		86.95	100
8460	8″ x 8″ x 3/4″		90	.267		122	9.45		131.45	149
8470	8″ x 8″ x 1″		90	.267		122	9.45		131.45	149
8480	8″ x 8″ x 1 1/2″		90	.267		122	9.45		131.45	149
8490	8″ x 8″ x 2″		90	.267		122	9.45		131.45	149
8500	Repair coupling 4″		100	.240		111	8.50		119.50	136
8520	6″ diameter		90	.267		135	9.45		144.45	164
8540	8″ diameter		50	.480		153	16.95		169.95	196
8560	10″ diameter		50	.480		178	16.95		194.95	224
8580	12″ diameter		50	.480		218	16.95		234.95	268
8600	Plug end 4″		100	.240		38	8.50		46.50	55.50
8620	6″ diameter		90	.267		73	9.45		82.45	95.50
8640	8″ diameter		50	.480		78.50	16.95		95.45	114
8660	10″ diameter		50	.480		125	16.95		141.95	166
8680	12″ diameter		50	.480		155	16.95		171.95	198
8700	PVC pipe, joint restraint									
8710	4″ diameter	B-20A	32	1	Ea.	34.50	38		72.50	99
8720	6″ diameter		25.60	1.250		42	48		90	122
8730	8″ diameter		21.33	1.500		62	57.50		119.50	160
8740	10″ diameter		18.28	1.751		117	67		184	236
8750	12″ diameter		16.84	1.900		123	72.50		195.50	251
8760	14″ diameter		16	2		196	76.50		272.50	340
8770	16″ diameter		11.64	2.749		264	105		369	460
8780	18″ diameter		11.03	2.901		325	111		436	535
8785	20″ diameter		9.14	3.501		405	134		539	660
8790	24″ diameter		7.53	4.250		470	163		633	775

33 11 13.35 Water Supply, HDPE		Crew	Daily Output	Labor-Hours	Unit	Material	Labor	Equipment	Total	Total Incl O&P
0010	**WATER SUPPLY, HDPE** R331113-80									
0011	Butt fusion joints, SDR 21,40′ lengths not including excavation or backfill									
0100	4″ diameter	B-22A	400	.095	L.F.	3.61	3.39	1.08	8.08	10.70
0200	6″ diameter		380	.100		10.60	3.57	1.14	15.31	18.70
0300	8″ diameter		320	.119		12.20	4.24	1.35	17.79	22
0400	10″ diameter		300	.127		18.90	4.52	1.45	24.87	30
0500	12″ diameter		260	.146		26.50	5.20	1.67	33.37	40
0600	14″ diameter		220	.173		32.50	6.15	1.97	40.62	47.50
0700	16″ diameter		180	.211		42	7.55	2.41	51.96	61
0800	18″ diameter		140	.271		53	9.70	3.10	65.80	77.50
0900	24″ diameter		100	.380		94.50	13.55	4.34	112.39	131
1000	Fittings									
1100	Elbows, 90 degrees									

33 11 13.35 Water Supply, HDPE		Crew	Daily Output	Labor-Hours	Unit	Material	2009 Bare Costs Labor	Equipment	Total	Total Incl O&P
1200	4″ diameter	B-22B	32	.500	Ea.	33.50	18.10	4.59	56.19	71
1300	6″ diameter		28	.571		103	20.50	5.25	128.75	153
1400	8″ diameter		24	.667		217	24	6.10	247.10	285
1500	10″ diameter		18	.889		350	32	8.15	390.15	445
1600	12″ diameter		12	1.333		525	48.50	12.25	585.75	670
1700	14″ diameter		9	1.778		630	64.50	16.30	710.80	815
1800	16″ diameter		6	2.667		760	96.50	24.50	881	1,025
1900	18″ diameter		4	4		1,050	145	36.50	1,231.50	1,425
2000	24″ diameter		3	5.333		2,100	193	49	2,342	2,700
2100	Tees									
2200	4″ diameter	B-22B	30	.533	Ea.	41	19.30	4.90	65.20	82.50
2300	6″ diameter		26	.615		122	22.50	5.65	150.15	177
2400	8″ diameter		22	.727		305	26.50	6.70	338.20	385
2500	10″ diameter		15	1.067		400	38.50	9.80	448.30	515
2600	12″ diameter		10	1.600		560	58	14.70	632.70	725
2700	14″ diameter		8	2		660	72.50	18.35	750.85	865
2800	16″ diameter		6	2.667		780	96.50	24.50	901	1,050
2900	18″ diameter		4	4		1,025	145	36.50	1,206.50	1,425
3000	24″ diameter		2	8		2,150	290	73.50	2,513.50	2,900

33 11 13.45 Water Supply, Copper Pipe		Crew	Daily Output	Labor-Hours	Unit	Material	Labor	Equipment	Total	Total Incl O&P
0010	**WATER SUPPLY, COPPER PIPE**									
0020	Not including excavation or backfill									
2000	Tubing, type K, 20′ joints, 3/4″ diameter	Q-1	400	.040	L.F.	8.75	1.76		10.51	12.35
2200	1″ diameter		320	.050		11.45	2.19		13.64	16
3000	1-1/2″ diameter		265	.060		18.55	2.65		21.20	24.50
3020	2″ diameter		230	.070		28.50	3.05		31.55	36
3040	2-1/2″ diameter		146	.110		42	4.81		46.81	53.50
3060	3″ diameter		134	.119		58.50	5.25		63.75	72.50
4012	4″ diameter		95	.168		96.50	7.40		103.90	117
4016	6″ diameter	Q-2	80	.300		245	13.65		258.65	290
5000	Tubing, type L									
5108	2″ diameter	Q-1	230	.070	L.F.	23.50	3.05		26.55	30
6010	3″ diameter		134	.119		46.50	5.25		51.75	59
6012	4″ diameter		95	.168		77	7.40		84.40	96
6016	6″ diameter	Q-2	80	.300		215	13.65		228.65	258
7020	Fittings, brass, corporation stops, 3/4″ diameter	1 Plum	19	.421	Ea.	33	20.50		53.50	68.50
7040	1″ diameter		16	.500		48.50	24.50		73	91.50
7060	1-1/2″ diameter		13	.615		142	30		172	203
7080	2″ diameter		11	.727		223	35.50		258.50	300
7100	Curb stops, 3/4″ diameter		19	.421		73.50	20.50		94	113
7120	1″ diameter		16	.500		89	24.50		113.50	136
7140	1-1/2″ diameter		13	.615		183	30		213	248
7160	2″ diameter		11	.727		305	35.50		340.50	390
7180	Curb box, cast iron, 1/2″ to 1″ curb stops		12	.667		79.50	32.50		112	138
7200	1-1/4″ to 2″ curb stops		8	1		115	49		164	203
7220	Saddles, 3/4″ diameter, add					121			121	133
7240	2″ diameter, add					119			119	131

33 12 13.15 Tapping, Crosses and Sleeves		Crew	Daily Output	Labor-Hours	Unit	2009 Bare Costs Material	Labor	Equipment	Total	Total Incl O&P
0010	**TAPPING, CROSSES AND SLEEVES**									
4000	Drill and tap pressurized main (labor only)									
4100	6" main, 1" to 2" service	Q-1	3	5.333	Ea.		234		234	365
4150	8" main, 1" to 2" service	"	2.75	5.818	"		255		255	395
4500	Tap and insert gate valve									
4600	8" main, 4" branch	B-21	3.20	8.750	Ea.		320	40.50	360.50	565
4650	6" branch		2.70	10.370			375	48	423	670
4700	10" Main, 4" branch		2.70	10.370			375	48	423	670
4750	6" branch		2.35	11.915			435	55	490	765
4800	12" main, 6" branch		2.35	11.915			435	55	490	765
4850	8" branch		2.35	11.915			435	55	490	765
7000	Tapping crosses, sleeves, valves; with rubber gaskets									
7020	Crosses, 4" x 4"	B-21	37	.757	Ea.	585	27.50	3.49	615.99	690
7060	8" x 6"		21	1.333		850	48.50	6.15	904.65	1,025
7080	8" x 8"		21	1.333		920	48.50	6.15	974.65	1,100
7100	10" x 6"		21	1.333		1,675	48.50	6.15	1,729.65	1,925
7160	12" x 12"		18	1.556		2,125	56.50	7.15	2,188.65	2,450
7180	14" x 6"		16	1.750		4,225	63.50	8.05	4,296.55	4,775
7240	16" x 10"		14	2		4,575	73	9.20	4,657.20	5,175
7280	18" x 6"		10	2.800		6,775	102	12.90	6,889.90	7,625
7320	18" x 18"		10	2.800		6,225	102	12.90	6,339.90	7,025
7340	20" x 6"		8	3.500		5,475	127	16.15	5,618.15	6,250
7360	20" x 12"		8	3.500		5,850	127	16.15	5,993.15	6,675
7420	24" x 12"		6	4.667		7,175	170	21.50	7,366.50	8,175
7440	24" x 18"		6	4.667		10,700	170	21.50	10,891.50	12,100
7600	Cut-in sleeves with rubber gaskets, 4"		18	1.556		210	56.50	7.15	273.65	330
7640	8"		10	2.800		365	102	12.90	479.90	580
7680	12"		9	3.111		600	113	14.35	727.35	860
7800	Cut-in valves with rubber gaskets, 4"		18	1.556		430	56.50	7.15	493.65	570
7840	8"		10	2.800		805	102	12.90	919.90	1,075
7880	12"		9	3.111		1,200	113	14.35	1,327.35	1,500
7900	Tapping Valve 4 inch, MJ, ductile iron		18	1.556		565	56.50	7.15	628.65	720
7920	Tapping Valve 6 inch, MJ, ductile iron		12	2.333		745	85	10.75	840.75	970
8000	Sleeves with rubber gaskets, 4" x 4"		37	.757		655	27.50	3.49	685.99	775
8030	8" x 4"		21	1.333		760	48.50	6.15	814.65	920
8040	8" x 6"		21	1.333		750	48.50	6.15	804.65	910
8060	8" x 8"		21	1.333		985	48.50	6.15	1,039.65	1,150
8070	10" x 4"		21	1.333		780	48.50	6.15	834.65	945
8080	10" x 6"		21	1.333		850	48.50	6.15	904.65	1,025
8090	10" x 8"		21	1.333		1,000	48.50	6.15	1,054.65	1,175
8140	12" x 12"		18	1.556		1,800	56.50	7.15	1,863.65	2,100
8160	14" x 6"		16	1.750		1,150	63.50	8.05	1,221.55	1,375
8220	16" x 10"		14	2		2,725	73	9.20	2,807.20	3,100
8260	18" x 6"		10	2.800		975	102	12.90	1,089.90	1,250
8300	18" x 18"		10	2.800		2,675	102	12.90	2,789.90	3,125
8320	20" x 6"		8	3.500		2,000	127	16.15	2,143.15	2,425
8340	20" x 12"		8	3.500		3,425	127	16.15	3,568.15	3,975
8400	24" x 12"		6	4.667		3,550	170	21.50	3,741.50	4,200
8420	24" x 18"		6	4.667		3,050	170	21.50	3,241.50	3,675
8800	Curb box, 6' long	B-20	20	1.200		282	42.50		324.50	380
8820	8' long		18	1.333		355	47		402	470
8830	Valve box w/ lid		14	1.714		148	60.50		208.50	263
8840	Valve box and large base w/ lid		14	1.714		249	60.50		309.50	375

33 12 16.10 Valves

		Crew	Daily Output	Labor-Hours	Unit	2009 Bare Costs Material	Labor	Equipment	Total	Total Incl O&P
0010	**VALVES**, water distribution									
0011	See Div. 22 05 23.20 and 22 05 23.60									
3000	Butterfly valves with boxes, cast iron, mech. jt.									
3100	4″ diameter	B-6	6	4	Ea.	1,000	136	49	1,185	1,375
3180	8″ diameter		6	4		1,650	136	49	1,835	2,100
3340	12″ diameter		6	4		2,850	136	49	3,035	3,425
3400	14″ diameter		4	6		3,575	204	73.50	3,852.50	4,325
3460	18″ diameter		4	6		5,975	204	73.50	6,252.50	6,975
3480	20″ diameter		4	6		6,825	204	73.50	7,102.50	7,925
3500	24″ diameter		4	6		7,425	204	73.50	7,702.50	8,575
3600	With lever operator									
3610	4″ diameter	B-6	6	4	Ea.	860	136	49	1,045	1,225
3616	8″ diameter		6	4		1,525	136	49	1,710	1,950
3620	12″ diameter		6	4		2,700	136	49	2,885	3,250
3624	16″ diameter		4	6		5,125	204	73.50	5,402.50	6,025
3630	24″ diameter		4	6		7,175	204	73.50	7,452.50	8,300
3700	Check valves, flanged									
3710	4″ diameter	B-6	6	4	Ea.	505	136	49	690	830
3714	6″ diameter		6	4		905	136	49	1,090	1,275
3716	8″ diameter		6	4		1,700	136	49	1,885	2,150
3720	12″ diameter		6	4		3,900	136	49	4,085	4,550
3724	16″ diameter		4	6		6,050	204	73.50	6,327.50	7,050
3726	18″ diameter		4	6		11,400	204	73.50	11,677.50	13,000
3730	24″ diameter		4	6		19,300	204	73.50	19,577.50	21,700
3800	Gate valves, C.I., 250 PSI, mechanical joint, w/boxes									
3810	4″ diameter	B-6	6	4	Ea.	600	136	49	785	935
3814	6″ diameter		6	4		685	136	49	870	1,025
3816	8″ diameter		6	4		995	136	49	1,180	1,375
3820	12″ diameter		6	4		2,175	136	49	2,360	2,650
3824	16″ diameter		4	6		8,875	204	73.50	9,152.50	10,200
3828	20″ diameter		4	6		14,600	204	73.50	14,877.50	16,500
3830	24″ diameter		4	6		19,600	204	73.50	19,877.50	21,900

33 12 16.20 Valves

		Crew	Daily Output	Labor-Hours	Unit	Material	Labor	Equipment	Total	Total Incl O&P
0010	**VALVES**									
0011	Special trim or use									
9000	Valves, gate valve, N.R.S. post type, 4″ diameter	B-6	6	4	Ea.	600	136	49	785	935
9040	8″ diameter		6	4		995	136	49	1,180	1,375
9080	12″ diameter		6	4		2,175	136	49	2,360	2,650
9120	O.S.&Y., 4″ diameter		6	4		525	136	49	710	850
9160	8″ diameter		6	4		1,375	136	49	1,560	1,775
9200	12″ diameter		6	4		2,800	136	49	2,985	3,350
9220	14″ diameter		4	6		6,625	204	73.50	6,902.50	7,700
9400	Check valves, rubber disc, 2-1/2″ diameter		6	4		340	136	49	525	650
9440	4″ diameter		6	4		505	136	49	690	830
9500	8″ diameter		6	4		1,700	136	49	1,885	2,150
9540	12″ diameter		6	4		3,900	136	49	4,085	4,550
9700	Detector check valves, reducing, 4″ diameter		6	4		1,275	136	49	1,460	1,675
9740	8″ diameter		6	4		2,525	136	49	2,710	3,050
9800	Galvanized, 4″ diameter		6	4		1,575	136	49	1,760	2,025
9840	8″ diameter		6	4		3,675	136	49	3,860	4,325

33 12 Water Utility Distribution Equipment

33 12 19 – Water Utility Distribution Fire Hydrants

33 12 19.10 Fire Hydrants		Crew	Daily Output	Labor-Hours	Unit	2009 Bare Costs Material	Labor	Equipment	Total	Total Incl O&P
0010	**FIRE HYDRANTS**									
0020	Mechanical joints unless otherwise noted									
1000	Fire hydrants, two way; excavation and backfill not incl.									
1100	4-1/2" valve size, depth 2'-0"	B-21	10	2.800	Ea.	1,425	102	12.90	1,539.90	1,750
1200	4'-6"		9	3.111		1,625	113	14.35	1,752.35	1,975
1260	6'-0"		7	4		1,775	146	18.45	1,939.45	2,200
1340	8'-0"		6	4.667		2,025	170	21.50	2,216.50	2,525
1420	10'-0"		5	5.600		2,125	204	26	2,355	2,700
2000	5-1/4" valve size, depth 2'-0"		10	2.800		1,500	102	12.90	1,614.90	1,825
2080	4'-0"		9	3.111		1,700	113	14.35	1,827.35	2,075
2160	6'-0"		7	4		1,900	146	18.45	2,064.45	2,325
2240	8'-0"		6	4.667		2,100	170	21.50	2,291.50	2,625
2320	10'-0"		5	5.600		2,300	204	26	2,530	2,875
2350	For threeway valves, add					7%				
2400	Lower barrel extensions with stems, 1'-0"	B-20	14	1.714		600	60.50		660.50	760
2440	2'-0"		13	1.846		735	65.50		800.50	910
2480	3'-0"		12	2		865	70.50		935.50	1,075
2520	4'-0"		10	2.400		1,000	85		1,085	1,250
5000	Indicator post									
5020	Adjustable, valve size 4" to 14", 4' bury	B-21	10	2.800	Ea.	760	102	12.90	874.90	1,025
5060	8' bury		7	4		970	146	18.45	1,134.45	1,325
5080	10' bury		6	4.667		995	170	21.50	1,186.50	1,400
5100	12' bury		5	5.600		1,500	204	26	1,730	2,000
5120	14' bury		4	7		1,625	255	32.50	1,912.50	2,250
5500	Non-adjustable, valve size 4" to 14", 3' bury		10	2.800		760	102	12.90	874.90	1,025
5520	3'-6" bury		10	2.800		760	102	12.90	874.90	1,025
5540	4' bury		9	3.111		760	113	14.35	887.35	1,025

33 16 Water Utility Storage Tanks

33 16 13 – Aboveground Water Utility Storage Tanks

33 16 13.13 Steel Water Storage Tanks		Crew	Daily Output	Labor-Hours	Unit	Material	Labor	Equipment	Total	Total Incl O&P
0010	**STEEL WATER STORAGE TANKS**									
0910	Steel, ground level, ht/diam. less than 1, not incl. fdn., 100,000 gallons				Ea.				173,000	190,000
1000	250,000 gallons								184,000	202,000
1200	500,000 gallons								280,000	308,000
1250	750,000 gallons								334,000	367,000
1300	1,000,000 gallons								509,000	560,000
1500	2,000,000 gallons								822,000	904,000
1600	4,000,000 gallons								1,277,000	1,405,000
1800	6,000,000 gallons								1,772,000	1,949,000
1850	8,000,000 gallons								2,320,000	2,126,500
1910	10,000,000 gallons								3,460,000	3,805,000
2100	Steel standpipes, hgt/diam. more than 1, 100' to overflow, no fdn									
2200	500,000 gallons				Ea.				367,000	404,000
2400	750,000 gallons								448,000	493,000
2500	1,000,000 gallons								545,000	600,000
2700	1,500,000 gallons								762,000	838,000
2800	2,000,000 gallons								936,000	1,030,000

33 16 Water Utility Storage Tanks

33 16 13 – Aboveground Water Utility Storage Tanks

33 16 13.16 Prestressed Conc. Abvgrnd Water Util. Stor. Tanks		Crew	Daily Output	Labor-Hours	Unit	2009 Bare Costs Material	Labor	Equipment	Total	Total Incl O&P
0010	**PRESTRESSED CONC. ABOVEGROUND WATER UTILITY STORAGE TANKS**									
0020	Not including pipe or pumps, 250,000 gallons				Ea.				270,000	297,000
0100	500,000 gallons								400,000	440,000
0300	1,000,000 gallons								580,000	638,000
0400	2,000,000 gallons								880,000	968,000
0600	4,000,000 gallons								1,400,000	1,540,000
0700	6,000,000 gallons								1,860,000	2,046,000
0750	8,000,000 gallons								2,400,000	2,640,000
0800	10,000,000 gallons				↓				2,900,000	3,190,000

33 16 13.23 Plastic-Coated Fabric Pillow Water Tanks

		Crew	Daily Output	Labor-Hours	Unit	Material	Labor	Equipment	Total	Total Incl O&P
0010	**PLASTIC-COATED FABRIC PILLOW WATER TANKS**									
7000	Water tanks, vinyl coated fabric pillow tanks, freestanding, 5,000 gallons	4 Clab	4	8	Ea.	9,500	253		9,753	10,900
7100	Supporting embankment not included, 25,000 gallons	6 Clab	2	24		14,900	760		15,660	17,700
7200	50,000 gallons	8 Clab	1.50	42.667		27,600	1,350		28,950	32,600
7300	100,000 gallons	9 Clab	.90	80		60,000	2,525		62,525	70,000
7400	150,000 gallons		.50	144		82,000	4,550		86,550	97,500
7500	200,000 gallons		.40	180		100,500	5,700		106,200	120,000
7600	250,000 gallons	↓	.30	240	↓	143,000	7,575		150,575	170,000

33 16 19 – Elevated Water Utility Storage Tanks

33 16 19.50 Elevated Water Storage Tanks

		Crew	Daily Output	Labor-Hours	Unit	Material	Labor	Equipment	Total	Total Incl O&P
0010	**ELEVATED WATER STORAGE TANKS**									
0011	Not incl pipe, pumps or foundation									
3000	Elevated water tanks, 100′ to bottom capacity line, incl painting									
3010	50,000 gallons				Ea.				222,000	242,000
3300	100,000 gallons								306,000	337,000
3400	250,000 gallons								422,000	464,000
3600	500,000 gallons								681,000	749,000
3700	750,000 gallons								940,000	1,034,000
3900	1,000,000 gallons				↓				944,000	1,038,000

33 21 Water Supply Wells

33 21 13 – Public Water Supply Wells

33 21 13.10 Wells and Accessories

		Crew	Daily Output	Labor-Hours	Unit	Material	Labor	Equipment	Total	Total Incl O&P
0010	**WELLS & ACCESSORIES**									
0011	Domestic									
0100	Drilled, 4″ to 6″ diameter	B-23	120	.333	L.F.		10.65	23	33.65	43
0200	8″ diameter	″	95.20	.420	″		13.45	29	42.45	54
0400	Gravel pack well, 40′ deep, incl. gravel & casing, complete									
0500	24″ diameter casing x 18″ diameter screen	B-23	.13	307	Total	30,600	9,850	21,400	61,850	73,500
0600	36″ diameter casing x 18″ diameter screen		.12	333	″	32,900	10,700	23,100	66,700	79,000
0800	Observation wells, 1-1/4″ riser pipe	↓	163	.245	V.L.F.	16.85	7.85	17.05	41.75	50.50
0900	For flush Buffalo roadway box, add	1 Skwk	16.60	.482	Ea.	46	19.70		65.70	82.50
1200	Test well, 2-1/2″ diameter, up to 50′ deep (15 to 50 GPM)	B-23	1.51	26.490	″	690	850	1,850	3,390	4,175
1300	Over 50′ deep, add	″	121.80	.328	L.F.	18.35	10.50	23	51.85	62.50
1500	Pumps, installed in wells to 100′ deep, 4″ submersible									
1510	1/2 H.P.	Q-1	3.22	4.969	Ea.	460	218		678	845
1520	3/4 H.P.		2.66	6.015		565	264		829	1,025
1600	1 H.P.	↓	2.29	6.987		615	305		920	1,150
1700	1-1/2 H.P.	Q-22	1.60	10		800	440	480	1,720	2,100
1800	2 H.P.	↓	1.33	12.030	↓	1,200	530	580	2,310	2,750

33 21 Water Supply Wells

33 21 13 – Public Water Supply Wells

33 21 13.10 Wells and Accessories		Crew	Daily Output	Labor-Hours	Unit	2009 Bare Costs Material	Labor	Equipment	Total	Total Incl O&P
1900	3 H.P.	Q-22	1.14	14.035	Ea.	1,975	615	675	3,265	3,875
2000	5 H.P.		1.14	14.035		2,775	615	675	4,065	4,750
2050	Remove and install motor only, 4 H.P.		1.14	14.035		2,375	615	675	3,665	4,300
3000	Pump, 6" submersible, 25' to 150' deep, 25 H.P., 249 to 297 GPM		.89	17.978		5,425	790	865	7,080	8,150
3100	25' to 500' deep, 30 H.P., 100 to 300 GPM		.73	21.918		6,300	960	1,050	8,310	9,575
8000	Steel well casing	B-23A	3020	.008	Lb.	.87	.28	.88	2.03	2.39
8110	Well screen assembly, stainless steel, 2" diameter		273	.088	L.F.	78	3.12	9.70	90.82	101
8120	3" diameter		253	.095		108	3.37	10.45	121.82	136
8130	4" diameter		200	.120		124	4.26	13.25	141.51	157
8140	5" diameter		168	.143		144	5.05	15.75	164.80	185
8150	6" diameter		126	.190		173	6.75	21	200.75	224
8160	8" diameter		98.50	.244		225	8.65	27	260.65	291
8170	10" diameter		73	.329		283	11.70	36.50	331.20	370
8180	12" diameter		62.50	.384		335	13.65	42.50	391.15	435
8190	14" diameter		54.30	.442		380	15.70	49	444.70	495
8200	16" diameter		48.30	.497		415	17.65	55	487.65	550
8210	18" diameter		39.20	.612		520	22	67.50	609.50	680
8220	20" diameter		31.20	.769		590	27.50	85	702.50	790
8230	24" diameter		23.80	1.008		730	36	111	877	985
8240	26" diameter		21	1.143		820	40.50	126	986.50	1,100
8300	Slotted PVC, 1-1/4" diameter		521	.046		1.93	1.64	5.10	8.67	10.35
8310	1-1/2" diameter		488	.049		2.79	1.75	5.45	9.99	11.85
8320	2" diameter		273	.088		3.87	3.12	9.70	16.69	20
8330	3" diameter		253	.095		4.11	3.37	10.45	17.93	21.50
8340	4" diameter		200	.120		4.74	4.26	13.25	22.25	26.50
8350	5" diameter		168	.143		4.98	5.05	15.75	25.78	31
8360	6" diameter		126	.190		6.90	6.75	21	34.65	41.50
8370	8" diameter		98.50	.244		10.45	8.65	27	46.10	55
8400	Artificial gravel pack, 2" screen, 6" casing	B-23B	174	.138		3.59	4.90	17	25.49	30.50
8405	8" casing		111	.216		4.91	7.70	26.50	39.11	47.50
8410	10" casing		74.50	.322		6.15	11.45	39.50	57.10	69
8415	12" casing		60	.400		8.35	14.20	49.50	72.05	86.50
8420	14" casing		50.20	.478		10.80	17	59	86.80	104
8425	16" casing		40.70	.590		14.85	21	72.50	108.35	130
8430	18" casing		36	.667		17.25	23.50	82	122.75	148
8435	20" casing		29.50	.814		19.90	29	100	148.90	179
8440	24" casing		25.70	.934		22	33	115	170	205
8445	26" casing		24.60	.976		24.50	34.50	120	179	215
8450	30" casing		20	1.200		28.50	42.50	148	219	263
8455	36" casing		16.40	1.463		30.50	52	180	262.50	315
8500	Develop well		8	3	Hr.	269	107	370	746	875
8550	Pump test well		8	3		71.50	107	370	548.50	655
8560	Standby well	B-23A	8	3		69.50	107	330	506.50	615
8570	Standby, drill rig		8	3			107	330	437	540
8580	Surface seal well, concrete filled		1	24	Ea.	740	850	2,650	4,240	5,125
8590	Well test pump, install & remove	B-23	1	40			1,275	2,775	4,050	5,150
8600	Well sterilization, chlorine	2 Clab	1	16		500	505		1,005	1,375
8610	Well water pressure switch	1 Clab	12	.667		66	21		87	108
9000	Minimum labor/equipment charge	B-21	1.80	15.556	Job		565	71.50	636.50	1,000
9950	See Div. 31 23 19.40 for wellpoints									
9960	See Div. 31 23 19.30 for drainage wells									

33 21 Water Supply Wells

33 21 13 – Public Water Supply Wells

33 21 13.20 Water Supply Wells, Pumps		Crew	Daily Output	Labor-Hours	Unit	Material	Labor	Equipment	Total	Total Incl O&P
						2009 Bare Costs				
0010	**WATER SUPPLY WELLS, PUMPS**									
0011	With pressure control									
1000	Deep well, jet, 42 gal. galvanized tank									
1040	3/4 HP	1 Plum	.80	10	Ea.	685	490		1,175	1,500
3000	Shallow well, jet, 30 gal. galvanized tank									
3040	1/2 HP	1 Plum	2	4	Ea.	495	195		690	845

33 31 Sanitary Utility Sewerage Piping

33 31 13 – Public Sanitary Utility Sewerage Piping

33 31 13.13 Sewage Collection, Vent Cast Iron Pipe		Crew	Daily Output	Labor-Hours	Unit	Material	Labor	Equipment	Total	Total Incl O&P
0010	**SEWAGE COLLECTION, VENT CAST IRON PIPE**									
0020	Not including excavation or backfill									
2022	Sewage vent cast iron, B & S, 4″ diameter	Q-1	66	.242	L.F.	11.35	10.65		22	29
2024	5″ diameter	Q-2	88	.273		16	12.40		28.40	37
2026	6″ diameter	″	84	.286		19.40	13		32.40	41.50
2028	8″ diameter	Q-3	70	.457		31.50	21		52.50	67.50
2030	10″ diameter		66	.485		51.50	22.50		74	92
2032	12″ diameter		57	.561		73.50	26		99.50	121
2034	15″ diameter		49	.653		105	30.50		135.50	163
8001	Fittings, bends and elbows									
8110	4″ diameter	Q-1	13	1.231	Ea.	39	54		93	127
8112	5″ diameter	Q-2	18	1.333		56	60.50		116.50	156
8114	6″ diameter	″	17	1.412		66.50	64		130.50	173
8116	8″ diameter	Q-3	11	2.909		187	135		322	415
8118	10″ diameter		10	3.200		276	149		425	535
8120	12″ diameter		9	3.556		360	165		525	655
8122	15″ diameter		7	4.571		1,025	212		1,237	1,450
8500	Wyes and tees									
8510	4″ diameter	Q-1	8	2	Ea.	57.50	88		145.50	200
8512	5″ diameter	Q-2	12	2		105	91		196	256
8514	6″ diameter	″	11	2.182		133	99.50		232.50	300
8516	8″ diameter	Q-3	7	4.571		315	212		527	680
8518	10″ diameter		6	5.333		505	248		753	940
8520	12″ diameter		4	8		935	370		1,305	1,600
8522	15″ diameter		3	10.667		1,600	495		2,095	2,525

33 31 13.15 Sewage Collection, Concrete Pipe	
0010	**SEWAGE COLLECTION, CONCRETE PIPE**
0020	See Div. 33 41 13.60 for sewage/drainage collection, concrete pipe

33 31 13.20 Sewage Collection, Plastic Pipe		Crew	Daily Output	Labor-Hours	Unit	Material	Labor	Equipment	Total	Total Incl O&P
0010	**SEWAGE COLLECTION, PLASTIC PIPE**									
0020	Not including excavation & backfill									
1100	Piping, DWV Sch 40 ABS, 4″ diameter	B-20	375	.064	L.F.	6.20	2.26		8.46	10.50
1110	6″ diameter		350	.069	″	14.05	2.42		16.47	19.50
1120	Fitting, 1/4 bend, 4″		19	1.263	Ea.	20.50	44.50		65	96.50
1130	6″		15	1.600		14.05	56.50		70.55	109
1140	Tee, 4″		12	2		20.50	70.50		91	139
3000	Piping, HDPE Corrugated Type S with watertight gaskets, 4″ diameter		425	.056	L.F.	.90	2		2.90	4.27
3020	6″ diameter		400	.060		2.05	2.12		4.17	5.75
3040	8″ diameter		380	.063		3.91	2.23		6.14	7.95
3060	10″ diameter		370	.065		5.40	2.29		7.69	9.70

33 31 13 – Public Sanitary Utility Sewerage Piping

33 31 13.20 Sewage Collection, Plastic Pipe		Crew	Daily Output	Labor-Hours	Unit	2009 Bare Costs Material	Labor	Equipment	Total	Total Incl O&P
3080	12″ diameter	B-20	340	.071	L.F.	7.35	2.50		9.85	12.20
3100	15″ diameter		300	.080		10	2.83		12.83	15.65
3120	18″ diameter	B-21	275	.102		14.25	3.70	.47	18.42	22
3140	24″ diameter		250	.112		22	4.07	.52	26.59	31.50
3160	30″ diameter		200	.140		35	5.10	.65	40.75	47.50
3180	36″ diameter		180	.156		44.50	5.65	.72	50.87	59
3200	42″ diameter		175	.160		55.50	5.80	.74	62.04	71.50
3220	48″ diameter		170	.165		72	6	.76	78.76	89.50
3240	54″ diameter		160	.175		111	6.35	.81	118.16	133
3260	60″ diameter		150	.187		129	6.80	.86	136.66	154
3300	Watertight elbows 12″ diameter	B-20	11	2.182	Ea.	61	77		138	194
3320	15″ diameter	″	9	2.667		94	94.50		188.50	259
3340	18″ diameter	B-21	9	3.111		155	113	14.35	282.35	370
3360	24″ diameter		9	3.111		330	113	14.35	457.35	565
3380	30″ diameter		8	3.500		525	127	16.15	668.15	805
3400	36″ diameter		8	3.500		680	127	16.15	823.15	970
3420	42″ diameter		6	4.667		855	170	21.50	1,046.50	1,250
3440	48″ diameter		6	4.667		1,300	170	21.50	1,491.50	1,725
3460	Watertight tee 12″ diameter	B-20	7	3.429		137	121		258	350
3480	15″ diameter	″	6	4		163	141		304	410
3500	18″ diameter	B-21	6	4.667		239	170	21.50	430.50	565
3520	24″ diameter		5	5.600		315	204	26	545	705
3540	30″ diameter		5	5.600		590	204	26	820	1,000
3560	36″ diameter		4	7		770	255	32.50	1,057.50	1,300
3580	42″ diameter		4	7		1,325	255	32.50	1,612.50	1,900
3600	48″ diameter		4	7		2,200	255	32.50	2,487.50	2,850
33 31 13.25 Sewage Collection, Polyvinyl Chloride Pipe										
0010	**SEWAGE COLLECTION, POLYVINYL CHLORIDE PIPE**									
0020	Not including excavation or backfill									
2000	20′ lengths, S.D.R. 35, B&S, 4″ diameter	B-20	375	.064	L.F.	1.66	2.26		3.92	5.55
2040	6″ diameter		350	.069		3.40	2.42		5.82	7.70
2080	13′ lengths , S.D.R. 35, B&S, 8″ diameter		335	.072		7.05	2.53		9.58	11.90
2120	10″ diameter	B-21	330	.085		11.15	3.09	.39	14.63	17.75
2160	12″ diameter		320	.088		12.70	3.18	.40	16.28	19.60
2200	15″ diameter		240	.117		12.10	4.24	.54	16.88	21
3040	Fittings, bends or elbows, 4″ diameter	2 Skwk	19	.842	Ea.	8.25	34.50		42.75	65
3080	6″ diameter		15	1.067		26	43.50		69.50	100
3120	Tees, 4″ diameter		12	1.333		14.35	54.50		68.85	104
3160	6″ diameter		10	1.600		43	65.50		108.50	153
3200	Wyes, 4″ diameter		12	1.333		18.15	54.50		72.65	108
3240	6″ diameter		10	1.600		57.50	65.50		123	170
4000	Piping, DWV PVC, no exc/bkfill, 10′ L, Sch 40, 4″ diameter	B-20	375	.064	L.F.	3.84	2.26		6.10	7.95
4010	6″ diameter		350	.069		7.65	2.42		10.07	12.40
4020	8″ diameter		335	.072		13.30	2.53		15.83	18.80

33 36 Utility Septic Tanks

33 36 13 – Utility Septic Tank and Effluent Wet Wells

33 36 13.10 Septic Tanks		Crew	Daily Output	Labor-Hours	Unit	Material	2009 Bare Costs Labor	Equipment	Total	Total Incl O&P
0010	**SEPTIC TANKS**									
0011	Not including excavation or piping									
0015	Septic tanks, not incl exc or piping, precast, 1,000 gal	B-21	8	3.500	Ea.	730	127	16.15	873.15	1,025
0020	1,250 gallon		8	3.500		950	127	16.15	1,093.15	1,275
0060	1,500 gallon		7	4		1,125	146	18.45	1,289.45	1,500
0100	2,000 gallon		5	5.600		1,625	204	26	1,855	2,150
0140	2,500 gallon		5	5.600		2,300	204	26	2,530	2,875
0180	4,000 gallon		4	7		5,675	255	32.50	5,962.50	6,675
0220	5,000 gal., 4 piece	B-13	3	18.667		9,075	640	263	9,978	11,300
0300	15,000 gallon, 4 piece	B-13B	1.70	32.941		20,500	1,125	660	22,285	25,200
0400	25,000 gallon, 4 piece		1.10	50.909		43,400	1,750	1,025	46,175	52,000
0500	40,000 gallon, 4 piece		.80	70		51,000	2,400	1,400	54,800	61,500
0520	50,000 gallon, 5 piece	B-13C	.60	93.333		59,000	3,200	2,675	64,875	72,500
0540	75,000 gallon, cast in place	C-14C	.25	448		71,500	17,100	103	88,703	107,000
0560	100,000 gallon	"	.15	746		88,500	28,600	172	117,272	145,000
0600	High density polyethylene, 1,000 gallon	B-21	6	4.667		1,125	170	21.50	1,316.50	1,525
0700	1,500 gallon		4	7		1,450	255	32.50	1,737.50	2,050
0900	Galley, 4′ x 4′ x 4′		16	1.750		270	63.50	8.05	341.55	410
1000	Distribution boxes, concrete, 7 outlets	2 Clab	16	1		75	31.50		106.50	135
1100	9 outlets	"	8	2		440	63		503	590
1150	Leaching field chambers, 13′ x 3′-7″ x 1′-4″, standard	B-13	16	3.500		510	120	49.50	679.50	810
1200	Heavy duty, 8′ x 4′ x 1′-6″		14	4		475	137	56.50	668.50	810
1300	13′ x 3′-9″ x 1′-6″		12	4.667		1,000	160	66	1,226	1,425
1350	20′ x 4′ x 1′-6″		5	11.200		1,050	385	158	1,593	1,950
1600	Leaching pit, 6′-6″ diameter, 6′ deep	B-21	5	5.600		1,325	204	26	1,555	1,825
1620	8′ deep		4	7		1,550	255	32.50	1,837.50	2,175
1700	8′ diameter, H-20 load, 6′ deep		4	7		2,200	255	32.50	2,487.50	2,875
1720	8′ deep		3	9.333		2,600	340	43	2,983	3,450
2000	Velocity reducing pit, precast conc., 6′ diameter, 3′ deep		4.70	5.957		1,475	217	27.50	1,719.50	2,000
2200	Excavation for septic tank, 3/4 C.Y. backhoe	B-12F	145	.110	C.Y.		4.09	4.13	8.22	11.10
2400	4′ trench for disposal field, 3/4 C.Y. backhoe	"	335	.048	L.F.		1.77	1.79	3.56	4.81
2600	Gravel fill, run of bank	B-6	150	.160	C.Y.	25	5.45	1.96	32.41	38.50
2800	Crushed stone, 3/4″	"	150	.160	"	29	5.45	1.96	36.41	42.50

33 41 Storm Utility Drainage Piping

33 41 13 – Public Storm Utility Drainage Piping

33 41 13.40 Piping, Storm Drainage, Corrugated Metal		Crew	Daily Output	Labor-Hours	Unit	Material	Labor	Equipment	Total	Total Incl O&P
0010	**PIPING, STORM DRAINAGE, CORRUGATED METAL**									
0020	Not including excavation or backfill									
2000	Corrugated metal pipe, galvanized									
2020	Bituminous coated with paved invert, 20′ lengths									
2040	8″ diameter, 16 ga.	B-14	330	.145	L.F.	12.65	4.83	.89	18.37	23
2060	10″ diameter, 16 ga.		260	.185		15.20	6.15	1.13	22.48	28
2080	12″ diameter, 16 ga.		210	.229		18.20	7.60	1.40	27.20	34
2100	15″ diameter, 16 ga.		200	.240		22	7.95	1.47	31.42	39
2120	18″ diameter, 16 ga.		190	.253		28.50	8.40	1.55	38.45	46.50
2140	24″ diameter, 14 ga.		160	.300		34.50	9.95	1.84	46.29	56.50
2160	30″ diameter, 14 ga.	B-13	120	.467		46	15.95	6.60	68.55	84
2180	36″ diameter, 12 ga.		120	.467		67.50	15.95	6.60	90.05	107
2200	48″ diameter, 12 ga.		100	.560		102	19.15	7.90	129.05	153
2220	60″ diameter, 10 ga.	B-13B	75	.747		132	25.50	15	172.50	204

33 41 13.40 Piping, Storm Drainage, Corrugated Metal		Crew	Daily Output	Labor-Hours	Unit	Material	Labor	Equipment	Total	Total Incl O&P
						2009 Bare Costs				
2240	72″ diameter, 8 ga.	B-13B	45	1.244	L.F.	198	42.50	25	265.50	315
2250	End sections, 8″ diameter, 16 ga.	B-14	20	2.400	Ea.	74	79.50	14.70	168.20	228
2255	10″ diameter, 16 ga.		20	2.400		84.50	79.50	14.70	178.70	239
2260	12″ diameter, 16 ga.		18	2.667		104	88.50	16.30	208.80	277
2265	15″ diameter, 16 ga.		18	2.667		129	88.50	16.30	233.80	305
2270	18″ diameter, 16 ga.		16	3		163	99.50	18.35	280.85	360
2275	24″ diameter, 16 ga.	B-13	16	3.500		232	120	49.50	401.50	505
2280	30″ diameter, 16 ga.		14	4		440	137	56.50	633.50	770
2285	36″ diameter, 14 ga.		14	4		580	137	56.50	773.50	925
2290	48″ diameter, 14 ga.		10	5.600		1,200	191	79	1,470	1,725
2292	60″ diameter, 14 ga.		6	9.333		1,650	320	132	2,102	2,475
2294	72″ diameter, 14 ga.	B-13B	5	11.200		2,875	385	225	3,485	4,050
2300	Bends or elbows, 8″ diameter	B-14	28	1.714		119	57	10.50	186.50	236
2320	10″ diameter		25	1.920		147	63.50	11.75	222.25	279
2340	12″ diameter, 16 ga.		30	1.600		171	53	9.80	233.80	286
2342	18″ diameter, 16 ga.		20	2.400		246	79.50	14.70	340.20	415
2344	24″ diameter, 14 ga.		16	3		345	99.50	18.35	462.85	565
2346	30″ diameter, 14 ga.		15	3.200		425	106	19.60	550.60	665
2348	36″ diameter, 14 ga.	B-13	15	3.733		575	128	52.50	755.50	900
2350	48″ diameter, 12 ga.	"	12	4.667		825	160	66	1,051	1,225
2352	60″ diameter, 10 ga.	B-13B	10	5.600		1,100	191	113	1,404	1,625
2354	72″ diameter, 10 ga.	"	6	9.333		1,375	320	188	1,883	2,225
2360	Wyes or tees, 8″ diameter	B-14	25	1.920		166	63.50	11.75	241.25	300
2380	10″ diameter		21	2.286		207	76	14	297	365
2400	12″ diameter, 16 ga.		22.48	2.135		246	71	13.05	330.05	400
2410	18″ diameter, 16 ga.		16	3		330	99.50	18.35	447.85	545
2412	24″ diameter, 14 ga.		16	3		495	99.50	18.35	612.85	730
2414	30″ diameter, 14 ga.	B-13	12	4.667		640	160	66	866	1,025
2416	36″ diameter, 14 ga.		11	5.091		790	174	72	1,036	1,225
2418	48″ diameter, 12 ga.		10	5.600		1,200	191	79	1,470	1,725
2420	60″ diameter, 10 ga.	B-13B	8	7		1,750	239	141	2,130	2,475
2422	72″ diameter, 10 ga.	"	5	11.200		2,100	385	225	2,710	3,175
2500	Galvanized, uncoated, 20′ lengths									
2520	8″ diameter, 16 ga.	B-14	355	.135	L.F.	10.35	4.49	.83	15.67	19.65
2540	10″ diameter, 16 ga.		280	.171		11.35	5.70	1.05	18.10	23
2560	12″ diameter, 16 ga.		220	.218		12.95	7.25	1.34	21.54	27.50
2580	15″ diameter, 16 ga.		220	.218		16.50	7.25	1.34	25.09	31.50
2600	18″ diameter, 16 ga.		205	.234		22	7.75	1.43	31.18	38.50
2620	24″ diameter, 14 ga.		175	.274		31.50	9.10	1.68	42.28	52
2640	30″ diameter, 14 ga.	B-13	130	.431		40.50	14.75	6.05	61.30	75
2660	36″ diameter, 12 ga.		130	.431		66	14.75	6.05	86.80	103
2680	48″ diameter, 12 ga.		110	.509		88.50	17.40	7.20	113.10	133
2690	60″ diameter, 10 ga.	B-13B	78	.718		139	24.50	14.40	177.90	209
2695	72″ diameter, 10 ga.	"	60	.933		176	32	18.75	226.75	266
2711	Bends or elbows, 12″ diameter, 16 ga.	B-14	30	1.600	Ea.	147	53	9.80	209.80	260
2712	15″ diameter, 16 ga.		25.04	1.917		182	63.50	11.75	257.25	315
2714	18″ diameter, 16 ga.		20	2.400		204	79.50	14.70	298.20	370
2716	24″ diameter, 14 ga.		16	3		298	99.50	18.35	415.85	515
2718	30″ diameter, 14 ga.		15	3.200		375	106	19.60	500.60	605
2720	36″ diameter, 14 ga.	B-13	15	3.733		510	128	52.50	690.50	825
2722	48″ diameter, 12 ga.		12	4.667		700	160	66	926	1,100
2724	60″ diameter, 10 ga.		10	5.600		1,075	191	79	1,345	1,575
2726	72″ diameter, 10 ga.		6	9.333		1,325	320	132	1,777	2,150

33 41 13.40 Piping, Storm Drainage, Corrugated Metal

	33 41 13.40 Piping, Storm Drainage, Corrugated Metal	Crew	Daily Output	Labor-Hours	Unit	2009 Bare Costs Material	Labor	Equipment	Total	Total Incl O&P
2728	Wyes or tees, 12" diameter, 16 ga.	B-14	22.48	2.135	Ea.	195	71	13.05	279.05	345
2730	18" diameter, 16 ga.		15	3.200		285	106	19.60	410.60	510
2732	24" diameter, 14 ga.		15	3.200		440	106	19.60	565.60	680
2734	30" diameter, 14 ga.		14	3.429		580	114	21	715	845
2736	36" diameter, 14 ga.	B-13	14	4		750	137	56.50	943.50	1,100
2738	48" diameter, 12 ga.		12	4.667		1,100	160	66	1,326	1,550
2740	60" diameter, 10 ga.		10	5.600		1,600	191	79	1,870	2,175
2742	72" diameter, 10 ga.		6	9.333		1,900	320	132	2,352	2,775
2780	End sections, 8" diameter	B-14	24	2		101	66.50	12.25	179.75	233
2785	10" diameter		22	2.182		104	72.50	13.35	189.85	247
2790	12" diameter		35	1.371		117	45.50	8.40	170.90	213
2800	18" diameter		30	1.600		139	53	9.80	201.80	251
2810	24" diameter	B-13	25	2.240		202	76.50	31.50	310	380
2820	30" diameter		25	2.240		385	76.50	31.50	493	585
2825	36" diameter		20	2.800		510	95.50	39.50	645	760
2830	48" diameter		10	5.600		1,050	191	79	1,320	1,550
2835	60" diameter	B-13B	5	11.200		1,375	385	225	1,985	2,400
2840	72" diameter	"	4	14		2,475	480	281	3,236	3,825
2850	Couplings, 12" diameter					22.50			22.50	25
2855	18" diameter					33			33	36.50
2860	24" diameter					44.50			44.50	49
2865	30" diameter					55			55	60
2870	36" diameter					76.50			76.50	84
2875	48" diameter					104			104	114
2880	60" diameter					163			163	180
2885	72" diameter					226			226	249

33 41 13.50 Piping, Drainage & Sewage, Corrug. HDPE Type S

		Crew	Daily Output	Labor-Hours	Unit	Material	Labor	Equipment	Total	Total Incl O&P
0010	**PIPING, DRAINAGE & SEWAGE, CORRUGATED HDPE TYPE S**									
0020	Not including excavation & backfill, bell & spigot									
1000	With gaskets, 4" diameter	B-20	425	.056	L.F.	.78	2		2.78	4.14
1010	6" diameter		400	.060		1.78	2.12		3.90	5.45
1020	8" diameter		380	.063		3.40	2.23		5.63	7.40
1030	10" diameter		370	.065		4.71	2.29		7	8.95
1040	12" diameter		340	.071		6.40	2.50		8.90	11.15
1050	15" diameter		300	.080		8.70	2.83		11.53	14.20
1060	18" diameter	B-21	275	.102		12.40	3.70	.47	16.57	20
1070	24" diameter		250	.112		19.20	4.07	.52	23.79	28
1080	30" diameter		200	.140		30.50	5.10	.65	36.25	42.50
1090	36" diameter		180	.156		38.50	5.65	.72	44.87	52.50
1100	42" diameter		175	.160		48	5.80	.74	54.54	63.50
1110	48" diameter		170	.165		62.50	6	.76	69.26	79.50
1120	54" diameter		160	.175		96.50	6.35	.81	103.66	117
1130	60" diameter		150	.187		112	6.80	.86	119.66	136
1135	Add 15% to material pipe cost for water tight connection bell & spigot									
1140	HDPE type s, elbows 12" diameter	B-20	11	2.182	Ea.	53	77		130	186
1150	15" diameter	"	9	2.667		82	94.50		176.50	245
1160	18" diameter	B-21	9	3.111		135	113	14.35	262.35	350
1170	24" diameter		9	3.111		287	113	14.35	414.35	515
1180	30" diameter		8	3.500		460	127	16.15	603.15	730
1190	36" diameter		8	3.500		590	127	16.15	733.15	875
1240	HDPE type s, Tee 12" diameter	B-20	7	3.429		119	121		240	330
1260	15" diameter	"	6	4		142	141		283	390

33 41 13.50 Piping, Drainage & Sewage, Corrug. HDPE Type S		Crew	Daily Output	Labor-Hours	Unit	2009 Bare Costs Material	Labor	Equipment	Total	Total Incl O&P
1280	18″ diameter	B-21	6	4.667	Ea.	208	170	21.50	399.50	530
1300	24″ diameter		5	5.600		274	204	26	504	660
1320	30″ diameter		5	5.600		515	204	26	745	925
1340	36″ diameter		4	7		670	255	32.50	957.50	1,175
1360	42″ diameter		4	7		1,150	255	32.50	1,437.50	1,725
1380	48″ diameter		4	7		1,900	255	32.50	2,187.50	2,550
1400	Add to basic installation cost for each split coupling joint									
1402	HDPE type s, split coupling, 12″ diameter	B-20	17	1.412	Ea.	5.85	50		55.85	88.50
1420	15″ diameter		15	1.600		6.50	56.50		63	100
1440	18″ diameter		13	1.846		11.25	65.50		76.75	119
1460	24″ diameter		12	2		16.40	70.50		86.90	134
1480	30″ diameter		10	2.400		27	85		112	169
1500	36″ diameter		9	2.667		34.50	94.50		129	193
1520	42″ diameter		8	3		41.50	106		147.50	220
1540	48″ diameter		8	3		53.50	106		159.50	233

33 41 13.60 Sewage/Drainage Collection, Concrete Pipe		Crew	Daily Output	Labor-Hours	Unit	Material	Labor	Equipment	Total	Total Incl O&P
0010	**SEWAGE/DRAINAGE COLLECTION, CONCRETE PIPE**									
0020	Not including excavation or backfill									
0100	Box culvert, precast, base price, 8′ long, 6′ x 3′	B-69	140	.343	L.F.	286	11.85	9.25	307.10	345
0150	6′ x 7′		125	.384		430	13.30	10.35	453.65	510
0200	8′ x 3′		133	.361		395	12.50	9.75	417.25	460
0250	8′ x 8′		100	.480		530	16.60	12.95	559.55	625
0300	10′ x 3′		110	.436		585	15.10	11.80	611.90	675
0350	10′ x 8′		80	.600		660	21	16.20	697.20	780
0400	12′ x 3′		100	.480		565	16.60	12.95	594.55	665
0450	12′ x 8′		67	.716		975	25	19.35	1,019.35	1,125
0500	Set up charge at plant, add to base price				Job	5,075			5,075	5,600
0510	Inserts and keyway, add				Ea.	475			475	525
0520	Sloped or skewed end, add				″	760			760	835
1000	Non-reinforced pipe, extra strength, B&S or T&G joints									
1010	6″ diameter	B-14	265.04	.181	L.F.	6.05	6	1.11	13.16	17.70
1020	8″ diameter		224	.214		6.70	7.10	1.31	15.11	20.50
1030	10″ diameter		216	.222		7.40	7.35	1.36	16.11	21.50
1040	12″ diameter		200	.240		9.10	7.95	1.47	18.52	24.50
1050	15″ diameter		180	.267		10.60	8.85	1.63	21.08	28
1060	18″ diameter		144	.333		13.05	11.05	2.04	26.14	34.50
1070	21″ diameter		112	.429		16.10	14.20	2.62	32.92	44
1080	24″ diameter		100	.480		19.75	15.95	2.94	38.64	50.50
1560	Reinforced culvert, class 2, no gaskets									
1590	27″ diameter	B-21	88	.318	L.F.	41	11.60	1.47	54.07	65.50
1592	30″ diameter	B-13	80	.700		44.50	24	9.85	78.35	98.50
1594	36″ diameter	″	72	.778		52	26.50	10.95	89.45	112
2000	Reinforced culvert, class 3, no gaskets									
2010	12″ diameter	B-14	150	.320	L.F.	10.20	10.60	1.96	22.76	30.50
2020	15″ diameter		150	.320		14.95	10.60	1.96	27.51	36
2030	18″ diameter		132	.364		17.45	12.05	2.23	31.73	41.50
2035	21″ diameter		120	.400		21.50	13.25	2.45	37.20	48
2040	24″ diameter		100	.480		29	15.95	2.94	47.89	60.50
2045	27″ diameter	B-13	92	.609		36.50	21	8.60	66.10	83.50
2050	30″ diameter		88	.636		42	22	8.95	72.95	91.50
2060	36″ diameter		72	.778		53.50	26.50	10.95	90.95	114
2070	42″ diameter	B-13B	72	.778		77.50	26.50	15.65	119.65	146

33 41 13.60 Sewage/Drainage Collection, Concrete Pipe		Crew	Daily Output	Labor-Hours	Unit	Material	2009 Bare Costs Labor	Equipment	Total	Total Incl O&P
2080	48″ diameter	B-13B	64	.875	L.F.	113	30	17.60	160.60	192
2090	60″ diameter		48	1.167		202	40	23.50	265.50	315
2100	72″ diameter		40	1.400		225	48	28	301	355
2120	84″ diameter		32	1.750		255	60	35	350	415
2140	96″ diameter		24	2.333		305	80	47	432	515
2200	With gaskets, class 3, 12″ diameter	B-21	168	.167		14.15	6.05	.77	20.97	26.50
2220	15″ diameter		160	.175		17.55	6.35	.81	24.71	30.50
2230	18″ diameter		152	.184		23.50	6.70	.85	31.05	38
2240	24″ diameter		136	.206		36	7.50	.95	44.45	52.50
2260	30″ diameter	B-13	88	.636		50.50	22	8.95	81.45	101
2270	36″ diameter	″	72	.778		64	26.50	10.95	101.45	126
2290	48″ diameter	B-13B	64	.875		129	30	17.60	176.60	210
2310	72″ diameter	″	40	1.400		251	48	28	327	385
2330	Flared ends, 6′-1″ long, 12″ diameter	B-21	190	.147		51.50	5.35	.68	57.53	66
2340	15″ diameter		155	.181		58.50	6.55	.83	65.88	76
2400	6′-2″ long, 18″ diameter		122	.230		61.50	8.35	1.06	70.91	82.50
2420	24″ diameter		88	.318		70.50	11.60	1.47	83.57	98
2440	36″ diameter	B-13	60	.933		128	32	13.15	173.15	208
2500	Class 4									
2510	12″ diameter	B-21	168	.167	L.F.	20	6.05	.77	26.82	33
2512	15″ diameter		160	.175		24	6.35	.81	31.16	37.50
2514	18″ diameter		152	.184		28	6.70	.85	35.55	43
2516	21″ diameter		144	.194		35	7.05	.90	42.95	51
2518	24″ diameter		136	.206		41	7.50	.95	49.45	58
2520	27″ diameter		120	.233		51.50	8.50	1.08	61.08	71.50
2522	30″ diameter	B-13	88	.636		56.50	22	8.95	87.45	107
2524	36″ diameter	″	72	.778		83.50	26.50	10.95	120.95	147
2600	Class 5									
2610	12″ diameter	B-21	168	.167	L.F.	20.50	6.05	.77	27.32	33.50
2612	15″ diameter		160	.175		24	6.35	.81	31.16	38
2614	18″ diameter		152	.184		29	6.70	.85	36.55	43.50
2616	21″ diameter		144	.194		36	7.05	.90	43.95	52
2618	24″ diameter		136	.206		42	7.50	.95	50.45	59
2620	27″ diameter		120	.233		54	8.50	1.08	63.58	74.50
2622	30″ diameter	B-13	88	.636		59.50	22	8.95	90.45	111
2624	36″ diameter	″	72	.778		89.50	26.50	10.95	126.95	154
2800	Add for rubber joints 12″- 36″ diameter					12%				
3080	Radius pipe, add to pipe prices, 12″ to 60″ diameter					50%				
3090	Over 60″ diameter, add					20%				
3500	Reinforced elliptical, 8′ lengths, C507 class 3									
3520	14″ x 23″ inside, round equivalent 18″ diameter	B-21	82	.341	L.F.	49.50	12.40	1.57	63.47	76.50
3530	24″ x 38″ inside, round equivalent 30″ diameter	B-13	58	.966		73.50	33	13.60	120.10	150
3540	29″ x 45″ inside, round equivalent 36″ diameter		52	1.077		122	37	15.20	174.20	211
3550	38″ x 60″ inside, round equivalent 48″ diameter		38	1.474		209	50.50	21	280.50	335
3560	48″ x 76″ inside, round equivalent 60″ diameter		26	2.154		294	73.50	30.50	398	480
3570	58″ x 91″ inside, round equivalent 72″ diameter		22	2.545		410	87	36	533	630
3780	Concrete slotted pipe, class 4 mortar joint									
3800	12″ diameter	B-21	168	.167	L.F.	18.80	6.05	.77	25.62	31.50
3840	18″ diameter	″	152	.184	″	29	6.70	.85	36.55	44
3900	Class 4 O-ring									
3940	12″ diameter	B-21	168	.167	L.F.	19.70	6.05	.77	26.52	32.50
3960	18″ diameter	″	152	.184	″	26.50	6.70	.85	34.05	41
6200	Gasket, conc. pipe joint, 12″				Ea.	3.95			3.95	4.35

33 41 Storm Utility Drainage Piping

33 41 13 – Public Storm Utility Drainage Piping

33 41 13.60 Sewage/Drainage Collection, Concrete Pipe		Crew	Daily Output	Labor-Hours	Unit	2009 Bare Costs Material	Labor	Equipment	Total	Total Incl O&P
6220	24″				Ea.	7.20			7.20	7.95
6240	36″					10.50			10.50	11.55
6260	48″					16.60			16.60	18.25
6265	54″					18.40			18.40	20
6270	60″					23.50			23.50	26
6280	72″				↓	26.50			26.50	29

33 42 Culverts

33 42 16 – Concrete Culverts

33 42 16.15 Oval Arch Culverts

		Crew	Daily Output	Labor-Hours	Unit	Material	Labor	Equipment	Total	Total Incl O&P
0010	**OVAL ARCH CULVERTS**									
3000	Corrugated galvanized or aluminum, coated & paved									
3020	17″ x 13″, 16 ga., 15″ equivalent	B-14	200	.240	L.F.	39.50	7.95	1.47	48.92	58
3040	21″ x 15″, 16 ga., 18″ equivalent		150	.320		51	10.60	1.96	63.56	75.50
3060	28″ x 20″, 14 ga., 24″ equivalent		125	.384		73	12.75	2.35	88.10	104
3080	35″ x 24″, 14 ga., 30″ equivalent	↓	100	.480		89	15.95	2.94	107.89	127
3100	42″ x 29″, 12 ga., 36″ equivalent	B-13	100	.560		133	19.15	7.90	160.05	186
3120	49″ x 33″, 12 ga., 42″ equivalent		90	.622		160	21.50	8.75	190.25	220
3140	57″ x 38″, 12 ga., 48″ equivalent	↓	75	.747	↓	178	25.50	10.55	214.05	248
3160	Steel, plain oval arch culverts, plain									
3180	17″ x 13″, 16 ga., 15″ equivalent	B-14	225	.213	L.F.	21.50	7.10	1.31	29.91	36.50
3200	21″ x 15″, 16 ga., 18″ equivalent		175	.274		25.50	9.10	1.68	36.28	45
3220	28″ x 20″, 14 ga., 24″ equivalent	↓	150	.320		40.50	10.60	1.96	53.06	64.50
3240	35″ x 24″, 14 ga., 30″ equivalent	B-13	108	.519		51	17.75	7.30	76.05	93
3260	42″ x 29″, 12 ga., 36″ equivalent		108	.519		84.50	17.75	7.30	109.55	130
3280	49″ x 33″, 12 ga., 42″ equivalent		92	.609		99.50	21	8.60	129.10	153
3300	57″ x 38″, 12 ga., 48″ equivalent		75	.747	↓	88	25.50	10.55	124.05	150
3320	End sections, 17″ x 13″		22	2.545	Ea.	117	87	36	240	310
3340	42″ x 29″	↓	17	3.294	″	480	113	46.50	639.50	765
3360	Multi-plate arch, steel	B-20	1690	.014	Lb.	1.34	.50		1.84	2.29

33 44 Storm Utility Water Drains

33 44 13 – Utility Area Drains

33 44 13.13 Catchbasins

		Crew	Daily Output	Labor-Hours	Unit	Material	Labor	Equipment	Total	Total Incl O&P
0010	**CATCHBASINS**									
0011	Not including footing & excavation									
1600	Frames & covers, C.I., 24″ square, 500 lb.	B-6	7.80	3.077	Ea.	298	105	37.50	440.50	540
1700	26″ D shape, 600 lb.		7	3.429		510	117	42	669	800
1800	Light traffic, 18″ diameter, 100 lb.		10	2.400		167	82	29.50	278.50	350
1900	24″ diameter, 300 lb.		8.70	2.759		259	94	34	387	475
2000	36″ diameter, 900 lb.		5.80	4.138		520	141	50.50	711.50	855
2100	Heavy traffic, 24″ diameter, 400 lb.		7.80	3.077		249	105	37.50	391.50	485
2200	36″ diameter, 1150 lb.		3	8		825	273	98	1,196	1,450
2300	Mass. State standard, 26″ diameter, 475 lb.		7	3.429		625	117	42	784	920
2400	30″ diameter, 620 lb.		7	3.429		395	117	42	554	670
2500	Watertight, 24″ diameter, 350 lb.		7.80	3.077		425	105	37.50	567.50	675
2600	26″ diameter, 500 lb.		7	3.429		400	117	42	559	675
2700	32″ diameter, 575 lb.	↓	6	4	↓	895	136	49	1,080	1,250
2800	3 piece cover & frame, 10″ deep,									

33 44 Storm Utility Water Drains

33 44 13 – Utility Area Drains

33 44 13.13 Catchbasins		Crew	Daily Output	Labor-Hours	Unit	2009 Bare Costs Material	Labor	Equipment	Total	Total Incl O&P
2900	1200 lbs., for heavy equipment	B-6	3	8	Ea.	1,275	273	98	1,646	1,950
3000	Raised for paving 1-1/4" to 2" high									
3100	4 piece expansion ring									
3200	20" to 26" diameter	1 Clab	3	2.667	Ea.	140	84.50		224.50	293
3300	30" to 36" diameter	"	3	2.667	"	195	84.50		279.50	355
3320	Frames and covers, existing, raised for paving, 2", including									
3340	row of brick, concrete collar, up to 12" wide frame	B-6	18	1.333	Ea.	43.50	45.50	16.30	105.30	139
3360	20" to 26" wide frame		11	2.182		66	74.50	26.50	167	223
3380	30" to 36" wide frame		9	2.667		82	91	32.50	205.50	273
3400	Inverts, single channel brick	D-1	3	5.333		92	194		286	415
3500	Concrete		5	3.200		72	116		188	267
3600	Triple channel, brick		2	8		139	291		430	625
3700	Concrete		3	5.333		123	194		317	450

33 46 Subdrainage

33 46 16 – Subdrainage Piping

33 46 16.20 Piping, Subdrainage, Concrete

		Crew	Daily Output	Labor-Hours	Unit	Material	Labor	Equipment	Total	Total Incl O&P
0010	**PIPING, SUBDRAINAGE, CONCRETE**									
0021	Not including excavation and backfill									
3000	Porous wall concrete underdrain, std. strength, 4" diameter	B-20	335	.072	L.F.	3.30	2.53		5.83	7.80
3020	6" diameter	"	315	.076		4.29	2.69		6.98	9.15
3040	8" diameter	B-21	310	.090		5.30	3.29	.42	9.01	11.65
3060	12" diameter		285	.098		11.20	3.57	.45	15.22	18.70
3080	15" diameter		230	.122		12.85	4.43	.56	17.84	22
3100	18" diameter		165	.170		17	6.15	.78	23.93	29.50
4000	Extra strength, 6" diameter	B-20	315	.076		3.97	2.69		6.66	8.80
4020	8" diameter	B-21	310	.090		5.95	3.29	.42	9.66	12.35
4040	10" diameter		295	.095		11.90	3.45	.44	15.79	19.25
4060	12" diameter		285	.098		12.90	3.57	.45	16.92	20.50
4080	15" diameter		230	.122		14.30	4.43	.56	19.29	23.50
4100	18" diameter		165	.170		21	6.15	.78	27.93	34

33 46 16.25 Piping, Subdrainage, Corrugated Metal

		Crew	Daily Output	Labor-Hours	Unit	Material	Labor	Equipment	Total	Total Incl O&P
0010	**PIPING, SUBDRAINAGE, CORRUGATED METAL**									
0021	Not including excavation and backfill									
2010	Aluminum, perforated									
2020	6" diameter, 18 ga.	B-14	380	.126	L.F.	4.70	4.19	.77	9.66	12.85
2200	8" diameter, 16 ga.		370	.130		7.15	4.30	.79	12.24	15.75
2220	10" diameter, 16 ga.		360	.133		8.90	4.42	.82	14.14	17.95
2240	12" diameter, 16 ga.		285	.168		10	5.60	1.03	16.63	21.50
2260	18" diameter, 16 ga.		205	.234		14.95	7.75	1.43	24.13	30.50
3000	Uncoated galvanized, perforated									
3020	6" diameter, 18 ga.	B-20	380	.063	L.F.	8.25	2.23		10.48	12.70
3200	8" diameter, 16 ga.	"	370	.065		11.35	2.29		13.64	16.20
3220	10" diameter, 16 ga.	B-21	360	.078		17	2.83	.36	20.19	23.50
3240	12" diameter, 16 ga.		285	.098		17.80	3.57	.45	21.82	26
3260	18" diameter, 16 ga.		205	.137		27	4.97	.63	32.60	39
4000	Steel, perforated, asphalt coated									
4020	6" diameter 18 ga.	B-20	380	.063	L.F.	6.60	2.23		8.83	10.90
4030	8" diameter 18 ga.	"	370	.065		10.30	2.29		12.59	15.10
4040	10" diameter 16 ga.	B-21	360	.078		11.85	2.83	.36	15.04	18
4050	12" diameter 16 ga.		285	.098		13.60	3.57	.45	17.62	21.50

33 46 Subdrainage

33 46 16 – Subdrainage Piping

33 46 16.25 Piping, Subdrainage, Corrugated Metal		Crew	Daily Output	Labor-Hours	Unit	2009 Bare Costs Material	Labor	Equipment	Total	Total Incl O&P
4060	18″ diameter 16 ga.	B-21	205	.137	L.F.	18.50	4.97	.63	24.10	29.50

33 46 16.30 Piping, Subdrainage, Plastic

		Crew	Daily Output	Labor-Hours	Unit	Material	Labor	Equipment	Total	Total Incl O&P
0010	**PIPING, SUBDRAINAGE, PLASTIC**									
0020	Not including excavation and backfill									
2100	Perforated PVC, 4″ diameter	B-14	314	.153	L.F.	1.09	5.05	.94	7.08	10.55
2110	6″ diameter		300	.160		2.02	5.30	.98	8.30	12
2120	8″ diameter		290	.166		2.29	5.50	1.01	8.80	12.65
2130	10″ diameter		280	.171		3.54	5.70	1.05	10.29	14.35
2140	12″ diameter		270	.178		4.91	5.90	1.09	11.90	16.25

33 46 26 – Geotextile Subsurface Drainage Filtration

33 46 26.10 Geotextiles for Subsurface Drainage

		Crew	Daily Output	Labor-Hours	Unit	Material	Labor	Equipment	Total	Total Incl O&P
0010	**GEOTEXTILES FOR SUBSURFACE DRAINAGE**									
0100	Fabric, laid in trench, polypropylene, ideal conditions	2 Clab	2400	.007	S.Y.	1.54	.21		1.75	2.04
0110	Adverse conditions		1600	.010	"	1.54	.32		1.86	2.21
0170	Fabric ply bonded to 3 dimen. nylon mat, .4″ thick, ideal conditions		2000	.008	S.F.	.26	.25		.51	.71
0180	Adverse conditions		1200	.013	"	.33	.42		.75	1.06
0185	Soil drainage mat on vertical wall, 0.44″ thick		265	.060	S.Y.	1.94	1.91		3.85	5.30
0188	0.25″ thick		300	.053	"	1.46	1.69		3.15	4.39
0190	0.8″ thick, ideal conditions		2400	.007	S.F.	.20	.21		.41	.57
0200	Adverse conditions		1600	.010	"	.33	.32		.65	.88
0300	Drainage material, 3/4″ gravel fill in trench	B-6	260	.092	C.Y.	30	3.15	1.13	34.28	39.50
0400	Pea stone	"	260	.092	"	31.50	3.15	1.13	35.78	41

33 49 Storm Drainage Structures

33 49 13 – Storm Drainage Manholes, Frames, and Covers

33 49 13.10 Storm Drainage Manholes, Frames and Covers

		Crew	Daily Output	Labor-Hours	Unit	Material	Labor	Equipment	Total	Total Incl O&P
0010	**STORM DRAINAGE MANHOLES, FRAMES & COVERS**									
0020	Excludes footing, excavation, backfill (See line items for frame & cover)									
0050	Brick, 4′ inside diameter, 4′ deep	D-1	1	16	Ea.	400	580		980	1,375
0100	6′ deep		.70	22.857		555	830		1,385	1,950
0150	8′ deep		.50	32		710	1,175		1,885	2,650
0200	For depths over 8′, add		4	4	V.L.F.	162	145		307	415
0400	Concrete blocks (radial), 4′ I.D., 4′ deep		1.50	10.667	Ea.	350	390		740	1,000
0500	6′ deep		1	16		460	580		1,040	1,450
0600	8′ deep		.70	22.857		575	830		1,405	1,975
0700	For depths over 8′, add		5.50	2.909	V.L.F.	59	106		165	236
0800	Concrete, cast in place, 4′ x 4′, 8″ thick, 4′ deep	C-14H	2	24	Ea.	560	945	12.95	1,517.95	2,175
0900	6′ deep		1.50	32		810	1,250	17.30	2,077.30	2,975
1000	8′ deep		1	48		1,150	1,900	26	3,076	4,425
1100	For depths over 8′, add		8	6	V.L.F.	132	236	3.24	371.24	540
1110	Precast, 4′ I.D., 4′ deep	B-22	4.10	7.317	Ea.	880	269	47.50	1,196.50	1,450
1120	6′ deep		3	10		1,100	370	64.50	1,534.50	1,900
1130	8′ deep		2	15		1,325	550	97	1,972	2,450
1140	For depths over 8′, add		16	1.875	V.L.F.	181	69	12.10	262.10	325
1150	5′ I.D., 4′ deep	B-6	3	8	Ea.	910	273	98	1,281	1,550
1160	6′ deep		2	12		1,225	410	147	1,782	2,175
1170	8′ deep		1.50	16		1,550	545	196	2,291	2,800
1180	For depths over 8′, add		12	2	V.L.F.	202	68	24.50	294.50	360
1190	6′ I.D., 4′ deep		2	12	Ea.	1,500	410	147	2,057	2,450
1200	6′ deep		1.50	16		1,925	545	196	2,666	3,225

33 49 Storm Drainage Structures

33 49 13 – Storm Drainage Manholes, Frames, and Covers

33 49 13.10 Storm Drainage Manholes, Frames and Covers		Crew	Daily Output	Labor-Hours	Unit	2009 Bare Costs Material	Labor	Equipment	Total	Total Incl O&P
1210	8′ deep	B-6	1	24	Ea.	2,375	820	294	3,489	4,275
1220	For depths over 8′, add		8	3	V.L.F.	310	102	36.50	448.50	550
1250	Slab tops, precast, 8″ thick									
1300	4′ diameter manhole	B-6	8	3	Ea.	209	102	36.50	347.50	435
1400	5′ diameter manhole		7.50	3.200		415	109	39	563	675
1500	6′ diameter manhole		7	3.429		595	117	42	754	890
3800	Steps, heavyweight cast iron, 7″ x 9″	1 Bric	40	.200		12.95	8.10		21.05	27.50
3900	8″ x 9″		40	.200		19.45	8.10		27.55	34.50
3928	12″ x 10-1/2″		40	.200		22	8.10		30.10	37.50
4000	Standard sizes, galvanized steel		40	.200		18.75	8.10		26.85	33.50
4100	Aluminum		40	.200		25.50	8.10		33.60	41

33 51 Natural-Gas Distribution

33 51 13 – Natural-Gas Piping

33 51 13.10 Piping, Gas Service and Distribution, P.E.

		Crew	Daily Output	Labor-Hours	Unit	Material	Labor	Equipment	Total	Total Incl O&P
0010	**PIPING, GAS SERVICE AND DISTRIBUTION, POLYETHYLENE**									
0020	Not including excavation or backfill									
1000	60 psi coils, compression coupling @ 100′, 1/2″ diameter, SDR 11	B-20A	608	.053	L.F.	1.61	2.01		3.62	4.98
1040	1-1/4″ diameter, SDR 11		544	.059		2.49	2.25		4.74	6.30
1100	2″ diameter, SDR 11		488	.066		3.09	2.51		5.60	7.40
1160	3″ diameter, SDR 11		408	.078		6.45	3		9.45	11.90
1500	60 PSI 40′ joints with coupling, 3″ diameter, SDR 11	B-21A	408	.098		6.45	3.83	1.48	11.76	14.80
1540	4″ diameter, SDR 11		352	.114		14.75	4.44	1.71	20.90	25
1600	6″ diameter, SDR 11		328	.122		46	4.77	1.84	52.61	60
1640	8″ diameter, SDR 11		272	.147		62.50	5.75	2.21	70.46	80.50
9000	Minimum labor/equipment charge	B-20	2	12	Job		425		425	695

33 51 13.20 Piping, Gas Service and Distribution, Steel

		Crew	Daily Output	Labor-Hours	Unit	Material	Labor	Equipment	Total	Total Incl O&P
0010	**PIPING, GAS SERVICE & DISTRIBUTION, STEEL**									
0020	Not including excavation or backfill, tar coated and wrapped									
4000	Pipe schedule 40, plain end									
4040	1″ diameter	Q-4	300	.107	L.F.	7.55	4.95	.19	12.69	16.20
4080	2″ diameter		280	.114		11.85	5.30	.20	17.35	21.50
4120	3″ diameter		260	.123		19.60	5.70	.21	25.51	30.50
4160	4″ diameter	B-35	255	.188		25.50	7.35	2.57	35.42	42.50
4200	5″ diameter		220	.218		37	8.50	2.97	48.47	57.50
4240	6″ diameter		180	.267		45	10.40	3.63	59.03	70.50
4280	8″ diameter		140	.343		71.50	13.40	4.67	89.57	106
4320	10″ diameter		100	.480		91	18.75	6.55	116.30	137
4360	12″ diameter		80	.600		95	23.50	8.20	126.70	152
4400	14″ diameter		75	.640		101	25	8.70	134.70	162
4440	16″ diameter		70	.686		111	27	9.35	147.35	175
4480	18″ diameter		65	.738		143	29	10.05	182.05	214
4520	20″ diameter		60	.800		222	31	10.90	263.90	305
4560	24″ diameter		50	.960		253	37.50	13.10	303.60	355
6000	Schedule 80, plain end									
6002	4″ diameter	B-35	144	.333	L.F.	33	13	4.54	50.54	62
6006	6″ diameter		126	.381		62.50	14.85	5.20	82.55	98
6008	8″ diameter		108	.444		83.50	17.35	6.05	106.90	126
6012	12″ diameter		72	.667		167	26	9.10	202.10	235
8008	Elbow, weld joint, standard weight									
8020	4″ diameter	Q-16	6.80	3.529	Ea.	68.50	161	8.20	237.70	335

33 51 Natural-Gas Distribution

33 51 13 – Natural-Gas Piping

33 51 13.20 Piping, Gas Service and Distribution, Steel		Crew	Daily Output	Labor-Hours	Unit	Material	2009 Bare Costs Labor	Equipment	Total	Total Incl O&P
8026	8″ diameter	Q-16	3.40	7.059	Ea.	282	320	16.40	618.40	830
8030	12″ diameter		2.30	10.435		630	475	24	1,129	1,450
8034	16″ diameter		1.50	16		1,275	730	37	2,042	2,575
8038	20″ diameter		1.20	20		2,300	910	46.50	3,256.50	3,975
8040	24″ diameter	↓	1.02	23.529	↓	3,325	1,075	54.50	4,454.50	5,375
8100	Extra heavy									
8102	4″ diameter	Q-16	5.30	4.528	Ea.	137	206	10.50	353.50	485
8108	8″ diameter		2.60	9.231		420	420	21.50	861.50	1,150
8112	12″ diameter		1.80	13.333		840	605	31	1,476	1,900
8116	16″ diameter		1.20	20		1,675	910	46.50	2,631.50	3,300
8120	20″ diameter		.94	25.532		3,050	1,150	59	4,259	5,225
8122	24″ diameter	↓	.80	30	↓	4,450	1,375	69.50	5,894.50	7,075
8200	Malleable, standard weight									
8202	4″ diameter	B-20	12	2	Ea.	335	70.50		405.50	485
8208	8″ diameter		6	4		915	141		1,056	1,225
8212	12″ diameter	↓	4	6	↓	1,000	212		1,212	1,450
8300	Extra heavy									
8302	4″ diameter	B-20	12	2	Ea.	670	70.50		740.50	855
8308	8″ diameter	B-21	6	4.667		1,150	170	21.50	1,341.50	1,575
8312	12″ diameter	″	4	7	↓	1,600	255	32.50	1,887.50	2,225
8500	Tee weld, standard weight									
8510	4″ diameter	Q-16	4.50	5.333	Ea.	109	243	12.35	364.35	510
8514	6″ diameter		3	8		195	365	18.55	578.55	800
8516	8″ diameter		2.30	10.435		340	475	24	839	1,125
8520	12″ diameter		1.50	16		950	730	37	1,717	2,225
8524	16″ diameter		1	24		1,850	1,100	55.50	3,005.50	3,800
8528	20″ diameter		.80	30		4,550	1,375	69.50	5,994.50	7,200
8530	24″ diameter	↓	.70	34.286	↓	5,875	1,550	79.50	7,504.50	9,000
8810	Malleable, standard weight									
8812	4″ diameter	B-20	8	3	Ea.	570	106		676	805
8818	8″ diameter	B-21	4	7		980	255	32.50	1,267.50	1,525
8822	12″ diameter	″	2.70	10.370	↓	1,875	375	48	2,298	2,725
8900	Extra heavy									
8902	4″ diameter	B-20	8	3	Ea.	855	106		961	1,125
8908	8″ diameter	B-21	4	7		1,225	255	32.50	1,512.50	1,800
8912	12″ diameter	″	2.70	10.370	↓	1,875	375	48	2,298	2,725

33 51 33 – Natural-Gas Metering

33 51 33.10 Piping, Valves and Meters, Gas Distribution		Crew	Daily Output	Labor-Hours	Unit	Material	Labor	Equipment	Total	Total Incl O&P
0010	**PIPING, VALVES & METERS, GAS DISTRIBUTION**									
0020	Not including excavation or backfill									
0100	Gas stops, with or without checks									
0140	1-1/4″ size	1 Plum	12	.667	Ea.	70	32.50		102.50	128
0180	1-1/2″ size		10	.800		94.50	39		133.50	165
0200	2″ size	↓	8	1	↓	142	49		191	232
0600	Pressure regulator valves, iron and bronze									
0640	1-1/2″ diameter	1 Plum	13	.615	Ea.	129	30		159	189
0680	2″ diameter	″	11	.727		245	35.50		280.50	325
0700	3″ diameter	Q-1	13	1.231		495	54		549	630
0740	4″ diameter	″	8	2	↓	1,675	88		1,763	1,950
2000	Lubricated semi-steel plug valve									
2040	3/4″ diameter	1 Plum	16	.500	Ea.	73	24.50		97.50	118
2080	1″ diameter	↓	14	.571	↓	93	28		121	145

33 51 Natural-Gas Distribution

33 51 33 – Natural-Gas Metering

33 51 33.10 Piping, Valves and Meters, Gas Distribution		Crew	Daily Output	Labor-Hours	Unit	2009 Bare Costs Material	Labor	Equipment	Total	Total Incl O&P
2100	1-1/4" diameter	1 Plum	12	.667	Ea.	121	32.50		153.50	185
2140	1-1/2" diameter		11	.727		119	35.50		154.50	186
2180	2" diameter		8	1		142	49		191	232
2300	2-1/2" diameter	Q-1	5	3.200		218	140		358	455
2340	3" diameter	"	4.50	3.556		268	156		424	535

33 52 Liquid Fuel Distribution

33 52 16 – Gasoline Distribution

33 52 16.13 Gasoline Piping

		Crew	Daily Output	Labor-Hours	Unit	Material	Labor	Equipment	Total	Total Incl O&P
0010	**GASOLINE PIPING**									
0020	Primary containment pipe, fiberglass-reinforced									
0030	Plastic pipe 15' & 30' lengths									
0040	2" diameter	Q-6	425	.056	L.F.	4.94	2.60		7.54	9.50
0050	3" diameter		400	.060		6.45	2.76		9.21	11.40
0060	4" diameter		375	.064		8.35	2.95		11.30	13.70
0100	Fittings									
0110	Elbows, 90° & 45°, bell-ends, 2"	Q-6	24	1	Ea.	50	46		96	127
0120	3" diameter		22	1.091		52.50	50.50		103	136
0130	4" diameter		20	1.200		69.50	55.50		125	162
0200	Tees, bell ends, 2"		21	1.143		60.50	52.50		113	148
0210	3" diameter		18	1.333		61	61.50		122.50	163
0220	4" diameter		15	1.600		83.50	73.50		157	206
0230	Flanges bell ends, 2"		24	1		20	46		66	93.50
0240	3" diameter		22	1.091		25	50.50		75.50	106
0250	4" diameter		20	1.200		34.50	55.50		90	124
0260	Sleeve couplings, 2"		21	1.143		12.75	52.50		65.25	95.50
0270	3" diameter		18	1.333		18.05	61.50		79.55	115
0280	4" diameter		15	1.600		25	73.50		98.50	142
0290	Threaded adapters 2"		21	1.143		16.75	52.50		69.25	100
0300	3" diameter		18	1.333		29.50	61.50		91	128
0310	4" diameter		15	1.600		40	73.50		113.50	158
0320	Reducers, 2"		27	.889		22.50	41		63.50	88
0330	3" diameter		22	1.091		26	50.50		76.50	107
0340	4" diameter		20	1.200		33.50	55.50		89	122
1010	Gas station product line for secondary containment (double wall)									
1100	Fiberglass reinforced plastic pipe 25' lengths									
1120	Pipe, plain end, 3" diameter	Q-6	375	.064	L.F.	8.50	2.95		11.45	13.90
1130	4" diameter		350	.069		13.90	3.16		17.06	20
1140	5" diameter		325	.074		18.05	3.40		21.45	25
1150	6" diameter		300	.080		18.60	3.69		22.29	26
1200	Fittings									
1230	Elbows, 90° & 45°, 3" diameter	Q-6	18	1.333	Ea.	61	61.50		122.50	163
1240	4" diameter		16	1.500		106	69		175	224
1250	5" diameter		14	1.714		247	79		326	395
1260	6" diameter		12	2		250	92		342	420
1270	Tees, 3" diameter		15	1.600		90.50	73.50		164	214
1280	4" diameter		12	2		133	92		225	289
1290	5" diameter		9	2.667		269	123		392	485
1300	6" diameter		6	4		281	184		465	595
1310	Couplings, 3" diameter		18	1.333		43	61.50		104.50	143
1320	4" diameter		16	1.500		111	69		180	229

33 52 Liquid Fuel Distribution

33 52 16 – Gasoline Distribution

33 52 16.13 Gasoline Piping		Crew	Daily Output	Labor-Hours	Unit	2009 Bare Costs Material	Labor	Equipment	Total	Total Incl O&P
1330	5″ diameter	Q-6	14	1.714	Ea.	230	79		309	375
1340	6″ diameter		12	2		238	92		330	405
1350	Cross-over nipples, 3″ diameter		18	1.333		9.80	61.50		71.30	106
1360	4″ diameter		16	1.500		11.50	69		80.50	120
1370	5″ diameter		14	1.714		17.05	79		96.05	141
1380	6″ diameter		12	2		17.85	92		109.85	163
1400	Telescoping, reducers, concentric 4″ x 3″		18	1.333		32.50	61.50		94	132
1410	5″ x 4″		17	1.412		85	65		150	195
1420	6″ x 5″		16	1.500		204	69		273	330

33 61 Hydronic Energy Distribution

33 61 13 – Underground Hydronic Energy Distribution

33 61 13.20 Pipe Conduit, Prefabricated/Preinsulated		Crew	Daily Output	Labor-Hours	Unit	Material	Labor	Equipment	Total	Total Incl O&P
0010	**PIPE CONDUIT, PREFABRICATED/PREINSULATED**									
0020	Does not include trenching, fittings or crane.									
0300	For cathodic protection, add 12 to 14%									
0310	of total built-up price (casing plus service pipe)									
0580	Polyurethane insulated system, 250° F. max. temp.									
0620	Black steel service pipe, standard wt., 1/2″ insulation									
0660	3/4″ diam. pipe size	Q-17	54	.296	L.F.	50	13.15	1.03	64.18	76.50
0670	1″ diam. pipe size		50	.320		55	14.20	1.11	70.31	83.50
0680	1-1/4″ diam. pipe size		47	.340		61	15.15	1.18	77.33	92.50
0690	1-1/2″ diam. pipe size		45	.356		66.50	15.80	1.24	83.54	99
0700	2″ diam. pipe size		42	.381		69	16.95	1.33	87.28	104
0710	2-1/2″ diam. pipe size		34	.471		70	21	1.64	92.64	111
0720	3″ diam. pipe size		28	.571		81	25.50	1.99	108.49	131
0730	4″ diam. pipe size		22	.727		102	32.50	2.53	137.03	165
0740	5″ diam. pipe size		18	.889		130	39.50	3.09	172.59	207
0750	6″ diam. pipe size	Q-18	23	1.043		152	48	2.42	202.42	244
0760	8″ diam. pipe size		19	1.263		222	58	2.93	282.93	340
0770	10″ diam. pipe size		16	1.500		293	69	3.48	365.48	430
0780	12″ diam. pipe size		13	1.846		365	85	4.28	454.28	535
0790	14″ diam. pipe size		11	2.182		405	101	5.05	511.05	605
0800	16″ diam. pipe size		10	2.400		465	111	5.55	581.55	685
0810	18″ diam. pipe size		8	3		535	138	6.95	679.95	810
0820	20″ diam. pipe size		7	3.429		595	158	7.95	760.95	910
0830	24″ diam. pipe size		6	4		730	184	9.30	923.30	1,100
0900	For 1″ thick insulation, add					10%				
0940	For 1-1/2″ thick insulation, add					13%				
0980	For 2″ thick insulation, add					20%				
1500	Gland seal for system, 3/4″ diam. pipe size	Q-17	32	.500	Ea.	675	22	1.74	698.74	780
1510	1″ diam. pipe size		32	.500		675	22	1.74	698.74	780
1540	1-1/4″ diam. pipe size		30	.533		725	23.50	1.86	750.36	835
1550	1-1/2″ diam. pipe size		30	.533		725	23.50	1.86	750.36	835
1560	2″ diam. pipe size		28	.571		860	25.50	1.99	887.49	985
1570	2-1/2″ diam. pipe size		26	.615		925	27.50	2.14	954.64	1,075
1580	3″ diam. pipe size		24	.667		995	29.50	2.32	1,026.82	1,150
1590	4″ diam. pipe size		22	.727		1,175	32.50	2.53	1,210.03	1,350
1600	5″ diam. pipe size		19	.842		1,450	37.50	2.93	1,490.43	1,650
1610	6″ diam. pipe size	Q-18	26	.923		1,525	42.50	2.14	1,569.64	1,775
1620	8″ diam. pipe size		25	.960		1,775	44	2.23	1,821.23	2,025

33 61 Hydronic Energy Distribution

33 61 13 – Underground Hydronic Energy Distribution

33 61 13.20 Pipe Conduit, Prefabricated/Preinsulated		Crew	Daily Output	Labor-Hours	Unit	Material	2009 Bare Costs Labor	Equipment	Total	Total Incl O&P
1630	10" diam. pipe size	Q-18	23	1.043	Ea.	2,175	48	2.42	2,225.42	2,450
1640	12" diam. pipe size		21	1.143		2,400	52.50	2.65	2,455.15	2,700
1650	14" diam. pipe size		19	1.263		2,675	58	2.93	2,735.93	3,050
1660	16" diam. pipe size		18	1.333		3,150	61.50	3.09	3,214.59	3,550
1670	18" diam. pipe size		16	1.500		3,375	69	3.48	3,447.48	3,800
1680	20" diam. pipe size		14	1.714		3,825	79	3.98	3,907.98	4,325
1690	24" diam. pipe size		12	2		4,250	92	4.64	4,346.64	4,825
2000	Elbow, 45° for system									
2020	3/4" diam. pipe size	Q-17	14	1.143	Ea.	425	51	3.98	479.98	555
2040	1" diam. pipe size		13	1.231		440	54.50	4.28	498.78	570
2050	1-1/4" diam. pipe size		11	1.455		495	64.50	5.05	564.55	650
2060	1-1/2" diam. pipe size		9	1.778		520	79	6.20	605.20	700
2070	2" diam. pipe size		6	2.667		545	118	9.30	672.30	795
2080	2-1/2" diam. pipe size		4	4		585	178	13.90	776.90	935
2090	3" diam. pipe size		3.50	4.571		680	203	15.90	898.90	1,075
2100	4" diam. pipe size		3	5.333		790	237	18.55	1,045.55	1,250
2110	5" diam. pipe size		2.80	5.714		1,025	254	19.90	1,298.90	1,550
2120	6" diam. pipe size	Q-18	4	6		1,150	276	13.90	1,439.90	1,725
2130	8" diam. pipe size		3	8		1,675	370	18.55	2,063.55	2,450
2140	10" diam. pipe size		2.40	10		2,125	460	23	2,608	3,100
2150	12" diam. pipe size		2	12		2,800	555	28	3,383	3,950
2160	14" diam. pipe size		1.80	13.333		3,475	615	31	4,121	4,825
2170	16" diam. pipe size		1.60	15		4,150	690	35	4,875	5,675
2180	18" diam. pipe size		1.30	18.462		5,200	850	43	6,093	7,100
2190	20" diam. pipe size		1	24		6,550	1,100	55.50	7,705.50	8,975
2200	24" diam. pipe size		.70	34.286		8,225	1,575	79.50	9,879.50	11,600
2260	For elbow, 90°, add					25%				
2300	For tee, straight, add					85%	30%			
2340	For tee, reducing, add					170%	30%			
2380	For weldolet, straight, add					50%				

33 63 Steam Energy Distribution

33 63 13 – Underground Steam and Condensate Distribution Piping

33 63 13.10 Calcium Silicate Insulated System		Crew	Daily Output	Labor-Hours	Unit	Material	Labor	Equipment	Total	Total Incl O&P
0010	**CALCIUM SILICATE INSULATED SYSTEM**									
0011	High temp. (1200 degrees F)									
2840	Steel casing with protective exterior coating									
2850	6-5/8" diameter	Q-18	52	.462	L.F.	96	21.50	1.07	118.57	139
2860	8-5/8" diameter		50	.480		105	22	1.11	128.11	151
2870	10-3/4" diameter		47	.511		123	23.50	1.18	147.68	173
2880	12-3/4" diameter		44	.545		134	25	1.27	160.27	187
2890	14" diameter		41	.585		150	27	1.36	178.36	209
2900	16" diameter		39	.615		162	28.50	1.43	191.93	224
2910	18" diameter		36	.667		178	30.50	1.55	210.05	245
2920	20" diameter		34	.706		201	32.50	1.64	235.14	273
2930	22" diameter		32	.750		279	34.50	1.74	315.24	360
2940	24" diameter		29	.828		320	38	1.92	359.92	410
2950	26" diameter		26	.923		360	42.50	2.14	404.64	470
2960	28" diameter		23	1.043		450	48	2.42	500.42	570
2970	30" diameter		21	1.143		480	52.50	2.65	535.15	610
2980	32" diameter		19	1.263		540	58	2.93	600.93	690

33 63 Steam Energy Distribution

33 63 13 – Underground Steam and Condensate Distribution Piping

33 63 13.10 Calcium Silicate Insulated System		Crew	Daily Output	Labor-Hours	Unit	2009 Bare Costs Material	Labor	Equipment	Total	Total Incl O&P
2990	34" diameter	Q-18	18	1.333	L.F.	545	61.50	3.09	609.59	700
3000	36" diameter		16	1.500		585	69	3.48	657.48	755
3040	For multi-pipe casings, add					10%				
3060	For oversize casings, add					2%				
3400	Steel casing gland seal, single pipe									
3420	6-5/8" diameter	Q-18	25	.960	Ea.	1,325	44	2.23	1,371.23	1,550
3440	8-5/8" diameter		23	1.043		1,550	48	2.42	1,600.42	1,800
3450	10-3/4" diameter		21	1.143		1,750	52.50	2.65	1,805.15	2,000
3460	12-3/4" diameter		19	1.263		2,100	58	2.93	2,160.93	2,400
3470	14" diameter		17	1.412		2,275	65	3.28	2,343.28	2,625
3480	16" diameter		16	1.500		2,675	69	3.48	2,747.48	3,050
3490	18" diameter		15	1.600		2,975	73.50	3.71	3,052.21	3,400
3500	20" diameter		13	1.846		3,300	85	4.28	3,389.28	3,750
3510	22" diameter		12	2		3,700	92	4.64	3,796.64	4,225
3520	24" diameter		11	2.182		4,150	101	5.05	4,256.05	4,725
3530	26" diameter		10	2.400		4,725	111	5.55	4,841.55	5,375
3540	28" diameter		9.50	2.526		5,350	116	5.85	5,471.85	6,050
3550	30" diameter		9	2.667		5,450	123	6.20	5,579.20	6,200
3560	32" diameter		8.50	2.824		6,125	130	6.55	6,261.55	6,950
3570	34" diameter		8	3		6,675	138	6.95	6,819.95	7,575
3580	36" diameter		7	3.429		7,100	158	7.95	7,265.95	8,075
3620	For multi-pipe casings, add					5%				
4000	Steel casing anchors, single pipe									
4020	6-5/8" diameter	Q-18	8	3	Ea.	1,200	138	6.95	1,344.95	1,550
4040	8-5/8" diameter		7.50	3.200		1,250	147	7.40	1,404.40	1,600
4050	10-3/4" diameter		7	3.429		1,650	158	7.95	1,815.95	2,050
4060	12-3/4" diameter		6.50	3.692		1,750	170	8.55	1,928.55	2,200
4070	14" diameter		6	4		2,050	184	9.30	2,243.30	2,575
4080	16" diameter		5.50	4.364		2,400	201	10.10	2,611.10	2,950
4090	18" diameter		5	4.800		2,700	221	11.15	2,932.15	3,325
4100	20" diameter		4.50	5.333		2,975	246	12.35	3,233.35	3,675
4110	22" diameter		4	6		3,325	276	13.90	3,614.90	4,100
4120	24" diameter		3.50	6.857		3,625	315	15.90	3,955.90	4,475
4130	26" diameter		3	8		4,100	370	18.55	4,488.55	5,100
4140	28" diameter		2.50	9.600		4,475	440	22.50	4,937.50	5,625
4150	30" diameter		2	12		4,775	555	28	5,358	6,125
4160	32" diameter		1.50	16		5,700	735	37	6,472	7,475
4170	34" diameter		1	24		6,400	1,100	55.50	7,555.50	8,825
4180	36" diameter		1	24		6,975	1,100	55.50	8,130.50	9,450
4220	For multi-pipe, add					5%	20%			
4800	Steel casing elbow									
4820	6-5/8" diameter	Q-18	15	1.600	Ea.	1,650	73.50	3.71	1,727.21	1,925
4830	8-5/8" diameter		15	1.600		1,750	73.50	3.71	1,827.21	2,050
4850	10-3/4" diameter		14	1.714		2,125	79	3.98	2,207.98	2,450
4860	12-3/4" diameter		13	1.846		2,500	85	4.28	2,589.28	2,875
4870	14" diameter		12	2		2,675	92	4.64	2,771.64	3,100
4880	16" diameter		11	2.182		2,925	101	5.05	3,031.05	3,375
4890	18" diameter		10	2.400		3,350	111	5.55	3,466.55	3,850
4900	20" diameter		9	2.667		3,575	123	6.20	3,704.20	4,125
4910	22" diameter		8	3		3,850	138	6.95	3,994.95	4,450
4920	24" diameter		7	3.429		4,225	158	7.95	4,390.95	4,900
4930	26" diameter		6	4		4,625	184	9.30	4,818.30	5,400
4940	28" diameter		5	4.800		5,000	221	11.15	5,232.15	5,825

33 63 13.10 Calcium Silicate Insulated System		Crew	Daily Output	Labor-Hours	Unit	2009 Bare Costs Material	Labor	Equipment	Total	Total Incl O&P
4950	30″ diameter	Q-18	4	6	Ea.	5,025	276	13.90	5,314.90	5,975
4960	32″ diameter		3	8		5,700	370	18.55	6,088.55	6,875
4970	34″ diameter		2	12		6,275	555	28	6,858	7,775
4980	36″ diameter		2	12		6,675	555	28	7,258	8,225
5500	Black steel service pipe, std. wt., 1″ thick insulation									
5510	3/4″ diameter pipe size	Q-17	54	.296	L.F.	42	13.15	1.03	56.18	67.50
5540	1″ diameter pipe size		50	.320		43.50	14.20	1.11	58.81	71
5550	1-1/4″ diameter pipe size		47	.340		49	15.15	1.18	65.33	78.50
5560	1-1/2″ diameter pipe size		45	.356		53	15.80	1.24	70.04	84.50
5570	2″ diameter pipe size		42	.381		59	16.95	1.33	77.28	93
5580	2-1/2″ diameter pipe size		34	.471		62	21	1.64	84.64	103
5590	3″ diameter pipe size		28	.571		71	25.50	1.99	98.49	120
5600	4″ diameter pipe size		22	.727		90.50	32.50	2.53	125.53	152
5610	5″ diameter pipe size		18	.889		123	39.50	3.09	165.59	199
5620	6″ diameter pipe size	Q-18	23	1.043		134	48	2.42	184.42	224
6000	Black steel service pipe, std. wt., 1-1/2″ thick insul.									
6010	3/4″ diameter pipe size	Q-17	54	.296	L.F.	42.50	13.15	1.03	56.68	68
6040	1″ diameter pipe size		50	.320		47.50	14.20	1.11	62.81	75
6050	1-1/4″ diameter pipe size		47	.340		53	15.15	1.18	69.33	83.50
6060	1-1/2″ diameter pipe size		45	.356		58	15.80	1.24	75.04	90
6070	2″ diameter pipe size		42	.381		63	16.95	1.33	81.28	97.50
6080	2-1/2″ diameter pipe size		34	.471		67	21	1.64	89.64	108
6090	3″ diameter pipe size		28	.571		75.50	25.50	1.99	102.99	125
6100	4″ diameter pipe size		22	.727		96	32.50	2.53	131.03	159
6110	5″ diameter pipe size		18	.889		123	39.50	3.09	165.59	199
6120	6″ diameter pipe size	Q-18	23	1.043		139	48	2.42	189.42	230
6130	8″ diameter pipe size		19	1.263		201	58	2.93	261.93	315
6140	10″ diameter pipe size		16	1.500		262	69	3.48	334.48	400
6150	12″ diameter pipe size		13	1.846		310	85	4.28	399.28	480
6190	For 2″ thick insulation, add					15%				
6220	For 2-1/2″ thick insulation, add					25%				
6260	For 3″ thick insulation, add					30%				
6800	Black steel service pipe, ex. hvy. wt., 1″ thick insul.									
6820	3/4″ diameter pipe size	Q-17	50	.320	L.F.	44	14.20	1.11	59.31	71.50
6840	1″ diameter pipe size		47	.340		47	15.15	1.18	63.33	76.50
6850	1-1/4″ diameter pipe size		44	.364		54	16.15	1.27	71.42	86
6860	1-1/2″ diameter pipe size		42	.381		57.50	16.95	1.33	75.78	91
6870	2″ diameter pipe size		40	.400		61	17.75	1.39	80.14	96
6880	2-1/2″ diameter pipe size		31	.516		75.50	23	1.80	100.30	120
6890	3″ diameter pipe size		27	.593		84.50	26.50	2.06	113.06	136
6900	4″ diameter pipe size		21	.762		112	34	2.65	148.65	178
6910	5″ diameter pipe size		17	.941		156	42	3.28	201.28	241
6920	6″ diameter pipe size	Q-18	22	1.091		144	50.50	2.53	197.03	239
7400	Black steel service pipe, ex. hvy. wt., 1-1/2″ thick insul.									
7420	3/4″ diameter pipe size	Q-17	50	.320	L.F.	36.50	14.20	1.11	51.81	63.50
7440	1″ diameter pipe size		47	.340		42	15.15	1.18	58.33	71
7450	1-1/4″ diameter pipe size		44	.364		48.50	16.15	1.27	65.92	80
7460	1-1/2″ diameter pipe size		42	.381		53	16.95	1.33	71.28	86.50
7470	2″ diameter pipe size		40	.400		53.50	17.75	1.39	72.64	88
7480	2-1/2″ diameter pipe size		31	.516		63.50	23	1.80	88.30	107
7490	3″ diameter pipe size		27	.593		75	26.50	2.06	103.56	126
7500	4″ diameter pipe size		21	.762		97.50	34	2.65	134.15	162
7510	5″ diameter pipe size		17	.941		135	42	3.28	180.28	217

33 63 Steam Energy Distribution

33 63 13 – Underground Steam and Condensate Distribution Piping

33 63 13.10 Calcium Silicate Insulated System		Crew	Daily Output	Labor-Hours	Unit	2009 Bare Costs Material	Labor	Equipment	Total	Total Incl O&P
7520	6" diameter pipe size	Q-18	22	1.091	L.F.	149	50.50	2.53	202.03	244
7530	8" diameter pipe size		18	1.333		223	61.50	3.09	287.59	345
7540	10" diameter pipe size		15	1.600		265	73.50	3.71	342.21	410
7550	12" diameter pipe size		13	1.846		325	85	4.28	414.28	495
7590	For 2" thick insulation, add					13%				
7640	For 2-1/2" thick insulation, add					18%				
7680	For 3" thick insulation, add					24%				

33 71 Electrical Utility Transmission and Distribution

33 71 16 – Electrical Utility Poles

33 71 16.33 Wood Electrical Utility Poles

		Crew	Daily Output	Labor-Hours	Unit	Material	Labor	Equipment	Total	Total Incl O&P
0010	**WOOD ELECTRICAL UTILITY POLES**									
0011	Excludes excavation, backfill and cast-in-place concrete									
6200	Electric & tel sitework, 20' high, treated wd, see Div. 26 56 13.10	R-3	3.10	6.452	Ea.	263	299	41.50	603.50	795
6400	25' high		2.90	6.897		278	320	44.50	642.50	850
6600	30' high		2.60	7.692		305	355	49.50	709.50	940
6800	35' high		2.40	8.333		400	385	54	839	1,100
7000	40' high		2.30	8.696		490	405	56	951	1,225
7200	45' high		1.70	11.765		600	545	76	1,221	1,575
7400	Cross arms with hardware & insulators									
7600	4' long	1 Elec	2.50	3.200	Ea.	125	150		275	370
7800	5' long		2.40	3.333		145	157		302	400
8000	6' long		2.20	3.636		167	171		338	445

33 71 19 – Electrical Underground Ducts and Manholes

33 71 19.15 Underground Ducts and Manholes

		Crew	Daily Output	Labor-Hours	Unit	Material	Labor	Equipment	Total	Total Incl O&P
0010	**UNDERGROUND DUCTS AND MANHOLES**									
0011	Not incl. excavation, backfill and concrete, in slab or duct bank									
1000	Direct burial									
1010	PVC, schedule 40, w/coupling, 1/2" diameter	1 Elec	340	.024	L.F.	.69	1.11		1.80	2.46
1020	3/4" diameter		290	.028		.92	1.30		2.22	3
1030	1" diameter		260	.031		1.37	1.45		2.82	3.73
1040	1-1/2" diameter		210	.038		2.23	1.79		4.02	5.20
1050	2" diameter		180	.044		2.92	2.09		5.01	6.40
1060	3" diameter	2 Elec	240	.067		5.60	3.13		8.73	10.95
1070	4" diameter		160	.100		7.85	4.70		12.55	15.85
1080	5" diameter		120	.133		11.40	6.25		17.65	22
1090	6" diameter		90	.178		15.05	8.35		23.40	29.50
1110	Elbows, 1/2" diameter	1 Elec	48	.167	Ea.	1.58	7.85		9.43	13.80
1120	3/4" diameter		38	.211		1.59	9.90		11.49	16.95
1130	1" diameter		32	.250		2.69	11.75		14.44	21
1140	1-1/2" diameter		21	.381		5.20	17.90		23.10	33
1150	2" diameter		16	.500		7.55	23.50		31.05	44.50
1160	3" diameter		12	.667		23	31.50		54.50	73.50
1170	4" diameter		9	.889		40	42		82	109
1180	5" diameter		8	1		70	47		117	150
1190	6" diameter		5	1.600		119	75		194	247
1210	Adapters, 1/2" diameter		52	.154		.51	7.25		7.76	11.65
1220	3/4" diameter		43	.186		.94	8.75		9.69	14.50
1230	1" diameter		39	.205		1.19	9.65		10.84	16.15
1240	1-1/2" diameter		35	.229		1.85	10.75		12.60	18.60

33 71 19.15 Underground Ducts and Manholes		Crew	Daily Output	Labor-Hours	Unit	Material	2009 Bare Costs Labor	Equipment	Total	Total Incl O&P
1250	2" diameter	1 Elec	26	.308	Ea.	2.66	14.45		17.11	25.50
1260	3" diameter		20	.400		6.75	18.80		25.55	36.50
1270	4" diameter		14	.571		11.80	27		38.80	54.50
1280	5" diameter		12	.667		23	31.50		54.50	73.50
1290	6" diameter		9	.889		28	42		70	95.50
1340	Bell end & cap, 1-1/2" diameter		35	.229		13.40	10.75		24.15	31.50
1350	Bell end & plug, 2" diameter		26	.308		13.90	14.45		28.35	38
1360	3" diameter		20	.400		17.30	18.80		36.10	48
1370	4" diameter		14	.571		21.50	27		48.50	65
1380	5" diameter		12	.667		32.50	31.50		64	84
1390	6" diameter		9	.889		37	42		79	105
1450	Base spacer, 2" diameter		56	.143		2.18	6.70		8.88	12.75
1460	3" diameter		46	.174		2.43	8.15		10.58	15.20
1470	4" diameter		41	.195		2.64	9.15		11.79	17
1480	5" diameter		37	.216		2.89	10.15		13.04	18.85
1490	6" diameter		34	.235		4.67	11.05		15.72	22
1550	Intermediate spacer, 2" diameter		60	.133		2.18	6.25		8.43	12.05
1560	3" diameter		46	.174		2.43	8.15		10.58	15.20
1570	4" diameter		41	.195		2.64	9.15		11.79	17
1580	5" diameter		37	.216		2.89	10.15		13.04	18.85
1590	6" diameter		34	.235		4.67	11.05		15.72	22
4010	PVC, schedule 80, w/coupling, 1/2" diameter		215	.037	L.F.	.84	1.75		2.59	3.61
4020	3/4" diameter		180	.044		1.13	2.09		3.22	4.45
4030	1" diameter		145	.055		1.61	2.59		4.20	5.75
4040	1-1/2" diameter		120	.067		2.68	3.13		5.81	7.75
4050	2" diameter		100	.080		3.69	3.76		7.45	9.85
4060	3" diameter	2 Elec	130	.123		7.55	5.80		13.35	17.20
4070	4" diameter		90	.178		10.95	8.35		19.30	25
4080	5" diameter		70	.229		15.80	10.75		26.55	34
4090	6" diameter		50	.320		19.75	15.05		34.80	44.50
4110	Elbows, 1/2" diameter	1 Elec	29	.276	Ea.	3.89	12.95		16.84	24
4120	3/4" diameter		23	.348		6.10	16.35		22.45	31.50
4130	1" diameter		20	.400		7.95	18.80		26.75	38
4140	1-1/2" diameter		16	.500		15.70	23.50		39.20	53.50
4150	2" diameter		12	.667		21.50	31.50		53	71.50
4160	3" diameter		9	.889		72.50	42		114.50	145
4170	4" diameter		7	1.143		128	53.50		181.50	224
4180	5" diameter		6	1.333		186	62.50		248.50	300
4190	6" diameter		4	2		365	94		459	545
4210	Adapter, 1/2" diameter		39	.205		.51	9.65		10.16	15.40
4220	3/4" diameter		33	.242		.94	11.40		12.34	18.60
4230	1" diameter		29	.276		1.19	12.95		14.14	21.50
4240	1-1/2" diameter		26	.308		1.85	14.45		16.30	24.50
4250	2" diameter		23	.348		2.66	16.35		19.01	28
4260	3" diameter		18	.444		6.75	21		27.75	39.50
4270	4" diameter		13	.615		11.80	29		40.80	57.50
4280	5" diameter		11	.727		23	34		57	78
4290	6" diameter		8	1		28	47		75	104
4310	Bell end & cap, 1-1/2" diameter		26	.308		13.40	14.45		27.85	37.50
4320	Bell end & plug, 2" diameter		23	.348		13.90	16.35		30.25	40.50
4330	3" diameter		18	.444		17.30	21		38.30	51
4340	4" diameter		13	.615		21.50	29		50.50	68
4350	5" diameter		11	.727		32.50	34		66.50	88.50

33 71 19.15 Underground Ducts and Manholes		Crew	Daily Output	Labor-Hours	Unit	Material	2009 Bare Costs Labor	Equipment	Total	Total Incl O&P
4360	6″ diameter	1 Elec	8	1	Ea.	37	47		84	113
4370	Base spacer, 2″ diameter		42	.190		2.18	8.95		11.13	16.15
4380	3″ diameter		33	.242		2.43	11.40		13.83	20
4390	4″ diameter		29	.276		2.64	12.95		15.59	23
4400	5″ diameter		26	.308		2.89	14.45		17.34	25.50
4410	6″ diameter		25	.320		4.67	15.05		19.72	28
4420	Intermediate spacer, 2″ diameter		45	.178		2.18	8.35		10.53	15.25
4430	3″ diameter		34	.235		2.43	11.05		13.48	19.65
4440	4″ diameter		31	.258		2.64	12.15		14.79	21.50
4450	5″ diameter		28	.286		2.89	13.45		16.34	23.50
4460	6″ diameter	↓	25	.320	↓	4.67	15.05		19.72	28

33 71 19.17 Electric and Telephone Underground		Crew	Daily Output	Labor-Hours	Unit	Material	Labor	Equipment	Total	Total Incl O&P
0010	**ELECTRIC AND TELEPHONE UNDERGROUND**									
0011	Not including excavation									
0200	backfill and cast in place concrete									
0400	Hand holes, precast concrete, with concrete cover									
0600	2′ x 2′ x 3′ deep	R-3	2.40	8.333	Ea.	385	385	54	824	1,075
0800	3′ x 3′ x 3′ deep		1.90	10.526		495	485	68	1,048	1,375
1000	4′ x 4′ x 4′ deep	↓	1.40	14.286	↓	995	660	92.50	1,747.50	2,225
1200	Manholes, precast with iron racks & pulling irons, C.I. frame									
1400	and cover, 4′ x 6′ x 7′ deep	B-13	2	28	Ea.	1,900	955	395	3,250	4,075
1600	6′ x 8′ x 7′ deep		1.90	29.474		2,350	1,000	415	3,765	4,675
1800	6′ x 10′ x 7′ deep	↓	1.80	31.111	↓	2,625	1,075	440	4,140	5,100
4200	Underground duct, banks ready for concrete fill, min. of 7.5″									
4400	between conduits, center to center									
4580	PVC, type EB, 1 @ 2″ diameter	2 Elec	480	.033	L.F.	.87	1.57		2.44	3.36
4600	2 @ 2″ diameter		240	.067		1.74	3.13		4.87	6.75
4800	4 @ 2″ diameter		120	.133		3.47	6.25		9.72	13.45
4900	1 @ 3″ diameter		400	.040		1.18	1.88		3.06	4.19
5000	2 @ 3″ diameter		200	.080		2.37	3.76		6.13	8.40
5200	4 @ 3″ diameter		100	.160		4.73	7.50		12.23	16.75
5300	1 @ 4″ diameter		320	.050		1.79	2.35		4.14	5.60
5400	2 @ 4″ diameter		160	.100		3.59	4.70		8.29	11.20
5600	4 @ 4″ diameter		80	.200		7.20	9.40		16.60	22.50
5800	6 @ 4″ diameter		54	.296		10.75	13.95		24.70	33.50
5810	1 @ 5″ diameter		260	.062		2.62	2.89		5.51	7.35
5820	2 @ 5″ diameter		130	.123		5.25	5.80		11.05	14.65
5840	4 @ 5″ diameter		70	.229		10.50	10.75		21.25	28
5860	6 @ 5″ diameter		50	.320		15.70	15.05		30.75	40.50
5870	1 @ 6″ diameter		200	.080		3.81	3.76		7.57	10
5880	2 @ 6″ diameter		100	.160		7.60	7.50		15.10	19.95
5900	4 @ 6″ diameter		50	.320		15.25	15.05		30.30	40
5920	6 @ 6″ diameter		30	.533		23	25		48	63.50
6200	Rigid galvanized steel, 2 @ 2″ diameter		180	.089		16.60	4.18		20.78	24.50
6400	4 @ 2″ diameter		90	.178		33	8.35		41.35	49.50
6800	2 @ 3″ diameter		100	.160		39	7.50		46.50	54
7000	4 @ 3″ diameter		50	.320		77.50	15.05		92.55	109
7200	2 @ 4″ diameter		70	.229		56	10.75		66.75	78
7400	4 @ 4″ diameter		34	.471		112	22		134	157
7600	6 @ 4″ diameter		22	.727		167	34		201	237
7620	2 @ 5″ diameter		60	.267		114	12.55		126.55	144
7640	4 @ 5″ diameter	↓	30	.533	↓	228	25		253	290

33 71 19.17 Electric and Telephone Underground		Crew	Daily Output	Labor-Hours	Unit	2009 Bare Costs Material	Labor	Equipment	Total	Total Incl O&P
7660	6 @ 5" diameter	2 Elec	18	.889	L.F.	340	42		382	440
7680	2 @ 6" diameter		40	.400		166	18.80		184.80	211
7700	4 @ 6" diameter		20	.800		330	37.50		367.50	425
7720	6 @ 6" diameter		14	1.143		495	53.50		548.50	630
7800	For Cast-in-place Concrete - Add									
7810	Under 1 C.Y.	C-6	16	3	C.Y.	162	99	3.24	264.24	345
7820	1 C.Y. - 5 C.Y.		19.20	2.500		146	82.50	2.70	231.20	299
7830	Over 5 C.Y.		24	2		121	66	2.16	189.16	243
7850	For Reinforcing Rods - Add									
7860	#4 to #7	2 Rodm	1.10	14.545	Ton	1,550	650		2,200	2,800
7870	#8 to #14	"	1.50	10.667	"	1,550	475		2,025	2,500
8000	Fittings, PVC type EB, elbow, 2" diameter	1 Elec	16	.500	Ea.	11.75	23.50		35.25	49
8200	3" diameter		14	.571		17.65	27		44.65	61
8400	4" diameter		12	.667		25	31.50		56.50	75.50
8420	5" diameter		10	.800		59	37.50		96.50	123
8440	6" diameter		9	.889		113	42		155	189
8500	Coupling, 2" diameter					1.17			1.17	1.29
8600	3" diameter					4.07			4.07	4.48
8700	4" diameter					6.40			6.40	7
8720	5" diameter					11.50			11.50	12.65
8740	6" diameter					35.50			35.50	39
8800	Adapter, 2" diameter	1 Elec	26	.308		2.09	14.45		16.54	25
9000	3" diameter		20	.400		5.35	18.80		24.15	35
9200	4" diameter		16	.500		7.15	23.50		30.65	44
9220	5" diameter		13	.615		19.25	29		48.25	65.50
9240	6" diameter		10	.800		25.50	37.50		63	86
9400	End bell, 2" diameter		16	.500		10.10	23.50		33.60	47
9600	3" diameter		14	.571		12.10	27		39.10	55
9800	4" diameter		12	.667		14.25	31.50		45.75	63.50
9810	5" diameter		10	.800		21.50	37.50		59	81.50
9820	6" diameter		8	1		42	47		89	119
9830	5° angle coupling, 2" diameter		26	.308		18.05	14.45		32.50	42.50
9840	3" diameter		20	.400		23	18.80		41.80	54
9850	4" diameter		16	.500		27	23.50		50.50	66
9860	5" diameter		13	.615		29.50	29		58.50	77
9870	6" diameter		10	.800		30.50	37.50		68	91.50
9880	Expansion joint, 2" diameter		16	.500		53	23.50		76.50	94
9890	3" diameter		18	.444		93.50	21		114.50	135
9900	4" diameter		12	.667		135	31.50		166.50	196
9910	5" diameter		10	.800		210	37.50		247.50	289
9920	6" diameter		8	1		261	47		308	360
9930	Heat bender, 2" diameter					485			485	535
9940	6" diameter					1,350			1,350	1,475
9950	Cement, quart					26.50			26.50	29.50
9960	Nylon polyethylene pull rope, 1/4"	2 Elec	2000	.008	L.F.	.11	.38		.49	.70
9990	Minimum labor/equipment charge	1 Elec	3.50	2.286	Job		107		107	165

33 81 Communications Structures

33 81 13 – Communications Transmission Towers

33 81 13.10 Radio Towers		Crew	Daily Output	Labor-Hours	Unit	2009 Bare Costs Material	Labor	Equipment	Total	Total Incl O&P
0010	**RADIO TOWERS**									
0020	Guyed, 50′ H, 40 lb. sec., 70 MPH basic wind spd.	2 Sswk	1	16	Ea.	3,275	715		3,990	4,925
0100	Wind load 90 MPH basic wind speed	"	1	16		3,275	715		3,990	4,925
0300	190′ high, 40 lb. section, wind load 70 MPH basic wind speed	K-2	.33	72.727		8,825	2,975	735	12,535	15,800
0400	200′ high, 70 lb. section, wind load 90 MPH basic wind speed		.33	72.727		17,800	2,975	735	21,510	25,600
0600	300′ high, 70 lb. section, wind load 70 MPH basic wind speed		.20	120		25,200	4,900	1,225	31,325	37,800
0700	270′ high, 90 lb. section, wind load 90 MPH basic wind speed		.20	120		29,100	4,900	1,225	35,225	42,100
0800	400′ high, 100 lb. section, wind load 70 MPH basic wind speed		.14	171		42,400	7,000	1,725	51,125	61,000
0900	Self-supporting, 60′ high, wind load 70 MPH basic wind speed		.80	30		6,825	1,225	305	8,355	10,000
0910	60′ high, wind load 90 MPH basic wind speed		.45	53.333		12,300	2,175	540	15,015	18,000
1000	120′ high, wind load 70 MPH basic wind speed		.40	60		16,700	2,450	605	19,755	23,500
1200	190′ high, wind load 90 MPH basic wind speed		.20	120		40,600	4,900	1,225	46,725	54,500
2000	For states west of Rocky Mountains, add for shipping					10%				

Estimating Tips

34 11 00 Rail Tracks

This subdivision includes items that may involve either repair of existing, or construction of new, railroad tracks. Additional preparation work, such as the roadbed earthwork, would be found in Division 31. Additional new construction siding and turnouts are found in Subdivision 34 72. Maintenance of railroads is found under 34 01 23 Operation and Maintenance of Railways.

34 41 13 Traffic Signals

This subdivision includes traffic signal systems. Other traffic control devices such as traffic signs are found in Subdivision 10 14 53 Traffic Signage.

34 71 13 Vehicle Barriers

This subdivision includes security vehicle barriers, guide and guard rails, crash barriers, and delineators. The actual maintenance and construction of concrete and asphalt pavement is found in Division 32.

Reference Numbers

Reference numbers are shown in shaded boxes at the beginning of some major classifications. These numbers refer to related items in the Reference Section. The reference information may be an estimating procedure, an alternate pricing method, or technical information.

Note: Not all subdivisions listed here necessarily appear in this publication.

34 01 Operation and Maintenance of Transportation

34 01 23 – Operation and Maintenance of Railways

34 01 23.51 Maintenance of Railroads		Crew	Daily Output	Labor-Hours	Unit	2009 Bare Costs Material	Labor	Equipment	Total	Total Incl O&P
0010	MAINTENANCE OF RAILROADS									
0400	Resurface and realign existing track	B-14	200	.240	L.F.		7.95	1.47	9.42	14.60
0600	For crushed stone ballast, add	"	500	.096	"	17.50	3.19	.59	21.28	25

34 11 Rail Tracks

34 11 13 – Track Rails

34 11 13.23 Heavy Rail Track		Crew	Daily Output	Labor-Hours	Unit	Material	Labor	Equipment	Total	Total Incl O&P
0010	HEAVY RAIL TRACK R347216-10									
1000	Rail, 100 lb. prime grade				L.F.	28.50			28.50	31.50
1500	Relay rail				"	14.35			14.35	15.75

34 11 33 – Track Cross Ties

34 11 33.13 Concrete Track Cross Ties		Crew	Daily Output	Labor-Hours	Unit	Material	Labor	Equipment	Total	Total Incl O&P
0010	CONCRETE TRACK CROSS TIES									
1400	Ties, concrete, 8'-6" long, 30" O.C.	B-14	80	.600	Ea.	108	19.90	3.67	131.57	156

34 11 33.16 Timber Track Cross Ties		Crew	Daily Output	Labor-Hours	Unit	Material	Labor	Equipment	Total	Total Incl O&P
0010	TIMBER TRACK CROSS TIES									
1600	Wood, pressure treated, 6" x 8" x 8'-6", C.L. lots	B-14	90	.533	Ea.	41	17.70	3.26	61.96	78
1700	L.C.L. lots		90	.533		43.50	17.70	3.26	64.46	80
1900	Heavy duty, 7" x 9" x 8'-6", C.L. lots		70	.686		45.50	23	4.20	72.70	91.50
2000	L.C.L. lots		70	.686		45.50	23	4.20	72.70	91.50

34 11 33.17 Timber Switch Ties		Crew	Daily Output	Labor-Hours	Unit	Material	Labor	Equipment	Total	Total Incl O&P
0010	TIMBER SWITCH TIES									
1200	Switch timber, for a #8 switch, pressure treated	B-14	3.70	12.973	M.B.F.	2,475	430	79.50	2,984.50	3,525
1300	Complete set of timbers, 3.7 M.B.F. for #8 switch	"	1	48	Total	9,525	1,600	294	11,419	13,400

34 11 93 – Track Appurtenances and Accessories

34 11 93.50 Track Accessories		Crew	Daily Output	Labor-Hours	Unit	Material	Labor	Equipment	Total	Total Incl O&P
0010	TRACK ACCESSORIES									
0020	Car bumpers, test	B-14	2	24	Ea.	3,200	795	147	4,142	4,975
0100	Heavy duty R347216-20		2	24		6,075	795	147	7,017	8,125
0200	Derails hand throw (sliding)		10	4.800		1,025	159	29.50	1,213.50	1,425
0300	Hand throw with standard timbers, open stand & target		8	6		1,100	199	36.50	1,335.50	1,575
2400	Wheel stops, fixed		18	2.667	Pr.	720	88.50	16.30	824.80	955
2450	Hinged		14	3.429	"	970	114	21	1,105	1,275

34 11 93.60 Track Material		Crew	Daily Output	Labor-Hours	Unit	Material	Labor	Equipment	Total	Total Incl O&P
0010	TRACK MATERIAL									
0020	Track bolts				Ea.	2.53			2.53	2.78
0100	Joint bars				Pr.	76.50			76.50	84
0200	Spikes				Ea.	.68			.68	.75
0300	Tie plates				"	12.75			12.75	14

34 41 Roadway Signaling and Control Equipment

34 41 13 – Traffic Signals

34 41 13.10 Traffic Signals Systems		Crew	Daily Output	Labor-Hours	Unit	2009 Bare Costs Material	Labor	Equipment	Total	Total Incl O&P
0010	**TRAFFIC SIGNALS SYSTEMS**									
0020	Mid block pedestrian crosswalk with pushbutton and mast arms	R-11	.30	186	Total	83,500	8,250	3,125	94,875	108,000
0600	Traffic signals, school flashing system, solar powered, remote controlled	"	1	56	Signal	19,100	2,475	940	22,515	25,900
1000	Intersection traffic signals, LED, mast, programmable, no lane control									
1010	Includes all labor, material and equip. for complete installation	R-11	1	56	Signal	195,500	2,475	940	198,915	220,000
1100	Intersection traffic signals, LED, mast, programmable, R/L lane control									
1110	Includes all labor, material and equip. for complete installation	R-11	1	56	Signal	261,000	2,475	940	264,415	292,000
1200	Add protective/permissive left turns to existing traffic light									
1210	Includes all labor, material and equip. for complete installation	R-11	1	56	Signal	78,000	2,475	940	81,415	91,000
1300	Replace existing light heads with LED heads wire hung									
1310	Includes all labor, material and equip. for complete installation	R-11	1	56	Signal	52,000	2,475	940	55,415	62,500
1400	Replace existing light heads with LED heads mast arm hung									
1410	Includes all labor, material and equip. for complete installation	R-11	1	56	Signal	104,500	2,475	940	107,915	120,000

34 43 Airfield Signaling and Control Equipment

34 43 13 – Airfield Signals

34 43 13.10 Airport Lighting		Crew	Daily Output	Labor-Hours	Unit	Material	Labor	Equipment	Total	Total Incl O&P
0010	**AIRPORT LIGHTING**									
0100	Runway centerline, bidir., semi-flush, 200 W, w/shallow insert base	R-22	12.40	3.006	Ea.	1,500	120		1,620	1,850
0120	Flush, 200 W, w/shallow insert base		12.40	3.006		1,475	120		1,595	1,825
0130	for mounting in base housing		18.64	2		775	80		855	980
0150	Touchdown zone light, unidirectional, 200 W, w/shallow insert base		12.40	3.006		1,200	120		1,320	1,525
0160	115 W		12.40	3.006		1,275	120		1,395	1,625
0180	Unidirectional, 200 W, for mounting in base housing		18.60	2.004		590	80		670	775
0190	115 W		18.60	2.004		635	80		715	825
0210	Runway edge & threshold light, bidir., 200 W, for base housing		9.36	3.983		835	159		994	1,175
0240	Threshold & approach light, unidir., 200 W, for base housing		9.36	3.983		485	159		644	785
0260	Runway edge, bi-directional, 2-115 W, for base housing		12.40	3.006		1,300	120		1,420	1,625
0280	Runway threshold & end, bidir., 2-115 W, for base housing		12.40	3.006		1,425	120		1,545	1,775
0370	45 W, flush, for mounting in base housing		18.64	2		700	80		780	895
0380	115 W		18.64	2		715	80		795	915

34 43 23 – Weather Observation Equipment

34 43 23.16 Airfield Wind Cones		Crew	Daily Output	Labor-Hours	Unit	Material	Labor	Equipment	Total	Total Incl O&P
1000	**AIRFIELD WIND CONES**									
1200	Wind cone, 12′ lighted assembly, rigid, w/obstruction light	R-21	1.36	24.118	Ea.	8,200	1,125	53	9,378	10,800
1210	Without obstruction light		1.52	21.579		6,875	1,025	47.50	7,947.50	9,175
1220	Unlighted assembly, w/obstruction light		1.68	19.524		7,500	915	43	8,458	9,700
1230	Without obstruction light		1.84	17.826		6,875	835	39	7,749	8,925
1240	Wind cone slip fitter, 2-1/2″ pipe		21.84	1.502		90	70.50	3.29	163.79	212
1250	Wind cone sock, 12′ x 3′, cotton		6.56	5		500	235	10.95	745.95	920
1260	Nylon		6.56	5		505	235	10.95	750.95	925

34 71 Roadway Construction

34 71 13 – Vehicle Barriers

34 71 13.17 Security Vehicle Barriers		Crew	Daily Output	Labor-Hours	Unit	Material (2009 Bare Costs)	Labor (2009 Bare Costs)	Equipment (2009 Bare Costs)	Total (2009 Bare Costs)	Total Incl O&P
0010	**SECURITY VEHICLE BARRIERS**									
0020	Security planters excludes filling material									
0100	Concrete security planter, exposed aggregate 36" dia x 30" high	B-11M	8	2	Ea.	450	73	43	566	660
0200	48" dia x 36" high		8	2		530	73	43	646	750
0300	53" dia x 18" high		8	2		605	73	43	721	830
0400	72" dia x 18" high with seats		8	2		1,200	73	43	1,316	1,500
0450	84" dia x 36" high		8	2		1,300	73	43	1,416	1,600
0500	36" x 36" x 24" high square		8	2		435	73	43	551	645
0600	36" x 36" x 30" high square		8	2		465	73	43	581	675
0700	48" L x 24" W x 30" H rectangle		8	2		450	73	43	566	660
0800	72" L x 24" W x 30" H rectangle		8	2		600	73	43	716	825
0900	96" L x 24" W x 30" H rectangle		8	2		730	73	43	846	970
0950	Decorative geometric concrete barrier, 96" L x 24" W x 36" H		8	2		950	73	43	1,066	1,225
1000	Concrete security planter, filling with washed sand or gravel °1 CY		8	2		33	73	43	149	201
1050	Concrete security planter, filling with washed sand or gravel °2 CY		6	2.667		65.50	97.50	57.50	220.50	292
1200	Jersey concrete barrier, 10' L x 2' by 0.5' W x 30" H	B-21B	16	2.500		400	85.50	48	533.50	630
1300	10' L x 2' by 0.5' W x 30" H, 10 or more same site		24	1.667		400	57	32	489	570
1400	Jersey concrete barrier, 10' L x 2' by 0.5' W x 32" H		16	2.500		400	85.50	48	533.50	630
1500	10' L x 2' by 6" W x 32" H, 10 or more same site		24	1.667		400	57	32	489	570
1600	Jersey concrete barrier, 20' L x 2' by 0.5' W x 30" H		12	3.333		625	114	64	803	940
1700	20' L x 2' by 0.5' W x 30" H, 10 or more same site		18	2.222		625	76	42.50	743.50	855
1800	Jersey concrete barrier, 20' L x 2' by 0.5' W x 32" H		12	3.333		900	114	64	1,078	1,250
1900	20' L x 2' by 0.5' W x 32" H, 10 or more same site		18	2.222		900	76	42.50	1,018.50	1,150
2000	GFRC decorative security barrier per 10 feet section including concrete		4	10		2,325	340	192	2,857	3,350
2100	GFRC decorative security barrier per 12 feet section including concrete		4	10		2,800	340	192	3,332	3,850
2210	GFRC decorative security barrier will stop 30 MPH, 4000 lb vehicle									
2300	High security barrier base prep per 12 feet section on bare ground	B-11C	4	4	Ea.	39.50	146	73.50	259	360
2310	GFRC barrier base prep does not include haul away of excavated matl									
2400	GFRC decorative high security barrier per 12 feet section w/concrete	B-21B	3	13.333	Ea.	5,825	455	256	6,536	7,425
2410	GFRC decorative high security barrier will stop 50 MPH, 15000 lb vehl									
2500	GFRC decorative impaler security barrier per 10 feet section w/prep	B-6	4	6	Ea.	2,275	204	73.50	2,552.50	2,900
2600	GFRC decorative impaler security barrier per 12 feet section w/prep	"	4	6	"	2,725	204	73.50	3,002.50	3,375
2610	Impaler barrier should stop 50 MPH, 15000 lb vehl w/some penetr									
2700	Pipe bollards, steel, concrete filled/painted, 8' L x 4' D hole, 8" diam.	B-6	10	2.400	Ea.	720	82	29.50	831.50	955
2710	Schedule 80 concrete bollards will stop 4000 lb vehicle @ 30 MPH									
2800	GFRC decorative jersey barrier cover per 10 feet section excludes soil	B-6	8	3	Ea.	950	102	36.50	1,088.50	1,250
2900	GFRC decorative jersey barrier cover per 12 feet section excludes soil		8	3		1,150	102	36.50	1,288.50	1,450
3000	GFRC decorative 8" diameter bollard cover		12	2		500	68	24.50	592.50	690
3100	GFRC decorative barrier face 10 foot section excludes earth backing		10	2.400		950	82	29.50	1,061.50	1,225
3200	Drop arm crash barrier, 15000 lb vehl @ 50 MPH									
3205	Includes all material, labor for complete installation									
3210	20' width or less				Ea.				70,000	77,000
3310	24' width								78,500	86,350
3410	Portable wedge crash barrier, 10' width								75,075	82,582
3510	12' width								76,000	83,600
3610	Sliding crash barrier, 12' width								138,000	151,800
3710	Sliding roller crash barrier, 20' width								158,000	173,800
3810	Sliding cantilever crash barrier, 20' width								158,000	173,800
3890	Note: Raised bollard crash barriers should be used w/tire shredders									
3910	Raised bollard crash barrier, 10' width				Ea.				35,000	38,500
4010	12' width								40,000	44,000
4110	Raised bollard crash barrier, 10' width,solar powered								42,000	46,200
4210	12' width								47,000	51,700

34 71 Roadway Construction

34 71 13 – Vehicle Barriers

34 71 13.17 Security Vehicle Barriers		Crew	Daily Output	Labor-Hours	Unit	2009 Bare Costs Material	Labor	Equipment	Total	Total Incl O&P
4310	In ground tire shredder, 16′ width				Ea.				40,000	44,000

34 71 13.26 Vehicle Guide Rails

		Crew	Daily Output	Labor-Hours	Unit	Material	Labor	Equipment	Total	Total Incl O&P
0010	**VEHICLE GUIDE RAILS**									
0012	Corrugated stl, galv. stl posts, 6′-3″ O.C.	B-80	850	.038	L.F.	22.50	1.27	.77	24.54	27.50
0200	End sections, galvanized, flared		50	.640	Ea.	93	21.50	13.10	127.60	151
0300	Wrap around end		50	.640	″	139	21.50	13.10	173.60	202
0400	Timber guide rail, 4″ x 8″ with 6″ x 8″ wood posts, treated		960	.033	L.F.	26	1.13	.68	27.81	31
0600	Cable guide rail, 3 at 3/4″ cables, steel posts, single face		900	.036		9.50	1.20	.73	11.43	13.20
0700	Wood posts		950	.034		13.15	1.14	.69	14.98	17.05
0900	Guide rail, steel box beam, 6″ x 6″		120	.267		32.50	9	5.45	46.95	56
1100	Median barrier, steel box beam, 6″ x 8″		215	.149		42.50	5.05	3.05	50.60	58.50
1400	Resilient guide fence and light shield, 6′ high	B-2	130	.308		36	9.85		45.85	56
1500	Concrete posts, individual, 6′-5″, triangular	B-80	110	.291	Ea.	66.50	9.85	5.95	82.30	95.50
1550	Square	″	110	.291	″	71	9.85	5.95	86.80	101
2000	Median, precast concrete, 3′-6″ high, 2′ wide, single face	B-29	380	.147	L.F.	52	5.05	2.52	59.57	68
2200	Double face	″	340	.165	″	59	5.65	2.82	67.47	77.50
2400	Speed bumps, thermoplastic, 10-1/2″ x 2-1/4″ x 48″ long	B-2	120	.333	Ea.	134	10.65		144.65	166
3030	Impact barrier, UTMCD, barrel type	B-16	30	1.067	″	500	34.50	17.75	552.25	625

34 71 19 – Vehicle Delineators

34 71 19.13 Fixed Vehicle Delineators

		Crew	Daily Output	Labor-Hours	Unit	Material	Labor	Equipment	Total	Total Incl O&P
0010	**FIXED VEHICLE DELINEATORS**									
0020	Crash barriers									
0100	Traffic channelizing pavement markers, layout only	A-7	2000	.012	Ea.		.52	.04	.56	.87
0110	13″ x 7-1/2″ x 2-1/2″ high, non-plowable install	2 Clab	96	.167		26	5.25		31.25	37
0200	8″ x 8″x 3-1/4″ high, non-plowable, install		96	.167		23.50	5.25		28.75	34.50
0230	4″ x 4″ x 3/4″ high, non-plowable, install		120	.133		3.50	4.21		7.71	10.80
0240	9-1/4″ x 5-7/8″ x 1/4″ high, plowable, concrete pav′t	A-2A	70	.343		20	10.75	4.79	35.54	45
0250	9-1/4″ x 5-7/8″ x 1/4″ high, plowable, asphalt pav′t	″	120	.200		4	6.30	2.80	13.10	17.80
0300	Barrier and curb delineators,reflectorized, 2″ x 4″	2 Clab	150	.107		2.25	3.37		5.62	8.05
0310	3″ x 5″	″	150	.107		4.50	3.37		7.87	10.50
0500	Rumble strip, polycarbonate									
0510	24″ x 3-1/2″ x 1/2″ high	2 Clab	50	.320	Ea.	8	10.10		18.10	25.50

34 72 Railway Construction

34 72 16 – Railway Siding

34 72 16.50 Railroad Sidings

		Crew	Daily Output	Labor-Hours	Unit	Material	Labor	Equipment	Total	Total Incl O&P
0010	**RAILROAD SIDINGS** R347216-10									
0800	Siding, yard spur, level grade									
0820	100 lb. new rail	B-14	57	.842	L.F.	112	28	5.15	145.15	175
1020	100 lb. new rail	″	22	2.182	″	157	72.50	13.35	242.85	305

34 72 16.60 Railroad Turnouts

		Crew	Daily Output	Labor-Hours	Unit	Material	Labor	Equipment	Total	Total Incl O&P
0010	**RAILROAD TURNOUTS**									
2200	Turnouts, #8, incl. 100 lb. rails, plates, bars, frog, switch pt.									
2300	Timbers and ballast 6″ below bottom of tie	B-14	.50	96	Ea.	11,500	3,175	590	15,265	18,400

Division Notes

		CREW	DAILY OUTPUT	LABOR-HOURS	UNIT	2009 BARE COSTS MAT.	LABOR	EQUIP.	TOTAL	TOTAL INCL O&P

Estimating Tips

35 20 16 Hydraulic Gates

This subdivision includes various types of gates that are commonly used in waterway and canal construction. Various earthwork items and structural support is found in Division 31, and concrete work in Division 3.

35 20 23 Dredging

This subdivision includes barge and shore dredging systems for rivers, canals, and channels.

35 31 00 Shoreline Protection

This subdivision includes breakwaters, bulkheads, and revetments for ocean and river inlets. Additional earthwork may be required from Division 31, and concrete work from Division 3.

35 49 00 Waterway Structures

This subdivision includes breakwaters and bulkheads for canals.

35 51 00 Floating Construction

This section includes floating piers, docks, and dock accessories. Fixed Pier Timber Construction is found in 06 13 33. Driven piles are found in Division 31, as well as sheet piling, cofferdams, and riprap.

Reference Numbers

Reference numbers are shown in shaded boxes at the beginning of some major classifications. These numbers refer to related items in the Reference Section. The reference information may be an estimating procedure, an alternate pricing method, or technical information.

Note: Not all subdivisions listed here necessarily appear in this publication.

35 20 16 – Hydraulic Gates

35 20 16.26 Hydraulic Sluice Gates		Crew	Daily Output	Labor-Hours	Unit	Material	2009 Bare Costs Labor	Equipment	Total	Total Incl O&P
0010	**HYDRAULIC SLUICE GATES**									
0100	Heavy duty, self contained w/crank oper. gate, 18" x 18"	L-5A	1.70	18.824	Ea.	7,400	840	420	8,660	10,100
0110	24" x 24"		1.20	26.667		10,000	1,200	595	11,795	13,800
0120	30" x 30"		1	32		11,500	1,425	715	13,640	15,900
0130	36" x 36"		.90	35.556		14,300	1,600	795	16,695	19,400
0140	42" x 42"		.80	40		16,500	1,775	895	19,170	22,300
0150	48" x 48"		.50	64		17,000	2,850	1,425	21,275	25,400
0160	54" x 54"		.40	80		21,900	3,575	1,800	27,275	32,500
0170	60" x 60"		.30	106		25,100	4,775	2,400	32,275	38,700
0180	66" x 66"		.30	106		30,800	4,775	2,400	37,975	45,000
0190	72" x 72"		.20	160		34,500	7,150	3,575	45,225	55,000
0200	78" x 78"		.20	160		47,000	7,150	3,575	57,725	68,500
0210	84" x 84"		.10	320		44,400	14,300	7,175	65,875	82,000
0220	90" x 90"	E-20	.30	213		53,500	9,325	4,425	67,250	80,000
0230	96" x 96"		.30	213		59,000	9,325	4,425	72,750	86,500
0240	108" x 108"		.20	320		67,000	14,000	6,625	87,625	106,000
0250	120" x 120"		.10	640		78,500	28,000	13,300	119,800	151,000
0260	132" x 132"		.10	640		138,500	28,000	13,300	179,800	217,000

35 20 16.63 Canal Gates		Crew	Daily Output	Labor-Hours	Unit	Material	Labor	Equipment	Total	Total Incl O&P
0010	**CANAL GATES**									
0011	Cast iron body, fabricated frame									
0100	12" diameter	L-5A	4.60	6.957	Ea.	720	310	156	1,186	1,525
0110	18" diameter		4	8		1,175	355	179	1,709	2,125
0120	24" diameter		3.50	9.143		2,400	410	205	3,015	3,575
0130	30" diameter		2.80	11.429		2,775	510	256	3,541	4,250
0140	36" diameter		2.30	13.913		3,400	620	310	4,330	5,200
0150	42" diameter		1.70	18.824		5,750	840	420	7,010	8,300
0160	48" diameter		1.20	26.667		7,650	1,200	595	9,445	11,200
0170	54" diameter		.90	35.556		13,200	1,600	795	15,595	18,200
0180	60" diameter		.50	64		15,700	2,850	1,425	19,975	24,000
0190	66" diameter		.50	64		11,500	2,850	1,425	15,775	19,400
0200	72" diameter		.40	80		21,500	3,575	1,800	26,875	32,000

35 20 16.66 Flap Gates		Crew	Daily Output	Labor-Hours	Unit	Material	Labor	Equipment	Total	Total Incl O&P
0010	**FLAP GATES**									
0100	Aluminum, 18" diameter	L-5A	5	6.400	Ea.	1,900	286	143	2,329	2,750
0110	24" diameter		4	8		2,475	355	179	3,009	3,525
0120	30" diameter		3.50	9.143		3,050	410	205	3,665	4,300
0130	36" diameter		2.80	11.429		3,975	510	256	4,741	5,575
0140	42" diameter		2.30	13.913		6,025	620	310	6,955	8,075
0150	48" diameter		1.70	18.824		7,075	840	420	8,335	9,750
0160	54" diameter		1.20	26.667		8,800	1,200	595	10,595	12,500
0170	60" diameter		.80	40		11,400	1,775	895	14,070	16,800
0180	66" diameter		.50	64		14,400	2,850	1,425	18,675	22,500
0190	72" diameter		.40	80		17,100	3,575	1,800	22,475	27,200

35 20 16.69 Knife Gates		Crew	Daily Output	Labor-Hours	Unit	Material	Labor	Equipment	Total	Total Incl O&P
0010	**KNIFE GATES**									
0100	Incl. handwheel operator for hub, 6" diameter	Q-23	7.70	3.117	Ea.	810	146	140	1,096	1,275
0110	8" diameter		7.20	3.333		1,075	157	150	1,382	1,600
0120	10" diameter		4.80	5		1,375	235	225	1,835	2,150
0130	12" diameter		3.60	6.667		2,075	315	300	2,690	3,100
0140	14" diameter		3.40	7.059		3,600	330	320	4,250	4,825
0150	16" diameter		3.20	7.500		4,000	350	340	4,690	5,325

35 20 Waterway and Marine Construction and Equipment

35 20 16 – Hydraulic Gates

35 20 16.69 Knife Gates

	35 20 16.69 Knife Gates	Crew	Daily Output	Labor-Hours	Unit	2009 Bare Costs Material	Labor	Equipment	Total	Total Incl O&P
0160	18″ diameter	Q-23	3	8	Ea.	5,300	375	360	6,035	6,800
0170	20″ diameter		2.70	8.889		6,475	415	400	7,290	8,225
0180	24″ diameter		2.40	10		7,850	470	450	8,770	9,850
0190	30″ diameter		1.80	13.333		18,500	625	600	19,725	22,000
0200	36″ diameter		1.20	20		24,100	940	900	25,940	28,900

35 20 16.73 Slide Gates

		Crew	Daily Output	Labor-Hours	Unit	Material	Labor	Equipment	Total	Total Incl O&P
0010	**SLIDE GATES**									
0100	Steel, self contained incl. anchor bolts and grout, 12″ x 12″	L-5A	4.60	6.957	Ea.	3,175	310	156	3,641	4,200
0110	18″ x 18″		4	8		3,500	355	179	4,034	4,675
0120	24″ x 24″		3.50	9.143		3,850	410	205	4,465	5,200
0130	30″ x 30″		2.80	11.429		4,250	510	256	5,016	5,875
0140	36″ x 36″		2.30	13.913		4,575	620	310	5,505	6,500
0150	42″ x 42″		1.70	18.824		5,225	840	420	6,485	7,700
0160	48″ x 48″		1.20	26.667		5,475	1,200	595	7,270	8,800
0170	54″ x 54″		.90	35.556		6,700	1,600	795	9,095	11,100
0180	60″ x 60″		.55	58.182		7,275	2,600	1,300	11,175	14,100
0190	72″ x 72″		.36	88.889		10,600	3,975	2,000	16,575	21,000

35 20 23 – Dredging

35 20 23.13 Mechanical Dredging

		Crew	Daily Output	Labor-Hours	Unit	Material	Labor	Equipment	Total	Total Incl O&P
0010	**MECHANICAL DREDGING**									
0020	Dredging mobilization and demobilization., add to below, minimum	B-8	.53	120	Total		4,225	5,375	9,600	12,700
0100	Maximum	″	.10	640	″		22,400	28,500	50,900	67,500
0300	Barge mounted clamshell excavation into scows									
0310	Dumped 20 miles at sea, minimum	B-57	310	.155	B.C.Y.		5.55	4.99	10.54	14.45
0400	Maximum	″	213	.225	″		8.10	7.25	15.35	21
0500	Barge mounted dragline or clamshell, hopper dumped,									
0510	pumped 1000′ to shore dump, minimum	B-57	340	.141	B.C.Y.		5.05	4.55	9.60	13.15
0525	All pumping uses 2000 gallons of water per cubic yard									
0600	Maximum	B-57	243	.198	B.C.Y.		7.10	6.35	13.45	18.40

35 20 23.23 Hydraulic Dredging

		Crew	Daily Output	Labor-Hours	Unit	Material	Labor	Equipment	Total	Total Incl O&P
0010	**HYDRAULIC DREDGING**									
1000	Hydraulic method, pumped 1000′ to shore dump, minimum	B-57	460	.104	B.C.Y.		3.74	3.36	7.10	9.70
1100	Maximum		310	.155			5.55	4.99	10.54	14.45
1400	Into scows dumped 20 miles, minimum		425	.113			4.05	3.64	7.69	10.50
1500	Maximum		243	.198			7.10	6.35	13.45	18.40
1600	For inland rivers and canals in South, deduct								30%	30%

35 31 Shoreline Protection

35 31 16 – Seawalls

35 31 16.13 Concrete Seawalls

		Crew	Daily Output	Labor-Hours	Unit	Material	Labor	Equipment	Total	Total Incl O&P
0010	**CONCRETE SEAWALLS**									
0011	Reinforced concrete									
0015	include footing and tie-backs									
0060	Maximum	C-17C	24.25	3.423	L.F.	81	141	20	242	340
0100	12′ high, minimum		20	4.150		131	171	24.50	326.50	450
0160	Maximum		18.50	4.486		152	185	26.50	363.50	495

35 31 16.19 Steel Sheet Piling Seawalls

		Crew	Daily Output	Labor-Hours	Unit	Material	Labor	Equipment	Total	Total Incl O&P
0010	**STEEL SHEET PILING SEAWALLS**									
0200	Steel sheeting, with 4′ x 4′ x 8″ concrete deadmen, @ 10′ O.C.									
0210	12′ high, shore driven	B-40	27	2.370	L.F.	82.50	94	123	299.50	380

35 31 Shoreline Protection

35 31 16 – Seawalls

35 31 16.19 Steel Sheet Piling Seawalls		Crew	Daily Output	Labor-Hours	Unit	2009 Bare Costs Material	Labor	Equipment	Total	Total Incl O&P
0260	Barge driven	B-76	15	4.800	L.F.	124	189	198	511	665

35 31 19 – Revetments

35 31 19.18 Revetments, Concrete

		Crew	Daily Output	Labor-Hours	Unit	Material	Labor	Equipment	Total	Total Incl O&P
0010	**REVETMENTS, CONCRETE**									
0100	Concrete revetment matt 8′ x20′ x 4 1/2 ″ excluding site preparation									
0110	Includes all labor, material and equip. for complete installation				Ea.	3,275			3,275	3,600

35 51 Floating Construction

35 51 13 – Floating Piers

35 51 13.23 Floating Wood Piers

		Crew	Daily Output	Labor-Hours	Unit	Material	Labor	Equipment	Total	Total Incl O&P
0010	**FLOATING WOOD PIERS**									
0020	Polyethylene encased polystyrene, no pilings included	F-3	330	.121	S.F.	38.50	4.91	2.33	45.74	52.50
0200	Pile supported, shore constructed, bare, 3″ decking		130	.308		26.50	12.45	5.90	44.85	56
0250	4″ decking		120	.333		26	13.50	6.40	45.90	57.50
0400	Floating, small boat, prefab, no shore facilities, minimum		250	.160		13.30	6.50	3.08	22.88	28.50
0500	Maximum		150	.267		44	10.80	5.15	59.95	72
0700	Per slip, minimum (180 S.F. each)		1.59	25.157	Ea.	2,900	1,025	485	4,410	5,375
0800	Maximum		1.40	28.571	″	8,625	1,150	550	10,325	12,000

Estimating Tips

Products such as conveyors, material handling cranes and hoists, as well as other items specified in this division, may require trained installers. The general contractor may not have any choice as to who will perform the installation or when it will be performed. Long lead times are often required for these products, making early decisions in purchasing and scheduling necessary.

The installation of this type of equipment may require the embedment of mounting hardware during construction of floors, structural walls, or interior walls/partitions. Electrical connections will require coordination with the electrical contractor.

Reference Numbers

Reference numbers are shown in shaded boxes at the beginning of some major classifications. These numbers refer to related items in the Reference Section. The reference information may be an estimating procedure, an alternate pricing method, or technical information.

Note: Not all subdivisions listed here necessarily appear in this publication.

41 21 Conveyors

41 21 23 – Piece Material Conveyors

41 21 23.16 Material Handling Conveyors		Crew	Daily Output	Labor-Hours	Unit	Material	2009 Bare Costs Labor	Equipment	Total	Total Incl O&P
0010	**MATERIAL HANDLING CONVEYORS**									
0020	Gravity fed, 2″ rollers, 3″ O.C.									
0050	10′ sections with 2 supports, 600 lb. capacity, 18″ wide				Ea.	450			450	495
0100	24″ wide					515			515	570
0150	1400 lb. capacity, 18″ wide					415			415	455
0200	24″ wide					460			460	505
0350	Horizontal belt, center drive and takeup, 60 fpm									
0400	16″ belt, 26.5′ length	2 Mill	.50	32	Ea.	3,150	1,325		4,475	5,575
0450	24″ belt, 41.5′ length		.40	40		4,675	1,675		6,350	7,750
0500	61.5′ length		.30	53.333		6,025	2,225		8,250	10,100
0600	Inclined belt, 10′ rise with horizontal loader and									
0620	End idler assembly, 27.5′ length, 18″ belt	2 Mill	.30	53.333	Ea.	9,200	2,225		11,425	13,600
0700	24″ belt	″	.15	106	″	10,300	4,425		14,725	18,400
3600	Monorail, overhead, manual, channel type									
3700	125 lb. per L.F.	1 Mill	26	.308	L.F.	10.25	12.80		23.05	31.50
3900	500 lb. per L.F.	″	21	.381	″	9.25	15.85		25.10	35
4000	Trolleys for above, 2 wheel, 125 lb. capacity				Ea.	77			77	84.50
4200	4 wheel, 250 lb. capacity					207			207	227
4300	8 wheel, 1,000 lb. capacity					410			410	450

41 22 Cranes and Hoists

41 22 13 – Cranes

41 22 13.10 Crane Rail

		Crew	Daily Output	Labor-Hours	Unit	Material	Labor	Equipment	Total	Total Incl O&P
0010	**CRANE RAIL**									
0020	Box beam bridge, no equipment included	E-4	3400	.009	Lb.	1.19	.43	.04	1.66	2.14
0210	Running track only, 104 lb per yard, 20′ piece	″	160	.200	L.F.	21.50	9.05	.84	31.39	41.50

41 22 13.13 Bridge Cranes

		Crew	Daily Output	Labor-Hours	Unit	Material	Labor	Equipment	Total	Total Incl O&P
0010	**BRIDGE CRANES**									
0100	1 girder, 20′ span, 3 ton	M-3	1	34	Ea.	23,800	1,550	175	25,525	28,800
0125	5 ton		1	34		26,000	1,550	175	27,725	31,200
0150	7.5 ton		1	34		31,000	1,550	175	32,725	36,700
0175	10 ton		.80	42.500		41,000	1,925	219	43,144	48,200
0200	15 ton		.80	42.500		52,500	1,925	219	54,644	61,000
0225	30′ span, 3 ton		1	34		24,700	1,550	175	26,425	29,800
0250	5 ton		1	34		27,200	1,550	175	28,925	32,500
0275	7.5 ton		1	34		32,600	1,550	175	34,325	38,400
0300	10 ton		.80	42.500		42,400	1,925	219	44,544	49,800
0325	15 ton		.80	42.500		55,000	1,925	219	57,144	63,500
0350	2 girder, 40′ span, 3 ton	M-4	.50	72		40,900	3,225	470	44,595	50,500
0375	5 ton		.50	72		42,800	3,225	470	46,495	52,500
0400	7.5 ton		.50	72		46,900	3,225	470	50,595	57,000
0425	10 ton		.40	90		55,000	4,050	585	59,635	67,500
0450	15 ton		.40	90		74,500	4,050	585	79,135	89,000
0475	25 ton		.30	120		88,000	5,400	780	94,180	106,500
0500	50′ span, 3 ton		.50	72		46,500	3,225	470	50,195	56,500
0525	5 ton		.50	72		48,400	3,225	470	52,095	58,500
0550	7.5 ton		.50	72		52,000	3,225	470	55,695	62,500
0575	10 ton		.40	90		59,500	4,050	585	64,135	72,500
0600	15 ton		.40	90		78,000	4,050	585	82,635	93,000
0625	25 ton		.30	120		92,500	5,400	780	98,680	111,000

41 22 23 – Hoists

41 22 23.10 Material Handling		Crew	Daily Output	Labor-Hours	Unit	Material	2009 Bare Costs Labor	Equipment	Total	Total Incl O&P
0010	**MATERIAL HANDLING**, cranes, hoists and lifts									
1500	Cranes, portable hydraulic, floor type, 2,000 lb capacity				Ea.	2,475			2,475	2,700
1600	4,000 lb capacity					4,150			4,150	4,575
1800	Movable gantry type, 12′ to 15′ range, 2,000 lb capacity					3,600			3,600	3,975
1900	6,000 lb capacity					5,225			5,225	5,750
2100	Hoists, electric overhead, chain, hook hung, 15′ lift, 1 ton cap.					2,050			2,050	2,250
2200	3 ton capacity					2,575			2,575	2,825
2500	5 ton capacity					6,625			6,625	7,275
2600	For hand-pushed trolley, add					15%				
2700	For geared trolley, add					30%				
2800	For motor trolley, add					75%				
3000	For lifts over 15′, 1 ton, add				L.F.	23.50			23.50	25.50
3100	5 ton, add				"	52			52	57
3110	Air powered hoist, 500 lb capacity	2 Mill	2.67	5.993	Ea.	2,800	249		3,049	3,475
3120	1000 lb capacity		2.46	6.504		3,200	271		3,471	3,950
3130	2000 lb capacity		2.29	6.987		4,150	291		4,441	5,025
3140	4000 lb capacity		2	8		4,775	335		5,110	5,775
3160	6000 lb capacity		1.88	8.511		6,775	355		7,130	8,000
3180	8000 lb capacity		1.68	9.524		6,325	395		6,720	7,575
3190	10000 lb capacity		1.60	10		7,625	415		8,040	9,025
3300	Lifts, scissor type, portable, electric, 36″ high, 2,000 lb					4,275			4,275	4,700
3400	48″ high, 4,000 lb					4,675			4,675	5,150

Division Notes

		CREW	DAILY OUTPUT	LABOR-HOURS	UNIT	2009 BARE COSTS MAT.	LABOR	EQUIP.	TOTAL	TOTAL INCL O&P

Estimating Tips

There are four basic types of pollution: Air, Noise, Water, and Solid.

These systems may be interrelated and care must be taken that the complete systems are estimated. For example, Air Pollution Equipment may include dust and air-entrained particles that have to be collected. The vacuum systems could be noisy, requiring silencers to reduce noise pollution, and the collected solids have to be disposed of to prevent Solid Pollution.

Water Treatment may be accomplished as comparatively small packaged plants providing Chemical, Biological or Thermal Treatment. These are usually priced by GPH or GPD.

Large plants and municipal treatment facilities are sized MGD. These facilities are usually priced in two steps. First, the large concrete structures used as settling ponds, etc.; and second, the equipment required. Storm water does not need as much treatment as sanitary wastewater.

The total cost of the system is greatly affected by the level of purification required by specifications and codes before release.

An additional cost may be encountered if there is a possibility that the water being treated may contain petroleum products, which have to be isolated and disposed of in an environmentally acceptable manner.

Reference Numbers

Reference numbers are shown in shaded boxes at the beginning of some major classifications. These numbers refer to related items in the Reference Section. The reference information may be an estimating procedure, an alternate pricing method, or technical information.

Note: Not all subdivisions listed here necessarily appear in this publication.

Division 44 - Pollution Control Equipment

44 11 16.10 Dust Collection Systems		Crew	Daily Output	Labor-Hours	Unit	Material (2009 Bare Costs)	Labor (2009 Bare Costs)	Equipment (2009 Bare Costs)	Total (2009 Bare Costs)	Total Incl O&P
0010	**DUST COLLECTION SYSTEMS** Commercial / industrial									
0120	Central vacuum units									
0130	Includes stand, filters and motorized shaker									
0200	500 CFM, 10" inlet, 2 HP	Q-20	2.40	8.333	Ea.	3,725	360		4,085	4,675
0220	1000 CFM, 10" inlet, 3 HP		2.20	9.091		3,900	395		4,295	4,925
0240	1500 CFM, 10" inlet, 5 HP		2	10		4,100	435		4,535	5,175
0260	3000 CFM, 13" inlet, 10 HP		1.50	13.333		10,400	580		10,980	12,400
0280	5000 CFM, 16" inlet, 2 @ 10 HP	↓	1	20	↓	11,000	870		11,870	13,500
1000	Vacuum tubing, galvanized									
1100	2-1/8" OD, 16 ga.	Q-9	440	.036	L.F.	2.45	1.54		3.99	5.15
1110	2-1/2" OD, 16 ga.		420	.038		2.82	1.62		4.44	5.65
1120	3" OD, 16 ga.		400	.040		3.23	1.70		4.93	6.25
1130	3-1/2" OD, 16 ga.		380	.042		4.67	1.79		6.46	8
1140	4" OD, 16 ga.		360	.044		5.25	1.89		7.14	8.75
1150	5" OD, 14 ga.		320	.050		9.40	2.12		11.52	13.75
1160	6" OD, 14 ga.		280	.057		10.90	2.43		13.33	15.80
1170	8" OD, 14 ga.		200	.080		16.40	3.40		19.80	23.50
1180	10" OD, 12 ga.		160	.100		16.40	4.25		20.65	25
1190	12" OD, 12 ga.		120	.133		28.50	5.65		34.15	40.50
1200	14" OD, 12 ga.	↓	80	.200	↓	34.50	8.50		43	51.50
1940	Hose, flexible wire reinforced rubber									
1956	3" dia.	Q-9	400	.040	L.F.	7.05	1.70		8.75	10.45
1960	4" dia.		360	.044		7.95	1.89		9.84	11.75
1970	5" dia.		320	.050		9.10	2.12		11.22	13.40
1980	6" dia.	↓	280	.057	↓	9.80	2.43		12.23	14.65
2000	90° Elbow, slip fit									
2110	2-1/8" dia.	Q-9	70	.229	Ea.	10.25	9.70		19.95	26.50
2120	2-1/2" dia.		65	.246		14.30	10.45		24.75	32.50
2130	3" dia.		60	.267		19.40	11.35		30.75	39.50
2140	3-1/2" dia.		55	.291		24	12.35		36.35	46
2150	4" dia.		50	.320		30	13.60		43.60	54
2160	5" dia.		45	.356		56.50	15.10		71.60	86
2170	6" dia.		40	.400		77.50	17		94.50	113
2180	8" dia.	↓	30	.533	↓	148	22.50		170.50	198
2400	45° Elbow, slip fit									
2410	2-1/8" dia.	Q-9	70	.229	Ea.	8.95	9.70		18.65	25.50
2420	2-1/2" dia.		65	.246		13.20	10.45		23.65	31
2430	3" dia.		60	.267		15.95	11.35		27.30	35.50
2440	3-1/2" dia.		55	.291		20	12.35		32.35	41.50
2450	4" dia.		50	.320		26	13.60		39.60	50
2460	5" dia.		45	.356		44	15.10		59.10	72
2470	6" dia.		40	.400		59	17		76	92
2480	8" dia.	↓	35	.457	↓	116	19.40		135.40	159
2800	90° TY, slip fit thru 6" dia									
2810	2-1/8" dia.	Q-9	42	.381	Ea.	20	16.20		36.20	47.50
2820	2-1/2" dia.		39	.410		26	17.45		43.45	56
2830	3" dia.		36	.444		34.50	18.90		53.40	68
2840	3-1/2" dia.		33	.485		44	20.50		64.50	80.50
2850	4" dia.		30	.533		63	22.50		85.50	105
2860	5" dia.		27	.593		120	25		145	172
2870	6" dia.	↓	24	.667		163	28.50		191.50	224
2880	8" dia., butt end					345			345	380
2890	10" dia., butt end				↓	625			625	685

44 11 Air Pollution Control Equipment

44 11 16 – Industrial Dust Collectors

44 11 16.10 Dust Collection Systems		Crew	Daily Output	Labor-Hours	Unit	2009 Bare Costs Material	Labor	Equipment	Total	Total Incl O&P
2900	12" dia., butt end				Ea.	625			625	685
2910	14" dia., butt end					1,125			1,125	1,250
2920	6" x 4" dia., butt end					99.50			99.50	109
2930	8" x 4" dia., butt end					188			188	207
2940	10" x 4" dia., butt end					247			247	271
2950	12" x 4" dia., butt end					283			283	310
3100	90° Elbow, butt end segmented									
3110	8" dia., butt end, segmented				Ea.	148			148	162
3120	10" dia., butt end, segmented					365			365	405
3130	12" dia., butt end, segmented					465			465	510
3140	14" dia., butt end, segmented					580			580	640
3200	45° Elbow, butt end segmented									
3210	8" dia., butt end, segmented				Ea.	116			116	128
3220	10" dia., butt end, segmented					265			265	292
3230	12" dia., butt end, segmented					272			272	299
3240	14" dia., butt end, segmented					335			335	365
3400	All butt end fittings require one coupling per joint.									
3410	Labor for fitting included with couplings.									
3460	Compression coupling, galvanized, neoprene gasket									
3470	2-1/8" dia.	Q-9	44	.364	Ea.	11.85	15.45		27.30	37.50
3480	2-1/2" dia.		44	.364		11.85	15.45		27.30	37.50
3490	3" dia.		38	.421		17.10	17.90		35	47.50
3500	3-1/2" dia.		35	.457		19.40	19.40		38.80	52.50
3510	4" dia.		33	.485		21	20.50		41.50	55.50
3520	5" dia.		29	.552		24	23.50		47.50	63.50
3530	6" dia.		26	.615		24	26		50	68
3540	8" dia.		22	.727		50	31		81	104
3550	10" dia.		20	.800		67.50	34		101.50	129
3560	12" dia.		18	.889		83.50	38		121.50	152
3570	14" dia.		16	1		107	42.50		149.50	186
3800	Air gate valves, galvanized									
3810	2-1/8" dia.	Q-9	30	.533	Ea.	173	22.50		195.50	226
3820	2-1/2" dia.		28	.571		191	24.50		215.50	249
3830	3" dia.		26	.615		209	26		235	272
3840	4" dia.		23	.696		263	29.50		292.50	335
3850	6" dia.		18	.889		355	38		393	450

44 41 Packaged Water Treatment

44 41 13 – Packaged Water Treatment Plants

44 41 13.16 Biological Packaged Water Treatment Plants		Crew	Daily Output	Labor-Hours	Unit	Material	Labor	Equipment	Total	Total Incl O&P
0010	**BIOLOGICAL PACKAGED WATER TREATMENT PLANTS**									
0011	Not including fencing or external piping									
0020	Steel packaged, blown air aeration plants									
0100	1,000 GPD				Gal.				300	345
0200	5,000 GPD								220	255
0300	15,000 GPD								110	127
0400	30,000 GPD								63.40	72.90
0500	50,000 GPD								53.75	61.80
0600	100,000 GPD								34.80	40
0700	200,000 GPD								32.30	37.10
0800	500,000 GPD								24.75	28.50

44 41 13 – Packaged Water Treatment Plants

44 41 13.16 Biological Packaged Water Treatment Plants		Crew	Daily Output	Labor-Hours	Unit	2009 Bare Costs Material	Labor	Equipment	Total	Total Incl O&P
1000	Concrete, extended aeration, primary and secondary treatment									
1010	10,000 GPD				Gal.				99.60	114.50
1100	30,000 GPD								57.65	66.30
1200	50,000 GPD								48.85	56.20
1400	100,000 GPD								31.65	36.40
1500	500,000 GPD				↓				22.50	25.90
1700	Municipal wastewater treatment facility									
1720	1.0 MGD				Gal.				9.20	10.60
1740	1.5 MGD								9	10.40
1760	2.0 MGD								7	8.10
1780	3.0 MGD								6	6.90
1800	5.0 MGD				↓				5.50	6.30
2000	Holding tank system, not incl. excavation or backfill									
2010	Recirculating chemical water closet	2 Plum	4	4	Ea.	965	195		1,160	1,350
2100	For voltage converter, add	"	16	1		255	49		304	355
2200	For high level alarm, add	1 Plum	7.80	1.026	↓	146	50		196	238

44 41 13.17 Wastewater Treatment System		Crew	Daily Output	Labor-Hours	Unit	Material	Labor	Equipment	Total	Total Incl O&P
0010	**WASTEWATER TREATMENT SYSTEM**									
0020	Fiberglass, 1,000 gallon	B-21	1.29	21.705	Ea.	3,300	790	100	4,190	5,000
0100	1,500 gallon	"	1.03	27.184	"	7,200	990	125	8,315	9,675

Assemblies Section

Table of Contents

How to Use the Assemblies Cost Tables

The following is a detailed explanation of a sample Assemblies Cost Table. Most Assembly Tables are separated into three parts: 1) an illustration of the system to be estimated; 2) the components and related costs of a typical system; and 3) the costs for similar systems with dimensional and/or size variations. For costs of the components that comprise these systems, or assemblies, refer to the Unit Price Section. Next to each bold number below is the described item with the appropriate component of the sample entry following in parentheses. In most cases, if the work is to be subcontracted, the general contractor will need to add an additional markup (RSMeans suggests using 10%) to the "Total" figures.

1 *System/Line Numbers (C1010 124 1250)*

Each Assemblies Cost Line has been assigned a unique identification number based on the UNIFORMAT II classification system.

UNIFORMAT II Major Group

C1010 124 1250

UNIFORMAT II Level 3

Means Major Classification

Means Individual Line Number

C10 Interior Construction

C1010 Partitions

1

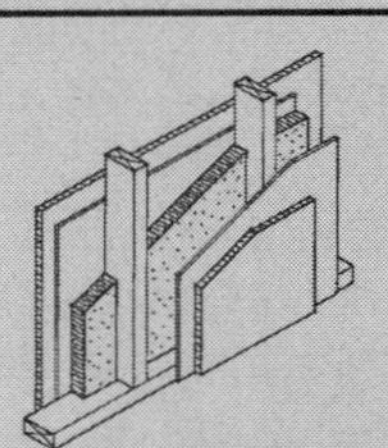

Wood Stud Framing

2

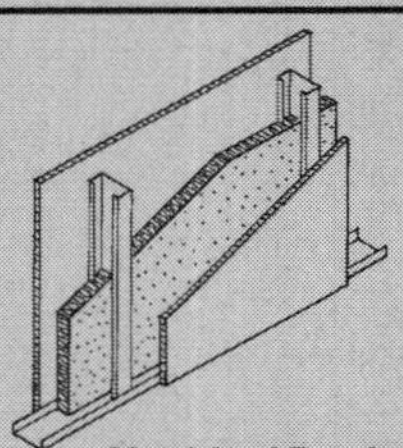

Metal Stud Framing

The Drywall Partitions/Stud Framing Systems are defined by type of drywall and number of layers, type and spacing of stud framing, and treatment on the opposite face. Components include taping and finishing.

Cost differences between regular and fire resistant drywall are negligible, and terminology is interchangeable. In some cases fiberglass insulation is included for additional sound deadening.

System Components	QUANTITY	UNIT	COST PER S.F. MAT.	INST.	TOTAL
SYSTEM C1010 124 1250					
DRYWALL PARTITION,5/8" F.R.1 SIDE,5/8" REG.1 SIDE,2"X4"STUDS,16" O.C.	5		6		
3 Gypsum plasterboard, nailed/screwed to studs, 5/8"F.R. fire resistant	1.000	S.F.	.40	.53	.93
Gypsum plasterboard, nailed/screwed to studs, 5/8" regular	1.000	S.F.	.37	.53	.90
Taping and finishing joints 4	2.000	S.F.	.10	1.06	1.16
Framing, 2 x 4 studs @ 16" O.C., 10' high	1.000	S.F.	.34	1.06	1.40
TOTAL			1.21	3.18	4.39

C1010 124	Drywall Partitions/Wood Stud Framing							
	FACE LAYER	BASE LAYER	FRAMING	OPPOSITE FACE	INSULATION	COST PER S.F. MAT.	INST.	TOTAL
1202	5/8" FR drywall	none	2 x 4, @ 16" O.C.	same	0	1.27	3.18	4.45
1250				5/8" reg. drywall	0	1.21	3.18	4.39
1302				nothing	0	.82	2.12	2.94
1402		1/4" SD gypsum	2 x 4 @ 16" O.C.	same	1-1/2" fiberglass	2.79	4.89	7.68
1452				5/8" FR drywall	1-1/2" fiberglass	2.38	4.30	6.68
1502				nothing	1-1/2" fiberglass	1.93	3.24	5.17
1602		resil. channels	2 x 4 @ 16", O.C.	same	1-1/2" fiberglass	2.32	6.20	8.52
1652				5/8" FR drywall	1-1/2" fiberglass	2.15	4.95	7.10
1702				nothing	1-1/2" fiberglass	1.70	3.89	5.59

7 8 9

2 *Illustration*

At the top of most assembly pages are an illustration, a brief description, and the design criteria used to develop the cost.

3 *System Components*

The components of a typical system are listed separately to show what has been included in the development of the total system price. The table below contains prices for other similar systems with dimensional and/or size variations.

4 *Quantity*

This is the number of line item units required for one system unit. For example, we assume that it will take 1 S.F. of 5/8″ fire-resistant gypsum board per square foot of wall area.

5 *Unit of Measure for Each Item*

The abbreviated designation indicates the unit of measure, as defined by industry standards, upon which the price of the component is based. For example, gypsum drywall is priced by S.F. (square foot).

6 *Unit of Measure for Each System (Cost per S.F.)*

Costs shown in the three right-hand columns have been adjusted by the component quantity and unit of measure for the entire system. In this example, "Cost per S.F." is the unit of measure for this system, or assembly.

7 *Materials (1.21)*

This column contains the Materials Cost of each component. These cost figures are bare costs plus 10% for profit.

8 *Installation (3.18)*

Installation includes labor and equipment plus the installing contractor's overhead and profit. Equipment costs are the bare rental costs plus 10% for profit. The labor overhead and profit are defined on the inside back cover of this book.

9 *Total (4.39)*

The figure in this column is the sum of the material and installation costs.

Material Cost	+	Installation Cost	=	Total
$1.21	+	$3.18	=	$4.39

A Substructure

A1010 Standard Foundations

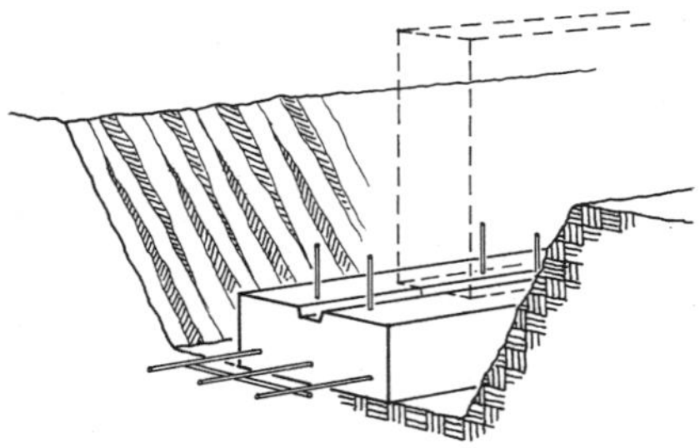

The Strip Footing System includes: excavation; hand trim; all forms needed for footing placement; forms for 2″ x 6″ keyway (four uses); dowels; and 3,000 p.s.i. concrete.

The footing size required varies for different soils. Soil bearing capacities are listed for 3 KSF and 6 KSF. Depths of the system range from 8″ to 24″. Widths range from 16″ to 96″. Smaller strip footings may not require reinforcement.

Please see the reference section for further design and cost information.

System Components	QUANTITY	UNIT	COST PER L.F.		
			MAT.	INST.	TOTAL
SYSTEM A1010 110 2500					
STRIP FOOTING, LOAD 5.1KLF, SOIL CAP. 3 KSF, 24″ WIDE X12″ DEEP, REINF.					
Trench excavation	.148	C.Y.		1.25	1.25
Hand trim	2.000	S.F.		1.76	1.76
Compacted backfill	.074	C.Y.		.25	.25
Formwork, 4 uses	2.000	S.F.	5.34	8.24	13.58
Keyway form, 4 uses	1.000	L.F.	.29	1.06	1.35
Reinforcing, fy = 60000 psi	3.000	Lb.	2.67	1.74	4.41
Dowels	2.000	Ea.	2.50	5.06	7.56
Concrete, f'c = 3000 psi	.074	C.Y.	8.21		8.21
Place concrete, direct chute	.074	C.Y.		1.63	1.63
Screed finish	2.000	S.F.		.68	.68
TOTAL			19.01	21.67	40.68

A1010 110	Strip Footings	COST PER L.F.		
		MAT.	INST.	TOTAL
2100	Strip footing, load 2.6KLF, soil capacity 3KSF, 16″ wide x 8″ deep plain	9.65	13	22.65
2300	Load 3.9 KLF, soil capacity, 3 KSF, 24″ wide x 8″ deep, plain	11.55	14.20	25.75
2500	Load 5.1KLF, soil capacity 3 KSF, 24″ wide x 12″ deep, reinf.	19	21.50	40.50
2700	Load 11.1KLF, soil capacity 6 KSF, 24″ wide x 12″ deep, reinf.	19	21.50	40.50
2900	Load 6.8 KLF, soil capacity 3 KSF, 32″ wide x 12″ deep, reinf.	23	23.50	46.50
3100	Load 14.8 KLF, soil capacity 6 KSF, 32″ wide x 12″ deep, reinf.	23	23.50	46.50
3300	Load 9.3 KLF, soil capacity 3 KSF, 40″ wide x 12″ deep, reinf.	26.50	25.50	52
3500	Load 18.4 KLF, soil capacity 6 KSF, 40″ wide x 12″ deep, reinf.	27	25.50	52.50
3700	Load 10.1KLF, soil capacity 3 KSF, 48″ wide x 12″ deep, reinf.	29	27.50	56.50
3900	Load 22.1KLF, soil capacity 6 KSF, 48″ wide x 12″ deep, reinf.	31.50	29.50	61
4100	Load 11.8KLF, soil capacity 3 KSF, 56″ wide x 12″ deep, reinf.	34	30.50	64.50
4300	Load 25.8KLF, soil capacity 6 KSF, 56″ wide x 12″ deep, reinf.	38	33	71
4500	Load 10KLF, soil capacity 3 KSF, 48″ wide x 16″ deep, reinf.	36	32	68
4700	Load 22KLF, soil capacity 6 KSF, 48″ wide, 16″ deep, reinf.	37.50	32.50	70
4900	Load 11.6KLF, soil capacity 3 KSF, 56″ wide x 16″ deep, reinf.	41	45.50	86.50
5100	Load 25.6KLF, soil capacity 6 KSF, 56″ wide x 16″ deep, reinf.	44	47.50	91.50
5300	Load 13.3KLF, soil capacity 3 KSF, 64″ wide x 16″ deep, reinf.	47	38	85
5500	Load 29.3KLF, soil capacity 6 KSF, 64″ wide x 16″ deep, reinf.	51.50	41	92.50
5700	Load 15KLF, soil capacity 3 KSF, 72″ wide x 20″ deep, reinf.	61.50	45.50	107
5900	Load 33KLF, soil capacity 6 KSF, 72″ wide x 20″ deep, reinf.	66.50	48.50	115
6100	Load 18.3KLF, soil capacity 3 KSF, 88″ wide x 24″ deep, reinf.	86	58	144
6300	Load 40.3KLF, soil capacity 6 KSF, 88″ wide x 24″ deep, reinf.	96	64.50	160.50
6500	Load 20KLF, soil capacity 3 KSF, 96″ wide x 24″ deep, reinf.	93.50	61.50	155
6700	Load 44 KLF, soil capacity 6 KSF, 96″ wide x 24″ deep, reinf.	101	66.50	167.50

A1010 Standard Foundations

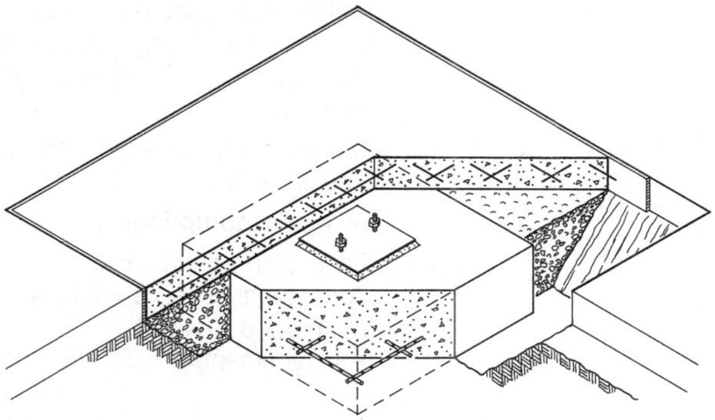

Column footing includes: Remove existing slab; excavation; forms for concrete; reinforcing steel; expansion joint; anchor bolts; grouting column base plate; and concrete placed and finished.

The footing size required varies for different soils.

Design assumptions:

Conc. slab f'c = 3.5 ksi
Conc. foundation f'c = 3.0 ksi
WWF 6 x 6 #10/#10
Reinforcement, fy = 60 ksi

System Components	QUANTITY	UNIT	COST EACH MAT.	COST EACH INST.	COST EACH TOTAL
SYSTEM A1010 260 0200					
CONC. COLUMN FOOTING IN EXISTING BUILDING, LOAD 10K, SOIL CAPACITY 3KSF, 3′ SQ X 12″ DEEP, 3,000 PSI					
Saw cut slab	16.000	S.F.	51.84	131.52	183.36
Break up slab, 6″	16.000	S.F.		212	212
Remove concrete; hand	.755	C.Y.		26.43	26.43
Excavate; hand	.592	C.Y.		61.57	61.57
Forms, 1 use	12.000	SFCA	47.52	231	278.52
Reinforcing steel, #4 bar	15.000	Lb.	13.35	8.70	22.05
Concrete, 3,000 psi, for footing, incl. premium delv. chg.	.333	C.Y.	51.73		51.73
Place conc. footing	.333	C.Y.		22.38	22.38
Backfill by hand	.311	C.Y.		9.33	9.33
Compaction; hand tamp	.311	C.Y.		6.38	6.38
Welded wire fabric	.160	C.S.F.	3.18	5.52	8.70
Premolded, bituminous fiber, 1/2″ x 6″	16.000	L.F.	7.04	22.56	29.60
Concrete, 3500 psi, for slab, incl. premium delv. chg.	.166	C.Y.	26.45		26.45
Place conc. slab, 6″	.166	C.Y.		3.72	3.72
Finish slab; steel trowel	16.000	S.F.		13.28	13.28
Anchor bolts, 3/4″ x 12″	2.000	Ea.	5.12	15.10	20.22
Grout level plate	1.000	S.F.	7.85	13.70	21.55
TOTAL			214.08	783.19	997.27
COST per C.Y. of FOOTING			642.88	2,351.92	2,994.80

A1020 Special Foundations

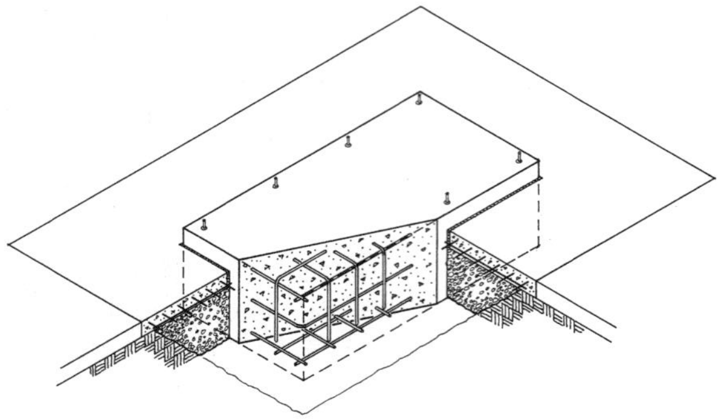

Equipment foundation includes: Remove existing slab; excavation; forms for concrete; reinforcing steel; expansion joint; anchor bolts; and concrete placed and screeded.

The foundation size required varies for different soils.

Design assumptions:

Conc. slab f'c = 3.5 ksi
Conc. foundation f'c = 3.0 ksi
WWF 6 x 6 #10/#10
Reinforcement, fy = 60 ksi

System Components	QUANTITY	UNIT	COST EACH		
			MAT.	INST.	TOTAL
SYSTEM A1020 810 0170					
CONC. EQUIPMENT FOUNDATION, 4'X8'X30" DEEP, 3,000 PSI.					
Saw cut slab	45.000	S.F.	90.72	230.16	320.88
Break up slab; 6"	45.000	S.F.		596.25	596.25
Remove concrete; hand	3.133	C.Y.		109.66	109.66
Excavate; hand	2.500	C.Y.		260	260
Forms, 1 use	60.000	SFCA	237.60	1,155	1,392.60
Reinforcing steel, #6 bar	318.000	Lb.	283.02	184.44	467.46
Concrete, 3000 psi, for foundation, incl. premium delv. chg.	3.000	C.Y.	460.43		460.43
Anchor bolts, 1" diam., with sleeves	8.000	Ea.	73.20	93.60	166.80
Place concrete foundation	3.000	C.Y.		91.17	91.17
Backfill by hand	.866	C.Y.		25.98	25.98
Compaction in 6" layers, hand tamp	.866	C.Y.		17.75	17.75
Premolded, bituminous fiber, 1/2"x6"	24.000	L.F.	10.56	33.84	44.40
Welded wire fabric; 6x6-#10/10	.130	C.S.F.	2.58	4.49	7.07
Concrete, 3500 psi, for slab, incl. premium delv. chg.	.241	C.Y.	38.42		38.42
Place conc. slab	.241	C.Y.		5.41	5.41
Finish slab; steel trowel	13.000	S.F.		10.79	10.79
Finish equip. foundation; screed	32.000	S.F.		9.28	9.28
TOTAL			1,196.53	2,827.82	4,024.35
COST per C.Y. of FOUNDATION			398.84	942.61	1,341.45

B Shell

B1010 Floor Construction

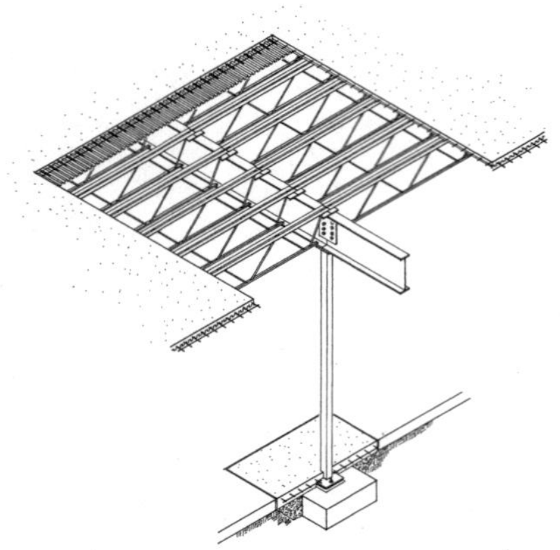

Mezzanine addition to existing building includes: Column footings; steel columns; structural steel; open web steel joists; uncoated 28 ga. steel slab forms; 2-1/2″ concrete slab reinforced with welded wire fabric; steel trowel finish.

Design assumptions:

Structural steel is A36, high strength bolted. Slab form is 28 gauge, galvanized.
WWF 6 x 6 #10/#10
Conc. slab f'c = 3 ksi

System Components	QUANTITY	UNIT	COST EACH MAT.	COST EACH INST.	COST EACH TOTAL
SYSTEM B1010 310 0142					
MEZZANINE ADDITION TO EXISTING BUILDING; 100 PSF SUPERIMPOSED LOAD, SLAB FORM DECK, MTL. JOISTS, 2.5″ CONC. SLAB, 3,000 PSI, 3,000 S.F.					
Saw cut existing slab, 6″ thick	192.000	L.F.	622.08	1,578.24	2,200.32
Cut out existing slab, 4′ x 4′ openings	192.000	S.F.		2,544	2,544
Remove concrete debris	9.060	C.Y.		317.10	317.10
Excavate by hand	7.104	C.Y.		738.82	738.82
Backfill by hand	3.732	C.Y.		111.96	111.96
Compaction of backfill	3.732	C.Y.		76.51	76.51
Forms for footing, 3′ x 3′ x 1′	144.000	SFCA	570.24	2,772	3,342.24
Reinforcing bar for footing, 3#5 E.W. x 30″ long	180.000	Lb.	160.20	104.40	264.60
Anchor bolts, 12″, 2 per column	24.000	Ea.	61.44	181.20	242.64
Concrete, 3000 psi, footing	5.592	C.Y.	620.71		620.71
Place concrete in footing, direct chute	5.592	C.Y.		268.64	268.64
Premolded, bituminous fiber, 1/2″ x 6″	192.000	L.F.	84.48	270.72	355.20
Reinforcing mesh for slab cutouts	1.920	C.S.F.	38.11	66.24	104.35
Concrete, 3500 psi, slab cutouts	2.784	C.Y.	317.38		317.38
Place concrete slab cutouts, direct chute	2.784	C.Y.		44.68	44.68
Machine trowel slab cutouts	192.000	S.F.		159.36	159.36
Grout for leveling plates	12.000	S.F.	94.20	164.40	258.60
Column, 4″ x 4″ x 1/4″ x12′	12.000	Ea.	3,264	1,296	4,560
Structural steel W 21x50	80.000	L.F.	10,920	952.80	11,872.80
W 16x31	80.000	L.F.	6,780	834	7,614
Open web joists, H or K series	6.800	Ton	22,950	7,282.80	30,232.80
Slab form, steel 28 gauge, galvanized	30.000	C.S.F.	5,670	2,130	7,800
Concrete 3,000 psi, 2 1/2″ slab, incl. premium delv. chg.	23.500	C.Y.	3,912.75		3,912.75
Welded wire fabric 6x6 #10/#10 (w1.4/w1.4)	30.000	C.S.F.	595.50	1,035	1,630.50
Place concrete	23.500	C.Y.		1,099.80	1,099.80
Monolithic steel trowel finish	30.000	C.S.F.		2,490	2,490
Curing with sprayed membrane curing compound	30.000	C.S.F.	172.50	264	436.50
TOTAL			56,833.59	26,782.67	83,616.26
COST PER S.F.			18.94	8.93	27.87

B1010 Floor Construction

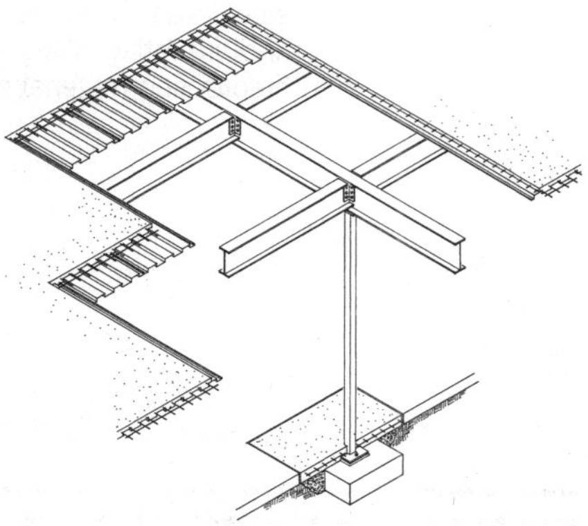

Mezzanine addition to existing building includes: Column footings; steel columns; structural steel; galvanized composite steel deck; 4″ concrete slab reinforced with welded wire fabric; steel trowel finish.

Design assumptions:

Structural steel is A36, high strength bolted.
Composite deck is 22 gauge, galvanized.
WWF 6 x 6 #10/#10
Conc. slab f'c = 3 ksi

System Components	QUANTITY	UNIT	COST EACH		
			MAT.	INST.	TOTAL
SYSTEM B1010 320 0152					
MEZZANINE ADDITION TO EXISTING BUILDING; 100 PSF SUPERIMPOSED LOAD					
COMPOSITE DECK, BEAMS, 4″ CONC. SLAB, 3,000 PSI, 3,000 S.F.					
Saw cut existing slab, 6″ thick	192.000	L.F.	622.08	1,578.24	2,200.32
Remove existing slab, 4′ x 4′ openings	192.000	S.F.		2,544	2,544
Remove concrete debris	9.060	C.Y.		317.10	317.10
Excavate by hand	7.104	C.Y.		738.82	738.82
Backfill by hand, no compaction, light soil	3.732	C.Y.		111.96	111.96
Compaction of backfill	3.732	C.Y.		76.51	76.51
Forms in place, footing, 3′ x 3′ x 1′	144.000	SFCA	570.24	2,772	3,342.24
Footings, #4 to #7 bars, 3#5 E.W. x 30″ long	180.000	Lb.	160.20	104.40	264.60
Anchor bolts, 12″ long, 2 per column	24.000	Ea.	61.44	181.20	242.64
Concrete, 3000 psi, footing	5.592	C.Y.	620.71		620.71
Place concrete in footings, direct chute	5.592	C.Y.		268.64	268.64
Premolded, bituminous fiber, 1/2″ x 6″	192.000	L.F.	84.48	270.72	355.20
Structural steel W 12x19	225.000	L.F.	9,000	1,599.75	10,599.75
Reinforcing mesh for slab cutouts	1.920	C.S.F.	38.11	66.24	104.35
Concrete, 3500 psi, for slab cutouts	2.784	C.Y.	317.38		317.38
W 14x30	375.000	L.F.	30,656.25	3,909.38	34,565.63
Place concrete, slab cutouts, direct chute	2.784	C.Y.		44.68	44.68
W 16x31	80.000	L.F.	6,780	834	7,614
Machine trowel slab cutouts	192.000	S.F.		159.36	159.36
Grout for leveling plates	12.000	S.F.	94.20	164.40	258.60
Column, 4″ x 4″ x 1/4″ x 12′	12.000	Ea.	3,264	1,296	4,560
W 18x50	80.000	L.F.	10,920	1,105.20	12,025.20
Metal deck, galvanized, 2″ deep, 22 gauge.	30.000	C.S.F.	8,040	2,190	10,230
Concrete, 3,000 p.s.i., 4″ slab, incl. premium delv. chg.	37.000	C.Y.	6,160.50		6,160.50
Welded wire fabric 6x6 #10/#10 (w1.4/w1.4)	30.000	C.S.F.	595.50	1,035	1,630.50
Place concrete	37.000	C.Y.		1,731.60	1,731.60
Monolithic steel trowel finish	30.000	C.S.F.		2,490	2,490
Curing with sprayed membrane curing compound	30.000	C.S.F.	172.50	264	436.50
TOTAL			78,157.59	25,853.20	104,010.79
COST PER S.F.			26.05	8.62	34.67

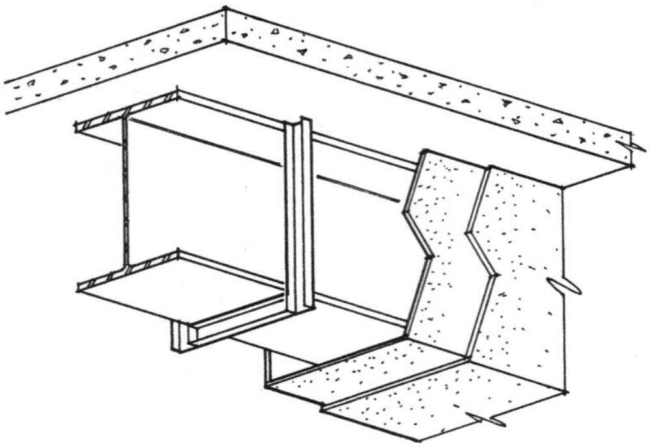

The table below lists fireproofing costs for steel beams by type, beam size, thickness and fire rating. Weights listed are for the fireproofing material only.

System Components	QUANTITY	UNIT	COST PER L.F. MAT.	COST PER L.F. INST.	COST PER L.F. TOTAL
SYSTEM B1010 710 1300					
FIREPROOFING, 5/8" F.R. GYP. BOARD, 12"X 4" BEAM, 2" THICK, 2 HR. F.R.					
Corner bead for drywall, 1-1/4" x 1-1/4", galvanized	.020	C.L.F.	.40	3.02	3.42
L bead for drywall, galvanized	.020	C.L.F.	.43	3.52	3.95
Furring, beams & columns, 3/4" galv. channels, 24" O.C.	2.330	S.F.	.54	5.66	6.20
Drywall on beam, no finish, 2 layers at 5/8" thick	3.000	S.F.	2.67	10.56	13.23
Drywall, taping and finishing joints, add	3.000	S.F.	.15	1.59	1.74
TOTAL			4.19	24.35	28.54

B1010 710 Steel Beam Fireproofing

	ENCASEMENT SYSTEM	BEAM SIZE (IN.)	THICKNESS (IN.)	FIRE RATING (HRS.)	WEIGHT (P.L.F.)	COST PER L.F. MAT.	COST PER L.F. INST.	COST PER L.F. TOTAL
0400	Concrete	12x4	1	1	77	7.65	26	33.65
0450	3000 PSI		1-1/2	2	93	8.80	28	36.80
0500			2	3	121	9.85	30.50	40.35
0550		14x5	1	1	100	10	31.50	41.50
0600			1-1/2	2	122	11.60	35	46.60
0650			2	3	142	13.20	38.50	51.70
0700		16x7	1	1	147	11.15	33.50	44.65
0750			1-1/2	2	169	11.80	34.50	46.30
0800			2	3	195	13.50	38	51.50
0850		18x7-1/2	1	1	172	13.10	39	52.10
0900			1-1/2	2	196	14.85	42.50	57.35
0950			2	3	225	16.65	46.50	63.15
1000		24x9	1	1	264	16.85	47	63.85
1050			1-1/2	2	295	18.25	49.50	67.75
1100			2	3	328	19.70	52.50	72.20
1150		30x10-1/2	1	1	366	22	59.50	81.50
1200			1-1/2	2	404	24	63	87
1250			2	3	449	25.50	66.50	92
1300	5/8" fire rated	12x4	2	2	15	4.19	24.50	28.69
1350	Gypsum board		2-5/8	3	24	5.85	30	35.85
1400		14x5	2	2	17	4.50	25.50	30
1450			2-5/8	3	27	6.45	32	38.45
1500		16x7	2	2	20	5.05	28	33.05
1550			2-5/8	3	31	6.50	28	34.50
1600		18x7-1/2	2	2	22	5.45	30	35.45
1650			2-5/8	3	34	7.75	38	45.75

B1010 Floor Construction

B1010 710 Steel Beam Fireproofing

	ENCASEMENT SYSTEM	BEAM SIZE (IN.)	THICKNESS (IN.)	FIRE RATING (HRS.)	WEIGHT (P.L.F.)	COST PER L.F.		
						MAT.	INST.	TOTAL
1700	5/8″ fire rated	24x9	2	2	27	6.75	37	43.75
1750	Gypsum board		2-5/8	3	42	9.55	47	56.55
1800		30x10-1/2	2	2	33	7.95	43.50	51.45
1850			2-5/8	3	51	11.25	55	66.25
1900	Gypsum	12x4	1-1/8	3	18	6	18.40	24.40
1950	Plaster on	14x5	1-1/8	3	21	6.80	21	27.80
2000	Metal lath	16x7	1-1/8	3	25	7.95	23.50	31.45
2050		18x7-1/2	1-1/8	3	32	9	26.50	35.50
2100		24x9	1-1/8	3	35	11.10	32.50	43.60
2150		30x10-1/2	1-1/8	3	44	13.35	38.50	51.85
2200	Perlite plaster	12x4	1-1/8	2	16	5.20	20.50	25.70
2250	On metal lath		1-1/4	3	20	5.50	21.50	27
2300			1-1/2	4	22	5.90	22.50	28.40
2350		14x5	1-1/8	2	18	6.75	25	31.75
2400			1-1/4	3	23	7.15	26	33.15
2450			1-1/2	4	25	7.90	28	35.90
2500		16x7	1-1/8	2	21	6.25	25.50	31.75
2550			1-1/4	3	26	6.55	26	32.55
2600			1-1/2	4	29	6.95	27.50	34.45
2650		18x7-1/2	1-1/8	2	26	7.75	29.50	37.25
2700			1-1/4	3	33	8.10	31	39.10
2750			1-1/2	4	36	8.90	33	41.90
2800		24x9	1-1/8	2	30	7.95	33.50	41.45
2850			1-1/4	3	38	8.30	34	42.30
2900			1-1/2	4	41	8.70	35.50	44.20
2950		30x10-1/2	1-1/8	2	37	10.30	41.50	51.80
3000			1-1/4	3	46	10.70	42.50	53.20
3050			1-1/2	4	51	11.50	44.50	56
3100	Sprayed Fiber	12x4	5/8	1	12	1.26	2.09	3.35
3150	(non asbestos)		1-1/8	2	21	2.26	3.75	6.01
3200			1-1/4	3	22	2.52	4.16	6.68
3250		14x5	5/8	1	14	1.26	2.09	3.35
3300			1-1/8	2	26	2.26	3.75	6.01
3350			1-1/4	3	29	2.52	4.16	6.68
3400		16x7	5/8	1	18	1.59	2.64	4.23
3450			1-1/8	2	32	2.86	4.74	7.60
3500			1-1/4	3	36	3.18	5.25	8.43
3550		18x7-1/2	5/8	1	20	1.76	2.91	4.67
3600			1-1/8	2	35	3.16	5.25	8.41
3650			1-1/4	3	39	3.49	5.80	9.29
3700		24x9	5/8	1	25	2.22	3.67	5.89
3750			1-1/8	2	45	4.02	6.65	10.67
3800			1-1/4	3	50	4.47	7.40	11.87
3850		30x10-1/2	5/8	1	31	2.69	4.46	7.15
3900			1-1/8	2	55	4.84	8	12.84
3950			1-1/4	3	61	5.40	8.90	14.30
4000	On Decking	Flat	1		6	.58	.60	1.18
4050	Per S.F.	Corrugated	1		7	.87	1.15	2.02
4100		Fluted	1		7	.87	1.15	2.02

B1010 Floor Construction

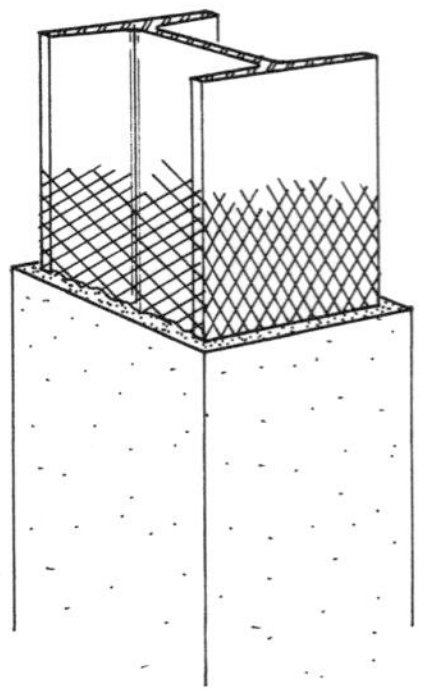

Listed below are costs per V.L.F. for fireproofing by material, column size, thickness and fire rating. Weights listed are for the fireproofing material only.

System Components	QUANTITY	UNIT	COST PER V.L.F. MAT.	COST PER V.L.F. INST.	COST PER V.L.F. TOTAL
SYSTEM B1010 720 3000					
CONCRETE FIREPROOFING, 8″ STEEL COLUMN, 1″ THICK, 1 HR. FIRE RATING					
Forms in place, columns, plywood, 4 uses	3.330	SFCA	2.16	26.64	28.80
Welded wire fabric, 2 x 2 #14 galv. 21 lb./C.S.F., column wrap	2.700	S.F.	1.97	5.05	7.02
Concrete ready mix, regular weight, 3000 psi	.621	C.F.	2.55		2.55
Place and vibrate concrete, 12″ sq./round columns, pumped	.621	C.F.		1.69	1.69
TOTAL			6.68	33.38	40.06

B1010 720 Steel Column Fireproofing

	ENCASEMENT SYSTEM	COLUMN SIZE (IN.)	THICKNESS (IN.)	FIRE RATING (HRS.)	WEIGHT (P.L.F.)	COST PER V.L.F. MAT.	COST PER V.L.F. INST.	COST PER V.L.F. TOTAL
3000	Concrete	8	1	1	110	6.70	33.50	40.20
3050			1-1/2	2	133	7.70	37	44.70
3100			2	3	145	8.55	40	48.55
3150		10	1	1	145	8.85	41.50	50.35
3200			1-1/2	2	168	9.75	44.50	54.25
3250			2	3	196	10.75	47.50	58.25
3600	Gypsum board	8	1	3	14	4.43	26	30.43
3650	1/2″ fire rated	10	1	3	17	4.76	27.50	32.26
3700	2 layers	14	1	3	22	4.93	28.50	33.43
3750	Gypsum board	8	1-1/2	3	23	5.90	33	38.90
3800	1/2″ fire rated	10	1-1/2	3	27	6.65	36.50	43.15
3850	3 layers	14	1-1/2	3	35	7.40	40	47.40
3900	Sprayed fiber	8	1-1/2	2	6.3	3.93	6.50	10.43
3950	Direct application		2	3	8.3	5.40	9	14.40
4000			2-1/2	4	10.4	7	11.65	18.65
4050		10	1-1/2	2	7.9	4.75	7.90	12.65
4100			2	3	10.5	6.50	10.75	17.25
4150			2-1/2	4	13.1	8.40	13.85	22.25
4500	Perlite plaster	8	1	2	18	8.15	29	37.15
4550	On metal lath		1-3/8	3	23	8.85	31.50	40.35
4600			1-3/4	4	35	10.40	38.50	48.90
4650	Perlite plaster	10	1	2	21	9.30	33.50	42.80
4700			1-3/8	3	27	10.45	38	48.45
4750			1-3/4	4	41	11.70	44	55.70

B1010 Floor Construction

B1010 720 Steel Column Fireproofing

	ENCASEMENT SYSTEM	COLUMN SIZE (IN.)	THICKNESS (IN.)	FIRE RATING (HRS.)	WEIGHT (P.L.F.)	COST PER V.L.F.		
						MAT.	INST.	TOTAL
5100	5/8" gypsum plaster	8	1	1-1/2	20	7.40	23.50	30.90
5150	On 3/8" gypsum lath	10	1	1-1/2	24	8.80	28	36.80
5200		14	1	1-1/2	33	10.75	34.50	45.25
5250	1"perlite plaster	8	1-3/8	2	23	8.10	25	33.10
5300	On 3/8" gypsum lath	10	1-3/8	2	28	9.30	28.50	37.80
5350		14	1-3/8	2	37	11.60	36.50	48.10
5400	1-3/8" perlite plaster	8	1-3/4	3	27	8.50	28	36.50
5450	On 3/8" gypsum lath	10	1-3/4	3	33	9.75	32.50	42.25
5500		14	1-3/4	3	43	12.15	40.50	52.65
5550	Concrete masonry	8	4-3/4	4	126	10.65	31	41.65
5600	Units 4" thick	10	4-3/4	4	166	13.35	39	52.35
5650	75% solid	14	4-3/4	4	262	16	46.50	62.50

B2010 Exterior Walls

This page illustrates masonry cleaning and restoration systems, including staging, cleaning and repointing.

System Components	QUANTITY	UNIT	COST PER S.F. MAT.	COST PER S.F. INST.	COST PER S.F. TOTAL
SYSTEM B2010 143 1300					
Repoint existing building, brick, common bond, high pressure cleaning,					
Water only, soft old mortar.		S.F.			
Scaffolding,steel tubular,bldg ext wall face, labor only, 1 to 5 stories	.010	S.F.		1.98	1.98
Cleaning, high pressure water only	1.100	S.F.		1.70	1.70
Repoint, brick common bond	1.000	S.F.	.58	5.45	6.03
TOTAL			.58	9.13	9.71

B2010 143	Masonry Restoration - Cleaning	COST PER S.F. MAT.	COST PER S.F. INST.	COST PER S.F. TOTAL
1000	Repoint existing building, high pressure cleaning, water only,			
1100	soft old mortar			
1200	For alternate masonry surfaces:			
1300	Brick, common bond	.58	9.15	9.73
1400	Flemish bond	.61	9.50	10.11
1500	English bond	.61	10.10	10.71
1700	Stone, 2′ x 2′ blocks	.77	6.95	7.72
1800	2′ x 4′ blocks	.58	6.15	6.73
2000	Add to above prices for alternate cleaning systems:			
2100	Chemical brush and wash	.04	.59	.63
2200	High pressure chemical and water	.04	.58	.62
2300	Sandblasting, wet system	.58	1.58	2.16
2400	Dry system	.54	.93	1.47
2500	Steam cleaning		.47	.47

B3010 Roof Coverings

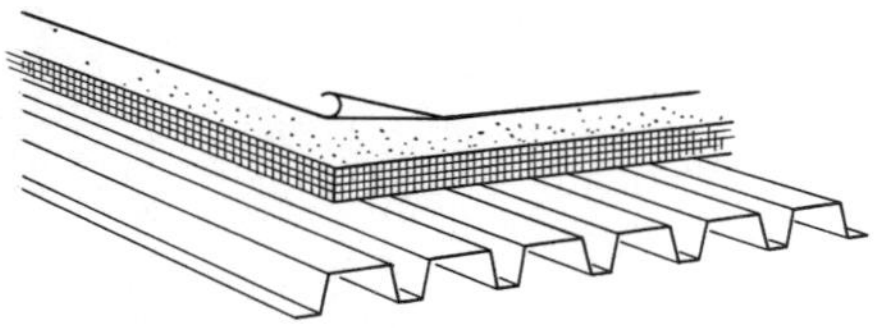

Fully Adhered

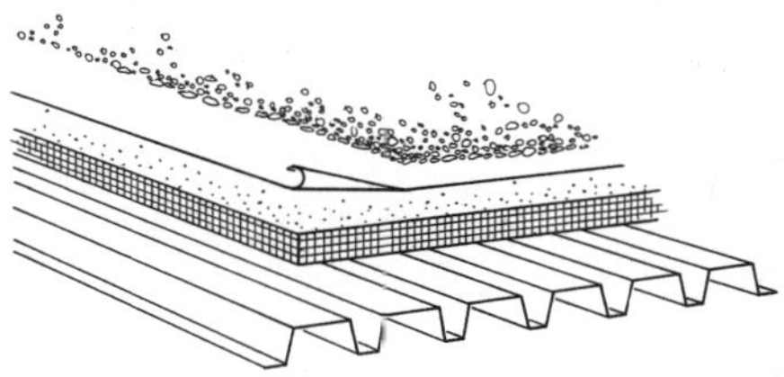

Ballasted

The systems listed below reflect only the cost for the single ply membrane.

For additional components see:

Insulation	B3010 320
Base Flashing	B3010 410
Roof Edge	B3010 420
Roof Openings	B3020 210

B3010 120	Single Ply Membrane	COST PER S.F.		
		MAT.	INST.	TOTAL
1000	CSPE (Chlorosulfonated polyethylene), 35 mils, fully adhered	2.04	.93	2.97
1100	Loosely laid with stone ballast	2.12	.48	2.60
2000	EPDM (Ethylene propylene diene monomer), 45 mils, fully adhered	1.02	.93	1.95
2100	Loosely laid with stone ballast	.81	.48	1.29
2200	Mechanically fastened with batten strips	.71	.69	1.40
3300	60 mils, fully adhered	1.22	.93	2.15
3400	Loosely laid with stone ballast	1.02	.48	1.50
3500	Mechanically fastened with batten strips	.91	.69	1.60
4000	Modified bit., SBS modified, granule surface cap sheet, mopped, 150 mils	.58	1.84	2.42
4100	Smooth surface cap sheet, mopped, 145 mils	.54	1.75	2.29
4500	APP modified, granule surface cap sheet, torched, 180 mils	.65	1.21	1.86
4600	Smooth surface cap sheet, torched, 170 mils	.61	1.15	1.76
5000	PIB (Polyisobutylene), 100 mils, fully adhered with contact cement	2.58	.93	3.51
5100	Loosely laid with stone ballast	1.99	.48	2.47
5200	Partially adhered with adhesive	2.50	.69	3.19
5300	Hot asphalt attachment	2.38	.69	3.07
6000	Reinforced PVC, 48 mils, loose laid and ballasted with stone	1.28	.48	1.76
6100	Partially adhered with mechanical fasteners	1.17	.69	1.86
6200	Fully adhered with adhesive	1.71	.93	2.64
6300	Reinforced PVC, 60 mils, loose laid and ballasted with stone	1.30	.48	1.78
6400	Partially adhered with mechanical fasteners	1.19	.69	1.88
6500	Fully adhered with adhesive	1.73	.93	2.66

C Interiors

C1010 Partitions

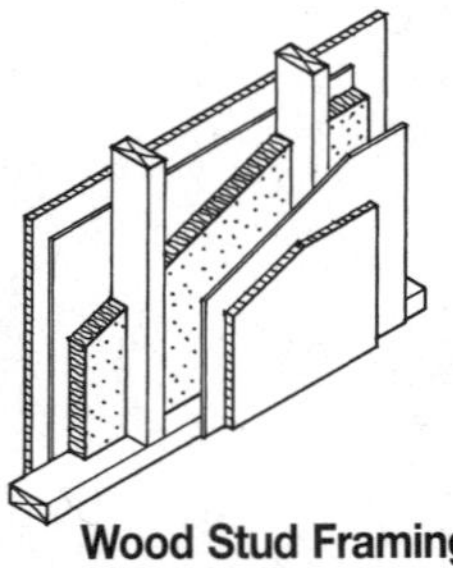

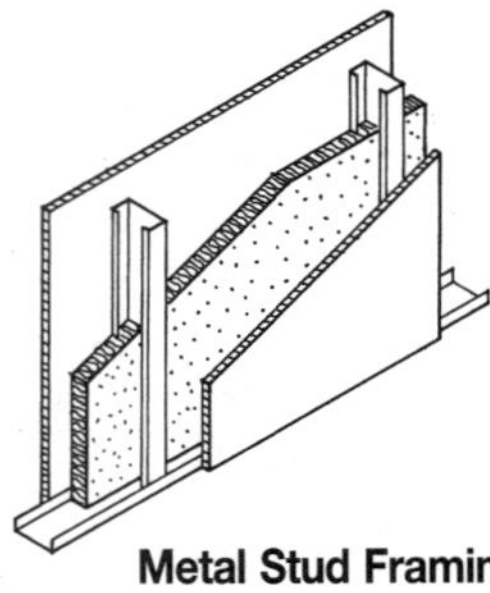

The Drywall Partitions/Stud Framing Systems are defined by type of drywall and number of layers, type and spacing of stud framing, and treatment on the opposite face. Components include taping and finishing.

Cost differences between regular and fire resistant drywall are negligible, and terminology is interchangeable. In some cases fiberglass insulation is included for additional sound deadening.

System Components	QUANTITY	UNIT	COST PER S.F. MAT.	INST.	TOTAL
SYSTEM C1010 124 1250					
DRYWALL PARTITION,5/8″ F.R.1 SIDE,5/8″ REG.1 SIDE,2″X4″STUDS,16″ O.C.					
Gypsum plasterboard, nailed/screwed to studs, 5/8″F.R. fire resistant	1.000	S.F.	.40	.53	.93
Gypsum plasterboard, nailed/screwed to studs, 5/8″ regular	1.000	S.F.	.37	.53	.90
Taping and finishing joints	2.000	S.F.	.10	1.06	1.16
Framing, 2 x 4 studs @ 16″ O.C., 10′ high	1.000	S.F.	.34	1.06	1.40
TOTAL			1.21	3.18	4.39

C1010 124 Drywall Partitions/Wood Stud Framing

	FACE LAYER	BASE LAYER	FRAMING	OPPOSITE FACE	INSULATION	COST PER S.F. MAT.	INST.	TOTAL
1202	5/8″ FR drywall	none	2 x 4, @ 16″ O.C.	same	0	1.27	3.18	4.45
1250				5/8″ reg. drywall	0	1.21	3.18	4.39
1302				nothing	0	.82	2.12	2.94
1402		1/4″ SD gypsum	2 x 4 @ 16″ O.C.	same	1-1/2″ fiberglass	2.79	4.89	7.68
1452				5/8″ FR drywall	1-1/2″ fiberglass	2.38	4.30	6.68
1502				nothing	1-1/2″ fiberglass	1.93	3.24	5.17
1602		resil. channels	2 x 4 @ 16″, O.C.	same	1-1/2″ fiberglass	2.32	6.20	8.52
1652				5/8″ FR drywall	1-1/2″ fiberglass	2.15	4.95	7.10
1702				nothing	1-1/2″ fiberglass	1.70	3.89	5.59
1802		5/8″ FR drywall	2 x 4 @ 24″ O.C.	same	0	1.98	4.03	6.01
1852				5/8″ FR drywall	0	1.58	3.50	5.08
1902				nothing	0	1.13	2.44	3.57
1952		5/8″ FR drywall	2 x 4, 16″ O.C.	same	0	2.07	4.24	6.31
1957				5/8″ FR drywall	0	1.67	3.71	5.38
2002				nothing	0	1.22	2.65	3.87
2012		5/8″ FR drywall	staggered, 6″ plate	same	0	2.44	5.30	7.74
2017				5/8″ FR drywall	0	2.04	4.78	6.82
2022				nothing	0	1.59	3.72	5.31
2202		5/8″ FR drywall	2 rows-2 x 4	same	2″ fiberglass	3.20	5.80	9
2252			16″O.C.	5/8″ FR drywall	2″ fiberglass	2.80	5.30	8.10
2302				nothing	2″ fiberglass	2.35	4.23	6.58
2402	5/8″ WR drywall	none	2 x 4, @ 16″ O.C.	same	0	1.51	3.18	4.69
2452				5/8″ FR drywall	0	1.39	3.18	4.57
2502				nothing	0	.94	2.12	3.06
2602		5/8″ FR drywall	2 x 4, @ 24″ O.C.	same	0	2.22	4.03	6.25
2652				5/8″ FR drywall	0	1.70	3.50	5.20
2702				nothing	0	1.25	2.44	3.69

C1010 Partitions

C1010 128	Drywall Components	COST PER S.F.		
		MAT.	INST.	TOTAL
0140	Metal studs, 24" O.C. including track, load bearing, 18 gage, 2-1/2"	.87	.99	1.86
0160	3-5/8"	1.03	1.01	2.04
0180	4"	.93	1.03	1.96
0200	6"	1.38	1.05	2.43
0220	16 gage, 2-1/2"	1	1.13	2.13
0240	3-5/8"	1.20	1.15	2.35
0260	4"	1.27	1.17	2.44
0280	6"	1.60	1.20	2.80
0300	Non load bearing, 25 gage, 1-5/8"	.25	.69	.94
0340	3-5/8"	.36	.71	1.07
0360	4"	.39	.71	1.10
0380	6"	.55	.73	1.28
0400	20 gage, 2-1/2"	.48	.88	1.36
0420	3-5/8"	.52	.89	1.41
0440	4"	.61	.89	1.50
0460	6"	.72	.91	1.63
0542	Wood studs including blocking, shoe and double top plate, 2"x4", 12"O.C.	.46	1.32	1.78
0562	16" O.C.	.37	1.06	1.43
0582	24" O.C.	.28	.85	1.13
0602	2"x6", 12" O.C.	.82	1.51	2.33
0622	16" O.C.	.66	1.17	1.83
0641	24" O.C.	.51	.92	1.43
0642	Furring one side only, steel channels, 3/4", 12" O.C.	.38	1.91	2.29
0644	16" O.C.	.34	1.70	2.04
0646	24" O.C.	.23	1.28	1.51
0647	1-1/2" , 12" O.C.	.51	2.14	2.65
0648	16" O.C.	.46	1.87	2.33
0649	24" O.C.	.30	1.47	1.77
0656	Wood strips, 1" x 3", on wood, 12" O.C.	.38	.96	1.34
0657	16" O.C.	.29	.72	1.01
0658	On masonry, 12" O.C.	.38	1.07	1.45
0659	16" O.C.	.29	.80	1.09
0660	On concrete, 12" O.C.	.38	2.03	2.41
0661	16" O.C.	.29	1.52	1.81
0662	Gypsum board, one face only, exterior sheathing, 1/2"	.67	1.01	1.68
0680	Interior, fire resistant, 1/2"	.35	.53	.88
0700	5/8"	.40	.53	.93
0720	Sound deadening board 1/4"	.41	.59	1
0740	Standard drywall 3/8"	.34	.53	.87
0760	1/2"	.34	.53	.87
0780	5/8"	.37	.53	.90
0800	Tongue & groove coreboard 1"	.97	2.20	3.17
0820	Water resistant, 1/2"	.53	.53	1.06
0840	5/8"	.52	.53	1.05
0860	Add for the following:, foil backing	.15		.15
0880	Fiberglass insulation, 3-1/2"	.63	.39	1.02
0900	6"	.85	.39	1.24
0920	Rigid insulation 1"	.54	.53	1.07
0940	Resilient furring @ 16" O.C.	.24	1.66	1.90
0960	Taping and finishing	.05	.53	.58
0980	Texture spray	.09	.56	.65
1000	Thin coat plaster	.09	.66	.75
1042	2"x4" staggered studs 2"x6" plates & blocking	.04	.12	.16

C1020 Interior Doors

Steel Door, Half Glass
Steel Frame

Steel Door, Flush
Steel Frame

Wood Door, Flush
Wood Frame

The Metal Door/Metal Frame Systems are defined as follows: door type, design and size, frame type and depth. Included in the components for each system is painting the door and frame. No hardware has been included in the systems.

System Components	QUANTITY	UNIT	COST EACH MAT.	INST.	TOTAL
SYSTEM C1020 114 1200					
STEEL DOOR, HOLLOW, 20 GA., HALF GLASS, 2′-8″X6′-8″, D.W.FRAME, 4-7/8″ DP					
Steel door, flush, hollow core, 1-3/4″ thick,half glass, 20 ga.,2′-8″x6′-8″	1.000	Ea.	475	53	528
Steel frame, KD, 16 ga., drywall, 4-7/8″ deep, 2′-8″x6′-8″, single	1.000	Ea.	122	70.50	192.50
Float glass, 3/16″ thick, clear, tempered	5.000	S.F.	33.25	38.25	71.50
Paint door and frame each side, primer	1.000	Ea.	5.16	90	95.16
Paint door and frame each side, 2 coats	1.000	Ea.	8.14	150	158.14
TOTAL			643.55	401.75	1,045.30

C1020 114 Metal Door/Metal Frame

	TYPE	DESIGN	SIZE	FRAME	DEPTH	COST EACH MAT.	INST.	TOTAL
1000	Flush-hollow	20 ga. full panel	2′-8″ x 6′-8″	drywall K.D.	4-7/8″	470	370	840
1020				butt welded	8-3/4″	580	415	995
1160			6′-0″ x 7′-0″	drywall K.D.	4-7/8″	920	595	1,515
1180				butt welded	8-3/4″	1,050	645	1,695
1200		20 ga. half glass	2′-8″ x 6′-8″	drywall K.D.	4-7/8″	645	400	1,045
1220				butt welded	8-3/4″	755	450	1,205
1360			6′-0″ x 7′-0″	drywall K.D.	4-7/8″	1,200	665	1,865
1380				butt welded	8-3/4″	1,325	715	2,040
1800		18 ga. full panel	2′-8″ x 6′-8″	drywall K.D.	4-7/8″	475	375	850
1820				butt welded	8-3/4″	585	420	1,005
1960			6′-0″ x 7′-0″	drywall K.D.	4-7/8″	1,000	595	1,595
1980				butt welded	8-3/4″	1,150	645	1,795
2000		18 ga. half glass	2′-8″ x 6′-8″	drywall K.D.	4-7/8″	675	405	1,080
2020				butt welded	8-3/4″	785	455	1,240
2160			6′-0″ x 7′-0″	drywall K.D.	4-7/8″	1,300	675	1,975
2180				butt welded	8-3/4″	1,425	720	2,145

C1020 Interior Doors

C1020 310	Hardware	COST EACH MAT.	COST EACH INST.	COST EACH TOTAL
0060	**HINGES**			
0080				
0100	Full mortise, low frequency, steel base, 4-1/2" x 4-1/2", USP	6.65		6.65
0120	5" x 5" USP	16.50		16.50
0140	6" x 6" USP	33		33
0160	Average frequency, steel base, 4-1/2" x 4-1/2" USP	13.50		13.50
0180	5" x 5" USP	22.50		22.50
0200	6" x 6" USP	48		48
0220	High frequency, steel base, 4-1/2" x 4-1/2", USP	32.50		32.50
0240	5" x 5" USP	30		30
0260	6" x 6" USP	73.50		73.50
0280				
0300	**LOCKSETS**			
0320				
0340	Heavy duty cylindrical, passage door			
0360	Non-keyed, passage	53.50	44	97.50
0380	Privacy	67	44	111
0400	Keyed, single cylinder function	92	53	145
0420	Hotel	132	66	198
0440				
0460	For re-core cylinder, add	37.50		37.50
0480				
0500	**CLOSERS**			
0520				
0540	Rack & pinion			
0560	Adjustable backcheck, 3 way mount, all sizes, regular arm	172	88	260
0580	Hold open arm	172	88	260
0600	Fusible link	143	81	224
0620	Non-sized, regular arm	162	88	250
0640	4 way mount, non-sized, regular arm	229	88	317
0660	Hold open arm	235	88	323
0680				
0700	**PUSH, PULL**			
0720				
0740	Push plate, aluminum	9.50	44	53.50
0760	Bronze	22.50	44	66.50
0780	Pull handle, push bar, aluminum	139	48	187
0800	Bronze	180	53	233
0810				
0820	**PANIC DEVICES**			
0840				
0860	Narrow stile, rim mounted, bar, exit only	680	88	768
0880	Outside key & pull	740	106	846
0900	Bar and vertical rod, exit only	700	106	806
0920	Outside key & pull	700	132	832
0940	Bar and concealed rod, exit only	815	176	991
0960	Mortise, bar, exit only	625	132	757
0980	Touch bar, exit only	910	132	1,042
1000				
1020	**WEATHERSTRIPPING**			
1040				
1060	Interlocking, 3' x 7', zinc	17.05	176	193.05
1080	Spring type, 3' x 7', bronze	39	176	215

C3010 Wall Finishes

C3010 230	Paint & Covering	COST PER S.F.		
		MAT.	INST.	TOTAL
0060	Painting, interior on plaster and drywall, brushwork, primer & 1 coat	.11	.69	.80
0080	Primer & 2 coats	.17	.92	1.09
0100	Primer & 3 coats	.24	1.13	1.37
0120	Walls & ceilings, roller work, primer & 1 coat	.11	.46	.57
0140	Primer & 2 coats	.17	.59	.76
0160	Woodwork incl. puttying, brushwork, primer & 1 coat	.11	1	1.11
0180	Primer & 2 coats	.17	1.32	1.49
0200	Primer & 3 coats	.24	1.80	2.04
0260	Cabinets and casework, enamel, primer & 1 coat	.12	1.13	1.25
0280	Primer & 2 coats	.18	1.38	1.56
0300	Masonry or concrete, latex, brushwork, primer & 1 coat	.23	.94	1.17
0320	Primer & 2 coats	.30	1.34	1.64
0340	Addition for block filler	.15	1.15	1.30
0380	Fireproof paints, intumescent, 1/8" thick 3/4 hour	2.14	.92	3.06
0420	7/16" thick 2 hour	6.15	3.21	9.36
0440	1-1/16" thick 3 hour	10.05	6.45	16.50
0500	Gratings, primer & 1 coat	.26	1.41	1.67
0600	Pipes over 12" diameter	.49	4.50	4.99
0720	Heavy framing 50-100 S.F./Ton	.08	.86	.94
0740	Spraywork, light framing 300-500 S.F./Ton	.09	.38	.47
0800	Varnish, interior wood trim, no sanding sealer & 1 coat	.08	1.13	1.21
0820	Hardwood floor, no sanding 2 coats	.17	.24	.41
0860	Maximum	.66	1.48	2.14
0880	Epoxy coatings, minimum	.41	.86	1.27
1080	Maximum	1.06	3.10	4.16
1100	High build epoxy 50 mil, minimum	.68	1.15	1.83
1120	Maximum	1.17	4.74	5.91
1140	Laminated epoxy with fiberglass minimum	.74	1.53	2.27
1160	Maximum	1.31	3.10	4.41
1180	Sprayed perlite or vermiculite 1/16" thick, minimum	.26	.15	.41
1200	Maximum	.75	.70	1.45
1260	Wall coatings, vinyl plastic, minimum	.33	.61	.94
1280	Maximum	.83	1.87	2.70
1300	Urethane on smooth surface, 2 coats, minimum	.26	.40	.66
1320	Maximum	.58	.68	1.26
1340	3 coats, minimum	.35	.54	.89
1360	Maximum	.79	.96	1.75
1380	Ceramic-like glazed coating, cementitious, minimum	.48	1.02	1.50
1420	Resin base, minimum	.33	.70	1.03
1440	Maximum	.54	1.36	1.90
1460	Wall coverings, aluminum foil	1.07	1.65	2.72
1500	Vinyl backing	5.70	1.89	7.59
1940	Ceramic tile, thin set, 4-1/4" x 4-1/4"	2.74	4.49	7.23
1960	12" x 12"	3.98	2.84	6.82

C3020 Floor Finishes

C3020 410	Tile & Covering	COST PER S.F.		
		MAT.	INST.	TOTAL
0060	Carpet tile, nylon, fusion bonded, 18" x 18" or 24" x 24", 24 oz.	3.55	.66	4.21
0080	35 oz.	4.16	.66	4.82
0100	42 oz.	5.55	.66	6.21
0140	Carpet, tufted, nylon, roll goods, 12' wide, 26 oz.	5.10	.75	5.85
0160	36 oz.	7.85	.75	8.60
0180	Woven, wool, 36 oz.	13.30	.75	14.05
0200	42 oz.	13.65	.75	14.40
0220	Padding, add to above, minimum	.56	.35	.91
0240	Maximum	.95	.35	1.30
0260	Composition flooring, acrylic, 1/4" thick	1.60	5.10	6.70
0280	3/8" thick	2.04	5.90	7.94
0300	Epoxy, minimum	2.74	3.92	6.66
0320	Maximum	3.32	5.40	8.72
0340	Epoxy terrazzo, minimum	5.80	5.70	11.50
0360	Maximum	8.50	7.55	16.05
0380	Mastic, hot laid, 1-1/2" thick, minimum	4.02	3.84	7.86
0400	Maximum	5.15	5.10	10.25
0420	Neoprene 1/4" thick, minimum	3.95	4.85	8.80
0440	Maximum	5.45	6.20	11.65
0460	Polyacrylate with ground granite 1/4", minimum	3.28	3.61	6.89
0480	Maximum	5.65	5.50	11.15
0500	Polyester with colored quart 2 chips 1/16", minimum	2.94	2.48	5.42
0520	Maximum	4.64	3.92	8.56
0540	Polyurethane with vinyl chips, minimum	7.70	2.48	10.18
0560	Maximum	11.10	3.08	14.18
0600	Concrete topping, granolithic concrete, 1/2" thick	.32	4.20	4.52
0620	1" thick	.63	4.30	4.93
0640	2" thick	1.26	4.95	6.21
0660	Heavy duty 3/4" thick, minimum	.43	6.50	6.93
0680	Maximum	.80	7.75	8.55
0700	For colors, add to above, minimum	.47	1.50	1.97
0720	Maximum	.78	1.65	2.43
0740	Exposed aggregate finish, minimum	.25	.77	1.02
0760	Maximum	.73	1.03	1.76
0780	Abrasives, .25 P.S.F. add to above, minimum	.48	.56	1.04
0800	Maximum	.67	.56	1.23
0820	Dust on coloring, add, minimum	.47	.37	.84
0840	Maximum	.78	.77	1.55
0860	Floor coloring using 0.6 psf powdered color, 1/2" integral, minimum	5.35	4.20	9.55
0880	Maximum	5.65	4.20	9.85
1020	Integral topping and finish, 1:1:2 mix, 3/16" thick	.11	2.47	2.58
1040	1/2" thick	.28	2.61	2.89
1060	3/4" thick	.43	2.91	3.34
1080	1" thick	.57	3.30	3.87
1100	Terrazzo, minimum	3.16	15.15	18.31
1120	Maximum	5.90	18.95	24.85
1280	Resilient, asphalt tile, 1/8" thick on concrete, minimum	1.24	1.19	2.43
1300	Maximum	1.36	1.19	2.55
1320	On wood, add for felt underlay	.22		.22

C3030 Ceiling Finishes

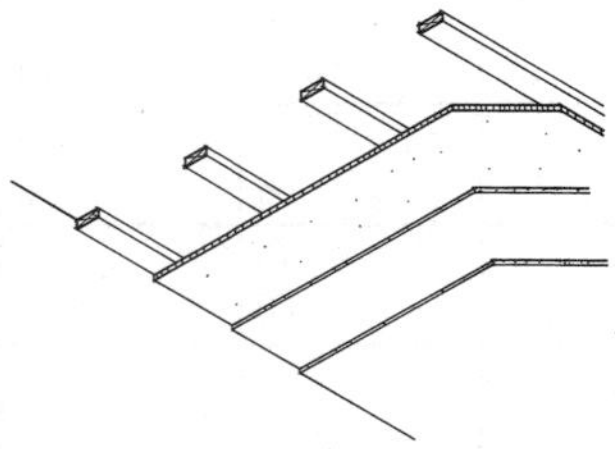

2 Coats of Plaster on Gypsum Lath on Wood Furring

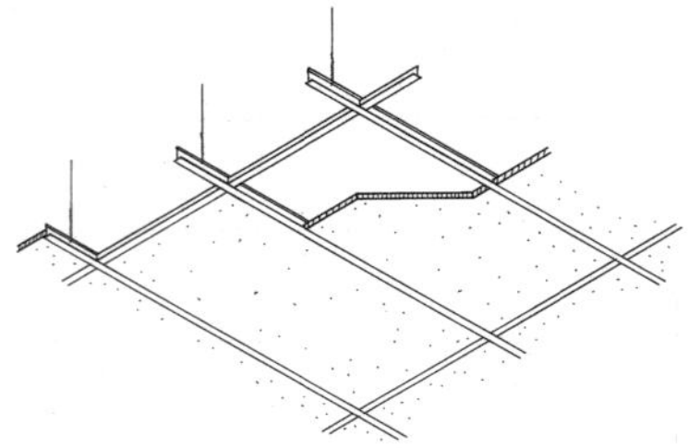

Fiberglass Board on Exposed Suspended Grid System

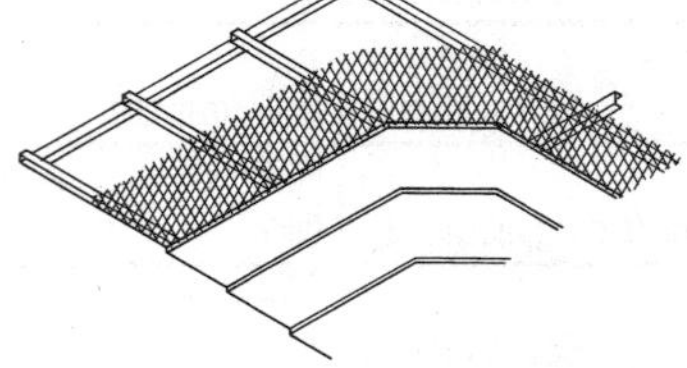

Plaster and Metal Lath on Metal Furring

System Components	QUANTITY	UNIT	COST PER S.F. MAT.	COST PER S.F. INST.	COST PER S.F. TOTAL
SYSTEM C3030 105 2400					
GYPSUM PLASTER, 2 COATS, 3/8″GYP. LATH, WOOD FURRING, FRAMING					
Gypsum plaster 2 coats no lath, on ceilings	.110	S.Y.	.46	2.83	3.29
Gypsum lath plain/perforated nailed, 3/8″ thick	.110	S.Y.	.67	.59	1.26
Add for ceiling installation	.110	S.Y.		.23	.23
Furring, 1″ x 3″ wood strips on wood joists	.750	L.F.	.29	1.13	1.42
Paint, primer and one coat	1.000	S.F.	.11	.46	.57
TOTAL			1.53	5.24	6.77

C3030 105 Plaster Ceilings

	TYPE	LATH	FURRING	SUPPORT		COST PER S.F. MAT.	COST PER S.F. INST.	COST PER S.F. TOTAL
2400	2 coat gypsum	3/8″ gypsum	1″x3″ wood, 16″ O.C.	wood		1.53	5.20	6.73
2500	Painted			masonry		1.53	5.35	6.88
2600				concrete		1.53	5.95	7.48
2700	3 coat gypsum	3.4# metal	1″x3″ wood, 16″ O.C.	wood		1.62	5.65	7.27
2800	Painted			masonry		1.62	5.75	7.37
2900				concrete		1.62	6.40	8.02
3000	2 coat perlite	3/8″ gypsum	1″x3″ wood, 16″ O.C.	wood		1.57	5.50	7.07
3100	Painted			masonry		1.57	5.60	7.17
3200				concrete		1.57	6.25	7.82
3300	3 coat perlite	3.4# metal	1″x3″ wood, 16″ O.C.	wood		1.46	5.60	7.06
3400	Painted			masonry		1.46	5.70	7.16
3500				concrete		1.46	6.35	7.81
3600	2 coat gypsum	3/8″ gypsum	3/4″ CRC, 12″ O.C.	1-1/2″ CRC, 48″O.C.		1.62	6.25	7.87
3700	Painted		3/4″ CRC, 16″ O.C.	1-1/2″ CRC, 48″O.C.		1.58	5.65	7.23
3800			3/4″ CRC, 24″ O.C.	1-1/2″ CRC, 48″O.C.		1.47	5.15	6.62
3900	2 coat perlite	3/8″ gypsum	3/4″ CRC, 12″ O.C.	1-1/2″ CRC, 48″O.C		1.66	6.75	8.41
4000	Painted		3/4″ CRC, 16″ O.C.	1-1/2″ CRC, 48″O.C.		1.62	6.15	7.77
4100			3/4″ CRC, 24″ O.C.	1-1/2″ CRC, 48″O.C.		1.51	5.65	7.16
4200	3 coat gypsum	3.4# metal	3/4″ CRC, 12″ O.C.	1-1/2″ CRC, 36″ O.C.		2.02	8.25	10.27
4300	Painted		3/4″ CRC, 16″ O.C.	1-1/2″ CRC, 36″ O.C.		1.98	7.65	9.63
4400			3/4″ CRC, 24″ O.C.	1-1/2″ CRC, 36″ O.C.		1.87	7.15	9.02
4500	3 coat perlite	3.4# metal	3/4″ CRC, 12″ O.C.	1-1/2″ CRC,36″ O.C.		2.13	9.05	11.18
4600	Painted		3/4″ CRC, 16″ O.C.	1-1/2″ CRC, 36″ O.C.		2.09	8.50	10.59
4700			3/4″ CRC, 24″ O.C.	1-1/2″ CRC, 36″ O.C.		1.98	8	9.98

D Services

D1010 Elevators and Lifts

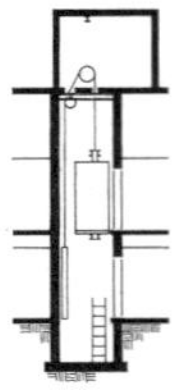

Geared traction elevators are the intermediate group both in cost and in operating areas. These are in buildings of four to fifteen floors and speed ranges from 200′ to 500′ per minute.

System Components

System Components	QUANTITY	UNIT	COST EACH MAT.	COST EACH INST.	COST EACH TOTAL
SYSTEM D1010 140 1600					
PASSENGER 2500 LB., 5 FLOORS, 200 FPM					
Passenger elevator, geared, 2000 lb. capacity, 4 stop, 200 FPM	1.000	Ea.	100,500	27,900	128,400
Over 40′ travel height, passenger elevator electric, add	10.000	V.L.F.	7,150	1,920	9,070
Passenger elevator, electric, 2500 lb. cap., add	1.000	Ea.	4,200		4,200
Over 4 stops, passenger elevator, electric, add	1.000	Stop	8,600	5,150	13,750
Hall lantern	5.000	Ea.	2,975	870	3,845
Maintenance agreement for passenger elevator, 9 months	1.000	Ea.	4,000		4,000
Position indicator at lobby	1.000	Ea.	685	43.50	728.50
TOTAL			128,110	35,883.50	163,993.50

D1010 140 Traction Geared Elevators

D1010 140	Traction Geared Elevators	COST EACH MAT.	COST EACH INST.	COST EACH TOTAL
1300	Passenger, 2000 Lb., 5 floors, 200 FPM	124,000	35,900	159,900
1500	15 floors, 350 FPM	300,500	108,500	409,000
1600	2500 Lb., 5 floors, 200 FPM	128,000	35,900	163,900
1800	15 floors, 350 FPM	305,000	108,500	413,500
1900	3000 Lb., 5 floors, 200 FPM	129,000	35,900	164,900
2100	15 floors, 350 FPM	305,500	108,500	414,000
2200	3500 Lb., 5 floors, 200 FPM	132,500	35,900	168,400
2400	15 floors, 350 FPM	309,000	108,500	417,500
2500	4000 Lb., 5 floors, 200 FPM	131,500	35,900	167,400
2700	15 floors, 350 FPM	308,500	108,500	417,000
2800	4500 Lb., 5 floors, 200 FPM	134,000	35,900	169,900
3000	15 floors, 350 FPM	310,500	108,500	419,000
3100	5000 Lb., 5 floors, 200 FPM	136,000	35,900	171,900
3300	15 floors, 350 FPM	312,500	108,500	421,000
4000	Hospital, 3500 Lb., 5 floors, 200 FPM	127,000	35,900	162,900
4200	15 floors, 350 FPM	301,500	108,500	410,000
4300	4000 Lb., 5 floors, 200 FPM	127,000	35,900	162,900
4500	15 floors, 350 FPM	301,500	108,500	410,000
4600	4500 Lb., 5 floors, 200 FPM	133,000	35,900	168,900
4800	15 floors, 350 FPM	307,500	108,500	416,000
4900	5000 Lb., 5 floors, 200 FPM	134,500	35,900	170,400
5100	15 floors, 350 FPM	309,000	108,500	417,500
6000	Freight, 4000 Lb., 5 floors, 50 FPM class 'B'	142,000	37,100	179,100
6200	15 floors, 200 FPM class 'B'	357,500	129,000	486,500
6300	8000 Lb., 5 floors, 50 FPM class 'B'	174,500	37,100	211,600
6500	15 floors, 200 FPM class 'B'	390,000	129,000	519,000
7000	10,000 Lb., 5 floors, 50 FPM class 'B'	192,500	37,100	229,600
7200	15 floors, 200 FPM class 'B'	556,000	129,000	685,000
8000	20,000 Lb., 5 floors, 50 FPM class 'B'	222,500	37,100	259,600
8200	15 floors, 200 FPM class 'B'	585,500	129,000	714,500

D1020 Escalators and Moving Walks

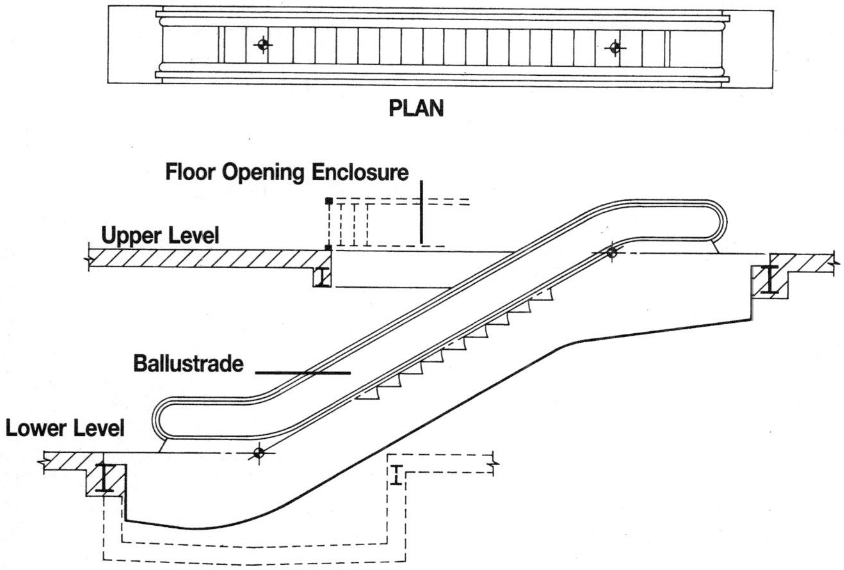

Moving stairs can be used for buildings where 600 or more people are to be carried to the second floor or beyond. Freight cannot be carried on escalators and at least one elevator must be available for this function.

Carrying capacity is 5000 to 8000 people per hour. Power requirement is 2 KW to 3 KW per hour and incline angle is 30°.

D1020 110 Moving Stairs

	TYPE	HEIGHT	WIDTH	BALUSTRADE MATERIAL		COST EACH		
						MAT.	INST.	TOTAL
0100	Escalator	10ft	32″	glass		99,500	38,700	138,200
0150				metal		105,500	38,700	144,200
0200			48″	glass		108,000	38,700	146,700
0250				metal		114,500	38,700	153,200
0300		15ft	32″	glass		105,500	45,200	150,700
0350				metal		114,500	45,200	159,700
0400			48″	glass		112,000	45,200	157,200
0450				metal		114,500	45,200	159,700
0500		20ft	32″	glass		111,000	55,500	166,500
0550				metal		118,000	55,500	173,500
0600			48″	glass		121,000	55,500	176,500
0650				metal		126,500	55,500	182,000
0700		25ft	32″	glass		117,500	67,500	185,000
0750				metal		126,500	67,500	194,000
0800			48″	glass		136,500	67,500	204,000
0850				metal		126,500	67,500	194,000

D1020 210 Moving Walks

	TYPE	STORY HEIGHT	DEGREE SLOPE	WIDTH	COST RANGE	COST PER L.F.		
						MAT.	INST.	TOTAL
1500	Ramp	10′-23′	12	3′-4″	minimum	2,525	760	3,285
1600					maximum	3,200	930	4,130
2000	Walk	0′	0	2′-0″	minimum	940	415	1,355
2500					maximum	1,300	615	1,915
3000					minimum	2,125	615	2,740
3500		0′	0	3′-4″	maximum	2,475	710	3,185

D2010 Plumbing Fixtures

Systems are complete with trim seat and rough-in (supply, waste and vent) for connection to supply branches and waste mains.

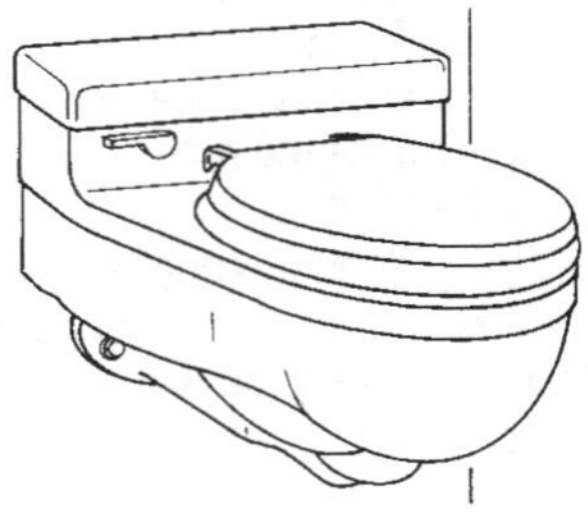

One Piece Wall Hung

Supply

Waste/Vent

Floor Mount

System Components	QUANTITY	UNIT	COST EACH MAT.	COST EACH INST.	COST EACH TOTAL
SYSTEM D2010 110 1880					
WATER CLOSET, VITREOUS CHINA, ELONGATED					
TANK TYPE, WALL HUNG, TWO PIECE		Ea.			
Water closet, tank type vit china wall hung 2 pc. w/seat supply & stop	1.000	Ea.	610	205	815
Pipe Steel galvanized, schedule 40, threaded, 2″ diam.	4.000	L.F.	45.60	68	113.60
Pipe, CI soil, no hub, cplg 10′ OC, hanger 5′ OC, 4″ diam.	2.000	L.F.	27.90	37.50	65.40
Pipe, coupling, standard coupling, CI soil, no hub, 4″ diam.	2.000	Ea.	45	66	111
Copper tubing type L solder joint, hangar 10′ O.C., 1/2″ diam.	6.000	L.F.	25.26	44.70	69.96
Wrought copper 90° elbow for solder joints 1/2″ diam.	2.000	Ea.	2.70	60	62.70
Wrought copper Tee for solder joints 1/2″ diam.	1.000	Ea.	2.32	46.50	48.82
Supports/carrier, water closet, siphon jet, horiz, single, 4″ waste	1.000	Ea.	550	113	663
TOTAL			1,308.78	640.70	1,949.48

D2010 110	Water Closet Systems		COST EACH MAT.	COST EACH INST.	COST EACH TOTAL
1800	Water closet, vitreous china, elongated				
1840	Tank type, wall hung				
1880	Close coupled two piece	R224000-30	1,300	640	1,940
1920	Floor mount, one piece		915	680	1,595
1960	One piece low profile		775	680	1,455
2000	Two piece close coupled		535	680	1,215
2040	Bowl only with flush valve				
2080	Wall hung		1,175	730	1,905
2120	Floor mount		720	695	1,415
2160	Floor mount, ADA compliant with 18″ high bowl		685	710	1,395

D2010 Plumbing Fixtures

Systems are complete with trim, flush valve and rough-in (supply, waste and vent) for connection to supply branches and waste mains.

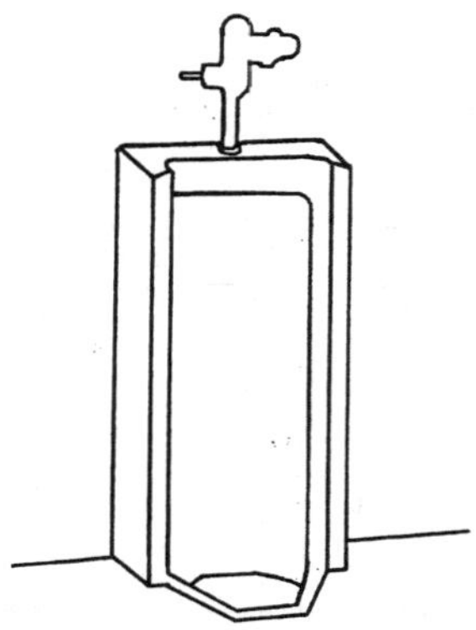

Stall Type

Supply

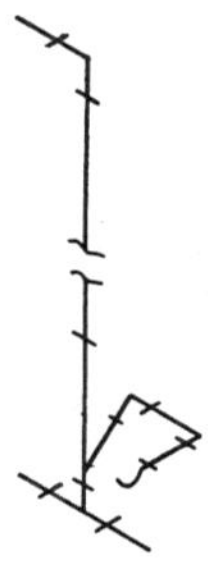

Waste/Vent

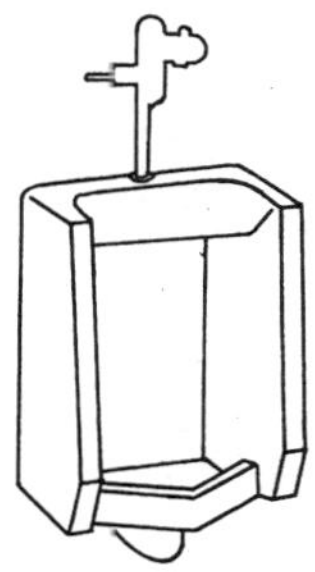

Wall Hung

System Components	QUANTITY	UNIT	COST EACH MAT.	COST EACH INST.	COST EACH TOTAL
SYSTEM D2010 210 2000					
URINAL, VITREOUS CHINA, WALL HUNG					
Urinal, wall hung, vitreous china, incl. hanger	1.000	Ea.	273	365	638
Pipe, steel, galvanized, schedule 40, threaded, 1-1/2" diam.	5.000	L.F.	42.75	68	110.75
Copper tubing type DWV, solder joint, hangers 10′ OC, 2" diam.	3.000	L.F.	61.50	41.25	102.75
Combination Y & 1/8 bend for CI soil pipe, no hub, 3" diam.	1.000	Ea.	15.65		15.65
Pipe, CI, no hub, cplg. 10′ OC, hanger 5′ OC, 3" diam.	4.000	L.F.	43.20	68	111.20
Pipe coupling standard, CI soil, no hub, 3" diam.	3.000	Ea.	38	57	95
Copper tubing type L, solder joint, hanger 10′ OC 3/4" diam.	5.000	L.F.	33	39.75	72.75
Wrought copper 90° elbow for solder joints 3/4" diam.	1.000	Ea.	3.04	32	35.04
Wrought copper Tee for solder joints, 3/4" diam.	1.000	Ea.	5.60	50.50	56.10
TOTAL			515.74	721.50	1,237.24

D2010 210	Urinal Systems		COST EACH MAT.	COST EACH INST.	COST EACH TOTAL
2000	Urinal, vitreous china, wall hung	R224000-30	515	720	1,235
2040	Stall type		1,025	860	1,885

D2010 Plumbing Fixtures

Systems are complete with trim and rough-in (supply, waste and vent) to connect to supply branches and waste mains.

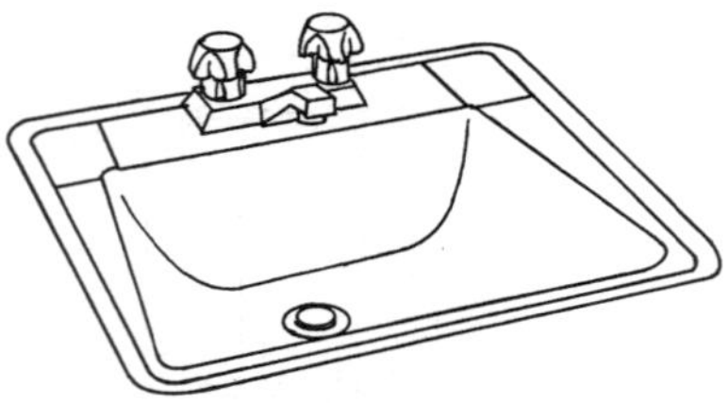

Vanity Top

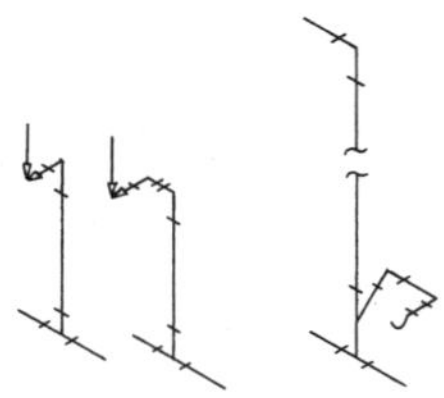

Supply Waste/Vent

Wall Hung

System Components	QUANTITY	UNIT	COST EACH		
			MAT.	INST.	TOTAL
SYSTEM D2010 310 1560					
LAVATORY W/TRIM, VANITY TOP, P.E. ON C.I., 20″ X 18″					
Lavatory w/trim, PE on CI, white, vanity top, 20″ x 18″ oval	1.000	Ea.	254	170	424
Pipe, steel, galvanized, schedule 40, threaded, 1-1/4″ diam.	4.000	L.F.	28.80	49	77.80
Copper tubing type DWV, solder joint, hanger 10′ OC 1-1/4″ diam.	4.000	L.F.	48	40.40	88.40
Wrought copper DWV, Tee, sanitary, 1-1/4″ diam.	1.000	Ea.	26	67	93
P trap w/cleanout, 20 ga., 1-1/4″ diam.	1.000	Ea.	79	33.50	112.50
Copper tubing type L, solder joint, hanger 10′ OC 1/2″ diam.	10.000	L.F.	42.10	74.50	116.60
Wrought copper 90° elbow for solder joints 1/2″ diam.	2.000	Ea.	2.70	60	62.70
Wrought copper Tee for solder joints, 1/2″ diam.	2.000	Ea.	4.64	93	97.64
Stop, chrome, angle supply, 1/2″ diam.	2.000	Ea.	16.20	55	71.20
TOTAL			501.44	642.40	1,143.84

D2010 310	Lavatory Systems	COST EACH		
		MAT.	INST.	TOTAL
1560	Lavatory w/trim, vanity top, PE on CI, 20″ x 18″, Vanity top by others.	500	640	1,140
1600	19″ x 16″ oval	420	640	1,060
1640	18″ round	485	640	1,125
1680	Cultured marble, 19″ x 17″	435	640	1,075
1720	25″ x 19″	445	640	1,085
1760	Stainless, self-rimming, 25″ x 22″	605	640	1,245
1800	17″ x 22″ R224000-30	595	640	1,235
1840	Steel enameled, 20″ x 17″	420	660	1,080
1880	19″ round	420	660	1,080
1920	Vitreous china, 20″ x 16″	530	675	1,205
1960	19″ x 16″	535	675	1,210
2000	22″ x 13″	540	675	1,215
2040	Wall hung, PE on CI, 18″ x 15″	760	710	1,470
2080	19″ x 17″	775	710	1,485
2120	20″ x 18″	725	710	1,435
2160	Vitreous china, 18″ x 15″	655	730	1,385
2200	19″ x 17″	590	730	1,320
2240	24″ x 20″	925	730	1,655

D2010 Plumbing Fixtures

Systems are complete with trim and rough-in (supply, waste and vent) to connect to supply branches and waste mains.

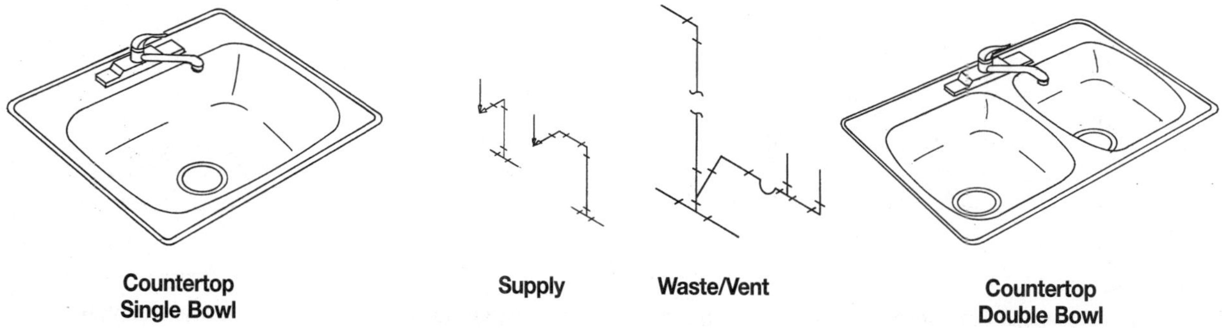

System Components	QUANTITY	UNIT	COST EACH MAT.	INST.	TOTAL
SYSTEM D2010 410 1720					
KITCHEN SINK W/TRIM, COUNTERTOP, P.E. ON C.I., 24″ X 21″, SINGLE BOWL					
Kitchen sink, countertop, PE on CI, 1 bowl, 24″ x 21″ OD	1.000	Ea.	299	194	493
Pipe, steel, galvanized, schedule 40, threaded, 1-1/4″ diam.	4.000	L.F.	28.80	49	77.80
Copper tubing, type DWV, solder, hangers 10′ OC 1-1/2″ diam.	6.000	L.F.	90.60	67.20	157.80
Wrought copper, DWV, Tee, sanitary, 1-1/2″ diam.	1.000	Ea.	32	75.50	107.50
P trap, standard, copper, 1-1/2″ diam.	1.000	Ea.	76.50	35.50	112
Copper tubing, type L, solder joints, hangers 10′ OC 1/2″ diam.	10.000	L.F.	42.10	74.50	116.60
Wrought copper 90° elbow for solder joints 1/2″ diam.	2.000	Ea.	2.70	60	62.70
Wrought copper Tee for solder joints, 1/2″ diam.	2.000	Ea.	4.64	93	97.64
Stop, angle supply, chrome, 1/2″ CTS	2.000	Ea.	16.20	55	71.20
TOTAL			592.54	703.70	1,296.24

D2010 410	Kitchen Sink Systems	COST EACH MAT.	INST.	TOTAL
1720	Kitchen sink w/trim, countertop, PE on CI, 24″x21″, single bowl	595	705	1,300
1760	30″ x 21″ single bowl	600	705	1,305
1800	32″ x 21″ double bowl	735	760	1,495
1840	42″ x 21″ double bowl	890	770	1,660
1880	Stainless steel, 19″ x 18″ single bowl	855	705	1,560
1920	25″ x 22″ single bowl	920	705	1,625
1960	33″ x 22″ double bowl	1,225	760	1,985
2000	43″ x 22″ double bowl	1,400	770	2,170
2040	44″ x 22″ triple bowl	1,425	800	2,225
2080	44″ x 24″ corner double bowl	1,075	770	1,845
2120	Steel, enameled, 24″ x 21″ single bowl	455	705	1,160
2160	32″ x 21″ double bowl	505	760	1,265
2240	Raised deck, PE on CI, 32″ x 21″, dual level, double bowl	660	965	1,625
2280	42″ x 21″ dual level, triple bowl	1,250	1,050	2,300

D2010 Plumbing Fixtures

Systems are complete with trim and rough-in (supply, waste and vent) to connect to supply branches and waste mains.

Recessed Bathtub

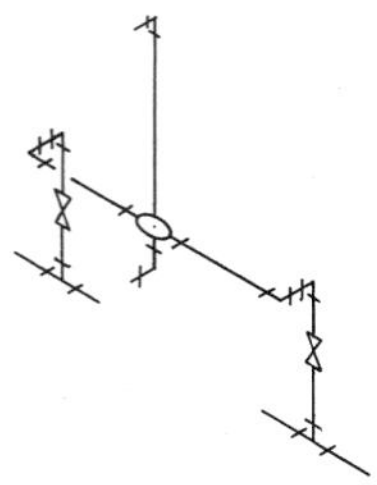
Supply

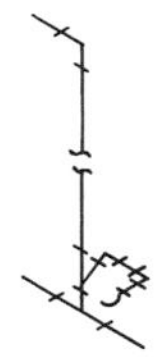
Waste/Vent

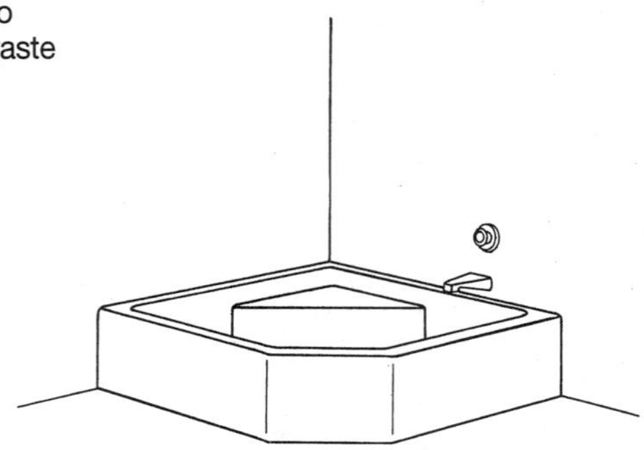
Corner Bathtub

System Components	QUANTITY	UNIT	COST EACH		
			MAT.	INST.	TOTAL
SYSTEM D2010 510 2000					
BATHTUB, RECESSED, PORCELAIN ENAMEL ON CAST IRON,, 48″ x 42″					
Bath tub, porcelain enamel on cast iron, w/fittings, 48″ x 42″	1.000	Ea.	1,975	272	2,247
Pipe, steel, galvanized, schedule 40, threaded, 1-1/4″ diam	4.000	L.F.	28.80	49	77.80
Pipe, CI no hub soil w/couplings 10′ OC, hangers 5′ OC, 4″ diam	3.000	L.F.	41.85	56.25	98.10
Combination Y and 1/8 bend for C.I. soil pipe, no hub, 4″ pipe size	1.000	Ea.	37.50		37.50
Drum trap, 3″ x 5″, copper, 1-1/2″ diam	1.000	Ea.	107	38	145
Copper tubing type L, solder joints, hangers 10′ OC 1/2″ diam	10.000	L.F.	42.10	74.50	116.60
Wrought copper 90° elbow, solder joints, 1/2″ diam	2.000	Ea.	2.70	60	62.70
Wrought copper Tee, solder joints, 1/2″ diam	2.000	Ea.	4.64	93	97.64
Stop, angle supply, 1/2″ diameter	2.000	Ea.	16.20	55	71.20
Copper tubing type DWV, solder joints, hanger 10′ OC 1-1/2″ diam	3.000	L.F.	45.30	33.60	78.90
Pipe coupling, standard, C.I. soil no hub, 4″ pipe size	2.000	Ea.	45	66	111
TOTAL			2,346.09	797.35	3,143.44

D2010 510	Bathtub Systems		COST EACH		
			MAT.	INST.	TOTAL
2000	Bathtub, recessed, P.E. on CI., 48″ x 42″	R224000 -30	2,350	795	3,145
2040	72″ x 36″		2,425	890	3,315
2080	Mat bottom, 5′ long		1,250	770	2,020
2120	5′-6″ long		1,750	795	2,545
2160	Corner, 48″ x 42″		2,350	770	3,120
2200	Formed steel, enameled, 4′-6″ long		770	715	1,485

D2010 Plumbing Fixtures

Systems are complete with trim, flush valve and rough-in (supply, waste and vent) for connection to supply branches and waste mains.

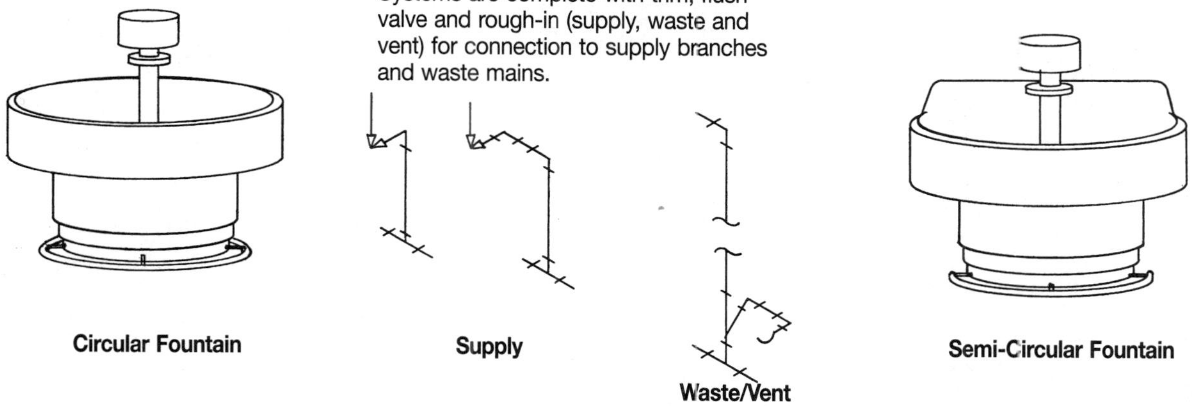

System Components	QUANTITY	UNIT	COST EACH MAT.	COST EACH INST.	COST EACH TOTAL
SYSTEM D2010 610 1760					
GROUP WASH FOUNTAIN, PRECAST TERRAZZO					
CIRCULAR, 36″ DIAMETER					
Wash fountain, group, precast terrazzo, foot control 36″ diam	1.000	Ea.	4,475	565	5,040
Copper tubing type DWV, solder joint, hanger 10′ OC, 2″ diam.	10.000	L.F.	205	137.50	342.50
P trap, standard, copper, 2″ diam.	1.000	Ea.	118	40.50	158.50
Wrought copper, Tee, sanitary, 2″ diam.	1.000	Ea.	37.50	86.50	124
Copper tubing type L, solder joint, hanger 10′ OC 1/2″ diam.	20.000	L.F.	84.20	149	233.20
Wrought copper 90° elbow for solder joints 1/2″ diam.	3.000	Ea.	4.05	90	94.05
Wrought copper Tee for solder joints, 1/2″ diam.	2.000	Ea.	4.64	93	97.64
TOTAL			4,928.39	1,161.50	6,089.89

D2010 610	Group Wash Fountain Systems	COST EACH MAT.	COST EACH INST.	COST EACH TOTAL
1740	Group wash fountain, precast terrazzo			
1760	Circular, 36″ diameter	4,925	1,150	6,075
1800	54″ diameter (R224000-30)	6,025	1,275	7,300
1840	Semi-circular, 36″ diameter	4,575	1,150	5,725
1880	54″ diameter	5,425	1,275	6,700
1960	Stainless steel, circular, 36″ diameter	4,825	1,075	5,900
2000	54″ diameter	6,125	1,200	7,325
2040	Semi-circular, 36″ diameter	4,275	1,075	5,350
2080	54″ diameter	5,350	1,200	6,550
2160	Thermoplastic, circular, 36″ diameter	3,725	880	4,605
2200	54″ diameter	4,275	1,025	5,300
2240	Semi-circular, 36″ diameter	3,475	880	4,355
2280	54″ diameter	4,150	1,025	5,175

D2010 Plumbing Fixtures

Systems are complete with trim and rough-in (supply, waste and vent) for connection to supply branches and waste mains.

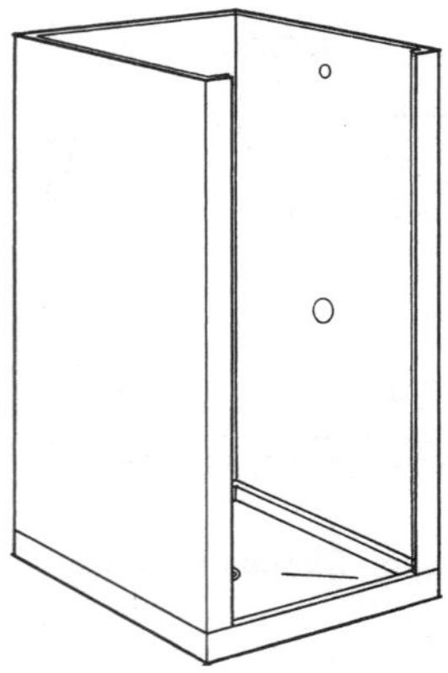

Three Wall

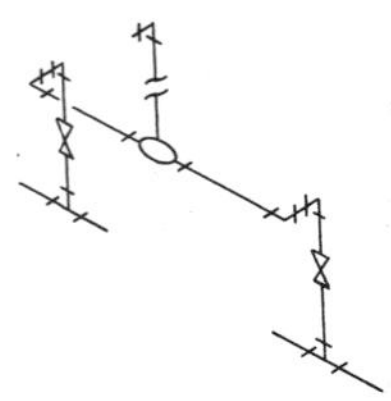

Supply

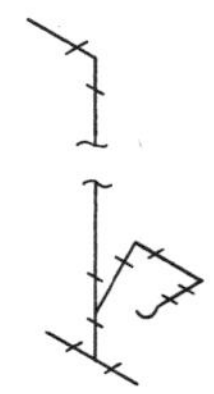

Waste/Vent

Corner Angle

System Components	QUANTITY	UNIT	COST EACH MAT.	COST EACH INST.	COST EACH TOTAL
SYSTEM D2010 710 1560					
SHOWER, STALL, BAKED ENAMEL, MOLDED STONE RECEPTOR, 30″ SQUARE					
Shower stall, enameled steel, molded stone receptor, 30″ square	1.000	Ea.	805	209	1,014
Copper tubing type DWV, solder joints, hangers 10′ OC, 2″ diam.	6.000	L.F.	90.60	67.20	157.80
Wrought copper DWV, Tee, sanitary, 2″ diam.	1.000	Ea.	32	75.50	107.50
Trap, standard, copper, 2″ diam.	1.000	Ea.	76.50	35.50	112
Copper tubing type L, solder joint, hanger 10′ OC 1/2″ diam.	16.000	L.F.	67.36	119.20	186.56
Wrought copper 90° elbow for solder joints 1/2″ diam.	3.000	Ea.	4.05	90	94.05
Wrought copper Tee for solder joints, 1/2″ diam.	2.000	Ea.	4.64	93	97.64
Stop and waste, straightway, bronze, solder joint 1/2″ diam	2.000	Ea.	6.18	50	56.18
TOTAL			1,086.33	739.40	1,825.73

D2010 710	Shower Systems		COST EACH MAT.	COST EACH INST.	COST EACH TOTAL
1560	Shower, stall, baked enamel, molded stone receptor, 30″ square		1,075	740	1,815
1600	32″ square		680	750	1,430
1640	Terrazzo receptor, 32″ square	R224000 -30	1,100	750	1,850
1680	36″ square		1,550	755	2,305
1720	36″ corner angle		1,675	755	2,430
1800	Fiberglass one piece, three walls, 32″ square		790	730	1,520
1840	36″ square		855	730	1,585
1880	Polypropylene, molded stone receptor, 30″ square		830	1,075	1,905
1920	32″ square		845	1,075	1,920
1960	Built-in head, arm, bypass, stops and handles		254	278	532

Systems are complete with trim and rough-in (supply, waste and vent) for connection to supply branches and waste mains.

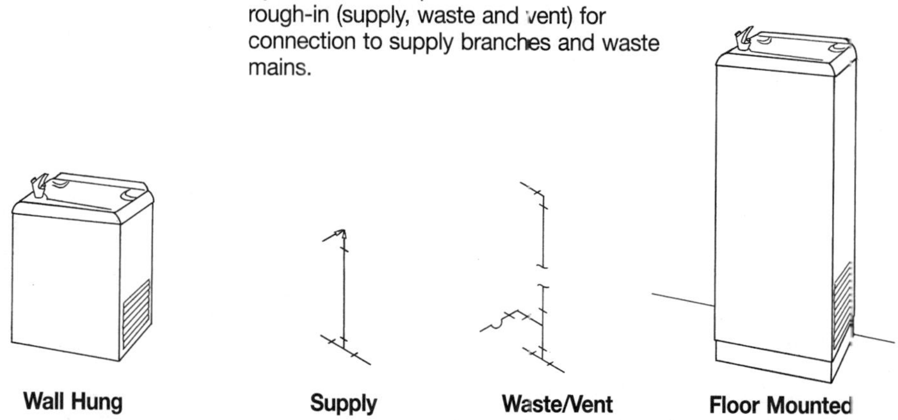

System Components	QUANTITY	UNIT	COST EACH MAT.	COST EACH INST.	COST EACH TOTAL
SYSTEM D2010 820 1840					
WATER COOLER, ELECTRIC, SELF CONTAINED, WALL HUNG, 8.2 GPH					
Water cooler, wall mounted, 8.2 GPH	1.000	Ea.	710	272	982
Copper tubing type DWV, solder joint, hanger 10′ OC 1-1/4″ diam.	4.000	L.F.	48	40.40	88.40
Wrought copper DWV, Tee, sanitary 1-1/4″ diam.	1.000	Ea.	26	67	93
P trap, copper drainage, 1-1/4″ diam.	1.000	Ea.	79	33.50	112.50
Copper tubing type L, solder joint, hanger 10′ OC 3/8″ diam.	5.000	L.F.	17.90	36	53.90
Wrought copper 90° elbow for solder joints 3/8″ diam.	1.000	Ea.	4.06	27.50	31.56
Wrought copper Tee for solder joints, 3/8″ diam.	1.000	Ea.	6.85	43	49.85
Stop and waste, straightway, bronze, solder, 3/8″ diam.	1.000	Ea.	6.75	25	31.75
TOTAL			898.56	544.40	1,442.96

D2010 820	Water Cooler Systems		COST EACH MAT.	COST EACH INST.	COST EACH TOTAL
1840	Water cooler, electric, wall hung, 8.2 GPH	R224000-30	900	545	1,445
1880	Dual height, 14.3 GPH		1,275	560	1,835
1920	Wheelchair type, 7.5 G.P.H.		2,000	545	2,545
1960	Semi recessed, 8.1 G.P.H.		1,250	545	1,795
2000	Full recessed, 8 G.P.H.		1,875	580	2,455
2040	Floor mounted, 14.3 G.P.H.		975	475	1,450
2080	Dual height, 14.3 G.P.H.		1,350	570	1,920
2120	Refrigerated compartment type, 1.5 G.P.H.		1,600	475	2,075

D2010 Plumbing Fixtures

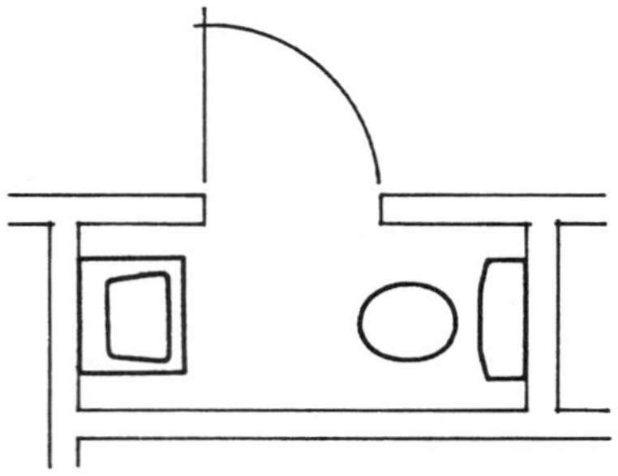

Two Fixture Bathroom Systems consisting of a lavatory, water closet, and rough-in service piping.

- Prices for plumbing and fixtures only.

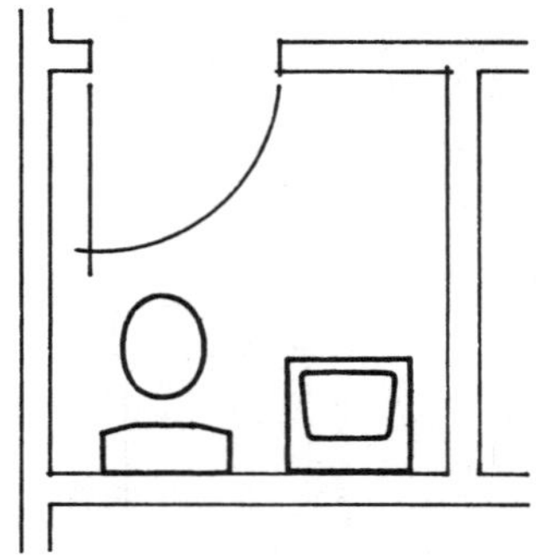

*Common wall is with an adjacent bathroom

System Components	QUANTITY	UNIT	COST EACH MAT.	COST EACH INST.	COST EACH TOTAL
SYSTEM D2010 920 1180					
BATHROOM, LAVATORY & WATER CLOSET, 2 WALL PLUMBING, STAND ALONE					
Water closet, 2 Pc. close cpld. vit .china flr. mntd. w/seat, supply & stop	1.000	Ea.	203	205	408
Water closet, rough-in waste & vent	1.000	Set	300	355	655
Lavatory w/ftngs., wall hung, white, PE on CI, 20″ x 18″	1.000	Ea.	325	136	461
Lavatory, rough-in waste & vent	1.000	Set	400	655	1,055
Copper tubing type L, solder joint, hanger 10′ OC 1/2″ diam.	10.000	L.F.	42.10	74.50	116.60
Pipe, steel, galvanized, schedule 40, threaded, 2″ diam.	12.000	L.F.	136.80	204	340.80
Pipe, CI soil, no hub, coupling 10′ OC, hanger 5′ OC, 4″ diam.	7.000	L.F.	88.55	138.60	227.15
TOTAL			1,495.45	1,768.10	3,263.55

D2010 920	Two Fixture Bathroom, Two Wall Plumbing	COST EACH MAT.	COST EACH INST.	COST EACH TOTAL
1180	Bathroom, lavatory & water closet, 2 wall plumbing, stand alone	1,500	1,775	3,275
1200	Share common plumbing wall*	1,350	1,525	2,875

D2010 922	Two Fixture Bathroom, One Wall Plumbing	COST EACH MAT.	COST EACH INST.	COST EACH TOTAL
2220	Bathroom, lavatory & water closet, one wall plumbing, stand alone	1,375	1,600	2,975
2240	Share common plumbing wall*	1,225	1,350	2,575

D3030 Cooling Generating Systems

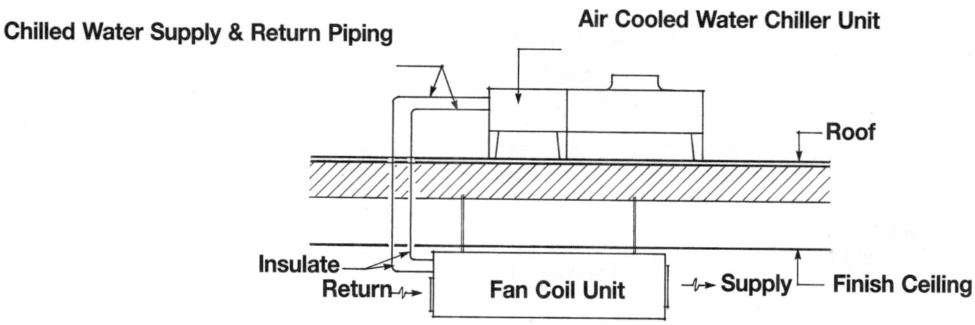

Design Assumptions: The chilled water, air cooled systems priced, utilize reciprocating hermetic compressors and propeller-type condenser fans. Piping with pumps and expansion tanks is included based on a two pipe system. No ducting is included and the fan-coil units are cooling only. Water treatment and balancing are not included. Chilled water piping is insulated. Area distribution is through the use of multiple fan coil units. Fewer but larger fan coil units with duct distribution would be approximately the same S.F. cost.

System Components	QUANTITY	UNIT	COST EACH MAT.	COST EACH INST.	COST EACH TOTAL
SYSTEM D3030 110 1200					
PACKAGED CHILLER, AIR COOLED, WITH FAN COIL UNIT					
APARTMENT CORRIDORS, 3,000 S.F., 5.50 TON					
Fan coil air conditioning unit, cabinet mounted & filters chilled water	1.000	Ea.	3,345.23	505.91	3,851.14
Water chiller, air conditioning unit, air cooler	1.000	Ea.	6,572.50	1,883.75	8,456.25
Chilled water unit coil connections	1.000	Ea.	980	1,375	2,355
Chilled water distribution piping	440.000	L.F.	10,340	18,700	29,040
TOTAL			21,237.73	22,464.66	43,702.39
COST PER S.F.			7.08	7.49	14.57

D3030 110	Chilled Water, Air Cooled Condenser Systems	COST PER S.F. MAT.	COST PER S.F. INST.	COST PER S.F. TOTAL
1180	Packaged chiller, air cooled, with fan coil unit			
1200	Apartment corridors, 3,000 S.F., 5.50 ton	7.07	7.50	14.57
1360	40,000 S.F., 73.33 ton (R262213 -27)	4.16	3.33	7.49
1440	Banks and libraries, 3,000 S.F., 12.50 ton	10.60	8.80	19.40
1560	20,000 S.F., 83.33 ton	7.30	4.55	11.85
1680	Bars and taverns, 3,000 S.F., 33.25 ton	17.30	10.45	27.75
1760	10,000 S.F., 110.83 ton	12.75	2.59	15.34
1920	Bowling alleys, 3,000 S.F., 17.00 ton	13.15	9.90	23.05
2040	20,000 S.F., 113.33 ton	8.40	4.59	12.99
2160	Department stores, 3,000 S.F., 8.75 ton	9.55	8.35	17.90
2320	40,000 S.F., 116.66 ton	5.45	3.44	8.89
2400	Drug stores, 3,000 S.F., 20.00 ton	14.85	10.20	25.05
2520	20,000 S.F., 133.33 ton	10.40	4.85	15.25
2640	Factories, 2,000 S.F., 10.00 ton	9.30	8.45	17.75
2800	40,000 S.F., 133.33 ton	5.95	3.53	9.48
2880	Food supermarkets, 3,000 S.F., 8.50 ton	9.40	8.30	17.70
3040	40,000 S.F., 113.33 ton	5.20	3.45	8.65
3120	Medical centers, 3,000 S.F., 7.00 ton	8.50	8.05	16.55
3280	40,000 S.F., 93.33 ton	4.68	3.38	8.06
3360	Offices, 3,000 S.F., 9.50 ton	8.65	8.25	16.90
3520	40,000 S.F., 126.66 ton	5.85	3.56	9.41
3600	Restaurants, 3,000 S.F., 15.00 ton	11.85	9.15	21
3720	20,000 S.F., 100.00 ton	8.45	4.80	13.25
3840	Schools and colleges, 3,000 S.F., 11.50 ton	10.10	8.65	18.75
3960	20,000 S.F., 76.66 ton	7.10	4.59	11.69

D3030 Cooling Generating Systems

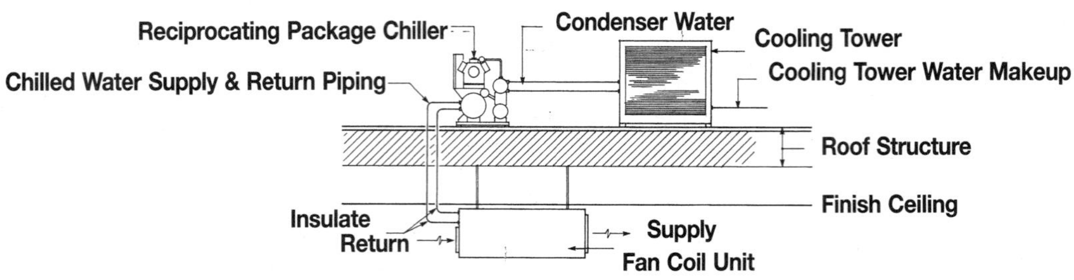

General: Water cooled chillers are available in the same sizes as air cooled units. They are also available in larger capacities.

Design Assumptions: The chilled water systems with water cooled condenser include reciprocating hermetic compressors, water cooling tower, pumps, piping and expansion tanks and are based on a two pipe system. Chilled water piping is insulated. No ducts are included and fan-coil units are cooling only. Area distribution is through use of multiple fan coil units. Fewer but larger fan coil units with duct distribution would be approximately the same S.F. cost. Water treatment and balancing are not included.

System Components	QUANTITY	UNIT	COST EACH MAT.	COST EACH INST.	COST EACH TOTAL
SYSTEM D3030 115 1320					
PACKAGED CHILLER, WATER COOLED, WITH FAN COIL UNIT					
APARTMENT CORRIDORS, 4,000 S.F., 7.33 TON					
Fan coil air conditioner unit, cabinet mounted & filters, chilled water	2.000	Ea.	4,458.48	674.27	5,132.75
Water chiller, water cooled, 1 compressor, hermetic scroll,	1.000	Ea.	5,358.60	3,297.60	8,656.20
Cooling tower, draw thru single flow, belt drive	1.000	Ea.	1,106.83	139.64	1,246.47
Cooling tower pumps & piping	1.000	System	553.42	329.85	883.27
Chilled water unit coil connections	2.000	Ea.	1,960	2,750	4,710
Chilled water distribution piping	520.000	L.F.	12,220	22,100	34,320
TOTAL			25,657.33	29,291.36	54,948.69
COST PER S.F.			6.41	7.32	13.73

D3030 115	Chilled Water, Cooling Tower Systems	COST PER S.F. MAT.	COST PER S.F. INST.	COST PER S.F. TOTAL
1300	Packaged chiller, water cooled, with fan coil unit			
1320	Apartment corridors, 4,000 S.F., 7.33 ton	6.43	7.33	13.76
1600	Banks and libraries, 4,000 S.F., 16.66 ton R262213 -27	10.05	8.05	18.10
1800	60,000 S.F., 250.00 ton	7.30	6.50	13.80
1880	Bars and taverns, 4,000 S.F., 44.33 ton	16.25	10.15	26.40
2000	20,000 S.F., 221.66 ton	17.80	8.45	26.25
2160	Bowling alleys, 4,000 S.F., 22.66 ton	11.75	8.85	20.60
2320	40,000 S.F., 226.66 ton	10.05	6.10	16.15
2440	Department stores, 4,000 S.F., 11.66 ton	7.25	7.95	15.20
2640	60,000 S.F., 175.00 ton	6.55	5.90	12.45
2720	Drug stores, 4,000 S.F., 26.66 ton	12.55	9.10	21.65
2880	40,000 S.F., 266.67 ton	9.90	6.90	16.80
3000	Factories, 4,000 S.F., 13.33 ton	8.80	7.65	16.45
3200	60,000 S.F., 200.00 ton	6.60	6.20	12.80
3280	Food supermarkets, 4,000 S.F., 11.33 ton	7.10	7.90	15
3480	60,000 S.F., 170.00 ton	6.50	5.90	12.40
3560	Medical centers, 4.000 S.F., 9.33 ton	6.25	7.25	13.50
3760	60,000 S.F., 140.00 ton	5.40	6	11.40
3840	Offices, 4,000 S.F., 12.66 ton	8.55	7.60	16.15
4040	60,000 S.F., 190.00 ton	6.40	6.15	12.55
4120	Restaurants, 4,000 S.F., 20.00 ton	10.60	8.20	18.80
4320	60,000 S.F., 300.00 ton	8.20	6.70	14.90
4400	Schools and colleges, 4,000 S.F., 15.33 ton	9.55	7.90	17.45
4600	60,000 S.F., 230.00 ton	6.80	6.20	13

D3050 Terminal & Package Units

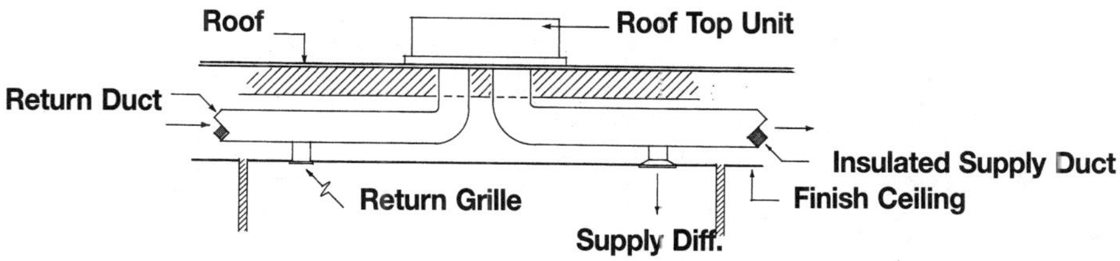

System Description: Rooftop single zone units are electric cooling and gas heat. Duct systems are low velocity, galvanized steel supply and return. Price variations between sizes are due to several factors. Jumps in the cost of the rooftop unit occur when the manufacturer shifts from the largest capacity unit on a small frame to the smallest capacity on the next larger frame, or changes from one compressor to two. As the unit capacity increases for larger areas the duct distribution grows in proportion. For most applications there is a tradeoff point where it is less expensive and more efficient to utilize smaller units with short simple distribution systems. Larger units also require larger initial supply and return ducts which can create a space problem. Supplemental heat may be desired in colder locations. The table below is based on one unit supplying the area listed. The 10,000 S.F. unit for bars and taverns is not listed because a nominal 110 ton unit would be required and this is above the normal single zone rooftop capacity.

System Components	QUANTITY	UNIT	COST EACH MAT.	COST EACH INST.	COST EACH TOTAL
SYSTEM D3050 150 1280					
ROOFTOP, SINGLE ZONE, AIR CONDITIONER					
APARTMENT CORRIDORS, 500 S.F., .92 TON					
Rooftop air-conditioner, 1 zone, electric cool, standard controls, curb	1.000	Ea.	1,621.50	724.50	2,346
Ductwork package for rooftop single zone units	1.000	System	289.80	1,035	1,324.80
TOTAL			1,911.30	1,759.50	3,670.80
COST PER S.F.			3.82	3.52	7.34

D3050 150	Rooftop Single Zone Unit Systems	COST PER S.F. MAT.	COST PER S.F. INST.	COST PER S.F. TOTAL
1260	Rooftop, single zone, air conditioner			
1280	Apartment corridors, 500 S.F., .92 ton	3.80	3.50	7.30
1480	10,000 S.F., 18.33 ton (R262213 -27)	2.61	2.38	4.99
1560	Banks or libraries, 500 S.F., 2.08 ton	8.65	7.95	16.60
1760	10,000 S.F., 41.67 ton	5.10	5.40	10.50
1840	Bars and taverns, 500 S.F. 5.54 ton	11.90	10.60	22.50
2000	5,000 S.F., 55.42 ton	11.30	8.05	19.35
2080	Bowling alleys, 500 S.F., 2.83 ton	7.60	8.95	16.55
2280	10,000 S.F., 56.67 ton	6.70	7.30	14
2360	Department stores, 500 S.F., 1.46 ton	6.05	5.60	11.65
2560	10,000 S.F., 29.17 ton	3.67	3.76	7.43
2640	Drug stores, 500 S.F., 3.33 ton	8.95	10.55	19.50
2840	10,000 S.F., 66.67 ton	7.85	8.60	16.45
2920	Factories, 500 S.F., 1.67 ton	6.90	6.35	13.25
3120	10,000 S.F., 33.33 ton	4.19	4.30	8.49
3200	Food supermarkets, 500 S.F., 1.42 ton	5.90	5.45	11.35
3400	10,000 S.F., 28.33 ton	3.56	3.66	7.22
3480	Medical centers, 500 S.F., 1.17 ton	4.85	4.47	9.32
3680	10,000 S.F., 23.33 ton	3.31	3.03	6.34
3760	Offices, 500 S.F., 1.58 ton	6.60	6.05	12.65
3960	10,000 S.F., 31.67 ton	3.99	4.09	8.08
4000	Restaurants, 500 S.F., 2.50 ton	10.40	9.55	19.95
4200	10,000 S.F., 50.00 ton	5.90	6.45	12.35
4240	Schools and colleges, 500 S.F., 1.92 ton	7.95	7.35	15.30
4440	10,000 S.F., 38.33 ton	4.68	4.95	9.63

D3050 Terminal & Package Units

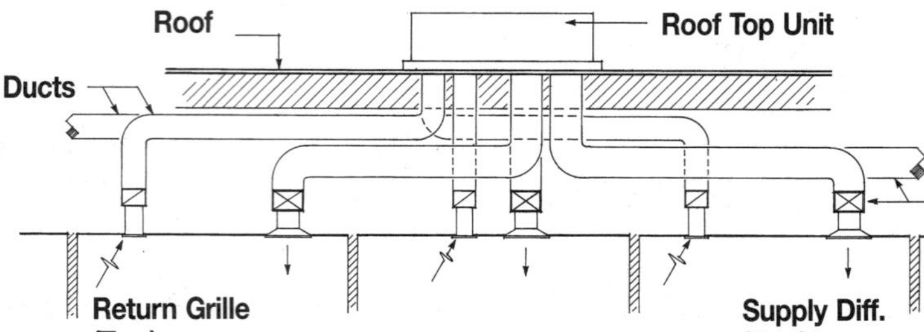

System Description: Rooftop units are multizone with up to 12 zones, and include electric cooling, gas heat, thermostats, filters, supply and return fans complete. Duct systems are low velocity, galvanized steel supply and return with insulated supplies.

Multizone units cost more per ton of cooling than single zone. However, they offer flexibility where load conditions are varied due to heat generating areas or exposure to radiational heating. For example, perimeter offices on the "sunny side" may require cooling at the same time "shady side" or central offices may require heating. It is possible to accomplish similar results using duct heaters in branches of the single zone unit. However, heater location could be a problem and total system operating energy efficiency could be lower.

System Components	QUANTITY	UNIT	COST EACH MAT.	COST EACH INST.	COST EACH TOTAL
SYSTEM D3050 155 1280					
ROOFTOP, MULTIZONE, AIR CONDITIONER					
APARTMENT CORRIDORS, 3,000 S.F., 5.50 TON					
Ductwork package for rooftop single zone units	1.000	Ea.	25,960	1,683	27,643
Ductwork package for rooftop multizone units	1.000	System	2,530	11,825	14,355
TOTAL			28,490	13,508	41,998
COST PER S.F.			9.50	4.50	14
Note A: Small single zone unit recommended					

D3050 155	Rooftop Multizone Unit Systems		COST PER S.F. MAT.	COST PER S.F. INST.	COST PER S.F. TOTAL
1240	Rooftop, multizone, air conditioner				
1260	Apartment corridors, 1,500 S.F., 2.75 ton. See Note A.				
1280	3,000 S.F., 5.50 ton	R262213 -27	9.50	4.50	14
1440	25,000 S.F., 45.80 ton		6.35	4.35	10.70
1520	Banks or libraries, 1,500 S.F., 6.25 ton		21.50	10.25	31.75
1640	15,000 S.F., 62.50 ton		10.80	9.80	20.60
1720	25,000 S.F., 104.00 ton		9.75	9.80	19.55
1800	Bars and taverns, 1,500 S.F., 16.62 ton		46	14.75	60.75
1840	3,000 S.F., 33.24 ton		38.50	14.20	52.70
1880	10,000 S.F., 110.83 ton		23.50	14.15	37.65
2080	Bowling alleys, 1,500 S.F., 8.50 ton		29.50	13.90	43.40
2160	10,000 S.F., 56.70 ton		19.60	13.45	33.05
2240	20,000 S.F., 113.00 ton		13.25	13.30	26.55
2640	Drug stores, 1,500 S.F., 10.00 ton		34.50	16.35	50.85
2680	3,000 S.F., 20.00 ton		24	15.80	39.80
2760	15,000 S.F., 100.00 ton		15.60	15.70	31.30
3760	Offices, 1,500 S.F., 4.75 ton, See Note A				
3880	15,000 S.F., 47.50 ton		10.95	7.50	18.45
3960	25,000 S.F., 79.16 ton		8.20	7.45	15.65
4000	Restaurants, 1,500 S.F., 7.50 ton		26	12.30	38.30
4080	10,000 S.F., 50.00 ton		17.30	11.85	29.15
4160	20,000 S.F., 100.00 ton		11.70	11.75	23.45
4240	Schools and colleges, 1,500 S.F., 5.75 ton		19.85	9.40	29.25
4360	15,000 S.F., 57.50 ton		9.95	9.05	19
4441	25,000 S.F., 95.83 ton		9.40	9.05	18.45

D3050 Terminal & Package Units

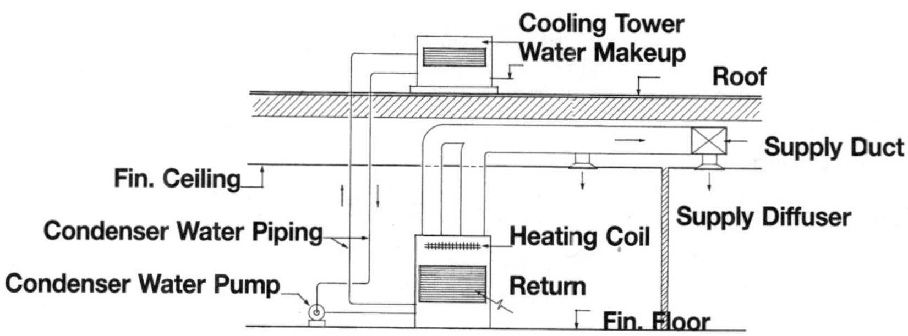

System Description: Self-contained, single package water cooled units include cooling tower, pump, piping allowance. Systems for 1000 S.F. and up include duct and diffusers to provide for even distribution of air. Smaller units distribute air through a supply air plenum, which is integral with the unit.

Returns are not ducted and supplies are not insulated.

Hot water or steam heating coils are included but piping to boiler and the boiler itself is not included.

Where local codes or conditions permit single pass cooling for the smaller units, deduct 10%.

System Components	QUANTITY	UNIT	COST EACH MAT.	COST EACH INST.	COST EACH TOTAL
SYSTEM D3050 160 1300					
SELF-CONTAINED, WATER COOLED UNIT					
APARTMENT CORRIDORS, 500 S.F., .92 TON					
Self contained, water cooled, single package air conditioner unit	1.000	Ea.	1,142.40	579.60	1,722
Ductwork package for water or air cooled packaged units	1.000	System	108.56	786.60	895.16
Cooling tower, draw thru single flow, belt drive	1.000	Ea.	138.92	17.53	156.45
Cooling tower pumps & piping	1.000	System	69.46	41.40	110.86
TOTAL			1,459.34	1,425.13	2,884.47
COST PER S.F.			2.92	2.85	5.77

D3050 160	Self-contained, Water Cooled Unit Systems	COST PER S.F. MAT.	COST PER S.F. INST.	COST PER S.F. TOTAL
1280	Self-contained, water cooled unit	2.92	2.85	5.77
1300	Apartment corridors, 500 S.F., .92 ton	2.90	2.85	5.75
1440	10,000 S.F., 18.33 ton (R262213-27)	3.50	1.98	5.48
1520	Banks or libraries, 500 S.F., 2.08 ton	6.10	2.90	9
1680	10,000 S.F., 41.66 ton	6.90	4.50	11.40
1760	Bars and taverns, 500 S.F., 5.54 ton	13.20	4.88	18.08
1920	10,000 S.F., 110.00 ton	16.95	7	23.95
2000	Bowling alleys, 500 S.F., 2.83 ton	8.35	3.94	12.29
2160	10,000 S.F., 56.66 ton	9.10	6	15.10
2200	Department stores, 500 S.F., 1.46 ton	4.30	2.04	6.34
2360	10,000 S.F., 29.17 ton	5	3.04	8.04
2440	Drug stores, 500 S.F., 3.33 ton	9.80	4.65	14.45
2600	10,000 S.F., 66.66 ton	10.75	7.10	17.85
2680	Factories, 500 S.F., 1.66 ton	4.90	2.31	7.21
2840	10,000 S.F., 33.33 ton	5.75	3.47	9.22
2920	Food supermarkets, 500 S.F., 1.42 ton	4.17	1.97	6.14
3080	10,000 S.F., 28.33 ton	4.88	2.95	7.83
3160	Medical centers, 500 S.F., 1.17 ton	3.43	1.61	5.04
3320	10,000 S.F., 23.33 ton	4.44	2.51	6.95
3400	Offices, 500 S.F., 1.58 ton	4.66	2.20	6.86
3560	10,000 S.F., 31.67 ton	5.45	3.30	8.75
3640	Restaurants, 500 S.F., 2.50 ton	7.35	3.49	10.84
3800	10,000 S.F., 50.00 ton	5.85	4.98	10.83
3880	Schools and colleges, 500 S.F., 1.92 ton	5.65	2.66	8.31
4040	10,000 S.F., 38.33 ton	6.35	4.13	10.48

D3050 Terminal & Package Units

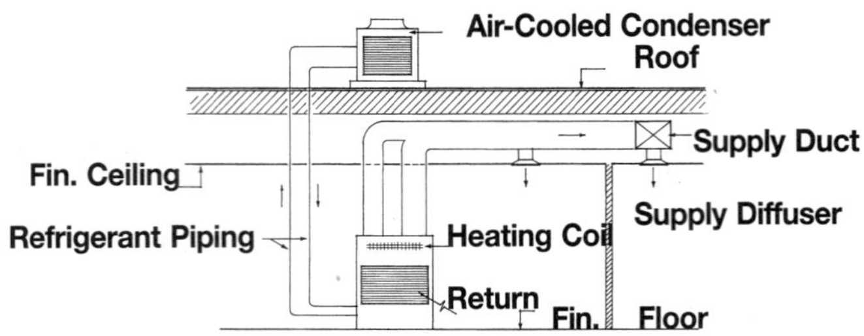

System Description: Self-contained air cooled units with remote air cooled condenser and interconnecting tubing. Systems for 1000 S.F. and up include duct and diffusers. Smaller units distribute air directly.

Returns are not ducted and supplies are not insulated.

Potential savings may be realized by using a single zone rooftop system or through-the-wall unit, especially in the smaller capacities, if the application permits.

Hot water or steam heating coils are included but piping to boiler and the boiler itself is not included.

Condenserless models are available for 15% less where remote refrigerant source is available.

System Components	QUANTITY	UNIT	COST EACH		
			MAT.	INST.	TOTAL
SYSTEM D3050 165 1320					
SELF-CONTAINED, AIR COOLED UNIT					
APARTMENT CORRIDORS, 500 S.F., .92 TON					
Air cooled, package unit	1.000	Ea.	1,198.50	374	1,572.50
Ductwork package for water or air cooled packaged units	1.000	System	108.56	786.60	895.16
Refrigerant piping	1.000	System	506	506	1,012
Air cooled condenser, direct drive, propeller fan	1.000	Ea.	348.75	142.60	491.35
TOTAL			2,161.81	1,809.20	3,971.01
COST PER S.F.			4.32	3.62	7.94

D3050 165	Self-contained, Air Cooled Unit Systems	COST PER S.F.		
		MAT.	INST.	TOTAL
1300	Self-contained, air cooled unit			
1320	Apartment corridors, 500 S.F., .92 ton	4.30	3.60	7.90
1480	10,000 S.F., 18.33 ton (R262213-27)	3.15	2.94	6.09
1560	Banks or libraries, 500 S.F., 2.08 ton	9.20	4.59	13.79
1720	10,000 S.F., 41.66 ton	8.30	6.55	14.85
1800	Bars and taverns, 500 S.F., 5.54 ton	19.30	10	29.30
1960	10,000 S.F., 110.00 ton	20.50	12.55	33.05
2040	Bowling alleys, 500 S.F., 2.83 ton	12.55	6.25	18.80
2200	10,000 S.F., 56.66 ton	10.70	8.85	19.55
2240	Department stores, 500 S.F., 1.46 ton	6.45	3.22	9.67
2400	10,000 S.F., 29.17 ton	5.10	4.55	9.65
2480	Drug stores, 500 S.F., 3.33 ton	14.75	7.35	22.10
2640	10,000 S.F., 66.66 ton	12.65	10.45	23.10
2720	Factories, 500 S.F., 1.66 ton	7.45	3.71	11.16
2880	10,000 S.F., 33.33 ton	5.85	5.20	11.05
3200	Medical centers, 500 S.F., 1.17 ton	5.20	2.59	7.79
3360	10,000 S.F., 23.33 ton	4.02	3.73	7.75
3440	Offices, 500 S.F., 1.58 ton	7	3.51	10.51
3600	10,000 S.F., 31.66 ton	5.60	4.96	10.56
3680	Restaurants, 500 S.F., 2.50 ton	11.05	5.50	16.55
3840	10,000 S.F., 50.00 ton	10.15	7.80	17.95
3920	Schools and colleges, 500 S.F., 1.92 ton	8.50	4.24	12.74
4080	10,000 S.F., 38.33 ton	7.65	6	13.65

D3050 Terminal & Package Units

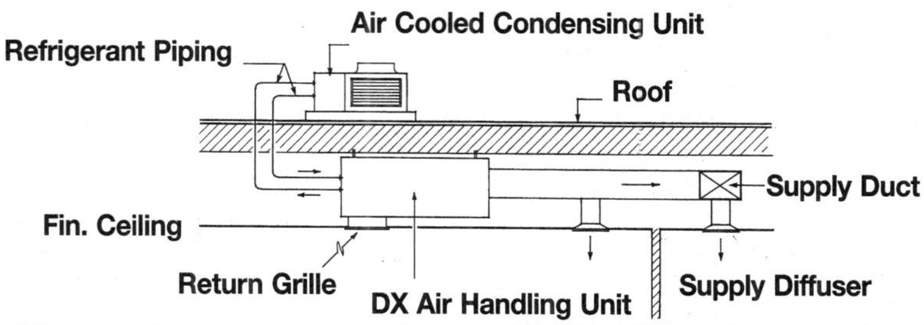

General: Split systems offer several important advantages which should be evaluated when a selection is to be made. They provide a greater degree of flexibility in component selection which permits an accurate match-up of the proper equipment size and type with the particular needs of the building. This allows for maximum use of modern energy saving concepts in heating and cooling. Outdoor installation of the air cooled condensing unit allows space savings in the building and also isolates the equipment operating sounds from building occupants.

Design Assumptions: The systems below are comprised of a direct expansion air handling unit and air cooled condensing unit with interconnecting copper tubing. Ducts and diffusers are also included for distribution of air. Systems are priced for cooling only. Heat can be added as desired either by putting hot water/steam coils into the air unit or into the duct supplying the particular area of need. Gas fired duct furnaces are also available. Refrigerant liquid line is insulated.

System Components	QUANTITY	UNIT	COST EACH MAT.	COST EACH INST.	COST EACH TOTAL
SYSTEM D3050 170 1280					
SPLIT SYSTEM, AIR COOLED CONDENSING UNIT					
APARTMENT CORRIDORS, 1,000 S.F., 1.80 TON					
Fan coil AC unit, cabinet mntd. & filters direct expansion air cool	1.000	Ea.	411.75	133.59	545.34
Ductwork package, for split system, remote condensing unit	1.000	System	101.57	750.30	851.87
Refrigeration piping	1.000	System	406.26	841.80	1,248.06
Condensing unit, air cooled, incls. compressor & standard controls	1.000	Ea.	1,525	536.80	2,061.80
TOTAL			2,444.58	2,262.49	4,707.07
COST PER S.F.			2.44	2.26	4.70

*Cooling requirements would lead to choosing a water cooled unit

D3050 170	Split Systems With Air Cooled Condensing Units		COST PER S.F. MAT.	COST PER S.F. INST.	COST PER S.F. TOTAL
1260	Split system, air cooled condensing unit				
1280	Apartment corridors, 1,000 S.F., 1.83 ton		2.45	2.25	4.70
1440	20,000 S.F., 36.66 ton	R262213 -27	2.58	3.17	5.75
1520	Banks and libraries, 1,000 S.F., 4.17 ton		4.02	5.20	9.22
1680	20,000 S.F., 83.32 ton		7.05	7.55	14.60
1760	Bars and taverns, 1,000 S.F., 11.08 ton		10.50	10.25	20.75
1880	10,000 S.F., 110.84 ton		16.05	12	28.05
2000	Bowling alleys, 1,000 S.F., 5.66 ton		5.70	9.75	15.45
2160	20,000 S.F., 113.32 ton		11.10	10.65	21.75
2320	Department stores, 1,000 S.F., 2.92 ton		2.86	3.58	6.44
2480	20,000 S.F., 58.33 ton		4.10	5.05	9.15
2560	Drug stores, 1,000 S.F., 6.66 ton		6.70	11.45	18.15
2720	20,000 S.F., 133.32 ton*				
2800	Factories, 1,000 S.F., 3.33 ton		3.25	4.07	7.32
2960	20,000 S.F., 66.66 ton		5.10	6.05	11.15
3040	Food supermarkets, 1,000 S.F., 2.83 ton		2.78	3.47	6.25
3200	20,000 S.F., 56.66 ton		3.98	4.92	8.90
3280	Medical centers, 1,000 S.F., 2.33 ton		2.46	2.81	5.27
3440	20,000 S.F., 46.66 ton		3.28	4.05	7.33
3520	Offices, 1,000 S.F., 3.17 ton		3.09	3.89	6.98
3680	20,000 S.F., 63.32 ton		4.85	5.75	10.60
3760	Restaurants, 1,000 S.F., 5.00 ton		5.05	8.60	13.65
3920	20,000 S.F., 100.00 ton		9.80	9.40	19.20
4000	Schools and colleges, 1,000 S.F., 3.83 ton		3.69	4.78	8.47
4160	20,000 S.F., 76.66 ton		5.90	6.90	12.80

D3050 Terminal & Package Units

Computer rooms impose special requirements on air conditioning systems. A prime requirement is reliability, due to the potential monetary loss that could be incurred by a system failure. A second basic requirement is the tolerance of control with which temperature and humidity are regulated, and dust eliminated. As the air conditioning system reliability is so vital, the additional cost of reserve capacity and redundant components is often justified.

System Descriptions: Computer areas may be environmentally controlled by one of three methods as follows:

1. **Self-contained Units**
   These are units built to higher standards of performance and reliability. They usually contain alarms and controls to indicate component operation failure, filter change, etc. It should be remembered that these units in the room will occupy space that is relatively expensive to build and that all alterations and service of the equipment will also have to be accomplished within the computer area.
2. **Decentralized Air Handling Units**
   In operation these are similar to the self-contained units except that their cooling capability comes from remotely located refrigeration equipment as refrigerant or chilled water. As no compressors or refrigerating equipment are required in the air units, they are smaller and require less service than self-contained units. An added plus for this type of system occurs if some of the computer components themselves also require chilled water for cooling.
3. **Central System Supply**
   Cooling is obtained from a central source which, since it is not located within the computer room, may have excess capacity and permit greater flexibility without interfering with the computer components. System performance criteria must still be met.

Note: The costs shown below do not include an allowance for ductwork or piping.

D3050 190	Computer Room Cooling Units	COST EACH		
		MAT.	INST.	TOTAL
0560	Computer room unit, air cooled, includes remote condenser			
0580	3 ton	17,200	2,200	19,400
0600	5 ton	18,400	2,450	20,850
0620	8 ton	34,500	4,075	38,575
0640	10 ton	36,100	4,400	40,500
0660	15 ton	39,700	5,000	44,700
0680	20 ton	48,200	7,150	55,350
0700	23 ton	59,500	8,175	67,675
0800	Chilled water, for connection to existing chiller system			
0820	5 ton	13,200	1,500	14,700
0840	8 ton	13,300	2,200	15,500
0860	10 ton	13,400	2,250	15,650
0880	15 ton	14,200	2,300	16,500
0900	20 ton	15,000	2,400	17,400
0920	23 ton	16,000	2,725	18,725
1000	Glycol system, complete except for interconnecting tubing			
1020	3 ton	21,700	2,750	24,450
1040	5 ton	23,700	2,900	26,600
1060	8 ton	38,100	4,800	42,900
1080	10 ton	40,500	5,250	45,750
1100	15 ton	50,000	6,600	56,600
1120	20 ton	55,500	7,150	62,650
1140	23 ton	58,000	7,450	65,450
1240	Water cooled, not including condenser water supply or cooling tower			
1260	3 ton	17,500	1,775	19,275
1280	5 ton	19,100	2,050	21,150
1300	8 ton	31,300	3,350	34,650
1320	15 ton	38,000	4,075	42,075
1340	20 ton	40,900	4,525	45,425
1360	23 ton	42,800	4,900	47,700

D4010 Sprinklers

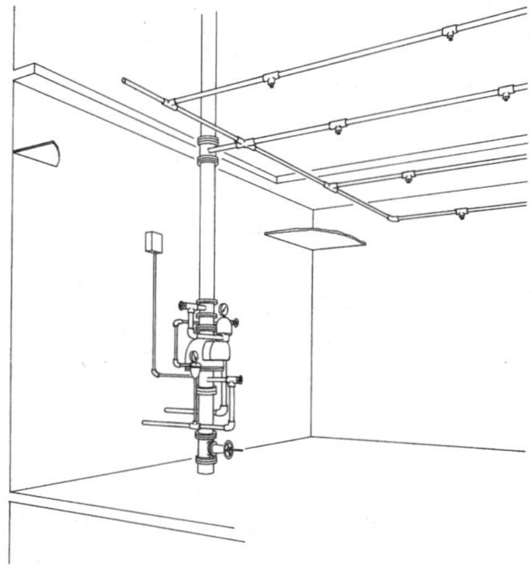

Dry Pipe System: A system employing automatic sprinklers attached to a piping system containing air under pressure, the release of which as from the opening of sprinklers permits the water pressure to open a valve known as a "dry pipe valve". The water then flows into the piping system and out the opened sprinklers.

All areas are assumed to be open.

System Components	QUANTITY	UNIT	COST EACH MAT.	COST EACH INST.	COST EACH TOTAL
SYSTEM D4010 310 0580					
DRY PIPE SPRINKLER, STEEL BLACK, SCH. 40 PIPE					
LIGHT HAZARD, ONE FLOOR, 2000 S.F.					
Valve, gate, iron body 125 lb., OS&Y, flanged, 4″ pipe size	1.000	Ea.	420	273.75	693.75
Valve, swing check, bronze, 125 lb., regrinding disc, 2-1/2″ pipe size	1.000	Ea.	420	54.38	474.38
Valve, angle, bronze, 150 lb., rising stem, threaded, 2″ pipe size	1.000	Ea.	393.75	41.25	435
*Alarm valve, 2-1/2″ pipe size	1.000	Ea.	1 012.50	270	1,282.50
Alarm, water motor, complete with gong	1.000	Ea.	251.25	111.75	363
Fire alarm horn, electric	1.000	Ea.	33.38	64.88	98.26
Valve swing check w/balldrip CI with brass trim, 4″ pipe size	1.000	Ea.	210	270	480
Pipe, steel, black, schedule 40, 4″ diam.	10.000	L.F.	149.25	233.70	382.95
Dry pipe valve, trim & gauges, 4″ pipe size	1.000	Ea.	1 668.75	806.25	2,475
Pipe, steel, black, schedule 40, threaded, cplg & hngr 10′OC 2-1/2″ diam.	20.000	L.F.	198.75	330	528.75
Pipe, steel, black, schedule 40, threaded, cplg & hngr 10′OC 2″ diam.	12.500	L.F.	79.22	159.38	238.60
Pipe, steel, black, schedule 40, threaded, cplg & hngr 10′OC 1-1/4″ diam.	37.500	L.F.	151.88	344.53	496.41
Pipe, steel, black, schedule 40, threaded, cplg & hngr 10′OC 1″ diam.	112.000	L.F.	357.84	957.60	1,315.44
Pipe Tee, malleable iron black, 150 lb. threaded, 4″ pipe size	2.000	Ea.	291	408	699
Pipe Tee, malleable iron black, 150 lb. threaded, 2-1/2″ pipe size	2.000	Ea.	81.75	181.50	263.25
Pipe Tee, malleable iron black, 150 lb. threaded, 2″ pipe size	1.000	Ea.	19.13	74.25	93.38
Pipe Tee, malleable iron black, 150 lb. threaded, 1-1/4″ pipe size	5.000	Ea.	44.63	292.50	337.13
Pipe Tee, malleable iron black, 150 lb. threaded, 1″ pipe size	4.000	Ea.	22.05	226.50	248.55
Pipe 90° elbow malleable iron black, 150 lb. threaded, 1″ pipe size	6.000	Ea.	21.15	209.25	230.40
Sprinkler head dry 1/2″ orifice 1″ NPT, 3″ to 4-3/4″ length	12.000	Ea.	1,194	510	1,704
Air compressor, 200 Gal sprinkler system capacity, 1/3 HP	1.000	Ea.	573.75	345	918.75
*Standpipe connection, wall, flush, brs. w/plug & chain 2-1/2″x2-1/2″	1.000	Ea.	101.25	161.25	262.50
Valve gate bronze, 300 psi, NRS, class 150, threaded, 1″ pipe size	1.000	Ea.	58.88	24	82.88
TOTAL			7,754.16	6,349.72	14,103.88
COST PER S.F.			3.88	3.17	7.05

D4010 310	Dry Pipe Sprinkler Systems	COST PER S.F. MAT.	COST PER S.F. INST.	COST PER S.F. TOTAL
0520	Dry pipe sprinkler systems, steel, black, sch. 40 pipe			
0530	Light hazard, one floor, 500 S.F.	7.25	5.40	12.65
0560	1000 S.F. R211313-20	4.06	3.13	7.19
0580	2000 S.F.	3.88	3.16	7.04
0600	5000 S.F.	2.05	2.11	4.16
0620	10,000 S.F.	1.47	1.78	3.25
0640	50,000 S.F.	1.17	1.58	2.75
0660	Each additional floor, 500 S.F.	1.64	2.63	4.27

D4010 Sprinklers

D4010 310	Dry Pipe Sprinkler Systems	COST PER S.F.		
		MAT.	INST.	TOTAL
0680	1000 S.F.	1.35	2.14	3.49
0700	2000 S.F.	1.32	1.98	3.30
0720	5000 S.F.	1.10	1.68	2.78
0740	10,000 S.F.	1.01	1.56	2.57
0760	50,000 S.F.	.94	1.40	2.34
1000	Ordinary hazard, one floor, 500 S.F.	7.40	5.45	12.85
1020	1000 S.F.	4.18	3.17	7.35
1040	2000 S.F.	3.98	3.30	7.28
1060	5000 S.F.	2.34	2.28	4.62
1080	10,000 S.F.	1.93	2.37	4.30
1100	50,000 S.F.	1.70	2.24	3.94
1140	Each additional floor, 500 S.F.	1.79	2.69	4.48
1160	1000 S.F.	1.58	2.41	3.99
1180	2000 S.F.	1.56	2.19	3.75
1200	5000 S.F.	1.44	1.89	3.33
1220	10,000 S.F.	1.33	1.86	3.19
1240	50,000 S.F.	1.24	1.61	2.85
1500	Extra hazard, one floor, 500 S.F.	10.15	6.80	16.95
1520	1000 S.F.	6.20	4.88	11.08
1540	2000 S.F.	4.47	4.26	8.73
1560	5000 S.F.	2.77	3.18	5.95
1580	10,000 S.F.	2.83	3.07	5.90
1600	50,000 S.F.	2.99	2.95	5.94
1660	Each additional floor, 500 S.F.	2.37	3.32	5.69
1680	1000 S.F.	2.32	3.15	5.47
1700	2000 S.F.	2.17	3.17	5.34
1720	5000 S.F.	1.87	2.75	4.62
1740	10,000 S.F.	2.14	2.53	4.67
1760	50,000 S.F.	2.15	2.42	4.57
2020	Grooved steel, black, sch. 40 pipe, light hazard, one floor, 2000 S.F.	3.73	2.70	6.43
2060	10,000 S.F.	1.49	1.56	3.05
2100	Each additional floor, 2000 S.F.	1.31	1.59	2.90
2150	10,000 S.F.	1.03	1.34	2.37
2200	Ordinary hazard, one floor, 2000 S.F.	3.95	2.88	6.83
2250	10,000 S.F.	1.89	1.99	3.88
2300	Each additional floor, 2000 S.F.	1.53	1.77	3.30
2350	10,000 S.F.	1.44	1.78	3.22
2400	Extra hazard, one floor, 2000 S.F.	4.47	3.64	8.11
2450	10,000 S.F.	2.56	2.57	5.13
2500	Each additional floor, 2000 S.F.	2.18	2.61	4.79
2550	10,000 S.F.	1.98	2.25	4.23
3050	Grooved steel black sch. 10 pipe, light hazard, one floor, 2000 S.F.	3.66	2.68	6.34
3100	10,000 S.F.	1.43	1.54	2.97
3150	Each additional floor, 2000 S.F.	1.24	1.56	2.80
3200	10,000 S.F.	.97	1.32	2.29
3250	Ordinary hazard, one floor, 2000 S.F.	3.88	2.86	6.74
3300	10,000 S.F.	1.75	1.95	3.70
3350	Each additional floor, 2000 S.F.	1.46	1.75	3.21
3400	10,000 S.F.	1.30	1.74	3.04
3450	Extra hazard, one floor, 2000 S.F.	4.41	3.61	8.02
3500	10,000 S.F.	2.42	2.52	4.94
3550	Each additional floor, 2000 S.F.	2.12	2.58	4.70
3600	10,000 S.F.	1.90	2.22	4.12
4050	Copper tubing, type M, light hazard, one floor, 2000 S.F.	4.56	2.69	7.25
4100	10,000 S.F.	2.26	1.55	3.81
4150	Each additional floor, 2000 S.F.	2.16	1.61	3.77
4200	10,000 S.F.	1.80	1.34	3.14
4250	Ordinary hazard, one floor, 2000 S.F.	4.94	2.98	7.92

D4010 Sprinklers

D4010 310	Dry Pipe Sprinkler Systems	COST PER S.F.		
		MAT.	INST.	TOTAL
4300	10,000 S.F.	2.82	1.84	4.66
4350	Each additional floor, 2000 S.F.	2.96	1.84	4.80
4400	10,000 S.F.	2.30	1.59	3.89
4450	Extra hazard, one floor, 2000 S.F.	5.75	3.63	9.38
4500	10,000 S.F.	5.50	2.76	8.26
4550	Each additional floor, 2000 S.F.	3.47	2.60	6.07
4600	10,000 S.F.	3.96	2.42	6.38
5050	Copper tubing, type M, T-drill system, light hazard, one floor			
5060	2000 S.F.	4.58	2.53	7.11
5100	10,000 S.F.	2.14	1.29	3.43
5150	Each additional floor, 2000 S.F.	2.18	1.45	3.63
5200	10,000 S.F.	1.68	1.08	2.76
5250	Ordinary hazard, one floor, 2000 S.F.	4.75	2.56	7.31
5300	10,000 S.F.	2.73	1.65	4.38
5350	Each additional floor, 2000 S.F.	2.33	1.45	3.78
5400	10,000 S.F.	2.15	1.37	3.52
5450	Extra hazard, one floor, 2000 S.F.	5.20	3.02	8.22
5500	10,000 S.F.	4.55	2.03	6.58
5550	Each additional floor, 2000 S.F.	2.91	1.99	4.90
5600	10,000 S.F.	3.03	1.69	4.72

D4010 Sprinklers

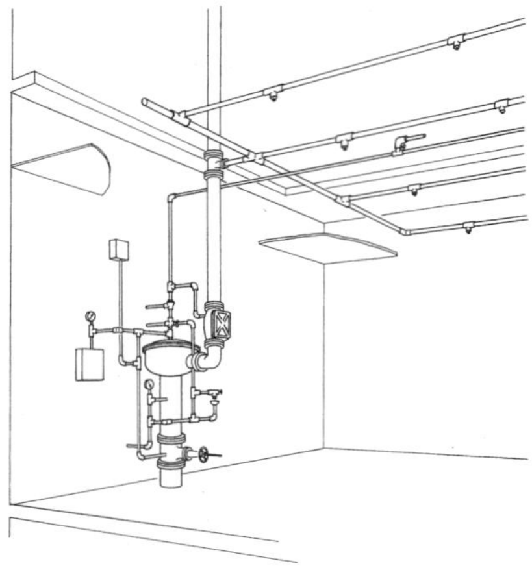

Pre-Action System: A system employing automatic sprinklers attached to a piping system containing air that may or may not be under pressure, with a supplemental heat responsive system of generally more sensitive characteristics than the automatic sprinklers themselves, installed in the same areas as the sprinklers. Actuation of the heat responsive system, as from a fire, opens a valve which permits water to flow into the sprinkler piping system and to be discharged from those sprinklers which were opened by heat from the fire.

All areas are assumed to be open.

System Components	QUANTITY	UNIT	COST EACH		
			MAT.	INST.	TOTAL
SYSTEM D4010 350 0580					
PREACTION SPRINKLER SYSTEM, STEEL BLACK SCH. 40 PIPE					
LIGHT HAZARD, 1 FLOOR, 2000 S.F.					
Valve, gate, iron body 125 lb., OS&Y, flanged, 4″ pipe size	1.000	Ea.	420	273.75	693.75
*Valve, swing check w/ball drip CI with brass trim 4″ pipe size	1.000	Ea.	210	270	480
Valve, swing check, bronze, 125 lb., regrinding disc, 2-1/2″ pipe size	1.000	Ea.	420	54.38	474.38
Valve, angle, bronze, 150 lb., rising stem, threaded, 2″ pipe size	1.000	Ea.	393.75	41.25	435
*Alarm valve, 2-1/2″ pipe size	1.000	Ea.	1,012.50	270	1,282.50
Alarm, water motor, complete with gong	1.000	Ea.	251.25	111.75	363
Fire alarm horn, electric	1.000	Ea.	33.38	64.88	98.26
Thermostatic release for release line	2.000	Ea.	825	45	870
Pipe, steel, black, schedule 40, 4″ diam.	10.000	L.F.	149.25	233.70	382.95
Dry pipe valve, trim & gauges, 4″ pipe size	1.000	Ea.	1,668.75	806.25	2,475
Pipe, steel, black, schedule 40, threaded, cplg. & hngr. 10′OC 2-1/2″ diam.	20.000	L.F.	198.75	330	528.75
Pipe steel black, schedule 40, threaded, cplg. & hngr.10′OC 2″ diam.	12.500	L.F.	79.22	159.38	238.60
Pipe, steel, black, schedule 40, threaded, cplg. & hngr. 10′OC 1-1/4″ diam.	37.500	L.F.	151.88	344.53	496.41
Pipe, steel, black, schedule 40, threaded, cplg. & hngr. 10′OC 1″ diam.	112.000	L.F.	357.84	957.60	1,315.44
Pipe, Tee, malleable iron, black, 150 lb. threaded, 4″ diam.	2.000	Ea.	291	408	699
Pipe, Tee, malleable iron, black, 150 lb. threaded, 2-1/2″ pipe size	2.000	Ea.	81.75	181.50	263.25
Pipe, Tee, malleable iron, black, 150 lb. threaded, 2″ pipe size	1.000	Ea.	19.13	74.25	93.38
Pipe, Tee, malleable iron, black, 150 lb. threaded, 1-1/4″ pipe size	5.000	Ea.	44.63	292.50	337.13
Pipe, Tee, malleable iron, black, 150 lb. threaded, 1″ pipe size	4.000	Ea.	22.05	226.50	248.55
Pipe, 90° elbow, malleable iron, blk., 150 lb. threaded, 1″ pipe size	6.000	Ea.	21.15	209.25	230.40
Sprinkler head, std. spray, brass 135°-286° F 1/2″ NPT, 3/8″ orifice	12.000	Ea.	151.20	450	601.20
Air compressor auto complete 200 Gal sprinkler sys. cap., 1/3 HP	1.000	Ea.	573.75	345	918.75
*Standpipe conn.,wall, flush, brass w/plug & chain 2-1/2″ x 2-1/2″	1.000	Ea.	101.25	161.25	262.50
Valve, gate, bronze, 300 psi, NRS, class 150, threaded, 1″ pipe size	1.000	Ea.	58.88	24	82.88
TOTAL			7,536.36	6,334.72	13,871.08
COST PER S.F.			3.77	3.17	6.94
*Not included in systems under 2000 S.F.					

D4010 350	Preaction Sprinkler Systems		COST PER S.F.		
			MAT.	INST.	TOTAL
0520	Preaction sprinkler systems, steel, black, sch. 40 pipe	R211313 -20			
0530	Light hazard, one floor, 500 S.F.		7.10	4.31	11.41
0560	1000 S.F.		4.10	3.20	7.30
0580	2000 S.F.		3.76	3.16	6.92
0600	5000 S.F.		1.96	2.11	4.07
0620	10,000 S.F.		1.39	1.78	3.17

D4010 350	Preaction Sprinkler Systems	COST PER S.F.		
		MAT.	INST.	TOTAL
0640	50,000 S.F.	1.10	1.57	2.67
0660	Each additional floor, 500 S.F.	1.81	2.34	4.15
0680	1000 S.F.	1.37	2.14	3.51
0700	2000 S.F.	1.34	1.98	3.32
0720	5000 S.F.	1.01	1.68	2.69
0740	10,000 S.F.	.93	1.56	2.49
0760	50,000 S.F.	.95	1.44	2.39
1000	Ordinary hazard, one floor, 500 S.F.	7.40	4.67	12.07
1020	1000 S.F.	4.07	3.16	7.23
1040	2000 S.F.	4.08	3.30	7.38
1060	5000 S.F.	2.15	2.27	4.42
1080	10,000 S.F.	1.70	2.36	4.06
1100	50,000 S.F.	1.45	2.21	3.66
1140	Each additional floor, 500 S.F.	2.09	2.70	4.79
1160	1000 S.F.	1.36	2.17	3.53
1180	2000 S.F.	1.25	2.17	3.42
1200	5000 S.F.	1.36	2.04	3.40
1220	10,000 S.F.	1.25	2.15	3.40
1240	50,000 S.F.	1.13	1.89	3.02
1500	Extra hazard, one floor, 500 S.F.	9.85	6	15.85
1520	1000 S.F.	5.65	4.46	10.11
1540	2000 S.F.	4.09	4.24	8.33
1560	5000 S.F.	2.66	3.43	6.09
1580	10,000 S.F.	2.50	3.40	5.90
1600	50,000 S.F.	2.59	3.27	5.86
1660	Each additional floor, 500 S.F.	2.40	3.32	5.72
1680	1000 S.F.	1.94	3.13	5.07
1700	2000 S.F.	1.79	3.15	4.94
1720	5000 S.F.	1.52	2.77	4.29
1740	10,000 S.F.	1.66	2.56	4.22
1760	50,000 S.F.	1.59	2.38	3.97
2020	Grooved steel, black, sch. 40 pipe, light hazard, one floor, 2000 S.F.	3.75	2.70	6.45
2060	10,000 S.F.	1.41	1.56	2.97
2100	Each additional floor of 2000 S.F.	1.33	1.59	2.92
2150	10,000 S.F.	.95	1.34	2.29
2200	Ordinary hazard, one floor, 2000 S.F.	3.84	2.87	6.71
2250	10,000 S.F.	1.66	1.98	3.64
2300	Each additional floor, 2000 S.F.	1.42	1.76	3.18
2350	10,000 S.F.	1.21	1.77	2.98
2400	Extra hazard, one floor, 2000 S.F.	4.09	3.62	7.71
2450	10,000 S.F.	2.09	2.55	4.64
2500	Each additional floor, 2000 S.F.	1.80	2.59	4.39
2550	10,000 S.F.	1.48	2.22	3.70
3050	Grooved steel, black, sch. 10 pipe light hazard, one floor, 2000 S.F.	3.68	2.68	6.36
3100	10,000 S.F.	1.35	1.54	2.89
3150	Each additional floor, 2000 S.F.	1.26	1.56	2.82
3200	10,000 S.F.	.89	1.32	2.21
3250	Ordinary hazard, one floor, 2000 S.F.	3.71	2.68	6.39
3300	10,000 S.F.	1.31	1.92	3.23
3350	Each additional floor, 2000 S.F.	1.35	1.74	3.09
3400	10,000 S.F.	1.07	1.73	2.80
3450	Extra hazard, one floor, 2000 S.F.	4.03	3.59	7.62
3500	10,000 S.F.	1.92	2.49	4.41
3550	Each additional floor, 2000 S.F.	1.74	2.56	4.30
3600	10,000 S.F.	1.40	2.19	3.59
4050	Copper tubing, type M, light hazard, one floor, 2000 S.F.	4.59	2.69	7.28
4100	10,000 S.F.	2.19	1.55	3.74
4150	Each additional floor, 2000 S.F.	2.19	1.61	3.80

D4010 Sprinklers

D4010 350	Preaction Sprinkler Systems	COST PER S.F. MAT.	COST PER S.F. INST.	COST PER S.F. TOTAL
4200	10,000 S.F.	1.51	1.32	2.83
4250	Ordinary hazard, one floor, 2000 S.F.	4.83	2.97	7.80
4300	10,000 S.F.	2.59	1.83	4.42
4350	Each additional floor, 2000 S.F.	2.20	1.63	3.83
4400	10,000 S.F.	1.85	1.46	3.31
4450	Extra hazard, one floor, 2000 S.F.	5.40	3.61	9.01
4500	10,000 S.F.	4.95	2.72	7.67
4550	Each additional floor, 2000 S.F.	3.09	2.58	5.67
4600	10,000 S.F.	3.46	2.39	5.85
5050	Copper tubing, type M, T-drill system, light hazard, one floor			
5060	2000 S.F.	4.60	2.53	7.13
5100	10,000 S.F.	2.06	1.29	3.35
5150	Each additional floor, 2000 S.F.	2.20	1.45	3.65
5200	10,000 S.F.	1.60	1.08	2.68
5250	Ordinary hazard, one floor, 2000 S.F.	4.64	2.55	7.19
5300	10,000 S.F.	2.50	1.64	4.14
5350	Each additional floor, 2000 S.F.	2.22	1.45	3.67
5400	10,000 S.F.	2.05	1.43	3.48
5450	Extra hazard, one floor, 2000 S.F.	4.82	3	7.82
5500	10,000 S.F.	4.02	1.99	6.01
5550	Each additional floor, 2000 S.F.	2.53	1.97	4.50
5600	10,000 S.F.	2.53	1.66	4.19

D4010 Sprinklers

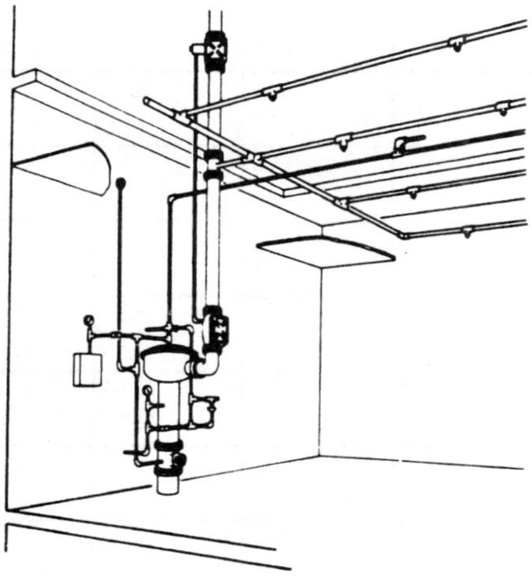

Deluge System: A system employing open sprinklers attached to a piping system connected to a water supply through a valve which is opened by the operation of a heat responsive system installed in the same areas as the sprinklers. When this valve opens, water flows into the piping system and discharges from all sprinklers attached thereto.

All areas are assumed to be open.

System Components	QUANTITY	UNIT	COST EACH MAT.	COST EACH INST.	COST EACH TOTAL
SYSTEM D4010 370 0580					
DELUGE SPRINKLER SYSTEM, STEEL BLACK SCH. 40 PIPE					
LIGHT HAZARD, 1 FLOOR, 2000 S.F.					
Valve, gate, iron body 125 lb., OS&Y, flanged, 4″ pipe size	1.000	Ea.	420	273.75	693.75
Valve, swing check w/ball drip, CI w/brass ftngs., 4″ pipe size	1.000	Ea.	210	270	480
Valve, swing check, bronze, 125 lb., regrinding disc, 2-1/2″ pipe size	1.000	Ea.	420	54.38	474.38
Valve, angle, bronze, 150 lb., rising stem, threaded, 2″ pipe size	1.000	Ea.	393.75	41.25	435
*Alarm valve, 2-1/2″ pipe size	1.000	Ea.	1,012.50	270	1,282.50
Alarm, water motor, complete with gong	1.000	Ea.	251.25	111.75	363
Fire alarm horn, electric	1.000	Ea.	33.38	64.88	98.26
Thermostatic release for release line	2.000	Ea.	825	45	870
Pipe, steel, black, schedule 40, 4″ diam.	10.000	L.F.	149.25	233.70	382.95
Deluge valve trim, pressure relief, emergency release, gauge, 4″ pipe size	1.000	Ea.	2 268.75	806.25	3,075
Deluge system pressured monitoring panel, 120V	1.000	Ea.	6 862.50	24.75	6,887.25
Pipe, steel, black, schedule 40, threaded, cplg & hngr 10′ OC 2-1/2″ diam.	20.000	L.F.	198.75	330	528.75
Pipe, steel, black, schedule 40, threaded, cplg & hngr 10′ OC 2″ diam.	12.500	L.F.	79.22	159.38	238.60
Pipe, steel, black, schedule 40, threaded, cplg & hngr 10′ OC 1-1/4″ diam.	37.500	L.F.	151.88	344.53	496.41
Pipe, steel, black, schedule 40, threaded, cplg & hngr 10′ OC 1″ diam.	112.000	L.F.	357.84	957.60	1,315.44
Pipe, Tee, malleable iron, black, 150 lb. threaded, 4″ pipe size	2.000	Ea.	291	408	699
Pipe, Tee, malleable iron, black, 150 lb. threaded, 2-1/2″ pipe size	2.000	Ea.	81.75	181.50	263.25
Pipe, Tee, malleable iron, black, 150 lb. threaded, 2″ pipe size	1.000	Ea.	19.13	74.25	93.38
Pipe, Tee, malleable iron, black, 150 lb. threaded, 1-1/4″ pipe size	5.000	Ea.	44.63	292.50	337.13
Pipe, Tee, malleable iron, black, 150 lb. threaded, 1″ pipe size	4.000	Ea.	22.05	226.50	248.55
Pipe, 90° elbow, malleable iron, black, 150 lb. threaded 1″ pipe size	6.000	Ea.	21.15	209.25	230.40
Sprinkler head, std spray, brass 135°-286° F 1/2″ NPT, 3/8″ orifice	9.720	Ea.	151.20	450	601.20
Air compressor, auto, complete, 200 Gal sprinkler sys. cap., 1/3 HP	1.000	Ea.	573.75	345	918.75
*Standpipe connection, wall, flush w/plug & chain 2-1/2″ x 2-1/2″	1.000	Ea.	101.25	161.25	262.50
Valve, gate, bronze, 300 psi, NRS, class 150, threaded, 1″ pipe size	1.000	Ea.	58.88	24	82.88
TOTAL			14,998.86	6,359.47	21,358.33
COST PER S.F.			7.50	3.18	10.68

*Not included in systems under 2000 S.F.

D4010 370	Deluge Sprinkler Systems		COST PER S.F. MAT.	COST PER S.F. INST.	COST PER S.F. TOTAL
0520	Deluge sprinkler systems, steel, black, sch. 40 pipe	R211313 -20			
0530	Light hazard, one floor, 500 S.F.		21	4.36	25.36

D4010 Sprinklers

D4010 370	Deluge Sprinkler Systems	COST PER S.F. MAT.	COST PER S.F. INST.	COST PER S.F. TOTAL
0560	1000 S.F.	11.05	3.07	14.12
0580	2000 S.F.	7.50	3.18	10.68
0600	5000 S.F.	3.45	2.11	5.56
0620	10,000 S.F.	2.14	1.78	3.92
0640	50,000 S.F.	1.26	1.57	2.83
0660	Each additional floor, 500 S.F.	1.81	2.34	4.15
0680	1000 S.F.	1.37	2.14	3.51
0700	2000 S.F.	1.34	1.98	3.32
0720	5000 S.F.	1.01	1.68	2.69
0740	10,000 S.F.	.93	1.56	2.49
0760	50,000 S.F.	.95	1.44	2.39
1000	Ordinary hazard, one floor, 500 S.F.	22	4.98	26.98
1020	1000 S.F.	11.05	3.18	14.23
1040	2000 S.F.	7.80	3.31	11.11
1060	5000 S.F.	3.64	2.27	5.91
1080	10,000 S.F.	2.45	2.36	4.81
1100	50,000 S.F.	1.64	2.24	3.88
1140	Each additional floor, 500 S.F.	2.09	2.70	4.79
1160	1000 S.F.	1.36	2.17	3.53
1180	2000 S.F.	1.25	2.17	3.42
1200	5000 S.F.	1.25	1.88	3.13
1220	10,000 S.F.	1.19	1.89	3.08
1240	50,000 S.F.	1.06	1.73	2.79
1500	Extra hazard, one floor, 500 S.F.	24	6.05	30.05
1520	1000 S.F.	12.95	4.63	17.58
1540	2000 S.F.	7.80	4.25	12.05
1560	5000 S.F.	3.87	3.16	7.03
1580	10,000 S.F.	3.13	3.11	6.24
1600	50,000 S.F.	3.07	3.01	6.08
1660	Each additional floor, 500 S.F.	2.40	3.32	5.72
1680	1000 S.F.	1.94	3.13	5.07
1700	2000 S.F.	1.79	3.15	4.94
1720	5000 S.F.	1.52	2.77	4.29
1740	10,000 S.F.	1.71	2.66	4.37
1760	50,000 S.F.	1.69	2.54	4.23
2000	Grooved steel, black, sch. 40 pipe, light hazard, one floor			
2020	2000 S.F.	7.50	2.71	10.21
2060	10,000 S.F.	2.18	1.58	3.76
2100	Each additional floor, 2,000 S.F.	1.33	1.59	2.92
2150	10,000 S.F.	.95	1.34	2.29
2200	Ordinary hazard, one floor, 2000 S.F.	3.84	2.87	6.71
2250	10,000 S.F.	2.41	1.98	4.39
2300	Each additional floor, 2000 S.F.	1.42	1.76	3.18
2350	10,000 S.F.	1.21	1.77	2.98
2400	Extra hazard, one floor, 2000 S.F.	7.80	3.63	11.43
2450	10,000 S.F.	2.86	2.56	5.42
2500	Each additional floor, 2000 S.F.	1.80	2.59	4.39
2550	10,000 S.F.	1.48	2.22	3.70
3000	Grooved steel, black, sch. 10 pipe, light hazard, one floor			
3050	2000 S.F.	6.95	2.57	9.52
3100	10,000 S.F.	2.10	1.54	3.64
3150	Each additional floor, 2000 S.F.	1.26	1.56	2.82
3200	10,000 S.F.	.89	1.32	2.21
3250	Ordinary hazard, one floor, 2000 S.F.	7.50	2.86	10.36
3300	10,000 S.F.	2.06	1.92	3.98
3350	Each additional floor, 2000 S.F.	1.35	1.74	3.09
3400	10,000 S.F.	1.07	1.73	2.80
3450	Extra hazard, one floor, 2000 S.F.	7.75	3.60	11.35

D4010 Sprinklers

D4010 370	Deluge Sprinkler Systems	COST PER S.F. MAT.	INST.	TOTAL
3500	10,000 S.F.	2.67	2.49	5.16
3550	Each additional floor, 2000 S.F.	1.74	2.56	4.30
3600	10,000 S.F.	1.40	2.19	3.59
4000	Copper tubing, type M, light hazard, one floor			
4050	2000 S.F.	8.30	2.70	11
4100	10,000 S.F.	2.93	1.55	4.48
4150	Each additional floor, 2000 S.F.	2.18	1.61	3.79
4200	10,000 S.F.	1.51	1.32	2.83
4250	Ordinary hazard, one floor, 2000 S.F.	8.55	2.98	11.53
4300	10,000 S.F.	3.34	1.83	5.17
4350	Each additional floor, 2000 S.F.	2.20	1.63	3.83
4400	10,000 S.F.	1.85	1.46	3.31
4450	Extra hazard, one floor, 2000 S.F.	9.10	3.62	12.72
4500	10,000 S.F.	5.75	2.74	8.49
4550	Each additional floor, 2000 S.F.	3.09	2.58	5.67
4600	10,000 S.F.	3.46	2.39	5.85
5000	Copper tubing, type M, T-drill system, light hazard, one floor			
5050	2000 S.F.	8.35	2.54	10.89
5100	10,000 S.F.	2.81	1.29	4.10
5150	Each additional floor, 2000 S.F.	2.11	1.47	3.58
5200	10,000 S.F.	1.60	1.08	2.68
5250	Ordinary hazard, one floor, 2000 S.F.	8.35	2.56	10.91
5300	10,000 S.F.	3.25	1.64	4.89
5350	Each additional floor, 2000 S.F.	2.22	1.44	3.66
5400	10,000 S.F.	2.05	1.43	3.48
5450	Extra hazard, one floor, 2000 S.F.	8.55	3.01	11.56
5500	10,000 S.F.	4.77	1.99	6.76
5550	Each additional floor, 2000 S.F.	2.53	1.97	4.50
5600	10,000 S.F.	2.53	1.66	4.19

D4010 Sprinklers

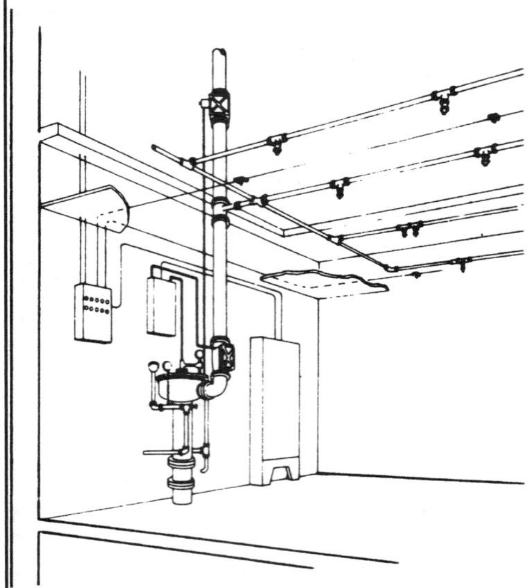

Firecycle is a fixed fire protection sprinkler system utilizing water as its extinguishing agent. It is a time delayed, recycling, preaction type which automatically shuts the water off when heat is reduced below the detector operating temperature and turns the water back on when that temperature is exceeded.

The system senses a fire condition through a closed circuit electrical detector system which controls water flow to the fire automatically. Batteries supply up to 90 hour emergency power supply for system operation. The piping system is dry (until water is required) and is monitored with pressurized air. Should any leak in the system piping occur, an alarm will sound, but water will not enter the system until heat is sensed by a Firecycle detector.

All areas are assumed to be open.

System Components	QUANTITY	UNIT	COST EACH MAT.	COST EACH INST.	COST EACH TOTAL
SYSTEM D4010 390 0580					
FIRECYCLE SPRINKLER SYSTEM, STEEL BLACK SCH. 40 PIPE					
LIGHT HAZARD, ONE FLOOR, 2000 S.F.					
Valve, gate, iron body 125 lb., OS&Y, flanged, 4″ pipe size	1.000	Ea.	420	273.75	693.75
Valve, angle, bronze, 150 lb., rising stem, threaded, 2″ pipe size	1.000	Ea.	393.75	41.25	435
Valve, swing check, bronze, 125 lb., regrinding disc, 2-1/2″ pipe size	1.000	Ea.	420	54.38	474.38
*Alarm valve, 2-1/2″ pipe size	1.000	Ea.	1,012.50	270	1,282.50
Alarm, water motor, complete with gong	1.000	Ea.	251.25	111.75	363
Pipe, steel, black, schedule 40, 4″ diam.	10.000	L.F.	149.25	233.70	382.95
Fire alarm, horn, electric	1.000	Ea.	33.38	64.88	98.26
Pipe, steel, black, schedule 40, threaded, cplg & hngr 10′ OC 2-1/2″ diam.	20.000	L.F.	198.75	330	528.75
Pipe, steel, black, schedule 40, threaded, cplg & hngr 10′ OC 2″ diam.	12.500	L.F.	79.22	159.38	238.60
Pipe, steel, black, schedule 40, threaded, cplg & hngr 10′ OC 1-1/4″ diam.	37.500	L.F.	151.88	344.53	496.41
Pipe, steel, black, schedule 40, threaded, cplg & hngr 10′ OC 1″ diam.	112.000	L.F.	357.84	957.60	1,315.44
Pipe, Tee, malleable iron, black, 150 lb. threaded, 4″ pipe size	2.000	Ea.	291	408	699
Pipe, Tee, malleable iron, black, 150 lb. threaded, 2-1/2″ pipe size	2.000	Ea.	81.75	181.50	263.25
Pipe, Tee, malleable iron, black, 150 lb. threaded, 2″ pipe size	1.000	Ea.	19.13	74.25	93.38
Pipe, Tee, malleable iron, black, 150 lb. threaded, 1-1/4″ pipe size	5.000	Ea.	44.63	292.50	337.13
Pipe, Tee, malleable iron, black, 150 lb. threaded, 1″ pipe size	4.000	Ea.	22.05	226.50	248.55
Pipe, 90° elbow, malleable iron, black, 150 lb. threaded, 1″ pipe size	6.000	Ea.	21.15	209.25	230.40
Sprinkler head std spray, brass 135°-286° F 1/2″ NPT, 3/8″ orifice	12.000	Ea.	151.20	450	601.20
Firecycle controls, incls panel, battery, solenoid valves, press switches	1.000	Ea.	8,925	1,706.25	10,631.25
Detector, firecycle system	2.000	Ea.	862.50	56.25	918.75
Firecycle pkg, swing check & flow control valves w/trim 4″ pipe size	1.000	Ea.	4,087.50	806.25	4,893.75
Air compressor, auto, complete, 200 Gal sprinkler sys. cap., 1/3 HP	1.000	Ea.	573.75	345	918.75
*Standpipe connection, wall, flush, brass w/plug & chain 2-1/2″x2-1/2″	1.000	Ea.	101.25	161.25	262.50
Valve, gate, bronze 300 psi, NRS, class 150, threaded, 1″ diam.	1.000	Ea.	58.88	24	82.88
TOTAL			18,707.61	7,782.22	26,489.83
COST PER S.F.			9.35	3.89	13.24

*Not included in systems under 2000 S.F.

D4010 390	Firecycle Sprinkler Systems	COST PER S.F.		
		MAT.	INST.	TOTAL
0520	Firecycle sprinkler systems, steel black sch. 40 pipe			
0530	Light hazard, one floor, 500 S.F. R211313 -20	30	8.45	38.45
0560	1000 S.F.	15.45	5.30	20.75
0580	2000 S.F.	9.35	3.89	13.24
0600	5000 S.F.	4.20	2.40	6.60
0620	10,000 S.F.	2.56	1.92	4.48
0640	50,000 S.F.	1.37	1.60	2.97
0660	Each additional floor of 500 S.F.	1.85	2.36	4.21
0680	1000 S.F.	1.39	2.15	3.54
0700	2000 S.F.	1.15	1.97	3.12
0720	5000 S.F.	1.02	1.68	2.70
0740	10,000 S.F.	.99	1.56	2.55
0760	50,000 S.F.	.99	1.44	2.43
1000	Ordinary hazard, one floor, 500 S.F.	30	8.85	38.85
1020	1000 S.F.	15.40	5.25	20.65
1040	2000 S.F.	9.45	4.02	13.47
1060	5000 S.F.	4.39	2.56	6.95
1080	10,000 S.F.	2.87	2.50	5.37
1100	50,000 S.F.	1.91	2.51	4.42
1140	Each additional floor, 500 S.F.	2.13	2.72	4.85
1160	1000 S.F.	1.38	2.18	3.56
1180	2000 S.F.	1.40	2	3.40
1200	5000 S.F.	1.26	1.88	3.14
1220	10,000 S.F.	1.16	1.85	3.01
1240	50,000 S.F.	1.07	1.65	2.72
1500	Extra hazard, one floor, 500 S.F.	32.50	10.15	42.65
1520	1000 S.F.	16.75	6.55	23.30
1540	2000 S.F.	9.70	4.97	14.67
1560	5000 S.F.	4.62	3.45	8.07
1580	10,000 S.F.	3.62	3.52	7.14
1600	50,000 S.F.	3.10	3.87	6.97
1660	Each additional floor, 500 S.F.	2.44	3.34	5.78
1680	1000 S.F.	1.96	3.14	5.10
1700	2000 S.F.	1.81	3.16	4.97
1720	5000 S.F.	1.53	2.77	4.30
1740	10,000 S.F.	1.72	2.56	4.28
1760	50,000 S.F.	1.70	2.46	4.16
2020	Grooved steel, black, sch. 40 pipe, light hazard, one floor			
2030	2000 S.F.	9.35	3.43	12.78
2060	10,000 S.F.	2.81	2.36	5.17
2100	Each additional floor, 2000 S.F.	1.35	1.60	2.95
2150	10,000 S.F.	1.01	1.34	2.35
2200	Ordinary hazard, one floor, 2000 S.F.	9.45	3.60	13.05
2250	10,000 S.F.	3.01	2.24	5.25
2300	Each additional floor, 2000 S.F.	1.44	1.77	3.21
2350	10,000 S.F.	1.27	1.77	3.04
2400	Extra hazard, one floor, 2000 S.F.	9.70	4.35	14.05
2450	10,000 S.F.	3.21	2.67	5.88
2500	Each additional floor, 2000 S.F.	1.82	2.60	4.42
2550	10,000 S.F.	1.54	2.22	3.76
3050	Grooved steel, black, sch. 10 pipe light hazard, one floor,			
3060	2000 S.F.	9.25	3.41	12.66
3100	10,000 S.F.	2.52	1.68	4.20
3150	Each additional floor, 2000 S.F.	1.28	1.57	2.85
3200	10,000 S.F.	.95	1.32	2.27
3250	Ordinary hazard, one floor, 2000 S.F.	9.35	3.58	12.93
3300	10,000 S.F.	2.69	2.08	4.77
3350	Each additional floor, 2000 S.F.	1.37	1.75	3.12

D4010 Sprinklers

D4010 390	Firecycle Sprinkler Systems	COST PER S.F.		
		MAT.	INST.	TOTAL
3400	10,000 S.F.	1.13	1.73	2.86
3450	Extra hazard, one floor, 2000 S.F.	9.60	4.32	13.92
3500	10,000 S.F.	3.07	2.62	5.69
3550	Each additional floor, 2000 S.F.	1.76	2.57	4.33
3600	10,000 S.F.	1.46	2.19	3.65
4060	Copper tubing, type M, light hazard, one floor, 2000 S.F.	10.15	3.42	13.57
4100	10,000 S.F.	3.35	1.69	5.04
4150	Each additional floor, 2000 S.F.	2.20	1.62	3.82
4200	10,000 S.F.	1.78	1.34	3.12
4250	Ordinary hazard, one floor, 2000 S.F.	10.40	3.70	14.10
4300	10,000 S.F.	3.76	1.97	5.73
4350	Each additional floor, 2000 S.F.	2.22	1.64	3.86
4400	10,000 S.F.	1.88	1.44	3.32
4450	Extra hazard, one floor, 2000 S.F.	10.95	4.34	15.29
4500	10,000 S.F.	6.20	2.89	9.09
4550	Each additional floor, 2000 S.F.	3.11	2.59	5.70
4600	10,000 S.F.	3.52	2.39	5.91
5060	Copper tubing, type M, T-drill system, light hazard, one floor 2000 S.F.	10.20	3.26	13.46
5100	10,000 S.F.	3.23	1.43	4.66
5150	Each additional floor, 2000 S.F.	2.35	1.54	3.89
5200	10,000 S.F.	1.66	1.08	2.74
5250	Ordinary hazard, one floor, 2000 S.F.	10.25	3.28	13.53
5300	10,000 S.F.	3.67	1.78	5.45
5350	Each additional floor, 2000 S.F.	2.24	1.45	3.69
5400	10,000 S.F.	2.11	1.43	3.54
5450	Extra hazard, one floor, 2000 S.F.	10.40	3.73	14.13
5500	10,000 S.F.	5.15	2.12	7.27
5550	Each additional floor, 2000 S.F.	2.55	1.98	4.53
5600	10,000 S.F.	2.59	1.66	4.25

D4010 Sprinklers

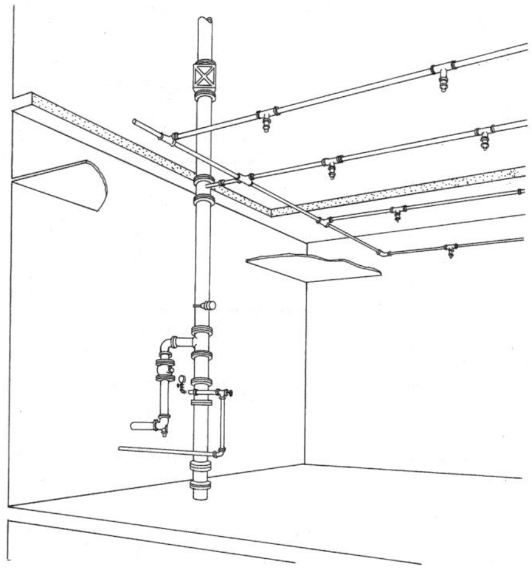

Wet Pipe System. A system employing automatic sprinklers attached to a piping system containing water and connected to a water supply so that water discharges immediately from sprinklers opened by heat from a fire.

All areas are assumed to be open.

System Components	QUANTITY	UNIT	COST EACH MAT.	COST EACH INST.	COST EACH TOTAL
SYSTEM D4010 410 0580					
WET PIPE SPRINKLER, STEEL, BLACK, SCH. 40 PIPE					
LIGHT HAZARD, ONE FLOOR, 2000 S.F.					
Valve, gate, iron body, 125 lb., OS&Y, flanged, 4″ diam.	1.000	Ea.	420	273.75	693.75
Valve, swing check, bronze, 125 lb., regrinding disc, 2-1/2″ pipe size	1.000	Ea.	420	54.38	474.38
Valve, angle, bronze, 150 lb., rising stem, threaded, 2″ diam.	1.000	Ea.	393.75	41.25	435
*Alarm valve, 2-1/2″ pipe size	1.000	Ea.	1,012.50	270	1,282.50
Alarm, water motor, complete with gong	1.000	Ea.	251.25	111.75	363
Valve, swing check, w/balldrip CI with brass trim 4″ pipe size	1.000	Ea.	210	270	480
Pipe, steel, black, schedule 40, 4″ diam.	10.000	L.F.	149.25	233.70	382.95
*Flow control valve, trim & gauges, 4″ pipe size	1.000	Set	2,325	611.25	2,936.25
Fire alarm horn, electric	1.000	Ea.	33.38	64.88	98.26
Pipe, steel, black, schedule 40, threaded, cplg & hngr 10′ OC, 2-1/2″ diam.	20.000	L.F.	198.75	330	528.75
Pipe, steel, black, schedule 40, threaded, cplg & hngr 10′ OC, 2″ diam.	12.500	L.F.	79.22	159.38	238.60
Pipe, steel, black, schedule 40, threaded, cplg & hngr 10′ OC, 1-1/4″ diam.	37.500	L.F.	151.88	344.53	496.41
Pipe steel, black, schedule 40, threaded cplg & hngr 10′ OC, 1″ diam.	112.000	L.F.	357.84	957.60	1,315.44
Pipe Tee, malleable iron black, 150 lb. threaded, 4″ pipe size	2.000	Ea.	291	408	699
Pipe Tee, malleable iron black, 150 lb. threaded, 2-1/2″ pipe size	2.000	Ea.	81.75	181.50	263.25
Pipe Tee, malleable iron black, 150 lb. threaded, 2″ pipe size	1.000	Ea.	19.13	74.25	93.38
Pipe Tee, malleable iron black, 150 lb. threaded, 1-1/4″ pipe size	5.000	Ea.	44.63	292.50	337.13
Pipe Tee, malleable iron black, 150 lb. threaded, 1″ pipe size	4.000	Ea.	22.05	226.50	248.55
Pipe 90° elbow, malleable iron black, 150 lb. threaded, 1″ pipe size	6.000	Ea.	21.15	209.25	230.40
Sprinkler head, standard spray, brass 135°-286° F 1/2″ NPT, 3/8″ orifice	12.000	Ea.	151.20	450	601.20
Valve, gate, bronze, NRS, class 150, threaded, 1″ pipe size	1.000	Ea.	58.88	24	82.88
*Standpipe connection, wall, single, flush w/plug & chain 2-1/2″x2-1/2″	1.000	Ea.	101.25	161.25	262.50
TOTAL			6,793.86	5,749.72	12,543.58
COST PER S.F.			3.40	2.87	6.27

*Not included in systems under 2000 S.F.

D4010 410	Wet Pipe Sprinkler Systems		COST PER S.F. MAT.	COST PER S.F. INST.	COST PER S.F. TOTAL
0520	Wet pipe sprinkler systems, steel, black, sch. 40 pipe				
0530	Light hazard, one floor, 500 S.F.		2.10	2.77	4.87
0560	1000 S.F.	R211313 -20	3.27	2.83	6.10
0580	2000 S.F.		3.40	2.86	6.26
0600	5000 S.F.	R211313 -40	1.65	1.98	3.63
0620	10,000 S.F.		1.10	1.71	2.81

D4010 Sprinklers

D4010 410	Wet Pipe Sprinkler Systems	COST PER S.F. MAT.	COST PER S.F. INST.	COST PER S.F. TOTAL
0640	50,000 S.F.	.85	1.54	2.39
0660	Each additional floor, 500 S.F.	1.01	2.35	3.36
0680	1000 S.F.	.98	2.18	3.16
0700	2000 S.F.	.93	1.96	2.89
0720	5000 S.F.	.68	1.66	2.34
0740	10,000 S.F.	.64	1.54	2.18
0760	50,000 S.F.	.57	1.21	1.78
1000	Ordinary hazard, one floor, 500 S.F.	2.32	2.96	5.28
1020	1000 S.F.	3.24	2.79	6.03
1040	2000 S.F.	3.50	3.01	6.51
1060	5000 S.F.	1.84	2.14	3.98
1080	10,000 S.F.	1.41	2.29	3.70
1100	50,000 S.F.	1.17	2.16	3.33
1140	Each additional floor, 500 S.F.	1.27	2.66	3.93
1160	1000 S.F.	.95	2.15	3.10
1180	2000 S.F.	1.04	2.16	3.20
1200	5000 S.F.	1.05	2.06	3.11
1220	10,000 S.F.	.96	2.13	3.09
1240	50,000 S.F.	.87	1.88	2.75
1500	Extra hazard, one floor, 500 S.F.	6.15	4.57	10.72
1520	1000 S.F.	4.09	3.95	8.04
1540	2000 S.F.	3.73	4.07	7.80
1560	5000 S.F.	2.55	3.50	6.05
1580	10,000 S.F.	2.19	3.36	5.55
1600	50,000 S.F.	2.37	3.23	5.60
1660	Each additional floor, 500 S.F.	1.58	3.28	4.86
1680	1000 S.F.	1.53	3.11	4.64
1700	2000 S.F.	1.38	3.13	4.51
1720	5000 S.F.	1.19	2.75	3.94
1740	10,000 S.F.	1.37	2.54	3.91
1760	50,000 S.F.	1.36	2.42	3.78
2020	Grooved steel, black sch. 40 pipe, light hazard, one floor, 2000 S.F.	3.38	2.42	5.80
2060	10,000 S.F.	1.36	1.55	2.91
2100	Each additional floor, 2000 S.F.	.92	1.57	2.49
2150	10,000 S.F.	.66	1.32	1.98
2200	Ordinary hazard, one floor, 2000 S.F.	3.47	2.59	6.06
2250	10,000 S.F.	1.37	1.91	3.28
2300	Each additional floor, 2000 S.F.	1.01	1.74	2.75
2350	10,000 S.F.	.92	1.75	2.67
2400	Extra hazard, one floor, 2000 S.F.	3.72	3.34	7.06
2450	10,000 S.F.	1.77	2.47	4.24
2500	Each additional floor, 2000 S.F.	1.39	2.57	3.96
2550	10,000 S.F.	1.19	2.20	3.39
3050	Grooved steel black sch. 10 pipe, light hazard, one floor, 2000 S.F.	3.31	2.40	5.71
3100	10,000 S.F.	1.06	1.47	2.53
3150	Each additional floor, 2000 S.F.	.85	1.54	2.39
3200	10,000 S.F.	.60	1.30	1.90
3250	Ordinary hazard, one floor, 2000 S.F.	3.40	2.57	5.97
3300	10,000 S.F.	1.23	1.87	3.10
3350	Each additional floor, 2000 S.F.	.94	1.72	2.66
3400	10,000 S.F.	.78	1.71	2.49
3450	Extra hazard, one floor, 2000 S.F.	3.66	3.31	6.97
3500	10,000 S.F.	1.63	2.42	4.05
3550	Each additional floor, 2000 S.F.	1.33	2.54	3.87
3600	10,000 S.F.	1.11	2.17	3.28
4050	Copper tubing, type M, light hazard, one floor, 2000 S.F.	4.21	2.41	6.62
4100	10,000 S.F.	1.89	1.48	3.37
4150	Each additional floor, 2000 S.F.	1.77	1.59	3.36

D4010 Sprinklers

D4010 410	Wet Pipe Sprinkler Systems	COST PER S.F. MAT.	COST PER S.F. INST.	COST PER S.F. TOTAL
4200	10,000 S.F.	1.43	1.32	2.75
4250	Ordinary hazard, one floor, 2000 S.F.	4.46	2.69	7.15
4300	10,000 S.F.	2.30	1.76	4.06
4350	Each additional floor, 2000 S.F.	2.09	1.75	3.84
4400	10,000 S.F.	1.78	1.56	3.34
4450	Extra hazard, one floor, 2000 S.F.	5	3.33	8.33
4500	10,000 S.F.	4.66	2.65	7.31
4550	Each additional floor, 2000 S.F.	2.68	2.56	5.24
4600	10,000 S.F.	3.17	2.37	5.54
5050	Copper tubing, type M, T-drill system, light hazard, one floor			
5060	2000 S.F.	4.23	2.25	6.48
5100	10,000 S.F.	1.77	1.22	2.99
5150	Each additional floor, 2000 S.F.	1.79	1.43	3.22
5200	10,000 S.F.	1.31	1.06	2.37
5250	Ordinary hazard, one floor, 2000 S.F.	4.27	2.27	6.54
5300	10,000 S.F.	2.21	1.57	3.78
5350	Each additional floor, 2000 S.F.	1.81	1.42	3.23
5400	10,000 S.F.	1.76	1.41	3.17
5450	Extra hazard, one floor, 2000 S.F.	4.45	2.72	7.17
5500	10,000 S.F.	3.73	1.92	5.65
5550	Each additional floor, 2000 S.F.	2.19	1.99	4.18
5600	10,000 S.F.	2.24	1.64	3.88

D4020 Standpipes

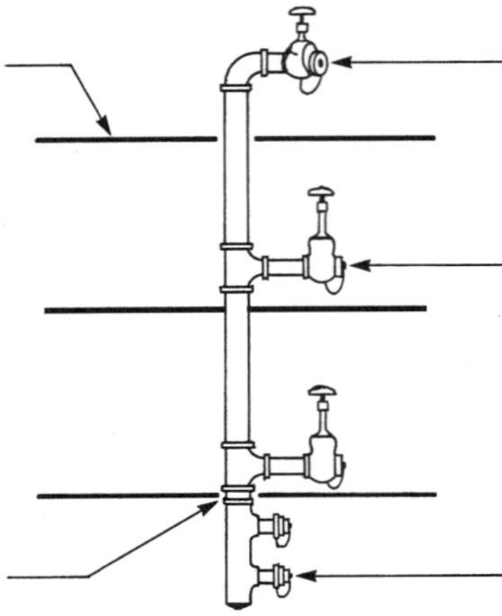

System Components	QUANTITY	UNIT	COST PER FLOOR MAT.	INST.	TOTAL
SYSTEM D4020 310 0560					
WET STANDPIPE RISER, CLASS I, STEEL, BLACK, SCH. 40 PIPE, 10′ HEIGHT					
4″ DIAMETER PIPE, ONE FLOOR					
Pipe, steel, black, schedule 40, threaded, 4″ diam.	20.000	L.F.	500	600	1,100
Pipe, Tee, malleable iron, black, 150 lb. threaded, 4″ pipe size	2.000	Ea.	388	544	932
Pipe, 90° elbow, malleable iron, black, 150 lb. threaded 4″ pipe size	1.000	Ea.	123	181	304
Pipe, nipple, steel, black, schedule 40, 2-1/2″ pipe size x 3″ long	2.000	Ea.	24.90	136	160.90
Fire valve, gate, 300 lb., brass w/handwheel, 2-1/2″ pipe size	1.000	Ea.	171	85.50	256.50
Fire valve, pressure restricting, adj., rgh. brs., 2-1/2″ pipe size	1.000	Ea.	304	171	475
Valve, swing check, w/ball drip, CI w/brs. ftngs., 4″ pipe size	1.000	Ea.	280	360	640
Standpipe conn wall dble. flush brs. w/plugs & chains 2-1/2″x2-1/2″x4″	1.000	Ea.	470	215	685
Valve, swing check, bronze, 125 lb., regrinding disc, 2-1/2″ pipe size	1.000	Ea.	560	72.50	632.50
Roof manifold, fire, w/valves & caps, horiz./vert. brs. 2-1/2″x2-1/2″x4″	1.000	Ea.	152	224	376
Fire, hydrolator, vent & drain, 2-1/2″ pipe size	1.000	Ea.	64.50	50	114.50
Valve, gate, iron body 125 lb., OS&Y, threaded, 4″ pipe size	1.000	Ea.	560	365	925
TOTAL			3,597.40	3,004	6,601.40

D4020 310	Wet Standpipe Risers, Class I		COST PER FLOOR MAT.	INST.	TOTAL
0550	Wet standpipe risers, Class I, steel black sch. 40, 10′ height				
0560	4″ diameter pipe, one floor		3,600	3,000	6,600
0580	Additional floors		855	930	1,785
0600	6″ diameter pipe, one floor		5,925	5,275	11,200
0620	Additional floors	R211226 -20	1,450	1,475	2,925
0640	8″ diameter pipe, one floor		8,850	6,375	15,225
0660	Additional floors		2,075	1,775	3,850
0680					

D4020 310	Wet Standpipe Risers, Class II	COST PER FLOOR MAT.	INST.	TOTAL
1030	Wet standpipe risers, Class II, steel black sch. 40, 10′ height			
1040	2″ diameter pipe, one floor	1,425	1,075	2,500
1060	Additional floors	465	420	885
1080	2-1/2″ diameter pipe, one floor	2,075	1,600	3,675
1100	Additional floors	535	490	1,025
1120				

D4020 Standpipes

D4020 310	Wet Standpipe Risers, Class III	COST PER FLOOR MAT.	COST PER FLOOR INST.	COST PER FLOOR TOTAL
1530	Wet standpipe risers, Class III, steel black sch. 40, 10′ height			
1540	4″ diameter pipe, one floor	3,700	3,000	6,700
1560	Additional floors	725	775	1,500
1580	6″ diameter pipe, one floor	6,025	5,275	11,300
1600	Additional floors	1,500	1,475	2,975
1620	8″ diameter pipe, one floor	8,950	6,375	15,325
1640	Additional floors	2,125	1,775	3,900

D4020 Standpipes

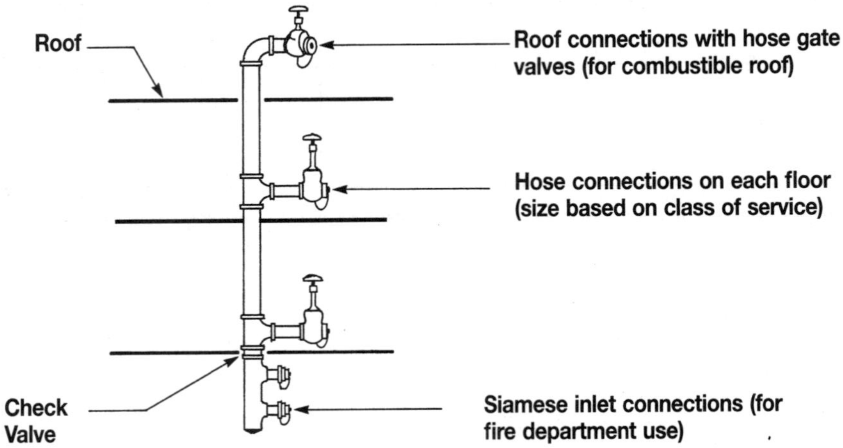

System Components	QUANTITY	UNIT	COST PER FLOOR MAT.	INST.	TOTAL
SYSTEM D4020 330 0540					
DRY STANDPIPE RISER, CLASS I, PIPE, STEEL, BLACK, SCH 40, 10′ HEIGHT					
4″ DIAMETER PIPE, ONE FLOOR					
Pipe, steel, black, schedule 40, threaded, 4″ diam.	20.000	L.F.	500	600	1,100
Pipe, Tee, malleable iron, black, 150 lb. threaded, 4″ pipe size	2.000	Ea.	388	544	932
Pipe, 90° elbow, malleable iron, black, 150 lb. threaded 4″ pipe size	1.000	Ea.	123	181	304
Pipe, nipple, steel, black, schedule 40, 2-1/2″ pipe size x 3″ long	2.000	Ea.	24.90	136	160.90
Fire valve gate NRS 300 lb., brass w/handwheel, 2-1/2″ pipe size	1.000	Ea.	171	85.50	256.50
Fire valve, pressure restricting, adj., rgh. brs., 2-1/2″ pipe size	1.000	Ea.	152	85.50	237.50
Standpipe conn wall dble. flush brs. w/plugs & chains 2-1/2″x2-1/2″x4″	1.000	Ea.	470	215	685
Valve swing check w/ball drip CI w/brs. ftngs., 4″pipe size	1.000	Ea.	280	360	640
Roof manifold, fire, w/valves & caps, horiz./vert. brs. 2-1/2″x2-1/2″x4″	1.000	Ea.	152	224	376
TOTAL			2,260.90	2,431	4,691.90

D4020 330	Dry Standpipe Risers, Class I	COST PER FLOOR MAT.	INST.	TOTAL
0530	Dry standpipe riser, Class I, steel black sch. 40, 10′ height			
0540	4″ diameter pipe, one floor	2,250	2,425	4,675
0560	Additional floors (R211226-20)	790	880	1,670
0580	6″ diameter pipe, one floor	4,325	4,200	8,525
0600	Additional floors	1,375	1,425	2,800
0620	8″ diameter pipe, one floor	6,500	5,100	11,600
0640	Additional floors	2,000	1,725	3,725
0660				

D4020 330	Dry Standpipe Risers, Class II	COST PER FLOOR MAT.	INST.	TOTAL
1030	Dry standpipe risers, Class II, steel black sch. 40, 10′ height			
1040	2″ diameter pipe, one floor	1,225	1,125	2,350
1060	Additional floor	400	370	770
1080	2-1/2″ diameter pipe, one floor	1,650	1,325	2,975
1100	Additional floors	470	440	910
1120				

D4020 Standpipes

D4020 330	Dry Standpipe Risers, Class III	COST PER FLOOR		
		MAT.	INST.	TOTAL
1530	Dry standpipe risers, Class III, steel black sch. 40, 10′ height			
1540	4″ diameter pipe, one floor	2,300	2,375	4,675
1560	Additional floors	670	795	1,465
1580	6″ diameter pipe, one floor	4,375	4,200	8,575
1600	Additional floors	1,425	1,425	2,850
1620	8″ diameter pipe, one floor	6,550	5,100	11,650
1640	Additional floor	2,050	1,725	3,775

D4020 410	Fire Hose Equipment	COST EACH		
		MAT.	INST.	TOTAL
0100	Adapters, reducing, 1 piece, FxM, hexagon, cast brass, 2-1/2" x 1-1/2"	50		50
0200	Pin lug, 1-1/2" x 1"	43.50		43.50
0250	3" x 2-1/2"	113		113
0300	For polished chrome, add 75% mat.			
0400	Cabinets, D.S. glass in door, recessed, steel box, not equipped			
0500	Single extinguisher, steel door & frame	95.50	134	229.50
0550	Stainless steel door & frame	204	134	338
0600	Valve, 2-1/2" angle, steel door & frame	123	89.50	212.50
0650	Aluminum door & frame	149	89.50	238.50
0700	Stainless steel door & frame	200	89.50	289.50
0750	Hose rack assy, 2-1/2" x 1-1/2" valve & 100' hose, steel door & frame	229	179	408
0800	Aluminum door & frame	335	179	514
0850	Stainless steel door & frame	450	179	629
0900	Hose rack assy,& extinguisher, 2-1/2"x1-1/2" valve & hose,steel door & fram	228	215	443
0950	Aluminum	430	215	645
1000	Stainless steel	785	215	1,000
1550	Compressor, air, dry pipe system, automatic, 200 gal., 3/4 H.P.	765	460	1,225
1600	520 gal., 1 H.P.	805	460	1,265
1650	Alarm, electric pressure switch (circuit closer)	265	23	288
2500	Couplings, hose, rocker lug, cast brass, 1-1/2"	48		48
2550	2-1/2"	63		63
3000	Escutcheon plate, for angle valves, polished brass, 1-1/2"	12.15		12.15
3050	2-1/2"	29		29
3500	Fire pump, electric, w/controller, fittings, relief valve			
3550	4" pump, 30 H.P., 500 G.P.M.	21,800	3,350	25,150
3600	5" pump, 40 H.P., 1000 G.P.M.	31,800	3,800	35,600
3650	5" pump, 100 H.P., 1000 G.P.M.	35,700	4,225	39,925
3700	For jockey pump system, add	3,900	540	4,440
3800	Fire pump, diesel, w/controller, 6" pump, 140 HP, 1500 GPM, 100 psi	93,500	5,625	99,125
3900	Fire pump, diesel, w/controller, 10" pump, 300 HP, 3500 GPM, 100 psi	162,500	10,200	172,700
4100	Fire pump, electric, w/controller, 6" pump, 139 HP, 1500 GPM, 1770 RPM	46,100	5,625	51,725
4200	Fire pump, electric, w/controller, 10" pump, 300 HP, 3500 GPM, 1770 RPM	88,000	9,475	97,475
5000	Hose, per linear foot, synthetic jacket, lined,			
5100	300 lb. test, 1-1/2" diameter	2.63	.41	3.04
5150	2-1/2" diameter	4.38	.49	4.87
5200	500 lb. test, 1-1/2" diameter	2.72	.41	3.13
5250	2-1/2" diameter	4.72	.49	5.21
5500	Nozzle, plain stream, polished brass, 1-1/2" x 10"	43.50		43.50
5550	2-1/2" x 15" x 13/16" or 1-1/2"	78.50		78.50
5600	Heavy duty combination adjustable fog and straight stream w/handle 1-1/2"	405		405
5650	2-1/2" direct connection	505		505
6000	Rack, for 1-1/2" diameter hose 100 ft. long, steel	54	54	108
6050	Brass	66.50	54	120.50
6500	Reel, steel, for 50 ft. long 1-1/2" diameter hose	118	77	195
6550	For 75 ft. long 2-1/2" diameter hose	192	77	269
7050	Siamese, w/plugs & chains, polished brass, sidewalk, 4" x 2-1/2" x 2-1/2"	520	430	950
7100	6" x 2-1/2" x 2-1/2"	710	540	1,250
7200	Wall type, flush, 4" x 2-1/2" x 2-1/2"	470	215	685
7250	6" x 2-1/2" x 2-1/2"	655	234	889
7300	Projecting, 4" x 2-1/2" x 2-1/2"	430	215	645
7350	6" x 2-1/2" x 2-1/2"	725	234	959
7400	For chrome plate, add 15% mat.			
8000	Valves, angle, wheel handle, 300 Lb., rough brass, 1-1/2"	63.50	50	113.50
8050	2-1/2"	66	85.50	151.50
8100	Combination pressure restricting, 1-1/2"	77	50	127
8150	2-1/2"	153	85.50	238.50
8200	Pressure restricting, adjustable, satin brass, 1-1/2"	109	50	159
8250	2-1/2"	152	85.50	237.50

D4020 Standpipes

D4020 410	Fire Hose Equipment	COST EACH		
		MAT.	INST.	TOTAL
8300	Hydrolator, vent and drain, rough brass, 1-1/2"	64.50	50	114.50
8350	2-1/2"	64.50	50	114.50
8400	Cabinet assy, incls. adapter, rack, hose, and nozzle	675	325	1,000

D4090 Other Fire Protection Systems

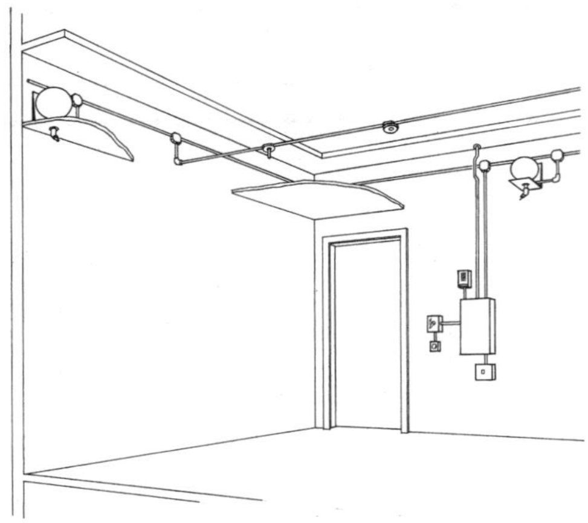

General: Automatic fire protection (suppression) systems other than water sprinklers may be desired for special environments, high risk areas, isolated locations or unusual hazards. Some typical applications would include:

Paint dip tanks
Securities vaults
Electronic data processing
Tape and data storage
Transformer rooms
Spray booths
Petroleum storage
High rack storage

Piping and wiring costs are dependent on the individual application and must be added to the component costs shown below.

All areas are assumed to be open.

D4090 910	Unit Components	COST EACH		
		MAT.	INST.	TOTAL
0020	Detectors with brackets			
0040	Fixed temperature heat detector	36	72.50	108.50
0060	Rate of temperature rise detector	42.50	72.50	115
0080	Ion detector (smoke) detector	96	93.50	189.50
0200	Extinguisher agent			
0240	200 lb FM200, container	6,700	272	6,972
0280	75 lb carbon dioxide cylinder	1,100	181	1,281
0320	Dispersion nozzle			
0340	FM200 1-1/2" dispersion nozzle	58	43	101
0380	Carbon dioxide 3" x 5" dispersion nozzle	58	33.50	91.50
0420	Control station			
0440	Single zone control station with batteries	1,475	580	2,055
0470	Multizone (4) control station with batteries	2,825	1,150	3,975
0490				
0500	Electric mechanical release	144	296	440
0520				
0550	Manual pull station	52	101	153
0570				
0640	Battery standby power 10" x 10" x 17"	885	145	1,030
0700				
0740	Bell signalling device	63.50	72.50	136

D4090 920	FM200 Systems	COST PER C.F.		
		MAT.	INST.	TOTAL
0820	Average FM200 system, minimum			1.52
0840	Maximum			3.03

D5010 Electrical Service/Distribution

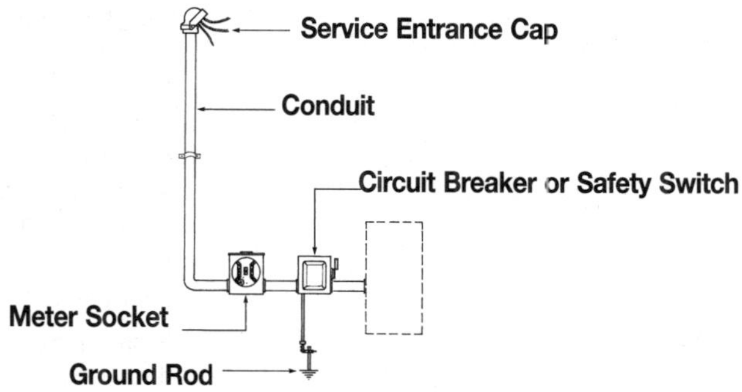

System Components	QUANTITY	UNIT	COST EACH MAT.	COST EACH INST.	COST EACH TOTAL
SYSTEM D5010 120 0220					
SERVICE INSTALLATION, INCLUDES BREAKERS, METERING, 20′ CONDUIT & WIRE					
3 PHASE, 4 WIRE, 60 A					
Circuit breaker, enclosed (NEMA 1), 600 volt, 3 pole, 60 A	1.000	Ea.	670	207	877
Meter socket, single position, 4 terminal, 100 A	1.000	Ea.	47.50	181	228.50
Rigid galvanized steel conduit, 3/4″, including fittings	20.000	L.F.	71	145	216
Wire, 600V type XHHW, copper stranded #6	.900	C.L.F.	110.70	80.10	190.80
Service entrance cap 3/4″ diameter	1.000	Ea.	11.55	44.50	56.05
Conduit LB fitting with cover, 3/4″ diameter	1.000	Ea.	15.30	44.50	59.80
Ground rod, copper clad, 8′ long, 3/4″ diameter	1.000	Ea.	30.50	109	139.50
Ground rod clamp, bronze, 3/4″ diameter	1.000	Ea.	7.35	18.10	25.45
Ground wire, bare armored, #6-1 conductor	.200	C.L.F.	28.80	64	92.80
TOTAL			992.70	893.20	1,885.90

D5010 120	Electric Service, 3 Phase - 4 Wire	COST EACH MAT.	COST EACH INST.	COST EACH TOTAL
0200	Service installation, includes breakers, metering, 20′ conduit & wire			
0220	3 phase, 4 wire, 120/208 volts, 60 A	995	895	1,890
0240	100 A	1,225	1,075	2,300
0280	200 A	1,875	1,650	3,525
0320	400 A	4,450	3,025	7,475
0360	600 A	8,400	4,100	12,500
0400	800 A	10,800	4,925	15,725
0440	1000 A	13,300	5,675	18,975
0480	1200 A	16,700	5,800	22,500
0520	1600 A	30,200	8,350	38,550
0560	2000 A	33,300	9,500	42,800
0570	Add 25% for 277/480 volt			
0580				
0610	1 phase, 3 wire, 120/240 volts, 100 A	525	970	1,495
0620	200 A	1,100	1,425	2,525

D5020 Lighting and Branch Wiring

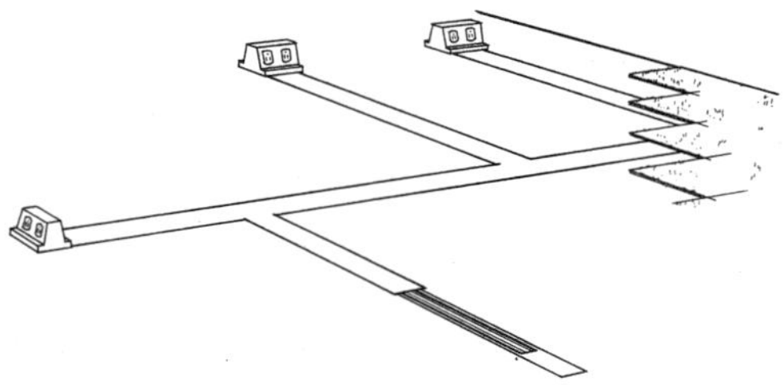

Underfloor Receptacle System

Description: Table D5020 115 includes installed costs of raceways and copper wire from panel to and including receptacle.

National Electrical Code prohibits use of undercarpet system in residential, school or hospital buildings. Can only be used with carpet squares.

Low density = (1) Outlet per 259 S.F. of floor area.

High density = (1) Outlet per 127 S.F. of floor area.

System Components	QUANTITY	UNIT	COST PER S.F. MAT.	INST.	TOTAL
SYSTEM D5020 115 0200					
RECEPTACLE SYSTEMS, UNDERFLOOR DUCT, 5′ ON CENTER, LOW DENSITY					
Underfloor duct 3-1/8″ x 7/8″ w/insert 24″ on center	.190	L.F.	3.55	1.57	5.12
Vertical elbow for underfloor duct, 3-1/8″, included					
Underfloor duct conduit adapter, 2″ x 1-1/4″, included					
Underfloor duct junction box, single duct, 3-1/8″	.003	Ea.	1.26	.44	1.70
Underfloor junction box carpet pan	.003	Ea.	1.05	.02	1.07
Underfloor duct outlet, high tension receptacle	.004	Ea.	.39	.29	.68
Wire 600V type THWN-THHN copper solid #12	.010	C.L.F.	.18	.53	.71
TOTAL			6.43	2.85	9.28

D5020 115	Receptacles, Floor	COST PER S.F. MAT.	INST.	TOTAL
0200	Receptacle systems, underfloor duct, 5′ on center, low density	6.45	2.85	9.30
0240	High density	7	3.66	10.66
0280	7′ on center, low density	5.10	2.44	7.54
0320	High density	5.65	3.25	8.90
0400	Poke thru fittings, low density	1.15	1.40	2.55
0440	High density	2.26	2.75	5.01
0520	Telepoles, using Romex, low density	1.14	.85	1.99
0560	High density	2.27	1.71	3.98
0600	Using EMT, low density	1.21	1.14	2.35
0640	High density	2.40	2.27	4.67
0720	Conduit system with floor boxes, low density	1.09	.98	2.07
0760	High density	2.19	1.96	4.15
0840	Undercarpet power system, 3 conductor with 5 conductor feeder, low density	1.46	.34	1.80
0880	High density	2.81	.72	3.53

D5020 Lighting and Branch Wiring

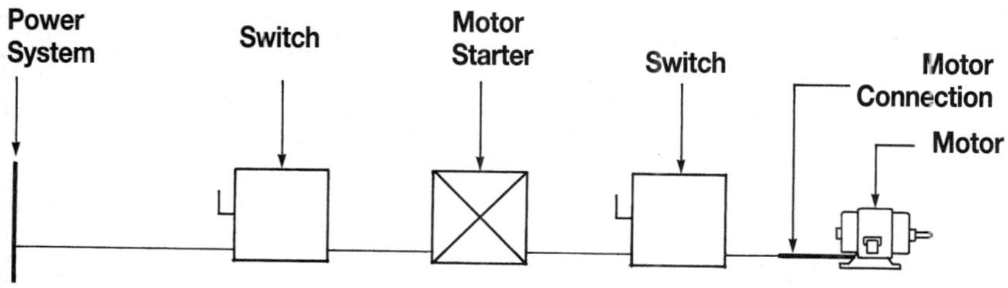

System D5020 145 installed cost of motor wiring as per Table D5010 170 using 50′ of rigid conduit and copper wire. **Cost and setting of motor not included.**

System Components	QUANTITY	UNIT	COST EACH MAT.	COST EACH INST.	COST EACH TOTAL
SYSTEM D5020 145 0200					
MOTOR INST., SINGLE PHASE, 115V, TO AND INCLUDING 1/3 HP MOTOR SIZE					
Wire 600V type THWN-THHN, copper solid #12	1.250	C.L.F.	21.88	65.63	87.51
Steel intermediate conduit, (IMC) 1/2″ diam.	50.000	L.F.	112	290	402
Magnetic FVNR, 115V, 1/3 HP, size 00 starter	1.000	Ea.	205	145	350
Safety switch, fused, heavy duty, 240V 2P 30 amp	1.000	Ea.	151	165	316
Safety switch, non fused, heavy duty, 600V, 3 phase, 30 A	1.000	Ea.	177	181	358
Flexible metallic conduit, Greenfield 1/2″ diam.	1.500	L.F.	.81	4.34	5.15
Connectors for flexible metallic conduit Greenfield 1/2″ diam.	1.000	Ea.	3.70	7.25	10.95
Coupling for Greenfield to conduit 1/2″ diam flexible metalic conduit	1.000	Ea.	2.20	11.55	13.75
Fuse cartridge nonrenewable, 250V 30 amp	1.000	Ea.	2.18	11.55	13.73
TOTAL			675.77	881.32	1,557.09

D5020 145	Motor Installation	COST EACH MAT.	COST EACH INST.	COST EACH TOTAL
0200	Motor installation, single phase, 115V, 1/3 HP motor size	675	880	1,555
0240	1 HP motor size	700	880	1,580
0280	2 HP motor size	750	935	1,685
0320	3 HP motor size	835	955	1,790
0360	230V, 1 HP motor size	680	895	1,575
0400	2 HP motor size	710	895	1,605
0440	3 HP motor size	790	960	1,750
0520	Three phase, 200V, 1-1/2 HP motor size	795	980	1,775
0560	3 HP motor size	845	1,075	1,920
0600	5 HP motor size	905	1,175	2,080
0640	7-1/2 HP motor size	940	1,200	2,140
0680	10 HP motor size	1,475	1,525	3,000
0720	15 HP motor size	1,900	1,700	3,600
0760	20 HP motor size	2,475	1,950	4,425
0800	25 HP motor size	2,525	1,950	4,475
0840	30 HP motor size	3,925	2,300	6,225
0880	40 HP motor size	5,050	2,725	7,775
0920	50 HP motor size	8,375	3,150	11,525
0960	60 HP motor size	8,825	3,350	12,175
1000	75 HP motor size	11,500	3,825	15,325
1040	100 HP motor size	27,100	4,525	31,625
1080	125 HP motor size	28,300	4,975	33,275
1120	150 HP motor size	33,600	5,825	39,425
1160	200 HP motor size	34,100	7,100	41,200
1240	230V, 1-1/2 HP motor size	760	970	1,730
1280	3 HP motor size	810	1,050	1,860
1320	5 HP motor size	865	1,175	2,040
1360	7-1/2 HP motor size	865	1,175	2,040
1400	10 HP motor size	1,325	1,425	2,750
1440	15 HP motor size	1,550	1,575	3,125
1480	20 HP motor size	2,275	1,900	4,175
1520	25 HP motor size	2,475	1,950	4,425

D5020 Lighting and Branch Wiring

D5020 145	Motor Installation	COST EACH		
		MAT.	INST.	TOTAL
1560	30 HP motor size	2,525	1,950	4,475
1600	40 HP motor size	4,625	2,650	7,275
1640	50 HP motor size	4,900	2,800	7,700
1680	60 HP motor size	8,375	3,175	11,550
1720	75 HP motor size	10,500	3,600	14,100
1760	100 HP motor size	11,800	4,000	15,800
1800	125 HP motor size	27,500	4,675	32,175
1840	150 HP motor size	30,200	5,300	35,500
1880	200 HP motor size	31,800	5,875	37,675
1960	460V, 2 HP motor size	930	980	1,910
2000	5 HP motor size	975	1,075	2,050
2040	10 HP motor size	1,000	1,175	2,175
2080	15 HP motor size	1,300	1,350	2,650
2120	20 HP motor size	1,350	1,450	2,800
2160	25 HP motor size	1,500	1,525	3,025
2200	30 HP motor size	1,875	1,625	3,500
2240	40 HP motor size	2,425	1,750	4,175
2280	50 HP motor size	2,750	1,950	4,700
2320	60 HP motor size	3,925	2,275	6,200
2360	75 HP motor size	4,650	2,525	7,175
2400	100 HP motor size	5,200	2,825	8,025
2440	125 HP motor size	8,675	3,175	11,850
2480	150 HP motor size	10,500	3,575	14,075
2520	200 HP motor size	12,600	4,025	16,625
2600	575V, 2 HP motor size	930	980	1,910
2640	5 HP motor size	975	1,075	2,050
2680	10 HP motor size	1,000	1,175	2,175
2720	20 HP motor size	1,300	1,350	2,650
2760	25 HP motor size	1,350	1,450	2,800
2800	30 HP motor size	1,875	1,625	3,500
2840	50 HP motor size	2,025	1,700	3,725
2880	60 HP motor size	3,875	2,275	6,150
2920	75 HP motor size	3,925	2,275	6,200
2960	100 HP motor size	4,650	2,525	7,175
3000	125 HP motor size	8,000	3,125	11,125
3040	150 HP motor size	8,675	3,175	11,850
3080	200 HP motor size	11,200	3,600	14,800

D5020 Lighting and Branch Wiring

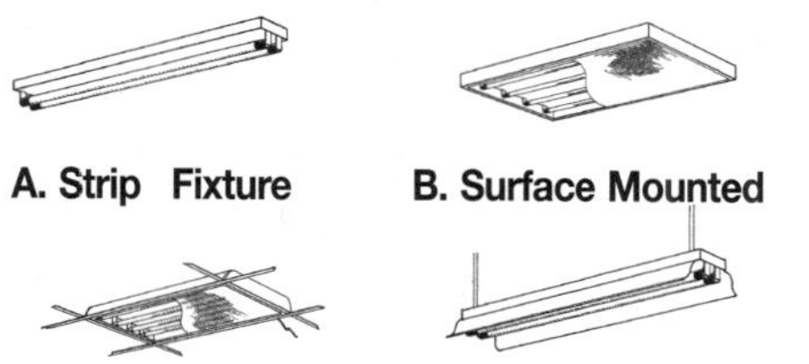

Design Assumptions:

1. A 100 footcandle average maintained level of illumination.
2. Ceiling heights range from 9′ to 11′.
3. Average reflectance values are assumed for ceilings, walls and floors.
4. Cool white (CW) flucrescent lamps with 3150 lumens for 40 watt lamps and 6300 lumens for 8′ slimline lamps.
5. Four 40 watt lamps per 4′ fixture and two 8′ lamps per 8′ fixture.
6. Average fixture efficiency values and spacing to mounting height ratios.
7. Installation labor is average U.S. rate as of January 1.

System Components	QUANTITY	UNIT	COST PER S.F. MAT.	COST PER S.F. INST.	COST PER S.F. TOTAL
SYSTEM D5020 208 0520					
FLUORESCENT FIXTURES MOUNTED 9′-11″ ABOVE FLOOR, 100 FC					
TYPE A, 8 FIXTURES PER 400 S.F.					
Steel intermediate conduit, (IMC) 1/2″ diam.	.404	L.F.	.90	2.34	3.24
Wire, 600V, type THWN-THHN, copper, solid, #12	.008	C.L.F.	.14	.42	.56
Fluorescent strip fixture 8′ long, surface mounted, two 75W SL	.020	Ea.	1.23	1.87	3.10
Steel outlet box 4″ concrete	.020	Ea.	.33	.58	.91
Steel outlet box plate with stud, 4″ concrete	.020	Ea.	.13	.15	.28
TOTAL			2.73	5.36	8.09

D5020 208	Fluorescent Fixtures (by Type)	COST PER S.F. MAT.	COST PER S.F. INST.	COST PER S.F. TOTAL
0520	Fluorescent fixtures, type A, 8 fixtures per 400 S.F.	2.73	5.35	8.08
0560	11 fixtures per 600 S.F.	2.60	5.20	7.80
0600	17 fixtures per 1000 S.F.	2.54	5.10	7.64
0640	23 fixtures per 1600 S.F.	2.33	4.84	7.17
0680	28 fixtures per 2000 S.F.	2.33	4.84	7.17
0720	41 fixtures per 3000 S.F.	2.29	4.83	7.12
0800	53 fixtures per 4000 S.F.	2.25	4.71	6.96
0840	64 fixtures per 5000 S.F.	2.25	4.71	6.96
0880	Type B, 11 fixtures per 400 S.F.	4.94	7.85	12.79
0920	15 fixtures per 600 S.F.	4.63	7.55	12.18
0960	24 fixtures per 1000 S.F.	4.53	7.50	12.03
1000	35 fixtures per 1600 S.F.	4.27	7.15	11.42
1040	42 fixtures per 2000 S.F.	4.19	7.15	11.34
1080	61 fixtures per 3000 S.F.	4.18	6.90	11.08
1160	80 fixtures per 4000 S.F.	4.09	7.05	11.14
1200	98 fixtures per 5000 S.F.	4.07	7	11.07
1240	Type C, 11 fixtures per 400 S.F.	4.05	8.25	12.30
1280	14 fixtures per 600 S.F.	3.67	7.70	11.37
1320	23 fixtures per 1000 S.F.	3.65	7.65	11.30
1360	34 fixtures per 1600 S.F.	3.56	7.55	11.11
1400	43 fixtures per 2000 S.F.	3.57	7.50	11.07
1440	63 fixtures per 3000 S.F.	3.49	7.40	10.89
1520	81 fixtures per 4000 S.F.	3.42	7.30	10.72
1560	101 fixtures per 5000 S.F.	3.42	7.30	10.72
1600	Type D, 8 fixtures per 400 S.F.	3.73	6.40	10.13
1640	12 fixtures per 600 S.F.	3.73	6.35	10.08
1680	19 fixtures per 1000 S.F.	3.60	6.20	9.80
1720	27 fixtures per 1600 S.F.	3.40	6.05	9.45
1760	34 fixtures per 2000 S.F.	3.38	6	9.38
1800	48 fixtures per 3000 S.F.	3.26	5.80	9.06
1880	64 fixtures per 4000 S.F.	3.26	5.80	9.06
1920	79 fixtures per 5000 S.F.	3.26	5.80	9.06

D5020 Lighting and Branch Wiring

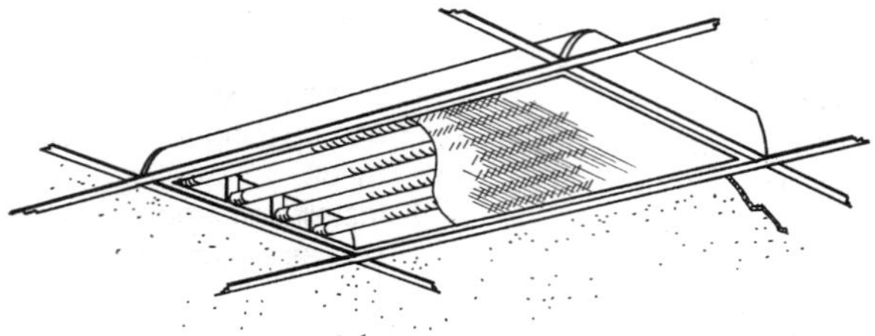

Type C. Recessed, mounted on grid ceiling suspension system, 2′ x 4′, four 40 watt lamps, acrylic prismatic diffusers.

5.3 watts per S.F. for 100 footcandles.

3 watts per S.F. for 57 footcandles.

System Components	QUANTITY	UNIT	COST PER S.F. MAT.	INST.	TOTAL
SYSTEM D5020 210 0200					
FLUORESCENT FIXTURES RECESS MOUNTED IN CEILING					
1 WATT PER S.F., 20 FC, 5 FIXTURES PER 1000 S.F.					
Steel intermediate conduit, (IMC) 1/2″ diam.	.128	L.F.	.29	.74	1.03
Wire, 600 volt, type THW, copper, solid, #12	.003	C.L.F.	.05	.16	.21
Fluorescent fixture, recessed, 2′x 4′, four 40W, w/ lens, for grid ceiling	.005	Ea.	.35	.62	.97
Steel outlet box 4″ square	.005	Ea.	.08	.15	.23
Fixture whip, Greenfield w/#12 THHN wire	.005	Ea.	.03	.04	.07
TOTAL			.80	1.71	2.51

D5020 210	Fluorescent Fixtures (by Wattage)	COST PER S.F. MAT.	INST.	TOTAL
0190	Fluorescent fixtures recess mounted in ceiling			
0195	T-12, standard 40 watt lamps			
0200	1 watt per S.F., 20 FC, 5 fixtures @40 watts per 1000 S.F.	.80	1.71	2.51
0240	2 watt per S.F., 40 FC, 10 fixtures @40 watt per 1000 S.F.	1.60	3.33	4.93
0280	3 watt per S.F., 60 FC, 15 fixtures @40 watt per 1000 S.F	2.40	5.05	7.45
0320	4 watt per S.F., 80 FC, 20 fixtures @40 watt per 1000 S.F.	3.20	6.70	9.90
0400	5 watt per S.F., 100 FC, 25 fixtures @40 watt per 1000 S.F.	4	8.40	12.40
0450	T-8, energy saver 32 watt lamps			
0500	0.8 watt per S.F., 20 FC, 5 fixtures @32 watt per 1000 S.F.	.88	1.71	2.59
0520	1.6 watt per S.F., 40 FC, 10 fixtures @32 watt per 1000 S.F.	1.76	3.33	5.09
0540	2.4 watt per S.F., 60 FC, 15 fixtures @ 32 watt per 1000 S.F	2.64	5.05	7.69
0560	3.2 watt per S.F., 80 FC, 20 fixtures @32 watt per 1000 S.F.	3.52	6.70	10.22
0580	4 watt per S.F., 100 FC, 25 fixtures @32 watt per 1000 S.F.	4.40	8.40	12.80

D5030 Communications and Security

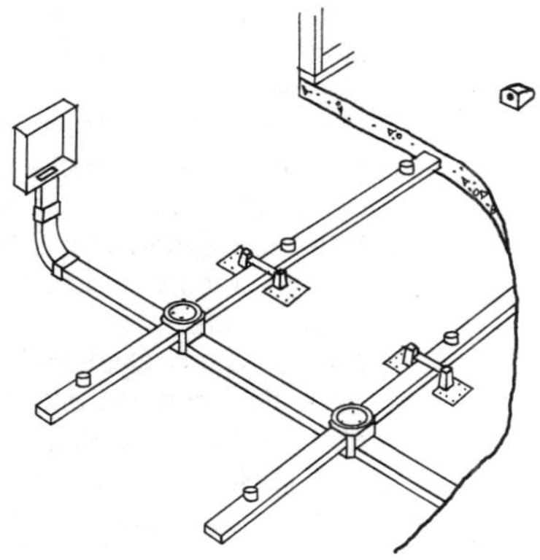

Description: System below includes telephone fitting installed. Does not include cable.

When poke thru fittings and telepoles are used for power, they can also be used for telephones at a negligible additional cost.

System Components	QUANTITY	UNIT	COST PER S.F. MAT.	COST PER S.F. INST.	COST PER S.F. TOTAL
SYSTEM D5030 310 0200					
TELEPHONE SYSTEMS, UNDERFLOOR DUCT, 5′ ON CENTER, LOW DENSITY					
Underfloor duct 7-1/4″ w/insert 2′ O.C. 1-3/8″ x 7-1/4″ super duct	.190	L.F.	6.18	2.19	8.37
Vertical elbow for underfloor superduct, 7-1/4″, included					
Underfloor duct conduit adapter, 2″ x 1-1/4″, included					
Underfloor duct junction box, single duct, 7-1/4″ x 3 1/8″	.003	Ea.	1.47	.44	1.91
Underfloor junction box carpet pan	.003	Ea.	1.05	.02	1.07
Underfloor duct outlet, low tension	.004	Ea.	.39	.29	.68
TOTAL			9.09	2.94	12.03

D5030 310	Telephone Systems	COST PER S.F. MAT.	COST PER S.F. INST.	COST PER S.F. TOTAL
0200	Telephone systems, underfloor duct, 5′ on center, low density	9.10	2.94	12.04
0240	5′ on center, high density	9.50	3.23	12.73
0280	7′ on center, low density	7.25	2.44	9.69
0320	7′ on center, high density	7.65	2.73	10.38
0400	Poke thru fittings, low density	1.06	1.05	2.11
0440	High density	2.12	2.10	4.22
0520	Telepoles, low density	1.08	.62	1.70
0560	High density	2.16	1.25	3.41
0640	Conduit system with floor boxes, low density	1.23	1.03	2.26
0680	High density	2.44	2.03	4.47
1020	Telephone wiring for offices & laboratories, 8 jacks/MSF	.56	2.51	3.07

D5030 Communications and Security

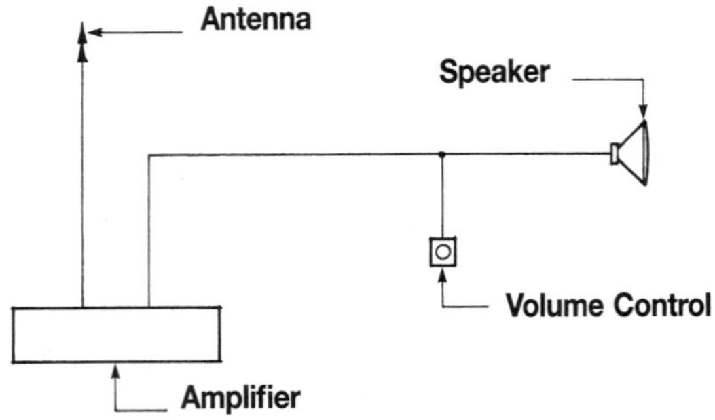

Sound System Includes AM–FM antenna, outlets, rigid conduit, and copper wire.
Fire Detection System Includes pull stations, signals, smoke and heat detectors, rigid conduit, and copper wire.
Intercom System Includes master and remote stations, rigid conduit, and copper wire.
Master Clock System Includes clocks, bells, rigid conduit, and copper wire.
Master TV Antenna Includes antenna, VHF–UHF reception and distribution, rigid conduit, and copper wire.

System Components	QUANTITY	UNIT	COST EACH MAT.	COST EACH INST.	COST EACH TOTAL
SYSTEM D5030 910 0220					
SOUND SYSTEM, INCLUDES OUTLETS, BOXES, CONDUIT & WIRE					
Steel intermediate conduit, (IMC) 1/2" diam	1200.000	L.F.	2,688	6,960	9,648
Wire sound shielded w/drain, #22-2 conductor	15.500	C.L.F.	465	1,123.75	1,588.75
Sound system speakers ceiling or wall	12.000	Ea.	1,452	870	2,322
Sound system volume control	12.000	Ea.	1,080	870	1,950
Sound system amplifier, 250 Watts	1.000	Ea.	1,800	580	2,380
Sound system antenna, AM FM	1.000	Ea.	225	145	370
Sound system monitor panel	1.000	Ea.	400	145	545
Sound system cabinet	1.000	Ea.	875	580	1,455
Steel outlet box 4" square	12.000	Ea.	31.44	348	379.44
Steel outlet box 4" plaster rings	12.000	Ea.	37.44	108.60	146.04
TOTAL			9,053.88	11,730.35	20,784.23

D5030 910	Communication & Alarm Systems	COST EACH MAT.	COST EACH INST.	COST EACH TOTAL
0200	Communication & alarm systems, includes outlets, boxes, conduit & wire			
0210	Sound system, 6 outlets	6,475	7,325	13,800
0220	12 outlets	9,050	11,700	20,750
0240	30 outlets	15,700	22,200	37,900
0280	100 outlets	47,800	74,000	121,800
0320	Fire detection systems, non-addressable, 12 detectors	3,125	6,075	9,200
0360	25 detectors	5,450	10,300	15,750
0400	50 detectors	10,700	20,400	31,100
0440	100 detectors	19,900	37,100	57,000
0450	Addressable type, 12 detectors	4,225	6,125	10,350
0452	25 detectors	7,675	10,400	18,075
0454	50 detectors	14,900	20,400	35,300
0456	100 detectors	28,600	37,400	66,000
0458	Fire alarm control panel, 8 zone	740	1,150	1,890
0459	12 zone	2,525	1,725	4,250
0460	Fire alarm command center, addressable without voice	2,600	1,000	3,600
0480	Intercom systems, 6 stations	3,625	5,000	8,625
0520	12 stations	7,375	10,000	17,375
0560	25 stations	12,700	19,200	31,900
0600	50 stations	24,500	36,100	60,600
0640	100 stations	48,300	70,500	118,800
0680	Master clock systems, 6 rooms	4,800	8,175	12,975
0720	12 rooms	7,425	13,900	21,325
0760	20 rooms	10,100	19,700	29,800
0800	30 rooms	17,100	36,400	53,500
0840	50 rooms	27,800	61,500	89,300

D5030 Communications and Security

D5030 910	Communication & Alarm Systems	COST EACH		
		MAT.	INST.	TOTAL
0880	100 rooms	54,000	121,500	175,500
0920	Master TV antenna systems, 6 outlets	3,250	5,175	8,425
0960	12 outlets	6,050	9,625	15,675
1000	30 outlets	12,400	22,300	34,700
1040	100 outlets	41,100	72,500	113,600

E Equipment & Furnishings

E2020 Moveable Furnishings

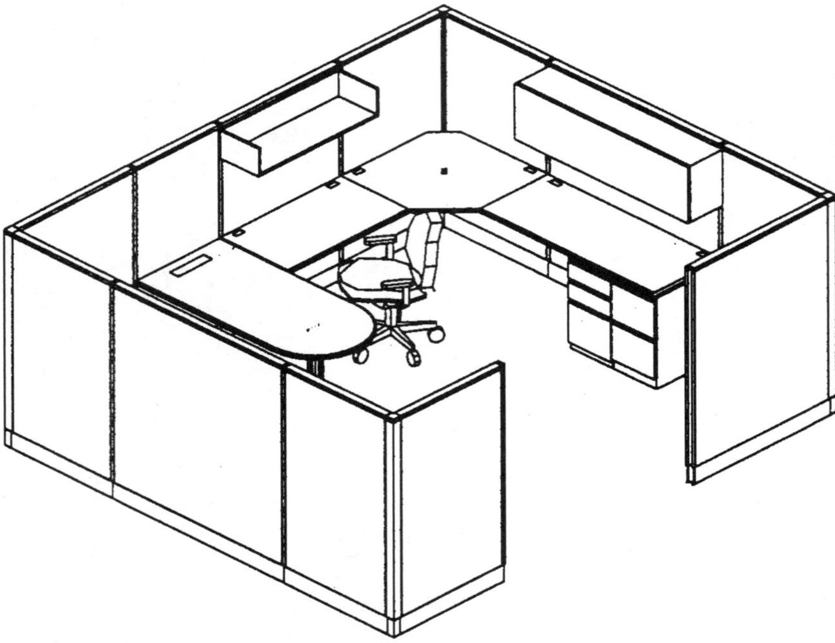

The work station shown here represents one of many types and sizes on the market. Costs can vary considerably due to the various parameters involved. For instance, wood systems can cost up to 40% more than metal or fabric systems. Generally, the cost per work station will decrease for "quantity" purchases.

Note: 4′ raceway included in panels (not pre-wired)

System Components	QUANTITY	UNIT	COST EACH MAT.	COST EACH INST.	COST EACH TOTAL
SYSTEM E2020 410 2600					
WORK STATION - MANAGER					
Acoustical panel, aluminum frame, fabric covered, 30″ wide, 64″ high	1.000	Ea.	335		335
36″ wide, 64″ high	6.000	Ea.	2,160		2,160
42″ wide, 64″ high	2.000	Ea.	830		830
48″ wide, 64″ high	1.000	Ea.	440		440
60″ wide, 64″ high	2.000	Ea.	970		970
Straight connector kit, 64″ high	7.000	Ea.	203		203
Ell connector kit, 64″ high, 90″ deep	4.000	Ea.	166		166
Panel end cover, 64″ high	2.000	Ea.	58		58
Worksurface, 24″ deep, 42″ wide	1.000	Ea.	170		170
Worksurface, 24″ deep, 60″ wide	1.000	Ea.	228		228
Corner worksurface, 24″ deep, 36″ wide	1.000	Ea.	335		335
Peninsula worksurface, 36″ wide, 66″ long	1.000	Ea.	435		435
Peninsula support column, 29-1/2″ high	1.000	Ea.	101		101
Cantilever bracket, 20″ deep, 24″ deep	3.000	Ea.	118.50		118.50
Pedestal spacer, 22″ deep, 15″ wide	2.000	Ea.	95		95
Pedestal, floorstanding, 2 box, 1 file	1.000	Ea.	330		330
Pedestal, floorstanding, 2 file, 22″ deep	1.000	Ea.	310		310
Overhead storage cabinet with door, 60″ wide	1.000	Ea.	415		415
Open bookshelf, 42″ wide	1.000	Ea.	123		123
Task light, recessed, 42″ - 48″ wide	1.000	Ea.	139		139
Task light, recessed, 60″ wide	1.000	Ea.	151		151
Electrical power harness, 36″ wide	2.000	Ea.	189		189
Electrical power harness, 42″ wide	1.000	Ea.	199		199
Electrical power harness, 60″ wide	1.000	Ea.	99.50		99.50
Base power in-feed cable	1.000	Ea.	105		105
Duplex receptacle, circuit 1	2.000	Ea.	29		29
Duplex receptacle, circuit 2	2.000	Ea.	29		29
Installation	1.000	Ea.		530	530
TOTAL			8,763	530	9,293

E2020 Moveable Furnishings

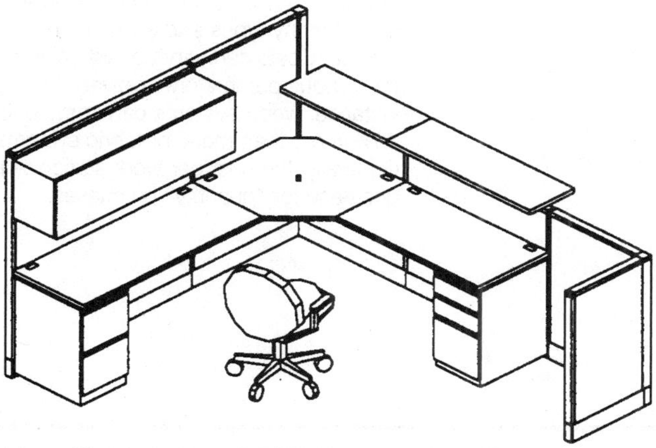

The work station shown here represents one of many types and sizes on the market. Costs can vary considerably due to the various parameters involved. For instance, wood systems can cost up to 40% more than metal or fabric systems. Generally, the cost per work station will decrease for "quantity" purchases.

System Components	QUANTITY	UNIT	COST EACH MAT.	COST EACH INST.	COST EACH TOTAL
SYSTEM E2020 420 2300					
WORK STATION - PROFESSIONAL/EXECUTIVE SECRETARY					
Acoustical panel, aluminum frame, fabric covered, 24″ wide x 43″ high	1.000	Ea.	253		253
30″ wide x 43″ high	1.000	Ea.	282		282
36″ wide x 43″ high	1.000	Ea.	310		310
48″ wide x 43″ high	1.000	Ea.	375		375
36″ wide x 64″ high	1.000	Ea.	360		360
60″ wide x 64″ high	1.000	Ea.	485		485
Straight connector kit, 43″ high	2.000	Ea.	58		58
Straight connector kit, 64″ high	1.000	Ea.	29		29
Ell connector kit, 43″ high, 90″ deep	1.000	Ea.	70.50		70.50
Ell connector kit, 64″ high	1.000	Ea.	41.50		41.50
Panel end cover, 43″ high	1.000	Ea.	29		29
Panel end cover, 64″ high	1.000	Ea.	29		29
Finish end cover, variable height, 2-way	1.000	Ea.	43.50		43.50
Corner worksurface, 24″ deep, 36″ wide	1.000	Ea.	335		335
Worksurface, 24″ deep, 48″ wide	1.000	Ea.	182		182
Worksurface, 24″ deep, 60″ wide	1.000	Ea.	228		228
Cantilever bracket, 20″ deep, 24″ deep	2.000	Ea.	79		79
Countertop brackets, 1 pair	2.000	Ea.	32.90		32.90
Countertop, 15″ deep, 36″ wide	1.000	Ea.	133		133
Countertop, 15″ deep, 48″ wide	1.000	Ea.	149		149
Pedestal spacer, 22″ deep, 15″ wide	2.000	Ea.	95		95
Pedestal, floorstanding, 2box, 1 file, 22″ deep	1.000	Ea.	330		330
Pedestal, floorstanding, 2 file, 22″ deep	1.000	Ea.	310		310
Overhead storage cabinet with door, 60″ wide	1.000	Ea.	415		415
Task light, recessed, 60″ wide	1.000	Ea.	151		151
Electrical power harness, 36″ wide	2.000	Ea.	189		189
Electrical power harness, 60″ wide	1.000	Ea.	99.50		99.50
Base power in-feed cable	1.000	Ea.	105		105
Duplex receptacle, circuit 1	2.000	Ea.	29		29
Duplex receptacle, circuit 2	1.000	Ea.	14.50		14.50
Installation	1.000	Ea.		320	320
TOTAL			5,242.40	320	5,562.40

E2020 Moveable Furnishings

The work station shown here represents one of many types and sizes on the market. Costs can vary considerably due to the various parameters involved. For instance, wood systems can cost up to 40% more than metal or fabric systems. Generally, the cost per work station will decrease for "quantity" purchases.

System Components	QUANTITY	UNIT	COST EACH		
			MAT.	INST.	TOTAL
SYSTEM E2020 430 1500					
WORK STATION - SECRETARY/CLERK					
Acoustical panel, aluminum frame, fabric covered, 30″ wide x 64″ high	2.000	Ea.	670		670
36″ wide x 64″ high	2.000	Ea.	720		720
42″ wide x 64″ high	1.000	Ea.	415		415
Straight connector kit, 64″ high	2.000	Ea.	58		58
Ell connector kit, 64″ high, 90″ deep	2.000	Ea.	83		83
Panel end cover, 64″ high	2.000	Ea.	58		58
Worksurface, 24″ deep, 42″ wide	1.000	Ea.	170		170
Worksurface, 72″ wide, 30″ deep	1.000	Ea.	290		290
Worksurface bracket kit, pair	2.000	Ea.	32.90		32.90
Flat bracket, 20″ deep, 24″ deep	1.000	Ea.	20.50		20.50
Pedestal spacer 22″ deep, 15″ wide	1.000	Ea.	47.50		47.50
Pedestal spacer, 28″ deep, 15″ wide	1.000	Ea.	57		57
Pedestal, 2 box, 1 file, 28″ deep	1.000	Ea.	345		345
Pedestal, 2 file, 22″ deep	1.000	Ea.	310		310
Overhead storage cabinet with door, 42″ wide	1.000	Ea.	274		274
Task light, recessed, 42″ - 48″ wide	1.000	Ea.	139		139
Installation	1.000	Ea.		211	211
TOTAL			3,689.90	211	3,900.90

E2020 Moveable Furnishings

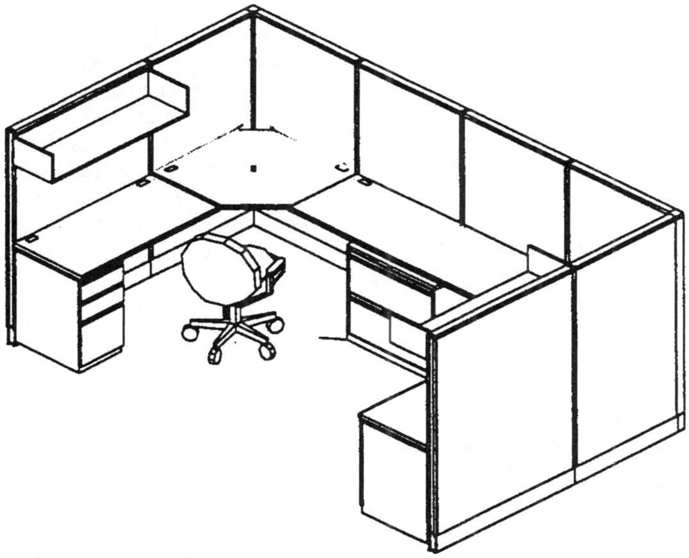

The work station shown here represents one of many types and sizes on the market. Costs can vary considerably due to the various parameters involved. For instance, wood systems can cost up to 40% more than metal or fabric systems. Generally, the cost per work station will decrease for "quantity" purchases.

System Components	QUANTITY	UNIT	COST EACH MAT.	INST.	TOTAL
SYSTEM E2020 440 1600					
WORK STATION - CLERICAL, TWO PERSON, UNIT					
Acoustical panel, aluminum frame, fabric covered, 36" wide, 64" high	6.000	Ea.	2,160		2,160
48" wide x 64" high	2.000	Ea.	880		880
Lateral file, 2 drawer, 30" wide	1.000	Ea.	435		435
Ell connector kit, 64" high, 90" deep	2.000	Ea.	83		83
Panel end cover, 64" high	2.000	Ea.	58		58
Corner worksurface, 36" wide x 24" deep	2.000	Ea.	670		670
Worksurface, 48" wide x 24" deep	2.000	Ea.	364		364
Worksurface, 72" wide x 24" deep	1.000	Ea.	257		257
Cantilever bracket, 20" deep, 24" deep	4.000	Ea.	158		158
Open bookshelf, 48" wide	2.000	Ea.	254		254
Pedestal spacer, 22" deep, 15" wide	2.000	Ea.	95		95
Pedestal, floorstanding, 2 box, 1 file, 22" deep	2.000	Ea.	660		660
Straight connector kit, 64" high	5.000	Ea.	145		145
Task light, recessed, 42" - 48" wide	2.000	Ea.	278		278
Electrical power harness, 36" wide	4.000	Ea.	378		378
Base power in-feed cable	1.000	Ea.	105		105
Duplex receptacle, circuit 1	2.000	Ea.	29		29
Duplex receptacle, circuit 2	2.000	Ea.	29		29
Installation	1.000	Ea.		420	420
TOTAL			7,038	420	7,458

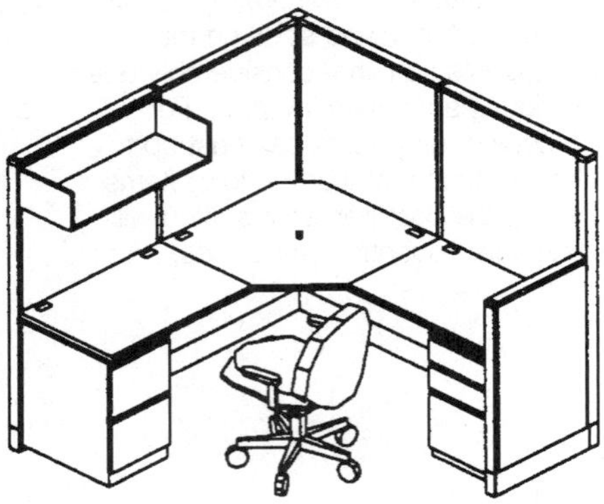

The work station shown here represents one of many types and sizes on the market. Costs can vary considerably due to the various parameters involved. For instance, wood systems can cost up to 40% more than metal or fabric systems. Generally, the cost per work station will decrease for "quantity" purchases.

System Components	QUANTITY	UNIT	COST EACH		
			MAT.	INST.	TOTAL
SYSTEM E2020 450 1600					
WORK STATION - SECRETARY/WORD PROCESSING					
Acoustical panel, aluminum frame, fabric covered, 43″ high, 24″ wide	1.000	Ea.	253		253
Acoustical panel, 64″ high, 36″ wide	4.000	Ea.	1,440		1,440
Straight connector kit, 64″ high	2.000	Ea.	58		58
Ell connector kit, 64″ high, 90″ deep	2.000	Ea.	83		83
Panel end cover, 43″ high	1.000	Ea.	29		29
Panel end cover, 64″ high	1.000	Ea.	29		29
Variable height finish end cover, 2-way	1.000	Ea.	43.50		43.50
Corner worksurface, 24″ deep, 36″ wide	1.000	Ea.	335		335
Worksurface, 24″ deep, 36″ wide	2.000	Ea.	276		276
Cantilever bracket, 20″ deep, 24″ deep	2.000	Ea.	79		79
Pedestal spacer, 22″ deep, 15″ wide	2.000	Ea.	95		95
Pedestal, floorstanding, 2 box, 1 file, 22″ deep	1.000	Ea.	330		330
Pedestal, floorstanding, 2 file, 22″ deep	1.000	Ea.	310		310
Open bookshelf, 36″ wide	1.000	Ea.	117		117
Task light, recessed, 30″ - 36″ wide	1.000	Ea.	129		129
Electrical power harness, 36″ wide	3.000	Ea.	283.50		283.50
Base power in-feed cable	1.000	Ea.	105		105
Duplex receptacle, circuit 1	2.000	Ea.	29		29
Duplex receptacle, circuit 2	1.000	Ea.	14.50		14.50
Installation	1.000	Ea.		264	264
TOTAL			4,038.50	264	4,302.50

F Special Construction

F1020 Integrated Construction

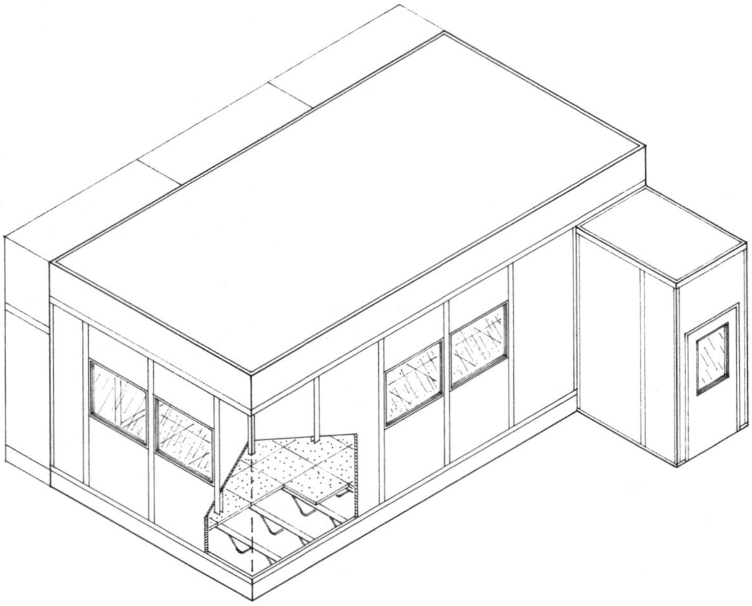

A clean room is a free standing enclosure designed to control particulate levels through air filtration and pressurization control, as well as temperature and humidity levels. These controlled environments are necessary for semi-conductor and electronics-related processes, precision machinery and metrology applications, laser and optic devices, pharmaceutical, biotechnology and medical device manufacture.

The class 10 and 100 prefab clean rooms include: 100% filtered ceiling coverage using air plenum modules, complete packaged HVAC with integrated electronic controls, fully insulated steel demountable modular wall partitions with integral utility chases, ante-room gown-up/airlock with interlocked doors and passthru.

All costs shown include factory pre-assembled and pre-tested components, field installed and tested, with only electrical power, water supply and drain connections required.

Costs are based on 288 S.F. net usable floor area.

System Components	QUANTITY	UNIT	COST EACH MAT.	COST EACH INST.	COST EACH TOTAL
CLASS 10 CLEAN ROOM SYSTEM					
Flooring, raised pedestal type w/steel perforated tiles, incl. steps	1.000	Ea.	6,175	2,050	8,225
Walls, modular baked enamel panels, insulated, all steel, incl. windows, doors, power outlets & light switches	1.000	Ea.	46,200	15,600	61,800
Ceiling, sealed plenum type, 99.9995% on .12 micron, HEPA, 100% coverage, 2″ gel grid tee bar incl. lights	1.000	Ea.	21,600	7,325	28,925
HVAC sys. incl. pressure, temp. and humidity control system and recirc. fan	1.000	Ea.	123,500	41,500	165,000
Gown-up, incl. floor, walls, ceiling, filters, lights, doors	1.000	Ea.	4,025	7,150	11,175
Test, balance and certification	1.000	Ea.	825	2,900	3,725
TOTAL			202,325	76,525	278,850

System Components	QUANTITY	UNIT	COST EACH MAT.	COST EACH INST.	COST EACH TOTAL
CLASS 100 CLEAN ROOM SYSTEM					
Flooring, seamless vinyl with coving	1.000	Ea.	4,150	2,050	6,200
Walls, modular baked enamel insul. panels, all steel incl. windows, doors, power outlets and light switches	1.000	Ea.	31,500	13,500	45,000
Ceiling, sealed plenum type, 99.99% on .3 micron HEPA, 100% coverage, 2″ gel grid tee bar incl. lights	1.000	Ea.	97,500	5,150	102,650
HVAC sys. incl. pressure, temp., and humidity control system, & recirc. fan	1.000	Ea.	82,500	36,400	118,900
Gown-up, incl. floor, walls, ceiling, filters, lights, doors with interlock	1.000	Ea.	17,700	7,150	24,850
Test, balance and certification	1.000	Ea.	8,350	2,900	11,250
TOTAL			241,700	67,150	308,850

F1020 Integrated Construction

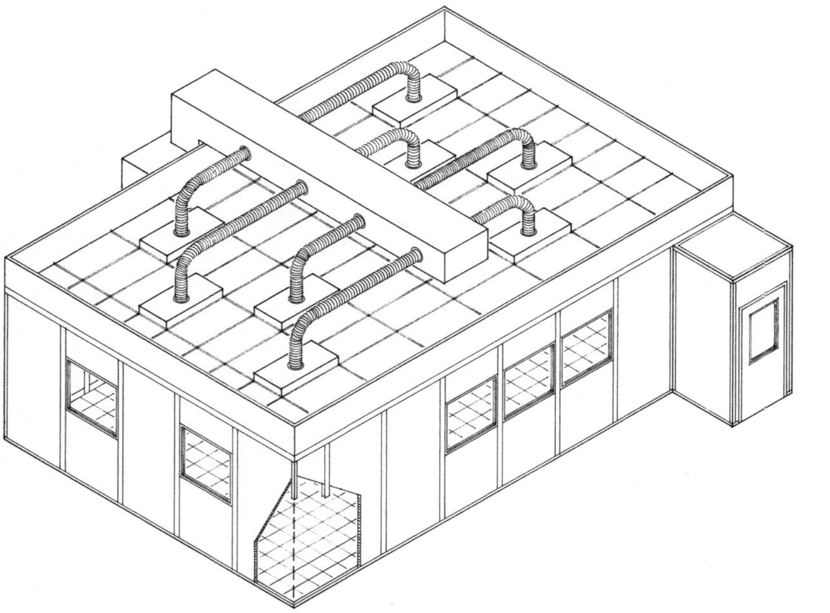

The Class 1000 through Class 100,000 prefab clean rooms use gasketed 2″ heavy duty tee-bar ceilings, ducted supply to and from the HVAC unit, sidewall returns, complete packaged HVAC and integrated electronic controls, fully insulated steel demountable modular wall partitions with integral utility chases, ante-room gown-up/airlock with interlocked doors and passthru.

All costs include factory pre-assembled and pre-tested components, field installation and certification, with the only connections required being electrical power, water supply, and drain.

Costs are based on 600 S.F. net usable floor area.

System Components	QUANTITY	UNIT	COST EACH MAT.	COST EACH INST.	COST EACH TOTAL
SYSTEM F1020 220 0310					
CLASS 1000 CLEAN ROOM SYSTEM					
Flooring, seamless vinyl with coving	1.000	Ea.	9,650	3,250	12,900
Walls, modular baked enamel insul panels, all steel incl. windows, doors, power outlets and light switch	1.000	Ea.	13,100	14,000	27,100
Ceiling,thermal ducted type 99.99% on .3 micron HEPA, steel blank tiles, 2″ gasketed grid tee bars including lights	1.000	Ea.	48,200	16,200	64,400
HVAC system incl. pressure, temp., and humidity control sys. and recirc fan	1.000	Ea.	96,500	32,500	129,000
Gown-up/air lock, incl. floor, walls, ceiling, filter, light, door	1.000	Ea.	3,000	1,025	4,025
Test, balance and certification	1.000	Ea.	445	6,180	6,625
TOTAL			170,895	73,155	244,050

F1020 220	Prefab Clean Room	COST EACH MAT.	COST EACH INST.	COST EACH TOTAL
0310	Prefab clean room, Class 1000	199,500	73,000	272,500
0320	Class 10,000	180,000	60,500	240,500
0330	Class 100,000	178,500	56,500	235,000

G Building Sitework

G1030 Site Earthwork

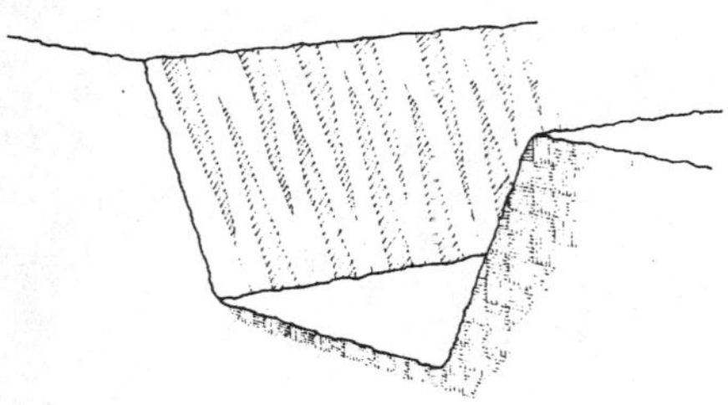

Trenching Systems are shown on a cost per linear foot basis. The systems include: excavation; backfill and removal of spoil; and compaction for various depths and trench bottom widths. The backfill has been reduced to accomodate a pipe of suitable diameter and bedding.

The slope for trench sides varies from none to 1:1.

The Expanded System Listing shows Trenching Systems that range from 2' to 12' in width. Depths range from 2' to 25'.

System Components	QUANTITY	UNIT	COST PER L.F. EQUIP.	COST PER L.F. LABOR	COST PER L.F. TOTAL
SYSTEM G1030 805 1310					
TRENCHING COMMON EARTH, NO SLOPE, 2′ WIDE, 2′ DP, 3/8 C.Y. BUCKET					
Excavation, trench, hyd. backhoe, track mtd., 3/8 C.Y. bucket	.148	B.C.Y.	.32	.93	1.25
Backfill and load spoil, from stockpile	.153	L.C.Y.	.10	.28	.38
Compaction by vibrating plate, 6″ lifts, 4 passes	.118	E.C.Y.	.03	.35	.38
Remove excess spoil, 8 C.Y. dump truck, 2 mile roundtrip	.040	L.C.Y.	.12	.14	.26
TOTAL			.57	1.70	2.27

G1030 805	Trenching Common Earth	COST PER L.F. EQUIP.	COST PER L.F. LABOR	COST PER L.F. TOTAL
1310	Trenching, common earth, no slope, 2′ wide, 2′ deep, 3/8 C.Y. bucket	.57	1.70	2.27
1320	3′ deep, 3/8 C.Y. bucket	.82	2.56	3.38
1330	4′ deep, 3/8 C.Y. bucket	1.06	3.41	4.47
1340	6′ deep, 3/8 C.Y. bucket	1.42	4.43	5.85
1350	8′ deep, 1/2 C.Y. bucket	1.88	5.90	7.78
1360	10′ deep, 1 C.Y. bucket	3.07	7	10.07
1400	4′ wide, 2′ deep, 3/8 C.Y. bucket	1.29	3.35	4.64
1410	3′ deep, 3/8 C.Y. bucket	1.76	5.05	6.81
1420	4′ deep, 1/2 C.Y. bucket	2.15	5.70	7.85
1430	6′ deep, 1/2 C.Y. bucket	3.44	9.05	12.49
1440	8′ deep, 1/2 C.Y. bucket	5.70	11.65	17.35
1450	10′ deep, 1 C.Y. bucket	6.95	14.45	21.40
1460	12′ deep, 1 C.Y. bucket	8.95	18.45	27.40
1470	15′ deep, 1-1/2 C.Y. bucket	7.70	16.45	24.15
1480	18′ deep, 2-1/2 C.Y. bucket	10.60	23	33.60
1520	6′ wide, 6′ deep, 5/8 C.Y. bucket w/trench box	7.65	13.40	21.05
1530	8′ deep, 3/4 C.Y. bucket	10.15	17.65	27.80
1540	10′ deep, 1 C.Y. bucket	9.75	18.25	28
1550	12′ deep, 1-1/2 C.Y. bucket	10.45	19.65	30.10
1560	16′ deep, 2-1/2 C.Y. bucket	14.75	24	38.75
1570	20′ deep, 3-1/2 C.Y. bucket	18.85	29.50	48.35
1580	24′ deep, 3-1/2 C.Y. bucket	22.50	35	57.50
1640	8′ wide, 12′ deep, 1-1/2 C.Y. bucket w/trench box	14.60	25	39.60
1650	15′ deep, 1-1/2 C.Y. bucket	19.10	32.50	51.60
1660	18′ deep, 2-1/2 C.Y. bucket	21.50	32	53.50
1680	24′ deep, 3-1/2 C.Y. bucket	30.50	45.50	76
1730	10′ wide, 20′ deep, 3-1/2 C.Y. bucket w/trench box	24.50	43	67.50
1740	24′ deep, 3-1/2 C.Y. bucket	37	52	89
1780	12′ wide, 20′ deep, 3-1/2 C.Y. bucket w/trench box	39	55	94
1790	25′ deep, bucket	48	70	118
1800	1/2 to 1 slope, 2′ wide, 2′ deep, 3/8 C.Y. bucket	.82	2.56	3.38
1810	3′ deep, 3/8 C.Y. bucket	1.36	4.49	5.85
1820	4′ deep, 3/8 C.Y. bucket	2.01	6.85	8.86
1840	6′ deep, 3/8 C.Y. bucket	3.43	11.15	14.58

G1030 Site Earthwork

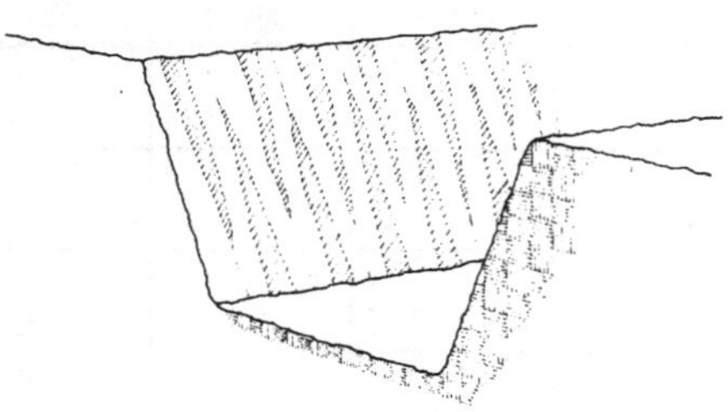

Trenching Systems are shown on a cost per linear foot basis. The systems include: excavation; backfill and removal of spoil; and compaction for various depths and trench bottom widths. The backfill has been reduced to accomodate a pipe of suitable diameter and bedding.

The slope for trench sides varies from none to 1:1.

The Expanded System Listing shows Trenching Systems that range from 2' to 12' in width. Depths range from 2' to 25'.

System Components	QUANTITY	UNIT	COST PER L.F. EQUIP.	LABOR	TOTAL
SYSTEM G1030 806 1310					
TRENCHING LOAM & SANDY CLAY, NO SLOPE, 2′ WIDE, 2′ DP, 3/8 C.Y. BUCKET					
Excavation, trench, hyd. backhoe, track mtd., 3/8 C.Y. bucket	.148	B.C.Y.	.30	.86	1.16
Backfill and load spoil, from stockpile	.165	L.C.Y.	.11	.30	.41
Compaction by vibrating plate 18″ wide, 6″ lifts, 4 passes	.118	E.C.Y.	.03	.35	.38
Remove excess spoil, 8 C.Y. dump truck, 2 mile roundtrip	.042	L.C.Y.	.13	.15	.28
TOTAL			.57	1.70	2.27

G1030 806	Trenching Loam & Sandy Clay	COST PER L.F. EQUIP.	LABOR	TOTAL
1310	Trenching, loam & sandy clay, no slope, 2′ wide, 2′ deep, 3/8 C.Y. bucket	.57	1.66	2.23
1320	3′ deep, 3/8 C.Y. bucket	.89	2.74	3.63
1330	4′ deep, 3/8 C.Y. bucket	1.04	3.34	4.38
1340	6′ deep, 3/8 C.Y. bucket	1.52	3.97	5.50
1350	8′ deep, 1/2 C.Y. bucket	2.02	5.25	7.25
1360	10′ deep, 1 C.Y. bucket	2.21	5.65	7.85
1400	4′ wide, 2′ deep, 3/8 C.Y. bucket	1.29	3.29	4.58
1410	3′ deep, 3/8 C.Y. bucket	1.75	4.97	6.70
1420	4′ deep, 1/2 C.Y. bucket	2.13	5.60	7.75
1430	6′ deep, 1/2 C.Y. bucket	3.66	8.10	11.75
1440	8′ deep, 1/2 C.Y. bucket	5.55	11.50	17.05
1450	10′ deep, 1 C.Y. bucket	5.25	11.75	17
1460	12′ deep, 1 C.Y. bucket	6.60	14.65	21.50
1470	15′ deep, 1-1/2 C.Y. bucket	8.05	17	25
1480	18′ deep, 2-1/2 C.Y. bucket	9.85	18.35	28
1520	6′ wide, 6′ deep, 5/8 C.Y. bucket w/trench box	7.35	13.15	20.50
1530	8′ deep, 3/4 C.Y. bucket	9.75	17.30	27
1540	10′ deep, 1 C.Y. bucket	8.90	17.55	26.50
1550	12′ deep, 1-1/2 C.Y. bucket	10.15	19.70	30
1560	16′ deep, 2-1/2 C.Y. bucket	14.30	24.50	39
1570	20′ deep, 3-1/2 C.Y. bucket	17.25	29.50	47
1580	24′ deep, 3-1/2 C.Y. bucket	21.50	35.50	57
1640	8′ wide, 12′ deep, 1-1/4 C.Y. bucket w/trench box	14.35	25	39.50
1650	15′ deep, 1-1/2 C.Y. bucket	17.55	31.50	49
1660	18′ deep, 2-1/2 C.Y. bucket	21.50	35	56.50
1680	24′ deep, 3-1/2 C.Y. bucket	29.50	46	75.50
1730	10′ wide, 20′ deep, 3-1/2 C.Y. bucket w/trench box	29.50	46.50	76
1740	24′ deep, 3-1/2 C.Y. bucket	37	57	94
1780	12′ wide, 20′ deep, 3-1/2 C.Y. bucket w/trench box	36	55	91
1790	25′ deep, bucket	46.50	71	118
1800	1/2:1 slope, 2′ wide, 2′ deep, 3/8 C.Y. BK	.80	2.50	3.30
1810	3′ deep, 3/8 C.Y. bucket	1.33	4.39	5.70
1820	4′ deep, 3/8 C.Y. bucket	1.98	6.70	8.70
1840	6′ deep, 3/8 C.Y. bucket	3.69	9.95	13.65
1860	8′ deep, 1/2 C.Y. bucket	5.85	15.90	22
1880	10′ deep, 1 C.Y. bucket	7.55	19.95	27.50

G1030 Site Earthwork

G1030 806	Trenching Loam & Sandy Clay	COST PER L.F.		
		EQUIP.	LABOR	TOTAL
2300	4' wide, 2' deep, 3/8 C.Y. bucket	1.52	4.13	5.65
2310	3' deep, 3/8 C.Y. bucket	2.28	6.85	9.15
2320	4' deep, 1/2 C.Y. bucket	2.99	8.50	11.50
2340	6' deep, 1/2 C.Y. bucket	6.15	14.40	20.50
2360	8' deep, 1/2 C.Y. bucket	10.75	23	34
2380	10' deep, 1 C.Y. bucket	11.50	27	38.50
2400	12' deep, 1 C.Y. bucket	19.30	44.50	64
2430	15' deep, 1-1/2 C.Y. bucket	23	50	73
2460	18' deep, 2-1/2 C.Y. bucket	39	75	114
2840	6' wide, 6' deep, 5/8 C.Y. bucket w/trench box	10.75	19.95	30.50
2860	8' deep, 3/4 C.Y. bucket	15.75	29.50	45.50
2880	10' deep, 1 C.Y. bucket	15.90	33	49
2900	12' deep, 1-1/2 C.Y. bucket	19.80	40.50	60.50
2940	16' deep, 2-1/2 C.Y. bucket	32.50	58	90.50
2980	20' deep, 3-1/2 C.Y. bucket	45.50	80	126
3020	24' deep, 3-1/2 C.Y. bucket	64.50	109	174
3100	8' wide, 12' deep, 1-1/2 C.Y. bucket w/trench box	24	45.50	69.50
3120	15' deep, 1-1/2 C.Y. bucket	32.50	64	96.50
3140	18' deep, 2-1/2 C.Y. bucket	45	77.50	123
3180	24' deep, 3-1/2 C.Y. bucket	72	120	192
3270	10' wide, 20' deep, 3-1/2 C.Y. bucket w/trench box	58	97	155
3280	24' deep, 3-1/2 C.Y. bucket	80	131	211
3320	12' wide, 20' deep, 3-1/2 C.Y. bucket w/trench box	64	105	169
3380	25' deep, 3-1/2 C.Y. bucket w/trench box	87.50	141	229
3500	1:1 slope, 2' wide, 2' deep, 3/8 C.Y. bucket	1.04	3.34	4.38
3520	3' deep, 3/8 C.Y. bucket	1.86	6.30	8.15
3540	4' deep, 3/8 C.Y. bucket	2.91	10.05	12.95
3560	6' deep, 3/8 C.Y. bucket	3.69	9.95	13.65
3580	8' deep, 1/2 C.Y. bucket	9.70	26.50	36
3600	10' deep, 1 C.Y. bucket	12.85	34.50	47.50
3800	2' deep, 3/8 C.Y. bucket	1.75	4.97	6.70
3820	4' wide, 3' deep, 3/8 C.Y. bucket	2.82	8.75	11.55
3840	4' deep, 1/2 C.Y. bucket	3.88	11.45	15.35
3860	6' deep, 1/2 C.Y. bucket	8.65	20.50	29
3880	8' deep, 1/2 C.Y. bucket	15.95	35	51
3900	10' deep, 1 C.Y. bucket	17.75	42	60
3920	12' deep, 1 C.Y. bucket	25.50	60	85.50
3940	15' deep, 1-1/2 C.Y. bucket	37.50	82.50	120
3960	18' deep, 2-1/2 C.Y. bucket	53.50	103	157
4030	6' wide, 6' deep, 5/8 C.Y. bucket w/trench box	14.15	26.50	40.50
4040	8' deep, 3/4 C.Y. bucket	22	41.50	63.50
4050	10' deep, 1 C.Y. bucket	23	48.50	71.50
4060	12' deep, 1-1/2 C.Y. bucket	29.50	61	90.50
4070	16' deep, 2-1/2 C.Y. bucket	51	91.50	143
4080	20' deep, 3-1/2 C.Y. bucket	73.50	130	204
4090	24' deep, 3-1/2 C.Y. bucket	107	183	290
4500	8' wide, 12' deep, 1-1/4 C.Y. bucket w/trench box	33.50	66	99.50
4550	15' deep, 1-1/2 C.Y. bucket	48	96	144
4600	18' deep, 2-1/2 C.Y. bucket	68	120	188
4650	24' deep, 3-1/2 C.Y. bucket	115	193	310
4800	10' wide, 20' deep, 3-1/2 C.Y. bucket w/trench box	86	147	233
4850	24' deep, 3-1/2 C.Y. bucket	123	204	325
4950	12' wide, 20' deep, 3-1/2 C.Y. bucket w/trench box	92	156	248
4980	25' deep, bucket	139	231	370

G1030 Site Earthwork

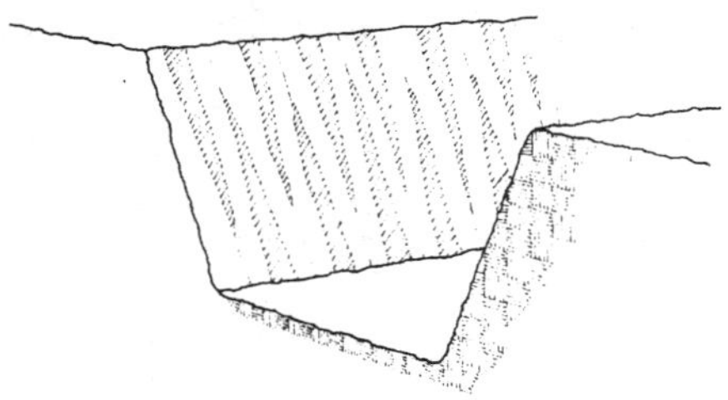

Trenching Systems are shown on a cost per linear foot basis. The systems include: excavation; backfill and removal of spoil; and compaction for various depths and trench bottom widths. The backfill has been reduced to accomodate a pipe of suitable diameter and bedding.

The slope for trench sides varies from none to 1:1.

The Expanced System Listing shows Trenching Systems that range from 2' to 12' in width. Depths range from 2' to 25'.

System Components	QUANTITY	UNIT	COST PER L.F. EQUIP.	LABOR	TOTAL
SYSTEM G1030 807 1310					
TRENCHING SAND & GRAVEL, NO SLOPE, 2′ WIDE, 2′ DEEP, 3/8 C.Y. BUCKET					
Excavation, trench, hyd. backhoe, track mtd., 3/8 C.Y. bucket	.148	B.C.Y.	.29	.84	1.13
Backfill and load spoil, from stockpile	.140	L.C.Y.	.09	.25	.34
Compaction by vibrating plate 18″ wide, 6″ lifts, 4 passes	.118	E.C.Y.	.03	.35	.38
Remove excess spoil, 8 C.Y. dump truck, 2 mile roundtrip	.035	L.C.Y.	.11	.12	.23
TOTAL			.52	1.56	2.08

G1030 807	Trenching Sand & Gravel	COST PER L.F. EQUIP.	LABOR	TOTAL
1310	Trenching, sand & gravel, no slope, 2′ wide, 2′ deep, 3/8 C.Y. bucket	.52	1.56	2.08
1320	3′ deep, 3/8 C.Y. bucket	.84	2.62	3.46
1330	4′ deep, 3/8 C.Y. bucket	.97	3.17	4.14
1340	6′ deep, 3/8 C.Y. bucket	1.45	3.78	5.25
1350	8′ deep, 1/2 C.Y. bucket	1.93	5	6.95
1360	10′ deep, 1 C.Y. bucket	2.06	5.30	7.35
1400	4′ wide, 2′ deep, 3/8 C.Y. bucket	1.18	3.10	4.28
1410	3′ deep, 3/8 C.Y. bucket	1.62	4.70	6.30
1420	4′ deep, 1/2 C.Y. bucket	1.97	5.30	7.25
1430	6′ deep, 1/2 C.Y. bucket	3.48	7.75	11.25
1440	8′ deep, 1/2 C.Y. bucket	5.25	10.90	16.15
1450	10′ deep, 1 C.Y. bucket	4.93	11.10	16.05
1460	12′ deep, 1 C.Y. bucket	6.20	13.85	20
1470	15′ deep, 1-1/2 C.Y. bucket	7.60	16.05	23.50
1480	18′ deep, 2-1/2 C.Y. bucket	9.30	17.25	26.50
1520	6′ wide, 6′ deep, 5/8 C.Y. bucket w/trench box	7	12.45	19.45
1530	8′ deep, 3/4 C.Y. bucket	9.25	16.45	25.50
1540	10′ deep, 1 C.Y. bucket	8.40	16.60	25
1550	12′ deep, 1-1/2 C.Y. bucket	9.60	18.55	28
1560	16′ deep, 2 C.Y. bucket	13.70	23	36.50
1570	20′ deep, 3-1/2 C.Y. bucket	16.40	27.50	44
1580	24′ deep, 3-1/2 C.Y. bucket	20.50	33.50	54
1640	8′ wide, 12′ deep, 1-1/2 C.Y. bucket w/trench box	13.50	23.50	37
1650	15′ deep, 1-1/2 C.Y. bucket	16.45	30	46.50
1660	18′ deep, 2-1/2 C.Y. bucket	20.50	33	53.50
1680	24′ deep, 3-1/2 C.Y. bucket	28	43.50	71.50
1730	10′ wide, 20′ deep, 3-1/2 C.Y. bucket w/trench box	28	43.50	71.50
1740	24′ deep, 3-1/2 C.Y. bucket	35	53.50	88.50
1780	12′ wide, 20′ deep, 3-1/2 C.Y. bucket w/trench box	33.50	51.50	85
1790	25′ deep, bucket	44	66.50	111
1800	1/2:1 slope, 2′ wide, 2′ deep, 3/8 CY bk	.75	2.37	3.12
1810	3′ deep, 3/8 C.Y. bucket	1.24	4.18	5.40
1820	4′ deep, 3/8 C.Y. bucket	1.85	6.40	8.25
1840	6′ deep, 3/8 C.Y. bucket	3.52	9.50	13
1860	8′ deep, 1/2 C.Y. bucket	5.60	15.20	21
1880	10′ deep, 1 C.Y. bucket	7.05	18.85	26

G1030 Site Earthwork

G1030 807	Trenching Sand & Gravel	COST PER L.F.		
		EQUIP.	LABOR	TOTAL
2300	4' wide, 2' deep, 3/8 C.Y. bucket	1.40	3.90	5.30
2310	3' deep, 3/8 C.Y. bucket	2.12	6.50	8.60
2320	4' deep, 1/2 C.Y. bucket	2.78	8.05	10.85
2340	6' deep, 1/2 C.Y. bucket	5.90	13.80	19.70
2360	8' deep, 1/2 C.Y. bucket	10.25	22	32.50
2380	10' deep, 1 C.Y. bucket	10.90	25.50	36.50
2400	12' deep, 1 C.Y. bucket	18.30	42.50	61
2430	15' deep, 1-1/2 C.Y. bucket	21.50	47	68.50
2460	18' deep, 2-1/2 C.Y. bucket	37	70.50	108
2840	6' wide, 6' deep, 5/8 C.Y. bucket w/trench box	10.25	18.90	29
2860	8' deep, 3/4 C.Y. bucket	15	28	43
2880	10' deep, 1 C.Y. bucket	15.10	31	46
2900	12' deep, 1-1/2 C.Y. bucket	18.80	38	57
2940	16' deep, 2 C.Y. bucket	31.50	54.50	86
2980	20' deep, 3-1/2 C.Y. bucket	43	75	118
3020	24' deep, 3-1/2 C.Y. bucket	61.50	103	165
3100	8' wide, 12' deep, 1-1/4 C.Y. bucket w/trench box	23	43	66
3120	15' deep, 1-1/2 C.Y. bucket	31	60.50	91.50
3140	18' deep, 2-1/2 C.Y. bucket	42.50	73	116
3180	24' deep, 3-1/2 C.Y. bucket	68.50	112	181
3270	10' wide, 20' deep, 3-1/2 C.Y. bucket w/trench box	55	91	146
3280	24' deep, 3-1/2 C.Y. bucket	76	123	199
3370	12' wide, 20' deep, 3-1/2 C.Y. bucket w/trench box	61	101	162
3380	25' deep, bucket	88	142	230
3500	1:1 slope, 2' wide, 2' deep, 3/8 CY bk	1.64	3.90	5.55
3520	3' deep, 3/8 C.Y. bucket	1.74	6	7.75
3540	4' deep, 3/8 C.Y. bucket	2.73	9.60	12.35
3560	6' deep, 3/8 C.Y. bucket	3.52	9.50	13
3580	8' deep, 1/2 C.Y. bucket	9.25	25.50	35
3600	10' deep, 1 C.Y. bucket	12.05	32.50	44.50
3800	4' wide, 2' deep, 3/8 C.Y. bucket	1.62	4.70	6.30
3820	3' deep, 3/8 C.Y. bucket	2.62	8.30	10.90
3840	4' deep, 1/2 C.Y. bucket	3.60	10.85	14.45
3860	6' deep, 1/2 C.Y. bucket	8.30	19.85	28
3880	8' deep, 1/2 C.Y. bucket	15.20	33.50	48.50
3900	10' deep, 1 C.Y. bucket	16.80	40	57
3920	12' deep, 1 C.Y. bucket	24.50	56.50	81
3940	15' deep, 1-1/2 C.Y. bucket	35.50	78	114
3960	18' deep, 2-1/2 C.Y. bucket	50.50	97	148
4030	6' wide, 6' deep, 5/8 C.Y. bucket w/trench box	13.45	25.50	39
4040	8' deep, 3/4 C.Y. bucket	21	39.50	60.50
4050	10' deep, 1 C.Y. bucket	21.50	46	67.50
4060	12' deep, 1-1/2 C.Y. bucket	28	57.50	85.50
4070	16' deep, 2 C.Y. bucket	49	86.50	136
4080	20' deep, 3-1/2 C.Y. bucket	70	122	192
4090	24' deep, 3-1/2 C.Y. bucket	102	172	274
4500	8' wide, 12' deep, 1-1/2 C.Y. bucket w/trench box	32	62.50	94.50
4550	15' deep, 1-1/2 C.Y. bucket	45	90.50	136
4600	18' deep, 2-1/2 C.Y. bucket	64.50	113	178
4650	24' deep, 3-1/2 C.Y. bucket	109	182	291
4800	10' wide, 20' deep, 3-1/2 C.Y. bucket w/trench box	81.50	138	220
4850	24' deep, 3-1/2 C.Y. bucket	117	192	310
4950	12' wide, 20' deep, 3-1/2 C.Y. bucket w/trench box	87.50	146	234
4980	25' deep, bucket	132	217	350

G2020 Parking Lots

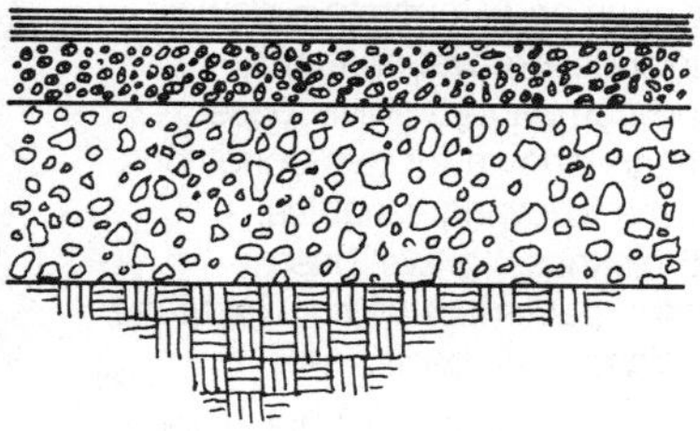

The Parking Lot System includes: compacted bank-run gravel; fine grading with a grader and roller; and bituminous concrete wearing course. All Parking Lot systems are on a cost per car basis. There are three basic types of systems: 90° angle, 60° angle, and 45° angle. The gravel base is compacted to 98%. Final stall design and lay-out of the parking lot with precast bumpers, sealcoating and white paint is also included.

The Expanded System Listing shows the three basic parking lot types with various depths of both gravel base and wearing course. The gravel base depths range from 6″ to 10″. The bituminous paving wearing course varies from a depth of 3″ to 6″.

System Components	QUANTITY	UNIT	COST PER CAR MAT.	COST PER CAR INST.	COST PER CAR TOTAL
SYSTEM G2020 220 1500					
PARKING LOT, 90° ANGLE PARKING, 3″ BITUMINOUS PAVING, 6″ GRAVEL BASE					
Surveying crew for layout, 4 man crew	.020	Day		45.04	45.04
Borrow, bank run gravel, haul 2 mi., spread w/dozer, no compaction	7.223	C.Y.	216.69	47.31	264
Grading, fine grade 3 passes with motor grader	43.333	S.Y.		84.70	84.70
Compact w/ vibrating plate, 8″ lifts, granular mat'l. to 98%	7.223	C.Y.		43.48	43.48
Bituminous paving, 3″ thick	43.333	S.Y.	487.50	74.53	562.03
Seal coating, petroleum resistant under 1,000 S.Y.	43.333	S.Y.	58.93	52.87	111.80
Painting lines on pavement, parking stall, white	1.000	Ea.	6.60	6.97	13.57
Precast concrete parking bar, 6″ x 10″ x 6′-0″	1.000	Ea.	64.50	17.60	82.10
TOTAL			834.22	372.50	1,206.72

G2020 220	Parking Lots	COST PER CAR MAT.	COST PER CAR INST.	COST PER CAR TOTAL
1500	Parking lot, 90° angle parking, 3″ bituminous paving, 6″ gravel base	835	375	1,210
1520	8″ gravel base	905	405	1,310
1540	10″ gravel base	980	430	1,410
1560	4″ bituminous paving, 6″ gravel base	960	370	1,330
1580	8″ gravel base	1,025	400	1,425
1600	10″ gravel base	1,100	430	1,530
1620	6″ bituminous paving, 6″ gravel base	1,225	390	1,615
1640	8″ gravel base	1,300	415	1,715
1661	10″ gravel base	1,375	445	1,820
1800	60° angle parking, 3″ bituminous paving, 6″ gravel base	835	375	1,210
1820	8″ gravel base	905	405	1,310
1840	10″ gravel base	980	430	1,410
1860	4″ bituminous paving, 6″ gravel base	960	370	1,330
1880	8″ gravel base	1,025	400	1,425
1900	10″ gravel base	1,100	430	1,530
1920	6″ bituminous paving, 6″ gravel base	1,225	390	1,615
1940	8″gravel base	1,300	415	1,715
1961	10″ gravel base	1,375	445	1,820
2200	45° angle parking, 3″ bituminous paving, 6″ gravel base	855	385	1,240
2220	8″ gravel base	930	415	1,345
2240	10″ gravel base	1,000	445	1,445
2260	4″ bituminous paving, 6″ gravel base	980	380	1,360
2280	8″ gravel base	1,050	415	1,465
2300	10″ gravel base	1,125	445	1,570
2320	6″ bituminous paving, 6″ gravel base	1,275	395	1,670
2340	8″ gravel base	1,350	430	1,780
2361	10″ gravel base	1,425	460	1,885

G2030 Pedestrian Paving

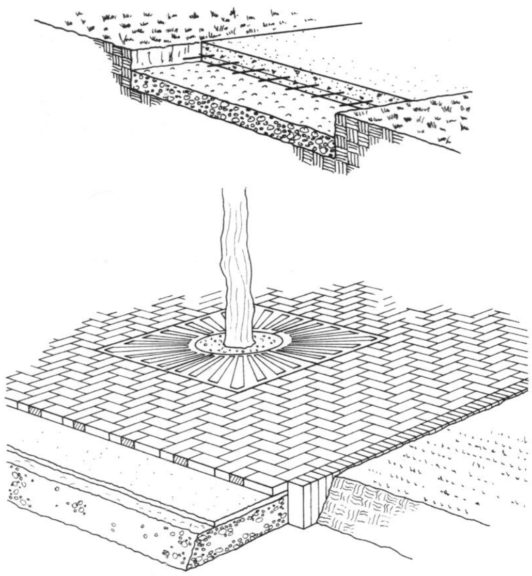

The Bituminous and Concrete Sidewalk Systems include excavation, hand grading and compacted gravel base. Pavements are shown for two conditions of each of the following variables; pavement thickness, gravel base thickness and pavement width. Costs are given on a linear foot basis.

The Plaza Systems listed include several brick and tile paving surfaces on two different bases: gravel and slab on grade. The type of bedding for the pavers depends on the base being used, and alternate bedding may be desirable. Also included in the paving costs are edging and precast grating costs and where concrete bases are involved, expansion joints. Costs are given on a square foot basis.

G2030 110	Bituminous Sidewalks	COST PER L.F.		
		MAT.	INST.	TOTAL
1580	Bituminous sidewalk, 1" thick paving, 4" gravel base, 3' width	2.22	3.90	6.12
1600	4' width	2.95	4.23	7.18
1640	6" gravel base, 3' width	2.79	4.10	6.89
1660	4' width	3.70	4.49	8.19
2120	2" thick paving, 4" gravel base, 3' width	3.32	4.74	8.06
2140	4' width	4.42	5.30	9.72
2180	6" gravel base, 3' width	3.89	4.94	8.83
2201	4' width	5.15	5.55	10.70

G2030 120	Concrete Sidewalks	COST PER L.F.		
		MAT.	INST.	TOTAL
1580	Concrete sidewalk, 4" thick, 4" gravel base, 3' wide	6.65	10.65	17.30
1600	4' wide	8.85	13.20	22.05
1640	6" gravel base, 3' wide	7.25	10.85	18.10
1660	4' wide	9.60	13.40	23
2120	6" thick concrete, 4" gravel base, 3' wide	9.75	12.20	21.95
2140	4' wide	13	15.15	28.15
2180	6" gravel base, 3' wide	10.30	12.40	22.70
2201	4' wide	13.75	16	29.75

G2030 150	Brick & Tile Plazas	COST PER S.F.		
		MAT.	INST.	TOTAL
2050	Brick pavers, 4" x 8" x 2-1/4", aggregate base, course washed sand bedding	8	5.40	13.40
2100	Slab on grade, asphalt bedding	10.45	7.85	18.30
3550	Concrete paving stone, 4" x 8" x 2-1/2", gravel base, sand bedding	4.62	3.76	8.38
3600	Slab on grade, asphalt bedding	6.55	5.70	12.25
4050	Concrete patio blocks, 8" x 16" x 2", gravel base, sand bedding	4.42	4.94	9.36
4100	Slab on grade, asphalt bedding	6.70	7.25	13.95
6050	Granite pavers, 3-1/2" x 3-1/2" x 3-1/2", gravel base, sand bedding	13.85	12.05	25.90
6101	Slab on grade, mortar bedding	16.15	16.10	32.25
7050	Limestone, 3" thick, gravel base, sand bedding	13.10	14.90	28
7100	Slab on grade, mortar bedding	15.40	18.90	34.30
8050	Slate flagging, 3/4" thick, gravel base, sand bedding	10.50	12.05	22.55
8100	Slab on grade, mastic bedding	22.50	17.25	39.75

G4020 Site Lighting

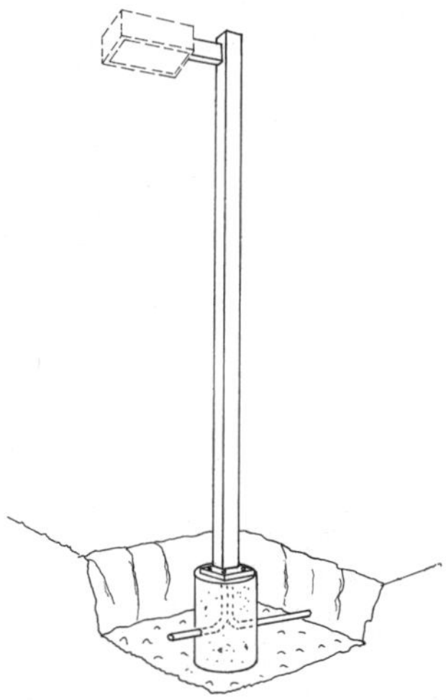

Table G4020 210 Procedure for Calculating Floodlights Required for Various Footcandles
Poles should not be spaced more than 4 times the fixture mounting height for good light distribution.

Estimating Chart
Select Lamp type.

Determine total square feet.

Chart will show quantity of fixtures to provide 1 footcandle initial, at intersection of lines. Multiply fixture quantity by desired footcandle level.

Chart based on use of wide beam luminaires in an area whose dimensions are large compared to mounting height and is approximate only.

To maintain 1 footcandle over a large area use these watts per square foot:

Incandescent	0.15
Metal Halide	0.032
Mercury Vapor	0.05
High Pressure Sodium	0.024

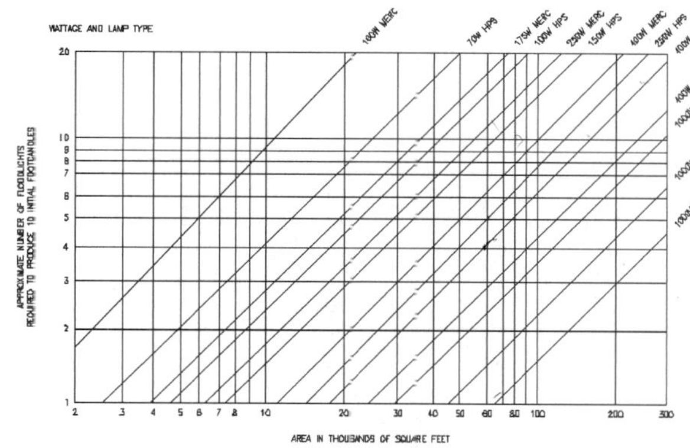

System Components	QUANTITY	UNIT	COST EACH MAT.	COST EACH INST.	COST EACH TOTAL
SYSTEM G4020 210 0200					
LIGHT POLES, ALUMINUM, 20′ HIGH, 1 ARM BRACKET					
Aluminum light pole, 20′, no concrete base	1.000	Ea.	950	544	1,494
Bracket arm for Aluminum light pole	1.000	Ea.	125	72.50	197.50
Excavation by hand, pits to 6′ deep, heavy soil or clay	2.368	C.Y.		246.27	246.27
Footing, concrete incl forms, reinforcing, spread, under 1 C.Y.	.465	C.Y.	80.91	86.07	166.98
Backfill by hand	1.903	C.Y.		72.31	72.31
Compaction vibrating plate	1.903	C.Y.		9.74	9.74
TOTAL			1,155.91	1,030.89	2,186.80

G4020 210	Light Pole (Installed)	COST EACH MAT.	COST EACH INST.	COST EACH TOTAL
0200	Light pole, aluminum, 20′ high, 1 arm bracket	1,150	1,025	2,175
0240	2 arm brackets	1,275	1,025	2,300
0280	3 arm brackets R265723-20	1,400	1,075	2,475
0320	4 arm brackets	1,525	1,075	2,600
0360	30′ high, 1 arm bracket	2,075	1,300	3,375
0400	2 arm brackets	2,200	1,300	3,500
0440	3 arm brackets	2,325	1,325	3,650
0480	4 arm brackets	2,450	1,325	3,775
0680	40′ high, 1 arm bracket	2,550	1,750	4,300
0720	2 arm brackets	2,675	1,750	4,425
0760	3 arm brackets	2,800	1,775	4,575
0800	4 arm brackets	2,925	1,775	4,700
0840	Steel, 20′ high, 1 arm bracket	1,400	1,075	2,475
0880	2 arm brackets	1,500	1,075	2,575
0920	3 arm brackets	1,525	1,125	2,650
0960	4 arm brackets	1,650	1,125	2,775
1000	30′ high, 1 arm bracket	1,600	1,400	3,000
1040	2 arm brackets	1,725	1,400	3,125
1080	3 arm brackets	1,725	1,425	3,150
1120	4 arm brackets	1,850	1,425	3,275
1320	40′ high, 1 arm bracket	2,100	1,875	3,975
1360	2 arm brackets	2,200	1,875	4,075
1400	3 arm brackets	2,225	1,900	4,125
1440	4 arm brackets	2,350	1,900	4,250

Reference Section

All the reference information is in one section, making it easy to find what you need to know . . . and easy to use the book on a daily basis. This section is visually identified by a vertical gray bar on the page edges.

In this Reference Section, we've included Equipment Rental Costs, a listing of rental and operating costs; Crew Listings, a full listing of all crews and equipment, and their costs; Historical Cost Indexes for cost comparisons over time; City Cost Indexes and Location Factors for adjusting costs to the region you are in; Reference Tables, where you will find explanations, estimating information and procedures, or technical data; Change Orders, information on pricing changes to contract documents; Square Foot Costs that allow you to make a rough estimate for the overall cost of a project; and an explanation of all the Abbreviations in the book.

Table of Contents

Estimating Tips

- This section contains the average costs to rent and operate hundreds of pieces of construction equipment. This is useful information when estimating the time and material requirements of any particular operation in order to establish a unit or total cost. Equipment costs include not only rental, but also operating costs for equipment under normal use.

Rental Costs

- Equipment rental rates are obtained from industry sources throughout North America–contractors, suppliers, dealers, manufacturers, and distributors.
- Rental rates vary throughout the country, with larger cities generally having lower rates. Lease plans for new equipment are available for periods in excess of six months, with a percentage of payments applying toward purchase.
- Monthly rental rates vary from 2% to 5% of the purchase price of the equipment depending on the anticipated life of the equipment and its wearing parts.
- Weekly rental rates are about 1/3 the monthly rates, and daily rental rates are about 1/3 the weekly rate.

Operating Costs

- The operating costs include parts and labor for routine servicing, such as repair and replacement of pumps, filters and worn lines. Normal operating expendables, such as fuel, lubricants, tires and electricity (where applicable), are also included.
- Extraordinary operating expendables with highly variable wear patterns, such as diamond bits and blades, are excluded. These costs can be found as material costs in the Unit Price section.
- The hourly operating costs listed do not include the operator's wages.

Equipment Cost/Day

- Any power equipment required by a crew is shown in the Crew Listings with a daily cost.
- The daily cost of equipment needed by a crew is based on dividing the weekly rental rate by 5 (number of working days in the week), and then adding the hourly operating cost times 8 (the number of hours in a day). This "Equipment Cost/Day" is shown in the far right column of the Equipment Rental pages.
- If equipment is needed for only one or two days, it is best to develop your own cost by including components for daily rent and hourly operating cost. This is important when the listed Crew for a task does not contain the equipment needed, such as a crane for lifting mechanical heating/cooling equipment up onto a roof.
- If the quantity of work is less than the crew's Daily Output shown for a Unit Price line item that includes a bare unit equipment cost, it is recommended to estimate one day's rental cost and operating cost for equipment shown in the Crew Listing for that line item.

Mobilization/ Demobilization

- The cost to move construction equipment from an equipment yard or rental company to the jobsite and back again is not included in equipment rental costs listed in the Reference section, nor in the bare equipment cost of any Unit Price line item, nor in any equipment costs shown in the Crew listings.
- Mobilization (to the site) and demobilization (from the site) costs can be found in the Unit Price section.
- If a piece of equipment is already at the jobsite, it is not appropriate to utilize mobil./ demob. costs again in an estimate.

01 54 33 | Equipment Rental

			UNIT	HOURLY OPER. COST	RENT PER DAY	RENT PER WEEK	RENT PER MONTH	EQUIPMENT COST/DAY	
10	0010	**CONCRETE EQUIPMENT RENTAL** without operators R015433-10							10
	0200	Bucket, concrete lightweight, 1/2 C.Y.	Ea.	.65	20	60	180	17.20	
	0300	1 C.Y.		.70	24.50	73	219	20.20	
	0400	1-1/2 C.Y.		.90	33	99	297	27	
	0500	2 C.Y.		1.00	40	120	360	32	
	0580	8 C.Y.		5.50	262	785	2,350	201	
	0600	Cart, concrete, self propelled, operator walking, 10 C.F.		2.65	56.50	170	510	55.20	
	0700	Operator riding, 18 C.F.		4.30	86.50	260	780	86.40	
	0800	Conveyer for concrete, portable, gas, 16" wide, 26' long		10.00	123	370	1,100	154	
	0900	46' long		10.40	150	450	1,350	173.20	
	1000	56' long		10.50	158	475	1,425	179	
	1100	Core drill, electric, 2-1/2 H.P., 1" to 8" bit diameter		1.72	65.50	196	590	52.95	
	1150	11 HP, 8" to 18" cores		5.15	115	345	1,025	110.20	
	1200	Finisher, concrete floor, gas, riding trowel, 96" wide		9.95	145	435	1,300	166.60	
	1300	Gas, walk-behind, 3 blade, 36" trowel		1.65	19.65	59	177	25	
	1400	4 blade, 48" trowel		3.20	28	84	252	42.40	
	1500	Float, hand-operated (Bull float) 48" wide		.08	13.65	41	123	8.85	
	1570	Curb builder, 14 H.P., gas, single screw		11.85	243	730	2,200	240.80	
	1590	Double screw		12.55	285	855	2,575	271.40	
	1600	Grinder, concrete and terrazzo, electric, floor		2.45	103	310	930	81.60	
	1700	Wall grinder		1.23	51.50	155	465	40.85	
	1800	Mixer, powered, mortar and concrete, gas, 6 C.F., 18 H.P.		6.95	118	355	1,075	126.60	
	1900	10 C.F., 25 H.P.		8.65	143	430	1,300	155.20	
	2000	16 C.F.		8.95	167	500	1,500	171.60	
	2100	Concrete, stationary, tilt drum, 2 C.Y.		6.10	232	695	2,075	187.80	
	2120	Pump, concrete, truck mounted 4" line 80' boom		22.85	925	2,775	8,325	737.80	
	2140	5" line, 110' boom		29.75	1,225	3,695	11,100	977	
	2160	Mud jack, 50 C.F. per hr.		6.10	130	390	1,175	126.80	
	2180	225 C.F. per hr.		7.95	147	440	1,325	151.60	
	2190	Shotcrete pump rig, 12 CY/hr		12.60	228	685	2,050	237.80	
	2600	Saw, concrete, manual, gas, 18 H.P.		5.15	36.50	110	330	63.20	
	2650	Self-propelled, gas, 30 H.P.		10.25	98.50	295	885	141	
	2700	Vibrators, concrete, electric, 60 cycle, 2 H.P.		.46	8.35	25	75	8.70	
	2800	3 H.P.		.64	10.35	31	93	11.30	
	2900	Gas engine, 5 H.P.		1.55	14.35	43	129	21	
	3000	8 H.P.		2.10	15.35	46	138	26	
	3050	Vibrating screed, gas engine, 8 H.P.		3.32	73	219	655	70.35	
	3120	Concrete transit mixer, 6 x 4, 250 H.P., 8 C.Y., rear discharge		47.15	575	1,720	5,150	721.20	
	3200	Front discharge		55.50	710	2,125	6,375	869	
	3300	6 x 6, 285 H.P., 12 C.Y., rear discharge		54.60	665	2,000	6,000	836.80	
	3400	Front discharge	↓	57.05	715	2,150	6,450	886.40	
20	0010	**EARTHWORK EQUIPMENT RENTAL** without operators R015433-10							20
	0040	Aggregate spreader, push type 8' to 12' wide	Ea.	2.45	25.50	77	231	35	
	0045	Tailgate type, 8' wide		2.35	33.50	100	300	38.80	
	0055	Earth auger, truck-mounted, for fence & sign posts, utility poles		13.50	505	1,520	4,550	412	
	0060	For borings and monitoring wells		38.20	670	2,005	6,025	706.60	
	0070	Portable, trailer mounted		2.65	26	78	234	36.80	
	0075	Truck-mounted, for caissons, water wells		85.35	3,075	9,250	27,800	2,533	
	0080	Horizontal boring machine, 12" to 36" diameter, 45 H.P.		20.30	202	605	1,825	283.40	
	0090	12" to 48" diameter, 65 H.P.		28.65	360	1,075	3,225	444.20	
	0095	Auger, for fence posts, gas engine, hand held		.50	5.65	17	51	7.40	
	0100	Excavator, diesel hydraulic, crawler mounted, 1/2 C.Y. cap.		20.95	380	1,135	3,400	394.60	
	0120	5/8 C.Y. capacity		25.40	525	1,570	4,700	517.20	
	0140	3/4 C.Y. capacity		29.70	600	1,805	5,425	598.60	
	0150	1 C.Y. capacity		36.90	625	1,870	5,600	669.20	
	0200	1-1/2 C.Y. capacity		45.10	840	2,525	7,575	865.80	
	0300	2 C.Y. capacity	↓	61.75	1,150	3,440	10,300	1,182	

01 54 33 | Equipment Rental

			UNIT	HOURLY OPER. COST	RENT PER DAY	RENT PER WEEK	RENT PER MONTH	EQUIPMENT COST/DAY	
20	0320	2-1/2 C.Y. capacity	Ea.	83.30	1 600	4,765	14,300	1,619	20
	0325	3-1/2 C.Y. capacity		111.30	2 125	6,360	19,100	2,162	
	0330	4-1/2 C.Y. capacity		134.65	2 600	7,790	23,400	2,635	
	0335	6 C.Y. capacity		168.45	2 875	8,660	26,000	3,080	
	0340	7 C.Y. capacity		173.20	3 125	9,350	28,100	3,256	
	0342	Excavator attachments, bucket thumbs		2.70	215	645	1,925	150.60	
	0345	Grapples		2.45	188	565	1,700	132.60	
	0347	Hydraulic hammer for boom mounting, 5000 ft-lb		12.10	405	1,220	3,650	340.80	
	0349	11,000 ft-lb		20.40	730	2,190	6,575	601.20	
	0350	Gradall type, truck mounted, 3 ton @ 15' radius, 5/8 C.Y.		49.60	935	2,810	8,425	958.80	
	0370	1 C.Y. capacity		55.10	1 150	3,445	10,300	1,130	
	0400	Backhoe-loader, 40 to 45 H.P., 5/8 C.Y. capacity		11.90	227	680	2,050	231.20	
	0450	45 H.P. to 60 H.P., 3/4 C.Y. capacity		15.85	278	835	2,500	293.80	
	0460	80 H.P., 1-1/4 C.Y. capacity		19.00	320	965	2,900	345	
	0470	112 H.P., 1-1/2 C.Y. capacity		25.45	465	1,400	4,200	483.60	
	0482	Backhoe-loader attachment, compactor, 20,000 lb		4.70	127	380	1,150	113.60	
	0485	Hydraulic hammer, 750 ft-lbs		2.80	90	270	810	76.40	
	0486	Hydraulic hammer, 1200 ft-lbs		5.25	185	555	1,675	153	
	0500	Brush chipper, gas engine, 6" cutter head, 35 H.P.		8.95	108	325	975	136.60	
	0550	12" cutter head, 130 H.P.		14.95	177	530	1,600	225.60	
	0600	15" cutter head, 165 H.P.		22.45	217	650	1,950	309.60	
	0750	Bucket, clamshell, general purpose, 3/8 C.Y.		1.10	36.50	110	330	30.80	
	0800	1/2 C.Y.		1.20	45	135	405	36.60	
	0850	3/4 C.Y.		1.35	55	165	495	43.80	
	0900	1 C.Y.		1.40	58.50	175	525	46.20	
	0950	1-1/2 C.Y.		2.25	80	240	720	66	
	1000	2 C.Y.		2.40	88.50	265	795	72.20	
	1010	Bucket, dragline, medium duty, 1/2 C.Y.		.65	23.50	70	210	19.20	
	1020	3/4 C.Y.		.65	24.50	74	222	20	
	1030	1 C.Y.		.70	26.50	80	240	21.60	
	1040	1-1/2 C.Y.		1.10	40	120	360	32.80	
	1050	2 C.Y.		1.15	45	135	405	36.20	
	1070	3 C.Y.		1.65	55	165	495	46.20	
	1200	Compactor, manually guided 2-drum vibratory smooth roller, 7.5 H.P.		5.65	167	500	1,500	145.20	
	1250	Rammer/tamper, gas, 8"		2.30	41.50	125	375	43.40	
	1260	15"		2.55	48.50	145	435	49.40	
	1300	Vibratory plate, gas, 18" plate, 3000 lb. blow		2.25	23	69	207	31.80	
	1350	21" plate, 5000 lb. blow		2.80	28	84	252	39.20	
	1370	Curb builder/extruder, 14 H.P., gas, single screw		11.85	243	730	2,200	240.80	
	1390	Double screw		12.55	285	855	2,575	271.40	
	1500	Disc harrow attachment, for tractor		.39	65.50	197	590	42.50	
	1810	Feller buncher, shearing & accumulating trees, 100 H.P.		28.55	555	1,665	5,000	561.40	
	1860	Grader, self-propelled, 25,000 lb.		24.85	450	1,345	4,025	467.80	
	1910	30,000 lb.		29.65	520	1,565	4,700	550.20	
	1920	40,000 lb.		45.50	995	2,985	8,950	961	
	1930	55,000 lb.		59.10	1.400	4,215	12,600	1,316	
	1950	Hammer, pavement breaker, self-propelled, diesel, 1000 to 1250 lb		23.55	340	1,025	3,075	393.40	
	2000	1300 to 1500 lb.		35.33	685	2,050	6,150	692.65	
	2050	Pile driving hammer, steam or air, 4150 ft.-lb. @ 225 BPM		9.35	470	1,405	4,225	355.80	
	2100	8750 ft.-lb. @ 145 BPM		11.35	650	1,955	5,875	481.80	
	2150	15,000 ft.-lb. @ 60 BPM		12.95	790	2,375	7,125	578.60	
	2200	24,450 ft.-lb. @ 111 BPM		13.95	885	2,655	7,975	642.60	
	2250	Leads, 60' high for pile driving hammers up to 20,000 ft.-lb.		2.85	78	234	700	69.60	
	2300	90' high for hammers over 20,000 ft.-lb.		4.35	140	420	1,250	118.80	
	2350	Diesel type hammer, 22,400 ft.-lb.		25.75	640	1,920	5,750	590	
	2400	41,300 ft.-lb.		32.70	695	2,080	6,250	677.60	
	2450	141,000 ft.-lb.		49.25	1,200	3,570	10,700	1,108	
	2500	Vib. elec. hammer/extractor, 200 KW diesel generator, 34 H.P.	↓	41.05	665	2,000	6,000	728.40	

01 54 33 | Equipment Rental

			UNIT	HOURLY OPER. COST	RENT PER DAY	RENT PER WEEK	RENT PER MONTH	EQUIPMENT COST/DAY	
20	2550	80 H.P.	Ea.	73.75	980	2,940	8,825	1,178	20
	2600	150 H.P.		139.15	1,900	5,720	17,200	2,257	
	2700	Extractor, steam or air, 700 ft.-lb.		17.15	510	1,530	4,600	443.20	
	2750	1000 ft.-lb.		19.55	625	1,875	5,625	531.40	
	2800	Log chipper, up to 22" diam, 600 H.P.		40.05	445	1,340	4,025	588.40	
	2850	Logger, for skidding & stacking logs, 150 H.P.		44.95	895	2,690	8,075	897.60	
	2860	Mulcher, diesel powered, trailer mounted		17.60	218	655	1,975	271.80	
	2900	Rake, spring tooth, with tractor		11.34	278	835	2,500	257.70	
	3000	Roller, vibratory, tandem, smooth drum, 20 H.P.		7.35	132	395	1,175	137.80	
	3050	35 H.P.		9.50	262	785	2,350	233	
	3100	Towed type vibratory compactor, smooth drum, 50 H.P.		22.75	380	1,145	3,425	411	
	3150	Sheepsfoot, 50 H.P.		25.05	440	1,325	3,975	465.40	
	3170	Landfill compactor, 220 HP		69.10	1,450	4,315	12,900	1,416	
	3200	Pneumatic tire roller, 80 H.P.		13.10	340	1,015	3,050	307.80	
	3250	120 H.P.		19.90	600	1,805	5,425	520.20	
	3300	Sheepsfoot vibratory roller, 240 H.P.		56.65	1,150	3,445	10,300	1,142	
	3320	340 H.P.		75.60	1,425	4,240	12,700	1,453	
	3350	Smooth drum vibratory roller, 75 H.P.		19.70	520	1,560	4,675	469.60	
	3400	125 H.P.		25.70	650	1,945	5,825	594.60	
	3410	Rotary mower, brush, 60", with tractor		15.30	252	755	2,275	273.40	
	3420	Rototiller, walk-behind, gas, 5 H.P.		2.43	62.50	188	565	57.05	
	3422	8 H.P.		3.89	100	300	900	91.10	
	3440	Scrapers, towed type, 7 CY capacity		4.90	105	315	945	102.20	
	3450	10 C.Y. capacity		6.20	173	520	1,550	153.60	
	3500	15 C.Y. capacity		6.85	190	570	1,700	168.80	
	3525	Self-propelled, single engine, 14 C.Y. capacity		83.90	1,400	4,190	12,600	1,509	
	3550	Dual engine, 21 C.Y. capacity		119.25	1,625	4,850	14,600	1,924	
	3600	31 C.Y. capacity		165.60	2,600	7,765	23,300	2,878	
	3640	44 C.Y. capacity		195.70	3,150	9,430	28,300	3,452	
	3650	Elevating type, single engine, 11 CY capacity		56.70	1,075	3,205	9,625	1,095	
	3700	22 C.Y. capacity		105.60	2,125	6,410	19,200	2,127	
	3710	Screening plant 110 H.P. w/ 5' x 10' screen		30.75	410	1,230	3,700	492	
	3720	5' x 16' screen		32.90	515	1,540	4,625	571.20	
	3850	Shovel, crawler-mounted, front-loading, 7 C.Y. capacity		171.90	2,850	8,575	25,700	3,090	
	3855	12 C.Y. capacity		244.40	3,650	10,965	32,900	4,148	
	3860	Shovel/backhoe bucket, 1/2 C.Y.		2.10	60	180	540	52.80	
	3870	3/4 C.Y.		2.15	68.50	205	615	58.20	
	3880	1 C.Y.		2.25	76.50	230	690	64	
	3890	1-1/2 C.Y.		2.40	91.50	275	825	74.20	
	3910	3 C.Y.		2.70	127	380	1,150	97.60	
	3950	Stump chipper, 18" deep, 30 H.P.		8.99	215	646	1,950	201.10	
	4110	Tractor, crawler, with bulldozer, torque converter, diesel 80 H.P.		21.90	375	1,120	3,350	399.20	
	4150	105 H.P.		31.70	580	1,735	5,200	600.60	
	4200	140 H.P.		36.50	650	1,950	5,850	682	
	4260	200 H.P.		55.85	1,050	3,175	9,525	1,082	
	4310	300 H.P.		71.45	1,425	4,255	12,800	1,423	
	4360	410 H.P.		95.30	1,800	5,435	16,300	1,849	
	4370	500 H.P.		127.20	2,400	7,230	21,700	2,464	
	4380	700 H.P.		204.15	4,325	12,995	39,000	4,232	
	4400	Loader, crawler, torque conv., diesel, 1-1/2 C.Y., 80 H.P.		20.00	345	1,035	3,100	367	
	4450	1-1/2 to 1-3/4 C.Y., 95 H.P.		23.75	455	1,365	4,100	463	
	4510	1-3/4 to 2-1/4 C.Y., 130 H.P.		35.05	810	2,435	7,300	767.40	
	4530	2-1/2 to 3-1/4 C.Y., 190 H.P.		47.75	1,025	3,040	9,125	990	
	4560	3-1/2 to 5 C.Y., 275 H.P.		68.30	1,500	4,520	13,600	1,450	
	4610	Front end loader, 4WD, articulated frame, 1 to 1-1/4 C.Y., 70 H.P.		14.25	208	625	1,875	239	
	4620	1-1/2 to 1-3/4 C.Y., 95 H.P.		18.90	295	885	2,650	328.20	
	4650	1-3/4 to 2 C.Y., 130 H.P.		21.60	345	1,035	3,100	379.80	
	4710	2-1/2 to 3-1/2 C.Y., 145 H.P.	↓	22.85	355	1,065	3,200	395.80	

01 54 33 | Equipment Rental

		01 54 33 Equipment Rental	UNIT	HOURLY OPER. COST	RENT PER DAY	RENT PER WEEK	RENT PER MONTH	EQUIPMENT COST/DAY	
20	4730	3 to 4-1/2 C.Y., 185 H.P.	Ea.	29.75	505	1,515	4,550	541	20
	4760	5-1/4 to 5-3/4 C.Y., 270 H.P.		48.05	785	2,350	7,050	854.40	
	4810	7 to 9 C.Y., 475 H.P.		81.85	1,400	4,210	12,600	1,497	
	4870	9 - 11 C.Y., 620 H.P.		112.20	2,200	6,635	19,900	2,225	
	4880	Skid steer loader, wheeled, 10 C.F., 30 H.P. gas		8.15	140	420	1,250	149.20	
	4890	1 C.Y., 78 H.P., diesel		14.95	233	700	2,100	259.60	
	4892	Skid-steer attachment, auger		.54	89.50	268	805	57.90	
	4893	Backhoe		.67	112	336	1,000	72.55	
	4894	Broom		.70	117	352	1,050	76	
	4895	Forks		.24	39.50	118	355	25.50	
	4896	Grapple		.57	95.50	287	860	61.95	
	4897	Concrete hammer		1.06	176	528	1,575	114.10	
	4898	Tree spade		.80	133	398	1,200	86	
	4899	Trencher		.77	128	385	1,150	83.15	
	4900	Trencher, chain, boom type, gas, operator walking, 12 H.P.		3.95	46.50	140	420	59.60	
	4910	Operator riding, 40 H.P.		15.50	295	885	2,650	301	
	5000	Wheel type, diesel, 4' deep, 12" wide		65.30	810	2,435	7,300	1,009	
	5100	6' deep, 20" wide		77.15	1,900	5,680	17,000	1,753	
	5150	Chain type, diesel, 5' deep, 8" wide		28.90	580	1,745	5,225	580.20	
	5200	Diesel, 8' deep, 16" wide		74.70	1,975	5,895	17,700	1,777	
	5210	Tree spade, self-propelled		16.78	267	800	2,400	294.25	
	5250	Truck, dump, 2-axle, 12 ton, 8 CY payload, 220 H.P.		24.55	217	650	1,950	326.40	
	5300	Three axle dump, 16 ton, 12 CY payload, 400 H.P.		43.25	310	935	2,800	533	
	5350	Dump trailer only, rear dump, 16-1/2 C.Y.		4.70	130	390	1,175	115.60	
	5400	20 C.Y.		5.15	148	445	1,325	130.20	
	5450	Flatbed, single axle, 1-1/2 ton rating		19.30	66.50	200	600	194.40	
	5500	3 ton rating		23.20	95	285	855	242.60	
	5550	Off highway rear dump, 25 ton capacity		59.95	1,325	4,000	12,000	1,280	
	5600	35 ton capacity		55.80	1,175	3,490	10,500	1,144	
	5610	50 ton capacity		76.40	1,625	4,855	14,600	1,582	
	5620	65 ton capacity		78.80	1,600	4,820	14,500	1,594	
	5630	100 ton capacity		101.35	2,050	6,180	18,500	2,047	
	6000	Vibratory plow, 25 H.P., walking	↓	6.65	61.50	185	555	90.20	
40	0010	**GENERAL EQUIPMENT RENTAL** without operators R015433-10							40
	0150	Aerial lift, scissor type, to 15' high, 1000 lb. cap., electric	Ea.	2.70	56.50	170	510	55.60	
	0160	To 25' high, 2000 lb. capacity		3.15	80	240	720	73.20	
	0170	Telescoping boom to 40' high, 500 lb. capacity, gas		17.10	320	960	2,875	328.80	
	0180	To 45' high, 500 lb. capacity		18.10	365	1,100	3,300	364.80	
	0190	To 60' high, 600 lb. capacity		20.45	505	1,510	4,525	465.60	
	0195	Air compressor, portable, 6.5 CFM, electric		.47	12.65	38	114	11.35	
	0196	Gasoline		.96	19	57	171	19.10	
	0200	Towed type, gas engine, 60 C.F.M.		11.70	48.50	145	435	122.60	
	0300	160 C.F.M.		13.60	50	150	450	138.80	
	0400	Diesel engine, rotary screw, 250 C.F.M.		12.80	100	300	900	162.40	
	0500	365 C.F.M.		17.15	123	370	1,100	211.20	
	0550	450 C.F.M.		21.75	153	460	1,375	266	
	0600	600 C.F.M.		38.05	217	650	1,950	434.40	
	0700	750 C.F.M.	↓	38.20	223	670	2,000	439.60	
	0800	For silenced models, small sizes, add to rent		3%	5%	5%	5%		
	0900	Large sizes, add to rent		5%	7%	7%	7%		
	0930	Air tools, breaker, pavement, 60 lb.	Ea.	.40	8.35	25	75	8.20	
	0940	80 lb.		.40	8.35	25	75	8.20	
	0950	Drills, hand (jackhammer) 65 lb.		.50	16	48	144	13.60	
	0960	Track or wagon, swing boom, 4" drifter		49.20	760	2,285	6,850	850.60	
	0970	5" drifter		60.00	825	2,470	7,400	974	
	0975	Track mounted quarry drill, 6" diameter drill		82.25	1,250	3,780	11,300	1,414	
	0980	Dust control per drill	↓	.99	19	57	171	19.30	

01 54 33 | Equipment Rental

			UNIT	HOURLY OPER. COST	RENT PER DAY	RENT PER WEEK	RENT PER MONTH	EQUIPMENT COST/DAY	
40	0990	Hammer, chipping, 12 lb.	Ea.	.45	22.50	67	201	17	40
	1000	Hose, air with couplings, 50' long, 3/4" diameter		.03	4.33	13	39	2.85	
	1100	1" diameter		.04	6.35	19	57	4.10	
	1200	1-1/2" diameter		.06	9.65	29	87	6.30	
	1300	2" diameter		.13	21.50	65	195	14.05	
	1400	2-1/2" diameter		.12	19.65	59	177	12.75	
	1410	3" diameter		.15	25	75	225	16.20	
	1450	Drill, steel, 7/8" x 2'		.05	8.35	25	75	5.40	
	1460	7/8" x 6'		.06	9.35	28	84	6.10	
	1520	Moil points		.02	3.67	11	33	2.35	
	1525	Pneumatic nailer w/accessories		.45	30	90	270	21.60	
	1530	Sheeting driver for 60 lb. breaker		.04	6	18	54	3.90	
	1540	For 90 lb. breaker		.12	8	24	72	5.75	
	1550	Spade, 25 lb.		.40	6	18	54	6.80	
	1560	Tamper, single, 35 lb.		.50	33.50	100	300	24	
	1570	Triple, 140 lb.		.75	50	150	450	36	
	1580	Wrenches, impact, air powered, up to 3/4" bolt		.30	8	24	72	7.20	
	1590	Up to 1-1/4" bolt		.45	21	63	189	16.20	
	1600	Barricades, barrels, reflectorized, 1 to 50 barrels		.03	4.60	13.80	41.50	3	
	1610	100 to 200 barrels		.02	3.53	10.60	32	2.30	
	1620	Barrels with flashers, 1 to 50 barrels		.03	5.25	15.80	47.50	3.40	
	1630	100 to 200 barrels		.03	4.20	12.60	38	2.75	
	1640	Barrels with steady burn type C lights		.04	7	21	63	4.50	
	1650	Illuminated board, trailer mounted, with generator		.70	120	360	1,075	77.60	
	1670	Portable barricade, stock, with flashers, 1 to 6 units		.03	5.25	15.80	47.50	3.40	
	1680	25 to 50 units		.03	4.90	14.70	44	3.20	
	1690	Butt fusion machine, electric		20.85	55	165	495	199.80	
	1695	Electro fusion machine		15.35	40	120	360	146.80	
	1700	Carts, brick, hand powered, 1000 lb. capacity		.38	63	189	565	40.85	
	1800	Gas engine, 1500 lb., 7-1/2' lift		4.73	119	357	1,075	109.25	
	1822	Dehumidifier, medium, 6 lb/hr, 150 CFM		.85	51.50	155	465	37.80	
	1824	Large, 18 lb/hr, 600 CFM		1.64	100	300	900	73.10	
	1830	Distributor, asphalt, trailer mtd, 2000 gal., 38 H.P. diesel		9.15	340	1,020	3,050	277.20	
	1840	3000 gal., 38 H.P. diesel		10.40	370	1,105	3,325	304.20	
	1850	Drill, rotary hammer, electric, 1-1/2" diameter		.70	22	66	198	18.80	
	1860	Carbide bit for above		.05	9	27	81	5.80	
	1865	Rotary, crawler, 250 H.P.		122.45	2,075	6,240	18,700	2,228	
	1870	Emulsion sprayer, 65 gal., 5 H.P. gas engine		3.20	99	297	890	85	
	1880	200 gal., 5 H.P. engine		6.85	165	495	1,475	153.80	
	1930	Floodlight, mercury vapor, or quartz, on tripod, 1000 watt		.34	13	39	117	10.50	
	1940	2000 watt		.59	24.50	73	219	19.30	
	1950	Floodlights, trailer mounted with generator, 1 - 300 watt light		3.00	70	210	630	66	
	1960	2 - 1000 watt lights		4.15	107	320	960	97.20	
	2000	4 - 300 watt lights		3.65	86.50	260	780	81.20	
	2005	Foam spray rig, incl. box trailer, compressor, generator, proportioner		24.90	480	1,445	4,325	488.20	
	2020	Forklift, straight mast, 12' lift, 5000 lb., 2 wheel drive, gas		22.00	198	595	1,775	295	
	2040	21' lift, 5000 lb., 4 wheel drive, diesel		16.65	247	740	2,225	281.20	
	2050	For rough terrain, 42' lift, 35' reach, 9000 lb., 110 HP		22.50	445	1,330	4,000	446	
	2060	For plant, 4 T. capacity, 80 H.P., 2 wheel drive, gas		13.55	105	315	945	171.40	
	2080	10 T. capacity, 120 H.P., 2 wheel drive, diesel		18.65	183	550	1,650	259.20	
	2100	Generator, electric, gas engine, 1.5 KW to 3 KW		3.10	10.35	31	93	31	
	2200	5 KW		3.95	14.35	43	129	40.20	
	2300	10 KW		7.45	33.50	100	300	79.60	
	2400	25 KW		9.10	65	195	585	111.80	
	2500	Diesel engine, 20 KW		9.25	73.50	220	660	118	
	2600	50 KW		16.60	98.50	295	885	191.80	
	2700	100 KW		31.90	120	360	1,075	327.20	
	2800	250 KW		63.05	232	695	2,075	643.40	

01 54 33 | Equipment Rental

			UNIT	HOURLY OPER. COST	RENT PER DAY	RENT PER WEEK	RENT PER MONTH	EQUIPMENT COST/DAY	
40	2850	Hammer, hydraulic, for mounting on boom, to 500 ft.-lb.	Ea.	2.15	68.50	205	615	58.20	40
	2860	1000 ft.-lb.		3.65	112	335	1,000	96.20	
	2900	Heaters, space, oil or electric, 50 MBH		2.50	6.65	20	60	24	
	3000	100 MBH		4.52	9	27	81	41.55	
	3100	300 MBH		14.52	36.50	110	330	138.15	
	3150	500 MBH		29.33	45	135	405	261.65	
	3200	Hose, water, suction with coupling, 20' long, 2" diameter		.02	3	9	27	1.95	
	3210	3" diameter		.03	4.67	14	42	3.05	
	3220	4" diameter		.03	5.35	16	48	3.45	
	3230	6" diameter		.11	18.35	55	165	11.90	
	3240	8" diameter		.20	33.50	100	300	21.60	
	3250	Discharge hose with coupling, 50' long, 2" diameter		.01	1.67	5	15	1.10	
	3260	3" diameter		.02	2.67	8	24	1.75	
	3270	4" diameter		.02	3.67	11	33	2.35	
	3280	6" diameter		.06	9.65	29	87	6.30	
	3290	8" diameter		.20	33.50	100	300	21.60	
	3295	Insulation blower		.22	6	18	54	5.35	
	3300	Ladders, extension type, 16' to 36' long		.14	23	69	207	14.90	
	3400	40' to 60' long		.20	33.50	100	300	21.60	
	3405	Lance for cutting concrete		2.97	104	313	940	86.35	
	3407	Lawn mower, rotary, 22", 5HP		2.48	65	195	585	58.85	
	3408	48" self propelled		4.48	128	385	1,150	112.85	
	3410	Level, laser type, for pipe and sewer leveling		1.46	97.50	292	875	70.10	
	3430	Electronic		.75	50	150	450	36	
	3440	Laser type, rotating beam for grade control		1.17	77.50	233	700	55.95	
	3460	Builders level with tripod and rod		.08	14	42	126	9.05	
	3500	Light towers, towable, with diesel generator, 2000 watt		3.65	86.50	260	780	81.20	
	3600	4000 watt		4.15	107	320	960	97.20	
	3700	Mixer, powered, plaster and mortar, 6 C.F., 7 H.P.		2.10	19.65	59	177	28.60	
	3800	10 C.F., 9 H.P.		2.25	31.50	95	285	37	
	3850	Nailer, pneumatic		.45	30	90	270	21.60	
	3900	Paint sprayers complete, 8 CFM		.67	44.50	134	400	32.15	
	4000	17 CFM		1.17	77.50	233	700	55.95	
	4020	Pavers, bituminous, rubber tires, 8' wide, 50 H.P., diesel		35.70	1,050	3,155	9,475	916.60	
	4030	10' wide, 150 H.P.		75.65	1,700	5,095	15,300	1,624	
	4050	Crawler, 8' wide, 100 H.P., diesel		75.65	1,775	5,355	16,100	1,676	
	4060	10' wide, 150 H.P.		84.80	2,175	6,505	19,500	1,979	
	4070	Concrete paver, 12' to 24' wide, 250 H.P.		81.00	1,575	4,755	14,300	1,599	
	4080	Placer-spreader-trimmer, 24' wide, 300 H.P.		116.90	2,600	7,825	23,500	2,500	
	4100	Pump, centrifugal gas pump, 1-1/2" diam., 65 GPM		3.35	45	135	405	53.80	
	4200	2" diameter, 130 GPM		4.55	50	150	450	66.40	
	4300	3" diameter, 250 GPM		4.80	51.50	155	465	69.40	
	4400	6" diameter, 1500 GPM		25.55	175	525	1,575	309.40	
	4500	Submersible electric pump, 1-1/4" diameter, 55 GPM		.39	15.65	47	141	12.50	
	4600	1-1/2" diameter, 83 GPM		.42	18	54	162	14.15	
	4700	2" diameter, 120 GPM		1.20	22.50	67	201	23	
	4800	3" diameter, 300 GPM		2.05	41.50	125	375	41.40	
	4900	4" diameter, 560 GPM		8.90	163	490	1,475	169.20	
	5000	6" diameter, 1590 GPM		13.05	222	665	2,000	237.40	
	5100	Diaphragm pump, gas, single, 1-1/2" diameter		1.24	45.50	136	410	37.10	
	5200	2" diameter		3.65	56.50	170	510	63.20	
	5300	3" diameter		3.65	56.50	170	510	63.20	
	5400	Double, 4" diameter		4.90	78.50	235	705	86.20	
	5450	Pressure Washer 5 Ga.Pm,3000 PSI		4.00	35	105	315	53	
	5500	Trash pump, self-priming, gas, 2" diameter		3.85	20	60	180	42.80	
	5600	Diesel, 4" diameter		10.25	58.50	175	525	117	
	5650	Diesel, 6" diameter		34.00	130	390	1,175	350	
	5655	Grout Pump	↓	13.80	93.50	280	840	166.40	

01 54 33 | Equipment Rental

			UNIT	HOURLY OPER. COST	RENT PER DAY	RENT PER WEEK	RENT PER MONTH	EQUIPMENT COST/DAY	
40	5700	Salamanders, L.P. gas fired, 100,000 BTU	Ea.	4.56	15.35	46	138	45.70	40
	5705	50,000 BTU		3.40	8.65	26	78	32.40	
	5720	Sandblaster, portable, open top, 3 C.F. capacity		.55	27	81	243	20.60	
	5730	6 C.F. capacity		.85	40	120	360	30.80	
	5740	Accessories for above		.12	19.65	59	177	12.75	
	5750	Sander, floor		.81	18.65	56	168	17.70	
	5760	Edger		.75	25.50	77	231	21.40	
	5800	Saw, chain, gas engine, 18" long		1.90	17.65	53	159	25.80	
	5900	Hydraulic powered, 36" long		.60	55	165	495	37.80	
	5950	60" long		.65	56.50	170	510	39.20	
	6000	Masonry, table mounted, 14" diameter, 5 H.P.		1.25	53	159	475	41.80	
	6050	Portable cut-off, 8 H.P.		2.05	28	84	252	33.20	
	6100	Circular, hand held, electric, 7-1/4" diameter		.19	4	12	36	3.90	
	6200	12" diameter		.26	7.35	22	66	6.50	
	6250	Wall saw, w/hydraulic power, 10 H.P		7.50	60	180	540	96	
	6275	Shot blaster, walk behind, 20" wide		4.60	300	905	2,725	217.80	
	6280	Sidewalk broom, walk-behind		2.31	57	171	515	52.70	
	6300	Steam cleaner, 100 gallons per hour		2.90	46.50	140	420	51.20	
	6310	200 gallons per hour		4.15	55	165	495	66.20	
	6340	Tar Kettle/Pot, 400 gallon		6.15	80	240	720	97.20	
	6350	Torch, cutting, acetylene-oxygen, 150' hose		.50	21.50	65	195	17	
	6360	Hourly operating cost includes tips and gas		9.45				75.60	
	6410	Toilet, portable chemical		.11	19	57	171	12.30	
	6420	Recycle flush type		.14	23	69	207	14.90	
	6430	Toilet, fresh water flush, garden hose,		.16	26.50	79	237	17.10	
	6440	Hoisted, non-flush, for high rise		.14	22.50	68	204	14.70	
	6450	Toilet, trailers, minimum		.24	39.50	118	355	25.50	
	6460	Maximum		.72	119	358	1,075	77.35	
	6465	Tractor, farm with attachment		14.40	262	785	2,350	272.20	
	6500	Trailers, platform, flush deck, 2 axle, 25 ton capacity		4.75	107	320	960	102	
	6600	40 ton capacity		6.20	150	450	1,350	139.60	
	6700	3 axle, 50 ton capacity		6.70	165	495	1,475	152.60	
	6800	75 ton capacity		8.35	218	655	1,975	197.80	
	6810	Trailer mounted cable reel for H.V. line work		4.84	231	692	2,075	177.10	
	6820	Trailer mounted cable tensioning rig		9.59	455	1,370	4,100	350.70	
	6830	Cable pulling rig		69.62	2,575	7,710	23,100	2,099	
	6900	Water tank, engine driven discharge, 5000 gallons		6.25	143	430	1,300	136	
	6925	10,000 gallons		8.50	202	605	1,825	189	
	6950	Water truck, off highway, 6000 gallons		66.95	775	2,320	6,950	999.60	
	7010	Tram car for H.V. line work, powered, 2 conductor		6.48	125	375	1,125	126.85	
	7020	Transit (builder's level) with tripod		.08	14	42	126	9.05	
	7030	Trench box, 3000 lbs. 6'x8'		.56	93	279	835	60.30	
	7040	7200 lbs. 6'x20'		1.05	175	525	1,575	113.40	
	7050	8000 lbs., 8' x 16'		.95	158	475	1,425	102.60	
	7060	9500 lbs., 8'x20'		1.16	194	581	1,750	125.50	
	7065	11,000 lbs., 8'x24'		1.27	212	637	1,900	137.55	
	7070	12,000 lbs., 10' x 20'		1.71	285	855	2,575	184.70	
	7100	Truck, pickup, 3/4 ton, 2 wheel drive		10.35	56.50	170	510	116.80	
	7200	4 wheel drive		10.65	71.50	215	645	128.20	
	7250	Crew carrier, 9 passenger		14.70	86.50	260	780	169.60	
	7290	Flat bed truck, 20,000 G.V.W.		15.55	122	365	1,100	197.40	
	7300	Tractor, 4 x 2, 220 H.P.		21.80	190	570	1,700	288.40	
	7410	330 H.P.		32.15	262	785	2,350	414.20	
	7500	6 x 4, 380 H.P.		36.85	305	920	2,750	478.80	
	7600	450 H.P.		44.90	370	1,110	3,325	581.20	
	7620	Vacuum truck, hazardous material, 2500 gallon		10.80	310	925	2,775	271.40	
	7625	5,000 gallon		16.90	435	1,300	3,900	395.20	
	7640	Tractor, with A frame, boom and winch, 225 H.P.	↓	24.35	267	800	2,400	354.80	

01 54 33 | Equipment Rental

				UNIT	HOURLY OPER. COST	RENT PER DAY	RENT PER WEEK	RENT PER MONTH	EQUIPMENT COST/DAY	
40	7650	Vacuum, H.E.P.A., 16 gal., wet/dry		Ea.	.82	18	54	162	17.35	40
	7655	55 gal, wet/dry			.83	27	81	243	22.85	
	7660	Water tank, portable			.17	28.50	85.50	257	18.45	
	7690	Sewer/catch basin vacuum, 14 CY, 1500 Gallon			21.45	650	1,950	5,850	561.60	
	7700	Welder, electric, 200 amp			3.63	17.65	53	159	39.65	
	7800	300 amp			5.36	21.50	64	192	55.70	
	7900	Gas engine, 200 amp			12.85	25.50	77	231	118.20	
	8000	300 amp			14.70	27.50	83	249	134.20	
	8100	Wheelbarrow, any size			.07	11.65	35	105	7.55	
	8200	Wrecking ball, 4000 lb.			2.10	73.50	220	660	60.80	
50	0010	**HIGHWAY EQUIPMENT RENTAL** without operators	R015433 -10							50
	0050	Asphalt batch plant, portable drum mixer, 100 ton/hr.		Ea.	63.70	1,450	4,330	13,000	1,376	
	0060	200 ton/hr.			71.10	1,525	4,565	13,700	1,482	
	0070	300 ton/hr.			82.55	1,800	5,395	16,200	1,739	
	0100	Backhoe attachment, long stick, up to 185 HP, 10.5' long			.32	21.50	64	192	15.35	
	0140	Up to 250 HP, 12' long			.35	23	69	207	16.60	
	0180	Over 250 HP, 15' long			.47	31	93	279	22.35	
	0200	Special dipper arm, up to 100 HP, 32' long			.96	63.50	191	575	45.90	
	0240	Over 100 HP, 33' long			1.19	79.50	238	715	57.10	
	0280	Catch basin/sewer cleaning truck, 3 ton, 9 CY, 1000 Gal			34.00	405	1,210	3,625	514	
	0300	Concrete batch plant, portable, electric, 200 CY/Hr			17.20	510	1,530	4,600	443.60	
	0520	Grader/dozer attachment, ripper/scarifier, rear mounted, up to 135 HP			3.10	65	195	585	63.80	
	0540	Up to 180 HP			3.70	83.50	250	750	79.60	
	0580	Up to 250 HP			4.05	95	285	855	89.40	
	0700	Pvmt. removal bucket, for hyd. excavator, up to 90 HP			1.65	48.50	145	435	42.20	
	0740	Up to 200 HP			1.85	70	210	630	56.80	
	0780	Over 200 HP			1.95	83.50	250	750	65.60	
	0900	Aggregate spreader, self-propelled, 187 HP			54.90	695	2,080	6,250	855.20	
	1000	Chemical spreader, 3 C.Y.			2.75	43.50	130	390	48	
	1900	Hammermill, traveling, 250 HP			92.70	1,850	5,540	16,600	1,850	
	2000	Horizontal borer, 3" diam, 13 HP gas driven			5.45	56.50	170	510	77.60	
	2150	Horizontal directional drill, 20,000 lb. thrust, 78 H.P. diesel			26.15	700	2,095	6,275	628.20	
	2160	30,000 lb. thrust, 115 H.P. diesel			32.45	1,050	3,185	9,550	896.60	
	2170	50,000 lb. thrust, 170 H.P. diesel			45.85	1,350	4,080	12,200	1,183	
	2190	Mud trailer for HDD, 1500 gallon, 175 H.P., gas			24.40	147	440	1,325	283.20	
	2200	Hydromulchers, gas power, 3000 gal., for truck mounting			17.00	258	775	2,325	291	
	2400	Joint & crack cleaner, walk behind, 25 HP			3.10	50	150	450	54.80	
	2500	Filler, trailer mounted, 400 gal., 20 HP			7.90	218	655	1,975	194.20	
	3000	Paint striper, self propelled, double line, 30 HP			6.25	162	485	1,450	147	
	3200	Post drivers, 6" I-Beam frame, for truck mounting			13.40	435	1,305	3,925	368.20	
	3400	Road sweeper, self propelled, 8' wide, 90 HP			31.95	575	1,730	5,200	601.60	
	3450	Road sweeper, vacuum assisted, 4 CY, 220 Gal			56.60	630	1,895	5,675	831.80	
	4000	Road mixer, self-propelled, 130 HP			39.70	770	2,315	6,950	780.60	
	4100	310 HP			71.00	2,225	6,640	19,900	1,896	
	4220	Cold mix paver, incl pug mill and bitumen tank, 165 HP			86.60	2,175	6,515	19,500	1,996	
	4250	Paver, asphalt, wheel or crawler, 130 H.P., diesel			83.50	2,125	6,340	19,000	1,936	
	4300	Paver, road widener, gas 1' to 6', 67 HP			40.10	825	2,480	7,450	816.80	
	4400	Diesel, 2' to 14', 88 HP			52.30	1,050	3,175	9,525	1,053	
	4600	Slipform pavers, curb and gutter, 2 track, 75 HP			36.80	765	2,300	6,900	754.40	
	4700	4 track, 165 HP			45.85	815	2,445	7,325	855.80	
	4800	Median barrier, 215 HP			46.50	845	2,535	7,600	879	
	4901	Trailer, low bed, 75 ton capacity			9.00	218	655	1,975	203	
	5000	Road planer, walk behind, 10" cutting width, 10 HP			2.90	30.50	91	273	41.40	
	5100	Self propelled, 12" cutting width, 64 HP			8.25	127	380	1,150	142	
	5120	Traffic line remover, metal ball blaster, truck mounted, 115 HP			46.45	760	2,285	6,850	828.60	
	5140	Grinder, truck mounted, 115 HP			49.55	805	2,420	7,250	880.40	
	5160	Walk-behind, 11 HP			3.25	46.50	140	420	54	
	5200	Pavement profiler, 4' to 6' wide, 450 HP			210.55	3,400	10,220	30,700	3,728	

01 54 33 | Equipment Rental

			UNIT	HOURLY OPER. COST	RENT PER DAY	RENT PER WEEK	RENT PER MONTH	EQUIPMENT COST/DAY	
50	5300	8' to 10' wide, 750 HP	Ea.	333.90	4,725	14,160	42,500	5,503	50
	5400	Roadway plate, steel, 1"x8'x20'		.07	11.35	34	102	7.35	
	5600	Stabilizer,self-propelled, 150 HP		39.45	610	1,830	5,500	681.60	
	5700	310 HP		66.45	1,350	4,060	12,200	1,344	
	5800	Striper, thermal, truck mounted 120 gal. paint, 150 H.P.		49.90	505	1,520	4,550	703.20	
	6000	Tar kettle, 330 gal., trailer mounted		5.77	61.50	185	555	83.15	
	7000	Tunnel locomotive, diesel, 8 to 12 ton		26.70	585	1,750	5,250	563.60	
	7005	Electric, 10 ton		23.25	665	2,000	6,000	586	
	7010	Muck cars, 1/2 C.Y. capacity		1.75	23	69	207	27.80	
	7020	1 C.Y. capacity		1.95	31.50	94	282	34.40	
	7030	2 C.Y. capacity		2.10	36.50	110	330	38.80	
	7040	Side dump, 2 C.Y. capacity		2.30	45	135	405	45.40	
	7050	3 C.Y. capacity		3.10	51.50	155	465	55.80	
	7060	5 C.Y. capacity		4.40	65	195	585	74.20	
	7100	Ventilating blower for tunnel, 7-1/2 H.P.		1.35	51.50	155	465	41.80	
	7110	10 H.P.		1.57	53.50	160	480	44.55	
	7120	20 H.P.		2.58	69.50	208	625	62.25	
	7140	40 H.P.		4.56	98.50	295	885	95.50	
	7160	60 H.P.		6.90	152	455	1,375	146.20	
	7175	75 H.P.		8.81	202	607	1,825	191.90	
	7180	200 H.P.		19.95	305	910	2,725	341.60	
	7800	Windrow loader, elevating		47.65	1,350	4,080	12,200	1,197	
60	0010	**LIFTING AND HOISTING EQUIPMENT RENTAL** without operators (R015433-10)							60
	0120	Aerial lift truck, 2 person, to 80'	Ea.	25.55	725	2,180	6,550	640.40	
	0140	Boom work platform, 40' snorkel (R015433-15)		15.40	260	780	2,350	279.20	
	0150	Crane, flatbed mntd, 3 ton cap.		13.50	195	585	1,750	225	
	0200	Crane, climbing, 106' jib, 6000 lb. capacity, 410 FPM (R312316-45)		37.40	1,475	4,400	13,200	1,179	
	0300	101' jib, 10,250 lb. capacity, 270 FPM		43.30	1,850	5,580	16,700	1,462	
	0500	Tower, static, 130' high, 106' jib, 6200 lb. capacity at 400 FPM		40.85	1,700	5,090	15,300	1,345	
	0600	Crawler mounted, lattice boom, 1/2 C.Y., 15 tons at 12' radius		30.66	615	1,850	5,550	615.30	
	0700	3/4 C.Y., 20 tons at 12' radius		40.88	770	2,310	6,925	789.05	
	0800	1 C.Y., 25 tons at 12' radius		54.50	1,025	3,080	9,250	1,052	
	0900	1-1/2 C.Y., 40 tons at 12' radius		54.50	1,025	3,095	9,275	1,055	
	1000	2 C.Y., 50 tons at 12' radius		54.20	1,125	3,380	10,100	1,110	
	1100	3 C.Y., 75 tons at 12' radius		72.95	1,525	4,560	13,700	1,496	
	1200	100 ton capacity, 60' boom		71.55	1,725	5,195	15,600	1,611	
	1300	165 ton capacity, 60' boom		89.60	1,925	5,750	17,300	1,867	
	1400	200 ton capacity, 70' boom		111.30	2,350	7,085	21,300	2,307	
	1500	350 ton capacity, 80' boom		158.35	3,775	11,335	34,000	3,534	
	1600	Truck mounted, lattice boom, 6 x 4, 20 tons at 10' radius		44.77	1,200	3,610	10,800	1,080	
	1700	25 tons at 10' radius		48.31	1,300	3,930	11,800	1,172	
	1800	30 tons at 10' radius		52.66	1,400	4,180	12,500	1,257	
	1900	40 tons at 12' radius		56.59	1,450	4,370	13,100	1,327	
	2000	60 tons at 15' radius		64.84	1,550	4,620	13,900	1,443	
	2050	82 tons at 15' radius		73.58	1,650	4,940	14,800	1,577	
	2100	90 tons at 15' radius		83.16	1,800	5,380	16,100	1,741	
	2200	115 tons at 15' radius		94.14	2,000	6,020	18,100	1,957	
	2300	150 tons at 18' radius		77.90	2,100	6,335	19,000	1,890	
	2350	165 tons at 18' radius		112.04	2,250	6,720	20,200	2,240	
	2400	Truck mounted, hydraulic, 12 ton capacity		49.10	625	1,880	5,650	768.80	
	2500	25 ton capacity		49.30	660	1,975	5,925	789.40	
	2550	33 ton capacity		50.35	695	2,080	6,250	818.80	
	2560	40 ton capacity		48.50	670	2,015	6,050	791	
	2600	55 ton capacity		72.00	915	2,745	8,225	1,125	
	2700	80 ton capacity		85.50	1,025	3,060	9,175	1,296	
	2720	100 ton capacity		104.90	1,550	4,685	14,100	1,776	
	2740	120 ton capacity		83.95	2,000	5,975	17,900	1,867	
	2760	150 ton capacity		89.85	2,175	6,490	19,500	2,017	

01 54 33 | Equipment Rental

			UNIT	HOURLY OPER. COST	RENT PER DAY	RENT PER WEEK	RENT PER MONTH	EQUIPMENT COST/DAY	
60	2800	Self-propelled, 4 x 4, with telescoping boom, 5 ton	Ea.	14.65	235	705	2,125	258.20	60
	2900	12-1/2 ton capacity		34.60	540	1,625	4,875	601.80	
	3000	15 ton capacity		32.30	575	1,725	5,175	603.40	
	3050	20 ton capacity		37.95	660	1,985	5,950	700.60	
	3100	25 ton capacity		38.85	675	2,030	6,100	716.80	
	3150	40 ton capacity		50.15	895	2,680	8,050	937.20	
	3200	Derricks, guy, 20 ton capacity, 60' boom, 75' mast		33.55	360	1,078	3,225	484	
	3300	100' boom, 115' mast		51.95	615	1,850	5,550	785.60	
	3400	Stiffleg, 20 ton capacity, 70' boom, 37' mast		35.80	465	1,400	4,200	566.40	
	3500	100' boom, 47' mast		54.68	745	2,240	6,725	885.45	
	3550	Helicopter, small, lift to 1250 lbs. maximum, w/pilot		105.67	2,900	8,710	26,100	2,587	
	3600	Hoists, chain type, overhead, manual, 3/4 ton		.10	1	3	9	1.40	
	3900	10 ton		.70	9.65	29	87	11.40	
	4000	Hoist and tower, 5000 lb. cap., portable electric, 40' high		4.65	207	621	1,875	161.40	
	4100	For each added 10' section, add		.10	16.35	49	147	10.60	
	4200	Hoist and single tubular tower, 5000 lb. electric, 100' high		6.26	289	867	2,600	223.50	
	4300	For each added 6'-6" section, add		.17	27.50	83	249	17.95	
	4400	Hoist and double tubular tower, 5000 lb., 100' high		6.70	320	955	2,875	244.60	
	4500	For each added 6'-6" section, add		.19	31	93	279	20.10	
	4550	Hoist and tower, mast type, 6000 lb., 100' high		7.26	330	990	2,975	256.10	
	4570	For each added 10' section, add		.12	19.65	59	177	12.75	
	4600	Hoist and tower, personnel, electric, 2000 lb., 100' @ 125 FPM		14.74	880	2,640	7,925	645.90	
	4700	3000 lb., 100' @ 200 FPM		16.88	995	2,990	8,975	733.05	
	4800	3000 lb., 150' @ 300 FPM		18.63	1,125	3,340	10,000	817.05	
	4900	4000 lb., 100' @ 300 FPM		19.36	1,125	3,410	10,200	836.90	
	5000	6000 lb., 100' @ 275 FPM		20.98	1,200	3,580	10,700	883.85	
	5100	For added heights up to 500', add	L.F.	.01	1.67	5	15	1.10	
	5200	Jacks, hydraulic, 20 ton	Ea.	.05	1.67	5	15	1.40	
	5500	100 ton		.35	10.65	32	96	9.20	
	6100	Jacks, hydraulic, climbing w/ 50' jackrods, control console, 30 ton cap		1.79	119	357	1,075	85.70	
	6150	For each added 10' jackrod section, add		.05	3.33	10	30	2.40	
	6300	50 ton capacity		2.87	191	574	1,725	137.75	
	6350	For each added 10' jackrod section, add		.06	4	12	36	2.90	
	6500	125 ton capacity		7.55	505	1,510	4,525	362.40	
	6550	For each added 10' jackrod section, add		.52	34.50	103	310	24.75	
	6600	Cable jack, 10 ton capacity with 200' cable		1.50	99.50	299	895	71.80	
	6650	For each added 50' of cable, add		.17	11.35	34	102	8.15	
70	0010	**WELLPOINT EQUIPMENT RENTAL** without operators R015433-10							70
	0020	Based on 2 months rental							
	0100	Combination jetting & wellpoint pump, 60 H.P. diesel	Ea.	21.79	295	884	2,650	351.10	
	0200	High pressure gas jet pump, 200 H.P., 300 psi	"	57.29	252	756	2,275	609.50	
	0300	Discharge pipe, 8" diameter	L.F.	.01	.48	1.44	4.32	.35	
	0350	12" diameter		.01	.71	2.12	6.35	.50	
	0400	Header pipe, flows up to 150 G.P.M., 4" diameter		.01	.43	1.29	3.87	.35	
	0500	400 G.P.M., 6" diameter		.01	.51	1.54	4.62	.40	
	0600	800 G.P.M., 8" diameter		.01	.71	2.12	6.35	.50	
	0700	1500 G.P.M., 10" diameter		.01	.74	2.22	6.65	.50	
	0800	2500 G.P.M., 12" diameter		.02	1.40	4.21	12.65	1	
	0900	4500 G.P.M., 16" diameter		.03	1.80	5.39	16.15	1.30	
	0950	For quick coupling aluminum and plastic pipe, add		.03	1.86	5.58	16.75	1.35	
	1100	Wellpoint, 25' long, with fittings & riser pipe, 1-1/2" or 2" diameter	Ea.	.06	3.71	11.13	33.50	2.70	
	1200	Wellpoint pump, diesel powered, 4" diameter, 20 H.P.		8.77	170	510	1,525	172.15	
	1300	6" diameter, 30 H.P.		12.22	211	632	1,900	224.15	
	1400	8" suction, 40 H.P.		16.47	289	867	2,600	305.15	
	1500	10" suction, 75 H.P.		26.59	340	1,013	3,050	415.30	
	1600	12" suction, 100 H.P.		37.34	540	1,620	4,850	622.70	
	1700	12" suction, 175 H.P.		57.89	590	1,770	5,300	817.10	

01 54 33 | Equipment Rental

				UNIT	HOURLY OPER. COST	RENT PER DAY	RENT PER WEEK	RENT PER MONTH	EQUIPMENT COST/DAY	
80	0010	**MARINE EQUIPMENT RENTAL** without operators	R015433 -10							80
	0200	Barge, 400 Ton, 30' wide x 90' long		Ea.	14.80	1,050	3,140	9,425	746.40	
	0240	800 Ton, 45' wide x 90' long			17.95	1,275	3,800	11,400	903.60	
	2000	Tugboat, diesel, 100 HP			27.75	195	585	1,750	339	
	2040	250 HP			56.05	360	1,075	3,225	663.40	
	2080	380 HP		↓	115.60	1,075	3,195	9,575	1,564	

Crews

Crew No.	Bare Costs		Incl. Subs O & P		Cost Per Labor-Hour	
Crew A-1	**Hr.**	**Daily**	**Hr.**	**Daily**	**Bare Costs**	**Incl. O&P**
1 Building Laborer	$31.60	$252.80	$52.15	$417.20	$31.60	$52.15
1 Concrete saw, gas manual		63.20		69.52	7.90	8.69
8 L.H., Daily Totals		$316.00		$486.72	$39.50	$60.84

Crew A-1A	Hr.	Daily	Hr.	Daily	Bare Costs	Incl. O&P
1 Skilled Worker	$40.85	$326.80	$66.50	$532.00	$40.85	$66.50
1 Shot Blaster, 20″		217.80		239.58	27.23	29.95
8 L.H., Daily Totals		$544.60		$771.58	$68.08	$96.45

Crew A-1B	Hr.	Daily	Hr.	Daily	Bare Costs	Incl. O&P
1 Building Laborer	$31.60	$252.80	$52.15	$417.20	$31.60	$52.15
1 Concrete Saw		141.00		155.10	17.63	19.39
8 L.H., Daily Totals		$393.80		$572.30	$49.23	$71.54

Crew A-1C	Hr.	Daily	Hr.	Daily	Bare Costs	Incl. O&P
1 Building Laborer	$31.60	$252.80	$52.15	$417.20	$31.60	$52.15
1 Chain saw, gas, 18″		25.80		28.38	3.23	3.55
8 L.H., Daily Totals		$278.60		$445.58	$34.83	$55.70

Crew A-1D	Hr.	Daily	Hr.	Daily	Bare Costs	Incl. O&P
1 Building Laborer	$31.60	$252.80	$52.15	$417.20	$31.60	$52.15
1 Vibrating plate, gas, 18″		31.80		34.98	3.98	4.37
8 L.H., Daily Totals		$284.60		$452.18	$35.58	$56.52

Crew A-1E	Hr.	Daily	Hr.	Daily	Bare Costs	Incl. O&P
1 Building Laborer	$31.60	$252.80	$52.15	$417.20	$31.60	$52.15
1 Vibratory Plate, Gas, 21″		39.20		43.12	4.90	5.39
8 L.H., Daily Totals		$292.00		$460.32	$36.50	$57.54

Crew A-1F	Hr.	Daily	Hr.	Daily	Bare Costs	Incl. O&P
1 Building Laborer	$31.60	$252.80	$52.15	$417.20	$31.60	$52.15
1 Rammer/tamper, gas, 8″		43.40		47.74	5.42	5.97
8 L.H., Daily Totals		$296.20		$464.94	$37.02	$58.12

Crew A-1G	Hr.	Daily	Hr.	Daily	Bare Costs	Incl. O&P
1 Building Laborer	$31.60	$252.80	$52.15	$417.20	$31.60	$52.15
1 Rammer/tamper, gas, 15″		49.40		54.34	6.17	6.79
8 L.H., Daily Totals		$302.20		$471.54	$37.77	$58.94

Crew A-1H	Hr.	Daily	Hr.	Daily	Bare Costs	Incl. O&P
1 Building Laborer	$31.60	$252.80	$52.15	$417.20	$31.60	$52.15
1 Exterior Steam Cleaner		51.20		56.32	6.40	7.04
8 L.H., Daily Totals		$304.00		$473.52	$38.00	$59.19

Crew A-1J	Hr.	Daily	Hr.	Daily	Bare Costs	Incl. O&P
1 Building Laborer	$31.60	$252.80	$52.15	$417.20	$31.60	$52.15
1 Cultivator, Walk-Behind, 5 H.P.		57.05		62.76	7.13	7.84
8 L.H., Daily Totals		$309.85		$479.95	$38.73	$59.99

Crew A-1K	Hr.	Daily	Hr.	Daily	Bare Costs	Incl. O&P
1 Building Laborer	$31.60	$252.80	$52.15	$417.20	$31.60	$52.15
1 Cultivator, Walk-Behind, 8 H.P.		91.10		100.21	11.39	12.53
8 L.H., Daily Totals		$343.90		$517.41	$42.99	$64.68

Crew A-1M	Hr.	Daily	Hr.	Daily	Bare Costs	Incl. O&P
1 Building Laborer	$31.60	$252.80	$52.15	$417.20	$31.60	$52.15
1 Snow Blower, Walk-Behind		52.70		57.97	6.59	7.25
8 L.H., Daily Totals		$305.50		$475.17	$38.19	$59.40

Crew No.	Bare Costs		Incl. Subs O & P		Cost Per Labor-Hour	
Crew A-2	**Hr.**	**Daily**	**Hr.**	**Daily**	**Bare Costs**	**Incl. O&P**
2 Laborers	$31.60	$505.60	$52.15	$834.40	$31.38	$51.68
1 Truck Driver (light)	30.95	247.60	50.75	406.00		
1 Flatbed Truck, Gas, 1.5 Ton		194.40		213.84	8.10	8.91
24 L.H., Daily Totals		$947.60		$1454.24	$39.48	$60.59

Crew A-2A	Hr.	Daily	Hr.	Daily	Bare Costs	Incl. O&P
2 Laborers	$31.60	$505.60	$52.15	$834.40	$31.38	$51.68
1 Truck Driver (light)	30.95	247.60	50.75	406.00		
1 Flatbed Truck, Gas, 1.5 Ton		194.40		213.84		
1 Concrete Saw		141.00		155.10	13.98	15.37
24 L.H., Daily Totals		$1088.60		$1609.34	$45.36	$67.06

Crew A-2B	Hr.	Daily	Hr.	Daily	Bare Costs	Incl. O&P
1 Truck Driver (light)	$30.95	$247.60	$50.75	$406.00	$30.95	$50.75
1 Flatbed Truck, Gas, 1.5 Ton		194.40		213.84	24.30	26.73
8 L.H., Daily Totals		$442.00		$619.84	$55.25	$77.48

Crew A-3A	Hr.	Daily	Hr.	Daily	Bare Costs	Incl. O&P
1 Truck Driver (light)	$30.95	$247.60	$50.75	$406.00	$30.95	$50.75
1 Pickup truck, 4 x 4, 3/4 ton		128.20		141.02	16.02	17.63
8 L.H., Daily Totals		$375.80		$547.02	$46.98	$68.38

Crew A-3B	Hr.	Daily	Hr.	Daily	Bare Costs	Incl. O&P
1 Equip. Oper. (medium)	$41.35	$330.80	$65.05	$520.40	$36.65	$58.70
1 Truck Driver (heavy)	31.95	255.60	52.35	418.80		
1 Dump Truck, 12 C.Y., 400 H.P.		533.00		586.30		
1 F.E. Loader, W.M.,2.5 C.Y.		395.80		435.38	58.05	63.85
16 L.H., Daily Totals		$1515.20		$1960.88	$94.70	$122.56

Crew A-3C	Hr.	Daily	Hr.	Daily	Bare Costs	Incl. O&P
1 Equip. Oper. (light)	$39.05	$312.40	$61.45	$491.60	$39.05	$61.45
1 Loader, Skid Steer, 78 H.P.		259.60		285.56	32.45	35.70
8 L.H., Daily Totals		$572.00		$777.16	$71.50	$97.14

Crew A-3D	Hr.	Daily	Hr.	Daily	Bare Costs	Incl. O&P
1 Truck Driver, Light	$30.95	$247.60	$50.75	$406.00	$30.95	$50.75
1 Pickup truck, 4 x 4, 3/4 ton		128.20		141.02		
1 Flatbed Trailer, 25 Ton		102.00		112.20	28.77	31.65
8 L.H., Daily Totals		$477.80		$659.22	$59.73	$82.40

Crew A-3E	Hr.	Daily	Hr.	Daily	Bare Costs	Incl. O&P
1 Equip. Oper. (crane)	$42.55	$340.40	$66.95	$535.60	$37.25	$59.65
1 Truck Driver (heavy)	31.95	255.60	52.35	418.80		
1 Pickup truck, 4 x 4, 3/4 ton		128.20		141.02	8.01	8.81
16 L.H., Daily Totals		$724.20		$1095.42	$45.26	$68.46

Crew A-3F	Hr.	Daily	Hr.	Daily	Bare Costs	Incl. O&P
1 Equip. Oper. (crane)	$42.55	$340.40	$66.95	$535.60	$37.25	$59.65
1 Truck Driver (heavy)	31.95	255.60	52.35	418.80		
1 Pickup truck, 4 x 4, 3/4 ton		128.20		141.02		
1 Truck Tractor, 6x4, 380 H.P.		478.80		526.68		
1 Lowbed Trailer, 75 Ton		203.00		223.30	50.63	55.69
16 L.H., Daily Totals		$1406.00		$1845.40	$87.88	$115.34

Crew No.	Bare Costs		Incl. Subs O & P		Cost Per Labor-Hour	
Crew A-3G	**Hr.**	**Daily**	**Hr.**	**Daily**	**Bare Costs**	**Incl. O&P**
1 Equip. Oper. (crane)	$42.55	$340.40	$66.95	$535.60	$37.25	$59.65
1 Truck Driver (heavy)	31.95	255.60	52.35	418.80		
1 Pickup truck, 4 x 4, 3/4 ton		128.20		141.02		
1 Truck Tractor, 6x4, 450 H.P.		581.20		639.32		
1 Lowbed Trailer, 75 Ton		203.00		223.30	57.02	62.73
16 L.H., Daily Totals		$1508.40		$1958.04	$94.28	$122.38

Crew A-3H	Hr.	Daily	Hr.	Daily	Bare Costs	Incl. O&P
1 Equip. Oper. (crane)	$42.55	$340.40	$66.95	$535.60	$42.55	$66.95
1 Hyd. crane, 12 Ton (daily)		1018.00		1119.80	127.25	139.97
8 L.H., Daily Totals		$1358.40		$1655.40	$169.80	$206.93

Crew A-3I	Hr.	Daily	Hr.	Daily	Bare Costs	Incl. O&P
1 Equip. Oper. (crane)	$42.55	$340.40	$66.95	$535.60	$42.55	$66.95
1 Hyd. crane, 25 Ton (daily)		1054.00		1159.40	131.75	144.93
8 L.H., Daily Totals		$1394.40		$1695.00	$174.30	$211.88

Crew A-3J	Hr.	Daily	Hr.	Daily	Bare Costs	Incl. O&P
1 Equip. Oper. (crane)	$42.55	$340.40	$66.95	$535.60	$42.55	$66.95
1 Hyd. crane, 40 Ton (daily)		1058.00		1163.80	132.25	145.47
8 L.H., Daily Totals		$1398.40		$1699.40	$174.80	$212.43

Crew A-3K	Hr.	Daily	Hr.	Daily	Bare Costs	Incl. O&P
1 Equip. Oper. (crane)	$42.55	$340.40	$66.95	$535.60	$39.67	$62.42
1 Equip. Oper. Oiler	36.80	294.40	57.90	463.20		
1 Hyd. crane, 55 Ton (daily)		1491.00		1640.10		
1 P/U Truck, 3/4 Ton (daily)		137.80		151.58	101.80	111.98
16 L.H., Daily Totals		$2263.60		$2790.48	$141.47	$174.41

Crew A-3L	Hr.	Daily	Hr.	Daily	Bare Costs	Incl. O&P
1 Equip. Oper. (crane)	$42.55	$340.40	$66.95	$535.60	$39.67	$62.42
1 Equip. Oper. Oiler	36.80	294.40	57.90	463.20		
1 Hyd. crane, 80 Ton (daily)		1704.00		1874.40		
1 P/U Truck, 3/4 Ton (daily)		137.80		151.58	115.11	126.62
16 L.H., Daily Totals		$2476.60		$3024.78	$154.79	$189.05

Crew A-3M	Hr.	Daily	Hr.	Daily	Bare Costs	Incl. O&P
1 Equip. Oper. (crane)	$42.55	$340.40	$66.95	$535.60	$39.67	$62.42
1 Equip. Oper. Oiler	36.80	294.40	57.90	463.20		
1 Hyd. crane, 100 Ton (daily)		2399.00		2638.90		
1 P/U Truck, 3/4 Ton (daily)		137.80		151.58	158.55	174.41
16 L.H., Daily Totals		$3171.60		$3789.28	$198.22	$236.83

Crew A-3N	Hr.	Daily	Hr.	Daily	Bare Costs	Incl. O&P
1 Equip. Oper. (crane)	$42.55	$340.40	$66.95	$535.60	$42.55	$66.95
1 Tower crane (monthly)		1022.00		1124.20	127.75	140.53
8 L.H., Daily Totals		$1362.40		$1659.80	$170.30	$207.47

Crew A-3P	Hr.	Daily	Hr.	Daily	Bare Costs	Incl. O&P
1 Equip. Oper., Light	$39.05	$312.40	$61.45	$491.60	$39.05	$61.45
1 A.T. Forklift, 42' lift		446.00		490.60	55.75	61.33
8 L.H., Daily Totals		$758.40		$982.20	$94.80	$122.78

Crew A-4	Hr.	Daily	Hr.	Daily	Bare Costs	Incl. O&P
2 Carpenters	$39.95	$639.20	$65.95	$1055.20	$38.37	$62.72
1 Painter, Ordinary	35.20	281.60	56.25	450.00		
24 L.H., Daily Totals		$920.80		$1505.20	$38.37	$62.72

Crew No.	Bare Costs		Incl. Subs O & P		Cost Per Labor-Hour	
Crew A-5	**Hr.**	**Daily**	**Hr.**	**Daily**	**Bare Costs**	**Incl. O&P**
2 Laborers	$31.60	$505.60	$52.15	$834.40	$31.53	$51.99
.25 Truck Driver (light)	30.95	61.90	50.75	101.50		
.25 Flatbed Truck, Gas, 1.5 Ton		48.60		53.46	2.70	2.97
18 L.H., Daily Totals		$616.10		$989.36	$34.23	$54.96

Crew A-6	Hr.	Daily	Hr.	Daily	Bare Costs	Incl. O&P
1 Instrument Man	$40.85	$326.80	$66.50	$532.00	$39.73	$64.35
1 Rodman/Chainman	38.60	308.80	62.20	497.60		
1 Laser Transit/Level		70.10		77.11	4.38	4.82
16 L.H., Daily Totals		$705.70		$1106.71	$44.11	$69.17

Crew A-7	Hr.	Daily	Hr.	Daily	Bare Costs	Incl. O&P
1 Chief Of Party	$50.20	$401.60	$79.65	$637.20	$43.22	$69.45
1 Instrument Man	40.85	326.80	66.50	532.00		
1 Rodman/Chainman	38.60	308.80	62.20	497.60		
1 Laser Transit/Level		70.10		77.11	2.92	3.21
24 L.H., Daily Totals		$1107.30		$1743.91	$46.14	$72.66

Crew A-8	Hr.	Daily	Hr.	Daily	Bare Costs	Incl. O&P
1 Chief of Party	$50.20	$401.60	$79.65	$637.20	$42.06	$67.64
1 Instrument Man	40.85	326.80	66.50	532.00		
2 Rodmen/Chainmen	38.60	617.60	62.20	995.20		
1 Laser Transit/Level		70.10		77.11	2.19	2.41
32 L.H., Daily Totals		$1416.10		$2241.51	$44.25	$70.05

Crew A-9	Hr.	Daily	Hr.	Daily	Bare Costs	Incl. O&P
1 Asbestos Foreman	$44.60	$356.80	$72.10	$576.80	$44.16	$71.40
7 Asbestos Workers	44.10	2469.60	71.30	3992.80		
64 L.H., Daily Totals		$2826.40		$4569.60	$44.16	$71.40

Crew A-10A	Hr.	Daily	Hr.	Daily	Bare Costs	Incl. O&P
1 Asbestos Foreman	$44.60	$356.80	$72.10	$576.80	$44.27	$71.57
2 Asbestos Workers	44.10	705.60	71.30	1140.80		
24 L.H., Daily Totals		$1062.40		$1717.60	$44.27	$71.57

Crew A-10B	Hr.	Daily	Hr.	Daily	Bare Costs	Incl. O&P
1 Asbestos Foreman	$44.60	$356.80	$72.10	$576.80	$44.23	$71.50
3 Asbestos Workers	44.10	1058.40	71.30	1711.20		
32 L.H., Daily Totals		$1415.20		$2288.00	$44.23	$71.50

Crew A-10C	Hr.	Daily	Hr.	Daily	Bare Costs	Incl. O&P
3 Asbestos Workers	$44.10	$1058.40	$71.30	$1711.20	$44.10	$71.30
1 Flatbed Truck, Gas, 1.5 Ton		194.40		213.84	8.10	8.91
24 L.H., Daily Totals		$1252.80		$1925.04	$52.20	$80.21

Crew A-10D	Hr.	Daily	Hr.	Daily	Bare Costs	Incl. O&P
2 Asbestos Workers	$44.10	$705.60	$71.30	$1140.80	$41.89	$66.86
1 Equip. Oper. (crane)	42.55	340.40	66.95	535.60		
1 Equip. Oper. Oiler	36.80	294.40	57.90	463.20		
1 Hydraulic Crane, 33 Ton		818.80		900.68	25.59	28.15
32 L.H., Daily Totals		$2159.20		$3040.28	$67.47	$95.01

Crew A-11	Hr.	Daily	Hr.	Daily	Bare Costs	Incl. O&P
1 Asbestos Foreman	$44.60	$356.80	$72.10	$576.80	$44.16	$71.40
7 Asbestos Workers	44.10	2469.60	71.30	3992.80		
2 Chipping Hammers, 12 Lb., Elec.		34.00		37.40	.53	.58
64 L.H., Daily Totals		$2860.40		$4607.00	$44.69	$71.98

Crew No.	Bare Costs		Incl. Subs O & P		Cost Per Labor-Hour	
Crew A-12	**Hr.**	**Daily**	**Hr.**	**Daily**	**Bare Costs**	**Incl. O&P**
1 Asbestos Foreman	$44.60	$356.80	$72.10	$576.80	$44.16	$71.40
7 Asbestos Workers	44.10	2469.60	71.30	3992.80		
1 Trk-mtd vac, 14 CY, 1500 Gal.		561.60		617.76		
1 Flatbed Truck, 20,000 GVW		197.40		217.14	11.86	13.05
64 L.H., Daily Totals		$3585.40		$5404.50	$56.02	$84.45

Crew A-13	Hr.	Daily	Hr.	Daily	Bare Costs	Incl. O&P
1 Equip. Oper. (light)	$39.05	$312.40	$61.45	$491.60	$39.05	$61.45
1 Trk-mtd vac, 14 CY, 1500 Gal.		561.60		617.76		
1 Flatbed Truck, 20,000 GVW		197.40		217.14	94.88	104.36
8 L.H., Daily Totals		$1071.40		$1326.50	$133.93	$165.81

Crew B-1	Hr.	Daily	Hr.	Daily	Bare Costs	Incl. O&P
1 Labor Foreman (outside)	$33.60	$268.80	$55.45	$443.60	$32.27	$53.25
2 Laborers	31.60	505.60	52.15	834.40		
24 L.H., Daily Totals		$774.40		$1278.00	$32.27	$53.25

Crew B-1A	Hr.	Daily	Hr.	Daily	Bare Costs	Incl. O&P
1 Laborer Foreman	$33.60	$268.80	$55.45	$443.60	$32.27	$53.25
2 Laborers	31.60	505.60	52.15	834.40		
2 Cutting Torches		34.00		37.40		
2 Gases		151.20		166.32	7.72	8.49
24 L.H., Daily Totals		$959.60		$1481.72	$39.98	$61.74

Crew B-1B	Hr.	Daily	Hr.	Daily	Bare Costs	Incl. O&P
1 Laborer Foreman	$33.60	$268.80	$55.45	$443.60	$34.84	$56.67
2 Laborers	31.60	505.60	52.15	834.40		
1 Equip. Oper. (crane)	42.55	340.40	66.95	535.60		
2 Cutting Torches		34.00		37.40		
2 Gases		151.20		166.32		
1 Hyd. Crane, 12 Ton		768.80		845.68	29.81	32.79
32 L.H., Daily Totals		$2068.80		$2863.00	$64.65	$89.47

Crew B-2	Hr.	Daily	Hr.	Daily	Bare Costs	Incl. O&P
1 Labor Foreman (outside)	$33.60	$268.80	$55.45	$443.60	$32.00	$52.81
4 Laborers	31.60	1011.20	52.15	1668.80		
40 L.H., Daily Totals		$1280.00		$2112.40	$32.00	$52.81

Crew B-3	Hr.	Daily	Hr.	Daily	Bare Costs	Incl. O&P
1 Labor Foreman (outside)	$33.60	$268.80	$55.45	$443.60	$33.67	$54.92
2 Laborers	31.60	505.60	52.15	834.40		
1 Equip. Oper. (med.)	41.35	330.80	65.05	520.40		
2 Truck Drivers (heavy)	31.95	511.20	52.35	837.60		
1 Crawler Loader, 3 C.Y.		990.00		1089.00		
2 Dump Trucks 12 C.Y., 400 H.P.		1066.00		1172.60	42.83	47.12
48 L.H., Daily Totals		$3672.40		$4897.60	$76.51	$102.03

Crew B-3A	Hr.	Daily	Hr.	Daily	Bare Costs	Incl. O&P
4 Laborers	$31.60	$1011.20	$52.15	$1668.80	$33.55	$54.73
1 Equip. Oper. (med.)	41.35	330.80	65.05	520.40		
1 Hyd. Excavator, 1.5 C.Y.		865.80		952.38	21.65	23.81
40 L.H., Daily Totals		$2207.80		$3141.58	$55.20	$78.54

Crew B-3B	Hr.	Daily	Hr.	Daily	Bare Costs	Incl. O&P
2 Laborers	$31.60	$505.60	$52.15	$834.40	$34.13	$55.42
1 Equip. Oper. (med.)	41.35	330.80	65.05	520.40		
1 Truck Driver (heavy)	31.95	255.60	52.35	418.80		
1 Backhoe Loader, 80 H.P.		345.00		379.50		
1 Dump Truck, 12 C.Y., 400 H.P.		533.00		586.30	27.44	30.18
32 L.H., Daily Totals		$1970.00		$2739.40	$61.56	$85.61

Crew No.	Bare Costs		Incl. Subs O & P		Cost Per Labor-Hour	
Crew B-3C	**Hr.**	**Daily**	**Hr.**	**Daily**	**Bare Costs**	**Incl. O&P**
3 Laborers	$31.60	$758.40	$52.15	$1251.60	$34.04	$55.38
1 Equip. Oper. (med.)	41.35	330.80	65.05	520.40		
1 Crawler Loader, 4 C.Y.		1450.00		1595.00	45.31	49.84
32 L.H., Daily Totals		$2539.20		$3367.00	$79.35	$105.22

Crew B-4	Hr.	Daily	Hr.	Daily	Bare Costs	Incl. O&P
1 Labor Foreman (outside)	$33.60	$268.80	$55.45	$443.60	$31.99	$52.73
4 Laborers	31.60	1011.20	52.15	1668.80		
1 Truck Driver (heavy)	31.95	255.60	52.35	418.80		
1 Truck Tractor, 220 H.P.		288.40		317.24		
1 Flatbed Trailer, 40 Ton		139.60		153.56	8.92	9.81
48 L.H., Daily Totals		$1963.60		$3002.00	$40.91	$62.54

Crew B-5	Hr.	Daily	Hr.	Daily	Bare Costs	Incl. O&P
1 Labor Foreman (outside)	$33.60	$268.80	$55.45	$443.60	$34.67	$56.31
4 Laborers	31.60	1011.20	52.15	1668.80		
2 Equip. Oper. (med.)	41.35	661.60	65.05	1040.80		
1 Air Compressor, 250 cfm		162.40		178.64		
2 Breakers, Pavement, 60 lb.		16.40		18.04		
2 -50' Air Hoses, 1.5"		12.60		13.86		
1 Crawler Loader, 3 C.Y.		990.00		1089.00	21.10	23.21
56 L.H., Daily Totals		$3123.00		$4452.74	$55.77	$79.51

Crew B-6	Hr.	Daily	Hr.	Daily	Bare Costs	Incl. O&P
2 Laborers	$31.60	$505.60	$52.15	$834.40	$34.08	$55.25
1 Equip. Oper. (light)	39.05	312.40	61.45	491.60		
1 Backhoe Loader, 48 H.P.		293.80		323.18	12.24	13.47
24 L.H., Daily Totals		$1111.80		$1649.18	$46.33	$68.72

Crew B-6B	Hr.	Daily	Hr.	Daily	Bare Costs	Incl. O&P
2 Labor Foremen (out)	$33.60	$537.60	$55.45	$887.20	$32.27	$53.25
4 Laborers	31.60	1011.20	52.15	1668.80		
1 S.P. Crane, 4x4, 5 Ton		258.20		284.02		
1 Flatbed Truck, Gas, 1.5 Ton		194.40		213.84		
1 Butt Fusion Machine		199.80		219.78	13.59	14.95
48 L.H., Daily Totals		$2201.20		$3273.64	$45.86	$68.20

Crew B-7	Hr.	Daily	Hr.	Daily	Bare Costs	Incl. O&P
1 Labor Foreman (outside)	$33.60	$268.80	$55.45	$443.60	$33.56	$54.85
4 Laborers	31.60	1011.20	52.15	1668.80		
1 Equip. Oper. (med.)	41.35	330.80	65.05	520.40		
1 Brush Chipper, 12", 130 H.P.		225.60		248.16		
1 Crawler Loader, 3 C.Y.		990.00		1089.00		
2 Chainsaws, Gas, 36" Long		75.60		83.16	26.90	29.59
48 L.H., Daily Totals		$2902.00		$4053.12	$60.46	$84.44

Crew B-7A	Hr.	Daily	Hr.	Daily	Bare Costs	Incl. O&P
2 Laborers	$31.60	$505.60	$52.15	$834.40	$34.08	$55.25
1 Equip. Oper. (light)	39.05	312.40	61.45	491.60		
1 Rake w/Tractor		257.70		283.47		
2 Chain Saws, gas, 18"		51.60		56.76	12.89	14.18
24 L.H., Daily Totals		$1127.30		$1666.23	$46.97	$69.43

Crews

Crew No.	Bare Costs		Incl. Subs O & P		Cost Per Labor-Hour	

Crew B-8	Hr.	Daily	Hr.	Daily	Bare Costs	Incl. O&P
1 Labor Foreman (outside)	$33.60	$268.80	$55.45	$443.60	$35.02	$56.56
2 Laborers	31.60	505.60	52.15	834.40		
2 Equip. Oper. (med.)	41.35	661.60	65.05	1040.80		
1 Equip. Oper. Oiler	36.80	294.40	57.90	463.20		
2 Truck Drivers (heavy)	31.95	511.20	52.35	837.60		
1 Hyd. Crane, 25 Ton		789.40		868.34		
1 Crawler Loader, 3 C.Y.		990.00		1089.00		
2 Dump Trucks, 12 C.Y., 400 H.P.		1066.00		1172.60	44.46	48.91
64 L.H., Daily Totals		$5087.00		$6749.54	$79.48	$105.46

Crew B-9	Hr.	Daily	Hr.	Daily	Bare Costs	Incl. O&P
1 Labor Foreman (outside)	$33.60	$268.80	$55.45	$443.60	$32.00	$52.81
4 Laborers	31.60	1011.20	52.15	1668.80		
1 Air Compressor, 250 cfm		162.40		178.64		
2 Breakers, Pavement, 60 lb.		16.40		18.04		
2 -50' Air Hoses, 1.5"		12.60		13.86	4.79	5.26
40 L.H., Daily Totals		$1471.40		$2322.94	$36.78	$58.07

Crew B-9A	Hr.	Daily	Hr.	Daily	Bare Costs	Incl. O&P
2 Laborers	$31.60	$505.60	$52.15	$834.40	$31.72	$52.22
1 Truck Driver (heavy)	31.95	255.60	52.35	418.80		
1 Water Tanker, 5000 Gal.		136.00		149.60		
1 Truck Tractor, 220 H.P.		288.40		317.24		
2 -50' Discharge Hoses, 3"		3.50		3.85	17.83	19.61
24 L.H., Daily Totals		$1189.10		$1723.89	$49.55	$71.83

Crew B-9B	Hr.	Daily	Hr.	Daily	Bare Costs	Incl. O&P
2 Laborers	$31.60	$505.60	$52.15	$834.40	$31.72	$52.22
1 Truck Driver (heavy)	31.95	255.60	52.35	418.80		
2 -50' Discharge Hoses, 3"		3.50		3.85		
1 Water Tanker, 5000 Gal.		136.00		149.60		
1 Truck Tractor, 220 H.P.		288.40		317.24		
1 Pressure Washer		53.00		58.30	20.04	22.04
24 L.H., Daily Totals		$1242.10		$1782.19	$51.75	$74.26

Crew B-9D	Hr.	Daily	Hr.	Daily	Bare Costs	Incl. O&P
1 Labor Foreman (Outside)	$33.60	$268.80	$55.45	$443.60	$32.00	$52.81
4 Common Laborers	31.60	1011.20	52.15	1668.80		
1 Air Compressor, 250 cfm		162.40		178.64		
2 -50' Air Hoses, 1.5"		12.60		13.86		
2 Air Powered Tampers		48.00		52.80	5.58	6.13
40 L.H., Daily Totals		$1503.00		$2357.70	$37.58	$58.94

Crew B-10	Hr.	Daily	Hr.	Daily	Bare Costs	Incl. O&P
1 Equip. Oper. (med.)	$41.35	$330.80	$65.05	$520.40	$38.10	$60.75
.5 Laborer	31.60	126.40	52.15	208.60		
12 L.H., Daily Totals		$457.20		$729.00	$38.10	$60.75

Crew B-10A	Hr.	Daily	Hr.	Daily	Bare Costs	Incl. O&P
1 Equip. Oper. (med.)	$41.35	$330.80	$65.05	$520.40	$38.10	$60.75
.5 Laborer	31.60	126.40	52.15	208.60		
1 Roller, 2-Drum, W.B., 7.5 H.P.		145.20		159.72	12.10	13.31
12 L.H., Daily Totals		$602.40		$888.72	$50.20	$74.06

Crew B-10B	Hr.	Daily	Hr.	Daily	Bare Costs	Incl. O&P
1 Equip. Oper. (med.)	$41.35	$330.80	$65.05	$520.40	$38.10	$60.75
.5 Laborer	31.60	126.40	52.15	208.60		
1 Dozer, 200 H.P.		1082.00		1190.20	90.17	99.18
12 L.H., Daily Totals		$1539.20		$1919.20	$128.27	$159.93

Crew B-10C	Hr.	Daily	Hr.	Daily	Bare Costs	Incl. O&P
1 Equip. Oper. (med.)	$41.35	$330.80	$65.05	$520.40	$38.10	$60.75
.5 Laborer	31.60	126.40	52.15	208.60		
1 Dozer, 200 H.P.		1082.00		1190.20		
1 Vibratory Roller, Towed, 23 Ton		411.00		452.10	124.42	136.86
12 L.H., Daily Totals		$1950.20		$2371.30	$162.52	$197.61

Crew B-10D	Hr.	Daily	Hr.	Daily	Bare Costs	Incl. O&P
1 Equip. Oper. (med.)	$41.35	$330.80	$65.05	$520.40	$38.10	$60.75
.5 Laborer	31.60	126.40	52.15	208.60		
1 Dozer, 200 H.P.		1082.00		1190.20		
1 Sheepsft. Roller, Towed		465.40		511.94	128.95	141.85
12 L.H., Daily Totals		$2004.60		$2431.14	$167.05	$202.60

Crew B-10E	Hr.	Daily	Hr.	Daily	Bare Costs	Incl. O&P
1 Equip. Oper. (med.)	$41.35	$330.80	$65.05	$520.40	$38.10	$60.75
.5 Laborer	31.60	126.40	52.15	208.60		
1 Tandem Roller, 5 Ton		137.80		151.58	11.48	12.63
12 L.H., Daily Totals		$595.00		$880.58	$49.58	$73.38

Crew B-10F	Hr.	Daily	Hr.	Daily	Bare Costs	Incl. O&P
1 Equip. Oper. (med.)	$41.35	$330.80	$65.05	$520.40	$38.10	$60.75
.5 Laborer	31.60	126.40	52.15	208.60		
1 Tandem Roller, 10 Ton		233.00		256.30	19.42	21.36
12 L.H., Daily Totals		$690.20		$985.30	$57.52	$82.11

Crew B-10G	Hr.	Daily	Hr.	Daily	Bare Costs	Incl. O&P
1 Equip. Oper. (med.)	$41.35	$330.80	$65.05	$520.40	$38.10	$60.75
.5 Laborer	31.60	126.40	52.15	208.60		
1 Sheepsft. Roll., 240 H.P.		1142.00		1256.20	95.17	104.68
12 L.H., Daily Totals		$1599.20		$1985.20	$133.27	$165.43

Crew B-10H	Hr.	Daily	Hr.	Daily	Bare Costs	Incl. O&P
1 Equip. Oper. (med.)	$41.35	$330.80	$65.05	$520.40	$38.10	$60.75
.5 Laborer	31.60	126.40	52.15	208.60		
1 Diaphragm Water Pump, 2"		63.20		69.52		
1 -20' Suction Hose, 2"		1.95		2.15		
2 -50' Discharge Hoses, 2"		2.20		2.42	5.61	6.17
12 L.H., Daily Totals		$524.55		$803.09	$43.71	$66.92

Crew B-10I	Hr.	Daily	Hr.	Daily	Bare Costs	Incl. O&P
1 Equip. Oper. (med.)	$41.35	$330.80	$65.05	$520.40	$38.10	$60.75
.5 Laborer	31.60	126.40	52.15	208.60		
1 Diaphragm Water Pump, 4"		86.20		94.82		
1 -20' Suction Hose, 4"		3.45		3.79		
2 -50' Discharge Hoses, 4"		4.70		5.17	7.86	8.65
12 L.H., Daily Totals		$551.55		$832.78	$45.96	$69.40

Crew B-10J	Hr.	Daily	Hr.	Daily	Bare Costs	Incl. O&P
1 Equip. Oper. (med.)	$41.35	$330.80	$65.05	$520.40	$38.10	$60.75
.5 Laborer	31.60	126.40	52.15	208.60		
1 Centrifugal Water Pump, 3"		69.40		76.34		
1 -20' Suction Hose, 3"		3.05		3.36		
2 -50' Discharge Hoses, 3"		3.50		3.85	6.33	6.96
12 L.H., Daily Totals		$533.15		$812.54	$44.43	$67.71

Crew No.	Bare Costs		Incl. Subs O & P		Cost Per Labor-Hour	
Crew B-10K	**Hr.**	**Daily**	**Hr.**	**Daily**	**Bare Costs**	**Incl. O&P**
1 Equip. Oper. (med.)	$41.35	$330.80	$65.05	$520.40	$38.10	$60.75
.5 Laborer	31.60	126.40	52.15	208.60		
1 Centr. Water Pump, 6″		309.40		340.34		
1 -20′ Suction Hose, 6″		11.90		13.09		
2 -50′ Discharge Hoses, 6″		12.60		13.86	27.82	30.61
12 L.H., Daily Totals		$791.10		$1096.29	$65.92	$91.36
Crew B-10L	**Hr.**	**Daily**	**Hr.**	**Daily**	**Bare Costs**	**Incl. O&P**
1 Equip. Oper. (med.)	$41.35	$330.80	$65.05	$520.40	$38.10	$60.75
.5 Laborer	31.60	126.40	52.15	208.60		
1 Dozer, 80 H.P.		399.20		439.12	33.27	36.59
12 L.H., Daily Totals		$856.40		$1168.12	$71.37	$97.34
Crew B-10M	**Hr.**	**Daily**	**Hr.**	**Daily**	**Bare Costs**	**Incl. O&P**
1 Equip. Oper. (med.)	$41.35	$330.80	$65.05	$520.40	$38.10	$60.75
.5 Laborer	31.60	126.40	52.15	208.60		
1 Dozer, 300 H.P.		1423.00		1565.30	118.58	130.44
12 L.H., Daily Totals		$1880.20		$2294.30	$156.68	$191.19
Crew B-10N	**Hr.**	**Daily**	**Hr.**	**Daily**	**Bare Costs**	**Incl. O&P**
1 Equip. Oper. (med.)	$41.35	$330.80	$65.05	$520.40	$38.10	$60.75
.5 Laborer	31.60	126.40	52.15	208.60		
1 F.E. Loader, T.M., 1.5 C.Y		367.00		403.70	30.58	33.64
12 L.H., Daily Totals		$824.20		$1132.70	$68.68	$94.39
Crew B-10O	**Hr.**	**Daily**	**Hr.**	**Daily**	**Bare Costs**	**Incl. O&P**
1 Equip. Oper. (med.)	$41.35	$330.80	$65.05	$520.40	$38.10	$60.75
.5 Laborer	31.60	126.40	52.15	208.60		
1 F.E. Loader, T.M., 2.25 C.Y.		767.40		844.14	63.95	70.34
12 L.H., Daily Totals		$1224.60		$1573.14	$102.05	$131.10
Crew B-10P	**Hr.**	**Daily**	**Hr.**	**Daily**	**Bare Costs**	**Incl. O&P**
1 Equip. Oper. (med.)	$41.35	$330.80	$65.05	$520.40	$38.10	$60.75
.5 Laborer	31.60	126.40	52.15	208.60		
1 Crawler Loader, 3 C.Y.		990.00		1089.00	82.50	90.75
12 L.H., Daily Totals		$1447.20		$1818.00	$120.60	$151.50
Crew B-10Q	**Hr.**	**Daily**	**Hr.**	**Daily**	**Bare Costs**	**Incl. O&P**
1 Equip. Oper. (med.)	$41.35	$330.80	$65.05	$520.40	$38.10	$60.75
.5 Laborer	31.60	126.40	52.15	208.60		
1 Crawler Loader, 4 C.Y.		1450.00		1595.00	120.83	132.92
12 L.H., Daily Totals		$1907.20		$2324.00	$158.93	$193.67
Crew B-10R	**Hr.**	**Daily**	**Hr.**	**Daily**	**Bare Costs**	**Incl. O&P**
1 Equip. Oper. (med.)	$41.35	$330.80	$65.05	$520.40	$38.10	$60.75
.5 Laborer	31.60	126.40	52.15	208.60		
1 F.E. Loader, W.M., 1 C.Y.		239.00		262.90	19.92	21.91
12 L.H., Daily Totals		$696.20		$991.90	$58.02	$82.66
Crew B-10S	**Hr.**	**Daily**	**Hr.**	**Daily**	**Bare Costs**	**Incl. O&P**
1 Equip. Oper. (med.)	$41.35	$330.80	$65.05	$520.40	$38.10	$60.75
.5 Laborer	31.60	126.40	52.15	208.60		
1 F.E. Loader, W.M., 1.5 C.Y.		328.20		361.02	27.35	30.09
12 L.H., Daily Totals		$785.40		$1090.02	$65.45	$90.83
Crew B-10T	**Hr.**	**Daily**	**Hr.**	**Daily**	**Bare Costs**	**Incl. O&P**
1 Equip. Oper. (med.)	$41.35	$330.80	$65.05	$520.40	$38.10	$60.75
.5 Laborer	31.60	126.40	52.15	208.60		
1 F.E. Loader, W.M.,2.5 C.Y.		395.80		435.38	32.98	36.28
12 L.H., Daily Totals		$853.00		$1164.38	$71.08	$97.03

Crew No.	Bare Costs		Incl. Subs O & P		Cost Per Labor-Hour	
Crew B-10U	**Hr.**	**Daily**	**Hr.**	**Daily**	**Bare Costs**	**Incl. O&P**
1 Equip. Oper. (med.)	$41.35	$330.80	$65.05	$520.40	$38.10	$60.75
.5 Laborer	31.60	126.40	52.15	208.60		
1 F.E. Loader, W.M., 5.5 C.Y.		854.40		939.84	71.20	78.32
12 L.H., Daily Totals		$1311.60		$1668.84	$109.30	$139.07
Crew B-10V	**Hr.**	**Daily**	**Hr.**	**Daily**	**Bare Costs**	**Incl. O&P**
1 Equip. Oper. (med.)	$41.35	$330.80	$65.05	$520.40	$38.10	$60.75
.5 Laborer	31.60	126.40	52.15	208.60		
1 Dozer, 700 H.P.		4232.00		4655.20	352.67	387.93
12 L.H., Daily Totals		$4689.20		$5384.20	$390.77	$448.68
Crew B-10W	**Hr.**	**Daily**	**Hr.**	**Daily**	**Bare Costs**	**Incl. O&P**
1 Equip. Oper. (med.)	$41.35	$330.80	$65.05	$520.40	$38.10	$60.75
.5 Laborer	31.60	126.40	52.15	208.60		
1 Dozer, 105 H.P.		600.60		660.66	50.05	55.06
12 L.H., Daily Totals		$1057.80		$1389.66	$88.15	$115.81
Crew B-10X	**Hr.**	**Daily**	**Hr.**	**Daily**	**Bare Costs**	**Incl. O&P**
1 Equip. Oper. (med.)	$41.35	$330.80	$65.05	$520.40	$38.10	$60.75
.5 Laborer	31.60	126.40	52.15	208.60		
1 Dozer, 410 H.P.		1849.00		2033.90	154.08	169.49
12 L.H., Daily Totals		$2306.20		$2762.90	$192.18	$230.24
Crew B-10Y	**Hr.**	**Daily**	**Hr.**	**Daily**	**Bare Costs**	**Incl. O&P**
1 Equip. Oper. (med.)	$41.35	$330.80	$65.05	$520.40	$38.10	$60.75
.5 Laborer	31.60	126.40	52.15	208.60		
1 Vibr. Roller, Towed, 12 Ton		469.60		516.56	39.13	43.05
12 L.H., Daily Totals		$926.80		$1245.56	$77.23	$103.80
Crew B-11A	**Hr.**	**Daily**	**Hr.**	**Daily**	**Bare Costs**	**Incl. O&P**
1 Equipment Oper. (med.)	$41.35	$330.80	$65.05	$520.40	$36.48	$58.60
1 Laborer	31.60	252.80	52.15	417.20		
1 Dozer, 200 H.P.		1082.00		1190.20	67.63	74.39
16 L.H., Daily Totals		$1665.60		$2127.80	$104.10	$132.99
Crew B-11B	**Hr.**	**Daily**	**Hr.**	**Daily**	**Bare Costs**	**Incl. O&P**
1 Equipment Oper. (light)	$39.05	$312.40	$61.45	$491.60	$35.33	$56.80
1 Laborer	31.60	252.80	52.15	417.20		
1 Air Powered Tamper		24.00		26.40		
1 Air Compressor, 365 cfm		211.20		232.32		
2 -50′ Air Hoses, 1.5″		12.60		13.86	15.49	17.04
16 L.H., Daily Totals		$813.00		$1181.38	$50.81	$73.84
Crew B-11C	**Hr.**	**Daily**	**Hr.**	**Daily**	**Bare Costs**	**Incl. O&P**
1 Equipment Oper. (med.)	$41.35	$330.80	$65.05	$520.40	$36.48	$58.60
1 Laborer	31.60	252.80	52.15	417.20		
1 Backhoe Loader, 48 H.P.		293.80		323.18	18.36	20.20
16 L.H., Daily Totals		$877.40		$1260.78	$54.84	$78.80
Crew B-11K	**Hr.**	**Daily**	**Hr.**	**Daily**	**Bare Costs**	**Incl. O&P**
1 Equipment Oper. (med.)	$41.35	$330.80	$65.05	$520.40	$36.48	$58.60
1 Laborer	31.60	252.80	52.15	417.20		
1 Trencher, Chain Type, 8′ D		1777.00		1954.70	111.06	122.17
16 L.H., Daily Totals		$2360.60		$2892.30	$147.54	$180.77
Crew B-11L	**Hr.**	**Daily**	**Hr.**	**Daily**	**Bare Costs**	**Incl. O&P**
1 Equipment Oper. (med.)	$41.35	$330.80	$65.05	$520.40	$36.48	$58.60
1 Laborer	31.60	252.80	52.15	417.20		
1 Grader, 30,000 Lbs.		550.20		605.22	34.39	37.83
16 L.H., Daily Totals		$1133.80		$1542.82	$70.86	$96.43

Crew No.	Bare Costs		Incl. Subs O & P		Cost Per Labor-Hour	
Crew B-11M	**Hr.**	**Daily**	**Hr.**	**Daily**	**Bare Costs**	**Incl. O&P**
1 Equipment Oper. (med.)	$41.35	$330.80	$65.05	$520.40	$36.48	$58.60
1 Laborer	31.60	252.80	52.15	417.20		
1 Backhoe Loader, 80 H.P.		345.00		379.50	21.56	23.72
16 L.H., Daily Totals		$928.60		$1317.10	$58.04	$82.32

Crew B-11S	Hr.	Daily	Hr.	Daily	Bare Costs	Incl. O&P
1 Equipment Operator (med.)	$41.35	$330.80	$65.05	$520.40	$38.10	$60.75
.5 Laborer	31.60	126.40	52.15	208.60		
1 Dozer, 300 H.P.		1423.00		1565.30		
1 Ripper, beam & 1 shank		79.60		87.56	125.22	137.74
12 L.H., Daily Totals		$1959.80		$2381.86	$163.32	$198.49

Crew B-11T	Hr.	Daily	Hr.	Daily	Bare Costs	Incl. O&P
1 Equipment Operator (med.)	$41.35	$330.80	$65.05	$520.40	$38.10	$60.75
.5 Laborer	31.60	126.40	52.15	208.60		
1 Dozer, 410 H.P.		1849.00		2033.90		
1 Ripper, beam & 2 shanks		89.40		98.34	161.53	177.69
12 L.H., Daily Totals		$2395.60		$2861.24	$199.63	$238.44

Crew B-11W	Hr.	Daily	Hr.	Daily	Bare Costs	Incl. O&P
1 Equipment Operator (med.)	$41.35	$330.80	$65.05	$520.40	$32.70	$53.39
1 Common Laborer	31.60	252.80	52.15	417.20		
10 Truck Drivers (hvy.)	31.95	2556.00	52.35	4188.00		
1 Dozer, 200 H.P.		1082.00		1190.20		
1 Vibratory Roller, Towed, 23 Ton		411.00		452.10		
10 Dump Trucks, 8 C.Y., 220 H.P.		3264.00		3590.40	49.55	54.51
96 L.H., Daily Totals		$7896.60		$10358.30	$82.26	$107.90

Crew B-11Y	Hr.	Daily	Hr.	Daily	Bare Costs	Incl. O&P
1 Labor Foreman (Outside)	$33.60	$268.80	$55.45	$443.60	$35.07	$56.82
5 Common Laborers	31.60	1264.00	52.15	2086.00		
3 Equipment Operators (med.)	41.35	992.40	65.05	1561.20		
1 Dozer, 80 H.P.		399.20		439.12		
2 Roller, 2-Drum, W.B., 7.5 H.P.		290.40		319.44		
4 Vibratory Plates, Gas, 21"		156.80		172.48	11.76	12.93
72 L.H., Daily Totals		$3371.60		$5021.84	$46.83	$69.75

Crew B-12A	Hr.	Daily	Hr.	Daily	Bare Costs	Incl. O&P
1 Equip. Oper. (crane)	$42.55	$340.40	$66.95	$535.60	$37.08	$59.55
1 Laborer	31.60	252.80	52.15	417.20		
1 Hyd. Excavator, 1 C.Y.		669.20		736.12	41.83	46.01
16 L.H., Daily Totals		$1262.40		$1688.92	$78.90	$105.56

Crew B-12B	Hr.	Daily	Hr.	Daily	Bare Costs	Incl. O&P
1 Equip. Oper. (crane)	$42.55	$340.40	$66.95	$535.60	$37.08	$59.55
1 Laborer	31.60	252.80	52.15	417.20		
1 Hyd. Excavator, 1.5 C.Y.		865.80		952.38	54.11	59.52
16 L.H., Daily Totals		$1459.00		$1905.18	$91.19	$119.07

Crew B-12C	Hr.	Daily	Hr.	Daily	Bare Costs	Incl. O&P
1 Equip. Oper. (crane)	$42.55	$340.40	$66.95	$535.60	$37.08	$59.55
1 Laborer	31.60	252.80	52.15	417.20		
1 Hyd. Excavator, 2 C.Y.		1182.00		1300.20	73.88	81.26
16 L.H., Daily Totals		$1775.20		$2253.00	$110.95	$140.81

Crew B-12D	Hr.	Daily	Hr.	Daily	Bare Costs	Incl. O&P
1 Equip. Oper. (crane)	$42.55	$340.40	$66.95	$535.60	$37.08	$59.55
1 Laborer	31.60	252.80	52.15	417.20		
1 Hyd. Excavator, 3.5 C.Y.		2162.00		2378.20	135.13	148.64
16 L.H., Daily Totals		$2755.20		$3331.00	$172.20	$208.19

Crew No.	Bare Costs		Incl. Subs O & P		Cost Per Labor-Hour	
Crew B-12E	**Hr.**	**Daily**	**Hr.**	**Daily**	**Bare Costs**	**Incl. O&P**
1 Equip. Oper. (crane)	$42.55	$340.40	$66.95	$535.60	$37.08	$59.55
1 Laborer	31.60	252.80	52.15	417.20		
1 Hyd. Excavator, .5 C.Y.		394.60		434.06	24.66	27.13
16 L.H., Daily Totals		$987.80		$1386.86	$61.74	$86.68

Crew B-12F	Hr.	Daily	Hr.	Daily	Bare Costs	Incl. O&P
1 Equip. Oper. (crane)	$42.55	$340.40	$66.95	$535.60	$37.08	$59.55
1 Laborer	31.60	252.80	52.15	417.20		
1 Hyd. Excavator, .75 C.Y.		598.60		658.46	37.41	41.15
16 L.H., Daily Totals		$1191.80		$1611.26	$74.49	$100.70

Crew B-12G	Hr.	Daily	Hr.	Daily	Bare Costs	Incl. O&P
1 Equip. Oper. (crane)	$42.55	$340.40	$66.95	$535.60	$37.08	$59.55
1 Laborer	31.60	252.80	52.15	417.20		
1 Crawler Crane, 15 Ton		615.30		676.83		
1 Clamshell Bucket, .5 C.Y.		36.60		40.26	40.74	44.82
16 L.H., Daily Totals		$1245.10		$1669.89	$77.82	$104.37

Crew B-12H	Hr.	Daily	Hr.	Daily	Bare Costs	Incl. O&P
1 Equip. Oper. (crane)	$42.55	$340.40	$66.95	$535.60	$37.08	$59.55
1 Laborer	31.60	252.80	52.15	417.20		
1 Crawler Crane, 25 Ton		1052.00		1157.20		
1 Clamshell Bucket, 1 C.Y.		46.20		50.82	68.64	75.50
16 L.H., Daily Totals		$1691.40		$2160.82	$105.71	$135.05

Crew B-12I	Hr.	Daily	Hr.	Daily	Bare Costs	Incl. O&P
1 Equip. Oper. (crane)	$42.55	$340.40	$66.95	$535.60	$37.08	$59.55
1 Laborer	31.60	252.80	52.15	417.20		
1 Crawler Crane, 20 Ton		789.05		867.96		
1 Dragline Bucket, .75 C.Y.		20.00		22.00	50.57	55.62
16 L.H., Daily Totals		$1402.25		$1842.76	$87.64	$115.17

Crew B-12J	Hr.	Daily	Hr.	Daily	Bare Costs	Incl. O&P
1 Equip. Oper. (crane)	$42.55	$340.40	$66.95	$535.60	$37.08	$59.55
1 Laborer	31.60	252.80	52.15	417.20		
1 Gradall, 5/8 C.Y.		958.80		1054.68	59.92	65.92
16 L.H., Daily Totals		$1552.00		$2007.48	$97.00	$125.47

Crew B-12K	Hr.	Daily	Hr.	Daily	Bare Costs	Incl. O&P
1 Equip. Oper. (crane)	$42.55	$340.40	$66.95	$535.60	$37.08	$59.55
1 Laborer	31.60	252.80	52.15	417.20		
1 Gradall, 3 Ton, 1 C.Y.		1130.00		1243.00	70.63	77.69
16 L.H., Daily Totals		$1723.20		$2195.80	$107.70	$137.24

Crew B-12L	Hr.	Daily	Hr.	Daily	Bare Costs	Incl. O&P
1 Equip. Oper. (crane)	$42.55	$340.40	$66.95	$535.60	$37.08	$59.55
1 Laborer	31.60	252.80	52.15	417.20		
1 Crawler Crane, 15 Ton		615.30		676.83		
1 F.E. Attachment, .5 C.Y.		52.80		58.08	41.76	45.93
16 L.H., Daily Totals		$1261.30		$1687.71	$78.83	$105.48

Crew B-12M	Hr.	Daily	Hr.	Daily	Bare Costs	Incl. O&P
1 Equip. Oper. (crane)	$42.55	$340.40	$66.95	$535.60	$37.08	$59.55
1 Laborer	31.60	252.80	52.15	417.20		
1 Crawler Crane, 20 Ton		789.05		867.96		
1 F.E. Attachment, .75 C.Y.		58.20		64.02	52.95	58.25
16 L.H., Daily Totals		$1440.45		$1884.78	$90.03	$117.80

Crew No.	Bare Costs		Incl. Subs O & P		Cost Per Labor-Hour	
Crew B-12N	**Hr.**	**Daily**	**Hr.**	**Daily**	**Bare Costs**	**Incl. O&P**
1 Equip. Oper. (crane)	$42.55	$340.40	$66.95	$535.60	$37.08	$59.55
1 Laborer	31.60	252.80	52.15	417.20		
1 Crawler Crane, 25 Ton		1052.00		1157.20		
1 F.E. Attachment, 1 C.Y.		64.00		70.40	69.75	76.72
16 L.H., Daily Totals		$1709.20		$2180.40	$106.83	$136.28

Crew B-12O	Hr.	Daily	Hr.	Daily	Bare Costs	Incl. O&P
1 Equip. Oper. (crane)	$42.55	$340.40	$66.95	$535.60	$37.08	$59.55
1 Laborer	31.60	252.80	52.15	417.20		
1 Crawler Crane, 40 Ton		1055.00		1160.50		
1 F.E. Attachment, 1.5 C.Y.		74.20		81.62	70.58	77.63
16 L.H., Daily Totals		$1722.40		$2194.92	$107.65	$137.18

Crew B-12P	Hr.	Daily	Hr.	Daily	Bare Costs	Incl. O&P
1 Equip. Oper. (crane)	$42.55	$340.40	$66.95	$535.60	$37.08	$59.55
1 Laborer	31.60	252.80	52.15	417.20		
1 Crawler Crane, 40 Ton		1055.00		1160.50		
1 Dragline Bucket, 1.5 C.Y.		32.80		36.08	67.99	74.79
16 L.H., Daily Totals		$1681.00		$2149.38	$105.06	$134.34

Crew B-12Q	Hr.	Daily	Hr.	Daily	Bare Costs	Incl. O&P
1 Equip. Oper. (crane)	$42.55	$340.40	$66.95	$535.60	$37.08	$59.55
1 Laborer	31.60	252.80	52.15	417.20		
1 Hyd. Excavator, 5/8 C.Y.		517.20		568.92	32.33	35.56
16 L.H., Daily Totals		$1110.40		$1521.72	$69.40	$95.11

Crew B-12S	Hr.	Daily	Hr.	Daily	Bare Costs	Incl. O&P
1 Equip. Oper. (crane)	$42.55	$340.40	$66.95	$535.60	$37.08	$59.55
1 Laborer	31.60	252.80	52.15	417.20		
1 Hyd. Excavator, 2.5 C.Y.		1619.00		1780.90	101.19	111.31
16 L.H., Daily Totals		$2212.20		$2733.70	$138.26	$170.86

Crew B-12T	Hr.	Daily	Hr.	Daily	Bare Costs	Incl. O&P
1 Equip. Oper. (crane)	$42.55	$340.40	$66.95	$535.60	$37.08	$59.55
1 Laborer	31.60	252.80	52.15	417.20		
1 Crawler Crane, 75 Ton		1496.00		1645.60		
1 F.E. Attachment, 3 C.Y.		97.60		107.36	99.60	109.56
16 L.H., Daily Totals		$2186.80		$2705.76	$136.68	$169.11

Crew B-12V	Hr.	Daily	Hr.	Daily	Bare Costs	Incl. O&P
1 Equip. Oper. (crane)	$42.55	$340.40	$66.95	$535.60	$37.08	$59.55
1 Laborer	31.60	252.80	52.15	417.20		
1 Crawler Crane, 75 Ton		1496.00		1645.60		
1 Dragline Bucket, 3 C.Y.		46.20		50.82	96.39	106.03
16 L.H., Daily Totals		$2135.40		$2649.22	$133.46	$165.58

Crew B-13	Hr.	Daily	Hr.	Daily	Bare Costs	Incl. O&P
1 Labor Foreman (outside)	$33.60	$268.80	$55.45	$443.60	$34.19	$55.56
4 Laborers	31.60	1011.20	52.15	1668.80		
1 Equip. Oper. (crane)	42.55	340.40	66.95	535.60		
1 Equip. Oper. Oiler	36.80	294.40	57.90	463.20		
1 Hyd. Crane, 25 Ton		789.40		868.34	14.10	15.51
56 L.H., Daily Totals		$2704.20		$3979.54	$48.29	$71.06

Crew No.	Bare Costs		Incl. Subs O & P		Cost Per Labor-Hour	
Crew B-13A	**Hr.**	**Daily**	**Hr.**	**Daily**	**Bare Costs**	**Incl. O&P**
1 Foreman	$33.60	$268.80	$55.45	$443.60	$34.77	$56.36
2 Laborers	31.60	505.60	52.15	834.40		
2 Equipment Operators	41.35	661.60	65.05	1040.80		
2 Truck Drivers (heavy)	31.95	511.20	52.35	837.60		
1 Crawler Crane, 75 Ton		1496.00		1645.60		
1 Crawler Loader, 4 C.Y.		1450.00		1595.00		
2 Dump Trucks, 8 C.Y., 220 H.P.		652.80		718.08	64.26	70.69
56 L.H., Daily Totals		$5546.00		$7115.08	$99.04	$127.06

Crew B-13B	Hr.	Daily	Hr.	Daily	Bare Costs	Incl. O&P
1 Labor Foreman (outside)	$33.60	$268.80	$55.45	$443.60	$34.19	$55.56
4 Laborers	31.60	1011.20	52.15	1668.80		
1 Equip. Oper. (crane)	42.55	340.40	66.95	535.60		
1 Equip. Oper. Oiler	36.80	294.40	57.90	463.20		
1 Hyd. Crane, 55 Ton		1125.00		1237.50	20.09	22.10
56 L.H., Daily Totals		$3039.80		$4348.70	$54.28	$77.66

Crew B-13C	Hr.	Daily	Hr.	Daily	Bare Costs	Incl. O&P
1 Labor Foreman (outside)	$33.60	$268.80	$55.45	$443.60	$34.19	$55.56
4 Laborers	31.60	1011.20	52.15	1668.80		
1 Equip. Oper. (crane)	42.55	340.40	66.95	535.60		
1 Equip. Oper. Oiler	36.80	294.40	57.90	463.20		
1 Crawler Crane, 100 Ton		1611.00		1772.10	28.77	31.64
56 L.H., Daily Totals		$3525.80		$4883.30	$62.96	$87.20

Crew B-13D	Hr.	Daily	Hr.	Daily	Bare Costs	Incl. O&P
1 Laborer	$31.60	$252.80	$52.15	$417.20	$37.08	$59.55
1 Equip. Oper. (crane)	42.55	340.40	66.95	535.60		
1 Hyd. Excavator, 1 C.Y.		669.20		736.12		
1 Trench Box		113.40		124.74	48.91	53.80
16 L.H., Daily Totals		$1375.80		$1813.66	$85.99	$113.35

Crew B-13E	Hr.	Daily	Hr.	Daily	Bare Costs	Incl. O&P
1 Laborer	$31.60	$252.80	$52.15	$417.20	$37.08	$59.55
1 Equip. Oper. (crane)	42.55	340.40	66.95	535.60		
1 Hyd. Excavator, 1.5 C.Y.		865.80		952.38		
1 Trench Box		113.40		124.74	61.20	67.32
16 L.H., Daily Totals		$1572.40		$2029.92	$98.28	$126.87

Crew B-13F	Hr.	Daily	Hr.	Daily	Bare Costs	Incl. O&P
1 Laborer	$31.60	$252.80	$52.15	$417.20	$37.08	$59.55
1 Equip. Oper. (crane)	42.55	340.40	66.95	535.60		
1 Hyd. Excavator, 3.5 C.Y.		2162.00		2378.20		
1 Trench Box		113.40		124.74	142.21	156.43
16 L.H., Daily Totals		$2868.60		$3455.74	$179.29	$215.98

Crew B-13G	Hr.	Daily	Hr.	Daily	Bare Costs	Incl. O&P
1 Laborer	$31.60	$252.80	$52.15	$417.20	$37.08	$59.55
1 Equip. Oper. (crane)	42.55	340.40	66.95	535.60		
1 Hyd. Excavator, .75 C.Y.		598.60		658.46		
1 Trench Box		113.40		124.74	44.50	48.95
16 L.H., Daily Totals		$1305.20		$1736.00	$81.58	$108.50

Crew B-13H	Hr.	Daily	Hr.	Daily	Bare Costs	Incl. O&P
1 Laborer	$31.60	$252.80	$52.15	$417.20	$37.08	$59.55
1 Equip. Oper. (crane)	42.55	340.40	66.95	535.60		
1 Gradall, 5/8 C.Y.		958.80		1054.68		
1 Trench Box		113.40		124.74	67.01	73.71
16 L.H., Daily Totals		$1665.40		$2132.22	$104.09	$133.26

Crew No.	Bare Costs Hr.	Bare Costs Daily	Incl. Subs O & P Hr.	Incl. Subs O & P Daily	Cost Per Labor-Hour Bare Costs	Cost Per Labor-Hour Incl. O&P
Crew B-13I	**Hr.**	**Daily**	**Hr.**	**Daily**	**Bare Costs**	**Incl. O&P**
1 Laborer	$31.60	$252.80	$52.15	$417.20	$37.08	$59.55
1 Equip. Oper. (crane)	42.55	340.40	66.95	535.60		
1 Gradall, 3 Ton, 1 C.Y.		1130.00		1243.00		
1 Trench Box		113.40		124.74	77.71	85.48
16 L.H., Daily Totals		$1836.60		$2320.54	$114.79	$145.03

Crew B-13J	Hr.	Daily	Hr.	Daily	Bare Costs	Incl. O&P
1 Laborer	$31.60	$252.80	$52.15	$417.20	$37.08	$59.55
1 Equip. Oper. (crane)	42.55	340.40	66.95	535.60		
1 Hyd. Excavator, 2.5 C.Y.		1619.00		1780.90		
1 Trench Box		113.40		124.74	108.28	119.10
16 L.H., Daily Totals		$2325.60		$2858.44	$145.35	$178.65

Crew B-14	Hr.	Daily	Hr.	Daily	Bare Costs	Incl. O&P
1 Labor Foreman (outside)	$33.60	$268.80	$55.45	$443.60	$33.17	$54.25
4 Laborers	31.60	1011.20	52.15	1668.80		
1 Equip. Oper. (light)	39.05	312.40	61.45	491.60		
1 Backhoe Loader, 48 H.P.		293.80		323.18	6.12	6.73
48 L.H., Daily Totals		$1886.20		$2927.18	$39.30	$60.98

Crew B-14A	Hr.	Daily	Hr.	Daily	Bare Costs	Incl. O&P
1 Equip. Oper. (crane)	$42.55	$340.40	$66.95	$535.60	$38.90	$62.02
.5 Laborer	31.60	126.40	52.15	208.60		
1 Hyd. Excavator, 4.5 C.Y.		2635.00		2898.50	219.58	241.54
12 L.H., Daily Totals		$3101.80		$3642.70	$258.48	$303.56

Crew B-14B	Hr.	Daily	Hr.	Daily	Bare Costs	Incl. O&P
1 Equip. Oper. (crane)	$42.55	$340.40	$66.95	$535.60	$38.90	$62.02
.5 Laborer	31.60	126.40	52.15	208.60		
1 Hyd. Excavator, 6 C.Y.		3080.00		3388.00	256.67	282.33
12 L.H., Daily Totals		$3546.80		$4132.20	$295.57	$344.35

Crew B-14C	Hr.	Daily	Hr.	Daily	Bare Costs	Incl. O&P
1 Equip. Oper. (crane)	$42.55	$340.40	$66.95	$535.60	$38.90	$62.02
.5 Laborer	31.60	126.40	52.15	208.60		
1 Hyd. Excavator, 7 C.Y.		3256.00		3581.60	271.33	298.47
12 L.H., Daily Totals		$3722.80		$4325.80	$310.23	$360.48

Crew B-14F	Hr.	Daily	Hr.	Daily	Bare Costs	Incl. O&P
1 Equip. Oper. (crane)	$42.55	$340.40	$66.95	$535.60	$38.90	$62.02
.5 Laborer	31.60	126.40	52.15	208.60		
1 Hyd. Shovel, 7 C.Y.		3090.00		3399.00	257.50	283.25
12 L.H., Daily Totals		$3556.80		$4143.20	$296.40	$345.27

Crew B-14G	Hr.	Daily	Hr.	Daily	Bare Costs	Incl. O&P
1 Equip. Oper. (crane)	$42.55	$340.40	$66.95	$535.60	$38.90	$62.02
.5 Laborer	31.60	126.40	52.15	208.60		
1 Hyd. Shovel, 12 C.Y.		4148.00		4562.80	345.67	380.23
12 L.H., Daily Totals		$4614.80		$5307.00	$384.57	$442.25

Crew B-14J	Hr.	Daily	Hr.	Daily	Bare Costs	Incl. O&P
1 Equip. Oper. (med.)	$41.35	$330.80	$65.05	$520.40	$38.10	$60.75
.5 Laborer	31.60	126.40	52.15	208.60		
1 F.E. Loader, 8 C.Y.		1497.00		1646.70	124.75	137.22
12 L.H., Daily Totals		$1954.20		$2375.70	$162.85	$197.97

Crew B-14K	Hr.	Daily	Hr.	Daily	Bare Costs	Incl. O&P
1 Equip. Oper. (med.)	$41.35	$330.80	$65.05	$520.40	$38.10	$60.75
.5 Laborer	31.60	126.40	52.15	208.60		
1 F.E. Loader, 10 C.Y.		2225.00		2447.50	185.42	203.96
12 L.H., Daily Totals		$2682.20		$3176.50	$223.52	$264.71

Crew B-15	Hr.	Daily	Hr.	Daily	Bare Costs	Incl. O&P
1 Equipment Oper. (med)	$41.35	$330.80	$65.05	$520.40	$34.59	$55.95
.5 Laborer	31.60	126.40	52.15	208.60		
2 Truck Drivers (heavy)	31.95	511.20	52.35	837.60		
2 Dump Trucks, 12 C.Y., 400 H.P.		1066.00		1172.60		
1 Dozer, 200 H.P.		1082.00		1190.20	76.71	84.39
28 L.H., Daily Totals		$3116.40		$3929.40	$111.30	$140.34

Crew B-16	Hr.	Daily	Hr.	Daily	Bare Costs	Incl. O&P
1 Labor Foreman (outside)	$33.60	$268.80	$55.45	$443.60	$32.19	$53.02
2 Laborers	31.60	505.60	52.15	834.40		
1 Truck Driver (heavy)	31.95	255.60	52.35	418.80		
1 Dump Truck, 12 C.Y., 400 H.P.		533.00		586.30	16.66	18.32
32 L.H., Daily Totals		$1563.00		$2283.10	$48.84	$71.35

Crew B-17	Hr.	Daily	Hr.	Daily	Bare Costs	Incl. O&P
2 Laborers	$31.60	$505.60	$52.15	$834.40	$33.55	$54.52
1 Equip. Oper. (light)	39.05	312.40	61.45	491.60		
1 Truck Driver (heavy)	31.95	255.60	52.35	418.80		
1 Backhoe Loader, 48 H.P.		293.80		323.18		
1 Dump Truck, 8 C.Y., 220 H.P.		326.40		359.04	19.38	21.32
32 L.H., Daily Totals		$1693.80		$2427.02	$52.93	$75.84

Crew B-17A	Hr.	Daily	Hr.	Daily	Bare Costs	Incl. O&P
2 Laborer Foremen	$33.60	$537.60	$55.45	$887.20	$34.05	$56.01
6 Laborers	31.60	1516.80	52.15	2503.20		
1 Skilled Worker Foreman	42.85	342.80	69.75	558.00		
1 Skilled Worker	40.85	326.80	66.50	532.00		
80 L.H., Daily Totals		$2724.00		$4480.40	$34.05	$56.01

Crew B-18	Hr.	Daily	Hr.	Daily	Bare Costs	Incl. O&P
1 Labor Foreman (outside)	$33.60	$268.80	$55.45	$443.60	$32.27	$53.25
2 Laborers	31.60	505.60	52.15	834.40		
1 Vibratory Plate, Gas, 21"		39.20		43.12	1.63	1.80
24 L.H., Daily Totals		$813.60		$1321.12	$33.90	$55.05

Crew B-19	Hr.	Daily	Hr.	Daily	Bare Costs	Incl. O&P
1 Pile Driver Foreman	$40.50	$324.00	$67.80	$542.40	$39.55	$64.67
4 Pile Drivers	38.50	1232.00	64.45	2062.40		
2 Equip. Oper. (crane)	42.55	680.80	66.95	1071.20		
1 Equip. Oper. Oiler	36.80	294.40	57.90	463.20		
1 Crawler Crane, 40 Ton		1055.00		1160.50		
1 Lead, 90' high		118.80		130.68		
1 Hammer, Diesel, 22k Ft-Lb		590.00		649.00	27.56	30.32
64 L.H., Daily Totals		$4295.00		$6079.38	$67.11	$94.99

Crew B-19A	Hr.	Daily	Hr.	Daily	Bare Costs	Incl. O&P
1 Pile Driver Foreman	$40.50	$324.00	$67.80	$542.40	$39.55	$64.67
4 Pile Drivers	38.50	1232.00	64.45	2062.40		
2 Equip. Oper. (crane)	42.55	680.80	66.95	1071.20		
1 Equip. Oper. Oiler	36.80	294.40	57.90	463.20		
1 Crawler Crane, 75 Ton		1496.00		1645.60		
1 Lead, 90' high		118.80		130.68		
1 Hammer, Diesel, 41k Ft-Lb		677.60		745.36	35.82	39.40
64 L.H., Daily Totals		$4823.60		$6660.84	$75.37	$104.08

Crew No.	Bare Costs		Incl. Subs O & P		Cost Per Labor-Hour	

Crew B-20	Hr.	Daily	Hr.	Daily	Bare Costs	Incl. O&P
1 Labor Foreman (out)	$33.60	$268.80	$55.45	$443.60	$35.35	$58.03
1 Skilled Worker	40.85	326.80	66.50	532.00		
1 Laborer	31.60	252.80	52.15	417.20		
24 L.H., Daily Totals		$848.40		$1392.80	$35.35	$58.03

Crew B-20A	Hr.	Daily	Hr.	Daily	Bare Costs	Incl. O&P
1 Labor Foreman	$33.60	$268.80	$55.45	$443.60	$38.24	$60.92
1 Laborer	31.60	252.80	52.15	417.20		
1 Plumber	48.75	390.00	75.60	604.80		
1 Plumber Apprentice	39.00	312.00	60.50	484.00		
32 L.H., Daily Totals		$1223.60		$1949.60	$38.24	$60.92

Crew B-21	Hr.	Daily	Hr.	Daily	Bare Costs	Incl. O&P
1 Labor Foreman (out)	$33.60	$268.80	$55.45	$443.60	$36.38	$59.31
1 Skilled Worker	40.85	326.80	66.50	532.00		
1 Laborer	31.60	252.80	52.15	417.20		
.5 Equip. Oper. (crane)	42.55	170.20	66.95	267.80		
.5 S.P. Crane, 4x4, 5 Ton		129.10		142.01	4.61	5.07
28 L.H., Daily Totals		$1147.70		$1802.61	$40.99	$64.38

Crew B-21A	Hr.	Daily	Hr.	Daily	Bare Costs	Incl. O&P
1 Labor Foreman	$33.60	$268.80	$55.45	$443.60	$39.10	$62.13
1 Laborer	31.60	252.80	52.15	417.20		
1 Plumber	48.75	390.00	75.60	604.80		
1 Plumber Apprentice	39.00	312.00	60.50	484.00		
1 Equip. Oper. (crane)	42.55	340.40	66.95	535.60		
1 S.P. Crane, 4x4, 12 Ton		601.80		661.98	15.05	16.55
40 L.H., Daily Totals		$2165.80		$3147.18	$54.15	$78.68

Crew B-21B	Hr.	Daily	Hr.	Daily	Bare Costs	Incl. O&P
1 Laborer Foreman	$33.60	$268.80	$55.45	$443.60	$34.19	$55.77
3 Laborers	31.60	758.40	52.15	1251.60		
1 Equip. Oper. (crane)	42.55	340.40	66.95	535.60		
1 Hyd. Crane, 12 Ton		768.80		845.68	19.22	21.14
40 L.H., Daily Totals		$2136.40		$3076.48	$53.41	$76.91

Crew B-21C	Hr.	Daily	Hr.	Daily	Bare Costs	Incl. O&P
1 Laborer Foreman	$33.60	$268.80	$55.45	$443.60	$34.19	$55.56
4 Laborers	31.60	1011.20	52.15	1668.80		
1 Equip. Oper. (crane)	42.55	340.40	66.95	535.60		
1 Equip. Oper. Oiler	36.80	294.40	57.90	463.20		
2 Cutting Torches		34.00		37.40		
2 Gases		151.20		166.32		
1 Lattice Boom Crane, 90 Ton		1741.00		1915.10	34.40	37.84
56 L.H., Daily Totals		$3841.00		$5230.02	$68.59	$93.39

Crew B-22	Hr.	Daily	Hr.	Daily	Bare Costs	Incl. O&P
1 Labor Foreman (out)	$33.60	$268.80	$55.45	$443.60	$36.79	$59.82
1 Skilled Worker	40.85	326.80	66.50	532.00		
1 Laborer	31.60	252.80	52.15	417.20		
.75 Equip. Oper. (crane)	42.55	255.30	66.95	401.70		
.75 S.P. Crane, 4x4, 5 Ton		193.65		213.01	6.46	7.10
30 L.H., Daily Totals		$1297.35		$2007.52	$43.24	$66.92

Crew No.	Bare Costs		Incl. Subs O & P		Cost Per Labor-Hour	

Crew B-22A	Hr.	Daily	Hr.	Daily	Bare Costs	Incl. O&P
1 Labor Foreman (out)	$33.60	$268.80	$55.45	$443.60	$35.70	$58.20
1 Skilled Worker	40.85	326.80	66.50	532.00		
2 Laborers	31.60	505.60	52.15	834.40		
.75 Equipment Oper. (crane)	42.55	255.30	66.95	401.70		
.75 S.P. Crane, 4x4, 5 Ton		193.65		213.01		
1 Generator, 5 kW		40.20		44.22		
1 Butt Fusion Machine		199.80		219.78	11.41	12.55
38 L.H., Daily Totals		$1790.15		$2688.72	$47.11	$70.76

Crew B-22B	Hr.	Daily	Hr.	Daily	Bare Costs	Incl. O&P
1 Skilled Worker	$40.85	$326.80	$66.50	$532.00	$36.23	$59.33
1 Laborer	31.60	252.80	52.15	417.20		
1 Electro Fusion Machine		146.80		161.48	9.18	10.09
16 L.H., Daily Totals		$726.40		$1110.68	$45.40	$69.42

Crew B-23	Hr.	Daily	Hr.	Daily	Bare Costs	Incl. O&P
1 Labor Foreman (outside)	$33.60	$268.80	$55.45	$443.60	$32.00	$52.81
4 Laborers	31.60	1011.20	52.15	1668.80		
1 Drill Rig, Truck-Mounted		2533.00		2786.30		
1 Flatbed Truck, Gas, 3 Ton		242.60		266.86	69.39	76.33
40 L.H., Daily Totals		$4055.60		$5165.56	$101.39	$129.14

Crew B-23A	Hr.	Daily	Hr.	Daily	Bare Costs	Incl. O&P
1 Labor Foreman (outside)	$33.60	$268.80	$55.45	$443.60	$35.52	$57.55
1 Laborer	31.60	252.80	52.15	417.20		
1 Equip. Operator (medium)	41.35	330.80	65.05	520.40		
1 Drill Rig, Truck-Mounted		2533.00		2786.30		
1 Pickup Truck, 3/4 Ton		116.80		128.48	110.41	121.45
24 L.H., Daily Totals		$3502.20		$4295.98	$145.93	$179.00

Crew B-23B	Hr.	Daily	Hr.	Daily	Bare Costs	Incl. O&P
1 Labor Foreman (outside)	$33.60	$268.80	$55.45	$443.60	$35.52	$57.55
1 Laborer	31.60	252.80	52.15	417.20		
1 Equip. Operator (medium)	41.35	330.80	65.05	520.40		
1 Drill Rig, Truck-Mounted		2533.00		2786.30		
1 Pickup Truck, 3/4 Ton		116.80		128.48		
1 Centr. Water Pump, 6"		309.40		340.34	123.30	135.63
24 L.H., Daily Totals		$3811.60		$4636.32	$158.82	$193.18

Crew B-24	Hr.	Daily	Hr.	Daily	Bare Costs	Incl. O&P
1 Cement Finisher	$38.30	$306.40	$59.90	$479.20	$36.62	$59.33
1 Laborer	31.60	252.80	52.15	417.20		
1 Carpenter	39.95	319.60	65.95	527.60		
24 L.H., Daily Totals		$878.80		$1424.00	$36.62	$59.33

Crew B-25	Hr.	Daily	Hr.	Daily	Bare Costs	Incl. O&P
1 Labor Foreman	$33.60	$268.80	$55.45	$443.60	$34.44	$55.97
7 Laborers	31.60	1769.60	52.15	2920.40		
3 Equip. Oper. (med.)	41.35	992.40	65.05	1561.20		
1 Asphalt Paver, 130 H.P.		1936.00		2129.60		
1 Tandem Roller, 10 Ton		233.00		256.30		
1 Roller, Pneum. Whl, 12 Ton		307.80		338.58	28.15	30.96
88 L.H., Daily Totals		$5507.60		$7649.68	$62.59	$86.93

Crew No.	Bare Costs		Incl. Subs O & P		Cost Per Labor-Hour	

Crew B-25B	Hr.	Daily	Hr.	Daily	Bare Costs	Incl. O&P
1 Labor Foreman	$33.60	$268.80	$55.45	$443.60	$35.02	$56.73
7 Laborers	31.60	1769.60	52.15	2920.40		
4 Equip. Oper. (medium)	41.35	1323.20	65.05	2081.60		
1 Asphalt Paver, 130 H.P.		1936.00		2129.60		
2 Tandem Rollers, 10 Ton		466.00		512.60		
1 Roller, Pneum. Whl, 12 Ton		307.80		338.58	28.23	31.05
96 L.H., Daily Totals		$6071.40		$8426.38	$63.24	$87.77

Crew B-25C	Hr.	Daily	Hr.	Daily	Bare Costs	Incl. O&P
1 Labor Foreman	$33.60	$268.80	$55.45	$443.60	$35.18	$57.00
3 Laborers	31.60	758.40	52.15	1251.60		
2 Equip. Oper. (medium)	41.35	661.60	65.05	1040.80		
1 Asphalt Paver, 130 H.P.		1936.00		2129.60		
1 Tandem Roller, 10 Ton		233.00		256.30	45.19	49.71
48 L.H., Daily Totals		$3857.80		$5121.90	$80.37	$106.71

Crew B-26	Hr.	Daily	Hr.	Daily	Bare Costs	Incl. O&P
1 Labor Foreman (outside)	$33.60	$268.80	$55.45	$443.60	$35.34	$57.66
6 Laborers	31.60	1516.80	52.15	2503.20		
2 Equip. Oper. (med.)	41.35	661.60	65.05	1040.80		
1 Rodman (reinf.)	44.55	356.40	75.95	607.60		
1 Cement Finisher	38.30	306.40	59.90	479.20		
1 Grader, 30,000 Lbs.		550.20		605.22		
1 Paving Mach. & Equip.		2500.00		2750.00	34.66	38.13
88 L.H., Daily Totals		$6160.20		$8429.62	$70.00	$95.79

Crew B-26A	Hr.	Daily	Hr.	Daily	Bare Costs	Incl. O&P
1 Labor Foreman (outside)	$33.60	$268.80	$55.45	$443.60	$35.34	$57.66
6 Laborers	31.60	1516.80	52.15	2503.20		
2 Equip. Oper. (med.)	41.35	661.60	65.05	1040.80		
1 Rodman (reinf.)	44.55	356.40	75.95	607.60		
1 Cement Finisher	38.30	306.40	59.90	479.20		
1 Grader, 30,000 Lbs.		550.20		605.22		
1 Paving Mach. & Equip.		2500.00		2750.00		
1 Concrete Saw		141.00		155.10	36.26	39.89
88 L.H., Daily Totals		$6301.20		$8584.72	$71.60	$97.55

Crew B-26B	Hr.	Daily	Hr.	Daily	Bare Costs	Incl. O&P
1 Labor Foreman (outside)	$33.60	$268.80	$55.45	$443.60	$35.84	$58.28
6 Laborers	31.60	1516.80	52.15	2503.20		
3 Equip. Oper. (med.)	41.35	992.40	65.05	1561.20		
1 Rodman (reinf.)	44.55	356.40	75.95	607.60		
1 Cement Finisher	38.30	306.40	59.90	479.20		
1 Grader, 30,000 Lbs.		550.20		605.22		
1 Paving Mach. & Equip.		2500.00		2750.00		
1 Concrete Pump, 110' Boom		977.00		1074.70	41.95	46.15
96 L.H., Daily Totals		$7468.00		$10024.72	$77.79	$104.42

Crew B-27	Hr.	Daily	Hr.	Daily	Bare Costs	Incl. O&P
1 Labor Foreman (outside)	$33.60	$268.80	$55.45	$443.60	$32.10	$52.98
3 Laborers	31.60	758.40	52.15	1251.60		
1 Berm Machine		271.40		298.54	8.48	9.33
32 L.H., Daily Totals		$1298.60		$1993.74	$40.58	$62.30

Crew B-28	Hr.	Daily	Hr.	Daily	Bare Costs	Incl. O&P
2 Carpenters	$39.95	$639.20	$65.95	$1055.20	$37.17	$61.35
1 Laborer	31.60	252.80	52.15	417.20		
24 L.H., Daily Totals		$892.00		$1472.40	$37.17	$61.35

Crew B-29	Hr.	Daily	Hr.	Daily	Bare Costs	Incl. O&P
1 Labor Foreman (outside)	$33.60	$268.80	$55.45	$443.60	$34.19	$55.56
4 Laborers	31.60	1011.20	52.15	1668.80		
1 Equip. Oper. (crane)	42.55	340.40	66.95	535.60		
1 Equip. Oper. Oiler	36.80	294.40	57.90	463.20		
1 Gradall, 5/8 C.Y.		958.80		1054.68	17.12	18.83
56 L.H., Daily Totals		$2873.60		$4165.88	$51.31	$74.39

Crew B-30	Hr.	Daily	Hr.	Daily	Bare Costs	Incl. O&P
1 Equip. Oper. (med.)	$41.35	$330.80	$65.05	$520.40	$35.08	$56.58
2 Truck Drivers (heavy)	31.95	511.20	52.35	837.60		
1 Hyd. Excavator, 1.5 C.Y.		865.80		952.38		
2 Dump Trucks, 12 C.Y., 400 H.P.		1066.00		1172.60	80.49	88.54
24 L.H., Daily Totals		$2773.80		$3482.98	$115.58	$145.12

Crew B-31	Hr.	Daily	Hr.	Daily	Bare Costs	Incl. O&P
1 Labor Foreman (outside)	$33.60	$268.80	$55.45	$443.60	$33.67	$55.57
3 Laborers	31.60	758.40	52.15	1251.60		
1 Carpenter	39.95	319.60	65.95	527.60		
1 Air Compressor, 250 cfm		162.40		178.64		
1 Sheeting Driver		5.75		6.33		
2 -50' Air Hoses, 1.5"		12.60		13.86	4.52	4.97
40 L.H., Daily Totals		$1527.55		$2421.63	$38.19	$60.54

Crew B-32	Hr.	Daily	Hr.	Daily	Bare Costs	Incl. O&P
1 Laborer	$31.60	$252.80	$52.15	$417.20	$38.91	$61.83
3 Equip. Oper. (med.)	41.35	992.40	65.05	1561.20		
1 Grader, 30,000 Lbs.		550.20		605.22		
1 Tandem Roller, 10 Ton		233.00		256.30		
1 Dozer, 200 H.P.		1082.00		1190.20	58.29	64.12
32 L.H., Daily Totals		$3110.40		$4030.12	$97.20	$125.94

Crew B-32A	Hr.	Daily	Hr.	Daily	Bare Costs	Incl. O&P
1 Laborer	$31.60	$252.80	$52.15	$417.20	$38.10	$60.75
2 Equip. Oper. (medium)	41.35	661.60	65.05	1040.80		
1 Grader, 30,000 Lbs.		550.20		605.22		
1 Roller, Vibratory, 25 Ton		594.60		654.06	47.70	52.47
24 L.H., Daily Totals		$2059.20		$2717.28	$85.80	$113.22

Crew B-32B	Hr.	Daily	Hr.	Daily	Bare Costs	Incl. O&P
1 Laborer	$31.60	$252.80	$52.15	$417.20	$38.10	$60.75
2 Equip. Oper. (medium)	41.35	661.60	65.05	1040.80		
1 Dozer, 200 H.P.		1082.00		1190.20		
1 Roller, Vibratory, 25 Ton		594.60		654.06	69.86	76.84
24 L.H., Daily Totals		$2591.00		$3302.26	$107.96	$137.59

Crew B-32C	Hr.	Daily	Hr.	Daily	Bare Costs	Incl. O&P
1 Labor Foreman	$33.60	$268.80	$55.45	$443.60	$36.81	$59.15
2 Laborers	31.60	505.60	52.15	834.40		
3 Equip. Oper. (medium)	41.35	992.40	65.05	1561.20		
1 Grader, 30,000 Lbs.		550.20		605.22		
1 Tandem Roller, 10 Ton		233.00		256.30		
1 Dozer, 200 H.P.		1082.00		1190.20	38.86	42.74
48 L.H., Daily Totals		$3632.00		$4890.92	$75.67	$101.89

Crew B-33A	Hr.	Daily	Hr.	Daily	Bare Costs	Incl. O&P
1 Equip. Oper. (med.)	$41.35	$330.80	$65.05	$520.40	$38.56	$61.36
.5 Laborer	31.60	126.40	52.15	208.60		
.25 Equip. Oper. (med.)	41.35	82.70	65.05	130.10		
1 Scraper, Towed, 7 C.Y.		102.20		112.42		
1.25 Dozers, 300 H.P.		1778.75		1956.63	134.35	147.79
14 L.H., Daily Totals		$2420.85		$2928.15	$172.92	$209.15

Crew No.	Bare Costs		Incl. Subs O & P		Cost Per Labor-Hour	
Crew B-33B	**Hr.**	**Daily**	**Hr.**	**Daily**	**Bare Costs**	**Incl. O&P**
1 Equip. Oper. (med.)	$41.35	$330.80	$65.05	$520.40	$38.56	$61.36
.5 Laborer	31.60	126.40	52.15	208.60		
.25 Equip. Oper. (med.)	41.35	82.70	65.05	130.10		
1 Scraper, Towed, 10 C.Y.		153.60		168.96		
1.25 Dozers, 300 H.P.		1778.75		1956.63	138.03	151.83
14 L.H., Daily Totals		$2472.25		$2984.68	$176.59	$213.19

Crew B-33C	Hr.	Daily	Hr.	Daily	Bare Costs	Incl. O&P
1 Equip. Oper. (med.)	$41.35	$330.80	$65.05	$520.40	$38.56	$61.36
.5 Laborer	31.60	126.40	52.15	208.60		
.25 Equip. Oper. (med.)	41.35	82.70	65.05	130.10		
1 Scraper, Towed, 15 C.Y.		168.80		185.68		
1.25 Dozers, 300 H.P.		1778.75		1956.63	139.11	153.02
14 L.H., Daily Totals		$2487.45		$3001.41	$177.68	$214.39

Crew B-33D	Hr.	Daily	Hr.	Daily	Bare Costs	Incl. O&P
1 Equip. Oper. (med.)	$41.35	$330.80	$65.05	$520.40	$38.56	$61.36
.5 Laborer	31.60	126.40	52.15	208.60		
.25 Equip. Oper. (med.)	41.35	82.70	65.05	130.10		
1 S.P. Scraper, 14 C.Y.		1509.00		1659.90		
.25 Dozer, 300 H.P.		355.75		391.32	133.20	146.52
14 L.H., Daily Totals		$2404.65		$2910.32	$171.76	$207.88

Crew B-33E	Hr.	Daily	Hr.	Daily	Bare Costs	Incl. O&P
1 Equip. Oper. (med.)	$41.35	$330.80	$65.05	$520.40	$38.56	$61.36
.5 Laborer	31.60	126.40	52.15	208.60		
.25 Equip. Oper. (med.)	41.35	82.70	65.05	130.10		
1 S.P. Scraper, 21 C.Y.		1924.00		2116.40		
.25 Dozer, 300 H.P.		355.75		391.32	162.84	179.12
14 L.H., Daily Totals		$2819.65		$3366.82	$201.40	$240.49

Crew B-33F	Hr.	Daily	Hr.	Daily	Bare Costs	Incl. O&P
1 Equip. Oper. (med.)	$41.35	$330.80	$65.05	$520.40	$38.56	$61.36
.5 Laborer	31.60	126.40	52.15	208.60		
.25 Equip. Oper. (med.)	41.35	82.70	65.05	130.10		
1 Elev. Scraper, 11 C.Y.		1095.00		1204.50		
.25 Dozer, 300 H.P.		355.75		391.32	103.63	113.99
14 L.H., Daily Totals		$1990.65		$2454.93	$142.19	$175.35

Crew B-33G	Hr.	Daily	Hr.	Daily	Bare Costs	Incl. O&P
1 Equip. Oper. (med.)	$41.35	$330.80	$65.05	$520.40	$38.56	$61.36
.5 Laborer	31.60	126.40	52.15	208.60		
.25 Equip. Oper. (med.)	41.35	82.70	65.05	130.10		
1 Elev. Scraper, 22 C.Y.		2127.00		2339.70		
.25 Dozer, 300 H.P.		355.75		391.32	177.34	195.07
14 L.H., Daily Totals		$3022.65		$3590.13	$215.90	$256.44

Crew B-33K	Hr.	Daily	Hr.	Daily	Bare Costs	Incl. O&P
1 Equipment Operator (med.)	$41.35	$330.80	$65.05	$520.40	$38.56	$61.36
.25 Equipment Operator (med.)	41.35	82.70	65.05	130.10		
.5 Laborer	31.60	126.40	52.15	208.60		
1 S.P. Scraper, 31 C.Y.		2878.00		3165.80		
.25 Dozer, 410 H.P.		462.25		508.48	238.59	262.45
14 L.H., Daily Totals		$3880.15		$4533.38	$277.15	$323.81

Crew B-34A	Hr.	Daily	Hr.	Daily	Bare Costs	Incl. O&P
1 Truck Driver (heavy)	$31.95	$255.60	$52.35	$418.80	$31.95	$52.35
1 Dump Truck, 8 C.Y., 220 H.P.		326.40		359.04	40.80	44.88
8 L.H., Daily Totals		$582.00		$777.84	$72.75	$97.23

Crew No.	Bare Costs		Incl. Subs O & P		Cost Per Labor-Hour	
Crew B-34B	**Hr.**	**Daily**	**Hr.**	**Daily**	**Bare Costs**	**Incl. O&P**
1 Truck Driver (heavy)	$31.95	$255.60	$52.35	$418.80	$31.95	$52.35
1 Dump Truck, 12 C.Y., 400 H.P.		533.00		586.30	66.63	73.29
8 L.H., Daily Totals		$788.60		$1005.10	$98.58	$125.64

Crew B-34C	Hr.	Daily	Hr.	Daily	Bare Costs	Incl. O&P
1 Truck Driver (heavy)	$31.95	$255.60	$52.35	$418.80	$31.95	$52.35
1 Truck Tractor, 6x4, 380 H.P.		478.80		526.68		
1 Dump Trailer, 16.5 C.Y.		115.60		127.16	74.30	81.73
8 L.H., Daily Totals		$850.00		$1072.64	$106.25	$134.08

Crew B-34D	Hr.	Daily	Hr.	Daily	Bare Costs	Incl. O&P
1 Truck Driver (heavy)	$31.95	$255.60	$52.35	$418.80	$31.95	$52.35
1 Truck Tractor, 6x4, 380 H.P.		478.80		526.68		
1 Dump Trailer, 20 C.Y.		130.20		143.22	76.13	83.74
8 L.H., Daily Totals		$864.60		$1088.70	$108.08	$136.09

Crew B-34E	Hr.	Daily	Hr.	Daily	Bare Costs	Incl. O&P
1 Truck Driver (heavy)	$31.95	$255.60	$52.35	$418.80	$31.95	$52.35
1 Dump Truck, Off Hwy., 25 Ton		1280.00		1408.00	160.00	176.00
8 L.H., Daily Totals		$1535.60		$1826.80	$191.95	$228.35

Crew B-34F	Hr.	Daily	Hr.	Daily	Bare Costs	Incl. O&P
1 Truck Driver (heavy)	$31.95	$255.60	$52.35	$418.80	$31.95	$52.35
1 Dump Truck, Off Hwy., 35 Ton		1144.00		1258.40	143.00	157.30
8 L.H., Daily Totals		$1399.60		$1677.20	$174.95	$209.65

Crew B-34G	Hr.	Daily	Hr.	Daily	Bare Costs	Incl. O&P
1 Truck Driver (heavy)	$31.95	$255.60	$52.35	$418.80	$31.95	$52.35
1 Dump Truck, Off Hwy., 50 Ton		1582.00		1740.20	197.75	217.53
8 L.H., Daily Totals		$1837.60		$2159.00	$229.70	$269.88

Crew B-34H	Hr.	Daily	Hr.	Daily	Bare Costs	Incl. O&P
1 Truck Driver (heavy)	$31.95	$255.60	$52.35	$418.80	$31.95	$52.35
1 Dump Truck, Off Hwy., 65 Ton		1594.00		1753.40	199.25	219.18
8 L.H., Daily Totals		$1849.60		$2172.20	$231.20	$271.52

Crew B-34J	Hr.	Daily	Hr.	Daily	Bare Costs	Incl. O&P
1 Truck Driver (heavy)	$31.95	$255.60	$52.35	$418.80	$31.95	$52.35
1 Dump Truck, Off Hwy., 100 Ton		2047.00		2251.70	255.88	281.46
8 L.H., Daily Totals		$2302.60		$2670.50	$287.82	$333.81

Crew B-34K	Hr.	Daily	Hr.	Daily	Bare Costs	Incl. O&P
1 Truck Driver (heavy)	$31.95	$255.60	$52.35	$418.80	$31.95	$52.35
1 Truck Tractor, 6x4, 450 H.P.		581.20		639.32		
1 Lowbed Trailer, 75 Ton		203.00		223.30	98.03	107.83
8 L.H., Daily Totals		$1039.80		$1281.42	$129.97	$160.18

Crew B-34L	Hr.	Daily	Hr.	Daily	Bare Costs	Incl. O&P
1 Equip. Oper. (light)	$39.05	$312.40	$61.45	$491.60	$39.05	$61.45
1 Flatbed Truck, Gas, 1.5 Ton		194.40		213.84	24.30	26.73
8 L.H., Daily Totals		$506.80		$705.44	$63.35	$88.18

Crew B-34N	Hr.	Daily	Hr.	Daily	Bare Costs	Incl. O&P
1 Truck Driver (heavy)	$31.95	$255.60	$52.35	$418.80	$31.95	$52.35
1 Dump Truck, 8 C.Y., 220 H.P.		326.40		359.04		
1 Flatbed Trailer, 40 Ton		139.60		153.56	58.25	64.08
8 L.H., Daily Totals		$721.60		$931.40	$90.20	$116.43

Crew B-34P	Hr.	Daily	Hr.	Daily	Bare Costs	Incl. O&P
	Bare Costs		Incl. Subs O & P		Cost Per Labor-Hour	
1 Pipe Fitter	$49.35	$394.80	$76.55	$612.40	$40.55	$64.12
1 Truck Driver (light)	30.95	247.60	50.75	406.00		
1 Equip. Oper. (medium)	41.35	330.80	65.05	520.40		
1 Flatbed Truck, Gas, 3 Ton		242.60		266.86		
1 Backhoe Loader, 48 H.P.		293.80		323.18	22.35	24.59
24 L.H., Daily Totals		$1509.60		$2128.84	$62.90	$88.70

Crew B-34Q	Hr.	Daily	Hr.	Daily	Bare Costs	Incl. O&P
1 Pipe Fitter	$49.35	$394.80	$76.55	$612.40	$40.95	$64.75
1 Truck Driver (light)	30.95	247.60	50.75	406.00		
1 Eqip. Oper. (crane)	42.55	340.40	66.95	535.60		
1 Flatbed Trailer, 25 Ton		102.00		112.20		
1 Dump Truck, 8 C.Y., 220 H.P.		326.40		359.04		
1 Hyd. Crane, 25 Ton		789.40		868.34	50.74	55.82
24 L.H., Daily Totals		$2200.60		$2893.58	$91.69	$120.57

Crew B-34R	Hr.	Daily	Hr.	Daily	Bare Costs	Incl. O&P
1 Pipe Fitter	$49.35	$394.80	$76.55	$612.40	$40.95	$64.75
1 Truck Driver (light)	30.95	247.60	50.75	406.00		
1 Eqip. Oper. (crane)	42.55	340.40	66.95	535.60		
1 Flatbed Trailer, 25 Ton		102.00		112.20		
1 Dump Truck, 8 C.Y., 220 H.P.		326.40		359.04		
1 Hyd. Crane, 25 Ton		789.40		868.34		
1 Hyd. Excavator, 1 C.Y.		669.20		736.12	78.63	86.49
24 L.H., Daily Totals		$2869.80		$3629.70	$119.58	$151.24

Crew B-34S	Hr.	Daily	Hr.	Daily	Bare Costs	Incl. O&P
2 Pipe Fitters	$49.35	$789.60	$76.55	$1224.80	$43.30	$68.10
1 Truck Driver (heavy)	31.95	255.60	52.35	418.80		
1 Eqip. Oper. (crane)	42.55	340.40	66.95	535.60		
1 Flatbed Trailer, 40 Ton		139.60		153.56		
1 Truck Tractor, 6x4, 380 H.P.		478.80		526.68		
1 Hyd. Crane, 80 Ton		1296.00		1425.60		
1 Hyd. Excavator, 2 C.Y.		1182.00		1300.20	96.76	106.44
32 L.H., Daily Totals		$4482.00		$5585.24	$140.06	$174.54

Crew B-34T	Hr.	Daily	Hr.	Daily	Bare Costs	Incl. O&P
2 Pipe Fitters	$49.35	$789.60	$76.55	$1224.80	$43.30	$68.10
1 Truck Driver (heavy)	31.95	255.60	52.35	418.80		
1 Eqip. Oper. (crane)	42.55	340.40	66.95	535.60		
1 Flatbed Trailer, 40 Ton		139.60		153.56		
1 Truck Tractor, 6x4, 380 H.P.		478.80		526.68		
1 Hyd. Crane, 80 Ton		1296.00		1425.60	59.83	65.81
32 L.H., Daily Totals		$3300.00		$4285.04	$103.13	$133.91

Crew B-35	Hr.	Daily	Hr.	Daily	Bare Costs	Incl. O&P
1 Laborer Foreman (out)	$33.60	$268.80	$55.45	$443.60	$39.02	$62.42
1 Skilled Worker	40.85	326.80	66.50	532.00		
1 Welder (plumber)	48.75	390.00	75.60	604.80		
1 Laborer	31.60	252.80	52.15	417.20		
1 Equip. Oper. (crane)	42.55	340.40	66.95	535.60		
1 Equip. Oper. Oiler	36.80	294.40	57.90	463.20		
1 Welder, electric, 300 amp		55.70		61.27		
1 Hyd. Excavator, .75 C.Y.		598.60		658.46	13.63	14.99
48 L.H., Daily Totals		$2527.50		$3716.13	$52.66	$77.42

Crew B-35A	Hr.	Daily	Hr.	Daily	Bare Costs	Incl. O&P
1 Laborer Foreman (out)	$33.60	$268.80	$55.45	$443.60	$37.96	$60.96
2 Laborers	31.60	505.60	52.15	834.40		
1 Skilled Worker	40.85	326.80	66.50	532.00		
1 Welder (plumber)	48.75	390.00	75.60	604.80		
1 Equip. Oper. (crane)	42.55	340.40	66.95	535.60		
1 Equip. Oper. Oiler	36.80	294.40	57.90	463.20		
1 Welder, gas engine, 300 amp		134.20		147.62		
1 Crawler Crane, 75 Ton		1496.00		1645.60	29.11	32.02
56 L.H., Daily Totals		$3756.20		$5206.82	$67.08	$92.98

Crew B-36	Hr.	Daily	Hr.	Daily	Bare Costs	Incl. O&P
1 Labor Foreman (outside)	$33.60	$268.80	$55.45	$443.60	$35.90	$57.97
2 Laborers	31.60	505.60	52.15	834.40		
2 Equip. Oper. (med.)	41.35	661.60	65.05	1040.80		
1 Dozer, 200 H.P.		1082.00		1190.20		
1 Aggregate Spreader		35.00		38.50		
1 Tandem Roller, 10 Ton		233.00		256.30	33.75	37.13
40 L.H., Daily Totals		$2786.00		$3803.80	$69.65	$95.09

Crew B-36A	Hr.	Daily	Hr.	Daily	Bare Costs	Incl. O&P
1 Labor Foreman (outside)	$33.60	$268.80	$55.45	$443.60	$37.46	$59.99
2 Laborers	31.60	505.60	52.15	834.40		
4 Equip. Oper. (med.)	41.35	1323.20	65.05	2081.60		
1 Dozer, 200 H.P.		1082.00		1190.20		
1 Aggregate Spreader		35.00		38.50		
1 Tandem Roller, 10 Ton		233.00		256.30		
1 Roller, Pneum. Whl, 12 Ton		307.80		338.58	29.60	32.56
56 L.H., Daily Totals		$3755.40		$5183.18	$67.06	$92.56

Crew B-36B	Hr.	Daily	Hr.	Daily	Bare Costs	Incl. O&P
1 Labor Foreman (outside)	$33.60	$268.80	$55.45	$443.60	$36.77	$59.04
2 Laborers	31.60	505.60	52.15	834.40		
4 Equip. Oper. (medium)	41.35	1323.20	65.05	2081.60		
1 Truck Driver, Heavy	31.95	255.60	52.35	418.80		
1 Grader, 30,000 Lbs.		550.20		605.22		
1 F.E. Loader, crl, 1.5 C.Y.		463.00		509.30		
1 Dozer, 300 H.P.		1423.00		1565.30		
1 Roller, Vibratory, 25 Ton		594.60		654.06		
1 Truck Tractor, 6x4, 450 H.P.		581.20		639.32		
1 Water Tanker, 5000 Gal.		136.00		149.60	58.56	64.42
64 L.H., Daily Totals		$6101.20		$7901.20	$95.33	$123.46

Crew B-36C	Hr.	Daily	Hr.	Daily	Bare Costs	Incl. O&P
1 Labor Foreman (outside)	$33.60	$268.80	$55.45	$443.60	$37.92	$60.59
3 Equip. Oper. (medium)	41.35	992.40	65.05	1561.20		
1 Truck Driver, Heavy	31.95	255.60	52.35	418.80		
1 Grader, 30,000 Lbs.		550.20		605.22		
1 Dozer, 300 H.P.		1423.00		1565.30		
1 Roller, Vibratory, 25 Ton		594.60		654.06		
1 Truck Tractor, 6x4, 450 H.P.		581.20		639.32		
1 Water Tanker, 5000 Gal.		136.00		149.60	82.13	90.34
40 L.H., Daily Totals		$4801.80		$6037.10	$120.05	$150.93

Crew B-37	Hr.	Daily	Hr.	Daily	Bare Costs	Incl. O&P
1 Labor Foreman (outside)	$33.60	$268.80	$55.45	$443.60	$33.17	$54.25
4 Laborers	31.60	1011.20	52.15	1668.80		
1 Equip. Oper. (light)	39.05	312.40	61.45	491.60		
1 Tandem Roller, 5 Ton		137.80		151.58	2.87	3.16
48 L.H., Daily Totals		$1730.20		$2755.58	$36.05	$57.41

Crew No.	Bare Costs		Incl. Subs O & P		Cost Per Labor-Hour	
Crew B-38	**Hr.**	**Daily**	**Hr.**	**Daily**	**Bare Costs**	**Incl. O&P**
1 Labor Foreman (outside)	$33.60	$268.80	$55.45	$443.60	$35.44	$57.25
2 Laborers	31.60	505.60	52.15	834.40		
1 Equip. Oper. (light)	39.05	312.40	61.45	491.60		
1 Equip. Oper. (medium)	41.35	330.80	65.05	520.40		
1 Backhoe Loader, 48 H.P.		293.80		323.18		
1 Hyd.Hammer, (1200 lb.)		153.00		168.30		
1 F.E. Loader, W.M., 4 C.Y.		541.00		595.10		
1 Pavt. Rem. Bucket		56.80		62.48	26.11	28.73
40 L.H., Daily Totals		$2462.20		$3439.06	$61.56	$85.98
Crew B-39	**Hr.**	**Daily**	**Hr.**	**Daily**	**Bare Costs**	**Incl. O&P**
1 Labor Foreman (outside)	$33.60	$268.80	$55.45	$443.60	$33.17	$54.25
4 Laborers	31.60	1011.20	52.15	1668.80		
1 Equip. Oper. (light)	39.05	312.40	61.45	491.60		
1 Air Compressor, 250 cfm		162.40		178.64		
2 Breakers, Pavement, 60 lb.		16.40		18.04		
2 -50' Air Hoses, 1.5"		12.60		13.86	3.99	4.39
48 L.H., Daily Totals		$1783.80		$2814.54	$37.16	$58.64
Crew B-40	**Hr.**	**Daily**	**Hr.**	**Daily**	**Bare Costs**	**Incl. O&P**
1 Pile Driver Foreman (out)	$40.50	$324.00	$67.80	$542.40	$39.55	$64.67
4 Pile Drivers	38.50	1232.00	64.45	2062.40		
2 Equip. Oper. (crane)	42.55	680.80	66.95	1071.20		
1 Equip. Oper. Oiler	36.80	294.40	57.90	463.20		
1 Crawler Crane, 40 Ton		1055.00		1160.50		
1 Vibratory Hammer & Gen.		2257.00		2482.70	51.75	56.92
64 L.H., Daily Totals		$5843.20		$7782.40	$91.30	$121.60
Crew B-40B	**Hr.**	**Daily**	**Hr.**	**Daily**	**Bare Costs**	**Incl. O&P**
1 Laborer Foreman	$33.60	$268.80	$55.45	$443.60	$34.63	$56.13
3 Laborers	31.60	758.40	52.15	1251.60		
1 Equip. Oper. (crane)	42.55	340.40	66.95	535.60		
1 Equip. Oper. Oiler	36.80	294.40	57.90	463.20		
1 Lattice Boom Crane, 40 Ton		1327.00		1459.70	27.65	30.41
48 L.H., Daily Totals		$2989.00		$4153.70	$62.27	$86.54
Crew B-41	**Hr.**	**Daily**	**Hr.**	**Daily**	**Bare Costs**	**Incl. O&P**
1 Labor Foreman (outside)	$33.60	$268.80	$55.45	$443.60	$32.70	$53.68
4 Laborers	31.60	1011.20	52.15	1668.80		
.25 Equip. Oper. (crane)	42.55	85.10	66.95	133.90		
.25 Equip. Oper. Oiler	36.80	73.60	57.90	115.80		
.25 Crawler Crane, 40 Ton		263.75		290.13	5.99	6.59
44 L.H., Daily Totals		$1702.45		$2652.22	$38.69	$60.28
Crew B-42	**Hr.**	**Daily**	**Hr.**	**Daily**	**Bare Costs**	**Incl. O&P**
1 Labor Foreman (outside)	$33.60	$268.80	$55.45	$443.60	$35.51	$58.96
4 Laborers	31.60	1011.20	52.15	1668.80		
1 Equip. Oper. (crane)	42.55	340.40	66.95	535.60		
1 Equip. Oper. Oiler	36.80	294.40	57.90	463.20		
1 Welder	44.70	357.60	82.80	662.40		
1 Hyd. Crane, 25 Ton		789.40		868.34		
1 Welder, gas engine, 300 amp		134.20		147.62		
1 Horz. Boring Csg. Mch.		444.20		488.62	21.37	23.51
64 L.H., Daily Totals		$3640.20		$5278.18	$56.88	$82.47

Crew No.	Bare Costs		Incl. Subs O & P		Cost Per Labor-Hour	
Crew B-43	**Hr.**	**Daily**	**Hr.**	**Daily**	**Bare Costs**	**Incl. O&P**
1 Labor Foreman (outside)	$33.60	$268.80	$55.45	$443.60	$34.63	$56.13
3 Laborers	31.60	758.40	52.15	1251.60		
1 Equip. Oper. (crane)	42.55	340.40	66.95	535.60		
1 Equip. Oper. Oiler	36.80	294.40	57.90	463.20		
1 Drill Rig, Truck-Mounted		2533.00		2786.30	52.77	58.05
48 L.H., Daily Totals		$4195.00		$5480.30	$87.40	$114.17
Crew B-44	**Hr.**	**Daily**	**Hr.**	**Daily**	**Bare Costs**	**Incl. O&P**
1 Pile Driver Foreman	$40.50	$324.00	$67.80	$542.40	$38.90	$63.96
4 Pile Drivers	38.50	1232.00	64.45	2062.40		
2 Equip. Oper. (crane)	42.55	680.80	66.95	1071.20		
1 Laborer	31.60	252.80	52.15	417.20		
1 Crawler Crane, 40 Ton		1055.00		1160.50		
1 Lead, 60' high		69.60		76.56		
1 Hammer, diesel, 15K Ft.-Lbs.		578.60		636.46	26.61	29.27
64 L.H., Daily Totals		$4192.80		$5966.72	$65.51	$93.23
Crew B-45	**Hr.**	**Daily**	**Hr.**	**Daily**	**Bare Costs**	**Incl. O&P**
1 Equip. Oper. (med.)	$41.35	$330.80	$65.05	$520.40	$36.65	$58.70
1 Truck Driver (heavy)	31.95	255.60	52.35	418.80		
1 Dist. Tanker, 3000 Gallon		304.20		334.62		
1 Truck Tractor, 6x4, 380 H.P.		478.80		526.68	48.94	53.83
16 L.H., Daily Totals		$1369.40		$1800.50	$85.59	$112.53
Crew B-46	**Hr.**	**Daily**	**Hr.**	**Daily**	**Bare Costs**	**Incl. O&P**
1 Pile Driver Foreman	$40.50	$324.00	$67.80	$542.40	$35.38	$58.86
2 Pile Drivers	38.50	616.00	64.45	1031.20		
3 Laborers	31.60	758.40	52.15	1251.60		
1 Chainsaw, Gas, 36" Long		37.80		41.58	.79	.87
48 L.H., Daily Totals		$1736.20		$2866.78	$36.17	$59.72
Crew B-47	**Hr.**	**Daily**	**Hr.**	**Daily**	**Bare Costs**	**Incl. O&P**
1 Blast Foreman	$33.60	$268.80	$55.45	$443.60	$34.75	$56.35
1 Driller	31.60	252.80	52.15	417.20		
1 Equip. Oper. (light)	39.05	312.40	61.45	491.60		
1 Air Track Drill, 4"		850.60		935.66		
1 Air Compressor, 600 cfm		434.40		477.84		
2 -50' Air Hoses, 3"		32.40		35.64	54.89	60.38
24 L.H., Daily Totals		$2151.40		$2801.54	$89.64	$116.73
Crew B-47A	**Hr.**	**Daily**	**Hr.**	**Daily**	**Bare Costs**	**Incl. O&P**
1 Drilling Foreman	$33.60	$268.80	$55.45	$443.60	$37.65	$60.10
1 Equip. Oper. (heavy)	42.55	340.40	66.95	535.60		
1 Oiler	36.80	294.40	57.90	463.20		
1 Air Track Drill, 5"		974.00		1071.40	40.58	44.64
24 L.H., Daily Totals		$1877.60		$2513.80	$78.23	$104.74
Crew B-47C	**Hr.**	**Daily**	**Hr.**	**Daily**	**Bare Costs**	**Incl. O&P**
1 Laborer	$31.60	$252.80	$52.15	$417.20	$35.33	$56.80
1 Equip. Oper. (light)	39.05	312.40	61.45	491.60		
1 Air Compressor, 750 cfm		439.60		483.56		
2 -50' Air Hoses, 3"		32.40		35.64		
1 Air Track Drill, 4"		850.60		935.66	82.66	90.93
16 L.H., Daily Totals		$1887.80		$2363.66	$117.99	$147.73
Crew B-47E	**Hr.**	**Daily**	**Hr.**	**Daily**	**Bare Costs**	**Incl. O&P**
1 Laborer Foreman	$33.60	$268.80	$55.45	$443.60	$32.10	$52.98
3 Laborers	31.60	758.40	52.15	1251.60		
1 Flatbed Truck, Gas, 3 Ton		242.60		266.86	7.58	8.34
32 L.H., Daily Totals		$1269.80		$1962.06	$39.68	$61.31

Crew No.	Bare Costs		Incl. Subs O & P		Cost Per Labor-Hour	
Crew B-47G	**Hr.**	**Daily**	**Hr.**	**Daily**	**Bare Costs**	**Incl. O&P**
1 Laborer Foreman	$33.60	$268.80	$55.45	$443.60	$33.96	$55.30
2 Laborers	31.60	505.60	52.15	834.40		
1 Equip. Oper. (light)	39.05	312.40	61.45	491.60		
1 Air Track Drill, 4″		850.60		935.66		
1 Air Compressor, 600 cfm		434.40		477.84		
2 -50′ Air Hoses, 3″		32.40		35.64		
1 Grout Pump		166.40		183.04	46.37	51.01
32 L.H., Daily Totals		$2570.60		$3401.78	$80.33	$106.31
Crew B-48	**Hr.**	**Daily**	**Hr.**	**Daily**	**Bare Costs**	**Incl. O&P**
1 Labor Foreman (outside)	$33.60	$268.80	$55.45	$443.60	$35.26	$56.89
3 Laborers	31.60	758.40	52.15	1251.60		
1 Equip. Oper. (crane)	42.55	340.40	66.95	535.60		
1 Equip. Oper. Oiler	36.80	294.40	57.90	463.20		
1 Equip. Oper. (light)	39.05	312.40	61.45	491.60		
1 Centr. Water Pump, 6″		309.40		340.34		
1 -20′ Suction Hose, 6″		11.90		13.09		
1 -50′ Discharge Hose, 6″		6.30		6.93		
1 Drill Rig, Truck-Mounted		2533.00		2786.30	51.08	56.19
56 L.H., Daily Totals		$4835.00		$6332.26	$86.34	$113.08
Crew B-49	**Hr.**	**Daily**	**Hr.**	**Daily**	**Bare Costs**	**Incl. O&P**
1 Labor Foreman (outside)	$33.60	$268.80	$55.45	$443.60	$36.65	$59.27
3 Laborers	31.60	758.40	52.15	1251.60		
2 Equip. Oper. (crane)	42.55	680.80	66.95	1071.20		
2 Equip. Oper. Oilers	36.80	588.80	57.90	926.40		
1 Equip. Oper. (light)	39.05	312.40	61.45	491.60		
2 Pile Drivers	38.50	616.00	64.45	1031.20		
1 Hyd. Crane, 25 Ton		789.40		868.34		
1 Centr. Water Pump, 6″		309.40		340.34		
1 -20′ Suction Hose, 6″		11.90		13.09		
1 -50′ Discharge Hose, 6″		6.30		6.93		
1 Drill Rig, Truck-Mounted		2533.00		2786.30	41.48	45.63
88 L.H., Daily Totals		$6875.20		$9230.60	$78.13	$104.89
Crew B-50	**Hr.**	**Daily**	**Hr.**	**Daily**	**Bare Costs**	**Incl. O&P**
2 Pile Driver Foremen	$40.50	$648.00	$67.80	$1084.80	$37.76	$62.18
6 Pile Drivers	38.50	1848.00	64.45	3093.60		
2 Equip. Oper. (crane)	42.55	680.80	66.95	1071.20		
1 Equip. Oper. Oiler	36.80	294.40	57.90	463.20		
3 Laborers	31.60	758.40	52.15	1251.60		
1 Crawler Crane, 40 Ton		1055.00		1160.50		
1 Lead, 60′ high		69.60		76.56		
1 Hammer, diesel, 15K Ft.-Lbs.		578.60		636.46		
1 Air Compressor, 600 cfm		434.40		477.84		
2 -50′ Air Hoses, 3″		32.40		35.64		
1 Chainsaw, Gas, 36″ Long		37.80		41.58	19.71	21.68
112 L.H., Daily Totals		$6437.40		$9392.98	$57.48	$83.87
Crew B-51	**Hr.**	**Daily**	**Hr.**	**Daily**	**Bare Costs**	**Incl. O&P**
1 Labor Foreman (outside)	$33.60	$268.80	$55.45	$443.60	$31.82	$52.47
4 Laborers	31.60	1011.20	52.15	1668.80		
1 Truck Driver (light)	30.95	247.60	50.75	406.00		
1 Flatbed Truck, Gas, 1.5 Ton		194.40		213.84	4.05	4.46
48 L.H., Daily Totals		$1722.00		$2732.24	$35.88	$56.92

Crew No.	Bare Costs		Incl. Subs O & P		Cost Per Labor-Hour	
Crew B-52	**Hr.**	**Daily**	**Hr.**	**Daily**	**Bare Costs**	**Incl. O&P**
1 Carpenter Foreman	$41.95	$335.60	$69.25	$554.00	$36.85	$60.29
1 Carpenter	39.95	319.60	65.95	527.60		
3 Laborers	31.60	758.40	52.15	1251.60		
1 Cement Finisher	38.30	306.40	59.90	479.20		
.5 Rodman (reinf.)	44.55	178.20	75.95	303.80		
.5 Equip. Oper. (med.)	41.35	165.40	65.05	260.20		
.5 Crawler Loader, 3 C.Y.		495.00		544.50	8.84	9.72
56 L.H., Daily Totals		$2558.60		$3920.90	$45.69	$70.02
Crew B-53	**Hr.**	**Daily**	**Hr.**	**Daily**	**Bare Costs**	**Incl. O&P**
1 Equip. Oper. (light)	$39.05	$312.40	$61.45	$491.60	$39.05	$61.45
1 Trencher, Chain, 12 H.P.		59.60		65.56	7.45	8.20
8 L.H., Daily Totals		$372.00		$557.16	$46.50	$69.64
Crew B-54	**Hr.**	**Daily**	**Hr.**	**Daily**	**Bare Costs**	**Incl. O&P**
1 Equip. Oper. (light)	$39.05	$312.40	$61.45	$491.60	$39.05	$61.45
1 Trencher, Chain, 40 H.P.		301.00		331.10	37.63	41.39
8 L.H., Daily Totals		$613.40		$822.70	$76.67	$102.84
Crew B-54A	**Hr.**	**Daily**	**Hr.**	**Daily**	**Bare Costs**	**Incl. O&P**
.17 Labor Foreman (outside)	$33.60	$45.70	$55.45	$75.41	$40.22	$63.66
1 Equipment Operator (med.)	41.35	330.80	65.05	520.40		
1 Wheel Trencher, 67 H.P.		1009.00		1109.90	107.80	118.58
9.36 L.H., Daily Totals		$1385.50		$1705.71	$148.02	$182.23
Crew B-54B	**Hr.**	**Daily**	**Hr.**	**Daily**	**Bare Costs**	**Incl. O&P**
.25 Labor Foreman (outside)	$33.60	$67.20	$55.45	$110.90	$39.80	$63.13
1 Equipment Operator (med.)	41.35	330.80	65.05	520.40		
1 Wheel Trencher, 150 H.P.		1753.00		1928.30	175.30	192.83
10 L.H., Daily Totals		$2151.00		$2559.60	$215.10	$255.96
Crew B-55	**Hr.**	**Daily**	**Hr.**	**Daily**	**Bare Costs**	**Incl. O&P**
2 Laborers	$31.60	$505.60	$52.15	$834.40	$31.38	$51.68
1 Truck Driver (light)	30.95	247.60	50.75	406.00		
1 Truck-mounted earth auger		706.60		777.26		
1 Flatbed Truck, Gas, 3 Ton		242.60		266.86	39.55	43.51
24 L.H., Daily Totals		$1702.40		$2284.52	$70.93	$95.19
Crew B-56	**Hr.**	**Daily**	**Hr.**	**Daily**	**Bare Costs**	**Incl. O&P**
1 Laborer	$31.60	$252.80	$52.15	$417.20	$35.33	$56.80
1 Equip. Oper. (light)	39.05	312.40	61.45	491.60		
1 Air Track Drill, 4″		850.60		935.66		
1 Air Compressor, 600 cfm		434.40		477.84		
1 -50′ Air Hose, 3″		16.20		17.82	81.33	89.46
16 L.H., Daily Totals		$1866.40		$2340.12	$116.65	$146.26
Crew B-57	**Hr.**	**Daily**	**Hr.**	**Daily**	**Bare Costs**	**Incl. O&P**
1 Labor Foreman (outside)	$33.60	$268.80	$55.45	$443.60	$35.87	$57.67
2 Laborers	31.60	505.60	52.15	834.40		
1 Equip. Oper. (crane)	42.55	340.40	66.95	535.60		
1 Equip. Oper. (light)	39.05	312.40	61.45	491.60		
1 Equip. Oper. Oiler	36.80	294.40	57.90	463.20		
1 Crawler Crane, 25 Ton		1052.00		1157.20		
1 Clamshell Bucket, 1 C.Y.		46.20		50.82		
1 Centr. Water Pump, 6″		309.40		340.34		
1 -20′ Suction Hose, 6″		11.90		13.09		
20 -50′ Discharge Hoses, 6″		126.00		138.60	32.20	35.42
48 L.H., Daily Totals		$3267.10		$4468.45	$68.06	$93.09

Crew No.	Bare Costs		Incl. Subs O & P		Cost Per Labor-Hour	

Crew B-58	Hr.	Daily	Hr.	Daily	Bare Costs	Incl. O&P
2 Laborers	$31.60	$505.60	$52.15	$834.40	$34.08	$55.25
1 Equip. Oper. (light)	39.05	312.40	61.45	491.60		
1 Backhoe Loader, 48 H.P.		293.80		323.18		
1 Small Helicopter, w/pilot		2587.00		2845.70	120.03	132.04
24 L.H., Daily Totals		$3698.80		$4494.88	$154.12	$187.29

Crew B-59	Hr.	Daily	Hr.	Daily	Bare Costs	Incl. O&P
1 Truck Driver (heavy)	$31.95	$255.60	$52.35	$418.80	$31.95	$52.35
1 Truck Tractor, 220 H.P.		288.40		317.24		
1 Water Tanker, 5000 Gal.		136.00		149.60	53.05	58.35
8 L.H., Daily Totals		$680.00		$885.64	$85.00	$110.71

Crew B-59A	Hr.	Daily	Hr.	Daily	Bare Costs	Incl. O&P
2 Laborers	$31.60	$505.60	$52.15	$834.40	$31.72	$52.22
1 Truck Driver (heavy)	31.95	255.60	52.35	418.80		
1 Water Tanker, 5000 Gal.		136.00		149.60		
1 Truck Tractor, 220 H.P.		288.40		317.24	17.68	19.45
24 L.H., Daily Totals		$1185.60		$1720.04	$49.40	$71.67

Crew B-60	Hr.	Daily	Hr.	Daily	Bare Costs	Incl. O&P
1 Labor Foreman (outside)	$33.60	$268.80	$55.45	$443.60	$36.32	$58.21
2 Laborers	31.60	505.60	52.15	834.40		
1 Equip. Oper. (crane)	42.55	340.40	66.95	535.60		
2 Equip. Oper. (light)	39.05	624.80	61.45	983.20		
1 Equip. Oper. Oiler	36.80	294.40	57.90	463.20		
1 Crawler Crane, 40 Ton		1055.00		1160.50		
1 Lead, 60' high		69.60		76.56		
1 Hammer, diesel, 15K Ft.-Lbs.		578.60		636.46		
1 Backhoe Loader, 48 H.P.		293.80		323.18	35.66	39.23
56 L.H., Daily Totals		$4031.00		$5456.70	$71.98	$97.44

Crew B-61	Hr.	Daily	Hr.	Daily	Bare Costs	Incl. O&P
1 Labor Foreman (outside)	$33.60	$268.80	$55.45	$443.60	$33.49	$54.67
3 Laborers	31.60	758.40	52.15	1251.60		
1 Equip. Oper. (light)	39.05	312.40	61.45	491.60		
1 Cement Mixer, 2 C.Y.		187.80		206.58		
1 Air Compressor, 160 cfm		138.80		152.68	8.16	8.98
40 L.H., Daily Totals		$1666.20		$2546.06	$41.66	$63.65

Crew B-62	Hr.	Daily	Hr.	Daily	Bare Costs	Incl. O&P
2 Laborers	$31.60	$505.60	$52.15	$834.40	$34.08	$55.25
1 Equip. Oper. (light)	39.05	312.40	61.45	491.60		
1 Loader, Skid Steer, 30 H.P., gas		149.20		164.12	6.22	6.84
24 L.H., Daily Totals		$967.20		$1490.12	$40.30	$62.09

Crew B-63	Hr.	Daily	Hr.	Daily	Bare Costs	Incl. O&P
4 Laborers	$31.60	$1011.20	$52.15	$1668.80	$33.09	$54.01
1 Equip. Oper. (light)	39.05	312.40	61.45	491.60		
1 Loader, Skid Steer, 30 H.P., gas		149.20		164.12	3.73	4.10
40 L.H., Daily Totals		$1472.80		$2324.52	$36.82	$58.11

Crew B-64	Hr.	Daily	Hr.	Daily	Bare Costs	Incl. O&P
1 Laborer	$31.60	$252.80	$52.15	$417.20	$31.27	$51.45
1 Truck Driver (light)	30.95	247.60	50.75	406.00		
1 Power Mulcher (small)		136.60		150.26		
1 Flatbed Truck, Gas, 1.5 Ton		194.40		213.84	20.69	22.76
16 L.H., Daily Totals		$831.40		$1187.30	$51.96	$74.21

Crew B-65	Hr.	Daily	Hr.	Daily	Bare Costs	Incl. O&P
1 Laborer	$31.60	$252.80	$52.15	$417.20	$31.27	$51.45
1 Truck Driver (light)	30.95	247.60	50.75	406.00		
1 Power Mulcher (large)		271.80		298.98		
1 Flatbed Truck, Gas, 1.5 Ton		194.40		213.84	29.14	32.05
16 L.H., Daily Totals		$966.60		$1336.02	$60.41	$83.50

Crew B-66	Hr.	Daily	Hr.	Daily	Bare Costs	Incl. O&P
1 Equip. Oper. (light)	$39.05	$312.40	$61.45	$491.60	$39.05	$61.45
1 Loader-Backhoe		231.20		254.32	28.90	31.79
8 L.H., Daily Totals		$543.60		$745.92	$67.95	$93.24

Crew B-67	Hr.	Daily	Hr.	Daily	Bare Costs	Incl. O&P
1 Millwright	$41.60	$332.80	$65.25	$522.00	$40.33	$63.35
1 Equip. Oper. (light)	39.05	312.40	61.45	491.60		
1 Forklift, R/T, 4,000 Lb.		281.20		309.32	17.57	19.33
16 L.H., Daily Totals		$926.40		$1322.92	$57.90	$82.68

Crew B-68	Hr.	Daily	Hr.	Daily	Bare Costs	Incl. O&P
2 Millwrights	$41.60	$665.60	$65.25	$1044.00	$40.75	$63.98
1 Equip. Oper. (light)	39.05	312.40	61.45	491.60		
1 Forklift, R/T, 4,000 Lb.		281.20		309.32	11.72	12.89
24 L.H., Daily Totals		$1259.20		$1844.92	$52.47	$76.87

Crew B-69	Hr.	Daily	Hr.	Daily	Bare Costs	Incl. O&P
1 Labor Foreman (outside)	$33.60	$268.80	$55.45	$443.60	$34.63	$56.13
3 Laborers	31.60	758.40	52.15	1251.60		
1 Equip Oper. (crane)	42.55	340.40	66.95	535.60		
1 Equip Oper. Oiler	36.80	294.40	57.90	463.20		
1 Hyd. Crane, 80 Ton		1296.00		1425.60	27.00	29.70
48 L.H., Daily Totals		$2958.00		$4119.60	$61.63	$85.83

Crew B-69A	Hr.	Daily	Hr.	Daily	Bare Costs	Incl. O&P
1 Labor Foreman	$33.60	$268.80	$55.45	$443.60	$34.67	$56.14
3 Laborers	31.60	758.40	52.15	1251.60		
1 Equip. Oper. (medium)	41.35	330.80	65.05	520.40		
1 Concrete Finisher	38.30	306.40	59.90	479.20		
1 Curb/Gutter Paver, 2-Track		754.40		829.84	15.72	17.29
48 L.H., Daily Totals		$2418.80		$3524.64	$50.39	$73.43

Crew B-69B	Hr.	Daily	Hr.	Daily	Bare Costs	Incl. O&P
1 Labor Foreman	$33.60	$268.80	$55.45	$443.60	$34.67	$56.14
3 Laborers	31.60	758.40	52.15	1251.60		
1 Equip. Oper. (medium)	41.35	330.80	65.05	520.40		
1 Cement Finisher	38.30	306.40	59.90	479.20		
1 Curb/Gutter Paver, 4-Track		855.80		941.38	17.83	19.61
48 L.H., Daily Totals		$2520.20		$3636.18	$52.50	$75.75

Crew B-70	Hr.	Daily	Hr.	Daily	Bare Costs	Incl. O&P
1 Labor Foreman (outside)	$33.60	$268.80	$55.45	$443.60	$36.06	$58.15
3 Laborers	31.60	758.40	52.15	1251.60		
3 Equip. Oper. (med.)	41.35	992.40	65.05	1561.20		
1 Grader, 30,000 Lbs.		550.20		605.22		
1 Ripper, beam & 1 shank		79.60		87.56		
1 Road Sweeper, S.P., 8' wide		601.60		661.76		
1 F.E. Loader, W.M., 1.5 C.Y.		328.20		361.02	27.85	30.64
56 L.H., Daily Totals		$3579.20		$4971.96	$63.91	$88.78

Crew No.	Bare Costs		Incl. Subs O & P		Cost Per Labor-Hour	
Crew B-71	**Hr.**	**Daily**	**Hr.**	**Daily**	**Bare Costs**	**Incl. O&P**
1 Labor Foreman (outside)	$33.60	$268.80	$55.45	$443.60	$36.06	$58.15
3 Laborers	31.60	758.40	52.15	1251.60		
3 Equip. Oper. (med.)	41.35	992.40	65.05	1561.20		
1 Pvmt. Profiler, 750 H.P.		5503.00		6053.30		
1 Road Sweeper, S.P., 8' wide		601.60		661.76		
1 F.E. Loader, W.M., 1.5 C.Y.		328.20		361.02	114.87	126.36
56 L.H., Daily Totals		$8452.40		$10332.48	$150.94	$184.51

Crew B-72	Hr.	Daily	Hr.	Daily	Bare Costs	Incl. O&P
1 Labor Foreman (outside)	$33.60	$268.80	$55.45	$443.60	$36.73	$59.01
3 Laborers	31.60	758.40	52.15	1251.60		
4 Equip. Oper. (med.)	41.35	1323.20	65.05	2081.60		
1 Pvmt. Profiler, 750 H.P.		5503.00		6053.30		
1 Hammermill, 250 H.P.		1850.00		2035.00		
1 Windrow Loader		1197.00		1316.70		
1 Mix Paver 165 H.P.		1996.00		2195.60		
1 Roller, Pneum. Whl, 12 Ton		307.80		338.58	169.59	186.55
64 L.H., Daily Totals		$13204.20		$15715.98	$206.32	$245.56

Crew B-73	Hr.	Daily	Hr.	Daily	Bare Costs	Incl. O&P
1 Labor Foreman (outside)	$33.60	$268.80	$55.45	$443.60	$37.94	$60.63
2 Laborers	31.60	505.60	52.15	834.40		
5 Equip. Oper. (med.)	41.35	1654.00	65.05	2602.00		
1 Road Mixer, 310 H.P.		1896.00		2085.60		
1 Tandem Roller, 10 Ton		233.00		256.30		
1 Hammermill, 250 H.P.		1850.00		2035.00		
1 Grader, 30,000 Lbs.		550.20		605.22		
.5 F.E. Loader, W.M., 1.5 C.Y.		164.10		180.51		
.5 Truck Tractor, 220 H.P.		144.20		158.62		
.5 Water Tanker, 5000 Gal.		68.00		74.80	76.65	84.31
64 L.H., Daily Totals		$7333.90		$9276.05	$114.59	$144.94

Crew B-74	Hr.	Daily	Hr.	Daily	Bare Costs	Incl. O&P
1 Labor Foreman (outside)	$33.60	$268.80	$55.45	$443.60	$36.81	$59.06
1 Laborer	31.60	252.80	52.15	417.20		
4 Equip. Oper. (med.)	41.35	1323.20	65.05	2081.60		
2 Truck Drivers (heavy)	31.95	511.20	52.35	837.60		
1 Grader, 30,000 Lbs.		550.20		605.22		
1 Ripper, beam & 1 shank		79.60		87.56		
2 Stabilizers, 310 H.P.		2688.00		2956.80		
1 Flatbed Truck, Gas, 3 Ton		242.60		266.86		
1 Chem. Spreader, Towed		48.00		52.80		
1 Roller, Vibratory, 25 Ton		594.60		654.06		
1 Water Tanker, 5000 Gal.		136.00		149.60		
1 Truck Tractor, 220 H.P.		288.40		317.24	72.30	79.53
64 L.H., Daily Totals		$6983.40		$8870.14	$109.12	$138.60

Crew B-75	Hr.	Daily	Hr.	Daily	Bare Costs	Incl. O&P
1 Labor Foreman (outside)	$33.60	$268.80	$55.45	$443.60	$37.51	$60.02
1 Laborer	31.60	252.80	52.15	417.20		
4 Equip. Oper. (med.)	41.35	1323.20	65.05	2081.60		
1 Truck Driver (heavy)	31.95	255.60	52.35	418.80		
1 Grader, 30,000 Lbs.		550.20		605.22		
1 Ripper, beam & 1 shank		79.60		87.56		
2 Stabilizers, 310 H.P.		2688.00		2956.80		
1 Dist. Tanker, 3000 Gallon		304.20		334.62		
1 Truck Tractor, 6x4, 380 H.P.		478.80		526.68		
1 Roller, Vibratory, 25 Ton		594.60		654.06	83.85	92.23
56 L.H., Daily Totals		$6795.80		$8526.14	$121.35	$152.25

Crew No.	Bare Costs		Incl. Subs O & P		Cost Per Labor-Hour	
Crew B-76	**Hr.**	**Daily**	**Hr.**	**Daily**	**Bare Costs**	**Incl. O&P**
1 Dock Builder Foreman	$40.50	$324.00	$67.80	$542.40	$39.43	$64.65
5 Dock Builders	38.50	1540.00	64.45	2578.00		
2 Equip. Oper. (crane)	42.55	680.80	66.95	1071.20		
1 Equip. Oper. Oiler	36.80	294.40	57.90	463.20		
1 Crawler Crane, 50 Ton		1110.00		1221.00		
1 Barge, 400 Ton		746.40		821.04		
1 Hammer, diesel, 15K Ft.-Lbs.		578.60		636.46		
1 Lead, 60' high		69.60		76.56		
1 Air Compressor, 600 cfm		434.40		477.84		
2 -50' Air Hoses, 3"		32.40		35.64	41.27	45.40
72 L.H., Daily Totals		$5810.60		$7923.34	$80.70	$110.05

Crew B-76A	Hr.	Daily	Hr.	Daily	Bare Costs	Incl. O&P
1 Laborer Foreman	$33.60	$268.80	$55.45	$443.60	$33.87	$55.13
5 Laborers	31.60	1264.00	52.15	2086.00		
1 Equip. Oper. (crane)	42.55	340.40	66.95	535.60		
1 Equip. Oper. Oiler	36.80	294.40	57.90	463.20		
1 Crawler Crane, 50 Ton		1110.00		1221.00		
1 Barge, 400 Ton		746.40		821.04	29.01	31.91
64 L.H., Daily Totals		$4024.00		$5570.44	$62.88	$87.04

Crew B-77	Hr.	Daily	Hr.	Daily	Bare Costs	Incl. O&P
1 Labor Foreman	$33.60	$268.80	$55.45	$443.60	$31.87	$52.53
3 Laborers	31.60	758.40	52.15	1251.60		
1 Truck Driver (light)	30.95	247.60	50.75	406.00		
1 Crack Cleaner, 25 H.P.		54.80		60.28		
1 Crack Filler, Trailer Mtd.		194.20		213.62		
1 Flatbed Truck, Gas, 3 Ton		242.60		266.86	12.29	13.52
40 L.H., Daily Totals		$1766.40		$2641.96	$44.16	$66.05

Crew B-78	Hr.	Daily	Hr.	Daily	Bare Costs	Incl. O&P
1 Labor Foreman	$33.60	$268.80	$55.45	$443.60	$31.82	$52.47
4 Laborers	31.60	1011.20	52.15	1668.80		
1 Truck Driver (light)	30.95	247.60	50.75	406.00		
1 Paint Striper, S.P.		147.00		161.70		
1 Flatbed Truck, Gas, 3 Ton		242.60		266.86		
1 Pickup Truck, 3/4 Ton		116.80		128.48	10.55	11.61
48 L.H., Daily Totals		$2034.00		$3075.44	$42.38	$64.07

Crew B-78A	Hr.	Daily	Hr.	Daily	Bare Costs	Incl. O&P
1 Equip. Oper. (light)	$39.05	$312.40	$61.45	$491.60	$39.05	$61.45
1 Line Rem. (metal balls) 115 H.P.		828.60		911.46	103.58	113.93
8 L.H., Daily Totals		$1141.00		$1403.06	$142.63	$175.38

Crew B-78B	Hr.	Daily	Hr.	Daily	Bare Costs	Incl. O&P
2 Laborers	$31.60	$505.60	$52.15	$834.40	$32.43	$53.18
.25 Equip. Oper. (light)	39.05	78.10	61.45	122.90		
1 Pickup Truck, 3/4 Ton		116.80		128.48		
1 Line Rem., 11 H.P., walk behind		54.00		59.40		
.25 Road Sweeper, S.P., 8' wide		150.40		165.44	17.84	19.63
18 L.H., Daily Totals		$904.90		$1310.62	$50.27	$72.81

Crew B-79	Hr.	Daily	Hr.	Daily	Bare Costs	Incl. O&P
1 Labor Foreman	$33.60	$268.80	$55.45	$443.60	$31.87	$52.53
3 Laborers	31.60	758.40	52.15	1251.60		
1 Truck Driver (light)	30.95	247.60	50.75	406.00		
1 Thermo. Striper, T.M.		703.20		773.52		
1 Flatbed Truck, Gas, 3 Ton		242.60		266.86		
2 Pickup Trucks, 3/4 Ton		233.60		256.96	29.48	32.43
40 L.H., Daily Totals		$2454.20		$3398.54	$61.35	$84.96

Crew No.	Bare Costs		Incl. Subs O & P		Cost Per Labor-Hour	
Crew B-79A	**Hr.**	**Daily**	**Hr.**	**Daily**	**Bare Costs**	**Incl. O&P**
1.5 Equip. Oper. (light)	$39.05	$468.60	$61.45	$737.40	$39.05	$61.45
.5 Line Remov. (grinder) 115 H.P.		440.20		484.22		
1 Line Rem. (metal balls) 115 H.P.		828.60		911.46	105.73	116.31
12 L.H., Daily Totals		$1737.40		$2133.08	$144.78	$177.76
Crew B-80	**Hr.**	**Daily**	**Hr.**	**Daily**	**Bare Costs**	**Incl. O&P**
1 Labor Foreman	$33.60	$268.80	$55.45	$443.60	$33.80	$54.95
1 Laborer	31.60	252.80	52.15	417.20		
1 Truck Driver (light)	30.95	247.60	50.75	406.00		
1 Equip. Oper. (light)	39.05	312.40	61.45	491.60		
1 Flatbed Truck, Gas, 3 Ton		242.60		266.86		
1 Earth Auger, Truck-Mtd.		412.00		453.20	20.46	22.50
32 L.H., Daily Totals		$1736.20		$2478.46	$54.26	$77.45
Crew B-80A	**Hr.**	**Daily**	**Hr.**	**Daily**	**Bare Costs**	**Incl. O&P**
3 Laborers	$31.60	$758.40	$52.15	$1251.60	$31.60	$52.15
1 Flatbed Truck, Gas, 3 Ton		242.60		266.86	10.11	11.12
24 L.H., Daily Totals		$1001.00		$1518.46	$41.71	$63.27
Crew B-80B	**Hr.**	**Daily**	**Hr.**	**Daily**	**Bare Costs**	**Incl. O&P**
3 Laborers	$31.60	$758.40	$52.15	$1251.60	$33.46	$54.48
1 Equip. Oper. (light)	39.05	312.40	61.45	491.60		
1 Crane, Flatbed Mounted, 3 Ton		225.00		247.50	7.03	7.73
32 L.H., Daily Totals		$1295.80		$1990.70	$40.49	$62.21
Crew B-80C	**Hr.**	**Daily**	**Hr.**	**Daily**	**Bare Costs**	**Incl. O&P**
2 Laborers	$31.60	$505.60	$52.15	$834.40	$31.38	$51.68
1 Truck Driver (light)	30.95	247.60	50.75	406.00		
1 Flatbed Truck, Gas, 1.5 Ton		194.40		213.84		
1 Manual fence post auger, gas		7.40		8.14	8.41	9.25
24 L.H., Daily Totals		$955.00		$1462.38	$39.79	$60.93
Crew B-81	**Hr.**	**Daily**	**Hr.**	**Daily**	**Bare Costs**	**Incl. O&P**
1 Laborer	$31.60	$252.80	$52.15	$417.20	$34.97	$56.52
1 Equip. Oper. (med.)	41.35	330.80	65.05	520.40		
1 Truck Driver (heavy)	31.95	255.60	52.35	418.80		
1 Hydromulcher, T.M.		291.00		320.10		
1 Truck Tractor, 220 H.P.		288.40		317.24	24.14	26.56
24 L.H., Daily Totals		$1418.60		$1993.74	$59.11	$83.07
Crew B-82	**Hr.**	**Daily**	**Hr.**	**Daily**	**Bare Costs**	**Incl. O&P**
1 Laborer	$31.60	$252.80	$52.15	$417.20	$35.33	$56.80
1 Equip. Oper. (light)	39.05	312.40	61.45	491.60		
1 Horiz. Borer, 6 H.P.		77.60		85.36	4.85	5.34
16 L.H., Daily Totals		$642.80		$994.16	$40.17	$62.13
Crew B-82A	**Hr.**	**Daily**	**Hr.**	**Daily**	**Bare Costs**	**Incl. O&P**
1 Laborer	$31.60	$252.80	$52.15	$417.20	$35.33	$56.80
1 Equip. Oper. (light)	39.05	312.40	61.45	491.60		
1 Flatbed Truck, Gas, 3 Ton		242.60		266.86		
1 Flatbed Trailer, 25 Ton		102.00		112.20		
1 Horiz. Dir. Drill, 20k lb. thrust		628.20		691.02	60.80	66.88
16 L.H., Daily Totals		$1538.00		$1978.88	$96.13	$123.68

Crew No.	Bare Costs		Incl. Subs O & P		Cost Per Labor-Hour	
Crew B-82B	**Hr.**	**Daily**	**Hr.**	**Daily**	**Bare Costs**	**Incl. O&P**
2 Laborers	$31.60	$505.60	$52.15	$834.40	$34.08	$55.25
1 Equip. Oper. (light)	39.05	312.40	61.45	491.60		
1 Flatbed Truck, Gas, 3 Ton		242.60		266.86		
1 Flatbed Trailer, 25 Ton		102.00		112.20		
1 Horiz. Dir. Drill, 30k lb. thrust		896.60		986.26	51.72	56.89
24 L.H., Daily Totals		$2059.20		$2691.32	$85.80	$112.14
Crew B-82C	**Hr.**	**Daily**	**Hr.**	**Daily**	**Bare Costs**	**Incl. O&P**
2 Laborers	$31.60	$505.60	$52.15	$834.40	$34.08	$55.25
1 Equip. Oper. (light)	39.05	312.40	61.45	491.60		
1 Flatbed Truck, Gas, 3 Ton		242.60		266.86		
1 Flatbed Trailer, 25 Ton		102.00		112.20		
1 Horiz. Dir. Drill, 50k lb. thrust		1183.00		1301.30	63.65	70.02
24 L.H., Daily Totals		$2345.60		$3006.36	$97.73	$125.27
Crew B-82D	**Hr.**	**Daily**	**Hr.**	**Daily**	**Bare Costs**	**Incl. O&P**
1 Equip. Oper. (light)	$39.05	$312.40	$61.45	$491.60	$39.05	$61.45
1 Mud Trailer for HDD, 1500 gallon		283.20		311.52	35.40	38.94
8 L.H., Daily Totals		$595.60		$803.12	$74.45	$100.39
Crew B-83	**Hr.**	**Daily**	**Hr.**	**Daily**	**Bare Costs**	**Incl. O&P**
1 Tugboat Captain	$41.35	$330.80	$65.05	$520.40	$36.48	$58.60
1 Tugboat Hand	31.60	252.80	52.15	417.20		
1 Tugboat, 250 H.P.		663.40		729.74	41.46	45.61
16 L.H., Daily Totals		$1247.00		$1667.34	$77.94	$104.21
Crew B-84	**Hr.**	**Daily**	**Hr.**	**Daily**	**Bare Costs**	**Incl. O&P**
1 Equip. Oper. (med.)	$41.35	$330.80	$65.05	$520.40	$41.35	$65.05
1 Rotary Mower/Tractor		273.40		300.74	34.17	37.59
8 L.H., Daily Totals		$604.20		$821.14	$75.53	$102.64
Crew B-85	**Hr.**	**Daily**	**Hr.**	**Daily**	**Bare Costs**	**Incl. O&P**
3 Laborers	$31.60	$758.40	$52.15	$1251.60	$33.62	$54.77
1 Equip. Oper. (med.)	41.35	330.80	65.05	520.40		
1 Truck Driver (heavy)	31.95	255.60	52.35	418.80		
1 Aerial Lift Truck, 80'		640.40		704.44		
1 Brush Chipper, 12", 130 H.P.		225.60		248.16		
1 Pruning Saw, Rotary		6.50		7.15	21.81	23.99
40 L.H., Daily Totals		$2217.30		$3150.55	$55.43	$78.76
Crew B-86	**Hr.**	**Daily**	**Hr.**	**Daily**	**Bare Costs**	**Incl. O&P**
1 Equip. Oper. (med.)	$41.35	$330.80	$65.05	$520.40	$41.35	$65.05
1 Stump Chipper, S.P.		201.10		221.21	25.14	27.65
8 L.H. Daily Totals		$531.90		$741.61	$66.49	$92.70
Crew B-86A	**Hr.**	**Daily**	**Hr.**	**Daily**	**Bare Costs**	**Incl. O&P**
1 Equip. Oper. (medium)	$41.35	$330.80	$65.05	$520.40	$41.35	$65.05
1 Grader, 30,000 Lbs.		550.20		605.22	68.78	75.65
8 L.H., Daily Totals		$881.00		$1125.62	$110.13	$140.70
Crew B-86B	**Hr.**	**Daily**	**Hr.**	**Daily**	**Bare Costs**	**Incl. O&P**
1 Equip. Oper. (medium)	$41.35	$330.80	$65.05	$520.40	$41.35	$65.05
1 Dozer, 200 H.P.		1082.00		1190.20	135.25	148.78
8 L.H., Daily Totals		$1412.80		$1710.60	$176.60	$213.82

Crew B-87	Bare Costs Hr.	Bare Costs Daily	Incl. Subs O & P Hr.	Incl. Subs O & P Daily	Cost Per Labor-Hour Bare Costs	Cost Per Labor-Hour Incl. O&P
1 Laborer	$31.60	$252.80	$52.15	$417.20	$39.40	$62.47
4 Equip. Oper. (med.)	41.35	1323.20	65.05	2081.60		
2 Feller Bunchers, 100 H.P.		1122.80		1235.08		
1 Log Chipper, 22" Tree		588.40		647.24		
1 Dozer, 105 H.P.		600.60		660.66		
1 Chainsaw, Gas, 36" Long		37.80		41.58	58.74	64.61
40 L.H., Daily Totals		$3925.60		$5083.36	$98.14	$127.08

Crew B-88	Hr.	Daily	Hr.	Daily	Bare Costs	Incl. O&P
1 Laborer	$31.60	$252.80	$52.15	$417.20	$39.96	$63.21
6 Equip. Oper. (med.)	41.35	1984.80	65.05	3122.40		
2 Feller Bunchers, 100 H.P.		1122.80		1235.08		
1 Log Chipper, 22" Tree		588.40		647.24		
2 Log Skidders, 50 H.P.		1795.20		1974.72		
1 Dozer, 105 H.P.		600.60		660.66		
1 Chainsaw, Gas, 36" Long		37.80		41.58	74.01	81.42
56 L.H., Daily Totals		$6382.40		$8098.88	$113.97	$144.62

Crew B-89	Hr.	Daily	Hr.	Daily	Bare Costs	Incl. O&P
1 Equip. Oper. (light)	$39.05	$312.40	$61.45	$491.60	$35.00	$56.10
1 Truck Driver (light)	30.95	247.60	50.75	406.00		
1 Flatbed Truck, Gas, 3 Ton		242.60		266.86		
1 Concrete Saw		141.00		155.10		
1 Water Tank, 65 Gal.		18.45		20.30	25.13	27.64
16 L.H., Daily Totals		$962.05		$1339.86	$60.13	$83.74

Crew B-89A	Hr.	Daily	Hr.	Daily	Bare Costs	Incl. O&P
1 Skilled Worker	$40.85	$326.80	$66.50	$532.00	$36.23	$59.33
1 Laborer	31.60	252.80	52.15	417.20		
1 Core Drill (large)		110.20		121.22	6.89	7.58
16 L.H., Daily Totals		$689.80		$1070.42	$43.11	$66.90

Crew B-89B	Hr.	Daily	Hr.	Daily	Bare Costs	Incl. O&P
1 Equip. Oper. (light)	$39.05	$312.40	$61.45	$491.60	$35.00	$56.10
1 Truck Driver, Light	30.95	247.60	50.75	406.00		
1 Wall Saw, Hydraulic, 10 H.P.		96.00		105.60		
1 Generator, Diesel, 100 kW		327.20		359.92		
1 Water Tank, 65 Gal.		18.45		20.30		
1 Flatbed Truck, Gas, 3 Ton		242.60		266.86	42.77	47.04
16 L.H., Daily Totals		$1244.25		$1650.28	$77.77	$103.14

Crew B-90	Hr.	Daily	Hr.	Daily	Bare Costs	Incl. O&P
1 Labor Foreman (outside)	$33.60	$268.80	$55.45	$443.60	$33.80	$54.94
3 Laborers	31.60	758.40	52.15	1251.60		
2 Equip. Oper. (light)	39.05	624.80	61.45	983.20		
2 Truck Drivers (heavy)	31.95	511.20	52.35	837.60		
1 Road Mixer, 310 H.P.		1896.00		2085.60		
1 Dist. Truck, 2000 Gal.		277.20		304.92	33.96	37.35
64 L.H., Daily Totals		$4336.40		$5906.52	$67.76	$92.29

Crew B-90A	Hr.	Daily	Hr.	Daily	Bare Costs	Incl. O&P
1 Labor Foreman	$33.60	$268.80	$55.45	$443.60	$37.46	$59.99
2 Laborers	31.60	505.60	52.15	834.40		
4 Equip. Oper. (medium)	41.35	1323.20	65.05	2081.60		
2 Graders, 30,000 Lbs.		1100.40		1210.44		
1 Tandem Roller, 10 Ton		233.00		256.30		
1 Roller, Pneum. Whl, 12 Ton		307.80		338.58	29.31	32.24
56 L.H., Daily Totals		$3738.80		$5164.92	$66.76	$92.23

Crew B-90B	Hr.	Daily	Hr.	Daily	Bare Costs	Incl. O&P
1 Labor Foreman	$33.60	$268.80	$55.45	$443.60	$36.81	$59.15
2 Laborers	31.60	505.60	52.15	834.40		
3 Equip. Oper. (medium)	41.35	992.40	65.05	1561.20		
1 Tandem Roller, 10 Ton		233.00		256.30		
1 Roller, Pneum. Whl, 12 Ton		307.80		338.58		
1 Road Mixer, 310 H.P.		1896.00		2085.60	50.77	55.84
48 L.H., Daily Totals		$4203.60		$5519.68	$87.58	$114.99

Crew B-91	Hr.	Daily	Hr.	Daily	Bare Costs	Incl. O&P
1 Labor Foreman (outside)	$33.60	$268.80	$55.45	$443.60	$36.77	$59.04
2 Laborers	31.60	505.60	52.15	834.40		
4 Equip. Oper. (med.)	41.35	1323.20	65.05	2081.60		
1 Truck Driver (heavy)	31.95	255.60	52.35	418.80		
1 Dist. Tanker, 3000 Gallon		304.20		334.62		
1 Truck Tractor, 6x4, 380 H.P.		478.80		526.68		
1 Aggreg. Spreader, S.P.		855.20		940.72		
1 Roller, Pneum. Whl, 12 Ton		307.80		338.58		
1 Tandem Roller, 10 Ton		233.00		256.30	34.05	37.45
64 L.H., Daily Totals		$4532.20		$6175.30	$70.82	$96.49

Crew B-92	Hr.	Daily	Hr.	Daily	Bare Costs	Incl. O&P
1 Labor Foreman (outside)	$33.60	$268.80	$55.45	$443.60	$32.10	$52.98
3 Laborers	31.60	758.40	52.15	1251.60		
1 Crack Cleaner, 25 H.P.		54.80		60.28		
1 Air Compressor, 60 cfm		122.60		134.86		
1 Tar Kettle, T.M.		83.15		91.47		
1 Flatbed Truck, Gas, 3 Ton		242.60		266.86	15.72	17.30
32 L.H., Daily Totals		$1530.35		$2248.67	$47.82	$70.27

Crew B-93	Hr.	Daily	Hr.	Daily	Bare Costs	Incl. O&P
1 Equip. Oper. (med.)	$41.35	$330.80	$65.05	$520.40	$41.35	$65.05
1 Feller Buncher, 100 H.P.		561.40		617.54	70.17	77.19
8 L.H., Daily Totals		$892.20		$1137.94	$111.53	$142.24

Crew B-94A	Hr.	Daily	Hr.	Daily	Bare Costs	Incl. O&P
1 Laborer	$31.60	$252.80	$52.15	$417.20	$31.60	$52.15
1 Diaphragm Water Pump, 2"		63.20		69.52		
1 -20' Suction Hose, 2"		1.95		2.15		
2 -50' Discharge Hoses, 2"		2.20		2.42	8.42	9.26
8 L.H., Daily Totals		$320.15		$491.29	$40.02	$61.41

Crew B-94B	Hr.	Daily	Hr.	Daily	Bare Costs	Incl. O&P
1 Laborer	$31.60	$252.80	$52.15	$417.20	$31.60	$52.15
1 Diaphragm Water Pump, 4"		86.20		94.82		
1 -20' Suction Hose, 4"		3.45		3.79		
2 -50' Discharge Hoses, 4"		4.70		5.17	11.79	12.97
8 L.H., Daily Totals		$347.15		$520.99	$43.39	$65.12

Crew B-94C	Hr.	Daily	Hr.	Daily	Bare Costs	Incl. O&P
1 Laborer	$31.60	$252.80	$52.15	$417.20	$31.60	$52.15
1 Centrifugal Water Pump, 3"		69.40		76.34		
1 -20' Suction Hose, 3"		3.05		3.36		
2 -50' Discharge Hoses, 3"		3.50		3.85	9.49	10.44
8 L.H., Daily Totals		$328.75		$500.75	$41.09	$62.59

Crew B-94D	Hr.	Daily	Hr.	Daily	Bare Costs	Incl. O&P
1 Laborer	$31.60	$252.80	$52.15	$417.20	$31.60	$52.15
1 Centr. Water Pump, 6"		309.40		340.34		
1 -20' Suction Hose, 6"		11.90		13.09		
2 -50' Discharge Hoses, 6"		12.60		13.86	41.74	45.91
8 L.H., Daily Totals		$586.70		$784.49	$73.34	$98.06

Crew No.	Bare Costs		Incl. Subs O & P		Cost Per Labor-Hour	
Crew C-1	**Hr.**	**Daily**	**Hr.**	**Daily**	**Bare Costs**	**Incl. O&P**
3 Carpenters	$39.95	$958.80	$65.95	$1582.80	$37.86	$62.50
1 Laborer	31.60	252.80	52.15	417.20		
32 L.H., Daily Totals		$1211.60		$2000.00	$37.86	$62.50
Crew C-2	**Hr.**	**Daily**	**Hr.**	**Daily**	**Bare Costs**	**Incl. O&P**
1 Carpenter Foreman (out)	$41.95	$335.60	$69.25	$554.00	$38.89	$64.20
4 Carpenters	39.95	1278.40	65.95	2110.40		
1 Laborer	31.60	252.80	52.15	417.20		
48 L.H., Daily Totals		$1866.80		$3081.60	$38.89	$64.20
Crew C-2A	**Hr.**	**Daily**	**Hr.**	**Daily**	**Bare Costs**	**Incl. O&P**
1 Carpenter Foreman (out)	$41.95	$335.60	$69.25	$554.00	$38.62	$63.19
3 Carpenters	39.95	958.80	65.95	1582.80		
1 Cement Finisher	38.30	306.40	59.90	479.20		
1 Laborer	31.60	252.80	52.15	417.20		
48 L.H., Daily Totals		$1853.60		$3033.20	$38.62	$63.19
Crew C-3	**Hr.**	**Daily**	**Hr.**	**Daily**	**Bare Costs**	**Incl. O&P**
1 Rodman Foreman	$46.55	$372.40	$79.35	$634.80	$40.88	$68.61
4 Rodmen (reinf.)	44.55	1425.60	75.95	2430.40		
1 Equip. Oper. (light)	39.05	312.40	61.45	491.60		
2 Laborers	31.60	505.60	52.15	834.40		
3 Stressing Equipment		27.60		30.36		
.5 Grouting Equipment		75.80		83.38	1.62	1.78
64 L.H., Daily Totals		$2719.40		$4504.94	$42.49	$70.39
Crew C-4	**Hr.**	**Daily**	**Hr.**	**Daily**	**Bare Costs**	**Incl. O&P**
1 Rodman Foreman	$46.55	$372.40	$79.35	$634.80	$45.05	$76.80
3 Rodmen (reinf.)	44.55	1069.20	75.95	1822.80		
3 Stressing Equipment		27.60		30.36	.86	.95
32 L.H., Daily Totals		$1469.20		$2487.96	$45.91	$77.75
Crew C-5	**Hr.**	**Daily**	**Hr.**	**Daily**	**Bare Costs**	**Incl. O&P**
1 Rodman Foreman	$46.55	$372.40	$79.35	$634.80	$43.44	$72.57
4 Rodmen (reinf.)	44.55	1425.60	75.95	2430.40		
1 Equip. Oper. (crane)	42.55	340.40	66.95	535.60		
1 Equip. Oper. Oiler	36.80	294.40	57.90	463.20		
1 Hyd. Crane, 25 Ton		789.40		868.34	14.10	15.51
56 L.H., Daily Totals		$3222.20		$4932.34	$57.54	$88.08
Crew C-6	**Hr.**	**Daily**	**Hr.**	**Daily**	**Bare Costs**	**Incl. O&P**
1 Labor Foreman (outside)	$33.60	$268.80	$55.45	$443.60	$33.05	$53.99
4 Laborers	31.60	1011.20	52.15	1668.80		
1 Cement Finisher	38.30	306.40	59.90	479.20		
2 Gas Engine Vibrators		52.00		57.20	1.08	1.19
48 L.H., Daily Totals		$1638.40		$2648.80	$34.13	$55.18
Crew C-7	**Hr.**	**Daily**	**Hr.**	**Daily**	**Bare Costs**	**Incl. O&P**
1 Labor Foreman (outside)	$33.60	$268.80	$55.45	$443.60	$34.23	$55.45
5 Laborers	31.60	1264.00	52.15	2086.00		
1 Cement Finisher	38.30	306.40	59.90	479.20		
1 Equip. Oper. (med.)	41.35	330.80	65.05	520.40		
1 Equip. Oper. (oiler)	36.80	294.40	57.90	463.20		
2 Gas Engine Vibrators		52.00		57.20		
1 Concrete Bucket, 1 C.Y.		20.20		22.22		
1 Hyd. Crane, 55 Ton		1125.00		1237.50	16.63	18.29
72 L.H., Daily Totals		$3661.60		$5309.32	$50.86	$73.74

Crew No.	Bare Costs		Incl. Subs O & P		Cost Per Labor-Hour	
Crew C-8	**Hr.**	**Daily**	**Hr.**	**Daily**	**Bare Costs**	**Incl. O&P**
1 Labor Foreman (outside)	$33.60	$268.80	$55.45	$443.60	$35.19	$56.68
3 Laborers	31.60	758.40	52.15	1251.60		
2 Cement Finishers	38.30	612.80	59.90	958.40		
1 Equip. Oper. (med.)	41.35	330.80	65.05	520.40		
1 Concrete Pump (small)		737.80		811.58	13.18	14.49
56 L.H., Daily Totals		$2708.60		$3985.58	$48.37	$71.17
Crew C-8A	**Hr.**	**Daily**	**Hr.**	**Daily**	**Bare Costs**	**Incl. O&P**
1 Labor Foreman (outside)	$33.60	$268.80	$55.45	$443.60	$34.17	$55.28
3 Laborers	31.60	758.40	52.15	1251.60		
2 Cement Finishers	38.30	612.80	59.90	958.40		
48 L.H., Daily Totals		$1640.00		$2653.60	$34.17	$55.28
Crew C-8B	**Hr.**	**Daily**	**Hr.**	**Daily**	**Bare Costs**	**Incl. O&P**
1 Labor Foreman (outside)	$33.60	$268.80	$55.45	$443.60	$33.95	$55.39
3 Laborers	31.60	758.40	52.15	1251.60		
1 Equip. Oper. (med.)	41.35	330.80	65.05	520.40		
1 Vibrating Power Screed		70.35		77.39		
1 Roller, Vibratory, 25 Ton		594.60		654.06		
1 Dozer, 200 H.P.		1082.00		1190.20	43.67	48.04
40 L.H., Daily Totals		$3104.95		$4137.24	$77.62	$103.43
Crew C-8C	**Hr.**	**Daily**	**Hr.**	**Daily**	**Bare Costs**	**Incl. O&P**
1 Labor Foreman (outside)	$33.60	$268.80	$55.45	$443.60	$34.67	$56.14
3 Laborers	31.60	758.40	52.15	1251.60		
1 Cement Finisher	38.30	306.40	59.90	479.20		
1 Equip. Oper. (med.)	41.35	330.80	65.05	520.40		
1 Shotcrete Rig, 12 C.Y./hr		237.80		261.58	4.95	5.45
48 L.H., Daily Totals		$1902.20		$2956.38	$39.63	$61.59
Crew C-8D	**Hr.**	**Daily**	**Hr.**	**Daily**	**Bare Costs**	**Incl. O&P**
1 Labor Foreman (outside)	$33.60	$268.80	$55.45	$443.60	$35.64	$57.24
1 Laborer	31.60	252.80	52.15	417.20		
1 Cement Finisher	38.30	306.40	59.90	479.20		
1 Equipment Oper. (light)	39.05	312.40	61.45	491.60		
1 Air Compressor, 250 cfm		162.40		178.64		
2 -50' Air Hoses, 1"		8.20		9.02	5.33	5.86
32 L.H., Daily Totals		$1311.00		$2019.26	$40.97	$63.10
Crew C-8E	**Hr.**	**Daily**	**Hr.**	**Daily**	**Bare Costs**	**Incl. O&P**
1 Labor Foreman (outside)	$33.60	$268.80	$55.45	$443.60	$35.64	$57.24
1 Laborer	31.60	252.80	52.15	417.20		
1 Cement Finisher	38.30	306.40	59.90	479.20		
1 Equipment Oper. (light)	39.05	312.40	61.45	491.60		
1 Air Compressor, 250 cfm		162.40		178.64		
2 -50' Air Hoses, 1"		8.20		9.02		
1 Concrete Pump (small)		737.80		811.58	28.39	31.23
32 L.H., Daily Totals		$2048.80		$2830.84	$64.03	$88.46
Crew C-10	**Hr.**	**Daily**	**Hr.**	**Daily**	**Bare Costs**	**Incl. O&P**
1 Laborer	$31.60	$252.80	$52.15	$417.20	$36.07	$57.32
2 Cement Finishers	38.30	612.80	59.90	958.40		
24 L.H., Daily Totals		$865.60		$1375.60	$36.07	$57.32
Crew C-10B	**Hr.**	**Daily**	**Hr.**	**Daily**	**Bare Costs**	**Incl. O&P**
3 Laborers	$31.60	$758.40	$52.15	$1251.60	$34.28	$55.25
2 Cement Finishers	38.30	612.80	59.90	958.40		
1 Concrete Mixer, 10 C.F.		155.20		170.72		
2 Trowels, 48" Walk-Behind		84.80		93.28	6.00	6.60
40 L.H., Daily Totals		$1611.20		$2474.00	$40.28	$61.85

Crew C-10C	Bare Costs Hr.	Bare Costs Daily	Incl. Subs O & P Hr.	Incl. Subs O & P Daily	Cost Per Labor-Hour Bare Costs	Cost Per Labor-Hour Incl. O&P
1 Laborer	$31.60	$252.80	$52.15	$417.20	$36.07	$57.32
2 Cement Finishers	38.30	612.80	59.90	958.40		
1 Trowel, 48" Walk-Behind		42.40		46.64	1.77	1.94
24 L.H., Daily Totals		$908.00		$1422.24	$37.83	$59.26

Crew C-10D	Hr.	Daily	Hr.	Daily	Bare Costs	Incl. O&P
1 Laborer	$31.60	$252.80	$52.15	$417.20	$36.07	$57.32
2 Cement Finishers	38.30	612.80	59.90	958.40		
1 Vibrating Power Screed		70.35		77.39		
1 Trowel, 48" Walk-Behind		42.40		46.64	4.70	5.17
24 L.H., Daily Totals		$978.35		$1499.63	$40.76	$62.48

Crew C-10E	Hr.	Daily	Hr.	Daily	Bare Costs	Incl. O&P
1 Laborer	$31.60	$252.80	$52.15	$417.20	$36.07	$57.32
2 Cement Finishers	38.30	612.80	59.90	958.40		
1 Vibrating Power Screed		70.35		77.39		
1 Cement Trowel, 96" Ride-On		166.60		183.26	9.87	10.86
24 L.H., Daily Totals		$1102.55		$1636.24	$45.94	$68.18

Crew C-11	Hr.	Daily	Hr.	Daily	Bare Costs	Incl. O&P
1 Struc. Steel Foreman	$46.70	$373.60	$86.50	$692.00	$43.81	$78.68
6 Struc. Steel Workers	44.70	2145.60	82.80	3974.40		
1 Equip. Oper. (crane)	42.55	340.40	66.95	535.60		
1 Equip. Oper. Oiler	36.80	294.40	57.90	463.20		
1 Lattice Boom Crane, 150 Ton		1890.00		2079.00	26.25	28.88
72 L.H., Daily Totals		$5044.00		$7744.20	$70.06	$107.56

Crew C-12	Hr.	Daily	Hr.	Daily	Bare Costs	Incl. O&P
1 Carpenter Foreman (out)	$41.95	$335.60	$69.25	$554.00	$39.33	$64.37
3 Carpenters	39.95	958.80	65.95	1582.80		
1 Laborer	31.60	252.80	52.15	417.20		
1 Equip. Oper. (crane)	42.55	340.40	66.95	535.60		
1 Hyd. Crane, 12 Ton		768.80		845.68	16.02	17.62
48 L.H., Daily Totals		$2656.40		$3935.28	$55.34	$81.98

Crew C-13	Hr.	Daily	Hr.	Daily	Bare Costs	Incl. O&P
1 Struc. Steel Worker	$44.70	$357.60	$82.80	$662.40	$43.12	$77.18
1 Welder	44.70	357.60	82.80	662.40		
1 Carpenter	39.95	319.60	65.95	527.60		
1 Welder, gas engine, 300 amp		134.20		147.62	5.59	6.15
24 L.H., Daily Totals		$1169.00		$2000.02	$48.71	$83.33

Crew C-14	Hr.	Daily	Hr.	Daily	Bare Costs	Incl. O&P
1 Carpenter Foreman (out)	$41.95	$335.60	$69.25	$554.00	$39.01	$64.22
5 Carpenters	39.95	1598.00	65.95	2638.00		
4 Laborers	31.60	1011.20	52.15	1668.80		
4 Rodmen (reinf.)	44.55	1425.60	75.95	2430.40		
2 Cement Finishers	38.30	612.80	59.90	958.40		
1 Equip. Oper. (crane)	42.55	340.40	66.95	535.60		
1 Equip. Oper. Oiler	36.80	294.40	57.90	463.20		
1 Hyd. Crane, 80 Ton		1296.00		1425.60	9.00	9.90
144 L.H., Daily Totals		$6914.00		$10674.00	$48.01	$74.13

Crew C-14A	Bare Costs Hr.	Bare Costs Daily	Incl. Subs O & P Hr.	Incl. Subs O & P Daily	Cost Per Labor-Hour Bare Costs	Cost Per Labor-Hour Incl. O&P
1 Carpenter Foreman (out)	$41.95	$335.60	$69.25	$554.00	$40.09	$66.30
16 Carpenters	39.95	5113.60	65.95	8441.60		
4 Rodmen (reinf.)	44.55	1425.60	75.95	2430.40		
2 Laborers	31.60	505.60	52.15	834.40		
1 Cement Finisher	38.30	306.40	59.90	479.20		
1 Equip. Oper. (med.)	41.35	330.80	65.05	520.40		
1 Gas Engine Vibrator		26.00		28.60		
1 Concrete Pump (small)		737.80		811.58	3.82	4.20
200 L.H., Daily Totals		$8781.40		$14100.18	$43.91	$70.50

Crew C-14B	Hr.	Daily	Hr.	Daily	Bare Costs	Incl. O&P
1 Carpenter Foreman (out)	$41.95	$335.60	$69.25	$554.00	$40.02	$66.05
16 Carpenters	39.95	5113.60	65.95	8441.60		
4 Rodmen (reinf.)	44.55	1425.60	75.95	2430.40		
2 Laborers	31.60	505.60	52.15	834.40		
2 Cement Finishers	38.30	612.80	59.90	958.40		
1 Equip. Oper. (med.)	41.35	330.80	65.05	520.40		
1 Gas Engine Vibrator		26.00		28.60		
1 Concrete Pump (small)		737.80		811.58	3.67	4.04
208 L.H., Daily Totals		$9087.80		$14579.38	$43.69	$70.09

Crew C-14C	Hr.	Daily	Hr.	Daily	Bare Costs	Incl. O&P
1 Carpenter Foreman (out)	$41.95	$335.60	$69.25	$554.00	$38.25	$63.24
6 Carpenters	39.95	1917.60	65.95	3165.60		
2 Rodmen (reinf.)	44.55	712.80	75.95	1215.20		
4 Laborers	31.60	1011.20	52.15	1668.80		
1 Cement Finisher	38.30	306.40	59.90	479.20		
1 Gas Engine Vibrator		26.00		28.60	.23	.26
112 L.H., Daily Totals		$4309.60		$7111.40	$38.48	$63.49

Crew C-14D	Hr.	Daily	Hr.	Daily	Bare Costs	Incl. O&P
1 Carpenter Foreman (out)	$41.95	$335.60	$69.25	$554.00	$39.72	$65.50
18 Carpenters	39.95	5752.80	65.95	9496.80		
2 Rodmen (reinf.)	44.55	712.80	75.95	1215.20		
2 Laborers	31.60	505.60	52.15	834.40		
1 Cement Finisher	38.30	306.40	59.90	479.20		
1 Equip. Oper. (med.)	41.35	330.80	65.05	520.40		
1 Gas Engine Vibrator		26.00		28.60		
1 Concrete Pump (small)		737.80		811.58	3.82	4.20
200 L.H., Daily Totals		$8707.80		$13940.18	$43.54	$69.70

Crew C-14E	Hr.	Daily	Hr.	Daily	Bare Costs	Incl. O&P
1 Carpenter Foreman (out)	$41.95	$335.60	$69.25	$554.00	$39.38	$65.57
2 Carpenters	39.95	639.20	65.95	1055.20		
4 Rodmen (reinf.)	44.55	1425.60	75.95	2430.40		
3 Laborers	31.60	758.40	52.15	1251.60		
1 Cement Finisher	38.30	306.40	59.90	479.20		
1 Gas Engine Vibrator		26.00		28.60	.30	.33
88 L.H., Daily Totals		$3491.20		$5799.00	$39.67	$65.90

Crew C-14F	Hr.	Daily	Hr.	Daily	Bare Costs	Incl. O&P
1 Laborer Foreman (out)	$33.60	$268.80	$55.45	$443.60	$36.29	$57.68
2 Laborers	31.60	505.60	52.15	834.40		
6 Cement Finishers	38.30	1838.40	59.90	2875.20		
1 Gas Engine Vibrator		26.00		28.60	.36	.40
72 L.H., Daily Totals		$2638.80		$4181.80	$36.65	$58.08

Crew No.	Bare Costs		Incl. Subs O & P		Cost Per Labor-Hour	
Crew C-14G	**Hr.**	**Daily**	**Hr.**	**Daily**	**Bare Costs**	**Incl. O&P**
1 Laborer Foreman (out)	$33.60	$268.80	$55.45	$443.60	$35.71	$57.05
2 Laborers	31.60	505.60	52.15	834.40		
4 Cement Finishers	38.30	1225.60	59.90	1916.80		
1 Gas Engine Vibrator		26.00		28.60	.46	.51
56 L.H., Daily Totals		$2026.00		$3223.40	$36.18	$57.56
Crew C-14H	**Hr.**	**Daily**	**Hr.**	**Daily**	**Bare Costs**	**Incl. O&P**
1 Carpenter Foreman (out)	$41.95	$335.60	$69.25	$554.00	$39.38	$64.86
2 Carpenters	39.95	639.20	65.95	1055.20		
1 Rodman (reinf.)	44.55	356.40	75.95	607.60		
1 Laborer	31.60	252.80	52.15	417.20		
1 Cement Finisher	38.30	306.40	59.90	479.20		
1 Gas Engine Vibrator		26.00		28.60	.54	.60
48 L.H., Daily Totals		$1916.40		$3141.80	$39.92	$65.45
Crew C-14L	**Hr.**	**Daily**	**Hr.**	**Daily**	**Bare Costs**	**Incl. O&P**
1 Carpenter Foreman (out)	$41.95	$335.60	$69.25	$554.00	$37.20	$61.12
6 Carpenters	39.95	1917.60	65.95	3165.60		
4 Laborers	31.60	1011.20	52.15	1668.80		
1 Cement Finisher	38.30	306.40	59.90	479.20		
1 Gas Engine Vibrator		26.00		28.60	.27	.30
96 L.H., Daily Totals		$3596.80		$5896.20	$37.47	$61.42
Crew C-15	**Hr.**	**Daily**	**Hr.**	**Daily**	**Bare Costs**	**Incl. O&P**
1 Carpenter Foreman (out)	$41.95	$335.60	$69.25	$554.00	$37.53	$61.48
2 Carpenters	39.95	639.20	65.95	1055.20		
3 Laborers	31.60	758.40	52.15	1251.60		
2 Cement Finishers	38.30	612.80	59.90	958.40		
1 Rodman (reinf.)	44.55	356.40	75.95	607.60		
72 L.H., Daily Totals		$2702.40		$4426.80	$37.53	$61.48
Crew C-16	**Hr.**	**Daily**	**Hr.**	**Daily**	**Bare Costs**	**Incl. O&P**
1 Labor Foreman (outside)	$33.60	$268.80	$55.45	$443.60	$37.27	$60.96
3 Laborers	31.60	758.40	52.15	1251.60		
2 Cement Finishers	38.30	612.80	59.90	958.40		
1 Equip. Oper. (med.)	41.35	330.80	65.05	520.40		
2 Rodmen (reinf.)	44.55	712.80	75.95	1215.20		
1 Concrete Pump (small)		737.80		811.58	10.25	11.27
72 L.H., Daily Totals		$3421.40		$5200.78	$47.52	$72.23
Crew C-17	**Hr.**	**Daily**	**Hr.**	**Daily**	**Bare Costs**	**Incl. O&P**
2 Skilled Worker Foremen	$42.85	$685.60	$69.75	$1116.00	$41.25	$67.15
8 Skilled Workers	40.85	2614.40	66.50	4256.00		
80 L.H., Daily Totals		$3300.00		$5372.00	$41.25	$67.15
Crew C-17A	**Hr.**	**Daily**	**Hr.**	**Daily**	**Bare Costs**	**Incl. O&P**
2 Skilled Worker Foremen	$42.85	$685.60	$69.75	$1116.00	$41.27	$67.15
8 Skilled Workers	40.85	2614.40	66.50	4256.00		
.125 Equip. Oper. (crane)	42.55	42.55	66.95	66.95		
.125 Hyd. Crane, 80 Ton		162.00		178.20	2.00	2.20
81 L.H., Daily Totals		$3504.55		$5617.15	$43.27	$69.35
Crew C-17B	**Hr.**	**Daily**	**Hr.**	**Daily**	**Bare Costs**	**Incl. O&P**
2 Skilled Worker Foremen	$42.85	$685.60	$69.75	$1116.00	$41.28	$67.15
8 Skilled Workers	40.85	2614.40	66.50	4256.00		
.25 Equip. Oper. (crane)	42.55	85.10	66.95	133.90		
.25 Hyd. Crane, 80 Ton		324.00		356.40		
.25 Trowel, 48" Walk-Behind		10.60		11.66	4.08	4.49
82 L.H., Daily Totals		$3719.70		$5873.96	$45.36	$71.63

Crew No.	Bare Costs		Incl. Subs O & P		Cost Per Labor-Hour	
Crew C-17C	**Hr.**	**Daily**	**Hr.**	**Daily**	**Bare Costs**	**Incl. O&P**
2 Skilled Worker Foremen	$42.85	$685.60	$69.75	$1116.00	$41.30	$67.14
8 Skilled Workers	40.85	2614.40	66.50	4256.00		
.375 Equip. Oper. (crane)	42.55	127.65	66.95	200.85		
.375 Hyd. Crane, 80 Ton		486.00		534.60	5.86	6.44
83 L.H., Daily Totals		$3913.65		$6107.45	$47.15	$73.58
Crew C-17D	**Hr.**	**Daily**	**Hr.**	**Daily**	**Bare Costs**	**Incl. O&P**
2 Skilled Worker Foremen	$42.85	$685.60	$69.75	$1116.00	$41.31	$67.14
8 Skilled Workers	40.85	2614.40	66.50	4256.00		
.5 Equip. Oper. (crane)	42.55	170.20	66.95	267.80		
.5 Hyd. Crane, 80 Ton		648.00		712.80	7.71	8.49
84 L.H., Daily Totals		$4118.20		$6352.60	$49.03	$75.63
Crew C-17E	**Hr.**	**Daily**	**Hr.**	**Daily**	**Bare Costs**	**Incl. O&P**
2 Skilled Worker Foremen	$42.85	$685.60	$69.75	$1116.00	$41.25	$67.15
8 Skilled Workers	40.85	2614.40	66.50	4256.00		
1 Hyd. Jack with Rods		85.70		94.27	1.07	1.18
80 L.H., Daily Totals		$3385.70		$5466.27	$42.32	$68.33
Crew C-18	**Hr.**	**Daily**	**Hr.**	**Daily**	**Bare Costs**	**Incl. O&P**
.125 Labor Foreman (out)	$33.60	$33.60	$55.45	$55.45	$31.82	$52.52
1 Laborer	31.60	252.80	52.15	417.20		
1 Concrete Cart, 10 C.F.		55.20		60.72	6.13	6.75
9 L.H., Daily Totals		$341.60		$533.37	$37.96	$59.26
Crew C-19	**Hr.**	**Daily**	**Hr.**	**Daily**	**Bare Costs**	**Incl. O&P**
.125 Labor Foreman (out)	$33.60	$33.60	$55.45	$55.45	$31.82	$52.52
1 Laborer	31.60	252.80	52.15	417.20		
1 Concrete Cart, 18 C.F.		86.40		95.04	9.60	10.56
9 L.H., Daily Totals		$372.80		$567.69	$41.42	$63.08
Crew C-20	**Hr.**	**Daily**	**Hr.**	**Daily**	**Bare Costs**	**Incl. O&P**
1 Labor Foreman (outside)	$33.60	$268.80	$55.45	$443.60	$33.91	$55.14
5 Laborers	31.60	1264.00	52.15	2086.00		
1 Cement Finisher	38.30	306.40	59.90	479.20		
1 Equip. Oper. (med.)	41.35	330.80	65.05	520.40		
2 Gas Engine Vibrators		52.00		57.20		
1 Concrete Pump (small)		737.80		811.58	12.34	13.57
64 L.H., Daily Totals		$2959.80		$4397.98	$46.25	$68.72
Crew C-21	**Hr.**	**Daily**	**Hr.**	**Daily**	**Bare Costs**	**Incl. O&P**
1 Labor Foreman (outside)	$33.60	$268.80	$55.45	$443.60	$33.91	$55.14
5 Laborers	31.60	1264.00	52.15	2086.00		
1 Cement Finisher	38.30	306.40	59.90	479.20		
1 Equip. Oper. (med.)	41.35	330.80	65.05	520.40		
2 Gas Engine Vibrators		52.00		57.20		
1 Concrete Conveyer		179.00		196.90	3.61	3.97
64 L.H., Daily Totals		$2401.00		$3783.30	$37.52	$59.11
Crew C-22	**Hr.**	**Daily**	**Hr.**	**Daily**	**Bare Costs**	**Incl. O&P**
1 Rodman Foreman	$46.55	$372.40	$79.35	$634.80	$44.70	$75.95
4 Rodmen (reinf.)	44.55	1425.60	75.95	2430.40		
.125 Equip. Oper. (crane)	42.55	42.55	66.95	66.95		
.125 Equip. Oper. Oiler	36.80	36.80	57.90	57.90		
.125 Hyd. Crane, 25 Ton		98.67		108.54	2.35	2.58
42 L.H., Daily Totals		$1976.03		$3298.59	$47.05	$78.54

Crew No.	Bare Costs		Incl. Subs O & P		Cost Per Labor-Hour	
Crew C-23	**Hr.**	**Daily**	**Hr.**	**Daily**	**Bare Costs**	**Incl. O&P**
2 Skilled Worker Foremen	$42.85	$685.60	$69.75	$1116.00	$41.02	$66.33
6 Skilled Workers	40.85	1960.80	66.50	3192.00		
1 Equip. Oper. (crane)	42.55	340.40	66.95	535.60		
1 Equip. Oper. Oiler	36.80	294.40	57.90	463.20		
1 Lattice Boom Crane, 90 Ton		1741.00		1915.10	21.76	23.94
80 L.H., Daily Totals		$5022.20		$7221.90	$62.78	$90.27

Crew C-24	**Hr.**	**Daily**	**Hr.**	**Daily**	**Bare Costs**	**Incl. O&P**
2 Skilled Worker Foremen	$42.85	$685.60	$69.75	$1116.00	$41.02	$66.33
6 Skilled Workers	40.85	1960.80	66.50	3192.00		
1 Equip. Oper. (crane)	42.55	340.40	66.95	535.60		
1 Equip. Oper. Oiler	36.80	294.40	57.90	463.20		
1 Lattice Boom Crane, 150 Ton		1890.00		2079.00	23.63	25.99
80 L.H., Daily Totals		$5171.20		$7385.80	$64.64	$92.32

Crew C-25	**Hr.**	**Daily**	**Hr.**	**Daily**	**Bare Costs**	**Incl. O&P**
2 Rodmen (reinf.)	$44.55	$712.80	$75.95	$1215.20	$34.95	$60.60
2 Rodmen Helpers	25.35	405.60	45.25	724.00		
32 L.H., Daily Totals		$1118.40		$1939.20	$34.95	$60.60

Crew C-27	**Hr.**	**Daily**	**Hr.**	**Daily**	**Bare Costs**	**Incl. O&P**
2 Cement Finishers	$38.30	$612.80	$59.90	$958.40	$38.30	$59.90
1 Concrete Saw		141.00		155.10	8.81	9.69
16 L.H., Daily Totals		$753.80		$1113.50	$47.11	$69.59

Crew C-28	**Hr.**	**Daily**	**Hr.**	**Daily**	**Bare Costs**	**Incl. O&P**
1 Cement Finisher	$38.30	$306.40	$59.90	$479.20	$38.30	$59.90
1 Portable Air Compressor, Gas		19.10		21.01	2.39	2.63
8 L.H., Daily Totals		$325.50		$500.21	$40.69	$62.53

Crew D-1	**Hr.**	**Daily**	**Hr.**	**Daily**	**Bare Costs**	**Incl. O&P**
1 Bricklayer	$40.50	$324.00	$65.50	$524.00	$36.33	$58.75
1 Bricklayer Helper	32.15	257.20	52.00	416.00		
16 L.H., Daily Totals		$581.20		$940.00	$36.33	$58.75

Crew D-2	**Hr.**	**Daily**	**Hr.**	**Daily**	**Bare Costs**	**Incl. O&P**
3 Bricklayers	$40.50	$972.00	$65.50	$1572.00	$37.41	$60.63
2 Bricklayer Helpers	32.15	514.40	52.00	832.00		
.5 Carpenter	39.95	159.80	65.95	263.80		
44 L.H., Daily Totals		$1646.20		$2667.80	$37.41	$60.63

Crew D-3	**Hr.**	**Daily**	**Hr.**	**Daily**	**Bare Costs**	**Incl. O&P**
3 Bricklayers	$40.50	$972.00	$65.50	$1572.00	$37.29	$60.38
2 Bricklayer Helpers	32.15	514.40	52.00	832.00		
.25 Carpenter	39.95	79.90	65.95	131.90		
42 L.H., Daily Totals		$1566.30		$2535.90	$37.29	$60.38

Crew D-4	**Hr.**	**Daily**	**Hr.**	**Daily**	**Bare Costs**	**Incl. O&P**
1 Bricklayer	$40.50	$324.00	$65.50	$524.00	$35.96	$57.74
2 Bricklayer Helpers	32.15	514.40	52.00	832.00		
1 Equip. Oper. (light)	39.05	312.40	61.45	491.60		
1 Grout Pump, 50 C.F./hr.		126.80		139.48	3.96	4.36
32 L.H., Daily Totals		$1277.60		$1987.08	$39.92	$62.10

Crew D-5	**Hr.**	**Daily**	**Hr.**	**Daily**	**Bare Costs**	**Incl. O&P**
1 Bricklayer	$40.50	$324.00	$65.50	$524.00	$40.50	$65.50
8 L.H., Daily Totals		$324.00		$524.00	$40.50	$65.50

Crew No.	Bare Costs		Incl. Subs O & P		Cost Per Labor-Hour	
Crew D-6	**Hr.**	**Daily**	**Hr.**	**Daily**	**Bare Costs**	**Incl. O&P**
3 Bricklayers	$40.50	$972.00	$65.50	$1572.00	$36.47	$59.04
3 Bricklayer Helpers	32.15	771.60	52.00	1248.00		
.25 Carpenter	39.95	79.90	65.95	131.90		
50 L.H., Daily Totals		$1823.50		$2951.90	$36.47	$59.04

Crew D-7	**Hr.**	**Daily**	**Hr.**	**Daily**	**Bare Costs**	**Incl. O&P**
1 Tile Layer	$38.10	$304.80	$59.60	$476.80	$34.08	$53.30
1 Tile Layer Helper	30.05	240.40	47.00	376.00		
16 L.H., Daily Totals		$545.20		$852.80	$34.08	$53.30

Crew D-8	**Hr.**	**Daily**	**Hr.**	**Daily**	**Bare Costs**	**Incl. O&P**
3 Bricklayers	$40.50	$972.00	$65.50	$1572.00	$37.16	$60.10
2 Bricklayer Helpers	32.15	514.40	52.00	832.00		
40 L.H., Daily Totals		$1486.40		$2404.00	$37.16	$60.10

Crew D-9	**Hr.**	**Daily**	**Hr.**	**Daily**	**Bare Costs**	**Incl. O&P**
3 Bricklayers	$40.50	$972.00	$65.50	$1572.00	$36.33	$58.75
3 Bricklayer Helpers	32.15	771.60	52.00	1248.00		
48 L.H., Daily Totals		$1743.60		$2820.00	$36.33	$58.75

Crew D-10	**Hr.**	**Daily**	**Hr.**	**Daily**	**Bare Costs**	**Incl. O&P**
1 Bricklayer Foreman	$42.50	$340.00	$68.70	$549.60	$39.42	$63.29
1 Bricklayer	40.50	324.00	65.50	524.00		
1 Bricklayer Helper	32.15	257.20	52.00	416.00		
1 Equip. Oper. (crane)	42.55	340.40	66.95	535.60		
1 S.P. Crane, 4x4, 12 Ton		601.80		661.98	18.81	20.69
32 L.H., Daily Totals		$1863.40		$2687.18	$58.23	$83.97

Crew D-11	**Hr.**	**Daily**	**Hr.**	**Daily**	**Bare Costs**	**Incl. O&P**
1 Bricklayer Foreman	$42.50	$340.00	$68.70	$549.60	$38.38	$62.07
1 Bricklayer	40.50	324.00	65.50	524.00		
1 Bricklayer Helper	32.15	257.20	52.00	416.00		
24 L.H., Daily Totals		$921.20		$1489.60	$38.38	$62.07

Crew D-12	**Hr.**	**Daily**	**Hr.**	**Daily**	**Bare Costs**	**Incl. O&P**
1 Bricklayer Foreman	$42.50	$340.00	$68.70	$549.60	$36.83	$59.55
1 Bricklayer	40.50	324.00	65.50	524.00		
2 Bricklayer Helpers	32.15	514.40	52.00	832.00		
32 L.H., Daily Totals		$1178.40		$1905.60	$36.83	$59.55

Crew D-13	**Hr.**	**Daily**	**Hr.**	**Daily**	**Bare Costs**	**Incl. O&P**
1 Bricklayer Foreman	$42.50	$340.00	$68.70	$549.60	$38.30	$61.85
1 Bricklayer	40.50	324.00	65.50	524.00		
2 Bricklayer Helpers	32.15	514.40	52.00	832.00		
1 Carpenter	39.95	319.60	65.95	527.60		
1 Equip. Oper. (crane)	42.55	340.40	66.95	535.60		
1 S.P. Crane, 4x4, 12 Ton		601.80		661.98	12.54	13.79
48 L.H., Daily Totals		$2440.20		$3630.78	$50.84	$75.64

Crew E-1	**Hr.**	**Daily**	**Hr.**	**Daily**	**Bare Costs**	**Incl. O&P**
1 Welder Foreman	$46.70	$373.60	$86.50	$692.00	$43.48	$76.92
1 Welder	44.70	357.60	82.80	662.40		
1 Equip. Oper. (light)	39.05	312.40	61.45	491.60		
1 Welder, gas engine, 300 amp		134.20		147.62	5.59	6.15
24 L.H., Daily Totals		$1177.80		$1993.62	$49.08	$83.07

Crew No.	Bare Costs		Incl. Subs O & P		Cost Per Labor-Hour	
Crew E-2	**Hr.**	**Daily**	**Hr.**	**Daily**	**Bare Costs**	**Incl. O&P**
1 Struc. Steel Foreman	$46.70	$373.60	$86.50	$692.00	$43.55	$77.51
4 Struc. Steel Workers	44.70	1430.40	82.80	2649.60		
1 Equip. Oper. (crane)	42.55	340.40	66.95	535.60		
1 Equip. Oper. Oiler	36.80	294.40	57.90	463.20		
1 Lattice Boom Crane, 90 Ton		1741.00		1915.10	31.09	34.20
56 L.H., Daily Totals		$4179.80		$6255.50	$74.64	$111.71
Crew E-3	**Hr.**	**Daily**	**Hr.**	**Daily**	**Bare Costs**	**Incl. O&P**
1 Struc. Steel Foreman	$46.70	$373.60	$86.50	$692.00	$45.37	$84.03
1 Struc. Steel Worker	44.70	357.60	82.80	662.40		
1 Welder	44.70	357.60	82.80	662.40		
1 Welder, gas engine, 300 amp		134.20		147.62	5.59	6.15
24 L.H., Daily Totals		$1223.00		$2164.42	$50.96	$90.18
Crew E-4	**Hr.**	**Daily**	**Hr.**	**Daily**	**Bare Costs**	**Incl. O&P**
1 Struc. Steel Foreman	$46.70	$373.60	$86.50	$692.00	$45.20	$83.72
3 Struc. Steel Workers	44.70	1072.80	82.80	1987.20		
1 Welder, gas engine, 300 amp		134.20		147.62	4.19	4.61
32 L.H., Daily Totals		$1580.60		$2826.82	$49.39	$88.34
Crew E-5	**Hr.**	**Daily**	**Hr.**	**Daily**	**Bare Costs**	**Incl. O&P**
2 Struc. Steel Foremen	$46.70	$747.20	$86.50	$1384.00	$44.09	$79.47
5 Struc. Steel Workers	44.70	1788.00	82.80	3312.00		
1 Equip. Oper. (crane)	42.55	340.40	66.95	535.60		
1 Welder	44.70	357.60	82.80	662.40		
1 Equip. Oper. Oiler	36.80	294.40	57.90	463.20		
1 Lattice Boom Crane, 90 Ton		1741.00		1915.10		
1 Welder, gas engine, 300 amp		134.20		147.62	23.44	25.78
80 L.H., Daily Totals		$5402.80		$8419.92	$67.53	$105.25
Crew E-6	**Hr.**	**Daily**	**Hr.**	**Daily**	**Bare Costs**	**Incl. O&P**
3 Struc. Steel Foremen	$46.70	$1120.80	$86.50	$2076.00	$44.09	$79.61
9 Struc. Steel Workers	44.70	3218.40	82.80	5961.60		
1 Equip. Oper. (crane)	42.55	340.40	66.95	535.60		
1 Welder	44.70	357.60	82.80	662.40		
1 Equip. Oper. Oiler	36.80	294.40	57.90	463.20		
1 Equip. Oper. (light)	39.05	312.40	61.45	491.60		
1 Lattice Boom Crane, 90 Ton		1741.00		1915.10		
1 Welder, gas engine, 300 amp		134.20		147.62		
1 Air Compressor, 160 cfm		138.80		152.68		
2 Impact Wrenches		32.40		35.64	15.99	17.59
128 L.H., Daily Totals		$7690.40		$12441.44	$60.08	$97.20
Crew E-7	**Hr.**	**Daily**	**Hr.**	**Daily**	**Bare Costs**	**Incl. O&P**
1 Struc. Steel Foreman	$46.70	$373.60	$86.50	$692.00	$44.09	$79.47
4 Struc. Steel Workers	44.70	1430.40	82.80	2649.60		
1 Equip. Oper. (crane)	42.55	340.40	66.95	535.60		
1 Equip. Oper. Oiler	36.80	294.40	57.90	463.20		
1 Welder Foreman	46.70	373.60	86.50	692.00		
2 Welders	44.70	715.20	82.80	1324.80		
1 Lattice Boom Crane, 90 Ton		1741.00		1915.10		
2 Welders, gas engine, 300 amp		268.40		295.24	25.12	27.63
80 L.H., Daily Totals		$5537.00		$8567.54	$69.21	$107.09

Crew No.	Bare Costs		Incl. Subs O & P		Cost Per Labor-Hour	
Crew E-8	**Hr.**	**Daily**	**Hr.**	**Daily**	**Bare Costs**	**Incl. O&P**
1 Struc. Steel Foreman	$46.70	$373.60	$86.50	$692.00	$43.80	$78.59
4 Struc. Steel Workers	44.70	1430.40	82.80	2649.60		
1 Welder Foreman	46.70	373.60	86.50	692.00		
4 Welders	44.70	1430.40	82.80	2649.60		
1 Equip. Oper. (crane)	42.55	340.40	66.95	535.60		
1 Equip. Oper. Oiler	36.80	294.40	57.90	463.20		
1 Equip. Oper. (light)	39.05	312.40	61.45	491.60		
1 Lattice Boom Crane, 90 Ton		1741.00		1915.10		
4 Welders, gas engine, 300 amp		536.80		590.48	21.90	24.09
104 L.H., Daily Totals		$6833.00		$10679.18	$65.70	$102.68
Crew E-9	**Hr.**	**Daily**	**Hr.**	**Daily**	**Bare Costs**	**Incl. O&P**
2 Struc. Steel Foremen	$46.70	$747.20	$86.50	$1384.00	$44.09	$79.61
5 Struc. Steel Workers	44.70	1788.00	82.80	3312.00		
1 Welder Foreman	46.70	373.60	86.50	692.00		
5 Welders	44.70	1788.00	82.80	3312.00		
1 Equip. Oper. (crane)	42.55	340.40	66.95	535.60		
1 Equip. Oper. Oiler	36.80	294.40	57.90	463.20		
1 Equip. Oper. (light)	39.05	312.40	61.45	491.60		
1 Lattice Boom Crane, 90 Ton		1741.00		1915.10		
5 Welders, gas engine, 300 amp		671.00		738.10	18.84	20.73
128 L.H., Daily Totals		$8056.00		$12843.60	$62.94	$100.34
Crew E-10	**Hr.**	**Daily**	**Hr.**	**Daily**	**Bare Costs**	**Incl. O&P**
1 Welder Foreman	$46.70	$373.60	$86.50	$692.00	$45.70	$84.65
1 Welder	44.70	357.60	82.80	662.40		
1 Welder, gas engine, 300 amp		134.20		147.62		
1 Flatbed Truck, Gas, 3 Ton		242.60		266.86	23.55	25.91
16 L.H., Daily Totals		$1108.00		$1768.88	$69.25	$110.56
Crew E-11	**Hr.**	**Daily**	**Hr.**	**Daily**	**Bare Costs**	**Incl. O&P**
2 Painters, Struc. Steel	$36.00	$576.00	$68.95	$1103.20	$35.66	$62.88
1 Building Laborer	31.60	252.80	52.15	417.20		
1 Equip. Oper. (light)	39.05	312.40	61.45	491.60		
1 Air Compressor, 250 cfm		162.40		178.64		
1 Sandblaster, portable, 3 C.F.		20.60		22.66		
1 Set Sand Blasting Accessories		12.75		14.03	6.12	6.73
32 L.H., Daily Totals		$1336.95		$2227.32	$41.78	$69.60
Crew E-12	**Hr.**	**Daily**	**Hr.**	**Daily**	**Bare Costs**	**Incl. O&P**
1 Welder Foreman	$46.70	$373.60	$86.50	$692.00	$42.88	$73.97
1 Equip. Oper. (light)	39.05	312.40	61.45	491.60		
1 Welder, gas engine, 300 amp		134.20		147.62	8.39	9.23
16 L.H. Daily Totals		$820.20		$1331.22	$51.26	$83.20
Crew E-13	**Hr.**	**Daily**	**Hr.**	**Daily**	**Bare Costs**	**Incl. O&P**
1 Welder Foreman	$46.70	$373.60	$86.50	$692.00	$44.15	$78.15
.5 Equip. Oper. (light)	39.05	156.20	61.45	245.80		
1 Welder, gas engine, 300 amp		134.20		147.62	11.18	12.30
12 L.H., Daily Totals		$664.00		$1085.42	$55.33	$90.45
Crew E-14	**Hr.**	**Daily**	**Hr.**	**Daily**	**Bare Costs**	**Incl. O&P**
1 Welder Foreman	$46.70	$373.60	$86.50	$692.00	$46.70	$86.50
1 Welder, gas engine, 300 amp		134.20		147.62	16.77	18.45
8 L.H., Daily Totals		$507.80		$839.62	$63.48	$104.95
Crew E-16	**Hr.**	**Daily**	**Hr.**	**Daily**	**Bare Costs**	**Incl. O&P**
1 Welder Foreman	$46.70	$373.60	$86.50	$692.00	$45.70	$84.65
1 Welder	44.70	357.60	82.80	662.40		
1 Welder, gas engine, 300 amp		134.20		147.62	8.39	9.23
16 L.H., Daily Totals		$865.40		$1502.02	$54.09	$93.88

Crew No.	Bare Costs		Incl. Subs O & P		Cost Per Labor-Hour	

Crew E-17	Hr.	Daily	Hr.	Daily	Bare Costs	Incl. O&P
1 Structural Steel Foreman	$46.70	$373.60	$86.50	$692.00	$45.70	$84.65
1 Structural Steel Worker	44.70	357.60	82.80	662.40		
16 L.H., Daily Totals		$731.20		$1354.40	$45.70	$84.65

Crew E-18	Hr.	Daily	Hr.	Daily	Bare Costs	Incl. O&P
1 Structural Steel Foreman	$46.70	$373.60	$86.50	$692.00	$44.43	$79.99
3 Structural Steel Workers	44.70	1072.80	82.80	1987.20		
1 Equipment Operator (med.)	41.35	330.80	65.05	520.40		
1 Lattice Boom Crane, 20 Ton		1080.00		1188.00	27.00	29.70
40 L.H., Daily Totals		$2857.20		$4387.60	$71.43	$109.69

Crew E-19	Hr.	Daily	Hr.	Daily	Bare Costs	Incl. O&P
1 Structural Steel Worker	$44.70	$357.60	$82.80	$662.40	$43.48	$76.92
1 Structural Steel Foreman	46.70	373.60	86.50	692.00		
1 Equip. Oper. (light)	39.05	312.40	61.45	491.60		
1 Lattice Boom Crane, 20 Ton		1080.00		1188.00	45.00	49.50
24 L.H., Daily Totals		$2123.60		$3034.00	$88.48	$126.42

Crew E-20	Hr.	Daily	Hr.	Daily	Bare Costs	Incl. O&P
1 Structural Steel Foreman	$46.70	$373.60	$86.50	$692.00	$43.69	$78.17
5 Structural Steel Workers	44.70	1788.00	82.80	3312.00		
1 Equip. Oper. (crane)	42.55	340.40	66.95	535.60		
1 Oiler	36.80	294.40	57.90	463.20		
1 Lattice Boom Crane, 40 Ton		1327.00		1459.70	20.73	22.81
64 L.H., Daily Totals		$4123.40		$6462.50	$64.43	$100.98

Crew E-22	Hr.	Daily	Hr.	Daily	Bare Costs	Incl. O&P
1 Skilled Worker Foreman	$42.85	$342.80	$69.75	$558.00	$41.52	$67.58
2 Skilled Workers	40.85	653.60	66.50	1064.00		
24 L.H., Daily Totals		$996.40		$1622.00	$41.52	$67.58

Crew E-24	Hr.	Daily	Hr.	Daily	Bare Costs	Incl. O&P
3 Structural Steel Workers	$44.70	$1072.80	$82.80	$1987.20	$43.86	$78.36
1 Equipment Operator (medium)	41.35	330.80	65.05	520.40		
1 Hyd. Crane, 25 Ton		789.40		868.34	24.67	27.14
32 L.H., Daily Totals		$2193.00		$3375.94	$68.53	$105.50

Crew E-25	Hr.	Daily	Hr.	Daily	Bare Costs	Incl. O&P
1 Welder Foreman	$46.70	$373.60	$86.50	$692.00	$46.70	$86.50
1 Cutting Torch		17.00		18.70		
1 Gases		75.60		83.16	11.57	12.73
8 L.H., Daily Totals		$466.20		$793.86	$58.27	$99.23

Crew F-3	Hr.	Daily	Hr.	Daily	Bare Costs	Incl. O&P
4 Carpenters	$39.95	$1278.40	$65.95	$2110.40	$40.47	$66.15
1 Equip. Oper. (crane)	42.55	340.40	66.95	535.60		
1 Hyd. Crane, 12 Ton		768.80		845.68	19.22	21.14
40 L.H., Daily Totals		$2387.60		$3491.68	$59.69	$87.29

Crew F-4	Hr.	Daily	Hr.	Daily	Bare Costs	Incl. O&P
4 Carpenters	$39.95	$1278.40	$65.95	$2110.40	$39.86	$64.78
1 Equip. Oper. (crane)	42.55	340.40	66.95	535.60		
1 Equip. Oper. Oiler	36.80	294.40	57.90	463.20		
1 Hyd. Crane, 55 Ton		1125.00		1237.50	23.44	25.78
48 L.H., Daily Totals		$3038.20		$4346.70	$63.30	$90.56

Crew F-5	Hr.	Daily	Hr.	Daily	Bare Costs	Incl. O&P
1 Carpenter Foreman	$41.95	$335.60	$69.25	$554.00	$40.45	$66.78
3 Carpenters	39.95	958.80	65.95	1582.80		
32 L.H., Daily Totals		$1294.40		$2136.80	$40.45	$66.78

Crew F-6	Hr.	Daily	Hr.	Daily	Bare Costs	Incl. O&P
2 Carpenters	$39.95	$639.20	$65.95	$1055.20	$37.13	$60.63
2 Building Laborers	31.60	505.60	52.15	834.40		
1 Equip. Oper. (crane)	42.55	340.40	66.95	535.60		
1 Hyd. Crane, 12 Ton		768.80		845.68	19.22	21.14
40 L.H., Daily Totals		$2254.00		$3270.88	$56.35	$81.77

Crew F-7	Hr.	Daily	Hr.	Daily	Bare Costs	Incl. O&P
2 Carpenters	$39.95	$639.20	$65.95	$1055.20	$35.77	$59.05
2 Building Laborers	31.60	505.60	52.15	834.40		
32 L.H., Daily Totals		$1144.80		$1889.60	$35.77	$59.05

Crew G-1	Hr.	Daily	Hr.	Daily	Bare Costs	Incl. O&P
1 Roofer Foreman	$36.25	$290.00	$64.70	$517.60	$31.99	$57.11
4 Roofers, Composition	34.25	1096.00	61.15	1956.80		
2 Roofer Helpers	25.35	405.60	45.25	724.00		
1 Application Equipment		173.20		190.52		
1 Tar Kettle/Pot		97.20		106.92		
1 Crew Truck		169.60		186.56	7.86	8.64
56 L.H., Daily Totals		$2231.60		$3682.40	$39.85	$65.76

Crew G-2	Hr.	Daily	Hr.	Daily	Bare Costs	Incl. O&P
1 Plasterer	$36.15	$289.20	$58.15	$465.20	$33.35	$54.08
1 Plasterer Helper	32.30	258.40	51.95	415.60		
1 Building Laborer	31.60	252.80	52.15	417.20		
1 Grout Pump, 50 C.F./hr.		126.80		139.48	5.28	5.81
24 L.H., Daily Totals		$927.20		$1437.48	$38.63	$59.90

Crew G-2A	Hr.	Daily	Hr.	Daily	Bare Costs	Incl. O&P
1 Roofer, composition	$34.25	$274.00	$61.15	$489.20	$30.40	$52.85
1 Roofer Helper	25.35	202.80	45.25	362.00		
1 Building Laborer	31.60	252.80	52.15	417.20		
1 Foam spray rig, trailer-mtd.		488.20		537.02		
1 Pickup Truck, 3/4 Ton		116.80		128.48	25.21	27.73
24 L.H., Daily Totals		$1334.60		$1933.90	$55.61	$80.58

Crew G-3	Hr.	Daily	Hr.	Daily	Bare Costs	Incl. O&P
2 Sheet Metal Workers	$47.20	$755.20	$75.00	$1200.00	$39.40	$63.58
2 Building Laborers	31.60	505.60	52.15	834.40		
32 L.H., Daily Totals		$1260.80		$2034.40	$39.40	$63.58

Crew G-4	Hr.	Daily	Hr.	Daily	Bare Costs	Incl. O&P
1 Labor Foreman (outside)	$33.60	$268.80	$55.45	$443.60	$32.27	$53.25
2 Building Laborers	31.60	505.60	52.15	834.40		
1 Flatbed Truck, Gas, 1.5 Ton		194.40		213.84		
1 Air Compressor, 160 cfm		138.80		152.68	13.88	15.27
24 L.H., Daily Totals		$1107.60		$1644.52	$46.15	$68.52

Crew G-5	Hr.	Daily	Hr.	Daily	Bare Costs	Incl. O&P
1 Roofer Foreman	$36.25	$290.00	$64.70	$517.60	$31.09	$55.50
2 Roofers, Composition	34.25	548.00	61.15	978.40		
2 Roofer Helpers	25.35	405.60	45.25	724.00		
1 Application Equipment		173.20		190.52	4.33	4.76
40 L.H., Daily Totals		$1416.80		$2410.52	$35.42	$60.26

Crew G-6A	Hr.	Daily	Hr.	Daily	Bare Costs	Incl. O&P
2 Roofers Composition	$34.25	$548.00	$61.15	$978.40	$34.25	$61.15
1 Small Compressor, Electric		11.35		12.48		
2 Pneumatic Nailers		43.20		47.52	3.41	3.75
16 L.H., Daily Totals		$602.55		$1038.41	$37.66	$64.90

Crew No.	Bare Costs		Incl. Subs O & P		Cost Per Labor-Hour	
Crew G-7	**Hr.**	**Daily**	**Hr.**	**Daily**	**Bare Costs**	**Incl. O&P**
1 Carpenter	$39.95	$319.60	$65.95	$527.60	$39.95	$65.95
1 Small Compressor, Electric		11.35		12.48		
1 Pneumatic Nailer		21.60		23.76	4.12	4.53
8 L.H., Daily Totals		$352.55		$563.85	$44.07	$70.48

Crew H-1	Hr.	Daily	Hr.	Daily	Bare Costs	Incl. O&P
2 Glaziers	$38.60	$617.60	$62.20	$995.20	$41.65	$72.50
2 Struc. Steel Workers	44.70	715.20	82.80	1324.80		
32 L.H., Daily Totals		$1332.80		$2320.00	$41.65	$72.50

Crew H-2	Hr.	Daily	Hr.	Daily	Bare Costs	Incl. O&P
2 Glaziers	$38.60	$617.60	$62.20	$995.20	$36.27	$58.85
1 Building Laborer	31.60	252.80	52.15	417.20		
24 L.H., Daily Totals		$870.40		$1412.40	$36.27	$58.85

Crew H-3	Hr.	Daily	Hr.	Daily	Bare Costs	Incl. O&P
1 Glazier	$38.60	$308.80	$62.20	$497.60	$34.45	$56.02
1 Helper	30.30	242.40	49.85	398.80		
16 L.H., Daily Totals		$551.20		$896.40	$34.45	$56.02

Crew J-1	Hr.	Daily	Hr.	Daily	Bare Costs	Incl. O&P
3 Plasterers	$36.15	$867.60	$58.15	$1395.60	$34.61	$55.67
2 Plasterer Helpers	32.30	516.80	51.95	831.20		
1 Mixing Machine, 6 C.F.		126.60		139.26	3.17	3.48
40 L.H., Daily Totals		$1511.00		$2366.06	$37.77	$59.15

Crew J-2	Hr.	Daily	Hr.	Daily	Bare Costs	Incl. O&P
3 Plasterers	$36.15	$867.60	$58.15	$1395.60	$34.77	$55.75
2 Plasterer Helpers	32.30	516.80	51.95	831.20		
1 Lather	35.55	284.40	56.15	449.20		
1 Mixing Machine, 6 C.F.		126.60		139.26	2.64	2.90
48 L.H., Daily Totals		$1795.40		$2815.26	$37.40	$58.65

Crew J-3	Hr.	Daily	Hr.	Daily	Bare Costs	Incl. O&P
1 Terrazzo Worker	$37.70	$301.60	$58.95	$471.60	$34.27	$53.60
1 Terrazzo Helper	30.85	246.80	48.25	386.00		
1 Terrazzo Grinder, Electric		81.60		89.76		
1 Terrazzo Mixer		171.60		188.76	15.82	17.41
16 L.H., Daily Totals		$801.60		$1136.12	$50.10	$71.01

Crew K-1	Hr.	Daily	Hr.	Daily	Bare Costs	Incl. O&P
1 Carpenter	$39.95	$319.60	$65.95	$527.60	$35.45	$58.35
1 Truck Driver (light)	30.95	247.60	50.75	406.00		
1 Flatbed Truck, Gas, 3 Ton		242.60		266.86	15.16	16.68
16 L.H., Daily Totals		$809.80		$1200.46	$50.61	$75.03

Crew K-2	Hr.	Daily	Hr.	Daily	Bare Costs	Incl. O&P
1 Struc. Steel Foreman	$46.70	$373.60	$86.50	$692.00	$40.78	$73.35
1 Struc. Steel Worker	44.70	357.60	82.80	662.40		
1 Truck Driver (light)	30.95	247.60	50.75	406.00		
1 Flatbed Truck, Gas, 3 Ton		242.60		266.86	10.11	11.12
24 L.H., Daily Totals		$1221.40		$2027.26	$50.89	$84.47

Crew L-1	Hr.	Daily	Hr.	Daily	Bare Costs	Incl. O&P
1 Electrician	$47.00	$376.00	$72.30	$578.40	$47.88	$73.95
1 Plumber	48.75	390.00	75.60	604.80		
16 L.H., Daily Totals		$766.00		$1183.20	$47.88	$73.95

Crew No.	Bare Costs		Incl. Subs O & P		Cost Per Labor-Hour	
Crew L-2	**Hr.**	**Daily**	**Hr.**	**Daily**	**Bare Costs**	**Incl. O&P**
1 Carpenter	$39.95	$319.60	$65.95	$527.60	$35.13	$57.90
1 Carpenter Helper	30.30	242.40	49.85	398.80		
16 L.H., Daily Totals		$562.00		$926.40	$35.13	$57.90

Crew L-3	Hr.	Daily	Hr.	Daily	Bare Costs	Incl. O&P
1 Carpenter	$39.95	$319.60	$65.95	$527.60	$43.52	$69.80
.5 Electrician	47.00	188.00	72.30	289.20		
.5 Sheet Metal Worker	47.20	188.80	75.00	300.00		
16 L.H., Daily Totals		$696.40		$1116.80	$43.52	$69.80

Crew L-3A	Hr.	Daily	Hr.	Daily	Bare Costs	Incl. O&P
1 Carpenter Foreman (outside)	$41.95	$335.60	$69.25	$554.00	$43.70	$71.17
.5 Sheet Metal Worker	47.20	188.80	75.00	300.00		
12 L.H., Daily Totals		$524.40		$854.00	$43.70	$71.17

Crew L-4	Hr.	Daily	Hr.	Daily	Bare Costs	Incl. O&P
2 Skilled Workers	$40.85	$653.60	$66.50	$1064.00	$37.33	$60.95
1 Helper	30.30	242.40	49.85	398.80		
24 L.H., Daily Totals		$896.00		$1462.80	$37.33	$60.95

Crew L-5	Hr.	Daily	Hr.	Daily	Bare Costs	Incl. O&P
1 Struc. Steel Foreman	$46.70	$373.60	$86.50	$692.00	$44.68	$81.06
5 Struc. Steel Workers	44.70	1788.00	82.80	3312.00		
1 Equip. Oper. (crane)	42.55	340.40	66.95	535.60		
1 Hyd. Crane, 25 Ton		789.40		868.34	14.10	15.51
56 L.H., Daily Totals		$3291.40		$5407.94	$58.77	$96.57

Crew L-5A	Hr.	Daily	Hr.	Daily	Bare Costs	Incl. O&P
1 Structural Steel Foreman	$46.70	$373.60	$86.50	$692.00	$44.66	$79.76
2 Structural Steel Workers	44.70	715.20	82.80	1324.80		
1 Equip. Oper. (crane)	42.55	340.40	66.95	535.60		
1 S.P. Crane, 4x4, 25 Ton		716.80		788.48	22.40	24.64
32 L.H., Daily Totals		$2146.00		$3340.88	$67.06	$104.40

Crew L-5B	Hr.	Daily	Hr.	Daily	Bare Costs	Incl. O&P
1 Structural Steel Foreman	$46.70	$373.60	$86.50	$692.00	$45.35	$74.96
2 Structural Steel Workers	44.70	715.20	82.80	1324.80		
2 Electricians	47.00	752.00	72.30	1156.80		
2 Steamfitters/Pipefitters	49.35	789.60	76.55	1224.80		
1 Equip. Oper. (crane)	42.55	340.40	66.95	535.60		
1 Equip. Oper. Oiler	36.80	294.40	57.90	463.20		
1 Hyd. Crane, 80 Ton		1296.00		1425.60	18.00	19.80
72 L.H., Daily Totals		$4561.20		$6822.80	$63.35	$94.76

Crew L-6	Hr.	Daily	Hr.	Daily	Bare Costs	Incl. O&P
1 Plumber	$48.75	$390.00	$75.60	$604.80	$48.17	$74.50
.5 Electrician	47.00	188.00	72.30	289.20		
12 L.H., Daily Totals		$578.00		$894.00	$48.17	$74.50

Crew L-7	Hr.	Daily	Hr.	Daily	Bare Costs	Incl. O&P
2 Carpenters	$39.95	$639.20	$65.95	$1055.20	$38.57	$62.91
1 Building Laborer	31.60	252.80	52.15	417.20		
.5 Electrician	47.00	188.00	72.30	289.20		
28 L.H., Daily Totals		$1080.00		$1761.60	$38.57	$62.91

Crew L-8	Hr.	Daily	Hr.	Daily	Bare Costs	Incl. O&P
2 Carpenters	$39.95	$639.20	$65.95	$1055.20	$41.71	$67.88
.5 Plumber	48.75	195.00	75.60	302.40		
20 L.H., Daily Totals		$834.20		$1357.60	$41.71	$67.88

Crew No.	Bare Costs		Incl. Subs O & P		Cost Per Labor-Hour	
Crew L-9	**Hr.**	**Daily**	**Hr.**	**Daily**	**Bare Costs**	**Incl. O&P**
1 Labor Foreman (inside)	$32.10	$256.80	$53.00	$424.00	$36.33	$61.39
2 Building Laborers	31.60	505.60	52.15	834.40		
1 Struc. Steel Worker	44.70	357.60	82.80	662.40		
.5 Electrician	47.00	188.00	72.30	289.20		
36 L.H., Daily Totals		$1308.00		$2210.00	$36.33	$61.39

Crew No.	Bare Costs		Incl. Subs O & P		Cost Per Labor-Hour	
Crew L-10	**Hr.**	**Daily**	**Hr.**	**Daily**	**Bare Costs**	**Incl. O&P**
1 Structural Steel Foreman	$46.70	$373.60	$86.50	$692.00	$44.65	$78.75
1 Structural Steel Worker	44.70	357.60	82.80	662.40		
1 Equip. Oper. (crane)	42.55	340.40	66.95	535.60		
1 Hyd. Crane, 12 Ton		768.80		845.68	32.03	35.24
24 L.H., Daily Totals		$1840.40		$2735.68	$76.68	$113.99

Crew No.	Bare Costs		Incl. Subs O & P		Cost Per Labor-Hour	
Crew L-11	**Hr.**	**Daily**	**Hr.**	**Daily**	**Bare Costs**	**Incl. O&P**
2 Wreckers	$31.60	$505.60	$57.30	$916.80	$36.20	$60.75
1 Equip. Oper. (crane)	42.55	340.40	66.95	535.60		
1 Equip. Oper. (light)	39.05	312.40	61.45	491.60		
1 Hyd. Excavator, 2.5 C.Y.		1619.00		1780.90		
1 Loader, Skid Steer, 78 H.P.		259.60		285.56	58.71	64.58
32 L.H., Daily Totals		$3037.00		$4010.46	$94.91	$125.33

Crew No.	Bare Costs		Incl. Subs O & P		Cost Per Labor-Hour	
Crew M-1	**Hr.**	**Daily**	**Hr.**	**Daily**	**Bare Costs**	**Incl. O&P**
3 Elevator Constructors	$56.60	$1358.40	$87.10	$2090.40	$53.77	$82.75
1 Elevator Apprentice	45.30	362.40	69.70	557.60		
5 Hand Tools		57.00		62.70	1.78	1.96
32 L.H., Daily Totals		$1777.80		$2710.70	$55.56	$84.71

Crew No.	Bare Costs		Incl. Subs O & P		Cost Per Labor-Hour	
Crew M-3	**Hr.**	**Daily**	**Hr.**	**Daily**	**Bare Costs**	**Incl. O&P**
1 Electrician Foreman (out)	$49.00	$392.00	$75.35	$602.80	$45.37	$70.72
1 Common Laborer	31.60	252.80	52.15	417.20		
.25 Equipment Operator, Medium	41.35	82.70	65.05	130.10		
1 Elevator Constructor	56.60	452.80	87.10	696.80		
1 Elevator Apprentice	45.30	362.40	69.70	557.60		
.25 S.P. Crane, 4x4, 20 Ton		175.15		192.66	5.15	5.67
34 L.H., Daily Totals		$1717.85		$2597.17	$50.52	$76.39

Crew No.	Bare Costs		Incl. Subs O & P		Cost Per Labor-Hour	
Crew M-4	**Hr.**	**Daily**	**Hr.**	**Daily**	**Bare Costs**	**Incl. O&P**
1 Electrician Foreman (out)	$49.00	$392.00	$75.35	$602.80	$44.96	$70.11
1 Common Laborer	31.60	252.80	52.15	417.20		
.25 Equipment Operator, Crane	42.55	85.10	66.95	133.90		
.25 Equipment Operator, Oiler	36.80	73.60	57.90	115.80		
1 Elevator Constructor	56.60	452.80	87.10	696.80		
1 Elevator Apprentice	45.30	362.40	69.70	557.60		
.25 S.P. Crane, 4x4, 40 Ton		234.30		257.73	6.51	7.16
36 L.H., Daily Totals		$1853.00		$2781.83	$51.47	$77.27

Crew No.	Bare Costs		Incl. Subs O & P		Cost Per Labor-Hour	
Crew Q-1	**Hr.**	**Daily**	**Hr.**	**Daily**	**Bare Costs**	**Incl. O&P**
1 Plumber	$48.75	$390.00	$75.60	$604.80	$43.88	$68.05
1 Plumber Apprentice	39.00	312.00	60.50	484.00		
16 L.H., Daily Totals		$702.00		$1088.80	$43.88	$68.05

Crew No.	Bare Costs		Incl. Subs O & P		Cost Per Labor-Hour	
Crew Q-1A	**Hr.**	**Daily**	**Hr.**	**Daily**	**Bare Costs**	**Incl. O&P**
.25 Plumber Foreman (out)	$50.75	$101.50	$78.70	$157.40	$49.15	$76.22
1 Plumber	48.75	390.00	75.60	604.80		
10 L.H., Daily Totals		$491.50		$762.20	$49.15	$76.22

Crew No.	Bare Costs		Incl. Subs O & P		Cost Per Labor-Hour	
Crew Q-1C	**Hr.**	**Daily**	**Hr.**	**Daily**	**Bare Costs**	**Incl. O&P**
1 Plumber	$48.75	$390.00	$75.60	$604.80	$43.03	$67.05
1 Plumber Apprentice	39.00	312.00	60.50	484.00		
1 Equip. Oper. (medium)	41.35	330.80	65.05	520.40		
1 Trencher, Chain Type, 8' D		1777.00		1954.70	74.04	81.45
24 L.H., Daily Totals		$2809.80		$3563.90	$117.08	$148.50

Crew No.	Bare Costs		Incl. Subs O & P		Cost Per Labor-Hour	
Crew Q-2	**Hr.**	**Daily**	**Hr.**	**Daily**	**Bare Costs**	**Incl. O&P**
2 Plumbers	$48.75	$780.00	$75.60	$1209.60	$45.50	$70.57
1 Plumber Apprentice	39.00	312.00	60.50	484.00	45.50	70.57
24 L.H., Daily Totals		$1092.00		$1693.60	$45.50	$70.57

Crew No.	Bare Costs		Incl. Subs O & P		Cost Per Labor-Hour	
Crew Q-3	**Hr.**	**Daily**	**Hr.**	**Daily**	**Bare Costs**	**Incl. O&P**
1 Plumber Foreman (inside)	$49.25	$394.00	$76.40	$611.20	$46.44	$72.03
2 Plumbers	48.75	780.00	75.60	1209.60		
1 Plumber Apprentice	39.00	312.00	60.50	484.00		
32 L.H., Daily Totals		$1486.00		$2304.80	$46.44	$72.03

Crew No.	Bare Costs		Incl. Subs O & P		Cost Per Labor-Hour	
Crew Q-4	**Hr.**	**Daily**	**Hr.**	**Daily**	**Bare Costs**	**Incl. O&P**
1 Plumber Foreman (inside)	$49.25	$394.00	$76.40	$611.20	$46.44	$72.03
1 Plumber	48.75	390.00	75.60	604.80		
1 Welder (plumber)	48.75	390.00	75.60	604.80		
1 Plumber Apprentice	39.00	312.00	60.50	484.00		
1 Welder, electric, 300 amp		55.70		61.27	1.74	1.91
32 L.H., Daily Totals		$1541.70		$2366.07	$48.18	$73.94

Crew No.	Bare Costs		Incl. Subs O & P		Cost Per Labor-Hour	
Crew Q-5	**Hr.**	**Daily**	**Hr.**	**Daily**	**Bare Costs**	**Incl. O&P**
1 Steamfitter	$49.35	$394.80	$76.55	$612.40	$44.42	$68.90
1 Steamfitter Apprentice	39.50	316.00	61.25	490.00		
16 L.H., Daily Totals		$710.80		$1102.40	$44.42	$68.90

Crew No.	Bare Costs		Incl. Subs O & P		Cost Per Labor-Hour	
Crew Q-6	**Hr.**	**Daily**	**Hr.**	**Daily**	**Bare Costs**	**Incl. O&P**
2 Steamfitters	$49.35	$789.60	$76.55	$1224.80	$46.07	$71.45
1 Steamfitter Apprentice	39.50	316.00	61.25	490.00	46.07	71.45
24 L.H., Daily Totals		$1105.60		$1714.80	$46.07	$71.45

Crew No.	Bare Costs		Incl. Subs O & P		Cost Per Labor-Hour	
Crew Q-7	**Hr.**	**Daily**	**Hr.**	**Daily**	**Bare Costs**	**Incl. O&P**
1 Steamfitter Foreman (inside)	$49.85	$398.80	$77.30	$618.40	$47.01	$72.91
2 Steamfitters	49.35	789.60	76.55	1224.80		
1 Steamfitter Apprentice	39.50	316.00	61.25	490.00		
32 L.H., Daily Totals		$1504.40		$2333.20	$47.01	$72.91

Crew No.	Bare Costs		Incl. Subs O & P		Cost Per Labor-Hour	
Crew Q-8	**Hr.**	**Daily**	**Hr.**	**Daily**	**Bare Costs**	**Incl. O&P**
1 Steamfitter Foreman (inside)	$49.85	$398.80	$77.30	$618.40	$47.01	$72.91
1 Steamfitter	49.35	394.80	76.55	612.40		
1 Welder (steamfitter)	49.35	394.80	76.55	612.40		
1 Steamfitter Apprentice	39.50	316.00	61.25	490.00		
1 Welder, electric, 300 amp		55.70		61.27	1.74	1.91
32 L.H., Daily Totals		$1560.10		$2394.47	$48.75	$74.83

Crew No.	Bare Costs		Incl. Subs O & P		Cost Per Labor-Hour	
Crew Q-9	**Hr.**	**Daily**	**Hr.**	**Daily**	**Bare Costs**	**Incl. O&P**
1 Sheet Metal Worker	$47.20	$377.60	$75.00	$600.00	$42.48	$67.50
1 Sheet Metal Apprentice	37.75	302.00	60.00	480.00		
16 L.H., Daily Totals		$679.60		$1080.00	$42.48	$67.50

Crew No.	Bare Costs		Incl. Subs O & P		Cost Per Labor-Hour	
Crew Q-10	**Hr.**	**Daily**	**Hr.**	**Daily**	**Bare Costs**	**Incl. O&P**
2 Sheet Metal Workers	$47.20	$755.20	$75.00	$1200.00	$44.05	$70.00
1 Sheet Metal Apprentice	37.75	302.00	60.00	480.00		
24 L.H., Daily Totals		$1057.20		$1680.00	$44.05	$70.00

Crews

Crew Q-11	Bare Costs Hr.	Bare Costs Daily	Incl. Subs O & P Hr.	Incl. Subs O & P Daily	Cost Per Labor-Hour Bare Costs	Cost Per Labor-Hour Incl. O&P
1 Sheet Metal Foreman (inside)	$47.70	$381.60	$75.80	$606.40	$44.96	$71.45
2 Sheet Metal Workers	47.20	755.20	75.00	1200.00		
1 Sheet Metal Apprentice	37.75	302.00	60.00	480.00		
32 L.H., Daily Totals		$1438.80		$2286.40	$44.96	$71.45

Crew Q-12	Hr.	Daily	Hr.	Daily	Bare Costs	Incl. O&P
1 Sprinkler Installer	$48.15	$385.20	$74.70	$597.60	$43.33	$67.20
1 Sprinkler Apprentice	38.50	308.00	59.70	477.60		
16 L.H., Daily Totals		$693.20		$1075.20	$43.33	$67.20

Crew Q-13	Hr.	Daily	Hr.	Daily	Bare Costs	Incl. O&P
1 Sprinkler Foreman (inside)	$48.65	$389.20	$75.45	$603.60	$45.86	$71.14
2 Sprinkler Installers	48.15	770.40	74.70	1195.20		
1 Sprinkler Apprentice	38.50	308.00	59.70	477.60		
32 L.H., Daily Totals		$1467.60		$2276.40	$45.86	$71.14

Crew Q-14	Hr.	Daily	Hr.	Daily	Bare Costs	Incl. O&P
1 Asbestos Worker	$44.10	$352.80	$71.30	$570.40	$39.70	$64.20
1 Asbestos Apprentice	35.30	282.40	57.10	456.80		
16 L.H., Daily Totals		$635.20		$1027.20	$39.70	$64.20

Crew Q-15	Hr.	Daily	Hr.	Daily	Bare Costs	Incl. O&P
1 Plumber	$48.75	$390.00	$75.60	$604.80	$43.88	$68.05
1 Plumber Apprentice	39.00	312.00	60.50	484.00		
1 Welder, electric, 300 amp		55.70		61.27	3.48	3.83
16 L.H., Daily Totals		$757.70		$1150.07	$47.36	$71.88

Crew Q-16	Hr.	Daily	Hr.	Daily	Bare Costs	Incl. O&P
2 Plumbers	$48.75	$780.00	$75.60	$1209.60	$45.50	$70.57
1 Plumber Apprentice	39.00	312.00	60.50	484.00		
1 Welder, electric, 300 amp		55.70		61.27	2.32	2.55
24 L.H., Daily Totals		$1147.70		$1754.87	$47.82	$73.12

Crew Q-17	Hr.	Daily	Hr.	Daily	Bare Costs	Incl. O&P
1 Steamfitter	$49.35	$394.80	$76.55	$612.40	$44.42	$68.90
1 Steamfitter Apprentice	39.50	316.00	61.25	490.00		
1 Welder, electric, 300 amp		55.70		61.27	3.48	3.83
16 L.H., Daily Totals		$766.50		$1163.67	$47.91	$72.73

Crew Q-17A	Hr.	Daily	Hr.	Daily	Bare Costs	Incl. O&P
1 Steamfitter	$49.35	$394.80	$76.55	$612.40	$43.80	$68.25
1 Steamfitter Apprentice	39.50	316.00	61.25	490.00		
1 Equip. Oper. (crane)	42.55	340.40	66.95	535.60		
1 Hyd. Crane, 12 Ton		768.80		845.68		
1 Welder, electric, 300 amp		55.70		61.27	34.35	37.79
24 L.H., Daily Totals		$1875.70		$2544.95	$78.15	$106.04

Crew Q-18	Hr.	Daily	Hr.	Daily	Bare Costs	Incl. O&P
2 Steamfitters	$49.35	$789.60	$76.55	$1224.80	$46.07	$71.45
1 Steamfitter Apprentice	39.50	316.00	61.25	490.00		
1 Welder, electric, 300 amp		55.70		61.27	2.32	2.55
24 L.H., Daily Totals		$1161.30		$1776.07	$48.39	$74.00

Crew Q-19	Hr.	Daily	Hr.	Daily	Bare Costs	Incl. O&P
1 Steamfitter	$49.35	$394.80	$76.55	$612.40	$45.28	$70.03
1 Steamfitter Apprentice	39.50	316.00	61.25	490.00		
1 Electrician	47.00	376.00	72.30	578.40		
24 L.H., Daily Totals		$1086.80		$1680.80	$45.28	$70.03

Crew Q-20	Hr.	Daily	Hr.	Daily	Bare Costs	Incl. O&P
1 Sheet Metal Worker	$47.20	$377.60	$75.00	$600.00	$43.38	$68.46
1 Sheet Metal Apprentice	37.75	302.00	60.00	480.00		
.5 Electrician	47.00	188.00	72.30	289.20		
20 L.H., Daily Totals		$867.60		$1369.20	$43.38	$68.46

Crew Q-21	Hr.	Daily	Hr.	Daily	Bare Costs	Incl. O&P
2 Steamfitters	$49.35	$789.60	$76.55	$1224.80	$46.30	$71.66
1 Steamfitter Apprentice	39.50	316.00	61.25	490.00		
1 Electrician	47.00	376.00	72.30	578.40		
32 L.H., Daily Totals		$1481.60		$2293.20	$46.30	$71.66

Crew Q-22	Hr.	Daily	Hr.	Daily	Bare Costs	Incl. O&P
1 Plumber	$48.75	$390.00	$75.60	$604.80	$43.88	$68.05
1 Plumber Apprentice	39.00	312.00	60.50	484.00		
1 Hyd. Crane, 12 Ton		768.80		845.68	48.05	52.85
16 L.H., Daily Totals		$1470.80		$1934.48	$91.92	$120.91

Crew Q-22A	Hr.	Daily	Hr.	Daily	Bare Costs	Incl. O&P
1 Plumber	$48.75	$390.00	$75.60	$604.80	$40.48	$63.80
1 Plumber Apprentice	39.00	312.00	60.50	484.00		
1 Laborer	31.60	252.80	52.15	417.20		
1 Equip. Oper. (crane)	42.55	340.40	66.95	535.60		
1 Hyd. Crane, 12 Ton		768.80		845.68	24.02	26.43
32 L.H., Daily Totals		$2064.00		$2887.28	$64.50	$90.23

Crew Q-23	Hr.	Daily	Hr.	Daily	Bare Costs	Incl. O&P
1 Plumber Foreman	$50.75	$406.00	$78.70	$629.60	$46.95	$73.12
1 Plumber	48.75	390.00	75.60	604.80		
1 Equip. Oper. (medium)	41.35	330.80	65.05	520.40		
1 Lattice Boom Crane, 20 Ton		1080.00		1188.00	45.00	49.50
24 L.H., Daily Totals		$2206.80		$2942.80	$91.95	$122.62

Crew R-1	Hr.	Daily	Hr.	Daily	Bare Costs	Incl. O&P
1 Electrician Foreman	$47.50	$380.00	$73.05	$584.40	$41.52	$64.94
3 Electricians	47.00	1128.00	72.30	1735.20		
2 Helpers	30.30	484.80	49.85	797.60		
48 L.H., Daily Totals		$1992.80		$3117.20	$41.52	$64.94

Crew R-1A	Hr.	Daily	Hr.	Daily	Bare Costs	Incl. O&P
1 Electrician	$47.00	$376.00	$72.30	$578.40	$38.65	$61.08
1 Helper	30.30	242.40	49.85	398.80		
16 L.H., Daily Totals		$618.40		$977.20	$38.65	$61.08

Crew R-2	Hr.	Daily	Hr.	Daily	Bare Costs	Incl. O&P
1 Electrician Foreman	$47.50	$380.00	$73.05	$584.40	$41.66	$65.23
3 Electricians	47.00	1128.00	72.30	1735.20		
2 Helpers	30.30	484.80	49.85	797.60		
1 Equip. Oper. (crane)	42.55	340.40	66.95	535.60		
1 S.P. Crane, 4x4, 5 Ton		258.20		284.02	4.61	5.07
56 L.H., Daily Totals		$2591.40		$3936.82	$46.27	$70.30

Crew R-3	Hr.	Daily	Hr.	Daily	Bare Costs	Incl. O&P
1 Electrician Foreman	$47.50	$380.00	$73.05	$584.40	$46.31	$71.53
1 Electrician	47.00	376.00	72.30	578.40		
.5 Equip. Oper. (crane)	42.55	170.20	66.95	267.80		
.5 S.P. Crane, 4x4, 5 Ton		129.10		142.01	6.46	7.10
20 L.H., Daily Totals		$1055.30		$1572.61	$52.77	$78.63

Crew No.	Bare Costs		Incl. Subs O & P		Cost Per Labor-Hour	
Crew R-4	**Hr.**	**Daily**	**Hr.**	**Daily**	**Bare Costs**	**Incl. O&P**
1 Struc. Steel Foreman	$46.70	$373.60	$86.50	$692.00	$45.56	$81.44
3 Struc. Steel Workers	44.70	1072.80	82.80	1987.20		
1 Electrician	47.00	376.00	72.30	578.40		
1 Welder, gas engine, 300 amp		134.20		147.62	3.36	3.69
40 L.H., Daily Totals		$1956.60		$3405.22	$48.91	$85.13

Crew R-5	Hr.	Daily	Hr.	Daily	Bare Costs	Incl. O&P
1 Electrician Foreman	$47.50	$380.00	$73.05	$584.40	$40.97	$64.20
4 Electrician Linemen	47.00	1504.00	72.30	2313.60		
2 Electrician Operators	47.00	752.00	72.30	1156.80		
4 Electrician Groundmen	30.30	969.60	49.85	1595.20		
1 Crew Truck		169.60		186.56		
1 Flatbed Truck, 20,000 GVW		197.40		217.14		
1 Pickup Truck, 3/4 Ton		116.80		128.48		
.2 Hyd. Crane, 55 Ton		225.00		247.50		
.2 Hyd. Crane, 12 Ton		153.76		169.14		
.2 Earth Auger, Truck-Mtd.		82.40		90.64		
1 Tractor w/Winch		354.80		390.28	14.77	16.25
88 L.H., Daily Totals		$4905.36		$7079.74	$55.74	$80.45

Crew R-6	Hr.	Daily	Hr.	Daily	Bare Costs	Incl. O&P
1 Electrician Foreman	$47.50	$380.00	$73.05	$584.40	$40.97	$64.20
4 Electrician Linemen	47.00	1504.00	72.30	2313.60		
2 Electrician Operators	47.00	752.00	72.30	1156.80		
4 Electrician Groundmen	30.30	969.60	49.85	1595.20		
1 Crew Truck		169.60		186.56		
1 Flatbed Truck, 20,000 GVW		197.40		217.14		
1 Pickup Truck, 3/4 Ton		116.80		128.48		
.2 Hyd. Crane, 55 Ton		225.00		247.50		
.2 Hyd. Crane, 12 Ton		153.76		169.14		
.2 Earth Auger, Truck-Mtd.		82.40		90.64		
1 Tractor w/Winch		354.80		390.28		
3 Cable Trailers		531.30		584.43		
.5 Tensioning Rig		175.35		192.88		
.5 Cable Pulling Rig		1049.50		1154.45	34.73	38.20
88 L.H., Daily Totals		$6661.51		$9011.50	$75.70	$102.40

Crew R-7	Hr.	Daily	Hr.	Daily	Bare Costs	Incl. O&P
1 Electrician Foreman	$47.50	$380.00	$73.05	$584.40	$33.17	$53.72
5 Electrician Groundmen	30.30	1212.00	49.85	1994.00		
1 Crew Truck		169.60		186.56	3.53	3.89
48 L.H., Daily Totals		$1761.60		$2764.96	$36.70	$57.60

Crew R-8	Hr.	Daily	Hr.	Daily	Bare Costs	Incl. O&P
1 Electrician Foreman	$47.50	$380.00	$73.05	$584.40	$41.52	$64.94
3 Electrician Linemen	47.00	1128.00	72.30	1735.20		
2 Electrician Groundmen	30.30	484.80	49.85	797.60		
1 Pickup Truck, 3/4 Ton		116.80		128.48		
1 Crew Truck		169.60		186.56	5.97	6.56
48 L.H., Daily Totals		$2279.20		$3432.24	$47.48	$71.50

Crew R-9	Hr.	Daily	Hr.	Daily	Bare Costs	Incl. O&P
1 Electrician Foreman	$47.50	$380.00	$73.05	$584.40	$38.71	$61.17
1 Electrician Lineman	47.00	376.00	72.30	578.40		
2 Electrician Operators	47.00	752.00	72.30	1156.80		
4 Electrician Groundmen	30.30	969.60	49.85	1595.20		
1 Pickup Truck, 3/4 Ton		116.80		128.48		
1 Crew Truck		169.60		186.56	4.47	4.92
64 L.H., Daily Totals		$2764.00		$4229.84	$43.19	$66.09

Crew R-10	Hr.	Daily	Hr.	Daily	Bare Costs	Incl. O&P
1 Electrician Foreman	$47.50	$380.00	$73.05	$584.40	$44.30	$68.68
4 Electrician Linemen	47.00	1504.00	72.30	2313.60		
1 Electrician Groundman	30.30	242.40	49.85	398.80		
1 Crew Truck		169.60		186.56		
3 Tram Cars		380.55		418.61	11.46	12.61
48 L.H., Daily Totals		$2676.55		$3901.97	$55.76	$81.29

Crew R-11	Hr.	Daily	Hr.	Daily	Bare Costs	Incl. O&P
1 Electrician Foreman	$47.50	$380.00	$73.05	$584.40	$44.24	$68.76
4 Electricians	47.00	1504.00	72.30	2313.60		
1 Equip. Oper. (crane)	42.55	340.40	66.95	535.60		
1 Common Laborer	31.60	252.80	52.15	417.20		
1 Crew Truck		169.60		186.56		
1 Hyd. Crane, 12 Ton		768.80		845.68	16.76	18.43
56 L.H., Daily Totals		$3415.60		$4883.04	$60.99	$87.20

Crew R-12	Hr.	Daily	Hr.	Daily	Bare Costs	Incl. O&P
1 Carpenter Foreman	$40.45	$323.60	$66.80	$534.40	$37.52	$62.46
4 Carpenters	39.95	1278.40	65.95	2110.40		
4 Common Laborers	31.60	1011.20	52.15	1668.80		
1 Equip. Oper. (med.)	41.35	330.80	65.05	520.40		
1 Steel Worker	44.70	357.60	82.80	662.40		
1 Dozer, 200 H.P.		1082.00		1190.20		
1 Pickup Truck, 3/4 Ton		116.80		128.48	13.62	14.98
88 L.H., Daily Totals		$4500.40		$6815.08	$51.14	$77.44

Crew R-13	Hr.	Daily	Hr.	Daily	Bare Costs	Incl. O&P
1 Electrician Foreman	$47.50	$380.00	$73.05	$584.40	$44.94	$69.45
3 Electricians	47.00	1128.00	72.30	1735.20		
.25 Equip. Oper. (crane)	42.55	85.10	66.95	133.90		
1 Equipment Oiler	36.80	294.40	57.90	463.20		
.25 Hydraulic Crane, 33 Ton		204.70		225.17	4.87	5.36
42 L.H., Daily Totals		$2092.20		$3141.87	$49.81	$74.81

Crew R-15	Hr.	Daily	Hr.	Daily	Bare Costs	Incl. O&P
1 Electrician Foreman	$47.50	$380.00	$73.05	$584.40	$45.76	$70.62
4 Electricians	47.00	1504.00	72.30	2313.60		
1 Equipment Oper. (Light)	39.05	312.40	61.45	491.60		
1 Aerial Lift Truck		328.80		361.68	6.85	7.54
48 L.H., Daily Totals		$2525.20		$3751.28	$52.61	$78.15

Crew R-18	Hr.	Daily	Hr.	Daily	Bare Costs	Incl. O&P
.25 Electrician Foreman	$47.50	$95.00	$73.05	$146.10	$36.76	$58.54
1 Electrician	47.00	376.00	72.30	578.40		
2 Helpers	30.30	484.80	49.85	797.60		
26 L.H., Daily Totals		$955.80		$1522.10	$36.76	$58.54

Crew R-19	Hr.	Daily	Hr.	Daily	Bare Costs	Incl. O&P
.5 Electrician Foreman	$47.50	$190.00	$73.05	$292.20	$47.10	$72.45
2 Electricians	47.00	752.00	72.30	1156.80		
20 L.H., Daily Totals		$942.00		$1449.00	$47.10	$72.45

Crew R-21	Hr.	Daily	Hr.	Daily	Bare Costs	Incl. O&P
1 Electrician Foreman	$47.50	$380.00	$73.05	$584.40	$46.98	$72.31
3 Electricians	47.00	1128.00	72.30	1735.20		
.1 Equip. Oper. (med.)	41.35	33.08	65.05	52.04		
.1 S.P. Crane, 4x4, 25 Ton		71.68		78.85	2.19	2.40
32.8 L.H., Daily Totals		$1612.76		$2450.49	$49.17	$74.71

Crew No.	Bare Costs		Incl. Subs O & P		Cost Per Labor-Hour	
Crew R-22	**Hr.**	**Daily**	**Hr.**	**Daily**	**Bare Costs**	**Incl. O&P**
.66 Electrician Foreman	$47.50	$250.80	$73.05	$385.70	$39.90	$62.77
2 Helpers	30.30	484.80	49.85	797.60		
2 Electricians	47.00	752.00	72.30	1156.80		
37.28 L.H., Daily Totals		$1487.60		$2340.10	$39.90	$62.77
Crew R-30	**Hr.**	**Daily**	**Hr.**	**Daily**	**Bare Costs**	**Incl. O&P**
.25 Electrician Foreman (out)	$49.00	$98.00	$75.35	$150.70	$37.68	$60.13
1 Electrician	47.00	376.00	72.30	578.40		
2 Laborers, (Semi-Skilled)	31.60	505.60	52.15	834.40		
26 L.H., Daily Totals		$979.60		$1563.50	$37.68	$60.13
Crew R-31	**Hr.**	**Daily**	**Hr.**	**Daily**	**Bare Costs**	**Incl. O&P**
1 Electrician	$47.00	$376.00	$72.30	$578.40	$47.00	$72.30
1 Core Drill, Electric, 2.5 H.P.		52.95		58.24	6.62	7.28
8 L.H., Daily Totals		$428.95		$636.64	$53.62	$79.58

Historical Cost Indexes

The table below lists both the RSMeans Historical Cost Index based on Jan. 1, 1993 = 100 as well as the computed value of an index based on Jan. 1, 2009 costs. Since the Jan. 1, 2009 figure is estimated, space is left to write in the actual index figures as they become available through either the quarterly "RSMeans Construction Cost Indexes" or as printed in the "Engineering News-Record." To compute the actual index based on Jan. 1, 2009 = 100, divide the Historical Cost Index for a particular year by the actual Jan. 1, 2009 Construction Cost Index. Space has been left to advance the index figures as the year progresses.

Year	Historical Cost Index Jan. 1, 1993 = 100		Current Index Based on Jan. 1, 2009 = 100	
	Est.	Actual	Est.	Actual
Oct 2009				
July 2009				
April 2009				
Jan 2009	185.9		100.0	100.0
July 2008		180.4	97.0	
2007		169.4	91.1	
2006		162.0	87.1	
2005		151.6	81.5	
2004		143.7	77.3	
2003		132.0	71.0	
2002		128.7	69.2	
2001		125.1	67.3	
2000		120.9	65.0	
1999		117.6	63.3	
1998		115.1	61.9	
1997		112.8	60.7	
1996		110.2	59.3	
1995		107.6	57.9	

Year	Historical Cost Index Jan. 1, 1993 = 100	Current Index Based on Jan. 1, 2009 = 100	
	Actual	Est.	Actual
July 1994	104.4	60.3	
1993	101.7	54.7	
1992	99.4	53.5	
1991	96.8	52.1	
1990	94.3	50.7	
1989	92.1	49.6	
1988	89.9	48.3	
1987	87.7	47.2	
1986	84.2	45.3	
1985	82.6	44.4	
1984	82.0	44.1	
1983	80.2	43.1	
1982	76.1	41.0	
1981	70.0	37.6	
1980	62.9	33.8	
1979	57.8	31.1	
1978	53.5	28.8	
1977	49.5	26.6	

Year	Historical Cost Index Jan. 1, 1993 = 100	Current Index Based on Jan. 1, 2009 = 100	
	Actual	Est.	Actual
July 1976	46.9	27.1	
1975	44.8	24.1	
1974	41.4	22.3	
1973	37.7	20.3	
1972	34.8	18.7	
1971	32.1	17.3	
1970	28.7	15.4	
1969	26.9	14.5	
1968	24.9	13.4	
1967	23.5	12.6	
1966	22.7	12.2	
1965	21.7	11.7	
1964	21.2	11.4	
1963	20.7	11.1	
1962	20.2	10.9	
1961	19.8	10.7	
1960	19.7	10.6	
1959	19.3	10.4	

Adjustments to Costs

The Historical Cost Index can be used to convert National Average building costs at a particular time to the approximate building costs for some other time.

Example:

Estimate and compare construction costs for different years in the same city.

To estimate the National Average construction cost of a building in 1970, knowing that it cost $900,000 in 2009:

INDEX in 1970 = 28.7

INDEX in 2009 = 185.9

Time Adjustment using the Historical Cost Indexes:

$$\frac{\text{Index for Year A}}{\text{Index for Year B}} \text{ x Cost in Year B = Cost in Year A}$$

$$\frac{\text{INDEX 1970}}{\text{INDEX 2009}} \text{ x Cost 2009 = Cost 1970}$$

$$\frac{28.7}{185.9} \text{ x \$900,000 = .154 x \$900,000 = \$138,600}$$

The construction cost of the building in 1970 is $138,600.

Note: The City Cost Indexes for Canada can be used to convert U.S. National averages to local costs in Canadian dollars.

How to Use the City Cost Indexes

What you should know before you begin

RSMeans City Cost Indexes (CCI) are an extremely useful tool to use when you want to compare costs from city to city and region to region.

This publication contains average construction cost indexes for 731 U.S. and Canadian cities covering over 930 three-digit zip code locations, as listed directly under each city.

Keep in mind that a City Cost Index number is a percentage ratio of a specific city's cost to the national average cost of the same item at a stated time period.

In other words, these index figures represent relative construction factors (or, if you prefer, multipliers) for Material and Installation costs, as well as the weighted average for Total In Place costs for each CSI MasterFormat division. Installation costs include both labor and equipment rental costs. When estimating equipment rental rates only, for a specific location, use 015433 CONTRACTOR EQUIPMENT index.

The 30 City Average Index is the average of 30 major U.S. cities and serves as a National Average.

Index figures for both material and installation are based on the 30 major city average of 100 and represent the cost relationship as of July 1, 2008. The index for each division is computed from representative material and labor quantities for that division. The weighted average for each city is a weighted total of the components listed above it, but does not include relative productivity between trades or cities.

As changes occur in local material prices, labor rates, and equipment rental rates, (including fuel costs) the impact of these changes should be accurately measured by the change in the City Cost Index for each particular city (as compared to the 30 City Average).

Therefore, if you know (or have estimated) building costs in one city today, you can easily convert those costs to expected building costs in another city.

In addition, by using the Historical Cost Index, you can easily convert National Average building costs at a particular time to the approximate building costs for some other time. The City Cost Indexes can then be applied to calculate the costs for a particular city.

Quick Calculations

Location Adjustment Using the City Cost Indexes:

$$\frac{\text{Index for City A}}{\text{Index for City B}} \times \text{Cost in City B} = \text{Cost in City A}$$

Time Adjustment for the National Average Using the Historical Cost Index:

$$\frac{\text{Index for Year A}}{\text{Index for Year B}} \times \text{Cost in Year B} = \text{Cost in Year A}$$

Adjustment from the National Average:

$$\frac{\text{Index for City A}}{100} \times \text{National Average Cost} = \text{Cost in City A}$$

Since each of the other RSMeans publications contains many different items, any *one* item multiplied by the particular city index may give incorrect results. However, the larger the number of items compiled, the closer the results should be to actual costs for that particular city.

The City Cost Indexes for Canadian cities are calculated using Canadian material and equipment prices and labor rates, in Canadian dollars. Therefore, indexes for Canadian cities can be used to convert U.S. National Average prices to local costs in Canadian dollars.

How to use this section

1. Compare costs from city to city.

In using the RSMeans Indexes, remember that an index number is not a fixed number but a ratio: It's a percentage ratio of a building component's cost at any stated time to the National Average cost of that same component at the same time period. Put in the form of an equation:

$$\frac{\text{Specific City Cost}}{\text{National Average Cost}} \times 100 = \text{City Index Number}$$

Therefore, when making cost comparisons between cities, do not subtract one city's index number from the index number of another city and read the result as a percentage difference. Instead, divide one city's index number by that of the other city. The resulting number may then be used as a multiplier to calculate cost differences from city to city.

The formula used to find cost differences between cities for the purpose of comparison is as follows:

$$\frac{\text{City A Index}}{\text{City B Index}} \times \text{City B Cost (Known)} = \text{City A Cost (Unknown)}$$

In addition, you can use RSMeans CCI to calculate and compare costs division by division between cities using the same basic formula. (Just be sure that you're comparing similar divisions.)

2. Compare a specific city's construction costs with the National Average.

When you're studying construction location feasibility, it's advisable to compare a prospective project's cost index with an index of the National Average cost.

For example, divide the weighted average index of construction costs of a specific city by that of the 30 City Average, which = 100.

$$\frac{\text{City Index}}{100} = \text{\% of National Average}$$

As a result, you get a ratio that indicates the relative cost of construction in that city in comparison with the National Average.

3. Convert U.S. National Average to actual costs in Canadian City.

$$\frac{\text{Index for Canadian City}}{100} \times \text{National Average Cost} = \text{Cost in Canadian City in \$ CAN}$$

4. Adjust construction cost data based on a National Average.

When you use a source of construction cost data which is based on a National Average (such as RSMeans cost data publications), it is necessary to adjust those costs to a specific location.

$$\frac{\text{City Index}}{100} \times \begin{matrix}\text{"Book" Cost Based on} \\ \text{National Average Costs}\end{matrix} = \begin{matrix}\text{City Cost} \\ \text{(Unknown)}\end{matrix}$$

5. When applying the City Cost Indexes to demolition projects, use the appropriate division installation index. For example, for removal of existing doors and windows, use Division 8 (Openings) index.

What you might like to know about how we developed the Indexes

The information presented in the CCI is organized according to the Construction Specifications Institute (CSI) MasterFormat 2004.

To create a reliable index, RSMeans researched the building type most often constructed in the United States and Canada. Because it was concluded that no one type of building completely represented the building construction industry, nine different types of buildings were combined to create a composite model.

The exact material, labor and equipment quantities are based on detailed analysis of these nine building types, then each quantity is weighted in proportion to expected usage. These various material items, labor hours, and equipment rental rates are thus combined to form a composite building representing as closely as possible the actual usage of materials, labor and equipment used in the North American Building Construction Industry.

The following structures were chosen to make up that composite model:

1. Factory, 1 story
2. Office, 2–4 story
3. Store, Retail
4. Town Hall, 2–3 story
5. High School, 2–3 story
6. Hospital, 4–8 story
7. Garage, Parking
8. Apartment, 1–3 story
9. Hotel/Motel, 2–3 story

For the purposes of ensuring the timeliness of the data, the components of the index for the composite model have been streamlined. They currently consist of:

- specific quantities of 66 commonly used construction materials;
- specific labor-hours for 21 building construction trades; and
- specific days of equipment rental for 6 types of construction equipment (normally used to install the 66 material items by the 21 trades.) Fuel costs and routine maintenance costs are included in the equipment cost.

A sophisticated computer program handles the updating of all costs for each city on a quarterly basis. Material and equipment price quotations are gathered quarterly from 731 cities in the United States and Canada. These prices and the latest negotiated labor wage rates for 21 different building trades are used to compile the quarterly update of the City Cost Index.

The 30 major U.S. cities used to calculate the National Average are:

Atlanta, GA	Memphis, TN
Baltimore, MD	Milwaukee, WI
Boston, MA	Minneapolis, MN
Buffalo, NY	Nashville, TN
Chicago, IL	New Orleans, LA
Cincinnati, OH	New York, NY
Cleveland, OH	Philadelphia, PA
Columbus, OH	Phoenix, AZ
Dallas, TX	Pittsburgh, PA
Denver, CO	St. Louis, MO
Detroit, MI	San Antonio, TX
Houston, TX	San Diego, CA
Indianapolis, IN	San Francisco, CA
Kansas City, MO	Seattle, WA
Los Angeles, CA	Washington, DC

What the CCI does not indicate

The weighted average for each city is a total of the divisional components weighted to reflect typical usage, but it does not include the productivity variations between trades or cities.

In addition, the CCI does not take into consideration factors such as the following:

- managerial efficiency
- competitive conditions
- automation
- restrictive union practices
- unique local requirements
- regional variations due to specific building codes

DIVISION		UNITED STATES			ALABAMA																	
		30 CITY AVERAGE			ANNISTON			BIRMINGHAM			BUTLER			DECATUR			DOTHAN					
					362			350 - 352			369			356			363					
		MAT.	INST.	TOTAL	MAT.	INST.	TOTAL	MAT.	INST.	TOTAL	MAT.	INST.	TOTAL	MAT.	INST.	TOTAL	MAT.	INST.	TOTAL			
015433	**CONTRACTOR EQUIPMENT**		100.0	100.0		100.4	100.4		100.5	100.5		98.9	98.9		100.4	100.4		98.9	98.9			
0241, 31 - 34	**SITE & INFRASTRUCTURE, DEMOLITION**	100.0	100.0	100.0	89.6	90.4	90.2	86.5	91.9	90.2	104.0	86.7	92.1	82.5	90.0	87.6	101.4	86.6	91.3			
0310	Concrete Forming & Accessories	100.0	100.0	100.0	90.6	48.9	54.6	91.0	72.9	75.3	87.0	49.3	54.4	94.9	51.9	57.8	95.8	48.6	55.0			
0320	Concrete Reinforcing	100.0	100.0	100.0	89.7	78.8	85.2	89.7	80.3	85.8	94.7	53.7	78.0	89.7	50.7	73.8	94.7	52.8	77.6			
0330	Cast-in-Place Concrete	100.0	100.0	100.0	92.6	49.0	75.9	99.2	71.4	88.6	90.3	61.2	79.2	92.1	59.0	79.4	90.3	51.3	75.4			
03	**CONCRETE**	100.0	100.0	100.0	93.8	56.3	76.8	92.8	74.8	84.6	94.9	56.1	77.2	89.7	56.0	74.4	94.3	52.2	75.1			
04	**MASONRY**	100.0	100.0	100.0	84.9	43.7	60.2	87.3	81.1	83.6	88.8	55.8	69.0	85.7	59.2	69.8	89.7	41.6	60.9			
05	**METALS**	100.0	100.0	100.0	99.2	87.6	95.8	99.9	91.5	97.4	98.1	80.2	92.9	101.3	79.7	95.0	98.2	78.5	92.4			
06	**WOOD, PLASTICS & COMPOSITES**	100.0	100.0	100.0	89.8	50.0	68.0	91.7	70.1	79.9	85.0	49.6	65.6	94.6	50.3	70.3	96.2	49.6	70.6			
07	**THERMAL & MOISTURE PROTECTION**	100.0	100.0	100.0	92.9	48.7	75.7	94.8	85.8	91.3	92.9	62.3	81.0	92.7	62.9	81.1	92.9	55.4	78.3			
08	**OPENINGS**	100.0	100.0	100.0	93.2	53.6	83.4	96.8	73.7	91.1	93.2	51.7	83.0	97.1	50.2	85.5	93.3	51.7	83.0			
0920	Plaster & Gypsum Board	100.0	100.0	100.0	97.0	48.7	65.6	103.6	69.6	81.5	95.5	48.3	64.8	100.7	49.0	67.1	103.3	48.3	67.6			
0950, 0980	Ceilings & Acoustic Treatment	100.0	100.0	100.0	94.4	48.7	66.1	97.1	69.6	80.0	94.4	48.3	65.9	98.5	49.0	67.9	94.4	48.3	65.9			
0960	Flooring	100.0	100.0	100.0	100.6	31.1	81.8	103.5	67.6	93.8	106.8	62.8	94.9	103.5	45.7	87.9	114.4	30.9	91.8			
0970, 0990	Wall Finishes & Painting/Coating	100.0	100.0	100.0	100.1	36.2	61.7	100.1	63.9	78.4	100.1	61.3	76.8	100.1	57.3	74.4	100.1	61.3	76.8			
09	**FINISHES**	100.0	100.0	100.0	95.4	44.7	68.6	97.7	70.6	83.4	99.0	53.0	74.7	97.3	51.3	73.0	102.1	46.5	72.7			
COVERS	**DIVS. 10 - 14, 25, 28, 41, 43, 44**	100.0	100.0	100.0	100.0	52.0	90.5	100.0	87.4	97.5	100.0	52.1	90.5	100.0	54.3	91.0	100.0	52.1	90.5			
21, 22, 23	**FIRE SUPPRESSION, PLUMBING & HVAC**	100.0	100.0	100.0	98.9	39.0	75.1	99.9	67.3	86.9	96.7	40.0	74.2	99.9	48.7	79.5	96.7	39.9	74.1			
26, 27, 3370	**ELECTRICAL, COMMUNICATIONS & UTIL.**	100.0	100.0	100.0	97.7	39.6	69.4	108.6	60.1	85.0	99.7	46.9	74.0	99.3	61.7	81.0	98.5	46.9	73.4			
MF2004	**WEIGHTED AVERAGE**	100.0	100.0	100.0	96.3	52.9	78.1	98.3	74.9	88.5	96.8	55.7	79.5	97.0	59.9	81.4	96.9	52.5	78.3			

DIVISION		ALABAMA																	
		EVERGREEN			GADSDEN			HUNTSVILLE			JASPER			MOBILE			MONTGOMERY		
		364			359			357 - 358			355			365 - 366			360 - 361		
		MAT.	INST.	TOTAL	MAT.	INST.	TOTAL	MAT.	INST.	TOTAL	MAT.	INST.	TOTAL	MAT.	INST.	TOTAL	MAT.	INST.	TOTAL
015433	**CONTRACTOR EQUIPMENT**		98.9	98.9		100.4	100.4		100.4	100.4		100.4	100.4		98.9	98.9		98.9	98.9
0241, 31 - 34	**SITE & INFRASTRUCTURE, DEMOLITION**	104.5	87.0	92.5	88.2	90.6	89.8	82.2	91.4	88.5	87.8	90.3	89.5	95.7	87.7	90.2	93.7	87.8	89.6
0310	Concrete Forming & Accessories	83.4	51.3	55.6	87.1	47.6	53.0	94.9	67.1	70.9	92.5	39.3	46.5	95.0	59.2	64.1	92.7	49.9	55.7
0320	Concrete Reinforcing	94.7	53.7	78.0	94.9	79.3	88.5	89.7	76.3	84.2	89.7	79.0	85.3	92.5	52.8	76.3	92.5	78.9	87.0
0330	Cast-in-Place Concrete	90.3	64.1	80.3	92.1	54.6	77.8	89.7	68.5	81.6	102.0	49.0	81.8	94.7	64.3	83.1	97.0	52.0	79.8
03	**CONCRETE**	95.0	57.9	78.1	93.9	57.7	77.4	88.6	70.6	80.4	96.8	52.1	76.5	91.5	61.3	77.8	92.5	57.8	76.7
04	**MASONRY**	88.8	61.1	72.2	84.3	49.5	63.5	86.8	66.0	74.3	81.1	43.1	58.4	87.5	60.7	71.4	86.0	41.3	59.2
05	**METALS**	98.2	80.2	92.9	99.3	89.3	96.4	101.3	90.4	98.1	99.3	88.1	96.0	100.1	80.0	94.2	99.9	88.7	96.6
06	**WOOD, PLASTICS & COMPOSITES**	80.8	49.6	63.7	85.6	45.4	63.5	94.6	67.4	79.7	91.7	37.0	61.7	94.8	59.1	75.2	91.3	49.6	68.4
07	**THERMAL & MOISTURE PROTECTION**	92.9	63.7	81.5	92.8	70.5	84.1	92.4	78.2	86.9	92.8	51.0	76.5	92.4	77.4	86.6	91.8	68.2	82.6
08	**OPENINGS**	93.3	51.7	83.0	93.2	52.3	83.1	97.1	64.9	89.1	93.2	51.1	82.8	97.1	56.5	87.0	97.1	57.2	87.2
0920	Plaster & Gypsum Board	93.3	48.3	64.1	95.1	43.9	61.8	98.3	66.7	77.8	96.9	35.2	56.9	100.7	58.1	73.0	103.6	48.3	67.7
0950, 0980	Ceilings & Acoustic Treatment	94.4	48.3	65.9	94.4	43.9	63.1	100.4	66.7	79.6	92.7	35.2	57.2	100.4	58.1	74.2	97.1	48.3	66.9
0960	Flooring	104.0	62.8	92.9	98.6	50.5	85.5	103.5	60.6	91.9	101.6	29.9	82.2	113.4	62.8	99.7	111.6	33.2	90.4
0970, 0990	Wall Finishes & Painting/Coating	100.1	61.3	76.8	100.1	69.8	81.9	100.1	60.5	76.4	100.1	44.3	66.6	104.2	62.4	79.1	100.1	61.3	76.8
09	**FINISHES**	97.9	54.3	74.9	94.4	49.3	70.6	97.4	64.8	80.2	95.1	37.4	64.6	102.5	59.9	79.9	101.0	46.9	72.4
COVERS	**DIVS. 10 - 14, 25, 28, 41, 43, 44**	100.0	54.0	90.9	100.0	78.8	95.8	100.0	83.7	96.8	100.0	49.9	90.1	100.0	81.4	96.3	100.0	78.7	95.8
21, 22, 23	**FIRE SUPPRESSION, PLUMBING & HVAC**	96.7	42.8	75.3	101.5	43.1	78.3	99.8	62.1	84.8	101.5	56.7	83.7	99.8	63.7	85.4	99.9	40.6	76.3
26, 27, 3370	**ELECTRICAL, COMMUNICATIONS & UTIL.**	97.1	46.9	72.6	99.4	60.1	80.2	100.5	67.3	84.3	98.9	33.7	67.2	100.5	58.4	80.0	103.4	71.0	87.6
MF2004	**WEIGHTED AVERAGE**	96.4	57.3	80.0	96.9	59.4	81.2	97.0	71.1	86.1	97.2	54.0	79.0	97.9	65.9	84.5	98.0	59.2	81.7

DIVISION		ALABAMA									ALASKA								
		PHENIX CITY			SELMA			TUSCALOOSA			ANCHORAGE			FAIRBANKS			JUNEAU		
		368			367			354			995 - 996			997			998		
		MAT.	INST.	TOTAL	MAT.	INST.	TOTAL	MAT.	INST.	TOTAL	MAT.	INST.	TOTAL	MAT.	INST.	TOTAL	MAT.	INST.	TOTAL
015433	**CONTRACTOR EQUIPMENT**		98.9	98.9		98.9	98.9		100.4	100.4		113.0	113.0		113.0	113.0		113.0	113.0
0241, 31 - 34	**SITE & INFRASTRUCTURE, DEMOLITION**	108.6	87.8	94.4	101.2	87.7	92.0	82.7	90.8	88.2	140.1	126.1	130.5	126.3	126.2	126.3	139.2	126.1	130.2
0310	Concrete Forming & Accessories	90.5	44.3	50.6	88.2	49.1	54.4	94.8	53.8	59.4	131.5	114.1	116.5	137.2	116.5	119.3	133.2	114.0	116.6
0320	Concrete Reinforcing	94.6	71.7	85.3	94.7	78.8	88.2	89.7	79.3	85.4	136.8	103.8	123.3	139.4	103.8	124.9	104.9	103.7	104.4
0330	Cast-in-Place Concrete	90.3	52.7	76.0	90.3	51.9	75.6	93.4	54.5	78.6	168.8	111.3	146.8	148.0	112.6	134.5	169.7	111.2	147.3
03	**CONCRETE**	97.4	54.3	77.8	93.8	57.4	77.2	90.4	60.4	76.8	138.1	110.7	125.6	125.3	112.1	119.3	132.2	110.6	122.4
04	**MASONRY**	88.8	47.3	63.9	92.3	41.8	62.0	85.9	52.9	66.1	207.3	116.1	152.7	216.8	117.7	157.5	198.8	116.1	149.3
05	**METALS**	98.1	86.4	94.7	98.1	87.8	95.1	100.5	89.2	97.2	124.0	100.9	117.3	124.1	101.1	117.4	124.1	100.8	117.3
06	**WOOD, PLASTICS & COMPOSITES**	89.5	41.5	63.1	86.6	49.6	66.3	94.6	52.0	71.2	114.3	113.6	113.9	118.5	115.7	117.0	114.3	113.6	113.9
07	**THERMAL & MOISTURE PROTECTION**	93.2	69.1	83.8	92.8	56.3	78.6	92.7	72.1	84.7	181.2	111.1	153.9	178.9	113.9	153.6	179.5	111.1	152.9
08	**OPENINGS**	93.2	50.9	82.8	93.2	57.2	84.3	97.1	63.8	88.8	125.0	109.8	121.3	122.2	111.0	119.4	122.2	109.8	119.1
0920	Plaster & Gypsum Board	98.1	39.9	60.3	96.6	48.3	65.2	100.7	50.8	68.3	139.3	113.8	122.7	156.6	116.0	130.2	139.3	113.8	122.7
0950, 0980	Ceilings & Acoustic Treatment	94.4	39.9	60.7	94.4	48.3	65.9	100.4	50.8	69.7	125.5	113.8	118.3	131.4	116.0	121.8	125.5	113.8	118.3
0960	Flooring	109.6	33.2	89.0	107.5	24.5	85.1	103.5	42.1	86.9	158.9	120.7	148.6	158.9	120.7	148.6	158.9	120.7	148.6
0970, 0990	Wall Finishes & Painting/Coating	100.1	61.3	76.8	100.1	61.3	76.8	100.1	52.8	71.7	155.7	112.9	130.0	155.7	121.0	134.8	155.7	112.9	130.0
09	**FINISHES**	100.6	42.4	69.9	99.1	45.4	70.7	97.7	50.8	72.9	148.5	115.6	131.1	150.3	118.2	133.3	147.3	115.6	130.5
COVERS	**DIVS. 10 - 14, 25, 28, 41, 43, 44**	100.0	78.2	95.7	100.0	52.1	90.5	100.0	80.9	96.2	100.0	110.2	102.0	100.0	111.0	102.2	100.0	110.2	102.0
21, 22, 23	**FIRE SUPPRESSION, PLUMBING & HVAC**	96.7	41.2	74.7	96.7	40.2	74.3	99.9	40.4	76.2	100.5	103.0	101.5	100.4	109.8	104.2	100.5	94.4	98.1
26, 27, 3370	**ELECTRICAL, COMMUNICATIONS & UTIL.**	99.1	78.5	89.1	98.2	46.9	73.2	100.0	60.1	80.5	140.0	109.5	125.1	148.2	109.5	129.4	141.5	109.5	125.9
MF2004	**WEIGHTED AVERAGE**	97.3	59.4	81.4	96.6	54.4	78.9	97.0	60.5	81.7	128.3	110.3	120.8	127.6	112.6	121.3	127.0	108.5	119.2

	DIVISION	ALASKA			ARIZONA															
		KETCHIKAN			CHAMBERS			FLAGSTAFF			GLOBE			KINGMAN			MESA/TEMPE		
		999			865			860			855			864			852		
		MAT.	INST.	TOTAL	MAT.	INST.	TOTAL	MAT.	INST.	TOTAL	MAT.	INST.	TOTAL	MAT.	INST.	TOTAL	MAT.	INST.	TOTAL
015433	CONTRACTOR EQUIPMENT		113.0	113.0		96.2	96.2		96.2	96.2		99.2	99.2		96.2	96.2		99.2	99.2
0241, 31 - 34	SITE & INFRASTRUCTURE, DEMOLITION	189.6	126.1	146.2	63.8	100.8	89.1	81.1	102.1	95.4	95.7	104.2	101.6	63.8	102.3	90.1	86.2	104.8	98.9
0310	Concrete Forming & Accessories	128.6	113.9	115.9	98.1	49.8	56.4	103.2	65.4	70.6	101.0	49.6	56.6	96.2	63.5	68.0	103.8	60.0	66.0
0320	Concrete Reinforcing	109.4	103.7	107.1	100.0	74.0	89.4	99.9	79.4	91.5	98.2	74.0	88.3	100.2	79.4	91.7	98.8	79.4	90.9
0330	Cast-in-Place Concrete	299.9	111.2	227.8	95.9	59.2	81.9	96.0	77.4	88.9	104.0	59.5	87.0	95.6	60.2	82.1	104.8	65.6	89.8
03	CONCRETE	208.8	110.5	164.1	101.4	58.0	81.7	121.0	72.3	98.9	117.2	58.1	90.3	101.1	65.6	84.9	108.5	65.9	89.1
04	MASONRY	225.6	116.1	160.1	98.1	45.1	66.4	97.8	61.8	76.3	110.5	45.1	71.4	98.1	57.2	73.6	110.7	47.6	72.9
05	METALS	124.2	100.7	117.4	94.8	69.0	87.3	95.3	74.9	89.3	93.2	69.5	86.3	95.4	74.7	89.4	93.5	75.4	88.2
06	WOOD, PLASTICS & COMPOSITES	109.3	113.6	111.7	99.7	47.9	71.3	105.5	65.2	83.4	92.9	48.0	68.3	96.1	65.2	79.1	96.2	65.3	79.3
07	THERMAL & MOISTURE PROTECTION	181.7	111.1	154.2	97.1	55.7	81.0	98.9	67.0	86.5	108.2	53.3	86.9	97.1	61.4	83.2	107.4	58.4	88.3
08	OPENINGS	122.6	109.8	119.4	100.6	53.1	88.9	100.8	68.9	92.9	96.2	53.2	85.6	100.8	66.7	92.4	96.3	66.7	89.0
0920	Plaster & Gypsum Board	144.4	113.8	124.5	82.9	46.0	58.9	85.7	64.0	71.6	83.1	46.0	59.0	79.0	64.0	69.2	84.9	64.0	71.3
0950, 0980	Ceilings & Acoustic Treatment	124.8	113.8	118.0	111.2	46.0	70.9	112.0	64.0	82.3	101.2	46.0	67.1	112.0	64.0	82.3	101.2	64.0	78.2
0960	Flooring	158.9	120.7	148.6	95.7	44.3	81.9	98.3	44.5	83.7	101.5	44.3	86.0	94.7	60.2	85.4	102.9	55.5	90.1
0970, 0990	Wall Finishes & Painting/Coating	155.7	112.9	130.0	90.4	42.4	61.6	90.4	55.1	69.2	94.0	42.4	63.1	90.4	55.1	69.2	94.0	55.1	70.7
09	FINISHES	151.5	115.6	132.5	98.0	47.0	71.0	101.0	59.8	79.2	100.5	47.1	72.3	97.4	61.3	78.3	100.2	58.1	78.0
COVERS	DIVS. 10 - 14, 25, 28, 41, 43, 44	100.0	110.2	102.0	100.0	77.8	95.6	100.0	81.5	96.3	100.0	78.1	95.7	100.0	79.8	96.0	100.0	76.7	95.4
21, 22, 23	FIRE SUPPRESSION, PLUMBING & HVAC	98.9	92.5	96.4	96.5	71.6	86.6	100.2	76.8	90.9	94.3	65.8	83.0	96.5	74.4	87.7	100.1	66.6	86.8
26, 27, 3370	ELECTRICAL, COMMUNICATIONS & UTIL.	148.2	109.5	129.4	102.7	69.8	86.7	101.6	61.3	81.9	95.9	66.4	81.5	102.7	61.2	82.5	92.5	61.3	77.3
MF2004	WEIGHTED AVERAGE	139.6	108.1	126.4	97.5	64.4	83.6	101.4	71.6	88.9	99.5	63.1	84.2	97.5	69.6	85.8	99.2	66.7	85.6

	DIVISION	ARIZONA												ARKANSAS					
		PHOENIX			PRESCOTT			SHOW LOW			TUCSON			BATESVILLE			CAMDEN		
		850,853			863			859			856 - 857			725			717		
		MAT.	INST.	TOTAL	MAT.	INST.	TOTAL	MAT.	INST.	TOTAL	MAT.	INST.	TOTAL	MAT.	INST.	TOTAL	MAT.	INST.	TOTAL
015433	CONTRACTOR EQUIPMENT		99.8	99.8		96.2	96.2		99.2	99.2		99.2	99.2		92.0	92.0		92.0	92.0
0241, 31 - 34	SITE & INFRASTRUCTURE, DEMOLITION	86.7	106.0	99.9	69.8	100.8	91.0	97.9	104.3	102.2	82.0	105.6	98.1	74.3	90.5	85.4	75.1	90.1	85.3
0310	Concrete Forming & Accessories	104.8	69.2	74.0	99.4	49.6	56.4	107.6	49.9	57.8	104.1	68.8	73.6	85.1	52.1	56.6	84.8	34.7	41.5
0320	Concrete Reinforcing	97.4	79.7	90.2	99.9	74.3	89.4	98.8	67.4	86.0	81.4	79.4	80.6	98.4	70.9	87.2	99.7	48.0	78.6
0330	Cast-in-Place Concrete	104.9	77.9	94.6	95.9	59.1	81.8	104.0	59.6	87.0	107.9	77.8	96.4	78.5	51.5	68.2	80.5	42.0	65.8
03	CONCRETE	108.1	74.3	92.7	106.7	57.9	84.5	119.4	57.1	91.1	106.3	74.0	91.6	81.8	56.2	70.1	83.4	41.1	64.1
04	MASONRY	97.8	64.9	78.1	98.1	50.9	69.8	110.5	46.8	72.4	95.8	61.8	75.4	102.3	47.2	69.3	110.6	35.8	65.8
05	METALS	95.1	76.6	89.7	95.3	69.2	87.7	93.0	67.5	85.5	94.3	75.5	88.8	100.6	70.8	91.9	100.7	62.6	89.6
06	WOOD, PLASTICS & COMPOSITES	97.1	70.0	82.2	101.1	47.9	71.9	100.5	48.0	71.7	96.4	70.0	81.9	86.7	53.3	68.4	86.5	34.8	58.1
07	THERMAL & MOISTURE PROTECTION	107.3	68.1	92.1	97.6	55.0	81.0	108.4	55.9	88.0	108.7	64.4	91.5	97.9	50.3	79.4	97.8	36.4	73.9
08	OPENINGS	97.3	71.5	90.9	100.8	53.1	89.0	95.3	51.5	84.4	92.6	71.5	87.4	96.3	51.4	85.2	92.5	40.1	79.6
0920	Plaster & Gypsum Board	87.3	68.8	75.2	83.1	46.0	59.0	86.8	46.0	60.3	90.4	68.8	76.3	82.1	52.1	62.6	82.1	33.0	50.2
0950, 0980	Ceilings & Acoustic Treatment	108.7	68.8	84.0	110.3	46.0	70.6	101.2	46.0	67.1	102.1	68.8	81.5	87.7	52.1	65.7	87.7	33.0	53.9
0960	Flooring	103.2	58.0	91.0	96.7	44.3	82.5	104.8	50.5	90.1	93.7	44.5	80.4	103.6	66.2	93.5	103.5	44.4	87.5
0970, 0990	Wall Finishes & Painting/Coating	94.0	57.2	71.9	90.4	42.4	61.6	94.0	42.4	63.1	94.0	55.1	70.7	98.5	47.3	67.8	98.5	51.7	70.4
09	FINISHES	102.4	65.3	82.8	98.6	47.0	71.3	102.2	48.2	73.6	98.3	62.6	79.4	91.0	54.2	71.6	91.0	38.1	63.0
COVERS	DIVS. 10 - 14, 25, 28, 41, 43, 44	100.0	82.4	96.5	100.0	77.8	95.6	100.0	78.2	95.7	100.0	82.4	96.5	100.0	49.7	90.0	100.0	44.5	89.0
21, 22, 23	FIRE SUPPRESSION, PLUMBING & HVAC	100.1	76.9	90.9	100.2	71.2	88.7	94.3	71.6	85.3	100.1	68.8	87.7	94.3	40.9	73.1	94.3	33.3	70.0
26, 27, 3370	ELECTRICAL, COMMUNICATIONS & UTIL.	100.3	64.5	82.9	101.3	61.2	81.8	92.9	66.4	80.0	94.8	58.8	77.3	98.4	47.6	73.7	95.0	42.5	69.4
MF2004	WEIGHTED AVERAGE	99.9	74.0	89.0	99.2	63.7	84.3	99.5	64.2	84.7	98.0	70.6	86.5	94.7	54.3	77.7	94.6	44.6	73.6

	DIVISION	ARKANSAS																	
		FAYETTEVILLE			FORT SMITH			HARRISON			HOT SPRINGS			JONESBORO			LITTLE ROCK		
		727			729			726			719			724			720 - 722		
		MAT.	INST.	TOTAL	MAT.	INST.	TOTAL	MAT.	INST.	TOTAL	MAT.	INST.	TOTAL	MAT.	INST.	TOTAL	MAT.	INST.	TOTAL
015433	CONTRACTOR EQUIPMENT		92.0	92.0		92.0	92.0		92.0	92.0		92.0	92.0		105.6	105.6		92.0	92.0
0241, 31 - 34	SITE & INFRASTRUCTURE, DEMOLITION	73.7	90.7	85.3	78.5	90.7	86.9	79.2	90.6	87.0	78.8	90.1	86.5	101.1	96.1	97.7	87.0	90.8	89.6
0310	Concrete Forming & Accessories	80.8	43.2	48.3	99.9	40.9	48.9	89.5	52.4	57.4	82.2	40.8	46.4	88.5	55.8	60.3	92.2	68.5	71.7
0320	Concrete Reinforcing	98.5	68.3	86.2	99.6	69.8	87.4	98.0	70.8	86.9	97.9	61.8	83.2	95.3	71.7	85.7	99.7	66.5	86.2
0330	Cast-in-Place Concrete	78.6	48.5	67.1	89.9	68.1	81.5	87.2	49.4	72.7	82.4	43.3	67.5	85.6	62.4	76.7	90.3	68.3	81.9
03	CONCRETE	81.5	50.8	67.5	89.0	56.8	74.4	88.1	55.5	73.3	86.2	46.7	68.2	85.8	62.5	75.2	88.8	68.4	79.5
04	MASONRY	92.2	44.1	63.4	97.4	57.0	73.2	102.7	45.2	68.3	82.8	29.6	51.0	93.6	48.6	66.6	94.5	57.0	72.0
05	METALS	100.6	70.2	91.8	102.8	72.3	93.9	101.6	70.8	92.6	100.6	67.3	90.9	96.8	83.8	93.0	98.6	71.7	90.7
06	WOOD, PLASTICS & COMPOSITES	82.4	41.6	60.0	103.9	36.4	66.9	92.2	53.3	70.8	83.6	41.8	60.7	90.3	56.5	71.8	98.0	73.2	84.4
07	THERMAL & MOISTURE PROTECTION	98.9	50.0	79.9	99.2	52.7	81.1	98.2	48.8	79.0	98.0	40.5	75.6	102.5	56.3	84.5	98.5	56.6	82.2
08	OPENINGS	96.3	49.0	84.6	97.2	44.6	84.2	97.2	52.2	86.1	92.5	45.6	80.9	99.1	60.4	89.5	97.2	65.1	89.3
0920	Plaster & Gypsum Board	80.2	40.1	54.1	86.9	34.7	53.0	85.8	52.1	63.9	81.0	40.3	54.6	98.2	55.1	70.2	83.1	72.7	76.4
0950, 0980	Ceilings & Acoustic Treatment	87.7	40.1	58.2	92.8	34.7	56.9	91.9	52.1	67.3	87.7	40.3	58.4	91.6	55.1	69.1	90.3	72.7	79.4
0960	Flooring	100.2	66.2	91.0	112.0	68.2	100.1	106.6	66.2	95.7	102.0	66.2	92.3	74.9	59.9	70.8	113.1	68.2	101.0
0970, 0990	Wall Finishes & Painting/Coating	98.5	33.0	59.2	98.5	61.9	76.5	98.5	47.3	67.8	98.5	41.0	64.0	87.4	55.6	68.3	98.5	63.5	77.5
09	FINISHES	89.8	45.5	66.4	95.6	47.1	70.0	93.7	54.3	72.9	90.8	45.1	66.6	88.2	56.3	71.4	95.0	68.9	81.2
COVERS	DIVS. 10 - 14, 25, 28, 41, 43, 44	100.0	60.0	92.1	100.0	69.7	94.0	100.0	61.8	92.4	100.0	45.8	89.3	100.0	56.6	91.4	100.0	74.0	94.9
21, 22, 23	FIRE SUPPRESSION, PLUMBING & HVAC	94.3	45.3	74.8	100.1	48.1	79.5	94.3	47.3	75.6	94.3	38.1	72.0	100.3	48.1	79.6	100.1	66.4	86.7
26, 27, 3370	ELECTRICAL, COMMUNICATIONS & UTIL.	92.4	59.7	76.5	95.8	75.3	85.8	97.0	43.3	70.8	97.1	44.8	71.6	103.8	52.2	78.7	100.5	76.6	88.9
MF2004	WEIGHTED AVERAGE	93.4	54.7	77.1	97.4	60.1	81.8	95.9	55.0	78.8	93.8	47.9	74.5	96.9	60.2	81.5	97.1	69.9	85.7

DIVISION		ARKANSAS												CALIFORNIA					
		PINE BLUFF			RUSSELLVILLE			TEXARKANA			WEST MEMPHIS			ALHAMBRA			ANAHEIM		
		716			728			718			723			917 - 918			928		
		MAT.	INST.	TOTAL	MAT.	INST.	TOTAL	MAT.	INST.	TOTAL	MAT.	INST.	TOTAL	MAT.	INST.	TOTAL	MAT.	INST.	TOTAL
015433	CONTRACTOR EQUIPMENT		92.0	92.0		92.0	92.0		90.7	90.7		105.6	105.6		99.1	99.1		102.2	102.2
0241, 31 - 34	SITE & INFRASTRUCTURE, DEMOLITION	80.6	90.7	87.5	75.7	90.6	85.9	95.9	87.8	90.4	109.0	96.2	100.3	103.1	110.6	108.2	101.8	109.4	107.0
0310	Concrete Forming & Accessories	81.9	68.2	70.1	85.8	54.4	58.7	88.6	46.1	51.9	93.8	55.7	60.9	119.8	112.9	113.8	106.2	126.7	123.9
0320	Concrete Reinforcing	99.7	66.4	86.1	99.1	63.7	84.7	99.2	69.7	87.2	95.3	63.3	82.3	101.6	118.0	108.3	93.2	118.2	103.4
0330	Cast-in-Place Concrete	82.4	68.2	77.0	82.3	51.5	70.5	89.9	49.0	74.2	89.8	62.3	79.3	102.9	120.7	109.7	101.2	123.5	109.7
03	CONCRETE	86.9	68.2	78.4	84.6	55.9	71.5	86.1	52.5	70.8	91.8	61.0	77.8	107.7	115.7	111.4	104.4	122.9	112.8
04	MASONRY	118.5	57.0	81.6	96.7	43.3	64.7	97.4	33.2	59.0	80.0	48.6	61.2	124.8	113.9	118.3	88.8	115.3	104.7
05	METALS	101.3	71.4	92.6	100.6	68.7	91.3	93.7	69.9	86.8	96.0	81.3	91.7	87.1	100.7	91.1	106.4	103.8	105.6
06	WOOD, PLASTICS & COMPOSITES	83.1	73.2	77.6	88.1	56.3	70.7	91.9	49.2	68.4	96.3	56.5	74.5	95.3	108.8	102.7	94.5	126.5	112.1
07	THERMAL & MOISTURE PROTECTION	98.0	56.6	81.9	99.1	48.9	79.6	98.9	45.9	78.3	102.9	56.3	84.8	101.5	109.6	104.6	102.5	116.7	108.1
08	OPENINGS	92.5	65.1	85.7	96.3	54.7	86.0	97.7	52.0	86.4	99.1	58.7	89.1	92.3	109.6	96.6	104.5	121.5	108.7
0920	Plaster & Gypsum Board	80.6	72.7	75.5	82.1	55.3	64.7	84.8	47.9	60.8	100.1	55.1	70.9	116.4	109.1	111.7	105.6	127.3	119.7
0950, 0980	Ceilings & Acoustic Treatment	88.6	72.7	78.8	87.7	55.3	67.7	94.5	47.9	65.7	89.8	55.1	68.4	99.7	109.1	105.5	114.8	127.3	122.6
0960	Flooring	101.8	68.2	92.7	103.2	66.2	93.2	104.4	57.2	91.6	77.6	59.9	72.8	102.5	99.9	101.8	114.2	113.6	114.1
0970, 0990	Wall Finishes & Painting/Coating	98.5	63.5	77.5	98.5	40.2	63.5	98.5	32.6	59.0	87.4	55.6	68.3	106.9	111.7	109.8	104.7	105.8	105.3
09	FINISHES	90.9	68.9	79.3	91.1	55.2	72.1	94.2	46.6	69.0	89.5	56.3	72.0	104.3	109.8	107.2	109.0	122.4	116.1
COVERS	DIVS. 10 - 14, 25, 28, 41, 43, 44	100.0	74.0	94.9	100.0	50.0	90.1	100.0	41.1	88.3	100.0	56.6	91.4	100.0	112.3	102.4	100.0	114.8	102.9
21, 22, 23	FIRE SUPPRESSION, PLUMBING & HVAC	100.1	50.8	80.5	94.3	51.4	77.3	100.1	35.6	74.5	94.5	48.1	76.0	94.4	102.1	97.5	100.1	115.5	106.3
26, 27, 3370	ELECTRICAL, COMMUNICATIONS & UTIL.	95.1	76.6	86.1	95.8	52.2	74.5	97.0	42.0	70.2	105.4	52.2	79.5	117.2	104.6	111.1	91.5	105.2	98.2
MF2004	WEIGHTED AVERAGE	96.9	66.6	84.2	94.5	56.7	78.6	95.9	48.8	76.1	96.0	59.6	80.8	100.2	107.9	103.4	101.5	114.7	107.0

DIVISION		CALIFORNIA																	
		BAKERSFIELD			BERKELEY			EUREKA			FRESNO			INGLEWOOD			LONG BEACH		
		932 - 933			947			955			936 - 938			903 - 905			906 - 908		
		MAT.	INST.	TOTAL	MAT.	INST.	TOTAL	MAT.	INST.	TOTAL	MAT.	INST.	TOTAL	MAT.	INST.	TOTAL	MAT.	INST.	TOTAL
015433	CONTRACTOR EQUIPMENT		100.7	100.7		104.6	104.6		100.6	100.6		100.7	100.7		99.1	99.1		99.1	99.1
0241, 31 - 34	SITE & INFRASTRUCTURE, DEMOLITION	108.0	107.8	107.9	134.7	105.6	114.8	116.4	105.0	108.6	107.6	107.2	107.4	91.7	108.0	102.8	99.3	108.0	105.2
0310	Concrete Forming & Accessories	102.1	126.1	122.8	120.8	135.7	133.6	116.2	125.5	124.2	102.5	126.4	123.2	106.8	112.8	112.0	101.6	112.9	111.4
0320	Concrete Reinforcing	105.0	118.3	110.4	96.0	118.7	105.3	101.9	118.5	108.7	88.9	118.5	101.0	104.4	118.0	110.0	103.5	118.1	109.4
0330	Cast-in-Place Concrete	108.3	122.4	113.7	140.2	114.6	130.4	109.8	116.5	112.4	106.7	117.0	110.6	99.9	121.5	108.1	113.8	121.5	116.7
03	CONCRETE	107.6	122.3	114.3	122.3	123.9	123.1	115.5	119.9	117.5	105.9	120.5	112.5	102.2	116.0	108.5	112.2	116.1	114.0
04	MASONRY	109.8	117.4	114.3	148.6	119.5	131.2	116.9	121.2	119.5	112.8	118.2	116.1	80.4	114.0	100.5	89.4	114.0	104.1
05	METALS	101.3	103.1	101.8	104.3	105.9	104.8	106.1	101.9	104.8	106.5	102.7	105.4	97.4	101.9	98.7	97.2	101.9	98.6
06	WOOD, PLASTICS & COMPOSITES	87.6	126.6	109.0	120.7	137.6	130.0	107.9	127.0	118.4	99.6	127.0	114.6	90.3	108.7	100.4	84.0	108.7	97.6
07	THERMAL & MOISTURE PROTECTION	100.5	109.8	104.1	109.8	122.6	114.8	106.6	110.3	108.0	96.4	112.5	102.7	104.0	111.1	106.7	104.2	112.4	107.4
08	OPENINGS	102.0	119.4	106.3	104.1	128.5	110.2	103.3	110.6	105.1	103.8	118.1	107.3	92.1	109.6	96.4	92.0	109.6	96.4
0920	Plaster & Gypsum Board	100.8	127.3	118.0	114.2	138.4	129.9	109.9	127.8	121.5	101.3	127.8	118.5	103.0	109.1	107.0	100.0	109.1	105.9
0950, 0980	Ceilings & Acoustic Treatment	116.4	127.3	123.2	117.0	138.4	130.2	119.8	127.8	124.7	119.5	127.8	124.6	113.6	109.1	110.8	113.6	109.1	110.8
0960	Flooring	114.2	87.9	107.1	123.2	116.2	121.3	119.0	110.8	116.8	121.1	133.7	124.5	104.1	99.9	103.0	100.8	99.9	100.5
0970, 0990	Wall Finishes & Painting/Coating	106.2	101.6	103.5	108.1	128.2	120.2	105.5	53.1	74.1	124.4	100.4	110.0	97.5	111.7	106.0	97.5	111.7	106.0
09	FINISHES	110.5	117.3	114.1	116.7	132.5	125.1	115.0	116.6	115.8	114.9	126.0	120.7	106.7	109.8	108.3	105.8	109.8	107.9
COVERS	DIVS. 10 - 14, 25, 28, 41, 43, 44	100.0	115.2	103.0	100.0	129.0	105.8	100.0	126.2	105.2	100.0	126.3	105.2	100.0	112.2	102.4	100.0	112.2	102.4
21, 22, 23	FIRE SUPPRESSION, PLUMBING & HVAC	100.1	99.9	100.0	94.5	129.3	108.3	94.4	107.5	99.6	100.3	111.6	104.8	94.5	102.1	97.5	94.5	102.1	97.5
26, 27, 3370	ELECTRICAL, COMMUNICATIONS & UTIL.	104.2	97.6	101.0	105.6	130.0	117.5	99.9	90.1	95.1	90.7	94.5	92.5	103.4	104.6	104.0	103.1	104.6	103.8
MF2004	WEIGHTED AVERAGE	103.3	109.4	105.8	108.3	123.8	114.8	104.7	109.6	106.7	103.2	112.6	107.2	97.6	107.9	101.9	99.3	108.0	102.9

DIVISION		CALIFORNIA																	
		LOS ANGELES			MARYSVILLE			MODESTO			MOJAVE			OAKLAND			OXNARD		
		900 - 902			959			953			935			946			930		
		MAT.	INST.	TOTAL	MAT.	INST.	TOTAL	MAT.	INST.	TOTAL	MAT.	INST.	TOTAL	MAT.	INST.	TOTAL	MAT.	INST.	TOTAL
015433	CONTRACTOR EQUIPMENT		100.3	100.3		100.6	100.6		100.6	100.6		100.7	100.7		104.6	104.6		100.0	100.0
0241, 31 - 34	SITE & INFRASTRUCTURE, DEMOLITION	98.5	110.1	106.4	112.4	106.1	108.1	106.0	107.1	106.8	100.6	107.8	105.5	141.8	105.6	117.1	108.0	106.3	106.9
0310	Concrete Forming & Accessories	104.2	126.7	123.6	105.6	126.0	123.2	101.5	126.4	123.0	113.6	111.6	111.9	108.4	142.5	137.8	105.7	126.6	123.8
0320	Concrete Reinforcing	104.4	118.6	110.2	101.9	118.0	108.5	105.6	118.1	110.7	107.0	117.8	111.4	98.2	119.2	106.8	105.0	117.9	110.3
0330	Cast-in-Place Concrete	108.8	122.0	113.9	122.8	116.6	120.4	109.9	117.2	112.7	92.3	122.3	103.7	132.9	120.7	128.2	107.5	122.7	113.3
03	CONCRETE	108.7	122.4	115.0	117.5	120.1	118.7	108.5	120.5	114.0	98.7	115.8	106.5	122.1	129.1	125.3	107.5	122.6	114.3
04	MASONRY	95.5	121.8	111.2	117.9	110.1	113.2	114.2	110.2	111.8	110.9	110.5	110.6	156.6	127.6	139.2	114.7	109.9	111.8
05	METALS	104.0	103.2	103.8	105.6	101.3	104.3	102.7	102.9	102.7	102.9	102.2	102.7	98.8	106.9	101.2	101.1	103.1	101.7
06	WOOD, PLASTICS & COMPOSITES	86.8	126.3	108.5	94.8	127.0	112.5	90.1	127.0	110.4	100.1	107.6	104.2	106.4	146.2	128.2	94.3	126.6	112.0
07	THERMAL & MOISTURE PROTECTION	104.2	118.8	109.9	106.1	110.4	107.7	105.7	110.8	107.7	101.6	105.1	103.0	107.3	125.6	114.4	105.8	112.5	108.4
08	OPENINGS	98.3	121.4	104.0	102.5	118.9	106.5	101.2	120.1	105.9	97.8	106.4	100.0	104.2	133.2	111.4	100.9	121.6	106.0
0920	Plaster & Gypsum Board	104.4	127.3	119.3	103.8	127.8	119.4	106.0	127.8	120.1	113.1	107.6	109.5	108.3	147.3	133.6	105.6	127.3	119.7
0950, 0980	Ceilings & Acoustic Treatment	126.1	127.3	126.9	119.0	127.8	124.4	114.8	127.8	122.8	119.0	107.6	112.0	120.5	147.3	137.0	121.4	127.3	125.1
0960	Flooring	102.3	113.6	105.3	113.6	109.6	112.5	114.2	105.4	111.8	121.5	87.9	112.4	117.6	116.2	117.2	112.7	113.6	113.0
0970, 0990	Wall Finishes & Painting/Coating	97.5	111.7	106.0	105.5	104.7	105.0	105.5	99.8	102.1	105.5	90.0	96.2	108.1	144.5	129.9	105.5	96.5	100.1
09	FINISHES	109.6	122.9	116.6	111.8	122.1	117.3	110.8	120.9	116.1	114.7	104.6	109.4	115.6	139.4	128.2	111.8	121.4	116.9
COVERS	DIVS. 10 - 14, 25, 28, 41, 43, 44	100.0	114.3	102.8	100.0	126.3	105.2	100.0	126.3	105.2	100.0	113.0	102.6	100.0	130.1	106.0	100.0	115.0	103.0
21, 22, 23	FIRE SUPPRESSION, PLUMBING & HVAC	100.3	115.5	106.3	94.4	103.5	98.0	100.2	111.6	104.7	94.4	99.7	96.5	100.3	143.0	117.3	100.2	115.6	106.3
26, 27, 3370	ELECTRICAL, COMMUNICATIONS & UTIL.	100.4	114.5	107.3	96.4	107.6	101.9	99.1	98.9	99.0	89.5	91.6	90.5	104.9	130.0	117.1	96.4	99.5	97.9
MF2004	WEIGHTED AVERAGE	102.2	116.7	108.3	104.0	111.2	107.0	103.4	111.8	106.9	99.6	104.3	101.6	108.8	129.5	117.5	102.9	112.8	107.0

	DIVISION	CALIFORNIA																	
		PALM SPRINGS			PALO ALTO			PASADENA			REDDING			RICHMOND			RIVERSIDE		
		922			943			910 - 912			960			948			925		
		MAT.	INST.	TOTAL	MAT.	INST.	TOTAL	MAT.	INST.	TOTAL	MAT.	INST.	TOTAL	MAT.	INST.	TOTAL	MAT.	INST.	TOTAL
015433	CONTRACTOR EQUIPMENT		101.4	101.4		104.6	104.6		99.1	99.1		100.6	100.6		104.6	104.6		101.4	101.4
0241, 31 - 34	SITE & INFRASTRUCTURE, DEMOLITION	92.9	107.8	103.1	129.8	105.6	113.2	99.7	110.6	107.1	112.2	106.2	108.1	141.3	105.6	116.9	100.1	107.8	105.4
0310	Concrete Forming & Accessories	102.5	113.0	111.6	106.2	135.7	131.6	108.7	112.9	112.3	105.4	126.0	123.2	123.7	142.0	139.5	106.4	126.5	123.7
0320	Concrete Reinforcing	107.2	117.9	111.5	96.0	118.7	105.2	102.5	118.0	108.8	101.9	118.0	108.5	96.0	118.7	105.3	104.1	118.0	109.8
0330	Cast-in-Place Concrete	96.5	123.2	106.7	118.5	120.4	119.2	97.6	120.7	106.4	121.8	116.7	119.8	136.3	120.5	130.3	105.0	123.4	112.0
03	CONCRETE	100.2	116.8	107.7	110.2	125.9	117.4	103.2	115.7	108.9	116.4	120.1	118.1	124.2	128.7	126.2	106.2	122.8	113.7
04	MASONRY	86.8	110.8	101.1	122.7	119.5	120.8	106.9	113.9	111.1	117.9	110.1	113.3	148.3	119.5	131.1	87.8	110.8	101.6
05	METALS	107.1	102.8	105.9	96.6	105.9	99.3	87.2	100.7	91.1	105.9	101.6	104.7	96.6	105.9	99.3	106.6	103.3	105.6
06	WOOD, PLASTICS & COMPOSITES	89.3	108.9	100.1	103.6	137.6	122.3	82.3	108.8	96.8	94.5	127.0	112.3	124.9	146.2	136.6	94.5	126.5	112.1
07	THERMAL & MOISTURE PROTECTION	102.6	108.9	105.0	107.0	123.9	113.5	101.2	109.6	104.4	106.2	108.9	107.3	107.6	123.4	113.7	102.8	111.8	106.3
08	OPENINGS	99.9	109.7	102.3	104.2	128.5	110.2	92.3	109.6	96.6	103.5	118.9	107.3	104.2	133.2	111.4	103.0	121.5	107.6
0920	Plaster & Gypsum Board	101.6	109.1	106.5	106.8	138.4	127.3	111.6	109.1	110.0	104.0	127.8	119.4	115.8	147.3	136.2	105.0	127.3	119.5
0950, 0980	Ceilings & Acoustic Treatment	112.3	109.1	110.4	118.6	138.4	130.8	99.7	109.1	105.5	126.4	127.8	127.2	118.6	147.3	136.3	118.9	127.3	124.1
0960	Flooring	116.2	99.9	111.8	116.7	113.3	115.8	96.7	99.9	97.6	113.5	109.6	112.4	125.2	116.2	122.8	118.4	113.6	117.1
0970, 0990	Wall Finishes & Painting/Coating	102.7	127.9	117.8	108.1	128.2	120.2	106.9	111.7	109.8	105.5	104.7	105.0	108.1	128.2	120.2	102.7	105.8	104.6
09	FINISHES	107.7	111.7	109.8	113.9	132.0	123.5	101.7	109.8	106.0	113.4	122.1	118.0	118.5	137.5	128.6	110.7	122.4	116.9
COVERS	DIVS. 10 - 14, 25, 28, 41, 43, 44	100.0	112.7	102.5	100.0	129.0	105.7	100.0	112.3	102.4	100.0	126.3	105.2	100.0	130.0	105.9	100.0	114.7	102.9
21, 22, 23	FIRE SUPPRESSION, PLUMBING & HVAC	94.4	102.2	97.5	94.5	128.4	108.0	94.4	102.1	97.5	100.2	103.5	101.5	94.5	125.6	106.9	100.1	115.5	106.2
26, 27, 3370	ELECTRICAL, COMMUNICATIONS & UTIL.	95.1	96.3	95.7	104.7	128.9	116.5	114.1	104.6	109.4	99.4	98.1	98.8	105.3	124.4	114.6	91.6	96.3	93.9
MF2004	WEIGHTED AVERAGE	99.2	106.8	102.4	103.5	123.7	112.0	98.0	107.9	102.2	105.8	109.8	107.5	107.4	123.9	114.3	101.6	112.7	106.3

	DIVISION	CALIFORNIA																	
		SACRAMENTO			SALINAS			SAN BERNARDINO			SAN DIEGO			SAN FRANCISCO			SAN JOSE		
		942,956 - 958			939			923 - 924			919 - 921			940 - 941			951		
		MAT.	INST.	TOTAL	MAT.	INST.	TOTAL	MAT.	INST.	TOTAL	MAT.	INST.	TOTAL	MAT.	INST.	TOTAL	MAT.	INST.	TOTAL
015433	CONTRACTOR EQUIPMENT		102.9	102.9		100.7	100.7		101.4	101.4		99.6	99.6		108.1	108.1		102.4	102.4
0241, 31 - 34	SITE & INFRASTRUCTURE, DEMOLITION	111.5	110.7	110.9	123.1	107.6	112.5	77.9	107.8	98.3	105.6	104.1	104.6	144.3	111.2	121.7	144.1	103.4	116.3
0310	Concrete Forming & Accessories	106.9	127.0	124.2	109.1	129.6	126.8	110.1	112.8	112.5	108.1	111.0	110.6	108.2	144.2	139.3	107.9	143.1	138.3
0320	Concrete Reinforcing	90.8	118.6	102.1	105.6	118.5	110.9	104.1	118.2	109.9	99.3	118.2	107.0	111.8	119.9	115.1	92.9	119.3	103.7
0330	Cast-in-Place Concrete	112.3	118.0	114.5	106.2	117.5	110.5	72.4	123.2	91.8	107.0	105.8	106.6	136.3	122.7	131.1	127.8	120.7	125.1
03	CONCRETE	110.9	121.1	115.6	117.0	122.1	119.3	80.3	116.7	96.8	108.5	109.9	109.1	126.4	130.8	128.4	114.6	129.5	121.4
04	MASONRY	125.0	118.3	121.0	110.5	116.4	114.0	93.8	117.6	108.1	98.7	111.5	106.4	157.2	136.0	144.5	148.2	128.9	136.6
05	METALS	94.5	103.4	97.1	105.7	103.9	105.2	106.6	103.1	105.6	103.5	103.1	103.4	104.9	110.5	106.5	100.5	109.6	103.2
06	WOOD, PLASTICS & COMPOSITES	100.3	127.2	115.1	99.1	130.5	116.3	98.6	108.9	104.3	95.3	107.6	102.0	106.4	146.4	128.3	101.9	145.9	126.1
07	THERMAL & MOISTURE PROTECTION	112.8	114.5	113.4	102.5	116.4	107.9	101.8	111.7	105.7	108.4	105.6	107.3	109.9	138.3	120.9	102.1	135.8	115.2
08	OPENINGS	117.4	120.2	118.1	102.3	124.6	107.8	100.0	109.7	102.4	99.9	109.8	102.3	108.6	135.8	115.3	93.9	135.5	104.2
0920	Plaster & Gypsum Board	103.4	127.8	119.2	106.7	131.4	122.8	106.8	109.1	108.3	117.3	107.6	111.0	110.7	147.3	134.5	100.9	147.3	131.0
0950, 0980	Ceilings & Acoustic Treatment	118.6	127.8	124.3	119.0	131.4	126.7	114.8	109.1	111.3	107.2	107.6	107.4	127.8	147.3	139.8	110.5	147.3	133.3
0960	Flooring	117.0	114.3	116.3	115.1	116.2	115.4	120.7	94.1	113.5	105.4	107.6	106.0	117.6	126.2	119.9	106.2	126.2	111.6
0970, 0990	Wall Finishes & Painting/Coating	105.2	114.3	110.7	106.6	146.2	130.4	102.7	102.2	102.4	103.8	111.7	108.5	108.1	156.9	137.4	106.2	146.2	130.2
09	FINISHES	112.2	124.3	118.6	113.9	130.0	122.4	109.1	107.7	108.4	107.6	110.1	108.9	117.8	142.9	131.1	108.6	141.4	126.0
COVERS	DIVS. 10 - 14, 25, 28, 41, 43, 44	100.0	126.9	105.3	100.0	126.8	105.3	100.0	101.8	100.4	100.0	111.7	102.3	100.0	130.7	106.1	100.0	129.5	105.8
21, 22, 23	FIRE SUPPRESSION, PLUMBING & HVAC	100.1	112.9	105.2	94.4	111.9	101.4	94.4	102.2	97.5	100.3	113.8	105.6	100.3	164.9	126.0	100.2	145.6	118.3
26, 27, 3370	ELECTRICAL, COMMUNICATIONS & UTIL.	99.6	103.7	101.6	90.7	113.0	101.5	95.1	98.8	96.9	97.7	104.1	100.8	104.9	162.2	132.8	103.1	141.0	121.6
MF2004	WEIGHTED AVERAGE	104.9	114.6	108.9	103.3	116.4	108.8	96.9	107.2	101.2	102.4	108.8	105.1	11[illegible].2	141.3	123.8	106.0	132.5	117.1

	DIVISION	CALIFORNIA																	
		SAN LUIS OBISPO			SAN MATEO			SAN RAFAEL			SANTA ANA			SANTA BARBARA			SANTA CRUZ		
		934			944			949			926 - 927			931			950		
		MAT.	INST.	TOTAL	MAT.	INST.	TOTAL	MAT.	INST.	TOTAL	MAT.	INST.	TOTAL	MAT.	INST.	TOTAL	MAT.	INST.	TOTAL
015433	CONTRACTOR EQUIPMENT		100.7	100.7		104.6	104.6		103.0	103.0		101.4	101.4		100.7	100.7		102.4	102.4
0241, 31 - 34	SITE & INFRASTRUCTURE, DEMOLITION	114.2	107.8	109.8	138.3	105.6	116.0	126.8	109.0	114.6	91.2	107.8	102.5	107.9	107.8	107.9	143.3	103.2	115.9
0310	Concrete Forming & Accessories	115.5	112.7	113.1	112.8	142.4	138.4	117.9	141.7	138.5	110.8	113.0	112.7	106.5	126.4	123.7	107.9	129.8	126.8
0320	Concrete Reinforcing	107.0	117.8	111.4	96.0	118.9	105.3	96.6	118.8	105.6	107.7	118.2	112.0	105.0	118.2	110.4	115.1	118.5	116.5
0330	Cast-in-Place Concrete	114.1	122.3	117.2	132.0	120.6	127.7	153.7	119.2	140.5	92.6	123.2	104.3	107.1	122.6	113.0	126.9	119.4	124.0
03	CONCRETE	115.4	116.3	115.8	120.3	129.0	124.3	144.1	128.0	136.7	97.8	116.8	106.4	107.3	122.4	114.2	118.7	123.1	120.7
04	MASONRY	112.6	110.4	111.3	148.0	122.5	132.7	122.1	122.5	122.3	82.8	117.5	103.6	111.1	114.7	113.2	152.6	116.6	131.0
05	METALS	103.6	102.2	103.2	96.5	106.5	99.4	97.7	102.7	99.2	106.7	103.2	105.7	101.5	103.2	102.0	108.4	107.4	108.1
06	WOOD, PLASTICS & COMPOSITES	102.3	109.0	106.0	112.0	146.2	130.8	110.4	145.9	129.9	100.2	108.9	105.0	94.3	126.6	112.0	101.9	130.6	117.7
07	THERMAL & MOISTURE PROTECTION	102.5	105.9	103.8	107.3	125.9	114.6	112.3	122.6	116.3	102.9	110.6	105.9	101.9	109.6	104.9	101.9	119.1	108.6
08	OPENINGS	100.1	107.3	101.9	104.2	133.2	111.3	114.6	133.0	119.1	99.2	109.7	101.8	102.0	121.6	106.8	95.2	124.7	102.5
0920	Plaster & Gypsum Board	114.2	109.1	110.9	110.9	147.3	134.5	113.4	147.3	135.4	107.1	109.1	108.4	105.6	127.3	119.7	114.0	131.4	125.3
0950, 0980	Ceilings & Acoustic Treatment	119.0	109.1	112.9	118.6	147.3	136.3	126.1	147.3	139.2	114.8	109.1	111.3	121.4	127.3	125.1	116.5	131.4	125.7
0960	Flooring	122.7	87.9	113.3	119.7	116.2	118.8	131.1	113.3	126.3	121.3	99.9	115.6	114.2	87.9	107.1	110.8	116.2	112.3
0970, 0990	Wall Finishes & Painting/Coating	105.5	96.5	100.1	108.1	128.2	120.2	104.3	128.2	118.6	102.7	102.2	102.4	105.5	96.5	100.1	106.6	146.2	130.4
09	FINISHES	116.3	105.9	110.8	116.0	137.5	127.4	119.0	136.8	128.5	110.6	108.9	109.7	112.5	117.0	114.8	113.2	130.1	122.1
COVERS	DIVS. 10 - 14, 25, 28, 41, 43, 44	100.0	113.2	102.6	100.0	130.1	106.0	100.0	129.3	105.8	100.0	112.7	102.5	100.0	115.0	103.0	100.0	127.1	105.4
21, 22, 23	FIRE SUPPRESSION, PLUMBING & HVAC	94.4	102.2	97.5	94.5	144.0	114.2	94.4	139.7	112.4	94.4	102.2	97.5	100.2	115.5	106.3	100.2	111.9	104.9
26, 27, 3370	ELECTRICAL, COMMUNICATIONS & UTIL.	89.5	93.0	91.2	104.7	132.3	118.1	99.4	107.6	103.4	95.1	98.3	96.7	88.4	100.3	94.2	102.1	113.0	107.4
MF2004	WEIGHTED AVERAGE	102.5	105.3	103.7	106.4	129.3	116.0	108.7	124.5	115.3	98.9	107.5	102.5	101.9	112.9	106.5	108.4	116.6	111.8

	DIVISION	CALIFORNIA															COLORADO		
		SANTA ROSA			STOCKTON			SUSANVILLE			VALLEJO			VAN NUYS			ALAMOSA		
		954			952			961			945			913 - 916			811		
		MAT.	INST.	TOTAL	MAT.	INST.	TOTAL	MAT.	INST.	TOTAL	MAT.	INST.	TOTAL	MAT.	INST.	TOTAL	MAT.	INST.	TOTAL
015433	CONTRACTOR EQUIPMENT		100.8	100.8		100.6	100.6		100.6	100.6		103.0	103.0		99.1	99.1		95.8	95.8
0241, 31 - 34	SITE & INFRASTRUCTURE, DEMOLITION	106.6	106.6	106.6	105.7	107.1	106.7	119.7	106.2	110.5	110.6	110.1	110.2	118.6	110.6	113.1	126.0	91.3	102.3
0310	Concrete Forming & Accessories	103.6	141.6	136.5	106.0	126.3	123.6	106.8	126.0	123.4	107.3	141.4	136.8	115.2	112.9	113.2	105.6	71.5	76.1
0320	Concrete Reinforcing	102.8	119.6	109.7	105.6	118.1	110.7	101.9	118.0	108.5	97.8	118.8	106.4	102.5	118.0	108.8	108.2	78.8	96.2
0330	Cast-in-Place Concrete	120.6	118.6	119.9	107.0	117.2	110.9	110.7	116.7	113.0	122.3	119.1	121.1	102.9	120.7	109.7	107.0	80.5	96.9
03	CONCRETE	117.2	128.0	122.1	107.4	120.5	113.4	117.5	120.1	118.7	117.0	127.9	122.0	117.1	115.7	116.5	121.1	76.4	100.8
04	MASONRY	115.3	134.7	126.9	114.3	110.2	111.9	116.8	110.1	112.8	88.9	121.4	108.3	124.8	113.9	118.3	131.3	73.8	96.8
05	METALS	107.0	106.1	106.7	102.9	102.9	102.9	104.9	101.6	104.0	97.7	104.4	99.6	86.3	100.7	90.5	97.0	83.1	93.0
06	WOOD, PLASTICS & COMPOSITES	90.6	145.7	120.8	96.1	127.0	113.1	96.5	127.0	113.2	98.0	145.9	124.3	90.3	108.8	100.4	98.9	73.2	84.8
07	THERMAL & MOISTURE PROTECTION	103.3	125.9	112.1	105.8	110.1	107.4	106.8	106.3	106.6	110.0	122.1	114.7	102.1	109.6	105.0	104.9	78.1	94.5
08	OPENINGS	100.9	135.4	109.4	101.2	120.1	105.9	103.4	118.9	107.2	117.0	135.5	121.6	92.1	109.6	96.5	92.1	78.9	88.8
0920	Plaster & Gypsum Board	102.7	147.3	131.6	106.0	127.8	120.1	104.9	127.8	119.7	108.2	147.3	133.6	114.7	109.1	111.1	76.5	72.0	73.5
0950, 0980	Ceilings & Acoustic Treatment	114.8	147.3	134.9	121.4	127.8	125.3	119.0	127.8	124.4	128.0	147.3	139.9	97.2	109.1	104.6	107.8	72.0	85.7
0960	Flooring	116.9	126.2	119.4	114.2	105.4	111.8	114.2	109.6	113.0	125.9	126.2	126.0	100.1	99.9	100.0	114.9	61.4	100.4
0970, 0990	Wall Finishes & Painting/Coating	102.7	128.2	118.0	105.5	105.1	105.3	105.5	118.2	113.1	105.1	128.2	119.0	106.9	111.7	109.8	108.9	27.1	59.8
09	FINISHES	109.6	138.9	125.1	112.3	121.4	117.2	113.0	123.6	118.6	115.6	139.1	128.0	103.8	109.8	107.0	107.1	64.7	84.7
COVERS	DIVS. 10 - 14, 25, 28, 41, 43, 44	100.0	128.3	105.6	100.0	126.3	105.2	100.0	126.3	105.2	100.0	128.9	105.7	100.0	112.3	102.4	100.0	87.3	97.5
21, 22, 23	FIRE SUPPRESSION, PLUMBING & HVAC	94.4	138.8	112.0	100.2	103.7	101.6	94.4	103.5	98.0	100.2	117.5	107.1	94.4	102.1	97.5	94.2	67.8	83.7
26, 27, 3370	ELECTRICAL, COMMUNICATIONS & UTIL.	95.4	107.6	101.4	99.1	111.7	105.2	99.8	98.1	99.0	95.6	116.5	105.8	114.1	104.6	109.4	95.7	77.6	86.9
MF2004	WEIGHTED AVERAGE	103.3	126.1	112.9	103.5	112.0	107.0	104.6	110.0	106.8	104.1	121.6	111.4	101.2	107.9	104.0	102.4	75.3	91.0

	DIVISION	COLORADO																	
		BOULDER			COLORADO SPRINGS			DENVER			DURANGO			FORT COLLINS			FORT MORGAN		
		803			808 - 809			800 - 802			813			805			807		
		MAT.	INST.	TOTAL	MAT.	INST.	TOTAL	MAT.	INST.	TOTAL	MAT.	INST.	TOTAL	MAT.	INST.	TOTAL	MAT.	INST.	TOTAL
015433	CONTRACTOR EQUIPMENT		96.0	96.0		94.5	94.5		98.3	98.3		95.8	95.8		96.0	96.0		96.0	96.0
0241, 31 - 34	SITE & INFRASTRUCTURE, DEMOLITION	88.4	97.1	94.4	89.9	94.0	92.7	89.0	102.2	98.1	120.0	91.3	100.4	99.8	96.8	97.8	91.0	96.9	95.1
0310	Concrete Forming & Accessories	103.0	76.4	80.0	93.6	80.7	82.5	99.9	82.1	84.5	111.3	71.6	77.0	100.5	74.6	78.2	103.3	75.8	79.5
0320	Concrete Reinforcing	107.8	78.9	96.0	106.9	79.6	95.8	106.9	79.7	95.8	108.2	78.8	96.2	107.9	77.7	95.6	108.0	78.8	96.1
0330	Cast-in-Place Concrete	99.0	80.3	91.9	101.8	82.9	94.5	96.3	84.2	91.7	123.1	80.5	106.8	111.9	78.1	98.9	97.2	79.4	90.4
03	CONCRETE	104.7	78.6	92.9	108.0	81.5	96.0	102.5	82.6	93.5	123.7	76.5	102.2	114.9	76.8	97.6	103.4	77.9	91.8
04	MASONRY	97.0	75.2	83.9	101.4	75.9	86.1	103.3	79.0	88.7	119.0	73.8	91.9	112.1	52.7	76.5	112.9	73.9	89.6
05	METALS	98.8	85.5	94.9	101.7	86.0	97.1	104.2	86.1	98.9	97.0	83.3	93.0	99.8	81.5	94.5	98.5	83.0	94.0
06	WOOD, PLASTICS & COMPOSITES	103.7	78.6	89.9	94.4	84.4	88.9	102.1	84.2	92.3	106.7	73.2	88.3	101.9	78.6	89.1	104.1	78.6	90.1
07	THERMAL & MOISTURE PROTECTION	99.3	79.0	91.4	100.0	80.2	92.3	98.9	76.6	90.2	104.7	78.1	94.4	99.7	64.4	86.0	99.3	74.4	89.6
08	OPENINGS	95.3	81.8	92.0	99.5	85.2	96.0	99.4	85.1	95.9	99.3	78.9	94.3	95.3	81.8	92.0	95.3	81.8	91.9
0920	Plaster & Gypsum Board	98.0	78.0	85.0	82.4	83.9	83.4	95.8	83.9	88.1	85.4	72.0	76.7	94.3	78.0	83.7	99.1	78.0	85.4
0950, 0980	Ceilings & Acoustic Treatment	100.5	78.0	86.6	107.2	83.9	92.8	106.3	83.9	92.4	107.8	72.0	85.7	100.5	78.0	86.6	100.5	78.0	86.6
0960	Flooring	113.0	86.7	105.9	104.0	76.5	96.6	109.2	88.8	103.7	120.0	61.4	104.1	109.0	61.4	96.2	113.6	61.4	99.5
0970, 0990	Wall Finishes & Painting/Coating	106.0	76.5	88.3	105.7	46.6	70.2	106.0	76.5	88.3	108.9	27.1	59.8	106.0	46.2	70.1	106.0	61.6	79.3
09	FINISHES	103.3	77.7	89.7	100.0	76.7	87.7	103.1	83.0	92.4	109.1	64.7	85.6	102.0	69.4	84.8	103.5	71.6	86.7
COVERS	DIVS. 10 - 14, 25, 28, 41, 43, 44	100.0	86.8	97.4	100.0	87.8	97.6	100.0	88.6	97.7	100.0	87.3	97.5	100.0	85.9	97.2	100.0	86.8	97.4
21, 22, 23	FIRE SUPPRESSION, PLUMBING & HVAC	94.2	80.4	88.7	100.1	75.1	90.2	100.0	86.7	94.7	94.2	70.2	84.7	100.0	79.1	91.7	94.2	80.3	88.7
26, 27, 3370	ELECTRICAL, COMMUNICATIONS & UTIL.	101.2	85.4	93.5	104.9	83.3	94.3	106.7	85.4	96.3	95.1	78.7	87.1	101.2	85.3	93.5	101.5	85.3	93.6
MF2004	WEIGHTED AVERAGE	98.6	82.0	91.6	101.5	81.0	92.9	101.8	85.5	95.0	102.8	76.0	91.6	102.2	77.3	91.7	99.2	80.6	91.4

	DIVISION	COLORADO																	
		GLENWOOD SPRINGS			GOLDEN			GRAND JUNCTION			GREELEY			MONTROSE			PUEBLO		
		816			804			815			806			814			810		
		MAT.	INST.	TOTAL	MAT.	INST.	TOTAL	MAT.	INST.	TOTAL	MAT.	INST.	TOTAL	MAT.	INST.	TOTAL	MAT.	INST.	TOTAL
015433	CONTRACTOR EQUIPMENT		98.9	98.9		96.0	96.0		98.9	98.9		96.0	96.0		97.4	97.4		95.8	95.8
0241, 31 - 34	SITE & INFRASTRUCTURE, DEMOLITION	134.6	99.9	110.9	100.2	97.3	98.2	119.3	99.5	105.7	87.3	95.5	92.9	128.3	95.2	105.7	111.1	93.0	98.7
0310	Concrete Forming & Accessories	102.7	75.6	79.3	96.1	75.8	78.5	110.1	75.0	79.8	98.5	47.2	54.2	102.3	75.5	79.1	107.5	81.0	84.6
0320	Concrete Reinforcing	107.2	78.8	95.6	108.0	78.8	96.1	107.5	77.6	95.3	107.8	77.2	95.3	107.1	78.7	95.5	103.6	79.6	93.8
0330	Cast-in-Place Concrete	107.0	79.8	96.6	97.3	79.4	90.4	118.5	79.0	103.4	93.5	56.1	79.2	107.0	79.8	96.6	106.2	84.0	97.7
03	CONCRETE	126.2	78.0	104.3	113.5	77.9	97.3	119.9	77.2	100.5	100.2	57.0	80.6	117.2	77.9	99.3	109.6	82.1	97.1
04	MASONRY	107.0	74.0	87.2	116.0	75.0	91.5	139.9	54.8	88.9	106.6	35.0	63.7	111.1	73.8	88.8	103.5	74.2	86.0
05	METALS	96.7	83.6	92.9	98.7	83.2	94.2	98.1	80.9	93.1	99.8	80.1	94.0	95.9	82.7	92.0	99.7	86.8	96.0
06	WOOD, PLASTICS & COMPOSITES	94.0	78.7	85.6	97.0	78.6	86.9	103.9	78.7	90.1	99.4	46.9	70.6	94.8	78.9	86.1	101.2	84.7	92.2
07	THERMAL & MOISTURE PROTECTION	104.7	73.8	92.7	100.2	72.3	89.3	103.6	61.1	87.1	99.1	54.1	81.6	104.8	78.7	94.7	103.2	79.6	94.0
08	OPENINGS	98.5	81.9	94.4	95.3	81.8	92.0	99.3	81.9	95.0	95.3	64.6	87.7	99.5	82.0	95.2	93.8	85.4	91.7
0920	Plaster & Gypsum Board	96.9	78.0	84.7	92.4	78.0	83.1	105.1	78.0	87.5	93.1	45.2	62.0	74.2	78.0	76.7	80.6	83.9	82.7
0950, 0980	Ceilings & Acoustic Treatment	106.2	78.0	88.8	100.5	78.0	86.6	106.2	78.0	88.8	100.5	45.2	66.3	107.8	78.0	89.4	117.8	83.9	96.8
0960	Flooring	113.9	56.6	98.4	106.6	61.4	94.4	119.4	61.4	103.7	107.8	61.4	95.2	117.3	50.9	99.3	115.8	88.8	108.5
0970, 0990	Wall Finishes & Painting/Coating	108.9	61.6	80.5	106.0	76.5	88.3	108.9	76.5	89.5	106.0	28.8	59.7	108.9	27.1	59.8	108.9	43.2	69.5
09	FINISHES	109.5	70.9	89.1	101.7	73.3	86.7	110.7	73.4	91.0	100.8	46.5	72.1	107.5	65.9	85.5	108.6	78.8	92.9
COVERS	DIVS. 10 - 14, 25, 28, 41, 43, 44	100.0	87.1	97.4	100.0	86.8	97.4	100.0	87.0	97.4	100.0	78.8	95.8	100.0	87.5	97.5	100.0	88.6	97.8
21, 22, 23	FIRE SUPPRESSION, PLUMBING & HVAC	94.2	70.0	84.6	94.2	80.1	88.6	99.9	65.2	86.1	100.0	74.1	89.7	94.2	70.0	84.6	99.9	68.0	87.2
26, 27, 3370	ELECTRICAL, COMMUNICATIONS & UTIL.	92.4	78.7	85.8	101.5	85.3	93.6	94.7	61.7	78.7	101.2	85.3	93.5	94.7	61.7	78.7	95.7	77.6	86.9
MF2004	WEIGHTED AVERAGE	102.4	77.7	92.1	100.7	80.8	92.4	104.9	72.0	91.1	99.7	67.0	86.0	101.4	74.4	90.1	101.3	79.0	91.9

	DIVISION	COLORADO			CONNECTICUT															
		SALIDA			BRIDGEPORT			BRISTOL			HARTFORD			MERIDEN			NEW BRITAIN		
		812			066			060			061			064			060		
		MAT.	INST.	TOTAL	MAT.	INST.	TOTAL	MAT.	INST.	TOTAL	MAT.	INST.	TOTAL	MAT.	INST.	TOTAL	MAT.	INST.	TOTAL
015433	CONTRACTOR EQUIPMENT		97.4	97.4		100.4	100.4		100.4	100.4		100.4	100.4		100.6	100.6		100.4	100.4
0241, 31 - 34	SITE & INFRASTRUCTURE, DEMOLITION	119.6	95.6	103.2	98.6	104.2	102.5	97.8	104.2	102.2	95.1	104.2	101.3	96.0	104.6	101.9	98.0	104.2	102.2
0310	Concrete Forming & Accessories	110.0	75.6	80.2	96.4	121.9	118.4	96.4	121.6	118.2	94.5	121.6	117.9	96.1	121.8	118.3	96.6	121.6	118.2
0320	Concrete Reinforcing	106.8	78.7	95.4	100.4	127.1	111.3	100.4	127.1	111.3	100.4	127.1	111.3	100.4	127.1	111.3	100.4	127.1	111.3
0330	Cast-in-Place Concrete	122.6	79.8	106.2	100.0	120.2	107.7	93.7	120.1	103.8	96.1	120.1	105.3	90.1	120.1	101.6	95.2	120.1	104.7
03	CONCRETE	118.7	77.9	100.1	107.4	122.0	114.0	104.5	121.8	112.4	105.5	121.8	112.9	102.8	121.9	111.5	105.2	121.8	112.7
04	MASONRY	141.1	73.8	100.8	106.6	132.0	121.8	99.2	132.0	118.8	99.4	132.0	118.9	98.8	132.0	118.7	101.2	132.0	119.6
05	METALS	95.6	82.8	91.8	94.9	123.3	103.2	94.9	123.1	103.1	99.5	123.1	106.4	92.4	123.2	101.4	91.6	123.1	100.8
06	WOOD, PLASTICS & COMPOSITES	102.5	78.9	89.6	101.4	119.6	111.4	101.4	119.6	111.4	99.9	119.6	110.7	101.4	119.6	111.4	101.4	119.6	111.4
07	THERMAL & MOISTURE PROTECTION	103.8	78.7	94.0	94.5	127.7	107.4	94.6	124.1	106.1	95.1	124.1	106.4	94.6	124.1	106.1	94.6	124.5	106.2
08	OPENINGS	92.2	82.0	89.6	104.9	129.4	110.9	104.9	127.4	110.4	105.6	127.4	111.0	107.6	129.4	113.0	104.9	127.4	110.4
0920	Plaster & Gypsum Board	76.1	78.0	77.4	104.8	119.5	114.4	104.8	119.5	114.4	106.1	119.5	114.8	106.6	119.5	115.0	104.8	119.5	114.4
0950, 0980	Ceilings & Acoustic Treatment	107.8	78.0	89.4	99.1	119.5	111.8	99.1	119.5	111.8	99.1	119.5	111.8	103.9	119.5	113.6	99.1	119.5	111.8
0960	Flooring	122.8	50.9	103.3	96.8	125.6	104.6	96.8	111.9	100.9	96.8	111.9	100.9	96.8	125.6	104.6	96.8	111.9	100.9
0970, 0990	Wall Finishes & Painting/Coating	108.9	27.1	59.8	91.7	116.7	106.7	91.7	116.7	106.7	91.7	116.7	106.7	91.7	116.7	106.7	91.7	116.7	106.7
09	FINISHES	108.4	65.9	85.9	100.5	121.9	111.8	100.5	119.5	110.5	100.2	119.5	110.4	101.9	121.9	112.4	100.5	119.5	110.5
COVERS	DIVS. 10 - 14, 25, 28, 41, 43, 44	100.0	87.5	97.5	100.0	111.7	102.3	100.0	111.7	102.3	100.0	111.7	102.3	100.0	111.7	102.3	100.0	111.7	102.3
21, 22, 23	FIRE SUPPRESSION, PLUMBING & HVAC	94.2	67.7	83.6	100.2	114.4	105.9	100.2	114.4	105.8	100.1	114.4	105.8	94.4	114.4	102.3	100.2	114.4	105.8
26, 27, 3370	ELECTRICAL, COMMUNICATIONS & UTIL.	95.0	77.6	86.5	97.0	109.9	103.3	97.0	109.6	103.2	96.9	110.5	103.5	97.0	109.6	103.1	97.1	109.6	103.2
MF2004	WEIGHTED AVERAGE	102.2	76.2	91.3	100.3	118.6	108.0	99.6	118.0	107.3	100.5	118.1	107.9	97.9	118.5	106.6	99.2	118.0	107.1

	DIVISION	CONNECTICUT																	
		NEW HAVEN			NEW LONDON			NORWALK			STAMFORD			WATERBURY			WILLIMANTIC		
		065			063			068			069			067			062		
		MAT.	INST.	TOTAL	MAT.	INST.	TOTAL	MAT.	INST.	TOTAL	MAT.	INST.	TOTAL	MAT.	INST.	TOTAL	MAT.	INST.	TOTAL
015433	CONTRACTOR EQUIPMENT		100.6	100.6		100.6	100.6		100.4	100.4		100.4	100.4		100.4	100.4		100.4	100.4
0241, 31 - 34	SITE & INFRASTRUCTURE, DEMOLITION	97.9	104.6	102.5	90.5	104.6	100.2	98.4	104.2	102.4	99.0	104.3	102.6	98.2	104.2	102.3	98.4	104.2	102.4
0310	Concrete Forming & Accessories	96.1	121.6	118.1	96.1	121.2	117.8	96.4	122.1	118.6	96.4	122.4	118.8	96.4	121.8	118.3	96.4	121.3	117.9
0320	Concrete Reinforcing	100.4	127.1	111.2	78.7	127.0	98.4	100.4	127.2	111.3	100.4	127.3	111.3	100.4	127.1	111.3	100.4	127.0	111.2
0330	Cast-in-Place Concrete	96.9	111.0	102.3	82.5	110.9	93.4	98.4	128.8	110.0	100.0	128.9	111.1	100.0	120.1	107.7	93.4	101.2	96.4
03	CONCRETE	120.8	118.7	119.8	91.9	118.4	103.9	106.7	125.1	115.0	107.4	125.3	115.5	107.4	121.9	114.0	104.3	115.1	109.2
04	MASONRY	99.4	132.0	118.9	97.7	132.0	118.2	98.8	133.7	119.7	99.7	133.7	120.0	99.7	132.0	119.0	99.1	132.0	118.8
05	METALS	91.8	123.0	100.9	91.6	122.6	100.6	94.9	123.6	103.3	94.9	123.9	103.4	94.9	123.2	103.2	94.7	122.8	102.9
06	WOOD, PLASTICS & COMPOSITES	101.4	119.6	111.4	101.4	119.6	111.4	101.4	119.6	111.4	101.4	119.6	111.4	101.4	119.6	111.4	101.4	119.6	111.4
07	THERMAL & MOISTURE PROTECTION	94.7	123.0	105.7	94.6	122.7	105.5	94.7	129.5	108.2	94.6	129.5	108.2	94.6	124.5	106.2	94.8	121.1	105.1
08	OPENINGS	104.9	129.4	110.9	108.0	118.0	110.5	104.9	129.4	110.9	104.9	129.4	110.9	104.9	129.4	110.9	108.1	129.4	113.4
0920	Plaster & Gypsum Board	104.8	119.5	114.4	104.8	119.5	114.4	104.8	119.5	114.4	104.8	119.5	114.4	104.8	119.5	114.4	104.8	119.5	114.4
0950, 0980	Ceilings & Acoustic Treatment	99.1	119.5	111.8	97.3	119.5	111.0	99.1	119.5	111.8	99.1	119.5	111.8	99.1	119.5	111.8	97.3	119.5	111.0
0960	Flooring	96.8	125.6	104.6	96.8	111.9	100.9	96.8	120.6	103.2	96.8	120.6	103.2	96.8	125.6	104.6	96.8	125.6	104.6
0970, 0990	Wall Finishes & Painting/Coating	91.7	116.7	106.7	91.7	116.7	106.7	91.7	116.7	106.7	91.7	116.7	106.7	91.7	116.7	106.7	91.7	116.7	106.7
09	FINISHES	100.5	121.9	111.8	99.6	119.5	110.1	100.5	121.0	111.3	100.6	121.0	111.4	100.4	121.9	111.7	100.2	121.9	111.6
COVERS	DIVS. 10 - 14, 25, 28, 41, 43, 44	100.0	111.7	102.3	100.0	111.7	102.3	100.0	111.7	102.3	100.0	111.9	102.3	100.0	111.7	102.3	100.0	111.7	102.3
21, 22, 23	FIRE SUPPRESSION, PLUMBING & HVAC	100.2	114.4	105.8	94.4	114.4	102.3	100.2	114.5	105.9	100.2	114.5	105.9	100.2	114.4	105.9	100.2	114.2	105.8
26, 27, 3370	ELECTRICAL, COMMUNICATIONS & UTIL.	97.0	109.6	103.1	94.1	109.6	101.6	97.0	108.3	102.5	97.0	154.5	125.0	96.6	109.9	103.1	97.0	110.5	103.6
MF2004	WEIGHTED AVERAGE	101.0	118.0	108.2	95.8	117.1	104.8	99.9	119.0	107.9	100.0	125.4	110.7	99.9	118.5	107.7	99.8	117.5	107.2

	DIVISION	D.C.			DELAWARE									FLORIDA					
		WASHINGTON			DOVER			NEWARK			WILMINGTON			DAYTONA BEACH			FORT LAUDERDALE		
		200 - 205			199			197			198			321			333		
		MAT.	INST.	TOTAL	MAT.	INST.	TOTAL	MAT.	INST.	TOTAL	MAT.	INST.	TOTAL	MAT.	INST.	TOTAL	MAT.	INST.	TOTAL
015433	CONTRACTOR EQUIPMENT		103.6	103.6		114.5	114.5		114.5	114.5		114.7	114.7		98.9	98.9		93.2	93.2
0241, 31 - 34	SITE & INFRASTRUCTURE, DEMOLITION	110.3	91.8	97.7	103.1	108.7	106.9	102.9	108.7	106.9	91.3	109.0	103.4	116.1	89.5	97.9	101.4	78.7	85.9
0310	Concrete Forming & Accessories	101.3	81.9	84.5	96.1	102.8	101.9	98.2	102.8	102.2	99.5	102.8	102.4	95.8	77.8	80.2	94.0	68.1	71.6
0320	Concrete Reinforcing	113.5	91.1	104.3	98.6	100.1	99.2	99.5	100.1	99.7	99.5	100.1	99.7	92.5	82.2	88.3	92.5	80.4	87.6
0330	Cast-in-Place Concrete	127.6	88.5	112.7	103.7	89.1	98.1	94.1	89.1	92.2	98.3	89.1	94.8	91.9	77.2	86.3	96.4	74.4	88.0
03	CONCRETE	121.1	87.3	105.8	104.9	98.5	102.0	100.8	98.5	99.7	102.8	98.5	100.8	90.4	79.4	85.4	92.4	74.0	84.0
04	MASONRY	101.7	81.9	89.8	104.2	93.8	98.0	102.5	93.8	97.3	107.6	93.8	99.3	86.3	72.3	77.9	86.6	68.2	75.6
05	METALS	98.0	109.3	101.3	102.8	114.7	106.3	102.8	114.7	106.3	99.6	114.7	104.0	101.7	94.5	99.6	101.6	94.4	99.5
06	WOOD, PLASTICS & COMPOSITES	100.9	81.2	90.1	94.5	104.2	99.9	97.4	104.2	101.2	99.8	104.2	102.2	95.7	81.0	87.6	91.6	65.6	77.4
07	THERMAL & MOISTURE PROTECTION	105.3	84.7	97.3	99.4	102.9	100.7	99.1	102.9	100.6	99.8	102.9	101.0	95.1	79.8	89.1	95.1	82.3	90.1
08	OPENINGS	107.0	92.7	103.4	95.5	108.6	98.7	95.5	108.6	98.7	95.2	108.6	98.5	99.7	74.9	93.6	97.1	66.1	89.4
0920	Plaster & Gypsum Board	111.0	80.6	91.2	109.4	104.3	106.1	111.4	104.3	106.8	111.0	104.3	106.6	100.7	80.8	87.8	100.3	64.9	77.3
0950, 0980	Ceilings & Acoustic Treatment	107.4	80.6	90.8	98.9	104.3	102.2	102.3	104.3	103.5	94.7	104.3	100.6	100.4	80.8	88.3	100.4	64.9	78.4
0960	Flooring	127.7	100.2	120.3	82.3	98.2	86.6	82.5	98.2	86.7	82.1	98.2	86.5	117.2	79.9	107.1	117.2	65.4	103.2
0970, 0990	Wall Finishes & Painting/Coating	125.2	87.4	102.5	95.8	98.0	97.1	95.8	98.0	97.1	95.8	98.0	97.1	115.3	76.7	92.1	110.4	54.2	76.7
09	FINISHES	111.0	85.3	97.5	101.1	101.4	101.2	102.4	101.4	101.8	100.5	101.4	101.0	106.3	78.4	91.6	103.9	65.1	83.4
COVERS	DIVS. 10 - 14, 25, 28, 41, 43, 44	100.0	98.9	99.8	100.0	105.5	101.1	100.0	105.5	101.1	100.0	105.5	101.1	100.0	83.5	96.7	100.0	86.3	97.3
21, 22, 23	FIRE SUPPRESSION, PLUMBING & HVAC	100.1	93.8	97.6	100.0	112.4	105.0	100.2	112.4	105.1	100.2	112.4	105.1	99.9	69.7	87.8	99.9	67.2	86.9
26, 27, 3370	ELECTRICAL, COMMUNICATIONS & UTIL.	103.4	101.5	102.5	102.1	108.1	105.0	99.7	108.1	103.8	105.3	108.1	106.6	99.2	62.9	81.6	99.2	75.5	87.7
MF2004	WEIGHTED AVERAGE	104.6	92.8	99.7	101.2	105.8	103.1	100.5	105.8	102.7	100.6	105.9	102.8	99.1	76.5	89.6	98.5	73.6	88.0

DIVISION		FLORIDA																	
		FORT MYERS			GAINESVILLE			JACKSONVILLE			LAKELAND			MELBOURNE			MIAMI		
		339,341			326,344			320,322			338			329			330 - 332,340		
		MAT.	INST.	TOTAL	MAT.	INST.	TOTAL	MAT.	INST.	TOTAL	MAT.	INST.	TOTAL	MAT.	INST.	TOTAL	MAT.	INST.	TOTAL
015433	**CONTRACTOR EQUIPMENT**		98.9	98.9		98.9	98.9		98.9	98.9		98.9	98.9		98.9	98.9		93.2	93.2
0241, 31 - 34	**SITE & INFRASTRUCTURE, DEMOLITION**	113.3	88.7	96.5	127.0	88.3	100.5	116.2	88.7	97.4	115.5	88.6	97.1	124.6	88.7	100.0	104.8	79.0	87.2
0310	Concrete Forming & Accessories	89.7	82.0	83.0	90.6	59.1	63.4	95.5	59.1	64.1	85.9	82.6	83.1	92.0	79.7	81.4	98.8	68.8	72.9
0320	Concrete Reinforcing	93.5	85.5	90.3	98.2	51.4	79.1	92.5	51.9	75.9	95.8	86.2	91.9	93.5	82.2	88.9	92.5	80.6	87.6
0330	Cast-in-Place Concrete	100.6	66.3	87.5	105.5	56.1	86.6	92.8	66.0	82.5	102.8	67.4	89.3	110.8	79.5	98.8	104.6	75.8	93.6
03	**CONCRETE**	93.3	78.2	86.4	101.1	58.3	81.6	90.8	61.8	77.6	95.2	78.9	87.8	101.5	81.1	92.2	96.5	74.8	86.6
04	**MASONRY**	80.3	61.9	69.3	99.1	49.9	69.6	86.2	58.6	69.7	95.0	73.0	81.8	83.7	76.3	79.2	85.9	72.0	77.6
05	**METALS**	104.3	95.3	101.6	100.8	80.6	94.9	100.2	81.2	94.7	104.2	95.4	101.6	110.7	94.9	106.1	107.0	94.5	103.4
06	**WOOD, PLASTICS & COMPOSITES**	88.6	86.4	87.4	89.7	57.9	72.2	95.7	57.9	74.9	84.3	86.4	85.5	91.4	81.0	85.7	97.6	65.6	80.1
07	**THERMAL & MOISTURE PROTECTION**	94.9	86.4	91.6	95.4	61.7	82.3	95.3	64.7	83.4	94.8	90.3	93.1	95.5	85.9	91.7	103.2	79.5	94.0
08	**OPENINGS**	98.4	77.3	93.2	97.8	51.5	86.4	99.7	52.1	88.0	98.3	75.4	92.7	98.9	80.9	94.5	97.1	66.4	89.5
0920	Plaster & Gypsum Board	96.6	86.4	90.0	96.6	56.9	70.8	100.7	56.9	72.2	94.3	86.4	89.2	97.0	80.8	86.5	96.3	64.9	75.9
0950, 0980	Ceilings & Acoustic Treatment	94.4	86.4	89.5	92.7	56.9	70.5	100.4	56.9	73.5	92.7	86.4	88.8	96.2	80.8	86.7	97.1	64.9	77.2
0960	Flooring	113.1	48.1	95.6	113.7	40.8	94.0	117.2	55.9	100.6	110.2	60.3	96.7	114.0	79.9	104.7	124.4	68.9	109.4
0970, 0990	Wall Finishes & Painting/Coating	115.3	61.6	83.1	115.3	44.4	72.8	115.3	49.6	75.9	115.3	61.6	83.1	115.3	105.7	109.5	110.4	54.2	76.7
09	**FINISHES**	102.9	74.2	87.7	103.9	53.5	77.3	106.3	57.2	80.4	101.4	76.7	88.4	104.5	82.7	93.0	105.0	66.2	84.5
COVERS	**DIVS. 10 - 14, 25, 28, 41, 43, 44**	100.0	80.3	96.1	100.0	81.2	96.3	100.0	79.0	95.8	100.0	80.3	96.1	100.0	85.0	97.0	100.0	86.9	97.4
21, 22, 23	**FIRE SUPPRESSION, PLUMBING & HVAC**	96.8	60.0	82.1	98.9	68.4	86.8	99.9	48.1	79.3	96.8	89.3	93.8	99.9	71.9	88.7	99.9	72.1	88.8
26, 27, 3370	**ELECTRICAL, COMMUNICATIONS & UTIL.**	101.5	41.3	72.2	99.5	73.1	86.7	98.9	68.2	84.0	99.5	41.8	71.4	100.5	72.9	87.1	107.1	76.3	92.1
MF2004	**WEIGHTED AVERAGE**	98.6	70.1	86.6	100.5	66.0	86.0	98.8	63.1	83.8	99.2	77.8	90.2	102.0	80.0	92.7	101.2	75.4	90.3

DIVISION		FLORIDA																	
		ORLANDO			PANAMA CITY			PENSACOLA			SARASOTA			ST. PETERSBURG			TALLAHASSEE		
		327 - 328,347			324			325			342			337			323		
		MAT.	INST.	TOTAL	MAT.	INST.	TOTAL	MAT.	INST.	TOTAL	MAT.	INST.	TOTAL	MAT.	INST.	TOTAL	MAT.	INST.	TOTAL
015433	**CONTRACTOR EQUIPMENT**		98.9	98.9		98.9	98.9		98.9	98.9		98.9	98.9		98.9	98.9		98.9	98.9
0241, 31 - 34	**SITE & INFRASTRUCTURE, DEMOLITION**	118.7	88.3	98.0	131.2	86.4	100.6	130.8	87.9	101.5	117.6	88.4	97.7	117.0	87.8	97.0	114.3	87.6	96.0
0310	Concrete Forming & Accessories	95.5	79.4	81.6	94.8	43.6	50.6	92.8	55.9	60.9	95.8	82.1	83.9	93.6	49.9	55.9	97.3	45.9	52.9
0320	Concrete Reinforcing	92.5	79.0	87.0	96.6	48.8	77.1	99.0	48.7	78.5	92.5	86.1	89.9	95.8	55.4	79.3	92.5	51.6	75.8
0330	Cast-in-Place Concrete	121.1	76.9	104.2	97.5	46.3	77.9	120.5	62.6	98.4	105.1	67.1	90.6	104.0	61.8	87.8	102.3	54.3	83.9
03	**CONCRETE**	105.3	79.5	93.5	98.7	47.1	75.2	109.1	58.6	86.1	96.5	78.5	88.4	96.6	56.9	78.5	95.3	51.9	75.6
04	**MASONRY**	90.3	72.3	79.5	90.5	44.5	62.9	109.4	53.0	75.7	87.3	73.0	78.7	130.7	48.6	81.5	89.0	47.1	63.9
05	**METALS**	107.8	92.8	103.4	101.5	69.5	92.1	102.6	80.0	96.0	104.7	94.9	101.8	105.0	81.7	98.2	94.7	80.7	90.6
06	**WOOD, PLASTICS & COMPOSITES**	95.4	83.9	89.1	94.6	43.7	66.7	93.0	57.6	73.5	95.7	86.4	90.6	93.1	49.4	69.1	93.5	44.4	66.6
07	**THERMAL & MOISTURE PROTECTION**	96.2	80.1	89.9	95.5	47.4	76.8	95.5	59.2	81.3	95.1	90.3	93.2	95.0	55.6	79.7	100.5	74.5	90.4
08	**OPENINGS**	99.7	73.0	93.1	97.4	36.6	82.3	97.3	53.8	86.6	99.7	77.1	94.1	98.3	50.1	86.4	100.2	45.6	86.7
0920	Plaster & Gypsum Board	109.7	83.8	92.9	98.8	42.2	62.0	100.0	56.5	71.8	100.7	86.4	91.4	98.0	48.0	65.6	114.3	42.9	67.9
0950, 0980	Ceilings & Acoustic Treatment	100.4	83.8	90.2	94.6	42.2	62.2	96.2	56.5	71.7	98.5	86.4	91.1	94.6	48.0	65.8	100.4	42.9	64.8
0960	Flooring	117.4	79.9	107.2	116.6	46.9	97.8	110.9	59.4	97.0	117.2	57.5	101.0	115.6	60.3	100.7	117.2	50.3	99.1
0970, 0990	Wall Finishes & Painting/Coating	115.3	58.7	81.4	115.3	41.0	70.7	115.3	49.0	75.5	115.3	61.6	83.1	115.3	43.3	72.1	115.3	43.8	72.4
09	**FINISHES**	107.6	78.1	92.0	105.9	44.0	73.2	104.4	55.7	78.7	105.9	76.2	90.2	104.1	50.7	75.9	107.8	45.5	74.9
COVERS	**DIVS. 10 - 14, 25, 28, 41, 43, 44**	100.0	83.7	96.8	100.0	46.9	89.5	100.0	48.0	89.7	100.0	80.3	96.1	100.0	56.8	91.4	100.0	71.7	94.4
21, 22, 23	**FIRE SUPPRESSION, PLUMBING & HVAC**	99.8	64.3	85.7	99.9	37.8	75.2	99.9	57.1	82.8	99.8	60.6	84.2	99.9	55.6	82.3	99.8	42.2	76.9
26, 27, 3370	**ELECTRICAL, COMMUNICATIONS & UTIL.**	107.0	40.4	74.6	97.9	41.8	70.6	101.9	58.2	80.6	99.2	37.3	69.1	99.5	47.0	73.9	106.8	36.7	72.6
MF2004	**WEIGHTED AVERAGE**	103.1	72.0	90.0	100.3	48.6	78.6	102.9	61.3	85.4	100.4	71.1	88.1	102.3	58.1	83.7	99.6	53.0	80.1

DIVISION		FLORIDA						GEORGIA											
		TAMPA			WEST PALM BEACH			ALBANY			ATHENS			ATLANTA			AUGUSTA		
		335 - 336,346			334,349			317,398			306			300 - 303,399			308 - 309		
		MAT.	INST.	TOTAL	MAT.	INST.	TOTAL	MAT.	INST.	TOTAL	MAT.	INST.	TOTAL	MAT.	INST.	TOTAL	MAT.	INST.	TOTAL
015433	**CONTRACTOR EQUIPMENT**		98.9	98.9		93.2	93.2		94.3	94.3		92.7	92.7		93.4	93.4		92.7	92.7
0241, 31 - 34	**SITE & INFRASTRUCTURE, DEMOLITION**	117.3	88.5	97.6	98.1	78.8	84.9	101.7	81.2	87.6	108.5	90.8	96.4	104.1	94.1	97.3	100.4	90.9	93.9
0310	Concrete Forming & Accessories	96.7	82.8	84.7	97.3	67.5	71.5	94.9	49.5	55.7	94.0	46.5	53.0	97.6	76.1	79.0	95.1	55.1	60.6
0320	Concrete Reinforcing	92.5	86.2	89.9	95.1	80.2	89.0	92.1	89.8	91.2	99.0	88.1	94.5	98.4	91.7	95.6	99.3	69.3	87.0
0330	Cast-in-Place Concrete	101.6	67.4	88.5	91.6	64.1	81.1	98.1	51.4	80.3	112.0	54.1	89.9	112.0	74.5	97.7	105.7	53.6	85.8
03	**CONCRETE**	94.9	79.0	87.7	89.7	70.1	80.8	93.1	59.2	77.7	103.8	57.7	82.8	101.6	78.6	91.1	97.3	57.9	79.4
04	**MASONRY**	87.5	73.0	78.8	86.2	66.8	74.6	89.4	45.4	63.1	74.2	62.0	66.9	87.7	73.7	79.3	87.9	45.1	62.3
05	**METALS**	103.9	95.5	101.5	100.9	93.7	98.8	101.0	89.0	97.5	92.8	76.4	88.0	93.6	83.5	90.7	92.4	71.5	86.3
06	**WOOD, PLASTICS & COMPOSITES**	97.0	86.4	91.2	96.2	65.6	79.4	94.6	44.5	67.1	93.3	40.2	64.2	97.3	77.4	86.4	94.5	55.6	73.1
07	**THERMAL & MOISTURE PROTECTION**	95.3	90.3	93.3	94.8	70.5	85.4	92.3	60.5	80.0	94.3	57.2	79.8	94.3	78.2	88.0	93.9	56.1	79.2
08	**OPENINGS**	99.7	77.8	94.3	96.2	66.1	88.8	97.1	51.0	85.7	93.3	48.8	82.3	99.1	75.4	93.2	93.3	54.8	83.8
0920	Plaster & Gypsum Board	100.7	86.4	91.4	103.0	64.9	78.2	98.0	43.0	62.3	117.4	38.5	66.1	118.8	77.0	91.7	117.7	54.3	76.6
0950, 0980	Ceilings & Acoustic Treatment	100.4	86.4	91.8	94.4	64.9	76.1	99.6	43.0	64.6	109.8	38.5	65.7	109.8	77.0	89.5	110.7	54.3	75.8
0960	Flooring	117.2	60.3	101.8	119.8	63.1	104.5	117.2	34.6	94.9	87.8	60.4	80.4	89.4	71.8	84.6	88.3	44.1	76.3
0970, 0990	Wall Finishes & Painting/Coating	115.3	61.6	83.1	110.4	48.1	73.0	110.4	46.5	72.1	102.8	36.9	63.3	102.8	86.2	92.9	102.8	42.0	66.3
09	**FINISHES**	106.4	76.7	90.7	103.4	63.9	82.5	103.3	44.8	72.4	100.4	46.8	72.1	100.7	76.2	87.7	100.2	51.2	74.3
COVERS	**DIVS. 10 - 14, 25, 28, 41, 43, 44**	100.0	84.7	97.0	100.0	86.2	97.3	100.0	81.4	96.3	100.0	49.1	89.9	100.0	85.7	97.2	100.0	59.5	92.0
21, 22, 23	**FIRE SUPPRESSION, PLUMBING & HVAC**	99.9	89.3	95.7	96.8	68.8	85.6	99.8	65.6	86.2	94.2	76.6	87.2	100.0	78.0	91.2	100.0	73.6	89.5
26, 27, 3370	**ELECTRICAL, COMMUNICATIONS & UTIL.**	99.1	47.0	73.8	100.5	54.2	78.0	94.5	58.3	76.9	99.0	76.4	88.0	98.7	76.7	88.0	99.6	68.7	84.5
MF2004	**WEIGHTED AVERAGE**	100.1	78.7	91.1	97.2	69.8	85.7	97.9	62.0	82.9	95.9	66.9	83.7	98.2	79.2	90.2	96.9	64.3	83.2

DIVISION		GEORGIA																	
		COLUMBUS			DALTON			GAINESVILLE			MACON			SAVANNAH			STATESBORO		
		318 - 319			307			305			310 - 312			313 - 314			304		
		MAT.	INST.	TOTAL	MAT.	INST.	TOTAL	MAT.	INST.	TOTAL	MAT.	INST.	TOTAL	MAT.	INST.	TOTAL	MAT.	INST.	TOTAL
015433	CONTRACTOR EQUIPMENT		94.3	94.3		104.5	104.5		92.7	92.7		102.3	102.3		95.0	95.0		95.0	95.0
0241, 31 - 34	SITE & INFRASTRUCTURE, DEMOLITION	101.6	81.9	88.1	109.3	94.5	99.2	107.9	91.3	96.6	102.1	94.0	96.6	104.1	81.4	88.6	109.8	78.1	88.2
0310	Concrete Forming & Accessories	94.8	62.4	66.8	86.6	49.6	54.6	97.3	49.0	55.5	94.2	59.4	64.1	95.0	54.7	60.2	81.2	44.9	49.9
0320	Concrete Reinforcing	92.5	90.9	91.9	98.5	70.5	87.1	98.8	87.3	94.1	93.7	90.2	92.2	93.4	69.5	83.6	98.1	48.1	77.7
0330	Cast-in-Place Concrete	97.8	61.1	83.7	108.7	41.5	83.0	117.7	56.2	94.2	96.4	53.5	80.0	114.0	52.5	90.5	111.7	48.5	87.6
03	CONCRETE	93.0	68.6	81.9	102.7	51.9	79.6	105.7	59.2	84.5	92.6	64.5	79.8	100.7	58.2	81.4	103.1	48.6	78.3
04	MASONRY	88.4	64.0	73.8	75.3	36.8	52.3	83.5	38.2	56.4	101.5	43.2	66.6	88.4	55.8	68.9	76.8	45.3	57.9
05	METALS	100.5	92.8	98.2	94.0	72.2	87.6	92.0	70.8	85.8	96.1	92.0	94.9	96.5	82.9	92.5	97.4	73.3	90.4
06	WOOD, PLASTICS & COMPOSITES	94.6	61.0	76.2	76.4	52.4	63.2	97.1	44.4	68.2	103.4	59.9	79.5	109.0	52.4	77.9	69.4	46.1	56.6
07	THERMAL & MOISTURE PROTECTION	92.5	66.4	82.3	95.9	46.7	76.7	94.2	49.7	76.9	91.1	66.2	81.4	95.1	57.8	80.6	94.7	43.1	74.7
08	OPENINGS	97.1	62.1	88.4	95.0	39.2	81.2	93.2	42.3	80.6	95.2	60.5	86.6	98.2	50.6	86.4	95.7	36.8	81.2
0920	Plaster & Gypsum Board	100.5	60.1	74.3	103.3	51.1	69.4	118.8	42.7	69.4	108.5	58.9	76.3	97.1	51.2	67.3	103.8	44.6	65.4
0950, 0980	Ceilings & Acoustic Treatment	99.6	60.1	75.2	118.2	51.1	76.7	109.8	42.7	68.3	95.7	58.9	73.0	100.4	51.2	70.0	116.4	44.6	72.0
0960	Flooring	117.2	56.5	100.8	88.7	14.5	68.6	89.3	14.5	69.1	91.2	40.7	77.5	118.0	52.0	100.1	104.1	50.3	89.6
0970, 0990	Wall Finishes & Painting/Coating	110.4	54.9	77.1	93.0	44.2	63.8	102.8	38.6	64.3	112.7	51.8	76.1	112.3	52.6	76.5	100.7	37.7	62.9
09	FINISHES	103.6	59.7	80.4	107.9	42.9	73.5	100.9	40.1	68.8	92.4	54.1	72.2	103.9	53.6	77.3	110.9	45.7	76.4
COVERS	DIVS. 10 - 14, 25, 28, 41, 43, 44	100.0	83.3	96.7	100.0	26.7	85.5	100.0	49.6	90.0	100.0	81.1	96.3	100.0	61.0	92.3	100.0	45.7	89.2
21, 22, 23	FIRE SUPPRESSION, PLUMBING & HVAC	99.9	55.9	82.4	94.2	38.1	71.9	94.2	78.0	87.7	99.9	65.7	86.3	99.9	59.6	83.9	94.6	38.9	72.4
26, 27, 3370	ELECTRICAL, COMMUNICATIONS & UTIL.	94.6	61.4	78.4	110.7	66.2	89.0	99.0	76.4	88.0	93.1	63.8	78.8	97.4	60.5	79.4	99.6	35.6	68.4
MF2004	WEIGHTED AVERAGE	97.8	66.8	84.8	98.0	52.3	78.8	96.5	63.2	82.5	96.5	66.6	83.9	98.8	62.0	83.3	97.9	48.1	77.0

DIVISION		GEORGIA						HAWAII									IDAHO		
		VALDOSTA			WAYCROSS			HILO			HONOLULU			STATES & POSS., GUAM			BOISE		
		316			315			967			968			969			836 - 837		
		MAT.	INST.	TOTAL	MAT.	INST.	TOTAL	MAT.	INST.	TOTAL	MAT.	INST.	TOTAL	MAT.	INST.	TOTAL	MAT.	INST.	TOTAL
015433	CONTRACTOR EQUIPMENT		94.3	94.3		94.3	94.3		101.7	101.7		101.7	101.7		166.8	166.8		101.1	101.1
0241, 31 - 34	SITE & INFRASTRUCTURE, DEMOLITION	112.2	81.1	90.9	108.9	79.5	88.8	132.7	107.0	115.2	137.0	107.0	116.5	175.5	104.4	126.9	77.2	101.1	93.6
0310	Concrete Forming & Accessories	84.2	47.4	52.4	86.1	48.5	53.6	108.3	134.3	130.8	116.5	134.3	131.9	110.8	65.8	71.9	101.1	75.5	79.0
0320	Concrete Reinforcing	94.4	48.1	75.5	94.4	84.9	90.5	106.0	112.4	108.6	105.0	112.4	108.1	185.6	35.3	124.3	102.2	72.7	90.2
0330	Cast-in-Place Concrete	96.0	54.8	80.3	108.5	53.8	87.6	205.0	120.7	172.8	174.6	120.7	154.0	177.7	117.2	154.6	96.8	81.3	90.9
03	CONCRETE	96.9	51.9	76.4	100.8	58.5	81.5	153.4	124.4	140.2	139.4	124.4	132.6	156.7	78.9	121.3	105.8	77.2	92.8
04	MASONRY	94.2	50.6	68.1	95.1	45.2	65.2	135.7	125.4	129.5	126.5	125.4	125.9	193.1	50.3	107.6	129.1	64.4	90.4
05	METALS	100.1	75.6	92.9	99.3	81.8	94.2	114.0	105.6	111.5	137.8	105.6	128.4	147.9	85.3	129.6	100.9	76.3	93.7
06	WOOD, PLASTICS & COMPOSITES	81.9	43.4	60.8	84.5	49.6	65.3	97.9	139.0	120.5	109.8	139.0	125.8	109.7	67.2	86.4	93.5	76.2	84.0
07	THERMAL & MOISTURE PROTECTION	92.7	63.3	81.2	92.5	45.8	74.3	105.8	122.7	112.4	115.3	122.7	118.2	123.5	73.1	103.9	95.4	71.0	85.9
08	OPENINGS	92.6	40.8	79.8	92.6	47.8	81.5	100.8	129.4	107.9	106.1	129.4	111.9	105.6	56.9	93.6	92.6	71.5	87.4
0920	Plaster & Gypsum Board	92.8	41.9	59.7	94.7	48.3	64.5	106.1	140.2	128.2	136.6	140.2	138.9	209.0	53.9	108.2	81.2	75.2	77.3
0950, 0980	Ceilings & Acoustic Treatment	94.6	41.9	62.0	92.7	48.3	65.2	116.5	140.2	131.1	119.7	140.2	132.3	244.1	53.9	126.5	112.2	75.2	89.3
0960	Flooring	109.3	43.4	91.5	110.7	33.5	89.8	140.7	125.5	136.6	164.8	125.5	154.2	164.5	52.0	134.1	100.3	71.3	92.5
0970, 0990	Wall Finishes & Painting/Coating	110.4	37.0	66.4	110.4	45.9	71.7	99.6	139.2	123.4	104.0	139.2	125.1	105.5	41.3	67.0	98.1	45.4	66.5
09	FINISHES	99.9	44.9	70.8	99.8	45.2	70.9	120.6	134.7	128.0	133.9	134.7	134.3	200.4	61.4	126.9	100.8	72.2	85.7
COVERS	DIVS. 10 - 14, 25, 28, 41, 43, 44	100.0	59.8	92.0	100.0	46.0	89.3	100.0	116.4	103.2	100.0	116.4	103.2	100.0	91.3	98.3	100.0	71.5	94.4
21, 22, 23	FIRE SUPPRESSION, PLUMBING & HVAC	99.9	40.8	76.4	96.1	48.5	77.2	100.2	107.3	103.0	100.2	107.3	103.0	102.8	44.9	79.8	99.9	69.2	87.7
26, 27, 3370	ELECTRICAL, COMMUNICATIONS & UTIL.	92.7	36.0	65.1	96.9	66.4	82.0	107.8	121.3	114.3	111.7	121.3	116.4	159.7	48.1	105.4	97.4	77.1	87.5
MF2004	WEIGHTED AVERAGE	97.8	51.0	78.2	97.7	57.2	80.7	114.2	118.7	116.1	118.7	118.7	118.7	138.1	64.3	107.1	100.6	74.8	89.8

DIVISION		IDAHO															ILLINOIS		
		COEUR D'ALENE			IDAHO FALLS			LEWISTON			POCATELLO			TWIN FALLS			BLOOMINGTON		
		838			834			835			832			833			617		
		MAT.	INST.	TOTAL	MAT.	INST.	TOTAL	MAT.	INST.	TOTAL	MAT.	INST.	TOTAL	MAT.	INST.	TOTAL	MAT.	INST.	TOTAL
015433	CONTRACTOR EQUIPMENT		97.4	97.4		101.1	101.1		97.4	97.4		101.1	101.1		101.1	101.1		100.5	100.5
0241, 31 - 34	SITE & INFRASTRUCTURE, DEMOLITION	78.3	99.9	93.1	75.5	100.1	92.3	84.6	100.5	95.4	78.5	101.1	94.0	84.9	99.1	94.6	94.6	94.0	94.2
0310	Concrete Forming & Accessories	114.5	61.8	69.0	95.2	45.4	52.2	119.6	63.6	71.2	101.2	75.2	78.8	102.0	32.3	41.8	86.9	115.4	111.5
0320	Concrete Reinforcing	109.4	95.3	103.7	104.1	70.8	90.5	109.4	95.6	103.8	102.6	73.0	90.5	104.4	71.2	90.8	102.1	107.9	104.5
0330	Cast-in-Place Concrete	104.7	88.4	98.5	92.1	56.9	78.7	108.8	90.1	101.7	99.6	81.1	92.5	102.1	43.4	79.7	97.3	99.3	98.0
03	CONCRETE	112.9	77.8	97.0	98.0	55.2	78.5	116.7	79.3	99.7	105.2	77.0	92.4	112.9	44.8	81.9	96.2	108.4	101.8
04	MASONRY	130.9	80.8	100.9	124.4	34.6	70.6	131.3	85.4	103.8	126.4	70.6	93.0	129.2	33.8	72.1	111.6	113.9	113.0
05	METALS	94.8	86.2	92.3	108.8	73.4	98.5	94.3	87.3	92.2	108.9	76.4	99.4	108.9	73.0	98.4	92.5	110.1	97.7
06	WOOD, PLASTICS & COMPOSITES	98.0	55.7	74.8	87.1	43.9	63.4	103.4	55.7	77.2	93.5	76.2	84.0	94.1	31.3	59.6	88.1	115.2	103.0
07	THERMAL & MOISTURE PROTECTION	146.0	74.2	118.1	95.0	47.9	76.7	146.2	78.1	119.7	95.5	66.3	84.1	96.2	41.7	75.0	98.5	109.2	102.7
08	OPENINGS	113.2	59.9	100.0	96.6	48.1	84.6	113.1	62.9	100.7	93.3	65.7	86.5	96.6	38.8	82.3	90.4	106.2	94.3
0920	Plaster & Gypsum Board	134.7	54.1	82.3	74.8	41.7	53.3	136.2	54.1	82.8	78.7	75.2	76.4	78.4	28.6	46.1	91.7	115.7	107.3
0950, 0980	Ceilings & Acoustic Treatment	143.8	54.1	88.3	108.7	41.7	67.3	143.8	54.1	88.3	117.8	75.2	91.5	111.2	28.6	60.1	87.6	115.7	105.0
0960	Flooring	139.4	51.4	115.6	100.0	48.3	86.0	142.5	89.7	128.2	103.6	71.3	94.8	104.9	48.3	89.6	92.9	110.2	97.6
0970, 0990	Wall Finishes & Painting/Coating	119.6	67.5	88.3	98.1	37.3	61.6	119.6	67.5	88.3	98.0	47.2	67.5	98.1	27.8	55.9	93.0	112.8	104.9
09	FINISHES	166.9	59.2	109.9	98.6	44.6	70.0	168.1	67.5	114.9	102.8	72.4	86.7	102.2	33.8	66.0	92.6	114.9	104.4
COVERS	DIVS. 10 - 14, 25, 28, 41, 43, 44	100.0	59.7	92.0	100.0	46.1	89.3	100.0	74.7	95.0	100.0	71.5	94.4	100.0	41.3	88.4	100.0	102.0	100.4
21, 22, 23	FIRE SUPPRESSION, PLUMBING & HVAC	100.0	80.9	92.4	100.7	61.3	85.0	101.0	83.4	94.0	99.9	69.2	87.7	99.9	57.6	83.1	94.1	108.3	99.8
26, 27, 3370	ELECTRICAL, COMMUNICATIONS & UTIL.	86.8	78.0	82.5	86.5	71.7	79.3	84.8	78.0	81.5	93.8	72.8	83.6	87.9	63.6	76.1	93.8	92.6	93.2
MF2004	WEIGHTED AVERAGE	108.3	77.4	95.3	99.9	59.8	83.0	109.0	80.4	97.0	101.6	74.5	90.2	102.5	54.0	82.1	95.1	106.4	99.8

DIVISION		ILLINOIS																	
		CARBONDALE			CENTRALIA			CHAMPAIGN			CHICAGO			DECATUR			EAST ST. LOUIS		
		629			628			618 - 619			606 - 608			625			620 - 622		
		MAT.	INST.	TOTAL	MAT.	INST.	TOTAL	MAT.	INST.	TOTAL	MAT.	INST.	TOTAL	MAT.	INST.	TOTAL	MAT.	INST.	TOTAL
015433	**CONTRACTOR EQUIPMENT**		108.1	108.1		108.1	108.1		101.3	101.3		94.1	94.1		101.3	101.3		108.1	108.1
0241, 31 - 34	**SITE & INFRASTRUCTURE, DEMOLITION**	102.4	96.6	98.4	102.7	97.4	99.1	104.0	96.2	98.7	98.1	95.3	96.2	88.3	95.1	93.0	104.6	98.0	100.1
0310	Concrete Forming & Accessories	95.7	97.0	96.9	97.8	104.7	103.8	93.2	114.4	111.5	100.5	153.2	146.1	98.1	108.8	107.4	93.2	110.3	107.9
0320	Concrete Reinforcing	101.0	82.4	93.4	101.0	97.7	99.7	102.1	95.3	99.3	106.3	157.6	127.2	99.0	96.9	98.1	101.0	105.9	103.0
0330	Cast-in-Place Concrete	91.0	79.0	86.4	91.5	106.0	97.0	112.8	109.9	111.7	105.9	145.2	120.9	99.4	108.5	102.9	93.0	107.7	98.6
03	**CONCRETE**	87.8	89.1	88.4	88.2	104.7	95.7	107.9	109.2	108.5	104.5	149.8	125.1	97.3	106.6	101.5	89.1	109.2	98.2
04	**MASONRY**	75.4	93.2	86.0	75.5	109.8	96.0	134.4	104.3	116.4	94.4	152.2	129.0	72.1	109.4	94.5	75.6	113.5	98.3
05	**METALS**	91.7	105.1	95.6	91.8	113.2	98.0	92.5	103.9	95.9	90.1	130.7	102.0	95.0	104.6	97.8	92.7	117.4	99.9
06	**WOOD, PLASTICS & COMPOSITES**	102.9	98.9	100.7	105.5	102.7	103.9	95.1	114.7	105.9	105.2	152.3	131.1	102.9	108.2	105.8	100.2	109.1	105.1
07	**THERMAL & MOISTURE PROTECTION**	93.5	88.7	91.6	93.6	101.9	96.8	99.1	110.1	103.4	100.7	142.7	117.1	99.2	102.3	100.4	93.6	108.4	99.3
08	**OPENINGS**	86.5	100.9	90.1	86.5	104.1	90.9	91.1	108.5	95.4	101.6	155.3	114.8	96.9	105.3	99.0	86.6	113.9	93.3
0920	Plaster & Gypsum Board	95.8	98.8	97.8	96.9	102.7	100.7	94.3	115.2	107.9	92.0	154.1	132.3	98.5	108.4	104.9	95.1	109.4	104.4
0950, 0980	Ceilings & Acoustic Treatment	87.6	98.8	94.5	87.6	102.7	96.9	87.6	115.2	104.7	101.5	154.1	134.0	93.4	108.4	102.7	87.6	109.4	101.1
0960	Flooring	110.9	87.2	104.5	112.0	110.7	111.7	96.2	107.5	99.2	92.0	149.5	107.5	100.2	111.2	103.2	109.6	118.3	111.9
0970, 0990	Wall Finishes & Painting/Coating	100.6	91.2	94.9	100.6	99.1	99.7	93.0	110.9	103.8	86.5	148.0	123.4	93.0	106.6	101.2	100.6	103.3	102.2
09	**FINISHES**	96.6	95.2	95.9	97.1	105.3	101.5	94.6	113.6	104.6	93.8	152.7	124.9	97.1	109.9	103.8	96.2	110.9	104.0
COVERS	**DIVS. 10 - 14, 25, 28, 41, 43, 44**	100.0	91.9	98.4	100.0	96.3	99.3	100.0	96.4	99.3	100.0	126.8	105.3	100.0	95.4	99.1	100.0	96.9	99.4
21, 22, 23	**FIRE SUPPRESSION, PLUMBING & HVAC**	94.1	89.0	92.1	94.1	91.3	93.0	94.1	104.6	98.3	99.9	133.8	113.4	99.9	101.2	100.4	99.9	96.5	98.5
26, 27, 3370	**ELECTRICAL, COMMUNICATIONS & UTIL.**	92.6	96.0	94.2	93.9	91.1	92.6	96.9	98.7	97.8	97.5	135.8	116.1	96.5	91.2	93.9	93.5	100.0	96.7
MF2004	**WEIGHTED AVERAGE**	92.1	94.1	92.9	92.4	100.5	95.8	98.4	105.0	101.2	97.9	138.3	114.9	96.2	102.5	98.8	93.9	105.7	98.8

DIVISION		ILLINOIS																	
		EFFINGHAM			GALESBURG			JOLIET			KANKAKEE			LA SALLE			NORTH SUBURBAN		
		624			614			604			609			613			600 - 603		
		MAT.	INST.	TOTAL	MAT.	INST.	TOTAL	MAT.	INST.	TOTAL	MAT.	INST.	TOTAL	MAT.	INST.	TOTAL	MAT.	INST.	TOTAL
015433	**CONTRACTOR EQUIPMENT**		101.3	101.3		100.5	100.5		92.0	92.0		92.0	92.0		100.5	100.5		92.0	92.0
0241, 31 - 34	**SITE & INFRASTRUCTURE, DEMOLITION**	95.8	94.9	95.2	97.0	93.9	94.9	97.9	93.6	94.9	92.1	91.6	91.7	96.4	94.5	95.1	97.2	92.5	94.0
0310	Concrete Forming & Accessories	102.6	103.2	103.1	93.1	108.2	106.1	102.5	134.2	129.9	95.6	103.4	102.3	106.9	111.6	111.0	101.7	125.5	122.3
0320	Concrete Reinforcing	102.1	91.8	97.9	101.6	107.5	104.0	106.3	146.9	122.9	107.2	100.6	104.5	101.7	103.7	102.5	106.3	150.9	124.5
0330	Cast-in-Place Concrete	99.1	100.5	99.6	100.2	96.4	98.7	105.8	138.0	118.1	98.6	122.2	107.6	100.0	103.4	101.3	105.9	135.7	117.3
03	**CONCRETE**	98.2	100.4	99.2	98.7	104.1	101.2	104.6	136.9	119.3	98.9	109.2	103.6	99.5	107.5	103.1	104.6	132.9	117.5
04	**MASONRY**	79.2	77.1	77.9	111.8	108.0	109.5	97.2	139.5	122.5	93.9	124.9	112.5	111.8	115.9	114.2	94.4	140.7	122.1
05	**METALS**	92.5	100.2	94.7	92.5	108.8	97.3	88.2	124.1	98.7	88.2	101.9	92.2	92.6	110.7	97.9	89.3	123.4	99.2
06	**WOOD, PLASTICS & COMPOSITES**	105.3	105.2	105.3	95.0	107.2	101.7	107.0	131.5	120.5	99.1	96.7	97.8	110.1	111.8	111.0	105.2	120.2	113.4
07	**THERMAL & MOISTURE PROTECTION**	98.8	93.2	96.6	98.7	100.5	99.4	100.6	134.6	113.8	99.9	122.4	108.7	98.8	107.6	102.2	101.1	130.2	112.4
08	**OPENINGS**	90.9	101.5	93.5	90.4	103.3	93.6	99.4	141.2	109.8	91.5	108.0	95.6	90.4	103.1	93.5	99.5	134.2	108.1
0920	Plaster & Gypsum Board	98.1	105.3	102.8	94.3	107.4	102.8	89.4	132.6	117.5	86.4	96.5	93.0	100.3	112.1	108.0	92.0	120.9	110.7
0950, 0980	Ceilings & Acoustic Treatment	87.6	105.3	98.6	87.6	107.4	99.9	101.5	132.6	120.8	101.5	96.5	98.4	87.6	112.1	102.8	101.5	120.9	113.5
0960	Flooring	101.5	103.6	102.0	96.1	105.5	98.6	91.6	138.4	104.2	88.6	95.9	90.6	103.7	106.9	104.6	92.0	138.4	104.5
0970, 0990	Wall Finishes & Painting/Coating	93.0	96.4	95.1	93.0	92.8	92.9	84.6	135.3	115.0	84.6	112.8	101.5	93.0	106.9	101.3	86.5	135.3	115.8
09	**FINISHES**	96.1	103.5	100.0	94.0	106.8	100.8	93.2	135.2	115.4	91.5	100.9	96.5	97.1	110.3	104.1	93.7	128.5	112.1
COVERS	**DIVS. 10 - 14, 25, 28, 41, 43, 44**	100.0	78.7	95.8	100.0	97.2	99.4	100.0	121.3	104.2	100.0	92.1	98.4	100.0	96.7	99.3	100.0	105.1	101.0
21, 22, 23	**FIRE SUPPRESSION, PLUMBING & HVAC**	94.1	96.9	95.2	94.1	98.0	95.7	100.0	125.8	110.3	94.2	96.8	95.2	94.1	113.4	101.8	99.9	122.6	109.0
26, 27, 3370	**ELECTRICAL, COMMUNICATIONS & UTIL.**	94.4	95.9	95.2	94.7	75.3	85.2	96.3	118.0	106.9	91.3	125.4	107.9	91.9	89.9	90.9	96.2	116.9	106.3
MF2004	**WEIGHTED AVERAGE**	94.3	95.9	95.0	95.7	99.0	97.1	97.4	126.9	109.8	93.6	107.0	99.2	95.9	106.3	100.3	97.4	123.7	108.5

DIVISION		ILLINOIS																	
		PEORIA			QUINCY			ROCK ISLAND			ROCKFORD			SOUTH SUBURBAN			SPRINGFIELD		
		615 - 616			623			612			610 - 611			605			626 - 627		
		MAT.	INST.	TOTAL	MAT.	INST.	TOTAL	MAT.	INST.	TOTAL	MAT.	INST.	TOTAL	MAT.	INST.	TOTAL	MAT.	INST.	TOTAL
015433	**CONTRACTOR EQUIPMENT**		100.5	100.5		101.3	101.3		100.5	100.5		100.5	100.5		92.0	92.0		101.3	101.3
0241, 31 - 34	**SITE & INFRASTRUCTURE, DEMOLITION**	97.2	94.2	95.2	94.6	94.4	94.5	95.1	93.6	94.1	96.3	95.6	95.8	97.2	92.5	94.0	95.5	95.2	95.3
0310	Concrete Forming & Accessories	96.5	109.8	108.0	100.4	108.3	107.3	94.8	102.8	101.7	100.6	119.9	117.3	101.7	125.6	122.3	98.7	108.7	107.4
0320	Concrete Reinforcing	99.0	107.5	102.4	101.7	86.0	95.3	101.6	104.2	102.7	93.5	127.0	107.2	106.3	150.9	124.5	99.0	100.1	99.4
0330	Cast-in-Place Concrete	97.1	103.1	99.4	99.3	92.5	96.7	98.0	99.8	98.7	99.4	116.6	106.0	105.9	135.7	117.3	88.4	103.8	94.3
03	**CONCRETE**	96.1	107.2	101.1	97.9	98.8	98.3	97.1	102.4	99.5	96.3	120.0	107.1	104.6	133.0	117.5	92.1	105.6	98.2
04	**MASONRY**	112.5	109.2	110.5	98.2	93.5	95.3	111.6	98.7	103.9	87.6	110.3	101.2	94.4	140.7	122.1	76.8	110.3	96.8
05	**METALS**	95.0	109.6	99.2	92.5	98.9	94.4	92.6	107.5	96.9	95.0	119.6	102.2	89.3	123.5	99.3	97.4	106.0	99.9
06	**WOOD, PLASTICS & COMPOSITES**	102.9	106.1	104.7	102.8	113.4	108.6	96.9	102.1	99.8	102.9	117.7	111.1	105.2	120.2	113.4	104.0	107.4	105.8
07	**THERMAL & MOISTURE PROTECTION**	99.3	107.5	102.5	98.8	95.0	97.3	98.7	99.4	98.9	102.0	116.7	107.7	101.1	130.2	112.5	99.0	105.8	101.6
08	**OPENINGS**	97.1	107.5	99.6	91.7	104.1	94.8	90.4	101.0	93.0	97.1	119.8	102.7	99.5	134.2	108.1	98.1	105.7	100.0
0920	Plaster & Gypsum Board	98.5	106.2	103.5	96.9	113.8	107.9	95.1	102.1	99.6	98.5	118.3	111.4	92.0	120.9	110.7	94.9	107.6	103.1
0950, 0980	Ceilings & Acoustic Treatment	93.4	106.2	101.4	87.6	113.8	103.8	87.6	102.1	96.6	93.4	118.3	108.8	101.5	120.9	113.5	91.8	107.6	101.6
0960	Flooring	100.2	110.2	102.9	100.2	98.4	99.7	97.5	101.4	98.5	100.2	113.6	103.8	92.0	138.4	104.5	100.2	102.7	100.9
0970, 0990	Wall Finishes & Painting/Coating	93.0	112.8	104.9	93.0	98.9	96.6	93.0	95.2	94.3	93.0	122.6	110.8	86.5	135.3	115.8	93.0	106.2	100.9
09	**FINISHES**	97.2	110.3	104.1	95.5	107.3	101.7	94.4	102.1	98.5	97.2	119.2	108.8	93.7	128.5	112.1	96.0	107.8	102.3
COVERS	**DIVS. 10 - 14, 25, 28, 41, 43, 44**	100.0	102.0	100.4	100.0	82.0	96.4	100.0	92.3	98.5	100.0	102.4	100.5	100.0	105.1	101.0	100.0	95.6	99.1
21, 22, 23	**FIRE SUPPRESSION, PLUMBING & HVAC**	99.9	105.9	102.3	94.1	92.4	93.4	94.1	98.1	95.7	100.0	111.9	104.7	99.9	122.6	109.0	99.8	102.1	100.7
26, 27, 3370	**ELECTRICAL, COMMUNICATIONS & UTIL.**	95.6	95.1	95.4	92.0	79.8	86.0	87.1	95.8	91.3	96.0	122.8	109.0	96.2	123.1	109.3	101.8	96.9	99.4
MF2004	**WEIGHTED AVERAGE**	98.2	104.9	101.0	94.9	94.8	94.9	94.7	99.5	96.7	97.1	115.0	104.7	97.4	124.5	108.8	97.0	103.4	99.7

DIVISION		INDIANA																	
		ANDERSON			BLOOMINGTON			COLUMBUS			EVANSVILLE			FORT WAYNE			GARY		
		460			474			472			476 - 477			467 - 468			463 - 464		
		MAT.	INST.	TOTAL	MAT.	INST.	TOTAL	MAT.	INST.	TOTAL	MAT.	INST.	TOTAL	MAT.	INST.	TOTAL	MAT.	INST.	TOTAL
015433	CONTRACTOR EQUIPMENT		96.9	96.9		87.5	87.5		87.5	87.5		113.5	113.5		96.9	96.9		96.9	96.9
0241, 31 - 34	SITE & INFRASTRUCTURE, DEMOLITION	83.4	94.7	91.1	74.8	94.8	88.5	71.9	94.7	87.5	79.4	121.3	108.0	84.5	94.5	91.3	83.9	97.6	93.2
0310	Concrete Forming & Accessories	94.9	78.7	80.9	99.9	76.1	79.4	94.5	73.7	76.6	93.9	81.8	83.4	93.6	74.1	76.7	95.0	106.7	105.1
0320	Concrete Reinforcing	95.1	76.5	87.5	86.1	76.5	82.2	86.5	76.1	82.3	94.4	78.7	88.0	95.1	75.8	87.2	95.1	91.8	93.8
0330	Cast-in-Place Concrete	95.0	79.6	89.1	95.9	76.5	88.5	95.5	71.7	86.4	91.7	89.6	90.9	100.9	78.3	92.3	99.2	107.2	102.3
03	CONCRETE	90.2	79.3	85.2	97.9	76.3	88.1	97.4	73.5	86.5	98.6	84.2	92.1	92.9	76.6	85.5	92.2	104.0	97.6
04	MASONRY	85.8	80.4	82.6	86.5	74.2	79.1	86.4	74.1	79.1	82.8	84.6	83.9	89.4	77.8	82.5	87.0	104.4	97.4
05	METALS	100.2	87.8	96.6	94.3	76.1	89.0	94.3	75.5	88.8	87.0	86.1	86.8	100.2	86.9	96.3	100.2	98.8	99.8
06	WOOD, PLASTICS & COMPOSITES	106.6	78.4	91.1	115.7	75.4	93.6	110.7	72.2	89.6	95.2	80.5	87.1	106.3	73.1	88.1	104.5	107.0	105.9
07	THERMAL & MOISTURE PROTECTION	96.2	78.8	89.4	92.8	79.0	87.4	92.5	78.6	87.1	96.8	85.3	92.3	96.0	81.7	90.4	95.2	104.4	98.8
08	OPENINGS	95.6	78.1	91.3	103.1	76.5	96.5	98.5	74.7	92.6	95.5	79.1	91.5	95.6	72.7	90.0	95.6	104.7	97.9
0920	Plaster & Gypsum Board	95.7	78.1	84.3	88.8	75.2	80.0	86.2	71.9	76.9	84.7	79.1	81.1	94.9	72.6	80.4	90.5	107.8	101.7
0950, 0980	Ceilings & Acoustic Treatment	91.1	78.1	83.1	79.9	75.2	77.0	79.9	71.9	75.0	85.3	79.1	81.5	91.1	72.6	79.7	91.1	107.8	101.4
0960	Flooring	92.9	87.0	91.3	104.8	84.1	99.2	99.4	84.1	95.3	98.3	82.6	94.0	92.9	76.2	88.4	92.9	112.5	98.2
0970, 0990	Wall Finishes & Painting/Coating	95.1	69.4	79.7	91.6	85.5	87.9	91.6	85.5	87.9	97.7	82.0	88.3	95.1	76.1	83.7	95.1	104.9	100.9
09	FINISHES	91.4	79.3	85.0	93.7	78.6	85.7	91.7	76.7	83.8	92.4	81.9	86.9	91.2	74.6	82.5	90.7	108.0	99.8
COVERS	DIVS. 10 - 14, 25, 28, 41, 43, 44	100.0	89.9	98.0	100.0	80.3	96.1	100.0	79.9	96.0	100.0	85.6	97.1	100.0	79.4	95.9	100.0	86.2	97.3
21, 22, 23	FIRE SUPPRESSION, PLUMBING & HVAC	99.9	74.0	89.6	99.7	78.3	91.2	93.9	80.2	88.4	99.9	80.6	92.2	99.9	73.7	89.5	99.9	100.5	100.2
26, 27, 3370	ELECTRICAL, COMMUNICATIONS & UTIL.	83.2	93.0	88.0	98.6	78.8	89.0	97.8	87.5	92.8	93.8	87.3	90.6	84.0	77.4	80.8	96.0	97.4	96.7
MF2004	WEIGHTED AVERAGE	94.8	82.5	89.6	96.9	78.8	89.3	94.7	79.5	88.3	94.4	86.5	91.1	95.3	78.4	88.2	96.3	101.5	98.5

DIVISION		INDIANA																	
		INDIANAPOLIS			KOKOMO			LAFAYETTE			LAWRENCEBURG			MUNCIE			NEW ALBANY		
		461 - 462			469			479			470			473			471		
		MAT.	INST.	TOTAL	MAT.	INST.	TOTAL	MAT.	INST.	TOTAL	MAT.	INST.	TOTAL	MAT.	INST.	TOTAL	MAT.	INST.	TOTAL
015433	CONTRACTOR EQUIPMENT		92.0	92.0		96.9	96.9		87.5	87.5		104.0	104.0		96.5	96.5		94.2	94.2
0241, 31 - 34	SITE & INFRASTRUCTURE, DEMOLITION	83.0	98.0	93.2	80.6	94.5	90.1	72.8	94.7	87.8	71.1	109.7	97.5	74.9	95.1	88.7	67.8	95.0	86.4
0310	Concrete Forming & Accessories	95.5	83.7	85.3	97.7	75.5	78.5	92.3	79.0	80.8	91.0	73.0	75.5	92.2	78.4	80.3	88.9	70.9	73.3
0320	Concrete Reinforcing	94.5	80.9	88.9	86.1	74.5	81.3	86.1	81.1	84.1	85.4	76.9	81.9	95.3	76.5	87.6	86.8	77.2	82.9
0330	Cast-in-Place Concrete	91.4	85.9	89.3	94.0	81.5	89.2	96.0	82.3	90.8	89.8	73.1	83.4	100.8	80.8	93.2	92.7	72.4	84.9
03	CONCRETE	91.7	83.8	88.1	86.9	78.1	82.9	97.5	80.4	89.7	91.2	74.6	83.6	97.5	79.5	89.3	96.1	73.1	85.6
04	MASONRY	93.5	83.6	87.6	85.4	78.8	81.5	91.5	77.9	83.3	72.6	73.5	73.2	88.5	80.4	83.6	79.4	69.0	73.2
05	METALS	102.2	80.7	95.9	96.9	86.5	93.9	92.9	77.8	88.5	89.3	87.0	88.6	95.9	87.5	93.5	91.4	82.4	88.8
06	WOOD, PLASTICS & COMPOSITES	103.5	83.3	92.4	109.7	74.4	90.3	107.8	78.4	91.7	94.2	72.2	82.1	109.8	78.2	92.5	95.9	71.1	82.3
07	THERMAL & MOISTURE PROTECTION	94.4	83.9	90.3	95.8	78.1	88.9	92.5	79.4	87.4	95.1	75.9	87.6	95.2	78.9	88.9	85.4	69.3	79.2
08	OPENINGS	103.6	83.4	98.6	90.2	75.4	86.5	97.0	79.4	92.6	97.0	74.9	91.5	97.9	78.0	93.0	95.1	73.8	89.8
0920	Plaster & Gypsum Board	94.0	83.0	86.8	99.9	73.9	83.0	82.8	78.3	79.9	70.6	71.7	71.3	84.5	78.1	80.4	83.6	70.4	75.0
0950, 0980	Ceilings & Acoustic Treatment	95.3	83.0	87.7	90.3	73.9	80.2	74.1	78.3	76.7	89.5	71.7	78.5	80.8	78.1	79.1	85.3	70.4	76.1
0960	Flooring	92.7	89.6	91.9	97.1	85.9	94.1	98.2	88.6	95.6	73.0	87.0	76.8	97.7	87.0	94.8	96.4	65.3	88.0
0970, 0990	Wall Finishes & Painting/Coating	95.1	90.8	92.5	95.1	82.5	87.5	91.6	83.7	86.8	93.1	73.2	81.2	91.6	69.4	78.3	97.7	67.3	79.5
09	FINISHES	92.5	85.6	88.9	93.0	78.0	85.1	89.6	81.2	85.2	83.8	75.9	79.6	90.9	79.1	84.7	91.9	69.6	80.1
COVERS	DIVS. 10 - 14, 25, 28, 41, 43, 44	100.0	91.7	98.4	100.0	80.7	96.2	100.0	89.5	97.9	100.0	50.9	90.3	100.0	89.3	97.9	100.0	50.5	90.2
21, 22, 23	FIRE SUPPRESSION, PLUMBING & HVAC	99.9	85.7	94.2	94.1	80.3	88.6	93.9	82.4	89.3	94.6	74.2	86.5	99.7	74.0	89.4	94.1	73.2	85.8
26, 27, 3370	ELECTRICAL, COMMUNICATIONS & UTIL.	98.3	93.0	95.7	87.8	79.1	83.6	97.3	83.0	90.4	91.5	67.4	79.8	89.1	79.7	84.5	92.3	75.0	83.9
MF2004	WEIGHTED AVERAGE	98.0	86.7	93.2	92.5	80.8	87.6	94.3	82.1	89.2	91.1	76.9	85.1	95.5	80.6	89.3	92.4	74.4	84.9

DIVISION		INDIANA									IOWA								
		SOUTH BEND			TERRE HAUTE			WASHINGTON			BURLINGTON			CARROLL			CEDAR RAPIDS		
		465 - 466			478			475			526			514			522 - 524		
		MAT.	INST.	TOTAL	MAT.	INST.	TOTAL	MAT.	INST.	TOTAL	MAT.	INST.	TOTAL	MAT.	INST.	TOTAL	MAT.	INST.	TOTAL
015433	CONTRACTOR EQUIPMENT		105.8	105.8		113.5	113.5		113.5	113.5		99.9	99.9		99.9	99.9		96.8	96.8
0241, 31 - 34	SITE & INFRASTRUCTURE, DEMOLITION	86.0	95.0	92.1	81.1	121.4	108.7	81.2	119.3	107.2	92.8	96.0	95.0	81.7	93.6	89.8	94.5	96.5	95.9
0310	Concrete Forming & Accessories	96.5	77.7	80.3	95.0	80.8	82.7	95.6	77.0	79.5	94.7	68.2	71.8	82.9	45.2	50.3	100.0	77.3	80.4
0320	Concrete Reinforcing	95.1	72.9	86.1	94.4	83.8	90.0	87.2	56.7	74.8	101.5	76.7	91.4	102.3	46.1	79.4	102.3	80.6	93.4
0330	Cast-in-Place Concrete	96.2	82.3	90.9	88.8	87.6	88.3	96.6	89.7	93.9	105.0	51.5	84.6	105.0	55.5	86.1	105.3	79.0	95.3
03	CONCRETE	87.4	79.9	84.0	101.0	84.0	93.3	105.8	77.9	93.1	100.5	65.1	84.4	99.5	50.5	77.2	100.6	79.3	90.9
04	MASONRY	86.0	79.0	81.8	89.4	81.2	84.5	82.9	81.8	82.3	96.2	55.3	71.7	98.8	41.5	64.5	102.5	75.8	86.5
05	METALS	100.2	98.0	99.5	87.7	88.6	88.0	82.9	74.5	80.4	85.6	89.9	86.9	85.7	71.3	81.5	87.7	91.0	88.7
06	WOOD, PLASTICS & COMPOSITES	103.7	77.1	89.1	97.3	80.2	87.9	97.6	76.3	85.9	102.2	74.3	86.9	88.3	46.7	65.4	108.4	76.4	90.8
07	THERMAL & MOISTURE PROTECTION	91.2	83.8	88.3	96.8	80.9	90.7	97.0	81.1	90.8	100.8	60.1	85.0	101.1	48.8	80.7	101.7	77.6	92.3
08	OPENINGS	88.9	76.2	85.8	96.1	80.9	92.3	92.8	70.4	87.3	92.8	68.3	86.7	98.0	42.9	84.4	99.0	79.8	94.2
0920	Plaster & Gypsum Board	88.6	76.7	80.9	84.7	78.8	80.9	83.7	74.8	77.9	95.1	73.4	81.0	89.9	44.8	60.6	98.5	75.8	83.8
0950, 0980	Ceilings & Acoustic Treatment	87.0	76.7	80.6	85.3	78.8	81.3	77.0	74.8	75.6	109.1	73.4	87.0	109.1	44.8	69.3	113.2	75.8	90.1
0960	Flooring	91.3	83.4	89.2	98.3	87.3	95.3	99.4	85.7	95.7	114.1	42.4	94.7	106.9	37.3	88.0	132.2	50.0	110.0
0970, 0990	Wall Finishes & Painting/Coating	94.6	81.4	86.7	97.7	84.1	89.6	97.7	85.5	90.4	102.0	70.6	83.1	102.0	35.7	62.3	105.6	73.3	86.2
09	FINISHES	89.0	79.1	83.8	92.4	82.3	87.1	90.9	80.1	85.2	107.3	64.1	84.5	102.9	42.5	71.0	114.3	71.0	91.4
COVERS	DIVS. 10 - 14, 25, 28, 41, 43, 44	100.0	80.2	96.1	100.0	92.8	98.6	100.0	84.4	96.9	100.0	78.9	95.8	100.0	74.8	95.0	100.0	84.7	97.0
21, 22, 23	FIRE SUPPRESSION, PLUMBING & HVAC	99.8	80.4	92.1	99.9	82.7	93.1	94.1	78.8	88.0	94.4	67.0	83.5	94.4	45.5	75.0	100.2	81.3	92.7
26, 27, 3370	ELECTRICAL, COMMUNICATIONS & UTIL.	103.7	82.7	93.5	92.1	83.7	88.0	92.6	83.5	88.2	98.8	71.5	85.5	99.5	36.8	69.0	96.5	78.6	87.7
MF2004	WEIGHTED AVERAGE	95.6	83.1	90.4	95.0	86.6	91.5	92.7	82.4	88.4	95.8	70.7	85.2	95.6	51.4	77.0	98.8	80.8	91.2

	DIVISION	IOWA																	
		COUNCIL BLUFFS			CRESTON			DAVENPORT			DECORAH			DES MOINES			DUBUQUE		
		515			508			527 - 528			521			500 - 503,509			520		
		MAT.	INST.	TOTAL	MAT.	INST.	TOTAL	MAT.	INST.	TOTAL	MAT.	INST.	TOTAL	MAT.	INST.	TOTAL	MAT.	INST.	TOTAL
015433	CONTRACTOR EQUIPMENT		96.1	96.1		99.9	99.9		99.9	99.9		99.9	99.9		101.2	101.2		95.9	95.9
0241, 31 - 34	SITE & INFRASTRUCTURE, DEMOLITION	101.6	92.4	95.3	81.7	94.6	90.5	92.6	99.8	97.5	91.4	94.1	93.3	85.1	99.7	95.1	92.5	93.6	93.3
0310	Concrete Forming & Accessories	82.1	54.4	58.2	83.4	66.0	68.4	99.7	87.9	89.5	92.1	42.4	49.2	100.7	72.6	76.4	83.5	68.8	70.8
0320	Concrete Reinforcing	104.3	75.4	92.5	101.7	79.0	92.4	102.3	92.5	98.3	101.5	46.2	79.0	102.3	77.1	92.0	100.9	80.4	92.6
0330	Cast-in-Place Concrete	109.5	77.0	97.1	104.7	57.7	86.8	101.4	91.7	97.7	102.2	52.5	83.2	97.2	92.5	95.4	103.1	89.1	97.8
03	CONCRETE	103.0	67.6	86.9	99.3	66.5	84.4	98.8	90.7	95.1	98.6	48.2	75.7	95.8	81.3	89.2	97.6	79.0	89.1
04	MASONRY	102.9	76.2	86.9	98.8	45.0	66.6	98.6	84.9	90.4	116.6	38.6	69.9	94.1	69.4	79.3	103.4	67.3	81.8
05	METALS	93.1	88.5	91.8	85.6	85.3	85.5	87.7	99.9	91.3	85.8	70.8	81.4	87.9	90.6	88.7	86.5	90.6	87.7
06	WOOD, PLASTICS & COMPOSITES	87.1	47.8	65.5	88.8	72.1	79.6	108.4	86.6	96.4	99.2	43.4	68.6	108.2	72.1	88.4	88.9	68.4	77.7
07	THERMAL & MOISTURE PROTECTION	101.1	64.4	86.9	101.9	53.0	82.8	101.2	86.7	95.6	101.0	43.8	78.8	102.4	74.4	91.6	101.4	68.4	88.6
08	OPENINGS	98.0	63.1	89.4	110.8	70.3	100.8	99.0	89.5	96.6	96.3	44.8	83.6	99.0	77.0	93.6	98.0	77.4	92.9
0920	Plaster & Gypsum Board	89.9	46.2	61.5	89.9	71.2	77.8	98.5	86.2	90.5	94.0	41.4	59.8	88.9	71.2	77.4	90.3	67.6	75.6
0950, 0980	Ceilings & Acoustic Treatment	109.1	46.2	70.2	109.1	71.2	85.6	113.2	86.2	96.5	109.1	41.4	67.2	109.1	71.2	85.6	109.1	67.6	83.4
0960	Flooring	105.8	40.8	88.2	107.1	37.3	88.2	116.8	93.1	110.4	113.4	52.9	97.1	116.5	37.3	95.1	121.3	34.8	97.9
0970, 0990	Wall Finishes & Painting/Coating	98.4	70.2	81.5	102.0	44.1	67.3	102.0	88.5	93.9	102.0	37.8	63.5	102.0	76.4	86.6	104.6	73.3	85.8
09	FINISHES	103.7	51.5	76.1	103.0	59.3	79.9	109.5	88.8	98.5	106.9	44.0	73.6	106.0	65.6	84.6	108.7	62.0	84.0
COVERS	DIVS. 10 - 14, 25, 28, 41, 43, 44	100.0	83.5	96.7	100.0	79.2	95.9	100.0	88.0	97.6	100.0	46.0	89.3	100.0	83.3	96.7	100.0	81.5	96.3
21, 22, 23	FIRE SUPPRESSION, PLUMBING & HVAC	100.2	77.8	91.3	94.4	47.4	75.7	100.2	88.0	95.4	94.4	39.4	72.5	100.1	70.5	88.4	100.2	73.0	89.4
26, 27, 3370	ELECTRICAL, COMMUNICATIONS & UTIL.	101.6	82.4	92.3	94.4	48.0	71.8	93.9	87.3	90.7	96.5	36.6	67.3	109.1	72.8	91.4	100.2	75.0	87.9
MF2004	WEIGHTED AVERAGE	99.6	74.6	89.1	96.2	61.2	81.5	97.7	90.2	94.5	96.5	48.8	76.5	98.2	76.7	89.2	97.9	75.8	88.6

	DIVISION	IOWA																	
		FORT DODGE			MASON CITY			OTTUMWA			SHENANDOAH			SIBLEY			SIOUX CITY		
		505			504			525			516			512			510 - 511		
		MAT.	INST.	TOTAL	MAT.	INST.	TOTAL	MAT.	INST.	TOTAL	MAT.	INST.	TOTAL	MAT.	INST.	TOTAL	MAT.	INST.	TOTAL
015433	CONTRACTOR EQUIPMENT		99.9	99.9		99.9	99.9		95.9	95.9		96.1	96.1		99.9	99.9		99.9	99.9
0241, 31 - 34	SITE & INFRASTRUCTURE, DEMOLITION	88.6	93.8	92.2	88.6	94.1	92.4	93.3	91.8	92.3	99.9	89.2	92.5	102.7	93.5	96.4	103.8	95.9	98.4
0310	Concrete Forming & Accessories	84.0	49.5	54.2	88.5	43.4	49.5	90.3	57.1	61.6	83.9	54.9	58.9	84.3	38.8	45.0	100.0	62.1	67.3
0320	Concrete Reinforcing	101.7	46.1	79.0	101.5	59.3	84.3	101.5	79.4	92.5	104.3	73.5	91.8	104.3	58.6	85.7	102.3	60.8	85.4
0330	Cast-in-Place Concrete	98.3	48.0	79.1	98.3	52.9	81.0	105.7	58.8	87.8	105.9	48.6	84.0	103.8	48.6	82.7	104.4	54.0	85.2
03	CONCRETE	95.1	49.8	74.5	95.3	51.3	75.3	100.1	63.2	83.3	100.7	57.5	81.0	99.7	47.6	76.0	100.0	60.3	81.9
04	MASONRY	97.0	34.7	59.7	109.7	43.6	70.1	100.1	46.3	67.9	102.5	31.4	60.0	121.7	35.3	70.0	96.2	53.0	70.3
05	METALS	85.7	71.6	81.6	85.7	77.2	83.2	85.6	89.0	86.6	92.3	81.1	89.0	85.8	75.6	82.9	87.7	81.0	85.8
06	WOOD, PLASTICS & COMPOSITES	89.4	52.9	69.4	94.3	43.4	66.4	96.7	59.0	76.0	88.9	61.8	74.0	89.8	38.7	61.7	108.4	62.4	83.1
07	THERMAL & MOISTURE PROTECTION	101.2	36.4	76.0	100.7	45.0	79.0	101.5	54.8	83.4	100.5	41.4	77.5	100.8	46.3	79.6	101.2	52.2	82.2
08	OPENINGS	103.3	44.6	88.8	93.8	48.3	82.5	98.0	64.3	89.7	87.6	59.9	80.8	94.1	44.3	81.8	99.0	59.9	89.3
0920	Plaster & Gypsum Board	90.3	51.2	64.9	91.1	41.4	58.8	92.5	57.8	70.0	90.3	60.7	71.1	90.3	36.5	55.4	98.5	61.1	74.2
0950, 0980	Ceilings & Acoustic Treatment	109.1	51.2	73.3	109.1	41.4	67.2	109.1	57.8	77.4	109.1	60.7	79.2	109.1	36.5	64.2	113.2	61.1	81.0
0960	Flooring	108.9	52.9	93.8	111.7	52.9	95.8	125.0	56.3	106.4	106.7	38.4	88.2	107.9	38.1	89.0	117.3	52.7	99.9
0970, 0990	Wall Finishes & Painting/Coating	102.0	16.2	50.5	102.0	34.6	61.6	104.6	79.1	89.3	98.4	24.5	54.1	102.0	35.7	62.3	103.0	61.0	77.8
09	FINISHES	104.8	47.0	74.3	105.8	43.6	72.9	110.2	58.6	82.9	103.9	49.6	75.2	106.2	38.0	70.1	111.2	60.2	84.2
COVERS	DIVS. 10 - 14, 25, 28, 41, 43, 44	100.0	75.5	95.2	100.0	46.9	89.5	100.0	76.5	95.3	100.0	75.1	95.1	100.0	56.1	91.3	100.0	80.5	96.1
21, 22, 23	FIRE SUPPRESSION, PLUMBING & HVAC	94.4	66.6	83.3	94.4	73.2	86.0	94.4	74.3	86.4	94.4	30.2	68.9	94.4	35.9	71.1	100.2	72.4	89.1
26, 27, 3370	ELECTRICAL, COMMUNICATIONS & UTIL.	101.2	48.8	75.7	100.2	64.9	83.0	98.6	71.5	85.4	96.5	29.1	63.7	96.5	36.8	67.4	96.5	71.5	84.3
MF2004	WEIGHTED AVERAGE	96.0	57.1	79.7	95.7	61.4	81.3	96.6	69.3	85.1	96.3	49.3	76.6	96.9	47.6	76.2	98.4	68.8	86.0

	DIVISION	IOWA						KANSAS											
		SPENCER			WATERLOO			BELLEVILLE			COLBY			DODGE CITY			EMPORIA		
		513			506 - 507			669			677			678			668		
		MAT.	INST.	TOTAL	MAT.	INST.	TOTAL	MAT.	INST.	TOTAL	MAT.	INST.	TOTAL	MAT.	INST.	TOTAL	MAT.	INST.	TOTAL
015433	CONTRACTOR EQUIPMENT		99.9	99.9		99.9	99.9		101.9	101.9		101.9	101.9		101.9	101.9		100.5	100.5
0241, 31 - 34	SITE & INFRASTRUCTURE, DEMOLITION	102.7	93.5	96.4	93.1	95.0	94.4	112.1	90.4	97.2	109.4	91.2	97.0	111.7	90.9	97.5	103.1	89.2	93.6
0310	Concrete Forming & Accessories	90.2	38.7	45.7	100.4	43.7	51.5	95.7	49.2	55.6	102.2	61.2	66.8	95.3	61.1	65.8	87.4	36.3	43.3
0320	Concrete Reinforcing	104.3	37.2	76.9	102.3	80.2	93.3	107.2	54.4	85.6	109.8	54.4	87.2	107.2	54.8	85.8	105.6	55.1	85.0
0330	Cast-in-Place Concrete	103.8	48.6	82.7	105.3	45.9	82.6	117.9	53.3	93.2	113.3	54.2	90.7	115.4	53.9	91.9	114.0	42.6	86.7
03	CONCRETE	100.1	43.6	74.4	100.6	53.0	79.0	115.4	53.2	87.1	112.9	58.9	88.3	113.8	58.8	88.8	108.4	44.1	79.1
04	MASONRY	121.7	35.3	70.0	97.6	55.6	72.5	97.0	45.1	65.9	98.3	44.0	65.8	104.7	51.3	72.8	102.4	54.4	73.6
05	METALS	85.8	66.8	80.3	87.7	89.0	88.1	90.9	77.1	86.9	91.2	79.8	87.9	92.4	79.3	88.6	90.6	80.6	87.7
06	WOOD, PLASTICS & COMPOSITES	96.6	38.7	64.8	109.0	43.4	73.0	96.9	52.0	72.3	104.1	67.8	84.2	96.3	67.8	80.6	88.3	35.2	59.1
07	THERMAL & MOISTURE PROTECTION	101.7	40.8	78.0	101.0	48.4	80.5	99.9	54.4	82.2	99.9	55.7	82.7	99.9	57.5	83.4	98.6	56.6	82.3
08	OPENINGS	107.4	38.1	90.3	94.7	56.2	85.2	96.8	49.3	85.0	96.9	57.9	87.2	96.8	57.9	87.2	94.4	40.1	81.0
0920	Plaster & Gypsum Board	92.5	36.5	56.2	98.5	41.4	61.4	96.4	50.2	66.4	100.1	66.5	78.3	95.6	66.5	76.7	92.6	32.7	53.7
0950, 0980	Ceilings & Acoustic Treatment	109.1	36.5	64.2	113.2	41.4	68.8	86.8	50.2	64.1	86.8	66.5	74.2	86.8	66.5	74.2	86.8	32.7	53.3
0960	Flooring	111.3	38.1	91.5	118.7	52.9	100.9	99.7	43.9	84.6	103.6	43.9	87.5	99.5	43.9	84.5	94.1	40.8	79.7
0970, 0990	Wall Finishes & Painting/Coating	102.0	35.7	62.3	103.0	32.0	60.4	93.0	44.7	64.0	93.0	44.7	64.0	93.0	44.7	64.0	93.0	44.7	64.0
09	FINISHES	107.5	38.0	70.7	110.1	43.3	74.8	96.3	47.7	70.6	97.8	57.0	76.2	96.1	57.0	75.4	93.2	36.6	63.3
COVERS	DIVS. 10 - 14, 25, 28, 41, 43, 44	100.0	73.9	94.8	100.0	74.6	95.0	100.0	44.1	88.9	100.0	45.9	89.3	100.0	46.0	89.3	100.0	39.0	87.9
21, 22, 23	FIRE SUPPRESSION, PLUMBING & HVAC	94.4	36.1	71.2	100.2	39.9	76.2	94.1	66.6	83.2	94.1	59.4	80.3	99.9	59.4	83.8	94.1	65.7	82.8
26, 27, 3370	ELECTRICAL, COMMUNICATIONS & UTIL.	98.1	36.8	68.3	96.5	64.9	81.1	99.8	64.9	82.8	99.8	65.0	82.8	96.7	71.8	84.5	97.0	72.1	84.9
MF2004	WEIGHTED AVERAGE	98.5	46.3	76.6	97.8	58.4	81.2	98.4	61.0	82.7	98.4	62.3	83.3	99.9	64.0	84.8	96.7	59.7	81.2

DIVISION		KANSAS																	
		FORT SCOTT			HAYS			HUTCHINSON			INDEPENDENCE			KANSAS CITY			LIBERAL		
		667			676			675			673			660 - 662			679		
		MAT.	INST.	TOTAL	MAT.	INST.	TOTAL	MAT.	INST.	TOTAL	MAT.	INST.	TOTAL	MAT.	INST.	TOTAL	MAT.	INST.	TOTAL
015433	CONTRACTOR EQUIPMENT		101.2	101.2		101.9	101.9		101.9	101.9		101.9	101.9		96.4	96.4		101.9	101.9
0241, 31 - 34	SITE & INFRASTRUCTURE, DEMOLITION	99.5	89.8	92.9	115.3	91.2	98.9	92.2	91.2	91.5	113.0	91.3	98.2	91.9	85.4	87.5	115.3	91.2	98.8
0310	Concrete Forming & Accessories	103.3	77.3	80.8	99.7	61.2	66.4	89.8	59.5	63.6	111.0	54.3	62.0	100.0	99.0	99.1	95.8	59.5	64.5
0320	Concrete Reinforcing	105.0	87.1	97.7	107.2	54.5	85.7	107.2	54.8	85.8	106.5	57.5	86.5	101.8	88.5	96.4	108.7	54.8	86.7
0330	Cast-in-Place Concrete	105.7	57.6	87.3	89.3	54.2	75.9	82.7	41.3	66.9	115.8	42.0	87.6	90.5	96.2	92.7	89.3	41.3	71.0
03	CONCRETE	103.7	73.1	89.8	104.1	58.9	83.6	88.6	53.7	72.7	115.0	52.2	86.4	95.7	96.3	96.0	106.2	53.7	82.3
04	MASONRY	102.5	55.7	74.5	105.6	45.4	69.5	97.7	47.1	67.4	95.5	48.3	67.3	105.5	94.8	99.1	102.6	47.1	69.3
05	METALS	90.6	91.6	90.9	90.8	79.9	87.6	90.7	79.1	87.3	90.6	80.5	87.6	97.2	96.9	97.1	91.1	79.1	87.6
06	WOOD, PLASTICS & COMPOSITES	106.4	86.1	95.2	101.3	67.8	82.9	90.9	67.8	78.2	114.1	59.5	84.2	102.4	100.4	101.3	96.9	67.8	80.9
07	THERMAL & MOISTURE PROTECTION	99.3	69.6	87.8	100.2	56.0	83.0	99.0	54.2	81.6	100.0	62.4	85.4	98.5	97.3	98.0	100.3	54.2	82.4
08	OPENINGS	94.4	79.8	90.7	96.8	57.9	87.1	96.7	57.9	87.1	94.4	54.1	84.4	95.8	92.5	95.0	96.8	57.9	87.2
0920	Plaster & Gypsum Board	98.6	85.5	90.1	98.6	66.5	77.7	94.1	66.5	76.2	105.3	58.0	74.5	92.1	100.3	97.4	96.4	66.5	77.0
0950, 0980	Ceilings & Acoustic Treatment	86.8	85.5	86.0	86.8	66.5	74.2	86.8	66.5	74.2	86.8	58.0	69.0	87.6	100.3	95.5	86.8	66.5	74.2
0960	Flooring	110.5	44.2	92.6	102.3	43.9	86.5	96.1	43.9	82.0	108.7	43.9	91.2	89.2	97.8	91.5	99.7	43.9	84.6
0970, 0990	Wall Finishes & Painting/Coating	95.0	44.7	64.9	93.0	44.7	64.0	93.0	44.7	64.0	93.0	44.7	64.0	100.6	77.8	86.9	93.0	44.7	64.0
09	FINISHES	99.2	68.9	83.2	97.8	57.0	76.2	93.4	56.0	73.6	100.2	51.4	74.4	94.0	95.9	95.0	96.9	56.0	75.3
COVERS	DIVS. 10 - 14, 25, 28, 41, 43, 44	100.0	54.0	90.9	100.0	45.9	89.3	100.0	44.5	89.0	100.0	44.0	88.9	100.0	74.5	94.9	100.0	44.6	89.0
21, 22, 23	FIRE SUPPRESSION, PLUMBING & HVAC	94.1	62.2	81.4	94.1	59.4	80.3	94.1	57.4	79.5	94.1	65.2	82.6	99.9	90.2	96.0	94.1	57.4	79.5
26, 27, 3370	ELECTRICAL, COMMUNICATIONS & UTIL.	96.3	72.1	84.5	98.7	65.0	82.3	93.8	65.0	79.7	95.9	72.1	84.3	101.8	98.7	100.3	96.7	71.8	84.5
MF2004	WEIGHTED AVERAGE	96.6	71.4	86.0	97.6	62.5	82.9	93.8	61.2	80.1	98.1	63.3	83.5	98.3	93.7	96.4	97.5	62.2	82.7

DIVISION		KANSAS									KENTUCKY								
		SALINA			TOPEKA			WICHITA			ASHLAND			BOWLING GREEN			CAMPTON		
		674			664 - 666			670 - 672			411 - 412			421 - 422			413 - 414		
		MAT.	INST.	TOTAL	MAT.	INST.	TOTAL	MAT.	INST.	TOTAL	MAT.	INST.	TOTAL	MAT.	INST.	TOTAL	MAT.	INST.	TOTAL
015433	CONTRACTOR EQUIPMENT		101.9	101.9		100.5	100.5		101.9	101.9		101.1	101.1		94.2	94.2		101.0	101.0
0241, 31 - 34	SITE & INFRASTRUCTURE, DEMOLITION	101.4	91.3	94.5	95.7	89.1	91.2	94.2	91.1	92.0	100.4	90.0	93.3	67.9	96.2	87.2	76.1	97.4	90.6
0310	Concrete Forming & Accessories	91.8	49.4	55.2	99.0	46.2	53.4	98.7	51.0	57.4	88.2	98.6	97.2	85.2	83.5	83.7	89.1	52.6	57.5
0320	Concrete Reinforcing	106.5	70.6	91.9	99.0	95.2	97.4	99.0	78.0	90.4	88.2	101.8	93.8	85.5	88.2	86.6	86.3	70.4	79.8
0330	Cast-in-Place Concrete	100.1	46.8	79.7	91.5	51.2	76.1	87.3	55.0	75.0	84.4	99.0	90.0	83.7	106.4	92.4	93.4	64.3	82.3
03	CONCRETE	101.7	54.2	80.1	93.6	58.8	77.8	91.6	58.9	76.8	92.3	100.0	95.8	90.5	92.4	91.4	94.2	61.0	79.1
04	MASONRY	119.4	45.0	74.9	99.6	58.9	75.2	93.5	58.3	72.4	87.7	102.7	96.7	89.1	76.0	81.3	88.1	39.0	58.7
05	METALS	92.3	85.6	90.3	97.4	96.0	97.0	97.4	87.2	94.4	90.5	108.2	95.7	92.0	87.2	90.6	91.4	78.4	87.6
06	WOOD, PLASTICS & COMPOSITES	92.8	52.0	70.4	98.2	43.7	68.3	99.8	49.8	72.4	77.6	96.8	88.1	91.0	82.5	86.4	89.6	50.9	68.4
07	THERMAL & MOISTURE PROTECTION	99.5	53.1	81.4	100.4	64.6	86.5	99.0	59.4	83.6	88.7	98.7	92.6	85.3	86.6	85.8	97.1	53.5	80.1
08	OPENINGS	96.7	51.9	85.6	97.1	59.6	87.8	97.1	57.1	87.2	93.6	94.3	93.8	95.1	80.4	91.5	96.7	61.5	88.0
0920	Plaster & Gypsum Board	94.9	50.2	65.8	93.8	41.5	59.8	90.9	47.9	62.9	56.9	96.8	82.8	80.3	82.2	81.5	80.3	48.4	59.6
0950, 0980	Ceilings & Acoustic Treatment	86.8	50.2	64.1	87.6	41.5	59.1	85.9	47.9	62.4	79.8	96.8	90.3	85.3	82.2	83.4	85.3	48.4	62.5
0960	Flooring	97.6	32.1	79.9	101.1	40.8	84.8	100.0	70.6	92.0	78.3	100.1	84.2	94.1	85.0	91.7	95.6	46.2	82.3
0970, 0990	Wall Finishes & Painting/Coating	93.0	44.7	64.0	93.0	60.4	73.4	93.0	55.1	70.3	100.7	96.8	98.4	97.7	72.4	82.5	97.7	39.7	62.9
09	FINISHES	94.7	45.1	68.5	95.3	44.8	68.6	94.2	54.6	73.3	80.1	98.8	90.0	90.7	82.5	86.3	91.3	48.8	68.8
COVERS	DIVS. 10 - 14, 25, 28, 41, 43, 44	100.0	56.1	91.3	100.0	59.3	91.9	100.0	58.5	91.8	100.0	86.9	97.4	100.0	69.7	94.0	100.0	61.3	92.3
21, 22, 23	FIRE SUPPRESSION, PLUMBING & HVAC	99.9	59.3	83.8	99.9	63.9	85.6	99.8	57.5	83.0	94.0	92.2	93.3	99.9	85.3	94.1	94.2	72.8	85.7
26, 27, 3370	ELECTRICAL, COMMUNICATIONS & UTIL.	96.4	71.8	84.4	108.4	72.2	90.7	103.2	71.8	87.9	89.9	91.7	90.8	92.8	79.4	86.2	90.2	77.1	83.8
MF2004	WEIGHTED AVERAGE	98.7	61.6	83.1	98.9	66.1	85.1	97.6	64.9	83.9	91.7	96.7	93.8	93.6	84.6	89.8	93.1	66.2	81.8

DIVISION		KENTUCKY																	
		CORBIN			COVINGTON			ELIZABETHTOWN			FRANKFORT			HAZARD			HENDERSON		
		407 - 409			410			427			406			417 - 418			424		
		MAT.	INST.	TOTAL	MAT.	INST.	TOTAL	MAT.	INST.	TOTAL	MAT.	INST.	TOTAL	MAT.	INST.	TOTAL	MAT.	INST.	TOTAL
015433	CONTRACTOR EQUIPMENT		101.0	101.0		104.0	104.0		94.2	94.2		101.0	101.0		101.0	101.0		113.5	113.5
0241, 31 - 34	SITE & INFRASTRUCTURE, DEMOLITION	73.7	97.4	89.9	72.3	112.7	99.9	63.3	95.9	85.6	76.9	98.3	91.5	74.2	99.2	91.3	70.8	120.7	104.9
0310	Concrete Forming & Accessories	86.3	52.1	56.8	85.2	105.5	102.7	80.7	81.9	81.7	96.7	63.7	68.2	86.3	45.4	51.0	92.1	81.4	82.9
0320	Concrete Reinforcing	85.9	70.4	79.6	85.0	97.6	90.1	85.9	93.8	89.1	87.2	86.3	86.8	86.8	71.1	80.4	85.6	86.4	85.9
0330	Cast-in-Place Concrete	88.4	64.1	79.1	89.3	96.5	92.1	75.7	78.7	76.8	84.5	71.4	79.5	89.8	93.0	91.0	73.9	90.4	80.2
03	CONCRETE	89.8	60.8	76.6	92.6	100.7	96.3	83.6	83.2	83.4	88.9	71.3	80.9	91.3	67.9	80.7	87.8	85.7	86.9
04	MASONRY	87.5	41.3	59.9	100.8	115.3	109.5	74.0	80.4	77.8	85.4	74.9	79.1	84.8	48.9	63.3	92.7	91.0	91.7
05	METALS	91.3	78.2	87.5	89.3	95.6	91.1	91.3	89.4	90.8	98.8	87.5	95.5	91.4	78.2	87.6	82.7	89.1	84.6
06	WOOD, PLASTICS & COMPOSITES	86.7	50.9	67.1	87.6	98.4	93.5	86.3	82.5	84.2	98.6	61.5	78.2	86.7	37.5	59.7	93.4	79.5	85.7
07	THERMAL & MOISTURE PROTECTION	96.9	55.1	80.7	95.3	100.6	97.3	85.0	80.7	83.3	97.0	74.0	88.0	97.0	52.9	79.8	96.5	93.8	95.4
08	OPENINGS	96.0	53.7	85.5	98.1	94.7	97.2	95.1	85.4	92.7	95.3	70.9	89.2	97.1	44.0	84.0	93.3	81.7	90.4
0920	Plaster & Gypsum Board	79.5	48.4	59.3	67.4	98.9	87.9	79.2	82.2	81.1	86.3	59.4	68.8	79.5	34.5	50.3	81.4	78.0	79.2
0950, 0980	Ceilings & Acoustic Treatment	85.3	48.4	62.5	87.0	98.9	94.4	85.3	82.2	83.4	86.1	59.4	69.6	85.3	34.5	53.9	77.0	78.0	77.6
0960	Flooring	93.8	46.2	80.9	70.5	91.2	76.1	91.3	67.7	84.9	98.7	61.0	88.5	93.8	41.0	79.5	97.4	91.1	95.7
0970, 0990	Wall Finishes & Painting/Coating	97.7	39.7	62.9	93.1	86.8	89.3	97.7	79.8	87.0	97.7	54.0	71.5	97.7	32.6	58.6	97.7	98.8	98.4
09	FINISHES	90.4	48.8	68.4	82.2	101.0	92.1	89.3	79.0	83.9	93.2	61.0	76.2	90.4	41.5	64.6	89.3	84.9	87.0
COVERS	DIVS. 10 - 14, 25, 28, 41, 43, 44	100.0	55.3	91.2	100.0	106.1	101.2	100.0	84.2	96.9	100.0	75.2	95.1	100.0	63.3	92.7	100.0	77.1	95.5
21, 22, 23	FIRE SUPPRESSION, PLUMBING & HVAC	94.1	72.2	85.4	94.6	99.2	96.5	94.5	83.4	90.1	100.0	73.8	89.6	94.2	47.4	75.6	94.5	78.7	88.2
26, 27, 3370	ELECTRICAL, COMMUNICATIONS & UTIL.	90.4	77.1	83.9	93.7	79.9	87.0	90.0	79.4	84.8	102.1	101.6	101.9	90.2	58.6	74.8	92.2	80.4	86.4
MF2004	WEIGHTED AVERAGE	92.3	65.8	81.2	92.8	99.3	95.5	90.1	83.6	87.3	96.3	78.9	89.0	92.4	58.7	78.3	90.7	86.8	89.0

DIVISION		KENTUCKY																	
		LEXINGTON			LOUISVILLE			OWENSBORO			PADUCAH			PIKEVILLE			SOMERSET		
		403 - 405			400 - 402			423			420			415 - 416			425 - 426		
		MAT.	INST.	TOTAL	MAT.	INST.	TOTAL	MAT.	INST.	TOTAL	MAT.	INST.	TOTAL	MAT.	INST.	TOTAL	MAT.	INST.	TOTAL
015433	CONTRACTOR EQUIPMENT		101.0	101.0		94.2	94.2		113.5	113.5		113.5	113.5		101.1	101.1		101.0	101.0
0241, 31 - 34	SITE & INFRASTRUCTURE, DEMOLITION	75.3	99.3	91.7	65.8	95.9	86.4	79.4	121.8	108.4	73.4	121.3	106.1	110.1	89.6	96.1	68.0	98.0	88.5
0310	Concrete Forming & Accessories	97.8	75.8	78.8	93.5	82.0	83.6	90.6	75.0	77.1	88.6	81.7	82.6	96.5	72.5	75.7	86.9	50.6	55.5
0320	Concrete Reinforcing	94.4	93.4	94.0	94.4	93.8	94.1	85.6	91.5	88.0	86.1	83.4	85.0	88.7	95.1	91.3	85.9	85.1	85.6
0330	Cast-in-Place Concrete	90.4	84.0	87.9	94.3	78.8	88.4	86.3	103.0	92.7	78.8	92.9	84.2	92.7	97.9	94.7	73.9	107.9	86.9
03	CONCRETE	93.1	82.3	88.2	94.7	83.3	89.5	98.8	88.2	93.9	91.9	86.1	89.3	104.5	86.8	96.5	79.1	77.9	78.5
04	MASONRY	86.6	68.3	75.6	87.8	80.4	83.4	86.1	74.3	79.0	89.2	91.7	90.7	86.3	70.0	76.5	80.6	59.1	67.7
05	METALS	93.4	90.4	92.5	100.8	89.6	97.5	83.9	90.0	85.7	81.4	87.5	83.2	90.4	100.8	93.5	91.3	84.6	89.4
06	WOOD, PLASTICS & COMPOSITES	101.2	75.1	86.9	98.6	82.5	89.8	91.4	71.3	80.4	89.3	80.2	84.3	86.7	64.4	74.4	87.4	43.2	63.1
07	THERMAL & MOISTURE PROTECTION	97.1	89.0	94.0	91.1	80.8	87.1	96.8	78.5	89.7	96.5	84.8	91.9	89.3	63.8	79.4	96.5	59.1	81.9
08	OPENINGS	96.1	80.0	92.1	96.1	85.4	93.4	93.3	78.1	89.6	92.4	79.6	89.3	94.4	73.9	89.3	96.0	61.0	87.3
0920	Plaster & Gypsum Board	87.3	73.5	78.4	83.8	82.2	82.8	80.3	69.6	73.4	79.6	78.8	79.1	59.9	63.2	62.0	79.9	40.5	54.3
0950, 0980	Ceilings & Acoustic Treatment	89.5	73.5	79.6	86.1	82.2	83.7	77.0	69.6	72.4	77.0	78.8	78.1	79.8	63.2	69.5	85.3	40.5	57.6
0960	Flooring	99.0	62.2	89.1	97.5	71.4	90.5	96.6	85.0	93.5	95.3	66.6	87.6	83.0	55.0	75.4	94.0	46.2	81.1
0970, 0990	Wall Finishes & Painting/Coating	97.7	72.2	82.4	97.7	79.8	87.0	97.7	92.7	94.7	97.7	80.2	87.2	100.7	93.4	96.3	97.7	54.8	71.9
09	FINISHES	94.1	72.1	82.5	91.9	79.6	85.4	89.4	78.0	83.4	88.6	78.4	83.2	82.6	70.3	76.1	90.0	48.2	67.9
COVERS	DIVS. 10 - 14, 25, 28, 41, 43, 44	100.0	93.4	98.7	100.0	92.9	98.6	100.0	89.3	97.9	100.0	74.2	94.9	100.0	66.3	93.3	100.0	65.5	93.2
21, 22, 23	FIRE SUPPRESSION, PLUMBING & HVAC	99.9	68.9	87.6	99.9	83.4	93.4	99.9	73.0	89.2	94.5	83.7	90.2	94.0	83.5	89.8	94.5	68.5	84.2
26, 27, 3370	ELECTRICAL, COMMUNICATIONS & UTIL.	93.2	76.6	85.1	104.3	79.4	92.2	92.2	80.4	86.5	94.5	84.7	89.7	92.8	85.6	89.3	90.6	76.6	83.8
MF2004	WEIGHTED AVERAGE	95.1	78.5	88.1	97.1	83.9	91.6	93.4	83.2	89.1	90.9	87.2	89.4	94.0	81.7	88.8	90.6	70.3	82.1

DIVISION		LOUISIANA																	
		ALEXANDRIA			BATON ROUGE			HAMMOND			LAFAYETTE			LAKE CHARLES			MONROE		
		713 - 714			707 - 708			704			705			706			712		
		MAT.	INST.	TOTAL	MAT.	INST.	TOTAL	MAT.	INST.	TOTAL	MAT.	INST.	TOTAL	MAT.	INST.	TOTAL	MAT.	INST.	TOTAL
015433	CONTRACTOR EQUIPMENT		90.7	90.7		89.6	89.6		90.9	90.9		90.9	90.9		89.6	89.6		90.7	90.7
0241, 31 - 34	SITE & INFRASTRUCTURE, DEMOLITION	103.3	88.4	93.1	117.8	86.5	96.4	118.3	88.9	98.2	118.9	88.8	98.3	119.6	86.2	96.8	103.3	87.9	92.8
0310	Concrete Forming & Accessories	84.2	39.7	45.8	100.0	63.7	68.7	84.1	53.2	57.4	102.2	49.1	56.4	103.4	54.1	60.8	83.6	40.4	46.3
0320	Concrete Reinforcing	101.1	63.1	85.6	100.9	63.3	85.6	99.5	61.1	83.9	100.9	62.9	85.4	100.9	60.4	84.4	100.0	63.1	84.9
0330	Cast-in-Place Concrete	93.9	42.6	74.3	91.5	62.6	80.4	89.2	48.9	73.8	88.7	49.3	73.7	93.4	62.6	81.6	93.9	60.7	81.2
03	CONCRETE	90.9	46.6	70.7	93.7	63.9	80.1	91.5	54.1	74.5	92.5	52.8	74.5	94.8	59.1	78.6	90.6	53.0	73.5
04	MASONRY	115.2	49.6	75.9	106.9	54.3	75.4	106.1	56.2	76.2	106.1	47.2	70.8	105.0	51.8	73.1	109.9	50.2	74.2
05	METALS	92.4	75.4	87.4	105.1	73.7	95.9	100.2	72.8	92.2	99.3	73.0	91.6	99.3	72.2	91.4	92.4	74.3	87.1
06	WOOD, PLASTICS & COMPOSITES	86.6	38.8	60.3	104.1	69.2	84.9	88.3	55.0	70.0	110.8	50.8	77.9	108.8	56.3	80.0	85.9	39.1	60.2
07	THERMAL & MOISTURE PROTECTION	99.3	49.9	80.1	101.3	59.7	85.1	103.2	59.7	86.3	103.7	53.6	84.2	102.9	57.1	85.1	99.3	53.5	81.5
08	OPENINGS	99.3	45.9	86.1	102.6	62.3	92.6	98.1	58.2	88.2	102.6	52.3	90.1	102.6	55.8	91.0	99.3	50.2	87.2
0920	Plaster & Gypsum Board	81.9	37.1	52.8	99.4	68.4	79.3	97.4	53.8	69.1	105.2	49.4	68.9	105.2	55.1	72.6	81.5	37.5	52.9
0950, 0980	Ceilings & Acoustic Treatment	92.0	37.1	58.0	98.6	68.4	80.0	98.5	53.8	70.9	96.9	49.4	67.5	97.8	55.1	71.4	92.0	37.5	58.3
0960	Flooring	102.0	64.0	91.7	107.2	68.7	96.8	97.0	51.9	84.8	107.4	38.9	88.8	107.4	64.0	95.6	101.5	64.0	91.4
0970, 0990	Wall Finishes & Painting/Coating	98.5	36.6	61.4	106.6	42.0	67.8	106.6	55.5	76.0	106.6	42.0	67.8	106.6	38.7	65.8	98.5	43.7	65.6
09	FINISHES	93.0	43.1	66.6	102.0	63.1	81.4	98.6	53.2	74.6	102.3	45.9	72.5	102.5	54.2	77.0	92.8	44.2	67.1
COVERS	DIVS. 10 - 14, 25, 28, 41, 43, 44	100.0	59.3	91.9	100.0	77.9	95.6	100.0	57.0	91.5	100.0	75.4	95.1	100.0	76.4	95.3	100.0	54.5	91.0
21, 22, 23	FIRE SUPPRESSION, PLUMBING & HVAC	100.1	55.4	82.3	99.9	54.4	81.8	94.2	51.4	77.2	100.0	55.4	82.2	100.0	55.9	82.5	100.1	49.1	79.8
26, 27, 3370	ELECTRICAL, COMMUNICATIONS & UTIL.	93.4	55.6	75.0	107.6	62.9	85.8	94.7	52.4	74.1	95.8	61.1	78.9	95.3	61.1	78.7	95.2	56.7	76.5
MF2004	WEIGHTED AVERAGE	97.0	56.0	79.8	102.2	63.8	86.1	97.6	58.4	81.2	100.0	58.5	82.6	100.1	61.1	83.7	96.9	56.0	79.7

DIVISION		LOUISIANA									MAINE								
		NEW ORLEANS			SHREVEPORT			THIBODAUX			AUGUSTA			BANGOR			BATH		
		700 - 701			710 - 711			703			043			044			045		
		MAT.	INST.	TOTAL	MAT.	INST.	TOTAL	MAT.	INST.	TOTAL	MAT.	INST.	TOTAL	MAT.	INST.	TOTAL	MAT.	INST.	TOTAL
015433	CONTRACTOR EQUIPMENT		91.0	91.0		90.7	90.7		90.9	90.9		100.4	100.4		100.4	100.4		100.4	100.4
0241, 31 - 34	SITE & INFRASTRUCTURE, DEMOLITION	122.9	89.8	100.2	105.0	87.8	93.2	121.3	88.9	99.1	81.6	100.5	94.5	80.8	98.1	92.6	79.0	100.5	93.7
0310	Concrete Forming & Accessories	104.8	66.4	71.7	99.2	42.6	50.3	97.1	61.7	66.5	95.0	88.6	89.5	90.4	83.8	84.7	86.4	88.7	88.4
0320	Concrete Reinforcing	100.9	64.9	86.2	99.6	63.2	84.8	99.5	61.0	83.8	82.0	99.0	88.9	81.7	99.2	88.8	80.8	99.0	88.2
0330	Cast-in-Place Concrete	92.5	71.9	84.6	93.3	48.9	76.3	95.8	52.1	79.1	83.5	58.8	74.0	77.5	59.5	70.6	77.5	58.8	70.4
03	CONCRETE	94.4	68.7	82.7	90.6	49.9	72.1	96.2	58.9	79.3	100.6	79.8	91.1	97.0	77.8	88.3	97.0	79.9	89.2
04	MASONRY	106.3	59.7	78.4	101.0	51.9	71.6	134.6	53.1	85.8	97.6	45.5	66.4	111.9	52.0	76.0	118.1	45.1	74.4
05	METALS	113.3	76.7	102.6	88.6	74.2	84.4	100.3	72.6	92.2	82.7	83.3	82.9	82.4	79.5	81.6	81.1	83.3	81.7
06	WOOD, PLASTICS & COMPOSITES	105.0	69.2	85.3	103.1	42.5	69.8	97.6	65.9	80.2	99.9	97.7	98.7	94.8	90.4	92.4	89.7	97.7	94.1
07	THERMAL & MOISTURE PROTECTION	103.7	62.8	87.8	98.6	52.4	80.7	103.4	57.2	85.4	94.4	54.1	78.7	94.8	54.4	79.1	94.8	57.9	80.4
08	OPENINGS	103.4	68.1	94.7	97.2	48.0	85.0	103.6	64.1	93.8	101.7	77.4	95.7	101.6	74.9	95.0	101.6	78.9	96.0
0920	Plaster & Gypsum Board	104.0	68.4	80.9	86.3	40.9	56.8	101.5	65.1	77.8	106.9	96.8	100.4	100.4	89.3	93.2	97.4	96.8	97.0
0950, 0980	Ceilings & Acoustic Treatment	96.9	68.4	79.3	90.3	40.9	59.8	98.5	65.1	77.8	90.0	96.8	94.2	90.8	89.3	89.9	88.9	96.8	93.8
0960	Flooring	107.7	68.7	97.2	111.4	68.7	99.9	104.0	49.4	89.3	96.8	37.7	80.8	94.6	47.0	81.7	92.6	37.7	77.8
0970, 0990	Wall Finishes & Painting/Coating	108.1	63.4	81.3	98.5	36.6	61.4	108.1	55.5	76.6	91.7	31.7	55.7	91.7	29.2	54.2	91.7	31.7	55.7
09	FINISHES	102.9	66.8	83.8	96.3	46.2	69.8	101.5	59.2	79.1	97.3	74.7	85.3	96.0	72.5	83.6	94.5	74.7	84.0
COVERS	DIVS. 10 - 14, 25, 28, 41, 43, 44	100.0	79.8	96.0	100.0	69.0	93.9	100.0	78.1	95.7	100.0	58.2	91.7	100.0	70.4	94.1	100.0	58.2	91.7
21, 22, 23	FIRE SUPPRESSION, PLUMBING & HVAC	100.0	66.1	86.5	100.0	55.3	82.3	94.2	59.5	80.4	100.0	69.1	87.7	100.2	67.9	87.3	94.4	69.1	84.3
26, 27, 3370	ELECTRICAL, COMMUNICATIONS & UTIL.	96.3	68.6	82.8	102.6	68.7	86.1	93.3	63.5	78.8	98.3	80.1	89.5	95.3	80.0	87.9	93.5	80.1	87.0
MF2004	WEIGHTED AVERAGE	102.9	69.6	88.9	96.8	59.2	81.0	100.4	63.7	85.0	96.1	74.3	86.9	95.8	73.7	86.5	94.2	74.4	85.9

DIVISION		MAINE																	
		HOULTON			KITTERY			LEWISTON			MACHIAS			PORTLAND			ROCKLAND		
		047			039			042			046			040 - 041			048		
		MAT.	INST.	TOTAL	MAT.	INST.	TOTAL	MAT.	INST.	TOTAL	MAT.	INST.	TOTAL	MAT.	INST.	TOTAL	MAT.	INST.	TOTAL
015433	CONTRACTOR EQUIPMENT		100.4	100.4		100.4	100.4		100.4	100.4		100.4	100.4		100.4	100.4		100.4	100.4
0241, 31 - 34	SITE & INFRASTRUCTURE, DEMOLITION	80.7	100.5	94.2	75.5	100.5	92.6	78.2	98.1	91.8	80.1	100.5	94.0	79.6	98.1	92.2	76.8	100.5	93.0
0310	Concrete Forming & Accessories	93.5	88.7	89.4	86.9	88.7	88.5	95.3	83.8	85.3	91.0	88.6	88.9	94.4	83.8	85.2	91.8	88.7	89.1
0320	Concrete Reinforcing	81.7	99.0	88.8	80.3	99.0	87.9	100.4	99.2	99.9	81.7	99.0	88.7	100.4	99.2	99.9	81.7	99.0	88.8
0330	Cast-in-Place Concrete	77.5	58.8	70.4	77.5	58.8	70.3	79.1	59.5	71.6	77.5	58.7	70.3	100.6	59.5	84.9	79.1	58.8	71.3
03	CONCRETE	98.0	79.9	89.8	92.3	79.9	86.7	97.6	77.8	88.6	97.5	79.8	89.4	107.6	77.8	94.0	94.4	79.9	87.8
04	MASONRY	95.8	45.1	65.4	108.8	45.1	70.6	96.5	52.0	69.8	95.8	45.1	65.4	96.8	52.0	70.0	90.0	46.3	63.8
05	METALS	81.2	83.3	81.9	81.0	83.3	81.7	85.3	79.5	83.6	81.3	83.2	81.8	86.5	79.5	84.5	81.1	83.4	81.8
06	WOOD, PLASTICS & COMPOSITES	98.6	97.7	98.1	90.2	97.7	94.3	100.6	90.4	95.0	95.7	97.7	96.8	99.8	90.4	94.6	96.6	97.7	97.2
07	THERMAL & MOISTURE PROTECTION	94.9	57.9	80.5	94.4	57.9	80.2	94.6	54.4	79.0	94.8	57.8	80.4	97.2	54.4	80.5	94.6	58.2	80.4
08	OPENINGS	101.7	78.9	96.0	101.6	78.9	96.0	104.9	74.9	97.5	101.7	76.9	95.6	104.9	74.9	97.5	101.6	76.9	95.5
0920	Plaster & Gypsum Board	102.2	96.8	98.7	97.4	96.8	97.0	105.5	89.3	95.0	101.1	96.8	98.3	111.1	89.3	96.9	101.1	96.8	98.3
0950, 0980	Ceilings & Acoustic Treatment	88.9	96.8	93.8	88.9	96.8	93.8	99.1	89.3	93.0	88.9	96.8	93.8	96.6	89.3	92.1	88.9	96.8	93.8
0960	Flooring	95.8	37.7	80.1	92.9	37.7	78.0	97.5	47.0	83.9	94.9	18.5	74.2	96.8	47.0	83.3	95.2	37.7	79.6
0970, 0990	Wall Finishes & Painting/Coating	91.7	31.7	55.7	91.7	31.7	55.7	91.7	29.2	54.2	91.7	31.7	55.7	91.7	29.2	54.2	91.7	31.7	55.7
09	FINISHES	96.2	74.7	84.8	94.2	74.7	83.9	99.1	72.5	85.1	95.7	71.3	82.8	99.0	72.5	85.0	95.4	74.7	84.4
COVERS	DIVS. 10 - 14, 25, 28, 41, 43, 44	100.0	58.2	91.7	100.0	58.2	91.7	100.0	70.4	94.1	100.0	58.2	91.7	100.0	70.4	94.1	100.0	58.2	91.7
21, 22, 23	FIRE SUPPRESSION, PLUMBING & HVAC	94.4	69.1	84.3	94.4	69.1	84.3	100.2	67.9	87.3	94.4	69.1	84.3	100.0	67.9	87.2	94.4	69.1	84.3
26, 27, 3370	ELECTRICAL, COMMUNICATIONS & UTIL.	97.1	80.1	88.8	93.5	80.1	87.0	97.2	80.0	88.8	97.1	80.1	88.8	103.7	80.0	92.2	97.1	80.1	88.8
MF2004	WEIGHTED AVERAGE	93.9	74.4	85.7	93.0	74.4	85.2	96.4	73.7	86.9	93.7	73.9	85.4	98.5	73.7	88.1	92.9	74.5	85.2

DIVISION		MAINE			MARYLAND														
		WATERVILLE			ANNAPOLIS			BALTIMORE			COLLEGE PARK			CUMBERLAND			EASTON		
		049			214			210 - 212			207 - 208			215			216		
		MAT.	INST.	TOTAL	MAT.	INST.	TOTAL	MAT.	INST.	TOTAL	MAT.	INST.	TOTAL	MAT.	INST.	TOTAL	MAT.	INST.	TOTAL
015433	CONTRACTOR EQUIPMENT		100.4	100.4		101.9	101.9		104.8	104.8		103.6	103.6		101.9	101.9		101.9	101.9
0241, 31 - 34	SITE & INFRASTRUCTURE, DEMOLITION	80.6	100.5	94.2	98.5	94.0	95.4	96.0	97.2	96.8	107.8	90.5	96.0	87.6	93.5	91.6	94.3	89.7	91.1
0310	Concrete Forming & Accessories	85.9	88.7	88.3	98.6	54.1	60.1	101.7	77.9	81.1	88.7	69.9	72.5	92.4	79.0	80.8	90.4	32.1	40.1
0320	Concrete Reinforcing	81.7	99.0	88.8	100.9	86.7	95.1	100.9	86.7	95.1	113.5	78.4	99.2	89.2	74.0	83.0	88.4	18.7	60.0
0330	Cast-in-Place Concrete	77.5	58.8	70.4	111.8	78.9	99.2	95.2	79.8	89.3	128.7	77.1	109.0	85.0	83.2	84.3	94.4	43.2	74.8
03	CONCRETE	98.7	79.9	90.1	108.9	70.5	91.4	101.3	81.3	92.2	121.1	75.4	100.3	90.4	80.6	85.9	97.6	35.8	69.5
04	MASONRY	105.0	45.5	69.4	95.7	71.5	81.2	96.9	71.5	81.7	115.0	66.8	86.2	94.5	73.3	81.8	108.6	33.6	63.7
05	METALS	81.2	83.3	81.8	96.2	97.1	96.5	97.3	97.9	97.5	85.1	95.7	88.2	93.7	92.3	93.3	93.9	63.7	85.1
06	WOOD, PLASTICS & COMPOSITES	89.2	97.7	93.9	94.9	48.0	69.1	97.9	79.9	88.0	85.7	71.9	78.1	88.0	78.2	82.6	85.3	32.2	56.2
07	THERMAL & MOISTURE PROTECTION	94.9	57.1	80.2	97.1	76.6	89.1	97.6	80.2	90.8	105.2	74.9	93.4	97.1	77.7	89.6	97.2	37.8	74.1
08	OPENINGS	101.7	77.4	95.7	91.7	67.8	85.8	93.7	85.8	91.8	101.1	75.7	94.8	92.8	80.3	89.7	91.0	29.1	75.7
0920	Plaster & Gypsum Board	97.4	96.8	97.0	102.9	46.4	66.2	104.6	79.4	88.2	102.4	70.9	82.0	100.9	77.8	85.9	99.4	30.1	54.4
0950, 0980	Ceilings & Acoustic Treatment	88.9	96.8	93.8	92.9	46.4	64.2	95.4	79.4	85.5	97.4	70.9	81.0	95.4	77.8	84.5	95.4	30.1	55.0
0960	Flooring	92.4	37.7	77.6	89.9	78.2	86.7	89.9	78.2	86.7	117.9	57.9	101.7	85.7	72.9	82.2	84.9	14.4	65.8
0970, 0990	Wall Finishes & Painting/Coating	91.7	31.7	55.7	94.6	81.0	86.4	94.6	81.0	86.4	125.2	80.8	98.6	94.6	82.6	87.4	94.6	23.3	51.8
09	FINISHES	94.6	74.7	84.1	95.2	59.0	76.1	96.0	77.8	86.4	104.5	68.4	85.4	93.6	78.0	85.4	93.6	28.6	59.3
COVERS	DIVS. 10 - 14, 25, 28, 41, 43, 44	100.0	58.2	91.7	100.0	81.9	96.4	100.0	86.0	97.2	100.0	61.6	92.4	100.0	88.5	97.7	100.0	44.3	89.0
21, 22, 23	FIRE SUPPRESSION, PLUMBING & HVAC	94.4	69.1	84.3	99.8	88.5	95.3	99.9	88.5	95.4	94.3	80.3	88.7	94.1	72.9	85.7	94.1	24.7	66.5
26, 27, 3370	ELECTRICAL, COMMUNICATIONS & UTIL.	97.1	80.1	88.8	99.2	89.8	94.6	98.2	91.7	95.0	102.8	89.6	96.4	95.6	81.4	88.7	95.1	60.8	78.4
MF2004	WEIGHTED AVERAGE	94.2	74.4	85.9	98.7	80.2	90.9	98.2	86.0	93.1	100.4	79.4	91.6	94.0	80.4	88.3	95.5	42.8	73.4

DIVISION		MARYLAND															MASSACHUSETTS		
		ELKTON			HAGERSTOWN			SALISBURY			SILVER SPRING			WALDORF			BOSTON		
		219			217			218			209			206			020 - 022, 024		
		MAT.	INST.	TOTAL	MAT.	INST.	TOTAL	MAT.	INST.	TOTAL	MAT.	INST.	TOTAL	MAT.	INST.	TOTAL	MAT.	INST.	TOTAL
015433	CONTRACTOR EQUIPMENT		101.9	101.9		101.9	101.9		101.9	101.9		99.3	99.3		99.3	99.3		105.7	105.7
0241, 31 - 34	SITE & INFRASTRUCTURE, DEMOLITION	82.0	90.2	87.6	85.7	94.6	91.8	94.2	89.2	90.8	94.3	88.3	90.2	101.8	87.7	92.1	85.3	106.8	100.0
0310	Concrete Forming & Accessories	96.4	58.3	63.5	91.6	79.4	81.1	104.3	30.4	40.4	97.4	56.0	61.6	104.2	49.2	56.7	99.3	137.7	132.5
0320	Concrete Reinforcing	88.4	54.7	74.7	89.2	73.9	83.0	88.4	54.5	74.6	112.3	78.0	98.3	113.1	54.6	89.2	99.3	146.1	118.4
0330	Cast-in-Place Concrete	76.4	48.7	65.8	81.0	60.1	73.0	94.4	41.9	74.3	131.8	64.3	106.0	147.7	57.8	113.4	100.0	144.7	117.1
03	CONCRETE	83.6	55.5	70.8	87.0	72.8	80.5	98.4	40.5	72.1	119.1	64.7	94.4	130.6	54.9	96.2	110.3	140.6	124.1
04	MASONRY	95.1	36.6	60.1	100.6	73.3	84.3	107.7	31.2	61.9	113.5	63.7	83.7	97.7	60.1	75.2	114.9	155.6	139.3
05	METALS	93.9	72.2	87.6	93.8	91.7	93.2	93.9	64.9	85.5	89.1	92.3	90.1	89.1	78.4	86.0	96.8	126.2	105.4
06	WOOD, PLASTICS & COMPOSITES	92.4	64.2	76.9	86.7	79.8	82.9	101.8	28.7	61.7	92.5	55.3	72.1	99.8	49.6	72.2	104.1	136.6	121.9
07	THERMAL & MOISTURE PROTECTION	96.9	45.7	77.0	96.9	68.5	85.8	97.4	55.5	81.1	106.6	76.8	95.0	107.0	71.6	93.3	94.8	148.7	115.8
08	OPENINGS	91.0	52.5	81.5	90.9	78.9	88.0	91.2	24.5	74.7	94.2	66.6	87.3	95.0	48.2	83.4	100.9	141.0	110.8
0920	Plaster & Gypsum Board	102.4	63.3	77.0	99.8	79.4	86.5	106.5	26.4	54.5	108.8	54.6	73.6	111.4	48.7	70.6	108.0	137.2	127.0
0950, 0980	Ceilings & Acoustic Treatment	95.4	63.3	75.5	96.3	79.4	85.9	95.4	26.4	52.7	107.4	54.6	74.7	107.4	48.7	71.1	98.2	137.2	122.3
0960	Flooring	87.4	54.9	78.6	85.4	72.9	82.0	91.4	54.9	81.6	125.9	57.9	107.5	131.0	54.9	110.4	97.3	167.4	116.2
0970, 0990	Wall Finishes & Painting/Coating	94.6	80.8	86.3	94.6	35.7	59.3	94.6	26.4	53.7	131.9	80.8	101.3	131.9	61.5	89.7	94.4	158.2	132.7
09	FINISHES	94.0	61.0	76.6	93.4	73.7	83.0	96.5	33.5	63.2	106.1	57.7	80.5	108.3	50.7	77.9	99.1	145.7	123.7
COVERS	DIVS. 10 - 14, 25, 28, 41, 43, 44	100.0	49.2	89.9	100.0	88.7	97.8	100.0	43.4	88.8	100.0	56.5	91.4	100.0	46.4	89.4	100.0	121.9	104.3
21, 22, 23	FIRE SUPPRESSION, PLUMBING & HVAC	94.1	76.8	87.2	99.9	87.9	95.1	94.1	54.2	78.2	94.3	78.2	87.9	94.3	76.3	87.2	100.2	131.5	112.6
26, 27, 3370	ELECTRICAL, COMMUNICATIONS & UTIL.	96.8	57.5	77.7	95.4	81.4	88.6	93.9	60.8	77.8	99.3	96.0	97.7	96.6	93.8	95.3	96.3	137.1	116.2
MF2004	WEIGHTED AVERAGE	93.1	62.9	80.4	94.9	81.6	89.3	95.8	50.3	76.7	99.7	75.5	89.5	100.5	69.5	87.5	100.5	135.9	115.4

City Cost Indexes

DIVISION		MASSACHUSETTS																	
		BROCKTON			BUZZARDS BAY			FALL RIVER			FITCHBURG			FRAMINGHAM			GREENFIELD		
		023			025			027			014			017			013		
		MAT.	INST.	TOTAL	MAT.	INST.	TOTAL	MAT.	INST.	TOTAL	MAT.	INST.	TOTAL	MAT.	INST.	TOTAL	MAT.	INST.	TOTAL
015433	CONTRACTOR EQUIPMENT		102.3	102.3		102.3	102.3		103.2	103.2		100.4	100.4		101.7	101.7		100.4	100.4
0241, 31 - 34	SITE & INFRASTRUCTURE, DEMOLITION	82.9	103.8	97.2	74.4	103.8	94.5	82.0	103.7	96.8	75.8	103.7	94.9	72.7	103.8	94.0	79.0	102.0	94.7
0310	Concrete Forming & Accessories	99.5	122.8	119.6	97.0	122.2	118.7	99.5	122.5	119.4	91.6	125.8	121.1	98.1	122.6	119.3	90.0	99.8	98.4
0320	Concrete Reinforcing	100.4	145.7	118.9	80.5	122.4	97.6	100.4	122.5	109.4	80.5	125.1	98.7	80.4	145.6	107.0	84.0	108.9	94.2
0330	Cast-in-Place Concrete	95.2	142.8	113.4	79.1	144.3	104.0	92.1	144.9	112.3	82.9	142.5	105.7	83.0	137.1	103.6	85.3	119.1	98.2
03	CONCRETE	107.9	133.2	119.4	90.8	129.1	108.3	106.4	129.5	116.9	88.9	130.4	107.8	91.9	131.2	109.8	92.9	107.8	99.7
04	MASONRY	111.4	146.4	132.4	102.0	148.4	129.7	111.2	148.3	133.4	98.2	146.5	127.1	104.1	143.8	127.9	102.7	118.2	112.0
05	METALS	93.8	123.2	102.4	89.2	114.0	96.4	93.8	114.5	99.9	89.1	112.2	95.8	89.1	122.8	98.9	91.1	101.2	94.0
06	WOOD, PLASTICS & COMPOSITES	102.9	120.1	112.3	100.0	120.0	111.0	102.9	120.3	112.5	96.4	125.4	112.3	102.5	119.8	112.0	94.4	97.2	96.0
07	THERMAL & MOISTURE PROTECTION	94.9	142.3	113.3	94.3	138.2	111.4	94.9	137.4	111.4	94.3	134.8	110.1	94.4	141.6	112.7	94.4	112.3	101.3
08	OPENINGS	98.9	129.0	106.3	94.4	117.7	100.1	98.9	117.9	103.6	103.5	123.6	108.5	94.8	126.0	102.5	103.6	102.1	103.3
0920	Plaster & Gypsum Board	103.1	120.0	114.1	99.0	120.0	112.7	103.1	120.0	114.1	100.4	125.5	116.7	102.6	120.0	113.9	101.5	96.4	98.1
0950, 0980	Ceilings & Acoustic Treatment	100.1	120.0	112.4	89.9	120.0	108.5	100.1	120.0	112.4	88.9	125.5	111.6	88.9	120.0	108.2	97.3	96.4	96.7
0960	Flooring	97.8	167.4	116.6	96.4	167.4	115.6	97.6	167.4	116.5	94.6	167.4	114.3	96.2	167.4	115.5	93.8	131.4	103.9
0970, 0990	Wall Finishes & Painting/Coating	93.9	130.3	115.7	93.9	130.3	115.7	93.9	130.3	115.7	91.7	130.3	114.9	92.8	127.1	113.4	91.7	104.5	99.4
09	FINISHES	99.0	131.6	116.3	95.1	131.6	114.4	99.0	131.8	116.4	94.4	134.8	115.7	95.1	131.1	114.1	96.6	105.9	101.5
COVERS	DIVS. 10 - 14, 25, 28, 41, 43, 44	100.0	118.1	103.6	100.0	118.1	103.6	100.0	118.8	103.7	100.0	101.5	100.3	100.0	117.5	103.5	100.0	95.0	99.0
21, 22, 23	FIRE SUPPRESSION, PLUMBING & HVAC	100.2	108.5	103.5	94.4	105.6	98.8	100.2	108.3	103.4	95.3	105.2	99.2	95.3	121.4	105.7	95.3	96.9	95.9
26, 27, 3370	ELECTRICAL, COMMUNICATIONS & UTIL.	95.4	99.5	97.4	92.9	99.4	96.1	95.2	99.4	97.3	97.6	95.7	96.7	94.4	125.4	109.5	97.6	90.8	94.3
MF2004	WEIGHTED AVERAGE	99.2	120.7	108.2	93.3	118.2	103.8	99.0	118.9	107.3	94.4	117.6	104.1	93.9	126.1	107.4	95.7	102.3	98.4

DIVISION		MASSACHUSETTS																	
		HYANNIS			LAWRENCE			LOWELL			NEW BEDFORD			PITTSFIELD			SPRINGFIELD		
		026			019			018			027			012			010 - 011		
		MAT.	INST.	TOTAL	MAT.	INST.	TOTAL	MAT.	INST.	TOTAL	MAT.	INST.	TOTAL	MAT.	INST.	TOTAL	MAT.	INST.	TOTAL
015433	CONTRACTOR EQUIPMENT		102.3	102.3		102.3	102.3		100.4	100.4		103.2	103.2		100.4	100.4		100.4	100.4
0241, 31 - 34	SITE & INFRASTRUCTURE, DEMOLITION	79.8	103.8	96.2	83.6	103.8	97.4	82.5	103.8	97.0	80.5	103.7	96.4	83.6	101.8	96.0	83.0	102.1	96.0
0310	Concrete Forming & Accessories	91.6	122.2	118.0	99.3	122.8	119.6	96.5	122.9	119.3	99.5	122.5	119.4	96.5	103.9	102.9	96.7	105.4	104.2
0320	Concrete Reinforcing	80.5	122.4	97.6	99.5	131.4	112.5	100.4	131.6	113.1	100.4	122.5	109.4	83.3	111.1	94.7	100.4	109.0	103.9
0330	Cast-in-Place Concrete	86.9	144.3	108.8	96.0	140.7	113.1	87.3	140.8	107.7	81.1	144.9	105.5	95.2	117.3	103.7	90.8	119.2	101.7
03	CONCRETE	97.5	129.2	111.9	108.0	129.9	118.0	99.1	129.8	113.1	101.3	129.5	114.1	99.4	109.4	103.9	100.8	110.3	105.1
04	MASONRY	110.2	148.4	133.0	110.8	143.8	130.6	97.1	146.5	126.7	110.3	148.3	133.0	97.7	114.9	108.0	97.4	118.2	109.9
05	METALS	90.5	114.1	97.4	91.4	117.7	99.1	91.4	115.6	98.5	93.8	114.5	99.9	91.2	102.0	94.4	93.8	101.3	96.0
06	WOOD, PLASTICS & COMPOSITES	93.7	120.0	108.2	102.9	120.1	112.3	102.3	120.1	112.1	102.9	120.3	112.5	102.3	104.7	103.6	102.3	104.7	103.6
07	THERMAL & MOISTURE PROTECTION	94.5	138.2	111.5	94.9	141.3	112.9	94.7	141.9	113.0	94.9	137.4	111.4	94.7	111.7	101.3	94.7	113.1	101.8
08	OPENINGS	95.0	117.7	100.6	98.9	122.4	104.7	104.9	122.4	109.2	98.9	117.9	103.6	104.9	106.7	105.3	104.9	106.1	105.2
0920	Plaster & Gypsum Board	96.1	120.0	111.6	105.5	120.0	115.0	105.5	120.0	115.0	103.1	120.0	114.1	105.5	104.1	104.6	105.5	104.1	104.6
0950, 0980	Ceilings & Acoustic Treatment	91.7	120.0	109.2	99.1	120.0	112.1	99.1	120.0	112.1	100.1	120.0	112.4	99.1	104.1	102.2	99.1	104.1	102.2
0960	Flooring	94.1	167.4	113.9	96.8	167.4	115.9	96.8	167.4	115.9	97.6	167.4	116.5	97.0	131.4	106.3	96.7	131.4	106.1
0970, 0990	Wall Finishes & Painting/Coating	93.9	127.1	113.8	91.8	130.3	114.9	91.7	127.1	113.0	93.9	130.3	115.7	91.7	104.5	99.4	93.3	104.5	100.0
09	FINISHES	94.8	131.3	114.1	98.7	131.6	116.1	98.7	131.3	115.9	98.9	131.8	116.3	98.7	109.4	104.4	98.8	110.3	104.8
COVERS	DIVS. 10 - 14, 25, 28, 41, 43, 44	100.0	118.1	103.6	100.0	118.2	103.6	100.0	118.2	103.6	100.0	118.8	103.7	100.0	94.7	98.9	100.0	95.9	99.2
21, 22, 23	FIRE SUPPRESSION, PLUMBING & HVAC	100.2	108.2	103.4	100.2	112.9	105.3	100.2	122.9	109.2	100.2	108.3	103.4	100.2	87.5	95.1	100.2	101.9	100.9
26, 27, 3370	ELECTRICAL, COMMUNICATIONS & UTIL.	93.5	99.4	96.4	96.7	125.4	110.7	97.1	125.4	110.9	96.3	99.4	97.8	97.1	90.8	94.1	97.2	90.8	94.1
MF2004	WEIGHTED AVERAGE	96.2	118.7	105.7	98.9	123.6	109.3	97.7	125.7	109.5	98.4	118.9	107.0	97.8	101.0	99.1	98.4	104.5	101.0

DIVISION		MASSACHUSETTS			MICHIGAN														
		WORCESTER			ANN ARBOR			BATTLE CREEK			BAY CITY			DEARBORN			DETROIT		
		015 - 016			481			490			487			481			482		
		MAT.	INST.	TOTAL	MAT.	INST.	TOTAL	MAT.	INST.	TOTAL	MAT.	INST.	TOTAL	MAT.	INST.	TOTAL	MAT.	INST.	TOTAL
015433	CONTRACTOR EQUIPMENT		100.4	100.4		111.3	111.3		101.4	101.4		111.3	111.3		111.3	111.3		98.5	98.5
0241, 31 - 34	SITE & INFRASTRUCTURE, DEMOLITION	82.9	103.7	97.1	77.1	96.5	90.4	79.7	82.4	81.6	69.4	94.8	86.8	76.9	96.7	90.4	90.9	98.3	95.9
0310	Concrete Forming & Accessories	97.0	126.0	122.0	98.2	110.1	108.5	97.7	89.9	90.9	98.3	86.9	88.4	98.1	119.5	116.6	99.1	119.6	116.8
0320	Concrete Reinforcing	100.4	141.6	117.2	96.1	128.4	109.3	94.4	85.8	90.9	96.1	127.4	108.9	96.1	128.7	109.4	95.4	128.7	109.0
0330	Cast-in-Place Concrete	90.3	142.6	110.3	85.0	114.5	96.3	90.3	104.8	95.8	81.4	93.7	86.1	83.1	118.0	96.4	91.3	118.0	101.5
03	CONCRETE	100.5	133.6	115.6	89.6	115.3	101.3	93.1	93.7	93.3	87.9	97.7	92.4	88.7	120.7	103.3	92.5	119.5	104.7
04	MASONRY	96.9	146.5	126.6	92.7	113.2	105.0	93.1	86.7	89.3	92.3	89.5	90.6	92.5	119.5	108.7	91.8	119.5	108.4
05	METALS	93.8	118.6	101.1	86.1	121.4	96.4	92.3	85.4	90.3	86.6	117.6	95.6	86.1	122.4	96.7	89.9	104.6	94.2
06	WOOD, PLASTICS & COMPOSITES	102.8	125.4	115.2	96.7	109.5	103.7	97.3	89.9	93.2	96.7	86.2	90.9	96.7	119.9	109.4	97.2	119.9	109.7
07	THERMAL & MOISTURE PROTECTION	94.7	134.8	110.3	102.5	111.9	106.2	92.7	87.5	90.7	100.2	87.9	95.4	101.1	121.3	109.0	99.3	121.3	107.8
08	OPENINGS	104.9	130.8	111.3	94.5	109.6	98.2	90.7	83.4	88.9	94.5	93.0	94.1	94.5	115.3	99.6	96.2	117.4	101.4
0920	Plaster & Gypsum Board	105.5	125.5	118.5	85.4	108.7	100.5	85.8	85.2	85.4	85.4	84.6	84.9	85.4	119.6	107.6	85.4	119.6	107.6
0950, 0980	Ceilings & Acoustic Treatment	99.1	125.5	115.4	88.7	108.7	101.1	89.5	85.2	86.8	89.7	84.6	86.5	88.7	119.6	107.8	89.7	119.6	108.2
0960	Flooring	96.8	163.3	114.8	92.7	120.3	100.2	97.9	95.1	97.1	92.5	79.9	89.1	92.4	126.2	101.5	92.4	126.2	101.6
0970, 0990	Wall Finishes & Painting/Coating	91.7	130.3	114.9	92.9	102.4	98.6	97.7	75.3	84.3	92.9	83.8	87.4	92.9	108.3	102.1	94.4	108.3	102.7
09	FINISHES	98.7	133.9	117.3	90.8	111.2	101.6	93.4	89.3	91.3	90.6	84.8	87.5	90.7	120.1	106.2	92.0	120.1	106.8
COVERS	DIVS. 10 - 14, 25, 28, 41, 43, 44	100.0	101.5	100.3	100.0	106.7	101.3	100.0	99.0	99.8	100.0	91.6	98.3	100.0	109.0	101.8	100.0	109.0	101.8
21, 22, 23	FIRE SUPPRESSION, PLUMBING & HVAC	100.2	105.4	102.3	99.9	99.5	99.7	99.9	87.8	95.1	99.9	84.5	93.8	99.9	110.7	104.2	99.9	112.8	105.0
26, 27, 3370	ELECTRICAL, COMMUNICATIONS & UTIL.	97.2	95.7	96.5	95.2	76.9	86.3	92.8	87.7	90.3	94.0	72.8	83.7	95.2	116.0	105.3	96.5	115.9	106.0
MF2004	WEIGHTED AVERAGE	98.3	118.9	107.0	93.7	104.5	98.2	94.5	88.2	91.9	93.1	90.0	91.8	93.5	115.5	102.7	95.3	114.2	103.2

DIVISION		MICHIGAN																	
		FLINT			GAYLORD			GRAND RAPIDS			IRON MOUNTAIN			JACKSON			KALAMAZOO		
		484 - 485			497			493,495			498 - 499			492			491		
		MAT.	INST.	TOTAL	MAT.	INST.	TOTAL	MAT.	INST.	TOTAL	MAT.	INST.	TOTAL	MAT.	INST.	TOTAL	MAT.	INST.	TOTAL
015433	CONTRACTOR EQUIPMENT		111.3	111.3		108.2	108.2		101.4	101.4		95.3	95.3		108.2	108.2		101.4	101.4
0241, 31 - 34	SITE & INFRASTRUCTURE, DEMOLITION	67.3	95.7	86.7	75.9	88.2	84.3	81.3	81.7	81.6	82.7	95.7	91.6	94.4	89.9	91.4	80.0	82.4	81.7
0310	Concrete Forming & Accessories	101.2	95.0	95.8	94.9	57.0	62.1	98.1	72.0	75.5	88.4	81.0	82.0	92.7	86.9	87.7	97.7	89.5	90.7
0320	Concrete Reinforcing	96.1	127.9	109.1	88.0	101.7	93.6	94.4	83.7	90.0	87.8	96.6	91.4	85.4	127.7	102.6	94.4	85.7	90.8
0330	Cast-in-Place Concrete	85.6	100.4	91.3	90.1	71.2	82.9	91.4	90.4	91.0	106.7	89.2	100.0	89.9	93.7	91.4	92.1	104.6	96.9
03	CONCRETE	90.1	103.6	96.2	89.3	71.9	81.4	93.6	80.4	87.6	98.3	87.1	93.2	84.5	97.6	90.5	96.0	93.5	94.9
04	MASONRY	92.7	101.3	97.8	103.2	74.6	86.1	88.8	47.0	63.8	90.2	87.9	88.9	84.7	96.4	91.7	91.9	86.7	88.8
05	METALS	86.1	119.0	95.7	93.8	104.7	97.0	93.8	80.7	90.0	93.3	91.3	92.7	94.0	116.4	100.5	92.3	85.1	90.2
06	WOOD, PLASTICS & COMPOSITES	99.9	93.6	96.4	90.0	53.8	70.1	94.9	73.7	83.2	85.6	79.7	82.4	88.9	85.0	86.7	97.3	89.9	93.2
07	THERMAL & MOISTURE PROTECTION	100.2	96.9	99.0	91.3	65.0	81.0	94.4	56.5	79.7	94.5	79.4	88.7	90.7	95.9	92.7	92.7	87.5	90.7
08	OPENINGS	94.5	97.0	95.1	92.3	59.5	84.2	94.1	66.1	87.2	97.2	71.1	90.8	91.3	89.6	90.9	90.7	83.4	88.9
0920	Plaster & Gypsum Board	86.9	92.2	90.3	85.2	50.7	62.8	84.3	68.4	74.0	56.4	79.5	71.4	84.5	83.0	83.5	85.8	85.2	85.4
0950, 0980	Ceilings & Acoustic Treatment	88.7	92.2	90.9	87.0	50.7	64.6	87.8	68.4	75.8	87.6	79.5	82.6	88.6	83.0	85.2	89.5	85.2	86.8
0960	Flooring	92.5	82.5	89.8	91.0	54.0	81.0	98.1	38.0	81.8	111.0	93.5	106.3	89.4	74.3	85.3	97.9	74.2	91.5
0970, 0990	Wall Finishes & Painting/Coating	92.9	88.8	90.4	93.1	49.7	67.1	97.7	37.1	61.4	116.3	40.5	70.8	93.1	82.5	86.8	97.7	75.3	84.3
09	FINISHES	90.2	91.4	90.8	93.2	54.1	72.5	93.1	61.8	76.5	96.0	79.0	87.0	94.2	83.4	88.5	93.4	85.2	89.1
COVERS	DIVS. 10 - 14, 25, 28, 41, 43, 44	100.0	94.3	98.9	100.0	84.6	96.9	100.0	93.8	98.8	100.0	87.5	97.5	100.0	92.9	98.6	100.0	98.9	99.8
21, 22, 23	FIRE SUPPRESSION, PLUMBING & HVAC	99.9	101.4	100.5	94.5	75.3	86.8	99.9	56.3	82.6	94.4	78.6	88.1	94.5	89.2	92.4	99.9	82.3	92.9
26, 27, 3370	ELECTRICAL, COMMUNICATIONS & UTIL.	95.2	99.3	97.2	90.9	53.8	72.8	103.2	53.6	79.1	97.2	77.7	87.7	94.7	76.9	86.0	92.6	81.8	87.3
MF2004	WEIGHTED AVERAGE	93.4	100.8	96.5	93.3	72.1	84.4	96.1	65.2	83.1	95.3	83.2	90.2	92.7	91.6	92.3	94.8	85.6	91.0

DIVISION		MICHIGAN															MINNESOTA		
		LANSING			MUSKEGON			ROYAL OAK			SAGINAW			TRAVERSE CITY			BEMIDJI		
		488 - 489			494			480,483			486			496			566		
		MAT.	INST.	TOTAL	MAT.	INST.	TOTAL	MAT.	INST.	TOTAL	MAT.	INST.	TOTAL	MAT.	INST.	TOTAL	MAT.	INST.	TOTAL
015433	CONTRACTOR EQUIPMENT		111.3	111.3		101.4	101.4		95.9	95.9		111.3	111.3		95.3	95.3		98.7	98.7
0241, 31 - 34	SITE & INFRASTRUCTURE, DEMOLITION	85.4	95.2	92.1	77.9	82.3	80.9	80.6	95.4	90.7	70.4	94.8	87.1	70.3	94.8	87.1	90.2	99.0	96.2
0310	Concrete Forming & Accessories	101.8	95.1	96.0	98.3	81.1	83.4	94.7	106.5	104.9	98.3	86.8	88.4	88.4	55.0	59.5	85.5	91.1	90.4
0320	Concrete Reinforcing	96.1	127.6	108.9	95.0	85.0	90.9	87.1	103.3	93.7	96.1	127.4	108.9	89.2	84.7	87.4	104.5	101.3	103.2
0330	Cast-in-Place Concrete	87.4	99.4	92.0	90.0	99.0	93.4	74.4	111.2	88.5	84.0	93.6	87.7	83.2	62.9	75.4	99.7	101.6	100.4
03	CONCRETE	90.9	103.3	96.5	91.6	87.7	89.8	77.7	106.5	90.8	89.2	97.6	93.0	81.8	64.4	73.9	96.3	97.8	96.9
04	MASONRY	86.5	96.5	92.5	90.5	73.9	80.6	87.6	115.3	104.2	94.0	89.5	91.3	88.4	74.8	80.3	97.0	105.2	101.9
05	METALS	84.7	118.2	94.5	90.1	83.7	88.2	89.2	91.2	89.8	86.2	117.5	95.3	93.3	86.8	91.4	88.9	116.6	97.0
06	WOOD, PLASTICS & COMPOSITES	99.3	94.1	96.4	93.7	79.6	86.0	92.2	105.2	99.4	92.6	86.2	89.1	85.6	52.0	67.1	76.9	87.9	82.9
07	THERMAL & MOISTURE PROTECTION	101.4	93.8	98.4	91.9	76.2	85.8	98.8	110.7	103.4	101.0	88.2	96.0	93.7	63.2	81.8	102.6	91.8	98.4
08	OPENINGS	94.5	97.3	95.2	89.9	76.2	86.5	94.5	100.3	96.0	92.4	93.0	92.6	97.2	54.9	86.7	98.8	108.3	101.1
0920	Plaster & Gypsum Board	83.8	92.8	89.6	72.6	74.6	73.9	83.1	104.3	96.9	85.4	84.6	84.9	56.4	50.7	52.7	101.0	87.8	92.4
0950, 0980	Ceilings & Acoustic Treatment	87.1	92.8	90.6	91.1	74.6	80.9	88.0	104.3	98.1	88.7	84.6	86.2	87.6	50.7	64.8	133.0	87.8	105.0
0960	Flooring	102.6	87.9	98.6	97.1	70.0	89.8	90.0	113.8	96.4	92.7	79.9	89.3	111.0	54.0	95.6	110.3	126.6	114.7
0970, 0990	Wall Finishes & Painting/Coating	105.2	91.7	97.1	97.0	53.8	71.1	94.4	102.3	99.2	92.9	83.8	87.4	116.3	49.7	76.3	98.4	92.5	94.9
09	FINISHES	94.8	93.3	94.0	91.2	75.6	83.0	89.8	107.6	99.2	90.5	84.8	87.5	95.1	52.8	72.7	112.0	97.7	104.4
COVERS	DIVS. 10 - 14, 25, 28, 41, 43, 44	100.0	94.5	98.9	100.0	97.5	99.5	100.0	106.0	101.2	100.0	91.6	98.3	100.0	80.0	96.0	100.0	89.7	98.0
21, 22, 23	FIRE SUPPRESSION, PLUMBING & HVAC	99.8	99.8	99.8	99.8	84.2	93.6	94.5	103.6	98.1	99.9	84.2	93.6	94.4	74.8	86.6	94.6	82.4	89.7
26, 27, 3370	ELECTRICAL, COMMUNICATIONS & UTIL.	101.2	77.2	89.5	93.2	79.1	86.3	97.3	103.3	100.2	92.9	72.8	83.1	92.7	53.1	73.4	103.1	88.8	96.1
MF2004	WEIGHTED AVERAGE	94.4	96.9	95.5	93.5	81.4	88.4	91.4	103.9	96.6	93.0	89.9	91.7	92.3	69.1	82.5	97.1	95.9	96.6

DIVISION		MINNESOTA																	
		BRAINERD			DETROIT LAKES			DULUTH			MANKATO			MINNEAPOLIS			ROCHESTER		
		564			565			556 - 558			560			553 - 555			559		
		MAT.	INST.	TOTAL	MAT.	INST.	TOTAL	MAT.	INST.	TOTAL	MAT.	INST.	TOTAL	MAT.	INST.	TOTAL	MAT.	INST.	TOTAL
015433	CONTRACTOR EQUIPMENT		100.8	100.8		98.7	98.7		101.1	101.1		100.8	100.8		104.1	104.1		101.1	101.1
0241, 31 - 34	SITE & INFRASTRUCTURE, DEMOLITION	88.9	103.2	98.6	88.3	99.4	95.9	93.3	102.7	99.7	85.4	102.5	97.1	91.9	107.7	102.7	90.9	101.7	98.3
0310	Concrete Forming & Accessories	87.8	93.4	92.6	82.7	95.0	93.3	97.1	114.3	112.0	95.5	103.6	102.5	99.4	135.9	131.0	99.8	104.9	104.2
0320	Concrete Reinforcing	103.2	101.3	102.5	104.5	101.3	103.2	102.0	101.7	101.9	103.1	122.0	110.8	102.2	122.7	110.6	102.0	122.2	110.3
0330	Cast-in-Place Concrete	108.4	105.5	107.3	96.8	104.7	99.8	107.7	104.9	106.6	99.8	108.6	103.2	111.9	121.2	115.4	109.7	101.3	106.5
03	CONCRETE	100.0	100.1	100.1	94.0	100.5	97.0	100.8	109.3	104.7	95.6	109.5	102.0	104.4	128.4	115.3	101.9	107.8	104.6
04	MASONRY	119.6	111.3	114.6	118.5	111.1	114.1	102.0	116.9	110.9	107.8	111.6	110.0	100.9	131.8	119.4	100.1	115.7	109.5
05	METALS	90.2	116.5	97.8	88.9	116.3	96.9	89.3	119.0	98.0	90.0	126.2	100.6	91.8	131.8	103.5	89.2	129.7	101.0
06	WOOD, PLASTICS & COMPOSITES	97.5	87.7	92.1	73.7	90.5	82.9	111.8	114.3	113.2	106.8	102.7	104.6	115.2	135.9	126.6	115.2	102.8	108.4
07	THERMAL & MOISTURE PROTECTION	101.0	103.2	101.9	102.4	96.0	99.9	102.3	115.3	107.3	101.3	95.7	99.1	102.5	133.8	114.7	102.6	102.3	102.5
08	OPENINGS	86.4	108.1	91.8	98.8	109.7	101.5	94.7	116.1	100.0	91.2	122.1	98.9	98.0	142.0	108.9	94.7	124.0	102.0
0920	Plaster & Gypsum Board	84.7	87.8	86.7	99.5	90.5	93.6	94.5	115.3	108.0	88.7	103.4	98.3	98.9	137.5	124.0	98.5	103.4	101.7
0950, 0980	Ceilings & Acoustic Treatment	69.9	87.8	80.9	133.0	90.5	106.7	96.3	115.3	108.1	69.9	103.4	90.6	99.7	137.5	123.1	98.0	103.4	101.4
0960	Flooring	109.1	126.6	113.9	109.0	126.6	113.7	107.5	123.6	111.9	111.4	122.1	114.3	105.3	126.6	111.0	107.7	92.5	103.6
0970, 0990	Wall Finishes & Painting/Coating	93.6	92.5	92.9	98.4	92.5	94.9	91.3	100.9	97.1	103.2	101.5	102.2	98.9	128.0	116.4	94.2	101.5	98.6
09	FINISHES	94.3	99.0	96.8	111.3	100.7	105.7	101.7	115.2	108.9	96.0	107.2	101.9	102.7	134.3	119.4	102.6	102.8	102.7
COVERS	DIVS. 10 - 14, 25, 28, 41, 43, 44	100.0	91.3	98.3	100.0	92.0	98.4	100.0	95.1	99.0	100.0	92.0	98.4	100.0	106.0	101.2	100.0	96.9	99.4
21, 22, 23	FIRE SUPPRESSION, PLUMBING & HVAC	93.7	85.4	90.4	94.6	85.1	90.8	99.8	102.4	100.9	93.7	92.0	93.0	99.8	117.1	106.7	99.8	98.0	99.1
26, 27, 3370	ELECTRICAL, COMMUNICATIONS & UTIL.	100.9	96.9	98.9	102.8	62.9	83.4	102.5	96.9	99.7	107.3	85.7	96.8	101.9	115.3	108.5	100.6	85.7	93.4
MF2004	WEIGHTED AVERAGE	95.9	99.5	97.4	97.7	94.6	96.4	98.2	108.2	102.4	96.0	103.2	99.0	99.3	124.2	109.8	98.1	104.7	100.9

DIVISION		MINNESOTA															MISSISSIPPI		
		SAINT PAUL			ST. CLOUD			THIEF RIVER FALLS			WILLMAR			WINDOM			BILOXI		
		550 - 551			563			567			562			561			395		
		MAT.	INST.	TOTAL	MAT.	INST.	TOTAL	MAT.	INST.	TOTAL	MAT.	INST.	TOTAL	MAT.	INST.	TOTAL	MAT.	INST.	TOTAL
015433	**CONTRACTOR EQUIPMENT**		101.1	101.1		100.8	100.8		98.7	98.7		100.8	100.8		100.8	100.8		99.5	99.5
0241, 31 - 34	**SITE & INFRASTRUCTURE, DEMOLITION**	95.7	103.5	101.0	83.7	105.3	98.5	89.0	99.0	95.8	83.4	98.9	94.0	77.1	97.7	91.2	103.7	89.6	94.1
0310	Concrete Forming & Accessories	93.5	125.6	121.2	85.1	123.3	118.1	86.3	90.8	90.2	84.9	53.6	57.9	89.1	50.1	55.4	93.6	61.8	66.1
0320	Concrete Reinforcing	98.4	122.6	108.3	103.3	122.2	111.0	104.9	101.3	103.4	102.9	121.5	110.5	102.9	120.6	110.1	92.5	63.2	80.6
0330	Cast-in-Place Concrete	113.9	120.4	116.4	95.6	118.8	104.5	98.8	90.3	95.6	97.1	75.3	88.8	84.0	55.4	73.1	106.8	60.3	89.0
03	**CONCRETE**	106.4	123.6	114.2	91.9	121.9	105.5	95.0	93.7	94.4	91.9	75.7	84.5	83.4	67.2	76.0	97.1	63.2	81.7
04	**MASONRY**	109.8	131.8	123.0	105.0	124.3	116.5	97.0	105.2	101.9	108.3	87.9	96.1	118.2	64.8	86.2	88.7	50.1	65.6
05	**METALS**	88.8	131.3	101.2	90.7	129.1	101.9	89.0	116.2	97.0	90.0	119.5	98.6	89.9	118.4	98.2	97.6	87.9	94.7
06	**WOOD, PLASTICS & COMPOSITES**	107.8	122.3	115.7	94.4	120.8	108.9	77.9	87.9	83.4	94.2	50.7	70.3	98.8	50.7	72.4	94.9	65.7	78.8
07	**THERMAL & MOISTURE PROTECTION**	103.6	131.8	114.6	101.1	113.7	106.0	103.6	92.8	99.4	101.0	57.6	84.1	100.9	48.8	80.7	92.6	62.2	80.7
08	**OPENINGS**	92.3	134.5	102.8	91.2	133.8	101.8	98.8	108.3	101.1	88.8	75.7	85.5	93.0	75.7	88.7	97.1	66.2	89.4
0920	Plaster & Gypsum Board	91.8	123.6	112.4	83.9	122.2	108.8	100.6	87.8	92.3	83.9	49.5	61.6	84.7	49.5	61.8	104.0	64.9	78.6
0950, 0980	Ceilings & Acoustic Treatment	93.0	123.6	111.9	69.9	122.2	102.2	133.0	87.8	105.0	69.9	49.5	57.3	69.9	49.5	57.3	100.4	64.9	78.5
0960	Flooring	101.2	126.6	108.0	105.0	126.6	110.9	110.0	126.6	114.5	106.9	122.1	111.0	109.8	122.1	113.1	117.2	59.2	101.5
0970, 0990	Wall Finishes & Painting/Coating	98.9	119.8	111.4	103.2	128.0	118.1	98.4	92.5	94.9	98.4	92.5	94.9	98.4	101.5	100.3	110.4	49.9	74.1
09	**FINISHES**	99.4	125.3	113.1	93.2	124.8	109.9	111.8	97.7	104.3	93.5	67.3	79.6	93.9	66.2	79.3	104.3	60.3	81.1
COVERS	**DIVS. 10 - 14, 25, 28, 41, 43, 44**	100.0	104.0	100.8	100.0	98.8	99.8	100.0	89.6	97.9	100.0	52.4	90.6	100.0	49.5	90.0	100.0	69.6	94.0
21, 22, 23	**FIRE SUPPRESSION, PLUMBING & HVAC**	99.8	112.5	104.8	99.5	115.8	106.0	94.6	82.0	89.6	93.7	91.2	92.7	93.7	75.3	86.4	99.9	50.0	80.0
26, 27, 3370	**ELECTRICAL, COMMUNICATIONS & UTIL.**	100.1	102.4	101.2	100.8	102.4	101.6	100.3	62.9	82.1	100.8	68.8	85.2	107.3	85.6	96.7	99.9	55.5	78.3
MF2004	**WEIGHTED AVERAGE**	98.6	118.7	107.0	95.8	117.5	104.9	96.7	91.7	94.6	94.3	82.8	89.5	94.8	77.6	87.6	98.5	62.6	83.4

DIVISION		MISSISSIPPI																	
		CLARKSDALE			COLUMBUS			GREENVILLE			GREENWOOD			JACKSON			LAUREL		
		386			397			387			389			390 - 392			394		
		MAT.	INST.	TOTAL	MAT.	INST.	TOTAL	MAT.	INST.	TOTAL	MAT.	INST.	TOTAL	MAT.	INST.	TOTAL	MAT.	INST.	TOTAL
015433	**CONTRACTOR EQUIPMENT**		99.5	99.5		99.5	99.5		99.5	99.5		99.5	99.5		99.5	99.5		99.5	99.5
0241, 31 - 34	**SITE & INFRASTRUCTURE, DEMOLITION**	101.1	89.4	93.1	102.4	89.7	93.7	107.6	89.7	95.4	104.4	89.0	93.9	98.9	89.7	92.6	109.2	89.5	95.7
0310	Concrete Forming & Accessories	84.4	57.9	61.5	82.5	60.0	63.1	81.0	74.8	75.6	93.5	57.8	62.7	93.0	67.4	70.9	82.6	62.1	64.9
0320	Concrete Reinforcing	99.6	48.7	78.9	99.0	49.6	78.9	100.3	70.6	88.1	99.6	61.5	84.1	92.5	61.7	79.9	99.6	45.8	77.7
0330	Cast-in-Place Concrete	104.1	61.3	87.7	108.9	63.3	91.5	107.2	63.7	90.6	112.0	60.9	92.4	99.2	63.7	85.6	106.5	63.2	89.9
03	**CONCRETE**	95.8	59.1	79.1	99.2	60.9	81.8	100.6	71.4	87.3	101.7	61.1	83.3	93.5	66.5	81.3	101.1	61.2	82.9
04	**MASONRY**	88.6	52.2	66.8	110.4	55.7	77.7	129.1	56.0	85.3	89.2	52.0	66.9	89.8	56.0	69.6	107.1	55.8	76.4
05	**METALS**	94.8	81.4	90.9	94.6	81.8	90.9	95.7	91.2	94.4	94.8	82.9	91.3	97.5	87.6	94.6	94.7	81.2	90.8
06	**WOOD, PLASTICS & COMPOSITES**	82.9	60.0	70.3	80.9	61.1	70.1	78.9	79.7	79.4	94.9	60.0	75.7	93.7	69.8	80.6	81.4	63.9	71.8
07	**THERMAL & MOISTURE PROTECTION**	92.4	54.3	77.6	92.5	57.6	78.9	92.7	68.8	83.4	92.7	54.2	77.7	94.4	67.7	84.0	92.6	61.2	80.4
08	**OPENINGS**	96.4	57.8	86.8	96.4	57.9	86.9	96.4	71.8	90.3	96.4	59.3	87.2	97.4	63.9	89.2	93.1	58.9	84.6
0920	Plaster & Gypsum Board	94.7	59.0	71.5	94.3	60.2	72.2	92.8	79.5	84.2	102.5	59.0	74.3	96.8	69.2	78.9	92.8	63.1	73.5
0950, 0980	Ceilings & Acoustic Treatment	92.7	59.0	71.9	92.7	60.2	72.6	94.6	79.5	85.2	92.7	59.0	71.9	97.1	69.2	79.8	92.7	63.1	74.4
0960	Flooring	110.3	59.2	96.5	109.0	59.2	95.5	108.0	59.2	94.8	117.2	59.2	101.5	117.6	59.2	101.8	106.9	59.2	94.0
0970, 0990	Wall Finishes & Painting/Coating	110.4	53.0	75.9	110.4	60.9	80.7	110.4	75.7	89.6	110.4	53.0	75.9	110.4	75.7	89.6	110.4	70.0	86.1
09	**FINISHES**	99.0	57.9	77.3	98.8	60.3	78.4	99.1	72.9	85.3	102.5	57.9	78.9	102.7	67.1	83.9	98.4	63.0	79.7
COVERS	**DIVS. 10 - 14, 25, 28, 41, 43, 44**	100.0	70.1	94.1	100.0	71.5	94.4	100.0	73.5	94.8	100.0	70.0	94.1	100.0	72.3	94.5	100.0	47.1	89.5
21, 22, 23	**FIRE SUPPRESSION, PLUMBING & HVAC**	97.4	49.8	78.5	96.7	45.2	76.2	99.9	57.4	83.0	97.4	47.4	77.5	99.9	63.4	85.4	96.8	55.2	80.2
26, 27, 3370	**ELECTRICAL, COMMUNICATIONS & UTIL.**	99.2	54.1	77.3	97.4	57.3	77.9	99.2	73.2	86.5	99.2	50.7	75.6	104.6	73.2	89.3	98.9	63.1	81.4
MF2004	**WEIGHTED AVERAGE**	96.6	60.4	81.4	97.7	61.0	82.3	100.0	70.9	87.8	97.8	59.9	81.9	98.5	69.9	86.5	97.8	63.7	83.5

DIVISION		MISSISSIPPI									MISSOURI								
		MCCOMB			MERIDIAN			TUPELO			BOWLING GREEN			CAPE GIRARDEAU			CHILLICOTHE		
		396			393			388			633			637			646		
		MAT.	INST.	TOTAL	MAT.	INST.	TOTAL	MAT.	INST.	TOTAL	MAT.	INST.	TOTAL	MAT.	INST.	TOTAL	MAT.	INST.	TOTAL
015433	**CONTRACTOR EQUIPMENT**		99.5	99.5		99.5	99.5		99.5	99.5		107.1	107.1		107.1	107.1		99.5	99.5
0241, 31 - 34	**SITE & INFRASTRUCTURE, DEMOLITION**	94.5	89.2	90.9	98.7	90.0	92.7	98.6	89.3	92.2	92.3	93.4	93.1	93.4	93.9	93.7	104.3	85.3	91.3
0310	Concrete Forming & Accessories	82.5	59.9	63.0	79.8	74.1	74.9	81.6	60.2	63.1	93.1	86.8	87.6	86.6	74.2	75.9	92.3	87.0	87.7
0320	Concrete Reinforcing	100.3	48.1	79.0	99.0	61.7	83.8	97.6	61.5	82.9	103.8	93.2	99.5	105.1	83.1	96.1	97.7	84.4	92.3
0330	Cast-in-Place Concrete	94.5	60.7	81.6	101.1	66.3	87.8	104.1	62.9	88.4	95.2	79.6	89.3	94.2	83.8	90.2	92.6	67.4	83.0
03	**CONCRETE**	89.7	59.6	76.0	93.6	70.4	83.1	95.2	62.9	80.5	94.3	86.8	90.9	93.6	80.9	87.8	97.8	80.2	89.8
04	**MASONRY**	112.4	51.4	75.9	88.4	60.6	71.8	118.8	55.5	80.9	121.8	95.0	105.7	119.3	77.3	94.1	101.4	85.1	91.7
05	**METALS**	94.9	79.4	90.4	95.6	87.8	93.3	94.7	83.2	91.3	93.3	110.0	98.2	94.2	104.9	97.4	92.1	91.4	91.9
06	**WOOD, PLASTICS & COMPOSITES**	80.9	63.2	71.2	77.8	76.3	77.0	79.5	61.1	69.4	95.7	87.4	91.1	88.5	72.1	79.5	92.7	93.7	93.2
07	**THERMAL & MOISTURE PROTECTION**	92.1	53.4	77.0	92.3	70.4	83.8	92.4	60.9	80.1	102.1	95.7	99.6	102.0	78.1	92.7	96.7	79.2	89.9
08	**OPENINGS**	96.4	59.7	87.3	96.4	73.9	90.8	96.4	59.2	87.2	98.9	83.4	95.0	98.8	72.5	92.3	90.3	79.7	87.7
0920	Plaster & Gypsum Board	94.3	62.4	73.6	92.8	75.9	81.8	92.8	60.2	71.6	98.1	86.9	90.8	95.2	71.0	79.5	91.5	93.2	92.6
0950, 0980	Ceilings & Acoustic Treatment	92.7	62.4	74.0	94.6	75.9	83.0	92.7	60.2	72.6	86.1	86.9	86.6	86.1	71.0	76.8	82.5	93.2	89.2
0960	Flooring	109.0	59.2	95.5	106.9	59.2	94.0	108.4	59.2	95.1	90.9	67.2	84.5	86.9	67.2	81.5	97.0	48.2	83.8
0970, 0990	Wall Finishes & Painting/Coating	110.4	48.6	73.3	110.4	87.5	96.7	110.4	67.5	84.6	98.3	105.1	102.4	98.3	77.8	86.0	95.1	51.5	69.0
09	**FINISHES**	98.2	59.1	77.5	98.1	73.4	85.0	98.2	61.0	78.5	92.5	84.9	88.5	90.8	72.3	81.1	95.9	78.3	86.6
COVERS	**DIVS. 10 - 14, 25, 28, 41, 43, 44**	100.0	74.3	94.9	100.0	74.7	95.0	100.0	71.5	94.4	100.0	63.0	92.7	100.0	60.5	92.2	100.0	60.8	92.2
21, 22, 23	**FIRE SUPPRESSION, PLUMBING & HVAC**	96.7	41.7	74.8	99.9	65.9	86.4	97.7	52.1	79.6	94.1	91.1	92.9	99.9	89.9	95.9	94.3	45.5	74.9
26, 27, 3370	**ELECTRICAL, COMMUNICATIONS & UTIL.**	95.9	57.7	77.3	98.9	58.9	79.4	98.9	57.5	78.8	100.9	85.3	93.3	100.9	108.3	104.5	92.8	38.4	66.3
MF2004	**WEIGHTED AVERAGE**	96.3	59.3	80.8	96.8	70.9	85.9	97.9	63.1	83.2	97.1	90.3	94.2	98.2	87.4	93.7	95.1	68.5	83.9

DIVISION		MISSOURI																	
		COLUMBIA			FLAT RIVER			HANNIBAL			HARRISONVILLE			JEFFERSON CITY			JOPLIN		
		652			636			634			647			650 - 651			648		
		MAT.	INST.	TOTAL	MAT.	INST.	TOTAL	MAT.	INST.	TOTAL	MAT.	INST.	TOTAL	MAT.	INST.	TOTAL	MAT.	INST.	TOTAL
015433	CONTRACTOR EQUIPMENT		108.1	108.1		107.1	107.1		107.1	107.1		99.5	99.5		108.1	108.1		105.9	105.9
0241, 31 - 34	SITE & INFRASTRUCTURE, DEMOLITION	102.8	95.7	97.9	95.0	94.0	94.3	89.6	92.7	91.7	95.4	89.9	91.6	104.5	95.1	98.1	104.2	98.8	100.5
0310	Concrete Forming & Accessories	88.2	66.7	69.6	98.7	82.9	85.1	91.4	63.8	67.5	89.3	98.7	97.4	98.5	59.3	64.6	104.3	69.2	74.0
0320	Concrete Reinforcing	102.9	113.6	107.3	105.1	103.0	104.2	103.3	80.0	93.8	97.3	104.9	100.4	101.6	108.3	104.3	100.8	70.9	88.6
0330	Cast-in-Place Concrete	89.0	85.3	87.6	98.3	79.4	91.1	90.1	73.2	83.6	94.9	98.2	96.1	89.0	69.1	81.4	100.3	69.9	88.7
03	CONCRETE	86.6	83.8	85.3	97.1	86.9	92.5	90.8	72.1	82.3	94.6	100.1	97.1	87.6	74.0	81.4	97.9	71.0	85.6
04	MASONRY	131.0	84.4	103.1	119.9	78.9	95.3	112.7	83.7	95.4	96.6	97.8	97.3	95.5	84.4	88.8	93.1	58.8	72.6
05	METALS	90.0	118.4	98.3	93.2	112.9	99.0	93.3	104.3	96.5	92.5	107.8	97.0	89.6	116.1	97.3	94.9	87.6	92.8
06	WOOD, PLASTICS & COMPOSITES	94.8	59.6	75.5	102.6	82.4	91.5	93.7	63.0	76.8	88.6	98.7	94.1	106.2	50.0	75.4	104.2	68.2	84.4
07	THERMAL & MOISTURE PROTECTION	93.7	81.5	88.9	102.3	90.8	97.8	102.0	83.5	94.8	96.0	99.2	97.2	93.4	78.1	87.5	96.6	64.8	84.2
08	OPENINGS	91.4	86.9	90.3	98.8	83.5	95.0	98.9	66.5	90.9	90.6	100.4	93.0	88.6	80.2	86.6	91.5	70.4	86.3
0920	Plaster & Gypsum Board	93.6	58.1	70.5	102.2	81.7	88.9	96.7	61.6	73.9	85.2	98.4	93.8	93.3	48.1	63.9	97.0	66.8	77.4
0950, 0980	Ceilings & Acoustic Treatment	87.6	58.1	69.3	86.1	81.7	83.4	86.1	61.6	71.0	82.5	98.4	92.4	85.1	48.1	62.2	85.0	66.8	73.8
0960	Flooring	106.5	88.5	101.7	94.2	67.2	86.9	89.9	67.2	83.7	92.4	97.0	93.6	114.7	88.5	107.6	122.7	45.2	101.7
0970, 0990	Wall Finishes & Painting/Coating	100.6	75.5	85.5	98.3	68.2	80.3	98.3	73.5	83.4	99.5	105.1	102.9	100.6	75.5	85.5	94.6	40.3	62.0
09	FINISHES	94.9	68.1	80.7	94.2	77.7	85.5	91.8	64.3	77.2	93.3	99.1	96.3	97.0	62.4	78.7	103.1	61.8	81.3
COVERS	DIVS. 10 - 14, 25, 28, 41, 43, 44	100.0	94.1	98.8	100.0	62.3	92.5	100.0	56.0	91.3	100.0	89.4	97.9	100.0	93.0	98.6	100.0	69.5	94.0
21, 22, 23	FIRE SUPPRESSION, PLUMBING & HVAC	99.9	91.3	96.5	94.1	91.0	92.8	94.1	84.7	90.4	94.2	94.8	94.4	99.8	95.4	98.0	100.1	52.6	81.2
26, 27, 3370	ELECTRICAL, COMMUNICATIONS & UTIL.	96.0	85.3	90.8	105.5	108.3	106.9	99.7	78.4	89.3	99.0	101.1	100.0	102.5	85.3	94.2	90.9	68.8	80.1
MF2004	WEIGHTED AVERAGE	96.3	88.1	92.9	98.1	91.1	95.1	95.9	80.1	89.3	94.7	98.4	96.2	95.4	86.0	91.5	97.1	68.0	84.9

DIVISION		MISSOURI																	
		KANSAS CITY			KIRKSVILLE			POPLAR BLUFF			ROLLA			SEDALIA			SIKESTON		
		640 - 641			635			639			654 - 655			653			638		
		MAT.	INST.	TOTAL	MAT.	INST.	TOTAL	MAT.	INST.	TOTAL	MAT.	INST.	TOTAL	MAT.	INST.	TOTAL	MAT.	INST.	TOTAL
015433	CONTRACTOR EQUIPMENT		100.9	100.9		95.8	95.8		100.3	100.3		108.1	108.1		96.4	96.4		100.3	100.3
0241, 31 - 34	SITE & INFRASTRUCTURE, DEMOLITION	96.7	93.5	94.5	90.5	82.2	84.8	76.9	90.5	86.2	102.3	93.1	96.0	93.9	84.0	87.1	80.4	90.6	87.4
0310	Concrete Forming & Accessories	103.3	107.7	107.1	84.4	54.8	58.8	84.5	69.6	71.7	95.6	81.7	83.6	93.0	61.1	65.4	85.5	69.8	71.9
0320	Concrete Reinforcing	95.8	114.7	103.5	103.9	79.2	93.8	107.1	75.6	94.2	103.4	52.3	82.5	101.8	96.0	99.4	106.4	67.2	90.4
0330	Cast-in-Place Concrete	93.2	107.9	98.8	98.2	64.2	85.2	75.6	62.1	70.4	91.0	75.2	85.0	95.1	67.9	84.7	80.8	70.1	76.7
03	CONCRETE	93.9	109.3	100.9	106.3	64.3	87.2	84.5	69.4	77.7	88.2	75.2	82.3	99.1	71.5	86.5	87.9	70.7	80.1
04	MASONRY	98.8	107.6	104.1	125.5	70.9	92.8	117.8	56.4	81.0	108.6	44.8	70.4	114.2	53.8	78.1	117.5	61.5	84.0
05	METALS	102.4	113.4	105.6	93.0	94.7	93.5	93.5	92.5	93.2	89.6	86.0	88.5	88.2	100.7	91.9	93.8	88.9	92.4
06	WOOD, PLASTICS & COMPOSITES	103.5	107.6	105.8	81.5	53.1	65.9	80.8	72.1	76.0	102.7	93.1	97.4	95.2	53.7	72.4	82.0	72.1	76.6
07	THERMAL & MOISTURE PROTECTION	96.3	106.2	100.1	108.2	76.4	95.8	107.4	72.0	93.6	93.8	58.6	80.1	99.0	76.9	90.4	107.5	66.1	91.4
08	OPENINGS	97.5	110.3	100.7	102.6	61.3	92.4	103.7	73.0	96.1	91.4	70.3	86.2	94.8	73.3	89.5	103.6	63.6	93.8
0920	Plaster & Gypsum Board	93.7	107.7	102.8	91.2	51.3	65.3	91.8	71.0	78.3	95.8	92.7	93.8	89.2	51.9	65.0	92.9	71.0	78.7
0950, 0980	Ceilings & Acoustic Treatment	90.8	107.7	101.3	83.6	51.3	63.7	86.1	71.0	76.8	87.6	92.7	90.8	86.8	51.9	65.2	86.1	71.0	76.8
0960	Flooring	98.9	99.9	99.2	70.4	49.4	64.8	82.6	55.9	75.4	110.8	42.1	92.2	85.6	97.0	88.7	83.0	45.2	72.8
0970, 0990	Wall Finishes & Painting/Coating	99.5	120.1	111.9	93.9	40.3	61.7	93.4	78.7	84.5	100.6	83.7	90.5	100.6	105.1	103.3	93.4	58.0	72.1
09	FINISHES	98.3	107.5	103.2	90.1	50.6	69.2	90.7	68.2	78.8	96.6	76.4	85.9	92.5	70.6	80.9	91.2	63.9	76.8
COVERS	DIVS. 10 - 14, 25, 28, 41, 43, 44	100.0	99.4	99.9	100.0	51.9	90.5	100.0	55.0	91.1	100.0	57.3	91.5	100.0	80.9	96.2	100.0	55.8	91.2
21, 22, 23	FIRE SUPPRESSION, PLUMBING & HVAC	100.0	104.1	101.6	94.1	83.2	89.8	94.1	72.4	85.5	94.1	64.4	82.3	94.0	89.9	92.4	94.1	73.3	85.8
26, 27, 3370	ELECTRICAL, COMMUNICATIONS & UTIL.	101.2	105.2	103.1	99.5	78.4	89.2	99.9	108.3	104.0	94.5	85.3	90.0	95.4	101.1	98.2	99.0	108.3	103.5
MF2004	WEIGHTED AVERAGE	99.2	106.1	102.1	98.7	73.3	88.0	95.6	77.7	88.1	94.1	73.0	85.2	95.4	81.7	89.6	96.1	77.2	88.2

DIVISION		MISSOURI									MONTANA								
		SPRINGFIELD			ST. JOSEPH			ST. LOUIS			BILLINGS			BUTTE			GREAT FALLS		
		656 - 658			644 - 645			630 - 631			590 - 591			597			594		
		MAT.	INST.	TOTAL	MAT.	INST.	TOTAL	MAT.	INST.	TOTAL	MAT.	INST.	TOTAL	MAT.	INST.	TOTAL	MAT.	INST.	TOTAL
015433	CONTRACTOR EQUIPMENT		101.2	101.2		99.5	99.5		108.0	108.0		99.1	99.1		98.7	98.7		98.7	98.7
0241, 31 - 34	SITE & INFRASTRUCTURE, DEMOLITION	95.9	91.6	92.9	98.2	88.6	91.7	93.2	97.5	96.1	91.0	97.7	95.6	99.9	96.7	97.7	103.6	97.5	99.4
0310	Concrete Forming & Accessories	101.2	71.7	75.7	103.1	79.4	82.6	98.8	105.3	104.4	96.9	65.3	69.6	85.4	69.7	71.8	96.8	66.7	70.8
0320	Concrete Reinforcing	99.0	113.3	104.8	94.6	104.7	98.7	96.1	109.4	101.5	102.3	70.6	89.4	111.0	70.7	94.6	102.3	70.5	89.3
0330	Cast-in-Place Concrete	96.6	62.9	83.8	93.2	103.3	97.0	94.2	112.3	101.1	108.7	69.2	93.6	119.7	70.9	101.0	126.5	68.4	104.3
03	CONCRETE	96.1	77.4	87.6	93.6	93.4	93.5	92.5	109.3	100.2	102.3	68.8	87.1	106.2	71.3	90.4	110.3	69.1	91.6
04	MASONRY	88.7	78.1	82.4	98.4	90.3	93.6	99.5	110.3	106.0	120.5	71.1	90.9	116.3	74.4	91.2	120.3	73.2	92.1
05	METALS	93.8	105.3	97.2	98.6	107.2	101.1	98.9	119.5	104.9	106.1	86.5	100.3	98.9	86.8	95.4	102.8	86.5	98.1
06	WOOD, PLASTICS & COMPOSITES	103.2	72.6	86.4	104.2	74.6	88.0	102.4	102.5	102.5	102.6	65.0	82.0	90.4	70.0	79.2	103.8	65.0	82.5
07	THERMAL & MOISTURE PROTECTION	98.2	73.3	88.5	96.6	92.3	94.9	101.9	107.8	104.2	101.3	69.3	88.9	101.2	73.8	90.5	101.7	68.2	88.7
08	OPENINGS	97.1	81.0	93.1	96.2	89.9	94.7	97.7	109.7	100.7	97.7	62.6	89.0	95.6	65.8	88.2	99.0	63.1	90.1
0920	Plaster & Gypsum Board	96.9	71.5	80.4	98.4	73.5	82.2	103.2	102.6	102.8	93.6	64.1	74.4	91.0	69.3	76.9	98.5	64.1	76.1
0950, 0980	Ceilings & Acoustic Treatment	87.6	71.5	77.7	90.0	73.5	79.8	91.1	102.6	98.2	106.7	64.1	80.3	111.6	69.3	85.4	113.2	64.1	82.8
0960	Flooring	109.2	45.2	91.9	102.0	97.0	100.7	94.0	102.4	96.3	110.1	61.4	96.9	106.9	83.2	100.5	115.5	83.2	106.7
0970, 0990	Wall Finishes & Painting/Coating	95.0	57.7	72.6	95.1	85.4	89.3	98.3	109.2	104.8	100.5	53.8	72.5	98.4	58.9	74.7	98.4	55.9	72.9
09	FINISHES	98.4	64.9	80.7	99.4	81.8	90.1	95.3	104.3	100.1	105.0	62.9	82.8	104.5	71.1	86.9	108.8	68.2	87.3
COVERS	DIVS. 10 - 14, 25, 28, 41, 43, 44	100.0	89.1	97.8	100.0	92.8	98.6	100.0	103.9	100.8	100.0	73.6	94.8	100.0	74.7	95.0	100.0	74.9	95.0
21, 22, 23	FIRE SUPPRESSION, PLUMBING & HVAC	99.9	68.3	87.4	100.1	90.1	96.1	99.9	108.7	103.4	100.1	76.1	90.6	100.2	69.9	88.2	100.2	74.3	89.9
26, 27, 3370	ELECTRICAL, COMMUNICATIONS & UTIL.	99.7	73.1	86.7	99.0	84.0	91.7	103.6	111.7	107.6	95.1	75.5	85.6	102.4	72.6	87.9	95.2	72.5	84.1
MF2004	WEIGHTED AVERAGE	97.3	77.6	89.1	98.3	90.2	94.9	98.6	108.8	102.9	101.9	74.6	90.4	101.6	75.0	90.4	103.1	74.8	91.2

DIVISION		MONTANA																	
		HAVRE			HELENA			KALISPELL			MILES CITY			MISSOULA			WOLF POINT		
		595			596			599			593			598			592		
		MAT.	INST.	TOTAL	MAT.	INST.	TOTAL	MAT.	INST.	TOTAL	MAT.	INST.	TOTAL	MAT.	INST.	TOTAL	MAT.	INST.	TOTAL
015433	CONTRACTOR EQUIPMENT		98.7	98.7		98.7	98.7		98.7	98.7		98.7	98.7		98.7	98.7		98.7	98.7
0241, 31 - 34	SITE & INFRASTRUCTURE, DEMOLITION	108.5	97.3	100.8	100.7	96.6	97.9	89.7	96.3	94.2	96.7	97.0	96.9	81.3	96.6	91.8	116.1	97.2	103.2
0310	Concrete Forming & Accessories	78.6	58.8	61.5	97.1	66.1	70.3	88.2	57.0	61.2	95.1	59.0	63.9	88.2	61.0	64.7	88.3	55.9	60.3
0320	Concrete Reinforcing	111.8	70.6	95.0	105.8	70.5	91.4	114.0	80.2	100.2	111.4	70.6	94.8	112.9	80.2	99.6	113.0	70.6	95.7
0330	Cast-in-Place Concrete	128.9	67.1	105.3	106.7	67.0	91.5	103.9	63.5	88.5	113.8	68.2	96.4	88.1	66.0	79.7	127.5	68.6	105.0
03	CONCRETE	114.0	65.2	91.8	101.8	68.4	86.6	96.9	64.9	82.3	103.4	65.7	86.2	86.0	67.5	77.6	117.5	64.4	93.3
04	MASONRY	117.3	70.7	89.4	113.0	71.6	88.2	115.3	68.8	87.5	122.2	69.3	90.5	139.5	69.1	97.3	123.3	68.8	90.7
05	METALS	95.6	86.6	93.0	102.0	86.4	97.5	95.5	90.4	94.0	94.9	86.7	92.5	95.9	90.8	94.4	95.0	86.6	92.5
06	WOOD, PLASTICS & COMPOSITES	82.4	55.4	67.5	103.9	65.0	82.6	94.0	55.1	72.6	100.9	57.0	76.8	94.0	60.6	75.6	93.1	52.4	70.7
07	THERMAL & MOISTURE PROTECTION	101.5	66.1	87.7	101.4	73.7	90.6	100.9	70.3	89.0	101.2	68.1	88.3	100.5	70.8	89.0	101.9	67.0	88.4
08	OPENINGS	95.7	57.8	86.3	98.5	61.8	89.4	95.7	60.4	86.9	95.1	58.7	86.1	95.6	63.1	87.6	95.2	55.9	85.5
0920	Plaster & Gypsum Board	87.3	54.1	65.7	91.2	64.1	73.6	91.7	53.7	67.0	97.4	55.8	70.4	91.7	59.4	70.8	93.7	51.0	65.9
0950, 0980	Ceilings & Acoustic Treatment	111.6	54.1	76.0	108.2	64.1	80.9	111.6	53.7	75.8	109.1	55.8	76.1	111.6	59.4	79.3	109.1	51.0	73.1
0960	Flooring	103.4	83.2	98.0	115.5	66.3	102.2	109.0	83.2	102.0	115.2	78.8	105.4	109.0	83.2	102.0	110.7	78.8	102.1
0970, 0990	Wall Finishes & Painting/Coating	98.4	55.9	72.9	98.4	49.7	69.2	98.4	43.0	65.2	98.4	51.6	70.3	98.4	54.7	72.2	98.4	55.9	72.9
09	FINISHES	103.6	61.9	81.6	106.4	64.2	84.1	104.7	59.4	80.7	107.2	61.0	82.7	104.2	63.8	82.8	106.6	58.9	81.4
COVERS	DIVS. 10 - 14, 25, 28, 41, 43, 44	100.0	84.8	97.0	100.0	86.3	97.3	100.0	83.3	96.7	100.0	84.1	96.8	100.0	83.7	96.8	100.0	83.8	96.8
21, 22, 23	FIRE SUPPRESSION, PLUMBING & HVAC	94.4	73.1	85.9	100.1	70.3	88.3	94.4	66.9	83.5	94.4	71.7	85.4	100.2	66.6	86.8	94.4	72.1	85.6
26, 27, 3370	ELECTRICAL, COMMUNICATIONS & UTIL.	95.2	72.5	84.1	101.0	72.5	87.1	99.3	76.7	88.3	95.2	75.5	85.6	100.3	76.7	88.8	95.2	70.3	83.1
MF2004	WEIGHTED AVERAGE	100.0	72.8	88.6	101.8	73.5	89.9	97.9	72.0	87.0	98.9	72.8	88.0	99.0	73.2	88.2	101.1	71.5	88.7

DIVISION		NEBRASKA																	
		ALLIANCE			COLUMBUS			GRAND ISLAND			HASTINGS			LINCOLN			MCCOOK		
		693			686			688			689			683 - 685			690		
		MAT.	INST.	TOTAL	MAT.	INST.	TOTAL	MAT.	INST.	TOTAL	MAT.	INST.	TOTAL	MAT.	INST.	TOTAL	MAT.	INST.	TOTAL
015433	CONTRACTOR EQUIPMENT		97.0	97.0		100.5	100.5		100.5	100.5		100.5	100.5		100.5	100.5		100.5	100.5
0241, 31 - 34	SITE & INFRASTRUCTURE, DEMOLITION	96.1	98.6	97.8	95.2	89.8	91.5	99.7	90.7	93.6	98.8	90.3	93.0	89.3	90.7	90.3	99.7	90.0	93.0
0310	Concrete Forming & Accessories	89.8	62.2	65.9	98.3	65.1	69.6	97.9	82.8	84.8	101.2	81.7	84.4	102.5	65.5	70.5	95.5	62.3	66.8
0320	Concrete Reinforcing	117.3	52.5	90.8	108.4	51.2	85.0	107.8	73.6	93.8	107.8	51.9	85.0	99.0	73.0	88.4	109.8	73.2	94.9
0330	Cast-in-Place Concrete	107.2	63.7	90.6	109.6	60.3	90.8	116.1	74.6	100.2	116.1	69.6	98.3	92.6	77.3	86.7	116.1	63.2	95.9
03	CONCRETE	117.5	61.5	92.0	103.9	62.0	84.8	108.2	78.8	94.8	108.4	72.7	92.1	94.3	72.1	84.2	108.5	65.8	89.1
04	MASONRY	109.4	73.3	87.8	114.4	78.5	92.9	107.5	84.8	93.9	116.0	97.2	104.7	96.6	74.6	83.4	104.9	73.4	86.0
05	METALS	96.4	69.9	88.6	88.6	79.5	86.0	90.1	87.7	89.4	91.1	81.1	88.2	95.0	86.8	92.6	91.3	86.3	89.9
06	WOOD, PLASTICS & COMPOSITES	88.6	61.1	73.5	100.4	66.5	81.8	99.8	84.6	91.4	103.5	84.6	93.1	102.7	61.5	80.1	97.6	61.2	77.6
07	THERMAL & MOISTURE PROTECTION	105.4	73.8	93.1	98.9	70.5	87.8	99.0	80.5	91.8	99.0	94.5	97.3	98.5	75.3	89.5	99.0	70.7	88.0
08	OPENINGS	91.0	57.1	82.6	90.7	62.0	83.6	90.7	77.9	87.5	90.7	71.9	86.1	96.2	61.0	87.5	90.8	62.5	83.8
0920	Plaster & Gypsum Board	87.6	59.7	69.5	95.2	65.1	75.7	94.9	83.9	87.8	96.4	83.9	88.3	91.4	60.0	71.0	94.9	59.7	72.0
0950, 0980	Ceilings & Acoustic Treatment	89.9	59.7	71.2	86.8	65.1	73.4	86.8	83.9	85.0	86.8	83.9	85.0	89.3	60.0	71.2	86.8	59.7	70.0
0960	Flooring	98.7	92.0	96.9	98.6	112.5	102.4	98.5	120.2	104.3	100.0	112.5	103.4	100.2	93.7	98.4	97.3	92.0	95.9
0970, 0990	Wall Finishes & Painting/Coating	158.3	63.0	101.1	93.0	73.3	81.2	93.0	77.7	83.8	93.0	73.3	81.2	93.0	89.5	90.9	93.0	54.9	70.1
09	FINISHES	97.1	66.9	81.1	95.1	73.8	83.8	95.2	88.6	91.7	95.9	86.1	90.7	95.5	72.6	83.4	94.9	65.9	79.6
COVERS	DIVS. 10 - 14, 25, 28, 41, 43, 44	100.0	79.0	95.8	100.0	82.7	96.6	100.0	88.2	97.7	100.0	87.3	97.5	100.0	85.5	97.1	100.0	78.2	95.7
21, 22, 23	FIRE SUPPRESSION, PLUMBING & HVAC	94.3	66.7	83.3	94.1	74.5	86.3	99.9	80.1	92.1	94.1	78.3	87.8	99.8	80.1	92.0	94.1	76.4	87.1
26, 27, 3370	ELECTRICAL, COMMUNICATIONS & UTIL.	91.4	81.0	86.3	91.1	89.9	90.5	89.9	74.0	82.1	89.3	97.5	93.3	103.8	74.0	89.3	93.5	81.0	87.4
MF2004	WEIGHTED AVERAGE	98.6	71.6	87.3	95.5	76.4	87.5	97.3	82.4	91.0	96.5	84.8	91.6	97.6	77.1	89.0	96.4	75.1	87.5

DIVISION		NEBRASKA												NEVADA					
		NORFOLK			NORTH PLATTE			OMAHA			VALENTINE			CARSON CITY			ELKO		
		687			691			680 - 681			692			897			898		
		MAT.	INST.	TOTAL	MAT.	INST.	TOTAL	MAT.	INST.	TOTAL	MAT.	INST.	TOTAL	MAT.	INST.	TOTAL	MAT.	INST.	TOTAL
015433	CONTRACTOR EQUIPMENT		91.9	91.9		100.5	100.5		91.9	91.9		94.8	94.8		101.1	101.1		101.1	101.1
0241, 31 - 34	SITE & INFRASTRUCTURE, DEMOLITION	79.0	90.4	86.8	100.6	90.2	93.5	80.1	90.7	87.3	83.7	95.4	91.7	68.7	103.0	92.2	57.0	102.7	88.2
0310	Concrete Forming & Accessories	84.8	80.9	81.4	97.8	81.6	83.8	94.4	72.4	75.4	86.0	61.7	65.0	101.7	92.8	94.0	108.7	79.7	83.6
0320	Concrete Reinforcing	108.5	54.1	86.3	109.2	74.3	95.0	103.6	73.2	91.2	109.9	50.9	85.8	107.5	113.2	109.8	109.2	100.9	105.8
0330	Cast-in-Place Concrete	110.4	68.8	94.5	116.1	71.2	98.9	99.3	79.2	91.6	102.2	60.9	86.5	109.5	91.2	102.5	102.5	75.0	92.0
03	CONCRETE	102.7	71.9	88.7	108.5	77.2	94.3	97.1	75.3	87.2	105.3	60.2	84.8	110.8	96.1	104.1	104.7	82.4	94.6
04	MASONRY	120.5	76.2	94.0	92.2	97.1	95.2	101.8	80.2	88.9	104.4	73.4	85.9	123.0	78.0	96.0	124.7	70.4	92.2
05	METALS	91.8	72.6	86.2	90.3	88.4	89.8	94.3	78.7	89.8	102.2	71.0	93.1	94.9	101.3	96.7	94.0	94.6	94.2
06	WOOD, PLASTICS & COMPOSITES	82.8	84.0	83.5	99.6	84.6	91.4	93.7	71.3	81.4	83.2	60.6	70.8	93.3	93.6	93.5	102.3	80.6	90.4
07	THERMAL & MOISTURE PROTECTION	98.0	80.4	91.2	99.0	87.9	94.6	94.4	82.9	89.9	98.7	73.6	89.0	99.4	87.1	94.6	100.7	73.3	90.0
08	OPENINGS	92.6	72.5	87.6	90.0	77.2	86.8	99.3	71.2	92.4	92.7	58.3	84.2	93.3	106.4	96.5	95.9	83.9	92.9
0920	Plaster & Gypsum Board	94.3	83.9	87.5	95.1	83.9	87.8	93.7	70.7	78.8	97.3	59.7	72.9	78.5	93.2	88.1	86.3	79.7	82.0
0950, 0980	Ceilings & Acoustic Treatment	101.3	83.9	90.5	87.6	83.9	85.3	102.1	70.7	82.7	104.3	59.7	76.7	107.8	93.2	98.8	111.2	79.7	91.7
0960	Flooring	122.8	112.5	120.0	98.4	112.5	102.2	128.1	54.5	108.2	125.5	89.5	115.8	104.9	57.3	92.0	110.7	57.3	96.3
0970, 0990	Wall Finishes & Painting/Coating	158.4	73.3	107.3	93.0	73.3	81.2	158.4	71.7	106.4	158.4	75.9	108.9	99.2	87.7	92.3	98.1	83.8	89.5
09	FINISHES	114.2	86.0	99.3	95.4	86.1	90.5	116.0	68.3	90.8	116.7	67.3	90.6	101.0	85.1	92.6	103.4	76.1	89.0
COVERS	DIVS. 10 - 14, 25, 28, 41, 43, 44	100.0	85.8	97.2	100.0	82.2	96.5	100.0	85.3	97.1	100.0	75.8	95.2	100.0	109.7	101.9	100.0	74.9	95.0
21, 22, 23	FIRE SUPPRESSION, PLUMBING & HVAC	93.9	78.1	87.6	99.9	78.2	91.3	99.7	80.0	91.8	93.8	64.6	82.2	99.9	87.3	94.9	97.2	71.4	86.9
26, 27, 3370	ELECTRICAL, COMMUNICATIONS & UTIL.	90.1	90.7	90.4	91.8	89.9	90.9	103.6	83.2	93.7	88.9	101.0	94.8	98.8	99.4	99.1	93.9	61.3	78.0
MF2004	WEIGHTED AVERAGE	97.2	80.4	90.1	96.8	85.0	91.8	99.5	78.8	90.8	98.7	73.6	88.2	99.9	93.1	97.1	98.2	77.6	89.6

	DIVISION	NEVADA									NEW HAMPSHIRE								
		ELY			LAS VEGAS			RENO			CHARLESTON			CLAREMONT			CONCORD		
		893			889 - 891			894 - 895			036			037			032 - 033		
		MAT.	INST.	TOTAL	MAT.	INST.	TOTAL	MAT.	INST.	TOTAL	MAT.	INST.	TOTAL	MAT.	INST.	TOTAL	MAT.	INST.	TOTAL
015433	CONTRACTOR EQUIPMENT		101.1	101.1		101.1	101.1		101.1	101.1		100.4	100.4		100.4	100.4		100.4	100.4
0241, 31 - 34	SITE & INFRASTRUCTURE, DEMOLITION	61.8	102.2	89.4	63.5	106.4	92.9	61.8	103.0	90.0	77.2	96.2	90.2	71.6	96.2	88.4	84.6	98.3	94.0
0310	Concrete Forming & Accessories	101.7	64.6	69.7	103.7	110.5	109.6	98.7	92.9	93.7	84.7	48.0	53.0	90.5	48.0	53.8	92.5	80.7	82.3
0320	Concrete Reinforcing	108.0	93.2	102.0	99.6	125.4	110.2	102.1	124.4	111.2	80.3	51.5	68.5	80.3	51.5	68.5	80.3	93.1	85.5
0330	Cast-in-Place Concrete	110.1	72.8	95.8	106.6	115.0	109.8	116.4	91.2	106.8	92.8	54.7	78.2	85.2	54.7	73.6	98.1	93.4	96.3
03	CONCRETE	113.3	73.6	95.3	108.0	114.4	110.9	112.8	98.2	106.1	101.5	51.9	78.9	93.2	51.9	74.5	102.3	87.5	95.6
04	MASONRY	130.0	62.9	89.8	117.2	103.7	109.1	123.8	78.0	96.4	91.8	41.5	61.7	92.1	41.5	61.8	94.8	87.9	90.7
05	METALS	93.9	91.5	93.2	101.1	110.0	103.7	95.3	105.6	98.3	88.6	66.7	82.2	88.6	66.7	82.2	89.2	90.0	89.4
06	WOOD, PLASTICS & COMPOSITES	93.2	64.2	77.3	93.3	108.8	101.8	89.4	93.6	91.7	88.3	49.6	67.1	94.9	49.6	70.0	98.9	80.4	88.8
07	THERMAL & MOISTURE PROTECTION	101.3	68.3	88.5	115.4	102.4	110.4	100.8	87.1	95.5	94.0	45.0	74.9	93.9	45.0	74.8	92.8	88.6	91.2
08	OPENINGS	95.8	72.9	90.2	94.8	117.7	100.5	93.6	106.6	96.8	101.5	44.9	87.5	102.7	44.9	88.4	102.7	77.3	96.4
0920	Plaster & Gypsum Board	82.5	62.7	69.7	84.5	109.0	100.4	78.7	93.2	88.2	97.0	47.0	64.5	99.3	47.0	65.3	101.7	78.9	86.9
0950, 0980	Ceilings & Acoustic Treatment	111.2	62.7	81.2	121.2	109.0	113.6	117.8	93.2	102.6	88.9	47.0	63.0	88.9	47.0	63.0	88.9	78.9	82.8
0960	Flooring	108.2	64.1	96.3	98.8	98.1	98.6	104.5	57.3	91.8	91.3	36.0	76.3	93.9	36.0	78.3	91.4	100.4	93.9
0970, 0990	Wall Finishes & Painting/Coating	98.1	86.2	91.0	100.4	118.4	111.2	98.1	87.7	91.9	91.7	52.6	68.2	91.7	52.6	68.2	91.7	96.7	94.7
09	FINISHES	102.6	66.3	83.4	102.4	108.7	105.7	102.4	85.1	93.2	92.9	46.5	68.4	93.5	46.5	68.7	94.5	86.0	90.0
COVERS	DIVS. 10 - 14, 25, 28, 41, 43, 44	100.0	62.8	92.6	100.0	104.5	100.9	100.0	109.7	101.9	100.0	45.0	89.1	100.0	45.0	89.1	100.0	69.8	94.0
21, 22, 23	FIRE SUPPRESSION, PLUMBING & HVAC	97.2	67.4	85.3	100.0	110.2	104.1	99.9	87.5	95.0	94.4	44.1	74.4	94.4	44.1	74.4	100.0	85.0	94.0
26, 27, 3370	ELECTRICAL, COMMUNICATIONS & UTIL.	94.1	61.3	78.1	99.0	128.0	113.1	94.5	99.4	96.9	94.8	37.5	66.9	94.8	37.5	66.9	96.9	65.2	81.5
MF2004	WEIGHTED AVERAGE	99.5	72.0	88.0	101.1	112.0	105.7	99.8	93.9	97.3	94.6	50.8	76.2	93.7	50.8	75.7	97.0	83.8	91.5

	DIVISION	NEW HAMPSHIRE															NEW JERSEY		
		KEENE			LITTLETON			MANCHESTER			NASHUA			PORTSMOUTH			ATLANTIC CITY		
		034			035			031			030			038			082,084		
		MAT.	INST.	TOTAL	MAT.	INST.	TOTAL	MAT.	INST.	TOTAL	MAT.	INST.	TOTAL	MAT.	INST.	TOTAL	MAT.	INST.	TOTAL
015433	CONTRACTOR EQUIPMENT		100.4	100.4		100.4	100.4		100.4	100.4		100.4	100.4		100.4	100.4		98.5	98.5
0241, 31 - 34	SITE & INFRASTRUCTURE, DEMOLITION	83.9	96.4	92.5	71.6	95.9	88.2	82.1	98.4	93.2	84.7	98.4	94.1	79.1	99.7	93.2	87.0	103.3	98.2
0310	Concrete Forming & Accessories	89.2	45.9	51.8	99.1	51.0	57.6	94.5	81.3	83.1	96.7	81.3	83.4	86.0	83.7	84.0	104.8	120.4	118.3
0320	Concrete Reinforcing	80.3	61.3	72.5	81.0	36.9	63.0	100.4	93.2	97.4	100.4	93.2	97.4	80.3	93.3	85.6	76.5	110.6	90.4
0330	Cast-in-Place Concrete	93.2	56.3	79.1	83.6	57.0	73.4	101.0	117.1	107.2	88.2	117.1	99.3	83.6	120.9	97.9	79.1	125.2	96.7
03	CONCRETE	101.1	53.4	79.4	92.6	51.3	73.9	107.8	96.0	102.4	101.9	96.0	99.2	92.7	98.5	95.4	94.6	119.2	105.8
04	MASONRY	95.7	43.8	64.6	104.6	41.2	66.7	98.0	87.9	91.9	97.8	87.9	91.8	93.3	93.1	93.2	99.4	120.3	111.9
05	METALS	89.1	71.0	83.8	89.2	60.8	80.9	93.8	90.6	92.9	93.8	90.6	92.8	90.4	93.6	91.4	88.9	101.1	92.5
06	WOOD, PLASTICS & COMPOSITES	93.3	44.3	66.4	104.1	50.6	74.7	98.9	80.4	88.7	102.3	80.4	90.3	89.6	80.4	84.6	112.7	120.8	117.2
07	THERMAL & MOISTURE PROTECTION	94.5	47.1	76.0	94.0	48.9	76.4	94.2	92.3	93.4	94.8	92.3	93.8	94.5	113.5	101.9	94.5	120.8	104.7
08	OPENINGS	100.0	50.1	87.7	103.8	42.5	88.6	104.9	77.3	98.1	104.9	77.3	98.1	105.5	72.2	97.3	102.4	116.5	105.9
0920	Plaster & Gypsum Board	98.5	41.5	61.5	110.4	48.1	69.9	108.7	78.9	89.4	105.5	78.9	88.3	97.0	78.9	85.3	108.6	120.8	116.5
0950, 0980	Ceilings & Acoustic Treatment	88.9	41.5	59.6	88.9	48.1	63.7	96.6	78.9	85.7	99.1	78.9	86.6	90.8	78.9	83.5	88.9	120.8	108.7
0960	Flooring	93.3	59.0	84.1	103.0	36.0	84.9	97.9	100.4	98.5	96.8	100.4	97.8	91.8	100.4	94.1	100.7	136.7	110.4
0970, 0990	Wall Finishes & Painting/Coating	91.7	52.6	68.2	91.7	68.9	78.0	91.7	96.7	94.7	91.7	96.7	94.7	91.7	96.7	94.7	91.7	124.2	111.2
09	FINISHES	94.8	48.7	70.4	97.7	50.2	72.6	98.7	86.0	92.0	99.1	86.0	92.2	94.1	87.2	90.5	98.8	124.6	112.5
COVERS	DIVS. 10 - 14, 25, 28, 41, 43, 44	100.0	61.5	92.4	100.0	47.1	89.5	100.0	69.9	94.0	100.0	69.9	94.0	100.0	71.7	94.4	100.0	110.2	102.0
21, 22, 23	FIRE SUPPRESSION, PLUMBING & HVAC	94.4	47.8	75.8	94.4	73.6	86.1	100.0	85.0	94.1	100.2	85.0	94.2	100.2	87.7	95.2	99.7	113.9	105.4
26, 27, 3370	ELECTRICAL, COMMUNICATIONS & UTIL.	94.8	37.5	66.9	95.8	65.1	80.9	96.7	83.4	90.2	97.0	83.4	90.4	95.2	83.4	89.5	92.8	134.0	112.9
MF2004	WEIGHTED AVERAGE	95.1	53.4	77.6	95.0	60.6	80.5	99.1	87.7	94.3	98.6	87.7	94.0	95.9	90.1	93.5	96.3	117.6	105.2

	DIVISION	NEW JERSEY																	
		CAMDEN			DOVER			ELIZABETH			HACKENSACK			JERSEY CITY			LONG BRANCH		
		081			078			072			076			073			077		
		MAT.	INST.	TOTAL	MAT.	INST.	TOTAL	MAT.	INST.	TOTAL	MAT.	INST.	TOTAL	MAT.	INST.	TOTAL	MAT.	INST.	TOTAL
015433	CONTRACTOR EQUIPMENT		98.5	98.5		100.4	100.4		100.4	100.4		100.4	100.4		98.5	98.5		98.1	98.1
0241, 31 - 34	SITE & INFRASTRUCTURE, DEMOLITION	87.9	103.7	98.7	96.9	104.6	102.2	100.8	104.6	103.4	97.6	104.5	102.3	87.9	104.4	99.2	92.6	104.3	100.6
0310	Concrete Forming & Accessories	96.7	120.5	117.2	93.3	121.2	117.4	104.7	121.1	118.9	93.4	121.1	117.3	96.7	121.2	117.9	97.2	121.0	117.7
0320	Concrete Reinforcing	100.4	114.2	106.0	77.3	119.5	94.5	77.3	119.5	94.5	77.3	119.5	94.5	100.4	119.5	108.2	77.3	119.5	94.5
0330	Cast-in-Place Concrete	76.7	123.8	94.7	98.4	125.2	108.6	84.5	125.1	100.0	96.1	125.1	107.2	76.7	125.2	95.2	85.3	125.1	100.5
03	CONCRETE	96.5	119.4	106.9	101.9	121.4	110.7	97.8	121.4	108.5	100.1	121.3	109.7	96.5	121.2	107.8	99.9	121.1	109.6
04	MASONRY	91.2	120.3	108.6	94.7	120.3	110.0	110.7	120.3	116.4	98.7	120.3	111.6	88.6	120.3	107.6	103.2	120.3	113.4
05	METALS	93.6	102.4	96.2	88.9	108.9	94.7	90.2	108.8	95.6	89.0	108.8	94.8	93.7	106.5	97.4	89.0	106.2	94.0
06	WOOD, PLASTICS & COMPOSITES	102.3	120.8	112.5	101.6	120.8	112.2	116.2	120.8	118.8	101.6	120.8	112.2	102.3	120.8	112.5	103.4	120.8	112.9
07	THERMAL & MOISTURE PROTECTION	94.4	120.6	104.6	94.9	125.8	106.9	95.1	125.8	107.0	94.7	120.7	104.8	94.4	125.8	106.6	94.5	125.5	106.6
08	OPENINGS	104.9	117.4	108.0	108.0	119.9	110.9	105.9	119.9	109.4	105.2	119.9	108.8	104.9	119.9	108.6	100.8	119.8	105.5
0920	Plaster & Gypsum Board	105.5	120.8	115.5	102.2	120.8	114.3	109.3	120.8	116.8	102.2	120.8	114.3	105.5	120.8	115.5	104.1	120.8	115.0
0950, 0980	Ceilings & Acoustic Treatment	99.1	120.8	112.5	88.9	120.8	108.7	90.8	120.8	109.4	88.9	120.8	108.7	99.1	120.8	112.5	88.9	120.8	108.7
0960	Flooring	96.8	136.7	107.6	95.7	151.8	110.9	101.1	151.8	114.8	95.7	151.8	110.9	96.8	151.8	111.7	97.0	151.8	111.8
0970, 0990	Wall Finishes & Painting/Coating	91.7	124.2	111.2	91.6	126.9	112.8	91.6	126.9	112.8	91.6	126.9	112.8	91.7	126.9	112.8	91.7	124.2	111.2
09	FINISHES	99.5	124.6	112.8	96.6	127.5	112.9	100.1	127.5	114.6	96.5	127.5	112.9	99.5	127.5	114.3	97.5	127.2	113.2
COVERS	DIVS. 10 - 14, 25, 28, 41, 43, 44	100.0	110.2	102.0	100.0	118.0	103.6	100.0	118.0	103.6	100.0	118.0	103.6	100.0	118.0	103.6	100.0	110.1	102.0
21, 22, 23	FIRE SUPPRESSION, PLUMBING & HVAC	100.2	113.9	105.6	99.8	122.8	108.9	100.2	120.6	108.3	99.8	122.8	108.9	100.2	122.8	109.2	99.8	121.9	108.6
26, 27, 3370	ELECTRICAL, COMMUNICATIONS & UTIL.	97.5	134.0	115.3	94.4	137.5	115.3	95.0	137.5	115.6	94.4	141.1	117.1	98.8	141.1	119.4	94.1	137.4	115.2
MF2004	WEIGHTED AVERAGE	97.8	117.8	106.2	97.6	122.0	107.8	98.6	121.5	108.2	97.3	122.3	107.8	97.8	122.2	108.0	97.1	121.2	107.2

	DIVISION	NEW JERSEY																	
		NEW BRUNSWICK			NEWARK			PATERSON			POINT PLEASANT			SUMMIT			TRENTON		
		088 - 089			070 - 071			074 - 075			087			079			085 - 086		
		MAT.	INST.	TOTAL	MAT.	INST.	TOTAL	MAT.	INST.	TOTAL	MAT.	INST.	TOTAL	MAT.	INST.	TOTAL	MAT.	INST.	TOTAL
015433	**CONTRACTOR EQUIPMENT**		98.1	98.1		100.4	100.4		100.4	100.4		98.1	98.1		100.4	100.4		98.1	98.1
0241, 31 - 34	**SITE & INFRASTRUCTURE, DEMOLITION**	99.9	104.3	102.9	103.4	104.6	104.2	99.1	104.5	102.8	101.5	104.3	103.4	98.5	104.6	102.6	87.6	104.3	99.0
0310	Concrete Forming & Accessories	99.6	121.1	118.2	93.3	121.2	117.4	95.5	121.1	117.6	94.4	120.9	117.3	96.2	121.1	117.7	93.9	120.8	117.1
0320	Concrete Reinforcing	77.3	119.5	94.5	100.4	119.5	108.2	100.4	119.5	108.2	77.3	119.5	94.5	77.3	119.5	94.5	100.4	113.2	105.6
0330	Cast-in-Place Concrete	97.8	125.1	108.2	98.1	125.1	108.4	97.8	125.1	108.2	97.8	125.1	108.2	81.7	125.1	98.3	87.3	125.0	101.7
03	**CONCRETE**	111.8	121.2	116.1	106.3	121.4	113.2	106.3	121.3	113.2	111.5	121.0	115.8	94.8	121.4	106.9	101.4	119.8	109.7
04	**MASONRY**	97.7	120.3	111.2	99.9	120.3	112.1	95.4	120.3	110.3	88.7	120.3	107.6	97.2	120.3	111.0	87.2	120.3	107.0
05	**METALS**	89.0	106.3	94.0	93.7	108.9	98.1	88.9	108.8	94.7	89.0	106.1	94.0	88.9	108.8	94.7	88.9	102.3	92.8
06	**WOOD, PLASTICS & COMPOSITES**	106.4	120.8	114.3	100.9	120.8	111.8	104.1	120.8	113.3	99.9	120.8	111.4	105.3	120.8	113.8	99.3	120.8	111.1
07	**THERMAL & MOISTURE PROTECTION**	94.7	125.5	106.7	94.6	125.8	106.7	95.0	120.7	105.0	94.8	125.5	106.7	95.3	125.8	107.1	90.6	120.7	102.3
08	**OPENINGS**	96.9	119.8	102.6	110.9	119.9	113.1	110.9	119.9	113.1	98.9	119.8	104.1	113.1	119.9	114.8	105.4	117.1	108.3
0920	Plaster & Gypsum Board	105.6	120.8	115.5	108.7	120.8	116.6	105.5	120.8	115.5	102.2	120.8	114.3	104.1	120.8	115.0	108.7	120.8	116.6
0950, 0980	Ceilings & Acoustic Treatment	88.9	120.8	108.7	96.6	120.8	111.6	99.1	120.8	112.5	88.9	120.8	108.7	88.9	120.8	108.7	96.6	120.8	111.6
0960	Flooring	98.3	151.8	112.8	97.0	151.8	111.8	96.8	151.8	111.7	95.7	136.7	106.8	97.1	151.8	111.9	97.2	151.8	111.9
0970, 0990	Wall Finishes & Painting/Coating	91.7	126.9	112.8	91.6	126.9	112.8	91.6	126.9	112.8	91.7	124.2	111.2	91.6	126.9	112.8	91.7	124.2	111.2
09	**FINISHES**	98.9	127.5	114.0	99.6	127.5	114.4	99.6	127.5	114.4	97.6	124.6	111.9	97.5	127.5	113.4	99.3	127.2	114.0
COVERS	**DIVS. 10 - 14, 25, 28, 41, 43, 44**	100.0	117.8	103.5	100.0	118.0	103.6	100.0	118.0	103.6	100.0	108.4	101.7	100.0	118.0	103.6	100.0	110.1	102.0
21, 22, 23	**FIRE SUPPRESSION, PLUMBING & HVAC**	99.7	122.0	108.6	100.2	120.6	108.3	100.2	122.8	109.2	99.7	121.9	108.6	99.8	120.6	108.1	100.3	121.8	108.8
26, 27, 3370	**ELECTRICAL, COMMUNICATIONS & UTIL.**	93.5	137.4	114.9	99.0	141.1	119.5	98.8	137.5	117.6	92.8	137.4	114.6	95.0	137.5	115.6	101.4	139.2	119.8
MF2004	**WEIGHTED AVERAGE**	98.1	121.4	107.9	100.5	122.0	109.5	99.3	121.8	108.8	97.7	120.8	107.4	97.6	121.5	107.6	97.6	120.6	107.3

	DIVISION	NEW JERSEY			NEW MEXICO														
		VINELAND			ALBUQUERQUE			CARRIZOZO			CLOVIS			FARMINGTON			GALLUP		
		080,083			870 - 872			883			881			874			873		
		MAT.	INST.	TOTAL	MAT.	INST.	TOTAL	MAT.	INST.	TOTAL	MAT.	INST.	TOTAL	MAT.	INST.	TOTAL	MAT.	INST.	TOTAL
015433	**CONTRACTOR EQUIPMENT**		98.5	98.5		112.1	112.1		112.1	112.1		112.1	112.1		112.1	112.1		112.1	112.1
0241, 31 - 34	**SITE & INFRASTRUCTURE, DEMOLITION**	91.4	103.7	99.8	78.2	108.0	98.6	98.7	108.0	105.1	87.0	108.0	101.3	84.5	108.0	100.6	92.1	108.0	103.0
0310	Concrete Forming & Accessories	92.0	120.5	116.6	99.8	69.1	73.3	99.8	69.1	73.3	99.8	68.7	72.9	99.8	69.1	73.3	99.8	69.1	73.3
0320	Concrete Reinforcing	76.5	110.6	90.4	102.8	70.9	89.8	110.2	70.9	94.2	111.4	50.9	86.7	112.3	70.9	95.4	107.5	70.9	92.5
0330	Cast-in-Place Concrete	85.3	125.2	100.5	108.9	77.6	96.9	101.3	77.6	92.3	101.2	77.4	92.1	109.9	77.6	97.6	103.4	77.6	93.5
03	**CONCRETE**	99.5	119.2	108.5	109.5	73.5	93.1	124.7	73.5	101.4	113.2	69.5	93.3	113.8	73.5	95.5	119.7	73.5	98.7
04	**MASONRY**	87.7	120.3	107.2	113.9	65.3	84.8	111.2	65.3	83.7	111.3	65.3	83.7	120.3	65.3	87.3	111.3	65.3	83.8
05	**METALS**	88.9	101.1	92.4	101.4	89.8	98.0	98.3	89.8	95.8	98.0	80.6	92.9	99.1	89.8	96.4	98.4	89.8	95.9
06	**WOOD, PLASTICS & COMPOSITES**	97.1	120.8	110.1	93.4	69.5	80.3	93.5	69.5	80.3	93.5	69.5	80.3	93.5	69.5	80.3	93.5	69.5	80.3
07	**THERMAL & MOISTURE PROTECTION**	94.3	120.8	104.6	100.3	74.8	90.4	102.1	74.8	91.5	100.8	74.8	90.7	100.6	74.8	90.6	101.7	74.8	91.2
08	**OPENINGS**	98.4	116.5	102.9	93.4	71.8	88.1	92.2	71.8	87.2	92.4	65.6	85.8	96.2	71.8	90.1	96.2	71.8	90.2
0920	Plaster & Gypsum Board	101.1	120.8	113.9	83.2	68.0	73.3	76.1	68.0	70.8	76.1	68.0	70.8	76.1	68.0	70.8	76.1	68.0	70.8
0950, 0980	Ceilings & Acoustic Treatment	88.9	120.8	108.7	112.4	68.0	84.9	107.8	68.0	83.2	107.8	68.0	83.2	107.8	68.0	83.2	107.8	68.0	83.2
0960	Flooring	94.8	136.7	106.1	102.1	69.0	93.2	104.5	69.0	94.9	104.5	69.0	94.9	104.5	69.0	94.9	104.5	69.0	94.9
0970, 0990	Wall Finishes & Painting/Coating	91.7	124.2	111.2	102.6	54.9	74.0	98.1	54.9	72.2	98.1	54.9	72.2	98.1	54.9	72.2	98.1	54.9	72.2
09	**FINISHES**	96.4	124.6	111.3	101.7	67.5	83.6	101.9	67.5	83.7	100.7	67.5	83.1	100.4	67.5	83.0	101.4	67.5	83.5
COVERS	**DIVS. 10 - 14, 25, 28, 41, 43, 44**	100.0	110.2	102.0	100.0	75.1	95.1	100.0	75.1	95.1	100.0	75.1	95.1	100.0	75.1	95.1	100.0	75.1	95.1
21, 22, 23	**FIRE SUPPRESSION, PLUMBING & HVAC**	99.7	113.9	105.4	100.0	72.3	89.0	96.6	72.3	86.9	96.6	71.7	86.7	99.9	72.3	88.9	96.4	72.3	86.8
26, 27, 3370	**ELECTRICAL, COMMUNICATIONS & UTIL.**	92.8	134.0	112.9	88.6	73.8	81.4	88.6	73.8	81.4	86.1	73.8	80.1	86.8	73.8	80.5	86.0	73.8	80.1
MF2004	**WEIGHTED AVERAGE**	95.7	117.6	104.9	99.8	76.0	89.8	100.7	76.0	90.3	98.5	74.2	88.3	100.3	76.0	90.1	99.9	76.0	89.9

	DIVISION	NEW MEXICO																	
		LAS CRUCES			LAS VEGAS			ROSWELL			SANTA FE			SOCORRO			TRUTH/CONSEQUENCES		
		880			877			882			875			878			879		
		MAT.	INST.	TOTAL	MAT.	INST.	TOTAL	MAT.	INST.	TOTAL	MAT.	INST.	TOTAL	MAT.	INST.	TOTAL	MAT.	INST.	TOTAL
015433	**CONTRACTOR EQUIPMENT**		90.9	90.9		112.1	112.1		112.1	112.1		112.1	112.1		112.1	112.1		90.9	90.9
0241, 31 - 34	**SITE & INFRASTRUCTURE, DEMOLITION**	89.2	91.1	90.5	84.1	108.0	100.4	89.0	108.0	102.0	84.0	108.0	100.4	80.5	108.0	99.3	100.6	91.1	94.1
0310	Concrete Forming & Accessories	97.0	67.4	71.4	99.8	69.1	73.3	99.8	68.8	73.1	98.9	69.1	73.2	99.8	69.1	73.3	96.9	67.5	71.5
0320	Concrete Reinforcing	106.4	50.6	83.7	109.2	70.8	93.6	111.4	50.9	86.7	110.2	70.9	94.2	111.4	70.9	94.9	105.1	50.7	82.9
0330	Cast-in-Place Concrete	95.6	69.2	85.5	106.9	77.6	95.7	101.3	77.5	92.2	125.5	77.6	107.2	104.7	77.6	94.4	114.9	69.3	97.5
03	**CONCRETE**	91.2	65.7	79.6	110.8	73.5	93.8	114.0	69.6	93.8	118.7	73.5	98.1	109.6	73.5	93.2	103.3	65.8	86.2
04	**MASONRY**	107.2	62.7	80.6	111.7	63.1	82.6	121.7	65.3	87.9	116.6	65.3	85.9	111.6	65.3	83.9	108.7	62.7	81.2
05	**METALS**	99.0	73.7	91.6	98.1	89.6	95.6	99.1	80.8	93.7	99.1	89.8	96.4	98.4	89.8	95.9	97.9	73.8	90.9
06	**WOOD, PLASTICS & COMPOSITES**	84.5	68.3	75.6	93.5	69.5	80.3	93.5	69.5	80.3	92.8	69.5	80.0	93.5	69.5	80.3	84.5	68.3	75.6
07	**THERMAL & MOISTURE PROTECTION**	86.5	69.1	79.7	100.3	74.2	90.1	100.9	74.8	90.7	100.2	74.8	90.3	100.2	74.8	90.3	87.1	69.1	80.1
08	**OPENINGS**	86.0	64.9	80.8	92.4	71.8	87.3	92.2	65.6	85.6	92.4	71.8	87.3	92.2	71.8	87.2	85.9	64.9	80.7
0920	Plaster & Gypsum Board	78.3	68.0	71.6	76.1	68.0	70.8	76.1	68.0	70.8	84.5	68.0	73.8	76.1	68.0	70.8	78.3	68.0	71.6
0950, 0980	Ceilings & Acoustic Treatment	101.3	68.0	80.7	107.8	68.0	83.2	107.8	68.0	83.2	107.8	68.0	83.2	107.8	68.0	83.2	101.3	68.0	80.7
0960	Flooring	138.2	65.2	118.4	104.5	69.0	94.9	104.5	69.0	94.9	104.5	69.0	94.9	104.5	69.0	94.9	138.2	65.2	118.4
0970, 0990	Wall Finishes & Painting/Coating	90.8	54.9	69.3	98.1	54.9	72.2	98.1	54.9	72.2	98.1	54.9	72.2	98.1	54.9	72.2	90.8	54.9	69.3
09	**FINISHES**	115.2	65.8	89.0	100.3	67.5	82.9	100.8	67.5	83.2	101.6	67.5	83.6	100.2	67.5	82.9	115.7	65.8	89.3
COVERS	**DIVS. 10 - 14, 25, 28, 41, 43, 44**	100.0	71.8	94.4	100.0	75.1	95.1	100.0	75.1	95.1	100.0	75.1	95.1	100.0	75.1	95.1	100.0	71.9	94.4
21, 22, 23	**FIRE SUPPRESSION, PLUMBING & HVAC**	100.2	71.7	88.9	96.4	72.3	86.8	99.9	72.0	88.8	99.9	72.3	88.9	96.4	72.3	86.8	96.3	71.7	86.5
26, 27, 3370	**ELECTRICAL, COMMUNICATIONS & UTIL.**	88.0	57.1	72.9	88.6	73.8	81.4	87.5	73.8	80.9	101.6	73.8	88.1	86.5	73.8	80.3	90.9	73.8	82.6
MF2004	**WEIGHTED AVERAGE**	97.0	68.5	85.0	98.4	75.8	88.9	100.3	74.3	89.3	102.0	76.0	91.1	97.9	76.0	88.7	98.1	70.9	86.7

DIVISION		NEW MEXICO			NEW YORK														
		TUCUMCARI			ALBANY			BINGHAMTON			BRONX			BROOKLYN			BUFFALO		
		884			120 - 122			137 - 139			104			112			140 - 142		
		MAT.	INST.	TOTAL	MAT.	INST.	TOTAL	MAT.	INST.	TOTAL	MAT.	INST.	TOTAL	MAT.	INST.	TOTAL	MAT.	INST.	TOTAL
015433	CONTRACTOR EQUIPMENT		112.1	112.1		112.9	112.9		117.7	117.7		109.7	109.7		109.6	109.6		98.1	98.1
0241, 31 - 34	SITE & INFRASTRUCTURE, DEMOLITION	86.7	108.0	101.2	73.4	105.3	95.2	95.6	96.0	95.9	121.0	119.8	120.1	122.4	122.2	122.2	95.5	99.3	98.1
0310	Concrete Forming & Accessories	99.8	68.7	72.9	96.1	87.4	88.6	101.0	79.0	82.0	102.5	181.2	170.5	107.3	180.6	170.6	97.1	113.7	111.4
0320	Concrete Reinforcing	109.2	50.9	85.5	99.1	94.3	97.1	98.1	94.4	96.6	100.8	184.6	135.0	99.4	184.5	134.1	98.4	102.3	100.0
0330	Cast-in-Place Concrete	101.2	77.4	92.1	83.4	101.9	90.5	102.9	90.2	98.1	96.6	160.0	120.8	104.8	158.2	125.2	110.7	116.6	112.9
03	CONCRETE	112.3	69.5	92.9	99.0	94.8	97.1	99.7	87.7	94.2	101.3	172.1	133.5	111.6	171.2	138.7	107.4	111.8	109.4
04	MASONRY	125.2	65.3	89.3	91.8	99.1	96.2	107.0	81.0	91.5	96.6	162.0	135.8	116.5	162.0	143.7	109.3	118.4	114.8
05	METALS	98.0	80.6	92.9	94.7	105.7	97.9	91.3	114.7	98.1	96.8	140.9	109.6	99.8	140.6	111.7	94.1	95.5	94.5
06	WOOD, PLASTICS & COMPOSITES	93.5	69.5	80.3	95.1	84.5	89.3	105.3	79.7	91.3	105.9	188.2	151.1	107.3	187.9	151.5	98.4	113.4	106.6
07	THERMAL & MOISTURE PROTECTION	100.8	74.8	90.6	89.5	91.8	90.4	103.1	80.5	94.3	109.0	163.5	130.2	106.9	159.7	127.4	99.1	109.0	102.9
08	OPENINGS	92.1	65.6	85.6	96.0	82.8	92.7	91.0	78.8	88.0	92.4	176.5	113.2	90.4	176.3	111.7	95.2	102.2	96.9
0920	Plaster & Gypsum Board	76.1	68.0	70.8	107.9	83.7	92.2	119.9	78.5	93.0	120.0	191.1	166.2	114.3	191.1	164.2	99.0	113.6	108.5
0950, 0980	Ceilings & Acoustic Treatment	107.8	68.0	83.2	92.4	83.7	87.0	96.6	78.5	85.4	85.8	191.1	150.9	80.0	191.1	148.7	98.5	113.6	107.8
0960	Flooring	104.5	69.0	94.9	83.8	86.7	84.6	94.4	81.9	91.0	91.8	169.6	112.9	100.9	169.6	119.5	89.1	117.2	96.7
0970, 0990	Wall Finishes & Painting/Coating	98.1	54.9	72.2	82.4	79.3	80.5	86.7	85.2	85.8	90.3	154.5	128.8	111.0	154.5	137.1	86.1	112.6	102.0
09	FINISHES	100.7	67.5	83.1	94.3	85.8	89.8	97.3	79.8	88.1	101.5	177.5	141.6	108.3	177.2	144.8	92.5	115.0	104.4
COVERS	DIVS. 10 - 14, 25, 28, 41, 43, 44	100.0	75.1	95.1	100.0	93.2	98.7	100.0	93.3	98.7	100.0	136.2	107.2	100.0	135.3	107.0	100.0	107.5	101.5
21, 22, 23	FIRE SUPPRESSION, PLUMBING & HVAC	96.6	71.7	86.7	100.0	93.0	97.3	100.4	79.0	91.9	100.2	159.6	123.8	99.8	159.5	123.5	99.9	97.1	98.8
26, 27, 3370	ELECTRICAL, COMMUNICATIONS & UTIL.	88.6	73.8	81.4	106.9	93.9	100.6	102.2	86.8	94.7	101.6	179.2	139.4	103.0	179.2	140.1	99.8	98.5	99.1
MF2004	WEIGHTED AVERAGE	99.3	74.2	88.8	97.4	94.7	96.3	98.1	86.9	93.4	100.0	162.0	126.0	103.1	161.9	127.8	99.0	104.8	101.5

DIVISION		NEW YORK																	
		ELMIRA			FAR ROCKAWAY			FLUSHING			GLENS FALLS			HICKSVILLE			JAMAICA		
		148 - 149			116			113			128			115,117,118			114		
		MAT.	INST.	TOTAL	MAT.	INST.	TOTAL	MAT.	INST.	TOTAL	MAT.	INST.	TOTAL	MAT.	INST.	TOTAL	MAT.	INST.	TOTAL
015433	CONTRACTOR EQUIPMENT		118.7	118.7		109.6	109.6		109.6	109.6		112.9	112.9		109.6	109.6		109.6	109.6
0241, 31 - 34	SITE & INFRASTRUCTURE, DEMOLITION	91.2	94.8	93.7	126.8	122.2	123.6	126.8	122.2	123.6	63.6	105.1	92.0	114.6	121.4	119.3	120.2	122.2	121.6
0310	Concrete Forming & Accessories	82.4	69.7	71.4	93.7	180.6	168.8	97.5	180.6	169.3	84.6	82.5	82.8	90.2	144.0	136.7	97.5	180.6	169.3
0320	Concrete Reinforcing	99.5	88.7	95.1	99.4	184.5	134.1	101.1	184.5	135.1	98.8	94.2	96.9	99.4	184.3	134.0	99.4	184.5	134.1
0330	Cast-in-Place Concrete	103.4	89.6	98.1	113.4	158.2	130.5	113.4	158.2	130.5	78.2	100.8	86.8	96.3	156.1	119.1	104.8	158.2	125.2
03	CONCRETE	101.4	82.6	92.9	117.9	171.2	142.2	118.5	171.2	142.5	90.4	92.2	91.2	103.5	154.3	126.6	111.0	171.2	138.4
04	MASONRY	105.5	81.2	91.0	120.6	162.0	145.4	114.3	162.0	142.8	92.9	97.2	95.5	110.9	159.6	140.0	118.7	162.0	144.6
05	METALS	89.7	115.5	97.2	99.8	140.6	111.7	99.8	140.6	111.7	91.2	105.2	95.3	101.1	138.4	112.0	99.8	140.6	111.7
06	WOOD, PLASTICS & COMPOSITES	82.7	68.2	74.7	91.2	187.9	144.3	95.8	187.9	146.3	82.2	79.5	80.7	87.5	140.7	116.7	95.8	187.9	146.3
07	THERMAL & MOISTURE PROTECTION	99.4	76.6	90.6	106.8	159.6	127.4	106.8	159.6	127.4	89.0	90.5	89.6	106.4	150.4	123.5	106.7	159.6	127.3
08	OPENINGS	96.2	71.2	90.0	88.8	176.3	110.4	88.8	176.3	110.4	89.5	80.0	87.2	88.8	150.6	104.1	88.8	176.3	110.4
0920	Plaster & Gypsum Board	94.0	66.7	76.3	108.0	191.1	162.0	109.9	191.1	162.6	105.6	78.4	88.0	107.3	142.2	129.9	109.9	191.1	162.6
0950, 0980	Ceilings & Acoustic Treatment	94.4	66.7	77.3	80.0	191.1	148.7	80.0	191.1	148.7	90.0	78.4	82.9	79.1	142.2	118.1	80.0	191.1	148.7
0960	Flooring	79.5	79.5	79.5	95.5	169.6	115.5	97.1	169.6	116.7	75.5	86.7	78.5	94.2	160.0	112.0	97.1	169.6	116.7
0970, 0990	Wall Finishes & Painting/Coating	84.6	83.5	84.0	111.0	154.5	137.1	111.0	154.5	137.1	82.4	79.3	80.5	111.0	154.5	137.1	111.0	154.5	137.1
09	FINISHES	87.7	72.1	79.5	106.3	177.2	143.8	107.0	177.2	144.1	90.1	82.3	86.0	104.7	147.1	127.1	106.6	177.2	143.9
COVERS	DIVS. 10 - 14, 25, 28, 41, 43, 44	100.0	80.2	96.1	100.0	135.3	107.0	100.0	135.3	107.0	100.0	61.0	92.3	100.0	128.9	105.7	100.0	135.3	107.0
21, 22, 23	FIRE SUPPRESSION, PLUMBING & HVAC	94.2	78.4	88.0	94.0	159.5	120.0	94.0	159.5	120.0	94.5	85.5	90.9	99.8	146.9	118.5	94.0	159.5	120.0
26, 27, 3370	ELECTRICAL, COMMUNICATIONS & UTIL.	98.6	88.2	93.5	110.1	179.2	143.7	110.1	179.2	143.7	97.7	93.9	95.8	102.2	149.3	125.1	101.2	179.2	139.2
MF2004	WEIGHTED AVERAGE	95.5	84.4	90.8	103.1	161.9	127.8	103.0	161.9	127.7	92.2	91.0	91.7	101.2	146.4	120.1	101.2	161.9	126.7

DIVISION		NEW YORK																	
		JAMESTOWN			KINGSTON			LONG ISLAND CITY			MONTICELLO			MOUNT VERNON			NEW ROCHELLE		
		147			124			111			127			105			108		
		MAT.	INST.	TOTAL	MAT.	INST.	TOTAL	MAT.	INST.	TOTAL	MAT.	INST.	TOTAL	MAT.	INST.	TOTAL	MAT.	INST.	TOTAL
015433	CONTRACTOR EQUIPMENT		95.3	95.3		109.6	109.6		109.6	109.6		109.6	109.6		109.7	109.7		109.7	109.7
0241, 31 - 34	SITE & INFRASTRUCTURE, DEMOLITION	92.5	95.9	94.8	122.1	118.2	119.5	123.8	122.2	122.7	117.1	118.3	117.9	130.4	114.7	119.6	128.9	114.7	119.2
0310	Concrete Forming & Accessories	82.4	78.6	79.1	86.2	104.0	101.6	101.8	180.6	169.9	93.2	104.2	102.7	92.7	128.8	123.9	107.6	128.8	125.9
0320	Concrete Reinforcing	99.7	80.3	91.8	99.1	123.3	109.0	99.4	184.5	134.1	98.4	123.3	108.5	99.7	183.6	133.9	99.8	183.6	134.0
0330	Cast-in-Place Concrete	107.3	100.7	104.8	105.6	132.6	115.9	108.2	158.2	127.3	98.9	132.7	111.8	108.0	133.8	117.9	108.0	133.8	117.9
03	CONCRETE	104.5	86.8	96.5	112.1	117.3	114.5	114.2	171.2	140.2	106.7	117.4	111.6	111.0	139.6	124.0	110.4	139.6	123.7
04	MASONRY	114.6	87.6	98.4	108.1	137.6	125.8	113.4	162.0	142.5	101.1	137.6	123.0	102.4	133.1	120.8	102.5	133.1	120.8
05	METALS	87.2	83.6	86.2	99.8	110.1	102.8	99.8	140.6	111.7	99.8	110.3	102.8	96.6	132.6	107.1	96.8	132.6	107.3
06	WOOD, PLASTICS & COMPOSITES	81.4	73.4	77.0	83.4	96.6	90.6	101.3	187.9	148.8	91.6	96.6	94.3	95.2	129.2	113.9	113.1	129.2	121.9
07	THERMAL & MOISTURE PROTECTION	99.0	87.2	94.4	108.5	135.7	119.1	106.8	159.6	127.4	108.2	135.7	118.9	109.9	141.0	122.0	110.0	141.0	122.1
08	OPENINGS	96.1	74.7	90.8	94.1	111.6	98.4	88.7	176.3	110.4	88.8	111.6	94.4	92.5	142.9	104.9	92.5	142.9	105.0
0920	Plaster & Gypsum Board	89.6	72.1	78.3	104.1	96.4	99.1	112.8	191.1	163.7	106.7	96.4	100.0	115.2	129.9	124.7	125.6	129.9	128.4
0950, 0980	Ceilings & Acoustic Treatment	91.0	72.1	79.3	75.0	96.4	88.2	80.0	191.1	148.7	75.0	96.4	88.2	83.3	129.9	112.1	83.3	129.9	112.1
0960	Flooring	82.1	82.1	82.1	88.3	81.3	86.4	98.7	169.6	117.8	91.1	81.3	88.5	83.7	164.6	105.6	91.7	164.6	111.4
0970, 0990	Wall Finishes & Painting/Coating	86.1	81.7	83.5	109.0	112.1	110.9	111.0	154.5	137.1	109.0	102.7	105.2	88.2	154.5	128.0	88.2	154.5	128.0
09	FINISHES	87.2	78.7	82.7	102.0	98.4	100.1	107.7	177.2	144.4	102.8	97.3	99.9	98.6	137.7	119.3	102.1	137.7	120.9
COVERS	DIVS. 10 - 14, 25, 28, 41, 43, 44	100.0	88.6	97.7	100.0	115.0	103.0	100.0	135.3	107.0	100.0	111.3	102.2	100.0	122.9	104.5	100.0	122.9	104.5
21, 22, 23	FIRE SUPPRESSION, PLUMBING & HVAC	94.1	70.9	84.8	94.4	96.9	95.4	99.8	159.5	123.5	94.4	113.3	101.9	94.5	127.3	107.6	94.5	127.3	107.6
26, 27, 3370	ELECTRICAL, COMMUNICATIONS & UTIL.	97.2	87.9	92.7	99.5	120.4	109.7	101.7	179.2	139.4	99.5	120.4	109.7	99.0	137.5	117.8	99.0	137.5	117.8
MF2004	WEIGHTED AVERAGE	95.6	82.5	90.1	100.8	112.5	105.7	102.9	161.9	127.6	99.3	115.7	106.2	99.8	132.8	113.6	100.1	132.8	113.8

	DIVISION	NEW YORK																	
		NEW YORK			NIAGARA FALLS			PLATTSBURGH			POUGHKEEPSIE			QUEENS			RIVERHEAD		
		100 - 102			143			129			125 - 126			110			119		
		MAT.	INST.	TOTAL	MAT.	INST.	TOTAL	MAT.	INST.	TOTAL	MAT.	INST.	TOTAL	MAT.	INST.	TOTAL	MAT.	INST.	TOTAL
015433	CONTRACTOR EQUIPMENT		110.2	110.2		95.3	95.3		98.5	98.5		109.6	109.6		109.6	109.6		109.6	109.6
0241, 31 - 34	SITE & INFRASTRUCTURE, DEMOLITION	130.7	121.2	124.2	94.8	97.1	96.4	86.7	102.8	97.7	118.1	118.3	118.2	118.3	122.2	120.9	115.6	121.4	119.6
0310	Concrete Forming & Accessories	107.0	184.8	174.2	82.4	113.2	109.0	88.9	77.9	79.4	86.2	166.9	155.9	90.4	180.6	168.3	94.9	144.0	137.3
0320	Concrete Reinforcing	106.7	190.7	140.9	98.4	96.5	97.6	103.1	93.8	99.3	99.1	123.5	109.1	101.1	184.5	135.1	101.2	184.3	135.2
0330	Cast-in-Place Concrete	108.8	168.8	131.7	111.0	108.1	109.9	95.6	96.8	96.1	102.3	132.1	113.7	99.6	158.2	122.0	97.9	156.1	120.1
03	CONCRETE	112.0	177.7	141.9	106.8	107.5	107.1	103.8	87.4	96.3	109.1	144.8	125.4	106.9	171.2	136.1	104.3	154.3	127.1
04	MASONRY	107.8	171.5	145.9	122.7	116.2	118.8	89.0	91.1	90.2	101.3	135.4	121.7	107.7	162.0	140.2	116.4	159.6	142.3
05	METALS	108.4	140.9	117.9	89.7	91.1	90.1	95.7	86.0	92.8	99.8	112.0	103.4	99.8	140.6	111.7	101.5	138.4	112.3
06	WOOD, PLASTICS & COMPOSITES	110.5	188.2	153.2	81.3	112.1	98.2	89.3	73.6	80.7	83.4	180.9	136.9	87.7	187.9	142.7	92.9	140.7	119.2
07	THERMAL & MOISTURE PROTECTION	109.3	167.8	132.1	99.1	105.7	101.7	101.3	85.1	95.0	108.5	143.6	122.1	106.5	159.6	127.2	107.0	150.4	123.9
08	OPENINGS	98.0	177.2	117.5	96.1	99.2	96.8	99.0	76.2	93.4	94.1	157.5	109.8	88.7	176.3	110.4	88.8	150.6	104.1
0920	Plaster & Gypsum Board	128.3	191.1	169.1	89.6	112.3	104.3	123.7	71.8	90.0	104.1	183.8	155.9	107.3	191.1	161.7	109.4	142.2	130.7
0950, 0980	Ceilings & Acoustic Treatment	109.1	191.1	159.8	91.0	112.3	104.2	100.3	71.8	82.7	75.0	183.8	142.3	80.0	191.1	148.7	81.6	142.2	119.1
0960	Flooring	93.3	169.6	113.9	82.0	108.1	89.1	97.3	93.6	96.3	88.3	145.4	103.7	94.2	169.6	114.6	95.5	160.0	112.9
0970, 0990	Wall Finishes & Painting/Coating	90.3	156.7	130.2	86.1	104.0	96.9	103.6	76.7	87.5	109.0	102.7	105.2	111.0	154.5	137.1	111.0	154.5	137.1
09	FINISHES	109.2	180.1	146.7	87.3	112.3	100.5	99.3	80.1	89.1	101.8	159.5	132.3	105.1	177.2	143.2	105.9	147.1	127.7
COVERS	DIVS. 10 - 14, 25, 28, 41, 43, 44	100.0	139.6	107.8	100.0	97.6	99.5	100.0	56.1	91.3	100.0	120.5	104.1	100.0	135.3	107.0	100.0	128.9	105.7
21, 22, 23	FIRE SUPPRESSION, PLUMBING & HVAC	100.2	164.6	125.8	94.1	96.2	94.9	94.4	85.3	90.8	94.4	97.0	95.5	99.8	159.5	123.5	99.9	146.9	118.6
26, 27, 3370	ELECTRICAL, COMMUNICATIONS & UTIL.	109.4	173.1	140.4	95.9	92.3	94.2	93.8	87.7	90.8	99.5	120.5	109.7	102.2	179.2	139.7	103.8	149.3	126.0
MF2004	WEIGHTED AVERAGE	106.1	164.6	130.7	96.7	101.5	98.7	96.7	85.9	92.2	100.0	127.3	111.5	101.3	161.9	126.7	102.0	146.4	120.6

	DIVISION	NEW YORK																	
		ROCHESTER			SCHENECTADY			STATEN ISLAND			SUFFERN			SYRACUSE			UTICA		
		144 - 146			123			103			109			130 - 132			133 - 135		
		MAT.	INST.	TOTAL	MAT.	INST.	TOTAL	MAT.	INST.	TOTAL	MAT.	INST.	TOTAL	MAT.	INST.	TOTAL	MAT.	INST.	TOTAL
015433	CONTRACTOR EQUIPMENT		114.6	114.6		112.9	112.9		109.7	109.7		109.7	109.7		112.9	112.9		112.9	112.9
0241, 31 - 34	SITE & INFRASTRUCTURE, DEMOLITION	72.0	105.3	94.8	73.4	105.4	95.3	135.4	119.8	124.7	125.3	115.3	118.4	94.4	104.9	101.6	71.6	102.8	92.9
0310	Concrete Forming & Accessories	98.0	92.4	93.2	100.7	88.1	89.8	92.2	181.2	169.1	101.4	124.3	121.2	99.9	88.9	90.4	101.2	85.6	87.8
0320	Concrete Reinforcing	99.2	89.2	95.1	97.7	94.3	96.3	100.8	184.6	135.0	99.8	123.3	109.4	99.1	94.7	97.3	99.1	91.4	96.0
0330	Cast-in-Place Concrete	99.5	99.1	99.4	94.5	102.9	97.7	108.0	160.0	127.9	104.7	125.3	112.6	95.5	96.9	96.0	87.4	90.1	88.4
03	CONCRETE	111.1	95.2	103.9	104.1	95.3	100.1	112.7	172.1	139.7	107.4	123.7	114.8	103.0	93.7	98.8	101.1	89.3	95.7
04	MASONRY	105.3	95.8	99.6	91.2	100.7	96.9	110.8	162.0	141.4	102.5	122.5	114.5	98.0	96.0	96.8	89.8	85.7	87.4
05	METALS	90.8	105.9	95.2	94.8	105.7	98.0	94.6	140.9	108.1	94.6	111.3	99.5	94.6	104.8	97.6	92.9	103.5	96.0
06	WOOD, PLASTICS & COMPOSITES	95.4	92.8	94.0	101.7	84.5	92.3	93.7	188.2	145.6	105.5	128.0	117.9	101.7	87.2	93.7	101.7	87.7	94.0
07	THERMAL & MOISTURE PROTECTION	94.9	95.6	95.2	90.1	92.6	91.1	109.3	163.5	130.4	109.9	133.0	118.9	100.2	94.7	98.0	89.9	90.2	90.0
08	OPENINGS	98.8	87.9	96.1	96.0	82.8	92.7	92.4	176.5	113.2	92.5	128.7	101.5	94.0	82.1	91.0	96.0	83.9	93.0
0920	Plaster & Gypsum Board	98.2	92.5	94.5	115.0	83.7	94.7	115.5	191.1	164.6	118.9	128.7	125.2	115.0	86.4	96.4	115.0	87.0	96.8
0950, 0980	Ceilings & Acoustic Treatment	97.8	92.5	94.5	96.6	83.7	88.6	85.8	191.1	150.9	83.3	128.7	111.4	96.6	86.4	90.3	96.6	87.0	90.7
0960	Flooring	79.3	101.1	85.2	83.4	86.7	84.3	87.4	169.6	109.6	87.4	160.0	107.1	85.1	88.4	86.0	83.4	93.3	86.1
0970, 0990	Wall Finishes & Painting/Coating	84.7	96.0	91.5	82.4	79.3	80.5	90.3	154.5	128.8	88.2	110.0	101.3	88.3	91.0	89.9	82.4	80.1	81.0
09	FINISHES	92.1	94.6	93.5	95.9	86.2	90.8	100.7	177.5	141.3	99.7	126.7	114.0	97.5	88.7	92.9	96.0	86.6	91.0
COVERS	DIVS. 10 - 14, 25, 28, 41, 43, 44	100.0	88.4	97.7	100.0	93.9	98.8	100.0	136.2	107.2	100.0	121.0	104.2	100.0	97.5	99.5	100.0	90.0	98.0
21, 22, 23	FIRE SUPPRESSION, PLUMBING & HVAC	99.9	89.6	95.8	100.2	93.9	97.7	100.2	159.6	123.8	94.5	116.7	103.3	100.2	86.9	94.9	100.2	81.7	92.9
26, 27, 3370	ELECTRICAL, COMMUNICATIONS & UTIL.	101.0	86.2	93.8	102.2	93.9	98.2	101.6	179.2	139.4	106.7	128.9	117.5	102.2	91.0	96.8	99.7	90.4	95.2
MF2004	WEIGHTED AVERAGE	98.4	94.1	96.6	97.7	95.3	96.7	101.9	162.0	127.1	99.9	121.8	109.1	98.7	93.1	96.4	96.7	89.5	93.6

	DIVISION	NEW YORK									NORTH CAROLINA								
		WATERTOWN			WHITE PLAINS			YONKERS			ASHEVILLE			CHARLOTTE			DURHAM		
		136			106			107			287 - 288			281 - 282			277		
		MAT.	INST.	TOTAL	MAT.	INST.	TOTAL	MAT.	INST.	TOTAL	MAT.	INST.	TOTAL	MAT.	INST.	TOTAL	MAT.	INST.	TOTAL
015433	CONTRACTOR EQUIPMENT		112.9	112.9		109.7	109.7		109.7	109.7		98.0	98.0		98.0	98.0		102.1	102.1
0241, 31 - 34	SITE & INFRASTRUCTURE, DEMOLITION	80.1	105.4	97.4	120.5	114.7	116.5	130.0	114.5	119.4	106.5	78.0	87.0	110.4	78.0	88.3	106.6	85.3	92.0
0310	Concrete Forming & Accessories	86.2	92.4	91.6	106.4	128.8	125.7	106.7	128.8	125.8	96.6	41.4	48.9	101.2	43.4	51.3	99.0	41.9	49.7
0320	Concrete Reinforcing	99.7	94.6	97.6	99.8	183.6	134.0	103.8	183.6	136.3	101.1	45.2	78.3	101.6	42.2	77.4	102.7	52.8	82.3
0330	Cast-in-Place Concrete	101.8	101.0	101.5	95.9	133.8	110.4	107.3	133.8	117.4	103.2	49.5	82.7	116.0	51.6	91.4	104.9	49.7	83.8
03	CONCRETE	113.8	96.6	106.0	100.8	139.6	118.4	110.8	139.6	123.9	106.2	47.1	79.3	111.9	48.1	82.9	106.8	48.8	80.4
04	MASONRY	91.0	98.5	95.5	101.8	133.1	120.6	107.3	133.1	122.7	85.3	41.9	59.3	92.1	45.9	64.4	90.9	37.7	59.0
05	METALS	92.9	104.8	96.4	95.9	132.6	106.6	104.7	132.6	112.8	88.3	77.6	85.2	92.5	76.7	87.9	102.1	80.7	95.8
06	WOOD, PLASTICS & COMPOSITES	84.0	91.2	87.9	111.6	129.2	121.3	111.4	129.2	121.2	96.4	40.8	65.9	101.7	43.0	69.5	98.9	42.3	67.8
07	THERMAL & MOISTURE PROTECTION	90.2	96.4	92.6	109.7	141.0	121.9	110.0	141.0	122.1	102.3	45.2	80.1	99.8	46.8	79.2	103.1	47.3	81.4
08	OPENINGS	96.0	88.3	94.1	92.5	142.9	105.0	96.0	143.5	107.7	94.2	40.5	80.9	98.3	42.1	84.4	98.3	44.8	85.1
0920	Plaster & Gypsum Board	107.9	90.6	96.7	121.9	129.9	127.1	127.9	129.9	129.2	106.6	38.4	62.3	108.1	40.7	64.4	110.6	40.0	64.8
0950, 0980	Ceilings & Acoustic Treatment	96.6	90.6	92.9	83.3	129.9	112.1	107.5	129.9	121.3	92.6	38.4	59.1	95.1	40.7	61.5	97.6	40.0	62.0
0960	Flooring	76.1	93.3	80.8	89.7	164.6	110.0	89.3	164.6	109.6	107.3	46.7	90.9	110.9	46.9	93.6	111.5	46.7	94.0
0970, 0990	Wall Finishes & Painting/Coating	82.4	89.0	86.4	88.2	154.5	128.0	88.2	154.5	128.0	114.4	36.8	67.8	114.4	42.0	71.0	114.4	38.6	68.9
09	FINISHES	93.5	92.3	92.8	100.3	137.7	120.1	107.4	137.7	123.4	101.8	41.6	70.0	103.5	43.7	71.9	104.8	42.1	71.6
COVERS	DIVS. 10 - 14, 25, 28, 41, 43, 44	100.0	94.9	99.0	100.0	122.9	104.5	100.0	122.9	104.5	100.0	75.0	95.1	100.0	75.6	95.2	100.0	70.2	94.1
21, 22, 23	FIRE SUPPRESSION, PLUMBING & HVAC	100.2	76.0	90.6	100.4	127.3	111.1	100.4	127.3	111.1	100.2	41.5	76.9	99.9	41.7	76.8	100.3	40.2	76.4
26, 27, 3370	ELECTRICAL, COMMUNICATIONS & UTIL.	102.2	90.9	96.7	99.0	137.5	117.8	106.8	137.5	121.7	97.7	40.2	69.7	100.3	52.3	76.9	98.6	50.9	75.4
MF2004	WEIGHTED AVERAGE	98.4	92.2	95.8	99.8	132.8	113.6	104.7	132.8	116.5	97.6	49.6	77.5	100.2	52.2	80.0	101.1	51.7	80.4

DIVISION		NORTH CAROLINA																	
		ELIZABETH CITY			FAYETTEVILLE			GASTONIA			GREENSBORO			HICKORY			KINSTON		
		279			283			280			270,272 - 274			286			285		
		MAT.	INST.	TOTAL	MAT.	INST.	TOTAL	MAT.	INST.	TOTAL	MAT.	INST.	TOTAL	MAT.	INST.	TOTAL	MAT.	INST.	TOTAL
015433	CONTRACTOR EQUIPMENT		105.8	105.8		102.1	102.1		98.0	98.0		102.1	102.1		102.1	102.1		102.1	102.1
0241, 31 - 34	SITE & INFRASTRUCTURE, DEMOLITION	112.1	85.2	93.7	105.4	85.5	91.8	106.0	78.0	86.8	106.5	85.5	92.1	105.7	85.4	91.8	104.3	85.2	91.2
0310	Concrete Forming & Accessories	85.2	36.0	42.7	96.1	42.7	49.9	103.3	42.4	50.7	98.8	42.6	50.2	92.5	40.4	47.5	88.9	34.1	41.6
0320	Concrete Reinforcing	100.6	43.9	77.5	104.9	53.0	83.7	101.6	45.2	78.6	101.6	52.7	81.6	101.1	20.5	68.3	100.6	19.9	67.7
0330	Cast-in-Place Concrete	105.1	37.6	79.3	108.2	52.9	87.1	100.9	57.2	84.2	104.1	51.1	83.8	103.1	50.7	83.1	99.6	40.9	77.1
03	CONCRETE	107.1	40.3	76.7	108.3	50.2	81.9	104.9	50.1	80.0	106.2	49.5	80.4	106.0	42.6	77.2	102.9	36.4	72.6
04	MASONRY	102.5	27.8	57.8	89.0	42.9	61.4	90.2	39.6	59.9	89.0	38.4	58.7	75.1	43.1	56.0	81.5	30.3	50.8
05	METALS	90.2	76.7	86.2	106.4	80.7	98.9	89.1	77.6	85.7	95.7	80.3	91.2	98.5	70.9	83.4	87.4	69.7	82.3
06	WOOD, PLASTICS & COMPOSITES	82.5	37.7	57.9	95.2	42.3	66.1	104.9	42.2	70.5	98.6	42.2	67.7	90.8	40.5	63.2	87.1	34.5	58.2
07	THERMAL & MOISTURE PROTECTION	102.4	33.1	75.4	102.2	45.8	80.2	102.6	45.5	80.4	102.8	44.8	80.3	102.6	44.1	79.8	102.5	33.6	75.7
08	OPENINGS	95.1	34.4	80.1	94.3	44.8	82.0	98.3	42.2	84.4	98.3	44.7	85.0	94.2	32.4	78.9	94.3	29.4	78.3
0920	Plaster & Gypsum Board	103.8	34.5	58.8	110.0	40.0	64.5	112.3	39.9	65.3	112.3	39.9	65.3	106.6	38.2	62.1	106.0	31.9	57.9
0950, 0980	Ceilings & Acoustic Treatment	97.6	34.5	58.6	93.4	40.0	60.4	97.6	39.9	61.9	97.6	39.9	61.9	92.6	38.2	58.9	97.6	31.9	57.0
0960	Flooring	100.7	15.1	77.6	107.4	46.7	91.0	111.5	46.7	94.0	111.5	43.2	93.0	107.1	36.0	87.9	103.8	20.6	81.3
0970, 0990	Wall Finishes & Painting/Coating	114.4	47.6	74.3	114.4	36.8	67.8	114.4	42.2	71.1	114.4	35.8	67.2	114.4	35.8	67.2	114.4	32.6	65.3
09	FINISHES	101.0	33.2	65.1	102.5	42.5	70.7	105.0	42.9	72.1	105.1	41.7	71.6	101.9	38.5	68.4	101.8	31.5	64.6
COVERS	DIVS. 10 - 14, 25, 28, 41, 43, 44	100.0	73.9	94.8	100.0	71.1	94.3	100.0	75.0	95.0	100.0	75.3	95.1	100.0	75.0	95.0	100.0	69.0	93.9
21, 22, 23	FIRE SUPPRESSION, PLUMBING & HVAC	94.4	32.3	69.7	100.1	41.4	76.8	100.2	41.1	76.7	100.2	41.5	76.8	94.4	34.2	70.4	94.4	27.6	67.8
26, 27, 3370	ELECTRICAL, COMMUNICATIONS & UTIL.	98.2	31.9	65.9	97.2	46.2	72.4	97.1	37.4	68.0	97.5	40.4	69.7	95.2	33.8	65.3	94.9	52.7	74.4
MF2004	WEIGHTED AVERAGE	97.7	43.0	74.7	101.1	52.1	80.5	98.5	49.6	78.0	99.8	50.7	79.2	95.5	45.9	74.7	95.2	43.3	73.4

DIVISION		NORTH CAROLINA															NORTH DAKOTA		
		MURPHY			RALEIGH			ROCKY MOUNT			WILMINGTON			WINSTON-SALEM			BISMARCK		
		289			275 - 276			278			284			271			585		
		MAT.	INST.	TOTAL	MAT.	INST.	TOTAL	MAT.	INST.	TOTAL	MAT.	INST.	TOTAL	MAT.	INST.	TOTAL	MAT.	INST.	TOTAL
015433	CONTRACTOR EQUIPMENT		98.0	98.0		102.1	102.1		102.1	102.1		98.0	98.0		102.1	102.1		98.7	98.7
0241, 31 - 34	SITE & INFRASTRUCTURE, DEMOLITION	108.2	77.7	87.3	110.6	85.5	93.5	110.0	85.0	92.9	107.6	78.0	87.3	106.8	85.4	92.2	96.1	97.3	96.9
0310	Concrete Forming & Accessories	103.8	26.4	36.9	99.3	43.5	51.1	91.5	31.1	39.3	98.0	41.9	49.5	100.0	41.7	49.6	94.7	45.5	52.2
0320	Concrete Reinforcing	100.7	19.8	67.7	101.6	53.0	81.7	100.6	29.0	71.4	101.9	51.1	81.2	101.6	41.9	77.2	111.0	69.7	94.1
0330	Cast-in-Place Concrete	106.7	37.1	80.1	118.9	56.5	95.1	102.8	34.5	76.7	102.8	50.8	82.9	106.7	48.5	84.4	101.0	53.0	82.7
03	CONCRETE	109.2	31.6	73.9	113.1	51.8	85.2	107.8	34.4	74.4	106.2	48.8	80.1	107.5	46.2	79.6	100.1	54.1	79.2
04	MASONRY	77.3	27.9	47.7	91.3	43.9	62.9	81.8	27.0	49.0	75.1	39.7	53.9	89.2	39.5	59.5	100.2	58.0	74.9
05	METALS	86.5	69.2	81.4	93.3	80.7	89.6	89.5	69.8	83.8	88.0	79.8	85.6	93.3	76.2	88.3	92.7	80.8	89.2
06	WOOD, PLASTICS & COMPOSITES	105.6	24.4	61.0	98.6	43.0	68.1	89.7	31.9	58.0	98.1	41.6	67.1	98.6	41.5	67.3	87.0	40.7	61.6
07	THERMAL & MOISTURE PROTECTION	102.5	32.0	75.1	102.0	46.1	80.2	102.8	29.0	74.1	102.3	45.2	80.1	102.8	44.4	80.1	101.6	50.6	81.8
08	OPENINGS	94.1	27.1	77.6	95.1	44.9	82.6	94.3	27.2	77.7	94.3	43.3	81.7	98.3	41.4	84.2	99.1	46.4	86.1
0920	Plaster & Gypsum Board	111.0	21.5	52.9	108.1	40.7	64.3	105.3	29.3	55.9	107.5	39.3	63.2	112.3	39.2	64.8	99.2	38.8	60.0
0950, 0980	Ceilings & Acoustic Treatment	92.6	21.5	48.6	95.1	40.7	61.5	93.4	29.3	53.8	93.4	39.3	59.9	97.6	39.2	61.5	139.6	38.8	77.3
0960	Flooring	111.9	22.0	87.6	111.5	46.7	94.0	105.4	14.5	80.8	108.2	46.7	91.6	111.5	46.7	94.0	115.9	67.4	102.8
0970, 0990	Wall Finishes & Painting/Coating	114.4	16.7	55.8	114.4	40.6	70.1	114.4	34.5	66.5	114.4	35.7	67.2	114.4	37.5	68.3	98.4	33.8	59.7
09	FINISHES	104.0	23.9	61.7	104.0	43.5	72.0	101.8	28.3	62.9	102.4	41.8	70.4	105.1	42.0	71.7	115.4	46.9	79.2
COVERS	DIVS. 10 - 14, 25, 28, 41, 43, 44	100.0	71.8	94.4	100.0	71.5	94.4	100.0	67.5	93.6	100.0	70.8	94.2	100.0	74.9	95.0	100.0	80.0	96.0
21, 22, 23	FIRE SUPPRESSION, PLUMBING & HVAC	94.4	20.5	65.0	100.0	39.6	76.0	94.4	35.2	70.9	100.2	41.2	76.7	100.2	40.1	76.3	100.3	60.9	84.6
26, 27, 3370	ELECTRICAL, COMMUNICATIONS & UTIL.	98.6	28.2	64.4	104.9	40.1	73.3	100.3	38.5	70.2	98.0	38.8	69.2	97.5	44.0	71.5	103.0	66.2	85.1
MF2004	WEIGHTED AVERAGE	96.3	35.7	70.9	100.7	51.3	80.0	96.9	41.6	73.7	97.2	49.6	77.2	99.5	50.0	78.7	100.1	62.9	84.5

DIVISION		NORTH DAKOTA																	
		DEVILS LAKE			DICKINSON			FARGO			GRAND FORKS			JAMESTOWN			MINOT		
		583			586			580 - 581			582			584			587		
		MAT.	INST.	TOTAL	MAT.	INST.	TOTAL	MAT.	INST.	TOTAL	MAT.	INST.	TOTAL	MAT.	INST.	TOTAL	MAT.	INST.	TOTAL
015433	CONTRACTOR EQUIPMENT		98.7	98.7		98.7	98.7		98.7	98.7		98.7	98.7		98.7	98.7		98.7	98.7
0241, 31 - 34	SITE & INFRASTRUCTURE, DEMOLITION	100.6	95.5	97.1	109.4	94.4	99.1	95.6	97.3	96.8	104.5	94.4	97.6	99.5	94.3	96.0	101.8	97.3	98.7
0310	Concrete Forming & Accessories	98.0	40.2	48.0	87.9	39.7	46.2	95.6	45.8	52.6	91.7	39.5	46.6	89.4	39.3	46.2	87.6	60.6	64.3
0320	Concrete Reinforcing	111.0	70.2	94.3	112.0	46.0	85.1	102.3	68.9	88.7	109.3	70.1	93.3	111.6	54.9	88.4	113.0	70.1	95.5
0330	Cast-in-Place Concrete	116.0	51.8	91.4	104.9	50.5	84.1	97.2	54.8	81.0	104.9	50.4	84.1	114.5	50.4	90.0	104.9	51.9	84.6
03	CONCRETE	109.8	51.4	83.3	109.5	46.2	80.7	105.6	54.8	82.5	106.5	50.4	81.0	108.6	47.6	80.8	105.5	60.4	85.0
04	MASONRY	108.3	65.6	82.7	110.9	59.7	80.2	101.6	43.4	66.8	102.7	65.2	80.2	120.8	38.0	71.2	103.0	65.2	80.4
05	METALS	92.7	80.6	89.2	92.6	68.8	85.7	94.9	79.7	90.4	92.7	77.0	88.1	92.7	70.4	86.1	92.9	81.6	89.6
06	WOOD, PLASTICS & COMPOSITES	90.8	37.5	61.6	79.5	37.5	56.4	86.9	41.3	61.8	83.7	37.5	58.3	81.4	37.5	57.3	79.2	60.8	69.1
07	THERMAL & MOISTURE PROTECTION	102.3	52.7	83.0	102.8	50.3	82.4	100.2	48.3	80.0	102.5	52.8	83.2	102.2	45.8	80.2	102.3	54.4	83.7
08	OPENINGS	99.1	40.9	84.7	99.1	35.3	83.3	99.1	46.8	86.1	99.1	40.9	84.7	99.1	37.7	83.9	99.3	57.4	88.9
0920	Plaster & Gypsum Board	112.7	35.5	62.6	104.9	35.5	59.8	101.6	39.4	61.2	106.4	35.5	60.4	106.0	35.5	60.2	104.9	59.7	75.5
0950, 0980	Ceilings & Acoustic Treatment	142.1	35.5	76.2	142.1	35.5	76.2	139.6	39.4	77.7	142.1	35.5	76.2	142.1	35.5	76.2	142.1	59.7	91.1
0960	Flooring	120.2	40.2	98.6	111.7	40.2	92.4	115.6	40.2	95.2	114.2	40.2	94.2	113.0	40.2	93.3	111.5	80.8	103.2
0970, 0990	Wall Finishes & Painting/Coating	98.4	25.8	54.9	98.4	25.8	54.9	98.4	69.7	81.2	98.4	25.8	54.9	98.4	25.8	54.9	98.4	28.3	56.4
09	FINISHES	119.4	37.4	76.1	116.7	37.4	74.8	115.7	45.8	78.7	117.1	37.4	75.0	116.3	37.4	74.6	115.9	60.8	86.8
COVERS	DIVS. 10 - 14, 25, 28, 41, 43, 44	100.0	40.8	88.3	100.0	40.9	88.3	100.0	80.1	96.1	100.0	40.9	88.3	100.0	77.6	95.6	100.0	82.4	96.5
21, 22, 23	FIRE SUPPRESSION, PLUMBING & HVAC	94.6	61.3	81.3	94.6	71.5	85.4	100.3	64.6	86.1	100.4	39.8	76.3	94.6	39.7	72.7	100.4	60.4	84.5
26, 27, 3370	ELECTRICAL, COMMUNICATIONS & UTIL.	94.0	41.8	68.6	102.5	67.4	85.4	104.9	66.3	86.1	97.4	62.5	80.4	94.0	41.7	68.5	100.6	67.4	84.4
MF2004	WEIGHTED AVERAGE	100.0	57.3	82.0	100.9	60.1	83.7	101.4	62.0	84.9	100.8	55.1	81.6	100.0	49.1	78.7	100.9	67.3	86.8

	DIVISION	NORTH DAKOTA			OHIO														
		WILLISTON			AKRON			ATHENS			CANTON			CHILLICOTHE			CINCINNATI		
		588			442 - 443			457			446 - 447			456			451 - 452		
		MAT.	INST.	TOTAL	MAT.	INST.	TOTAL	MAT.	INST.	TOTAL	MAT.	INST.	TOTAL	MAT.	INST.	TOTAL	MAT.	INST.	TOTAL
015433	**CONTRACTOR EQUIPMENT**		98.7	98.7		95.4	95.4		90.5	90.5		95.4	95.4		100.3	100.3		100.2	100.2
0241, 31 - 34	**SITE & INFRASTRUCTURE, DEMOLITION**	102.7	94.4	97.0	89.0	102.8	98.4	85.9	91.3	89.6	89.1	102.8	98.5	76.3	105.5	96.3	73.9	105.2	95.3
0310	Concrete Forming & Accessories	93.3	39.7	47.0	101.0	96.3	96.9	92.5	87.5	88.2	101.0	85.4	87.5	94.3	92.9	93.1	96.2	85.4	86.8
0320	Concrete Reinforcing	114.2	46.0	86.4	98.0	93.3	96.1	87.4	89.6	88.3	98.0	77.1	89.5	84.6	85.9	85.1	89.7	84.8	87.7
0330	Cast-in-Place Concrete	104.9	50.5	84.1	93.2	100.7	96.1	97.6	97.6	97.6	94.1	98.3	95.7	88.6	93.8	90.6	81.6	87.3	83.8
03	**CONCRETE**	106.8	46.2	79.2	95.0	96.4	95.6	97.4	90.5	94.3	95.4	87.8	92.0	92.4	91.9	92.2	87.8	86.1	87.0
04	**MASONRY**	97.9	59.7	75.0	90.0	94.7	92.8	69.0	92.5	83.1	90.7	87.0	88.4	76.1	94.3	87.0	75.7	92.4	85.7
05	**METALS**	92.8	68.8	85.8	92.1	82.3	89.2	104.0	73.8	95.2	92.1	75.5	87.2	95.4	87.9	93.2	97.5	88.1	94.8
06	**WOOD, PLASTICS & COMPOSITES**	85.0	37.5	58.9	96.9	96.8	96.9	79.8	87.2	83.9	97.2	85.5	90.8	91.6	91.6	91.6	93.8	82.8	87.8
07	**THERMAL & MOISTURE PROTECTION**	102.5	50.3	82.2	104.4	97.5	101.7	99.6	95.0	97.8	105.3	93.7	100.8	100.7	93.4	97.8	98.8	93.5	96.8
08	**OPENINGS**	99.2	35.3	83.4	107.6	96.0	104.7	98.6	81.5	94.4	101.7	75.2	95.1	89.9	87.1	89.2	97.8	82.9	94.1
0920	Plaster & Gypsum Board	106.4	35.5	60.4	90.7	96.4	94.4	90.0	86.4	87.7	91.5	84.6	87.0	91.2	91.6	91.4	92.6	82.5	86.0
0950, 0980	Ceilings & Acoustic Treatment	142.1	35.5	76.2	92.3	96.4	94.8	103.8	86.4	93.0	92.3	84.6	87.5	98.4	91.6	94.2	99.2	82.5	88.9
0960	Flooring	115.4	40.2	95.1	102.6	88.2	98.7	128.2	94.8	119.2	102.8	80.9	96.9	105.4	93.9	102.3	106.6	93.6	103.1
0970, 0990	Wall Finishes & Painting/Coating	98.4	25.8	54.9	105.7	108.0	107.1	110.0	47.9	72.7	105.7	81.4	91.1	106.9	92.5	98.3	106.9	88.9	96.1
09	**FINISHES**	117.4	37.4	75.1	99.9	96.3	98.0	102.3	85.4	93.4	100.1	84.2	91.7	98.8	93.3	95.9	99.4	86.8	92.7
COVERS	**DIVS. 10 - 14, 25, 28, 41, 43, 44**	100.0	40.9	88.3	100.0	88.8	97.8	100.0	63.5	92.8	100.0	68.1	93.7	100.0	85.5	97.1	100.0	91.9	98.4
21, 22, 23	**FIRE SUPPRESSION, PLUMBING & HVAC**	94.6	71.5	85.4	99.9	95.5	98.2	94.2	61.4	81.2	99.9	81.4	92.6	94.8	90.1	92.9	100.0	87.0	94.9
26, 27, 3370	**ELECTRICAL, COMMUNICATIONS & UTIL.**	97.8	67.4	83.0	97.5	87.7	92.7	96.8	51.1	74.6	96.8	88.2	92.6	95.5	85.5	90.6	94.7	80.0	87.5
MF2004	**WEIGHTED AVERAGE**	99.4	60.1	82.9	97.8	93.7	96.1	96.7	75.9	88.0	97.3	85.0	92.1	93.8	91.4	92.8	95.3	88.1	92.3

	DIVISION	OHIO																	
		CLEVELAND			COLUMBUS			DAYTON			HAMILTON			LIMA			LORAIN		
		441			430 - 432			453 - 454			450			458			440		
		MAT.	INST.	TOTAL	MAT.	INST.	TOTAL	MAT.	INST.	TOTAL	MAT.	INST.	TOTAL	MAT.	INST.	TOTAL	MAT.	INST.	TOTAL
015433	**CONTRACTOR EQUIPMENT**		96.1	96.1		95.3	95.3		95.1	95.1		100.3	100.3		93.2	93.2		95.4	95.4
0241, 31 - 34	**SITE & INFRASTRUCTURE, DEMOLITION**	89.0	104.5	99.6	88.5	100.1	96.4	73.0	104.9	94.8	73.0	105.1	95.0	81.2	92.5	88.9	88.4	102.3	97.9
0310	Concrete Forming & Accessories	101.1	103.3	103.0	98.3	83.6	85.6	96.1	80.6	82.7	96.2	84.3	85.9	92.4	83.4	84.6	101.1	95.1	95.9
0320	Concrete Reinforcing	98.6	93.8	96.6	92.8	86.2	90.1	89.7	80.5	85.9	89.7	84.8	87.7	87.4	80.7	84.6	98.0	93.7	96.2
0330	Cast-in-Place Concrete	91.4	109.9	98.5	89.5	93.0	90.8	75.9	90.4	81.5	81.4	85.8	83.1	89.7	97.2	92.6	88.7	105.6	95.2
03	**CONCRETE**	94.3	102.9	98.2	92.6	87.1	90.1	85.2	83.8	84.6	87.7	85.1	86.5	91.3	87.5	89.6	92.9	97.7	95.1
04	**MASONRY**	94.4	107.2	102.1	93.1	93.2	93.1	75.2	88.1	82.9	75.6	89.7	84.1	93.8	87.3	89.9	86.7	100.9	95.2
05	**METALS**	93.5	85.7	91.3	91.9	83.4	89.4	96.8	79.1	91.6	96.9	88.0	94.3	104.0	82.2	97.7	92.6	83.9	90.1
06	**WOOD, PLASTICS & COMPOSITES**	96.0	100.8	98.6	98.1	81.2	88.8	95.0	78.1	85.7	93.8	82.8	87.8	79.7	81.5	80.7	96.9	93.9	95.2
07	**THERMAL & MOISTURE PROTECTION**	103.2	110.2	106.0	100.2	93.8	97.7	104.3	89.3	98.5	100.8	91.8	97.3	99.2	98.8	99.0	105.2	104.2	104.8
08	**OPENINGS**	97.8	98.2	97.9	100.2	81.4	95.6	98.0	78.3	93.1	95.5	82.9	92.4	98.6	79.4	93.9	101.7	94.4	99.9
0920	Plaster & Gypsum Board	89.9	100.5	96.8	88.0	80.5	83.1	92.6	77.6	82.9	92.6	82.5	86.0	90.0	80.5	83.8	90.7	93.3	92.4
0950, 0980	Ceilings & Acoustic Treatment	90.6	100.5	96.8	91.1	80.5	84.5	100.2	77.6	86.2	99.2	82.5	88.9	102.9	80.5	89.0	92.3	93.3	92.9
0960	Flooring	102.4	103.3	102.6	93.6	91.5	93.0	109.3	86.2	103.0	106.6	90.5	102.2	127.4	93.1	118.1	102.8	108.5	104.3
0970, 0990	Wall Finishes & Painting/Coating	105.7	112.3	109.7	97.3	94.2	95.4	106.9	85.5	94.1	106.9	86.9	94.9	110.0	83.3	94.0	105.7	112.3	109.7
09	**FINISHES**	99.5	104.2	102.0	93.6	85.6	89.4	100.4	81.6	90.5	99.3	85.4	92.0	101.5	84.6	92.6	100.0	99.4	99.7
COVERS	**DIVS. 10 - 14, 25, 28, 41, 43, 44**	100.0	103.5	100.7	100.0	93.2	98.6	100.0	90.5	98.1	100.0	90.9	98.2	100.0	87.2	97.5	100.0	100.2	100.0
21, 22, 23	**FIRE SUPPRESSION, PLUMBING & HVAC**	99.9	105.5	102.1	99.7	89.8	95.8	101.0	85.2	94.7	100.6	85.7	94.7	94.2	84.7	90.4	99.9	84.1	93.6
26, 27, 3370	**ELECTRICAL, COMMUNICATIONS & UTIL.**	97.3	104.2	100.7	98.1	85.1	91.8	93.1	83.3	88.3	93.4	70.0	82.0	97.1	80.3	88.9	97.0	91.6	94.4
MF2004	**WEIGHTED AVERAGE**	97.2	102.6	99.5	96.3	88.6	93.0	95.2	85.5	91.1	95.0	85.8	91.1	97.0	85.3	92.1	96.9	93.6	95.5

	DIVISION	OHIO																	
		MANSFIELD			MARION			SPRINGFIELD			STEUBENVILLE			TOLEDO			YOUNGSTOWN		
		448 - 449			433			455			439			434 - 436			444 - 445		
		MAT.	INST.	TOTAL	MAT.	INST.	TOTAL	MAT.	INST.	TOTAL	MAT.	INST.	TOTAL	MAT.	INST.	TOTAL	MAT.	INST.	TOTAL
015433	**CONTRACTOR EQUIPMENT**		95.4	95.4		94.5	94.5		95.1	95.1		99.1	99.1		97.4	97.4		95.4	95.4
0241, 31 - 34	**SITE & INFRASTRUCTURE, DEMOLITION**	85.2	103.4	97.6	84.5	98.3	93.9	73.3	103.2	93.8	126.9	108.4	114.2	87.7	99.5	95.8	88.9	103.4	98.8
0310	Concrete Forming & Accessories	91.1	93.0	92.8	95.1	79.0	81.2	96.1	82.2	84.1	96.7	89.2	90.3	98.3	102.8	102.2	101.1	89.9	91.4
0320	Concrete Reinforcing	89.2	77.7	84.5	85.6	86.2	85.8	89.7	85.8	88.1	83.2	85.9	84.3	92.8	93.1	93.0	98.0	88.2	94.0
0330	Cast-in-Place Concrete	86.3	93.8	89.2	81.7	90.1	84.9	78.0	89.9	82.5	88.7	95.4	91.3	89.5	106.1	95.8	92.3	102.3	96.1
03	**CONCRETE**	87.5	89.8	88.5	85.0	84.0	84.6	86.1	85.3	85.8	88.5	90.1	89.3	92.6	101.5	96.7	94.6	93.2	94.0
04	**MASONRY**	89.1	96.9	93.8	95.1	92.8	93.7	75.4	87.3	82.5	81.6	92.3	88.0	102.1	100.3	101.0	90.2	93.6	92.3
05	**METALS**	92.8	77.4	88.3	91.0	80.4	87.9	96.8	80.8	92.1	87.5	80.2	85.4	91.7	89.1	91.0	92.1	80.6	88.8
06	**WOOD, PLASTICS & COMPOSITES**	85.7	93.9	90.2	94.4	78.6	85.7	96.3	80.9	87.9	90.5	88.7	89.5	98.1	103.6	101.1	96.9	88.6	92.4
07	**THERMAL & MOISTURE PROTECTION**	104.7	95.5	101.1	99.6	85.5	94.1	104.2	89.3	98.4	111.5	93.2	104.4	101.8	107.2	103.9	105.5	94.8	101.3
08	**OPENINGS**	100.9	85.4	97.0	94.5	79.1	90.7	95.9	81.1	92.2	95.2	85.8	92.9	98.1	98.5	98.2	101.7	89.7	98.7
0920	Plaster & Gypsum Board	86.2	93.3	90.8	86.5	77.8	80.8	92.6	80.5	84.7	87.3	87.7	87.6	88.0	103.7	98.2	90.7	87.9	88.9
0950, 0980	Ceilings & Acoustic Treatment	94.8	93.3	93.9	91.1	77.8	82.9	100.2	80.5	88.0	88.2	87.7	87.9	91.1	103.7	98.9	92.3	87.9	89.5
0960	Flooring	97.8	107.1	100.3	91.9	78.2	88.2	109.3	86.2	103.0	119.5	89.2	111.3	92.8	98.8	94.4	102.8	93.1	100.2
0970, 0990	Wall Finishes & Painting/Coating	105.7	88.3	95.3	97.3	54.6	71.7	106.9	85.5	94.1	112.2	94.5	101.6	97.3	104.5	101.6	105.7	93.8	98.5
09	**FINISHES**	98.2	95.7	96.9	92.6	76.0	83.8	100.4	83.0	91.2	109.6	89.6	99.0	93.4	102.6	98.2	100.1	90.7	95.1
COVERS	**DIVS. 10 - 14, 25, 28, 41, 43, 44**	100.0	91.9	98.4	100.0	63.2	92.7	100.0	90.5	98.1	100.0	91.7	98.4	100.0	91.3	98.3	100.0	87.7	97.6
21, 22, 23	**FIRE SUPPRESSION, PLUMBING & HVAC**	94.1	82.0	89.3	93.9	85.2	90.4	101.0	84.8	94.6	94.1	84.1	90.2	99.7	98.4	99.2	99.9	86.1	94.5
26, 27, 3370	**ELECTRICAL, COMMUNICATIONS & UTIL.**	94.8	80.7	87.9	92.2	79.9	86.2	93.1	82.2	87.8	86.0	105.6	95.5	98.0	105.0	101.4	97.0	86.0	91.6
MF2004	**WEIGHTED AVERAGE**	94.4	88.3	91.9	92.6	83.5	88.8	95.1	85.7	91.2	94.1	91.6	93.0	96.5	99.7	97.9	97.2	89.7	94.1

City Cost Indexes

DIVISION		OHIO			OKLAHOMA															
		ZANESVILLE			ARDMORE			CLINTON			DURANT			ENID			GUYMON			
		437 - 438			734			736			747			737			739			
		MAT.	INST.	TOTAL	MAT.	INST.	TOTAL	MAT.	INST.	TOTAL	MAT.	INST.	TOTAL	MAT.	INST.	TOTAL	MAT.	INST.	TOTAL	
015433	CONTRACTOR EQUIPMENT		94.5	94.5		82.1	82.1		81.5	81.5		81.5	81.5		81.5	81.5		81.5	81.5	
0241, 31 - 34	SITE & INFRASTRUCTURE, DEMOLITION	87.2	99.7	95.7	106.4	92.2	96.7	108.0	91.1	96.5	100.6	90.9	93.9	109.5	91.1	96.9	113.3	89.5	97.1	
0310	Concrete Forming & Accessories	92.1	79.9	81.6	94.0	43.1	50.0	92.8	37.2	44.8	85.6	43.2	49.0	96.3	36.5	44.6	100.1	23.1	33.6	
0320	Concrete Reinforcing	85.1	85.5	85.2	99.4	77.2	90.3	100.0	77.2	90.7	100.4	60.2	84.0	99.3	77.2	90.3	100.0	30.0	71.4	
0330	Cast-in-Place Concrete	86.1	91.0	88.0	95.9	46.7	77.1	92.6	46.6	75.1	89.6	46.6	73.2	92.6	49.2	76.0	92.7	32.7	69.8	
03	CONCRETE	88.3	84.6	86.6	92.4	51.2	73.7	91.7	48.6	72.1	87.7	48.1	69.7	92.1	49.2	72.6	94.5	28.7	64.6	
04	MASONRY	92.0	85.7	88.2	94.9	55.6	71.3	121.5	55.6	82.0	93.4	69.7	79.2	103.5	55.6	74.8	97.5	21.8	52.2	
05	METALS	92.4	81.5	89.2	94.6	70.0	87.4	94.7	69.9	87.4	94.6	64.0	85.7	95.9	70.0	88.3	95.1	44.2	80.2	
06	WOOD, PLASTICS & COMPOSITES	90.8	78.6	84.1	97.9	42.2	67.3	96.9	34.3	62.5	88.5	42.2	63.1	100.9	33.3	63.8	104.9	22.2	59.5	
07	THERMAL & MOISTURE PROTECTION	99.7	91.1	96.4	99.3	58.9	83.6	99.4	58.0	83.3	99.0	60.8	84.2	99.5	57.9	83.3	99.8	28.0	71.9	
08	OPENINGS	94.5	79.9	90.9	95.5	51.5	84.6	95.5	47.2	83.6	95.5	47.2	83.5	95.5	46.5	83.4	95.6	23.6	77.8	
0920	Plaster & Gypsum Board	84.6	77.8	80.2	83.8	40.8	55.8	83.4	32.6	50.4	80.4	40.8	54.7	84.5	31.6	50.2	85.2	20.1	42.9	
0950, 0980	Ceilings & Acoustic Treatment	91.1	77.8	82.9	82.7	40.8	56.8	82.7	32.6	51.7	82.7	40.8	56.8	83.6	31.6	51.5	85.2	20.1	44.9	
0960	Flooring	89.9	82.4	87.9	108.6	48.4	92.3	107.1	46.1	90.6	102.8	66.0	92.9	109.3	46.1	92.2	111.6	27.3	88.8	
0970, 0990	Wall Finishes & Painting/Coating	97.3	77.8	85.6	98.5	58.9	74.7	98.5	58.9	74.7	98.5	58.9	74.7	98.5	58.9	74.7	98.5	17.5	49.9	
09	FINISHES	91.9	79.6	85.4	93.6	44.2	67.5	93.3	39.2	64.7	91.1	48.0	68.3	94.4	38.6	64.9	96.0	23.1	57.4	
COVERS	DIVS. 10 - 14, 25, 28, 41, 43, 44	100.0	81.6	96.4	100.0	66.1	93.3	100.0	65.1	93.1	100.0	66.2	93.3	100.0	65.0	93.1	100.0	61.9	92.5	
21, 22, 23	FIRE SUPPRESSION, PLUMBING & HVAC	93.9	80.2	88.5	94.3	63.6	82.1	94.3	63.6	82.1	94.3	63.8	82.2	100.1	63.6	85.6	94.3	24.6	66.6	
26, 27, 3370	ELECTRICAL, COMMUNICATIONS & UTIL.	92.5	74.7	83.8	94.5	66.6	80.9	95.5	66.6	81.4	97.1	56.7	77.5	95.5	66.6	81.4	97.1	18.0	58.6	
MF2004	WEIGHTED AVERAGE	93.1	82.5	88.7	95.2	61.1	80.9	96.6	59.7	81.1	94.4	60.4	80.2	97.4	59.7	81.6	96.4	32.0	69.4	

DIVISION		OKLAHOMA																	
		LAWTON			MCALESTER			MIAMI			MUSKOGEE			OKLAHOMA CITY			PONCA CITY		
		735			745			743			744			730 - 731			746		
		MAT.	INST.	TOTAL	MAT.	INST.	TOTAL	MAT.	INST.	TOTAL	MAT.	INST.	TOTAL	MAT.	INST.	TOTAL	MAT.	INST.	TOTAL
015433	CONTRACTOR EQUIPMENT		82.1	82.1		81.5	81.5		90.7	90.7		90.7	90.7		82.3	82.3		81.5	81.5
0241, 31 - 34	SITE & INFRASTRUCTURE, DEMOLITION	104.6	92.2	96.1	92.8	91.2	91.7	94.9	89.3	91.0	95.0	87.8	90.1	104.0	92.7	96.3	101.3	91.1	94.3
0310	Concrete Forming & Accessories	99.8	49.3	56.1	83.5	44.8	50.0	96.5	52.8	58.7	101.0	32.8	42.0	98.9	42.7	50.4	92.0	43.4	50.0
0320	Concrete Reinforcing	99.6	77.2	90.4	100.0	44.8	77.5	98.5	77.2	89.8	99.3	34.3	72.8	99.6	77.2	90.4	99.3	77.1	90.3
0330	Cast-in-Place Concrete	89.6	49.2	74.2	78.5	49.3	67.4	82.3	48.4	69.4	83.3	38.8	66.3	92.3	52.0	76.9	92.0	41.5	72.7
03	CONCRETE	88.9	54.8	73.4	79.5	46.9	64.7	83.2	56.9	71.3	84.8	36.7	62.9	90.1	52.9	73.2	89.4	49.6	71.3
04	MASONRY	97.4	55.6	72.3	111.5	56.2	78.4	96.0	56.8	72.5	113.5	48.9	74.8	97.1	57.5	73.4	88.2	56.2	69.0
05	METALS	99.5	70.1	90.9	94.6	57.0	83.6	94.5	82.3	91.0	95.8	60.3	85.4	101.5	70.0	92.3	94.5	69.7	87.3
06	WOOD, PLASTICS & COMPOSITES	103.9	50.5	74.6	85.9	45.7	63.8	100.8	54.6	75.4	106.0	33.0	65.9	103.1	40.7	68.8	96.1	42.8	66.8
07	THERMAL & MOISTURE PROTECTION	99.3	59.7	83.9	98.7	58.1	82.9	99.1	59.9	83.8	99.1	43.7	77.6	96.8	59.5	82.3	99.2	58.0	83.2
08	OPENINGS	97.2	55.8	87.0	95.5	44.4	82.8	95.5	58.8	86.4	95.5	32.0	79.8	97.2	50.5	85.7	95.5	51.8	84.7
0920	Plaster & Gypsum Board	86.9	49.4	62.6	79.3	44.4	56.6	84.5	53.5	64.4	88.4	31.1	51.2	83.1	39.3	54.6	83.4	41.4	56.1
0950, 0980	Ceilings & Acoustic Treatment	92.8	49.4	66.0	82.7	44.4	59.0	82.7	53.5	64.6	92.8	31.1	54.7	90.3	39.3	58.7	82.7	41.4	57.2
0960	Flooring	112.0	46.1	94.2	101.5	46.1	86.6	110.2	66.0	98.3	113.2	37.2	92.6	112.0	46.1	94.2	106.6	46.1	90.2
0970, 0990	Wall Finishes & Painting/Coating	98.5	58.9	74.7	98.5	43.5	65.5	98.5	76.1	85.0	98.5	31.6	58.3	98.5	58.9	74.7	98.5	58.9	74.7
09	FINISHES	97.2	48.7	71.5	90.0	44.3	65.8	93.2	57.3	74.2	97.0	33.7	63.5	95.9	43.4	68.2	92.8	44.3	67.2
COVERS	DIVS. 10 - 14, 25, 28, 41, 43, 44	100.0	67.0	93.5	100.0	66.7	93.4	100.0	68.3	93.7	100.0	64.2	92.9	100.0	66.6	93.4	100.0	66.3	93.3
21, 22, 23	FIRE SUPPRESSION, PLUMBING & HVAC	100.1	63.6	85.6	94.3	33.9	70.3	94.3	61.8	81.4	100.1	25.8	70.6	100.0	64.6	86.0	94.3	61.4	81.2
26, 27, 3370	ELECTRICAL, COMMUNICATIONS & UTIL.	97.1	66.6	82.3	95.6	64.0	80.2	97.0	64.0	80.9	95.2	30.7	63.8	103.0	66.6	85.3	95.2	66.5	81.2
MF2004	WEIGHTED AVERAGE	97.8	62.5	83.0	93.8	52.4	76.5	94.1	64.5	81.7	96.9	41.6	73.7	98.7	61.7	83.1	94.4	60.4	80.1

DIVISION		OKLAHOMA												OREGON					
		POTEAU			SHAWNEE			TULSA			WOODWARD			BEND			EUGENE		
		749			748			740 - 741			738			977			974		
		MAT.	INST.	TOTAL	MAT.	INST.	TOTAL	MAT.	INST.	TOTAL	MAT.	INST.	TOTAL	MAT.	INST.	TOTAL	MAT.	INST.	TOTAL
015433	CONTRACTOR EQUIPMENT		92.0	92.0		81.5	81.5		90.7	90.7		81.5	81.5		100.7	100.7		100.7	100.7
0241, 31 - 34	SITE & INFRASTRUCTURE, DEMOLITION	76.7	91.3	86.7	104.7	91.0	95.4	101.6	89.3	93.2	108.6	91.1	96.6	119.3	104.9	109.5	106.9	104.9	105.5
0310	Concrete Forming & Accessories	89.7	45.9	51.9	85.5	42.0	48.0	100.8	42.8	50.7	92.9	37.1	44.7	109.3	100.1	101.4	105.7	100.0	100.8
0320	Concrete Reinforcing	100.5	77.2	91.0	99.3	49.2	78.9	99.6	77.1	90.4	99.3	77.2	90.3	98.1	95.5	97.1	102.3	95.5	99.5
0330	Cast-in-Place Concrete	82.3	48.0	69.2	95.0	46.1	76.3	90.8	48.1	74.5	92.6	46.6	75.1	108.3	102.9	106.2	104.9	102.8	104.1
03	CONCRETE	85.1	53.8	70.9	91.0	45.4	70.3	89.5	52.4	72.6	91.9	48.6	72.2	114.3	100.0	107.8	105.7	100.0	103.1
04	MASONRY	96.7	56.2	72.5	112.8	54.9	78.1	96.8	57.5	73.3	91.8	55.6	70.1	117.0	98.1	105.7	114.5	98.1	104.7
05	METALS	94.6	82.2	90.9	94.5	58.6	84.0	98.9	81.5	93.8	94.7	70.0	87.5	90.6	96.0	92.2	91.3	95.9	92.6
06	WOOD, PLASTICS & COMPOSITES	93.1	45.8	67.2	88.3	41.9	62.8	104.5	41.9	70.2	97.1	34.2	62.5	98.5	100.0	99.3	94.2	100.0	97.3
07	THERMAL & MOISTURE PROTECTION	99.2	58.4	83.3	99.2	57.9	83.1	99.1	56.8	82.7	99.5	58.0	83.3	106.1	94.3	101.5	105.4	91.5	100.0
08	OPENINGS	95.5	54.0	85.2	95.5	41.3	82.1	97.2	50.8	85.7	95.5	47.1	83.5	98.8	101.3	99.4	99.2	101.3	99.7
0920	Plaster & Gypsum Board	82.7	44.4	57.8	80.4	40.5	54.5	86.9	40.4	56.7	84.1	32.5	50.6	104.2	99.7	101.3	102.3	99.7	100.7
0950, 0980	Ceilings & Acoustic Treatment	82.7	44.4	59.0	82.7	40.5	56.6	92.8	40.4	60.4	85.2	32.5	52.6	102.4	99.7	100.7	103.3	99.7	101.1
0960	Flooring	105.8	66.0	95.0	102.7	36.7	84.9	111.8	48.0	94.6	107.1	48.4	91.2	114.3	99.0	110.2	112.4	99.0	108.8
0970, 0990	Wall Finishes & Painting/Coating	98.5	58.9	74.7	98.5	39.2	62.9	98.5	46.6	67.3	98.5	58.9	74.7	110.1	67.7	84.7	110.1	67.7	84.7
09	FINISHES	90.8	50.1	69.3	91.3	39.6	64.0	96.6	43.1	68.3	94.1	39.5	65.2	109.1	96.5	102.4	107.4	96.5	101.6
COVERS	DIVS. 10 - 14, 25, 28, 41, 43, 44	100.0	67.0	93.5	100.0	65.8	93.2	100.0	66.9	93.5	100.0	65.1	93.1	100.0	101.3	100.3	100.0	101.3	100.3
21, 22, 23	FIRE SUPPRESSION, PLUMBING & HVAC	94.3	61.5	81.3	94.3	62.9	81.8	100.1	53.1	81.4	94.3	63.6	82.1	94.4	99.6	96.4	100.2	99.5	99.9
26, 27, 3370	ELECTRICAL, COMMUNICATIONS & UTIL.	95.3	64.0	80.1	97.2	66.6	82.3	97.1	40.1	69.4	97.0	66.6	82.2	100.6	95.1	97.9	99.3	95.1	97.2
MF2004	WEIGHTED AVERAGE	93.5	62.8	80.6	95.9	57.8	79.9	97.6	56.3	80.3	95.4	59.8	80.4	101.0	98.5	100.0	100.7	98.4	99.7

	DIVISION	OREGON																	
		KLAMATH FALLS			MEDFORD			PENDLETON			PORTLAND			SALEM			VALE		
		976			975			978			970 - 972			973			979		
		MAT.	INST.	TOTAL	MAT.	INST.	TOTAL	MAT.	INST.	TOTAL	MAT.	INST.	TOTAL	MAT.	INST.	TOTAL	MAT.	INST.	TOTAL
015433	**CONTRACTOR EQUIPMENT**		100.7	100.7		100.7	100.7		99.3	99.3		100.7	100.7		100.7	100.7		99.3	99.3
0241, 31 - 34	**SITE & INFRASTRUCTURE, DEMOLITION**	124.1	104.9	111.0	116.1	104.9	108.4	113.4	99.6	104.0	109.7	104.9	106.5	103.5	104.9	104.5	99.3	99.6	99.5
0310	Concrete Forming & Accessories	102.3	100.0	100.3	101.4	100.0	100.2	103.1	100.2	100.6	107.1	100.4	101.3	104.5	100.2	100.7	109.5	99.4	100.8
0320	Concrete Reinforcing	98.1	95.5	97.0	99.8	95.5	98.0	97.4	95.8	96.7	103.1	95.8	100.1	103.2	95.8	100.2	95.0	95.6	95.3
0330	Cast-in-Place Concrete	108.3	101.2	105.6	108.3	102.8	106.2	109.1	102.6	106.6	107.8	103.0	106.0	94.7	102.9	97.9	85.8	102.3	92.1
03	**CONCRETE**	117.2	99.4	109.1	111.8	100.0	106.4	97.8	100.1	98.8	107.3	100.2	104.1	101.1	100.1	100.7	82.4	99.5	90.2
04	**MASONRY**	133.4	98.1	112.3	111.6	98.1	103.5	121.9	103.3	110.8	115.5	103.2	108.1	119.7	103.2	109.8	120.0	103.3	110.0
05	**METALS**	90.7	95.8	92.2	90.9	95.9	92.3	97.4	96.8	97.2	92.3	96.6	93.6	91.7	96.4	93.1	97.3	96.0	96.9
06	**WOOD, PLASTICS & COMPOSITES**	89.6	100.0	95.3	88.4	100.0	94.7	91.7	100.1	96.3	95.5	100.0	97.9	91.6	100.0	96.2	100.3	100.1	100.2
07	**THERMAL & MOISTURE PROTECTION**	106.3	93.5	101.3	106.0	93.5	101.1	97.2	94.9	96.3	105.2	99.9	103.1	102.8	95.6	100.0	96.7	92.4	95.0
08	**OPENINGS**	98.8	101.3	99.4	101.8	101.3	101.7	94.3	101.3	96.1	96.8	101.3	97.9	98.7	101.3	99.3	94.3	94.9	94.4
0920	Plaster & Gypsum Board	101.6	99.7	100.4	100.9	99.7	100.1	83.2	99.7	93.9	107.8	99.7	102.6	95.4	99.7	98.2	88.4	99.7	95.8
0950, 0980	Ceilings & Acoustic Treatment	112.3	99.7	104.6	118.9	99.7	107.0	70.6	99.7	88.6	107.0	99.7	102.5	104.7	99.7	101.6	70.6	99.7	88.6
0960	Flooring	110.7	99.0	107.5	110.0	99.0	107.0	78.0	99.0	83.7	109.9	99.0	106.9	112.6	99.0	108.9	80.1	99.0	85.2
0970, 0990	Wall Finishes & Painting/Coating	110.1	62.1	81.3	110.1	62.1	81.3	95.2	72.7	81.7	109.6	72.7	87.5	110.1	67.7	84.7	95.2	67.7	78.7
09	**FINISHES**	110.5	95.9	102.8	110.8	95.9	102.9	75.4	97.2	86.9	108.3	97.1	102.3	106.5	96.5	101.2	75.7	96.6	86.8
COVERS	**DIVS. 10 - 14, 25, 28, 41, 43, 44**	100.0	101.3	100.3	100.0	101.3	100.3	100.0	101.5	100.3	100.0	101.4	100.3	100.0	101.3	100.3	100.0	101.6	100.3
21, 22, 23	**FIRE SUPPRESSION, PLUMBING & HVAC**	94.4	98.6	96.1	100.2	99.5	99.9	95.7	102.0	98.2	100.1	104.3	101.8	100.0	99.6	99.9	95.7	75.9	87.8
26, 27, 3370	**ELECTRICAL, COMMUNICATIONS & UTIL.**	99.4	85.0	92.4	102.8	85.0	94.1	90.9	92.8	91.8	99.6	99.8	99.7	107.7	95.1	101.6	90.9	72.9	82.1
MF2004	**WEIGHTED AVERAGE**	102.2	96.7	99.9	102.3	97.0	100.1	96.1	99.0	97.3	101.1	101.0	101.0	101.0	99.1	100.2	93.8	90.3	92.3

	DIVISION	PENNSYLVANIA																	
		ALLENTOWN			ALTOONA			BEDFORD			BRADFORD			BUTLER			CHAMBERSBURG		
		181			166			155			167			160			172		
		MAT.	INST.	TOTAL	MAT.	INST.	TOTAL	MAT.	INST.	TOTAL	MAT.	INST.	TOTAL	MAT.	INST.	TOTAL	MAT.	INST.	TOTAL
015433	**CONTRACTOR EQUIPMENT**		112.9	112.9		112.9	112.9		111.0	111.0		112.9	112.9		112.9	112.9		112.1	112.1
0241, 31 - 34	**SITE & INFRASTRUCTURE, DEMOLITION**	93.2	104.4	100.9	98.1	104.1	102.2	109.7	102.2	104.6	92.8	103.7	100.2	88.8	105.2	100.0	87.3	101.6	97.0
0310	Concrete Forming & Accessories	99.3	112.3	110.5	84.5	80.5	81.0	85.6	82.3	82.8	86.7	79.9	80.9	86.0	92.6	91.7	85.2	81.7	82.2
0320	Concrete Reinforcing	99.1	103.5	100.9	96.1	92.7	94.7	95.6	79.1	88.9	98.1	92.8	96.0	96.7	104.6	100.0	96.5	100.6	98.2
0330	Cast-in-Place Concrete	86.5	100.6	91.9	96.4	84.4	91.8	99.8	76.5	90.9	92.1	88.7	90.8	85.2	91.7	87.6	93.5	76.3	86.9
03	**CONCRETE**	97.8	107.3	102.1	93.1	85.7	89.7	99.3	81.2	91.1	99.6	87.0	93.9	85.2	95.9	90.0	105.3	84.9	96.0
04	**MASONRY**	94.9	99.5	97.7	97.0	74.6	83.6	95.0	84.7	88.8	94.1	81.0	86.2	99.1	90.4	93.9	95.2	87.3	90.5
05	**METALS**	95.2	119.3	102.2	89.1	112.0	95.8	92.6	104.6	96.1	93.3	112.1	98.8	88.9	119.2	97.7	93.3	114.4	99.5
06	**WOOD, PLASTICS & COMPOSITES**	101.1	116.0	109.3	80.0	82.7	81.5	79.4	82.7	81.2	86.5	79.2	82.5	81.6	93.2	88.0	89.8	82.7	85.9
07	**THERMAL & MOISTURE PROTECTION**	100.2	114.4	105.7	99.4	88.0	95.0	102.5	86.0	96.1	100.2	90.4	96.4	99.1	96.2	98.0	99.5	74.4	89.8
08	**OPENINGS**	94.0	110.9	98.2	88.4	88.3	88.3	89.3	84.6	88.2	94.4	90.4	93.4	88.3	102.6	91.9	90.8	84.5	89.2
0920	Plaster & Gypsum Board	112.6	116.3	115.0	106.3	81.8	90.4	92.4	81.8	85.5	107.1	78.2	88.3	106.6	92.7	97.6	110.1	81.8	91.7
0950, 0980	Ceilings & Acoustic Treatment	87.4	116.3	105.3	91.6	81.8	85.5	85.5	81.8	83.2	90.0	78.2	82.7	92.5	92.7	92.6	90.0	81.8	84.9
0960	Flooring	85.1	94.7	87.7	79.2	51.1	71.6	90.0	51.9	79.7	78.8	99.5	84.4	80.0	90.8	82.9	82.7	51.8	74.3
0970, 0990	Wall Finishes & Painting/Coating	88.3	81.1	84.0	84.2	105.7	97.1	97.0	77.7	85.4	88.3	89.4	89.0	84.2	105.5	97.0	88.3	66.4	75.2
09	**FINISHES**	95.0	105.5	100.6	93.7	76.8	84.8	93.4	75.4	83.9	93.0	83.0	87.7	93.7	92.9	93.2	93.7	74.3	83.4
COVERS	**DIVS. 10 - 14, 25, 28, 41, 43, 44**	100.0	103.4	100.7	100.0	94.0	98.8	100.0	96.4	99.3	100.0	95.5	99.1	100.0	98.5	99.7	100.0	88.7	97.8
21, 22, 23	**FIRE SUPPRESSION, PLUMBING & HVAC**	100.2	109.1	103.8	99.8	84.8	93.8	94.1	88.2	91.8	94.5	82.6	89.8	94.0	85.3	90.6	94.5	90.7	93.0
26, 27, 3370	**ELECTRICAL, COMMUNICATIONS & UTIL.**	101.3	96.3	98.9	90.6	98.0	94.2	93.8	98.0	95.8	95.0	98.0	96.5	91.2	97.3	94.2	93.5	84.3	89.0
MF2004	**WEIGHTED AVERAGE**	97.7	106.5	101.4	94.3	89.4	92.2	95.1	89.3	92.6	95.4	90.7	93.4	91.9	96.3	93.8	95.5	88.9	92.7

	DIVISION	PENNSYLVANIA																	
		DOYLESTOWN			DUBOIS			ERIE			GREENSBURG			HARRISBURG			HAZLETON		
		189			158			164 - 165			156			170 - 171			182		
		MAT.	INST.	TOTAL	MAT.	INST.	TOTAL	MAT.	INST.	TOTAL	MAT.	INST.	TOTAL	MAT.	INST.	TOTAL	MAT.	INST.	TOTAL
015433	**CONTRACTOR EQUIPMENT**		95.4	95.4		111.0	111.0		112.9	112.9		111.0	111.0		112.1	112.1		112.9	112.9
0241, 31 - 34	**SITE & INFRASTRUCTURE, DEMOLITION**	106.9	91.8	96.6	115.7	102.4	106.6	94.4	104.8	101.5	105.0	103.8	104.2	83.6	102.8	96.7	86.3	104.2	98.5
0310	Concrete Forming & Accessories	83.4	130.1	123.7	85.0	83.5	83.7	98.8	88.7	90.0	92.0	92.7	92.6	92.5	84.3	85.4	81.3	85.1	84.6
0320	Concrete Reinforcing	95.9	135.5	112.0	95.0	93.0	94.2	98.1	93.8	96.4	95.0	104.6	98.9	99.1	100.9	99.8	96.2	104.7	99.6
0330	Cast-in-Place Concrete	81.7	92.6	85.9	96.2	89.6	93.7	94.8	86.3	91.6	92.6	91.4	92.1	96.1	93.8	95.2	81.7	88.5	84.3
03	**CONCRETE**	93.5	117.5	104.4	100.6	88.8	95.2	92.1	90.2	91.2	94.6	95.7	95.1	100.1	92.2	96.5	90.5	91.5	91.0
04	**MASONRY**	98.1	128.5	116.3	95.0	85.7	89.4	87.2	90.9	89.4	104.5	90.3	96.0	95.0	87.8	90.7	107.0	94.5	99.5
05	**METALS**	92.7	121.8	101.2	92.6	111.5	98.1	89.3	112.6	96.1	92.6	118.2	100.0	97.0	117.6	103.0	94.9	118.7	101.9
06	**WOOD, PLASTICS & COMPOSITES**	81.9	133.1	110.0	78.4	82.7	80.7	97.3	87.8	92.1	86.5	93.2	90.2	94.9	83.0	88.4	80.5	82.9	81.8
07	**THERMAL & MOISTURE PROTECTION**	98.6	127.0	109.6	102.7	93.2	99.0	99.6	94.5	97.6	102.4	95.7	99.8	105.4	105.4	105.4	99.7	103.4	101.2
08	**OPENINGS**	95.7	137.9	106.1	89.3	92.3	90.1	88.5	89.5	88.7	89.3	101.8	92.4	94.0	90.8	93.2	94.7	92.1	94.0
0920	Plaster & Gypsum Board	104.7	134.1	123.8	91.7	81.8	85.2	112.6	87.1	96.0	94.0	92.7	93.2	109.7	82.1	91.8	105.2	82.0	90.1
0950, 0980	Ceilings & Acoustic Treatment	86.7	134.1	116.0	85.5	81.8	83.2	87.4	87.1	87.2	84.6	92.7	89.6	89.9	82.1	85.1	88.3	82.0	84.4
0960	Flooring	69.5	92.9	75.9	89.6	99.5	92.3	86.5	76.1	83.7	93.3	76.3	88.7	85.7	90.0	86.8	76.0	93.2	80.7
0970, 0990	Wall Finishes & Painting/Coating	87.8	79.3	82.7	97.0	105.5	102.1	94.5	89.4	91.4	97.0	105.5	102.1	88.3	87.7	88.0	88.3	98.0	94.1
09	**FINISHES**	86.3	118.4	103.3	93.7	86.7	90.0	96.4	86.0	90.9	94.1	91.1	92.5	94.1	85.2	89.4	91.0	86.3	88.5
COVERS	**DIVS. 10 - 14, 25, 28, 41, 43, 44**	100.0	83.4	96.7	100.0	96.4	99.3	100.0	98.7	99.8	100.0	98.5	99.7	100.0	88.9	97.8	100.0	97.8	99.6
21, 22, 23	**FIRE SUPPRESSION, PLUMBING & HVAC**	94.0	123.8	105.9	94.1	88.5	91.9	99.8	86.4	94.5	94.1	89.8	92.4	100.1	91.6	96.7	94.5	84.1	90.3
26, 27, 3370	**ELECTRICAL, COMMUNICATIONS & UTIL.**	94.1	116.2	104.9	94.4	98.0	96.1	92.4	85.6	89.1	94.4	98.0	96.1	100.0	84.4	92.4	95.9	82.2	89.2
MF2004	**WEIGHTED AVERAGE**	94.4	118.6	104.5	95.5	93.2	94.5	94.2	92.0	93.3	95.0	96.8	95.8	97.9	93.1	95.9	94.9	92.4	93.9

DIVISION		PENNSYLVANIA																	
		INDIANA			JOHNSTOWN			KITTANNING			LANCASTER			LEHIGH VALLEY			MONTROSE		
		157			159			162			175 - 176			180			188		
		MAT.	INST.	TOTAL	MAT.	INST.	TOTAL	MAT.	INST.	TOTAL	MAT.	INST.	TOTAL	MAT.	INST.	TOTAL	MAT.	INST.	TOTAL
015433	**CONTRACTOR EQUIPMENT**		111.0	111.0		111.0	111.0		112.9	112.9		112.1	112.1		112.9	112.9		112.9	112.9
0241, 31 - 34	**SITE & INFRASTRUCTURE, DEMOLITION**	103.0	103.0	103.0	110.3	103.2	105.5	91.6	105.1	100.9	78.8	102.8	95.2	90.6	104.4	100.0	89.5	102.6	98.4
0310	Concrete Forming & Accessories	86.3	84.7	85.0	85.0	83.6	83.8	86.0	92.9	91.9	87.1	83.8	84.3	93.0	112.3	109.7	82.1	84.7	84.3
0320	Concrete Reinforcing	94.3	104.7	98.5	95.6	104.3	99.2	96.7	104.7	100.0	96.2	100.8	98.1	96.2	103.5	99.2	100.8	104.5	102.3
0330	Cast-in-Place Concrete	90.8	91.4	91.0	100.7	88.2	95.9	88.5	91.7	89.7	79.5	93.8	85.0	88.5	100.5	93.1	86.8	87.4	87.0
03	**CONCRETE**	92.4	92.2	92.3	100.0	90.5	95.7	87.7	96.0	91.5	93.0	92.0	92.6	96.8	107.3	101.6	95.7	90.7	93.4
04	**MASONRY**	91.5	91.9	91.7	92.3	85.1	88.0	102.7	91.9	96.2	100.4	88.4	93.2	94.9	99.4	97.6	94.9	94.5	94.6
05	**METALS**	92.7	118.1	100.1	92.6	116.3	99.5	88.9	119.5	97.9	93.3	116.9	100.2	94.9	119.2	102.0	93.4	114.7	99.6
06	**WOOD, PLASTICS & COMPOSITES**	80.2	82.7	81.5	78.4	82.7	80.7	81.6	93.3	88.0	92.5	83.0	87.3	93.4	116.0	105.8	81.2	83.4	82.4
07	**THERMAL & MOISTURE PROTECTION**	102.3	93.9	99.0	102.5	91.9	98.4	99.2	96.7	98.2	99.1	98.6	98.9	100.1	106.3	102.5	99.7	93.4	97.3
08	**OPENINGS**	89.3	95.8	90.9	89.3	91.7	89.9	88.4	101.6	91.6	90.8	92.2	91.1	94.7	110.9	98.7	91.0	92.1	91.3
0920	Plaster & Gypsum Board	92.8	81.8	85.6	91.4	81.8	85.2	106.6	92.8	97.6	111.9	82.1	92.6	108.5	116.3	113.6	105.6	82.5	90.6
0950, 0980	Ceilings & Acoustic Treatment	85.5	81.8	83.2	84.6	81.8	82.9	92.5	92.8	92.7	90.0	82.1	85.1	88.3	116.3	105.6	90.0	82.5	85.4
0960	Flooring	90.4	99.5	92.9	89.6	65.1	83.0	80.0	99.5	85.3	83.7	90.0	85.4	81.6	94.7	85.1	76.7	64.9	73.5
0970, 0990	Wall Finishes & Painting/Coating	97.0	105.5	102.1	97.0	105.7	102.2	84.2	105.5	97.0	88.3	65.7	74.7	88.3	76.4	81.1	88.3	98.0	94.1
09	**FINISHES**	93.1	88.3	90.5	93.0	81.5	86.9	93.8	94.5	94.2	93.7	81.1	87.0	93.5	105.0	99.6	92.0	82.0	86.7
COVERS	**DIVS. 10 - 14, 25, 28, 41, 43, 44**	100.0	97.2	99.5	100.0	96.6	99.3	100.0	98.5	99.7	100.0	88.9	97.8	100.0	103.5	100.7	100.0	98.0	99.6
21, 22, 23	**FIRE SUPPRESSION, PLUMBING & HVAC**	94.1	89.8	92.4	94.1	88.8	92.0	94.0	93.8	94.0	94.5	91.8	93.4	94.5	109.1	100.3	94.5	88.2	92.0
26, 27, 3370	**ELECTRICAL, COMMUNICATIONS & UTIL.**	94.4	98.0	96.1	94.4	98.0	96.1	90.6	98.0	94.2	94.9	51.4	73.7	95.9	134.6	114.7	95.0	95.8	95.4
MF2004	**WEIGHTED AVERAGE**	93.9	95.6	94.6	95.1	93.2	94.3	92.4	98.6	95.0	94.2	87.9	91.6	95.5	111.4	102.2	94.4	93.7	94.1

DIVISION		PENNSYLVANIA																	
		NEW CASTLE			NORRISTOWN			OIL CITY			PHILADELPHIA			PITTSBURGH			POTTSVILLE		
		161			194			163			190 - 191			150 - 152			179		
		MAT.	INST.	TOTAL	MAT.	INST.	TOTAL	MAT.	INST.	TOTAL	MAT.	INST.	TOTAL	MAT.	INST.	TOTAL	MAT.	INST.	TOTAL
015433	**CONTRACTOR EQUIPMENT**		112.9	112.9		97.3	97.3		112.9	112.9		96.8	96.8		111.7	111.7		112.1	112.1
0241, 31 - 34	**SITE & INFRASTRUCTURE, DEMOLITION**	89.3	105.2	100.1	95.8	98.3	97.5	87.6	104.0	98.8	103.0	97.5	99.3	108.6	105.1	106.2	81.9	102.7	96.1
0310	Concrete Forming & Accessories	86.0	92.3	91.4	84.5	127.7	121.8	86.0	82.0	82.6	100.8	135.8	131.0	99.7	93.0	93.9	79.0	84.7	83.9
0320	Concrete Reinforcing	95.5	95.2	95.4	98.2	140.2	115.3	96.7	95.0	96.0	101.3	140.6	117.3	96.1	104.8	99.7	95.4	97.4	96.2
0330	Cast-in-Place Concrete	85.9	91.5	88.1	88.9	127.4	103.6	83.5	94.5	87.7	109.1	127.5	116.1	96.2	91.5	94.4	84.7	81.2	83.4
03	**CONCRETE**	85.5	93.9	89.3	95.4	129.5	110.9	84.0	90.3	86.9	108.2	133.1	119.6	97.7	95.9	96.9	96.8	87.4	92.5
04	**MASONRY**	98.3	90.4	93.5	109.5	127.6	120.3	98.8	89.7	93.4	96.6	134.8	119.5	86.6	93.2	90.5	94.6	90.2	91.9
05	**METALS**	88.9	115.0	96.5	98.5	127.0	106.8	88.9	113.2	96.0	102.3	127.7	109.7	93.9	118.6	101.1	93.5	115.3	99.9
06	**WOOD, PLASTICS & COMPOSITES**	81.6	93.3	88.0	81.3	126.5	106.1	81.6	79.2	80.3	100.0	137.2	120.5	96.0	93.2	94.5	82.9	82.9	82.9
07	**THERMAL & MOISTURE PROTECTION**	99.1	95.0	97.5	98.9	130.9	111.3	99.0	94.1	97.1	99.5	133.9	112.9	102.6	96.6	100.3	99.2	103.3	100.8
08	**OPENINGS**	88.4	98.5	90.9	89.5	135.6	100.9	88.4	82.4	86.9	97.7	141.5	108.5	91.8	102.6	94.5	90.8	89.7	90.5
0920	Plaster & Gypsum Board	106.6	92.8	97.6	96.8	127.4	116.6	106.6	78.2	88.1	103.8	138.5	126.3	99.1	92.8	95.0	107.5	82.0	90.9
0950, 0980	Ceilings & Acoustic Treatment	92.5	92.8	92.7	96.4	127.4	115.6	92.5	78.2	83.6	96.4	138.5	122.4	85.5	92.8	90.0	90.0	82.0	85.1
0960	Flooring	80.0	64.4	75.8	76.2	136.6	92.6	80.0	99.5	85.3	84.3	136.6	98.5	97.3	99.5	97.9	79.6	90.0	82.4
0970, 0990	Wall Finishes & Painting/Coating	84.2	105.7	97.1	97.3	144.5	125.6	84.2	105.5	97.0	97.3	144.5	125.6	97.0	105.7	102.2	88.3	98.0	94.1
09	**FINISHES**	93.7	88.0	90.7	96.7	131.0	114.8	93.6	85.8	89.5	100.4	137.3	119.9	96.3	94.8	95.5	92.1	86.9	89.3
COVERS	**DIVS. 10 - 14, 25, 28, 41, 43, 44**	100.0	98.5	99.7	100.0	117.1	103.4	100.0	97.5	99.5	100.0	120.3	104.0	100.0	98.5	99.7	100.0	89.7	98.0
21, 22, 23	**FIRE SUPPRESSION, PLUMBING & HVAC**	94.0	85.2	90.5	94.2	127.6	107.5	94.0	85.8	90.7	100.0	127.6	111.0	99.9	93.8	97.5	94.5	87.7	91.8
26, 27, 3370	**ELECTRICAL, COMMUNICATIONS & UTIL.**	91.2	96.3	93.7	94.2	137.5	115.3	93.1	97.3	95.2	98.7	137.5	117.6	96.7	98.0	97.3	93.1	88.0	90.6
MF2004	**WEIGHTED AVERAGE**	91.9	94.7	93.1	96.2	127.4	109.3	91.9	92.9	92.3	101.0	130.0	113.2	96.9	98.7	97.6	94.1	92.3	93.4

DIVISION		PENNSYLVANIA																	
		READING			SCRANTON			STATE COLLEGE			STROUDSBURG			SUNBURY			UNIONTOWN		
		195 - 196			184 - 185			168			183			178			154		
		MAT.	INST.	TOTAL	MAT.	INST.	TOTAL	MAT.	INST.	TOTAL	MAT.	INST.	TOTAL	MAT.	INST.	TOTAL	MAT.	INST.	TOTAL
015433	**CONTRACTOR EQUIPMENT**		114.5	114.5		112.9	112.9		112.1	112.1		112.9	112.9		112.9	112.9		111.0	111.0
0241, 31 - 34	**SITE & INFRASTRUCTURE, DEMOLITION**	100.1	107.9	105.4	93.7	104.3	100.9	84.2	102.7	96.9	88.1	102.7	98.1	92.5	103.9	100.3	103.6	103.7	103.7
0310	Concrete Forming & Accessories	100.2	84.7	86.8	99.4	82.7	85.0	84.3	80.4	81.0	87.5	85.4	85.7	90.8	83.8	84.7	79.4	92.5	90.7
0320	Concrete Reinforcing	99.5	97.5	98.7	99.1	104.7	101.4	97.3	93.1	95.6	99.4	103.0	100.9	98.1	100.9	99.2	95.0	104.6	98.9
0330	Cast-in-Place Concrete	79.7	96.6	86.1	90.4	88.9	89.8	87.1	73.3	81.8	85.1	77.8	82.4	92.7	93.3	92.9	90.8	91.3	91.0
03	**CONCRETE**	94.1	92.8	93.5	99.6	90.5	95.5	99.7	81.9	91.6	94.4	87.4	91.2	100.3	91.8	96.4	92.1	95.6	93.7
04	**MASONRY**	97.8	91.1	93.8	95.2	94.8	95.0	100.4	81.2	88.9	91.8	99.0	96.1	94.1	87.9	90.4	106.5	90.3	96.8
05	**METALS**	98.8	116.2	103.9	97.1	119.1	103.5	93.2	112.4	98.8	95.0	113.6	100.4	93.3	116.7	100.1	92.4	117.9	99.9
06	**WOOD, PLASTICS & COMPOSITES**	100.2	82.0	90.2	101.1	78.9	88.9	88.7	82.7	85.4	87.4	83.4	85.2	90.9	83.0	86.6	71.8	93.2	83.5
07	**THERMAL & MOISTURE PROTECTION**	99.1	107.7	102.5	100.1	98.2	99.4	99.4	91.5	96.3	99.9	83.4	93.5	100.0	100.3	100.1	102.2	95.9	99.7
08	**OPENINGS**	95.2	90.6	94.1	94.0	89.9	93.0	90.7	88.3	90.1	94.7	84.0	92.1	90.9	90.8	90.9	89.3	102.6	92.5
0920	Plaster & Gypsum Board	107.4	81.2	90.4	115.0	77.8	90.9	108.7	81.8	91.2	106.6	82.5	90.9	107.5	82.1	91.0	88.8	92.7	91.4
0950, 0980	Ceilings & Acoustic Treatment	86.3	81.2	83.2	96.6	77.8	85.0	87.5	81.8	84.0	86.7	82.5	84.1	87.5	82.1	84.2	84.6	92.7	89.6
0960	Flooring	82.5	91.3	84.9	85.1	107.4	91.1	82.1	55.8	75.0	79.3	57.6	73.4	80.6	47.6	71.7	86.8	71.8	82.8
0970, 0990	Wall Finishes & Painting/Coating	95.8	99.0	97.7	88.3	98.0	94.1	88.3	105.7	98.7	88.3	74.2	79.9	88.3	98.0	94.1	97.0	105.7	102.2
09	**FINISHES**	98.0	86.1	91.7	97.5	88.0	92.5	92.5	77.8	84.7	92.0	78.8	85.0	93.1	79.0	85.6	91.3	89.5	90.3
COVERS	**DIVS. 10 - 14, 25, 28, 41, 43, 44**	100.0	96.7	99.4	100.0	97.5	99.5	100.0	86.3	97.3	100.0	72.9	94.6	100.0	88.9	97.8	100.0	98.5	99.7
21, 22, 23	**FIRE SUPPRESSION, PLUMBING & HVAC**	100.2	105.6	102.3	100.2	88.8	95.7	94.5	87.6	91.8	94.5	90.2	92.8	94.5	91.4	93.3	94.1	84.6	90.3
26, 27, 3370	**ELECTRICAL, COMMUNICATIONS & UTIL.**	100.9	88.0	94.6	101.3	95.8	98.7	94.0	98.0	95.9	95.9	143.1	118.8	93.6	85.2	89.5	91.4	98.0	94.6
MF2004	**WEIGHTED AVERAGE**	98.5	97.6	98.1	98.5	95.2	97.1	94.9	90.1	92.9	94.8	98.8	96.5	95.0	92.2	93.8	94.1	95.5	94.7

DIVISION		PENNSYLVANIA																	
		WASHINGTON			WELLSBORO			WESTCHESTER			WILKES-BARRE			WILLIAMSPORT			YORK		
		153			169			193			186 - 187			177			173 - 174		
		MAT.	INST.	TOTAL	MAT.	INST.	TOTAL	MAT.	INST.	TOTAL	MAT.	INST.	TOTAL	MAT.	INST.	TOTAL	MAT.	INST.	TOTAL
015433	CONTRACTOR EQUIPMENT		111.0	111.0		112.9	112.9		97.3	97.3		112.9	112.9		112.9	112.9		112.1	112.1
0241, 31 - 34	SITE & INFRASTRUCTURE, DEMOLITION	103.6	103.8	103.7	97.0	102.5	100.8	102.4	96.4	98.3	86.1	104.2	98.5	83.2	103.0	96.7	82.4	102.8	96.4
0310	Concrete Forming & Accessories	86.4	92.9	92.0	86.2	79.7	80.6	91.2	130.9	125.5	90.0	85.6	86.2	87.5	65.6	68.6	82.3	84.5	84.2
0320	Concrete Reinforcing	95.0	104.7	99.0	97.3	104.4	100.2	97.2	107.9	101.6	98.1	104.7	100.8	97.3	74.9	88.1	98.1	100.9	99.3
0330	Cast-in-Place Concrete	90.8	91.5	91.1	91.3	85.1	89.0	98.4	127.1	109.4	81.7	88.6	84.3	78.4	77.8	78.2	85.2	94.1	88.6
03	CONCRETE	92.5	95.8	94.0	102.0	87.7	95.5	103.2	124.8	113.0	91.4	91.7	91.5	88.3	73.6	81.6	98.1	92.5	95.5
04	MASONRY	91.5	91.9	91.7	100.0	88.2	92.9	104.0	127.7	118.2	107.7	94.4	99.8	86.4	89.6	88.3	95.1	88.4	91.1
05	METALS	92.4	118.4	100.0	93.3	115.0	99.6	98.5	112.6	102.6	93.4	118.9	100.8	93.3	104.8	96.6	94.7	117.6	101.4
06	WOOD, PLASTICS & COMPOSITES	80.3	93.2	87.4	85.9	79.0	82.1	88.9	133.0	113.1	90.0	83.2	86.3	87.4	59.7	72.2	86.7	83.0	84.7
07	THERMAL & MOISTURE PROTECTION	102.3	96.3	100.0	100.4	90.6	96.5	99.2	126.5	109.8	99.7	103.4	101.2	99.6	98.0	99.0	99.3	105.7	101.8
08	OPENINGS	89.3	102.6	92.5	94.4	89.8	93.3	89.5	124.9	98.3	91.0	93.6	91.6	90.9	62.6	83.9	90.8	90.8	90.8
0920	Plaster & Gypsum Board	92.6	92.7	92.7	106.4	78.0	87.9	99.0	134.1	121.8	107.8	82.4	91.3	107.5	58.0	75.3	108.9	82.1	91.5
0950, 0980	Ceilings & Acoustic Treatment	84.6	92.7	89.6	87.5	78.0	81.6	96.4	134.1	119.7	90.0	82.4	85.3	90.0	58.0	70.2	89.1	82.1	84.8
0960	Flooring	90.5	99.5	92.9	78.6	56.9	72.7	79.8	136.6	95.2	80.3	93.2	83.8	79.3	55.8	72.9	81.1	90.0	83.5
0970, 0990	Wall Finishes & Painting/Coating	97.0	105.7	102.2	88.3	98.0	94.1	97.3	144.5	125.6	88.3	98.0	94.1	88.3	98.0	94.1	88.3	87.7	88.0
09	FINISHES	92.9	94.6	93.8	92.7	76.7	84.2	98.5	132.8	116.7	93.1	87.6	90.2	92.6	65.5	78.3	92.5	85.3	88.7
COVERS	DIVS. 10 - 14, 25, 28, 41, 43, 44	100.0	98.5	99.7	100.0	88.4	97.7	100.0	117.7	103.5	100.0	97.8	99.6	100.0	86.3	97.3	100.0	89.1	97.8
21, 22, 23	FIRE SUPPRESSION, PLUMBING & HVAC	94.1	93.8	94.0	94.5	90.8	93.0	94.2	125.7	106.7	94.5	84.0	90.3	94.5	91.5	93.3	100.2	91.9	96.9
26, 27, 3370	ELECTRICAL, COMMUNICATIONS & UTIL.	93.7	98.0	95.8	95.0	85.9	90.6	94.1	104.3	99.1	95.9	82.2	89.2	94.1	54.4	74.7	94.9	84.4	89.8
MF2004	WEIGHTED AVERAGE	93.8	98.4	95.7	96.0	90.7	93.8	97.2	119.9	106.8	94.7	92.7	93.8	92.9	81.1	87.9	96.1	93.3	94.9

DIVISION		PUERTO RICO			RHODE ISLAND						SOUTH CAROLINA								
		SAN JUAN			NEWPORT			PROVIDENCE			AIKEN			BEAUFORT			CHARLESTON		
		009			028			029			298			299			294		
		MAT.	INST.	TOTAL	MAT.	INST.	TOTAL	MAT.	INST.	TOTAL	MAT.	INST.	TOTAL	MAT.	INST.	TOTAL	MAT.	INST.	TOTAL
015433	CONTRACTOR EQUIPMENT		85.9	85.9		102.0	102.0		102.0	102.0		101.7	101.7		101.7	101.7		101.7	101.7
0241, 31 - 34	SITE & INFRASTRUCTURE, DEMOLITION	118.7	82.9	94.2	78.2	103.3	95.4	81.6	103.3	96.4	126.1	86.7	99.1	120.6	86.0	96.9	101.6	85.5	90.5
0310	Concrete Forming & Accessories	90.3	18.4	28.2	99.5	112.2	110.4	98.4	112.2	110.3	100.8	67.3	71.8	99.5	43.5	51.1	98.5	45.4	52.6
0320	Concrete Reinforcing	180.4	10.8	111.2	100.4	113.9	105.9	100.4	113.9	105.9	102.8	65.2	87.4	101.8	30.1	72.6	101.6	64.9	86.6
0330	Cast-in-Place Concrete	100.0	31.3	73.7	77.4	117.2	92.6	94.2	117.2	103.0	81.4	70.1	77.1	81.4	53.4	70.7	95.5	54.1	79.7
03	CONCRETE	114.5	22.7	72.8	99.5	113.8	106.1	107.3	113.8	110.3	111.3	69.2	92.1	108.3	46.7	80.3	102.2	53.8	80.2
04	MASONRY	88.1	15.1	44.4	103.4	123.6	115.5	107.4	123.6	117.1	79.5	59.2	67.3	95.3	38.1	61.1	96.8	42.6	64.4
05	METALS	107.3	37.5	86.9	93.8	108.6	98.1	93.8	108.6	98.1	87.2	86.1	86.9	87.2	74.5	83.5	88.9	82.3	87.0
06	WOOD, PLASTICS & COMPOSITES	95.1	18.6	53.1	102.9	110.1	106.8	102.0	110.1	106.4	101.8	68.5	83.5	99.9	44.3	69.4	98.6	44.3	68.8
07	THERMAL & MOISTURE PROTECTION	131.3	21.7	88.6	94.8	114.7	102.5	93.0	114.7	101.4	103.6	66.3	89.1	103.3	41.4	79.2	102.5	46.4	80.6
08	OPENINGS	152.2	15.3	118.4	98.9	112.2	102.2	99.5	112.2	102.6	94.2	65.5	87.1	94.2	42.9	81.5	98.3	48.3	85.9
0920	Plaster & Gypsum Board	119.4	15.5	51.9	102.7	109.7	107.2	103.3	109.7	107.5	109.5	67.2	82.0	113.1	42.1	66.9	114.8	42.1	67.5
0950, 0980	Ceilings & Acoustic Treatment	219.3	15.5	93.2	96.5	109.7	104.7	95.9	109.7	104.4	92.6	67.2	76.9	97.6	42.1	63.2	97.6	42.1	63.2
0960	Flooring	199.5	14.7	149.6	97.6	126.8	105.5	97.2	126.8	105.2	110.1	64.3	97.7	112.0	57.4	97.2	111.5	65.3	99.0
0970, 0990	Wall Finishes & Painting/Coating	198.5	13.7	87.6	93.9	120.7	110.0	93.9	120.7	110.0	114.4	69.5	87.4	114.4	39.9	69.7	114.4	45.0	72.8
09	FINISHES	198.7	17.4	102.9	98.1	115.9	107.5	98.0	115.9	107.5	105.9	67.3	85.5	107.7	46.1	75.1	105.8	48.6	75.6
COVERS	DIVS. 10 - 14, 25, 28, 41, 43, 44	100.0	19.5	84.1	100.0	98.2	99.6	100.0	98.2	99.6	100.0	74.0	94.9	100.0	54.4	91.0	100.0	68.9	93.8
21, 22, 23	FIRE SUPPRESSION, PLUMBING & HVAC	104.1	13.7	68.1	100.2	107.2	103.0	100.0	107.2	102.9	94.4	66.6	83.3	94.4	28.5	68.2	100.2	44.8	78.1
26, 27, 3370	ELECTRICAL, COMMUNICATIONS & UTIL.	122.6	13.8	69.6	96.3	94.5	95.4	95.9	94.5	95.2	95.2	66.7	81.3	98.8	35.3	67.9	97.2	103.6	100.3
MF2004	WEIGHTED AVERAGE	120.1	23.9	79.7	97.7	109.1	102.5	98.8	109.1	103.2	97.1	70.0	85.7	97.9	46.2	76.2	98.4	62.2	83.2

DIVISION		SOUTH CAROLINA															SOUTH DAKOTA		
		COLUMBIA			FLORENCE			GREENVILLE			ROCK HILL			SPARTANBURG			ABERDEEN		
		290 - 292			295			296			297			293			574		
		MAT.	INST.	TOTAL	MAT.	INST.	TOTAL	MAT.	INST.	TOTAL	MAT.	INST.	TOTAL	MAT.	INST.	TOTAL	MAT.	INST.	TOTAL
015433	CONTRACTOR EQUIPMENT		101.7	101.7		101.7	101.7		101.7	101.7		101.7	101.7		101.7	101.7		98.7	98.7
0241, 31 - 34	SITE & INFRASTRUCTURE, DEMOLITION	102.8	85.4	90.9	113.3	85.4	94.2	107.6	85.0	92.2	105.1	84.9	91.3	107.4	85.0	92.1	90.7	95.3	93.8
0310	Concrete Forming & Accessories	100.5	44.3	52.0	86.2	44.5	50.2	98.2	44.3	51.7	96.7	41.7	49.2	101.5	44.3	52.1	94.9	43.0	50.0
0320	Concrete Reinforcing	101.6	64.7	86.5	101.2	64.8	86.4	101.1	51.7	81.0	102.0	51.5	81.4	101.1	51.7	81.0	109.5	44.4	82.9
0330	Cast-in-Place Concrete	102.3	54.7	84.1	81.4	53.9	70.9	81.4	53.8	70.8	81.4	48.5	68.8	81.4	53.8	70.8	100.0	49.9	80.8
03	CONCRETE	105.4	53.6	81.8	102.5	53.4	80.1	100.8	50.9	78.1	98.9	47.9	75.7	101.0	50.9	78.2	99.9	47.2	75.9
04	MASONRY	91.5	40.3	60.9	79.6	42.6	57.4	77.6	42.6	56.6	102.9	37.4	63.7	79.6	42.6	57.4	104.3	60.4	78.0
05	METALS	88.9	81.5	86.7	87.9	81.8	86.1	87.9	77.3	84.8	87.2	76.8	84.2	87.9	77.3	84.8	99.1	70.6	90.8
06	WOOD, PLASTICS & COMPOSITES	100.9	43.8	69.5	83.8	43.8	61.8	98.4	43.8	68.4	96.6	41.6	66.4	102.5	43.8	70.2	101.5	42.2	68.9
07	THERMAL & MOISTURE PROTECTION	101.6	45.0	79.6	102.7	46.3	80.8	102.7	46.3	80.8	102.6	43.7	79.7	102.7	46.3	80.8	101.1	53.0	82.4
08	OPENINGS	98.3	48.1	85.9	94.3	48.1	82.8	94.2	45.0	82.0	94.2	43.9	81.8	94.2	45.0	82.0	94.8	43.0	82.0
0920	Plaster & Gypsum Board	108.1	41.5	64.9	102.7	41.5	63.0	108.1	41.5	64.8	107.3	39.3	63.1	109.9	41.5	65.5	100.3	40.4	61.4
0950, 0980	Ceilings & Acoustic Treatment	95.1	41.5	62.0	93.4	41.5	61.3	92.6	41.5	61.0	92.6	39.3	59.7	92.6	41.5	61.0	112.0	40.4	67.7
0960	Flooring	107.9	48.9	92.0	101.4	48.9	87.2	108.7	64.1	96.6	107.6	49.8	92.0	110.4	64.1	97.9	116.2	56.4	100.1
0970, 0990	Wall Finishes & Painting/Coating	111.9	45.0	71.7	114.4	45.0	72.8	114.4	45.0	72.8	114.4	45.0	72.8	114.4	45.0	72.8	98.4	42.7	65.0
09	FINISHES	102.8	45.0	72.3	101.3	45.5	71.8	103.6	48.1	74.3	102.9	43.3	71.4	104.4	48.1	74.6	108.6	45.0	75.0
COVERS	DIVS. 10 - 14, 25, 28, 41, 43, 44	100.0	68.6	93.8	100.0	68.6	93.8	100.0	68.6	93.8	100.0	67.6	93.6	100.0	68.6	93.8	100.0	51.0	90.3
21, 22, 23	FIRE SUPPRESSION, PLUMBING & HVAC	99.9	39.6	76.0	100.2	39.7	76.1	100.2	38.0	75.5	94.4	37.0	71.5	100.2	38.0	75.5	100.1	47.9	79.4
26, 27, 3370	ELECTRICAL, COMMUNICATIONS & UTIL.	101.3	53.0	77.8	95.2	50.4	73.4	97.3	54.5	76.4	97.3	54.5	76.4	97.3	54.5	76.4	97.9	59.7	79.3
MF2004	WEIGHTED AVERAGE	98.6	53.3	79.6	96.6	53.3	78.4	96.7	52.9	78.3	96.1	50.9	77.1	96.9	52.9	78.4	99.9	56.3	81.6

	DIVISION	SOUTH DAKOTA																	
		MITCHELL			MOBRIDGE			PIERRE			RAPID CITY			SIOUX FALLS			WATERTOWN		
		573			576			575			577			570 - 571			572		
		MAT.	INST.	TOTAL	MAT.	INST.	TOTAL	MAT.	INST.	TOTAL	MAT.	INST.	TOTAL	MAT.	INST.	TOTAL	MAT.	INST.	TOTAL
015433	CONTRACTOR EQUIPMENT		98.7	98.7		98.7	98.7		98.7	98.7		98.7	98.7		99.4	99.4		98.7	98.7
0241, 31 - 34	SITE & INFRASTRUCTURE, DEMOLITION	87.8	95.3	92.9	87.7	95.3	92.9	89.8	95.3	93.6	88.9	95.4	93.3	90.5	96.6	94.7	87.6	95.3	92.9
0310	Concrete Forming & Accessories	94.1	43.4	50.3	85.6	43.1	48.9	93.5	43.9	50.7	101.2	41.7	49.8	94.3	46.7	53.2	82.3	42.9	48.3
0320	Concrete Reinforcing	108.8	54.2	86.5	111.6	44.4	84.2	109.0	55.9	87.3	102.3	55.7	83.3	102.3	55.9	83.4	105.8	44.4	80.8
0330	Cast-in-Place Concrete	97.1	48.5	78.6	97.1	49.9	79.1	92.0	48.0	75.2	96.4	48.6	78.1	91.9	49.0	75.5	97.1	49.9	79.1
03	CONCRETE	97.8	48.5	75.4	97.8	47.3	74.8	95.4	49.1	74.3	96.5	48.3	74.6	93.0	50.7	73.8	96.5	47.2	74.1
04	MASONRY	93.1	59.8	73.1	101.1	60.4	76.7	99.8	60.0	76.0	101.1	55.9	74.0	97.2	58.1	73.8	125.4	60.1	86.3
05	METALS	98.3	70.9	90.3	98.3	70.5	90.2	99.1	75.7	92.3	100.8	75.7	93.4	100.9	76.2	93.7	98.3	70.9	90.3
06	WOOD, PLASTICS & COMPOSITES	100.5	42.8	68.8	90.3	42.4	64.0	99.3	43.0	68.4	105.2	39.1	68.9	100.2	46.0	70.5	86.6	42.2	62.2
07	THERMAL & MOISTURE PROTECTION	100.9	51.0	81.5	101.0	53.0	82.3	100.9	51.3	81.6	101.5	52.9	82.6	102.6	54.7	84.0	100.8	52.9	82.1
08	OPENINGS	93.3	43.6	81.0	96.2	42.5	82.9	98.0	47.0	85.4	99.0	44.9	85.6	99.2	48.6	86.7	93.3	42.7	80.7
0920	Plaster & Gypsum Board	98.9	41.0	61.3	94.3	40.7	59.5	94.3	41.3	59.9	100.2	37.2	59.3	99.8	44.4	63.8	92.2	40.4	58.5
0950, 0980	Ceilings & Acoustic Treatment	109.5	41.0	67.1	112.0	40.7	67.9	107.0	41.3	66.4	116.2	37.2	67.3	109.5	44.4	69.2	109.5	40.4	66.8
0960	Flooring	115.8	56.4	99.7	110.1	56.4	95.6	115.5	40.2	95.1	115.5	71.0	103.5	115.5	76.7	105.0	108.4	56.4	94.4
0970, 0990	Wall Finishes & Painting/Coating	98.4	46.7	67.4	98.4	47.9	68.1	98.4	51.2	70.1	98.4	51.2	70.1	98.4	51.2	70.1	98.4	42.7	65.0
09	FINISHES	107.5	45.8	74.9	105.8	45.8	74.0	106.2	43.3	73.0	109.2	47.4	76.5	107.7	52.4	78.5	104.4	45.0	73.0
COVERS	DIVS. 10 - 14, 25, 28, 41, 43, 44	100.0	47.3	89.6	100.0	51.0	90.3	100.0	75.8	95.2	100.0	75.7	95.2	100.0	76.2	95.3	100.0	51.0	90.3
21, 22, 23	FIRE SUPPRESSION, PLUMBING & HVAC	94.3	47.5	75.7	94.3	46.3	75.2	100.0	47.5	79.2	100.1	48.0	79.4	100.0	46.2	78.6	94.3	48.1	75.9
26, 27, 3370	ELECTRICAL, COMMUNICATIONS & UTIL.	96.3	51.3	74.4	97.9	52.3	75.7	100.9	51.8	77.0	94.6	51.8	73.8	102.9	69.1	86.5	95.5	52.3	74.5
MF2004	WEIGHTED AVERAGE	97.1	55.2	79.5	97.7	55.0	79.8	99.5	56.5	81.4	99.7	56.5	81.5	99.9	60.1	83.2	98.1	55.3	80.1

	DIVISION	TENNESSEE																	
		CHATTANOOGA			COLUMBIA			COOKEVILLE			JACKSON			JOHNSON CITY			KNOXVILLE		
		373 - 374			384			385			383			376			377 - 379		
		MAT.	INST.	TOTAL	MAT.	INST.	TOTAL	MAT.	INST.	TOTAL	MAT.	INST.	TOTAL	MAT.	INST.	TOTAL	MAT.	INST.	TOTAL
015433	CONTRACTOR EQUIPMENT		103.7	103.7		98.9	98.9		98.9	98.9		103.8	103.8		98.8	98.8		98.8	98.8
0241, 31 - 34	SITE & INFRASTRUCTURE, DEMOLITION	102.0	95.5	97.6	92.0	86.7	88.4	98.5	86.8	90.5	100.7	93.9	96.1	111.5	86.6	94.5	88.0	86.6	87.1
0310	Concrete Forming & Accessories	94.6	42.0	49.2	81.5	47.0	51.7	81.6	40.2	45.8	88.1	34.0	41.4	82.1	42.7	48.1	93.7	42.7	49.6
0320	Concrete Reinforcing	84.7	59.6	74.5	87.3	60.2	76.3	87.3	60.2	76.3	87.3	37.9	67.2	85.3	58.6	74.4	84.7	58.6	74.1
0330	Cast-in-Place Concrete	97.3	46.6	78.0	92.8	51.2	76.9	105.2	48.9	83.7	102.7	40.9	79.1	78.3	50.5	67.7	91.4	46.8	74.4
03	CONCRETE	90.6	49.1	71.7	92.5	53.2	74.6	101.7	49.3	77.9	93.9	39.5	69.2	95.8	50.5	75.2	88.1	49.2	70.4
04	MASONRY	97.7	38.4	62.2	109.0	48.5	72.7	104.5	48.8	71.2	109.4	29.9	61.8	108.7	38.3	66.5	75.1	38.3	53.0
05	METALS	103.4	83.8	97.7	96.0	88.1	93.7	96.1	87.3	93.5	98.1	72.8	90.7	100.4	83.4	95.4	104.0	83.6	98.1
06	WOOD, PLASTICS & COMPOSITES	99.3	42.3	68.0	70.3	48.8	58.5	70.5	39.4	53.5	86.0	33.9	57.4	74.0	43.3	57.1	87.6	43.3	63.3
07	THERMAL & MOISTURE PROTECTION	95.4	54.6	79.6	89.9	55.5	76.5	90.3	50.1	74.6	92.0	36.3	70.3	90.2	54.7	76.4	88.4	54.2	75.1
08	OPENINGS	99.4	48.4	86.8	91.5	49.4	81.1	91.5	45.9	80.3	100.5	35.4	84.4	95.3	49.3	83.9	92.2	49.3	81.6
0920	Plaster & Gypsum Board	82.6	40.8	55.4	90.7	47.4	62.6	90.7	37.7	56.3	90.6	32.0	52.5	96.2	41.8	60.8	102.7	41.8	63.1
0950, 0980	Ceilings & Acoustic Treatment	97.1	40.8	62.3	86.0	47.4	62.2	86.0	37.7	56.2	95.1	32.0	56.1	93.5	41.8	61.5	94.3	41.8	61.8
0960	Flooring	99.5	44.3	84.6	91.1	23.0	72.7	91.3	21.5	72.4	91.7	21.5	72.7	93.8	45.6	80.7	99.2	45.6	84.7
0970, 0990	Wall Finishes & Painting/Coating	108.8	41.3	68.3	98.9	38.4	62.6	98.9	41.9	64.7	100.4	27.9	56.9	106.1	48.4	71.5	106.1	48.4	71.5
09	FINISHES	93.8	41.9	66.4	93.2	41.6	65.9	93.8	35.6	63.0	94.0	31.1	60.7	95.6	43.3	68.0	90.7	43.3	65.6
COVERS	DIVS. 10 - 14, 25, 28, 41, 43, 44	100.0	46.6	89.4	100.0	52.4	90.6	100.0	51.5	90.4	100.0	50.4	90.2	100.0	53.0	90.7	100.0	53.0	90.7
21, 22, 23	FIRE SUPPRESSION, PLUMBING & HVAC	100.0	39.7	76.0	97.0	46.5	76.9	97.0	60.1	82.3	99.9	48.8	79.6	99.7	52.2	80.8	99.7	52.2	80.8
26, 27, 3370	ELECTRICAL, COMMUNICATIONS & UTIL.	106.5	69.3	88.4	96.1	62.2	79.6	97.8	62.2	80.4	104.1	45.8	75.7	95.0	53.9	75.0	100.8	58.3	80.1
MF2004	WEIGHTED AVERAGE	99.4	55.0	80.8	95.6	57.0	79.4	97.0	58.1	80.6	99.0	47.9	77.5	98.4	55.4	80.3	95.8	55.8	79.0

	DIVISION	TENNESSEE									TEXAS								
		MCKENZIE			MEMPHIS			NASHVILLE			ABILENE			AMARILLO			AUSTIN		
		382			375,380 - 381			370 - 372			795 - 796			790 - 791			786 - 787		
		MAT.	INST.	TOTAL	MAT.	INST.	TOTAL	MAT.	INST.	TOTAL	MAT.	INST.	TOTAL	MAT.	INST.	TOTAL	MAT.	INST.	TOTAL
015433	CONTRACTOR EQUIPMENT		98.9	98.9		101.9	101.9		104.9	104.9		90.7	90.7		90.7	90.7		90.0	90.0
0241, 31 - 34	SITE & INFRASTRUCTURE, DEMOLITION	97.9	86.5	90.2	90.4	91.5	91.1	96.9	98.6	98.1	102.0	88.4	92.7	103.2	89.3	93.7	91.2	88.8	89.5
0310	Concrete Forming & Accessories	89.4	33.9	41.5	94.8	60.5	65.2	94.0	64.0	68.1	97.9	39.2	47.2	98.2	49.2	55.9	91.9	54.1	59.3
0320	Concrete Reinforcing	87.5	31.4	64.6	91.0	65.6	80.6	91.0	64.5	80.2	96.6	53.3	78.9	96.6	48.0	76.8	97.4	46.2	76.5
0330	Cast-in-Place Concrete	102.9	43.7	80.3	86.6	66.5	78.9	90.5	67.4	81.7	94.2	42.3	74.4	93.3	47.5	75.8	91.5	48.0	74.9
03	CONCRETE	100.3	39.4	72.6	87.1	65.3	77.2	89.3	66.9	79.1	90.3	44.4	69.4	90.0	49.6	71.6	84.5	51.4	69.5
04	MASONRY	107.6	47.8	71.8	87.8	68.0	76.0	83.4	62.6	71.0	100.4	50.3	70.4	100.1	45.8	67.6	94.0	48.8	66.9
05	METALS	96.1	75.4	90.1	105.1	91.0	101.0	105.1	89.4	100.5	102.8	70.5	93.4	102.8	68.4	92.8	98.1	65.3	88.5
06	WOOD, PLASTICS & COMPOSITES	79.7	31.9	53.5	91.7	61.1	74.9	97.1	64.9	79.4	100.6	38.3	66.4	99.0	51.7	73.0	96.5	58.1	75.4
07	THERMAL & MOISTURE PROTECTION	90.3	47.6	73.7	90.1	67.5	81.3	93.4	64.6	82.2	99.2	46.2	78.6	99.4	45.1	78.3	87.0	51.3	73.1
08	OPENINGS	91.5	32.0	76.8	98.4	63.3	89.7	97.6	65.6	89.7	93.2	42.3	80.6	93.2	46.2	81.6	101.2	54.7	89.7
0920	Plaster & Gypsum Board	94.4	29.9	52.5	91.4	60.1	71.1	98.7	64.0	76.2	86.9	36.6	54.3	82.7	50.5	61.8	89.9	57.1	68.6
0950, 0980	Ceilings & Acoustic Treatment	86.0	29.9	51.3	98.7	60.1	74.8	96.1	64.0	76.3	92.8	36.6	58.1	90.3	50.5	65.7	86.3	57.1	68.3
0960	Flooring	94.7	44.1	81.0	94.2	40.5	79.6	101.6	71.6	93.5	112.0	64.7	99.2	111.8	57.3	97.1	94.6	43.9	80.9
0970, 0990	Wall Finishes & Painting/Coating	98.9	48.8	68.9	101.3	57.0	74.7	108.4	66.4	83.2	97.1	49.2	68.4	97.1	34.1	59.3	92.7	38.3	60.0
09	FINISHES	95.2	36.3	64.1	90.8	55.8	72.3	99.3	65.7	81.5	96.6	44.5	69.0	95.7	49.1	71.1	89.5	50.6	68.9
COVERS	DIVS. 10 - 14, 25, 28, 41, 43, 44	100.0	31.9	86.5	100.0	79.0	95.9	100.0	79.9	96.0	100.0	73.3	94.7	100.0	66.5	93.4	100.0	67.0	93.5
21, 22, 23	FIRE SUPPRESSION, PLUMBING & HVAC	97.0	63.9	83.8	99.9	66.6	86.6	99.9	78.9	91.5	100.1	39.9	76.2	100.0	48.3	79.5	100.0	55.5	82.3
26, 27, 3370	ELECTRICAL, COMMUNICATIONS & UTIL.	97.5	63.5	81.0	103.2	68.7	86.4	103.0	65.4	84.7	97.3	43.8	71.2	101.9	68.6	85.7	104.2	67.6	86.4
MF2004	WEIGHTED AVERAGE	97.1	55.2	79.5	97.5	70.1	86.0	98.5	73.6	88.0	98.2	50.8	78.3	98.6	56.8	81.0	96.5	59.1	80.8

DIVISION		TEXAS																	
		BEAUMONT			BROWNWOOD			BRYAN			CHILDRESS			CORPUS CHRISTI			DALLAS		
		776 - 777			768			778			792			783 - 784			752 - 753		
		MAT.	INST.	TOTAL	MAT.	INST.	TOTAL	MAT.	INST.	TOTAL	MAT.	INST.	TOTAL	MAT.	INST.	TOTAL	MAT.	INST.	TOTAL
015433	CONTRACTOR EQUIPMENT		92.0	92.0		90.7	90.7		92.0	92.0		90.7	90.7		96.8	96.8		99.4	99.4
0241, 31 - 34	SITE & INFRASTRUCTURE, DEMOLITION	98.8	89.0	92.1	109.2	86.8	93.9	88.8	90.6	90.0	114.9	87.4	96.1	122.7	82.5	95.2	119.7	88.1	98.1
0310	Concrete Forming & Accessories	104.7	48.2	55.9	96.2	26.6	36.1	83.4	40.1	46.0	96.4	47.5	54.2	98.0	35.7	44.1	94.8	59.3	64.2
0320	Concrete Reinforcing	96.4	39.4	73.2	96.9	23.6	67.0	98.7	46.2	77.3	96.7	36.7	72.3	96.7	44.8	75.5	96.7	52.9	78.8
0330	Cast-in-Place Concrete	94.3	52.5	78.3	101.6	37.8	77.2	75.5	67.6	72.5	96.6	45.6	77.1	98.9	45.0	78.3	92.3	55.4	78.2
03	CONCRETE	92.7	49.3	72.9	96.9	31.6	67.2	78.7	52.3	66.7	97.5	45.9	74.1	90.0	42.8	68.5	88.7	58.4	75.0
04	MASONRY	101.3	56.1	74.2	128.2	44.3	78.0	135.5	57.8	89.0	104.8	44.2	68.5	83.8	48.5	62.7	102.2	58.6	76.1
05	METALS	102.3	65.8	91.6	100.6	54.0	87.0	102.4	69.2	92.7	100.4	60.8	88.8	97.6	76.1	91.3	102.5	82.2	96.6
06	WOOD, PLASTICS & COMPOSITES	111.4	48.3	76.8	99.2	25.6	58.8	77.6	32.6	52.9	100.0	51.7	73.5	115.1	34.9	71.0	98.7	60.2	77.6
07	THERMAL & MOISTURE PROTECTION	104.5	55.7	85.5	99.4	38.0	75.5	97.7	55.9	81.4	99.8	42.4	77.5	95.3	44.3	75.5	93.2	63.4	81.6
08	OPENINGS	94.9	44.3	82.4	89.1	25.9	73.5	96.9	42.0	83.3	90.2	43.9	78.7	108.6	36.1	90.7	103.4	54.0	91.2
0920	Plaster & Gypsum Board	96.3	47.1	64.3	85.8	23.4	45.3	86.5	30.8	50.3	85.8	50.5	62.9	93.9	32.9	54.3	90.5	59.2	70.1
0950, 0980	Ceilings & Acoustic Treatment	104.4	47.1	69.0	87.7	23.4	48.0	98.3	30.8	56.5	87.7	50.5	64.7	87.9	32.9	53.9	94.2	59.2	72.5
0960	Flooring	111.1	70.9	100.2	110.7	36.5	90.7	82.5	58.7	76.1	109.7	38.2	90.4	107.1	43.4	89.9	100.6	53.2	87.8
0970, 0990	Wall Finishes & Painting/Coating	93.9	46.0	65.1	97.1	27.4	55.3	89.1	61.2	72.4	97.1	36.2	60.6	105.5	53.5	74.3	103.5	53.1	73.2
09	FINISHES	95.0	51.8	72.2	95.5	28.2	59.9	82.8	43.9	62.2	95.8	45.1	69.0	96.3	38.2	65.6	98.4	57.5	76.7
COVERS	DIVS. 10 - 14, 25, 28, 41, 43, 44	100.0	80.4	96.1	100.0	40.4	88.2	100.0	67.8	93.6	100.0	65.8	93.2	100.0	77.8	95.6	100.0	79.2	95.9
21, 22, 23	FIRE SUPPRESSION, PLUMBING & HVAC	100.0	59.1	83.7	94.3	26.3	67.3	94.2	71.7	85.2	94.3	47.1	75.6	99.9	40.6	76.3	99.9	63.8	85.5
26, 27, 3370	ELECTRICAL, COMMUNICATIONS & UTIL.	95.0	65.0	80.4	93.4	25.7	60.5	93.9	68.4	81.5	97.3	35.1	67.0	97.6	50.1	74.5	97.3	69.2	83.6
MF2004	WEIGHTED AVERAGE	98.4	60.1	82.3	98.0	37.2	72.5	95.4	62.6	81.6	97.6	49.6	77.4	98.4	50.4	78.3	99.4	66.2	85.4

DIVISION		TEXAS																	
		DEL RIO			DENTON			EASTLAND			EL PASO			FORT WORTH			GALVESTON		
		788			762			764			798 - 799,885			760 - 761			775		
		MAT.	INST.	TOTAL	MAT.	INST.	TOTAL	MAT.	INST.	TOTAL	MAT.	INST.	TOTAL	MAT.	INST.	TOTAL	MAT.	INST.	TOTAL
015433	CONTRACTOR EQUIPMENT		90.0	90.0		96.4	96.4		90.7	90.7		90.7	90.7		90.7	90.7		100.1	100.1
0241, 31 - 34	SITE & INFRASTRUCTURE, DEMOLITION	110.5	86.3	94.0	109.1	79.3	88.7	112.6	86.9	95.1	100.2	87.7	91.7	104.7	89.0	93.9	125.6	85.8	98.4
0310	Concrete Forming & Accessories	92.5	23.8	33.2	102.7	34.8	44.1	97.1	31.1	40.1	97.3	43.2	50.6	93.7	58.5	63.3	92.0	65.8	69.3
0320	Concrete Reinforcing	97.4	16.7	64.4	98.2	46.8	77.2	97.1	38.1	73.1	96.6	43.5	75.0	96.6	52.7	78.7	98.2	58.0	81.8
0330	Cast-in-Place Concrete	106.5	33.9	78.7	78.6	48.1	66.9	107.4	38.4	81.0	86.4	41.0	69.1	98.2	51.5	80.3	100.3	68.8	88.2
03	CONCRETE	103.8	27.7	69.2	78.3	43.9	62.7	101.0	36.3	71.6	86.7	43.7	67.1	92.0	55.9	75.6	94.3	67.1	81.9
04	MASONRY	97.5	40.8	63.6	136.4	47.8	83.3	98.8	43.5	65.6	98.0	46.2	67.0	94.7	58.5	73.0	96.4	57.9	73.3
05	METALS	96.6	49.9	83.0	100.2	76.9	93.4	100.5	56.4	87.6	102.6	63.5	91.1	99.5	70.4	91.0	104.1	87.2	99.2
06	WOOD, PLASTICS & COMPOSITES	95.8	23.8	56.3	110.1	35.0	68.9	106.1	31.8	65.3	100.2	45.8	70.3	101.1	60.1	78.6	94.4	66.7	79.2
07	THERMAL & MOISTURE PROTECTION	92.3	32.3	69.0	98.2	44.4	77.3	99.7	38.2	75.8	95.7	52.4	78.8	100.4	52.1	81.6	96.9	66.3	85.0
08	OPENINGS	101.9	20.0	81.7	104.7	38.7	88.4	77.5	28.9	65.5	93.2	41.3	80.3	87.1	53.9	78.9	102.1	63.6	92.6
0920	Plaster & Gypsum Board	90.0	21.6	45.6	90.1	33.0	53.0	85.8	29.9	49.5	85.2	44.4	58.7	86.2	59.2	68.6	92.3	65.9	75.2
0950, 0980	Ceilings & Acoustic Treatment	84.6	21.6	45.7	92.7	33.0	55.8	87.7	29.9	51.9	90.3	44.4	61.9	92.8	59.2	72.0	100.8	65.9	79.2
0960	Flooring	91.7	17.5	71.6	103.1	42.6	86.8	144.2	26.3	112.3	112.0	62.4	98.6	145.6	42.6	117.7	97.2	58.7	86.8
0970, 0990	Wall Finishes & Painting/Coating	92.7	24.1	51.5	107.9	27.7	59.8	98.5	28.6	56.6	97.1	32.7	58.5	98.5	48.6	68.6	101.4	58.2	75.5
09	FINISHES	89.6	23.1	54.4	92.5	35.0	62.1	105.9	30.2	65.9	95.7	46.0	69.4	107.1	54.7	79.4	92.9	63.6	77.4
COVERS	DIVS. 10 - 14, 25, 28, 41, 43, 44	100.0	38.4	87.8	100.0	41.9	88.5	100.0	40.8	88.3	100.0	64.9	93.0	100.0	78.9	95.8	100.0	86.0	97.2
21, 22, 23	FIRE SUPPRESSION, PLUMBING & HVAC	94.0	17.6	63.7	94.3	41.8	73.5	94.3	26.5	67.4	100.0	34.9	74.1	100.0	53.6	81.6	94.1	68.4	83.9
26, 27, 3370	ELECTRICAL, COMMUNICATIONS & UTIL.	99.2	18.3	59.8	96.2	41.7	69.7	93.3	32.1	63.5	98.6	51.9	75.9	98.1	60.3	79.7	95.3	61.5	78.8
MF2004	WEIGHTED AVERAGE	97.6	31.9	70.1	97.6	48.1	76.8	96.9	39.4	72.8	97.5	49.8	77.5	98.0	60.7	82.4	98.2	69.1	86.0

DIVISION		TEXAS																	
		GIDDINGS			GREENVILLE			HOUSTON			HUNTSVILLE			LAREDO			LONGVIEW		
		789			754			770 - 772			773			780			756		
		MAT.	INST.	TOTAL	MAT.	INST.	TOTAL	MAT.	INST.	TOTAL	MAT.	INST.	TOTAL	MAT.	INST.	TOTAL	MAT.	INST.	TOTAL
015433	CONTRACTOR EQUIPMENT		90.0	90.0		97.2	97.2		100.1	100.1		92.0	92.0		90.0	90.0		92.3	92.3
0241, 31 - 34	SITE & INFRASTRUCTURE, DEMOLITION	96.8	86.7	89.9	111.7	82.8	91.9	123.0	85.7	97.5	108.9	87.8	94.5	90.5	88.6	89.2	106.7	92.3	96.8
0310	Concrete Forming & Accessories	90.9	31.1	39.2	86.9	26.4	34.7	94.1	65.9	69.8	90.5	32.3	40.2	92.6	35.9	43.6	83.4	27.3	34.9
0320	Concrete Reinforcing	97.9	38.9	73.8	97.2	23.8	67.2	98.0	58.2	81.8	99.0	25.9	69.2	97.4	44.9	75.9	96.3	21.7	65.9
0330	Cast-in-Place Concrete	90.2	38.4	70.4	84.0	37.6	66.3	97.2	68.9	86.4	103.9	44.5	81.2	76.2	59.8	69.9	97.7	35.2	73.9
03	CONCRETE	86.3	36.4	63.6	82.6	32.3	59.7	92.2	67.2	80.8	99.6	37.0	71.1	81.2	47.3	65.8	95.2	30.6	65.8
04	MASONRY	104.6	45.2	69.0	161.6	46.6	92.7	96.1	60.8	75.0	136.0	30.4	72.8	91.7	49.6	66.5	157.0	41.5	87.9
05	METALS	96.1	56.2	84.5	100.0	65.4	89.9	106.9	87.5	101.2	102.3	56.3	88.9	99.1	64.5	89.0	92.8	53.2	81.3
06	WOOD, PLASTICS & COMPOSITES	94.9	31.8	60.2	90.0	25.7	54.7	96.8	66.7	80.3	86.6	33.8	57.6	95.8	34.7	62.3	83.7	27.7	52.9
07	THERMAL & MOISTURE PROTECTION	93.0	38.9	71.9	92.9	39.4	72.1	96.8	67.4	85.4	98.6	36.2	74.3	91.4	49.4	75.1	93.5	33.8	70.3
08	OPENINGS	101.1	28.9	83.3	102.1	26.0	83.3	104.5	63.6	94.4	96.8	28.8	80.0	101.9	36.6	85.7	92.1	25.0	75.5
0920	Plaster & Gypsum Board	88.7	29.9	50.5	86.6	23.4	45.6	94.8	65.9	76.0	90.6	32.0	52.5	90.9	32.9	53.3	83.4	25.7	45.9
0950, 0980	Ceilings & Acoustic Treatment	83.8	29.9	50.4	90.8	23.4	49.2	105.8	65.9	81.1	98.3	32.0	57.3	87.9	32.9	53.9	85.8	25.7	48.6
0960	Flooring	91.3	26.3	73.7	96.0	42.6	81.6	98.5	58.7	87.7	86.4	26.3	70.1	91.6	43.4	78.6	101.8	37.5	84.4
0970, 0990	Wall Finishes & Painting/Coating	92.7	28.6	54.2	103.5	23.6	55.5	101.4	61.2	77.3	89.1	29.4	53.3	92.7	50.0	67.1	95.2	31.4	57.0
09	FINISHES	88.3	30.0	57.5	95.2	28.8	60.1	99.9	64.0	80.9	85.5	31.2	56.8	89.2	37.7	62.0	99.5	29.5	62.5
COVERS	DIVS. 10 - 14, 25, 28, 41, 43, 44	100.0	39.3	88.0	100.0	40.2	88.2	100.0	86.0	97.2	100.0	39.4	88.0	100.0	71.7	94.4	100.0	23.4	84.8
21, 22, 23	FIRE SUPPRESSION, PLUMBING & HVAC	94.1	26.5	67.2	94.0	25.8	66.9	99.9	71.9	88.8	94.2	25.0	66.7	99.9	36.6	74.7	94.0	25.5	66.7
26, 27, 3370	ELECTRICAL, COMMUNICATIONS & UTIL.	96.0	32.1	64.9	93.8	25.8	60.7	97.2	68.5	83.2	93.9	31.4	63.5	99.3	64.9	82.6	93.8	50.1	72.5
MF2004	WEIGHTED AVERAGE	94.9	39.5	71.7	98.7	38.4	73.4	100.7	71.2	88.3	98.8	37.9	73.2	95.8	51.7	77.3	98.0	39.9	73.6

DIVISION		TEXAS																	
		LUBBOCK			LUFKIN			MCALLEN			MCKINNEY			MIDLAND			ODESSA		
		793 - 794			759			785			750			797			797		
		MAT.	INST.	TOTAL	MAT.	INST.	TOTAL	MAT.	INST.	TOTAL	MAT.	INST.	TOTAL	MAT.	INST.	TOTAL	MAT.	INST.	TOTAL
015433	**CONTRACTOR EQUIPMENT**		98.4	98.4		92.3	92.3		96.8	96.8		97.2	97.2		98.4	98.4		90.7	90.7
0241, 31 - 34	**SITE & INFRASTRUCTURE, DEMOLITION**	130.8	84.3	99.0	100.7	92.3	95.0	127.1	82.5	96.6	107.1	83.6	91.0	134.9	84.3	100.3	102.2	88.8	93.0
0310	Concrete Forming & Accessories	96.9	38.6	46.5	86.5	37.7	44.4	97.0	33.7	42.3	85.9	39.7	46.0	100.7	36.9	45.5	97.9	36.7	45.0
0320	Concrete Reinforcing	97.7	53.1	79.5	97.9	28.9	69.7	96.9	44.7	75.6	97.2	46.8	76.6	98.7	53.0	80.1	96.6	53.0	78.8
0330	Cast-in-Place Concrete	94.3	47.7	76.5	87.4	40.0	69.3	107.5	43.5	83.1	78.8	49.1	67.4	100.3	45.2	79.2	94.2	44.1	75.0
03	**CONCRETE**	89.5	46.7	70.0	88.8	38.4	65.9	95.1	41.4	70.7	78.8	46.3	64.0	93.6	45.1	71.5	90.3	43.8	69.2
04	**MASONRY**	99.7	44.4	66.6	119.7	35.5	69.3	97.4	48.5	68.1	174.1	50.7	100.2	117.7	44.8	74.0	100.4	44.7	67.0
05	**METALS**	106.5	82.3	99.4	100.1	59.4	88.2	96.5	75.9	90.5	99.9	75.9	92.9	104.9	81.8	98.1	102.1	69.5	92.6
06	**WOOD, PLASTICS & COMPOSITES**	100.4	38.4	66.4	92.2	39.1	63.1	113.5	32.5	69.1	88.6	39.8	61.8	105.6	36.6	67.7	100.6	36.5	65.4
07	**THERMAL & MOISTURE PROTECTION**	89.9	46.8	73.1	93.3	38.5	72.0	95.3	43.3	75.1	92.8	47.6	75.2	90.1	43.3	71.9	99.2	42.5	77.2
08	**OPENINGS**	104.2	40.7	88.5	73.7	35.8	64.3	106.9	35.3	89.2	102.1	41.4	87.1	103.1	39.4	87.4	93.2	39.3	79.9
0920	Plaster & Gypsum Board	87.4	36.6	54.4	83.3	37.6	53.6	93.9	30.5	52.7	85.5	38.1	54.7	89.7	34.7	54.0	86.9	34.7	53.0
0950, 0980	Ceilings & Acoustic Treatment	94.5	36.6	58.7	78.4	37.6	53.1	87.9	30.5	52.4	90.8	38.1	58.2	91.0	34.7	56.2	92.8	34.7	56.9
0960	Flooring	104.0	36.1	85.6	136.5	35.0	109.1	106.5	64.7	95.2	95.6	42.6	81.3	105.6	35.7	86.7	112.0	35.7	91.4
0970, 0990	Wall Finishes & Painting/Coating	108.4	30.6	61.7	95.2	35.4	59.4	105.5	27.1	58.4	103.5	44.2	67.9	108.4	30.2	61.5	97.1	30.6	57.2
09	**FINISHES**	98.8	36.3	65.7	107.8	37.0	70.4	96.4	38.2	65.7	94.6	40.4	66.0	99.2	35.1	65.3	96.6	35.1	64.1
COVERS	**DIVS. 10 - 14, 25, 28, 41, 43, 44**	100.0	73.1	94.7	100.0	65.5	93.2	100.0	71.6	94.4	100.0	43.4	88.8	100.0	64.5	93.0	100.0	64.1	92.9
21, 22, 23	**FIRE SUPPRESSION, PLUMBING & HVAC**	99.6	45.9	78.2	94.0	29.6	68.4	94.1	28.8	68.1	94.0	43.4	73.9	93.8	33.4	69.8	100.1	33.4	73.6
26, 27, 3370	**ELECTRICAL, COMMUNICATIONS & UTIL.**	95.8	41.2	69.2	95.2	43.9	70.2	96.8	32.5	65.5	93.9	50.1	72.6	95.8	38.4	67.9	97.4	38.3	68.7
MF2004	**WEIGHTED AVERAGE**	100.1	51.1	79.5	95.7	43.9	74.0	98.0	45.1	75.8	98.7	51.5	78.8	100.0	47.4	77.9	98.1	46.3	76.3

DIVISION		TEXAS																	
		PALESTINE			SAN ANGELO			SAN ANTONIO			TEMPLE			TEXARKANA			TYLER		
		758			769			781 - 782			765			755			757		
		MAT.	INST.	TOTAL	MAT.	INST.	TOTAL	MAT.	INST.	TOTAL	MAT.	INST.	TOTAL	MAT.	INST.	TOTAL	MAT.	INST.	TOTAL
015433	**CONTRACTOR EQUIPMENT**		92.3	92.3		90.7	90.7		91.8	91.8		90.7	90.7		92.3	92.3		92.3	92.3
0241, 31 - 34	**SITE & INFRASTRUCTURE, DEMOLITION**	109.2	91.9	97.4	104.9	89.2	94.2	90.2	92.3	91.6	91.7	87.5	88.8	93.8	93.1	93.3	106.2	93.2	97.3
0310	Concrete Forming & Accessories	78.4	31.1	37.5	96.6	34.8	43.2	92.7	55.6	60.6	100.0	33.1	42.2	93.1	33.3	41.4	88.1	32.3	39.9
0320	Concrete Reinforcing	95.5	38.7	72.3	96.8	53.0	78.9	104.4	49.3	81.9	96.9	45.5	76.0	95.4	52.5	77.9	96.3	52.7	78.5
0330	Cast-in-Place Concrete	79.9	37.7	63.8	95.8	42.4	75.4	74.7	67.5	71.9	78.5	47.3	66.6	80.5	43.6	66.4	96.0	42.7	75.6
03	**CONCRETE**	88.9	36.1	64.9	92.7	42.2	69.7	82.0	59.5	71.7	80.4	41.8	62.8	83.0	42.0	64.4	94.7	41.3	70.5
04	**MASONRY**	115.2	42.8	71.9	124.9	44.2	76.6	91.5	60.6	73.0	135.5	47.0	82.5	173.6	44.8	98.5	167.6	48.6	96.4
05	**METALS**	99.9	55.6	86.9	100.8	66.6	90.8	99.7	69.4	90.9	100.5	63.8	89.8	92.7	69.1	85.8	99.8	69.6	91.0
06	**WOOD, PLASTICS & COMPOSITES**	82.8	31.7	54.8	99.4	34.9	64.0	95.8	54.0	72.9	109.2	31.8	66.7	94.7	31.3	59.9	93.8	29.6	58.5
07	**THERMAL & MOISTURE PROTECTION**	93.7	39.4	72.5	99.2	40.9	76.5	91.4	64.9	81.1	98.8	43.3	77.2	93.1	47.6	75.4	93.5	52.9	77.7
08	**OPENINGS**	73.7	29.5	62.7	89.2	38.4	76.6	103.9	53.6	91.4	75.4	33.7	65.1	92.0	37.1	78.4	73.6	36.1	64.4
0920	Plaster & Gypsum Board	79.2	29.9	47.2	85.8	33.1	51.6	90.9	52.9	66.2	87.3	29.9	50.0	85.9	29.4	49.6	83.3	27.6	47.2
0950, 0980	Ceilings & Acoustic Treatment	78.4	29.9	48.4	87.7	33.1	53.9	87.9	52.9	66.3	87.7	29.9	51.9	85.0	29.4	50.6	78.4	27.6	47.0
0960	Flooring	127.2	26.3	99.9	110.9	31.8	89.5	91.6	66.5	84.8	147.1	43.7	119.2	110.5	58.3	96.4	138.4	38.8	111.5
0970, 0990	Wall Finishes & Painting/Coating	95.2	28.6	55.3	97.1	31.6	57.8	92.7	50.0	67.1	98.5	31.6	58.3	95.2	36.9	60.2	95.2	36.9	60.2
09	**FINISHES**	105.1	29.8	65.3	95.2	33.2	62.4	89.2	55.9	71.6	105.7	34.1	67.8	101.7	37.2	67.6	108.7	32.2	68.3
COVERS	**DIVS. 10 - 14, 25, 28, 41, 43, 44**	100.0	40.7	88.3	100.0	63.2	92.7	100.0	78.0	95.7	100.0	40.8	88.3	100.0	71.8	94.4	100.0	71.5	94.4
21, 22, 23	**FIRE SUPPRESSION, PLUMBING & HVAC**	94.0	26.5	67.2	94.3	38.5	72.1	99.9	67.0	86.8	94.3	31.5	69.4	94.0	34.8	70.5	94.0	58.5	79.9
26, 27, 3370	**ELECTRICAL, COMMUNICATIONS & UTIL.**	91.2	36.3	64.5	97.3	30.6	64.8	99.8	64.9	82.8	94.5	71.8	83.5	95.1	65.0	80.4	93.8	56.0	75.4
MF2004	**WEIGHTED AVERAGE**	95.0	40.3	72.0	97.6	45.3	75.7	96.2	65.4	83.3	95.6	48.7	75.9	97.6	50.8	77.9	98.8	54.2	80.1

DIVISION		TEXAS															UTAH		
		VICTORIA			WACO			WAXAHACKIE			WHARTON			WICHITA FALLS			LOGAN		
		779			766 - 767			751			774			763			843		
		MAT.	INST.	TOTAL	MAT.	INST.	TOTAL	MAT.	INST.	TOTAL	MAT.	INST.	TOTAL	MAT.	INST.	TOTAL	MAT.	INST.	TOTAL
015433	**CONTRACTOR EQUIPMENT**		99.3	99.3		90.7	90.7		97.2	97.2		100.1	100.1		90.7	90.7		101.4	101.4
0241, 31 - 34	**SITE & INFRASTRUCTURE, DEMOLITION**	131.8	81.8	97.6	100.8	88.9	92.7	109.6	84.0	92.1	139.5	83.0	100.9	101.6	88.5	92.6	88.6	101.7	97.5
0310	Concrete Forming & Accessories	92.7	37.7	45.2	98.5	38.8	46.9	85.9	44.3	50.0	86.9	34.6	41.7	98.5	39.6	47.6	105.2	55.4	62.2
0320	Concrete Reinforcing	94.3	28.9	67.6	96.6	46.1	76.0	97.2	46.8	76.6	98.1	56.3	81.1	96.6	47.3	76.5	102.6	74.0	90.9
0330	Cast-in-Place Concrete	112.6	41.6	85.5	85.0	52.9	72.7	83.2	41.7	67.3	115.7	48.3	90.0	90.8	47.3	74.2	92.3	71.4	84.3
03	**CONCRETE**	101.1	40.0	73.3	86.1	46.5	68.1	81.9	45.9	65.5	105.0	45.8	78.1	88.8	45.3	69.0	112.9	65.1	91.2
04	**MASONRY**	111.5	35.6	66.1	97.0	55.3	72.0	162.3	49.0	94.5	97.6	31.2	57.8	97.5	56.3	72.8	113.2	57.9	80.1
05	**METALS**	102.4	74.8	94.4	102.7	67.4	92.4	100.0	77.5	93.4	104.1	80.5	97.2	102.6	71.3	93.5	100.8	75.6	93.5
06	**WOOD, PLASTICS & COMPOSITES**	98.4	39.3	65.9	106.9	34.0	66.9	88.6	46.8	65.7	88.3	35.7	59.4	106.9	38.3	69.2	85.2	52.3	67.1
07	**THERMAL & MOISTURE PROTECTION**	100.2	40.2	76.9	99.7	47.2	79.3	92.9	44.9	74.2	97.3	46.1	77.4	99.7	51.0	80.8	99.5	63.9	85.7
08	**OPENINGS**	102.0	37.4	86.0	87.1	35.0	74.2	102.1	45.2	88.0	102.1	39.6	86.7	87.1	42.2	76.0	88.2	54.8	79.9
0920	Plaster & Gypsum Board	91.1	37.6	56.3	86.9	32.1	51.3	85.7	45.4	59.5	89.4	33.8	53.3	86.9	36.6	54.2	76.3	50.4	59.5
0950, 0980	Ceilings & Acoustic Treatment	101.6	37.6	62.0	92.8	32.1	55.3	91.7	45.4	63.0	100.8	33.8	59.4	92.8	36.6	58.0	108.7	50.4	72.6
0960	Flooring	96.0	35.0	79.5	145.6	33.1	115.2	95.6	35.0	79.2	94.3	26.3	75.9	146.7	75.0	127.3	104.5	53.5	90.7
0970, 0990	Wall Finishes & Painting/Coating	101.6	35.4	61.9	98.5	31.6	58.3	103.5	35.4	62.6	101.4	29.4	58.2	102.1	50.0	70.8	98.1	43.1	65.1
09	**FINISHES**	91.1	37.1	62.6	106.9	35.5	69.1	95.0	41.8	66.9	92.5	32.5	60.7	107.5	46.6	75.3	101.5	52.9	75.8
COVERS	**DIVS. 10 - 14, 25, 28, 41, 43, 44**	100.0	43.1	88.7	100.0	75.8	95.2	100.0	44.5	89.0	100.0	41.8	88.5	100.0	64.9	93.0	100.0	64.0	92.9
21, 22, 23	**FIRE SUPPRESSION, PLUMBING & HVAC**	94.2	29.6	68.5	100.1	54.1	81.9	94.0	42.4	73.5	94.1	24.7	66.5	100.1	45.6	78.5	99.9	69.5	87.8
26, 27, 3370	**ELECTRICAL, COMMUNICATIONS & UTIL.**	100.1	44.0	72.8	97.4	71.9	85.0	93.8	69.2	81.8	98.8	31.5	66.0	99.7	60.7	80.7	94.5	66.8	81.0
MF2004	**WEIGHTED AVERAGE**	100.1	44.3	76.7	97.8	56.8	80.6	98.6	54.2	80.0	100.2	42.2	75.8	98.5	55.3	80.3	100.3	67.3	86.4

		UTAH												VERMONT					
	DIVISION	OGDEN			PRICE			PROVO			SALT LAKE CITY			BELLOWS FALLS			BENNINGTON		
		842,844			845			846 - 847			840 - 841			051			052		
		MAT.	INST.	TOTAL	MAT.	INST.	TOTAL	MAT.	INST.	TOTAL	MAT.	INST.	TOTAL	MAT.	INST.	TOTAL	MAT.	INST.	TOTAL
015433	CONTRACTOR EQUIPMENT		101.4	101.4		100.7	100.7		100.7	100.7		101.4	101.4		100.4	100.4		100.4	100.4
0241, 31 - 34	SITE & INFRASTRUCTURE, DEMOLITION	77.5	101.7	94.0	86.5	99.0	95.1	85.2	100.4	95.6	77.3	101.7	93.9	74.0	96.7	89.5	73.4	97.5	89.9
0310	Concrete Forming & Accessories	105.3	55.4	62.2	107.3	30.1	40.6	106.5	55.5	62.5	104.7	55.5	62.2	96.5	43.6	50.8	94.1	54.7	60.1
0320	Concrete Reinforcing	102.2	74.0	90.7	110.0	69.1	93.3	110.9	74.0	95.9	104.6	74.0	92.1	80.3	69.8	76.0	80.3	69.7	76.0
0330	Cast-in-Place Concrete	93.7	71.4	85.2	92.4	45.4	74.4	92.4	71.4	84.4	102.4	71.4	90.6	89.9	63.0	79.7	89.9	68.8	81.9
03	CONCRETE	102.6	65.1	85.6	114.6	44.1	82.5	113.1	65.2	91.3	122.1	65.2	96.2	97.6	56.2	78.8	97.5	63.1	81.8
04	MASONRY	106.9	57.9	77.6	118.5	37.1	69.8	118.6	57.9	82.3	120.8	57.9	83.2	95.5	34.3	58.8	104.6	51.3	72.7
05	METALS	101.3	75.6	93.8	98.2	72.4	90.6	99.0	75.7	92.2	106.0	75.7	97.1	88.8	73.5	84.3	88.8	72.6	84.0
06	WOOD, PLASTICS & COMPOSITES	85.2	52.3	67.1	87.8	28.1	55.0	86.6	52.3	67.7	86.9	52.3	67.9	102.3	42.4	69.4	99.4	49.5	72.0
07	THERMAL & MOISTURE PROTECTION	98.3	63.9	84.9	101.3	50.4	81.5	101.3	63.9	86.8	101.8	63.9	87.1	94.0	51.9	77.6	94.0	52.6	77.9
08	OPENINGS	88.2	54.8	79.9	92.4	39.3	79.3	92.5	54.8	83.1	90.1	54.8	81.3	101.6	45.4	87.7	101.5	47.1	88.1
0920	Plaster & Gypsum Board	76.3	50.4	59.5	77.8	25.3	43.7	76.7	50.4	59.6	78.7	50.4	60.3	103.4	39.5	61.9	101.5	46.9	66.0
0950, 0980	Ceilings & Acoustic Treatment	108.7	50.4	72.6	108.7	25.3	57.1	108.7	50.4	72.6	103.1	50.4	70.5	88.9	39.5	58.4	88.9	46.9	62.9
0960	Flooring	102.2	53.5	89.0	105.7	39.8	87.9	105.3	53.5	91.3	103.3	53.5	89.9	96.8	34.2	79.9	95.6	34.2	79.0
0970, 0990	Wall Finishes & Painting/Coating	98.1	43.1	65.1	98.1	43.1	65.1	98.1	60.7	75.7	100.7	60.7	76.7	91.7	30.7	55.1	91.7	30.7	55.1
09	FINISHES	99.7	52.9	75.0	102.6	31.9	65.2	102.2	54.9	77.2	99.7	54.9	76.0	95.1	39.1	65.5	94.5	47.6	69.7
COVERS	DIVS. 10 - 14, 25, 28, 41, 43, 44	100.0	64.0	92.9	100.0	48.0	89.7	100.0	64.0	92.9	100.0	64.0	92.9	100.0	44.0	88.9	100.0	50.9	90.3
21, 22, 23	FIRE SUPPRESSION, PLUMBING & HVAC	99.9	69.5	87.8	97.0	35.6	72.6	99.9	69.5	87.8	100.0	69.5	87.9	94.4	78.3	88.0	94.4	85.7	90.9
26, 27, 3370	ELECTRICAL, COMMUNICATIONS & UTIL.	94.9	66.8	81.2	100.0	39.7	70.7	95.1	69.7	82.8	98.2	69.7	84.3	96.9	83.1	90.2	96.9	62.8	80.3
MF2004	WEIGHTED AVERAGE	98.4	67.3	85.3	100.7	46.5	78.0	100.8	67.8	87.0	102.9	67.9	88.2	94.8	64.0	81.9	95.2	66.8	83.2

		VERMONT																	
	DIVISION	BRATTLEBORO			BURLINGTON			GUILDHALL			MONTPELIER			RUTLAND			ST. JOHNSBURY		
		053			054			059			056			057			058		
		MAT.	INST.	TOTAL	MAT.	INST.	TOTAL	MAT.	INST.	TOTAL	MAT.	INST.	TOTAL	MAT.	INST.	TOTAL	MAT.	INST.	TOTAL
015433	CONTRACTOR EQUIPMENT		100.4	100.4		100.4	100.4		100.4	100.4		100.4	100.4		100.4	100.4		100.4	100.4
0241, 31 - 34	SITE & INFRASTRUCTURE, DEMOLITION	74.8	97.7	90.5	76.5	96.8	90.3	73.2	96.0	88.8	77.6	96.8	90.7	76.6	96.8	90.4	73.3	96.2	89.0
0310	Concrete Forming & Accessories	96.7	49.8	56.2	93.5	55.8	60.9	94.1	47.9	54.2	98.1	56.3	62.0	97.0	56.3	61.8	92.3	48.0	54.1
0320	Concrete Reinforcing	79.5	69.8	75.5	100.4	81.1	92.5	81.0	41.3	64.8	79.9	81.2	80.4	100.4	81.2	92.5	79.5	47.1	66.3
0330	Cast-in-Place Concrete	92.8	69.2	83.8	94.2	107.0	99.1	87.2	59.6	76.7	99.0	107.2	102.1	88.1	107.2	95.4	87.2	59.9	76.8
03	CONCRETE	99.9	61.1	82.2	104.5	78.9	92.9	95.3	51.7	75.5	103.0	79.2	92.2	101.9	79.2	91.6	94.9	52.8	75.8
04	MASONRY	104.6	51.4	72.8	107.1	53.8	75.2	104.9	39.1	65.5	90.2	53.8	68.4	85.9	53.8	66.7	129.9	39.2	75.6
05	METALS	88.8	73.3	84.3	95.4	79.7	90.8	88.8	62.3	81.1	88.8	80.3	86.3	93.8	80.3	89.8	88.8	62.8	81.2
06	WOOD, PLASTICS & COMPOSITES	102.6	42.4	69.6	97.3	55.9	74.6	98.6	49.5	71.7	101.6	55.9	76.5	102.8	55.9	77.1	94.8	49.5	69.9
07	THERMAL & MOISTURE PROTECTION	94.1	52.7	78.0	94.1	54.6	78.7	93.8	42.5	73.9	93.5	63.0	81.6	94.1	63.0	82.0	93.8	42.5	73.8
08	OPENINGS	101.5	45.4	87.7	104.9	55.5	92.6	101.6	42.1	86.9	101.5	55.5	90.1	104.9	55.5	92.6	101.6	42.1	86.9
0920	Plaster & Gypsum Board	103.4	39.5	61.9	105.1	53.6	71.6	108.2	46.9	68.4	114.2	53.6	74.8	104.0	53.6	71.2	111.5	46.9	69.5
0950, 0980	Ceilings & Acoustic Treatment	88.9	39.5	58.4	92.5	53.6	68.4	88.9	46.9	62.9	88.9	53.6	67.1	93.3	53.6	68.7	88.9	46.9	62.9
0960	Flooring	96.9	34.2	80.0	97.0	61.8	87.5	100.2	34.2	82.3	103.9	61.8	92.5	96.8	61.8	87.4	103.9	34.2	85.1
0970, 0990	Wall Finishes & Painting/Coating	91.7	30.7	55.1	91.7	35.7	58.1	91.7	23.8	51.0	91.7	35.7	58.1	91.7	35.7	58.1	91.7	25.0	51.7
09	FINISHES	95.2	43.4	67.8	96.4	54.0	74.0	96.7	42.7	68.2	99.0	54.0	75.2	96.3	54.0	74.0	98.3	42.9	69.0
COVERS	DIVS. 10 - 14, 25, 28, 41, 43, 44	100.0	50.2	90.1	100.0	81.9	96.4	100.0	45.0	89.1	100.0	81.9	96.4	100.0	81.9	96.4	100.0	45.0	89.1
21, 22, 23	FIRE SUPPRESSION, PLUMBING & HVAC	94.4	87.2	91.5	100.0	64.9	86.1	94.4	63.2	82.0	94.2	65.0	82.6	100.2	65.0	86.2	94.4	63.2	82.0
26, 27, 3370	ELECTRICAL, COMMUNICATIONS & UTIL.	96.9	83.1	90.2	99.2	62.8	81.5	96.9	62.8	80.3	100.4	62.8	82.1	96.9	62.8	80.3	96.9	62.8	80.3
MF2004	WEIGHTED AVERAGE	95.6	69.0	84.4	99.3	67.7	86.1	95.1	56.9	79.0	95.9	68.1	84.2	97.5	68.1	85.2	96.3	57.1	79.9

		VERMONT			VIRGINIA														
	DIVISION	WHITE RIVER JCT.			ALEXANDRIA			ARLINGTON			BRISTOL			CHARLOTTESVILLE			CULPEPER		
		050			223			222			242			229			227		
		MAT.	INST.	TOTAL	MAT.	INST.	TOTAL	MAT.	INST.	TOTAL	MAT.	INST.	TOTAL	MAT.	INST.	TOTAL	MAT.	INST.	TOTAL
015433	CONTRACTOR EQUIPMENT		100.4	100.4		102.9	102.9		102.1	102.1		102.1	102.1		105.8	105.8		102.1	102.1
0241, 31 - 34	SITE & INFRASTRUCTURE, DEMOLITION	76.9	96.1	90.1	116.5	89.1	97.8	128.2	86.7	99.8	111.8	86.1	94.2	116.9	87.0	96.5	114.8	87.1	95.9
0310	Concrete Forming & Accessories	91.6	46.2	52.4	94.1	76.1	78.6	93.4	74.1	76.7	89.1	48.4	53.9	87.5	55.2	59.6	85.0	73.1	74.7
0320	Concrete Reinforcing	80.3	41.4	64.4	90.2	82.2	86.9	101.8	78.7	92.4	101.9	64.9	86.8	101.2	71.0	88.9	101.8	72.4	89.8
0330	Cast-in-Place Concrete	92.8	59.4	80.0	106.8	82.9	97.7	103.9	80.3	94.9	103.5	53.5	84.4	107.6	62.2	90.3	106.4	72.5	93.4
03	CONCRETE	101.9	50.9	78.7	108.2	80.8	95.7	113.7	77.9	97.4	109.8	55.2	85.0	110.4	62.3	88.6	107.2	73.7	92.0
04	MASONRY	116.6	41.4	71.6	91.6	72.3	80.1	104.3	69.2	83.3	92.6	45.5	64.4	118.0	52.6	78.8	107.1	67.8	83.6
05	METALS	88.8	62.7	81.2	97.2	96.7	97.0	95.9	86.6	93.2	94.8	86.7	92.5	95.1	90.9	93.9	95.2	86.1	92.5
06	WOOD, PLASTICS & COMPOSITES	96.4	47.6	69.6	97.7	75.7	85.6	94.4	75.7	84.2	86.8	46.8	64.8	85.0	53.3	67.6	83.8	75.7	79.4
07	THERMAL & MOISTURE PROTECTION	94.1	38.2	72.4	101.5	80.6	93.4	103.2	72.5	91.3	102.8	52.7	83.3	102.5	58.7	85.4	102.6	67.3	88.9
08	OPENINGS	101.6	43.6	87.2	98.3	78.3	93.3	96.2	72.3	90.3	99.6	50.8	87.5	97.6	57.4	87.6	97.9	68.3	90.6
0920	Plaster & Gypsum Board	100.4	44.9	64.4	112.3	74.7	87.9	109.0	74.7	86.7	105.1	44.7	65.8	104.3	50.6	69.5	104.9	74.7	85.3
0950, 0980	Ceilings & Acoustic Treatment	88.9	44.9	61.7	97.6	74.7	83.4	93.4	74.7	81.8	92.6	44.7	63.0	92.6	50.6	66.7	93.4	74.7	81.8
0960	Flooring	94.6	34.2	78.3	111.5	88.9	105.4	109.5	61.7	96.6	105.1	41.9	88.0	103.4	49.5	88.8	103.4	61.7	92.1
0970, 0990	Wall Finishes & Painting/Coating	91.7	19.0	48.1	127.6	83.3	101.0	127.6	83.3	101.0	114.4	47.8	74.4	114.4	67.6	86.3	127.6	60.1	87.1
09	FINISHES	94.3	40.9	66.1	105.9	78.9	91.6	105.1	73.4	88.4	101.5	46.5	72.4	100.9	54.7	76.5	101.6	70.7	85.3
COVERS	DIVS. 10 - 14, 25, 28, 41, 43, 44	100.0	44.5	89.0	100.0	89.2	97.9	100.0	80.8	96.2	100.0	68.0	93.7	100.0	78.6	95.8	100.0	68.7	93.8
21, 22, 23	FIRE SUPPRESSION, PLUMBING & HVAC	94.4	62.7	81.8	100.2	88.5	95.5	100.2	82.4	93.1	94.4	58.1	79.9	94.4	63.8	82.2	94.4	70.5	84.9
26, 27, 3370	ELECTRICAL, COMMUNICATIONS & UTIL.	96.9	62.8	80.3	97.6	98.2	97.9	95.1	93.4	94.3	97.2	41.3	70.0	97.1	69.8	83.8	99.7	60.8	80.8
MF2004	WEIGHTED AVERAGE	96.3	56.6	79.7	100.7	86.0	94.5	101.5	80.8	92.8	98.7	57.4	81.4	100.0	66.6	85.9	99.3	72.0	87.9

DIVISION		VIRGINIA																	
		FAIRFAX			FARMVILLE			FREDERICKSBURG			GRUNDY			HARRISONBURG			LYNCHBURG		
		220 - 221			239			224 - 225			246			228			245		
		MAT.	INST.	TOTAL	MAT.	INST.	TOTAL	MAT.	INST.	TOTAL	MAT.	INST.	TOTAL	MAT.	INST.	TOTAL	MAT.	INST.	TOTAL
015433	**CONTRACTOR EQUIPMENT**		102.1	102.1		105.8	105.8		102.1	102.1		102.1	102.1		102.1	102.1		102.1	102.1
0241, 31 - 34	**SITE & INFRASTRUCTURE, DEMOLITION**	127.2	87.3	99.9	112.6	86.4	94.7	114.3	87.2	95.8	109.4	84.9	92.6	123.7	86.3	98.2	110.2	86.3	93.9
0310	Concrete Forming & Accessories	87.7	74.4	76.2	99.5	50.5	57.2	87.7	54.6	59.1	92.1	42.2	49.0	83.9	48.6	53.4	89.2	69.4	72.1
0320	Concrete Reinforcing	101.8	82.1	93.8	101.9	58.4	84.1	102.6	81.2	93.9	100.5	59.0	83.6	101.8	53.0	81.9	101.2	65.6	86.7
0330	Cast-in-Place Concrete	103.9	80.9	95.1	107.6	60.3	89.5	105.5	58.6	87.5	103.5	54.1	84.6	103.9	65.6	89.3	103.5	65.3	88.9
03	**CONCRETE**	113.3	79.3	97.8	109.4	57.1	85.6	107.0	62.7	86.8	108.4	51.2	82.4	111.1	57.6	86.7	108.4	68.7	90.3
04	**MASONRY**	104.2	69.4	83.4	102.4	51.9	72.1	105.8	50.4	72.6	97.0	54.9	71.8	102.8	57.1	75.5	110.8	59.4	80.0
05	**METALS**	95.2	95.5	95.3	95.1	82.9	91.5	95.2	94.3	94.9	94.8	76.6	89.5	95.1	89.1	93.4	95.0	88.2	93.0
06	**WOOD, PLASTICS & COMPOSITES**	86.8	75.7	80.7	99.8	51.5	73.3	86.8	52.6	68.0	90.1	41.0	63.1	82.7	45.6	62.3	86.8	73.3	79.4
07	**THERMAL & MOISTURE PROTECTION**	103.1	72.9	91.4	102.5	53.2	83.3	102.6	55.7	84.4	102.8	52.4	83.2	103.0	66.3	88.7	102.6	61.1	86.5
08	**OPENINGS**	96.2	77.6	91.6	97.9	50.3	86.1	97.6	59.5	88.2	99.6	40.3	84.9	97.9	57.3	87.9	97.9	65.2	89.8
0920	Plaster & Gypsum Board	105.3	74.7	85.4	111.0	48.8	70.6	105.3	50.6	69.8	105.8	38.7	62.2	104.3	43.5	64.8	105.1	72.1	83.7
0950, 0980	Ceilings & Acoustic Treatment	93.4	74.7	81.8	92.6	48.8	65.5	93.4	50.6	67.0	92.6	38.7	59.3	92.6	43.5	62.2	92.6	72.1	79.9
0960	Flooring	105.4	61.7	93.6	111.5	67.6	99.6	105.4	83.0	99.4	106.8	37.8	88.2	103.0	83.0	97.6	105.1	46.8	89.3
0970, 0990	Wall Finishes & Painting/Coating	127.6	83.3	101.0	114.4	44.5	72.5	127.6	73.8	95.3	114.4	40.4	70.0	127.6	58.3	86.0	114.4	47.8	74.4
09	**FINISHES**	103.3	73.4	87.5	104.0	53.3	77.2	102.2	61.1	80.5	102.0	40.7	69.6	101.9	55.0	77.1	101.3	63.4	81.3
COVERS	**DIVS. 10 - 14, 25, 28, 41, 43, 44**	100.0	80.8	96.2	100.0	57.2	91.5	100.0	68.9	93.8	100.0	54.9	91.1	100.0	67.7	93.6	100.0	72.0	94.4
21, 22, 23	**FIRE SUPPRESSION, PLUMBING & HVAC**	94.4	82.6	89.7	94.4	41.2	73.2	94.4	79.8	88.6	94.4	46.2	75.2	94.4	60.3	80.8	94.4	63.2	82.0
26, 27, 3370	**ELECTRICAL, COMMUNICATIONS & UTIL.**	98.3	93.4	95.9	94.3	54.8	75.1	95.3	93.4	94.4	97.2	52.1	75.2	97.4	91.2	94.4	98.3	61.4	80.3
MF2004	**WEIGHTED AVERAGE**	100.2	82.2	92.6	99.0	56.9	81.3	98.8	73.8	88.3	98.8	54.1	80.0	99.6	68.2	86.4	99.4	68.1	86.2

DIVISION		VIRGINIA																	
		NEWPORT NEWS			NORFOLK			PETERSBURG			PORTSMOUTH			PULASKI			RICHMOND		
		236			233 - 235			238			237			243			230 - 232		
		MAT.	INST.	TOTAL	MAT.	INST.	TOTAL	MAT.	INST.	TOTAL	MAT.	INST.	TOTAL	MAT.	INST.	TOTAL	MAT.	INST.	TOTAL
015433	**CONTRACTOR EQUIPMENT**		105.8	105.8		106.2	106.2		105.8	105.8		105.8	105.8		102.1	102.1		105.8	105.8
0241, 31 - 34	**SITE & INFRASTRUCTURE, DEMOLITION**	110.7	87.7	95.0	111.7	88.4	95.8	115.4	88.2	96.8	109.2	87.0	94.0	103.7	84.9	92.4	111.5	88.1	95.5
0310	Concrete Forming & Accessories	98.7	72.5	76.1	100.4	72.6	76.4	92.3	65.5	69.2	88.3	59.5	63.4	92.2	42.3	49.1	98.2	65.5	70.0
0320	Concrete Reinforcing	101.6	71.3	89.2	101.6	71.4	89.3	101.2	71.8	89.2	101.2	71.3	89.0	100.5	59.0	83.6	101.6	71.8	89.4
0330	Cast-in-Place Concrete	104.6	65.6	89.7	115.0	65.6	96.1	111.1	60.2	91.6	103.6	65.0	88.8	103.5	54.2	84.6	106.2	60.2	88.6
03	**CONCRETE**	106.4	71.2	90.4	111.4	71.2	93.1	111.9	66.3	91.1	105.3	65.1	87.0	103.4	51.3	82.4	107.1	66.3	88.5
04	**MASONRY**	97.7	58.4	74.2	103.1	58.4	76.3	109.9	60.9	80.6	102.5	58.0	75.9	89.3	54.9	68.7	96.0	60.9	75.0
05	**METALS**	97.2	91.7	95.6	96.2	91.7	94.9	95.2	92.4	94.4	96.2	89.8	94.3	94.9	76.7	89.6	99.3	92.4	97.3
06	**WOOD, PLASTICS & COMPOSITES**	98.6	77.4	87.0	100.5	77.4	87.8	90.5	68.4	78.4	86.4	60.1	72.0	90.1	41.0	63.1	97.4	68.4	81.5
07	**THERMAL & MOISTURE PROTECTION**	102.5	61.3	86.4	99.9	61.3	84.8	102.5	61.8	86.7	102.5	58.7	85.5	102.8	52.4	83.2	101.5	61.8	86.0
08	**OPENINGS**	98.3	69.5	91.2	98.3	69.5	91.2	97.6	65.7	89.7	98.4	60.1	88.9	99.6	40.3	84.9	98.3	65.7	90.2
0920	Plaster & Gypsum Board	112.3	75.6	88.5	108.8	75.6	87.2	106.0	66.4	80.3	105.3	57.7	74.4	105.8	38.7	62.2	117.0	66.4	84.1
0950, 0980	Ceilings & Acoustic Treatment	97.6	75.6	84.0	97.6	75.6	84.0	93.4	66.4	76.7	97.6	57.7	72.9	92.6	38.7	59.3	100.9	66.4	79.5
0960	Flooring	111.5	56.7	96.7	110.9	56.7	96.3	106.4	79.8	99.2	102.5	71.7	94.2	105.8	37.8	88.2	110.9	79.8	102.5
0970, 0990	Wall Finishes & Painting/Coating	114.4	48.3	74.8	114.4	64.6	84.5	114.4	67.6	86.3	114.4	64.6	84.5	114.4	40.4	70.0	114.4	67.6	86.3
09	**FINISHES**	105.1	67.6	85.3	104.4	69.4	85.9	102.2	68.7	84.5	101.5	61.9	80.6	102.0	40.7	69.6	106.2	68.7	86.4
COVERS	**DIVS. 10 - 14, 25, 28, 41, 43, 44**	100.0	80.8	96.2	100.0	80.7	96.2	100.0	79.0	95.8	100.0	71.9	94.4	100.0	54.9	91.1	100.0	79.0	95.8
21, 22, 23	**FIRE SUPPRESSION, PLUMBING & HVAC**	100.2	66.9	86.9	100.0	65.9	86.4	94.4	61.7	81.4	100.2	65.9	86.5	94.4	46.5	75.3	100.0	61.7	84.8
26, 27, 3370	**ELECTRICAL, COMMUNICATIONS & UTIL.**	97.2	63.8	80.9	105.5	60.4	83.5	97.4	69.8	84.0	95.4	60.4	78.4	97.2	56.2	77.2	104.2	69.8	87.5
MF2004	**WEIGHTED AVERAGE**	100.5	70.9	88.1	101.9	70.5	88.7	99.9	70.1	87.4	99.8	67.4	86.2	93.4	54.7	80.1	101.7	70.1	88.4

DIVISION		VIRGINIA									WASHINGTON								
		ROANOKE			STAUNTON			WINCHESTER			CLARKSTON			EVERETT			OLYMPIA		
		240 - 241			244			226			994			982			985		
		MAT.	INST.	TOTAL	MAT.	INST.	TOTAL	MAT.	INST.	TOTAL	MAT.	INST.	TOTAL	MAT.	INST.	TOTAL	MAT.	INST.	TOTAL
015433	**CONTRACTOR EQUIPMENT**		102.1	102.1		105.8	105.8		102.1	102.1		95.9	95.9		101.3	101.3		101.3	101.3
0241, 31 - 34	**SITE & INFRASTRUCTURE, DEMOLITION**	108.2	86.2	93.2	113.2	87.0	95.3	122.3	86.4	97.8	103.1	94.6	97.3	93.0	110.8	106.7	101.5	111.1	108.0
0310	Concrete Forming & Accessories	98.5	69.0	73.0	91.8	57.7	62.3	86.2	54.1	58.5	117.7	76.2	81.9	111.8	100.6	102.1	101.5	102.4	102.2
0320	Concrete Reinforcing	101.6	65.7	86.9	101.2	57.2	83.3	101.2	81.4	93.1	103.7	86.1	96.5	113.0	93.3	105.0	113.6	93.3	105.3
0330	Cast-in-Place Concrete	117.6	65.2	97.6	107.6	58.3	88.8	103.9	61.6	87.7	110.9	81.4	99.7	101.0	100.9	101.0	94.4	108.5	99.8
03	**CONCRETE**	112.5	68.5	92.4	109.5	59.6	86.8	110.3	63.4	89.0	103.7	80.0	92.9	93.0	98.8	98.3	95.5	102.2	98.5
04	**MASONRY**	98.2	61.2	76.1	107.9	59.6	79.0	100.2	50.2	70.2	115.8	80.4	94.6	133.6	96.3	113.3	128.5	100.8	111.9
05	**METALS**	97.0	88.0	94.3	95.1	87.7	92.9	95.2	91.8	94.2	90.9	83.3	88.7	109.5	88.6	103.4	103.8	88.7	99.4
06	**WOOD, PLASTICS & COMPOSITES**	98.6	73.3	84.7	90.1	58.3	72.6	85.3	52.6	67.3	96.6	74.7	84.6	103.9	102.1	102.9	90.5	102.1	96.9
07	**THERMAL & MOISTURE PROTECTION**	102.4	60.0	85.9	102.4	57.3	84.9	103.1	62.4	87.2	150.3	77.4	121.9	105.1	93.7	100.6	104.7	94.7	100.8
08	**OPENINGS**	98.3	65.2	90.1	97.9	56.3	87.6	99.7	59.6	89.8	110.9	76.8	102.5	103.7	97.9	102.3	103.6	97.9	102.2
0920	Plaster & Gypsum Board	112.3	72.1	86.2	105.8	55.8	73.4	105.3	50.6	69.8	140.1	73.5	96.9	112.7	102.1	105.8	108.2	102.1	104.3
0950, 0980	Ceilings & Acoustic Treatment	97.6	72.1	81.8	92.6	55.8	69.9	93.4	50.6	67.0	104.9	73.5	85.5	105.1	102.1	103.3	97.4	102.1	100.3
0960	Flooring	111.5	43.9	93.2	106.1	44.3	89.4	104.7	83.0	98.8	107.4	52.8	92.6	122.4	102.7	117.1	114.9	102.9	111.6
0970, 0990	Wall Finishes & Painting/Coating	114.4	47.8	74.4	114.4	33.3	65.8	127.6	51.3	81.8	109.0	67.3	84.0	109.3	81.6	92.7	109.3	89.6	97.5
09	**FINISHES**	105.0	62.9	82.7	101.7	53.3	76.1	102.6	58.6	79.3	120.9	70.4	94.2	110.2	99.4	104.5	105.7	101.4	103.4
COVERS	**DIVS. 10 - 14, 25, 28, 41, 43, 44**	100.0	71.9	94.4	100.0	71.6	94.4	100.0	64.2	92.9	100.0	93.6	98.7	100.0	98.2	99.6	100.0	101.4	100.3
21, 22, 23	**FIRE SUPPRESSION, PLUMBING & HVAC**	100.2	53.2	81.5	94.4	62.2	81.6	94.4	75.6	86.9	94.7	81.5	89.4	100.2	96.4	98.7	100.0	98.8	99.5
26, 27, 3370	**ELECTRICAL, COMMUNICATIONS & UTIL.**	97.2	44.4	71.5	96.0	48.2	72.8	95.7	50.1	73.5	100.2	90.7	95.6	103.3	90.3	97.0	106.2	97.1	101.8
MF2004	**WEIGHTED AVERAGE**	101.2	63.7	85.4	99.3	62.8	84.0	99.4	66.5	85.6	102.9	82.2	94.2	105.0	96.7	101.5	103.1	99.5	101.6

DIVISION		WASHINGTON																	
		RICHLAND			SEATTLE			SPOKANE			TACOMA			VANCOUVER			WENATCHEE		
		993			980 - 981,987			990 - 992			983 - 984			986			988		
		MAT.	INST.	TOTAL	MAT.	INST.	TOTAL	MAT.	INST.	TOTAL	MAT.	INST.	TOTAL	MAT.	INST.	TOTAL	MAT.	INST.	TOTAL
015433	**CONTRACTOR EQUIPMENT**		95.9	95.9		101.8	101.8		95.9	95.9		101.3	101.3		98.2	98.2		101.3	101.3
0241, 31 - 34	**SITE & INFRASTRUCTURE, DEMOLITION**	104.1	96.0	98.5	102.3	110.1	107.6	103.4	96.0	98.3	101.4	111.1	108.0	113.5	100.9	104.9	113.2	109.3	110.5
0310	Concrete Forming & Accessories	117.8	80.6	85.7	103.4	102.7	102.8	123.3	80.6	86.4	103.4	102.5	102.6	104.4	91.6	93.4	104.7	75.8	79.7
0320	Concrete Reinforcing	99.5	86.1	94.0	111.7	93.3	104.2	100.2	86.2	94.5	111.7	93.3	104.2	112.6	92.9	104.6	112.5	86.2	101.8
0330	Cast-in-Place Concrete	111.2	86.8	101.9	106.1	108.7	107.1	115.6	86.8	104.6	103.9	108.6	105.7	116.1	98.0	109.2	106.1	82.5	97.1
03	**CONCRETE**	103.1	83.8	94.3	100.7	102.5	101.5	105.6	83.8	95.7	99.7	102.3	100.9	109.1	94.0	102.2	106.6	80.2	94.6
04	**MASONRY**	115.6	81.3	95.1	133.2	100.8	113.8	116.3	83.7	96.8	133.0	100.8	113.7	133.4	93.0	109.2	136.5	78.4	101.7
05	**METALS**	91.2	83.8	89.0	111.2	90.3	105.1	93.7	83.9	90.8	111.2	88.8	104.7	108.4	89.5	102.9	108.8	83.9	101.6
06	**WOOD, PLASTICS & COMPOSITES**	96.8	79.9	87.5	94.8	102.1	98.8	105.7	79.9	91.5	93.7	102.1	98.3	86.4	91.6	89.3	94.6	74.4	83.5
07	**THERMAL & MOISTURE PROTECTION**	151.3	80.0	123.6	105.0	97.3	102.0	148.0	80.9	121.9	104.7	96.4	101.5	105.3	86.7	98.1	105.2	76.8	94.2
08	**OPENINGS**	111.6	74.8	102.5	106.0	97.9	104.0	112.2	74.8	103.0	104.4	97.9	102.8	102.2	91.4	99.6	103.9	71.7	95.9
0920	Plaster & Gypsum Board	140.1	78.9	100.4	107.3	102.1	103.9	137.2	78.9	99.4	109.7	102.1	104.8	107.8	91.5	97.2	110.6	73.4	86.4
0950, 0980	Ceilings & Acoustic Treatment	111.4	78.9	91.3	108.2	102.1	104.4	106.7	78.9	89.5	108.4	102.1	104.5	103.5	91.5	96.1	101.5	73.4	84.1
0960	Flooring	107.6	38.8	89.0	114.7	102.9	111.5	108.0	70.2	97.8	115.3	102.9	111.9	121.2	71.4	107.7	118.5	70.2	105.5
0970, 0990	Wall Finishes & Painting/Coating	109.0	67.3	84.0	109.3	89.6	97.5	108.9	67.3	83.9	109.3	89.6	97.5	115.7	64.1	84.7	109.3	67.3	84.1
09	**FINISHES**	122.6	70.8	95.2	108.1	101.4	104.5	121.3	77.2	98.0	108.5	101.4	104.8	107.0	84.7	95.2	109.0	73.2	90.1
COVERS	**DIVS. 10 - 14, 25, 28, 41, 43, 44**	100.0	76.1	95.3	100.0	101.4	100.3	100.0	76.0	95.3	100.0	101.4	100.3	100.0	72.3	94.5	100.0	92.4	98.5
21, 22, 23	**FIRE SUPPRESSION, PLUMBING & HVAC**	100.6	96.4	98.9	100.2	107.0	102.9	100.5	82.1	93.2	100.2	98.8	99.7	100.3	100.9	100.6	94.4	78.1	87.9
26, 27, 3370	**ELECTRICAL, COMMUNICATIONS & UTIL.**	97.0	90.8	93.9	103.1	99.4	101.3	95.3	76.3	86.1	103.1	97.1	100.2	109.7	97.6	103.8	103.9	81.9	93.2
MF2004	**WEIGHTED AVERAGE**	104.1	85.7	96.4	105.4	101.7	103.9	104.5	81.8	95.0	105.1	99.6	102.8	106.5	93.8	101.1	104.8	81.4	95.0

DIVISION		WASHINGTON			WEST VIRGINIA														
		YAKIMA			BECKLEY			BLUEFIELD			BUCKHANNON			CHARLESTON			CLARKSBURG		
		989			258 - 259			247 - 248			262			250 - 253			263 - 264		
		MAT.	INST.	TOTAL	MAT.	INST.	TOTAL	MAT.	INST.	TOTAL	MAT.	INST.	TOTAL	MAT.	INST.	TOTAL	MAT.	INST.	TOTAL
015433	**CONTRACTOR EQUIPMENT**		101.3	101.3		102.1	102.1		102.1	102.1		102.1	102.1		102.1	102.1		102.1	102.1
0241, 31 - 34	**SITE & INFRASTRUCTURE, DEMOLITION**	104.4	109.4	107.8	103.1	88.3	93.0	102.8	88.3	92.9	109.6	88.0	94.8	103.9	89.4	94.0	110.2	88.0	95.0
0310	Concrete Forming & Accessories	103.7	94.9	96.1	88.0	84.9	85.3	89.4	83.8	84.5	88.7	83.4	84.1	104.3	84.5	87.2	86.3	83.3	83.7
0320	Concrete Reinforcing	112.2	85.7	101.4	100.2	81.0	92.3	100.2	62.1	84.6	100.7	88.1	95.6	101.6	81.1	93.2	100.7	83.9	93.9
0330	Cast-in-Place Concrete	111.1	82.4	100.1	101.2	98.5	100.2	101.2	98.3	100.1	100.9	96.3	99.1	99.4	99.8	99.5	110.6	96.3	105.1
03	**CONCRETE**	104.2	88.5	97.1	103.9	89.7	97.5	104.0	85.7	95.7	107.3	89.6	99.3	104.3	90.0	97.8	111.2	88.8	101.0
04	**MASONRY**	125.2	63.4	88.2	95.9	87.6	90.9	90.8	87.6	88.9	102.3	81.5	89.8	94.7	89.4	91.5	103.8	81.5	90.4
05	**METALS**	109.4	84.2	102.0	95.1	95.8	95.3	95.1	88.7	93.2	95.2	98.6	96.2	97.2	96.7	97.1	95.2	97.1	95.8
06	**WOOD, PLASTICS & COMPOSITES**	94.0	102.1	98.4	86.8	85.5	86.1	88.5	85.5	86.8	87.7	82.8	85.0	104.8	82.8	92.7	84.6	82.8	83.6
07	**THERMAL & MOISTURE PROTECTION**	104.8	76.6	93.9	102.5	87.2	96.5	102.6	82.1	94.6	102.8	83.4	95.3	99.7	87.7	95.0	102.8	83.4	95.2
08	**OPENINGS**	103.8	84.5	99.1	98.9	79.0	94.0	99.8	74.7	93.6	99.9	79.2	94.8	99.5	77.5	94.1	99.9	82.1	95.5
0920	Plaster & Gypsum Board	109.1	102.1	104.6	102.9	84.8	91.1	103.3	84.8	91.3	103.7	82.0	89.6	107.9	82.0	91.1	102.3	82.0	89.1
0950, 0980	Ceilings & Acoustic Treatment	103.4	102.1	102.6	90.1	84.8	86.8	90.1	84.8	86.8	91.8	82.0	85.7	94.3	82.0	86.7	91.8	82.0	85.7
0960	Flooring	116.5	58.3	100.8	101.9	105.0	102.7	102.5	105.0	103.2	102.1	94.5	100.0	111.5	105.0	109.7	100.7	94.5	99.1
0970, 0990	Wall Finishes & Painting/Coating	109.3	67.3	84.1	114.4	40.2	69.9	114.4	40.2	69.9	114.4	91.3	100.5	114.4	87.1	98.0	114.4	91.3	100.5
09	**FINISHES**	107.8	86.0	96.3	99.0	84.4	91.3	99.3	84.4	91.4	100.1	86.2	92.8	103.5	88.5	95.6	99.4	86.2	92.4
COVERS	**DIVS. 10 - 14, 25, 28, 41, 43, 44**	100.0	95.7	99.2	100.0	63.9	92.9	100.0	63.8	92.8	100.0	83.0	96.6	100.0	83.9	96.8	100.0	83.0	96.6
21, 22, 23	**FIRE SUPPRESSION, PLUMBING & HVAC**	100.2	93.0	97.4	94.4	81.2	89.1	94.4	59.4	80.5	94.4	90.8	93.0	100.0	83.0	93.3	94.4	86.2	91.1
26, 27, 3370	**ELECTRICAL, COMMUNICATIONS & UTIL.**	106.1	90.8	98.6	93.4	88.0	90.8	96.1	69.0	82.9	97.5	92.8	95.2	102.5	88.0	95.4	97.5	92.8	95.2
MF2004	**WEIGHTED AVERAGE**	105.3	88.0	98.0	97.3	86.0	92.6	97.5	77.3	89.0	98.9	88.9	94.7	100.4	87.8	95.1	99.3	87.8	94.5

DIVISION		WEST VIRGINIA																	
		GASSAWAY			HUNTINGTON			LEWISBURG			MARTINSBURG			MORGANTOWN			PARKERSBURG		
		266			255 - 257			249			254			265			261		
		MAT.	INST.	TOTAL	MAT.	INST.	TOTAL	MAT.	INST.	TOTAL	MAT.	INST.	TOTAL	MAT.	INST.	TOTAL	MAT.	INST.	TOTAL
015433	**CONTRACTOR EQUIPMENT**		102.1	102.1		102.1	102.1		102.1	102.1		102.1	102.1		102.1	102.1		102.1	102.1
0241, 31 - 34	**SITE & INFRASTRUCTURE, DEMOLITION**	106.7	88.5	94.2	107.3	90.3	95.7	120.2	88.3	98.4	107.0	85.9	92.6	103.8	88.3	93.2	112.4	89.4	96.7
0310	Concrete Forming & Accessories	88.1	82.4	83.2	99.7	89.4	90.8	86.6	84.5	84.8	88.0	79.3	80.5	86.6	83.3	83.8	90.5	83.3	84.3
0320	Concrete Reinforcing	100.7	77.8	91.4	101.6	86.3	95.4	100.7	62.2	85.0	100.2	71.3	88.4	100.7	83.9	93.9	100.2	88.1	95.2
0330	Cast-in-Place Concrete	105.6	98.4	102.9	110.4	104.1	108.0	101.3	98.3	100.1	106.0	91.0	100.3	100.9	96.3	99.1	103.1	95.7	100.3
03	**CONCRETE**	107.5	88.1	98.7	109.2	94.6	102.6	114.3	86.1	101.5	107.6	82.7	96.2	103.8	88.8	97.0	109.2	89.4	100.2
04	**MASONRY**	105.0	86.9	94.2	97.6	86.6	91.0	96.4	87.6	91.1	98.1	71.2	82.0	124.3	81.5	98.7	81.7	83.9	83.1
05	**METALS**	95.2	95.6	95.3	97.3	98.2	97.5	95.2	89.4	93.5	95.4	86.1	92.7	95.3	97.1	95.8	95.8	98.7	96.6
06	**WOOD, PLASTICS & COMPOSITES**	87.0	81.2	83.8	98.6	87.8	92.7	84.9	85.5	85.2	86.8	83.2	84.8	84.9	82.8	83.7	88.3	81.2	84.4
07	**THERMAL & MOISTURE PROTECTION**	102.6	85.8	96.1	102.7	88.3	97.1	103.4	87.1	97.1	102.8	66.2	88.5	102.6	83.4	95.1	102.5	84.7	95.6
08	**OPENINGS**	97.8	75.9	92.4	98.3	81.5	94.1	99.8	74.7	93.6	101.2	70.0	93.5	101.3	82.1	96.5	98.9	78.3	93.8
0920	Plaster & Gypsum Board	103.4	80.4	88.4	111.0	87.1	95.5	102.3	84.8	90.9	103.4	82.4	89.7	102.3	82.0	89.1	105.2	80.4	89.1
0950, 0980	Ceilings & Acoustic Treatment	91.8	80.4	84.7	92.6	87.1	89.2	91.8	84.8	87.4	91.8	82.4	86.0	91.8	82.0	85.7	91.8	80.4	84.7
0960	Flooring	101.9	105.0	102.7	111.3	93.6	106.5	100.9	105.0	102.0	101.9	60.2	90.6	100.9	94.5	99.1	105.7	95.6	103.0
0970, 0990	Wall Finishes & Painting/Coating	114.4	87.1	98.0	114.4	87.1	98.0	114.4	40.2	69.9	114.4	47.5	74.3	114.4	91.3	100.5	114.4	87.2	98.1
09	**FINISHES**	99.7	86.9	93.0	103.5	89.9	96.3	100.5	84.4	92.0	99.7	73.6	85.9	99.0	86.2	92.3	101.5	85.7	93.1
COVERS	**DIVS. 10 - 14, 25, 28, 41, 43, 44**	100.0	82.8	96.6	100.0	85.7	97.2	100.0	63.8	92.8	100.0	61.9	92.5	100.0	74.5	95.0	100.0	83.8	96.8
21, 22, 23	**FIRE SUPPRESSION, PLUMBING & HVAC**	94.4	93.0	93.8	100.2	84.9	94.1	94.4	81.0	89.0	94.4	70.7	85.0	94.4	90.6	92.9	100.1	87.3	95.0
26, 27, 3370	**ELECTRICAL, COMMUNICATIONS & UTIL.**	97.5	88.0	92.9	97.2	95.8	96.5	93.4	69.0	81.5	99.3	79.6	89.7	97.7	92.8	95.3	97.7	94.3	96.0
MF2004	**WEIGHTED AVERAGE**	98.7	88.8	94.5	100.6	90.3	96.3	99.3	82.0	92.0	98.9	76.5	89.5	99.4	88.5	94.8	99.6	88.7	95.0

City Cost Indexes

DIVISION		WEST VIRGINIA									WISCONSIN								
		PETERSBURG			ROMNEY			WHEELING			BELOIT			EAU CLAIRE			GREEN BAY		
		268			267			260			535			547			541 - 543		
		MAT.	INST.	TOTAL	MAT.	INST.	TOTAL	MAT.	INST.	TOTAL	MAT.	INST.	TOTAL	MAT.	INST.	TOTAL	MAT.	INST.	TOTAL
015433	CONTRACTOR EQUIPMENT		102.1	102.1		102.1	102.1		102.1	102.1		100.4	100.4		101.0	101.0		98.7	98.7
0241, 31 - 34	SITE & INFRASTRUCTURE, DEMOLITION	103.1	88.7	93.2	106.1	88.7	94.2	113.1	88.9	96.6	91.2	106.2	101.5	90.8	104.0	99.8	94.1	99.5	97.8
0310	Concrete Forming & Accessories	89.7	85.0	85.6	85.9	84.7	84.8	92.2	83.6	84.8	97.0	93.9	94.3	97.0	94.4	94.7	105.0	93.6	95.2
0320	Concrete Reinforcing	100.2	71.9	88.6	100.7	68.3	87.5	99.5	84.0	93.1	100.6	108.4	103.8	100.8	100.0	100.5	98.9	85.6	93.5
0330	Cast-in-Place Concrete	100.9	96.0	99.0	105.6	95.9	101.9	103.1	96.4	100.6	99.4	95.2	97.8	97.8	94.4	96.5	101.2	96.2	99.3
03	CONCRETE	103.9	87.3	96.3	107.4	86.4	97.8	109.2	89.0	100.0	98.0	97.3	97.7	97.7	95.8	96.8	98.3	93.4	96.1
04	MASONRY	98.3	81.5	88.2	96.8	67.5	79.3	105.3	81.5	91.0	103.3	94.9	98.3	89.5	95.1	92.8	119.3	94.2	104.3
05	METALS	95.3	92.6	94.5	95.4	90.4	93.9	95.9	97.4	96.4	98.5	103.5	100.0	91.9	101.6	94.8	94.1	95.7	94.5
06	WOOD, PLASTICS & COMPOSITES	88.7	86.0	87.2	83.9	86.0	85.0	90.1	82.8	86.1	110.1	94.7	101.7	113.5	94.8	103.3	115.8	94.8	104.3
07	THERMAL & MOISTURE PROTECTION	102.7	73.1	91.2	102.7	70.2	90.1	102.9	83.8	95.5	99.5	93.7	97.2	100.1	85.2	94.3	102.0	83.5	94.8
08	OPENINGS	101.3	77.2	95.3	101.2	76.6	95.1	99.8	82.1	95.4	103.3	99.5	102.4	100.7	93.5	98.9	98.8	91.8	97.1
0920	Plaster & Gypsum Board	103.7	85.3	91.8	101.5	85.3	91.0	105.6	82.0	90.3	95.6	95.0	95.2	100.4	95.0	96.9	92.7	95.0	94.2
0950, 0980	Ceilings & Acoustic Treatment	91.8	85.3	87.8	91.8	85.3	87.8	91.8	82.0	85.7	83.4	95.0	90.6	99.9	95.0	96.9	92.5	95.0	94.0
0960	Flooring	103.0	94.5	100.7	100.6	67.8	91.8	106.8	94.5	103.5	104.5	111.6	106.4	97.0	106.7	99.6	116.7	106.7	114.0
0970, 0990	Wall Finishes & Painting/Coating	114.4	86.8	97.9	114.4	86.8	97.9	114.4	91.3	100.5	93.6	93.5	93.5	89.2	91.8	90.8	101.0	74.5	85.1
09	FINISHES	99.9	87.6	93.4	99.1	82.2	90.1	101.9	86.2	93.6	96.7	97.5	97.2	99.4	96.8	98.0	103.8	94.6	99.0
COVERS	DIVS. 10 - 14, 25, 28, 41, 43, 44	100.0	74.9	95.0	100.0	83.4	96.7	100.0	93.5	98.7	100.0	85.9	97.2	100.0	87.3	97.5	100.0	86.6	97.3
21, 22, 23	FIRE SUPPRESSION, PLUMBING & HVAC	94.4	84.9	90.6	94.4	84.7	90.5	100.2	90.6	96.4	99.9	88.3	95.3	100.1	80.1	92.2	100.4	83.8	93.8
26, 27, 3370	ELECTRICAL, COMMUNICATIONS & UTIL.	101.0	79.7	90.6	100.2	79.7	90.2	94.8	92.8	93.8	98.2	85.9	92.2	101.9	87.7	94.9	96.6	83.4	90.2
MF2004	WEIGHTED AVERAGE	98.6	84.6	92.7	98.8	82.3	91.9	100.6	89.2	95.8	99.3	94.6	97.3	97.9	92.0	95.4	99.7	90.4	95.8

DIVISION		WISCONSIN																	
		KENOSHA			LA CROSSE			LANCASTER			MADISON			MILWAUKEE			NEW RICHMOND		
		531			546			538			537			530,532			540		
		MAT.	INST.	TOTAL	MAT.	INST.	TOTAL	MAT.	INST.	TOTAL	MAT.	INST.	TOTAL	MAT.	INST.	TOTAL	MAT.	INST.	TOTAL
015433	CONTRACTOR EQUIPMENT		98.3	98.3		101.0	101.0		100.4	100.4		100.4	100.4		89.4	89.4		100.8	100.8
0241, 31 - 34	SITE & INFRASTRUCTURE, DEMOLITION	97.0	103.1	101.1	84.4	104.1	97.9	90.7	106.4	101.4	91.0	106.5	101.6	91.9	97.4	95.7	86.1	103.4	97.9
0310	Concrete Forming & Accessories	104.1	104.4	104.4	84.4	94.9	93.5	96.3	94.2	94.4	95.9	95.7	95.7	99.6	113.8	111.8	92.0	103.7	102.1
0320	Concrete Reinforcing	100.4	102.3	101.1	100.5	88.2	95.5	102.0	88.2	96.3	100.6	88.6	95.7	100.6	102.7	101.5	97.9	99.9	98.7
0330	Cast-in-Place Concrete	108.2	100.4	105.2	87.9	94.5	90.4	98.8	97.6	98.3	97.3	98.2	97.6	97.1	108.5	101.5	101.8	87.2	96.2
03	CONCRETE	102.5	102.6	102.5	89.7	93.8	91.6	97.8	94.4	96.3	96.9	95.4	96.2	97.2	109.0	102.6	95.6	97.4	96.4
04	MASONRY	100.4	107.7	104.8	88.7	98.0	94.2	103.5	100.5	101.7	101.2	104.0	102.9	103.2	117.5	111.8	113.8	95.3	102.7
05	METALS	99.3	102.6	100.3	91.8	96.5	93.2	96.3	94.0	95.6	101.1	95.7	99.5	101.8	96.2	100.2	92.1	100.9	94.6
06	WOOD, PLASTICS & COMPOSITES	112.2	103.9	107.6	98.3	94.8	96.4	109.3	94.7	101.3	107.0	94.7	100.3	112.3	113.6	113.0	102.2	108.2	105.5
07	THERMAL & MOISTURE PROTECTION	99.8	97.3	98.8	99.6	85.2	94.0	99.2	81.2	92.2	95.5	92.9	94.5	97.8	112.1	103.4	101.1	94.9	98.7
08	OPENINGS	98.4	105.9	100.3	100.7	82.9	96.3	98.4	83.8	94.8	103.3	95.2	101.3	105.6	111.3	107.0	88.4	100.8	91.5
0920	Plaster & Gypsum Board	85.0	104.4	97.6	94.8	95.0	94.9	94.3	95.0	94.7	92.1	95.0	94.0	96.5	114.3	108.1	85.5	109.1	100.8
0950, 0980	Ceilings & Acoustic Treatment	82.6	104.4	96.1	98.3	95.0	96.2	78.5	95.0	88.7	85.3	95.0	91.3	86.8	114.3	103.8	67.4	109.1	93.2
0960	Flooring	122.8	112.2	119.9	89.5	108.7	94.7	103.7	105.0	104.1	98.8	105.0	100.5	107.4	120.5	110.9	109.0	106.7	108.4
0970, 0990	Wall Finishes & Painting/Coating	102.8	101.8	102.2	89.2	71.5	78.6	93.6	76.3	83.2	93.6	93.5	93.5	95.6	113.1	106.1	103.2	91.8	96.4
09	FINISHES	101.3	106.3	103.9	95.6	95.6	95.6	95.1	94.0	94.5	94.8	97.6	96.3	98.6	115.5	107.5	94.3	102.8	98.8
COVERS	DIVS. 10 - 14, 25, 28, 41, 43, 44	100.0	102.3	100.5	100.0	88.3	97.7	100.0	56.8	91.4	100.0	87.9	97.6	100.0	105.1	101.0	100.0	88.3	97.7
21, 22, 23	FIRE SUPPRESSION, PLUMBING & HVAC	100.1	88.5	95.5	100.1	81.3	92.7	94.1	81.1	88.9	99.7	89.3	95.6	99.9	101.4	100.5	93.7	79.6	88.1
26, 27, 3370	ELECTRICAL, COMMUNICATIONS & UTIL.	98.6	92.9	95.8	102.1	87.7	95.1	97.9	87.6	92.9	101.3	88.9	95.3	99.6	104.3	101.9	100.1	87.7	94.0
MF2004	WEIGHTED AVERAGE	100.0	99.3	99.7	96.3	91.3	94.2	96.9	90.3	94.1	99.5	95.0	97.6	100.3	106.4	102.9	95.5	93.6	94.7

DIVISION		WISCONSIN																	
		OSHKOSH			PORTAGE			RACINE			RHINELANDER			SUPERIOR			WAUSAU		
		549			539			534			545			548			544		
		MAT.	INST.	TOTAL	MAT.	INST.	TOTAL	MAT.	INST.	TOTAL	MAT.	INST.	TOTAL	MAT.	INST.	TOTAL	MAT.	INST.	TOTAL
015433	CONTRACTOR EQUIPMENT		98.7	98.7		100.4	100.4		100.4	100.4		98.7	98.7		100.8	100.8		98.7	98.7
0241, 31 - 34	SITE & INFRASTRUCTURE, DEMOLITION	86.0	99.5	95.2	81.5	106.4	98.5	91.0	107.3	102.1	98.8	99.6	99.4	82.8	103.5	96.9	81.7	99.5	93.9
0310	Concrete Forming & Accessories	89.0	93.3	92.7	89.5	95.1	94.4	97.4	104.6	103.6	86.7	93.9	92.9	90.6	94.3	93.8	88.5	93.2	92.6
0320	Concrete Reinforcing	99.0	85.6	93.5	102.1	88.3	96.5	100.6	102.3	101.3	99.2	86.2	93.9	97.9	87.0	93.5	99.2	88.1	94.7
0330	Cast-in-Place Concrete	93.8	96.0	94.7	84.8	100.4	90.8	97.5	100.2	98.5	106.4	95.3	102.1	96.0	95.9	95.9	87.4	96.0	90.7
03	CONCRETE	91.0	93.2	92.0	86.7	95.9	90.9	97.1	102.6	99.6	102.3	93.3	98.2	90.5	93.8	92.0	86.5	93.6	89.7
04	MASONRY	101.3	94.2	97.0	102.4	100.9	101.5	103.3	107.7	105.9	118.7	94.7	104.3	113.1	104.0	107.6	100.8	94.2	96.9
05	METALS	92.3	95.3	93.2	96.9	94.9	96.3	100.2	102.8	100.9	92.2	95.0	93.0	92.9	96.9	94.1	92.1	96.2	93.3
06	WOOD, PLASTICS & COMPOSITES	96.6	94.8	95.6	100.6	94.7	97.4	110.3	103.9	106.8	93.8	94.8	94.4	100.5	93.8	96.8	95.9	94.8	95.3
07	THERMAL & MOISTURE PROTECTION	101.2	84.9	94.9	98.8	91.1	95.8	99.5	98.6	99.1	102.0	80.4	93.6	100.9	95.9	98.9	101.1	78.5	92.3
08	OPENINGS	94.1	91.8	93.5	98.6	91.4	96.8	103.3	105.9	103.9	94.1	85.0	91.9	88.3	90.8	88.9	94.3	92.5	93.9
0920	Plaster & Gypsum Board	83.7	95.0	91.0	90.1	95.0	93.3	95.6	104.4	101.3	83.0	95.0	90.8	85.3	94.2	91.1	83.4	95.0	90.9
0950, 0980	Ceilings & Acoustic Treatment	92.5	95.0	94.0	79.3	95.0	89.0	83.4	104.4	96.4	92.5	95.0	94.0	68.2	94.2	84.3	92.5	95.0	94.0
0960	Flooring	106.9	106.7	106.9	99.7	112.1	103.1	104.5	112.2	106.6	106.0	102.4	105.0	110.4	121.1	113.3	106.7	106.7	106.7
0970, 0990	Wall Finishes & Painting/Coating	98.4	74.5	84.1	93.6	76.3	83.2	93.5	101.8	98.5	98.4	69.4	81.0	93.6	101.6	98.4	98.4	74.5	84.1
09	FINISHES	98.9	93.3	96.0	93.0	96.2	94.6	96.7	106.3	101.8	99.6	93.4	96.3	94.0	100.4	97.3	98.6	94.6	96.5
COVERS	DIVS. 10 - 14, 25, 28, 41, 43, 44	100.0	74.4	94.9	100.0	74.0	94.8	100.0	102.3	100.5	100.0	63.5	92.8	100.0	87.0	97.4	100.0	86.5	97.3
21, 22, 23	FIRE SUPPRESSION, PLUMBING & HVAC	94.6	83.7	90.3	94.1	89.1	92.1	99.9	93.7	97.5	94.6	85.4	90.9	93.7	93.7	93.7	94.6	78.8	88.3
26, 27, 3370	ELECTRICAL, COMMUNICATIONS & UTIL.	100.6	83.7	92.4	101.7	92.2	97.1	98.0	96.2	97.1	100.0	77.4	89.0	104.9	88.3	96.8	101.8	77.4	89.9
MF2004	WEIGHTED AVERAGE	95.5	89.9	93.2	95.6	94.3	95.0	99.5	101.3	100.2	98.0	88.7	94.1	95.4	95.7	95.5	94.9	88.5	92.2

	DIVISION	WYOMING																	
		CASPER			CHEYENNE			NEWCASTLE			RAWLINS			RIVERTON			ROCK SPRINGS		
		826			820			827			823			825			829 - 831		
		MAT.	INST.	TOTAL	MAT.	INST.	TOTAL	MAT.	INST.	TOTAL	MAT.	INST.	TOTAL	MAT.	INST.	TOTAL	MAT.	INST.	TOTAL
015433	**CONTRACTOR EQUIPMENT**		101.1	101.1		101.1	101.1		101.1	101.1		101.1	101.1		101.1	101.1		101.1	101.1
0241, 31 - 34	**SITE & INFRASTRUCTURE, DEMOLITION**	92.6	100.1	97.7	85.8	100.2	95.6	78.8	98.9	92.5	92.8	98.9	97.0	86.3	98.9	94.9	82.2	98.9	93.6
0310	Concrete Forming & Accessories	103.9	45.0	53.0	105.5	61.3	67.3	96.0	40.4	48.0	100.2	40.4	48.6	94.9	40.4	47.8	101.8	40.4	48.7
0320	Concrete Reinforcing	109.3	46.0	83.5	103.0	46.9	80.1	110.8	46.7	84.7	110.4	46.7	84.4	111.4	46.7	85.0	111.4	46.7	85.0
0330	Cast-in-Place Concrete	102.7	76.1	92.5	98.0	76.3	89.7	99.0	56.8	82.9	99.1	56.8	83.0	99.1	56.8	82.9	99.0	56.8	82.9
03	**CONCRETE**	108.1	57.1	84.9	105.3	64.5	86.7	106.3	48.4	80.0	119.7	48.4	87.3	114.3	48.4	84.3	106.7	48.4	80.2
04	**MASONRY**	107.3	39.3	66.6	104.0	54.2	74.2	100.9	48.6	69.6	100.9	48.6	69.6	100.9	48.6	69.6	162.7	48.6	94.4
05	**METALS**	98.3	64.7	88.5	99.5	66.3	89.8	94.7	64.2	85.8	94.8	64.2	85.8	94.9	64.1	85.9	95.5	64.1	86.3
06	**WOOD, PLASTICS & COMPOSITES**	97.7	42.4	67.3	99.3	63.7	79.8	89.0	39.1	61.6	93.4	39.1	63.6	87.9	39.1	61.1	96.1	39.1	64.8
07	**THERMAL & MOISTURE PROTECTION**	97.6	51.9	79.8	99.4	58.1	83.3	100.6	49.9	80.9	102.4	49.9	81.9	101.7	49.9	81.5	100.7	49.9	80.9
08	**OPENINGS**	92.7	41.9	80.2	94.8	53.5	84.6	98.7	40.1	84.2	98.4	40.1	84.0	98.6	40.1	84.1	99.1	40.1	84.5
0920	Plaster & Gypsum Board	86.7	40.2	56.5	82.1	62.2	69.2	77.0	36.7	50.8	78.5	36.7	51.4	77.0	36.7	50.8	84.1	36.7	53.3
0950, 0980	Ceilings & Acoustic Treatment	112.1	40.2	67.6	106.0	62.2	78.9	108.8	36.7	64.2	108.8	36.7	64.2	108.8	36.7	64.2	108.8	36.7	64.2
0960	Flooring	103.5	37.0	85.5	108.0	59.4	94.9	101.5	52.3	88.2	104.5	52.3	90.4	100.8	52.3	87.6	107.0	52.3	92.2
0970, 0990	Wall Finishes & Painting/Coating	98.0	49.5	68.9	101.2	49.5	70.2	97.3	31.7	57.9	97.3	31.7	57.9	97.3	31.7	57.9	97.3	31.7	57.9
09	**FINISHES**	104.7	42.9	72.1	104.8	60.0	81.1	99.7	40.4	68.4	102.3	40.4	69.6	100.4	40.4	68.7	102.3	40.4	69.6
COVERS	**DIVS. 10 - 14, 25, 28, 41, 43, 44**	100.0	79.4	95.9	100.0	82.0	96.4	100.0	65.4	93.1	100.0	65.4	93.1	100.0	65.3	93.1	100.0	65.3	93.1
21, 22, 23	**FIRE SUPPRESSION, PLUMBING & HVAC**	100.0	60.7	84.4	99.9	59.9	84.0	97.0	57.3	81.2	97.0	57.3	81.2	97.0	57.3	81.2	99.9	57.3	83.0
26, 27, 3370	**ELECTRICAL, COMMUNICATIONS & UTIL.**	101.2	59.3	80.8	97.6	72.7	85.5	95.1	73.4	84.5	95.1	73.4	84.5	95.1	60.4	78.2	93.3	60.4	77.3
MF2004	**WEIGHTED AVERAGE**	100.6	58.4	82.9	100.0	65.9	85.7	97.9	58.4	81.3	100.2	58.4	82.6	99.2	56.5	81.3	102.0	56.5	82.9

	DIVISION	WYOMING												CANADA					
		SHERIDAN			WHEATLAND			WORLAND			YELLOWSTONE NAT'L PA			BARRIE, ONTARIO			BATHURST, NEW BRUNSWICK		
		828			822			824			821								
		MAT.	INST.	TOTAL	MAT.	INST.	TOTAL	MAT.	INST.	TOTAL	MAT.	INST.	TOTAL	MAT.	INST.	TOTAL	MAT.	INST.	TOTAL
015433	**CONTRACTOR EQUIPMENT**		101.1	101.1		101.1	101.1		101.1	101.1		101.1	101.1		102.2	102.2		103.3	103.3
0241, 31 - 34	**SITE & INFRASTRUCTURE, DEMOLITION**	86.4	100.1	95.8	83.1	98.9	93.9	80.6	98.9	93.1	80.7	99.3	93.4	116.5	100.6	105.6	96.1	97.8	97.3
0310	Concrete Forming & Accessories	103.8	45.3	53.2	98.0	40.8	48.6	98.1	40.4	48.2	98.1	42.8	50.3	123.3	87.5	92.4	102.1	62.6	68.0
0320	Concrete Reinforcing	111.4	46.8	85.1	110.8	46.7	84.7	111.4	46.7	85.0	113.4	46.7	86.2	179.6	84.7	140.9	147.9	57.2	110.9
0330	Cast-in-Place Concrete	102.4	76.2	92.4	103.4	57.3	85.8	99.0	56.8	82.9	99.0	60.3	84.2	167.7	86.9	136.8	138.3	62.0	109.1
03	**CONCRETE**	114.3	57.4	88.4	111.1	48.8	82.7	106.5	48.4	80.1	106.9	50.7	81.3	154.3	87.0	123.7	136.2	62.3	102.6
04	**MASONRY**	101.2	53.5	72.7	101.3	49.6	70.3	100.9	48.6	69.6	101.0	55.1	73.5	169.3	96.0	125.4	159.7	63.7	102.2
05	**METALS**	98.4	65.9	88.9	94.7	64.2	85.8	94.9	64.1	85.9	95.5	64.1	86.3	107.7	90.9	102.8	105.2	74.6	96.3
06	**WOOD, PLASTICS & COMPOSITES**	100.6	42.4	68.6	90.8	39.1	62.4	90.9	39.1	62.4	90.9	39.1	62.4	118.1	85.9	100.4	94.0	62.5	76.7
07	**THERMAL & MOISTURE PROTECTION**	101.6	55.7	83.8	101.0	50.9	81.5	100.7	49.9	80.9	100.3	53.3	82.0	110.8	87.6	101.8	104.2	61.7	87.6
08	**OPENINGS**	99.4	42.9	85.4	97.4	40.1	83.2	98.9	40.1	84.4	92.3	40.1	79.4	91.0	84.5	89.4	86.2	54.7	78.4
0920	Plaster & Gypsum Board	95.4	40.2	59.5	77.0	36.7	50.8	77.0	36.7	50.8	77.5	36.7	51.0	153.5	85.3	109.2	165.9	61.0	97.7
0950, 0980	Ceilings & Acoustic Treatment	111.6	40.2	67.4	108.8	36.7	64.2	108.8	36.7	64.2	110.4	36.7	64.8	95.4	85.3	89.1	96.2	61.0	74.4
0960	Flooring	103.6	52.3	89.8	103.0	52.3	89.3	103.0	52.3	89.3	103.0	52.3	89.3	132.1	94.8	122.0	113.9	45.2	95.4
0970, 0990	Wall Finishes & Painting/Coating	99.3	49.5	69.4	97.3	31.7	57.9	97.3	31.7	57.9	97.3	31.7	57.9	110.4	86.8	96.2	109.6	49.8	73.8
09	**FINISHES**	106.5	45.7	74.3	100.5	40.7	68.9	100.2	40.4	68.6	100.6	42.1	69.7	119.0	88.9	103.1	113.9	58.0	84.3
COVERS	**DIVS. 10 - 14, 25, 28, 41, 43, 44**	100.0	79.4	95.9	100.0	65.7	93.2	100.0	65.3	93.1	100.0	67.7	93.6	140.0	80.0	128.1	140.0	70.6	126.3
21, 22, 23	**FIRE SUPPRESSION, PLUMBING & HVAC**	97.0	59.8	82.2	97.0	57.8	81.4	97.0	57.3	81.2	97.0	60.7	82.6	102.5	101.1	102.0	102.4	71.1	90.0
26, 27, 3370	**ELECTRICAL, COMMUNICATIONS & UTIL.**	97.6	60.4	79.5	95.1	72.7	84.2	95.1	60.4	78.2	94.0	60.4	77.6	123.1	92.7	108.3	126.2	63.4	95.6
MF2004	**WEIGHTED AVERAGE**	100.7	60.5	83.8	98.6	58.6	81.8	98.1	56.5	80.6	97.5	58.6	81.2	119.2	93.0	108.2	114.7	67.8	95.0

	DIVISION	CANADA																	
		BRANDON, MANITOBA			BRANTFORD, ONTARIO			BRIDGEWATER, NOVA SCOTIA			CALGARY, ALBERTA			CAP-DE-LA-MADELEINE, QUEBEC			CHARLESBOURG, QUEBEC		
		MAT.	INST.	TOTAL	MAT.	INST.	TOTAL	MAT.	INST.	TOTAL	MAT.	INST.	TOTAL	MAT.	INST.	TOTAL	MAT.	INST.	TOTAL
015433	**CONTRACTOR EQUIPMENT**		104.6	104.6		102.2	102.2		102.9	102.9		106.4	106.4		104.0	104.0		104.0	104.0
0241, 31 - 34	**SITE & INFRASTRUCTURE, DEMOLITION**	111.9	99.7	103.6	114.2	101.1	105.3	98.9	99.4	99.2	124.3	104.8	111.0	94.1	101.1	98.9	94.1	101.1	98.9
0310	Concrete Forming & Accessories	122.6	71.1	78.1	123.2	94.8	98.6	96.1	70.3	73.8	127.0	90.2	95.2	129.0	86.2	92.0	129.0	86.2	92.0
0320	Concrete Reinforcing	161.1	53.8	117.3	174.5	83.4	137.3	147.9	47.3	106.8	147.9	63.2	113.4	147.9	79.7	120.1	147.9	79.7	120.1
0330	Cast-in-Place Concrete	151.2	75.8	122.4	163.6	107.1	142.0	168.2	71.6	131.3	186.7	101.0	153.9	132.0	95.6	118.1	132.0	95.6	118.1
03	**CONCRETE**	136.5	70.4	106.4	151.4	97.0	126.7	147.9	67.4	111.3	157.0	89.4	126.3	131.6	88.6	112.0	131.6	88.6	112.0
04	**MASONRY**	163.8	66.3	105.4	165.8	100.5	126.7	160.8	70.5	106.7	213.5	91.4	140.4	160.7	85.4	115.6	160.7	85.4	115.6
05	**METALS**	107.9	79.0	99.4	107.1	92.0	102.7	106.2	76.9	97.7	151.1	88.1	132.7	105.5	89.4	100.8	105.5	89.4	100.8
06	**WOOD, PLASTICS & COMPOSITES**	113.9	71.7	90.7	116.1	93.2	103.5	86.0	69.3	76.9	114.3	89.5	100.7	127.6	85.7	104.6	127.6	85.7	104.6
07	**THERMAL & MOISTURE PROTECTION**	104.5	70.8	91.4	110.6	92.9	103.7	105.5	69.0	91.4	123.0	87.9	109.3	104.6	88.3	98.3	104.6	88.3	98.3
08	**OPENINGS**	92.0	63.4	85.0	90.8	90.4	90.7	82.8	63.4	78.0	92.0	79.9	89.0	92.0	79.2	88.9	92.0	79.2	88.9
0920	Plaster & Gypsum Board	126.1	70.2	89.8	153.5	92.8	114.1	160.1	68.1	100.3	162.0	88.6	114.3	186.5	84.9	120.5	186.5	84.9	120.5
0950, 0980	Ceilings & Acoustic Treatment	95.4	70.2	79.8	95.4	92.8	93.8	95.4	68.1	78.5	144.8	88.6	110.0	95.4	84.9	88.9	95.4	84.9	88.9
0960	Flooring	132.0	67.3	114.5	132.0	94.8	122.0	110.1	64.5	97.8	134.4	88.8	122.1	132.0	95.4	122.1	132.0	95.4	122.1
0970, 0990	Wall Finishes & Painting/Coating	109.7	56.7	77.9	109.6	95.3	101.0	109.6	61.9	81.0	109.6	100.7	104.3	109.6	88.6	97.0	109.6	88.6	97.0
09	**FINISHES**	114.3	69.0	90.4	117.8	95.0	105.8	111.5	68.3	88.7	132.7	91.6	111.0	121.6	88.2	104.0	121.6	88.2	104.0
COVERS	**DIVS. 10 - 14, 25, 28, 41, 43, 44**	140.0	72.9	126.7	140.0	82.2	128.6	140.0	71.8	126.5	140.0	102.3	132.5	140.0	91.6	130.4	140.0	91.6	130.4
21, 22, 23	**FIRE SUPPRESSION, PLUMBING & HVAC**	102.4	85.7	95.8	102.4	104.1	103.1	102.4	86.0	95.9	100.1	87.6	95.1	103.0	92.7	98.9	103.0	92.7	98.9
26, 27, 3370	**ELECTRICAL, COMMUNICATIONS & UTIL.**	125.4	71.4	99.1	123.1	92.2	108.0	128.7	66.0	98.2	125.2	84.7	105.5	121.8	74.4	98.7	121.8	74.4	98.7
MF2004	**WEIGHTED AVERAGE**	116.4	76.1	99.5	118.4	96.9	109.4	116.1	75.0	98.9	130.7	89.9	113.6	115.3	87.8	103.8	115.3	87.8	103.8

DIVISION		CANADA																	
		CHARLOTTETOWN, PRINCE EDWARD ISLAND			CHICOUTIMI, QUEBEC			CORNER BROOK, NEWFOUNDLAND			CORNWALL, ONTARIO			DALHOUSIE, NEW BRUNSWICK			DARTMOUTH, NOVA SCOTIA		
		MAT.	INST.	TOTAL	MAT.	INST.	TOTAL	MAT.	INST.	TOTAL	MAT.	INST.	TOTAL	MAT.	INST.	TOTAL	MAT.	INST.	TOTAL
015433	CONTRACTOR EQUIPMENT		102.8	102.8		104.0	104.0		104.8	104.8		102.2	102.2		103.3	103.3		102.9	102.9
0241, 31 - 34	SITE & INFRASTRUCTURE, DEMOLITION	106.9	96.9	100.0	93.5	100.7	98.4	115.1	98.5	103.8	112.1	100.6	104.3	96.1	97.8	97.2	106.9	99.4	101.8
0310	Concrete Forming & Accessories	96.4	57.2	62.6	129.0	93.4	98.3	103.8	60.6	66.5	120.9	87.6	92.2	102.1	62.6	68.0	96.2	70.3	73.8
0320	Concrete Reinforcing	158.2	47.3	113.0	111.6	93.9	104.4	147.9	49.1	107.6	174.5	83.1	137.2	147.9	57.2	110.9	154.5	47.3	110.8
0330	Cast-in-Place Concrete	159.4	61.2	121.9	130.2	100.5	118.9	172.4	70.2	133.4	147.1	97.9	128.3	138.3	62.0	109.1	170.0	71.6	132.4
03	CONCRETE	153.9	57.8	110.2	123.5	96.1	111.0	161.4	63.0	116.7	143.5	90.7	119.5	136.2	62.3	102.6	150.1	67.4	112.5
04	MASONRY	161.0	61.5	101.4	160.8	93.6	120.6	161.1	62.9	102.3	164.5	92.2	121.2	157.7	63.7	101.4	172.0	70.5	111.2
05	METALS	106.3	70.8	95.9	105.1	94.4	102.0	108.2	76.2	98.9	107.1	90.7	102.3	105.2	74.6	96.3	108.4	76.9	99.2
06	WOOD, PLASTICS & COMPOSITES	86.3	56.5	70.0	127.6	93.7	109.0	95.1	59.5	75.5	114.9	86.7	99.4	94.0	62.5	76.7	86.0	69.3	76.9
07	THERMAL & MOISTURE PROTECTION	105.2	59.2	87.3	104.6	96.4	101.4	107.5	61.3	89.5	110.4	86.5	101.1	107.8	61.7	89.9	105.5	69.0	91.4
08	OPENINGS	94.3	48.9	83.1	91.6	80.4	88.8	98.2	55.9	87.8	92.0	84.2	90.1	86.2	54.7	78.4	82.8	63.4	78.0
0920	Plaster & Gypsum Board	164.0	54.8	93.0	186.5	93.2	125.9	166.1	57.7	95.7	228.3	86.1	135.9	165.9	61.0	97.7	162.7	68.1	101.2
0950, 0980	Ceilings & Acoustic Treatment	108.7	54.8	75.4	95.4	93.2	94.0	97.1	57.7	72.7	98.7	86.1	90.9	96.2	61.0	74.4	105.4	68.1	82.3
0960	Flooring	110.1	60.3	96.7	132.0	95.4	122.1	114.8	54.9	98.6	132.0	93.4	121.6	113.9	45.2	95.4	110.1	64.5	97.8
0970, 0990	Wall Finishes & Painting/Coating	109.6	41.5	68.8	109.6	104.2	106.4	109.6	59.7	79.7	109.6	88.8	97.1	109.6	49.8	73.8	109.6	61.9	81.0
09	FINISHES	116.0	56.3	84.4	121.6	95.6	107.9	115.3	59.3	85.7	128.9	88.9	107.8	113.9	58.0	84.3	114.2	68.3	89.9
COVERS	DIVS. 10 - 14, 25, 28, 41, 43, 44	140.0	70.1	126.2	140.0	93.5	130.8	140.0	71.0	126.3	140.0	79.6	128.0	140.0	70.6	126.3	140.0	71.8	126.5
21, 22, 23	FIRE SUPPRESSION, PLUMBING & HVAC	102.4	65.0	87.6	102.4	94.3	99.2	102.4	73.2	90.8	103.0	102.0	102.6	102.4	71.1	90.0	102.4	86.0	95.9
26, 27, 3370	ELECTRICAL, COMMUNICATIONS & UTIL.	122.5	53.6	89.0	121.8	86.1	104.4	121.7	59.7	91.6	123.6	93.2	108.8	125.4	63.4	95.2	126.5	66.0	97.0
MF2004	WEIGHTED AVERAGE	117.8	63.3	94.9	114.0	93.5	105.4	119.7	68.2	98.0	118.5	93.3	107.9	114.6	67.8	95.0	117.5	75.0	99.7

DIVISION		CANADA																	
		EDMONTON, ALBERTA			FORT MCMURRAY, ALBERTA			FREDERICTON, NEW BRUNSWICK			GATINEAU, QUEBEC			GRANBY, QUEBEC			HALIFAX, NOVA SCOTIA		
		MAT.	INST.	TOTAL	MAT.	INST.	TOTAL	MAT.	INST.	TOTAL	MAT.	INST.	TOTAL	MAT.	INST.	TOTAL	MAT.	INST.	TOTAL
015433	CONTRACTOR EQUIPMENT		106.4	106.4		104.2	104.2		103.3	103.3		104.0	104.0		104.0	104.0		102.9	102.9
0241, 31 - 34	SITE & INFRASTRUCTURE, DEMOLITION	154.6	104.8	120.6	120.2	102.4	108.0	95.5	97.8	97.1	93.9	101.1	98.8	94.4	101.1	99.0	99.3	99.2	99.2
0310	Concrete Forming & Accessories	128.4	90.2	95.4	121.1	90.3	94.5	120.9	62.8	70.7	129.0	86.1	91.9	129.0	86.0	91.9	96.4	76.4	79.1
0320	Concrete Reinforcing	147.9	63.2	113.4	161.1	63.2	121.2	148.8	57.3	111.5	156.6	79.7	125.2	156.6	79.7	125.2	154.5	61.2	116.4
0330	Cast-in-Place Concrete	208.9	101.0	167.6	217.5	100.2	172.6	131.6	62.0	105.0	130.2	95.5	116.9	134.5	95.5	119.6	171.3	75.7	134.7
03	CONCRETE	167.5	89.4	131.9	173.7	89.1	135.2	134.5	62.4	101.7	132.5	88.5	112.5	134.5	88.5	113.6	150.7	74.0	115.8
04	MASONRY	202.0	91.4	135.8	204.7	91.3	136.8	177.8	65.2	110.4	160.6	85.4	115.5	160.9	85.4	115.7	173.5	82.9	119.2
05	METALS	151.6	88.0	133.0	133.4	88.1	120.2	105.1	75.0	96.3	105.5	89.3	100.8	105.5	89.2	100.7	120.4	81.4	109.0
06	WOOD, PLASTICS & COMPOSITES	110.9	89.5	99.2	110.7	89.4	99.0	116.9	62.5	87.0	127.6	85.7	104.6	127.6	85.7	104.6	86.1	75.5	80.3
07	THERMAL & MOISTURE PROTECTION	123.7	87.9	109.8	119.1	87.1	106.7	108.1	62.6	90.4	104.6	88.3	98.3	104.6	86.8	97.7	106.0	76.2	94.4
08	OPENINGS	92.0	79.9	89.0	92.0	79.8	89.0	86.1	53.6	78.0	92.0	74.7	87.7	92.0	74.7	87.7	82.8	68.9	79.4
0920	Plaster & Gypsum Board	158.2	88.6	113.0	154.1	88.6	111.5	181.4	61.0	103.2	140.0	84.9	104.2	142.5	84.9	105.1	163.1	74.4	105.5
0950, 0980	Ceilings & Acoustic Treatment	148.1	88.6	111.3	105.4	88.6	95.0	102.9	61.0	77.0	95.4	84.9	88.9	95.4	84.9	88.9	105.4	74.4	86.2
0960	Flooring	135.6	88.8	122.9	132.0	88.8	120.3	127.3	45.2	105.1	132.0	95.4	122.1	132.0	95.4	122.1	110.1	64.5	97.8
0970, 0990	Wall Finishes & Painting/Coating	109.7	92.2	99.2	109.7	92.2	99.2	109.6	63.5	82.0	109.6	88.6	97.0	109.6	88.6	97.0	109.6	74.4	88.5
09	FINISHES	135.2	90.7	111.6	121.6	90.6	105.2	121.6	59.5	88.8	115.2	88.2	100.9	115.5	88.2	101.1	114.2	74.2	93.1
COVERS	DIVS. 10 - 14, 25, 28, 41, 43, 44	140.0	102.3	132.5	140.0	102.1	132.5	140.0	70.6	126.3	140.0	91.6	130.4	140.0	91.6	130.4	140.0	73.8	126.9
21, 22, 23	FIRE SUPPRESSION, PLUMBING & HVAC	100.1	87.6	95.2	103.0	100.6	102.0	102.4	80.8	93.8	103.0	92.7	98.9	102.4	92.7	98.6	99.2	76.1	90.0
26, 27, 3370	ELECTRICAL, COMMUNICATIONS & UTIL.	118.5	84.7	102.1	115.6	87.6	102.0	134.3	79.0	107.4	121.8	74.4	98.7	122.7	74.4	99.2	128.4	74.6	102.2
MF2004	WEIGHTED AVERAGE	131.8	89.8	114.1	127.7	92.6	112.9	117.1	72.4	98.3	114.9	87.6	103.4	115.1	87.6	103.5	119.0	78.0	101.8

DIVISION		CANADA																	
		HAMILTON, ONTARIO			HULL, QUEBEC			JOLIETTE, QUEBEC			KAMLOOPS, BRITISH COLUMBIA			KINGSTON, ONTARIO			KITCHENER, ONTARIO		
		MAT.	INST.	TOTAL	MAT.	INST.	TOTAL	MAT.	INST.	TOTAL	MAT.	INST.	TOTAL	MAT.	INST.	TOTAL	MAT.	INST.	TOTAL
015433	CONTRACTOR EQUIPMENT		107.3	107.3		104.0	104.0		104.0	104.0		106.7	106.7		104.0	104.0		103.8	103.8
0241, 31 - 34	SITE & INFRASTRUCTURE, DEMOLITION	114.9	111.1	112.3	93.9	101.1	98.8	94.6	101.1	99.0	117.4	104.1	108.3	112.1	103.8	106.4	101.3	104.7	103.6
0310	Concrete Forming & Accessories	129.6	95.4	100.1	129.0	86.1	91.9	129.0	86.2	92.0	121.0	88.9	93.3	121.1	87.7	92.2	117.4	87.8	91.8
0320	Concrete Reinforcing	161.2	90.3	132.3	156.6	79.7	125.2	147.9	79.7	120.1	116.2	77.6	100.5	174.5	83.1	137.2	110.2	90.2	102.0
0330	Cast-in-Place Concrete	151.9	101.4	132.6	130.2	95.5	116.9	135.6	95.6	120.3	117.3	99.3	110.4	147.1	97.9	128.3	140.8	82.6	118.6
03	CONCRETE	143.6	96.6	122.2	132.5	88.5	112.5	133.3	88.6	113.0	137.1	90.7	116.0	143.5	90.7	119.5	121.6	86.7	105.7
04	MASONRY	181.0	101.5	133.4	160.6	85.4	115.5	161.0	85.4	115.7	167.0	93.0	122.7	170.8	92.2	123.8	161.9	97.8	123.5
05	METALS	127.9	94.2	118.1	105.5	89.3	100.8	105.5	89.4	100.8	107.7	89.2	102.3	103.7	90.6	103.4	116.1	93.8	109.6
06	WOOD, PLASTICS & COMPOSITES	115.5	94.6	104.0	127.6	85.7	104.6	127.6	85.7	104.6	98.2	87.1	92.1	114.9	86.8	99.5	109.3	86.1	96.6
07	THERMAL & MOISTURE PROTECTION	120.1	96.5	110.9	104.6	88.3	98.2	104.6	88.3	98.2	120.0	85.4	106.6	110.4	87.6	101.5	110.0	93.2	103.5
08	OPENINGS	92.0	91.4	91.9	92.0	74.7	87.7	92.0	79.2	88.9	87.9	83.7	86.9	92.0	83.9	90.0	83.3	85.2	83.7
0920	Plaster & Gypsum Board	195.8	94.3	129.9	140.0	84.9	104.2	186.5	84.9	120.5	137.7	86.0	104.1	232.0	86.2	137.3	152.4	85.5	109.0
0950, 0980	Ceilings & Acoustic Treatment	120.6	94.3	104.3	95.4	84.9	88.9	95.4	84.9	88.9	95.4	86.0	89.6	112.9	86.2	96.4	105.4	85.5	93.1
0960	Flooring	134.4	94.8	123.7	132.0	95.4	122.1	132.0	95.4	122.1	129.7	53.3	109.1	132.0	93.4	121.6	127.8	94.8	118.9
0970, 0990	Wall Finishes & Painting/Coating	109.6	100.5	104.2	109.6	88.6	97.0	109.6	88.6	97.0	109.6	81.1	92.5	109.6	82.0	93.1	109.6	89.7	97.7
09	FINISHES	130.8	95.8	112.3	115.2	88.2	100.9	121.6	88.2	104.0	116.9	81.7	98.3	132.8	88.2	109.2	118.0	89.0	102.6
COVERS	DIVS. 10 - 14, 25, 28, 41, 43, 44	140.0	106.6	133.4	140.0	91.6	130.4	140.0	91.6	130.4	140.0	102.6	132.6	140.0	79.6	128.0	140.0	104.6	133.0
21, 22, 23	FIRE SUPPRESSION, PLUMBING & HVAC	100.0	91.9	96.8	102.4	92.7	98.6	102.4	92.7	98.6	102.4	96.1	99.9	103.0	102.1	102.7	99.2	88.9	95.1
26, 27, 3370	ELECTRICAL, COMMUNICATIONS & UTIL.	128.8	96.1	112.9	124.5	74.4	100.1	122.7	74.4	99.2	127.0	84.0	106.1	123.6	91.8	108.1	124.1	93.4	109.1
MF2004	WEIGHTED AVERAGE	123.4	96.9	112.3	115.0	87.6	103.5	115.5	87.8	103.9	117.2	90.8	106.1	119.4	93.4	108.5	114.5	92.2	105.1

		CANADA																	
DIVISION		LAVAL, QUEBEC			LETHBRIDGE, ALBERTA			LLOYDMINSTER, ALBERTA			LONDON, ONTARIO			MEDICINE HAT, ALBERTA			MONCTON, NEW BRUNSWICK		
		MAT.	INST.	TOTAL	MAT.	INST.	TOTAL	MAT.	INST.	TOTAL	MAT.	INST.	TOTAL	MAT.	INST.	TOTAL	MAT.	INST.	TOTAL
015433	**CONTRACTOR EQUIPMENT**		104.0	104.0		104.2	104.2		104.2	104.2		104.1	104.1		104.2	104.2		103.3	103.3
0241, 31 - 34	**SITE & INFRASTRUCTURE, DEMOLITION**	94.4	101.1	99.0	111.8	102.4	105.4	111.6	101.8	104.9	114.5	104.9	108.0	110.3	101.8	104.5	95.5	98.0	97.2
0310	Concrete Forming & Accessories	129.2	86.1	91.9	122.4	90.3	94.7	120.5	80.8	86.2	129.7	89.0	94.5	122.4	80.7	86.4	102.1	62.9	68.2
0320	Concrete Reinforcing	156.6	79.7	125.2	161.1	63.2	121.2	161.1	63.2	121.2	130.4	88.9	113.5	161.1	63.2	121.2	147.9	60.2	112.1
0330	Cast-in-Place Concrete	134.5	95.5	119.6	163.0	100.2	139.0	151.2	96.7	130.3	149.3	99.4	130.2	151.2	96.6	130.3	133.1	62.3	106.1
03	**CONCRETE**	134.5	88.5	113.6	148.3	89.1	121.4	142.7	83.8	115.9	136.2	92.8	116.5	142.8	83.7	115.9	133.8	63.2	101.7
04	**MASONRY**	160.9	85.4	115.7	178.3	91.3	126.2	161.0	85.0	115.5	182.4	99.1	132.5	161.0	85.0	115.5	159.3	63.8	102.1
05	**METALS**	105.5	89.3	100.8	127.3	88.1	115.9	107.9	87.9	102.0	125.3	92.1	115.6	107.9	87.9	102.0	105.2	77.6	97.2
06	**WOOD, PLASTICS & COMPOSITES**	127.9	85.7	104.8	114.0	89.4	100.5	110.7	79.9	93.8	115.5	87.0	99.9	114.0	79.9	95.3	94.0	62.5	76.7
07	**THERMAL & MOISTURE PROTECTION**	105.1	88.3	98.5	116.4	87.1	105.0	113.3	82.9	101.5	120.9	93.2	110.1	118.8	82.9	104.8	107.8	63.4	90.6
08	**OPENINGS**	92.0	74.7	87.7	92.0	79.8	89.0	92.0	74.6	87.7	93.2	85.4	91.2	92.0	74.6	87.7	86.2	56.3	78.8
0920	Plaster & Gypsum Board	145.7	84.9	106.2	136.2	88.6	105.3	131.9	78.8	97.4	199.5	86.4	126.1	133.8	78.8	98.0	165.9	61.0	97.7
0950, 0980	Ceilings & Acoustic Treatment	95.4	84.9	88.9	104.6	88.6	94.7	95.4	78.8	85.1	138.5	86.4	106.3	95.4	78.8	85.1	96.2	61.0	74.4
0960	Flooring	132.0	95.4	122.1	132.0	88.8	120.3	132.0	88.8	120.3	134.6	94.8	123.9	132.0	88.8	120.3	113.9	55.7	98.2
0970, 0990	Wall Finishes & Painting/Coating	109.6	88.6	97.0	109.6	100.7	104.3	109.7	78.4	91.0	109.6	97.0	102.0	109.6	78.4	90.9	109.6	49.8	73.8
09	**FINISHES**	116.0	88.2	101.3	118.0	91.5	104.0	115.4	81.9	97.7	135.6	90.6	111.8	115.5	81.9	97.7	113.9	60.1	85.5
COVERS	**DIVS. 10 - 14, 25, 28, 41, 43, 44**	140.0	91.6	130.4	140.0	102.1	132.5	140.0	98.7	131.8	140.0	105.2	133.1	140.0	98.7	131.8	140.0	70.6	126.3
21, 22, 23	**FIRE SUPPRESSION, PLUMBING & HVAC**	99.2	92.7	96.6	102.8	97.2	100.5	103.0	97.3	100.8	100.0	89.0	95.6	102.4	93.9	99.0	102.4	71.2	90.0
26, 27, 3370	**ELECTRICAL, COMMUNICATIONS & UTIL.**	123.5	74.4	99.6	117.7	87.6	103.0	114.4	87.6	101.4	123.3	93.4	108.8	114.4	87.6	101.4	130.9	63.4	98.0
MF2004	**WEIGHTED AVERAGE**	114.5	87.6	103.2	121.9	92.0	109.3	116.4	88.8	104.8	122.0	93.3	110.0	116.4	88.1	104.5	115.0	68.7	95.5

		CANADA																	
DIVISION		MONTREAL, QUEBEC			MOOSE JAW, SASKATCHEWAN			NEW GLASGOW, NOVA SCOTIA			NEWCASTLE, NEW BRUNSWICK			NORTH BAY, ONTARIO			OSHAWA, ONTARIO		
		MAT.	INST.	TOTAL	MAT.	INST.	TOTAL	MAT.	INST.	TOTAL	MAT.	INST.	TOTAL	MAT.	INST.	TOTAL	MAT.	INST.	TOTAL
015433	**CONTRACTOR EQUIPMENT**		105.6	105.6		101.2	101.2		102.9	102.9		103.3	103.3		102.2	102.2		103.8	103.8
0241, 31 - 34	**SITE & INFRASTRUCTURE, DEMOLITION**	101.1	101.0	101.0	111.5	96.8	101.5	99.2	99.4	99.3	96.1	97.8	97.3	114.2	100.2	104.6	114.1	104.0	107.2
0310	Concrete Forming & Accessories	135.9	93.7	99.5	105.8	59.5	65.8	96.1	70.3	73.8	102.1	62.6	68.0	123.2	85.2	90.4	123.3	87.7	92.6
0320	Concrete Reinforcing	146.2	93.9	124.9	113.7	62.6	92.9	147.9	47.3	106.8	147.9	57.2	110.9	174.5	82.7	137.0	174.5	85.3	138.1
0330	Cast-in-Place Concrete	174.7	101.9	146.9	146.6	70.3	117.5	170.0	71.6	132.4	138.3	62.0	109.1	163.6	85.3	133.7	162.9	87.0	133.9
03	**CONCRETE**	151.6	96.7	126.6	124.2	64.8	97.2	148.7	67.4	111.7	136.2	62.3	102.6	151.4	85.2	121.3	151.0	87.4	122.1
04	**MASONRY**	167.3	93.7	123.2	159.2	63.1	101.7	160.9	70.5	106.8	159.7	63.7	102.2	165.8	88.2	119.3	165.4	93.8	122.5
05	**METALS**	123.5	94.8	115.1	104.5	76.3	96.3	106.2	76.9	97.7	105.2	74.6	96.3	107.1	90.5	102.2	107.1	92.7	102.9
06	**WOOD, PLASTICS & COMPOSITES**	127.9	93.9	109.3	95.6	57.8	74.8	86.0	69.3	76.9	94.0	62.5	76.7	116.1	85.3	99.2	116.4	85.9	99.6
07	**THERMAL & MOISTURE PROTECTION**	117.1	96.9	109.3	104.0	63.4	88.2	105.5	69.0	91.4	107.8	61.7	89.9	110.6	84.2	100.3	110.9	86.4	101.4
08	**OPENINGS**	92.0	82.1	89.6	87.0	55.6	79.2	82.8	63.4	78.0	86.2	54.7	78.4	90.8	81.8	88.6	90.8	85.8	89.6
0920	Plaster & Gypsum Board	152.5	93.2	114.0	131.4	56.2	82.5	160.1	68.1	100.3	165.9	61.0	97.7	153.5	84.7	108.8	155.3	85.3	109.8
0950, 0980	Ceilings & Acoustic Treatment	102.0	93.2	96.5	95.4	56.2	71.1	95.4	68.1	78.5	96.2	61.0	74.4	95.4	84.7	88.8	99.6	85.3	90.7
0960	Flooring	133.9	95.4	123.5	119.8	60.7	103.8	110.1	64.5	97.8	113.9	45.2	95.4	132.0	93.4	121.6	132.0	97.2	122.6
0970, 0990	Wall Finishes & Painting/Coating	109.6	104.2	106.4	109.6	64.5	82.6	109.6	61.9	81.0	109.6	49.8	73.8	109.6	88.1	96.7	109.6	104.9	106.8
09	**FINISHES**	119.2	95.8	106.9	111.5	59.8	84.2	111.5	68.3	88.7	113.9	58.0	84.3	117.8	87.2	101.7	119.0	91.1	104.2
COVERS	**DIVS. 10 - 14, 25, 28, 41, 43, 44**	140.0	94.1	130.9	140.0	69.7	126.1	140.0	71.8	126.5	140.0	70.6	126.3	140.0	78.2	127.8	140.0	104.7	133.0
21, 22, 23	**FIRE SUPPRESSION, PLUMBING & HVAC**	100.1	94.4	97.8	103.0	77.6	92.9	102.4	86.0	95.9	102.4	71.1	90.0	102.4	100.0	101.4	99.2	103.2	100.8
26, 27, 3370	**ELECTRICAL, COMMUNICATIONS & UTIL.**	129.8	86.1	108.5	125.9	64.4	95.9	122.2	66.0	94.8	125.4	63.4	95.2	123.1	93.1	108.5	125.3	92.7	109.4
MF2004	**WEIGHTED AVERAGE**	121.7	93.8	110.0	113.5	69.9	95.2	115.5	75.0	98.5	114.7	67.8	95.0	118.4	91.2	107.0	117.9	94.7	108.2

		CANADA																	
DIVISION		OTTAWA, ONTARIO			OWEN SOUND, ONTARIO			PETERBOROUGH, ONTARIO			PORTAGE LA PRAIRIE, MANITOBA			PRINCE ALBERT, SASKATCHEWAN			PRINCE GEORGE, BRITISH COLUMBIA		
		MAT.	INST.	TOTAL	MAT.	INST.	TOTAL	MAT.	INST.	TOTAL	MAT.	INST.	TOTAL	MAT.	INST.	TOTAL	MAT.	INST.	TOTAL
015433	**CONTRACTOR EQUIPMENT**		103.8	103.8		102.2	102.2		102.2	102.2		104.6	104.6		101.2	101.2		106.7	106.7
0241, 31 - 34	**SITE & INFRASTRUCTURE, DEMOLITION**	111.7	104.6	106.9	116.5	100.4	105.5	114.2	100.6	104.9	112.3	99.7	103.7	106.2	97.0	99.9	121.2	104.1	109.5
0310	Concrete Forming & Accessories	130.2	91.7	97.0	123.3	83.7	89.1	123.2	86.2	91.3	122.6	71.0	78.0	105.8	59.3	65.6	111.1	88.9	92.0
0320	Concrete Reinforcing	160.1	88.8	131.0	179.6	84.7	140.9	174.5	83.2	137.2	161.1	53.8	117.3	118.8	62.6	95.8	116.2	77.6	100.5
0330	Cast-in-Place Concrete	153.0	98.5	132.2	167.7	81.3	134.7	163.6	86.9	134.3	151.2	75.8	122.3	132.8	70.3	108.9	147.2	99.3	128.9
03	**CONCRETE**	143.9	93.7	121.1	154.3	83.4	122.1	151.4	86.3	121.8	136.5	70.4	106.4	118.8	64.6	94.1	150.5	90.7	123.3
04	**MASONRY**	179.7	97.2	130.3	169.3	93.7	124.0	165.8	94.8	123.3	163.8	66.3	105.4	158.2	63.1	101.3	169.4	93.0	123.7
05	**METALS**	126.2	93.7	116.7	107.7	90.7	102.8	107.1	90.9	102.4	107.9	79.0	99.4	104.6	76.1	96.3	107.7	89.2	102.3
06	**WOOD, PLASTICS & COMPOSITES**	115.2	90.9	101.9	118.1	82.1	98.3	116.1	84.0	98.5	113.9	71.7	90.7	95.6	57.8	74.8	98.2	87.1	92.1
07	**THERMAL & MOISTURE PROTECTION**	123.4	92.7	111.5	110.8	85.0	100.8	110.6	89.5	102.4	104.5	70.8	91.4	104.0	62.3	87.8	114.9	85.4	103.5
08	**OPENINGS**	92.0	88.4	91.1	91.0	81.1	88.6	90.8	83.5	89.0	92.0	63.4	85.0	85.6	55.6	78.1	87.9	83.7	86.9
0920	Plaster & Gypsum Board	234.7	90.5	141.1	153.5	81.4	106.7	153.5	83.4	107.9	126.1	70.2	89.8	131.4	56.2	82.5	135.2	86.0	103.3
0950, 0980	Ceilings & Acoustic Treatment	131.4	90.5	106.1	95.4	81.4	86.7	95.4	83.4	88.0	95.4	70.2	79.8	95.4	56.2	71.1	95.4	86.0	89.6
0960	Flooring	134.4	93.4	123.4	132.1	94.8	122.0	132.0	93.4	121.6	132.0	67.3	114.5	119.8	60.7	103.8	126.6	53.3	106.8
0970, 0990	Wall Finishes & Painting/Coating	109.6	91.4	98.7	110.4	86.8	96.2	109.6	90.3	98.0	109.7	56.7	77.9	109.6	55.0	76.9	109.6	81.1	92.5
09	**FINISHES**	138.2	92.0	113.8	119.0	86.1	101.6	117.8	87.9	102.0	114.4	69.0	90.4	111.5	58.7	83.6	115.6	81.7	97.7
COVERS	**DIVS. 10 - 14, 25, 28, 41, 43, 44**	140.0	103.1	132.7	140.0	78.7	127.9	140.0	79.8	128.1	140.0	72.9	126.7	140.0	69.7	126.1	140.0	102.6	132.6
21, 22, 23	**FIRE SUPPRESSION, PLUMBING & HVAC**	100.2	89.4	95.9	102.5	99.9	101.5	102.4	103.5	102.9	102.4	85.7	95.8	103.0	69.9	89.9	102.4	96.1	99.9
26, 27, 3370	**ELECTRICAL, COMMUNICATIONS & UTIL.**	119.7	94.4	107.4	124.7	91.8	108.7	123.1	92.7	108.3	125.4	61.5	94.3	125.9	64.4	95.9	123.1	84.0	104.1
MF2004	**WEIGHTED AVERAGE**	122.7	93.9	110.6	119.4	91.2	107.6	118.4	93.2	107.8	116.5	74.7	98.9	112.5	68.1	93.9	118.3	90.8	106.7

DIVISION		CANADA																	
		QUEBEC, QUEBEC			RED DEER, ALBERTA			REGINA, SASKATCHEWAN			RIMOUSKI, QUEBEC			ROUYN-NORANDA, QUEBEC			SAINT HYACINTHE, QUEBEC		
		MAT.	INST.	TOTAL	MAT.	INST.	TOTAL	MAT.	INST.	TOTAL	MAT.	INST.	TOTAL	MAT.	INST.	TOTAL	MAT.	INST.	TOTAL
015433	CONTRACTOR EQUIPMENT		106.0	106.0		104.2	104.2		101.2	101.2		104.0	104.0		104.0	104.0		104.0	104.0
0241, 31 - 34	SITE & INFRASTRUCTURE, DEMOLITION	99.7	101.1	100.7	110.3	101.8	104.5	113.0	96.8	101.9	94.3	100.7	98.7	93.9	101.1	98.8	94.4	101.1	99.0
0310	Concrete Forming & Accessories	137.1	94.0	99.8	136.8	80.7	88.4	106.0	59.5	65.8	129.0	93.4	98.3	129.0	86.1	91.9	129.0	86.1	91.9
0320	Concrete Reinforcing	140.1	94.0	121.3	161.1	63.2	121.2	125.7	62.6	100.0	111.6	93.9	104.4	156.6	79.7	125.2	156.6	79.7	125.2
0330	Cast-in-Place Concrete	148.3	102.3	130.7	151.2	96.6	130.3	156.3	70.3	123.4	137.0	100.5	123.0	130.2	95.5	116.9	134.5	95.5	119.6
03	CONCRETE	138.1	96.9	119.4	143.7	83.7	116.4	131.1	64.8	100.9	126.7	96.1	112.7	132.5	88.5	112.5	134.5	88.5	113.6
04	MASONRY	174.3	93.7	126.0	161.0	85.0	115.5	166.4	63.6	104.8	160.0	93.6	120.3	160.6	85.4	115.5	160.9	85.4	115.7
05	METALS	125.2	95.1	116.4	107.9	87.9	102.0	105.5	76.4	97.0	105.1	94.4	102.0	105.5	89.3	100.8	105.5	89.3	100.8
06	WOOD, PLASTICS & COMPOSITES	128.6	94.0	109.6	114.0	79.9	95.3	95.8	57.8	74.9	127.6	93.7	109.0	127.6	85.7	104.6	127.6	85.7	104.6
07	THERMAL & MOISTURE PROTECTION	118.6	97.1	110.2	127.2	82.9	110.0	106.0	63.5	89.5	104.6	96.4	101.4	104.6	88.3	98.2	104.8	88.3	98.4
08	OPENINGS	92.0	89.6	91.4	92.0	74.6	87.7	87.0	55.6	79.2	91.6	80.4	88.8	92.0	74.7	87.7	92.0	74.7	87.7
0920	Plaster & Gypsum Board	191.0	93.2	127.4	133.8	78.8	98.0	165.1	56.2	94.3	185.9	93.2	125.6	139.4	84.9	104.0	144.3	84.9	105.7
0950, 0980	Ceilings & Acoustic Treatment	93.6	93.2	93.3	95.4	78.8	85.1	119.5	56.2	80.3	92.9	93.2	93.1	92.9	84.9	88.0	92.9	84.9	88.0
0960	Flooring	134.4	95.4	123.9	134.4	88.8	122.1	119.8	60.7	103.8	132.0	95.4	122.1	132.0	95.4	122.1	132.0	95.4	122.1
0970, 0990	Wall Finishes & Painting/Coating	110.2	104.2	106.6	109.6	78.4	90.9	109.6	64.5	82.6	109.6	104.2	106.4	109.6	88.6	97.0	109.6	88.6	97.0
09	FINISHES	123.1	95.9	108.7	116.2	81.9	98.1	122.0	59.8	89.1	121.0	95.6	107.6	114.5	88.2	100.6	115.2	88.2	100.9
COVERS	DIVS. 10 - 14, 25, 28, 41, 43, 44	140.0	94.3	130.9	140.0	98.7	131.8	140.0	69.7	126.1	140.0	93.5	130.8	140.0	91.6	130.4	140.0	91.6	130.4
21, 22, 23	FIRE SUPPRESSION, PLUMBING & HVAC	99.9	94.4	97.7	102.4	93.9	99.0	99.4	77.6	90.8	102.4	94.3	99.2	102.4	92.7	98.6	97.9	92.7	95.8
26, 27, 3370	ELECTRICAL, COMMUNICATIONS & UTIL.	121.7	86.1	104.4	114.4	87.6	101.4	126.7	64.4	96.4	122.7	86.1	104.9	122.7	74.4	99.2	123.5	74.4	99.6
MF2004	WEIGHTED AVERAGE	120.1	94.2	109.2	116.9	88.1	104.8	115.1	69.9	96.1	114.4	93.5	105.6	114.8	87.6	103.4	114.2	87.6	103.0

DIVISION		CANADA																	
		SAINT JOHN, NEW BRUNSWICK			SARNIA, ONTARIO			SASKATOON, SASKATCHEWAN			SAULT STE MARIE, ONTARIO			SHERBROOKE, QUEBEC			SOREL, QUEBEC		
		MAT.	INST.	TOTAL	MAT.	INST.	TOTAL	MAT.	INST.	TOTAL	MAT.	INST.	TOTAL	MAT.	INST.	TOTAL	MAT.	INST.	TOTAL
015433	CONTRACTOR EQUIPMENT		103.3	103.3		102.2	102.2		101.2	101.2		102.2	102.2		104.0	104.0		104.0	104.0
0241, 31 - 34	SITE & INFRASTRUCTURE, DEMOLITION	96.2	98.8	98.0	112.6	100.7	104.4	107.4	97.0	100.3	102.2	100.2	100.9	94.4	101.1	99.0	94.6	101.1	99.0
0310	Concrete Forming & Accessories	120.9	62.2	70.2	121.9	93.2	97.1	106.0	59.4	65.7	111.4	87.1	90.4	129.0	86.1	91.9	129.0	86.2	92.0
0320	Concrete Reinforcing	147.9	60.4	112.2	122.9	84.5	107.3	118.8	62.6	95.9	112.1	83.1	100.2	156.6	79.7	125.2	147.9	79.7	120.1
0330	Cast-in-Place Concrete	136.2	62.3	107.9	150.9	99.4	131.2	142.0	70.3	114.6	135.6	85.5	116.4	134.5	95.5	119.6	135.6	95.6	120.3
03	CONCRETE	136.5	62.9	103.0	135.0	93.8	116.3	123.1	64.7	96.5	119.1	86.2	104.1	134.5	88.5	113.6	133.3	88.6	113.0
04	MASONRY	177.2	65.5	110.3	176.0	97.5	129.0	165.9	63.6	104.6	161.8	93.8	121.1	160.9	85.4	115.7	161.0	85.4	115.7
05	METALS	105.1	78.1	97.2	107.0	91.2	102.4	105.5	76.3	97.0	106.4	92.0	102.2	105.5	89.3	100.8	105.5	89.4	100.8
06	WOOD, PLASTICS & COMPOSITES	116.9	61.3	86.3	115.2	92.1	102.6	94.2	57.8	74.2	102.5	87.4	94.2	127.6	85.7	104.6	127.6	85.7	104.6
07	THERMAL & MOISTURE PROTECTION	108.1	63.3	90.7	110.6	92.9	103.7	104.9	62.5	88.4	109.5	87.4	100.9	104.6	88.3	98.3	104.6	88.3	98.2
08	OPENINGS	86.1	54.5	78.3	93.2	87.2	91.7	86.0	55.6	78.5	83.7	84.8	84.0	92.0	74.7	87.7	92.0	79.2	88.9
0920	Plaster & Gypsum Board	181.4	59.6	102.3	180.2	91.8	122.8	145.5	56.2	87.5	141.2	86.9	105.9	141.8	84.9	104.9	185.9	84.9	120.3
0950, 0980	Ceilings & Acoustic Treatment	102.9	59.6	76.1	99.6	91.8	94.7	119.5	56.2	80.3	95.4	86.9	90.1	92.9	84.9	88.0	92.9	84.9	88.0
0960	Flooring	127.3	55.7	108.0	132.0	98.7	123.0	119.8	60.7	103.8	123.3	93.4	115.2	132.0	95.4	122.1	132.0	95.4	122.1
0970, 0990	Wall Finishes & Painting/Coating	109.6	69.1	85.3	109.6	102.1	105.1	109.6	55.0	76.9	109.6	94.6	100.6	109.6	88.6	97.0	109.6	88.6	97.0
09	FINISHES	121.6	61.5	89.8	122.5	95.4	108.2	119.1	58.7	87.2	112.7	89.2	100.3	114.8	88.2	100.8	121.0	88.2	103.7
COVERS	DIVS. 10 - 14, 25, 28, 41, 43, 44	140.0	70.4	126.2	140.0	81.3	128.4	140.0	69.7	126.1	140.0	103.6	132.8	140.0	91.6	130.4	140.0	91.6	130.4
21, 22, 23	FIRE SUPPRESSION, PLUMBING & HVAC	102.4	77.3	92.4	102.4	109.8	105.3	99.3	77.6	90.7	102.4	96.6	100.1	103.0	92.7	98.9	102.4	92.7	98.6
26, 27, 3370	ELECTRICAL, COMMUNICATIONS & UTIL.	133.9	78.8	107.1	127.1	95.0	111.5	127.0	64.4	96.5	125.3	93.1	109.6	122.7	74.4	99.2	122.7	74.4	99.2
MF2004	WEIGHTED AVERAGE	117.3	72.4	98.4	117.9	97.5	109.3	113.6	69.8	95.2	112.9	92.6	104.4	115.2	87.6	103.6	115.4	87.8	103.8

DIVISION		CANADA																	
		ST CATHARINES, ONTARIO			ST JEROME, QUEBEC			ST JOHNS, NEWFOUNDLAND			SUDBURY, ONTARIO			SUMMERSIDE, PRINCE EDWARD ISLAND			SYDNEY, NOVA SCOTIA		
		MAT.	INST.	TOTAL	MAT.	INST.	TOTAL	MAT.	INST.	TOTAL	MAT.	INST.	TOTAL	MAT.	INST.	TOTAL	MAT.	INST.	TOTAL
015433	CONTRACTOR EQUIPMENT		102.2	102.2		104.0	104.0		104.8	104.8		102.2	102.2		102.8	102.8		102.9	102.9
0241, 31 - 34	SITE & INFRASTRUCTURE, DEMOLITION	102.1	101.9	102.0	93.9	101.1	98.8	115.7	99.5	104.7	102.2	101.5	101.7	107.2	96.9	100.1	94.1	99.4	97.7
0310	Concrete Forming & Accessories	115.3	92.6	95.7	129.0	86.1	91.9	103.8	64.3	69.7	111.4	88.5	91.6	96.4	57.2	62.6	96.1	70.3	73.8
0320	Concrete Reinforcing	111.1	90.2	102.6	156.6	79.7	125.2	163.8	57.7	120.5	112.1	88.5	102.5	147.9	47.3	106.8	147.9	47.3	106.8
0330	Cast-in-Place Concrete	134.4	101.2	121.7	130.2	95.5	116.9	172.4	77.6	136.2	135.6	97.6	121.0	161.3	61.2	123.0	130.9	71.6	108.2
03	CONCRETE	118.6	95.3	108.0	132.5	88.5	112.5	164.6	68.7	121.0	119.1	91.9	106.8	152.7	57.8	109.6	130.5	67.4	101.8
04	MASONRY	161.4	101.5	125.5	160.6	85.4	115.5	164.6	65.1	105.1	161.5	96.6	122.6	160.2	61.5	101.1	157.8	70.5	105.5
05	METALS	106.4	93.7	102.7	105.5	89.3	100.8	108.2	79.1	99.7	106.4	93.5	102.6	106.3	70.8	95.9	106.2	76.9	97.7
06	WOOD, PLASTICS & COMPOSITES	106.9	91.5	98.4	127.6	85.7	104.6	95.2	64.3	78.3	102.5	87.4	94.2	86.3	56.5	70.0	86.0	69.3	76.9
07	THERMAL & MOISTURE PROTECTION	110.0	93.4	103.5	104.6	88.3	98.2	109.3	63.4	91.5	109.5	91.3	102.4	105.1	59.2	87.3	105.5	69.0	91.4
08	OPENINGS	82.7	88.5	84.2	92.0	74.7	87.7	98.2	60.2	88.8	83.7	84.8	84.0	94.3	48.9	83.1	82.8	63.4	78.0
0920	Plaster & Gypsum Board	137.2	91.1	107.2	139.4	84.9	104.0	168.5	62.7	99.8	141.2	86.9	105.9	160.5	54.8	91.8	160.1	68.1	100.3
0950, 0980	Ceilings & Acoustic Treatment	99.6	91.1	94.3	92.9	84.9	88.0	104.6	62.7	78.7	95.4	86.9	90.1	95.4	54.8	70.3	95.4	68.1	78.5
0960	Flooring	126.0	94.8	117.6	132.0	95.4	122.1	114.8	54.9	98.6	123.3	93.4	115.2	110.1	60.3	96.7	110.1	64.5	97.8
0970, 0990	Wall Finishes & Painting/Coating	109.6	100.5	104.2	109.6	88.6	97.0	109.6	64.5	82.6	109.6	90.8	98.4	109.6	41.5	68.8	109.6	61.9	81.0
09	FINISHES	114.0	94.0	103.4	114.5	88.2	100.6	117.4	62.5	88.3	112.7	89.4	100.4	112.4	56.3	82.8	111.5	68.3	88.7
COVERS	DIVS. 10 - 14, 25, 28, 41, 43, 44	140.0	81.2	128.4	140.0	91.6	130.4	140.0	71.2	126.4	140.0	104.6	133.0	140.0	70.1	126.2	140.0	71.8	126.5
21, 22, 23	FIRE SUPPRESSION, PLUMBING & HVAC	99.2	90.6	95.8	102.4	92.7	98.6	99.6	64.2	85.5	102.5	89.8	97.5	102.4	65.0	87.6	102.4	86.0	95.9
26, 27, 3370	ELECTRICAL, COMMUNICATIONS & UTIL.	126.3	94.5	110.8	123.6	74.4	99.6	122.9	65.8	95.1	123.1	94.5	109.1	120.8	53.6	88.1	122.2	66.0	94.8
MF2004	WEIGHTED AVERAGE	112.3	94.2	104.7	114.8	87.6	103.4	119.9	69.2	98.6	112.7	92.9	104.4	117.2	63.3	94.6	113.1	75.0	97.1

	DIVISION	CANADA																	
		THUNDER BAY, ONTARIO			TIMMINS, ONTARIO			TORONTO, ONTARIO			TROIS RIVIERES, QUEBEC			TRURO, NOVA SCOTIA			VANCOUVER, BRITISH COLUMBIA		
		MAT.	INST.	TOTAL	MAT.	INST.	TOTAL	MAT.	INST.	TOTAL	MAT.	INST.	TOTAL	MAT.	INST.	TOTAL	MAT.	INST.	TOTAL
015433	**CONTRACTOR EQUIPMENT**		102.2	102.2		102.2	102.2		104.1	104.1		104.0	104.0		102.9	102.9		112.7	112.7
0241, 31 - 34	**SITE & INFRASTRUCTURE, DEMOLITION**	107.6	101.8	103.7	114.2	100.2	104.6	139.0	105.5	116.1	94.6	101.1	99.0	99.2	99.4	99.3	121.8	107.4	111.9
0310	Concrete Forming & Accessories	123.3	93.0	97.1	123.2	85.2	90.4	131.1	100.7	104.8	129.0	86.2	92.0	96.1	70.3	73.8	124.8	78.0	84.4
0320	Concrete Reinforcing	99.3	89.6	95.4	174.5	82.7	137.0	157.0	91.0	130.1	147.9	79.7	120.1	147.9	47.3	106.8	156.7	72.3	122.3
0330	Cast-in-Place Concrete	148.0	99.8	129.6	163.6	85.3	133.7	147.4	109.5	132.9	135.6	95.6	120.3	170.0	71.6	132.4	163.0	92.2	135.9
03	**CONCRETE**	126.3	94.8	112.0	151.4	85.2	121.3	140.8	101.8	123.1	133.3	88.6	113.0	148.7	67.4	111.7	153.2	82.7	121.1
04	**MASONRY**	162.1	101.3	125.7	165.8	88.2	119.3	193.0	106.7	141.4	161.0	85.4	115.7	160.9	70.5	106.8	177.4	81.1	119.8
05	**METALS**	106.2	92.7	102.3	107.1	90.5	102.2	127.7	94.9	118.2	105.5	89.4	100.8	106.2	77.0	97.7	165.9	89.2	143.5
06	**WOOD, PLASTICS & COMPOSITES**	116.4	91.6	102.8	116.1	85.3	99.2	116.4	99.4	107.1	127.6	85.7	104.6	86.0	69.3	76.9	114.3	75.1	92.8
07	**THERMAL & MOISTURE PROTECTION**	110.2	94.8	104.2	110.6	84.2	100.3	121.0	101.2	113.3	104.6	88.3	98.2	105.5	69.0	91.4	132.4	82.1	112.8
08	**OPENINGS**	81.7	87.9	83.3	90.8	81.8	88.6	90.8	96.0	92.1	92.0	79.2	88.9	82.8	63.4	78.0	93.8	75.8	89.3
0920	Plaster & Gypsum Board	166.5	91.2	117.6	153.5	84.7	108.8	177.3	99.3	126.6	185.9	84.9	120.3	160.1	68.1	100.3	154.9	73.5	102.0
0950, 0980	Ceilings & Acoustic Treatment	95.4	91.2	92.8	95.4	84.7	88.8	132.6	99.3	112.0	92.9	84.9	88.0	95.4	68.1	78.5	141.0	73.5	99.3
0960	Flooring	132.0	54.6	111.1	132.0	93.4	121.6	135.6	100.5	126.1	132.0	95.4	122.1	110.1	64.5	97.8	134.8	92.0	123.2
0970, 0990	Wall Finishes & Painting/Coating	109.6	93.2	99.8	109.6	88.1	96.7	111.5	104.9	107.5	109.6	88.6	97.0	109.6	61.9	81.0	109.6	88.7	97.1
09	**FINISHES**	119.2	86.7	102.0	117.8	87.2	101.7	133.2	101.2	116.3	121.0	88.2	103.7	111.5	68.3	88.7	130.7	81.2	104.5
COVERS	**DIVS. 10 - 14, 25, 28, 41, 43, 44**	140.0	81.6	128.4	140.0	78.2	127.8	140.0	108.5	133.8	140.0	91.6	130.4	140.0	71.8	126.5	140.0	99.9	132.1
21, 22, 23	**FIRE SUPPRESSION, PLUMBING & HVAC**	99.2	90.7	95.8	102.4	100.0	101.4	99.9	95.9	98.3	102.4	92.7	98.6	102.4	86.0	95.9	99.9	73.3	89.3
26, 27, 3370	**ELECTRICAL, COMMUNICATIONS & UTIL.**	124.1	92.7	108.8	125.3	93.1	109.6	122.0	96.8	109.8	122.7	74.4	99.2	122.2	66.0	94.8	124.7	75.0	100.5
MF2004	**WEIGHTED AVERAGE**	113.5	92.9	104.9	118.6	91.2	107.1	123.6	99.8	113.6	115.4	87.8	103.8	115.5	75.0	98.5	131.2	82.0	110.5

	DIVISION	CANADA																	
		VICTORIA, BRITISH COLUMBIA			WHITEHORSE, YUKON			WINDSOR, ONTARIO			WINNIPEG, MANITOBA			YARMOUTH, NOVA SCOTIA			YELLOWKNIFE, NWT		
		MAT.	INST.	TOTAL	MAT.	INST.	TOTAL	MAT.	INST.	TOTAL	MAT.	INST.	TOTAL	MAT.	INST.	TOTAL	MAT.	INST.	TOTAL
015433	**CONTRACTOR EQUIPMENT**		109.0	109.0		103.0	103.0		102.2	102.2		106.4	106.4		102.9	102.9		103.0	103.0
0241, 31 - 34	**SITE & INFRASTRUCTURE, DEMOLITION**	121.2	106.6	111.2	108.0	97.8	101.1	97.2	101.5	100.1	119.3	100.9	106.7	98.9	99.4	99.2	131.1	102.4	111.5
0310	Concrete Forming & Accessories	111.1	77.1	81.8	106.0	58.6	65.1	123.3	90.0	94.5	130.3	65.5	74.3	96.1	70.3	73.8	106.0	79.9	83.5
0320	Concrete Reinforcing	116.2	72.2	98.2	118.8	60.7	95.1	108.9	88.9	100.7	147.9	56.0	110.4	147.9	47.3	106.8	118.8	64.4	96.6
0330	Cast-in-Place Concrete	147.2	91.3	125.8	146.6	69.8	117.3	137.7	100.8	123.6	183.5	72.3	141.0	168.2	71.6	131.3	243.5	87.5	183.9
03	**CONCRETE**	150.5	81.7	119.2	125.2	63.9	97.3	120.2	93.7	108.2	155.5	67.1	115.3	147.9	67.4	111.3	170.4	80.2	129.4
04	**MASONRY**	168.5	81.1	116.2	158.6	61.6	100.5	161.5	99.9	124.6	180.0	62.1	109.4	160.8	70.5	106.7	167.3	74.7	111.8
05	**METALS**	107.7	84.8	101.0	105.7	76.3	97.1	106.3	93.4	102.5	159.9	78.6	136.1	106.2	76.9	97.7	106.4	82.0	99.3
06	**WOOD, PLASTICS & COMPOSITES**	98.2	75.0	85.5	95.6	57.0	74.4	116.4	88.3	101.0	114.2	66.0	87.7	86.0	69.3	76.9	95.6	81.1	87.6
07	**THERMAL & MOISTURE PROTECTION**	114.9	80.4	101.5	103.9	61.5	87.4	110.0	93.6	103.6	114.1	67.3	95.9	105.5	69.0	91.4	107.4	78.3	96.1
08	**OPENINGS**	87.9	72.2	84.0	85.6	54.4	77.8	81.5	86.6	82.7	92.0	60.7	84.3	82.8	63.4	78.0	85.6	68.7	81.4
0920	Plaster & Gypsum Board	135.2	73.5	95.2	116.8	55.2	76.8	156.7	87.8	111.9	157.0	64.1	96.7	160.1	68.1	100.3	165.6	80.2	110.1
0950, 0980	Ceilings & Acoustic Treatment	95.4	73.5	81.9	95.4	55.2	70.6	95.4	87.8	90.7	121.9	64.1	86.2	95.4	68.1	78.5	132.0	80.2	100.0
0960	Flooring	126.6	44.4	104.4	119.8	58.8	103.3	132.0	95.5	122.1	133.3	67.3	115.5	110.1	64.5	97.8	119.8	90.2	111.8
0970, 0990	Wall Finishes & Painting/Coating	109.6	75.7	89.2	109.6	54.1	76.3	109.6	92.8	99.5	109.7	48.1	72.7	109.6	61.9	81.0	109.6	78.9	91.2
09	**FINISHES**	115.6	71.5	92.3	109.4	57.8	82.1	117.5	91.3	103.6	125.4	64.2	93.1	111.5	68.3	88.7	124.8	81.0	101.6
COVERS	**DIVS. 10 - 14, 25, 28, 41, 43, 44**	140.0	75.0	127.1	140.0	69.1	126.0	140.0	80.6	128.2	140.0	71.7	126.5	140.0	71.8	126.5	140.0	73.5	126.8
21, 22, 23	**FIRE SUPPRESSION, PLUMBING & HVAC**	102.4	71.4	90.1	102.4	76.2	92.0	99.2	90.2	95.6	99.9	65.6	86.3	102.4	86.0	95.9	102.9	97.4	100.7
26, 27, 3370	**ELECTRICAL, COMMUNICATIONS & UTIL.**	123.7	74.8	99.9	125.9	63.8	95.6	130.2	94.4	112.8	127.9	67.0	98.3	122.2	66.0	94.8	126.8	88.8	108.3
MF2004	**WEIGHTED AVERAGE**	118.3	78.8	101.7	113.2	68.9	94.6	113.0	93.2	104.7	129.6	69.6	104.4	115.4	75.0	98.5	121.4	85.7	106.4

Location Factors

Costs shown in RSMeans cost data publications are based on national averages for materials and installation. To adjust these costs to a specific location, simply multiply the base cost by the factor and divide by 100 for that city. The data is arranged alphabetically by state and postal zip code numbers. For a city not listed, use the factor for a nearby city with similar economic characteristics.

STATE/ZIP	CITY	MAT.	INST.	TOTAL
ALABAMA				
350-352	Birmingham	98.3	74.9	88.5
354	Tuscaloosa	97.0	60.5	81.7
355	Jasper	97.2	54.0	79.0
356	Decatur	97.0	59.9	81.4
357-358	Huntsville	97.0	71.1	86.1
359	Gadsden	96.9	59.4	81.2
360-361	Montgomery	98.0	59.2	81.7
362	Anniston	96.3	52.9	78.1
363	Dothan	96.9	52.5	78.3
364	Evergreen	96.4	57.3	80.0
365-366	Mobile	97.9	65.9	84.5
367	Selma	96.6	54.4	78.9
368	Phenix City	97.3	59.4	81.4
369	Butler	96.8	55.7	79.5
ALASKA				
995-996	Anchorage	128.3	110.3	120.8
997	Fairbanks	127.6	112.6	121.3
998	Juneau	127.0	108.5	119.2
999	Ketchikan	139.6	108.1	126.4
ARIZONA				
850,853	Phoenix	99.9	74.0	89.0
852	Mesa/Tempe	99.2	66.7	85.6
855	Globe	99.5	63.1	84.2
856-857	Tucson	98.0	70.6	86.5
859	Show Low	99.5	64.2	84.7
860	Flagstaff	101.4	71.6	88.9
863	Prescott	99.2	63.7	84.3
864	Kingman	97.5	69.6	85.8
865	Chambers	97.5	64.4	83.6
ARKANSAS				
716	Pine Bluff	96.9	66.6	84.2
717	Camden	94.6	44.6	73.6
718	Texarkana	95.9	48.8	76.1
719	Hot Springs	93.8	47.9	74.5
720-722	Little Rock	97.1	69.9	85.7
723	West Memphis	96.0	59.6	80.8
724	Jonesboro	96.9	60.2	81.5
725	Batesville	94.7	54.3	77.7
726	Harrison	95.9	55.0	78.8
727	Fayetteville	93.4	54.7	77.1
728	Russellville	94.5	56.7	78.6
729	Fort Smith	97.4	60.1	81.8
CALIFORNIA				
900-902	Los Angeles	102.2	116.7	108.3
903-905	Inglewood	97.6	107.9	101.9
906-908	Long Beach	99.3	108.0	102.9
910-912	Pasadena	98.0	107.9	102.2
913-916	Van Nuys	101.2	107.9	104.0
917-918	Alhambra	100.2	107.9	103.4
919-921	San Diego	102.4	108.8	105.1
922	Palm Springs	99.2	106.8	102.4
923-924	San Bernardino	96.9	107.2	101.2
925	Riverside	101.6	112.7	106.3
926-927	Santa Ana	98.9	107.5	102.5
928	Anaheim	101.5	114.7	107.0
930	Oxnard	102.9	112.8	107.0
931	Santa Barbara	101.9	112.9	106.5
932-933	Bakersfield	103.3	109.4	105.8
934	San Luis Obispo	102.5	105.3	103.7
935	Mojave	99.6	104.3	101.6
936-938	Fresno	103.2	112.6	107.2
939	Salinas	103.3	116.4	108.8
940-941	San Francisco	111.2	141.3	123.8
942,956-958	Sacramento	104.9	114.6	108.9
943	Palo Alto	103.5	123.7	112.0
944	San Mateo	106.4	129.3	116.0
945	Vallejo	104.1	121.6	111.4
946	Oakland	108.8	129.5	117.5
947	Berkeley	108.3	123.8	114.8
948	Richmond	107.4	123.9	114.3
949	San Rafael	108.7	124.5	115.3
950	Santa Cruz	108.4	116.6	111.8
CALIFORNIA (CONT'D)				
951	San Jose	106.0	132.5	117.1
952	Stockton	103.5	112.0	107.0
953	Modesto	103.4	111.8	106.9
954	Santa Rosa	103.3	126.1	112.9
955	Eureka	104.7	109.6	106.7
959	Marysville	104.0	111.2	107.0
960	Redding	105.8	109.8	107.5
961	Susanville	104.6	110.0	106.8
COLORADO				
800-802	Denver	101.8	85.5	95.0
803	Boulder	98.6	82.0	91.6
804	Golden	100.7	80.8	92.4
805	Fort Collins	102.2	77.3	91.7
806	Greeley	99.7	67.0	86.0
807	Fort Morgan	99.2	80.6	91.4
808-809	Colorado Springs	101.5	81.0	92.9
810	Pueblo	101.3	79.0	91.9
811	Alamosa	102.4	75.3	91.0
812	Salida	102.2	76.2	91.3
813	Durango	102.8	76.0	91.6
814	Montrose	101.4	74.4	90.1
815	Grand Junction	104.9	72.0	91.1
816	Glenwood Springs	102.4	77.7	92.1
CONNECTICUT				
060	New Britain	99.2	118.0	107.1
061	Hartford	100.5	118.1	107.9
062	Willimantic	99.8	117.5	107.2
063	New London	95.8	117.1	104.8
064	Meriden	97.9	118.5	106.6
065	New Haven	101.0	118.0	108.2
066	Bridgeport	100.3	118.6	108.0
067	Waterbury	99.9	118.5	107.7
068	Norwalk	99.9	119.0	107.9
069	Stamford	100.0	125.4	110.7
D.C.				
200-205	Washington	104.6	92.8	99.7
DELAWARE				
197	Newark	100.5	105.8	102.7
198	Wilmington	100.6	105.9	102.8
199	Dover	101.2	105.8	103.1
FLORIDA				
320,322	Jacksonville	98.8	63.1	83.8
321	Daytona Beach	99.1	76.5	89.6
323	Tallahassee	99.6	53.0	80.1
324	Panama City	100.3	48.6	78.6
325	Pensacola	102.9	61.3	85.4
326,344	Gainesville	100.5	66.0	86.0
327-328,347	Orlando	103.1	72.0	90.0
329	Melbourne	102.0	80.0	92.7
330-332,340	Miami	101.2	75.4	90.3
333	Fort Lauderdale	98.5	73.6	88.0
334,349	West Palm Beach	97.2	69.8	85.7
335-336,346	Tampa	100.1	78.7	91.1
337	St. Petersburg	102.3	58.1	83.7
338	Lakeland	99.2	77.8	90.2
339,341	Fort Myers	98.6	70.1	86.6
342	Sarasota	100.4	71.1	88.1
GEORGIA				
300-303,399	Atlanta	98.2	79.2	90.2
304	Statesboro	97.9	48.1	77.0
305	Gainesville	96.5	63.2	82.5
306	Athens	95.9	66.9	83.7
307	Dalton	98.0	52.3	78.8
308-309	Augusta	96.9	64.3	83.2
310-312	Macon	96.5	66.6	83.9
313-314	Savannah	98.8	62.0	83.3
315	Waycross	97.7	57.2	80.7
316	Valdosta	97.8	51.0	78.2
317,398	Albany	97.9	62.0	82.9
318-319	Columbus	97.8	66.8	84.8

STATE/ZIP	CITY	MAT.	INST.	TOTAL
HAWAII				
967	Hilo	114.2	118.7	116.1
968	Honolulu	118.7	118.7	118.7
STATES & POSS.				
969	Guam	138.1	64.3	107.1
IDAHO				
832	Pocatello	101.6	74.5	90.2
833	Twin Falls	102.5	54.0	82.1
834	Idaho Falls	99.9	59.8	83.0
835	Lewiston	109.0	80.4	97.0
836-837	Boise	100.6	74.8	89.8
838	Coeur d'Alene	108.3	77.4	95.3
ILLINOIS				
600-603	North Suburban	97.4	123.7	108.5
604	Joliet	97.4	126.9	109.8
605	South Suburban	97.4	124.5	108.8
606-608	Chicago	97.9	138.3	114.9
609	Kankakee	93.6	107.0	99.2
610-611	Rockford	97.1	115.0	104.7
612	Rock Island	94.7	99.5	96.7
613	La Salle	95.9	106.3	100.3
614	Galesburg	95.7	99.0	97.1
615-616	Peoria	98.2	104.9	101.0
617	Bloomington	95.1	106.4	99.8
618-619	Champaign	98.4	105.0	101.2
620-622	East St. Louis	93.9	105.7	98.8
623	Quincy	94.9	94.8	94.9
624	Effingham	94.3	95.9	95.0
625	Decatur	96.2	102.5	98.8
626-627	Springfield	97.0	103.4	99.7
628	Centralia	92.4	100.5	95.8
629	Carbondale	92.1	94.1	92.9
INDIANA				
460	Anderson	94.8	82.5	89.6
461-462	Indianapolis	98.0	86.7	93.2
463-464	Gary	96.3	101.5	98.5
465-466	South Bend	95.6	83.1	90.4
467-468	Fort Wayne	95.3	78.4	88.2
469	Kokomo	92.5	80.8	87.6
470	Lawrenceburg	91.1	76.9	85.1
471	New Albany	92.4	74.4	84.9
472	Columbus	94.7	79.5	88.3
473	Muncie	95.5	80.6	89.3
474	Bloomington	96.9	78.8	89.3
475	Washington	92.7	82.4	88.4
476-477	Evansville	94.4	86.5	91.1
478	Terre Haute	95.0	86.6	91.5
479	Lafayette	94.3	82.1	89.2
IOWA				
500-503,509	Des Moines	98.2	76.7	89.2
504	Mason City	95.7	61.4	81.3
505	Fort Dodge	96.0	57.1	79.7
506-507	Waterloo	97.8	58.4	81.2
508	Creston	96.2	61.2	81.5
510-511	Sioux City	98.4	68.8	86.0
512	Sibley	96.9	47.6	76.2
513	Spencer	98.5	46.3	76.6
514	Carroll	95.6	51.4	77.0
515	Council Bluffs	99.6	74.6	89.1
516	Shenandoah	96.3	49.3	76.6
520	Dubuque	97.9	75.8	88.6
521	Decorah	96.5	48.8	76.5
522-524	Cedar Rapids	98.8	80.8	91.2
525	Ottumwa	96.6	69.3	85.1
526	Burlington	95.8	70.7	85.2
527-528	Davenport	97.7	90.2	94.5
KANSAS				
660-662	Kansas City	98.3	93.7	96.4
664-666	Topeka	98.9	66.1	85.1
667	Fort Scott	96.6	71.4	86.0
668	Emporia	96.7	59.7	81.2
669	Belleville	98.4	61.0	82.7
670-672	Wichita	97.6	64.9	83.9
673	Independence	98.1	63.3	83.5
674	Salina	98.7	61.6	83.1
675	Hutchinson	93.8	61.2	80.1
676	Hays	97.6	62.5	82.9
677	Colby	98.4	62.3	83.3

STATE/ZIP	CITY	MAT.	INST.	TOTAL
KANSAS (CONT'D)				
678	Dodge City	99.9	64.0	84.8
679	Liberal	97.5	62.2	82.7
KENTUCKY				
400-402	Louisville	97.1	83.9	91.6
403-405	Lexington	95.1	78.5	88.1
406	Frankfort	96.3	78.9	89.0
407-409	Corbin	92.3	65.8	81.2
410	Covington	92.8	99.3	95.5
411-412	Ashland	91.7	96.7	93.8
413-414	Campton	93.1	66.2	81.8
415-416	Pikeville	94.0	81.7	88.8
417-418	Hazard	92.4	58.7	78.3
420	Paducah	90.9	87.2	89.4
421-422	Bowling Green	93.6	84.6	89.8
423	Owensboro	93.4	83.2	89.1
424	Henderson	90.7	86.8	89.0
425-426	Somerset	90.6	70.3	82.1
427	Elizabethtown	90.1	83.6	87.3
LOUISIANA				
700-701	New Orleans	102.9	69.6	88.9
703	Thibodaux	100.4	63.7	85.0
704	Hammond	97.6	58.4	81.2
705	Lafayette	100.0	58.5	82.6
706	Lake Charles	100.1	61.1	83.7
707-708	Baton Rouge	102.2	63.8	86.1
710-711	Shreveport	96.8	59.2	81.0
712	Monroe	96.9	56.0	79.7
713-714	Alexandria	97.0	56.0	79.8
MAINE				
039	Kittery	93.0	74.4	85.2
040-041	Portland	98.5	73.7	88.1
042	Lewiston	96.4	73.7	86.9
043	Augusta	96.1	74.3	86.9
044	Bangor	95.8	73.7	86.5
045	Bath	94.2	74.4	85.9
046	Machias	93.7	73.9	85.4
047	Houlton	93.9	74.4	85.7
048	Rockland	92.9	74.5	85.2
049	Waterville	94.2	74.4	85.9
MARYLAND				
206	Waldorf	100.5	69.5	87.5
207-208	College Park	100.4	79.4	91.6
209	Silver Spring	99.7	75.5	89.5
210-212	Baltimore	98.2	86.0	93.1
214	Annapolis	98.7	80.2	90.9
215	Cumberland	94.0	80.4	88.3
216	Easton	95.5	42.8	73.4
217	Hagerstown	94.9	81.6	89.3
218	Salisbury	95.8	50.3	76.7
219	Elkton	93.1	62.9	80.4
MASSACHUSETTS				
010-011	Springfield	98.4	104.5	101.0
012	Pittsfield	97.8	101.0	99.1
013	Greenfield	95.7	102.3	98.4
014	Fitchburg	94.4	117.6	104.1
015-016	Worcester	98.3	118.9	107.0
017	Framingham	93.9	126.1	107.4
018	Lowell	97.7	125.7	109.5
019	Lawrence	98.9	123.6	109.3
020-022, 024	Boston	100.5	135.9	115.4
023	Brockton	99.2	120.7	108.2
025	Buzzards Bay	93.3	118.2	103.8
026	Hyannis	96.2	118.7	105.7
027	New Bedford	98.4	118.9	107.0
MICHIGAN				
480,483	Royal Oak	91.4	103.9	96.6
481	Ann Arbor	93.7	104.5	98.2
482	Detroit	95.3	114.2	103.2
484-485	Flint	93.4	100.8	96.5
486	Saginaw	93.0	89.9	91.7
487	Bay City	93.1	90.0	91.8
488-489	Lansing	94.4	96.9	95.5
490	Battle Creek	94.5	88.2	91.9
491	Kalamazoo	94.8	85.6	91.0
492	Jackson	92.7	91.6	92.3
493,495	Grand Rapids	96.1	65.2	83.1
494	Muskegon	93.5	81.4	88.4

Location Factors

STATE/ZIP	CITY	MAT.	INST.	TOTAL
MICHIGAN (CONT'D)				
496	Traverse City	92.3	69.1	82.5
497	Gaylord	93.3	72.1	84.4
498-499	Iron Mountain	95.3	83.2	90.2
MINNESOTA				
550-551	Saint Paul	98.6	118.7	107.0
553-555	Minneapolis	99.3	124.2	109.8
556-558	Duluth	98.2	108.2	102.4
MINNESOTA (CONT'd)				
559	Rochester	98.1	104.7	100.9
560	Mankato	96.0	103.2	99.0
561	Windom	94.8	77.6	87.6
562	Willmar	94.3	82.8	89.5
563	St. Cloud	95.8	117.5	104.9
564	Brainerd	95.9	99.5	97.4
565	Detroit Lakes	97.7	94.6	96.4
566	Bemidji	97.1	95.9	96.6
567	Thief River Falls	96.7	91.7	94.6
MISSISSIPPI				
386	Clarksdale	96.6	60.4	81.4
387	Greenville	100.0	70.9	87.8
388	Tupelo	97.9	63.1	83.2
389	Greenwood	97.8	59.9	81.9
390-392	Jackson	98.5	69.9	86.5
393	Meridian	96.8	70.9	85.9
394	Laurel	97.8	63.7	83.5
395	Biloxi	98.5	62.6	83.4
396	Mccomb	96.3	59.3	80.8
397	Columbus	97.7	61.0	82.3
MISSOURI				
630-631	St. Louis	98.6	108.8	102.9
633	Bowling Green	97.1	90.3	94.2
634	Hannibal	95.9	80.1	89.3
635	Kirksville	98.7	73.3	88.0
636	Flat River	98.1	91.1	95.1
637	Cape Girardeau	98.2	87.4	93.7
638	Sikeston	96.1	77.2	88.2
639	Poplar Bluff	95.6	77.7	88.1
640-641	Kansas City	99.2	106.1	102.1
644-645	St. Joseph	98.3	90.2	94.9
646	Chillicothe	95.1	68.5	83.9
647	Harrisonville	94.7	98.4	96.2
648	Joplin	97.1	68.0	84.9
650-651	Jefferson City	95.4	86.0	91.5
652	Columbia	96.3	88.1	92.9
653	Sedalia	95.4	81.7	89.6
654-655	Rolla	94.1	73.0	85.2
656-658	Springfield	97.3	77.6	89.1
MONTANA				
590-591	Billings	101.9	74.6	90.4
592	Wolf Point	101.1	71.5	88.7
593	Miles City	98.9	72.8	88.0
594	Great Falls	103.1	74.8	91.2
595	Havre	100.0	72.8	88.6
596	Helena	101.8	73.5	89.9
597	Butte	101.6	75.0	90.4
598	Missoula	99.0	73.2	88.2
599	Kalispell	97.9	72.0	87.0
NEBRASKA				
680-681	Omaha	99.5	78.8	90.8
683-685	Lincoln	97.6	77.1	89.0
686	Columbus	95.5	76.4	87.5
687	Norfolk	97.2	80.4	90.1
688	Grand Island	97.3	82.4	91.0
689	Hastings	96.5	84.8	91.6
690	Mccook	96.4	75.1	87.5
691	North Platte	96.8	85.0	91.8
692	Valentine	98.7	73.6	88.2
693	Alliance	98.6	71.6	87.3
NEVADA				
889-891	Las Vegas	101.1	112.0	105.7
893	Ely	99.5	72.0	88.0
894-895	Reno	99.8	93.9	97.3
897	Carson City	99.9	93.1	97.1
898	Elko	98.2	77.6	89.6
NEW HAMPSHIRE				
030	Nashua	98.6	87.7	94.0
031	Manchester	99.1	87.7	94.3

STATE/ZIP	CITY	MAT.	INST.	TOTAL
NEW HAMPSHIRE (CONT'D)				
032-033	Concord	97.0	83.8	91.5
034	Keene	95.1	53.4	77.6
035	Littleton	95.0	60.6	80.5
036	Charleston	94.6	50.8	76.2
037	Claremont	93.7	50.8	75.7
038	Portsmouth	95.9	90.1	93.5
NEW JERSEY				
070-071	Newark	100.5	122.0	109.5
072	Elizabeth	98.6	121.5	108.2
073	Jersey City	97.8	122.2	108.0
074-075	Paterson	99.3	121.8	108.8
076	Hackensack	97.3	122.3	107.8
077	Long Branch	97.1	121.2	107.2
078	Dover	97.6	122.0	107.8
079	Summit	97.6	121.5	107.6
080,083	Vineland	95.7	117.6	104.9
081	Camden	97.8	117.8	106.2
082,084	Atlantic City	96.3	117.6	105.2
085-086	Trenton	97.6	120.6	107.3
087	Point Pleasant	97.7	120.8	107.4
088-089	New Brunswick	98.1	121.4	107.9
NEW MEXICO				
870-872	Albuquerque	99.8	76.0	89.8
873	Gallup	99.9	76.0	89.9
874	Farmington	100.3	76.0	90.1
875	Santa Fe	102.0	76.0	91.1
877	Las Vegas	98.4	75.8	88.9
878	Socorro	97.9	76.0	88.7
879	Truth/Consequences	98.1	70.9	86.7
880	Las Cruces	97.0	68.5	85.0
881	Clovis	98.5	74.2	88.3
882	Roswell	100.3	74.3	89.3
883	Carrizozo	100.7	76.0	90.3
884	Tucumcari	99.3	74.2	88.8
NEW YORK				
100-102	New York	106.1	164.6	130.7
103	Staten Island	101.9	162.0	127.1
104	Bronx	100.0	162.0	126.0
105	Mount Vernon	99.8	132.8	113.6
106	White Plains	99.8	132.8	113.6
107	Yonkers	104.7	132.8	116.5
108	New Rochelle	100.1	132.8	113.8
109	Suffern	99.9	121.8	109.1
110	Queens	101.3	161.9	126.7
111	Long Island City	102.9	161.9	127.6
112	Brooklyn	103.1	161.9	127.8
113	Flushing	103.0	161.9	127.7
114	Jamaica	101.2	161.9	126.7
115,117,118	Hicksville	101.2	146.4	120.1
116	Far Rockaway	103.1	161.9	127.8
119	Riverhead	102.0	146.4	120.6
120-122	Albany	97.4	94.7	96.3
123	Schenectady	97.7	95.3	96.7
124	Kingston	100.8	112.5	105.7
125-126	Poughkeepsie	100.0	127.3	111.5
127	Monticello	99.3	115.7	106.2
128	Glens Falls	92.2	91.0	91.7
129	Plattsburgh	96.7	85.9	92.2
130-132	Syracuse	98.7	93.1	96.4
133-135	Utica	96.7	89.5	93.6
136	Watertown	98.4	92.2	95.8
137-139	Binghamton	98.1	86.9	93.4
140-142	Buffalo	99.0	104.8	101.5
143	Niagara Falls	96.7	101.5	98.7
144-146	Rochester	98.4	94.1	96.6
147	Jamestown	95.6	82.5	90.1
148-149	Elmira	95.5	84.4	90.8
NORTH CAROLINA				
270,272-274	Greensboro	99.8	50.7	79.2
271	Winston-Salem	99.5	50.0	78.7
275-276	Raleigh	100.7	51.3	80.0
277	Durham	101.1	51.7	80.4
278	Rocky Mount	96.9	41.6	73.7
279	Elizabeth City	97.7	43.0	74.7
280	Gastonia	98.5	49.6	78.0
281-282	Charlotte	100.2	52.2	80.0
283	Fayetteville	101.1	52.1	80.5
284	Wilmington	97.2	49.6	77.2
285	Kinston	95.2	43.3	73.4

STATE/ZIP	CITY	MAT.	INST.	TOTAL
NORTH CAROLINA (CONT'D)				
286	Hickory	95.5	45.9	74.7
287-288	Asheville	97.6	49.6	77.5
289	Murphy	96.3	35.7	70.9
NORTH DAKOTA				
580-581	Fargo	101.4	62.0	84.9
582	Grand Forks	100.8	55.1	81.6
583	Devils Lake	100.0	57.3	82.0
584	Jamestown	100.0	49.1	78.7
585	Bismarck	100.1	62.9	84.5
586	Dickinson	100.9	60.1	83.7
587	Minot	100.9	67.3	86.8
588	Williston	99.4	60.1	82.9
OHIO				
430-432	Columbus	96.3	88.6	93.0
433	Marion	92.6	83.5	88.8
434-436	Toledo	96.5	99.7	97.9
437-438	Zanesville	93.1	82.5	88.7
439	Steubenville	94.1	91.6	93.0
440	Lorain	96.9	93.6	95.5
441	Cleveland	97.2	102.6	99.5
442-443	Akron	97.8	93.7	96.1
444-445	Youngstown	97.2	89.7	94.1
446-447	Canton	97.3	85.0	92.1
448-449	Mansfield	94.4	88.3	91.9
450	Hamilton	95.0	85.8	91.1
451-452	Cincinnati	95.3	88.1	92.3
453-454	Dayton	95.2	85.5	91.1
455	Springfield	95.1	85.7	91.2
456	Chillicothe	93.8	91.4	92.8
457	Athens	96.7	75.9	88.0
458	Lima	97.0	85.3	92.1
OKLAHOMA				
730-731	Oklahoma City	98.7	61.7	83.1
734	Ardmore	95.2	61.1	80.9
735	Lawton	97.8	62.5	83.0
736	Clinton	96.6	59.7	81.1
737	Enid	97.4	59.7	81.6
738	Woodward	95.4	59.8	80.4
739	Guymon	96.4	32.0	69.4
740-741	Tulsa	97.6	56.3	80.3
743	Miami	94.1	64.5	81.7
744	Muskogee	96.9	41.6	73.7
745	Mcalester	93.8	52.4	76.5
746	Ponca City	94.4	60.4	80.1
747	Durant	94.4	60.4	80.2
748	Shawnee	95.9	57.8	79.9
749	Poteau	93.5	62.8	80.6
OREGON				
970-972	Portland	101.1	101.0	101.0
973	Salem	101.0	99.1	100.2
974	Eugene	100.7	98.4	99.7
975	Medford	102.3	97.0	100.1
976	Klamath Falls	102.2	96.7	99.9
977	Bend	101.0	98.5	100.0
978	Pendleton	96.1	99.0	97.3
979	Vale	93.8	90.3	92.3
PENNSYLVANIA				
150-152	Pittsburgh	96.9	98.7	97.6
153	Washington	93.8	98.4	95.7
154	Uniontown	94.1	95.5	94.7
155	Bedford	95.1	89.3	92.6
156	Greensburg	95.0	96.8	95.8
157	Indiana	93.9	95.6	94.6
158	Dubois	95.5	93.2	94.5
159	Johnstown	95.1	93.2	94.3
160	Butler	91.9	96.3	93.8
161	New Castle	91.9	94.7	93.1
162	Kittanning	92.4	98.6	95.0
163	Oil City	91.9	92.9	92.3
164-165	Erie	94.2	92.0	93.3
166	Altoona	94.3	89.4	92.2
167	Bradford	95.4	90.7	93.4
168	State College	94.9	90.1	92.9
169	Wellsboro	96.0	90.7	93.8
170-171	Harrisburg	97.9	93.1	95.9
172	Chambersburg	95.5	88.9	92.7
173-174	York	96.1	93.3	94.9
175-176	Lancaster	94.2	87.9	91.6

STATE/ZIP	CITY	MAT.	INST.	TOTAL
PENNSYLVANIA (CONT'D)				
177	Williamsport	92.9	81.1	87.9
178	Sunbury	95.0	92.2	93.8
179	Pottsville	94.1	92.3	93.4
180	Lehigh Valley	95.5	111.4	102.2
181	Allentown	97.7	106.5	101.4
182	Hazleton	94.9	92.4	93.9
183	Stroudsburg	94.8	98.8	96.5
184-185	Scranton	98.5	95.2	97.1
186-187	Wilkes-Barre	94.7	92.7	93.8
188	Montrose	94.4	93.7	94.1
189	Doylestown	94.4	118.6	104.5
190-191	Philadelphia	101.0	130.0	113.2
193	Westchester	97.2	119.9	106.8
194	Norristown	96.2	127.4	109.3
195-196	Reading	98.5	97.6	98.1
PUERTO RICO				
009	San Juan	120.1	23.9	79.7
RHODE ISLAND				
028	Newport	97.7	109.1	102.5
029	Providence	98.8	109.1	103.2
SOUTH CAROLINA				
290-292	Columbia	98.6	53.3	79.6
293	Spartanburg	96.9	52.9	78.4
294	Charleston	98.4	62.2	83.2
295	Florence	96.6	53.3	78.4
296	Greenville	96.7	52.9	78.3
297	Rock Hill	96.1	50.9	77.1
298	Aiken	97.1	70.0	85.7
299	Beaufort	97.9	46.2	76.2
SOUTH DAKOTA				
570-571	Sioux Falls	99.9	60.1	83.2
572	Watertown	98.1	55.3	80.1
573	Mitchell	97.1	55.2	79.5
574	Aberdeen	99.9	56.3	81.6
575	Pierre	99.5	56.5	81.4
576	Mobridge	97.7	55.0	79.8
577	Rapid City	99.7	56.5	81.5
TENNESSEE				
370-372	Nashville	98.5	73.6	88.0
373-374	Chattanooga	99.4	55.0	80.8
375,380-381	Memphis	97.5	70.1	86.0
376	Johnson City	98.4	55.4	80.3
377-379	Knoxville	95.8	55.8	79.0
382	McKenzie	97.1	55.2	79.5
383	Jackson	99.0	47.9	77.5
384	Columbia	95.6	57.0	79.4
385	Cookeville	97.0	58.1	80.6
TEXAS				
750	McKinney	98.7	51.5	78.8
751	Waxahackie	98.6	54.2	80.0
752-753	Dallas	99.4	66.2	85.4
754	Greenville	98.7	38.4	73.4
755	Texarkana	97.6	50.8	77.9
756	Longview	98.0	39.9	73.6
757	Tyler	98.8	54.2	80.1
758	Palestine	95.0	40.3	72.0
759	Lufkin	95.7	43.9	74.0
760-761	Fort Worth	98.0	60.7	82.4
762	Denton	97.6	48.1	76.8
763	Wichita Falls	98.5	55.3	80.3
764	Eastland	96.9	39.4	72.8
765	Temple	95.6	48.7	75.9
766-767	Waco	97.8	56.8	80.6
768	Brownwood	98.0	37.2	72.5
769	San Angelo	97.6	45.3	75.7
770-772	Houston	100.7	71.2	88.3
773	Huntsville	98.8	37.9	73.2
774	Wharton	100.2	42.2	75.8
775	Galveston	98.2	69.1	86.0
776-777	Beaumont	98.4	60.1	82.3
778	Bryan	95.4	62.6	81.6
779	Victoria	100.1	44.3	76.7
780	Laredo	95.8	51.7	77.3
781-782	San Antonio	96.2	65.4	83.3
783-784	Corpus Christi	98.4	50.4	78.3
785	McAllen	98.0	45.1	75.8
786-787	Austin	96.5	59.1	80.8

Location Factors

STATE/ZIP	CITY	MAT.	INST.	TOTAL
TEXAS (CONT'D)				
788	Del Rio	97.6	31.9	70.1
789	Giddings	94.9	39.5	71.7
790-791	Amarillo	98.6	56.8	81.0
792	Childress	97.6	49.6	77.4
793-794	Lubbock	100.1	51.1	79.5
795-796	Abilene	98.2	50.8	78.3
797	Midland	100.0	47.4	77.9
798-799,885	El Paso	97.5	49.8	77.5
UTAH				
840-841	Salt Lake City	102.9	67.9	88.2
842,844	Ogden	98.4	67.3	85.3
843	Logan	100.3	67.3	86.4
845	Price	100.7	46.5	78.0
846-847	Provo	100.8	67.8	87.0
VERMONT				
050	White River Jct.	96.3	56.6	79.7
051	Bellows Falls	94.8	64.0	81.9
052	Bennington	95.2	66.8	83.2
053	Brattleboro	95.6	69.0	84.4
054	Burlington	99.3	67.7	86.1
056	Montpelier	95.9	68.1	84.2
057	Rutland	97.5	68.1	85.2
058	St. Johnsbury	96.3	57.1	79.9
059	Guildhall	95.1	56.9	79.0
VIRGINIA				
220-221	Fairfax	100.2	82.2	92.6
222	Arlington	101.5	80.8	92.8
223	Alexandria	100.7	86.0	94.5
224-225	Fredericksburg	98.8	73.8	88.3
226	Winchester	99.4	66.5	85.6
227	Culpeper	99.3	72.0	87.9
228	Harrisonburg	99.6	68.2	86.4
229	Charlottesville	100.0	66.6	85.9
230-232	Richmond	101.7	70.1	88.4
233-235	Norfolk	101.9	70.5	88.7
236	Newport News	100.5	70.9	88.1
237	Portsmouth	99.8	67.4	86.2
238	Petersburg	99.9	70.1	87.4
239	Farmville	99.0	56.9	81.3
240-241	Roanoke	101.2	63.7	85.4
242	Bristol	98.7	57.4	81.4
243	Pulaski	98.4	54.7	80.1
244	Staunton	99.3	62.8	84.0
245	Lynchburg	99.4	68.1	86.2
246	Grundy	98.8	54.1	80.0
WASHINGTON				
980-981,987	Seattle	105.4	101.7	103.9
982	Everett	105.0	96.7	101.5
983-984	Tacoma	105.1	99.6	102.8
985	Olympia	103.1	99.5	101.6
986	Vancouver	106.5	93.8	101.1
988	Wenatchee	104.8	81.4	95.0
989	Yakima	105.3	88.0	98.0
990-992	Spokane	104.5	81.8	95.0
993	Richland	104.1	85.7	96.4
994	Clarkston	102.9	82.2	94.2
WEST VIRGINIA				
247-248	Bluefield	97.5	77.3	89.0
249	Lewisburg	99.3	82.0	92.0
250-253	Charleston	100.4	87.8	95.1
254	Martinsburg	98.9	76.5	89.5
255-257	Huntington	100.6	90.3	96.3
258-259	Beckley	97.3	86.0	92.6
260	Wheeling	100.6	89.2	95.8
261	Parkersburg	99.6	88.7	95.0
262	Buckhannon	98.9	88.9	94.7
263-264	Clarksburg	99.3	87.8	94.5
265	Morgantown	99.4	88.5	94.8
266	Gassaway	98.7	88.8	94.5
267	Romney	98.8	82.3	91.9
268	Petersburg	98.6	84.6	92.7
WISCONSIN				
530,532	Milwaukee	100.3	106.4	102.9
531	Kenosha	100.0	99.3	99.7
534	Racine	99.5	101.3	100.2
535	Beloit	99.3	94.6	97.3
537	Madison	99.5	95.0	97.6

STATE/ZIP	CITY	MAT.	INST.	TOTAL
WISCONSIN (CONT'D)				
538	Lancaster	96.9	90.3	94.1
539	Portage	95.6	94.3	95.0
540	New Richmond	95.5	93.6	94.7
541-543	Green Bay	99.7	90.4	95.8
544	Wausau	94.9	88.5	92.2
545	Rhinelander	98.0	88.7	94.1
546	La Crosse	96.3	91.3	94.2
547	Eau Claire	97.9	92.0	95.4
548	Superior	95.4	95.7	95.5
549	Oshkosh	95.5	89.9	93.2
WYOMING				
820	Cheyenne	100.0	65.9	85.7
821	Yellowstone Nat'l Park	97.5	58.6	81.2
822	Wheatland	98.6	58.6	81.8
823	Rawlins	100.2	58.4	82.6
824	Worland	98.1	56.5	80.6
825	Riverton	99.2	56.5	81.3
826	Casper	100.6	58.4	82.9
827	Newcastle	97.9	58.4	81.3
828	Sheridan	100.7	60.5	83.8
829-831	Rock Springs	102.0	56.5	82.9
CANADIAN FACTORS (reflect Canadian currency)				
ALBERTA				
	Calgary	130.7	89.9	113.6
	Edmonton	131.8	89.8	114.1
	Fort McMurray	127.7	92.6	112.9
	Lethbridge	121.9	92.0	109.3
	Lloydminster	116.4	88.8	104.8
	Medicine Hat	116.4	88.1	104.5
	Red Deer	116.9	88.1	104.8
BRITISH COLUMBIA				
	Kamloops	117.2	90.8	106.1
	Prince George	118.3	90.8	106.7
	Vancouver	131.2	82.0	110.5
	Victoria	118.3	78.8	101.7
MANITOBA				
	Brandon	116.4	76.1	99.5
	Portage la Prairie	116.5	74.7	98.9
	Winnipeg	129.6	69.6	104.4
NEW BRUNSWICK				
	Bathurst	114.7	67.8	95.0
	Dalhousie	114.6	67.8	95.0
	Fredericton	117.1	72.4	98.3
	Moncton	115.0	68.7	95.5
	Newcastle	114.7	67.8	95.0
	St. John	117.3	72.4	98.4
NEWFOUNDLAND				
	Corner Brook	119.7	68.2	98.0
	St. Johns	119.9	69.2	98.6
NORTHWEST TERRITORIES				
	Yellowknife	121.4	85.7	106.4
NOVA SCOTIA				
	Bridgewater	116.1	75.0	98.9
	Dartmouth	117.5	75.0	99.7
	Halifax	119.0	78.0	101.8
	New Glasgow	115.5	75.0	98.5
	Sydney	113.1	75.0	97.1
	Truro	115.5	75.0	98.5
	Yarmouth	115.4	75.0	98.5
ONTARIO				
	Barrie	119.2	93.0	108.2
	Brantford	118.4	96.9	109.4
	Cornwall	118.5	93.3	107.9
	Hamilton	123.4	96.9	112.3
	Kingston	119.4	93.4	108.5
	Kitchener	114.5	92.2	105.1
	London	122.0	93.3	110.0
	North Bay	118.4	91.2	107.0
	Oshawa	117.9	94.7	108.2
	Ottawa	122.7	93.9	110.6
	Owen Sound	119.4	91.2	107.6
	Peterborough	118.4	93.2	107.8
	Sarnia	117.9	97.5	109.3

STATE/ZIP	CITY	MAT.	INST.	TOTAL
ONTARIO (CONT'D)				
	Sault Ste Marie	112.9	92.6	104.4
	St. Catharines	112.3	94.2	104.7
	Sudbury	112.7	92.9	104.4
	Thunder Bay	113.5	92.9	104.9
	Timmins	118.6	91.2	107.1
	Toronto	123.6	99.8	113.6
	Windsor	113.0	93.2	104.7
PRINCE EDWARD ISLAND				
	Charlottetown	117.8	63.3	94.9
	Summerside	117.2	63.3	94.6
QUEBEC				
	Cap-de-la-Madeleine	115.3	87.8	103.8
	Charlesbourg	115.3	87.8	103.8
	Chicoutimi	114.0	93.5	105.4
	Gatineau	114.9	87.6	103.4
	Granby	115.1	87.6	103.5
	Hull	115.0	87.6	103.5
	Joliette	115.5	87.8	103.9
	Laval	114.5	87.6	103.2
	Montreal	121.7	93.8	110.0
	Quebec	120.1	94.2	109.2
	Rimouski	114.4	93.5	105.6
	Rouyn-Noranda	114.8	87.6	103.4
	Saint Hyacinthe	114.2	87.6	103.0
	Sherbrooke	115.2	87.6	103.6
	Sorel	115.4	87.8	103.8
	St. Jerome	114.8	87.6	103.4
	Trois Rivieres	115.4	87.8	103.8
SASKATCHEWAN				
	Moose Jaw	113.5	69.9	95.2
	Prince Albert	112.5	68.1	93.9
	Regina	115.1	69.9	96.1
	Saskatoon	113.6	69.8	95.2
YUKON				
	Whitehorse	113.2	68.9	94.6

R011105-05 Tips for Accurate Estimating

1. Use pre-printed or columnar forms for orderly sequence of dimensions and locations and for recording telephone quotations.
2. Use only the front side of each paper or form except for certain pre-printed summary forms.
3. Be consistent in listing dimensions: For example, length x width x height. This helps in rechecking to ensure that, the total length of partitions is appropriate for the building area.
4. Use printed (rather than measured) dimensions where given.
5. Add up multiple printed dimensions for a single entry where possible.
6. Measure all other dimensions carefully.
7. Use each set of dimensions to calculate multiple related quantities.
8. Convert foot and inch measurements to decimal feet when listing. Memorize decimal equivalents to .01 parts of a foot (1/8″ equals approximately .01′).
9. Do not "round off" quantities until the final summary.
10. Mark drawings with different colors as items are taken off.
11. Keep similar items together, different items separate.
12. Identify location and drawing numbers to aid in future checking for completeness.
13. Measure or list everything on the drawings or mentioned in the specifications.
14. It may be necessary to list items not called for to make the job complete.
15. Be alert for: Notes on plans such as N.T.S. (not to scale); changes in scale throughout the drawings; reduced size drawings; discrepancies between the specifications and the drawings.
16. Develop a consistent pattern of performing an estimate. For example:
    a. Start the quantity takeoff at the lower floor and move to the next higher floor.
    b. Proceed from the main section of the building to the wings.
    c. Proceed from south to north or vice versa, clockwise or counterclockwise.
    d. Take off floor plan quantities first, elevations next, then detail drawings.
17. List all gross dimensions that can be either used again for different quantities, or used as a rough check of other quantities for verification (exterior perimeter, gross floor area, individual floor areas, etc.).
18. Utilize design symmetry or repetition (repetitive floors, repetitive wings, symmetrical design around a center line, similar room layouts, etc.). Note: Extreme caution is needed here so as not to omit or duplicate an area.
19. Do not convert units until the final total is obtained. For instance, when estimating concrete work, keep all units to the nearest cubic foot, then summarize and convert to cubic yards.
20. When figuring alternatives, it is best to total all items involved in the basic system, then total all items involved in the alternates. Therefore you work with positive numbers in all cases. When adds and deducts are used, it is often confusing whether to add or subtract a portion of an item; especially on a complicated or involved alternate.

R011110-10 Architectural Fees

Tabulated below are typical percentage fees by project size, for good professional architectural service. Fees may vary from those listed depending upon degree of design difficulty and economic conditions in any particular area.

Rates can be interpolated horizontally and vertically. Various portions of the same project requiring different rates should be adjusted proportionately. For alterations, add 50% to the fee for the first $500,000 of project cost and add 25% to the fee for project cost over $500,000.

Architectural fees tabulated below include Structural, Mechanical, and Electrical Engineering Fees. They do not include the fees for special consultants such as kitchen planning, security, acoustical, interior design, etc.

Civil Engineering fees are included in the Architectural fee for project sites requiring minimal design such as city sites. However, separate Civil Engineering fees must be added when utility connections require design, drainage calculations are needed, stepped foundations are required, or provisions are required to protect adjacent wetlands.

	Total Project Size in Thousands of Dollars						
Building Types	100	250	500	1,000	5,000	10,000	50,000
Factories, garages, warehouses, repetitive housing	9.0%	8.0%	7.0%	6.2%	5.3%	4.9%	4.5%
Apartments, banks, schools, libraries, offices, municipal buildings	12.2	12.3	9.2	8.0	7.0	6.6	6.2
Churches, hospitals, homes, laboratories, museums, research	15.0	13.6	12.7	11.9	9.5	8.8	8.0
Memorials, monumental work, decorative furnishings	—	16.0	14.5	13.1	10.0	9.0	8.3

R011110-30 Engineering Fees

Typical **Structural Engineering Fees** based on type of construction and total project size. These fees are included in Architectural Fees.

Type of Construction	Total Project Size (in thousands of dollars)			
	$500	$500-$1,000	$1,000-$5,000	Over $5000
Industrial buildings, factories & warehouses	Technical payroll times 2.0 to 2.5	1.60%	1.25%	1.00%
Hotels, apartments, offices, dormitories, hospitals, public buildings, food stores	↓	2.00%	1.70%	1.20%
Museums, banks, churches and cathedrals	↓	2.00%	1.75%	1.25%
Thin shells, prestressed concrete, earthquake resistive	↓	2.00%	1.75%	1.50%
Parking ramps, auditoriums, stadiums, convention halls, hangars & boiler houses	↓	2.50%	2.00%	1.75%
Special buildings, major alterations, underpinning & future expansion	↓	Add to above 0.5%	Add to above 0.5%	Add to above 0.5%

For complex reinforced concrete or unusually complicated structures, add 20% to 50%.

Typical **Mechanical and Electrical Engineering Fees** are based on the size of the subcontract. The fee structure for both are shown below. These fees are included in Architectural Fees.

Type of Construction	Subcontract Size							
	$25,000	$50,000	$100,000	$225,000	$350,000	$500,000	$750,000	$1,000,000
Simple structures	6.4%	5.7%	4.8%	4.5%	4.4%	4.3%	4.2%	4.1%
Intermediate structures	8.0	7.3	6.5	5.6	5.1	5.0	4.9	4.8
Complex structures	10.1	9.0	9.0	8.0	7.5	7.5	7.0	7.0

For renovations, add 15% to 25% to applicable fee.

General Requirements R0121 Allowances

R012153-10 Repair and Remodeling

Cost figures are based on new construction utilizing the most cost-effective combination of labor, equipment, and material with the work scheduled in proper sequence to allow the various trades to accomplish their work in an efficient manner.

The costs for repair and remodeling work must be modified due to the following factors that may be present in any given repair and remodeling project:

1. Equipment usage curtailment due to the physical limitations of the project, with only hand-operated equipment being used.
2. Increased requirement for shoring and bracing to hold up the building while structural changes are being made and to allow for temporary storage of construction materials on above-grade floors.
3. Material handling becomes more costly due to having to move within the confines of an enclosed building. For multi-story construction, low capacity elevators and stairwells may be the only access to the upper floors.
4. Large amount of cutting and patching and attempting to match the existing construction is required. It is often more economical to remove entire walls rather than create many new door and window openings. This sort of trade-off has to be carefully analyzed.
5. Cost of protection of completed work is increased since the usual sequence of construction usually cannot be accomplished.
6. Economies of scale usually associated with new construction may not be present. If small quantities of components must be custom fabricated due to job requirements, unit costs will naturally increase. Also, if only small work areas are available at a given time, job scheduling between trades becomes difficult and subcontractor quotations may reflect the excessive start-up and shut-down phases of the job.
7. Work may have to be done on other than normal shifts and may have to be done around an existing production facility which has to stay in production during the course of the repair and remodeling.
8. Dust and noise protection of adjoining non-construction areas can involve substantial special protection and alter usual construction methods.
9. Job may be delayed due to unexpected conditions discovered during demolition or removal. These delays ultimately increase construction costs.
10. Piping and ductwork runs may not be as simple as for new construction. Wiring may have to be snaked through walls and floors.
11. Matching "existing construction" may be impossible because materials may no longer be manufactured. Substitutions may be expensive.
12. Weather protection of existing structure requires additional temporary structures to protect building at openings.
13. On small projects, because of local conditions, it may be necessary to pay a tradesman for a minimum of four hours for a task that is completed in one hour.

All of the above areas can contribute to increased costs for a repair and remodeling project. Each of the above factors should be considered in the planning, bidding and construction stage in order to minimize the increased costs associated with repair and remodeling jobs.

General Requirements — R0121 Allowances

R012157-20 Construction Time Requirements

Table at left is average construction time in months for different types of building projects. Table at right is the construction time in months for different size projects. Design time runs 25% to 40% of construction time.

Building Type	Size S.F.
Industrial/Warehouse	100,000
	500,000
	1,000,000
Offices/Retail	50,000
	250,000
	500,000
Institutional/Hospitals/Laboratory	200,000
	500,000
	750,000
	1,000,000

Project Value	Construction Duration
$8,000,000	14 months
$32,000,000	19 months
$75,000,000	21 months
$7,000,000	15 months
$28,000,000	23 months
$58,000,000	34 months
$45,000,000	31 months
$110,000,000	52 months
$160,000,000	55 months
$210,000,000	60 months

General Requirements — R0129 Payment Procedures

R012909-80 Sales Tax by State

State sales tax on materials is tabulated below (5 states have no sales tax). Many states allow local jurisdictions, such as a county or city, to levy additional sales tax. Some projects may be sales tax exempt, particularly those constructed with public funds.

State	Tax (%)	State	Tax (%)	State	Tax (%)	State	Tax (%)
Alabama	4	Illinois	6.25	Montana	0	Rhode Island	7
Alaska	0	Indiana	6	Nebraska	5.5	South Carolina	6
Arizona	5.6	Iowa	5	Nevada	6.5	South Dakota	4
Arkansas	6	Kansas	5.3	New Hampshire	0	Tennessee	7
California	7.25	Kentucky	6	New Jersey	7	Texas	6.25
Colorado	2.9	Louisiana	4	New Mexico	5	Utah	4.65
Connecticut	6	Maine	5	New York	4	Vermont	6
Delaware	0	Maryland	6	North Carolina	4.25	Virginia	5
District of Columbia	5.75	Massachusetts	5	North Dakota	5	Washington	6.5
Florida	6	Michigan	6	Ohio	5.5	West Virginia	6
Georgia	4	Minnesota	6.5	Oklahoma	4.5	Wisconsin	5
Hawaii	4	Mississippi	7	Oregon	0	Wyoming	4
Idaho	6	Missouri	4.225	Pennsylvania	6	Average	4.91 %

Sales Tax by Province (Canada)

GST - a value-added tax, which the government imposes on most goods and services provided in or imported into Canada. PST - a retail sales tax, which five of the provinces impose on the price of most goods and some services. QST - a value-added tax, similar to the federal GST, which Quebec imposes. HST - Three provinces have combined their retail sales tax with the federal GST into one harmonized tax.

Province	PST (%)	QST (%)	GST(%)	HST(%)
Alberta	0	0	5	0
British Columbia	7	0	5	0
Manitoba	7	0	5	0
New Brunswick	0	0	0	13
Newfoundland	0	0	0	13
Northwest Territories	0	0	5	0
Nova Scotia	0	0	0	13
Ontario	8	0	5	0
Prince Edward Island	10	0	5	0
Quebec	0	7.5	5	0
Saskatchewan	5	0	5	0
Yukon	0	0	5	0

General Requirements — R0129 Payment Procedures

R012909-85 Unemployment Taxes and Social Security Taxes

State Unemployment Tax rates vary not only from state to state, but also with the experience rating of the contractor. The Federal Unemployment Tax rate is 6.2% of the first $7,000 of wages. This is reduced by a credit of up to 5.4% for timely payment to the state. The minimum Federal Unemployment Tax is 0.8% after all credits.

Social Security (FICA) for 2009 is estimated at time of publication to be 7.65% of wages up to $102,000.

R012909-90 Overtime

One way to improve the completion date of a project or eliminate negative float from a schedule is to compress activity duration times. This can be achieved by increasing the crew size or working overtime with the proposed crew.

To determine the costs of working overtime to compress activity duration times, consider the following examples. Below is an overtime efficiency and cost chart based on a five, six, or seven day week with an eight through twelve hour day. Payroll percentage increases for time and one half and double time are shown for the various working days.

Days per Week	Hours per Day	Production Efficiency					Payroll Cost Factors	
		1st Week	2nd Week	3rd Week	4th Week	Average 4 Weeks	@ 1-1/2 Times	@ 2 Times
5	8	100%	100%	100%	100%	100%	100%	100%
	9	100	100	95	90	96.25	105.6	111.1
	10	100	95	90	85	91.25	110.0	120.0
	11	95	90	75	65	81.25	113.6	127.3
	12	90	85	70	60	76.25	116.7	133.3
6	8	100	100	95	90	96.25	108.3	116.7
	9	100	95	90	85	92.50	113.0	125.9
	10	95	90	85	80	87.50	116.7	133.3
	11	95	85	70	65	78.75	119.7	139.4
	12	90	80	65	60	73.75	122.2	144.4
7	8	100	95	85	75	88.75	114.3	128.6
	9	95	90	80	70	83.75	118.3	136.5
	10	90	85	75	65	78.75	121.4	142.9
	11	85	80	65	60	72.50	124.0	148.1
	12	85	75	60	55	68.75	126.2	152.4

General Requirements — R0131 Project Management & Coordination

R013113-40 Builder's Risk Insurance

Builder's Risk Insurance is insurance on a building during construction. Premiums are paid by the owner or the contractor. Blasting, collapse, and underground insurance would raise total insurance costs above those listed. Floater policy for materials delivered to the job runs $.75 to $1.25 per $100 value. Contractor equipment insurance runs $.50 to $1.50 per $100 value. Insurance for miscellaneous tools to $1,500 value runs from $3.00 to $7.50 per $100 value.

Tabulated below are New England Builder's Risk insurance rates in dollars per $100 value for $1,000 deductible. For $25,000 deductible, rates can be reduced 13% to 34%. On contracts over $1,000,000, rates may be lower than those tabulated. Policies are written annually for the total completed value in place. For "all risk" insurance (excluding flood, earthquake, and certain other perils) add $.025 to total rates below.

Coverage	Frame Construction (Class 1)				Brick Construction (Class 4)				Fire Resistive (Class 6)			
	Range			Average	Range			Average	Range			Average
Fire Insurance	$.350	to	$ .850	$.600	$.158	to	$.189	$.174	$.052	to	$.080	$.070
Extended Coverage	.115	to	.200	.158	.080	to	.105	.101	.081	to	.105	.100
Vandalism	.012	to	.016	.014	.008	to	.011	.011	.008	to	.011	.010
Total Annual Rate	$.477	to	$1.066	$.772	$.246	to	$.305	$.286	$.141	to	$.196	$.180

R013113-50 General Contractor's Overhead

There are two distinct types of overhead on a construction project: Project Overhead and Main Office Overhead. Project Overhead includes those costs at a construction site not directly associated with the installation of construction materials. Examples of Project Overhead costs include the following:

1. Superintendent
2. Construction office and storage trailers
3. Temporary sanitary facilities
4. Temporary utilities
5. Security fencing
6. Photographs
7. Clean up
8. Performance and payment bonds

The above Project Overhead items are also referred to as General Requirements and therefore are estimated in Division 1. Division 1 is the first division listed in the CSI MasterFormat but it is usually the last division estimated. The sum of the costs in Divisions 1 through 49 is referred to as the sum of the direct costs.

All construction projects also include indirect costs. The primary components of indirect costs are the contractor's Main Office Overhead and Profit. The amount of the Main Office Overhead expense varies depending on the the following:

1. Owner's compensation
2. Project managers and estimator's wages
3. Clerical support wages
4. Office rent and utilities
5. Corporate legal and accounting costs
6. Advertising
7. Automobile expenses
8. Association dues
9. Travel and entertainment expenses

These costs are usually calculated as a percentage of annual sales volume. This percentage can range from 35% for a small contractor doing less than $500.000 to 5% for a large contractor with sales in excess of $100 million.

R013113-60 Workers' Compensation Insurance Rates by Trade

The table below tabulates the national averages for Workers' Compensation insurance rates by trade and type of building. The average "Insurance Rate" is multiplied by the "% of Building Cost" for each trade. This produces the "Workers' Compensation Cost" by % of total labor cost, to be added for each trade by building type to determine the weighted average Workers' Compensation rate for the building types analyzed.

Trade	Insurance Rate (% Labor Cost) Range	Insurance Rate (% Labor Cost) Average	% of Building Cost Office Bldgs.	% of Building Cost Schools & Apts.	% of Building Cost Mfg.	Workers' Compensation Office Bldgs.	Workers' Compensation Schools & Apts.	Workers' Compensation Mfg.
Excavation, Grading, etc.	4.2 % to 17.5%	10.0%	4.8%	4.9%	4.5%	0.48%	0.49%	0.45%
Piles & Foundations	5.9 to 40.3	20.1	7.1	5.2	8.7	1.43	1.05	1.75
Concrete	5.1 to 26.8	14.6	5.0	14.8	3.7	0.73	2.16	0.54
Masonry	4.8 to 43.3	14.4	6.9	7.5	1.9	0.99	1.08	0.27
Structural Steel	5.9 to 104.1	37.9	10.7	3.9	17.6	4.06	1.48	6.67
Miscellaneous & Ornamental Metals	3.4 to 22.8	10.7	2.8	4.0	3.6	0.30	0.43	0.39
Carpentry & Millwork	5.9 to 53.2	17.8	3.7	4.0	0.5	0.66	0.71	0.09
Metal or Composition Siding	5.9 to 38	16.6	2.3	0.3	4.3	0.38	0.05	0.71
Roofing	5.9 to 77.1	31.2	2.3	2.6	3.1	0.72	0.81	0.97
Doors & Hardware	4.9 to 32	11.6	0.9	1.4	0.4	0.10	0.16	0.05
Sash & Glazing	5.9 to 32.4	13.9	3.5	4.0	1.0	0.49	0.56	0.14
Lath & Plaster	3.3 to 43.9	13.6	3.3	6.9	0.8	0.45	0.94	0.11
Tile, Marble & Floors	3.1 to 17.9	9.1	2.6	3.0	0.5	0.24	0.27	0.05
Acoustical Ceilings	2.6 to 45.6	10.6	2.4	0.2	0.3	0.25	0.02	0.03
Painting	4.7 to 29.6	12.5	1.5	1.6	1.6	0.19	0.20	0.20
Interior Partitions	5.9 to 53.2	17.8	3.9	4.3	4.4	0.69	0.77	0.78
Miscellaneous Items	2.1 to 168.2	16.0	5.2	3.7	9.7	0.83	0.59	1.55
Elevators	2.8 to 11.8	6.6	2.1	1.1	2.2	0.14	0.07	0.15
Sprinklers	2.5 to 14.3	7.8	0.5	—	2.0	0.04	—	0.16
Plumbing	2.9 to 12.4	7.8	4.9	7.2	5.2	0.38	0.56	0.41
Heat., Vent., Air Conditioning	4.3 to 23	11.6	13.5	11.0	12.9	1.57	1.28	1.50
Electrical	2.8 to 11.5	6.5	10.1	8.4	11.1	0.66	0.55	0.72
Total	2.1 % to 168.2%	—	100.0%	100.0%	100.0%	15.78%	14.23%	17.69%
Overall Weighted Average 15.90%								

Workers' Compensation Insurance Rates by States

The table below lists the weighted average Workers' Compensation base rate for each state with a factor comparing this with the national average of 15.5%.

State	Weighted Average	Factor	State	Weighted Average	Factor	State	Weighted Average	Factor
Alabama	24.2%	156	Kentucky	18.4%	119	North Dakota	13.7%	88
Alaska	21.7	140	Louisiana	28.3	183	Ohio	14.8	95
Arizona	9.5	61	Maine	15.4	99	Oklahoma	14.6	94
Arkansas	13.2	85	Maryland	16.7	108	Oregon	13.5	87
California	19.6	126	Massachusetts	12.5	81	Pennsylvania	13.8	89
Colorado	13.2	85	Michigan	17.3	112	Rhode Island	21.2	137
Connecticut	21.0	135	Minnesota	25.7	166	South Carolina	18.3	118
Delaware	15.1	97	Mississippi	19.1	123	South Dakota	19.1	123
District of Columbia	13.7	88	Missouri	17.2	111	Tennessee	15.1	97
Florida	13.8	89	Montana	16.6	107	Texas	12.2	79
Georgia	26.2	169	Nebraska	23.0	148	Utah	12.3	79
Hawaii	17.0	110	Nevada	11.6	75	Vermont	18.0	116
Idaho	10.8	70	New Hampshire	20.0	129	Virginia	11.7	75
Illinois	21.6	139	New Jersey	13.4	86	Washington	9.8	63
Indiana	5.9	38	New Mexico	18.5	119	West Virginia	9.2	59
Iowa	11.5	74	New York	12.6	81	Wisconsin	15.0	97
Kansas	8.9	57	North Carolina	18.9	122	Wyoming	6.5	42
Weighted Average for U.S. is 15.9% of payroll = 100%								

Rates in the following table are the base or manual costs per $100 of payroll for Workers' Compensation in each state. Rates are usually applied to straight time wages only and not to premium time wages and bonuses.

The weighted average skilled worker rate for 35 trades is 15.5%. For bidding purposes, apply the full value of Workers' Compensation directly to total labor costs, or if labor is 38%, materials 42% and overhead and profit 20% of total cost, carry 38/80 x 15.5% =7.4% of cost (before overhead and profit) into overhead. Rates vary not only from state to state but also with the experience rating of the contractor.

Rates are the most current available at the time of publication.

R013113-060 Workers' Compensation Insurance Rates by Trade and State (cont.)

State	Carpentry — 3 stories or less 5651	Carpentry — interior cab. work 5437	Carpentry — general 5403	Concrete Work — NOC 5213	Concrete Work — flat (flr., sdwk.) 5221	Electrical Wiring — inside 5190	Excavation — earth NOC 6217	Excavation — rock 6217	Glaziers 5462	Insulation Work 5479	Lathing 5443	Masonry 5022	Painting & Decorating 5474	Pile Driving 6003	Plastering 5480	Plumbing 5183	Roofing 5551	Sheet Metal Work (HVAC) 5538	Steel Erection — door & sash 5102	Steel Erection — inter., ornam. 5102	Steel Erection — structure 5040	Steel Erection — NOC 5057	Tile Work — (interior ceramic) 5348	Waterproofing 9014	Wrecking 5701
AL	37.98	22.82	32.46	16.20	11.27	10.68	11.16	11.16	21.26	16.28	11.93	27.00	25.81	32.65	18.85	12.27	58.14	22.96	11.01	11.01	55.68	25.37	13.31	6.56	55.68
AK	14.73	12.27	17.44	11.07	9.71	8.72	12.14	12.14	32.36	27.78	9.74	43.27	21.95	32.40	43.85	9.97	35.12	8.61	9.39	9.39	43.01	33.00	6.76	6.61	43.01
AZ	11.60	6.45	15.40	8.56	4.45	4.59	4.90	4.90	6.93	12.48	6.61	6.51	6.33	11.45	5.78	5.24	13.38	7.38	10.66	10.66	21.10	15.34	3.17	2.59	21.10
AR	14.82	7.45	16.26	12.17	6.48	5.02	7.76	7.76	9.82	16.49	5.93	9.93	11.50	16.48	16.10	5.15	23.09	13.09	6.77	6.77	32.38	25.50	6.08	3.74	32.38
CA	31.95	31.95	31.95	11.89	11.89	7.98	13.79	13.79	15.74	12.50	11.24	16.21	17.08	15.27	20.53	12.03	44.43	15.60	12.46	12.46	17.17	20.88	8.54	17.08	20.88
CO	13.57	8.92	12.33	11.57	7.87	5.28	10.34	10.34	9.75	16.84	6.02	14.01	9.02	14.31	9.58	7.43	26.09	11.61	9.31	9.31	32.62	17.58	7.06	5.03	17.58
CT	15.67	14.66	28.76	24.79	10.09	8.01	12.91	12.91	22.98	15.94	21.94	22.24	17.61	22.88	14.48	11.16	41.95	15.61	18.57	18.57	42.72	23.45	11.58	5.41	42.72
DE	15.17	15.17	11.46	11.51	9.49	5.89	9.32	9.32	13.11	11.46	13.11	12.77	26.77	18.75	13.11	7.87	26.71	9.78	12.28	12.28	26.77	12.28	9.29	12.77	26.77
DC	11.01	12.78	10.19	9.66	9.14	6.01	12.54	12.54	17.02	8.18	9.40	11.97	7.51	12.11	12.48	10.38	19.21	8.93	10.61	10.61	37.21	16.49	17.86	4.00	37.21
FL	13.05	10.47	14.39	15.31	6.97	6.64	8.30	8.30	9.91	9.24	5.98	11.32	9.73	38.24	18.70	6.75	22.19	12.68	8.85	8.85	27.89	14.56	6.55	5.17	27.89
GA	35.68	20.54	25.67	16.85	13.26	10.82	17.38	17.38	18.77	23.64	15.09	21.06	27.77	27.36	22.83	11.4	54.71	20.76	19.93	19.93	64.78	42.21	12.42	8.50	64.78
HI	17.38	11.62	28.20	13.74	12.27	7.10	7.73	7.73	20.55	21.19	10.94	17.87	11.33	20.15	15.61	6.06	33.24	8.02	11.11	11.11	31.71	21.70	9.78	11.54	31.71
ID	9.16	6.04	10.35	11.25	6.13	3.70	5.72	5.72	8.70	7.67	8.89	7.59	8.45	12.13	13.37	5.36	30.27	8.89	5.72	5.72	29.16	10.28	9.79	4.75	29.16
IL	24.93	14.46	20.78	26.59	12.10	9.62	10.06	10.06	19.18	17.48	15.59	19.27	11.62	34.85	15.88	11.07	31.06	14.96	16.05	16.05	72.39	25.69	14.02	4.84	72.39
IN	9.65	4.92	7.29	5.10	3.22	2.84	4.87	4.87	6.28	7.44	2.62	4.84	4.72	6.44	3.72	2.91	10.88	4.45	3.43	3.43	12.12	7.04	3.09	2.47	12.12
IA	10.43	10.02	10.34	12.12	7.00	4.96	6.27	6.27	9.42	7.06	5.58	9.61	6.65	9.78	7.33	6.46	22.58	7.52	6.15	6.15	30.21	35.21	7.42	3.87	35.21
KS	10.05	7.85	9.89	7.19	6.42	3.37	4.41	4.41	9.22	7.45	4.68	6.91	7.57	11.31	6.25	4.85	15.06	7.14	4.82	4.82	26.19	13.21	6.35	3.46	13.21
KY	18.72	14.07	25.30	13.39	8.00	5.43	9.15	9.15	17.37	17.59	11.24	9.12	13.18	27.72	14.41	7.06	38.25	19.58	12.32	12.32	54.03	22.56	13.67	5.74	56.16
LA	22.02	24.93	53.17	26.43	15.61	9.88	17.49	17.49	20.14	21.43	24.51	28.33	29.58	31.25	22.37	8.64	77.12	21.20	18.98	18.98	51.76	24.21	13.77	13.78	66.41
ME	13.79	10.40	28.55	20.94	8.74	6.31	9.05	9.05	18.14	11.82	7.90	14.76	12.73	16.70	11.79	10.56	22.87	11.61	9.73	9.73	31.67	23.11	7.00	7.12	31.67
MD	15.05	11.06	14.11	15.67	6.67	9.24	9.20	9.20	20.03	15.02	8.83	11.43	7.48	20.09	15.02	6.11	34.74	13.54	9.51	9.51	59.03	28.60	7.60	5.36	28.60
MA	6.80	5.60	11.46	19.51	6.57	3.20	4.19	4.19	8.96	10.06	6.35	10.81	4.79	14.68	5.08	3.98	32.80	5.15	7.61	7.61	45.71	36.69	6.21	2.09	25.88
MI	18.40	11.39	20.64	19.45	8.95	4.72	10.20	10.20	12.41	14.75	13.59	15.54	13.64	40.27	15.18	6.84	30.67	10.00	10.16	10.16	40.27	22.21	11.11	4.87	40.27
MN	19.53	19.80	41.48	14.65	15.73	7.53	14.95	14.95	15.68	11.90	22.05	19.18	16.95	24.63	22.05	10.88	70.18	14.23	10.58	10.58	104.13	33.30	14.55	6.68	36.03
MS	21.07	11.11	22.29	15.49	9.26	6.28	11.91	11.91	13.07	14.85	7.91	14.09	13.14	35.34	31.72	9.80	44.11	16.91	14.23	14.23	36.11	25.03	9.84	4.18	36.11
MO	29.99	11.16	13.98	18.91	10.39	7.64	9.17	9.17	11.13	15.17	7.93	15.01	12.53	19.12	17.02	10.05	29.98	13.19	11.31	11.31	34.56	35.84	11.18	6.51	34.56
MT	17.12	11.15	22.57	12.37	12.79	6.39	15.54	15.54	10.67	32.03	11.06	13.75	9.61	25.18	10.02	8.75	37.61	11.53	8.70	8.70	27.45	15.73	7.53	7.46	15.73
NE	24.65	17.47	21.80	25.07	15.13	10.38	16.90	16.90	23.70	31.50	12.20	22.22	20.07	22.50	21.67	12.35	35.22	19.42	16.05	16.05	45.40	35.27	11.10	6.38	47.05
NV	15.25	7.14	11.58	10.76	8.23	7.17	9.77	9.77	10.19	8.12	5.27	7.50	7.97	13.33	7.30	5.98	14.08	18.10	8.67	8.67	25.10	21.54	5.39	5.55	21.54
NH	25.92	13.36	18.90	26.79	14.17	6.94	14.52	14.52	13.07	21.36	9.63	24.34	7.77	18.99	11.71	10.13	49.23	13.59	15.27	15.27	52.24	18.83	11.37	6.09	52.24
NJ	14.34	8.94	14.34	16.01	9.99	4.76	8.35	8.35	8.55	12.65	11.64	13.09	11.92	16.03	11.64	6.17	37.77	6.49	9.46	9.46	23.72	13.59	6.99	5.72	22.31
NM	18.91	7.15	20.11	17.36	10.05	7.51	11.20	11.20	21.26	13.72	8.39	15.34	11.64	21.86	13.44	8.13	36.37	13.47	13.69	13.69	55.78	38.81	6.59	6.29	55.78
NY	12.72	6.45	12.67	16.02	11.64	6.03	8.29	8.29	10.97	7.98	12.10	16.02	9.89	12.77	9.03	6.88	27.98	10.46	8.84	8.84	23.45	12.72	6.66	6.04	8.73
NC	18.76	14.56	15.36	18.15	6.88	11.50	11.70	11.70	15.21	16.00	15.28	12.51	12.90	17.50	15.68	9.16	29.70	15.90	10.44	10.44	79.66	26.21	9.28	5.55	79.66
ND	11.17	11.17	11.17	6.42	6.42	3.63	6.64	6.64	11.17	11.17	7.51	7.71	6.61	22.78	7.51	5.22	23.53	5.22	22.78	22.78	22.78	22.78	11.17	23.53	10.98
OH	11.04	8.52	11.68	12.33	9.99	6.14	8.70	8.70	7.86	16.94	45.64	12.71	12.37	19.06	3.28	6.99	31.63	9.71	7.59	7.59	29.23	16.59	9.60	6.42	16.59
OK	13.42	9.17	11.76	11.76	6.34	5.73	10.83	10.83	16.07	19.47	8.45	10.02	8.47	20.25	12.19	7.09	21.13	9.05	13.76	13.76	39.72	23.97	6.50	5.71	39.72
OR	15.43	8.10	15.69	14.22	8.97	5.14	9.33	9.33	15.22	11.80	7.60	14.07	10.63	16.17	12.73	5.50	26.30	9.41	7.93	7.93	26.94	18.78	11.04	4.54	26.94
PA	13.18	13.18	11.38	14.21	10.46	5.95	8.27	8.27	11.47	11.38	11.47	12.22	13.31	15.78	11.47	7.33	27.42	7.84	14.30	14.30	21.31	14.30	8.02	12.22	21.37
RI	19.53	11.65	18.07	18.23	16.24	4.43	10.38	10.38	12.85	22.78	11.97	25.11	24.13	37.66	17.25	8.52	33.92	10.29	14.07	14.07	59.49	37.50	14.36	7.70	59.49
SC	22.54	15.41	20.91	14.80	7.92	10.92	13.16	13.16	16.12	14.69	11.32	12.88	16.52	19.48	16.02	11.47	45.24	12.95	12.99	12.99	32.57	26.39	9.62	6.68	47.05
SD	27.77	8.13	29.58	25.56	9.55	5.52	11.01	11.01	12.16	12.05	7.70	11.31	9.53	21.32	12.44	11.75	20.73	12.48	8.98	8.98	80.86	40.98	8.74	4.98	80.86
TN	15.68	13.11	16.06	13.89	7.97	7.35	13.68	13.68	13.09	11.80	7.30	15.60	12.63	24.89	16.02	7.66	22.25	12.77	7.17	7.17	29.18	23.47	10.15	4.65	29.18
TX	11.25	8.89	11.25	9.00	7.26	6.15	8.54	8.54	9.65	12.87	6.36	10.91	8.98	22.43	8.98	6.75	19.59	12.84	9.09	9.09	29.68	13.37	7.07	6.69	9.96
UT	15.74	8.88	12.97	9.81	8.47	4.57	11.59	11.59	14.70	12.49	6.35	11.25	8.37	12.47	7.26	6.18	24.99	12.64	7.16	7.16	25.67	14.67	6.80	4.27	18.37
VT	15.70	10.75	17.37	17.27	7.21	6.98	12.59	12.59	19.75	17.78	8.33	15.82	11.62	16.74	10.78	10.44	36.43	11.59	12.51	12.51	44.95	42.80	11.24	7.28	44.95
VA	11.05	8.02	9.16	11.05	5.31	5.52	7.57	7.57	10.25	9.90	14.34	9.62	8.99	10.75	10.22	5.83	22.52	7.65	8.37	8.37	33.52	19.66	4.74	3.04	33.52
WA	8.48	8.48	8.48	7.32	7.32	2.81	6.69	6.69	12.27	9.77	8.43	9.09	10.93	16.12	9.95	4.17	16.36	4.27	7.60	7.60	7.60	7.60	8.45	16.36	7.60
WV	10.65	7.30	10.23	8.24	4.41	4.70	6.22	6.22	7.89	7.25	6.11	7.98	7.90	11.90	7.93	4.90	18.05	5.74	5.58	5.58	24.17	13.20	5.65	2.66	17.34
WI	9.99	11.82	14.45	11.15	9.69	6.02	8.19	8.19	10.44	13.58	6.25	15.31	13.03	16.47	9.96	6.24	37.36	7.21	9.36	9.36	28.23	44.93	13.80	4.86	28.23
WY	5.87	5.87	5.87	5.87	5.87	5.87	5.87	5.87	5.87	5.87	5.87	5.87	5.87	5.87	5.87	5.87	5.87	5.87	5.87	5.87	5.87	5.87	5.87	5.87	5.87
AVG.	16.63	11.62	17.80	14.58	9.14	6.46	10.01	10.01	13.89	14.44	10.63	14.37	12.49	20.09	13.60	7.84	31.18	11.57	10.74	10.74	37.94	23.15	9.13	6.69	34.13

R013113-060 Workers' Compensation (cont.) (Canada in Canadian dollars)

Province		Alberta	British Columbia	Manitoba	Ontario	New Brunswick	Newfndld. & Labrador	Northwest Territories	Nova Scotia	Prince Edward Island	Quebec	Saskat-chewan	Yukon
Carpentry—3 stories or less	Rate	7.32	4.74	4.36	4.35	3.79	10.08	3.84	7.82	5.96	15.05	6.23	8.89
	Code	42143	721028	40102	723	4226	4226	4-41	4226	401	80110	B1317	202
Carpentry—interior cab. work	Rate	2.18	4.90	4.36	4.35	4.40	4.69	3.84	4.52	3.71	15.05	3.51	8.89
	Code	42133	721021	40102	723	4279	4270	4-41	4274	402	80110	B11-27	202
CARPENTRY—general	Rate	7.32	4.74	4.36	4.35	3.79	4.69	3.84	7.82	5.96	15.05	6.23	8.89
	Code	42143	721028	40102	723	4226	4299	4-41	4226	401	80110	B1317	202
CONCRETE WORK—NOC	Rate	4.41	4.81	7.46	16.02	3.79	10.08	3.84	4.52	5.96	17.99	6.24	4.67
	Code	42104	721010	40110	748	4224	4224	4-41	4224	401	80100	B13-14	203
CONCRETE WORK—flat (flr. sidewalk)	Rate	4.41	4.81	7.46	16.02	3.79	10.08	3.84	4.52	5.96	17.99	6.24	4.67
	Code	42104	721010	40110	748	4224	4224	4-41	4224	401	80100	B13-14	203
ELECTRICAL Wiring—inside	Rate	2.13	1.77	2.51	3.79	2.17	3.01	3.52	2.31	3.71	6.12	3.51	4.67
	Code	42124	721019	40203	704	4261	4261	4-46	4261	402	80170	B11-05	206
EXCAVATION—earth NOC	Rate	2.51	3.56	3.76	4.55	2.62	4.37	3.64	3.34	3.82	7.13	3.93	4.67
	Code	40604	721031	40706	711	4214	4214	4-43	4214	404	80030	R11-06	207
EXCAVATION—rock	Rate	2.51	3.56	3.76	4.55	2.62	4.37	3.64	3.34	3.82	7.13	3.93	4.67
	Code	40604	721031	40706	711	4214	4214	4-43	4214	404	80030	R11-06	207
GLAZIERS	Rate	3.16	3.62	4.36	8.90	4.55	6.81	3.84	5.21	3.71	17.08	6.23	4.67
	Code	42121	715020	40109	751	4233	4233	4-41	4233	402	80150	B13-04	212
INSULATION WORK	Rate	2.57	8.42	4.36	8.90	4.55	6.81	3.84	5.21	5.96	15.05	5.98	8.89
	Code	42184	721029	40102	751	4234	4234	4-41	4234	401	80110	B12-07	202
LATHING	Rate	5.59	6.78	4.36	4.35	4.40	4.69	3.84	4.52	3.71	15.05	6.23	8.89
	Code	42135	721033	40102	723	4273	4279	4-41	4271	402	80110	B13-16	202
MASONRY	Rate	4.41	6.78	4.36	11.15	4.55	6.81	3.84	5.21	5.96	17.99	6.23	8.89
	Code	42102	721037	40102	741	4231	4231	4-41	4231	401	80100	B13-18	202
PAINTING & DECORATING	Rate	4.35	4.94	3.18	6.75	4.40	4.69	3.84	4.52	3.71	15.05	5.98	8.89
	Code	42111	721041	40105	719	4275	4275	4-41	4275	402	80110	B12-01	202
PILE DRIVING	Rate	4.41	5.18	3.76	6.34	3.79	9.78	3.64	4.64	5.96	7.13	6.23	8.89
	Code	42159	722004	40706	732	4221	4221	4-43	4221	401	80030	B13-10	202
PLASTERING	Rate	5.59	6.78	5.08	6.75	4.40	4.69	3.84	4.52	3.71	15.05	5.98	8.89
	Code	42135	721042	40108	719	4271	4271	4-41	4271	402	80110	B12-21	202
PLUMBING	Rate	2.13	3.11	3.05	4.02	2.32	3.37	3.52	2.43	3.71	6.93	3.51	4.67
	Code	42122	721043	40204	707	4241	4241	4-46	4241	402	80160	B11-01	214
ROOFING	Rate	8.12	8.68	7.02	12.98	4.55	10.08	3.84	10.57	5.96	23.01	6.23	8.89
	Code	42118	721036	40403	728	4236	4236	4-41	4236	401	80130	B13-20	202
SHEET METAL WORK (HVAC)	Rate	2.13	3.11	7.02	4.02	2.32	3.37	3.52	2.43	3.71	6.93	3.51	4.67
	Code	42117	721043	40402	707	4244	4244	4-46	4244	402	80160	B11-07	208
STEEL ERECTION—door & sash	Rate	2.57	14.07	11.32	16.02	3.79	10.08	3.84	5.21	5.96	25.32	6.23	8.89
	Code	42106	722005	40502	748	4227	4227	4-41	4227	401	80080	B13-22	202
STEEL ERECTION—inter., ornam.	Rate	2.57	14.07	11.32	16.02	3.79	10.08	3.84	5.21	5.96	25.32	6.23	8.89
	Code	42106	722005	40502	748	4227	4227	4-41	4227	401	80080	B13-22	202
STEEL ERECTION—structure	Rate	2.57	14.07	11.32	16.02	3.79	10.08	3.84	5.21	5.96	25.32	6.23	8.89
	Code	42106	722005	40502	748	4227	4227	4-41	4227	401	80080	B13-22	202
STEEL ERECTION—NOC	Rate	2.57	14.07	11.32	16.02	3.79	10.08	3.84	5.21	5.96	25.32	6.23	8.89
	Code	42106	722005	40502	748	4227	4227	4-41	4227	401	80080	B13-22	202
TILE WORK—inter. (ceramic)	Rate	3.63	5.62	1.97	6.75	4.40	10.08	3.84	4.52	3.71	15.05	6.23	8.89
	Code	42113	721054	40103	719	4276	4276	4-41	4276	402	80110	B13-01	202
WATERPROOFING	Rate	4.35	4.93	4.36	4.35	4.55	4.69	3.84	5.21	3.71	23.01	5.98	8.89
	Code	42139	721016	40102	723	4239	4299	4-41	4239	402	80130	B12-17	202
WRECKING	Rate	2.51	4.59	6.59	16.02	2.62	4.39	3.64	3.34	5.96	15.05	6.23	8.89
	Code	40604	721005	40106	748	4211	4211	4-43	4211	401	80110	B13-09	202

R013113-80 Performance Bond

This table shows the cost of a Performance Bond for a construction job scheduled to be completed in 12 months. Add 1% of the premium cost per month for jobs requiring more than 12 months to complete. The rates are "standard" rates offered to contractors that the bonding company considers financially sound and capable of doing the work. Preferred rates are offered by some bonding companies based upon financial strength of the contractor. Actual rates vary from contractor to contractor and from bonding company to bonding company. Contractors should prequalify through a bonding agency before submitting a bid on a contract that requires a bond.

Contract Amount	Building Construction Class B Projects	Highways & Bridges: Class A New Construction	Highways & Bridges: Class A-1 Highway Resurfacing
First $ 100,000 bid	$25.00 per M	$15.00 per M	$9.40 per M
Next 400,000 bid	$ 2,500 plus $15.00 per M	$ 1,500 plus $10.00 per M	$ 940 plus $7.20 per M
Next 2,000,000 bid	8,500 plus 10.00 per M	5,500 plus 7.00 per M	3,820 plus 5.00 per M
Next 2,500,000 bid	28,500 plus 7.50 per M	19,500 plus 5.50 per M	15,820 plus 4.50 per M
Next 2,500,000 bid	47,250 plus 7.00 per M	33,250 plus 5.00 per M	28,320 plus 4.50 per M
Over 7,500,000 bid	64,750 plus 6.00 per M	45,750 plus 4.50 per M	39,570 plus 4.00 per M

R015423-10 Steel Tubular Scaffolding

On new construction, tubular scaffolding is efficient up to 60′ high or five stories. Above this it is usually better to use a hung scaffolding if construction permits. Swing scaffolding operations may interfere with tenants. In this case, the tubular is more practical at all heights.

In repairing or cleaning the front of an existing building the cost of tubular scaffolding per S.F. of building front increases as the height increases above the first tier. The first tier cost is relatively high due to leveling and alignment.

The minimum efficient crew for erecting and dismantling is three workers. They can set up and remove 18 frame sections per day up to 5 stories high. For 6 to 12 stories high, a crew of four is most efficient. Use two or more on top and two on the bottom for handing up or hoisting. They can also set up and remove 18 frame sections per day. At 7' horizontal spacing, this will run about 800 S.F. per day of erecting and dismantling. Time for placing and removing planks must be added to the above. A crew of three can place and remove 72 planks per day up to 5 stories. For over 5 stories, a crew of four can place and remove 80 planks per day.

The table below shows the number of pieces required to erect tubular steel scaffolding for 1000 S.F. of building frontage. This area is made up of a scaffolding system that is 12 frames (11 bays) long by 2 frames high.

For jobs under twenty-five frames, add 50% to rental cost. Rental rates will be lower for jobs over three months duration. Large quantities for long periods can reduce rental rates by 20%.

Description of Component	Number of Pieces for 1000 S.F. of Building Front	Unit
5′ Wide Standard Frame, 6′-4″ High	24	Ea.
Leveling Jack & Plate	24	
Cross Brace	44	
Side Arm Bracket, 21″	12	
Guardrail Post	12	
Guardrail, 7′ section	22	
Stairway Section	2	
Stairway Starter Bar	1	
Stairway Inside Handrail	2	
Stairway Outside Handrail	2	
Walk-Thru Frame Guardrail	2	↓

Scaffolding is often used as falsework over 15′ high during construction of cast-in-place concrete beams and slabs. Two foot wide scaffolding is generally used for heavy beam construction. The span between frames depends upon the load to be carried with a maximum span of 5′.

Heavy duty shoring frames with a capacity of 10,000#/leg can be spaced up to 10′ O.C. depending upon form support design and loading.

Scaffolding used as horizontal shoring requires less than half the material required with conventional shoring.

On new construction, erection is done by carpenters.

Rolling towers supporting horizontal shores can reduce labor and speed the job. For maintenance work, catwalks with spans up to 70′ can be supported by the rolling towers.

R015423-20 Pump Staging

Pump staging is generally not available for rent. The table below shows the number of pieces required to erect pump staging for 2400 S.F. of building frontage. This area is made up of a pump jack system that is 3 poles (2 bays) wide by 2 poles high.

Item	Number of Pieces for 2400 S.F. of Building Front	Unit
Aluminum pole section, 24′ long	6	Ea.
Aluminum splice joint, 6′ long	3	
Aluminum foldable brace	3	
Aluminum pump jack	3	
Aluminum support for workbench/back safety rail	3	
Aluminum scaffold plank/workbench, 14″ wide x 24′ long	4	
Safety net, 22′ long	2	
Aluminum plank end safety rail	2	↓

The cost in place for this 2400 S.F. will depend on how many uses are realized during the life of the equipment.

R015433-10 Contractor Equipment

Rental Rates shown elsewhere in the book pertain to late model high quality machines in excellent working condition, rented from equipment dealers. Rental rates from contractors may be substantially lower than the rental rates from equipment dealers depending upon economic conditions; for older, less productive machines, reduce rates by a maximum of 15%. Any overtime must be added to the base rates. For shift work, rates are lower. Usual rule of thumb is 150% of one shift rate for two shifts; 200% for three shifts.

For periods of less than one week, operated equipment is usually more economical to rent than renting bare equipment and hiring an operator.

Costs to move equipment to a job site (mobilization) or from a job site (demobilization) are not included in rental rates, nor in any Equipment costs on any Unit Price line items or crew listings. These costs can be found elsewhere. If a piece of equipment is already at a job site, it is not appropriate to utilize mob/demob costs in an estimate again.

Rental rates vary throughout the country with larger cities generally having lower rates. Lease plans for new equipment are available for periods in excess of six months with a percentage of payments applying toward purchase.

Monthly rental rates vary from 2% to 5% of the cost of the equipment depending on the anticipated life of the equipment and its wearing parts. Weekly rates are about 1/3 the monthly rates and daily rental rates about 1/3 the weekly rate.

The hourly operating costs for each piece of equipment include costs to the user such as fuel, oil, lubrication, normal expendables for the equipment, and a percentage of mechanic's wages chargeable to maintenance. The hourly operating costs listed do not include the operator's wages.

The daily cost for equipment used in the standard crews is figured by dividing the weekly rate by five, then adding eight times the hourly operating cost to give the total daily equipment cost, not including the operator. This figure is in the right hand column of the Equipment listings under Equipment Cost/Day.

Pile Driving rates shown for pile hammer and extractor do not include leads, crane, boiler or compressor. Vibratory pile driving requires an added field specialist during set-up and pile driving operation for the electric model. The hydraulic model requires a field specialist for set-up only. Up to 125 reuses of sheet piling are possible using vibratory drivers. For normal conditions, crane capacity for hammer type and size are as follows.

Crane Capacity	Hammer Type and Size		
	Air or Steam	Diesel	Vibratory
25 ton	to 8,750 ft.-lb.		70 H.P.
40 ton	15,000 ft.-lb.	to 32,000 ft.-lb.	170 H.P.
60 ton	25,000 ft.-lb.		300 H.P.
100 ton		112,000 ft.-lb.	

Cranes should be specified for the job by size, building and site characteristics, availability, performance characteristics, and duration of time required.

Backhoes & Shovels rent for about the same as equivalent size cranes but maintenance and operating expense is higher. Crane operators rate must be adjusted for high boom heights. Average adjustments: for 150′ boom add 2% per hour; over 185′, add 4% per hour; over 210′, add 6% per hour; over 250′, add 8% per hour and over 295′, add 12% per hour.

Tower Cranes of the climbing or static type have jibs from 50′ to 200′ and capacities at maximum reach range from 4,000 to 14,000 pounds. Lifting capacities increase up to maximum load as the hook radius decreases.

Typical rental rates, based on purchase price are about 2% to 3% per month.

Erection and dismantling runs between 500 and 2000 labor hours. Climbing operation takes 10 labor hours per 20′ climb. Crane dead time is about 5 hours per 40′ climb. If crane is bolted to side of the building add cost of ties and extra mast sections. Climbing cranes have from 80′ to 180′ of mast while static cranes have 80′ to 800′ of mast.

Truck Cranes can be converted to tower cranes by using tower attachments. Mast heights over 400′ have been used.

A single 100′ high material **Hoist and Tower** can be erected and dismantled in about 400 labor hours; a double 100′ high hoist and tower in about 600 labor hours. Erection times for additional heights are 3 and 4 labor hours per vertical foot respectively up to 150′, and 4 to 5 labor hours per vertical foot over 150′ high. A 40′ high portable Buck hoist takes about 160 labor hours to erect and dismantle. Additional heights take 2 labor hours per vertical foot to 80′ and 3 labor hours per vertical foot for the next 100′. Most material hoists do not meet local code requirements for carrying personnel.

A 150′ high **Personnel Hoist** requires about 500 to 800 labor hours to erect and dismantle. Budget erection time at 5 labor hours per vertical foot for all trades. Local code requirements or labor scarcity requiring overtime can add up to 50% to any of the above erection costs.

Earthmoving Equipment: The selection of earthmoving equipment depends upon the type and quantity of material, moisture content, haul distance, haul road, time available, and equipment available. Short haul cut and fill operations may require dozers only, while another operation may require excavators, a fleet of trucks, and spreading and compaction equipment. Stockpiled material and granular material are easily excavated with front end loaders. Scrapers are most economically used with hauls between 300′ and 1-1/2 miles if adequate haul roads can be maintained. Shovels are often used for blasted rock and any material where a vertical face of 8′ or more can be excavated. Special conditions may dictate the use of draglines, clamshells, or backhoes. Spreading and compaction equipment must be matched to the soil characteristics, the compaction required, and the rate the fill is being supplied.

R015433-15 Heavy Lifting

Hydraulic Climbing Jacks

The use of hydraulic heavy lift systems is an alternative to conventional type crane equipment. The lifting, lowering, pushing, or pulling mechanism is a hydraulic climbing jack moving on a square steel jackrod from 1-5/8″ to 4″ square, or a steel cable. The jackrod or cable can be vertical or horizontal, stationary or movable, depending on the individual application. When the jackrod is stationary, the climbing jack will climb the rod and push or pull the load along with itself. When the climbing jack is stationary, the jackrod is movable with the load attached to the end and the climbing jack will lift or lower the jackrod with the attached load. The heavy lift system is normally operated by a single control lever located at the hydraulic pump.

The system is flexible in that one or more climbing jacks can be applied wherever a load support point is required, and the rate of lift synchronized.

Economic benefits have been demonstrated on projects such as: erection of ground assembled roofs and floors, complete bridge spans, girders and trusses, towers, chimney liners and steel vessels, storage tanks, and heavy machinery. Other uses are raising and lowering offshore work platforms, caissons, tunnel sections and pipelines.

R019313-10 Facility Maintenance - Frequency Table

The following table lists "average" frequency data for selected facility maintenance activities. The frequencies given are for a normal standard of maintenance under average conditions.

Activity	Average Frequency	Notes
Acoustical tile, cleaning		
Heavy smoking	2-3 years	
Non-smoking	10 years	
Carpet, cleaning		Frequency depends on the type of carpet and the occupancy in the building.
Heavy traffic	Weekly	
Light traffic	Every 6 weeks	
Carpet, vacuuming		
Heavy traffic	Daily	6-8 passes by machine in key areas.
Light traffic	Twice weekly	
Offices	Weekly	
Corridor and lobby, policing		Includes picking up loose trash, removing cigarette butts from in and around jardinieres and/or sand urns. Frequency depends on occupancy in the building.
Main	4 times daily	
Secondary	Daily	
Elevator, cleaning		The frequency of elevator cleaning is dependent upon the amount of traffic, size of the elevator car and the occupancy in the building.
Passenger	Daily	
Freight	Weekly	
Elevator lobby, cleaning	Daily	Includes sweeping, mopping and rinsing.
Escalator, cleaning	Daily	Frequency given is for a normal standard of cleaning under average conditions. Approximately 40% of escalator treads are exposed when escalator is stopped.
Floors, mopping		Frequency given is for a normal standard of cleaning under average conditions.
Main corridors	Daily	
Secondary	Weekly	
Floors, sweeping	Daily	Frequency given is for a normal standard of cleaning under average conditions.
Floors, waxing and polishing		Frequency given is for a normal standard of cleaning under average conditions.
Office area	Every 9 weeks	
Open area	Every 9 weeks	
Flower beds		Frequency depends on geographic location.
Fall clean-up	Every year	
Fertilize	2 times per year	
Mulch	Every year	
Police-up	30 times per year	
Weed		
With mulch	15 times per year	
No mulch	25 times per year	
Lawn, mowing	30 times per year	Frequency depends on geographic location.
Fertilize	2 times per year	
Sweep	3 times per year	
Weed control	2 times per year	
Edge-trim		
Walks	30 times per year	
Shrub	10 times per year	
Light fixtures		Frequency depends on occupancy in the building.
Dusting	Every month	
Washing	Every 3 months	
Shrub Areas		Frequency depends on geographic location.
Fertilize	Every year	
Mulch	2 times per year	
Police-up	30 times per year	
Prune	5 times per year	
Weed	10 times per year	
Stairway		Frequency is dependent upon weather conditions, type of building, type of occupancy, and amount of traffic.
Sweeping and dusting	Daily	
Mopping or scrubbing	Weekly	
Toilet, cleaning	Daily	Includes toilet cleaning, collect waste, cleaning wash basins, urinals, water closets, partitions, walls and floors.
Trash collection	Daily	
Trees		Frequency depends on geographic location.
Fertilize	Every year	
Prune	2 times per year	
Pest control		
Spray	3 times per year	
Systemic	Every year	
Urn and Jardiniere cleaning	Daily	Includes the removal of refuse and debris, cleaning and polishing.
Walks, sweeping	30 times per year	
Walls, periodic cleaning	Every 6 months	Damp wipe, spot removal.
Windows, washing	Every month	Frequency given is for a normal standard of cleaning under average conditions.

R019313-20 Facility Maintenance Labor-Hours

This section lists minimum and maximum cleaning times per unit.
For more information, see *Means Facilities Maintenance Standards* book.

	Unit	
	Minimum	**Maximum**
Blast Cleaning		
White-metal	100 S.F./Hr	
Near-white	175 S.F./Hr.	
Commercial	370 S.F./Hr.	
Brush-off	870 S.F./Hr.	
Paint Application		
Brushing	125 S.F./Hr.	
Rolling	125 S.F./Hr	
Spraying	500 S.F./Hr.	
Plaster Cleaning		
Wall dusting	2 sec./S.F.	3 sec./S.F.
Vacuuming	4 sec./S.F.	5 sec./S.F.
Spot washing	125 S.F./Hr.	175 S.F./Hr.
Thorough cleaning	275 S.F./Hr.	
Plaster Repair		
Gypsum and lime repair	5 S.Y./Hr.	10 S.Y./Hr.
Ceramic Tile Repair		
General	7 S.F./Hr.	10 S.F./Hr.
Adhesive tile setting	9S.F./Hr	12 S.F./Hr.
Pointing tile joints	10 S.F./Hr.	15 S.F./Hr.
Floor Cleaning		
Manual sweeping	10 min./1000 S.F.	25 min./1000 S.F.
Dust mopping	5 min./1000 S.F.	20 min./1000 S.F.
Buffing	15 min./1000 S.F.	40 min./1000 S.F.
Spray buffing	20 min./1000 S.F.	50 min./1000 S.F.
Damp mopping	15 min./1000 S.F.	30 min./1000 S.F.
Wet mopping	30 min./1000 S.F.	50 min./1000 S.F.
Scrubbing	50 min./1000 S.F.	140 min./1000 S.F.
Scrubbing using electric floor machine		
General	15 min./1000 S.F.	30 min./1000 S.F.
Stripping	100 min./1000 S.F.	200 min./1000 S.F.
Waxing and Buffing using power maching		
Rewaxing	15 min./1000 S.F.	30 min./1000 S.F.
Stripping and rewaxing (two coats)	100 min./1000 S.F.	300 min./1000 S.F.
Waxing and buffing (one coat)	30 min./1000 S.F.	70 min./1000 S.F.
Carpets		
Dry vacuuming	15 min./1000 S.F.	40 min./1000 S.F.
Wet vacuuming	30 min./1000 S.F.	50 min./1000 S.F.
Carpet mopping	20 min./1000 S.F.	40 min./1000 S.F.
Shampooing	175 min./1000 S.F.	250 min./1000 S.F.
Resilient Floor Repair		
Grinding	50 S.F./Hr.	80 S.F./Hr.
Floor Replacement		
Removal (by hand)		
Tiles	100 S.F./Hr.	130 S.F./Hr.
Sheet	120 S.F./Hr	160 S.F./Hr.
Hardwood	40 S.F./Hr.	60 S.F./Hr.
Replacement		
Ceramic	10 S.F./Hr.	20 S.F./Hr.
Resilient	40 S.F./Hr.	70 S.F./Hr.
Hardwood	25 S.F./Hr.	35 S.F./Hr.
Add for related items:		
Replace wood subfloor	80 S.F./Hr.	100 S.F./Hr.
Replace underlayment	75 S.F./Hr.	90 S.F./Hr.
Replace floor moulding	10 S.F./Hr.	30 S.F./Hr.

Unit		
	Minimum	**Maximum**
Seamless Floor Repair	5 S.F./Hr.	15 S.F./Hr.
Wood Floors		
Sanding	40 S.F./Hr.	60 S.F./Hr.
Sealing	200 S.F./Hr.	300 S.F./Hr.
Waxing*		
Wood Floor Repair		
Loose boards or tiles	50 S.F./Hr.	250S.F./Hr.
Wood strip floor replacement	30 S.F./Hr.	60 S.F./Hr.
Window Washing	300 S.F./Hr.	450 S.F./Hr.
Venetian Blinds, cleaning	15 min./set of blinds	

This section lists average cleaning times per each, or square foot	
	Average
Floor Operations	
Sweeping	
Halls and Corridors	15min./1000 S.F.
General Rooms	30 min./100 S.F.
Dust Mop (unobstructed)	10 min./1000 S.F.
Dust Mop (obstructed)	15 min./1000 S.F.
Damp Mop (unobstructed)	20 min./1000 S.F.
Damp Mop (obstructed)	40 min./1000 S.F.
Wet Mop and Rinse	100 min./1000 S.F.
Hand Scrubbing 12″ brush	300 min./1000 S.F.
Deck Scrubbing	100 min./1000 S.F.
Machine Scrubbing	
12″ diameter	50 min./1000 S.F.
14″ diameter	40 min./1000 S.F.
16″ diameter	35 min./1000 S.F.
18″ diameter	31 min./1000 S.F.
19″ diameter	28 min./1000 S.F.
21″ diameter	25 min./1000 S.F.
23″ diameter	23 min./1000 S.F.
24″ diameter	20 min./1000 S.F.
32″ diameter	18 min./1000 S.F.
36″ diameter	15 min./1000 S.F.
Automatic Scrub Machine (24″)	5 min./1000 S.F.
Vacuum (unobstructed)	20 min./1000 S.F.
Vacuum (obstructed)	30 min./1000 S.F.
Waxing	30 min./1000 S.F.
Machine Polish (19″ machine)	15 min./1000 S.F.
Rectangular Machine (48″ plate)	5 min./1000 S.F.
Buff with steel wool	20 min./1000 S.F.
Strip and Rewax	150 min./1000 S.F.
Dry Strip and Rewax	120 min./1000 S.F.
Spray Buffing (unobstructed)	30 min./1000 S.F.
Spray Buffing (obstructed)	45 min./1000 S.F.
Carpeting	
Vacuuming (unobstructed)	20 min./1000 S.F.
Vacuuming (obstructed)	30 min./1000 S.F.
Spot Vacuuming	15 min./1000 S.F.
Shampoo (dry foam)	60 min. 1000 S.F.
Pile Lift	30 min 1000 S.F.
Lockers	.20 min
Radiators	.30 min.
Tables (medium)	.50 min.
Telephones	.15 min.
Towel dispensers	.12 min.

*See waxing and buffing using electric power machine.

R019313-20 Facility Maintenance Labor-Hours (cont.)

	Average
Towel Disposal Cans	.40 min.
Typewriter and Stand	.50 min
Wash Basin (office)	.60 min
Waste Basin (office)	.50 min.
Window Sill	.20 min.
Venetian Blinds, std. size	3.50 min.
Washrooms	
Cleaning commode	4 min.
Door (spot wash both sides)	1 min.
Mirrors	1 min.
Sanitary napkin dispenser	.50 min.
Urinals	3 min.
Wash basin-soap dispenser	3 min.
General cleaning	120 min./1000 S.F.
Wall Washing	
Painted walls (manual)	240 min./1000 S.F.
Painted walls (machine)	150 min./1000 S.F.
Marble walls (manual)	90 min./1000 S.F.
Ceiling Washing	
Ceiling washing (manual)	300 min./1000 S.F.
Ceiling washing (machine)	180 min./1000 S.F.
Window Washing	
Single pane	125 min./1000 S.F.
Multi-pane	170 min./1000 S.F.
Frosted single pane	190 min./1000 S.F.
Opaque glass	50 min./1000 S.F.
Plate glass	35 min./1000 S.F.
Office partitions (glass)	110 min./1000 S.F.
Dusting Lamps and Lighting Fixtures	
Wall fluorescent fixtures	.15 min.
Desk fluorescent lamp	.30 min.
Table lamp and shade	.60 min.
Floor lamp and shade	.60 min.
Washing Fluorescent Light Fixtures	
Ceiling fixtures (egg crate) 4′ ea.	9 min.
Ceiling fixtures (egg crate) 8′ ea.	12 min.
Dusting	
Air conditioners	.30 min.
Ash trays (desk)	.25 min.
Book cases (3-tier set)	.30 min.
Chairs	.30 min.
Cigarette stands	.40 min.

	Average
Couch	.25 min.
Desks	.80 min.
Desk Trays	.15 min.
File cabinets (4 drawer)	.40 min.
Fabric Upholstery Cleaning	
Whisk or vacuum armless chair	.50 min.
Armchair	1 min.
Couch	2 min.
Shampooing armless chair	4 min.
Armchair	7 min.
Couch	20 min.
Stairway Cleaning	
Sweep and dust, 1 flight, 15 steps	6 min.
Damp mop, 1 flight, 15 steps	5 min.
Scrubbing (hand)	20 min.
Carpentry	
Repair door surface closer	1-1/2 Hrs.
Repair concealed door closer	4 Hrs.
Repair door damage at shop	4 Hrs.
Repair and replace screens	1 Hr.
Repair broken glass	1-1/2 Hrs.
Repair and replace ceiling tile	2 Hrs.
Repair small drywall damage	4 Hrs.
Repair and replace sash balance	4 Hrs.
Change lock on door	1-1/2 Hrs.
Painting	
Repaint 20′ x 15′ room, 1 coat	34 Hrs.
Repaint small bathroom, 1 coat	11-1/2 Hrs.
Repaint exterior window	2-1/2 Hrs.
Plumbing and Steamfitting	
Clear stopped water closet	3 Hrs.
Clear stopped basin	1-1/2 Hrs.
Replace leaking radiator valve	6 Hrs.
Clear external sewer stoppage	9 Hrs.
Install replacement valve, faucet, or trap	10 Hrs.
Electrical	
Replace fluorescent lamp ballast	2 Hrs.
Replace fractional HP motor	10 Hrs.
Replace blown fuse or reset	
Circuit breaker	1-1/2 Hrs.
Repair exterior light damage	4 Hrs.
Control circuit problems	5 Hrs.

R019313-30 A Review of Major Building Materials–Advantages and Disadvantages

Material	Advantages	Disadvantages
Aluminum	Lightweight High resistance to corrosion Low electrical resistance Good conductor	Softness Limited strength for structural uses Low stiffness High rate of thermal expansion Low rate of fire resistance Relatively high cost
Concrete	High compressive resistance Durable Resistance to moisture, rot, insects, fire, and wear Watertight (depending on water-cement ratio)	Workability Lack of tensile strength Lack of resistance to many types of chemical exposure, such as salt Hard to remove stains
Copper	High resistance to corrosion Good electrical conductivity Workable Forms its own surface protection Thermal contraction and expansion not high	Costly Strength varies with treatment and mechanical working
Epoxy	Can have varied properties depending on composition Liquid use very applicable Controllable Excellent strength properties Small creep during curing Hard Tough Resistant to abrasion Resistant to corrosion, salts, acids, petroleum products, solvents, and other chemicals Adhesion to surfaces good	Color Form
Glass	Considerably strong Non-corrosive	Brittle, subject to shattering under shock Transmits energy at a rapid rate Large sizes expensive
Lead	Good resistance to corrosion	Heavy High coefficient of thermal expansion Difficult to hold in place
Masonry	Available in small units Appearance (available in many textures, sizes, and colors) Good insulator	Stains hard to remove Porous (absorption rate high) Shrinkage of mortar Thermal and expansion cracking Others — see concrete
Paper	Readily available Low in cost Used in conjunction with other materials	Susceptible to water damage Susceptible to rot Relatively weak Highly combustible
Plastics — general	Applicable to many uses Relatively low in cost Workable into many shapes High strength Lightweight Non-corrosive Others — see specific plastic entry	Lack of resistance to fire Low stiffness High rate of thermal expansion Low thermal conductivity Some cases of chemical or physical instability with time Non-salvageable Others — see specific plastic entry

R019313-30 A Review of Major Building Materials—Advantages and Disadvantages (cont.)

Material	Advantages	Disadvantages
Plastics — specific		
Acrylics	Transparent Hard Weather resistant Shatter resistant	Easily scratched
Alkyds	Water resistant Touch Good adhesive properties	
Melamines	Hard Durable Abrasive resistant Chemical and heat resistant	
Polyamides (nylon)	Hard Tough Wear resistant	Costly
Polyesters	Weather and chemical resistant Stiff Hard	
Polyethylene	Flexible Tough Translucent Low cost	Easily scratched
Polystyrene	Hard Clear Water and chemical resistant Low cost	Brittle
Vinyls	Tough Wear resistant Stain resistant	
Plywood	Stronger than standard lumber Durable High resistance to impact and load Not as affected by moisture changes as standard wood Workability See Wood for others	Dimensional stability not as good as that of standard structural lumber Poor quality glues are possible Thermal expansion and contraction more of a problem
Steel	Strong Most resistant to aging Most reliable in quality, non-combustible, non-rotting, dimensionally stable with time and moisture change Resistant to staining	Costly Resultant loss of strength when exposed to intense heat, heat, rapid heat gain and loss Corrosive when exposed to moisture and air or other corrosive conditions
Tin	Extremely workable Very resistant to corrosion	Heavy Loss of strength when exposed to heat
Wood	Readily available Relatively low in cost Simple to work with Available in many shapes and forms Good insulating properties	Paintability changes with moisture content Combustible Susceptible to rot and insect infestation Soft and easily damaged (porous) Dimensional changes due to changes in temperature and moisture Strength changes with changing moisture content
Zinc	Fairly high resistance to corrosion Forms its own protective surface Workable	Brittle

R024119-10 Demolition Defined

Whole Building Demolition - Demolition of the whole building with no concern for any particular building element, component, or material type being demolished. This type of demolition is accomplished with large pieces of construction equipment that break up the structure, load it into trucks, and haul it to a disposal site, but disposal or dump fees are not included. Demolition of below-grade foundation elements, such as footings, foundation walls, grade beams, slabs on grade, etc., is not included. Certain mechanical equipment containing flammable liquids or ozone-depleting refrigerants, electric lighting elements, communication equipment components, and other building elements may contain hazardous waste, and must be removed, either selectively or carefully, as hazardous waste before the building can be demolished.

Foundation Demolition - Demolition of below-grade foundation footings, foundation walls, grade beams, and slabs on grade. This type of demolition is accomplished by hand or pneumatic hand tools, and does not include saw cutting, or handling, loading, hauling, or disposal of the debris.

Gutting - Removal of building interior finishes and electrical/mechanical systems down to the load-bearing and sub-floor elements of the rough building frame, with no concern for any particular building element, component, or material type being demolished. This type of demolition is accomplished by hand or pneumatic hand tools, and includes loading into trucks, but not hauling, disposal or dump fees, scaffolding, or shoring. Certain mechanical equipment containing flammable liquids or ozone-depleting refrigerants, electric lighting elements, communication equipment components, and other building elements may contain hazardous waste, and must be removed, either selectively or carefully, as hazardous waste, before the building is gutted.

Selective Demolition - Demolition of a selected building element, component, or finish, with some concern for surrounding or adjacent elements, components, or finishes (see the first Subdivision(s) at the beginning of appropriate Divisions). This type of demolition is accomplished by hand or pneumatic hand tools, and does not include handling, loading, storing, hauling, or disposal of the debris, scaffolding, or shoring. "Gutting" methods may be used in order to save time, but damage that is caused to surrounding or adjacent elements, components, or finishes may have to be repaired at a later time.

Careful Removal - Removal of a piece of service equipment, building element or component, or material type, with great concern for both the removed item and surrounding or adjacent elements, components or finishes. The purpose of careful removal may be to protect the removed item for later re-use, preserve a higher salvage value of the removed item, or replace an item while taking care to protect surrounding or adjacent elements, components, connections, or finishes from cosmetic and/or structural damage. An approximation of the time required to perform this type of removal is 1/3 to 1/2 the time it would take to install a new item of like kind (see Reference Number R220105-10). This type of removal is accomplished by hand or pneumatic hand tools, and does not include loading, hauling, or storing the removed item, scaffolding, shoring, or lifting equipment.

Cutout Demolition - Demolition of a small quantity of floor, wall, roof, or other assembly, with concern for the appearance and structural integrity of the surrounding materials. This type of demolition is accomplished by hand or pneumatic hand tools, and does not include saw cutting, handling, loading, hauling, or disposal of debris, scaffolding, or shoring.

Rubbish Handling - Work activities that involve handling, loading or hauling of debris. Generally, the cost of rubbish handling must be added to the cost of all types of demolition, with the exception of whole building demolition.

Minor Site Demolition - Demolition of site elements outside the footprint of a building. This type of demolition is accomplished by hand or pneumatic hand tools, or with larger pieces of construction equipment, and may include loading a removed item onto a truck (check the Crew for equipment used). It does not include saw cutting, hauling, or disposal of debris, and, sometimes, handling or loading.

R024119-20 Dumpsters

Dumpster rental costs on construction sites are presented in two ways.

The cost per week rental includes the delivery of the dumpster, its pulling or emptying once per week, and its final removal. The assumption is made that the dumpster contractor could choose to empty a dumpster by simply bringing in an empty unit and removing the full one. These costs also include the disposal of the materials in the dumpster.

The Alternate Pricing can be used when actual planned conditions are not approximated by the weekly numbers. For example, these lines can be used when a dumpster is needed for 4 weeks and will need to be emptied 2 or 3 times per week. Conversely the Alternate Pricing lines can be used when a dumpster will be rented for several weeks or months but needs to be emptied only a few times over this period.

Existing Conditions R0265 Underground Storage Tank Removal

R026510-20 Underground Storage Tank Removal

Underground Storage Tank Removal can be divided into two categories: Non-Leaking and Leaking. Prior to removing an underground storage tank, tests should be made, with the proper authorities present, to determine whether a tank has been leaking or the surrounding soil has been contaminated.

To safely remove Liquid Underground Storage Tanks:

1. Excavate to the top of the tank.
2. Disconnect all piping.
3. Open all tank vents and access ports.
4. Remove all liquids and/or sludge.
5. Purge the tank with an inert gas.
6. Provide access to the inside of the tank and clean out the interior using proper personal protective equipment (PPE).
7. Excavate soil surrounding the tank using proper PPE for on-site personnel.
8. Pull and properly dispose of the tank.
9. Clean up the site of all contaminated material.
10. Install new tanks or close the excavation.

Existing Conditions — R0282 Asbestos Remediation

R028213-20 Asbestos Removal Process

Asbestos removal is accomplished by a specialty contractor who understands the federal and state regulations regarding the handling and disposal of the material. The process of asbestos removal is divided into many individual steps. An accurate estimate can be calculated only after all the steps have been priced.

The steps are generally as follows:

1. Obtain an asbestos abatement plan from an industrial hygienist.
2. Monitor the air quality in and around the removal area and along the path of travel between the removal area and transport area. This establishes the background contamination.
3. Construct a two part decontamination chamber at entrance to removal area.
4. Install a HEPA filter to create a negative pressure in the removal area.
5. Install wall, floor, and ceiling protection as required by the plan, usually 2 layers of fireproof 6 mil polyethylene.
6. Industrial hygienist visually inspects work area to verify compliance with plan.
7. Provide temporary supports for conduit and piping affected by the removal process.
8. Proceed with asbestos removal and bagging process. Monitor air quality as described in Step #2. Discontinue operations when contaminate levels exceed applicable standards.
9. Document the legal disposal of materials in accordance with EPA standards.
10. Thoroughly clean removal area including all ledges, crevices, and surfaces.
11. Post abatement inspection by industrial hygienist to verify plan compliance.
12. Provide a certificate from a licensed industrial hygienist attesting that contaminate levels are within acceptable standards before returning area to regular use.

Existing Conditions — R0283 Lead Remediation

R028319-60 Lead Paint Remediation Methods

Lead paint remediation can be accomplished by the following methods:

1. Abrasive blast
2. Chemical stripping
3. Power tool cleaning with vacuum collection system
4. Encapsulation
5. Remove and replace
6. Enclosure

Each of these methods has strengths and weakness depending on the specific circumstances of the project. The following is an overview of each method.

1. **Abrasive blasting** is usually accomplished with sand or recyclable metallic blast. Before work can begin, the area must be contained to ensure the blast material with lead does not escape to the atmosphere. The use of vacuum blast greatly reduces the containment requirements. Lead abatement equipment that may be associated with this work includes a negative air machine. In addition, it is necessary to have an industrial hygienist monitor the project on a continual basis. When the work is complete, the spent blast sand with lead must be disposed of as a hazardous material. If metallic shot was used, the lead is separated from the shot and disposed of as hazardous material. Worker protection includes disposable clothing and respiratory protection.
2. **Chemical stripping** requires strong chemicals be applied to the surface to remove the lead paint. Before the work can begin, the area under/adjacent to the work area must be covered to catch the chemical and removed lead. After the chemical is applied to the painted surface it is usually covered with paper. The chemical is left in place for the specified period, then the paper with lead paint is pulled or scraped off. The process may require several chemical applications. The paper with chemicals and lead paint adhered to it, plus the containment and loose scrapings collected by a HEPA (High Efficiency Particulate Air Filter) vac, must be disposed of as a hazardous material. The chemical stripping process usually requires a neutralizing agent and several wash downs after the paint is removed. Worker protection includes a neoprene or other compatible protective clothing and respiratory protection with face shield. An industrial hygienist is required intermittently during the process.
3. **Power tool cleaning** is accomplished using shrouded needle blasting guns. The shrouding with different end configurations is held up against the surface to be cleaned. The area is blasted with hardened needles and the shroud captures the lead with a HEPA vac and deposits it in a holding tank. An industrial hygienist monitors the project, protective clothing and a respirator is required until air samples prove otherwise. When the work is complete the lead must be disposed of as a hazardous material.
4. **Encapsulation** is a method that leaves the well bonded lead paint in place after the peeling paint has been removed. Before the work can begin, the area under/adjacent to the work must be covered to catch the scrapings. The scraped surface is then washed with a detergent and rinsed. The prepared surface is covered with approximately 10 mils of paint. A reinforcing fabric can also be embedded in the paint covering. The scraped paint and containment must be disposed of as a hazardous material. Workers must wear protective clothing and respirators.
5. **Remove and replace** is an effective way to remove lead paint from windows, gypsum walls and concrete masonry surfaces. The painted materials are removed and new materials are installed. Workers should wear a respirator and tyvek suit. The demolished materials must be disposed of as hazardous waste if it fails the TCLP (Toxicity Characteristic Leachate Process) test.
6. **Enclosure** is the process that permanently seals lead painted materials in place. This process has many applications such as covering lead painted drywall with new drywall, covering exterior construction with tyvek paper then residing, or covering lead painted structural members with aluminum or plastic. The seams on all enclosing materials must be securely sealed. An industrial hygienist monitors the project, and protective clothing and a respirator is required until air samples prove otherwise.

All the processes require clearance monitoring and wipe testing as required by the hygienist.

R031113-60 Formwork Labor-Hours

Item	Hours Required				Total Hours	Multiple Use		
	Unit	Fabricate	Erect & Strip	Clean & Move	1 Use	2 Use	3 Use	4 Use
Beam and Girder, interior beams, 12″ wide	100 S.F.	6.4	8.3	1.3	16.0	13.3	12.4	12.0
Hung from steel beams		5.8	7.7	1.3	14.8	12.4	11.6	11.2
Beam sides only, 36″ high		5.8	7.2	1.3	14.3	11.9	11.1	10.7
Beam bottoms only, 24″ wide		6.6	13.0	1.3	20.9	18.1	17.2	16.7
Box out for openings		9.9	10.0	1.1	21.0	16.6	15.1	14.3
Buttress forms, to 8′ high		6.0	6.5	1.2	13.7	11.2	10.4	10.0
Centering, steel, 3/4″ rib lath			1.0		1.0			
3/8″ rib lath or slab form			0.9		0.9			
Chamfer strip or keyway	100 L.F.		1.5		1.5	1.5	1.5	1.5
Columns, fiber tube 8″ diameter			20.6		20.6			
12″			21.3		21.3			
16″			22.9		22.9			
20″			23.7		23.7			
24″			24.6		24.6			
30″			25.6		25.6			
Columns, round steel, 12″ diameter			22.0		22.0	22.0	22.0	22.0
16″			25.6		25.6	25.6	25.6	25.6
20″			30.5		30.5	30.5	30.5	30.5
24″			37.7		37.7	37.7	37.7	37.7
Columns, plywood 8″ x 8″	100 S.F.	7.0	11.0	1.2	19.2	16.2	15.2	14.7
12″ x 12″		6.0	10.5	1.2	17.7	15.2	14.4	14.0
16″ x 16″		5.9	10.0	1.2	17.1	14.7	13.8	13.4
24″ x 24″		5.8	9.8	1.2	16.8	14.4	13.6	13.2
Columns, steel framed plywood 8″ x 8″			10.0	1.0	11.0	11.0	11.0	11.0
12″ x 12″			9.3	1.0	10.3	10.3	10.3	10.3
16″ x 16″			8.5	1.0	9.5	9.5	9.5	9.5
24″ x 24″			7.8	1.0	8.8	8.8	8.8	8.8
Drop head forms, plywood		9.0	12.5	1.5	23.0	19.0	17.7	17.0
Coping forms		8.5	15.0	1.5	25.0	21.3	20.0	19.4
Culvert, box			14.5	4.3	18.8	18.8	18.8	18.8
Curb forms, 6″ to 12″ high, on grade		5.0	8.5	1.2	14.7	12.7	12.1	11.7
On elevated slabs		6.0	10.8	1.2	18.0	15.5	14.7	14.3
Edge forms to 6″ high, on grade	100 L.F.	2.0	3.5	0.6	6.1	5.6	5.4	5.3
7″ to 12″ high	100 S.F.	2.5	5.0	1.0	8.5	7.8	7.5	7.4
Equipment foundations		10.0	18.0	2.0	30.0	25.5	24.0	23.3
Flat slabs, including drops		3.5	6.0	1.2	10.7	9.5	9.0	8.8
Hung from steel		3.0	5.5	1.2	9.7	8.7	8.4	8.2
Closed deck for domes		3.0	5.8	1.2	10.0	9.0	8.7	8.5
Open deck for pans		2.2	5.3	1.0	8.5	7.9	7.7	7.6
Footings, continuous, 12″ high		3.5	3.5	1.5	8.5	7.3	6.8	6.6
Spread, 12″ high		4.7	4.2	1.6	10.5	8.7	8.0	7.7
Pile caps, square or rectangular		4.5	5.0	1.5	11.0	9.3	8.7	8.4
Grade beams, 24″ deep		2.5	5.3	1.2	9.0	8.3	8.0	7.9
Lintel or Sill forms		8.0	17.0	2.0	27.0	23.5	22.3	21.8
Spandrel beams, 12″ wide		9.0	11.2	1.3	21.5	17.5	16.2	15.5
Stairs			25.0	4.0	29.0	29.0	29.0	29.0
Trench forms in floor		4.5	14.0	1.5	20.0	18.3	17.7	17.4
Walls, Plywood, at grade, to 8′ high		5.0	6.5	1.5	13.0	11.0	9.7	9.5
8′ to 16′		7.5	8.0	1.5	17.0	13.8	12.7	12.1
16′ to 20′		9.0	10.0	1.5	20.5	16.5	15.2	14.5
Foundation walls, to 8′ high		4.5	6.5	1.0	12.0	10.3	9.7	9.4
8′ to 16′ high		5.5	7.5	1.0	14.0	11.8	11.0	10.6
Retaining wall to 12′ high, battered		6.0	8.5	1.5	16.0	13.5	12.7	12.3
Radial walls to 12′ high, smooth		8.0	9.5	2.0	19.5	16.0	14.8	14.3
2′ chords		7.0	8.0	1.5	16.5	13.5	12.5	12.0
Prefabricated modular, to 8′ high		—	4.3	1.0	5.3	5.3	5.3	5.3
Steel, to 8′ high		—	6.8	1.2	8.0	8.0	8.0	8.0
8′ to 16′ high		—	9.1	1.5	10.6	10.3	10.2	10.2
Steel framed plywood to 8′ high		—	6.8	1.2	8.0	7.5	7.3	7.2
8′ to 16′ high		—	9.3	1.2	10.5	9.5	9.2	9.0

Concrete — R0321 Reinforcing Steel

R032110-10 Reinforcing Steel Weights and Measures

Bar Designation No.**	Nominal Weight Lb./Ft.	U.S. Customary Units: Nominal Dimensions* Diameter in.	Cross Sectional Area, in.^{2}	Perimeter in.	SI Units: Nominal Weight kg/m	Nominal Dimensions* Diameter mm	Cross Sectional Area, cm^2	Perimeter mm
3	.376	.375	.11	1.178	.560	9.52	.71	29.9
4	.668	.500	.20	1.571	.994	12.70	1.29	39.9
5	1.043	.625	.31	1.963	1.552	15.88	2.00	49.9
6	1.502	.750	.44	2.356	2.235	19.05	2.84	59.8
7	2.044	.875	.60	2.749	3.042	22.22	3.87	69.8
8	2.670	1.000	.79	3.142	3.973	25.40	5.10	79.8
9	3.400	1.128	1.00	3.544	5.059	28.65	6.45	90.0
10	4.303	1.270	1.27	3.990	6.403	32.26	8.19	101.4
11	5.313	1.410	1.56	4.430	7.906	35.81	10.06	112.5
14	7.650	1.693	2.25	5.320	11.384	43.00	14.52	135.1
18	13.600	2.257	4.00	7.090	20.238	57.33	25.81	180.1

* The nominal dimensions of a deformed bar are equivalent to those of a plain round bar having the same weight per foot as the deformed bar.
** Bar numbers are based on the number of eighths of an inch included in the nominal diameter of the bars.

R032110-80 Shop-Fabricated Reinforcing Steel

The material prices for reinforcing, shown in the unit cost sections of the book, are for 50 tons or more of shop-fabricated reinforcing steel and include:

1. Mill base price of reinforcing steel
2. Mill grade/size/length extras
3. Mill delivery to the fabrication shop
4. Shop storage and handling
5. Shop drafting/detailing
6. Shop shearing and bending
7. Shop listing
8. Shop delivery to the job site

Both material and installation costs can be considerably higher for small jobs consisting primarily of smaller bars, while material costs may be slightly lower for larger jobs.

Concrete — R0330 Cast-In-Place Concrete

R033053-50 Industrial Chimneys

Foundation requirements in C.Y. of concrete for various sized chimneys.

Size Chimney	2 Ton Soil	3 Ton Soil	Size Chimney	2 Ton Soil	3 Ton Soil	Size Chimney	2 Ton Soil	3 Ton Soil
75′ x 3′-0″	13 C.Y.	11 C.Y.	160′ x 6′-6″	86 C.Y.	76 C.Y.	300′ x 10′-0″	325 C.Y.	245 C.Y.
85′ x 5′-6″	19	16	175′ x 7′-0″	108	95	350′ x 12′-0″	422	320
100′ x 5′-0″	24	20	200′ x 6′-0″	125	105	400′ x 14′-0″	520	400
125′ x 5′-6″	43	36	250′ x 8′-0″	230	175	500′ x 18′-0″	725	575

Concrete — R0331 Structural Concrete

R033105-70 Placing Ready-Mixed Concrete

For ground pours allow for 5% waste when figuring quantities.

Prices in the front of the book assume normal deliveries. If deliveries are made before 8 A.M. or after 5 P.M. or on Saturday afternoons add 30%. Negotiated discounts for large volumes are not included in prices in front of book.

For the lower floors without truck access, concrete may be wheeled in rubber-tired buggies, conveyer handled, crane handled or pumped. Pumping is economical if there is top steel. Conveyers are more efficient for thick slabs.

At higher floors the rubber-tired buggies may be hoisted by a hoisting tower and wheeled to location. Placement by a conveyer is limited to three floors and is best for high-volume pours. Pumped concrete is best when building has no crane access. Concrete may be pumped directly as high as thirty-six stories using special pumping techniques. Normal maximum height is about fifteen stories.

Best pumping aggregate is screened and graded bank gravel rather than crushed stone.

Pumping downward is more difficult than pumping upward. Horizontal distance from pump to pour may increase preparation time prior to pour. Placing by cranes, either mobile, climbing or tower types, continues as the most efficient method for high-rise concrete buildings.

R033105-80 Slab on Grade

General: Ground slabs are classified on the basis of use. Thickness is generally controlled by the heaviest concentrated load supported. If load area is greater than 80 sq. in., soil bearing may be important. The base granular fill must be a uniformly compacted material of limited capillarity, such as gravel or crushed rock. Concrete is placed on this surface of the vapor barrier on top of base.

Ground slabs are either single or two course floors. Single course are widely used. Two course floors have a subsequent wear resistant topping.

Reinforcement is provided to maintain tightly closed cracks.

Control joints limit crack locations and provide for differential horizontal movement only. Isolation joints allow both horizontal and vertical differential movement.

Use of Table: Determine appropriate type of slab (A, B, C, or D) by considering type of use or amount of abrasive wear of traffic type.

Determine thickness by maximum allowable wheel load or uniform load, opposite 1st column, thickness. Increase the controlling thickness if details require, and select either plain or reinforced slab thickness and type.

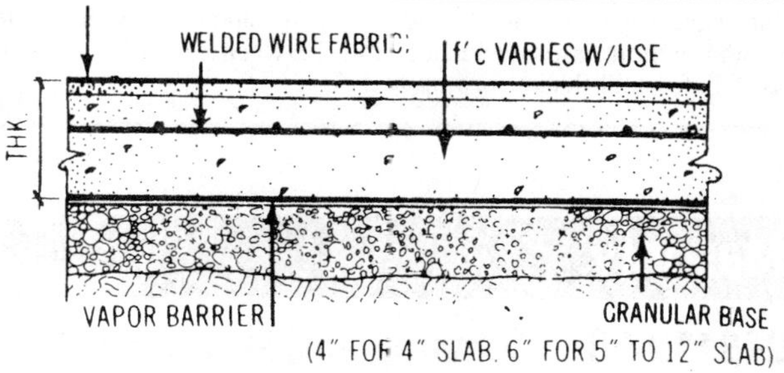

Slab on Grade

Thickness and Loading Assumptions by Type of Use

SLAB THICKNESS (IN.)	TYPE	A	B	C	D	◄ Slab I.D.
		Non	Light	Normal	Heavy	◄ Industrial
		Little	Light	Moderate	Severe	◄ Abrasion
		Foot Only	Pneumatic Wheels	Solid Rubber Wheels	Steel Tires	◄ Type of Traffic
		Load* (K)	Load* (K)	Load* (K)	Load* (K)	Max. Uniform Load to Slab ▼ (PSF)
4"	Reinf. Plain	4K				100
5"	Reinf. Plain	6K	4K			200
6"	Reinf. Plain		8K	6K	6K	500 to 800
7"	Reinf. Plain			9K	8K	1,500
8"	Reinf. Plain				11K	* Max. Wheel Load in Kips (incl. impact)
10"	Reinf. Plain				14K	
12"	Reinf. Plain					
DESIGN ASSUMPTIONS	Concrete, Chuted	f'c = 3.5 KSI	4 KSI	4.5 KSI	Slab @ 3.5 KSI	ASSUMPTIONS BY SLAB TYPE
	Toppings			1" Integral	1" Bonded	
	Finish	Steel Trowel	Steel Trowel	Steel Trowel	Screed & Steel Trowel	
	Compacted Granular Base	4" deep for 4" slab thickness 6" deep for 5" slab thickness & greater				ASSUMPTIONS FOR ALL SLAB TYPES
	Vapor Barrier	6 mil polyethylene				
	Forms & Joints	Allowances included				
	Rein-forcement	WWF as required ≥ 60,000 psi				

R040519-50 Masonry Reinforcing

Horizontal joint reinforcing helps prevent wall cracks where wall movement may occur and in many locations is required by code. Horizontal joint reinforcing is generally not considered to be structural reinforcing and an unreinforced wall may still contain joint reinforcing.

Reinforcing strips come in 10′ and 12′ lengths and in truss and ladder shapes, with and without drips. Field labor runs between 2.7 to 5.3 hours per 1000 L.F. for wall thicknesses up to 12″.

The wire meets ASTM A82 for cold drawn steel wire and the typical size is 9 ga. sides and ties with 3/16″ diameter also available. Typical finish is mill galvanized with zinc coating at .10 oz. per S.F. Class I (.40 oz. per S.F.) and Class III (.80 oz per S.F.) are also available, as is hot dipped galvanizing at 1.50 oz. per S.F.

R042110-20 Common and Face Brick

Common building brick manufactured according to ASTM C62 and facing brick manufactured according to ASTM C216 are the two standard bricks available for general building use.

Building brick is made in three grades; SW, where high resistance to damage caused by cyclic freezing is required; MW, where moderate resistance to cyclic freezing is needed; and NW, where little resistance to cyclic freezing is needed. Facing brick is made in only the two grades SW and MW. Additionally, facing brick is available in three types; FBS, for general use; FBX, for general use where a higher degree of precision and lower permissible variation in size than FBS is needed; and FBA, for general use to produce characteristic architectural effects resulting from non-uniformity in size and texture of the units.

In figuring the material cost of brickwork, an allowance of 25% mortar waste and 3% brick breakage was included. If bricks are delivered palletized with 280 to 300 per pallet, or packaged, allow only 1-1/2% for breakage. Packaged or palletized delivery is practical when a job is big enough to have a crane or other equipment available to handle a package of brick. This is so on all industrial work but not always true on small commercial buildings.

The use of buff and gray face is increasing, and there is a continuing trend to the Norman, Roman, Jumbo and SCR brick.

Common red clay brick for backup is not used that often. Concrete block is the most usual backup material with occasional use of sand lime or cement brick. Building brick is commonly used in solid walls for strength and as a fire stop.

Brick panels built on the ground and then crane erected to the upper floors have proven to be economical. This allows the work to be done under cover and without scaffolding.

R042110-50 Brick, Block & Mortar Quantities

Running Bond						For Other Bonds Standard Size Add to S.F. Quantities in Table to Left		
Number of Brick per S.F. of Wall - Single Wythe with 3/8″ Joints				C.F. of Mortar per M Bricks, Waste Included				
Type Brick	Nominal Size (incl. mortar) L H W	Modular Coursing	Number of Brick per S.F.	3/8″ Joint	1/2″ Joint	Bond Type	Description	Factor
Standard	8 x 2-2/3 x 4	3C=8″	6.75	10.3	12.9	Common	full header every fifth course	+20%
Economy	8 x 4 x 4	1C=4″	4.50	11.4	14.6		full header every sixth course	+16.7%
Engineer	8 x 3-1/5 x 4	5C=16″	5.63	10.6	13.6	English	full header every second course	+50%
Fire	9 x 2-1/2 x 4-1/2	2C=5″	6.40	550 # Fireclay	—	Flemish	alternate headers every course	+33.3%
Jumbo	12 x 4 x 6 or 8	1C=4″	3.00	23.8	30.8		every sixth course	+5.6%
Norman	12 x 2-2/3 x 4	3C=8″	4.50	14.0	17.9	Header = W x H exposed		+100%
Norwegian	12 x 3-1/5 x 4	5C=16″	3.75	14.6	18.6	Rowlock = H x W exposed		+100%
Roman	12 x 2 x 4	2C=4″	6.00	13.4	17.0	Rowlock stretcher = L x W exposed		+33.3%
SCR	12 x 2-2/3 x 6	3C=8″	4.50	21.8	28.0	Soldier = H x L exposed		—
Utility	12 x 4 x 4	1C=4″	3.00	15.4	19.6	Sailor = W x L exposed		-33.3%

Concrete Blocks Nominal Size	Approximate Weight per S.F.		Blocks per 100 S.F.	Mortar per M block, waste included	
	Standard	Lightweight		Partitions	Back up
2″ x 8″ x 16″	20 PSF	15 PSF	113	27 C.F.	36 C.F.
4″ ↓	30 ↓	20 ↓	↓	41 ↓	51 ↓
6″	42	30		56	66
8″	55	38		72	82
10″	70	47		87	97
12″	85	55		102	112

R042210-20 Concrete Block

The material cost of special block such as corner, jamb and head block can be figured at the same price as ordinary block of same size. Labor on specials is about the same as equal-sized regular block.

Bond beam and 16″ high lintel blocks are more expensive than regular units of equal size. Lintel blocks are 8″ long and either 8″ or 16″ high.

Use of motorized mortar spreader box will speed construction of continuous walls.

Hollow non-load-bearing units are made according to ASTM C129 and hollow load-bearing units according to ASTM C90.

Metals | R0505 Common Work Results for Metals

R050516-30 Coating Structural Steel

On field-welded jobs, the shop-applied primer coat is necessarily omitted. All painting must be done in the field and usually consists of red oxide rust inhibitive paint or an aluminum paint. The table below shows paint coverage and daily production for field painting.

See Division 09 97 13.23 for hot-dipped galvanizing and for field-applied cold galvanizing and other paints and protective coatings.

See Division 05 01 10.51 for steel surface preparation treatments such as wire brushing, pressure washing and sand blasting.

Type Construction	Surface Area per Ton	Coat	One Gallon Covers		In 8 Hrs. Person Covers		Average per Ton Spray	
			Brush	Spray	Brush	Spray	Gallons	Labor-hours
Light Structural	300 S.F. to 500 S.F.	1st	500 S.F.	455 S.F.	640 S.F.	2000 S.F.	0.9 gals.	1.6 L.H.
		2nd	450	410	800	2400	1.0	1.3
		3rd	450	410	960	3200	1.0	1.0
Medium	150 S.F. to 300 S.F.	All	400	365	1600	3200	0.6	0.6
Heavy Structural	50 S.F. to 150 S.F.	1st	400	365	1920	4000	0.2	0.2
		2nd	400	365	2000	4000	0.2	0.2
		3rd	400	365	2000	4000	0.2	0.2
Weighted Average	225 S.F.	All	400	365	1350	3000	0.6	0.6

R050521-20 Welded Structural Steel

Usual weight reductions with welded design run 10% to 20% compared with bolted or riveted connections. This amounts to about the same total cost compared with bolted structures since field welding is more expensive than bolts. For normal spans of 18′ to 24′ figure 6 to 7 connections per ton.

Trusses — For welded trusses add 4% to weight of main members for connections. Up to 15% less steel can be expected in a welded truss compared to one that is shop bolted. Cost of erection is the same whether shop bolted or welded.

General — Typical electrodes for structural steel welding are E6010, E6011, E60T and E70T. Typical buildings vary between 2# to 8# of weld rod per ton of steel. Buildings utilizing continuous design require about three times as much welding as conventional welded structures. In estimating field erection by welding, it is best to use the average linear feet of weld per ton to arrive at the welding cost per ton. The type, size and position of the weld will have a direct bearing on the cost per linear foot. A typical field welder will deposit 1.8# to 2# of weld rod per hour manually. Using semiautomatic methods can increase production by as much as 50% to 75%.

R050523-10 High Strength Bolts

Common bolts (A307) are usually used in secondary connections (see Division 05 05 23.10).

High strength bolts (A325 and A490) are usually specified for primary connections such as column splices, beam and girder connections to columns, column bracing, connections for supports of operating equipment or of other live loads which produce impact or reversal of stress, and in structures carrying cranes of over 5-ton capacity.

Allow 20 field bolts per ton of steel for a 6 story office building, apartment house or light industrial building. For 6 to 12 stories allow 25 bolts per ton, and above 12 stories, 30 bolts per ton. On power stations, 20 to 25 bolts per ton are needed.

R051223-10 Structural Steel

The bare material prices for structural steel, shown in the unit cost sections of the book, are for 100 tons of shop-fabricated structural steel and include:

1. Mill base price of structural steel
2. Mill scrap/grade/size/length extras
3. Mill delivery to a metals service center (warehouse)
4. Service center storage and handling
5. Service center delivery to a fabrication shop
6. Shop storage and handling
7. Shop drafting/detailing
8. Shop fabrication
9. Shop coat of primer paint
10. Shop listing
11. Shop delivery to the job site

In unit cost sections of the book that contain items for field fabrication of steel components, the bare material cost of steel includes:

1. Mill base price of structural steel
2. Mill scrap/grade/size/length extras
3. Mill delivery to a metals service center (warehouse)
4. Service center storage and handling
5. Service center delivery to the job site

R051223-15 Structural Steel Estimating for Repair and Remodeling Projects

The correct approach to estimating structural steel is dependent upon the amount of steel required for the particular project. If the project is a sizable addition to a building, the data can be used directly from the cost data book. This is not the case however if the project requires a small amount of steel, for instance to reinforce existing roof or floor structural systems. To better understand this, please refer to the unit price line for a W16x31 beam with bolted connections. Assume your project requires the reinforcement of the structural members of the roof system of a 3 story building required for the installation a new piece of HVAC equipment. The project will need 4 pieces of W16x31 that are 30′ long. After pricing this 120 L.F. job using the unit prices for a W16x31, an analysis will reveal that the price is wholly inadequate.

The first problem is apparent if you examine the amount of steel that can be installed per day. The unit price line indicates 900 linear feet per day can be installed by a 5 man crew with an 90 ton crane. This productivity is correct for new construction but certainly is not for repair and remodeling work. Installation of new structural steel considers that each member is installed from the foundation to the roof of the structure in a planned and systematic manner with a crane having unrestricted access to all parts of the project. Additionally each connection is planned and detailed with full field access for fit-up and final bolting. The erection is planned and progresses such that interferences and conflicts with other structural members are minimized, if not completely eliminated. All of these assumptions are clearly not the case with a repair and remodeling job, and a significant decrease in the stated productivity will be observed.

A crane will certainly be needed to lift the members into the general area of the project but in most cases will not be able to place the beams into their final position. An opening in the existing roof may not be large enough to permit the beams to pass through and it may be necessary to bring them into the building through existing windows or doors. Moving the beams to the actual area where they will be installed may involve hand labor and the use of dollies. Finally, hoists and/or jacks may be needed for final positioning.

The connection of new members to existing can often be accomplished by field bolting with accurate field measurements and good planning but in many cases access to both sides of existing members is not possible and field welding becomes the only alternative. In addition to the cost of the actual welding, protection of existing finishes, systems and structure and fire protection must be considered.

New beams can never be installed tight to the existing decks or floors which they must support and the use of shims and tack welding becomes necessary. Additionally, further planning and cost is involved in assuring that existing loads are minimized during the installation and shimming process.

It is apparent that installation of structural steel as part of a repair and remodeling project involves more than simply installing the members and estimating the cost in the same manner as new construction. The best procedure for estimating the total cost is adequate planning and coordination of each process and activity that will be needed. Unit costs for the materials, labor and equipment can then be attached to each needed activity and a final, complete price can be determined.

R051223-20 Steel Estimating Quantities

One estimate on erection is that a crane can handle 35 to 60 pieces per day. Say the average is 45. With usual sizes of beams, girders, and columns, this would amount to about 20 tons per day. The type of connection greatly affects the speed of erection. Moment connections for continuous design slow down production and increase erection costs.

Short open web bar joists can be set at the rate of 75 to 80 per day, with 50 per day being the average for setting long span joists.

After main members are calculated, add the following for usual allowances: base plates 2% to 3%; column splices 4% to 5%; and miscellaneous details 4% to 5%, for a total of 10% to 13% in addition to main members.

The ratio of column to beam tonnage varies depending on type of steels used, typical spans, story heights and live loads.

It is more economical to keep the column size constant and to vary the strength of the column by using high strength steels. This also saves floor space. Buildings have recently gone as high as ten stories with 8″ high strength columns. For light columns under W8X31 lb. sections, concrete filled steel columns are economical.

High strength steels may be used in columns and beams to save floor space and to meet head room requirements. High strength steels in some sizes sometimes require long lead times.

Round, square and rectangular columns, both plain and concrete filled, are readily available and save floor area, but are higher in cost per pound than rolled columns. For high unbraced columns, tube columns may be less expensive.

Below are average minimum figures for the weights of the structural steel frame for different types of buildings using A36 steel, rolled shapes and simple joints. For economy in domes, rise to span ratio = .13. Open web joist framing systems will reduce weights by 10% to 40%. Composite design can reduce steel weight by up to 25% but additional concrete floor slab thickness may be required. Continuous design can reduce the weights up to 20%. There are many building codes with different live load requirements and different structural requirements, such as hurricane and earthquake loadings, which can alter the figures.

Structural Steel Weights per S.F. of Floor Area

Type of Building	No. of Stories	Avg. Spans	L.L. #/S.F.	Lbs. Per S.F.	Type of Building	No. of Stories	Avg. Spans	L.L. #/S.F.	Lbs. Per S.F.
		20′x20′		8	Apartments	2-8	20′x20′	40	8
Steel Frame Mfg.	1	30′x30′	40	13		9-25			14
		40′x40′		18		to 10			10
Parking garage	4	Various	80	8.5	Office	20	Various	80	18
Domes	1	200′	30	10		30			26
(Schwedler)*		300′		15		over 50			35

R051223-25 Common Structural Steel Specifications

ASTM A992 (formerly A36, then A572 Grade 50) is the all-purpose carbon grade steel widely used in building and bridge construction.

The other high-strength steels listed below may each have certain advantages over ASTM A992 stuctural carbon steel, depending on the application. They have proven to be economical choices where, due to lighter members, the reduction of dead load and the associated savings in shipping cost can be significant.

ASTM A588 atmospheric weathering, high-strength low-alloy steels can be used in the bare (uncoated) condition, where exposure to normal atmosphere causes a tightly adherant oxide to form on the surface protecting the steel from further oxidation. ASTM A242 corrosion-resistant, high-strength low-alloy steels have enhanced atmospheric corrosion resistance of at least two times that of carbon structural steels with copper, or four times that of carbon structural steels without copper. The reduction or elimination of maintenance resulting from the use of these steels often offsets their higher initial cost.

Steel Type	ASTM Designation	Minimum Yield Stress in KSI	Shapes Available
Carbon	A36	36	All structural shape groups, and plates & bars up thru 8″ thick
	A529	50	Structural shape group 1, and plates & bars up thru 2″ thick
High-Strength Low-Alloy Quenched & Self-Tempered	A913	50	All structural shape groups
		60	
		65	
		70	
High-Strength Low-Alloy Columbium-Vanadium	A572	42	All structural shape groups, and plates & bars up thru 6″ thick
		50	All structural shape groups, and plates & bars up thru 4″ thick
		55	Structural shape groups 1 & 2, and plates & bars up thru 2″ thick
		60	Structural shape groups 1 & 2, and plates & bars up thru 1-1/4″ thick
		65	Structural shape group 1, and plates & bars up thru 1-1/4″ thick
High-Strength Low-Alloy Columbium-Vanadium	A992	50	All structural shape groups
Weathering High-Strength Low-Alloy	A242	42	Structural shape groups 4 & 5, and plates & bars over 1-1/2″ up thru 4″ thick
		46	Structural shape group 3, and plates & bars over 3/4″ up thru 1-1/2″ thick
		50	Structural shape groups 1 & 2, and plates & bars up thru 3/4″ thick
Weathering High-Strength Low-Alloy	A588	42	Plates & bars over 5″ up thru 8″ thick
		46	Plates & bars over 4″ up thru 5″ thick
		50	All structural shape groups, and plates & bars up thru 4″ thick
Quenched and Tempered Low-Alloy	A852	70	Plates & bars up thru 4″ thick
Quenched and Tempered Alloy	A514	90	Plates & bars over 2-1/2″ up thru 6″ thick
		100	Plates & bars up thru 2-1/2″ thick

R051223-30 High Strength Steels

The mill price of high strength steels may be higher than A992 carbon steel but their proper use can achieve overall savings thru total reduced weights. For columns with L/r over 100, A992 steel is best; under 100, high strength steels are economical. For heavy columns, high strength steels are economical when cover plates are eliminated. There is no economy using high strength steels for clip angles or supports or for beams where deflection governs. Thinner members are more economical than thick.

The per ton erection and fabricating costs of the high strength steels will be higher than for A992 since the same number of pieces, but less weight, will be installed.

R051223-35 Common Steel Sections

The upper portion of this table shows the name, shape, common designation, and basic characteristics of commonly used steel sections. The lower portion explains how to read the designations used for the above illustrated common sections.

Shape & Designation	Name & Characteristics	Shape & Designation	Name & Characteristics
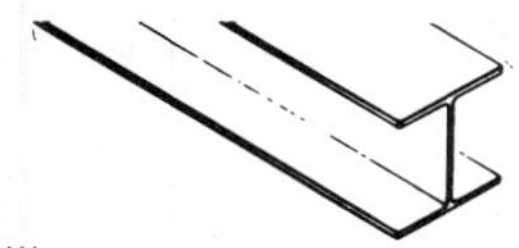 W	W Shape Parallel flange surfaces	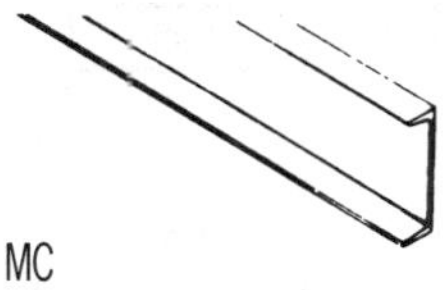 MC	Miscellaneous Channel Infrequently rolled by some producers
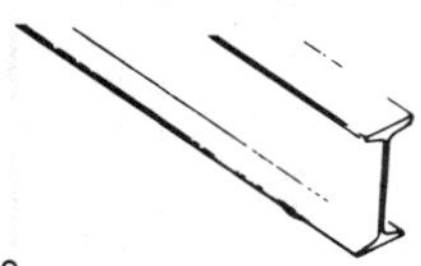 S	American Standard Beam (I Beam) Sloped inner flange	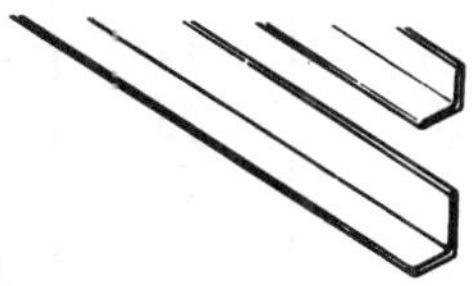 L	Angle Equal or unequal legs, constant thickness
 M	Miscellaneous Beams Cannot be classified as W, HP or S; infrequently rolled by some producers	 T	Structural Tee Cut from W, M or S on center of web
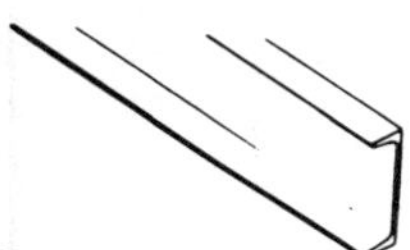 C	American Standard Channel Sloped inner flange	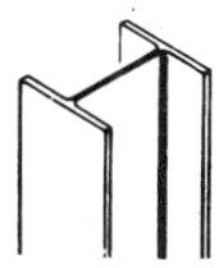 HP	Bearing Pile Parallel flanges and equal flange and web thickness

Common drawing designations follow:

W Shape
W 18 x 35
Weight in Pounds per Foot
Nominal Depth in Inches (Actual 17-3/4″)

American Standard Beam
S 12 x 31.8
Weight in Pounds per Foot
Depth in Inches

Miscellaneous Beam
M 8 x 6.5
Weight in Pounds per Foot
Depth in Inches

American Standard Channel
C 8 x 11.5
Weight in Pounds per Foot
Depth in Inches

Miscellaneous Channel
MC 8 x 22.8
Weight in Pounds per Foot
Depth in Inches

Angle
Length of Long Leg in Inches
L 6 x 3-1/2 x 3/8 Thickness of Each Leg in Inches
Length of Other Leg in Inches

Tee Cut from W16 x 100
WT 8 x 50
Weight in Pounds per Foot
Nominal Depth in Inches (Actual 8-1/2″)

Tee Cut from S12 x 35
ST 6 x 17.5
Weight in Pounds per Foot
Depth in Inches (Actual 6-1/4″)

Tee Cut from M10 x 9
MT 5 x 4.5
Weight in Pounds per Foot
Depth in Inches

Bearing Pile
HP 12 x 84
Weight in Pounds per Foot
Nominal Depth in Inches (Actual 12-1/4″)

R051223-45 Installation Time for Structural Steel Building Components

The following tables show the expected average installation times for various structural steel shapes. Table A presents installation times for columns, Table B for beams, Table C for light framing and bolts, and Table D for structural steel for various project types.

Table A

Description	Labor-Hours	Unit
Columns		
Steel, Concrete Filled		
3-1/2″ Diameter	.933	Ea.
6-5/8″ Diameter	1.120	Ea.
Steel Pipe		
3″ Diameter	.933	Ea.
8″ Diameter	1.120	Ea.
12″ Diameter	1.244	Ea.
Structural Tubing		
4″ x 4″	.966	Ea.
8″ x 8″	1.120	Ea.
12″ x 8″	1.167	Ea.
W Shape 2 Tier		
W8 x 31	.052	L.F.
W8 x 67	.057	L.F.
W10 x 45	.054	L.F.
W10 x 112	.058	L.F.
W12 x 50	.054	L.F.
W12 x 190	.061	L.F.
W14 x 74	.057	L.F.
W14 x 176	.061	L.F.

Table B

Description	Labor-Hours	Unit	Labor-Hours	Unit
Beams, W Shape				
W6 x 9	.949	Ea.	.093	L.F.
W10 x 22	1.037	Ea.	.085	L.F.
W12 x 26	1.037	Ea.	.064	L.F.
W14 x 34	1.333	Ea.	.069	L.F.
W16 x 31	1.333	Ea.	.062	L.F.
W18 x 50	2.162	Ea.	.088	L.F.
W21 x 62	2.222	Ea.	.077	L.F.
W24 x 76	2.353	Ea.	.072	L.F.
W27 x 94	2.581	Ea.	.067	L.F.
W30 x 108	2.857	Ea.	.067	L.F.
W33 x 130	3.200	Ea.	.071	L.F.
W36 x 300	3.810	Ea.	.077	L.F.

Table C

Description	Labor-Hours	Unit
Light Framing		
Angles 4″ and Larger	.055	lbs.
Less than 4″	.091	lbs.
Channels 8″ and Larger	.048	lbs.
Less than 8″	.072	lbs.
Cross Bracing Angles	.055	lbs.
Rods	.034	lbs.
Hanging Lintels	.069	lbs.
High Strength Bolts in Place		
3/4″ Bolts	.070	Ea.
7/8″ Bolts	.076	Ea.

Table D

Description	Labor-Hours	Unit	Labor-Hours	Unit
Apartments, Nursing Homes, etc.				
1-2 Stories	4.211	Piece	7.767	Ton
3-6 Stories	4.444	Piece	7.921	Ton
7-15 Stories	4.923	Piece	9.014	Ton
Over 15 Stories	5.333	Piece	9.209	Ton
Offices, Hospitals, etc.				
1-2 Stories	4.211	Piece	7.767	Ton
3-6 Stories	4.741	Piece	8.889	Ton
7-15 Stories	4.923	Piece	9.014	Ton
Over 15 Stories	5.120	Piece	9.209	Ton
Industrial Buildings				
1 Story	3.478	Piece	6.202	Ton

R051223-50 Subpurlins

Bulb tee subpurlins are structural members designed to support and reinforce a variety of roof deck systems such as precast cement fiber roof deck tiles, monolithic roof deck systems, and gypsum or lightweight concrete over formboard. Other uses include interstitial service ceiling systems, wall panel systems, and joist anchoring in bond beams. See Unit Price section for pricing on a square foot basis at 32-5/8″ O.C. Maximum span is based on a 3-span condition with a total allowable vertical load of 40 psf.

R051223-80 Dimensions and Weights of Sheet Steel

Gauge No.	Approximate Thickness				Weight		
	Inches (in fractions)	Inches (in decimal parts)		Millimeters			
	Wrought Iron	Wrought Iron	Steel	Steel	per S.F. in Ounces	per S.F. in Lbs.	per Square Meter in Kg.
0000000	1/2″	.5	.4782	12.146	320	20.000	97.650
000000	15/32″	.46875	.4484	11.389	300	18.750	91.550
00000	7/16″	.4375	.4185	10.630	280	17.500	85.440
0000	13/32″	.40625	.3886	9.870	260	16.250	79.330
000	3/8″	.375	.3587	9.111	240	15.000	73.240
00	11/32″	.34375	.3288	8.352	220	13.750	67.130
0	5/16″	.3125	.2989	7.592	200	12.500	61.030
1	9/32″	.28125	.2690	6.833	180	11.250	54.930
2	17/64″	.265625	.2541	6.454	170	10.625	51.880
3	1/4″	.25	.2391	6.073	160	10.000	48.820
4	15/64″	.234375	.2242	5.695	150	9.375	45.770
5	7/32″	.21875	.2092	5.314	140	8.750	42.720
6	13/64″	.203125	.1943	4.935	130	8.125	39.670
7	3/16″	.1875	.1793	4.554	120	7.500	36.320
8	11/64″	.171875	.1644	4.176	110	6.875	33.570
9	5/32″	.15625	.1495	3.797	100	6.250	30.520
10	9/64″	.140625	.1345	3.416	90	5.625	27.460
11	1/8″	.125	.1196	3.038	80	5.000	24.410
12	7/64″	.109375	.1046	2.657	70	4.375	21.360
13	3/32″	.09375	.0897	2.278	60	3.750	18.310
14	5/64″	.078125	.0747	1.897	50	3.125	15.260
15	9/128″	.0713125	.0673	1.709	45	2.813	13.730
16	1/16″	.0625	.0598	1.519	40	2.500	12.210
17	9/160″	.05625	.0538	1.367	36	2.250	10.990
18	1/20″	.05	.0478	1.214	32	2.000	9.765
19	7/160″	.04375	.0418	1.062	28	1.750	8.544
20	3/80″	.0375	.0359	.912	24	1.500	7.324
21	11/320″	.034375	.0329	.836	22	1.375	6.713
22	1/32″	.03125	.0299	.759	20	1.250	6.103
23	9/320″	.028125	.0269	.683	18	1.125	5.490
24	1/40″	.025	.0239	.607	16	1.000	4.882
25	7/320″	.021875	.0209	.531	14	.875	4.272
26	3/160″	.01875	.0179	.455	12	.750	3.662
27	11/640″	.0171875	.0164	.417	11	.688	3.357
28	1/64″	.015625	.0149	.378	10	.625	3.052

R053100-10 Decking Descriptions

General - All Deck Products

Steel deck is made by cold forming structural grade sheet steel into a repeating pattern of parallel ribs. The strength and stiffness of the panels are the result of the ribs and the material properties of the steel. Deck lengths can be varied to suit job conditions, but because of shipping considerations, are usually less than 40 feet. Standard deck width varies with the product used but full sheets are usually 12″, 18″, 24″, 30″, or 36″. Deck is typically furnished in a standard width with the ends cut square. Any cutting for width, such as at openings or for angular fit, is done at the job site.

Deck is typically attached to the building frame with arc puddle welds, self-drilling screws, or powder or pneumatically driven pins. Sheet-to-sheet fastening is done with screws, button punching (crimping), or welds.

Composite Floor Deck

After installation and adequate fastening, floor deck serves several purposes. It (a) acts as a working platform, (b) stabilizes the frame, (c) serves as a concrete form for the slab, and (d) reinforces the slab to carry the design loads applied during the life of the building. Composite decks are distinguished by the presence of shear connector devices as part of the deck. These devices are designed to mechanically lock the concrete and deck together so that the concrete and the deck work together to carry subsequent floor loads. These shear connector devices can be rolled-in embossments, lugs, holes, or wires welded to the panels. The deck profile can also be used to interlock concrete and steel.

Composite deck finishes are either galvanized (zinc coated) or phosphatized/painted. Galvanized deck has a zinc coating on both the top and bottom surfaces. The phosphatized/painted deck has a bare (phosphatized) top surface that will come into contact with the concrete. This bare top surface can be expected to develop rust before the concrete is placed. The bottom side of the deck has a primer coat of paint.

Composite floor deck is normally installed so the panel ends do not overlap on the supporting beams. Shear lugs or panel profile shape often prevent a tight metal to metal fit if the panel ends overlap; the air gap caused by overlapping will prevent proper fusion with the structural steel supports when the panel end laps are shear stud welded.

Adequate end bearing of the deck must be obtained as shown on the drawings. If bearing is actually less in the field than shown on the drawings, further investigation is required.

Roof Deck

Roof deck is not designed to act compositely with other materials. Roof deck acts alone in transferring horizontal and vertical loads into the building frame. Roof deck rib openings are usually narrower than floor deck rib openings. This provides adequate support of rigid thermal insulation board.

Roof deck is typically installed to endlap approximately 2″ over supports. However, it can be butted (or lapped more than 2″) to solve field fit problems. Since designers frequently use the installed deck system as part of the horizontal bracing system (the deck as a diaphragm), any fastening substitution or change should be approved by the designer. Continuous perimeter support of the deck is necessary to limit edge deflection in the finished roof and may be required for diaphragm shear transfer.

Standard roof deck finishes are galvanized or primer painted. The standard factory applied paint for roof deck is a primer paint and is not intended to weather for extended periods of time. Field painting or touching up of abrasions and deterioration of the primer coat or other protective finishes is the responsibility of the contractor.

Cellular Deck

Cellular deck is made by attaching a bottom steel sheet to a roof deck or composite floor deck panel. Cellular deck can be used in the same manner as floor deck. Electrical, telephone, and data wires are easily run through the chase created between the deck panel and the bottom sheet.

When used as part of the electrical distribution system, the cellular deck must be installed so that the ribs line up and create a smooth cell transition at abutting ends. The joint that occurs at butting cell ends must be taped or otherwise sealed to prevent wet concrete from seeping into the cell. Cell interiors must be free of welding burrs, or other sharp intrusions, to prevent damage to wires.

When used as a roof deck, the bottom flat plate is usually left exposed to view. Care must be maintained during erection to keep good alignment and prevent damage.

Cellular deck is sometimes used with the flat plate on the top side to provide a flat working surface. Installation of the deck for this purpose requires special methods for attachment to the frame because the flat plate, now on the top, can prevent direct access to the deck material that is bearing on the structural steel. It may be advisable to treat the flat top surface to prevent slipping.

Cellular deck is always furnished galvanized or painted over galvanized.

Form Deck

Form deck can be any floor or roof deck product used as a concrete form. Connections to the frame are by the same methods used to anchor floor and roof deck. Welding washers are recommended when welding deck that is less than 20 gauge thickness.

Form deck is furnished galvanized, prime painted, or uncoated. Galvanized deck must be used for those roof deck systems where form deck is used to carry a lightweight insulating concrete fill.

Wood, Plastics & Comp. R0611 Wood Framing

R061110-30 Lumber Product Material Prices

The price of forest products fluctuates widely from location to location and from season to season depending upon economic conditions. The bare material prices in the unit cost sections of the book show the National Average material prices in effect Jan. 1 of this book year. It must be noted that lumber prices in general may change significantly during the year.

Availability of certain items depends upon geographic location and must be checked prior to firm-price bidding.

Wood, Plastics & Comp. R0616 Sheathing

R061636-20 Plywood

There are two types of plywood used in construction: interior, which is moisture-resistant but not waterproofed, and exterior, which is waterproofed.

The grade of the exterior surface of the plywood sheets is designated by the first letter: A, for smooth surface with patches allowed; B, for solid surface with patches and plugs allowed; C, which may be surface plugged or may have knot holes up to 1″ wide; and D, which is used only for interior type plywood and may have knot holes up to 2-1/2″ wide. "Structural Grade" is specifically designed for engineered applications such as box beams. All CC & DD grades have roof and floor spans marked on them.

Underlayment-grade plywood runs from 1/4″ to 1-1/4″ thick. Thicknesses 5/8″ and over have optional tongue and groove joints which eliminate the need for blocking the edges. Underlayment 19/32″ and over may be referred to as Sturd-i-Floor.

The price of plywood can fluctuate widely due to geographic and economic conditions.

Typical uses for various plywood grades are as follows:

AA-AD Interior — cupboards, shelving, paneling, furniture

BB Plyform — concrete form plywood

CDX — wall and roof sheathing

Structural — box beams, girders, stressed skin panels

AA-AC Exterior — fences, signs, siding, soffits, etc.

Underlayment — base for resilient floor coverings

Overlaid HDO — high density for concrete forms & highway signs

Overlaid MDO — medium density for painting, siding, soffits & signs

303 Siding — exterior siding, textured, striated, embossed, etc.

Thermal & Moist. Protec. R0752 Modified Bituminous Membrane Roofing

R075213-30 Modified Bitumen Roofing

The cost of modified bitumen roofing is highly dependent on the type of installation that is planned. Installation is based on the type of modifier used in the bitumen. The two most popular modifiers are atactic polypropylene (APP) and styrene butadiene styrene (SBS). The modifiers are added to heated bitumen during the manufacturing process to change its characteristics. A polyethylene, polyester or fiberglass reinforcing sheet is then sandwiched between layers of this bitumen. When completed, the result is a pre-assembled, built-up roof that has increased elasticity and weatherablility. Some manufacturers include a surfacing material such as ceramic or mineral granules, metal particles or sand.

The preferred method of adhering SBS-modified bitumen roofing to the substrate is with hot-mopped asphalt (much the same as built-up roofing). This installation method requires a tar kettle/pot to heat the asphalt, as well as the labor, tools and equipment necessary to distribute and spread the hot asphalt.

The alternative method for applying APP and SBS modified bitumen is as follows. A skilled installer uses a torch to melt a small pool of bitumen off the membrane. This pool must form across the entire roll for proper adhesion. The installer must unroll the roofing at a pace slow enough to melt the bitumen, but fast enough to prevent damage to the rest of the membrane.

Modified bitumen roofing provides the advantages of both built-up and single-ply roofing. Labor costs are reduced over those of built-up roofing because only a single ply is necessary. The elasticity of single-ply roofing is attained with the reinforcing sheet and polymer modifiers. Modifieds have some self-healing characteristics and because of their multi-layer construction, they offer the reliability and safety of built-up roofing.

R078413-30 Firestopping

Firestopping is the sealing of structural, mechanical, electrical, and other penetrations through fire-rated assemblies. The basic components of firestop systems are safing insulation and firestop sealant on both sides of wall penetrations and the top side of floor penetrations.

Pipe penetrations are assumed to be through concrete, grout, or joint compound and can be sleeved or unsleeved. Costs for the penetrations and sleeves are not included. An annular space of 1″ is assumed. Escutcheons are not included.

Metallic pipe is assumed to be copper, aluminum, cast iron, or similar metallic material. Insulated metallic pipe is assumed to be covered with a thermal insulating jacket of varying thickness and materials.

Non-metallic pipe is assumed to be PVC, CPVC, FR Polypropylene, or similar plastic piping material. Intumescent firestop sealant or wrap strips are included. Collars on both sides of wall penetrations and a sheet metal plate on the underside of floor penetrations are included.

Ductwork is assumed to be sheet metal, stainless steel or similar metallic material. Duct penetrations are assumed to be through concrete, grout or joint compound. Costs for penetrations and sleeves are not included. An annular space of 1/2″ is assumed.

Multi-trade openings include costs for sheet metal forms, firestop mortar, wrap strips, collars, and sealants as necessary.

Structural penetrations joints are assumed to be 1/2″ or less. CMU walls are assumed to be within 1-1/2″ of metal deck. Drywall walls are assumed to be tight to the underside of metal decking.

Metal panel, glass, or curtain wall systems include a spandrel area of 5′ filled with mineral wool foil-faced insulation. Fasteners and stiffeners are included.

Openings — R0813 Metal Doors

R081313-20 Steel Door Selection Guide

Standard steel doors are classified into four levels, as recommended by the Steel Door Institute in the chart below. Each of the four levels offers a range of construction models and designs, to meet architectural requirements for preference and appearance, including full flush, seamless, and stile & rail. Recommended minimum gauge requirements are also included.

For complete standard steel door construction specifications and available sizes, refer to the Steel Door Institute Technical Data Series, ANSI A250.8-98 (SDI-100), and ANSI A250.4-94 Test Procedure and Acceptance Criteria for Physical Endurance of Steel Door and Hardware Reinforcements.

Level		Model	Construction	For Full Flush or Seamless Min. Gauge	Thickness (in)	Thickness (mm)
I	Standard Duty	1 2	Full Flush Seamless	20	0.032	0.8
II	Heavy Duty	1 2	Full Flush Seamless	18	0.042	1.0
III	Extra Heavy Duty	1 2 3	Full Flush Seamless *Stile & Rail	16	0.053	1.3
IV	Maximum Duty	1 2	Full Flush Seamless	14	0.067	1.6

*Stiles & rails are 16 gauge; flush panels, when specified, are 18 gauge

R092910-10 Levels of Gypsum Drywall Finish

In the past, contract documents often used phrases such as "industry standard" and "workmanlike finish" to specify the expected quality of gypsum board wall and ceiling installations. The vagueness of these descriptions led to unacceptable work and disputes.

In order to resolve this problem, four major trade associations concerned with the manufacture, erection, finish and decoration of gypsum board wall and ceiling systems have developed an industry-wide *Recommended Levels of Gypsum Board Finish*.

The finish of gypsum board walls and ceilings for specific final decoration is dependent on a number of factors. A primary consideration is the location of the surface and the degree of decorative treatment desired. Painted and unpainted surfaces in warehouses and other areas where appearance is normally not critical may simply require the taping of wallboard joints and 'spotting' of fastener heads. Blemish-free, smooth, monolithic surfaces often intended for painted and decorated walls and ceilings in habitated structures, ranging from single-family dwellings through monumental buildings, require additional finishing prior to the application of the final decoration.

Other factors to be considered in determining the level of finish of the gypsum board surface are (1) the type of angle of surface illumination (both natural and artificial lighting), and (2) the paint and method of application or the type and finish of wallcovering specified as the final decoration. Critical lighting conditions, gloss paints, and thin wallcoverings require a higher level of gypsum board finish than do heavily textured surfaces which are subsequently painted or surfaces which are to be decorated with heavy grade wallcoverings.

The following descriptions were developed jointly by the Association of the Wall and Ceiling Industries-International (AWCI), Ceiling & Interior Systems Construction Association (CISCA), Gypsum Association (GA), and Painting and Decorating Contractors of America (PDCA) as a guide.

Level 0: No taping, finishing, or accessories required. This level of finish may be useful in temporary construction or whenever the final decoration has not been determined.

Level 1: All joints and interior angles shall have tape set in joint compound. Surface shall be free of excess joint compound. Tool marks and ridges are acceptable. Frequently specified in plenum areas above ceilings, in attics, in areas where the assembly would generally be concealed or in building service corridors, and other areas not normally open to public view.

Level 2: All joints and interior angles shall have tape embedded in joint compound and wiped with a joint knife leaving a thin coating of joint compound over all joints and interior angles. Fastener heads and accessories shall be covered with a coat of joint compound. Surface shall be free of excess joint compound. Tool marks and ridges are acceptable. Joint compound applied over the body of the tape at the time of tape embedment shall be considered a separate coat of joint compound and shall satisfy the conditions of this level. Specified where water-resistant gypsum backing board is used as a substrate for tile; may be specified in garages, warehouse storage, or other similar areas where surface appearance is not of primary concern.

Level 3: All joints and interior angles shall have tape embedded in joint compound and one additional coat of joint compound applied over all joints and interior angles. Fastener heads and accessories shall be covered with two separate coats of joint compound. All joint compound shall be smooth and free of tool marks and ridges. Typically specified in appearance areas which are to receive heavy- or medium-texture (spray or hand applied) finishes before final painting, or where heavy-grade wallcoverings are to be applied as the final decoration. This level of finish is not recommended where smooth painted surfaces or light to medium wallcoverings are specified.

Level 4: All joints and interior angles shall have tape embedded in joint compound and two separate coats of joint compound applied over all flat joints and one separate coat of joint compound applied over interior angles. Fastener heads and accessories shall be covered with three separate coats of joint compound. All joint compound shall be smooth and free of tool marks and ridges. This level should be specified where flat paints, light textures, or wallcoverings are to be applied. In critical lighting areas, flat paints applied over light textures tend to reduce joint photographing. Gloss, semi-gloss, and enamel paints are not recommended over this level of finish. The weight, texture, and sheen level of wallcoverings applied over this level of finish should be carefully evaluated. Joints and fasteners must be adequately concealed if the wallcovering material is lightweight, contains limited pattern, has a gloss finish, or any combination of these finishes is present. Unbacked vinyl wallcoverings are not recommended over this level of finish.

Level 5: All joints and interior angles shall have tape embedded in joint compound and two separate coats of joint compound applied over all flat joints and one separate coat of joint compound applied over interior angles. Fastener heads and accessories shall be covered with three separate coats of joint compound. A thin skim coat of joint compound or a material manufactured especially for this purpose, shall be applied to the entire surface. The surface shall be smooth and free of tool marks and ridges. This level of finish is highly recommended where gloss, semi-gloss, enamel, or nontextured flat paints are specified or where severe lighting conditions occur. This highest quality finish is the most effective method to provide a uniform surface and minimize the possibility of joint photographing and of fasteners showing through the final decoration.

R099100-10 Painting Estimating Techniques

Proper estimating methodology is needed to obtain an accurate painting estimate. There is no known reliable shortcut or square foot method. The following steps should be followed:

- List all surfaces to be painted, with an accurate quantity (area) of each. Items having similar surface condition, finish, application method and accessibility may be grouped together.
- List all the tasks required for each surface to be painted, including surface preparation, masking, and protection of adjacent surfaces. Surface preparation may include minor repairs, washing, sanding and puttying.
- Select the proper Means line for each task. Review and consider all adjustments to labor and materials for type of paint and location of work. Apply the height adjustment carefully. For instance, when applying the adjustment for work over 8' high to a wall that is 12' high, apply the adjustment only to the area between 8' and 12' high, and not to the entire wall.

When applying more than one percent (%) adjustment, apply each to the base cost of the data, rather than applying one percentage adjustment on top of the other.

When estimating the cost of painting walls and ceilings remember to add the brushwork for all cut-ins at inside corners and around windows and doors as a LF measure. One linear foot of cut-in with brush equals one square foot of painting.

All items for spray painting include the labor for roll-back.

Deduct for openings greater than 100 SF, or openings that extend from floor to ceiling and are greater than 5' wide. Do not deduct small openings.

The cost of brushes, rollers, ladders and spray equipment are considered to be part of a painting contractor's overhead, and should not be added to the estimate. The cost of rented equipment such as scaffolding and swing staging should be added to the estimate.

R099100-20 Painting

Item	Coat	One Gallon Covers			In 8 Hours a Laborer Covers			Labor-Hours per 100 S.F.		
		Brush	Roller	Spray	Brush	Roller	Spray	Brush	Roller	Spray
Paint wood siding	prime	250 S.F.	225 S.F.	290 S.F.	1150 S.F.	1300 S.F.	2275 S.F.	.695	.615	.351
	others	270	250	290	1300	1625	2600	.615	.492	.307
Paint exterior trim	prime	400	—	—	650	—	—	1.230	—	—
	1st	475	—	—	800	—	—	1.000	—	—
	2nd	520	—	—	975	—	—	.820	—	—
Paint shingle siding	prime	270	255	300	650	975	1950	1.230	.820	.410
	others	360	340	380	800	1150	2275	1.000	.695	.351
Stain shingle siding	1st	180	170	200	750	1125	2250	1.068	.711	.355
	2nd	270	250	290	900	1325	2600	.888	.603	.307
Paint brick masonry	prime	180	135	160	750	800	1800	1.066	1.000	.444
	1st	270	225	290	815	975	2275	.981	.820	.351
	2nd	340	305	360	815	1150	2925	.981	.695	.273
Paint interior plaster or drywall	prime	400	380	495	1150	2000	3250	.695	.400	.246
	others	450	425	495	1300	2300	4000	.615	.347	.200
Paint interior doors and windows	prime	400	—	—	650	—	—	1.230	—	—
	1st	425	—	—	800	—	—	1.000	—	—
	2nd	450	—	—	975	—	—	.820	—	—

R133419-10 Pre-Engineered Steel Buildings

These buildings are manufactured by many companies and normally erected by franchised dealers throughout the U.S. The four basic types are: Rigid Frames, Truss type, Post and Beam and the Sloped Beam type. Most popular roof slope is low pitch of 1″ in 12″. The minimum economical area of these buildings is about 3000 S.F. of floor area. Bay sizes are usually 20′ to 24′ but can go as high as 30′ with heavier girts and purlins. Eave heights are usually 12′ to 24′ with 18′ to 20′ most typical.

Material prices shown in the Unit Price section are bare costs for the building shell only and do not include floors, foundations, anchor bolts, interior finishes or utilities. Costs assume at least three bays of 24′ each, a 1″ in 12″ roof slope, and they are based on 30 psf roof load and 20 psf wind load (wind load is a function of wind speed, building height, and terrain characteristics; this should be determined by a Registered Structural Engineer) and no unusual requirements. Costs include the structural frame, 26 ga. non-insulated colored corrugated or ribbed roofing and siding panels, fasteners, closures, trim and flashing but no allowance for insulation, doors, windows, skylights, gutters or downspouts. Very large projects would generally cost less for materials than the prices shown. For roof panel substitutions and wall panel substitutions, see appropriate Unit Price sections.

Conditions at the site, weather, shape and size of the building, and labor availability will affect the erection cost of the building.

R142000-10 Freight Elevators

Capacities run from 2,000 lbs. to over 100,000 lbs. with 3,000 lbs. to 10,000 lbs. most common. Travel speeds are generally lower and control less intricate than on passenger elevators. Costs in the Unit Price sections are for hydraulic and geared elevators.

R142000-20 Elevator Selective Costs See R142000-40 for cost development.

	Passenger		Freight		Hospital	
A. Base Unit	Hydraulic	Electric	Hydraulic	Electric	Hydraulic	Electric
Capacity	1,500 lb.	2,000 lb.	2,000 lb.	4,000 lb.	4,000 lb.	4,000 lb.
Speed	100 F.P.M.	200 F.P.M.	100 F.P.M.	200 F.P.M.	100 F.P.M.	200 F.P.M.
#Stops/Travel Ft.	2/12	4/40	2/20	4/40	2/20	4/40
Push Button Oper.	Yes	Yes	Yes	Yes	Yes	Yes
Telephone Box & Wire	"	"	"	"	"	"
Emergency Lighting	"	"	No	No	"	"
Cab	Plastic Lam. Walls	Plastic Lam. Walls	Painted Steel	Painted Steel	Plastic Lam. Walls	Plastic Lam. Walls
Cove Lighting	Yes	Yes	No	No	Yes	Yes
Floor	V.C.T.	V.C.T.	Wood w/Safety Treads	Wood w/Safety Treads	V.C.T.	V.C.T.
Doors, & Speedside Slide	Yes	Yes	Yes	Yes	Yes	Yes
Gates, Manual	No	No	No	No	No	No
Signals, Lighted Buttons	Car and Hall	Car and Hall	Car and Hall	Car and Hall	Car and Hall	Car and Hall
O.H. Geared Machine	N.A.	Yes	N.A.	Yes	N.A.	Yes
Variable Voltage Contr.	"	"	N.A.	"	"	"
Emergency Alarm	Yes	"	Yes	"	Yes	"
Class "A" Loading	N.A.	N.A.	"	"	N.A.	N.A.

R142000-30 Passenger Elevators

Electric elevators are used generally but hydraulic elevators can be used for lifts up to 70′ and where large capacities are required. Hydraulic speeds are limited to 200 F.P.M. but cars are self leveling at the stops. On low rises, hydraulic installation runs about 15% less than standard electric types but on higher rises this installation cost advantage is reduced. Maintenance of hydraulic elevators is about the same as electric type but underground portion is not included in the maintenance contract.

In electric elevators there are several control systems available, the choice of which will be based upon elevator use, size, speed and cost criteria. The two types of drives are geared for low speeds and gearless for 450 F.P.M. and over.

The tables above illustrate typical installed costs of the various types of elevators available.

R142000-40 Elevator Cost Development

To price a new car or truck from the factory, you must start with the manufacturer's basic model, then add or exchange optional equipment and features. The same is true for pricing elevators.

Example:

Requirement: One-passenger elevator, five-story hydraulic, 2,500 lb. capacity, 12′ floor to floor, speed 150 F.P.M., emergency power switching and maintenance contract.

Description	Adjustment
A. Base Elevator: Hydraulic Passenger, 1500 lb. Capacity, 100 fpm, 2 Stops, Standard Finish	1 Ea.
B. Capacity Adjustment (2,500 lb.)	1 Ea.
C. Excess Travel Adjustment: 48′ Total Travel (4 x 12′) minus 12′ Base Unit Travel =	36 V.L.F.
D. Stops Adjustment: 5 Total Stops minus 2 Stops (Base Unit) =	3 Stops
E. Speed Adjustment (150 F.P.M.)	1 Ea.
F. Options:	
1. Intercom Service	1 Ea.
2. Emergency Power Switching, Automatic	1 Ea.
3. Stainless Steel Entrance Doors	5 Ea.
4. Maintenance Contract (12 Months)	1 Ea.
5. Position Indicator for main floor level (none indicated in Base Unit)	1 Ea.

Conveying Equipment — R1431 Escalators

R143110-10 Escalators

Moving stairs can be used for buildings where 600 or more people are to be carried to the second floor or beyond. Freight cannot be carried on escalators and at least one elevator must be available for this function.

Carrying capacity is 5,000 to 8,000 people per hour. Power requirement is 2 to 3 KW per hour and incline angle is 30°.

Conveying Equipment — R1432 Moving Walks

R143210-20 Moving Ramps and Walks

These are a specialized form of conveyor 3′ to 6′ wide with capacities of 3,600 to 18,000 persons per hour. Maximum speed is 140 F.P.M. and normal incline is 0° to 15°.

Local codes will determine the maximum angle. Outdoor units would require additional weather protection.

Fire Suppression — R2112 Fire-Suppression Standpipes

R211226-10 Standpipe Systems

The basis for standpipe system design is National Fire Protection Association NFPA 14, however, the authority having jurisdiction should be consulted for special conditions, local requirements and approval.

Standpipe systems, properly designed and maintained, are an effective and valuable time saving aid for extinquishing fires, especially in the upper stories of tall buildings, the interior of large commercial or industrial malls, or other areas where construction features or access make the laying of temporary hose lines time consuming and/or hazardous. Standpipes are frequently installed with automatic sprinkler systems for maximum protection.

There are three general classes of service for standpipe systems:
Class I for use by fire departments and personnel with special training for heavy streams (2-1/2″ hose connections).
Class II for use by building occupants until the arrival of the fire department (1-1/2″ hose connector with hose).
Class III for use by either fire departments and trained personnel or by the building occupants (both 2-1/2″ and 1-1/2″ hose connections or one 2-1/2″ hose valve with an easily removable 2-1/2″ by 1-1/2″ adapter).

Standpipe systems are also classified by the way water is supplied to the system. The four basic types are:
Type 1: Wet standpipe system having supply valve open and water pressure maintained at all times.
Type 2: Standpipe system so arranged through the use of approved devices as to admit water to the system automatically by opening a hose valve.
Type 3: Standpipe system arranged to admit water to the system through manual operation of approved remote control devices located at each hose station.
Type 4: Dry standpipe having no permanent water supply.

R211226-20 NFPA 14 Basic Standpipe Design

Class	Design-Use	Pipe Size Minimums	Water Supply Minimums
Class I	2 1/2″ hose connection on each floor All areas within 150′ of an exit in every exit stairway Fire Department Trained Personnel	Height to 100′, 4″ dia. Heights above 100′, 6″ dia. (275′ max. except with pressure regulators 400′ max.)	For each standpipe riser 500 GPM flow For common supply pipe allow 500 GPM for first standpipe plus 250 GPM for each additional standpipe (2500 GPM max. total) 30 min. duration 65 PSI at 500 GPM
Class II	1 1/2″ hose connection with hose on each floor All areas within 130′ of hose connection measured along path of hose travel Occupant personnel	Height to 50′, 2″ dia. Height above 50′, 2 1/2″ dia.	For each standpipe riser 100 GPM flow For multiple riser common supply pipe 100 GPM 300 min. duration, 65 PSI at 100 GPM
Class III	Both of above. Class I valved connections will meet Class III with additional 2 1/2″ by 1 1/2″ adapter and 1 1/2″ hose.	Same as Class I	Same as Class I

*Note: Where 2 or more standpipes are installed in the same building or section of building they shall be interconnected at the bottom.

Combined Systems

Combined systems are systems where the risers supply both automatic sprinklers and 2-1/2″ hose connection outlets for fire department use. In such a system the sprinkler spacing pattern shall be in accordance with NFPA 13 while the risers and supply piping will be sized in accordance with NFPA 14. When the building is completely sprinklered the risers may be sized by hydraulic calculation.The minimum size riser for buildings not completely sprinklered is 6″.

The minimum water supply of a completely sprinklered, light hazard, high-rise occupancy building will be 500 GPM while the supply required for other types of completely sprinklered high-rise buildings is 1000 GPM.

General System Requirements

1. Approved valves will be provided at the riser for controlling branch lines to hose outlets.
2. A hose valve will be provided at each outlet for attachment of hose.
3. Where pressure at any standpipe outlet exceeds 100 PSI a pressure reducer must be installed to limit the pressure to 100 PSI. Note that the pressure head due to gravity in 100′ of riser is 43.4 PSI. This must be overcome by city pressure, fire pumps, or gravity tanks to provide adequate pressure at the top of the riser.
4. Each hose valve on a wet system having linen hose shall have an automatic drip connection to prevent valve leakage from entering the hose.
5. Each riser will have a valve to isolate it from the rest of the system.
6. One or more fire department connections as an auxiliary supply shall be provided for each Class I or Class III standpipe system. In buildings having two or more zones, a connection will be provided for each zone.
7. There will be no shutoff valve in the fire department connection, but a check valve will be located in the line before it joins the system.
8. All hose connections street side will be identified on a cast plate or fitting as to purpose.

R211313-10 Sprinkler Systems (Automatic)

Sprinkler systems may be classified by type as follows:

1. **Wet Pipe System.** A system employing automatic sprinklers attached to a piping system containing water and connected to a water supply so that water discharges immediately from sprinklers opened by a fire.
2. **Dry Pipe System.** A system employing automatic sprinklers attached to a piping system containing air under pressure, the release of which as from the opening of sprinklers permits the water pressure to open a valve known as a "dry pipe valve". The water then flows into the piping system and out the opened sprinklers.
3. **Pre-Action System.** A system employing automatic sprinklers attached to a piping system containing air that may or may not be under pressure, with a supplemental heat responsive system of generally more sensitive characteristics than the automatic sprinklers themselves, installed in the same areas as the sprinklers; actuation of the heat responsive system, as from a fire, opens a valve which permits water to flow into the sprinkler piping system and to be discharged from any sprinklers which may be open.
4. **Deluge System.** A system employing open sprinklers attached to a piping system connected to a water supply through a valve which is opened by the operation of a heat responsive system installed in the same areas as the sprinklers. When this valve opens, water flows into the piping system and discharges from all sprinklers attached thereto.
5. **Combined Dry Pipe and Pre-Action Sprinkler System.** A system employing automatic sprinklers attached to a piping system containing air under pressure with a supplemental heat responsive system of generally more sensitive characteristics than the automatic sprinklers themselves, installed in the same areas as the sprinklers; operation of the heat responsive system, as from a fire, actuates tripping devices which open dry pipe valves simultaneously and without loss of air pressure in the system. Operation of the heat responsive system also opens approved air exhaust valves at the end of the feed main which facilitates the filling of the system with water which usually precedes the opening of sprinklers. The heat responsive system also serves as an automatic fire alarm system.
6. **Limited Water Supply System.** A system employing automatic sprinklers and conforming to these standards but supplied by a pressure tank of limited capacity.
7. **Chemical Systems.** Systems using halon, carbon dioxide, dry chemical or high expansion foam as selected for special requirements. Agent may extinguish flames by chemically inhibiting flame propagation, suffocate flames by excluding oxygen, interrupting chemical action of oxygen uniting with fuel or sealing and cooling the combustion center.
8. **Firecycle System.** Firecycle is a fixed fire protection sprinkler system utilizing water as its extinguishing agent. It is a time delayed, recycling, preaction type which automatically shuts the water off when heat is reduced below the detector operating temperature and turns the water back on when that temperature is exceeded. The system senses a fire condition through a closed circuit electrical detector system which controls water flow to the fire automatically. Batteries supply up to 90 hour emergency power supply for system operation. The piping system is dry (until water is required) and is monitored with pressurized air. Should any leak in the system piping occur, an alarm will sound, but water will not enter the system until heat is sensed by a firecycle detector.

Area coverage sprinkler systems may be laid out and fed from the supply in any one of several patterns as shown below. It is desirable, if possible, to utilize a central feed and achieve a shorter flow path from the riser to the furthest sprinkler. This permits use of the smallest sizes of pipe possible with resulting savings.

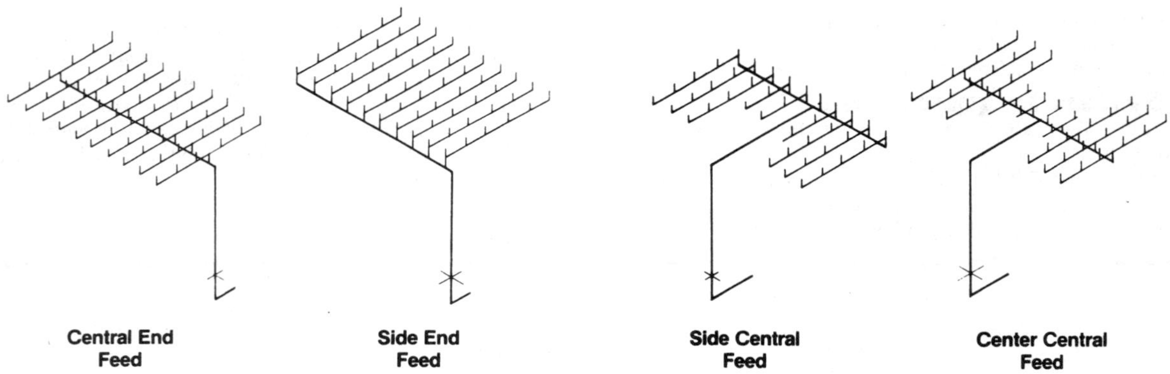

R211313-20 System Classification

System Classification

Rules for installation of sprinkler systems vary depending on the classification of occupancy falling into one of three categories as follows:

Light Hazard Occupancy

The protection area allotted per sprinkler should not exceed 225 S.F., with the maximum distance between lines and sprinklers on lines being 15′. The sprinklers do not need to be staggered. Branch lines should not exceed eight sprinklers on either side of a cross main. Each large area requiring more than 100 sprinklers and without a sub-dividing partition should be supplied by feed mains or risers sized for ordinary hazard occupancy.
Maximum system area = 52,000 S.F.

Included in this group are:

Churches
Clubs
Educational
Hospitals
Institutional
Libraries
(except large stack rooms)
Museums
Nursing Homes
Offices
Residential
Restaurants
Theaters and Auditoriums
(except stages and prosceniums)
Unused Attics

Ordinary Hazard Occupancy

The protection area allotted per sprinkler shall not exceed 130 S.F. of noncombustible ceiling and 130 S.F. of combustible ceiling. The maximum allowable distance between sprinkler lines and sprinklers on line is 15′. Sprinklers shall be staggered if the distance between heads exceeds 12′. Branch lines should not exceed eight sprinklers on either side of a cross main.
Maximum system area = 52,000 S.F.

Included in this group are:

Group 1
Automotive Parking and Showrooms
Bakeries
Beverage manufacturing
Canneries
Dairy Products Manufacturing/Processing
Electronic Plans
Glass and Glass Products Manufacturing
Laundries
Restaurant Service Areas

Group 2
Cereal Mills
Chemical Plants—Ordinary
Confectionery Products
Distilleries
Dry Cleaners
Feed Mills
Horse Stables
Leather Goods Manufacturing
Libraries—Large Stack Room Areas
Machine Shops
Metal Working
Mercantile
Paper and Pulp Mills
Paper Process Plants
Piers and Wharves
Post Offices
Printing and Publishing
Repair Garages
Stages
Textile Manufacturing
Tire Manufacturing
Tobacco Products Manufacturing
Wood Machining
Wood Product Assembly

Extra Hazard Occupancy

The protection area allotted per sprinkler shall not exceed 100 S.F. of noncombustible ceiling and 100 S.F. of combustible ceiling. The maximum allowable distance between lines and between sprinklers on lines is 12′. Sprinklers on alternate lines shall be staggered if the distance between sprinklers on lines exceeds 8′. Branch lines should not exceed six sprinklers on either side of a cross main.
Maximum system area:
Design by pipe schedule = 25,000 S.F.
Design by hydraulic calculation = 40,000 S.F.

Included in this group are:

Group 1
Aircraft hangars
Combustible Hydraulic Fluid Use Area
Die Casting
Metal Extruding
Plywood/Particle Board Manufacturing
Printing (inks with flash points < 100 degrees F)
Rubber Reclaiming, Compounding, Drying, Milling, Vulcanizing
Saw Mills
Textile Picking, Opening, Blending, Garnetting, Carding, Combing of Cotton, Synthetics, Wood Shoddy, or Burlap
Upholstering with Plastic Foams

Group 2
Asphalt Saturating
Flammable Liquids Spraying
Flow Coating
Manufactured/Modular Home Building Assemblies (where finished enclosure is present and has combustible interiors)
Open Oil Quenching
Plastics Processing
Solvent Cleaning
Varnish and Paint Dipping

R211313-30 Sprinkler Quantities for Various Sizes and Types of Pipe

Sprinkler Quantities: The table below lists the usual maximum number of sprinkler heads for each size of copper and steel pipe for both wet and dry systems. These quantities may be adjusted to meet individual structural needs or local code requirements. Maximum area on any one floor for one system is: light hazard and ordinary hazard 52,000 S.F., extra hazardous 25,000 S.F.

Pipe Size	Light Hazard Occupancy		Ordinary Hazard Occupancy		Extra Hazard Occupancy	
Diameter	Steel Pipe	Copper Pipe	Steel Pipe	Copper Pipe	Steel Pipe	Copper Pipe
1"	2 sprinklers	2 sprinklers	2 sprinklers	2 sprinklers	1 sprinklers	1 sprinklers
1-1/4"	3	3	3	3	2	2
1-1/2"	5	5	5	5	5	5
2"	10	12	10	12	8	8
2-1/2"	30	40	20	25	15	20
3"	60	65	40	45	27	30
3-1/2"	100	115	65	75	40	45
4"			100	115	55	65
5"			160	180	90	100
6"			275	300	150	170

Dry Pipe Systems: A dry pipe system should be installed where a wet pipe system is impractical, as in rooms or buildings that cannot be properly heated.

The use of an approved dry pipe system is more desirable than shutting off the water supply during cold weather.

Not more than 750 gallons of system capacity should be controlled by one dry pipe valve. Where two or more dry pipe valves are used, systems should preferably be divided horizontally.

R211313-40 Adjustment for Sprinkler/Standpipe Installations

Quality/Complexity Multiplier (For all installations)

Economy installation, add .... 0 to 5%
Good quality, medium complexity, add .... 5 to 15%
Above average quality and complexity, add .... 15 to 25%

Plumbing — R2201 Operation & Maintenance of Plumbing

R220105-10 Demolition (Selective vs. Removal for Replacement)

Demolition can be divided into two basic categories.

One type of demolition involves the removal of material with no concern for its replacement. The labor-hours to estimate this work are found under "Selective Demolition" in the Fire Protection, Plumbing and HVAC Divisions. It is selective in that individual items or all the material installed as a system or trade grouping such as plumbing or heating systems are removed. This may be accomplished by the easiest way possible, such as sawing, torch cutting, or sledge hammer as well as simple unbolting.

The second type of demolition is the removal of some item for repair or replacement. This removal may involve careful draining, opening of unions, disconnecting and tagging of electrical connections, capping of pipes/ducts to prevent entry of debris or leakage of the material contained as well as transport of the item away from its in-place location to a truck/dumpster. An approximation of the time required to accomplish this type of demolition is to use half of the time indicated as necessary to install a new unit. For example; installation of a new pump might be listed as requiring 6 labor-hours so if we had to estimate the removal of the old pump we would allow an additional 3 hours for a total of 9 hours. That is, the complete replacement of a defective pump with a new pump would be estimated to take 9 labor-hours.

Plumbing — R2205 Common Work Results for Plumbing

R220523-90 Valve Selection Considerations

INTRODUCTION: In any piping application, valve performance is critical. Valves should be selected to give the best performance at the lowest cost.

The following is a list of performance characteristics generally expected of valves.

1. Stopping flow or starting it.
2. Throttling flow (Modulation).
3. Flow direction changing.
4. Checking backflow (Permitting flow in only one direction).
5. Relieving or regulating pressure.

In order to properly select the right valve, some facts must be determined.

A. What liquid or gas will flow through the valve?

B. Does the fluid contain suspended particles?

C. Does the fluid remain in liquid form at all times?

D. Which metals does fluid corrode?

E. What are the pressure and temperature limits? (As temperature and pressure rise, so will the price of the valve.)

F. Is there constant line pressure?

G. Is the valve merely an on-off valve?

H. Will checking of backflow be required?

I. Will the valve operate frequently or infrequently?

Valves are classified by design type into such classifications as Gate, Globe, Angle, Check, Ball, Butterfly and Plug. They are also classified by end connection, stem, pressure restrictions and material such as bronze, cast iron, etc. Each valve has a specific use. A quality valve used correctly will provide a lifetime of trouble-free service, but a high quality valve installed in the wrong service may require frequent attention.

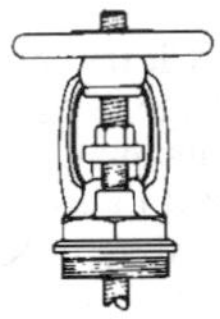

STEM TYPES

(OS & Y)—Rising Stem-Outside Screw and Yoke

Offers a visual indication of whether the valve is open or closed. Recommended where high temperatures, corrosives, and solids in the line might cause damage to inside-valve stem threads. The stem threads are engaged by the yoke bushing so the stem rises through the hand wheel as it is turned.

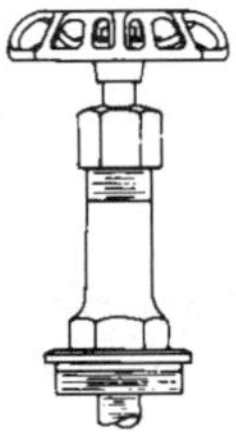

(R.S.)—Rising Stem-Inside Screw

Adequate clearance for operation must be provided because both the hand wheel and the stem rise.
The valve wedge position is indicated by the position of the stem and hand wheel.

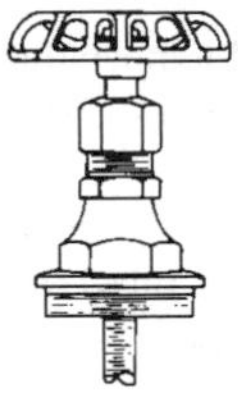

(N.R.S.)—Non-Rising Stem-Inside Screw

A minimum clearance is required for operating this type of valve. Excessive wear or damage to stem threads inside the valve may be caused by heat, corrosion, and solids. Because the hand wheel and stem do not rise, wedge position cannot be visually determined.

VALVE TYPES

Gate Valves

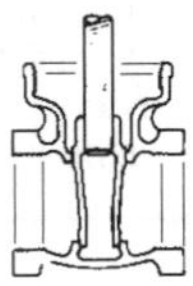

Provide full flow, minute pressure drop, minimum turbulence and minimum fluid trapped in the line.
They are normally used where operation is infrequent.

Globe Valves

Globe valves are designed for throttling and/or frequent operation with positive shut-off. Particular attention must be paid to the several types of seating materials available to avoid unnecessary wear. The seats must be compatible with the fluid in service and may be composition or metal. The configuration of the Globe valve opening causes turbulence which results in increased resistance. Most bronze Globe valves are rising stem-inside screw, but they are also available on O.S. & Y.

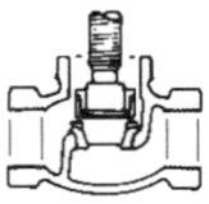

Angle Valves

The fundamental difference between the Angle valve and the Globe valve is the fluid flow through the Angle valve. It makes a 90° turn and offers less resistance to flow than the Globe valve while replacing an elbow. An Angle valve thus reduces the number of joints and installation time.

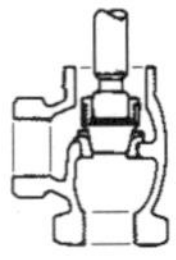

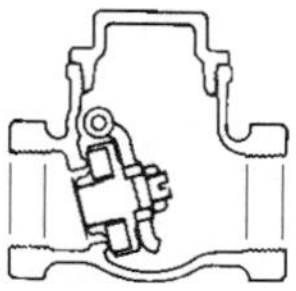

Check Valves

Check valves are designed to prevent backflow by automatically seating when the direction of fluid is reversed.
Swing Check valves are generally installed with Gate-valves, as they provide comparable full flow. Usually recommended for lines where flow velocities are low and should not be used on lines with pulsating flow. Recommended for horizontal installation, or in vertical lines only where flow is upward.

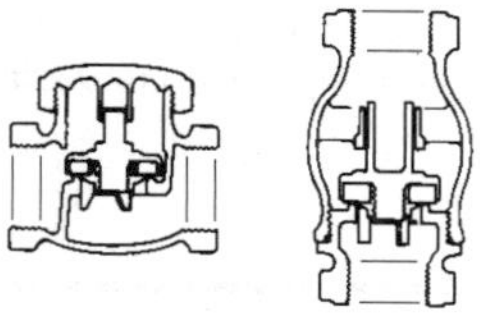

Lift Check Valves

These are commonly used with Globe and Angle valves since they have similar diaphragm seating arrangements and are recommended for preventing backflow of steam, air, gas and water, and on vapor lines with high flow velocities. For horizontal lines, horizontal lift checks should be used and vertical lift checks for vertical lines.

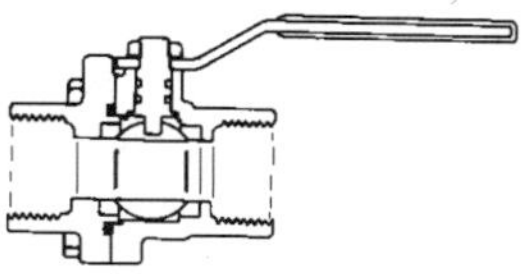

Ball Valves

Ball valves are light and easily installed, yet because of modern elastomeric seats, provide tight closure. Flow is controlled by rotating up to 90° a drilled ball which fits tightly against resilient seals. This ball seats with flow in either direction, and valve handle indicates the degree of opening. Recommended for frequent operation readily adaptable to automation, ideal for installation where space is limited.

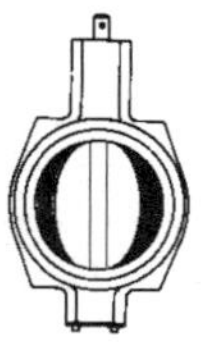

Butterfly Valves

Butterfly valves provide bubble-tight closure with excellent throttling characteristics. They can be used for full-open, closed and for throttling applications.

The Butterfly valve consists of a disc within the valve body which is controlled by a shaft. In its closed position, the valve disc seals against a resilient seat. The disc position throughout the full 90° rotation is visually indicated by the position of the operator.

A Butterfly valve is only a fraction of the weight of a Gate valve and requires no gaskets between flanges in most cases. Recommended for frequent operation and adaptable to automation where space is limited.

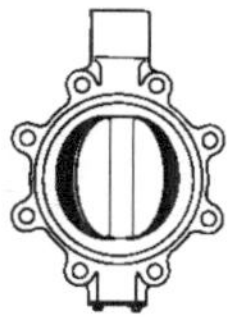

Wafer and Lug type bodies when installed between two pipe flanges, can be easily removed from the line. The pressure of the bolted flanges holds the valve in place.
Locating lugs makes installation easier.

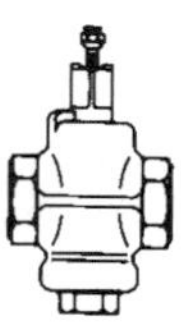

Plug Valves

Lubricated plug valves, because of the wide range of service to which they are adapted, may be classified as all purpose valves. They can be safely used at all pressure and vacuums, and at all temperatures up to the limits of available lubricants. They are the most satisfactory valves for the handling of gritty suspensions and many other destructive, erosive, corrosive and chemical solutions.

R221113-50 Pipe Material Considerations

1. Malleable fittings should be used for gas service.
2. Malleable fittings are used where there are stresses/strains due to expansion and vibration.
3. Cast fittings may be broken as an aid to disassembling of heating lines frozen by long use, temperature and minerals.
4. Cast iron pipe is extensively used for underground and submerged service.
5. Type M (light wall) copper tubing is available in hard temper only and is used for nonpressure and less severe applications than K and L.
6. Type L (medium wall) copper tubing, available hard or soft for interior service.
7. Type K (heavy wall) copper tubing, available in hard or soft temper for use where conditions are severe. For underground and interior service.
8. Hard drawn tubing requires fewer hangers or supports but should not be bent. Silver brazed fittings are recommended, however soft solder is normally used.
9. Type DMV (very light wall) copper tubing designed for drainage, waste and vent plus other non-critical pressure services.

Domestic/Imported Pipe and Fittings Cost

The prices shown in this publication for steel/cast iron pipe and steel, cast iron, malleable iron fittings are based on domestic production sold at the normal trade discounts. The above listed items of foreign manufacture may be available at prices of 1/3 to 1/2 those shown. Some imported items after minor machining or finishing operations are being sold as domestic to further complicate the system.

Caution: Most pipe prices in this book also include a coupling and pipe hangers, which, for the larger sizes, can add significantly to the per-foot cost and should be taken into account when comparing "book cost" with quoted supplier's cost.

R224000-30 Minimum Plumbing Fixture Requirements

Minimum Plumbing Fixture Requirements

Type of Building or Occupancy (2)	Water Closets (14) (Fixtures per Person)		Urinals (5,10) (Fixtures per Person)		Lavatories (Fixtures per Person)		Bathtubs or Showers (Fixtures per Person)	Drinking Fountains (Fixtures per Person) (3, 13)
	Male	Female	Male	Female	Male	Female		
Assembly Places- Theatres, Auditoriums, Convention Halls, etc.-for permanent employee use	1: 1 - 15 2: 16 - 35 3: 36 - 55 Over 55, add 1 fixture for each additional 40 persons	1: 1 - 15 2: 16 - 35 3: 36 - 55	0: 1 - 9 1: 10 - 50 Add one fixture for each additional 50 males		1 per 40	1 per 40		
Assembly Places- Theatres, Auditoriums, Convention Halls, etc. - for public use	1: 1 - 100 2: 101 - 200 3: 201 - 400 Over 400, add 1 fixture for each additional 500 males and 1 for each additional 125 females	3: 1 - 50 4: 51 - 100 8: 101 - 200 11: 201 - 400	1: 1 - 100 2: 101 - 200 3: 201 - 400 4: 401 - 600 Over 600, add 1 fixture for each additional 300 males		1: 1 - 200 2: 201 - 400 3: 401 - 750 Over 750, add 1 fixture for each additional 500 persons	1: 1 - 200 2: 201 - 400 3: 401 - 750		1: 1 - 150 2: 151 - 400 3: 401 - 750 Over 750, add one fixture for each additional 500 persons
Dormitories (9) School or Labor	1 per 10 Add 1 fixture for each additional 25 males (over 10) and 1 for each additional 20 females (over 8)	1 per 8	1 per 25 Over 150, add 1 fixture for each additional 50 males		1 per 12 Over 12 add 1 fixture for each additional 20 males and 1 for each 15 additional females	1 per 12	1 per 8 For females add 1 bathtub per 30. Over 150, add 1 per 20	1 per 150 (12)
Dormitories- for Staff Use	1: 1 - 15 2: 16 - 35 3: 36 - 55 Over 55, add 1 fixture for each additional 40 persons	1: 1 - 15 3: 16 - 35 4: 36 - 55	1 per 50		1 per 40	1 per 40	1 per 8	
Dwellings: Single Dwelling Multiple Dwelling or Apartment House	1 per dwelling 1 per dwelling or apartment unit				1 per dwelling 1 per dwelling or apartment unit		1 per dwelling 1 per dwelling or apartment unit	
Hospital Waiting rooms	1 per room				1 per room			1 per 150 (12)
Hospitals- for employee use	1: 1 - 15 2: 16 - 35 3: 36 - 55 Over 55, add 1 fixture for each additional 40 persons	1: 1 - 15 3: 16 - 35 4: 36 - 55	0: 1 - 9 1: 10 - 50 Add 1 fixture for each additional 50 males		1 per 40	1 per 40		
Hospitals: Individual Room Ward Room	1 per room 1 per 8 patients				1 per room 1 per 10 patients		1 per room 1 per 20 patients	1 per 150 (12)
Industrial (6) Warehouses Workshops, Foundries and similar establishments- for employee use	1: 1 -10 2: 11 - 25 3: 26 - 50 4: 51 - 75 5: 76 - 100 Over 100, add 1 fixture for each additional 30 persons	1: 1 -10 2: 11 - 25 3: 26 - 50 4: 51 - 75 5: 76 - 100			Up to 100, per 10 persons Over 100, 1 per 15 persons (7, 8)		1 shower for each 15 persons exposed to excessive heat or to skin contamination with poisonous, infectious or irritating material	1 per 150 (12)
Institutional - Other than Hospitals or Penal Institutions (on each occupied floor)	1 per 25	1 per 20	0: 1 - 9 1: 10 - 50 Add 1 fixture for each additional 50 males		1 per 10	1 per 10	1 per 8	1 per 150 (12)
Institutional - Other than Hospitals or Penal Institutions (on each occupied floor)- for employee use	1: 1 - 15 2: 16 - 35 3: 36 - 55 Over 55, add 1 fixture for each additional 40 persons	1: 1 - 15 3: 16 - 35 4: 36 - 55	0: 1 - 9 1: 10 - 50 Add 1 fixture for each additional 50 males		1 per 40	1 per 40	1 per 8	1 per 150 (12)
Office or Public Buildings	1: 1 - 100 2: 101 - 200 3: 201 - 400 Over 400, add 1 fixture for each additional 500 males and 1 for each additional 150 females	3: 1 - 50 4: 51 - 100 8: 101 - 200 11: 201 - 400	1: 1 - 100 2: 101 - 200 3: 201 - 400 4: 401 - 600 Over 600, add 1 fixture for each additional 300 males		1: 1 - 200 2: 201 - 400 3: 401 - 750 Over 750, add 1 fixture for each additional 500 persons	1: 1 - 200 2: 201 - 400 3: 401 - 750		1 per 150 (12)
Office or Public Buildings - for employee use	1: 1 - 15 2: 16 - 35 3: 36 - 55 Over 55, add 1 fixture for each additional 40 persons	1: 1 - 15 3: 16 - 35 4: 36 - 55	0: 1 - 9 1: 10 - 50 Add 1 fixture for each additional 50 males		1 per 40	1 per 40		

R224000-30 Minimum Plumbing Fixture Requirements (cont.)

Minimum Plumbing Fixture Requirements

Type of Building or Occupancy	Water Closets (14) (Fixtures per Person)		Urinals (5, 10) (Fixtures per Person)		Lavatories (Fixtures per Person)		Bathtubs or Showers (Fixtures per Person)	Drinking Fountains (Fixtures per Person) (3, 13)
	Male	Female	Male	Female	Male	Female		
Penal Institutions - for employee use	1: 1 - 15 2: 16 - 35 3: 36 - 55	1: 1 - 15 3: 16 - 35 4: 36 - 55	0: 1 - 9 1: 10 - 50		1 per 40	1 per 40		1 per 150 (12)
	Over 55, add 1 fixture for each additional 40 persons		Add 1 fixture for each additional 50 males					
Penal Institutions - for prison use								
Cell	1 per cell				1 per cell			1 per cellblock floor
Exercise room	1 per exercise room		1 per exercise room		1 per exercise room			1 per exercise room
Restaurants, Pubs and Lounges (11)	1: 1 - 50 2: 51 - 150 3: 151 - 300	1: 1 - 50 2: 51 - 150 4: 151 - 300	1: 1 - 150		1: 1 - 150 2: 151 - 200 3: 201 - 400	1: 1 - 150 2: 151 - 200 3: 201 - 400		
	Over 300, add 1 fixture for each additional 200 persons		Over 150, add 1 fixture for each additional 150 males		Over 400, add 1 fixture for each additional 400 persons			
Schools - for staff use All Schools	1: 1 - 15 2: 16 - 35 3: 36 - 55	1: 1 - 15 3: 16 - 35 4: 36 - 55	1 per 50		1 per 40	1 per 40		
	Over 55, add 1 fixture for each additional 40 persons							
Schools - for student use:								
Nursery	1: 1 - 20 2: 21 - 50	1: 1 - 20 2: 21 - 50			1: 1 - 25 2: 26 - 50	1: 1 - 25 2: 26 - 50		1 per 150 (12)
	Over 50, add 1 fixture for each additional 50 persons				Over 50, add 1 fixture for each additional 50 persons			
Elementary	1 per 30	1 per 25	1 per 75		1 per 35	1 per 35		1 per 150 (12)
Secondary	1 per 40	1 per 30	1 per 35		1 per 40	1 per 40		1 per 150 (12)
Others (Colleges, Universities, Adult Centers, etc.	1 per 40	1 per 30	1 per 35		1 per 40	1 per 40		1 per 150 (12)
Worship Places Educational and Activities Unit	1 per 150	1 per 75	1 per 150		1 per 2 water closets			1 per 150 (12)
Worship Places Principal Assembly Place	1 per 150	1 per 75	1 per 150		1 per 2 water closets			1 per 150 (12)

Notes:

1. The figures shown are based upon one (1) fixture being the minimum required for the number of persons indicated or any fraction thereof.
2. Building categories not shown on this table shall be considered separately by the Administrative Authority.
3. Drinking fountains shall not be installed in toilet rooms.
4. Laundry trays. One (1) laundry tray or one (1) automatic washer standpipe for each dwelling unit or one (1) laundry trays or one (1) automatic washer standpipes, or combination thereof, for each twelve (12) apartments. Kitchen sinks, one (1) for each dwelling or apartment unit.
5. For each urinal added in excess of the minimum required, one water closet may be deducted. The number of water closets shall not be reduced to less than two-thirds (2/3) of the minimum requirement.
6. As required by ANSI Z4.1-1968, Sanitation in Places of Employment.
7. Where there is exposure to skin contamination with poisonous, infectious, or irritating materials, provide one (1) lavatory for each five (5) persons.
8. Twenty-four (24) lineal inches of wash sink or eighteen (18) inches of a circular basin, when provided with water outlets for such space shall be considered equivalent to one (1) lavatory.
9. Laundry trays, one (1) for each fifty (50) persons. Service sinks, one (1) for each hundred (100) persons.
10. General. In applying this schedule of facilities, consideration shall be given to the accessibility of the fixtures. Conformity purely on a numerical basis may not result in an installation suited to the need of the individual establishment. For example, schools should be provided with toilet facilities on each floor having classrooms.
    a. Surrounding materials, wall and floor space to a point two (2) feet in front of urinal lip and four (4) feet above the floor, and at least two (2) feet to each side of the urinal shall be lined with non-absorbent materials.
    b. Trough urinals shall be prohibited.
11. A restaurant is defined as a business which sells food to be consumed on the premises.
    a. The number of occupants for a drive-in restaurant shall be considered as equal to the number of parking stalls.
    b. Employee toilet facilities shall not to be included in the above restaurant requirements. Hand washing facilities shall be available in the kitchen for employees.
12. Where food is consumed indoors, water stations may be substituted for drinking fountains. Offices, or public buildings for use by more than six (6) persons shall have one (1) drinking fountain for the first one hundred fifty (150) persons and one additional fountain for each three hundred (300) persons thereafter.
13. There shall be a minimum of one (1) drinking fountain per occupied floor in schools, theaters, auditoriums, dormitories, offices of public building.
14. The total number of water closets for females shall be at least equal to the total number of water closets and urinals required for males.

R233100-40 Sheet Metal Calculator (Weight in Lb./Ft. of Length)

Gauge	26	24	22	20	18	16
Wt.-Lb./S.F.	.906	1.156	1.406	1.656	2.156	2.656
SMACNA Max. Dimension – Long Side		30″	54″	84″	85″ Up	
Sum-2 sides						
2	.3	.40	.50	.60	.80	.90
3	.5	.65	.80	.90	1.1	1.4
4	.7	.85	1.0	1.2	1.5	1.8
5	.8	1.1	1.3	1.5	1.9	2.3
6	1.0	1.3	1.5	1.7	2.3	2.7
7	1.2	1.5	1.8	2.0	2.7	3.2
8	1.3	1.7	2.0	2.3	3.0	3.6
9	1.5	1.9	2.3	2.6	3.4	4.1
10	1.7	2.2	2.5	2.9	3.8	4.5
11	1.8	2.4	2.8	3.2	4.2	5.0
12	2.0	2.6	3.0	3.5	4.6	5.4
13	2.2	2.8	3.3	3.8	4.9	5.9
14	2.3	3.0	3.5	4.1	5.3	6.3
15	2.5	3.2	3.8	4.4	5.7	6.8
16	2.7	3.4	4.0	4.6	6.1	7.2
17	2.8	3.7	4.3	4.9	6.5	7.7
18	3.0	3.9	4.5	5.2	6.8	8.1
19	3.2	4.1	4.8	5.5	7.2	8.6
20	3.3	4.3	5.0	5.8	7.6	9.0
21	3.5	4.5	5.3	6.1	8.0	9.5
22	3.7	4.7	5.5	6.4	8.4	9.9
23	3.8	5.0	5.8	6.7	8.7	10.4
24	4.0	5.2	6.0	7.0	9.1	10.8
25	4.2	5.4	6.3	7.3	9.5	11.3
26	4.3	5.6	6.5	7.5	9.9	11.7
27	4.5	5.8	6.8	7.8	10.3	12.2
28	4.7	6.0	7.0	8.1	10.6	12.6
29	4.8	6.2	7.3	8.4	11.0	13.1
30	5.0	6.5	7.5	8.7	11.4	13.5
31	5.2	6.7	7.8	9.0	11.8	14.0
32	5.3	6.9	8.0	9.3	12.2	14.4
33	5.5	7.1	8.3	9.6	12.5	14.9
34	5.7	7.3	8.5	9.9	12.9	15.3
35	5.8	7.5	8.8	10.2	13.3	15.8
36	6.0	7.8	9.0	10.4	13.7	16.2
37	6.2	8.0	9.3	10.7	14.1	16.7
38	6.3	8.2	9.5	11.0	14.4	17.1
39	6.5	8.4	9.8	11.3	14.8	17.6
40	6.7	8.6	10.0	11.6	15.2	18.0
41	6.8	8.8	10.3	11.9	15.6	18.5
42	7.0	9.0	10.5	12.2	16.0	18.9
43	7.2	9.2	10.8	12.5	16.3	19.4
44	7.3	9.5	11.0	12.8	16.7	19.8
45	7.5	9.7	11.3	13.1	17.1	20.3
46	7.7	9.9	11.5	13.3	17.5	20.7
47	7.8	10.1	11.8	13.6	17.9	21.2
48	8.0	10.3	12.0	13.9	18.2	21.6
49	8.2	10.5	12.3	14.2	18.6	22.1
50	8.3	10.7	12.5	14.5	19.0	22.5
51	8.5	11.0	12.8	14.8	19.4	23.0
52	8.7	11.2	13.0	15.1	19.8	23.4
53	8.8	11.4	13.3	15.4	20.1	23.9
54	9.0	11.6	13.5	15.7	20.5	24.3
55	9.2	11.8	13.8	16.0	20.9	24.8
56	9.3	12.0	14.0	16.2	21.3	25.2
57	9.5	12.3	14.3	16.5	21.7	25.7
58	9.7	12.5	14.5	16.8	22.0	26.1
59	9.8	12.7	14.8	17.1	22.4	26.6
60	10.0	12.9	15.0	17.4	22.8	27.0
61	10.2	13.1	15.3	17.7	23.2	27.5
62	10.3	13.3	15.5	18.0	23.6	27.9
63	10.5	13.5	15.8	18.3	24.0	28.4
64	10.7	13.7	16.0	18.6	24.3	28.8
65	10.8	13.9	16.3	18.9	24.7	29.3
66	11.0	14.1	16.5	19.1	25.1	29.7
67	11.2	14.3	16.8	19.4	25.5	30.2
68	11.3	14.6	17.0	19.7	25.8	30.6
69	11.5	14.8	17.3	20.0	26.2	31.1
70	11.7	15.0	17.5	20.3	26.6	31.5
71	11.8	15.2	17.8	20.6	27.0	32.0
72	12.0	15.4	18.0	20.9	27.4	32.4
73	12.2	15.6	18.3	21.2	27.7	32.9
74	12.3	15.8	18.5	21.5	28.1	33.3
75	12.5	16.1	18.8	21.8	28.5	33.8
76	12.7	16.3	19.0	22.0	28.9	34.2
77	12.8	16.5	19.3	22.3	29.3	34.7
78	13.0	16.7	19.5	22.6	29.6	35.1
79	13.2	16.9	19.8	22.9	30.0	35.6
80	13.3	17.1	20.0	23.2	30.4	36.0
81	13.5	17.3	20.3	23.5	30.8	36.5
82	13.7	17.5	20.5	23.8	31.2	36.9
83	13.8	17.8	20.8	24.1	31.5	37.4
84	14.0	18.0	21.0	24.4	31.9	37.8
85	14.2	18.2	21.3	24.7	32.3	38.3
86	14.3	18.4	21.5	24.9	32.7	38.7
87	14.5	18.6	21.8	25.2	33.1	39.2
88	14.7	18.8	22.0	25.5	33.4	39.6
89	14.8	19.0	22.3	25.8	33.8	40.1
90	15.0	19.3	22.5	26.1	34.2	40.5
91	15.2	19.5	22.8	26.4	34.6	41.0
92	15.3	19.7	23.0	26.7	35.0	41.4
93	15.5	19.9	23.3	27.0	35.3	41.9
94	15.7	20.1	23.5	27.3	35.7	42.3
95	15.8	20.3	23.8	27.6	36.1	42.8
96	16.0	20.5	24.0	27.8	36.5	43.2
97	16.2	20.8	24.3	28.1	36.9	43.7
98	16.3	21.0	24.5	28.4	37.2	44.1
99	16.5	21.2	24.8	28.7	37.6	44.6
100	16.7	21.4	25.0	29.0	38.0	45.0
101	16.8	21.6	25.3	29.3	38.4	45.5
102	17.0	21.8	25.5	29.6	38.8	45.9
103	17.2	22.0	25.8	29.9	39.1	46.4
104	17.3	22.3	26.0	30.2	39.5	46.8
105	17.5	22.5	26.3	30.5	39.9	47.3
106	17.7	22.7	26.5	30.7	40.3	47.7
107	17.8	22.9	26.8	31.0	40.7	48.2
108	18.0	23.1	27.0	31.3	41.0	48.6
109	18.2	23.3	27.3	31.6	41.4	49.1
110	18.3	23.5	27.5	31.9	41.8	49.5

Example: If duct is 34″ x 20″ x 15′ long, 34″ is greater than 30″ maximum, for 24 ga. so must be 22 ga. 34″ + 20″ = 54″ going across from 54″ find 13.5 lb. per foot. 13.5 x 15′ = 202.5 lbs. For S.F. of surface area 202.5 ÷ 1.406 = 144 S.F.

Note: Figures include an allowance for scrap.

R233400-10 Recommended Ventilation Air Changes

Table below lists range of time in minutes per change for various types of facilities.

Assembly Halls	2-10	Dance Halls	2-10	Laundries	1-3
Auditoriums	2-10	Dining Rooms	3-10	Markets	2-10
Bakeries	2-3	Dry Cleaners	1-5	Offices	2-10
Banks	3-10	Factories	2-5	Pool Rooms	2-5
Bars	2-5	Garages	2-10	Recreation Rooms	2-10
Beauty Parlors	2-5	Generator Rooms	2-5	Sales Rooms	2-10
Boiler Rooms	1-5	Gymnasiums	2-10	Theaters	2-8
Bowling Alleys	2-10	Kitchens-Hospitals	2-5	Toilets	2-5
Churches	5-10	Kitchens-Restaurant	1-3	Transformer Rooms	1-5

CFM air required for changes = Volume of room in cubic feet ÷ Minutes per change.

R235000-50 Factor for Determining Heat Loss for Various Types of Buildings

General: While the most accurate estimates of heating requirements would naturally be based on detailed information about the building being considered, it is possible to arrive at a reasonable approximation using the following procedure:

1. Calculate the cubic volume of the room or building.
2. Select the appropriate factor from Table 1 below. Note that the factors apply only to inside temperatures listed in the first column and to 0° F outside temperature.
3. If the building has bad north and west exposures, multiply the heat loss factor by 1.1.
4. If the outside design temperature is other than 0° F, multiply the factor from Table 1 by the factor from Table 2.
5. Multiply the cubic volume by the factor selected from Table 1. This will give the estimated BTUH heat loss which must be made up to maintain inside temperature.

Table 1 — Building Type	Conditions	Qualifications	Loss Factor*
Factories & Industrial Plants General Office Areas at 70° F	One Story	Skylight in Roof	6.2
		No Skylight in Roof	5.7
	Multiple Story	Two Story	4.6
		Three Story	4.3
		Four Story	4.1
		Five Story	3.9
		Six Story	3.6
	All Walls Exposed	Flat Roof	6.9
		Heated Space Above	5.2
	One Long Warm Common Wall	Flat Roof	6.3
		Heated Space Above	4.7
	Warm Common Walls on Both Long Sides	Flat Roof	5.8
		Heated Space Above	4.1
Warehouses at 60° F	All Walls Exposed	Skylights in Roof	5.5
		No Skylight in Roof	5.1
		Heated Space Above	4.0
	One Long Warm Common Wall	Skylight in Roof	5.0
		No Skylight in Roof	4.9
		Heated Space Above	3.4
	Warm Common Walls on Both Long Sides	Skylight in Roof	4.7
		No Skylight in Roof	4.4
		Heated Space Above	3.0

*Note: This table tends to be conservative particularly for new buildings designed for minimum energy consumption.

Table 2 — Outside Design Temperature Correction Factor (for Degrees Fahrenheit)									
Outside Design Temperature	50	40	30	20	10	0	-10	-20	-30
Correction Factor	.29	.43	.57	.72	.86	1.00	1.14	1.28	1.43

R236000-30 Psychrometric Table

Dewpoint or Saturation Temperature (F)

Relative humidity (%)															
100	32	35	40	45	50	55	60	65	70	75	80	85	90	95	100
90	30	33	37	42	47	52	57	62	67	72	77	82	87	92	97
80	27	30	34	39	44	49	54	58	64	68	73	78	83	88	93
70	24	27	31	36	40	45	50	55	60	64	69	74	79	84	88
60	20	24	28	32	36	41	46	51	55	60	65	69	74	79	83
50	16	20	24	28	33	36	41	46	50	55	60	64	69	73	78
40	12	15	18	23	27	31	35	40	45	49	53	58	62	67	71
30	8	10	14	18	21	25	29	33	37	42	46	50	54	59	62
20	6	7	8	9	13	16	20	24	28	31	35	40	43	48	52
10	4	4	5	5	6	8	9	10	13	17	20	24	27	30	34
	32	**35**	**40**	**45**	**50**	**55**	**60**	**65**	**70**	**75**	**80**	**85**	**90**	**95**	**100**

Dry bulb temperature (F)

This table shows the relationship between RELATIVE HUMIDITY, DRY BULB TEMPERATURE AND DEWPOINT.

As an example, assume that the thermometer in a room reads 75° F, and we know that the relative humidity is 50%. The chart shows the dewpoint temperature to be 55° F. That is, any surface colder than 55° F will "sweat" or collect condensing moisture. This surface could be the outside of an uninsulated chilled water pipe in the summertime, or the inside surface of a wall or deck in the wintertime. After determining the extreme ambient parameters, the table at the left is useful in determining which surfaces need insulation or vapor barrier protection.

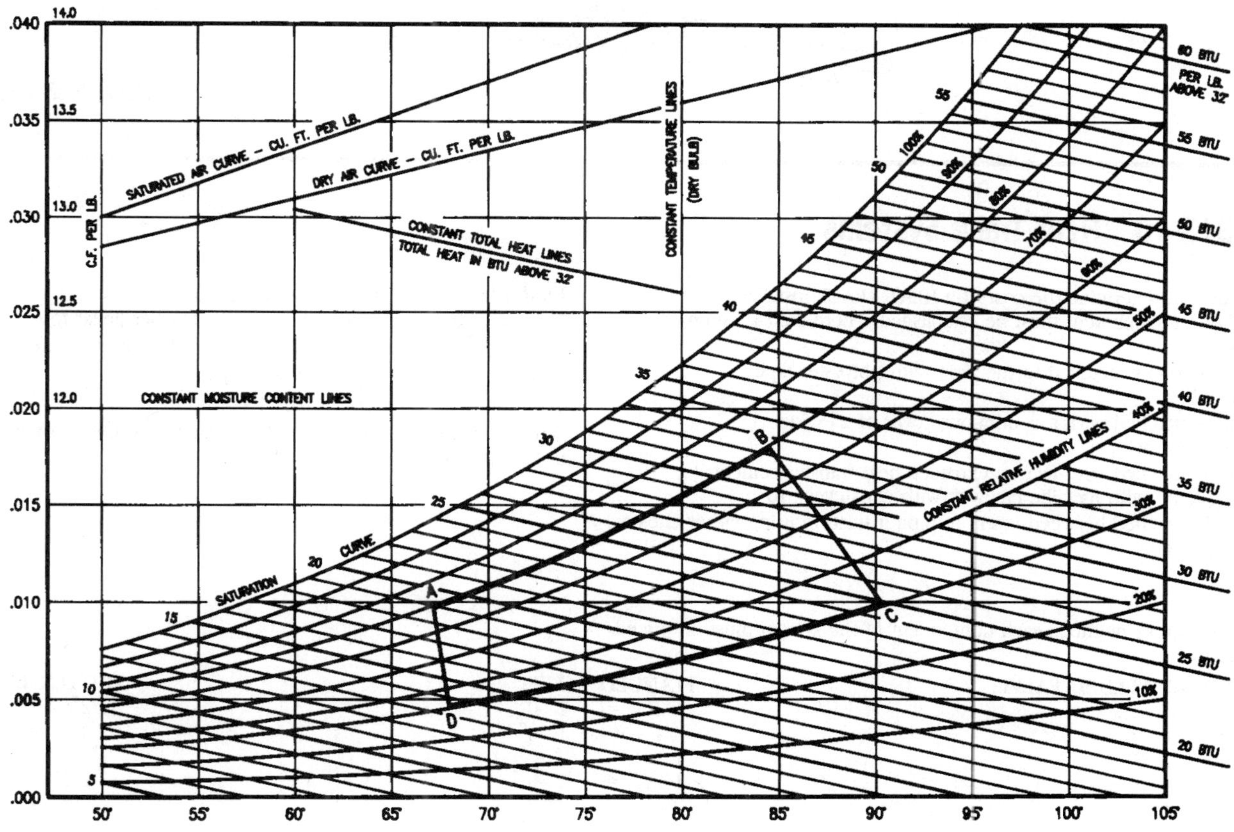

TEMPERATURE DEGREES FAHRENHEIT
TOTAL PRESSURE = 14.696 LB. PER SQ. IN. ABS.

Psychrometric chart showing different variables based on one pound of dry air. Space marked A B C D is temperature-humidity range which is most comfortable for majority of people.

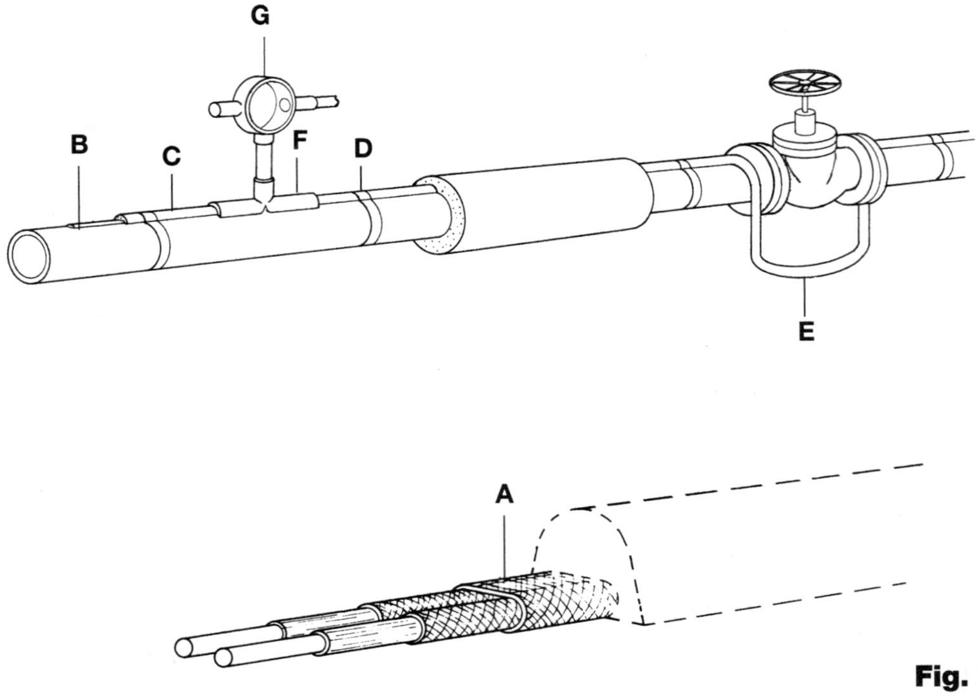

Fig. R238313-11

R238313-10 Heat Trace Systems

Before you can determine the cost of a HEAT TRACE installation the method of attachment must be established. There are (4) common methods:

1. Cable is simply attached to the pipe with polyester tape every 12′.
2. Cable is attached with a continuous cover of 2″ wide aluminum tape.
3. Cable is attached with factory extruded heat transfer cement and covered with metallic raceway with clips every 10′.
4. Cable is attached between layers of pipe insulation using either clips or polyester tape.

Example: Components for method 3 must include:

A. Heat trace cable by voltage and watts per linear foot.
B. Heat transfer cement, 1 gallon per 60 linear feet of cover.
C. Metallic raceway by size and type.
D. Raceway clips by size of pipe.

When taking off linear foot lengths of cable add the following for each valve in the system. (E)

In all of the above methods each component of the system must be priced individually.

SCREWED OR WELDED VALVE:

1/2″	=	6″
3/4″	=	9″
1″	=	1′ -0″
1-1/2″	=	1′ -6″
2″	=	2′
2-1/2″	=	2′ -6″
3″	=	2′ -6″
4″	=	4′ -0″
6″	=	7′ -0″
8″	=	9′ -6″
10″	=	12′ -6″
12″	=	15′ -0″
14″	=	18′ -0″
16″	=	21′ -6″
18″	=	25′ -6″
20″	=	28′ -6″
24″	=	34′ -0″
30″	=	40′ -0″

FLANGED VALVE:

1/2″	=	1′ -0″
3/4″	=	1′ -6″
1″	=	2′ -0″
1-1/2″	=	2′ -6″
2″	=	2′ -6″
2-1/2″	=	3′ -0″
3″	=	3′ -6″
4″	=	4′ -0″
6″	=	8′ -0″
8″	=	11′ -0″
10″	=	14′ -0″
12″	=	16′ -6″
14″	=	19′ -6″
16″	=	23′ -0″
18″	=	27′ -0″
20″	=	30′ -0″
24″	=	36′ -0″
30″	=	42′ -0″

BUTTERFLY VALVE:

1/2″	=	0′
3/4″	=	0′
1″	=	1′ -0″
1-1/2″	=	1′ -6″
2″	=	2′ -0″
2-1/2″	=	2′ -6″
3″	=	2′ -6″
4″	=	3′ -0″
6″	=	3′ -6″
8″	=	4′ -0″
10″	=	4′ -0″
12″	=	5′ -0″
14″	=	5′ -6″
16″	=	6′ -0″
18″	=	6′ -6″
20″	=	7′ -0″
24″	=	8′ -0″
30″	=	10′ -0″

R238313-10 Heat Trace Systems (cont.)

Add the following quantities of heat transfer cement to linear foot totals for each valve:

Nominal Valve Size	Gallons of Cement per Valve
1/2″	0.14
3/4″	0.21
1″	0.29
1-1/2″	0.36
2″	0.43
2-1/2″	0.70
3″	0.71
4″	1.00
6″	1.43
8″	1.48
10″	1.50
12″	1.60
14″	1.75
16″	2.00
18″	2.25
20″	2.50
24″	3.00
30″	3.75

The following must be added to the list of components to accurately price HEAT TRACE systems:

1. Expediter fitting and clamp fasteners (F)
2. Junction box and nipple connected to expediter fitting (G)
3. Field installed terminal blocks within junction box
4. Ground lugs
5. Piping from power source to expediter fitting
6. Controls
7. Thermostats
8. Branch wiring
9. Cable splices
10. End of cable terminations
11. Branch piping fittings and boxes

Deduct the following percentages from labor if cable lengths in the same area exceed:

150′ to 250′	10%	351′ to 500′	20%
251′ to 350′	15%	Over 500′	25%

Add the following percentages to labor for elevated installations:

15′ to 20′ high	10%	31′ to 35′ high	40%
21′ to 25′ high	20%	36′ to 40′ high	50%
26′ to 30′ high	30%	Over 40′ high	60%

R238313-20 Spiral-Wrapped Heat Trace Cable (Pitch Table)

In order to increase the amount of heat, occasionally heat trace cable is wrapped in a spiral fashion around a pipe; increasing the number of feet of heater cable per linear foot of pipe.

Engineers first determine the heat loss per foot of pipe (based on the insulating material, its thickness, and the temperature differential across it). A ratio is then calculated by the formula:

$$\text{Feet of Heat Trace per Foot of Pipe} = \frac{\text{Watts/Foot of Heat Loss}}{\text{Watts/Foot of the Cable}}$$

The linear distance between wraps (pitch) is then taken from a chart or table. Generally, the pitch is listed on a drawing leaving the estimator to calculate the total length of heat tape required. An approximation may be taken from this table.

Feet of Heat Trace Per Foot of Pipe																
	Nominal Pipe Size in Inches															
Pitch In Inches	1	1¼	1½	2	2½	3	4	6	8	10	12	14	16	18	20	24
3.5	1.80															
4	1.65															
5	1.46	1.60	1.80													
6	1.34	1.45	1.55	1.75												
7	1.25	1.35	1.43	1.57	1.75											
8	1.20	1.28	1.34	1.45	1.60	1.80										
9	1.16	1.23	1.28	1.37	1.51	1.68										
10	1.13	1.19	1.24	1.32	1.44	1.57	1.82									
15	1.06	1.08	1.10	1.15	1.21	1.29	1.42	1.78								
20	1.04	1.05	1.06	1.08	1.13	1.17	1.25	1.49	1.73							
25		1.04	1.04	1.06	1.08	1.11	1.17	1.33	1.51	1.72						
30				1.04	1.05	1.07	1.12	1.24	1.37	1.54	1.70	1.80				
35					1.06	1.06	1.09	1.17	1.28	1.42	1.54	1.64	1.78			
40						1.05	1.07	1.14	1.22	1.33	1.44	1.52	1.64	1.75		
50							1.05	1.09	1.15	1.22	1.29	1.35	1.44	1.53	1.64	1.83
60								1.06	1.11	1.16	1.21	1.25	1.31	1.39	1.46	1.62
70								1.05	1.08	1.12	1.17	1.19	1.24	1.30	1.35	1.47
80									1.06	1.09	1.13	1.15	1.19	1.24	1.28	1.38
90									1.04	1.06	1.10	1.13	1.16	1.19	1.23	1.32
100										1.05	1.08	1.10	1.13	1.15	1.19	1.23

Note: Common practice would normally limit the lower end of the table to 5% of additional heat and above 80% an engineer would likely opt for two (2) parallel cables.

R260105-30 Electrical Demolition (Removal for Replacement)

The purpose of this reference number is to provide a guide to users for electrical "removal for replacement" by applying the rule of thumb: 1/3 of new installation time (typical range from 20% to 50%) for removal. Remember to use reasonable judgment when applying the suggested percentage factor. For example:

Contractors have been requested to remove an existing fluorescent lighting fixture and replace with a new fixture utilizing energy saver lamps and electronic ballast.

In order to fully understand the extent of the project, contractors should visit the job site and estimate the time to perform the renovation work in accordance with applicable national, state and local regulation and codes.

The contractor may need to add extra labor hours to his estimate if he discovers unknown concealed conditions such as: contaminated asbestos ceiling, broken acoustical ceiling tile and need to repair, patch and touch-up paint all the damaged or disturbed areas, tasks normally assigned to general contractors. In addition, the owner could request that the contractors salvage the materials removed and turn over the materials to the owner or dispose of the materials to a reclamation station. The normal removal item is 0.5 labor hour for a lighting fixture and 1.5 labor-hours for new installation time. Revise the estimate times from 2 labor-hours work up to a minimum 4 labor-hours work for just fluorescent lighting fixture.

For removal of large concentrations of lighting fixtures in the same area, apply an "economy of scale" to reduce estimating labor-hours.

R260519-80 Undercarpet Systems

Takeoff Procedure for Power Systems: List components for each fitting type, tap, splice, and bend on your quantity takeoff sheet. Each component must be priced separately. Start at the power supply transition fittings and survey each circuit for the components needed. List the quantities of each component under a specific circuit number. Use the floor plan layout scale to get cable footage.

Reading across the list, combine the totals of each component in each circuit and list the total quantity in the last column. Calculate approximately 5% for scrap for items such as cable, top shield, tape, and spray adhesive. Also provide for final variations that may occur on-site.

Suggested guidelines are:
1. Equal amounts of cable and top shield should be priced.
2. For each roll of cable, price a set of cable splices.
3. For every 1 ft. of cable, price 2-1/2 ft. of hold-down tape.
4. For every 3 rolls of hold-down tape, price 1 can of spray adhesive.

Adjust final figures wherever possible to accommodate standard packaging of the product. This information is available from the distributor.

Each transition fitting requires:
1. 1 base
2. 1 cover
3. 1 transition block

Each floor fitting requires:
1. 1 frame/base kit
2. 1 transition block
3. 2 covers (duplex/blank)

Each tap requires:
1. 1 tap connector for each conductor
2. 1 pair insulating patches
3. 2 top shield connectors

Each splice requires:
1. 1 splice connector for each conductor
2. 1 pair insulating patches
3. 3 top shield connectors

Each cable bend requires:
1. 2 top shield connectors

Each cable dead end (outside of transition block) requires:
1. 1 pair insulating patches

Labor does not include:
1. Patching or leveling uneven floors
2. Filling in holes or removing projections from concrete slabs.
3. Sealing porous floors
4. Sweeping and vacuuming floors
5. Removal of existing carpeting
6. Carpet square cut-outs
7. Installation of carpet squares

Takeoff Procedures for Telephone Systems: After reviewing floor plans identify each transition. Number or letter each cable run from that fitting. Start at the transition fitting and survey each circuit for the components needed. List the cable type, terminations, cable length, and floor fitting type under the specific circuit number. Use the floor plan layout scale to get the cable footage. Add some extra length (next higher increment of 5 feet) to preconnectorized cable.

Transition fittings require:
1. 1 base plate
2. 1 cover
3. 1 transition block

Floor fittings require:
1. 1 frame/base kit
2. 2 covers
3. Modular jacks

Reading across the list, combine the list of components in each circuit and list the total quantity in the last column. Calculate the necessary scrap factors for such items as tape, bottom shield and spray adhesive. Also provide for final variations that may occur on-site.

Adjust final figures whenever possible to accommodate standard packaging. Check that items such as transition fittings, floor boxes, and floor fittings that are to utilize both power and telephone have been priced as combination fittings, so as to avoid duplication.

Make sure to include marking of floors and drilling of fasteners if fittings specified are not the adhesive type.

Labor does not include:
1. Conduit or raceways before transition of floor boxes
2. Telephone cable before transition boxes
3. Terminations before transition boxes
4. Floor preparation as described in power section

Be sure to include all cable folds when pricing labor.

Takeoff Procedure for Data Systems: Start at the transition fittings and take off quantities in the same manner as the telephone system, keeping in mind that data cable does not require top or bottom shields.

The data cable is simply cross-taped on the cable run to the floor fitting.

Data cable can be purchased in either bulk form in which case coaxial connector material and labor must be priced, or in preconnectorized cut lengths.

Data cable cannot be folded and must be notched at 1 inch intervals. A count of all turns must be added to the labor portion of the estimate. (Note: Some manufacturers have prenotched cable.)

Notching required:
1. 90 degree turn requires 8 notches per side
2. 180 degree turn requires 16 notches per side

Floor boxes, transition boxes, and fittings are the same as described in the power and telephone procedures.

Since undercarpet systems require special hand tools, be sure to include this cost in proportion to number of crews involved in the installation.

Job Conditions: Productivity is based on new construction in an unobstructed area. Staging area is assumed to be within 200′ of work being performed.

R260519-90 Wire

Wire quantities are taken off by either measuring each cable run or by extending the conduit and raceway quantities times the number of conductors in the raceway. Ten percent should be added for waste and tie-ins. Keep in mind that the unit of measure of wire is C.L.F. not L.F. as in raceways so the formula would read:

$$\frac{\text{(L.F. Raceway x No. of Conductors) x 1.10}}{100} = \text{C.L.F.}$$

Price per C.L.F. of wire includes:
1. Set up wire coils or spools on racks
2. Attaching wire to pull in means
3. Measuring and cutting wire
4. Pulling wire into a raceway
5. Identifying and tagging

Price does not include:
1. Connections to breakers, panelboards, or equipment
2. Splices

Job Conditions: Productivity is based on new construction to a height of 15′ using rolling staging in an unobstructed area. Material staging is assumed to be within 100′ of work being performed.

Economy of Scale: If more than three wires at a time are being pulled, deduct the following percentages from the labor of that grouping:

4-5 wires	25%
6-10 wires	30%
11-15 wires	35%
over 15 wires	40%

If a wire pull is less than 100′ in length and is interrupted several times by boxes, lighting outlets, etc., it may be necessary to add the following lengths to each wire being pulled:

Junction box to junction box	2 L.F.
Lighting panel to junction box	6 L.F.
Distribution panel to sub panel	8 L.F.
Switchboard to distribution panel	12 L.F.
Switchboard to motor control center	20 L.F.
Switchboard to cable tray	40 L.F.

Measure of Drops and Riser: It is important when taking off wire quantities to include the wire for drops to electrical equipment. If heights of electrical equipment are not clearly stated, use the following guide:

	Bottom A.F.F.	Top A.F.F.	Inside Cabinet
Safety switch to 100A	5′	6′	2′
Safety switch 400 to 600A	4′	6′	3′
100A panel 12 to 30 circuit	4′	6′	3′
42 circuit panel	3′	6′	4′
Switch box	3′	3′6″	1′
Switchgear	0′	8′	8′
Motor control centers	0′	8′	8′
Transformers - wall mount	4′	8′	2′
Transformers - floor mount	0′	12′	4′

R260519-91 Maximum Circuit Length (approximate) for Various Power Requirements Assuming THW, Copper Wire @ 75°C, Based Upon a 4% Voltage Drop

Maximum Circuit Length: Table R260519-91 indicates typical maximum installed length a circuit can have and still maintain an adequate voltage level at the point of use. The circuit length is similar to the conduit length.

If the circuit length for an ampere load and a copper wire size exceeds the length obtained from Table R260519-91, use the next largest wire size to compensate voltage drop.

Example: A 130 ampere load at 480 volts, 3 phase, 3 wire with No. 1 wire can be run a maximum of 555 L.F. and provide satisfactory operation. If the same load is to be wired at the end of a 625 L.F. circuit, then a larger wire must be used.

		Maximum Circuit Length in Feet				
	Wire	2 Wire, 1 Phase		3 Wire, 3 Phase		
Amperes	Size	120V	240V	240V	480V	600V
15	14*	50	105	120	240	300
	14	50	100	120	235	295
20	12*	60	125	145	290	360
	12	60	120	140	280	350
30	10*	65	130	155	305	380
	10	65	130	150	300	375
50	8	60	125	145	285	355
65	6	75	150	175	345	435
85	4	90	185	210	425	530
115	2	110	215	250	500	620
130	1	120	240	275	555	690
150	1/0	130	260	305	605	760
175	2/0	140	285	330	655	820
200	3/0	155	315	360	725	904
230	4/0	170	345	395	795	990
255	250	185	365	420	845	1055
285	300	195	395	455	910	1140
310	350	210	420	485	975	1220
380	500	245	490	565	1130	1415

*Solid Conductor

Note: The circuit length is the one-way distance between the origin and the load.

R260519-92 Minimum Copper and Aluminum Wire Size Allowed for Various Types of Insulation

Minimum Wire Sizes									
	Copper		Aluminum			Copper		Aluminum	
Amperes	THW THWN or XHHW	THHN XHHW *	THW XHHW	THHN XHHW *	Amperes	THW THWN or XHHW	THHN XHHW *	THW XHHW	THHN XHHW *
15A	#14	#14	#12	#12	195	3/0	2/0	250kcmil	4/0
20	#12	#12	#10	#10	200	3/0	3/0	250kcmil	4/0
25	#10	#10	#10	#10	205	4/0	3/0	250kcmil	4/0
30	#10	#10	# 8	# 8	225	4/0	3/0	300kcmil	250kcmil
40	# 8	# 8	# 8	# 8	230	4/0	4/0	300kcmil	250kcmil
45	# 8	# 8	# 6	# 8	250	250kcmil	4/0	350kcmil	300kcmil
50	# 8	# 8	# 6	# 6	255	250kcmil	4/0	400kcmil	300kcmil
55	# 6	# 8	# 4	# 6	260	300kcmil	4/0	400kcmil	350kcmil
60	# 6	# 6	# 4	# 6	270	300kcmil	250kcmil	400kcmil	350kcmil
65	# 6	# 6	# 4	# 4	280	300kcmil	250kcmil	500kcmil	350kcmil
75	# 4	# 6	# 3	# 4	285	300kcmil	250kcmil	500kcmil	400kcmil
85	# 4	# 4	# 2	# 3	290	350kcmil	250kcmil	500kcmil	400kcmil
90	# 3	# 4	# 2	# 2	305	350kcmil	300kcmil	500kcmil	400kcmil
95	# 3	# 4	# 1	# 2	310	350kcmil	300kcmil	500kcmil	500kcmil
100	# 3	# 3	# 1	# 2	320	400kcmil	300kcmil	600kcmil	500kcmil
110	# 2	# 3	1/0	# 1	335	400kcmil	350kcmi	600kcmil	500kcmil
115	# 2	# 2	1/0	# 1	340	500kcmil	350kcmi	600kcmil	500kcmil
120	# 1	# 2	1/0	1/0	350	500kcmil	350kcmi	700kcmil	500kcmil
130	# 1	# 2	2/0	1/0	375	500kcmil	400kcmi	700kcmil	600kcmil
135	1/0	# 1	2/0	1/0	380	500kcmil	400kcmi	750kcmil	600kcmil
150	1/0	# 1	3/0	2/0	385	600kcmil	500kcmi	750kcmil	600kcmil
155	2/0	1/0	3/0	3/0	420	600kcmil	500kcmi		700kcmil
170	2/0	1/0	4/0	3/0	430		500kcmi		750kcmil
175	2/0	2/0	4/0	3/0	435		600kcmi		750kcmil
180	3/0	2/0	4/0	4/0	475		600kcmi		

*Dry Locations Only

Notes:

1. Size #14 to 4/0 is in AWG units (American Wire Gauge).
2. Size 250 to 750 is in kcmil units (Thousand Circular Mils).
3. Use next higher ampere value if exact value is not listed in table.
4. For loads that operate continuously increase ampere value by 25% to obtain proper wire size.
5. Refer to Table R260519-91 for the maximum circuit length for the various size wires.
6. Table R260519-92 has been written for estimating purpose only, based on ambient temperature of 30°C (86° F); for ambient temperature other than 30°C (86° F), ampacity correction factors will be applied.

R260533-20 Conduit To 15′ High

List conduit by quantity, size, and type. Do not deduct for lengths occupied by fittings, since this will be allowance for scrap. Example:

A. Aluminum — size
B. Rigid galvanized — size
C. Steel intermediate (IMC) — size
D. Rigid steel, plastic-coated 20 Mil. — size
E. Rigid steel, plastic-coated 40 Mil. — size
F. Electric metallic tubing (EMT) — size
G. PVC Schedule 40 — size

Types (A) thru (E) listed above contain the following per 100 L.F.:

1. (11) Threaded couplings
2. (11) Beam-type hangers
3. (2) Factory sweeps
4. (2) Fiber bushings
5. (4) Locknuts
6. (2) Field threaded pipe terminations
7. (2) Removal of concentric knockouts

Type (F) contains per 100 L.F.:

1. (11) Set screw couplings
2. (11) Beam clamps
3. (2) Field bends on 1/2″ and 3/4″ diameter
4. (2) Factory sweeps for 1″ and above
5. (2) Set screw steel connectors
6. (2) Removal of concentric knockouts

Type (G) contains per 100 L.F.:

1. (11) Field cemented couplings
2. (34) Beam clamps
3. (2) Factory sweeps
4. (2) Adapters
5. (2) Locknuts
6. (2) Removal of concentric knockouts

Labor-hours for all conduit to 15′ include:

1. Unloading by hand
2. Hauling by hand to an area up to 200′ from loading dock
3. Setup of rolling staging
4. Installation of conduit and fittings as described in Conduit models (A) thru (G)

Not included in the material and labor are:

1. Staging rental or purchase
2. Structural modifications
3. Wire
4. Junction boxes
5. Fittings in excess of those described in conduit models (A) thru (G)
6. Painting of conduit

Fittings

Only those fittings listed above are included in the linear foot totals, although they should be listed separately from conduit lengths, without prices, to ensure proper quantities for material procurement.

If the fittings required exceed the quantities included in the model conduit runs, then material and labor costs must be added to the difference. If actual needs per 100 L.F. of conduit are: (2) sweeps, (4) LBs and (1) field bend, then, (4) LBs and (1) field bend must be priced additionally.

R260533-21 Hangers

It is sometimes desirable to substitute an alternate style of hanger if the support being used is not the type described in the conduit models.

One approach is the substitution method:

1. Find the cost of the type hanger described in the conduit model.
2. Calculate the cost of the desired type hanger (it may be necessary to calculate individual components such as drilling, expansion shields, etc.).
3. Calculate the cost difference (delta) between the two types of hangers.
4. Multiply the cost delta by the number of hangers in the model.
5. Divide the total delta cost for hangers in the model by the length of the model to find the delta cost per L.F. for that model.
6. Modify the given unit costs per L.F. for the model by the delta cost per L.F.

Another approach to hanger configurations would be to start with the conduit only and add all the supports and any other items as separate lines. This procedure is most useful if the project involves racking many runs of conduit on a single hanger, for instance, a trapeze type hanger.

Example: Five (5) 2″ RGS conduits, 50 L.F. each, are to be run on trapeze hangers from one pull box to another. The run includes one 90° bend.

1. List the hangers' components to create an assembly cost for each 2′-wide trapeze.
2. List the components for the 50′ conduit run, noting that 6 trapeze supports will be required.

Job Conditions: Productivities are based on new construction to 15′ high, using scaffolding in an unobstructed area. Material storage is assumed to be within 100′ of work being performed.

Add to labor for elevated installations:

15′ to 20′ High–10%	30′ to 35 High–30%
20′ to 25′ High–20%	35′ to 40′ High–35%
25′ to 30′ High–25%	Over 40′ High–40%

Add these percentages to the L.F. labor cost, but not to fittings. Add these percentages only to quantities exceeding the different height levels, rather than the total conduit quantities.

Linear foot price for labor does not include penetrations in walls or floors and must be added to the estimate.

R260533-22 Conductors in Conduit

Table below lists maximum number of conductors for various sized conduit using THW, TW or THWN insulations.

Copper Wire Size	Conduit Size																										
	1/2″			3/4″			1″			1-1/4″			1-1/2″			2″			2-1/2″			3″		3-1/2″		4″	
	TW	THW	THWN	TW	THW	THWN	TW	THW	THWN	TW	THW	THWN	TW	THW	THWN	TW	THW	THWN	TW	THW	THWN	THW	THWN	THW	THWN	THW	THWN
#14	9	6	13	15	10	24	25	16	39	44	29	69	60	40	94	99	65	154	142	93		143		192			
#12	7	4	10	12	8	18	19	13	29	35	24	51	47	32	70	78	53	114	111	76	164	117		157			
#10	5	4	6	9	6	11	15	11	18	26	19	32	36	26	44	60	43	73	85	61	104	95	160	127		163	
#8	2	1	3	4	3	5	7	5	9	12	10	16	17	13	22	28	22	36	40	32	51	49	79	66	106	85	136
#6		1	1		2	4		4	6		7	11		10	15		16	26		23	37	36	57	48	76	62	98
#4		1	1		1	2		3	4		5	7		7	9		12	16		17	22	27	35	36	47	47	60
#3		1	1		1	1		2	3		4	6		6	8		10	13		15	19	23	29	31	39	40	51
#2		1	1		1	1		2	3		4	5		5	7		9	11		13	16	20	25	27	33	34	43
#1					1	1		1	1		3	3		4	5		6	8		9	12	14	18	19	25	25	32
1/0					1	1		1	1		2	3		3	4		5	7		8	10	12	15	16	21	21	27
2/0					1	1		1	1		1	2		3	3		5	6		7	8	10	13	14	17	18	22
3/0					1	1		1	1		1	1		2	3		4	5		6	7	9	11	12	14	15	18
4/0						1		1	1		1	1		1	2		3	4		5	6	7	9	10	12	13	15
250 kcmil								1	1		1	1		1	1		2	3		4	4	6	7	8	10	10	12
300								1	1		1	1		1	1		2	3		3	4	5	6	7	8	9	11
350									1		1	1		1	1		1	2		3	3	4	5	6	7	8	9
400											1	1		1	1		1	1		2	3	4	5	5	6	7	8
500											1	1		1	1		1	1		1	2	3	4	4	5	6	7
600												1		1	1		1	1		1	1	3	3	4	4	5	5
700														1	1		1	1		1	1	2	3	3	4	4	5
750														1	1		1	1		1	1	2	2	3	3	4	4

R262213-27 Central Air Conditioning Watts per S.F., BTUs per Hour per S.F. of Floor Area and S.F. per Ton of Air Conditioning

Type Building	Watts per S.F.	BTUH per S.F.	S.F. per Ton	Type Building	Watts per S.F.	BTUH per S.F.	S.F. per Ton	Type Building	Watts per S.F.	BTUH per S.F.	S.F. per Ton
Apartments, Individual	3	26	450	Dormitory, Rooms	4.5	40	300	Libraries	5.7	50	240
Corridors	2.5	22	550	Corridors	3.4	30	400	Low Rise Office, Ext.	4.3	38	320
Auditoriums & Theaters	3.3	40	300/18*	Dress Shops	4.9	43	280	Interior	3.8	33	360
Banks	5.7	50	240	Drug Stores	9	80	150	Medical Centers	3.2	28	425
Barber Shops	5.5	48	250	Factories	4.5	40	300	Motels	3.2	28	425
Bars & Taverns	15	133	90	High Rise Off.-Ext. Rms.	5.2	46	263	Office (small suite)	4.9	43	280
Beauty Parlors	7.6	66	180	Interior Rooms	4.2	37	325	Post Office, Int. Office	4.9	42	285
Bowling Alleys	7.8	68	175	Hospitals, Core	4.9	43	280	Central Area	5.3	46	260
Churches	3.3	36	330/20*	Perimeter	5.3	46	260	Residences	2.3	20	600
Cocktail Lounges	7.8	68	175	Hotels, Guest Rooms	5	44	275	Restaurants	6.8	60	200
Computer Rooms	16	141	85	Public Spaces	6.2	55	220	Schools & Colleges	5.3	46	260
Dental Offices	6	52	230	Corridors	3.4	30	400	Shoe Stores	6.2	55	220
Dept. Stores, Basement	4	34	350	Industrial Plants, Offices	4.3	38	320	Shop'g. Ctrs., Sup. Mkts.	4	34	350
Main Floor	4.5	40	300	General Offices	4	34	350	Retail Stores	5.5	48	250
Upper Floor	3.4	30	400	Plant Areas	4.5	40	300	Specialty Shops	6.8	60	200

*Persons per ton

12,000 BTUH = 1 ton of air conditioning

Electrical — R2624 Switchboards & Panelboards

R262416-50 Load Centers and Panelboards

When pricing Load Centers list panels by size and type. List Breakers in a separate column of the "Quantity Sheet," and define by phase and ampere rating.

Material and Labor prices include breakers; for example: a 100A, 3-Wire, 102/240V, 18 circuit panel w/ main breaker, as described in the unit cost section, contains 18 single pole 20A breakers.

If you do not choose to include a full panel of single pole breakers, use the following method to adjust material and labor costs.

Example: In an 18 circuit panel only 16 single pole breakers are desired, requiring that the cost of 2 breakers be subtracted from the panel cost.

1. Go to the appropriate unit cost section of the book to find the unit prices of the given circuit breaker type.
2. Modify those costs as follows:
   Bare material price x 0.50
   Bare labor cost x 0.60.
3. Multiply those modified bare costs by 2 (breakers in this example).
4. Subtract the modified costs for the 2 breakers from the given cost of the panel.

Labor-hours for Load Center installation includes:

1. Unloading, uncrating, and handling enclosures 200′ from unloading area
2. Measuring and marking
3. Drilling (4) lead anchor type fasteners using a hammer drill
4. Mounting and leveling panel to a height of 6′
5. Preparation and termination of feeder cable to lugs or main breaker
6. Branch circuit identification
7. Lacing using tie wraps
8. Testing and load balancing
9. Marking panel directory

Not included in the material and labor are:

1. Modifications to enclosure
2. Structural supports
3. Additional lugs
4. Plywood backboards
5. Painting or lettering

Note: Knockouts are included in the price of terminating pipe runs and need not be added to the Load Center costs.

Job Conditions: Productivity is based on new construction to a height of 6′, in an unobstructed area. Material staging area is assumed to be within 100′ of work being performed.

Electrical — R2627 Low Voltage Distribution Equipment

R262726-90 Wiring Devices

Wiring devices should be priced on a separate takeoff form which includes boxes, covers, conduit and wire.

Labor-hours for devices include:

1. Stripping of wire
2. Attaching wire to device using terminators on the device itself, lugs, set screws etc.
3. Mounting of device in box

Labor-hours do not include:

1. Conduit
2. Wire
3. Boxes
4. Plates

Economy of Scale – for large concentrations of devices in the same area deduct the following percentages from labor-hours:

1	to	10	0%
11	to	25	20%
26	to	50	25%
51	to	100	30%
	over	100	35%

R262726-90 Wiring Devices (cont.)

NEMA No.	15 R	20 R	30 R	50 R	60 R
1 125V 2 Pole, 2 Wire	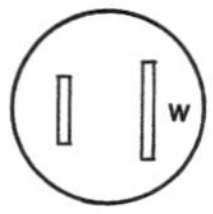				
2 250V 2 Pole, 2 Wire		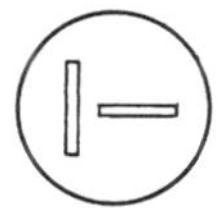			
5 125V 2 Pole, 3 Wire		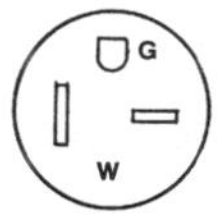	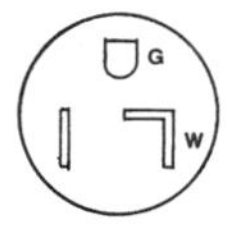	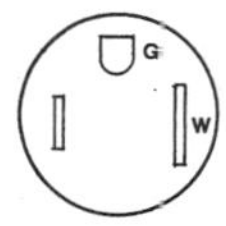	
6 250V 2 Pole, 3 Wire		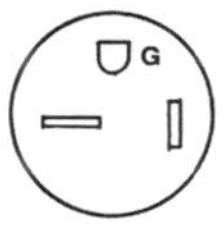	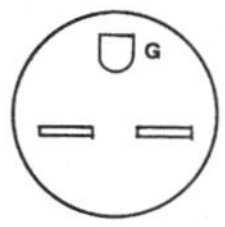	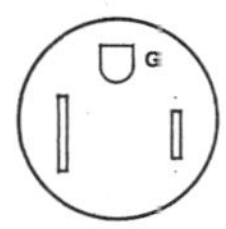	
7 277V, AC 2 Pole, 3 Wire		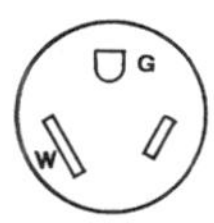			
10 125/250V 3 Pole, 3 Wire		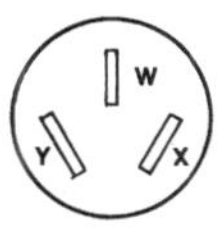		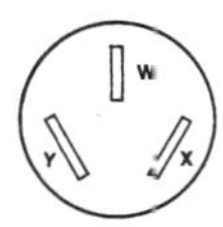	
11 3 Phase 250V 3 Pole, 3 Wire	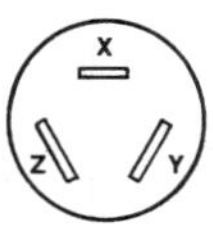	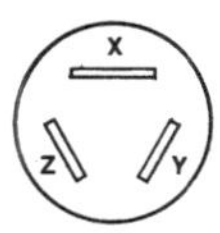	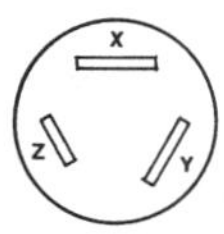	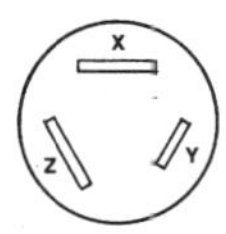	
14 125/250V 3 Pole, 4 Wire	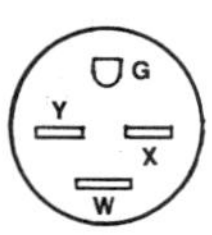	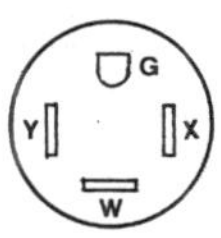	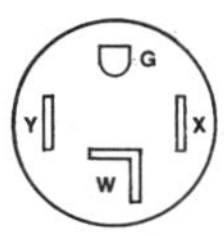	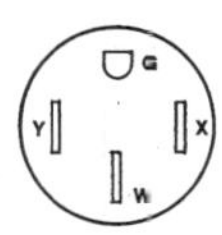	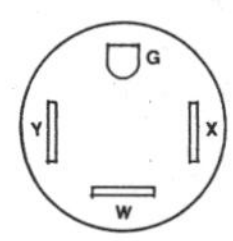
15 3 Phase 250V 3 Pole, 4 Wire			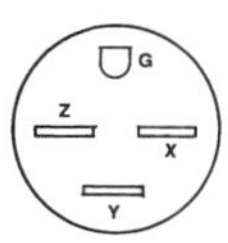	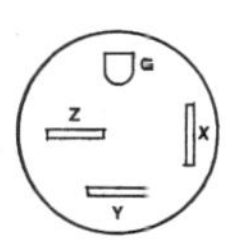	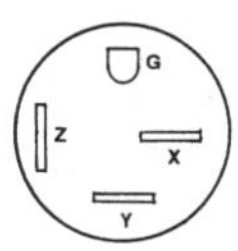
18 3 Phase 208Y/120V 4 Pole, 4 Wire	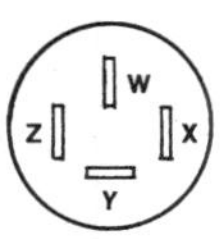	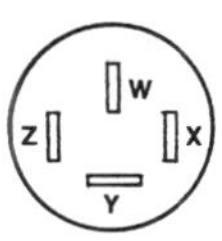	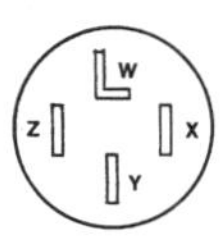	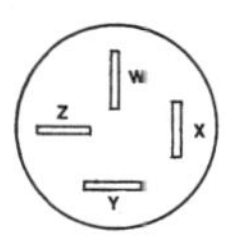	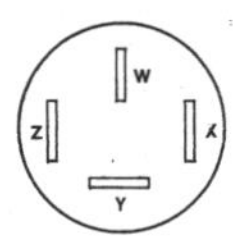

R262726-90 Wiring Devices (cont.)

NEMA No.	15 R	20 R	30 R	NEMA No.	15 R	20 R	30 R
L 1 125V 2 Pole, 2 Wire				L 13 3 Phase 600V 3 Pole, 3 Wire			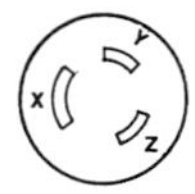
L 2 250V 2 Pole, 2 Wire				L 14 125/250V 3 Pole, 4 Wire		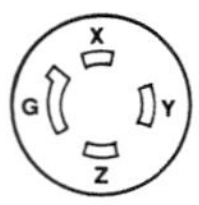	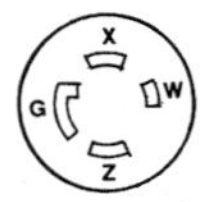
L 5 125 V 2 Pole, 3 Wire	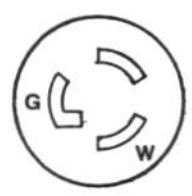			L 15 3 Phase 250 V 3 Pole, 4 Wire		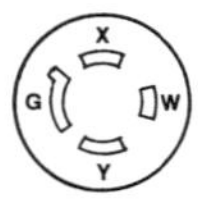	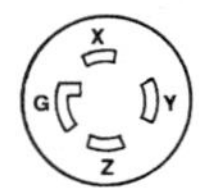
L 6 250 V 2 Pole, 3 Wire			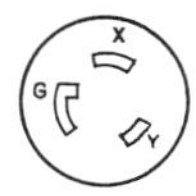	L 16 3 Phase 480V 3 Pole, 4 Wire		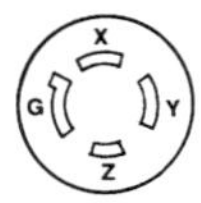	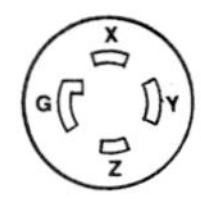
L 7 227 V, AC 2 Pole, 3 Wire		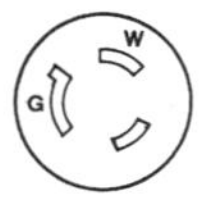		L 17 3 Phase 600 V 3 Pole, 4 Wire			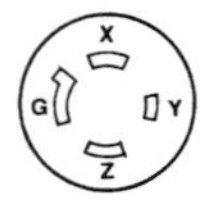
L 8 480 V 2 Pole, 3 Wire			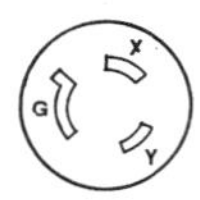	L 18 3 Phase 208Y/120V 4 Pole, 4 Wire		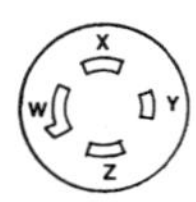	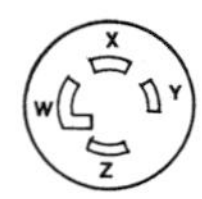
L 9 600 V 2 Pole, 3 Wire		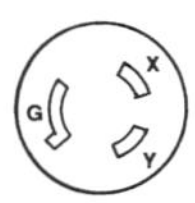	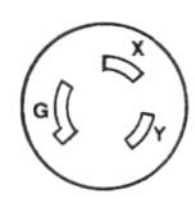	L 19 3 Phase 480Y/277V 4 Pole, 4 Wire		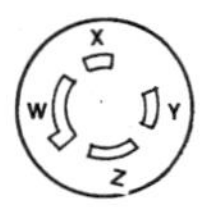	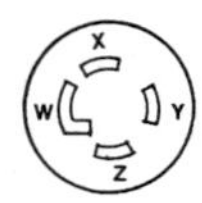
L 10 125 /250V 3 Pole, 3 Wire			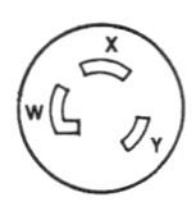	L 20 3 Phase 600Y/347V 4 Pole, 4 Wire		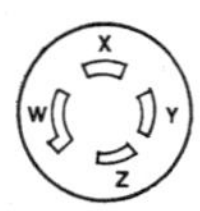	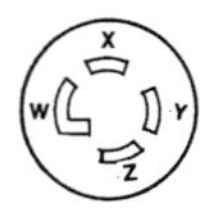
L 11 3 Phase 250 V 3 Pole, 3 Wire		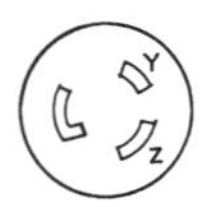	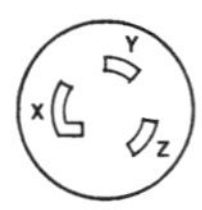	L 21 3 Phase 208Y/120V 4 Pole, 5 Wire			
L 12 3 Phase 480 V 3 Pole, 3 Wire		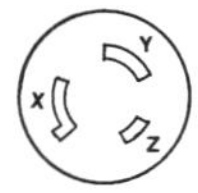	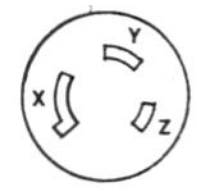	L 22 3 Phase 480Y/277V 4 Pole, 5 Wire			
				L 23 3 Phase 600Y/347V 4 Pole, 5 Wire			

R265113-40 Interior Lighting Fixtures

When taking off interior lighting fixtures, it is advisable to set up your quantity work sheet to conform to the lighting schedule as it appears on the print. Include the alpha-numeric code plus the symbol on your work sheet.

Take off a particular section or floor of the building and count each type of fixture before going on to another type. It would also be advantageous to include on the same work sheet the pipe, wire, fittings and circuit number associated with each type of lighting fixture. This will help you identify the costs associated with any particular lighting system and in turn make material purchases more specific as to when and how much to order under the classification of lighting.

By taking off lighting first you can get a complete "WALK THRU" of the job. This will become helpful when doing other phases of the project.

Materials for a recessed fixture include:
1. Fixture
2. Lamps
3. 6′ of jack chain
4. (2) S hooks
5. (2) Wire nuts

Labor for interior recessed fixtures include:
1. Unloading by hand
2. Hauling by hand to an area up to 200′ from loading dock
3. Uncrating
4. Layout
5. Installing fixture
6. Attaching jack chain & S hooks
7. Connecting circuit power
8. Reassembling fixture
9. Installing lamps
10. Testing

Material for surface mounted fixtures includes:
1. Fixture
2. Lamps
3. Either (4) lead type anchors, (4) toggle bolts, or (4) ceiling grid clips
4. (2) Wire nuts

Material for pendent mounted fixtures includes:
1. Fixture
2. Lamps
3. (2) Wire nuts
4. Rigid pendents as required by type of fixtures
5. Canopies as required by type of fixture

Labor hours include the following for both surface and pendent fixtures:
1. Unloading by hand
2. Hauling by hand to an area up to 200′ from loading dock
3. Uncrating
4. Layout and marking
5. Drilling (4) holes for either lead anchors or toggle bolts using a hammer drill
6. Installing fixture
7. Leveling fixture
8. Connecting circuit power
9. Installing lamps
10. Testing

Labor for surface or pendent fixtures does not include:
1. Conduit
2. Boxes or covers
3. Connectors
4. Fixture whips
5. Special support
6. Switching
7. Wire

Economy of Scale: For large concentrations of lighting fixtures in the same area deduct the following percentages from labor:

25	to	50	fixtures	15%
51	to	75	fixtures	20%
76	to	100	fixtures	25%
101 and over				30%

Job Conditions: Productivity is based on new construction in a unobstructed first floor location, using rolling staging to 15′ high.

Material staging is assumed to be within 100′ of work being performed.

Add the following percentages to labor for elevated installations:

15′	to	20′	high	10%
21′	to	25′	high	20%
26′	to	30′	high	30%
31′	to	35′	high	40%
36′	to	40′	high	50%
41′ and over				60%

R265723-05 Comparison - Operation of High Intensity Discharge Lamps

Lamp Type	Wattage	Life (Hours)	1 Circuit Wattage	Average Initial Lumens	2 L.L.D.	3 Mean Lumens
M.V.	100 DX	24,000	125	4,000	61%	2,440
L.P.S.	SOX-35	18,000	65	4,800	100%	4,800
H.P.S.	LU-70	12,000	84	5,800	90%	5,220
M.H.	No Equivalent					
M.V.	175 DX	24,000	210	8,500	66%	5,676
M.V.	250 DX	24,000	295	13,000	66%	7,986
L.P.S.	SOX-55	18,000	82	8,000	100%	8,000
H.P.S.	LU-100	12,000	120	9,500	90%	8,550
M.H.	No Equivalent					
M.V.	400 DX	24,000	465	24,000	64%	14,400
L.P.S.	SOX-90	18,000	141	13,500	100%	13,500
H.P.S.	LU-150	16,000	188	16,000	90%	14,400
M.H.	MH-175	7,500	210	14,000	73%	10,200
M.V.	No Equivalent					
L.P.S.	SOX-135	18,000	147	22,500	100%	22,500
H.P.S.	LU-250	20,000	310	25,500	92%	23,205
M.H.	MH-250	7,500	295	20,500	78%	16,000
M.V.	1000 DX	24,000	1,085	63,000	61%	37,820
L.P.S.	SOX-180	18,000	248	33,000	100%	33,000
H.P.S.	LU-400	20,000	480	50,000	90%	45,000
M.H.	MH-400	15,000	465	34,000	72%	24,600
H.P.S.	LU-1000	15,000	1,100	140,000	91%	127,400

1. Includes ballast losses and average lamp watts
2. Lamp lumen depreciation (% of initial light output at 70% rated life)
3. Lamp lumen output at 70% rated life (L.L.D. x initial)
4. Cost of yearly operation = Cicuit Wattage/1000 x annual operating hours x average cost per KWH

M.V. = Mercury Vapor
L.P.S. = Low pressure sodium
H.P.S. = High pressure sodium
L.H. = Metal halide

R265723-20 Lamp Comparison Chart with Enclosed Floodlight, Ballast & Lamp for Pole Mounting

Type	Watts	Initial Lumens	Lumens per Watt	Lumens @ 40% Life	Life (Hours)
Incandescent	150	2,880	19	85	750
	300	6,360	21	84	750
	500	10,850	22	80	1,000
	1,000	23,740	24	80	1,000
	1,500	34,400	23	80	1,000
Tungsten Halogen	500	10,950	22	97	2,000
	1,500	35,800	24	97	2,000
Fluorescent Cool White	40	3,150	79	88	20,000
	110	9,200	84	87	12,000
	215	16,000	74	81	12,000
Deluxe Mercury	250	12,100	48	86	24,000
	400	22,500	56	85	24,000
	1,000	63,000	63	75	24,000
Metal Halide	175	14,000	80	77	7,500
	400	34,000	85	75	15,000
	1,000	100,000	100	83	10,000
	1,500	155,000	103	92	1,500
High Pressure Sodium	70	5,800	83	90	20,000
	100	9,500	95	90	20,000
	150	16,000	107	90	24,000
	400	50,000	125	90	24,000
	1,000	140,000	140	90	24,000
Low Pressure Sodium	55	4,600	131	98	18,000
	90	12,750	142	98	18,000
	180	33,000	183	98	18,000

Color: High Pressure Sodium — Slightly Yellow
Low Pressure Sodium — Yellow
Mercury Vapor — Green-Blue
Metal Halide — Blue White

Note: Pole not included.

Communications — R2713 Communications Backbone Cabling

R271323-40 Fiber Optics

Fiber optic systems use optical fiber such as plastic, glass, or fused silica, a transparent material, to transmit radiant power (i.e., light) for control, communication, and signaling applications. The types of fiber optic cables can be nonconductive, conductive, or composite. The composite cables contain fiber optics and current-carrying electrical conductors. The configuration for one of the fiber optic systems is as follows:

The transceiver module acts as transmitting and receiving equipment in a common house, which converts electrical energy to light energy or vice versa.

Pricing the fiber optic system is not an easy task. The performance of the whole system will affect the cost significantly. New specialized tools and techniques decrease the installing cost tremendously. In the fiber optic section of Means Electrical Cost Data, a benchmark for labor-hours and material costs is set up so that users can adjust their costs according to unique project conditions.

Units for Measure: Fiber optic cable is measured in hundred linear feet (C.L.F.) or industry units of measure - meter (m) or kilometer (km). The connectors are counted as units (EA.)

Material Units: Generally, the material costs include only the cable. All the accessories shall be priced separately.

Labor Units: The following procedures are generally included for the installation of fiber optic cables:

- Receiving
- Material handling
- Setting up pulling equipment
- Measuring and cutting cable
- Pulling cable

These additional items are listed and extended: Terminations

Takeoff Procedure: Cable should be taken off by type, size, number of fibers, and number of terminations required. List the lengths of each type of cable on the takeoff sheets. Total and add 10% for waste. Transfer the figures to a cost analysis sheet and extend.

Communications — R2715 Communications Horizontal Cabling

R271513-75 High Performance Cable

There are several categories used to describe high performance cable. The following information includes a description of categories CAT 3, 5, 5e, 6, and 7, and details classifications of frequency and specific standards. The category standards have evolved under the sponsorship of organizations such as the Telecommunication Industry Association (TIA), the Electronic Industries Alliance (EIA), the American National Standards Institute (ANSI), the International Organization for Standardization (ISO), and the International Electrotechnical Commission (IEC), all of which have catered to the increasing complexities of modern network technology. For network cabling, users must comply with national or international standards. A breakdown of these categories is as follows:

Category 3: Designed to handle frequencies up to 16 MHz.

Category 5: (TIA/EIA 568A) Designed to handle frequencies up to 100 MHz.

Category 5e: Additional transmission performance to exceed Category 5.

Category 6 (draft): Development by TIA and other international groups to handle frequencies of 250 MHz.

Category 7 (draft): Under development to handle a frequency range from 1 to 600 MHz.

Exterior Improvements — R3292 Turf & Grasses

R329219-50 Seeding

The type of grass is determined by light, shade and moisture content of soil plus intended use. Fertilizer should be disked 4″ before seeding. For steep slopes disk five tons of mulch and lay two tons of hay or straw on surface per acre after seeding. Surface mulch can be staked, lightly disked or tar emulsion sprayed. Material for mulch can be wood chips, peat moss, partially rotted hay or straw, wood fibers, and sprayed emulsions. Hemp seed blankets with fertilizer are also available. For spring seeding, watering is necessary. Late fall seeding may have to be reseeded in the spring. Hydraulic seeding, power mulching, and aerial seeding can be used on large areas.

R331113-80 Piping Designations

There are several systems currently in use to describe pipe and fittings. The following paragraphs will help to identify and clarify classifications of piping systems used for water distribution.

Piping may be classified by schedule. Piping schedules include 5S, 10S, 10, 20, 30, Standard, 40, 60, Extra Strong, 80, 100, 120, 140, 160, and Double Extra Strong. These schedules are dependent upon the pipe wall thickness. The wall thickness of a particular schedule may vary with pipe size.

Ductile iron pipe for water distribution is classified by Pressure Classes such as Class 150, 200, 250, 300, and 350. These classes are actually the rated water working pressure of the pipe in pounds per square inch (psi). The pipe in these pressure classes is designed to withstand the rated water working pressure plus a surge allowance of 100 psi.

The American Water Works Association (AWWA) provides standards for various types of **plastic pipe**. C-900 is the specification for polyvinyl chloride (PVC) piping used for water distribution in sizes ranging from 4″ through 12″. C-901 is the specification for polyethylene (PE) pressure pipe, tubing and fittings used for water distribution in sizes ranging from 1/2″ through 3″. C-905 is the specification for PVC piping sizes 14″ and greater.

PVC pressure-rated pipe is identified using the standard dimensional ratio (SDR) method. This method is defined by the American Society for Testing and Materials (ASTM) Standard D 2241. This pipe is available in SDR numbers 64, 41, 32.5, 26, 21, 17, and 13.5. Pipe with an SDR of 64 will have the thinnest wall while pipe with an SDR of 13.5 will have the thickest wall. When the pressure rating (PR) of a pipe is given in psi, it is based on a line supplying water at 73 degrees F.

The National Sanitation Foundation (NSF) seal of approval is applied to products that can be used with potable water. These products have been tested to ANSI/NSF Standard 14.

Valves and strainers are classified by American National Standards Institute (ANSI) Classes. These Classes are 125, 150, 200, 250, 300, 400, 600, 900, 1500 and 2500. Within each class there is an operating pressure range dependent upon temperature. Design parameters should be compared to the appropriate material dependent, pressure-temperature rating chart for accurate valve selection.

R337119-30 Concrete for Conduit Encasement

Table below lists C.Y. of concrete for 100 L.F. of trench. Conduits separation center to center should meet 7.5″ (N.E.C.).

Number of Conduits	1	2	3	4	6	8	9	Number of Conduits
Trench Dimension	11.5″ x 11.5″	11.5″ x 19″	11.5″ x 27″	19″ x 19″	19″ x 27″	19″ x 38″	27″ x 27″	Trench Dimension
Conduit Diameter 2.0″	3.29	5.39	7.64	8.83	12.51	17.66	17.72	Conduit Diameter 2.0″
2.5″	3.23	5.29	7.49	8.62	12.19	17.23	17.25	2.5″
3.0″	3.15	5.13	7.24	8.29	11.71	16.59	16.52	3.0″
3.5″	3.08	4.97	7.02	7.99	11.26	15.98	15.84	3.5″
4.0″	2.99	4.80	6.76	7.65	10.74	15.30	15.07	4.0″
5.0″	2.78	4.37	6.11	6.78	9.44	13.57	13.12	5.0″
6.0″	2.52	3.84	5.33	5.74	7.87	11.48	10.77	6.0″
					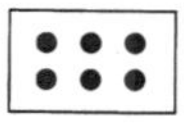	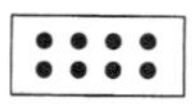		

R347216-10 Single Track R.R. Siding

The costs for a single track RR siding in the Unit Price section include the components shown in the table below.

Description of Component	Qty. per L.F. of Track	Unit
Ballast, 1-1/2" crushed stone	.667	C.Y.
6" x 8" x 8'-6" Treated timber ties, 22" O.C.	.545	Ea.
Tie plates, 2 per tie	1.091	Ea.
Track rail	2.000	L.F.
Spikes, 6", 4 per tie	2.182	Ea.
Splice bars w/ bolts, lock washers & nuts, @ 33' O.C.	.061	Pair
Crew B-14 @ 57 L.F./Day	.018	Day

R347216-20 Single Track, Steel Ties, Concrete Bed

The costs for a R.R. siding with steel ties and a concrete bed in the Unit Price section include the components shown in the table below.

Description of Component	Qty. per L.F. of Track	Unit
Concrete bed, 9' wide, 10" thick	.278	C.Y.
Ties, W6x16 x 6'-6" long, @ 30" O.C.	.400	Ea.
Tie plates, 4 per tie	1.600	Ea.
Track rail	2.000	L.F.
Tie plate bolts, 1", 8 per tie	3.200	Ea.
Splice bars w/bolts, lock washers & nuts, @ 33' O.C.	.061	Pair
Crew B-14 @ 22 L.F./Day	.045	Day

Change Orders

Change Order Considerations

A Change Order is a written document, usually prepared by the design professional, and signed by the owner, the architect/engineer and the contractor. A change order states the agreement of the parties to: an addition, deletion, or revision in the work; an adjustment in the contract sum, if any; or an adjustment in the contract time, if any. Change orders, or "extras" in the construction process occur after execution of the construction contract and impact architects/engineers, contractors and owners.

Change orders that are properly recognized and managed can ensure orderly, professional and profitable progress for all who are involved in the project. There are many causes for change orders and change order requests. In all cases, change orders or change order requests should be addressed promptly and in a precise and prescribed manner. The following paragraphs include information regarding change order pricing and procedures.

The Causes of Change Orders

Reasons for issuing change orders include:

- Unforeseen field conditions that require a change in the work
- Correction of design discrepancies, errors or omissions in the contract documents
- Owner-requested changes, either by design criteria, scope of work, or project objectives
- Completion date changes for reasons unrelated to the construction process
- Changes in building code interpretations, or other public authority requirements that require a change in the work
- Changes in availability of existing or new materials and products

Procedures

Properly written contract documents must include the correct change order procedures for all parties—owners, design professionals and contractors—to follow in order to avoid costly delays and litigation.

Being "in the right" is not always a sufficient or acceptable defense. The contract provisions requiring notification and documentation must be adhered to within a defined or reasonable time frame.

The appropriate method of handling change orders is by a written proposal and acceptance by all parties involved. Prior to starting work on a project, all parties should identify their authorized agents who may sign and accept change orders, as well as any limits placed on their authority.

Time may be a critical factor when the need for a change arises. For such cases, the contractor might be directed to proceed on a "time and materials" basis, rather than wait for all paperwork to be processed—a delay that could impede progress. In this situation, the contractor must still follow the prescribed change order procedures, including but not limited to, notification and documentation.

All forms used for change orders should be dated and signed by the proper authority. Lack of documentation can be very costly, especially if legal judgments are to be made and if certain field personnel are no longer available. For time and material change orders, the contractor should keep accurate daily records of all labor and material allocated to the change. Forms that can be used to document change order work are available in *Means Forms for Building Construction Professionals.*

Owners or awarding authorities who do considerable and continual building construction (such as the federal government) realize the inevitability of change orders for numerous reasons, both predictable and unpredictable. As a result, the federal government, the American Institute of Architects (AIA), the Engineers Joint Contract Documents Committee (EJCDC) and other contractor, legal and technical organizations have developed standards and procedures to be followed by all parties to achieve contract continuance and timely completion, while being financially fair to all concerned.

In addition to the change order standards put forth by industry associations, there are also many books available on the subject.

Pricing Change Orders

When pricing change orders, regardless of their cause, the most significant factor is when the change occurs. The need for a change may be perceived in the field or requested by the architect/engineer *before* any of the actual installation has begun, or may evolve or appear *during* construction when the item of work in question is partially installed. In the latter cases, the original sequence of construction is disrupted, along with all contiguous and supporting systems. Change orders cause the greatest impact when they occur *after* the installation has been completed and must be uncovered, or even replaced. Post-completion changes may be caused by necessary design changes, product failure, or changes in the owner's requirements that are not discovered until the building or the systems begin to function.

Specified procedures of notification and record keeping must be adhered to and enforced regardless of the stage of construction: *before, during,* or *after* installation. Some bidding documents anticipate change orders by requiring that unit prices including overhead and profit percentages—for additional as well as deductible changes—be listed. Generally these unit prices do not fully take into account the ripple effect, or impact on other trades, and should be used for general guidance only.

When pricing change orders, it is important to classify the time frame in which the change occurs. There are two basic time frames for change orders: *pre-installation change orders,* which occur before the start of construction, and *post-installation change orders,* which involve reworking after the original installation. Change orders that occur between these stages may be priced according to the extent of work completed using a combination of techniques developed for pricing *pre-* and *post-installation* changes.

The following factors are the basis for a check list to use when preparing a change order estimate.

Factors To Consider When Pricing Change Orders

As an estimator begins to prepare a change order, the following questions should be reviewed to determine their impact on the final price.

General

- Is the change order work *pre-installation* or *post-installation?*

  Change order work costs vary according to how much of the installation has been completed. Once workers have the project scoped in their mind, even though they have not started, it can be difficult to refocus. Consequently they may spend more than the normal amount of time understanding the change. Also, modifications to work in place such as trimming or refitting usually take more time than was initially estimated. The greater the amount of work in place, the more reluctant workers are to change it. Psychologically they may resent the change and as a result the rework takes longer than normal. Post-installation change order estimates must include demolition of existing work as required to accomplish the change. If the work is performed at a later time, additional obstacles such as building finishes may be present which must be protected. Regardless of whether the change occurs pre-installation or post-installation, attempt to isolate the identifiable factors and price them separately. For example, add shipping costs that may be required pre-installation or any demolition required post-installation. Then analyze the potential impact on productivity of psychological and/or learning curve factors and adjust the output rates accordingly. One approach is to break down the typical workday into segments and quantify the impact on each segment. The following chart may be useful as a guide:

Activities (Productivity) Expressed as Percentages of a Workday			
Task	Means Mechanical Cost Data (for New Construction)	Pre-Installation Change Orders	Post-Installation Change Orders
1. Study plans	3%	6%	6%
2. Material procurement	3%	3%	3%
3. Receiving and storing	3%	3%	3%
4. Mobilization	5%	5%	5%
5. Site movement	5%	5%	8%
6. Layout and marking	8%	10%	12%
7. Actual installation	64%	59%	54%
8. Clean-up	3%	3%	3%
9. Breaks—non-productive	6%	6%	6%
Total	100%	100%	100%

Change Order Installation Efficiency

The labor-hours expressed (for new construction) are based on average installation time, using an efficiency level of approximately 60-65%. For change order situations, adjustments to this efficiency level should reflect the daily labor-hour allocation for that particular occurrence.

If any of the specific percentages expressed in the above chart do not apply to a particular project situation, then those percentage points should be reallocated to the appropriate task(s). Example: Using data for new construction, assume there is no new material being utilized. The percentages for Tasks 2 and 3 would therefore be reallocated to other tasks. If the time required for Tasks 2 and 3 can now be applied to installation, we can add the time allocated for *Material Procurement* and *Receiving and Storing* to the *Actual Installation* time for new construction, thereby increasing the Actual Installation percentage.

This chart shows that, due to reduced productivity, labor costs will be higher than those for new construction by 5% to 15% for pre-installation change orders and by 15% to 25% for post-installation change orders. Each job and change order is unique and must be examined individually. Many factors, covered elsewhere in this section, can each have a significant impact on productivity and change order costs. All such factors should be considered in every case.

- Will the change substantially delay the original completion date?

  A significant change in the project may cause the original completion date to be extended. The extended schedule may subject the contractor to new wage rates dictated by relevant labor contracts. Project supervision and other project overhead must also be extended beyond the original completion date. The schedule extension may also put installation into a new weather season. For example, underground piping scheduled for October installation was delayed until January. As a result, frost penetrated the trench area, thereby changing the degree of difficulty of the task. Changes and delays may have a ripple effect throughout the project. This effect must be analyzed and negotiated with the owner.

- What is the net effect of a deduct change order?

  In most cases, change orders resulting in a deduction or credit reflect only bare costs. The contractor may retain the overhead and profit based on the original bid.

Materials

- Will you have to pay more or less for the new material, required by the change order, than you paid for the original purchase?

  The same material prices or discounts will usually apply to materials purchased for change orders as new construction. In some instances, however, the contractor may forfeit the advantages of competitive pricing for change orders. Consider the following example:

  A contractor purchased over $20,000 worth of fan coil units for an installation, and obtained the maximum discount. Some time later it was determined the project required an additional matching unit. The contractor has to purchase this unit from the original supplier to ensure a match. The supplier at this time may not discount the unit because of the small quantity, and the fact that he is no longer in a competitive situation. The impact of quantity on purchase can add between 0% and 25% to material prices and/or subcontractor quotes.

- If materials have been ordered or delivered to the job site, will they be subject to a cancellation charge or restocking fee?

  Check with the supplier to determine if ordered materials are subject to a cancellation charge. Delivered materials not used as result of a change order may be subject to a restocking fee if returned to the supplier. Common restocking charges run between 20% and 40%. Also, delivery charges to return the goods to the supplier must be added.

Labor

- How efficient is the existing crew at the actual installation?

  Is the same crew that performed the initial work going to do the change order? Possibly the change consists of the installation of a unit identical to one already installed; therefore the change should take less time. Be sure to consider this potential productivity increase and modify the productivity rates accordingly.

- If the crew size is increased, what impact will that have on supervision requirements?

  Under most bargaining agreements or management practices, there is a point at which a working foreman is replaced by a nonworking foreman. This replacement increases project overhead by adding a nonproductive worker. If additional workers are added to accelerate the project or to perform changes while maintaining the schedule, be sure to add additional supervision time if warranted. Calculate the hours involved and the additional cost directly if possible.

- What are the other impacts of increased crew size?

  The larger the crew, the greater the potential for productivity to decrease. Some of the factors that cause this productivity loss are: overcrowding (producing restrictive conditions in the working space), and possibly a shortage of any special tools and equipment required. Such factors affect not only the crew working on the elements directly involved in the change order, but other crews whose movement may also be hampered.

  As the crew increases, check its basic composition for changes by the addition or deletion of apprentices or nonworking foreman and quantify the potential effects of equipment shortages or other logistical factors.

- As new crews, unfamiliar with the project, are brought onto the site, how long will it take them to become oriented to the project requirements?

  The orientation time for a new crew to become 100% effective varies with the site and type of project. Orientation is easiest at a new construction site, and most difficult at existing, very restrictive renovation sites. The type of work also affects orientation time. When all elements of the work are exposed, such as concrete or masonry work, orientation is decreased. When the work is concealed or less visible, such as existing electrical systems, orientation takes longer. Usually orientation can be accomplished in one day or less. Costs for added orientation should be itemized and added to the total estimated cost.

- How much actual production can be gained by working overtime?

  Short term overtime can be used effectively to accomplish more work in a day. However, as overtime is scheduled to run beyond several weeks, studies have shown marked decreases in output. The following chart shows the effect of long term overtime on worker efficiency. If the anticipated change requires extended overtime to keep the job on schedule, these factors can be used as a guide to predict the impact on time and cost. Add project overhead, particularly supervision, that may also be incurred.

Days per Week	Hours per Day	Production Efficiency					Payroll Cost Factors	
		1 Week	2 Weeks	3 Weeks	4 Weeks	Average 4 Weeks	@ 1-1/2 Times	@ 2 Times
5	8	100%	100%	100%	100%	100%	100%	100%
	9	100	100	95	90	96.25	105.6	111.1
	10	100	95	90	85	91.25	110.0	120.0
	11	95	90	75	65	81.25	113.6	127.3
	12	90	85	70	60	76.25	116.7	133.3
6	8	100	100	95	90	96.25	108.3	116.7
	9	100	95	90	85	92.50	113.0	125.9
	10	95	90	85	80	87.50	116.7	133.3
	11	95	85	70	65	78.75	119.7	139.4
	12	90	80	65	60	73.75	122.2	144.4
7	8	100	95	85	75	88.75	114.3	128.6
	9	95	90	80	70	83.75	118.3	136.5
	10	90	85	75	65	78.75	121.4	142.9
	11	85	80	65	60	72.50	124.0	148.1
	12	85	75	60	55	68.75	126.2	152.4

Effects of Overtime

Caution: Under many labor agreements, Sundays and holidays are paid at a higher premium than the normal overtime rate.

The use of long-term overtime is counterproductive on almost any construction job; that is, the longer the period of overtime, the lower the actual production rate. Numerous studies have been conducted, and while they have resulted in slightly different numbers, all reach the same conclusion. The figure above tabulates the effects of overtime work on efficiency.

As illustrated, there can be a difference between the *actual* payroll cost per hour and the *effective* cost per hour for overtime work. This is due to the reduced production efficiency with the increase in weekly hours beyond 40. This difference between actual and effective cost results from overtime work over a prolonged period. Short-term overtime work does not result in as great a reduction in efficiency, and in such cases, effective cost may not vary significantly from the actual payroll cost. As the total hours per week are increased on a regular basis, more time is lost because of fatigue, lowered morale, and an increased accident rate.

As an example, assume a project where workers are working 6 days a week, 10 hours per day. From the figure above (based on productivity studies), the average effective productive hours over a four-week period are:

$$0.875 \times 60 = 52.5$$

Depending upon the locale and day of week, overtime hours may be paid at time and a half or double time. For time and a half, the overall (average) *actual* payroll cost (including regular and overtime hours) is determined as follows:

$$\frac{40 \text{ reg. hrs.} + (20 \text{ overtime hrs.} \times 1.5)}{60 \text{ hrs.}} = 1.167$$

Based on 60 hours, the payroll cost per hour will be 116.7% of the normal rate at 40 hours per week. However, because the effective production (efficiency) for 60 hours is reduced to the equivalent of 52.5 hours, the effective cost of overtime is calculated as follows:

For time and a half:

$$\frac{40 \text{ reg. hrs.} + (20 \text{ overtime hrs.} \times 1.5)}{52.5 \text{ hrs.}} = 1.33$$

Installed cost will be 133% of the normal rate (for labor).

Thus, when figuring overtime, the actual cost per unit of work will be higher than the apparent overtime payroll dollar increase, due to the reduced productivity of the longer workweek. These efficiency calculations are true only for those cost factors determined by hours worked. Costs that are applied weekly or monthly, such as equipment rentals, will not be similarly affected.

Equipment

- What equipment is required to complete the change order?

  Change orders may require extending the rental period of equipment already on the job site, or the addition of special equipment brought in to accomplish the change work. In either case, the additional rental charges and operator labor charges must be added.

Summary

The preceding considerations and others you deem appropriate should be analyzed and applied to a change order estimate. The impact of each should be quantified and listed on the estimate to form an audit trail.

Change orders that are properly identified, documented, and managed help to ensure the orderly, professional and profitable progress of the work. They also minimize potential claims or disputes at the end of the project.

Estimating Tips

- The cost figures in this Square Foot Cost section were derived from approximately 11,200 projects contained in the RSMeans database of completed construction projects. They include the contractor's overhead and profit, but do not generally include architectural fees or land costs. The figures have been adjusted to January of the current year. New projects are added to our files each year, and outdated projects are discarded. For this reason, certain costs may not show a uniform annual progression. In no case are all subdivisions of a project listed.
- These projects were located throughout the U.S. and reflect a tremendous variation in square foot (S.F.) and cubic foot (C.F.) costs. This is due to differences, not only in labor and material costs, but also in individual owners' requirements. For instance, a bank in a large city would have different features than one in a rural area. This is true of all the different types of buildings analyzed. Therefore, caution should be exercised when using these Square Foot costs. For example, for court houses, costs in the database are local court house costs and will not apply to the larger, more elaborate federal court houses. As a general rule, the projects in the 1/4 column do not include any site work or equipment, while the projects in the 3/4 column may include both equipment and site work. The median figures do not generally include site work.
- None of the figures "go with" any others. All individual cost items were computed and tabulated separately. Thus, the sum of the median figures for Plumbing, HVAC and Electrical will not normally total up to the total Mechanical and Electrical costs arrived at by separate analysis and tabulation of the projects.
- Each building was analyzed as to total and component costs and percentages. The figures were arranged in ascending order with the results tabulated as shown. The 1/4 column shows that 25% of the projects had lower costs and 75% had higher. The 3/4 column shows that 75% of the projects had lower costs and 25% had higher. The median column shows that 50% of the projects had lower costs and 50% had higher.
- There are two times when square foot costs are useful. The first is in the conceptual stage when no details are available. Then square foot costs make a useful starting point. The second is after the bids are in and the costs can be worked back into their appropriate units for information purposes. As soon as details become available in the project design, the square foot approach should be discontinued and the project priced as to its particular components. When more precision is required, or for estimating the replacement cost of specific buildings, the current edition of *RSMeans Square Foot Costs* should be used.
- In using the figures in this section, it is recommended that the median column be used for preliminary figures if no additional information is available. The median figures, when multiplied by the total city construction cost index figures (see City Cost Indexes) and then multiplied by the project size modifier at the end of this section, should present a fairly accurate base figure, which would then have to be adjusted in view of the estimator's experience, local economic conditions, code requirements, and the owner's particular requirements. There is no need to factor the percentage figures, as these should remain constant from city to city. All tabulations mentioning air conditioning had at least partial air conditioning.
- The editors of this book would greatly appreciate receiving cost figures on one or more of your recent projects, which would then be included in the averages for next year. All cost figures received will be kept confidential, except that they will be averaged with other similar projects to arrive at Square Foot cost figures for next year's book. See the last page of the book for details and the discount available for submitting one or more of your projects.

50 17 00 | S.F. Costs

			Unit	Unit Costs 1/4	Unit Costs Median	Unit Costs 3/4	% of Total 1/4	% of Total Median	% of Total 3/4
01	0010	**APARTMENTS Low Rise (1 to 3 story)**	S.F.	67	84.50	112			
	0020	Total project cost	C.F.	6	8	9.85			
	0100	Site work	S.F.	5.75	7.85	13.75	6.05%	10.55%	14.05%
	0500	Masonry		1.32	3.07	5.30	1.54%	3.67%	6.35%
	1500	Finishes		7.10	9.75	12.10	9.05%	10.75%	12.85%
	1800	Equipment		2.19	3.32	4.94	2.73%	4.03%	5.95%
	2720	Plumbing		5.20	6.70	8.55	6.65%	8.95%	10.05%
	2770	Heating, ventilating, air conditioning		3.33	4.10	6.05	4.20%	5.60%	7.60%
	2900	Electrical		3.88	5.15	6.90	5.20%	6.65%	8.40%
	3100	Total: Mechanical & Electrical		13.45	17.10	21.50	15.90%	18.05%	23%
	9000	Per apartment unit, total cost	Apt.	62,500	95,500	140,500			
	9500	Total: Mechanical & Electrical	"	11,800	18,600	24,300			
02	0010	**APARTMENTS Mid Rise (4 to 7 story)**	S.F.	89	107	133			
	0020	Total project costs	C.F.	6.95	9.60	13.10			
	0100	Site work	S.F.	3.56	7.05	12.65	5.25%	6.70%	9.15%
	0500	Masonry		5.90	8.15	11.15	5.10%	7.25%	10.50%
	1500	Finishes		11.20	14.60	18.40	10.55%	13.45%	17.70%
	1800	Equipment		2.58	3.90	5.05	2.54%	3.48%	4.31%
	2500	Conveying equipment		1.91	2.44	2.96	1.94%	2.27%	2.69%
	2720	Plumbing		5.20	8.35	8.85	5.70%	7.20%	8.95%
	2900	Electrical		5.85	7.95	9.65	6.65%	7.20%	8.95%
	3100	Total: Mechanical & Electrical		18.75	23.50	28.50	18.50%	21%	23%
	9000	Per apartment unit, total cost	Apt.	100,500	118,500	196,500			
	9500	Total: Mechanical & Electrical	"	19,000	22,000	27,800			
03	0010	**APARTMENTS High Rise (8 to 24 story)**	S.F.	101	116	139			
	0020	Total project costs	C.F.	9.80	11.40	14.55			
	0100	Site work	S.F.	3.66	5.90	8.30	2.58%	4.84%	6.15%
	0500	Masonry		5.85	10.60	13.20	4.74%	9.65%	11.05%
	1500	Finishes		11.20	14	16.50	9.75%	11.80%	13.70%
	1800	Equipment		3.24	3.99	5.30	2.78%	3.49%	4.35%
	2500	Conveying equipment		2.29	3.48	4.73	2.23%	2.78%	3.37%
	2720	Plumbing		7.40	8.75	12.25	6.80%	7.20%	10.45%
	2900	Electrical		6.95	8.75	11.85	6.45%	7.65%	8.80%
	3100	Total: Mechanical & Electrical		21	26.50	32	17.95%	22.50%	24.50%
	9000	Per apartment unit, total cost	Apt.	105,000	115,500	160,000			
	9500	Total: Mechanical & Electrical	"	22,700	25,900	27,400			
04	0010	**AUDITORIUMS**	S.F.	104	141	203			
	0020	Total project costs	C.F.	6.55	9.15	13.10			
	2720	Plumbing	S.F.	6.35	9.20	11.15	5.85%	7.20%	8.70%
	2900	Electrical		8.10	11.70	15.75	6.80%	8.95%	11.30%
	3100	Total: Mechanical & Electrical		23	46	56	24.50%	30.50%	31.50%
05	0010	**AUTOMOTIVE SALES**	S.F.	77.50	105	130			
	0020	Total project costs	C.F.	5.10	6.15	7.95			
	2720	Plumbing	S.F.	3.54	6.15	6.70	2.89%	6.05%	6.50%
	2770	Heating, ventilating, air conditioning		5.45	8.35	9	4.61%	10%	10.35%
	2900	Electrical		6.25	9.80	13.30	7.25%	9.80%	12.15%
	3100	Total: Mechanical & Electrical		19.60	28	33.50	19.15%	20.50%	22%
06	0010	**BANKS**	S.F.	151	189	240			
	0020	Total project costs	C.F.	10.85	14.75	19.50			
	0100	Site work	S.F.	17.35	26.50	38	7.75%	12.45%	17%
	0500	Masonry		7.90	15.15	26.50	3.36%	6.95%	10.35%
	1500	Finishes		13.45	19.50	24	5.80%	8.45%	11.25%
	1800	Equipment		5.70	12.60	26	1.34%	5.95%	10.65%
	2720	Plumbing		4.76	6.80	9.95	2.82%	3.90%	4.93%
	2770	Heating, ventilating, air conditioning		9.05	12.10	16.10	4.86%	7.15%	8.50%
	2900	Electrical		14.35	19.20	25	8.20%	10.20%	12.15%
	3100	Total: Mechanical & Electrical		33.50	45.50	53	16.25%	19.40%	23%
	3500	See also Divisions 11020 & 11030 (MF2004 11 16 00 & 11 17 00)							

50 17 00 | S.F. Costs

		50 17 00 | S.F. Costs	UNIT	UNIT COSTS 1/4	UNIT COSTS MEDIAN	UNIT COSTS 3/4	% OF TOTAL 1/4	% OF TOTAL MEDIAN	% OF TOTAL 3/4	
13	0010	**CHURCHES**	S.F.	102	130	168				13
	0020	Total project costs	C.F.	6.35	8	10.55				
	1800	Equipment	S.F.	1.22	2.93	6.25	.95%	2.11%	4.50%	
	2720	Plumbing		3.99	5.55	8.20	3.51%	4.96%	6.25%	
	2770	Heating, ventilating, air conditioning		9.30	12.15	17.20	7.50%	10%	12%	
	2900	Electrical		8.60	11.80	15.90	7.35%	8.80%	11%	
	3100	Total: Mechanical & Electrical		26.50	34.50	46	18.30%	22%	24.50%	
	3500	See also Division 11040 (MF2004 11 91 00)								
15	0010	**CLUBS, COUNTRY**	S.F.	110	132	167				15
	0020	Total project costs	C.F.	8.85	10.80	14.90				
	2720	Plumbing	S.F.	6.65	9.85	22.50	5.60%	7.90%	10%	
	2900	Electrical		8.65	11.85	15.45	7%	8.95%	11%	
	3100	Total: Mechanical & Electrical		46	57.50	60.50	19%	26.50%	29.50%	
17	0010	**CLUBS, SOCIAL Fraternal**	S.F.	87.50	126	169				17
	0020	Total project costs	C.F.	5.50	8.30	9.90				
	2720	Plumbing	S.F.	5.50	6.85	10.40	5.60%	6.90%	8.55%	
	2770	Heating, ventilating, air conditioning		7.95	9.60	12.35	8.20%	9.25%	14.40%	
	2900	Electrical		6.60	10.85	12.40	6.50%	9.50%	10.55%	
	3100	Total: Mechanical & Electrical		19.45	37	47	21%	23%	23.50%	
18	0010	**CLUBS, Y.M.C.A.**	S.F.	110	143	179				18
	0020	Total project costs	C.F.	5.05	8.50	12.65				
	2720	Plumbing	S.F.	6.95	13.85	15.55	5.65%	7.60%	10.85%	
	2900	Electrical		8.80	11	15.15	6.05%	7.60%	9.25%	
	3100	Total: Mechanical & Electrical		33.50	37.50	51.50	18.40%	21.50%	28.50%	
19	0010	**COLLEGES Classrooms & Administration**	S.F.	111	152	201				19
	0020	Total project costs	C.F.	8.10	11.80	18.20				
	0500	Masonry	S.F.	8.15	15.35	18.85	5.65%	8.25%	10.50%	
	2720	Plumbing		5.65	11.60	20.50	5.10%	6.60%	8.95%	
	2900	Electrical		9.25	14.05	19.20	7.70%	9.85%	12%	
	3100	Total: Mechanical & Electrical		37	51.50	61.50	24%	28%	31.50%	
21	0010	**COLLEGES Science, Engineering, Laboratories**	S.F.	207	242	280				21
	0020	Total project costs	C.F.	11.85	17.30	19.65				
	1800	Equipment	S.F.	11.50	26	28.50	2%	6.45%	12.65%	
	2900	Electrical		17.05	23.50	37	7.10%	9.40%	12.10%	
	3100	Total: Mechanical & Electrical		63.50	75	116	28.50%	31.50%	41%	
	3500	See also Division 11600 (MF2004 11 53 00)								
23	0010	**COLLEGES Student Unions**	S.F.	132	179	216				23
	0020	Total project costs	C.F.	7.35	9.65	11.90				
	3100	Total: Mechanical & Electrical	S.F.	49.50	53.50	63.50	23.50%	26%	29%	
25	0010	**COMMUNITY CENTERS**	S.F.	109	134	181				25
	0020	Total project costs	C.F.	7.10	10.15	13.15				
	1800	Equipment	S.F.	2.64	4.46	7.10	1.48%	3.01%	5.45%	
	2720	Plumbing		5.15	9	12.30	4.85%	7%	8.95%	
	2770	Heating, ventilating, air conditioning		8.25	12.10	17.20	6.80%	10.35%	12.90%	
	2900	Electrical		8.90	11.90	17.05	7.30%	9%	10.45%	
	3100	Total: Mechanical & Electrical		31	38	54	20%	25%	31%	
28	0010	**COURT HOUSES**	S.F.	157	180	233				28
	0020	Total project costs	C.F.	12.05	14.40	18.15				
	2720	Plumbing	S.F.	7.45	10.45	15	5.95%	7.45%	8.20%	
	2900	Electrical		16.70	18.50	27	8.90%	10.45%	11.55%	
	3100	Total: Mechanical & Electrical		42.50	58	64	22.50%	27.50%	30.50%	
30	0010	**DEPARTMENT STORES**	S.F.	58	78.50	99				30
	0020	Total project costs	C.F.	3.11	4.03	5.50				
	2720	Plumbing	S.F.	1.81	2.28	3.47	1.82%	4.21%	5.90%	
	2770	Heating, ventilating, air conditioning		5.30	8.15	12.25	8.20%	9.10%	14.80%	

	50 17 00 | S.F. Costs		UNIT	UNIT COSTS 1/4	UNIT COSTS MEDIAN	UNIT COSTS 3/4	% OF TOTAL 1/4	% OF TOTAL MEDIAN	% OF TOTAL 3/4
30	2900	Electrical	S.F.	6.65	9.15	10.80	9.05%	12.15%	14.95%
	3100	Total: Mechanical & Electrical		11.75	15	26.50	13.20%	21.50%	50%
31	0010	**DORMITORIES Low Rise (1 to 3 story)**	S.F.	109	143	185			
	0020	Total project costs	C.F.	6.20	10.05	15.05			
	2720	Plumbing	S.F.	6.60	8.85	11.20	8.05%	9%	9.65%
	2770	Heating, ventilating, air conditioning		7	8.40	11.15	4.61%	8.05%	10%
	2900	Electrical		7.15	10.95	14.85	6.25%	8.65%	9.50%
	3100	Total: Mechanical & Electrical		36.50	40.50	63.50	21.50%	24%	27%
	9000	Per bed, total cost	Bed	46,700	52,000	111,000			
32	0010	**DORMITORIES Mid Rise (4 to 8 story)**	S.F.	135	176	218			
	0020	Total project costs	C.F.	14.90	16.35	19.60			
	2900	Electrical	S.F.	14.35	16.30	22	8.20%	10.20%	11.95%
	3100	Total: Mechanical & Electrical	″	40	41.50	81.50	25.50%	34.50%	37.50%
	9000	Per bed, total cost	Bed	19,200	43,900	111,000			
34	0010	**FACTORIES**	S.F.	51	76.50	117			
	0020	Total project costs	C.F.	3.28	4.89	8.10			
	0100	Site work	S.F.	5.85	10.65	16.85	6.95%	11.45%	17.95%
	2720	Plumbing		2.76	5.15	8.50	3.73%	6.05%	8.10%
	2770	Heating, ventilating, air conditioning		5.35	7.70	10.40	5.25%	8.45%	11.35%
	2900	Electrical		6.35	10.05	15.35	8.10%	10.50%	14.20%
	3100	Total: Mechanical & Electrical		18.15	24.50	37	21%	28.50%	35.50%
36	0010	**FIRE STATIONS**	S.F.	102	139	182			
	0020	Total project costs	C.F.	5.95	8.15	10.85			
	0500	Masonry	S.F.	15	27	34.50	8.60%	11.65%	16.45%
	1140	Roofing		3.31	8.95	10.20	1.90%	4.94%	5.05%
	1580	Painting		2.57	3.84	3.93	1.37%	1.57%	2.07%
	1800	Equipment		1.31	2.57	4.47	.74%	1.98%	3.54%
	2720	Plumbing		5.70	9.15	13.25	5.85%	7.35%	9.45%
	2770	Heating, ventilating, air conditioning		5.60	9.10	14.10	5.15%	7.40%	9.40%
	2900	Electrical		7.20	12.55	16.95	6.85%	8.65%	10.60%
	3100	Total: Mechanical & Electrical		35.50	45.50	52.50	18.75%	23%	27%
37	0010	**FRATERNITY HOUSES & Sorority Houses**	S.F.	102	131	179			
	0020	Total project costs	C.F.	10.10	10.55	12.70			
	2720	Plumbing	S.F.	7.65	8.80	16.10	6.80%	8%	10.85%
	2900	Electrical		6.70	14.45	17.70	6.60%	9.90%	10.65%
	3100	Total: Mechanical & Electrical		17.90	26	31		15.10%	15.90%
38	0010	**FUNERAL HOMES**	S.F.	107	146	265			
	0020	Total project costs	C.F.	10.95	12.15	23.50			
	2900	Electrical	S.F.	4.73	8.65	9.50	3.58%	4.44%	5.95%
	3100	Total: Mechanical & Electrical		16.75	24.50	34	12.90%	12.90%	12.90%
39	0010	**GARAGES, COMMERCIAL (Service)**	S.F.	60.50	93.50	129			
	0020	Total project costs	C.F.	3.99	5.90	8.55			
	1800	Equipment	S.F.	3.42	7.70	12	2.21%	4.62%	6.80%
	2720	Plumbing		4.20	6.45	11.80	5.45%	7.85%	10.65%
	2730	Heating & ventilating		5.50	7.30	9.90	5.25%	6.85%	8.20%
	2900	Electrical		5.75	8.80	12.65	7.15%	9.25%	10.85%
	3100	Total: Mechanical & Electrical		12.70	24.50	36	12.35%	17.40%	26%
40	0010	**GARAGES, MUNICIPAL (Repair)**	S.F.	89	119	167			
	0020	Total project costs	C.F.	5.55	7.05	12.15			
	0500	Masonry	S.F.	8.35	16.35	25.50	5.60%	9.15%	12.50%
	2720	Plumbing		4	7.65	14.45	3.59%	6.70%	7.95%
	2730	Heating & ventilating		6.85	9.90	19.10	6.15%	7.45%	13.50%
	2900	Electrical		6.60	10.35	14.90	6.65%	8.15%	11.15%
	3100	Total: Mechanical & Electrical		30.50	42	61	21.50%	25.50%	28.50%

50 17 00 | S.F. Costs

		50 17 00 | S.F. Costs	UNIT	UNIT COSTS 1/4	UNIT COSTS MEDIAN	UNIT COSTS 3/4	% OF TOTAL 1/4	% OF TOTAL MEDIAN	% OF TOTAL 3/4	
41	0010	**GARAGES, PARKING**	S.F.	34.50	50.50	87				41
	0020	Total project costs	C.F.	3.26	4.42	6.45				
	2720	Plumbing	S.F.	.98	1.52	2.35	1.72%	2.70%	3.85%	
	2900	Electrical		1.90	2.33	3.67	4.33%	5.20%	6.30%	
	3100	Total: Mechanical & Electrical		3.89	5.40	6.75	7%	8.90%	11.05%	
	3200									
	9000	Per car, total cost	Car	14,600	18,400	23,400				
43	0010	**GYMNASIUMS**	S.F.	96.50	129	174				43
	0020	Total project costs	C.F.	4.81	6.50	8				
	1800	Equipment	S.F.	2.29	4.30	8.25	1.81%	3.30%	6.70%	
	2720	Plumbing		6.10	7.25	9.35	4.65%	6.40%	7.75%	
	2770	Heating, ventilating, air conditioning		6.55	10	20	5.15%	9.05%	11.10%	
	2900	Electrical		7.40	10.10	13.30	6.60%	8.30%	10.30%	
	3100	Total: Mechanical & Electrical		27	36.50	43.50	19.75%	23.50%	29%	
	3500	See also Division 11480 (MF2004 11 67 00)								
46	0010	**HOSPITALS**	S.F.	185	228	315				46
	0020	Total project costs	C.F.	14.05	17.45	25				
	1800	Equipment	S.F.	4.70	9.05	15.60	.96%	2.63%	5%	
	2720	Plumbing		15.95	22.50	28.50	7.60%	9.10%	10.85%	
	2770	Heating, ventilating, air conditioning		23.50	30	40.50	7.80%	12.95%	16.65%	
	2900	Electrical		20	26.50	39	9.85%	11.55%	13.90%	
	3100	Total: Mechanical & Electrical		57	81.50	123	27%	33.50%	36.50%	
	9000	Per bed or person, total cost	Bed	214,500	295,500	340,500				
	9900	See also Division 11700 (MF2004 11 71 00)								
48	0010	**HOUSING For the Elderly**	S.F.	91	115	142				48
	0020	Total project costs	C.F.	6.50	9	11.50				
	0100	Site work	S.F.	6.35	9.85	14.45	5.05%	7.90%	12.10%	
	0500	Masonry		2.77	10.35	15.15	1.30%	6.05%	11%	
	1800	Equipment		2.20	3.03	4.82	1.88%	3.23%	4.43%	
	2510	Conveying systems		2.21	2.98	4.03	1.78%	2.20%	2.81%	
	2720	Plumbing		6.75	8.60	10.85	8.15%	9.55%	10.50%	
	2730	Heating, ventilating, air conditioning		3.47	4.91	7.35	3.30%	5.60%	7.25%	
	2900	Electrical		6.80	9.20	11.80	7.30%	8.50%	10.25%	
	3100	Total: Mechanical & Electrical		23.50	28	37	18.10%	22.50%	29%	
	9000	Per rental unit, total cost	Unit	84,500	99,000	110,500				
	9500	Total: Mechanical & Electrical	"	18,800	21,700	25,300				
50	0010	**HOUSING Public (Low Rise)**	S.F.	76.50	106	138				50
	0020	Total project costs	C.F.	6.80	8.50	10.55				
	0100	Site work	S.F.	9.75	14.05	22.50	8.35%	11.75%	16.50%	
	1800	Equipment		2.08	3.39	5.15	2.26%	3.03%	4.24%	
	2720	Plumbing		5.50	7.30	9.25	7.15%	9.05%	11.60%	
	2730	Heating, ventilating, air conditioning		2.77	5.40	5.90	4.26%	6.05%	6.45%	
	2900	Electrical		4.63	6.90	9.55	5.10%	6.55%	8.25%	
	3100	Total: Mechanical & Electrical		22	28.50	31.50	14.50%	17.55%	26.50%	
	9000	Per apartment, total cost	Apt.	84,000	95,500	120,000				
	9500	Total: Mechanical & Electrical	"	17,900	22,100	24,500				
51	0010	**ICE SKATING RINKS**	S.F.	65.50	153	168				51
	0020	Total project costs	C.F.	4.81	4.92	5.65				
	2720	Plumbing	S.F.	2.44	4.58	4.69	3.12%	3.23%	5.65%	
	2900	Electrical		7	10.75	11.35	6.30%	10.15%	15.05%	
	3100	Total: Mechanical & Electrical		11.65	16.50	20.50	18.95%	18.95%	18.95%	
52	0010	**JAILS**	S.F.	199	257	330				52
	0020	Total project costs	C.F.	17.95	25	29.50				
	1800	Equipment	S.F.	7.75	23	39	2.80%	5.55%	10.35%	
	2720	Plumbing		19.15	25.50	34	7%	8.90%	13.35%	
	2770	Heating, ventilating, air conditioning		17.95	24	46.50	7.50%	9.45%	17.75%	
	2900	Electrical		20.50	27.50	35	8.20%	11.55%	14.95%	

50 17 00 | S.F. Costs

				UNIT COSTS			% OF TOTAL		
			UNIT	1/4	MEDIAN	3/4	1/4	MEDIAN	3/4
52	3100	Total: Mechanical & Electrical	S.F.	54	99	117	28%	30%	34%
53	0010	**LIBRARIES**	S.F.	126	161	207			
	0020	Total project costs	C.F.	8.60	10.80	13.75			
	0500	Masonry	S.F.	9.85	17.40	29	5.80%	7.60%	11.80%
	1800	Equipment		1.72	4.63	7	.37%	1.50%	4.16%
	2720	Plumbing		4.65	6.75	9.15	3.38%	4.60%	5.70%
	2770	Heating, ventilating, air conditioning		10.30	17.40	22.50	7.80%	10.95%	12.80%
	2900	Electrical		12.85	16.65	21	8.30%	10.30%	11.95%
	3100	Total: Mechanical & Electrical		38	48	58.50	19.65%	23%	26.50%
54	0010	**LIVING, ASSISTED**	S.F.	116	137	162			
	0020	Total project costs	C.F.	9.80	11.45	13			
	0500	Masonry	S.F.	3.42	4.08	4.79	2.37%	3.16%	3.86%
	1800	Equipment		2.65	3.08	3.94	2.12%	2.45%	2.66%
	2720	Plumbing		9.75	13.05	13.55	6.05%	8.15%	10.60%
	2770	Heating, ventilating, air conditioning		11.60	12.10	13.25	7.95%	9.35%	9.70%
	2900	Electrical		11.40	12.60	14.55	9%	10%	10.70%
	3100	Total: Mechanical & Electrical		32	37.50	43	26%	29%	31.50%
55	0010	**MEDICAL CLINICS**	S.F.	118	146	185			
	0020	Total project costs	C.F.	8.70	11.25	14.95			
	1800	Equipment	S.F.	3.20	6.70	10.45	1.05%	2.94%	6.35%
	2720	Plumbing		7.85	11.10	14.80	6.15%	8.40%	10.10%
	2770	Heating, ventilating, air conditioning		9.40	12.30	18.10	6.65%	8.85%	11.35%
	2900	Electrical		10.20	14.45	18.90	8.10%	10%	12.25%
	3100	Total: Mechanical & Electrical		32.50	44	60.50	22.50%	27%	33.50%
	3500	See also Division 11700 (MF2004 11 71 00)							
57	0010	**MEDICAL OFFICES**	S.F.	112	138	169			
	0020	Total project costs	C.F.	8.30	11.25	15.25			
	1800	Equipment	S.F.	3.68	7.25	10.35	.70%	5.10%	7.05%
	2720	Plumbing		6.15	9.50	12.80	5.60%	6.80%	8.50%
	2770	Heating, ventilating, air conditioning		7.45	10.75	14.20	6.10%	8%	9.70%
	2900	Electrical		9.10	13.05	18.10	7.65%	9.80%	11.70%
	3100	Total: Mechanical & Electrical		26	35.50	51.50	19.35%	23%	29%
59	0010	**MOTELS**	S.F.	70.50	102	133			
	0020	Total project costs	C.F.	6.25	8.40	13.70			
	2720	Plumbing	S.F.	7.15	9.05	10.80	9.45%	10.60%	12.55%
	2770	Heating, ventilating, air conditioning		4.34	6.50	11.60	5.60%	5.60%	10%
	2900	Electrical		6.65	8.40	10.45	7.45%	9.05%	10.45%
	3100	Total: Mechanical & Electrical		22.50	28.50	48.50	18.50%	24%	25.50%
	5000								
	9000	Per rental unit, total cost	Unit	35,800	68,000	73,500			
	9500	Total: Mechanical & Electrical	"	6,975	10,600	12,300			
60	0010	**NURSING HOMES**	S.F.	110	142	174			
	0020	Total project costs	C.F.	8.65	10.80	14.80			
	1800	Equipment	S.F.	3.47	4.60	7.70	2.02%	3.62%	4.99%
	2720	Plumbing		9.45	14.30	17.25	8.75%	10.10%	12.70%
	2770	Heating, ventilating, air conditioning		9.95	15.10	20	9.70%	11.45%	11.80%
	2900	Electrical		10.90	13.60	18.55	9.40%	10.55%	12.45%
	3100	Total: Mechanical & Electrical		26	36.50	61	26%	29.50%	30.50%
	9000	Per bed or person, total cost	Bed	48,900	61,000	79,000			
61	0010	**OFFICES Low Rise (1 to 4 story)**	S.F.	93	120	155			
	0020	Total project costs	C.F.	6.65	9.15	12.05			
	0100	Site work	S.F.	7.20	12.60	18.85	5.90%	9.70%	13.60%
	0500	Masonry		3.46	7.25	13.15	2.62%	5.50%	8.45%
	1800	Equipment		.99	1.93	5.25	.69%	1.50%	3.63%
	2720	Plumbing		3.32	5.15	7.50	3.66%	4.50%	6.10%
	2770	Heating, ventilating, air conditioning		7.35	10.25	15	7.20%	10.30%	11.70%
	2900	Electrical		7.55	10.75	15.20	7.45%	9.60%	11.35%

	50 17 00 | S.F. Costs		UNIT	UNIT COSTS 1/4	UNIT COSTS MEDIAN	UNIT COSTS 3/4	% OF TOTAL 1/4	% OF TOTAL MEDIAN	% OF TOTAL 3/4	
61	3100	Total: Mechanical & Electrical	S.F.	20.50	27.50	41	18%	22%	27%	61
62	0010	**OFFICES Mid Rise (5 to 10 story)**	S.F.	98.50	119	157				62
	0020	Total project costs	C.F.	6.95	8.90	12.60				
	2720	Plumbing	S.F.	2.97	4.61	6.60	2.83%	3.74%	4.50%	
	2770	Heating, ventilating, air conditioning		7.45	10.65	17.05	7.65%	9.40%	11%	
	2900	Electrical		7.30	9.35	12.95	6.35%	7.80%	10%	
	3100	Total: Mechanical & Electrical	↓	18.90	24.50	46.50	19.15%	21.50%	27.50%	
63	0010	**OFFICES High Rise (11 to 20 story)**	S.F.	121	152	187				63
	0020	Total project costs	C.F.	8.45	10.55	15.15				
	2900	Electrical	S.F.	7.35	8.95	13.30	5.80%	7.85%	10.50%	
	3100	Total: Mechanical & Electrical	↓	23.50	31.50	53.50	16.90%	23.50%	34%	
64	0010	**POLICE STATIONS**	S.F.	145	190	240				64
	0020	Total project costs	C.F.	11.55	14.15	19.35				
	0500	Masonry	S.F.	13.60	24	30	6.70%	9.10%	11.35%	
	1800	Equipment		2.30	10.10	16.05	.98%	3.35%	6.70%	
	2720	Plumbing		8.10	16.15	20	5.65%	6.90%	10.75%	
	2770	Heating, ventilating, air conditioning		12.65	16.80	23	5.85%	10.55%	11.70%	
	2900	Electrical		15.85	22.50	30	9.80%	11.85%	14.80%	
	3100	Total: Mechanical & Electrical	↓	52	62.50	84.50	25%	31.50%	32.50%	
65	0010	**POST OFFICES**	S.F.	114	141	180				65
	0020	Total project costs	C.F.	6.90	8.70	9.90				
	2720	Plumbing	S.F.	5.15	6.40	8.05	4.24%	5.30%	5.60%	
	2770	Heating, ventilating, air conditioning		8.05	9.95	11.05	6.65%	7.15%	9.35%	
	2900	Electrical		9.45	13.30	15.75	7.25%	9%	11%	
	3100	Total: Mechanical & Electrical	↓	27.50	35.50	40.50	16.25%	18.80%	22%	
66	0010	**POWER PLANTS**	S.F.	795	1,000	1,925				66
	0020	Total project costs	C.F.	22	47.50	102				
	2900	Electrical	S.F.	56	119	177	9.30%	12.75%	21.50%	
	8100	Total: Mechanical & Electrical	↓	140	455	1,025	32.50%	32.50%	52.50%	
67	0010	**RELIGIOUS EDUCATION**	S.F.	92	120	148				67
	0020	Total project costs	C.F.	5.10	7.35	9.15				
	2720	Plumbing	S.F.	3.85	5.45	7.70	4.40%	5.30%	7.10%	
	2770	Heating, ventilating, air conditioning		9.75	11	15.55	10.05%	11.45%	12.35%	
	2900	Electrical		7.30	10.30	13.65	7.60%	9.05%	10.35%	
	3100	Total: Mechanical & Electrical	↓	29.50	39	47.50	22%	23%	27%	
69	0010	**RESEARCH Laboratories & Facilities**	S.F.	138	198	288				69
	0020	Total project costs	C.F.	10.40	20	24				
	1800	Equipment	S.F.	6.10	12.05	29.50	.94%	4.58%	8.80%	
	2720	Plumbing		13.85	17.70	28.50	6.15%	8.30%	10.80%	
	2770	Heating, ventilating, air conditioning		12.40	41.50	49.50	7.25%	16.50%	17.50%	
	2900	Electrical		16.15	26.50	43.50	9.55%	11.50%	15.40%	
	3100	Total: Mechanical & Electrical	↓	49	92	132	29.50%	37%	42%	
70	0010	**RESTAURANTS**	S.F.	134	172	224				70
	0020	Total project costs	C.F.	11.25	14.75	19.35				
	1800	Equipment	S.F.	8.65	21	32	6.10%	13%	15.65%	
	2720	Plumbing		10.60	12.85	16.85	6.10%	8.15%	9%	
	2770	Heating, ventilating, air conditioning		13.45	18.60	24	9.20%	12%	12.40%	
	2900	Electrical		14.15	17.45	22.50	8.35%	10.55%	11.55%	
	3100	Total: Mechanical & Electrical	↓	44	47	60.50	21%	24.50%	29.50%	
	9000	Per seat unit, total cost	Seat	4,900	6,550	7,750				
	9500	Total: Mechanical & Electrical	"	1,225	1,625	1,925				
72	0010	**RETAIL STORES**	S.F.	62.50	84	111				72
	0020	Total project costs	C.F.	4.23	6.05	8.40				
	2720	Plumbing	S.F.	2.26	3.78	6.45	3.26%	4.60%	6.80%	
	2770	Heating, ventilating, air conditioning		4.89	6.70	10.05	6.75%	8.75%	10.15%	
	2900	Electrical		5.65	7.70	11.10	7.25%	9.90%	11.65%	
	3100	Total: Mechanical & Electrical	↓	14.95	19.20	26.50	17.05%	21%	23.50%	

		50 17 00 | S.F. Costs	UNIT	UNIT COSTS 1/4	UNIT COSTS MEDIAN	UNIT COSTS 3/4	% OF TOTAL 1/4	% OF TOTAL MEDIAN	% OF TOTAL 3/4
74	0010	**SCHOOLS Elementary**	S.F.	101	125	155			
	0020	Total project costs	C.F.	6.65	8.55	11.05			
	0500	Masonry	S.F.	9.10	15.70	23.50	5.80%	11%	14.95%
	1800	Equipment		2.78	4.70	8.75	1.89%	3.32%	4.71%
	2720	Plumbing		5.85	8.30	11.05	5.70%	7.15%	9.35%
	2730	Heating, ventilating, air conditioning		8.80	14	19.55	8.15%	10.80%	14.90%
	2900	Electrical		9.60	12.70	15.95	8.40%	10.05%	11.70%
	3100	Total: Mechanical & Electrical	↓	34	43	52.50	25%	27.50%	30%
	9000	Per pupil, total cost	Ea.	11,700	17,400	50,000			
	9500	Total: Mechanical & Electrical	"	3,300	4,175	14,500			
76	0010	**SCHOOLS Junior High & Middle**	S.F.	103	129	157			
	0020	Total project costs	C.F.	6.65	8.65	9.70			
	0500	Masonry	S.F.	13.10	16.90	19.75	8%	11.40%	14.30%
	1800	Equipment		3.36	5.40	8.15	1.81%	3.26%	4.86%
	2720	Plumbing		6.10	7.55	9.35	5.30%	6.80%	7.25%
	2770	Heating, ventilating, air conditioning		12.20	14.85	26	8.90%	11.55%	14.20%
	2900	Electrical		10.30	12.40	15.95	7.90%	9.35%	10.60%
	3100	Total: Mechanical & Electrical	↓	33.50	43	53	23.50%	27%	29.50%
	9000	Per pupil, total cost	Ea.	13,300	17,500	23,500			
78	0010	**SCHOOLS Senior High**	S.F.	108	133	167			
	0020	Total project costs	C.F.	6.60	9.70	15.65			
	1800	Equipment	S.F.	2.87	6.75	9.50	1.88%	2.98%	4.80%
	2720	Plumbing		6.15	9.20	16.80	5.60%	6.90%	8.30%
	2770	Heating, ventilating, air conditioning		12.50	14.35	27.50	8.95%	11.60%	15%
	2900	Electrical		10.95	14.25	21	8.65%	10.15%	11.95%
	3100	Total: Mechanical & Electrical	↓	36.50	42.50	71.50	23.50%	26.50%	28.50%
	9000	Per pupil, total cost	Ea.	10,300	21,000	26,200			
80	0010	**SCHOOLS Vocational**	S.F.	88.50	128	159			
	0020	Total project costs	C.F.	5.50	7.90	10.90			
	0500	Masonry	S.F.	5.20	12.80	19.60	3.53%	6.70%	10.95%
	1800	Equipment		2.77	6.90	9.60	1.24%	3.10%	4.26%
	2720	Plumbing		5.65	8.45	12.40	5.40%	6.90%	8.55%
	2770	Heating, ventilating, air conditioning		7.90	14.75	24.50	8.60%	11.90%	14.65%
	2900	Electrical		9.20	12.05	16.60	8.45%	10.95%	13.20%
	3100	Total: Mechanical & Electrical	↓	32	35.50	61	23.50%	27.50%	31%
	9000	Per pupil, total cost	Ea.	12,300	33,000	49,200			
83	0010	**SPORTS ARENAS**	S.F.	77.50	103	159			
	0020	Total project costs	C.F.	4.20	7.50	9.70			
	2720	Plumbing	S.F.	4.49	6.80	14.35	4.35%	6.35%	9.40%
	2770	Heating, ventilating, air conditioning		9.65	11.40	15.85	8.80%	10.20%	13.55%
	2900	Electrical		8.05	10.95	14.15	8.60%	9.90%	12.25%
	3100	Total: Mechanical & Electrical	↓	20	35	47	21.50%	25%	27.50%
85	0010	**SUPERMARKETS**	S.F.	71.50	83	97			
	0020	Total project costs	C.F.	3.98	4.81	7.30			
	2720	Plumbing	S.F.	3.99	5.05	5.85	5.40%	6%	7.45%
	2770	Heating, ventilating, air conditioning		5.85	7.80	9.50	8.60%	8.65%	9.60%
	2900	Electrical		8.95	10.30	12.15	10.40%	12.45%	13.60%
	3100	Total: Mechanical & Electrical	↓	23	25	32.50	20.50%	26.50%	31%
86	0010	**SWIMMING POOLS**	S.F.	116	194	415			
	0020	Total project costs	C.F.	9.25	11.55	12.60			
	2720	Plumbing	S.F.	10.70	12.20	16.60	4.80%	9.70%	20.50%
	2900	Electrical		8.70	14.10	20.50	5.75%	6.95%	7.60%
	3100	Total: Mechanical & Electrical	↓	21	55	73.50	11.15%	14.10%	23.50%
87	0010	**TELEPHONE EXCHANGES**	S.F.	154	225	285			
	0020	Total project costs	C.F.	9.55	15.35	21			
	2720	Plumbing	S.F.	6.50	10	14.65	4.52%	5.80%	6.90%
	2770	Heating, ventilating, air conditioning	↓	15.05	30.50	37.50	11.80%	16.05%	18.40%

	50 17 00 | S.F. Costs		UNIT	UNIT COSTS 1/4	UNIT COSTS MEDIAN	UNIT COSTS 3/4	% OF TOTAL 1/4	% OF TOTAL MEDIAN	% OF TOTAL 3/4	
87	2900	Electrical	S.F.	15.65	25	44	10.90%	14%	17.85%	87
	3100	Total: Mechanical & Electrical	↓	46	87.50	124	29.50%	33.50%	44.50%	
91	0010	**THEATERS**	S.F.	96.50	120	183				91
	0020	Total project costs	C.F.	4.46	6.60	9.70				
	2720	Plumbing	S.F.	3.22	3.49	14.25	2.92%	4.70%	6.80%	
	2770	Heating, ventilating, air conditioning		9.40	11.35	14.05	8%	12.25%	13.40%	
	2900	Electrical		8.45	11.40	23	8.05%	9.95%	12.25%	
	3100	Total: Mechanical & Electrical	↓	21.50	32.50	66.50	23%	26.50%	27.50%	
94	0010	**TOWN HALLS City Halls & Municipal Buildings**	S.F.	108	137	179				94
	0020	Total project costs	C.F.	9.85	11.80	16.60				
	2720	Plumbing	S.F.	4.50	8.40	15.50	4.31%	5.95%	7.95%	
	2770	Heating, ventilating, air conditioning		8.15	16.15	23.50	7.05%	9.05%	13.45%	
	2900	Electrical		10.25	14.60	19.95	8.05%	9.45%	11.65%	
	3100	Total: Mechanical & Electrical	↓	35.50	45	69	22%	26.50%	31%	
97	0010	**WAREHOUSES & Storage Buildings**	S.F.	40.50	60	86				97
	0020	Total project costs	C.F.	2.11	3.30	5.45				
	0100	Site work	S.F.	4.15	8.25	12.45	6.05%	12.95%	19.85%	
	0500	Masonry		2.42	5.70	12.35	3.73%	7.40%	12.30%	
	1800	Equipment		.65	1.39	7.80	.91%	1.82%	5.55%	
	2720	Plumbing		1.34	2.41	4.50	2.90%	4.80%	6.55%	
	2730	Heating, ventilating, air conditioning		1.53	4.32	5.80	2.41%	5%	8.90%	
	2900	Electrical		2.38	4.48	7.40	5.15%	7.20%	10.10%	
	3100	Total: Mechanical & Electrical	↓	6.65	10.20	20	12.75%	18.90%	26%	
99	0010	**WAREHOUSE & OFFICES Combination**	S.F.	49.50	66	90.50				99
	0020	Total project costs	C.F.	2.53	3.67	5.45				
	1800	Equipment	S.F.	.86	1.66	2.47	.52%	1.20%	2.40%	
	2720	Plumbing		1.91	3.39	4.96	3.74%	4.76%	6.30%	
	2770	Heating, ventilating, air conditioning		3.02	4.72	6.60	5%	5.65%	10.05%	
	2900	Electrical		3.31	4.93	7.75	5.75%	8%	10%	
	3100	Total: Mechanical & Electrical	↓	9.25	14.25	22.50	14.40%	19.95%	24.50%	

Square Foot Project Size Modifier

One factor that affects the S.F. cost of a particular building is the size. In general, for buildings built to the same specifications in the same locality, the larger building will have the lower S.F. cost. This is due mainly to the decreasing contribution of the exterior walls plus the economy of scale usually achievable in larger buildings. The Area Conversion Scale shown below will give a factor to convert costs for the typical size building to an adjusted cost for the particular project.

The Square Foot Base Size lists the median costs, most typical project size in our accumulated data, and the range in size of the projects.

The Size Factor for your project is determined by dividing your project area in S.F. by the typical project size for the particular Building Type. With this factor, enter the Area Conversion Scale at the appropriate Size Factor and determine the appropriate cost multiplier for your building size.

Example: Determine the cost per S.F. for a 100,000 S.F. Mid-rise apartment building.

$$\frac{\text{Proposed building area} = 100{,}000 \text{ S.F.}}{\text{Typical size from below} = 50{,}000 \text{ S.F.}} = 2.00$$

Enter Area Conversion scale at 2.0, intersect curve, read horizontally the appropriate cost multiplier of .94. Size adjusted cost becomes .94 x $107.00 = $101.00 based on national average costs.

Note: For Size Factors less than .50, the Cost Multiplier is 1.1
For Size Factors greater than 3.5, the Cost Multiplier is .90

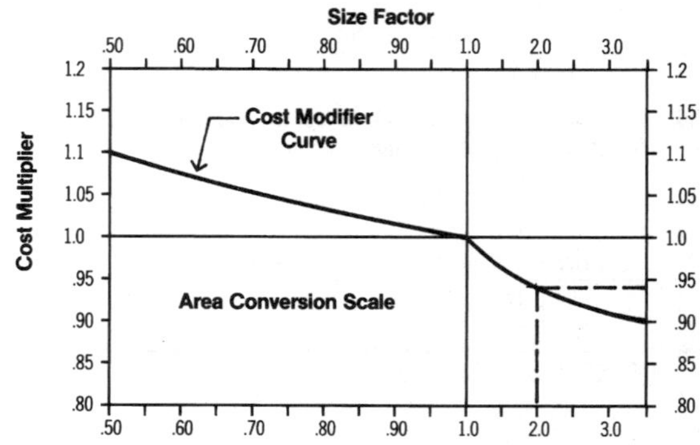

Square Foot Base Size

Building Type	Median Cost per S.F.	Typical Size Gross S.F.	Typical Range Gross S.F.	Building Type	Median Cost per S.F.	Typical Size Gross S.F.	Typical Range Gross S.F.
Apartments, Low Rise	$ 84.50	21,000	9,700 - 37,200	Jails	$ 257.00	40,000	5,500 - 145,000
Apartments, Mid Rise	107.00	50,000	32,000 - 100,000	Libraries	161.00	12,000	7,000 - 31,000
Apartments, High Rise	116.00	145,000	95,000 - 600,000	Living, Assisted	137.00	32,300	23,500 - 50,300
Auditoriums	141.00	25,000	7,600 - 39,000	Medical Clinics	146.00	7,200	4,200 - 15,700
Auto Sales	105.00	20,000	10,800 - 28,600	Medical Offices	138.00	6,000	4,000 - 15,000
Banks	189.00	4,200	2,500 - 7,500	Motels	102.00	40,000	15,800 - 120,000
Churches	130.00	17,000	2,000 - 42,000	Nursing Homes	142.00	23,000	15,000 - 37,000
Clubs, Country	132.00	6,500	4,500 - 15,000	Offices, Low Rise	120.00	20,000	5,000 - 80,000
Clubs, Social	126.00	10,000	6,000 - 13,500	Offices, Mid Rise	119.00	120,000	20,000 - 300,000
Clubs, YMCA	143.00	28,300	12,800 - 39,400	Offices, High Rise	152.00	260,000	120,000 - 800,000
Colleges (Class)	152.00	50,000	15,000 - 150,000	Police Stations	190.00	10,500	4,000 - 19,000
Colleges (Science Lab)	242.00	45,600	16,600 - 80,000	Post Offices	141.00	12,400	6,800 - 30,000
College (Student Union)	179.00	33,400	16,000 - 85,000	Power Plants	1000.00	7,500	1,000 - 20,000
Community Center	134.00	9,400	5,300 - 16,700	Religious Education	120.00	9,000	6,000 - 12,000
Court Houses	180.00	32,400	17,800 - 106,000	Research	198.00	19,000	6,300 - 45,000
Dept. Stores	78.50	90,000	44,000 - 122,000	Restaurants	172.00	4,400	2,800 - 6,000
Dormitories, Low Rise	143.00	25,000	10,000 - 95,000	Retail Stores	84.00	7,200	4,000 - 17,600
Dormitories, Mid Rise	176.00	85,000	20,000 - 200,000	Schools, Elementary	125.00	41,000	24,500 - 55,000
Factories	76.50	26,400	12,900 - 50,000	Schools, Jr. High	129.00	92,000	52,000 - 119,000
Fire Stations	139.00	5,800	4,000 - 8,700	Schools, Sr. High	133.00	101,000	50,500 - 175,000
Fraternity Houses	131.00	12,500	8,200 - 14,800	Schools, Vocational	128.00	37,000	20,500 - 82,000
Funeral Homes	146.00	10,000	4,000 - 20,000	Sports Arenas	103.00	15,000	5,000 - 40,000
Garages, Commercial	93.50	9,300	5,000 - 13,600	Supermarkets	83.00	44,000	12,000 - 60,000
Garages, Municipal	119.00	8,300	4,500 - 12,600	Swimming Pools	194.00	20,000	10,000 - 32,000
Garages, Parking	50.50	163,000	76,400 - 225,300	Telephone Exchange	225.00	4,500	1,200 - 10,600
Gymnasiums	129.00	19,200	11,600 - 41,000	Theaters	120.00	10,500	8,800 - 17,500
Hospitals	228.00	55,000	27,200 - 125,000	Town Halls	137.00	10,800	4,800 - 23,400
House (Elderly)	115.00	37,000	21,000 - 66,000	Warehouses	60.00	25,000	8,000 - 72,000
Housing (Public)	106.00	36,000	14,400 - 74,400	Warehouse & Office	66.00	25,000	8,000 - 72,000
Ice Rinks	153.00	29,000	27,200 - 33,600				

Abbreviations

A	Area Square Feet; Ampere
ABS	Acrylonitrile Butadiene Stryrene; Asbestos Bonded Steel
A.C.	Alternating Current; Air-Conditioning; Asbestos Cement; Plywood Grade A & C
A.C.I.	American Concrete Institute
AD	Plywood, Grade A & D
Addit.	Additional
Adj.	Adjustable
af	Audio-frequency
A.G.A.	American Gas Association
Agg.	Aggregate
A.H.	Ampere Hours
A hr.	Ampere-hour
A.H.U.	Air Handling Unit
A.I.A.	American Institute of Architects
AIC	Ampere Interrupting Capacity
Allow.	Allowance
alt.	Altitude
Alum.	Aluminum
a.m.	Ante Meridiem
Amp.	Ampere
Anod.	Anodized
Approx.	Approximate
Apt.	Apartment
Asb.	Asbestos
A.S.B.C.	American Standard Building Code
Asbe.	Asbestos Worker
ASCE.	American Society of Civil Engineers
A.S.H.R.A.E.	American Society of Heating, Refrig. & AC Engineers
A.S.M.E.	American Society of Mechanical Engineers
A.S.T.M.	American Society for Testing and Materials
Attchmt.	Attachment
Avg., Ave.	Average
A.W.G.	American Wire Gauge
AWWA	American Water Works Assoc.
Bbl.	Barrel
B&B	Grade B and Better; Balled & Burlapped
B.&S.	Bell and Spigot
B.&W.	Black and White
b.c.c.	Body-centered Cubic
B.C.Y.	Bank Cubic Yards
BE	Bevel End
B.F.	Board Feet
Bg. cem.	Bag of Cement
BHP	Boiler Horsepower; Brake Horsepower
B.I.	Black Iron
Bit., Bitum.	Bituminous
Bit., Conc.	Bituminous Concrete
Bk.	Backed
Bkrs.	Breakers
Bldg.	Building
Blk.	Block
Bm.	Beam
Boil.	Boilermaker
B.P.M.	Blows per Minute
BR	Bedroom
Brg.	Bearing
Brhe.	Bricklayer Helper
Bric.	Bricklayer
Brk.	Brick
Brng.	Bearing
Brs.	Brass
Brz.	Bronze
Bsn.	Basin
Btr.	Better
Btu	British Thermal Unit
BTUH	BTU per Hour
B.U.R.	Built-up Roofing
BX	Interlocked Armored Cable
°C	degree centegrade
c	Conductivity, Copper Sweat
C	Hundred; Centigrade
C/C	Center to Center, Cedar on Cedar
C-C	Center to Center
Cab.	Cabinet
Cair.	Air Tool Laborer
Calc	Calculated
Cap.	Capacity
Carp.	Carpenter
C.B.	Circuit Breaker
C.C.A.	Chromate Copper Arsenate
C.C.F.	Hundred Cubic Feet
cd	Candela
cd/sf	Candela per Square Foot
CD	Grade of Plywood Face & Back
CDX	Plywood, Grade C & D, exterior glue
Cefi.	Cement Finisher
Cem.	Cement
CF	Hundred Feet
C.F.	Cubic Feet
CFM	Cubic Feet per Minute
c.g.	Center of Gravity
CHW	Chilled Water; Commercial Hot Water
C.I.	Cast Iron
C.I.P.	Cast in Place
Circ.	Circuit
C.L.	Carload Lot
Clab.	Common Laborer
Clam	Common maintenance laborer
C.L.F.	Hundred Linear Feet
CLF	Current Limiting Fuse
CLP	Cross Linked Polyethylene
cm	Centimeter
CMP	Corr. Metal Pipe
C.M.U.	Concrete Masonry Unit
CN	Change Notice
Col.	Column
CO_2	Carbon Dioxide
Comb.	Combination
Compr.	Compressor
Conc.	Concrete
Cont.	Continuous; Continued, Container
Corr.	Corrugated
Cos	Cosine
Cot	Cotangent
Cov.	Cover
C/P	Cedar on Paneling
CPA	Control Point Adjustment
Cplg.	Coupling
C.P.M.	Critical Path Method
CPVC	Chlorinated Polyvinyl Chloride
C.Pr.	Hundred Pair
CRC	Cold Rolled Channel
Creos.	Creosote
Crpt.	Carpet & Linoleum Layer
CRT	Cathode-ray Tube
CS	Carbon Steel, Constant Shear Bar Joist
Csc	Cosecant
C.S.F.	Hundred Square Feet
CSI	Construction Specifications Institute
C.T.	Current Transformer
CTS	Copper Tube Size
Cu	Copper, Cubic
Cu. Ft.	Cubic Foot
cw	Continuous Wave
C.W.	Cool White; Cold Water
Cwt.	100 Pounds
C.W.X.	Cool White Deluxe
C.Y.	Cubic Yard (27 cubic feet)
C.Y./Hr.	Cubic Yard per Hour
Cyl.	Cylinder
d	Penny (nail size)
D	Deep; Depth; Discharge
Dis., Disch.	Discharge
Db.	Decibel
Dbl.	Double
DC	Direct Current
DDC	Direct Digital Control
Demob.	Demobilization
d.f.u.	Drainage Fixture Units
D.H.	Double Hung
DHW	Domestic Hot Water
DI	Ductile Iron
Diag.	Diagonal
Diam., Dia	Diameter
Distrib.	Distribution
Div.	Division
Dk.	Deck
D.L.	Dead Load; Diesel
DLH	Deep Long Span Bar Joist
Do.	Ditto
Dp.	Depth
D.P.S.T.	Double Pole, Single Throw
Dr.	Drive
Drink.	Drinking
D.S.	Double Strength
D.S.A.	Double Strength A Grade
D.S.B.	Double Strength B Grade
Dty.	Duty
DWV	Drain Waste Vent
DX	Deluxe White, Direct Expansion
dyn	Dyne
e	Eccentricity
E	Equipment Only; East
Ea.	Each
E.B.	Encased Burial
Econ.	Economy
E.C.Y	Embankment Cubic Yards
EDP	Electronic Data Processing
EIFS	Exterior Insulation Finish System
E.D.R.	Equiv. Direct Radiation
Eq.	Equation
EL	elevation
Elec.	Electrician; Electrical
Elev.	Elevator; Elevating
EMT	Electrical Metallic Conduit; Thin Wall Conduit
Eng.	Engine, Engineered
EPDM	Ethylene Propylene Diene Monomer
EPS	Expanded Polystyrene
Eqhv.	Equip. Oper., Heavy
Eqlt.	Equip. Oper., Light
Eqmd.	Equip. Oper., Medium
Eqmm.	Equip. Oper., Master Mechanic
Eqol.	Equip. Oper., Oilers
Equip.	Equipment
ERW	Electric Resistance Welded
E.S.	Energy Saver
Est.	Estimated
esu	Electrostatic Units
E.W.	Each Way
EWT	Entering Water Temperature
Excav.	Excavation
Exp.	Expansion, Exposure
Ext.	Exterior
Extru.	Extrusion
f.	Fiber stress
F	Fahrenheit; Female; Fill
Fab.	Fabricated

FBGS	Fiberglass
F.C.	Footcandles
f.c.c.	Face-centered Cubic
f'c.	Compressive Stress in Concrete; Extreme Compressive Stress
F.E.	Front End
FEP	Fluorinated Ethylene Propylene (Teflon)
F.G.	Flat Grain
F.H.A.	Federal Housing Administration
Fig.	Figure
Fin.	Finished
Fixt.	Fixture
Fl. Oz.	Fluid Ounces
Flr.	Floor
F.M.	Frequency Modulation; Factory Mutual
Fmg.	Framing
Fdn.	Foundation
Fori.	Foreman, Inside
Foro.	Foreman, Outside
Fount.	Fountain
fpm	Feet per Minute
FPT	Female Pipe Thread
Fr.	Frame
F.R.	Fire Rating
FRK	Foil Reinforced Kraft
FRP	Fiberglass Reinforced Plastic
FS	Forged Steel
FSC	Cast Body; Cast Switch Box
Ft.	Foot; Feet
Ftng.	Fitting
Ftg.	Footing
Ft lb.	Foot Pound
Furn.	Furniture
FVNR	Full Voltage Non-Reversing
FXM	Female by Male
Fy.	Minimum Yield Stress of Steel
g	Gram
G	Gauss
Ga.	Gauge
Gal., gal.	Gallon
gpm, GPM	Gallon per Minute
Galv.	Galvanized
Gen.	General
G.F.I.	Ground Fault Interrupter
Glaz.	Glazier
GPD	Gallons per Day
GPH	Gallons per Hour
GPM	Gallons per Minute
GR	Grade
Gran.	Granular
Grnd.	Ground
H	High Henry
H.C.	High Capacity
H.D.	Heavy Duty; High Density
H.D.O.	High Density Overlaid
H.D.P.E.	high density polyethelene
Hdr.	Header
Hdwe.	Hardware
Help.	Helper Average
HEPA	High Efficiency Particulate Air Filter
Hg	Mercury
HIC	High Interrupting Capacity
HM	Hollow Metal
HMWPE	high molecular weight polyethylene
H.O.	High Output
Horiz.	Horizontal
H.P.	Horsepower; High Pressure
H.P.F.	High Power Factor
Hr.	Hour
Hrs./Day	Hours per Day
HSC	High Short Circuit
Ht.	Height
Htg.	Heating
Htrs.	Heaters
HVAC	Heating, Ventilation & Air-Conditioning
Hvy.	Heavy
HW	Hot Water
Hyd.;Hydr.	Hydraulic
Hz.	Hertz (cycles)
I.	Moment of Inertia
IBC	International Building Code
I.C.	Interrupting Capacity
ID	Inside Diameter
I.D.	Inside Dimension; Identification
I.F.	Inside Frosted
I.M.C.	Intermediate Metal Conduit
In.	Inch
Incan.	Incandescent
Incl.	Included; Including
Int.	Interior
Inst.	Installation
Insul.	Insulation/Insulated
I.P.	Iron Pipe
I.P.S.	Iron Pipe Size
I.P.T.	Iron Pipe Threaded
I.W.	Indirect Waste
J	Joule
J.I.C.	Joint Industrial Council
K	Thousand; Thousand Pounds; Heavy Wall Copper Tubing, Kelvin
K.A.H.	Thousand Amp. Hours
KCMIL	Thousand Circular Mils
KD	Knock Down
K.D.A.T.	Kiln Dried After Treatment
kg	Kilogram
kG	Kilogauss
kgf	Kilogram Force
kHz	Kilohertz
Kip.	1000 Pounds
KJ	Kiljoule
K.L.	Effective Length Factor
K.L.F.	Kips per Linear Foot
Km	Kilometer
K.S.F.	Kips per Square Foot
K.S.I.	Kips per Square Inch
kV	Kilovolt
kVA	Kilovolt Ampere
K.V.A.R.	Kilovar (Reactance)
KW	Kilowatt
KWh	Kilowatt-hour
L	Labor Only; Length; Long; Medium Wall Copper Tubing
Lab.	Labor
lat	Latitude
Lath.	Lather
Lav.	Lavatory
lb.; #	Pound
L.B.	Load Bearing; L Conduit Body
L. & E.	Labor & Equipment
lb./hr.	Pounds per Hour
lb./L.F.	Pounds per Linear Foot
lbf/sq.in.	Pound-force per Square Inch
L.C.L.	Less than Carload Lot
L.C.Y.	Loose Cubic Yard
Ld.	Load
LE	Lead Equivalent
LED	Light Emitting Diode
L.F.	Linear Foot
L.F. Nose	Linear Foot of Stair Nosing
L.F. Rsr	Linear Foot of Stair Riser
Lg.	Long; Length; Large
L & H	Light and Heat
LH	Long Span Bar Joist
L.H.	Labor Hours
L.L.	Live Load
L.L.D.	Lamp Lumen Depreciation
lm	Lumen
lm/sf	Lumen per Square Foot
lm/W	Lumen per Watt
L.O.A.	Length Over All
log	Logarithm
L-O-L	Lateralolet
long.	longitude
L.P.	Liquefied Petroleum; Low Pressu
L.P.F.	Low Power Factor
LR	Long Radius
L.S.	Lump Sum
Lt.	Light
Lt. Ga.	Light Gauge
L.T.L.	Less than Truckload Lot
Lt. Wt.	Lightweight
L.V.	Low Voltage
M	Thousand; Material; Male; Light Wall Copper Tubing
M^2CA	Meters Squared Contact Area
m/hr.; M.H.	Man-hour
mA	Milliampere
Mach.	Machine
Mag. Str.	Magnetic Starter
Maint.	Maintenance
Marb.	Marble Setter
Mat; Mat'l.	Material
Max.	Maximum
MBF	Thousand Board Feet
MBH	Thousand BTU's per hr.
MC	Metal Clad Cable
M.C.F.	Thousand Cubic Feet
M.C.F.M.	Thousand Cubic Feet per Minute
M.C.M.	Thousand Circular Mils
M.C.P.	Motor Circuit Protector
MD	Medium Duty
M.D.O.	Medium Density Overlaid
Med.	Medium
MF	Thousand Feet
M.F.B.M.	Thousand Feet Board Measure
Mfg.	Manufacturing
Mfrs.	Manufacturers
mg	Milligram
MGD	Million Gallons per Day
MGPH	Thousand Gallons per Hour
MH, M.H.	Manhole; Metal Halide; Man-Hour
MHz	Megahertz
Mi.	Mile
MI	Malleable Iron; Mineral Insulated
mm	Millimeter
Mill.	Millwright
Min., min.	Minimum, minute
Misc.	Miscellaneous
ml	Milliliter, Mainline
M.L.F.	Thousand Linear Feet
Mo.	Month
Mobil.	Mobilization
Mog.	Mogul Base
MPH	Miles per Hour
MPT	Male Pipe Thread

Abbreviations

MRT	Mile Round Trip
ms	Millisecond
M.S.F.	Thousand Square Feet
Mstz.	Mosaic & Terrazzo Worker
M.S.Y.	Thousand Square Yards
Mtd., mtd.	Mounted
Mthe.	Mosaic & Terrazzo Helper
Mtng.	Mounting
Mult.	Multi; Multiply
M.V.A.	Million Volt Amperes
M.V.A.R.	Million Volt Amperes Reactance
MV	Megavolt
MW	Megawatt
MXM	Male by Male
MYD	Thousand Yards
N	Natural; North
nA	Nanoampere
NA	Not Available; Not Applicable
N.B.C.	National Building Code
NC	Normally Closed
N.E.M.A.	National Electrical Manufacturers Assoc.
NEHB	Bolted Circuit Breaker to 600V.
N.L.B.	Non-Load-Bearing
NM	Non-Metallic Cable
nm	Nanometer
No.	Number
NO	Normally Open
N.O.C.	Not Otherwise Classified
Nose.	Nosing
N.P.T.	National Pipe Thread
NQOD	Combination Plug-on/Bolt on Circuit Breaker to 240V.
N.R.C.	Noise Reduction Coefficient/ Nuclear Regulator Commission
N.R.S.	Non Rising Stem
ns	Nanosecond
nW	Nanowatt
OB	Opposing Blade
OC	On Center
OD	Outside Diameter
O.D.	Outside Dimension
ODS	Overhead Distribution System
O.G.	Ogee
O.H.	Overhead
O&P	Overhead and Profit
Oper.	Operator
Opng.	Opening
Orna.	Ornamental
OSB	Oriented Strand Board
O.S.&Y.	Outside Screw and Yoke
Ovhd.	Overhead
OWG	Oil, Water or Gas
Oz.	Ounce
P.	Pole; Applied Load; Projection
p.	Page
Pape.	Paperhanger
P.A.P.R.	Powered Air Purifying Respirator
PAR	Parabolic Reflector
Pc., Pcs.	Piece, Pieces
P.C.	Portland Cement; Power Connector
P.C.F.	Pounds per Cubic Foot
P.C.M.	Phase Contrast Microscopy
P.E.	Professional Engineer; Porcelain Enamel; Polyethylene; Plain End
Perf.	Perforated
PEX	Cross linked polyethylene
Ph.	Phase
P.I.	Pressure Injected
Pile.	Pile Driver
Pkg.	Package
Pl.	Plate
Plah.	Plasterer Helper
Plas.	Plasterer
Pluh.	Plumbers Helper
Plum.	Plumber
Ply.	Plywood
p.m.	Post Meridiem
Pntd.	Painted
Pord.	Painter, Ordinary
pp	Pages
PP, PPL	Polypropylene
P.P.M.	Parts per Million
Pr.	Pair
P.E.S.B.	Pre-engineered Steel Building
Prefab.	Prefabricated
Prefin.	Prefinished
Prop.	Propelled
PSF, psf	Pounds per Square Foot
PSI, psi	Pounds per Square Inch
PSIG	Pounds per Square Inch Gauge
PSP	Plastic Sewer Pipe
Pspr.	Painter, Spray
Psst.	Painter, Structural Steel
P.T.	Potential Transformer
P. & T.	Pressure & Temperature
Ptd.	Painted
Ptns.	Partitions
Pu	Ultimate Load
PVC	Polyvinyl Chloride
Pvmt.	Pavement
Pwr.	Power
Q	Quantity Heat Flow
Qt.	Quart
Quan., Qty.	Quantity
Q.C.	Quick Coupling
r	Radius of Gyration
R	Resistance
R.C.P.	Reinforced Concrete Pipe
Rect.	Rectangle
Reg.	Regular
Reinf.	Reinforced
Req'd.	Required
Res.	Resistant
Resi.	Residential
Rgh.	Rough
RGS	Rigid Galvanized Steel
R.H.W.	Rubber, Heat & Water Resistant; Residential Hot Water
rms	Root Mean Square
Rnd.	Round
Rodm.	Rodman
Rofc.	Roofer, Composition
Rofp.	Roofer, Precast
Rohe.	Roofer Helpers (Composition)
Rots.	Roofer, Tile & Slate
R.O.W.	Right of Way
RPM	Revolutions per Minute
R.S.	Rapid Start
Rsr	Riser
RT	Round Trip
S.	Suction; Single Entrance; South
SC	Screw Cover
SCFM	Standard Cubic Feet per Minute
Scaf.	Scaffold
Sch., Sched.	Schedule
S.C.R.	Modular Brick
S.D.	Sound Deadening
S.D.R.	Standard Dimension Ratio
S.E.	Surfaced Edge
Sel.	Select
S.E.R., S.E.U.	Service Entrance Cable
S.F.	Square Foot
S.F.C.A.	Square Foot Contact Area
S.F. Flr.	Square Foot of Floor
S.F.G.	Square Foot of Ground
S.F. Hor.	Square Foot Horizontal
S.F.R.	Square Feet of Radiation
S.F. Shlf.	Square Foot of Shelf
S4S	Surface 4 Sides
Shee.	Sheet Metal Worker
Sin.	Sine
Skwk.	Skilled Worker
SL	Saran Lined
S.L.	Slimline
Sldr.	Solder
SLH	Super Long Span Bar Joist
S.N.	Solid Neutral
S-O-L	Socketolet
sp	Standpipe
S.P.	Static Pressure; Single Pole; Self-Propelled
Spri.	Sprinkler Installer
spwg	Static Pressure Water Gauge
S.P.D.T.	Single Pole, Double Throw
SPF	Spruce Pine Fir
S.P.S.T.	Single Pole, Single Throw
SPT	Standard Pipe Thread
Sq.	Square; 100 Square Feet
Sq. Hd.	Square Head
Sq. In.	Square Inch
S.S.	Single Strength; Stainless Steel
S.S.B.	Single Strength B Grade
sst, ss	Stainless Steel
Sswk.	Structural Steel Worker
Sswl.	Structural Steel Welder
St.;Stl.	Steel
S.T.C.	Sound Transmission Coefficient
Std.	Standard
Stg.	Staging
STK	Select Tight Knot
STP	Standard Temperature & Pressure
Stpi.	Steamfitter, Pipefitter
Str.	Strength; Starter; Straight
Strd.	Stranded
Struct.	Structural
Sty.	Story
Subj.	Subject
Subs.	Subcontractors
Surf.	Surface
Sw.	Switch
Swbd.	Switchboard
S.Y.	Square Yard
Syn.	Synthetic
S.Y.P.	Southern Yellow Pine
Sys.	System
t.	Thickness
T	Temperature; Ton
Tan	Tangent
T.C.	Terra Cotta
T & C	Threaded and Coupled
T.D.	Temperature Difference
Tdd	Telecommunications Device for the Deaf
T.E.M.	Transmission Electron Microscopy
TFE	Tetrafluoroethylene (Teflon)
T. & G.	Tongue & Groove; Tar & Gravel
Th., Thk.	Thick
Thn.	Thin
Thrded	Threaded
Tilf.	Tile Layer, Floor
Tilh.	Tile Layer, Helper
THHN	Nylon Jacketed Wire
THW.	Insulated Strand Wire

Abbreviations

THWN	Nylon Jacketed Wire
T.L.	Truckload
T.M.	Track Mounted
Tot.	Total
T-O-L	Threadolet
T.S.	Trigger Start
Tr.	Trade
Transf.	Transformer
Trhv.	Truck Driver, Heavy
Trlr	Trailer
Trlt.	Truck Driver, Light
TTY	Teletypewriter
TV	Television
T.W.	Thermoplastic Water Resistant Wire
UCI	Uniform Construction Index
UF	Underground Feeder
UGND	Underground Feeder
U.H.F.	Ultra High Frequency
U.I.	United Inch
U.L.	Underwriters Laboratory
Uld.	unloading
Unfin.	Unfinished
URD	Underground Residential Distribution
US	United States
USP	United States Primed
UTP	Unshielded Twisted Pair
V	Volt
V.A.	Volt Amperes
V.C.T.	Vinyl Composition Tile
VAV	Variable Air Volume
VC	Veneer Core
Vent.	Ventilation
Vert.	Vertical
V.F.	Vinyl Faced
V.G.	Vertical Grain
V.H.F.	Very High Frequency
VHO	Very High Output
Vib.	Vibrating
V.L.F.	Vertical Linear Foot
Vol.	Volume
VRP	Vinyl Reinforced Polyester
W	Wire; Watt; Wide; West
w/	With
W.C.	Water Column; Water Closet
W.F.	Wide Flange
W.G.	Water Gauge
Wldg.	Welding
W. Mile	Wire Mile
W-O-L	Weldolet
W.R.	Water Resistant
Wrck.	Wrecker
W.S.P.	Water, Steam, Petroleum
WT., Wt.	Weight
WWF	Welded Wire Fabric
XFER	Transfer
XFMR	Transformer
XHD	Extra Heavy Duty
XHHW, XLPE	Cross-Linked Polyethylene Wire Insulation
XLP	Cross-linked Polyethylene
Y	Wye
yd	Yard
yr	Year
Δ	Delta
%	Percent
~	Approximately
Ø	Phase; diameter
@	At
#	Pound; Number
<	Less Than
>	Greater Than
Z	zone

A

B

C

F

H

I

J

K

L

T

U

V

W

X

Y

Z

Notes

Reed Construction Data/RSMeans—
a tradition of excellence in construction cost information and services since 1942

Table of Contents

For more information visit the RSMeans websi at www.rsmeans.com

Book Selection Guide

The following table provides definitive information on the content of each cost data publication. The number of lines of data provided in each unit price or assemblies division, as well as the number of reference tables and crews, is listed for each book. The presence of other elements such as equipment rental costs, historical cost indexes, city cost indexes, square foot models, or cross-referenced indexes is also indicated. You can use the table to h select the RSMeans book that has the quantity and type of information you most need in your work.

Unit Cost Divisions	Building Construction Costs	Mechanical	Electrical	Repair & Remodel	Square Foot	Site Work Landsc.	Assemblies	Interior	Concrete Masonry	Open Shop	Heavy Construction	Light Commercial	Facilities Construction	Plumbing	Residenti
1	553	372	385	474		489		301	437	552	483	220	967	386	15
2	555	239	60	544		782		333	189	553	680	409	1022	262	21
3	1471	193	179	830		1321		266	1803	1467	1459	304	1386	162	30
4	878	24	0	677		683		557	1082	854	576	469	1125	0	38
5	1847	197	152	951		796		1023	751	1816	1045	809	1837	275	72
6	1870	78	80	1458		461		1191	310	1849	395	1562	1484	22	207
7	1395	164	71	1418		506		513	439	1393	0	1091	1451	175	85
8	1941	61	10	1969		298		1677	673	1906	0	1389	2134	0	131
9	1800	72	26	1606		292		1899	392	1761	0	1513	2035	54	135
10	978	17	10	582		226		800	156	978	29	465	1072	233	21
11	998	209	160	476		126		848	28	983	0	223	1012	170	11
12	547	0	2	303		248		1691	12	536	19	349	1736	23	31
13	730	119	111	247		341		255	69	702	253	80	740	74	7
14	281	36	0	230		28		260	0	280	0	12	300	21	
21	78	0	16	31		0		252	0	78	0	60	370	375	
22	1090	6955	154	1049		1409		596	20	1069	1634	751	6824	8656	57
23	1189	7103	609	776		159		682	38	1103	110	649	5143	1860	42
26	1246	493	9825	991		743		1053	55	1234	527	1060	9761	438	55
27	75	0	200	34		9		67	0	75	35	48	198	0	
28	87	59	94	64		0		69	0	72	0	42	109	45	2
31	1904	1162	1031	1225		2731		7	1635	1848	2804	1027	1964	1087	103
32	737	46	0	603		3849		360	271	710	1366	353	1370	158	41
33	514	1020	429	199		1957		0	225	512	1883	114	1547	1189	11
34	99	0	20	4		134		0	31	55	113	0	119	0	
35	18	0	0	0		166		0	0	18	281	0	83	0	
41	52	0	0	24		7		22	0	52	30	0	59	14	
44	98	90	0	0		32		0	0	23	32	0	98	105	
Totals	23031	18709	13264	16765		17793		14722	8616	22479	13754	12999	45946	15784	1126

Assembly Divisions	Building Construction Costs	Mechanical	Elec-trical	Repair & Remodel	Square Foot	Site Work Landscape	Assemblies	Interior	Concrete Masonry	Open Shop	Heavy Construction	Light Commercial	Facilities Construction	Plumbing	Asm Div	Residenti
A		15	0	182	150	530	598	0	536		570	154	24	0	1	37
B		0	0	808	2471	0	5577	329	1914		368	2014	143	0	2	21
C		0	0	630	864	0	1229	1567	146		0	765	242	0	3	58
D		1066	810	694	1779	0	2408	765	0		0	1270	1011	1068	4	86
E		0	0	85	260	0	297	5	0		0	257	5	0	5	39
F		0	0	0	123	0	124	0	0		0	123	3	0	6	35
G		528	294	311	202	2206	668	0	535		1194	200	293	685	7	30
															8	76
															9	8
															10	
															11	
															12	
Totals		1609	1104	2710	5849	2736	10901	2666	3131		2132	4783	1721	1753		393

Reference Section	Building Construction Costs	Mechanical	Electrical	Repair & Remodel	Square Foot	Site Work Landscape	Assemblies	Interior	Concrete Masonry	Open Shop	Heavy Construction	Light Commercial	Facilities Construction	Plumbing	Residenti
Reference Tables	yes	yes	yes	yes	no	yes	yes	yes	yes	yes	yes	yes	yes	yes	ye
Models					105							46			32
Crews	482	482	482	462		482		482	482	460	482	460	462	482	46
Equipment Rental Costs	yes	yes	yes	yes		yes		yes	yes	yes	yes	yes	yes	yes	yes
Historical Cost Indexes	yes	yes	yes	yes	yes	yes	yes	yes	yes	yes	yes	yes	yes	yes	no
City Cost Indexes	yes	yes	yes	yes	yes	yes	yes	yes	yes	yes	yes	yes	yes	yes	yes

For more information
visit the RSMeans website
at www.rsmeans.com

Annual Cost Guides

RSMeans Square Foot Costs 2009

It's Accurate and Easy To Use!

- **Updated price information** based on nationwide figures from suppliers, estimators, labor experts, and contractors
- "How-to-Use" sections, with **clear examples** of commercial, residential, industrial, and institutional structures
- Realistic graphics, offering true-to-life illustrations of building projects
- Extensive information on using square foot cost data, including sample estimates and alternate pricing methods

$168.95 per copy
Available Oct. 2008
Catalog no. 60059

RSMeans Repair & Remodeling Cost Data 2009

Commercial/Residential

Use this valuable tool to estimate commercial and residential renovation and remodeling.

Includes: New costs for hundreds of unique methods, materials, and conditions that only come up in repair and remodeling, PLUS:

- Unit costs for over 15,000 constructic components
- Installed costs for over 90 assemblies
- Over 930 "on-site" localization factoı for the U.S. and Canada.

Unit prices organized according to MasterFormat 200

$132.95 per copy
Available Nov. 2008
Catalog no. 60049

RSMeans Electrical Cost Data 2009

Pricing information for every part of electrical cost planning. More than 13,000 unit and systems costs with design tables; clear specifications and drawings; engineering guides; illustrated estimating procedures; complete labor-hour and materials costs for better scheduling and procurement; and the latest electrical products and construction methods.

- A variety of special electrical systems including cathodic protection
- Costs for maintenance, demolition, HVAC/mechanical, specialties, equipment, and more

$154.95 per copy
Available Oct. 2008
Catalog no. 60039

Unit prices organized according to MasterFormat 2004!

RSMeans Electrical Change Order Cost Data 2009

RSMeans Electrical Change Order Cost Data provides you with electrical unit prices exclusively for pricing change orders—based on the recent, direct experience of contractors and suppliers. Analyze and check your own change order estimates against the experience others have had doing the same work. It also covers productivity analysis and change order cost justifications. With useful information for calculating the effects of change orders and dealing witl their administration.

$154.95 per copy
Available Dec. 2008
Catalog no. 60239

₹SMeans Assemblies ;ost Data 2009

SMeans Assemblies Cost Data takes the guesswork ıt of preliminary or conceptual estimates. Now you ›n't have to try to calculate the assembled cost by orking up individual component costs. We've done l the work for you.

·esents detailed illustrations, descriptions, ·ecifications, and costs for every conceivable ıilding assembly—240 types in all—arranged in the .sy-to-use UNIFORMAT II system. Each illustrated ıssembled" cost includes a complete grouping of aterials and associated installation costs, including .e installing contractor's overhead and profit.

₹45.95 per copy
vailable Oct. 2008
atalog no. 60069

RSMeans Open Shop Building Construction Cost Data 2009

The latest costs for accurate budgeting and estimating of new commercial and residential construction. . . renovation work. . . change orders. . . cost engineering.

RSMeans Open Shop "BCCD" will assist you to:

- Develop benchmark prices for change orders
- Plug gaps in preliminary estimates and budgets
- Estimate complex projects
- Substantiate invoices on contracts
- Price ADA-related renovations

Unit prices organized according to MasterFormat 2004!

$154.95 per copy
Available Dec. 2008
Catalog no. 60159

₹SMeans Residential ;ost Data 2009

›ntains square foot costs for 30 basic home models ith the look of today, plus hundreds of custom lditions and modifications you can quote right f the page. With costs for the 100 residential stems you're most likely to use in the year ahead. ›mplete with blank estimating forms, sample timates, and step-by-step instructions.

ow contains line items for cultured stone and ick, PVC trim lumber, and TPO roofing.

132.95 per copy
vailable Oct. 2008
atalog no. 60179

Unit prices organized according to MasterFormat 2004!

RSMeans Site Work & Landscape Cost Data 2009

Includes unit and assemblies costs for earthwork, sewerage, piped utilities, site improvements, drainage, paving, trees & shrubs, street openings/ repairs, underground tanks, and more. Contains 57 tables of assemblies costs for accurate conceptual estimates.

Includes:

- Estimating for infrastructure improvements
- Environmentally-oriented construction
- ADA-mandated handicapped access
- Hazardous waste line items

$154.95 per copy
Available Nov. 2008
Catalog no. 60289

₹SMeans Facilities Maintenance & Repair Cost Data 2009

SMeans Facilities Maintenance & Repair Cost ata gives you a complete system to manage and an your facility repair and maintenance costs ıd budget efficiently. Guidelines for auditing a cility and developing an annual maintenance plan. ıdgeting is included, along with reference tables on ›st and management, and information on frequency ıd productivity of maintenance operations.

he only nationally recognized source of ıaintenance and repair costs. Developed in ›operation with the Civil Engineering Research aboratory (CERL) of the Army Corps of Engineers.

336.95 per copy
vailable Dec. 2008
atalog no. 60309

RSMeans Light Commercial Cost Data 2009

Specifically addresses the light commercial market, which is a specialized niche in the construction industry. Aids you, the owner/designer/contractor, in preparing all types of estimates—from budgets to detailed bids. Includes new advances in methods and materials.

Assemblies Section allows you to evaluate alternatives in the early stages of design/planning.

Over 11,000 unit costs ensure that you have the prices you need. . . when you need them.

Unit prices organized according to MasterFormat 2004!

$132.95 per copy
Available Nov. 2008
Catalog no. 60189

For more information
visit the RSMeans website
at www.rsmeans.com

Annual Cost Guides

RSMeans Concrete & Masonry Cost Data 2009

Provides you with cost facts for virtually all concrete/masonry estimating needs, from complicated formwork to various sizes and face finishes of brick and block—all in great detail. The comprehensive Unit Price Section contains more than 7,500 selected entries. Also contains an Assemblies [Cost] Section, and a detailed Reference Section that supplements the cost data.

Unit prices organized according to MasterFormat 2004!

$139.95 per copy
Available Dec. 2008
Catalog no. 60119

RSMeans Labor Rates for the Construction Industry 2009

Complete information for estimating labor costs, making comparisons, and negotiating wage rates by trade for over 300 U.S. and Canadian cities. With 46 construction trades listed by local union numbe in each city, and historical wage rates included for comparison. Each city chart lists the county and is alphabetically arranged with handy visual flip tabs for quick reference.

$336.95 per copy
Available Dec. 2008
Catalog no. 60129

RSMeans Construction Cost Indexes 2009

What materials and labor costs will change unexpectedly this year? By how much?

- Breakdowns for 316 major cities
- National averages for 30 key cities
- Expanded five major city indexes
- Historical construction cost indexes

$335.00 per year (subscription)
Catalog no. 50149

$83.25 individual quarters
Catalog no. 60149 A,B,C,D

RSMeans Interior Cost Data 2009

Provides you with prices and guidance needed to make accurate interior work estimates. Contains costs on materials, equipment, hardware, custom installations, furnishings, and labor costs. . . for ne and remodel commercial and industrial interior construction, including updated information on office furnishings, and reference information.

Unit prices organized according to MasterFormat 2004!

$154.95 per copy
Available Nov. 2008
Catalog no. 60099

RSMeans Heavy Construction Cost Data 2009

A comprehensive guide to heavy construction costs. Includes costs for highly specialized projects such as tunnels, dams, highways, airports, and waterways. Information on labor rates, equipment, and materials costs is included. Features unit price costs, systems costs, and numerous reference tables for costs and design.

$154.95 per copy
Available Dec. 2008
Catalog no. 60169

Unit prices organized according to MasterFormat 2004!

Reference Books

For more information
visit the RSMeans website
at www.rsmeans.com

alue Engineering: ractical Applications

Design, Construction, Maintenance & Operations

Alphonse Dell'Isola, PE

:ool for immediate application—for engineers, hitects, facility managers, owners, and contractors. :ludes making the case for VE—the management efing; integrating VE into planning, budgeting, d design; conducting life cycle costing; using VE thodology in design review and consultant selection; e studies; VE workbook; and a life cycle costing gram on disk.

.95 per copy
er 450 pages, illustrated, softcover
talog no. 67319A

Facilities Operations & Engineering Reference

by the Association for Facilities Engineering and RSMeans

An all-in-one technical reference for planning and managing facility projects and solving day-to-day operations problems. Selected as the official certified plant engineer reference, this handbook covers financial analysis, maintenance, HVAC and energy efficiency, and more.

$54.98 per copy
Over 700 pages, illustrated, hardcover
Catalog no. 67318

he Building Professional's uide to Contract Documents

d Edition

Waller S. Poage, AIA, CSI, CVS

comprehensive reference for owners, design fessionals, contractors, and students

Structure your documents for maximum efficiency.
Effectively communicate construction requirements.
Understand the roles and responsibilities of construction professionals.
Improve methods of project delivery.

4.95 per copy, 400 pages
agrams and construction forms, hardcover
talog no. 67261A

Complete Book of Framing

by Scot Simpson

This straightforward easy-to-learn method will help framers, carpenters, and handy homeowners build their skills in rough carpentry and framing. Shows how to frame all the parts of a house: floors, walls, roofs, door & window openings, and stairs—with hundreds of color photographs and drawings that show every detail.

The book gives beginners all the basics they need to go from zero framing knowledge to a journeyman level.

$29.95 per copy
352 pages, softcover
Catalog no. 67353

ost Planning & Estimating or Facilities Maintenance

this unique book, a team of facilities management thorities shares their expertise on:

Evaluating and budgeting maintenance operations
Maintaining and repairing key building components
Applying *RSMeans Facilities Maintenance & Repair Cost Data* to your estimating

vers special maintenance requirements of the ten ajor building types.

.95 per copy
er 475 pages, hardcover
talog no. 67314

Life Cycle Costing for Facilities

by Alphonse Dell'Isola and Dr. Steven Kirk

Guidance for achieving higher quality design and construction projects at lower costs! Cost-cutting efforts often sacrifice quality to yield the cheapest product. Life cycle costing enables building designers and owners to achieve both. The authors of this book show how LCC can work for a variety of projects — from roads to HVAC upgrades to different types of buildings.

$99.95 per copy
450 pages, hardcover
Catalog no. 67341

lanning & Managing Interior rojects 2nd Edition

Carol E. Farren, CFM

pert guidance on managing renovation & relocation ojects

is book guides you through every step in relocating a new space or renovating an old one. From initial eeting through design and construction, to post-project ministration, it helps you get the most for your mpany or client. Includes sample forms, spec lists, reements, drawings, and much more!

4.98 per copy
0 pages, softcover
talog no. 67245A

Construction Business Management

by Nick Ganaway

Only 43% of construction firms stay in business after four years. Make sure your company thrives with valuable guidance from a pro with 25 years of success as a commercial contractor. Find out what it takes to build all aspects of a business that is profitable, enjoyable, and enduring. With a bonus chapter on retail construction.

$49.95 per copy
200 pages, softcover
Catalog no. 67352

Interior Home Improvement Costs 9th Edition

Updated estimates for the most popular remodeling and repair projects—from small, do-it-yourself jobs to major renovations and new construction. Includes: kitchens & baths; new living space from your attic, basement, or garage; new floors, paint, and wallpaper; tearing out or building new walls; closets, stairs, and fireplaces; new energy-saving improvements, home theaters, and more!

$24.95 per copy
250 pages, illustrated, softcover
Catalog no. 67308E

Exterior Home Improvement Costs 9th Edition

Updated estimates for the most popular remodeling and repair projects—from small, do-it-yourself jobs, to major renovations and new construction. Includes: curb appeal projects—landscaping, patios, porches, driveways, and walkways; new windows and doors; decks, greenhouses, a sunrooms; room additions and garages; roofing, siding, anc painting; "green" improvements to save energy & water.

$24.95 per copy
Over 275 pages, illustrated, softcover
Catalog no. 67309E

Builder's Essentials: Plan Reading & Material Takeoff

For Residential and Light Commercial Construction

by Wayne J. DelPico

A valuable tool for understanding plans and specs, and accurately calculating material quantities. Step-by-step instructions and takeoff procedures based on a full set of working drawings.

$35.95 per copy
Over 420 pages, softcover
Catalog no. 67307

Means Unit Price Estimating Methods

New 4th Edition

This new edition includes up-to-date cost data and estimating examples, updated to reflect changes to the CS numbering system and new features of RSMeans cost data It describes the most productive, universally accepted way to estimate, and uses checklists and forms to illustrate shortcuts and timesavers. A model estimate demonstrates procedures. A new chapter explores computer estimating alternatives.

$59.95 per copy
Over 350 pages, illustrated, softcover
Catalog no. 67303B

Total Productive Facilities Management

by Richard W. Sievert, Jr.

Today, facilities are viewed as strategic resources. . . elevating the facility manager to the role of asset manager supporting the organization's overall business goals. Now, Richard Sievert Jr., in this well-articulated guidebook, sets forth a new operational standard for the facility manager's emerging role. . . a comprehensive program for managing facilities as a true profit center.

$29.98 per copy
275 pages, softcover
Catalog no. 67321

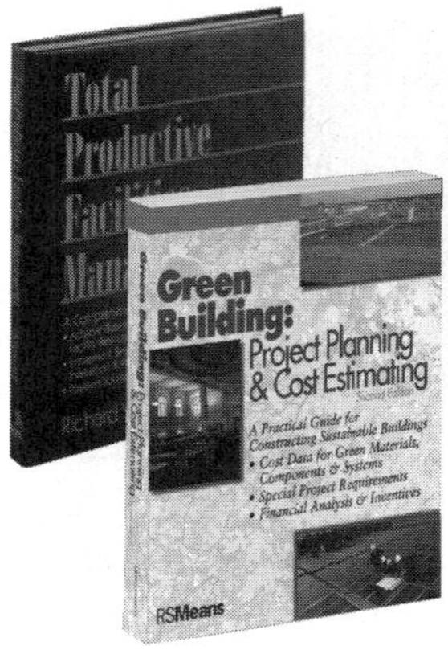

Green Building: Project Planning & Cost Estimating 2nd Edition

This new edition has been completely updated with the latest in green building technologies, design concepts, standards, and costs. Now includes a 2009 Green Building *CostWorks* CD with more than 300 green building assemblies and over 5,000 unit price line items for sustainable building. The new edition is also full-color with all new case studies—plus a new chapter on deconstruction, a key aspect of green building.

$129.95 per copy
350 pages, softcover
Catalog no. 67338A

Concrete Repair and Maintenance Illustrated

by Peter Emmons

Hundreds of illustrations show users how to analyze, repair, clean, and maintain concrete structures for optimal performance and cost effectiveness. From parking garages to roads and bridges to structural concrete, this comprehensive book describes the causes, effects, and remedies for concrete wear and failure. Invaluable for planning jobs, selecting materials, and training employees, this book is a must-have for concrete specialists, general contractors, facility managers, civil and structural engineers, and architects.

$69.95 per copy
300 pages, illustrated, softcover
Catalog no. 67146

For more information
visit the RSMeans website
at www.rsmeans.com

ɪeans Illustrated Construction ictionary Condensed, 2nd Edition

.e best portable dictionary for office or field use—an ential tool for contractors, architects, insurance and l estate personnel, facility managers, homeowners, and yone who needs quick, clear definitions for construction ms. The second edition has been further enhanced with dates and hundreds of new terms and illustrations . . . keeping with the most recent developments in the nstruction industry.

ɔw with a quick-reference Spanish section. Includes tools d equipment, materials, tasks, and more!

).95 per copy
er 500 pages, softcover
talog no. 67282A

Means Repair & Remodeling Estimating 4th Edition

By Edward B. Wetherill and RSMeans

This important reference focuses on the unique problems of estimating renovations of existing structures, and helps you determine the true costs of remodeling through careful evaluation of architectural details and a site visit.

New section on disaster restoration costs.

$69.95 per copy
Over 450 pages, illustrated, hardcover
Catalog no. 67265B

acilities Planning & elocation

y David D. Owen

complete system for planning space needs and managing ocations. Includes step-by-step manual, over 50 forms, d extensive reference section on materials and nishings.

w lower price and user-friendly format.

).98 per copy
er 450 pages, softcover
talog no. 67301

Means Square Foot & UNIFORMAT Assemblies Estimating Methods 3rd Edition

Develop realistic square foot and assemblies costs for budgeting and construction funding. The new edition features updated guidance on square foot and assemblies estimating using UNIFORMAT II. An essential reference for anyone who performs conceptual estimates.

$69.95 per copy
Over 300 pages, illustrated, softcover
Catalog no. 67145B

ɪeans Electrical Estimating ɪethods 3rd Edition

panded edition includes sample estimates and cost ormation in keeping with the latest version of the CSI asterFormat and UNIFORMAT II. Complete coverage fiber optic and uninterruptible power supply electrical stems, broken down by components, and explained in tail. Includes a new chapter on computerized estimating ethods. A practical companion to *RSMeans Electrical ɔst Data.*

ɪ.95 per copy
er 325 pages, hardcover
talog no. 67230B

Means Mechanical Estimating Methods 4th Edition

Completely updated, this guide assists you in making a review of plans, specs, and bid packages, with suggestions for takeoff procedures, listings, substitutions, and pre-bid scheduling for all components of HVAC. Includes suggestions for budgeting labor and equipment usage. Compares materials and construction methods to allow you to select the best option.

$64.95 per copy
Over 350 pages, illustrated, softcover
Catalog no. 67294B

ɪeans ADA Compliance Pricing uide 2nd Edition

y Adaptive Environments and SMeans

ɔmpletely updated and revised to the new 2004 *Americans th Disabilities Act Accessibility Guidelines,* is book features more than 70 of the most commonly eded modifications for ADA compliance. Projects range ɔm installing ramps and walkways, widening ɔrways and entryways, and installing and refitting evators, to relocating light switches and signage.

9.95 per copy
er 350 pages, illustrated, softcover
talog no. 67310A

Project Scheduling & Management for Construction 3rd Edition

by David R. Pierce, Jr.

A comprehensive yet easy-to-follow guide to construction project scheduling and control—from vital project management principles through the latest scheduling, tracking, and controlling techniques. The author is a leading authority on scheduling, with years of field and teaching experience at leading academic institutions. Spend a few hours with this book and come away with a solid understanding of this essential management topic.

$64.95 per copy
Over 300 pages, illustrated, hardcover
Catalog no. 67247B

The Practice of Cost Segregation Analysis

by Bruce A. Desrosiers and Wayne J. DelPico

This expert guide walks you through the practice of cost segregation analysis, which enables property owners to defer taxes and benefit from "accelerated cost recovery" through depreciation deductions on assets that are properly identified and classified.

With a glossary of terms, sample cost segregation estimates for various building types, key information resources, and updates via a dedicated website, this book is a critical resource for anyone involved in cost segregation analysis.

$99.95 per copy
Over 225 pages
Catalog no. 67345

Preventive Maintenance fo Multi-Family Housing

by John C. Maciha

Prepared by one of the nation's leading experts on multi-family housing.

This complete PM system for apartment and condominium communities features expert guidance, checklists for buildings and grounds maintenance tasks and their frequencies, a reusable wall chart to track maintenance, and a dedicated website featuring customizable electronic forms. A must-have for anyor involved with multi-family housing maintenance and upkeep.

$89.95 per copy
225 pages
Catalog no. 67346

How to Estimate with Means Data & CostWorks

New 3rd Edition

by RSMeans and Saleh A. Mubarak, Ph.D.

New 3rd Edition—fully updated with new chapters, plus new CD with updated *CostWorks* cost data and MasterFormat organization. Includes all major construction items—with more than 300 exercises and two sets of plans that show how to estimate for a broad range of construction items and systems—including general conditions and equipment costs.

$59.95 per copy
272 pages, softcover
Includes CostWorks CD
Catalog no. 67324B

Job Order Contracting

Expediting Construction Project Delivery

by Allen Henderson

Expert guidance to help you implement JOC—fast becoming the preferred project delivery method for repair and renovation, minor new construction, and maintenance projects in the public sector and in many states and municipalities. The author, a leading JOC expert and practitioner, shows how to:

- Establish a JOC program
- Evaluate proposals and award contracts
- Handle general requirements and estimating
- Partner for maximum benefits

$89.95 per copy
192 pages, illustrated, hardcover
Catalog no. 67348

Builder's Essentials: Estimating Building Costs

For the Residential & Light Commercial Contractor

by Wayne J. DelPico

Step-by-step estimating methods for residential and light commercial contractors. Includes a detailed look at every construction specialty—explaining all the components, takeoff units, and labor needed for well-organized, complete estimates. Covers correctly interpreting plans and specifications, and developing accurate and complete labor and material costs.

$29.95 per copy
Over 400 pages, illustrated, softcover
Catalog no. 67343

Building & Renovating Schools

This all-inclusive guide covers every step of the schoo construction process—from initial planning, needs assessment, and design, right through moving into the new facility. A must-have resource for anyone concerned with new school construction or renovatior With square foot cost models for elementary, middle, and high school facilities, and real-life case studies of recently completed school projects.

The contributors to this book—architects, constructio project managers, contractors, and estimators who specialize in school construction—provide start-to-finish, expert guidance on the process.

$69.98 per copy
Over 425 pages, hardcover
Catalog no. 67342

Reference Books

For more information visit the RSMeans website at www.rsmeans.com

'he Homeowner's Guide to 1old By Michael Pugliese

pert guidance to protect your health and your home.

old, whether caused by leaks, humidity or flooding, is eal health and financial issue—for homeowners and ntractors. This full-color book explains:
- Construction and maintenance practices to prevent mold
- How to inspect for and remove mold
- Mold remediation procedures and costs
- What to do after a flood
- How to deal with insurance companies

1.95 per copy
4 pages, softcover
talog no. 67344

Means Illustrated Construction Dictionary
Unabridged 3rd Edition, with CD-ROM

Long regarded as the industry's finest, *Means Illustrated Construction Dictionary* is now even better. With the addition of over 1,000 new terms and hundreds of new illustrations, it is the clear choice for the most comprehensive and current information. The companion CD-ROM that comes with this new edition adds many extra features: larger graphics, expanded definitions, and links to both CSI MasterFormat numbers and product information.

$99.95 per copy
Over 790 pages, illustrated, hardcover
Catalog no. 67292A

1eans Landscape :stimating Methods
ew 5th Edition

iswers questions about preparing competitive landscape nstruction estimates, with up-to-date cost estimates and e new MasterFormat classification system. Expanded d revised to address the latest materials and methods, cluding new coverage on approaches to green building. cludes:
- Step-by-step explanation of the estimating process
- Sample forms and worksheets that save time and prevent errors

4.95 per copy
er 350 pages, softcover
talog no. 67295C

Means Plumbing Estimating Methods 3rd Edition
by Joseph Galeno and Sheldon Greene

Updated and revised! This practical guide walks you through a plumbing estimate, from basic materials and installation methods through change order analysis. *Plumbing Estimating Methods* covers residential, commercial, industrial, and medical systems, and features sample takeoff and estimate forms and detailed illustrations of systems and components.

$29.98 per copy
330+ pages, softcover
Catalog no. 67283B

Inderstanding & Negotiating :onstruction Contracts
y Kit Werremeyer

ke advantage of the author's 30 years' experience in iall-to-large (including international) construction ojects. Learn how to identify, understand, and evaluate gh risk terms and conditions typically found in all nstruction contracts—then negotiate to lower or minate the risk, improve terms of payment, and duce exposure to claims and disputes. The author oids "legalese" and gives real-life examples from actual ojects.

9.95 per copy
0 pages, softcover
talog no. 67350

Means Estimating Handbook
2nd Edition

Updated Second Edition answers virtually any estimating technical question—all organized by CSI MasterFormat. This comprehensive reference covers the full spectrum of technical data required to estimate construction costs. The book includes information on sizing, productivity, equipment requirements, code-mandated specifications, design standards, and engineering factors.

$99.95 per copy
Over 900 pages, hardcover
Catalog no. 67276A

1eans Spanish/English Construction Dictionary 2nd Edition
y RSMeans and the iternational Code Council

iis expanded edition features thousands of the most mmon words and useful phrases in the construction dustry with easy-to-follow pronunciations in both Spanish d English. Over 800 new terms, phrases, and illustrations ve been added. It also features a new stand-alone "Safety & nergencies" section, with colored pages for quick access.

nique to this dictionary are the systems illustrations owing the relationship of components in the most mmon building systems for all major trades.

3.95 per copy
er 400 pages
talog no. 67327A

How Your House Works
by Charlie Wing

A must-have reference for every homeowner, handyman, and contractor—for repair, remodeling, and new construction. This book uncovers the mysteries behind just about every major appliance and building element in your house—from electrical, heating and AC, to plumbing, framing, foundations and appliances. Clear, "exploded" drawings show exactly how things should be put together and how they function—what to check if they don't work, and what you can do that might save you having to call in a professional.

$21.95 per copy
160 pages, softcover
Catalog no. 67351

Professional Development

Means CostWorks® Training

This one-day course helps users become more familiar with the functionality of *Means CostWorks* program. Each menu, icon, screen, and function found in the program is explained in depth. Time is devoted to hands-on estimating exercises.

Some of what you'll learn:
- Search the database utilizing all navigation methods
- Export RSMeans Data to your preferred spreadsheet format
- View crews, assembly components, and much more!
- Automatically regionalize the database

You are required to bring your own laptop computer for this course.

When you register for this course you will receive an outline for your laptop requirements.

Unit Price Estimating

This interactive *two-day* seminar teaches attendees how to interpret project information and process it into final, detailed estimates with the greatest accuracy level.

The most important credential an estimator can take to the job is the ability to visualize construction, and estimate accurately.

Some of what you'll learn:
- Interpreting the design in terms of cost
- The most detailed, time-tested methodology for accurate pricing
- Key cost drivers—material, labor, equipment, staging, and subcontracts
- Understanding direct and indirect costs for accurate job cost accounting and change order management

Who should attend: Corporate and government estimators and purchasers, architects, engineers... and others needing to produce accurate project estimates.

Square Foot & Assemblies Estimating

This *two-day* course teaches attendees how to quickly deliver accurate square foot estimates using limited budget and design information.

Some of what you'll learn:
- How square foot costing gets the estimate done faster
- Taking advantage of a "systems" or "assemblies" format
- The RSMeans "building assemblies/square foot cost approach"
- How to create a reliable preliminary systems estimate using bare-bones design information

Who should attend: Facilities managers, facilities engineers, estimators, planners, developers, construction finance professionals... and others needing to make quick, accurate construction cost estimates at commercial, government, educational, and medical facilities.

Repair & Remodeling Estimating

This *two-day* seminar emphasizes all the underlying considerations unique to repair/remodeling estimating and presents the correct methods for generating accurate, reliable R&R project costs using the unit price and assemblies methods.

Some of what you'll learn:
- Estimating considerations—like labor-hours, building code compliance, working within existing structures, purchasing materials in smaller quantities, unforeseen deficiencies
- Identifying problems and providing solutions to estimating building alterations
- Rules for factoring in minimum labor costs, accurate productivity estimates, and allowances for project contingencies
- R&R estimating examples calculated using unit price and assemblies data

Who should attend: Facilities managers, plant engineers, architects, contractors, estimators, builders... and others who are concerned with the proper preparation and/or evaluation of repair and remodeling estimates.

Mechanical & Electrical Estimating

This *two-day* course teaches attendees how to prepare more accurate and complete mechanical/electrical estimates, avoiding the pitfalls of omission and double-counting, while understanding the composition and rationale within the RSMeans Mechanical/Electrical database.

Some of what you'll learn:
- The unique way mechanical and electrical systems are interrelated
- M&E estimates–conceptual, planning, budgeting, and bidding stages
- Order of magnitude, square foot, assemblies, and unit price estimating
- Comparative cost analysis of equipment and design alternatives

Who should attend: Architects, engineers, facilities managers, mechanical and electrical contractors... and others needing a highly reliable method for developing, understanding, and evaluating mechanical and electrical contracts.

Green Building Planning & Construction

In this *two-day* course, learn about tailoring a building and its placement on the site to the local climate, site conditions, culture, and community. Includes information to reduce resource consumption and augment resource supply.

Some of what you'll learn:
- Green technologies, materials, systems, and standards
- Energy efficiencies with energy modeling tools
- Cost vs. value of green products over their life cycle
- Low-cost strategies and economic incentives and funding
- Health, comfort, and productivity goals and techniques

Who should attend: Contractors, project managers, building owners, building officials, healthcare and insurance professionals.

Facilities Maintenance & Repair Estimating

This *two-day* course teaches attendees how to plan, budget, and estimate the cost of ongoing and preventive maintenance and repair for existing buildings and grounds.

Some of what you'll learn:
- The most financially favorable maintenance, repair, and replacement scheduling and estimating
- Auditing and value engineering facilities
- Preventive planning and facilities upgrading
- Determining both in-house and contract-out service costs
- Annual, asset-protecting M&R plan

Who should attend: Facility managers, maintenance supervisors, buildings and grounds superintendents, plant managers, planners, estimators... and others involved in facilities planning and budgeting.

Practical Project Management for Construction Professionals

In this *two-day* course, acquire the essential knowledge and develop the skills to effectively and efficiently execute the day-to-day responsibilities of the construction project manager.

Covers:
- General conditions of the construction contract
- Contract modifications: change orders and construction change directives
- Negotiations with subcontractors and vendors
- Effective writing: notification and communications
- Dispute resolution: claims and liens

Who should attend: Architects, engineers, owner's representatives, project managers.

Facilities Repair & Remodeling Estimating

In this *two-day* course, professionals working in facilities management can get help with their daily challenges to establish budgets for all phase of a project

Some of what you'll learn:
- Determine the full scope of a project
- Identify the scope of risks & opportunities
- Creative solutions to estimating issues
- Organizing estimates for presentation & discussion
- Special techniques for repair/remodel and maintenance projects
- Negotiating project change orders

Who should attend: Facility managers, engineers, contractors, facility trades-people, planners & project managers

Professional Development

For more information visit the RSMeans website at www.rsmeans.com

cheduling and Project Management

'his *two-day* course teaches attendees the most urrent and proven scheduling and management echniques needed to bring projects in on time and n budget.

ome of what you'll learn:

- Crucial phases of planning and scheduling
- How to establish project priorities and develop realistic schedules and management techniques
- Critical Path and Precedence Methods
- Special emphasis on cost control

Vho should attend: Construction project ianagers, supervisors, engineers, estimators, ontractors... and others who want to improve heir project planning, scheduling, and management kills.

Understanding & Negotiating Construction Contracts

In this *two-day* course, learn how to protect the assets of your company by justifying or eliminating commercial risk through negotiation of a contract's terms and conditions.

Some of what you'll learn:

- Myths and paradigms of contracts
- Scope of work, terms of payment, and scheduling
- Dispute resolution
- Why clients love CLAIMS!
- Negotiating issues and much more

Who should attend: Contractors & subcontractors, material suppliers, project managers, risk & insurance managers, procurement managers, owners and facility managers, corporate executives.

Site Work & Heavy Construction Estimating

This *two-day* course teaches attendees how to estimate earthwork, site utilities, foundations, and site improvements, using the assemblies and the unit price methods.

Some of what you'll learn:

- Basic site work and heavy construction estimating skills
- Estimating foundations, utilities, earthwork, and site improvements
- Correct equipment usage, quality control, and site investigation for estimating purposes

Who should attend: Project managers, design engineers, estimators, and contractors doing site work and heavy construction.

ssessing Scope of Work for Facility Construction Estimating

'his *two-day* course is a practical training rogram that addresses the vital importance of nderstanding the SCOPE of projects in order to roduce accurate cost estimates in a facility repair nd remodeling environment.

ome of what you'll learn:

- Discussions of site visits, plans/specs, record drawings of facilities, and site-specific lists
- Review of CSI divisions, including means, methods, materials, and the challenges of scoping each topic
- Exercises in SCOPE identification and SCOPE writing for accurate estimating of projects
- Hands-on exercises that require SCOPE, take-off, and pricing

Vho should attend: Corporate and government stimators, planners, facility managers, and others vho need to produce accurate project estimates.

2009 RSMeans Seminar Schedule

Note: call for exact dates and details.

Location	Dates
Las Vegas, NV	March
Washington, DC	April
Denver, CO	May
San Francisco, CA	June
Washington, DC	September
Dallas, TX	September
Las Vegas, NV	October
Atlantic City, NJ	October
Orlando, FL	November
San Diego, CA	December

1-800-334-3509, ext. 5115

For more information visit the RSMeans website at www.rsmeans.com

Professional Development

Registration Information

Register early... Save up to $100! Register 30 days before the start date of a seminar and save $100 off your total fee. *Note: This discount can be applied only once per order. It cannot be applied to team discount registrations or any other special offer.*

How to register Register by phone today! The RSMeans toll-free number for making reservations is **1-800-334-3509, ext. 5115.**

Individual seminar registration fee - $935. *Means CostWorks®* training registration fee - $375. To register by mail, complete the registration form and return, with your full fee, to: RSMeans Seminars, 63 Smiths Lane, Kingston, MA 02364.

Government pricing All federal government employees save off the regular seminar price. Other promotional discounts cannot be combined with the government discount.

Team discount program for two to four seminar registrations. Call for pricing: 1-800-334-3509, ext. 5115

Multiple course discounts When signing up for two or more courses, call for pricing.

Refund policy Cancellations will be accepted up to ten business days prior to the seminar start. There are no refunds for cancellations received later than ten working days prior to the first day of the seminar. A $150 processing fee will be applied for all cancellations. Written notice of cancellation is required. Substitutions can be made at any time before the session starts. **No-shows are subject to the full seminar fee.**

AACE approved courses Many seminars described and offered here have been approved for 14 hours (1.4 recertification credits) of credit by the AACE International Certification Board toward meeting the continuing education requirements for recertification as a Certified Cost Engineer/Certified Cost Consultant.

AIA Continuing Education We are registered with the AIA Continuing Education System (AIA/CES) and are committed to developing quality learning activities in accordance with the CES criteria. Many seminars meet the AIA/CES criteria for Quality Level 2. AIA members may receive (14) learning units (LUs) f each two-day RSMeans course.

NASBA CPE sponsor credits We are part of the National Registry of CPE sponsors. Attendees may be eligible for (16) CPE credits.

Daily course schedule The first day of each seminar session begins at 8:30 a.m. and ends at 4:30 p.m. The second day begins at 8:00 a.m. and ends at 4:00 p. Participants are urged to bring a hand-held calculator, since many actual problem will be worked out in each session.

Continental breakfast Your registration includes the cost of a continental breakfast, and a morning and afternoon refreshment break. These informal segments allow you to discuss topics of mutual interest with other seminar attendees. (You are free to make your own lunch and dinner arrangements.)

Hotel/transportation arrangements RSMeans arranges to hold a block of rooms at most host hotels. To take advantage of special group rates when making your reservation, be sure to mention that you are attending the RSMean seminar. You are, of course, free to stay at the lodging place of your choice. **(Hote reservations and transportation arrangements should be made directly by semina attendees.)**

Important Class sizes are limited, so please register as soon as possible.

Note: *Pricing subject to change.*

Registration Form

ADDS-100

Call 1-800-334-3509, ext. 5115 to register or FAX this form 1-800-632-6732. Visit our Web site: www.rsmeans.com

Please register the following people for the RSMeans construction seminars as shown here. We understand that we must make our own hotel reservations if overnight stays are necessary.

☐ Full payment of $_______________ enclosed.

☐ Bill me

Please print name of registrant(s)

(To appear on certificate of completion)

____

____

____

____

P.O. #: ______________________________

GOVERNMENT AGENCIES MUST SUPPLY PURCHASE ORDER NUMBER OR TRAINING FORM.

Firm name______________________________

Address______________________________

City/State/Zip______________________________

Telephone no.________________ Fax no.________________

E-mail address______________________________

Charge registration(s) to: ☐ MasterCard ☐ VISA ☐ American Express

Account no.________________ Exp. date________________

Cardholder's signature______________________________

Seminar name______________________________

Seminar City ______________________________

Please mail check to: RSMeans Seminars, 63 Smiths Lane, P.O. Box 800, Kingston, MA 02364 USA

Harness the full power of RSMeans online with *MeansCostWorks.com*

MeansCostWorks.com

Manage your estimates in a secure environment... available 24/7 from a web-enabled computer.

To Learn MORE call 1.800.334.3509 or go to: www.MeansCostWorks.com
View the online demo...Get a FREE TRIAL Subscription!

MeansCostWorks.com from RSMeans is web-based software for estimating construction costs on commercial and residential construction projects. It provides fast and accurate construction estimating using the latest construction cost data available from RSMeans. *It's an invaluable tool for construction, engineering and design professionals.*

Introducing *MeansCostsWorks.com*

Now you can access and use the world's largest construction cost database for estimating interior projects... verifying bids, managing change orders, budgeting, designing, and more.

MeansCostWorks.com is current, comprehensive, and very easy to use. It combines exhaustive cost data for virtually every construction component and task with state-of-the-art estimating and robust customization.

- Estimate faster and more accurately than you ever thought possible.
- Collaborate seamlessly with authorized team members.
- Easily print, e-mail, or export your work product to your Microsoft EXCEL program.

MeansData™

Software Integration

CONSTRUCTION COSTS FOR SOFTWARE APPLICATIONS

Your construction estimating software is only as good as your cost data.

BECK TECHNOLOGY

D Profiler

Have you ever wondered why so many projects are over budget, over scope, and take longer to design and build than expected?

DProfiler with RSMeans is the solution that reduces your costs, the risks to the owner, and gives you a competitive advantage.

Use DProfiler with RSMeans cost data at the start of a project to establish the project scope and budget.

www.beck-technology.com

SECURE INTERNET ORDER SITE
www.rsmeans.com
RSMeans

The Gypsum Construction Handbook by USG

An invaluable reference of construction procedures for gypsum drywall, cement board, veneer plaster, and conventional plaster. This book includes the newest product developments, installation methods, fire- and sound-rated construction information, illustrated framing-to-finish application instructions, estimating and planning information, and more. Great for architects and engineers; contractors, builders, and dealers; apprentices and training programs; building inspectors and code officials; and anyone interested in gypsum construction. Features information on tools and safety practices, a glossary of construction terms, and a list of important agencies and associations. Also available in Spanish.

$29.95
Over 500 pages, illustrated, softcover
Catalog no. 67357

BIM Handbook
A Guide to Building Information Modeling for Owners, Managers, Designers, Engineers and Contractors
by Chuck Eastman, Paul Teicholz, Rafael Sacks, Kathleen Liston

Building Information Modeling (BIM) is a new approach to design, construction, and facility management where a digital depiction of the building process is used to facilitate the exchange information in digital format.

The BIM Handbook provides an in-depth understanding of BIM technologies, the business and organizational issues associated with its implementation, and the profound advantages tha effective use of BIM can provide to all members o project team. Also includes a rich set of informati BIM case studies.

$85.00
504 pages
Catalog no. 67356

Universal Design Ideas for Style, Comfort & Safety
by RSMeans and Lexicon Consulting, Inc.

Incorporating universal design when building or remodeling helps people of any age and physical ability more fully and safely enjoy their living spaces. This book shows how universal design can be artfully blended into the most attractive homes. It discusses specialized products like adjustable countertops and chair lifts, as well as simple ways to enhance a home's safety and comfort. With color photos and expert guidance, every area of the home is covered. Includes budget estimates that give an idea how much projects will cost.

$21.95
160 pages, illustrated, softcover
Catalog no. 67354

Green Home Improvement
by Daniel D. Chiras, PhD

With energy costs rising and environmental awareness increasing, many people are looking to make their homes greener. This book makes it easy, with 65 projects and actual costs and project savings that will help homeowners prioritize thei green improvements.

Projects range from simple water savers that cost only a few dollars, to bigger-ticket items such as efficient appliances and HVAC systems, green roofing, flooring, landscaping, and much more. With color photos and cost estimates, each projec compares options and describes the work involve the benefits, and the savings.

$24.95
320 pages, illustrated, softcover
Catalog no. 67355

Residential & Light Commercial Construction Standards, 3rd Edition, Updated
by RSMeans and contributing authors

This book provides authoritative requirements and recommendations compiled from leading professional associations, industry publications, and building code organizations. This all-in-one reference helps establish a standard for workmanship, quickly resolve disputes, and avoid defect claims. Updated third edition includes new coverage of green building, seismic, hurricane, and mold-resistant construction, plus updated national building code requirements.

Covers all areas of building construction, including paving; concrete; framing; finish carpentry; insulation; roofing and siding; doors and windows; drywall; ceilings and floors; wall coverings; plumbing; HVAC; electrical; and more!

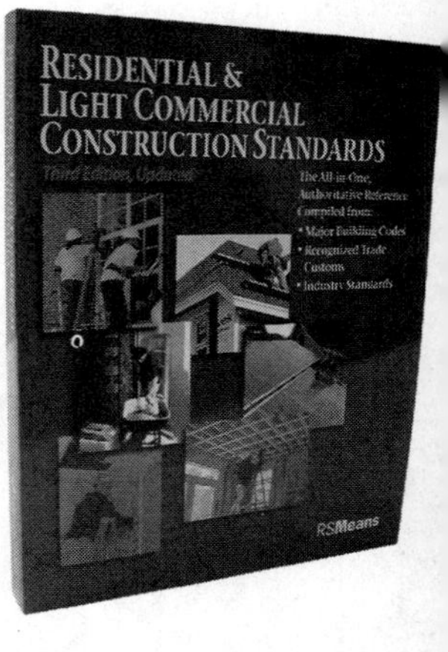

$59.95
Over 550 pages, illustrated, softcover
Catalog no. 67322B

RDER TOLL FREE 1-800-334-3509
R FAX 1-800-632-6732

2009 Order Form

Qty.	Book no.	COST ESTIMATING BOOKS	Unit Price	Total
	60069	Assemblies Cost Data 2009	$245.95	
	60019	Building Construction Cost Data 2009	154.95	
	60119	Concrete & Masonry Cost Data 2009	136.95	
	50149	Construction Cost Indexes 2009 (subscription)	335.00	
	60149A	Construction Cost Index–January 2009	83.75	
	60149B	Construction Cost Index–April 2009	83.75	
	60149C	Construction Cost Index–July 2009	83.75	
	60149D	Construction Cost Index–October 2009	83.75	
	60349	Contr. Pricing Guide: Resid. R & R Costs 2009	39.95	
	60339	Contr. Pricing Guide: Resid. Detailed 2009	39.95	
	60329	Contr. Pricing Guide: Resid. Sq. Ft. 2009	39.95	
	60239	Electrical Change Order Cost Data 2009	154.95	
	60039	Electrical Cost Data 2009	154.95	
	60209	Facilities Construction Cost Data 2009	368.95	
	60309	Facilities Maintenance & Repair Cost Data 2009	336.95	
	60169	Heavy Construction Cost Data 2009	154.95	
	60099	Interior Cost Data 2009	154.95	
	60129	Labor Rates for the Const. Industry 2009	336.95	
	60189	Light Commercial Cost Data 2009	132.95	
	60029	Mechanical Cost Data 2009	154.95	
	60159	Open Shop Building Const. Cost Data 2009	154.95	
	60219	Plumbing Cost Data 2009	154.95	
	60049	Repair & Remodeling Cost Data 2009	132.95	
	60179	Residential Cost Data 2009	132.95	
	60289	Site Work & Landscape Cost Data 2009	154.95	
	60059	Square Foot Costs 2009	168.95	
	62018	Yardsticks for Costing (2008)	149.95	
	62019	Yardsticks for Costing (2009)	154.95	
		REFERENCE BOOKS		
	67310A	ADA Compliance Pricing Guide, 2nd Ed.	79.95	
	67330	Bldrs Essentials: Adv. Framing Methods	12.98	
	67329	Bldrs Essentials: Best Bus. Practices for Bldrs	29.95	
	67298A	Bldrs Essentials: Framing/Carpentry 2nd Ed.	15.98	
	67298AS	Bldrs Essentials: Framing/Carpentry Spanish	15.98	
	67307	Bldrs Essentials: Plan Reading & Takeoff	35.95	
	67342	Building & Renovating Schools	69.98	
	67339	Building Security: Strategies & Costs	44.98	
	67353	Complete Book of Framing	29.95	
	67146	Concrete Repair & Maintenance Illustrated	69.95	
	67352	Construction Business Management	49.95	
	67314	Cost Planning & Est. for Facil. Maint.	89.95	
	67328A	Designing & Building with the IBC, 2nd Ed.	99.95	
	67230B	Electrical Estimating Methods, 3rd Ed.	64.95	
	64777A	Environmental Remediation Est. Methods, 2nd Ed.	49.98	
	67343	Estimating Bldg. Costs for Resi. & Lt. Comm.	29.95	
	67276A	Estimating Handbook, 2nd Ed.	$99.95	
	67318	Facilities Operations & Engineering Reference	54.98	
	67301	Facilities Planning & Relocation	49.98	
	67338A	Green Building: Proj. Planning & Cost Est., 2nd Ed.	129.95	
	67355	Green Home Improvement	24.95	

Qty.	Book no.	REFERENCE BOOKS (Cont.)	Unit Price	Total
	67357	The Gypsum Construction Handbook	29.95	
	67357S	The Gypsum Construction Handbook (Spanish)	29.95	
	67323	Historic Preservation: Proj. Planning & Est.	49.98	
	67349	Home Addition & Renovation Project Costs	19.98	
	67308E	Home Improvement Costs–Int. Projects, 9th Ed.	24.95	
	67309E	Home Improvement Costs–Ext. Projects, 9th Ed.	24.95	
	67344	Homeowner's Guide to Mold	21.95	
	67324B	How to Est.w/Means Data & CostWorks, 3rd Ed.	59.95	
	67351	How Your House Works	21.95	
	67282A	Illustrated Const. Dictionary, Condensed, 2nd Ed.	59.95	
	67292A	Illustrated Const. Dictionary, w/CD-ROM, 3rd Ed.	99.95	
	67348	Job Order Contracting	89.95	
	67347	Kitchen & Bath Project Costs	19.98	
	67295C	Landscape Estimating Methods, 5th Ed.	64.95	
	67341	Life Cycle Costing for Facilities	99.95	
	67294B	Mechanical Estimating Methods, 4th Ed.	64.95	
	67245A	Planning & Managing Interior Projects, 2nd Ed.	34.98	
	67283B	Plumbing Estimating Methods, 3rd Ed.	29.98	
	67345	Practice of Cost Segregation Analysis	99.95	
	67337	Preventive Maint. for Higher Education Facilities	149.95	
	67346	Preventive Maint. for Multi-Family Housing	89.95	
	67326	Preventive Maint. Guidelines for School Facil.	149.95	
	67247B	Project Scheduling & Management for Constr. 3rd Ed.	64.95	
	67265B	Repair & Remodeling Estimating Methods, 4th Ed.	69.95	
	67322B	Resi. & Light Commercial Const. Stds., 3rd Ed.	59.95	
	67327A	Spanish/English Construction Dictionary, 2nd Ed.	23.95	
	67145B	Sq. Ft. & Assem. Estimating Methods, 3rd Ed.	69.95	
	67321	Total Productive Facilities Management	29.98	
	67350	Understanding and Negotiating Const. Contracts	69.95	
	67303B	Unit Price Estimating Methods, 4th Ed.	59.95	
	67354	Universal Design	21.95	
	67319A	Value Engineering: Practical Applications	79.95	

MA residents add 5% state sales tax	
Shipping & Handling**	
Total (U.S. Funds)*	

Prices are subject to change and are for U.S. delivery only. *Canadian customers may call for current prices. **Shipping & handling charges: Add 7% of total order for check and credit card payments. Add 9% of total order for invoiced orders.

Send order to: **ADDV-1000**

Name (please print) ______________________________

Company ______________________________

☐ **Company**

☐ **Home Address** ______________________________

____

City/State/Zip ______________________________

Phone # ______________________ P.O. # ______________________

(Must accompany all orders being billed)

ail to: RSMeans, P.O. Box 800, Kingston, MA 02364

RSMeans Project Cost Report

By filling out this report, your project data will contribute to the database that supports the RSMeans Project Cost Square Foot Data. When you fill out this form, RSMeans will provide a $30 discount off one of the RSMeans products advertised in the preceding pages. Please complete the form including all items where you have cost data, and all the items marked (✔).

$30.00 Discount per product for each report you submit.

Project Description (No remodeling projects, please.)

✔ Building Use (Office, School...) ____________

✔ Address (City, State) ____________

____

✔ Building Area (SF) ____________

✔ Frame (Wood, Steel...) ____________

✔ Exterior Wall (Brick, Tilt-up...) ____________

✔ Basement: (check one) ☐ Full ☐ Partial ☐ None

✔ Number Stories ____________

✔ Floor-to-Floor Height ____________

✔ Volume (C.F.) ____________

% Air Conditioned ____________ Tons ____________

Total Project Cost $ ____________

Owner ____________

Architect ____________

General Contractor ____________

✔ Bid Date ____________

✔ Typical Bay Size ____________

✔ Capacity ____________

✔ Labor Force: _____ % Union _____ % Non-Union

✔ Project Description (Circle one number in each line)

1. Economy 2. Average 3. Custom 4. Luxury
1. Square 2. Rectangular 3. Irregular 4. Very Irregular

Comments ____________

____

A	✔	General Conditions	$
B	✔	Site Work	$
C	✔	Concrete	$
D	✔	Masonry	$
E	✔	Metals	$
F	✔	Wood & Plastics	$
G	✔	Thermal & Moisture Protection	$
GR		Roofing & Flashing	$
H	✔	Doors and Windows	$
J	✔	Finishes	$
JP		Painting & Wall Covering	$

K	✔	Specialties	$
L	✔	Equipment	$
M	✔	Furnishings	$
N	✔	Special Construction	$
P	✔	Conveying Systems	$
Q	✔	Mechanical	$
QP		Plumbing	$
QB		HVAC	$
R	✔	Electrical	$
S	✔	Mech./Elec. Combined	$

Please specify the RSMeans product you wish to receive. Complete the address information.

Product Name ____________

Product Number ____________

Your Name ____________

Title ____________

Company ____________

☐ Company

☐ Home Street Address ____________

City, State, Zip ____________

-n by mail or Fax 888-492-6770.

Method of Payment:

Credit Card # ____________

Expiration Date ____________

Check ____________

Purchase Order ____________

R.S. Means Company, Inc.
Square Foot Costs Department
P.O. Box 800
Kingston, MA 02364-9988